# Handbook of Truly Concurrent Process Algebra

# Handbook of Truly Concurrent Process Algebra

**Yong Wang**
Department of Computer Science and Technology
Faculty of Information Technology
Beijing University of Technology
Beijing, China

MK
MORGAN KAUFMANN PUBLISHERS
ELSEVIER AN IMPRINT OF ELSEVIER

Morgan Kaufmann is an imprint of Elsevier
50 Hampshire Street, 5th Floor, Cambridge, MA 02139, United States



**Notices**

Knowledge and best practice in this field are constantly changing. As new research and experience broaden our understanding, changes in research methods, professional practices, or medical treatment may become necessary.

Practitioners and researchers must always rely on their own experience and knowledge in evaluating and using any information, methods, compounds, or experiments described herein. In using such information or methods they should be mindful of their own safety and the safety of others, including parties for whom they have a professional responsibility.

To the fullest extent of the law, neither the Publisher nor the authors, contributors, or editors, assume any liability for any injury and/or damage to persons or property as a matter of products liability, negligence or otherwise, or from any use or operation of any methods, products, instructions, or ideas contained in the material herein.

ISBN: 978-0-443-21515-5

For information on all Morgan Kaufmann publications
visit our website at https://www.elsevier.com/books-and-journals

*Publisher:* Mara Conner
*Acquisitions Editor:* Chris Katsaropoulos
*Editorial Project Manager:* Rafael Guilherme Trombaco
*Production Project Manager:* Neena S. Maheen
*Cover Designer:* Mark Rogers

Typeset by VTeX

# Contents

## 38. Composition of patterns

Chapter 1

# Introduction to algebraic theory for reversible computing

Reversible computing [6] [7] [8] is an interesting topic in Computer Science. Reversible computing has been used in many area, such as quantum computing, transactions in databases, and business transactions. In these application areas, there are two aspects of reversible computing: one is that there is a corresponding reverse atomic action for each atomic action, so it is reversible at the level of atomic action; the other is in program logic, for the forward logic is important and the reverse logic is important too. The relation between the forward logic and reverse logic is the main content of reversible computing.

In the following, we take an example of business transactions. Traditional transactions with ACID (Atomicity, Consistency, Isolation and Durability) properties are supported by most of the current database management systems. They are usually implemented by middle states and a commit–rollback mechanism before being permanently stored or aborted. LLT (Long Living Transaction) or LRBT (Long Running Business Transaction) is a transaction surviving a long period of time. Compared to LLTs, LRBTs must satisfy the heterogeneity of applications located in different organizations, not only a long period of running time. More importantly, they are usually interactive, that is, LRBTs and their inner contained business activities or nested LRBTs are interacting with applications or humans. A LRBT may contain nested LRBTs or atomic business activities and finally be consisted of business activities. A business activity is usually located in one organization domain and survives for a shorter period of time. Hence, business activities are like traditional transactions and can be processed by use of traditional transaction process mechanisms. However, the abortion or failure of a LRBT cannot be settled by middle states to be rollbacked or permanently stored finally like traditional transaction processing, and the failure or abortion processing of a LRBT must adopt a special kind of mechanism, unlike rollback. Every step of a LRBT produces permanent effects without middle states to rollback, so a compensation operation corresponding to this step seems to be the only choice to cancel this step's effects when the LRBT fails or is aborted. Corresponding to a LLT, with the running of a LRBT, every business activity with ACID attributes is finished, so when a LRBT is aborted or fails, a compensation activity according to each finished business activity must be executed to eliminate the permanent effect produced by the business activity.

The functions of compensation mechanisms in LRBTs have already been accepted by the industry. Since Web Service is one kind of solution to solve cross-organizational application integrations, a compensation mechanism was adopted in Web Service orchestration specifications, such as the early WSFL (Web Service Flow Language) and XLANG. The late OASIS's WS-BPEL (Web Services Business Process Execution Language) also adopts compensation as the mechanism to abort a LRBT defined in a composite Web Service.

An example of an LRBT called TravelAgent is illustrated in Fig. 1.1a. TravelAgent LRBT is provided for the users to plan their traveling. It has eight business activities:

- SDC (Select Destination City): this is used for a user to select a destination traveling city.
- SH (Select Hotel): when the city is selected, SH is used to select a hotel for future traveling.
- BA (By Air): if the user decides to travel by air, he/she uses BA business activity to book an airline ticket.
- BT (By Train): if the user decides to travel by train, he/she uses BT business activity to book a train ticket.
- BB (By Bus): if the user decides to travel by bus, he/she uses BB business activity to book a bus ticket.
- PFA (Pay For Air): this activity is provided to pay for the ticket if BA is selected.
- PFT (Pay For Train): this activity is provided to pay for the ticket if BT selected.
- PFB (Pay For Bus): this activity is provided to pay for the ticket if BB is selected.

The process of TravelAgent LRBT is as follows:

1. First, a user selects a destination city by use of SDC business activity.
2. Then, he/she selects a hotel for the traveling through SH business activity.
3. Then, the travel means to the destination city is selected and there are three candidate means to be selected: by air through BA activity, by train through BT activity, and by bus through BB activity.

Handbook of Truly Concurrent Process Algebra. https://doi.org/10.1016/B978-0-44-321515-5.00005-6

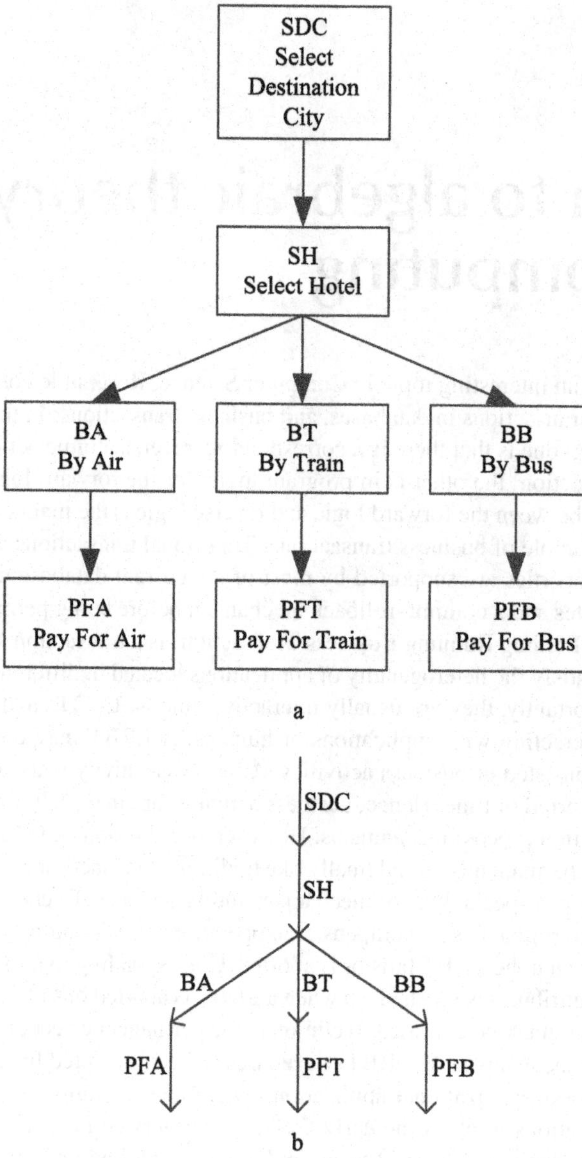


**FIGURE 1.1** TravelAgent example.

4. Finally, the user pays for the ticket booking according to the selected travel means. That is, he/she pays for the ticket booking through PFA activity if BA is selected, and selects PFT activity if BT is selected, and selects PFB activity if BB is selected.

The process graph of TravelAgent example is as Fig. 1.1b illustrates. It is described by process algebra ACP [3] as the following process term:

$SDC \cdot SH \cdot BA \cdot PFA + SDC \cdot SH \cdot BT \cdot PFT + SDC \cdot SH \cdot BB \cdot PFB$

Each business activity has its corresponding compensation activity implemented at the same time. In the compensation process of TravelAgent LRBT, there are eight compensation activities corresponding to the eight business activities and they are named as cSDC, cSH, cBA, cBT, cBB, cPFA, cPFT, and cPFB.

The compensation process can be described by ACP by use of the following process term:

$cSDC \cdot cSH \cdot cBA \cdot cPFA + cSDC \cdot cSH \cdot cBT \cdot cPFT + cSDC \cdot cSH \cdot cBB \cdot cPFB$

At runtime, there is only one execution path selected from the three candidate execution paths in the process graph as Fig. 1.1b shows. Without loss of generality, one path for selection of airline way is described by ACP through the following process term:

$SDC \cdot SH \cdot BA \cdot PFA$

The corresponding compensation process is represented by the following ACP process term:

$cSDC \cdot cSH \cdot cBA \cdot cPFA$

There exist several studies on reversible computing based on process algebra. In [6], an algebraic way of reversible computing used a communication key was proposed. In [6], a reversible version of process algebra CCS [1] [2] was presented, however, the properties of this reversible CCS based on the so-called forward–reverse interleaving bisimulation semantics were not discussed. In the studies of reversible computing, the whole algebraic architecture for reversible computing is still void, just because the interleaving concurrency cannot be reversible.

Phillips et al. [7] [8] researched the relation of concurrency and reversibility, and pointed out that true concurrency is reversible. We have done some work on truly concurrent process algebras, including CTC (Calculus for True Concurrency) [9], which is a generalization of CCS [1] [2] for true concurrency, APTC (Algebra of Parallelism for True Concurrency) [9], which is a generalization of ACP [3] for true concurrency and $\pi_{tc}$ [9] ($\pi$ calculus for True Concurrency), which is a generalization of $\pi$ calculus [4] [5] for true concurrency. In this book, we will discuss reversible computing based on these truly concurrent process algebras.

This book first gives the whole concrete algebraic system for reversible computing based on our previous work on truly concurrent process algebras, including calculus for fully reversible computing in Chapter 3, algebraic axiomatization for fully reversible computing in Chapter 4, calculus for partially reversible computing in Chapter 5, the sound and complete algebraic axiomatization for partially reversible computing in Chapter 6, and mobile calculus for partially reversible computing in Chapter 7.

Chapter 2

# Backgrounds

In this chapter, we introduce some preliminaries for self-sufficiency.

## 2.1 Operational semantics

**Definition 2.1** (Congruence). Let $\Sigma$ be a signature. An equivalence relation $R$ on $\mathcal{T}(\Sigma)$ is a congruence if for each $f \in \Sigma$, if $s_i R t_i$ for $i \in \{1, \cdots, ar(f)\}$, then $f(s_1, \cdots, s_{ar(f)}) R f(t_1, \cdots, t_{ar(f)})$.

**Definition 2.2** (Conservative extension). Let $T_0$ and $T_1$ be TSSs (transition system specifications) over signatures $\Sigma_0$ and $\Sigma_1$, respectively. The TSS $T_0 \oplus T_1$ is a conservative extension of $T_0$ if the LTSs (labeled transition systems) generated by $T_0$ and $T_0 \oplus T_1$ contain exactly the same transitions $t \xrightarrow{a} t'$ and $tP$ with $t \in \mathcal{T}(\Sigma_0)$.

**Definition 2.3** (Source dependency). The source-dependent variables in a transition rule of $\rho$ are defined inductively as follows: (1) all variables in the source of $\rho$ are source dependent; (2) if $t \xrightarrow{a} t'$ is a premise of $\rho$ and all variables in $t$ are source dependent, then all variables in $t'$ are source dependent. A transition rule is source dependent if all its variables are. A TSS is source dependent if all its rules are.

**Definition 2.4** (Freshness). Let $T_0$ and $T_1$ be TSSs over signatures $\Sigma_0$ and $\Sigma_1$, respectively. A term in $\mathbb{T}(T_0 \oplus T_1)$ is said to be fresh if it contains a function symbol from $\Sigma_1 \setminus \Sigma_0$. Similarly, a transition label or predicate symbol in $T_1$ is fresh if it does not occur in $T_0$.

**Theorem 2.5** (Conservative extension). *Let $T_0$ and $T_1$ be TSSs over signatures $\Sigma_0$ and $\Sigma_1$, respectively, where $T_0$ and $T_0 \oplus T_1$ are positive after reduction. Under the following conditions, $T_0 \oplus T_1$ is a conservative extension of $T_0$. (1) $T_0$ is source dependent. (2) For each $\rho \in T_1$, either the source of $\rho$ is fresh, or $\rho$ has a premise of the form $t \xrightarrow{a} t'$ or $tP$, where $t \in \mathbb{T}(\Sigma_0)$, all variables in $t$ occur in the source of $\rho$ and $t'$, $a$ or $P$ is fresh.*

## 2.2 Proof techniques

In this section, we introduce the concepts and conclusions about elimination, which is very important in the proof of the completeness theorem.

**Definition 2.6** (Elimination property). Let a process algebra with a defined set of basic terms be a subset of the set of closed terms over the process algebra. Then, the process algebra has the elimination to basic terms property if for every closed term $s$ of the algebra, there exists a basic term $t$ of the algebra such that the algebra $\vdash s = t$.

**Definition 2.7** (Strongly normalizing). A term $s_0$ is called strongly normalizing if there is no infinite series of reductions beginning in $s_0$.

**Definition 2.8.** We write $s >_{lpo} t$ if $s \rightarrow^{+} t$, where $\rightarrow^{+}$ is the transitive closure of the reduction relation defined by the transition rules of an algebra.

**Theorem 2.9** (Strong normalization). *Let a term rewriting (TRS) system with finitely many rewriting rules and let $>$ be a well-founded ordering on the signature of the corresponding algebra. If $s >_{lpo} t$ for each rewriting rule $s \rightarrow t$ in the TRS, then the term rewriting system is strongly normalizing.*

## 2.3 CTC

CTC [9] is a calculus of truly concurrent systems. It includes syntax and semantics:

**Handbook of Truly Concurrent Process Algebra. https://doi.org/10.1016/B978-0-44-321515-5.00006-8**

1. Its syntax includes actions, process constants, and operators acting between actions, like Prefix, Summation, Composition, Restriction, Relabeling.
2. Its semantics is based on labeled transition systems, Prefix, Summation, Composition, Restriction, Relabeling have their transition rules. CTC has good semantic properties based on the truly concurrent bisimulations. These properties include monoid laws, static laws, new expansion laws for strongly truly concurrent bisimulations, $\tau$ laws for weakly truly concurrent bisimulations, and full congruences for strongly and weakly truly concurrent bisimulations, and also a unique solution for recursion.

CTC can be used widely in verification of computer systems with a truly concurrent flavor.

## 2.4 $\pi_{tc}$

$\pi_{tc}$ [9] is an extension of $CTC$ and a generalization of $\pi$ calculus.

1. It treats names, variables, and substitutions more carefully, since names may be free or bound.
2. Names are mobile by references, rather than by values.
3. There are three kinds of prefixes, $\tau$ prefix $\tau.P$, output prefix $\overline{x}y.P$, and input prefix $x(y).P$, which are most distinctive to $CTC$.
4. $\pi_{tc}$ has good semantic properties based on the truly concurrent bisimulations. These properties include summation laws, identity laws, restriction laws, parallel laws, new expansion laws for strongly truly concurrent bisimulations, and full congruences for truly concurrent bisimulations, and also a unique solution for recursion.

## 2.5 APTC

APTC eliminates the differences of structures of transition system, event structure, etc., and discusses their behavioral equivalences. It considers that there are two kinds of causality relations: the chronological order modeled by the sequential composition and the causal order between different parallel branches modeled by the communication merge. It also considers that there exist two kinds of confliction relations: the structural confliction modeled by the alternative composition and the conflictions in different parallel branches that should be eliminated. Based on conservative extension, there are four modules in APTC: BATC (Basic Algebra for True Concurrency), APTC (Algebra for Parallelism in True Concurrency), recursion, and abstraction.

### 2.5.1 Basic algebra for true concurrency

BATC has sequential composition $\cdot$ and alternative composition $+$ to capture the chronological ordered causality and the structural confliction. The constants are ranged over $A$, the set of atomic actions. The algebraic laws on $\cdot$ and $+$ are sound and complete modulo truly concurrent bisimulation equivalences (including pomset bisimulation, step bisimulation, hp-bisimulation, and hhp-bisimulation).

**Definition 2.10** (Prime event structure with silent event). Let $\Lambda$ be a fixed set of labels, ranged over $a, b, c, \cdots$ and $\tau$. A ($\Lambda$-labeled) prime event structure with silent event $\tau$ is a tuple $\mathcal{E} = \langle \mathbb{E}, \leq, \sharp, \lambda \rangle$, where $\mathbb{E}$ is a denumerable set of events, including the silent event $\tau$. Let $\hat{\mathbb{E}} = \mathbb{E} \backslash \{\tau\}$, exactly excluding $\tau$, it is obvious that $\hat{\tau^*} = \epsilon$, where $\epsilon$ is the empty event. Let $\lambda : \mathbb{E} \to \Lambda$ be a labeling function and let $\lambda(\tau) = \tau$. Also, $\leq$, $\sharp$ are binary relations on $\mathbb{E}$, called causality and conflict, respectively, such that:

1. $\leq$ is a partial order and $\lceil e \rceil = \{e' \in \mathbb{E} | e' \leq e\}$ is finite for all $e \in \mathbb{E}$. It is easy to see that $e \leq \tau^* \leq e' = e \leq \tau \leq \cdots \leq \tau \leq e'$, then $e \leq e'$.
2. $\sharp$ is irreflexive, symmetric, and hereditary with respect to $\leq$, that is, for all $e, e', e'' \in \mathbb{E}$, if $e \sharp e' \leq e''$, then $e \sharp e''$.

Then, the concepts of consistency and concurrency can be drawn from the above definition:

1. $e, e' \in \mathbb{E}$ are consistent, denoted as $e \frown e'$, if $\neg(e \sharp e')$. A subset $X \subseteq \mathbb{E}$ is called consistent, if $e \frown e'$ for all $e, e' \in X$.
2. $e, e' \in \mathbb{E}$ are concurrent, denoted as $e \parallel e'$, if $\neg(e \leq e')$, $\neg(e' \leq e)$, and $\neg(e \sharp e')$.

**Definition 2.11** (Configuration). Let $\mathcal{E}$ be a PES. A (finite) configuration in $\mathcal{E}$ is a (finite) consistent subset of events $C \subseteq \mathcal{E}$, closed with respect to causality (i.e., $\lceil C \rceil = C$). The set of finite configurations of $\mathcal{E}$ is denoted by $\mathcal{C}(\mathcal{E})$. We let $\hat{C} = C \backslash \{\tau\}$.

A consistent subset of $X \subseteq \mathbb{E}$ of events can be seen as a pomset. Given $X, Y \subseteq \mathbb{E}$, $\hat{X} \sim \hat{Y}$ if $\hat{X}$ and $\hat{Y}$ are isomorphic as pomsets. In the remainder of the chapter, when we say $C_1 \sim C_2$, we mean $\hat{C_1} \sim \hat{C_2}$.

**TABLE 2.1** Axioms of BATC.

| No. | Axiom |
|---|---|
| $A1$ | $x+y=y+x$ |
| $A2$ | $(x+y)+z=x+(y+z)$ |
| $A3$ | $x+x=x$ |
| $A4$ | $(x+y)\cdot z=x\cdot z+y\cdot z$ |
| $A5$ | $(x\cdot y)\cdot z=x\cdot(y\cdot z)$ |

**TABLE 2.2** Transition rules of BATC.

$$\frac{}{e\xrightarrow{e}\surd}$$

$$\frac{x\xrightarrow{e}\surd}{x+y\xrightarrow{e}\surd}\quad\frac{x\xrightarrow{e}x'}{x+y\xrightarrow{e}x'}\quad\frac{y\xrightarrow{e}\surd}{x+y\xrightarrow{e}\surd}\quad\frac{y\xrightarrow{e}y'}{x+y\xrightarrow{e}y'}$$

$$\frac{x\xrightarrow{e}\surd}{x\cdot y\xrightarrow{e}y}\quad\frac{x\xrightarrow{e}x'}{x\cdot y\xrightarrow{e}x'\cdot y}$$

**Definition 2.12** (Pomset transitions and step). Let $\mathcal{E}$ be a PES and let $C\in\mathcal{C}(\mathcal{E})$, and $\emptyset\neq X\subseteq\mathbb{E}$, if $C\cap X=\emptyset$ and $C'=C\cup X\in\mathcal{C}(\mathcal{E})$, then $C\xrightarrow{X}C'$ is called a pomset transition from $C$ to $C'$. When the events in $X$ are pairwise concurrent, we say that $C\xrightarrow{X}C'$ is a step.

**Definition 2.13** (Pomset, step bisimulation). Let $\mathcal{E}_1$, $\mathcal{E}_2$ be PESs. A pomset bisimulation is a relation $R\subseteq\mathcal{C}(\mathcal{E}_1)\times\mathcal{C}(\mathcal{E}_2)$, such that if $(C_1,C_2)\in R$, and $C_1\xrightarrow{X_1}C_1'$, then $C_2\xrightarrow{X_2}C_2'$, with $X_1\subseteq\mathbb{E}_1$, $X_2\subseteq\mathbb{E}_2$, $X_1\sim X_2$, and $(C_1',C_2')\in R$, and vice versa. We say that $\mathcal{E}_1$, $\mathcal{E}_2$ are pomset bisimilar, written $\mathcal{E}_1\sim_p\mathcal{E}_2$, if there exists a pomset bisimulation $R$, such that $(\emptyset,\emptyset)\in R$. By replacing pomset transitions with steps, we can obtain the definition of step bisimulation. When PESs $\mathcal{E}_1$ and $\mathcal{E}_2$ are step bisimilar, we write $\mathcal{E}_1\sim_s\mathcal{E}_2$.

**Definition 2.14** (Posetal product). Given two PESs $\mathcal{E}_1$, $\mathcal{E}_2$, the posetal product of their configurations, denoted $\mathcal{C}(\mathcal{E}_1)\overline{\times}\mathcal{C}(\mathcal{E}_2)$, is defined as

$$\{(C_1,f,C_2)|C_1\in\mathcal{C}(\mathcal{E}_1),C_2\in\mathcal{C}(\mathcal{E}_2),f:C_1\to C_2\text{ isomorphism}\}$$

A subset $R\subseteq\mathcal{C}(\mathcal{E}_1)\overline{\times}\mathcal{C}(\mathcal{E}_2)$ is called a posetal relation. We say that $R$ is downward closed when for any $(C_1,f,C_2)$, $(C_1',f',C_2')\in\mathcal{C}(\mathcal{E}_1)\overline{\times}\mathcal{C}(\mathcal{E}_2)$, if $(C_1,f,C_2)\subseteq(C_1',f',C_2')$ pointwise and $(C_1',f',C_2')\in R$, then $(C_1,f,C_2)\in R$.

For $f:X_1\to X_2$, we define $f[x_1\mapsto x_2]:X_1\cup\{x_1\}\to X_2\cup\{x_2\}$, $z\in X_1\cup\{x_1\}$, (1) $f[x_1\mapsto x_2](z)=x_2$, if $z=x_1$; (2) $f[x_1\mapsto x_2](z)=f(z)$, otherwise, where $X_1\subseteq\mathbb{E}_1$, $X_2\subseteq\mathbb{E}_2$, $x_1\in\mathbb{E}_1$, $x_2\in\mathbb{E}_2$.

**Definition 2.15** ((Hereditary) history-preserving bisimulation). A history-preserving (hp-) bisimulation is a posetal relation $R\subseteq\mathcal{C}(\mathcal{E}_1)\overline{\times}\mathcal{C}(\mathcal{E}_2)$ such that if $(C_1,f,C_2)\in R$, and $C_1\xrightarrow{e_1}C_1'$, then $C_2\xrightarrow{e_2}C_2'$, with $(C_1',f[e_1\mapsto e_2],C_2')\in R$, and vice versa. $\mathcal{E}_1$, $\mathcal{E}_2$ are history-preserving (hp-)bisimilar and are written $\mathcal{E}_1\sim_{hp}\mathcal{E}_2$ if there exists a hp-bisimulation $R$ such that $(\emptyset,\emptyset,\emptyset)\in R$.

A hereditary history-preserving (hhp-)bisimulation is a downward closed hp-bisimulation. $\mathcal{E}_1$, $\mathcal{E}_2$ are hereditary history-preserving (hhp-)bisimilar and are written as $\mathcal{E}_1\sim_{hhp}\mathcal{E}_2$.

In the following, let $e_1,e_2,e_1',e_2'\in\mathbb{E}$, and let variables $x,y,z$ range over the set of terms for true concurrency, $p,q,s$ range over the set of closed terms. The set of axioms of BATC consists of the laws given in Table 2.1.

We give the operational transition rules of operators $\cdot$ and $+$ as Table 2.2 shows. The predicate $\xrightarrow{e}\surd$ represents successful termination after execution of the event $e$.

**Theorem 2.16** (Soundness of BATC modulo truly concurrent bisimulation equivalences). *The axiomatization of BATC is sound modulo truly concurrent bisimulation equivalences $\sim_p$, $\sim_s$, $\sim_{hp}$, and $\sim_{hhp}$. That is,*

1. *let $x$ and $y$ be BATC terms. If BATC $\vdash x=y$, then $x\sim_p y$;*
2. *let $x$ and $y$ be BATC terms. If BATC $\vdash x=y$, then $x\sim_s y$;*

3. *let $x$ and $y$ be BATC terms. If BATC $\vdash x = y$, then $x \sim_{hp} y$;*
4. *let $x$ and $y$ be BATC terms. If BATC $\vdash x = y$, then $x \sim_{hhp} y$.*

**Theorem 2.17** (Completeness of BATC modulo truly concurrent bisimulation equivalences). *The axiomatization of BATC is complete modulo truly concurrent bisimulation equivalences $\sim_p$, $\sim_s$, $\sim_{hp}$, and $\sim_{hhp}$. That is,*

1. *let $p$ and $q$ be closed BATC terms, if $p \sim_p q$ then $p = q$;*
2. *let $p$ and $q$ be closed BATC terms, if $p \sim_s q$ then $p = q$;*
3. *let $p$ and $q$ be closed BATC terms, if $p \sim_{hp} q$ then $p = q$;*
4. *let $p$ and $q$ be closed BATC terms, if $p \sim_{hhp} q$ then $p = q$.*

### 2.5.2 Algebra for parallelism in true concurrency

APTC uses the whole parallel operator $\between$, the auxiliary binary parallel $\parallel$ to model parallelism, and the communication merge $\mid$ to model communications among different parallel branches, and also the unary conflict elimination operator $\Theta$ and the binary unless operator $\triangleleft$ to eliminate conflicts among different parallel branches. Since a communication may be blocked, a new constant called deadlock $\delta$ is extended to $A$, and also a new unary encapsulation operator $\partial_H$ is introduced to eliminate $\delta$, which may exist in the processes. The algebraic laws on these operators are also sound and complete modulo truly concurrent bisimulation equivalences (including pomset bisimulation, step bisimulation, hp-bisimulation, but not hhp-bisimulation). Note that the parallel operator $\parallel$ in a process cannot be eliminated by deductions on the process using axioms of APTC, but other operators can eventually be steadied by $\cdot$, $+$ and $\parallel$, this is also why truly concurrent bisimulations are called *truly concurrent* semantics.

We design the axioms of APTC in Table 2.3, including algebraic laws of the parallel operator $\parallel$, communication operator $\mid$, conflict elimination operator $\Theta$, unless operator $\triangleleft$, encapsulation operator $\partial_H$, the deadlock constant $\delta$, and also the whole parallel operator $\between$.

We give the transition rules of APTC in Table 2.4, it is suitable for all truly concurrent behavioral equivalence, including pomset bisimulation, step bisimulation, hp-bisimulation, and hhp-bisimulation.

**Theorem 2.18** (Soundness of APTC modulo truly concurrent bisimulation equivalences). *The axiomatization of APTC is sound modulo truly concurrent bisimulation equivalences $\sim_p$, $\sim_s$, and $\sim_{hp}$. That is,*

1. *let $x$ and $y$ be APTC terms. If APTC $\vdash x = y$, then $x \sim_p y$;*
2. *let $x$ and $y$ be APTC terms. If APTC $\vdash x = y$, then $x \sim_s y$;*
3. *let $x$ and $y$ be APTC terms. If APTC $\vdash x = y$, then $x \sim_{hp} y$.*

**Theorem 2.19** (Completeness of APTC modulo truly concurrent bisimulation equivalences). *The axiomatization of APTC is complete modulo truly concurrent bisimulation equivalences $\sim_p$, $\sim_s$, and $\sim_{hp}$. That is,*

1. *let $p$ and $q$ be closed APTC terms, if $p \sim_p q$ then $p = q$;*
2. *let $p$ and $q$ be closed APTC terms, if $p \sim_s q$ then $p = q$;*
3. *let $p$ and $q$ be closed APTC terms, if $p \sim_{hp} q$ then $p = q$.*

### 2.5.3 Recursion

To model infinite computation, recursion is introduced into APTC. In order to obtain a sound and complete theory, guarded recursion and linear recursion are needed. The corresponding axioms are RSP (Recursive Specification Principle) and RDP (Recursive Definition Principle), RDP states that the solutions of a recursive specification can represent the behaviors of the specification, while RSP states that a guarded recursive specification has only one solution, they are sound with respect to APTC with guarded recursion modulo several truly concurrent bisimulation equivalences (including pomset bisimulation, step bisimulation, and hp-bisimulation), and they are complete with respect to APTC with linear recursion modulo several truly concurrent bisimulation equivalences (including pomset bisimulation, step bisimulation, and hp-bisimulation). In the following, $E, F, G$ are recursion specifications, $X, Y, Z$ are recursive variables.

For a guarded recursive specifications $E$ with the form

$$X_1 = t_1(X_1, \cdots, X_n)$$

$$\cdots$$

$$X_n = t_n(X_1, \cdots, X_n)$$

**TABLE 2.3** Axioms of APTC.

| No. | Axiom |
|---|---|
| $A6$ | $x+\delta=x$ |
| $A7$ | $\delta\cdot x=\delta$ |
| $P1$ | $x \between y=x\parallel y+x\mid y$ |
| $P2$ | $x\parallel y=y\parallel x$ |
| $P3$ | $(x\parallel y)\parallel z=x\parallel(y\parallel z)$ |
| $P4$ | $e_1\parallel(e_2\cdot y)=(e_1\parallel e_2)\cdot y$ |
| $P5$ | $(e_1\cdot x)\parallel e_2=(e_1\parallel e_2)\cdot x$ |
| $P6$ | $(e_1\cdot x)\parallel(e_2\cdot y)=(e_1\parallel e_2)\cdot(x\between y)$ |
| $P7$ | $(x+y)\parallel z=(x\parallel z)+(y\parallel z)$ |
| $P8$ | $x\parallel(y+z)=(x\parallel y)+(x\parallel z)$ |
| $P9$ | $\delta\parallel x=\delta$ |
| $P10$ | $x\parallel\delta=\delta$ |
| $C11$ | $e_1\mid e_2=\gamma(e_1,e_2)$ |
| $C12$ | $e_1\mid(e_2\cdot y)=\gamma(e_1,e_2)\cdot y$ |
| $C13$ | $(e_1\cdot x)\mid e_2=\gamma(e_1,e_2)\cdot x$ |
| $C14$ | $(e_1\cdot x)\mid(e_2\cdot y)=\gamma(e_1,e_2)\cdot(x\between y)$ |
| $C15$ | $(x+y)\mid z=(x\mid z)+(y\mid z)$ |
| $C16$ | $x\mid(y+z)=(x\mid y)+(x\mid z)$ |
| $C17$ | $\delta\mid x=\delta$ |
| $C18$ | $x\mid\delta=\delta$ |
| $CE19$ | $\Theta(e)=e$ |
| $CE20$ | $\Theta(\delta)=\delta$ |
| $CE21$ | $\Theta(x+y)=\Theta(x)+\Theta(y)$ |
| $CE22$ | $\Theta(x\cdot y)=\Theta(x)\cdot\Theta(y)$ |
| $CE23$ | $\Theta(x\parallel y)=((\Theta(x)\triangleleft y)\parallel y)+((\Theta(y)\triangleleft x)\parallel x)$ |
| $CE24$ | $\Theta(x\mid y)=((\Theta(x)\triangleleft y)\mid y)+((\Theta(y)\triangleleft x)\mid x)$ |
| $U25$ | $(\sharp(e_1,e_2))\quad e_1\triangleleft e_2=\tau$ |
| $U26$ | $(\sharp(e_1,e_2),e_2\leq e_3)\quad e_1\triangleleft e_3=\tau$ |
| $U27$ | $(\sharp(e_1,e_2),e_2\leq e_3)\quad e3\triangleleft e_1=\tau$ |
| $U28$ | $e\triangleleft\delta=e$ |
| $U29$ | $\delta\triangleleft e=\delta$ |
| $U30$ | $(x+y)\triangleleft z=(x\triangleleft z)+(y\triangleleft z)$ |
| $U31$ | $(x\cdot y)\triangleleft z=(x\triangleleft z)\cdot(y\triangleleft z)$ |
| $U32$ | $(x\parallel y)\triangleleft z=(x\triangleleft z)\parallel(y\triangleleft z)$ |
| $U33$ | $(x\mid y)\triangleleft z=(x\triangleleft z)\mid(y\triangleleft z)$ |
| $U34$ | $x\triangleleft(y+z)=(x\triangleleft y)\triangleleft z$ |
| $U35$ | $x\triangleleft(y\cdot z)=(x\triangleleft y)\triangleleft z$ |
| $U36$ | $x\triangleleft(y\parallel z)=(x\triangleleft y)\triangleleft z$ |
| $U37$ | $x\triangleleft(y\mid z)=(x\triangleleft y)\triangleleft z$ |
| $D1$ | $e\notin H\quad\partial_H(e)=e$ |
| $D2$ | $e\in H\quad\partial_H(e)=\delta$ |
| $D3$ | $\partial_H(\delta)=\delta$ |
| $D4$ | $\partial_H(x+y)=\partial_H(x)+\partial_H(y)$ |
| $D5$ | $\partial_H(x\cdot y)=\partial_H(x)\cdot\partial_H(y)$ |
| $D6$ | $\partial_H(x\parallel y)=\partial_H(x)\parallel\partial_H(y)$ |

**TABLE 2.4** Transition rules of APTC.

| | |
|---|---|
| $\frac{x\xrightarrow{e_1}\surd\quad y\xrightarrow{e_2}\surd}{x\parallel y\xrightarrow{\{e_1,e_2\}}\surd}$ | $\frac{x\xrightarrow{e_1}x'\quad y\xrightarrow{e_2}\surd}{x\parallel y\xrightarrow{\{e_1,e_2\}}x'}$ |
| $\frac{x\xrightarrow{e_1}\surd\quad y\xrightarrow{e_2}y'}{x\parallel y\xrightarrow{\{e_1,e_2\}}y'}$ | $\frac{x\xrightarrow{e_1}x'\quad y\xrightarrow{e_2}y'}{x\parallel y\xrightarrow{\{e_1,e_2\}}x'\between y'}$ |
| $\frac{x\xrightarrow{e_1}\surd\quad y\xrightarrow{e_2}\surd}{x\mid y\xrightarrow{\gamma(e_1,e_2)}\surd}$ | $\frac{x\xrightarrow{e_1}x'\quad y\xrightarrow{e_2}\surd}{x\mid y\xrightarrow{\gamma(e_1,e_2)}x'}$ |
| $\frac{x\xrightarrow{e_1}\surd\quad y\xrightarrow{e_2}y'}{x\mid y\xrightarrow{\gamma(e_1,e_2)}y'}$ | $\frac{x\xrightarrow{e_1}x'\quad y\xrightarrow{e_2}y'}{x\mid y\xrightarrow{\gamma(e_1,e_2)}x'\between y'}$ |
| $\frac{x\xrightarrow{e_1}\surd\quad (\sharp(e_1,e_2))}{\Theta(x)\xrightarrow{e_1}\surd}$ | $\frac{x\xrightarrow{e_2}\surd\quad (\sharp(e_1,e_2))}{\Theta(x)\xrightarrow{e_2}\surd}$ |
| $\frac{x\xrightarrow{e_1}x'\quad (\sharp(e_1,e_2))}{\Theta(x)\xrightarrow{e_1}\Theta(x')}$ | $\frac{x\xrightarrow{e_2}x'\quad (\sharp(e_1,e_2))}{\Theta(x)\xrightarrow{e_2}\Theta(x')}$ |
| $\frac{x\xrightarrow{e_1}\surd\quad y\nrightarrow^{e_2}\quad (\sharp(e_1,e_2))}{x\triangleleft y\xrightarrow{\tau}\surd}$ | $\frac{x\xrightarrow{e_1}x'\quad y\nrightarrow^{e_2}\quad (\sharp(e_1,e_2))}{x\triangleleft y\xrightarrow{\tau}x'}$ |
| $\frac{x\xrightarrow{e_1}\surd\quad y\nrightarrow^{e_3}\quad (\sharp(e_1,e_2),e_2\leq e_3)}{x\triangleleft y\xrightarrow{\tau}\surd}$ | $\frac{x\xrightarrow{e_1}x'\quad y\nrightarrow^{e_3}\quad (\sharp(e_1,e_2),e_2\leq e_3)}{x\triangleleft y\xrightarrow{\tau}x'}$ |
| $\frac{x\xrightarrow{e_3}\surd\quad y\nrightarrow^{e_2}\quad (\sharp(e_1,e_2),e_1\leq e_3)}{x\triangleleft y\xrightarrow{\tau}\surd}$ | $\frac{x\xrightarrow{e_3}x'\quad y\nrightarrow^{e_2}\quad (\sharp(e_1,e_2),e_1\leq e_3)}{x\triangleleft y\xrightarrow{\tau}x'}$ |
| $\frac{x\xrightarrow{e}\surd}{\partial_H(x)\xrightarrow{e}\surd}\quad (e\notin H)$ | $\frac{x\xrightarrow{e}x'}{\partial_H(x)\xrightarrow{e}\partial_H(x')}\quad (e\notin H)$ |

**TABLE 2.5** Transition rules of guarded recursion.

| |
|---|
| $\frac{t_i(\langle X_1\mid E\rangle,\cdots,\langle X_n\mid E\rangle)\xrightarrow{\{e_1,\cdots,e_k\}}\surd}{\langle X_i\mid E\rangle\xrightarrow{\{e_1,\cdots,e_k\}}\surd}$ |
| $\frac{t_i(\langle X_1\mid E\rangle,\cdots,\langle X_n\mid E\rangle)\xrightarrow{\{e_1,\cdots,e_k\}}y}{\langle X_i\mid E\rangle\xrightarrow{\{e_1,\cdots,e_k\}}y}$ |

**TABLE 2.6** Recursive definition and specification principle.

| No. | Axiom |
|---|---|
| $RDP$ | $\langle X_i\mid E\rangle = t_i(\langle X_1\mid E,\cdots,X_n\mid E\rangle)\quad (i\in\{1,\cdots,n\})$ |
| $RSP$ | if $y_i=t_i(y_1,\cdots,y_n)$ for $i\in\{1,\cdots,n\}$, then $y_i=\langle X_i\mid E\rangle\quad (i\in\{1,\cdots,n\})$ |

The behavior of the solution $\langle X_i|E\rangle$ for the recursion variable $X_i$ in $E$, where $i\in\{1,\cdots,n\}$, is exactly the behavior of their right-hand sides $t_i(X_1,\cdots,X_n)$, which is captured by the two transition rules in Table 2.5.

The $RDP$ (Recursive Definition Principle) and the $RSP$ (Recursive Specification Principle) are shown in Table 2.6.

**Theorem 2.20** (Soundness of $APTC$ with guarded recursion). *Let $x$ and $y$ be $APTC$ with guarded recursion terms. If $APTC$ with guarded recursion $\vdash x=y$, then*

1. $x\sim_s y$;
2. $x\sim_p y$;
3. $x\sim_{hp} y$.

**Theorem 2.21** (Completeness of $APTC$ with linear recursion). *Let $p$ and $q$ be closed $APTC$ with linear recursion terms, then,*

1. *if $p \sim_s q$ then $p = q$;*
2. *if $p \sim_p q$ then $p = q$;*
3. *if $p \sim_{hp} q$ then $p = q$.*

### 2.5.4 Abstraction

To abstract away internal implementations from the external behaviors, a new constant $\tau$ called a silent step is added to $A$, and also a new unary abstraction operator $\tau_I$ is used to rename actions in $I$ into $\tau$ (the resulted APTC with silent step and abstraction operator is called $\text{APTC}_\tau$). The recursive specification is adapted to guarded linear recursion to prevent infinite $\tau$-loops specifically. The axioms of $\tau$ and $\tau_I$ are sound modulo rooted branching truly concurrent bisimulation equivalences (several kinds of weakly truly concurrent bisimulation equivalences, including rooted branching pomset bisimulation, rooted branching step bisimulation, and rooted branching hp-bisimulation). To eliminate infinite $\tau$-loops caused by $\tau_I$ and obtain completeness, CFAR (Cluster Fair Abstraction Rule) is used to prevent infinite $\tau$-loops in a constructible way.

**Definition 2.22** (Weak pomset transitions and weak step). Let $\mathcal{E}$ be a PES and let $C \in \mathcal{C}(\mathcal{E})$, and $\emptyset \neq X \subseteq \hat{\mathbb{E}}$, if $C \cap X = \emptyset$ and $\hat{C'} = \hat{C} \cup X \in \mathcal{C}(\mathcal{E})$, then $C \overset{X}{\Rightarrow} C'$ is called a weak pomset transition from $C$ to $C'$, where we define $\overset{e}{\Rightarrow} \triangleq \overset{\tau^*}{\longrightarrow} \overset{e}{\rightarrow} \overset{\tau^*}{\longrightarrow}$. Also, $\overset{X}{\Rightarrow} \triangleq \overset{\tau^*}{\longrightarrow} \overset{e}{\rightarrow} \overset{\tau^*}{\longrightarrow}$, for every $e \in X$. When the events in $X$ are pairwise concurrent, we say that $C \overset{X}{\Rightarrow} C'$ is a weak step.

**Definition 2.23** (Branching pomset, step bisimulation). Assume a special termination predicate $\downarrow$, and let $\surd$ represent a state with $\surd \downarrow$. Let $\mathcal{E}_1$, $\mathcal{E}_2$ be PESs. A branching pomset bisimulation is a relation $R \subseteq \mathcal{C}(\mathcal{E}_1) \times \mathcal{C}(\mathcal{E}_2)$, such that:

1. if $(C_1, C_2) \in R$, and $C_1 \overset{X}{\rightarrow} C_1'$, then
   - either $X \equiv \tau^*$, and $(C_1', C_2) \in R$;
   - or there is a sequence of (zero or more) $\tau$-transitions $C_2 \overset{\tau^*}{\longrightarrow} C_2^0$, such that $(C_1, C_2^0) \in R$ and $C_2^0 \overset{X}{\Rightarrow} C_2'$ with $(C_1', C_2') \in R$;
2. if $(C_1, C_2) \in R$, and $C_2 \overset{X}{\rightarrow} C_2'$, then
   - either $X \equiv \tau^*$, and $(C_1, C_2') \in R$;
   - or there is a sequence of (zero or more) $\tau$-transitions $C_1 \overset{\tau^*}{\longrightarrow} C_1^0$, such that $(C_1^0, C_2) \in R$ and $C_1^0 \overset{X}{\Rightarrow} C_1'$ with $(C_1', C_2') \in R$;
3. if $(C_1, C_2) \in R$ and $C_1 \downarrow$, then there is a sequence of (zero or more) $\tau$-transitions $C_2 \overset{\tau^*}{\longrightarrow} C_2^0$ such that $(C_1, C_2^0) \in R$ and $C_2^0 \downarrow$;
4. if $(C_1, C_2) \in R$ and $C_2 \downarrow$, then there is a sequence of (zero or more) $\tau$-transitions $C_1 \overset{\tau^*}{\longrightarrow} C_1^0$ such that $(C_1^0, C_2) \in R$ and $C_1^0 \downarrow$.

We say that $\mathcal{E}_1$, $\mathcal{E}_2$ are branching pomset bisimilar, written $\mathcal{E}_1 \approx_{bp} \mathcal{E}_2$, if there exists a branching pomset bisimulation $R$, such that $(\emptyset, \emptyset) \in R$.

By replacing pomset transitions with steps, we can obtain the definition of branching step bisimulation. When PESs $\mathcal{E}_1$ and $\mathcal{E}_2$ are branching step bisimilar, we write $\mathcal{E}_1 \approx_{bs} \mathcal{E}_2$.

**Definition 2.24** (Rooted branching pomset, step bisimulation). Assume a special termination predicate $\downarrow$, and let $\surd$ represent a state with $\surd \downarrow$. Let $\mathcal{E}_1$, $\mathcal{E}_2$ be PESs. A branching pomset bisimulation is a relation $R \subseteq \mathcal{C}(\mathcal{E}_1) \times \mathcal{C}(\mathcal{E}_2)$, such that:

1. if $(C_1, C_2) \in R$, and $C_1 \overset{X}{\rightarrow} C_1'$, then $C_2 \overset{X}{\rightarrow} C_2'$ with $C_1' \approx_{bp} C_2'$;
2. if $(C_1, C_2) \in R$, and $C_2 \overset{X}{\rightarrow} C_2'$, then $C_1 \overset{X}{\rightarrow} C_1'$ with $C_1' \approx_{bp} C_2'$;
3. if $(C_1, C_2) \in R$ and $C_1 \downarrow$, then $C_2 \downarrow$;
4. if $(C_1, C_2) \in R$ and $C_2 \downarrow$, then $C_1 \downarrow$.

We say that $\mathcal{E}_1$, $\mathcal{E}_2$ are rooted branching pomset bisimilar, written $\mathcal{E}_1 \approx_{rbp} \mathcal{E}_2$, if there exists a rooted branching pomset bisimulation $R$, such that $(\emptyset, \emptyset) \in R$.

By replacing pomset transitions with steps, we can obtain the definition of rooted branching step bisimulation. When PESs $\mathcal{E}_1$ and $\mathcal{E}_2$ are rooted branching step bisimilar, we write $\mathcal{E}_1 \approx_{rbs} \mathcal{E}_2$.

**Definition 2.25** (Branching (hereditary) history-preserving bisimulation). Assume a special termination predicate $\downarrow$, and let $\surd$ represent a state with $\surd \downarrow$. A branching history-preserving (hp-) bisimulation is a weakly posetal relation $R \subseteq \mathcal{C}(\mathcal{E}_1)\overline{\times}\mathcal{C}(\mathcal{E}_2)$ such that:

1. if $(C_1, f, C_2) \in R$, and $C_1 \xrightarrow{e_1} C_1'$, then
   - either $e_1 \equiv \tau$, and $(C_1', f[e_1 \mapsto \tau], C_2) \in R$;
   - or there is a sequence of (zero or more) $\tau$-transitions $C_2 \xrightarrow{\tau^*} C_2^0$, such that $(C_1, f, C_2^0) \in R$ and $C_2^0 \xrightarrow{e_2} C_2'$ with $(C_1', f[e_1 \mapsto e_2], C_2') \in R$;
2. if $(C_1, f, C_2) \in R$, and $C_2 \xrightarrow{e_2} C_2'$, then
   - either $X \equiv \tau$, and $(C_1, f[e_2 \mapsto \tau], C_2') \in R$;
   - or there is a sequence of (zero or more) $\tau$-transitions $C_1 \xrightarrow{\tau^*} C_1^0$, such that $(C_1^0, f, C_2) \in R$ and $C_1^0 \xrightarrow{e_1} C_1'$ with $(C_1', f[e_2 \mapsto e_1], C_2') \in R$;
3. if $(C_1, f, C_2) \in R$ and $C_1 \downarrow$, then there is a sequence of (zero or more) $\tau$-transitions $C_2 \xrightarrow{\tau^*} C_2^0$ such that $(C_1, f, C_2^0) \in R$ and $C_2^0 \downarrow$;
4. if $(C_1, f, C_2) \in R$ and $C_2 \downarrow$, then there is a sequence of (zero or more) $\tau$-transitions $C_1 \xrightarrow{\tau^*} C_1^0$ such that $(C_1^0, f, C_2) \in R$ and $C_1^0 \downarrow$.

$\mathcal{E}_1, \mathcal{E}_2$ are branching history-preserving (hp-)bisimilar and are written $\mathcal{E}_1 \approx_{bhp} \mathcal{E}_2$ if there exists a branching hp-bisimulation $R$ such that $(\emptyset, \emptyset, \emptyset) \in R$.

A branching hereditary history-preserving (hhp-)bisimulation is a downward closed branching hhp-bisimulation. $\mathcal{E}_1, \mathcal{E}_2$ are branching hereditary history-preserving (hhp-)bisimilar and are written as $\mathcal{E}_1 \approx_{bhhp} \mathcal{E}_2$.

**Definition 2.26** (Rooted branching (hereditary) history-preserving bisimulation). Assume a special termination predicate $\downarrow$, and let $\surd$ represent a state with $\surd \downarrow$. A rooted branching history-preserving (hp-) bisimulation is a weakly posetal relation $R \subseteq \mathcal{C}(\mathcal{E}_1)\overline{\times}\mathcal{C}(\mathcal{E}_2)$ such that:

1. if $(C_1, f, C_2) \in R$, and $C_1 \xrightarrow{e_1} C_1'$, then $C_2 \xrightarrow{e_2} C_2'$ with $C_1' \approx_{bhp} C_2'$;
2. if $(C_1, f, C_2) \in R$, and $C_2 \xrightarrow{e_2} C_1'$, then $C_1 \xrightarrow{e_1} C_2'$ with $C_1' \approx_{bhp} C_2'$;
3. if $(C_1, f, C_2) \in R$ and $C_1 \downarrow$, then $C_2 \downarrow$;
4. if $(C_1, f, C_2) \in R$ and $C_2 \downarrow$, then $C_1 \downarrow$.

$\mathcal{E}_1, \mathcal{E}_2$ are rooted branching history-preserving (hp-)bisimilar and are written $\mathcal{E}_1 \approx_{rbhp} \mathcal{E}_2$ if there exists rooted a branching hp-bisimulation $R$ such that $(\emptyset, \emptyset, \emptyset) \in R$.

A rooted branching hereditary history-preserving (hhp-)bisimulation is a downward closed rooted branching hhp-bisimulation. $\mathcal{E}_1, \mathcal{E}_2$ are rooted branching hereditary history-preserving (hhp-)bisimilar and are written as $\mathcal{E}_1 \approx_{rbhhp} \mathcal{E}_2$.

The axioms and transition rules of $\text{APTC}_\tau$ are shown in Table 2.7 and Table 2.8.

**Theorem 2.27** (Soundness of $APTC_\tau$ with guarded linear recursion). *Let $x$ and $y$ be $APTC_\tau$ with guarded linear recursion terms. If $APTC_\tau$ with guarded linear recursion $\vdash x = y$, then*

1. $x \approx_{rbs} y$;
2. $x \approx_{rbp} y$;
3. $x \approx_{rbhp} y$.

**Theorem 2.28** (Soundness of $CFAR$). *$CFAR$ is sound modulo rooted branching truly concurrent bisimulation equivalences $\approx_{rbs}$, $\approx_{rbp}$, and $\approx_{rbhp}$.*

**Theorem 2.29** (Completeness of $APTC_\tau$ with guarded linear recursion and $CFAR$). *Let $p$ and $q$ be closed $APTC_\tau$ with guarded linear recursion and $CFAR$ terms, then,*

1. *if $p \approx_{rbs} q$, then $p = q$;*
2. *if $p \approx_{rbp} q$, then $p = q$;*
3. *if $p \approx_{rbhp} q$, then $p = q$.*

**TABLE 2.7** Axioms of $APTC_\tau$.

| No. | Axiom |
|---|---|
| $B1$ | $e \cdot \tau = e$ |
| $B2$ | $e \cdot (\tau \cdot (x+y) + x) = e \cdot (x+y)$ |
| $B3$ | $x \parallel \tau = x$ |
| $TI1$ | $e \notin I \quad \tau_I(e) = e$ |
| $TI2$ | $e \in I \quad \tau_I(e) = \tau$ |
| $TI3$ | $\tau_I(\delta) = \delta$ |
| $TI4$ | $\tau_I(x+y) = \tau_I(x) + \tau_I(y)$ |
| $TI5$ | $\tau_I(x \cdot y) = \tau_I(x) \cdot \tau_I(y)$ |
| $TI6$ | $\tau_I(x \parallel y) = \tau_I(x) \parallel \tau_I(y)$ |
| $CFAR$ | If $X$ is in a cluster for $I$ with exits $\{(a_{11} \parallel \cdots \parallel a_{1i})Y_1, \cdots, (a_{m1} \parallel \cdots \parallel a_{mi})Y_m, b_{11} \parallel \cdots \parallel b_{1j}, \cdots, b_{n1} \parallel \cdots \parallel b_{nj}\}$, then $\tau \cdot \tau_I(\langle X\|E\rangle) = \tau \cdot \tau_I((a_{11} \parallel \cdots \parallel a_{1i})\langle Y_1\|E\rangle + \cdots + (a_{m1} \parallel \cdots \parallel a_{mi})\langle Y_m\|E\rangle + b_{11} \parallel \cdots \parallel b_{1j} + \cdots + b_{n1} \parallel \cdots \parallel b_{nj})$ |

**TABLE 2.8** Transition rule of $APTC_\tau$.

$$\frac{}{\tau \xrightarrow{\tau} \surd}$$

$$\frac{x \xrightarrow{e} \surd}{\tau_I(x) \xrightarrow{e} \surd}\ e \notin I \qquad \frac{x \xrightarrow{e} x'}{\tau_I(x) \xrightarrow{e} \tau_I(x')}\ e \notin I$$

$$\frac{x \xrightarrow{e} \surd}{\tau_I(x) \xrightarrow{\tau} \surd}\ e \in I \qquad \frac{x \xrightarrow{e} x'}{\tau_I(x) \xrightarrow{\tau} \tau_I(x')}\ e \in I$$

**TABLE 2.9** Transition rules of left parallel operator $\leftmerge$.

$$\frac{x \xrightarrow{e_1} \surd \quad y \xrightarrow{e_2} \surd \quad (e_1 \le e_2)}{x \leftmerge y \xrightarrow{\{e_1,e_2\}} \surd} \qquad \frac{x \xrightarrow{e_1} x' \quad y \xrightarrow{e_2} \surd \quad (e_1 \le e_2)}{x \leftmerge y \xrightarrow{\{e_1,e_2\}} x'}$$

$$\frac{x \xrightarrow{e_1} \surd \quad y \xrightarrow{e_2} y' \quad (e_1 \le e_2)}{x \leftmerge y \xrightarrow{\{e_1,e_2\}} y'} \qquad \frac{x \xrightarrow{e_1} x' \quad y \xrightarrow{e_2} y' \quad (e_1 \le e_2)}{x \leftmerge y \xrightarrow{\{e_1,e_2\}} x' \between y'}$$

### 2.5.5 Axiomatization for hhp-bisimilarity

Since hhp-bisimilarity is a downward closed hp-bisimilarity and can be downward closed to a single atomic event, this implies bisimilarity. As Moller [15] has proven, there is no finite sound and complete axiomatization for parallelism $\parallel$ modulo bisimulation equivalence, so there is no finite sound and complete axiomatization for parallelism $\parallel$ modulo hhp-bisimulation equivalence either. Inspired by the way of left merge to modeling the full merge for bisimilarity, we introduce a left parallel composition $\leftmerge$ to model the full parallelism $\parallel$ for hhp-bisimilarity.

In the following section, we add left parallel composition $\leftmerge$ to the whole theory. As the resulting theory is similar to the former, we only list the significant differences, and all proofs of the conclusions are left to the reader.

The transition rules of left parallel composition $\leftmerge$ are shown in Table 2.9. With a little modification, we extend the causal order relation $\le$ on $\mathbb{E}$ to include the original partial order (denoted by $<$) and concurrency (denoted by $=$).

The new axioms for parallelism are listed in Table 2.10.

**Definition 2.30** (Basic terms of $APTC$ with left parallel composition). The set of basic terms of $APTC$, $\mathcal{B}(APTC)$, is inductively defined as follows:

1. $\mathbb{E} \subset \mathcal{B}(APTC)$;
2. if $e \in \mathbb{E}, t \in \mathcal{B}(APTC)$, then $e \cdot t \in \mathcal{B}(APTC)$;
3. if $t, s \in \mathcal{B}(APTC)$, then $t + s \in \mathcal{B}(APTC)$;
4. if $t, s \in \mathcal{B}(APTC)$, then $t \leftmerge s \in \mathcal{B}(APTC)$.

**TABLE 2.10 Axioms of parallelism with left parallel composition.**

| No. | Axiom |
|---|---|
| $A6$ | $x + \delta = x$ |
| $A7$ | $\delta \cdot x = \delta$ |
| $P1$ | $x \between y = x \parallel y + x \mid y$ |
| $P2$ | $x \parallel y = y \parallel x$ |
| $P3$ | $(x \parallel y) \parallel z = x \parallel (y \parallel z)$ |
| $P4$ | $x \parallel y = x \llfloor y + y \llfloor x$ |
| $P5$ | $(e_1 \leq e_2) \quad e_1 \llfloor (e_2 \cdot y) = (e_1 \llfloor e_2) \cdot y$ |
| $P6$ | $(e_1 \leq e_2) \quad (e_1 \cdot x) \llfloor e_2 = (e_1 \llfloor e_2) \cdot x$ |
| $P7$ | $(e_1 \leq e_2) \quad (e_1 \cdot x) \llfloor (e_2 \cdot y) = (e_1 \llfloor e_2) \cdot (x \between y)$ |
| $P8$ | $(x + y) \llfloor z = (x \llfloor z) + (y \llfloor z)$ |
| $P9$ | $\delta \llfloor x = \delta$ |
| $C10$ | $e_1 \mid e_2 = \gamma(e_1, e_2)$ |
| $C11$ | $e_1 \mid (e_2 \cdot y) = \gamma(e_1, e_2) \cdot y$ |
| $C12$ | $(e_1 \cdot x) \mid e_2 = \gamma(e_1, e_2) \cdot x$ |
| $C13$ | $(e_1 \cdot x) \mid (e_2 \cdot y) = \gamma(e_1, e_2) \cdot (x \between y)$ |
| $C14$ | $(x + y) \mid z = (x \mid z) + (y \mid z)$ |
| $C15$ | $x \mid (y + z) = (x \mid y) + (x \mid z)$ |
| $C16$ | $\delta \mid x = \delta$ |
| $C17$ | $x \mid \delta = \delta$ |
| $CE18$ | $\Theta(e) = e$ |
| $CE19$ | $\Theta(\delta) = \delta$ |
| $CE20$ | $\Theta(x + y) = \Theta(x) + \Theta(y)$ |
| $CE21$ | $\Theta(x \cdot y) = \Theta(x) \cdot \Theta(y)$ |
| $CE22$ | $\Theta(x \llfloor y) = ((\Theta(x) \triangleleft y) \llfloor y) + ((\Theta(y) \triangleleft x) \llfloor x)$ |
| $CE23$ | $\Theta(x \mid y) = ((\Theta(x) \triangleleft y) \mid y) + ((\Theta(y) \triangleleft x) \mid x)$ |
| $U24$ | $(\sharp(e_1, e_2)) \quad e_1 \triangleleft e_2 = \tau$ |
| $U25$ | $(\sharp(e_1, e_2), e_2 \leq e_3) \quad e_1 \triangleleft e_3 = \tau$ |
| $U26$ | $(\sharp(e_1, e_2), e_2 \leq e_3) \quad e3 \triangleleft e_1 = \tau$ |
| $U27$ | $e \triangleleft \delta = e$ |
| $U28$ | $\delta \triangleleft e = \delta$ |
| $U29$ | $(x + y) \triangleleft z = (x \triangleleft z) + (y \triangleleft z)$ |
| $U30$ | $(x \cdot y) \triangleleft z = (x \triangleleft z) \cdot (y \triangleleft z)$ |
| $U31$ | $(x \llfloor y) \triangleleft z = (x \triangleleft z) \llfloor (y \triangleleft z)$ |
| $U32$ | $(x \mid y) \triangleleft z = (x \triangleleft z) \mid (y \triangleleft z)$ |
| $U33$ | $x \triangleleft (y + z) = (x \triangleleft y) \triangleleft z$ |
| $U34$ | $x \triangleleft (y \cdot z) = (x \triangleleft y) \triangleleft z$ |
| $U35$ | $x \triangleleft (y \llfloor z) = (x \triangleleft y) \triangleleft z$ |
| $U36$ | $x \triangleleft (y \mid z) = (x \triangleleft y) \triangleleft z$ |

**Theorem 2.31** (Generalization of the algebra for left parallelism with respect to $BATC$). *The algebra for left parallelism is a generalization of $BATC$.*

**Theorem 2.32** (Congruence theorem of $APTC$ with left parallel composition). *Truly concurrent bisimulation equivalences $\sim_p$, $\sim_s$, $\sim_{hp}$, and $\sim_{hhp}$ are all congruences with respect to $APTC$ with left parallel composition.*

**TABLE 2.11** Axioms of encapsulation operator with left parallel composition.

| No. | Axiom |
|---|---|
| $D1$ | $e \notin H \quad \partial_H(e) = e$ |
| $D2$ | $e \in H \quad \partial_H(e) = \delta$ |
| $D3$ | $\partial_H(\delta) = \delta$ |
| $D4$ | $\partial_H(x + y) = \partial_H(x) + \partial_H(y)$ |
| $D5$ | $\partial_H(x \cdot y) = \partial_H(x) \cdot \partial_H(y)$ |
| $D6$ | $\partial_H(x \llcorner y) = \partial_H(x) \llcorner \partial_H(y)$ |

**Theorem 2.33** (Elimination theorem of parallelism with left parallel composition). *Let $p$ be a closed $APTC$ with left parallel composition term. Then, there is a basic $APTC$ term $q$ such that $APTC \vdash p = q$.*

**Theorem 2.34** (Soundness of parallelism with left parallel composition modulo truly concurrent bisimulation equivalences). *Let $x$ and $y$ be $APTC$ with left parallel composition terms. If $APTC \vdash x = y$, then*

1. $x \sim_s y$;
2. $x \sim_p y$;
3. $x \sim_{hp} y$;
4. $x \sim_{hhp} y$.

**Theorem 2.35** (Completeness of parallelism with left parallel composition modulo truly concurrent bisimulation equivalences). *Let $x$ and $y$ be $APTC$ terms.*

1. *If $x \sim_s y$, then $APTC \vdash x = y$;*
2. *if $x \sim_p y$, then $APTC \vdash x = y$;*
3. *if $x \sim_{hp} y$, then $APTC \vdash x = y$;*
4. *if $x \sim_{hhp} y$, then $APTC \vdash x = y$.*

The transition rules of the encapsulation operator are the same, and the its axioms are shown in Table 2.11.

**Theorem 2.36** (Conservativity of $APTC$ with respect to the algebra for parallelism with left parallel composition). *$APTC$ is a conservative extension of the algebra for parallelism with left parallel composition.*

**Theorem 2.37** (Congruence theorem of encapsulation operator $\partial_H$). *Truly concurrent bisimulation equivalences $\sim_p$, $\sim_s$, $\sim_{hp}$, and $\sim_{hhp}$ are all congruences with respect to encapsulation operator $\partial_H$.*

**Theorem 2.38** (Elimination theorem of $APTC$). *Let $p$ be a closed $APTC$ term including the encapsulation operator $\partial_H$. Then, there is a basic $APTC$ term $q$ such that $APTC \vdash p = q$.*

**Theorem 2.39** (Soundness of $APTC$ modulo truly concurrent bisimulation equivalences). *Let $x$ and $y$ be $APTC$ terms including encapsulation operator $\partial_H$. If $APTC \vdash x = y$, then*

1. $x \sim_s y$;
2. $x \sim_p y$;
3. $x \sim_{hp} y$;
4. $x \sim_{hhp} y$.

**Theorem 2.40** (Completeness of $APTC$ modulo truly concurrent bisimulation equivalences). *Let $p$ and $q$ be closed $APTC$ terms including encapsulation operator $\partial_H$,*

1. *if $p \sim_s q$, then $p = q$;*
2. *if $p \sim_p q$, then $p = q$;*
3. *if $p \sim_{hp} q$, then $p = q$;*
4. *if $p \sim_{hhp} q$, then $p = q$.*

**Definition 2.41** (Recursive specification). A recursive specification is a finite set of recursive equations

$$X_1 = t_1(X_1, \cdots, X_n)$$

$$\cdots$$

$$X_n = t_n(X_1, \cdots, X_n)$$

where the left-hand sides of $X_i$ are called recursion variables, and the right-hand sides $t_i(X_1, \cdots, X_n)$ are process terms in $APTC$ with possible occurrences of the recursion variables $X_1, \cdots, X_n$.

**Definition 2.42** (Solution). Processes $p_1, \cdots, p_n$ are a solution for a recursive specification $\{X_i = t_i(X_1, \cdots, X_n)|i \in \{1, \cdots, n\}\}$ (with respect to truly concurrent bisimulation equivalences $\sim_s(\sim_p, \sim_{hp}, \sim_{hhp})$) if $p_i \sim_s (\sim_p, \sim_{hp}, \sim hhp)t_i(p_1, \cdots, p_n)$ for $i \in \{1, \cdots, n\}$.

**Definition 2.43** (Guarded recursive specification). A recursive specification

$$X_1 = t_1(X_1, \cdots, X_n)$$

$$\cdots$$

$$X_n = t_n(X_1, \cdots, X_n)$$

is guarded if the right-hand sides of its recursive equations can be adapted to the form by applications of the axioms in $APTC$ and replacing recursion variables by the right-hand sides of their recursive equations,

$$(a_{11} \between \cdots \between a_{1i_1}) \cdot s_1(X_1, \cdots, X_n) + \cdots + (a_{k1} \between \cdots \between a_{ki_k}) \cdot s_k(X_1, \cdots, X_n) + (b_{11} \between \cdots \between b_{1j_1}) + \cdots + (b_{1j_1} \between \cdots \between b_{lj_l})$$

where $a_{11}, \cdots, a_{1i_1}, a_{k1}, \cdots, a_{ki_k}, b_{11}, \cdots, b_{1j_1}, b_{1j_1}, \cdots, b_{lj_l} \in \mathbb{E}$, and the sum above is allowed to be empty, in which case it represents the deadlock $\delta$.

**Definition 2.44** (Linear recursive specification). A recursive specification is linear if its recursive equations are of the form

$$(a_{11} \between \cdots \between a_{1i_1})X_1 + \cdots + (a_{k1} \between \cdots \between a_{ki_k})X_k + (b_{11} \between \cdots \between b_{1j_1}) + \cdots + (b_{1j_1} \between \cdots \between b_{lj_l})$$

where $a_{11}, \cdots, a_{1i_1}, a_{k1}, \cdots, a_{ki_k}, b_{11}, \cdots, b_{1j_1}, b_{1j_1}, \cdots, b_{lj_l} \in \mathbb{E}$, and the sum above is allowed to be empty, in which case it represents the deadlock $\delta$.

**Theorem 2.45** (Conservativity of $APTC$ with guarded recursion). *$APTC$ with guarded recursion is a conservative extension of $APTC$.*

**Theorem 2.46** (Congruence theorem of $APTC$ with guarded recursion). *Truly concurrent bisimulation equivalences $\sim_p$, $\sim_s$, $\sim_{hp}$, and $\sim_{hhp}$ are all congruences with respect to $APTC$ with guarded recursion.*

**Theorem 2.47** (Elimination theorem of $APTC$ with linear recursion). *Each process term in $APTC$ with linear recursion is equal to a process term $\langle X_1|E\rangle$ with $E$ a linear recursive specification.*

**Theorem 2.48** (Soundness of $APTC$ with guarded recursion). *Let $x$ and $y$ be $APTC$ with guarded recursion terms. If $APTC$ with guarded recursion $\vdash x = y$, then*

1. $x \sim_s y$;
2. $x \sim_p y$;
3. $x \sim_{hp} y$;
4. $x \sim_{hhp} y$.

**Theorem 2.49** (Completeness of $APTC$ with linear recursion). *Let $p$ and $q$ be closed $APTC$ with linear recursion terms, then*

1. *if $p \sim_s q$, then $p = q$;*
2. *if $p \sim_p q$, then $p = q$;*
3. *if $p \sim_{hp} q$, then $p = q$;*
4. *if $p \sim_{hhp} q$, then $p = q$.*

**Definition 2.50** (Guarded linear recursive specification). A recursive specification is linear if its recursive equations are of the form

$$(a_{11} \between \cdots \between a_{1i_1})X_1 + \cdots + (a_{k1} \between \cdots \between a_{ki_k})X_k + (b_{11} \between \cdots \between b_{1j_1}) + \cdots + (b_{1j_1} \between \cdots \between b_{lj_l})$$

**TABLE 2.12** Axioms of silent step.

| No. | Axiom |
|---|---|
| $B1$ | $e\cdot\tau=e$ |
| $B2$ | $e\cdot(\tau\cdot(x+y)+x)=e\cdot(x+y)$ |
| $B3$ | $x\leftmerge\tau=x$ |

where $a_{11},\cdots,a_{1i_1},a_{k1},\cdots,a_{ki_k},b_{11},\cdots,b_{1j_1},b_{1j_1},\cdots,b_{lj_l}\in\mathbb{E}\cup\{\tau\}$, and the sum above is allowed to be empty, in which case it represents the deadlock $\delta$.

A linear recursive specification $E$ is guarded if there does not exist an infinite sequence of $\tau$-transitions $\langle X|E\rangle\xrightarrow{\tau}\langle X'|E\rangle\xrightarrow{\tau}\langle X''|E\rangle\xrightarrow{\tau}\cdots$.

The transition rules of $\tau$ are the same, and the axioms of $\tau$ are as Table 2.12 shows.

**Theorem 2.51** (Conservativity of $APTC$ with silent step and guarded linear recursion). *$APTC$ with silent step and guarded linear recursion is a conservative extension of $APTC$ with linear recursion.*

**Theorem 2.52** (Congruence theorem of $APTC$ with silent step and guarded linear recursion). *Rooted branching truly concurrent bisimulation equivalences $\approx_{rbp}$, $\approx_{rbs}$, $\approx_{rbhp}$, and $\approx_{rbhhp}$ are all congruences with respect to $APTC$ with silent step and guarded linear recursion.*

**Theorem 2.53** (Elimination theorem of $APTC$ with silent step and guarded linear recursion). *Each process term in $APTC$ with silent step and guarded linear recursion is equal to a process term $\langle X_1|E\rangle$ with $E$ a guarded linear recursive specification.*

**Theorem 2.54** (Soundness of $APTC$ with silent step and guarded linear recursion). *Let $x$ and $y$ be $APTC$ with silent step and guarded linear recursion terms. If $APTC$ with silent step and guarded linear recursion $\vdash x=y$, then*

1. $x\approx_{rbs} y$;
2. $x\approx_{rbp} y$;
3. $x\approx_{rbhp} y$;
4. $x\approx_{rbhhp} y$.

**Theorem 2.55** (Completeness of $APTC$ with silent step and guarded linear recursion). *Let $p$ and $q$ be closed $APTC$ with silent step and guarded linear recursion terms, then*

1. *if* $p\approx_{rbs} q$, *then* $p=q$;
2. *if* $p\approx_{rbp} q$, *then* $p=q$;
3. *if* $p\approx_{rbhp} q$, *then* $p=q$;
4. *if* $p\approx_{rbhhp} q$, *then* $p=q$.

The transition rules of $\tau_I$ are the same, and the axioms are shown in Table 2.13.

**Theorem 2.56** (Conservativity of $APTC_\tau$ with guarded linear recursion). *$APTC_\tau$ with guarded linear recursion is a conservative extension of $APTC$ with silent step and guarded linear recursion.*

**Theorem 2.57** (Congruence theorem of $APTC_\tau$ with guarded linear recursion). *Rooted branching truly concurrent bisimulation equivalences $\approx_{rbp}$, $\approx_{rbs}$, $\approx_{rbhp}$, and $\approx_{rbhhp}$ are all congruences with respect to $APTC_\tau$ with guarded linear recursion.*

**Theorem 2.58** (Soundness of $APTC_\tau$ with guarded linear recursion). *Let $x$ and $y$ be $APTC_\tau$ with guarded linear recursion terms. If $APTC_\tau$ with guarded linear recursion $\vdash x=y$, then*

1. $x\approx_{rbs} y$;
2. $x\approx_{rbp} y$;
3. $x\approx_{rbhp} y$;
4. $x\approx_{rbhhp} y$.

**TABLE 2.13** Axioms of abstraction operator.

| No. | Axiom |
|---|---|
| $TI1$ | $e \notin I \quad \tau_I(e) = e$ |
| $TI2$ | $e \in I \quad \tau_I(e) = \tau$ |
| $TI3$ | $\tau_I(\delta) = \delta$ |
| $TI4$ | $\tau_I(x + y) = \tau_I(x) + \tau_I(y)$ |
| $TI5$ | $\tau_I(x \cdot y) = \tau_I(x) \cdot \tau_I(y)$ |
| $TI6$ | $\tau_I(x \between y) = \tau_I(x) \between \tau_I(y)$ |

**TABLE 2.14** Cluster fair abstraction rule.

| No. | Axiom |
|---|---|
| $CFAR$ | If $X$ is in a cluster for $I$ with exits $\{(a_{11}\between\cdots\between a_{1i})Y_1, \cdots, (a_{m1}\between\cdots\between a_{mi})Y_m, b_{11}\between\cdots\between b_{1j}, \cdots, b_{n1}\between\cdots\between b_{nj}\}$, then $\tau \cdot \tau_I(\langle X\|E\rangle) =$ $\tau \cdot \tau_I((a_{11}\between\cdots\between a_{1i})\langle Y_1\|E\rangle + \cdots + (a_{m1}\between\cdots\between a_{mi})\langle Y_m\|E\rangle + b_{11}\between\cdots\between b_{1j} + \cdots + b_{n1}\between\cdots\between b_{nj})$ |

**TABLE 2.15** Transition rule of the shadow constant.

| |
|---|
| $\overline{Ⓢ \rightarrow \surd}$ |

**Definition 2.59** (Cluster). Let $E$ be a guarded linear recursive specification, and $I \subseteq \mathbb{E}$. Two recursion variables $X$ and $Y$ in $E$ are in the same cluster for $I$ iff there exist sequences of transitions $\langle X|E\rangle \xrightarrow{\{b_{11},\cdots,b_{1i}\}} \cdots \xrightarrow{\{b_{m1},\cdots,b_{mi}\}} \langle Y|E\rangle$ and $\langle Y|E\rangle \xrightarrow{\{c_{11},\cdots,c_{1j}\}} \cdots \xrightarrow{\{c_{n1},\cdots,c_{nj}\}} \langle X|E\rangle$, where $b_{11}, \cdots, b_{mi}, c_{11}, \cdots, c_{nj} \in I \cup \{\tau\}$.

$a_1 \between \cdots \between a_k$ or $(a_1 \between \cdots \between a_k)X$ is an exit for the cluster $C$ iff: (1) $a_1 \between \cdots \between a_k$ or $(a_1 \between \cdots \between a_k)X$ is a summand at the right-hand side of the recursive equation for a recursion variable in $C$, and (2) in the case of $(a_1 \between \cdots \between a_k)X$, either $a_l \notin I \cup \{\tau\} (l \in \{1, 2, \cdots, k\})$ or $X \notin C$. The CFAR please see Table 2.14.

**Theorem 2.60** (Soundness of $CFAR$). *$CFAR$ is sound modulo rooted branching truly concurrent bisimulation equivalences $\approx_{rbs}$, $\approx_{rbp}$, $\approx_{rbhp}$, and $\approx_{rbhhp}$.*

**Theorem 2.61** (Completeness of $APTC_\tau$ with guarded linear recursion and $CFAR$). *Let $p$ and $q$ be closed $APTC_\tau$ with guarded linear recursion and $CFAR$ terms, then*

1. *if $p \approx_{rbs} q$, then $p = q$;*
2. *if $p \approx_{rbp} q$, then $p = q$;*
3. *if $p \approx_{rbhp} q$, then $p = q$;*
4. *if $p \approx_{rbhhp} q$, then $p = q$.*

### 2.5.6 Placeholder

We introduce a constant called shadow constant Ⓢ to act for the placeholder that we always used to deal with entanglement in quantum process algebra. The transition rule of the shadow constant Ⓢ is shown in Table 2.15. The rule says that Ⓢ can terminate successfully without executing any action.

We need to adjust the definition of guarded linear recursive specification to the following one.

**Definition 2.62** (Guarded linear recursive specification). A linear recursive specification $E$ is guarded if there does not exist an infinite sequence of $\tau$-transitions $\langle X|E\rangle \xrightarrow{\tau} \langle X'|E\rangle \xrightarrow{\tau} \langle X''|E\rangle \xrightarrow{\tau} \cdots$, and there does not exist an infinite sequence of Ⓢ-transitions $\langle X|E\rangle \rightarrow \langle X'|E\rangle \rightarrow \langle X''|E\rangle \rightarrow \cdots$.

**Theorem 2.63** (Conservativity of $APTC$ with respect to the shadow constant). *$APTC_\tau$ with guarded linear recursion and shadow constant is a conservative extension of $APTC_\tau$ with guarded linear recursion.*

**TABLE 2.16** Axioms of shadow constant.

| No. | Axiom |
|---|---|
| $SC1$ | $\circledS \cdot x = x$ |
| $SC2$ | $x \cdot \circledS = x$ |
| $SC3$ | $\circledS^e \parallel e = e$ |
| $SC4$ | $e \parallel (\circledS^e \cdot y) = e \cdot y$ |
| $SC5$ | $\circledS^e \parallel (e \cdot y) = e \cdot y$ |
| $SC6$ | $(e \cdot x) \parallel \circledS^e = e \cdot x$ |
| $SC7$ | $(\circledS^e \cdot x) \parallel e = e \cdot x$ |
| $SC8$ | $(e \cdot x) \parallel (\circledS^e \cdot y) = e \cdot (x \between y)$ |
| $SC9$ | $(\circledS^e \cdot x) \parallel (e \cdot y) = e \cdot (x \between y)$ |

We design the axioms for the shadow constant $\circledS$ in Table 2.16. For $\circledS^e_i$, we add superscript $e$ to denote $\circledS$ is belonging to $e$ and subscript $i$ to denote that it is the $i$th shadow of $e$. Also, we extend the set $\mathbb{E}$ to the set $\mathbb{E} \cup \{\tau\} \cup \{\delta\} \cup \{\circledS^e_i\}$.

The mismatch of action and its shadows in parallelism will cause deadlock, that is, $e \parallel \circledS^{e'} = \delta$ with $e \neq e'$. We must make all shadows $\circledS^e_i$ distinct to ensure $f$ in hp-bisimulation is an isomorphism.

**Theorem 2.64** (Soundness of the shadow constant). *Let $x$ and $y$ be $APTC_\tau$ with guarded linear recursion and the shadow constant terms. If $APTC_\tau$ with guarded linear recursion and the shadow constant $\vdash x = y$, then*

1. $x \approx_{rbs} y$;
2. $x \approx_{rbp} y$;
3. $x \approx_{rbhp} y$.

**Theorem 2.65** (Completeness of the shadow constant). *Let $p$ and $q$ be closed $APTC_\tau$ with guarded linear recursion and $CFAR$ and the shadow constant terms, then*

1. *if $p \approx_{rbs} q$, then $p = q$;*
2. *if $p \approx_{rbp} q$, then $p = q$;*
3. *if $p \approx_{rbhp} q$, then $p = q$.*

With the shadow constant, we have

$$\begin{aligned}\partial_H((a \cdot r_b) \between w_b) &= \partial_H((a \cdot r_b) \between (\circledS^a_1 \cdot w_b)) \\ &= a \cdot c_b\end{aligned}$$

with $H = \{r_b, w_b\}$ and $\gamma(r_b, w_b) \triangleq c_b$.

Also, we see the following example:

$$\begin{aligned}a \between b &= a \parallel b + a \mid b \\ &= a \parallel b + a \parallel b + a \parallel b + a \mid b \\ &= a \parallel (\circledS^a_1 \cdot b) + (\circledS^b_1 \cdot a) \parallel b + a \parallel b + a \mid b \\ &= (a \parallel \circledS^a_1) \cdot b + (\circledS^b_1 \parallel b) \cdot a + a \parallel b + a \mid b \\ &= a \cdot b + b \cdot a + a \parallel b + a \mid b\end{aligned}$$

What do we see? Yes, the parallelism contains both interleaving and true concurrency. This may be why true concurrency is called ***true** concurrency*.

## 2.6 Forward–reverse truly concurrent bisimulations

**Definition 2.66** (Forward–reverse (FR) pomset transitions and forward–reverse (FR) step). Let $\mathcal{E}$ be a PES and let $C \in \mathcal{C}(\mathcal{E})$, $\emptyset \neq X \subseteq \mathbb{E}$, $\mathcal{K} \subseteq \mathbb{N}$, and $X[\mathcal{K}]$ denotes that for each $e \in X$, there is $e[m] \in X[\mathcal{K}]$, where $(m \in \mathcal{K})$, which is called the past of $e$, and we extend $\mathbb{E}$ to $\mathbb{E} \cup \{\tau\} \cup \mathbb{E}[\mathcal{K}]$. If $C \cap X[\mathcal{K}] = \emptyset$ and $C' = C \cup X[\mathcal{K}]$, $X \in \mathcal{C}(\mathcal{E})$, then $C \xrightarrow{X} C'$ is called a forward pomset transition from $C$ to $C'$, and $C' \xrightarrow{X[\mathcal{K}]} C$ is called a reverse pomset transition from $C'$ to $C$. When the events in $X$ are pairwise concurrent, we say that $C \xrightarrow{X} C'$ is a forward step and $C' \xrightarrow{X[\mathcal{K}]} C$ is a reverse step.

**Definition 2.67** (Weak forward–reverse (FR) pomset transitions and weak forward–reverse (FR) step). Let $\mathcal{E}$ be a PES and let $C \in \mathcal{C}(\mathcal{E})$, and $\emptyset \neq X \subseteq \hat{\mathbb{E}}$, $\mathcal{K} \subseteq \mathbb{N}$, and $X[\mathcal{K}]$ denotes that for each $e \in X$, there is $e[m] \in X[\mathcal{K}]$, where $(m \in \mathcal{K})$, which is called the past of $e$. If $C \cap X[\mathcal{K}] = \emptyset$ and $\hat{C'} = \hat{C} \cup X[\mathcal{K}]$, $X \in \mathcal{C}(\mathcal{E})$, then $C \xRightarrow{X} C'$ is called a weak forward pomset transition from $C$ to $C'$, where we define $\xRightarrow{e} \triangleq \xrightarrow{\tau^*}\xrightarrow{e}\xrightarrow{\tau^*}$ and $\xRightarrow{X} \triangleq \xrightarrow{\tau^*}\xrightarrow{e}\xrightarrow{\tau^*}$, for every $e \in X$. Also, $C' \xRightarrow{X[\mathcal{K}]} C$ is called a weak reverse pomset transition from $C'$ to $C$, where we define $\xRightarrow{e[m]} \triangleq \xrightarrow{\tau^*}\xrightarrow{e[m]}\xrightarrow{\tau^*}$, $\xRightarrow{X[\mathcal{K}]} \triangleq \xrightarrow{\tau^*}\xrightarrow{e[m]}\xrightarrow{\tau^*}$, for every $e \in X$ and $m \in \mathcal{K}$. When the events in $X$ are pairwise concurrent, we say that $C \xRightarrow{X} C'$ is a weak forward step and $C' \xRightarrow{X[\mathcal{K}]} C$ is a weak reverse step.

We will also suppose that all the PESs in this chapter are image finite, that is, for any PES $\mathcal{E}$ and $C \in \mathcal{C}(\mathcal{E})$, and $a \in \Lambda$, $\{e \in \mathbb{E} | C \xrightarrow{e} C' \wedge \lambda(e) = a\}$ and $\{e \in \hat{\mathbb{E}} | C \xRightarrow{e} C' \wedge \lambda(e) = a\}$, and $a \in \Lambda$, $\{e \in \mathbb{E} | C' \xrightarrow{e[m]} C \wedge \lambda(e) = a\}$ and $\{e \in \hat{\mathbb{E}} | C' \xRightarrow{e[m]} C \wedge \lambda(e) = a\}$ are finite.

**Definition 2.68** (Forward–reverse (FR) pomset, step bisimulation). Let $\mathcal{E}_1$, $\mathcal{E}_2$ be PESs. An FR pomset bisimulation is a relation $R \subseteq \mathcal{C}(\mathcal{E}_1) \times \mathcal{C}(\mathcal{E}_2)$, such that (1) if $(C_1, C_2) \in R$, and $C_1 \xrightarrow{X_1} C_1'$ then $C_2 \xrightarrow{X_2} C_2'$, with $X_1 \subseteq \mathbb{E}_1$, $X_2 \subseteq \mathbb{E}_2$, $X_1 \sim X_2$ and $(C_1', C_2') \in R$, and vice versa; (2) if $(C_1', C_2') \in R$, and $C_1' \xrightarrow{X_1[\mathcal{K}_1]} C_1$ then $C_2' \xrightarrow{X_2[\mathcal{K}_2]} C_2$, with $X_1 \subseteq \mathbb{E}_1$, $X_2 \subseteq \mathbb{E}_2$, $\mathcal{K}_1, \mathcal{K}_2 \subseteq \mathbb{N}$, $X_1 \sim X_2$ and $(C_1, C_2) \in R$, and vice versa. We say that $\mathcal{E}_1$, $\mathcal{E}_2$ are FR pomset bisimilar, written $\mathcal{E}_1 \sim_p^{fr} \mathcal{E}_2$, if there exists an FR pomset bisimulation $R$, such that $(\emptyset, \emptyset) \in R$. By replacing FR pomset transitions with FR steps, we can obtain the definition of FR step bisimulation. When PESs $\mathcal{E}_1$ and $\mathcal{E}_2$ are FR step bisimilar, we write $\mathcal{E}_1 \sim_s^{fr} \mathcal{E}_2$.

**Definition 2.69** (Weak forward–reverse (FR) pomset, step bisimulation). Let $\mathcal{E}_1$, $\mathcal{E}_2$ be PESs. A weak FR pomset bisimulation is a relation $R \subseteq \mathcal{C}(\mathcal{E}_1) \times \mathcal{C}(\mathcal{E}_2)$, such that (1) if $(C_1, C_2) \in R$, and $C_1 \xRightarrow{X_1} C_1'$, then $C_2 \xRightarrow{X_2} C_2'$, with $X_1 \subseteq \hat{\mathbb{E}}_1$, $X_2 \subseteq \hat{\mathbb{E}}_2$, $X_1 \sim X_2$ and $(C_1', C_2') \in R$, and vice versa; (2) if $(C_1', C_2') \in R$, and $C_1' \xRightarrow{X_1[\mathcal{K}_1]} C_1$, then $C_2' \xRightarrow{X_2[\mathcal{K}_2]} C_2$, with $X_1 \subseteq \hat{\mathbb{E}}_1$, $X_2 \subseteq \hat{\mathbb{E}}_2$, $\mathcal{K}_1, \mathcal{K}_2 \subseteq \mathbb{N}$, $X_1 \sim X_2$ and $(C_1, C_2) \in R$, and vice versa. We say that $\mathcal{E}_1$, $\mathcal{E}_2$ are weak FR pomset bisimilar, written $\mathcal{E}_1 \approx_p^{fr} \mathcal{E}_2$, if there exists a weak FR pomset bisimulation $R$, such that $(\emptyset, \emptyset) \in R$. By replacing weak FR pomset transitions with weak FR steps, we can obtain the definition of weak FR step bisimulation. When PESs $\mathcal{E}_1$ and $\mathcal{E}_2$ are weak FR step bisimilar, we write $\mathcal{E}_1 \approx_s^{fr} \mathcal{E}_2$.

**Definition 2.70** (Forward–reverse (FR) (hereditary) history-preserving bisimulation). An FR history-preserving (hp-) bisimulation is a posetal relation $R \subseteq \mathcal{C}(\mathcal{E}_1)\overline{\times}\mathcal{C}(\mathcal{E}_2)$ such that (1) if $(C_1, f, C_2) \in R$, and $C_1 \xrightarrow{e_1} C_1'$, then $C_2 \xrightarrow{e_2} C_2'$, with $(C_1', f[e_1 \mapsto e_2], C_2') \in R$, and vice versa, (2) if $(C_1', f', C_2') \in R$, and $C_1' \xrightarrow{e_1[m]} C_1$, then $C_2' \xrightarrow{e_2[n]} C_2$, with $(C_1, f'[e_1[m] \mapsto e_2[n]], C_2) \in R$, and vice versa. $\mathcal{E}_1, \mathcal{E}_2$ are FR history-preserving (hp-) bisimilar and are written $\mathcal{E}_1 \sim_{hp}^{fr} \mathcal{E}_2$ if there exists an FR hp-bisimulation $R$ such that $(\emptyset, \emptyset, \emptyset) \in R$.

An FR hereditary history-preserving (hhp-)bisimulation is a downward closed FR hp-bisimulation. $\mathcal{E}_1, \mathcal{E}_2$ are FR hereditary history-preserving (hhp-)bisimilar and are written $\mathcal{E}_1 \sim_{hhp}^{fr} \mathcal{E}_2$.

**Definition 2.71** (Weak forward–reverse (FR) (hereditary) history-preserving bisimulation). A weak FR history-preserving (hp-) bisimulation is a weakly posetal relation $R \subseteq \mathcal{C}(\mathcal{E}_1)\overline{\times}\mathcal{C}(\mathcal{E}_2)$ such that (1) if $(C_1, f, C_2) \in R$, and $C_1 \xRightarrow{e_1} C_1'$, then $C_2 \xRightarrow{e_2} C_2'$, with $(C_1', f[e_1 \mapsto e_2], C_2') \in R$, and vice versa, (2) if $(C_1', f', C_2') \in R$, and $C_1' \xRightarrow{e_1[m]} C_1$, then $C_2' \xRightarrow{e_2[n]} C_2$, with $(C_1, f'[e_1[m] \mapsto e_2[n]], C_2) \in R$, and vice versa. $\mathcal{E}_1, \mathcal{E}_2$ are weak FR history-preserving (hp-) bisimilar and are written $\mathcal{E}_1 \approx_{hp}^{fr} \mathcal{E}_2$ if there exists a weak FR hp-bisimulation $R$ such that $(\emptyset, \emptyset, \emptyset) \in R$.

A weak FR hereditary history-preserving (hhp-) bisimulation is a downward closed weak FR hp-bisimulation. $\mathcal{E}_1, \mathcal{E}_2$ are weak FR hereditary history-preserving (hhp-) bisimilar and are written $\mathcal{E}_1 \approx_{hhp}^{fr} \mathcal{E}_2$.

**Definition 2.72** (Branching forward–reverse pomset, step bisimulation). Assume a special termination predicate $\downarrow$, and let $\surd$ represent a state with $\surd \downarrow$. Let $\mathcal{E}_1$, $\mathcal{E}_2$ be PESs. A branching FR pomset bisimulation is a relation $R \subseteq \mathcal{C}(\mathcal{E}_1) \times \mathcal{C}(\mathcal{E}_2)$, such that:

1. if $(C_1, C_2) \in R$, and $C_1 \xrightarrow{X} C_1'$, then
   - either $X \equiv \tau^*$, and $(C_1', C_2) \in R$;
   - or there is a sequence of (zero or more) $\tau$-transitions $C_2 \xrightarrow{\tau^*} C_2^0$, such that $(C_1, C_2^0) \in R$ and $C_2^0 \xRightarrow{X} C_2'$ with $(C_1', C_2') \in R$;
2. if $(C_1, C_2) \in R$, and $C_2 \xrightarrow{X} C_2'$, then
   - either $X \equiv \tau^*$, and $(C_1, C_2') \in R$;
   - or there is a sequence of (zero or more) $\tau$-transitions $C_1 \xrightarrow{\tau^*} C_1^0$, such that $(C_1^0, C_2) \in R$ and $C_1^0 \xRightarrow{X} C_1'$ with $(C_1', C_2') \in R$;
3. if $(C_1, C_2) \in R$ and $C_1 \downarrow$, then there is a sequence of (zero or more) $\tau$-transitions $C_2 \xrightarrow{\tau^*} C_2^0$ such that $(C_1, C_2^0) \in R$ and $C_2^0 \downarrow$;
4. if $(C_1, C_2) \in R$ and $C_2 \downarrow$, then there is a sequence of (zero or more) $\tau$-transitions $C_1 \xrightarrow{\tau^*} C_1^0$ such that $(C_1^0, C_2) \in R$ and $C_1^0 \downarrow$;
5. if $(C_1', C_2') \in R$, and $C_1' \xtwoheadrightarrow{X[\mathcal{K}]} C_1$, then
   - either $X[\mathcal{K}] \equiv \tau^*$, and $(C_1, C_2') \in R$;
   - or there is a sequence of (zero or more) $\tau$-transitions $C_2' \xtwoheadrightarrow{\tau^*} C_2'^0$, such that $(C_1', C_2'^0) \in R$ and $C_2'^0 \xRightarrow{X[\mathcal{K}]} C_2$ with $(C_1, C_2) \in R$;
6. if $(C_1', C_2') \in R$, and $C_2' \xtwoheadrightarrow{X} C_2$, then
   - either $X[\mathcal{K}] \equiv \tau^*$, and $(C_1', C_2) \in R$;
   - or there is a sequence of (zero or more) $\tau$-transitions $C_1' \xtwoheadrightarrow{\tau^*} C_1'^0$, such that $(C_1'^0, C_2') \in R$ and $C_1'^0 \xRightarrow{X[\mathcal{K}]} C_1$ with $(C_1, C_2) \in R$;
7. if $(C_1', C_2') \in R$ and $C_1' \downarrow$, then there is a sequence of (zero or more) $\tau$-transitions $C_2' \xtwoheadrightarrow{\tau^*} C_2'^0$ such that $(C_1', C_2'^0) \in R$ and $C_2'^0 \downarrow$;
8. if $(C_1', C_2') \in R$ and $C_2' \downarrow$, then there is a sequence of (zero or more) $\tau$-transitions $C_1' \xtwoheadrightarrow{\tau^*} C_1'^0$ such that $(C_1'^0, C_2') \in R$ and $C_1'^0 \downarrow$.

We say that $\mathcal{E}_1, \mathcal{E}_2$ are branching FR pomset bisimilar, written $\mathcal{E}_1 \approx_{bp}^{fr} \mathcal{E}_2$, if there exists a branching FR pomset bisimulation $R$, such that $(\emptyset, \emptyset) \in R$.

By replacing FR pomset transitions with FR steps, we can obtain the definition of branching FR step bisimulation. When PESs $\mathcal{E}_1$ and $\mathcal{E}_2$ are branching FR step bisimilar, we write $\mathcal{E}_1 \approx_{bs}^{fr} \mathcal{E}_2$.

**Definition 2.73** (Rooted branching forward–reverse (FR) pomset, step bisimulation). Assume a special termination predicate $\downarrow$, and let $\surd$ represent a state with $\surd \downarrow$. Let $\mathcal{E}_1, \mathcal{E}_2$ be PESs. A rooted branching FR pomset bisimulation is a relation $R \subseteq \mathcal{C}(\mathcal{E}_1) \times \mathcal{C}(\mathcal{E}_2)$, such that:

1. if $(C_1, C_2) \in R$, and $C_1 \xrightarrow{X} C_1'$, then $C_2 \xrightarrow{X} C_2'$ with $C_1' \approx_{bp}^{fr} C_2'$;
2. if $(C_1, C_2) \in R$, and $C_2 \xrightarrow{X} C_2'$, then $C_1 \xrightarrow{X} C_1'$ with $C_1' \approx_{bp}^{fr} C_2'$;
3. if $(C_1', C_2') \in R$, and $C_1' \xtwoheadrightarrow{X[\mathcal{K}]} C_1$, then $C_2' \xtwoheadrightarrow{X[\mathcal{K}]} C_2$ with $C_1 \approx_{bp}^{fr} C_2$;
4. if $(C_1', C_2') \in R$, and $C_2' \xtwoheadrightarrow{X[\mathcal{K}]} C_2$, then $C_1' \xtwoheadrightarrow{X[\mathcal{K}]} C_1$ with $C_1 \approx_{bp}^{fr} C_2$;
5. if $(C_1, C_2) \in R$ and $C_1 \downarrow$, then $C_2 \downarrow$;
6. if $(C_1, C_2) \in R$ and $C_2 \downarrow$, then $C_1 \downarrow$.

We say that $\mathcal{E}_1, \mathcal{E}_2$ are rooted branching FR pomset bisimilar, written $\mathcal{E}_1 \approx_{rbp}^{fr} \mathcal{E}_2$, if there exists a rooted branching FR pomset bisimulation $R$, such that $(\emptyset, \emptyset) \in R$.

By replacing FR pomset transitions with FR steps, we can obtain the definition of rooted branching FR step bisimulation. When PESs $\mathcal{E}_1$ and $\mathcal{E}_2$ are rooted branching FR step bisimilar, we write $\mathcal{E}_1 \approx_{rbs}^{fr} \mathcal{E}_2$.

**Definition 2.74** (Branching forward–reverse (FR) (hereditary) history-preserving bisimulation). Assume a special termination predicate $\downarrow$, and let $\surd$ represent a state with $\surd\downarrow$. A branching FR history-preserving (hp-) bisimulation is a weakly posetal relation $R \subseteq \mathcal{C}(\mathcal{E}_1)\overline{\times}\mathcal{C}(\mathcal{E}_2)$ such that:

1. if $(C_1, f, C_2) \in R$, and $C_1 \xrightarrow{e_1} C_1'$, then
   - either $e_1 \equiv \tau$, and $(C_1', f[e_1 \mapsto \tau^{e_1}], C_2) \in R$;
   - or there is a sequence of (zero or more) $\tau$-transitions $C_2 \xrightarrow{\tau^*} C_2^0$, such that $(C_1, f, C_2^0) \in R$ and $C_2^0 \xrightarrow{e_2} C_2'$ with $(C_1', f[e_1 \mapsto e_2], C_2') \in R$;
2. if $(C_1, f, C_2) \in R$, and $C_2 \xrightarrow{e_2} C_2'$, then
   - either $e_2 \equiv \tau$, and $(C_1, f[e_2 \mapsto \tau^{e_2}], C_2') \in R$;
   - or there is a sequence of (zero or more) $\tau$-transitions $C_1 \xrightarrow{\tau^*} C_1^0$, such that $(C_1^0, f, C_2) \in R$ and $C_1^0 \xrightarrow{e_1} C_1'$ with $(C_1', f[e_2 \mapsto e_1], C_2') \in R$;
3. if $(C_1, f, C_2) \in R$ and $C_1 \downarrow$, then there is a sequence of (zero or more) $\tau$-transitions $C_2 \xrightarrow{\tau^*} C_2^0$ such that $(C_1, f, C_2^0) \in R$ and $C_2^0 \downarrow$;
4. if $(C_1, f, C_2) \in R$ and $C_2 \downarrow$, then there is a sequence of (zero or more) $\tau$-transitions $C_1 \xrightarrow{\tau^*} C_1^0$ such that $(C_1^0, f, C_2) \in R$ and $C_1^0 \downarrow$;
5. if $(C_1', f', C_2') \in R$, and $C_1' \xtwoheadrightarrow{e_1[m]} C_1$, then
   - either $e_1[m] \equiv \tau$, and $(C_1, f'[e_1[m] \mapsto \tau^{e_1[m]}], C_2') \in R$;
   - or there is a sequence of (zero or more) $\tau$-transitions $C_2' \xtwoheadrightarrow{\tau^*} C_2'^0$, such that $(C_1', f', C_2'^0) \in R$ and $C_2'^0 \xtwoheadrightarrow{e_2[n]} C_2$ with $(C_1, f'[e_1[m] \mapsto e_2[n]], C_2) \in R$;
6. if $(C_1', f', C_2') \in R$, and $C_2' \xtwoheadrightarrow{e_2[n]} C_2$, then
   - either $e_2[n] \equiv \tau$, and $(C_1', f'[e_2[n] \mapsto \tau^{e_2[n]}], C_2) \in R$;
   - or there is a sequence of (zero or more) $\tau$-transitions $C_1' \xtwoheadrightarrow{\tau^*} C_1'^0$, such that $(C_1'^0, f', C_2') \in R$ and $C_1'^0 \xtwoheadrightarrow{e_1[m]} C_1$ with $(C_1, f[e_2[n] \mapsto e_1[m]], C_2) \in R$;
7. if $(C_1', f', C_2') \in R$ and $C_1' \downarrow$, then there is a sequence of (zero or more) $\tau$-transitions $C_2' \xtwoheadrightarrow{\tau^*} C_2'^0$ such that $(C_1', f', C_2'^0) \in R$ and $C_2'^0 \downarrow$;
8. if $(C_1', f', C_2') \in R$ and $C_2' \downarrow$, then there is a sequence of (zero or more) $\tau$-transitions $C_1' \xtwoheadrightarrow{\tau^*} C_1'^0$ such that $(C_1'^0, f', C_2') \in R$ and $C_1'^0 \downarrow$.

$\mathcal{E}_1, \mathcal{E}_2$ are branching FR history-preserving (hp-)bisimilar and are written $\mathcal{E}_1 \approx_{bhp}^{fr} \mathcal{E}_2$ if there exists a branching FR hp-bisimulation $R$ such that $(\emptyset, \emptyset, \emptyset) \in R$.

A branching FR hereditary history-preserving (hhp-)bisimulation is a downward closed branching FR hp-bisimulation. $\mathcal{E}_1, \mathcal{E}_2$ are branching FR hereditary history-preserving (hhp-)bisimilar and are written $\mathcal{E}_1 \approx_{bhhp}^{fr} \mathcal{E}_2$.

**Definition 2.75** (Rooted branching forward–reverse (FR) (hereditary) history-preserving bisimulation). Assume a special termination predicate $\downarrow$, and let $\surd$ represent a state with $\surd\downarrow$. A rooted branching FR history-preserving (hp-) bisimulation is a weakly posetal relation $R \subseteq \mathcal{C}(\mathcal{E}_1)\overline{\times}\mathcal{C}(\mathcal{E}_2)$ such that:

1. if $(C_1, f, C_2) \in R$, and $C_1 \xrightarrow{e_1} C_1'$, then $C_2 \xrightarrow{e_2} C_2'$ with $C_1' \approx_{bhp}^{fr} C_2'$;
2. if $(C_1, f, C_2) \in R$, and $C_2 \xrightarrow{e_2} C_2'$, then $C_1 \xrightarrow{e_1} C_1'$ with $C_1' \approx_{bhp}^{fr} C_2'$;
3. if $(C_1', f', C_2') \in R$, and $C_1' \xtwoheadrightarrow{e_1[m]} C_1$, then $C_2' \xtwoheadrightarrow{e_2[n]} C_2$ with $C_1 \approx_{bhp}^{fr} C_2$;
4. if $(C_1', f', C_2') \in R$, and $C_2' \xtwoheadrightarrow{e_2[n]} C_2$, then $C_1' \xtwoheadrightarrow{e_1[m]} C_1$ with $C_1 \approx_{bhp}^{fr} C_2$;
5. if $(C_1, f, C_2) \in R$ and $C_1 \downarrow$, then $C_2 \downarrow$;
6. if $(C_1, f, C_2) \in R$ and $C_2 \downarrow$, then $C_1 \downarrow$.

$\mathcal{E}_1, \mathcal{E}_2$ are rooted branching FR history-preserving (hp-)bisimilar and are written as $\mathcal{E}_1 \approx^{fr}_{rbhp} \mathcal{E}_2$ if there exists a rooted branching FR hp-bisimulation $R$ such that $(\emptyset, \emptyset, \emptyset) \in R$.

A rooted branching FR hereditary history-preserving (hhp-)bisimulation is a downward closed rooted branching FR hp-bisimulation. $\mathcal{E}_1, \mathcal{E}_2$ are rooted branching FR hereditary history-preserving (hhp-)bisimilar and are written as $\mathcal{E}_1 \approx^{fr}_{rbhhp} \mathcal{E}_2$.

Chapter 3

# Reversible calculus

In this chapter, we introduce reversible computing in CTC, which is called RCTC. This chapter is organized as follows. We give the syntax and operational semantics of RCTC in Section 3.1. We discuss the properties of RCTC based on strongly forward–reverse truly concurrent bisimilarities in Section 3.2, and the properties of RCTC based on weakly forward–reverse truly concurrent bisimilarities in Section 3.3.

## 3.1 Syntax and operational semantics

We assume an infinite set $\mathcal{N}$ of (action or event) names, and use $a, b, c, \cdots$ to range over $\mathcal{N}$. We denote by $\overline{\mathcal{N}}$ the set of co-names and let $\overline{a}, \overline{b}, \overline{c}, \cdots$ range over $\overline{\mathcal{N}}$. Then, we set $\mathcal{L} = \mathcal{N} \cup \overline{\mathcal{N}}$ as the set of labels, and use $l, \overline{l}$ to range over $\mathcal{L}$. We extend complementation to $\mathcal{L}$ such that $\overline{\overline{a}} = a$. Let $\tau$ denote the silent step (internal action or event) and define $Act = \mathcal{L} \cup \{\tau\} \cup \mathcal{L}[\mathcal{K}]$ to be the set of actions, $\alpha, \beta$ range over $Act$. Also, $K, L$ are used to stand for subsets of $\mathcal{L}$ and $\overline{L}$ is used for the set of complements of labels in $L$. A relabeling function $f$ is a function from $\mathcal{L}$ to $\mathcal{L}$ such that $f(\overline{l}) = \overline{f(l)}$. By defining $f(\tau) = \tau$, we extend $f$ to $Act$. We write $\mathcal{P}$ for the set of processes. Sometimes, we use $I, J$ to stand for an indexing set, and we write $E_i : i \in I$ for a family of expressions indexed by $I$. $Id_D$ is the identity function or relation over set $D$.

For each process constant schema $A$, a defining equation of the form

$$A \overset{\text{def}}{=} P$$

is assumed, where $P$ is a process.

### 3.1.1 Syntax

We use the Prefix . to model the causality relation $\leq$ in true concurrency, the Summation $+$ to model the conflict relation $\sharp$ in true concurrency, and the Composition $\parallel$ to explicitly model concurrent relation in true concurrency. Also, we follow the conventions of process algebra.

**Definition 3.1** (Syntax). Reversible truly concurrent processes RCTC are defined inductively by the following formation rules:

1. $A \in \mathcal{P}$;
2. $\mathbf{nil} \in \mathcal{P}$;
3. if $P \in \mathcal{P}$, then the Prefix $\alpha.P \in \mathcal{P}$ and $P.\alpha[m] \in \mathcal{P}$, for $\alpha \in Act$ and $m \in \mathcal{K}$;
4. if $P, Q \in \mathcal{P}$, then the Summation $P + Q \in \mathcal{P}$;
5. if $P, Q \in \mathcal{P}$, then the Composition $P \parallel Q \in \mathcal{P}$;
6. if $P \in \mathcal{P}$, then the Prefix $(\alpha_1 \parallel \cdots \parallel \alpha_n).P \in \mathcal{P} \quad (n \in I)$ and $P.(\alpha_1[m] \parallel \cdots \parallel \alpha_n[m]) \in \mathcal{P} \quad (n \in I)$, for $\alpha_1, \cdots, \alpha_n \in Act$ and $m \in \mathcal{K}$;
7. if $P \in \mathcal{P}$, then the Restriction $P \setminus L \in \mathcal{P}$ with $L \in \mathcal{L}$;
8. if $P \in \mathcal{P}$, then the Relabeling $P[f] \in \mathcal{P}$.

The standard BNF grammar of syntax of RCTC can be summarized as follows:
$P ::= A \mid \mathbf{nil} \mid \alpha.P \mid P.\alpha[m] \mid P+P \mid P \parallel P \mid (\alpha_1 \parallel \cdots \parallel \alpha_n).P \mid P.(\alpha_1[m] \parallel \cdots \parallel \alpha_n[m]) \mid P \setminus L \mid P[f]$

### 3.1.2 Operational semantics

The operational semantics is defined by LTSs (labeled transition systems), and it is detailed by the following definition.

Handbook of Truly Concurrent Process Algebra. https://doi.org/10.1016/B978-0-44-321515-5.00007-X

**TABLE 3.1 Forward transition rules of Prefix and Summation.**

| | |
|---|---|
| $\frac{}{\alpha\xrightarrow{\alpha}\alpha[m]}$ | |
| $\frac{P\xrightarrow{\alpha}\alpha[m]\quad \alpha\notin Q}{P+Q\xrightarrow{\alpha}\alpha[m]+Q}$ | $\frac{P\xrightarrow{\alpha}P'\quad \alpha\notin Q}{P+Q\xrightarrow{\alpha}P'+Q}$ |
| $\frac{Q\xrightarrow{\alpha}\alpha[m]\quad \alpha\notin P}{P+Q\xrightarrow{\alpha}P+\alpha[m]}$ | $\frac{Q\xrightarrow{\alpha}Q'\quad \alpha\notin P}{P+Q\xrightarrow{\alpha}P+Q'}$ |
| $\frac{P\xrightarrow{\alpha}\alpha[m]\quad Q\xrightarrow{\alpha}\alpha[m]}{P+Q\xrightarrow{\alpha}\alpha[m]+\alpha[m]}$ | $\frac{P\xrightarrow{\alpha}P'\quad Q\xrightarrow{\alpha}\alpha[m]}{P+Q\xrightarrow{\alpha}P'+\alpha[m]}$ |
| $\frac{P\xrightarrow{\alpha}\alpha[m]\quad Q\xrightarrow{\alpha}Q'}{P+Q\xrightarrow{\alpha}\alpha[m]+Q'}$ | $\frac{P\xrightarrow{\alpha}P'\quad Q\xrightarrow{\alpha}Q'}{P+Q\xrightarrow{\alpha}P'+Q'}$ |
| $\frac{P\xrightarrow{\alpha}\alpha[m]\quad \text{Std}(Q)}{P.Q\xrightarrow{\alpha}\alpha[m].Q}$ | $\frac{P\xrightarrow{\alpha}P'\quad \text{Std}(Q)}{P.Q\xrightarrow{\alpha}P'.Q}$ |
| $\frac{Q\xrightarrow{\beta}\beta[n]\quad \text{NStd}(P)}{P.Q\xrightarrow{\beta}P.\beta[n]}$ | $\frac{Q\xrightarrow{\beta}Q'\quad \text{NStd}(P)}{P.Q\xrightarrow{\beta}P.Q'}$ |

**TABLE 3.2 Reverse transition rules of Prefix and Summation.**

| | |
|---|---|
| $\frac{}{\alpha[m]\overset{\alpha[m]}{\twoheadrightarrow}\alpha}$ | |
| $\frac{P\overset{\alpha[m]}{\twoheadrightarrow}\alpha\quad \alpha\notin Q}{P+Q\overset{\alpha[m]}{\twoheadrightarrow}\alpha+Q}$ | $\frac{P\overset{\alpha[m]}{\twoheadrightarrow}P'\quad \alpha\notin Q}{P+Q\overset{\alpha[m]}{\twoheadrightarrow}P'+Q}$ |
| $\frac{Q\overset{\alpha[m]}{\twoheadrightarrow}\alpha\quad \alpha\notin P}{P+Q\overset{\alpha[m]}{\twoheadrightarrow}P+\alpha}$ | $\frac{Q\overset{\alpha[m]}{\twoheadrightarrow}Q'\quad \alpha\notin P}{P+Q\overset{\alpha[m]}{\twoheadrightarrow}P+Q'}$ |
| $\frac{P\overset{\alpha[m]}{\twoheadrightarrow}\alpha\quad Q\overset{\alpha[m]}{\twoheadrightarrow}\alpha}{P+Q\overset{\alpha[m]}{\twoheadrightarrow}\alpha+\alpha}$ | $\frac{P\overset{\alpha[m]}{\twoheadrightarrow}P'\quad Q\overset{\alpha[m]}{\twoheadrightarrow}\alpha}{P+Q\overset{\alpha[m]}{\twoheadrightarrow}P'+\alpha}$ |
| $\frac{P\overset{\alpha[m]}{\twoheadrightarrow}\alpha\quad Q\overset{\alpha[m]}{\twoheadrightarrow}Q'}{P+Q\overset{\alpha[m]}{\twoheadrightarrow}\alpha+Q'}$ | $\frac{P\overset{\alpha[m]}{\twoheadrightarrow}P'\quad Q\overset{\alpha[m]}{\twoheadrightarrow}Q'}{P+Q\overset{\alpha[m]}{\twoheadrightarrow}P'+Q'}$ |
| $\frac{P\overset{\alpha[m]}{\twoheadrightarrow}\alpha\quad \text{Std}(Q)}{P.Q\overset{\alpha[m]}{\twoheadrightarrow}\alpha.Q}$ | $\frac{P\overset{\alpha[m]}{\twoheadrightarrow}P'\quad \text{Std}(Q)}{P.Q\overset{\alpha[m]}{\twoheadrightarrow}P'.Q}$ |
| $\frac{Q\overset{\beta[n]}{\twoheadrightarrow}\beta\quad \text{NStd}(P)}{P.Q\overset{\beta[n]}{\twoheadrightarrow}P.\beta}$ | $\frac{Q\overset{\beta[n]}{\twoheadrightarrow}Q'\quad \text{NStd}(P)}{P.Q\overset{\beta[n]}{\twoheadrightarrow}P.Q'}$ |

**Definition 3.2** (Semantics). The operational semantics of CTC corresponding to the syntax in Definition 3.1 is defined by a series of transition rules, they are shown in Tables 3.1, 3.2, 3.3, 3.4, 3.5, 3.6, 3.7, and 3.8. Also, the predicate $\xrightarrow{\alpha}\alpha[m]$ represents successful forward termination after execution of the action $\alpha$, the predicate $\overset{\alpha[m]}{\twoheadrightarrow}\alpha$ represents successful reverse termination after execution of the event $\alpha[m]$, the predicate Std(P) represents that $p$ is a standard process containing no past events, the predicate NStd(P) represents that $P$ is a process full of past events.

The forward transition rules for Prefix and Summation are shown in Table 3.1.

The reverse transition rules for Prefix and Summation are shown in Table 3.2.

The forward and reverse pomset transition rules of Prefix and Summation are shown in Table 3.3 and Table 3.4, which are different from single event transition rules in Table 3.1 and Table 3.2, the forward and reverse pomset transition rules are labeled by pomsets, which are defined by causality . and conflict $+$.

**TABLE 3.3 Forward pomset transition rules of Prefix and Summation.**

| | |
|---|---|
| $\frac{}{p\xrightarrow{p}p[\mathcal{K}]}$ | |
| $\frac{P\xrightarrow{p}p[\mathcal{K}]\quad p\nsubseteq Q}{P+Q\xrightarrow{p}p[\mathcal{K}]+Q}$ | $\frac{P\xrightarrow{p}P'\quad p\nsubseteq Q}{P+Q\xrightarrow{p}P'+Q}$ |
| $\frac{Q\xrightarrow{q}q[\mathcal{J}]\quad q\nsubseteq P}{P+Q\xrightarrow{q}P+q[\mathcal{J}]}$ | $\frac{Q\xrightarrow{q}Q'\quad q\nsubseteq P}{P+Q\xrightarrow{q}P+Q'}$ |
| $\frac{P\xrightarrow{p}p[\mathcal{K}]\quad Q\xrightarrow{p}p[\mathcal{K}]}{P+Q\xrightarrow{p}p[\mathcal{K}]+p[\mathcal{K}]}$ | $\frac{P\xrightarrow{p}P'\quad Q\xrightarrow{p}p[\mathcal{K}]}{P+Q\xrightarrow{p}P'+p[\mathcal{K}]}$ |
| $\frac{P\xrightarrow{p}p[\mathcal{K}]\quad Q\xrightarrow{p}Q'}{P+Q\xrightarrow{p}p[\mathcal{K}]+Q'}$ | $\frac{P\xrightarrow{p}P'\quad Q\xrightarrow{p}Q'}{P+Q\xrightarrow{p}P'+Q'}$ |
| $\frac{P\xrightarrow{p}p[\mathcal{K}]\quad \text{Std}(Q)}{P.Q\xrightarrow{p}p[\mathcal{K}].Q}(p\subseteq P)$ | $\frac{P\xrightarrow{p}P'\quad \text{Std}(Q)}{P.Q\xrightarrow{p}P'.Q}(p\subseteq P)$ |
| $\frac{Q\xrightarrow{q}q[\mathcal{J}]\quad \text{NStd}(P)}{P.Q\xrightarrow{q}P.q[\mathcal{J}]}(q\subseteq Q)$ | $\frac{Q\xrightarrow{q}Q'\quad \text{NStd}(P)}{P.Q\xrightarrow{q}P.Q'}(q\subseteq Q)$ |

**TABLE 3.4 Reverse pomset transition rules of Prefix and Summation.**

| | |
|---|---|
| $\frac{}{p[\mathcal{K}]\xtwoheadrightarrow{p[\mathcal{K}]}p}$ | |
| $\frac{P\xtwoheadrightarrow{p[\mathcal{K}]}p\quad p\nsubseteq Q}{P+Q\xtwoheadrightarrow{p[\mathcal{K}]}p+Q}$ | $\frac{P\xtwoheadrightarrow{p[\mathcal{K}]}P'\quad p\nsubseteq Q}{P+Q\xtwoheadrightarrow{p[\mathcal{K}]}P'+Q}$ |
| $\frac{Q\xtwoheadrightarrow{q[\mathcal{J}]}q\quad q\nsubseteq P}{P+Q\xtwoheadrightarrow{q[\mathcal{J}]}P+q}$ | $\frac{Q\xtwoheadrightarrow{q[\mathcal{J}]}Q'\quad q\nsubseteq P}{P+Q\xtwoheadrightarrow{q[\mathcal{J}]}P+Q'}$ |
| $\frac{P\xtwoheadrightarrow{p[\mathcal{K}]}p\quad Q\xtwoheadrightarrow{p[\mathcal{K}]}p}{P+Q\xtwoheadrightarrow{p[\mathcal{K}]}p+p}$ | $\frac{P\xtwoheadrightarrow{p[\mathcal{K}]}P'\quad Q\xtwoheadrightarrow{p[\mathcal{K}]}p}{P+Q\xtwoheadrightarrow{p[\mathcal{K}]}P'+p}$ |
| $\frac{P\xtwoheadrightarrow{p[\mathcal{K}]}p\quad Q\xtwoheadrightarrow{p[\mathcal{K}]}Q'}{P+Q\xtwoheadrightarrow{p[\mathcal{K}]}p+Q'}$ | $\frac{P\xtwoheadrightarrow{p[\mathcal{K}]}P'\quad Q\xtwoheadrightarrow{p[\mathcal{K}]}Q'}{P+Q\xtwoheadrightarrow{p[\mathcal{K}]}P'+Q'}$ |
| $\frac{P\xtwoheadrightarrow{p[\mathcal{K}]}p\quad \text{Std}(Q)}{P.Q\xtwoheadrightarrow{p[\mathcal{K}]}p.Q}(p\subseteq P)$ | $\frac{P\xtwoheadrightarrow{p[\mathcal{K}]}P'\quad \text{Std}(Q)}{P.Q\xtwoheadrightarrow{p[\mathcal{K}]}P'.Q}(p\subseteq P)$ |
| $\frac{Q\xtwoheadrightarrow{q[\mathcal{J}]}q\quad \text{NStd}(P)}{P.Q\xtwoheadrightarrow{q[\mathcal{J}]}P.q}(q\subseteq Q)$ | $\frac{Q\xtwoheadrightarrow{q[\mathcal{J}]}Q'\quad \text{NStd}(P)}{P\cdot Q\xtwoheadrightarrow{q[\mathcal{J}]}P.Q'}(q\subseteq Q)$ |

The forward transition rules for Composition are shown in Table 3.5.
The reverse transition rules for Composition are shown in Table 3.6.
The forward transition rules for Restriction, Relabeling, and Constants are shown in Table 3.7.
The reverse transition rules for Restriction, Relabeling, and Constants are shown in Table 3.8.

### 3.1.3 Properties of transitions

**Definition 3.3** (Sorts). Given the sorts $\mathcal{L}(A)$ and $\mathcal{L}(X)$ of constants and variables, we define $\mathcal{L}(P)$ inductively as follows:

1. $\mathcal{L}(l.P)=\{l\}\cup\mathcal{L}(P)$;
2. $\mathcal{L}(P.l[m])=\{l\}\cup\mathcal{L}(P)$;

**TABLE 3.5** Forward transition rules of Composition.

| |
|---|
| $\dfrac{P \xrightarrow{\alpha} P' \quad Q \nrightarrow}{P \parallel Q \xrightarrow{\alpha} P' \parallel Q}$ |
| $\dfrac{Q \xrightarrow{\alpha} Q' \quad P \nrightarrow}{P \parallel Q \xrightarrow{\alpha} P \parallel Q'}$ |
| $\dfrac{P \xrightarrow{\alpha} P' \quad Q \xrightarrow{\beta} Q'}{P \parallel Q \xrightarrow{\{\alpha,\beta\}} P' \parallel Q'} \quad (\beta \neq \overline{\alpha})$ |
| $\dfrac{P \xrightarrow{l} P' \quad Q \xrightarrow{\overline{l}} Q'}{P \parallel Q \xrightarrow{\tau} P' \parallel Q'}$ |

**TABLE 3.6** Reverse transition rules of Composition.

| |
|---|
| $\dfrac{P \xtwoheadrightarrow{\alpha[m]} P' \quad Q \not\twoheadrightarrow}{P \parallel Q \xtwoheadrightarrow{\alpha[m]} P' \parallel Q}$ |
| $\dfrac{Q \xtwoheadrightarrow{\alpha[m]} Q' \quad P \not\twoheadrightarrow}{P \parallel Q \xtwoheadrightarrow{\alpha[m]} P \parallel Q'}$ |
| $\dfrac{P \xtwoheadrightarrow{\alpha[m]} P' \quad Q \xtwoheadrightarrow{\beta[m]} Q'}{P \parallel Q \xtwoheadrightarrow{\{\alpha[m],\beta[m]\}} P' \parallel Q'} \quad (\beta \neq \overline{\alpha})$ |
| $\dfrac{P \xtwoheadrightarrow{l[m]} P' \quad Q \xtwoheadrightarrow{\overline{l}[m]} Q'}{P \parallel Q \xtwoheadrightarrow{\tau} P' \parallel Q'}$ |

**TABLE 3.7** Forward transition rules of Restriction, Relabeling, and Constants.

| |
|---|
| $\dfrac{P \xrightarrow{\alpha} P'}{P \setminus L \xrightarrow{\alpha} P' \setminus L} \quad (\alpha, \overline{\alpha} \notin L)$ |
| $\dfrac{P \xrightarrow{\{\alpha_1,\cdots,\alpha_n\}} P'}{P \setminus L \xrightarrow{\{\alpha_1,\cdots,\alpha_n\}} P' \setminus L} \quad (\alpha_1, \overline{\alpha_1}, \cdots, \alpha_n, \overline{\alpha_n} \notin L)$ |
| $\dfrac{P \xrightarrow{\alpha} P'}{P[f] \xrightarrow{f(\alpha)} P'[f]}$ |
| $\dfrac{P \xrightarrow{\{\alpha_1,\cdots,\alpha_n\}} P'}{P[f] \xrightarrow{\{f(\alpha_1),\cdots,f(\alpha_n)\}} P'[f]}$ |
| $\dfrac{P \xrightarrow{\alpha} P'}{A \xrightarrow{\alpha} P'} \quad (A \stackrel{\text{def}}{=} P)$ |
| $\dfrac{P \xrightarrow{\{\alpha_1,\cdots,\alpha_n\}} P'}{A \xrightarrow{\{\alpha_1,\cdots,\alpha_n\}} P'} \quad (A \stackrel{\text{def}}{=} P)$ |

3. $\mathcal{L}((l_1 \parallel \cdots \parallel l_n).P) = \{l_1, \cdots, l_n\} \cup \mathcal{L}(P)$;
4. $\mathcal{L}(P.(l_1[m] \parallel \cdots \parallel l_n[m])) = \{l_1, \cdots, l_n\} \cup \mathcal{L}(P)$;
5. $\mathcal{L}(\tau.P) = \mathcal{L}(P)$;
6. $\mathcal{L}(P + Q) = \mathcal{L}(P) \cup \mathcal{L}(Q)$;
7. $\mathcal{L}(P \parallel Q) = \mathcal{L}(P) \cup \mathcal{L}(Q)$;
8. $\mathcal{L}(P \setminus L) = \mathcal{L}(P) - (L \cup \overline{L})$;

**TABLE 3.8 Reverse transition rules of Restriction, Relabeling, and Constants.**

| |
|---|
| $\dfrac{P \xrightarrow{\alpha[m]} P'}{P \setminus L \xrightarrow{\alpha[m]} P' \setminus L} \quad (\alpha, \overline{\alpha} \notin L)$ |
| $\dfrac{P \xrightarrow{\{\alpha_1[m], \cdots, \alpha_n[m]\}} P'}{P \setminus L \xrightarrow{\{\alpha_1[m], \cdots, \alpha_n[m]\}} P' \setminus L} \quad (\alpha_1, \overline{\alpha_1}, \cdots, \alpha_n, \overline{\alpha_n} \notin L)$ |
| $\dfrac{P \xrightarrow{\alpha[m]} P'}{P[f] \xrightarrow{f(\alpha[m])} P'[f]}$ |
| $\dfrac{P \xrightarrow{\{\alpha_1[m], \cdots, \alpha_n[m]\}} P'}{P[f] \xrightarrow{\{f(\alpha_1)[m], \cdots, f(\alpha_n)[m]\}} P'[f]}$ |
| $\dfrac{P \xrightarrow{\alpha[m]} P'}{A \xrightarrow{\alpha[m]} P'} \quad (A \overset{\text{def}}{=} P)$ |
| $\dfrac{P \xrightarrow{\{\alpha_1[m], \cdots, \alpha_n[m]\}} P'}{A \xrightarrow{\{\alpha_1[m], \cdots, \alpha_n[m]\}} P'} \quad (A \overset{\text{def}}{=} P)$ |

9. $\mathcal{L}(P[f]) = \{f(l) : l \in \mathcal{L}(P)\}$;
10. for $A \overset{\text{def}}{=} P$, $\mathcal{L}(P) \subseteq \mathcal{L}(A)$.

Now, we present some properties of the transition rules defined in Definition 3.2.

**Proposition 3.4.** *If $P \xrightarrow{\alpha} P'$, then*

1. $\alpha \in \mathcal{L}(P) \cup \{\tau\}$;
2. $\mathcal{L}(P') \subseteq \mathcal{L}(P)$.

*If $P \xrightarrow{\{\alpha_1, \cdots, \alpha_n\}} P'$, then*

1. $\alpha_1, \cdots, \alpha_n \in \mathcal{L}(P) \cup \{\tau\}$;
2. $\mathcal{L}(P') \subseteq \mathcal{L}(P)$.

*Proof.* By induction on the inference of $P \xrightarrow{\alpha} P'$ and $P \xrightarrow{\{\alpha_1, \cdots, \alpha_n\}} P'$, there are several cases corresponding to the forward transition rules in Definition 3.2, however, we omit them here. □

**Proposition 3.5.** *If $P \xrightarrow{\alpha[m]} P'$, then*

1. $\alpha \in \mathcal{L}(P) \cup \{\tau\}$;
2. $\mathcal{L}(P') \subseteq \mathcal{L}(P)$.

*If $P \xrightarrow{\{\alpha_1[m], \cdots, \alpha_n[m]\}} P'$, then*

1. $\alpha_1, \cdots, \alpha_n \in \mathcal{L}(P) \cup \{\tau\}$;
2. $\mathcal{L}(P') \subseteq \mathcal{L}(P)$.

*Proof.* By induction on the inference of $P \xrightarrow{\alpha} P'$ and $P \xrightarrow{\{\alpha_1, \cdots, \alpha_n\}} P'$, there are several cases corresponding to the forward transition rules in Definition 3.2, however, we omit them here. □

## 3.2 Strongly forward–reverse truly concurrent bisimulations

Based on the concepts of strongly FR truly concurrent bisimulation equivalences, we obtain the following laws.

**Proposition 3.6** (Monoid laws for strongly FR pomset bisimulation). *The monoid laws for strongly FR pomset bisimulation are as follows:*

1. $P+Q\sim_p^{fr} Q+P$;
2. $P+(Q+R)\sim_p^{fr} (P+Q)+R$;
3. $P+P\sim_p^{fr} P$;
4. $P+\textbf{nil}\sim_p^{fr} P$.

*Proof.* **1.** $P+Q\sim_p^{fr} Q+P$. There are several cases, however, we will not enumerate them all here. By the forward transition rules of Summation in Table 3.3, we obtain

$$\frac{P\xrightarrow{p}P' \quad p\nsubseteq Q}{P+Q\xrightarrow{p}P'+Q}(p\subseteq P) \qquad \frac{P\xrightarrow{p}P' \quad p\nsubseteq Q}{Q+P\xrightarrow{p}Q+P'}(p\subseteq P)$$

$$\frac{Q\xrightarrow{q}Q' \quad q\nsupseteq P}{P+Q\xrightarrow{q}P+Q'}(q\subseteq Q) \qquad \frac{Q\xrightarrow{q}Q' \quad q\nsubseteq P}{Q+P\xrightarrow{q}Q'+P}(q\subseteq Q)$$

By the reverse transition rules of Summation in Table 3.4, we obtain

$$\frac{P\overset{p[\mathcal{K}]}{\twoheadrightarrow}P' \quad p\nsubseteq Q}{P+Q\overset{p[\mathcal{K}]}{\twoheadrightarrow}P'+Q}(p\subseteq P) \qquad \frac{P\overset{p[\mathcal{K}]}{\twoheadrightarrow}P' \quad p\nsubseteq Q}{Q+P\overset{p[\mathcal{K}]}{\twoheadrightarrow}Q+P'}(p\subseteq P)$$

$$\frac{Q\overset{q[\mathcal{J}]}{\twoheadrightarrow}Q' \quad q\nsubseteq P}{P+Q\overset{q[\mathcal{J}]}{\twoheadrightarrow}P+Q'}(q\subseteq Q) \qquad \frac{Q\overset{q[\mathcal{J}]}{\twoheadrightarrow}Q' \quad q\nsubseteq P}{Q+P\overset{q[\mathcal{J}]}{\twoheadrightarrow}Q'+P}(q\subseteq Q)$$

With the assumptions $P'+Q\sim_p^{fr} Q+P'$ and $P+Q'\sim_p^{fr} Q'+P$, hence, $P+Q\sim_p^{fr} Q+P$, as desired.

**2.** $P+(Q+R)\sim_p^{fr}(P+Q)+R$. There are several cases, however, we will not enumerate them all here. By the forward transition rules of Summation in Table 3.3, we obtain

$$\frac{P\xrightarrow{p}P' \quad p\nsubseteq Q \quad p\nsubseteq R}{P+(Q+R)\xrightarrow{p}P'+(Q+R)}(p\subseteq P) \qquad \frac{P\xrightarrow{p}P' \quad p\nsubseteq Q \quad p\nsubseteq R}{(P+Q)+R\xrightarrow{p}(P'+Q)+R}(p\subseteq P)$$

$$\frac{Q\xrightarrow{q}Q' \quad q\nsubseteq P \quad q\nsubseteq R}{P+(Q+R)\xrightarrow{q}P+(Q'+R)}(q\subseteq Q) \qquad \frac{Q\xrightarrow{q}Q' \quad q\nsubseteq P \quad q\nsubseteq R}{(P+Q)+R\xrightarrow{q}(P+Q')+R}(q\subseteq Q)$$

$$\frac{R\xrightarrow{r}R' \quad r\nsubseteq P \quad r\nsubseteq Q}{P+(Q+R)\xrightarrow{r}P+(Q+R')}(r\subseteq R) \qquad \frac{R\xrightarrow{r}R' \quad r\nsubseteq P \quad r\nsubseteq Q}{(P+Q)+R\xrightarrow{r}(P+Q)+R'}(r\subseteq R)$$

By the reverse transition rules of Summation in Table 3.4, we obtain

$$\frac{P\overset{p[\mathcal{K}]}{\twoheadrightarrow}P' \quad p\nsubseteq Q \quad p\nsubseteq R}{P+(Q+R)\overset{p[\mathcal{K}]}{\twoheadrightarrow}P'+(Q+R)}(p\subseteq P) \qquad \frac{P\overset{p[\mathcal{K}]}{\twoheadrightarrow}P' \quad p\nsubseteq Q \quad p\nsubseteq R}{(P+Q)+R\overset{p[\mathcal{K}]}{\twoheadrightarrow}(P'+Q)+R}(p\subseteq P)$$

$$\frac{Q\overset{q[\mathcal{J}]}{\twoheadrightarrow}Q' \quad q\nsubseteq P \quad q\nsubseteq R}{P+(Q+R)\overset{q[\mathcal{J}]}{\twoheadrightarrow}P+(Q'+R)}(q\subseteq Q) \qquad \frac{Q\overset{q[\mathcal{J}]}{\twoheadrightarrow}Q' \quad q\nsubseteq P \quad q\nsubseteq R}{(P+Q)+R\overset{q[\mathcal{J}]}{\twoheadrightarrow}(P+Q')+R}(q\subseteq Q)$$

$$\frac{R\overset{r[\mathcal{I}]}{\twoheadrightarrow}R' \quad r\nsubseteq P \quad r\nsubseteq Q}{P+(Q+R)\overset{r[\mathcal{I}]}{\twoheadrightarrow}P+(Q+R')}(r\subseteq R) \qquad \frac{R\overset{r[\mathcal{I}]}{\twoheadrightarrow}R' \quad r\nsubseteq P \quad r\nsubseteq Q}{(P+Q)+R\overset{r[\mathcal{I}]}{\twoheadrightarrow}(P+Q)+R'}(r\subseteq R)$$

With the assumptions $P'+(Q+R)\sim_p^{fr}(P'+Q)+R$, $P+(Q'+R)\sim_p^{fr}(P+Q')+R$ and $P+(Q+R')\sim_p^{fr}(P+Q)+R'$, hence, $P+(Q+R)\sim_p^{fr}(P+Q)+R$, as desired.

3. $P + P \sim_p^{fr} P$. By the forward transition rules of Summation, we obtain

$$\frac{P \xrightarrow{p} p[\mathcal{K}]}{P + P \xrightarrow{p} p[\mathcal{K}] + p[\mathcal{K}]}(p \subseteq P) \quad \frac{P \xrightarrow{p} p[\mathcal{K}]}{P \xrightarrow{p} p[\mathcal{K}]}(p \subseteq P)$$

$$\frac{P \xrightarrow{p} P'}{P + P \xrightarrow{p} P' + P'}(p \subseteq P) \quad \frac{P \xrightarrow{p} P'}{P \xrightarrow{p} P'}(p \subseteq P)$$

By the reverse transition rules of Summation, we obtain

$$\frac{P \xrightarrow{p[\mathcal{K}]} p}{P + P \xrightarrow{p[\mathcal{K}]} p + p}(p \subseteq P) \quad \frac{P \xrightarrow{p[\mathcal{K}]} p}{P \xrightarrow{p[\mathcal{K}]} p}(p \subseteq P)$$

$$\frac{P \xrightarrow{p[\mathcal{K}]} P'}{P + P \xrightarrow{p[\mathcal{K}]} P' + P'}(p \subseteq P) \quad \frac{P \xrightarrow{p[\mathcal{K}]} P'}{P \xrightarrow{p[\mathcal{K}]} P'}(p \subseteq P)$$

With the assumptions $p[\mathcal{K}] + p[\mathcal{K}] \sim_p^{fr} p[\mathcal{K}]$, $p + p \sim_p^{fr} p$ and $P' + P' \sim_p^{fr} P'$, hence, $P + P \sim_p^{fr} P$, as desired.

4. $P + \mathbf{nil} \sim_p^{fr} P$. There are several cases, however, we will not enumerate them all here. By the forward transition rules of Summation in Table 3.3, we obtain

$$\frac{P \xrightarrow{p} P'}{P + \mathbf{nil} \xrightarrow{p} P'}(p \subseteq P) \quad \frac{P \xrightarrow{p} P'}{P \xrightarrow{p} P'}(p \subseteq P)$$

By the reverse transition rules of Summation in Table 3.4, we obtain

$$\frac{P \xrightarrow{p[\mathcal{K}]} P'}{P + \mathbf{nil} \xrightarrow{p[\mathcal{K}]} P'}(p \subseteq P) \quad \frac{P \xrightarrow{p[\mathcal{K}]} P'}{P \xrightarrow{p[\mathcal{K}]} P'}(p \subseteq P)$$

Since $P' \sim_p^{fr} P'$, $P + \mathbf{nil} \sim_p^{fr} P$, as desired. □

**Proposition 3.7** (Monoid laws for strongly FR step bisimulation). *The monoid laws for strongly FR step bisimulation are as follows:*

1. $P + Q \sim_s^{fr} Q + P$;
2. $P + (Q + R) \sim_s^{fr} (P + Q) + R$;
3. $P + P \sim_s^{fr} P$;
4. $P + \mathit{nil} \sim_s^{fr} P$.

*Proof.* 1. $P + Q \sim_s^{fr} Q + P$. There are several cases, however, we will not enumerate them all here. By the forward transition rules of Summation in Table 3.3, we obtain

$$\frac{P \xrightarrow{p} P' \quad p \not\subseteq Q}{P + Q \xrightarrow{p} P' + Q}(p \subseteq P, \forall\alpha, \beta \in p, \text{ are pairwise concurrent})$$

$$\frac{P \xrightarrow{p} P' \quad p \not\subseteq Q}{Q + P \xrightarrow{p} Q + P'}(p \subseteq P, \forall\alpha, \beta \in p, \text{ are pairwise concurrent})$$

$$\frac{Q \xrightarrow{q} Q' \quad q \not\supseteq P}{P + Q \xrightarrow{q} P + Q'}(q \subseteq Q, \forall\alpha, \beta \in q, \text{ are pairwise concurrent})$$

$$\frac{Q \xrightarrow{q} Q' \quad q \not\subseteq P}{Q + P \xrightarrow{q} Q' + P}(q \subseteq Q, \forall\alpha, \beta \in q, \text{ are pairwise concurrent})$$

By the reverse transition rules of Summation in Table 3.4, we obtain

$$\frac{P \xrightarrow{p[\mathcal{K}]} P' \quad p \nsubseteq Q}{P + Q \xrightarrow{p[\mathcal{K}]} P' + Q} (p \subseteq P, \forall \alpha, \beta \in p, \text{ are pairwise concurrent})$$

$$\frac{P \xrightarrow{p[\mathcal{K}]} P' \quad p \nsubseteq Q}{Q + P \xrightarrow{p[\mathcal{K}]} Q + P'} (p \subseteq P, \forall \alpha, \beta \in p, \text{ are pairwise concurrent})$$

$$\frac{Q \xrightarrow{q[\mathcal{J}]} Q' \quad q \nsubseteq P}{P + Q \xrightarrow{q[\mathcal{J}]} P + Q'} (q \subseteq Q, \forall \alpha, \beta \in q, \text{ are pairwise concurrent})$$

$$\frac{Q \xrightarrow{q[\mathcal{J}]} Q' \quad q \nsubseteq P}{Q + P \xrightarrow{q[\mathcal{J}]} Q' + P} (q \subseteq Q, \forall \alpha, \beta \in q, \text{ are pairwise concurrent})$$

With the assumptions $P' + Q \sim_s^{fr} Q + P'$ and $P + Q' \sim_s^{fr} Q' + P$, hence, $P + Q \sim_s^{fr} Q + P$, as desired.

2. $P + (Q + R) \sim_s^{fr} (P + Q) + R$. There are several cases, however, we will not enumerate them all here. By the forward transition rules of Summation in Table 3.3, we obtain

$$\frac{P \xrightarrow{p} P' \quad p \nsubseteq Q \quad p \nsubseteq R}{P + (Q + R) \xrightarrow{p} P' + (Q + R)} (p \subseteq P, \forall \alpha, \beta \in p, \text{ are pairwise concurrent})$$

$$\frac{P \xrightarrow{p} P' \quad p \nsubseteq Q \quad p \nsubseteq R}{(P + Q) + R \xrightarrow{p} (P' + Q) + R} (p \subseteq P, \forall \alpha, \beta \in p, \text{ are pairwise concurrent})$$

$$\frac{Q \xrightarrow{q} Q' \quad q \nsubseteq P \quad q \nsubseteq R}{P + (Q + R) \xrightarrow{q} P + (Q' + R)} (q \subseteq Q, \forall \alpha, \beta \in q, \text{ are pairwise concurrent})$$

$$\frac{Q \xrightarrow{q} Q' \quad q \nsubseteq P \quad q \nsubseteq R}{(P + Q) + R \xrightarrow{q} (P + Q') + R} (q \subseteq Q, \forall \alpha, \beta \in q, \text{ are pairwise concurrent})$$

$$\frac{R \xrightarrow{r} R' \quad r \nsubseteq P \quad r \nsubseteq Q}{P + (Q + R) \xrightarrow{r} P + (Q + R')} (r \subseteq R, \forall \alpha, \beta \in r, \text{ are pairwise concurrent})$$

$$\frac{R \xrightarrow{r} R' \quad r \nsubseteq P \quad r \nsubseteq Q}{(P + Q) + R \xrightarrow{r} (P + Q) + R'} (r \subseteq R, \forall \alpha, \beta \in r, \text{ are pairwise concurrent})$$

By the reverse transition rules of Summation in Table 3.4, we obtain

$$\frac{P \xrightarrow{p[\mathcal{K}]} P' \quad p \nsubseteq Q \quad p \nsubseteq R}{P + (Q + R) \xrightarrow{p[\mathcal{K}]} P' + (Q + R)} (p \subseteq P, \forall \alpha, \beta \in p, \text{ are pairwise concurrent})$$

$$\frac{P \xrightarrow{p[\mathcal{K}]} P' \quad p \nsubseteq Q \quad p \nsubseteq R}{(P + Q) + R \xrightarrow{p[\mathcal{K}]} (P' + Q) + R} (p \subseteq P, \forall \alpha, \beta \in p, \text{ are pairwise concurrent})$$

$$\frac{Q \xrightarrow{q[\mathcal{J}]} Q' \quad q \nsubseteq P \quad q \nsubseteq R}{P + (Q + R) \xrightarrow{q[\mathcal{J}]} P + (Q' + R)} (q \subseteq Q, \forall \alpha, \beta \in q, \text{ are pairwise concurrent})$$

$$\frac{Q \xrightarrow{q[\mathcal{J}]} Q' \quad q \nsubseteq P \quad q \nsubseteq R}{(P + Q) + R \xrightarrow{q[\mathcal{J}]} (P + Q') + R} (q \subseteq Q, \forall \alpha, \beta \in q, \text{ are pairwise concurrent})$$

$$\frac{R \xrightarrow{r[\mathcal{I}]} R' \quad r \not\subseteq P \quad r \not\subseteq Q}{P + (Q + R) \xrightarrow{r[\mathcal{I}]} P + (Q + R')} (r \subseteq R, \forall \alpha, \beta \in r, \text{ are pairwise concurrent})$$

$$\frac{R \xrightarrow{r[\mathcal{I}]} R' \quad r \not\subseteq P \quad r \not\subseteq Q}{(P + Q) + R \xrightarrow{r[\mathcal{I}]} (P + Q) + R'} (r \subseteq R, \forall \alpha, \beta \in r, \text{ are pairwise concurrent})$$

With the assumptions $P' + (Q + R) \sim_s^{fr} (P' + Q) + R$, $P + (Q' + R) \sim_s^{fr} (P + Q') + R$ and $P + (Q + R') \sim_s^{fr} (P + Q) + R'$, hence, $P + (Q + R) \sim_s^{fr} (P + Q) + R$, as desired.

**3.** $P + P \sim_s^{fr} P$. By the forward transition rules of Summation, we obtain

$$\frac{P \xrightarrow{p} p[\mathcal{K}]}{P + P \xrightarrow{p} p[\mathcal{K}] + p[\mathcal{K}]} (p \subseteq P, \forall \alpha, \beta \in p, \text{ are pairwise concurrent})$$

$$\frac{P \xrightarrow{p} p[\mathcal{K}]}{P \xrightarrow{p} p[\mathcal{K}]} (p \subseteq P, \forall \alpha, \beta \in p, \text{ are pairwise concurrent})$$

$$\frac{P \xrightarrow{p} P'}{P + P \xrightarrow{p} P' + P'} (p \subseteq P, \forall \alpha, \beta \in p, \text{ are pairwise concurrent})$$

$$\frac{P \xrightarrow{p} P'}{P \xrightarrow{p} P'} (p \subseteq P, \forall \alpha, \beta \in p, \text{ are pairwise concurrent})$$

By the reverse transition rules of Summation, we obtain

$$\frac{P \xrightarrow{p[\mathcal{K}]} p}{P + P \xrightarrow{p[\mathcal{K}]} p + p} (p \subseteq P, \forall \alpha, \beta \in p, \text{ are pairwise concurrent})$$

$$\frac{P \xrightarrow{p[\mathcal{K}]} p}{P \xrightarrow{p[\mathcal{K}]} p} (p \subseteq P, \forall \alpha, \beta \in p, \text{ are pairwise concurrent})$$

$$\frac{P \xrightarrow{p[\mathcal{K}]} P'}{P + P \xrightarrow{p[\mathcal{K}]} P' + P'} (p \subseteq P, \forall \alpha, \beta \in p, \text{ are pairwise concurrent})$$

$$\frac{P \xrightarrow{p[\mathcal{K}]} P'}{P \xrightarrow{p[\mathcal{K}]} P'} (p \subseteq P, \forall \alpha, \beta \in p, \text{ are pairwise concurrent})$$

With the assumptions $p[\mathcal{K}] + p[\mathcal{K}] \sim_s^{fr} p[\mathcal{K}]$, $p + p \sim_s^{fr} p$ and $P' + P' \sim_s^{fr} P'$, hence, $P + P \sim_s^{fr} P$, as desired.

**4.** $P + \mathbf{nil} \sim_s^{fr} P$. There are several cases, however, we will not enumerate them all here. By the forward transition rules of Summation in Table 3.3, we obtain

$$\frac{P \xrightarrow{p} P'}{P + \mathbf{nil} \xrightarrow{p} P'} (p \subseteq P, \forall \alpha, \beta \in p, \text{ are pairwise concurrent})$$

$$\frac{P \xrightarrow{p} P'}{P \xrightarrow{p} P'} (p \subseteq P, \forall \alpha, \beta \in p, \text{ are pairwise concurrent})$$

By the reverse transition rules of Summation in Table 3.4, we obtain

$$\frac{P \xrightarrow{p[\mathcal{K}]} P'}{P + \mathbf{nil} \xrightarrow{p[\mathcal{K}]} P'} (p \subseteq P, \forall \alpha, \beta \in p, \text{ are pairwise concurrent})$$

$$\frac{P \xrightarrow{p[\mathcal{K}]} P'}{P \xrightarrow{p[\mathcal{K}]} P'} (p \subseteq P, \forall \alpha, \beta \in p, \text{ are pairwise concurrent})$$

Since $P' \sim_s^{fr} P'$, $P + \mathbf{nil} \sim_s^{fr} P$, as desired. □

**Proposition 3.8** (Monoid laws for strongly FR hp-bisimulation). *The monoid laws for strongly FR hp-bisimulation are as follows:*

1. $P + Q \sim_{hp}^{fr} Q + P$;
2. $P + (Q + R) \sim_{hp}^{fr} (P + Q) + R$;
3. $P + P \sim_{hp}^{fr} P$;
4. $P + \mathbf{nil} \sim_{hp}^{fr} P$.

*Proof.* **1.** $P + Q \sim_{hp}^{fr} Q + P$. There are several cases, however, we will not enumerate them all here. By the forward transition rules of Summation in Table 3.1, we obtain

$$\frac{P \xrightarrow{\alpha} P' \quad \alpha \notin Q}{P + Q \xrightarrow{\alpha} P' + Q} (\alpha \in P) \qquad \frac{P \xrightarrow{\alpha} P' \quad \alpha \notin Q}{Q + P \xrightarrow{\alpha} Q + P'} (\alpha \in P)$$

$$\frac{Q \xrightarrow{\beta} Q' \quad \beta \notin P}{P + Q \xrightarrow{\beta} P + Q'} (\beta \in Q) \qquad \frac{Q \xrightarrow{\beta} Q' \quad \beta \notin P}{Q + P \xrightarrow{\beta} Q' + P} (\beta \in Q)$$

By the reverse transition rules of Summation in Table 3.2, we obtain

$$\frac{P \xrightarrow{\alpha[m]} P' \quad \alpha \notin Q}{P + Q \xrightarrow{\alpha[m]} P' + Q} (\alpha \in P) \qquad \frac{P \xrightarrow{\alpha[m]} P' \quad \alpha \notin Q}{Q + P \xrightarrow{\alpha[m]} Q + P'} (\alpha P)$$

$$\frac{Q \xrightarrow{\beta[n]} Q' \quad \beta \notin P}{P + Q \xrightarrow{\beta[n]} P + Q'} (\beta \in Q) \qquad \frac{Q \xrightarrow{\beta[n]} Q' \quad \beta \notin P}{Q + P \xrightarrow{\beta[n]} Q' + P} (\beta \in Q)$$

Since $(C(P + Q), f, C(Q + P)) \in \sim_{hp}^{fr}$, $(C((P + Q)'), f[\alpha \mapsto \alpha], C((Q + P)')) \in \sim_{hp}^{fr}$ and $(C((P + Q)'), f[\beta \mapsto \beta], C((Q + P)')) \in \sim_{hp}^{fr}$, $P + Q \sim_{hp}^{fr} Q + P$, as desired.

**2.** $P + (Q + R) \sim_{hp}^{fr} (P + Q) + R$. There are several cases, however, we will not enumerate them all here. By the forward transition rules of Summation in Table 3.1, we obtain

$$\frac{P \xrightarrow{\alpha} P' \quad \alpha \notin Q \quad \alpha \notin R}{P + (Q + R) \xrightarrow{\alpha} P' + (Q + R)} (\alpha \in P) \qquad \frac{P \xrightarrow{\alpha} P' \quad \alpha \notin Q \quad \alpha \notin R}{(P + Q) + R \xrightarrow{\alpha} (P' + Q) + R} (\alpha \in P)$$

$$\frac{Q \xrightarrow{\beta} Q' \quad \beta \notin P \quad \beta \notin R}{P + (Q + R) \xrightarrow{\beta} P + (Q' + R)} (\beta \in Q) \qquad \frac{Q \xrightarrow{\beta} Q' \quad \beta \notin P \quad \beta \notin R}{(P + Q) + R \xrightarrow{\beta} (P + Q') + R} (\beta \in Q)$$

$$\frac{R \xrightarrow{\gamma} R' \quad \gamma \notin P \quad \gamma \notin Q}{P + (Q + R) \xrightarrow{\gamma} P + (Q + R')} (\gamma \in R) \qquad \frac{R \xrightarrow{\gamma} R' \quad \gamma \notin P \quad \gamma \notin Q}{(P + Q) + R \xrightarrow{\gamma} (P + Q) + R'} (\gamma \in R)$$

By the reverse transition rules of Summation in Table 3.2, we obtain

$$\frac{P \xrightarrow{\alpha[m]} P' \quad \alpha \notin Q \quad \alpha \notin R}{P + (Q + R) \xrightarrow{\alpha[m]} P' + (Q + R)} (\alpha \in P) \qquad \frac{P \xrightarrow{\alpha[m]} P' \quad \alpha \notin Q \quad \alpha \notin R}{(P + Q) + R \xrightarrow{\alpha[m]} (P' + Q) + R} (\alpha \in P)$$

$$\frac{Q \xrightarrow{\beta[n]} Q' \quad \beta \notin P \quad \beta \notin R}{P + (Q + R) \xrightarrow{\beta[n]} P + (Q' + R)} (\beta \in Q) \qquad \frac{Q \xrightarrow{\beta[n]} Q' \quad \beta \notin P \quad \beta \notin R}{(P + Q) + R \xrightarrow{\beta[n]} (P + Q') + R} (\beta Q)$$

$$\frac{R \xrightarrow{\gamma[k]} R' \quad \gamma \notin P \quad \gamma \notin Q}{P + (Q + R) \xrightarrow{\gamma[k]} P + (Q + R')} (\gamma \in R) \qquad \frac{R \xrightarrow{\gamma[k]} R' \quad \gamma \notin P \quad \gamma \notin Q}{(P + Q) + R \xrightarrow{\gamma[k]} (P + Q) + R'} (\gamma \in R)$$

Since $(C(P + (Q + R)), f, C((P + Q) + R)) \in \sim_{hp}^{fr}$, $(C(P + (Q + R))', f[\alpha \mapsto \alpha], C((P + Q) + R)') \in \sim_{hp}^{fr}$, $(C(P + (Q + R))', f[\beta \mapsto \beta], C((P + Q) + R)') \in \sim_{hp}^{fr}$ and $(C((P + (Q + R))'), f[\gamma \mapsto \gamma], C((P + Q) + R)') \in \sim_{hp}^{fr}$, $P + (Q + R) \sim_{hp}^{fr} (P + Q) + R$, as desired.

3. $P + P \sim_{hp}^{fr} P$. By the forward transition rules of Summation, we obtain

$$\frac{P \xrightarrow{\alpha} \alpha[m]}{P + P \xrightarrow{\alpha} \alpha[m] + \alpha[m]} (\alpha \in P) \qquad \frac{P \xrightarrow{\alpha} \alpha[m]}{P \xrightarrow{\alpha} \alpha[m]} (\alpha \in P)$$

$$\frac{P \xrightarrow{\alpha} P'}{P + P \xrightarrow{\alpha} P' + P'} (p \subseteq P) \qquad \frac{P \xrightarrow{\alpha} P'}{P \xrightarrow{\alpha} P'} (\alpha \in P)$$

By the reverse transition rules of Summation, we obtain

$$\frac{P \xrightarrow{\alpha[m]} \alpha}{P + P \xrightarrow{\alpha[m]} \alpha + \alpha} (\alpha \in P) \qquad \frac{P \xrightarrow{\alpha[m]} \alpha}{P \xrightarrow{\alpha[m]} \alpha} (\alpha \in P)$$

$$\frac{P \xrightarrow{\alpha[m]} P'}{P + P \xrightarrow{\alpha[m]} P' + P'} (\alpha \in P) \qquad \frac{P \xrightarrow{\alpha[m]} P'}{P \xrightarrow{\alpha[m} P'} (\alpha \in P)$$

Since $(C(P + P), f, C(P)) \in \sim_{hp}^{fr}$, $(C(P + P)', f[\alpha \mapsto \alpha], C(P)') \in \sim_{hp}^{fr}$, $P + P \sim_{hp}^{fr} P$, as desired.

4. $P + \mathbf{nil} \sim_{hp}^{fr} P$. There are several cases, however, we will not enumerate them all here. By the forward transition rules of Summation in Table 3.1, we obtain

$$\frac{P \xrightarrow{\alpha} P'}{P + \mathbf{nil} \xrightarrow{\alpha} P'} (\alpha P) \qquad \frac{P \xrightarrow{\alpha} P'}{P \xrightarrow{\alpha} P'} (\alpha \in P)$$

By the reverse transition rules of Summation in Table 3.2, we obtain

$$\frac{P \xrightarrow{\alpha[m]} P'}{P + \mathbf{nil} \xrightarrow{\alpha[m]} P'} (\alpha \in P) \qquad \frac{P \xrightarrow{\alpha[m]} P'}{P \xrightarrow{\alpha[m]} P'} (\alpha \in P)$$

Since $(C(P + \mathbf{nil}), f, C(P)) \in \sim_{hp}^{fr}$, $(C(P + \mathbf{nil})', f[\alpha \mapsto \alpha], C(P)') \in \sim_{hp}^{fr}$, $P + \mathbf{nil} \sim_{hp}^{fr} P$, as desired. □

**Proposition 3.9** (Monoid laws for strongly FR hhp-bisimulation). *The monoid laws for strongly FR hhp-bisimulation are as follows:*

1. $P + Q \sim_{hhp}^{fr} Q + P$;
2. $P + (Q + R) \sim_{hhp}^{fr} (P + Q) + R$;
3. $P + P \sim_{hhp}^{fr} P$;
4. $P + \mathbf{nil} \sim_{hhp}^{fr} P$.

*Proof.* **1.** $P + Q \sim^{fr}_{hhp} Q + P$. There are several cases, however, we will not enumerate them all here. By the forward transition rules of Summation in Table 3.1, we obtain

$$\frac{P \xrightarrow{\alpha} P' \quad \alpha \notin Q}{P + Q \xrightarrow{\alpha} P' + Q}(\alpha \in P) \qquad \frac{P \xrightarrow{\alpha} P' \quad \alpha \notin Q}{Q + P \xrightarrow{\alpha} Q + P'}(\alpha \in P)$$

$$\frac{Q \xrightarrow{\beta} Q' \quad \beta \notin P}{P + Q \xrightarrow{\beta} P + Q'}(\beta \in Q) \qquad \frac{Q \xrightarrow{\beta} Q' \quad \beta \notin P}{Q + P \xrightarrow{\beta} Q' + P}(\beta \in Q)$$

By the reverse transition rules of Summation in Table 3.2, we obtain

$$\frac{P \xrightarrow{\alpha[m]} P' \quad \alpha \notin Q}{P + Q \xrightarrow{\alpha[m]} P' + Q}(\alpha \in P) \qquad \frac{P \xrightarrow{\alpha[m]} P' \quad \alpha \notin Q}{Q + P \xrightarrow{\alpha[m]} Q + P'}(\alpha P)$$

$$\frac{Q \xrightarrow{\beta[n]} Q' \quad \beta \notin P}{P + Q \xrightarrow{\beta[n]} P + Q'}(\beta \in Q) \qquad \frac{Q \xrightarrow{\beta[n]} Q' \quad \beta \notin P}{Q + P \xrightarrow{\beta[n]} Q' + P}(\beta \in Q)$$

Since $(C(P+Q), f, C(Q+P)) \in \sim^{fr}_{hhp}$, $(C(P+Q)', f[\alpha \mapsto \alpha], C(Q+P)') \in \sim^{fr}_{hhp}$ and $(C((P+Q)'), f[\beta \mapsto \beta], C(Q+P)') \in \sim^{fr}_{hhp}$, $P + Q \sim^{fr}_{hhp} Q + P$, as desired.

**2.** $P + (Q + R) \sim^{fr}_{hhp} (P + Q) + R$. There are several cases, however, we will not enumerate them all here. By the forward transition rules of Summation in Table 3.1, we obtain

$$\frac{P \xrightarrow{\alpha} P' \quad \alpha \notin Q \quad \alpha \notin R}{P + (Q + R) \xrightarrow{\alpha} P' + (Q + R)}(\alpha \in P) \qquad \frac{P \xrightarrow{\alpha} P' \quad \alpha \notin Q \quad \alpha \notin R}{(P + Q) + R \xrightarrow{\alpha} (P' + Q) + R}(\alpha \in P)$$

$$\frac{Q \xrightarrow{\beta} Q' \quad \beta \notin P \quad \beta \notin R}{P + (Q + R) \xrightarrow{\beta} P + (Q' + R)}(\beta \in Q) \qquad \frac{Q \xrightarrow{\beta} Q' \quad \beta \notin P \quad \beta \notin R}{(P + Q) + R \xrightarrow{\beta} (P + Q') + R}(\beta \in Q)$$

$$\frac{R \xrightarrow{\gamma} R' \quad \gamma \notin P \quad \gamma \notin Q}{P + (Q + R) \xrightarrow{\gamma} P + (Q + R')}(\gamma \in R) \qquad \frac{R \xrightarrow{\gamma} R' \quad \gamma \notin P \quad \gamma \notin Q}{(P + Q) + R \xrightarrow{\gamma} (P + Q) + R'}(\gamma \in R)$$

By the reverse transition rules of Summation in Table 3.2, we obtain

$$\frac{P \xrightarrow{\alpha[m]} P' \quad \alpha \notin Q \quad \alpha \notin R}{P + (Q + R) \xrightarrow{\alpha[m]} P' + (Q + R)}(\alpha \in P) \qquad \frac{P \xrightarrow{\alpha[m]} P' \quad \alpha \notin Q \quad \alpha \notin R}{(P + Q) + R \xrightarrow{\alpha[m]} (P' + Q) + R}(\alpha \in P)$$

$$\frac{Q \xrightarrow{\beta[n]} Q' \quad \beta \notin P \quad \beta \notin R}{P + (Q + R) \xrightarrow{\beta[n]} P + (Q' + R)}(\beta \in Q) \qquad \frac{Q \xrightarrow{\beta[n]} Q' \quad \beta \notin P \quad \beta \notin R}{(P + Q) + R \xrightarrow{\beta[n]} (P + Q') + R}(\beta Q)$$

$$\frac{R \xrightarrow{\gamma[k]} R' \quad \gamma \notin P \quad \gamma \notin Q}{P + (Q + R) \xrightarrow{\gamma[k]} P + (Q + R')}(\gamma \in R) \qquad \frac{R \xrightarrow{\gamma[k]} R' \quad \gamma \notin P \quad \gamma \notin Q}{(P + Q) + R \xrightarrow{\gamma[k]} (P + Q) + R'}(\gamma \in R)$$

Since $(C(P+(Q+R)), f, C((P+Q)+R)) \in \sim^{fr}_{hhp}$, $(C(P+(Q+R))', f[\alpha \mapsto \alpha], C((P+Q)+R)') \in \sim^{fr}_{hhp}$, $(C(P+(Q+R))', f[\beta \mapsto \beta], C((P+Q)+R)') \in \sim^{fr}_{hhp}$ and $(C(P+(Q+R))', f[\gamma \mapsto \gamma], C((P+Q)+R)') \in \sim^{fr}_{hhp}$, $P + (Q + R) \sim^{fr}_{hhp} (P + Q) + R$, as desired.

**3.** $P + P \sim^{fr}_{hhp} P$. By the forward transition rules of Summation, we obtain

$$\frac{P \xrightarrow{\alpha} \alpha[m]}{P + P \xrightarrow{\alpha} \alpha[m] + \alpha[m]}(\alpha \in P) \qquad \frac{P \xrightarrow{\alpha} \alpha[m]}{P \xrightarrow{\alpha} \alpha[m]}(\alpha \in P)$$

$$\frac{P \xrightarrow{\alpha} P'}{P + P \xrightarrow{\alpha} P' + P'}(p \subseteq P) \quad \frac{P \xrightarrow{\alpha} P'}{P \xrightarrow{\alpha} P'}(\alpha \in P)$$

By the reverse transition rules of Summation, we obtain

$$\frac{P \xrightarrow{\alpha[m]} \alpha}{P + P \xrightarrow{\alpha[m]} \alpha + \alpha}(\alpha \in P) \quad \frac{P \xrightarrow{\alpha[m]} \alpha}{P \xrightarrow{\alpha[m]} \alpha}(\alpha \in P)$$

$$\frac{P \xrightarrow{\alpha[m]} P'}{P + P \xrightarrow{\alpha[m]} P' + P'}(\alpha \in P) \quad \frac{P \xrightarrow{\alpha[m]} P'}{P \xrightarrow{\alpha[m} P'}(\alpha \in P)$$

Since $(C(P+P), f, C(P)) \in \sim_{hhp}^{fr}$, $(C(P+P)', f[\alpha \mapsto \alpha], C(P)') \in \sim_{hhp}^{fr}$, $P + P \sim_{hhp}^{fr} P$, as desired.

4. $P + \mathbf{nil} \sim_{hhp}^{fr} P$. There are several cases, however, we will not enumerate them all here. By the forward transition rules of Summation in Table 3.1, we obtain

$$\frac{P \xrightarrow{\alpha} P'}{P + \mathbf{nil} \xrightarrow{\alpha} P'}(\alpha P) \quad \frac{P \xrightarrow{\alpha} P'}{P \xrightarrow{\alpha} P'}(\alpha \in P)$$

By the reverse transition rules of Summation in Table 3.2, we obtain

$$\frac{P \xrightarrow{\alpha[m]} P'}{P + \mathbf{nil} \xrightarrow{\alpha[m]} P'}(\alpha \in P) \quad \frac{P \xrightarrow{\alpha[m]} P'}{P \xrightarrow{\alpha[m]} P'}(\alpha \in P)$$

Since $(C(P + \mathbf{nil}), f, C(P)) \in \sim_{hhp}^{fr}$, $(C(P + \mathbf{nil})', f[\alpha \mapsto \alpha], C(P)') \in \sim_{hhp}^{fr}$, $P + \mathbf{nil} \sim_{hhp}^{fr} P$, as desired. □

**Proposition 3.10** (Static laws for strongly FR step bisimulation). *The static laws for strongly FR step bisimulation are as follows:*

1. $P \parallel Q \sim_s^{fr} Q \parallel P$;
2. $P \parallel (Q \parallel R) \sim_s^{fr} (P \parallel Q) \parallel R$;
3. $P \parallel \mathbf{nil} \sim_s^{fr} P$;
4. $P \setminus L \sim_s^{fr} P$, *if* $\mathcal{L}(P) \cap (L \cup \overline{L}) = \emptyset$;
5. $P \setminus K \setminus L \sim_s^{fr} P \setminus (K \cup L)$;
6. $P[f] \setminus L \sim_s^{fr} P \setminus f^{-1}(L)[f]$;
7. $(P \parallel Q) \setminus L \sim_s^{fr} P \setminus L \parallel Q \setminus L$, *if* $\mathcal{L}(P) \cap \overline{\mathcal{L}(Q)} \cap (L \cup \overline{L}) = \emptyset$;
8. $P[Id] \sim_s^{fr} P$;
9. $P[f] \sim_s^{fr} P[f']$, *if* $f \restriction \mathcal{L}(P) = f' \restriction \mathcal{L}(P)$;
10. $P[f][f'] \sim_s^{fr} P[f' \circ f]$;
11. $(P \parallel Q)[f] \sim_s^{fr} P[f] \parallel Q[f]$, *if* $f \restriction (L \cup \overline{L})$ *is one-to-one, where* $L = \mathcal{L}(P) \cup \mathcal{L}(Q)$.

*Proof.* Although transition rules in Tables 3.5, 3.6, 3.7, and 3.8 are defined in the flavor of a single event, they can be modified into a step (a set of events within which each event is pairwise concurrent), however, we omit them here. If we treat a single event as a step containing just one event, the proof of the static laws does not pose any problem, hence, we use this way and still use the transition rules in Tables 3.5, 3.6, 3.7, and 3.8.

1. $P \parallel Q \sim_s^{fr} Q \parallel P$. By the forward transition rules of Composition, we obtain

$$\frac{P \xrightarrow{\alpha} P' \quad Q \nrightarrow}{P \parallel Q \xrightarrow{\alpha} P' \parallel Q} \quad \frac{P \xrightarrow{\alpha} P' \quad Q \nrightarrow}{Q \parallel P \xrightarrow{\alpha} Q \parallel P'}$$

$$\frac{Q \xrightarrow{\beta} Q' \quad P \nrightarrow}{P \parallel Q \xrightarrow{\beta} P \parallel Q'} \quad \frac{Q \xrightarrow{\beta} Q' \quad P \nrightarrow}{Q \parallel P \xrightarrow{\beta} Q' \parallel P}$$

$$\frac{P \xrightarrow{\alpha} P' \quad Q \xrightarrow{\beta} Q'}{P \parallel Q \xrightarrow{\{\alpha,\beta\}} P' \parallel Q'} (\beta \neq \overline{\alpha}) \quad \frac{P \xrightarrow{\alpha} P' \quad Q \xrightarrow{\beta} Q'}{Q \parallel P \xrightarrow{\{\alpha,\beta\}} Q' \parallel P'} (\beta \neq \overline{\alpha})$$

$$\frac{P \xrightarrow{l} P' \quad Q \xrightarrow{\overline{l}} Q'}{P \parallel Q \xrightarrow{\tau} P' \parallel Q'} \quad \frac{P \xrightarrow{l} P' \quad Q \xrightarrow{\overline{l}} Q'}{Q \parallel P \xrightarrow{\tau} Q' \parallel P'}$$

By the reverse transition rules of Composition, we obtain

$$\frac{P \xrightarrow{\alpha[m]} P' \quad Q \not\twoheadrightarrow}{P \parallel Q \xrightarrow{\alpha[m]} P' \parallel Q} \quad \frac{P \xrightarrow{\alpha[m]} P' \quad Q \not\twoheadrightarrow}{Q \parallel P \xrightarrow{\alpha[m]} Q \parallel P'}$$

$$\frac{Q \xrightarrow{\beta[n]} Q' \quad P \not\twoheadrightarrow}{P \parallel Q \xrightarrow{\beta[n]} P \parallel Q'} \quad \frac{Q \xrightarrow{\beta[n]} Q' \quad P \not\twoheadrightarrow}{Q \parallel P \xrightarrow{\beta[n]} Q' \parallel P}$$

$$\frac{P \xrightarrow{\alpha[m]} P' \quad Q \xrightarrow{\beta[m]} Q'}{P \parallel Q \xrightarrow{\{\alpha[m],\beta[m]\}} P' \parallel Q'} (\beta \neq \overline{\alpha}) \quad \frac{P \xrightarrow{\alpha[m]} P' \quad Q \xrightarrow{\beta[m]} Q'}{Q \parallel P \xrightarrow{\{\alpha[m],\beta[m]\}} Q' \parallel P'} (\beta \neq \overline{\alpha})$$

$$\frac{P \xrightarrow{l[m]} P' \quad Q \xrightarrow{\overline{l}[m]} Q'}{P \parallel Q \xrightarrow{\tau} P' \parallel Q'} \quad \frac{P \xrightarrow{l[m]} P' \quad Q \xrightarrow{\overline{l}[m]} Q'}{Q \parallel P \xrightarrow{\tau} Q' \parallel P'}$$

Hence, with the assumptions $P' \parallel Q \sim_s^{fr} Q \parallel P'$, $P \parallel Q' \sim_s^{fr} Q' \parallel P$ and $P' \parallel Q' \sim_s^{fr} Q' \parallel P'$, $P \parallel Q \sim_s^{fr} Q \parallel P$, as desired.

**2.** $P \parallel (Q \parallel R) \sim_s^{fr} (P \parallel Q) \parallel R$. By the forward transition rules of Composition, we obtain

$$\frac{P \xrightarrow{\alpha} P' \quad Q \nrightarrow \quad R \nrightarrow}{P \parallel (Q \parallel R) \xrightarrow{\alpha} P' \parallel (Q \parallel R)} \quad \frac{P \xrightarrow{\alpha} P' \quad Q \nrightarrow \quad R \nrightarrow}{(P \parallel Q) \parallel R \xrightarrow{\alpha} (P' \parallel Q) \parallel R}$$

$$\frac{Q \xrightarrow{\beta} Q' \quad P \nrightarrow \quad R \nrightarrow}{P \parallel (Q \parallel R) \xrightarrow{\beta} P \parallel (Q' \parallel R)} \quad \frac{Q \xrightarrow{\beta} Q' \quad P \nrightarrow \quad R \nrightarrow}{(P \parallel Q) \parallel R \xrightarrow{\beta} (P \parallel Q') \parallel R}$$

$$\frac{R \xrightarrow{\gamma} R' \quad P \nrightarrow \quad Q \nrightarrow}{P \parallel (Q \parallel R) \xrightarrow{\gamma} P \parallel (Q \parallel R')} \quad \frac{R \xrightarrow{\gamma} R' \quad P \nrightarrow \quad Q \nrightarrow}{(P \parallel Q) \parallel R \xrightarrow{\gamma} (P \parallel Q) \parallel R'}$$

$$\frac{P \xrightarrow{\alpha} P' \quad Q \xrightarrow{\beta} Q' \quad R \nrightarrow}{P \parallel (Q \parallel R) \xrightarrow{\{\alpha,\beta\}} P' \parallel (Q' \parallel R)} (\beta \neq \overline{\alpha}) \quad \frac{P \xrightarrow{\alpha} P' \quad Q \xrightarrow{\beta} Q' \quad R \nrightarrow}{(P \parallel Q) \parallel R \xrightarrow{\{\alpha,\beta\}} (P' \parallel Q') \parallel R} (\beta \neq \overline{\alpha})$$

$$\frac{P \xrightarrow{\alpha} P' \quad R \xrightarrow{\gamma} R' \quad Q \nrightarrow}{P \parallel (Q \parallel R) \xrightarrow{\{\alpha,\gamma\}} P' \parallel (Q \parallel R')} (\gamma \neq \overline{\alpha}) \quad \frac{P \xrightarrow{\alpha} P' \quad R \xrightarrow{\gamma} R' \quad Q \nrightarrow}{(P \parallel Q) \parallel R \xrightarrow{\{\alpha,\gamma\}} (P' \parallel Q) \parallel R]} (\gamma \neq \overline{\alpha})$$

$$\frac{Q \xrightarrow{\beta} P' \quad R \xrightarrow{\gamma} R' \quad P \nrightarrow}{P \parallel (Q \parallel R) \xrightarrow{\{\beta,\gamma\}} P \parallel (Q' \parallel R')} (\gamma \neq \overline{\beta}) \quad \frac{Q \xrightarrow{\beta} Q' \quad R \xrightarrow{\gamma} R' \quad P \nrightarrow}{(P \parallel Q) \parallel R \xrightarrow{\{\beta,\gamma\}} (P \parallel Q') \parallel R'} (\gamma \neq \overline{\beta})$$

$$\frac{P \xrightarrow{\alpha} P' \quad Q \xrightarrow{\beta} Q' \quad R \xrightarrow{\gamma} R'}{P \parallel (Q \parallel R) \xrightarrow{\{\alpha,\beta,\gamma\}} P' \parallel (Q' \parallel R')} (\beta \neq \overline{\alpha}, \gamma \neq \overline{\alpha}, \gamma \neq \overline{\beta})$$

$$\frac{P \xrightarrow{\alpha} P' \quad Q \xrightarrow{\beta} Q' \quad R \xrightarrow{\gamma} R'}{(P \parallel Q) \parallel R \xrightarrow{\{\alpha,\beta,\gamma\}} (P' \parallel Q') \parallel R'} (\beta \neq \overline{\alpha}, \gamma \neq \overline{\alpha}, \gamma \neq \overline{\beta})$$

$$\frac{P \xrightarrow{l} P' \quad Q \xrightarrow{\bar{l}} Q' \quad R \nrightarrow}{P \parallel (Q \parallel R) \xrightarrow{\tau} P' \parallel (Q' \parallel R)} \quad \frac{P \xrightarrow{l} P' \quad Q \xrightarrow{\bar{l}} Q' \quad R \nrightarrow}{(P \parallel Q) \parallel R \xrightarrow{\tau} (P' \parallel Q') \parallel R}$$

$$\frac{P \xrightarrow{l} P' \quad R \xrightarrow{\bar{l}} R' \quad Q \nrightarrow}{P \parallel (Q \parallel R) \xrightarrow{\tau} P' \parallel (Q \parallel R')} \quad \frac{P \xrightarrow{l} P' \quad R \xrightarrow{\bar{l}} R' \quad Q \nrightarrow}{(P \parallel Q) \parallel R \xrightarrow{\tau} (P' \parallel Q) \parallel R]}$$

$$\frac{Q \xrightarrow{l} P' \quad R \xrightarrow{\bar{l}} R' \quad P \nrightarrow}{P \parallel (Q \parallel R) \xrightarrow{\tau} P \parallel (Q' \parallel R')} \quad \frac{Q \xrightarrow{l} Q' \quad R \xrightarrow{\bar{l}} R' \quad P \nrightarrow}{(P \parallel Q) \parallel R \xrightarrow{\tau} (P \parallel Q') \parallel R'}$$

$$\frac{P \xrightarrow{l} P' \quad Q \xrightarrow{\bar{l}} Q' \quad R \xrightarrow{\gamma} R'}{P \parallel (Q \parallel R) \xrightarrow{\{\tau,\gamma\}} P' \parallel (Q' \parallel R')} \quad \frac{P \xrightarrow{l} P' \quad Q \xrightarrow{\bar{l}} Q' \quad R \xrightarrow{\gamma} R'}{(P \parallel Q) \parallel R \xrightarrow{\{\tau,\gamma\}} (P' \parallel Q') \parallel R'}$$

$$\frac{P \xrightarrow{l} P' \quad R \xrightarrow{\bar{l}} R' \quad Q \xrightarrow{\beta} Q'}{P \parallel (Q \parallel R) \xrightarrow{\{\tau,\beta\}} P' \parallel (Q' \parallel R')} \quad \frac{P \xrightarrow{l} P' \quad R \xrightarrow{\bar{l}} R' \quad Q \xrightarrow{\beta} Q'}{(P \parallel Q) \parallel R \xrightarrow{\{\tau,\beta\}} (P' \parallel Q') \parallel R]}$$

$$\frac{Q \xrightarrow{l} Q' \quad R \xrightarrow{\bar{l}} R' \quad P \xrightarrow{\alpha} P'}{P \parallel (Q \parallel R) \xrightarrow{\{\tau,\alpha\}} P' \parallel (Q' \parallel R')} \quad \frac{Q \xrightarrow{l} Q' \quad R \xrightarrow{\bar{l}} R' \quad P \xrightarrow{\alpha} P'}{(P \parallel Q) \parallel R \xrightarrow{\{\tau,\alpha\}} (P' \parallel Q') \parallel R'}$$

By the reverse transition rules of Composition, we obtain

$$\frac{P \overset{\alpha[m]}{\twoheadrightarrow} P' \quad Q \not\twoheadrightarrow \quad R \not\twoheadrightarrow}{P \parallel (Q \parallel R) \overset{\alpha[m]}{\twoheadrightarrow} P' \parallel (Q \parallel R)} \quad \frac{P \overset{\alpha[m]}{\twoheadrightarrow} P' \quad Q \not\twoheadrightarrow \quad R \not\twoheadrightarrow}{(P \parallel Q) \parallel R \overset{\alpha[m]}{\twoheadrightarrow} (P' \parallel Q) \parallel R}$$

$$\frac{Q \overset{\beta[n]}{\twoheadrightarrow} Q' \quad P \not\twoheadrightarrow \quad R \not\twoheadrightarrow}{P \parallel (Q \parallel R) \overset{\beta[n]}{\twoheadrightarrow} P \parallel (Q' \parallel R)} \quad \frac{Q \overset{\beta[n]}{\twoheadrightarrow} Q' \quad P \not\twoheadrightarrow \quad R \not\twoheadrightarrow}{(P \parallel Q) \parallel R \overset{\beta[n]}{\twoheadrightarrow} (P \parallel Q') \parallel R}$$

$$\frac{R \overset{\gamma[k]}{\twoheadrightarrow} R' \quad P \not\twoheadrightarrow \quad Q \not\twoheadrightarrow}{P \parallel (Q \parallel R) \overset{\gamma[k]}{\twoheadrightarrow} P \parallel (Q \parallel R')} \quad \frac{R \overset{\gamma[k]}{\twoheadrightarrow} R' \quad P \not\twoheadrightarrow \quad Q \not\twoheadrightarrow}{(P \parallel Q) \parallel R \overset{\gamma[k]}{\twoheadrightarrow} (P \parallel Q) \parallel R'}$$

$$\frac{P \overset{\alpha[m]}{\twoheadrightarrow} P' \quad Q \overset{\beta[m]}{\twoheadrightarrow} Q' \quad R \not\twoheadrightarrow}{P \parallel (Q \parallel R) \overset{\{\alpha[m],\beta[m]\}}{\twoheadrightarrow} P' \parallel (Q' \parallel R)} (\beta \neq \overline{\alpha}) \quad \frac{P \overset{\alpha[m]}{\twoheadrightarrow} P' \quad Q \overset{\beta[m]}{\twoheadrightarrow} Q' \quad R \not\twoheadrightarrow}{(P \parallel Q) \parallel R \overset{\{\alpha[m],\beta[m]\}}{\twoheadrightarrow} (P' \parallel Q') \parallel R} (\beta \neq \overline{\alpha})$$

$$\frac{P \overset{\alpha[m]}{\twoheadrightarrow} P' \quad R \overset{\gamma[m]}{\twoheadrightarrow} R' \quad Q \not\twoheadrightarrow}{P \parallel (Q \parallel R) \overset{\{\alpha[m],\gamma[m]\}}{\twoheadrightarrow} P' \parallel (Q \parallel R')} (\gamma \neq \overline{\alpha}) \quad \frac{P \overset{\alpha[m]}{\twoheadrightarrow} P' \quad R \overset{\gamma[m]}{\twoheadrightarrow} R' \quad Q \not\twoheadrightarrow}{(P \parallel Q) \parallel R \overset{\{\alpha[m],\gamma[m]\}}{\twoheadrightarrow} (P' \parallel Q) \parallel R]} (\gamma \neq \overline{\alpha})$$

$$\frac{Q \overset{\beta[m]}{\twoheadrightarrow} P' \quad R \overset{\gamma[m]}{\twoheadrightarrow} R' \quad P \not\twoheadrightarrow}{P \parallel (Q \parallel R) \overset{\{\beta[m],\gamma[m]\}}{\twoheadrightarrow} P \parallel (Q' \parallel R')} (\gamma \neq \overline{\beta}) \quad \frac{Q \overset{\beta[m]}{\twoheadrightarrow} Q' \quad R \overset{\gamma[m]}{\twoheadrightarrow} R' \quad P \not\twoheadrightarrow}{(P \parallel Q) \parallel R \overset{\{\beta[m],\gamma[m]\}}{\twoheadrightarrow} (P \parallel Q') \parallel R'} (\gamma \neq \overline{\beta})$$

$$\frac{P \overset{\alpha[m]}{\twoheadrightarrow} P' \quad Q \overset{\beta[m]}{\twoheadrightarrow} Q' \quad R \overset{\gamma[m]}{\twoheadrightarrow} R'}{P \parallel (Q \parallel R) \overset{\{\alpha[m],\beta[m],\gamma[m]\}}{\twoheadrightarrow} P' \parallel (Q' \parallel R')} (\beta \neq \overline{\alpha}, \gamma \neq \overline{\alpha}, \gamma \neq \overline{\beta})$$

$$\frac{P \overset{\alpha[m]}{\twoheadrightarrow} P' \quad Q \overset{\beta[m]}{\twoheadrightarrow} Q' \quad R \overset{\gamma[m]}{\twoheadrightarrow} R'}{(P \parallel Q) \parallel R \overset{\{\alpha[m],\beta[m],\gamma[m]\}}{\twoheadrightarrow} (P' \parallel Q') \parallel R'} (\beta \neq \overline{\alpha}, \gamma \neq \overline{\alpha}, \gamma \neq \overline{\beta})$$

$$\frac{P \overset{l[m]}{\twoheadrightarrow} P' \quad Q \overset{\bar{l}[m]}{\twoheadrightarrow} Q' \quad R \not\twoheadrightarrow}{P \parallel (Q \parallel R) \overset{\tau}{\twoheadrightarrow} P' \parallel (Q' \parallel R)} \quad \frac{P \overset{l[m]}{\twoheadrightarrow} P' \quad Q \overset{\bar{l}[m]}{\twoheadrightarrow} Q' \quad R \not\twoheadrightarrow}{(P \parallel Q) \parallel R \overset{\tau}{\twoheadrightarrow} (P' \parallel Q') \parallel R}$$

$$\frac{P\overset{l[m]}{\twoheadrightarrow}P' \quad R\overset{\bar{l}[m]}{\twoheadrightarrow}R' \quad Q\not\twoheadrightarrow}{P\parallel(Q\parallel R)\overset{\tau}{\twoheadrightarrow}P'\parallel(Q\parallel R')} \quad \frac{P\overset{l[m]}{\twoheadrightarrow}P' \quad R\overset{\bar{l}[m]}{\twoheadrightarrow}R' \quad Q\not\twoheadrightarrow}{(P\parallel Q)\parallel R\overset{\tau}{\twoheadrightarrow}(P'\parallel Q)\parallel R]}$$

$$\frac{Q\overset{l[m]}{\twoheadrightarrow}P' \quad R\overset{\bar{l}[m]}{\twoheadrightarrow}R' \quad P\not\twoheadrightarrow}{P\parallel(Q\parallel R)\overset{\tau}{\twoheadrightarrow}P\parallel(Q'\parallel R')} \quad \frac{Q\overset{l[m]}{\twoheadrightarrow}Q' \quad R\overset{\bar{l}[m]}{\twoheadrightarrow}R' \quad P\not\twoheadrightarrow}{(P\parallel Q)\parallel R\overset{\tau}{\twoheadrightarrow}(P\parallel Q')\parallel R'}$$

$$\frac{P\overset{l[m]}{\twoheadrightarrow}P' \quad Q\overset{\bar{l}[m]}{\twoheadrightarrow}Q' \quad R\overset{\gamma[m]}{\twoheadrightarrow}R'}{P\parallel(Q\parallel R)\overset{\{\tau,\gamma[m]\}}{\twoheadrightarrow}P'\parallel(Q'\parallel R')} \quad \frac{P\overset{l[m]}{\twoheadrightarrow}P' \quad Q\overset{\bar{l}[m]}{\twoheadrightarrow}Q' \quad R\overset{\gamma[m]}{\twoheadrightarrow}R'}{(P\parallel Q)\parallel R\overset{\{\tau,\gamma[m]\}}{\twoheadrightarrow}(P'\parallel Q')\parallel R'}$$

$$\frac{P\overset{l[m]}{\twoheadrightarrow}P' \quad R\overset{\bar{l}[m]}{\twoheadrightarrow}R' \quad Q\overset{\beta[m]}{\twoheadrightarrow}Q'}{P\parallel(Q\parallel R)\overset{\{\tau,\beta[m]\}}{\twoheadrightarrow}P'\parallel(Q'\parallel R')} \quad \frac{P\overset{l[m]}{\twoheadrightarrow}P' \quad R\overset{\bar{l}[m]}{\twoheadrightarrow}R' \quad Q\overset{\beta[m]}{\twoheadrightarrow}Q'}{(P\parallel Q)\parallel R\overset{\{\tau,\beta[m]\}}{\twoheadrightarrow}(P'\parallel Q')\parallel R]}$$

$$\frac{Q\overset{l[m]}{\twoheadrightarrow}Q' \quad R\overset{\bar{l}[m]}{\twoheadrightarrow}R' \quad P\overset{\alpha[m]}{\twoheadrightarrow}P'}{P\parallel(Q\parallel R)\overset{\{\tau,\alpha[m]\}}{\twoheadrightarrow}P'\parallel(Q'\parallel R')} \quad \frac{Q\overset{l[m]}{\twoheadrightarrow}Q' \quad R\overset{\bar{l}[m]}{\twoheadrightarrow}R' \quad P\overset{\alpha[m]}{\twoheadrightarrow}P'}{(P\parallel Q)\parallel R\overset{\{\tau,\alpha[m]\}}{\twoheadrightarrow}(P'\parallel Q')\parallel R'}$$

Hence, with the assumptions $P'\parallel(Q\parallel R)\sim_s^{fr}(P'\parallel Q)\parallel R$, $P\parallel(Q'\parallel R)\sim_s^{fr}(P\parallel Q')\parallel R$, $P\parallel(Q\parallel R')\sim_s^{fr}(P\parallel Q)\parallel R'$, $P'\parallel(Q'\parallel R)\sim_s^{fr}(P'\parallel Q')\parallel R$, $P'\parallel(Q\parallel R')\sim_s^{fr}(P'\parallel Q)\parallel R'$, $P\parallel(Q'\parallel R')\sim_s^{fr}(P\parallel Q')\parallel R'$ and $P'\parallel(Q'\parallel R')\sim_s^{fr}(P'\parallel Q')\parallel R'$, $P\parallel(Q\parallel R)\sim_s^{fr}(P\parallel Q)\parallel R$, as desired.

**3.** $P\parallel\mathbf{nil}\sim_s^{fr}P$. By the forward transition rules of Composition, we obtain

$$\frac{P\xrightarrow{\alpha}P'}{P\parallel\mathbf{nil}\xrightarrow{\alpha}P'} \quad \frac{P\xrightarrow{\alpha}P'}{P\xrightarrow{\alpha}P'}$$

By the reverse transition rules of Composition, we obtain

$$\frac{P\overset{\alpha[m]}{\twoheadrightarrow}P'}{P\parallel\mathbf{nil}\overset{\alpha[m]}{\twoheadrightarrow}P'} \quad \frac{P\overset{\alpha[m]}{\twoheadrightarrow}P'}{P\overset{\alpha[m]}{\twoheadrightarrow}P'}$$

Since $P'\sim_s^{fr}P'$, $P\parallel\mathbf{nil}\sim_s^{fr}P$, as desired.

**4.** $P\setminus L\sim_s^{fr}P$, if $\mathcal{L}(P)\cap(L\cup\overline{L})=\emptyset$. By the forward transition rules of Restriction, we obtain

$$\frac{P\xrightarrow{\alpha}P'}{P\setminus L\xrightarrow{\alpha}P'\setminus L}(\mathcal{L}(P)\cap(L\cup\overline{L})=\emptyset) \quad \frac{P\xrightarrow{\alpha}P'}{P\xrightarrow{\alpha}P'}$$

By the reverse transition rules of Restriction, we obtain

$$\frac{P\overset{\alpha[m]}{\twoheadrightarrow}P'}{P\setminus L\overset{\alpha[m]}{\twoheadrightarrow}P'\setminus L}(\mathcal{L}(P)\cap(L\cup\overline{L})=\emptyset) \quad \frac{P\overset{\alpha[m]}{\twoheadrightarrow}P'}{P\overset{\alpha[m]}{\twoheadrightarrow}P'}$$

Since $P'\sim_s^{fr}P'$, and with the assumption $P'\setminus L\sim_s^{fr}P'$, $P\setminus L\sim_s^{fr}P$, if $\mathcal{L}(P)\cap(L\cup\overline{L})=\emptyset$, as desired.

**5.** $P\setminus K\setminus L\sim_s^{fr}P\setminus(K\cup L)$. By the forward transition rules of Restriction, we obtain

$$\frac{P\xrightarrow{\alpha}P'}{P\setminus K\setminus L\xrightarrow{\alpha}P'\setminus K\setminus L} \quad \frac{P\xrightarrow{\alpha}P'}{P\setminus(K\cup L)\xrightarrow{\alpha}P'\setminus(K\cup L)}$$

By the reverse transition rules of Restriction, we obtain

$$\frac{P\overset{\alpha[m]}{\twoheadrightarrow}P'}{P\setminus K\setminus L\overset{\alpha[m]}{\twoheadrightarrow}P'\setminus K\setminus L} \quad \frac{P\overset{\alpha[m]}{\twoheadrightarrow}P'}{P\setminus(K\cup L)\overset{\alpha[m]}{\twoheadrightarrow}P'\setminus(K\cup L)}$$

Since $P' \sim^{fr}_s P'$, and with the assumption $P' \setminus K \setminus L \sim^{fr}_s P' \setminus (K \cup L)$, $P \setminus K \setminus L \sim^{fr}_s P \setminus (K \cup L)$, as desired.

**6.** $P[f] \setminus L \sim^{fr}_s P \setminus f^{-1}(L)[f]$. By the forward transition rules of Restriction and Relabeling, we obtain

$$\frac{P \xrightarrow{\alpha} P'}{P[f] \setminus L \xrightarrow{f(\alpha)} P'[f] \setminus L} \quad \frac{P \xrightarrow{\alpha} P'}{P \setminus f^{-1}(L)[f] \xrightarrow{f(\alpha)} P' \setminus f^{-1}(L)[f]}$$

By the reverse transition rules of Restriction and Relabeling, we obtain

$$\frac{P \xtwoheadrightarrow{\alpha[m]} P'}{P[f] \setminus L \xtwoheadrightarrow{f(\alpha)[m]} P'[f] \setminus L} \quad \frac{P \xtwoheadrightarrow{\alpha[m]} P'}{P \setminus f^{-1}(L)[f] \xtwoheadrightarrow{f(\alpha)[m]} P' \setminus f^{-1}(L)[f]}$$

Hence, with the assumption $P'[f] \setminus L \sim^{fr}_s P' \setminus f^{-1}(L)[f]$, $P[f] \setminus L \sim^{fr}_s P \setminus f^{-1}(L)[f]$, as desired.

**7.** $(P \parallel Q) \setminus L \sim^{fr}_s P \setminus L \parallel Q \setminus L$, if $\mathcal{L}(P) \cap \overline{\mathcal{L}(Q)} \cap (L \cup \overline{L}) = \emptyset$. By the forward transition rules of Composition and Restriction, we obtain

$$\frac{P \xrightarrow{\alpha} P' \quad Q \nrightarrow}{(P \parallel Q) \setminus L \xrightarrow{\alpha} (P' \parallel Q) \setminus L} (\mathcal{L}(P) \cap \overline{\mathcal{L}(Q)} \cap (L \cup \overline{L}) = \emptyset)$$

$$\frac{P \xrightarrow{\alpha} P' \quad Q \nrightarrow}{P \setminus L \parallel Q \setminus L \xrightarrow{\alpha} P' \setminus L \parallel Q \setminus L} (\mathcal{L}(P) \cap \overline{\mathcal{L}(Q)} \cap (L \cup \overline{L}) = \emptyset)$$

$$\frac{Q \xrightarrow{\beta} Q' \quad P \nrightarrow}{(P \parallel Q) \setminus L \xrightarrow{\beta} (P \parallel Q') \setminus L} (\mathcal{L}(P) \cap \overline{\mathcal{L}(Q)} \cap (L \cup \overline{L}) = \emptyset)$$

$$\frac{Q \xrightarrow{\beta} Q' \quad P \nrightarrow}{P \setminus L \parallel Q \setminus L \xrightarrow{\beta} P \setminus L \parallel Q' \setminus L} (\mathcal{L}(P) \cap \overline{\mathcal{L}(Q)} \cap (L \cup \overline{L}) = \emptyset)$$

$$\frac{P \xrightarrow{\alpha} P' \quad Q \xrightarrow{\beta} Q'}{(P \parallel Q) \setminus L \xrightarrow{\{\alpha,\beta\}} (P' \parallel Q') \setminus L} (\mathcal{L}(P) \cap \overline{\mathcal{L}(Q)} \cap (L \cup \overline{L}) = \emptyset)$$

$$\frac{P \xrightarrow{\alpha} P' \quad Q \xrightarrow{\beta} Q'}{P \setminus L \parallel Q \setminus L \xrightarrow{\{\alpha,\beta\}} (P' \parallel Q') \setminus L} (\mathcal{L}(P) \cap \overline{\mathcal{L}(Q)} \cap (L \cup \overline{L}) = \emptyset)$$

$$\frac{P \xrightarrow{l} P' \quad Q \xrightarrow{\overline{l}} Q'}{(P \parallel Q) \setminus L \xrightarrow{\tau} (P' \parallel Q') \setminus L} (\mathcal{L}(P) \cap \overline{\mathcal{L}(Q)} \cap (L \cup \overline{L}) = \emptyset)$$

$$\frac{P \xrightarrow{l} P' \quad Q \xrightarrow{\overline{l}} Q'}{(P \setminus L \parallel Q \setminus L \xrightarrow{\tau} P' \setminus L \parallel Q' \setminus L} (\mathcal{L}(P) \cap \overline{\mathcal{L}(Q)} \cap (L \cup \overline{L}) = \emptyset)$$

By the reverse transition rules of Composition and Restriction, we obtain

$$\frac{P \xtwoheadrightarrow{\alpha[m]} P' \quad Q \not\twoheadrightarrow}{(P \parallel Q) \setminus L \xtwoheadrightarrow{\alpha[m]} (P' \parallel Q) \setminus L} (\mathcal{L}(P) \cap \overline{\mathcal{L}(Q)} \cap (L \cup \overline{L}) = \emptyset)$$

$$\frac{P \xtwoheadrightarrow{\alpha[m]} P' \quad Q \not\twoheadrightarrow}{P \setminus L \parallel Q \setminus L \xtwoheadrightarrow{\alpha[m]} P' \setminus L \parallel Q \setminus L} (\mathcal{L}(P) \cap \overline{\mathcal{L}(Q)} \cap (L \cup \overline{L}) = \emptyset)$$

$$\frac{Q \xtwoheadrightarrow{\beta[n]} Q' \quad P \not\twoheadrightarrow}{(P \parallel Q) \setminus L \xtwoheadrightarrow{\beta[n]} (P \parallel Q') \setminus L} (\mathcal{L}(P) \cap \overline{\mathcal{L}(Q)} \cap (L \cup \overline{L}) = \emptyset)$$

$$\frac{Q\overset{\beta[n]}{\twoheadrightarrow}Q'\quad P\not\twoheadrightarrow}{P\setminus L\parallel Q\setminus L\overset{\beta[n]}{\twoheadrightarrow}P\setminus L\parallel Q'\setminus L}(\mathcal{L}(P)\cap\overline{\mathcal{L}(Q)}\cap(L\cup\overline{L})=\emptyset)$$

$$\frac{P\overset{\alpha[m]}{\twoheadrightarrow}P'\quad Q\overset{\beta[m]}{\twoheadrightarrow}Q'}{(P\parallel Q)\setminus L\overset{\{\alpha[m],\beta[m]\}}{\twoheadrightarrow}(P'\parallel Q')\setminus L}(\mathcal{L}(P)\cap\overline{\mathcal{L}(Q)}\cap(L\cup\overline{L})=\emptyset)$$

$$\frac{P\overset{\alpha[m]}{\twoheadrightarrow}P'\quad Q\overset{\beta[m]}{\twoheadrightarrow}Q'}{P\setminus L\parallel Q\setminus L\overset{\{\alpha[m],\beta[m]\}}{\twoheadrightarrow}(P'\parallel Q')\setminus L}(\mathcal{L}(P)\cap\overline{\mathcal{L}(Q)}\cap(L\cup\overline{L})=\emptyset)$$

$$\frac{P\overset{l[m]}{\twoheadrightarrow}P'\quad Q\overset{\overline{l}[m]}{\twoheadrightarrow}Q'}{(P\parallel Q)\setminus L\overset{\tau}{\twoheadrightarrow}(P'\parallel Q')\setminus L}(\mathcal{L}(P)\cap\overline{\mathcal{L}(Q)}\cap(L\cup\overline{L})=\emptyset)$$

$$\frac{P\overset{l[m]}{\twoheadrightarrow}P'\quad Q\overset{\overline{l}[m]}{\twoheadrightarrow}Q'}{(P\setminus L\parallel Q\setminus L\overset{\tau}{\twoheadrightarrow}P'\setminus L\parallel Q'\setminus L}(\mathcal{L}(P)\cap\overline{\mathcal{L}(Q)}\cap(L\cup\overline{L})=\emptyset)$$

Since $(P'\parallel Q)\setminus L\sim_s^{fr}P'\setminus L\parallel Q\setminus L$, $(P\parallel Q')\setminus L\sim_s^{fr}P\setminus L\parallel Q'\setminus L$ and $(P'\parallel Q')\setminus L\sim_s^{fr}P'\setminus L\parallel Q'\setminus L$, $(P\parallel Q)\setminus L\sim_s^{fr}P\setminus L\parallel Q\setminus L$, if $\mathcal{L}(P)\cap\overline{\mathcal{L}(Q)}\cap(L\cup\overline{L})=\emptyset$, as desired.

**8.** $P[Id]\sim_s^{fr}P$. By the forward transition rules Relabeling, we obtain

$$\frac{P\xrightarrow{\alpha}P'}{P[Id]\xrightarrow{Id(\alpha)}P'[Id]}\qquad\frac{P\xrightarrow{\alpha}P'}{P\xrightarrow{\alpha}P'}$$

By the reverse transition rules Relabeling, we obtain

$$\frac{P\overset{\alpha[m]}{\twoheadrightarrow}P'}{P[Id]\overset{Id(\alpha[m])}{\twoheadrightarrow}P'[Id]}\qquad\frac{P\overset{\alpha[m]}{\twoheadrightarrow}P'}{P\overset{\alpha[m]}{\twoheadrightarrow}P'}$$

Hence, with the assumption $P'[Id]\sim_s^{fr}P'$ and $Id(\alpha)=\alpha$, $P[Id]\sim_s^{fr}P$, as desired.

**9.** $P[f]\sim_s^{fr}P[f']$, if $f\restriction\mathcal{L}(P)=f'\restriction\mathcal{L}(P)$. By the forward transition rules of Relabeling, we obtain

$$\frac{P\xrightarrow{\alpha}P'}{P[f]\xrightarrow{f(\alpha)}P'[f]}\qquad\frac{P\xrightarrow{\alpha}P'}{P[f']\xrightarrow{f'(\alpha)}P'[f']}$$

By the reverse transition rules of Relabeling, we obtain

$$\frac{P\overset{\alpha[m]}{\twoheadrightarrow}P'}{P[f]\overset{f(\alpha)[m]}{\twoheadrightarrow}P'[f]}\qquad\frac{P\overset{\alpha[m]}{\twoheadrightarrow}P'}{P[f']\overset{f'(\alpha)[m]}{\twoheadrightarrow}P'[f']}$$

Hence, with the assumption $P'[f]\sim_s^{fr}P'[f']$ and $f(\alpha)=f'(\alpha)$, if $f\restriction\mathcal{L}(P)=f'\restriction\mathcal{L}(P)$, $P[f]\sim_s^{fr}P[f']$, as desired.

**10.** $P[f][f']\sim_s^{fr}P[f'\circ f]$. By the forward transition rules of Relabeling, we obtain

$$\frac{P\xrightarrow{\alpha}P'}{P[f][f']\xrightarrow{f'(f(\alpha))}P'[f][f']}\qquad\frac{P\xrightarrow{\alpha}P'}{P[f'\circ f]\xrightarrow{f'(f(\alpha))}P'[f'\circ f]}$$

By the reverse transition rules of Relabeling, we obtain

$$\frac{P\overset{\alpha[m]}{\twoheadrightarrow}P'}{P[f][f']\xrightarrow{f'(f(\alpha))[m]}P'[f][f']}\qquad\frac{P\overset{\alpha[m]}{\twoheadrightarrow}P'}{P[f'\circ f]\overset{f'(f(\alpha))[m]}{\twoheadrightarrow}P'[f'\circ f]}$$

Hence, with the assumption $P'[f][f'] \sim_s^{fr} P'[f' \circ f]$, $P[f][f'] \sim_s^{fr} P[f' \circ f]$, as desired.

**11.** $(P \parallel Q)[f] \sim_s^{fr} P[f] \parallel Q[f]$, if $f \upharpoonright (L \cup \overline{L})$ is one-to-one, where $L = \mathcal{L}(P) \cup \mathcal{L}(Q)$. By the forward transition rules of Composition and Relabeling, we obtain

$$\frac{P \xrightarrow{\alpha} P' \quad Q \nrightarrow}{(P \parallel Q)[f] \xrightarrow{f(\alpha)} (P' \parallel Q)[f]}\text{(if } f \upharpoonright (L \cup \overline{L}) \text{ is one-to-one, where } L = \mathcal{L}(P) \cup \mathcal{L}(Q))$$

$$\frac{P \xrightarrow{\alpha} P' \quad Q \nrightarrow}{P[f] \parallel Q[f] \xrightarrow{\alpha} P'[f] \parallel Q[f]}\text{(if } f \upharpoonright (L \cup \overline{L}) \text{ is one-to-one, where } L = \mathcal{L}(P) \cup \mathcal{L}(Q))$$

$$\frac{Q \xrightarrow{\beta} Q' \quad P \nrightarrow}{(P \parallel Q)[f] \xrightarrow{f(\beta)} (P \parallel Q')[f]}\text{(if } f \upharpoonright (L \cup \overline{L}) \text{ is one-to-one, where } L = \mathcal{L}(P) \cup \mathcal{L}(Q))$$

$$\frac{Q \xrightarrow{\beta} Q' \quad P \nrightarrow}{P[f] \parallel Q[f] \xrightarrow{\beta} P[f] \parallel Q'[f]}\text{(if } f \upharpoonright (L \cup \overline{L}) \text{ is one-to-one, where } L = \mathcal{L}(P) \cup \mathcal{L}(Q))$$

$$\frac{P \xrightarrow{\alpha} P' \quad Q \xrightarrow{\beta} Q'}{(P \parallel Q)[f] \xrightarrow{\{f(\alpha), f(\beta)\}} (P' \parallel Q')[f]}\text{(if } f \upharpoonright (L \cup \overline{L}) \text{ is one-to-one, where } L = \mathcal{L}(P) \cup \mathcal{L}(Q))$$

$$\frac{P \xrightarrow{\alpha} P' \quad Q \xrightarrow{\beta} Q'}{P[f] \parallel Q[f] \xrightarrow{\{f(\alpha), f(\beta)\}} P'[f] \parallel Q'[f]}\text{(if } f \upharpoonright (L \cup \overline{L}) \text{ is one-to-one, where } L = \mathcal{L}(P) \cup \mathcal{L}(Q))$$

$$\frac{P \xrightarrow{l} P' \quad Q \xrightarrow{\overline{l}} Q'}{(P \parallel Q)[f] \xrightarrow{\tau} (P' \parallel Q')[f]}\text{(if } f \upharpoonright (L \cup \overline{L}) \text{ is one-to-one, where } L = \mathcal{L}(P) \cup \mathcal{L}(Q))$$

$$\frac{P \xrightarrow{l} P' \quad Q \xrightarrow{\overline{l}} Q'}{(P[f] \parallel Q[f] \xrightarrow{\tau} P'[f] \parallel Q'[f]}\text{(if } f \upharpoonright (L \cup \overline{L}) \text{ is one-to-one, where } L = \mathcal{L}(P) \cup \mathcal{L}(Q))$$

By the reverse transition rules of Composition and Relabeling, we obtain

$$\frac{P \overset{\alpha[m]}{\twoheadrightarrow} P' \quad Q \not\twoheadrightarrow}{(P \parallel Q)[f] \overset{f(\alpha)[m]}{\twoheadrightarrow} (P' \parallel Q)[f]}\text{(if } f \upharpoonright (L \cup \overline{L}) \text{ is one-to-one, where } L = \mathcal{L}(P) \cup \mathcal{L}(Q))$$

$$\frac{P \overset{\alpha[m]}{\twoheadrightarrow} P' \quad Q \not\twoheadrightarrow}{P[f] \parallel Q[f] \overset{f(\alpha)[m]}{\twoheadrightarrow} P'[f] \parallel Q[f]}\text{(if } f \upharpoonright (L \cup \overline{L}) \text{ is one-to-one, where } L = \mathcal{L}(P) \cup \mathcal{L}(Q))$$

$$\frac{Q \overset{\beta[n]}{\twoheadrightarrow} Q' \quad P \not\twoheadrightarrow}{(P \parallel Q)[f] \overset{f(\beta)[n]}{\twoheadrightarrow} (P \parallel Q')[f]}\text{(if } f \upharpoonright (L \cup \overline{L}) \text{ is one-to-one, where } L = \mathcal{L}(P) \cup \mathcal{L}(Q))$$

$$\frac{Q \overset{\beta[n]}{\twoheadrightarrow} Q' \quad P \not\twoheadrightarrow}{P[f] \parallel Q[f] \overset{f(\beta)[n]}{\twoheadrightarrow} P[f] \parallel Q'[f]}\text{(if } f \upharpoonright (L \cup \overline{L}) \text{ is one-to-one, where } L = \mathcal{L}(P) \cup \mathcal{L}(Q))$$

$$\frac{P \overset{\alpha[m]}{\twoheadrightarrow} P' \quad Q \overset{\beta[m]}{\twoheadrightarrow} Q'}{(P \parallel Q)[f] \overset{\{f(\alpha)[m], f(\beta)[m]\}}{\twoheadrightarrow} (P' \parallel Q')[f]}\text{(if } f \upharpoonright (L \cup \overline{L}) \text{ is one-to-one, where } L = \mathcal{L}(P) \cup \mathcal{L}(Q))$$

$$\frac{P \overset{\alpha[m]}{\twoheadrightarrow} P' \quad Q \overset{\beta[m]}{\twoheadrightarrow} Q'}{P[f] \parallel Q[f] \overset{\{f(\alpha)[m], f(\beta)[m]\}}{\twoheadrightarrow} P'[f] \parallel Q'[f]}\text{(if } f \upharpoonright (L \cup \overline{L}) \text{ is one-to-one, where } L = \mathcal{L}(P) \cup \mathcal{L}(Q))$$

$$\frac{P \xrightarrow{l[m]} P' \quad Q \xrightarrow{\bar{l}[m]} Q'}{(P \parallel Q)[f] \xrightarrow{\tau} (P' \parallel Q')[f]} (\text{if } f \restriction (L \cup \overline{L}) \text{ is one-to-one, where } L = \mathcal{L}(P) \cup \mathcal{L}(Q))$$

$$\frac{P \xrightarrow{l[m]} P' \quad Q \xrightarrow{\bar{l}[m]} Q'}{(P[f] \parallel Q[f] \xrightarrow{\tau} P'[f] \parallel Q'[f]} (\text{if } f \restriction (L \cup \overline{L}) \text{ is one-to-one, where } L = \mathcal{L}(P) \cup \mathcal{L}(Q))$$

Hence, with the assumptions $(P' \parallel Q)[f] \sim_s^{fr} P'[f] \parallel Q[f]$, $(P \parallel Q')[f] \sim_s^{fr} P[f] \parallel Q'[f]$ and $(P' \parallel Q')[f] \sim_s^{fr} P'[f] \parallel Q'[f]$, $(P \parallel Q)[f] \sim_s^{fr} P[f] \parallel Q[f]$, if $f \restriction (L \cup \overline{L})$ is one-to-one, where $L = \mathcal{L}(P) \cup \mathcal{L}(Q)$, as desired. □

**Proposition 3.11** (Static laws for strongly FR pomset bisimulation). *The static laws for strongly FR pomset bisimulation are as follows:*

1. $P \parallel Q \sim_p^{fr} Q \parallel P$;
2. $P \parallel (Q \parallel R) \sim_p^{fr} (P \parallel Q) \parallel R$;
3. $P \parallel \textbf{nil} \sim_p^{fr} P$;
4. $P \setminus L \sim_p^{fr} P$, *if* $\mathcal{L}(P) \cap (L \cup \overline{L}) = \emptyset$;
5. $P \setminus K \setminus L \sim_p^{fr} P \setminus (K \cup L)$;
6. $P[f] \setminus L \sim_p^{fr} P \setminus f^{-1}(L)[f]$;
7. $(P \parallel Q) \setminus L \sim_p^{fr} P \setminus L \parallel Q \setminus L$, *if* $\mathcal{L}(P) \cap \overline{\mathcal{L}(Q)} \cap (L \cup \overline{L}) = \emptyset$;
8. $P[Id] \sim_p^{fr} P$;
9. $P[f] \sim_p^{fr} P[f']$, *if* $f \restriction \mathcal{L}(P) = f' \restriction \mathcal{L}(P)$;
10. $P[f][f'] \sim_p^{fr} P[f' \circ f]$;
11. $(P \parallel Q)[f] \sim_p^{fr} P[f] \parallel Q[f]$, *if* $f \restriction (L \cup \overline{L})$ *is one-to-one, where* $L = \mathcal{L}(P) \cup \mathcal{L}(Q)$.

*Proof.* From the definition of strongly FR pomset bisimulation (see Definition 2.68), we know that strongly FR pomset bisimulation is defined by FR pomset transitions, which are labeled by pomsets. In an FR pomset transition, the events in the pomset are either within causality relations (defined by the prefix .) or in concurrency (implicitly defined by . and +, and explicitly defined by ∥), of course, they are pairwise consistent (without conflicts). In Proposition 3.10, we have already proven the case that all events are pairwise concurrent, hence, we only need to prove the case of events in causality. Without loss of generality, we take a pomset of $p = \{\alpha, \beta : \alpha.\beta\}$. Then, the FR pomset transition labeled by the above $p$ is just composed of one single event transition labeled by $\alpha$ succeeded by another single event transition labeled by $\beta$, that is, $\xrightarrow{p} = \xrightarrow{\alpha}\xrightarrow{\beta}$ and $\xrightarrow{p[\mathcal{K}]} = \xrightarrow{\beta[n]}\xrightarrow{\alpha[m]}$.

Similarly to the proof of static laws for strongly FR step bisimulation (see Proposition 3.10), we can prove that the static laws hold for strongly FR pomset bisimulation, however, we omit them here. □

**Proposition 3.12** (Static laws for strongly FR hp-bisimulation). *The static laws for strongly FR hp-bisimulation are as follows:*

1. $P \parallel Q \sim_{hp}^{fr} Q \parallel P$;
2. $P \parallel (Q \parallel R) \sim_{hp}^{fr} (P \parallel Q) \parallel R$;
3. $P \parallel \textbf{nil} \sim_{hp}^{fr} P$;
4. $P \setminus L \sim_{hp}^{fr} P$, *if* $\mathcal{L}(P) \cap (L \cup \overline{L}) = \emptyset$;
5. $P \setminus K \setminus L \sim_{hp}^{fr} P \setminus (K \cup L)$;
6. $P[f] \setminus L \sim_{hp}^{fr} P \setminus f^{-1}(L)[f]$;
7. $(P \parallel Q) \setminus L \sim_{hp}^{fr} P \setminus L \parallel Q \setminus L$, *if* $\mathcal{L}(P) \cap \overline{\mathcal{L}(Q)} \cap (L \cup \overline{L}) = \emptyset$;
8. $P[Id] \sim_{hp}^{fr} P$;
9. $P[f] \sim_{hp}^{fr} P[f']$, *if* $f \restriction \mathcal{L}(P) = f' \restriction \mathcal{L}(P)$;
10. $P[f][f'] \sim_{hp}^{fr} P[f' \circ f]$;
11. $(P \parallel Q)[f] \sim_{hp}^{fr} P[f] \parallel Q[f]$, *if* $f \restriction (L \cup \overline{L})$ *is one-to-one, where* $L = \mathcal{L}(P) \cup \mathcal{L}(Q)$.

*Proof.* From the definition of strongly FR hp-bisimulation (see Definition 2.70), we know that strongly FR hp-bisimulation is defined on the posetal product $(C_1, f, C_2), f : C_1 \to C_2$ isomorphism. Two processes $P$ related to $C_1$ and $Q$ related to $C_2$, and $f : C_1 \to C_2$ isomorphism. Initially, $(C_1, f, C_2) = (\emptyset, \emptyset, \emptyset)$, and $(\emptyset, \emptyset, \emptyset) \in \sim_{hp}^{fr}$. When $P \xrightarrow{\alpha} P'$ $(C_1 \xrightarrow{\alpha} C_1')$, there will be $Q \xrightarrow{\alpha} Q'$ $(C_2 \xrightarrow{\alpha} C_2')$, and we define $f' = f[\alpha \mapsto \alpha]$. Also, when $P \xrightarrow{\alpha[m]} P'$ $(C_1 \xrightarrow{\alpha[m]} C_1')$, there will be $Q \xrightarrow{\alpha[m]} Q'$ $(C_2 \xrightarrow{\alpha[m]} C_2')$, and we define $f' = f[\alpha[m] \mapsto \alpha[m]]$. Then, if $(C_1, f, C_2) \in \sim_{hp}^{fr}$, then $(C_1', f', C_2') \in \sim_{hp}^{fr}$.

Similarly to the proof of static laws for strongly FR pomset bisimulation (see Proposition 3.11), we can prove that static laws hold for strongly FR hp-bisimulation, we just need additionally to check the above conditions on FR hp-bisimulation, however, we omit them here. □

**Proposition 3.13** (Static laws for strongly FR hhp-bisimulation). *The static laws for strongly FR hhp-bisimulation are as follows:*

1. $P \parallel Q \sim_{hhp}^{fr} Q \parallel P$;
2. $P \parallel (Q \parallel R) \sim_{hhp}^{fr} (P \parallel Q) \parallel R$;
3. $P \parallel \mathbf{nil} \sim_{hhp}^{fr} P$;
4. $P \setminus L \sim_{hhp}^{fr} P$, *if* $\mathcal{L}(P) \cap (L \cup \overline{L}) = \emptyset$;
5. $P \setminus K \setminus L \sim_{hhp}^{fr} P \setminus (K \cup L)$;
6. $P[f] \setminus L \sim_{hhp}^{fr} P \setminus f^{-1}(L)[f]$;
7. $(P \parallel Q) \setminus L \sim_{hhp}^{fr} P \setminus L \parallel Q \setminus L$, *if* $\mathcal{L}(P) \cap \overline{\mathcal{L}(Q)} \cap (L \cup \overline{L}) = \emptyset$;
8. $P[Id] \sim_{hhp}^{fr} P$;
9. $P[f] \sim_{hhp}^{fr} P[f']$, *if* $f \upharpoonright \mathcal{L}(P) = f' \upharpoonright \mathcal{L}(P)$;
10. $P[f][f'] \sim_{hhp}^{fr} P[f' \circ f]$;
11. $(P \parallel Q)[f] \sim_{hhp}^{fr} P[f] \parallel Q[f]$, *if* $f \upharpoonright (L \cup \overline{L})$ *is one-to-one, where* $L = \mathcal{L}(P) \cup \mathcal{L}(Q)$.

*Proof.* From the definition of strongly FR hhp-bisimulation (see Definition 2.70), we know that strongly FR hhp-bisimulation is downward closed for strongly FR hp-bisimulation.

Similarly to the proof of static laws for strongly FR hp-bisimulation (see Proposition 3.12), we can prove that static laws hold for strongly FR hhp-bisimulation, that is, they are downward closed for strongly FR hp-bisimulation, however, we omit them here. □

**Proposition 3.14** (Milner's expansion law for strongly FR truly concurrent bisimulations). *Milner's expansion law does not hold any longer for any strongly FR truly concurrent bisimulation, that is,*

1. $\alpha \parallel \beta \nsim_p^{fr} \alpha.\beta + \beta.\alpha$;
2. $\alpha \parallel \beta \nsim_s^{fr} \alpha.\beta + \beta.\alpha$;
3. $\alpha \parallel \beta \nsim_{hp}^{fr} \alpha.\beta + \beta.\alpha$;
4. $\alpha \parallel \beta \nsim_{hhp}^{fr} \alpha.\beta + \beta.\alpha$.

*Proof.* In nature, it is caused by $\alpha \parallel \beta$ and $\alpha.\beta + \beta.\alpha$ having a different causality structure. By the transition rules of Composition and Prefix, we have

$$\alpha \parallel \beta \xrightarrow{\{\alpha,\beta\}} \mathbf{nil}$$

while

$$\alpha.\beta + \beta.\alpha \nrightarrow^{\{\alpha,\beta\}} .$$

Also,

$$\alpha[m] \parallel \beta[m] \xrightarrow{\{\alpha[m],\beta[m]\}} \mathbf{nil}$$

while

$$\alpha[m].\beta[m] + \beta[m].\alpha[m] \not\xrightarrow{\{\alpha[m],\beta[m]\}} .$$

□

**Proposition 3.15** (New expansion law for strongly FR step bisimulation). *Let* $P \equiv (P_1[f_1] \parallel \cdots \parallel P_n[f_n]) \setminus L$, *with* $n \geq 1$. *Then,*

$$\begin{aligned} P \sim_s^f \{(f_1(\alpha_1) \parallel \cdots \parallel f_n(\alpha_n)).(P_1'[f_1] \parallel \cdots \parallel P_n'[f_n]) \setminus L : \\ P_i \xrightarrow{\alpha_i} P_i', i \in \{1, \cdots, n\}, f_i(\alpha_i) \notin L \cup \overline{L}\} \\ + \sum \{\tau.(P_1[f_1] \parallel \cdots \parallel P_i'[f_i] \parallel \cdots \parallel P_j'[f_j] \parallel \cdots \parallel P_n[f_n]) \setminus L : \\ P_i \xrightarrow{l_1} P_i', P_j \xrightarrow{l_2} P_j', f_i(l_1) = \overline{f_j(l_2)}, i < j\} \end{aligned} \tag{3.1}$$

$$\begin{aligned} P \sim_s^r \{(P_1'[f_1] \parallel \cdots \parallel P_n'[f_n]).(f_1(\alpha_1[m]) \parallel \cdots \parallel f_n(\alpha_n)[m]) \setminus L : \\ P_i \xrightarrow{\alpha_i[m]} P_i', i \in \{1, \cdots, n\}, f_i(\alpha_i) \notin L \cup \overline{L}\} \\ + \sum \{(P_1[f_1] \parallel \cdots \parallel P_i'[f_i] \parallel \cdots \parallel P_j'[f_j] \parallel \cdots \parallel P_n[f_n]).\tau \setminus L : \\ P_i \xrightarrow{l_1[m]} P_i', P_j \xrightarrow{l_2[m]} P_j', f_i(l_1) = \overline{f_j(l_2)}, i < j\} \end{aligned} \tag{3.2}$$

*Proof.* Although transition rules in Definition 3.2 are defined in the flavor of a single event, they can be modified into a step (a set of events within which each event is pairwise concurrent), however, we omit them here. If we treat a single event as a step containing just one event, the proof of the new expansion law poses no problem, hence, we use this way and still use the transition rules in Definition 3.2.

(1) The case of strongly forward step bisimulation.

First, we consider the case without Restriction and Relabeling. That is, we suffice to prove the following case by induction on the size $n$.

For $P \equiv P_1 \parallel \cdots \parallel P_n$, with $n \geq 1$, we need to prove

$$\begin{aligned} P \sim_s \{(\alpha_1 \parallel \cdots \parallel \alpha_n).(P_1' \parallel \cdots \parallel P_n') : P_i \xrightarrow{\alpha_i} P_i', i \in \{1, \cdots, n\} \\ + \sum \{\tau.(P_1 \parallel \cdots \parallel P_i' \parallel \cdots \parallel P_j' \parallel \cdots \parallel P_n) : P_i \xrightarrow{l} P_i', P_j \xrightarrow{\overline{l}} P_j', i < j\} \end{aligned}$$

For $n = 1$, $P_1 \sim_s^f \alpha_1.P_1' : P_1 \xrightarrow{\alpha_1} P_1'$ is obvious. Then, with a hypothesis $n$, we consider $R \equiv P \parallel P_{n+1}$. By the forward transition rules of Composition, we can obtain

$$\begin{aligned} R \sim_s^f \{(p \parallel \alpha_{n+1}).(P' \parallel P_{n+1}') : P \xrightarrow{p} P', P_{n+1} \xrightarrow{\alpha_{n+1}} P_{n+1}', p \subseteq P\} \\ + \sum \{\tau.(P' \parallel P_{n+1}') : P \xrightarrow{l} P', P_{n+1} \xrightarrow{\overline{l}} P_{n+1}'\} \end{aligned}$$

Now with the induction assumption $P \equiv P_1 \parallel \cdots \parallel P_n$, the right-hand side can be reformulated as follows:

$$\begin{aligned} \{(\alpha_1 \parallel \cdots \parallel \alpha_n \parallel \alpha_{n+1}).(P_1' \parallel \cdots \parallel P_n' \parallel P_{n+1}') : \\ P_i \xrightarrow{\alpha_i} P_i', i \in \{1, \cdots, n+1\} \\ + \sum \{\tau.(P_1 \parallel \cdots \parallel P_i' \parallel \cdots \parallel P_j' \parallel \cdots \parallel P_n \parallel P_{n+1}) : \\ P_i \xrightarrow{l} P_i', P_j \xrightarrow{\overline{l}} P_j', i < j\} \\ + \sum \{\tau.(P_1 \parallel \cdots \parallel P_i' \parallel \cdots \parallel P_j \parallel \cdots \parallel P_n \parallel P_{n+1}') : \\ P_i \xrightarrow{l} P_i', P_{n+1} \xrightarrow{\overline{l}} P_{n+1}', i \in \{1, \cdots, n\}\} \end{aligned}$$

Hence,

$$\begin{aligned} R \sim_s^f \{(\alpha_1 \parallel \cdots \parallel \alpha_n \parallel \alpha_{n+1}).(P_1' \parallel \cdots \parallel P_n' \parallel P_{n+1}') : \\ P_i \xrightarrow{\alpha_i} P_i', i \in \{1, \cdots, n+1\} \\ + \sum \{\tau.(P_1 \parallel \cdots \parallel P_i' \parallel \cdots \parallel P_j' \parallel \cdots \parallel P_n) : \end{aligned}$$

$$P_i \xrightarrow{l} P_i', P_j \xrightarrow{\bar{l}} P_j', 1 \leq i < j \geq n+1\}$$

Then, we can easily add the full conditions with Restriction and Relabeling.

(2) The case of strongly reverse step bisimulation.

First, we consider the case without Restriction and Relabeling. That is, we suffice to prove the following case by induction on the size $n$.

For $P \equiv P_1 \parallel \cdots \parallel P_n$, with $n \geq 1$, we need to prove

$$P \sim_s^r \{(P_1' \parallel \cdots \parallel P_n').(\alpha_1[m] \parallel \cdots \parallel \alpha_n[m]) : P_i \xrightarrow{\alpha_i[m]} P_i', i \in \{1, \cdots, n\}$$
$$+\sum\{(P_1 \parallel \cdots \parallel P_i' \parallel \cdots \parallel P_j' \parallel \cdots \parallel P_n).\tau : P_i \xrightarrow{l[m]} P_i', P_j \xrightarrow{\bar{l}[m]} P_j', i < j\}$$

For $n = 1$, $P_1 \sim_s^r P_1'.\alpha_1[m] : P_1 \xrightarrow{\alpha_1[m]} P_1'$ is obvious. Then, with a hypothesis $n$, we consider $R \equiv P \parallel P_{n+1}$. By the reverse transition rules of Composition, we can obtain

$$R \sim_s^r \{(P' \parallel P_{n+1}').(p[m] \parallel \alpha_{n+1}[m]) : P \xrightarrow{p[m]} P', P_{n+1} \xrightarrow{\alpha_{n+1}[m]} P_{n+1}', p \subseteq P\}$$
$$+\sum\{(P' \parallel P_{n+1}').\tau : P \xrightarrow{l[m]} P', P_{n+1} \xrightarrow{\bar{l}[m]} P_{n+1}'\}$$

Now, with the induction assumption $P \equiv P_1 \parallel \cdots \parallel P_n$, the right-hand side can be reformulated as follows:

$$\{(P_1' \parallel \cdots \parallel P_n' \parallel P_{n+1}).(\alpha_1[m] \parallel \cdots \parallel \alpha_n[m] \parallel \alpha_{n+1}[m]) :$$
$$P_i \xrightarrow{\alpha_i[m]} P_i', i \in \{1, \cdots, n+1\}$$
$$+\sum\{(P_1 \parallel \cdots \parallel P_i' \parallel \cdots \parallel P_j' \parallel \cdots \parallel P_n \parallel P_{n+1}).\tau :$$
$$P_i \xrightarrow{l[m]} P_i', P_j \xrightarrow{\bar{l}[m]} P_j', i < j\}$$
$$+\sum\{(P_1 \parallel \cdots \parallel P_i' \parallel \cdots \parallel P_j' \parallel \cdots \parallel P_n \parallel P_{n+1}).\tau :$$
$$P_i \xrightarrow{l[m]} P_i', P_{n+1} \xrightarrow{\bar{l}[m]} P_{n+1}', i \in \{1, \cdots, n\}\}$$

Hence,

$$R \sim_s^r \{(P_1' \parallel \cdots \parallel P_n' \parallel P_{n+1}').(\alpha_1[m] \parallel \cdots \parallel \alpha_n[m] \parallel \alpha_{n+1}[m]) :$$
$$P_i \xrightarrow{\alpha_i[m]} P_i', i \in \{1, \cdots, n+1\}$$
$$+\sum\{(P_1 \parallel \cdots \parallel P_i' \parallel \cdots \parallel P_j' \parallel \cdots \parallel P_n).\tau :$$
$$P_i \xrightarrow{l[m]} P_i', P_j \xrightarrow{\bar{l}[m]} P_j', 1 \leq i < j \geq n+1\}$$

Then, we can easily add the full conditions with Restriction and Relabeling. □

**Proposition 3.16** (New expansion law for strongly FR pomset bisimulation). *Let* $P \equiv (P_1[f_1] \parallel \cdots \parallel P_n[f_n]) \setminus L$, *with* $n \geq 1$. *Then,*

$$P \sim_p^f \{(f_1(\alpha_1) \parallel \cdots \parallel f_n(\alpha_n)).(P_1'[f_1] \parallel \cdots \parallel P_n'[f_n]) \setminus L :$$
$$P_i \xrightarrow{\alpha_i} P_i', i \in \{1, \cdots, n\}, f_i(\alpha_i) \notin L \cup \bar{L}\}$$
$$+\sum\{\tau.(P_1[f_1] \parallel \cdots \parallel P_i'[f_i] \parallel \cdots \parallel P_j'[f_j] \parallel \cdots \parallel P_n[f_n]) \setminus L :$$
$$P_i \xrightarrow{l_1} P_i', P_j \xrightarrow{l_2} P_j', f_i(l_1) = \overline{f_j(l_2)}, i < j\} \quad (3.3)$$
$$P \sim_p^r \{(P_1'[f_1] \parallel \cdots \parallel P_n'[f_n]).(f_1(\alpha_1[m]) \parallel \cdots \parallel f_n(\alpha_n)[m]) \setminus L :$$

$$P_i \xrightarrow{\alpha_i[m]} P'_i, i \in \{1, \cdots, n\}, f_i(\alpha_i) \notin L \cup \overline{L}\}$$
$$+\sum\{(P_1[f_1] \parallel \cdots \parallel P'_i[f_i] \parallel \cdots \parallel P'_j[f_j] \parallel \cdots \parallel P_n[f_n]).\tau \setminus L :$$
$$P_i \xrightarrow{l_1[m]} P'_i, P_j \xrightarrow{l_2[m]} P'_j, f_i(l_1) = \overline{f_j(l_2)}, i < j\} \quad (3.4)$$

*Proof.* From the definition of strongly FR pomset bisimulation (see Definition 2.68), we know that strongly FR pomset bisimulation is defined by FR pomset transitions, which are labeled by pomsets. In an FR pomset transition, the events in the pomset are either within causality relations (defined by the prefix .) or in concurrency (implicitly defined by . and +, and explicitly defined by $\parallel$), of course, they are pairwise consistent (without conflicts). In Proposition 3.15, we have already proven the case that all events are pairwise concurrent, hence, we only need to prove the case of events in causality. Without loss of generality, we take a pomset of $p = \{\alpha, \beta : \alpha.\beta\}$. Then, the FR pomset transition labeled by the above $p$ is just composed of one single event transition labeled by $\alpha$ succeeded by another single event transition labeled by $\beta$, that is, $\xrightarrow{p} = \xrightarrow{\alpha}\xrightarrow{\beta}$ and $\xrightarrow{p[\mathcal{K}]} = \xrightarrow{\beta[n]}\xrightarrow{\alpha[m]}$.

Similarly to the proof of the new expansion law for strongly FR step bisimulation (see Proposition 3.15), we can prove that the new expansion law holds for strongly FR pomset bisimulation, however, we omit this proof here. □

**Proposition 3.17** (New expansion law for strongly FR hp-bisimulation). *Let $P \equiv (P_1[f_1] \parallel \cdots \parallel P_n[f_n]) \setminus L$, with $n \geq 1$. Then,*

$$P \sim^f_{hp} \{(f_1(\alpha_1) \parallel \cdots \parallel f_n(\alpha_n)).(P'_1[f_1] \parallel \cdots \parallel P'_n[f_n]) \setminus L :$$
$$P_i \xrightarrow{\alpha_i} P'_i, i \in \{1, \cdots, n\}, f_i(\alpha_i) \notin L \cup \overline{L}\}$$
$$+\sum\{\tau.(P_1[f_1] \parallel \cdots \parallel P'_i[f_i] \parallel \cdots \parallel P'_j[f_j] \parallel \cdots \parallel P_n[f_n]) \setminus L :$$
$$P_i \xrightarrow{l_1} P'_i, P_j \xrightarrow{l_2} P'_j, f_i(l_1) = \overline{f_j(l_2)}, i < j\} \quad (3.5)$$
$$P \sim^r_{hp} \{(P'_1[f_1] \parallel \cdots \parallel P'_n[f_n]).(f_1(\alpha_1[m]) \parallel \cdots \parallel f_n(\alpha_n)[m]) \setminus L :$$
$$P_i \xrightarrow{\alpha_i[m]} P'_i, i \in \{1, \cdots, n\}, f_i(\alpha_i) \notin L \cup \overline{L}\}$$
$$+\sum\{(P_1[f_1] \parallel \cdots \parallel P'_i[f_i] \parallel \cdots \parallel P'_j[f_j] \parallel \cdots \parallel P_n[f_n]).\tau \setminus L :$$
$$P_i \xrightarrow{l_1[m]} P'_i, P_j \xrightarrow{l_2[m]} P'_j, f_i(l_1) = \overline{f_j(l_2)}, i < j\} \quad (3.6)$$

*Proof.* From the definition of strongly FR hp-bisimulation (see Definition 2.70), we know that strongly FR hp-bisimulation is defined on the posetal product $(C_1, f, C_2)$, $f : C_1 \to C_2$ isomorphism. Two processes $P$ related to $C_1$ and $Q$ related to $C_2$, and $f : C_1 \to C_2$ isomorphism. Initially, $(C_1, f, C_2) = (\emptyset, \emptyset, \emptyset)$, and $(\emptyset, \emptyset, \emptyset) \in \sim_{hp}$. When $P \xrightarrow{\alpha} P'$ $(C_1 \xrightarrow{\alpha} C'_1)$, there will be $Q \xrightarrow{\alpha} Q'$ $(C_2 \xrightarrow{\alpha} C'_2)$, and we define $f' = f[\alpha \mapsto \alpha]$. Also, when $P \xrightarrow{\alpha[m]} P'$ $(C_1 \xrightarrow{\alpha[m]} C'_1)$, there will be $Q \xrightarrow{\alpha[m]} Q'$ $(C_2 \xrightarrow{\alpha[m]} C'_2)$, and we define $f' = f[\alpha[m] \mapsto \alpha[m]]$. Then, if $(C_1, f, C_2) \in \sim^{fr}_{hp}$, then $(C'_1, f', C'_2) \in \sim^{fr}_{hp}$.

Similarly to the proof of the new expansion law for strongly FR pomset bisimulation (see Proposition 3.16), we can prove that the new expansion law holds for strongly FR hp-bisimulation, we just need additionally to check the above conditions on FR hp-bisimulation, however, we omit them here. □

**Proposition 3.18** (New expansion law for strongly FR hhp-bisimulation). *Let $P \equiv (P_1[f_1] \parallel \cdots \parallel P_n[f_n]) \setminus L$, with $n \geq 1$. Then,*

$$P \sim^f_{hhp} \{(f_1(\alpha_1) \parallel \cdots \parallel f_n(\alpha_n)).(P'_1[f_1] \parallel \cdots \parallel P'_n[f_n]) \setminus L :$$
$$P_i \xrightarrow{\alpha_i} P'_i, i \in \{1, \cdots, n\}, f_i(\alpha_i) \notin L \cup \overline{L}\}$$
$$+\sum\{\tau.(P_1[f_1] \parallel \cdots \parallel P'_i[f_i] \parallel \cdots \parallel P'_j[f_j] \parallel \cdots \parallel P_n[f_n]) \setminus L :$$
$$P_i \xrightarrow{l_1} P'_i, P_j \xrightarrow{l_2} P'_j, f_i(l_1) = \overline{f_j(l_2)}, i < j\} \quad (3.7)$$
$$P \sim^r_{hhp} \{(P'_1[f_1] \parallel \cdots \parallel P'_n[f_n]).(f_1(\alpha_1[m]) \parallel \cdots \parallel f_n(\alpha_n)[m]) \setminus L :$$

$$P_i \xrightarrow{\alpha_i[m]} P_i', i \in \{1, \cdots, n\}, f_i(\alpha_i) \notin L \cup \overline{L}\}$$

$$+\sum\{(P_1[f_1] \parallel \cdots \parallel P_i'[f_i] \parallel \cdots \parallel P_j'[f_j] \parallel \cdots \parallel P_n[f_n]).\tau \setminus L :$$

$$P_i \xrightarrow{l_1[m]} P_i', P_j \xrightarrow{l_2[m]} P_j', f_i(l_1) = \overline{f_j(l_2)}, i < j\} \tag{3.8}$$

*Proof.* From the definition of strongly FR hhp-bisimulation (see Definition 2.70), we know that strongly FR hhp-bisimulation is downward closed for strongly FR hp-bisimulation.

Similarly to the proof of the new expansion law for strongly FR hp-bisimulation (see Proposition 3.17), we can prove that the new expansion law holds for strongly FR hhp-bisimulation, that is, they are downward closed for strongly FR hp-bisimulation, however, we omit them here. □

**Theorem 3.19** (Congruence for strongly FR step bisimulation). *We can enjoy the congruence for strongly FR step bisimulation as follows:*

1. *If $A \overset{\text{def}}{=} P$, then $A \sim_s^{fr} P$;*
2. *Let $P_1 \sim_s^{fr} P_2$. Then,*
   (a) $\alpha.P_1 \sim_s^f \alpha.P_2$;
   (b) $(\alpha_1 \parallel \cdots \parallel \alpha_n).P_1 \sim_s^f (\alpha_1 \parallel \cdots \parallel \alpha_n).P_2$;
   (c) $P_1.\alpha[m] \sim_s^r P_2.\alpha[m]$;
   (d) $P_1.(\alpha_1[m] \parallel \cdots \parallel \alpha_n[m]) \sim_s^r P_2.(\alpha_1[m] \parallel \cdots \parallel \alpha_n[m]).$;
   (e) $P_1 + Q \sim_s^{fr} P_2 + Q$;
   (f) $P_1 \parallel Q \sim_s^{fr} P_2 \parallel Q$;
   (g) $P_1 \setminus L \sim_s^{fr} P_2 \setminus L$;
   (h) $P_1[f] \sim_s^{fr} P_2[f]$.

*Proof.* Although transition rules in Definition 3.2 are defined in the flavor of a single event, they can be modified into a step (a set of events within which each event is pairwise concurrent), however, we omit them here. If we treat a single event as a step containing just one event, the proof of the congruence does not pose any problem, hence, we use this way and still use the transition rules in Definition 3.2.

1. If $A \overset{\text{def}}{=} P$, then $A \sim_s^{fr} P$. This is obvious.
2. Let $P_1 \sim_s^{fr} P_2$. Then,
   (a) $\alpha.P_1 \sim_s^f \alpha.P_2$. By the forward transition rules of Prefix, we can obtain

$$\alpha.P_1 \xrightarrow{\alpha} \alpha[m]P_1$$

$$\alpha.P_2 \xrightarrow{\alpha} \alpha[m]P_2$$

   Since $P_1 \sim_s^{fr} P_2$, we obtain $\alpha.P_1 \sim_s^f \alpha.P_2$, as desired.
   (b) $(\alpha_1 \parallel \cdots \parallel \alpha_n).P_1 \sim_s^f (\alpha_1 \parallel \cdots \parallel \alpha_n).P_2$. By the transition rules of Prefix, we can obtain

$$(\alpha_1 \parallel \cdots \parallel \alpha_n).P_1 \xrightarrow{\{\alpha_1, \cdots, \alpha_n\}} (\alpha_1[m] \parallel \cdots \parallel \alpha_n[m]).P_1$$

$$(\alpha_1 \parallel \cdots \parallel \alpha_n).P_2 \xrightarrow{\{\alpha_1, \cdots, \alpha_n\}} (\alpha_1[m] \parallel \cdots \parallel \alpha_n[m]).P_2$$

   Since $P_1 \sim_s^{fr} P_2$, we obtain $(\alpha_1 \parallel \cdots \parallel \alpha_n).P_1 \sim_s^f (\alpha_1 \parallel \cdots \parallel \alpha_n).P_2$, as desired.
   (c) $P_1.\alpha[m] \sim_s^r P_2.\alpha[m]$. By the reverse transition rules of Prefix, we can obtain

$$P_1.\alpha[m] \xrightarrow{\alpha[m]} P_1.\alpha$$

$$P_2.\alpha[m] \xrightarrow{\alpha[m]} P_2.\alpha$$

   Since $P_1 \sim_s^{fr} P_2$, we obtain $P_1.\alpha[m] \sim_s^r P_2.\alpha[m]$, as desired.

**(d)** $P_1.(\alpha_1[m] \parallel \cdots \parallel \alpha_n[m]) \sim_s^r P_2.(\alpha_1[m] \parallel \cdots \parallel \alpha_n[m])$. By the reverse transition rules of Prefix, we can obtain

$$P_1.(\alpha_1[m] \parallel \cdots \parallel \alpha_n[m]) \xrightarrow{\{\alpha_1[m],\cdots,\alpha_n[m]\}} \!\!\!\!\!\twoheadrightarrow P_1.(\alpha_1 \parallel \cdots \parallel \alpha_n)$$

$$P_2.(\alpha_1[m] \parallel \cdots \parallel \alpha_n[m]) \xrightarrow{\{\alpha_1[m],\cdots,\alpha_n[m]\}} \!\!\!\!\!\twoheadrightarrow P_2.(\alpha_1 \parallel \cdots \parallel \alpha_n)$$

Since $P_1 \sim_s^{fr} P_2$, we obtain $(\alpha_1 \parallel \cdots \parallel \alpha_n).P_1 \sim_s^r (\alpha_1 \parallel \cdots \parallel \alpha_n).P_2$, as desired.

**(e)** $P_1 + Q \sim_s^{fr} P_2 + Q$. There are several cases, however, we will not enumerate them all here. By the forward transition rules of Summation, we can obtain

$$\frac{P_1 \xrightarrow{\alpha} P_1'}{P_2 \xrightarrow{\alpha} P_2'}(P_1' \sim_s^{fr} P_2')$$

$$\frac{P_1 \xrightarrow{\alpha} P_1' \quad \alpha \notin Q}{P_1 + Q \xrightarrow{\alpha} P_1' + Q} \qquad \frac{P_2 \xrightarrow{\alpha} P_2' \quad \alpha \notin Q}{P_2 + Q \xrightarrow{\alpha} P_2' + Q}$$

$$\frac{Q \xrightarrow{\beta} Q' \quad \beta \notin P_1}{P_1 + Q \xrightarrow{\beta} P_1 + Q'} \qquad \frac{Q \xrightarrow{\beta} Q' \quad \beta \notin P_2}{P_2 + Q \xrightarrow{\beta} P_2 + Q'}$$

By the reverse transition rules of Summation, we can obtain

$$\frac{P_1 \xrightarrow{\alpha[m]} \!\!\!\!\!\twoheadrightarrow P_1'}{P_2 \xrightarrow{\alpha[m]} \!\!\!\!\!\twoheadrightarrow P_2'}(P_1' \sim_s^{fr} P_2')$$

$$\frac{P_1 \xrightarrow{\alpha[m]} \!\!\!\!\!\twoheadrightarrow P_1' \quad \alpha \notin Q}{P_1 + Q \xrightarrow{\alpha[m]} \!\!\!\!\!\twoheadrightarrow P_1' + Q} \qquad \frac{P_2 \xrightarrow{\alpha[m]} \!\!\!\!\!\twoheadrightarrow P_2' \quad \alpha \notin Q}{P_2 + Q \xrightarrow{\alpha[m]} \!\!\!\!\!\twoheadrightarrow P_2' + Q}$$

$$\frac{Q \xrightarrow{\beta[n]} \!\!\!\!\!\twoheadrightarrow Q' \quad \beta \notin P_1}{P_1 + Q \xrightarrow{\beta[n]} \!\!\!\!\!\twoheadrightarrow P_1 + Q'} \qquad \frac{Q \xrightarrow{\beta[n]} \!\!\!\!\!\twoheadrightarrow Q' \quad \beta \notin P_2}{P_2 + Q \xrightarrow{\beta[n]} \!\!\!\!\!\twoheadrightarrow P_2 + Q'}$$

With the assumptions $P_1' + \sim_s^{fr} P_2' + Q$ and $P_1 + Q' \sim_s^{fr} P_2 + Q'$, we obtain $P_1 + Q \sim_s^{fr} P_2 + Q$, as desired.

**(f)** $P_1 \parallel Q \sim_s^{fr} P_2 \parallel Q$. There are several cases, however, we will not enumerate them all here. By the forward transition rules of Composition, we can obtain

$$\frac{P_1 \xrightarrow{\alpha} P_1'}{P_2 \xrightarrow{\alpha} P_2'}(P_1' \sim_s^{fr} P_2')$$

$$\frac{P_1 \xrightarrow{\alpha} P_1' \quad Q \nrightarrow}{P_1 \parallel Q \xrightarrow{\alpha} P_1' \parallel Q} \qquad \frac{P_2 \xrightarrow{\alpha} P_2' \quad Q \nrightarrow}{P_2 \parallel Q \xrightarrow{\alpha} P_2' \parallel Q}$$

$$\frac{Q \xrightarrow{\beta} Q' \quad P_1 \nrightarrow}{P_1 \parallel Q \xrightarrow{\beta} P_1 \parallel Q'} \qquad \frac{Q \xrightarrow{\beta} P_2' \quad P_2 \nrightarrow}{P_2 \parallel Q \xrightarrow{\beta} P_2 \parallel Q'}$$

$$\frac{P_1 \xrightarrow{\alpha} P_1' \quad Q \xrightarrow{\beta} Q'}{P_1 \parallel Q \xrightarrow{\{\alpha,\beta\}} P_1' \parallel Q'}(\beta \neq \overline{\alpha}) \qquad \frac{P_2 \xrightarrow{\alpha} P_2' \quad Q \xrightarrow{\beta} Q'}{P_2 \parallel Q \xrightarrow{\{\alpha,\beta\}} P_2' \parallel Q'}(\beta \neq \overline{\alpha})$$

$$\frac{P_1 \xrightarrow{l} P_1' \quad Q \xrightarrow{\overline{l}} Q'}{P_1 \parallel Q \xrightarrow{\tau} P_1' \parallel Q'} \qquad \frac{P_2 \xrightarrow{l} P_2' \quad Q \xrightarrow{\overline{l}} Q'}{P_2 \parallel Q \xrightarrow{\tau} P_2' \parallel Q'}$$

By the reverse transition rules of Composition, we can obtain

$$\frac{P_1 \xrightarrow{\alpha[m]} P_1'}{P_2 \xrightarrow{\alpha[m]} P_2'} (P_1' \sim_s^{fr} P_2')$$

$$\frac{P_1 \xrightarrow{\alpha[m]} P_1' \quad Q \nrightarrow}{P_1 \parallel Q \xrightarrow{\alpha[m]} P_1' \parallel Q} \quad \frac{P_2 \xrightarrow{\alpha[m]} P_2' \quad Q \nrightarrow}{P_2 \parallel Q \xrightarrow{\alpha[m]} P_2' \parallel Q}$$

$$\frac{Q \xrightarrow{\beta[n]} Q' \quad P_1 \nrightarrow}{P_1 \parallel Q \xrightarrow{\beta[n]} P_1 \parallel Q'} \quad \frac{Q \xrightarrow{\beta[n]} P_2' \quad P_2 \nrightarrow}{P_2 \parallel Q \xrightarrow{\beta[n]} P_2 \parallel Q'}$$

$$\frac{P_1 \xrightarrow{\alpha[m]} P_1' \quad Q \xrightarrow{\beta[m]} Q'}{P_1 \parallel Q \xrightarrow{\{\alpha[m],\beta[m]\}} P_1' \parallel Q'} (\beta \neq \overline{\alpha}) \quad \frac{P_2 \xrightarrow{\alpha[m]} P_2' \quad Q \xrightarrow{\beta[m]} Q'}{P_2 \parallel Q \xrightarrow{\{\alpha[m],\beta[m]\}} P_2' \parallel Q'} (\beta \neq \overline{\alpha})$$

$$\frac{P_1 \xrightarrow{l[m]} P_1' \quad Q \xrightarrow{\overline{l}[m]} Q'}{P_1 \parallel Q \xrightarrow{\tau} P_1' \parallel Q'} \quad \frac{P_2 \xrightarrow{l[m]} P_2' \quad Q \xrightarrow{\overline{l}[m]} Q'}{P_2 \parallel Q \xrightarrow{\tau} P_2' \parallel Q'}$$

Since $P_1' \sim_s^{fr} P_2'$ and $Q' \sim_s^{fr} Q'$, and with the assumptions $P_1' \parallel Q \sim_s^{fr} P_2' \parallel Q$, $P_1 \parallel Q' \sim_s^{fr} P_2 \parallel Q'$ and $P_1' \parallel Q' \sim_s^{fr} P_2' \parallel Q'$, we obtain $P_1 \parallel Q \sim_s^{fr} P_2 \parallel Q$, as desired.

**(g)** $P_1 \setminus L \sim_s^{fr} P_2 \setminus L$. There are several cases, however, we will not enumerate them all here. By the forward transition rules of Restriction, we obtain

$$\frac{P_1 \xrightarrow{\alpha} P_1'}{P_2 \xrightarrow{\alpha} P_2'} (P_1' \sim_s^{fr} P_2')$$

$$\frac{P_1 \xrightarrow{\alpha} P_1'}{P_1 \setminus L \xrightarrow{\alpha} P_1' \setminus L}$$

$$\frac{P_2 \xrightarrow{\alpha} P_2'}{P_2 \setminus L \xrightarrow{\alpha} P_2' \setminus L}$$

By the reverse transition rules of Restriction, we obtain

$$\frac{P_1 \xrightarrow{\alpha[m]} P_1'}{P_2 \xrightarrow{\alpha[m]} P_2'} (P_1' \sim_s^{fr} P_2')$$

$$\frac{P_1 \xrightarrow{\alpha[m]} P_1'}{P_1 \setminus L \xrightarrow{\alpha[m]} P_1' \setminus L}$$

$$\frac{P_2 \xrightarrow{\alpha[m]} P_2'}{P_2 \setminus L \xrightarrow{\alpha[m]} P_2' \setminus L}$$

Since $P_1' \sim_s^{fr} P_2'$, and with the assumption $P_1' \setminus L \sim_s^{fr} P_2' \setminus L$, we obtain $P_1 \setminus L \sim_s^{fr} P_2 \setminus L$, as desired.

**(h)** $P_1[f] \sim_s^{fr} P_2[f]$. By the forward transition rules of Relabeling, we obtain

$$\frac{P_1 \xrightarrow{\alpha} P_1'}{P_2 \xrightarrow{\alpha} P_2'} (P_1' \sim_s^{fr} P_2')$$

$$\frac{P_1 \xrightarrow{\alpha} P_1'}{P_1[f] \xrightarrow{f(\alpha)} P_1'[f]}$$

$$\frac{P_2 \xrightarrow{\alpha} P_2'}{P_2[f] \xrightarrow{f(\alpha)} P_2'[f]}$$

By the reverse transition rules of Relabeling, we obtain

$$\frac{P_1 \xrightarrow{\alpha[m]} P_1'}{P_2 \xrightarrow{\alpha[m]} P_2'}(P_1' \sim_s^{fr} P_2')$$

$$\frac{P_1 \xrightarrow{\alpha[m]} P_1'}{P_1[f] \xrightarrow{f(\alpha)[m]} P_1'[f]}$$

$$\frac{P_2 \xrightarrow{\alpha[m]} P_2'}{P_2[f] \xrightarrow{f(\alpha)[m]} P_2'[f]}$$

Since $P_1' \sim_s^{fr} P_2'$, and with the assumption $P_1'[f] \sim_s^{fr} P_2'[f]$, we obtain $P_1[f] \sim_s^{fr} P_2[f]$, as desired. □

**Theorem 3.20** (Congruence for strongly FR pomset bisimulation). *We can enjoy the congruence for strongly FR pomset bisimulation as follows:*

1. *If $A \stackrel{def}{=} P$, then $A \sim_p^{fr} P$;*
2. *Let $P_1 \sim_p^{fr} P_2$. Then,*
   (a) $\alpha.P_1 \sim_p^{f} \alpha.P_2$;
   (b) $(\alpha_1 \parallel \cdots \parallel \alpha_n).P_1 \sim_p^{f} (\alpha_1 \parallel \cdots \parallel \alpha_n).P_2$;
   (c) $P_1.\alpha[m] \sim_p^{r} P_2.\alpha[m]$;
   (d) $P_1.(\alpha_1[m] \parallel \cdots \parallel \alpha_n[m]) \sim_p^{r} P_2.(\alpha_1[m] \parallel \cdots \parallel \alpha_n[m]).$;
   (e) $P_1 + Q \sim_p^{fr} P_2 + Q$;
   (f) $P_1 \parallel Q \sim_p^{fr} P_2 \parallel Q$;
   (g) $P_1 \setminus L \sim_p^{fr} P_2 \setminus L$;
   (h) $P_1[f] \sim_p^{fr} P_2[f]$.

*Proof.* From the definition of strongly FR pomset bisimulation (see Definition 2.68), we know that strongly FR pomset bisimulation is defined by FR pomset transitions, which are labeled by pomsets. In an FR pomset transition, the events in the pomset are either within causality relations (defined by the prefix .) or in concurrency (implicitly defined by . and +, and explicitly defined by $\parallel$), of course, they are pairwise consistent (without conflicts). In Theorem 3.19, we have already proven the case that all events are pairwise concurrent, hence, we only need to prove the case of events in causality. Without loss of generality, we take a pomset of $p = \{\alpha, \beta : \alpha.\beta\}$. Then, the FR pomset transition labeled by the above $p$ is just composed of one single event transition labeled by $\alpha$ succeeded by another single event transition labeled by $\beta$, that is, $\xrightarrow{p} = \xrightarrow{\alpha}\xrightarrow{\beta}$ and $\xrightarrow{p[\mathcal{K}]} = \xrightarrow{\beta[n]}\xrightarrow{\alpha[m]}$.

Similarly to the proof of congruence for strongly FR step bisimulation (see Theorem 3.19), we can prove that the congruence holds for strongly FR pomset bisimulation, however, we omit this proof here. □

**Theorem 3.21** (Congruence for strongly FR hp-bisimulation). *We can enjoy the congruence for strongly FR hp-bisimulation as follows:*

1. *If $A \stackrel{def}{=} P$, then $A \sim_{hp}^{fr} P$;*
2. *Let $P_1 \sim_{hp}^{fr} P_2$. Then,*
   (a) $\alpha.P_1 \sim_{hp}^{f} \alpha.P_2$;

**(b)** $(\alpha_1 \parallel \cdots \parallel \alpha_n).P_1 \sim^f_{hp} (\alpha_1 \parallel \cdots \parallel \alpha_n).P_2$;
**(c)** $P_1.\alpha[m] \sim^r_{hp} P_2.\alpha[m]$;
**(d)** $P_1.(\alpha_1[m] \parallel \cdots \parallel \alpha_n[m]) \sim^r_{hp} P_2.(\alpha_1[m] \parallel \cdots \parallel \alpha_n[m])$.;
**(e)** $P_1 + Q \sim^{fr}_{hp} P_2 + Q$;
**(f)** $P_1 \parallel Q \sim^{fr}_{hp} P_2 \parallel Q$;
**(g)** $P_1 \setminus L \sim^{fr}_{hp} P_2 \setminus L$;
**(h)** $P_1[f] \sim^{fr}_{hp} P_2[f]$.

*Proof.* From the definition of strongly FR hp-bisimulation (see Definition 2.70), we know that strongly FR hp-bisimulation is defined on the posetal product $(C_1, f, C_2)$, $f : C_1 \to C_2$ isomorphism. Two processes $P$ related to $C_1$ and $Q$ related to $C_2$, and $f : C_1 \to C_2$ isomorphism. Initially, $(C_1, f, C_2) = (\emptyset, \emptyset, \emptyset)$, and $(\emptyset, \emptyset, \emptyset) \in \sim^{fr}_{hp}$. When $P \xrightarrow{\alpha} P'$ $(C_1 \xrightarrow{\alpha} C_1')$, there will be $Q \xrightarrow{\alpha} Q'$ $(C_2 \xrightarrow{\alpha} C_2')$, and we define $f' = f[\alpha \mapsto \alpha]$. Also, when $P \xrightarrow{\alpha[m]} P'$ $(C_1 \xrightarrow{\alpha[m]} C_1')$, there will be $Q \xrightarrow{\alpha[m]} Q'$ $(C_2 \xrightarrow{\alpha[m]} C_2')$, and we define $f' = f[\alpha[m] \mapsto \alpha[m]]$. Then, if $(C_1, f, C_2) \in \sim^{fr}_{hp}$, then $(C_1', f', C_2') \in \sim^{fr}_{hp}$.

Similarly to the proof of congruence for strongly FR pomset bisimulation (see Theorem 3.20), we can prove that the congruence holds for strongly FR hp-bisimulation, we just need additionally to check the above conditions on FR hp-bisimulation, however, we omit them here. □

**Theorem 3.22** (Congruence for strongly FR hhp-bisimulation). *We can enjoy the congruence for strongly FR hhp-bisimulation as follows:*

1. *If* $A \stackrel{def}{=} P$, *then* $A \sim^{fr}_{hhp} P$;
2. *Let* $P_1 \sim^{fr}_{hhp} P_2$. *Then,*

**(a)** $\alpha.P_1 \sim^f_{hhp} \alpha.P_2$;
**(b)** $(\alpha_1 \parallel \cdots \parallel \alpha_n).P_1 \sim^f_{hhp} (\alpha_1 \parallel \cdots \parallel \alpha_n).P_2$;
**(c)** $P_1.\alpha[m] \sim^r_{hhp} P_2.\alpha[m]$;
**(d)** $P_1.(\alpha_1[m] \parallel \cdots \parallel \alpha_n[m]) \sim^r_{hhp} P_2.(\alpha_1[m] \parallel \cdots \parallel \alpha_n[m])$.;
**(e)** $P_1 + Q \sim^{fr}_{hhp} P_2 + Q$;
**(f)** $P_1 \parallel Q \sim^{fr}_{hhp} P_2 \parallel Q$;
**(g)** $P_1 \setminus L \sim^{fr}_{hhp} P_2 \setminus L$;
**(h)** $P_1[f] \sim^{fr}_{hhp} P_2[f]$.

*Proof.* From the definition of strongly FR hhp-bisimulation (see Definition 2.70), we know that strongly FR hhp-bisimulation is downward closed for strongly FR hp-bisimulation.

Similarly to the proof of congruence for strongly FR hp-bisimulation (see Theorem 3.21), we can prove that the congruence holds for strongly FR hhp-bisimulation, however, we omit this proof here. □

**Definition 3.23** (Weakly guarded recursive expression). $X$ is weakly guarded in $E$ if each occurrence of $X$ is with some subexpression $\alpha.F$ or $(\alpha_1 \parallel \cdots \parallel \alpha_n).F$ or $F.\alpha[m]$ or $F.(\alpha_1[m] \parallel \cdots \parallel \alpha_n[m])$ of $E$.

**Proposition 3.24.** *If the variables* $\widetilde{X}$ *are weakly guarded in E, and* $E\{\widetilde{P}/\widetilde{X}\} \xrightarrow{\{\alpha_1,\cdots,\alpha_n\}} P'$, *or* $E\{\widetilde{P}/\widetilde{X}\} \xrightarrow{\{\alpha_1[m],\cdots,\alpha_n[m]\}} P'$, *then* $P'$ *cannot take the form* $E'\{\widetilde{P}/\widetilde{X}\}$ *for some expression* $E'$.

*Proof.* It needs to induct on the depth of the inference of $E\{\widetilde{P}/\widetilde{X}\} \xrightarrow{\{\alpha_1,\cdots,\alpha_n\}} P'$ or $E\{\widetilde{P}/\widetilde{X}\} \xrightarrow{\{\alpha_1[m],\cdots,\alpha_n[m]\}} P'$. We consider $E\{\widetilde{P}/\widetilde{X}\} \xrightarrow{\{\alpha_1,\cdots,\alpha_n\}} P'$.

Case $E \equiv E_1 + E_2$. We may have $E_1 \xrightarrow{e_1} e_1[m] \cdot E_1'$ $e_1 \notin E_2$, $E_1 + E_2 \xrightarrow{e_1} e_1[m] \cdot E_1' + E_2$, $e_1[m] \cdot E_1' + E_2$ cannot take the form $E'\{\widetilde{P}/\widetilde{X}\}$ for some expression $E'$.

Hence, there may be no recursive expression for strongly FR truly concurrent bisimulations. For the same reason, there also may be no recursive expression for weakly FR truly concurrent bisimulations. □

**TABLE 3.9** Forward and reverse transition rules of $\tau$.

| |
|---|
| $\dfrac{}{\tau \xrightarrow{\tau} \surd}$ |
| $\dfrac{}{\tau \overset{\tau}{\twoheadrightarrow} \surd}$ |

## 3.3 Weakly forward–reverse truly concurrent bisimulations

Remembering that $\tau$ can neither be restricted nor relabeled, we know that the monoid laws, the static laws, and the new expansion law in Section 3.2 still holds with respect to the corresponding weakly FR truly concurrent bisimulations. Also, we can enjoy the congruence of Prefix, Summation, Composition, Restriction, Relabeling, and Constants with respect to corresponding weakly FR truly concurrent bisimulations. We will not retype these laws, and just give the $\tau$-specific laws.

The forward and reverse transition rules of $\tau$ are shown in Table 3.9, where $\xrightarrow{\tau} \surd$ is a predicate that represents a successful termination after execution of the silent step $\tau$.

**Proposition 3.25** ($\tau$ laws for weakly FR step bisimulation). *The $\tau$ laws for weakly FR step bisimulation are as follows:*

1. $P \approx_s^f \tau.P$;
2. $P \approx_s^r P.\tau$;
3. $\alpha.\tau.P \approx_s^f \alpha.P$;
4. $P.\tau.\alpha[m] \approx_s^r P.\alpha[m]$;
5. $(\alpha_1 \parallel \cdots \parallel \alpha_n).\tau.P \approx_s^f (\alpha_1 \parallel \cdots \parallel \alpha_n).P$;
6. $P.\tau.(\alpha_1[m] \parallel \cdots \parallel \alpha_n[m]) \approx_s^r P.(\alpha_1[m] \parallel \cdots \parallel \alpha_n[m])$;
7. $P + \tau.P \approx_s^f \tau.P$;
8. $P + P.\tau \approx_s^r P.\tau$;
9. $\alpha.(\tau.(P+Q)+P) \approx_s^f \alpha.(P+Q)$;
10. $((P+Q).\tau + P).\alpha[m] \approx_s^r (P+Q).\alpha[m]$;
11. $(\alpha_1 \parallel \cdots \parallel \alpha_n).(\tau.(P+Q)+P) \approx_s^f (\alpha_1 \parallel \cdots \parallel \alpha_n).(P+Q)$;
12. $((P+Q).\tau + P).(\alpha_1[m] \parallel \cdots \parallel \alpha_n[m]) \approx_s^r (P+Q).(\alpha_1[m] \parallel \cdots \parallel \alpha_n[m])$;
13. $P \approx_s^{fr} \tau \parallel P$.

*Proof.* Although transition rules in Definition 3.2 are defined in the flavor of a single event, they can be modified into a step (a set of events within which each event is pairwise concurrent), however, we omit them here. If we treat a single event as a step containing just one event, the proof of $\tau$ laws does not pose any problem, hence, we use this way and still use the transition rules in Definition 3.2.

1. $P \approx_s^f \tau.P$. By the forward transition rules of Prefix, we obtain

$$\frac{P \overset{\alpha}{\Rightarrow} P'}{P \overset{\alpha}{\Rightarrow} P'} \quad \frac{P \overset{\alpha}{\Rightarrow} P'}{\tau.P \overset{\alpha}{\Rightarrow} P'}$$

Since $P' \approx_s^f P'$, we obtain $P \approx_s^f \tau.P$, as desired.

2. $P \approx_s^r P.\tau$. By the reverse transition rules of Prefix, we obtain

$$\frac{P \overset{\alpha[m]}{\Rrightarrow} P'}{P \overset{\alpha[m]}{\Rrightarrow} P'} \quad \frac{P \overset{\alpha[m]}{\Rrightarrow} P'}{\tau.P \overset{\alpha[m]}{\Rrightarrow} P'}$$

Since $P' \approx_s^r P'$, we obtain $P \approx_s^r P.\tau$, as desired.

3. $\alpha.\tau.P \approx_s^f \alpha.P$. By the forward transition rules of Prefix, we obtain

$$\frac{}{\alpha.\tau.P \overset{\alpha}{\Rightarrow} \alpha[m].P} \quad \frac{}{\alpha.P \overset{\alpha}{\Rightarrow} \alpha[m].P}$$

Since $\alpha[m].P \approx_s^f \alpha[m].P$, we obtain $\alpha.\tau.P \approx_s^f \alpha.P$, as desired.

4. $P.\tau.\alpha[m] \approx_s^r P.\alpha[m]$. By the reverse transition rules of Prefix, we obtain

$$\frac{}{P.\tau.\alpha[m] \xrightarrow{\alpha[m]} P.\alpha} \quad \frac{}{P.\alpha[m] \xrightarrow{\alpha[m]} P.\alpha}$$

Since $P.\alpha \approx_s^r P.\alpha$, we obtain $P.\tau.\alpha[m] \approx_s^r P.\alpha[m]$, as desired.

5. $(\alpha_1 \parallel \cdots \parallel \alpha_n).\tau.P \approx_s^f (\alpha_1 \parallel \cdots \parallel \alpha_n).P$. By the forward transition rules of Prefix, we obtain

$$\frac{}{(\alpha_1 \parallel \cdots \parallel \alpha_n).\tau.P \xrightarrow{\{\alpha_1,\cdots,\alpha_n\}} (\alpha_1[m] \parallel \cdots \parallel \alpha_n[m]).P}$$

$$\frac{}{(\alpha_1 \parallel \cdots \parallel \alpha_n).P \xrightarrow{\{\alpha_1,\cdots,\alpha_n\}} (\alpha_1[m] \parallel \cdots \parallel \alpha_n[m]).P}$$

Since $(\alpha_1[m] \parallel \cdots \parallel \alpha_n[m]).P \approx_s^f (\alpha_1[m] \parallel \cdots \parallel \alpha_n[m]).P$, we obtain $(\alpha_1 \parallel \cdots \parallel \alpha_n).\tau.P \approx_s^f (\alpha_1 \parallel \cdots \parallel \alpha_n).P$, as desired.

6. $P.\tau.(\alpha_1[m] \parallel \cdots \parallel \alpha_n[m]) \approx_s^r P.(\alpha_1[m] \parallel \cdots \parallel \alpha_n[m])$. By the reverse transition rules of Prefix, we obtain

$$\frac{}{P.\tau.(\alpha_1[m] \parallel \cdots \parallel \alpha_n[m]) \xrightarrow{\{\alpha_1[m],\cdots,\alpha_n[m]\}} (P.(\alpha_1 \parallel \cdots \parallel \alpha_n)}$$

$$\frac{}{P.(\alpha_1[m] \parallel \cdots \parallel \alpha_n[m]) \xrightarrow{\{\alpha_1[m],\cdots,\alpha_n[m]\}} (P.(\alpha_1 \parallel \cdots \parallel \alpha_n)}$$

Since $P.(\alpha_1 \parallel \cdots \parallel \alpha_n) \approx_s^r P.(\alpha_1 \parallel \cdots \parallel \alpha_n)$, we obtain $P.\tau.(\alpha_1[m] \parallel \cdots \parallel \alpha_n[m]) \approx_s^r P.(\alpha_1[m] \parallel \cdots \parallel \alpha_n[m])$, as desired.

7. $P + \tau.P \approx_s^f \tau.P$. By the forward transition rules of Summation, we obtain

$$\frac{P \xrightarrow{\alpha} P'}{P + \tau.P \xrightarrow{\alpha} P' + P'} \quad \frac{P \xrightarrow{\alpha} P'}{\tau.P \xrightarrow{\alpha} P'}$$

Since $P' + P' \approx_s^f P'$, we obtain $P + \tau.P \approx_s^f \tau.P$, as desired.

8. $P + P.\tau \approx_s^r P.\tau$. By the reverse transition rules of Summation, we obtain

$$\frac{P \xrightarrow{\alpha[m]} P'}{P + P.\tau \xrightarrow{\alpha[m]} P' + P'} \quad \frac{P \xrightarrow{\alpha[m]} P'}{P.\tau \xrightarrow{\alpha[m]} P'}$$

Since $P' + P' \approx_s^r P'$, we obtain $P + P.\tau \approx_s^r P.\tau$, as desired.

9. $\alpha.(\tau.(P + Q) + P) \approx_s^f \alpha.(P + Q)$. By the forward transition rules of Prefix and Summation, we obtain

$$\frac{}{\alpha.(\tau.(P + Q) + P) \xrightarrow{\alpha} \alpha[m].(P + Q + P)} \quad \frac{}{\alpha.(P + Q) \xrightarrow{\alpha} \alpha[m].(P + Q)}$$

Since $\alpha[m].(P + Q + P) \approx_s^f \alpha[m].(P + Q)$, we obtain $\alpha.(\tau.(P + Q) + P) \approx_s^f \alpha.(P + Q)$, as desired.

10. $((P + Q).\tau + P).\alpha[m] \approx_s^r (P + Q).\alpha[m]$. By the reverse transition rules of Prefix and Summation, we obtain

$$\frac{}{((P + Q).\tau + P).\alpha[m] \xrightarrow{\alpha[m]} (P + Q + P).\alpha} \quad \frac{}{(P + Q).\alpha[m] \xrightarrow{\alpha[m]} (P + Q).\alpha}$$

Since $(P + Q + P).\alpha \approx_s^r (P + Q).\alpha$, we obtain $((P + Q).\tau + P).\alpha[m] \approx_s^r (P + Q).\alpha[m]$, as desired.

11. $(\alpha_1 \parallel \cdots \parallel \alpha_n).(\tau.(P + Q) + P) \approx_s^f (\alpha_1 \parallel \cdots \parallel \alpha_n).(P + Q)$. By the forward transition rules of Prefix and Summation, we obtain

$$\frac{}{(\alpha_1 \parallel \cdots \parallel \alpha_n).(\tau.(P + Q) + P) \xrightarrow{\{\alpha_1,\cdots,\alpha_n\}} (\alpha_1[m] \parallel \cdots \parallel \alpha_n[m]).(P + Q + P)}$$

$$\frac{}{(\alpha_1 \parallel \cdots \parallel \alpha_n).(P+Q) \xRightarrow{\{\alpha_1,\cdots,\alpha_n\}} (\alpha_1[m] \parallel \cdots \parallel \alpha_n[m]).(P+Q)}$$

Since $(\alpha_1[m] \parallel \cdots \parallel \alpha_n[m]).(P+Q+P) \approx_s^f (\alpha_1[m] \parallel \cdots \parallel \alpha_n[m]).(P+Q)$, we obtain $(\alpha_1 \parallel \cdots \parallel \alpha_n).(\tau.(P+Q)+P) \approx_s^f (\alpha_1 \parallel \cdots \parallel \alpha_n).(P+Q)$, as desired.

12. $((P+Q).\tau+P).(\alpha_1[m] \parallel \cdots \parallel \alpha_n[m]) \approx_s^r (P+Q).(\alpha_1[m] \parallel \cdots \parallel \alpha_n[m])$. By the reverse transition rules of Prefix and Summation, we obtain

$$\frac{}{((P+Q).\tau+P).(\alpha_1[m] \parallel \cdots \parallel \alpha_n[m]) \xRightarrow{\{\alpha_1[m],\cdots,\alpha_n[m]\}} (P+Q+P).(\alpha_1 \parallel \cdots \parallel \alpha_n)}$$

$$\frac{}{(P+Q).(\alpha_1[m] \parallel \cdots \parallel \alpha_n[m]) \xRightarrow{\{\alpha_1[m],\cdots,\alpha_n[m]\}} (P+Q).(\alpha_1 \parallel \cdots \parallel \alpha_n)}$$

Since $(P+Q+P).(\alpha_1 \parallel \cdots \parallel \alpha_n) \approx_s^r (P+Q).(\alpha_1 \parallel \cdots \parallel \alpha_n)$, we obtain $((P+Q).\tau+P).(\alpha_1[m] \parallel \cdots \parallel \alpha_n[m]) \approx_s^r (P+Q).(\alpha_1[m] \parallel \cdots \parallel \alpha_n[m])$, as desired.

13. $P \approx_s^{fr} \tau \parallel P$. By the forward transition rules of Composition, we obtain

$$\frac{P \xRightarrow{\alpha} P'}{P \xRightarrow{\alpha} P'} \qquad \frac{P \xRightarrow{\alpha} P'}{\tau \parallel P \xRightarrow{\alpha} P'}$$

By the reverse transition rules of Composition, we obtain

$$\frac{P \xRightarrow{\alpha} P'}{P \xRightarrow{\alpha} P'} \qquad \frac{P \xRightarrow{\alpha} P'}{\tau \parallel P \xRightarrow{\alpha} P'}$$

Since $P' \approx_s^{fr} P'$, we obtain $P \approx_s^{fr} \tau \parallel P$, as desired. □

**Proposition 3.26** ($\tau$ laws for weakly FR pomset bisimulation). *The $\tau$ laws for weakly FR pomset bisimulation are as follows:*

1. $P \approx_p^f \tau.P$;
2. $P \approx_p^r P.\tau$;
3. $\alpha.\tau.P \approx_p^f \alpha.P$;
4. $P.\tau.\alpha[m] \approx_p^r P.\alpha[m]$;
5. $(\alpha_1 \parallel \cdots \parallel \alpha_n).\tau.P \approx_p^f (\alpha_1 \parallel \cdots \parallel \alpha_n).P$;
6. $P.\tau.(\alpha_1[m] \parallel \cdots \parallel \alpha_n[m]) \approx_p^r P.(\alpha_1[m] \parallel \cdots \parallel \alpha_n[m])$;
7. $P+\tau.P \approx_p^f \tau.P$;
8. $P+P.\tau \approx_p^r P.\tau$;
9. $\alpha.(\tau.(P+Q)+P) \approx_s^f \alpha.(P+Q)$;
10. $((P+Q).\tau+P).\alpha[m] \approx_s^r (P+Q).\alpha[m]$;
11. $(\alpha_1 \parallel \cdots \parallel \alpha_n).(\tau.(P+Q)+P) \approx_s^f (\alpha_1 \parallel \cdots \parallel \alpha_n).(P+Q)$;
12. $((P+Q).\tau+P).(\alpha_1[m] \parallel \cdots \parallel \alpha_n[m]) \approx_s^r (P+Q).(\alpha_1[m] \parallel \cdots \parallel \alpha_n[m])$;
13. $P \approx_p^{fr} \tau \parallel P$.

*Proof.* From the definition of weakly FR pomset bisimulation $\approx_p^{fr}$ (see Definition 2.69), we know that weakly FR pomset bisimulation $\approx_p^{fr}$ is defined by weakly FR pomset transitions, which are labeled by pomsets with $\tau$. In a weakly FR pomset transition, the events in the pomset are either within causality relations (defined by .) or in concurrency (implicitly defined by . and +, and explicitly defined by $\parallel$), of course, they are pairwise consistent (without conflicts). In Proposition 3.25, we have already proven the case that all events are pairwise concurrent, hence, we only need to prove the case of events in causality. Without loss of generality, we take a pomset of $p = \{\alpha, \beta : \alpha.\beta\}$. Then, the weakly forward pomset transition labeled by the above $p$ is just composed of one single event transition labeled by $\alpha$ succeeded by another single event transition labeled by $\beta$, that is, $\xRightarrow{p} = \xRightarrow{\alpha}\xRightarrow{\beta}$ and $\xRightarrow{p[\mathcal{K}]} = \xRightarrow{\beta[n]}\xRightarrow{\alpha[m]}$.

Similarly to the proof of $\tau$ laws for weakly FR step bisimulation $\approx_s^{fr}$ (Proposition 3.25), we can prove that $\tau$ laws hold for weakly FR pomset bisimulation $\approx_p^{fr}$, however, we omit them here. □

**Proposition 3.27** ($\tau$ laws for weakly FR hp-bisimulation). *The $\tau$ laws for weakly FR hp-bisimulation are as follows:*

1. $P \approx_{hp}^{f} \tau.P$;
2. $P \approx_{hp}^{r} P.\tau$;
3. $\alpha.\tau.P \approx_{hp}^{f} \alpha.P$;
4. $P.\tau.\alpha[m] \approx_{hp}^{r} P.\alpha[m]$;
5. $(\alpha_1 \parallel \cdots \parallel \alpha_n).\tau.P \approx_{hp}^{f} (\alpha_1 \parallel \cdots \parallel \alpha_n).P$;
6. $P.\tau.(\alpha_1[m] \parallel \cdots \parallel \alpha_n[m]) \approx_{hp}^{r} P.(\alpha_1[m] \parallel \cdots \parallel \alpha_n[m])$;
7. $P + \tau.P \approx_{hp}^{f} \tau.P$;
8. $P + P.\tau \approx_{hp}^{r} P.\tau$;
9. $\alpha.(\tau.(P+Q)+P) \approx_s^{f} \alpha.(P+Q)$;
10. $((P+Q).\tau + P).\alpha[m] \approx_s^{r} (P+Q).\alpha[m]$;
11. $(\alpha_1 \parallel \cdots \parallel \alpha_n).(\tau.(P+Q)+P) \approx_s^{f} (\alpha_1 \parallel \cdots \parallel \alpha_n).(P+Q)$;
12. $((P+Q).\tau + P).(\alpha_1[m] \parallel \cdots \parallel \alpha_n[m]) \approx_s^{r} (P+Q).(\alpha_1[m] \parallel \cdots \parallel \alpha_n[m])$;
13. $P \approx_{hp}^{fr} \tau \parallel P$.

*Proof.* From the definition of weakly FR hp-bisimulation $\approx_{hp}^{fr}$ (see Definition 2.71), we know that weakly FR hp-bisimulation $\approx_{hp}^{fr}$ is defined on the weakly posetal product $(C_1, f, C_2)$, $f : \hat{C_1} \to \hat{C_2}$ isomorphism. Two processes $P$ related to $C_1$ and $Q$ related to $C_2$, and $f : \hat{C_1} \to \hat{C_2}$ isomorphism. Initially, $(C_1, f, C_2) = (\emptyset, \emptyset, \emptyset)$, and $(\emptyset, \emptyset, \emptyset) \in \approx_{hp}^{fr}$. When $P \stackrel{\alpha}{\Rightarrow} P'$ ($C_1 \stackrel{\alpha}{\Rightarrow} C_1'$), there will be $Q \stackrel{\alpha}{\Rightarrow} Q'$ ($C_2 \stackrel{\alpha}{\Rightarrow} C_2'$), and we define $f' = f[\alpha \mapsto \alpha]$. Also, when $P \stackrel{\alpha[m]}{\Longrightarrow} P'$ ($C_1 \stackrel{\alpha[m]}{\Longrightarrow} C_1'$), there will be $Q \stackrel{\alpha[m]}{\Longrightarrow} Q'$ ($C_2 \stackrel{\alpha[m]}{\Longrightarrow} C_2'$), and we define $f' = f[\alpha[m] \mapsto \alpha[m]]$. Then, if $(C_1, f, C_2) \in \approx_{hp}^{fr}$, then $(C_1', f', C_2') \in \approx_{hp}^{fr}$.

Similarly to the proof of $\tau$ laws for weakly FR pomset bisimulation (Proposition 3.26), we can prove that $\tau$ laws hold for weakly FR hp-bisimulation, we just need additionally to check the above conditions on weakly FR hp-bisimulation, however, we omit them here. □

**Proposition 3.28** ($\tau$ laws for weakly FR hhp-bisimulation). *The $\tau$ laws for weakly FR hhp-bisimulation are as follows:*

1. $P \approx_{hhp}^{f} \tau.P$;
2. $P \approx_{hhp}^{r} P.\tau$;
3. $\alpha.\tau.P \approx_{hhp}^{f} \alpha.P$;
4. $P.\tau.\alpha[m] \approx_{hhp}^{r} P.\alpha[m]$;
5. $(\alpha_1 \parallel \cdots \parallel \alpha_n).\tau.P \approx_{hhp}^{f} (\alpha_1 \parallel \cdots \parallel \alpha_n).P$;
6. $P.\tau.(\alpha_1[m] \parallel \cdots \parallel \alpha_n[m]) \approx_{hhp}^{r} P.(\alpha_1[m] \parallel \cdots \parallel \alpha_n[m])$;
7. $P + \tau.P \approx_{hhp}^{f} \tau.P$;
8. $P + P.\tau \approx_{hhp}^{r} P.\tau$;
9. $\alpha.(\tau.(P+Q)+P) \approx_s^{f} \alpha.(P+Q)$;
10. $((P+Q).\tau + P).\alpha[m] \approx_s^{r} (P+Q).\alpha[m]$;
11. $(\alpha_1 \parallel \cdots \parallel \alpha_n).(\tau.(P+Q)+P) \approx_s^{f} (\alpha_1 \parallel \cdots \parallel \alpha_n).(P+Q)$;
12. $((P+Q).\tau + P).(\alpha_1[m] \parallel \cdots \parallel \alpha_n[m]) \approx_s^{r} (P+Q).(\alpha_1[m] \parallel \cdots \parallel \alpha_n[m])$;
13. $P \approx_{hhp}^{fr} \tau \parallel P$.

*Proof.* From the definition of weakly FR hhp-bisimulation (see Definition 2.71), we know that weakly FR hhp-bisimulation is downward closed for weakly FR hp-bisimulation.

Similarly to the proof of $\tau$ laws for weakly FR hp-bisimulation (see Proposition 3.27), we can prove that the $\tau$ laws hold for weakly FR hhp-bisimulation, however, we omit them here. □

Chapter 4

# Algebraic laws for reversible computing

We found algebraic laws for true concurrency called $APTC$ [9], we now design a reversible version of $APTC$ called $RAPTC$. This chapter is organized as follows. We introduce Basic Reversible Algebra for True Concurrency ($BRATC$) in Section 4.1, Reversible Algebra for Parallelism in True Concurrency ($RAPTC$) in Section 4.2, and abstraction in Section 4.3.

## 4.1 Basic reversible algebra for true concurrency

In this section, we will discuss the algebraic laws for prime event structure $\mathcal{E}$, exactly for causality $\leq$ and conflict $\sharp$, with a reversible flavor. The resulted algebra is called Basic Reversible Algebra for True Concurrency, abbreviated $BRATC$.

### 4.1.1 Axiom system of $BRATC$

In the following, let $e_1, e_2, e_1', e_2' \in \mathbb{E}$, and let variables $x, y, z$ range over the set of terms for true concurrency, $p, q, s$ range over the set of closed terms, the predicate Std(p) represents that $p$ is a standard process containing no past events, the predicate NStd(p) represents that $p$ is a process full of past events. The set of axioms of $BRATC$ consists of the laws given in Table 4.1.

Intuitively, the axiom $A1$ says that the binary operator $+$ satisfies the commutative law. The axiom $A2$ says that $+$ satisfies associativity. $A3$ says that $+$ satisfies idempotency. The axiom $A4$ is the left distributivity of the binary operator $\cdot$ to $+$. The axiom $RA4$ is the right distributivity of the binary operator $\cdot$ to $+$, and $A5$ is the associativity of $\cdot$.

### 4.1.2 Properties of $BRATC$

**Definition 4.1** (Basic terms of $BRATC$). The set of basic terms of $BRATC$, $\mathcal{B}(BRATC)$, is inductively defined as follows:

1. $\mathbb{E} \subset \mathcal{B}(BRATC)$;
2. if $e \in \mathbb{E}, t \in \mathcal{B}(BRATC)$, then $e \cdot t \in \mathcal{B}(BRATC)$;
3. if $t, s \in \mathcal{B}(BRATC)$, then $t + s \in \mathcal{B}(BRATC)$.

**Theorem 4.2** (Elimination theorem of $BRATC$). *Let $p$ be a closed $BRATC$ term. Then, there is a basic $BRATC$ term $q$ such that $BRATC \vdash p = q$.*

*Proof.* (1) First, suppose that the following ordering on the signature of $BRATC$ is defined: $\cdot > +$ and the symbol $\cdot$ is given the lexicographical status for the first argument, then for each rewrite rule $p \to q$ (or $p \twoheadrightarrow q$) in Table 4.2 relation $p >_{lpo} q$ can easily be proved. We obtain that the term rewrite system shown in Table 4.2 is strongly normalizing, for it

**TABLE 4.1** Axioms of $BRATC$.

| No. | Axiom |
|---|---|
| $A1$ | $x + y = y + x$ |
| $A2$ | $(x + y) + z = x + (y + z)$ |
| $A3$ | $x + x = x$ |
| $A4$ | $x \cdot (y + z) = x \cdot y + x \cdot z(\text{Std}(x), \text{Std}(y), \text{Std}(z))$ |
| $RA4$ | $(x + y) \cdot z = x \cdot z + y \cdot z(\text{NStd}(x), \text{NStd}(y), \text{NStd}(z))$ |
| $A5$ | $(x \cdot y) \cdot z = x \cdot (y \cdot z)$ |

Handbook of Truly Concurrent Process Algebra. https://doi.org/10.1016/B978-0-44-321515-5.00008-1

**TABLE 4.2** Term rewrite system of $BRATC$.

| No. | Rewriting Rule |
|---|---|
| $RRA3$ | $x + x \to x$ |
| $RRA4$ | $x \cdot (y + z) \to x \cdot y + x \cdot z$ |
| $RRRA4$ | $(x + y) \cdot z \to x \cdot z + y \cdot z$ |
| $RRA5$ | $(x \cdot y) \cdot z \to x \cdot (y \cdot z)$ |

has finitely many rewriting rules, and $>$ is a well-founded ordering on the signature of $BRATC$, and if $s >_{lpo} t$, for each rewriting rule $s \to t$ is in Table 4.2 (see Theorem 2.9).

(2) Then, we prove that the normal forms of closed $BRATC$ terms are basic $BRATC$ terms.

Suppose that $p$ is a normal form of some closed $BRATC$ term and suppose that $p$ is not a basic term. Let $p'$ denote the smallest sub-term of $p$ that is not a basic term. This implies that each sub-term of $p'$ is a basic term. Then, we prove that $p$ is not a term in normal form. It is sufficient to induct on the structure of $p'$:

- Case $p' \equiv eore[m], e \in \mathbb{E}$. $p'$ is a basic term, which contradicts the assumption that $p'$ is not a basic term, so this case should not occur.
- Case $p' \equiv p_1 \cdot p_2$. By induction on the structure of the basic term $p_1$:
  - Subcase $p_1 \in \mathbb{E}$. $p'$ would be a basic term, which contradicts the assumption that $p'$ is not a basic term;
  - Subcase $p_1 \equiv e \cdot p_1'$. $RRA5$ rewriting rule can be applied. Hence, $p$ is not a normal form;
  - Subcase $p_1 \equiv p_1' \cdot e[m]$. $RRA5$ rewriting rule can be applied. Hence, $p$ is not a normal form;
  - Subcase $p_1 \equiv p_1' + p_1''$. $RRA4$ or $RRRA4$ rewriting rule can be applied. Hence, $p$ is not a normal form.
- Case $p' \equiv p_1 + p_2$. By induction on the structure of the basic terms both $p_1$ and $p_2$, all subcases will lead to that $p'$ would be a basic term, which contradicts the assumption that $p'$ is not a basic term. □

### 4.1.3 Structured operational semantics of $BRATC$

In this section, we will define a term-deduction system that gives the operational semantics of $BRATC$. We give the operational transition rules for operators $\cdot$ and $+$ as Table 4.3 and Table 4.4 show. Also, the predicate $\xrightarrow{\alpha} \alpha[m]$ represents successful forward termination after execution of the action $\alpha$, the predicate $\xrightarrow{\alpha[m]} \alpha$ represents successful reverse termination after execution of the event $\alpha[m]$.

The pomset transition rules are shown in Table 4.5 and Table 4.6, which are different from the single event transition rules in Table 4.3 and Table 4.4, the pomset transition rules are labeled by pomsets, which are defined by causality $\cdot$ and conflict $+$.

**Theorem 4.3** (Congruence of $BRATC$ with respect to FR pomset bisimulation equivalence). *FR pomset bisimulation equivalence $\sim_p^{fr}$ is a congruence with respect to $BRATC$.*

*Proof.* It is easy to see that FR pomset bisimulation is an equivalent relation on $BRATC$ terms, we only need to prove that $\sim_p^{fr}$ is preserved by the operators $\cdot$ and $+$.

- Causality operator $\cdot$. Let $x_1, x_2$ and $y_1, y_2$ be $BRATC$ processes, and $x_1 \sim_p^{fr} y_1$, $x_2 \sim_p^{fr} y_2$, it is sufficient to prove that $x_1 \cdot x_2 \sim_p^{fr} y_1 \cdot y_2$.

  By the definition of FR pomset bisimulation $\sim_p^{fr}$ (Definition 2.68), $x_1 \sim_p^{fr} y_1$ means that

$$x_1 \xrightarrow{X_1} x_1' \quad y_1 \xrightarrow{Y_1} y_1'$$

$$x_1 \xrightarrow{X_1[\mathcal{K}]} x_1'' \quad y_1 \xrightarrow{Y_1[\mathcal{J}]} y_1''$$

  with $X_1 \subseteq x_1$, $Y_1 \subseteq y_1$, $X_1 \sim Y_1$, $x_1' \sim_p^{fr} y_1'$, and $x_1'' \sim_p^{fr} y_1''$. The meaning of $x_2 \sim_p^{fr} y_2$ is similar.

**TABLE 4.3** Forward single event transition rules of $BRATC$.

| | |
|---|---|
| $\frac{}{e \xrightarrow{e} e[m]}$ | |
| $\frac{x \xrightarrow{e} e[m] \quad e \notin y}{x + y \xrightarrow{e} e[m] + y}$ | $\frac{x \xrightarrow{e} x' \quad e \notin y}{x + y \xrightarrow{e} x' + y}$ |
| $\frac{y \xrightarrow{e} e[m] \quad e \notin x}{x + y \xrightarrow{e} x + e[m]}$ | $\frac{y \xrightarrow{e} y' \quad e \notin x}{x + y \xrightarrow{e} x + y'}$ |
| $\frac{x \xrightarrow{e} e[m] \quad y \xrightarrow{e} e[m]}{x + y \xrightarrow{e} e[m] + e[m]}$ | $\frac{x \xrightarrow{e} x' \quad y \xrightarrow{e} e[m]}{x + y \xrightarrow{e} x' + e[m]}$ |
| $\frac{x \xrightarrow{e} e[m] \quad y \xrightarrow{e} y'}{x + y \xrightarrow{e} e[m] + y'}$ | $\frac{x \xrightarrow{e} x' \quad y \xrightarrow{e} y'}{x + y \xrightarrow{e} x' + y'}$ |
| $\frac{x \xrightarrow{e} e[m] \quad \mathrm{Std}(y)}{x \cdot y \xrightarrow{e} e[m] \cdot y}$ | $\frac{x \xrightarrow{e} x' \quad \mathrm{Std}(y)}{x \cdot y \xrightarrow{e} x' \cdot y}$ |
| $\frac{y \xrightarrow{e'} e'[n] \quad \mathrm{NStd}(x)}{x \cdot y \xrightarrow{e'} x \cdot e'[n]}$ | $\frac{y \xrightarrow{e'} y' \quad \mathrm{NStd}(x)}{x \cdot y \xrightarrow{e'} x \cdot y'}$ |

**TABLE 4.4** Reverse single event transition rules of $BRATC$.

| | |
|---|---|
| $\frac{}{e[m] \overset{e[m]}{\twoheadrightarrow} e}$ | |
| $\frac{x \overset{e[m]}{\twoheadrightarrow} e \quad e \notin y}{x + y \overset{e[m]}{\twoheadrightarrow} e + y}$ | $\frac{x \overset{e[m]}{\twoheadrightarrow} x' \quad e \notin y}{x + y \overset{e[m]}{\twoheadrightarrow} x' + y}$ |
| $\frac{y \overset{e[m]}{\twoheadrightarrow} e \quad e \notin x}{x + y \overset{e[m]}{\twoheadrightarrow} x + e}$ | $\frac{y \overset{e[m]}{\twoheadrightarrow} y' \quad e \notin x}{x + y \overset{e[m]}{\twoheadrightarrow} x + y'}$ |
| $\frac{x \overset{e[m]}{\twoheadrightarrow} e \quad y \overset{e[m]}{\twoheadrightarrow} e}{x + y \overset{e[m]}{\twoheadrightarrow} e + e}$ | $\frac{x \overset{e[m]}{\twoheadrightarrow} x' \quad y \overset{e[m]}{\twoheadrightarrow} e}{x + y \overset{e[m]}{\twoheadrightarrow} x' + e}$ |
| $\frac{x \overset{e[m]}{\twoheadrightarrow} e \quad y \overset{e[m]}{\twoheadrightarrow} y'}{x + y \overset{e[m]}{\twoheadrightarrow} e + y'}$ | $\frac{x \overset{e[m]}{\twoheadrightarrow} x' \quad y \overset{e[m]}{\twoheadrightarrow} y'}{x + y \overset{e[m]}{\twoheadrightarrow} x' + y'}$ |
| $\frac{x \overset{e[m]}{\twoheadrightarrow} e \quad \mathrm{Std}(y)}{x \cdot y \overset{e[m]}{\twoheadrightarrow} e \cdot y}$ | $\frac{x \overset{e[m]}{\twoheadrightarrow} x' \quad \mathrm{Std}(y)}{x \cdot y \overset{e[m]}{\twoheadrightarrow} x' \cdot y}$ |
| $\frac{y \overset{e'[n]}{\twoheadrightarrow} e' \quad \mathrm{NStd}(x)}{x \cdot y \overset{e'[n]}{\twoheadrightarrow} x \cdot e'}$ | $\frac{y \overset{e'[n]}{\twoheadrightarrow} y' \quad \mathrm{NStd}(x)}{x \cdot y \overset{e'[n]}{\twoheadrightarrow} x \cdot y'}$ |

By the FR pomset transition rules for causality operator $\cdot$ in Table 4.5 and Table 4.6, we can obtain

$$x_1 \cdot x_2 \xrightarrow{X_1} X_1[\mathcal{K}] \cdot x_2 \quad y_1 \cdot y_2 \xrightarrow{Y_1} Y_1[\mathcal{J}] \cdot y_2$$

$$x_1 \cdot x_2 \overset{X_2[\mathcal{K}]}{\twoheadrightarrow} x_1 \cdot X_2 \quad y_1 \cdot y_2 \overset{Y_2[\mathcal{J}]}{\twoheadrightarrow} y_1 \cdot Y_2$$

with $X_1 \subseteq x_1$, $Y_1 \subseteq y_1$, $X_2 \subseteq x_2$, $Y_2 \subseteq y_2$, $X_1 \sim Y_1$, $X_2 \sim Y_2$, and the assumptions $X_1[\mathcal{K}] \cdot x_2 \sim_p^{fr} Y_1[\mathcal{J}] \cdot y_2$ and $x_1 \cdot X_2 \sim_p^{fr} y_1 \cdot Y_2$, hence, we obtain $x_1 \cdot x_2 \sim_p^{fr} y_1 \cdot y_2$, as desired.
Or, we can obtain

$$x_1 \cdot x_2 \xrightarrow{X_1} x_1' \cdot x_2 \quad y_1 \cdot y_2 \xrightarrow{Y_1} y_1' \cdot y_2$$

**TABLE 4.5** Forward pomset transition rules of *BRATC*.

| | |
|---|---|
| $\frac{}{X \xrightarrow{X} X[\mathcal{K}]}$ | |
| $\frac{x \xrightarrow{X} X[\mathcal{K}] \quad X \nsubseteq y}{x+y \xrightarrow{X} X[\mathcal{K}]+y}$ | $\frac{x \xrightarrow{X} x' \quad X \nsubseteq y}{x+y \xrightarrow{X} x'+y}$ |
| $\frac{y \xrightarrow{Y} Y[\mathcal{J}] \quad Y \nsubseteq x}{x+y \xrightarrow{Y} x+Y[\mathcal{J}]}$ | $\frac{y \xrightarrow{Y} y' \quad Y \nsubseteq x}{x+y \xrightarrow{Y} x+y'}$ |
| $\frac{x \xrightarrow{X} X[\mathcal{K}] \quad y \xrightarrow{X} X[\mathcal{K}]}{x+y \xrightarrow{X} X[\mathcal{K}]+X[\mathcal{K}]}$ | $\frac{x \xrightarrow{X} x' \quad y \xrightarrow{X} X[\mathcal{K}]}{x+y \xrightarrow{X} x'+X[\mathcal{K}]}$ |
| $\frac{x \xrightarrow{X} X[\mathcal{K}] \quad y \xrightarrow{X} y'}{x+y \xrightarrow{X} X[\mathcal{K}]+y'}$ | $\frac{x \xrightarrow{X} x' \quad y \xrightarrow{X} y'}{x+y \xrightarrow{X} x'+y'}$ |
| $\frac{x \xrightarrow{X} X[\mathcal{K}] \quad Std(y)}{x \cdot y \xrightarrow{X} X[\mathcal{K}] \cdot y}(X \subseteq x)$ | $\frac{x \xrightarrow{X} x' \quad Std(y)}{x \cdot y \xrightarrow{X} x' \cdot y}(X \subseteq x)$ |
| $\frac{y \xrightarrow{Y} Y[\mathcal{J}] \quad NStd(x)}{x \cdot y \xrightarrow{Y} x \cdot Y[\mathcal{J}]}(Y \subseteq y)$ | $\frac{y \xrightarrow{Y} y' \quad NStd(x)}{x \cdot y \xrightarrow{Y} x \cdot y'}(Y \subseteq y)$ |

**TABLE 4.6** Reverse pomset transition rules of *BRATC*.

| | |
|---|---|
| $\frac{}{X[\mathcal{K}] \overset{X[\mathcal{K}]}{\twoheadrightarrow} X}$ | |
| $\frac{x \overset{X[\mathcal{K}]}{\twoheadrightarrow} X \quad X \nsubseteq y}{x+y \overset{X[\mathcal{K}]}{\twoheadrightarrow} X+y}$ | $\frac{x \overset{X[\mathcal{K}]}{\twoheadrightarrow} x' \quad X \nsubseteq y}{x+y \overset{X[\mathcal{K}]}{\twoheadrightarrow} x'+y}$ |
| $\frac{y \overset{Y[\mathcal{J}]}{\twoheadrightarrow} Y \quad Y \nsubseteq x}{x+y \overset{Y[\mathcal{J}]}{\twoheadrightarrow} x+Y}$ | $\frac{y \overset{Y[\mathcal{J}]}{\twoheadrightarrow} y' \quad Y \nsubseteq x}{x+y \overset{Y[\mathcal{J}]}{\twoheadrightarrow} x+y'}$ |
| $\frac{x \overset{X[\mathcal{K}]}{\twoheadrightarrow} X \quad y \overset{X[\mathcal{K}]}{\twoheadrightarrow} X}{x+y \overset{X[\mathcal{K}]}{\twoheadrightarrow} X+X}$ | $\frac{x \overset{X[\mathcal{K}]}{\twoheadrightarrow} x' \quad y \overset{X[\mathcal{K}]}{\twoheadrightarrow} X}{x+y \overset{X[\mathcal{K}]}{\twoheadrightarrow} x'+X}$ |
| $\frac{x \overset{X[\mathcal{K}]}{\twoheadrightarrow} X \quad y \overset{X[\mathcal{K}]}{\twoheadrightarrow} y'}{x+y \overset{X[\mathcal{K}]}{\twoheadrightarrow} X+y'}$ | $\frac{x \overset{X[\mathcal{K}]}{\twoheadrightarrow} x' \quad y \overset{X[\mathcal{K}]}{\twoheadrightarrow} y'}{x+y \overset{X[\mathcal{K}]}{\twoheadrightarrow} x'+y'}$ |
| $\frac{x \overset{X[\mathcal{K}]}{\twoheadrightarrow} X \quad Std(y)}{x \cdot y \overset{X[\mathcal{K}]}{\twoheadrightarrow} X \cdot y}(X \subseteq x)$ | $\frac{x \overset{X[\mathcal{K}]}{\twoheadrightarrow} x' \quad Std(y)}{x \cdot y \overset{X[\mathcal{K}]}{\twoheadrightarrow} x' \cdot y}(X \subseteq x)$ |
| $\frac{y \overset{Y[\mathcal{J}]}{\twoheadrightarrow} Y \quad NStd(x)}{x \cdot y \overset{Y[\mathcal{J}]}{\twoheadrightarrow} x \cdot Y}(Y \subseteq y)$ | $\frac{y \overset{Y[\mathcal{J}]}{\twoheadrightarrow} y' \quad NStd(x)}{x \cdot y \overset{Y[\mathcal{J}]}{\twoheadrightarrow} x \cdot y'}(Y \subseteq y)$ |

$$x_1 \cdot x_2 \overset{X_2[\mathcal{K}]}{\twoheadrightarrow} x_1 \cdot x_2' \quad y_1 \cdot y_2 \overset{Y_2[\mathcal{J}]}{\twoheadrightarrow} y_1 \cdot y_2'$$

with $X_1 \subseteq x_1$, $Y_1 \subseteq y_1$, $X_2 \subseteq x_2$, $Y_2 \subseteq y_2$, $X_1 \sim Y_1$, $X_2 \sim Y_2$, and the assumptions $x_1' \cdot x_2 \sim_p^{fr} y_1' \cdot y_2$, $x_1 \cdot x_2' \sim_p^{fr} y_1 \cdot y_2'$, hence, we obtain $x_1 \cdot x_2 \sim_p^{fr} y_1 \cdot y_2$, as desired.

- Conflict operator $+$. Let $x_1, x_2$ and $y_1, y_2$ be *BRATC* processes, and $x_1 \sim_p^{fr} y_1$, $x_2 \sim_p^{fr} y_2$, it is sufficient to prove that $x_1 + x_2 \sim_p^{fr} y_1 + y_2$. The meanings of $x_1 \sim_p^{fr} y_1$ and $x_2 \sim_p^{fr} y_2$ are the same as the above case, according to the definition of FR pomset bisimulation $\sim_p^{fr}$ in Definition 2.68.

By the FR pomset transition rules for conflict operator $+$ in Table 4.5 and Table 4.6, we can obtain several cases:

$$x_1 + x_2 \xrightarrow{X_1} X_1[\mathcal{K}] + x_2 \quad y_1 + y_2 \xrightarrow{Y_1} Y_1[\mathcal{J}] + y_2$$

$$x_1 + x_2 \xtwoheadrightarrow{X_1[\mathcal{K}]} X_1 + x_2 \quad y_1 + y_2 \xtwoheadrightarrow{Y_1[\mathcal{J}]} Y_1 + y_2$$

with $X_1 \subseteq x_1$, $Y_1 \subseteq y_1$, $X_1 \sim Y_1$, and the assumptions $X_1[\mathcal{K}] + x_2 \sim_p^{fr} Y_1[\mathcal{J}] + y_2$ and $X_1 + x_2 \sim_p^{fr} Y_1 + y_2$, hence, we obtain $x_1 + x_2 \sim_p^{fr} y_1 + y_2$, as desired.
Or, we can obtain

$$x_1 + x_2 \xrightarrow{X_1} x_1' + x_2 \quad y_1 + y_2 \xrightarrow{Y_1} y_1' + y_2$$

$$x_1 + x_2 \xtwoheadrightarrow{X_1[\mathcal{K}]} x_1' + x_2 \quad y_1 + y_2 \xtwoheadrightarrow{Y_1[\mathcal{J}]} y_1' + y_2$$

with $X_1 \subseteq x_1$, $Y_1 \subseteq y_1$, $X_1 \sim Y_1$, and $x_1' + x_2 \sim_p^{fr} y_1' + y_2$, hence, we obtain $x_1 + x_2 \sim_p^{fr} y_1 + y_2$, as desired.
Or, we can obtain

$$x_1 + x_2 \xrightarrow{X_2} x_1 + X_2[\mathcal{K}] \quad y_1 + y_2 \xrightarrow{Y_2} y_1 + Y_2[\mathcal{J}]$$

$$x_1 + x_2 \xrightarrow{X_2[\mathcal{K}]} x_1 + X_2 \quad y_1 + y_2 \xrightarrow{Y_2[\mathcal{J}]} y_1 + Y_2$$

with $X_2 \subseteq x_2$, $Y_2 \subseteq y_2$, $X_2 \sim Y_2$, and the assumptions $x_1 + X_2[\mathcal{K}] \sim_p^{fr} y_1 + Y_2[\mathcal{J}]$ and $x_1 + X_2 \sim_p^{fr} y_1 + Y_2$, hence, we obtain $x_1 + x_2 \sim_p^{fr} y_1 + y_2$, as desired.
Or, we can obtain

$$x_1 + x_2 \xrightarrow{X_2} x_1 + x_2' \quad y_1 + y_2 \xrightarrow{Y_2} y_1 + y_2'$$

$$x_1 + x_2 \xtwoheadrightarrow{X_2[\mathcal{K}]} x_1 + x_2' \quad y_1 + y_2 \xtwoheadrightarrow{Y_2[\mathcal{J}]} y_1 + y_2'$$

with $X_2 \subseteq x_2$, $Y_2 \subseteq y_2$, $X_2 \sim Y_2$, and the assumption $x_1 + x_2' \sim_p^{fr} y_1 + y_2'$, hence, we obtain $x_1 + x_2 \sim_p^{fr} y_1 + y_2$, as desired.
Or, we can obtain

$$x_1 + x_2 \xrightarrow{X} x_1' + x_2' \quad y_1 + y_2 \xrightarrow{Y} y_1' + y_2'$$

$$x_1 + x_2 \xtwoheadrightarrow{X[\mathcal{K}]} x_1' + x_2' \quad y_1 + y_2 \xtwoheadrightarrow{Y[\mathcal{J}]} y_1' + y_2'$$

with $X \subseteq x_1$, $Y \subseteq y_1$, $X \subseteq x_2$, $Y \subseteq y_2$, $X \sim Y$, and the assumption $x_1' + x_2' \sim_p^{fr} y_1' + y_2'$, hence, we obtain $x_1 + x_2 \sim_p^{fr} y_1 + y_2$, as desired. □

**Theorem 4.4** (Soundness of $BRATC$ modulo FR pomset bisimulation equivalence). *Let $x$ and $y$ be $BRATC$ terms. If $BRATC \vdash x = y$, then $x \sim_p^{fr} y$.*

*Proof.* Since FR pomset bisimulation $\sim_p^{fr}$ is both an equivalent and a congruent relation, we only need to check if each axiom in Table 4.1 is sound modulo FR pomset bisimulation equivalence.

- **Axiom** $A1$. Let $p, q$ be $BRATC$ processes, and $p + q = q + p$, it is sufficient to prove that $p + q \sim_p^{fr} q + p$. By the forward pomset transition rules for operator $+$ in Table 4.5, we obtain

$$\frac{p \xrightarrow{P} P[\mathcal{K}]}{p + q \xrightarrow{P} P[\mathcal{K}] + q}(P \subseteq p, P \nsubseteq q) \quad \frac{p \xrightarrow{P} P[\mathcal{K}]}{q + p \xrightarrow{P} q + P[\mathcal{K}]}(P \subseteq p, P \nsubseteq q)$$

$$\frac{p \xrightarrow{P} p'}{p + q \xrightarrow{P} p' + q}(P \subseteq p, P \nsubseteq q) \quad \frac{p \xrightarrow{P} p'}{q + p \xrightarrow{P} q + p'}(P \subseteq p, P \nsubseteq q)$$

$$\frac{q \xrightarrow{Q} Q[\mathcal{J}]}{p+q \xrightarrow{Q} p+Q[\mathcal{J}]}(Q \subseteq q, Q \nsubseteq p) \quad \frac{q \xrightarrow{Q} Q[\mathcal{J}]}{q+p \xrightarrow{Q} Q[\mathcal{J}]+p}(Q \subseteq q, Q \nsubseteq p)$$

$$\frac{q \xrightarrow{Q} q'}{p+q \xrightarrow{Q} p+q'}(Q \subseteq q, Q \nsubseteq p) \quad \frac{q \xrightarrow{Q} q'}{q+p \xrightarrow{Q} q'+p}(Q \subseteq q, Q \nsubseteq p)$$

$$\frac{p \xrightarrow{P} p' \quad q \xrightarrow{P} q'}{p+q \xrightarrow{P} p'+q'}(P \subseteq p, P \subseteq q) \quad \frac{p \xrightarrow{P} p' \quad q \xrightarrow{P} q'}{q+p \xrightarrow{P} q'+p'}(P \subseteq p, P \subseteq q)$$

By the reverse pomset transition rules for operator $+$ in Table 4.6, we obtain

$$\frac{p \xrightarrow{P[\mathcal{K}]} P \quad P \nsubseteq q}{p+q \xrightarrow{P[\mathcal{K}]} P+q}(P \subseteq p) \quad \frac{p \xrightarrow{P[\mathcal{K}]} P \quad P \nsubseteq q}{q+p \xrightarrow{P[\mathcal{K}]} q+P}(P \subseteq p)$$

$$\frac{p \xrightarrow{P[\mathcal{K}]} p' \quad P \nsubseteq q}{p+q \xrightarrow{P[\mathcal{K}]} p'+q}(P \subseteq p) \quad \frac{p \xrightarrow{P[\mathcal{K}]} p' \quad P \nsubseteq q}{q+p \xrightarrow{P[\mathcal{K}]} q+p'}(P \subseteq p)$$

$$\frac{q \xrightarrow{Q[\mathcal{J}]} Q \quad Q \nsubseteq p}{p+q \xrightarrow{Q[\mathcal{J}]} p+Q}(Q \subseteq q) \quad \frac{q \xrightarrow{Q[\mathcal{J}]} Q \quad Q \nsubseteq p}{q+p \xrightarrow{Q[\mathcal{J}]} Q+p}(Q \subseteq q)$$

$$\frac{q \xrightarrow{Q[\mathcal{J}]} q' \quad Q \nsubseteq p}{p+q \xrightarrow{Q[\mathcal{J}]} p+q'}(Q \subseteq q) \quad \frac{q \xrightarrow{Q[\mathcal{J}]} q' \quad Q \nsubseteq p}{q+p \xrightarrow{Q[\mathcal{J}]} q'+p}(Q \subseteq q)$$

$$\frac{p \xrightarrow{P[\mathcal{K}]} p' \quad q \xrightarrow{P[\mathcal{K}]} q'}{p+q \xrightarrow{P[\mathcal{K}]} p'+q'}(P \subseteq p, P \subseteq q) \quad \frac{p \xrightarrow{P[\mathcal{K}]} p' \quad q \xrightarrow{P[\mathcal{K}]} q'}{q+p \xrightarrow{P[\mathcal{K}]} q'+p'}(P \subseteq p, P \subseteq q)$$

With the assumptions $P[\mathcal{K}]+q \sim_p^{fr} q+P[\mathcal{K}]$, $P+q \sim_p^{fr} q+P$, $p+Q[\mathcal{J}] \sim_p^{fr} Q[\mathcal{J}]+p$, $p'+q \sim_p^{fr} q+p'$, $p+q' \sim_p^{fr} q'+p$, and $p'+q' \sim_p^{fr} q'+p'$, hence, $p+q \sim_p^{fr} q+p$, as desired.

- **Axiom** $A2$. Let $p,q,s$ be $BRATC$ processes, and $(p+q)+s = p+(q+s)$, it is sufficient to prove that $(p+q)+s \sim_p^{fr} p+(q+s)$. By the forward pomset transition rules for operator $+$ in Table 4.5, we obtain

$$\frac{p \xrightarrow{P} P[\mathcal{K}] \quad P \nsubseteq q \quad P \nsubseteq s}{(p+q)+s \xrightarrow{P} (P[\mathcal{K}]+q)+s}(P \subseteq p) \quad \frac{p \xrightarrow{P} P[\mathcal{K}] \quad P \nsubseteq q \quad P \nsubseteq s}{p+(q+s) \xrightarrow{P} P[\mathcal{K}]+(q+s)}(P \subseteq p)$$

$$\frac{p \xrightarrow{P} p' \quad P \nsubseteq q \quad P \nsubseteq s}{(p+q)+s \xrightarrow{P} (p'+q)+s}(P \subseteq p) \quad \frac{p \xrightarrow{P} p' \quad P \nsubseteq q \quad P \nsubseteq s}{p+(q+s) \xrightarrow{P} p'+(q+s)}(P \subseteq p)$$

$$\frac{q \xrightarrow{Q} Q[\mathcal{J}] \quad Q \nsubseteq p \quad Q \nsubseteq s}{(p+q)+s \xrightarrow{Q} (p+Q[\mathcal{J}])+s}(Q \subseteq q) \quad \frac{q \xrightarrow{Q} Q[\mathcal{J}] \quad Q \nsubseteq p \quad Q \nsubseteq s}{p+(q+s) \xrightarrow{Q} p+(Q[\mathcal{J}]+s)}(Q \subseteq q)$$

$$\frac{q \xrightarrow{Q} q' \quad Q \nsubseteq p \quad Q \nsubseteq s}{(p+q)+s \xrightarrow{Q} (p+q')+s}(Q \subseteq q) \quad \frac{q \xrightarrow{Q} q' \quad Q \nsubseteq p \quad Q \nsubseteq s}{p+(q+s) \xrightarrow{Q} p+(q'+s)}(Q \subseteq q)$$

$$\frac{s \xrightarrow{S} S[\mathcal{I}] \quad S \nsubseteq p \quad S \nsubseteq q}{(p+q)+s \xrightarrow{S} (p+q)+S[\mathcal{I}]}(S \subseteq s) \quad \frac{s \xrightarrow{S} S[\mathcal{I}] \quad S \nsubseteq p \quad S \nsubseteq q}{p+(q+s) \xrightarrow{S} p+(q+S[\mathcal{I}])}(S \subseteq s)$$

$$\frac{s \xrightarrow{S} s' \quad S \nsubseteq p \quad S \nsubseteq q}{(p+q)+s \xrightarrow{S} (p+q)+s'}(S \subseteq s) \quad \frac{s \xrightarrow{S} s' \quad S \nsubseteq p \quad S \nsubseteq q}{p+(q+s) \xrightarrow{S} p+(q+s')}(S \subseteq s)$$

$$\frac{p \xrightarrow{P} p' \quad q \xrightarrow{P} q' \quad s \xrightarrow{P} s'}{(p+q)+s \xrightarrow{P} (p'+q')+s'}(P \subseteq p, P \subseteq q, P \subseteq s) \quad \frac{p \xrightarrow{P} p' \quad q \xrightarrow{P} q' \quad s \xrightarrow{P} s'}{p+(q+s) \xrightarrow{P} p'+(q'+s')}(P \subseteq p, P \subseteq q, P \subseteq s)$$

By the reverse pomset transition rules for operator + in Table 4.6, we obtain

$$\frac{p \xrightarrow{P[\mathcal{K}]} P \quad P \nsubseteq q \quad P \nsubseteq s}{(p+q)+s \xrightarrow{P[\mathcal{K}]} (P+q)+s}(P \subseteq p) \quad \frac{p \xrightarrow{P[\mathcal{K}]} P \quad P \nsubseteq q \quad P \nsubseteq s}{p+(q+s) \xrightarrow{P[\mathcal{K}]} P+(q+s)}(P \subseteq p)$$

$$\frac{p \xrightarrow{P[\mathcal{K}]} p' \quad P \nsubseteq q \quad P \nsubseteq s}{(p+q)+s \xrightarrow{P[\mathcal{K}]} (p'+q)+s}(P \subseteq p) \quad \frac{p \xrightarrow{P[\mathcal{K}]} p' \quad P \nsubseteq q \quad P \nsubseteq s}{p+(q+s) \xrightarrow{P[\mathcal{K}]} p'+(q+s)}(P \subseteq p)$$

$$\frac{q \xrightarrow{Q[\mathcal{J}]} Q \quad Q \nsubseteq p \quad Q \nsubseteq s}{(p+q)+s \xrightarrow{Q[\mathcal{J}]} (p+Q)+s}(Q \subseteq q) \quad \frac{q \xrightarrow{Q[\mathcal{J}]} Q \quad Q \nsubseteq p \quad Q \nsubseteq s}{p+(q+s) \xrightarrow{Q[\mathcal{J}]} p+(Q+s)}(Q \subseteq q)$$

$$\frac{q \xrightarrow{Q[\mathcal{J}]} q' \quad Q \nsubseteq p \quad Q \nsubseteq s}{(p+q)+s \xrightarrow{Q[\mathcal{J}]} (p+q')+s}(Q \subseteq q) \quad \frac{q \xrightarrow{Q[\mathcal{J}]} q' \quad Q \nsubseteq p \quad Q \nsubseteq s}{p+(q+s) \xrightarrow{Q[\mathcal{J}]} p+(q'+s)}(Q \subseteq q)$$

$$\frac{s \xrightarrow{S[\mathcal{I}]} S \quad S \nsubseteq p \quad S \nsubseteq q}{(p+q)+s \xrightarrow{S[\mathcal{I}]} (p+q)+S}(S \subseteq s) \quad \frac{s \xrightarrow{S[\mathcal{I}]} S \quad S \nsubseteq p \quad S \nsubseteq q}{p+(q+s) \xrightarrow{S[\mathcal{I}]} p+(q+S)}(S \subseteq s)$$

$$\frac{s \xrightarrow{S[\mathcal{I}]} s' \quad S \nsubseteq p \quad S \nsubseteq q}{(p+q)+s \xrightarrow{S[\mathcal{I}]} (p+q)+s'}(S \subseteq s) \quad \frac{s \xrightarrow{S[\mathcal{I}]} s' \quad S \nsubseteq p \quad S \nsubseteq q}{p+(q+s) \xrightarrow{S[\mathcal{I}]} p+(q+s')}(S \subseteq s)$$

$$\frac{p \xrightarrow{P[\mathcal{K}]} p' \quad q \xrightarrow{P[\mathcal{K}]} q' \quad s \xrightarrow{P[\mathcal{K}]} s'}{(p+q)+s \xrightarrow{P[\mathcal{K}]} (p'+q')+s'}(P \subseteq p, P \subseteq q, P \subseteq s)$$

$$\frac{p \xrightarrow{P[\mathcal{K}]} p' \quad q \xrightarrow{P[\mathcal{K}]} q' \quad s \xrightarrow{P[\mathcal{K}]} s'}{p+(q+s) \xrightarrow{P[\mathcal{K}]} p'+(q'+s')}(P \subseteq p, P \subseteq q, P \subseteq s)$$

With the assumptions $(P[\mathcal{K}]+q)+s \sim_p^{fr} P[\mathcal{K}]+(q+s)$, $(P+q)+s \sim_p^{fr} P+(q+s)$, $(p+Q[\mathcal{J}])+s \sim_p^{fr} p+(Q[\mathcal{J}]+s)$, $(p+Q)+s \sim_p^{fr} p+(Q+s)$, $(p+q)+S[\mathcal{I}] \sim_p^{fr} p+(q+S[\mathcal{I}])$, $(p+q)+S \sim_p^{fr} p+(q+S)$, $(p'+q)+s \sim_p^{fr} p'+(q+s)$, $(p+q')+s \sim_p^{fr} p+(q'+s)$, $(p+q)+s' \sim_p^{fr} p+(q+s')$, and $(p'+q')+s' \sim_p^{fr} p'+(q'+s')$, hence, $(p+q)+s \sim_p^{fr} p+(q+s)$, as desired.

- **Axiom** $A3$. Let $p$ be a $BRATC$ process, and $p+p=p$, it is sufficient to prove that $p+p \sim_p^{fr} p$. By the forward pomset transition rules for operator + in Table 4.5, we obtain

$$\frac{p \xrightarrow{P} P[\mathcal{K}]}{p+p \xrightarrow{P} P[\mathcal{K}]+P[\mathcal{K}]}(P \subseteq p) \quad \frac{p \xrightarrow{P} P[\mathcal{K}]}{p \xrightarrow{P} P[\mathcal{K}]}(P \subseteq p)$$

$$\frac{p \xrightarrow{P} p'}{p+p \xrightarrow{P} p'+p'}(P \subseteq p) \quad \frac{p \xrightarrow{P} p'}{p \xrightarrow{P} p'}(P \subseteq p)$$

By the reverse pomset transition rules for operator + in Table 4.6, we obtain

$$\frac{p \xrightarrow{P[\mathcal{K}]} P}{p+p \xrightarrow{P[\mathcal{K}]} P+P}(P \subseteq p) \quad \frac{p \xrightarrow{P[\mathcal{K}]} P}{p \xrightarrow{P[\mathcal{K}]} P}(P \subseteq p)$$

$$\frac{p\xrightarrow{P[\mathcal{K}]}p'}{p+p\xrightarrow{P[\mathcal{K}]}p'+p'}(P\subseteq p)\quad\frac{p\xrightarrow{P[\mathcal{K}]}p'}{p\xrightarrow{P[\mathcal{K}]}p'}(P\subseteq p)$$

With the assumptions $P[\mathcal{K}]+P[\mathcal{K}]\sim_p^{fr}P[\mathcal{K}]$, $P+P\sim_p^{fr}P$, and $p'+p'\sim_p^{fr}p'$, hence, $p+p\sim_p^{fr}p$, as desired.

- **Axiom** $A4$. Let $p,q,s$ be $BRATC$ processes, and $p\cdot(q+s)=p\cdot q+p\cdot s(\text{Std}(p),\text{Std}(q),\text{Std}(s))$, it is sufficient to prove that $p\cdot(q+s)\sim_p^{fr}p\cdot q+p\cdot s$. By the pomset transition rules for operators $+$ and $\cdot$ in Table 4.5, we obtain

$$\frac{p\xrightarrow{P}P[\mathcal{K}]}{p\cdot(q+s)\xrightarrow{P}P[\mathcal{K}]\cdot(q+s)}(P\subseteq p)\quad\frac{p\xrightarrow{P}P[\mathcal{K}]}{p\cdot q+p\cdot s\xrightarrow{P}P[\mathcal{K}]\cdot q+P[\mathcal{K}]\cdot s}(P\subseteq p)$$

$$\frac{p\xrightarrow{P}p'}{p\cdot(q+s)\xrightarrow{P}p'\cdot(q+s)}(P\subseteq p)\quad\frac{p\xrightarrow{P}p'}{p\cdot q+p\cdot s\xrightarrow{P}p'\cdot q+p'\cdot s}(P\subseteq p)$$

By the reverse transition rules for operators $+$ and $\cdot$ in Table 4.6, there are no transitions.
With the assumptions $P[\mathcal{K}]\cdot(q+s)\sim_p^{fr}P[\mathcal{K}]\cdot q+P[\mathcal{K}]\cdot s$, $p'\cdot(q+s)\sim_p^{fr}p'\cdot q+p'\cdot s$, hence, $p\cdot(q+s)\sim_p^{fr}p\cdot q+p\cdot s(\text{Std}(p),\text{Std}(q),\text{Std}(s))$, as desired.

- **Axiom** $RA4$. Let $p,q,s$ be $BRATC$ processes, and $(q+s)\cdot p=q\cdot p+s\cdot p(\text{NStd}(p),\text{NStd}(q),\text{NStd}(s))$, it is sufficient to prove that $(q+s)\cdot p\sim_p^{fr}q\cdot p+s\cdot p$. By the pomset transition rules for operators $+$ and $\cdot$ in Table 4.5, there are no transitions.
By the reverse transition rules for operators $+$ and $\cdot$ in Table 4.6, we obtain

$$\frac{p\xrightarrow{P[\mathcal{K}]}P}{(q+s)\cdot p\xrightarrow{P[\mathcal{K}]}(q+s)\cdot P}(P\subseteq p)\quad\frac{p\xrightarrow{P[\mathcal{K}]}P}{q\cdot p+s\cdot p\xrightarrow{P[\mathcal{K}]}q\cdot P+s\cdot P}(P\subseteq p)$$

$$\frac{p\xrightarrow{P[\mathcal{K}]}p'}{(q+s)\cdot p\xrightarrow{P[\mathcal{K}]}(q+s)\cdot p'}(P\subseteq p)\quad\frac{p\xrightarrow{P[\mathcal{K}]}p'}{q\cdot p+s\cdot p\xrightarrow{P[\mathcal{K}]}q\cdot p'+s\cdot p'}(P\subseteq p)$$

With the assumptions $(q+s)\cdot P\sim_p^{fr}q\cdot P+s\cdot P$, $(q+s)\cdot p'\sim_p^{fr}q\cdot p'+s\cdot p'$, hence, $(q+s)\cdot p\sim_p^{fr}q\cdot p+s\cdot p(\text{NStd}(p),\text{NStd}(q),\text{NStd}(s))$, as desired.

- **Axiom** $A5$. Let $p,q,s$ be $BRATC$ processes, and $(p\cdot q)\cdot s=p\cdot(q\cdot s)$, it is sufficient to prove that $(p\cdot q)\cdot s\sim_p^{fr}p\cdot(q\cdot s)$. By the forward pomset transition rules for operator $\cdot$ in Table 4.5, we obtain

$$\frac{p\xrightarrow{P}P[\mathcal{K}]}{(p\cdot q)\cdot s\xrightarrow{P}(P[\mathcal{K}]\cdot q)\cdot s}(P\subseteq p)\quad\frac{p\xrightarrow{P}P[\mathcal{K}]}{p\cdot(q\cdot s)\xrightarrow{P}P[\mathcal{K}]\cdot(q\cdot s)}(P\subseteq p)$$

$$\frac{p\xrightarrow{P}p'}{(p\cdot q)\cdot s\xrightarrow{P}(p'\cdot q)\cdot s}(P\subseteq p)\quad\frac{p\xrightarrow{P}p'}{p\cdot(q\cdot s)\xrightarrow{P}p'\cdot(q\cdot s)}(P\subseteq p)$$

By the reverse pomset transition rules for operator $\cdot$ in Table 4.6, we obtain

$$\frac{s\xrightarrow{S[\mathcal{I}]}S}{(p\cdot q)\cdot s\xrightarrow{S[\mathcal{I}]}(p\cdot q)\cdot S}(S\subseteq s)\quad\frac{s\xrightarrow{S[\mathcal{I}]}S}{p\cdot(q\cdot s)\xrightarrow{S[\mathcal{I}]}p\cdot(q\cdot S)}(S\subseteq s)$$

$$\frac{s\xrightarrow{S[\mathcal{I}]}s'}{(p\cdot q)\cdot s\xrightarrow{S[\mathcal{I}]}(p\cdot q)\cdot s'}(S\subseteq s)\quad\frac{s\xrightarrow{S[\mathcal{I}]}s'}{p\cdot(q\cdot s)\xrightarrow{S[\mathcal{I}]}p\cdot(q\cdot s')}(S\subseteq s)$$

With assumptions $(P[\mathcal{K}]\cdot q)\cdot s\sim_p^{fr}P[\mathcal{K}]\cdot(q\cdot s)$, $(p'\cdot q)\cdot s\sim_p^{fr}p'\cdot(q\cdot s)$, $(p\cdot q)\cdot S\sim_p^{fr}p\cdot(q\cdot S)$, $(p\cdot q)\cdot s'\sim_p^{fr}p\cdot(q\cdot s')$, hence, $(p\cdot q)\cdot s\sim_p^{fr}p\cdot(q\cdot s)$, as desired. □

**Proposition 4.5** (About completeness of $BRATC$ modulo FR truly concurrent bisimulation equivalence). *Let p and q be closed $BRATC$ terms, if $p \sim_p^{fr} q$, then there may be $p \neq q$.*

*Proof.* First, by the elimination theorem of $BRATC$, we know that for each closed $BRATC$ term $p$, there exists a closed basic $BRATC$ term $p'$, such that $BRATC \vdash p = p'$, hence, we only need to consider closed basic $BRATC$ terms.

The basic terms (see Definition 4.1) modulo associativity and commutativity (AC) of conflict $+$ (defined by axioms $A1$ and $A2$ in Table 4.1), and this equivalence is denoted by $=_{AC}$. Then, each equivalence class $s$ modulo AC of $+$ has the following normal form

$$s_1 + \cdots + s_k$$

with each $s_i$ either an atomic event or of the form $t_1 \cdot t_2$, and each $s_i$ is called the summand of $s$.

Now, we try to prove that for normal forms $n$ and $n'$, if $n \sim_p^{fr} n'$, then $n =_{AC} n'$. It is sufficient to induct on the sizes of $n$ and $n'$.

Consider a summand $e_1 \cdot e_2 \cdot e_3$ of $n$. Then, $n \xrightarrow{e_1}\xrightarrow{e_2}\xrightarrow{e_3} e_1[1] \cdot e_2[2] \cdot e_3[3]$, $n'$ should also have $n' \xrightarrow{e_1}\xrightarrow{e_2}\xrightarrow{e_3} n''$, however, maybe $n'' = e_1[1] \cdot e_2[2] \cdot e_3[3]$, or maybe $n'' = e_1[1] \cdot e_3[3] + e_2[2] \cdot e_4$ according to the transition rules of $+$. Note that in the reversible version of $APTC$, the choice $+$ is different from that alternative composition $+$ in $APTC$. Although we define in $+$, if one branch forward or reverse executes successfully, then $+$ forward or reverse executes successfully, the above situation still stands.

That is, we cannot obtain $n =_{AC} n'$. Hence, we cannot give the completeness of $BRATC$ modulo FR pomset bisimulation equivalence. Similarly, we cannot give the completeness of $BRATC$ modulo any FR truly concurrent bisimulation equivalence. Also, in Section 4.2, since $BRATC$ is an embedding of $RAPTC$, we also cannot give the completeness of $RAPTC$ modulo any FR truly concurrent bisimulation equivalence. Furthermore, in Section 11.4, since $RAPTC_\tau$ is a conservative extension of $RAPTC$, we also cannot give the completeness of $RAPTC_\tau$ modulo any weakly FR truly concurrent bisimulation equivalence. □

The FR step transition rules are defined in Table 4.7 and Table 4.8, which are different from the FR pomset transition rules, the FR step transition rules are labeled by steps, in which every event is pairwise concurrent.

**Theorem 4.6** (Congruence of $BRATC$ with respect to FR step bisimulation equivalence). *FR step bisimulation equivalence $\sim_s^{fr}$ is a congruence with respect to $BRATC$.*

*Proof.* It is easy to see that FR step bisimulation is an equivalent relation on $BRATC$ terms, we only need to prove that $\sim_s^{fr}$ is preserved by the operators $\cdot$ and $+$.

- Causality operator $\cdot$. Let $x_1, x_2$ and $y_1, y_2$ be $BRATC$ processes, and $x_1 \sim_s^{fr} y_1$, $x_2 \sim_s^{fr} y_2$, it is sufficient to prove that $x_1 \cdot x_2 \sim_s^{fr} y_1 \cdot y_2$.

  By the definition of FR step bisimulation $\sim_s^{fr}$ (Definition 2.68), $x_1 \sim_s^{fr} y_1$ means that

$$x_1 \xrightarrow{X_1} x_1' \quad y_1 \xrightarrow{Y_1} y_1'$$

$$x_1 \xtwoheadrightarrow{X_1[\mathcal{K}]} x_1'' \quad y_1 \xtwoheadrightarrow{Y_1[\mathcal{J}]} y_1''$$

  with $X_1 \subseteq x_1$, $\forall e_1, e_2 \in X_1$ are pairwise concurrent, $Y_1 \subseteq y_1$, $\forall e_1, e_2 \in Y_1$ are pairwise concurrent, $X_1 \sim Y_1$, $x_1' \sim_s^{fr} y_1'$ and $x_1'' \sim_s^{fr} y_1''$. The meaning of $x_2 \sim_s^{fr} y_2$ is similar.

  By the FR step transition rules for causality operator $\cdot$ in Table 4.5 and Table 4.6, we can obtain

$$x_1 \cdot x_2 \xrightarrow{X_1} X_1[\mathcal{K}] \cdot x_2 \quad y_1 \cdot y_2 \xrightarrow{Y_1} Y_1[\mathcal{J}] \cdot y_2$$

$$x_1 \cdot x_2 \xtwoheadrightarrow{X_2[\mathcal{K}]} x_1 \cdot X_2 \quad y_1 \cdot y_2 \xtwoheadrightarrow{Y_2[\mathcal{J}]} y_1 \cdot Y_2$$

  where $X_1 \subseteq x_1$, $\forall e_1, e_2 \in X_1$ are pairwise concurrent, $Y_1 \subseteq y_1$, $\forall e_1, e_2 \in Y_1$ are pairwise concurrent, $X_2 \subseteq x_2$, $\forall e_1, e_2 \in X_2$ are pairwise concurrent, $Y_2 \subseteq y_2$, $\forall e_1, e_2 \in Y_2$ are pairwise concurrent, $X_1 \sim Y_1$, $X_2 \sim Y_2$, and the assumptions $X_1[\mathcal{K}] \cdot x_2 \sim_s^{fr} Y_1[\mathcal{J}] \cdot y_2$ and $x_1 \cdot X_2 \sim_s^{fr} y_1 \cdot Y_2$, hence, we obtain $x_1 \cdot x_2 \sim_s^{fr} y_1 \cdot y_2$, as desired.

  Or, we can obtain

$$x_1 \cdot x_2 \xrightarrow{X_1} x_1' \cdot x_2 \quad y_1 \cdot y_2 \xrightarrow{Y_1} y_1' \cdot y_2$$

**TABLE 4.7 Forward step transition rules of $BRATC$.**

$$\frac{}{X\xrightarrow{X}X[\mathcal{K}]}(\forall e_1,e_2\in X \text{ are pairwise concurrent})$$

$$\frac{x\xrightarrow{X}X[\mathcal{K}]\quad X\not\subseteq y}{x+y\xrightarrow{X}X[\mathcal{K}]+y}(\forall e_1,e_2\in X \text{ are pairwise concurrent})$$

$$\frac{x\xrightarrow{X}x'\quad X\not\subseteq y}{x+y\xrightarrow{X}x'+y}(\forall e_1,e_2\in X \text{ are pairwise concurrent})$$

$$\frac{y\xrightarrow{Y}Y[\mathcal{J}]\quad Y\not\subseteq x}{x+y\xrightarrow{Y}x+Y[\mathcal{J}]}(\forall e_1,e_2\in Y \text{ are pairwise concurrent})$$

$$\frac{y\xrightarrow{Y}y'\quad Y\not\subseteq x}{x+y\xrightarrow{Y}x+y'}(\forall e_1,e_2\in Y \text{ are pairwise concurrent})$$

$$\frac{x\xrightarrow{X}X[\mathcal{K}]\quad y\xrightarrow{X}X[\mathcal{K}]}{x+y\xrightarrow{X}X[\mathcal{K}]+X[\mathcal{K}]}(\forall e_1,e_2\in X \text{ are pairwise concurrent})$$

$$\frac{x\xrightarrow{X}x'\quad y\xrightarrow{X}X[\mathcal{K}]}{x+y\xrightarrow{X}x'+X[\mathcal{K}]}(\forall e_1,e_2\in X \text{ are pairwise concurrent})$$

$$\frac{x\xrightarrow{X}X[\mathcal{K}]\quad y\xrightarrow{X}y'}{x+y\xrightarrow{X}X[\mathcal{K}]+y'}(\forall e_1,e_2\in X \text{ are pairwise concurrent})$$

$$\frac{x\xrightarrow{X}x'\quad y\xrightarrow{X}y'}{x+y\xrightarrow{X}x'+y'}(\forall e_1,e_2\in X \text{ are pairwise concurrent})$$

$$\frac{x\xrightarrow{X}X[\mathcal{K}]\quad \text{Std}(y)}{x\cdot y\xrightarrow{X}X[\mathcal{K}]\cdot y}(X\subseteq x,\forall e_1,e_2\in X \text{ are pairwise concurrent})$$

$$\frac{x\xrightarrow{X}x'\quad \text{Std}(y)}{x\cdot y\xrightarrow{X}x'\cdot y}(X\subseteq x,\forall e_1,e_2\in X \text{ are pairwise concurrent})$$

$$\frac{y\xrightarrow{Y}Y[\mathcal{J}]\quad \text{NStd}(x)}{x\cdot y\xrightarrow{Y}x\cdot Y[\mathcal{J}]}(Y\subseteq y,\forall e_1,e_2\in Y \text{ are pairwise concurrent})$$

$$\frac{y\xrightarrow{Y}y'\quad \text{NStd}(x)}{x\cdot y\xrightarrow{Y}x\cdot y'}(Y\subseteq y,\forall e_1,e_2\in Y \text{ are pairwise concurrent})$$

$$x_1\cdot x_2\overset{X_2[\mathcal{K}]}{\twoheadrightarrow}x_1\cdot x_2' \quad y_1\cdot y_2\overset{Y_2[\mathcal{J}]}{\twoheadrightarrow}y_1\cdot y_2'$$

where $X_1\subseteq x_1$, $\forall e_1,e_2\in X_1$ are pairwise concurrent, $Y_1\subseteq y_1$, $\forall e_1,e_2\in Y_1$ are pairwise concurrent, $X_2\subseteq x_2$, $\forall e_1,e_2\in X_2$ are pairwise concurrent, $Y_2\subseteq y_2$, $\forall e_1,e_2\in Y_2$ are pairwise concurrent, $X_1\sim Y_1$, $X_2\sim Y_2$, and the assumptions $x_1'\cdot x_2\sim_s^{fr}y_1'\cdot y_2$, $x_1\cdot x_2'\sim_s^{fr}y_1\cdot y_2'$, hence, we obtain $x_1\cdot x_2\sim_s^{fr}y_1\cdot y_2$, as desired.

- Conflict operator $+$. Let $x_1,x_2$ and $y_1,y_2$ be $BRATC$ processes, and $x_1\sim_s^{fr}y_1$, $x_2\sim_s^{fr}y_2$, it is sufficient to prove that $x_1+x_2\sim_s^{fr}y_1+y_2$. The meanings of $x_1\sim_s^{fr}y_1$ and $x_2\sim_s^{fr}y_2$ are the same as the above case, according to the definition of FR step bisimulation $\sim_s^{fr}$ in Definition 2.68.
  By the FR step transition rules for conflict operator $+$ in Table 4.5 and Table 4.6, we can obtain several cases:

$$x_1+x_2\xrightarrow{X_1}X_1[\mathcal{K}]+x_2 \quad y_1+y_2\xrightarrow{Y_1}Y_1[\mathcal{J}]+y_2$$

$$x_1+x_2\overset{X_1[\mathcal{K}]}{\twoheadrightarrow}X_1+x_2 \quad y_1+y_2\overset{Y_1[\mathcal{J}]}{\twoheadrightarrow}Y_1+y_2$$

where $X_1\subseteq x_1$, $\forall e_1,e_2\in X_1$ are pairwise concurrent, $Y_1\subseteq y_1$, $\forall e_1,e_2\in Y_1$ are pairwise concurrent, $X_1\sim Y_1$, and the assumptions $X_1[\mathcal{K}]+x_2\sim_s^{fr}Y_1[\mathcal{J}]+y_2$ and $X_1+x_2\sim_s^{fr}Y_1+y_2$, hence, we obtain $x_1+x_2\sim_s^{fr}y_1+y_2$, as desired.

**TABLE 4.8 Reverse step transition rules of $BRATC$.**

| |
|---|
| $\frac{}{X[\mathcal{K}]\xrightarrow{X[\mathcal{K}]}X}(\forall e_1,e_2\in X \text{ are pairwise concurrent})$ |
| $\frac{x\xrightarrow{X[\mathcal{K}]}X \quad X\not\subseteq y}{x+y\xrightarrow{X[\mathcal{K}]}X+y}(\forall e_1,e_2\in X \text{ are pairwise concurrent})$ |
| $\frac{x\xrightarrow{X[\mathcal{K}]}x' \quad X\not\subseteq y}{x+y\xrightarrow{X[\mathcal{K}]}x'+y}(\forall e_1,e_2\in X \text{ are pairwise concurrent})$ |
| $\frac{y\xrightarrow{Y[\mathcal{J}]}Y \quad Y\not\subseteq x}{x+y\xrightarrow{Y[\mathcal{J}]}x+Y}(\forall e_1,e_2\in Y \text{ are pairwise concurrent})$ |
| $\frac{y\xrightarrow{Y[\mathcal{J}]}y' \quad Y\not\subseteq x}{x+y\xrightarrow{Y[\mathcal{J}]}x+y'}(\forall e_1,e_2\in Y \text{ are pairwise concurrent})$ |
| $\frac{x\xrightarrow{X[\mathcal{K}]}X \quad y\xrightarrow{X[\mathcal{K}]}X}{x+y\xrightarrow{X[\mathcal{K}]}X+X}(\forall e_1,e_2\in X \text{ are pairwise concurrent})$ |
| $\frac{x\xrightarrow{X[\mathcal{K}]}x' \quad y\xrightarrow{X[\mathcal{K}]}X}{x+y\xrightarrow{X[\mathcal{K}]}x'+X}(\forall e_1,e_2\in X \text{ are pairwise concurrent})$ |
| $\frac{x\xrightarrow{X[\mathcal{K}]}X \quad y\xrightarrow{X[\mathcal{K}]}y'}{x+y\xrightarrow{X[\mathcal{K}]}X+y'}(\forall e_1,e_2\in X \text{ are pairwise concurrent})$ |
| $\frac{x\xrightarrow{X[\mathcal{K}]}x' \quad y\xrightarrow{X[\mathcal{K}]}y'}{x+y\xrightarrow{X[\mathcal{K}]}x'+y'}(\forall e_1,e_2\in X \text{ are pairwise concurrent})$ |
| $\frac{x\xrightarrow{X[\mathcal{K}]}X \quad \text{Std}(y)}{x\cdot y\xrightarrow{X[\mathcal{K}]}X\cdot y}(X\subseteq x,\forall e_1,e_2\in X \text{ are pairwise concurrent})$ |
| $\frac{x\xrightarrow{X[\mathcal{K}]}x' \quad \text{Std}(y)}{x\cdot y\xrightarrow{X[\mathcal{K}]}x'\cdot y}(X\subseteq x,\forall e_1,e_2\in X \text{ are pairwise concurrent})$ |
| $\frac{y\xrightarrow{Y[\mathcal{J}]}Y \quad \text{NStd}(x)}{x\cdot y\xrightarrow{Y[\mathcal{J}]}x\cdot Y}(Y\subseteq y,\forall e_1,e_2\in Y \text{ are pairwise concurrent})$ |
| $\frac{y\xrightarrow{Y[\mathcal{J}]}y' \quad \text{NStd}(x)}{x\cdot y\xrightarrow{Y[\mathcal{J}]}x\cdot y'}(Y\subseteq y,\forall e_1,e_2\in Y \text{ are pairwise concurrent})$ |

Or, we can obtain

$$x_1+x_2\xrightarrow{X_1}x_1'+x_2 \quad y_1+y_2\xrightarrow{Y_1}y_1'+y_2$$

$$x_1+x_2\xrightarrow{X_1[\mathcal{K}]}x_1'+x_2 \quad y_1+y_2\xrightarrow{Y_1[\mathcal{J}]}y_1'+y_2$$

where $X_1\subseteq x_1$, $\forall e_1,e_2\in X_1$ are pairwise concurrent, $Y_1\subseteq y_1$, $\forall e_1,e_2\in Y_1$ are pairwise concurrent, $X_1\sim Y_1$, and $x_1'+x_2\sim_s^{fr}y_1'+y_2$, hence, we obtain $x_1+x_2\sim_s^{fr}y_1+y_2$, as desired.

Or, we can obtain

$$x_1+x_2\xrightarrow{X_2}x_1+X_2[\mathcal{K}] \quad y_1+y_2\xrightarrow{Y_2}y_1+Y_2[\mathcal{J}]$$

$$x_1+x_2\xrightarrow{X_2[\mathcal{K}]}x_1+X_2 \quad y_1+y_2\xrightarrow{Y_2[\mathcal{J}]}y_1+Y_2$$

where $X_2 \subseteq x_2$, $\forall e_1, e_2 \in X_2$ are pairwise concurrent, $Y_2 \subseteq y_2$, $\forall e_1, e_2 \in Y_2$ are pairwise concurrent, $X_2 \sim Y_2$, and the assumptions $x_1 + X_2[\mathcal{K}] \sim_s^{fr} y_1 + Y_2[\mathcal{J}]$ and $x_1 + X_2 \sim_s^{fr} y_1 + Y_2$, hence, we obtain $x_1 + x_2 \sim_s^{fr} y_1 + y_2$, as desired. Or, we can obtain

$$x_1 + x_2 \xrightarrow{X_2} x_1 + x_2' \quad y_1 + y_2 \xrightarrow{Y_2} y_1 + y_2'$$

$$x_1 + x_2 \xtwoheadrightarrow{X_2[\mathcal{K}]} x_1 + x_2' \quad y_1 + y_2 \xtwoheadrightarrow{Y_2[\mathcal{J}]} y_1 + y_2'$$

where $X_2 \subseteq x_2$, $\forall e_1, e_2 \in X_2$ are pairwise concurrent, $Y_2 \subseteq y_2$, $X_2 \sim Y_2$, $\forall e_1, e_2 \in Y_2$ are pairwise concurrent, and the assumption $x_1 + x_2' \sim_s^{fr} y_1 + y_2'$, hence, we obtain $x_1 + x_2 \sim_s^{fr} y_1 + y_2$, as desired. Or, we can obtain

$$x_1 + x_2 \xrightarrow{X} x_1' + x_2' \quad y_1 + y_2 \xrightarrow{Y} y_1' + y_2'$$

$$x_1 + x_2 \xtwoheadrightarrow{X[\mathcal{K}]} x_1' + x_2' \quad y_1 + y_2 \xtwoheadrightarrow{Y[\mathcal{J}]} y_1' + y_2'$$

where $X \subseteq x_1$, $Y \subseteq y_1$, $X \subseteq x_2$, $Y \subseteq y_2$, $\forall e_1, e_2 \in X$ are pairwise concurrent, $\forall e_1, e_2 \in Y$ are pairwise concurrent, $X \sim Y$, and the assumption $x_1' + x_2' \sim_s^{fr} y_1' + y_2'$, hence, we obtain $x_1 + x_2 \sim_s^{fr} y_1 + y_2$, as desired. □

**Theorem 4.7** (Soundness of $BRATC$ modulo FR step bisimulation equivalence). *Let $x$ and $y$ be $BRATC$ terms. If $BRATC \vdash x = y$, then $x \sim_s^{fr} y$.*

*Proof.* Since FR step bisimulation $\sim_s^{fr}$ is both an equivalent and a congruent relation, we only need to check if each axiom in Table 4.1 is sound modulo FR step bisimulation equivalence.

- **Axiom** $A1$. Let $p, q$ be $BRATC$ processes, and $p + q = q + p$, it is sufficient to prove that $p + q \sim_s^{fr} q + p$. By the forward step transition rules for operator $+$ in Table 4.7, we obtain

$$\frac{p \xrightarrow{P} P[\mathcal{K}]}{p+q \xrightarrow{P} P[\mathcal{K}]+q} (P \subseteq p, P \nsubseteq q, \forall e_1, e_2 \in P \text{ are pairwise concurrent})$$

$$\frac{p \xrightarrow{P} P[\mathcal{K}]}{q+p \xrightarrow{P} q+P[\mathcal{K}]} (P \subseteq p, P \nsubseteq q, \forall e_1, e_2 \in P \text{ are pairwise concurrent})$$

$$\frac{p \xrightarrow{P} p'}{p+q \xrightarrow{P} p'+q} (P \subseteq p, P \nsubseteq q, \forall e_1, e_2 \in P \text{ are pairwise concurrent})$$

$$\frac{p \xrightarrow{P} p'}{q+p \xrightarrow{P} q+p'} (P \subseteq p, P \nsubseteq q, \forall e_1, e_2 \in P \text{ are pairwise concurrent})$$

$$\frac{q \xrightarrow{Q} Q[\mathcal{J}]}{p+q \xrightarrow{Q} p+Q[\mathcal{J}]} (Q \subseteq q, Q \nsubseteq p, \forall e_1, e_2 \in Q \text{ are pairwise concurrent})$$

$$\frac{q \xrightarrow{Q} Q[\mathcal{J}]}{q+p \xrightarrow{Q} Q[\mathcal{J}]+p} (Q \subseteq q, Q \nsubseteq p, \forall e_1, e_2 \in Q \text{ are pairwise concurrent})$$

$$\frac{q \xrightarrow{Q} q'}{p+q \xrightarrow{Q} p+q'} (Q \subseteq q, Q \nsubseteq p, \forall e_1, e_2 \in Q \text{ are pairwise concurrent})$$

$$\frac{q \xrightarrow{Q} q'}{q+p \xrightarrow{Q} q'+p} (Q \subseteq q, Q \nsubseteq p, \forall e_1, e_2 \in Q \text{ are pairwise concurrent})$$

$$\frac{p \xrightarrow{P} p' \quad q \xrightarrow{P} q'}{p+q \xrightarrow{P} p'+q'} (P \subseteq p, P \subseteq q, \forall e_1, e_2 \in P \text{ are pairwise concurrent})$$

$$\frac{p \xrightarrow{P} p' \quad q \xrightarrow{P} q'}{q+p \xrightarrow{P} q'+p'}(P \subseteq p, P \subseteq q, \forall e_1, e_2 \in P \text{ are pairwise concurrent})$$

By the reverse step transition rules for operator $+$ in Table 4.8, we obtain

$$\frac{p \xrightarrow{P[\mathcal{K}]} P \quad P \nsubseteq q}{p+q \xrightarrow{P[\mathcal{K}]} P+q}(P \subseteq p, \forall e_1, e_2 \in P \text{ are pairwise concurrent})$$

$$\frac{p \xrightarrow{P[\mathcal{K}]} P \quad P \nsubseteq q}{q+p \xrightarrow{P[\mathcal{K}]} q+P}(P \subseteq p, \forall e_1, e_2 \in P \text{ are pairwise concurrent})$$

$$\frac{p \xrightarrow{P[\mathcal{K}]} p' \quad P \nsubseteq q}{p+q \xrightarrow{P[\mathcal{K}]} p'+q}(P \subseteq p, \forall e_1, e_2 \in P \text{ are pairwise concurrent})$$

$$\frac{p \xrightarrow{P[\mathcal{K}]} p' \quad P \nsubseteq q}{q+p \xrightarrow{P[\mathcal{K}]} q+p'}(P \subseteq p, \forall e_1, e_2 \in P \text{ are pairwise concurrent})$$

$$\frac{q \xrightarrow{Q[\mathcal{J}]} Q \quad Q \nsubseteq p}{p+q \xrightarrow{Q[\mathcal{J}]} p+Q}(Q \subseteq q, \forall e_1, e_2 \in Q \text{ are pairwise concurrent})$$

$$\frac{q \xrightarrow{Q[\mathcal{J}]} Q \quad Q \nsubseteq p}{q+p \xrightarrow{Q[\mathcal{J}]} Q+p}(Q \subseteq q, \forall e_1, e_2 \in Q \text{ are pairwise concurrent})$$

$$\frac{q \xrightarrow{Q[\mathcal{J}]} q' \quad Q \nsubseteq p}{p+q \xrightarrow{Q[\mathcal{J}]} p+q'}(Q \subseteq q, \forall e_1, e_2 \in Q \text{ are pairwise concurrent})$$

$$\frac{q \xrightarrow{Q[\mathcal{J}]} q' \quad Q \nsubseteq p}{q+p \xrightarrow{Q[\mathcal{J}]} q'+p}(Q \subseteq q, \forall e_1, e_2 \in Q \text{ are pairwise concurrent})$$

$$\frac{p \xrightarrow{P[\mathcal{K}]} p' \quad q \xrightarrow{P[\mathcal{K}]} q'}{p+q \xrightarrow{P[\mathcal{K}]} p'+q'}(P \subseteq p, P \subseteq q, \forall e_1, e_2 \in P \text{ are pairwise concurrent})$$

$$\frac{p \xrightarrow{P[\mathcal{K}]} p' \quad q \xrightarrow{P[\mathcal{K}]} q'}{q+p \xrightarrow{P[\mathcal{K}]} q'+p'}(P \subseteq p, P \subseteq q, \forall e_1, e_2 \in P \text{ are pairwise concurrent})$$

With the assumptions $P[\mathcal{K}]+q \sim_s^{fr} q+P[\mathcal{K}]$, $P+q \sim_s^{fr} q+P$, $p+Q[\mathcal{J}] \sim_s^{fr} Q[\mathcal{J}]+p$, $p'+q \sim_s^{fr} q+p'$, $p+q' \sim_s^{fr} q'+p$, and $p'+q' \sim_s^{fr} q'+p'$, hence, $p+q \sim_s^{fr} q+p$, as desired.

- **Axiom** $A2$. Let $p,q,s$ be $BRATC$ processes, and $(p+q)+s = p+(q+s)$, it is sufficient to prove that $(p+q)+s \sim_s^{fr} p+(q+s)$. By the forward step transition rules for operator $+$ in Table 4.7, we obtain

$$\frac{p \xrightarrow{P} P[\mathcal{K}] \quad P \nsubseteq q \quad P \nsubseteq s}{(p+q)+s \xrightarrow{P} (P[\mathcal{K}]+q)+s}(P \subseteq p, \forall e_1, e_2 \in P \text{ are pairwise concurrent})$$

$$\frac{p \xrightarrow{P} P[\mathcal{K}] \quad P \nsubseteq q \quad P \nsubseteq s}{p+(q+s) \xrightarrow{P} P[\mathcal{K}]+(q+s)}(P \subseteq p, \forall e_1, e_2 \in P \text{ are pairwise concurrent})$$

$$\frac{p \xrightarrow{P} p' \quad P \nsubseteq q \quad P \nsubseteq s}{(p+q)+s \xrightarrow{P} (p'+q)+s} (P \subseteq p, \forall e_1, e_2 \in P \text{ are pairwise concurrent})$$

$$\frac{p \xrightarrow{P} p' \quad P \nsubseteq q \quad P \nsubseteq s}{p+(q+s) \xrightarrow{P} p'+(q+s)} (P \subseteq p, \forall e_1, e_2 \in P \text{ are pairwise concurrent})$$

$$\frac{q \xrightarrow{Q} Q[\mathcal{J}] \quad Q \nsubseteq p \quad Q \nsubseteq s}{(p+q)+s \xrightarrow{Q} (p+Q[\mathcal{J}])+s} (Q \subseteq q, \forall e_1, e_2 \in Q \text{ are pairwise concurrent})$$

$$\frac{q \xrightarrow{Q} Q[\mathcal{J}] \quad Q \nsubseteq p \quad Q \nsubseteq s}{p+(q+s) \xrightarrow{Q} p+(Q[\mathcal{J}]+s)} (Q \subseteq q, \forall e_1, e_2 \in Q \text{ are pairwise concurrent})$$

$$\frac{q \xrightarrow{Q} q' \quad Q \nsubseteq p \quad Q \nsubseteq s}{(p+q)+s \xrightarrow{Q} (p+q')+s} (Q \subseteq q, \forall e_1, e_2 \in Q \text{ are pairwise concurrent})$$

$$\frac{q \xrightarrow{Q} q' \quad Q \nsubseteq p \quad Q \nsubseteq s}{p+(q+s) \xrightarrow{Q} p+(q'+s)} (Q \subseteq q, \forall e_1, e_2 \in Q \text{ are pairwise concurrent})$$

$$\frac{s \xrightarrow{S} S[\mathcal{I}] \quad S \nsubseteq p \quad S \nsubseteq q}{(p+q)+s \xrightarrow{S} (p+q)+S[\mathcal{I}]} (S \subseteq s, \forall e_1, e_2 \in S \text{ are pairwise concurrent})$$

$$\frac{s \xrightarrow{S} S[\mathcal{I}] \quad S \nsubseteq p \quad S \nsubseteq q}{p+(q+s) \xrightarrow{S} p+(q+S[\mathcal{I}])} (S \subseteq s, \forall e_1, e_2 \in S \text{ are pairwise concurrent})$$

$$\frac{s \xrightarrow{S} s' \quad S \nsubseteq p \quad S \nsubseteq q}{(p+q)+s \xrightarrow{S} (p+q)+s'} (S \subseteq s, \forall e_1, e_2 \in S \text{ are pairwise concurrent})$$

$$\frac{s \xrightarrow{S} s' \quad S \nsubseteq p \quad S \nsubseteq q}{p+(q+s) \xrightarrow{S} p+(q+s')} (S \subseteq s, \forall e_1, e_2 \in S \text{ are pairwise concurrent})$$

$$\frac{p \xrightarrow{P} p' \quad q \xrightarrow{P} q' \quad s \xrightarrow{P} s'}{(p+q)+s \xrightarrow{P} (p'+q')+s'} (P \subseteq p, P \subseteq q, P \subseteq s, \forall e_1, e_2 \in P \text{ are pairwise concurrent})$$

$$\frac{p \xrightarrow{P} p' \quad q \xrightarrow{P} q' \quad s \xrightarrow{P} s'}{p+(q+s) \xrightarrow{P} p'+(q'+s')} (P \subseteq p, P \subseteq q, P \subseteq s, \forall e_1, e_2 \in P \text{ are pairwise concurrent})$$

By the reverse step transition rules for operator $+$ in Table 4.8, we obtain

$$\frac{p \xtwoheadrightarrow{P[\mathcal{K}]} P \quad P \nsubseteq q \quad P \nsubseteq s}{(p+q)+s \xtwoheadrightarrow{P[\mathcal{K}]} (P+q)+s} (P \subseteq p, \forall e_1, e_2 \in P \text{ are pairwise concurrent})$$

$$\frac{p \xtwoheadrightarrow{P[\mathcal{K}]} P \quad P \nsubseteq q \quad P \nsubseteq s}{p+(q+s) \xtwoheadrightarrow{P[\mathcal{K}]} P+(q+s)} (P \subseteq p, \forall e_1, e_2 \in P \text{ are pairwise concurrent})$$

$$\frac{p \xtwoheadrightarrow{P[\mathcal{K}]} p' \quad P \nsubseteq q \quad P \nsubseteq s}{(p+q)+s \xtwoheadrightarrow{P[\mathcal{K}]} (p'+q)+s} (P \subseteq p, \forall e_1, e_2 \in P \text{ are pairwise concurrent})$$

$$\frac{p \xrightarrow{P[\mathcal{K}]} p' \quad P \not\subseteq q \quad P \not\subseteq s}{p+(q+s) \xrightarrow{P[\mathcal{K}]} p'+(q+s)} (P \subseteq p, \forall e_1, e_2 \in P \text{ are pairwise concurrent})$$

$$\frac{q \xrightarrow{Q[\mathcal{J}]} Q \quad Q \not\subseteq p \quad Q \not\subseteq s}{(p+q)+s \xrightarrow{Q[\mathcal{J}]} (p+Q)+s} (Q \subseteq q, \forall e_1, e_2 \in Q \text{ are pairwise concurrent})$$

$$\frac{q \xrightarrow{Q[\mathcal{J}]} Q \quad Q \not\subseteq p \quad Q \not\subseteq s}{p+(q+s) \xrightarrow{Q[\mathcal{J}]} p+(Q+s)} (Q \subseteq q, \forall e_1, e_2 \in Q \text{ are pairwise concurrent})$$

$$\frac{q \xrightarrow{Q[\mathcal{J}]} q' \quad Q \not\subseteq p \quad Q \not\subseteq s}{(p+q)+s \xrightarrow{Q[\mathcal{J}]} (p+q')+s} (Q \subseteq q, \forall e_1, e_2 \in Q \text{ are pairwise concurrent})$$

$$\frac{q \xrightarrow{Q[\mathcal{J}]} q' \quad Q \not\subseteq p \quad Q \not\subseteq s}{p+(q+s) \xrightarrow{Q[\mathcal{J}]} p+(q'+s)} (Q \subseteq q, \forall e_1, e_2 \in Q \text{ are pairwise concurrent})$$

$$\frac{s \xrightarrow{S[\mathcal{I}]} S \quad S \not\subseteq p \quad S \not\subseteq q}{(p+q)+s \xrightarrow{S[\mathcal{I}]} (p+q)+S} (S \subseteq s, \forall e_1, e_2 \in S \text{ are pairwise concurrent})$$

$$\frac{s \xrightarrow{S[\mathcal{I}]} S \quad S \not\subseteq p \quad S \not\subseteq q}{p+(q+s) \xrightarrow{S[\mathcal{I}]} p+(q+S)} (S \subseteq s, \forall e_1, e_2 \in S \text{ are pairwise concurrent})$$

$$\frac{s \xrightarrow{S[\mathcal{I}]} s' \quad S \not\subseteq p \quad S \not\subseteq q}{(p+q)+s \xrightarrow{S[\mathcal{I}]} (p+q)+s'} (S \subseteq s, \forall e_1, e_2 \in S \text{ are pairwise concurrent})$$

$$\frac{s \xrightarrow{S[\mathcal{I}]} s' \quad S \not\subseteq p \quad S \not\subseteq q}{p+(q+s) \xrightarrow{S[\mathcal{I}]} p+(q+s')} (S \subseteq s, \forall e_1, e_2 \in S \text{ are pairwise concurrent})$$

$$\frac{p \xrightarrow{P[\mathcal{K}]} p' \quad q \xrightarrow{P[\mathcal{K}]} q' \quad s \xrightarrow{P[\mathcal{K}]} s'}{(p+q)+s \xrightarrow{P[\mathcal{K}]} (p'+q')+s'} (P \subseteq p, P \subseteq q, P \subseteq s, \forall e_1, e_2 \in P \text{ are pairwise concurrent})$$

$$\frac{p \xrightarrow{P[\mathcal{K}]} p' \quad q \xrightarrow{P[\mathcal{K}]} q' \quad s \xrightarrow{P[\mathcal{K}]} s'}{p+(q+s) \xrightarrow{P[\mathcal{K}]} p'+(q'+s')} (P \subseteq p, P \subseteq q, P \subseteq s, \forall e_1, e_2 \in P \text{ are pairwise concurrent})$$

With the assumptions $(P[\mathcal{K}]+q)+s \sim_s^{fr} P[\mathcal{K}]+(q+s)$, $(P+q)+s \sim_s^{fr} P+(q+s)$, $(p+Q[\mathcal{J}])+s \sim_s^{fr} p+(Q[\mathcal{J}]+s)$, $(p+Q)+s \sim_s^{fr} p+(Q+s)$, $(p+q)+S[\mathcal{I}] \sim_s^{fr} p+(q+S[\mathcal{I}])$, $(p+q)+S \sim_s^{fr} p+(q+S)$, $(p'+q)+s \sim_s^{fr} p'+(q+s)$, $(p+q')+s \sim_s^{fr} p+(q'+s)$, $(p+q)+s' \sim_s^{fr} p+(q+s')$ and $(p'+q')+s' \sim_s^{fr} p'+(q'+s')$, hence, $(p+q)+s \sim_s^{fr} p+(q+s)$, as desired.

- **Axiom** $A3$. Let $p$ be a $BRATC$ process, and $p+p=p$, it is sufficient to prove that $p+p \sim_s^{fr} p$. By the forward step transition rules for operator $+$ in Table 4.7, we obtain

$$\frac{p \xrightarrow{P} P[\mathcal{K}]}{p+p \xrightarrow{P} P[\mathcal{K}]+P[\mathcal{K}]} (P \subseteq p, \forall e_1, e_2 \in P \text{ are pairwise concurrent})$$

$$\frac{p \xrightarrow{P} P[\mathcal{K}]}{p \xrightarrow{P} P[\mathcal{K}]} (P \subseteq p, \forall e_1, e_2 \in P \text{ are pairwise concurrent})$$

$$\frac{p \xrightarrow{P} p'}{p+p \xrightarrow{P} p'+p'}(P \subseteq p, \forall e_1, e_2 \in P \text{ are pairwise concurrent})$$

$$\frac{p \xrightarrow{P} p'}{p \xrightarrow{P} p'}(P \subseteq p, \forall e_1, e_2 \in P \text{ are pairwise concurrent})$$

By the reverse step transition rules for operator $+$ in Table 4.8, we obtain

$$\frac{p \xtwoheadrightarrow{P[\mathcal{K}]} P}{p+p \xtwoheadrightarrow{P[\mathcal{K}]} P+P}(P \subseteq p, \forall e_1, e_2 \in P \text{ are pairwise concurrent})$$

$$\frac{p \xtwoheadrightarrow{P[\mathcal{K}]} P}{p \xtwoheadrightarrow{P[\mathcal{K}]} P}(P \subseteq p, \forall e_1, e_2 \in P \text{ are pairwise concurrent})$$

$$\frac{p \xtwoheadrightarrow{P[\mathcal{K}]} p'}{p+p \xtwoheadrightarrow{P[\mathcal{K}]} p'+p'}(P \subseteq p, \forall e_1, e_2 \in P \text{ are pairwise concurrent})$$

$$\frac{p \xtwoheadrightarrow{P[\mathcal{K}]} p'}{p \xtwoheadrightarrow{P[\mathcal{K}]} p'}(P \subseteq p, \forall e_1, e_2 \in P \text{ are pairwise concurrent})$$

With the assumptions $P[\mathcal{K}]+P[\mathcal{K}] \sim_s^{fr} P[\mathcal{K}]$, $P+P \sim_s^{fr} P$, and $p'+p' \sim_s^{fr} p'$, hence, $p+p \sim_s^{fr} p$, as desired.

- **Axiom** $A4$. Let $p, q, s$ be $BRATC$ processes, and $p \cdot (q+s) = p \cdot q + p \cdot s(\text{Std}(p), \text{Std}(q), \text{Std}(s))$, it is sufficient to prove that $p \cdot (q+s) \sim_p^{fr} p \cdot q + p \cdot s$. By the pomset transition rules for operators $+$ and $\cdot$ in Table 4.5, we obtain

$$\frac{p \xrightarrow{P} P[\mathcal{K}]}{p \cdot (q+s) \xrightarrow{P} P[\mathcal{K}] \cdot (q+s)}(P \subseteq p, \forall e_1, e_2 \in P \text{ are pairwise concurrent})$$

$$\frac{p \xrightarrow{P} P[\mathcal{K}]}{p \cdot q + p \cdot s \xrightarrow{P} P[\mathcal{K}] \cdot q + P[\mathcal{K}] \cdot s}(P \subseteq p, \forall e_1, e_2 \in P \text{ are pairwise concurrent})$$

$$\frac{p \xrightarrow{P} p'}{p \cdot (q+s) \xrightarrow{P} p' \cdot (q+s)}(P \subseteq p, \forall e_1, e_2 \in P \text{ are pairwise concurrent})$$

$$\frac{p \xrightarrow{P} p'}{p \cdot q + p \cdot s \xrightarrow{P} p' \cdot q + p' \cdot s}(P \subseteq p, \forall e_1, e_2 \in P \text{ are pairwise concurrent})$$

By the reverse transition rules for operators $+$ and $\cdot$ in Table 4.6, there are no transitions.
With the assumptions $P[\mathcal{K}] \cdot (q+s) \sim_p^{fr} P[\mathcal{K}] \cdot q + P[\mathcal{K}] \cdot s$, $p' \cdot (q+s) \sim_p^{fr} p' \cdot q + p' \cdot s$, hence, $p \cdot (q+s) \sim_p^{fr} p \cdot q + p \cdot s(\text{Std}(p), \text{Std}(q), \text{Std}(s))$, as desired.

- **Axiom** $RA4$. Let $p, q, s$ be $BRATC$ processes, and $(q+s) \cdot p = q \cdot p + s \cdot p(\text{NStd}(p), \text{NStd}(q), \text{NStd}(s))$, it is sufficient to prove that $(q+s) \cdot p \sim_p^{fr} q \cdot p + s \cdot p$. By the pomset transition rules for operators $+$ and $\cdot$ in Table 4.5, there are no transitions.
By the reverse transition rules for operators $+$ and $\cdot$ in Table 4.6, we obtain

$$\frac{p \xtwoheadrightarrow{P[\mathcal{K}]} P}{(q+s) \cdot p \xtwoheadrightarrow{P[\mathcal{K}]} (q+s) \cdot P}(P \subseteq p, \forall e_1, e_2 \in P \text{ are pairwise concurrent})$$

$$\frac{p \xtwoheadrightarrow{P[\mathcal{K}]} P}{q \cdot p + s \cdot p \xtwoheadrightarrow{P[\mathcal{K}]} q \cdot P + s \cdot P}(P \subseteq p, \forall e_1, e_2 \in P \text{ are pairwise concurrent})$$

$$\frac{p \xrightarrow{P[\mathcal{K}]} p'}{(q+s)\cdot p \xrightarrow{P[\mathcal{K}]} (q+s)\cdot p'}(P \subseteq p, \forall e_1, e_2 \in P \text{ are pairwise concurrent})$$

$$\frac{p \xrightarrow{P[\mathcal{K}]} p'}{q\cdot p + s\cdot p \xrightarrow{P[\mathcal{K}]} q\cdot p' + s\cdot p'}(P \subseteq p, \forall e_1, e_2 \in P \text{ are pairwise concurrent})$$

With the assumptions $(q+s)\cdot P \sim_p^{fr} q\cdot P + s\cdot P$, $(q+s)\cdot p' \sim_p^{fr} q\cdot p' + s\cdot p'$, hence, $(q+s)\cdot p \sim_p^{fr} q\cdot p + s\cdot p(\text{NStd}(p), \text{NStd}(q), \text{NStd}(s))$, as desired.

- **Axiom** $A5$. Let $p, q, s$ be $BRATC$ processes, and $(p\cdot q)\cdot s = p\cdot(q\cdot s)$, it is sufficient to prove that $(p\cdot q)\cdot s \sim_s^{fr} p\cdot(q\cdot s)$. By the forward step transition rules for operator $\cdot$ in Table 4.7, we obtain

$$\frac{p \xrightarrow{P} P[\mathcal{K}]}{(p\cdot q)\cdot s \xrightarrow{P} (P[\mathcal{K}]\cdot q)\cdot s}(P \subseteq p, \forall e_1, e_2 \in P \text{ are pairwise concurrent})$$

$$\frac{p \xrightarrow{P} P[\mathcal{K}]}{p\cdot(q\cdot s) \xrightarrow{P} P[\mathcal{K}]\cdot(q\cdot s)}(P \subseteq p, \forall e_1, e_2 \in P \text{ are pairwise concurrent})$$

$$\frac{p \xrightarrow{P} p'}{(p\cdot q)\cdot s \xrightarrow{P} (p'\cdot q)\cdot s}(P \subseteq p, \forall e_1, e_2 \in P \text{ are pairwise concurrent})$$

$$\frac{p \xrightarrow{P} p'}{p\cdot(q\cdot s) \xrightarrow{P} p'\cdot(q\cdot s)}(P \subseteq p, \forall e_1, e_2 \in P \text{ are pairwise concurrent})$$

By the reverse step transition rules for operator $\cdot$ in Table 4.8, we obtain

$$\frac{s \xrightarrow{S[\mathcal{I}]} S}{(p\cdot q)\cdot s \xrightarrow{S[\mathcal{I}]} (p\cdot q)\cdot S}(S \subseteq s, \forall e_1, e_2 \in S \text{ are pairwise concurrent})$$

$$\frac{s \xrightarrow{S[\mathcal{I}]} S}{p\cdot(q\cdot s) \xrightarrow{S[\mathcal{I}]} p\cdot(q\cdot S)}(S \subseteq s, \forall e_1, e_2 \in S \text{ are pairwise concurrent})$$

$$\frac{s \xrightarrow{S[\mathcal{I}]} s'}{(p\cdot q)\cdot s \xrightarrow{S[\mathcal{I}]} (p\cdot q)\cdot s'}(S \subseteq s, \forall e_1, e_2 \in S \text{ are pairwise concurrent})$$

$$\frac{s \xrightarrow{S[\mathcal{I}]} s'}{p\cdot(q\cdot s) \xrightarrow{S[\mathcal{I}]} p\cdot(q\cdot s')}(S \subseteq s, \forall e_1, e_2 \in S \text{ are pairwise concurrent})$$

With assumptions $(P[\mathcal{K}]\cdot q)\cdot s \sim_s^{fr} P[\mathcal{K}]\cdot(q\cdot s)$, $(p'\cdot q)\cdot s \sim_s^{fr} p'\cdot(q\cdot s)$, $(p\cdot q)\cdot S \sim_s^{fr} p\cdot(q\cdot S)$, $(p\cdot q)\cdot s' \sim_s^{fr} p\cdot(q\cdot s')$, hence, $(p\cdot q)\cdot s \sim_s^{fr} p\cdot(q\cdot s)$, as desired. □

The transition rules for FR (hereditary) hp-bisimulation of $BRATC$ are the same as the single event transition rules in Table 4.3.

**Theorem 4.8** (Congruence of $BRATC$ with respect to FR hp-bisimulation equivalence). *FR hp-bisimulation equivalence* $\sim_{hp}^{fr}$ *is a congruence with respect to* $BRATC$.

*Proof.* It is easy to see that FR history-preserving bisimulation is an equivalent relation on $BRATC$ terms, we only need to prove that $\sim_{hp}^{fr}$ is preserved by the operators $\cdot$ and $+$.

- Causality operator $\cdot$. Let $x_1, x_2$ and $y_1, y_2$ be $BRATC$ processes, and $x_1 \sim_{hp}^{fr} y_1$, $x_2 \sim_{hp}^{fr} y_2$, it is sufficient to prove that $x_1 \cdot x_2 \sim_{hp}^{fr} y_1 \cdot y_2$.
  By the definition of FR hp-bisimulation $\sim_{hp}^{fr}$ (Definition 2.70), $x_1 \sim_{hp}^{fr} y_1$ means that there is a posetal relation $(C(x_1), f, C(y_1)) \in \sim_{hp}^{fr}$, and

$$x_1 \xrightarrow{e_1} x_1' \quad y_1 \xrightarrow{e_2} y_1'$$

$$x_1 \xrightarrow{e_1[m]} x_1' \quad y_1 \xrightarrow{e_2[m]} y_1'$$

  with $(C(x_1'), f[e_1 \mapsto e_2], C(y_1')) \in \sim_{hp}^{fr}$. The meaning of $x_2 \sim_{hp}^{fr} y_2$ is similar.
  By the FR hp-transition rules for causality operator $\cdot$ in Table 4.3 and Table 4.4, we can obtain

$$x_1 \cdot x_2 \xrightarrow{e_1} e_1[m] \cdot x_2 \quad y_1 \cdot y_2 \xrightarrow{e_2} e_2[n] \cdot y_2$$

$$x_1 \cdot x_2 \xrightarrow{e_1'[m} x_1 \cdot e_1' \quad y_1 \cdot y_2 \xrightarrow{e_2'[n} y_1 \cdot e_2'$$

  With the assumptions $(C(e_1[m] \cdot x_2), f[e_1 \mapsto e_2], C(e_2[n] \cdot y_2)) \in \sim_{hp}^{fr}$ and $(C(x_1 \cdot e_1'), f[e_1' \mapsto e_2'], C(y_1 \cdot e_2')) \in \sim_{hp}^{fr}$, hence, we obtain $x_1 \cdot x_2 \sim_{hp}^{fr} y_1 \cdot y_2$, as desired.
  Or, we can obtain

$$x_1 \cdot x_2 \xrightarrow{e_1} x_1' \cdot x_2 \quad y_1 \cdot y_2 \xrightarrow{e_2} y_1' \cdot y_2$$

$$x_1 \cdot x_2 \xrightarrow{e1'[m]} x_1 \cdot x_2' \quad y_1 \cdot y_2 \xrightarrow{e_2'[n]} y_1 \cdot y_2'$$

  With the assumptions $(C(x_1' \cdot x_2), f[e_1 \mapsto e_2], C(y_1' \cdot y_2)) \in \sim_{hp}^{fr}$ and $(C(x_1 \cdot x_2'), f[e_1' \mapsto e_2'], C(y_1 \cdot y_2')) \in \sim_{hp}^{fr}$, hence, we obtain $x_1 \cdot x_2 \sim_{hp}^{fr} y_1 \cdot y_2$, as desired.
- Conflict operator $+$. Let $x_1, x_2$ and $y_1, y_2$ be $BRATC$ processes, and $x_1 \sim_{hp}^{fr} y_1$, $x_2 \sim_{hp}^{fr} y_2$, it is sufficient to prove that $x_1 + x_2 \sim_{hp}^{fr} y_1 + y_2$. The meanings of $x_1 \sim_{hp}^{fr} y_1$ and $x_2 \sim_{hp}^{fr} y_2$ are the same as the above case, according to the definition of FR hp-bisimulation $\sim_{hp}^{fr}$ in Definition 2.70.
  By the FR hp-transition rules for conflict operator $+$ in Table 4.3 and Table 4.4, we can obtain several cases:

$$x_1 + x_2 \xrightarrow{e_1} e_1[m] + x_2 \quad y_1 + y_2 \xrightarrow{e_2} e_2[n] + y_2$$

$$x_1 + x_2 \xrightarrow{e_1[m]} e_1 + x_2 \quad y_1 + y_2 \xrightarrow{e_2[n]} e_2 + y_2$$

  With the assumptions $(C(e_1[m] + x_2), f[e_1 \mapsto e_2], C(e_2[n] + y_2)) \in \sim_{hp}^{fr}$ and $(C(e_1 + x_2), f[e_1 \mapsto e_2], C(e_2 + y_2)) \in \sim_{hp}^{fr}$, hence, we obtain $x_1 + x_2 \sim_{hp}^{fr} y_1 + y_2$, as desired.
  Or, we can obtain

$$x_1 + x_2 \xrightarrow{e_1} x_1' + x_2 \quad y_1 + y_2 \xrightarrow{e_2} y_1' + y_2$$

$$x_1 + x_2 \xrightarrow{e_1[m]} x_1' + x_2 \quad y_1 + y_2 \xrightarrow{e_2[n]} y_1' + y_2$$

  With the assumptions $(C(x_1' + x_2), f[e_1 \mapsto e_2], C(y_1' + y_2)) \in \sim_{hp}^{fr}$, hence, we obtain $x_1 + x_2 \sim_{hp}^{fr} y_1 + y_2$, as desired.
  Or, we can obtain

$$x_1 + x_2 \xrightarrow{e_1'} x_1 + e_1'[m] \quad y_1 + y_2 \xrightarrow{e_2'} y_1 + e_2'[n]$$

$$x_1 + x_2 \xrightarrow{e_1'[m]} x_1 + e_1' \quad y_1 + y_2 \xrightarrow{e_2'[n]} y_1 + e_2'$$

  With the assumptions $(C(x_1 + e_1'[m]), f[e_1' \mapsto e_2'], C(y_1 + e_2'[n])) \in \sim_{hp}^{fr}$ and $(C(x_1 + e_1'), f[e_1' \mapsto e_2'], C(y_1 + e_2')) \in \sim_{hp}^{fr}$, hence, we obtain $x_1 + x_2 \sim_{hp}^{fr} y_1 + y_2$, as desired.

Or, we can obtain

$$x_1 + x_2 \xrightarrow{e_1'} x_1 + x_2' \quad y_1 + y_2 \xrightarrow{e_2'} y_1 + y_2'$$

$$x_1 + x_2 \xtwoheadrightarrow{e_1'[m]} x_1 + x_2' \quad y_1 + y_2 \xtwoheadrightarrow{e_2'[n]} y_1 + y_2'$$

With the assumptions $(C(x_1 + x_2'), f[e_1' \mapsto e_2'], C(y_1 + y_2')) \in \sim_{hp}^{fr}$, hence, we obtain $x_1 + x_2 \sim_{hp}^{fr} y_1 + y_2$, as desired. Or, we can obtain

$$x_1 + x_2 \xrightarrow{e_1} x_1' + x_2' \quad y_1 + y_2 \xrightarrow{e_2} y_1' + y_2'$$

$$x_1 + x_2 \xtwoheadrightarrow{e_1[m]} x_1' + x_2' \quad y_1 + y_2 \xtwoheadrightarrow{e_2[n]} y_1' + y_2'$$

With the assumptions $(C(x_1' + x_2'), f[e_1 \mapsto e_2], C(y_1' + y_2')) \in \sim_{hp}^{fr}$, hence, we obtain $x_1 + x_2 \sim_{hp}^{fr} y_1 + y_2$, as desired. □

**Theorem 4.9** (Soundness of $BRATC$ modulo FR hp-bisimulation equivalence). *Let $x$ and $y$ be $BRATC$ terms. If $BRATC \vdash x = y$, then $x \sim_{hp}^{fr} y$.*

*Proof.* Since FR hp-bisimulation $\sim_{hp}^{fr}$ is both an equivalent and a congruent relation, we only need to check if each axiom in Table 4.1 is sound modulo FR hp-bisimulation equivalence.

- **Axiom** $A1$. Let $p, q$ be $BRATC$ processes, and $p + q = q + p$, it is sufficient to prove that $p + q \sim_{hp}^{fr} q + p$. By the forward hp-transition rules for operator $+$ in Table 4.3, we obtain

$$\frac{p \xrightarrow{e_1} e_1[m]}{p + q \xrightarrow{e_1} e_1[m] + q} (e_1 \in p, e_1 \notin q) \qquad \frac{p \xrightarrow{e_1} e_1[m]}{q + p \xrightarrow{e_1} q + e_1[m]} (e_1 \in p, e_1 \notin q)$$

$$\frac{p \xrightarrow{e_1} p'}{p + q \xrightarrow{e_1} p' + q} (e_1 \in p, e_1 \notin q) \qquad \frac{p \xrightarrow{e_1} p'}{q + p \xrightarrow{e_1} q + p'} (e_1 \in p, e_1 \notin q)$$

$$\frac{q \xrightarrow{e_2} e_2[n]}{p + q \xrightarrow{e_2} p + e_2[n]} (e_2 \in q, e_2 \notin p) \qquad \frac{q \xrightarrow{e_2} e_2[n]}{q + p \xrightarrow{e_2} e_2[n] + p} (e_2 \in q, e_2 \notin p)$$

$$\frac{q \xrightarrow{e_2} q'}{p + q \xrightarrow{e_2} p + q'} (e_2 \in q, e_2 \notin p) \qquad \frac{q \xrightarrow{e_2} q'}{q + p \xrightarrow{e_2} q' + p} (e_2 \in q, e_2 \notin p)$$

$$\frac{p \xrightarrow{e_1} p' \quad q \xrightarrow{e_1} q'}{p + q \xrightarrow{e_1} p' + q'} (e_1 \in p, e_1 \in q) \qquad \frac{p \xrightarrow{e_1} p' \quad q \xrightarrow{e_1} q'}{q + p \xrightarrow{e_1} q' + p'} (e_1 \in p, e_1 \in q)$$

  By the reverse hp-transition rules for operator $+$ in Table 4.4, we obtain

$$\frac{p \xtwoheadrightarrow{e_1[m]} e_1 \quad e_1 \notin q}{p + q \xtwoheadrightarrow{e_1[m]} e_1 + q} (e_1 \in p) \qquad \frac{p \xtwoheadrightarrow{e_1[m]} e_1 \quad e_1 \notin q}{q + p \xtwoheadrightarrow{e_1[m]} q + e_1} (e_1 \in p)$$

$$\frac{p \xtwoheadrightarrow{e_1[m]} p' \quad e_1 \notin q}{p + q \xtwoheadrightarrow{e_1[m]} p' + q} (e_1 \in p) \qquad \frac{p \xtwoheadrightarrow{e_1[m]} p' \quad e_1 \notin q}{q + p \xtwoheadrightarrow{e_1[m]} q + p'} (e_1 \in p)$$

$$\frac{q \xtwoheadrightarrow{e_2[n]} e_2 \quad e_2 \notin p}{p + q \xtwoheadrightarrow{e_2[n]} p + e_2} (e_2 \in q) \qquad \frac{q \xtwoheadrightarrow{e_2[n]} e_2 \quad e_2 \notin p}{q + p \xtwoheadrightarrow{e_2[n]} e_2 + p} (e_2 \in q)$$

$$\frac{q \xtwoheadrightarrow{e_2[n]} q' \quad e_2 \notin p}{p + q \xtwoheadrightarrow{e_2[n]} p + q'} (e_2 \in q) \qquad \frac{q \xtwoheadrightarrow{e_2[n]} q' \quad e_2 \notin p}{q + p \xtwoheadrightarrow{e_2[n]} q' + p} (e_2 \in q)$$

$$\frac{p \xrightarrow{e_1[m]} p' \quad q \xrightarrow{e_1[m]} q'}{p+q \xrightarrow{e_1[m]} p'+q'}(e_1 \in p, e_1 \in q) \qquad \frac{p \xrightarrow{e_1[m]} p' \quad q \xrightarrow{e_1[m]} q'}{q+p \xrightarrow{e_1[m]} q'+p'}(e_1 \in p, e_1 \in q)$$

With the assumptions $(C(e_1[m]+q), f[e_1 \mapsto e_1], C(q+e_1[m])) \in \sim^{fr}_{hp}$, $C((e_1+q), f[e_1 \mapsto e_1], C(q+e_1)) \in \sim^{fr}_{hp}$, $(C(p+e_2[n]), f[e_2 \mapsto e_2], C(e_2[n]+p)) \in \sim^{fr}_{hp}$, $(C(p'+q), f[e_1 \mapsto e_1], C(q+p')) \in \sim^{fr}_{hp}$, $(C(p+q'), f[e_2 \mapsto e_2], C(q'+p)) \in \sim^{fr}_{hp}$, and $(C(p'+q'), f[e_1 \mapsto e_1], C(q'+p')) \in \sim^{fr}_{hp}$ hence, $p+q \sim^{fr}_{hp} q+p$, as desired.

- **Axiom** $A2$. Let $p, q, s$ be $BRATC$ processes, and $(p+q)+s = p+(q+s)$, it is sufficient to prove that $(p+q)+s \sim^{fr}_{hp} p+(q+s)$. By the forward hp-transition rules for operator $+$ in Table 4.3, we obtain

$$\frac{p \xrightarrow{e_1} e_1[m] \quad e_1 \notin q \quad e_1 \notin s}{(p+q)+s \xrightarrow{e_1} (e_1[m]+q)+s}(e_1 \in p) \qquad \frac{p \xrightarrow{e_1} e_1[m] \quad e_1 \notin q \quad e_1 \notin s}{p+(q+s) \xrightarrow{e_1} e_1[m]+(q+s)}(e_1 \in p)$$

$$\frac{p \xrightarrow{e_1} p' \quad e_1 \notin q \quad e_1 \notin s}{(p+q)+s \xrightarrow{e_1} (p'+q)+s}(e_1 \in p) \qquad \frac{p \xrightarrow{e_1} p' \quad e_1 \notin q \quad e_1 \notin s}{p+(q+s) \xrightarrow{e_1} p'+(q+s)}(e_1 \in p)$$

$$\frac{q \xrightarrow{e_2} e_2[n] \quad e_2 \notin p \quad e_2 \notin s}{(p+q)+s \xrightarrow{e_2} (p+e_2[n])+s}(e_2 \in q) \qquad \frac{q \xrightarrow{e_2} e_2[n] \quad e_2 \notin p \quad e_2 \notin s}{p+(q+s) \xrightarrow{e_2} e_1+(e_2[n]+s)}(e_2 \in q)$$

$$\frac{q \xrightarrow{e_2} q' \quad e_2 \notin p \quad e_2 \notin s}{(p+q)+s \xrightarrow{e_2} (p+q')+s}(e_2 \in q) \qquad \frac{q \xrightarrow{e_2} q' \quad e_2 \notin p \quad e_2 \notin s}{p+(q+s) \xrightarrow{e_2} p+(q'+s)}(e_2 \in q)$$

$$\frac{s \xrightarrow{e_3} e_3[l] \quad e_3 \notin p \quad e_3 \notin q}{(p+q)+s \xrightarrow{e_3} (p+q)+e_3[l]}(e_3 \in s) \qquad \frac{s \xrightarrow{e_3} e_3[l] \quad e_3 \notin p \quad e_3 \notin q}{p+(q+s) \xrightarrow{e_3} p+(q+e_3[l])}(e_3 \in s)$$

$$\frac{s \xrightarrow{e_3} s' \quad e_3 \notin p \quad e_3 \notin q}{(p+q)+s \xrightarrow{e_3} (p+q)+s'}(e_3 \in s) \qquad \frac{s \xrightarrow{e_3} s' \quad e_3 \notin p \quad e_3 \notin q}{p+(q+s) \xrightarrow{e_3} p+(q+s')}(e_3 \in s)$$

$$\frac{p \xrightarrow{e_1} p' \quad q \xrightarrow{e_1} q' \quad s \xrightarrow{e_1} s'}{(p+q)+s \xrightarrow{e_1} (p'+q')+s'}(e_1 \in p, e_1 \in q, e_1 \in s) \qquad \frac{p \xrightarrow{e_1} p' \quad q \xrightarrow{e_1} q' \quad s \xrightarrow{e_1} s'}{p+(q+s) \xrightarrow{e_1} p'+(q'+s')}(e_1 \in p, e_1 \in q, e_1 \in s)$$

By the reverse hp-transition rules for operator $+$ in Table 4.4, we obtain

$$\frac{p \xrightarrow{e_1[m]} e_1 \quad e_1 \notin q \quad e_1 \notin s}{(p+q)+s \xrightarrow{e_1[m]} (e_1+q)+s}(e_1 \in p) \qquad \frac{p \xrightarrow{e_1[m]} e_1 \quad e_1 \notin q \quad e_1 \notin s}{p+(q+s) \xrightarrow{e_1[m]} e_1+(q+s)}(e_1 \in p)$$

$$\frac{p \xrightarrow{e_1[m]} p' \quad e_1 \notin q \quad e_1 \notin s}{(p+q)+s \xrightarrow{e_1[m]} (p'+q)+s}(e_1 \in p) \qquad \frac{p \xrightarrow{e_1[m]} p' \quad e_1 \notin q \quad e_1 \notin s}{p+(q+s) \xrightarrow{e_1[m]} p'+(q+s)}(e_1 \in p)$$

$$\frac{q \xrightarrow{e_2[n]} e_2 \quad e_2 \notin p \quad e_2 \notin s}{(p+q)+s \xrightarrow{e_2[n]} (p+e_2)+s}(e_2 \in q) \qquad \frac{q \xrightarrow{e_2[n]} e_2 \quad e_2 \notin p \quad e_2 \notin s}{p+(q+s) \xrightarrow{e_2[n]} p+(e_2+s)}(e_2 \in q)$$

$$\frac{q \xrightarrow{e_2[n]} q' \quad e_2 \notin p \quad e_2 \notin s}{(p+q)+s \xrightarrow{e_2[n]} (p+q')+s}(e_2 \in q) \qquad \frac{q \xrightarrow{e_2[n]} q' \quad e_2 \notin p \quad e_2 \notin s}{p+(q+s) \xrightarrow{e_2[n]} p+(q'+s)}(e_2 \in q)$$

$$\frac{s \xrightarrow{e_3[l]} e_3 \quad e_3 \notin p \quad e_3 \notin q}{(p+q)+s \xrightarrow{e_3[l]} (p+q)+e_3}(e_3 \in s) \qquad \frac{s \xrightarrow{e_3[l]} e_3 \quad e_3 \notin p \quad e_3 \notin q}{p+(q+s) \xrightarrow{e_3[l]} p+(q+e_3)}(e_3 \in s)$$

$$\frac{s \xrightarrow{e_3[l]} s' \quad e_3 \notin p \quad e_3 \notin q}{(p+q)+s \xrightarrow{e_3[l]} (p+q)+s'}(e_3 \in s) \qquad \frac{s \xrightarrow{e_3[l]} s' \quad e_3 \notin p \quad e_3 \notin q}{p+(q+s) \xrightarrow{e_3[l]} p+(q+s')}(e_3 \in s)$$

$$\frac{p \xrightarrow{e_1[m]} p' \quad q \xrightarrow{e_1[m]} q' \quad s \xrightarrow{e_1[m]} s'}{(p+q)+s \xrightarrow{e_1[m]} (p'+q')+s'}(e_1 \in p, e_1 \in q, e_1 \in s)$$

$$\frac{p \xrightarrow{e_1[m]} p' \quad q \xrightarrow{e_1[m]} q' \quad s \xrightarrow{e_1[m]} s'}{p+(q+s) \xrightarrow{e_1[m]} p'+(q'+s')}(e_1 \in p, e_1 \in q, e_1 \in s)$$

With the assumptions $(C((e_1[m]+q)+s), f[e_1 \mapsto e_1], C(e_1[m]+(q+s))) \in \sim_{hp}^{fr}$, $(C((e_1+q)+s), f[e_1 \mapsto e_1], C(e_1+(q+s))) \in \sim_{hp}^{fr}$, $(C((p+e_2[n])+s), f[e_2 \mapsto e_2], C(p+(e_2[n]+s))) \in \sim_{hp}^{fr}$, $(C((p+e_2)+s), f[e_2 \mapsto e_2], C(p+(e_2+s))) \in \sim_{hp}^{fr}$, $(C((p+q)+e_3[l]), f[e_3 \mapsto e_3], C(p+(q+e_3[l]))) \in \sim_{hp}^{fr}$, $(C((p+q)+e_3 \sim_{hp}^{fr} p+(q+e_3)$, $(p'+q)+s \sim_{hp}^{fr} p'+(q+s)$, $(C((p+q')+s), f[e_2 \mapsto e_2], C(p+(q'+s))) \in \sim_{hp}^{fr}$, $(C((p+q)+s'), f[e_3 \mapsto e_3], C(p+(q+s'))) \in \sim_{hp}^{fr}$ and $(C((p'+q')+s'), f[e_1 \mapsto e_1], C(p'+(q'+s'))) \in \sim_{hp}^{fr}$, hence, $(p+q)+s \sim_{hp}^{fr} p+(q+s)$, as desired.

- **Axiom** $A3$. Let $p$ be a $BRATC$ process, and $p+p=p$, it is sufficient to prove that $p+p \sim_{hp}^{fr} p$. By the forward hp-transition rules for operator $+$ in Table 4.3, we obtain

$$\frac{p \xrightarrow{e_1} e_1[m]}{p+p \xrightarrow{e_1} e_1[m]+e_1[m]}(e_1 \in p) \quad \frac{p \xrightarrow{e_1} e_1[m]}{p \xrightarrow{e_1} e_1[m]}(e_1 \in p)$$

$$\frac{p \xrightarrow{e_1} p'}{p+p \xrightarrow{e_1} p'+p'}(e_1 \in p) \quad \frac{p \xrightarrow{e_1} p'}{p \xrightarrow{e_1} p'}(e_1 \in p)$$

By the reverse hp-transition rules for operator $+$ in Table 4.4, we obtain

$$\frac{p \xrightarrow{e_1[m]} e_1}{p+p \xrightarrow{e_1[m]} e_1+e_1}(e_1 \in p) \quad \frac{p \xrightarrow{e_1[m]} e_1}{p \xrightarrow{e_1[m]} e_1}(e_1 \in p)$$

$$\frac{p \xrightarrow{e_1[m]} p'}{p+p \xrightarrow{e_1[m]} p'+p'}(e_1 \in p) \quad \frac{p \xrightarrow{e_1[m]} p'}{p \xrightarrow{e_1[m]} p'}(e_1 \in p)$$

With the assumptions $(C(e_1[m]+e_1[m]), f[e_1 \mapsto e_1], C(e_1[m])) \in \sim_{hp}^{fr}$, $(C(e_1+e_1), f[e_1 \mapsto e_1], C(e_1)) \in \sim_{hp}^{fr}$ and $(C(p'+p'), f[e_1 \mapsto e_1], C(p')) \in \sim_{hp}^{fr}$, hence, $p+p \sim_{hp}^{fr} p$, as desired.

- **Axiom** $A4$. Let $p, q, s$ be $BRATC$ processes, and $p \cdot (q+s) = p \cdot q + p \cdot s(\text{Std}(p), \text{Std}(q), \text{Std}(s))$, it is sufficient to prove that $p \cdot (q+s) \sim_{hp}^{fr} p \cdot q + p \cdot s$. By the hp-transition rules for operators $+$ and $\cdot$ in Table 4.3, we obtain

$$\frac{p \xrightarrow{e_1} e_1[m]}{p \cdot (q+s) \xrightarrow{e_1} e_1[m] \cdot (q+s)}(e_1 \in p) \quad \frac{p \xrightarrow{e_1} e_1[m]}{p \cdot q + p \cdot s \xrightarrow{e_1} e_1[m] \cdot q + e_1[m] \cdot s}(e_1 \in p)$$

$$\frac{p \xrightarrow{e_1} p'}{p \cdot (q+s) \xrightarrow{e_1} p' \cdot (q+s)}(e_1 \in p) \quad \frac{p \xrightarrow{e_1} p'}{p \cdot q + p \cdot s \xrightarrow{e_1} p' \cdot q + p' \cdot s}(e_1 \in p)$$

By the reverse transition rules for operators $+$ and $\cdot$ in Table 4.4, there are no transition rules.
With the assumptions $(C(e_1[m] \cdot (q+s)), f[e_1 \mapsto e_1], C(e_1[m] \cdot q + e_1[m] \cdot s)) \in \sim_{hp}^{fr}$, $(C(p' \cdot (q+s)), f[e_1 \mapsto e_1], C(p' \cdot q + p' \cdot s)) \in \sim_{hp}^{fr}$, hence, $p \cdot (q+s) \sim_{hp}^{fr} p \cdot q + p \cdot s(\text{Std}(p), \text{Std}(q), \text{Std}(s))$, as desired.

- **Axiom** $RA4$. Let $p, q, s$ be $BRATC$ processes, and $(q+s) \cdot p = q \cdot p + s \cdot p(\text{NStd}(p), \text{NStd}(q), \text{NStd}(s))$, it is sufficient to prove that $(q+s) \cdot p \sim_{hp}^{fr} q \cdot p + s \cdot p$. By the hp-transition rules for operators $+$ and $\cdot$ in Table 4.3, there are no transition rules.

By the reverse transition rules for operators $+$ and $\cdot$ in Table 4.4, we obtain

$$\frac{p \xrightarrow{e_1[m]} e_1}{(q+s)\cdot p \xrightarrow{e_1[m]} (q+s)\cdot e_1}(e_1 \in p) \quad \frac{p \xrightarrow{e_1[m]} e_1}{q\cdot p+s\cdot p \xrightarrow{e_1[m]} q\cdot e_1+s\cdot e_1}(e_1 \in p)$$

$$\frac{p \xrightarrow{e_1[m]} p'}{(q+s)\cdot p \xrightarrow{e_1[m]} (q+s\cdot p')}(e_1 \in p) \quad \frac{p \xrightarrow{e_1[m]} p'}{q\cdot p+s\cdot p \xrightarrow{e_1[m]} q\cdot p'+s\cdot p'}(e_1 \in p)$$

With the assumptions $(C((q+s)\cdot e_1), f[e_1 \mapsto e_1], C(q\cdot e_1+s\cdot e_1)) \in \sim_{hp}^{fr}$, $(C((q+s)\cdot p'), f[e_1 \mapsto e_1], C(q\cdot p'+s\cdot p')) \in \sim_{hp}^{fr}$, hence, $(q+s)\cdot p \sim_{hp}^{fr} q\cdot p+s\cdot p(\mathrm{NStd}(p), \mathrm{NStd}(q), \mathrm{NStd}(s))$, as desired.

- **Axiom** *A*5. Let $p, q, s$ be $BRATC$ processes, and $(p\cdot q)\cdot s = p\cdot(q\cdot s)$, it is sufficient to prove that $(p\cdot q)\cdot s \sim_{hp}^{fr} p\cdot(q\cdot s)$. By the forward hp-transition rules for operator $\cdot$ in Table 4.3, we obtain

$$\frac{p \xrightarrow{e_1} e_1[m]}{(p\cdot q)\cdot s \xrightarrow{e_1} (e_1[m]\cdot q)\cdot s}(e_1 \in p) \quad \frac{p \xrightarrow{e_1} e_1[m]}{p\cdot(q\cdot s) \xrightarrow{e_1} e_1[m]\cdot(q\cdot s)}(e_1 \in p)$$

$$\frac{p \xrightarrow{e_1} p'}{(p\cdot q)\cdot s \xrightarrow{e_1} (p'\cdot q)\cdot s}(e_1 \in p) \quad \frac{p \xrightarrow{e_1} p'}{p\cdot(q\cdot s) \xrightarrow{e_1} p'\cdot(q\cdot s)}(e_1 \in p)$$

By the reverse hp-transition rules for operator $\cdot$ in Table 4.4, we obtain

$$\frac{s \xrightarrow{e_3[l]} e_3}{(p\cdot q)\cdot s \xrightarrow{e_3[l]} (p\cdot q)\cdot e_3}(e_3 \in s) \quad \frac{s \xrightarrow{e_3[l]} e_3}{p\cdot(q\cdot s) \xrightarrow{e_3[l]} p\cdot(q\cdot e_3)}(e_3 \in s)$$

$$\frac{s \xrightarrow{e_3[l]} s'}{(p\cdot q)\cdot s \xrightarrow{e_3[l]} (p\cdot q)\cdot s'}(e_3 \in s) \quad \frac{s \xrightarrow{e_3[l]} s'}{p\cdot(q\cdot s) \xrightarrow{e_3[l]} p\cdot(q\cdot s')}(e_3 \in s)$$

with assumptions $(C((e_1[m]\cdot q)\cdot s), f[e_1 \mapsto e_1], C(e_1[m]\cdot(q\cdot s))) \in \sim_{hp}^{fr}$, $(C((p'\cdot q)\cdot s), f[e_1 \mapsto e_1], C(p'\cdot(q\cdot s))) \in \sim_{hp}^{fr}$, $(C((p\cdot q)\cdot e_3), f[e_3 \mapsto e_3], C(p\cdot(q\cdot e_3))) \in \sim_{hp}^{fr}$, $(C((p\cdot q)\cdot s'), f[e_3 \mapsto e_3], C(p\cdot(q\cdot s'))) \in \sim_{hp}^{fr}$, hence, $(p\cdot q)\cdot s \sim_{hp}^{fr} p\cdot(q\cdot s)$, as desired. □

**Theorem 4.10** (Congruence of $BRATC$ with respect to FR hhp-bisimulation equivalence). *FR hhp-bisimulation equivalence $\sim_{hhp}^{fr}$ is a congruence with respect to $BRATC$.*

*Proof.* It is easy to see that FR hhp-bisimulation is an equivalent relation on $BRATC$ terms, we only need to prove that $\sim_{hhp}^{fr}$ is preserved by the operators $\cdot$ and $+$.

- Causality operator $\cdot$. Let $x_1, x_2$ and $y_1, y_2$ be $BRATC$ processes, and $x_1 \sim_{hhp}^{fr} y_1$, $x_2 \sim_{hhp}^{fr} y_2$, it is sufficient to prove that $x_1\cdot x_2 \sim_{hhp}^{fr} y_1\cdot y_2$.
  By the definition of FR hhp-bisimulation $\sim_{hhp}^{fr}$ (Definition 2.70), $x_1 \sim_{hhp}^{fr} y_1$ means that there is a posetal relation $(C(x_1), f, C(y_1)) \in \sim_{hhp}^{fr}$, and

$$x_1 \xrightarrow{e_1} x_1' \quad y_1 \xrightarrow{e_2} y_1'$$

$$x_1 \xrightarrow{e_1[m]} x_1' \quad y_1 \xrightarrow{e_2[m]} y_1'$$

  with $(C(x_1'), f[e_1 \mapsto e_2], C(y_1')) \in \sim_{hhp}^{fr}$. The meaning of $x_2 \sim_{hhp}^{fr} y_2$ is similar.
  By the FR hhp-transition rules for causality operator $\cdot$ in Table 4.3 and Table 4.4, we can obtain

$$x_1\cdot x_2 \xrightarrow{e_1} e_1[m]\cdot x_2 \quad y_1\cdot y_2 \xrightarrow{e_2} e_2[n]\cdot y_2$$

$$x_1 \cdot x_2 \xrightarrow{e_1'[m]} x_1 \cdot e_1' \quad y_1 \cdot y_2 \xrightarrow{e_2'[n]} y_1 \cdot e_2'$$

With the assumptions $(C(e_1[m] \cdot x_2), f[e_1 \mapsto e_2], C(e_2[n] \cdot y_2)) \in \sim_{hhp}^{fr}$ and $(C(x_1 \cdot e_1'), f[e_1' \mapsto e_2'], C(y_1 \cdot e_2')) \in \sim_{hhp}^{fr}$, hence, we obtain $x_1 \cdot x_2 \sim_{hhp}^{fr} y_1 \cdot y_2$, as desired.
Or, we can obtain

$$x_1 \cdot x_2 \xrightarrow{e_1} x_1' \cdot x_2 \quad y_1 \cdot y_2 \xrightarrow{e_2} y_1' \cdot y_2$$

$$x_1 \cdot x_2 \xrightarrow{e1'[m]} x_1 \cdot x_2' \quad y_1 \cdot y_2 \xrightarrow{e_2'[n]} y_1 \cdot y_2'$$

With the assumptions $(C(x_1' \cdot x_2), f[e_1 \mapsto e_2], C(y_1' \cdot y_2)) \in \sim_{hhp}^{fr}$ and $(C(x_1 \cdot x_2'), f[e_1' \mapsto e_2'], C(y_1 \cdot y_2')) \in \sim_{hhp}^{fr}$, hence, we obtain $x_1 \cdot x_2 \sim_{hhp}^{fr} y_1 \cdot y_2$, as desired.

- Conflict operator $+$. Let $x_1, x_2$ and $y_1, y_2$ be $BRATC$ processes, and $x_1 \sim_{hhp}^{fr} y_1$, $x_2 \sim_{hhp}^{fr} y_2$, it is sufficient to prove that $x_1 + x_2 \sim_{hhp}^{fr} y_1 + y_2$. The meanings of $x_1 \sim_{hhp}^{fr} y_1$ and $x_2 \sim_{hhp}^{fr} y_2$ are the same as the above case, according to the definition of FR hhp-bisimulation $\sim_{hp}^{fr}$ in Definition 2.70.
By the FR hhp-transition rules for conflict operator $+$ in Table 4.3 and Table 4.4, we can obtain several cases:

$$x_1 + x_2 \xrightarrow{e_1} e_1[m] + x_2 \quad y_1 + y_2 \xrightarrow{e_2} e_2[n] + y_2$$

$$x_1 + x_2 \xrightarrow{e_1[m]} e_1 + x_2 \quad y_1 + y_2 \xrightarrow{e_2[n]} e_2 + y_2$$

With the assumptions $(C(e_1[m] + x_2), f[e_1 \mapsto e_2], C(e_2[n] + y_2)) \in \sim_{hhp}^{fr}$ and $(C(e_1 + x_2), f[e_1 \mapsto e_2], C(e_2 + y_2)) \in \sim_{hhp}^{fr}$, hence, we obtain $x_1 + x_2 \sim_{hhp}^{fr} y_1 + y_2$, as desired.
Or, we can obtain

$$x_1 + x_2 \xrightarrow{e_1} x_1' + x_2 \quad y_1 + y_2 \xrightarrow{e_2} y_1' + y_2$$

$$x_1 + x_2 \xrightarrow{e_1[m]} x_1' + x_2 \quad y_1 + y_2 \xrightarrow{e_2[n]} y_1' + y_2$$

With the assumptions $(C(x_1' + x_2), f[e_1 \mapsto e_2], C(y_1' + y_2)) \in \sim_{hhp}^{fr}$, hence, we obtain $x_1 + x_2 \sim_{hhp}^{fr} y_1 + y_2$, as desired.
Or, we can obtain

$$x_1 + x_2 \xrightarrow{e_1'} x_1 + e_1'[m] \quad y_1 + y_2 \xrightarrow{e_2'} y_1 + e_2'[n]$$

$$x_1 + x_2 \xrightarrow{e_1'[m]} x_1 + e_1' \quad y_1 + y_2 \xrightarrow{e_2'[n]} y_1 + e_2'$$

With the assumptions $(C(x_1 + e_1'[m]), f[e_1' \mapsto e_2'], C(y_1 + e_2'[n])) \in \sim_{hhp}^{fr}$ and $(C(x_1 + e_1'), f[e_1' \mapsto e_2'], C(y_1 + e_2')) \in \sim_{hhp}^{fr}$, hence, we obtain $x_1 + x_2 \sim_{hhp}^{fr} y_1 + y_2$, as desired.
Or, we can obtain

$$x_1 + x_2 \xrightarrow{e_1'} x_1 + x_2' \quad y_1 + y_2 \xrightarrow{e_2'} y_1 + y_2'$$

$$x_1 + x_2 \xrightarrow{e_1'[m]} x_1 + x_2' \quad y_1 + y_2 \xrightarrow{e_2'[n]} y_1 + y_2'$$

With the assumptions $(C(x_1 + x_2'), f[e_1' \mapsto e_2'], C(y_1 + y_2')) \in \sim_{hhp}^{fr}$, hence, we obtain $x_1 + x_2 \sim_{hhp}^{fr} y_1 + y_2$, as desired.
Or, we can obtain

$$x_1 + x_2 \xrightarrow{e_1} x_1' + x_2' \quad y_1 + y_2 \xrightarrow{e_2} y_1' + y_2'$$

$$x_1 + x_2 \xrightarrow{e_1[m]} x_1' + x_2' \quad y_1 + y_2 \xrightarrow{e_2[n]} y_1' + y_2'$$

With the assumptions $(C(x_1' + x_2'), f[e_1 \mapsto e_2], C(y_1' + y_2')) \in \sim_{hhp}^{fr}$, hence, we obtain $x_1 + x_2 \sim_{hhp}^{fr} y_1 + y_2$, as desired.
□

**Theorem 4.11** (Soundness of $BRATC$ modulo FR hhp-bisimulation equivalence). *Let $x$ and $y$ be $BRATC$ terms. If $BRATC \vdash x = y$, then $x \sim^{fr}_{hhp} y$.*

*Proof.* Since FR hhp-bisimulation $\sim^{fr}_{hhp}$ is both an equivalent and a congruent relation, we only need to check if each axiom in Table 4.1 is sound modulo FR hhp-bisimulation equivalence.

- **Axiom** $A1$. Let $p, q$ be $BRATC$ processes, and $p + q = q + p$, it is sufficient to prove that $p + q \sim^{fr}_{hhp} q + p$. By the forward hhp-transition rules for operator $+$ in Table 4.3, we obtain

$$\frac{p \xrightarrow{e_1} e_1[m]}{p+q \xrightarrow{e_1} e_1[m]+q}(e_1 \in p, e_1 \notin q) \quad \frac{p \xrightarrow{e_1} e_1[m]}{q+p \xrightarrow{e_1} q+e_1[m]}(e_1 \in p, e_1 \notin q)$$

$$\frac{p \xrightarrow{e_1} p'}{p+q \xrightarrow{e_1} p'+q}(e_1 \in p, e_1 \notin q) \quad \frac{p \xrightarrow{e_1} p'}{q+p \xrightarrow{e_1} q+p'}(e_1 \in p, e_1 \notin q)$$

$$\frac{q \xrightarrow{e_2} e_2[n]}{p+q \xrightarrow{e_2} p+e_2[n]}(e_2 \in q, e_2 \notin p) \quad \frac{q \xrightarrow{e_2} e_2[n]}{q+p \xrightarrow{e_2} e_2[n]+p}(e_2 \in q, e_2 \notin p)$$

$$\frac{q \xrightarrow{e_2} q'}{p+q \xrightarrow{e_2} p+q'}(e_2 \in q, e_2 \notin p) \quad \frac{q \xrightarrow{e_2} q'}{q+p \xrightarrow{e_2} q'+p}(e_2 \in q, e_2 \notin p)$$

$$\frac{p \xrightarrow{e_1} p' \quad q \xrightarrow{e_1} q'}{p+q \xrightarrow{e_1} p'+q'}(e_1 \in p, e_1 \in q) \quad \frac{p \xrightarrow{e_1} p' \quad q \xrightarrow{e_1} q'}{q+p \xrightarrow{e_1} q'+p'}(e_1 \in p, e_1 \in q)$$

By the reverse hhp-transition rules for operator $+$ in Table 4.4, we obtain

$$\frac{p \xrightarrow{e_1[m]} e_1 \quad e_1 \notin q}{p+q \xrightarrow{e_1[m]} e_1+q}(e_1 \in p) \quad \frac{p \xrightarrow{e_1[m]} e_1 \quad e_1 \notin q}{q+p \xrightarrow{e_1[m]} q+e_1}(e_1 \in p)$$

$$\frac{p \xrightarrow{e_1[m]} p' \quad e_1 \notin q}{p+q \xrightarrow{e_1[m]} p'+q}(e_1 \in p) \quad \frac{p \xrightarrow{e_1[m]} p' \quad e_1 \notin q}{q+p \xrightarrow{e_1[m]} q+p'}(e_1 \in p)$$

$$\frac{q \xrightarrow{e_2[n]} e_2 \quad e_2 \notin p}{p+q \xrightarrow{e_2[n]} p+e_2}(e_2 \in q) \quad \frac{q \xrightarrow{e_2[n]} e_2 \quad e_2 \notin p}{q+p \xrightarrow{e_2[n]} e_2+p}(e_2 \in q)$$

$$\frac{q \xrightarrow{e_2[n]} q' \quad e_2 \notin p}{p+q \xrightarrow{e_2[n]} p+q'}(e_2 \in q) \quad \frac{q \xrightarrow{e_2[n]} q' \quad e_2 \notin p}{q+p \xrightarrow{e_2[n]} q'+p}(e_2 \in q)$$

$$\frac{p \xrightarrow{e_1[m]} p' \quad q \xrightarrow{e_1[m]} q'}{p+q \xrightarrow{e_1[m]} p'+q'}(e_1 \in p, e_1 \in q) \quad \frac{p \xrightarrow{e_1[m]} p' \quad q \xrightarrow{e_1[m]} q'}{q+p \xrightarrow{e_1[m]} q'+p'}(e_1 \in p, e_1 \in q)$$

With the assumptions $(C(e_1[m]+q), f[e_1 \mapsto e_1], C(q+e_1[m])) \in \sim^{fr}_{hhp}$, $C((e_1+q), f[e_1 \mapsto e_1], C(q+e_1)) \in \sim^{fr}_{hhp}$, $(C(p+e_2[n]), f[e_2 \mapsto e_2], C(e_2[n]+p)) \in \sim^{fr}_{hhp}$, $(C(p'+q), f[e_1 \mapsto e_1], C(q+p')) \in \sim^{fr}_{hhp}$, $(C(p+q'), f[e_2 \mapsto e_2], C(q'+p)) \in \sim^{fr}_{hhp}$ and $(C(p'+q'), f[e_1 \mapsto e_1], C(q'+p')) \in \sim^{fr}_{hhp}$, hence, $p+q \sim^{fr}_{hhp} q+p$, as desired.

- **Axiom** $A2$. Let $p, q, s$ be $BRATC$ processes, and $(p+q)+s = p+(q+s)$, it is sufficient to prove that $(p+q)+s \sim^{fr}_{hhp} p+(q+s)$. By the forward hhp-transition rules for operator $+$ in Table 4.3, we obtain

$$\frac{p \xrightarrow{e_1} e_1[m] \quad e_1 \notin q \quad e_1 \notin s}{(p+q)+s \xrightarrow{e_1} (e_1[m]+q)+s}(e_1 \in p) \quad \frac{p \xrightarrow{e_1} e_1[m] \quad e_1 \notin q \quad e_1 \notin s}{p+(q+s) \xrightarrow{e_1} e_1[m]+(q+s)}(e_1 \in p)$$

$$\frac{p \xrightarrow{e_1} p' \quad e_1 \notin q \quad e_1 \notin s}{(p+q)+s \xrightarrow{e_1} (p'+q)+s}(e_1 \in p) \qquad \frac{p \xrightarrow{e_1} p' \quad e_1 \notin q \quad e_1 \notin s}{p+(q+s) \xrightarrow{e_1} p'+(q+s)}(e_1 \in p)$$

$$\frac{q \xrightarrow{e_2} e_2[n] \quad e_2 \notin p \quad e_2 \notin s}{(p+q)+s \xrightarrow{e_2} (p+e_2[n])+s}(e_2 \in q) \qquad \frac{q \xrightarrow{e_2} e_2[n] \quad e_2 \notin p \quad e_2 \notin s}{p+(q+s) \xrightarrow{e_2} e_1+(e_2[n]+s)}(e_2 \in q)$$

$$\frac{q \xrightarrow{e_2} q' \quad e_2 \notin p \quad e_2 \notin s}{(p+q)+s \xrightarrow{e_2} (p+q')+s}(e_2 \in q) \qquad \frac{q \xrightarrow{e_2} q' \quad e_2 \notin p \quad e_2 \notin s}{p+(q+s) \xrightarrow{e_2} p+(q'+s)}(e_2 \in q)$$

$$\frac{s \xrightarrow{e_3} e_3[l] \quad e_3 \notin p \quad e_3 \notin q}{(p+q)+s \xrightarrow{e_3} (p+q)+e_3[l]}(e_3 \in s) \qquad \frac{s \xrightarrow{e_3} e_3[l] \quad e_3 \notin p \quad e_3 \notin q}{p+(q+s) \xrightarrow{e_3} p+(q+e_3[l])}(e_3 \in s)$$

$$\frac{s \xrightarrow{e_3} s' \quad e_3 \notin p \quad e_3 \notin q}{(p+q)+s \xrightarrow{e_3} (p+q)+s'}(e_3 \in s) \qquad \frac{s \xrightarrow{e_3} s' \quad e_3 \notin p \quad e_3 \notin q}{p+(q+s) \xrightarrow{e_3} p+(q+s')}(e_3 \in s)$$

$$\frac{p \xrightarrow{e_1} p' \quad q \xrightarrow{e_1} q' \quad s \xrightarrow{e_1} s'}{(p+q)+s \xrightarrow{e_1} (p'+q')+s'}(e_1 \in p, e_1 \in q, e_1 \in s) \qquad \frac{p \xrightarrow{e_1} p' \quad q \xrightarrow{e_1} q' \quad s \xrightarrow{e_1} s'}{p+(q+s) \xrightarrow{e_1} p'+(q'+s')}(e_1 \in p, e_1 \in q, e_1 \in s)$$

By the reverse hhp-transition rules for operator $+$ in Table 4.4, we obtain

$$\frac{p \xrightarrow{e_1[m]} e_1 \quad e_1 \notin q \quad e_1 \notin s}{(p+q)+s \xrightarrow{e_1[m]} (e_1+q)+s}(e_1 \in p) \qquad \frac{p \xrightarrow{e_1[m]} e_1 \quad e_1 \notin q \quad e_1 \notin s}{p+(q+s) \xrightarrow{e_1[m]} e_1+(q+s)}(e_1 \in p)$$

$$\frac{p \xrightarrow{e_1[m]} p' \quad e_1 \notin q \quad e_1 \notin s}{(p+q)+s \xrightarrow{e_1[m]} (p'+q)+s}(e_1 \in p) \qquad \frac{p \xrightarrow{e_1[m]} p' \quad e_1 \notin q \quad e_1 \notin s}{p+(q+s) \xrightarrow{e_1[m]} p'+(q+s)}(e_1 \in p)$$

$$\frac{q \xrightarrow{e_2[n]} e_2 \quad e_2 \notin p \quad e_2 \notin s}{(p+q)+s \xrightarrow{e_2[n]} (p+e_2)+s}(e_2 \in q) \qquad \frac{q \xrightarrow{e_2[n]} e_2 \quad e_2 \notin p \quad e_2 \notin s}{p+(q+s) \xrightarrow{e_2[n]} p+(e_2+s)}(e_2 \in q)$$

$$\frac{q \xrightarrow{e_2[n]} q' \quad e_2 \notin p \quad e_2 \notin s}{(p+q)+s \xrightarrow{e_2[n]} (p+q')+s}(e_2 \in q) \qquad \frac{q \xrightarrow{e_2[n]} q' \quad e_2 \notin p \quad e_2 \notin s}{p+(q+s) \xrightarrow{e_2[n]} p+(q'+s)}(e_2 \in q)$$

$$\frac{s \xrightarrow{e_3[l]} e_3 \quad e_3 \notin p \quad e_3 \notin q}{(p+q)+s \xrightarrow{e_3[l]} (p+q)+e_3}(e_3 \in s) \qquad \frac{s \xrightarrow{e_3[l]} e_3 \quad e_3 \notin p \quad e_3 \notin q}{p+(q+s) \xrightarrow{e_3[l]} p+(q+e_3)}(e_3 \in s)$$

$$\frac{s \xrightarrow{e_3[l]} s' \quad e_3 \notin p \quad e_3 \notin q}{(p+q)+s \xrightarrow{e_3[l]} (p+q)+s'}(e_3 \in s) \qquad \frac{s \xrightarrow{e_3[l]} s' \quad e_3 \notin p \quad e_3 \notin q}{p+(q+s) \xrightarrow{e_3[l]} p+(q+s')}(e_3 \in s)$$

$$\frac{p \xrightarrow{e_1[m]} p' \quad q \xrightarrow{e_1[m]} q' \quad s \xrightarrow{e_1[m]} s'}{(p+q)+s \xrightarrow{e_1[m]} (p'+q')+s'}(e_1 \in p, e_1 \in q, e_1 \in s)$$

$$\frac{p \xrightarrow{e_1[m]} p' \quad q \xrightarrow{e_1[m]} q' \quad s \xrightarrow{e_1[m]} s'}{p+(q+s) \xrightarrow{e_1[m]} p'+(q'+s')}(e_1 \in p, e_1 \in q, e_1 \in s)$$

With the assumptions $(C((e_1[m]+q)+s), f[e_1 \mapsto e_1], C(e_1[m]+(q+s))) \in \sim^{fr}_{hhp}$, $(C((e_1+q)+s), f[e_1 \mapsto e_1], C(e_1+(q+s))) \in \sim^{fr}_{hhp}$, $(C((p+e_2[n])+s), f[e_2 \mapsto e_2], C(p+(e_2[n]+s))) \in \sim^{fr}_{hhp}$, $(C((p+e_2)+s), f[e_2 \mapsto e_2], C(p+(e_2+s))) \in \sim^{fr}_{hhp}$, $(C((p+q)+e_3[l]), f[e_3 \mapsto e_3], C(p+(q+e_3[l]))) \in \sim^{fr}_{hhp}$, $(C((p+q)+e_3 \sim^{fr}_{hhp} p+(q+e_3)$, $(p'+q)+s \sim^{fr}_{hhp} p'+(q+s)$, $(C((p+q')+s), f[e_2 \mapsto e_2], C(p+(q'+s))) \in \sim^{fr}_{hhp}$, $(C((p+$

$q)+s'), f[e_3 \mapsto e_3], C(p+(q+s'))) \in \sim_{hhp}^{fr}$, and $(C((p'+q')+s'), f[e_1 \mapsto e_1], C(p'+(q'+s'))) \in \sim_{hhp}^{fr}$, hence, $(p+q)+s \sim_{hhp}^{fr} p+(q+s)$, as desired.

- **Axiom** $A3$. Let $p$ be a $BRATC$ process, and $p+p=p$, it is sufficient to prove that $p+p \sim_{hhp}^{fr} p$. By the forward hhp-transition rules for operator $+$ in Table 4.3, we obtain

$$\frac{p \xrightarrow{e_1} e_1[m]}{p+p \xrightarrow{e_1} e_1[m]+e_1[m]}(e_1 \in p) \qquad \frac{p \xrightarrow{e_1} e_1[m]}{p \xrightarrow{e_1} e_1[m]}(e_1 \in p)$$

$$\frac{p \xrightarrow{e_1} p'}{p+p \xrightarrow{e_1} p'+p'}(e_1 \in p) \qquad \frac{p \xrightarrow{e_1} p'}{p \xrightarrow{e_1} p'}(e_1 \in p)$$

By the reverse hhp-transition rules for operator $+$ in Table 4.4, we obtain

$$\frac{p \xrightarrow{e_1[m]} e_1}{p+p \xrightarrow{e_1[m]} e_1+e_1}(e_1 \in p) \qquad \frac{p \xrightarrow{e_1[m]} e_1}{p \xrightarrow{e_1[m]} e_1}(e_1 \in p)$$

$$\frac{p \xrightarrow{e_1[m]} p'}{p+p \xrightarrow{e_1[m]} p'+p'}(e_1 \in p) \qquad \frac{p \xrightarrow{e_1[m]} p'}{p \xrightarrow{e_1[m]} p'}(e_1 \in p)$$

With the assumptions $(C(e_1[m]+e_1[m]), f[e_1 \mapsto e_1], C(e_1[m])) \in \sim_{hhp}^{fr}$, $(C(e_1+e_1), f[e_1 \mapsto e_1], C(e_1)) \in \sim_{hhp}^{fr}$ and $(C(p'+p'), f[e_1 \mapsto e_1], C(p')) \in \sim_{hhp}^{fr}$, hence, $p+p \sim_{hhp}^{fr} p$, as desired.

- **Axiom** $A4$. Let $p,q,s$ be $BRATC$ processes, and $p\cdot(q+s)=p\cdot q+p\cdot s(\text{Std}(p), \text{Std}(q), \text{Std}(s))$, it is sufficient to prove that $p\cdot(q+s) \sim_{hhp}^{fr} p\cdot q+p\cdot s$. By the hhp-transition rules for operators $+$ and $\cdot$ in Table 4.3, we obtain

$$\frac{p \xrightarrow{e_1} e_1[m]}{p\cdot(q+s) \xrightarrow{e_1} e_1[m]\cdot(q+s)}(e_1 \in p) \qquad \frac{p \xrightarrow{e_1} e_1[m]}{p\cdot q+p\cdot s \xrightarrow{e_1} e_1[m]\cdot q+e_1[m]\cdot s}(e_1 \in p)$$

$$\frac{p \xrightarrow{e_1} p'}{p\cdot(q+s) \xrightarrow{e_1} p'\cdot(q+s)}(e_1 \in p) \qquad \frac{p \xrightarrow{e_1} p'}{p\cdot q+p\cdot s \xrightarrow{e_1} p'\cdot q+p'\cdot s}(e_1 \in p)$$

By the reverse transition rules for operators $+$ and $\cdot$ in Table 4.4, there are no transition rules.
With the assumptions $(C(e_1[m]\cdot(q+s)), f[e_1 \mapsto e_1], C(e_1[m]\cdot q+e_1[m]\cdot s)) \in \sim_{hhp}^{fr}$, $(C(p'\cdot(q+s)), f[e_1 \mapsto e_1], C(p'\cdot q+p'\cdot s)) \in \sim_{hhp}^{fr}$, hence, $p\cdot(q+s) \sim_{hhp}^{fr} p\cdot q+p\cdot s(\text{Std}(p), \text{Std}(q), \text{Std}(s))$, as desired.

- **Axiom** $RA4$. Let $p,q,s$ be $BRATC$ processes, and $(q+s)\cdot p=q\cdot p+s\cdot p(\text{NStd}(p), \text{NStd}(q), \text{NStd}(s))$, it is sufficient to prove that $(q+s)\cdot p \sim_{hhp}^{fr} q\cdot p+s\cdot p$. By the hhp-transition rules for operators $+$ and $\cdot$ in Table 4.3, there are no transition rules.
By the reverse transition rules for operators $+$ and $\cdot$ in Table 4.4, we obtain

$$\frac{p \xrightarrow{e_1[m]} e_1}{(q+s)\cdot p \xrightarrow{e_1[m]} (q+s)\cdot e_1}(e_1 \in p) \qquad \frac{p \xrightarrow{e_1[m]} e_1}{q\cdot p+s\cdot p \xrightarrow{e_1[m]} q\cdot e_1+s\cdot e_1}(e_1 \in p)$$

$$\frac{p \xrightarrow{e_1[m]} p'}{(q+s)\cdot p \xrightarrow{e_1[m]} (q+s\cdot p')}(e_1 \in p) \qquad \frac{p \xrightarrow{e_1[m]} p'}{q\cdot p+s\cdot p \xrightarrow{e_1[m]} q\cdot p'+s\cdot p'}(e_1 \in p)$$

With the assumptions $(C((q+s)\cdot e_1), f[e_1 \mapsto e_1], C(q\cdot e_1+s\cdot e_1)) \in \sim_{hhp}^{fr}$, $(C((q+s)\cdot p'), f[e_1 \mapsto e_1], C(q\cdot p'+s\cdot p')) \in \sim_{hhp}^{fr}$, hence, $(q+s)\cdot p \sim_{hhp}^{fr} q\cdot p+s\cdot p(\text{NStd}(p), \text{NStd}(q), \text{NStd}(s))$, as desired.

**TABLE 4.9** Forward transition rules of parallel operator $\parallel$.

| | |
|---|---|
| $\frac{x\xrightarrow{e_1}e_1[m]\quad y\xrightarrow{e_2}e_2[m]}{x\parallel y\xrightarrow{\{e_1,e_2\}}e_1[m]\parallel e_2[m]}$ | $\frac{x\xrightarrow{e_1}x'\quad y\xrightarrow{e_2}e_2[m]}{x\parallel y\xrightarrow{\{e_1,e_2\}}x'\parallel e_2[m]}$ |
| $\frac{x\xrightarrow{e_1}e_1[m]\quad y\xrightarrow{e_2}y'}{x\parallel y\xrightarrow{\{e_1,e_2\}}e_1[m]\parallel y'}$ | $\frac{x\xrightarrow{e_1}x'\quad y\xrightarrow{e_2}y'}{x\parallel y\xrightarrow{\{e_1,e_2\}}x'\between y'}$ |

**TABLE 4.10** Reverse transition rules of parallel operator $\parallel$.

| | |
|---|---|
| $\frac{x\xrightarrow{e_1[m]}\!\!\!\!\rightarrow e_1\quad y\xrightarrow{e_2[m]}\!\!\!\!\rightarrow e_2}{x\parallel y\xrightarrow{\{e_1[m],e_2[m]\}}\!\!\!\!\rightarrow e_1\parallel e_2}$ | $\frac{x\xrightarrow{e_1[m]}\!\!\!\!\rightarrow x'\quad y\xrightarrow{e_2[m]}\!\!\!\!\rightarrow e_2}{x\parallel y\xrightarrow{\{e_1[m],e_2[m]\}}\!\!\!\!\rightarrow x'\parallel e_2}$ |
| $\frac{x\xrightarrow{e_1[m]}\!\!\!\!\rightarrow e_1\quad y\xrightarrow{e_2[m]}\!\!\!\!\rightarrow y'}{x\parallel y\xrightarrow{\{e_1[m],e_2[m]\}}\!\!\!\!\rightarrow e_1\parallel y'}$ | $\frac{x\xrightarrow{e_1[m]}\!\!\!\!\rightarrow x'\quad y\xrightarrow{e_2[m]}\!\!\!\!\rightarrow y'}{x\parallel y\xrightarrow{\{e_1[m],e_2[m]\}}\!\!\!\!\rightarrow x'\between y'}$ |

- **Axiom** $A5$. Let $p,q,s$ be $BRATC$ processes, and $(p\cdot q)\cdot s=p\cdot(q\cdot s)$, it is sufficient to prove that $(p\cdot q)\cdot s\sim^{fr}_{hhp}p\cdot(q\cdot s)$. By the forward hhp-transition rules for operator $\cdot$ in Table 4.3, we obtain

$$\frac{p\xrightarrow{e_1}e_1[m]}{(p\cdot q)\cdot s\xrightarrow{e_1}(e_1[m]\cdot q)\cdot s}(e_1\in p)\qquad\frac{p\xrightarrow{e_1}e_1[m]}{p\cdot(q\cdot s)\xrightarrow{e_1}e_1[m]\cdot(q\cdot s)}(e_1\in p)$$

$$\frac{p\xrightarrow{e_1}p'}{(p\cdot q)\cdot s\xrightarrow{e_1}(p'\cdot q)\cdot s}(e_1\in p)\qquad\frac{p\xrightarrow{e_1}p'}{p\cdot(q\cdot s)\xrightarrow{e_1}p'\cdot(q\cdot s)}(e_1\in p)$$

By the reverse hhp-transition rules for operator $\cdot$ in Table 4.4, we obtain

$$\frac{s\xrightarrow{e_3[l]}\!\!\!\!\rightarrow e_3}{(p\cdot q)\cdot s\xrightarrow{e_3[l]}\!\!\!\!\rightarrow(p\cdot q)\cdot e_3}(e_3\in s)\qquad\frac{s\xrightarrow{e_3[l]}\!\!\!\!\rightarrow e_3}{p\cdot(q\cdot s)\xrightarrow{e_3[l]}\!\!\!\!\rightarrow p\cdot(q\cdot e_3)}(e_3\in s)$$

$$\frac{s\xrightarrow{e_3[l]}\!\!\!\!\rightarrow s'}{(p\cdot q)\cdot s\xrightarrow{e_3[l]}\!\!\!\!\rightarrow(p\cdot q)\cdot s'}(e_3\in s)\qquad\frac{s\xrightarrow{e_3[l]}\!\!\!\!\rightarrow s'}{p\cdot(q\cdot s)\xrightarrow{e_3[l]}\!\!\!\!\rightarrow p\cdot(q\cdot s')}(e_3\in s)$$

With the assumptions $(C((e_1[m]\cdot q)\cdot s),f[e_1\mapsto e_1],C(e_1[m]\cdot(q\cdot s)))\in\sim^{fr}_{hhp}$, $(C((p'\cdot q)\cdot s),f[e_1\mapsto e_1],C(p'\cdot(q\cdot s)))\in\sim^{fr}_{hhp}$, $(C((p\cdot q)\cdot e_3),f[e_3\mapsto e_3],C(p\cdot(q\cdot e_3)))\in\sim^{fr}_{hhp}$, $(C((p\cdot q)\cdot s'),f[e_3\mapsto e_3],C(p\cdot(q\cdot s')))\in\sim^{fr}_{hhp}$, hence, $(p\cdot q)\cdot s\sim^{fr}_{hhp}p\cdot(q\cdot s)$, as desired. □

## 4.2 Reversible algebra for parallelism in true concurrency

In this section, we will discuss reversible parallelism in true concurrency. The resulted algebra is called Reversible Algebra for Parallelism in True Concurrency, abbreviated $RAPTC$.

### 4.2.1 Parallelism

The forward transition rules for parallelism $\parallel$ are shown in Table 4.9, and the reverse transition rules for $\parallel$ are shown in Table 4.10.

The forward and reverse transition rules of communication $\mid$ are shown in Table 4.11 and Table 4.12.

The conflict elimination is also captured by two auxiliary operators, the unary conflict elimination operator $\Theta$ and the binary unless operator $\triangleleft$. The forward and reverse transition rules for $\Theta$ and $\triangleleft$ are expressed by the ten transition rules in Table 4.13 and Table 4.14.

**TABLE 4.11** Forward transition rules of communication operator $\mid$.

| | |
|---|---|
| $\frac{x \xrightarrow{e_1} e_1[m] \quad y \xrightarrow{e_2} e_2[m]}{x \mid y \xrightarrow{\gamma(e_1,e_2)} \gamma(e_1,e_2)[m]}$ | $\frac{x \xrightarrow{e_1} x' \quad y \xrightarrow{e_2} e_2[m]}{x \mid y \xrightarrow{\gamma(e_1,e_2)} \gamma(e_1,e_2)[m] \cdot x'}$ |
| $\frac{x \xrightarrow{e_1} e_1[m] \quad y \xrightarrow{e_2} y'}{x \mid y \xrightarrow{\gamma(e_1,e_2)} \gamma(e_1,e_2)[m] \cdot y'}$ | $\frac{x \xrightarrow{e_1} x' \quad y \xrightarrow{e_2} y'}{x \mid y \xrightarrow{\gamma(e_1,e_2)} \gamma(e_1,e_2)[m] \cdot x' \between y'}$ |

**TABLE 4.12** Reverse transition rules of communication operator $\mid$.

| | |
|---|---|
| $\frac{x \overset{e_1[m]}{\twoheadrightarrow} e_1 \quad y \overset{e_2[m]}{\twoheadrightarrow} e_2}{x \mid y \overset{\gamma(e_1,e_2)[m]}{\twoheadrightarrow} \gamma(e_1,e_2)}$ | $\frac{x \overset{e_1[m]}{\twoheadrightarrow} x' \quad y \overset{e_2[m]}{\twoheadrightarrow} e_2}{x \mid y \overset{\gamma(e_1,e_2)[m]}{\twoheadrightarrow} \gamma(e_1,e_2) \cdot x'}$ |
| $\frac{x \overset{e_1[m]}{\twoheadrightarrow} e_1 \quad y \overset{e_2[m]}{\twoheadrightarrow} y'}{x \mid y \overset{\gamma(e_1,e_2)[m]}{\twoheadrightarrow} \gamma(e_1,e_2) \cdot y'}$ | $\frac{x \overset{e_1[m]}{\twoheadrightarrow} x' \quad y \overset{e_2[m]}{\twoheadrightarrow} y'}{x \mid y \overset{\gamma(e_1,e_2)[m]}{\twoheadrightarrow} \gamma(e_1,e_2) \cdot x' \between y'}$ |

**TABLE 4.13** Forward transition rules of conflict elimination.

| | |
|---|---|
| $\frac{x \xrightarrow{e_1} e_1[m] \quad (\sharp(e_1,e_2))}{\Theta(x) \xrightarrow{e_1} e_1[m]}$ | $\frac{x \xrightarrow{e_2} e_2[n] \quad (\sharp(e_1,e_2))}{\Theta(x) \xrightarrow{e_2} e_2[n]}$ |
| $\frac{x \xrightarrow{e_1} x' \quad (\sharp(e_1,e_2))}{\Theta(x) \xrightarrow{e_1} \Theta(x')}$ | $\frac{x \xrightarrow{e_2} x' \quad (\sharp(e_1,e_2))}{\Theta(x) \xrightarrow{e_2} \Theta(x')}$ |
| $\frac{x \xrightarrow{e_1} e_1[m] \quad y \nrightarrow^{e_2} \quad (\sharp(e_1,e_2))}{x \triangleleft y \xrightarrow{\tau} \surd}$ | $\frac{x \xrightarrow{e_1} x' \quad y \nrightarrow^{e_2} \quad (\sharp(e_1,e_2))}{x \triangleleft y \xrightarrow{\tau} x'}$ |
| $\frac{x \xrightarrow{e_1} e_1[m] \quad y \nrightarrow^{e_3} \quad (\sharp(e_1,e_2), e_2 \leq e_3)}{x \triangleleft y \xrightarrow{\tau} \surd}$ | $\frac{x \xrightarrow{e_1} x' \quad y \nrightarrow^{e_3} \quad (\sharp(e_1,e_2), e_2 \leq e_3)}{x \triangleleft y \xrightarrow{\tau} x'}$ |
| $\frac{x \xrightarrow{e_3} e_3[l] \quad y \nrightarrow^{e_2} \quad (\sharp(e_1,e_2), e_1 \leq e_3)}{x \triangleleft y \xrightarrow{\tau} \surd}$ | $\frac{x \xrightarrow{e_3} x' \quad y \nrightarrow^{e_2} \quad (\sharp(e_1,e_2), e_1 \leq e_3)}{x \triangleleft y \xrightarrow{\tau} x'}$ |

**TABLE 4.14** Reverse transition rules of conflict elimination.

| | |
|---|---|
| $\frac{x \overset{e_1[m]}{\twoheadrightarrow} e_1 \quad (\sharp(e_1,e_2))}{\Theta(x) \overset{e_1[m]}{\twoheadrightarrow} e_1}$ | $\frac{x \overset{e_2[n]}{\twoheadrightarrow} e_2 \quad (\sharp(e_1,e_2))}{\Theta(x) \overset{e_2[n]}{\twoheadrightarrow} e_2}$ |
| $\frac{x \overset{e_1[m]}{\twoheadrightarrow} x' \quad (\sharp(e_1,e_2))}{\Theta(x) \overset{e_1[m]}{\twoheadrightarrow} \Theta(x')}$ | $\frac{x \overset{e_2[n]}{\twoheadrightarrow} x' \quad (\sharp(e_1,e_2))}{\Theta(x) \overset{e_2[n]}{\twoheadrightarrow} \Theta(x')}$ |
| $\frac{x \overset{e_1[m]}{\twoheadrightarrow} e_1 \quad y \overset{e_2[n]}{\not\twoheadrightarrow} \quad (\sharp(e_1,e_2))}{x \triangleleft y \overset{\tau}{\twoheadrightarrow} \surd}$ | $\frac{x \overset{e_1[m]}{\twoheadrightarrow} x' \quad y \overset{e_2[n]}{\not\twoheadrightarrow} \quad (\sharp(e_1,e_2))}{x \triangleleft y \overset{\tau}{\twoheadrightarrow} x'}$ |
| $\frac{x \overset{e_1[m]}{\twoheadrightarrow} e_1 \quad y \overset{e_3[l]}{\not\twoheadrightarrow} \quad (\sharp(e_1,e_2), e_2 \geq e_3)}{x \triangleleft y \overset{\tau}{\twoheadrightarrow} \surd}$ | $\frac{x \overset{e_1[m]}{\twoheadrightarrow} x' \quad y \overset{e_3[l]}{\not\twoheadrightarrow} \quad (\sharp(e_1,e_2), e_2 \geq e_3)}{x \triangleleft y \overset{\tau}{\twoheadrightarrow} x'}$ |
| $\frac{x \overset{e_3[l]}{\twoheadrightarrow} e_3 \quad y \overset{e_2[n]}{\not\twoheadrightarrow} \quad (\sharp(e_1,e_2), e_1 \geq e_3)}{x \triangleleft y \overset{\tau}{\twoheadrightarrow} \surd}$ | $\frac{x \overset{e_3[l]}{\twoheadrightarrow} x' \quad y \overset{e_2[n]}{\not\twoheadrightarrow} \quad (\sharp(e_1,e_2), e_1 \geq e_3)}{x \triangleleft y \overset{\tau}{\twoheadrightarrow} x'}$ |

**Theorem 4.12** (Congruence theorem of $RAPTC$). *FR truly concurrent bisimulation equivalences* $\sim_p^{fr}$, $\sim_s^{fr}$, $\sim_{hp}^{fr}$, *and* $\sim_{hhp}^{fr}$ *are all congruences with respect to* $RAPTC$.

*Proof.* (1) Case FR pomset bisimulation equivalence $\sim_p^{fr}$.

- Case parallel operator $\parallel$. Let $x_1, x_2$ and $y_1, y_2$ be $RAPTC$ processes, and $x_1 \sim_p^{fr} y_1$, $x_2 \sim_p^{fr} y_2$, it is sufficient to prove that $x_1 \parallel x_2 \sim_p^{fr} y_1 \parallel y_2$.
  By the definition of FR pomset bisimulation $\sim_p^{fr}$ (Definition 2.68), $x_1 \sim_p^{fr} y_1$ means that
  $$x_1 \xrightarrow{X_1} x_1' \quad y_1 \xrightarrow{Y_1} y_1'$$
  with $X_1 \subseteq x_1$, $Y_1 \subseteq y_1$, $X_1 \sim Y_1$, and $x_1' \sim_p^{fr} y_1'$. The meaning of $x_2 \sim_p^{fr} y_2$ is similar.
  By the forward transition rules for parallel operator $\parallel$ in Table 4.9, we can obtain
  $$x_1 \parallel x_2 \xrightarrow{\{X_1,X_2\}} X_1[\mathcal{K}] \parallel X_2[\mathcal{K}] \quad y_1 \parallel y_2 \xrightarrow{\{Y_1,Y_2\}} Y_1[\mathcal{J}] \parallel Y_2[\mathcal{J}]$$
  $$x_1 \parallel x_2 \xtwoheadrightarrow{\{X_1[\mathcal{K}],X_2[\mathcal{K}]\}} X_1 \parallel X_2 \quad y_1 \parallel y_2 \xtwoheadrightarrow{\{Y_1[\mathcal{J}],Y_2[\mathcal{J}]\}} Y_1 \parallel Y_2$$
  With $X_1 \subseteq x_1$, $Y_1 \subseteq y_1$, $X_2 \subseteq x_2$, $Y_2 \subseteq y_2$, $X_1 \sim Y_1$ and $X_2 \sim Y_2$, and the assumptions $X_1[\mathcal{K} \parallel X_2[\mathcal{K}]] \sim_p^{fr} Y_1[\mathcal{J}] \parallel Y_2[\mathcal{J}]$, and $X_1 \parallel X_2 \sim_p^{fr} Y_1 \parallel Y_2$, hence, we obtain $x_1 \parallel x_2 \sim_p^{fr} y_1 \parallel y_2$, as desired.
  Or, we can obtain
  $$x_1 \parallel x_2 \xrightarrow{\{X_1,X_2\}} x_1' \parallel X_2[\mathcal{K}] \quad y_1 \parallel y_2 \xrightarrow{\{Y_1,Y_2\}} y_1' \parallel Y_2[\mathcal{J}]$$
  $$x_1 \parallel x_2 \xtwoheadrightarrow{\{X_1[\mathcal{K}],X_2[\mathcal{K}]\}} x_1' \parallel X_2 \quad y_1 \parallel y_2 \xtwoheadrightarrow{\{Y_1[\mathcal{J}],Y_2[\mathcal{J}]\}} y_1' \parallel Y_2$$
  With $X_1 \subseteq x_1$, $Y_1 \subseteq y_1$, $X_2 \subseteq x_2$, $Y_2 \subseteq y_2$, $X_1 \sim Y_1$, $X_2 \sim Y_2$, and the assumptions $x_1' \parallel X_2[\mathcal{K}]] \sim_p^{fr} y_1' \parallel Y_2[\mathcal{J}]$ and $x_1' \parallel X_2 \sim_p^{fr} y_1' \parallel Y_2$, hence, we obtain $x_1 \parallel x_2 \sim_p^{fr} y_1 \parallel y_2$, as desired.
  Or, we can obtain
  $$x_1 \parallel x_2 \xrightarrow{\{X_1,X_2\}} X_1[\mathcal{K}] \parallel x_2' \quad y_1 \parallel y_2 \xrightarrow{\{Y_1,Y_2\}} Y_1[\mathcal{J}] \parallel y_2'$$
  $$x_1 \parallel x_2 \xtwoheadrightarrow{\{X_1[\mathcal{K}],X_2[\mathcal{K}]\}} X_1 \parallel x_2' \quad y_1 \parallel y_2 \xtwoheadrightarrow{\{Y_1[\mathcal{J}],Y_2[\mathcal{J}]\}} Y_1 \parallel y_2'$$
  With $X_1 \subseteq x_1$, $Y_1 \subseteq y_1$, $X_2 \subseteq x_2$, $Y_2 \subseteq y_2$, $X_1 \sim Y_1$, $X_2 \sim Y_2$, and the assumptions $X_1[\mathcal{K} \parallel x_2' \sim_p^{fr} Y_1[\mathcal{J}] \parallel y_2'$ and $X_1 \parallel x_2' \sim_p^{fr} Y_1 \parallel y_2'$, hence, we obtain $x_1 \parallel x_2 \sim_p^{fr} y_1 \parallel y_2$, as desired.
  Or, we can obtain
  $$x_1 \parallel x_2 \xrightarrow{\{X_1,X_2\}} x_1' \between x_2' \quad y_1 \parallel y_2 \xrightarrow{\{Y_1,Y_2\}} y_1' \between y_2'$$
  $$x_1 \parallel x_2 \xtwoheadrightarrow{\{X_1[\mathcal{K}],X_2[\mathcal{K}]\}} x_1' \between x_2' \quad y_1 \parallel y_2 \xtwoheadrightarrow{\{Y_1[\mathcal{J}],Y_2[\mathcal{J}]\}} y_1' \between y_2'$$
  With $X_1 \subseteq x_1$, $Y_1 \subseteq y_1$, $X_2 \subseteq x_2$, $Y_2 \subseteq y_2$, $X_1 \sim Y_1$, $X_2 \sim Y_2$, and the assumption $x_1' \between x_2' \sim_p^{fr} y_1' \between y_2'$, hence, we obtain $x_1 \parallel x_2 \sim_p^{fr} y_1 \parallel y_2$, as desired.
- Case communication operator $\mid$. It can be proved similarly to the case of parallel operator $\parallel$, hence, we omit it here. Note that a communication is defined between two single communicating events.
- Case conflict elimination operator $\Theta$. It can be proved similarly to the above cases, hence, we omit it here. Note that the conflict elimination operator $\Theta$ is a unary operator.
- Case unless operator $\triangleleft$. It can be proved similarly to the case of parallel operator $\parallel$, hence, we omit it here. Note that a conflict relation is defined between two single events.

(2) The cases of FR step bisimulation $\sim_s^{fr}$, FR hp-bisimulation $\sim_{hp}^{fr}$, and FR hhp-bisimulation $\sim_{hhp}^{fr}$ can be proven similarly, hence, we omit them here. □

### 4.2.2 Axiom system of parallelism

**Definition 4.13** (Basic terms of $RAPTC$). The set of basic terms of $RAPTC$, $\mathcal{B}(RAPTC)$, is inductively defined as follows:

1. $\mathbb{E} \subset \mathcal{B}(RAPTC)$;
2. if $e \in \mathbb{E}, t \in \mathcal{B}(RAPTC)$, then $e \cdot t \in \mathcal{B}(RAPTC)$;
3. if $t, s \in \mathcal{B}(RAPTC)$, then $t + s \in \mathcal{B}(RAPTC)$;
4. if $t, s \in \mathcal{B}(RAPTC)$, then $t \parallel s \in \mathcal{B}(RAPTC)$.

We design the axioms of parallelism in Table 4.15, including algebraic laws for parallel operator $\parallel$, communication operator $\mid$, conflict elimination operator $\Theta$, unless operator $\triangleleft$, and also the whole parallel operator $\between$. Since the communication between two communicating events in different parallel branches may cause deadlock (a state of inactivity), which is caused by mismatch of two communicating events or the imperfectness of the communication channel, we introduce a new constant $\delta$ to denote the deadlock, and let the atomic event $e \in \mathbb{E} \cup \{\delta\}$.

Based on the definition of basic terms for $RAPTC$ (see Definition 4.13) and axioms of parallelism (see Table 4.15), we can prove the elimination theorem of parallelism.

**Theorem 4.14** (Elimination theorem of FR parallelism). *Let $p$ be a closed $RAPTC$ term. Then, there is a basic $RAPTC$ term $q$ such that $RAPTC \vdash p = q$.*

*Proof.* (1) First, suppose that the following ordering on the signature of $RAPTC$ is defined: $\parallel > \cdot > +$ and the symbol $\cdot$ is given the lexicographical status for the first argument, then for each rewrite rule $p \rightarrow q$ in Table 4.16 relation $p >_{lpo} q$ can easily be proved. We obtain that the term rewrite system shown in Table 4.16 is strongly normalizing, for it has finitely many rewriting rules, and $>$ is a well-founded ordering on the signature of $RAPTC$, and if $s >_{lpo} t$, for each rewriting rule $s \rightarrow t$ is in Table 4.16 (see Theorem 2.9).

(2) Then, we prove that the normal forms of closed $RAPTC$ terms are basic $RAPTC$ terms.

Suppose that $p$ is a normal form of some closed $RAPTC$ term and suppose that $p$ is not a basic $RAPTC$ term. Let $p'$ denote the smallest sub-term of $p$ that is not a basic $RAPTC$ term. This implies that each sub-term of $p'$ is a basic $RAPTC$ term. Then, we prove that $p$ is not a term in normal form. It is sufficient to induct on the structure of $p'$:

- Case $p' \equiv e$ or $e[m]$, $e \in \mathbb{E}$. $p'$ is a basic $RAPTC$ term, which contradicts the assumption that $p'$ is not a basic $RAPTC$ term, hence, this case should not occur.
- Case $p' \equiv p_1 \cdot p_2$. By induction on the structure of the basic $RAPTC$ term $p_1$:
  - Subcase $p_1 \in \mathbb{E}$. $p'$ would be a basic $RAPTC$ term, which contradicts the assumption that $p'$ is not a basic $RAPTC$ term;
  - Subcase $p_1 \equiv e \cdot p_1'$. $RRA5$ rewriting rule in Table 4.2 can be applied. Hence, $p$ is not a normal form;
  - Subcase $p_1 \equiv p_1' \cdot e[m]$. $RRA5$ rewriting rule in Table 4.2 can be applied. Hence, $p$ is not a normal form;
  - Subcase $p_1 \equiv p_1' + p_1''$. $RRA4$ rewriting rule in Table 4.2 can be applied. Hence, $p$ is not a normal form;
  - Subcase $p_1 \equiv p_1' \parallel p_1''$. $p'$ would be a basic $RAPTC$ term, which contradicts the assumption that $p'$ is not a basic $RAPTC$ term;
  - Subcase $p_1 \equiv p_1' \mid p_1''$. $RRC11$ and $RRRC11$ rewrite rules in Table 4.16 can be applied. Hence, $p$ is not a normal form;
  - Subcase $p_1 \equiv \Theta(p_1')$. $RRCE19$, $RRRCE19$, and $RRCE20$ rewrite rules in Table 4.16 can be applied. Hence, $p$ is not a normal form.
- Case $p' \equiv p_1 + p_2$. By induction on the structure of the basic $RAPTC$ terms both $p_1$ and $p_2$, all subcases will lead to that $p'$ would be a basic $RAPTC$ term, which contradicts the assumption that $p'$ is not a basic $RAPTC$ term.
- Case $p' \equiv p_1 \parallel p_2$. By induction on the structure of the basic $RAPTC$ terms both $p_1$ and $p_2$, all subcases will lead to that $p'$ would be a basic $RAPTC$ term, which contradicts the assumption that $p'$ is not a basic $RAPTC$ term.
- Case $p' \equiv p_1 \mid p_2$. By induction on the structure of the basic $RAPTC$ terms both $p_1$ and $p_2$, all subcases will lead to that $p'$ would be a basic $RAPTC$ term, which contradicts the assumption that $p'$ is not a basic $RAPTC$ term.
- Case $p' \equiv \Theta(p_1)$. By induction on the structure of the basic $RAPTC$ term $p_1$, $RRCE19 - RRCE24$ rewrite rules in Table 4.16 can be applied. Hence, $p$ is not a normal form.
- Case $p' \equiv p_1 \triangleleft p_2$. By induction on the structure of the basic $RAPTC$ terms both $p_1$ and $p_2$, all subcases will lead to that $p'$ would be a basic $RAPTC$ term, which contradicts the assumption that $p'$ is not a basic $RAPTC$ term. □

### 4.2.3 Structured operational semantics of parallelism

**Theorem 4.15** (Generalization of the reversible algebra for parallelism with respect to $BRATC$). *The algebra for parallelism is a generalization of $BRATC$.*

**TABLE 4.15** Axioms of parallelism.

| No. | Axiom |
|---|---|
| $A6$ | $x + \delta = x$ |
| $A7$ | $\delta \cdot x = \delta(\text{Std(x)})$ |
| $RA7$ | $x \cdot \delta = \delta(\text{NStd(x)})$ |
| $P1$ | $x \between y = x \parallel y + x \mid y$ |
| $P2$ | $x \parallel y = y \parallel x$ |
| $P3$ | $(x \parallel y) \parallel z = x \parallel (y \parallel z)$ |
| $P4$ | $e_1 \parallel (e_2 \cdot y) = (e_1 \parallel e_2) \cdot y$ |
| $RP4$ | $e_1[m] \parallel (y \cdot e_2[m]) = y \cdot (e_1[m] \parallel e_2[m])$ |
| $P5$ | $(e_1 \cdot x) \parallel e_2 = (e_1 \parallel e_2) \cdot x$ |
| $RP5$ | $(x \cdot e_1[m]) \parallel e_2[m] = x \cdot (e_1[m] \parallel e_2[m])$ |
| $P6$ | $(e_1 \cdot x) \parallel (e_2 \cdot y) = (e_1 \parallel e_2) \cdot (x \between y)$ |
| $RP6$ | $(x \cdot e_1[m]) \parallel (y \cdot e_2[m]) = (x \between y) \cdot (e_1[m] \parallel e_2[m])$ |
| $P7$ | $(x + y) \parallel z = (x \parallel z) + (y \parallel z)$ |
| $P8$ | $x \parallel (y + z) = (x \parallel y) + (x \parallel z)$ |
| $P9$ | $\delta \parallel x = \delta$ |
| $P10$ | $x \parallel \delta = \delta$ |
| $C11$ | $e_1 \mid e_2 = \gamma(e_1, e_2)$ |
| $RC11$ | $e_1[m] \mid e_2[m] = \gamma(e_1, e_2)[m]$ |
| $C12$ | $e_1 \mid (e_2 \cdot y) = \gamma(e_1, e_2) \cdot y$ |
| $RC12$ | $e_1[m] \mid (y \cdot e_2[m]) = y \cdot \gamma(e_1, e_2)[m]$ |
| $C13$ | $(e_1 \cdot x) \mid e_2 = \gamma(e_1, e_2) \cdot x$ |
| $RC13$ | $(x \cdot e_1[m]) \mid e_2[m] = x \cdot \gamma(e_1, e_2)[m]$ |
| $C14$ | $(e_1 \cdot x) \mid (e_2 \cdot y) = \gamma(e_1, e_2) \cdot (x \between y)$ |
| $RC14$ | $(x \cdot e_1[m]) \mid (y \cdot e_2[m]) = (x \between y) \cdot \gamma(e_1, e_2)[m]$ |
| $C15$ | $(x + y) \mid z = (x \mid z) + (y \mid z)$ |
| $C16$ | $x \mid (y + z) = (x \mid y) + (x \mid z)$ |
| $C17$ | $\delta \mid x = \delta$ |
| $C18$ | $x \mid \delta = \delta$ |
| $CE19$ | $\Theta(e) = e$ |
| $RCE19$ | $\Theta(e[m]) = e[m]$ |
| $CE20$ | $\Theta(\delta) = \delta$ |
| $CE21$ | $\Theta(x + y) = \Theta(x) + \Theta(y)$ |
| $CE22$ | $\Theta(x \cdot y) = \Theta(x) \cdot \Theta(y)$ |
| $CE23$ | $\Theta(x \parallel y) = ((\Theta(x) \triangleleft y) \parallel y) + ((\Theta(y) \triangleleft x) \parallel x)$ |
| $CE24$ | $\Theta(x \mid y) = ((\Theta(x) \triangleleft y) \mid y) + ((\Theta(y) \triangleleft x) \mid x)$ |
| $U25$ | $(\sharp(e_1, e_2)) \quad e_1 \triangleleft e_2 = \tau$ |
| $RU25$ | $(\sharp(e_1[m], e_2[n])) \quad e_1[m] \triangleleft e_2[n] = \tau$ |
| $U26$ | $(\sharp(e_1, e_2), e_2 \leq e_3) \quad e_1 \triangleleft e_3 = \tau$ |
| $RU26$ | $(\sharp(e_1[m], e_2[n]), e_2[n] \geq e_3[l]) \quad e_1[m] \triangleleft e_3[l] = \tau$ |
| $U27$ | $(\sharp(e_1, e_2), e_2 \leq e_3) \quad e3 \triangleleft e_1 = \tau$ |
| $RU27$ | $(\sharp(e_1[m], e_2[n]), e_2[n] \geq e_3[l]) \quad e3[l] \triangleleft e_1[m] = \tau$ |
| $U28$ | $e \triangleleft \delta = e$ |
| $U29$ | $\delta \triangleleft e = \delta$ |

*continued on next page*

**TABLE 4.15** *(continued)*

| No. | Axiom |
|---|---|
| $U30$ | $(x+y)\triangleleft z=(x\triangleleft z)+(y\triangleleft z)$ |
| $U31$ | $(x\cdot y)\triangleleft z=(x\triangleleft z)\cdot(y\triangleleft z)$ |
| $U32$ | $(x\parallel y)\triangleleft z=(x\triangleleft z)\parallel(y\triangleleft z)$ |
| $U33$ | $(x\mid y)\triangleleft z=(x\triangleleft z)\mid(y\triangleleft z)$ |
| $U34$ | $x\triangleleft(y+z)=(x\triangleleft y)\triangleleft z$ |
| $U35$ | $x\triangleleft(y\cdot z)=(x\triangleleft y)\triangleleft z$ |
| $U36$ | $x\triangleleft(y\parallel z)=(x\triangleleft y)\triangleleft z$ |
| $U37$ | $x\triangleleft(y\mid z)=(x\triangleleft y)\triangleleft z$ |

*Proof.* It follows from the following two facts (see Theorem 2.5):

1. The transition rules of $BRATC$ in Section 4.1 are all source dependent;
2. The sources of the transition rules for the algebra for parallelism contain an occurrence of $\between$, or $\parallel$, or $\mid$, or $\Theta$, or $\triangleleft$;
3. The transition rules of $RAPTC$ are all source dependent.

Hence, the reversible algebra for parallelism is a generalization of $BRATC$, as desired. □

**Theorem 4.16** (Soundness of parallelism modulo FR step bisimulation equivalence). *Let $x$ and $y$ be $RAPTC$ terms. If $RAPTC\vdash x=y$, then $x\sim_s^{fr} y$.*

*Proof.* Since FR step bisimulation $\sim_s^{fr}$ is both an equivalent and a congruent relation with respect to the operators $\between$, $\parallel$, $\mid$, $\Theta$, and $\triangleleft$, we only need to check if each axiom in Table 4.15 is sound modulo FR step bisimulation equivalence.

Although transition rules in Tables 4.9, 4.11, 4.13, 4.10, 4.12, and 4.14 are defined in the flavor of single event, they can be modified into a step (a set of events within which each event is pairwise concurrent), hence, we omit them here. If we treat a single event as a step containing just one event, the proof of this soundness theorem does not pose any problem, hence, we use this way and still use the transition rules in Tables 4.9, 4.11, 4.13, 4.10, 4.12, and 4.14.

We omit the defining axioms, and the trivial axioms related to $\delta$, in the following, we only prove the soundness of the non-trivial axioms, including axioms $P2-P8$, $C12-C16$, $CE21-CE24$, and $U30-U37$.

- **Axiom** $P2$. Let $p,q$ be $RAPTC$ processes, and $p\parallel q=q\parallel p$, it is sufficient to prove that $p\parallel q\sim_s^{fr} q\parallel p$. By the forward transition rules for operator $\parallel$ in Table 4.9, we obtain

$$\frac{p\xrightarrow{e_1}e_1[m]\quad q\xrightarrow{e_2}e_2[m]}{p\parallel q\xrightarrow{\{e_1,e_2\}}e_1[m]\parallel e_2[m]}\qquad\frac{p\xrightarrow{e_1}e_1[m]\quad q\xrightarrow{e_2}e_2[m]}{q\parallel p\xrightarrow{\{e_1,e_2\}}e_2[m]\parallel e_1[m]}$$

$$\frac{p\xrightarrow{e_1}p'\quad q\xrightarrow{e_2}e_2[m]}{p\parallel q\xrightarrow{\{e_1,e_2\}}p'\parallel e_2[m]}\qquad\frac{p\xrightarrow{e_1}p'\quad q\xrightarrow{e_2}e_2[m]}{q\parallel p\xrightarrow{\{e_1,e_2\}}e_2[m]\parallel p'}$$

$$\frac{p\xrightarrow{e_1}e_1[m]\quad q\xrightarrow{e_2}q'}{p\parallel q\xrightarrow{\{e_1,e_2\}}e_1[m]\parallel q'}\qquad\frac{p\xrightarrow{e_1}e_1[m]\quad q\xrightarrow{e_2}q'}{q\parallel p\xrightarrow{\{e_1,e_2\}}q'\parallel e_1[m]}$$

$$\frac{p\xrightarrow{e_1}p'\quad q\xrightarrow{e_2}q'}{p\parallel q\xrightarrow{\{e_1,e_2\}}p'\between q'}\qquad\frac{p\xrightarrow{e_1}p'\quad q\xrightarrow{e_2}q'}{q\parallel p\xrightarrow{\{e_1,e_2\}}q'\between p'}$$

By the reverse transition rules for operator $\parallel$ in Table 4.10, we obtain

$$\frac{p\overset{e_1[m]}{\twoheadrightarrow}e_1\quad q\overset{e_2[m]}{\twoheadrightarrow}e_2}{p\parallel q\overset{\{e_1[m],e_2[m]\}}{\twoheadrightarrow}e_1\parallel e_2}\qquad\frac{p\overset{e_1[m]}{\twoheadrightarrow}e_1\quad q\overset{e_2[m]}{\twoheadrightarrow}e_2}{q\parallel p\overset{\{e_1[m],e_2[m]\}}{\twoheadrightarrow}e_2\parallel e_1}$$

$$\frac{p\overset{e_1[m]}{\twoheadrightarrow}p'\quad q\overset{e_2[m]}{\twoheadrightarrow}e_2}{p\parallel q\overset{\{e_1[m],e_2[m]\}}{\twoheadrightarrow}p'\parallel e_2}\qquad\frac{p\overset{e_1[m]}{\twoheadrightarrow}p'\quad q\overset{e_2[m]}{\twoheadrightarrow}e_2}{q\parallel p\overset{\{e_1[m],e_2[m]\}}{\twoheadrightarrow}e_2\parallel p'}$$

**TABLE 4.16** Term rewrite system of $RAPTC$.

| No. | Rewriting Rule |
|---|---|
| $RRA6$ | $x + \delta \rightarrow x$ |
| $RRA7$ | $\delta \cdot x \rightarrow \delta$ |
| $RRRA7$ | $x \cdot \delta \rightarrow \delta$ |
| $RRP1$ | $x \between y \rightarrow x \parallel y + x \mid y$ |
| $RRP2$ | $x \parallel y \rightarrow y \parallel x$ |
| $RRP3$ | $(x \parallel y) \parallel z \rightarrow x \parallel (y \parallel z)$ |
| $RRP4$ | $e_1 \parallel (e_2 \cdot y) \rightarrow (e_1 \parallel e_2) \cdot y$ |
| $RRRP4$ | $e_1[m] \parallel (y \cdot e_2[m]) \rightarrow y \cdot (e_1[m] \parallel e_2[m])$ |
| $RRP5$ | $(e_1 \cdot x) \parallel e_2 \rightarrow (e_1 \parallel e_2) \cdot x$ |
| $RRRP5$ | $(x \cdot e_1[m]) \parallel e_2[m] \rightarrow x \cdot (e_1[m] \parallel e_2[m])$ |
| $RRP6$ | $(e_1 \cdot x) \parallel (e_2 \cdot y) \rightarrow (e_1 \parallel e_2) \cdot (x \between y)$ |
| $RRRP6$ | $(x \cdot e_1[m]) \parallel (y \cdot e_2[m]) \rightarrow (x \between y) \cdot (e_1[m] \parallel e_2[m])$ |
| $RRP7$ | $(x + y) \parallel z \rightarrow (x \parallel z) + (y \parallel z)$ |
| $RRP8$ | $x \parallel (y + z) \rightarrow (x \parallel y) + (x \parallel z)$ |
| $RRP9$ | $\delta \parallel x \rightarrow \delta$ |
| $RRP10$ | $x \parallel \delta \rightarrow \delta$ |
| $RRC11$ | $e_1 \mid e_2 \rightarrow \gamma(e_1, e_2)$ |
| $RRRC11$ | $e_1[m] \mid e_2[m] \rightarrow \gamma(e_1, e_2)[m]$ |
| $RRC12$ | $e_1 \mid (e_2 \cdot y) \rightarrow \gamma(e_1, e_2) \cdot y$ |
| $RRRC12$ | $e_1[m] \mid (y \cdot e_2[m]) \rightarrow y \cdot \gamma(e_1, e_2)[m]$ |
| $RRC13$ | $(e_1 \cdot x) \mid e_2 \rightarrow \gamma(e_1, e_2) \cdot x$ |
| $RRRC13$ | $(x \cdot e_1[m]) \mid e_2[m] \rightarrow x \cdot \gamma(e_1, e_2)[m]$ |
| $RRC14$ | $(e_1 \cdot x) \mid (e_2 \cdot y) \rightarrow \gamma(e_1, e_2) \cdot (x \between y)$ |
| $RRRC14$ | $(x \cdot e_1[m]) \mid (y \cdot e_2[m]) \rightarrow (x \between y) \cdot \gamma(e_1, e_2)[m]$ |
| $RRC15$ | $(x + y) \mid z \rightarrow (x \mid z) + (y \mid z)$ |
| $RRC16$ | $x \mid (y + z) \rightarrow (x \mid y) + (x \mid z)$ |
| $RRC17$ | $\delta \mid x \rightarrow \delta$ |
| $RRC18$ | $x \mid \delta \rightarrow \delta$ |
| $RRCE19$ | $\Theta(e) \rightarrow e$ |
| $RRRCE19$ | $\Theta(e[m]) \rightarrow e[m]$ |
| $RRCE20$ | $\Theta(\delta) \rightarrow \delta$ |
| $RRCE21$ | $\Theta(x + y) \rightarrow \Theta(x) + \Theta(y)$ |
| $RRCE22$ | $\Theta(x \cdot y) \rightarrow \Theta(x) \cdot \Theta(y)$ |
| $RRCE23$ | $\Theta(x \parallel y) \rightarrow ((\Theta(x) \triangleleft y) \parallel y) + ((\Theta(y) \triangleleft x) \parallel x)$ |
| $RRCE24$ | $\Theta(x \mid y) \rightarrow ((\Theta(x) \triangleleft y) \mid y) + ((\Theta(y) \triangleleft x) \mid x)$ |
| $RRU25$ | $(\sharp(e_1, e_2)) \quad e_1 \triangleleft e_2 \rightarrow \tau$ |
| $RRRU25$ | $(\sharp(e_1[m], e_2[n])) \quad e_1[m] \triangleleft e_2[n] \rightarrow \tau$ |
| $RRU26$ | $(\sharp(e_1, e_2), e_2 \leq e_3) \quad e_1 \triangleleft e_3 \rightarrow \tau$ |
| $RRRU26$ | $(\sharp(e_1[m], e_2[n]), e_2[n] \geq e_3[l]) \quad e_1[m] \triangleleft e_3[l] \rightarrow \tau$ |
| $RRU27$ | $(\sharp(e_1, e_2), e_2 \leq e_3) \quad e3 \triangleleft e_1 \rightarrow \tau$ |
| $RRRU27$ | $(\sharp(e_1[m], e_2[n]), e_2[n] \geq e_3[l]) \quad e3[l] \triangleleft e_1[m] \rightarrow \tau$ |
| $RRU28$ | $e \triangleleft \delta \rightarrow e$ |
| $RRU29$ | $\delta \triangleleft e \rightarrow \delta$ |

continued on next page

**TABLE 4.16** *(continued)*

| No. | Rewriting Rule |
|---|---|
| $RRU30$ | $(x+y)\triangleleft z \rightarrow (x\triangleleft z)+(y\triangleleft z)$ |
| $RRU31$ | $(x\cdot y)\triangleleft z \rightarrow (x\triangleleft z)\cdot(y\triangleleft z)$ |
| $RRU32$ | $(x\parallel y)\triangleleft z \rightarrow (x\triangleleft z)\parallel(y\triangleleft z)$ |
| $RRU33$ | $(x\mid y)\triangleleft z \rightarrow (x\triangleleft z)\mid(y\triangleleft z)$ |
| $RRU34$ | $x\triangleleft(y+z) \rightarrow (x\triangleleft y)\triangleleft z$ |
| $RRU35$ | $x\triangleleft(y\cdot z) \rightarrow (x\triangleleft y)\triangleleft z$ |
| $RRU36$ | $x\triangleleft(y\parallel z) \rightarrow (x\triangleleft y)\triangleleft z$ |
| $RRU37$ | $x\triangleleft(y\mid z) \rightarrow (x\triangleleft y)\triangleleft z$ |

$$\frac{p\xrightarrow{e_1[m]}\!\!\twoheadrightarrow e_1\quad q\xrightarrow{e_2[m]}\!\!\twoheadrightarrow q'}{p\parallel q\xrightarrow{\{e_1[m],e_2[m]\}}\!\!\twoheadrightarrow e_1\parallel q'}\qquad\frac{p\xrightarrow{e_1[m]}\!\!\twoheadrightarrow e_1\quad q\xrightarrow{e_2[m]}\!\!\twoheadrightarrow q'}{q\parallel p\xrightarrow{\{e_1[m],e_2[m]\}}\!\!\twoheadrightarrow q'\parallel e_1}$$

$$\frac{p\xrightarrow{e_1[m]}\!\!\twoheadrightarrow p'\quad q\xrightarrow{e_2[m]}\!\!\twoheadrightarrow q'}{p\parallel q\xrightarrow{\{e_1[m],e_2[m]\}}\!\!\twoheadrightarrow p'\between q'}\qquad\frac{p\xrightarrow{e_1[m]}\!\!\twoheadrightarrow p'\quad q\xrightarrow{e_2[m]}\!\!\twoheadrightarrow q'}{q\parallel p\xrightarrow{\{e_1[m],e_2[m]\}}\!\!\twoheadrightarrow q'\between p'}$$

Hence, with the assumption $e_1[m]\parallel e_2[m]\sim_s^{fr} e_2[m]\parallel e_1[m]$, $e_1\parallel e_2\sim_s^{fr} e_2\parallel e_1$, $p'\parallel e_2[m]\sim_s^{fr} e_2[m]\parallel p'$, $p'\parallel e_2\sim_s^{fr} e_2\parallel p'$, $e_1[m]\parallel q'\sim_s^{fr} q'\parallel e_1[m]$, $e_1\parallel q'\sim_s^{fr} q'\parallel e_1$, $p'\between q'\sim_s^{fr} q'\between p'$, $p\parallel q\sim_s^{fr} q\parallel p$, as desired.

- **Axiom** $P3$. Let $p,q,r$ be $RAPTC$ processes, and $(p\parallel q)\parallel r=p\parallel(q\parallel r)$, it is sufficient to prove that $(p\parallel q)\parallel r\sim_s^{fr} p\parallel(q\parallel r)$. By the forward transition rules for operator $\parallel$ in Table 4.9, we obtain

$$\frac{p\xrightarrow{e_1}e_1[m]\quad q\xrightarrow{e_2}e_2[m]\quad r\xrightarrow{e_3}e_3[m]}{(p\parallel q)\parallel r\xrightarrow{\{e_1,e_2,e_3\}}(e_1[m]\parallel e_2[m])\parallel e_3[m]}\qquad\frac{p\xrightarrow{e_1}e_1[m]\quad q\xrightarrow{e_2}e_2[m]\quad r\xrightarrow{e_3}e_3[m]}{p\parallel(q\parallel r)\xrightarrow{\{e_1,e_2,e_3\}}e_1[m]\parallel(e_2[m]\parallel e_3[m])}$$

$$\frac{p\xrightarrow{e_1}p'\quad q\xrightarrow{e_2}e_2[m]\quad r\xrightarrow{e_3}e_3[m]}{(p\parallel q)\parallel r\xrightarrow{\{e_1,e_2,e_3\}}(p'\parallel e_2[m])\parallel e_3[m]}\qquad\frac{p\xrightarrow{e_1}p'\quad q\xrightarrow{e_2}e_2[m]\quad r\xrightarrow{e_3}e_3[m]}{p\parallel(q\parallel r)\xrightarrow{\{e_1,e_2,e_3\}}p'\parallel(e_2[m]\parallel e_3[m])}$$

There are also two other cases, however, we omit them here.

$$\frac{p\xrightarrow{e_1}p'\quad q\xrightarrow{e_2}q'\quad r\xrightarrow{e_3}e_3[m]}{(p\parallel q)\parallel r\xrightarrow{\{e_1,e_2,e_3\}}(p'\between q')\between e_3[m]}\qquad\frac{p\xrightarrow{e_1}p'\quad q\xrightarrow{e_2}q'\quad r\xrightarrow{e_3}e_3[m]}{p\parallel(q\parallel r)\xrightarrow{\{e_1,e_2,e_3\}}p'\between(q'\between e_3[m])}$$

There are also other cases, however, we also omit them here.

$$\frac{p\xrightarrow{e_1}p'\quad q\xrightarrow{e_2}q'\quad r\xrightarrow{e_3}r'}{(p\parallel q)\parallel r'\xrightarrow{\{e_1,e_2,e_3\}}(p'\between q')\between r'}\qquad\frac{p\xrightarrow{e_1}p'\quad q\xrightarrow{e_2}q'\quad r\xrightarrow{e_3}r'}{p\parallel(q\parallel r)\xrightarrow{\{e_1,e_2,e_3\}}p'\between(q'\between r')}$$

By the reverse transition rules for operator $\parallel$ in Table 4.10, we obtain

$$\frac{p\xrightarrow{e_1[m]}\!\!\twoheadrightarrow e_1\quad q\xrightarrow{e_2[m]}\!\!\twoheadrightarrow e_2\quad r\xrightarrow{e_3[m]}\!\!\twoheadrightarrow e_3}{(p\parallel q)\parallel r\xrightarrow{\{e_1[m],e_2[m],e_3[m]\}}\!\!\twoheadrightarrow(e_1\parallel e_2)\parallel e_3}\qquad\frac{p\xrightarrow{e_1[m]}\!\!\twoheadrightarrow e_1\quad q\xrightarrow{e_2[m]}\!\!\twoheadrightarrow e_2\quad r\xrightarrow{e_3[m]}\!\!\twoheadrightarrow e_3}{p\parallel(q\parallel r)\xrightarrow{\{e_1[m],e_2[m],e_3[m]\}}\!\!\twoheadrightarrow e_1\parallel(e_2\parallel e_3)}$$

$$\frac{p\xrightarrow{e_1[m]}\!\!\twoheadrightarrow p'\quad q\xrightarrow{e_2[m]}\!\!\twoheadrightarrow e_2\quad r\xrightarrow{e_3[m]}\!\!\twoheadrightarrow e_3}{(p\parallel q)\parallel r\xrightarrow{\{e_1[m],e_2[m],e_3[m]\}}\!\!\twoheadrightarrow(p'\parallel e_2)\parallel e_3}\qquad\frac{p\xrightarrow{e_1[m]}\!\!\twoheadrightarrow p'\quad q\xrightarrow{e_2[m]}\!\!\twoheadrightarrow e_2\quad r\xrightarrow{e_3[m]}\!\!\twoheadrightarrow e_3}{p\parallel(q\parallel r)\xrightarrow{\{e_1[m],e_2[m],e_3[m]\}}\!\!\twoheadrightarrow p'\parallel(e_2\parallel e_3)}$$

There are also two other cases, however, we omit them here.

$$\frac{p \xrightarrow{e_1[m]} p' \quad q \xrightarrow{e_2[m]} q' \quad r \xrightarrow{e_3[m]} e_3}{(p \parallel q) \parallel r \xrightarrow{\{e_1[m],e_2[m],e_3[m]\}} (p' \between q') \between e_3} \quad \frac{p \xrightarrow{e_1[m]} p' \quad q \xrightarrow{e_2[m]} q' \quad r \xrightarrow{e_3[m]} e_3}{p \parallel (q \parallel r) \xrightarrow{\{e_1[m],e_2[m],e_3[m]\}} p' \between (q' \between e_3)}$$

There are also two other cases, however, we omit them here.

$$\frac{p \xrightarrow{e_1[m]} p' \quad q \xrightarrow{e_2[m]} q' \quad r \xrightarrow{e_3[m]} r'}{(p \parallel q) \parallel r' \xrightarrow{\{e_1[m],e_2[m],e_3[m]\}} (p' \between q') \between r'} \quad \frac{p \xrightarrow{e_1[m]} p' \quad q \xrightarrow{e_2[m]} q' \quad r \xrightarrow{e_3[m]} r'}{p \parallel (q \parallel r) \xrightarrow{\{e_1[m],e_2[m],e_3[m]\}} p' \between (q' \between r')}$$

Hence, with the assumption $(e_1[m] \parallel e_2[m]) \parallel e_3[m] \sim_s^{fr} e_1[m] \parallel (e_2[m] \parallel e_3[m])$, $(e_1 \parallel e_2) \parallel e_3 \sim_s^{fr} e_1 \parallel (e_2 \parallel e_3)$, $(p' \parallel e_2[m]) \parallel e_3[m] \sim_s^{fr} p' \parallel (e_2[m] \parallel e_3[m])$, $(p' \parallel e_2) \parallel e_3 \sim_s^{fr} p' \parallel (e_2 \parallel e_3)$, $(p' \between q') \between e_3[m] \sim_s^{fr} p' \between (q' \between e_3[m])$, $(p' \between q') \between e_3 \sim_s^{fr} p' \between (q' \between e_3)$, $(p' \between q') \between r' \sim_s^{fr} p' \between (q' \between r')$, $(p \parallel q) \parallel r \sim_s^{fr} p \parallel (q \parallel r)$, as desired.

- **Axiom** $P4$. Let $q$ be an $RAPTC$ process, and $e_1 \parallel (e_2 \cdot q) = (e_1 \parallel e_2) \cdot q$, it is sufficient to prove that $e_1 \parallel (e_2 \cdot q) \sim_s^{fr} (e_1 \parallel e_2) \cdot q$. By the forward transition rules for operator $\parallel$ in Table 4.9, we obtain

$$\frac{e_1 \xrightarrow{e_1} e_1[m] \quad e_2 \xrightarrow{e_2} e_2[m]}{e_1 \parallel (e_2 \cdot q) \xrightarrow{\{e_1,e_2\}} e_1[m] \parallel (e_2[m] \cdot q)}$$

$$\frac{e_1 \xrightarrow{e_1} e_1[m] \quad e_2 \xrightarrow{e_2} e_2[m]}{(e_1 \parallel e_2) \cdot q \xrightarrow{\{e_1,e_2\}} (e_1[m] \parallel e_2[m]) \cdot q}$$

By the reverse transition rules for operator $\parallel$ in Table 4.10, there are no transitions.
Hence, with the assumption $e_1[m] \parallel (e_2[m] \cdot q) \sim_s^{fr} (e_1[m] \parallel e_2[m]) \cdot q$, $e_1 \parallel (e_2 \cdot q) \sim_s^{fr} (e_1 \parallel e_2) \cdot q$, as desired.

- **Axiom** $RP4$. Let $q$ be an $RAPTC$ process, and $e_1[m] \parallel (q \cdot e_2[m]) = q \cdot (e_1[m] \parallel e_2[m])$, it is sufficient to prove that $e_1[m] \parallel (q \cdot e_2[m]) \sim_s^{fr} q \cdot (e_1[m] \parallel e_2[m])$. By the forward transition rules for operator $\parallel$ in Table 4.9, there are no transitions.
By the reverse transition rules for operator $\parallel$ in Table 4.10, we obtain

$$\frac{e_1[m] \xrightarrow{e_1[m]} e_1 \quad e_2[m] \xrightarrow{e_2[m]} e_2}{e_1[m] \parallel (q \cdot e_2[m]) \xrightarrow{\{e_1[m],e_2[m]\}} e_1 \parallel (q \cdot e_2)}$$

$$\frac{e_1[m] \xrightarrow{e_1[m]} e_1 \quad e_2[m] \xrightarrow{e_2[m]} e_2}{q \cdot (e_1[m] \parallel e_2[m]) \xrightarrow{\{e_1[m],e_2[m]\}} q \cdot (e_1 \parallel e_2)}$$

Hence, with the assumption $e_1 \parallel (q \cdot e_2) \sim_s^{fr} q \cdot (e_1 \parallel e_2)$, $e_1[m] \parallel (q \cdot e_2[m]) \sim_s^{fr} q \cdot (e_1[m] \parallel e_2[m])$, as desired.

- **Axiom** $P5$. Let $p$ be an $RAPTC$ process, and $(e_1 \cdot p) \parallel e_2 = (e_1 \parallel e_2) \cdot p$, it is sufficient to prove that $(e_1 \cdot p) \parallel e_2 \sim_s^{fr} (e_1 \parallel e_2) \cdot p$. By the forward transition rules for operator $\parallel$ in Table 4.9, we obtain

$$\frac{e_1 \xrightarrow{e_1} e_1[m] \quad e_2 \xrightarrow{e_2} e_2[m]}{(e_1 \cdot p) \parallel e_2 \xrightarrow{\{e_1,e_2\}} (e_1[m] \cdot p) \parallel e_2[m]}$$

$$\frac{e_1 \xrightarrow{e_1} e_1[m] \quad e_2 \xrightarrow{e_2} e_2[m]}{(e_1 \parallel e_2) \cdot p \xrightarrow{\{e_1,e_2\}} (e_1[m] \parallel e_2[m]) \cdot p}$$

By the reverse transition rules for operator $\parallel$ in Table 4.10, there are no transitions.
Hence, with the assumption $(e_1[m] \cdot p) \parallel e_2[m] \sim_s^{fr} (e_1[m] \parallel e_2[m]) \cdot p$, $(e_1 \cdot p) \parallel e_2 \sim_s^{fr} (e_1 \parallel e_2) \cdot p$, as desired.

- **Axiom** $RP5$. Let $p$ be an $RAPTC$ process, and $(p \cdot e_1[m]) \parallel e_2[m] = p \cdot (e_1[m] \parallel e_2[m])$, it is sufficient to prove that $(p \cdot e_1[m]) \parallel e_2[m] \sim_s^{fr} p \cdot (e_1[m] \parallel e_2[m])$. By the forward transition rules for operator $\parallel$ in Table 4.9, there are no transitions.

By the reverse transition rules for operator $\parallel$ in Table 4.10, we obtain

$$\frac{e_1[m]\xrightarrow{e_1[m]}\!\!\!\!\twoheadrightarrow e_1\quad e_2[m]\xrightarrow{e_2[m]}\!\!\!\!\twoheadrightarrow e_2}{(p\cdot e_1[m])\parallel e_2[m]\xrightarrow{\{e_1[m],e_2[m]\}}\!\!\!\!\twoheadrightarrow (p\cdot e_1)\parallel e_2}$$

$$\frac{e_1[m]\xrightarrow{e_1[m]}\!\!\!\!\twoheadrightarrow e_1\quad e_2[m]\xrightarrow{e_2[m]}\!\!\!\!\twoheadrightarrow e_2}{p\cdot(e_1[m]\parallel e_2[m])\xrightarrow{\{e_1[m],e_2[m]\}}\!\!\!\!\twoheadrightarrow p\cdot(e_1\parallel e_2)}$$

Hence, with the assumptions $(p\cdot e_1)\parallel e_2\sim_s^{fr} p\cdot(e_1\parallel e_2)$, $(p\cdot e_1[m])\parallel e_2[m]\sim_s^{fr} p\cdot(e_1[m]\parallel e_2[m])$, as desired.

- **Axiom** $P6$. Let $p,q$ be $RAPTC$ processes, and $(e_1\cdot p)\parallel(e_2\cdot q)=(e_1\parallel e_2)\cdot(p\between q)$, it is sufficient to prove that $(e_1\cdot p)\parallel(e_2\cdot q)\sim_s^{fr}(e_1\parallel e_2)\cdot(p\between q)$. By the forward transition rules for operator $\parallel$ in Table 4.9, we obtain

$$\frac{e_1\xrightarrow{e_1}e_1[m]\quad e_2\xrightarrow{e_2}e_2[m]}{(e_1\cdot p)\parallel(e_2\cdot q)\xrightarrow{\{e_1,e_2\}}(e_1[m]\cdot p)\parallel(e_2[m]\cdot q)}$$

$$\frac{e_1\xrightarrow{e_1}e_1[m]\quad e_2\xrightarrow{e_2}e_2[m]}{(e_1\parallel e_2)\cdot(p\between q)\xrightarrow{\{e_1,e_2\}}(e_1[m]\parallel e_2[m])\cdot(p\between q)}$$

By the reverse transition rules for operator $\parallel$ in Table 4.10, there are no transitions.
Hence, with the assumptions $(e_1[m]\cdot p)\parallel(e_1[m]\cdot q)\sim_s^{fr}(e_1[m]\parallel e_2[m])\cdot(p\between q)$, $(e_1\cdot p)\parallel(e_2\cdot q)\sim_s^{fr}(e_1\parallel e_2)\cdot(p\between q)$, as desired.

- **Axiom** $RP6$. Let $p,q$ be $RAPTC$ processes, and $(p\cdot e_1[m])\parallel(q\cdot e_2[m])=(p\between q)\cdot(e_1[m]\parallel e_2[m])$, it is sufficient to prove that $(p\cdot e_1[m])\parallel(q\cdot e_2[m])\sim_s^{fr}(p\between q)\cdot(e_1[m]\parallel e_2[m])$. By the forward transition rules for operator $\parallel$ in Table 4.9, there are no transitions.
By the reverse transition rules for operator $\parallel$ in Table 4.10, we obtain

$$\frac{e_1[m]\xrightarrow{e_1[m]}\!\!\!\!\twoheadrightarrow e_1\quad e_2[m]\xrightarrow{e_2[m]}\!\!\!\!\twoheadrightarrow e_2}{(p\cdot e_1[m])\parallel(q\cdot e_2[m])\xrightarrow{\{e_1[m],e_2[m]\}}\!\!\!\!\twoheadrightarrow (p\cdot e_1)\parallel(q\cdot e_2)}$$

$$\frac{e_1\xrightarrow{e_1[m]}\!\!\!\!\twoheadrightarrow e_1\quad e_2\xrightarrow{e_2[m]}\!\!\!\!\twoheadrightarrow e_2}{(p\between q)\cdot(e_1[m]\parallel e_2[m])\xrightarrow{\{e_1[m],e_2[m]\}}\!\!\!\!\twoheadrightarrow (p\between q)\cdot(e_1\parallel e_2)}$$

Hence, with the assumptions $(p\cdot e_1)\parallel(q\cdot e_2)\sim_s^{fr}(p\between q)\cdot(e_1\parallel e_2)$, $(p\cdot e_1[m])\parallel(q\cdot e_2[m])\sim_s^{fr}(p\between q)\cdot(e_1[m]\parallel e_2[m])$, as desired.

- **Axiom** $P7$. Let $p,q,r$ be $RAPTC$ processes, and $(p+q)\parallel r=(p\parallel r)+(q\parallel r)$, it is sufficient to prove that $(p+q)\parallel r\sim_s^{fr}(p\parallel r)+(q\parallel r)$. There are several cases, however, we will not enumerate them all here. By the forward transition rules for operators $+$ and $\parallel$ in Tables 4.3 and 4.9, we obtain

$$\frac{p\xrightarrow{e_1}e_1[m]\quad q\xrightarrow{e_1}e_1[m]\quad r\xrightarrow{e_2}e_2[m]}{(p+q)\parallel r\xrightarrow{\{e_1,e_2\}}(e_1[m]+e_1[m])\parallel e_2[m]}$$

$$\frac{p\xrightarrow{e_1}e_1[m]\quad q\xrightarrow{e_1}e_1[m]\quad r\xrightarrow{e_2}e_2[m]}{(p\parallel r)+(q\parallel r)\xrightarrow{\{e_1,e_2\}}(e_1[m]\parallel e_2[m])+(e_1[m]\parallel e_2[m])}$$

$$\frac{p\xrightarrow{e_1}p'\quad q\xrightarrow{e_1}q'\quad r\xrightarrow{e_2}r'}{(p+q)\parallel r\xrightarrow{\{e_1,e_2\}}(p'+q')\parallel r'}\qquad\frac{p\xrightarrow{e_1}p'\quad q\xrightarrow{e_1}q'\quad r\xrightarrow{e_2}r'}{(p\parallel r)+(q\parallel r)\xrightarrow{\{e_1,e_2\}}(p'\parallel r')+(q'\parallel r')}$$

By the reverse transition rules for operators $+$ and $\parallel$ in Tables 4.4 and 4.10, we obtain

$$\frac{p\xrightarrow{e_1[m]}\!\!\!\!\rightarrow e_1\quad q\xrightarrow{e_1[m]}\!\!\!\!\rightarrow e_1\quad r\xrightarrow{e_2[m]}\!\!\!\!\rightarrow e_2}{(p+q)\parallel r\xrightarrow{\{e_1[m],e_2[m]\}}\!\!\!\!\rightarrow (e_1+e_1)\parallel e_2}\quad \frac{p\xrightarrow{e_1[m]}\!\!\!\!\rightarrow e_1\quad q\xrightarrow{e_1[m]}\!\!\!\!\rightarrow e_1\quad r\xrightarrow{e_2[m]}\!\!\!\!\rightarrow e_2}{(p\parallel r)+(q\parallel r)\xrightarrow{\{e_1[m],e_2[m]\}}\!\!\!\!\rightarrow (e_1\parallel e_2)+(e_1\parallel e_2)}$$

$$\frac{p\xrightarrow{e_1[m]}\!\!\!\!\rightarrow p'\quad q\xrightarrow{e_1[m]}\!\!\!\!\rightarrow q'\quad r\xrightarrow{e_2[m]}\!\!\!\!\rightarrow r'}{(p+q)\parallel r\xrightarrow{\{e_1[m],e_2[m]\}}\!\!\!\!\rightarrow (p'+q')\parallel r'}\quad \frac{p\xrightarrow{e_1[m]}\!\!\!\!\rightarrow p'\quad q\xrightarrow{e_1[m]}\!\!\!\!\rightarrow q'\quad r\xrightarrow{e_2[m]}\!\!\!\!\rightarrow r'}{(p\parallel r)+(q\parallel r)\xrightarrow{\{e_1[m],e_2[m]\}}\!\!\!\!\rightarrow (p'\parallel r')+(q'\parallel r')}$$

Hence, with the assumptions $(e_1[m]+e_1[m])\parallel e_2[m]\sim_s^{fr}(e_1[m]\parallel e_2[m])+(e_1[m]\parallel e_2[m])$, $(e_1+e_1)\parallel e_2\sim_s^{fr}(e_1\parallel e_2)+(e_1\parallel e_2)$, $(p'+q')\parallel r'\sim_s^{fr}(p'\parallel r')+(q'\parallel r')$, $(p+q)\parallel r\sim_s^{fr}(p\parallel r)+(q\parallel r)$, as desired.

- **Axiom** $P8$. Let $p,q,r$ be $RAPTC$ processes, and $p\parallel(q+r)=(p\parallel q)+(p\parallel r)$, it is sufficient to prove that $p\parallel(q+r)\sim_s^{fr}(p\parallel q)+(p\parallel r)$. There are several cases, however, we will not enumerate them all here. By the forward transition rules for operators $+$ and $\parallel$ in Tables 4.3 and 4.9, we obtain

$$\frac{p\xrightarrow{e_1}e_1[m]\quad q\xrightarrow{e_2}e_2[m]\quad r\xrightarrow{e_2}e_2[m]}{p\parallel(q+r)\xrightarrow{\{e_1,e_2\}}e_1[m]\parallel(e_2[m]+e_2[m])}$$

$$\frac{p\xrightarrow{e_1}e_1[m]\quad q\xrightarrow{e_2}e_2[m]\quad r\xrightarrow{e_2}e_2[m]}{(p\parallel q)+(p\parallel r)\xrightarrow{\{e_1,e_2\}}(e_1[m]\parallel e_2[m])+(e_1[m]+e_2[m])}$$

$$\frac{p\xrightarrow{e_1}p'\quad q\xrightarrow{e_2}q'\quad r\xrightarrow{e_2}r'}{p\parallel(q+r)\xrightarrow{\{e_1,e_2\}}p'\parallel(q'+r')}\quad \frac{p\xrightarrow{e_1}p'\quad q\xrightarrow{e_2}q'\quad r\xrightarrow{e_2}r'}{(p\parallel q)+(p\parallel r)\xrightarrow{\{e_1,e_2\}}(p'\parallel q')+(p'+r')}$$

By the reverse transition rules for operators $+$ and $\parallel$ in Tables 4.4 and 4.10, we obtain

$$\frac{p\xrightarrow{e_1[m]}\!\!\!\!\rightarrow e_1\quad q\xrightarrow{e_2[m]}\!\!\!\!\rightarrow e_2\quad r\xrightarrow{e_2[m]}\!\!\!\!\rightarrow e_2}{p\parallel(q+r)\xrightarrow{\{e_1[m],e_2[m]\}}\!\!\!\!\rightarrow e_1\parallel(e_2+e_2)}\quad \frac{p\xrightarrow{e_1[m]}\!\!\!\!\rightarrow e_1\quad q\xrightarrow{e_2[m]}\!\!\!\!\rightarrow e_2\quad r\xrightarrow{e_2[m]}\!\!\!\!\rightarrow e_2}{(p\parallel q)+(p\parallel r)\xrightarrow{\{e_1[m],e_2[m]\}}\!\!\!\!\rightarrow (e_1\parallel e_2)+(e_1+e_2)}$$

$$\frac{p\xrightarrow{e_1[m]}\!\!\!\!\rightarrow p'\quad q\xrightarrow{e_2[m]}\!\!\!\!\rightarrow q'\quad r\xrightarrow{e_2[m]}\!\!\!\!\rightarrow r'}{p\parallel(q+r)\xrightarrow{\{e_1[m],e_2[m]\}}\!\!\!\!\rightarrow p'\parallel(q'+r')}\quad \frac{p\xrightarrow{e_1[m]}\!\!\!\!\rightarrow p'\quad q\xrightarrow{e_2[m]}\!\!\!\!\rightarrow q'\quad r\xrightarrow{e_2[m]}\!\!\!\!\rightarrow r'}{(p\parallel q)+(p\parallel r)\xrightarrow{\{e_1[m],e_2[m]\}}\!\!\!\!\rightarrow (p'\parallel q')+(p'+r')}$$

Hence, with the assumptions $e_1[m]\parallel(e_2[m]+e_2[m])\sim_s^{fr}(e_1[m]\parallel e_2[m])+(e_1[m]\parallel e_2[m])$, $e_1\parallel(e_2+e_2)\sim_s^{fr}(e_1\parallel e_2)+(e_1\parallel e_2)$, $p'\parallel(q'+r')\sim_s^{fr}(p'\parallel q')+(p'\parallel r')$, $p\parallel(q+r)\sim_s^{fr}(p\parallel q)+(p\parallel r)$, as desired.

- **Axiom** $C12$. Let $q$ be an $RAPTC$ process, and $e_1\mid(e_2\cdot q)=\gamma(e_1,e_2)\cdot q$, it is sufficient to prove that $e_1\mid(e_2\cdot q)\sim_s^{fr}\gamma(e_1,e_2)\cdot q$. By the forward transition rules for operator $\mid$ in Table 4.11, we obtain

$$\frac{e_1\xrightarrow{e_1}e_1[m]\quad e_2\xrightarrow{e_2}e_2[m]}{e_1\mid(e_2\cdot q)\xrightarrow{\gamma(e_1,e_2)}e_1[m]\mid(e_2[m]\cdot q)}$$

$$\frac{e_1\xrightarrow{e_1}e_1[m]\quad e_2\xrightarrow{e_2}e_2[m]}{\gamma(e_1,e_2)\cdot q\xrightarrow{\gamma(e_1,e_2)}\gamma(e_1,e_2)[m]\cdot q}$$

By the reverse transition rules for operator $\mid$ in Table 4.12, there are no transitions.
Hence, with the assumptions $e_1[m]\mid(e_2[m]\cdot q)\sim_s^{fr}\gamma(e_1,e_2)[m]\cdot q$, $e_1\mid(e_2\cdot q)\sim_s^{fr}\gamma(e_1,e_2)\cdot q$, as desired.

- **Axiom** $RC12$. Let $q$ be an $RAPTC$ process, and $e_1[m]\mid(q\cdot e_2[m])=q\cdot\gamma(e_1,e_2)[m]$, it is sufficient to prove that $e_1[m]\mid(q\cdot e_2[m])\sim_s^{fr}q\cdot\gamma(e_1,e_2)[m]$. By the forward transition rules for operator $\mid$ in Table 4.11, there are no transitions. By the reverse transition rules for operator $\mid$ in Table 4.12, we obtain

$$\frac{e_1\xrightarrow{e_1[m]}\!\!\!\!\rightarrow e_1\quad e_2\xrightarrow{e_2[m]}\!\!\!\!\rightarrow e_2}{e_1[m]\mid(q\cdot e_2[m])\xrightarrow{\gamma(e_1,e_2)[m]}\!\!\!\!\rightarrow e_1\mid(q\cdot e_2)}$$

$$\frac{e_1 \xrightarrow{e_1[m]} e_1 \quad e_2 \xrightarrow{e_2[m]} e_2}{q \cdot \gamma(e_1, e_2)[m] \xrightarrow{\gamma(e_1,e_2)[m]} q \cdot \gamma(e_1, e_2)}$$

Hence, with the assumptions $e_1 \mid (q \cdot e_2) \sim_s^{fr} q \cdot \gamma(e_1, e_2)$, $e_1[m] \mid (q \cdot e_2[m]) \sim_s^{fr} q \cdot \gamma(e_1, e_2)[m]$, as desired.

- **Axiom** $C13$. Let $p$ be an $RAPTC$ process, and $(e_1 \cdot p) \mid e_2 = \gamma(e_1, e_2) \cdot p$, it is sufficient to prove that $(e_1 \cdot p) \mid e_2 \sim_s^{fr} \gamma(e_1, e_2) \cdot p$. By the forward transition rules for operator $\mid$ in Table 4.11, we obtain

$$\frac{e_1 \xrightarrow{e_1} e_1[m] \quad e_2 \xrightarrow{e_2} e_2[m]}{(e_1 \cdot p) \mid e_2 \xrightarrow{\gamma(e_1,e_2)} (e_1[m] \cdot p) \mid e_2[m]}$$

$$\frac{e_1 \xrightarrow{e_1} e_1[m] \quad e_2 \xrightarrow{e_2} e_2[m]}{\gamma(e_1, e_2) \cdot p \xrightarrow{\gamma(e_1,e_2)} \gamma(e_1, e_2)[m] \cdot p}$$

By the reverse transition rules for operator $\mid$ in Table 4.12, there are no transitions.
Hence, with the assumptions $(e_1[m] \cdot p) \mid e_2[m] \sim_s^{fr} \gamma(e_1, e_2)[m] \cdot p$, $(e_1 \cdot p) \mid e_2 \sim_s^{fr} \gamma(e_1, e_2) \cdot p$, as desired.

- **Axiom** $RC13$. Let $p$ be an $RAPTC$ process, and $(p \cdot e_1[m]) \mid e_2[m] = p \cdot \gamma(e_1, e_2)[m]$, it is sufficient to prove that $(p \cdot e_1[m]) \mid e_2[m] \sim_s^{fr} p \cdot \gamma(e_1, e_2)[m]$. By the forward transition rules for operator $\mid$ in Table 4.11, there are no transitions.
By the reverse transition rules for operator $\mid$ in Table 4.12, we obtain

$$\frac{e_1 \xrightarrow{e_1[m]} e_1 \quad e_2 \xrightarrow{e_2[m]} e_2}{(p \cdot e_1[m]) \mid e_2[m] \xrightarrow{\gamma(e_1,e_2)[m]} (p \cdot e_1) \mid e_2}$$

$$\frac{e_1 \xrightarrow{e_1[m]} e_1 \quad e_2 \xrightarrow{e_2[m]} e_2}{p \cdot \gamma(e_1, e_2)[m] \xrightarrow{\gamma(e_1,e_2)[m]} p \cdot \gamma(e_1, e_2)}$$

Hence, with the assumptions $(p \cdot e_1) \mid e_2 \sim_s^{fr} p \cdot \gamma(e_1, e_2)$, $(p \cdot e_1[m]) \mid e_2[m] \sim_s^{fr} p \cdot \gamma(e_1, e_2)[m]$, as desired.

- **Axiom** $C14$. Let $p, q$ be $RAPTC$ processes, and $(e_1 \cdot p) \mid (e_2 \cdot q) = \gamma(e_1, e_2) \cdot (p \between q)$, it is sufficient to prove that $(e_1 \cdot p) \mid (e_2 \cdot q) \sim_s^{fr} \gamma(e_1, e_2) \cdot (p \between q)$. By the forward transition rules for operator $\mid$ in Table 4.11, we obtain

$$\frac{e_1 \xrightarrow{e_1} e_1[m] \quad e_2 \xrightarrow{e_2} e_2[m]}{(e_1 \cdot p) \mid (e_2 \cdot q) \xrightarrow{\gamma(e_1,e_2)} (e_1[m] \cdot p) \between (e_2[m] \cdot q)}$$

$$\frac{e_1 \xrightarrow{e_1} e_1[m] \quad e_2 \xrightarrow{e_2} e_2[m]}{\gamma(e_1, e_2) \cdot (p \between q) \xrightarrow{\gamma(e_1,e_2)} \gamma(e_1, e_2)[m] \cdot (p \between q)}$$

By the reverse transition rules for operator $\mid$ in Table 4.12, there are no transitions.
Hence, with the assumptions $(e_1[m] \cdot p) \mid (e_2[m] \cdot q) \sim_s^{fr} \gamma(e_1, e_2)[m] \cdot (p \between q)$, $(e_1 \cdot p) \mid (e_2 \cdot q) \sim_s^{fr} \gamma(e_1, e_2) \cdot (p \between q)$, as desired.

- **Axiom** $RC14$. Let $p, q$ be $RAPTC$ processes, and $(p \cdot e_1[m]) \mid (q \cdot e_2[m]) = (p \between q) \cdot \gamma(e_1, e_2)[m]$, it is sufficient to prove that $(p \cdot e_1[m]) \mid (q \cdot e_2[m]) \sim_s^{fr} (p \between q) \cdot \gamma(e_1, e_2)[m]$. By the forward transition rules for operator $\mid$ in Table 4.11, there are no transitions.
By the reverse transition rules for operator $\mid$ in Table 4.12, we obtain

$$\frac{e_1 \xrightarrow{e_1[m]} e_1 \quad e_2 \xrightarrow{e_2[m]} e_2}{(p \cdot e_1[m]) \mid (q \cdot e_2[m]) \xrightarrow{\gamma(e_1,e_2)[m]} (p \cdot e_1) \mid (q \cdot e_2)}$$

$$\frac{e_1 \xrightarrow{e_1[m]} e_1 \quad e_2 \xrightarrow{e_2[m]} e_2}{(p \between q) \cdot \gamma(e_1, e_2)[m] \xrightarrow{\gamma(e_1,e_2)[m]} (p \between q) \cdot \gamma(e_1, e_2)}$$

Hence, with the assumptions $(p \cdot e_1) \mid (q \cdot e_2) \sim_s^{fr} (p \between q) \cdot \gamma(e_1, e_2)$, $(p \cdot e_1[m]) \mid (q \cdot e_2[m]) \sim_s^{fr} (p \between q) \cdot \gamma(e_1, e_2)[m]$, as desired.

- **Axiom** $C15$. Let $p, q, r$ be $RAPTC$ processes, and $(p+q) \mid r = (p \mid r) + (q \mid r)$, it is sufficient to prove that $(p+q) \mid r \sim_s^{fr} (p \mid r) + (q \mid r)$. There are several cases, however, we will not enumerate them all here. By the forward transition rules for operators $+$ and $\mid$ in Tables 4.3 and 4.11, we obtain

$$\frac{p \xrightarrow{e_1} e_1[m] \quad q \xrightarrow{e_1} e_1[m] \quad r \xrightarrow{e_2} e_2[m]}{(p+q) \mid r \xrightarrow{\gamma(e_1,e_2)} (e_1[m] + e_1[m]) \mid e_2[m]}$$

$$\frac{p \xrightarrow{e_1} e_1[m] \quad q \xrightarrow{e_1} e_1[m] \quad r \xrightarrow{e_2} e_2[m]}{(p \mid r) + (q \mid r) \xrightarrow{\gamma(e_1,e_2)} (e_1[m] \mid e_2[m]) + (e_1[m] \mid e_2[m])}$$

$$\frac{p \xrightarrow{e_1} p' \quad q \xrightarrow{e_1} q' \quad r \xrightarrow{e_2} r'}{(p+q) \mid r \xrightarrow{\gamma(e_1,e_2)} (p'+q') \mid r'} \quad \frac{p \xrightarrow{e_1} p' \quad q \xrightarrow{e_1} q' \quad r \xrightarrow{e_2} r'}{(p \mid r) + (q \mid r) \xrightarrow{\gamma(e_1,e_2)} (p' \mid r') + (q' \mid r')}$$

By the reverse transition rules for operators $+$ and $\mid$ in Tables 4.4 and 4.12, we obtain

$$\frac{p \xrightarrow{e_1[m]}\!\!\!\!\rightarrow e_1 \quad q \xrightarrow{e_1[m]}\!\!\!\!\rightarrow e_1 \quad r \xrightarrow{e_2[m]}\!\!\!\!\rightarrow e_2}{(p+q) \mid r \xrightarrow{\gamma(e_1,e_2)[m]}\!\!\!\!\rightarrow (e_1 + e_1) \mid e_2} \quad \frac{p \xrightarrow{e_1[m]}\!\!\!\!\rightarrow e_1 \quad q \xrightarrow{e_1[m]}\!\!\!\!\rightarrow e_1 \quad r \xrightarrow{e_2[m]}\!\!\!\!\rightarrow e_2}{(p \mid r) + (q \mid r) \xrightarrow{\gamma(e_1,e_2)[m]}\!\!\!\!\rightarrow (e_1 \mid e_2) + (e_1 \mid e_2)}$$

$$\frac{p \xrightarrow{e_1[m]}\!\!\!\!\rightarrow p' \quad q \xrightarrow{e_1[m]}\!\!\!\!\rightarrow q' \quad r \xrightarrow{e_2[m]}\!\!\!\!\rightarrow r'}{(p+q) \mid r \xrightarrow{\gamma(e_1,e_2)[m]}\!\!\!\!\rightarrow (p'+q') \mid r'} \quad \frac{p \xrightarrow{e_1[m]}\!\!\!\!\rightarrow p' \quad q \xrightarrow{e_1[m]}\!\!\!\!\rightarrow q' \quad r \xrightarrow{e_2[m]}\!\!\!\!\rightarrow r'}{(p \mid r) + (q \mid r) \xrightarrow{\gamma(e_1,e_2)[m]}\!\!\!\!\rightarrow (p' \mid r') + (q' \mid r')}$$

Hence, with the assumptions $(e_1[m] + e_1[m]) \mid e_2[m] \sim_s^{fr} (e_1[m] \mid e_2[m]) + (e_1[m] \mid e_2[m])$, $(e_1 + e_1) \mid e_2 \sim_s^{fr} (e_1 \mid e_2) + (e_1 \mid e_2)$, $(p'+q') \mid r' \sim_s^{fr} (p' \mid r') + (q' \mid r')$, $(p+q) \mid r \sim_s^{fr} (p \mid r) + (q \mid r)$, as desired.

- **Axiom** $C16$. Let $p, q, r$ be $RAPTC$ processes, and $p \mid (q+r) = (p \mid q) + (p \mid r)$, it is sufficient to prove that $p \mid (q+r) \sim_s^{fr} (p \mid q) + (p \mid r)$. There are several cases, however, we will not enumerate them all here. By the forward transition rules for operators $+$ and $\mid$ in Tables 4.3 and 4.11, we obtain

$$\frac{p \xrightarrow{e_1} e_1[m] \quad q \xrightarrow{e_2} e_2[m] \quad r \xrightarrow{e_2} e_2[m]}{p \mid (q+r) \xrightarrow{\gamma(e_1,e_2)} e_1[m] \parallel (e_2[m] + e_2[m])}$$

$$\frac{p \xrightarrow{e_1} e_1[m] \quad q \xrightarrow{e_2} e_2[m] \quad r \xrightarrow{e_2} e_2[m]}{(p \mid q) + (p \mid r) \xrightarrow{\gamma(e_1,e_2)} (e_1[m] \mid e_2[m]) + (e_1[m] \mid e_2[m])}$$

$$\frac{p \xrightarrow{e_1} p' \quad q \xrightarrow{e_2} q' \quad r \xrightarrow{e_2} r'}{p \mid (q+r) \xrightarrow{\gamma(e_1,e_2)} p' \parallel (q'+r')} \quad \frac{p \xrightarrow{e_1} p' \quad q \xrightarrow{e_2} q' \quad r \xrightarrow{e_2} r'}{(p \mid q) + (p \mid r) \xrightarrow{\gamma(e_1,e_2)} (p' \mid q') + (p' \mid r')}$$

By the reverse transition rules for operators $+$ and $\mid$ in Tables 4.4 and 4.12, we obtain

$$\frac{p \xrightarrow{e_1[m]}\!\!\!\!\rightarrow e_1 \quad q \xrightarrow{e_2[m]}\!\!\!\!\rightarrow e_2 \quad r \xrightarrow{e_2[m]}\!\!\!\!\rightarrow e_2}{p \mid (q+r) \xrightarrow{\gamma(e_1,e_2)[m]}\!\!\!\!\rightarrow e_1 \parallel (e_2 + e_2)} \quad \frac{p \xrightarrow{e_1[m]}\!\!\!\!\rightarrow e_1 \quad q \xrightarrow{e_2[m]}\!\!\!\!\rightarrow e_2 \quad r \xrightarrow{e_2[m]}\!\!\!\!\rightarrow e_2}{(p \mid q) + (p \mid r) \xrightarrow{\gamma(e_1,e_2)[m]}\!\!\!\!\rightarrow (e_1 \mid e_2) + (e_1 \mid e_2)}$$

$$\frac{p \xrightarrow{e_1[m]}\!\!\!\!\rightarrow p' \quad q \xrightarrow{e_2[m]}\!\!\!\!\rightarrow q' \quad r \xrightarrow{e_2[m]}\!\!\!\!\rightarrow r'}{p \mid (q+r) \xrightarrow{\gamma(e_1,e_2)[m]}\!\!\!\!\rightarrow p' \parallel (q'+r')} \quad \frac{p \xrightarrow{e_1[m]}\!\!\!\!\rightarrow p' \quad q \xrightarrow{e_2[m]}\!\!\!\!\rightarrow q' \quad r \xrightarrow{e_2[m]}\!\!\!\!\rightarrow r'}{(p \mid q) + (p \mid r) \xrightarrow{\gamma(e_1,e_2)[m]}\!\!\!\!\rightarrow (p' \mid q') + (p' \mid r')}$$

Hence, with the assumptions $e_1[m] \mid (e_2[m] + e_2[m]) \sim_s^{fr} (e_1[m] \mid e_2[m]) + (e_1[m] \mid e_2[m])$, $e_1 \mid (e_2 + e_2) \sim_s^{fr} (e_1 \mid e_2) + (e_1 \mid e_2)$, $p' \mid (q'+r') \sim_s^{fr} (p' \mid q') + (p' \mid r')$, $p \mid (q+r) \sim_s^{fr} (p \mid q) + (p \mid r)$, as desired.

- **Axiom** $CE21$. Let $p,q$ be $RAPTC$ processes, and $\Theta(p+q)=\Theta(p)+\Theta(q)$, it is sufficient to prove that $\Theta(p+q)\sim_s^{fr}\Theta(p)+\Theta(q)$. By the forward transition rules for operators $+$ in Table 4.3, and $\Theta$ and $\triangleleft$ in Table 4.13, we obtain

$$\frac{p\xrightarrow{e_1}e_1[m](\sharp(e_1,e_2))}{\Theta(p+q)\xrightarrow{e_1}\Theta(e_1[m]+q)}\quad\frac{p\xrightarrow{e_1}e_1[m](\sharp(e_1,e_2))}{\Theta(p)+\Theta(q)\xrightarrow{e_1}\Theta(e_1[m])+\Theta(q)}$$

$$\frac{q\xrightarrow{e_2}e_2[m](\sharp(e_1,e_2))}{\Theta(p+q)\xrightarrow{e_2}\Theta(p+e_2[m])}\quad\frac{q\xrightarrow{e_2}e_2[m](\sharp(e_1,e_2))}{\Theta(p)+\Theta(q)\xrightarrow{e_2}\Theta(p)+\Theta(e_2[m])}$$

$$\frac{p\xrightarrow{e_1}p'(\sharp(e_1,e_2))}{\Theta(p+q)\xrightarrow{e_1}\Theta(p'+q)}\quad\frac{p\xrightarrow{e_1}p'(\sharp(e_1,e_2))}{\Theta(p)+\Theta(q)\xrightarrow{e_1}\Theta(p')+\Theta(q)}$$

$$\frac{q\xrightarrow{e_2}q'(\sharp(e_1,e_2))}{\Theta(p+q)\xrightarrow{e_2}\Theta(p+q')}\quad\frac{q\xrightarrow{e_2}q'(\sharp(e_1,e_2))}{\Theta(p)+\Theta(q)\xrightarrow{e_2}\Theta(p)+\Theta(q')}$$

By the reverse transition rules for operators $+$ in Table 4.4, and $\Theta$ and $\triangleleft$ in Table 4.14, we obtain

$$\frac{p\xrightarrow{e_1[m]}\!\!\!\!\rightarrow e_1(\sharp(e_1,e_2))}{\Theta(p+q)\xrightarrow{e_1[m]}\!\!\!\!\rightarrow\Theta(e_1+q)}\quad\frac{p\xrightarrow{e_1[m]}\!\!\!\!\rightarrow e_1(\sharp(e_1,e_2))}{\Theta(p)+\Theta(q)\xrightarrow{e_1[m]}\!\!\!\!\rightarrow\Theta(e_1)+\Theta(q)}$$

$$\frac{q\xrightarrow{e_2[m]}\!\!\!\!\rightarrow e_2(\sharp(e_1,e_2))}{\Theta(p+q)\xrightarrow{e_2[m]}\!\!\!\!\rightarrow\Theta(p+e_2)}\quad\frac{q\xrightarrow{e_2[m]}\!\!\!\!\rightarrow e_2(\sharp(e_1,e_2))}{\Theta(p)+\Theta(q)\xrightarrow{e_2[m]}\!\!\!\!\rightarrow\Theta(p)+\Theta(e_2)}$$

$$\frac{p\xrightarrow{e_1[m]}\!\!\!\!\rightarrow p'(\sharp(e_1,e_2))}{\Theta(p+q)\xrightarrow{e_1[m]}\!\!\!\!\rightarrow\Theta(p'+q)}\quad\frac{p\xrightarrow{e_1[m]}\!\!\!\!\rightarrow p'(\sharp(e_1,e_2))}{\Theta(p)+\Theta(q)\xrightarrow{e_1[m]}\!\!\!\!\rightarrow\Theta(p')+\Theta(q)}$$

$$\frac{q\xrightarrow{e_2[m]}\!\!\!\!\rightarrow q'(\sharp(e_1,e_2))}{\Theta(p+q)\xrightarrow{e_2[m]}\!\!\!\!\rightarrow\Theta(p+q')}\quad\frac{q\xrightarrow{e_2[m]}\!\!\!\!\rightarrow q'(\sharp(e_1,e_2))}{\Theta(p)+\Theta(q)\xrightarrow{e_2[m]}\!\!\!\!\rightarrow\Theta(p)+\Theta(q')}$$

Hence, with the assumptions $\Theta(e_1[m]+q)\sim_s^{fr}\Theta(e_1[m])+\Theta(q)$, $\Theta(e_1+q)\sim_s^{fr}\Theta(e_1)+\Theta(q)$, $\Theta(p+e_2[m])\sim_s^{fr}\Theta(p)+\Theta(e_2[m])$, $\Theta(p+e_2)\sim_s^{fr}\Theta(p)+\Theta(e_2)$, $\Theta(p+q')\sim_s^{fr}\Theta(p)+\Theta(q')$, $\Theta(p'+q)\sim_s^{fr}\Theta(p')+\Theta(q)$, $\Theta(p+q)\sim_s^{fr}\Theta(p)+\Theta(q)$, as desired.

- **Axiom** $CE22$. Let $p,q$ be $RAPTC$ processes, and $\Theta(p\cdot q)=\Theta(p)\cdot\Theta(q)$, it is sufficient to prove that $\Theta(p\cdot q)\sim_s^{fr}\Theta(p)\cdot\Theta(q)$. There are several cases, however, we will not enumerate them all here. By the forward transition rules for operators $\cdot$ in Table 4.3, and $\Theta$ in Table 4.13, we obtain

$$\frac{p\xrightarrow{e_1}e_1[m]}{\Theta(p\cdot q)\xrightarrow{e_1}\Theta(e_1[m]\cdot q)}\quad\frac{p\xrightarrow{e_1}e_1[m]}{\Theta(p)\cdot\Theta(q)\xrightarrow{e_1}\Theta(e_1[m])\cdot\Theta(q)}$$

$$\frac{p\xrightarrow{e_1}p'}{\Theta(p\cdot q)\xrightarrow{e_1}\Theta(p'\cdot q)}\quad\frac{p\xrightarrow{e_1}p'}{\Theta(p)\cdot\Theta(q)\xrightarrow{e_1}\Theta(p')\cdot\Theta(q)}$$

By the reverse transition rules for operators $\cdot$ in Table 4.4, and $\Theta$ in Table 4.14, we obtain

$$\frac{q\xrightarrow{e_2[m]}\!\!\!\!\rightarrow e_1}{\Theta(p\cdot q)\xrightarrow{e_2[m]}\!\!\!\!\rightarrow\Theta(p\cdot e_2)}\quad\frac{q\xrightarrow{e_2[m]}\!\!\!\!\rightarrow e_2}{\Theta(p)\cdot\Theta(q)\xrightarrow{e_2[m]}\!\!\!\!\rightarrow\Theta(p)\cdot\Theta(e_2)}$$

$$\frac{q\xrightarrow{e_2[m]}\!\!\!\!\rightarrow q'}{\Theta(p\cdot q)\xrightarrow{e_2[m]}\!\!\!\!\rightarrow\Theta(p\cdot q')}\quad\frac{q\xrightarrow{e_2[m]}\!\!\!\!\rightarrow q'}{\Theta(p)\cdot\Theta(q)\xrightarrow{e_2[m]}\!\!\!\!\rightarrow\Theta(p)\cdot\Theta(q')}$$

Hence, with the assumptions $\Theta(e_1[m]\cdot q)\sim_s^{fr}\Theta(e_1[m])\cdot\Theta(q)$, $\Theta(p\cdot e_2)\sim_s^{fr}\Theta(p)\cdot\Theta(e_2)$, $\Theta(p'\cdot q)\sim_s^{fr}\Theta(p')\cdot\Theta(q)$, $\Theta(p\cdot q')\sim_s^{fr}\Theta(p)\cdot\Theta(q')$, $\Theta(p\cdot q)\sim_s^{fr}\Theta(p)\cdot\Theta(q)$, as desired.

- **Axiom** $CE23$. Let $p,q$ be $RAPTC$ processes, and $\Theta(p\parallel q)=((\Theta(p)\triangleleft q)\parallel q)+((\Theta(q)\triangleleft p)\parallel p)$, it is sufficient to prove that $\Theta(p\parallel q)\sim_s^{fr}((\Theta(p)\triangleleft q)\parallel q)+((\Theta(q)\triangleleft p)\parallel p)$. By the forward transition rules for operators $+$ in Table 4.3, and $\Theta$ and $\triangleleft$ in Table 4.13, and $\parallel$ in Table 4.9 we obtain

$$\frac{p\xrightarrow{e_1}e_1[m]\quad q\xrightarrow{e_2}e_2[m]}{\Theta(p\parallel q)\xrightarrow{\{e_1,e_2\}}\Theta(e_1[m]\parallel e_2[m])}$$

$$\frac{p\xrightarrow{e_1}e_1[m]\quad q\xrightarrow{e_2}e_2[m]}{((\Theta(p)\triangleleft q)\parallel q)+((\Theta(q)\triangleleft p)\parallel p)\xrightarrow{\{e_1,e_2\}}((\Theta(e_1[m])\triangleleft e_2[m])\parallel e_2[m])+((\Theta(e_2[m])\triangleleft e_1[m])\parallel e_1[m])}$$

$$\frac{p\xrightarrow{e_1}p'\quad q\xrightarrow{e_2}e_2[m]}{\Theta(p\parallel q)\xrightarrow{\{e_1,e_2\}}\Theta(p'\parallel e_2[m])}$$

$$\frac{p\xrightarrow{e_1}p'\quad q\xrightarrow{e_2}e_2[m]}{((\Theta(p)\triangleleft q)\parallel q)+((\Theta(q)\triangleleft p)\parallel p)\xrightarrow{\{e_1,e_2\}}((\Theta(p)\triangleleft e_2[m])\parallel e_2[m])+((\Theta(e_2[m])\triangleleft p)\parallel p)}$$

$$\frac{p\xrightarrow{e_1}e_1[m]\quad q\xrightarrow{e_2}q'}{\Theta(p\parallel q)\xrightarrow{\{e_1,e_2\}}\Theta(e_1[m]\parallel q')}$$

$$\frac{p\xrightarrow{e_1}e_1[m]\quad q\xrightarrow{e_2}q'}{((\Theta(p)\triangleleft q)\parallel q)+((\Theta(q)\triangleleft p)\parallel p)\xrightarrow{\{e_1,e_2\}}((\Theta(e_1[m])\triangleleft q)\parallel q)+((\Theta(q)\triangleleft e_1[m])\parallel e_1[m])}$$

$$\frac{p\xrightarrow{e_1}p'\quad q\xrightarrow{e_2}q'}{\Theta(p\parallel q)\xrightarrow{\{e_1,e_2\}}\Theta(p'\between q')}$$

$$\frac{p\xrightarrow{e_1}p'\quad q\xrightarrow{e_2}q'}{((\Theta(p)\triangleleft q)\parallel q)+((\Theta(q)\triangleleft p)\parallel p)\xrightarrow{\{e_1,e_2\}}((\Theta(p')\triangleleft q')\between q')+((\Theta(q')\triangleleft p')\between p')}$$

By the reverse transition rules for operators $+$ in Table 4.4, and $\Theta$ and $\triangleleft$ in Table 4.14, and $\parallel$ in Table 4.10 we obtain

$$\frac{p\xrightarrow{e_1[m]}\!\!\!\!\!\rightarrow e_1\quad q\xrightarrow{e_2[m]}\!\!\!\!\!\rightarrow e_2}{\Theta(p\parallel q)\xrightarrow{\{e_1[m],e_2[m]\}}\!\!\!\!\!\rightarrow\Theta(e_1\parallel e_2)}$$

$$\frac{p\xrightarrow{e_1[m]}\!\!\!\!\!\rightarrow e_1\quad q\xrightarrow{e_2[m]}\!\!\!\!\!\rightarrow e_2}{((\Theta(p)\triangleleft q)\parallel q)+((\Theta(q)\triangleleft p)\parallel p)\xrightarrow{\{e_1[m],e_2[m]\}}\!\!\!\!\!\rightarrow((\Theta(e_1)\triangleleft e_2)\parallel e_2)+((\Theta(e_2)\triangleleft e_1)\parallel e_1)}$$

$$\frac{p\xrightarrow{e_1[m]}\!\!\!\!\!\rightarrow p'\quad q\xrightarrow{e_2[m]}\!\!\!\!\!\rightarrow e_2}{\Theta(p\parallel q)\xrightarrow{\{e_1[m],e_2[m]\}}\!\!\!\!\!\rightarrow\Theta(p'\parallel e_2)}$$

$$\frac{p\xrightarrow{e_1[m]}\!\!\!\!\!\rightarrow p'\quad q\xrightarrow{e_2[m]}\!\!\!\!\!\rightarrow e_2}{((\Theta(p)\triangleleft q)\parallel q)+((\Theta(q)\triangleleft p)\parallel p)\xrightarrow{\{e_1[m],e_2[m]\}}\!\!\!\!\!\rightarrow((\Theta(p)\triangleleft e_2)\parallel e_2)+((\Theta(e_2)\triangleleft p)\parallel p)}$$

$$\frac{p\xrightarrow{e_1[m]}\!\!\!\!\!\rightarrow e_1\quad q\xrightarrow{e_2[m]}\!\!\!\!\!\rightarrow q'}{\Theta(p\parallel q)\xrightarrow{\{e_1[m],e_2[m]\}}\!\!\!\!\!\rightarrow\Theta(e_1\parallel q')}$$

$$\frac{p\xrightarrow{e_1[m]}\!\!\!\!\!\rightarrow e_1\quad q\xrightarrow{e_2[m]}\!\!\!\!\!\rightarrow q'}{((\Theta(p)\triangleleft q)\parallel q)+((\Theta(q)\triangleleft p)\parallel p)\xrightarrow{\{e_1[m],e_2[m]\}}\!\!\!\!\!\rightarrow((\Theta(e_1)\triangleleft q)\parallel q)+((\Theta(q)\triangleleft e_1)\parallel e_1)}$$

$$\frac{p\xrightarrow{e_1[m]}\!\!\!\!\rightarrow p' \quad q\xrightarrow{e_2[m]}\!\!\!\!\rightarrow q'}{\Theta(p\parallel q)\xrightarrow{\{e_1[m],e_2[m]\}}\!\!\!\!\rightarrow \Theta(p'\between q')}$$

$$\frac{p\xrightarrow{e_1[m]}\!\!\!\!\rightarrow p' \quad q\xrightarrow{e_2[m]}\!\!\!\!\rightarrow q'}{((\Theta(p)\triangleleft q)\parallel q)+((\Theta(q)\triangleleft p)\parallel p)\xrightarrow{\{e_1[m],e_2[m]\}}\!\!\!\!\rightarrow ((\Theta(p')\triangleleft q')\between q')+((\Theta(q')\triangleleft p')\between p')}$$

Hence, with the assumptions $\Theta(e_1[m]\parallel e_2[m])\sim_s^{fr}((\Theta(e_1[m])\triangleleft e_2[m])\parallel e_2[m])+((\Theta(e_2[m])\triangleleft e_1[m])\parallel e_1[m])$, $\Theta(e_1\parallel e_2)\sim_s^{fr}((\Theta(e_1)\triangleleft e_2)\parallel e_2)+((\Theta(e_2)\triangleleft e_1)\parallel e_1)$, $\Theta(e_1[m]\parallel q')\sim_s^{fr}((\Theta(e_1[m])\triangleleft q')\parallel q')+((\Theta(q')\triangleleft e_1[m])\parallel e_1[m])$, $\Theta(e_1\parallel q')\sim_s^{fr}((\Theta(e_1)\triangleleft q')\parallel q')+((\Theta(q')\triangleleft e_1)\parallel e_1)$, $\Theta(p'\parallel e_2[m])\sim_s^{fr}((\Theta(p')\triangleleft e_2[m])\parallel e_2[m])+((\Theta(e_2[m])\triangleleft p')\parallel p')$, $\Theta(p'\parallel e_2)\sim_s^{fr}((\Theta(p')\triangleleft e_2)\parallel e_2)+((\Theta(e_2)\triangleleft p')\parallel p')$, $\Theta(p'\between q')\sim_s^{fr}((\Theta(p')\triangleleft q')\between q')+((\Theta(q')\triangleleft p')\between p')$, $\Theta(p\parallel q)\sim_s^{fr}((\Theta(p)\triangleleft q)\parallel q)+((\Theta(q)\triangleleft p)\parallel p)$, as desired.

- **Axiom** $CE24$. Let $p,q$ be $RAPTC$ processes, and $\Theta(p\mid q)=((\Theta(p)\triangleleft q)\mid q)+((\Theta(q)\triangleleft p)\mid p)$, it is sufficient to prove that $\Theta(p\mid q)\sim_s^{fr}((\Theta(p)\triangleleft q)\mid q)+((\Theta(q)\triangleleft p)\mid p)$. By the forward transition rules for operators $+$ in Table 4.3, and $\Theta$ and $\triangleleft$ in Table 4.13, and $\mid$ in Table 4.11 we obtain

$$\frac{p\xrightarrow{e_1}e_1[m]\quad q\xrightarrow{e_2}e_2[m]}{\Theta(p\mid q)\xrightarrow{\gamma(e_1,e_2)}\Theta(e_1[m]\mid e_2[m])}$$

$$\frac{p\xrightarrow{e_1}e_1[m]\quad q\xrightarrow{e_2}e_2[m]}{((\Theta(p)\triangleleft q)\mid q)+((\Theta(q)\triangleleft p)\mid p)\xrightarrow{\gamma(e_1,e_2)}((\Theta(e_1[m])\triangleleft e_2[m])\mid e_2[m])+((\Theta(e_2[m])\triangleleft e_1[m])\mid e_1[m]}$$

$$\frac{p\xrightarrow{e_1}p'\quad q\xrightarrow{e_2}e_2[m]}{\Theta(p\mid q)\xrightarrow{\gamma(e_1,e_2)}\Theta(p'\mid e_2[m])}$$

$$\frac{p\xrightarrow{e_1}p'\quad q\xrightarrow{e_2}e_2[m]}{((\Theta(p)\triangleleft q)\mid q)+((\Theta(q)\triangleleft p)\mid p)\xrightarrow{\gamma(e_1,e_2)}((\Theta(p')\triangleleft e_2[m])\mid e_2[m])+((\Theta(e_2[m])\triangleleft p')\mid p'}$$

$$\frac{p\xrightarrow{e_1}e_1[m]\quad q\xrightarrow{e_2}q'}{\Theta(p\mid q)\xrightarrow{\gamma(e_1,e_2)}\Theta(e_1[m]\mid q')}$$

$$\frac{p\xrightarrow{e_1}e_1[m]\quad q\xrightarrow{e_2}q'}{((\Theta(p)\triangleleft q)\mid q)+((\Theta(q)\triangleleft p)\mid p)\xrightarrow{\gamma(e_1,e_2)}((\Theta(e_1[m])\triangleleft q')\mid q')+((\Theta(q')\triangleleft e_1[m])\mid e_1[m]}$$

$$\frac{p\xrightarrow{e_1}p'\quad q\xrightarrow{e_2}q'}{\Theta(p\mid q)\xrightarrow{\gamma(e_1,e_2)}\Theta(p'\between q')}$$

$$\frac{p\xrightarrow{e_1}p'\quad q\xrightarrow{e_2}q'}{((\Theta(p)\triangleleft q)\mid q)+((\Theta(q)\triangleleft p)\mid p)\xrightarrow{\gamma(e_1,e_2)}((\Theta(p')\triangleleft q')\between q')+((\Theta(q')\triangleleft p')\between p')}$$

By the reverse transition rules for operators $+$ in Table 4.4, and $\Theta$ and $\triangleleft$ in Table 4.14, and $\mid$ in Table 4.12 we obtain

$$\frac{p\xrightarrow{e_1[m]}\!\!\!\!\rightarrow e_1\quad q\xrightarrow{e_2[m]}\!\!\!\!\rightarrow e_2}{\Theta(p\mid q)\xrightarrow{\gamma(e_1,e_2)[m]}\!\!\!\!\rightarrow\Theta(e_1\mid e_2)}$$

$$\frac{p\xrightarrow{e_1[m]}\!\!\!\!\rightarrow e_1\quad q\xrightarrow{e_2[m]}\!\!\!\!\rightarrow e_2}{((\Theta(p)\triangleleft q)\mid q)+((\Theta(q)\triangleleft p)\mid p)\xrightarrow{\gamma(e_1,e_2)[m]}\!\!\!\!\rightarrow((\Theta(e_1)\triangleleft e_2)\mid e_2)+((\Theta(e_2)\triangleleft e_1)\mid e_1}$$

$$\frac{p\xrightarrow{e_1[m]}\!\!\!\!\rightarrow p'\quad q\xrightarrow{e_2[m]}\!\!\!\!\rightarrow e_2[m]}{\Theta(p\mid q)\xrightarrow{\gamma(e_1,e_2)[m]}\!\!\!\!\rightarrow\Theta(p'\mid e_2)}$$

$$\frac{p \overset{e_1[m]}{\twoheadrightarrow} p' \quad q \overset{e_2[m]}{\twoheadrightarrow} e_2}{((\Theta(p) \triangleleft q) \mid q) + ((\Theta(q) \triangleleft p) \mid p) \overset{\gamma(e_1,e_2)[m]}{\twoheadrightarrow} ((\Theta(p') \triangleleft e_2) \mid e_2) + ((\Theta(e_2) \triangleleft p') \mid p'}$$

$$\frac{p \overset{e_1[m]}{\twoheadrightarrow} e_1 \quad q \overset{e_2[m]}{\twoheadrightarrow} q'}{\Theta(p \mid q) \overset{\gamma(e_1,e_2)[m]}{\twoheadrightarrow} \Theta(e_1 \mid q')}$$

$$\frac{p \overset{e_1[m]}{\twoheadrightarrow} e_1 \quad q \overset{e_2[m]}{\twoheadrightarrow} q'}{((\Theta(p) \triangleleft q) \mid q) + ((\Theta(q) \triangleleft p) \mid p) \overset{\gamma(e_1,e_2)[m]}{\twoheadrightarrow} ((\Theta(e_1) \triangleleft q') \mid q') + ((\Theta(q') \triangleleft e_1) \mid e_1}$$

$$\frac{p \overset{e_1[m]}{\twoheadrightarrow} p' \quad q \overset{e_2[m]}{\twoheadrightarrow} q'}{\Theta(p \mid q) \overset{\gamma(e_1,e_2)[m]}{\twoheadrightarrow} \Theta(p' \between q')}$$

$$\frac{p \overset{e_1[m]}{\twoheadrightarrow} p' \quad q \overset{e_2[m]}{\twoheadrightarrow} q'}{((\Theta(p) \triangleleft q) \mid q) + ((\Theta(q) \triangleleft p) \mid p) \overset{\gamma(e_1,e_2)[m]}{\twoheadrightarrow} ((\Theta(p') \triangleleft q') \between q') + ((\Theta(q') \triangleleft p') \between p')}$$

Hence, with the assumptions $\Theta(e_1[m] \mid e_2[m]) \sim_s^{fr} ((\Theta(e_1[m]) \triangleleft e_2[m]) \mid e_2[m]) + ((\Theta(e_2[m]) \triangleleft e_1[m]) \mid e_1[m])$, $\Theta(e_1 \mid e_2) \sim_s^{fr} ((\Theta(e_1) \triangleleft e_2) \mid e_2) + ((\Theta(e_2) \triangleleft e_1) \mid e_1)$, $\Theta(e_1[m] \mid q') \sim_s^{fr} ((\Theta(e_1[m]) \triangleleft q') \mid q') + ((\Theta(q') \triangleleft e_1[m]) \mid e_1[m])$, $\Theta(e_1 \mid q') \sim_s^{fr} ((\Theta(e_1) \triangleleft q') \mid q') + ((\Theta(q') \triangleleft e_1) \mid e_1)$, $\Theta(p' \mid e_2[m]) \sim_s^{fr} ((\Theta(p') \triangleleft e_2[m]) \mid e_2[m]) + ((\Theta(e_2[m]) \triangleleft p') \mid p')$, $\Theta(p' \mid e_2) \sim_s^{fr} ((\Theta(p') \triangleleft e_2) \mid e_2) + ((\Theta(e_2) \triangleleft p') \mid p')$, $\Theta(p' \between q') \sim_s^{fr} ((\Theta(p') \triangleleft q') \between q') + ((\Theta(q') \triangleleft p') \between p')$, $\Theta(p \mid q) \sim_s^{fr} ((\Theta(p) \triangleleft q) \mid q) + ((\Theta(q) \triangleleft p) \mid p)$, as desired.

- **Axiom** $U30$. Let $p, q, r$ be $RAPTC$ processes, and $(p+q) \triangleleft r = (p \triangleleft r) + (q \triangleleft r)$, it is sufficient to prove that $(p+q) \triangleleft r \sim_s^{fr} (p \triangleleft r) + (q \triangleleft r)$. By the forward transition rules for operators $+$ and $\triangleleft$ in Table 4.3 and 4.13, we obtain

$$\frac{p \xrightarrow{e_1} e_1[m] \quad e_1 \notin q}{(p+q) \triangleleft r \xrightarrow{e_1} (e_1[m]+q) \triangleleft r} \qquad \frac{p \xrightarrow{e_1} e_1[m] \quad e_2 \notin q}{(p \triangleleft r) + (q \triangleleft r) \xrightarrow{e_1} (e_1[m] \triangleleft r) + (q \triangleleft r)}$$

$$\frac{q \xrightarrow{e_2} e_2[m] \quad e_2 \notin p}{(p+q) \triangleleft r \xrightarrow{e_2} (p+e_2[m]) \triangleleft r} \qquad \frac{q \xrightarrow{e_2} e_2[m] \quad e_2 \notin p}{(p \triangleleft r) + (q \triangleleft r) \xrightarrow{e_2} (p \triangleleft r) + (e_2[m] \triangleleft r)}$$

$$\frac{p \xrightarrow{e_1} p' \quad e_1 \notin q}{(p+q) \triangleleft r \xrightarrow{e_1} (p'+q) \triangleleft r} \qquad \frac{p \xrightarrow{e_1} p' \quad e_1 \notin q}{(p \triangleleft r) + (q \triangleleft r) \xrightarrow{e_1} (p' \triangleleft r) + (q \triangleleft r)}$$

$$\frac{q \xrightarrow{e_2} q' \quad e_2 \notin p}{(p+q) \triangleleft r \xrightarrow{e_2} (p+q') \triangleleft r} \qquad \frac{q \xrightarrow{e_2} q' \quad e_2 \notin p}{(p \triangleleft r) + (q \triangleleft r) \xrightarrow{e_2} (p \triangleleft r) + (q' \triangleleft r)}$$

$$\frac{p \xrightarrow{e_1} p' \quad q \xrightarrow{e_1} q'}{(p+q) \triangleleft r \xrightarrow{e_1} (p'+q') \triangleleft r} \qquad \frac{p \xrightarrow{e_1} p' \quad q \xrightarrow{e_1} q'}{(p \triangleleft r) + (q \triangleleft r) \xrightarrow{e_1} (p' \triangleleft r) + (q' \triangleleft r)}$$

By the reverse transition rules for operators $+$ and $\triangleleft$ in Table 4.4 and 4.14, we obtain

$$\frac{p \overset{e_1[m]}{\twoheadrightarrow} e_1 \quad e_1 \notin q}{(p+q) \triangleleft r \overset{e_1[m]}{\twoheadrightarrow} (e_1+q) \triangleleft r} \qquad \frac{p \overset{e_1[m]}{\twoheadrightarrow} e_1 \quad e_2 \notin q}{(p \triangleleft r) + (q \triangleleft r) \overset{e_1[m]}{\twoheadrightarrow} (e_1 \triangleleft r) + (q \triangleleft r)}$$

$$\frac{q \overset{e_2}{\twoheadrightarrow} e_2[m] \quad e_2 \notin p}{(p+q) \triangleleft r \overset{e_2[m]}{\twoheadrightarrow} (p+e_2) \triangleleft r} \qquad \frac{q \overset{e_2[m]}{\twoheadrightarrow} e_2 \quad e_2 \notin p}{(p \triangleleft r) + (q \triangleleft r) \overset{e_2[m]}{\twoheadrightarrow} (p \triangleleft r) + (e_2 \triangleleft r)}$$

$$\frac{p \overset{e_1[m]}{\twoheadrightarrow} p' \quad e_1 \notin q}{(p+q) \triangleleft r \overset{e_1[m]}{\twoheadrightarrow} (p'+q) \triangleleft r} \qquad \frac{p \overset{e_1[m]}{\twoheadrightarrow} p' \quad e_1 \notin q}{(p \triangleleft r) + (q \triangleleft r) \overset{e_1[m]}{\twoheadrightarrow} (p' \triangleleft r) + (q \triangleleft r)}$$

$$\frac{q \xrightarrow{e_2[m]} q' \quad e_2 \notin p}{(p+q) \triangleleft r \xrightarrow{e_2[m]} (p+q') \triangleleft r} \quad \frac{q \xrightarrow{e_2[m]} q' \quad e_2 \notin p}{(p \triangleleft r)+(q \triangleleft r) \xrightarrow{e_2[m]} (p \triangleleft r)+(q' \triangleleft r)}$$

$$\frac{p \xrightarrow{e_1[m]} p' \quad q \xrightarrow{e_1[m]} q'}{(p+q) \triangleleft r \xrightarrow{e_1[m]} (p'+q') \triangleleft r} \quad \frac{p \xrightarrow{e_1[m]} p' \quad q \xrightarrow{e_1[m]} q'}{(p \triangleleft r)+(q \triangleleft r) \xrightarrow{e_1[m]} (p' \triangleleft r)+(q' \triangleleft r)}$$

Hence, with the assumptions $(e_1[m]+q) \triangleleft r \sim_s^{fr} (e_1[m] \triangleleft r)+(q \triangleleft r)$, $(e_1+q) \triangleleft r \sim_s^{fr} (e_1 \triangleleft r)+(q \triangleleft r)$, $(p+e_2[m]) \triangleleft r \sim_s^{fr} (p \triangleleft r)+(e_2[m] \triangleleft r)$, $(p'+q) \triangleleft r \sim_s^{fr} (p' \triangleleft r)+(q \triangleleft r)$, $(p+q') \triangleleft r \sim_s^{fr} (p \triangleleft r)+(q' \triangleleft r)$, $(p'+q') \triangleleft r \sim_s^{fr} (p' \triangleleft r)+(q' \triangleleft r)$, $(p+q) \triangleleft r \sim_s^{fr} (p \triangleleft r)+(q \triangleleft r)$, as desired.

- **Axiom** $U31$. Let $p, q, r$ be $RAPTC$ processes, and $(p \cdot q) \triangleleft r = (p \triangleleft r) \cdot (q \triangleleft r)$, it is sufficient to prove that $(p \cdot q) \triangleleft r \sim_s^{fr} (p \triangleleft r) \cdot (q \triangleleft r)$. By the forward transition rules for operators $\cdot$ and $\triangleleft$ in Table 4.3 and 4.13, we obtain

$$\frac{p \xrightarrow{e_1} e_1[m]}{(p \cdot q) \triangleleft r \xrightarrow{e_1} (e_1[m] \cdot q) \triangleleft r} \quad \frac{p \xrightarrow{e_1} e_1[m]}{(p \triangleleft r) \cdot (q \triangleleft r) \xrightarrow{e_1} (e_1[m] \triangleleft r) \cdot (q \triangleleft r)}$$

$$\frac{p \xrightarrow{e_1} p'}{(p \cdot q) \triangleleft r \xrightarrow{e_1} (p' \cdot q) \triangleleft r} \quad \frac{p \xrightarrow{e_1} p'}{(p \triangleleft r) \cdot (q \triangleleft r) \xrightarrow{e_1} (p' \triangleleft r) \cdot (q \triangleleft r)}$$

By the reverse transition rules for operators $\cdot$ and $\triangleleft$ in Tables 4.4 and 4.14, we obtain

$$\frac{q \xrightarrow{e_2[m]} e_2}{(p \cdot q) \triangleleft r \xrightarrow{e_2[m]} (p \cdot e_2) \triangleleft r} \quad \frac{q \xrightarrow{e_2[m]} e_2}{(p \triangleleft r) \cdot (q \triangleleft r) \xrightarrow{e_2[m]} (p \triangleleft r) \cdot (e_2 \triangleleft r)}$$

$$\frac{q \xrightarrow{e_2[m]} q'}{(p \cdot q) \triangleleft r \xrightarrow{e_2[m]} (p \cdot q') \triangleleft r} \quad \frac{q \xrightarrow{e_2[m]} q'}{(p \triangleleft r) \cdot (q \triangleleft r) \xrightarrow{e_2[m]} (p \triangleleft r) \cdot (q' \triangleleft r)}$$

With the assumptions $(e_1[m] \cdot q) \triangleleft r \sim_s^{fr} (e_1[m] \triangleleft r) \cdot (q \triangleleft r)$, $(e_1 \cdot q) \triangleleft r \sim_s^{fr} (e_1 \triangleleft r) \cdot (q \triangleleft r)$, $(p' \cdot q) \triangleleft r = (p' \triangleleft r) \cdot (q \triangleleft r)$, $(p \cdot e_2[m]) \triangleleft r \sim_s^{fr} (p \triangleleft r) \cdot (e_2[m] \triangleleft r)$, $(p \cdot e_2) \triangleleft r \sim_s^{fr} (p \triangleleft r) \cdot (e_2 \triangleleft r)$, $(p \cdot q') \triangleleft r = (p \triangleleft r) \cdot (q' \triangleleft r)$, hence, $(p \cdot q) \triangleleft r \sim_s^{fr} (p \triangleleft r) \cdot (q \triangleleft r)$, as desired.

- **Axiom** $U32$. Let $p, q, r$ be $RAPTC$ processes, and $(p \parallel q) \triangleleft r = (p \triangleleft r) \parallel (q \triangleleft r)$, it is sufficient to prove that $(p \parallel q) \triangleleft r \sim_s^{fr} (p \triangleleft r) \parallel (q \triangleleft r)$. By the forward transition rules for operators $\parallel$ and $\triangleleft$ in Tables 4.9 and 4.13, we obtain

$$\frac{p \xrightarrow{e_1} e_1[m] \quad q \xrightarrow{e_2} e_2[m]}{(p \parallel q) \triangleleft r \xrightarrow{\{e_1,e_2\}} (e_1[m] \parallel e_2[m]) \triangleleft r} \quad \frac{p \xrightarrow{e_1} e_1[m] \quad q \xrightarrow{e_2} e_2[m]}{(p \triangleleft r) \parallel (q \triangleleft r) \xrightarrow{\{e_1,e_2\}} (e_1[m] \triangleleft r) \parallel (e_2[m] \triangleleft r)}$$

$$\frac{p \xrightarrow{e_1} p' \quad q \xrightarrow{e_2} e_2[m]}{(p \parallel q) \triangleleft r \xrightarrow{\{e_1,e_2\}} (p' \parallel e_2[m]) \triangleleft r} \quad \frac{p \xrightarrow{e_1} p' \quad q \xrightarrow{e_2} e_2[m]}{(p \triangleleft r) \parallel (q \triangleleft r) \xrightarrow{\{e_1,e_2\}} (p' \triangleleft r) \parallel (e_2[m] \triangleleft r)}$$

$$\frac{p \xrightarrow{e_1} e_1[m] \quad q \xrightarrow{e_2} q'}{(p \parallel q) \triangleleft r \xrightarrow{\{e_1,e_2\}} (e_1[m] \parallel q') \triangleleft r} \quad \frac{p \xrightarrow{e_1} e_1[m] \quad q \xrightarrow{e_2} q'}{(p \triangleleft r) \parallel (q \triangleleft r) \xrightarrow{\{e_1,e_2\}} (e_1[m] \triangleleft r) \parallel (q' \triangleleft r)}$$

$$\frac{p \xrightarrow{e_1} p' \quad q \xrightarrow{e_2} q'}{(p \parallel q) \triangleleft r \xrightarrow{\{e_1,e_2\}} (p' \between q') \triangleleft r} \quad \frac{p \xrightarrow{e_1} p' \quad q \xrightarrow{e_2} q'}{(p \triangleleft r) \parallel (q \triangleleft r) \xrightarrow{\{e_1,e_2\}} (p' \triangleleft r) \between (q' \triangleleft r)}$$

By the reverse transition rules for operators $\parallel$ and $\triangleleft$ in Tables 4.10 and 4.14, we obtain

$$\frac{p \xrightarrow{e_1[m]} e_1 \quad q \xrightarrow{e_2[m]} e_2}{(p \parallel q) \triangleleft r \xrightarrow{\{e_1[m],e_2[m]\}} (e_1 \parallel e_2) \triangleleft r} \quad \frac{p \xrightarrow{e_1[m]} e_1 \quad q \xrightarrow{e_2[m]} e_2}{(p \triangleleft r) \parallel (q \triangleleft r) \xrightarrow{\{e_1[m],e_2[m]\}} (e_1 \triangleleft r) \parallel (e_2 \triangleleft r)}$$

$$\frac{p\xrightarrow{e_1[m]}\!\!\!\!\twoheadrightarrow p' \quad q\xrightarrow{e_2[m]}\!\!\!\!\twoheadrightarrow e_2}{(p\parallel q)\triangleleft r\xrightarrow{\{e_1[m],e_2[m]\}}\!\!\!\!\twoheadrightarrow (p'\parallel e_2)\triangleleft r} \quad \frac{p\xrightarrow{e_1[m]}\!\!\!\!\twoheadrightarrow p' \quad q\xrightarrow{e_2[m]}\!\!\!\!\twoheadrightarrow e_2}{(p\triangleleft r)\parallel (q\triangleleft r)\xrightarrow{\{e_1[m],e_2[m]\}}\!\!\!\!\twoheadrightarrow (p'\triangleleft r)\parallel (e_2\triangleleft r)}$$

$$\frac{p\xrightarrow{e_1[m]}\!\!\!\!\twoheadrightarrow e_1 \quad q\xrightarrow{e_2[m]}\!\!\!\!\twoheadrightarrow q'}{(p\parallel q)\triangleleft r\xrightarrow{\{e_1[m],e_2[m]\}}\!\!\!\!\twoheadrightarrow (e_1\parallel q')\triangleleft r} \quad \frac{p\xrightarrow{e_1[m]}\!\!\!\!\twoheadrightarrow e_1 \quad q\xrightarrow{e_2[m]}\!\!\!\!\twoheadrightarrow q'}{(p\triangleleft r)\parallel (q\triangleleft r)\xrightarrow{\{e_1[m],e_2[m]\}}\!\!\!\!\twoheadrightarrow (e_1\triangleleft r)\parallel (q'\triangleleft r)}$$

$$\frac{p\xrightarrow{e_1[m]}\!\!\!\!\twoheadrightarrow p' \quad q\xrightarrow{e_2[m]}\!\!\!\!\twoheadrightarrow q'}{(p\parallel q)\triangleleft r\xrightarrow{\{e_1[m],e_2[m]\}}\!\!\!\!\twoheadrightarrow (p'\between q')\triangleleft r} \quad \frac{p\xrightarrow{e_1[m]}\!\!\!\!\twoheadrightarrow p' \quad q\xrightarrow{e_2[m]}\!\!\!\!\twoheadrightarrow q'}{(p\triangleleft r)\parallel (q\triangleleft r)\xrightarrow{\{e_1[m],e_2[m]\}}\!\!\!\!\twoheadrightarrow (p'\triangleleft r)\between (q'\triangleleft r)}$$

With the assumptions $(e_1[m]\parallel e_2[m])\triangleleft r\sim_s^{fr}(e_1[m]\triangleleft r)\parallel(e_2[m]\triangleleft r)$, $(e_1\parallel e_2)\triangleleft r\sim_s^{fr}(e_1\triangleleft r)\parallel(e_2\triangleleft r)$, $(e_1[m]\parallel q')\triangleleft r\sim_s^{fr}(e_1[m]\triangleleft r)\parallel(q'\triangleleft r)$, $(e_1\parallel q')\triangleleft r\sim_s^{fr}(e_1\triangleleft r)\parallel(q'\triangleleft r)$, $(p'\parallel e_2[m])\triangleleft r\sim_s^{fr}(p'\triangleleft r)\parallel(e_2[m]\triangleleft r)$, $(p'\parallel e_2)\triangleleft r\sim_s^{fr}(p'\triangleleft r)\parallel(e_2\triangleleft r)$, $(p'\between q')\triangleleft r=(p'\triangleleft r)\between(q'\triangleleft r)$, hence, $(p\parallel q)\triangleleft r\sim_s^{fr}(p\triangleleft r)\parallel(q\triangleleft r)$, as desired.

- **Axiom** $U33$. Let $p,q,r$ be $RAPTC$ processes, and $(p\mid q)\triangleleft r=(p\triangleleft r)\mid(q\triangleleft r)$, it is sufficient to prove that $(p\mid q)\triangleleft r\sim_s^{fr}(p\triangleleft r)\mid(q\triangleleft r)$. By the forward transition rules for operators $\mid$ and $\triangleleft$ in Tables 4.11 and 4.13, we obtain

$$\frac{p\xrightarrow{e_1}e_1[m] \quad q\xrightarrow{e_2}e_2[m]}{(p\mid q)\triangleleft r\xrightarrow{\gamma(e_1,e_2)}(e_1[m]\mid e_2[m])\triangleleft r} \quad \frac{p\xrightarrow{e_1}e_1[m] \quad q\xrightarrow{e_2}e_2[m]}{(p\triangleleft r)\mid(q\triangleleft r)\xrightarrow{\gamma(e_1,e_2)}(e_1[m]\triangleleft r)\mid(e_2[m]\triangleleft r)}$$

$$\frac{p\xrightarrow{e_1}p' \quad q\xrightarrow{e_2}e_2[m]}{(p\mid q)\triangleleft r\xrightarrow{\gamma(e_1,e_2)}(p'\mid e_2[m])\triangleleft r} \quad \frac{p\xrightarrow{e_1}p' \quad q\xrightarrow{e_2}e_2[m]}{(p\triangleleft r)\mid(q\triangleleft r)\xrightarrow{\gamma(e_1,e_2)}(p'\triangleleft r)\mid(e_2[m]\triangleleft r)}$$

$$\frac{p\xrightarrow{e_1}e_1[m] \quad q\xrightarrow{e_2}q'}{(p\mid q)\triangleleft r\xrightarrow{\gamma(e_1,e_2)}(e_1[m]\mid q')\triangleleft r} \quad \frac{p\xrightarrow{e_1}e_1[m] \quad q\xrightarrow{e_2}q'}{(p\triangleleft r)\mid(q\triangleleft r)\xrightarrow{\gamma(e_1,e_2)}(e_1[m]\triangleleft r)\mid(q'\triangleleft r)}$$

$$\frac{p\xrightarrow{e_1}p' \quad q\xrightarrow{e_2}q'}{(p\mid q)\triangleleft r\xrightarrow{\gamma(e_1,e_2)}(p'\between q')\triangleleft r} \quad \frac{p\xrightarrow{e_1}p' \quad q\xrightarrow{e_2}q'}{(p\triangleleft r)\mid(q\triangleleft r)\xrightarrow{\gamma(e_1,e_2)}(p'\triangleleft r)\between(q'\triangleleft r)}$$

By the reverse transition rules for operators $\mid$ and $\triangleleft$ in Tables 4.12 and 4.14, we obtain

$$\frac{p\xrightarrow{e_1[m]}\!\!\!\!\twoheadrightarrow e_1 \quad q\xrightarrow{e_2[m]}\!\!\!\!\twoheadrightarrow e_2}{(p\mid q)\triangleleft r\xrightarrow{\gamma(e_1,e_2)[m]}\!\!\!\!\twoheadrightarrow (e_1\mid e_2)\triangleleft r} \quad \frac{p\xrightarrow{e_1[m]}\!\!\!\!\twoheadrightarrow e_1 \quad q\xrightarrow{e_2[m]}\!\!\!\!\twoheadrightarrow e_2}{(p\triangleleft r)\mid(q\triangleleft r)\xrightarrow{\gamma(e_1,e_2)[m]}\!\!\!\!\twoheadrightarrow (e_1\triangleleft r)\mid(e_2\triangleleft r)}$$

$$\frac{p\xrightarrow{e_1[m]}\!\!\!\!\twoheadrightarrow p' \quad q\xrightarrow{e_2[m]}\!\!\!\!\twoheadrightarrow e_2}{(p\mid q)\triangleleft r\xrightarrow{\gamma(e_1,e_2)[m]}\!\!\!\!\twoheadrightarrow (p'\mid e_2)\triangleleft r} \quad \frac{p\xrightarrow{e_1[m]}\!\!\!\!\twoheadrightarrow p' \quad q\xrightarrow{e_2[m]}\!\!\!\!\twoheadrightarrow e_2}{(p\triangleleft r)\mid(q\triangleleft r)\xrightarrow{\gamma(e_1,e_2)[m]}\!\!\!\!\twoheadrightarrow (p'\triangleleft r)\mid(e_2\triangleleft r)}$$

$$\frac{p\xrightarrow{e_1[m]}\!\!\!\!\twoheadrightarrow e_1 \quad q\xrightarrow{e_2[m]}\!\!\!\!\twoheadrightarrow q'}{(p\mid q)\triangleleft r\xrightarrow{\gamma(e_1,e_2)[m]}\!\!\!\!\twoheadrightarrow (e_1\mid q')\triangleleft r} \quad \frac{p\xrightarrow{e_1[m]}\!\!\!\!\twoheadrightarrow e_1 \quad q\xrightarrow{e_2[m]}\!\!\!\!\twoheadrightarrow q'}{(p\triangleleft r)\mid(q\triangleleft r)\xrightarrow{\gamma(e_1,e_2)[m]}\!\!\!\!\twoheadrightarrow (e_1\triangleleft r)\mid(q'\triangleleft r)}$$

$$\frac{p\xrightarrow{e_1[m]}\!\!\!\!\twoheadrightarrow p' \quad q\xrightarrow{e_2[m]}\!\!\!\!\twoheadrightarrow q'}{(p\mid q)\triangleleft r\xrightarrow{\gamma(e_1,e_2)[m]}\!\!\!\!\twoheadrightarrow (p'\between q')\triangleleft r} \quad \frac{p\xrightarrow{e_1[m]}\!\!\!\!\twoheadrightarrow p' \quad q\xrightarrow{e_2[m]}\!\!\!\!\twoheadrightarrow q'}{(p\triangleleft r)\mid(q\triangleleft r)\xrightarrow{\gamma(e_1,e_2)[m]}\!\!\!\!\twoheadrightarrow (p'\triangleleft r)\between(q'\triangleleft r)}$$

With the assumptions $(e_1[m]\mid e_2[m])\triangleleft r\sim_s^{fr}(e_1[m]\triangleleft r)\mid(e_2[m]\triangleleft r)$, $(e_1[m]\mid q')\triangleleft r\sim_s^{fr}(e_1[m]\triangleleft r)\mid(q'\triangleleft r)$, $(e_1\mid q')\triangleleft r\sim_s^{fr}(e_1\triangleleft r)\mid(q'\triangleleft r)$, $(p'\mid e_2[m])\triangleleft r\sim_s^{fr}(p'\triangleleft r)\mid(e_2[m]\triangleleft r)$, $(p'\mid e_2)\triangleleft r\sim_s^{fr}(p'\triangleleft r)\mid(e_2\triangleleft r)$, $(p'\between q')\triangleleft r=(p'\triangleleft r)\between(q'\triangleleft r)$, hence, $(p\mid q)\triangleleft r\sim_s^{fr}(p\triangleleft r)\mid(q\triangleleft r)$, as desired.

- **Axiom** $U34$. Let $p,q,r$ be $RAPTC$ processes, and $p\triangleleft(q+r)=(p\triangleleft q)\triangleleft r$, it is sufficient to prove that $p\triangleleft(q+r)\sim_s^{fr}(p\triangleleft q)\triangleleft r$. By the forward transition rules for operators $+$ and $\triangleleft$ in Tables 4.3 and 4.13, we obtain

$$\frac{p\xrightarrow{e_1}e_1[m]}{p\triangleleft(q+r)\xrightarrow{e_1}e_1[m]\triangleleft(q+r)}\quad\frac{p\xrightarrow{e_1}e_1[m]}{(p\triangleleft q)\triangleleft r\xrightarrow{e_1}(e_1[m]\triangleleft q)\triangleleft r}$$

$$\frac{p\xrightarrow{e_1}p'}{p\triangleleft(q+r)\xrightarrow{e_1}p'\triangleleft(q+r)}\quad\frac{p\xrightarrow{e_1}p'}{(p\triangleleft q)\triangleleft r\xrightarrow{e_1}(p'\triangleleft q)\triangleleft r}$$

By the reverse transition rules for operators $+$ and $\triangleleft$ in Tables 4.4 and 4.14, we obtain

$$\frac{p\xrightarrow{e_1[m]}\!\!\!\!\rightarrow e_1}{p\triangleleft(q+r)\xrightarrow{e_1[m]}\!\!\!\!\rightarrow e_1\triangleleft(q+r)}\quad\frac{p\xrightarrow{e_1[m]}\!\!\!\!\rightarrow e_1}{(p\triangleleft q)\triangleleft r\xrightarrow{e_1[m]}\!\!\!\!\rightarrow(e_1\triangleleft q)\triangleleft r}$$

$$\frac{p\xrightarrow{e_1[m]}\!\!\!\!\rightarrow p'}{p\triangleleft(q+r)\xrightarrow{e_1[m]}\!\!\!\!\rightarrow p'\triangleleft(q+r)}\quad\frac{p\xrightarrow{e_1[m]}\!\!\!\!\rightarrow p'}{(p\triangleleft q)\triangleleft r\xrightarrow{e_1[m]}\!\!\!\!\rightarrow(p'\triangleleft q)\triangleleft r}$$

With the assumptions $e_1[m]\triangleleft(q+r)\sim_s^{fr}(e_1[m]\triangleleft q)\triangleleft r$, $e_1\triangleleft(q+r)\sim_s^{fr}(e_1\triangleleft q)\triangleleft r$, $p'\triangleleft(q+r)\sim_s^{fr}(p'\triangleleft q)\triangleleft r$, hence, $p\triangleleft(q+r)\sim_s^{fr}(p\triangleleft q)\triangleleft r$, as desired.
- **Axiom** $U35$. Let $p,q,r$ be $RAPTC$ processes, and $p\triangleleft(q\cdot r)=(p\triangleleft q)\triangleleft r$, it is sufficient to prove that $p\triangleleft(q\cdot r)\sim_s^{fr}(p\triangleleft q)\triangleleft r$. By the forward transition rules for operators $\cdot$ and $\triangleleft$ in Tables 4.3 and 4.13, we obtain

$$\frac{p\xrightarrow{e_1}e_1[m]}{p\triangleleft(q\cdot r)\xrightarrow{e_1}e_1[m]\triangleleft(q\cdot r)}\quad\frac{p\xrightarrow{e_1}e_1[m]}{(p\triangleleft q)\triangleleft r\xrightarrow{e_1}(e_1[m]\triangleleft q)\triangleleft r}$$

$$\frac{p\xrightarrow{e_1}p'}{p\triangleleft(q\cdot r)\xrightarrow{e_1}p'\triangleleft(q\cdot r)}\quad\frac{p\xrightarrow{e_1}p'}{(p\triangleleft q)\triangleleft r\xrightarrow{e_1}(p'\triangleleft q)\triangleleft r}$$

By the reverse transition rules for operators $\cdot$ and $\triangleleft$ in Tables 4.4 and 4.14, we obtain

$$\frac{p\xrightarrow{e_1[m]}\!\!\!\!\rightarrow e_1}{p\triangleleft(q\cdot r)\xrightarrow{e_1[m]}\!\!\!\!\rightarrow e_1\triangleleft(q\cdot r)}\quad\frac{p\xrightarrow{e_1[m]}\!\!\!\!\rightarrow e_1}{(p\triangleleft q)\triangleleft r\xrightarrow{e_1[m]}\!\!\!\!\rightarrow(e_1\triangleleft q)\triangleleft r}$$

$$\frac{p\xrightarrow{e_1[m]}\!\!\!\!\rightarrow p'}{p\triangleleft(q\cdot r)\xrightarrow{e_1[m]}\!\!\!\!\rightarrow p'\triangleleft(q\cdot r)}\quad\frac{p\xrightarrow{e_1[m]}\!\!\!\!\rightarrow p'}{(p\triangleleft q)\triangleleft r\xrightarrow{e_1[m]}\!\!\!\!\rightarrow(p'\triangleleft q)\triangleleft r}$$

With the assumptions $e_1[m]\triangleleft(q\cdot r)\sim_s^{fr}(e_1[m]\triangleleft q)\triangleleft r$, $e_1\triangleleft(q\cdot r)\sim_s^{fr}(e_1\triangleleft q)\triangleleft r$, $p'\triangleleft(q\cdot r)\sim_s^{fr}(p'\triangleleft q)\triangleleft r$, hence, $p\triangleleft(q\cdot r)\sim_s^{fr}(p\triangleleft q)\triangleleft r$, as desired.
- **Axiom** $U36$. Let $p,q,r$ be $RAPTC$ processes, and $p\triangleleft(q\parallel r)=(p\triangleleft q)\triangleleft r$, it is sufficient to prove that $p\triangleleft(q\parallel r)\sim_s^{fr}(p\triangleleft q)\triangleleft r$. By the forward transition rules for operators $\parallel$ and $\triangleleft$ in Tables 4.9 and 4.13, we obtain

$$\frac{p\xrightarrow{e_1}e_1[m]}{p\triangleleft(q\parallel r)\xrightarrow{e_1}e_1[m]\triangleleft(q\parallel r)}\quad\frac{p\xrightarrow{e_1}e_1[m]}{(p\triangleleft q)\triangleleft r\xrightarrow{e_1}(e_1[m]\triangleleft q)\triangleleft r}$$

$$\frac{p\xrightarrow{e_1}p'}{p\triangleleft(q\parallel r)\xrightarrow{e_1}p'\triangleleft(q\parallel r)}\quad\frac{p\xrightarrow{e_1}p'}{(p\triangleleft q)\triangleleft r\xrightarrow{e_1}(p'\triangleleft q)\triangleleft r}$$

By the reverse transition rules for operators $\parallel$ and $\triangleleft$ in Tables 4.10 and 4.14, we obtain

$$\frac{p\xrightarrow{e_1[m]}\!\!\!\!\rightarrow e_1}{p\triangleleft(q\parallel r)\xrightarrow{e_1[m]}\!\!\!\!\rightarrow e_1\triangleleft(q\parallel r)}\quad\frac{p\xrightarrow{e_1[m]}\!\!\!\!\rightarrow e_1}{(p\triangleleft q)\triangleleft r\xrightarrow{e_1[m]}\!\!\!\!\rightarrow(e_1\triangleleft q)\triangleleft r}$$

$$\frac{p \xrightarrow{e_1[m]} p'}{p \triangleleft (q \parallel r) \xrightarrow{e_1[m]} p' \triangleleft (q \parallel r)} \quad \frac{p \xrightarrow{e_1[m]} p'}{(p \triangleleft q) \triangleleft r \xrightarrow{e_1[m]} (p' \triangleleft q) \triangleleft r}$$

With the assumptions $e_1[m] \triangleleft (q \parallel r) \sim_s^{fr} (e_1[m] \triangleleft q) \triangleleft r$, $e_1 \triangleleft (q \parallel r) \sim_s^{fr} (e_1 \triangleleft q) \triangleleft r$, $p' \triangleleft (q \parallel r) \sim_s^{fr} (p' \triangleleft q) \triangleleft r$, hence, $p \triangleleft (q \parallel r) \sim_s^{fr} (p \triangleleft q) \triangleleft r$, as desired.

- **Axiom** $U37$. Let $p, q, r$ be $RAPTC$ processes, and $p \triangleleft (q \mid r) = (p \triangleleft q) \triangleleft r$, it is sufficient to prove that $p \triangleleft (q \mid r) \sim_s^{fr} (p \triangleleft q) \triangleleft r$. By the forward transition rules for operators $\mid$ and $\triangleleft$ in Tables 4.11 and 4.13, we obtain

$$\frac{p \xrightarrow{e_1} e_1[m]}{p \triangleleft (q \mid r) \xrightarrow{e_1} e_1[m] \triangleleft (q \mid r)} \quad \frac{p \xrightarrow{e_1} e_1[m]}{(p \triangleleft q) \triangleleft r \xrightarrow{e_1} e_1[m] \triangleleft (q \mid r)}$$

$$\frac{p \xrightarrow{e_1} p'}{p \triangleleft (q \mid r) \xrightarrow{e_1} p' \triangleleft (q \mid r)} \quad \frac{p \xrightarrow{e_1} p'}{(p \triangleleft q) \triangleleft r \xrightarrow{e_1} (p' \triangleleft q) \triangleleft r}$$

By the reverse transition rules for operators $\mid$ and $\triangleleft$ in Tables 4.12 and 4.14, we obtain

$$\frac{p \xrightarrow{e_1[m]} e_1}{p \triangleleft (q \mid r) \xrightarrow{e_1[m]} e_1 \triangleleft (q \mid r)} \quad \frac{p \xrightarrow{e_1[m]} e_1}{(p \triangleleft q) \triangleleft r \xrightarrow{e_1[m]} e_1 \triangleleft (q \mid r)}$$

$$\frac{p \xrightarrow{e_1[m]} p'}{p \triangleleft (q \mid r) \xrightarrow{e_1[m]} p' \triangleleft (q \mid r)} \quad \frac{p \xrightarrow{e_1[m]} p'}{(p \triangleleft q) \triangleleft r \xrightarrow{e_1[m]} (p' \triangleleft q) \triangleleft r}$$

With the assumptions $e_1[m] \triangleleft (q \mid r) \sim_s^{fr} (e_1[m] \triangleleft q) \triangleleft r$, $e_1 \triangleleft (q \mid r) \sim_s^{fr} (e_1 \triangleleft q) \triangleleft r$, $p' \triangleleft (q \mid r) \sim_s^{fr} (p' \triangleleft q) \triangleleft r$, hence, $p \triangleleft (q \mid r) \sim_s^{fr} (p \triangleleft q) \triangleleft r$, as desired. □

**Theorem 4.17** (Soundness of parallelism modulo FR pomset bisimulation equivalence). *Let x and y be $RAPTC$ terms. If $RAPTC \vdash x = y$, then $x \sim_p^{fr} y$.*

*Proof.* Since FR pomset bisimulation $\sim_p^{fr}$ is both an equivalent and a congruent relation with respect to the operators $\between$, $\parallel$, $\mid$, $\Theta$, and $\triangleleft$, we only need to check if each axiom in Table 4.15 is sound modulo FR pomset bisimulation equivalence.

From the definition of FR pomset bisimulation (see Definition 2.68), we know that FR pomset bisimulation is defined by pomset transitions, which are labeled by pomsets. In a pomset transition, the events in the pomset are either within causality relations (defined by $\cdot$) or in concurrency (implicitly defined by $\cdot$ and $+$, and explicitly defined by $\between$), of course, they are pairwise consistent (without conflicts). In Theorem 4.16, we have already proven the case that all events are pairwise concurrent, hence, we only need to prove the case of events in causality. Without loss of generality, we take a pomset of $P = \{e_1, e_2 : e_1 \cdot e_2\}$. Then, the pomset transition labeled by the above $P$ is just composed of one single event transition labeled by $e_1$ succeeded by another single event transition labeled by $e_2$, that is, $\xrightarrow{P} = \xrightarrow{e_1}\xrightarrow{e_2}$ or $\xrightarrow{P} = \xrightarrow{e_2[n]}\xrightarrow{e_1[m]}$.

Similarly to the proof of soundness of parallelism modulo FR step bisimulation equivalence (see Theorem 4.16), we can prove that each axiom in Table 4.15 is sound modulo FR pomset bisimulation equivalence, however, we omit these proofs here. □

**Theorem 4.18** (Soundness of parallelism modulo FR hp-bisimulation equivalence). *Let x and y be $RAPTC$ terms. If $RAPTC \vdash x = y$, then $x \sim_{hp}^{fr} y$.*

*Proof.* Since FR hp-bisimulation $\sim_{hp}^{fr}$ is both an equivalent and a congruent relation with respect to the operators $\between$, $\parallel$, $\mid$, $\Theta$, and $\triangleleft$, we only need to check if each axiom in Table 4.15 is sound modulo FR hp-bisimulation equivalence.

From the definition of FR hp-bisimulation (see Definition 2.70), we know that FR hp-bisimulation is defined on the posetal product $(C_1, f, C_2)$, $f : C_1 \to C_2$ isomorphism. Two process terms $s$ related to $C_1$ and $t$ related to $C_2$, and $f : C_1 \to C_2$ isomorphism. Initially, $(C_1, f, C_2) = (\emptyset, \emptyset, \emptyset)$, and $(\emptyset, \emptyset, \emptyset) \in \sim_{hp}^{fr}$. When $s \xrightarrow{e} s'$ $(C_1 \xrightarrow{e} C_1')$, there will be $t \xrightarrow{e} t'$ $(C_2 \xrightarrow{e} C_2')$, and we define $f' = f[e \mapsto e]$. Also, when $s \xrightarrow{e[m]} s'$ $(C_1 \xrightarrow{e[m]} C_1')$, there will be $t \xrightarrow{e[m]} t'$ $(C_2 \xrightarrow{e[m]} C_2')$, and we define $f' = f[e[m] \mapsto e[m]]$. Then, if $(C_1, f, C_2) \in \sim_{hp}^{fr}$, then $(C_1', f', C_2') \in \sim_{hp}^{fr}$.

**TABLE 4.17** Forward transition rules of encapsulation operator $\partial_H$.

$$\frac{x \xrightarrow{e} e[m]}{\partial_H(x) \xrightarrow{e} \partial_H(e[m])} \quad (e \notin H) \qquad \frac{x \xrightarrow{e} x'}{\partial_H(x) \xrightarrow{e} \partial_H(x')} \quad (e \notin H)$$

**TABLE 4.18** Reverse transition rules of encapsulation operator $\partial_H$.

$$\frac{x \xrightarrow{e[m]} e}{\partial_H(x) \xrightarrow{e[m]} e} \quad (e \notin H) \qquad \frac{x \xrightarrow{e} x'}{\partial_H(x) \xrightarrow{e} \partial_H(x')} \quad (e \notin H)$$

**TABLE 4.19** Axioms of encapsulation operator.

| No. | Axiom |
|---|---|
| $D1$ | $e \notin H \quad \partial_H(e) = e$ |
| $RD1$ | $e \notin H \quad \partial_H(e[m]) = e[m]$ |
| $D2$ | $e \in H \quad \partial_H(e) = \delta$ |
| $RD2$ | $e \in H \quad \partial_H(e[m]) = \delta$ |
| $D3$ | $\partial_H(\delta) = \delta$ |
| $D4$ | $\partial_H(x + y) = \partial_H(x) + \partial_H(y)$ |
| $D5$ | $\partial_H(x \cdot y) = \partial_H(x) \cdot \partial_H(y)$ |
| $D6$ | $\partial_H(x \parallel y) = \partial_H(x) \parallel \partial_H(y)$ |

Similarly to the proof of soundness of parallelism modulo FR pomset bisimulation equivalence (see Theorem 4.17), we can prove that each axiom in Table 4.15 is sound modulo FR hp-bisimulation equivalence, we just need additionally to check the above conditions on FR hp-bisimulation, however, we omit these checks here. □

### 4.2.4 Encapsulation

The mismatch of two communicating events in different parallel branches can cause deadlock, so the deadlocks in the concurrent processes should be eliminated. Like $APTC$ [9], we also introduce the unary encapsulation operator $\partial_H$ for set $H$ of atomic events, which renames all atomic events in $H$ into $\delta$. The whole algebra including parallelism for true concurrency in the above sections, deadlock $\delta$ and encapsulation operator $\partial_H$, is called Reversible Algebra for Parallelism in True Concurrency, abbreviated $RAPTC$.

The forward transition rules of encapsulation operator $\partial_H$ are shown in Table 4.17, and the reverse transition rules of encapsulation operator $\partial_H$ are shown in Table 4.18.

Based on the transition rules for encapsulation operator $\partial_H$ in Table 4.17 and Table 4.18, we design the axioms as Table 4.19 shows.

**Theorem 4.19** (Conservativity of $RAPTC$ with respect to the reversible algebra for parallelism). *$RAPTC$ is a conservative extension of the reversible algebra for parallelism.*

*Proof.* It follows from the following two facts (see Theorem 2.5):

1. The transition rules of the reversible algebra for parallelism in the above sections are all source dependent;
2. The sources of the transition rules for the encapsulation operator contain an occurrence of $\partial_H$.

Hence, $RAPTC$ is a conservative extension of the reversible algebra for parallelism, as desired. □

**Theorem 4.20** (Congruence theorem of encapsulation operator $\partial_H$). *FR truly concurrent bisimulation equivalences $\sim_p^{fr}$, $\sim_s^{fr}$, $\sim_{hp}^{fr}$, and $\sim_{hhp}^{fr}$ are all congruences with respect to encapsulation operator $\partial_H$.*

*Proof.* (1) Case FR pomset bisimulation equivalence $\sim_p^{fr}$.

**TABLE 4.20** Term rewrite system of encapsulation operator $\partial_H$.

| No. | Rewriting Rule |
|---|---|
| $RRD1$ | $e \notin H \quad \partial_H(e) \to e$ |
| $RRRD1$ | $e \notin H \quad \partial_H(e[m]) \to e[m]$ |
| $RRD2$ | $e \in H \quad \partial_H(e) \to \delta$ |
| $RRRD2$ | $e \in H \quad \partial_H(e[m]) \to \delta$ |
| $RRD3$ | $\partial_H(\delta) \to \delta$ |
| $RRD4$ | $\partial_H(x+y) \to \partial_H(x)+\partial_H(y)$ |
| $RRD5$ | $\partial_H(x \cdot y) \to \partial_H(x) \cdot \partial_H(y)$ |
| $RRD6$ | $\partial_H(x \parallel y) \to \partial_H(x) \parallel \partial_H(y)$ |

Let $x$ and $y$ be $RAPTC$ processes, and $x \sim_p^{fr} y$, it is sufficient to prove that $\partial_H(x) \sim_p^{fr} \partial_H(y)$.

By the definition of FR pomset bisimulation $\sim_p^{fr}$ (Definition 2.68), $x \sim_p^{fr} y$ means that

$$x \xrightarrow{X} x' \quad y \xrightarrow{Y} y'$$

$$x \xrightarrow{X[\mathcal{K}]} x' \quad y \xrightarrow{Y[\mathcal{J}]} y'$$

with $X \subseteq x$, $Y \subseteq y$, $X \sim Y$ and $x' \sim_p^{fr} y'$.

By the FR pomset transition rules for encapsulation operator $\partial_H$ in Table 4.17 and Table 4.18, we can obtain

$$\partial_H(x) \xrightarrow{X} \partial_H(X[\mathcal{K}])(X \not\subseteq H) \quad \partial_H(y) \xrightarrow{Y} \partial_H(Y[\mathcal{J}])(Y \not\subseteq H)$$

$$\partial_H(x) \xrightarrow{X[\mathcal{K}]} \partial_H(X)(X \not\subseteq H) \quad \partial_H(y) \xrightarrow{Y[\mathcal{J}]} \partial_H(Y)(Y \not\subseteq H)$$

with $X \subseteq x$, $Y \subseteq y$, and $X \sim Y$, and the assumptions $\partial_H(X[\mathcal{K}]) \sim_p^{fr} \partial_H(Y[\mathcal{J}])$, $\partial_H(X) \sim_p^{fr} \partial_H(Y)$ hence, we obtain $\partial_H(x) \sim_p^{fr} \partial_H(y)$, as desired.

Or, we can obtain

$$\partial_H(x) \xrightarrow{X} \partial_H(x')(X \not\subseteq H) \quad \partial_H(y) \xrightarrow{Y} \partial_H(y')(Y \not\subseteq H)$$

$$\partial_H(x) \xrightarrow{X} \partial_H(x')(X \not\subseteq H) \quad \partial_H(y) \xrightarrow{Y} \partial_H(y')(Y \not\subseteq H)$$

With $X \subseteq x$, $Y \subseteq y$, $X \sim Y$, $x' \sim_p^{fr} y'$ and the assumption $\partial_H(x') \sim_p^{fr} \partial_H(y')$, hence, we obtain $\partial_H(x) \sim_p^{fr} \partial_H(y)$, as desired.

(2) The cases of FR step bisimulation $\sim_s^{fr}$, FR hp-bisimulation $\sim_{hp}^{fr}$ and FR hhp-bisimulation $\sim_{hhp}^{fr}$ can be proven similarly, however, we omit them here. □

**Theorem 4.21** (Elimination theorem of $RAPTC$). *Let $p$ be a closed $RAPTC$ term including the encapsulation operator $\partial_H$. Then, there is a basic $RAPTC$ term $q$ such that $RAPTC \vdash p = q$.*

*Proof.* (1) First, suppose that the following ordering on the signature of $RAPTC$ is defined: $\parallel > \cdot > +$ and the symbol $\cdot$ is given the lexicographical status for the first argument, then for each rewrite rule $p \to q$ in Table 4.20 relation $p >_{lpo} q$ can easily be proved. We obtain that the term rewrite system shown in Table 4.20 is strongly normalizing, for it has finitely many rewriting rules, and $>$ is a well-founded ordering on the signature of $RAPTC$, and if $s >_{lpo} t$, for each rewriting rule $s \to t$ is in Table 4.20 (see Theorem 2.9).

(2) Then, we prove that the normal forms of closed $RAPTC$ terms including encapsulation operator $\partial_H$ are basic $RAPTC$ terms.

Suppose that $p$ is a normal form of some closed $RAPTC$ term and suppose that $p$ is not a basic $RAPTC$ term. Let $p'$ denote the smallest sub-term of $p$ that is not a basic $RAPTC$ term. This implies that each sub-term of $p'$ is a basic $RAPTC$ term. Then, we prove that $p$ is not a term in normal form. It is sufficient to induct on the structure of $p'$, following from Theorem 4.14, we only prove the new case $p' \equiv \partial_H(p_1)$:

- Case $p_1 \equiv e$. The transition rules $RRD1$ or $RRD2$ can be applied, hence, $p$ is not a normal form;
- Case $p_1 \equiv e[m]$. The transition rules $RRRD1$ or $RRRD2$ can be applied, hence, $p$ is not a normal form;
- Case $p_1 \equiv \delta$. The transition rules $RRD3$ can be applied, hence, $p$ is not a normal form;
- Case $p_1 \equiv p_1' + p_1''$. The transition rule $RRD4$ can be applied, hence, $p$ is not a normal form;
- Case $p_1 \equiv p_1' \cdot p_1''$. The transition rule $RRD5$ can be applied, hence, $p$ is not a normal form;
- Case $p_1 \equiv p_1' \parallel p_1''$. The transition rule $RRD6$ can be applied, hence, $p$ is not a normal form. □

**Theorem 4.22** (Soundness of $RAPTC$ modulo FR step bisimulation equivalence). *Let $x$ and $y$ be $RAPTC$ terms including encapsulation operator $\partial_H$. If $RAPTC \vdash x = y$, then $x \sim_s^{fr} y$.*

*Proof.* Since FR step bisimulation $\sim_s^{fr}$ is both an equivalent and a congruent relation with respect to the operator $\partial_H$, we only need to check if each axiom in Table 4.19 is sound modulo FR step bisimulation equivalence.

Although transition rules in Table 4.17 and Table 4.18 are defined in the flavor of a single event, they can be modified into a step (a set of events within which each event is pairwise concurrent), however, we omit them here. If we treat a single event as a step containing just one event, the proof of this soundness theorem does not pose any problem, hence, we use this way and still use the transition rules in Table 4.17 and Table 4.18.

We omit the defining axioms, including axioms $D1 - D3$, and we only prove the soundness of the non-trivial axioms, including axioms $D4 - D6$.

- **Axiom** $D4$. Let $p, q$ be $RAPTC$ processes, and $\partial_H(p+q) = \partial_H(p) + \partial_H(q)$, it is sufficient to prove that $\partial_H(p+q) \sim_s^{fr} \partial_H(p) + \partial_H(q)$. By the forward transition rules for operator $+$ in Table 4.3 and $\partial_H$ in Table 4.17, we obtain

$$\frac{p \xrightarrow{e_1} e_1[m] \quad e_1 \notin q \quad (e_1 \notin H)}{\partial_H(p+q) \xrightarrow{e_1} \partial_H(e_1[m]+q)} \qquad \frac{p \xrightarrow{e_1} e_1[m] \quad e_1 \notin q \quad (e_1 \notin H)}{\partial_H(p)+\partial_H(q) \xrightarrow{e_1} \partial_H(e_1[m])+\partial_H(q)}$$

$$\frac{q \xrightarrow{e_2} e_2[m] \quad e_2 \notin p \quad (e_2 \notin H)}{\partial_H(p+q) \xrightarrow{e_2} \partial_H(p+e_2[m])} \qquad \frac{q \xrightarrow{e_2} e_2[m] \quad e_2 \notin p \quad (e_2 \notin H)}{\partial_H(p)+\partial_H(q) \xrightarrow{e_2} \partial_H(p)+\partial_H(e_2[m])}$$

$$\frac{p \xrightarrow{e_1} p' \quad e_1 \notin q \quad (e_1 \notin H)}{\partial_H(p+q) \xrightarrow{e_1} \partial_H(p'+q)} \qquad \frac{p \xrightarrow{e_1} p' \quad e_1 \notin q \quad (e_1 \notin H)}{\partial_H(p)+\partial_H(q) \xrightarrow{e_1} \partial_H(p')+\partial(q)}$$

$$\frac{q \xrightarrow{e_2} q' \quad e_2 \notin p \quad (e_2 \notin H)}{\partial_H(p+q) \xrightarrow{e_2} \partial_H(p+q')} \qquad \frac{q \xrightarrow{e_2} q' \quad e_2 \notin p \quad (e_2 \notin H)}{\partial_H(p)+\partial_H(q) \xrightarrow{e_2} \partial_H(p)+\partial_H(q')}$$

$$\frac{q \xrightarrow{e_2} q' \quad e_2 \notin p \quad (e_2 \notin H)}{\partial_H(p+q) \xrightarrow{e_2} \partial_H(p+q')} \qquad \frac{q \xrightarrow{e_2} q' \quad e_2 \notin p \quad (e_2 \notin H)}{\partial_H(p)+\partial_H(q) \xrightarrow{e_2} \partial_H(p)+\partial_H(q')}$$

$$\frac{p \xrightarrow{e_1} p' \quad q \xrightarrow{e_1} q' \quad (e_1 \notin H)}{\partial_H(p+q) \xrightarrow{e_1} \partial_H(p'+q')} \qquad \frac{p \xrightarrow{e_1} p' \quad q \xrightarrow{e_1} q' \quad (e_1 \notin H)}{\partial_H(p)+\partial_H(q) \xrightarrow{e_1} \partial_H(p')+\partial(q')}$$

By the reverse transition rules for operator $+$ in Table 4.4 and $\partial_H$ in Table 4.18, we obtain

$$\frac{p \xrightarrow{e_1[m]} e_1 \quad e_1 \notin q \quad (e_1 \notin H)}{\partial_H(p+q) \xrightarrow{e_1[m]} \partial_H(e_1+q)} \qquad \frac{p \xrightarrow{e_1[m]} e_1 \quad e_1 \notin q \quad (e_1 \notin H)}{\partial_H(p)+\partial_H(q) \xrightarrow{e_1[m]} \partial_H(e_1)+\partial_H(q)}$$

$$\frac{q \xrightarrow{e_2[m]} e_2 \quad e_2 \notin p \quad (e_2 \notin H)}{\partial_H(p+q) \xrightarrow{e_2[m]} \partial_H(p+e_2)} \qquad \frac{q \xrightarrow{e_2[m]} e_2 \quad e_2 \notin p \quad (e_2 \notin H)}{\partial_H(p)+\partial_H(q) \xrightarrow{e_2[m]} \partial_H(p)+\partial_H(e_2)}$$

$$\frac{p \xrightarrow{e_1[m]} p' \quad e_1 \notin q \quad (e_1 \notin H)}{\partial_H(p+q) \xrightarrow{e_1[m]} \partial_H(p'+q)} \qquad \frac{p \xrightarrow{e_1[m]} p' \quad e_1 \notin q \quad (e_1 \notin H)}{\partial_H(p)+\partial_H(q) \xrightarrow{e_1[m]} \partial_H(p')+\partial(q)}$$

$$\frac{q \xrightarrow{e_2[m]} q' \quad e_2 \notin p \quad (e_2 \notin H)}{\partial_H(p+q) \xrightarrow{e_2[m]} \partial_H(p+q')} \qquad \frac{q \xrightarrow{e_2[m]} q' \quad e_2 \notin p \quad (e_2 \notin H)}{\partial_H(p)+\partial_H(q) \xrightarrow{e_2[m]} \partial_H(p)+\partial_H(q')}$$

$$\frac{q \xrightarrow{e_2[m]} q' \quad e_2 \notin p \quad (e_2 \notin H)}{\partial_H(p+q) \xrightarrow{e_2[m]} \partial_H(p+q')} \qquad \frac{q \xrightarrow{e_2[m]} q' \quad e_2 \notin p \quad (e_2 \notin H)}{\partial_H(p)+\partial_H(q) \xrightarrow{e_2[m]} \partial_H(p)+\partial_H(q')}$$

$$\frac{p \xrightarrow{e_1[m]} p' \quad q \xrightarrow{e_1[m]} q' \quad (e_1 \notin H)}{\partial_H(p+q) \xrightarrow{e_1[m]} \partial_H(p'+q')} \qquad \frac{p \xrightarrow{e_1[m]} p' \quad q \xrightarrow{e_1[m]} q' \quad (e_1 \notin H)}{\partial_H(p)+\partial_H(q) \xrightarrow{e_1[m]} \partial_H(p')+\partial(q')}$$

Hence, with the assumptions $\partial_H(e_1[m]+q) \sim_s^{fr} \partial_H(e_1[m])+\partial_H(q)$, $\partial_H(e_1+q) \sim_s^{fr} \partial_H(e_1)+\partial_H(q)$, $\partial_H(p+e_2[m]) \sim_s^{fr} \partial_H(p)+\partial_H(e_2[m])$, $\partial_H(p+e_2) \sim_s^{fr} \partial_H(p)+\partial_H(e_2)$, $\partial_H(p'+q) \sim_s^{fr} \partial_H(p')+\partial_H(q)$, $\partial_H(p+q') \sim_s^{fr} \partial_H(p)+\partial_H(q')$, $\partial_H(p'+q') \sim_s^{fr} \partial_H(p')+\partial_H(q')$, $\partial_H(p+q) \sim_s^{fr} \partial_H(p)+\partial_H(q)$, as desired.

- **Axiom** $D5$. Let $p, q$ be $RAPTC$ processes, and $\partial_H(p\cdot q) = \partial_H(p)\cdot\partial_H(q)$, it is sufficient to prove that $\partial_H(p\cdot q) \sim_s^{fr} \partial_H(p)\cdot\partial_H(q)$. By the forward transition rules for operator $\cdot$ in Table 4.3 and $\partial_H$ in Table 4.17, we obtain

$$\frac{p \xrightarrow{e_1} e_1[m] \quad (e_1 \notin H)}{\partial_H(p\cdot q) \xrightarrow{e_1} \partial_H(e_1[m]\cdot q)} \qquad \frac{p \xrightarrow{e_1} e_1[m] \quad (e_1 \notin H)}{\partial_H(p)\cdot\partial_H(q) \xrightarrow{e_1} \partial_H(e_1[m])\cdot\partial_H(q)}$$

$$\frac{p \xrightarrow{e_1} p' \quad (e_1 \notin H)}{\partial_H(p\cdot q) \xrightarrow{e_1} \partial_H(p'\cdot q)} \qquad \frac{p \xrightarrow{e_1} p' \quad (e_1 \notin H)}{\partial_H(p)\cdot\partial_H(q) \xrightarrow{e_1} \partial_H(p')\cdot\partial_H(q)}$$

By the reverse transition rules for operator $\cdot$ in Table 4.4 and $\partial_H$ in Table 4.18, we obtain

$$\frac{q \xrightarrow{e_2[m]} e_2 \quad (e_2 \notin H)}{\partial_H(p\cdot q) \xrightarrow{e_2[m]} \partial_H(p\cdot e_2)} \qquad \frac{q \xrightarrow{e_2[m]} e_2 \quad (e_2 \notin H)}{\partial_H(p)\cdot\partial_H(q) \xrightarrow{e_2[m]} \partial_H(p)\cdot\partial_H(e_2)}$$

$$\frac{q \xrightarrow{e_2[m]} q' \quad (e_2 \notin H)}{\partial_H(p\cdot q) \xrightarrow{e_2[m]} \partial_H(p\cdot q')} \qquad \frac{q \xrightarrow{e_2[m]} q' \quad (e_2 \notin H)}{\partial_H(p)\cdot\partial_H(q) \xrightarrow{e_2[m]} \partial_H(p)\cdot\partial_H(q')}$$

Hence, with the assumptions $\partial_H(e_1[m]\cdot q) \sim_s^{fr} \partial_H(e_1[m])\cdot\partial_H(q)$, $\partial_H(p\cdot e_2) \sim_s^{fr} \partial_H(p)\cdot\partial_H(e_2)$, $\partial_H(p'\cdot q) \sim_s^{fr} \partial_H(p')\cdot\partial_H(q)$, $\partial_H(p\cdot q') \sim_s^{fr} \partial_H(p)\cdot\partial_H(q')$, $\partial_H(p\cdot q) \sim_s^{fr} \partial_H(p)\cdot\partial_H(q)$, as desired.

- **Axiom** $D6$. Let $p, q$ be $RAPTC$ processes, and $\partial_H(p \parallel q) = \partial_H(p) \parallel \partial_H(q)$, it is sufficient to prove that $\partial_H(p \parallel q) \sim_s^{fr} \partial_H(p) \parallel \partial_H(q)$. By the forward transition rules for operator $\parallel$ in Table 4.9 and $\partial_H$ in Table 4.17, we obtain

$$\frac{p \xrightarrow{e_1} e_1[m] \quad q \xrightarrow{e_2} e_2[m] \quad (e_1, e_2 \notin H)}{\partial_H(p \parallel q) \xrightarrow{\{e_1,e_2\}} \partial_H(e_1[m] \parallel e_2[m])} \qquad \frac{p \xrightarrow{e_1} e_1[m] \quad q \xrightarrow{e_2} e_2[m] \quad (e_1, e_2 \notin H)}{\partial_H(p) \parallel \partial_H(q) \xrightarrow{\{e_1,e_2\}} \partial_H(e_1[m]) \parallel \partial_H(e_2[m])}$$

$$\frac{p \xrightarrow{e_1} p' \quad q \xrightarrow{e_2} e_2[m] \quad (e_1, e_2 \notin H)}{\partial_H(p \parallel q) \xrightarrow{\{e_1,e_2\}} \partial_H(p' \parallel e_2[m])} \qquad \frac{p \xrightarrow{e_1} p' \quad q \xrightarrow{e_2} e_2[m] \quad (e_1, e_2 \notin H)}{\partial_H(p) \parallel \partial_H(q) \xrightarrow{\{e_1,e_2\}} \partial_H(p') \parallel \partial_H(e_2[m])}$$

$$\frac{p \xrightarrow{e_1} e_1[m] \quad q \xrightarrow{e_2} q' \quad (e_1, e_2 \notin H)}{\partial_H(p \parallel q) \xrightarrow{\{e_1,e_2\}} \partial_H(e_1[m] \parallel q')} \qquad \frac{p \xrightarrow{e_1} e_1[m] \quad q \xrightarrow{e_2} q' \quad (e_1, e_2 \notin H)}{\partial_H(p) \parallel \partial_H(q) \xrightarrow{\{e_1,e_2\}} \partial_H(e_1[m])\partial_H(q')}$$

$$\frac{p \xrightarrow{e_1} p' \quad q \xrightarrow{e_2} q' \quad (e_1, e_2 \notin H)}{\partial_H(p \parallel q) \xrightarrow{\{e_1,e_2\}} \partial_H(p' \between q')} \qquad \frac{p \xrightarrow{e_1} p' \quad q \xrightarrow{e_2} q' \quad (e_1, e_2 \notin H)}{\partial_H(p) \parallel \partial_H(q) \xrightarrow{\{e_1,e_2\}} \partial_H(p') \between \partial_H(q')}$$

By the reverse transition rules for operator $\parallel$ in Table 4.10 and $\partial_H$ in Table 4.18, we obtain

$$\frac{p \xrightarrow{e_1[m]} e_1 \quad q \xrightarrow{e_2[m]} e_2 \quad (e_1, e_2 \notin H)}{\partial_H(p \parallel q) \xrightarrow{\{e_1[m],e_2[m]\}} \partial_H(e_1 \parallel e_2)} \qquad \frac{p \xrightarrow{e_1[m]} e_1 \quad q \xrightarrow{e_2[m]} e_2 \quad (e_1, e_2 \notin H)}{\partial_H(p) \parallel \partial_H(q) \xrightarrow{\{e_1[m],e_2[m]\}} \partial_H(e_1) \parallel \partial_H(e_2)}$$

$$\frac{p\xrightarrow{e_1[m]}p'\quad q\xrightarrow{e_2[m]}e_2\quad (e_1,e_2\notin H)}{\partial_H(p\parallel q)\xrightarrow{\{e_1[m],e_2[m]\}}\partial_H(p'\parallel e_2)}\qquad\frac{p\xrightarrow{e_1[m]}p'\quad q\xrightarrow{e_2[m]}e_2\quad (e_1,e_2\notin H)}{\partial_H(p)\parallel\partial_H(q)\xrightarrow{\{e_1[m],e_2[m]\}}\partial_H(p')\parallel\partial_H(e_2)}$$

$$\frac{p\xrightarrow{e_1[m]}e_1\quad q\xrightarrow{e_2[m]}q'\quad (e_1,e_2\notin H)}{\partial_H(p\parallel q)\xrightarrow{\{e_1[m],e_2[m]\}}\partial_H(e_1\parallel q')}\qquad\frac{p\xrightarrow{e_1[m]}e_1\quad q\xrightarrow{e_2[m]}q'\quad (e_1,e_2\notin H)}{\partial_H(p)\parallel\partial_H(q)\xrightarrow{\{e_1[m],e_2[m]\}}\partial_H(e_1)\partial_H(q')}$$

$$\frac{p\xrightarrow{e_1[m]}p'\quad q\xrightarrow{e_2[m]}q'\quad (e_1,e_2\notin H)}{\partial_H(p\parallel q)\xrightarrow{\{e_1[m],e_2[m]\}}\partial_H(p'\between q')}\qquad\frac{p\xrightarrow{e_1[m]}p'\quad q\xrightarrow{e_2[m]}q'\quad (e_1,e_2\notin H)}{\partial_H(p)\parallel\partial_H(q)\xrightarrow{\{e_1[m],e_2[m]\}}\partial_H(p')\between\partial_H(q')}$$

Hence, with the assumptions $\partial_H(e_1[m]\parallel q')\sim_s^{fr}\partial_H(e_1[m])\parallel\partial_H(q')$, $\partial_H(e_1\parallel q')\sim_s^{fr}\partial_H(e_1)\parallel\partial_H(q')$, $\partial_H(p'\parallel e_2[m])\sim_s^{fr}\partial_H(p')\parallel\partial_H(e_2[m])$, $\partial_H(p'\parallel e_2)\sim_s^{fr}\partial_H(p')\parallel\partial_H(e_2)$, $\partial_H(p'\between q')\sim_s^{fr}\partial_H(p')\between\partial_H(q')$, $\partial_H(p\parallel q)\sim_s^{fr}\partial_H(p)\parallel\partial_H(q)$, as desired. □

**Theorem 4.23** (Soundness of $RAPTC$ modulo FR pomset bisimulation equivalence). *Let $x$ and $y$ be $RAPTC$ terms including encapsulation operator $\partial_H$. If $RAPTC\vdash x=y$, then $x\sim_p^{fr}y$.*

*Proof.* Since FR pomset bisimulation $\sim_p^{fr}$ is both an equivalent and a congruent relation with respect to the operator $\partial_H$, we only need to check if each axiom in Table 4.19 is sound modulo FR pomset bisimulation equivalence.

From the definition of FR pomset bisimulation (see Definition 2.68), we know that FR pomset bisimulation is defined by pomset transitions, which are labeled by pomsets. In a pomset transition, the events in the pomset are either within causality relations (defined by $\cdot$) or in concurrency (implicitly defined by $\cdot$ and $+$, and explicitly defined by $\between$), of course, they are pairwise consistent (without conflicts). In Theorem 4.22, we have already proven the case that all events are pairwise concurrent, hence, we only need to prove the case of events in causality. Without loss of generality, we take a pomset of $P=\{e_1,e_2:e_1\cdot e_2\}$. Then, the pomset transition labeled by the above $P$ is just composed of one single event transition labeled by $e_1$ succeeded by another single event transition labeled by $e_2$, that is, $\xrightarrow{P}=\xrightarrow{e_1}\xrightarrow{e_2}$ or $\xrightarrow{P}=\xrightarrow{e_2[n]}\xrightarrow{e_1[m]}$.

Similarly to the proof of soundness of $RAPTC$ modulo FR step bisimulation equivalence (see Theorem 4.22), we can prove that each axiom in Table 4.19 is sound modulo FR pomset bisimulation equivalence, however, we omit these proofs here. □

**Theorem 4.24** (Soundness of $RAPTC$ modulo FR hp-bisimulation equivalence). *Let $x$ and $y$ be $RAPTC$ terms including encapsulation operator $\partial_H$. If $RAPTC\vdash x=y$, then $x\sim_{hp}^{fr}y$.*

*Proof.* Since FR hp-bisimulation $\sim_{hp}^{fr}$ is both an equivalent and a congruent relation with respect to the operator $\partial_H$, we only need to check if each axiom in Table 4.19 is sound modulo FR hp-bisimulation equivalence.

From the definition of FR hp-bisimulation (see Definition 2.70), we know that FR hp-bisimulation is defined on the posetal product $(C_1,f,C_2)$, $f:C_1\to C_2$ isomorphism. Two process terms $s$ related to $C_1$ and $t$ related to $C_2$, and $f:C_1\to C_2$ isomorphism. Initially, $(C_1,f,C_2)=(\emptyset,\emptyset,\emptyset)$, and $(\emptyset,\emptyset,\emptyset)\in\sim_{hp}^{fr}$. When $s\xrightarrow{e}s'$ $(C_1\xrightarrow{e}C_1')$, there will be $t\xrightarrow{e}t'$ $(C_2\xrightarrow{e}C_2')$, and we define $f'=f[e\mapsto e]$. Also, when $s\xrightarrow{e[m]}s'$ $(C_1\xrightarrow{e[m]}C_1')$, there will be $t\xrightarrow{e[m]}t'$ $(C_2\xrightarrow{e[m]}C_2')$, and we define $f'=f[e[m]\mapsto e[m]]$. Then, if $(C_1,f,C_2)\in\sim_{hp}^{fr}$, then $(C_1',f',C_2')\in\sim_{hp}^{fr}$.

Similarly to the proof of soundness of $RAPTC$ modulo FR pomset bisimulation equivalence (see Theorem 4.23), we can prove that each axiom in Table 4.19 is sound modulo FR hp-bisimulation equivalence, we just need additionally to check the above conditions on FR hp-bisimulation, however, we omit them here. □

### 4.2.5 Recursion

**Definition 4.25** (Weakly guarded recursive expression). $X$ is weakly guarded in $E$ if each occurrence of $X$ is with some subexpression $\alpha.F$ or $(\alpha_1\parallel\cdots\parallel\alpha_n).F$ or $F.\alpha[m]$ or $F.(\alpha_1[m]\parallel\cdots\parallel\alpha_n[m])$ of $E$.

**Proposition 4.26.** *If the variables $\widetilde{X}$ are weakly guarded in E, and $E\{\widetilde{P}/\widetilde{X}\}\xrightarrow{\{\alpha_1,\cdots,\alpha_n\}}P'$, or $E\{\widetilde{P}/\widetilde{X}\}\xrightarrow{\{\alpha_1[m],\cdots,\alpha_n[m]\}}P'$, then $P'$ cannot take the form $E'\{\widetilde{P}/\widetilde{X}\}$ for some expression $E'$.*

TABLE 4.21 Transition rule of the silent step.

| |
|---|
| $\frac{}{\tau \xrightarrow{\tau} \surd}$ |
| $\frac{}{\tau \xtwoheadrightarrow{\tau} \surd}$ |

TABLE 4.22 Axioms of silent step.

| No. | Axiom |
|---|---|
| $B1$ | $e \cdot \tau = e$ |
| $RB1$ | $\tau \cdot e[m] = e[m]$ |
| $B2$ | $e \cdot (\tau \cdot (x + y) + x) = e \cdot (x + y)$ |
| $RB2$ | $((x + y) \cdot \tau + x) \cdot e[m] = (x + y) \cdot e[m]$ |
| $B3$ | $x \parallel \tau = x$ |

*Proof.* It needs to induct on the depth of the inference of $E\{\widetilde{P}/\widetilde{X}\} \xrightarrow{\{\alpha_1,\cdots,\alpha_n\}} P'$ or $E\{\widetilde{P}/\widetilde{X}\} \xtwoheadrightarrow{\{\alpha_1[m],\cdots,\alpha_n[m]\}} P'$. We consider $E\{\widetilde{P}/\widetilde{X}\} \xrightarrow{\{\alpha_1,\cdots,\alpha_n\}} P'$.

Case $E \equiv E_1 + E_2$. We may have $E_1 \xrightarrow{e_1} e_1[m] \cdot E_1'$ $e_1 \notin E_2$, $E_1 + E_2 \xrightarrow{e_1} e_1[m] \cdot E_1' + E_2$, $e_1[m] \cdot E_1' + E_2$ cannot take the form $E'\{\widetilde{P}/\widetilde{X}\}$ for some expression $E'$.

Hence, there may be no recursive expression for strongly FR truly concurrent bisimulations. For the same reason, there also may be no recursive expression for weakly FR truly concurrent bisimulations. □

## 4.3 Abstraction

To abstract away from the internal implementations of a program, and verify that the program exhibits the desired external behaviors, the silent step $\tau$ and abstraction operator $\tau_I$ are introduced, where $I \subseteq \mathbb{E}$ denotes the internal events. The transition rule of $\tau$ is shown in Table 4.21. In the following, let the atomic event $e$ range over $\mathbb{E} \cup \{\delta\} \cup \{\tau\}$, and let the communication function $\gamma : \mathbb{E} \cup \{\tau\} \times \mathbb{E} \cup \{\tau\} \to \mathbb{E} \cup \{\delta\}$, with each communication involved $\tau$ resulting in $\delta$.

**Theorem 4.27** (Conservativity of $RAPTC$ with silent step). *$RAPTC$ with silent step is a conservative extension of $RAPTC$.*

*Proof.* Since the transition rules of $RAPTC$ are source dependent, and the transition rules for silent step in Table 4.21 contain only a fresh constant $\tau$ in their source, the transition rules of $RAPTC$ with silent step is a conservative extension of those of $RAPTC$. □

**Theorem 4.28** (Congruence theorem of $RAPTC$ with silent step). *Rooted branching FR truly concurrent bisimulation equivalences $\approx_{rbp}^{fr}$, $\approx_{rbs}^{fr}$, and $\approx_{rbhp}^{fr}$ are all congruences with respect to $RAPTC$ with silent step.*

*Proof.* It follows from the following two facts:

1. FR truly concurrent bisimulation equivalences $\sim_p^{fr}$, $\sim_s^{fr}$, and $\sim_{hp}^{fr}$ are all congruences with respect to all operators of $RAPTC$, while FR truly concurrent bisimulation equivalences $\sim_p^{fr}$, $\sim_s^{fr}$, and $\sim_{hp}^{fr}$ imply the corresponding rooted branching FR truly concurrent bisimulation $\approx rbp^{fr}$, $\approx_{rbs}^{fr}$, and $\approx_{rbhp}^{fr}$, hence, rooted branching FR truly concurrent bisimulation $\approx rbp^{fr}$, $\approx_{rbs}^{fr}$, and $\approx_{rbhp}^{fr}$ are all congruences with respect to all operators of $RAPTC$;
2. While $\mathbb{E}$ is extended to $\mathbb{E} \cup \{\tau\}$, it can be proved that rooted branching FR truly concurrent bisimulation $\approx rbp^{fr}$, $\approx_{rbs}^{fr}$, and $\approx_{rbhp}^{fr}$ are all congruences with respect to all operators of $RAPTC$, however, we omit these proofs here. □

### 4.3.1 Algebraic laws for the silent step

We design the axioms for the silent step $\tau$ in Table 4.22.

**Theorem 4.29** (Soundness of $RAPTC$ with silent step). *Let $x$ and $y$ be $RAPTC$ with silent step terms. If $RAPTC$ with silent step $\vdash x = y$, then*

1. $x \approx^{fr}_{rbs} y$;
2. $x \approx^{fr}_{rbp} y$;
3. $x \approx^{fr}_{rbhp} y$.

*Proof.* (1) Soundness of $RAPTC$ with silent step with respect to rooted branching FR step bisimulation $\approx^{fr}_{rbs}$.

Since rooted branching FR step bisimulation $\approx^{fr}_{rbs}$ is both an equivalent and a congruent relation with respect to $RAPTC$ with silent step, we only need to check if each axiom in Table 4.22 is sound modulo rooted branching FR step bisimulation equivalence.

Although transition rules in Table 4.21 are defined in the flavor of a single event, they can be modified into a step (a set of events within which each event is pairwise concurrent), however, we omit them here. If we treat a single event as a step containing just one event, the proof of this soundness theorem does not pose any problem, hence, we use this way and still use the transition rules in Table 4.21.

- **Axiom** $B1$. Assume that $e \cdot \tau = e$, it is sufficient to prove that $e \cdot \tau \approx^{fr}_{rbs} e$. By the forward transition rules for operator $\cdot$ in Table 4.7 and $\tau$ in Table 4.21, we obtain

$$\frac{e \xrightarrow{e} e[m]}{e \cdot \tau \xrightarrow{e} \xrightarrow{\tau} e[m]}$$

$$\frac{e \xrightarrow{e} e[m]}{e \xrightarrow{e} e[m]}$$

By the reverse transition rules for operator $\cdot$ in Table 4.8 and $\tau$ in Table 4.21, there are no transitions.
Hence, $e \cdot \tau \approx^{fr}_{rbs} e$, as desired.

- **Axiom** $RB1$. Assume that $\tau \cdot e[m] = e[m]$, it is sufficient to prove that $\tau \cdot e[m] \approx^{fr}_{rbs} e[m]$. By the forward transition rules for operator $\cdot$ in Table 4.7 and $\tau$ in Table 4.21, there are no transitions.
By the reverse transition rules for operator $\cdot$ in Table 4.8 and $\tau$ in Table 4.21, we obtain

$$\frac{e[m] \xrightarrow{e[m]}\!\!\!\!\twoheadrightarrow e}{\tau \cdot e[m] \xrightarrow{e[m]}\!\!\!\!\twoheadrightarrow \xrightarrow{\tau}\!\!\!\!\twoheadrightarrow e}$$

$$\frac{e[m] \xrightarrow{e[m]}\!\!\!\!\twoheadrightarrow e}{e[m] \xrightarrow{e[m]}\!\!\!\!\twoheadrightarrow e}$$

Hence, $\tau \cdot e[m] \approx^{fr}_{rbs} e[m]$, as desired.

- **Axiom** $B2$. Let $p$ and $q$ be $RAPTC$ with silent step processes, and assume that $e \cdot (\tau \cdot (p+q) + p) = e \cdot (p+q)$, it is sufficient to prove that $e \cdot (\tau \cdot (p+q) + p) \approx^{fr}_{rbs} e \cdot (p+q)$. There are several cases, however, we will not enumerate them all here. By the forward transition rules for operators $\cdot$ and $+$ in Table 4.7 and $\tau$ in Table 4.21, we obtain

$$\frac{e \xrightarrow{e} e[m] \quad p \xrightarrow{e_1} p' \quad q \xrightarrow{e_1} q'}{e \cdot (\tau \cdot (p+q) + p) \xrightarrow{e} \xrightarrow{\tau} \xrightarrow{e_1} e[m] \cdot ((p' + q') + p')}$$

$$\frac{e \xrightarrow{e} e[m] \quad p \xrightarrow{e_1} p'}{e \cdot (p+q) \xrightarrow{e} \xrightarrow{e_1} e[m] \cdot (p' + q')}$$

By the reverse transition rules for operators $\cdot$ and $+$ in Table 4.8 and $\tau$ in Table 4.21, there are no transitions.
Hence, $e \cdot (\tau \cdot (p+q) + p) \approx^{fr}_{rbs} e \cdot (p+q)$, as desired.

- **Axiom** $RB2$. Let $p$ and $q$ be $RAPTC$ with silent step processes, and assume that $((x+y) \cdot \tau + x) \cdot e[m] = (x+y) \cdot e[m]$, it is sufficient to prove that $((x+y) \cdot \tau + x) \cdot e[m] \approx^{fr}_{rbs} (x+y) \cdot e[m]$. There are several cases, however, we will not

enumerate them all here. By the forward transition rules for operators $\cdot$ and $+$ in Table 4.7 and $\tau$ in Table 4.21, there are no transitions.
By the reverse transition rules for operators $\cdot$ and $+$ in Table 4.8 and $\tau$ in Table 4.21, we obtain

$$\frac{e[m] \xrightarrow{e[m]} e \quad p \xrightarrow{e_1[n]} p' \quad q \xrightarrow{e_1[n]} q'}{((p+q)\cdot\tau+p)\cdot e[m] \xrightarrow{e[m]}\xrightarrow{\tau}\xrightarrow{e_1[n]} ((p'+q')+p')\cdot e}$$

$$\frac{e[m] \xrightarrow{e[m]} e \quad p \xrightarrow{e_1[n]} p'}{(p+q)\cdot e[m] \xrightarrow{e[m]}\xrightarrow{e_1[n]} (p'+q'\cdot e)}$$

Hence, $((p+q)\cdot\tau+p)\cdot e[m] \approx_{rbs}^{fr} (p+q)\cdot e[m]$, as desired.

- **Axiom** $B3$. Let $p$ be an $RAPTC$ with silent step, and assume that $p \parallel \tau = p$, it is sufficient to prove that $p \parallel \tau \approx_{rbs}^{fr} p$. By the forward transition rules for operator $\parallel$ in Table 4.9 and $\tau$ in Table 4.21, we obtain

$$\frac{p \xrightarrow{e} e[m]}{p \parallel \tau \xRightarrow{e} e[m]}$$

$$\frac{p \xrightarrow{e} p'}{p \parallel \tau \xRightarrow{e} p'}$$

By the reverse transition rules for operator $\parallel$ in Table 4.10 and $\tau$ in Table 4.21, we obtain

$$\frac{p \xrightarrow{e[m]} e}{p \parallel \tau \xRightarrow{e[m]} e}$$

$$\frac{p \xrightarrow{e[m]} p'}{p \parallel \tau \xRightarrow{e[m]} p'}$$

Hence, $p \parallel \tau \approx_{rbs}^{fr} p$, as desired.

(2) Soundness of $RAPTC$ with silent step with respect to rooted branching FR pomset bisimulation $\approx_{rbp}^{fr}$.

Since rooted branching FR pomset bisimulation $\approx_{rbp}^{fr}$ is both an equivalent and a congruent relation with respect to $RAPTC$ with silent step, we only need to check if each axiom in Table 4.22 is sound modulo rooted branching FR pomset bisimulation $\approx_{rbp}^{fr}$.

From the definition of rooted branching FR pomset bisimulation $\approx_{rbp}^{fr}$ (see Definition 2.73), we know that rooted branching FR pomset bisimulation $\approx_{rbp}^{fr}$ is defined by weak pomset transitions, which are labeled by pomsets with $\tau$. In a weak pomset transition, the events in the pomset are either within causality relations (defined by $\cdot$) or in concurrency (implicitly defined by $\cdot$ and $+$, and explicitly defined by $\between$), of course, they are pairwise consistent (without conflicts). In (1), we have already proven the case that all events are pairwise concurrent, hence, we only need to prove the case of events in causality. Without loss of generality, we take a pomset of $P = \{e_1, e_2 : e_1 \cdot e_2\}$. Then, the weak pomset transition labeled by the above $P$ is just composed of one single event transition labeled by $e_1$ succeeded by another single event transition labeled by $e_2$, that is, $\xRightarrow{P} = \xRightarrow{e_1}\xRightarrow{e_2}$ or $\xRightarrow{P} = \xRightarrow{e_2}\xRightarrow{e_1}$.

Similarly to the proof of soundness of $RAPTC$ with silent step modulo rooted branching FR step bisimulation $\approx_{rbs}^{fr}$ (1), we can prove that each axiom in Table 4.22 is sound modulo rooted branching FR pomset bisimulation $\approx_{rbp}^{fr}$, however, we omit these proofs here.

(3) Soundness of $RAPTC$ with silent step with respect to rooted branching FR hp-bisimulation $\approx_{rbhp}^{fr}$.

Since rooted branching FR hp-bisimulation $\approx_{rbhp}^{fr}$ is both an equivalent and a congruent relation with respect to $RAPTC$ with silent step, we only need to check if each axiom in Table 4.22 is sound modulo rooted branching FR hp-bisimulation $\approx_{rbhp}^{fr}$.

**TABLE 4.23 Transition rule of the abstraction operator.**

| | |
|---|---|
| $\dfrac{x \xrightarrow{e} \surd}{\tau_I(x) \xrightarrow{e} \surd} \quad e \notin I$ | $\dfrac{x \xrightarrow{e} x'}{\tau_I(x) \xrightarrow{e} \tau_I(x')} \quad e \notin I$ |
| $\dfrac{x \xrightarrow{e} \surd}{\tau_I(x) \xrightarrow{\tau} \surd} \quad e \in I$ | $\dfrac{x \xrightarrow{e} x'}{\tau_I(x) \xrightarrow{\tau} \tau_I(x')} \quad e \in I$ |
| $\dfrac{x \xtwoheadrightarrow{e[m]} e}{\tau_I(x) \xtwoheadrightarrow{e[m]} e} \quad e[m] \notin I$ | $\dfrac{x \xtwoheadrightarrow{e[m]} x'}{\tau_I(x) \xtwoheadrightarrow{e[m]} \tau_I(x')} \quad e[m] \notin I$ |
| $\dfrac{x \xtwoheadrightarrow{e[m]} \surd}{\tau_I(x) \xtwoheadrightarrow{\tau} \surd} \quad e[m] \in I$ | $\dfrac{x \xtwoheadrightarrow{e[m]} x'}{\tau_I(x) \xtwoheadrightarrow{\tau} \tau_I(x')} \quad e[m] \in I$ |

From the definition of rooted branching FR hp-bisimulation $\approx^{fr}_{rbhp}$ (see Definition 2.75), we know that rooted branching FR hp-bisimulation $\approx^{fr}_{rbhp}$ is defined on the weakly posetal product $(C_1, f, C_2), f : \hat{C_1} \to \hat{C_2}$ isomorphism. Two process terms $s$ related to $C_1$ and $t$ related to $C_2$, and $f : \hat{C_1} \to \hat{C_2}$ isomorphism. Initially, $(C_1, f, C_2) = (\emptyset, \emptyset, \emptyset)$, and $(\emptyset, \emptyset, \emptyset) \in \approx^{fr}_{rbhp}$. When $s \xrightarrow{e} s'$ $(C_1 \xrightarrow{e} C_1')$, there will be $t \xRightarrow{e} t'$ $(C_2 \xRightarrow{e} C_2')$, and we define $f' = f[e \mapsto e]$. Also, when $s \xRrightarrow{e[m]} s'$ $(C_1 \xRrightarrow{e[m]} C_1')$, there will be $t \xRrightarrow{e[m]} t'$ $(C_2 \xRrightarrow{e[m]} C_2')$, and we define $f' = f[e[m] \mapsto e[m]$. Then, if $(C_1, f, C_2) \in \approx^{fr}_{rbhp}$, then $(C_1', f', C_2') \in \approx^{fr}_{rbhp}$.

Similarly to the proof of soundness of $RAPTC$ with silent step modulo rooted branching FR pomset bisimulation equivalence (2), we can prove that each axiom in Table 4.22 is sound modulo rooted branching FR hp-bisimulation equivalence, we just need additionally to check the above conditions on rooted branching FR hp-bisimulation, however, we omit them here. □

### 4.3.2 Abstraction

The unary abstraction operator $\tau_I$ $(I \subseteq \mathbb{E})$ renames all atomic events in $I$ into $\tau$. $RAPTC$ with silent step and abstraction operator is called $RAPTC_\tau$. The transition rules of operator $\tau_I$ are shown in Table 4.23.

**Theorem 4.30** (Conservativity of $RAPTC_\tau$). *$RAPTC_\tau$ is a conservative extension of $RAPTC$ with silent step.*

*Proof.* Since the transition rules of $RAPTC$ with silent step are source dependent, and the transition rules for abstraction operator in Table 4.23 contain only a fresh operator $\tau_I$ in their source, the transition rules of $RAPTC_\tau$ is a conservative extension of those of $RAPTC$ with silent step. □

**Theorem 4.31** (Congruence theorem of $RAPTC_\tau$). *Rooted branching FR truly concurrent bisimulation equivalences $\approx^{fr}_{rbp}$, $\approx^{fr}_{rbs}$, and $\approx^{fr}_{rbhp}$ are all congruences with respect to $RAPTC_\tau$.*

*Proof.* (1) Case rooted branching FR pomset bisimulation equivalence $\approx^{fr}_{rbp}$.

Let $x$ and $y$ be $RAPTC_\tau$ processes, and $x \approx^{fr}_{rbp} y$, it is sufficient to prove that $\tau_I(x) \approx^{fr}_{rbp} \tau_I(y)$.

By the transition rules for operator $\tau_I$ in Table 4.23, we can obtain

$$\tau_I(x) \xrightarrow{X} X[\mathcal{K}](X \not\subseteq I) \quad \tau_I(y) \xrightarrow{Y} Y[\mathcal{J}](Y \not\subseteq I)$$

$$\tau_I(x) \xtwoheadrightarrow{X[\mathcal{K}]} X(X \not\subseteq I) \quad \tau_I(y) \xtwoheadrightarrow{Y[\mathcal{J}]} Y(Y \not\subseteq I)$$

with $X \subseteq x$, $Y \subseteq y$, and $X \sim Y$.

Or, we can obtain

$$\tau_I(x) \xrightarrow{X} \tau_I(x')(X \not\subseteq I) \quad \tau_I(y) \xrightarrow{Y} \tau_I(y')(Y \not\subseteq I)$$

$$\tau_I(x) \xtwoheadrightarrow{X[\mathcal{K}]} \tau_I(x')(X \not\subseteq I) \quad \tau_I(y) \xtwoheadrightarrow{Y[\mathcal{J}]} \tau_I(y')(Y \not\subseteq I)$$

**TABLE 4.24** Axioms of abstraction operator.

| No. | Axiom |
|---|---|
| $TI1$ | $e \notin I \quad \tau_I(e) = e$ |
| $RTI1$ | $e[m] \notin I \quad \tau_I(e[m]) = e[m]$ |
| $TI2$ | $e \in I \quad \tau_I(e) = \tau$ |
| $RTI2$ | $e[m] \in I \quad \tau_I(e[m]) = \tau$ |
| $TI3$ | $\tau_I(\delta) = \delta$ |
| $TI4$ | $\tau_I(x + y) = \tau_I(x) + \tau_I(y)$ |
| $TI5$ | $\tau_I(x \cdot y) = \tau_I(x) \cdot \tau_I(y)$ |
| $TI6$ | $\tau_I(x \parallel y) = \tau_I(x) \parallel \tau_I(y)$ |

with $X \subseteq x$, $Y \subseteq y$, and $X \sim Y$ and the hypothesis $\tau_I(x') \approx_{rbp}^{fr} \tau_I(y')$.

Or, we can obtain

$$\tau_I(x) \xrightarrow{\tau^*} \surd(X \subseteq I) \quad \tau_I(y) \xrightarrow{\tau^*} \surd(Y \subseteq I)$$

$$\tau_I(x) \overset{\tau^*}{\twoheadrightarrow} \surd(X \subseteq I) \quad \tau_I(y) \overset{\tau^*}{\twoheadrightarrow} \surd(Y \subseteq I)$$

with $X \subseteq x$, $Y \subseteq y$, and $X \sim Y$.

Or, we can obtain

$$\tau_I(x) \xrightarrow{\tau^*} \tau_I(x')(X \subseteq I) \quad \tau_I(y) \xrightarrow{\tau^*} \tau_I(y')(Y \subseteq I)$$

$$\tau_I(x) \overset{\tau^*}{\twoheadrightarrow} \tau_I(x')(X \subseteq I) \quad \tau_I(y) \overset{\tau^*}{\twoheadrightarrow} \tau_I(y')(Y \subseteq I)$$

with $X \subseteq x$, $Y \subseteq y$, and $X \sim Y$ and the hypothesis $\tau_I(x') \approx_{rbp}^{fr} \tau_I(y')$.

Hence, we obtain $\tau_I(x) \approx_{rbp}^{fr} \tau_I(y)$, as desired

(2) The cases of rooted branching FR step bisimulation $\approx_{rbs}^{fr}$, rooted branching FR hp-bisimulation $\approx_{rbhp}^{fr}$ can be proven similarly, however, we omit them here. □

We design the axioms for the abstraction operator $\tau_I$ in Table 4.24.

**Theorem 4.32** (Soundness of $RAPTC_\tau$). *Let $x$ and $y$ be $RAPTC_\tau$ terms. If $RAPTC_\tau \vdash x = y$, then*

1. $x \approx_{rbs}^{fr} y$;
2. $x \approx_{rbp}^{fr} y$;
3. $x \approx_{rbhp}^{fr} y$.

*Proof.* (1) Soundness of $RAPTC_\tau$ with respect to rooted branching FR step bisimulation $\approx_{rbs}^{fr}$.

Since rooted branching FR step bisimulation $\approx_{rbs}^{fr}$ is both an equivalent and a congruent relation with respect to $RAPTC_\tau$, we only need to check if each axiom in Table 4.24 is sound modulo rooted branching FR step bisimulation equivalence.

Although transition rules in Table 4.23 are defined in the flavor of a single event, they can be modified into a step (a set of events within which each event is pairwise concurrent), however, we omit them here. If we treat a single event as a step containing just one event, the proof of this soundness theorem does not pose any problem, hence, we use this way and still use the transition rules in Table 4.24.

We only prove soundness of the non-trivial axioms $TI4 - TI6$, and omit the defining axioms $TI1 - TI3$.

- **Axiom** $TI4$. Let $p, q$ be $RAPTC_\tau$ processes, and $\tau_I(p+q) = \tau_I(p) + \tau_I(q)$, it is sufficient to prove that $\tau_I(p+q) \approx_{rbs}^{fr} \tau_I(p) + \tau_I(q)$. By the forward transition rules for operator $+$ in Table 4.7 and $\tau_I$ in Table 4.23, we obtain

$$\frac{p \xrightarrow{e_1} e_1[m] \quad e_1 \notin q \quad (e_1 \notin I)}{\tau_I(p+q) \xrightarrow{e_1} \tau_I(e_1[m] + q)} \qquad \frac{p \xrightarrow{e_1} e_1[m] \quad (e_1 \notin I)}{\tau_I(p) + \tau_I(q) \xrightarrow{e_1} \tau_I(e_1[m]) + \tau_I(q)}$$

$$\frac{q \xrightarrow{e_2} e_2[m] \quad e_2 \notin p \quad (e_2 \notin I)}{\tau_I(p+q) \xrightarrow{e_2} \tau_I(p+e_2[m])} \quad \frac{q \xrightarrow{e_2} e_2[m] \quad e_2 \notin p \quad (e_2 \notin I)}{\tau_I(p)+\tau_I(q) \xrightarrow{e_2} \tau_I(p)+\tau_I(e_1[m])}$$

$$\frac{p \xrightarrow{e_1} p' \quad e_1 \notin q \quad (e_1 \notin I)}{\tau_I(p+q) \xrightarrow{e_1} \tau_I(p'+q)} \quad \frac{p \xrightarrow{e_1} p' \quad e_1 \notin q \quad (e_1 \notin I)}{\tau_I(p)+\tau_I(q) \xrightarrow{e_1} \tau_I(p')+\tau_I(q)}$$

$$\frac{q \xrightarrow{e_2} q' \quad e_2 \notin p \quad (e_2 \notin I)}{\tau_I(p+q) \xrightarrow{e_2} \tau_I(p+q')} \quad \frac{q \xrightarrow{e_2} q' \quad e_2 \notin p \quad (e_2 \notin I)}{\tau_I(p)+\tau_I(q) \xrightarrow{e_2} \tau_I(p)+\tau_I(q')}$$

$$\frac{p \xrightarrow{e_1} p' \quad q \xrightarrow{e_1} q' \quad (e_1 \notin I)}{\tau_I(p+q) \xrightarrow{e_1} \tau_I(p'+q')} \quad \frac{p \xrightarrow{e_1} p' \quad q \xrightarrow{e_1} q' \quad (e_1 \notin I)}{\tau_I(p)+\tau_I(q) \xrightarrow{e_1} \tau_I(p')+\tau_I(q')}$$

$$\frac{p \xrightarrow{e_1} e_1[m] \quad e_1 \notin q \quad (e_1 \in I)}{\tau_I(p+q) \xrightarrow{\tau} \tau_I(q)} \quad \frac{p \xrightarrow{e_1} e_1[m] \quad e_1 \notin q \quad (e_1 \in I)}{\tau_I(p)+\tau_I(q) \xrightarrow{\tau} \tau_I(q)}$$

$$\frac{q \xrightarrow{e_2} e_2[m] \quad e_2 \notin p \quad (e_2 \in I)}{\tau_I(p+q) \xrightarrow{\tau} \tau_I(p)} \quad \frac{q \xrightarrow{e_2} e_2[m] \quad e_2 \notin p \quad (e_2 \in I)}{\tau_I(p)+\tau_I(q) \xrightarrow{\tau} \tau_I(p)}$$

$$\frac{p \xrightarrow{e_1} p' \quad e_1 \notin q \quad (e_1 \in I)}{\tau_I(p+q) \xrightarrow{\tau} \tau_I(p'+q)} \quad \frac{p \xrightarrow{e_1} p' \quad e_1 \notin q \quad (e_1 \in I)}{\tau_I(p)+\tau_I(q) \xrightarrow{\tau} \tau_I(p')+\tau_I(q)}$$

$$\frac{q \xrightarrow{e_2} q' \quad e_2 \notin p \quad (e_2 \in I)}{\tau_I(p+q) \xrightarrow{\tau} \tau_I(p+q')} \quad \frac{q \xrightarrow{e_2} q' \quad e_2 \notin p \quad (e_2 \in I)}{\tau_I(p)+\tau_I(q) \xrightarrow{\tau} \tau_I(p)+\tau_I(q')}$$

$$\frac{p \xrightarrow{e_1} p' \quad q \xrightarrow{e_1} q' \quad (e_1 \in I)}{\tau_I(p+q) \xrightarrow{\tau} \tau_I(p'+q')} \quad \frac{p \xrightarrow{e_1} p' \quad q \xrightarrow{e_1} q' \quad (e_1 \in I)}{\tau_I(p)+\tau_I(q) \xrightarrow{\tau} \tau_I(p')+\tau_I(q')}$$

By the reverse transition rules for operator $+$ in Table 4.8 and $\tau_I$ in Table 4.23, we obtain

$$\frac{p \overset{e_1[m]}{\twoheadrightarrow} e_1 \quad e_1[m] \notin q \quad (e_1 \notin I)}{\tau_I(p+q) \overset{e_1[m]}{\twoheadrightarrow} \tau_I(e_1+q)} \quad \frac{p \overset{e_1[m]}{\twoheadrightarrow} e_1 \quad (e_1[m] \notin I)}{\tau_I(p)+\tau_I(q) \overset{e_1[m]}{\twoheadrightarrow} \tau_I(e_1)+\tau_I(q)}$$

$$\frac{q \overset{e_2[m]}{\twoheadrightarrow} e_2 \quad e_2 \notin p \quad (e_2[m] \notin I)}{\tau_I(p+q) \overset{e_2[m]}{\twoheadrightarrow} \tau_I(p+e_2)} \quad \frac{q \overset{e_2[m]}{\twoheadrightarrow} e_2 \quad e_2 \notin p \quad (e_2[m] \notin I)}{\tau_I(p)+\tau_I(q) \overset{e_2[m]}{\twoheadrightarrow} \tau_I(p)+\tau_I(e_1)}$$

$$\frac{p \overset{e_1[m]}{\twoheadrightarrow} p' \quad e_1 \notin q \quad (e_1[m] \notin I)}{\tau_I(p+q) \overset{e_1[m]}{\twoheadrightarrow} \tau_I(p'+q)} \quad \frac{p \overset{e_1[m]}{\twoheadrightarrow} p' \quad e_1 \notin q \quad (e_1[m] \notin I)}{\tau_I(p)+\tau_I(q) \overset{e_1[m]}{\twoheadrightarrow} \tau_I(p')+\tau_I(q)}$$

$$\frac{q \overset{e_2[m]}{\twoheadrightarrow} q' \quad e_2 \notin p \quad (e_2[m] \notin I)}{\tau_I(p+q) \overset{e_2[m]}{\twoheadrightarrow} \tau_I(p+q')} \quad \frac{q \overset{e_2[m]}{\twoheadrightarrow} q' \quad e_2 \notin p \quad (e_2[m] \notin I)}{\tau_I(p)+\tau_I(q) \overset{e_2[m]}{\twoheadrightarrow} \tau_I(p)+\tau_I(q')}$$

$$\frac{p \overset{e_1[m]}{\twoheadrightarrow} p' \quad q \overset{e_1}{\twoheadrightarrow} q' \quad (e_1[m] \notin I)}{\tau_I(p+q) \overset{e_1[m]}{\twoheadrightarrow} \tau_I(p'+q')} \quad \frac{p \overset{e_1[m]}{\twoheadrightarrow} p' \quad q \overset{e_1}{\twoheadrightarrow} q' \quad (e_1[m] \notin I)}{\tau_I(p)+\tau_I(q) \overset{e_1[m]}{\twoheadrightarrow} \tau_I(p')+\tau_I(q')}$$

$$\frac{p \overset{e_1[m]}{\twoheadrightarrow} e_1 \quad e_1 \notin q \quad (e_1[m] \in I)}{\tau_I(p+q) \overset{\tau}{\twoheadrightarrow} \tau_I(q)} \quad \frac{p \overset{e_1[m]}{\twoheadrightarrow} e_1 \quad e_1 \notin q \quad (e_1[m] \in I)}{\tau_I(p)+\tau_I(q) \overset{\tau}{\twoheadrightarrow} \tau_I(q)}$$

$$\frac{q \overset{e_2[m]}{\twoheadrightarrow} e_2 \quad e_2 \notin p \quad (e_2[m] \in I)}{\tau_I(p+q) \overset{\tau}{\twoheadrightarrow} \tau_I(p)} \quad \frac{q \overset{e_2[m]}{\twoheadrightarrow} e_2 \quad e_2 \notin p \quad (e_2[m] \in I)}{\tau_I(p)+\tau_I(q) \overset{\tau}{\twoheadrightarrow} \tau_I(p)}$$

$$\frac{p \xrightarrow{e_1[m]} p' \quad e_1 \notin q \quad (e_1[m] \in I)}{\tau_I(p+q) \xrightarrow{\tau} \tau_I(p'+q)} \quad \frac{p \xrightarrow{e_1} p' \quad e_1 \notin q \quad (e_1[m] \in I)}{\tau_I(p)+\tau_I(q) \xrightarrow{\tau} \tau_I(p')+\tau_I(q)}$$

$$\frac{q \xrightarrow{e_2[m]} q' \quad e_2 \notin p \quad (e_2[m] \in I)}{\tau_I(p+q) \xrightarrow{\tau} \tau_I(p+q')} \quad \frac{q \xrightarrow{e_2} q' \quad e_2 \notin p \quad (e_2[m] \in I)}{\tau_I(p)+\tau_I(q) \xrightarrow{\tau} \tau_I(p)+\tau_I(q')}$$

$$\frac{p \xrightarrow{e_1[m]} p' \quad q \xrightarrow{e_1} q' \quad (e_1[m] \in I)}{\tau_I(p+q) \xrightarrow{\tau} \tau_I(p'+q')} \quad \frac{p \xrightarrow{e_1[m]} p' \quad q \xrightarrow{e_1} q' \quad (e_1[m] \in I)}{\tau_I(p)+\tau_I(q) \xrightarrow{\tau} \tau_I(p')+\tau_I(q')}$$

Hence, with the assumptions $\tau_I(e_1[m]+q) \approx_{rbs}^{fr} \tau_I(e_1[m])+\tau_I(q)$, $\tau_I(e_1+q) \approx_{rbs}^{fr} \tau_I(e_1)+\tau_I(q)$, $\tau_I(p+e_2[m]) \approx_{rbs}^{fr} \tau_I(p)+\tau_I(e_2[m])$, $\tau_I(p+e_2) \approx_{rbs}^{fr} \tau_I(p)+\tau_I(e_2)$, $\tau_I(p'+q) \approx_{rbs}^{fr} \tau_I(p')+\tau_I(q)$, $\tau_I(p+q') \approx_{rbs}^{fr} \tau_I(p)+\tau_I(q')$, $\tau_I(p'+q') \approx_{rbs}^{fr} \tau_I(p')+\tau_I(q')$ $\tau_I(p+q) \approx_{rbs}^{fr} \tau_I(p)+\tau_I(q)$, as desired.

- **Axiom** $TI5$. Let $p,q$ be $RAPTC_\tau$ processes, and $\tau_I(p\cdot q)=\tau_I(p)\cdot\tau_I(q)$, it is sufficient to prove that $\tau_I(p\cdot q) \approx_{rbs}^{fr} \tau_I(p)\cdot\tau_I(q)$. By the forward transition rules for operator $\cdot$ in Table 4.7 and $\tau_I$ in Table 4.23, we obtain

$$\frac{p \xrightarrow{e_1} e_1[m] \quad (e_1 \notin I)}{\tau_I(p\cdot q) \xrightarrow{e_1} \tau_I(e_1[m]\cdot q)} \quad \frac{p \xrightarrow{e_1} e_1[m] \quad (e_1 \notin I)}{\tau_I(p)\cdot\tau_I(q) \xrightarrow{e_1} \tau_I(e_1[m])\cdot\tau_I(q)}$$

$$\frac{p \xrightarrow{e_1} p' \quad (e_1 \notin I)}{\tau_I(p\cdot q) \xrightarrow{e_1} \tau_I(p'\cdot q)} \quad \frac{p \xrightarrow{e_1} p' \quad (e_1 \notin I)}{\tau_I(p)\cdot\tau_I(q) \xrightarrow{e_1} \tau_I(p')\cdot\tau_I(q)}$$

$$\frac{p \xrightarrow{e_1} e_1[m] \quad (e_1 \in I)}{\tau_I(p\cdot q) \xrightarrow{\tau} \tau_I(q)} \quad \frac{p \xrightarrow{e_1} e_1[m] \quad (e_1 \in I)}{\tau_I(p)\cdot\tau_I(q) \xrightarrow{\tau} \tau_I(q)}$$

$$\frac{p \xrightarrow{e_1} p' \quad (e_1 \in I)}{\tau_I(p\cdot q) \xrightarrow{\tau} \tau_I(p'\cdot q)} \quad \frac{p \xrightarrow{e_1} p' \quad (e_1 \in I)}{\tau_I(p)\cdot\tau_I(q) \xrightarrow{\tau} \tau_I(p')\cdot\tau_I(q)}$$

By the reverse transition rules for operator $\cdot$ in Table 4.8 and $\tau_I$ in Table 4.23, we obtain

$$\frac{q \xrightarrow{e_2[m]} e_2 \quad (e_2 \notin I)}{\tau_I(p\cdot q) \xrightarrow{e_2[m]} \tau_I(p\cdot e_2)} \quad \frac{q \xrightarrow{e_2[m]} e_2 \quad (e_2 \notin I)}{\tau_I(p)\cdot\tau_I(q) \xrightarrow{e_2[m]} \tau_I(p)\cdot\tau_I(e_2)}$$

$$\frac{q \xrightarrow{e_2[m]} q' \quad (e_2 \notin I)}{\tau_I(p\cdot q) \xrightarrow{e_2[m]} \tau_I(p\cdot q')} \quad \frac{q \xrightarrow{e_2[m]} q' \quad (e_2 \notin I)}{\tau_I(p)\cdot\tau_I(q) \xrightarrow{e_2[m]} \tau_I(p)\cdot\tau_I(q')}$$

$$\frac{q \xrightarrow{e_2[m]} e_2 \quad (e_2[m] \in I)}{\tau_I(p\cdot q) \xrightarrow{\tau} \tau_I(p)} \quad \frac{q \xrightarrow{e_1[m]} e_2 \quad (e_2[m] \in I)}{\tau_I(p)\cdot\tau_I(q) \xrightarrow{\tau} \tau_I(p)}$$

$$\frac{q \xrightarrow{e_2[m]} q' \quad (e_2[m] \in I)}{\tau_I(p\cdot q) \xrightarrow{\tau} \tau_I(p\cdot q')} \quad \frac{q \xrightarrow{e_1[m]} q' \quad (e_2[m] \in I)}{\tau_I(p)\cdot\tau_I(q) \xrightarrow{\tau} \tau_I(p)\cdot\tau_I('q)}$$

Hence, with the assumptions $\tau_I(e_1[m]\cdot q) \approx_{rbs}^{fr} \tau_I(e_1[m])\cdot\tau_I(q)$, $\tau_I(p\cdot e_2) \approx_{rbs}^{fr} \tau_I(p)\cdot\tau_I(e_2)$, $\tau_I(p'\cdot q)=\tau_I(p')\cdot\tau_I(q)$, $\tau_I(p\cdot q')=\tau_I(p)\cdot\tau_I(q')$, $\tau_I(p\cdot q) \approx_{rbs}^{fr} \tau_I(p)\cdot\tau_I(q)$, as desired.

- **Axiom** $TI6$. Let $p,q$ be $RAPTC_\tau$ processes, and $\tau_I(p \parallel q)=\tau_I(p) \parallel \tau_I(q)$, it is sufficient to prove that $\tau_I(p \parallel q) \approx_{rbs}^{fr} \tau_I(p) \parallel \tau_I(q)$. By the forward transition rules for operator $\parallel$ in Table 4.9 and $\tau_I$ in Table 4.23, we obtain

$$\frac{p \xrightarrow{e_1} e_1[m] \quad q \xrightarrow{e_2} e_2[m] \quad (e_1,e_2 \notin I)}{\tau_I(p \parallel q) \xrightarrow{\{e_1,e_2\}} \tau_I(e_1[m] \parallel e_2[m])} \quad \frac{p \xrightarrow{e_1} e_1[m] \quad q \xrightarrow{e_2} e_2[m] \quad (e_1,e_2 \notin I)}{\tau_I(p) \parallel \tau_I(q) \xrightarrow{\{e_1,e_2\}} \tau_I(e_1[m]) \parallel \tau_I(e_2[m])}$$

$$\frac{p\xrightarrow{e_1}p'\quad q\xrightarrow{e_2}e_2[m]\quad (e_1,e_2\notin I)}{\tau_I(p\parallel q)\xrightarrow{\{e_1,e_2\}}\tau_I(p'\parallel e_2[m])}\quad \frac{p\xrightarrow{e_1}p'\quad q\xrightarrow{e_2}e_2[m]\quad (e_1,e_2\notin I)}{\tau_I(p)\parallel\tau_I(q)\xrightarrow{\{e_1,e_2\}}\tau_I(p')\parallel\tau_I(e_2[m])}$$

$$\frac{p\xrightarrow{e_1}e_1[m]\quad q\xrightarrow{e_2}q'\quad (e_1,e_2\notin I)}{\tau_I(p\parallel q)\xrightarrow{\{e_1,e_2\}}\tau_I(e_1[m]\parallel q')}\quad \frac{p\xrightarrow{e_1}e_1[m]\quad q\xrightarrow{e_2}q'\quad (e_1,e_2\notin I)}{\tau_I(p)\parallel\tau_I(q)\xrightarrow{\{e_1,e_2\}}\tau_I(e_1[m])\parallel\tau_I(q')}$$

$$\frac{p\xrightarrow{e_1}p'\quad q\xrightarrow{e_2}q'\quad (e_1,e_2\notin I)}{\tau_I(p\parallel q)\xrightarrow{\{e_1,e_2\}}\tau_I(p'\between q')}\quad \frac{p\xrightarrow{e_1}p'\quad q\xrightarrow{e_2}q'\quad (e_1,e_2\notin I)}{\tau_I(p)\parallel\tau_I(q)\xrightarrow{\{e_1,e_2\}}\tau_I(p')\between\tau_I(q')}$$

$$\frac{p\xrightarrow{e_1}e_1[m]\quad q\xrightarrow{e_2}e_2[m]\quad (e_1\notin I,e_2\in I)}{\tau_I(p\parallel q)\xRightarrow{e_1}\tau_I(e_1[m])}\quad \frac{p\xrightarrow{e_1}e_1[m]\quad q\xrightarrow{e_2}e_2[m]\quad (e_1\notin I,e_2\in I)}{\tau_I(p)\parallel\tau_I(q)\xRightarrow{e_1}\tau_I(e_1[m])}$$

$$\frac{p\xrightarrow{e_1}p'\quad q\xrightarrow{e_2}e_2[m]\quad (e_1\notin I,e_2\in I)}{\tau_I(p\parallel q)\xRightarrow{e_1}\tau_I(p')}\quad \frac{p\xrightarrow{e_1}p'\quad q\xrightarrow{e_2}e_2[m]\quad (e_1\notin I,e_2\in I)}{\tau_I(p)\parallel\tau_I(q)\xRightarrow{e_1}\tau_I(p')}$$

$$\frac{p\xrightarrow{e_1}e_1[m]\quad q\xrightarrow{e_2}q'\quad (e_1\notin I,e_2\in I)}{\tau_I(p\parallel q)\xRightarrow{e_1}\tau_I(e_1[m]\parallel q')}\quad \frac{p\xrightarrow{e_1}e_1[m]\quad q\xrightarrow{e_2}q'\quad (e_1\notin I,e_2\in I)}{\tau_I(p)\parallel\tau_I(q)\xRightarrow{e_1}\tau_I(e_1[m])\parallel\tau_I(q')}$$

$$\frac{p\xrightarrow{e_1}p'\quad q\xrightarrow{e_2}q'\quad (e_1\notin I,e_2\in I)}{\tau_I(p\parallel q)\xRightarrow{e_1}\tau_I(p'\between q')}\quad \frac{p\xrightarrow{e_1}p'\quad q\xrightarrow{e_2}q'\quad (e_1\notin I,e_2\in I)}{\tau_I(p)\parallel\tau_I(q)\xRightarrow{e_1}\tau_I(p')\between\tau_I(q')}$$

$$\frac{p\xrightarrow{e_1}e_1[m]\quad q\xrightarrow{e_2}e_2[m]\quad (e_1\in I,e_2\notin I)}{\tau_I(p\parallel q)\xRightarrow{e_2}\tau_I(e_2[m])}\quad \frac{p\xrightarrow{e_1}e_1[m]\quad q\xrightarrow{e_2}e_2[m]\quad (e_1\in I,e_2\notin I)}{\tau_I(p)\parallel\tau_I(q)\xRightarrow{e_2}\tau_I(e_2[m])}$$

$$\frac{p\xrightarrow{e_1}p'\quad q\xrightarrow{e_2}e_2[m]\quad (e_1\in I,e_2\notin I)}{\tau_I(p\parallel q)\xRightarrow{e_2}\tau_I(p'\parallel e_2[m])}\quad \frac{p\xrightarrow{e_1}p'\quad q\xrightarrow{e_2}e_2[m]\quad (e_1\in I,e_2\notin I)}{\tau_I(p)\parallel\tau_I(q)\xRightarrow{e_2}\tau_I(p')\parallel\tau_I(e_2[m])}$$

$$\frac{p\xrightarrow{e_1}e_1[m]\quad q\xrightarrow{e_2}q'\quad (e_1\in I,e_2\notin I)}{\tau_I(p\parallel q)\xRightarrow{e_2}\tau_I(q')}\quad \frac{p\xrightarrow{e_1}e_1[m]\quad q\xrightarrow{e_2}q'\quad (e_1\in I,e_2\notin I)}{\tau_I(p)\parallel\tau_I(q)\xRightarrow{e_2}\tau_I(q')}$$

$$\frac{p\xrightarrow{e_1}p'\quad q\xrightarrow{e_2}q'\quad (e_1\in I,e_2\notin I)}{\tau_I(p\parallel q)\xRightarrow{e_2}\tau_I(p'\between q')}\quad \frac{p\xrightarrow{e_1}p'\quad q\xrightarrow{e_2}q'\quad (e_1\in I,e_2\notin I)}{\tau_I(p)\parallel\tau_I(q)\xRightarrow{e_2}\tau_I(p')\between\tau_I(q')}$$

$$\frac{p\xrightarrow{e_1}e_1[m]\quad q\xrightarrow{e_2}e_2[m]\quad (e_1,e_2\in I)}{\tau_I(p\parallel q)\xrightarrow{\tau^*}\surd}\quad \frac{p\xrightarrow{e_1}e_1[m]\quad q\xrightarrow{e_2}e_2[m]\quad (e_1,e_2\in I)}{\tau_I(p)\parallel\tau_I(q)\xrightarrow{\tau^*}\surd}$$

$$\frac{p\xrightarrow{e_1}p'\quad q\xrightarrow{e_2}e_2[m]\quad (e_1,e_2\in I)}{\tau_I(p\parallel q)\xrightarrow{\tau^*}\tau_I(p')}\quad \frac{p\xrightarrow{e_1}p'\quad q\xrightarrow{e_2}e_2[m]\quad (e_1,e_2\in I)}{\tau_I(p)\parallel\tau_I(q)\xrightarrow{\tau^*}\tau_I(p')}$$

$$\frac{p\xrightarrow{e_1}e_1[m]\quad q\xrightarrow{e_2}q'\quad (e_1,e_2\in I)}{\tau_I(p\parallel q)\xrightarrow{\tau^*}\tau_I(q')}\quad \frac{p\xrightarrow{e_1}e_1[m]\quad q\xrightarrow{e_2}q'\quad (e_1,e_2\in I)}{\tau_I(p)\parallel\tau_I(q)\xrightarrow{\tau^*}\tau_I(q')}$$

$$\frac{p\xrightarrow{e_1}p'\quad q\xrightarrow{e_2}q'\quad (e_1,e_2\in I)}{\tau_I(p\parallel q)\xrightarrow{\tau^*}\tau_I(p'\between q')}\quad \frac{p\xrightarrow{e_1}p'\quad q\xrightarrow{e_2}q'\quad (e_1,e_2\in I)}{\tau_I(p)\parallel\tau_I(q)\xrightarrow{\tau^*}\tau_I(p')\between\tau_I(q')}$$

By the reverse transition rules for operator $\parallel$ in Table 4.10 and $\tau_I$ in Table 4.23, we obtain

$$\frac{p\xrightarrow{e_1[m]}\!\!\!\!\rightarrow e_1\quad q\xrightarrow{e_2[m]}\!\!\!\!\rightarrow e_2\quad (e_1[m],e_2[m]\notin I)}{\tau_I(p\parallel q)\xrightarrow{\{e_1[m],e_2[m]\}}\!\!\!\!\rightarrow\tau_I(e_1\parallel e_2)}\quad \frac{p\xrightarrow{e_1[m]}\!\!\!\!\rightarrow e_1\quad q\xrightarrow{e_2[m]}\!\!\!\!\rightarrow e_2\quad (e_1[m],e_2[m]\notin I)}{\tau_I(p)\parallel\tau_I(q)\xrightarrow{\{e_1[m],e_2[m]\}}\!\!\!\!\rightarrow\tau_I(e_1)\parallel\tau_I(e_2)}$$

$$\frac{p\xrightarrow{e_1[m]}\!\!\!\!\rightarrow p'\quad q\xrightarrow{e_2[m]}\!\!\!\!\rightarrow e_2\quad (e_1[m],e_2[m]\notin I)}{\tau_I(p\parallel q)\xrightarrow{\{e_1[m],e_2[m]\}}\!\!\!\!\rightarrow\tau_I(p'\parallel e_2)}$$

$$\frac{p\xrightarrow{e_1[m]}\!\!\!\!\rightarrow e_1\quad q\xrightarrow{e_2[m]}\!\!\!\!\rightarrow q'\quad (e_1[m],e_2[m]\notin I)}{\tau_I(p\parallel q)\xrightarrow{\{e_1[m],e_2[m]\}}\!\!\!\!\rightarrow\tau_I(e_1\parallel q')}$$

$$\frac{p\xrightarrow{e_1[m]}\!\!\!\!\rightarrow p'\quad q\xrightarrow{e_2[m]}\!\!\!\!\rightarrow q'\quad (e_1[m],e_2[m]\notin I)}{\tau_I(p\parallel q)\xrightarrow{\{e_1[m],e_2[m]\}}\!\!\!\!\rightarrow\tau_I(p'\between q')}$$

$$\frac{p\xrightarrow{e_1[m]}\!\!\!\!\rightarrow e_1\quad q\xrightarrow{e_2[m]}\!\!\!\!\rightarrow e_2\quad (e_1[m]\notin I,e_2[m]\in I)}{\tau_I(p\parallel q)\overset{e_1[m]}{\Rrightarrow}\tau_I(e_1)}$$

$$\frac{p\xrightarrow{e_1[m]}\!\!\!\!\rightarrow p'\quad q\xrightarrow{e_2[m]}\!\!\!\!\rightarrow e_2\quad (e_1[m]\notin I,e_2[m]\in I)}{\tau_I(p\parallel q)\overset{e_1[m]}{\Rrightarrow}\tau_I(p')}$$

$$\frac{p\xrightarrow{e_1[m]}\!\!\!\!\rightarrow e_1\quad q\xrightarrow{e_2[m]}\!\!\!\!\rightarrow q'\quad (e_1[m]\notin I,e_2[m]\in I)}{\tau_I(p\parallel q)\overset{e_1[m]}{\Rrightarrow}\tau_I(e_1\parallel q')}$$

$$\frac{p\xrightarrow{e_1[m]}\!\!\!\!\rightarrow p'\quad q\xrightarrow{e_2[m]}\!\!\!\!\rightarrow q'\quad (e_1[m]\notin I,e_2[m]\in I)}{\tau_I(p\parallel q)\overset{e_1[m]}{\Rrightarrow}\tau_I(p'\between q')}$$

$$\frac{p\xrightarrow{e_1[m]}\!\!\!\!\rightarrow e_1\quad q\xrightarrow{e_2[m]}\!\!\!\!\rightarrow e_2\quad (e_1[m]\in I,e_2[m]\notin I)}{\tau_I(p\parallel q)\overset{e_2[m]}{\Rrightarrow}\tau_I(e_2)}$$

$$\frac{p\xrightarrow{e_1[m]}\!\!\!\!\rightarrow p'\quad q\xrightarrow{e_2[m]}\!\!\!\!\rightarrow e_2\quad (e_1[m]\in I,e_2[m]\notin I)}{\tau_I(p\parallel q)\overset{e_2[m]}{\Rrightarrow}\tau_I(p'\parallel e_2)}$$

$$\frac{p\xrightarrow{e_1[m]}\!\!\!\!\rightarrow e_1\quad q\xrightarrow{e_2[m]}\!\!\!\!\rightarrow q'\quad (e_1[m]\in I,e_2[m]\notin I)}{\tau_I(p\parallel q)\overset{e_2[m]}{\Rrightarrow}\tau_I(q')}$$

$$\frac{p\xrightarrow{e_1[m]}\!\!\!\!\rightarrow p'\quad q\xrightarrow{e_2[m]}\!\!\!\!\rightarrow q'\quad (e_1[m]\in I,e_2[m]\notin I)}{\tau_I(p\parallel q)\overset{e_2[m]}{\Rrightarrow}\tau_I(p'\between q')}$$

$$\frac{p\xrightarrow{e_1[m]}\!\!\!\!\rightarrow e_1\quad q\xrightarrow{e_2[m]}\!\!\!\!\rightarrow e_2\quad (e_1[m],e_2[m]\in I)}{\tau_I(p\parallel q)\xrightarrow{\tau^*}\!\!\!\!\rightarrow\surd}$$

$$\frac{p\xrightarrow{e_1[m]}\!\!\!\!\rightarrow p'\quad q\xrightarrow{e_2[m]}\!\!\!\!\rightarrow e_2\quad (e_1[m],e_2[m]\in I)}{\tau_I(p\parallel q)\xrightarrow{\tau^*}\!\!\!\!\rightarrow\tau_I(p')}$$

$$\frac{p\xrightarrow{e_1[m]}\!\!\!\!\rightarrow e_1\quad q\xrightarrow{e_2[m]}\!\!\!\!\rightarrow q'\quad (e_1[m],e_2[m]\in I)}{\tau_I(p\parallel q)\xrightarrow{\tau^*}\!\!\!\!\rightarrow\tau_I(q')}$$

$$\frac{p\xrightarrow{e_1[m]}\!\!\!\!\rightarrow p'\quad q\xrightarrow{e_2[m]}\!\!\!\!\rightarrow q'\quad (e_1[m],e_2[m]\in I)}{\tau_I(p\parallel q)\xrightarrow{\tau^*}\!\!\!\!\rightarrow\tau_I(p'\between q')}$$

$$\frac{p\xrightarrow{e_1[m]}\!\!\!\!\rightarrow p'\quad q\xrightarrow{e_2[m]}\!\!\!\!\rightarrow e_2\quad (e_1[m],e_2[m]\notin I)}{\tau_I(p)\parallel\tau_I(q)\xrightarrow{\{e_1[m],e_2[m]\}}\!\!\!\!\rightarrow\tau_I(p')\parallel\tau_I(e_2)}$$

$$\frac{p\xrightarrow{e_1[m]}\!\!\!\!\rightarrow e_1\quad q\xrightarrow{e_2[m]}\!\!\!\!\rightarrow q'\quad (e_1[m],e_2[m]\notin I)}{\tau_I(p)\parallel\tau_I(q)\xrightarrow{\{e_1[m],e_2[m]\}}\!\!\!\!\rightarrow\tau_I(e_1)\parallel\tau_I(q')}$$

$$\frac{p\xrightarrow{e_1[m]}\!\!\!\!\rightarrow p'\quad q\xrightarrow{e_2[m]}\!\!\!\!\rightarrow q'\quad (e_1[m],e_2[m]\notin I)}{\tau_I(p)\parallel\tau_I(q)\xrightarrow{\{e_1[m],e_2[m]\}}\!\!\!\!\rightarrow\tau_I(p')\between\tau_I(q')}$$

$$\frac{p\xrightarrow{e_1[m]}\!\!\!\!\rightarrow e_1\quad q\xrightarrow{e_2[m]}\!\!\!\!\rightarrow e_2\quad (e_1[m]\notin I,e_2[m]\in I)}{\tau_I(p)\parallel\tau_I(q)\overset{e_1[m]}{\Rrightarrow}\tau_I(e_1)}$$

$$\frac{p\xrightarrow{e_1[m]}\!\!\!\!\rightarrow p'\quad q\xrightarrow{e_2[m]}\!\!\!\!\rightarrow e_2\quad (e_1[m]\notin I,e_2[m]\in I)}{\tau_I(p)\parallel\tau_I(q)\overset{e_1[m]}{\Rrightarrow}\tau_I(p')}$$

$$\frac{p\xrightarrow{e_1[m]}\!\!\!\!\rightarrow e_1\quad q\xrightarrow{e_2[m]}\!\!\!\!\rightarrow q'\quad (e_1[m]\notin I,e_2[m]\in I)}{\tau_I(p)\parallel\tau_I(q)\overset{e_1[m]}{\Rrightarrow}\tau_I(e_1)\parallel\tau_I(q')}$$

$$\frac{p\xrightarrow{e_1[m]}\!\!\!\!\rightarrow p'\quad q\xrightarrow{e_2[m]}\!\!\!\!\rightarrow q'\quad (e_1[m]\notin I,e_2[m]\in I)}{\tau_I(p)\parallel\tau_I(q)\overset{e_1[m]}{\Rrightarrow}\tau_I(p')\between\tau_I(q')}$$

$$\frac{p\xrightarrow{e_1[m]}\!\!\!\!\rightarrow e_1\quad q\xrightarrow{e_2[m]}\!\!\!\!\rightarrow e_2\quad (e_1[m]\in I,e_2[m]\notin I)}{\tau_I(p)\parallel\tau_I(q)\overset{e_2[m]}{\Rrightarrow}\tau_I(e_2)}$$

$$\frac{p\xrightarrow{e_1[m]}\!\!\!\!\rightarrow p'\quad q\xrightarrow{e_2[m]}\!\!\!\!\rightarrow e_2\quad (e_1[m]\in I,e_2[m]\notin I)}{\tau_I(p)\parallel\tau_I(q)\overset{e_2[m]}{\Rrightarrow}\tau_I(p')\parallel\tau_I(e_2)}$$

$$\frac{p\xrightarrow{e_1[m]}\!\!\!\!\rightarrow e_1\quad q\xrightarrow{e_2[m]}\!\!\!\!\rightarrow q'\quad (e_1[m]\in I,e_2[m]\notin I)}{\tau_I(p)\parallel\tau_I(q)\overset{e_2[m]}{\Rrightarrow}\tau_I(q')}$$

$$\frac{p\xrightarrow{e_1[m]}\!\!\!\!\rightarrow p'\quad q\xrightarrow{e_2[m]}\!\!\!\!\rightarrow q'\quad (e_1[m]\in I,e_2[m]\notin I)}{\tau_I(p)\parallel\tau_I(q)\overset{e_2[m]}{\Rrightarrow}\tau_I(p')\between\tau_I(q')}$$

$$\frac{p\xrightarrow{e_1[m]}\!\!\!\!\rightarrow e_1\quad q\xrightarrow{e_2[m]}\!\!\!\!\rightarrow e_2\quad (e_1[m],e_2[m]\in I)}{\tau_I(p)\parallel\tau_I(q)\xrightarrow{\tau^*}\!\!\!\!\rightarrow\surd}$$

$$\frac{p\xrightarrow{e_1[m]}\!\!\!\!\rightarrow p'\quad q\xrightarrow{e_2[m]}\!\!\!\!\rightarrow e_2\quad (e_1[m],e_2[m]\in I)}{\tau_I(p)\parallel\tau_I(q)\xrightarrow{\tau^*}\!\!\!\!\rightarrow\tau_I(p')}$$

$$\frac{p\xrightarrow{e_1[m]}\!\!\!\!\rightarrow e_1\quad q\xrightarrow{e_2[m]}\!\!\!\!\rightarrow q'\quad (e_1[m],e_2[m]\in I)}{\tau_I(p)\parallel\tau_I(q)\xrightarrow{\tau^*}\!\!\!\!\rightarrow\tau_I(q')}$$

$$\frac{p\xrightarrow{e_1[m]}\!\!\!\!\rightarrow p'\quad q\xrightarrow{e_2[m]}\!\!\!\!\rightarrow q'\quad (e_1[m],e_2[m]\in I)}{\tau_I(p)\parallel\tau_I(q)\xrightarrow{\tau^*}\!\!\!\!\rightarrow\tau_I(p')\between\tau_I(q')}$$

Hence, with the assumptions $\tau_I(p' \between q') = \tau_I(p') \between \tau_I(q')$, $\tau_I(p \parallel q) \approx^{fr}_{rbs} \tau_I(p) \parallel \tau_I(q)$, as desired.

(2) Soundness of $RAPTC_\tau$ with respect to rooted branching FR pomset bisimulation $\approx^{fr}_{rbp}$.

Since rooted branching FR pomset bisimulation $\approx^{fr}_{rbp}$ is both an equivalent and a congruent relation with respect to $RAPTC_\tau$, we only need to check if each axiom in Table 4.24 is sound modulo rooted branching FR pomset bisimulation $\approx^{fr}_{rbp}$.

From the definition of rooted branching FR pomset bisimulation $\approx^{fr}_{rbp}$ (see Definition 2.73), we know that rooted branching FR pomset bisimulation $\approx^{fr}_{rbp}$ is defined by weak pomset transitions, which are labeled by pomsets with $\tau$. In a weak pomset transition, the events in the pomset are either within causality relations (defined by $\cdot$) or in concurrency (implicitly defined by $\cdot$ and $+$, and explicitly defined by $\between$), of course, they are pairwise consistent (without conflicts). In (1), we have already proven the case that all events are pairwise concurrent, hence, we only need to prove the case of events in causality. Without loss of generality, we take a pomset of $P = \{e_1, e_2 : e_1 \cdot e_2\}$. Then, the weak pomset transition labeled by the above $P$ is just composed of one single event transition labeled by $e_1$ succeeded by another single event transition labeled by $e_2$, that is, $\overset{P}{\Rightarrow} = \overset{e_1}{\Rightarrow}\overset{e_2}{\Rightarrow}$ or $\overset{P}{\rightrightarrows} = \overset{e_2}{\rightrightarrows}\overset{e_1}{\rightrightarrows}$.

Similarly to the proof of soundness of $RAPTC_\tau$ modulo rooted branching FR step bisimulation $\approx^{fr}_{rbs}$ (1), we can prove that each axiom in Table 4.24 is sound modulo rooted branching FR pomset bisimulation $\approx^{fr}_{rbp}$, however, we omit them here.

(3) Soundness of $RAPTC_\tau$ with respect to rooted branching FR hp-bisimulation $\approx^{fr}_{rbhp}$.

Since rooted branching FR hp-bisimulation $\approx^{fr}_{rbhp}$ is both an equivalent and a congruent relation with respect to $RAPTC_\tau$, we only need to check if each axiom in Table 4.24 is sound modulo rooted branching FR hp-bisimulation $\approx^{fr}_{rbhp}$.

From the definition of rooted branching FR hp-bisimulation $\approx^{fr}_{rbhp}$ (see Definition 2.75), we know that rooted branching FR hp-bisimulation $\approx^{fr}_{rbhp}$ is defined on the weakly posetal product $(C_1, f, C_2)$, $f : \hat{C_1} \to \hat{C_2}$ isomorphism. Two process terms $s$ related to $C_1$ and $t$ related to $C_2$, and $f : \hat{C_1} \to \hat{C_2}$ isomorphism. Initially, $(C_1, f, C_2) = (\emptyset, \emptyset, \emptyset)$, and $(\emptyset, \emptyset, \emptyset) \in \approx^{fr}_{rbhp}$. When $s \xrightarrow{e} s'$ $(C_1 \xrightarrow{e} C_1')$, there will be $t \overset{e}{\Rightarrow} t'$ $(C_2 \overset{e}{\Rightarrow} C_2')$, and we define $f' = f[e \mapsto e]$. Also, when $s \overset{e[m]}{\rightrightarrows} s'$ $(C_1 \overset{e[m]}{\rightrightarrows} C_1')$, there will be $t \overset{e[m]}{\rightrightarrows} t'$ $(C_2 \overset{e[m]}{\rightrightarrows} C_2')$, and we define $f' = f[e[m] \mapsto e[m]]$. Then, if $(C_1, f, C_2) \in \approx^{fr}_{rbhp}$, then $(C_1', f', C_2') \in \approx^{fr}_{rbhp}$.

Similarly to the proof of soundness of $RAPTC_\tau$ modulo rooted branching FR pomset bisimulation equivalence (2), we can prove that each axiom in Table 4.24 is sound modulo rooted branching FR hp-bisimulation equivalence, we just need additionally to check the above conditions on rooted branching FR hp-bisimulation, however, we omit them here. □

Chapter 5

# Partially reversible calculus

Based on CTC, we also did some work on reversible algebra called RCTC. However, RCTC is imperfect, it is sound, but it has no recursion theory. The main reason for this is that the existence of a multichoice operator means that a recursion theory cannot be established. In this chapter, we try to use an alternative operator to replace a multichoice operator and we obtain a perfect partially reversible calculus. The main reason for using an alternative operator is that when an alternative branch is forward executing, the reverse branch is also determined and other branches have no necessity to remain. However, when a process is reversed, the other branches disappear. We call the reversible algebra using an alternative operator partially reversible calculus. This chapter is organized as follows. We give the syntax and operational semantics of RCTC in Section 5.1. We discuss the properties of RCTC based on strongly forward–reverse truly concurrent bisimilarities in Section 5.2, and the properties of RCTC based on weakly forward–reverse truly concurrent bisimilarities in Section 5.3.

## 5.1 Syntax and operational semantics

We assume an infinite set $\mathcal{N}$ of (action or event) names, and use $a, b, c, \cdots$ to range over $\mathcal{N}$. We denote by $\overline{\mathcal{N}}$ the set of co-names and let $\overline{a}, \overline{b}, \overline{c}, \cdots$ range over $\overline{\mathcal{N}}$. Then, we set $\mathcal{L} = \mathcal{N} \cup \overline{\mathcal{N}}$ as the set of labels, and use $l, \overline{l}$ to range over $\mathcal{L}$. We extend complementation to $\mathcal{L}$ such that $\overline{\overline{a}} = a$. Let $\tau$ denote the silent step (internal action or event) and define $Act = \mathcal{L} \cup \{\tau\} \cup \mathcal{L}[\mathcal{K}]$ to be the set of actions, $\alpha, \beta$ range over $Act$. Also, $K, L$ are used to stand for subsets of $\mathcal{L}$ and $\overline{L}$ is used for the set of complements of labels in $L$. A relabeling function $f$ is a function from $\mathcal{L}$ to $\mathcal{L}$ such that $f(\overline{l}) = \overline{f(l)}$. By defining $f(\tau) = \tau$, we extend $f$ to $Act$. We write $\mathcal{P}$ for the set of processes. Sometimes, we use $I, J$ to stand for an indexing set, and we write $E_i : i \in I$ for a family of expressions indexed by $I$. $Id_D$ is the identity function or relation over set $D$.

For each process constant schema $A$, a defining equation of the form

$$A \overset{\text{def}}{=} P$$

is assumed, where $P$ is a process.

### 5.1.1 Syntax

We use the Prefix . to model the causality relation $\leq$ in true concurrency, the Summation $+$ to model the conflict relation $\sharp$ in true concurrency, and the Composition $\parallel$ to explicitly model concurrent relation in true concurrency. Also, we follow the conventions of process algebra.

**Definition 5.1** (Syntax). Reversible truly concurrent processes RCTC are defined inductively by the following formation rules:

1. $A \in \mathcal{P}$;
2. $\mathbf{nil} \in \mathcal{P}$;
3. if $P \in \mathcal{P}$, then the Prefix $\alpha.P \in \mathcal{P}$ and $P.\alpha[m] \in \mathcal{P}$, for $\alpha \in Act$ and $m \in \mathcal{K}$;
4. if $P, Q \in \mathcal{P}$, then the Summation $P + Q \in \mathcal{P}$;
5. if $P, Q \in \mathcal{P}$, then the Composition $P \parallel Q \in \mathcal{P}$;
6. if $P \in \mathcal{P}$, then the Prefix $(\alpha_1 \parallel \cdots \parallel \alpha_n).P \in \mathcal{P} \quad (n \in I)$ and $P.(\alpha_1[m] \parallel \cdots \parallel \alpha_n[m]) \in \mathcal{P} \quad (n \in I)$, for $\alpha_1, \cdots, \alpha_n \in Act$ and $m \in \mathcal{K}$;
7. if $P \in \mathcal{P}$, then the Restriction $P \setminus L \in \mathcal{P}$ with $L \in \mathcal{L}$;
8. if $P \in \mathcal{P}$, then the Relabeling $P[f] \in \mathcal{P}$.

The standard BNF grammar of syntax of RCTC can be summarized as follows:

$P ::= A \ | \ \mathbf{nil} \ | \ \alpha.P \ | \ P.\alpha[m] \ | \ P+P \ | \ P \parallel P \ | \ (\alpha_1 \parallel \cdots \parallel \alpha_n).P \ | \ P.(\alpha_1[m] \parallel \cdots \parallel \alpha_n[m]) \ | \ P \setminus L \ | \ P[f]$

Handbook of Truly Concurrent Process Algebra. https://doi.org/10.1016/B978-0-44-321515-5.00009-3

**TABLE 5.1 Forward transition rules of Prefix and Summation.**

$$\frac{}{\alpha \xrightarrow{\alpha} \alpha[m]}$$

$$\frac{P \xrightarrow{\alpha} \alpha[m]}{P+Q \xrightarrow{\alpha} \alpha[m]} \quad \frac{P \xrightarrow{\alpha} P'}{P+Q \xrightarrow{\alpha} P'}$$

$$\frac{Q \xrightarrow{\alpha} \alpha[m]}{P+Q \xrightarrow{\alpha} \alpha[m]} \quad \frac{Q \xrightarrow{\alpha} Q'}{P+Q \xrightarrow{\alpha} Q'}$$

$$\frac{P \xrightarrow{\alpha} \alpha[m] \quad \mathrm{Std}(Q)}{P.Q \xrightarrow{\alpha} \alpha[m].Q} \quad \frac{P \xrightarrow{\alpha} P' \quad \mathrm{Std}(Q)}{P.Q \xrightarrow{\alpha} P'.Q}$$

$$\frac{Q \xrightarrow{\beta} \beta[n] \quad \mathrm{NStd}(P)}{P.Q \xrightarrow{\beta} P.\beta[n]} \quad \frac{Q \xrightarrow{\beta} Q' \quad \mathrm{NStd}(P)}{P.Q \xrightarrow{\beta} P.Q'}$$

**TABLE 5.2 Reverse transition rules of Prefix and Summation.**

$$\frac{}{\alpha[m] \overset{\alpha[m]}{\twoheadrightarrow} \alpha}$$

$$\frac{P \overset{\alpha[m]}{\twoheadrightarrow} \alpha}{P+Q \overset{\alpha[m]}{\twoheadrightarrow} \alpha} \quad \frac{P \overset{\alpha[m]}{\twoheadrightarrow} P'}{P+Q \overset{\alpha[m]}{\twoheadrightarrow} P'}$$

$$\frac{Q \overset{\alpha[m]}{\twoheadrightarrow} \alpha}{P+Q \overset{\alpha[m]}{\twoheadrightarrow} \alpha} \quad \frac{Q \overset{\alpha[m]}{\twoheadrightarrow} Q'}{P+Q \overset{\alpha[m]}{\twoheadrightarrow} Q'}$$

$$\frac{P \overset{\alpha[m]}{\twoheadrightarrow} \alpha \quad \mathrm{Std}(Q)}{P.Q \overset{\alpha[m]}{\twoheadrightarrow} \alpha.Q} \quad \frac{P \overset{\alpha[m]}{\twoheadrightarrow} P' \quad \mathrm{Std}(Q)}{P.Q \overset{\alpha[m]}{\twoheadrightarrow} P'.Q}$$

$$\frac{Q \overset{\beta[n]}{\twoheadrightarrow} \beta \quad \mathrm{NStd}(P)}{P.Q \overset{\beta[n]}{\twoheadrightarrow} P.\beta} \quad \frac{Q \overset{\beta[n]}{\twoheadrightarrow} Q' \quad \mathrm{NStd}(P)}{P.Q \overset{\beta[n]}{\twoheadrightarrow} P.Q'}$$

### 5.1.2 Operational semantics

The operational semantics is defined by LTSs (labeled transition systems), and it is detailed by the following definition.

**Definition 5.2** (Semantics). The operational semantics of CTC corresponding to the syntax in Definition 5.1 is defined by a series of transition rules, which are shown in Tables 5.1, 5.2, 5.3, 5.4, 5.5, 5.6, 5.7, and 5.8. Also, the predicate $\xrightarrow{\alpha} \alpha[m]$ represents successful forward termination after execution of the action $\alpha$, the predicate $\overset{\alpha[m]}{\twoheadrightarrow} \alpha$ represents successful reverse termination after execution of the event $\alpha[m]$, the predicate Std(P) represents that $p$ is a standard process containing no past events, the predicate NStd(P) represents that $P$ is a process full of past events.

The forward transition rules for Prefix and Summation are shown in Table 5.1.

The reverse transition rules for Prefix and Summation are shown in Table 5.2.

The forward and reverse pomset transition rules of Prefix and Summation are shown in Table 5.3 and Table 5.4, which are different from the single event transition rules in Table 5.1 and Table 5.2, the forward and reverse pomset transition rules are labeled by pomsets, which are defined by causality . and conflict $+$.

The forward transition rules for Composition are shown in Table 5.5.

The reverse transition rules for Composition are shown in Table 5.6.

The forward transition rules for Restriction, Relabeling, and Constants are shown in Table 5.7.

The reverse transition rules for Restriction, Relabeling, and Constants are shown in Table 5.8.

### 5.1.3 Properties of transitions

**Definition 5.3** (Sorts). Given the sorts $\mathcal{L}(A)$ and $\mathcal{L}(X)$ of constants and variables, we define $\mathcal{L}(P)$ inductively as follows:

**TABLE 5.3** Forward pomset transition rules of Prefix and Summation.

| |
|---|
| $\dfrac{}{p \xrightarrow{p} p[\mathcal{K}]}$ |
| $\dfrac{P \xrightarrow{p} p[\mathcal{K}]}{P + Q \xrightarrow{p} p[\mathcal{K}]}$ $\quad \dfrac{P \xrightarrow{p} P'}{P + Q \xrightarrow{p} P'}$ |
| $\dfrac{Q \xrightarrow{q} q[\mathcal{J}]}{P + Q \xrightarrow{q} q[\mathcal{J}]}$ $\quad \dfrac{Q \xrightarrow{q} Q'}{P + Q \xrightarrow{q} Q'}$ |
| $\dfrac{P \xrightarrow{p} p[\mathcal{K}] \quad \mathrm{Std}(Q)}{P.Q \xrightarrow{p} p[\mathcal{K}].Q}(p \subseteq P)$ $\quad \dfrac{P \xrightarrow{p} P' \quad \mathrm{Std}(Q)}{P.Q \xrightarrow{p} P'.Q}(p \subseteq P)$ |
| $\dfrac{Q \xrightarrow{q} q[\mathcal{J}] \quad \mathrm{NStd}(P)}{P.Q \xrightarrow{q} P.q[\mathcal{J}]}(q \subseteq Q)$ $\quad \dfrac{Q \xrightarrow{q} Q' \quad \mathrm{NStd}(P)}{P.Q \xrightarrow{q} P.Q'}(q \subseteq Q)$ |

**TABLE 5.4** Reverse pomset transition rules of Prefix and Summation.

| |
|---|
| $\dfrac{}{p[\mathcal{K}] \overset{p[\mathcal{K}]}{\twoheadrightarrow} p}$ |
| $\dfrac{P \overset{p[\mathcal{K}]}{\twoheadrightarrow} p}{P + Q \overset{p[\mathcal{K}]}{\twoheadrightarrow} p}$ $\quad \dfrac{P \overset{p[\mathcal{K}]}{\twoheadrightarrow} P'}{P + Q \overset{p[\mathcal{K}]}{\twoheadrightarrow} P'}$ |
| $\dfrac{Q \overset{q[\mathcal{J}]}{\twoheadrightarrow} q}{P + Q \overset{q[\mathcal{J}]}{\twoheadrightarrow} q}$ $\quad \dfrac{Q \overset{q[\mathcal{J}]}{\twoheadrightarrow} Q'}{P + Q \overset{q[\mathcal{J}]}{\twoheadrightarrow} Q'}$ |
| $\dfrac{P \overset{p[\mathcal{K}]}{\twoheadrightarrow} p \quad \mathrm{Std}(Q)}{P.Q \overset{p[\mathcal{K}]}{\twoheadrightarrow} p.Q}(p \subseteq P)$ $\quad \dfrac{P \overset{p[\mathcal{K}]}{\twoheadrightarrow} P' \quad \mathrm{Std}(Q)}{P.Q \overset{p[\mathcal{K}]}{\twoheadrightarrow} P'.Q}(p \subseteq P)$ |
| $\dfrac{Q \overset{q[\mathcal{J}]}{\twoheadrightarrow} q \quad \mathrm{NStd}(P)}{P.Q \overset{q[\mathcal{J}]}{\twoheadrightarrow} P.q}(q \subseteq Q)$ $\quad \dfrac{Q \overset{q[\mathcal{J}]}{\twoheadrightarrow} Q' \quad \mathrm{NStd}(P)}{P \cdot Q \overset{q[\mathcal{J}]}{\twoheadrightarrow} P.Q'}(q \subseteq Q)$ |

**TABLE 5.5** Forward transition rules of Composition.

| |
|---|
| $\dfrac{P \xrightarrow{\alpha} P' \quad Q \nrightarrow}{P \parallel Q \xrightarrow{\alpha} P' \parallel Q}$ |
| $\dfrac{Q \xrightarrow{\alpha} Q' \quad P \nrightarrow}{P \parallel Q \xrightarrow{\alpha} P \parallel Q'}$ |
| $\dfrac{P \xrightarrow{\alpha} P' \quad Q \xrightarrow{\beta} Q'}{P \parallel Q \xrightarrow{\{\alpha,\beta\}} P' \parallel Q'} \quad (\beta \neq \overline{\alpha})$ |
| $\dfrac{P \xrightarrow{l} P' \quad Q \xrightarrow{\overline{l}} Q'}{P \parallel Q \xrightarrow{\tau} P' \parallel Q'}$ |

1. $\mathcal{L}(l.P) = \{l\} \cup \mathcal{L}(P)$;
2. $\mathcal{L}(P.l[m]) = \{l\} \cup \mathcal{L}(P)$;
3. $\mathcal{L}((l_1 \parallel \cdots \parallel l_n).P) = \{l_1, \cdots, l_n\} \cup \mathcal{L}(P)$;
4. $\mathcal{L}(P.(l_1[m] \parallel \cdots \parallel l_n[m])) = \{l_1, \cdots, l_n\} \cup \mathcal{L}(P)$;
5. $\mathcal{L}(\tau.P) = \mathcal{L}(P)$;
6. $\mathcal{L}(P + Q) = \mathcal{L}(P) \cup \mathcal{L}(Q)$;
7. $\mathcal{L}(P \parallel Q) = \mathcal{L}(P) \cup \mathcal{L}(Q)$;

**TABLE 5.6** Reverse transition rules of Composition.

| |
|---|
| $\dfrac{P\overset{\alpha[m]}{\twoheadrightarrow}P' \quad Q\not\twoheadrightarrow}{P\parallel Q\overset{\alpha[m]}{\twoheadrightarrow}P'\parallel Q}$ |
| $\dfrac{Q\overset{\alpha[m]}{\twoheadrightarrow}Q' \quad P\not\twoheadrightarrow}{P\parallel Q\overset{\alpha[m]}{\twoheadrightarrow}P\parallel Q'}$ |
| $\dfrac{P\overset{\alpha[m]}{\twoheadrightarrow}P' \quad Q\overset{\beta[m]}{\twoheadrightarrow}Q'}{P\parallel Q\overset{\{\alpha[m],\beta[m]\}}{\twoheadrightarrow}P'\parallel Q'}\quad(\beta\neq\overline{\alpha})$ |
| $\dfrac{P\overset{l[m]}{\twoheadrightarrow}P' \quad Q\overset{\overline{l}[m]}{\twoheadrightarrow}Q'}{P\parallel Q\overset{\tau}{\twoheadrightarrow}P'\parallel Q'}$ |

**TABLE 5.7** Forward transition rules of Restriction, Relabeling, and Constants.

| |
|---|
| $\dfrac{P\xrightarrow{\alpha}P'}{P\setminus L\xrightarrow{\alpha}P'\setminus L}\quad(\alpha,\overline{\alpha}\notin L)$ |
| $\dfrac{P\xrightarrow{\{\alpha_1,\cdots,\alpha_n\}}P'}{P\setminus L\xrightarrow{\{\alpha_1,\cdots,\alpha_n\}}P'\setminus L}\quad(\alpha_1,\overline{\alpha_1},\cdots,\alpha_n,\overline{\alpha_n}\notin L)$ |
| $\dfrac{P\xrightarrow{\alpha}P'}{P[f]\xrightarrow{f(\alpha)}P'[f]}$ |
| $\dfrac{P\xrightarrow{\{\alpha_1,\cdots,\alpha_n\}}P'}{P[f]\xrightarrow{\{f(\alpha_1),\cdots,f(\alpha_n)\}}P'[f]}$ |
| $\dfrac{P\xrightarrow{\alpha}P'}{A\xrightarrow{\alpha}P'}\quad(A\overset{\text{def}}{=}P)$ |
| $\dfrac{P\xrightarrow{\{\alpha_1,\cdots,\alpha_n\}}P'}{A\xrightarrow{\{\alpha_1,\cdots,\alpha_n\}}P'}\quad(A\overset{\text{def}}{=}P)$ |

**TABLE 5.8** Reverse transition rules of Restriction, Relabeling, and Constants.

| |
|---|
| $\dfrac{P\overset{\alpha[m]}{\twoheadrightarrow}P'}{P\setminus L\overset{\alpha[m]}{\twoheadrightarrow}P'\setminus L}\quad(\alpha,\overline{\alpha}\notin L)$ |
| $\dfrac{P\overset{\{\alpha_1[m],\cdots,\alpha_n[m]\}}{\twoheadrightarrow}P'}{P\setminus L\overset{\{\alpha_1[m],\cdots,\alpha_n[m]\}}{\twoheadrightarrow}P'\setminus L}\quad(\alpha_1,\overline{\alpha_1},\cdots,\alpha_n,\overline{\alpha_n}\notin L)$ |
| $\dfrac{P\overset{\alpha[m]}{\twoheadrightarrow}P'}{P[f]\overset{f(\alpha[m])}{\twoheadrightarrow}P'[f]}$ |
| $\dfrac{P\overset{\{\alpha_1[m],\cdots,\alpha_n[m]\}}{\twoheadrightarrow}P'}{P[f]\overset{\{f(\alpha_1)[m],\cdots,f(\alpha_n)[m]\}}{\twoheadrightarrow}P'[f]}$ |
| $\dfrac{P\overset{\alpha[m]}{\twoheadrightarrow}P'}{A\overset{\alpha[m]}{\twoheadrightarrow}P'}\quad(A\overset{\text{def}}{=}P)$ |
| $\dfrac{P\overset{\{\alpha_1[m],\cdots,\alpha_n[m]\}}{\twoheadrightarrow}P'}{A\overset{\{\alpha_1[m],\cdots,\alpha_n[m]\}}{\twoheadrightarrow}P'}\quad(A\overset{\text{def}}{=}P)$ |

8. $\mathcal{L}(P \setminus L) = \mathcal{L}(P) - (L \cup \overline{L})$;
9. $\mathcal{L}(P[f]) = \{f(l) : l \in \mathcal{L}(P)\}$;
10. for $A \stackrel{\text{def}}{=} P$, $\mathcal{L}(P) \subseteq \mathcal{L}(A)$.

Now, we present some properties of the transition rules defined in Definition 5.2.

**Proposition 5.4.** *If $P \xrightarrow{\alpha} P'$, then*

1. $\alpha \in \mathcal{L}(P) \cup \{\tau\}$;
2. $\mathcal{L}(P') \subseteq \mathcal{L}(P)$.

*If $P \xrightarrow{\{\alpha_1, \cdots, \alpha_n\}} P'$, then*

1. $\alpha_1, \cdots, \alpha_n \in \mathcal{L}(P) \cup \{\tau\}$;
2. $\mathcal{L}(P') \subseteq \mathcal{L}(P)$.

*Proof.* By induction on the inference of $P \xrightarrow{\alpha} P'$ and $P \xrightarrow{\{\alpha_1, \cdots, \alpha_n\}} P'$, there are several cases corresponding to the forward transition rules in Definition 5.2, however, we omit them here. □

**Proposition 5.5.** *If $P \xrightarrow{\alpha[m]} P'$, then*

1. $\alpha \in \mathcal{L}(P) \cup \{\tau\}$;
2. $\mathcal{L}(P') \subseteq \mathcal{L}(P)$.

*If $P \xrightarrow{\{\alpha_1[m], \cdots, \alpha_n[m]\}} P'$, then*

1. $\alpha_1, \cdots, \alpha_n \in \mathcal{L}(P) \cup \{\tau\}$;
2. $\mathcal{L}(P') \subseteq \mathcal{L}(P)$.

*Proof.* By induction on the inference of $P \xrightarrow{\alpha} P'$ and $P \xrightarrow{\{\alpha_1, \cdots, \alpha_n\}} P'$, there are several cases corresponding to the forward transition rules in Definition 5.2, however, we omit them here. □

## 5.2 Strongly forward–reverse truly concurrent bisimulations

### 5.2.1 Laws and congruence

Based on the concepts of strongly FR truly concurrent bisimulation equivalences, we obtain the following laws:

**Proposition 5.6** (Monoid laws for strongly FR pomset bisimulation). *The monoid laws for strongly FR pomset bisimulation are as follows:*

1. $P + Q \sim_p^{fr} Q + P$;
2. $P + (Q + R) \sim_p^{fr} (P + Q) + R$;
3. $P + P \sim_p^{fr} P$;
4. $P + \mathbf{nil} \sim_p^{fr} P$.

*Proof.* 1. $P + Q \sim_p^{fr} Q + P$. There are several cases, however, we will not enumerate them all here. By the forward transition rules of Summation in Table 5.3, we obtain

$$\frac{P \xrightarrow{p} P'}{P + Q \xrightarrow{p} P'} (p \subseteq P) \quad \frac{P \xrightarrow{p} P'}{Q + P \xrightarrow{p} P'} (p \subseteq P)$$

$$\frac{Q \xrightarrow{q} Q'}{P + Q \xrightarrow{q} Q'} (q \subseteq Q) \quad \frac{Q \xrightarrow{q} Q'}{Q + P \xrightarrow{q} Q'} (q \subseteq Q)$$

By the reverse transition rules of Summation in Table 5.4, we obtain

$$\frac{P \xrightarrow{p[\mathcal{K}]} P'}{P + Q \xrightarrow{p[\mathcal{K}]} P'} (p \subseteq P) \quad \frac{P \xrightarrow{p[\mathcal{K}]} P'}{Q + P \xrightarrow{p[\mathcal{K}]} P'} (p \subseteq P)$$

$$\frac{Q\xrightarrow{q[\mathcal{J}]}\!\!\!\!\rightarrow Q'}{P+Q\xrightarrow{q[\mathcal{J}]}\!\!\!\!\rightarrow Q'}(q\subseteq Q)\quad\frac{Q\xrightarrow{q[\mathcal{J}]}\!\!\!\!\rightarrow Q'}{Q+P\xrightarrow{q[\mathcal{J}]}\!\!\!\!\rightarrow Q'}(q\subseteq Q)$$

Hence, $P+Q\sim_p^{fr} Q+P$, as desired.

2. $P+(Q+R)\sim_p^{fr}(P+Q)+R$. There are several cases, however, we will not enumerate them all here. By the forward transition rules of Summation in Table 5.3, we obtain

$$\frac{P\xrightarrow{p}P'}{P+(Q+R)\xrightarrow{p}P'}(p\subseteq P)\quad\frac{P\xrightarrow{p}P'}{(P+Q)+R\xrightarrow{p}P'}(p\subseteq P)$$

$$\frac{Q\xrightarrow{q}Q'}{P+(Q+R)\xrightarrow{q}Q'}(q\subseteq Q)\quad\frac{Q\xrightarrow{q}Q'}{(P+Q)+R\xrightarrow{q}Q'}(q\subseteq Q)$$

$$\frac{R\xrightarrow{r}R'}{P+(Q+R)\xrightarrow{r}R'}(r\subseteq R)\quad\frac{R\xrightarrow{r}R'}{(P+Q)+R\xrightarrow{r}R'}(r\subseteq R)$$

By the reverse transition rules of Summation in Table 5.4, we obtain

$$\frac{P\xrightarrow{p[\mathcal{K}]}\!\!\!\!\rightarrow P'}{P+(Q+R)\xrightarrow{p[\mathcal{K}]}\!\!\!\!\rightarrow P'}(p\subseteq P)\quad\frac{P\xrightarrow{p[\mathcal{K}]}\!\!\!\!\rightarrow P'}{(P+Q)+R\xrightarrow{p[\mathcal{K}]}\!\!\!\!\rightarrow P'}(p\subseteq P)$$

$$\frac{Q\xrightarrow{q[\mathcal{J}]}\!\!\!\!\rightarrow Q'}{P+(Q+R)\xrightarrow{q[\mathcal{J}]}\!\!\!\!\rightarrow Q'}(q\subseteq Q)\quad\frac{Q\xrightarrow{q[\mathcal{J}]}\!\!\!\!\rightarrow Q'}{(P+Q)+R\xrightarrow{q[\mathcal{J}]}\!\!\!\!\rightarrow Q'}(q\subseteq Q)$$

$$\frac{R\xrightarrow{r[\mathcal{I}]}\!\!\!\!\rightarrow R'}{P+(Q+R)\xrightarrow{r[\mathcal{I}]}\!\!\!\!\rightarrow R'}(r\subseteq R)\quad\frac{R\xrightarrow{r[\mathcal{I}]}\!\!\!\!\rightarrow R'}{(P+Q)+R\xrightarrow{r[\mathcal{I}]}\!\!\!\!\rightarrow R'}(r\subseteq R)$$

Hence, $P+(Q+R)\sim_p^{fr}(P+Q)+R$, as desired.

3. $P+P\sim_p^{fr}P$. By the forward transition rules of Summation, we obtain

$$\frac{P\xrightarrow{p}p[\mathcal{K}]}{P+P\xrightarrow{p}p[\mathcal{K}]}(p\subseteq P)\quad\frac{P\xrightarrow{p}p[\mathcal{K}]}{P\xrightarrow{p}p[\mathcal{K}]}(p\subseteq P)$$

$$\frac{P\xrightarrow{p}P'}{P+P\xrightarrow{p}P'}(p\subseteq P)\quad\frac{P\xrightarrow{p}P'}{P\xrightarrow{p}P'}(p\subseteq P)$$

By the reverse transition rules of Summation, we obtain

$$\frac{P\xrightarrow{p[\mathcal{K}]}\!\!\!\!\rightarrow p}{P+P\xrightarrow{p[\mathcal{K}]}\!\!\!\!\rightarrow p}(p\subseteq P)\quad\frac{P\xrightarrow{p[\mathcal{K}]}\!\!\!\!\rightarrow p}{P\xrightarrow{p[\mathcal{K}]}\!\!\!\!\rightarrow p}(p\subseteq P)$$

$$\frac{P\xrightarrow{p[\mathcal{K}]}\!\!\!\!\rightarrow P'}{P+P\xrightarrow{p[\mathcal{K}]}\!\!\!\!\rightarrow P'}(p\subseteq P)\quad\frac{P\xrightarrow{p[\mathcal{K}]}\!\!\!\!\rightarrow P'}{P\xrightarrow{p[\mathcal{K}]}\!\!\!\!\rightarrow P'}(p\subseteq P)$$

Hence, $P+P\sim_p^{fr}P$, as desired.

4. $P+\mathbf{nil}\sim_p^{fr}P$. There are several cases, however, we will not enumerate them all here. By the forward transition rules of Summation in Table 5.3, we obtain

$$\frac{P\xrightarrow{p}P'}{P+\mathbf{nil}\xrightarrow{p}P'}(p\subseteq P)\quad\frac{P\xrightarrow{p}P'}{P\xrightarrow{p}P'}(p\subseteq P)$$

By the reverse transition rules of Summation in Table 5.4, we obtain

$$\frac{P \xrightarrow{p[\mathcal{K}]} P'}{P + \mathbf{nil} \xrightarrow{p[\mathcal{K}]} P'}(p \subseteq P) \quad \frac{P \xrightarrow{p[\mathcal{K}]} P'}{P \xrightarrow{p[\mathcal{K}]} P'}(p \subseteq P)$$

Since $P' \sim_p^{fr} P'$, $P + \mathbf{nil} \sim_p^{fr} P$, as desired. □

**Proposition 5.7** (Monoid laws for strongly FR step bisimulation). *The monoid laws for strongly FR step bisimulation are as follows:*

1. $P + Q \sim_s^{fr} Q + P$;
2. $P + (Q + R) \sim_s^{fr} (P + Q) + R$;
3. $P + P \sim_s^{fr} P$;
4. $P + \boldsymbol{nil} \sim_s^{fr} P$.

*Proof.* 1. $P + Q \sim_s^{fr} Q + P$. There are several cases, however, we will not enumerate them all here. By the forward transition rules of Summation in Table 5.3, we obtain

$$\frac{P \xrightarrow{p} P'}{P + Q \xrightarrow{p} P'}(p \subseteq P, \forall\alpha, \beta \in p, \text{ are pairwise concurrent})$$

$$\frac{P \xrightarrow{p} P'}{Q + P \xrightarrow{p} P'}(p \subseteq P, \forall\alpha, \beta \in p, \text{ are pairwise concurrent})$$

$$\frac{Q \xrightarrow{q} Q'}{P + Q \xrightarrow{q} Q'}(q \subseteq Q, \forall\alpha, \beta \in q, \text{ are pairwise concurrent})$$

$$\frac{Q \xrightarrow{q} Q'}{Q + P \xrightarrow{q} Q'}(q \subseteq Q, \forall\alpha, \beta \in q, \text{ are pairwise concurrent})$$

By the reverse transition rules of Summation in Table 5.4, we obtain

$$\frac{P \xrightarrow{p[\mathcal{K}]} P'}{P + Q \xrightarrow{p[\mathcal{K}]} P'}(p \subseteq P, \forall\alpha, \beta \in p, \text{ are pairwise concurrent})$$

$$\frac{P \xrightarrow{p[\mathcal{K}]} P'}{Q + P \xrightarrow{p[\mathcal{K}]} P'}(p \subseteq P, \forall\alpha, \beta \in p, \text{ are pairwise concurrent})$$

$$\frac{Q \xrightarrow{q[\mathcal{J}]} Q'}{P + Q \xrightarrow{q[\mathcal{J}]} Q'}(q \subseteq Q, \forall\alpha, \beta \in q, \text{ are pairwise concurrent})$$

$$\frac{Q \xrightarrow{q[\mathcal{J}]} Q'}{Q + P \xrightarrow{q[\mathcal{J}]} Q'}(q \subseteq Q, \forall\alpha, \beta \in q, \text{ are pairwise concurrent})$$

Hence, $P + Q \sim_s^{fr} Q + P$, as desired.

2. $P + (Q + R) \sim_s^{fr} (P + Q) + R$. There are several cases, however, we will not enumerate them all here. By the forward transition rules of Summation in Table 5.3, we obtain

$$\frac{P \xrightarrow{p} P'}{P + (Q + R) \xrightarrow{p} P'}(p \subseteq P, \forall\alpha, \beta \in p, \text{ are pairwise concurrent})$$

$$\frac{P \xrightarrow{p} P'}{(P+Q)+R \xrightarrow{p} P'}(p \subseteq P, \forall \alpha, \beta \in p, \text{ are pairwise concurrent})$$

$$\frac{Q \xrightarrow{q} Q'}{P+(Q+R) \xrightarrow{q} Q'}(q \subseteq Q, \forall \alpha, \beta \in q, \text{ are pairwise concurrent})$$

$$\frac{Q \xrightarrow{q} Q'}{(P+Q)+R \xrightarrow{q} Q'}(q \subseteq Q, \forall \alpha, \beta \in q, \text{ are pairwise concurrent})$$

$$\frac{R \xrightarrow{r} R'}{P+(Q+R) \xrightarrow{r} R'}(r \subseteq R, \forall \alpha, \beta \in r, \text{ are pairwise concurrent})$$

$$\frac{R \xrightarrow{r} R'}{(P+Q)+R \xrightarrow{r} R'}(r \subseteq R, \forall \alpha, \beta \in r, \text{ are pairwise concurrent})$$

By the reverse transition rules of Summation in Table 5.4, we obtain

$$\frac{P \xrightarrow{p[\mathcal{K}]} P'}{P+(Q+R) \xrightarrow{p[\mathcal{K}]} P'}(p \subseteq P, \forall \alpha, \beta \in p, \text{ are pairwise concurrent})$$

$$\frac{P \xrightarrow{p[\mathcal{K}]} P'}{(P+Q)+R \xrightarrow{p[\mathcal{K}]} P'}(p \subseteq P, \forall \alpha, \beta \in p, \text{ are pairwise concurrent})$$

$$\frac{Q \xrightarrow{q[\mathcal{J}]} Q'}{P+(Q+R) \xrightarrow{q[\mathcal{J}]} Q'}(q \subseteq Q, \forall \alpha, \beta \in q, \text{ are pairwise concurrent})$$

$$\frac{Q \xrightarrow{q[\mathcal{J}]} Q'}{(P+Q)+R \xrightarrow{q[\mathcal{J}]} Q'}(q \subseteq Q, \forall \alpha, \beta \in q, \text{ are pairwise concurrent})$$

$$\frac{R \xrightarrow{r[\mathcal{I}]} R'}{P+(Q+R) \xrightarrow{r[\mathcal{I}]} R'}(r \subseteq R, \forall \alpha, \beta \in r, \text{ are pairwise concurrent})$$

$$\frac{R \xrightarrow{r[\mathcal{I}]} R'}{(P+Q)+R \xrightarrow{r[\mathcal{I}]} R'}(r \subseteq R, \forall \alpha, \beta \in r, \text{ are pairwise concurrent})$$

Hence, $P+(Q+R) \sim_s^{fr} (P+Q)+R$, as desired.

3. $P+P \sim_s^{fr} P$. By the forward transition rules of Summation, we obtain

$$\frac{P \xrightarrow{p} p[\mathcal{K}]}{P+P \xrightarrow{p} p[\mathcal{K}]}(p \subseteq P, \forall \alpha, \beta \in p, \text{ are pairwise concurrent})$$

$$\frac{P \xrightarrow{p} p[\mathcal{K}]}{P \xrightarrow{p} p[\mathcal{K}]}(p \subseteq P, \forall \alpha, \beta \in p, \text{ are pairwise concurrent})$$

$$\frac{P \xrightarrow{p} P'}{P+P \xrightarrow{p} P'}(p \subseteq P, \forall \alpha, \beta \in p, \text{ are pairwise concurrent})$$

$$\frac{P \xrightarrow{p} P'}{P \xrightarrow{p} P'}(p \subseteq P, \forall \alpha, \beta \in p, \text{ are pairwise concurrent})$$

By the reverse transition rules of Summation, we obtain

$$\frac{P \xrightarrow{p[\mathcal{K}]} p}{P + P \xrightarrow{p[\mathcal{K}]} p}(p \subseteq P, \forall \alpha, \beta \in p, \text{ are pairwise concurrent})$$

$$\frac{P \xrightarrow{p[\mathcal{K}]} p}{P \xrightarrow{p[\mathcal{K}]} p}(p \subseteq P, \forall \alpha, \beta \in p, \text{ are pairwise concurrent})$$

$$\frac{P \xrightarrow{p[\mathcal{K}]} P'}{P + P \xrightarrow{p[\mathcal{K}]} P'}(p \subseteq P, \forall \alpha, \beta \in p, \text{ are pairwise concurrent})$$

$$\frac{P \xrightarrow{p[\mathcal{K}]} P'}{P \xrightarrow{p[\mathcal{K}]} P'}(p \subseteq P, \forall \alpha, \beta \in p, \text{ are pairwise concurrent})$$

Hence, $P + P \sim_s^{fr} P$, as desired.

4. $P + \mathbf{nil} \sim_s^{fr} P$. There are several cases, however, we will not enumerate them all here. By the forward transition rules of Summation in Table 5.3, we obtain

$$\frac{P \xrightarrow{p} P'}{P + \mathbf{nil} \xrightarrow{p} P'}(p \subseteq P, \forall \alpha, \beta \in p, \text{ are pairwise concurrent})$$

$$\frac{P \xrightarrow{p} P'}{P \xrightarrow{p} P'}(p \subseteq P, \forall \alpha, \beta \in p, \text{ are pairwise concurrent})$$

By the reverse transition rules of Summation in Table 5.4, we obtain

$$\frac{P \xrightarrow{p[\mathcal{K}]} P'}{P + \mathbf{nil} \xrightarrow{p[\mathcal{K}]} P'}(p \subseteq P, \forall \alpha, \beta \in p, \text{ are pairwise concurrent})$$

$$\frac{P \xrightarrow{p[\mathcal{K}]} P'}{P \xrightarrow{p[\mathcal{K}]} P'}(p \subseteq P, \forall \alpha, \beta \in p, \text{ are pairwise concurrent})$$

Since $P' \sim_s^{fr} P'$, $P + \mathbf{nil} \sim_s^{fr} P$, as desired. □

**Proposition 5.8** (Monoid laws for strongly FR hp-bisimulation). *The monoid laws for strongly FR hp-bisimulation are as follows:*

1. $P + Q \sim_{hp}^{fr} Q + P$;
2. $P + (Q + R) \sim_{hp}^{fr} (P + Q) + R$;
3. $P + P \sim_{hp}^{fr} P$;
4. $P + \mathbf{nil} \sim_{hp}^{fr} P$.

*Proof.* 1. $P + Q \sim_{hp}^{fr} Q + P$. There are several cases, however, we will not enumerate them all here. By the forward transition rules of Summation in Table 5.1, we obtain

$$\frac{P \xrightarrow{\alpha} P'}{P + Q \xrightarrow{\alpha} P'}(\alpha \in P) \quad \frac{P \xrightarrow{\alpha} P'}{Q + P \xrightarrow{\alpha} P'}(\alpha \in P)$$

$$\frac{Q \xrightarrow{\beta} Q'}{P + Q \xrightarrow{\beta} Q'}(\beta \in Q) \quad \frac{Q \xrightarrow{\beta} Q'}{Q + P \xrightarrow{\beta} Q'}(\beta \in Q)$$

By the reverse transition rules of Summation in Table 5.2, we obtain

$$\frac{P \xrightarrow{\alpha[m]} P'}{P+Q \xrightarrow{\alpha[m]} P'}(\alpha \in P) \quad \frac{P \xrightarrow{\alpha[m]} P'}{Q+P \xrightarrow{\alpha[m]} P'}(\alpha P)$$

$$\frac{Q \xrightarrow{\beta[n]} Q'}{P+Q \xrightarrow{\beta[n]} Q'}(\beta \in Q) \quad \frac{Q \xrightarrow{\beta[n]} Q'}{Q+P \xrightarrow{\beta[n]} Q'}(\beta \in Q)$$

Hence, $P+Q \sim_{hp}^{fr} Q+P$, as desired.

2. $P+(Q+R) \sim_{hp}^{fr} (P+Q)+R$. There are several cases, however, we will not enumerate them all here. By the forward transition rules of Summation in Table 5.1, we obtain

$$\frac{P \xrightarrow{\alpha} P'}{P+(Q+R) \xrightarrow{\alpha} P'}(\alpha \in P) \quad \frac{P \xrightarrow{\alpha} P'}{(P+Q)+R \xrightarrow{\alpha} P'}(\alpha \in P)$$

$$\frac{Q \xrightarrow{\beta} Q'}{P+(Q+R) \xrightarrow{\beta} Q'}(\beta \in Q) \quad \frac{Q \xrightarrow{\beta} Q'}{(P+Q)+R \xrightarrow{\beta} Q'}(\beta \in Q)$$

$$\frac{R \xrightarrow{\gamma} R'}{P+(Q+R) \xrightarrow{\gamma} R'}(\gamma \in R) \quad \frac{R \xrightarrow{\gamma} R'}{(P+Q)+R \xrightarrow{\gamma} R'}(\gamma \in R)$$

By the reverse transition rules of Summation in Table 5.2, we obtain

$$\frac{P \xrightarrow{\alpha[m]} P'}{P+(Q+R) \xrightarrow{\alpha[m]} P'}(\alpha \in P) \quad \frac{P \xrightarrow{\alpha[m]} P'}{(P+Q)+R \xrightarrow{\alpha[m]} P'}(\alpha \in P)$$

$$\frac{Q \xrightarrow{\beta[n]} Q'}{P+(Q+R) \xrightarrow{\beta[n]} Q'}(\beta \in Q) \quad \frac{Q \xrightarrow{\beta[n]} Q'}{(P+Q)+R \xrightarrow{\beta[n]} Q'}(\beta Q)$$

$$\frac{R \xrightarrow{\gamma[k]} R'}{P+(Q+R) \xrightarrow{\gamma[k]} R'}(\gamma \in R) \quad \frac{R \xrightarrow{\gamma[k]} R'}{(P+Q)+R \xrightarrow{\gamma[k]} R'}(\gamma \in R)$$

Hence, $P+(Q+R) \sim_{hp}^{fr} (P+Q)+R$, as desired.

3. $P+P \sim_{hp}^{fr} P$. By the forward transition rules of Summation, we obtain

$$\frac{P \xrightarrow{\alpha} \alpha[m]}{P+P \xrightarrow{\alpha} \alpha[m]}(\alpha \in P) \quad \frac{P \xrightarrow{\alpha} \alpha[m]}{P \xrightarrow{\alpha} \alpha[m]}(\alpha \in P)$$

$$\frac{P \xrightarrow{\alpha} P'}{P+P \xrightarrow{\alpha} P'}(p \subseteq P) \quad \frac{P \xrightarrow{\alpha} P'}{P \xrightarrow{\alpha} P'}(\alpha \in P)$$

By the reverse transition rules of Summation, we obtain

$$\frac{P \xrightarrow{\alpha[m]} \alpha}{P+P \xrightarrow{\alpha[m]} \alpha}(\alpha \in P) \quad \frac{P \xrightarrow{\alpha[m]} \alpha}{P \xrightarrow{\alpha[m]} \alpha}(\alpha \in P)$$

$$\frac{P \xrightarrow{\alpha[m]} P'}{P+P \xrightarrow{\alpha[m]} P'}(\alpha \in P) \quad \frac{P \xrightarrow{\alpha[m]} P'}{P \xrightarrow{\alpha[m} P'}(\alpha \in P)$$

Hence, $P+P \sim_{hp}^{fr} P$, as desired.

4. $P + \mathbf{nil} \sim_{hp}^{fr} P$. There are several cases, however, we will not enumerate them all here. By the forward transition rules of Summation in Table 5.1, we obtain

$$\frac{P \xrightarrow{\alpha} P'}{P + \mathbf{nil} \xrightarrow{\alpha} P'}(\alpha P) \quad \frac{P \xrightarrow{\alpha} P'}{P \xrightarrow{\alpha} P'}(\alpha \in P)$$

By the reverse transition rules of Summation in Table 5.2, we obtain

$$\frac{P \xrightarrow{\alpha[m]} P'}{P + \mathbf{nil} \xrightarrow{\alpha[m]} P'}(\alpha \in P) \quad \frac{P \xrightarrow{\alpha[m]} P'}{P \xrightarrow{\alpha[m]} P'}(\alpha \in P)$$

Since $(C(P + \mathbf{nil}), f, C(P)) \in \sim_{hp}^{fr}$, $(C(P + \mathbf{nil})', f[\alpha \mapsto \alpha], C(P)') \in \sim_{hp}^{fr}$, $P + \mathbf{nil} \sim_{hp}^{fr} P$, as desired. □

**Proposition 5.9** (Monoid laws for strongly FR hhp-bisimulation). *The monoid laws for strongly FR hhp-bisimulation are as follows:*

1. $P + Q \sim_{hhp}^{fr} Q + P$;
2. $P + (Q + R) \sim_{hhp}^{fr} (P + Q) + R$;
3. $P + P \sim_{hhp}^{fr} P$;
4. $P + \mathbf{nil} \sim_{hhp}^{fr} P$.

*Proof.* 1. $P + Q \sim_{hhp}^{fr} Q + P$. There are several cases, however, we will not enumerate them all here. By the forward transition rules of Summation in Table 5.1, we obtain

$$\frac{P \xrightarrow{\alpha} P'}{P + Q \xrightarrow{\alpha} P'}(\alpha \in P) \quad \frac{P \xrightarrow{\alpha} P'}{Q + P \xrightarrow{\alpha} P'}(\alpha \in P)$$

$$\frac{Q \xrightarrow{\beta} Q'}{P + Q \xrightarrow{\beta} Q'}(\beta \in Q) \quad \frac{Q \xrightarrow{\beta} Q'}{Q + P \xrightarrow{\beta} Q'}(\beta \in Q)$$

By the reverse transition rules of Summation in Table 5.2, we obtain

$$\frac{P \xrightarrow{\alpha[m]} P'}{P + Q \xrightarrow{\alpha[m]} P'}(\alpha \in P) \quad \frac{P \xrightarrow{\alpha[m]} P'}{Q + P \xrightarrow{\alpha[m]} P'}(\alpha P)$$

$$\frac{Q \xrightarrow{\beta[n]} Q'}{P + Q \xrightarrow{\beta[n]} Q'}(\beta \in Q) \quad \frac{Q \xrightarrow{\beta[n]} Q'}{Q + P \xrightarrow{\beta[n]} Q'}(\beta \in Q)$$

Hence, $P + Q \sim_{hhp}^{fr} Q + P$, as desired.

2. $P + (Q + R) \sim_{hhp}^{fr} (P + Q) + R$. There are several cases, however, we will not enumerate them all here. By the forward transition rules of Summation in Table 5.1, we obtain

$$\frac{P \xrightarrow{\alpha} P'}{P + (Q + R) \xrightarrow{\alpha} P'}(\alpha \in P) \quad \frac{P \xrightarrow{\alpha} P'}{(P + Q) + R \xrightarrow{\alpha} P'}(\alpha \in P)$$

$$\frac{Q \xrightarrow{\beta} Q'}{P + (Q + R) \xrightarrow{\beta} Q'}(\beta \in Q) \quad \frac{Q \xrightarrow{\beta} Q'}{(P + Q) + R \xrightarrow{\beta} Q'}(\beta \in Q)$$

$$\frac{R \xrightarrow{\gamma} R'}{P + (Q + R) \xrightarrow{\gamma} R'}(\gamma \in R) \quad \frac{R \xrightarrow{\gamma} R'}{(P + Q) + R \xrightarrow{\gamma} R'}(\gamma \in R)$$

By the reverse transition rules of Summation in Table 5.2, we obtain

$$\frac{P \xrightarrow{\alpha[m]} P'}{P+(Q+R) \xrightarrow{\alpha[m]} P'}(\alpha \in P) \quad \frac{P \xrightarrow{\alpha[m]} P'}{(P+Q)+R \xrightarrow{\alpha[m]} P'}(\alpha \in P)$$

$$\frac{Q \xrightarrow{\beta[n]} Q'}{P+(Q+R) \xrightarrow{\beta[n]} Q'}(\beta \in Q) \quad \frac{Q \xrightarrow{\beta[n]} Q'}{(P+Q)+R \xrightarrow{\beta[n]} Q'}(\beta Q)$$

$$\frac{R \xrightarrow{\gamma[k]} R'}{P+(Q+R) \xrightarrow{\gamma[k]} R'}(\gamma \in R) \quad \frac{R \xrightarrow{\gamma[k]} R'}{(P+Q)+R \xrightarrow{\gamma[k]} R'}(\gamma \in R)$$

Hence, $P+(Q+R) \sim^{fr}_{hhp} (P+Q)+R$, as desired.

3. $P+P \sim^{fr}_{hhp} P$. By the forward transition rules of Summation, we obtain

$$\frac{P \xrightarrow{\alpha} \alpha[m]}{P+P \xrightarrow{\alpha} \alpha[m]}(\alpha \in P) \quad \frac{P \xrightarrow{\alpha} \alpha[m]}{P \xrightarrow{\alpha} \alpha[m]}(\alpha \in P)$$

$$\frac{P \xrightarrow{\alpha} P'}{P+P \xrightarrow{\alpha} P'}(p \subseteq P) \quad \frac{P \xrightarrow{\alpha} P'}{P \xrightarrow{\alpha} P'}(\alpha \in P)$$

By the reverse transition rules of Summation, we obtain

$$\frac{P \xrightarrow{\alpha[m]} \alpha}{P+P \xrightarrow{\alpha[m]} \alpha}(\alpha \in P) \quad \frac{P \xrightarrow{\alpha[m]} \alpha}{P \xrightarrow{\alpha[m]} \alpha}(\alpha \in P)$$

$$\frac{P \xrightarrow{\alpha[m]} P'}{P+P \xrightarrow{\alpha[m]} P'}(\alpha \in P) \quad \frac{P \xrightarrow{\alpha[m]} P'}{P \xrightarrow{\alpha[m} P'}(\alpha \in P)$$

Since $(C(P+P), f, C(P)) \in \sim^{fr}_{hhp}$, $(C((P+P)'), f[\alpha \mapsto \alpha], C((P)')) \in \sim^{fr}_{hhp}$, $P+P \sim^{fr}_{hhp} P$, as desired.

4. $P+\mathbf{nil} \sim^{fr}_{hhp} P$. There are several cases, however, we will not enumerate them all here. By the forward transition rules of Summation in Table 5.1, we obtain

$$\frac{P \xrightarrow{\alpha} P'}{P+\mathbf{nil} \xrightarrow{\alpha} P'}(\alpha P) \quad \frac{P \xrightarrow{\alpha} P'}{P \xrightarrow{\alpha} P'}(\alpha \in P)$$

By the reverse transition rules of Summation in Table 5.2, we obtain

$$\frac{P \xrightarrow{\alpha[m]} P'}{P+\mathbf{nil} \xrightarrow{\alpha[m]} P'}(\alpha \in P) \quad \frac{P \xrightarrow{\alpha[m]} P'}{P \xrightarrow{\alpha[m]} P'}(\alpha \in P)$$

Since $(C(P+\mathbf{nil}), f, C(P)) \in \sim^{fr}_{hhp}$, $(C(P+\mathbf{nil})', f[\alpha \mapsto \alpha], C(P)') \in \sim^{fr}_{hhp}$, $P+\mathbf{nil} \sim^{fr}_{hhp} P$, as desired. □

**Proposition 5.10** (Static laws for strongly FR step bisimulation). *The static laws for strongly FR step bisimulation are as follows:*

1. $P \parallel Q \sim^{fr}_{s} Q \parallel P$;
2. $P \parallel (Q \parallel R) \sim^{fr}_{s} (P \parallel Q) \parallel R$;
3. $P \parallel \mathbf{nil} \sim^{fr}_{s} P$;
4. $P \setminus L \sim^{fr}_{s} P$, *if* $\mathcal{L}(P) \cap (L \cup \overline{L}) = \emptyset$;
5. $P \setminus K \setminus L \sim^{fr}_{s} P \setminus (K \cup L)$;

6. $P[f]\setminus L \sim_s^{fr} P\setminus f^{-1}(L)[f]$;
7. $(P \parallel Q)\setminus L \sim_s^{fr} P\setminus L \parallel Q\setminus L$, *if* $\mathcal{L}(P)\cap\overline{\mathcal{L}(Q)}\cap(L\cup\overline{L})=\emptyset$;
8. $P[Id] \sim_s^{fr} P$;
9. $P[f] \sim_s^{fr} P[f']$, *if* $f\upharpoonright\mathcal{L}(P)=f'\upharpoonright\mathcal{L}(P)$;
10. $P[f][f'] \sim_s^{fr} P[f'\circ f]$;
11. $(P \parallel Q)[f] \sim_s^{fr} P[f] \parallel Q[f]$, *if* $f\upharpoonright(L\cup\overline{L})$ *is one-to-one, where* $L=\mathcal{L}(P)\cup\mathcal{L}(Q)$.

*Proof.* Although transition rules in Tables 5.5, 5.6, 5.7, and 5.8 are defined in the flavor of a single event, they can be modified into a step (a set of events within which each event is pairwise concurrent), however, we omit them here. If we treat a single event as a step containing just one event, the proof of the static laws does not pose any problem, so we use this way and still use the transition rules in Tables 5.5, 5.6, 5.7, and 5.8.

1. $P \parallel Q \sim_s^{fr} Q \parallel P$. By the forward transition rules of Composition, we obtain

$$\frac{P\xrightarrow{\alpha}P'\quad Q\nrightarrow}{P\parallel Q\xrightarrow{\alpha}P'\parallel Q}\qquad\frac{P\xrightarrow{\alpha}P'\quad Q\nrightarrow}{Q\parallel P\xrightarrow{\alpha}Q\parallel P'}$$

$$\frac{Q\xrightarrow{\beta}Q'\quad P\nrightarrow}{P\parallel Q\xrightarrow{\beta}P\parallel Q'}\qquad\frac{Q\xrightarrow{\beta}Q'\quad P\nrightarrow}{Q\parallel P\xrightarrow{\beta}Q'\parallel P}$$

$$\frac{P\xrightarrow{\alpha}P'\quad Q\xrightarrow{\beta}Q'}{P\parallel Q\xrightarrow{\{\alpha,\beta\}}P'\parallel Q'}(\beta\neq\overline{\alpha})\qquad\frac{P\xrightarrow{\alpha}P'\quad Q\xrightarrow{\beta}Q'}{Q\parallel P\xrightarrow{\{\alpha,\beta\}}Q'\parallel P'}(\beta\neq\overline{\alpha})$$

$$\frac{P\xrightarrow{l}P'\quad Q\xrightarrow{\overline{l}}Q'}{P\parallel Q\xrightarrow{\tau}P'\parallel Q'}\qquad\frac{P\xrightarrow{l}P'\quad Q\xrightarrow{\overline{l}}Q'}{Q\parallel P\xrightarrow{\tau}Q'\parallel P'}$$

By the reverse transition rules of Composition, we obtain

$$\frac{P\xrightarrow{\alpha[m]}\!\!\!\!\rightarrow P'\quad Q\not\twoheadrightarrow}{P\parallel Q\xrightarrow{\alpha[m]}\!\!\!\!\rightarrow P'\parallel Q}\qquad\frac{P\xrightarrow{\alpha[m]}\!\!\!\!\rightarrow P'\quad Q\not\twoheadrightarrow}{Q\parallel P\xrightarrow{\alpha[m]}\!\!\!\!\rightarrow Q\parallel P'}$$

$$\frac{Q\xrightarrow{\beta[n]}\!\!\!\!\rightarrow Q'\quad P\not\twoheadrightarrow}{P\parallel Q\xrightarrow{\beta[n]}\!\!\!\!\rightarrow P\parallel Q'}\qquad\frac{Q\xrightarrow{\beta[n]}\!\!\!\!\rightarrow Q'\quad P\not\twoheadrightarrow}{Q\parallel P\xrightarrow{\beta[n]}\!\!\!\!\rightarrow Q'\parallel P}$$

$$\frac{P\xrightarrow{\alpha[m]}\!\!\!\!\rightarrow P'\quad Q\xrightarrow{\beta[m]}\!\!\!\!\rightarrow Q'}{P\parallel Q\xrightarrow{\{\alpha[m],\beta[m]\}}\!\!\!\!\rightarrow P'\parallel Q'}(\beta\neq\overline{\alpha})\qquad\frac{P\xrightarrow{\alpha[m]}\!\!\!\!\rightarrow P'\quad Q\xrightarrow{\beta[m]}\!\!\!\!\rightarrow Q'}{Q\parallel P\xrightarrow{\{\alpha[m],\beta[m]\}}\!\!\!\!\rightarrow Q'\parallel P'}(\beta\neq\overline{\alpha})$$

$$\frac{P\xrightarrow{l[m]}\!\!\!\!\rightarrow P'\quad Q\xrightarrow{\overline{l}[m]}\!\!\!\!\rightarrow Q'}{P\parallel Q\xrightarrow{\tau}\!\!\!\!\rightarrow P'\parallel Q'}\qquad\frac{P\xrightarrow{l[m]}\!\!\!\!\rightarrow P'\quad Q\xrightarrow{\overline{l}[m]}\!\!\!\!\rightarrow Q'}{Q\parallel P\xrightarrow{\tau}\!\!\!\!\rightarrow Q'\parallel P'}$$

Hence, with the assumptions $P'\parallel Q\sim_s^{fr}Q\parallel P'$, $P\parallel Q'\sim_s^{fr}Q'\parallel P$ and $P'\parallel Q'\sim_s^{fr}Q'\parallel P'$, $P\parallel Q\sim_s^{fr}Q\parallel P$, as desired.

2. $P\parallel(Q\parallel R)\sim_s^{fr}(P\parallel Q)\parallel R$. By the forward transition rules of Composition, we obtain

$$\frac{P\xrightarrow{\alpha}P'\quad Q\nrightarrow\quad R\nrightarrow}{P\parallel(Q\parallel R)\xrightarrow{\alpha}P'\parallel(Q\parallel R)}\qquad\frac{P\xrightarrow{\alpha}P'\quad Q\nrightarrow\quad R\nrightarrow}{(P\parallel Q)\parallel R\xrightarrow{\alpha}(P'\parallel Q)\parallel R}$$

$$\frac{Q\xrightarrow{\beta}Q'\quad P\nrightarrow\quad R\nrightarrow}{P\parallel(Q\parallel R)\xrightarrow{\beta}P\parallel(Q'\parallel R)}\qquad\frac{Q\xrightarrow{\beta}Q'\quad P\nrightarrow\quad R\nrightarrow}{(P\parallel Q)\parallel R\xrightarrow{\beta}(P\parallel Q')\parallel R}$$

$$\frac{R\xrightarrow{\gamma}R' \quad P\nrightarrow \quad Q\nrightarrow}{P\parallel(Q\parallel R)\xrightarrow{\gamma}P\parallel(Q\parallel R')} \quad \frac{R\xrightarrow{\gamma}R' \quad P\nrightarrow \quad Q\nrightarrow}{(P\parallel Q)\parallel R\xrightarrow{\gamma}(P\parallel Q)\parallel R'}$$

$$\frac{P\xrightarrow{\alpha}P' \quad Q\xrightarrow{\beta}Q' \quad R\nrightarrow}{P\parallel(Q\parallel R)\xrightarrow{\{\alpha,\beta\}}P'\parallel(Q'\parallel R)}(\beta\neq\overline{\alpha}) \quad \frac{P\xrightarrow{\alpha}P' \quad Q\xrightarrow{\beta}Q' \quad R\nrightarrow}{(P\parallel Q)\parallel R\xrightarrow{\{\alpha,\beta\}}(P'\parallel Q')\parallel R}(\beta\neq\overline{\alpha})$$

$$\frac{P\xrightarrow{\alpha}P' \quad R\xrightarrow{\gamma}R' \quad Q\nrightarrow}{P\parallel(Q\parallel R)\xrightarrow{\{\alpha,\gamma\}}P'\parallel(Q\parallel R')}(\gamma\neq\overline{\alpha}) \quad \frac{P\xrightarrow{\alpha}P' \quad R\xrightarrow{\gamma}R' \quad Q\nrightarrow}{(P\parallel Q)\parallel R\xrightarrow{\{\alpha,\gamma\}}(P'\parallel Q)\parallel R]}(\gamma\neq\overline{\alpha})$$

$$\frac{Q\xrightarrow{\beta}P' \quad R\xrightarrow{\gamma}R' \quad P\nrightarrow}{P\parallel(Q\parallel R)\xrightarrow{\{\beta,\gamma\}}P\parallel(Q'\parallel R')}(\gamma\neq\overline{\beta}) \quad \frac{Q\xrightarrow{\beta}Q' \quad R\xrightarrow{\gamma}R' \quad P\nrightarrow}{(P\parallel Q)\parallel R\xrightarrow{\{\beta,\gamma\}}(P\parallel Q')\parallel R'}(\gamma\neq\overline{\beta})$$

$$\frac{P\xrightarrow{\alpha}P' \quad Q\xrightarrow{\beta}Q' \quad R\xrightarrow{\gamma}R'}{P\parallel(Q\parallel R)\xrightarrow{\{\alpha,\beta,\gamma\}}P'\parallel(Q'\parallel R')}(\beta\neq\overline{\alpha},\gamma\neq\overline{\alpha},\gamma\neq\overline{\beta})$$

$$\frac{P\xrightarrow{\alpha}P' \quad Q\xrightarrow{\beta}Q' \quad R\xrightarrow{\gamma}R'}{(P\parallel Q)\parallel R\xrightarrow{\{\alpha,\beta,\gamma\}}(P'\parallel Q')\parallel R'}(\beta\neq\overline{\alpha},\gamma\neq\overline{\alpha},\gamma\neq\overline{\beta})$$

$$\frac{P\xrightarrow{l}P' \quad Q\xrightarrow{\overline{l}}Q' \quad R\nrightarrow}{P\parallel(Q\parallel R)\xrightarrow{\tau}P'\parallel(Q'\parallel R)} \quad \frac{P\xrightarrow{l}P' \quad Q\xrightarrow{\overline{l}}Q' \quad R\nrightarrow}{(P\parallel Q)\parallel R\xrightarrow{\tau}(P'\parallel Q')\parallel R}$$

$$\frac{P\xrightarrow{l}P' \quad R\xrightarrow{\overline{l}}R' \quad Q\nrightarrow}{P\parallel(Q\parallel R)\xrightarrow{\tau}P'\parallel(Q\parallel R')} \quad \frac{P\xrightarrow{l}P' \quad R\xrightarrow{\overline{l}}R' \quad Q\nrightarrow}{(P\parallel Q)\parallel R\xrightarrow{\tau}(P'\parallel Q)\parallel R]}$$

$$\frac{Q\xrightarrow{l}P' \quad R\xrightarrow{\overline{l}}R' \quad P\nrightarrow}{P\parallel(Q\parallel R)\xrightarrow{\tau}P\parallel(Q'\parallel R')} \quad \frac{Q\xrightarrow{l}Q' \quad R\xrightarrow{\overline{l}}R' \quad P\nrightarrow}{(P\parallel Q)\parallel R\xrightarrow{\tau}(P\parallel Q')\parallel R'}$$

$$\frac{P\xrightarrow{l}P' \quad Q\xrightarrow{\overline{l}}Q' \quad R\xrightarrow{\gamma}R'}{P\parallel(Q\parallel R)\xrightarrow{\{\tau,\gamma\}}P'\parallel(Q'\parallel R')} \quad \frac{P\xrightarrow{l}P' \quad Q\xrightarrow{\overline{l}}Q' \quad R\xrightarrow{\gamma}R'}{(P\parallel Q)\parallel R\xrightarrow{\{\tau,\gamma\}}(P'\parallel Q')\parallel R'}$$

$$\frac{P\xrightarrow{l}P' \quad R\xrightarrow{\overline{l}}R' \quad Q\xrightarrow{\beta}Q'}{P\parallel(Q\parallel R)\xrightarrow{\{\tau,\beta\}}P'\parallel(Q'\parallel R')} \quad \frac{P\xrightarrow{l}P' \quad R\xrightarrow{\overline{l}}R' \quad Q\xrightarrow{\beta}Q'}{(P\parallel Q)\parallel R\xrightarrow{\{\tau,\beta\}}(P'\parallel Q')\parallel R]}$$

$$\frac{Q\xrightarrow{l}Q' \quad R\xrightarrow{\overline{l}}R' \quad P\xrightarrow{\alpha}P'}{P\parallel(Q\parallel R)\xrightarrow{\{\tau,\alpha\}}P'\parallel(Q'\parallel R')} \quad \frac{Q\xrightarrow{l}Q' \quad R\xrightarrow{\overline{l}}R' \quad P\xrightarrow{\alpha}P'}{(P\parallel Q)\parallel R\xrightarrow{\{\tau,\alpha\}}(P'\parallel Q')\parallel R'}$$

By the reverse transition rules of Composition, we obtain

$$\frac{P\xrightarrow{\alpha[m]}\!\!\!\!\rightarrow P' \quad Q\not\twoheadrightarrow \quad R\not\twoheadrightarrow}{P\parallel(Q\parallel R)\xrightarrow{\alpha[m]}\!\!\!\!\rightarrow P'\parallel(Q\parallel R)} \quad \frac{P\xrightarrow{\alpha[m]}\!\!\!\!\rightarrow P' \quad Q\not\twoheadrightarrow \quad R\not\twoheadrightarrow}{(P\parallel Q)\parallel R\xrightarrow{\alpha[m]}\!\!\!\!\rightarrow(P'\parallel Q)\parallel R}$$

$$\frac{Q\xrightarrow{\beta[n]}\!\!\!\!\rightarrow Q' \quad P\not\twoheadrightarrow \quad R\not\twoheadrightarrow}{P\parallel(Q\parallel R)\xrightarrow{\beta[n]}\!\!\!\!\rightarrow P\parallel(Q'\parallel R)} \quad \frac{Q\xrightarrow{\beta[n]}\!\!\!\!\rightarrow Q' \quad P\not\twoheadrightarrow \quad R\not\twoheadrightarrow}{(P\parallel Q)\parallel R\xrightarrow{\beta[n]}\!\!\!\!\rightarrow(P\parallel Q')\parallel R}$$

$$\frac{R\xrightarrow{\gamma[k]}\!\!\!\!\rightarrow R' \quad P\not\twoheadrightarrow \quad Q\not\twoheadrightarrow}{P\parallel(Q\parallel R)\xrightarrow{\gamma[k]}\!\!\!\!\rightarrow P\parallel(Q\parallel R')} \quad \frac{R\xrightarrow{\gamma[k]}\!\!\!\!\rightarrow R' \quad P\not\twoheadrightarrow \quad Q\not\twoheadrightarrow}{(P\parallel Q)\parallel R\xrightarrow{\gamma[k]}\!\!\!\!\rightarrow(P\parallel Q)\parallel R'}$$

$$\frac{P \xrightarrow{\alpha[m]} P' \quad Q \xrightarrow{\beta[m]} Q' \quad R \not\twoheadrightarrow}{P \parallel (Q \parallel R) \xrightarrow{\{\alpha[m],\beta[m]\}} P' \parallel (Q' \parallel R)} (\beta \neq \overline{\alpha}) \qquad \frac{P \xrightarrow{\alpha[m]} P' \quad Q \xrightarrow{\beta[m]} Q' \quad R \not\twoheadrightarrow}{(P \parallel Q) \parallel R \xrightarrow{\{\alpha[m],\beta[m]\}} (P' \parallel Q') \parallel R} (\beta \neq \overline{\alpha})$$

$$\frac{P \xrightarrow{\alpha[m]} P' \quad R \xrightarrow{\gamma[m]} R' \quad Q \not\twoheadrightarrow}{P \parallel (Q \parallel R) \xrightarrow{\{\alpha[m],\gamma[m]\}} P' \parallel (Q \parallel R')} (\gamma \neq \overline{\alpha}) \qquad \frac{P \xrightarrow{\alpha[m]} P' \quad R \xrightarrow{\gamma[m]} R' \quad Q \not\twoheadrightarrow}{(P \parallel Q) \parallel R \xrightarrow{\{\alpha[m],\gamma[m]\}} (P' \parallel Q) \parallel R]} (\gamma \neq \overline{\alpha})$$

$$\frac{Q \xrightarrow{\beta[m]} P' \quad R \xrightarrow{\gamma[m]} R' \quad P \not\twoheadrightarrow}{P \parallel (Q \parallel R) \xrightarrow{\{\beta[m],\gamma[m]\}} P \parallel (Q' \parallel R')} (\gamma \neq \overline{\beta}) \qquad \frac{Q \xrightarrow{\beta[m]} Q' \quad R \xrightarrow{\gamma[m]} R' \quad P \not\twoheadrightarrow}{(P \parallel Q) \parallel R \xrightarrow{\{\beta[m],\gamma[m]\}} (P \parallel Q') \parallel R'} (\gamma \neq \overline{\beta})$$

$$\frac{P \xrightarrow{\alpha[m]} P' \quad Q \xrightarrow{\beta[m]} Q' \quad R \xrightarrow{\gamma[m]} R'}{P \parallel (Q \parallel R) \xrightarrow{\{\alpha[m],\beta[m],\gamma[m]\}} P' \parallel (Q' \parallel R')} (\beta \neq \overline{\alpha}, \gamma \neq \overline{\alpha}, \gamma \neq \overline{\beta})$$

$$\frac{P \xrightarrow{\alpha[m]} P' \quad Q \xrightarrow{\beta[m]} Q' \quad R \xrightarrow{\gamma[m]} R'}{(P \parallel Q) \parallel R \xrightarrow{\{\alpha[m],\beta[m],\gamma[m]\}} (P' \parallel Q') \parallel R'} (\beta \neq \overline{\alpha}, \gamma \neq \overline{\alpha}, \gamma \neq \overline{\beta})$$

$$\frac{P \xrightarrow{l[m]} P' \quad Q \xrightarrow{\overline{l}[m]} Q' \quad R \not\twoheadrightarrow}{P \parallel (Q \parallel R) \xrightarrow{\tau} P' \parallel (Q' \parallel R)} \qquad \frac{P \xrightarrow{l[m]} P' \quad Q \xrightarrow{\overline{l}[m]} Q' \quad R \not\twoheadrightarrow}{(P \parallel Q) \parallel R \xrightarrow{\tau} (P' \parallel Q') \parallel R}$$

$$\frac{P \xrightarrow{l[m]} P' \quad R \xrightarrow{\overline{l}[m]} R' \quad Q \not\twoheadrightarrow}{P \parallel (Q \parallel R) \xrightarrow{\tau} P' \parallel (Q \parallel R')} \qquad \frac{P \xrightarrow{l[m]} P' \quad R \xrightarrow{\overline{l}[m]} R' \quad Q \not\twoheadrightarrow}{(P \parallel Q) \parallel R \xrightarrow{\tau} (P' \parallel Q) \parallel R]}$$

$$\frac{Q \xrightarrow{l[m]} P' \quad R \xrightarrow{\overline{l}[m]} R' \quad P \not\twoheadrightarrow}{P \parallel (Q \parallel R) \xrightarrow{\tau} P \parallel (Q' \parallel R')} \qquad \frac{Q \xrightarrow{l[m]} Q' \quad R \xrightarrow{\overline{l}[m]} R' \quad P \not\twoheadrightarrow}{(P \parallel Q) \parallel R \xrightarrow{\tau} (P \parallel Q') \parallel R'}$$

$$\frac{P \xrightarrow{l[m]} P' \quad Q \xrightarrow{\overline{l}[m]} Q' \quad R \xrightarrow{\gamma[m]} R'}{P \parallel (Q \parallel R) \xrightarrow{\{\tau,\gamma[m]\}} P' \parallel (Q' \parallel R')} \qquad \frac{P \xrightarrow{l[m]} P' \quad Q \xrightarrow{\overline{l}[m]} Q' \quad R \xrightarrow{\gamma[m]} R'}{(P \parallel Q) \parallel R \xrightarrow{\{\tau,\gamma[m]\}} (P' \parallel Q') \parallel R'}$$

$$\frac{P \xrightarrow{l[m]} P' \quad R \xrightarrow{\overline{l}[m]} R' \quad Q \xrightarrow{\beta[m]} Q'}{P \parallel (Q \parallel R) \xrightarrow{\{\tau,\beta[m]\}} P' \parallel (Q' \parallel R')} \qquad \frac{P \xrightarrow{l[m]} P' \quad R \xrightarrow{\overline{l}[m]} R' \quad Q \xrightarrow{\beta[m]} Q'}{(P \parallel Q) \parallel R \xrightarrow{\{\tau,\beta[m]\}} (P' \parallel Q') \parallel R]}$$

$$\frac{Q \xrightarrow{l[m]} Q' \quad R \xrightarrow{\overline{l}[m]} R' \quad P \xrightarrow{\alpha[m]} P'}{P \parallel (Q \parallel R) \xrightarrow{\{\tau,\alpha[m]\}} P' \parallel (Q' \parallel R')} \qquad \frac{Q \xrightarrow{l[m]} Q' \quad R \xrightarrow{\overline{l}[m]} R' \quad P \xrightarrow{\alpha[m]} P'}{(P \parallel Q) \parallel R \xrightarrow{\{\tau,\alpha[m]\}} (P' \parallel Q') \parallel R'}$$

Hence, with the assumptions $P' \parallel (Q \parallel R) \sim_s^{fr} (P' \parallel Q) \parallel R$, $P \parallel (Q' \parallel R) \sim_s^{fr} (P \parallel Q') \parallel R$, $P \parallel (Q \parallel R') \sim_s^{fr} (P \parallel Q) \parallel R'$, $P' \parallel (Q' \parallel R) \sim_s^{fr} (P' \parallel Q') \parallel R$, $P' \parallel (Q \parallel R') \sim_s^{fr} (P' \parallel Q) \parallel R'$, $P \parallel (Q' \parallel R') \sim_s^{fr} (P \parallel Q') \parallel R'$ and $P' \parallel (Q' \parallel R') \sim_s^{fr} (P' \parallel Q') \parallel R'$, $P \parallel (Q \parallel R) \sim_s^{fr} (P \parallel Q) \parallel R$, as desired.

**3.** $P \parallel \mathbf{nil} \sim_s^{fr} P$. By the forward transition rules of Composition, we obtain

$$\frac{P \xrightarrow{\alpha} P'}{P \parallel \mathbf{nil} \xrightarrow{\alpha} P'} \qquad \frac{P \xrightarrow{\alpha} P'}{P \xrightarrow{\alpha} P'}$$

By the reverse transition rules of Composition, we obtain

$$\frac{P \xrightarrow{\alpha[m]} P'}{P \parallel \mathbf{nil} \xrightarrow{\alpha[m]} P'} \qquad \frac{P \xrightarrow{\alpha[m]} P'}{P \xrightarrow{\alpha[m]} P'}$$

Since $P' \sim_s^{fr} P'$, $P \parallel \mathbf{nil} \sim_s^{fr} P$, as desired.

4. $P \setminus L \sim_s^{fr} P$, if $\mathcal{L}(P) \cap (L \cup \overline{L}) = \emptyset$. By the forward transition rules of Restriction, we obtain

$$\frac{P \xrightarrow{\alpha} P'}{P \setminus L \xrightarrow{\alpha} P' \setminus L}(\mathcal{L}(P) \cap (L \cup \overline{L}) = \emptyset) \quad \frac{P \xrightarrow{\alpha} P'}{P \xrightarrow{\alpha} P'}$$

By the reverse transition rules of Restriction, we obtain

$$\frac{P \xrightarrow{\alpha[m]} P'}{P \setminus L \xrightarrow{\alpha[m]} P' \setminus L}(\mathcal{L}(P) \cap (L \cup \overline{L}) = \emptyset) \quad \frac{P \xrightarrow{\alpha[m]} P'}{P \xrightarrow{\alpha[m]} P'}$$

Since $P' \sim_s^{fr} P'$, and with the assumptions $P' \setminus L \sim_s^{fr} P'$, $P \setminus L \sim_s^{fr} P$, if $\mathcal{L}(P) \cap (L \cup \overline{L}) = \emptyset$, as desired.

5. $P \setminus K \setminus L \sim_s^{fr} P \setminus (K \cup L)$. By the forward transition rules of Restriction, we obtain

$$\frac{P \xrightarrow{\alpha} P'}{P \setminus K \setminus L \xrightarrow{\alpha} P' \setminus K \setminus L} \quad \frac{P \xrightarrow{\alpha} P'}{P \setminus (K \cup L) \xrightarrow{\alpha} P' \setminus (K \cup L)}$$

By the reverse transition rules of Restriction, we obtain

$$\frac{P \xrightarrow{\alpha[m]} P'}{P \setminus K \setminus L \xrightarrow{\alpha[m]} P' \setminus K \setminus L} \quad \frac{P \xrightarrow{\alpha[m]} P'}{P \setminus (K \cup L) \xrightarrow{\alpha[m]} P' \setminus (K \cup L)}$$

Since $P' \sim_s^{fr} P'$, and with the assumptions $P' \setminus K \setminus L \sim_s^{fr} P' \setminus (K \cup L)$, $P \setminus K \setminus L \sim_s^{fr} P \setminus (K \cup L)$, as desired.

6. $P[f] \setminus L \sim_s^{fr} P \setminus f^{-1}(L)[f]$. By the forward transition rules of Restriction and Relabeling, we obtain

$$\frac{P \xrightarrow{\alpha} P'}{P[f] \setminus L \xrightarrow{f(\alpha)} P'[f] \setminus L} \quad \frac{P \xrightarrow{\alpha} P'}{P \setminus f^{-1}(L)[f] \xrightarrow{f(\alpha)} P' \setminus f^{-1}(L)[f]}$$

By the reverse transition rules of Restriction and Relabeling, we obtain

$$\frac{P \xrightarrow{\alpha[m]} P'}{P[f] \setminus L \xrightarrow{f(\alpha)[m]} P'[f] \setminus L} \quad \frac{P \xrightarrow{\alpha[m]} P'}{P \setminus f^{-1}(L)[f] \xrightarrow{f(\alpha)[m]} P' \setminus f^{-1}(L)[f]}$$

Hence, with the assumptions $P'[f] \setminus L \sim_s^{fr} P' \setminus f^{-1}(L)[f]$, $P[f] \setminus L \sim_s^{fr} P \setminus f^{-1}(L)[f]$, as desired.

7. $(P \parallel Q) \setminus L \sim_s^{fr} P \setminus L \parallel Q \setminus L$, if $\mathcal{L}(P) \cap \overline{\mathcal{L}(Q)} \cap (L \cup \overline{L}) = \emptyset$. By the forward transition rules of Composition and Restriction, we obtain

$$\frac{P \xrightarrow{\alpha} P' \quad Q \nrightarrow}{(P \parallel Q) \setminus L \xrightarrow{\alpha} (P' \parallel Q) \setminus L}(\mathcal{L}(P) \cap \overline{\mathcal{L}(Q)} \cap (L \cup \overline{L}) = \emptyset)$$

$$\frac{P \xrightarrow{\alpha} P' \quad Q \nrightarrow}{P \setminus L \parallel Q \setminus L \xrightarrow{\alpha} P' \setminus L \parallel Q \setminus L}(\mathcal{L}(P) \cap \overline{\mathcal{L}(Q)} \cap (L \cup \overline{L}) = \emptyset)$$

$$\frac{Q \xrightarrow{\beta} Q' \quad P \nrightarrow}{(P \parallel Q) \setminus L \xrightarrow{\beta} (P \parallel Q') \setminus L}(\mathcal{L}(P) \cap \overline{\mathcal{L}(Q)} \cap (L \cup \overline{L}) = \emptyset)$$

$$\frac{Q \xrightarrow{\beta} Q' \quad P \nrightarrow}{P \setminus L \parallel Q \setminus L \xrightarrow{\beta} P \setminus L \parallel Q' \setminus L}(\mathcal{L}(P) \cap \overline{\mathcal{L}(Q)} \cap (L \cup \overline{L}) = \emptyset)$$

$$\frac{P \xrightarrow{\alpha} P' \quad Q \xrightarrow{\beta} Q'}{(P \parallel Q) \setminus L \xrightarrow{\{\alpha,\beta\}} (P' \parallel Q') \setminus L}(\mathcal{L}(P) \cap \overline{\mathcal{L}(Q)} \cap (L \cup \overline{L}) = \emptyset)$$

$$\frac{P \xrightarrow{\alpha} P' \quad Q \xrightarrow{\beta} Q'}{P \setminus L \parallel Q \setminus L \xrightarrow{\{\alpha,\beta\}} (P' \parallel Q') \setminus L} (\mathcal{L}(P) \cap \overline{\mathcal{L}(Q)} \cap (L \cup \overline{L}) = \emptyset)$$

$$\frac{P \xrightarrow{l} P' \quad Q \xrightarrow{\overline{l}} Q'}{(P \parallel Q) \setminus L \xrightarrow{\tau} (P' \parallel Q') \setminus L} (\mathcal{L}(P) \cap \overline{\mathcal{L}(Q)} \cap (L \cup \overline{L}) = \emptyset)$$

$$\frac{P \xrightarrow{l} P' \quad Q \xrightarrow{\overline{l}} Q'}{(P \setminus L \parallel Q \setminus L \xrightarrow{\tau} P' \setminus L \parallel Q' \setminus L} (\mathcal{L}(P) \cap \overline{\mathcal{L}(Q)} \cap (L \cup \overline{L}) = \emptyset)$$

By the reverse transition rules of Composition and Restriction, we obtain

$$\frac{P \overset{\alpha[m]}{\twoheadrightarrow} P' \quad Q \not\twoheadrightarrow}{(P \parallel Q) \setminus L \overset{\alpha[m]}{\twoheadrightarrow} (P' \parallel Q) \setminus L} (\mathcal{L}(P) \cap \overline{\mathcal{L}(Q)} \cap (L \cup \overline{L}) = \emptyset)$$

$$\frac{P \overset{\alpha[m]}{\twoheadrightarrow} P' \quad Q \not\twoheadrightarrow}{P \setminus L \parallel Q \setminus L \overset{\alpha[m]}{\twoheadrightarrow} P' \setminus L \parallel Q \setminus L} (\mathcal{L}(P) \cap \overline{\mathcal{L}(Q)} \cap (L \cup \overline{L}) = \emptyset)$$

$$\frac{Q \overset{\beta[n]}{\twoheadrightarrow} Q' \quad P \not\twoheadrightarrow}{(P \parallel Q) \setminus L \overset{\beta[n]}{\twoheadrightarrow} (P \parallel Q') \setminus L} (\mathcal{L}(P) \cap \overline{\mathcal{L}(Q)} \cap (L \cup \overline{L}) = \emptyset)$$

$$\frac{Q \overset{\beta[n]}{\twoheadrightarrow} Q' \quad P \not\twoheadrightarrow}{P \setminus L \parallel Q \setminus L \overset{\beta[n]}{\twoheadrightarrow} P \setminus L \parallel Q' \setminus L} (\mathcal{L}(P) \cap \overline{\mathcal{L}(Q)} \cap (L \cup \overline{L}) = \emptyset)$$

$$\frac{P \overset{\alpha[m]}{\twoheadrightarrow} P' \quad Q \overset{\beta[m]}{\twoheadrightarrow} Q'}{(P \parallel Q) \setminus L \overset{\{\alpha[m],\beta[m]\}}{\twoheadrightarrow} (P' \parallel Q') \setminus L} (\mathcal{L}(P) \cap \overline{\mathcal{L}(Q)} \cap (L \cup \overline{L}) = \emptyset)$$

$$\frac{P \overset{\alpha[m]}{\twoheadrightarrow} P' \quad Q \overset{\beta[m]}{\twoheadrightarrow} Q'}{P \setminus L \parallel Q \setminus L \overset{\{\alpha[m],\beta[m]\}}{\twoheadrightarrow} (P' \parallel Q') \setminus L} (\mathcal{L}(P) \cap \overline{\mathcal{L}(Q)} \cap (L \cup \overline{L}) = \emptyset)$$

$$\frac{P \overset{l[m]}{\twoheadrightarrow} P' \quad Q \overset{\overline{l}[m]}{\twoheadrightarrow} Q'}{(P \parallel Q) \setminus L \overset{\tau}{\twoheadrightarrow} (P' \parallel Q') \setminus L} (\mathcal{L}(P) \cap \overline{\mathcal{L}(Q)} \cap (L \cup \overline{L}) = \emptyset)$$

$$\frac{P \overset{l[m]}{\twoheadrightarrow} P' \quad Q \overset{\overline{l}[m]}{\twoheadrightarrow} Q'}{(P \setminus L \parallel Q \setminus L \overset{\tau}{\twoheadrightarrow} P' \setminus L \parallel Q' \setminus L} (\mathcal{L}(P) \cap \overline{\mathcal{L}(Q)} \cap (L \cup \overline{L}) = \emptyset)$$

Since $(P' \parallel Q) \setminus L \sim_s^{fr} P' \setminus L \parallel Q \setminus L$, $(P \parallel Q') \setminus L \sim_s^{fr} P \setminus L \parallel Q' \setminus L$ and $(P' \parallel Q') \setminus L \sim_s^{fr} P' \setminus L \parallel Q' \setminus L$, $(P \parallel Q) \setminus L \sim_s^{fr} P \setminus L \parallel Q \setminus L$, if $\mathcal{L}(P) \cap \overline{\mathcal{L}(Q)} \cap (L \cup \overline{L}) = \emptyset$, as desired.

**8.** $P[Id] \sim_s^{fr} P$. By the forward transition rules Relabeling, we obtain

$$\frac{P \xrightarrow{\alpha} P'}{P[Id] \xrightarrow{Id(\alpha)} P'[Id]} \qquad \frac{P \xrightarrow{\alpha} P'}{P \xrightarrow{\alpha} P'}$$

By the reverse transition rules Relabeling, we obtain

$$\frac{P \overset{\alpha[m]}{\twoheadrightarrow} P'}{P[Id] \overset{Id(\alpha[m])}{\twoheadrightarrow} P'[Id]} \qquad \frac{P \overset{\alpha[m]}{\twoheadrightarrow} P'}{P \overset{\alpha[m]}{\twoheadrightarrow} P'}$$

Hence, with the assumptions $P'[Id] \sim_s^{fr} P'$ and $Id(\alpha) = \alpha$, $P[Id] \sim_s^{fr} P$, as desired.

9. $P[f]\sim_s^{fr} P[f']$, if $f\upharpoonright\mathcal{L}(P)=f'\upharpoonright\mathcal{L}(P)$. By the forward transition rules of Relabeling, we obtain

$$\frac{P\xrightarrow{\alpha}P'}{P[f]\xrightarrow{f(\alpha)}P'[f]}\quad\frac{P\xrightarrow{\alpha}P'}{P[f']\xrightarrow{f'(\alpha)}P'[f']}$$

By the reverse transition rules of Relabeling, we obtain

$$\frac{P\xtwoheadrightarrow{\alpha[m]}P'}{P[f]\xtwoheadrightarrow{f(\alpha)[m]}P'[f]}\quad\frac{P\xtwoheadrightarrow{\alpha[m]}P'}{P[f']\xtwoheadrightarrow{f'(\alpha)[m]}P'[f']}$$

Hence, with the assumptions $P'[f]\sim_s^{fr} P'[f']$ and $f(\alpha)=f'(\alpha)$, if $f\upharpoonright\mathcal{L}(P)=f'\upharpoonright\mathcal{L}(P)$, $P[f]\sim_s^{fr} P[f']$, as desired.

10. $P[f][f']\sim_s^{fr} P[f'\circ f]$. By the forward transition rules of Relabeling, we obtain

$$\frac{P\xrightarrow{\alpha}P'}{P[f][f']\xrightarrow{f'(f(\alpha))}P'[f][f']}\quad\frac{P\xrightarrow{\alpha}P'}{P[f'\circ f]\xrightarrow{f'(f(\alpha))}P'[f'\circ f]}$$

By the reverse transition rules of Relabeling, we obtain

$$\frac{P\xtwoheadrightarrow{\alpha[m]}P'}{P[f][f']\xrightarrow{f'(f(\alpha))[m]}P'[f][f']}\quad\frac{P\xtwoheadrightarrow{\alpha[m]}P'}{P[f'\circ f]\xtwoheadrightarrow{f'(f(\alpha))[m]}P'[f'\circ f]}$$

Hence, with the assumptions $P'[f][f']\sim_s^{fr} P'[f'\circ f]$, $P[f][f']\sim_s^{fr} P[f'\circ f]$, as desired.

11. $(P\parallel Q)[f]\sim_s^{fr} P[f]\parallel Q[f]$, if $f\upharpoonright(L\cup\overline{L})$ is one-to-one, where $L=\mathcal{L}(P)\cup\mathcal{L}(Q)$. By the forward transition rules of Composition and Relabeling, we obtain

$$\frac{P\xrightarrow{\alpha}P'\quad Q\nrightarrow}{(P\parallel Q)[f]\xrightarrow{f(\alpha)}(P'\parallel Q)[f]}(\text{if } f\upharpoonright(L\cup\overline{L}) \text{ is one-to-one, where } L=\mathcal{L}(P)\cup\mathcal{L}(Q))$$

$$\frac{P\xrightarrow{\alpha}P'\quad Q\nrightarrow}{P[f]\parallel Q[f]\xrightarrow{\alpha}P'[f]\parallel Q[f]}(\text{if } f\upharpoonright(L\cup\overline{L}) \text{ is one-to-one, where } L=\mathcal{L}(P)\cup\mathcal{L}(Q))$$

$$\frac{Q\xrightarrow{\beta}Q'\quad P\nrightarrow}{(P\parallel Q)[f]\xrightarrow{f(\beta)}(P\parallel Q')[f]}(\text{if } f\upharpoonright(L\cup\overline{L}) \text{ is one-to-one, where } L=\mathcal{L}(P)\cup\mathcal{L}(Q))$$

$$\frac{Q\xrightarrow{\beta}Q'\quad P\nrightarrow}{P[f]\parallel Q[f]\xrightarrow{\beta}P[f]\parallel Q'[f]}(\text{if } f\upharpoonright(L\cup\overline{L}) \text{ is one-to-one, where } L=\mathcal{L}(P)\cup\mathcal{L}(Q))$$

$$\frac{P\xrightarrow{\alpha}P'\quad Q\xrightarrow{\beta}Q'}{(P\parallel Q)[f]\xrightarrow{\{f(\alpha),f(\beta)\}}(P'\parallel Q')[f]}(\text{if } f\upharpoonright(L\cup\overline{L}) \text{ is one-to-one, where } L=\mathcal{L}(P)\cup\mathcal{L}(Q))$$

$$\frac{P\xrightarrow{\alpha}P'\quad Q\xrightarrow{\beta}Q'}{P[f]\parallel Q[f]\xrightarrow{\{f(\alpha),f(\beta)\}}P'[f]\parallel Q'[f]}(\text{if } f\upharpoonright(L\cup\overline{L}) \text{ is one-to-one, where } L=\mathcal{L}(P)\cup\mathcal{L}(Q))$$

$$\frac{P\xrightarrow{l}P'\quad Q\xrightarrow{\overline{l}}Q'}{(P\parallel Q)[f]\xrightarrow{\tau}(P'\parallel Q')[f]}(\text{if } f\upharpoonright(L\cup\overline{L}) \text{ is one-to-one, where } L=\mathcal{L}(P)\cup\mathcal{L}(Q))$$

$$\frac{P\xrightarrow{l}P'\quad Q\xrightarrow{\overline{l}}Q'}{(P[f]\parallel Q[f]\xrightarrow{\tau}P'[f]\parallel Q'[f]}(\text{if } f\upharpoonright(L\cup\overline{L}) \text{ is one-to-one, where } L=\mathcal{L}(P)\cup\mathcal{L}(Q))$$

By the reverse transition rules of Composition and Relabeling, we obtain

$$\frac{P\xrightarrow{\alpha[m]}\!\!\!\!\rightarrow P'\quad Q\not\twoheadrightarrow}{(P\parallel Q)[f]\xrightarrow{f(\alpha)[m]}\!\!\!\!\rightarrow (P'\parallel Q)[f]}(\text{if } f\restriction(L\cup\overline{L}) \text{ is one-to-one, where } L=\mathcal{L}(P)\cup\mathcal{L}(Q))$$

$$\frac{P\xrightarrow{\alpha[m]}\!\!\!\!\rightarrow P'\quad Q\not\twoheadrightarrow}{P[f]\parallel Q[f]\xrightarrow{f(\alpha)[m]}\!\!\!\!\rightarrow P'[f]\parallel Q[f]}(\text{if } f\restriction(L\cup\overline{L}) \text{ is one-to-one, where } L=\mathcal{L}(P)\cup\mathcal{L}(Q))$$

$$\frac{Q\xrightarrow{\beta[n]}\!\!\!\!\rightarrow Q'\quad P\not\twoheadrightarrow}{(P\parallel Q)[f]\xrightarrow{f(\beta)[n]}\!\!\!\!\rightarrow (P\parallel Q')[f]}(\text{if } f\restriction(L\cup\overline{L}) \text{ is one-to-one, where } L=\mathcal{L}(P)\cup\mathcal{L}(Q))$$

$$\frac{Q\xrightarrow{\beta[n]}\!\!\!\!\rightarrow Q'\quad P\not\twoheadrightarrow}{P[f]\parallel Q[f]\xrightarrow{f(\beta)[n]}\!\!\!\!\rightarrow P[f]\parallel Q'[f]}(\text{if } f\restriction(L\cup\overline{L}) \text{ is one-to-one, where } L=\mathcal{L}(P)\cup\mathcal{L}(Q))$$

$$\frac{P\xrightarrow{\alpha[m]}\!\!\!\!\rightarrow P'\quad Q\xrightarrow{\beta[m]}\!\!\!\!\rightarrow Q'}{(P\parallel Q)[f]\xrightarrow{\{f(\alpha)[m],f(\beta)[m]\}}\!\!\!\!\rightarrow (P'\parallel Q')[f]}(\text{if } f\restriction(L\cup\overline{L}) \text{ is one-to-one, where } L=\mathcal{L}(P)\cup\mathcal{L}(Q))$$

$$\frac{P\xrightarrow{\alpha[m]}\!\!\!\!\rightarrow P'\quad Q\xrightarrow{\beta[m]}\!\!\!\!\rightarrow Q'}{P[f]\parallel Q[f]\xrightarrow{\{f(\alpha)[m],f(\beta)[m]\}}\!\!\!\!\rightarrow P'[f]\parallel Q'[f]}(\text{if } f\restriction(L\cup\overline{L}) \text{ is one-to-one, where } L=\mathcal{L}(P)\cup\mathcal{L}(Q))$$

$$\frac{P\xrightarrow{l[m]}\!\!\!\!\rightarrow P'\quad Q\xrightarrow{\overline{l}[m]}\!\!\!\!\rightarrow Q'}{(P\parallel Q)[f]\xrightarrow{\tau}\!\!\!\!\rightarrow (P'\parallel Q')[f]}(\text{if } f\restriction(L\cup\overline{L}) \text{ is one-to-one, where } L=\mathcal{L}(P)\cup\mathcal{L}(Q))$$

$$\frac{P\xrightarrow{l[m]}\!\!\!\!\rightarrow P'\quad Q\xrightarrow{\overline{l}[m]}\!\!\!\!\rightarrow Q'}{(P[f]\parallel Q[f]\xrightarrow{\tau}\!\!\!\!\rightarrow P'[f]\parallel Q'[f]}(\text{if } f\restriction(L\cup\overline{L}) \text{ is one-to-one, where } L=\mathcal{L}(P)\cup\mathcal{L}(Q))$$

Hence, with the assumptions $(P'\parallel Q)[f]\sim_s^{fr} P'[f]\parallel Q[f]$, $(P\parallel Q')[f]\sim_s^{fr} P[f]\parallel Q'[f]$ and $(P'\parallel Q')[f]\sim_s^{fr} P'[f]\parallel Q'[f]$, $(P\parallel Q)[f]\sim_s^{fr} P[f]\parallel Q[f]$, if $f\restriction(L\cup\overline{L})$ is one-to-one, where $L=\mathcal{L}(P)\cup\mathcal{L}(Q)$, as desired. □

**Proposition 5.11** (Static laws for strongly FR pomset bisimulation). *The static laws for strongly FR pomset bisimulation are as follows:*

1. $P\parallel Q\sim_p^{fr} Q\parallel P$;
2. $P\parallel(Q\parallel R)\sim_p^{fr}(P\parallel Q)\parallel R$;
3. $P\parallel \boldsymbol{nil}\sim_p^{fr} P$;
4. $P\setminus L\sim_p^{fr} P$, *if* $\mathcal{L}(P)\cap(L\cup\overline{L})=\emptyset$;
5. $P\setminus K\setminus L\sim_p^{fr} P\setminus(K\cup L)$;
6. $P[f]\setminus L\sim_p^{fr} P\setminus f^{-1}(L)[f]$;
7. $(P\parallel Q)\setminus L\sim_p^{fr} P\setminus L\parallel Q\setminus L$, *if* $\mathcal{L}(P)\cap\overline{\mathcal{L}(Q)}\cap(L\cup\overline{L})=\emptyset$;
8. $P[Id]\sim_p^{fr} P$;
9. $P[f]\sim_p^{fr} P[f']$, *if* $f\restriction\mathcal{L}(P)=f'\restriction\mathcal{L}(P)$;
10. $P[f][f']\sim_p^{fr} P[f'\circ f]$;
11. $(P\parallel Q)[f]\sim_p^{fr} P[f]\parallel Q[f]$, *if* $f\restriction(L\cup\overline{L})$ *is one-to-one, where* $L=\mathcal{L}(P)\cup\mathcal{L}(Q)$.

*Proof.* From the definition of strongly FR pomset bisimulation (see Definition 2.68), we know that strongly FR pomset bisimulation is defined by FR pomset transitions, which are labeled by pomsets. In an FR pomset transition, the events in the pomset are either within causality relations (defined by the prefix .) or in concurrency (implicitly defined by . and +, and explicitly defined by ∥), of course, they are pairwise consistent (without conflicts). In Proposition 5.10, we have already proven the case that all events are pairwise concurrent, hence, we only need to prove the case of events in causality.

Without loss of generality, we take a pomset of $p = \{\alpha, \beta : \alpha.\beta\}$. Then, the FR pomset transition labeled by the above $p$ is just composed of one single event transition labeled by $\alpha$ succeeded by another single event transition labeled by $\beta$, that is, $\xrightarrow{p} = \xrightarrow{\alpha}\xrightarrow{\beta}$ and $\xrightarrow{p[\mathcal{K}]}\!\!\twoheadrightarrow = \xrightarrow{\beta[n]}\!\!\twoheadrightarrow\xrightarrow{\alpha[m]}\!\!\twoheadrightarrow$.

Similarly to the proof of static laws for strongly FR step bisimulation (see Proposition 5.10), we can prove that the static laws hold for strongly FR pomset bisimulation, however, we omit them here. □

**Proposition 5.12** (Static laws for strongly FR hp-bisimulation). *The static laws for strongly FR hp-bisimulation are as follows:*

1. $P \parallel Q \sim^{fr}_{hp} Q \parallel P$;
2. $P \parallel (Q \parallel R) \sim^{fr}_{hp} (P \parallel Q) \parallel R$;
3. $P \parallel \mathbf{nil} \sim^{fr}_{hp} P$;
4. $P \setminus L \sim^{fr}_{hp} P$, *if* $\mathcal{L}(P) \cap (L \cup \overline{L}) = \emptyset$;
5. $P \setminus K \setminus L \sim^{fr}_{hp} P \setminus (K \cup L)$;
6. $P[f] \setminus L \sim^{fr}_{hp} P \setminus f^{-1}(L)[f]$;
7. $(P \parallel Q) \setminus L \sim^{fr}_{hp} P \setminus L \parallel Q \setminus L$, *if* $\mathcal{L}(P) \cap \overline{\mathcal{L}(Q)} \cap (L \cup \overline{L}) = \emptyset$;
8. $P[Id] \sim^{fr}_{hp} P$;
9. $P[f] \sim^{fr}_{hp} P[f']$, *if* $f \restriction \mathcal{L}(P) = f' \restriction \mathcal{L}(P)$;
10. $P[f][f'] \sim^{fr}_{hp} P[f' \circ f]$;
11. $(P \parallel Q)[f] \sim^{fr}_{hp} P[f] \parallel Q[f]$, *if* $f \restriction (L \cup \overline{L})$ *is one-to-one, where* $L = \mathcal{L}(P) \cup \mathcal{L}(Q)$.

*Proof.* From the definition of strongly FR hp-bisimulation (see Definition 2.70), we know that strongly FR hp-bisimulation is defined on the posetal product $(C_1, f, C_2), f : C_1 \to C_2$ isomorphism. Two processes $P$ related to $C_1$ and $Q$ related to $C_2$, and $f : C_1 \to C_2$ isomorphism. Initially, $(C_1, f, C_2) = (\emptyset, \emptyset, \emptyset)$, and $(\emptyset, \emptyset, \emptyset) \in \sim_{hp}$. When $P \xrightarrow{\alpha} P'$ $(C_1 \xrightarrow{\alpha} C_1')$, there will be $Q \xrightarrow{\alpha} Q'$ $(C_2 \xrightarrow{\alpha} C_2')$, and we define $f' = f[\alpha \mapsto \alpha]$. Also, when $P \xrightarrow{\alpha[m]}\!\!\twoheadrightarrow P'$ $(C_1 \xrightarrow{\alpha[m]}\!\!\twoheadrightarrow C_1')$, there will be $Q \xrightarrow{\alpha[m]}\!\!\twoheadrightarrow Q'$ $(C_2 \xrightarrow{\alpha[m]}\!\!\twoheadrightarrow C_2')$, and we define $f' = f[\alpha[m] \mapsto \alpha[m]]$. Then, if $(C_1, f, C_2) \in \sim^{fr}_{hp}$, then $(C_1', f', C_2') \in \sim^{fr}_{hp}$.

Similarly to the proof of static laws for strongly FR pomset bisimulation (see Proposition 5.11), we can prove that static laws hold for strongly FR hp-bisimulation, we just need additionally to check the above conditions on FR hp-bisimulation, however, we omit them here. □

**Proposition 5.13** (Static laws for strongly FR hhp-bisimulation). *The static laws for strongly FR hhp-bisimulation are as follows:*

1. $P \parallel Q \sim^{fr}_{hhp} Q \parallel P$;
2. $P \parallel (Q \parallel R) \sim^{fr}_{hhp} (P \parallel Q) \parallel R$;
3. $P \parallel \mathbf{nil} \sim^{fr}_{hhp} P$;
4. $P \setminus L \sim^{fr}_{hhp} P$, *if* $\mathcal{L}(P) \cap (L \cup \overline{L}) = \emptyset$;
5. $P \setminus K \setminus L \sim^{fr}_{hhp} P \setminus (K \cup L)$;
6. $P[f] \setminus L \sim^{fr}_{hhp} P \setminus f^{-1}(L)[f]$;
7. $(P \parallel Q) \setminus L \sim^{fr}_{hhp} P \setminus L \parallel Q \setminus L$, *if* $\mathcal{L}(P) \cap \overline{\mathcal{L}(Q)} \cap (L \cup \overline{L}) = \emptyset$;
8. $P[Id] \sim^{fr}_{hhp} P$;
9. $P[f] \sim^{fr}_{hhp} P[f']$, *if* $f \restriction \mathcal{L}(P) = f' \restriction \mathcal{L}(P)$;
10. $P[f][f'] \sim^{fr}_{hhp} P[f' \circ f]$;
11. $(P \parallel Q)[f] \sim^{fr}_{hhp} P[f] \parallel Q[f]$, *if* $f \restriction (L \cup \overline{L})$ *is one-to-one, where* $L = \mathcal{L}(P) \cup \mathcal{L}(Q)$.

*Proof.* From the definition of strongly FR hhp-bisimulation (see Definition 2.70), we know that strongly FR hhp-bisimulation is downward closed for strongly FR hp-bisimulation.

Similarly to the proof of static laws for strongly FR hp-bisimulation (see Proposition 5.12), we can prove that static laws hold for strongly FR hhp-bisimulation, that is, they are downward closed for strongly FR hp-bisimulation, however, we omit them here. □

**Proposition 5.14** (Milner's expansion law for strongly FR truly concurrent bisimulations). *Milner's expansion law does not hold any longer for any strongly FR truly concurrent bisimulation, that is,*

1. $\alpha \parallel \beta \nsim_p^{fr} \alpha.\beta + \beta.\alpha$;
2. $\alpha \parallel \beta \nsim_s^{fr} \alpha.\beta + \beta.\alpha$;
3. $\alpha \parallel \beta \nsim_{hp}^{fr} \alpha.\beta + \beta.\alpha$;
4. $\alpha \parallel \beta \nsim_{hhp}^{fr} \alpha.\beta + \beta.\alpha$.

*Proof.* In nature, this is caused by $\alpha \parallel \beta$ and $\alpha.\beta + \beta.\alpha$ having different causality structures. By the transition rules of Composition and Prefix, we have

$$\alpha \parallel \beta \xrightarrow{\{\alpha,\beta\}} \mathbf{nil}$$

while

$$\alpha.\beta + \beta.\alpha \nrightarrow^{\{\alpha,\beta\}} .$$

Also,

$$\alpha[m] \parallel \beta[m] \xrightarrow{\{\alpha[m],\beta[m]\}} \mathbf{nil}$$

while

$$\alpha[m].\beta[m] + \beta[m].\alpha[m] \stackrel{\{\alpha[m],\beta[m]\}}{\not\twoheadrightarrow} .$$

□

**Proposition 5.15** (New expansion law for strongly FR step bisimulation). *Let $P \equiv (P_1[f_1] \parallel \cdots \parallel P_n[f_n]) \setminus L$, with $n \geq 1$. Then,*

$$\begin{aligned} P \sim_s^f &\{(f_1(\alpha_1) \parallel \cdots \parallel f_n(\alpha_n)).(P_1'[f_1] \parallel \cdots \parallel P_n'[f_n]) \setminus L : \\ &P_i \xrightarrow{\alpha_i} P_i', i \in \{1, \cdots, n\}, f_i(\alpha_i) \notin L \cup \overline{L}\} \\ &+\sum\{\tau.(P_1[f_1] \parallel \cdots \parallel P_i'[f_i] \parallel \cdots \parallel P_j'[f_j] \parallel \cdots \parallel P_n[f_n]) \setminus L : \\ &P_i \xrightarrow{l_1} P_i', P_j \xrightarrow{l_2} P_j', f_i(l_1) = \overline{f_j(l_2)}, i < j\} \end{aligned} \quad (5.1)$$

$$\begin{aligned} P \sim_s^r &\{(P_1'[f_1] \parallel \cdots \parallel P_n'[f_n]).(f_1(\alpha_1[m]) \parallel \cdots \parallel f_n(\alpha_n)[m]) \setminus L : \\ &P_i \xrightarrow{\alpha_i[m]} P_i', i \in \{1, \cdots, n\}, f_i(\alpha_i) \notin L \cup \overline{L}\} \\ &+\sum\{(P_1[f_1] \parallel \cdots \parallel P_i'[f_i] \parallel \cdots \parallel P_j'[f_j] \parallel \cdots \parallel P_n[f_n]).\tau \setminus L : \\ &P_i \xrightarrow{l_1[m]} P_i', P_j \xrightarrow{l_2[m]} P_j', f_i(l_1) = \overline{f_j(l_2)}, i < j\} \end{aligned} \quad (5.2)$$

*Proof.* Although transition rules in Definition 5.2 are defined in the flavor of a single event, they can be modified into a step (a set of events within which each event is pairwise concurrent), however, we omit them here. If we treat a single event as a step containing just one event, the proof of the new expansion law does not pose any problem, hence, we use this way and still use the transition rules in Definition 5.2.

(1) The case of strongly forward step bisimulation.

First, we consider the case without Restriction and Relabeling. That is, we suffice to prove the following case by induction on the size $n$.

For $P \equiv P_1 \parallel \cdots \parallel P_n$, with $n \geq 1$, we need to prove

$$P \sim_s \{(\alpha_1 \parallel \cdots \parallel \alpha_n).(P_1' \parallel \cdots \parallel P_n') : P_i \xrightarrow{\alpha_i} P_i', i \in \{1, \cdots, n\}$$

$$+\sum\{\tau.(P_1 \parallel \cdots \parallel P_i' \parallel \cdots \parallel P_j' \parallel \cdots \parallel P_n) : P_i \xrightarrow{l} P_i', P_j \xrightarrow{\bar{l}} P_j', i < j\}$$

For $n = 1$, $P_1 \sim_s^f \alpha_1.P_1' : P_1 \xrightarrow{\alpha_1} P_1'$ is obvious. Then, with a hypothesis $n$, we consider $R \equiv P \parallel P_{n+1}$. By the forward transition rules of Composition, we can obtain

$$R \sim_s^f \{(p \parallel \alpha_{n+1}).(P' \parallel P_{n+1}') : P \xrightarrow{p} P', P_{n+1} \xrightarrow{\alpha_{n+1}} P_{n+1}', p \subseteq P\}$$
$$+\sum\{\tau.(P' \parallel P_{n+1}') : P \xrightarrow{l} P', P_{n+1} \xrightarrow{\bar{l}} P_{n+1}'\}$$

Now, with the induction assumption $P \equiv P_1 \parallel \cdots \parallel P_n$, the right-hand side can be reformulated as follows:

$$\{(\alpha_1 \parallel \cdots \parallel \alpha_n \parallel \alpha_{n+1}).(P_1' \parallel \cdots \parallel P_n' \parallel P_{n+1}') :$$
$$P_i \xrightarrow{\alpha_i} P_i', i \in \{1, \cdots, n+1\}$$
$$+\sum\{\tau.(P_1 \parallel \cdots \parallel P_i' \parallel \cdots \parallel P_j' \parallel \cdots \parallel P_n \parallel P_{n+1}) :$$
$$P_i \xrightarrow{l} P_i', P_j \xrightarrow{\bar{l}} P_j', i < j\}$$
$$+\sum\{\tau.(P_1 \parallel \cdots \parallel P_i' \parallel \cdots \parallel P_j \parallel \cdots \parallel P_n \parallel P_{n+1}') :$$
$$P_i \xrightarrow{l} P_i', P_{n+1} \xrightarrow{\bar{l}} P_{n+1}', i \in \{1, \cdots, n\}\}$$

Hence,

$$R \sim_s^f \{(\alpha_1 \parallel \cdots \parallel \alpha_n \parallel \alpha_{n+1}).(P_1' \parallel \cdots \parallel P_n' \parallel P_{n+1}') :$$
$$P_i \xrightarrow{\alpha_i} P_i', i \in \{1, \cdots, n+1\}$$
$$+\sum\{\tau.(P_1 \parallel \cdots \parallel P_i' \parallel \cdots \parallel P_j' \parallel \cdots \parallel P_n) :$$
$$P_i \xrightarrow{l} P_i', P_j \xrightarrow{\bar{l}} P_j', 1 \leq i < j \geq n+1\}$$

Then, we can easily add the full conditions with Restriction and Relabeling.

(2) The case of strongly reverse step bisimulation.

First, we consider the case without Restriction and Relabeling. That is, we suffice to prove the following case by induction on the size $n$.

For $P \equiv P_1 \parallel \cdots \parallel P_n$, with $n \geq 1$, we need to prove

$$P \sim_s^r \{(P_1' \parallel \cdots \parallel P_n').(\alpha_1[m] \parallel \cdots \parallel \alpha_n[m]) : P_i \xrightarrow{\alpha_i[m]} P_i', i \in \{1, \cdots, n\}$$
$$+\sum\{(P_1 \parallel \cdots \parallel P_i' \parallel \cdots \parallel P_j' \parallel \cdots \parallel P_n).\tau : P_i \xrightarrow{l[m]} P_i', P_j \xrightarrow{\bar{l}[m]} P_j', i < j\}$$

For $n = 1$, $P_1 \sim_s^r P_1'.\alpha_1[m] : P_1 \xrightarrow{\alpha_1[m]} P_1'$ is obvious. Then, with a hypothesis $n$, we consider $R \equiv P \parallel P_{n+1}$. By the reverse transition rules of Composition, we can obtain

$$R \sim_s^r \{(P' \parallel P_{n+1}').(p[m] \parallel \alpha_{n+1}[m]) : P \xrightarrow{p[m]} P', P_{n+1} \xrightarrow{\alpha_{n+1}[m]} P_{n+1}', p \subseteq P\}$$
$$+\sum\{(P' \parallel P_{n+1}').\tau : P \xrightarrow{l[m]} P', P_{n+1} \xrightarrow{\bar{l}[m]} P_{n+1}'\}$$

Now, with the induction assumption $P \equiv P_1 \parallel \cdots \parallel P_n$, the right-hand side can be reformulated as follows:

$$\{(P_1' \parallel \cdots \parallel P_n' \parallel P_{n+1}).(\alpha_1[m] \parallel \cdots \parallel \alpha_n[m] \parallel \alpha_{n+1}[m]) :$$
$$P_i \xrightarrow{\alpha_i[m]} P_i', i \in \{1, \cdots, n+1\}$$
$$+\sum\{(P_1 \parallel \cdots \parallel P_i' \parallel \cdots \parallel P_j' \parallel \cdots \parallel P_n \parallel P_{n+1}).\tau :$$

$$P_i \xrightarrow{l[m]} P_i', P_j \xrightarrow{\overline{l}[m]} P_j', i < j\}$$

$$+\sum\{(P_1 \parallel \cdots \parallel P_i' \parallel \cdots \parallel P_j' \parallel \cdots \parallel P_n \parallel P_{n+1}).\tau :$$

$$P_i \xrightarrow{l[m]} P_i', P_{n+1} \xrightarrow{\overline{l}[m]} P_{n+1}', i \in \{1, \cdots, n\}\}$$

Hence,

$$R \sim_s^r \{(P_1' \parallel \cdots \parallel P_n' \parallel P_{n+1}').(\alpha_1[m] \parallel \cdots \parallel \alpha_n[m] \parallel \alpha_{n+1}[m]) :$$

$$P_i \xrightarrow{\alpha_i[m]} P_i', i \in \{1, \cdots, n+1\}$$

$$+\sum\{(P_1 \parallel \cdots \parallel P_i' \parallel \cdots \parallel P_j' \parallel \cdots \parallel P_n).\tau :$$

$$P_i \xrightarrow{l[m]} P_i', P_j \xrightarrow{\overline{l}[m]} P_j', 1 \leq i < j \geq n+1\}$$

Then, we can easily add the full conditions with Restriction and Relabeling. □

**Proposition 5.16** (New expansion law for strongly FR pomset bisimulation). *Let $P \equiv (P_1[f_1] \parallel \cdots \parallel P_n[f_n]) \setminus L$, with $n \geq 1$. Then,*

$$P \sim_p^f \{(f_1(\alpha_1) \parallel \cdots \parallel f_n(\alpha_n)).(P_1'[f_1] \parallel \cdots \parallel P_n'[f_n]) \setminus L :$$

$$P_i \xrightarrow{\alpha_i} P_i', i \in \{1, \cdots, n\}, f_i(\alpha_i) \notin L \cup \overline{L}\}$$

$$+\sum\{\tau.(P_1[f_1] \parallel \cdots \parallel P_i'[f_i] \parallel \cdots \parallel P_j'[f_j] \parallel \cdots \parallel P_n[f_n]) \setminus L :$$

$$P_i \xrightarrow{l_1} P_i', P_j \xrightarrow{l_2} P_j', f_i(l_1) = \overline{f_j(l_2)}, i < j\} \quad (5.3)$$

$$P \sim_p^r \{(P_1'[f_1] \parallel \cdots \parallel P_n'[f_n]).(f_1(\alpha_1[m]) \parallel \cdots \parallel f_n(\alpha_n)[m]) \setminus L :$$

$$P_i \xrightarrow{\alpha_i[m]} P_i', i \in \{1, \cdots, n\}, f_i(\alpha_i) \notin L \cup \overline{L}\}$$

$$+\sum\{(P_1[f_1] \parallel \cdots \parallel P_i'[f_i] \parallel \cdots \parallel P_j'[f_j] \parallel \cdots \parallel P_n[f_n]).\tau \setminus L :$$

$$P_i \xrightarrow{l_1[m]} P_i', P_j \xrightarrow{l_2[m]} P_j', f_i(l_1) = \overline{f_j(l_2)}, i < j\} \quad (5.4)$$

*Proof.* From the definition of strongly FR pomset bisimulation (see Definition 2.68), we know that strongly FR pomset bisimulation is defined by FR pomset transitions, which are labeled by pomsets. In an FR pomset transition, the events in the pomset are either within causality relations (defined by the prefix .) or in concurrency (implicitly defined by . and +, and explicitly defined by ∥), of course, they are pairwise consistent (without conflicts). In Proposition 5.15, we have already proven the case that all events are pairwise concurrent, hence, we only need to prove the case of events in causality. Without loss of generality, we take a pomset of $p = \{\alpha, \beta : \alpha.\beta\}$. Then, the FR pomset transition labeled by the above $p$ is just composed of one single event transition labeled by $\alpha$ succeeded by another single event transition labeled by $\beta$, that is, $\xrightarrow{p} = \xrightarrow{\alpha}\xrightarrow{\beta}$ and $\xrightarrow{p[\mathcal{K}]} = \xrightarrow{\beta[n]}\xrightarrow{\alpha[m]}$.

Similarly to the proof of the new expansion law for strongly FR step bisimulation (see Proposition 5.15), we can prove that the new expansion law holds for strongly FR pomset bisimulation, however, we omit this proof here. □

**Proposition 5.17** (New expansion law for strongly FR hp-bisimulation). *Let $P \equiv (P_1[f_1] \parallel \cdots \parallel P_n[f_n]) \setminus L$, with $n \geq 1$. Then,*

$$P \sim_{hp}^f \{(f_1(\alpha_1) \parallel \cdots \parallel f_n(\alpha_n)).(P_1'[f_1] \parallel \cdots \parallel P_n'[f_n]) \setminus L :$$

$$P_i \xrightarrow{\alpha_i} P_i', i \in \{1, \cdots, n\}, f_i(\alpha_i) \notin L \cup \overline{L}\}$$

$$+\sum\{\tau.(P_1[f_1] \parallel \cdots \parallel P_i'[f_i] \parallel \cdots \parallel P_j'[f_j] \parallel \cdots \parallel P_n[f_n]) \setminus L :$$

$$P_i \xrightarrow{l_1} P_i', P_j \xrightarrow{l_2} P_j', f_i(l_1) = \overline{f_j(l_2)}, i < j\} \quad (5.5)$$

$$P \sim^r_{hp} \{(P'_1[f_1] \parallel \cdots \parallel P'_n[f_n]).(f_1(\alpha_1[m]) \parallel \cdots \parallel f_n(\alpha_n)[m]) \setminus L : \\ P_i \xrightarrow{\alpha_i[m]} P'_i, i \in \{1, \cdots, n\}, f_i(\alpha_i) \notin L \cup \overline{L}\} \\ + \sum \{(P_1[f_1] \parallel \cdots \parallel P'_i[f_i] \parallel \cdots \parallel P'_j[f_j] \parallel \cdots \parallel P_n[f_n]).\tau \setminus L : \\ P_i \xrightarrow{l_1[m]} P'_i, P_j \xrightarrow{l_2[m]} P'_j, f_i(l_1) = \overline{f_j(l_2)}, i < j\} \quad (5.6)$$

*Proof.* From the definition of strongly FR hp-bisimulation (see Definition 2.70), we know that strongly FR hp-bisimulation is defined on the posetal product $(C_1, f, C_2), f : C_1 \to C_2$ isomorphism. Two processes $P$ related to $C_1$ and $Q$ related to $C_2$, and $f : C_1 \to C_2$ isomorphism. Initially, $(C_1, f, C_2) = (\emptyset, \emptyset, \emptyset)$, and $(\emptyset, \emptyset, \emptyset) \in \sim^{fr}_{hp}$. When $P \xrightarrow{\alpha} P'$ $(C_1 \xrightarrow{\alpha} C'_1)$, there will be $Q \xrightarrow{\alpha} Q'$ $(C_2 \xrightarrow{\alpha} C'_2)$, and we define $f' = f[\alpha \mapsto \alpha]$. Also, when $P \xrightarrow{\alpha[m]} P'$ $(C_1 \xrightarrow{\alpha[m]} C'_1)$, there will be $Q \xrightarrow{\alpha[m]} Q'$ $(C_2 \xrightarrow{\alpha[m]} C'_2)$, and we define $f' = f[\alpha[m] \mapsto \alpha[m]]$. Then, if $(C_1, f, C_2) \in \sim^{fr}_{hp}$, then $(C'_1, f', C'_2) \in \sim^{fr}_{hp}$.

Similarly to the proof of the new expansion law for strongly FR pomset bisimulation (see Proposition 5.16), we can prove that the new expansion law holds for strongly FR hp-bisimulation, we just need additionally to check the above conditions on FR hp-bisimulation, however, we omit them here. □

**Proposition 5.18** (New expansion law for strongly FR hhp-bisimulation). *Let $P \equiv (P_1[f_1] \parallel \cdots \parallel P_n[f_n]) \setminus L$, with $n \geq 1$. Then,*

$$P \sim^f_{hhp} \{(f_1(\alpha_1) \parallel \cdots \parallel f_n(\alpha_n)).(P'_1[f_1] \parallel \cdots \parallel P'_n[f_n]) \setminus L : \\ P_i \xrightarrow{\alpha_i} P'_i, i \in \{1, \cdots, n\}, f_i(\alpha_i) \notin L \cup \overline{L}\} \\ + \sum \{\tau.(P_1[f_1] \parallel \cdots \parallel P'_i[f_i] \parallel \cdots \parallel P'_j[f_j] \parallel \cdots \parallel P_n[f_n]) \setminus L : \\ P_i \xrightarrow{l_1} P'_i, P_j \xrightarrow{l_2} P'_j, f_i(l_1) = \overline{f_j(l_2)}, i < j\} \quad (5.7)$$

$$P \sim^r_{hhp} \{(P'_1[f_1] \parallel \cdots \parallel P'_n[f_n]).(f_1(\alpha_1[m]) \parallel \cdots \parallel f_n(\alpha_n)[m]) \setminus L : \\ P_i \xrightarrow{\alpha_i[m]} P'_i, i \in \{1, \cdots, n\}, f_i(\alpha_i) \notin L \cup \overline{L}\} \\ + \sum \{(P_1[f_1] \parallel \cdots \parallel P'_i[f_i] \parallel \cdots \parallel P'_j[f_j] \parallel \cdots \parallel P_n[f_n]).\tau \setminus L : \\ P_i \xrightarrow{l_1[m]} P'_i, P_j \xrightarrow{l_2[m]} P'_j, f_i(l_1) = \overline{f_j(l_2)}, i < j\} \quad (5.8)$$

*Proof.* From the definition of strongly FR hhp-bisimulation (see Definition 2.70), we know that strongly FR hhp-bisimulation is downward closed for strongly FR hp-bisimulation.

Similarly to the proof of the new expansion law for strongly FR hp-bisimulation (see Proposition 5.17), we can prove that the new expansion law holds for strongly FR hhp-bisimulation, that is, they are downward closed for strongly FR hp-bisimulation, however, we omit this proof here. □

**Theorem 5.19** (Congruence for strongly FR step bisimulation). *We can enjoy the congruence for strongly FR step bisimulation as follows:*

1. *If $A \stackrel{def}{=} P$, then $A \sim^{fr}_s P$;*
2. *Let $P_1 \sim^{fr}_s P_2$. Then,*
   (a) $\alpha.P_1 \sim^f_s \alpha.P_2$;
   (b) $(\alpha_1 \parallel \cdots \parallel \alpha_n).P_1 \sim^f_s (\alpha_1 \parallel \cdots \parallel \alpha_n).P_2$;
   (c) $P_1.\alpha[m] \sim^r_s P_2.\alpha[m]$;
   (d) $P_1.(\alpha_1[m] \parallel \cdots \parallel \alpha_n[m]) \sim^r_s P_2.(\alpha_1[m] \parallel \cdots \parallel \alpha_n[m])$.;
   (e) $P_1 + Q \sim^{fr}_s P_2 + Q$;
   (f) $P_1 \parallel Q \sim^{fr}_s P_2 \parallel Q$;
   (g) $P_1 \setminus L \sim^{fr}_s P_2 \setminus L$;
   (h) $P_1[f] \sim^{fr}_s P_2[f]$.

*Proof.* Although transition rules in Definition 5.2 are defined in the flavor of a single event, they can be modified into a step (a set of events within which each event is pairwise concurrent), however, we omit them here. If we treat a single event as a step containing just one event, the proof of the congruence does not pose any problem, hence, we use this way and still use the transition rules in Definition 5.2.

1. If $A \stackrel{\text{def}}{=} P$, then $A \sim_s^{fr} P$. This is obvious.
2. Let $P_1 \sim_s^{fr} P_2$. Then,
   (a) $\alpha.P_1 \sim_s^f \alpha.P_2$. By the forward transition rules of Prefix, we can obtain

$$\alpha.P_1 \xrightarrow{\alpha} \alpha[m]P_1$$

$$\alpha.P_2 \xrightarrow{\alpha} \alpha[m]P_2$$

   Since $P_1 \sim_s^{fr} P_2$, we obtain $\alpha.P_1 \sim_s^f \alpha.P_2$, as desired.

   (b) $(\alpha_1 \parallel \cdots \parallel \alpha_n).P_1 \sim_s^f (\alpha_1 \parallel \cdots \parallel \alpha_n).P_2$. By the transition rules of Prefix, we can obtain

$$(\alpha_1 \parallel \cdots \parallel \alpha_n).P_1 \xrightarrow{\{\alpha_1,\cdots,\alpha_n\}} (\alpha_1[m] \parallel \cdots \parallel \alpha_n[m]).P_1$$

$$(\alpha_1 \parallel \cdots \parallel \alpha_n).P_2 \xrightarrow{\{\alpha_1,\cdots,\alpha_n\}} (\alpha_1[m] \parallel \cdots \parallel \alpha_n[m]).P_2$$

   Since $P_1 \sim_s^{fr} P_2$, we obtain $(\alpha_1 \parallel \cdots \parallel \alpha_n).P_1 \sim_s^f (\alpha_1 \parallel \cdots \parallel \alpha_n).P_2$, as desired.

   (c) $P_1.\alpha[m] \sim_s^r P_2.\alpha[m]$. By the reverse transition rules of Prefix, we can obtain

$$P_1.\alpha[m] \xrightarrow{\alpha[m]} P_1.\alpha$$

$$P_2.\alpha[m] \xrightarrow{\alpha[m]} P_2.\alpha$$

   Since $P_1 \sim_s^{fr} P_2$, we obtain $P_1.\alpha[m] \sim_s^r P_2.\alpha[m]$, as desired.

   (d) $P_1.(\alpha_1[m] \parallel \cdots \parallel \alpha_n[m]) \sim_s^r P_2.(\alpha_1[m] \parallel \cdots \parallel \alpha_n[m])$. By the reverse transition rules of Prefix, we can obtain

$$P_1.(\alpha_1[m] \parallel \cdots \parallel \alpha_n[m]) \xrightarrow{\{\alpha_1[m],\cdots,\alpha_n[m]\}} P_1.(\alpha_1 \parallel \cdots \parallel \alpha_n)$$

$$P_2.(\alpha_1[m] \parallel \cdots \parallel \alpha_n[m]) \xrightarrow{\{\alpha_1[m],\cdots,\alpha_n[m]\}} P_2.(\alpha_1 \parallel \cdots \parallel \alpha_n)$$

   Since $P_1 \sim_s^{fr} P_2$, we obtain $(\alpha_1 \parallel \cdots \parallel \alpha_n).P_1 \sim_s^r (\alpha_1 \parallel \cdots \parallel \alpha_n).P_2$, as desired.

   (e) $P_1 + Q \sim_s^{fr} P_2 + Q$. There are several cases, however, we will not enumerate them all here. By the forward transition rules of Summation, we can obtain

$$\frac{P_1 \xrightarrow{\alpha} P_1'}{P_2 \xrightarrow{\alpha} P_2'} (P_1' \sim_s^{fr} P_2')$$

$$\frac{P_1 \xrightarrow{\alpha} P_1'}{P_1 + Q \xrightarrow{\alpha} P_1'} \quad \frac{P_2 \xrightarrow{\alpha} P_2'}{P_2 + Q \xrightarrow{\alpha} P_2'}$$

$$\frac{Q \xrightarrow{\beta} Q'}{P_1 + Q \xrightarrow{\beta} Q'} \quad \frac{Q \xrightarrow{\beta} Q'}{P_2 + Q \xrightarrow{\beta} Q'}$$

   By the reverse transition rules of Summation, we can obtain

$$\frac{P_1 \xrightarrow{\alpha[m]} P_1'}{P_2 \xrightarrow{\alpha[m]} P_2'} (P_1' \sim_s^{fr} P_2')$$

$$\frac{P_1 \xrightarrow{\alpha[m]} P_1'}{P_1 + Q \xrightarrow{\alpha[m]} P_1'} \quad \frac{P_2 \xrightarrow{\alpha[m]} P_2'}{P_2 + Q \xrightarrow{\alpha[m]} P_2'}$$

$$\frac{Q \xrightarrow{\beta[n]} Q'}{P_1 + Q \xrightarrow{\beta[n]} Q'} \quad \frac{Q \xrightarrow{\beta[n]} Q'}{P_2 + Q \xrightarrow{\beta[n]} Q'}$$

Hence, we obtain $P_1 + Q \sim_s^{fr} P_2 + Q$, as desired.

**(f)** $P_1 \parallel Q \sim_s^{fr} P_2 \parallel Q$. There are several cases, however, we will not enumerate them all here. By the forward transition rules of Composition, we can obtain

$$\frac{P_1 \xrightarrow{\alpha} P_1'}{P_2 \xrightarrow{\alpha} P_2'} (P_1' \sim_s^{fr} P_2')$$

$$\frac{P_1 \xrightarrow{\alpha} P_1' \quad Q \nrightarrow}{P_1 \parallel Q \xrightarrow{\alpha} P_1' \parallel Q} \quad \frac{P_2 \xrightarrow{\alpha} P_2' \quad Q \nrightarrow}{P_2 \parallel Q \xrightarrow{\alpha} P_2' \parallel Q}$$

$$\frac{Q \xrightarrow{\beta} Q' \quad P_1 \nrightarrow}{P_1 \parallel Q \xrightarrow{\beta} P_1 \parallel Q'} \quad \frac{Q \xrightarrow{\beta} P_2' \quad P_2 \nrightarrow}{P_2 \parallel Q \xrightarrow{\beta} P_2 \parallel Q'}$$

$$\frac{P_1 \xrightarrow{\alpha} P_1' \quad Q \xrightarrow{\beta} Q'}{P_1 \parallel Q \xrightarrow{\{\alpha,\beta\}} P_1' \parallel Q'} (\beta \neq \overline{\alpha}) \quad \frac{P_2 \xrightarrow{\alpha} P_2' \quad Q \xrightarrow{\beta} Q'}{P_2 \parallel Q \xrightarrow{\{\alpha,\beta\}} P_2' \parallel Q'} (\beta \neq \overline{\alpha})$$

$$\frac{P_1 \xrightarrow{l} P_1' \quad Q \xrightarrow{\overline{l}} Q'}{P_1 \parallel Q \xrightarrow{\tau} P_1' \parallel Q'} \quad \frac{P_2 \xrightarrow{l} P_2' \quad Q \xrightarrow{\overline{l}} Q'}{P_2 \parallel Q \xrightarrow{\tau} P_2' \parallel Q'}$$

By the reverse transition rules of Composition, we can obtain

$$\frac{P_1 \xrightarrow{\alpha[m]} P_1'}{P_2 \xrightarrow{\alpha[m]} P_2'} (P_1' \sim_s^{fr} P_2')$$

$$\frac{P_1 \xrightarrow{\alpha[m]} P_1' \quad Q \nrightarrow}{P_1 \parallel Q \xrightarrow{\alpha[m]} P_1' \parallel Q} \quad \frac{P_2 \xrightarrow{\alpha[m]} P_2' \quad Q \nrightarrow}{P_2 \parallel Q \xrightarrow{\alpha[m]} P_2' \parallel Q}$$

$$\frac{Q \xrightarrow{\beta[n]} Q' \quad P_1 \nrightarrow}{P_1 \parallel Q \xrightarrow{\beta[n]} P_1 \parallel Q'} \quad \frac{Q \xrightarrow{\beta[n]} P_2' \quad P_2 \nrightarrow}{P_2 \parallel Q \xrightarrow{\beta[n]} P_2 \parallel Q'}$$

$$\frac{P_1 \xrightarrow{\alpha[m]} P_1' \quad Q \xrightarrow{\beta[m]} Q'}{P_1 \parallel Q \xrightarrow{\{\alpha[m],\beta[m]\}} P_1' \parallel Q'} (\beta \neq \overline{\alpha}) \quad \frac{P_2 \xrightarrow{\alpha[m]} P_2' \quad Q \xrightarrow{\beta[m]} Q'}{P_2 \parallel Q \xrightarrow{\{\alpha[m],\beta[m]\}} P_2' \parallel Q'} (\beta \neq \overline{\alpha})$$

$$\frac{P_1 \xrightarrow{l[m]} P_1' \quad Q \xrightarrow{\overline{l}[m]} Q'}{P_1 \parallel Q \xrightarrow{\tau} P_1' \parallel Q'} \quad \frac{P_2 \xrightarrow{l[m]} P_2' \quad Q \xrightarrow{\overline{l}[m]} Q'}{P_2 \parallel Q \xrightarrow{\tau} P_2' \parallel Q'}$$

Since $P_1' \sim_s^{fr} P_2'$ and $Q' \sim_s^{fr} Q'$, and with the assumptions $P_1' \parallel Q \sim_s^{fr} P_2' \parallel Q$, $P_1 \parallel Q' \sim_s^{fr} P_2 \parallel Q'$ and $P_1' \parallel Q' \sim_s^{fr} P_2' \parallel Q'$, we obtain $P_1 \parallel Q \sim_s^{fr} P_2 \parallel Q$, as desired.

**(g)** $P_1 \setminus L \sim_s^{fr} P_2 \setminus L$. There are several cases, however, we will not enumerate them all here. By the forward transition rules of Restriction, we obtain

$$\frac{P_1 \xrightarrow{\alpha} P_1'}{P_2 \xrightarrow{\alpha} P_2'} (P_1' \sim_s^{fr} P_2')$$

$$\frac{P_1 \xrightarrow{\alpha} P_1'}{P_1 \setminus L \xrightarrow{\alpha} P_1' \setminus L}$$

$$\frac{P_2 \xrightarrow{\alpha} P_2'}{P_2 \setminus L \xrightarrow{\alpha} P_2' \setminus L}$$

By the reverse transition rules of Restriction, we obtain

$$\frac{P_1 \xrightarrow{\alpha[m]} P_1'}{P_2 \xrightarrow{\alpha[m]} P_2'} (P_1' \sim_s^{fr} P_2')$$

$$\frac{P_1 \xrightarrow{\alpha[m]} P_1'}{P_1 \setminus L \xrightarrow{\alpha[m]} P_1' \setminus L}$$

$$\frac{P_2 \xrightarrow{\alpha[m]} P_2'}{P_2 \setminus L \xrightarrow{\alpha[m]} P_2' \setminus L}$$

Since $P_1' \sim_s^{fr} P_2'$, and with the assumptions $P_1' \setminus L \sim_s^{fr} P_2' \setminus L$, we obtain $P_1 \setminus L \sim_s^{fr} P_2 \setminus L$, as desired.

**(h)** $P_1[f] \sim_s^{fr} P_2[f]$. By the forward transition rules of Relabeling, we obtain

$$\frac{P_1 \xrightarrow{\alpha} P_1'}{P_2 \xrightarrow{\alpha} P_2'} (P_1' \sim_s^{fr} P_2')$$

$$\frac{P_1 \xrightarrow{\alpha} P_1'}{P_1[f] \xrightarrow{f(\alpha)} P_1'[f]}$$

$$\frac{P_2 \xrightarrow{\alpha} P_2'}{P_2[f] \xrightarrow{f(\alpha)} P_2'[f]}$$

By the reverse transition rules of Relabeling, we obtain

$$\frac{P_1 \xrightarrow{\alpha[m]} P_1'}{P_2 \xrightarrow{\alpha[m]} P_2'} (P_1' \sim_s^{fr} P_2')$$

$$\frac{P_1 \xrightarrow{\alpha[m]} P_1'}{P_1[f] \xrightarrow{f(\alpha)[m]} P_1'[f]}$$

$$\frac{P_2 \xrightarrow{\alpha[m]} P_2'}{P_2[f] \xrightarrow{f(\alpha)[m]} P_2'[f]}$$

Since $P_1' \sim_s^{fr} P_2'$, and with the assumptions $P_1'[f] \sim_s^{fr} P_2'[f]$, we obtain $P_1[f] \sim_s^{fr} P_2[f]$, as desired. □

**Theorem 5.20** (Congruence for strongly FR pomset bisimulation). *We can enjoy the congruence for strongly FR pomset bisimulation as follows:*

1. *If $A \overset{def}{=} P$, then $A \sim_p^{fr} P$;*
2. *Let $P_1 \sim_p^{fr} P_2$. Then,*
   (a) $\alpha.P_1 \sim_p^{f} \alpha.P_2$;

(b) $(\alpha_1 \parallel \cdots \parallel \alpha_n).P_1 \sim_p^f (\alpha_1 \parallel \cdots \parallel \alpha_n).P_2$;
(c) $P_1.\alpha[m] \sim_p^r P_2.\alpha[m]$;
(d) $P_1.(\alpha_1[m] \parallel \cdots \parallel \alpha_n[m]) \sim_p^r P_2.(\alpha_1[m] \parallel \cdots \parallel \alpha_n[m])$.;
(e) $P_1 + Q \sim_p^{fr} P_2 + Q$;
(f) $P_1 \parallel Q \sim_p^{fr} P_2 \parallel Q$;
(g) $P_1 \setminus L \sim_p^{fr} P_2 \setminus L$;
(h) $P_1[f] \sim_p^{fr} P_2[f]$.

*Proof.* From the definition of strongly FR pomset bisimulation (see Definition 2.68), we know that strongly FR pomset bisimulation is defined by FR pomset transitions, which are labeled by pomsets. In an FR pomset transition, the events in the pomset are either within causality relations (defined by the prefix .) or in concurrency (implicitly defined by . and +, and explicitly defined by $\parallel$), of course, they are pairwise consistent (without conflicts). In Theorem 5.19, we have already proven the case that all events are pairwise concurrent, hence, we only need to prove the case of events in causality. Without loss of generality, we take a pomset of $p = \{\alpha, \beta : \alpha.\beta\}$. Then, the FR pomset transition labeled by the above $p$ is just composed of one single event transition labeled by $\alpha$ succeeded by another single event transition labeled by $\beta$, that is, $\xrightarrow{p} = \xrightarrow{\alpha}\xrightarrow{\beta}$ and $\xrightarrow{p[\mathcal{K}]} = \xrightarrow{\beta[n]}\xrightarrow{\alpha[m]}$.

Similarly to the proof of congruence for strongly FR step bisimulation (see Theorem 5.19), we can prove that the congruence holds for strongly FR pomset bisimulation, however, we omit this here. □

**Theorem 5.21** (Congruence for strongly FR hp-bisimulation). *We can enjoy the congruence for strongly FR hp-bisimulation as follows:*

1. *If $A \stackrel{def}{=} P$, then $A \sim_{hp}^{fr} P$;*
2. *Let $P_1 \sim_{hp}^{fr} P_2$. Then,*
   (a) $\alpha.P_1 \sim_{hp}^f \alpha.P_2$;
   (b) $(\alpha_1 \parallel \cdots \parallel \alpha_n).P_1 \sim_{hp}^f (\alpha_1 \parallel \cdots \parallel \alpha_n).P_2$;
   (c) $P_1.\alpha[m] \sim_{hp}^r P_2.\alpha[m]$;
   (d) $P_1.(\alpha_1[m] \parallel \cdots \parallel \alpha_n[m]) \sim_{hp}^r P_2.(\alpha_1[m] \parallel \cdots \parallel \alpha_n[m])$.;
   (e) $P_1 + Q \sim_{hp}^{fr} P_2 + Q$;
   (f) $P_1 \parallel Q \sim_{hp}^{fr} P_2 \parallel Q$;
   (g) $P_1 \setminus L \sim_{hp}^{fr} P_2 \setminus L$;
   (h) $P_1[f] \sim_{hp}^{fr} P_2[f]$.

*Proof.* From the definition of strongly FR hp-bisimulation (see Definition 2.70), we know that strongly FR hp-bisimulation is defined on the posetal product $(C_1, f, C_2)$, $f : C_1 \to C_2$ isomorphism. Two processes $P$ related to $C_1$ and $Q$ related to $C_2$, and $f : C_1 \to C_2$ isomorphism. Initially, $(C_1, f, C_2) = (\emptyset, \emptyset, \emptyset)$, and $(\emptyset, \emptyset, \emptyset) \in \sim_{hp}^{fr}$. When $P \xrightarrow{\alpha} P'$ $(C_1 \xrightarrow{\alpha} C_1')$, there will be $Q \xrightarrow{\alpha} Q'$ $(C_2 \xrightarrow{\alpha} C_2')$, and we define $f' = f[\alpha \mapsto \alpha]$. Also, when $P \xrightarrow{\alpha[m]} P'$ $(C_1 \xrightarrow{\alpha[m]} C_1')$, there will be $Q \xrightarrow{\alpha[m]} Q'$ $(C_2 \xrightarrow{\alpha[m]} C_2')$, and we define $f' = f[\alpha[m] \mapsto \alpha[m]]$. Then, if $(C_1, f, C_2) \in \sim_{hp}^{fr}$, then $(C_1', f', C_2') \in \sim_{hp}^{fr}$.

Similarly to the proof of congruence for strongly FR pomset bisimulation (see Theorem 5.20), we can prove that the congruence holds for strongly FR hp-bisimulation, we just need additionally to check the above conditions on FR hp-bisimulation, however, we omit them here. □

**Theorem 5.22** (Congruence for strongly FR hhp-bisimulation). *We can enjoy the congruence for strongly FR hhp-bisimulation as follows:*

1. *If $A \stackrel{def}{=} P$, then $A \sim_{hhp}^{fr} P$;*
2. *Let $P_1 \sim_{hhp}^{fr} P_2$. Then,*
   (a) $\alpha.P_1 \sim_{hhp}^f \alpha.P_2$;
   (b) $(\alpha_1 \parallel \cdots \parallel \alpha_n).P_1 \sim_{hhp}^f (\alpha_1 \parallel \cdots \parallel \alpha_n).P_2$;

(c) $P_1.\alpha[m] \sim^r_{hhp} P_2.\alpha[m]$;
(d) $P_1.(\alpha_1[m] \parallel \cdots \parallel \alpha_n[m]) \sim^r_{hhp} P_2.(\alpha_1[m] \parallel \cdots \parallel \alpha_n[m])$.;
(e) $P_1 + Q \sim^{fr}_{hhp} P_2 + Q$;
(f) $P_1 \parallel Q \sim^{fr}_{hhp} P_2 \parallel Q$;
(g) $P_1 \setminus L \sim^{fr}_{hhp} P_2 \setminus L$;
(h) $P_1[f] \sim^{fr}_{hhp} P_2[f]$.

*Proof.* From the definition of strongly FR hhp-bisimulation (see Definition 2.70), we know that strongly FR hhp-bisimulation is downward closed for strongly FR hp-bisimulation.

Similarly to the proof of congruence for strongly FR hp-bisimulation (see Theorem 5.21), we can prove that the congruence holds for strongly FR hhp-bisimulation, however, we omit this here. □

### 5.2.2 Recursion

**Definition 5.23** (Weakly guarded recursive expression). $X$ is weakly guarded in $E$ if each occurrence of $X$ is with some subexpression $\alpha.F$ or $(\alpha_1 \parallel \cdots \parallel \alpha_n).F$ or $F.\alpha[m]$ or $F.(\alpha_1[m] \parallel \cdots \parallel \alpha_n[m])$ of $E$.

**Lemma 5.24.** *If the variables $\widetilde{X}$ are weakly guarded in $E$, and $E\{\widetilde{P}/\widetilde{X}\} \xrightarrow{\{\alpha_1,\cdots,\alpha_n\}} P'$ or $E\{\widetilde{P}/\widetilde{X}\} \xrightarrow{\{\alpha_1[m],\cdots,\alpha_n[m]\}} P'$, then $P'$ takes the form $E'\{\widetilde{P}/\widetilde{X}\}$ for some expression $E'$, and moreover, for any $\widetilde{Q}$, $E\{\widetilde{Q}/\widetilde{X}\} \xrightarrow{\{\alpha_1,\cdots,\alpha_n\}} E'\{\widetilde{Q}/\widetilde{X}\}$ or $E\{\widetilde{Q}/\widetilde{X}\} \xrightarrow{\{\alpha_1[m],\cdots,\alpha_n[m]\}} E'\{\widetilde{Q}/\widetilde{X}\}$.*

*Proof.* We only prove the case of forward transition. It needs to induct on the depth of the inference of $E\{\widetilde{P}/\widetilde{X}\} \xrightarrow{\{\alpha_1,\cdots,\alpha_n\}} P'$.

1. Case $E \equiv Y$, a variable. Then, $Y \notin \widetilde{X}$. Since $\widetilde{X}$ are weakly guarded, $Y\{\widetilde{P}/\widetilde{X} \equiv Y\} \nrightarrow$, this case is impossible.
2. Case $E \equiv \beta.F$. Then, we must have $\alpha = \beta$, and $P' \equiv F\{\widetilde{P}/\widetilde{X}\}$, and $E\{\widetilde{Q}/\widetilde{X}\} \equiv \beta.F\{\widetilde{Q}/\widetilde{X}\} \xrightarrow{\beta} F\{\widetilde{Q}/\widetilde{X}\}$, then, let $E'$ be $F$, as desired.
3. Case $E \equiv (\beta_1 \parallel \cdots \parallel \beta_n).F$. Then, we must have $\alpha_i = \beta_i$ for $1 \leq i \leq n$, and $P' \equiv F\{\widetilde{P}/\widetilde{X}\}$, and $E\{\widetilde{Q}/\widetilde{X}\} \equiv (\beta_1 \parallel \cdots \parallel \beta_n).F\{\widetilde{Q}/\widetilde{X}\} \xrightarrow{\{\beta_1,\cdots,\beta_n\}} F\{\widetilde{Q}/\widetilde{X}\}$, then, let $E'$ be $F$, as desired.
4. Case $E \equiv E_1 + E_2$. Then, either $E_1\{\widetilde{P}/\widetilde{X}\} \xrightarrow{\{\alpha_1,\cdots,\alpha_n\}} P'$ or $E_2\{\widetilde{P}/\widetilde{X}\} \xrightarrow{\{\alpha_1,\cdots,\alpha_n\}} P'$, then, we can apply this lemma in either case, as desired.
5. Case $E \equiv E_1 \parallel E_2$. There are four possibilities:
   (a) We may have $E_1\{\widetilde{P}/\widetilde{X}\} \xrightarrow{\alpha} P_1'$ and $E_2\{\widetilde{P}/\widetilde{X}\} \nrightarrow$ with $P' \equiv P_1' \parallel (E_2\{\widetilde{P}/\widetilde{X}\})$, then by applying this lemma, $P_1'$ is of the form $E_1'\{\widetilde{P}/\widetilde{X}\}$, and for any $Q$, $E_1\{\widetilde{Q}/\widetilde{X}\} \xrightarrow{\alpha} E_1'\{\widetilde{Q}/\widetilde{X}\}$. Hence, $P'$ is of the form $E_1' \parallel E_2\{\widetilde{P}/\widetilde{X}\}$, and for any $Q$, $E\{\widetilde{Q}/\widetilde{X}\} \equiv E_1\{\widetilde{Q}/\widetilde{X}\} \parallel E_2\{\widetilde{Q}/\widetilde{X}\} \xrightarrow{\alpha} (E_1' \parallel E_2)\{\widetilde{Q}/\widetilde{X}\}$, then, let $E'$ be $E_1' \parallel E_2$, as desired.
   (b) We may have $E_2\{\widetilde{P}/\widetilde{X}\} \xrightarrow{\alpha} P_2'$ and $E_1\{\widetilde{P}/\widetilde{X}\} \nrightarrow$ with $P' \equiv P_2' \parallel (E_1\{\widetilde{P}/\widetilde{X}\})$, this case can be prove similarly to the above subcase, as desired.
   (c) We may have $E_1\{\widetilde{P}/\widetilde{X}\} \xrightarrow{\alpha} P_1'$ and $E_2\{\widetilde{P}/\widetilde{X}\} \xrightarrow{\beta} P_2'$ with $\alpha \neq \overline{\beta}$ and $P' \equiv P_1' \parallel P_2'$, then by applying this lemma, $P_1'$ is of the form $E_1'\{\widetilde{P}/\widetilde{X}\}$, and for any $Q$, $E_1\{\widetilde{Q}/\widetilde{X}\} \xrightarrow{\alpha} E_1'\{\widetilde{Q}/\widetilde{X}\}$; $P_2'$ is of the form $E_2'\{\widetilde{P}/\widetilde{X}\}$, and for any $Q$, $E_2\{\widetilde{Q}/\widetilde{X}\} \xrightarrow{\alpha} E_2'\{\widetilde{Q}/\widetilde{X}\}$. Hence, $P'$ is of the form $E_1' \parallel E_2'\{\widetilde{P}/\widetilde{X}\}$, and for any $Q$, $E\{\widetilde{Q}/\widetilde{X}\} \equiv E_1\{\widetilde{Q}/\widetilde{X}\} \parallel E_2\{\widetilde{Q}/\widetilde{X}\} \xrightarrow{\{\alpha,\beta\}} (E_1' \parallel E_2')\{\widetilde{Q}/\widetilde{X}\}$, then, let $E'$ be $E_1' \parallel E_2'$, as desired.
   (d) We may have $E_1\{\widetilde{P}/\widetilde{X}\} \xrightarrow{l} P_1'$ and $E_2\{\widetilde{P}/\widetilde{X}\} \xrightarrow{\overline{l}} P_2'$ with $P' \equiv P_1' \parallel P_2'$, then by applying this lemma, $P_1'$ is of the form $E_1'\{\widetilde{P}/\widetilde{X}\}$, and for any $Q$, $E_1\{\widetilde{Q}/\widetilde{X}\} \xrightarrow{l} E_1'\{\widetilde{Q}/\widetilde{X}\}$; $P_2'$ is of the form $E_2'\{\widetilde{P}/\widetilde{X}\}$, and for any $Q$, $E_2\{\widetilde{Q}/\widetilde{X}\} \xrightarrow{\overline{l}} E_2'\{\widetilde{Q}/\widetilde{X}\}$. Hence, $P'$ is of the form $E_1' \parallel E_2'\{\widetilde{P}/\widetilde{X}\}$, and for any $Q$, $E\{\widetilde{Q}/\widetilde{X}\} \equiv E_1\{\widetilde{Q}/\widetilde{X}\} \parallel E_2\{\widetilde{Q}/\widetilde{X}\} \xrightarrow{\tau} (E_1' \parallel E_2')\{\widetilde{Q}/\widetilde{X}\}$, then, let $E'$ be $E_1' \parallel E_2'$, as desired.
6. Case $E \equiv F[R]$ and $E \equiv F \setminus L$. These cases can be proven similarly to the above case.
7. Case $E \equiv C$, an agent constant defined by $C \stackrel{\text{def}}{=} R$. Then, there is no $X \in \widetilde{X}$ occurring in $E$, hence, $C \xrightarrow{\{\alpha_1,\cdots,\alpha_n\}} P'$, let $E'$ be $P'$, as desired.

For the case of reverse transition, it can be proven similarly, however, we omit it here. □

**Theorem 5.25** (Unique solution of equations for strongly FR step bisimulation). *Let the recursive expressions $E_i(i \in I)$ contain at most the variables $X_i(i \in I)$, and let each $X_j(j \in I)$ be weakly guarded in each $E_i$. Then,*

*If $\widetilde{P} \sim_s^{fr} \widetilde{E}\{\widetilde{P}/\widetilde{X}\}$ and $\widetilde{Q} \sim_s^{fr} \widetilde{E}\{\widetilde{Q}/\widetilde{X}\}$, then $\widetilde{P} \sim_s^{fr} \widetilde{Q}$.*

*Proof.* We only prove the case of forward transition. It is sufficient to induct on the depth of the inference of $E\{\widetilde{P}/\widetilde{X}\} \xrightarrow{\{\alpha_1,\cdots,\alpha_n\}} P'$.

1. Case $E \equiv X_i$. Then, we have $E\{\widetilde{P}/\widetilde{X}\} \equiv P_i \xrightarrow{\{\alpha_1,\cdots,\alpha_n\}} P'$, since $P_i \sim_s^{fr} E_i\{\widetilde{P}/\widetilde{X}\}$, we have $E_i\{\widetilde{P}/\widetilde{X}\} \xrightarrow{\{\alpha_1,\cdots,\alpha_n\}} P'' \sim_s^{fr} P'$. Since $\widetilde{X}$ are weakly guarded in $E_i$, by Lemma 5.24, $P'' \equiv E'\{\widetilde{P}/\widetilde{X}\}$ and $E_i\{\widetilde{P}/\widetilde{X}\} \xrightarrow{\{\alpha_1,\cdots,\alpha_n\}} E'\{\widetilde{P}/\widetilde{X}\}$. Since $E\{\widetilde{Q}/\widetilde{X}\} \equiv X_i\{\widetilde{Q}/\widetilde{X}\} \equiv Q_i \sim_s^{fr} E_i\{\widetilde{Q}/\widetilde{X}\}$, $E\{\widetilde{Q}/\widetilde{X}\} \xrightarrow{\{\alpha_1,\cdots,\alpha_n\}} Q' \sim_s^{fr} E'\{\widetilde{Q}/\widetilde{X}\}$. Hence, $P' \sim_s^{fr} Q'$, as desired.
2. Case $E \equiv \alpha.F$. This case can be proven similarly.
3. Case $E \equiv (\alpha_1 \parallel \cdots \parallel \alpha_n).F$. This case can be proven similarly.
4. Case $E \equiv E_1 + E_2$. We have $E_i\{\widetilde{P}/\widetilde{X}\} \xrightarrow{\{\alpha_1,\cdots,\alpha_n\}} P'$, $E_i\{\widetilde{Q}/\widetilde{X}\} \xrightarrow{\{\alpha_1,\cdots,\alpha_n\}} Q'$, then $P' \sim_s^{fr} Q'$, as desired.
5. Case $E \equiv E_1 \parallel E_2$, $E \equiv F[R]$ and $E \equiv F \setminus L$, $E \equiv C$. These cases can be proven similarly to the above case.

For the case of reverse transition, it can be proven similarly, however, we omit it here. □

**Theorem 5.26** (Unique solution of equations for strongly FR pomset bisimulation). *Let the recursive expressions $E_i(i \in I)$ contain at most the variables $X_i(i \in I)$, and let each $X_j(j \in I)$ be weakly guarded in each $E_i$. Then,*

*If $\widetilde{P} \sim_p^{fr} \widetilde{E}\{\widetilde{P}/\widetilde{X}\}$ and $\widetilde{Q} \sim_p^{fr} \widetilde{E}\{\widetilde{Q}/\widetilde{X}\}$, then $\widetilde{P} \sim_p^{fr} \widetilde{Q}$.*

*Proof.* From the definition of strongly FR pomset bisimulation (see Definition 2.68), we know that strongly FR pomset bisimulation is defined by FR pomset transitions, which are labeled by pomsets. In an FR pomset transition, the events in the pomset are either within causality relations (defined by the prefix .) or in concurrency (implicitly defined by . and +, and explicitly defined by $\parallel$), of course, they are pairwise consistent (without conflicts). In Theorem 5.19, we have already proven the case that all events are pairwise concurrent, hence, we only need to prove the case of events in causality. Without loss of generality, we take a pomset of $p = \{\alpha, \beta : \alpha.\beta\}$. Then, the FR pomset transition labeled by the above $p$ is just composed of one single event transition labeled by $\alpha$ succeeded by another single event transition labeled by $\beta$, that is, $\xrightarrow{p} = \xrightarrow{\alpha}\xrightarrow{\beta}$ and $\xtwoheadrightarrow{p[\mathcal{K}]} = \xtwoheadrightarrow{\beta[n]}\xtwoheadrightarrow{\alpha[m]}$.

Similarly to the proof of the unique solution of equations for strongly FR step bisimulation (see Theorem 5.25), we can prove that the unique solution of equations holds for strongly FR pomset bisimulation, however, we omit this here. □

**Theorem 5.27** (Unique solution of equations for strongly FR hp-bisimulation). *Let the recursive expressions $E_i(i \in I)$ contain at most the variables $X_i(i \in I)$, and let each $X_j(j \in I)$ be weakly guarded in each $E_i$. Then,*

*If $\widetilde{P} \sim_{hp}^{fr} \widetilde{E}\{\widetilde{P}/\widetilde{X}\}$ and $\widetilde{Q} \sim_{hp}^{fr} \widetilde{E}\{\widetilde{Q}/\widetilde{X}\}$, then $\widetilde{P} \sim_{hp}^{fr} \widetilde{Q}$.*

*Proof.* From the definition of strongly FR hp-bisimulation (see Definition 2.70), we know that strongly FR hp-bisimulation is defined on the posetal product $(C_1, f, C_2)$, $f : C_1 \to C_2$ isomorphism. Two processes $P$ related to $C_1$ and $Q$ related to $C_2$, and $f : C_1 \to C_2$ isomorphism. Initially, $(C_1, f, C_2) = (\emptyset, \emptyset, \emptyset)$, and $(\emptyset, \emptyset, \emptyset) \in \sim_{hp}^{fr}$. When $P \xrightarrow{\alpha} P'$ $(C_1 \xrightarrow{\alpha} C_1')$, there will be $Q \xrightarrow{\alpha} Q'$ $(C_2 \xrightarrow{\alpha} C_2')$, and we define $f' = f[\alpha \mapsto \alpha]$. Also, when $P \xtwoheadrightarrow{\alpha[m]} P'$ $(C_1 \xtwoheadrightarrow{\alpha[m]} C_1')$, there will be $Q \xtwoheadrightarrow{\alpha[m]} Q'$ $(C_2 \xtwoheadrightarrow{\alpha[m]} C_2')$, and we define $f' = f[\alpha[m] \mapsto \alpha[m]]$. Then, if $(C_1, f, C_2) \in \sim_{hp}^{fr}$, then $(C_1', f', C_2') \in \sim_{hp}^{fr}$.

Similarly to the proof of the unique solution of equations for strongly FR pomset bisimulation (see Theorem 5.26), we can prove that the unique solution of equations holds for strongly FR hp-bisimulation, we just need additionally to check the above conditions on hp-bisimulation, however, we omit them here. □

**Theorem 5.28** (Unique solution of equations for strongly FR hhp-bisimulation). *Let the recursive expressions $E_i(i \in I)$ contain at most the variables $X_i(i \in I)$, and let each $X_j(j \in I)$ be weakly guarded in each $E_i$. Then,*

*If $\widetilde{P} \sim_{hhp}^{fr} \widetilde{E}\{\widetilde{P}/\widetilde{X}\}$ and $\widetilde{Q} \sim_{hhp}^{fr} \widetilde{E}\{\widetilde{Q}/\widetilde{X}\}$, then $\widetilde{P} \sim_{hhp}^{fr} \widetilde{Q}$.*

*Proof.* From the definition of strongly FR hhp-bisimulation (see Definition 2.70), we know that strongly FR hhp-bisimulation is downward closed for strongly FR hp-bisimulation.

Similarly to the proof of the unique solution of equations for strongly FR hp-bisimulation (see Theorem 5.27), we can prove that the unique solution of equations holds for strongly FR hhp-bisimulation, however, we omit this here. □

**TABLE 5.9** Forward and reverse transition rules of $\tau$.

| |
|---|
| $\dfrac{}{\tau \xrightarrow{\tau} \surd}$ |
| $\dfrac{}{\tau \overset{\tau}{\twoheadrightarrow} \surd}$ |

## 5.3 Weakly forward–reverse truly concurrent bisimulations

### 5.3.1 Laws

Remembering that $\tau$ can neither be restricted nor relabeled, we know that the monoid laws, the static laws, and the new expansion law in Section 5.2 still hold with respect to the corresponding weakly FR truly concurrent bisimulations. Also, we can enjoy the congruence of Prefix, Summation, Composition, Restriction, Relabeling, and Constants with respect to corresponding weakly FR truly concurrent bisimulations. We will not retype these laws, and just give the $\tau$-specific laws. The forward and reverse transition rules of $\tau$ are shown in Table 5.9, where $\xrightarrow{\tau} \surd$ is a predicate that represents a successful termination after execution of the silent step $\tau$.

**Proposition 5.29** ($\tau$ laws for weakly FR step bisimulation). *The $\tau$ laws for weakly FR step bisimulation are as follows:*

1. $P \approx_s^f \tau.P$;
2. $P \approx_s^r P.\tau$;
3. $\alpha.\tau.P \approx_s^f \alpha.P$;
4. $P.\tau.\alpha[m] \approx_s^r P.\alpha[m]$;
5. $(\alpha_1 \parallel \cdots \parallel \alpha_n).\tau.P \approx_s^f (\alpha_1 \parallel \cdots \parallel \alpha_n).P$;
6. $P.\tau.(\alpha_1[m] \parallel \cdots \parallel \alpha_n[m]) \approx_s^r P.(\alpha_1[m] \parallel \cdots \parallel \alpha_n[m])$;
7. $P + \tau.P \approx_s^f \tau.P$;
8. $P + P.\tau \approx_s^r P.\tau$;
9. $\alpha.(\tau.(P+Q)+P) \approx_s^f \alpha.(P+Q)$;
10. $((P+Q).\tau + P).\alpha[m] \approx_s^r (P+Q).\alpha[m]$;
11. $(\alpha_1 \parallel \cdots \parallel \alpha_n).(\tau.(P+Q)+P) \approx_s^f (\alpha_1 \parallel \cdots \parallel \alpha_n).(P+Q)$;
12. $((P+Q).\tau + P).(\alpha_1[m] \parallel \cdots \parallel \alpha_n[m]) \approx_s^r (P+Q).(\alpha_1[m] \parallel \cdots \parallel \alpha_n[m])$;
13. $P \approx_s^{fr} \tau \parallel P$.

*Proof.* Although the transition rules in Definition 5.2 are defined in the flavor of a single event, they can be modified into a step (a set of events within which each event is pairwise concurrent), however, we omit them here. If we treat a single event as a step containing just one event, the proof of $\tau$ laws does not pose any problem, hence, we use this way and still use the transition rules in Definition 5.2.

1. $P \approx_s^f \tau.P$. By the forward transition rules of Prefix, we obtain

$$\frac{P \xRightarrow{\alpha} P'}{P \xRightarrow{\alpha} P'} \quad \frac{P \xRightarrow{\alpha} P'}{\tau.P \xRightarrow{\alpha} P'}$$

Since $P' \approx_s^f P'$, we obtain $P \approx_s^f \tau.P$, as desired.

2. $P \approx_s^r P.\tau$. By the reverse transition rules of Prefix, we obtain

$$\frac{P \overset{\alpha[m]}{\Rrightarrow} P'}{P \overset{\alpha[m]}{\Rrightarrow} P'} \quad \frac{P \overset{\alpha[m]}{\Rrightarrow} P'}{\tau.P \overset{\alpha[m]}{\Rrightarrow} P'}$$

Since $P' \approx_s^r P'$, we obtain $P \approx_s^r P.\tau$, as desired.

3. $\alpha.\tau.P \approx_s^f \alpha.P$. By the forward transition rules of Prefix, we obtain

$$\frac{}{\alpha.\tau.P \xRightarrow{\alpha} \alpha[m].P} \quad \frac{}{\alpha.P \xRightarrow{\alpha} \alpha[m].P}$$

Since $\alpha[m].P \approx_s^f \alpha[m].P$, we obtain $\alpha.\tau.P \approx_s^f \alpha.P$, as desired.

4. $P.\tau.\alpha[m] \approx_s^r P.\alpha[m]$. By the reverse transition rules of Prefix, we obtain

$$\frac{}{P.\tau.\alpha[m] \xRightarrow{\alpha[m]} P.\alpha} \quad \frac{}{P.\alpha[m] \xRightarrow{\alpha[m]} P.\alpha}$$

Since $P.\alpha \approx_s^r P.\alpha$, we obtain $P.\tau.\alpha[m] \approx_s^r P.\alpha[m]$, as desired.

5. $(\alpha_1 \parallel \cdots \parallel \alpha_n).\tau.P \approx_s^f (\alpha_1 \parallel \cdots \parallel \alpha_n).P$. By the forward transition rules of Prefix, we obtain

$$\frac{}{(\alpha_1 \parallel \cdots \parallel \alpha_n).\tau.P \xRightarrow{\{\alpha_1,\cdots,\alpha_n\}} (\alpha_1[m] \parallel \cdots \parallel \alpha_n[m]).P}$$

$$\frac{}{(\alpha_1 \parallel \cdots \parallel \alpha_n).P \xRightarrow{\{\alpha_1,\cdots,\alpha_n\}} (\alpha_1[m] \parallel \cdots \parallel \alpha_n[m]).P}$$

Since $(\alpha_1[m] \parallel \cdots \parallel \alpha_n[m]).P \approx_s^f (\alpha_1[m] \parallel \cdots \parallel \alpha_n[m]).P$, we obtain $(\alpha_1 \parallel \cdots \parallel \alpha_n).\tau.P \approx_s^f (\alpha_1 \parallel \cdots \parallel \alpha_n).P$, as desired.

6. $P.\tau.(\alpha_1[m] \parallel \cdots \parallel \alpha_n[m]) \approx_s^r P.(\alpha_1[m] \parallel \cdots \parallel \alpha_n[m])$. By the reverse transition rules of Prefix, we obtain

$$\frac{}{P.\tau.(\alpha_1[m] \parallel \cdots \parallel \alpha_n[m]) \xRightarrow{\{\alpha_1[m],\cdots,\alpha_n[m]\}} (P.(\alpha_1 \parallel \cdots \parallel \alpha_n)}$$

$$\frac{}{P.(\alpha_1[m] \parallel \cdots \parallel \alpha_n[m]) \xRightarrow{\{\alpha_1[m],\cdots,\alpha_n[m]\}} (P.(\alpha_1 \parallel \cdots \parallel \alpha_n)}$$

Since $P.(\alpha_1 \parallel \cdots \parallel \alpha_n) \approx_s^r P.(\alpha_1 \parallel \cdots \parallel \alpha_n)$, we obtain $P.\tau.(\alpha_1[m] \parallel \cdots \parallel \alpha_n[m]) \approx_s^r P.(\alpha_1[m] \parallel \cdots \parallel \alpha_n[m])$, as desired.

7. $P + \tau.P \approx_s^f \tau.P$. By the forward transition rules of Summation, we obtain

$$\frac{P \xRightarrow{\alpha} P'}{P + \tau.P \xRightarrow{\alpha} P'} \quad \frac{P \xRightarrow{\alpha} P'}{\tau.P \xRightarrow{\alpha} P'}$$

Hence, we obtain $P + \tau.P \approx_s^f \tau.P$, as desired.

8. $P + P.\tau \approx_s^r P.\tau$. By the reverse transition rules of Summation, we obtain

$$\frac{P \xRightarrow{\alpha[m]} P'}{P + P.\tau \xRightarrow{\alpha[m]} P'} \quad \frac{P \xRightarrow{\alpha[m]} P'}{P.\tau \xRightarrow{\alpha[m]} P'}$$

Hence, we obtain $P + P.\tau \approx_s^r P.\tau$, as desired.

9. $\alpha.(\tau.(P + Q) + P) \approx_s^f \alpha.(P + Q)$. By the forward transition rules of Prefix and Summation, we obtain

$$\frac{}{\alpha.(\tau.(P + Q) + P) \xRightarrow{\alpha} \alpha[m].(P + Q)} \quad \frac{}{\alpha.(P + Q) \xRightarrow{\alpha} \alpha[m].(P + Q)}$$

Hence, we obtain $\alpha.(\tau.(P + Q) + P) \approx_s^f \alpha.(P + Q)$, as desired.

10. $((P + Q).\tau + P).\alpha[m] \approx_s^r (P + Q).\alpha[m]$. By the reverse transition rules of Prefix and Summation, we obtain

$$\frac{}{((P + Q).\tau + P).\alpha[m] \xRightarrow{\alpha[m]} (P + Q).\alpha} \quad \frac{}{(P + Q).\alpha[m] \xRightarrow{\alpha[m]} (P + Q).\alpha}$$

Hence, we obtain $((P + Q).\tau + P).\alpha[m] \approx_s^r (P + Q).\alpha[m]$, as desired.

11. $(\alpha_1 \parallel \cdots \parallel \alpha_n).(\tau.(P+Q)+P) \approx_s^f (\alpha_1 \parallel \cdots \parallel \alpha_n).(P+Q)$. By the forward transition rules of Prefix and Summation, we obtain

$$\frac{}{(\alpha_1 \parallel \cdots \parallel \alpha_n).(\tau.(P + Q) + P) \xRightarrow{\{\alpha_1,\cdots,\alpha_n\}} (\alpha_1[m] \parallel \cdots \parallel \alpha_n[m]).(P + Q)}$$

$$\frac{}{(\alpha_1 \parallel \cdots \parallel \alpha_n).(P+Q) \xrightarrow{\{\alpha_1,\cdots,\alpha_n\}} (\alpha_1[m] \parallel \cdots \parallel \alpha_n[m]).(P+Q)}$$

Hence, we obtain $(\alpha_1 \parallel \cdots \parallel \alpha_n).(\tau.(P+Q)+P) \approx_s^f (\alpha_1 \parallel \cdots \parallel \alpha_n).(P+Q)$, as desired.

**12.** $((P+Q).\tau+P).(\alpha_1[m] \parallel \cdots \parallel \alpha_n[m]) \approx_s^r (P+Q).(\alpha_1[m] \parallel \cdots \parallel \alpha_n[m])$. By the reverse transition rules of Prefix and Summation, we obtain

$$\frac{}{((P+Q).\tau+P).(\alpha_1[m] \parallel \cdots \parallel \alpha_n[m]) \overset{\{\alpha_1[m],\cdots,\alpha_n[m]\}}{\rightrightarrows} (P+Q).(\alpha_1 \parallel \cdots \parallel \alpha_n)}$$

$$\frac{}{(P+Q).(\alpha_1[m] \parallel \cdots \parallel \alpha_n[m]) \overset{\{\alpha_1[m],\cdots,\alpha_n[m]\}}{\rightrightarrows} (P+Q).(\alpha_1 \parallel \cdots \parallel \alpha_n)}$$

Hence, we obtain $((P+Q).\tau+P).(\alpha_1[m] \parallel \cdots \parallel \alpha_n[m]) \approx_s^r (P+Q).(\alpha_1[m] \parallel \cdots \parallel \alpha_n[m])$, as desired.

**13.** $P \approx_s^{fr} \tau \parallel P$. By the forward transition rules of Composition, we obtain

$$\frac{P \overset{\alpha}{\Rightarrow} P'}{P \overset{\alpha}{\Rightarrow} P'} \quad \frac{P \overset{\alpha}{\Rightarrow} P'}{\tau \parallel P \overset{\alpha}{\Rightarrow} P'}$$

By the reverse transition rules of Composition, we obtain

$$\frac{P \overset{\alpha}{\rightrightarrows} P'}{P \overset{\alpha}{\rightrightarrows} P'} \quad \frac{P \overset{\alpha}{\rightrightarrows} P'}{\tau \parallel P \overset{\alpha}{\rightrightarrows} P'}$$

Since $P' \approx_s^{fr} P'$, we obtain $P \approx_s^{fr} \tau \parallel P$, as desired. □

**Proposition 5.30** ($\tau$ laws for weakly FR pomset bisimulation). *The $\tau$ laws for weakly FR pomset bisimulation are as follows:*

1. $P \approx_p^f \tau.P$;
2. $P \approx_p^r P.\tau$;
3. $\alpha.\tau.P \approx_p^f \alpha.P$;
4. $P.\tau.\alpha[m] \approx_p^r P.\alpha[m]$;
5. $(\alpha_1 \parallel \cdots \parallel \alpha_n).\tau.P \approx_p^f (\alpha_1 \parallel \cdots \parallel \alpha_n).P$;
6. $P.\tau.(\alpha_1[m] \parallel \cdots \parallel \alpha_n[m]) \approx_p^r P.(\alpha_1[m] \parallel \cdots \parallel \alpha_n[m])$;
7. $P+\tau.P \approx_p^f \tau.P$;
8. $P+P.\tau \approx_p^r P.\tau$;
9. $\alpha.(\tau.(P+Q)+P) \approx_s^f \alpha.(P+Q)$;
10. $((P+Q).\tau+P).\alpha[m] \approx_s^r (P+Q).\alpha[m]$;
11. $(\alpha_1 \parallel \cdots \parallel \alpha_n).(\tau.(P+Q)+P) \approx_s^f (\alpha_1 \parallel \cdots \parallel \alpha_n).(P+Q)$;
12. $((P+Q).\tau+P).(\alpha_1[m] \parallel \cdots \parallel \alpha_n[m]) \approx_s^r (P+Q).(\alpha_1[m] \parallel \cdots \parallel \alpha_n[m])$;
13. $P \approx_p^{fr} \tau \parallel P$.

*Proof.* From the definition of weakly FR pomset bisimulation $\approx_p^{fr}$ (see Definition 2.69), we know that weakly FR pomset bisimulation $\approx_p^{fr}$ is defined by weakly FR pomset transitions, which are labeled by pomsets with $\tau$. In a weakly FR pomset transition, the events in the pomset are either within causality relations (defined by .) or in concurrency (implicitly defined by . and +, and explicitly defined by $\parallel$), of course, they are pairwise consistent (without conflicts). In Proposition 5.29, we have already proven the case that all events are pairwise concurrent, hence, we only need to prove the case of events in causality. Without loss of generality, we take a pomset of $p = \{\alpha, \beta : \alpha.\beta\}$. Then, the weakly forward pomset transition labeled by the above $p$ is just composed of one single event transition labeled by $\alpha$ succeeded by another single event transition labeled by $\beta$, that is, $\overset{p}{\Rightarrow} = \overset{\alpha}{\Rightarrow}\overset{\beta}{\Rightarrow}$ and $\overset{p[\mathcal{K}]}{\rightrightarrows} = \overset{\beta[n]}{\rightrightarrows}\overset{\alpha[m]}{\rightrightarrows}$.

Similarly to the proof of $\tau$ laws for weakly FR step bisimulation $\approx_s^{fr}$ (Proposition 5.29), we can prove that $\tau$ laws hold for weakly FR pomset bisimulation $\approx_p^{fr}$, however, we omit this here. □

**Proposition 5.31** ($\tau$ laws for weakly FR hp-bisimulation). *The $\tau$ laws for weakly FR hp-bisimulation are as follows:*

1. $P \approx_{hp}^{f} \tau.P$;
2. $P \approx_{hp}^{r} P.\tau$;
3. $\alpha.\tau.P \approx_{hp}^{f} \alpha.P$;
4. $P.\tau.\alpha[m] \approx_{hp}^{r} P.\alpha[m]$;
5. $(\alpha_1 \parallel \cdots \parallel \alpha_n).\tau.P \approx_{hp}^{f} (\alpha_1 \parallel \cdots \parallel \alpha_n).P$;
6. $P.\tau.(\alpha_1[m] \parallel \cdots \parallel \alpha_n[m]) \approx_{hp}^{r} P.(\alpha_1[m] \parallel \cdots \parallel \alpha_n[m])$;
7. $P+\tau.P \approx_{hp}^{f} \tau.P$;
8. $P+P.\tau \approx_{hp}^{r} P.\tau$;
9. $\alpha.(\tau.(P+Q)+P) \approx_{s}^{f} \alpha.(P+Q)$;
10. $((P+Q).\tau+P).\alpha[m] \approx_{s}^{r} (P+Q).\alpha[m]$;
11. $(\alpha_1 \parallel \cdots \parallel \alpha_n).(\tau.(P+Q)+P) \approx_{s}^{f} (\alpha_1 \parallel \cdots \parallel \alpha_n).(P+Q)$;
12. $((P+Q).\tau+P).(\alpha_1[m] \parallel \cdots \parallel \alpha_n[m]) \approx_{s}^{r} (P+Q).(\alpha_1[m] \parallel \cdots \parallel \alpha_n[m])$;
13. $P \approx_{hp}^{fr} \tau \parallel P$.

*Proof.* From the definition of weakly FR hp-bisimulation $\approx_{hp}^{fr}$ (see Definition 2.71), we know that weakly FR hp-bisimulation $\approx_{hp}^{fr}$ is defined on the weakly posetal product $(C_1, f, C_2), f : \hat{C_1} \to \hat{C_2}$ isomorphism. Two processes $P$ related to $C_1$ and $Q$ related to $C_2$, and $f : \hat{C_1} \to \hat{C_2}$ isomorphism. Initially, $(C_1, f, C_2) = (\emptyset, \emptyset, \emptyset)$, and $(\emptyset, \emptyset, \emptyset) \in \approx_{hp}^{fr}$. When $P \xRightarrow{\alpha} P'$ ($C_1 \xRightarrow{\alpha} C_1'$), there will be $Q \xRightarrow{\alpha} Q'$ ($C_2 \xRightarrow{\alpha} C_2'$), and we define $f' = f[\alpha \mapsto \alpha]$. Also, when $P \xRightarrow{\alpha[m]} P'$ ($C_1 \xRightarrow{\alpha[m]} C_1'$), there will be $Q \xRightarrow{\alpha[m]} Q'$ ($C_2 \xRightarrow{\alpha[m]} C_2'$), and we define $f' = f[\alpha[m] \mapsto \alpha[m]]$. Then, if $(C_1, f, C_2) \in \approx_{hp}^{fr}$, then $(C_1', f', C_2') \in \approx_{hp}^{fr}$.

Similarly to the proof of $\tau$ laws for weakly FR pomset bisimulation (Proposition 5.30), we can prove that $\tau$ laws hold for weakly FR hp-bisimulation, we just need additionally to check the above conditions on weakly FR hp-bisimulation, however, we omit them here. □

**Proposition 5.32** ($\tau$ laws for weakly FR hhp-bisimulation). *The $\tau$ laws for weakly FR hhp-bisimulation are as follows:*

1. $P \approx_{hhp}^{f} \tau.P$;
2. $P \approx_{hhp}^{r} P.\tau$;
3. $\alpha.\tau.P \approx_{hhp}^{f} \alpha.P$;
4. $P.\tau.\alpha[m] \approx_{hhp}^{r} P.\alpha[m]$;
5. $(\alpha_1 \parallel \cdots \parallel \alpha_n).\tau.P \approx_{hhp}^{f} (\alpha_1 \parallel \cdots \parallel \alpha_n).P$;
6. $P.\tau.(\alpha_1[m] \parallel \cdots \parallel \alpha_n[m]) \approx_{hhp}^{r} P.(\alpha_1[m] \parallel \cdots \parallel \alpha_n[m])$;
7. $P+\tau.P \approx_{hhp}^{f} \tau.P$;
8. $P+P.\tau \approx_{hhp}^{r} P.\tau$;
9. $\alpha.(\tau.(P+Q)+P) \approx_{s}^{f} \alpha.(P+Q)$;
10. $((P+Q).\tau+P).\alpha[m] \approx_{s}^{r} (P+Q).\alpha[m]$;
11. $(\alpha_1 \parallel \cdots \parallel \alpha_n).(\tau.(P+Q)+P) \approx_{s}^{f} (\alpha_1 \parallel \cdots \parallel \alpha_n).(P+Q)$;
12. $((P+Q).\tau+P).(\alpha_1[m] \parallel \cdots \parallel \alpha_n[m]) \approx_{s}^{r} (P+Q).(\alpha_1[m] \parallel \cdots \parallel \alpha_n[m])$;
13. $P \approx_{hhp}^{fr} \tau \parallel P$.

*Proof.* From the definition of weakly FR hhp-bisimulation (see Definition 2.71), we know that weakly FR hhp-bisimulation is downward closed for weakly FR hp-bisimulation.

Similarly to the proof of $\tau$ laws for weakly FR hp-bisimulation (see Proposition 5.31), we can prove that the $\tau$ laws hold for weakly FR hhp-bisimulation, however, we omit this here. □

### 5.3.2 Recursion

**Definition 5.33** (Sequential). $X$ is sequential in $E$ if every subexpression of $E$ that contains $X$, apart from $X$ itself, is of the form $\alpha.F$ or $F.\alpha[m]$, or $(\alpha_1 \parallel \cdots \parallel \alpha_n).F$ or $F.(\alpha_1[m] \parallel \cdots \parallel \alpha_n[m])$, or $\sum \widetilde{F}$.

**Definition 5.34** (Guarded recursive expression). $X$ is guarded in $E$ if each occurrence of $X$ is with some subexpression $l.F$ or $F.l[m]$, or $(l_1 \parallel \cdots \parallel l_n).F$ or $F.(l_1[m] \parallel \cdots \parallel l_n[m])$ of $E$.

**Lemma 5.35.** *Let $G$ be guarded and sequential, $Vars(G) \subseteq \widetilde{X}$, and let $G\{\widetilde{P}/\widetilde{X}\} \xrightarrow{\{\alpha_1,\cdots,\alpha_n\}} P'$ or $G\{\widetilde{P}/\widetilde{X}\} \xtwoheadrightarrow{\{\alpha_1[m],\cdots,\alpha_n[m]\}} P'$. Then, there is an expression $H$ such that $G \xrightarrow{\{\alpha_1,\cdots,\alpha_n\}} H$ or $G \xtwoheadrightarrow{\{\alpha_1[m],\cdots,\alpha_n[m]\}} H$, $P' \equiv H\{\widetilde{P}/\widetilde{X}\}$, and for any $\widetilde{Q}$, $G\{\widetilde{Q}/\widetilde{X}\} \xrightarrow{\{\alpha_1,\cdots,\alpha_n\}} H\{\widetilde{Q}/\widetilde{X}\}$ or $G\{\widetilde{Q}/\widetilde{X}\} \xtwoheadrightarrow{\{\alpha_1[m],\cdots,\alpha_n[m]\}} H\{\widetilde{Q}/\widetilde{X}\}$. Moreover, $H$ is sequential, $Vars(H) \subseteq \widetilde{X}$, and if $\alpha_1 = \cdots = \alpha_n = \alpha_1[m] = \cdots = \alpha_n[m] = \tau$, then $H$ is also guarded.*

*Proof.* We only prove the case of forward transition. We need to induct on the structure of $G$.

If $G$ is a Constant, a Composition, a Restriction or a Relabeling, then it contains no variables, since $G$ is sequential and guarded, then $G \xrightarrow{\{\alpha_1,\cdots,\alpha_n\}} P'$, then let $H \equiv P'$, as desired.

$G$ cannot be a variable, since it is guarded.

If $G \equiv G_1 + G_2$. Then, either $G_1\{\widetilde{P}/\widetilde{X}\} \xrightarrow{\{\alpha_1,\cdots,\alpha_n\}} P'$ or $G_2\{\widetilde{P}/\widetilde{X}\} \xrightarrow{\{\alpha_1,\cdots,\alpha_n\}} P'$, then, we can apply this lemma in either case, as desired.

If $G \equiv \beta.H$. Then, we must have $\alpha = \beta$, and $P' \equiv H\{\widetilde{P}/\widetilde{X}\}$, and $G\{\widetilde{Q}/\widetilde{X}\} \equiv \beta.H\{\widetilde{Q}/\widetilde{X}\} \xrightarrow{\beta} H\{\widetilde{Q}/\widetilde{X}\}$, then, let $G'$ be $H$, as desired.

If $G \equiv (\beta_1 \parallel \cdots \parallel \beta_n).H$. Then, we must have $\alpha_i = \beta_i$ for $1 \le i \le n$, and $P' \equiv H\{\widetilde{P}/\widetilde{X}\}$, and $G\{\widetilde{Q}/\widetilde{X}\} \equiv (\beta_1 \parallel \cdots \parallel \beta_n).H\{\widetilde{Q}/\widetilde{X}\} \xrightarrow{\{\beta_1,\cdots,\beta_n\}} H\{\widetilde{Q}/\widetilde{X}\}$, then, let $G'$ be $H$, as desired.

If $G \equiv \tau.H$. Then, we must have $\tau = \tau$, and $P' \equiv H\{\widetilde{P}/\widetilde{X}\}$, and $G\{\widetilde{Q}/\widetilde{X}\} \equiv \tau.H\{\widetilde{Q}/\widetilde{X}\} \xrightarrow{\tau} H\{\widetilde{Q}/\widetilde{X}\}$, then, let $G'$ be $H$, as desired.

For the case of reverse transition, it can be proven similarly, however, we omit it here. □

**Theorem 5.36** (Unique solution of equations for weakly FR step bisimulation). *Let the guarded and sequential expressions $\widetilde{E}$ contain free variables $\subseteq \widetilde{X}$, then,*

*If $\widetilde{P} \approx_s^{fr} \widetilde{E}\{\widetilde{P}/\widetilde{X}\}$ and $\widetilde{Q} \approx_s^{fr} \widetilde{E}\{\widetilde{Q}/\widetilde{X}\}$, then $\widetilde{P} \approx_s^{fr} \widetilde{Q}$.*

*Proof.* We only prove the case of forward transition. Like the corresponding theorem in CCS, without loss of generality, we only consider a single equation $X = E$. Hence, we assume $P \approx_s^{fr} E(P)$, $Q \approx_s^{fr} E(Q)$, then $P \approx_s^{fr} Q$.

We will prove $\{(H(P), H(Q)) : H\}$ sequential, if $H(P) \xrightarrow{\{\alpha_1,\cdots,\alpha_n\}} P'$, then for some $Q'$, $H(Q) \xRightarrow{\{\alpha_1,\cdots,\alpha_n\}} Q'$ and $P' \approx_s^{fr} Q'$.

Let $H(P) \xrightarrow{\{\alpha_1,\cdot,\alpha_n\}} P'$, then $H(E(P)) \xRightarrow{\{\alpha_1,\cdots,\alpha_n\}} P''$ and $P' \approx_s^{fr} P''$.

By Lemma 5.35, we know there is a sequential $H'$ such that $H(E(P)) \xRightarrow{\{\alpha_1,\cdots,\alpha_n\}} H'(P) \Rightarrow P'' \approx_s^{fr} P'$.

Also, $H(E(Q)) \xRightarrow{\{\alpha_1,\cdots,\alpha_n\}} H'(Q) \Rightarrow Q''$ and $P'' \approx_s^{fr} Q''$. Also, $H(Q) \xrightarrow{\{\alpha_1,\cdots,\alpha_n\}} Q' \approx_s^{fr} Q''$. Hence, $P' \approx_s^{fr} Q'$, as desired.

For the case of reverse transition, it can be proven similarly, however, we omit it here. □

**Theorem 5.37** (Unique solution of equations for weakly FR pomset bisimulation). *Let the guarded and sequential expressions $\widetilde{E}$ contain free variables $\subseteq \widetilde{X}$, then*

*If $\widetilde{P} \approx_p^{fr} \widetilde{E}\{\widetilde{P}/\widetilde{X}\}$ and $\widetilde{Q} \approx_p^{fr} \widetilde{E}\{\widetilde{Q}/\widetilde{X}\}$, then $\widetilde{P} \approx_p^{fr} \widetilde{Q}$.*

*Proof.* From the definition of weakly FR pomset bisimulation $\approx_p^{fr}$ (see Definition 2.69), we know that weakly FR pomset bisimulation $\approx_p^{fr}$ is defined by weakly FR pomset transitions, which are labeled by pomsets with $\tau$. In a weakly FR pomset transition, the events in the pomset are either within causality relations (defined by .) or in concurrency (implicitly defined by . and +, and explicitly defined by $\parallel$), of course, they are pairwise consistent (without conflicts). In Proposition 5.29, we have already proven the case that all events are pairwise concurrent, hence, we only need to prove the case of events in causality. Without loss of generality, we take a pomset of $p = \{\alpha, \beta : \alpha.\beta\}$. Then, the weakly forward pomset transition labeled by the above $p$ is just composed of one single event transition labeled by $\alpha$ succeeded by another single event transition labeled by $\beta$, that is, $\xRightarrow{p} = \xRightarrow{\alpha}\xRightarrow{\beta}$ and $\overset{p[\mathcal{K}]}{\Rrightarrow} = \overset{\beta[n]}{\Rrightarrow}\overset{\alpha[m]}{\Rrightarrow}$.

Similarly to the proof of the unique solution of equations for weakly FR step bisimulation $\approx_s^{fr}$ (Theorem 5.36), we can prove that the unique solution of equations holds for weakly FR pomset bisimulation $\approx_p^{fr}$, however, we omit this here. □

**Theorem 5.38** (Unique solution of equations for weakly FR hp-bisimulation). *Let the guarded and sequential expressions $\widetilde{E}$ contain free variables $\subseteq \widetilde{X}$, then*

*If $\widetilde{P} \approx_{hp}^{fr} \widetilde{E}\{\widetilde{P}/\widetilde{X}\}$ and $\widetilde{Q} \approx_{hp}^{fr} \widetilde{E}\{\widetilde{Q}/\widetilde{X}\}$, then $\widetilde{P} \approx_{hp}^{fr} \widetilde{Q}$.*

*Proof.* From the definition of weakly FR hp-bisimulation $\approx_{hp}^{fr}$ (see Definition 2.71), we know that weakly FR hp-bisimulation $\approx_{hp}^{fr}$ is defined on the weakly posetal product $(C_1, f, C_2), f : \hat{C_1} \rightarrow \hat{C_2}$ isomorphism. Two processes $P$ related to $C_1$ and $Q$ related to $C_2$, and $f : \hat{C_1} \rightarrow \hat{C_2}$ isomorphism. Initially, $(C_1, f, C_2) = (\emptyset, \emptyset, \emptyset)$, and $(\emptyset, \emptyset, \emptyset) \in \approx_{hp}^{fr}$. When $P \xRightarrow{\alpha} P'$ $(C_1 \xRightarrow{\alpha} C_1')$, there will be $Q \xRightarrow{\alpha} Q'$ $(C_2 \xRightarrow{\alpha} C_2')$, and we define $f' = f[\alpha \mapsto \alpha]$. Also, when $P \xRightarrow{\alpha[m]} P'$ $(C_1 \xRightarrow{\alpha[m]} C_1')$, there will be $Q \xRightarrow{\alpha[m]} Q'$ $(C_2 \xRightarrow{\alpha[m]} C_2')$, and we define $f' = f[\alpha[m] \mapsto \alpha[m]]$. Then, if $(C_1, f, C_2) \in \approx_{hp}^{fr}$, then $(C_1', f', C_2') \in \approx_{hp}^{fr}$.

Similarly to the proof of the unique solution of equations for weakly FR pomset bisimulation (Theorem 5.37), we can prove that unique solution of equations holds for weakly FR hp-bisimulation, we just need additionally to check the above conditions on weakly FR hp-bisimulation, however, we omit them here. □

**Theorem 5.39** (Unique solution of equations for weakly FR hhp-bisimulation). *Let the guarded and sequential expressions $\widetilde{E}$ contain free variables $\subseteq \widetilde{X}$, then,*

*If $\widetilde{P} \approx_{hhp}^{fr} \widetilde{E}\{\widetilde{P}/\widetilde{X}\}$ and $\widetilde{Q} \approx_{hhp}^{fr} \widetilde{E}\{\widetilde{Q}/\widetilde{X}\}$, then $\widetilde{P} \approx_{hhp}^{fr} \widetilde{Q}$.*

*Proof.* From the definition of weakly FR hhp-bisimulation (see Definition 2.71), we know that weakly FR hhp-bisimulation is downward closed for weakly FR hp-bisimulation.

Similarly to the proof of the unique solution of equations for weakly FR hp-bisimulation (see Theorem 5.38), we can prove that the unique solution of equations holds for weakly FR hhp-bisimulation, however, we omit this here. □

Chapter 6

# Algebraic laws for partially reversible computing

Based on APTC, we also did some work on reversible algebra called RAPTC. However, the axiomatization of RAPTC is imperfect, it is sound, but not complete. The main reason for this is that the existence of a multichoice operator makes a sound and complete axiomatization cannot be established. In this chapter, we try to use an alternative operator to replace a multichoice operator and we obtain a sound and complete axiomatization for reversible computing. The main reason for using an alternative operator is that when an alternative branch is forward executing, the reverse branch is also determined and other branches have no necessity to remain. However, when a process is reversed, the other branches disappear. We call the reversible algebra using an alternative operator partially reversible algebra.

This chapter is organized as follows. We introduce the whole sound and complete axiomatization in Sections 6.1, 6.2, 6.3, and 6.4.

## 6.1 Basic algebra for reversible true concurrency

In this section, we will discuss the algebraic laws of the confliction $+$ and causal relation $\cdot$ based on reversible truly concurrent bisimulations. The resulted algebra is called Basic Algebra for Reversible True Concurrency, abbreviated BARTC.

### 6.1.1 Axiom system of BARTC

In the following, let $e_1, e_2, e'_1, e'_2 \in \mathbb{E}$, and let variables $x, y, z$ range over the set of terms for true concurrency, $p, q, s$ range over the set of closed terms. The predicate $Std(x)$ denotes that $x$ contains only standard events (no histories of events) and $NStd(x)$ means that $x$ only contains histories of events. The set of axioms of BARTC consists of the laws given in Table 6.1.

### 6.1.2 Properties of BARTC

**Definition 6.1** (Basic terms of BARTC). The set of basic terms of BARTC, $\mathcal{B}(BARTC)$, is inductively defined as follows:

1. $\mathbb{E} \subset \mathcal{B}(BARTC)$;
2. if $e \in \mathbb{E}, t \in \mathcal{B}(BARTC)$, then $e \cdot t \in \mathcal{B}(BARTC)$;
3. if $e[m] \in \mathbb{E}, t \in \mathcal{B}(BARTC)$, then $t \cdot e[m] \in \mathcal{B}(BARTC)$;
4. if $t, s \in \mathcal{B}(BARTC)$, then $t + s \in \mathcal{B}(BARTC)$.

**Theorem 6.2** (Elimination theorem of BARTC). *Let p be a closed BARTC term. Then, there is a basic BARTC term q such that* $BARTC \vdash p = q$.

**TABLE 6.1** Axioms of BARTC.

| No. | Axiom | |
|---|---|---|
| $A1$ | $x + y = y + x$ | |
| $A2$ | $(x + y) + z = x + (y + z)$ | |
| $A3$ | $x + x = x$ | |
| $A41$ | $(x + y) \cdot z = x \cdot z + y \cdot z$ | $Std(x), Std(y), Std(z)$ |
| $A42$ | $x \cdot (y + z) = x \cdot y + x \cdot z$ | $NStd(x), NStd(y), NStd(z)$ |
| $A5$ | $(x \cdot y) \cdot z = x \cdot (y \cdot z)$ | |

Handbook of Truly Concurrent Process Algebra. https://doi.org/10.1016/B978-0-44-321515-5.00010-X

**TABLE 6.2** Term rewrite system of BARTC.

| No. | Rewriting Rule |
|---|---|
| $RA3$ | $x+x \to x$ |
| $RA41$ | $(x+y)\cdot z \to x\cdot z+y\cdot z$ |
| $RA42$ | $x\cdot(y+z) \to x\cdot y+x\cdot z$ |
| $RA5$ | $(x\cdot y)\cdot z \to x\cdot(y\cdot z)$ |

*Proof.* (1) First, suppose that the following ordering on the signature of BARTC is defined: $\cdot > +$ and the symbol $\cdot$ is given the lexicographical status for the first argument, then for each rewrite rule $p \to q$ in Table 6.2 relation $p >_{lpo} q$ can easily be proved. We obtain that the term rewrite system shown in Table 6.2 is strongly normalizing, for it has finitely many rewriting rules, and $>$ is a well-founded ordering on the signature of BARTC, and if $s >_{lpo} t$, for each rewriting rule $s \to t$ is in Table 6.2.

(2) Then, we prove that the normal forms of closed BARTC terms are basic BARTC terms.

Suppose that $p$ is a normal form of some closed BARTC term and suppose that $p$ is not a basic term. Let $p'$ denote the smallest sub-term of $p$ that is not a basic term. This implies that each sub-term of $p'$ is a basic term. Then, we prove that $p$ is not a term in normal form. It is sufficient to induct on the structure of $p'$:

- Case $p' \equiv e, e \in \mathbb{E}$. $p'$ is a basic term, which contradicts the assumption that $p'$ is not a basic term, hence, this case should not occur.
- Case $p' \equiv p_1 \cdot p_2$. By induction on the structure of the basic term $p_1$:
  - Subcase $p_1 \in \mathbb{E}$. $p'$ would be a basic term, which contradicts the assumption that $p'$ is not a basic term;
  - Subcase $p_1 \equiv e \cdot p_1'$. $RA5$ rewriting rule can be applied. Hence, $p$ is not a normal form;
  - Subcase $p_1 \equiv p_1' \cdot e[m]$. $RA5$ rewriting rule can be applied. Hence, $p$ is not a normal form;
  - Subcase $p_1 \equiv p_1' + p_1''$. $RA41$ and $RA42$ rewriting rules can be applied. Hence, $p$ is not a normal form.
- Case $p' \equiv p_1 + p_2$. By induction on the structure of the basic terms both $p_1$ and $p_2$, all subcases will lead to that $p'$ would be a basic term, which contradicts the assumption that $p'$ is not a basic term. □

### 6.1.3 Structured operational semantics of BARTC

In this section, we will define a term-deduction system that gives the operational semantics of BARTC. We give the forward operational transition rules of operators $\cdot$ and $+$ as Table 6.3 shows, and the reverse rules of operators $\cdot$ and $+$ as Table 6.4 shows. Also, the predicate $\xrightarrow{e} e[m]$ represents successful forward termination after forward execution of the event $e$, the predicate $\xrightarrow{e[m]} e$ represents successful reverse termination after reverse execution of the event history $e[m]$.

The forward pomset transition rules are shown in Table 6.5, and the reverse pomset transition rules are shown in Table 6.6, which are different from the single event transition rules, the pomset transition rules are labeled by pomsets, which are defined by causality $\cdot$ and conflict $+$.

**Theorem 6.3** (Congruence of BARTC with respect to FR pomset bisimulation equivalence). *FR pomset bisimulation equivalence $\sim_p^{fr}$ is a congruence with respect to BARTC.*

*Proof.* It is easy to see that FR pomset bisimulation is an equivalent relation on BARTC terms, we only need to prove that $\sim_p^{fr}$ is preserved by the operators $\cdot$ and $+$.

- Causality operator $\cdot$. Let $x_1, x_2$ and $y_1, y_2$ be BARTC processes, and $x_1 \sim_p^{fr} y_1$, $x_2 \sim_p^{fr} y_2$, it is sufficient to prove that $x_1 \cdot x_2 \sim_p^{fr} y_1 \cdot y_2$.

  By the definition of FR pomset bisimulation $\sim_p^{fr}$ (Definition 2.68), $x_1 \sim_p^{fr} y_1$ means that

$$x_1 \xrightarrow{X_1} x_1' \quad y_1 \xrightarrow{Y_1} y_1'$$

$$x_1 \xrightarrow{X_1[\mathcal{K}]} x_1' \quad y_1 \xrightarrow{Y_1[\mathcal{L}]} y_1'$$

  with $X_1 \subseteq x_1$, $Y_1 \subseteq y_1$, $X_1 \sim Y_1$ and $x_1' \sim_p^{fr} y_1'$. The meaning of $x_2 \sim_p^{fr} y_2$ is similar.

**TABLE 6.3** Forward single event transition rules of BARTC.

| | | | |
|---|---|---|---|
| $\frac{}{e \xrightarrow{e} e[m]}$ | | | |
| $\frac{x \xrightarrow{e} e[m]}{x+y \xrightarrow{e} e[m]}$ | $\frac{x \xrightarrow{e} x'}{x+y \xrightarrow{e} x'}$ | $\frac{y \xrightarrow{e} e[m]}{x+y \xrightarrow{e} e[m]}$ | $\frac{y \xrightarrow{e} y'}{x+y \xrightarrow{e} y'}$ |
| $\frac{x \xrightarrow{e} e[m]}{x \cdot y \xrightarrow{e} e[m] \cdot y}$ | $\frac{x \xrightarrow{e} x'}{x \cdot y \xrightarrow{e} x' \cdot y}$ | | |

**TABLE 6.4** Reverse single event transition rules of BARTC.

| | | | |
|---|---|---|---|
| $\frac{}{e[m] \overset{e[m]}{\twoheadrightarrow} e}$ | | | |
| $\frac{x \overset{e[m]}{\twoheadrightarrow} e}{x+y \overset{e[m]}{\twoheadrightarrow} e}$ | $\frac{x \overset{e[m]}{\twoheadrightarrow} x'}{x+y \overset{e[m]}{\twoheadrightarrow} x'}$ | $\frac{y \overset{e[m]}{\twoheadrightarrow} e}{x+y \overset{e[m]}{\twoheadrightarrow} e}$ | $\frac{y \overset{e[m]}{\twoheadrightarrow} y'}{x+y \overset{e[m]}{\twoheadrightarrow} y'}$ |
| $\frac{y \overset{e[m]}{\twoheadrightarrow} e}{x \cdot y \overset{e[m]}{\twoheadrightarrow} x \cdot e}$ | $\frac{y \overset{e[m]}{\twoheadrightarrow} y'}{x \cdot y \overset{e[m]}{\twoheadrightarrow} x \cdot y'}$ | | |

**TABLE 6.5** Forward pomset transition rules of BARTC.

| | | | |
|---|---|---|---|
| $\frac{}{X \xrightarrow{X} X[\mathcal{K}]}$ | | | |
| $\frac{x \xrightarrow{X} X[\mathcal{K}]}{x+y \xrightarrow{X} X[\mathcal{K}]}(X \subseteq x)$ | $\frac{x \xrightarrow{X} x'}{x+y \xrightarrow{X} x'}(X \subseteq x)$ | $\frac{y \xrightarrow{Y} Y[\mathcal{K}]}{x+y \xrightarrow{Y} Y[\mathcal{K}]}(Y \subseteq y)$ | $\frac{y \xrightarrow{Y} y'}{x+y \xrightarrow{Y} y'}(Y \subseteq y)$ |
| $\frac{x \xrightarrow{X} X[\mathcal{K}]}{x \cdot y \xrightarrow{X} X[\mathcal{K}] \cdot y}(X \subseteq x)$ | $\frac{x \xrightarrow{X} x'}{x \cdot y \xrightarrow{X} x' \cdot y}(X \subseteq x)$ | | |

**TABLE 6.6** Reverse pomset transition rules of BARTC.

| | | | |
|---|---|---|---|
| $\frac{}{X[\mathcal{K}] \overset{X[\mathcal{K}]}{\twoheadrightarrow} X}$ | | | |
| $\frac{x \overset{X[\mathcal{K}]}{\twoheadrightarrow} X}{x+y \overset{X[\mathcal{K}]}{\twoheadrightarrow} X}(X \subseteq x)$ | $\frac{x \overset{X[\mathcal{K}]}{\twoheadrightarrow} x'}{x+y \overset{X[\mathcal{K}]}{\twoheadrightarrow} x'}(X \subseteq x)$ | $\frac{y \overset{Y[\mathcal{K}]}{\twoheadrightarrow} Y}{x+y \overset{Y[\mathcal{K}]}{\twoheadrightarrow} Y}(Y \subseteq y)$ | $\frac{y \overset{Y[\mathcal{K}]}{\twoheadrightarrow} y'}{x+y \overset{Y[\mathcal{K}]}{\twoheadrightarrow} y'}(Y \subseteq y)$ |
| $\frac{y \overset{Y[\mathcal{K}]}{\twoheadrightarrow} Y}{x \cdot y \overset{Y[\mathcal{K}]}{\twoheadrightarrow} x \cdot Y}(Y \subseteq y)$ | $\frac{y \overset{Y[\mathcal{K}]}{\twoheadrightarrow} y'}{x \cdot y \overset{Y[\mathcal{K}]}{\twoheadrightarrow} x \cdot y'}(Y \subseteq y)$ | | |

By the pomset transition rules for causality operator $\cdot$ in Table 6.5 and Table 6.6, we can obtain

$$x_1 \cdot x_2 \xrightarrow{X_1} X_1[\mathcal{K}] \cdot x_2 \quad y_1 \cdot y_2 \xrightarrow{Y_1} Y_1[\mathcal{L}] \cdot y_2$$

$$x_1 \cdot x_2 \overset{X_2[\mathcal{K}]}{\twoheadrightarrow} x_1 \cdot X_2 \quad y_1 \cdot y_2 \overset{Y_2[\mathcal{L}]}{\twoheadrightarrow} y_1 \cdot Y_2$$

with $X_1 \subseteq x_1$, $Y_1 \subseteq y_1$, $X_1 \sim Y_1$, and $x_2 \sim_p^{fr} y_2$; $X_2 \subseteq x_2$, $Y_2 \subseteq y_2$, $X_2 \sim Y_2$, and $x_1 \sim_p^{fr} y_1$, hence, we obtain $x_1 \cdot x_2 \sim_p^{fr} y_1 \cdot y_2$, as desired.

Or, we can obtain

$$x_1 \cdot x_2 \xrightarrow{X_1} x_1' \cdot x_2 \quad y_1 \cdot y_2 \xrightarrow{Y_1} y_1' \cdot y_2$$

$$x_1 \cdot x_2 \overset{X_2[\mathcal{K}]}{\twoheadrightarrow} x_1 \cdot x_2' \quad y_1 \cdot y_2 \overset{Y_2[\mathcal{L}]}{\twoheadrightarrow} y_1 \cdot y_2'$$

with $X_1 \subseteq x_1$, $Y_1 \subseteq y_1$, $X_1 \sim Y_1$, and $x_1' \sim_p^{fr} y_1'$, $x_2 \sim_p^{fr} y_2$; $X_2 \subseteq x_2$, $Y_2 \subseteq y_2$, $X_2 \sim Y_2$, and $x_2' \sim_p^{fr} y_2'$, $x_1 \sim_p^{fr} y_1$, hence, we obtain $x_1 \cdot x_2 \sim_p^{fr} y_1 \cdot y_2$, as desired.

- **Conflict operator $+$.** Let $x_1, x_2$ and $y_1, y_2$ be BARTC processes, and $x_1 \sim_p^{fr} y_1$, $x_2 \sim_p^{fr} y_2$, it is sufficient to prove that $x_1 + x_2 \sim_p^{fr} y_1 + y_2$. The meanings of $x_1 \sim_p^{fr} y_1$ and $x_2 \sim_p^{fr} y_2$ are the same as the above case, according to the definition of FR pomset bisimulation $\sim_p^{fr}$ in Definition 2.68.
By the pomset transition rules for conflict operator $+$ in Table 6.5 and Table 6.6, we can obtain four cases:

$$x_1 + x_2 \xrightarrow{X_1} X_1[\mathcal{K}] \quad y_1 + y_2 \xrightarrow{Y_1} Y_1[\mathcal{L}]$$

$$x_1 + x_2 \xtwoheadrightarrow{X_1[\mathcal{K}]} X_1 \quad y_1 + y_2 \xtwoheadrightarrow{Y_1[\mathcal{L}]} Y_1$$

with $X_1 \subseteq x_1$, $Y_1 \subseteq y_1$, $X_1 \sim Y_1$, hence, we obtain $x_1 + x_2 \sim_p^{fr} y_1 + y_2$, as desired.
Or, we can obtain

$$x_1 + x_2 \xrightarrow{X_1} x_1' \quad y_1 + y_2 \xrightarrow{Y_1} y_1'$$

$$x_1 + x_2 \xtwoheadrightarrow{X_1[\mathcal{K}]} x_1' \quad y_1 + y_2 \xtwoheadrightarrow{Y_1[\mathcal{L}]} y_1'$$

with $X_1 \subseteq x_1$, $Y_1 \subseteq y_1$, $X_1 \sim Y_1$, and $x_1' \sim_p^{fr} y_1'$, hence, we obtain $x_1 + x_2 \sim_p^{fr} y_1 + y_2$, as desired.
Or, we can obtain

$$x_1 + x_2 \xrightarrow{X_2} X_2[\mathcal{K}] \quad y_1 + y_2 \xrightarrow{Y_2} Y_2[\mathcal{L}]$$

$$x_1 + x_2 \xtwoheadrightarrow{X_2[\mathcal{K}]} X_2 \quad y_1 + y_2 \xtwoheadrightarrow{Y_2[\mathcal{L}]} Y_2$$

with $X_2 \subseteq x_2$, $Y_2 \subseteq y_2$, $X_2 \sim Y_2$, hence, we obtain $x_1 + x_2 \sim_p^{fr} y_1 + y_2$, as desired.
Or, we can obtain

$$x_1 + x_2 \xrightarrow{X_2} x_2' \quad y_1 + y_2 \xrightarrow{Y_2} y_2'$$

$$x_1 + x_2 \xtwoheadrightarrow{X_2[\mathcal{K}]} x_2' \quad y_1 + y_2 \xtwoheadrightarrow{Y_2[\mathcal{L}]} y_2'$$

with $X_2 \subseteq x_2$, $Y_2 \subseteq y_2$, $X_2 \sim Y_2$, and $x_2' \sim_p^{fr} y_2'$, hence, we obtain $x_1 + x_2 \sim_p^{fr} y_1 + y_2$, as desired. □

**Theorem 6.4** (Soundness of BARTC modulo FR pomset bisimulation equivalence). *Let $x$ and $y$ be BARTC terms. If $BARTC \vdash x = y$, then $x \sim_p^{fr} y$.*

*Proof.* Since FR pomset bisimulation $\sim_p^{fr}$ is both an equivalent and a congruent relation, we only need to check if each axiom in Table 6.1 is sound modulo FR pomset bisimulation equivalence.

- **Axiom $A1$.** Let $p, q$ be BARTC processes, and $p + q = q + p$, it is sufficient to prove that $p + q \sim_p^{fr} q + p$. By the pomset transition rules for operator $+$ in Table 6.5 and Table 6.6, we obtain

$$\frac{p \xrightarrow{P} P[\mathcal{K}]}{p + q \xrightarrow{P} P[\mathcal{K}]}(P \subseteq p) \quad \frac{p \xrightarrow{P} P[\mathcal{K}]}{q + p \xrightarrow{P} P[\mathcal{K}]}(P \subseteq p)$$

$$\frac{p \xtwoheadrightarrow{P[\mathcal{K}]} P}{p + q \xtwoheadrightarrow{P[\mathcal{K}]} P}(P \subseteq p) \quad \frac{p \xtwoheadrightarrow{P[\mathcal{K}]} P}{q + p \xtwoheadrightarrow{P[\mathcal{K}]} P}(P \subseteq p)$$

$$\frac{p \xrightarrow{P} p'}{p + q \xrightarrow{P} p'}(P \subseteq p) \quad \frac{p \xrightarrow{P} p'}{q + p \xrightarrow{P} p'}(P \subseteq p)$$

$$\frac{p \xtwoheadrightarrow{P[\mathcal{K}]} p'}{p + q \xtwoheadrightarrow{P[\mathcal{K}]} p'}(P \subseteq p) \quad \frac{p \xtwoheadrightarrow{P[\mathcal{K}]} p'}{q + p \xtwoheadrightarrow{P[\mathcal{K}]} p'}(P \subseteq p)$$

$$\frac{q\xrightarrow{Q}Q[\mathcal{L}]}{p+q\xrightarrow{Q}Q[\mathcal{L}]}(Q\subseteq q)\quad\frac{q\xrightarrow{Q}Q[\mathcal{L}]}{q+p\xrightarrow{Q}Q[\mathcal{L}]}(Q\subseteq q)$$

$$\frac{q\xrightarrow{Q[\mathcal{L}]}\mathrel{\mkern-14mu\rightarrow} Q}{p+q\xrightarrow{Q[\mathcal{L}]}\mathrel{\mkern-14mu\rightarrow} Q}(Q\subseteq q)\quad\frac{q\xrightarrow{Q[\mathcal{L}]}\mathrel{\mkern-14mu\rightarrow} Q}{q+p\xrightarrow{Q[\mathcal{L}]}\mathrel{\mkern-14mu\rightarrow} Q}(Q\subseteq q)$$

$$\frac{q\xrightarrow{Q}q'}{p+q\xrightarrow{Q}q'}(Q\subseteq q)\quad\frac{q\xrightarrow{Q}q'}{q+p\xrightarrow{Q}q'}(Q\subseteq q)$$

$$\frac{q\xrightarrow{Q[\mathcal{L}]}\mathrel{\mkern-14mu\rightarrow} q'}{p+q\xrightarrow{Q[\mathcal{L}]}\mathrel{\mkern-14mu\rightarrow} q'}(Q\subseteq q)\quad\frac{q\xrightarrow{Q[\mathcal{L}]}\mathrel{\mkern-14mu\rightarrow} q'}{q+p\xrightarrow{Q[\mathcal{L}]}\mathrel{\mkern-14mu\rightarrow} q'}(Q\subseteq q)$$

Hence, $p+q\sim_p^{fr}q+p$, as desired.

- **Axiom** $A2$. Let $p,q,s$ be BARTC processes, and $(p+q)+s=p+(q+s)$, it is sufficient to prove that $(p+q)+s\sim_p^{fr}p+(q+s)$. By the pomset transition rules for operator $+$ in Table 6.5 and Table 6.6, we obtain

$$\frac{p\xrightarrow{P}P[\mathcal{K}]}{(p+q)+s\xrightarrow{P}P[\mathcal{K}]}(P\subseteq p)\quad\frac{p\xrightarrow{P}P[\mathcal{K}]}{p+(q+s)\xrightarrow{P}P[\mathcal{K}]}(P\subseteq p)$$

$$\frac{p\xrightarrow{P[\mathcal{K}]}\mathrel{\mkern-14mu\rightarrow} P}{(p+q)+s\xrightarrow{P[\mathcal{K}]}\mathrel{\mkern-14mu\rightarrow} P}(P\subseteq p)\quad\frac{p\xrightarrow{P[\mathcal{K}]}\mathrel{\mkern-14mu\rightarrow} P}{p+(q+s)\xrightarrow{P[\mathcal{K}]}\mathrel{\mkern-14mu\rightarrow} P}(P\subseteq p)$$

$$\frac{p\xrightarrow{P}p'}{(p+q)+s\xrightarrow{P}p'}(P\subseteq p)\quad\frac{p\xrightarrow{P}p'}{p+(q+s)\xrightarrow{P}p'}(P\subseteq p)$$

$$\frac{p\xrightarrow{P[\mathcal{K}]}\mathrel{\mkern-14mu\rightarrow} p'}{(p+q)+s\xrightarrow{P[\mathcal{K}]}\mathrel{\mkern-14mu\rightarrow} p'}(P\subseteq p)\quad\frac{p\xrightarrow{P[\mathcal{K}]}\mathrel{\mkern-14mu\rightarrow} p'}{p+(q+s)\xrightarrow{P[\mathcal{K}]}\mathrel{\mkern-14mu\rightarrow} p'}(P\subseteq p)$$

$$\frac{q\xrightarrow{Q}Q[\mathcal{L}]}{(p+q)+s\xrightarrow{Q}Q[\mathcal{L}]}(Q\subseteq q)\quad\frac{q\xrightarrow{Q}Q[\mathcal{L}]}{p+(q+s)\xrightarrow{Q}Q[\mathcal{L}]}(Q\subseteq q)$$

$$\frac{q\xrightarrow{Q[\mathcal{L}]}\mathrel{\mkern-14mu\rightarrow} Q}{(p+q)+s\xrightarrow{Q[\mathcal{L}]}\mathrel{\mkern-14mu\rightarrow} Q}(Q\subseteq q)\quad\frac{q\xrightarrow{Q[\mathcal{L}]}\mathrel{\mkern-14mu\rightarrow} Q}{p+(q+s)\xrightarrow{Q[\mathcal{L}]}\mathrel{\mkern-14mu\rightarrow} Q}(Q\subseteq q)$$

$$\frac{q\xrightarrow{Q}q'}{(p+q)+s\xrightarrow{Q}q'}(Q\subseteq q)\quad\frac{q\xrightarrow{Q}q'}{p+(q+s)\xrightarrow{Q}q'}(Q\subseteq q)$$

$$\frac{q\xrightarrow{Q[\mathcal{L}]}\mathrel{\mkern-14mu\rightarrow} q'}{(p+q)+s\xrightarrow{Q[\mathcal{L}]}\mathrel{\mkern-14mu\rightarrow} q'}(Q\subseteq q)\quad\frac{q\xrightarrow{Q[\mathcal{L}]}\mathrel{\mkern-14mu\rightarrow} q'}{p+(q+s)\xrightarrow{Q[\mathcal{L}]}\mathrel{\mkern-14mu\rightarrow} q'}(Q\subseteq q)$$

$$\frac{s\xrightarrow{S}S[\mathcal{M}]}{(p+q)+s\xrightarrow{S}S[\mathcal{M}]}(S\subseteq s)\quad\frac{s\xrightarrow{S}S[\mathcal{M}]}{p+(q+s)\xrightarrow{S}S[\mathcal{M}]}(S\subseteq s)$$

$$\frac{s\xrightarrow{S[\mathcal{M}]}\mathrel{\mkern-14mu\rightarrow} S}{(p+q)+s\xrightarrow{S[\mathcal{M}]}\mathrel{\mkern-14mu\rightarrow} S}(S\subseteq s)\quad\frac{s\xrightarrow{S[\mathcal{M}]}\mathrel{\mkern-14mu\rightarrow} S}{p+(q+s)\xrightarrow{S[\mathcal{M}]}\mathrel{\mkern-14mu\rightarrow} S}(S\subseteq s)$$

$$\frac{s \xrightarrow{S} s'}{(p+q)+s \xrightarrow{S} s'}(S \subseteq s) \quad \frac{s \xrightarrow{S} s'}{p+(q+s) \xrightarrow{S} s'}(S \subseteq s)$$

$$\frac{s \xrightarrow{S[\mathcal{M}]} s'}{(p+q)+s \xrightarrow{S[\mathcal{M}]} s'}(S \subseteq s) \quad \frac{s \xrightarrow{S[\mathcal{M}]} s'}{p+(q+s) \xrightarrow{S[\mathcal{M}]} s'}(S \subseteq s)$$

Hence, $(p+q)+s \sim_p^{fr} p+(q+s)$, as desired.

- **Axiom** $A3$. Let $p$ be a BARTC process, and $p+p=p$, it is sufficient to prove that $p+p \sim_p^{fr} p$. By the pomset transition rules for operator $+$ in Table 6.5 and Table 6.6, we obtain

$$\frac{p \xrightarrow{P} P[\mathcal{K}]}{p+p \xrightarrow{P} P[\mathcal{K}]}(P \subseteq p) \quad \frac{p \xrightarrow{P} P[\mathcal{K}]}{p \xrightarrow{P} P[\mathcal{K}]}(P \subseteq p)$$

$$\frac{p \xrightarrow{P[\mathcal{K}]} P}{p+p \xrightarrow{P[\mathcal{K}]} P}(P \subseteq p) \quad \frac{p \xrightarrow{P[\mathcal{K}]} P}{p \xrightarrow{P[\mathcal{K}]} P}(P \subseteq p)$$

$$\frac{p \xrightarrow{P} p'}{p+p \xrightarrow{P} p'}(P \subseteq p) \quad \frac{p \xrightarrow{P} p'}{p \xrightarrow{P} p'}(P \subseteq p)$$

$$\frac{p \xrightarrow{P[\mathcal{K}]} p'}{p+p \xrightarrow{P[\mathcal{K}]} p'}(P \subseteq p) \quad \frac{p \xrightarrow{P[\mathcal{K}]} p'}{p \xrightarrow{P[\mathcal{K}]} p'}(P \subseteq p)$$

Hence, $p+p \sim_p^{fr} p$, as desired.

- **Axiom** $A41$. Let $p, q, s$ be BARTC processes, $Std(p)$, $Std(q)$, $Std(s)$, and $(p+q)\cdot s = p\cdot s + q\cdot s$, it is sufficient to prove that $(p+q)\cdot s \sim_p^{fr} p\cdot s + q\cdot s$. By the pomset transition rules for operators $+$ and $\cdot$ in Table 6.5, we obtain

$$\frac{p \xrightarrow{P} P[\mathcal{K}]}{(p+q)\cdot s \xrightarrow{P} P[\mathcal{K}]\cdot s}(P \subseteq p) \quad \frac{p \xrightarrow{P} P[\mathcal{K}]}{p\cdot s+q\cdot s \xrightarrow{P} P[\mathcal{K}]\cdot s}(P \subseteq p)$$

$$\frac{p \xrightarrow{P} p'}{(p+q)\cdot s \xrightarrow{P} p'\cdot s}(P \subseteq p) \quad \frac{p \xrightarrow{P} p'}{p\cdot s+q\cdot s \xrightarrow{P} p'\cdot s}(P \subseteq p)$$

$$\frac{q \xrightarrow{Q} Q[\mathcal{L}]}{(p+q)\cdot s \xrightarrow{Q} Q[\mathcal{K}]\cdot s}(Q \subseteq q) \quad \frac{q \xrightarrow{Q} Q[\mathcal{L}]}{p\cdot s+q\cdot s \xrightarrow{Q} Q[\mathcal{K}]\cdot s}(Q \subseteq q)$$

$$\frac{q \xrightarrow{Q} q'}{(p+q)\cdot s \xrightarrow{Q} q'\cdot s}(Q \subseteq q) \quad \frac{q \xrightarrow{Q} q'}{p\cdot s+q\cdot s \xrightarrow{Q} q'\cdot s}(Q \subseteq q)$$

Hence, $(p+q)\cdot s \sim_p^{fr} p\cdot s+q\cdot s$, as desired.

- **Axiom** $A42$. Let $p, q, s$ be BARTC processes, $NStd(p)$, $NStd(q)$, $NStd(s)$, and $p\cdot(q+s) = p\cdot q + p\cdot s$, it is sufficient to prove that $p\cdot(q+s) \sim_p^{fr} p\cdot q + p\cdot s$. By the pomset transition rules for operators $+$ and $\cdot$ in Table 6.6, we obtain

$$\frac{q \xrightarrow{Q[\mathcal{L}]} Q}{p\cdot(q+s) \xrightarrow{Q[\mathcal{L}]} p\cdot Q}(Q \subseteq q) \quad \frac{q \xrightarrow{Q[\mathcal{L}]} Q}{p\cdot q+p\cdot s \xrightarrow{Q[\mathcal{L}]} p\cdot Q}(Q \subseteq q)$$

$$\frac{q \xrightarrow{Q[\mathcal{L}]} q'}{p\cdot(q+s) \xrightarrow{Q[\mathcal{L}]} p\cdot q'}(Q \subseteq q) \quad \frac{q \xrightarrow{Q[\mathcal{L}]} q'}{p\cdot q+p\cdot s \xrightarrow{Q[\mathcal{L}]} p\cdot q'}(Q \subseteq q)$$

$$\frac{s \xrightarrow{S[\mathcal{M}]} S}{p\cdot(q+s) \xrightarrow{S[\mathcal{M}]} p\cdot S}(S\subseteq s) \quad \frac{s \xrightarrow{S[\mathcal{M}]} S}{p\cdot q + p\cdot s \xrightarrow{S[\mathcal{M}]} p\cdot S}(S\subseteq s)$$

$$\frac{s \xrightarrow{S[\mathcal{M}]} s'}{p\cdot(q+s) \xrightarrow{S[\mathcal{M}]} p\cdot s'}(S\subseteq s) \quad \frac{s \xrightarrow{S[\mathcal{M}]} s'}{p\cdot q + p\cdot s \xrightarrow{S[\mathcal{M}]} p\cdot s'}(S\subseteq s)$$

Hence, $p\cdot(q+s) \sim_p^{fr} p\cdot q + p\cdot s$, as desired.

- **Axiom** $A5$. Let $p, q, s$ be BARTC processes, and $(p\cdot q)\cdot s = p\cdot(q\cdot s)$, it is sufficient to prove that $(p\cdot q)\cdot s \sim_p^{fr} p\cdot(q\cdot s)$. By the pomset transition rules for operator $\cdot$ in Table 6.5 and Table 6.6, we obtain

$$\frac{p \xrightarrow{P} P[\mathcal{K}]}{(p\cdot q)\cdot s \xrightarrow{P} (P[\mathcal{K}]\cdot q)\cdot s}(P\subseteq p) \quad \frac{p \xrightarrow{P} P[\mathcal{K}]}{p\cdot(q\cdot s) \xrightarrow{P} P[\mathcal{K}]\cdot(q\cdot s)}(P\subseteq p)$$

$$\frac{p \xrightarrow{P} p'}{(p\cdot q)\cdot s \xrightarrow{P} (p'\cdot q)\cdot s}(P\subseteq p) \quad \frac{p \xrightarrow{P} p'}{p\cdot(q\cdot s) \xrightarrow{P} p'\cdot(q\cdot s)}(P\subseteq p)$$

$$\frac{s \xrightarrow{S[\mathcal{M}} S]}{(p\cdot q)\cdot s \xrightarrow{S[\mathcal{M}]} (p\cdot q)\cdot S}(S\subseteq s) \quad \frac{s \xrightarrow{S[\mathcal{M}]} S}{p\cdot(q\cdot s) \xrightarrow{S[\mathcal{K}]} p\cdot(q\cdot S)}(S\subseteq s)$$

$$\frac{s \xrightarrow{S[\mathcal{M}]} s'}{(p\cdot q)\cdot s \xrightarrow{S[\mathcal{M}]} (p\cdot q)\cdot s'}(S\subseteq s) \quad \frac{s \xrightarrow{S[\mathcal{M}]} s'}{p\cdot(q\cdot s) \xrightarrow{S[\mathcal{M}]} p\cdot(q\cdot s')}(S\subseteq s)$$

With assumptions $(p\cdot q)\cdot S = p\cdot(q\cdot S)$ and $(p\cdot q)\cdot s' = p\cdot(q\cdot s')$, hence, $(p\cdot q)\cdot s \sim_p^{fr} p\cdot(q\cdot s)$, as desired. □

**Theorem 6.5** (Completeness of BARTC modulo FR pomset bisimulation equivalence). *Let $p$ and $q$ be closed BARTC terms, if $p \sim_p^{fr} q$, then $p = q$.*

*Proof.* First, by the elimination theorem of BARTC, we know that for each closed BARTC term $p$, there exists a closed basic $BARTC$ term $p'$, such that $BARTC \vdash p = p'$, hence, we only need to consider closed basic $BARTC$ terms.

The basic terms (see Definition 6.1) modulo associativity and commutativity (AC) of conflict $+$ (defined by axioms $A1$ and $A2$ in Table 6.1), and this equivalence is denoted by $=_{AC}$. Then, each equivalence class $s$ modulo AC of $+$ has the following normal form

$$s_1 + \cdots + s_k$$

with each $s_i$ either an atomic event or of the form $t_1 \cdot t_2$, and each $s_i$ is called the summand of $s$.

Now, we prove that for normal forms $n$ and $n'$, if $n \sim_p^{fr} n'$, then $n =_{AC} n'$. It is sufficient to induct on the sizes of $n$ and $n'$.

- Consider a summand $e$ of $n$. Then, $n \xrightarrow{e} e[m]$, hence, $n \sim_p^{fr} n'$ implies $n' \xrightarrow{e} e[m]$, meaning that $n'$ also contains the summand $e$.
- Consider a summand $e[m]$ of $n$. Then, $n \xrightarrow{e[m]} e$, hence, $n \sim_p^{fr} n'$ implies $n' \xrightarrow{e[m]} e$, meaning that $n'$ also contains the summand $e[m]$.
- Consider a summand $t_1 \cdot t_2$ of $n$. Then, $n \xrightarrow{t_1} t_1[\mathcal{K}]\cdot t_2$, hence, $n \sim_p^{fr} n'$ implies $n' \xrightarrow{t_1} t_1[\mathcal{K}]\cdot t_2'$ with $t_1[\mathcal{K}]\cdot t_2 \sim_p^{fr} t_1[\mathcal{K}]\cdot t_2'$, meaning that $n'$ contains a summand $t_1 \cdot t_2'$. Since $t_2$ and $t_2'$ are normal forms and have sizes no greater than $n$ and $n'$, by the induction hypotheses $t_2 \sim_p^{fr} t_2'$ implies $t_2 =_{AC} t_2'$.
- Consider a summand $t_1 \cdot t_2[\mathcal{L}]$ of $n$. Then, $n \xrightarrow{t_2[\mathcal{L}]} t_1 \cdot t_2$, hence, $n \sim_p^{fr} n'$ implies $n' \xrightarrow{t_2[\mathcal{L}]} t_1' \cdot t_2$ with $t_1 \cdot t_2[\mathcal{L}] \sim_p^{fr} t_1' \cdot t_2[\mathcal{L}]$, meaning that $n'$ contains a summand $t_1' \cdot t_2[\mathcal{L}]$. Since $t_21$ and $t_1'$ are normal forms and have sizes no greater than $n$ and $n'$, by the induction hypotheses $t_1 \sim_p^{fr} t_1'$ implies $t_1 =_{AC} t_1'$.

Hence, we obtain $n =_{AC} n'$.

Finally, let $s$ and $t$ be basic terms, and $s \sim_p^{fr} t$, there are normal forms $n$ and $n'$, such that $s = n$ and $t = n'$. The soundness theorem of BARTC modulo FR pomset bisimulation equivalence (see Theorem 6.4) yields $s \sim_p^{fr} n$ and $t \sim_p^{fr} n'$, hence, $n \sim_p^{fr} s \sim_p^{fr} t \sim_p^{fr} n'$. Since if $n \sim_p^{fr} n'$, then $n =_{AC} n'$, $s = n =_{AC} n' = t$, as desired. □

The step transition rules are almost the same as the transition rules in Table 6.5 and Table 6.6, the difference is that events in the transition pomset are pairwise concurrent for the step transition rules, hence we omit them here.

**Theorem 6.6** (Congruence of BARTC with respect to FR step bisimulation equivalence). *Step bisimulation equivalence $\sim_s^{fr}$ is a congruence with respect to BARTC.*

*Proof.* It is easy to see that FR step bisimulation is an equivalent relation on BARTC terms, we only need to prove that $\sim_s^{fr}$ is preserved by the operators $\cdot$ and $+$. The proof is almost the same as the proof of congruence of BARTC with respect to FR pomset bisimulation equivalence, the difference is that events in the transition pomset are pairwise concurrent for FR step bisimulation equivalence, hence we omit it here. □

**Theorem 6.7** (Soundness of BARTC modulo FR step bisimulation equivalence). *Let $x$ and $y$ be BARTC terms. If $BARTC \vdash x = y$, then $x \sim_s^{fr} y$.*

*Proof.* Since FR step bisimulation $\sim_s^{fr}$ is both an equivalent and a congruent relation, we only need to check if each axiom in Table 6.1 is sound modulo FR step bisimulation equivalence. The soundness proof is almost the same as the soundness proof of BARTC modulo FR pomset bisimulation equivalence, the difference is that events in the transition pomset are pairwise concurrent, hence we omit it here. □

**Theorem 6.8** (Completeness of BARTC modulo FR step bisimulation equivalence). *Let $p$ and $q$ be closed BARTC terms, if $p \sim_s^{fr} q$, then $p = q$.*

*Proof.* The proof of completeness is almost the same as the proof of BARTC modulo FR pomset bisimulation equivalence, the only difference is that events in the transition pomset are pairwise concurrent, hence we omit it here. □

The transition rules for (hereditary) FR hp-bisimulation of BARTC are the same as the single event transition rules in Table 6.3 and Table 6.4.

**Theorem 6.9** (Congruence of BARTC with respect to FR hp-bisimulation equivalence). *Hp-bisimulation equivalence $\sim_{hp}^{fr}$ is a congruence with respect to BARTC.*

*Proof.* It is easy to see that history-preserving bisimulation is an equivalent relation on BARTC terms, we only need to prove that $\sim_{hp}^{fr}$ is preserved by the operators $\cdot$ and $+$.

The proof is similar to the proof of congruence of BARTC with respect to FR pomset bisimulation equivalence, hence, we omit it here. □

**Theorem 6.10** (Soundness of BARTC modulo FR hp-bisimulation equivalence). *Let $x$ and $y$ be BARTC terms. If $BARTC \vdash x = y$, then $x \sim_{hp}^{fr} y$.*

*Proof.* Since FR hp-bisimulation $\sim_{hp}^{fr}$ is both an equivalent and a congruent relation, we only need to check if each axiom in Table 6.1 is sound modulo FR hp-bisimulation equivalence.

The proof is similar to the proof of soundness of BARTC modulo FR pomset and step bisimulation equivalences, hence, we omit it here. □

**Theorem 6.11** (Completeness of BARTC modulo FR hp-bisimulation equivalence). *Let $p$ and $q$ be closed BARTC terms, if $p \sim_{hp}^{fr} q$ then $p = q$.*

*Proof.* The proof is similar to the proof of completeness of BARTC modulo FR pomset and step bisimulation equivalences, hence, we omit it here. □

**Theorem 6.12** (Congruence of BARTC with respect to FR hhp-bisimulation equivalence). *Hhp-bisimulation equivalence $\sim_{hhp}^{fr}$ is a congruence with respect to BARTC.*

**TABLE 6.7 Forward transition rules of parallel operator $\parallel$.**

| | |
|---|---|
| $\dfrac{x\xrightarrow{e_1}e_1[m]\quad y\xrightarrow{e_2}e_2[m]}{x\parallel y\xrightarrow{\{e_1,e_2\}}e_1[m]\parallel e_2[m]}$ | $\dfrac{x\xrightarrow{e_1}x'\quad y\xrightarrow{e_2}e_2[m]}{x\parallel y\xrightarrow{\{e_1,e_2\}}x'\parallel e_2[m]}$ |
| $\dfrac{x\xrightarrow{e_1}e_1[m]\quad y\xrightarrow{e_2}y'}{x\parallel y\xrightarrow{\{e_1,e_2\}}e_1[m]\parallel y'}$ | $\dfrac{x\xrightarrow{e_1}x'\quad y\xrightarrow{e_2}y'}{x\parallel y\xrightarrow{\{e_1,e_2\}}x'\between y'}$ |

**TABLE 6.8 Reverse transition rules of parallel operator $\parallel$.**

| | |
|---|---|
| $\dfrac{x\overset{e_1[m]}{\twoheadrightarrow}e_1\quad y\overset{e_2[m]}{\twoheadrightarrow}e_2}{x\parallel y\overset{\{e_1[m],e_2[m]\}}{\twoheadrightarrow}e_1\parallel e_2}$ | $\dfrac{x\overset{e_1[m]}{\twoheadrightarrow}x'\quad y\overset{e_2[m]}{\twoheadrightarrow}e_2}{x\parallel y\overset{\{e_1[m],e_2[m]\}}{\twoheadrightarrow}x'\parallel e_2}$ |
| $\dfrac{x\overset{e_1[m]}{\twoheadrightarrow}e_1\quad y\overset{e_2[m]}{\twoheadrightarrow}y'}{x\parallel y\overset{\{e_1[m],e_2[m]\}}{\twoheadrightarrow}e_1\parallel y'}$ | $\dfrac{x\overset{e_1[m]}{\twoheadrightarrow}x'\quad y\overset{e_2[m]}{\twoheadrightarrow}y'}{x\parallel y\overset{\{e_1[m],e_2[m]\}}{\twoheadrightarrow}x'\between y'}$ |

*Proof.* It is easy to see that FR hhp-bisimulation is an equivalent relation on BARTC terms, we only need to prove that $\sim^{fr}_{hhp}$ is preserved by the operators $\cdot$ and $+$.

The proof is similar to the proof of congruence of BARTC with respect to FR hp-bisimulation equivalence, hence, we omit it here. □

**Theorem 6.13** (Soundness of BARTC modulo FR hhp-bisimulation equivalence). *Let $x$ and $y$ be BARTC terms. If $BARTC \vdash x = y$, then $x \sim^{fr}_{hhp} y$.*

*Proof.* Since FR hhp-bisimulation $\sim^{fr}_{hhp}$ is both an equivalent and a congruent relation, we only need to check if each axiom in Table 6.1 is sound modulo FR hhp-bisimulation equivalence.

The proof is similar to the proof of soundness of BARTC modulo FR hp-bisimulation equivalence, hence, we omit it here. □

**Theorem 6.14** (Completeness of BARTC modulo FR hhp-bisimulation equivalence). *Let $p$ and $q$ be closed BARTC terms, if $p \sim^{fr}_{hhp} q$, then $p = q$.*

*Proof.* The proof is similar to the proof of BARTC modulo FR hp-bisimulation equivalence, hence, we omit it here. □

## 6.2 Algebra for parallelism in reversible true concurrency

In this section, we will discuss parallelism in reversible true concurrency. The resulted algebra is called Algebra for Parallelism in Reversible True Concurrency, abbreviated APRTC.

### 6.2.1 Parallelism

The forward transition rules for parallelism $\parallel$ are shown in Table 6.7, and the reverse transition rules for $\parallel$ are shown in Table 6.8.

The forward and reverse transition rules of communication $\mid$ are shown in Table 6.9 and Table 6.10.

The conflict elimination is also captured by two auxiliary operators, the unary conflict elimination operator $\Theta$ and the binary unless operator $\triangleleft$. The forward and reverse transition rules for $\Theta$ and $\triangleleft$ are expressed by the ten transition rules in Table 6.11 and Table 6.12.

**Theorem 6.15** (Congruence theorem of APRTC). *FR truly concurrent bisimulation equivalences $\sim^{fr}_{p}$, $\sim^{fr}_{s}$, $\sim^{fr}_{hp}$, and $\sim^{fr}_{hhp}$ are all congruences with respect to APRTC.*

*Proof.* (1) Case FR pomset bisimulation equivalence $\sim^{fr}_{p}$.

**TABLE 6.9** Forward transition rules of communication operator $\mid$.

| | |
|---|---|
| $\dfrac{x\xrightarrow{e_1}e_1[m]\quad y\xrightarrow{e_2}e_2[m]}{x\mid y\xrightarrow{\gamma(e_1,e_2)}\gamma(e_1,e_2)[m]}$ | $\dfrac{x\xrightarrow{e_1}x'\quad y\xrightarrow{e_2}e_2[m]}{x\mid y\xrightarrow{\gamma(e_1,e_2)}\gamma(e_1,e_2)[m]\cdot x'}$ |
| $\dfrac{x\xrightarrow{e_1}e_1[m]\quad y\xrightarrow{e_2}y'}{x\mid y\xrightarrow{\gamma(e_1,e_2)}\gamma(e_1,e_2)[m]\cdot y'}$ | $\dfrac{x\xrightarrow{e_1}x'\quad y\xrightarrow{e_2}y'}{x\mid y\xrightarrow{\gamma(e_1,e_2)}\gamma(e_1,e_2)[m]\cdot x'\between y'}$ |

**TABLE 6.10** Reverse transition rules of communication operator $\mid$.

| | |
|---|---|
| $\dfrac{x\overset{e_1[m]}{\twoheadrightarrow}e_1\quad y\overset{e_2[m]}{\twoheadrightarrow}e_2}{x\mid y\overset{\gamma(e_1,e_2)[m]}{\twoheadrightarrow}\gamma(e_1,e_2)}$ | $\dfrac{x\overset{e_1[m]}{\twoheadrightarrow}x'\quad y\overset{e_2[m]}{\twoheadrightarrow}e_2}{x\mid y\overset{\gamma(e_1,e_2)[m]}{\twoheadrightarrow}\gamma(e_1,e_2)\cdot x'}$ |
| $\dfrac{x\overset{e_1[m]}{\twoheadrightarrow}e_1\quad y\overset{e_2[m]}{\twoheadrightarrow}y'}{x\mid y\overset{\gamma(e_1,e_2)[m]}{\twoheadrightarrow}\gamma(e_1,e_2)\cdot y'}$ | $\dfrac{x\overset{e_1[m]}{\twoheadrightarrow}x'\quad y\overset{e_2[m]}{\twoheadrightarrow}y'}{x\mid y\overset{\gamma(e_1,e_2)[m]}{\twoheadrightarrow}\gamma(e_1,e_2)\cdot x'\between y'}$ |

**TABLE 6.11** Forward transition rules of conflict elimination.

| | |
|---|---|
| $\dfrac{x\xrightarrow{e_1}e_1[m]\quad(\sharp(e_1,e_2))}{\Theta(x)\xrightarrow{e_1}e_1[m]}$ | $\dfrac{x\xrightarrow{e_2}e_2[n]\quad(\sharp(e_1,e_2))}{\Theta(x)\xrightarrow{e_2}e_2[n]}$ |
| $\dfrac{x\xrightarrow{e_1}x'\quad(\sharp(e_1,e_2))}{\Theta(x)\xrightarrow{e_1}\Theta(x')}$ | $\dfrac{x\xrightarrow{e_2}x'\quad(\sharp(e_1,e_2))}{\Theta(x)\xrightarrow{e_2}\Theta(x')}$ |
| $\dfrac{x\xrightarrow{e_1}e_1[m]\quad y\nrightarrow^{e_2}\quad(\sharp(e_1,e_2))}{x\triangleleft y\xrightarrow{\tau}\surd}$ | $\dfrac{x\xrightarrow{e_1}x'\quad y\nrightarrow^{e_2}\quad(\sharp(e_1,e_2))}{x\triangleleft y\xrightarrow{\tau}x'}$ |
| $\dfrac{x\xrightarrow{e_1}e_1[m]\quad y\nrightarrow^{e_3}\quad(\sharp(e_1,e_2),e_2\leq e_3)}{x\triangleleft y\xrightarrow{\tau}\surd}$ | $\dfrac{x\xrightarrow{e_1}x'\quad y\nrightarrow^{e_3}\quad(\sharp(e_1,e_2),e_2\leq e_3)}{x\triangleleft y\xrightarrow{\tau}x'}$ |
| $\dfrac{x\xrightarrow{e_3}e_3[l]\quad y\nrightarrow^{e_2}\quad(\sharp(e_1,e_2),e_1\leq e_3)}{x\triangleleft y\xrightarrow{\tau}\surd}$ | $\dfrac{x\xrightarrow{e_3}x'\quad y\nrightarrow^{e_2}\quad(\sharp(e_1,e_2),e_1\leq e_3)}{x\triangleleft y\xrightarrow{\tau}x'}$ |

**TABLE 6.12** Reverse transition rules of conflict elimination.

| | |
|---|---|
| $\dfrac{x\overset{e_1[m]}{\twoheadrightarrow}e_1\quad(\sharp(e_1,e_2))}{\Theta(x)\overset{e_1[m]}{\twoheadrightarrow}e_1}$ | $\dfrac{x\overset{e_2[n]}{\twoheadrightarrow}e_2\quad(\sharp(e_1,e_2))}{\Theta(x)\overset{e_2[n]}{\twoheadrightarrow}e_2}$ |
| $\dfrac{x\overset{e_1[m]}{\twoheadrightarrow}x'\quad(\sharp(e_1,e_2))}{\Theta(x)\overset{e_1[m]}{\twoheadrightarrow}\Theta(x')}$ | $\dfrac{x\overset{e_2[n]}{\twoheadrightarrow}x'\quad(\sharp(e_1,e_2))}{\Theta(x)\overset{e_2[n]}{\twoheadrightarrow}\Theta(x')}$ |
| $\dfrac{x\overset{e_1[m]}{\twoheadrightarrow}e_1\quad y\overset{e_2[n]}{\not\twoheadrightarrow}\quad(\sharp(e_1,e_2))}{x\triangleleft y\overset{\tau}{\twoheadrightarrow}\surd}$ | $\dfrac{x\overset{e_1[m]}{\twoheadrightarrow}x'\quad y\overset{e_2[n]}{\not\twoheadrightarrow}\quad(\sharp(e_1,e_2))}{x\triangleleft y\overset{\tau}{\twoheadrightarrow}x'}$ |
| $\dfrac{x\overset{e_1[m]}{\twoheadrightarrow}e_1\quad y\overset{e_3[l]}{\not\twoheadrightarrow}\quad(\sharp(e_1,e_2),e_2\geq e_3)}{x\triangleleft y\overset{\tau}{\twoheadrightarrow}\surd}$ | $\dfrac{x\overset{e_1[m]}{\twoheadrightarrow}x'\quad y\overset{e_3[l]}{\not\twoheadrightarrow}\quad(\sharp(e_1,e_2),e_2\geq e_3)}{x\triangleleft y\overset{\tau}{\twoheadrightarrow}x'}$ |
| $\dfrac{x\overset{e_3[l]}{\twoheadrightarrow}e_3\quad y\overset{e_2[n]}{\not\twoheadrightarrow}\quad(\sharp(e_1,e_2),e_1\geq e_3)}{x\triangleleft y\overset{\tau}{\twoheadrightarrow}\surd}$ | $\dfrac{x\overset{e_3[l]}{\twoheadrightarrow}x'\quad y\overset{e_2[n]}{\not\twoheadrightarrow}\quad(\sharp(e_1,e_2),e_1\geq e_3)}{x\triangleleft y\overset{\tau}{\twoheadrightarrow}x'}$ |

- Case parallel operator $\parallel$. Let $x_1, x_2$ and $y_1, y_2$ be APRTC processes, and $x_1 \sim_p^{fr} y_1$, $x_2 \sim_p^{fr} y_2$, it is sufficient to prove that $x_1 \parallel x_2 \sim_p^{fr} y_1 \parallel y_2$.

By the definition of FR pomset bisimulation $\sim_p^{fr}$ (Definition 2.68), $x_1 \sim_p^{fr} y_1$ means that

$$x_1 \xrightarrow{X_1} x_1' \quad y_1 \xrightarrow{Y_1} y_1'$$

with $X_1 \subseteq x_1$, $Y_1 \subseteq y_1$, $X_1 \sim Y_1$ and $x_1' \sim_p^{fr} y_1'$. The meaning of $x_2 \sim_p^{fr} y_2$ is similar.
By the forward transition rules for parallel operator $\parallel$ in Table 6.7, we can obtain

$$x_1 \parallel x_2 \xrightarrow{\{X_1,X_2\}} X_1[\mathcal{K}] \parallel X_2[\mathcal{K}] \quad y_1 \parallel y_2 \xrightarrow{\{Y_1,Y_2\}} Y_1[\mathcal{J}] \parallel Y_2[\mathcal{J}]$$

$$x_1 \parallel x_2 \xtwoheadrightarrow{\{X_1[\mathcal{K}],X_2[\mathcal{K}]\}} X_1 \parallel X_2 \quad y_1 \parallel y_2 \xtwoheadrightarrow{\{Y_1[\mathcal{J}],Y_2[\mathcal{J}]\}} Y_1 \parallel Y_2$$

with $X_1 \subseteq x_1$, $Y_1 \subseteq y_1$, $X_2 \subseteq x_2$, $Y_2 \subseteq y_2$, $X_1 \sim Y_1$, and $X_2 \sim Y_2$, and the assumptions $X_1[\mathcal{K} \parallel X_2[\mathcal{K}]] \sim_p^{fr} Y_1[\mathcal{J}] \parallel Y_2[\mathcal{J}]$, and $X_1 \parallel X_2 \sim_p^{fr} Y_1 \parallel Y_2$, hence, we obtain $x_1 \parallel x_2 \sim_p^{fr} y_1 \parallel y_2$, as desired.
Or, we can obtain

$$x_1 \parallel x_2 \xrightarrow{\{X_1,X_2\}} x_1' \parallel X_2[\mathcal{K}] \quad y_1 \parallel y_2 \xrightarrow{\{Y_1,Y_2\}} y_1' \parallel Y_2[\mathcal{J}]$$

$$x_1 \parallel x_2 \xtwoheadrightarrow{\{X_1[\mathcal{K}],X_2[\mathcal{K}]\}} x_1' \parallel X_2 \quad y_1 \parallel y_2 \xtwoheadrightarrow{\{Y_1[\mathcal{J}],Y_2[\mathcal{J}]\}} y_1' \parallel Y_2$$

with $X_1 \subseteq x_1$, $Y_1 \subseteq y_1$, $X_2 \subseteq x_2$, $Y_2 \subseteq y_2$, $X_1 \sim Y_1$, $X_2 \sim Y_2$, and the assumptions $x_1' \parallel X_2[\mathcal{K}]] \sim_p^{fr} y_1' \parallel Y_2[\mathcal{J}]$ and $x_1' \parallel X_2 \sim_p^{fr} y_1' \parallel Y_2$, hence, we obtain $x_1 \parallel x_2 \sim_p^{fr} y_1 \parallel y_2$, as desired.
Or, we can obtain

$$x_1 \parallel x_2 \xrightarrow{\{X_1,X_2\}} X_1[\mathcal{K}] \parallel x_2' \quad y_1 \parallel y_2 \xrightarrow{\{Y_1,Y_2\}} Y_1[\mathcal{J}] \parallel y_2'$$

$$x_1 \parallel x_2 \xtwoheadrightarrow{\{X_1[\mathcal{K}],X_2[\mathcal{K}]\}} X_1 \parallel x_2' \quad y_1 \parallel y_2 \xtwoheadrightarrow{\{Y_1[\mathcal{J}],Y_2[\mathcal{J}]\}} Y_1 \parallel y_2'$$

with $X_1 \subseteq x_1$, $Y_1 \subseteq y_1$, $X_2 \subseteq x_2$, $Y_2 \subseteq y_2$, $X_1 \sim Y_1$, $X_2 \sim Y_2$, and the assumptions $X_1[\mathcal{K} \parallel x_2' \sim_p^{fr} Y_1[\mathcal{J}] \parallel y_2'$ and $X_1 \parallel x_2' \sim_p^{fr} Y_1 \parallel y_2'$, hence, we obtain $x_1 \parallel x_2 \sim_p^{fr} y_1 \parallel y_2$, as desired.
Or, we can obtain

$$x_1 \parallel x_2 \xrightarrow{\{X_1,X_2\}} x_1' \between x_2' \quad y_1 \parallel y_2 \xrightarrow{\{Y_1,Y_2\}} y_1' \between y_2'$$

$$x_1 \parallel x_2 \xtwoheadrightarrow{\{X_1[\mathcal{K}],X_2[\mathcal{K}]\}} x_1' \between x_2' \quad y_1 \parallel y_2 \xtwoheadrightarrow{\{Y_1[\mathcal{J}],Y_2[\mathcal{J}]\}} y_1' \between y_2'$$

with $X_1 \subseteq x_1$, $Y_1 \subseteq y_1$, $X_2 \subseteq x_2$, $Y_2 \subseteq y_2$, $X_1 \sim Y_1$, $X_2 \sim Y_2$, and the assumption $x_1' \between x_2' \sim_p^{fr} y_1' \between y_2'$, hence, we obtain $x_1 \parallel x_2 \sim_p^{fr} y_1 \parallel y_2$, as desired.

- Case communication operator $\mid$. It can be proved similarly to the case of parallel operator $\parallel$, hence, we omit it here. Note that a communication is defined between two single communicating events.
- Case conflict elimination operator $\Theta$. It can be proved similarly to the above cases, hence, we omit it here. Note that the conflict elimination operator $\Theta$ is a unary operator.
- Case unless operator $\triangleleft$. It can be proved similarly to the case of parallel operator $\parallel$, hence, we omit it here. Note that a conflict relation is defined between two single events.

(2) The cases of FR step bisimulation $\sim_s^{fr}$, FR hp-bisimulation $\sim_{hp}^{fr}$, and FR hhp-bisimulation $\sim_{hhp}^{fr}$ can be proven similarly, however, we omit them here. □

### 6.2.2 Axiom system of parallelism

**Definition 6.16** (Basic terms of APRTC). The set of basic terms of APRTC, $\mathcal{B}(APRTC)$, is inductively defined as follows:

1. $\mathbb{E} \subset \mathcal{B}(APRTC)$;
2. if $e \in \mathbb{E}, t \in \mathcal{B}(APRTC)$ then $e \cdot t \in \mathcal{B}(APRTC)$;
3. if $e[m] \in \mathbb{E}, t \in \mathcal{B}(APRTC)$ then $t \cdot e[m] \in \mathcal{B}(APRTC)$;
4. if $t, s \in \mathcal{B}(APRTC)$ then $t + s \in \mathcal{B}(APRTC)$;

5. if $t, s \in \mathcal{B}(APRTC)$ then $t \parallel s \in \mathcal{B}(APRTC)$.

We design the axioms of parallelism in Table 6.13, including algebraic laws for parallel operator $\parallel$, communication operator $\mid$, conflict elimination operator $\Theta$, unless operator $\triangleleft$, and also the whole parallel operator $\between$. Since the communication between two communicating events in different parallel branches may cause deadlock (a state of inactivity), which is caused by mismatch of two communicating events or the imperfectness of the communication channel, we introduce a new constant $\delta$ to denote the deadlock, and let the atomic event $e \in \mathbb{E} \cup \{\delta\}$.

Based on the definition of basic terms for APRTC (see Definition 6.16) and axioms of parallelism (see Table 6.13), we can prove the elimination theorem of parallelism.

**Theorem 6.17** (Elimination theorem of FR parallelism). *Let $p$ be a closed APRTC term. Then, there is a basic APRTC term $q$ such that $APRTC \vdash p = q$.*

*Proof.* (1) First, suppose that the following ordering on the signature of APRTC is defined: $\parallel > \cdot > +$ and the symbol $\cdot$ is given the lexicographical status for the first argument, then for each rewrite rule $p \to q$ in Table 6.14 relation $p >_{lpo} q$ can easily be proved. We obtain that the term rewrite system shown in Table 6.14 is strongly normalizing, for it has finitely many rewriting rules, and $>$ is a well-founded ordering on the signature of APRTC, and if $s >_{lpo} t$, for each rewriting rule $s \to t$ is in Table 6.14.

(2) Then, we prove that the normal forms of closed APRTC terms are basic APRTC terms.

Suppose that $p$ is a normal form of some closed APRTC term and suppose that $p$ is not a basic APRTC term. Let $p'$ denote the smallest sub-term of $p$ that is not a basic APRTC term. It implies that each sub-term of $p'$ is a basic APRTC term. Then, we prove that $p$ is not a term in normal form. It is sufficient to induct on the structure of $p'$:

- Case $p' \equiv e$ or $e[m], e \in \mathbb{E}$. $p'$ is a basic APRTC term, which contradicts the assumption that $p'$ is not a basic APRTC term, hence, this case should not occur.
- Case $p' \equiv p_1 \cdot p_2$. By induction on the structure of the basic APRTC term $p_1$:
  - Subcase $p_1 \in \mathbb{E}$. $p'$ would be a basic APRTC term, which contradicts the assumption that $p'$ is not a basic APRTC term;
  - Subcase $p_1 \equiv e \cdot p_1'$. $RR5$ rewriting rule in Table 6.2 can be applied. Hence, $p$ is not a normal form;
  - Subcase $p_1 \equiv p_1' \cdot e[m]$. $RA5$ rewriting rule in Table 6.2 can be applied. Hence, $p$ is not a normal form;
  - Subcase $p_1 \equiv p_1' + p_1''$. $RA4$ rewriting rule in Table 6.2 can be applied. Hence, $p$ is not a normal form;
  - Subcase $p_1 \equiv p_1' \parallel p_1''$. $p'$ would be a basic APRTC term, which contradicts the assumption that $p'$ is not a basic APRTC term;
  - Subcase $p_1 \equiv p_1' \mid p_1''$. $RC11$ and $RRC11$ rewrite rules in Table 6.14 can be applied. Hence, $p$ is not a normal form;
  - Subcase $p_1 \equiv \Theta(p_1')$. $RCE19$, $RRCE19$ and $RCE20$ rewrite rules in Table 6.14 can be applied. Hence, $p$ is not a normal form.
- Case $p' \equiv p_1 + p_2$. By induction on the structure of the basic APRTC terms both $p_1$ and $p_2$, all subcases will lead to that $p'$ would be a basic APRTC term, which contradicts the assumption that $p'$ is not a basic APRTC term.
- Case $p' \equiv p_1 \parallel p_2$. By induction on the structure of the basic APRTC terms both $p_1$ and $p_2$, all subcases will lead to that $p'$ would be a basic APRTC term, which contradicts the assumption that $p'$ is not a basic APRTC term.
- Case $p' \equiv p_1 \mid p_2$. By induction on the structure of the basic APRTC terms both $p_1$ and $p_2$, all subcases will lead to that $p'$ would be a basic APRTC term, which contradicts the assumption that $p'$ is not a basic APRTC term.
- Case $p' \equiv \Theta(p_1)$. By induction on the structure of the basic APRTC term $p_1$, $RCE19 - RCE24$ rewrite rules in Table 6.14 can be applied. Hence, $p$ is not a normal form.
- Case $p' \equiv p_1 \triangleleft p_2$. By induction on the structure of the basic APRTC terms both $p_1$ and $p_2$, all subcases will lead to that $p'$ would be a basic APRTC term, which contradicts the assumption that $p'$ is not a basic APRTC term. □

### 6.2.3 Structured operational semantics of parallelism

**Theorem 6.18** (Generalization of the algebra for parallelism with respect to BARTC). *The algebra for parallelism is a generalization of BARTC.*

*Proof.* This follows from the following three facts:

1. The transition rules of BARTC in Section 6.1 are all source dependent;
2. The sources of the transition rules for the algebra for parallelism contain an occurrence of $\between$, or $\parallel$, or $\mid$, or $\Theta$, or $\triangleleft$;

**TABLE 6.13** Axioms of parallelism.

| No. | Axiom |
|---|---|
| $A6$ | $x+\delta=x$ |
| $A7$ | $\delta\cdot x=\delta(\text{Std(x)})$ |
| $RA7$ | $x\cdot\delta=\delta(\text{NStd(x)})$ |
| $P1$ | $x\between y=x\parallel y+x\mid y$ |
| $P2$ | $x\parallel y=y\parallel x$ |
| $P3$ | $(x\parallel y)\parallel z=x\parallel(y\parallel z)$ |
| $P4$ | $e_1\parallel(e_2\cdot y)=(e_1\parallel e_2)\cdot y$ |
| $RP4$ | $e_1[m]\parallel(y\cdot e_2[m])=y\cdot(e_1[m]\parallel e_2[m])$ |
| $P5$ | $(e_1\cdot x)\parallel e_2=(e_1\parallel e_2)\cdot x$ |
| $RP5$ | $(x\cdot e_1[m])\parallel e_2[m]=x\cdot(e_1[m]\parallel e_2[m])$ |
| $P6$ | $(e_1\cdot x)\parallel(e_2\cdot y)=(e_1\parallel e_2)\cdot(x\between y)$ |
| $RP6$ | $(x\cdot e_1[m])\parallel(y\cdot e_2[m])=(x\between y)\cdot(e_1[m]\parallel e_2[m])$ |
| $P7$ | $(x+y)\parallel z=(x\parallel z)+(y\parallel z)$ |
| $P8$ | $x\parallel(y+z)=(x\parallel y)+(x\parallel z)$ |
| $P9$ | $\delta\parallel x=\delta$ |
| $P10$ | $x\parallel\delta=\delta$ |
| $C11$ | $e_1\mid e_2=\gamma(e_1,e_2)$ |
| $RC11$ | $e_1[m]\mid e_2[m]=\gamma(e_1,e_2)[m]$ |
| $C12$ | $e_1\mid(e_2\cdot y)=\gamma(e_1,e_2)\cdot y$ |
| $RC12$ | $e_1[m]\mid(y\cdot e_2[m])=y\cdot\gamma(e_1,e_2)[m]$ |
| $C13$ | $(e_1\cdot x)\mid e_2=\gamma(e_1,e_2)\cdot x$ |
| $RC13$ | $(x\cdot e_1[m])\mid e_2[m]=x\cdot\gamma(e_1,e_2)[m]$ |
| $C14$ | $(e_1\cdot x)\mid(e_2\cdot y)=\gamma(e_1,e_2)\cdot(x\between y)$ |
| $RC14$ | $(x\cdot e_1[m])\mid(y\cdot e_2[m])=(x\between y)\cdot\gamma(e_1,e_2)[m]$ |
| $C15$ | $(x+y)\mid z=(x\mid z)+(y\mid z)$ |
| $C16$ | $x\mid(y+z)=(x\mid y)+(x\mid z)$ |
| $C17$ | $\delta\mid x=\delta$ |
| $C18$ | $x\mid\delta=\delta$ |
| $CE19$ | $\Theta(e)=e$ |
| $RCE19$ | $\Theta(e[m])=e[m]$ |
| $CE20$ | $\Theta(\delta)=\delta$ |
| $CE21$ | $\Theta(x+y)=\Theta(x)+\Theta(y)$ |
| $CE22$ | $\Theta(x\cdot y)=\Theta(x)\cdot\Theta(y)$ |
| $CE23$ | $\Theta(x\parallel y)=((\Theta(x)\triangleleft y)\parallel y)+((\Theta(y)\triangleleft x)\parallel x)$ |
| $CE24$ | $\Theta(x\mid y)=((\Theta(x)\triangleleft y)\mid y)+((\Theta(y)\triangleleft x)\mid x)$ |
| $U25$ | $(\sharp(e_1,e_2))\quad e_1\triangleleft e_2=\tau$ |
| $RU25$ | $(\sharp(e_1[m],e_2[n]))\quad e_1[m]\triangleleft e_2[n]=\tau$ |
| $U26$ | $(\sharp(e_1,e_2),e_2\leq e_3)\quad e_1\triangleleft e_3=\tau$ |
| $RU26$ | $(\sharp(e_1[m],e_2[n]),e_2[n]\geq e_3[l])\quad e_1[m]\triangleleft e_3[l]=\tau$ |
| $U27$ | $(\sharp(e_1,e_2),e_2\leq e_3)\quad e3\triangleleft e_1=\tau$ |
| $RU27$ | $(\sharp(e_1[m],e_2[n]),e_2[n]\geq e_3[l])\quad e3[l]\triangleleft e_1[m]=\tau$ |
| $U28$ | $e\triangleleft\delta=e$ |
| $U29$ | $\delta\triangleleft e=\delta$ |

*continued on next page*

**TABLE 6.13** *(continued)*

| No. | Axiom |
|---|---|
| $U30$ | $(x+y) \triangleleft z = (x \triangleleft z) + (y \triangleleft z)$ |
| $U31$ | $(x \cdot y) \triangleleft z = (x \triangleleft z) \cdot (y \triangleleft z)$ |
| $U32$ | $(x \parallel y) \triangleleft z = (x \triangleleft z) \parallel (y \triangleleft z)$ |
| $U33$ | $(x \mid y) \triangleleft z = (x \triangleleft z) \mid (y \triangleleft z)$ |
| $U34$ | $x \triangleleft (y+z) = (x \triangleleft y) \triangleleft z$ |
| $U35$ | $x \triangleleft (y \cdot z) = (x \triangleleft y) \triangleleft z$ |
| $U36$ | $x \triangleleft (y \parallel z) = (x \triangleleft y) \triangleleft z$ |
| $U37$ | $x \triangleleft (y \mid z) = (x \triangleleft y) \triangleleft z$ |

**3.** The transition rules of APRTC are all source dependent.

Hence, the algebra for parallelism is a generalization of BARTC, that is, BARTC is an embedding of the algebra for parallelism, as desired. □

**Theorem 6.19** (Soundness of parallelism modulo FR step bisimulation equivalence). *Let $x$ and $y$ be APRTC terms. If $APRTC \vdash x = y$, then $x \sim_s^{fr} y$.*

*Proof.* Since FR step bisimulation $\sim_s^{fr}$ is both an equivalent and a congruent relation with respect to the operators $\between$, $\parallel$, $\mid$, $\Theta$, and $\triangleleft$, we only need to check if each axiom in Table 6.13 is sound modulo FR step bisimulation equivalence.

The proof is similar to the proof of soundness of BARTC modulo FR step bisimulation equivalence, hence, we omit it here. □

**Theorem 6.20** (Completeness of parallelism modulo FR step bisimulation equivalence). *Let $p$ and $q$ be closed APRTC terms, if $p \sim_s^{fr} q$, then $p = q$.*

*Proof.* First, by the elimination theorem of APRTC (see Theorem 6.17), we know that for each closed APRTC term $p$, there exists a closed basic APRTC term $p'$, such that $APRTC \vdash p = p'$, hence, we only need to consider closed basic APRTC terms.

The basic terms modulo associativity and commutativity (AC) of conflict $+$ (defined by axioms $A1$ and $A2$ in Table 6.1) and associativity and commutativity (AC) of parallel $\parallel$ (defined by axioms $P2$ and $P3$ in Table 6.13), and these equivalences are denoted by $=_{AC}$. Then, each equivalence class $s$ modulo AC of $+$ and $\parallel$ has the following normal form

$$s_1 + \cdots + s_k$$

with each $s_i$ either an atomic event or of the form

$$t_1 \cdot \cdots \cdot t_m$$

with each $t_j$ either an atomic event or of the form

$$u_1 \parallel \cdots \parallel u_n$$

with each $u_l$ an atomic event, and each $s_i$ is called the summand of $s$.

Now, we prove that for normal forms $n$ and $n'$, if $n \sim_s^{fr} n'$, then $n =_{AC} n'$. It is sufficient to induct on the sizes of $n$ and $n'$.

- Consider a summand $e$ of $n$. Then, $n \xrightarrow{e} e[m]$, hence, $n \sim_s^{fr} n'$ implies $n' \xrightarrow{e} e[m]$, meaning that $n'$ also contains the summand $e$.
- Consider a summand $e[m]$ of $n$. Then, $n \xtwoheadrightarrow{e[m]} e$, hence, $n \sim_s^{fr} n'$ implies $n' \xtwoheadrightarrow{e[m]} e$, meaning that $n'$ also contains the summand $e[m]$.
- Consider a summand $t_1 \cdot t_2$ of $n$,
  - if $t_1 \equiv e'$, then $n \xrightarrow{e'} e'[m] \cdot t_2$, hence, $n \sim_s^{fr} n'$ implies $n' \xrightarrow{e'} e'[m] \cdot t_2'$ with $e'[m] \cdot t_2 \sim_s^{fr} e'[m] \cdot t_2'$, meaning that $n'$ contains a summand $e' \cdot t_2'$. Since $t_2$ and $t_2'$ are normal forms and have sizes smaller than $n$ and $n'$, by the induction hypotheses if $t_2 \sim_s^{fr} t_2'$, then $t_2 =_{AC} t_2'$;

**TABLE 6.14** Term rewrite system of APRTC.

| No. | Rewriting Rule |
|---|---|
| $RA6$ | $x + \delta \rightarrow x$ |
| $RA7$ | $\delta \cdot x \rightarrow \delta$ |
| $RRA7$ | $x \cdot \delta \rightarrow \delta$ |
| $RP1$ | $x \between y \rightarrow x \parallel y + x \mid y$ |
| $RP2$ | $x \parallel y \rightarrow y \parallel x$ |
| $RP3$ | $(x \parallel y) \parallel z \rightarrow x \parallel (y \parallel z)$ |
| $RP4$ | $e_1 \parallel (e_2 \cdot y) \rightarrow (e_1 \parallel e_2) \cdot y$ |
| $RRP4$ | $e_1[m] \parallel (y \cdot e_2[m]) \rightarrow y \cdot (e_1[m] \parallel e_2[m])$ |
| $RP5$ | $(e_1 \cdot x) \parallel e_2 \rightarrow (e_1 \parallel e_2) \cdot x$ |
| $RRP5$ | $(x \cdot e_1[m]) \parallel e_2[m] \rightarrow x \cdot (e_1[m] \parallel e_2[m])$ |
| $RP6$ | $(e_1 \cdot x) \parallel (e_2 \cdot y) \rightarrow (e_1 \parallel e_2) \cdot (x \between y)$ |
| $RP6$ | $(x \cdot e_1[m]) \parallel (y \cdot e_2[m]) \rightarrow (x \between y) \cdot (e_1[m] \parallel e_2[m])$ |
| $RP7$ | $(x + y) \parallel z \rightarrow (x \parallel z) + (y \parallel z)$ |
| $RP8$ | $x \parallel (y + z) \rightarrow (x \parallel y) + (x \parallel z)$ |
| $RP9$ | $\delta \parallel x \rightarrow \delta$ |
| $RP10$ | $x \parallel \delta \rightarrow \delta$ |
| $RC11$ | $e_1 \mid e_2 \rightarrow \gamma(e_1, e_2)$ |
| $RRC11$ | $e_1[m] \mid e_2[m] \rightarrow \gamma(e_1, e_2)[m]$ |
| $RC12$ | $e_1 \mid (e_2 \cdot y) \rightarrow \gamma(e_1, e_2) \cdot y$ |
| $RRC12$ | $e_1[m] \mid (y \cdot e_2[m]) \rightarrow y \cdot \gamma(e_1, e_2)[m]$ |
| $RC13$ | $(e_1 \cdot x) \mid e_2 \rightarrow \gamma(e_1, e_2) \cdot x$ |
| $RRC13$ | $(x \cdot e_1[m]) \mid e_2[m] \rightarrow x \cdot \gamma(e_1, e_2)[m]$ |
| $RC14$ | $(e_1 \cdot x) \mid (e_2 \cdot y) \rightarrow \gamma(e_1, e_2) \cdot (x \between y)$ |
| $RRC14$ | $(x \cdot e_1[m]) \mid (y \cdot e_2[m]) \rightarrow (x \between y) \cdot \gamma(e_1, e_2)[m]$ |
| $RC15$ | $(x + y) \mid z \rightarrow (x \mid z) + (y \mid z)$ |
| $RC16$ | $x \mid (y + z) \rightarrow (x \mid y) + (x \mid z)$ |
| $RC17$ | $\delta \mid x \rightarrow \delta$ |
| $RC18$ | $x \mid \delta \rightarrow \delta$ |
| $RCE19$ | $\Theta(e) \rightarrow e$ |
| $RRCE19$ | $\Theta(e[m]) \rightarrow e[m]$ |
| $RCE20$ | $\Theta(\delta) \rightarrow \delta$ |
| $RCE21$ | $\Theta(x + y) \rightarrow \Theta(x) + \Theta(y)$ |
| $RCE22$ | $\Theta(x \cdot y) \rightarrow \Theta(x) \cdot \Theta(y)$ |
| $RCE23$ | $\Theta(x \parallel y) \rightarrow ((\Theta(x) \triangleleft y) \parallel y) + ((\Theta(y) \triangleleft x) \parallel x)$ |
| $RCE24$ | $\Theta(x \mid y) \rightarrow ((\Theta(x) \triangleleft y) \mid y) + ((\Theta(y) \triangleleft x) \mid x)$ |
| $RU25$ | $(\sharp(e_1, e_2)) \quad e_1 \triangleleft e_2 \rightarrow \tau$ |
| $RRU25$ | $(\sharp(e_1[m], e_2[n])) \quad e_1[m] \triangleleft e_2[n] \rightarrow \tau$ |
| $RU26$ | $(\sharp(e_1, e_2), e_2 \leq e_3) \quad e_1 \triangleleft e_3 \rightarrow \tau$ |
| $RRU26$ | $(\sharp(e_1[m], e_2[n]), e_2[n] \geq e_3[l]) \quad e_1[m] \triangleleft e_3[l] \rightarrow \tau$ |
| $RU27$ | $(\sharp(e_1, e_2), e_2 \leq e_3) \quad e3 \triangleleft e_1 \rightarrow \tau$ |
| $RRU27$ | $(\sharp(e_1[m], e_2[n]), e_2[n] \geq e_3[l]) \quad e3[l] \triangleleft e_1[m] \rightarrow \tau$ |
| $RU28$ | $e \triangleleft \delta \rightarrow e$ |
| $RU29$ | $\delta \triangleleft e \rightarrow \delta$ |

continued on next page

**TABLE 6.14** *(continued)*

| No. | Rewriting Rule |
|---|---|
| $RU30$ | $(x+y)\triangleleft z \to (x\triangleleft z)+(y\triangleleft z)$ |
| $RU31$ | $(x\cdot y)\triangleleft z \to (x\triangleleft z)\cdot(y\triangleleft z)$ |
| $RU32$ | $(x \parallel y)\triangleleft z \to (x\triangleleft z)\parallel(y\triangleleft z)$ |
| $RU33$ | $(x \mid y)\triangleleft z \to (x\triangleleft z)\mid(y\triangleleft z)$ |
| $RU34$ | $x\triangleleft(y+z) \to (x\triangleleft y)\triangleleft z$ |
| $RU35$ | $x\triangleleft(y\cdot z) \to (x\triangleleft y)\triangleleft z$ |
| $RU36$ | $x\triangleleft(y\parallel z) \to (x\triangleleft y)\triangleleft z$ |
| $RU37$ | $x\triangleleft(y\mid z) \to (x\triangleleft y)\triangleleft z$ |

- if $t_2 \equiv e'[m]$, then $n \xrightarrow{e'[m]}\!\!\!\!\rightarrow t_1\cdot e'$, hence, $n \sim_s^{fr} n'$ implies $n' \xrightarrow{e'[m]}\!\!\!\!\rightarrow t_1'\cdot e'$ with $t_1\cdot e' \sim_s^{fr} t_1'\cdot e'$, meaning that $n'$ contains a summand $t_1'\cdot e'$. Since $t_1$ and $t_1'$ are normal forms and have sizes smaller than $n$ and $n'$, by the induction hypotheses if $t_1 \sim_s^{fr} t_1'$, then $t_1 =_{AC} t_1'$;
- if $t_1 \equiv e_1 \parallel \cdots \parallel e_n$, then $n \xrightarrow{\{e_1,\cdots,e_n\}} (e_1[m] \parallel \cdots \parallel e_n[m])\cdot t_2$, hence, $n \sim_s^{fr} n'$ implies $n' \xrightarrow{\{e_1,\cdots,e_n\}} (e_1[m] \parallel \cdots \parallel e_n[m])\cdot t_2'$ with $t_2 \sim_s^{fr} t_2'$, meaning that $n'$ contains a summand $(e_1 \parallel \cdots \parallel e_n)\cdot t_2'$. Since $t_2$ and $t_2'$ are normal forms and have sizes smaller than $n$ and $n'$, by the induction hypotheses if $t_2 \sim_s^{fr} t_2'$, then $t_2 =_{AC} t_2'$.
- if $t_2 \equiv e_1[m] \parallel \cdots \parallel e_n[m]$, then $n \xrightarrow{\{e_1[m],\cdots,e_n[m]\}}\!\!\!\!\rightarrow (t_1\cdot e_1 \parallel \cdots \parallel e_n)$, hence, $n \sim_s^{fr} n'$ implies $n' \xrightarrow{\{e_1[m],\cdots,e_n[m]\}}\!\!\!\!\rightarrow t_1'\cdot(e_1 \parallel \cdots \parallel e_n)$ with $t_1 \sim_s^{fr} t_1'$, meaning that $n'$ contains a summand $t_1'\cdot(e_1[m] \parallel \cdots \parallel e_n[m])$. Since $t_1$ and $t_1'$ are normal forms and have sizes smaller than $n$ and $n'$, by the induction hypotheses if $t_1 \sim_s^{fr} t_1'$, then $t_1 =_{AC} t_1'$.

Hence, we obtain $n =_{AC} n'$.

Finally, let $s$ and $t$ be basic APRTC terms, and $s \sim_s^{fr} t$, there are normal forms $n$ and $n'$, such that $s = n$ and $t = n'$. The soundness theorem of parallelism modulo FR step bisimulation equivalence (see Theorem 6.19) yields $s \sim_s^{fr} n$ and $t \sim_s^{fr} n'$, hence, $n \sim_s^{fr} s \sim_s^{fr} t \sim_s^{fr} n'$. Since, if $n \sim_s^{fr} n'$, then $n =_{AC} n'$, $s = n =_{AC} n' = t$, as desired. □

**Theorem 6.21** (Soundness of parallelism modulo FR pomset bisimulation equivalence). *Let $x$ and $y$ be APRTC terms. If $APRTC \vdash x = y$, then $x \sim_p^{fr} y$.*

*Proof.* Since FR pomset bisimulation $\sim_p^{fr}$ is both an equivalent and a congruent relation with respect to the operators $\between$, $\parallel$, $\mid$, $\Theta$, and $\triangleleft$, we only need to check if each axiom in Table 6.13 is sound modulo FR pomset bisimulation equivalence.

From the definition of FR pomset bisimulation (see Definition 2.68), we know that FR pomset bisimulation is defined by pomset transitions, that are labeled by pomsets. In a pomset transition, the events in the pomset are either within causality relations (defined by $\cdot$) or in concurrency (implicitly defined by $\cdot$ and $+$, and explicitly defined by $\between$), of course, they are pairwise consistent (without conflicts). In Theorem 6.19, we have already proven the case that all events are pairwise concurrent, hence, we only need to prove the case of events in causality. Without loss of generality, we take a pomset of $P = \{e_1, e_2 : e_1 \cdot e_2\}$. Then, the pomset transition labeled by the above $P$ is just composed of one single event transition labeled by $e_1$ succeeded by another single event transition labeled by $e_2$, that is, $\xrightarrow{P} = \xrightarrow{e_1}\xrightarrow{e_2}$ or $\xrightarrow{P} = \xrightarrow{e_2[n]}\!\!\!\!\rightarrow\xrightarrow{e_1[m]}\!\!\!\!\rightarrow$.

Similarly to the proof of soundness of parallelism modulo FR step bisimulation equivalence (see Theorem 6.19), we can prove that each axiom in Table 6.13 is sound modulo FR pomset bisimulation equivalence, however, we omit them here. □

**Theorem 6.22** (Completeness of parallelism modulo FR pomset bisimulation equivalence). *Let $p$ and $q$ be closed APRTC terms, if $p \sim_p^{fr} q$, then $p = q$.*

*Proof.* The proof is similar to the proof of completeness of parallelism modulo FR step bisimulation equivalence, hence, we omit it here. □

**Theorem 6.23** (Soundness of parallelism modulo FR hp-bisimulation equivalence). *Let $x$ and $y$ be APRTC terms. If $APRTC \vdash x = y$, then $x \sim_{hp}^{fr} y$.*

**TABLE 6.15** Forward transition rules of encapsulation operator $\partial_H$.

$$\frac{x \xrightarrow{e} e[m]}{\partial_H(x) \xrightarrow{e} \partial_H(e[m])} \quad (e \notin H) \qquad \frac{x \xrightarrow{e} x'}{\partial_H(x) \xrightarrow{e} \partial_H(x')} \quad (e \notin H)$$

**TABLE 6.16** Reverse transition rules of encapsulation operator $\partial_H$.

$$\frac{x \xrightarrow{e[m]} e}{\partial_H(x) \xrightarrow{e[m]} e} \quad (e \notin H) \qquad \frac{x \xrightarrow{e} x'}{\partial_H(x) \xrightarrow{e} \partial_H(x')} \quad (e \notin H)$$

*Proof.* Since FR hp-bisimulation $\sim^{fr}_{hp}$ is both an equivalent and a congruent relation with respect to the operators $\between$, $\parallel$, $\mid$, $\Theta$, and $\triangleleft$, we only need to check if each axiom in Table 6.13 is sound modulo FR hp-bisimulation equivalence.

From the definition of FR hp-bisimulation (see Definition 2.70), we know that FR hp-bisimulation is defined on the posetal product $(C_1, f, C_2), f : C_1 \to C_2$ isomorphism. Two process terms $s$ related to $C_1$ and $t$ related to $C_2$, and $f : C_1 \to C_2$ isomorphism. Initially, $(C_1, f, C_2) = (\emptyset, \emptyset, \emptyset)$, and $(\emptyset, \emptyset, \emptyset) \in \sim^{fr}_{hp}$. When $s \xrightarrow{e} s'$ $(C_1 \xrightarrow{e} C_1')$, there will be $t \xrightarrow{e} t'$ $(C_2 \xrightarrow{e} C_2')$, and we define $f' = f[e \mapsto e]$. Also, when $s \xrightarrow{e[m]} s'$ $(C_1 \xrightarrow{e[m]} C_1')$, there will be $t \xrightarrow{e[m]} t'$ $(C_2 \xrightarrow{e[m]} C_2')$, and we define $f' = f[e[m] \mapsto e[m]]$. Then, if $(C_1, f, C_2) \in \sim^{fr}_{hp}$, then $(C_1', f', C_2') \in \sim^{fr}_{hp}$.

Similarly to the proof of soundness of parallelism modulo FR pomset bisimulation equivalence (see Theorem 6.21), we can prove that each axiom in Table 6.13 is sound modulo FR hp-bisimulation equivalence, we just need additionally to check the above conditions on FR hp-bisimulation, however, we omit them here. □

**Theorem 6.24** (Completeness of parallelism modulo FR hp-bisimulation equivalence). *Let $p$ and $q$ be closed APRTC terms, if $p \sim^{fr}_{hp} q$, then $p = q$.*

*Proof.* The proof is similar to the proof of completeness of parallelism modulo FR pomset bisimulation equivalence, hence, we omit it here. □

### 6.2.4 Encapsulation

The mismatch of two communicating events in different parallel branches can cause deadlock, hence, the deadlocks in the concurrent processes should be eliminated. Like *APTC* [9], we also introduce the unary encapsulation operator $\partial_H$ for set $H$ of atomic events, which renames all atomic events in $H$ into $\delta$. The whole algebra including parallelism for true concurrency in the above sections, deadlock $\delta$ and encapsulation operator $\partial_H$, is called Reversible Algebra for Parallelism in True Concurrency, abbreviated APRTC.

The forward transition rules of encapsulation operator $\partial_H$ are shown in Table 6.15, and the reverse transition rules of encapsulation operator $\partial_H$ are shown in Table 6.16.

Based on the transition rules for encapsulation operator $\partial_H$ in Table 6.15 and Table 6.16, we design the axioms as Table 6.17 shows.

**Theorem 6.25** (Conservativity of APRTC with respect to the algebra for parallelism). *APRTC is a conservative extension of the algebra for parallelism.*

*Proof.* This follows from the following two facts:

1. The transition rules of the algebra for parallelism in the above sections are all source dependent;
2. The sources of the transition rules for the encapsulation operator contain an occurrence of $\partial_H$.

Hence, APRTC is a conservative extension of the algebra for parallelism, as desired. □

**Theorem 6.26** (Congruence theorem of encapsulation operator $\partial_H$). *Truly concurrent bisimulation equivalences $\sim^{fr}_p$, $\sim^{fr}_s$, $\sim^{fr}_{hp}$, and $\sim^{fr}_{hhp}$ are all congruences with respect to encapsulation operator $\partial_H$.*

*Proof.* (1) Case FR pomset bisimulation equivalence $\sim^{fr}_p$.

**TABLE 6.17** Axioms of encapsulation operator.

| No. | Axiom |
|---|---|
| $D1$ | $e \notin H \quad \partial_H(e) = e$ |
| $RD1$ | $e \notin H \quad \partial_H(e[m]) = e[m]$ |
| $D2$ | $e \in H \quad \partial_H(e) = \delta$ |
| $RD2$ | $e \in H \quad \partial_H(e[m]) = \delta$ |
| $D3$ | $\partial_H(\delta) = \delta$ |
| $D4$ | $\partial_H(x + y) = \partial_H(x) + \partial_H(y)$ |
| $D5$ | $\partial_H(x \cdot y) = \partial_H(x) \cdot \partial_H(y)$ |
| $D6$ | $\partial_H(x \parallel y) = \partial_H(x) \parallel \partial_H(y)$ |

Let $x$ and $y$ be APRTC processes, and $x \sim_p^{fr} y$, it is sufficient to prove that $\partial_H(x) \sim_p^{fr} \partial_H(y)$.
By the definition of FR pomset bisimulation $\sim_p^{fr}$ (Definition 2.68), $x \sim_p^{fr} y$ means that

$$x \xrightarrow{X} x' \quad y \xrightarrow{Y} y'$$

$$x \xrightarrow{X[\mathcal{K}]} x' \quad y \xrightarrow{Y[\mathcal{J}]} y'$$

with $X \subseteq x$, $Y \subseteq y$, $X \sim Y$ and $x' \sim_p^{fr} y'$.

By the FR pomset transition rules for encapsulation operator $\partial_H$ in Table 6.15 and Table 6.16, we can obtain

$$\partial_H(x) \xrightarrow{X} \partial_H(X[\mathcal{K}])(X \not\subseteq H) \quad \partial_H(y) \xrightarrow{Y} \partial_H(Y[\mathcal{J}])(Y \not\subseteq H)$$

$$\partial_H(x) \xrightarrow{X[\mathcal{K}]} \partial_H(X)(X \not\subseteq H) \quad \partial_H(y) \xrightarrow{Y[\mathcal{J}]} \partial_H(Y)(Y \not\subseteq H)$$

with $X \subseteq x$, $Y \subseteq y$, and $X \sim Y$, and the assumptions $\partial_H(X[\mathcal{K}]) \sim_p^{fr} \partial_H(Y[\mathcal{J}])$, $\partial_H(X) \sim_p^{fr} \partial_H(Y)$, hence, we obtain $\partial_H(x) \sim_p^{fr} \partial_H(y)$, as desired.

Or, we can obtain

$$\partial_H(x) \xrightarrow{X} \partial_H(x')(X \not\subseteq H) \quad \partial_H(y) \xrightarrow{Y} \partial_H(y')(Y \not\subseteq H)$$

$$\partial_H(x) \xrightarrow{X} \partial_H(x')(X \not\subseteq H) \quad \partial_H(y) \xrightarrow{Y} \partial_H(y')(Y \not\subseteq H)$$

with $X \subseteq x$, $Y \subseteq y$, $X \sim Y$, $x' \sim_p^{fr} y'$ and the assumption $\partial_H(x') \sim_p^{fr} \partial_H(y')$, hence, we obtain $\partial_H(x) \sim_p^{fr} \partial_H(y)$, as desired.

(2) The cases of FR step bisimulation $\sim_s^{fr}$, FR hp-bisimulation $\sim_{hp}^{fr}$ and FR hhp-bisimulation $\sim_{hhp}^{fr}$ can be proven similarly, however, we omit them here. □

**Theorem 6.27** (Elimination theorem of APRTC). *Let $p$ be a closed APRTC term including the encapsulation operator $\partial_H$. Then, there is a basic APRTC term $q$ such that $APRTC \vdash p = q$.*

*Proof.* (1) First, suppose that the following ordering on the signature of APRTC is defined: $\parallel > \cdot > +$ and the symbol $\cdot$ is given the lexicographical status for the first argument, then for each rewrite rule $p \to q$ in Table 6.18 relation $p >_{lpo} q$ can easily be proved. We obtain that the term rewrite system shown in Table 6.18 is strongly normalizing, for it has finitely many rewriting rules, and $>$ is a well-founded ordering on the signature of APRTC, and if $s >_{lpo} t$, for each rewriting rule $s \to t$ is in Table 6.18.

(2) Then, we prove that the normal forms of closed APRTC terms including encapsulation operator $\partial_H$ are basic APRTC terms.

Suppose that $p$ is a normal form of some closed APRTC term and suppose that $p$ is not a basic APRTC term. Let $p'$ denote the smallest sub-term of $p$ that is not a basic APRTC term. It implies that each sub-term of $p'$ is a basic APRTC term. Then, we prove that $p$ is not a term in normal form. It is sufficient to induct on the structure of $p'$, following from Theorem 6.17, we only prove the new case $p' \equiv \partial_H(p_1)$:

**TABLE 6.18** Term rewrite system of encapsulation operator $\partial_H$.

| No. | Rewriting Rule |
|---|---|
| $RD1$ | $e \notin H \quad \partial_H(e) \rightarrow e$ |
| $RRD1$ | $e \notin H \quad \partial_H(e[m]) \rightarrow e[m]$ |
| $RD2$ | $e \in H \quad \partial_H(e) \rightarrow \delta$ |
| $RRD2$ | $e \in H \quad \partial_H(e[m]) \rightarrow \delta$ |
| $RD3$ | $\partial_H(\delta) \rightarrow \delta$ |
| $RD4$ | $\partial_H(x+y) \rightarrow \partial_H(x) + \partial_H(y)$ |
| $RD5$ | $\partial_H(x \cdot y) \rightarrow \partial_H(x) \cdot \partial_H(y)$ |
| $RD6$ | $\partial_H(x \parallel y) \rightarrow \partial_H(x) \parallel \partial_H(y)$ |

- Case $p_1 \equiv e$. The transition rules $RD1$ or $RD2$ can be applied, hence, $p$ is not a normal form;
- Case $p_1 \equiv e[m]$. The transition rules $RRD1$ or $RRD2$ can be applied, hence, $p$ is not a normal form;
- Case $p_1 \equiv \delta$. The transition rule $RD3$ can be applied, hence, $p$ is not a normal form;
- Case $p_1 \equiv p_1' + p_1''$. The transition rule $RD4$ can be applied, hence, $p$ is not a normal form;
- Case $p_1 \equiv p_1' \cdot p_1''$. The transition rule $RD5$ can be applied, hence, $p$ is not a normal form;
- Case $p_1 \equiv p_1' \parallel p_1''$. The transition rule $RD6$ can be applied, hence, $p$ is not a normal form. □

**Theorem 6.28** (Soundness of APRTC modulo FR step bisimulation equivalence). *Let $x$ and $y$ be APRTC terms including encapsulation operator $\partial_H$. If $APRTC \vdash x = y$, then $x \sim_s^{fr} y$.*

*Proof.* Since FR step bisimulation $\sim_s^{fr}$ is both an equivalent and a congruent relation with respect to the operator $\partial_H$, we only need to check if each axiom in Table 6.17 is sound modulo FR step bisimulation equivalence.

The proof is similar to the proof of soundness of the algebra of parallelism modulo FR step bisimulation equivalence, hence, we omit it here. □

**Theorem 6.29** (Completeness of APRTC modulo FR step bisimulation equivalence). *Let $p$ and $q$ be closed APRTC terms including encapsulation operator $\partial_H$, if $p \sim_s^{fr} q$, then $p = q$.*

*Proof.* First, by the elimination theorem of APRTC (see Theorem 6.27), we know that the normal form of APRTC does not contain $\partial_H$, and for each closed APRTC term $p$, there exists a closed basic APRTC term $p'$, such that $APRTC \vdash p = p'$, hence, we only need to consider closed basic APRTC terms.

Similarly to Theorem 6.20, we can prove that for normal forms $n$ and $n'$, if $n \sim_s^{fr} n'$, then $n =_{AC} n'$.

Finally, let $s$ and $t$ be basic APRTC terms, and $s \sim_s^{fr} t$, there are normal forms $n$ and $n'$, such that $s = n$ and $t = n'$. The soundness theorem of APRTC modulo FR step bisimulation equivalence (see Theorem 6.28) yields $s \sim_s^{fr} n$ and $t \sim_s^{fr} n'$, hence, $n \sim_s^{fr} s \sim_s^{fr} t \sim_s^{fr} n'$. Since if $n \sim_s^{fr} n'$, then $n =_{AC} n'$, $s = n =_{AC} n' = t$, as desired. □

**Theorem 6.30** (Soundness of APRTC modulo FR pomset bisimulation equivalence). *Let $x$ and $y$ be APRTC terms including encapsulation operator $\partial_H$. If $APRTC \vdash x = y$, then $x \sim_p^{fr} y$.*

*Proof.* Since FR pomset bisimulation $\sim_p^{fr}$ is both an equivalent and a congruent relation with respect to the operator $\partial_H$, we only need to check if each axiom in Table 6.17 is sound modulo FR pomset bisimulation equivalence.

From the definition of FR pomset bisimulation (see Definition 2.68), we know that FR pomset bisimulation is defined by pomset transitions, which are labeled by pomsets. In a pomset transition, the events in the pomset are either within causality relations (defined by $\cdot$) or in concurrency (implicitly defined by $\cdot$ and $+$, and explicitly defined by $\between$), of course, they are pairwise consistent (without conflicts). In Theorem 6.28, we have already proven the case that all events are pairwise concurrent, hence, we only need to prove the case of events in causality. Without loss of generality, we take a pomset of $P = \{e_1, e_2 : e_1 \cdot e_2\}$. Then, the pomset transition labeled by the above $P$ is just composed of one single event transition labeled by $e_1$ succeeded by another single event transition labeled by $e_2$, that is, $\xrightarrow{P} = \xrightarrow{e_1}\xrightarrow{e_2}$ or $\xrightarrow{P} = \xrightarrow{e_2[n]}\!\!\!\!\rightarrow\xrightarrow{e_1[m]}\!\!\!\!\rightarrow$.

Similarly to the proof of soundness of APRTC modulo FR step bisimulation equivalence (see Theorem 6.28), we can prove that each axiom in Table 6.17 is sound modulo FR pomset bisimulation equivalence, however, we omit them here. □

**TABLE 6.19 Transition rules of guarded recursion.**

$$\frac{t_i(\langle X_1|E\rangle,\cdots,\langle X_n|E\rangle)\xrightarrow{e}\surd}{\langle X_i|E\rangle\xrightarrow{e}\surd}$$

$$\frac{t_i(\langle X_1|E\rangle,\cdots,\langle X_n|E\rangle)\xrightarrow{e}y}{\langle X_i|E\rangle\xrightarrow{e}y}$$

**Theorem 6.31** (Completeness of APRTC modulo FR pomset bisimulation equivalence). *Let $p$ and $q$ be closed APRTC terms including encapsulation operator $\partial_H$, if $p\sim_p^{fr} q$, then $p=q$.*

*Proof.* The proof can be proven similarly to the proof of completeness of APRTC modulo FR step bisimulation equivalence, hence, we omit it here. □

**Theorem 6.32** (Soundness of APRTC modulo FR hp-bisimulation equivalence). *Let $x$ and $y$ be APRTC terms including encapsulation operator $\partial_H$. If $APRTC\vdash x=y$, then $x\sim_{hp}^{fr} y$.*

*Proof.* Since FR hp-bisimulation $\sim_{hp}^{fr}$ is both an equivalent and a congruent relation with respect to the operator $\partial_H$, we only need to check if each axiom in Table 6.17 is sound modulo FR hp-bisimulation equivalence.

From the definition of FR hp-bisimulation (see Definition 2.70), we know that FR hp-bisimulation is defined on the posetal product $(C_1,f,C_2), f:C_1\rightarrow C_2$ isomorphism. Two process terms $s$ related to $C_1$ and $t$ related to $C_2$, and $f: C_1\rightarrow C_2$ isomorphism. Initially, $(C_1,f,C_2)=(\emptyset,\emptyset,\emptyset)$, and $(\emptyset,\emptyset,\emptyset)\in\sim_{hp}^{fr}$. When $s\xrightarrow{e}s'$ $(C_1\xrightarrow{e}C_1')$, there will be $t\xrightarrow{e}t'$ $(C_2\xrightarrow{e}C_2')$, and we define $f'=f[e\mapsto e]$. Also, when $s\xtwoheadrightarrow{e[m]}s'$ $(C_1\xtwoheadrightarrow{e[m]}C_1')$, there will be $t\xtwoheadrightarrow{e[m]}t'$ $(C_2\xtwoheadrightarrow{e[m]}C_2')$, and we define $f'=f[e[m]\mapsto e[m]]$. Then, if $(C_1,f,C_2)\in\sim_{hp}^{fr}$, then $(C_1',f',C_2')\in\sim_{hp}^{fr}$.

Similarly to the proof of soundness of APRTC modulo FR pomset bisimulation equivalence (see Theorem 6.30), we can prove that each axiom in Table 6.17 is sound modulo FR hp-bisimulation equivalence, we just need additionally to check the above conditions on FR hp-bisimulation, however, we omit them here. □

**Theorem 6.33** (Completeness of APRTC modulo FR hp-bisimulation equivalence). *Let $p$ and $q$ be closed APRTC terms including encapsulation operator $\partial_H$, if $p\sim_{hp}^{fr} q$, then $p=q$.*

*Proof.* The proof is similar to the proof of completeness of APRTC modulo FR pomset bisimulation equivalence, hence, we omit it here. □

## 6.3 Recursion

In this section, we introduce recursion to capture infinite processes based on APRTC. In the following, $E, F, G$ are recursion specifications, $X, Y, Z$ are recursive variables.

The behavior of the solution $\langle X_i|E\rangle$ for the recursion variable $X_i$ in $E$, where $i\in\{1,\cdots,n\}$, is exactly the behavior of their right-hand sides $t_i(X_1,\cdots,X_n)$, which is captured by the two transition rules in Table 6.19.

**Theorem 6.34** (Conservativity of APRTC with guarded recursion). *APRTC with guarded recursion is a conservative extension of APRTC.*

*Proof.* Since the transition rules of APRTC are source dependent, and the transition rules for guarded recursion in Table 6.19 contain only a fresh constant in their source, the transition rules of APRTC with guarded recursion are a conservative extension of those of APRTC. □

**Theorem 6.35** (Congruence theorem of APRTC with guarded recursion). *Truly concurrent bisimulation equivalences $\sim_p^{fr}$, $\sim_s^{fr}$, and $\sim_{hp}^{fr}$ are all congruences with respect to APRTC with guarded recursion.*

*Proof.* This follows from the following two facts:

1. In a guarded recursive specification, the right-hand sides of its recursive equations can be adapted to the form by applications of the axioms in APRTC and replacing recursion variables by the right-hand sides of their recursive equations;
2. Truly concurrent bisimulation equivalences $\sim_p^{fr}$, $\sim_s^{fr}$, and $\sim_{hp}^{fr}$ are all congruences with respect to all operators of APRTC. □

**TABLE 6.20** Recursive definition and specification principle.

| No. | Axiom |
|---|---|
| RDP | $\langle X_i\|E\rangle = t_i(\langle X_1\|E,\cdots,X_n\|E\rangle)\quad (i\in\{1,\cdots,n\})$ |
| RSP | if $y_i = t_i(y_1,\cdots,y_n)$ for $i\in\{1,\cdots,n\}$, then $y_i=\langle X_i\|E\rangle\quad (i\in\{1,\cdots,n\})$ |

### 6.3.1 Recursive definition and specification principles

The RDP (Recursive Definition Principle) and the RSP (Recursive Specification Principle) are shown in Table 6.20.

**Theorem 6.36** (Elimination theorem of APRTC with linear recursion). *Each process term in APRTC with linear recursion is equal to a process term $\langle X_1|E\rangle$ with E a linear recursive specification.*

*Proof.* By applying structural induction with respect to term size, each process term $t_1$ in APRTC with linear recursion generates a process that can be expressed in the form of equations

$t_i = (a_{i11}\parallel\cdots\parallel a_{i1i_1})t_{i1}+\cdots+(a_{ik_i1}\parallel\cdots\parallel a_{ik_ii_k})t_{ik_i}+(b_{i11}\parallel\cdots\parallel b_{i1i_1})+\cdots+(b_{il_i1}\parallel\cdots\parallel b_{il_ii_l})$

for $i\in\{1,\cdots,n\}$. Or,

$t_i = t_{i1}(a_{i11}[m_{i1}]\parallel\cdots\parallel a_{i1i_1}[m_{i1}])+\cdots+t_{ik_i}(a_{ik_i1}[m_{ik}]\parallel\cdots\parallel a_{ik_ii_k}[m_{ik}])+(b_{i11}[n_{i1}]\parallel\cdots\parallel b_{i1i_1})[n_{i1}]+\cdots+(b_{il_i1}[n_{il}]\parallel\cdots\parallel b_{il_ii_l}[n_{il}])$

Let the linear recursive specification $E$ consist of the recursive equations

$X_i = (a_{i11}\parallel\cdots\parallel a_{i1i_1})X_{i1}+\cdots+(a_{ik_i1}\parallel\cdots\parallel a_{ik_ii_k})X_{ik_i}+(b_{i11}\parallel\cdots\parallel b_{i1i_1})+\cdots+(b_{il_i1}\parallel\cdots\parallel b_{il_ii_l})$

or the equations,

$X_i = X_{i1}(a_{i11}[m_{i1}]\parallel\cdots\parallel a_{i1i_1}[m_{i1}])+\cdots+X_{ik_i}(a_{ik_i1}[m_{ik}]\parallel\cdots\parallel a_{ik_ii_k}[m_{ik}])+(b_{i11}[n_{i1}]\parallel\cdots\parallel b_{i1i_1}[n_{i1}])+\cdots+(b_{il_i1}[n_{il}]\parallel\cdots\parallel b_{il_ii_l}[n_{il}])$

for $i\in\{1,\cdots,n\}$. Replacing $X_i$ by $t_i$ for $i\in\{1,\cdots,n\}$ is a solution for $E$, RSP yields $t_1=\langle X_1|E\rangle$. □

**Theorem 6.37** (Soundness of APRTC with guarded recursion). *Let x and y be APRTC with guarded recursion terms. If $APRTC$ with guarded recursion $\vdash x = y$, then*

1. $x \sim_s^{fr} y$;
2. $x \sim_p^{fr} y$;
3. $x \sim_{hp}^{fr} y$.

*Proof.* (1) Soundness of APRTC with guarded recursion with respect to FR step bisimulation $\sim_s^{fr}$.

Since FR step bisimulation $\sim_s^{fr}$ is both an equivalent and a congruent relation with respect to APRTC with guarded recursion, we only need to check if each axiom in Table 6.20 is sound modulo FR step bisimulation equivalence.

This can be proven similarly to the proof of soundness of APRTC modulo FR step bisimulation equivalence, however, we omit this here.

(2) Soundness of APRTC with guarded recursion with respect to FR pomset bisimulation $\sim_p^{fr}$.

Since FR pomset bisimulation $\sim_p^{fr}$ is both an equivalent and a congruent relation with respect to the guarded recursion, we only need to check if each axiom in Table 6.20 is sound modulo FR pomset bisimulation equivalence.

From the definition of FR pomset bisimulation (see Definition 2.68), we know that FR pomset bisimulation is defined by pomset transitions, which are labeled by pomsets. In a pomset transition, the events in the pomset are either within causality relations (defined by $\cdot$) or in concurrency (implicitly defined by $\cdot$ and $+$, and explicitly defined by $\between$), of course, they are pairwise consistent (without conflicts). We have already proven the case that all events are pairwise concurrent, hence, we only need to prove the case of events in causality. Without loss of generality, we take a pomset of $P=\{e_1,e_2: e_1\cdot e_2\}$. Then, the pomset transition labeled by the above $P$ is just composed of one single event transition labeled by $e_1$ succeeded by another single event transition labeled by $e_2$, that is, $\xrightarrow{P}=\xrightarrow{e_1}\xrightarrow{e_2}$ or $\xrightarrow{P}=\xrightarrow{e_2[n]}\!\!\!\!\rightarrow\xrightarrow{e_1[m]}\!\!\!\!\rightarrow$.

Similarly to the proof of soundness of APRTC with guarded recursion modulo FR step bisimulation equivalence (1), we can prove that each axiom in Table 6.20 is sound modulo FR pomset bisimulation equivalence, however, we omit them here.

(3) Soundness of APRTC with guarded recursion with respect to FR hp-bisimulation $\sim_{hp}^{fr}$.

Since FR hp-bisimulation $\sim_{hp}^{fr}$ is both an equivalent and a congruent relation with respect to guarded recursion, we only need to check if each axiom in Table 6.20 is sound modulo FR hp-bisimulation equivalence.

From the definition of FR hp-bisimulation (see Definition 2.70), we know that FR hp-bisimulation is defined on the posetal product $(C_1, f, C_2), f : C_1 \rightarrow C_2$ isomorphism. Two process terms $s$ related to $C_1$ and $t$ related to $C_2$, and $f : C_1 \rightarrow C_2$ isomorphism. Initially, $(C_1, f, C_2) = (\emptyset, \emptyset, \emptyset)$, and $(\emptyset, \emptyset, \emptyset) \in \sim_{hp}^{fr}$. When $s \xrightarrow{e} s'$ $(C_1 \xrightarrow{e} C_1')$, there will be $t \xrightarrow{e} t'$ $(C_2 \xrightarrow{e} C_2')$, and we define $f' = f[e \mapsto e]$. Also, when $s \xtwoheadrightarrow{e[m]} s'$ $(C_1 \xtwoheadrightarrow{e[m]} C_1')$, there will be $t \xtwoheadrightarrow{e[m]} t'$ $(C_2 \xtwoheadrightarrow{e[m]} C_2')$, and we define $f' = f[e[m] \mapsto e[m]]$. Then, if $(C_1, f, C_2) \in \sim_{hp}^{fr}$, then $(C_1', f', C_2') \in \sim_{hp}^{fr}$.

Similarly to the proof of soundness of APRTC with guarded recursion modulo FR pomset bisimulation equivalence (2), we can prove that each axiom in Table 6.20 is sound modulo FR hp-bisimulation equivalence, we just need additionally to check the above conditions on FR hp-bisimulation, however, we omit them here. □

**Theorem 6.38** (Completeness of APRTC with linear recursion). *Let $p$ and $q$ be closed APRTC with linear recursion terms, then,*

1. *if $p \sim_s^{fr} q$, then $p = q$;*
2. *if $p \sim_p^{fr} q$, then $p = q$;*
3. *if $p \sim_{hp}^{fr} q$, then $p = q$.*

*Proof.* First, by the elimination theorem of APRTC with guarded recursion (see Theorem 6.36), we know that each process term in APRTC with linear recursion is equal to a process term $\langle X_1|E\rangle$ with $E$ a linear recursive specification.

It remains to prove the following cases:

(1) If $\langle X_1|E_1\rangle \sim_s^{fr} \langle Y_1|E_2\rangle$ for linear recursive specification $E_1$ and $E_2$, then $\langle X_1|E_1\rangle = \langle Y_1|E_2\rangle$.

Let $E_1$ consist of recursive equations $X = t_X$ for $X \in \mathcal{X}$ and $E_2$ consist of recursion equations $Y = t_Y$ for $Y \in \mathcal{Y}$. Let the linear recursive specification $E$ consist of recursion equations $Z_{XY} = t_{XY}$, and $\langle X|E_1\rangle \sim_s^{fr} \langle Y|E_2\rangle$, and $t_{XY}$ consist of the following summands:

1. $t_{XY}$ contains a summand $(a_1 \parallel \cdots \parallel a_m)Z_{X'Y'}$ iff $t_X$ contains the summand $(a_1 \parallel \cdots \parallel a_m)X'$ and $t_Y$ contains the summand $(a_1 \parallel \cdots \parallel a_m)Y'$ such that $\langle X'|E_1\rangle \sim_s^{fr} \langle Y'|E_2\rangle$;
2. $t_{XY}$ contains a summand $Z_{X'Y'}(a_1[m] \parallel \cdots \parallel a_m[m])$ iff $t_X$ contains the summand $X'(a_1[m] \parallel \cdots \parallel a_m[m])$ and $t_Y$ contains the summand $Y'(a_1[m] \parallel \cdots \parallel a_m[m])$ such that $\langle X'|E_1\rangle \sim_s^{fr} \langle Y'|E_2\rangle$;
3. $t_{XY}$ contains a summand $b_1 \parallel \cdots \parallel b_n$ iff $t_X$ contains the summand $b_1 \parallel \cdots \parallel b_n$ and $t_Y$ contains the summand $b_1 \parallel \cdots \parallel b_n$;
4. $t_{XY}$ contains a summand $b_1[n] \parallel \cdots \parallel b_n[n]$ iff $t_X$ contains the summand $b_1[n] \parallel \cdots \parallel b_n[n]$ and $t_Y$ contains the summand $b_1[n] \parallel \cdots \parallel b_n[n]$.

Let $\sigma$ map recursion variable $X$ in $E_1$ to $\langle X|E_1\rangle$, and let $\psi$ map recursion variable $Z_{XY}$ in $E$ to $\langle X|E_1\rangle$. Hence, $\sigma((a_1 \parallel \cdots \parallel a_m)X') \equiv (a_1 \parallel \cdots \parallel a_m)\langle X'|E_1\rangle \equiv \psi((a_1 \parallel \cdots \parallel a_m)Z_{X'Y'})$, or $\sigma(X'(a_1[m] \parallel \cdots \parallel a_m[m])) \equiv \langle X'|E_1\rangle(a_1[m] \parallel \cdots \parallel a_m[m]) \equiv \psi(Z_{X'Y'}(a_1[m] \parallel \cdots \parallel a_m[m]))$, hence, by RDP, we obtain $\langle X|E_1\rangle = \sigma(t_X) = \psi(t_{XY})$. Then, by RSP, $\langle X|E_1\rangle = \langle Z_{XY}|E\rangle$, particularly, $\langle X_1|E_1\rangle = \langle Z_{X_1Y_1}|E\rangle$. Similarly, we can obtain $\langle Y_1|E_2\rangle = \langle Z_{X_1Y_1}|E\rangle$. Finally, $\langle X_1|E_1\rangle = \langle Z_{X_1Y_1}|E\rangle = \langle Y_1|E_2\rangle$, as desired.

(2) If $\langle X_1|E_1\rangle \sim_p^{fr} \langle Y_1|E_2\rangle$ for linear recursive specifications $E_1$ and $E_2$, then $\langle X_1|E_1\rangle = \langle Y_1|E_2\rangle$.

It can be proven similarly to (1), hence, we omit it here.

(3) If $\langle X_1|E_1\rangle \sim_{hp}^{fr} \langle Y_1|E_2\rangle$ for linear recursive specifications $E_1$ and $E_2$, then $\langle X_1|E_1\rangle = \langle Y_1|E_2\rangle$.

It can be proven similarly to (1), hence, we omit it here. □

## 6.4 Abstraction

To abstract away from the internal implementations of a program, and verify that the program exhibits the desired external behaviors, the silent step $\tau$ and abstraction operator $\tau_I$ are introduced, where $I \subseteq \mathbb{E}$ denotes the internal events. The transition rule of $\tau$ is shown in Table 6.21. In the following, let the atomic event $e$ range over $\mathbb{E} \cup \{\delta\} \cup \{\tau\}$, and let the communication function $\gamma : \mathbb{E} \cup \{\tau\} \times \mathbb{E} \cup \{\tau\} \rightarrow \mathbb{E} \cup \{\delta\}$, with each communication involved $\tau$ resulting in $\delta$.

**Theorem 6.39** (Conservativity of $RAPTC$ with silent step). *$RAPTC$ with silent step is a conservative extension of $RAPTC$.*

*Proof.* Since the transition rules of $RAPTC$ are source dependent, and the transition rules for silent step in Table 6.21 contain only a fresh constant $\tau$ in their source, the transition rules of $RAPTC$ with silent step is a conservative extension of those of $RAPTC$. □

**TABLE 6.21** Transition rule of the silent step.

| |
|---|
| $\frac{}{\tau \xrightarrow{\tau} \surd}$ |
| $\frac{}{\tau \xtwoheadrightarrow{\tau} \surd}$ |

**TABLE 6.22** Axioms of silent step.

| No. | Axiom |
|---|---|
| $B1$ | $e \cdot \tau = e$ |
| $RB1$ | $\tau \cdot e[m] = e[m]$ |
| $B2$ | $e \cdot (\tau \cdot (x + y) + x) = e \cdot (x + y)$ |
| $RB2$ | $((x+y) \cdot \tau + x) \cdot e[m] = (x+y) \cdot e[m]$ |
| $B3$ | $x \parallel \tau = x$ |

**Theorem 6.40** (Congruence theorem of $RAPTC$ with silent step). *Rooted branching FR truly concurrent bisimulation equivalences* $\approx^{fr}_{rbp}$, $\approx^{fr}_{rbs}$, *and* $\approx^{fr}_{rbhp}$ *are all congruences with respect to* $RAPTC$ *with silent step.*

*Proof.* This follows from the following two facts:

1. FR truly concurrent bisimulation equivalences $\sim^{fr}_{p}$, $\sim^{fr}_{s}$, and $\sim^{fr}_{hp}$ are all congruences with respect to all operators of $RAPTC$, while FR truly concurrent bisimulation equivalences $\sim^{fr}_{p}$, $\sim^{fr}_{s}$, and $\sim^{fr}_{hp}$ imply the corresponding rooted branching FR truly concurrent bisimulation $\approx rbp^{fr}$, $\approx^{fr}_{rbs}$, and $\approx^{fr}_{rbhp}$, hence, rooted branching FR truly concurrent bisimulation $\approx rbp^{fr}$, $\approx^{fr}_{rbs}$, and $\approx^{fr}_{rbhp}$ are all congruences with respect to all operators of $RAPTC$;
2. While $\mathbb{E}$ is extended to $\mathbb{E} \cup \{\tau\}$, it can be proved that rooted branching FR truly concurrent bisimulation $\approx rbp^{fr}$, $\approx^{fr}_{rbs}$, and $\approx^{fr}_{rbhp}$ are all congruences with respect to all operators of $RAPTC$, hence, we omit it here. □

### 6.4.1 Algebraic laws for the silent step

We design the axioms for the silent step $\tau$ in Table 6.22.

**Theorem 6.41** (Elimination theorem of APRTC with silent step and guarded linear recursion). *Each process term in APRTC with silent step and guarded linear recursion is equal to a process term* $\langle X_1|E\rangle$ *with* $E$ *a guarded linear recursive specification.*

*Proof.* By applying structural induction with respect to term size, each process term $t_1$ in APRTC with silent step and guarded linear recursion generates a process that can be expressed in the form of equations:

$t_i = (a_{i11} \parallel \cdots \parallel a_{i1i_1})t_{i1} + \cdots + (a_{ik_i1} \parallel \cdots \parallel a_{ik_ii_k})t_{ik_i} + (b_{i11} \parallel \cdots \parallel b_{i1i_1}) + \cdots + (b_{il_i1} \parallel \cdots \parallel b_{il_ii_l})$

or,

$t_i = t_{i1}(a_{i11}[m_{i1}] \parallel \cdots \parallel a_{i1i_1}[m_{i1}]) + \cdots + t_{ik_i}(a_{ik_i1}[m_{ik}] \parallel \cdots \parallel a_{ik_ii_k}[m_{ik}]) + (b_{i11}[n_{i1}] \parallel \cdots \parallel b_{i1i_1}[n_{i1}]) + \cdots + (b_{il_i1}[n_{il}] \parallel \cdots \parallel b_{il_ii_l}[n_{il}])$

for $i \in \{1, \cdots, n\}$. Let the linear recursive specification $E$ consist of the recursive equations

$$X_i = (a_{i11} \parallel \cdots \parallel a_{i1i_1})X_{i1} + \cdots + (a_{ik_i1} \parallel \cdots \parallel a_{ik_ii_k})X_{ik_i} + (b_{i11} \parallel \cdots \parallel b_{i1i_1}) + \cdots + (b_{il_i1} \parallel \cdots \parallel b_{il_ii_l})$$

or,

$X_i = X_{i1}(a_{i11}[m_{i1}] \parallel \cdots \parallel a_{i1i_1}[m_{i1}]) + \cdots + X_{ik_i}(a_{ik_i1}[m_{ik}] \parallel \cdots \parallel a_{ik_ii_k}[m_{ik}]) + (b_{i11}[n_{i1}] \parallel \cdots \parallel b_{i1i_1}[n_{i1}]) + \cdots + (b_{il_i1}[n_{il}] \parallel \cdots \parallel b_{il_ii_l}[n_{il}])$

for $i \in \{1, \cdots, n\}$. Replacing $X_i$ by $t_i$ for $i \in \{1, \cdots, n\}$ is a solution for $E$, RSP yields $t_1 = \langle X_1|E\rangle$. □

**Theorem 6.42** (Soundness of APRTC with silent step and guarded linear recursion). *Let* $x$ *and* $y$ *be APRTC with silent step and guarded linear recursion terms. If APRTC with silent step and guarded linear recursion* $\vdash x = y$, *then*

1. $x \approx_{rbs}^{fr} y$;
2. $x \approx_{rbp}^{fr} y$;
3. $x \approx_{rbhp}^{fr} y$.

*Proof.* (1) Soundness of APRTC with silent step and guarded linear recursion with respect to rooted branching FR step bisimulation $\approx_{rbs}^{fr}$.

Since rooted branching FR step bisimulation $\approx_{rbs}^{fr}$ is both an equivalent and a congruent relation with respect to APRTC with silent step and guarded linear recursion, we only need to check if each axiom in Table 6.22 is sound modulo rooted branching FR step bisimulation equivalence.

Although transition rules in Table 6.21 are defined in the flavor of a single event, they can be modified into a step (a set of events within which each event is pairwise concurrent), however, we omit them here. If we treat a single event as a step containing just one event, the proof of this soundness theorem does not pose any problem, hence, we use this way and still use the transition rules in Table 6.21.

- **Axiom** $B1$. Assume that $e \cdot \tau = e$, it is sufficient to prove that $e \cdot \tau \approx_{rbs}^{fr} e$. By the forward transition rules for operator $\cdot$ in Table 6.3 and $\tau$ in Table 6.21, we obtain

$$\frac{e \xrightarrow{e} e[m]}{e \cdot \tau \xrightarrow{e} \xrightarrow{\tau} e[m]}$$

$$\frac{e \xrightarrow{e} e[m]}{e \xrightarrow{e} e[m]}$$

  By the reverse transition rules for operator $\cdot$ in Table 6.4 and $\tau$ in Table 6.21, there are no transitions.
  Hence, $e \cdot \tau \approx_{rbs}^{fr} e$, as desired.
- **Axiom** $RB1$. Assume that $\tau \cdot e[m] = e[m]$, it is sufficient to prove that $\tau \cdot e[m] \approx_{rbs}^{fr} e[m]$. By the forward transition rules for operator $\cdot$ in Table 6.3 and $\tau$ in Table 6.21, there are no transitions.
  By the reverse transition rules for operator $\cdot$ in Table 6.4 and $\tau$ in Table 6.21, we obtain

$$\frac{e[m] \xrightarrow{e[m]} e}{\tau \cdot e[m] \xrightarrow{e[m]} \xrightarrow{\tau} e}$$

$$\frac{e[m] \xrightarrow{e[m]} e}{e[m] \xrightarrow{e[m]} e}$$

  Hence, $\tau \cdot e[m] \approx_{rbs}^{fr} e[m]$, as desired.
- **Axiom** $B2$. Let $p$ and $q$ be $RAPTC$ with silent step processes, and assume that $e \cdot (\tau \cdot (p+q) + p) = e \cdot (p+q)$, it is sufficient to prove that $e \cdot (\tau \cdot (p+q) + p) \approx_{rbs}^{fr} e \cdot (p+q)$. There are several cases, however, we will not enumerate them all here. By the forward transition rules for operators $\cdot$ and $+$ in Table 6.3 and $\tau$ in Table 6.21, we obtain

$$\frac{e \xrightarrow{e} e[m] \quad p \xrightarrow{e_1} p' \quad q \xrightarrow{e_1} q'}{e \cdot (\tau \cdot (p+q) + p) \xrightarrow{e} \xrightarrow{\tau} \xrightarrow{e_1} e[m] \cdot ((p'+q') + p')}$$

$$\frac{e \xrightarrow{e} e[m] \quad p \xrightarrow{e_1} p'}{e \cdot (p+q) \xrightarrow{e} \xrightarrow{e_1} e[m] \cdot (p'+q')}$$

  By the reverse transition rules for operators $\cdot$ and $+$ in Table 6.4 and $\tau$ in Table 6.21, there are no transitions.
  Hence, $e \cdot (\tau \cdot (p+q) + p) \approx_{rbs}^{fr} e \cdot (p+q)$, as desired.
- **Axiom** $RB2$. Let $p$ and $q$ be $RAPTC$ with silent step processes, and assume that $((x+y) \cdot \tau + x) \cdot e[m] = (x+y) \cdot e[m]$, it is sufficient to prove that $((x+y) \cdot \tau + x) \cdot e[m] \approx_{rbs}^{fr} (x+y) \cdot e[m]$. There are several cases, however, we will not enumerate them all here. By the forward transition rules for operators $\cdot$ and $+$ in Table 6.3 and $\tau$ in Table 6.21, there are no transitions.

By the reverse transition rules for operators $\cdot$ and $+$ in Table 6.4 and $\tau$ in Table 6.21, we obtain

$$\frac{e[m]\xrightarrow{e[m]}\!\!\!\twoheadrightarrow e\quad p\xrightarrow{e_1[n]}\!\!\!\twoheadrightarrow p'\quad q\xrightarrow{e_1[n]}\!\!\!\twoheadrightarrow q'}{((p+q)\cdot\tau+p)\cdot e[m]\xrightarrow{e[m]}\!\!\!\twoheadrightarrow\xrightarrow{\tau}\!\!\!\twoheadrightarrow\xrightarrow{e_1[n]}\!\!\!\twoheadrightarrow((p'+q')+p')\cdot e}$$

$$\frac{e[m]\xrightarrow{e[m]}\!\!\!\twoheadrightarrow e\quad p\xrightarrow{e_1[n]}\!\!\!\twoheadrightarrow p'}{(p+q)\cdot e[m]\xrightarrow{e[m]}\!\!\!\twoheadrightarrow\xrightarrow{e_1[n]}\!\!\!\twoheadrightarrow(p'+q'\cdot e)}$$

Hence, $((p+q)\cdot\tau+p)\cdot e[m]\approx_{rbs}^{fr}(p+q)\cdot e[m]$, as desired.

- **Axiom** $B3$. Let $p$ be an $RAPTC$ with silent step, and assume that $p\parallel\tau=p$, it is sufficient to prove that $p\parallel\tau\approx_{rbs}^{fr}p$. By the forward transition rules for operator $\parallel$ in Table 6.7 and $\tau$ in Table 6.21, we obtain

$$\frac{p\xrightarrow{e}e[m]}{p\parallel\tau\xRightarrow{e}e[m]}$$

$$\frac{p\xrightarrow{e}p'}{p\parallel\tau\xRightarrow{e}p'}$$

By the reverse transition rules for operator $\parallel$ in Table 6.8 and $\tau$ in Table 6.21, we obtain

$$\frac{p\xrightarrow{e[m]}\!\!\!\twoheadrightarrow e}{p\parallel\tau\overset{e[m]}{\Longrightarrow}e}$$

$$\frac{p\xrightarrow{e[m]}\!\!\!\twoheadrightarrow p'}{p\parallel\tau\overset{e[m]}{\Longrightarrow}p'}$$

Hence, $p\parallel\tau\approx_{rbs}^{fr}p$, as desired.

(2) Soundness of APRTC with silent step and guarded linear recursion with respect to rooted branching FR pomset bisimulation $\approx_{rbp}^{fr}$.

Since rooted branching FR pomset bisimulation $\approx_{rbp}^{fr}$ is both an equivalent and a congruent relation with respect to APRTC with silent step and guarded linear recursion, we only need to check if each axiom in Table 6.22 is sound modulo rooted branching FR pomset bisimulation $\approx_{rbp}^{fr}$.

From the definition of rooted branching FR pomset bisimulation $\approx_{rbp}^{fr}$ (see Definition 2.73), we know that rooted branching FR pomset bisimulation $\approx_{rbp}^{fr}$ is defined by weak pomset transitions, which are labeled by pomsets with $\tau$. In a weak pomset transition, the events in the pomset are either within causality relations (defined by $\cdot$) or in concurrency (implicitly defined by $\cdot$ and $+$, and explicitly defined by $\between$), of course, they are pairwise consistent (without conflicts). In (1), we have already proven the case that all events are pairwise concurrent, hence, we only need to prove the case of events in causality. Without loss of generality, we take a pomset of $P=\{e_1,e_2:e_1\cdot e_2\}$. Then, the weak pomset transition labeled by the above $P$ is just composed of one single event transition labeled by $e_1$ succeeded by another single event transition labeled by $e_2$, that is, $\xRightarrow{P}=\xRightarrow{e_1}\xRightarrow{e_2}$ or $\overset{P}{\Longrightarrow}=\overset{e_2}{\Longrightarrow}\overset{e_1}{\Longrightarrow}$.

Similarly to the proof of soundness of APRTC with silent step and guarded linear recursion modulo rooted branching FR step bisimulation $\approx_{rbs}^{fr}$ (1), we can prove that each axiom in Table 6.22 is sound modulo rooted branching FR pomset bisimulation $\approx_{rbp}^{fr}$, however, we omit this here.

(3) Soundness of APRTC with silent step and guarded linear recursion with respect to rooted branching FR hp-bisimulation $\approx_{rbhp}^{fr}$.

Since rooted branching FR hp-bisimulation $\approx_{rbhp}^{fr}$ is both an equivalent and a congruent relation with respect to APRTC with silent step and guarded linear recursion, we only need to check if each axiom in Table 6.22 is sound modulo rooted branching FR hp-bisimulation $\approx_{rbhp}^{fr}$.

From the definition of rooted branching FR hp-bisimulation $\approx^{fr}_{rbhp}$ (see Definition 2.75), we know that rooted branching FR hp-bisimulation $\approx^{fr}_{rbhp}$ is defined on the weakly posetal product $(C_1, f, C_2), f : \hat{C_1} \to \hat{C_2}$ isomorphism. Two process terms $s$ related to $C_1$ and $t$ related to $C_2$, and $f : \hat{C_1} \to \hat{C_2}$ isomorphism. Initially, $(C_1, f, C_2) = (\emptyset, \emptyset, \emptyset)$, and $(\emptyset, \emptyset, \emptyset) \in \approx^{fr}_{rbhp}$. When $s \xrightarrow{e} s'$ $(C_1 \xrightarrow{e} C_1')$, there will be $t \xRightarrow{e} t'$ $(C_2 \xRightarrow{e} C_2')$, and we define $f' = f[e \mapsto e]$. Also, when $s \overset{e[m]}{\twoheadrightarrow} s'$ $(C_1 \overset{e[m]}{\twoheadrightarrow} C_1')$, there will be $t \overset{e[m]}{\twoheadrightarrow} t'$ $(C_2 \overset{e[m]}{\twoheadrightarrow} C_2')$, and we define $f' = f[e[m] \mapsto e[m]$. Then, if $(C_1, f, C_2) \in \approx^{fr}_{rbhp}$, then $(C_1', f', C_2') \in \approx^{fr}_{rbhp}$.

Similarly to the proof of soundness of APRTC with silent step and guarded linear recursion modulo rooted branching FR pomset bisimulation equivalence (2), we can prove that each axiom in Table 6.22 is sound modulo rooted branching FR hp-bisimulation equivalence, we just need additionally to check the above conditions on rooted branching FR hp-bisimulation, however, we omit them here. □

**Theorem 6.43** (Completeness of APRTC with silent step and guarded linear recursion). *Let $p$ and $q$ be closed APRTC with silent step and guarded linear recursion terms, then*

1. *if $p \approx^{fr}_{rbs} q$, then $p = q$;*
2. *if $p \approx^{fr}_{rbp} q$, then $p = q$;*
3. *if $p \approx^{fr}_{rbhp} q$, then $p = q$.*

*Proof.* First, by the elimination theorem of APRTC with silent step and guarded linear recursion (see Theorem 6.41), we know that each process term in APRTC with silent step and guarded linear recursion is equal to a process term $\langle X_1|E\rangle$ with $E$ a guarded linear recursive specification.

It remains to prove the following cases:

(1) If $\langle X_1|E_1\rangle \approx^{fr}_{rbs} \langle Y_1|E_2\rangle$ for guarded linear recursive specification $E_1$ and $E_2$, then $\langle X_1|E_1\rangle = \langle Y_1|E_2\rangle$.

First, the recursive equation $W = \tau + \cdots + \tau$ with $W \not\equiv X_1$ in $E_1$ and $E_2$, can be removed, and the corresponding summands $aW$ are replaced by $a$, to obtain $E_1'$ and $E_2'$, by use of the axioms RDP, $A3$ and $B1$, $RB1$, and $\langle X|E_1\rangle = \langle X|E_1'\rangle$, $\langle Y|E_2\rangle = \langle Y|E_2'\rangle$.

Let $E_1$ consist of recursive equations $X = t_X$ for $X \in \mathcal{X}$ and $E_2$ consist of recursion equations $Y = t_Y$ for $Y \in \mathcal{Y}$, and are not the form $\tau + \cdots + \tau$. Let the guarded linear recursive specification $E$ consist of recursion equations $Z_{XY} = t_{XY}$, and $\langle X|E_1\rangle \approx^{fr}_{rbs} \langle Y|E_2\rangle$, and $t_{XY}$ consist of the following summands:

1. $t_{XY}$ contains a summand $(a_1 \parallel \cdots \parallel a_m)Z_{X'Y'}$ iff $t_X$ contains the summand $(a_1 \parallel \cdots \parallel a_m)X'$ and $t_Y$ contains the summand $(a_1 \parallel \cdots \parallel a_m)Y'$ such that $\langle X'|E_1\rangle \approx^{fr}_{rbs} \langle Y'|E_2\rangle$;
2. $t_{XY}$ contains a summand $Z_{X'Y'}(a_1[m] \parallel \cdots \parallel a_m[m])$ iff $t_X$ contains the summand $X'(a_1[m] \parallel \cdots \parallel a_m[m])$ and $t_Y$ contains the summand $Y'(a_1[m] \parallel \cdots \parallel a_m[m])$ such that $\langle X'|E_1\rangle \approx^{fr}_{rbs} \langle Y'|E_2\rangle$;
3. $t_{XY}$ contains a summand $b_1 \parallel \cdots \parallel b_n$ iff $t_X$ contains the summand $b_1 \parallel \cdots \parallel b_n$ and $t_Y$ contains the summand $b_1 \parallel \cdots \parallel b_n$;
4. $t_{XY}$ contains a summand $b_1[n] \parallel \cdots \parallel b_n[n]$ iff $t_X$ contains the summand $b_1[n] \parallel \cdots \parallel b_n[n]$ and $t_Y$ contains the summand $b_1[n] \parallel \cdots \parallel b_n[n]$;
5. $t_{XY}$ contains a summand $\tau Z_{X'Y}$ iff $XY \not\equiv X_1Y_1$, $t_X$ contains the summand $\tau X'$, and $\langle X'|E_1\rangle \approx^{fr}_{rbs} \langle Y|E_2\rangle$;
6. $t_{XY}$ contains a summand $Z_{X'Y}\tau$ iff $XY \not\equiv X_1Y_1$, $t_X$ contains the summand $X'\tau$, and $\langle X'|E_1\rangle \approx^{fr}_{rbs} \langle Y|E_2\rangle$;
7. $t_{XY}$ contains a summand $\tau Z_{XY'}$ iff $XY \not\equiv X_1Y_1$, $t_Y$ contains the summand $\tau Y'$, and $\langle X|E_1\rangle \approx^{fr}_{rbs} \langle Y'|E_2\rangle$;
8. $t_{XY}$ contains a summand $Z_{XY'}\tau$ iff $XY \not\equiv X_1Y_1$, $t_Y$ contains the summand $Y'\tau$, and $\langle X|E_1\rangle \approx^{fr}_{rbs} \langle Y'|E_2\rangle$.

Since $E_1$ and $E_2$ are guarded, $E$ is guarded. Constructing the process term $u_{XY}$ consists of the following summands:

1. $u_{XY}$ contains a summand $(a_1 \parallel \cdots \parallel a_m)\langle X'|E_1\rangle$ iff $t_X$ contains the summand $(a_1 \parallel \cdots \parallel a_m)X'$ and $t_Y$ contains the summand $(a_1 \parallel \cdots \parallel a_m)Y'$ such that $\langle X'|E_1\rangle \approx^{fr}_{rbs} \langle Y'|E_2\rangle$;
2. $u_{XY}$ contains a summand $\langle X'|E_1\rangle(a_1[m] \parallel \cdots \parallel a_m[m])$ iff $t_X$ contains the summand $X'(a_1[m] \parallel \cdots \parallel a_m[m])$ and $t_Y$ contains the summand $Y'(a_1[m] \parallel \cdots \parallel a_m[m])$ such that $\langle X'|E_1\rangle \approx^{fr}_{rbs} \langle Y'|E_2\rangle$;
3. $u_{XY}$ contains a summand $b_1 \parallel \cdots \parallel b_n$ iff $t_X$ contains the summand $b_1 \parallel \cdots \parallel b_n$ and $t_Y$ contains the summand $b_1 \parallel \cdots \parallel b_n$;
4. $u_{XY}$ contains a summand $b_1[n] \parallel \cdots \parallel b_n[n]$ iff $t_X$ contains the summand $b_1[n] \parallel \cdots \parallel b_n[n]$ and $t_Y$ contains the summand $b_1[n] \parallel \cdots \parallel b_n[n]$;

**TABLE 6.23 Transition rule of the abstraction operator.**

| | |
|---|---|
| $\dfrac{x \xrightarrow{e} \surd}{\tau_I(x) \xrightarrow{e} \surd} \quad e \notin I$ | $\dfrac{x \xrightarrow{e} x'}{\tau_I(x) \xrightarrow{e} \tau_I(x')} \quad e \notin I$ |
| $\dfrac{x \xrightarrow{e} \surd}{\tau_I(x) \xrightarrow{\tau} \surd} \quad e \in I$ | $\dfrac{x \xrightarrow{e} x'}{\tau_I(x) \xrightarrow{\tau} \tau_I(x')} \quad e \in I$ |
| $\dfrac{x \xtwoheadrightarrow{e[m]} e}{\tau_I(x) \xtwoheadrightarrow{e[m]} e} \quad e[m] \notin I$ | $\dfrac{x \xtwoheadrightarrow{e[m]} x'}{\tau_I(x) \xtwoheadrightarrow{e[m]} \tau_I(x')} \quad e[m] \notin I$ |
| $\dfrac{x \xtwoheadrightarrow{e[m]} \surd}{\tau_I(x) \xtwoheadrightarrow{\tau} \surd} \quad e[m] \in I$ | $\dfrac{x \xtwoheadrightarrow{e[m]} x'}{\tau_I(x) \xtwoheadrightarrow{\tau} \tau_I(x')} \quad e[m] \in I$ |

5. $u_{XY}$ contains a summand $\tau\langle X'|E_1\rangle$ iff $XY \not\equiv X_1Y_1$, $t_X$ contains the summand $\tau X'$, and $\langle X'|E_1\rangle \approx_{rbs}^{fr} \langle Y|E_2\rangle$;
6. $u_{XY}$ contains a summand $\langle X'|E_1\rangle\tau$ iff $XY \not\equiv X_1Y_1$, $t_X$ contains the summand $X'\tau$, and $\langle X'|E_1\rangle \approx_{rbs}^{fr} \langle Y|E_2\rangle$.

Let the process term $s_{XY}$ be defined as follows:

1. $s_{XY} \triangleq \tau\langle X|E_1\rangle + u_{XY}$ iff $XY \not\equiv X_1Y_1$, $t_Y$ contains the summand $\tau Y'$, and $\langle X|E_1\rangle \approx_{rbs}^{fr} \langle Y'|E_2\rangle$;
2. $s_{XY} \triangleq \langle X|E_1\rangle\tau + u_{XY}$ iff $XY \not\equiv X_1Y_1$, $t_Y$ contains the summand $Y'\tau$, and $\langle X|E_1\rangle \approx_{rbs}^{fr} \langle Y'|E_2\rangle$;
3. $s_{XY} \triangleq \langle X|E_1\rangle$, otherwise.

Hence, $\langle X|E_1\rangle = \langle X|E_1\rangle + u_{XY}$, and $(a_1 \parallel \cdots \parallel a_m)(\tau\langle X|E_1\rangle + u_{XY}) = (a_1 \parallel \cdots \parallel a_m)((\tau\langle X|E_1\rangle + u_{XY}) + u_{XY}) = (a_1 \parallel \cdots \parallel a_m)(\langle X|E_1\rangle + u_{XY}) = (a_1 \parallel \cdots \parallel a_m)\langle X|E_1\rangle$, or $((\langle X|E_1\rangle\tau + u_{XY})(a_1[m] \parallel \cdots \parallel a_m[m]) = (((\langle X|E_1\rangle\tau + u_{XY}) + u_{XY})(a_1[m] \parallel \cdots \parallel a_m[m]) = ((\langle X|E_1\rangle + u_{XY})(a_1[m] \parallel \cdots \parallel a_m[m]) = \langle X|E_1\rangle(a_1[m] \parallel \cdots \parallel a_m[m])$, hence, $s_{XY}(a_1 \parallel \cdots \parallel a_m) = (a_1[m] \parallel \cdots \parallel a_m[m])\langle X|E_1\rangle$.

Let $\sigma$ map recursion variable $X$ in $E_1$ to $\langle X|E_1\rangle$, and let $\psi$ map recursion variable $Z_{XY}$ in $E$ to $s_{XY}$. It is sufficient to prove $s_{XY} = \psi(t_{XY})$ for recursion variables $Z_{XY}$ in $E$. Either $XY \equiv X_1Y_1$ or $XY \not\equiv X_1Y_1$, hence, we can obtain $s_{XY} = \psi(t_{XY})$. Hence, $s_{XY} = \langle Z_{XY}|E\rangle$ for recursive variables $Z_{XY}$ in $E$ is a solution for $E$. Then, by RSP, particularly, $\langle X_1|E_1\rangle = \langle Z_{X_1Y_1}|E\rangle$. Similarly, we can obtain $\langle Y_1|E_2\rangle = \langle Z_{X_1Y_1}|E\rangle$. Finally, $\langle X_1|E_1\rangle = \langle Z_{X_1Y_1}|E\rangle = \langle Y_1|E_2\rangle$, as desired.

(2) If $\langle X_1|E_1\rangle \approx_{rbp}^{fr} \langle Y_1|E_2\rangle$ for guarded linear recursive specification $E_1$ and $E_2$, then $\langle X_1|E_1\rangle = \langle Y_1|E_2\rangle$.
It can be proven similarly to (1), hence, we omit it here.

(3) If $\langle X_1|E_1\rangle \approx_{rbhb} \langle Y_1|E_2\rangle$ for guarded linear recursive specification $E_1$ and $E_2$, then $\langle X_1|E_1\rangle = \langle Y_1|E_2\rangle$.
It can be proven similarly to (1), hence, we omit it here. □

### 6.4.2 Abstraction

The unary abstraction operator $\tau_I$ ($I \subseteq \mathbb{E}$) renames all atomic events in $I$ into $\tau$. APRTC with silent step and abstraction operator is called $APRTC_\tau$. The transition rules of operator $\tau_I$ are shown in Table 6.23.

**Theorem 6.44** (Conservativity of $APRTC_\tau$). *$APRTC_\tau$ is a conservative extension of APRTC with silent step.*

*Proof.* Since the transition rules of APRTC with silent step are source dependent, and the transition rules for abstraction operator in Table 6.23 contain only a fresh operator $\tau_I$ in their source, hence, the transition rules of $APRTC_\tau$ is a conservative extension of those of $RAPTC$ with silent step. □

**Theorem 6.45** (Congruence theorem of $APRTC_\tau$). *Rooted branching FR truly concurrent bisimulation equivalences $\approx_{rbp}^{fr}$, $\approx_{rbs}^{fr}$, and $\approx_{rbhp}^{fr}$ are all congruences with respect to $APRTC_\tau$.*

*Proof.* (1) Case rooted branching FR pomset bisimulation equivalence $\approx_{rbp}^{fr}$.

Let $x$ and $y$ be $APRTC_\tau$ processes, and $x \approx_{rbp}^{fr} y$, it is sufficient to prove that $\tau_I(x) \approx_{rbp}^{fr} \tau_I(y)$.
By the transition rules for operator $\tau_I$ in Table 6.23, we can obtain

$$\tau_I(x) \xrightarrow{X} X[\mathcal{K}](X \nsubseteq I) \quad \tau_I(y) \xrightarrow{Y} Y[\mathcal{J}](Y \nsubseteq I)$$

**TABLE 6.24 Axioms of abstraction operator.**

| No. | Axiom |
|---|---|
| $TI1$ | $e \notin I \quad \tau_I(e) = e$ |
| $RTI1$ | $e[m] \notin I \quad \tau_I(e[m]) = e[m]$ |
| $TI2$ | $e \in I \quad \tau_I(e) = \tau$ |
| $RTI2$ | $e[m] \in I \quad \tau_I(e[m]) = \tau$ |
| $TI3$ | $\tau_I(\delta) = \delta$ |
| $TI4$ | $\tau_I(x + y) = \tau_I(x) + \tau_I(y)$ |
| $TI5$ | $\tau_I(x \cdot y) = \tau_I(x) \cdot \tau_I(y)$ |
| $TI6$ | $\tau_I(x \parallel y) = \tau_I(x) \parallel \tau_I(y)$ |

$$\tau_I(x) \xrightarrow{X[\mathcal{K}]} X(X \not\subseteq I) \quad \tau_I(y) \xrightarrow{Y[\mathcal{J}]} Y(Y \not\subseteq I)$$

with $X \subseteq x$, $Y \subseteq y$, and $X \sim Y$.

Or, we can obtain

$$\tau_I(x) \xrightarrow{X} \tau_I(x')(X \not\subseteq I) \quad \tau_I(y) \xrightarrow{Y} \tau_I(y')(Y \not\subseteq I)$$

$$\tau_I(x) \xrightarrow{X[\mathcal{K}]} \tau_I(x')(X \not\subseteq I) \quad \tau_I(y) \xrightarrow{Y[\mathcal{J}]} \tau_I(y')(Y \not\subseteq I)$$

with $X \subseteq x$, $Y \subseteq y$, and $X \sim Y$ and the hypothesis $\tau_I(x') \approx^{fr}_{rbp} \tau_I(y')$.

Or, we can obtain

$$\tau_I(x) \xrightarrow{\tau^*} \surd(X \subseteq I) \quad \tau_I(y) \xrightarrow{\tau^*} \surd(Y \subseteq I)$$

$$\tau_I(x) \xrightarrow{\tau^*} \surd(X \subseteq I) \quad \tau_I(y) \xrightarrow{\tau^*} \surd(Y \subseteq I)$$

with $X \subseteq x$, $Y \subseteq y$, and $X \sim Y$.

Or, we can obtain

$$\tau_I(x) \xrightarrow{\tau^*} \tau_I(x')(X \subseteq I) \quad \tau_I(y) \xrightarrow{\tau^*} \tau_I(y')(Y \subseteq I)$$

$$\tau_I(x) \xrightarrow{\tau^*} \tau_I(x')(X \subseteq I) \quad \tau_I(y) \xrightarrow{\tau^*} \tau_I(y')(Y \subseteq I)$$

with $X \subseteq x$, $Y \subseteq y$, and $X \sim Y$ and the hypothesis $\tau_I(x') \approx^{fr}_{rbp} \tau_I(y')$.

Hence, we obtain $\tau_I(x) \approx^{fr}_{rbp} \tau_I(y)$, as desired

(2) The cases of rooted branching FR step bisimulation $\approx^{fr}_{rbs}$, rooted branching FR hp-bisimulation $\approx^{fr}_{rbhp}$ can be proven similarly, however, we omit them here. □

We design the axioms for the abstraction operator $\tau_I$ in Table 6.24.

**Theorem 6.46** (Soundness of $APRTC_\tau$ with guarded linear recursion). *Let x and y be $APRTC_\tau$ with guarded linear recursion terms. If $APRTC_\tau$ with guarded linear recursion $\vdash x = y$, then*

1. $x \approx^{fr}_{rbs} y$;
2. $x \approx^{fr}_{rbp} y$;
3. $x \approx^{fr}_{rbhp} y$.

*Proof.* (1) Soundness of $APRTC_\tau$ with guarded linear recursion with respect to rooted branching FR step bisimulation $\approx^{fr}_{rbs}$.

Since rooted branching FR step bisimulation $\approx^{fr}_{rbs}$ is both an equivalent and a congruent relation with respect to $APRTC_\tau$ with guarded linear recursion, we only need to check if each axiom in Table 6.24 is sound modulo rooted branching FR step bisimulation equivalence.

The proof is similar to the proof of soundness of APRTC with silent step and guarded linear recursion, however, we omit this here.

(2) Soundness of $APRTC_\tau$ with guarded linear recursion with respect to rooted branching FR pomset bisimulation $\approx^{fr}_{rbp}$.

Since rooted branching FR pomset bisimulation $\approx^{fr}_{rbp}$ is both an equivalent and a congruent relation with respect to $APRTC_\tau$ with guarded linear recursion, we only need to check if each axiom in Table 6.24 is sound modulo rooted branching FR pomset bisimulation $\approx^{fr}_{rbp}$.

From the definition of rooted branching FR pomset bisimulation $\approx^{fr}_{rbp}$ (see Definition 2.73), we know that rooted branching FR pomset bisimulation $\approx^{fr}_{rbp}$ is defined by weak pomset transitions, which are labeled by pomsets with $\tau$. In a weak pomset transition, the events in the pomset are either within causality relations (defined by $\cdot$) or in concurrency (implicitly defined by $\cdot$ and $+$, and explicitly defined by $\between$), of course, they are pairwise consistent (without conflicts). In (1), we have already proven the case that all events are pairwise concurrent, hence, we only need to prove the case of events in causality. Without loss of generality, we take a pomset of $P = \{e_1, e_2 : e_1 \cdot e_2\}$. Then, the weak pomset transition labeled by the above $P$ is just composed of one single event transition labeled by $e_1$ succeeded by another single event transition labeled by $e_2$, that is, $\xRightarrow{P} = \xRightarrow{e_1}\xRightarrow{e_2}$ or $\overset{P}{\Rrightarrow} = \overset{e_2}{\Rrightarrow}\overset{e_1}{\Rrightarrow}$.

Similarly to the proof of soundness of $APRTC_\tau$ with guarded linear recursion modulo rooted branching FR step bisimulation $\approx^{fr}_{rbs}$ (1), we can prove that each axiom in Table 6.24 is sound modulo rooted branching FR pomset bisimulation $\approx^{fr}_{rbp}$, however, we omit them here.

(3) Soundness of $APRTC_\tau$ with guarded linear recursion with respect to rooted branching FR hp-bisimulation $\approx^{fr}_{rbhp}$.

Since rooted branching FR hp-bisimulation $\approx^{fr}_{rbhp}$ is both an equivalent and a congruent relation with respect to $APRTC_\tau$ with guarded linear recursion, we only need to check if each axiom in Table 6.24 is sound modulo rooted branching FR hp-bisimulation $\approx^{fr}_{rbhp}$.

From the definition of rooted branching FR hp-bisimulation $\approx^{fr}_{rbhp}$ (see Definition 2.75), we know that rooted branching FR hp-bisimulation $\approx^{fr}_{rbhp}$ is defined on the weakly posetal product $(C_1, f, C_2), f : \hat{C_1} \to \hat{C_2}$ isomorphism. Two process terms $s$ related to $C_1$ and $t$ related to $C_2$, and $f : \hat{C_1} \to \hat{C_2}$ isomorphism. Initially, $(C_1, f, C_2) = (\emptyset, \emptyset, \emptyset)$, and $(\emptyset, \emptyset, \emptyset) \in \approx^{fr}_{rbhp}$. When $s \xrightarrow{e} s'$ $(C_1 \xrightarrow{e} C_1')$, there will be $t \xRightarrow{e} t'$ $(C_2 \xRightarrow{e} C_2')$, and we define $f' = f[e \mapsto e]$. Also, when $s \overset{e[m]}{\rightrightarrows} s'$ $(C_1 \overset{e[m]}{\rightrightarrows} C_1')$, there will be $t \overset{e[m]}{\Rrightarrow} t'$ $(C_2 \overset{e[m]}{\Rrightarrow} C_2')$, and we define $f' = f[e[m] \mapsto e[m]$. Then, if $(C_1, f, C_2) \in \approx^{fr}_{rbhp}$, then $(C_1', f', C_2') \in \approx^{fr}_{rbhp}$.

Similarly to the proof of soundness of $APRTC_\tau$ with guarded linear recursion modulo rooted branching FR pomset bisimulation equivalence (2), we can prove that each axiom in Table 6.24 is sound modulo rooted branching FR hp-bisimulation equivalence, we just need additionally to check the above conditions on rooted branching FR hp-bisimulation, however, we omit them here. □

Although $\tau$-loops are prohibited in guarded linear recursive specifications in a specifiable way, they can be constructed using the abstraction operator, for example, there exist $\tau$-loops in the process term $\tau_{\{a\}}(\langle X|X = aX\rangle)$. To avoid $\tau$-loops caused by $\tau_I$ and ensure fairness, the concepts of cluster and CFAR (Cluster Fair Abstraction Rule) are still valid in true concurrency, hence, we introduce them below.

**Definition 6.47** (Cluster). Let $E$ be a guarded linear recursive specification, and $I \subseteq \mathbb{E}$. Two recursion variables $X$ and $Y$ in $E$ are in the same cluster for $I$ iff there exist sequences of transitions $\langle X|E\rangle \xrightarrow{\{b_{11},\cdots,b_{1i}\}} \cdots \xrightarrow{\{b_{m1},\cdots,b_{mi}\}} \langle Y|E\rangle$ and $\langle Y|E\rangle \xrightarrow{\{c_{11},\cdots,c_{1j}\}} \cdots \xrightarrow{\{c_{n1},\cdots,c_{nj}\}} \langle X|E\rangle$, or $\langle X|E\rangle \xtwoheadrightarrow{\{b_{11}[m],\cdots,b_{1i}[m]\}} \cdots \xtwoheadrightarrow{\{b_{m1}[m],\cdots,b_{mi}[m]\}} \langle Y|E\rangle$ and $\langle Y|E\rangle \xtwoheadrightarrow{\{c_{11}[n],\cdots,c_{1j}[n]\}} \cdots \xtwoheadrightarrow{\{c_{n1}[n],\cdots,c_{nj}[n]\}} \langle X|E\rangle$, where $b_{11}, \cdots, b_{mi}, c_{11}, \cdots, c_{nj}, b_{11}[m], \cdots, b_{mi}[m], c_{11}[n], \cdots, c_{nj}[n] \in I \cup \{\tau\}$.

$a_1 \parallel \cdots \parallel a_k$, or $(a_1 \parallel \cdots \parallel a_k)X$, or $a_1[m] \parallel \cdots \parallel a_k[m]$, or $X(a_1[m] \parallel \cdots \parallel a_k[m])$ is an exit for the cluster $C$ iff: (1) $a_1 \parallel \cdots \parallel a_k$, or $(a_1 \parallel \cdots \parallel a_k)X$, or $a_1[m] \parallel \cdots \parallel a_k[m]$, or $X(a_1[m] \parallel \cdots \parallel a_k[m])$ is a summand at the right-hand side of the recursive equation for a recursion variable in $C$, and (2) in the case of $(a_1 \parallel \cdots \parallel a_k)X$, and $X(a_1[m] \parallel \cdots \parallel a_k[m])$ either $a_l, a_l[m] \notin I \cup \{\tau\} (l \in \{1, 2, \cdots, k\})$ or $X \notin C$.

**Theorem 6.48** (Soundness of CFAR). *CFAR is sound modulo rooted branching FR truly concurrent bisimulation equivalences* $\approx^{fr}_{rbs}$, $\approx^{fr}_{rbp}$, *and* $\approx^{fr}_{rbhp}$.

**TABLE 6.25 Cluster fair abstraction rule.**

| No. | Axiom |
|---|---|
| CFAR | If $X$ is in a cluster for $I$ with exits $\{(a_{11} \parallel \cdots \parallel a_{1i})Y_1, \cdots, (a_{m1} \parallel \cdots \parallel a_{mi})Y_m, b_{11} \parallel \cdots \parallel b_{1j}, \cdots, b_{n1} \parallel \cdots \parallel b_{nj}\}$, then $\tau \cdot \tau_I(\langle X\|E\rangle) =$ $\tau \cdot \tau_I((a_{11} \parallel \cdots \parallel a_{1i})\langle Y_1\|E\rangle + \cdots + (a_{m1} \parallel \cdots \parallel a_{mi})\langle Y_m\|E\rangle + b_{11} \parallel \cdots \parallel b_{1j} + \cdots + b_{n1} \parallel \cdots \parallel b_{nj})$ Or exists, $\{Y_1(a_{11}[m] \parallel \cdots \parallel a_{1i}[m1]), \cdots, Y_m(a_{m1}[mm] \parallel \cdots \parallel a_{mi}[mm]), b_{11}[n1] \parallel \cdots \parallel b_{1j}[n1], \cdots, b_{n1}[nn] \parallel \cdots \parallel b_{nj}[nn]\}$, then $\tau_I(\langle X\|E\rangle) \cdot \tau =$ $\tau_I(\langle Y_1\|E\rangle(a_{11}[m1] \parallel \cdots \parallel a_{1i}[m1]) + \cdots + \langle Y_m\|E\rangle(a_{m1}[mm] \parallel \cdots \parallel a_{mi}[mm]) + b_{11}[n1] \parallel \cdots \parallel b_{1j}[n1] + \cdots + b_{n1}[nn] \parallel \cdots \parallel b_{nj}[nn]) \cdot \tau$ |

*Proof.* (1) Soundness of CFAR with respect to rooted branching FR step bisimulation $\approx^{fr}_{rbs}$.

Let $X$ be in a cluster for $I$ with exits $\{(a_{11} \parallel \cdots \parallel a_{1i})Y_1, \cdots, (a_{m1} \parallel \cdots \parallel a_{mi})Y_m, b_{11} \parallel \cdots \parallel b_{1j}, \cdots, b_{n1} \parallel \cdots \parallel b_{nj}\}$ and $\{Y_1(a_{11}[m1] \parallel \cdots \parallel a_{1i}[m1]), \cdots, Y_m(a_{m1}[mm] \parallel \cdots \parallel a_{mi}[mm]), b_{11}[n1] \parallel \cdots \parallel b_{1j}[n1], \cdots, b_{n1}[nn] \parallel \cdots \parallel b_{nj}[nn]\}$. Then, $\langle X|E\rangle$ can execute a string of atomic events from $I \cup \{\tau\}$ inside the cluster of $X$, followed by an exit $(a_{i'1} \parallel \cdots \parallel a_{i'i})Y_{i'}$ for $i' \in \{1, \cdots, m\}$ or $b_{j'1} \parallel \cdots \parallel b_{j'j}$ for $j' \in \{1, \cdots, n\}$, or $Y_{i'}(a_{i'1}[m1] \parallel \cdots \parallel a_{i'i}[mi'])$ for $i' \in \{1, \cdots, m\}$ or $b_{j'1}[n1] \parallel \cdots \parallel b_{j'j}[nj']$ for $j' \in \{1, \cdots, n\}$. Hence, $\tau_I(\langle X|E\rangle)$ can execute a string of $\tau^*$ inside the cluster of $X$, followed by an exit $\tau_I((a_{i'1} \parallel \cdots \parallel a_{i'i})\langle Y_{i'}|E\rangle)$ for $i' \in \{1, \cdots, m\}$ or $\tau_I(b_{j'1} \parallel \cdots \parallel b_{j'j})$ for $j' \in \{1, \cdots, n\}$, or $\tau_I(\langle Y_{i'}|E\rangle(a_{i'1}[m1] \parallel \cdots \parallel a_{i'i}[mi']))$ for $i' \in \{1, \cdots, m\}$ or $\tau_I(b_{j'1}[n1] \parallel \cdots \parallel b_{j'j}[nj'])$ for $j' \in \{1, \cdots, n\}$. Also, these $\tau^*$ are non-initial in $\tau\tau_I(\langle X|E\rangle)$ and $\tau_I(\langle X|E\rangle)\tau$, hence, they are truly silent by the axiom $B1$ and $RB1$, we obtain $\tau\tau_I(\langle X|E\rangle) \approx^{fr}_{rbs} \tau \cdot \tau_I((a_{11} \parallel \cdots \parallel a_{1i})\langle Y_1|E\rangle + \cdots + (a_{m1} \parallel \cdots \parallel a_{mi})\langle Y_m|E\rangle + b_{11} \parallel \cdots \parallel b_{1j} + \cdots + b_{n1} \parallel \cdots \parallel b_{nj})$, and $\tau_I(\langle X|E\rangle)\tau \approx^{fr}_{rbs} \tau_I(\langle Y_1|E\rangle(a_{11}[m1] \parallel \cdots \parallel a_{1i}[m1]) + \cdots + \langle Y_m|E\rangle(a_{m1}[mm] \parallel \cdots \parallel a_{mi}[mm]) + b_{11}[n1] \parallel \cdots \parallel b_{1j}[n1] + \cdots + b_{n1}[nn] \parallel \cdots \parallel b_{nj}[nn]) \cdot \tau$ as desired.

(2) Soundness of CFAR with respect to rooted branching FR pomset bisimulation $\approx^{fr}_{rbp}$.

From the definition of rooted branching FR pomset bisimulation $\approx^{fr}_{rbp}$ (see Definition 2.73), we know that rooted branching FR pomset bisimulation $\approx^{fr}_{rbp}$ is defined by weak pomset transitions, which are labeled by pomsets with $\tau$. In a weak pomset transition, the events in the pomset are either within causality relations (defined by $\cdot$) or in concurrency (implicitly defined by $\cdot$ and $+$, and explicitly defined by $\between$), of course, they are pairwise consistent (without conflicts). In (1), we have already proven the case that all events are pairwise concurrent, hence, we only need to prove the case of events in causality. Without loss of generality, we take a pomset of $P = \{e_1, e_2 : e_1 \cdot e_2\}$. Then, the weak pomset transition labeled by the above $P$ is just composed of one single event transition labeled by $e_1$ succeeded by another single event transition labeled by $e_2$, that is, $\stackrel{P}{\Rightarrow} = \stackrel{e_1}{\Rightarrow}\stackrel{e_2}{\Rightarrow}$ or $\stackrel{P}{\Rrightarrow} = \stackrel{e_2}{\Rrightarrow}\stackrel{e_1}{\Rrightarrow}$.

Similarly to the proof of soundness of CFAR modulo rooted branching FR step bisimulation $\approx^{fr}_{rbs}$ (1), we can prove that CFAR in Table 6.25 is sound modulo rooted branching FR pomset bisimulation $\approx^{fr}_{rbp}$, however, we omit this here.

(3) Soundness of CFAR with respect to rooted branching FR hp-bisimulation $\approx^{fr}_{rbhp}$.

From the definition of rooted branching FR hp-bisimulation $\approx^{fr}_{rbhp}$ (see Definition 2.75), we know that rooted branching FR hp-bisimulation $\approx^{fr}_{rbhp}$ is defined on the weakly posetal product $(C_1, f, C_2), f : \hat{C_1} \to \hat{C_2}$ isomorphism. Two process terms $s$ related to $C_1$ and $t$ related to $C_2$, and $f : \hat{C_1} \to \hat{C_2}$ isomorphism. Initially, $(C_1, f, C_2) = (\emptyset, \emptyset, \emptyset)$, and $(\emptyset, \emptyset, \emptyset) \in \approx^{fr}_{rbhp}$. When $s \xrightarrow{e} s'$ $(C_1 \xrightarrow{e} C_1')$, there will be $t \stackrel{e}{\Rightarrow} t'$ $(C_2 \stackrel{e}{\Rightarrow} C_2')$, and we define $f' = f[e \mapsto e]$. Also, when $s \stackrel{e[m]}{\Rrightarrow} s'$ $(C_1 \stackrel{e[m]}{\Rrightarrow} C_1')$, there will be $t \stackrel{e[m]}{\Rrightarrow} t'$ $(C_2 \stackrel{e[m]}{\Rrightarrow} C_2')$, and we define $f' = f[e[m] \mapsto e[m]]$. Then, if $(C_1, f, C_2) \in \approx^{fr}_{rbhp}$, then $(C_1', f', C_2') \in \approx^{fr}_{rbhp}$.

Similarly to the proof of soundness of CFAR modulo rooted branching FR pomset bisimulation equivalence (2), we can prove that CFAR in Table 6.25 is sound modulo rooted branching FR hp-bisimulation equivalence, we just need additionally to check the above conditions on rooted branching FR hp-bisimulation, however, we omit them here. □

**Theorem 6.49** (Completeness of $APRTC_\tau$ with guarded linear recursion and CFAR). *Let $p$ and $q$ be closed $APRTC_\tau$ with guarded linear recursion and CFAR terms, then*

**1.** *if* $p \approx_{rbs}^{fr} q$, *then* $p = q$;
**2.** *if* $p \approx_{rbp}^{fr} q$, *then* $p = q$;
**3.** *if* $p \approx_{rbhp}^{fr} q$, *then* $p = q$.

*Proof.* (1) For the case of rooted branching FR step bisimulation, the proof is the following.

First, in the proof the Theorem 6.43, we know that each process term $p$ in APRTC with silent step and guarded linear recursion is equal to a process term $\langle X_1|E\rangle$ with $E$ a guarded linear recursive specification. Also, we prove if $\langle X_1|E_1\rangle \approx_{rbs}^{fr} \langle Y_1|E_2\rangle$, then $\langle X_1|E_1\rangle = \langle Y_1|E_2\rangle$

The only new case is $p \equiv \tau_I(q)$. Let $q = \langle X|E\rangle$ with $E$ a guarded linear recursive specification, hence, $p = \tau_I(\langle X|E\rangle)$. Then, the collection of recursive variables in $E$ can be divided into its clusters $C_1, \cdots, C_N$ for $I$. Let

$(a_{1i1} \parallel \cdots \parallel a_{k_{i1}i1})Y_{i1} + \cdots + (a_{1im_i} \parallel \cdots \parallel a_{k_{im_i}im_i})Y_{im_i} + b_{1i1} \parallel \cdots \parallel b_{l_{i1}i1} + \cdots + b_{1im_i} \parallel \cdots \parallel b_{l_{im_i}im_i}$

or,

$Y_{i1}(a_{1i1}[m1] \parallel \cdots \parallel a_{k_{i1}i1}[m1]) + \cdots + Y_{im_i}(a_{1im_i}[mm] \parallel \cdots \parallel a_{k_{im_i}im_i}[mm]) + b_{1i1}[n1] \parallel \cdots \parallel b_{l_{i1}i1}[n1] + \cdots + b_{1im_i}[nn] \parallel \cdots \parallel b_{l_{im_i}im_i[nn]}$

be the conflict composition of exits for the cluster $C_i$, with $i \in \{1, \cdots, N\}$.

For $Z, Z' \in C_i$ with $i \in \{1, \cdots, N\}$, we define

$s_Z \triangleq (\hat{a_{1i1}} \parallel \cdots \parallel \hat{a_{k_{i1}i1}})\tau_I(\langle Y_{i1}|E\rangle) + \cdots + (\hat{a_{1im_i}} \parallel \cdots \parallel \hat{a_{k_{im_i}im_i}})\tau_I(\langle Y_{im_i}|E\rangle) + \hat{b_{1i1}} \parallel \cdots \parallel \hat{b_{l_{i1}i1}} + \cdots + \hat{b_{1im_i}} \parallel \cdots \parallel \hat{b_{l_{im_i}im_i}}$

and

$s'_Z \triangleq \tau_I(\langle Y_{i1}|E\rangle)(\hat{a_{1i1}}[m1] \parallel \cdots \parallel \hat{a_{k_{i1}i1}}[m1]) + \cdots + (\hat{a_{1im_i}}[mm] \parallel \cdots \parallel \hat{a_{k_{im_i}im_i}}[mm])\tau_I(\langle Y_{im_i}|E\rangle) + \hat{b_{1i1}}[n1] \parallel \cdots \parallel \hat{b_{l_{i1}i1}}[n1] + \cdots + \hat{b_{1im_i}}[nn] \parallel \cdots \parallel \hat{b_{l_{im_i}im_i}}[nn]$

For $Z, Z' \in C_i$ and $a_1, \cdots, a_j \in \mathbb{E} \cup \{\tau\}$ with $j \in \mathbb{N}$, we have

$(a_1 \parallel \cdots \parallel a_j)\tau_I(\langle Z|E\rangle)$

$= (a_1 \parallel \cdots \parallel a_j)\tau_I((a_{1i1} \parallel \cdots \parallel a_{k_{i1}i1})\langle Y_{i1}|E\rangle + \cdots + (a_{1im_i} \parallel \cdots \parallel a_{k_{im_i}im_i})\langle Y_{im_i}|E\rangle + b_{1i1} \parallel \cdots \parallel b_{l_{i1}i1} + \cdots + b_{1im_i} \parallel \cdots \parallel b_{l_{im_i}im_i})$

$= (a_1 \parallel \cdots \parallel a_j)s_Z$

$\tau_I(\langle Z|E\rangle)(a_1[m] \parallel \cdots \parallel a_j[m])$

$= \tau_I(\langle Y_{i1}|E\rangle(a_{1i1}[m1] \parallel \cdots \parallel a_{k_{i1}i1}[m1]) + \cdots + \langle Y_{im_i}|E\rangle(a_{1im_i}[mm] \parallel \cdots \parallel a_{k_{im_i}im_i}[mm]) + b_{1i1}[n1] \parallel \cdots \parallel b_{l_{i1}i1}[n1] + \cdots + b_{1im_i}[nn] \parallel \cdots \parallel b_{l_{im_i}im_i}[nn])(a_1[m] \parallel \cdots \parallel a_j[m])$

$= (a_1 \parallel \cdots \parallel a_j)s'_Z$

Let the linear recursive specification $F$ contain the same recursive variables as $E$, for $Z, Z' \in C_i$, $F$ contains the following recursive equation

$Z = (\hat{a_{1i1}} \parallel \cdots \parallel \hat{a_{k_{i1}i1}})Y_{i1} + \cdots + (\hat{a_{1im_i}} \parallel \cdots \parallel \hat{a_{k_{im_i}im_i}})Y_{im_i} + \hat{b_{1i1}} \parallel \cdots \parallel \hat{b_{l_{i1}i1}} + \cdots + \hat{b_{1im_i}} \parallel \cdots \parallel \hat{b_{l_{im_i}im_i}}$

Let the linear recursive specification $F'$ contain the same recursive variables as $E$, for $Z, Z' \in C_i$, $F$ contains the following recursive equation

$Z' = Y_{i1}(\hat{a_{1i1}}[m1] \parallel \cdots \parallel \hat{a_{k_{i1}i1}}[m1]) + \cdots + Y_{im_i}(\hat{a_{1im_i}}[mm] \parallel \cdots \parallel \hat{a_{k_{im_i}im_i}}[mm]) + \hat{b_{1i1}}[n1] \parallel \cdots \parallel \hat{b_{l_{i1}i1}}[n1] + \cdots + \hat{b_{1im_i}}[nn] \parallel \cdots \parallel \hat{b_{l_{im_i}im_i}}[nn]$

It is easy to see that there is no sequence of one or more $\tau$-transitions from $\langle Z|F\rangle$ and $\langle Z'|F'\rangle$ to itself, hence, $F$ and $F'$ is guarded.

As

$s_Z = (\hat{a_{1i1}} \parallel \cdots \parallel \hat{a_{k_{i1}i1}})Y_{i1} + \cdots + (\hat{a_{1im_i}} \parallel \cdots \parallel \hat{a_{k_{im_i}im_i}})Y_{im_i} + \hat{b_{1i1}} \parallel \cdots \parallel \hat{b_{l_{i1}i1}} + \cdots + \hat{b_{1im_i}} \parallel \cdots \parallel \hat{b_{l_{im_i}im_i}}$

is a solution for $F$, hence, $(a_1 \parallel \cdots \parallel a_j)\tau_I(\langle Z|E\rangle) = (a_1 \parallel \cdots \parallel a_j)s_Z = (a_1 \parallel \cdots \parallel a_j)\langle Z|F\rangle$.

Hence,

$\langle Z|F\rangle = (\hat{a_{1i1}} \parallel \cdots \parallel \hat{a_{k_{i1}i1}})\langle Y_{i1}|F\rangle + \cdots + (\hat{a_{1im_i}} \parallel \cdots \parallel \hat{a_{k_{im_i}im_i}})\langle Y_{im_i}|F\rangle + \hat{b_{1i1}} \parallel \cdots \parallel \hat{b_{l_{i1}i1}} + \cdots + \hat{b_{1im_i}} \parallel \cdots \parallel \hat{b_{l_{im_i}im_i}}$

As

$s'_Z = Y_{i1}(\hat{a_{1i1}}[m1] \parallel \cdots \parallel \hat{a_{k_{i1}i1}}[m1]) + \cdots + Y_{im_i}(\hat{a_{1im_i}}[mm] \parallel \cdots \parallel \hat{a_{k_{im_i}im_i}}[mm]) + \hat{b_{1i1}}[n1] \parallel \cdots \parallel \hat{b_{l_{i1}i1}}[n1] + \cdots + \hat{b_{1im_i}}[nn] \parallel \cdots \parallel \hat{b_{l_{im_i}im_i}}[nn]$

is a solution for $F'$, hence, $\tau_I(\langle Z'|E\rangle)(a_1[m] \parallel \cdots \parallel a_j[m]) = s'_Z(a_1[m] \parallel \cdots \parallel a_j[m]) = \langle Z'|F'\rangle(a_1[m] \parallel \cdots \parallel a_j[m])$.

Hence,

$\langle Z'|F'\rangle = \langle Y_{i1}|F\rangle(\hat{a_{1i1}}[m1] \parallel \cdots \parallel \hat{a_{k_{i1}i1}}[m1]) + \cdots + \langle Y_{im_i}|F\rangle(\hat{a_{1im_i}}[mm] \parallel \cdots \parallel \hat{a_{k_{im_i}im_i}}[mm]) + \hat{b_{1i1}}[n1] \parallel \cdots \parallel \hat{b_{l_{i1}i1}}[n1] + \cdots + \hat{b_{1im_i}}[nn] \parallel \cdots \parallel \hat{b_{l_{im_i}im_i}}[nn]$

Hence, $\tau_I(\langle X|E\rangle = \langle Z|F\rangle)$, as desired.
(2) For the case of rooted branching FR pomset bisimulation, it can be proven similarly to (1), hence, we omit it here.
(3) For the case of rooted branching FR hp-bisimulation, it can be proven similarly to (1), hence, we omit it here. □

Chapter 7

# Partially reversible $\pi_{tc}$

In this chapter, we design $\pi_{tc}$ with partial reversibility. This chapter is organized as follows. In Section 7.1, we introduce the truly concurrent operational semantics. Then, we introduce the syntax and operational semantics, laws modulo strongly truly concurrent bisimulations, and algebraic theory of $\pi_{tc}$ with reversibility in Sections 7.2, 7.3, and 7.4, respectively.

## 7.1 Operational semantics

First, in this section, we introduce concepts of FR (strongly) truly concurrent bisimilarities, including FR pomset bisimilarity, FR step bisimilarity, FR history-preserving (hp-)bisimilarity, and FR hereditary history-preserving (hhp-)bisimilarity. In contrast to traditional FR truly concurrent bisimilarities, these versions in $\pi_{ptc}$ must take care of actions with bound objects. Note that these FR truly concurrent bisimilarities are defined as late bisimilarities, but not early bisimilarities, as defined in $\pi$-calculus [4] [5]. Note that here, a PES $\mathcal{E}$ is deemed as a process.

**Definition 7.1** (Prime event structure with silent event). Let $\Lambda$ be a fixed set of labels, ranged over $a, b, c, \cdots$ and $\tau$. A ($\Lambda$-labeled) prime event structure with silent event $\tau$ is a tuple $\mathcal{E} = \langle \mathbb{E}, \leq, \sharp, \lambda \rangle$, where $\mathbb{E}$ is a denumerable set of events, including the silent event $\tau$. Let $\hat{\mathbb{E}} = \mathbb{E} \backslash \{\tau\}$, exactly excluding $\tau$, it is obvious that $\hat{\tau^*} = \epsilon$, where $\epsilon$ is the empty event. Let $\lambda : \mathbb{E} \to \Lambda$ be a labeling function and let $\lambda(\tau) = \tau$. Also, $\leq$, $\sharp$ are binary relations on $\mathbb{E}$, called causality and conflict, respectively, such that:

1. $\leq$ is a partial order and $\lceil e \rceil = \{e' \in \mathbb{E} | e' \leq e\}$ is finite for all $e \in \mathbb{E}$. It is easy to see that $e \leq \tau^* \leq e' = e \leq \tau \leq \cdots \leq \tau \leq e'$, then $e \leq e'$.
2. $\sharp$ is irreflexive, symmetric, and hereditary with respect to $\leq$, that is, for all $e, e', e'' \in \mathbb{E}$, if $e \sharp e' \leq e''$, then $e \sharp e''$.

Then, the concepts of consistency and concurrency can be drawn from the above definition:

1. $e, e' \in \mathbb{E}$ are consistent, denoted as $e \frown e'$, if $\neg(e \sharp e')$. A subset $X \subseteq \mathbb{E}$ is called consistent, if $e \frown e'$ for all $e, e' \in X$.
2. $e, e' \in \mathbb{E}$ are concurrent, denoted as $e \parallel e'$, if $\neg(e \leq e')$, $\neg(e' \leq e)$, and $\neg(e \sharp e')$.

**Definition 7.2** (Configuration). Let $\mathcal{E}$ be a PES. A (finite) configuration in $\mathcal{E}$ is a (finite) consistent subset of events $C \subseteq \mathcal{E}$, closed with respect to causality (i.e., $\lceil C \rceil = C$). The set of finite configurations of $\mathcal{E}$ is denoted by $\mathcal{C}(\mathcal{E})$. We let $\hat{C} = C \backslash \{\tau\}$.

A consistent subset of $X \subseteq \mathbb{E}$ of events can be seen as a pomset. Given $X, Y \subseteq \mathbb{E}$, $\hat{X} \sim \hat{Y}$ if $\hat{X}$ and $\hat{Y}$ are isomorphic as pomsets. In the remainder of the chapter, when we say $C_1 \sim C_2$, we mean $\hat{C_1} \sim \hat{C_2}$.

**Definition 7.3** (Pomset transitions and step). Let $\mathcal{E}$ be a PES and let $C \in \mathcal{C}(\mathcal{E})$, and $\emptyset \neq X \subseteq \mathbb{E}$, if $C \cap X = \emptyset$ and $C' = C \cup X \in \mathcal{C}(\mathcal{E})$, then $C \xrightarrow{X} C'$ is called a pomset transition from $C$ to $C'$. When the events in $X$ are pairwise concurrent, we say that $C \xrightarrow{X} C'$ is a step.

**Definition 7.4** (FR pomset transitions and step). Let $\mathcal{E}$ be a PES and let $C \in \mathcal{C}(\mathcal{E})$, and $\emptyset \neq X \subseteq \mathbb{E}$, if $C \cap X = \emptyset$ and $C' = C \cup X \in \mathcal{C}(\mathcal{E})$, then $C \xrightarrow{X} C'$ is called a forward pomset transition from $C$ to $C'$ and $C' \overset{X[\mathcal{K}]}{\twoheadrightarrow} C$ is called a reverse pomset transition from $C'$ to $C$. When the events in $X$ and $X[\mathcal{K}]$ are pairwise concurrent, we say that $C \xrightarrow{X} C'$ is a forward step and $C' \xrightarrow{X[\mathcal{K}]} C$ is a reverse step. It is obvious that $\rightarrow^* \xrightarrow{X} \rightarrow^* = \xrightarrow{X}$ and $\rightarrow^* \xrightarrow{e} \rightarrow^* = \xrightarrow{e}$ for any $e \in \mathbb{E}$ and $X \subseteq \mathbb{E}$.

**Definition 7.5** (FR strongly pomset, step bisimilarity). Let $\mathcal{E}_1, \mathcal{E}_2$ be PESs. A FR strongly pomset bisimulation is a relation $R \subseteq \mathcal{C}(\mathcal{E}_1) \times \mathcal{C}(\mathcal{E}_2)$, such that (1) if $(C_1, C_2) \in R$, and $C_1 \xrightarrow{X_1} C_1'$ (with $\mathcal{E}_1 \xrightarrow{X_1} \mathcal{E}_1'$), then $C_2 \xrightarrow{X_2} C_2'$ (with $\mathcal{E}_2 \xrightarrow{X_2} \mathcal{E}_2'$), with $X_1 \subseteq \mathbb{E}_1$, $X_2 \subseteq \mathbb{E}_2$, $X_1 \sim X_2$ and $(C_1', C_2') \in R$:

1. for each fresh action $\alpha \in X_1$, if $C_1'' \xrightarrow{\alpha} C_1'''$ (with $\mathcal{E}_1'' \xrightarrow{\alpha} \mathcal{E}_1'''$), then for some $C_2''$ and $C_2'''$, $C_2'' \xrightarrow{\alpha} C_2'''$ (with $\mathcal{E}_2'' \xrightarrow{\alpha} \mathcal{E}_2'''$), such that if $(C_1'', C_2'') \in R$, then $(C_1''', C_2''') \in R$;

Handbook of Truly Concurrent Process Algebra. https://doi.org/10.1016/B978-0-44-321515-5.00011-1

2. for each $x(y) \in X_1$ with $(y \notin n(\mathcal{E}_1, \mathcal{E}_2))$, if $C_1'' \xrightarrow{x(y)} C_1'''$ (with $\mathcal{E}_1'' \xrightarrow{x(y)} \mathcal{E}_1'''\{w/y\}$) for all $w$, then for some $C_2''$ and $C_2'''$, $C_2'' \xrightarrow{x(y)} C_2'''$ (with $\mathcal{E}_2'' \xrightarrow{x(y)} \mathcal{E}_2'''\{w/y\}$) for all $w$, such that if $(C_1'', C_2'') \in R$, then $(C_1''', C_2''') \in R$;
3. for each two $x_1(y), x_2(y) \in X_1$ with $(y \notin n(\mathcal{E}_1, \mathcal{E}_2))$, if $C_1'' \xrightarrow{\{x_1(y), x_2(y)\}} C_1'''$ (with $\mathcal{E}_1'' \xrightarrow{\{x_1(y), x_2(y)\}} \mathcal{E}_1'''\{w/y\}$) for all $w$, then for some $C_2''$ and $C_2'''$, $C_2'' \xrightarrow{\{x_1(y), x_2(y)\}} C_2'''$ (with $\mathcal{E}_2'' \xrightarrow{\{x_1(y), x_2(y)\}} \mathcal{E}_2'''\{w/y\}$) for all $w$, such that if $(C_1'', C_2'') \in R$, then $(C_1''', C_2''') \in R$;
4. for each $\overline{x}(y) \in X_1$ with $y \notin n(\mathcal{E}_1, \mathcal{E}_2)$, if $C_1'' \xrightarrow{\overline{x}(y)} C_1'''$ (with $\mathcal{E}_1'' \xrightarrow{\overline{x}(y)} \mathcal{E}_1'''$), then for some $C_2''$ and $C_2'''$, $C_2'' \xrightarrow{\overline{x}(y)} C_2'''$ (with $\mathcal{E}_2'' \xrightarrow{\overline{x}(y)} \mathcal{E}_2'''$), such that if $(C_1'', C_2'') \in R$, then $(C_1''', C_2''') \in R$,

and vice versa; (2) if $(C_1, C_2) \in R$, and $C_1 \xtwoheadrightarrow{X_1[\mathcal{K}_1]} C_1'$ (with $\mathcal{E}_1 \xtwoheadrightarrow{X_1[\mathcal{K}_1]} \mathcal{E}_1'$), then $C_2 \xtwoheadrightarrow{X_2[\mathcal{K}_2]} C_2'$ (with $\mathcal{E}_2 \xtwoheadrightarrow{X_2[\mathcal{K}_2]} \mathcal{E}_2'$), with $X_1 \subseteq \mathbb{E}_1$, $X_2 \subseteq \mathbb{E}_2$, $X_1 \sim X_2$ and $(C_1', C_2') \in R$:

1. for each fresh action $\alpha \in X_1$, if $C_1'' \xtwoheadrightarrow{\alpha[m]} C_1'''$ (with $\mathcal{E}_1'' \xtwoheadrightarrow{\alpha[m]} \mathcal{E}_1'''$), then for some $C_2''$ and $C_2'''$, $C_2'' \xtwoheadrightarrow{\alpha[m]} C_2'''$ (with $\mathcal{E}_2'' \xtwoheadrightarrow{\alpha[m]} \mathcal{E}_2'''$), such that if $(C_1'', C_2'') \in R$, then $(C_1''', C_2''') \in R$;
2. for each $x(y) \in X_1$ with $(y \notin n(\mathcal{E}_1, \mathcal{E}_2))$, if $C_1'' \xtwoheadrightarrow{x(y)[m]} C_1'''$ (with $\mathcal{E}_1'' \xtwoheadrightarrow{x(y)[m]} \mathcal{E}_1'''\{w/y\}$) for all $w$, then for some $C_2''$ and $C_2'''$, $C_2'' \xtwoheadrightarrow{x(y)[m]} C_2'''$ (with $\mathcal{E}_2'' \xtwoheadrightarrow{x(y)[m]} \mathcal{E}_2'''\{w/y\}$) for all $w$, such that if $(C_1'', C_2'') \in R$, then $(C_1''', C_2''') \in R$;
3. for each two $x_1(y), x_2(y) \in X_1$ with $(y \notin n(\mathcal{E}_1, \mathcal{E}_2))$, if $C_1'' \xtwoheadrightarrow{\{x_1(y)[m], x_2(y)[n]\}} C_1'''$ (with $\mathcal{E}_1'' \xtwoheadrightarrow{\{x_1(y)[m], x_2(y)[n]\}} \mathcal{E}_1'''\{w/y\}$) for all $w$, then for some $C_2''$ and $C_2'''$, $C_2'' \xtwoheadrightarrow{\{x_1(y)[m], x_2(y)[n]\}} C_2'''$ (with $\mathcal{E}_2'' \xtwoheadrightarrow{\{x_1(y)[m], x_2(y)[n]\}} \mathcal{E}_2'''\{w/y\}$) for all $w$, such that if $(C_1'', C_2'') \in R$, then $(C_1''', C_2''') \in R$;
4. for each $\overline{x}(y) \in X_1$ with $y \notin n(\mathcal{E}_1, \mathcal{E}_2)$, if $C_1'' \xtwoheadrightarrow{\overline{x}(y)[m]} C_1'''$ (with $\mathcal{E}_1'' \xtwoheadrightarrow{\overline{x}(y)[m]} \mathcal{E}_1'''$), then for some $C_2''$ and $C_2'''$, $C_2'' \xtwoheadrightarrow{\overline{x}(y)[m]} C_2'''$ (with $\mathcal{E}_2'' \xtwoheadrightarrow{\overline{x}(y)[m]} \mathcal{E}_2'''$), such that if $(C_1'', C_2'') \in R$, then $(C_1''', C_2''') \in R$,

and vice versa.

We say that $\mathcal{E}_1, \mathcal{E}_2$ are FR strongly pomset bisimilar, written $\mathcal{E}_1 \sim_p^{fr} \mathcal{E}_2$, if there exists a FR strongly pomset bisimulation $R$, such that $(\emptyset, \emptyset) \in R$. By replacing FR pomset transitions with steps, we can obtain the definition of FR strongly step bisimulation. When PESs $\mathcal{E}_1$ and $\mathcal{E}_2$ are FR strongly step bisimilar, we write $\mathcal{E}_1 \sim_s^{fr} \mathcal{E}_2$.

**Definition 7.6** (Posetal product). Given two PESs $\mathcal{E}_1, \mathcal{E}_2$, the posetal product of their configurations, denoted $\mathcal{C}(\mathcal{E}_1)\overline{\times}\mathcal{C}(\mathcal{E}_2)$, is defined as

$$\{(C_1, f, C_2) | C_1 \in \mathcal{C}(\mathcal{E}_1), C_2 \in \mathcal{C}(\mathcal{E}_2), f : C_1 \to C_2 \text{ isomorphism}\}$$

A subset $R \subseteq \mathcal{C}(\mathcal{E}_1)\overline{\times}\mathcal{C}(\mathcal{E}_2)$ is called a posetal relation. We say that $R$ is downward closed when for any $(C_1, f, C_2)$, $(C_1', f', C_2') \in \mathcal{C}(\mathcal{E}_1)\overline{\times}\mathcal{C}(\mathcal{E}_2)$, if $(C_1, f, C_2) \subseteq (C_1', f', C_2')$ pointwise and $(C_1', f', C_2') \in R$, then $(C_1, f, C_2) \in R$.

For $f : X_1 \to X_2$, we define $f[x_1 \mapsto x_2] : X_1 \cup \{x_1\} \to X_2 \cup \{x_2\}$, $z \in X_1 \cup \{x_1\}$, (1) $f[x_1 \mapsto x_2](z) = x_2$, if $z = x_1$; (2) $f[x_1 \mapsto x_2](z) = f(z)$, otherwise, where $X_1 \subseteq \mathbb{E}_1$, $X_2 \subseteq \mathbb{E}_2$, $x_1 \in \mathbb{E}_1$, $x_2 \in \mathbb{E}_2$.

**Definition 7.7** (FR strongly (hereditary) history-preserving bisimilarity). A FR strongly history-preserving (hp-) bisimulation is a posetal relation $R \subseteq \mathcal{C}(\mathcal{E}_1)\overline{\times}\mathcal{C}(\mathcal{E}_2)$ such that (1) if $(C_1, f, C_2) \in R$, and

1. for $e_1 = \alpha$ a fresh action, if $C_1 \xrightarrow{\alpha} C_1'$ (with $\mathcal{E}_1 \xrightarrow{\alpha} \mathcal{E}_1'$), then for some $C_2'$ and $e_2 = \alpha$, $C_2 \xrightarrow{\alpha} C_2'$ (with $\mathcal{E}_2 \xrightarrow{\alpha} \mathcal{E}_2'$), such that $(C_1', f[e_1 \mapsto e_2], C_2') \in R$;
2. for $e_1 = x(y)$ with $(y \notin n(\mathcal{E}_1, \mathcal{E}_2))$, if $C_1 \xrightarrow{x(y)} C_1'$ (with $\mathcal{E}_1 \xrightarrow{x(y)} \mathcal{E}_1'\{w/y\}$) for all $w$, then for some $C_2'$ and $e_2 = x(y)$, $C_2 \xrightarrow{x(y)} C_2'$ (with $\mathcal{E}_2 \xrightarrow{x(y)} \mathcal{E}_2'\{w/y\}$) for all $w$, such that $(C_1', f[e_1 \mapsto e_2], C_2') \in R$;
3. for $e_1 = \overline{x}(y)$ with $y \notin n(\mathcal{E}_1, \mathcal{E}_2)$, if $C_1 \xrightarrow{\overline{x}(y)} C_1'$ (with $\mathcal{E}_1 \xrightarrow{\overline{x}(y)} \mathcal{E}_1'$), then for some $C_2'$ and $e_2 = \overline{x}(y)$, $C_2 \xrightarrow{\overline{x}(y)} C_2'$ (with $\mathcal{E}_2 \xrightarrow{\overline{x}(y)} \mathcal{E}_2'$), such that $(C_1', f[e_1 \mapsto e_2], C_2') \in R$,

and vice versa; (2) if $(C_1, f, C_2) \in R$, and

1. for $e_1 = \alpha$ a fresh action, if $C_1 \xtwoheadrightarrow{\alpha[m]} C_1'$ (with $\mathcal{E}_1 \xtwoheadrightarrow{\alpha[m]} \mathcal{E}_1'$), then for some $C_2'$ and $e_2 = \alpha$, $C_2 \xtwoheadrightarrow{\alpha[m]} C_2'$ (with $\mathcal{E}_2 \xtwoheadrightarrow{\alpha[m]} \mathcal{E}_2'$), such that $(C_1', f[e_1 \mapsto e_2], C_2') \in R$;

2. for $e_1 = x(y)$ with $(y \notin n(\mathcal{E}_1, \mathcal{E}_2))$, if $C_1 \xrightarrow{x(y)[m]} C_1'$ (with $\mathcal{E}_1 \xrightarrow{x(y)[m]} \mathcal{E}_1'\{w/y\}$) for all $w$, then for some $C_2'$ and $e_2 = x(y)$, $C_2 \xrightarrow{x(y)[m]} C_2'$ (with $\mathcal{E}_2 \xrightarrow{x(y)[m]} \mathcal{E}_2'\{w/y\}$) for all $w$, such that $(C_1', f[e_1 \mapsto e_2], C_2') \in R$;
3. for $e_1 = \overline{x}(y)$ with $y \notin n(\mathcal{E}_1, \mathcal{E}_2)$, if $C_1 \xrightarrow{\overline{x}(y)[m]} C_1'$ (with $\mathcal{E}_1 \xrightarrow{\overline{x}(y)[m]} \mathcal{E}_1'$), then for some $C_2'$ and $e_2 = \overline{x}(y)$, $C_2 \xrightarrow{\overline{x}(y)[m]} C_2'$ (with $\mathcal{E}_2 \xrightarrow{\overline{x}(y)[m]} \mathcal{E}_2'$), such that $(C_1', f[e_1 \mapsto e_2], C_2') \in R$,

and vice versa. $\mathcal{E}_1, \mathcal{E}_2$ are FR strongly history-preserving (hp-)bisimilar and are written $\mathcal{E}_1 \sim_{hp}^{fr} \mathcal{E}_2$ if there exists a FR strongly hp-bisimulation $R$ such that $(\emptyset, \emptyset, \emptyset) \in R$.

A FR strongly hereditary history-preserving (hhp-)bisimulation is a downward closed FR strongly hp-bisimulation. $\mathcal{E}_1, \mathcal{E}_2$ are FR strongly hereditary history-preserving (hhp-)bisimilar and are written $\mathcal{E}_1 \sim_{hhp}^{fr} \mathcal{E}_2$.

## 7.2 Syntax and operational semantics

We assume an infinite set $\mathcal{N}$ of (action or event) names, and use $a, b, c, \cdots$ to range over $\mathcal{N}$, use $x, y, z, w, u, v$ as meta-variables over names. We denote by $\overline{\mathcal{N}}$ the set of co-names and let $\overline{a}, \overline{b}, \overline{c}, \cdots$ range over $\overline{\mathcal{N}}$. Then, we set $\mathcal{L} = \mathcal{N} \cup \overline{\mathcal{N}}$ as the set of labels, and use $l, \overline{l}$ to range over $\mathcal{L}$. We extend complementation to $\mathcal{L}$ such that $\overline{\overline{a}} = a$. Let $\tau$ denote the silent step (internal action or event) and define $Act = \mathcal{L} \cup \{\tau\}$ to be the set of actions, $\alpha, \beta$ range over $Act$. Also, $K, L$ are used to stand for subsets of $\mathcal{L}$ and $\overline{L}$ is used for the set of complements of labels in $L$.

Further, we introduce a set $\mathcal{X}$ of process variables, and a set $\mathcal{K}$ of process constants, and let $X, Y, \cdots$ range over $\mathcal{X}$, and $A, B, \cdots$ range over $\mathcal{K}$. For each process constant $A$, a nonnegative arity $ar(A)$ is assigned to it. Let $\widetilde{x} = x_1, \cdots, x_{ar(A)}$ be a tuple of distinct name variables, then $A(\widetilde{x})$ is called a process constant. $\widetilde{X}$ is a tuple of distinct process variables, and also $E, F, \cdots$ range over the recursive expressions. We write $\mathcal{P}$ for the set of processes. Sometimes, we use $I, J$ to stand for an indexing set, and we write $E_i : i \in I$ for a family of expressions indexed by $I$. $Id_D$ is the identity function or relation over set $D$. The symbol $\equiv_\alpha$ denotes equality under standard alpha-convertibility, note that the subscript $\alpha$ has no relation to the action $\alpha$.

### 7.2.1 Syntax

We use the Prefix . to model the causality relation $\leq$ in true concurrency, the Summation $+$ to model the conflict relation $\sharp$ in true concurrency, and the Composition $\parallel$ to explicitly model concurrent relation in true concurrency. Also, we follow the conventions of process algebra.

**Definition 7.8** (Syntax). A truly concurrent process $\pi_{tc}$ with reversibility is defined inductively by the following formation rules:

1. $A(\widetilde{x}) \in \mathcal{P}$;
2. $\textbf{nil} \in \mathcal{P}$;
3. if $P \in \mathcal{P}$, then the Prefix $\tau.P \in \mathcal{P}$, for $\tau \in Act$ is the silent action;
4. if $P \in \mathcal{P}$, then the Output $\overline{x}y.P \in \mathcal{P}$, for $x, y \in Act$;
5. if $P \in \mathcal{P}$, then the Output $P.\overline{x}y[m] \in \mathcal{P}$, for $x, y \in Act$;
6. if $P \in \mathcal{P}$, then the Input $x(y).P \in \mathcal{P}$, for $x, y \in Act$;
7. if $P \in \mathcal{P}$, then the Input $P.x(y)[m] \in \mathcal{P}$, for $x, y \in Act$;
8. if $P \in \mathcal{P}$, then the Restriction $(x)P \in \mathcal{P}$, for $x \in Act$;
9. if $P, Q \in \mathcal{P}$, then the Summation $P + Q \in \mathcal{P}$;
10. if $P, Q \in \mathcal{P}$, then the Composition $P \parallel Q \in \mathcal{P}$.

The standard BNF grammar of syntax of $\pi_{tc}$ with reversibility can be summarized as follows:

$$P ::= A(\widetilde{x})|\textbf{nil}|\tau.P|\overline{x}y.P|x(y).P|P.\overline{x}y[m]|P.x(y)[m]|(x)P|P + P|P \parallel P$$

In $\overline{x}y$, $x(y)$ and $\overline{x}(y)$, $x$ is called the subject, $y$ is called the object and it may be free or bound.

**Definition 7.9** (Free variables). The free names of a process $P$, $fn(P)$, are defined as follows:

1. $fn(A(\widetilde{x})) \subseteq \{\widetilde{x}\}$;
2. $fn(\textbf{nil}) = \emptyset$;
3. $fn(\tau.P) = fn(P)$;

4. $fn(\overline{x}y.P) = fn(P) \cup \{x\} \cup \{y\}$;
5. $fn(P.\overline{x}y[m]) = fn(P) \cup \{x\} \cup \{y\}$;
6. $fn(x(y).P) = fn(P) \cup \{x\} - \{y\}$;
7. $fn(P.x(y)[m]) = fn(P) \cup \{x\} - \{y\}$;
8. $fn((x)P) = fn(P) - \{x\}$;
9. $fn(P + Q) = fn(P) \cup fn(Q)$;
10. $fn(P \parallel Q) = fn(P) \cup fn(Q)$.

**Definition 7.10** (Bound variables). Let $n(P)$ be the names of a process $P$, then the bound names $bn(P) = n(P) - fn(P)$.

For each process constant schema $A(\widetilde{x})$, a defining equation of the form

$$A(\widetilde{x}) \overset{\text{def}}{=} P$$

is assumed, where $P$ is a process with $fn(P) \subseteq \{\widetilde{x}\}$.

**Definition 7.11** (Substitutions). A substitution is a function $\sigma : \mathcal{N} \to \mathcal{N}$. For $x_i\sigma = y_i$ with $1 \leq i \leq n$, we write $\{y_1/x_1, \cdots, y_n/x_n\}$ or $\{\widetilde{y}/\widetilde{x}\}$ for $\sigma$. For a process $P \in \mathcal{P}$, $P\sigma$ is defined inductively as follows:

1. if $P$ is a process constant $A(\widetilde{x}) = A(x_1, \cdots, x_n)$, then $P\sigma = A(x_1\sigma, \cdots, x_n\sigma)$;
2. if $P = \textbf{nil}$, then $P\sigma = \textbf{nil}$;
3. if $P = \tau.P'$, then $P\sigma = \tau.P'\sigma$;
4. if $P = \overline{x}y.P'$, then $P\sigma = \overline{x\sigma}y\sigma.P'\sigma$;
5. if $P = P'.\overline{x}y[m]$, then $P\sigma = P'\sigma.\overline{x\sigma}y\sigma[m]$;
6. if $P = x(y).P'$, then $P\sigma = x\sigma(y).P'\sigma$;
7. if $P = P'.x(y)[m]$, then $P\sigma = P'\sigma.x\sigma(y)[m]$;
8. if $P = (x)P'$, then $P\sigma = (x\sigma)P'\sigma$;
9. if $P = P_1 + P_2$, then $P\sigma = P_1\sigma + P_2\sigma$;
10. if $P = P_1 \parallel P_2$, then $P\sigma = P_1\sigma \parallel P_2\sigma$.

### 7.2.2 Operational semantics

The operational semantics is defined by LTSs (labeled transition systems), and it is detailed by the following definition.

**Definition 7.12** (Semantics). The operational semantics of $\pi_{tc}$ with reversibility corresponding to the syntax in Definition 7.8 is defined by a series of transition rules, named **ACT**, **SUM**, **IDE**, **PAR**, **COM**, **CLOSE**, **RES**, **OPEN** that indicate that the rules are associated respectively with Prefix, Summation, Identity, Parallel Composition, Communication, and Restriction in Definition 7.8. They are shown in Table 7.1. See also Table 7.2.

### 7.2.3 Properties of transitions

**Proposition 7.13.** 1. *If $P \xrightarrow{\alpha} P'$, then*

   (a) $fn(\alpha) \subseteq fn(P)$;
   (b) $fn(P') \subseteq fn(P) \cup bn(\alpha)$;

2. *If $P \xrightarrow{\{\alpha_1, \cdots, \alpha_n\}} P'$, then*

   (a) $fn(\alpha_1) \cup \cdots \cup fn(\alpha_n) \subseteq fn(P)$;
   (b) $fn(P') \subseteq fn(P) \cup bn(\alpha_1) \cup \cdots \cup bn(\alpha_n)$.

*Proof.* By induction on the depth of inference. □

**Proposition 7.14.** *Suppose that $P \xrightarrow{\alpha(y)} P'$, where $\alpha = x$ or $\alpha = \overline{x}$, and $x \notin n(P)$, then there exists some $P'' \equiv_\alpha P'\{z/y\}$, $P \xrightarrow{\alpha(z)} P''$.*

*Proof.* By induction on the depth of inference. □

**Proposition 7.15.** *If $P \xrightarrow{\alpha} P'$, $bn(\alpha) \cap fn(P'\sigma) = \emptyset$, and $\sigma \lceil bn(\alpha) = id$, then there exists some $P'' \equiv_\alpha P'\sigma$, $P\sigma \xrightarrow{\alpha\sigma} P''$.*

**TABLE 7.1** Forward transition rules.

| Rule | Transition | Condition |
|---|---|---|
| TAU-ACT | $\dfrac{}{\tau.P \xrightarrow{\tau} P}$ | |
| OUTPUT-ACT | $\dfrac{}{\overline{x}y.P \xrightarrow{\overline{x}y} \overline{x}y[m].P}$ | |
| INPUT-ACT | $\dfrac{}{x(z).P \xrightarrow{x(w)} x(z)[m].P\{w/z\}}$ | $(w \notin fn((z)P))$ |
| PAR$_1$ | $\dfrac{P \xrightarrow{\alpha} P' \quad Q \nrightarrow}{P \parallel Q \xrightarrow{\alpha} P' \parallel Q}$ | $(bn(\alpha) \cap fn(Q) = \emptyset)$ |
| PAR$_2$ | $\dfrac{Q \xrightarrow{\alpha} Q' \quad P \nrightarrow}{P \parallel Q \xrightarrow{\alpha} P \parallel Q'}$ | $(bn(\alpha) \cap fn(P) = \emptyset)$ |
| PAR$_3$ | $\dfrac{P \xrightarrow{\alpha} P' \quad Q \xrightarrow{\beta} Q'}{P \parallel Q \xrightarrow{\{\alpha,\beta\}} P' \parallel Q'}$ | $(\beta \neq \overline{\alpha}, bn(\alpha) \cap bn(\beta) = \emptyset, bn(\alpha) \cap fn(Q) = \emptyset, bn(\beta) \cap fn(P) = \emptyset)$ |
| PAR$_4$ | $\dfrac{P \xrightarrow{x_1(z)} P' \quad Q \xrightarrow{x_2(z)} Q'}{P \parallel Q \xrightarrow{\{x_1(w), x_2(w)\}} P'\{w/z\} \parallel Q'\{w/z\}}$ | $(w \notin fn((z)P) \cup fn((z)Q))$ |
| COM | $\dfrac{P \xrightarrow{\overline{x}y} P' \quad Q \xrightarrow{x(z)} Q'}{P \parallel Q \xrightarrow{\tau} P' \parallel Q'\{y/z\}}$ | |
| CLOSE | $\dfrac{P \xrightarrow{\overline{x}(w)} P' \quad Q \xrightarrow{x(w)} Q'}{P \parallel Q \xrightarrow{\tau} (w)(P' \parallel Q')}$ | |
| SUM$_1$ | $\dfrac{P \xrightarrow{\alpha} P'}{P + Q \xrightarrow{\alpha} P'}$ | |
| SUM$_2$ | $\dfrac{P \xrightarrow{\{\alpha_1, \cdots, \alpha_n\}} P'}{P + Q \xrightarrow{\{\alpha_1, \cdots, \alpha_n\}} P'}$ | |
| IDE$_1$ | $\dfrac{P\{\widetilde{y}/\widetilde{x}\} \xrightarrow{\alpha} P'}{A(\widetilde{y}) \xrightarrow{\alpha} P'}$ | $(A(\widetilde{x}) \overset{\text{def}}{=} P)$ |
| IDE$_2$ | $\dfrac{P\{\widetilde{y}/\widetilde{x}\} \xrightarrow{\{\alpha_1, \cdots, \alpha_n\}} P'}{A(\widetilde{y}) \xrightarrow{\{\alpha_1, \cdots, \alpha_n\}} P'}$ | $(A(\widetilde{x}) \overset{\text{def}}{=} P)$ |
| RES$_1$ | $\dfrac{P \xrightarrow{\alpha} P'}{(y)P \xrightarrow{\alpha} (y)P'}$ | $(y \notin n(\alpha))$ |
| RES$_2$ | $\dfrac{P \xrightarrow{\{\alpha_1, \cdots, \alpha_n\}} P'}{(y)P \xrightarrow{\{\alpha_1, \cdots, \alpha_n\}} (y)P'}$ | $(y \notin n(\alpha_1) \cup \cdots \cup n(\alpha_n))$ |
| OPEN$_1$ | $\dfrac{P \xrightarrow{\overline{x}y} P'}{(y)P \xrightarrow{\overline{x}(w)} P'\{w/y\}}$ | $(y \neq x, w \notin fn((y)P'))$ |
| OPEN$_2$ | $\dfrac{P \xrightarrow{\{\overline{x}_1 y, \cdots, \overline{x}_n y\}} P'}{(y)P \xrightarrow{\{\overline{x}_1(w), \cdots, \overline{x}_n(w)\}} P'\{w/y\}}$ | $(y \neq x_1 \neq \cdots \neq x_n, w \notin fn((y)P'))$ |

*Proof.* By the definition of substitution (Definition 7.11) and induction on the depth of inference. □

**Proposition 7.16.** **1.** *If* $P\{w/z\} \xrightarrow{\alpha} P'$, *where* $w \notin fn(P)$ *and* $bn(\alpha) \cap fn(P, w) = \emptyset$, *then there exist some* $Q$ *and* $\beta$ *with* $Q\{w/z\} \equiv_\alpha P'$ *and* $\beta\sigma = \alpha$, $P \xrightarrow{\beta} Q$;

**2.** *If* $P\{w/z\} \xrightarrow{\{\alpha_1, \cdots, \alpha_n\}} P'$, *where* $w \notin fn(P)$ *and* $bn(\alpha_1) \cap \cdots \cap bn(\alpha_n) \cap fn(P, w) = \emptyset$, *then there exist some* $Q$ *and* $\beta_1, \cdots, \beta_n$ *with* $Q\{w/z\} \equiv_\alpha P'$ *and* $\beta_1\sigma = \alpha_1, \cdots, \beta_n\sigma = \alpha_n$, $P \xrightarrow{\{\beta_1, \cdots, \beta_n\}} Q$.

**TABLE 7.2 Reverse transition rules.**

| | | |
|---|---|---|
| RTAU-ACT | $\dfrac{}{\tau.P \overset{\tau}{\twoheadrightarrow} P}$ | |
| ROUTPUT-ACT | $\dfrac{}{P.\overline{x}y[m] \overset{\overline{x}y[m]}{\twoheadrightarrow} P.\overline{x}y}$ | |
| RINPUT-ACT | $\dfrac{}{P.x(z)[m] \overset{x(w)[m]}{\twoheadrightarrow} P\{w/z\}.x(z)}$ | $(w \notin fn((z)P))$ |
| $\text{RPAR}_1$ | $\dfrac{P \overset{\alpha[m]}{\twoheadrightarrow} P' \quad Q \not\twoheadrightarrow}{P \parallel Q \overset{\alpha[m]}{\twoheadrightarrow} P' \parallel Q}$ | $(bn(\alpha) \cap fn(Q) = \emptyset)$ |
| $\text{RPAR}_2$ | $\dfrac{Q \overset{\alpha[m]}{\twoheadrightarrow} Q' \quad P \not\twoheadrightarrow}{P \parallel Q \overset{\alpha[m]}{\twoheadrightarrow} P \parallel Q'}$ | $(bn(\alpha) \cap fn(P) = \emptyset)$ |
| $\text{RPAR}_3$ | $\dfrac{P \overset{\alpha[m]}{\twoheadrightarrow} P' \quad Q \overset{\beta[m]}{\twoheadrightarrow} Q'}{P \parallel Q \overset{\{\alpha[m],\beta[m]\}}{\twoheadrightarrow} P' \parallel Q'}$ | $(\beta \neq \overline{\alpha}, bn(\alpha) \cap bn(\beta) = \emptyset, bn(\alpha) \cap fn(Q) = \emptyset, bn(\beta) \cap fn(P) = \emptyset)$ |
| $\text{RPAR}_4$ | $\dfrac{P \overset{x_1(z)[m]}{\twoheadrightarrow} P' \quad Q \overset{x_2(z)[m]}{\twoheadrightarrow} Q'}{P \parallel Q \overset{\{x_1(w)[m],x_2(w)[m]\}}{\twoheadrightarrow} P'\{w/z\} \parallel Q'\{w/z\}}$ | $(w \notin fn((z)P) \cup fn((z)Q))$ |
| RCOM | $\dfrac{P \overset{\overline{x}y[m]}{\twoheadrightarrow} P' \quad Q \overset{x(z)[m]}{\twoheadrightarrow} Q'}{P \parallel Q \overset{\tau}{\twoheadrightarrow} P' \parallel Q'\{y/z\}}$ | |
| RCLOSE | $\dfrac{P \overset{\overline{x}(w)[m]}{\twoheadrightarrow} P' \quad Q \overset{x(w)[m]}{\twoheadrightarrow} Q'}{P \parallel Q \overset{\tau}{\twoheadrightarrow} (w)(P' \parallel Q')}$ | |
| $\text{RSUM}_1$ | $\dfrac{P \overset{\alpha[m]}{\twoheadrightarrow} P'}{P + Q \overset{\alpha[m]}{\twoheadrightarrow} P'}$ | |
| $\text{RSUM}_2$ | $\dfrac{P \overset{\{\alpha_1[m],\cdots,\alpha_n[m]\}}{\twoheadrightarrow} P'}{P + Q \overset{\{\alpha_1[m],\cdots,\alpha_n[m]\}}{\twoheadrightarrow} P'}$ | |
| $\text{RIDE}_1$ | $\dfrac{P\{\widetilde{y}/\widetilde{x}\} \overset{\alpha[m]}{\twoheadrightarrow} P'}{A(\widetilde{y}) \overset{\alpha[m]}{\twoheadrightarrow} P'}$ | $(A(\widetilde{x}) \overset{\text{def}}{=} P)$ |
| $\text{RIDE}_2$ | $\dfrac{P\{\widetilde{y}/\widetilde{x}\} \overset{\{\alpha_1[m],\cdots,\alpha_n[m]\}}{\twoheadrightarrow} P'}{A(\widetilde{y}) \overset{\{\alpha_1[m],\cdots,\alpha_n[m]\}}{\twoheadrightarrow} P'}$ | $(A(\widetilde{x}) \overset{\text{def}}{=} P)$ |
| $\text{RRES}_1$ | $\dfrac{P \overset{\alpha[m]}{\twoheadrightarrow} P'}{(y)P \overset{\alpha[m]}{\twoheadrightarrow} (y)P'}$ | $(y \notin n(\alpha))$ |
| $\text{RRES}_2$ | $\dfrac{P \overset{\{\alpha_1[m],\cdots,\alpha_n[m]\}}{\twoheadrightarrow} P'}{(y)P \overset{\{\alpha_1[m],\cdots,\alpha_n[m]\}}{\twoheadrightarrow} (y)P'}$ | $(y \notin n(\alpha_1) \cup \cdots \cup n(\alpha_n))$ |
| $\text{ROPEN}_1$ | $\dfrac{P \overset{\overline{x}y[m]}{\twoheadrightarrow} P'}{(y)P \overset{\overline{x}(w)[m]}{\twoheadrightarrow} P'\{w/y\}}$ | $(y \neq x, w \notin fn((y)P'))$ |
| $\text{ROPEN}_2$ | $\dfrac{P \overset{\{\overline{x}_1 y[m],\cdots,\overline{x}_n y[m]\}}{\twoheadrightarrow} P'}{(y)P \overset{\{\overline{x}_1(w)[m],\cdots,\overline{x}_n(w)[m]\}}{\twoheadrightarrow} P'\{w/y\}}$ | $(y \neq x_1 \neq \cdots \neq x_n, w \notin fn((y)P'))$ |

*Proof.* By the definition of substitution (Definition 7.11) and induction on the depth of inference. □

**Proposition 7.17. 1.** *If* $P \overset{\alpha[m]}{\twoheadrightarrow} P'$, *then*

**(a)** $fn(\alpha[m]) \subseteq fn(P)$;

(b) $fn(P') \subseteq fn(P) \cup bn(\alpha[m])$;

2. If $P \xrightarrow{\{\alpha_1[m],\cdots,\alpha_n[m]\}} P'$, then
   (a) $fn(\alpha_1[m]) \cup \cdots \cup fn(\alpha_n[m]) \subseteq fn(P)$;
   (b) $fn(P') \subseteq fn(P) \cup bn(\alpha_1[m]) \cup \cdots \cup bn(\alpha_n[m])$.

*Proof.* By induction on the depth of inference. □

**Proposition 7.18.** *Suppose that $P \xrightarrow{\alpha(y)[m]} P'$, where $\alpha = x$ or $\alpha = \overline{x}$, and $x \notin n(P)$, then there exists some $P'' \equiv_\alpha P'\{z/y\}$, $P \xrightarrow{\alpha(z)[m]} P''$.*

*Proof.* By induction on the depth of inference. □

**Proposition 7.19.** *If $P \xrightarrow{\alpha[m]} P'$, $bn(\alpha[m]) \cap fn(P'\sigma) = \emptyset$, and $\sigma\lceil bn(\alpha[m]) = id$, then there exists some $P'' \equiv_\alpha P'\sigma$, $P\sigma \xrightarrow{\alpha[m]\sigma} P''$.*

*Proof.* By the definition of substitution (Definition 7.11) and induction on the depth of inference. □

**Proposition 7.20.** 1. *If $P\{w/z\} \xrightarrow{\alpha[m]} P'$, where $w \notin fn(P)$ and $bn(\alpha) \cap fn(P, w) = \emptyset$, then there exist some $Q$ and $\beta$ with $Q\{w/z\} \equiv_\alpha P'$ and $\beta\sigma[m] = \alpha[m]$, $P \xrightarrow{\beta[m]} Q$;*

2. *If $P\{w/z\} \xrightarrow{\{\alpha_1[m],\cdots,\alpha_n[m]\}} P'$, where $w \notin fn(P)$ and $bn(\alpha_1[m]) \cap \cdots \cap bn(\alpha_n[m]) \cap fn(P, w) = \emptyset$, then there exist some $Q$ and $\beta_1[m], \cdots, \beta_n[m]$ with $Q\{w/z\} \equiv_\alpha P'$ and $\beta_1\sigma[m] = \alpha_1[m], \cdots, \beta_n\sigma[m] = \alpha_n[m]$, $P \xrightarrow{\{\beta_1[m],\cdots,\beta_n[m]\}} Q$.*

*Proof.* By the definition of substitution (Definition 7.11) and induction on the depth of inference. □

## 7.3 Strong bisimilarities

### 7.3.1 Laws and congruence

**Theorem 7.21.** *$\equiv_\alpha$ are FR strongly truly concurrent bisimulations. That is, if $P \equiv_\alpha Q$, then,*

1. $P \sim_p^{fr} Q$;
2. $P \sim_s^{fr} Q$;
3. $P \sim_{hp}^{fr} Q$;
4. $P \sim_{hhp}^{fr} Q$.

*Proof.* By induction on the depth of inference, we can obtain the following facts:

1. If $\alpha$ is a free action and $P \xrightarrow{\alpha} P'$, then equally for some $Q'$ with $P' \equiv_\alpha Q'$, $Q \xrightarrow{\alpha} Q'$;
2. If $P \xrightarrow{a(y)} P'$ with $a = x$ or $a = \overline{x}$ and $z \notin n(Q)$, then equally for some $Q'$ with $P'\{z/y\} \equiv_\alpha Q'$, $Q \xrightarrow{a(z)} Q'$;
3. If $\alpha[m]$ is a free action and $P \xrightarrow{\alpha[m]} P'$, then equally for some $Q'$ with $P' \equiv_\alpha Q'$, $Q \xrightarrow{\alpha[m]} Q'$;
4. If $P \xrightarrow{a(y)[m]} P'$ with $a = x$ or $a = \overline{x}$ and $z \notin n(Q)$, then equally for some $Q'$ with $P'\{z/y\} \equiv_\alpha Q'$, $Q \xrightarrow{a(z)[m]} Q'$.

Then, we can obtain:

1. by the definition of FR strongly pomset bisimilarity, $P \sim_p^{fr} Q$;
2. by the definition of FR strongly step bisimilarity, $P \sim_s^{fr} Q$;
3. by the definition of FR strongly hp-bisimilarity, $P \sim_{hp}^{fr} Q$;
4. by the definition of FR strongly hhp-bisimilarity, $P \sim_{hhp}^{fr} Q$. □

**Proposition 7.22** (Summation laws for FR strongly pomset bisimulation). *The Summation laws for FR strongly pomset bisimulation are as follows:*

1. $P + Q \sim_p^{fr} Q + P$;
2. $P + (Q + R) \sim_p^{fr} (P + Q) + R$;

3. $P + P \sim_p^{fr} P$;
4. $P + \mathbf{nil} \sim_p^{fr} P$.

*Proof.* 1. $P + Q \sim_p^{fr} Q + P$. It is sufficient to prove the relation $R = \{(P + Q, Q + P)\} \cup \mathbf{Id}$ is a FR strongly pomset bisimulation, however, we omit it here;
2. $P + (Q + R) \sim_p^{fr} (P + Q) + R$. It is sufficient to prove the relation $R = \{(P + (Q + R), (P + Q) + R)\} \cup \mathbf{Id}$ is a FR strongly pomset bisimulation, however, we omit it here;
3. $P + P \sim_p^{fr} P$. It is sufficient to prove the relation $R = \{(P + P, P)\} \cup \mathbf{Id}$ is a FR strongly pomset bisimulation, however, we omit it here;
4. $P + \mathbf{nil} \sim_p^{fr} P$. It is sufficient to prove the relation $R = \{(P + \mathbf{nil}, P)\} \cup \mathbf{Id}$ is a FR strongly pomset bisimulation, however, we omit it here. □

**Proposition 7.23** (Summation laws for FR strongly step bisimulation). *The Summation laws for FR strongly step bisimulation are as follows:*

1. $P + Q \sim_s^{fr} Q + P$;
2. $P + (Q + R) \sim_s^{fr} (P + Q) + R$;
3. $P + P \sim_s^{fr} P$;
4. $P + \mathbf{nil} \sim_s^{fr} P$.

*Proof.* 1. $P + Q \sim_s^{fr} Q + P$. It is sufficient to prove the relation $R = \{(P + Q, Q + P)\} \cup \mathbf{Id}$ is a FR strongly step bisimulation, however, we omit it here;
2. $P + (Q + R) \sim_s^{fr} (P + Q) + R$. It is sufficient to prove the relation $R = \{(P + (Q + R), (P + Q) + R)\} \cup \mathbf{Id}$ is a FR strongly step bisimulation, however, we omit it here;
3. $P + P \sim_s^{fr} P$. It is sufficient to prove the relation $R = \{(P + P, P)\} \cup \mathbf{Id}$ is a FR strongly step bisimulation, however, we omit it here;
4. $P + \mathbf{nil} \sim_s^{fr} P$. It is sufficient to prove the relation $R = \{(P + \mathbf{nil}, P)\} \cup \mathbf{Id}$ is a FR strongly step bisimulation, however, we omit it here. □

**Proposition 7.24** (Summation laws for FR strongly hp-bisimulation). *The Summation laws for FR strongly hp-bisimulation are as follows:*

1. $P + Q \sim_{hp}^{fr} Q + P$;
2. $P + (Q + R) \sim_{hp}^{fr} (P + Q) + R$;
3. $P + P \sim_{hp}^{fr} P$;
4. $P + \mathbf{nil} \sim_{hp}^{fr} P$.

*Proof.* 1. $P + Q \sim_{hp}^{fr} Q + P$. It is sufficient to prove the relation $R = \{(P + Q, Q + P)\} \cup \mathbf{Id}$ is a FR strongly hp-bisimulation, however, we omit it here;
2. $P + (Q + R) \sim_{hp}^{fr} (P + Q) + R$. It is sufficient to prove the relation $R = \{(P + (Q + R), (P + Q) + R)\} \cup \mathbf{Id}$ is a FR strongly hp-bisimulation, however, we omit it here;
3. $P + P \sim_{hp}^{fr} P$. It is sufficient to prove the relation $R = \{(P + P, P)\} \cup \mathbf{Id}$ is a FR strongly hp-bisimulation, however, we omit it here;
4. $P + \mathbf{nil} \sim_{hp}^{fr} P$. It is sufficient to prove the relation $R = \{(P + \mathbf{nil}, P)\} \cup \mathbf{Id}$ is a FR strongly hp-bisimulation, however, we omit it here. □

**Proposition 7.25** (Summation laws for FR strongly hhp-bisimulation). *The Summation laws for FR strongly hhp-bisimulation are as follows:*

1. $P + Q \sim_{hhp}^{fr} Q + P$;
2. $P + (Q + R) \sim_{hhp}^{fr} (P + Q) + R$;
3. $P + P \sim_{hhp}^{fr} P$;
4. $P + \mathbf{nil} \sim_{hhp}^{fr} P$.

*Proof.* **1.** $P + Q \sim^{fr}_{hhp} Q + P$. It is sufficient to prove the relation $R = \{(P + Q, Q + P)\} \cup \mathbf{Id}$ is a FR strongly hhp-bisimulation, however, we omit it here;

**2.** $P + (Q + R) \sim^{fr}_{hhp} (P + Q) + R$. It is sufficient to prove the relation $R = \{(P + (Q + R), (P + Q) + R)\} \cup \mathbf{Id}$ is a FR strongly hhp-bisimulation, however, we omit it here;

**3.** $P + P \sim^{fr}_{hhp} P$. It is sufficient to prove the relation $R = \{(P + P, P)\} \cup \mathbf{Id}$ is a FR strongly hhp-bisimulation, however, we omit it here;

**4.** $P + \mathbf{nil} \sim^{fr}_{hhp} P$. It is sufficient to prove the relation $R = \{(P + \mathbf{nil}, P)\} \cup \mathbf{Id}$ is a FR strongly hhp-bisimulation, however, we omit it here. □

**Theorem 7.26** (Identity law for FR strongly truly concurrent bisimilarities). *If* $A(\widetilde{x}) \stackrel{def}{=} P$, *then*

1. $A(\widetilde{y}) \sim^{fr}_{p} P\{\widetilde{y}/\widetilde{x}\}$;
2. $A(\widetilde{y}) \sim^{fr}_{s} P\{\widetilde{y}/\widetilde{x}\}$;
3. $A(\widetilde{y}) \sim^{fr}_{hp} P\{\widetilde{y}/\widetilde{x}\}$;
4. $A(\widetilde{y}) \sim^{fr}_{hhp} P\{\widetilde{y}/\widetilde{x}\}$.

*Proof.* **1.** $A(\widetilde{y}) \sim^{fr}_{p} P\{\widetilde{y}/\widetilde{x}\}$. It is sufficient to prove the relation $R = \{(A(\widetilde{y}), P\{\widetilde{y}/\widetilde{x}\})\} \cup \mathbf{Id}$ is a FR strongly pomset bisimulation, however, we omit it here;

**2.** $A(\widetilde{y}) \sim^{fr}_{s} P\{\widetilde{y}/\widetilde{x}\}$. It is sufficient to prove the relation $R = \{(A(\widetilde{y}), P\{\widetilde{y}/\widetilde{x}\})\} \cup \mathbf{Id}$ is a FR strongly step bisimulation, however, we omit it here;

**3.** $A(\widetilde{y}) \sim^{fr}_{hp} P\{\widetilde{y}/\widetilde{x}\}$. It is sufficient to prove the relation $R = \{(A(\widetilde{y}), P\{\widetilde{y}/\widetilde{x}\})\} \cup \mathbf{Id}$ is a FR strongly hp-bisimulation, however, we omit it here;

**4.** $A(\widetilde{y}) \sim^{fr}_{hhp} P\{\widetilde{y}/\widetilde{x}\}$. It is sufficient to prove the relation $R = \{(A(\widetilde{y}), P\{\widetilde{y}/\widetilde{x}\})\} \cup \mathbf{Id}$ is a FR strongly hhp-bisimulation, however, we omit it here. □

**Theorem 7.27** (Restriction laws for FR strongly pomset bisimilarity). *The restriction laws for FR strongly pomset bisimilarity are as follows:*

1. $(y)P \sim^{fr}_{p} P$, *if* $y \notin fn(P)$;
2. $(y)(z)P \sim^{fr}_{p} (z)(y)P$;
3. $(y)(P + Q) \sim^{fr}_{p} (y)P + (y)Q$;
4. $(y)\alpha.P \sim^{fr}_{p} \alpha.(y)P$ *if* $y \notin n(\alpha)$;
5. $(y)\alpha.P \sim^{fr}_{p} \mathbf{nil}$ *if* $y$ *is the subject of* $\alpha$.

*Proof.* **1.** $(y)P \sim^{fr}_{p} P$, if $y \notin fn(P)$. It is sufficient to prove the relation $R = \{((y)P, P)\} \cup \mathbf{Id}$, if $y \notin fn(P)$, is a FR strongly pomset bisimulation, however, we omit it here;

**2.** $(y)(z)P \sim^{fr}_{p} (z)(y)P$. It is sufficient to prove the relation $R = \{((y)(z)P, (z)(y)P)\} \cup \mathbf{Id}$ is a FR strongly pomset bisimulation, however, we omit it here;

**3.** $(y)(P + Q) \sim^{fr}_{p} (y)P + (y)Q$. It is sufficient to prove the relation $R = \{((y)(P + Q), (y)P + (y)Q)\} \cup \mathbf{Id}$ is a FR strongly pomset bisimulation, however, we omit it here;

**4.** $(y)\alpha.P \sim^{fr}_{p} \alpha.(y)P$ if $y \notin n(\alpha)$. It is sufficient to prove the relation $R = \{((y)\alpha.P, \alpha.(y)P)\} \cup \mathbf{Id}$, if $y \notin n(\alpha)$, is a FR strongly pomset bisimulation, however, we omit it here;

**5.** $(y)\alpha.P \sim^{fr}_{p} \mathbf{nil}$ if $y$ is the subject of $\alpha$. It is sufficient to prove the relation $R = \{((y)\alpha.P, \mathbf{nil})\} \cup \mathbf{Id}$, if $y$ is the subject of $\alpha$, is a FR strongly pomset bisimulation, however, we omit it here. □

**Theorem 7.28** (Restriction laws for FR strongly step bisimilarity). *The restriction laws for FR strongly step bisimilarity are as follows:*

1. $(y)P \sim^{fr}_{s} P$, *if* $y \notin fn(P)$;
2. $(y)(z)P \sim^{fr}_{s} (z)(y)P$;
3. $(y)(P + Q) \sim^{fr}_{s} (y)P + (y)Q$;
4. $(y)\alpha.P \sim^{fr}_{s} \alpha.(y)P$ *if* $y \notin n(\alpha)$;
5. $(y)\alpha.P \sim^{fr}_{s} \mathbf{nil}$ *if* $y$ *is the subject of* $\alpha$.

*Proof.* **1.** $(y)P \sim_s^{fr} P$, if $y \notin fn(P)$. It is sufficient to prove the relation $R = \{((y)P, P)\} \cup \mathbf{Id}$, if $y \notin fn(P)$, is a FR strongly step bisimulation, however, we omit it here;
**2.** $(y)(z)P \sim_s^{fr} (z)(y)P$. It is sufficient to prove the relation $R = \{((y)(z)P, (z)(y)P)\} \cup \mathbf{Id}$ is a FR strongly step bisimulation, however, we omit it here;
**3.** $(y)(P+Q) \sim_s^{fr} (y)P + (y)Q$. It is sufficient to prove the relation $R = \{((y)(P+Q), (y)P + (y)Q)\} \cup \mathbf{Id}$ is a FR strongly step bisimulation, however, we omit it here;
**4.** $(y)\alpha.P \sim_s^{fr} \alpha.(y)P$ if $y \notin n(\alpha)$. It is sufficient to prove the relation $R = \{((y)\alpha.P, \alpha.(y)P)\} \cup \mathbf{Id}$, if $y \notin n(\alpha)$, is a FR strongly step bisimulation, however, we omit it here;
**5.** $(y)\alpha.P \sim_s^{fr} \mathbf{nil}$ if $y$ is the subject of $\alpha$. It is sufficient to prove the relation $R = \{((y)\alpha.P, \mathbf{nil})\} \cup \mathbf{Id}$, if $y$ is the subject of $\alpha$, is a FR strongly step bisimulation, however, we omit it here. □

**Theorem 7.29** (Restriction laws for FR strongly hp-bisimilarity). *The restriction laws for FR strongly hp-bisimilarity are as follows:*

1. $(y)P \sim_{hp}^{fr} P$, *if* $y \notin fn(P)$;
2. $(y)(z)P \sim_{hp}^{fr} (z)(y)P$;
3. $(y)(P+Q) \sim_{hp}^{fr} (y)P + (y)Q$;
4. $(y)\alpha.P \sim_{hp}^{fr} \alpha.(y)P$ *if* $y \notin n(\alpha)$;
5. $(y)\alpha.P \sim_{hp}^{fr} \mathbf{nil}$ *if* $y$ *is the subject of* $\alpha$.

*Proof.* **1.** $(y)P \sim_{hp}^{fr} P$, if $y \notin fn(P)$. It is sufficient to prove the relation $R = \{((y)P, P)\} \cup \mathbf{Id}$, if $y \notin fn(P)$, is a FR strongly hp-bisimulation, however, we omit it here;
**2.** $(y)(z)P \sim_{hp}^{fr} (z)(y)P$. It is sufficient to prove the relation $R = \{((y)(z)P, (z)(y)P)\} \cup \mathbf{Id}$ is a FR strongly hp-bisimulation, however, we omit it here;
**3.** $(y)(P+Q) \sim_{hp}^{fr} (y)P + (y)Q$. It is sufficient to prove the relation $R = \{((y)(P+Q), (y)P + (y)Q)\} \cup \mathbf{Id}$ is a FR strongly hp-bisimulation, however, we omit it here;
**4.** $(y)\alpha.P \sim_{hp}^{fr} \alpha.(y)P$ if $y \notin n(\alpha)$. It is sufficient to prove the relation $R = \{((y)\alpha.P, \alpha.(y)P)\} \cup \mathbf{Id}$, if $y \notin n(\alpha)$, is a FR strongly hp-bisimulation, however, we omit it here;
**5.** $(y)\alpha.P \sim_{hp}^{fr} \mathbf{nil}$ if $y$ is the subject of $\alpha$. It is sufficient to prove the relation $R = \{((y)\alpha.P, \mathbf{nil})\} \cup \mathbf{Id}$, if $y$ is the subject of $\alpha$, is a FR strongly hp-bisimulation, however, we omit it here. □

**Theorem 7.30** (Restriction laws for FR strongly hhp-bisimilarity). *The restriction laws for FR strongly hhp-bisimilarity are as follows:*

1. $(y)P \sim_{hhp}^{fr} P$, *if* $y \notin fn(P)$;
2. $(y)(z)P \sim_{hhp}^{fr} (z)(y)P$;
3. $(y)(P+Q) \sim_{hhp}^{fr} (y)P + (y)Q$;
4. $(y)\alpha.P \sim_{hhp}^{fr} \alpha.(y)P$ *if* $y \notin n(\alpha)$;
5. $(y)\alpha.P \sim_{hhp}^{fr} \mathbf{nil}$ *if* $y$ *is the subject of* $\alpha$.

*Proof.* **1.** $(y)P \sim_{hhp}^{fr} P$, if $y \notin fn(P)$. It is sufficient to prove the relation $R = \{((y)P, P)\} \cup \mathbf{Id}$, if $y \notin fn(P)$, is a FR strongly hhp-bisimulation, however, we omit it here;
**2.** $(y)(z)P \sim_{hhp}^{fr} (z)(y)P$. It is sufficient to prove the relation $R = \{((y)(z)P, (z)(y)P)\} \cup \mathbf{Id}$ is a FR strongly hhp-bisimulation, however, we omit it here;
**3.** $(y)(P+Q) \sim_{hhp}^{fr} (y)P + (y)Q$. It is sufficient to prove the relation $R = \{((y)(P+Q), (y)P + (y)Q)\} \cup \mathbf{Id}$ is a FR strongly hhp-bisimulation, however, we omit it here;
**4.** $(y)\alpha.P \sim_{hhp}^{fr} \alpha.(y)P$ if $y \notin n(\alpha)$. It is sufficient to prove the relation $R = \{((y)\alpha.P, \alpha.(y)P)\} \cup \mathbf{Id}$, if $y \notin n(\alpha)$, is a FR strongly hhp-bisimulation, however, we omit it here;
**5.** $(y)\alpha.P \sim_{hhp}^{fr} \mathbf{nil}$ if $y$ is the subject of $\alpha$. It is sufficient to prove the relation $R = \{((y)\alpha.P, \mathbf{nil})\} \cup \mathbf{Id}$, if $y$ is the subject of $\alpha$, is a FR strongly hhp-bisimulation, however, we omit it here. □

**Theorem 7.31** (Parallel laws for FR strongly pomset bisimilarity). *The parallel laws for FR strongly pomset bisimilarity are as follows:*

1. $P \parallel \mathbf{nil} \sim_p^{fr} P$;
2. $P_1 \parallel P_2 \sim_p^{fr} P_2 \parallel P_1$;
3. $(P_1 \parallel P_2) \parallel P_3 \sim_p^{fr} P_1 \parallel (P_2 \parallel P_3)$;
4. $(y)(P_1 \parallel P_2) \sim_p^{fr} (y)P_1 \parallel (y)P_2$, *if* $y \notin fn(P_1) \cap fn(P_2)$.

*Proof.* 1. $P \parallel \mathbf{nil} \sim_p^{fr} P$. It is sufficient to prove the relation $R = \{(P \parallel \mathbf{nil}, P)\} \cup \mathbf{Id}$ is a FR strongly pomset bisimulation, however, we omit it here;
2. $P_1 \parallel P_2 \sim_p^{fr} P_2 \parallel P_1$. It is sufficient to prove the relation $R = \{(P_1 \parallel P_2, P_2 \parallel P_1)\} \cup \mathbf{Id}$ is a FR strongly pomset bisimulation, however, we omit it here;
3. $(P_1 \parallel P_2) \parallel P_3 \sim_p^{fr} P_1 \parallel (P_2 \parallel P_3)$. It is sufficient to prove the relation $R = \{((P_1 \parallel P_2) \parallel P_3, P_1 \parallel (P_2 \parallel P_3))\} \cup \mathbf{Id}$ is a FR strongly pomset bisimulation, however, we omit it here;
4. $(y)(P_1 \parallel P_2) \sim_p^{fr} (y)P_1 \parallel (y)P_2$, if $y \notin fn(P_1) \cap fn(P_2)$. It is sufficient to prove the relation $R = \{((y)(P_1 \parallel P_2), (y)P_1 \parallel (y)P_2)\} \cup \mathbf{Id}$, if $y \notin fn(P_1) \cap fn(P_2)$, is a FR strongly pomset bisimulation, however, we omit it here. □

**Theorem 7.32** (Parallel laws for FR strongly step bisimilarity). *The parallel laws for FR strongly step bisimilarity are as follows:*

1. $P \parallel \mathbf{nil} \sim_s^{fr} P$;
2. $P_1 \parallel P_2 \sim_s^{fr} P_2 \parallel P_1$;
3. $(P_1 \parallel P_2) \parallel P_3 \sim_s^{fr} P_1 \parallel (P_2 \parallel P_3)$;
4. $(y)(P_1 \parallel P_2) \sim_s^{fr} (y)P_1 \parallel (y)P_2$, *if* $y \notin fn(P_1) \cap fn(P_2)$.

*Proof.* 1. $P \parallel \mathbf{nil} \sim_s^{fr} P$. It is sufficient to prove the relation $R = \{(P \parallel \mathbf{nil}, P)\} \cup \mathbf{Id}$ is a FR strongly step bisimulation, however, we omit it here;
2. $P_1 \parallel P_2 \sim_s^{fr} P_2 \parallel P_1$. It is sufficient to prove the relation $R = \{(P_1 \parallel P_2, P_2 \parallel P_1)\} \cup \mathbf{Id}$ is a FR strongly step bisimulation, however, we omit it here;
3. $(P_1 \parallel P_2) \parallel P_3 \sim_s^{fr} P_1 \parallel (P_2 \parallel P_3)$. It is sufficient to prove the relation $R = \{((P_1 \parallel P_2) \parallel P_3, P_1 \parallel (P_2 \parallel P_3))\} \cup \mathbf{Id}$ is a FR strongly step bisimulation, however, we omit it here;
4. $(y)(P_1 \parallel P_2) \sim_s^{fr} (y)P_1 \parallel (y)P_2$, if $y \notin fn(P_1) \cap fn(P_2)$. It is sufficient to prove the relation $R = \{((y)(P_1 \parallel P_2), (y)P_1 \parallel (y)P_2)\} \cup \mathbf{Id}$, if $y \notin fn(P_1) \cap fn(P_2)$, is a FR strongly step bisimulation, however, we omit it here. □

**Theorem 7.33** (Parallel laws for FR strongly hp-bisimilarity). *The parallel laws for FR strongly hp-bisimilarity are as follows:*

1. $P \parallel \mathbf{nil} \sim_{hp}^{fr} P$;
2. $P_1 \parallel P_2 \sim_{hp}^{fr} P_2 \parallel P_1$;
3. $(P_1 \parallel P_2) \parallel P_3 \sim_{hp}^{fr} P_1 \parallel (P_2 \parallel P_3)$;
4. $(y)(P_1 \parallel P_2) \sim_{hp}^{fr} (y)P_1 \parallel (y)P_2$, *if* $y \notin fn(P_1) \cap fn(P_2)$.

*Proof.* 1. $P \parallel \mathbf{nil} \sim_{hp}^{fr} P$. It is sufficient to prove the relation $R = \{(P \parallel \mathbf{nil}, P)\} \cup \mathbf{Id}$ is a FR strongly hp-bisimulation, however, we omit it here;
2. $P_1 \parallel P_2 \sim_{hp}^{fr} P_2 \parallel P_1$. It is sufficient to prove the relation $R = \{(P_1 \parallel P_2, P_2 \parallel P_1)\} \cup \mathbf{Id}$ is a FR strongly hp-bisimulation, however, we omit it here;
3. $(P_1 \parallel P_2) \parallel P_3 \sim_{hp}^{fr} P_1 \parallel (P_2 \parallel P_3)$. It is sufficient to prove the relation $R = \{((P_1 \parallel P_2) \parallel P_3, P_1 \parallel (P_2 \parallel P_3))\} \cup \mathbf{Id}$ is a FR strongly hp-bisimulation, however, we omit it here;
4. $(y)(P_1 \parallel P_2) \sim_{hp}^{fr} (y)P_1 \parallel (y)P_2$, if $y \notin fn(P_1) \cap fn(P_2)$. It is sufficient to prove the relation $R = \{((y)(P_1 \parallel P_2), (y)P_1 \parallel (y)P_2)\} \cup \mathbf{Id}$, if $y \notin fn(P_1) \cap fn(P_2)$, is a FR strongly hp-bisimulation, however, we omit it here. □

**Theorem 7.34** (Parallel laws for FR strongly hhp-bisimilarity). *The parallel laws for FR strongly hhp-bisimilarity are as follows:*

1. $P \parallel \mathbf{nil} \sim_{hhp}^{fr} P$;
2. $P_1 \parallel P_2 \sim_{hhp}^{fr} P_2 \parallel P_1$;
3. $(P_1 \parallel P_2) \parallel P_3 \sim_{hhp}^{fr} P_1 \parallel (P_2 \parallel P_3)$;

4. $(y)(P_1 \parallel P_2) \sim^{fr}_{hhp} (y)P_1 \parallel (y)P_2$, *if* $y \notin fn(P_1) \cap fn(P_2)$.

*Proof.* 1. $P \parallel \mathbf{nil} \sim^{fr}_{hhp} P$. It is sufficient to prove the relation $R = \{(P \parallel \mathbf{nil}, P)\} \cup \mathbf{Id}$ is a FR strongly hhp-bisimulation, however, we omit it here;

2. $P_1 \parallel P_2 \sim^{fr}_{hhp} P_2 \parallel P_1$. It is sufficient to prove the relation $R = \{(P_1 \parallel P_2, P_2 \parallel P_1)\} \cup \mathbf{Id}$ is a FR strongly hhp-bisimulation, however, we omit it here;
3. $(P_1 \parallel P_2) \parallel P_3 \sim^{fr}_{hhp} P_1 \parallel (P_2 \parallel P_3)$. It is sufficient to prove the relation $R = \{((P_1 \parallel P_2) \parallel P_3, P_1 \parallel (P_2 \parallel P_3))\} \cup \mathbf{Id}$ is a FR strongly hhp-bisimulation, however, we omit it here;
4. $(y)(P_1 \parallel P_2) \sim^{fr}_{hhp} (y)P_1 \parallel (y)P_2$, if $y \notin fn(P_1) \cap fn(P_2)$. It is sufficient to prove the relation $R = \{((y)(P_1 \parallel P_2), (y)P_1 \parallel (y)P_2)\} \cup \mathbf{Id}$, if $y \notin fn(P_1) \cap fn(P_2)$, is a FR strongly hhp-bisimulation, however, we omit it here. □

**Theorem 7.35** (Expansion law for truly concurrent bisimilarities). *Let* $P \equiv \sum_i \alpha_i.P_i$ *and* $Q \equiv \sum_j \beta_j.Q_j$, *where* $bn(\alpha_i) \cap fn(Q) = \emptyset$ *for all* $i$, *and* $bn(\beta_j) \cap fn(P) = \emptyset$ *for all* $j$. *Then,*

1. $P \parallel Q \sim^{fr}_p \sum_i \sum_j (\alpha_i \parallel \beta_j).(P_i \parallel Q_j) + \sum_{\alpha_i\ comp\ \beta_j} \tau.R_{ij}$;
2. $P \parallel Q \sim^{fr}_s \sum_i \sum_j (\alpha_i \parallel \beta_j).(P_i \parallel Q_j) + \sum_{\alpha_i\ comp\ \beta_j} \tau.R_{ij}$;
3. $P \parallel Q \sim^{fr}_{hp} \sum_i \sum_j (\alpha_i \parallel \beta_j).(P_i \parallel Q_j) + \sum_{\alpha_i\ comp\ \beta_j} \tau.R_{ij}$;
4. $P \parallel Q \nsim_{phhp} \sum_i \sum_j (\alpha_i \parallel \beta_j).(P_i \parallel Q_j) + \sum_{\alpha_i\ comp\ \beta_j} \tau.R_{ij}$,

*where* $\alpha_i\ comp\ \beta_j$ *and* $R_{ij}$ *are defined as follows:*

1. $\alpha_i$ *is* $\overline{x}u$ *and* $\beta_j$ *is* $x(v)$, *then* $R_{ij} = P_i \parallel Q_j\{u/v\}$;
2. $\alpha_i$ *is* $\overline{x}(u)$ *and* $\beta_j$ *is* $x(v)$, *then* $R_{ij} = (w)(P_i\{w/u\} \parallel Q_j\{w/v\})$, *if* $w \notin fn((u)P_i) \cup fn((v)Q_j)$;
3. $\alpha_i$ *is* $x(v)$ *and* $\beta_j$ *is* $\overline{x}u$, *then* $R_{ij} = P_i\{u/v\} \parallel Q_j$;
4. $\alpha_i$ *is* $x(v)$ *and* $\beta_j$ *is* $\overline{x}(u)$, *then* $R_{ij} = (w)(P_i\{w/v\} \parallel Q_j\{w/u\})$, *if* $w \notin fn((v)P_i) \cup fn((u)Q_j)$.

*Let* $P \equiv \sum_i P_i.\alpha_i[m]$ *and* $Q \equiv \sum_i Q_j.\beta_j[m]$, *where* $bn(\alpha_i[m]) \cap fn(Q) = \emptyset$ *for all* $i$, *and* $bn(\beta_j[m]) \cap fn(P) = \emptyset$ *for all* $j$. *Then,*

1. $P \parallel Q \sim^{fr}_p \sum_i \sum_j (P_i \parallel Q_j).(\alpha_i[m] \parallel \beta_j[m]) + \sum_{\alpha_i\ comp\ \beta_j} R_{ij}.\tau$;
2. $P \parallel Q \sim^{fr}_s \sum_i \sum_j (P_i \parallel Q_j).(\alpha_i[m] \parallel \beta_j[m]) + \sum_{\alpha_i\ comp\ \beta_j} R_{ij}.\tau$;
3. $P \parallel Q \sim^{fr}_{hp} \sum_i \sum_j (P_i \parallel Q_j).(\alpha_i[m] \parallel \beta_j[m]) + \sum_{\alpha_i\ comp\ \beta_j} R_{ij}.\tau$;
4. $P \parallel Q \nsim_{phhp} \sum_i \sum_j (P_i \parallel Q_j).(\alpha_i[m] \parallel \beta_j[m]) + \sum_{\alpha_i\ comp\ \beta_j} R_{ij}.\tau$,

*where* $\alpha_i\ comp\ \beta_j$ *and* $R_{ij}$ *are defined as follows:*

1. $\alpha_i[m]$ *is* $\overline{x}u$ *and* $\beta_j[m]$ *is* $x(v)$, *then* $R_{ij} = P_i \parallel Q_j\{u/v\}$;
2. $\alpha_i[m]$ *is* $\overline{x}(u)$ *and* $\beta_j[m]$ *is* $x(v)$, *then* $R_{ij} = (w)(P_i\{w/u\} \parallel Q_j\{w/v\})$, *if* $w \notin fn((u)P_i) \cup fn((v)Q_j)$;
3. $\alpha_i[m]$ *is* $x(v)$ *and* $\beta_j[m]$ *is* $\overline{x}u$, *then* $R_{ij} = P_i\{u/v\} \parallel Q_j$;
4. $\alpha_i[m]$ *is* $x(v)$ *and* $\beta_j[m]$ *is* $\overline{x}(u)$, *then* $R_{ij} = (w)(P_i\{w/v\} \parallel Q_j\{w/u\})$, *if* $w \notin fn((v)P_i) \cup fn((u)Q_j)$.

*Proof.* According to the definition of FR strongly truly concurrent bisimulations, we can easily prove the above equations, however, we omit the proof here. □

**Theorem 7.36** (Equivalence and congruence for FR strongly pomset bisimilarity). *We can enjoy the full congruence modulo FR strongly pomset bisimilarity:*

1. $\sim^{fr}_p$ *is an equivalence relation;*
2. *If* $P \sim^{fr}_p Q$, *then*
   (a) $\alpha.P \sim^f_p \alpha.Q$, $\alpha$ *is a free action;*
   (b) $P.\alpha[m] \sim^r_p Q.\alpha[m]$, $\alpha[m]$ *is a free action;*
   (c) $P + R \sim^{fr}_p Q + R$;
   (d) $P \parallel R \sim^{fr}_p Q \parallel R$;
   (e) $(w)P \sim^{fr}_p (w)Q$;
   (f) $x(y).P \sim^f_p x(y).Q$;
   (g) $P.x(y)[m] \sim^r_p Q.x(y)[m]$.

*Proof.* **1.** $\sim_p^{fr}$ is an equivalence relation, it is obvious that;

**2.** If $P \sim_p^{fr} Q$, then

**(a)** $\alpha.P \sim_p^f \alpha.Q$, $\alpha$ is a free action. It is sufficient to prove the relation $R = \{(\alpha.P, \alpha.Q)\} \cup \mathbf{Id}$ is a F strongly pomset bisimulation, however, we omit it here;

**(b)** $P.\alpha[m] \sim_p^r Q.\alpha[m]$, $\alpha[m]$ is a free action. It is sufficient to prove the relation $R = \{(P.\alpha[m], Q.\alpha[m])\} \cup \mathbf{Id}$ is a R strongly pomset bisimulation, however, we omit it here;

**(c)** $P + R \sim_p^{fr} Q + R$. It is sufficient to prove the relation $R = \{(P + R, Q + R)\} \cup \mathbf{Id}$ is a FR strongly pomset bisimulation, however, we omit it here;

**(d)** $P \parallel R \sim_p^{fr} Q \parallel R$. It is sufficient to prove the relation $R = \{(P \parallel R, Q \parallel R)\} \cup \mathbf{Id}$ is a FR strongly pomset bisimulation, however, we omit it here;

**(e)** $(w)P \sim_p^{fr} (w)Q$. It is sufficient to prove the relation $R = \{((w)P, (w)Q)\} \cup \mathbf{Id}$ is a FR strongly pomset bisimulation, however, we omit it here;

**(f)** $x(y).P \sim_p^f x(y).Q$. It is sufficient to prove the relation $R = \{(x(y).P, x(y).Q)\} \cup \mathbf{Id}$ is a F strongly pomset bisimulation, however, we omit it here;

**(g)** $P.x(y)[m] \sim_p^r Q.x(y)[m]$. It is sufficient to prove the relation $R = \{(P.x(y)[m], Q.x(y)[m])\} \cup \mathbf{Id}$ is a R strongly pomset bisimulation, however, we omit it here. □

**Theorem 7.37** (Equivalence and congruence for FR strongly step bisimilarity). *We can enjoy the full congruence modulo FR strongly step bisimilarity:*

**1.** $\sim_s^{fr}$ *is an equivalence relation;*

**2.** *If* $P \sim_s^{fr} Q$*, then*

**(a)** $\alpha.P \sim_s^f \alpha.Q$, $\alpha$ *is a free action;*

**(b)** $P.\alpha[m] \sim_s^r Q.\alpha[m]$, $\alpha[m]$ *is a free action;*

**(c)** $P + R \sim_s^{fr} Q + R$;

**(d)** $P \parallel R \sim_s^{fr} Q \parallel R$;

**(e)** $(w)P \sim_s^{fr} (w)Q$;

**(f)** $x(y).P \sim_s^f x(y).Q$;

**(g)** $P.x(y)[m] \sim_s^r Q.x(y)[m]$.

*Proof.* **1.** $\sim_s^{fr}$ is an equivalence relation, it is obvious that;

**2.** If $P \sim_s^{fr} Q$, then

**(a)** $\alpha.P \sim_s^f \alpha.Q$, $\alpha$ is a free action. It is sufficient to prove the relation $R = \{(\alpha.P, \alpha.Q)\} \cup \mathbf{Id}$ is a F strongly step bisimulation, however, we omit it here;

**(b)** $P.\alpha[m] \sim_s^r Q.\alpha[m]$, $\alpha[m]$ is a free action. It is sufficient to prove the relation $R = \{(P.\alpha[m], Q.\alpha[m])\} \cup \mathbf{Id}$ is a R strongly step bisimulation, however, we omit it here;

**(c)** $P + R \sim_s^{fr} Q + R$. It is sufficient to prove the relation $R = \{(P + R, Q + R)\} \cup \mathbf{Id}$ is a FR strongly step bisimulation, however, we omit it here;

**(d)** $P \parallel R \sim_s^{fr} Q \parallel R$. It is sufficient to prove the relation $R = \{(P \parallel R, Q \parallel R)\} \cup \mathbf{Id}$ is a FR strongly step bisimulation, however, we omit it here;

**(e)** $(w)P \sim_s^{fr} (w)Q$. It is sufficient to prove the relation $R = \{((w)P, (w)Q)\} \cup \mathbf{Id}$ is a FR strongly step bisimulation, however, we omit it here;

**(f)** $x(y).P \sim_s^f x(y).Q$. It is sufficient to prove the relation $R = \{(x(y).P, x(y).Q)\} \cup \mathbf{Id}$ is a F strongly step bisimulation, however, we omit it here;

**(g)** $P.x(y)[m] \sim_s^r Q.x(y)[m]$. It is sufficient to prove the relation $R = \{(P.x(y)[m], Q.x(y)[m])\} \cup \mathbf{Id}$ is a R strongly step bisimulation, however, we omit it here. □

**Theorem 7.38** (Equivalence and congruence for FR strongly hp-bisimilarity). *We can enjoy the full congruence modulo FR strongly hp-bisimilarity:*

**1.** $\sim_{hp}^{fr}$ *is an equivalence relation;*

**2.** *If* $P \sim_{hp}^{fr} Q$*, then*

**(a)** $\alpha.P \sim_{hp}^f \alpha.Q$, $\alpha$ *is a free action;*

**(b)** $P.\alpha[m] \sim_{hp}^r Q.\alpha[m]$, $\alpha[m]$ *is a free action;*

(c) $P+R \sim_{hp}^{fr} Q+R$;
(d) $P \parallel R \sim_{hp}^{fr} Q \parallel R$;
(e) $(w)P \sim_{hp}^{fr} (w)Q$;
(f) $x(y).P \sim_{hp}^{f} x(y).Q$;
(g) $P.x(y)[m] \sim_{hp}^{r} Q.x(y)[m]$.

*Proof.* **1.** $\sim_{hp}^{fr}$ is an equivalence relation, it is obvious that;
**2.** If $P \sim_{hp}^{fr} Q$, then
   (a) $\alpha.P \sim_{hp}^{f} \alpha.Q$, $\alpha$ is a free action. It is sufficient to prove the relation $R = \{(\alpha.P, \alpha.Q)\} \cup \mathbf{Id}$ is a F strongly hp-bisimulation, however, we omit it here;
   (b) $P.\alpha[m] \sim_{hp}^{r} Q.\alpha[m]$, $\alpha[m]$ is a free action. It is sufficient to prove the relation $R = \{(P.\alpha[m], Q.\alpha[m])\} \cup \mathbf{Id}$ is a R strongly hp-bisimulation, however, we omit it here;
   (c) $P+R \sim_{hp}^{fr} Q+R$. It is sufficient to prove the relation $R = \{(P+R, Q+R)\} \cup \mathbf{Id}$ is a FR strongly hp-bisimulation, however, we omit it here;
   (d) $P \parallel R \sim_{hp}^{fr} Q \parallel R$. It is sufficient to prove the relation $R = \{(P \parallel R, Q \parallel R)\} \cup \mathbf{Id}$ is a FR strongly hp-bisimulation, however, we omit it here;
   (e) $(w)P \sim_{hp}^{fr} (w)Q$. It is sufficient to prove the relation $R = \{((w)P, (w)Q)\} \cup \mathbf{Id}$ is a FR strongly hp-bisimulation, however, we omit it here;
   (f) $x(y).P \sim_{hp}^{f} x(y).Q$. It is sufficient to prove the relation $R = \{(x(y).P, x(y).Q)\} \cup \mathbf{Id}$ is a F strongly hp-bisimulation, however, we omit it here;
   (g) $P.x(y)[m] \sim_{hp}^{r} Q.x(y)[m]$. It is sufficient to prove the relation $R = \{(P.x(y)[m], Q.x(y)[m])\} \cup \mathbf{Id}$ is a R strongly hp-bisimulation, however, we omit it here. □

**Theorem 7.39** (Equivalence and congruence for FR strongly hhp-bisimilarity). *We can enjoy the full congruence modulo FR strongly hhp-bisimilarity:*

**1.** $\sim_{hhp}^{fr}$ *is an equivalence relation;*
**2.** *If* $P \sim_{hhp}^{fr} Q$, *then*
   (a) $\alpha.P \sim_{hhp}^{f} \alpha.Q$, *$\alpha$ is a free action;*
   (b) $P.\alpha[m] \sim_{hhp}^{r} Q.\alpha[m]$, *$\alpha[m]$ is a free action;*
   (c) $P+R \sim_{hhp}^{fr} Q+R$;
   (d) $P \parallel R \sim_{hhp}^{fr} Q \parallel R$;
   (e) $(w)P \sim_{hhp}^{fr} (w)Q$;
   (f) $x(y).P \sim_{hhp}^{f} x(y).Q$;
   (g) $P.x(y)[m] \sim_{hhp}^{r} Q.x(y)[m]$.

*Proof.* **1.** $\sim_{hhp}^{fr}$ is an equivalence relation, it is obvious that;
**2.** If $P \sim_{hhp}^{fr} Q$, then
   (a) $\alpha.P \sim_{hhp}^{f} \alpha.Q$, $\alpha$ is a free action. It is sufficient to prove the relation $R = \{(\alpha.P, \alpha.Q)\} \cup \mathbf{Id}$ is a F strongly hhp-bisimulation, however, we omit it here;
   (b) $P.\alpha[m] \sim_{hhp}^{r} Q.\alpha[m]$, $\alpha[m]$ is a free action. It is sufficient to prove the relation $R = \{(P.\alpha[m], Q.\alpha[m])\} \cup \mathbf{Id}$ is a R strongly hhp-bisimulation, however, we omit it here;
   (c) $P+R \sim_{hhp}^{fr} Q+R$. It is sufficient to prove the relation $R = \{(P+R, Q+R)\} \cup \mathbf{Id}$ is a FR strongly hhp-bisimulation, however, we omit it here;
   (d) $P \parallel R \sim_{hhp}^{fr} Q \parallel R$. It is sufficient to prove the relation $R = \{(P \parallel R, Q \parallel R)\} \cup \mathbf{Id}$ is a FR strongly hhp-bisimulation, however, we omit it here;
   (e) $(w)P \sim_{hhp}^{fr} (w)Q$. It is sufficient to prove the relation $R = \{((w)P, (w)Q)\} \cup \mathbf{Id}$ is a FR strongly hhp-bisimulation, however, we omit it here;
   (f) $x(y).P \sim_{hhp}^{f} x(y).Q$. It is sufficient to prove the relation $R = \{(x(y).P, x(y).Q)\} \cup \mathbf{Id}$ is a F strongly hhp-bisimulation, however, we omit it here;

**(g)** $P.x(y)[m] \sim^r_{hhp} Q.x(y)[m]$. It is sufficient to prove the relation $R = \{(P.x(y)[m], Q.x(y)[m])\} \cup \mathbf{Id}$ is a R strongly hhp-bisimulation, however, we omit it here. □

### 7.3.2 Recursion

**Definition 7.40.** Let $X$ have arity $n$, and let $\widetilde{x} = x_1, \cdots, x_n$ be distinct names, and $fn(P) \subseteq \{x_1, \cdots, x_n\}$. The replacement of $X(\widetilde{x})$ by $P$ in $E$, written $E\{X(\widetilde{x}) := P\}$, means the result of replacing each sub-term $X(\widetilde{y})$ in $E$ by $P\{\widetilde{y}/\widetilde{x}\}$.

**Definition 7.41.** Let $E$ and $F$ be two process expressions containing only $X_1, \cdots, X_m$ with associated name sequences $\widetilde{x}_1, \cdots, \widetilde{x}_m$. Then,

1. $E \sim^{fr}_p F$ means $E(\widetilde{P}) \sim^{fr}_p F(\widetilde{P})$;
2. $E \sim^{fr}_s F$ means $E(\widetilde{P}) \sim^{fr}_s F(\widetilde{P})$;
3. $E \sim^{fr}_{hp} F$ means $E(\widetilde{P}) \sim^{fr}_{hp} F(\widetilde{P})$;
4. $E \sim^{fr}_{hhp} F$ means $E(\widetilde{P}) \sim^{fr}_{hhp} F(\widetilde{P})$;

for all $\widetilde{P}$ such that $fn(P_i) \subseteq \widetilde{x}_i$ for each $i$.

**Definition 7.42.** A term or identifier is weakly guarded in $P$ if it lies within some sub-term $\alpha.Q$ or $Q.\alpha[m]$ or $(\alpha_1 \parallel \cdots \parallel \alpha_n).Q$ or $Q.(\alpha_1[m] \parallel \cdots \parallel \alpha_n[m])$ of $P$.

**Theorem 7.43.** *Assume that $\widetilde{E}$ and $\widetilde{F}$ are expressions containing only $X_i$ with $\widetilde{x}_i$, and $\widetilde{A}$ and $\widetilde{B}$ are identifiers with $A_i$, $B_i$. Then, for all i,*

1. $E_i \sim^{fr}_s F_i$, $A_i(\widetilde{x}_i) \overset{def}{=} E_i(\widetilde{A})$, $B_i(\widetilde{x}_i) \overset{def}{=} F_i(\widetilde{B})$, *then* $A_i(\widetilde{x}_i) \sim^{fr}_s B_i(\widetilde{x}_i)$;
2. $E_i \sim^{fr}_p F_i$, $A_i(\widetilde{x}_i) \overset{def}{=} E_i(\widetilde{A})$, $B_i(\widetilde{x}_i) \overset{def}{=} F_i(\widetilde{B})$, *then* $A_i(\widetilde{x}_i) \sim^{fr}_p B_i(\widetilde{x}_i)$;
3. $E_i \sim^{fr}_{hp} F_i$, $A_i(\widetilde{x}_i) \overset{def}{=} E_i(\widetilde{A})$, $B_i(\widetilde{x}_i) \overset{def}{=} F_i(\widetilde{B})$, *then* $A_i(\widetilde{x}_i) \sim^{fr}_{hp} B_i(\widetilde{x}_i)$;
4. $E_i \sim^{fr}_{hhp} F_i$, $A_i(\widetilde{x}_i) \overset{def}{=} E_i(\widetilde{A})$, $B_i(\widetilde{x}_i) \overset{def}{=} F_i(\widetilde{B})$, *then* $A_i(\widetilde{x}_i) \sim^{fr}_{hhp} B_i(\widetilde{x}_i)$.

*Proof.* **1.** $E_i \sim^{fr}_s F_i$, $A_i(\widetilde{x}_i) \overset{\text{def}}{=} E_i(\widetilde{A})$, $B_i(\widetilde{x}_i) \overset{\text{def}}{=} F_i(\widetilde{B})$, then $A_i(\widetilde{x}_i) \sim^{fr}_s B_i(\widetilde{x}_i)$.

We will consider the case $I = \{1\}$ with loss of generality, and show the following relation $R$ is a FR strongly step bisimulation:

$$R = \{(G(A), G(B)) : G \text{ has only identifier } X\}$$

By choosing $G \equiv X(\widetilde{y})$, it follows that $A(\widetilde{y}) \sim^{fr}_s B(\widetilde{y})$. It is sufficient to prove the following:

**(a)** If $G(A) \xrightarrow{\{\alpha_1,\cdots,\alpha_n\}} P'$, where $\alpha_i (1 \leq i \leq n)$ is a free action or bound output action with $bn(\alpha_1) \cap \cdots \cap bn(\alpha_n) \cap n(G(A), G(B)) = \emptyset$, then $G(B) \xrightarrow{\{\alpha_1,\cdots,\alpha_n\}} Q''$ such that $P' \sim^{fr}_s Q''$;

**(b)** If $G(A) \xrightarrow{x(y)} P'$ with $x \notin n(G(A), G(B))$, then $G(B) \xrightarrow{x(y)} Q''$, such that for all $u$, $P'\{u/y\} \sim^{fr}_s Q''\{u/y\}$;

**(c)** If $G(A) \xtwoheadrightarrow{\{\alpha_1[m],\cdots,\alpha_n[m]\}} P'$, where $\alpha_i[m] (1 \leq i \leq n)$ is a free action or bound output action with $bn(\alpha_1[m]) \cap \cdots \cap bn(\alpha_n[m]) \cap n(G(A), G(B)) = \emptyset$, then $G(B) \xtwoheadrightarrow{\{\alpha_1[m],\cdots,\alpha_n[m]\}} Q''$ such that $P' \sim^{fr}_s Q''$;

**(d)** If $G(A) \xtwoheadrightarrow{x(y)[m]} P'$ with $x \notin n(G(A), G(B))$, then $G(B) \xtwoheadrightarrow{x(y)[m]} Q''$, such that for all $u$, $P'\{u/y\} \sim^{fr}_s Q''\{u/y\}$.

To prove the above properties, it is sufficient to induct on the depth of inference and it is quite routine, hence, we omit it here.

**2.** $E_i \sim^{fr}_p F_i$, $A_i(\widetilde{x}_i) \overset{\text{def}}{=} E_i(\widetilde{A})$, $B_i(\widetilde{x}_i) \overset{\text{def}}{=} F_i(\widetilde{B})$, then $A_i(\widetilde{x}_i) \sim^{fr}_p B_i(\widetilde{x}_i)$. It can be proven similarly to the above case.

**3.** $E_i \sim^{fr}_{hp} F_i$, $A_i(\widetilde{x}_i) \overset{\text{def}}{=} E_i(\widetilde{A})$, $B_i(\widetilde{x}_i) \overset{\text{def}}{=} F_i(\widetilde{B})$, then $A_i(\widetilde{x}_i) \sim^{fr}_{hp} B_i(\widetilde{x}_i)$. It can be proven similarly to the above case.

**4.** $E_i \sim^{fr}_{hhp} F_i$, $A_i(\widetilde{x}_i) \overset{\text{def}}{=} E_i(\widetilde{A})$, $B_i(\widetilde{x}_i) \overset{\text{def}}{=} F_i(\widetilde{B})$, then $A_i(\widetilde{x}_i) \sim^{fr}_{hhp} B_i(\widetilde{x}_i)$. It can be proven similarly to the above case. □

**Theorem 7.44** (Unique solution of equations). *Assume $\widetilde{E}$ are expressions containing only $X_i$ with $\widetilde{x}_i$, and each $X_i$ is weakly guarded in each $E_j$. Assume that $\widetilde{P}$ and $\widetilde{Q}$ are processes such that $fn(P_i) \subseteq \widetilde{x}_i$ and $fn(Q_i) \subseteq \widetilde{x}_i$. Then, for all i,*

1. *if* $P_i \sim^{fr}_p E_i(\widetilde{P})$, $Q_i \sim^{fr}_p E_i(\widetilde{Q})$, *then* $P_i \sim^{fr}_p Q_i$;

2. *if* $P_i \sim_s^{fr} E_i(\widetilde{P})$, $Q_i \sim_s^{fr} E_i(\widetilde{Q})$, *then* $P_i \sim_s^{fr} Q_i$;
3. *if* $P_i \sim_{hp}^{fr} E_i(\widetilde{P})$, $Q_i \sim_{hp}^{fr} E_i(\widetilde{Q})$, *then* $P_i \sim_{hp}^{fr} Q_i$;
4. *if* $P_i \sim_{hhp}^{fr} E_i(\widetilde{P})$, $Q_i \sim_{hhp}^{fr} E_i(\widetilde{Q})$, *then* $P_i \sim_{hhp}^{fr} Q_i$.

*Proof.* **1.** It is similar to the proof of unique solution of equations for FR strongly pomset bisimulation in CTC, please refer to Chapter 5 for details, however, we omit it here;

**2.** It is similar to the proof of unique solution of equations for FR strongly step bisimulation in CTC, please refer to Chapter 5 for details, however, we omit it here;

**3.** It is similar to the proof of unique solution of equations for FR strongly hp-bisimulation in CTC, please refer to Chapter 5 for details, however, we omit it here;

**4.** It is similar to the proof of unique solution of equations for FR strongly hhp-bisimulation in CTC, please refer to Chapter 5 for details, however, we omit it here. □

## 7.4 Algebraic theory

**Definition 7.45** (STC). The theory **STC** is consisted of the following axioms and inference rules:

1. Alpha-conversion **A**.

$$\text{if } P \equiv Q, \text{ then } P = Q$$

2. Congruence **C**. If $P = Q$, then

$$\tau.P = \tau.Q \quad \overline{x}y.P = \overline{x}y.Q \quad P.\overline{x}y[m] = Q.\overline{x}y[m]$$
$$P + R = Q + R \quad P \parallel R = Q \parallel R$$
$$(x)P = (x)Q \quad x(y).P = x(y).Q \quad P.x(y)[m] = Q.x(y)[m]$$

3. Summation **S**.

$$\mathbf{S0} \quad P + \mathbf{nil} = P$$
$$\mathbf{S1} \quad P + P = P$$
$$\mathbf{S2} \quad P + Q = Q + P$$
$$\mathbf{S3} \quad P + (Q + R) = (P + Q) + R$$

4. Restriction **R**.

$$\mathbf{R0} \quad (x)P = P \quad \text{if } x \notin fn(P)$$
$$\mathbf{R1} \quad (x)(y)P = (y)(x)P$$
$$\mathbf{R2} \quad (x)(P + Q) = (x)P + (x)Q$$
$$\mathbf{R3} \quad (x)\alpha.P = \alpha.(x)P \quad \text{if } x \notin n(\alpha)$$
$$\mathbf{R4} \quad (x)\alpha.P = \mathbf{nil} \quad \text{if } x \text{ is the subject of } \alpha$$

5. Expansion **E**. Let $P \equiv \sum_i \alpha_i.P_i$ and $Q \equiv \sum_j \beta_j.Q_j$, where $bn(\alpha_i) \cap fn(Q) = \emptyset$ for all $i$, and $bn(\beta_j) \cap fn(P) = \emptyset$ for all $j$. Then,
   - **(a)** $P \parallel Q \sim_p^{fr} \sum_i \sum_j (\alpha_i \parallel \beta_j).(P_i \parallel Q_j) + \sum_{\alpha_i \text{ comp } \beta_j} \tau.R_{ij}$;
   - **(b)** $P \parallel Q \sim_s^{fr} \sum_i \sum_j (\alpha_i \parallel \beta_j).(P_i \parallel Q_j) + \sum_{\alpha_i \text{ comp } \beta_j} \tau.R_{ij}$;
   - **(c)** $P \parallel Q \sim_{hp}^{fr} \sum_i \sum_j (\alpha_i \parallel \beta_j).(P_i \parallel Q_j) + \sum_{\alpha_i \text{ comp } \beta_j} \tau.R_{ij}$;
   - **(d)** $P \parallel Q \nsim_{phhp} \sum_i \sum_j (\alpha_i \parallel \beta_j).(P_i \parallel Q_j) + \sum_{\alpha_i \text{ comp } \beta_j} \tau.R_{ij}$,

   where $\alpha_i$ comp $\beta_j$ and $R_{ij}$ are defined as follows:
   - **(a)** $\alpha_i$ is $\overline{x}u$ and $\beta_j$ is $x(v)$, then $R_{ij} = P_i \parallel Q_j\{u/v\}$;
   - **(b)** $\alpha_i$ is $\overline{x}(u)$ and $\beta_j$ is $x(v)$, then $R_{ij} = (w)(P_i\{w/u\} \parallel Q_j\{w/v\})$, if $w \notin fn((u)P_i) \cup fn((v)Q_j)$;
   - **(c)** $\alpha_i$ is $x(v)$ and $\beta_j$ is $\overline{x}u$, then $R_{ij} = P_i\{u/v\} \parallel Q_j$;

**(d)** $\alpha_i$ is $x(v)$ and $\beta_j$ is $\overline{x}(u)$, then $R_{ij} = (w)(P_i\{w/v\} \parallel Q_j\{w/u\})$, if $w \notin fn((v)P_i) \cup fn((u)Q_j)$.
Let $P \equiv \sum_i P_i.\alpha_i[m]$ and $Q \equiv \sum_l Q_j.\beta_j[m]$, where $bn(\alpha_i[m]) \cap fn(Q) = \emptyset$ for all $i$, and $bn(\beta_j[m]) \cap fn(P) = \emptyset$ for all $j$. Then,
**(a)** $P \parallel Q \sim_p^{fr} \sum_i \sum_j (P_i \parallel Q_j).(\alpha_i[m] \parallel \beta_j[m]) + \sum_{\alpha_i \text{ comp } \beta_j} R_{ij}.\tau$;
**(b)** $P \parallel Q \sim_s^{fr} \sum_i \sum_j (P_i \parallel Q_j).(\alpha_i[m] \parallel \beta_j[m]) + \sum_{\alpha_i \text{ comp } \beta_j} R_{ij}.\tau$;
**(c)** $P \parallel Q \sim_{hp}^{fr} \sum_i \sum_j (P_i \parallel Q_j).(\alpha_i[m] \parallel \beta_j[m]) + \sum_{\alpha_i \text{ comp } \beta_j} R_{ij}.\tau$;
**(d)** $P \parallel Q \nsim_{phhp} \sum_i \sum_j (P_i \parallel Q_j).(\alpha_i[m] \parallel \beta_j[m]) + \sum_{\alpha_i \text{ comp } \beta_j} R_{ij}.\tau$,
where $\alpha_i$ comp $\beta_j$ and $R_{ij}$ are defined as follows:
**(a)** $\alpha_i[m]$ is $\overline{x}u$ and $\beta_j[m]$ is $x(v)$, then $R_{ij} = P_i \parallel Q_j\{u/v\}$;
**(b)** $\alpha_i[m]$ is $\overline{x}(u)$ and $\beta_j[m]$ is $x(v)$, then $R_{ij} = (w)(P_i\{w/u\} \parallel Q_j\{w/v\})$, if $w \notin fn((u)P_i) \cup fn((v)Q_j)$;
**(c)** $\alpha_i[m]$ is $x(v)$ and $\beta_j[m]$ is $\overline{x}u$, then $R_{ij} = P_i\{u/v\} \parallel Q_j$;
**(d)** $\alpha_i[m]$ is $x(v)$ and $\beta_j[m]$ is $\overline{x}(u)$, then $R_{ij} = (w)(P_i\{w/v\} \parallel Q_j\{w/u\})$, if $w \notin fn((v)P_i) \cup fn((u)Q_j)$.

6. Identifier **I**.

$$\text{If } A(\widetilde{x}) \overset{\text{def}}{=} P, \text{ then } A(\widetilde{y}) = P\{\widetilde{y}/\widetilde{x}\}.$$

**Theorem 7.46** (Soundness). *If* ***STC*** $\vdash P = Q$, *then*

1. $P \sim_p^{fr} Q$;
2. $P \sim_s^{fr} Q$;
3. $P \sim_{hp}^{fr} Q$;
4. $P \sim_{hhp}^{fr} Q$.

*Proof.* The soundness of these laws modulo strongly truly concurrent bisimilarities was already proven in Section 7.3. □

**Definition 7.47.** The agent identifier $A$ is weakly guardedly defined if every agent identifier is weakly guarded in the right-hand side of the definition of $A$.

**Definition 7.48** (Head normal form). A Process $P$ is in head normal form if it is a sum of the prefixes:

$$P \equiv \sum_i (\alpha_{i1} \parallel \cdots \parallel \alpha_{in}).P_i \quad P \equiv \sum_i P_i.(\alpha_{i1}[m] \parallel \cdots \parallel \alpha_{in}[m])$$

**Proposition 7.49.** *If every agent identifier is weakly guardedly defined, then for any process P, there is a head normal form H such that*

$$\textbf{\textit{STC}} \vdash P = H$$

*Proof.* It is sufficient to induct on the structure of $P$ and it is quite obvious. □

**Theorem 7.50** (Completeness). *For all processes P and Q,*

1. *if* $P \sim_p^{fr} Q$, *then* ***STC*** $\vdash P = Q$;
2. *if* $P \sim_s^{fr} Q$, *then* ***STC*** $\vdash P = Q$;
3. *if* $P \sim_{hp}^{fr} Q$, *then* ***STC*** $\vdash P = Q$.

*Proof.* **1.** if $P \sim_s^{fr} Q$, then **STC** $\vdash P = Q$.
For the forward transition case.
Since $P$ and $Q$ all have head normal forms, let $P \equiv \sum_{i=1}^k \alpha_i.P_i$ and $Q \equiv \sum_{i=1}^k \beta_i.Q_i$. Then, the depth of $P$, denoted as $d(P) = 0$, if $k = 0$; $d(P) = 1 + max\{d(P_i)\}$ for $1 \le j, i \le k$. The depth $d(Q)$ can be defined similarly.
It is sufficient to induct on $d = d(P) + d(Q)$. When $d = 0$, $P \equiv$ **nil** and $Q \equiv$ **nil**, $P = Q$, as desired.
Suppose $d > 0$.

- If $(\alpha_1 \parallel \cdots \parallel \alpha_n).M$ with $\alpha_i (1 \le i \le n)$ free actions is a summand of $P$, then $P \xrightarrow{\{\alpha_1, \cdots, \alpha_n\}} M$. Since $Q$ is in head normal form and has a summand $(\alpha_1 \parallel \cdots \parallel \alpha_n).N$ such that $M \sim_s^{fr} N$, by the induction hypothesis **STC** $\vdash M = N$, **STC** $\vdash (\alpha_1 \parallel \cdots \parallel \alpha_n).M = (\alpha_1 \parallel \cdots \parallel \alpha_n).N$;

- If $x(y).M$ is a summand of $P$, then for $z \notin n(P, Q)$, $P \xrightarrow{x(z)} M' \equiv M\{z/y\}$. Since $Q$ is in head normal form and has a summand $x(w).N$ such that for all $v$, $M'\{v/z\} \sim_s^{fr} N'\{v/z\}$, where $N' \equiv N\{z/w\}$, by the induction hypothesis **STC** $\vdash M'\{v/z\} = N'\{v/z\}$, by the axioms **C** and **A**, **STC** $\vdash x(y).M = x(w).N$;
- If $\overline{x}(y).M$ is a summand of $P$, then for $z \notin n(P, Q)$, $P \xrightarrow{\overline{x}(z)} M' \equiv M\{z/y\}$. Since $Q$ is in head normal form and has a summand $\overline{x}(w).N$ such that $M' \sim_s^{fr} N'$, where $N' \equiv N\{z/w\}$, by the induction hypothesis **STC** $\vdash M' = N'$, by the axioms **A** and **C**, **STC** $\vdash \overline{x}(y).M = \overline{x}(w).N$.

For the reverse transition case, it can be proven similarly, hence, we omit it here.

2. if $P \sim_p^{fr} Q$, then **STC** $\vdash P = Q$. It can be proven similarly to the above case.
3. if $P \sim_{hp}^{fr} Q$, then **STC** $\vdash P = Q$. It can be proven similarly to the above case. □

Chapter 8

# Introduction to probabilistic process algebra for true concurrency

Probabilistic phenomena exist in nature and computer science. Probabilistic aspects should be considered in several cases in computer science: (1) for an unreliable system, the failures are probabilistic in nature, and a whole system can be modeled as a probabilistic process; (2) in distributed algorithms, the randomness can be modeled as probability; (3) in closed quantum systems, quantum measurement is a kind of basic quantum operation, so probability is unavoidable.

There are several efforts to formalize probability in computer science, among them is the effort to introduce probability into process algebra. The well-known process algebras, such as CCS [1] [2], ACP [3], and $\pi$-calculus [4] [5], capture the interleaving concurrency based on bisimilarity semantics. The representative of introduction probability into traditional process algebra is Andova's probabilistic process algebra [10] [11] [12], which is an extension of process algebra ACP with probability.

We have done some work on truly concurrent process algebras, such as CTC (Calculus for True Concurrency) [9] which is a generalization of CCS [1] [2] for true concurrency, APTC (Algebra of Parallelism for True Concurrency) [9] that is a generalization of ACP [3] for true concurrency and $\pi_{tc}$ ($\pi$ Calculus for True Concurrency) [9] that is a generalization of $\pi$ calculus [4] [5] for true concurrency, they capture the so-called true concurrency based on truly concurrent bisimilarities, such as pomset bisimilarity, step bisimilarity, history-preserving (hp-) bisimilarity, and hereditary history-preserving (hhp-) bisimilarity. Truly concurrent process algebras are generalizations of the corresponding traditional process algebras.

In this book, we introduce probability into truly concurrent process algebras, based on the work on probabilistic process algebra [10] [11] [12].

We introduce the preliminaries in Chapter 9, including guards and probabilistic operational semantics, about the introductions on operational semantics, proof techniques, truly concurrent process algebras [9], which are based on truly concurrent operational semantics, please refer to Chapter 2.

We design a Calculus for Probabilistic True Concurrency abbreviated $CPTC$ in Chapter 10, which is an extension of CTC with probability, including the syntax and operational semantics of $CPTC$, its properties modulo strongly probabilistic truly concurrent bisimulations and its properties modulo weakly probabilistic truly concurrent bisimulations.

We introduce Algebra of Probabilistic Processes for True Concurrency abbreviated $APPTC$ in Chapter 11, which is an extension of APTC with probability, including $BAPTC$ (Basic Algebra for Probabilistic True Concurrency), $APPTC$ (Algebra for Parallelism in Probabilistic True Concurrency), recursion, and abstraction.

We design a calculus for mobile processes called $\pi_{ptc}$ in Chapter 12, which is an extension of $\pi_{tc}$ with probability, including the syntax and operational semantics of $\pi_{ptc}$, its properties modulo strongly probabilistic truly concurrent bisimulations, and its axiomatization.

We introduce guards $APPTC_G$ in Chapter 13, which is an extension of guards in APTC with probability, including the operational semantics of guards, $BAPTC$ with Guards, $APPTC$ with Guards, recursion, abstraction, and Hoare Logic for $APPTC_G$.

We design CTC with probability and guards in Chapter 14, which is an extension of CPTC with guards, including the syntax and operational semantics, its properties modulo strong bisimulations, and its properties modulo weak bisimulations.

We design $\pi_{tc}$ with probability and guards in Chapter 15, which is an extension of $\pi_{ptc}$ with guards, including its syntax and operational semantics, its properties modulo strong bisimulations, and its axiomatization.

Handbook of Truly Concurrent Process Algebra. https://doi.org/10.1016/B978-0-44-321515-5.00012-3

Chapter 9

# Backgrounds

## 9.1 Guards

To have the ability of data manipulation, we introduce guards into APTC in this section.

### 9.1.1 Operational semantics

In this section, we extend truly concurrent bisimilarities to the ones containing data states.

**Definition 9.1** (Prime event structure with silent event and empty event). Let $\Lambda$ be a fixed set of labels, ranged over $a, b, c, \cdots$ and $\tau, \epsilon$. A ($\Lambda$-labeled) prime event structure with silent event $\tau$ and empty event $\epsilon$ is a tuple $\mathcal{E} = \langle \mathbb{E}, \leq, \sharp, \lambda \rangle$, where $\mathbb{E}$ is a denumerable set of events, including the silent event $\tau$ and empty event $\epsilon$. Let $\hat{\mathbb{E}} = \mathbb{E} \backslash \{\tau, \epsilon\}$, exactly excluding $\tau$ and $\epsilon$, it is obvious that $\hat{\tau^*} = \epsilon$. Let $\lambda : \mathbb{E} \to \Lambda$ be a labeling function and let $\lambda(\tau) = \tau$ and $\lambda(\epsilon) = \epsilon$. Also, $\leq$, $\sharp$ are binary relations on $\mathbb{E}$, called causality and conflict, respectively, such that:

1. $\leq$ is a partial order and $\lceil e \rceil = \{e' \in \mathbb{E} | e' \leq e\}$ is finite for all $e \in \mathbb{E}$. It is easy to see that $e \leq \tau^* \leq e' = e \leq \tau \leq \cdots \leq \tau \leq e'$, then $e \leq e'$.
2. $\sharp$ is irreflexive, symmetric, and hereditary with respect to $\leq$, that is, for all $e, e', e'' \in \mathbb{E}$, if $e \sharp e' \leq e''$, then $e \sharp e''$.

Then, the concepts of consistency and concurrency can be drawn from the above definition:

1. $e, e' \in \mathbb{E}$ are consistent, denoted as $e \frown e'$, if $\neg(e \sharp e')$. A subset $X \subseteq \mathbb{E}$ is called consistent, if $e \frown e'$ for all $e, e' \in X$.
2. $e, e' \in \mathbb{E}$ are concurrent, denoted as $e \parallel e'$, if $\neg(e \leq e')$, $\neg(e' \leq e)$, and $\neg(e \sharp e')$.

**Definition 9.2** (Configuration). Let $\mathcal{E}$ be a PES. A (finite) configuration in $\mathcal{E}$ is a (finite) consistent subset of events $C \subseteq \mathcal{E}$, closed with respect to causality (i.e., $\lceil C \rceil = C$), and a data state $s \in S$ with $S$ the set of all data states, denoted $\langle C, s \rangle$. The set of finite configurations of $\mathcal{E}$ is denoted by $\langle \mathcal{C}(\mathcal{E}), S \rangle$. We let $\hat{C} = C \backslash \{\tau\} \cup \{\epsilon\}$.

A consistent subset of $X \subseteq \mathbb{E}$ of events can be seen as a pomset. Given $X, Y \subseteq \mathbb{E}$, $\hat{X} \sim \hat{Y}$ if $\hat{X}$ and $\hat{Y}$ are isomorphic as pomsets. In the remainder of the chapter, when we say $C_1 \sim C_2$, we mean $\hat{C_1} \sim \hat{C_2}$.

**Definition 9.3** (Pomset transitions and step). Let $\mathcal{E}$ be a PES and let $C \in \mathcal{C}(\mathcal{E})$, and $\emptyset \neq X \subseteq \mathbb{E}$, if $C \cap X = \emptyset$ and $C' = C \cup X \in \mathcal{C}(\mathcal{E})$, then $\langle C, s \rangle \xrightarrow{X} \langle C', s' \rangle$ is called a pomset transition from $\langle C, s \rangle$ to $\langle C', s' \rangle$. When the events in $X$ are pairwise concurrent, we say that $\langle C, s \rangle \xrightarrow{X} \langle C', s' \rangle$ is a step. It is obvious that $\to^* \xrightarrow{X} \to^* = \xrightarrow{X}$ and $\to^* \xrightarrow{e} \to^* = \xrightarrow{e}$ for any $e \in \mathbb{E}$ and $X \subseteq \mathbb{E}$.

**Definition 9.4** (Weak pomset transitions and weak step). Let $\mathcal{E}$ be a PES and let $C \in \mathcal{C}(\mathcal{E})$, and $\emptyset \neq X \subseteq \hat{\mathbb{E}}$, if $C \cap X = \emptyset$ and $\hat{C'} = \hat{C} \cup X \in \mathcal{C}(\mathcal{E})$, then $\langle C, s \rangle \xRightarrow{X} \langle C', s' \rangle$ is called a weak pomset transition from $\langle C, s \rangle$ to $\langle C', s' \rangle$, where we define $\xRightarrow{e} \triangleq \xrightarrow{\tau^*} \xrightarrow{e} \xrightarrow{\tau^*}$. Also, $\xRightarrow{X} \triangleq \xrightarrow{\tau^*} \xrightarrow{e} \xrightarrow{\tau^*}$, for every $e \in X$. When the events in $X$ are pairwise concurrent, we say that $\langle C, s \rangle \xRightarrow{X} \langle C', s' \rangle$ is a weak step.

We will also suppose that all the PESs in this chapter are image finite, that is, for any PES $\mathcal{E}$ and $C \in \mathcal{C}(\mathcal{E})$ and $a \in \Lambda$, $\{e \in \mathbb{E} | \langle C, s \rangle \xrightarrow{e} \langle C', s' \rangle \wedge \lambda(e) = a\}$ and $\{e \in \hat{\mathbb{E}} | \langle C, s \rangle \xRightarrow{e} \langle C', s' \rangle \wedge \lambda(e) = a\}$ is finite.

**Definition 9.5** (Pomset, step bisimulation). Let $\mathcal{E}_1$, $\mathcal{E}_2$ be PESs. A pomset bisimulation is a relation $R \subseteq \langle \mathcal{C}(\mathcal{E}_1), S \rangle \times \langle \mathcal{C}(\mathcal{E}_2), S \rangle$, such that if $(\langle C_1, s \rangle, \langle C_2, s \rangle) \in R$, and $\langle C_1, s \rangle \xrightarrow{X_1} \langle C_1', s' \rangle$, then $\langle C_2, s \rangle \xrightarrow{X_2} \langle C_2', s' \rangle$, with $X_1 \subseteq \mathbb{E}_1$, $X_2 \subseteq \mathbb{E}_2$, $X_1 \sim X_2$ and $(\langle C_1', s' \rangle, \langle C_2', s' \rangle) \in R$ for all $s, s' \in S$, and vice versa. We say that $\mathcal{E}_1$, $\mathcal{E}_2$ are pomset bisimilar, written $\mathcal{E}_1 \sim_p \mathcal{E}_2$, if there exists a pomset bisimulation $R$, such that $(\langle \emptyset, \emptyset \rangle, \langle \emptyset, \emptyset \rangle) \in R$. By replacing pomset transitions with steps, we can obtain the definition of step bisimulation. When PESs $\mathcal{E}_1$ and $\mathcal{E}_2$ are step bisimilar, we write $\mathcal{E}_1 \sim_s \mathcal{E}_2$.

Handbook of Truly Concurrent Process Algebra. https://doi.org/10.1016/B978-0-44-321515-5.00013-5

**Definition 9.6** (Weak pomset, step bisimulation). Let $\mathcal{E}_1$, $\mathcal{E}_2$ be PESs. A weak pomset bisimulation is a relation $R \subseteq \langle\mathcal{C}(\mathcal{E}_1), S\rangle \times \langle\mathcal{C}(\mathcal{E}_2), S\rangle$, such that if $(\langle C_1, s\rangle, \langle C_2, s\rangle) \in R$, and $\langle C_1, s\rangle \xRightarrow{X_1} \langle C_1', s'\rangle$, then $\langle C_2, s\rangle \xRightarrow{X_2} \langle C_2', s'\rangle$, with $X_1 \subseteq \hat{\mathbb{E}_1}$, $X_2 \subseteq \hat{\mathbb{E}_2}$, $X_1 \sim X_2$ and $(\langle C_1', s'\rangle, \langle C_2', s'\rangle) \in R$ for all $s, s' \in S$, and vice versa. We say that $\mathcal{E}_1$, $\mathcal{E}_2$ are weak pomset bisimilar, written $\mathcal{E}_1 \approx_p \mathcal{E}_2$, if there exists a weak pomset bisimulation $R$, such that $(\langle\emptyset, \emptyset\rangle, \langle\emptyset, \emptyset\rangle) \in R$. By replacing weak pomset transitions with weak steps, we can obtain the definition of weak step bisimulation. When PESs $\mathcal{E}_1$ and $\mathcal{E}_2$ are weak step bisimilar, we write $\mathcal{E}_1 \approx_s \mathcal{E}_2$.

**Definition 9.7** (Posetal product). Given two PESs $\mathcal{E}_1$, $\mathcal{E}_2$, the posetal product of their configurations, denoted $\langle\mathcal{C}(\mathcal{E}_1), S\rangle\overline{\times}\langle\mathcal{C}(\mathcal{E}_2), S\rangle$, is defined as

$$\{(\langle C_1, s\rangle, f, \langle C_2, s\rangle)|C_1 \in \mathcal{C}(\mathcal{E}_1), C_2 \in \mathcal{C}(\mathcal{E}_2), f : C_1 \to C_2 \text{ isomorphism}\}$$

A subset $R \subseteq \langle\mathcal{C}(\mathcal{E}_1), S\rangle\overline{\times}\langle\mathcal{C}(\mathcal{E}_2), S\rangle$ is called a posetal relation. We say that $R$ is downward closed when for any $(\langle C_1, s\rangle, f, \langle C_2, s\rangle), (\langle C_1', s'\rangle, f', \langle C_2', s'\rangle) \in \langle\mathcal{C}(\mathcal{E}_1), S\rangle\overline{\times}\langle\mathcal{C}(\mathcal{E}_2), S\rangle$, if $(\langle C_1, s\rangle, f, \langle C_2, s\rangle) \subseteq (\langle C_1', s'\rangle, f', \langle C_2', s'\rangle)$ pointwise and $(\langle C_1', s'\rangle, f', \langle C_2', s'\rangle) \in R$, then $(\langle C_1, s\rangle, f, \langle C_2, s\rangle) \in R$.

For $f : X_1 \to X_2$, we define $f[x_1 \mapsto x_2] : X_1 \cup \{x_1\} \to X_2 \cup \{x_2\}$, $z \in X_1 \cup \{x_1\}$, (1) $f[x_1 \mapsto x_2](z) = x_2$, if $z = x_1$; (2) $f[x_1 \mapsto x_2](z) = f(z)$, otherwise, where $X_1 \subseteq \mathbb{E}_1$, $X_2 \subseteq \mathbb{E}_2$, $x_1 \in \mathbb{E}_1$, $x_2 \in \mathbb{E}_2$.

**Definition 9.8** (Weakly posetal product). Given two PESs $\mathcal{E}_1$, $\mathcal{E}_2$, the weakly posetal product of their configurations, denoted $\langle\mathcal{C}(\mathcal{E}_1), S\rangle\overline{\times}\langle\mathcal{C}(\mathcal{E}_2), S\rangle$, is defined as

$$\{(\langle C_1, s\rangle, f, \langle C_2, s\rangle)|C_1 \in \mathcal{C}(\mathcal{E}_1), C_2 \in \mathcal{C}(\mathcal{E}_2), f : \hat{C_1} \to \hat{C_2} \text{ isomorphism}\}$$

A subset $R \subseteq \langle\mathcal{C}(\mathcal{E}_1), S\rangle\overline{\times}\langle\mathcal{C}(\mathcal{E}_2), S\rangle$ is called a weakly posetal relation. We say that $R$ is downward closed when for any $(\langle C_1, s\rangle, f, \langle C_2, s\rangle), (\langle C_1', s'\rangle, f, \langle C_2', s'\rangle) \in \langle\mathcal{C}(\mathcal{E}_1), S\rangle\overline{\times}\langle\mathcal{C}(\mathcal{E}_2), S\rangle$, if $(\langle C_1, s\rangle, f, \langle C_2, s\rangle) \subseteq (\langle C_1', s'\rangle, f', \langle C_2', s'\rangle)$ pointwise and $(\langle C_1', s'\rangle, f', \langle C_2', s'\rangle) \in R$, then $(\langle C_1, s\rangle, f, \langle C_2, s\rangle) \in R$.

For $f : X_1 \to X_2$, we define $f[x_1 \mapsto x_2] : X_1 \cup \{x_1\} \to X_2 \cup \{x_2\}$, $z \in X_1 \cup \{x_1\}$, (1) $f[x_1 \mapsto x_2](z) = x_2$, if $z = x_1$; (2) $f[x_1 \mapsto x_2](z) = f(z)$, otherwise, where $X_1 \subseteq \hat{\mathbb{E}_1}$, $X_2 \subseteq \hat{\mathbb{E}_2}$, $x_1 \in \hat{\mathbb{E}_1}$, $x_2 \in \hat{\mathbb{E}_2}$. Also, we define $f(\tau^*) = f(\tau^*)$.

**Definition 9.9** ((Hereditary) history-preserving bisimulation). A history-preserving (hp-) bisimulation is a posetal relation $R \subseteq \langle\mathcal{C}(\mathcal{E}_1), S\rangle\overline{\times}\langle\mathcal{C}(\mathcal{E}_2), S\rangle$ such that if $(\langle C_1, s\rangle, f, \langle C_2, s\rangle) \in R$, and $\langle C_1, s\rangle \xrightarrow{e_1} \langle C_1', s'\rangle$, then $\langle C_2, s\rangle \xrightarrow{e_2} \langle C_2', s'\rangle$, with $(\langle C_1', s'\rangle, f[e_1 \mapsto e_2], \langle C_2', s'\rangle) \in R$ for all $s, s' \in S$, and vice versa. $\mathcal{E}_1, \mathcal{E}_2$ are history-preserving (hp-)bisimilar and are written $\mathcal{E}_1 \sim_{hp} \mathcal{E}_2$ if there exists a hp-bisimulation $R$ such that $(\langle\emptyset, \emptyset\rangle, \emptyset, \langle\emptyset, \emptyset\rangle) \in R$.

A hereditary history-preserving (hhp-)bisimulation is a downward closed hp-bisimulation. $\mathcal{E}_1, \mathcal{E}_2$ are hereditary history-preserving (hhp-)bisimilar and are written $\mathcal{E}_1 \sim_{hhp} \mathcal{E}_2$.

**Definition 9.10** (Weak (hereditary) history-preserving bisimulation). A weak history-preserving (hp-) bisimulation is a weakly posetal relation $R \subseteq \langle\mathcal{C}(\mathcal{E}_1), S\rangle\overline{\times}\langle\mathcal{C}(\mathcal{E}_2), S\rangle$ such that if $(\langle C_1, s\rangle, f, \langle C_2, s\rangle) \in R$, and $\langle C_1, s\rangle \xRightarrow{e_1} \langle C_1', s'\rangle$, then $\langle C_2, s\rangle \xRightarrow{e_2} \langle C_2', s'\rangle$, with $(\langle C_1', s'\rangle, f[e_1 \mapsto e_2], \langle C_2', s'\rangle) \in R$ for all $s, s' \in S$, and vice versa. $\mathcal{E}_1, \mathcal{E}_2$ are weak history-preserving (hp-)bisimilar and are written $\mathcal{E}_1 \approx_{hp} \mathcal{E}_2$ if there exists a weak hp-bisimulation $R$ such that $(\langle\emptyset, \emptyset\rangle, \emptyset, \langle\emptyset, \emptyset\rangle) \in R$.

A weakly hereditary history-preserving (hhp-)bisimulation is a downward closed weak hp-bisimulation. $\mathcal{E}_1, \mathcal{E}_2$ are weakly hereditary history-preserving (hhp-)bisimilar and are written $\mathcal{E}_1 \approx_{hhp} \mathcal{E}_2$.

### 9.1.2 *BATC* with guards

In this section, we will discuss the guards for $BATC$, which are denoted as $BATC_G$. Let $\mathbb{E}$ be the set of atomic events (actions), and we assume that there is a data set $\Delta$ and data $D_1, \cdots, D_n \in \Delta$, the data variables $d_1, \cdots, d_n$ range over $\Delta$, and $d_i$ has the same data type as $D_i$ and can have a substitution $D_i/d_i$, for process $x$, $x[D_i/d_i]$ denotes that all occurrences of $d_i$ in $x$ are replaced by $D_i$. Also, the atomic action $e$ may manipulate on data and has the form $e(d_1, \cdots, d_n)$ or $e(D_1, \cdots, D_n)$. Let $G_{at}$ be the set of atomic guards, $\delta$ be the deadlock constant, and $\epsilon$ be the empty event. We extend $G_{at}$ to the set of basic guards $G$ with element $\phi, \psi, \cdots$, which is generated by the following formation rules:

$$\phi ::= \delta|\epsilon|\neg\phi|\psi \in G_{at}|\phi + \psi|\phi \cdot \psi$$

In the following, let $e_1, e_2, e_1', e_2' \in \mathbb{E}$, $\phi, \psi \in G$ and let variables $x, y, z$ range over the set of terms for true concurrency, $p, q, s$ range over the set of closed terms. The predicate $test(\phi, s)$ represents that $\phi$ holds in the state $s$, and $test(\epsilon, s)$ holds

**TABLE 9.1** Axioms of $BATC_G$.

| No. | Axiom |
|---|---|
| $A1$ | $x + y = y + x$ |
| $A2$ | $(x + y) + z = x + (y + z)$ |
| $A3$ | $x + x = x$ |
| $A4$ | $(x + y) \cdot z = x \cdot z + y \cdot z$ |
| $A5$ | $(x \cdot y) \cdot z = x \cdot (y \cdot z)$ |
| $A6$ | $x + \delta = x$ |
| $A7$ | $\delta \cdot x = \delta$ |
| $A8$ | $\epsilon \cdot x = x$ |
| $A9$ | $x \cdot \epsilon = x$ |
| $G1$ | $\phi \cdot \neg\phi = \delta$ |
| $G2$ | $\phi + \neg\phi = \epsilon$ |
| $G3$ | $\phi\delta = \delta$ |
| $G4$ | $\phi(x + y) = \phi x + \phi y$ |
| $G5$ | $\phi(x \cdot y) = \phi x \cdot y$ |
| $G6$ | $(\phi + \psi)x = \phi x + \psi x$ |
| $G7$ | $(\phi \cdot \psi) \cdot x = \phi \cdot (\psi \cdot x)$ |
| $G8$ | $\phi = \epsilon$ if $\forall s \in S.test(\phi, s)$ |
| $G9$ | $\phi_0 \cdot \cdots \cdot \phi_n = \delta$ if $\forall s \in S, \exists i \leq n.test(\neg\phi_i, s)$ |
| $G10$ | $wp(e, \phi)e\phi = wp(e, \phi)e$ |
| $G11$ | $\neg wp(e, \phi)e\neg\phi = \neg wp(e, \phi)e$ |

and $test(\delta, s)$ does not hold. $effect(e, s) \in S$ denotes $s'$ in $s \xrightarrow{e} s'$. The predicate weakest precondition $wp(e, \phi)$ denotes that $\forall s \in S, test(\phi, effect(e, s))$ holds.

The set of axioms of $BATC_G$ consists of the laws given in Table 9.1.

Note that by eliminating atomic event from the process terms, the axioms in Table 9.1 will lead to a Boolean Algebra. Also, $G9$ is a precondition of $e$ and $\phi$, $G10$ is the weakest precondition of $e$ and $\phi$. A data environment with $effect$ function is sufficiently deterministic, and it is obvious that if the weakest precondition is expressible and $G9$, $G10$ are sound, then the related data environment is sufficiently deterministic.

**Definition 9.11** (Basic terms of $BATC_G$). The set of basic terms of $BATC_G$, $\mathcal{B}(BATC_G)$, is inductively defined as follows:

1. $\mathbb{E} \subset \mathcal{B}(BATC_G)$;
2. $G \subset \mathcal{B}(BATC_G)$;
3. if $e \in \mathbb{E}, t \in \mathcal{B}(BATC_G)$, then $e \cdot t \in \mathcal{B}(BATC_G)$;
4. if $\phi \in G, t \in \mathcal{B}(BATC_G)$, then $\phi \cdot t \in \mathcal{B}(BATC_G)$;
5. if $t, s \in \mathcal{B}(BATC_G)$, then $t + s \in \mathcal{B}(BATC_G)$.

**Theorem 9.12** (Elimination theorem of $BATC_G$). *Let $p$ be a closed $BATC_G$ term. Then, there is a basic $BATC_G$ term $q$ such that $BATC_G \vdash p = q$.*

We will define a term-deduction system that gives the operational semantics of $BATC_G$. We give the operational transition rules for $\epsilon$, atomic guard $\phi \in G_{at}$, atomic event $e \in \mathbb{E}$, operators $\cdot$ and $+$ as Table 9.2 shows. Also, the predicate $\xrightarrow{e} \surd$ represents successful termination after execution of the event $e$.

Note that we replace the single atomic event $e \in \mathbb{E}$ by $X \subseteq \mathbb{E}$, we can obtain the pomset transition rules of $BATC_G$, however, we omit them here.

**Theorem 9.13** (Congruence of $BATC_G$ with respect to truly concurrent bisimulation equivalences). *(1) Pomset bisimulation equivalence $\sim_p$ is a congruence with respect to $BATC_G$;*

*(2) Step bisimulation equivalence $\sim_s$ is a congruence with respect to $BATC_G$;*

**TABLE 9.2** Single event transition rules of $BATC_G$.

| | |
|---|---|
| $\frac{}{\langle\epsilon, s\rangle \rightarrow \langle\surd, s\rangle}$ | |
| $\frac{}{\langle e, s\rangle \xrightarrow{e} \langle\surd, s'\rangle}$ if $s' \in effect(e, s)$ | |
| $\frac{}{\langle\phi, s\rangle \rightarrow \langle\surd, s\rangle}$ if $test(\phi, s)$ | |
| $\frac{\langle x, s\rangle \xrightarrow{e} \langle\surd, s'\rangle}{\langle x+y, s\rangle \xrightarrow{e} \langle\surd, s'\rangle}$ | $\frac{\langle x, s\rangle \xrightarrow{e} \langle x', s'\rangle}{\langle x+y, s\rangle \xrightarrow{e} \langle x', s'\rangle}$ |
| $\frac{\langle y, s\rangle \xrightarrow{e} \langle\surd, s'\rangle}{\langle x+y, s\rangle \xrightarrow{e} \langle\surd, s'\rangle}$ | $\frac{\langle y, s\rangle \xrightarrow{e} \langle y', s'\rangle}{\langle x+y, s\rangle \xrightarrow{e} \langle y', s'\rangle}$ |
| $\frac{\langle x, s\rangle \xrightarrow{e} \langle\surd, s'\rangle}{\langle x \cdot y, s\rangle \xrightarrow{e} \langle y, s'\rangle}$ | $\frac{\langle x, s\rangle \xrightarrow{e} \langle x', s'\rangle}{\langle x \cdot y, s\rangle \xrightarrow{e} \langle x' \cdot y, s'\rangle}$ |

*(3) Hp-bisimulation equivalence $\sim_{hp}$ is a congruence with respect to $BATC_G$;*
*(4) Hhp-bisimulation equivalence $\sim_{hhp}$ is a congruence with respect to $BATC_G$.*

**Theorem 9.14** (Soundness of $BATC_G$ modulo truly concurrent bisimulation equivalences). *(1) Let x and y be $BATC_G$ terms. If $BATC \vdash x = y$, then $x \sim_p y$;*
*(2) Let x and y be $BATC_G$ terms. If $BATC \vdash x = y$, then $x \sim_s y$;*
*(3) Let x and y be $BATC_G$ terms. If $BATC \vdash x = y$, then $x \sim_{hp} y$;*
*(4) Let x and y be $BATC_G$ terms. If $BATC \vdash x = y$, then $x \sim_{hhp} y$.*

**Theorem 9.15** (Completeness of $BATC_G$ modulo truly concurrent bisimulation equivalences). *(1) Let p and q be closed $BATC_G$ terms, if $p \sim_p q$, then $p = q$;*
*(2) Let p and q be closed $BATC_G$ terms, if $p \sim_s q$, then $p = q$;*
*(3) Let p and q be closed $BATC_G$ terms, if $p \sim_{hp} q$, then $p = q$;*
*(4) Let p and q be closed $BATC_G$ terms, if $p \sim_{hhp} q$, then $p = q$.*

**Theorem 9.16** (Sufficient determinacy). *All related data environments with respect to $BATC_G$ can be sufficiently deterministic.*

### 9.1.3 $APTC$ with guards

In this section, we will extend $APTC$ with guards, which is abbreviated $APTC_G$. The set of basic guards $G$ with element $\phi, \psi, \cdots$, which is extended by the following formation rules:

$$\phi ::= \delta|\epsilon|\neg\phi|\psi \in G_{at}|\phi + \psi|\phi \cdot \psi|\phi \parallel \psi$$

The set of axioms of $APTC_G$, including the axioms of $BATC_G$ in Table 9.1, is shown in Table 9.3.

**Definition 9.17** (Basic terms of $APTC_G$). The set of basic terms of $APTC_G$, $\mathcal{B}(APTC_G)$, is inductively defined as follows:

1. $\mathbb{E} \subset \mathcal{B}(APTC_G)$;
2. $G \subset \mathcal{B}(APTC_G)$;
3. if $e \in \mathbb{E}, t \in \mathcal{B}(APTC_G)$, then $e \cdot t \in \mathcal{B}(APTC_G)$;
4. if $\phi \in G, t \in \mathcal{B}(APTC_G)$, then $\phi \cdot t \in \mathcal{B}(APTC_G)$;
5. if $t, s \in \mathcal{B}(APTC_G)$, then $t + s \in \mathcal{B}(APTC_G)$;
6. if $t, s \in \mathcal{B}(APTC_G)$, then $t \parallel s \in \mathcal{B}(APTC_G)$.

Based on the definition of basic terms for $APTC_G$ (see Definition 9.17) and axioms of $APTC_G$, we can prove the elimination theorem of $APTC_G$.

**Theorem 9.18** (Elimination theorem of $APTC_G$). *Let p be a closed $APTC_G$ term. Then, there is a basic $APTC_G$ term q such that $APTC_G \vdash p = q$.*

**TABLE 9.3** Axioms of $APTC_G$.

| No. | Axiom |
|---|---|
| $P1$ | $x \between y = x \parallel y + x \mid y$ |
| $P2$ | $e_1 \parallel (e_2 \cdot y) = (e_1 \parallel e_2) \cdot y$ |
| $P3$ | $(e_1 \cdot x) \parallel e_2 = (e_1 \parallel e_2) \cdot x$ |
| $P4$ | $(e_1 \cdot x) \parallel (e_2 \cdot y) = (e_1 \parallel e_2) \cdot (x \between y)$ |
| $P5$ | $(x + y) \parallel z = (x \parallel z) + (y \parallel z)$ |
| $P6$ | $x \parallel (y + z) = (x \parallel y) + (x \parallel z)$ |
| $P7$ | $\delta \parallel x = \delta$ |
| $P8$ | $x \parallel \delta = \delta$ |
| $P9$ | $\epsilon \parallel x = x$ |
| $P10$ | $x \parallel \epsilon = x$ |
| $C1$ | $e_1 \mid e_2 = \gamma(e_1, e_2)$ |
| $C2$ | $e_1 \mid (e_2 \cdot y) = \gamma(e_1, e_2) \cdot y$ |
| $C3$ | $(e_1 \cdot x) \mid e_2 = \gamma(e_1, e_2) \cdot x$ |
| $C4$ | $(e_1 \cdot x) \mid (e_2 \cdot y) = \gamma(e_1, e_2) \cdot (x \between y)$ |
| $C5$ | $(x + y) \mid z = (x \mid z) + (y \mid z)$ |
| $C6$ | $x \mid (y + z) = (x \mid y) + (x \mid z)$ |
| $C7$ | $\delta \mid x = \delta$ |
| $C8$ | $x \mid \delta = \delta$ |
| $C9$ | $\epsilon \mid x = \delta$ |
| $C10$ | $x \mid \epsilon = \delta$ |
| $CE1$ | $\Theta(e) = e$ |
| $CE2$ | $\Theta(\delta) = \delta$ |
| $CE3$ | $\Theta(\epsilon) = \epsilon$ |
| $CE4$ | $\Theta(x + y) = \Theta(x) + \Theta(y)$ |
| $CE5$ | $\Theta(x \cdot y) = \Theta(x) \cdot \Theta(y)$ |
| $CE6$ | $\Theta(x \parallel y) = ((\Theta(x) \triangleleft y) \parallel y) + ((\Theta(y) \triangleleft x) \parallel x)$ |
| $CE7$ | $\Theta(x \mid y) = ((\Theta(x) \triangleleft y) \mid y) + ((\Theta(y) \triangleleft x) \mid x)$ |
| $U1$ | $(\sharp(e_1, e_2)) \quad e_1 \triangleleft e_2 = \tau$ |
| $U2$ | $(\sharp(e_1, e_2), e_2 \leq e_3) \quad e_1 \triangleleft e_3 = \tau$ |
| $U3$ | $(\sharp(e_1, e_2), e_2 \leq e_3) \quad e3 \triangleleft e_1 = \tau$ |
| $U4$ | $e \triangleleft \delta = e$ |
| $U5$ | $\delta \triangleleft e = \delta$ |
| $U6$ | $e \triangleleft \epsilon = e$ |
| $U7$ | $\epsilon \triangleleft e = e$ |
| $U8$ | $(x + y) \triangleleft z = (x \triangleleft z) + (y \triangleleft z)$ |
| $U9$ | $(x \cdot y) \triangleleft z = (x \triangleleft z) \cdot (y \triangleleft z)$ |
| $U10$ | $(x \parallel y) \triangleleft z = (x \triangleleft z) \parallel (y \triangleleft z)$ |
| $U11$ | $(x \mid y) \triangleleft z = (x \triangleleft z) \mid (y \triangleleft z)$ |
| $U12$ | $x \triangleleft (y + z) = (x \triangleleft y) \triangleleft z$ |
| $U13$ | $x \triangleleft (y \cdot z) = (x \triangleleft y) \triangleleft z$ |
| $U14$ | $x \triangleleft (y \parallel z) = (x \triangleleft y) \triangleleft z$ |
| $U15$ | $x \triangleleft (y \mid z) = (x \triangleleft y) \triangleleft z$ |

*continued on next page*

**TABLE 9.3** *(continued)*

| No. | Axiom |
|---|---|
| $D1$ | $e \notin H \quad \partial_H(e) = e$ |
| $D2$ | $e \in H \quad \partial_H(e) = \delta$ |
| $D3$ | $\partial_H(\delta) = \delta$ |
| $D4$ | $\partial_H(x + y) = \partial_H(x) + \partial_H(y)$ |
| $D5$ | $\partial_H(x \cdot y) = \partial_H(x) \cdot \partial_H(y)$ |
| $D6$ | $\partial_H(x \parallel y) = \partial_H(x) \parallel \partial_H(y)$ |
| $G12$ | $\phi(x \parallel y) = \phi x \parallel \phi y$ |
| $G13$ | $\phi(x \mid y) = \phi x \mid \phi y$ |
| $G14$ | $\phi \parallel \delta = \delta$ |
| $G15$ | $\delta \parallel \phi = \delta$ |
| $G16$ | $\phi \mid \delta = \delta$ |
| $G17$ | $\delta \mid \phi = \delta$ |
| $G18$ | $\phi \parallel \epsilon = \phi$ |
| $G19$ | $\epsilon \parallel \phi = \phi$ |
| $G20$ | $\phi \mid \epsilon = \delta$ |
| $G21$ | $\epsilon \mid \phi = \delta$ |
| $G22$ | $\phi \parallel \neg\phi = \delta$ |
| $G23$ | $\Theta(\phi) = \phi$ |
| $G24$ | $\partial_H(\phi) = \phi$ |
| $G25$ | $\phi_0 \parallel \cdots \parallel \phi_n = \delta$ if $\forall s_0, \cdots, s_n \in S, \exists i \leq n.test(\neg\phi_i, s_0 \cup \cdots \cup s_n)$ |

We will define a term-deduction system that gives the operational semantics of $APTC_G$. Two atomic events $e_1$ and $e_2$ are in race condition, which are denoted $e_1\%e_2$. See Table 9.4.

**Theorem 9.19** (Generalization of $APTC_G$ with respect to $BATC_G$). *$APTC_G$ is a generalization of $BATC_G$.*

**Theorem 9.20** (Congruence of $APTC_G$ with respect to truly concurrent bisimulation equivalences). *(1) Pomset bisimulation equivalence $\sim_p$ is a congruence with respect to $APTC_G$;*

*(2) Step bisimulation equivalence $\sim_s$ is a congruence with respect to $APTC_G$;*

*(3) Hp-bisimulation equivalence $\sim_{hp}$ is a congruence with respect to $APTC_G$;*

*(4) Hhp-bisimulation equivalence $\sim_{hhp}$ is a congruence with respect to $APTC_G$.*

**Theorem 9.21** (Soundness of $APTC_G$ modulo truly concurrent bisimulation equivalences). *(1) Let $x$ and $y$ be $APTC_G$ terms. If $APTC \vdash x = y$, then $x \sim_p y$;*

*(2) Let $x$ and $y$ be $APTC_G$ terms. If $APTC \vdash x = y$, then $x \sim_s y$;*

*(3) Let $x$ and $y$ be $APTC_G$ terms. If $APTC \vdash x = y$, then $x \sim_{hp} y$.*

**Theorem 9.22** (Completeness of $APTC_G$ modulo truly concurrent bisimulation equivalences). *(1) Let $p$ and $q$ be closed $APTC_G$ terms, if $p \sim_p q$, then $p = q$;*

*(2) Let $p$ and $q$ be closed $APTC_G$ terms, if $p \sim_s q$, then $p = q$;*

*(3) Let $p$ and $q$ be closed $APTC_G$ terms, if $p \sim_{hp} q$, then $p = q$.*

**Theorem 9.23** (Sufficient determinacy). *All related data environments with respect to $APTC_G$ can be sufficiently deterministic.*

### 9.1.4 Recursion

In this section, we introduce recursion to capture infinite processes based on $APTC_G$. In the following, $E, F, G$ are recursion specifications, $X, Y, Z$ are recursive variables.

**TABLE 9.4** Transition rules of $APTC_G$.

| | |
|---|---|
| $\frac{}{\langle e_1\parallel\cdots\parallel e_n,s\rangle\xrightarrow{\{e_1,\cdots,e_n\}}\langle\surd,s'\rangle}$ if $s'\in effect(e_1,s)\cup\cdots\cup effect(e_n,s)$ | |
| $\frac{}{\langle\phi_1\parallel\cdots\parallel\phi_n,s\rangle\rightarrow\langle\surd,s\rangle}$ if $test(\phi_1,s),\cdots,test(\phi_n,s)$ | |
| $\frac{\langle x,s\rangle\xrightarrow{e_1}\langle\surd,s'\rangle\quad\langle y,s\rangle\xrightarrow{e_2}\langle\surd,s''\rangle}{\langle x\parallel y,s\rangle\xrightarrow{\{e_1,e_2\}}\langle\surd,s'\cup s''\rangle}$ | $\frac{\langle x,s\rangle\xrightarrow{e_1}\langle x',s'\rangle\quad\langle y,s\rangle\xrightarrow{e_2}\langle\surd,s''\rangle}{\langle x\parallel y,s\rangle\xrightarrow{\{e_1,e_2\}}\langle x',s'\cup s''\rangle}$ |
| $\frac{\langle x,s\rangle\xrightarrow{e_1}\langle\surd,s'\rangle\quad\langle y,s\rangle\xrightarrow{e_2}\langle y',s''\rangle}{\langle x\parallel y,s\rangle\xrightarrow{\{e_1,e_2\}}\langle y',s'\cup s''\rangle}$ | $\frac{\langle x,s\rangle\xrightarrow{e_1}\langle x',s'\rangle\quad\langle y,s\rangle\xrightarrow{e_2}\langle y',s''\rangle}{\langle x\parallel y,s\rangle\xrightarrow{\{e_1,e_2\}}\langle x'\between y',s'\cup s''\rangle}$ |
| $\frac{\langle x,s\rangle\xrightarrow{e_1}\langle\surd,s'\rangle\quad\langle y,s\rangle\overset{e_2}{\nrightarrow}\quad(e_1\%e_2)}{\langle x\parallel y,s\rangle\xrightarrow{e_1}\langle y,s'\rangle}$ | $\frac{\langle x,s\rangle\xrightarrow{e_1}\langle x',s'\rangle\quad\langle y,s\rangle\overset{e_2}{\nrightarrow}\quad(e_1\%e_2)}{\langle x\parallel y,s\rangle\xrightarrow{e_1}\langle x'\between y,s'\rangle}$ |
| $\frac{\langle x,s\rangle\overset{e_1}{\nrightarrow}\quad\langle y,s\rangle\xrightarrow{e_2}\langle\surd,s''\rangle\quad(e_1\%e_2)}{\langle x\parallel y,s\rangle\xrightarrow{e_2}\langle x,s''\rangle}$ | $\frac{\langle x,s\rangle\overset{e_1}{\nrightarrow}\quad\langle y,s\rangle\xrightarrow{e_2}\langle y',s''\rangle\quad(e_1\%e_2)}{\langle x\parallel y,s\rangle\xrightarrow{e_2}\langle x\between y',s''\rangle}$ |
| $\frac{\langle x,s\rangle\xrightarrow{e_1}\langle\surd,s'\rangle\quad\langle y,s\rangle\xrightarrow{e_2}\langle\surd,s''\rangle}{\langle x\mid y,s\rangle\xrightarrow{\gamma(e_1,e_2)}\langle\surd,effect(\gamma(e_1,e_2),s)\rangle}$ | $\frac{\langle x,s\rangle\xrightarrow{e_1}\langle x',s'\rangle\quad\langle y,s\rangle\xrightarrow{e_2}\langle\surd,s''\rangle}{\langle x\mid y,s\rangle\xrightarrow{\gamma(e_1,e_2)}\langle x',effect(\gamma(e_1,e_2),s)\rangle}$ |
| $\frac{\langle x,s\rangle\xrightarrow{e_1}\langle\surd,s'\rangle\quad\langle y,s\rangle\xrightarrow{e_2}\langle y',s''\rangle}{\langle x\mid y,s\rangle\xrightarrow{\gamma(e_1,e_2)}\langle y',effect(\gamma(e_1,e_2),s)\rangle}$ | $\frac{\langle x,s\rangle\xrightarrow{e_1}\langle x',s'\rangle\quad\langle y,s\rangle\xrightarrow{e_2}\langle y',s''\rangle}{\langle x\mid y,s\rangle\xrightarrow{\gamma(e_1,e_2)}\langle x'\between y',effect(\gamma(e_1,e_2),s)\rangle}$ |
| $\frac{\langle x,s\rangle\xrightarrow{e_1}\langle\surd,s'\rangle\quad(\sharp(e_1,e_2))}{\langle\Theta(x),s\rangle\xrightarrow{e_1}\langle\surd,s'\rangle}$ | $\frac{\langle x,s\rangle\xrightarrow{e_2}\langle\surd,s''\rangle\quad(\sharp(e_1,e_2))}{\langle\Theta(x),s\rangle\xrightarrow{e_2}\langle\surd,s''\rangle}$ |
| $\frac{\langle x,s\rangle\xrightarrow{e_1}\langle x',s'\rangle\quad(\sharp(e_1,e_2))}{\langle\Theta(x),s\rangle\xrightarrow{e_1}\langle\Theta(x'),s'\rangle}$ | $\frac{\langle x,s\rangle\xrightarrow{e_2}\langle x'',s''\rangle\quad(\sharp(e_1,e_2))}{\langle\Theta(x),s\rangle\xrightarrow{e_2}\langle\Theta(x''),s''\rangle}$ |
| $\frac{\langle x,s\rangle\xrightarrow{e_1}\langle\surd,s'\rangle\quad\langle y,s\rangle\nrightarrow^{e_2}\quad(\sharp(e_1,e_2))}{\langle x\triangleleft y,s\rangle\xrightarrow{\tau}\langle\surd,s'\rangle}$ | $\frac{\langle x,s\rangle\xrightarrow{e_1}\langle x',s'\rangle\quad\langle y,s\rangle\nrightarrow^{e_2}\quad(\sharp(e_1,e_2))}{\langle x\triangleleft y,s\rangle\xrightarrow{\tau}\langle x',s'\rangle}$ |
| $\frac{\langle x,s\rangle\xrightarrow{e_1}\langle\surd,s\rangle\quad\langle y,s\rangle\nrightarrow^{e_3}\quad(\sharp(e_1,e_2),e_2\le e_3)}{\langle x\triangleleft y,s\rangle\xrightarrow{\tau}\langle\surd,s'\rangle}$ | $\frac{\langle x,s\rangle\xrightarrow{e_1}\langle x',s'\rangle\quad\langle y,s\rangle\nrightarrow^{e_3}\quad(\sharp(e_1,e_2),e_2\le e_3)}{\langle x\triangleleft y,s\rangle\xrightarrow{\tau}\langle x',s'\rangle}$ |
| $\frac{\langle x,s\rangle\xrightarrow{e_3}\langle\surd,s'\rangle\quad\langle y,s\rangle\nrightarrow^{e_2}\quad(\sharp(e_1,e_2),e_1\le e_3)}{\langle x\triangleleft y,s\rangle\xrightarrow{\tau}\langle\surd,s'\rangle}$ | $\frac{\langle x,s\rangle\xrightarrow{e_3}\langle x',s'\rangle\quad\langle y,s\rangle\nrightarrow^{e_2}\quad(\sharp(e_1,e_2),e_1\le e_3)}{\langle x\triangleleft y,s\rangle\xrightarrow{\tau}\langle x',s'\rangle}$ |
| $\frac{\langle x,s\rangle\xrightarrow{e}\langle\surd,s'\rangle}{\langle\partial_H(x),s\rangle\xrightarrow{e}\langle\surd,s'\rangle}\quad(e\notin H)$ | $\frac{\langle x,s\rangle\xrightarrow{e}\langle x',s'\rangle}{\langle\partial_H(x),s\rangle\xrightarrow{e}\langle\partial_H(x'),s'\rangle}\quad(e\notin H)$ |

**Definition 9.24** (Guarded recursive specification). A recursive specification

$$X_1 = t_1(X_1,\cdots,X_n)$$

$$\cdots$$

$$X_n = t_n(X_1,\cdots,X_n)$$

is guarded if the right-hand sides of its recursive equations can be adapted to the form by applications of the axioms in $APTC$ and replacing recursion variables by the right-hand sides of their recursive equations,

$$(a_{11}\parallel\cdots\parallel a_{1i_1})\cdot s_1(X_1,\cdots,X_n)+\cdots+(a_{k1}\parallel\cdots\parallel a_{ki_k})\cdot s_k(X_1,\cdots,X_n)+(b_{11}\parallel\cdots\parallel b_{1j_1})+\cdots+(b_{1j_1}\parallel\cdots\parallel b_{lj_l})$$

where $a_{11},\cdots,a_{1i_1},a_{k1},\cdots,a_{ki_k},b_{11},\cdots,b_{1j_1},b_{1j_1},\cdots,b_{lj_l}\in\mathbb{E}$, and the sum above is allowed to be empty, in which case it represents the deadlock $\delta$. Also, there does not exist an infinite sequence of $\epsilon$-transitions $\langle X|E\rangle\rightarrow\langle X'|E\rangle\rightarrow\langle X''|E\rangle\rightarrow\cdots$. The transition rules of guarded recursions please see Table 9.5.

**TABLE 9.5** Transition rules of guarded recursion.

| |
|---|
| $$\frac{\langle t_i(\langle X_1\|E\rangle,\cdots,\langle X_n\|E\rangle),s\rangle\xrightarrow{\{e_1,\cdots,e_k\}}\langle\surd,s'\rangle}{\langle\langle X_i\|E\rangle,s\rangle\xrightarrow{\{e_1,\cdots,e_k\}}\langle\surd,s'\rangle}$$ |
| $$\frac{\langle t_i(\langle X_1\|E\rangle,\cdots,\langle X_n\|E\rangle),s\rangle\xrightarrow{\{e_1,\cdots,e_k\}}\langle y,s'\rangle}{\langle\langle X_i\|E\rangle,s\rangle\xrightarrow{\{e_1,\cdots,e_k\}}\langle y,s'\rangle}$$ |

**TABLE 9.6** Transition rule of the silent step.

| |
|---|
| $$\frac{}{\langle\tau_\phi,s\rangle\rightarrow\langle\surd,s\rangle}\text{ if } test(\tau_\phi,s)$$ |
| $$\frac{}{\langle\tau,s\rangle\xrightarrow{\tau}\langle\surd,\tau(s)\rangle}$$ |

**Theorem 9.25** (Conservativity of $APTC_G$ with guarded recursion). *$APTC_G$ with guarded recursion is a conservative extension of $APTC_G$.*

**Theorem 9.26** (Congruence theorem of $APTC_G$ with guarded recursion). *Truly concurrent bisimulation equivalences $\sim_p$, $\sim_s$, and $\sim_{hp}$ are all congruences with respect to $APTC_G$ with guarded recursion.*

**Theorem 9.27** (Elimination theorem of $APTC_G$ with linear recursion). *Each process term in $APTC_G$ with linear recursion is equal to a process term $\langle X_1|E\rangle$ with $E$ a linear recursive specification.*

**Theorem 9.28** (Soundness of $APTC_G$ with guarded recursion). *Let $x$ and $y$ be $APTC_G$ with guarded recursion terms. If $APTC_G$ with guarded recursion $\vdash x=y$, then (1) $x\sim_s y$;*
*(2) $x\sim_p y$;*
*(3) $x\sim_{hp} y$.*

**Theorem 9.29** (Completeness of $APTC_G$ with linear recursion). *Let $p$ and $q$ be closed $APTC_G$ with linear recursion terms, then (1) if $p\sim_s q$, then $p=q$;*
*(2) if $p\sim_p q$, then $p=q$;*
*(3) if $p\sim_{hp} q$, then $p=q$.*

### 9.1.5 Abstraction

To abstract away from the internal implementations of a program, and verify that the program exhibits the desired external behaviors, the silent step $\tau$ and abstraction operator $\tau_I$ are introduced, where $I\subseteq\mathbb{E}\cup G_{at}$ denotes the internal events or guards. The silent step $\tau$ represents the internal events, and $\tau_\phi$ for internal guards, when we consider the external behaviors of a process, $\tau$ steps can be removed, that is, $\tau$ steps must remain silent. The transition rule of $\tau$ is shown in Table 9.6. In the following, let the atomic event $e$ range over $\mathbb{E}\cup\{\epsilon\}\cup\{\delta\}\cup\{\tau\}$, and $\phi$ range over $G\cup\{\tau\}$, and let the communication function $\gamma:\mathbb{E}\cup\{\tau\}\times\mathbb{E}\cup\{\tau\}\rightarrow\mathbb{E}\cup\{\delta\}$, with each communication involved $\tau$ resulting in $\delta$. We use $\tau(s)$ to denote $effect(\tau,s)$, for the fact that $\tau$ only changes the state of internal data environment, that is, for the external data environments, $s=\tau(s)$.

We have introduced $\tau$ into the event structure, and also give the concept of weakly true concurrency. In this section, we give the concepts of rooted branching truly concurrent bisimulation equivalences, based on these concepts, we can design the axiom system of the silent step $\tau$ and the abstraction operator $\tau_I$.

**Definition 9.30** (Branching pomset, step bisimulation). Assume a special termination predicate $\downarrow$, and let $\surd$ represent a state with $\surd\downarrow$. Let $\mathcal{E}_1$, $\mathcal{E}_2$ be PESs. A branching pomset bisimulation is a relation $R\subseteq\langle\mathcal{C}(\mathcal{E}_1),S\rangle\times\langle\mathcal{C}(\mathcal{E}_2),S\rangle$, such that:

1. if $(\langle C_1,s\rangle,\langle C_2,s\rangle)\in R$, and $\langle C_1,s\rangle\xrightarrow{X}\langle C_1',s'\rangle$, then
   - either $X\equiv\tau^*$, and $(\langle C_1',s'\rangle,\langle C_2,s\rangle)\in R$ with $s'\in\tau(s)$;
   - or there is a sequence of (zero or more) $\tau$-transitions $\langle C_2,s\rangle\xrightarrow{\tau^*}\langle C_2^0,s^0\rangle$, such that $(\langle C_1,s\rangle,\langle C_2^0,s^0\rangle)\in R$ and $\langle C_2^0,s^0\rangle\xRightarrow{X}\langle C_2',s'\rangle$ with $(\langle C_1',s'\rangle,\langle C_2',s'\rangle)\in R$;

2. if $(\langle C_1, s\rangle, \langle C_2, s\rangle) \in R$, and $\langle C_2, s\rangle \xrightarrow{X} \langle C_2', s'\rangle$, then
   - either $X \equiv \tau^*$, and $(\langle C_1, s\rangle, \langle C_2', s'\rangle) \in R$;
   - or there is a sequence of (zero or more) $\tau$-transitions $\langle C_1, s\rangle \xrightarrow{\tau^*} \langle C_1^0, s^0\rangle$, such that $(\langle C_1^0, s^0\rangle, \langle C_2, s\rangle) \in R$ and $\langle C_1^0, s^0\rangle \stackrel{X}{\Rightarrow} \langle C_1', s'\rangle$ with $(\langle C_1', s'\rangle, \langle C_2', s'\rangle) \in R$;
3. if $(\langle C_1, s\rangle, \langle C_2, s\rangle) \in R$ and $\langle C_1, s\rangle \downarrow$, then there is a sequence of (zero or more) $\tau$-transitions $\langle C_2, s\rangle \xrightarrow{\tau^*} \langle C_2^0, s^0\rangle$ such that $(\langle C_1, s\rangle, \langle C_2^0, s^0\rangle) \in R$ and $\langle C_2^0, s^0\rangle \downarrow$;
4. if $(\langle C_1, s\rangle, \langle C_2, s\rangle) \in R$ and $\langle C_2, s\rangle \downarrow$, then there is a sequence of (zero or more) $\tau$-transitions $\langle C_1, s\rangle \xrightarrow{\tau^*} \langle C_1^0, s^0\rangle$ such that $(\langle C_1^0, s^0\rangle, \langle C_2, s\rangle) \in R$ and $\langle C_1^0, s^0\rangle \downarrow$.

We say that $\mathcal{E}_1, \mathcal{E}_2$ are branching pomset bisimilar, written $\mathcal{E}_1 \approx_{bp} \mathcal{E}_2$, if there exists a branching pomset bisimulation $R$, such that $(\langle\emptyset, \emptyset\rangle, \langle\emptyset, \emptyset\rangle) \in R$.

By replacing pomset transitions with steps, we can obtain the definition of branching step bisimulation. When PESs $\mathcal{E}_1$ and $\mathcal{E}_2$ are branching step bisimilar, we write $\mathcal{E}_1 \approx_{bs} \mathcal{E}_2$.

**Definition 9.31** (Rooted branching pomset, step bisimulation). Assume a special termination predicate $\downarrow$, and let $\surd$ represent a state with $\surd \downarrow$. Let $\mathcal{E}_1, \mathcal{E}_2$ be PESs. A rooted branching pomset bisimulation is a relation $R \subseteq \langle \mathcal{C}(\mathcal{E}_1), S\rangle \times \langle \mathcal{C}(\mathcal{E}_2), S\rangle$, such that:

1. if $(\langle C_1, s\rangle, \langle C_2, s\rangle) \in R$, and $\langle C_1, s\rangle \xrightarrow{X} \langle C_1', s'\rangle$, then $\langle C_2, s\rangle \xrightarrow{X} \langle C_2', s'\rangle$ with $\langle C_1', s'\rangle \approx_{bp} \langle C_2', s'\rangle$;
2. if $(\langle C_1, s\rangle, \langle C_2, s\rangle) \in R$, and $\langle C_2, s\rangle \xrightarrow{X} \langle C_2', s'\rangle$, then $\langle C_1, s\rangle \xrightarrow{X} \langle C_1', s'\rangle$ with $\langle C_1', s'\rangle \approx_{bp} \langle C_2', s'\rangle$;
3. if $(\langle C_1, s\rangle, \langle C_2, s\rangle) \in R$ and $\langle C_1, s\rangle \downarrow$, then $\langle C_2, s\rangle \downarrow$;
4. if $(\langle C_1, s\rangle, \langle C_2, s\rangle) \in R$ and $\langle C_2, s\rangle \downarrow$, then $\langle C_1, s\rangle \downarrow$.

We say that $\mathcal{E}_1, \mathcal{E}_2$ are rooted branching pomset bisimilar, written $\mathcal{E}_1 \approx_{rbp} \mathcal{E}_2$, if there exists a rooted branching pomset bisimulation $R$, such that $(\langle\emptyset, \emptyset\rangle, \langle\emptyset, \emptyset\rangle) \in R$.

By replacing pomset transitions with steps, we can obtain the definition of rooted branching step bisimulation. When PESs $\mathcal{E}_1$ and $\mathcal{E}_2$ are rooted branching step bisimilar, we write $\mathcal{E}_1 \approx_{rbs} \mathcal{E}_2$.

**Definition 9.32** (Branching (hereditary) history-preserving bisimulation). Assume a special termination predicate $\downarrow$, and let $\surd$ represent a state with $\surd \downarrow$. A branching history-preserving (hp-) bisimulation is a weakly posetal relation $R \subseteq \langle \mathcal{C}(\mathcal{E}_1), S\rangle \overline{\times} \langle \mathcal{C}(\mathcal{E}_2), S\rangle$ such that:

1. if $(\langle C_1, s\rangle, f, \langle C_2, s\rangle) \in R$, and $\langle C_1, s\rangle \xrightarrow{e_1} \langle C_1', s'\rangle$, then
   - either $e_1 \equiv \tau$, and $(\langle C_1', s'\rangle, f[e_1 \mapsto \tau], \langle C_2, s\rangle) \in R$;
   - or there is a sequence of (zero or more) $\tau$-transitions $\langle C_2, s\rangle \xrightarrow{\tau^*} \langle C_2^0, s^0\rangle$, such that $(\langle C_1, s\rangle, f, \langle C_2^0, s^0\rangle) \in R$ and $\langle C_2^0, s^0\rangle \xrightarrow{e_2} \langle C_2', s'\rangle$ with $(\langle C_1', s'\rangle, f[e_1 \mapsto e_2], \langle C_2', s'\rangle) \in R$;
2. if $(\langle C_1, s\rangle, f, \langle C_2, s\rangle) \in R$, and $\langle C_2, s\rangle \xrightarrow{e_2} \langle C_2', s'\rangle$, then
   - either $e_2 \equiv \tau$, and $(\langle C_1, s\rangle, f[e_2 \mapsto \tau], \langle C_2', s'\rangle) \in R$;
   - or there is a sequence of (zero or more) $\tau$-transitions $\langle C_1, s\rangle \xrightarrow{\tau^*} \langle C_1^0, s^0\rangle$, such that $(\langle C_1^0, s^0\rangle, f, \langle C_2, s\rangle) \in R$ and $\langle C_1^0, s^0\rangle \xrightarrow{e_1} \langle C_1', s'\rangle$ with $(\langle C_1', s'\rangle, f[e_2 \mapsto e_1], \langle C_2', s'\rangle) \in R$;
3. if $(\langle C_1, s\rangle, f, \langle C_2, s\rangle) \in R$ and $\langle C_1, s\rangle \downarrow$, then there is a sequence of (zero or more) $\tau$-transitions $\langle C_2, s\rangle \xrightarrow{\tau^*} \langle C_2^0, s^0\rangle$ such that $(\langle C_1, s\rangle, f, \langle C_2^0, s^0\rangle) \in R$ and $\langle C_2^0, s^0\rangle \downarrow$;
4. if $(\langle C_1, s\rangle, f, \langle C_2, s\rangle) \in R$ and $\langle C_2, s\rangle \downarrow$, then there is a sequence of (zero or more) $\tau$-transitions $\langle C_1, s\rangle \xrightarrow{\tau^*} \langle C_1^0, s^0\rangle$ such that $(\langle C_1^0, s^0\rangle, f, \langle C_2, s\rangle) \in R$ and $\langle C_1^0, s^0\rangle \downarrow$.

$\mathcal{E}_1, \mathcal{E}_2$ are branching history-preserving (hp-)bisimilar and are written $\mathcal{E}_1 \approx_{bhp} \mathcal{E}_2$ if there exists a branching hp-bisimulation $R$ such that $(\langle\emptyset, \emptyset\rangle, \emptyset, \langle\emptyset, \emptyset\rangle) \in R$.

A branching hereditary history-preserving (hhp-)bisimulation is a downward closed branching hp-bisimulation. $\mathcal{E}_1, \mathcal{E}_2$ are branching hereditary history-preserving (hhp-)bisimilar and are written $\mathcal{E}_1 \approx_{bhhp} \mathcal{E}_2$.

**TABLE 9.7 Axioms of silent step.**

| No. | Axiom |
|---|---|
| $B1$ | $e \cdot \tau = e$ |
| $B2$ | $e \cdot (\tau \cdot (x + y) + x) = e \cdot (x + y)$ |
| $B3$ | $x \parallel \tau = x$ |
| $G26$ | $\tau_\phi \cdot x = x$ |
| $G27$ | $x \cdot \tau_\phi = x$ |
| $G28$ | $x \parallel \tau_\phi = x$ |

**Definition 9.33** (Rooted branching (hereditary) history-preserving bisimulation). Assume a special termination predicate $\downarrow$, and let $\surd$ represent a state with $\surd\downarrow$. A rooted branching history-preserving (hp-) bisimulation is a weakly posetal relation $R \subseteq \langle \mathcal{C}(\mathcal{E}_1), S\rangle \overline{\times} \langle \mathcal{C}(\mathcal{E}_2), S\rangle$ such that:

1. if $(\langle C_1, s\rangle, f, \langle C_2, s\rangle) \in R$, and $\langle C_1, s\rangle \xrightarrow{e_1} \langle C_1', s'\rangle$, then $\langle C_2, s\rangle \xrightarrow{e_2} \langle C_2', s'\rangle$ with $\langle C_1', s'\rangle \approx_{bhp} \langle C_2', s'\rangle$;
2. if $(\langle C_1, s\rangle, f, \langle C_2, s\rangle) \in R$, and $\langle C_2, s\rangle \xrightarrow{e_2} \langle C_2', s'\rangle$, then $\langle C_1, s\rangle \xrightarrow{e_1} \langle C_1', s'\rangle$ with $\langle C_1', s'\rangle \approx_{bhp} \langle C_2', s'\rangle$;
3. if $(\langle C_1, s\rangle, f, \langle C_2, s\rangle) \in R$ and $\langle C_1, s\rangle \downarrow$, then $\langle C_2, s\rangle \downarrow$;
4. if $(\langle C_1, s\rangle, f, \langle C_2, s\rangle) \in R$ and $\langle C_2, s\rangle \downarrow$, then $\langle C_1, s\rangle \downarrow$.

$\mathcal{E}_1, \mathcal{E}_2$ are rooted branching history-preserving (hp-)bisimilar and are written $\mathcal{E}_1 \approx_{rbhp} \mathcal{E}_2$ if there exists a rooted branching hp-bisimulation $R$ such that $(\langle\emptyset, \emptyset\rangle, \emptyset, \langle\emptyset, \emptyset\rangle) \in R$.

A rooted branching hereditary history-preserving (hhp-)bisimulation is a downward closed rooted branching hp-bisimulation. $\mathcal{E}_1, \mathcal{E}_2$ are rooted branching hereditary history-preserving (hhp-)bisimilar and are written $\mathcal{E}_1 \approx_{rbhhp} \mathcal{E}_2$.

**Definition 9.34** (Guarded linear recursive specification). A linear recursive specification $E$ is guarded if there does not exist an infinite sequence of $\tau$-transitions $\langle X|E\rangle \xrightarrow{\tau} \langle X'|E\rangle \xrightarrow{\tau} \langle X''|E\rangle \xrightarrow{\tau} \cdots$, and there does not exist an infinite sequence of $\epsilon$-transitions $\langle X|E\rangle \rightarrow \langle X'|E\rangle \rightarrow \langle X''|E\rangle \rightarrow \cdots$.

**Theorem 9.35** (Conservativity of $APTC_G$ with silent step and guarded linear recursion). *$APTC_G$ with silent step and guarded linear recursion is a conservative extension of $APTC_G$ with linear recursion.*

**Theorem 9.36** (Congruence theorem of $APTC_G$ with silent step and guarded linear recursion). *Rooted branching truly concurrent bisimulation equivalences $\approx_{rbp}$, $\approx_{rbs}$, and $\approx_{rbhp}$ are all congruences with respect to $APTC_G$ with silent step and guarded linear recursion.*

We design the axioms for the silent step $\tau$ in Table 9.7.

**Theorem 9.37** (Elimination theorem of $APTC_G$ with silent step and guarded linear recursion). *Each process term in $APTC_G$ with silent step and guarded linear recursion is equal to a process term $\langle X_1|E\rangle$ with E a guarded linear recursive specification.*

**Theorem 9.38** (Soundness of $APTC_G$ with silent step and guarded linear recursion). *Let x and y be $APTC_G$ with silent step and guarded linear recursion terms. If $APTC_G$ with silent step and guarded linear recursion $\vdash x = y$, then*

*(1) $x \approx_{rbs} y$;*
*(2) $x \approx_{rbp} y$;*
*(3) $x \approx_{rbhp} y$.*

**Theorem 9.39** (Completeness of $APTC_G$ with silent step and guarded linear recursion). *Let p and q be closed $APTC_G$ with silent step and guarded linear recursion terms, then*

*(1) if $p \approx_{rbs} q$, then $p = q$;*
*(2) if $p \approx_{rbp} q$, then $p = q$;*
*(3) if $p \approx_{rbhp} q$, then $p = q$.*

The unary abstraction operator $\tau_I$ ($I \subseteq \mathbb{E} \cup G_{at}$) renames all atomic events or atomic guards in $I$ into $\tau$. $APTC_G$ with silent step and abstraction operator is called $APTC_{G_\tau}$. The transition rules of operator $\tau_I$ are shown in Table 9.8.

**Theorem 9.40** (Conservativity of $APTC_{G_\tau}$ with guarded linear recursion). *$APTC_{G_\tau}$ with guarded linear recursion is a conservative extension of $APTC_G$ with silent step and guarded linear recursion.*

**TABLE 9.8** Transition rule of the abstraction operator.

$$\frac{\langle x,s\rangle \xrightarrow{e} \langle\surd,s'\rangle}{\langle\tau_I(x),s\rangle \xrightarrow{e} \langle\surd,s'\rangle}\ e\notin I \qquad \frac{\langle x,s\rangle \xrightarrow{e} \langle x',s'\rangle}{\langle\tau_I(x),s\rangle \xrightarrow{e} \langle\tau_I(x'),s'\rangle}\ e\notin I$$

$$\frac{\langle x,s\rangle \xrightarrow{e} \langle\surd,s'\rangle}{\langle\tau_I(x),s\rangle \xrightarrow{\tau} \langle\surd,\tau(s)\rangle}\ e\in I \qquad \frac{\langle x,s\rangle \xrightarrow{e} \langle x',s'\rangle}{\langle\tau_I(x),s\rangle \xrightarrow{\tau} \langle\tau_I(x'),\tau(s)\rangle}\ e\in I$$

**TABLE 9.9** Axioms of abstraction operator.

| No. | Axiom |
|---|---|
| $TI1$ | $e\notin I \quad \tau_I(e)=e$ |
| $TI2$ | $e\in I \quad \tau_I(e)=\tau$ |
| $TI3$ | $\tau_I(\delta)=\delta$ |
| $TI4$ | $\tau_I(x+y)=\tau_I(x)+\tau_I(y)$ |
| $TI5$ | $\tau_I(x\cdot y)=\tau_I(x)\cdot\tau_I(y)$ |
| $TI6$ | $\tau_I(x\parallel y)=\tau_I(x)\parallel\tau_I(y)$ |
| $G29$ | $\phi\notin I \quad \tau_I(\phi)=\phi$ |
| $G30$ | $\phi\in I \quad \tau_I(\phi)=\tau_\phi$ |

**Theorem 9.41** (Congruence theorem of $APTC_{G_\tau}$ with guarded linear recursion). *Rooted branching truly concurrent bisimulation equivalences $\approx_{rbp}$, $\approx_{rbs}$, and $\approx_{rbhp}$ are all congruences with respect to $APTC_{G_\tau}$ with guarded linear recursion.*

We design the axioms for the abstraction operator $\tau_I$ in Table 9.9.

**Theorem 9.42** (Soundness of $APTC_{G_\tau}$ with guarded linear recursion). *Let $x$ and $y$ be $APTC_{G_\tau}$ with guarded linear recursion terms. If $APTC_{G_\tau}$ with guarded linear recursion $\vdash x=y$, then*
*(1) $x\approx_{rbs} y$;*
*(2) $x\approx_{rbp} y$;*
*(3) $x\approx_{rbhp} y$.*

Although $\tau$-loops are prohibited in guarded linear recursive specifications (see Definition 9.34) in a specifiable way, they can be constructed using the abstraction operator, for example, there exist $\tau$-loops in the process term $\tau_{\{a\}}(\langle X|X=aX\rangle)$. To avoid $\tau$-loops caused by $\tau_I$ and ensure fairness, the concept of cluster and $CFAR$ (Cluster Fair Abstraction Rule) [17] are still needed.

**Theorem 9.43** (Completeness of $APTC_{G_\tau}$ with guarded linear recursion and $CFAR$). *Let $p$ and $q$ be closed $APTC_{G_\tau}$ with guarded linear recursion and $CFAR$ terms, then*
*(1) if $p\approx_{rbs} q$, then $p=q$;*
*(2) if $p\approx_{rbp} q$, then $p=q$;*
*(3) if $p\approx_{rbhp} q$, then $p=q$.*

## 9.2 Probabilistic operational semantics for true concurrency

In the following, the variables $x,x',y,y',z,z'$ range over the collection of process terms, $s,s',t,t',u,u'$ are closed terms, $\tau$ is the special constant silent step, $\delta$ is the special constant deadlock, $A$ is the collection of atomic actions, atomic actions $a,b\in A$, $A_\delta=A\cup\{\delta\}$, $A_\tau=A\cup\{\tau\}$. $\rightsquigarrow$ denotes probabilistic transition, and an action transition labeled by an atomic action $a\in A$, $\xrightarrow{a}$ and $\xrightarrow{a}\surd$. $x\xrightarrow{a}p$ means that by performing action $a$ process $x$ evolves into $p$; while $x\xrightarrow{a}\surd$ means that $x$ performs an $a$ action and then terminates. $p\rightsquigarrow x$ denotes that process $p$ chooses to behave like process $x$ with a non-zero probability $\pi>0$.

**Definition 9.44** (Probabilistic prime event structure with silent event). Let $\Lambda$ be a fixed set of labels, ranged over $a,b,c,\cdots$ and $\tau$. A ($\Lambda$-labeled) prime event structure with silent event $\tau$ is a quintuple $\mathcal{E}=\langle\mathbb{E},\leq,\sharp,\sharp_\pi,\lambda\rangle$, where $\mathbb{E}$ is a denumerable set of events, including the silent event $\tau$. Let $\hat{\mathbb{E}} = \mathbb{E}\backslash\{\tau\}$, exactly excluding $\tau$, it is obvious that $\hat{\tau^*} = \epsilon$, where $\epsilon$ is the empty event. Let $\lambda : \mathbb{E} \to \Lambda$ be a labeling function and let $\lambda(\tau) = \tau$. Also, $\leq$, $\sharp$, $\sharp_\pi$ are binary relations on $\mathbb{E}$, called causality, conflict, and probabilistic conflict, respectively, such that:

1. $\leq$ is a partial order and $\lceil e \rceil = \{e' \in \mathbb{E} | e' \leq e\}$ is finite for all $e \in \mathbb{E}$. It is easy to see that $e \leq \tau^* \leq e' = e \leq \tau \leq \cdots \leq \tau \leq e'$, then $e \leq e'$;
2. $\sharp$ is irreflexive, symmetric, and hereditary with respect to $\leq$, that is, for all $e, e', e'' \in \mathbb{E}$, if $e \sharp e' \leq e''$, then $e \sharp e''$;
3. $\sharp_\pi$ is irreflexive, symmetric, and hereditary with respect to $\leq$, that is, for all $e, e', e'' \in \mathbb{E}$, if $e \sharp_\pi e' \leq e''$, then $e \sharp_\pi e''$.

Then, the concepts of consistency and concurrency can be drawn from the above definition:

1. $e, e' \in \mathbb{E}$ are consistent, denoted as $e \frown e'$, if $\neg(e \sharp e')$ and $\neg(e \sharp_\pi e')$. A subset $X \subseteq \mathbb{E}$ is called consistent, if $e \frown e'$ for all $e, e' \in X$;
2. $e, e' \in \mathbb{E}$ are concurrent, denoted as $e \parallel e'$, if $\neg(e \leq e')$, $\neg(e' \leq e)$, and $\neg(e \sharp e')$ and $\neg(e \sharp_\pi e')$.

**Definition 9.45** (Configuration). Let $\mathcal{E}$ be a PES. A (finite) configuration in $\mathcal{E}$ is a (finite) consistent subset of events $C \subseteq \mathcal{E}$, closed with respect to causality (i.e., $\lceil C \rceil = C$). The set of finite configurations of $\mathcal{E}$ is denoted by $\mathcal{C}(\mathcal{E})$. We let $\hat{C} = C\backslash\{\tau\}$.

A consistent subset of $X \subseteq \mathbb{E}$ of events can be seen as a pomset. Given $X, Y \subseteq \mathbb{E}$, $\hat{X} \sim \hat{Y}$ if $\hat{X}$ and $\hat{Y}$ are isomorphic as pomsets. In the remainder of the chapter, when we say $C_1 \sim C_2$, we mean $\hat{C_1} \sim \hat{C_2}$.

**Definition 9.46** (Pomset transitions and step). Let $\mathcal{E}$ be a PES and let $C \in \mathcal{C}(\mathcal{E})$, and $\emptyset \neq X \subseteq \mathbb{E}$, if $C \cap X = \emptyset$ and $C' = C \cup X \in \mathcal{C}(\mathcal{E})$, then $C \xrightarrow{X} C'$ is called a pomset transition from $C$ to $C'$. When the events in $X$ are pairwise concurrent, we say that $C \xrightarrow{X} C'$ is a step.

**Definition 9.47** (Probabilistic transitions). Let $\mathcal{E}$ be a PES and let $C \in \mathcal{C}(\mathcal{E})$, the transition $C \xrightarrow{\pi} C^\pi$ is called a probabilistic transition from $C$ to $C^\pi$.

**Definition 9.48** (Weak pomset transitions and weak step). Let $\mathcal{E}$ be a PES and let $C \in \mathcal{C}(\mathcal{E})$, and $\emptyset \neq X \subseteq \hat{\mathbb{E}}$, if $C \cap X = \emptyset$ and $\hat{C'} = \hat{C} \cup X \in \mathcal{C}(\mathcal{E})$, then $C \xRightarrow{X} C'$ is called a weak pomset transition from $C$ to $C'$, where we define $\xRightarrow{e} \triangleq \xrightarrow{\tau^*}\xrightarrow{e}\xrightarrow{\tau^*}$. Also, $\xRightarrow{X} \triangleq \xrightarrow{\tau^*}\xrightarrow{e}\xrightarrow{\tau^*}$, for every $e \in X$. When the events in $X$ are pairwise concurrent, we say that $C \xRightarrow{X} C'$ is a weak step.

We will also suppose that all the PESs in this book are image finite, that is, for any PES $\mathcal{E}$ and $C \in \mathcal{C}(\mathcal{E})$ and $a \in \Lambda$, $\{\langle C, s\rangle \xrightarrow{\pi} \langle C^\pi, s\rangle\}$, $\{e \in \mathbb{E} | \langle C, s\rangle \xrightarrow{e} \langle C', s'\rangle \wedge \lambda(e) = a\}$ and $\{e \in \hat{\mathbb{E}} | \langle C, s\rangle \xRightarrow{e} \langle C', s'\rangle \wedge \lambda(e) = a\}$ is finite.

A probability distribution function (PDF) $\mu$ is a map $\mu : \mathcal{C} \times \mathcal{C} \to [0, 1]$ and $\mu^*$ is the cumulative probability distribution function (cPDF).

**Definition 9.49** (Probabilistic pomset, step bisimulation). Let $\mathcal{E}_1$, $\mathcal{E}_2$ be PESs. A probabilistic pomset bisimulation is a relation $R \subseteq \mathcal{C}(\mathcal{E}_1) \times \mathcal{C}(\mathcal{E}_2)$, such that (1) if $(C_1, C_2) \in R$, and $C_1 \xrightarrow{X_1} C_1'$, then $C_2 \xrightarrow{X_2} C_2'$, with $X_1 \subseteq \mathbb{E}_1$, $X_2 \subseteq \mathbb{E}_2$, $X_1 \sim X_2$ and $(C_1', C_2') \in R$, and vice versa; (2) if $(C_1, C_2) \in R$, and $C_1 \xrightarrow{\pi} C_1^\pi$, then $C_2 \xrightarrow{\pi} C_2^\pi$ and $(C_1^\pi, C_2^\pi) \in R$, and vice versa; (3) if $(C_1, C_2) \in R$, then $\mu(C_1, C) = \mu(C_2, C)$ for each $C \in \mathcal{C}(\mathcal{E})/R$; (4) $[\surd]_R = \{\surd\}$. We say that $\mathcal{E}_1$, $\mathcal{E}_2$ are probabilistic pomset bisimilar, written $\mathcal{E}_1 \sim_{pp} \mathcal{E}_2$, if there exists a probabilistic pomset bisimulation $R$, such that $(\emptyset, \emptyset) \in R$. By replacing probabilistic pomset transitions with steps, we can obtain the definition of probabilistic step bisimulation. When PESs $\mathcal{E}_1$ and $\mathcal{E}_2$ are probabilistic step bisimilar, we write $\mathcal{E}_1 \sim_{ps} \mathcal{E}_2$.

**Definition 9.50** (Posetal product). Given two PESs $\mathcal{E}_1$, $\mathcal{E}_2$, the posetal product of their configurations, denoted $\mathcal{C}(\mathcal{E}_1)\overline{\times}\mathcal{C}(\mathcal{E}_2)$, is defined as

$$\{(C_1, f, C_2) | C_1 \in \mathcal{C}(\mathcal{E}_1), C_2 \in \mathcal{C}(\mathcal{E}_2), f : C_1 \to C_2 \text{ isomorphism}\}$$

A subset $R \subseteq \mathcal{C}(\mathcal{E}_1)\overline{\times}\mathcal{C}(\mathcal{E}_2)$ is called a posetal relation. We say that $R$ is downward closed when for any $(C_1, f, C_2)$, $(C_1', f', C_2') \in \mathcal{C}(\mathcal{E}_1)\overline{\times}\mathcal{C}(\mathcal{E}_2)$, if $(C_1, f, C_2) \subseteq (C_1', f', C_2')$ pointwise and $(C_1', f', C_2') \in R$, then $(C_1, f, C_2) \in R$.

For $f : X_1 \to X_2$, we define $f[x_1 \mapsto x_2] : X_1 \cup \{x_1\} \to X_2 \cup \{x_2\}$, $z \in X_1 \cup \{x_1\}$, (1) $f[x_1 \mapsto x_2](z) = x_2$, if $z = x_1$; (2) $f[x_1 \mapsto x_2](z) = f(z)$, otherwise, where $X_1 \subseteq \mathbb{E}_1$, $X_2 \subseteq \mathbb{E}_2$, $x_1 \in \mathbb{E}_1$, $x_2 \in \mathbb{E}_2$.

**Definition 9.51** (Probabilistic (hereditary) history-preserving bisimulation). A probabilistic history-preserving (hp-) bisimulation is a posetal relation $R \subseteq \mathcal{C}(\mathcal{E}_1)\overline{\times}\mathcal{C}(\mathcal{E}_2)$ such that (1) if $(C_1, f, C_2) \in R$, and $C_1 \xrightarrow{e_1} C_1'$, then $C_2 \xrightarrow{e_2} C_2'$, with

$(C_1', f[e_1 \mapsto e_2], C_2') \in R$, and vice versa; (2) if $(C_1, f, C_2) \in R$, and $C_1 \xrightarrow{\pi} C_1^{\pi}$, then $C_2 \xrightarrow{\pi} C_2^{\pi}$ and $(C_1^{\pi}, f, C_2^{\pi}) \in R$, and vice versa; (3) if $(C_1, f, C_2) \in R$, then $\mu(C_1, C) = \mu(C_2, C)$ for each $C \in \mathcal{C}(\mathcal{E})/R$; (4) $[\surd]_R = \{\surd\}$. $\mathcal{E}_1, \mathcal{E}_2$ are probabilistic history-preserving (hp-)bisimilar and are written $\mathcal{E}_1 \sim_{php} \mathcal{E}_2$ if there exists a probabilistic hp-bisimulation $R$ such that $(\emptyset, \emptyset, \emptyset) \in R$.

A probabilistic hereditary history-preserving (hhp-)bisimulation is a downward closed probabilistic hp-bisimulation. $\mathcal{E}_1, \mathcal{E}_2$ are probabilistic hereditary history-preserving (hhp-)bisimilar and are written $\mathcal{E}_1 \sim_{phhp} \mathcal{E}_2$.

**Definition 9.52** (Weakly probabilistic pomset, step bisimulation). Let $\mathcal{E}_1$, $\mathcal{E}_2$ be PESs. A weakly probabilistic pomset bisimulation is a relation $R \subseteq \mathcal{C}(\mathcal{E}_1) \times \mathcal{C}(\mathcal{E}_2)$, such that (1) if $(C_1, C_2) \in R$, and $C_1 \xRightarrow{X_1} C_1'$, then $C_2 \xRightarrow{X_2} C_2'$, with $X_1 \subseteq \hat{\mathbb{E}}_1$, $X_2 \subseteq \hat{\mathbb{E}}_2$, $X_1 \sim X_2$ and $(C_1', C_2') \in R$, and vice versa; (2) if $(C_1, C_2) \in R$, and $C_1 \xrightarrow{\pi} C_1^{\pi}$, then $C_2 \xrightarrow{\pi} C_2^{\pi}$ and $(C_1^{\pi}, C_2^{\pi}) \in R$, and vice versa; (3) if $(C_1, C_2) \in R$, then $\mu(C_1, C) = \mu(C_2, C)$ for each $C \in \mathcal{C}(\mathcal{E})/R$; (4) $[\surd]_R = \{\surd\}$. We say that $\mathcal{E}_1$, $\mathcal{E}_2$ are weakly pomset bisimilar, written $\mathcal{E}_1 \approx_{pp} \mathcal{E}_2$, if there exists a weak probabilistic pomset bisimulation $R$, such that $(\emptyset, \emptyset) \in R$. By replacing weakly probabilistic pomset transitions with weak steps, we can obtain the definition of weakly probabilistic step bisimulation. When PESs $\mathcal{E}_1$ and $\mathcal{E}_2$ are weakly probabilistic step bisimilar, we write $\mathcal{E}_1 \approx_{ps} \mathcal{E}_2$.

**Definition 9.53** (Weakly posetal product). Given two PESs $\mathcal{E}_1$, $\mathcal{E}_2$, the weakly posetal product of their configurations, denoted $\mathcal{C}(\mathcal{E}_1)\overline{\times}\mathcal{C}(\mathcal{E}_2)$, is defined as

$$\{(C_1, f, C_2) | C_1 \in \mathcal{C}(\mathcal{E}_1), C_2 \in \mathcal{C}(\mathcal{E}_2), f : \hat{C}_1 \to \hat{C}_2 \text{ isomorphism}\}$$

A subset $R \subseteq \mathcal{C}(\mathcal{E}_1)\overline{\times}\mathcal{C}(\mathcal{E}_2)$ is called a weakly posetal relation. We say that $R$ is downward closed when for any $(C_1, f, C_2), (C_1', f, C_2') \in \mathcal{C}(\mathcal{E}_1)\overline{\times}\mathcal{C}(\mathcal{E}_2)$, if $(C_1, f, C_2) \subseteq (C_1', f', C_2')$ pointwise and $(C_1', f', C_2') \in R$, then $(C_1, f, C_2) \in R$.

For $f : X_1 \to X_2$, we define $f[x_1 \mapsto x_2] : X_1 \cup \{x_1\} \to X_2 \cup \{x_2\}$, $z \in X_1 \cup \{x_1\}$, (1) $f[x_1 \mapsto x_2](z) = x_2$, if $z = x_1$; (2) $f[x_1 \mapsto x_2](z) = f(z)$, otherwise, where $X_1 \subseteq \hat{\mathbb{E}}_1$, $X_2 \subseteq \hat{\mathbb{E}}_2$, $x_1 \in \hat{\mathbb{E}}_1$, $x_2 \in \hat{\mathbb{E}}_2$. Also, we define $f(\tau^*) = f(\tau^*)$.

**Definition 9.54** (Weakly probabilistic (hereditary) history-preserving bisimulation). A weakly probabilistic history-preserving (hp-) bisimulation is a weakly posetal relation $R \subseteq \mathcal{C}(\mathcal{E}_1)\overline{\times}\mathcal{C}(\mathcal{E}_2)$ such that (1) if $(C_1, f, C_2) \in R$, and $C_1 \xRightarrow{e_1} C_1'$, then $C_2 \xRightarrow{e_2} C_2'$, with $(C_1', f[e_1 \mapsto e_2], C_2') \in R$, and vice versa; (2) if $(C_1, f, C_2) \in R$, and $C_1 \xrightarrow{\pi} C_1^{\pi}$, then $C_2 \xrightarrow{\pi} C_2^{\pi}$ and $(C_1^{\pi}, f, C_2^{\pi}) \in R$, and vice versa; (3) if $(C_1, f, C_2) \in R$, then $\mu(C_1, C) = \mu(C_2, C)$ for each $C \in \mathcal{C}(\mathcal{E})/R$; (4) $[\surd]_R = \{\surd\}$. $\mathcal{E}_1, \mathcal{E}_2$ are weakly probabilistic history-preserving (hp-)bisimilar and are written $\mathcal{E}_1 \approx_{php} \mathcal{E}_2$ if there exists a weakly probabilistic hp-bisimulation $R$ such that $(\emptyset, \emptyset, \emptyset) \in R$.

A weakly probabilistic hereditary history-preserving (hhp-)bisimulation is a downward closed weakly probabilistic hp-bisimulation. $\mathcal{E}_1, \mathcal{E}_2$ are weakly probabilistic hereditary history-preserving (hhp-)bisimilar and are written $\mathcal{E}_1 \approx_{phhp} \mathcal{E}_2$.

**Definition 9.55** (Probabilistic branching pomset, step bisimulation). Assume a special termination predicate $\downarrow$, and let $\surd$ represent a state with $\surd\downarrow$. Let $\mathcal{E}_1$, $\mathcal{E}_2$ be PESs. A probabilistic branching pomset bisimulation is a relation $R \subseteq \mathcal{C}(\mathcal{E}_1) \times \mathcal{C}(\mathcal{E}_2)$, such that:

1. if $(C_1, C_2) \in R$, and $C_1 \xrightarrow{X} C_1'$, then
   - either $X \equiv \tau^*$, and $(C_1', C_2) \in R$;
   - or there is a sequence of (zero or more) probabilistic transitions and $\tau$-transitions $C_2 \rightsquigarrow^* \xrightarrow{\tau^*} C_2^0$, such that $(C_1, C_2^0) \in R$ and $C_2^0 \xRightarrow{X} C_2'$ with $(C_1', C_2') \in R$;
2. if $(C_1, C_2) \in R$, and $C_2 \xrightarrow{X} C_2'$, then
   - either $X \equiv \tau^*$, and $(C_1, C_2') \in R$;
   - or there is a sequence of (zero or more) probabilistic transitions and $\tau$-transitions $C_1 \rightsquigarrow^* \xrightarrow{\tau^*} C_1^0$, such that $(C_1^0, C_2) \in R$ and $C_1^0 \xRightarrow{X} C_1'$ with $(C_1', C_2') \in R$;
3. if $(C_1, C_2) \in R$ and $C_1 \downarrow$, then there is a sequence of (zero or more) probabilistic transitions and $\tau$-transitions $C_2 \rightsquigarrow^* \xrightarrow{\tau^*} C_2^0$ such that $(C_1, C_2^0) \in R$ and $C_2^0 \downarrow$;
4. if $(C_1, C_2) \in R$ and $C_2 \downarrow$, then there is a sequence of (zero or more) probabilistic transitions and $\tau$-transitions $C_1 \rightsquigarrow^* \xrightarrow{\tau^*} C_1^0$ such that $(C_1^0, C_2) \in R$ and $C_1^0 \downarrow$;

5. if $(C_1, C_2) \in R$, then $\mu(C_1, C) = \mu(C_2, C)$ for each $C \in \mathcal{C}(\mathcal{E})/R$;
6. $[\surd]_R = \{\surd\}$.

We say that $\mathcal{E}_1, \mathcal{E}_2$ are probabilistic branching pomset bisimilar, written $\mathcal{E}_1 \approx_{pbp} \mathcal{E}_2$, if there exists a probabilistic branching pomset bisimulation $R$, such that $(\emptyset, \emptyset) \in R$.

By replacing probabilistic branching pomset transitions with steps, we can obtain the definition of probabilistic branching step bisimulation. When PESs $\mathcal{E}_1$ and $\mathcal{E}_2$ are probabilistic branching step bisimilar, we write $\mathcal{E}_1 \approx_{pbs} \mathcal{E}_2$.

**Definition 9.56** (Probabilistic rooted branching pomset, step bisimulation). Assume a special termination predicate $\downarrow$, and let $\surd$ represent a state with $\surd\downarrow$. Let $\mathcal{E}_1, \mathcal{E}_2$ be PESs. A branching pomset bisimulation is a relation $R \subseteq \mathcal{C}(\mathcal{E}_1) \times \mathcal{C}(\mathcal{E}_2)$, such that:

1. if $(C_1, C_2) \in R$, and $C_1 \rightsquigarrow C_1^\pi \xrightarrow{X} C_1'$, then $C_2 \rightsquigarrow C_2^\pi \xrightarrow{X} C_2'$ with $C_1' \approx_{pbp} C_2'$;
2. if $(C_1, C_2) \in R$, and $C_2 \rightsquigarrow C_2^\pi \xrightarrow{X} C_2'$, then $C_1 \rightsquigarrow C_1^\pi \xrightarrow{X} C_1'$ with $C_1' \approx_{pbp} C_2'$;
3. if $(C_1, C_2) \in R$ and $C_1 \downarrow$, then $C_2 \downarrow$;
4. if $(C_1, C_2) \in R$ and $C_2 \downarrow$, then $C_1 \downarrow$.

We say that $\mathcal{E}_1, \mathcal{E}_2$ are probabilistic rooted branching pomset bisimilar, written $\mathcal{E}_1 \approx_{prbp} \mathcal{E}_2$, if there exists a probabilistic rooted branching pomset bisimulation $R$, such that $(\emptyset, \emptyset) \in R$.

By replacing probabilistic pomset transitions with steps, we can obtain the definition of probabilistic rooted branching step bisimulation. When PESs $\mathcal{E}_1$ and $\mathcal{E}_2$ are probabilistic rooted branching step bisimilar, we write $\mathcal{E}_1 \approx_{prbs} \mathcal{E}_2$.

**Definition 9.57** (Probabilistic branching (hereditary) history-preserving bisimulation). Assume a special termination predicate $\downarrow$, and let $\surd$ represent a state with $\surd\downarrow$. A probabilistic branching history-preserving (hp-) bisimulation is a weakly posetal relation $R \subseteq \mathcal{C}(\mathcal{E}_1)\overline{\times}\mathcal{C}(\mathcal{E}_2)$ such that:

1. if $(C_1, f, C_2) \in R$, and $C_1 \xrightarrow{e_1} C_1'$, then
   - either $e_1 \equiv \tau$, and $(C_1', f[e_1 \mapsto \tau], C_2) \in R$;
   - or there is a sequence of (zero or more) probabilistic transitions and $\tau$-transitions $C_2 \rightsquigarrow^* \xrightarrow{\tau^*} C_2^0$, such that $(C_1, f, C_2^0) \in R$ and $C_2^0 \xrightarrow{e_2} C_2'$ with $(C_1', f[e_1 \mapsto e_2], C_2') \in R$;
2. if $(C_1, f, C_2) \in R$, and $C_2 \xrightarrow{e_2} C_2'$, then
   - either $X \equiv \tau$, and $(C_1, f[e_2 \mapsto \tau], C_2') \in R$;
   - or there is a sequence of (zero or more) probabilistic transitions and $\tau$-transitions $C_1 \rightsquigarrow^* \xrightarrow{\tau^*} C_1^0$, such that $(C_1^0, f, C_2) \in R$ and $C_1^0 \xrightarrow{e_1} C_1'$ with $(C_1', f[e_2 \mapsto e_1], C_2') \in R$;
3. if $(C_1, f, C_2) \in R$ and $C_1 \downarrow$, then there is a sequence of (zero or more) probabilistic transitions and $\tau$-transitions $C_2 \rightsquigarrow^* \xrightarrow{\tau^*} C_2^0$ such that $(C_1, f, C_2^0) \in R$ and $C_2^0 \downarrow$;
4. if $(C_1, f, C_2) \in R$ and $C_2 \downarrow$, then there is a sequence of (zero or more) probabilistic transitions and $\tau$-transitions $C_1 \rightsquigarrow^* \xrightarrow{\tau^*} C_1^0$ such that $(C_1^0, f, C_2) \in R$ and $C_1^0 \downarrow$;
5. if $(C_1, C_2) \in R$, then $\mu(C_1, C) = \mu(C_2, C)$ for each $C \in \mathcal{C}(\mathcal{E})/R$;
6. $[\surd]_R = \{\surd\}$.

$\mathcal{E}_1, \mathcal{E}_2$ are probabilistic branching history-preserving (hp-)bisimilar and are written $\mathcal{E}_1 \approx_{pbhp} \mathcal{E}_2$ if there exists a probabilistic branching hp-bisimulation $R$ such that $(\emptyset, \emptyset, \emptyset) \in R$.

A probabilistic branching hereditary history-preserving (hhp-)bisimulation is a downward closed probabilistic branching hhp-bisimulation. $\mathcal{E}_1, \mathcal{E}_2$ are probabilistic branching hereditary history-preserving (hhp-)bisimilar and are written $\mathcal{E}_1 \approx_{pbhhp} \mathcal{E}_2$.

**Definition 9.58** (Probabilistic rooted branching (hereditary) history-preserving bisimulation). Assume a special termination predicate $\downarrow$, and let $\surd$ represent a state with $\surd\downarrow$. A probabilistic rooted branching history-preserving (hp-) bisimulation is a posetal relation $R \subseteq \mathcal{C}(\mathcal{E}_1)\overline{\times}\mathcal{C}(\mathcal{E}_2)$ such that:

1. if $(C_1, f, C_2) \in R$, and $C_1 \rightsquigarrow C_1^\pi \xrightarrow{e_1} C_1'$, then $C_2 \rightsquigarrow C_2^\pi \xrightarrow{e_2} C_2'$ with $C_1' \approx_{pbhp} C_2'$;
2. if $(C_1, f, C_2) \in R$, and $C_2 \rightsquigarrow C_2^\pi \xrightarrow{e_2} C_1'$, then $C_1 \rightsquigarrow C_1^\pi \xrightarrow{e_1} C_2'$ with $C_1' \approx_{pbhp} C_2'$;
3. if $(C_1, f, C_2) \in R$ and $C_1 \downarrow$, then $C_2 \downarrow$;
4. if $(C_1, f, C_2) \in R$ and $C_2 \downarrow$, then $C_1 \downarrow$.

$\mathcal{E}_1, \mathcal{E}_2$ are probabilistic rooted branching history-preserving (hp-)bisimilar and are written $\mathcal{E}_1 \approx_{prbhp} \mathcal{E}_2$ if there exists a probabilistic rooted branching hp-bisimulation $R$ such that $(\emptyset, \emptyset, \emptyset) \in R$.

A probabilistic rooted branching hereditary history-preserving (hhp-)bisimulation is a downward closed probabilistic rooted branching hhp-bisimulation. $\mathcal{E}_1, \mathcal{E}_2$ are probabilistic rooted branching hereditary history-preserving (hhp-)bisimilar and are written $\mathcal{E}_1 \approx_{prbhhp} \mathcal{E}_2$.

Chapter 10

# A calculus for probabilistic true concurrency

In this chapter, we design a calculus for probabilistic true concurrency (CPTC). This chapter is organized as follows. We introduce the syntax and operational semantics in Section 10.1, strongly probabilistic truly concurrent bisimulations in Section 10.2, and its properties for weakly probabilistic truly concurrent bisimulations in Section 10.3.

## 10.1 Syntax and operational semantics

We assume an infinite set $\mathcal{N}$ of (action or event) names, and use $a, b, c, \cdots$ to range over $\mathcal{N}$. We denote by $\overline{\mathcal{N}}$ the set of co-names and let $\overline{a}, \overline{b}, \overline{c}, \cdots$ range over $\overline{\mathcal{N}}$. Then, we set $\mathcal{L} = \mathcal{N} \cup \overline{\mathcal{N}}$ as the set of labels, and use $l, \overline{l}$ to range over $\mathcal{L}$. We extend complementation to $\mathcal{L}$ such that $\overline{\overline{a}} = a$. Let $\tau$ denote the silent step (internal action or event) and define $Act = \mathcal{L} \cup \{\tau\}$ to be the set of actions, $\alpha, \beta$ range over $Act$. Also, $K, L$ are used to stand for subsets of $\mathcal{L}$ and $\overline{L}$ is used for the set of complements of labels in $L$. A relabeling function $f$ is a function from $\mathcal{L}$ to $\mathcal{L}$ such that $f(\overline{l}) = \overline{f(l)}$. By defining $f(\tau) = \tau$, we extend $f$ to $Act$.

Further, we introduce a set $\mathcal{X}$ of process variables, and a set $\mathcal{K}$ of process constants, and let $X, Y, \cdots$ range over $\mathcal{X}$, and $A, B, \cdots$ range over $\mathcal{K}$, $\widetilde{X}$ is a tuple of distinct process variables, and also $E, F, \cdots$ range over the recursive expressions. We write $\mathcal{P}$ for the set of processes. Sometimes, we use $I, J$ to stand for an indexing set, and we write $E_i : i \in I$ for a family of expressions indexed by $I$. $Id_D$ is the identity function or relation over set $D$.

For each process constant schema $A$, a defining equation of the form

$$A \stackrel{\text{def}}{=} P$$

is assumed, where $P$ is a process.

### 10.1.1 Syntax

We use the Prefix . to model the causality relation $\leq$ in true concurrency, the Summation $+$ to model the conflict relation $\sharp$, and the Box-Summation $\boxplus_{\pi}$ to model the probabilistic conflict relation $\sharp_{\pi}$ in true concurrency, and the Composition $\parallel$ to explicitly model concurrent relation in true concurrency. Also, we follow the conventions of process algebra.

**Definition 10.1** (Syntax). Truly concurrent processes are defined inductively by the following formation rules:

1. $A \in \mathcal{P}$;
2. $\mathbf{nil} \in \mathcal{P}$;
3. if $P \in \mathcal{P}$, then the Prefix $\alpha.P \in \mathcal{P}$, for $\alpha \in Act$;
4. if $P, Q \in \mathcal{P}$, then the Summation $P + Q \in \mathcal{P}$;
5. if $P, Q \in \mathcal{P}$, then the Box-Summation $P \boxplus_{\pi} Q \in \mathcal{P}$;
6. if $P, Q \in \mathcal{P}$, then the Composition $P \parallel Q \in \mathcal{P}$;
7. if $P \in \mathcal{P}$, then the Prefix $(\alpha_1 \parallel \cdots \parallel \alpha_n).P \in \mathcal{P} \quad (n \in I)$, for $\alpha_1, \cdots, \alpha_n \in Act$;
8. if $P \in \mathcal{P}$, then the Restriction $P \setminus L \in \mathcal{P}$ with $L \in \mathcal{L}$;
9. if $P \in \mathcal{P}$, then the Relabeling $P[f] \in \mathcal{P}$.

The standard BNF grammar of syntax of CPTC can be summarized as follows:

$$P ::= A \quad | \quad \mathbf{nil} \quad | \quad \alpha.P \quad | \quad P + P \quad | \quad P \boxplus_{\pi} P \quad | \quad P \parallel P \quad | \quad (\alpha_1 \parallel \cdots \parallel \alpha_n).P \quad | \quad P \setminus L \quad | \quad P[f]$$

Handbook of Truly Concurrent Process Algebra. https://doi.org/10.1016/B978-0-44-321515-5.00014-7

**TABLE 10.1** Probabilistic transition rules of CPTC.

| | |
|---|---|
| **PAct$_1$** | $\frac{}{\alpha.P \rightsquigarrow \breve{\alpha}.P}$ |
| **PSum** | $\frac{P \rightsquigarrow P' \quad Q \rightsquigarrow Q'}{P+Q \rightsquigarrow P'+Q'}$ |
| **PBox-Sum** | $\frac{P \rightsquigarrow P'}{P \boxplus_\pi Q \rightsquigarrow P'} \quad \frac{Q \rightsquigarrow Q'}{P \boxplus_\pi Q \rightsquigarrow Q'}$ |
| **PCom** | $\frac{P \rightsquigarrow P' \quad Q \rightsquigarrow Q'}{P \parallel Q \rightsquigarrow P'+Q'}$ |
| **PAct$_2$** | $\frac{}{(\alpha_1 \parallel \cdots \parallel \alpha_n).P \rightsquigarrow (\breve{\alpha_1} \parallel \cdots \parallel \breve{\alpha_n}).P}$ |
| **PRes** | $\frac{P \rightsquigarrow P'}{P \setminus L \rightsquigarrow P' \setminus L}$ |
| **PRel** | $\frac{P \rightsquigarrow P'}{P[f] \rightsquigarrow P'[f]}$ |
| **PCon** | $\frac{P \rightsquigarrow P'}{A \rightsquigarrow P'} \quad (A \stackrel{\text{def}}{=} P)$ |

### 10.1.2 Operational semantics

The operational semantics is defined by LTSs (labeled transition systems), and it is detailed by the following definition.

**Definition 10.2** (Semantics). The operational semantics of CPTC corresponding to the syntax in Definition 10.1 is defined by a series of transition rules, named **PAct**, **PSum**, **PBox-Sum**, **Com**, **Res**, **Rel**, and **Con** and named **Act**, **Sum**, **Com**, **Res**, **Rel**, and **Con** indicate that the rules are associated respectively with Prefix, Summation, Composition, Restriction, Relabeling, and Constants in Definition 10.1. They are shown in Tables 10.1 and 10.2.

### 10.1.3 Properties of transitions

**Definition 10.3** (Sorts). Given the sorts $\mathcal{L}(A)$ and $\mathcal{L}(X)$ of constants and variables, we define $\mathcal{L}(P)$ inductively as follows:

1. $\mathcal{L}(l.P) = \{l\} \cup \mathcal{L}(P)$;
2. $\mathcal{L}((l_1 \parallel \cdots \parallel l_n).P) = \{l_1, \cdots, l_n\} \cup \mathcal{L}(P)$;
3. $\mathcal{L}(\tau.P) = \mathcal{L}(P)$;
4. $\mathcal{L}(P+Q) = \mathcal{L}(P) \cup \mathcal{L}(Q)$;
5. $\mathcal{L}(P \boxplus_\pi Q) = \mathcal{L}(P) \cup \mathcal{L}(Q)$;
6. $\mathcal{L}(P \parallel Q) = \mathcal{L}(P) \cup \mathcal{L}(Q)$;
7. $\mathcal{L}(P \setminus L) = \mathcal{L}(P) - (L \cup \overline{L})$;
8. $\mathcal{L}(P[f]) = \{f(l) : l \in \mathcal{L}(P)\}$;
9. for $A \stackrel{\text{def}}{=} P$, $\mathcal{L}(P) \subseteq \mathcal{L}(A)$.

Now, we present some properties of the transition rules defined in Table 10.2.

**Proposition 10.4.** *If $P \xrightarrow{\alpha} P'$, then*

1. $\alpha \in \mathcal{L}(P) \cup \{\tau\}$;
2. $\mathcal{L}(P') \subseteq \mathcal{L}(P)$.

*If $P \xrightarrow{\{\alpha_1,\cdots,\alpha_n\}} P'$, then*

1. $\alpha_1, \cdots, \alpha_n \in \mathcal{L}(P) \cup \{\tau\}$;
2. $\mathcal{L}(P') \subseteq \mathcal{L}(P)$.

*Proof.* By induction on the inference of $P \xrightarrow{\alpha} P'$ and $P \xrightarrow{\{\alpha_1,\cdots,\alpha_n\}} P'$, there are fourteen cases corresponding to the transition rules named **Act**$_{1,2}$, **Sum**$_{1,2}$, **Com**$_{1,2,3,4}$, **Res**$_{1,2}$, **Rel**$_{1,2}$, and **Con**$_{1,2}$ in Table 10.2, we just prove the one case **Act**$_1$ and **Act**$_2$, and omit the others.

Case **Act**$_1$: by **Act**$_1$, with $P \equiv \alpha.P'$. Then, by Definition 10.3, we have (1) $\mathcal{L}(P) = \{\alpha\} \cup \mathcal{L}(P')$ if $\alpha \neq \tau$; (2) $\mathcal{L}(P) = \mathcal{L}(P')$ if $\alpha = \tau$. Hence, $\alpha \in \mathcal{L}(P) \cup \{\tau\}$, and $\mathcal{L}(P') \subseteq \mathcal{L}(P)$, as desired.

**TABLE 10.2** Action transition rules of CPTC.

| | | |
|---|---|---|
| **Act$_1$** | $\dfrac{}{\breve{\alpha}.P \xrightarrow{\alpha} P}$ | |
| **Sum$_1$** | $\dfrac{P \xrightarrow{\alpha} P'}{P+Q \xrightarrow{\alpha} P'}$ | |
| **Com$_1$** | $\dfrac{P \xrightarrow{\alpha} P' \quad Q \nrightarrow}{P \parallel Q \xrightarrow{\alpha} P' \parallel Q}$ | |
| **Com$_2$** | $\dfrac{Q \xrightarrow{\alpha} Q' \quad P \nrightarrow}{P \parallel Q \xrightarrow{\alpha} P \parallel Q'}$ | |
| **Com$_3$** | $\dfrac{P \xrightarrow{\alpha} P' \quad Q \xrightarrow{\beta} Q'}{P \parallel Q \xrightarrow{\{\alpha,\beta\}} P' \parallel Q'}$ | $(\beta \neq \overline{\alpha})$ |
| **Com$_4$** | $\dfrac{P \xrightarrow{l} P' \quad Q \xrightarrow{\overline{l}} Q'}{P \parallel Q \xrightarrow{\tau} P' \parallel Q'}$ | |
| **Act$_2$** | $\dfrac{}{(\breve{\alpha_1} \parallel \cdots \parallel \breve{\alpha_n}).P \xrightarrow{\{\alpha_1,\cdots,\alpha_n\}} P}$ | $(\alpha_i \neq \overline{\alpha_j} \quad i,j \in \{1,\cdots,n\})$ |
| **Sum$_2$** | $\dfrac{P \xrightarrow{\{\alpha_1,\cdots,\alpha_n\}} P'}{P+Q \xrightarrow{\{\alpha_1,\cdots,\alpha_n\}} P'}$ | |
| **Res$_1$** | $\dfrac{P \xrightarrow{\alpha} P'}{P \setminus L \xrightarrow{\alpha} P' \setminus L}$ | $(\alpha, \overline{\alpha} \notin L)$ |
| **Res$_2$** | $\dfrac{P \xrightarrow{\{\alpha_1,\cdots,\alpha_n\}} P'}{P \setminus L \xrightarrow{\{\alpha_1,\cdots,\alpha_n\}} P' \setminus L}$ | $(\alpha_1, \overline{\alpha_1}, \cdots, \alpha_n, \overline{\alpha_n} \notin L)$ |
| **Rel$_1$** | $\dfrac{P \xrightarrow{\alpha} P'}{P[f] \xrightarrow{f(\alpha)} P'[f]}$ | |
| **Rel$_2$** | $\dfrac{P \xrightarrow{\{\alpha_1,\cdots,\alpha_n\}} P'}{P[f] \xrightarrow{\{f(\alpha_1),\cdots,f(\alpha_n)\}} P'[f]}$ | |
| **Con$_1$** | $\dfrac{P \xrightarrow{\alpha} P'}{A \xrightarrow{\alpha} P'}$ | $(A \overset{\text{def}}{=} P)$ |
| **Con$_2$** | $\dfrac{P \xrightarrow{\{\alpha_1,\cdots,\alpha_n\}} P'}{A \xrightarrow{\{\alpha_1,\cdots,\alpha_n\}} P'}$ | $(A \overset{\text{def}}{=} P)$ |

Case **Act$_2$**: by **Act$_2$**, with $P \equiv (\alpha_1 \parallel \cdots \parallel \alpha_n).P'$. Then, by Definition 10.3, we have (1) $\mathcal{L}(P) = \{\alpha_1, \cdots, \alpha_n\} \cup \mathcal{L}(P')$ if $\alpha_i \neq \tau$ for $i \leq n$; (2) $\mathcal{L}(P) = \mathcal{L}(P')$ if $\alpha_1, \cdots, \alpha_n = \tau$. Hence, $\alpha_1, \cdots, \alpha_n \in \mathcal{L}(P) \cup \{\tau\}$, and $\mathcal{L}(P') \subseteq \mathcal{L}(P)$, as desired. □

## 10.2 Strongly probabilistic truly concurrent bisimulations

### 10.2.1 Laws and congruence

Based on the concepts of strongly probabilistic truly concurrent bisimulation equivalences, we obtain the following laws.

**Proposition 10.5** (Monoid laws for strongly probabilistic pomset bisimulation). *The monoid laws for strongly probabilistic pomset bisimulation are as follows:*

1. $P + Q \sim_{pp} Q + P$;
2. $P + (Q + R) \sim_{pp} (P + Q) + R$;
3. $P + P \sim_{pp} P$;
4. $P + \mathbf{nil} \sim_{pp} P$.

*Proof.* According to the definition of strongly probabilistic pomset bisimulation, we can easily prove the above equations, however, we omit the proof here. □

**Proposition 10.6** (Monoid laws for strongly probabilistic step bisimulation). *The monoid laws for strongly probabilistic step bisimulation are as follows:*

1. $P + Q \sim_{ps} Q + P$;
2. $P + (Q + R) \sim_{ps} (P + Q) + R$;
3. $P + P \sim_{ps} P$;
4. $P + \mathbf{nil} \sim_{ps} P$.

*Proof.* According to the definition of strongly probabilistic step bisimulation, we can easily prove the above equations, however, we omit the proof here. □

**Proposition 10.7** (Monoid laws for strongly probabilistic hp-bisimulation). *The monoid laws for strongly probabilistic hp-bisimulation are as follows:*

1. $P + Q \sim_{php} Q + P$;
2. $P + (Q + R) \sim_{php} (P + Q) + R$;
3. $P + P \sim_{php} P$;
4. $P + \mathbf{nil} \sim_{php} P$.

*Proof.* According to the definition of strongly probabilistic hp-bisimulation, we can easily prove the above equations, however, we omit the proof here. □

**Proposition 10.8** (Monoid laws for strongly probabilistic hhp-bisimulation). *The monoid laws for strongly probabilistic hhp-bisimulation are as follows:*

1. $P + Q \sim_{phhp} Q + P$;
2. $P + (Q + R) \sim_{phhp} (P + Q) + R$;
3. $P + P \sim_{phhp} P$;
4. $P + \mathbf{nil} \sim_{phhp} P$.

*Proof.* According to the definition of strongly probabilistic hhp-bisimulation, we can easily prove the above equations, however, we omit the proof here. □

**Proposition 10.9** (Monoid laws 2 for strongly probabilistic pomset bisimulation). *The monoid laws 2 for strongly probabilistic pomset bisimulation are as follows:*

1. $P \boxplus_{\pi} Q \sim_{pp} Q \boxplus_{1-\pi} P$;
2. $P \boxplus_{\pi} (Q \boxplus_{\rho} R) \sim_{pp} (P \boxplus_{\frac{\pi}{\pi+\rho-\pi\rho}} Q) \boxplus_{\pi+\rho-\pi\rho} R$;
3. $P \boxplus_{\pi} P \sim_{pp} P$;
4. $P \boxplus_{\pi} \mathbf{nil} \sim_{pp} P$.

*Proof.* According to the definition of strongly probabilistic pomset bisimulation, we can easily prove the above equations, however, we omit the proof here. □

**Proposition 10.10** (Monoid laws 2 for strongly probabilistic step bisimulation). *The monoid laws 2 for strongly probabilistic step bisimulation are as follows:*

1. $P \boxplus_{\pi} Q \sim_{ps} Q \boxplus_{1-\pi} P$;
2. $P \boxplus_{\pi} (Q \boxplus_{\rho} R) \sim_{ps} (P \boxplus_{\frac{\pi}{\pi+\rho-\pi\rho}} Q) \boxplus_{\pi+\rho-\pi\rho} R$;
3. $P \boxplus_{\pi} P \sim_{ps} P$;
4. $P \boxplus_{\pi} \mathbf{nil} \sim_{ps} P$.

*Proof.* According to the definition of strongly probabilistic step bisimulation, we can easily prove the above equations, however, we omit the proof here. □

**Proposition 10.11** (Monoid laws 2 for strongly probabilistic hp-bisimulation). *The monoid laws 2 for strongly probabilistic hp-bisimulation are as follows:*

1. $P \boxplus_{\pi} Q \sim_{php} Q \boxplus_{1-\pi} P$;

2. $P \boxplus_{\pi} (Q \boxplus_{\rho} R) \sim_{php} (P \boxplus_{\frac{\pi}{\pi+\rho-\pi\rho}} Q) \boxplus_{\pi+\rho-\pi\rho} R$;
3. $P \boxplus_{\pi} P \sim_{php} P$;
4. $P \boxplus_{\pi} \textbf{nil} \sim_{php} P$.

*Proof.* According to the definition of strongly probabilistic hp-bisimulation, we can easily prove the above equations, however, we omit the proof here. □

**Proposition 10.12** (Monoid laws 2 for strongly probabilistic hhp-bisimulation). *The monoid laws 2 for strongly probabilistic hhp-bisimulation are as follows:*

1. $P \boxplus_{\pi} Q \sim_{phhp} Q \boxplus_{1-\pi} P$;
2. $P \boxplus_{\pi} (Q \boxplus_{\rho} R) \sim_{phhp} (P \boxplus_{\frac{\pi}{\pi+\rho-\pi\rho}} Q) \boxplus_{\pi+\rho-\pi\rho} R$;
3. $P \boxplus_{\pi} P \sim_{phhp} P$;
4. $P \boxplus_{\pi} \textbf{nil} \sim_{phhp} P$.

*Proof.* According to the definition of strongly probabilistic hhp-bisimulation, we can easily prove the above equations, however, we omit the proof here. □

**Proposition 10.13** (Static laws for strongly probabilistic pomset bisimulation). *The static laws for strongly probabilistic pomset bisimulation are as follows:*

1. $P \parallel Q \sim_{pp} Q \parallel P$;
2. $P \parallel (Q \parallel R) \sim_{pp} (P \parallel Q) \parallel R$;
3. $P \parallel \textbf{nil} \sim_{pp} P$;
4. $P \setminus L \sim_{pp} P$, *if* $\mathcal{L}(P) \cap (L \cup \overline{L}) = \emptyset$;
5. $P \setminus K \setminus L \sim_{pp} P \setminus (K \cup L)$;
6. $P[f] \setminus L \sim_{pp} P \setminus f^{-1}(L)[f]$;
7. $(P \parallel Q) \setminus L \sim_{pp} P \setminus L \parallel Q \setminus L$, *if* $\mathcal{L}(P) \cap \overline{\mathcal{L}(Q)} \cap (L \cup \overline{L}) = \emptyset$;
8. $P[Id] \sim_{pp} P$;
9. $P[f] \sim_{pp} P[f']$, *if* $f \restriction \mathcal{L}(P) = f' \restriction \mathcal{L}(P)$;
10. $P[f][f'] \sim_{pp} P[f' \circ f]$;
11. $(P \parallel Q)[f] \sim_{pp} P[f] \parallel Q[f]$, *if* $f \restriction (L \cup \overline{L})$ *is one-to-one, where* $L = \mathcal{L}(P) \cup \mathcal{L}(Q)$.

*Proof.* According to the definition of strongly probabilistic pomset bisimulation, we can easily prove the above equations, however, we omit the proof here. □

**Proposition 10.14** (Static laws for strongly probabilistic step bisimulation). *The static laws for strongly probabilistic step bisimulation are as follows:*

1. $P \parallel Q \sim_{ps} Q \parallel P$;
2. $P \parallel (Q \parallel R) \sim_{ps} (P \parallel Q) \parallel R$;
3. $P \parallel \textbf{nil} \sim_{ps} P$;
4. $P \setminus L \sim_{ps} P$, *if* $\mathcal{L}(P) \cap (L \cup \overline{L}) = \emptyset$;
5. $P \setminus K \setminus L \sim_{ps} P \setminus (K \cup L)$;
6. $P[f] \setminus L \sim_{ps} P \setminus f^{-1}(L)[f]$;
7. $(P \parallel Q) \setminus L \sim_{ps} P \setminus L \parallel Q \setminus L$, *if* $\mathcal{L}(P) \cap \overline{\mathcal{L}(Q)} \cap (L \cup \overline{L}) = \emptyset$;
8. $P[Id] \sim_{ps} P$;
9. $P[f] \sim_{ps} P[f']$, *if* $f \restriction \mathcal{L}(P) = f' \restriction \mathcal{L}(P)$;
10. $P[f][f'] \sim_{ps} P[f' \circ f]$;
11. $(P \parallel Q)[f] \sim_{ps} P[f] \parallel Q[f]$, *if* $f \restriction (L \cup \overline{L})$ *is one-to-one, where* $L = \mathcal{L}(P) \cup \mathcal{L}(Q)$.

*Proof.* According to the definition of strongly probabilistic step bisimulation, we can easily prove the above equations, however, we omit the proof here. □

**Proposition 10.15** (Static laws for strongly probabilistic hp-bisimulation). *The static laws for strongly probabilistic hp-bisimulation are as follows:*

1. $P \parallel Q \sim_{php} Q \parallel P$;
2. $P \parallel (Q \parallel R) \sim_{php} (P \parallel Q) \parallel R$;

3. $P \parallel \mathbf{nil} \sim_{php} P$;
4. $P \setminus L \sim_{php} P$, *if* $\mathcal{L}(P) \cap (L \cup \overline{L}) = \emptyset$;
5. $P \setminus K \setminus L \sim_{php} P \setminus (K \cup L)$;
6. $P[f] \setminus L \sim_{php} P \setminus f^{-1}(L)[f]$;
7. $(P \parallel Q) \setminus L \sim_{php} P \setminus L \parallel Q \setminus L$, *if* $\mathcal{L}(P) \cap \overline{\mathcal{L}(Q)} \cap (L \cup \overline{L}) = \emptyset$;
8. $P[Id] \sim_{php} P$;
9. $P[f] \sim_{php} P[f']$, *if* $f \restriction \mathcal{L}(P) = f' \restriction \mathcal{L}(P)$;
10. $P[f][f'] \sim_{php} P[f' \circ f]$;
11. $(P \parallel Q)[f] \sim_{php} P[f] \parallel Q[f]$, *if* $f \restriction (L \cup \overline{L})$ *is one-to-one, where* $L = \mathcal{L}(P) \cup \mathcal{L}(Q)$.

*Proof.* According to the definition of strongly probabilistic hp-bisimulation, we can easily prove the above equations, however, we omit the proof here. □

**Proposition 10.16** (Static laws for strongly probabilistic hhp-bisimulation). *The static laws for strongly probabilistic hhp-bisimulation are as follows:*

1. $P \parallel Q \sim_{phhp} Q \parallel P$;
2. $P \parallel (Q \parallel R) \sim_{phhp} (P \parallel Q) \parallel R$;
3. $P \parallel \mathbf{nil} \sim_{phhp} P$;
4. $P \setminus L \sim_{phhp} P$, *if* $\mathcal{L}(P) \cap (L \cup \overline{L}) = \emptyset$;
5. $P \setminus K \setminus L \sim_{phhp} P \setminus (K \cup L)$;
6. $P[f] \setminus L \sim_{phhp} P \setminus f^{-1}(L)[f]$;
7. $(P \parallel Q) \setminus L \sim_{phhp} P \setminus L \parallel Q \setminus L$, *if* $\mathcal{L}(P) \cap \overline{\mathcal{L}(Q)} \cap (L \cup \overline{L}) = \emptyset$;
8. $P[Id] \sim_{phhp} P$;
9. $P[f] \sim_{phhp} P[f']$, *if* $f \restriction \mathcal{L}(P) = f' \restriction \mathcal{L}(P)$;
10. $P[f][f'] \sim_{phhp} P[f' \circ f]$;
11. $(P \parallel Q)[f] \sim_{phhp} P[f] \parallel Q[f]$, *if* $f \restriction (L \cup \overline{L})$ *is one-to-one, where* $L = \mathcal{L}(P) \cup \mathcal{L}(Q)$.

*Proof.* According to the definition of strongly probabilistic hhp-bisimulation, we can easily prove the above equations, however, we omit the proof here. □

**Proposition 10.17** (Expansion law for strongly probabilistic step bisimulation). *Let* $P \equiv (P_1[f_1] \parallel \cdots \parallel P_n[f_n]) \setminus L$, *with* $n \geq 1$. *Then,*

$$\begin{aligned} P \sim_{ps} \{(f_1(\alpha_1) \parallel \cdots \parallel f_n(\alpha_n)).(P_1'[f_1] \parallel \cdots \parallel P_n'[f_n]) \setminus L :\\ P_i \rightsquigarrow \xrightarrow{\alpha_i} P_i', i \in \{1, \cdots, n\}, f_i(\alpha_i) \notin L \cup \overline{L}\} \\ + \sum \{\tau.(P_1[f_1] \parallel \cdots \parallel P_i'[f_i] \parallel \cdots \parallel P_j'[f_j] \parallel \cdots \parallel P_n[f_n]) \setminus L :\\ P_i \rightsquigarrow \xrightarrow{l_1} P_i', P_j \rightsquigarrow \xrightarrow{l_2} P_j', f_i(l_1) = \overline{f_j(l_2)}, i < j\} \end{aligned}$$

*Proof.* Although transition rules in Table 10.2 are defined in the flavor of a single event, they can be modified into a step (a set of events within which each event is pairwise concurrent), however, we omit them here. If we treat a single event as a step containing just one event, the proof of the new expansion law poses no problem, so we use this way and still use the transition rules in Table 10.2.

First, we consider the case without Restriction and Relabeling. That is, we suffice to prove the following case by induction on the size $n$.

For $P \equiv P_1 \parallel \cdots \parallel P_n$, with $n \geq 1$, we need to prove

$$\begin{aligned} P \sim_{ps} \{(\alpha_1 \parallel \cdots \parallel \alpha_n).(P_1' \parallel \cdots \parallel P_n') : P_i \rightsquigarrow \xrightarrow{\alpha_i} P_i', i \in \{1, \cdots, n\} \\ + \sum \{\tau.(P_1 \parallel \cdots \parallel P_i' \parallel \cdots \parallel P_j' \parallel \cdots \parallel P_n) : P_i \rightsquigarrow \xrightarrow{l} P_i', P_j \rightsquigarrow \xrightarrow{\overline{l}} P_j', i < j\} \end{aligned}$$

For $n = 1$, $P_1 \sim_{ps} \alpha_1.P_1' : P_1 \rightsquigarrow \xrightarrow{\alpha_1} P_1'$ is obvious. Then, with a hypothesis $n$, we consider $R \equiv P \parallel P_{n+1}$. By the transition rules $\mathbf{Com}_{1,2,3,4}$, we can obtain

$$R \sim_{ps} \{(p \parallel \alpha_{n+1}).(P' \parallel P_{n+1}') : P \rightsquigarrow \xrightarrow{p} P', P_{n+1} \rightsquigarrow \xrightarrow{\alpha_{n+1}} P_{n+1}', p \subseteq P\}$$

$$+\sum\{\tau.(P' \parallel P'_{n+1}) : P \rightsquigarrow\xrightarrow{l} P', P_{n+1} \rightsquigarrow\xrightarrow{\bar{l}} P'_{n+1}\}$$

Now, with the induction assumption $P \equiv P_1 \parallel \cdots \parallel P_n$, the right-hand side can be reformulated as follows:

$$\begin{aligned}
&\{(\alpha_1 \parallel \cdots \parallel \alpha_n \parallel \alpha_{n+1}).(P'_1 \parallel \cdots \parallel P'_n \parallel P'_{n+1}) :\\
&\quad P_i \rightsquigarrow\xrightarrow{\alpha_i} P'_i, i \in \{1, \cdots, n+1\}\\
&+\sum\{\tau.(P_1 \parallel \cdots \parallel P'_i \parallel \cdots \parallel P'_j \parallel \cdots \parallel P_n \parallel P_{n+1}) :\\
&\quad P_i \rightsquigarrow\xrightarrow{l} P'_i, P_j \rightsquigarrow\xrightarrow{\bar{l}} P'_j, i < j\}\\
&+\sum\{\tau.(P_1 \parallel \cdots \parallel P'_i \parallel \cdots \parallel P_j \parallel \cdots \parallel P_n \parallel P'_{n+1}) :\\
&\quad P_i \xrightarrow{l} P'_i, P_{n+1} \rightsquigarrow\xrightarrow{\bar{l}} P'_{n+1}, i \in \{1, \cdots, n\}\}
\end{aligned}$$

Hence,

$$\begin{aligned}
R \sim_{ps} &\{(\alpha_1 \parallel \cdots \parallel \alpha_n \parallel \alpha_{n+1}).(P'_1 \parallel \cdots \parallel P'_n \parallel P'_{n+1}) :\\
&\quad P_i \rightsquigarrow\xrightarrow{\alpha_i} P'_i, i \in \{1, \cdots, n+1\}\\
&+\sum\{\tau.(P_1 \parallel \cdots \parallel P'_i \parallel \cdots \parallel P'_j \parallel \cdots \parallel P_n) :\\
&\quad P_i \rightsquigarrow\xrightarrow{l} P'_i, P_j \rightsquigarrow\xrightarrow{\bar{l}} P'_j, 1 \leq i < j \geq n+1\}
\end{aligned}$$

Then, we can easily add the full conditions with Restriction and Relabeling. □

**Proposition 10.18** (Expansion law for strongly probabilistic pomset bisimulation). *Let $P \equiv (P_1[f_1] \parallel \cdots \parallel P_n[f_n]) \setminus L$, with $n \geq 1$. Then,*

$$\begin{aligned}
P \sim_{pp} &\{(f_1(\alpha_1) \parallel \cdots \parallel f_n(\alpha_n)).(P'_1[f_1] \parallel \cdots \parallel P'_n[f_n]) \setminus L :\\
&\quad P_i \rightsquigarrow\xrightarrow{\alpha_i} P'_i, i \in \{1, \cdots, n\}, f_i(\alpha_i) \notin L \cup \overline{L}\}\\
&+\sum\{\tau.(P_1[f_1] \parallel \cdots \parallel P'_i[f_i] \parallel \cdots \parallel P'_j[f_j] \parallel \cdots \parallel P_n[f_n]) \setminus L :\\
&\quad P_i \rightsquigarrow\xrightarrow{l_1} P'_i, P_j \rightsquigarrow\xrightarrow{l_2} P'_j, f_i(l_1) = \overline{f_j(l_2)}, i < j\}
\end{aligned}$$

*Proof.* Similarly to the proof of expansion law for strongly probabilistic step bisimulation (see Proposition 10.17), we can prove that the new expansion law holds for strongly probabilistic pomset bisimulation, however, we omit it here. □

**Proposition 10.19** (Expansion law for strongly probabilistic hp-bisimulation). *Let $P \equiv (P_1[f_1] \parallel \cdots \parallel P_n[f_n]) \setminus L$, with $n \geq 1$. Then,*

$$\begin{aligned}
P \sim_{php} &\{(f_1(\alpha_1) \parallel \cdots \parallel f_n(\alpha_n)).(P'_1[f_1] \parallel \cdots \parallel P'_n[f_n]) \setminus L :\\
&\quad P_i \rightsquigarrow\xrightarrow{\alpha_i} P'_i, i \in \{1, \cdots, n\}, f_i(\alpha_i) \notin L \cup \overline{L}\}\\
&+\sum\{\tau.(P_1[f_1] \parallel \cdots \parallel P'_i[f_i] \parallel \cdots \parallel P'_j[f_j] \parallel \cdots \parallel P_n[f_n]) \setminus L :\\
&\quad P_i \rightsquigarrow\xrightarrow{l_1} P'_i, P_j \rightsquigarrow\xrightarrow{l_2} P'_j, f_i(l_1) = \overline{f_j(l_2)}, i < j\}
\end{aligned}$$

*Proof.* Similarly to the proof of the expansion law for strongly probabilistic pomset bisimulation (see Proposition 10.18), we can prove that the expansion law holds for strongly probabilistic hp-bisimulation, we just need additionally to check the above conditions on hp-bisimulation, however, we omit these here. □

**Proposition 10.20** (Expansion law for strongly probabilistic hhp-bisimulation). *Let $P \equiv (P_1[f_1] \parallel \cdots \parallel P_n[f_n]) \setminus L$, with $n \geq 1$. Then,*

$$P \sim_{phhp} \{(f_1(\alpha_1) \parallel \cdots \parallel f_n(\alpha_n)).(P'_1[f_1] \parallel \cdots \parallel P'_n[f_n]) \setminus L :$$

$$P_i \overset{\alpha_i}{\rightsquigarrow} P_i', i \in \{1, \cdots, n\}, f_i(\alpha_i) \notin L \cup \overline{L}\}$$

$$+\sum\{\tau.(P_1[f_1] \parallel \cdots \parallel P_i'[f_i] \parallel \cdots \parallel P_j'[f_j] \parallel \cdots \parallel P_n[f_n]) \setminus L :$$

$$P_i \overset{l_1}{\rightsquigarrow} P_i', P_j \overset{l_2}{\rightsquigarrow} P_j', f_i(l_1) = \overline{f_j(l_2)}, i < j\}$$

*Proof.* From the definition of strongly probabilistic hhp-bisimulation (see Definition 2.15), we know that strongly hhp-bisimulation is downward closed for strongly probabilistic hp-bisimulation.

Similarly to the proof of the expansion law for strongly probabilistic hp-bisimulation (see Proposition 10.19), we can prove that the expansion law holds for strongly probabilistic hhp-bisimulation, that is, they are downward closed for strongly probabilistic hp-bisimulation, however, we omit it here. □

**Theorem 10.21** (Congruence for strongly probabilistic pomset bisimulation). *We can enjoy the full congruence for strongly probabilistic pomset bisimulation as follows:*

1. *If* $A \overset{def}{=} P$, *then* $A \sim_{pp} P$;
2. *Let* $P_1 \sim_{pp} P_2$. *Then,*
   (a) $\alpha.P_1 \sim_{pp} \alpha.P_2$;
   (b) $(\alpha_1 \parallel \cdots \parallel \alpha_n).P_1 \sim_{pp} (\alpha_1 \parallel \cdots \parallel \alpha_n).P_2$;
   (c) $P_1 + Q \sim_{pp} P_2 + Q$;
   (d) $P_1 \boxplus_\pi Q \sim_{pp} P_2 \boxplus_\pi Q$;
   (e) $P_1 \parallel Q \sim_{pp} P_2 \parallel Q$;
   (f) $P_1 \setminus L \sim_{pp} P_2 \setminus L$;
   (g) $P_1[f] \sim_{pp} P_2[f]$.

*Proof.* According to the definition of strongly probabilistic pomset bisimulation, we can easily prove the above equations, however, we omit the proof here. □

**Theorem 10.22** (Congruence for strongly probabilistic step bisimulation). *We can enjoy the full congruence for strongly probabilistic step bisimulation as follows:*

1. *If* $A \overset{def}{=} P$, *then* $A \sim_{ps} P$;
2. *Let* $P_1 \sim_{ps} P_2$. *Then,*
   (a) $\alpha.P_1 \sim_{ps} \alpha.P_2$;
   (b) $(\alpha_1 \parallel \cdots \parallel \alpha_n).P_1 \sim_{ps} (\alpha_1 \parallel \cdots \parallel \alpha_n).P_2$;
   (c) $P_1 + Q \sim_{ps} P_2 + Q$;
   (d) $P_1 \boxplus_\pi Q \sim_{ps} P_2 \boxplus_\pi Q$;
   (e) $P_1 \parallel Q \sim_{ps} P_2 \parallel Q$;
   (f) $P_1 \setminus L \sim_{ps} P_2 \setminus L$;
   (g) $P_1[f] \sim_{ps} P_2[f]$.

*Proof.* According to the definition of strongly probabilistic step bisimulation, we can easily prove the above equations, however, we omit the proof here. □

**Theorem 10.23** (Congruence for strongly probabilistic hp-bisimulation). *We can enjoy the full congruence for strongly probabilistic hp-bisimulation as follows:*

1. *If* $A \overset{def}{=} P$, *then* $A \sim_{php} P$;
2. *Let* $P_1 \sim_{php} P_2$. *Then,*
   (a) $\alpha.P_1 \sim_{php} \alpha.P_2$;
   (b) $(\alpha_1 \parallel \cdots \parallel \alpha_n).P_1 \sim_{php} (\alpha_1 \parallel \cdots \parallel \alpha_n).P_2$;
   (c) $P_1 + Q \sim_{php} P_2 + Q$;
   (d) $P_1 \boxplus_\pi Q \sim_{php} P_2 \boxplus_\pi Q$;
   (e) $P_1 \parallel Q \sim_{php} P_2 \parallel Q$;
   (f) $P_1 \setminus L \sim_{php} P_2 \setminus L$;
   (g) $P_1[f] \sim_{php} P_2[f]$.

*Proof.* According to the definition of strongly probabilistic hp-bisimulation, we can easily prove the above equations, however, we omit the proof here. □

**Theorem 10.24** (Congruence for strongly probabilistic hhp-bisimulation). *We can enjoy the full congruence for strongly probabilistic hhp-bisimulation as follows:*

1. *If* $A \stackrel{def}{=} P$, *then* $A \sim_{phhp} P$;
2. *Let* $P_1 \sim_{phhp} P_2$. *Then,*
   (a) $\alpha.P_1 \sim_{phhp} \alpha.P_2$;
   (b) $(\alpha_1 \parallel \cdots \parallel \alpha_n).P_1 \sim_{phhp} (\alpha_1 \parallel \cdots \parallel \alpha_n).P_2$;
   (c) $P_1 + Q \sim_{phhp} P_2 + Q$;
   (d) $P_1 \boxplus_\pi Q \sim_{phhp} P_2 \boxplus_\pi Q$;
   (e) $P_1 \parallel Q \sim_{phhp} P_2 \parallel Q$;
   (f) $P_1 \setminus L \sim_{phhp} P_2 \setminus L$;
   (g) $P_1[f] \sim_{phhp} P_2[f]$.

*Proof.* According to the definition of strongly probabilistic hhp-bisimulation, we can easily prove the above equations, however, we omit the proof here. □

### 10.2.2 Recursion

**Definition 10.25** (Weakly guarded recursive expression). $X$ is weakly guarded in $E$ if each occurrence of $X$ is with some subexpression $\alpha.F$ or $(\alpha_1 \parallel \cdots \parallel \alpha_n).F$ of $E$.

**Lemma 10.26.** *If the variables* $\widetilde{X}$ *are weakly guarded in* $E$, *and* $E\{\widetilde{P}/\widetilde{X}\} \rightsquigarrow \xrightarrow{\{\alpha_1,\cdots,\alpha_n\}} P'$, *then* $P'$ *takes the form* $E'\{\widetilde{P}/\widetilde{X}\}$ *for some expression* $E'$, *and moreover, for any* $\widetilde{Q}$, $E\{\widetilde{Q}/\widetilde{X}\} \rightsquigarrow \xrightarrow{\{\alpha_1,\cdots,\alpha_n\}} E'\{\widetilde{Q}/\widetilde{X}\}$.

*Proof.* It needs to induct on the depth of the inference of $E\{\widetilde{P}/\widetilde{X}\} \rightsquigarrow \xrightarrow{\{\alpha_1,\cdots,\alpha_n\}} P'$.

1. Case $E \equiv Y$, a variable. Then, $Y \notin \widetilde{X}$. Since $\widetilde{X}$ are weakly guarded, $Y\{\widetilde{P}/\widetilde{X} \equiv Y\} \nrightarrow$, this case is impossible.
2. Case $E \equiv \beta.F$. Then, we must have $\alpha = \beta$, and $P' \equiv F\{\widetilde{P}/\widetilde{X}\}$, and $E\{\widetilde{Q}/\widetilde{X}\} \equiv \beta.F\{\widetilde{Q}/\widetilde{X}\} \rightsquigarrow \xrightarrow{\beta} F\{\widetilde{Q}/\widetilde{X}\}$, then, let $E'$ be $F$, as desired.
3. Case $E \equiv (\beta_1 \parallel \cdots \parallel \beta_n).F$. Then, we must have $\alpha_i = \beta_i$ for $1 \leq i \leq n$, and $P' \equiv F\{\widetilde{P}/\widetilde{X}\}$, and $E\{\widetilde{Q}/\widetilde{X}\} \equiv (\beta_1 \parallel \cdots \parallel \beta_n).F\{\widetilde{Q}/\widetilde{X}\} \rightsquigarrow \xrightarrow{\{\beta_1,\cdots,\beta_n\}} F\{\widetilde{Q}/\widetilde{X}\}$, then, let $E'$ be $F$, as desired.
4. Case $E \equiv E_1 + E_2$. Then, either $E_1\{\widetilde{P}/\widetilde{X}\} \rightsquigarrow \xrightarrow{\{\alpha_1,\cdots,\alpha_n\}} P'$ or $E_2\{\widetilde{P}/\widetilde{X}\} \rightsquigarrow \xrightarrow{\{\alpha_1,\cdots,\alpha_n\}} P'$, then, we can apply this lemma in either case, as desired.
5. Case $E \equiv E_1 \boxplus_\pi E_2$. Then, either $E_1\{\widetilde{P}/\widetilde{X}\} \rightsquigarrow P'$ or $E_2\{\widetilde{P}/\widetilde{X}\} \rightsquigarrow P'$, then, we can apply this lemma in either case, as desired.
6. Case $E \equiv E_1 \parallel E_2$. There are four possibilities:
   (a) We may have $E_1\{\widetilde{P}/\widetilde{X}\} \rightsquigarrow \xrightarrow{\alpha} P_1'$ and $E_2\{\widetilde{P}/\widetilde{X}\} \nrightarrow$ with $P' \equiv P_1' \parallel (E_2\{\widetilde{P}/\widetilde{X}\})$, then by applying this lemma, $P_1'$ is of the form $E_1'\{\widetilde{P}/\widetilde{X}\}$, and for any $Q$, $E_1\{\widetilde{Q}/\widetilde{X}\} \rightsquigarrow \xrightarrow{\alpha} E_1'\{\widetilde{Q}/\widetilde{X}\}$. Hence, $P'$ is of the form $E_1' \parallel E_2\{\widetilde{P}/\widetilde{X}\}$, and for any $Q$, $E\{\widetilde{Q}/\widetilde{X}\} \equiv E_1\{\widetilde{Q}/\widetilde{X}\} \parallel E_2\{\widetilde{Q}/\widetilde{X}\} \rightsquigarrow \xrightarrow{\alpha} (E_1' \parallel E_2)\{\widetilde{Q}/\widetilde{X}\}$, then, let $E'$ be $E_1' \parallel E_2$, as desired.
   (b) We may have $E_2\{\widetilde{P}/\widetilde{X}\} \rightsquigarrow \xrightarrow{\alpha} P_2'$ and $E_1\{\widetilde{P}/\widetilde{X}\} \nrightarrow$ with $P' \equiv P_2' \parallel (E_1\{\widetilde{P}/\widetilde{X}\})$, this case can be prove similarly to the above subcase, as desired.
   (c) We may have $E_1\{\widetilde{P}/\widetilde{X}\} \rightsquigarrow \xrightarrow{\alpha} P_1'$ and $E_2\{\widetilde{P}/\widetilde{X}\} \rightsquigarrow \xrightarrow{\beta} P_2'$ with $\alpha \neq \overline{\beta}$ and $P' \equiv P_1' \parallel P_2'$, then by applying this lemma, $P_1'$ is of the form $E_1'\{\widetilde{P}/\widetilde{X}\}$, and for any $Q$, $E_1\{\widetilde{Q}/\widetilde{X}\} \rightsquigarrow \xrightarrow{\alpha} E_1'\{\widetilde{Q}/\widetilde{X}\}$; $P_2'$ is of the form $E_2'\{\widetilde{P}/\widetilde{X}\}$, and for any $Q$, $E_2\{\widetilde{Q}/\widetilde{X}\} \rightsquigarrow \xrightarrow{\alpha} E_2'\{\widetilde{Q}/\widetilde{X}\}$. Hence, $P'$ is of the form $E_1' \parallel E_2'\{\widetilde{P}/\widetilde{X}\}$, and for any $Q$, $E\{\widetilde{Q}/\widetilde{X}\} \equiv E_1\{\widetilde{Q}/\widetilde{X}\} \parallel E_2\{\widetilde{Q}/\widetilde{X}\} \rightsquigarrow \xrightarrow{\{\alpha,\beta\}} (E_1' \parallel E_2')\{\widetilde{Q}/\widetilde{X}\}$, then, let $E'$ be $E_1' \parallel E_2'$, as desired.
   (d) We may have $E_1\{\widetilde{P}/\widetilde{X}\} \rightsquigarrow \xrightarrow{l} P_1'$ and $E_2\{\widetilde{P}/\widetilde{X}\} \rightsquigarrow \xrightarrow{\overline{l}} P_2'$ with $P' \equiv P_1' \parallel P_2'$, then by applying this lemma, $P_1'$ is of the form $E_1'\{\widetilde{P}/\widetilde{X}\}$, and for any $Q$, $E_1\{\widetilde{Q}/\widetilde{X}\} \rightsquigarrow \xrightarrow{l} E_1'\{\widetilde{Q}/\widetilde{X}\}$; $P_2'$ is of the form $E_2'\{\widetilde{P}/\widetilde{X}\}$, and for any $Q$, $E_2\{\widetilde{Q}/\widetilde{X}\} \rightsquigarrow \xrightarrow{\overline{l}} E_2'\{\widetilde{Q}/\widetilde{X}\}$. Hence, $P'$ is of the form $E_1' \parallel E_2'\{\widetilde{P}/\widetilde{X}\}$, and for any $Q$, $E\{\widetilde{Q}/\widetilde{X}\} \equiv E_1\{\widetilde{Q}/\widetilde{X}\} \parallel E_2\{\widetilde{Q}/\widetilde{X}\} \rightsquigarrow \xrightarrow{\tau} (E_1' \parallel E_2')\{\widetilde{Q}/\widetilde{X}\}$, then, let $E'$ be $E_1' \parallel E_2'$, as desired.

7. Case $E \equiv F[R]$ and $E \equiv F \setminus L$. These cases can be proven similarly to the above case.
8. Case $E \equiv C$, an agent constant defined by $C \overset{\text{def}}{=} R$. Then, there is no $X \in \widetilde{X}$ occurring in $E$, hence, $C \xrightarrow{\{\alpha_1,\cdots,\alpha_n\}} P'$, let $E'$ be $P'$, as desired. □

**Theorem 10.27** (Unique solution of equations for strongly probabilistic step bisimulation). *Let the recursive expressions $E_i (i \in I)$ contain at most the variables $X_i (i \in I)$, and let each $X_j (j \in I)$ be weakly guarded in each $E_i$. Then,*
*If $\widetilde{P} \sim_{ps} \widetilde{E}\{\widetilde{P}/\widetilde{X}\}$ and $\widetilde{Q} \sim_{ps} \widetilde{E}\{\widetilde{Q}/\widetilde{X}\}$, then $\widetilde{P} \sim_{ps} \widetilde{Q}$.*

*Proof.* It is sufficient to induct on the depth of the inference of $E\{\widetilde{P}/\widetilde{X}\} \rightsquigarrow \xrightarrow{\{\alpha_1,\cdots,\alpha_n\}} P'$.

1. Case $E \equiv X_i$. Then, we have $E\{\widetilde{P}/\widetilde{X}\} \equiv P_i \rightsquigarrow \xrightarrow{\{\alpha_1,\cdots,\alpha_n\}} P'$, since $P_i \sim_{ps} E_i\{\widetilde{P}/\widetilde{X}\}$, we have $E_i\{\widetilde{P}/\widetilde{X}\} \rightsquigarrow \xrightarrow{\{\alpha_1,\cdots,\alpha_n\}} P'' \sim_{ps} P'$. Since $\widetilde{X}$ are weakly guarded in $E_i$, by Lemma 10.26, $P'' \equiv E'\{\widetilde{P}/\widetilde{X}\}$ and $E_i\{\widetilde{P}/\widetilde{X}\} \rightsquigarrow \xrightarrow{\{\alpha_1,\cdots,\alpha_n\}} E'\{\widetilde{P}/\widetilde{X}\}$. Since $E\{\widetilde{Q}/\widetilde{X}\} \equiv X_i\{\widetilde{Q}/\widetilde{X}\} \equiv Q_i \sim_{ps} E_i\{\widetilde{Q}/\widetilde{X}\}$, $E\{\widetilde{Q}/\widetilde{X}\} \rightsquigarrow \xrightarrow{\{\alpha_1,\cdots,\alpha_n\}} Q' \sim_{ps} E'\{\widetilde{Q}/\widetilde{X}\}$. Hence, $P' \sim_{ps} Q'$, as desired.
2. Case $E \equiv \alpha.F$. This case can be proven similarly.
3. Case $E \equiv (\alpha_1 \parallel \cdots \parallel \alpha_n).F$. This case can be proven similarly.
4. Case $E \equiv E_1 + E_2$. We have $E_i\{\widetilde{P}/\widetilde{X}\} \rightsquigarrow \xrightarrow{\{\alpha_1,\cdots,\alpha_n\}} P'$, $E_i\{\widetilde{Q}/\widetilde{X}\} \rightsquigarrow \xrightarrow{\{\alpha_1,\cdots,\alpha_n\}} Q'$, then, $P' \sim_{ps} Q'$, as desired.
5. Case $E \equiv E_1 \boxplus_\pi E_2$. We have $E_i\{\widetilde{P}/\widetilde{X}\} \rightsquigarrow P'$, $E_i\{\widetilde{Q}/\widetilde{X}\} \rightsquigarrow Q'$, then, $P' \sim_{ps} Q'$, as desired.
6. Case $E \equiv E_1 \parallel E_2$, $E \equiv F[R]$ and $E \equiv F \setminus L$, $E \equiv C$. These cases can be proven similarly to the above case. □

**Theorem 10.28** (Unique solution of equations for strongly probabilistic pomset bisimulation). *Let the recursive expressions $E_i (i \in I)$ contain at most the variables $X_i (i \in I)$, and let each $X_j (j \in I)$ be weakly guarded in each $E_i$. Then,*
*If $\widetilde{P} \sim_{pp} \widetilde{E}\{\widetilde{P}/\widetilde{X}\}$ and $\widetilde{Q} \sim_{pp} \widetilde{E}\{\widetilde{Q}/\widetilde{X}\}$, then $\widetilde{P} \sim_{pp} \widetilde{Q}$.*

*Proof.* Similarly to the proof of the unique solution of equations for strongly probabilistic step bisimulation (see Theorem 10.27), we can prove that the unique solution of equations holds for strongly probabilistic pomset bisimulation, however, we omit the proof here. □

**Theorem 10.29** (Unique solution of equations for strongly probabilistic hp-bisimulation). *Let the recursive expressions $E_i (i \in I)$ contain at most the variables $X_i (i \in I)$, and let each $X_j (j \in I)$ be weakly guarded in each $E_i$. Then,*
*If $\widetilde{P} \sim_{php} \widetilde{E}\{\widetilde{P}/\widetilde{X}\}$ and $\widetilde{Q} \sim_{php} \widetilde{E}\{\widetilde{Q}/\widetilde{X}\}$, then $\widetilde{P} \sim_{php} \widetilde{Q}$.*

*Proof.* Similarly to the proof of the unique solution of equations for strongly probabilistic pomset bisimulation (see Theorem 10.28), we can prove that the unique solution of equations holds for strongly probabilistic hp-bisimulation, we just need additionally to check the above conditions on hp-bisimulation, however, we omit these here. □

**Theorem 10.30** (Unique solution of equations for strongly probabilistic hhp-bisimulation). *Let the recursive expressions $E_i (i \in I)$ contain at most the variables $X_i (i \in I)$, and let each $X_j (j \in I)$ be weakly guarded in each $E_i$. Then,*
*If $\widetilde{P} \sim_{phhp} \widetilde{E}\{\widetilde{P}/\widetilde{X}\}$ and $\widetilde{Q} \sim_{phhp} \widetilde{E}\{\widetilde{Q}/\widetilde{X}\}$, then $\widetilde{P} \sim_{phhp} \widetilde{Q}$.*

*Proof.* Similarly to the proof of the unique solution of equations for strongly probabilistic hp-bisimulation (see Theorem 10.29), we can prove that the unique solution of equations holds for strongly probabilistic hhp-bisimulation, however, we omit the proof here. □

## 10.3 Weakly probabilistic truly concurrent bisimulations

The weak probabilistic transition rules of CPTC are the same as the strong one in Table 10.1. Also, the weak action transition rules of CPTC are listed in Table 10.3.

### 10.3.1 Laws and congruence

Remembering that $\tau$ can neither be restricted nor relabeled, we know that the monoid laws, the monoid laws 2, the static laws, and the expansion law in Section 10.2 still hold with respect to the corresponding weakly probabilistic truly concurrent bisimulations. Also, we can enjoy the full congruence of Prefix, Summation, Composition, Restriction, Relabeling, and Constants with respect to the corresponding weakly probabilistic truly concurrent bisimulations. We will not retype these laws, and just give the $\tau$-specific laws.

**TABLE 10.3** Weak action transition rules of CPTC.

| | |
|---|---|
| **WAct₁** | $\dfrac{}{\breve{\alpha}.P \xRightarrow{\alpha} P}$ |
| **WSum₁** | $\dfrac{P \xRightarrow{\alpha} P'}{P+Q \xRightarrow{\alpha} P'}$ |
| **WCom₁** | $\dfrac{P \xRightarrow{\alpha} P' \quad Q \nrightarrow}{P \parallel Q \xRightarrow{\alpha} P' \parallel Q}$ |
| **WCom₂** | $\dfrac{Q \xRightarrow{\alpha} Q' \quad P \nrightarrow}{P \parallel Q \xRightarrow{\alpha} P \parallel Q'}$ |
| **WCom₃** | $\dfrac{P \xRightarrow{\alpha} P' \quad Q \xRightarrow{\beta} Q'}{P \parallel Q \xRightarrow{\{\alpha,\beta\}} P' \parallel Q'} \quad (\beta \neq \overline{\alpha})$ |
| **WCom₄** | $\dfrac{P \xRightarrow{l} P' \quad Q \xRightarrow{\overline{l}} Q'}{P \parallel Q \xRightarrow{\tau} P' \parallel Q'}$ |
| **WAct₂** | $\dfrac{}{(\breve{\alpha}_1 \parallel \cdots \parallel \breve{\alpha}_n).P \xRightarrow{\{\alpha_1,\cdots,\alpha_n\}} P} \quad (\alpha_i \neq \overline{\alpha_j} \quad i,j \in \{1,\cdots,n\})$ |
| **WSum₂** | $\dfrac{P \xRightarrow{\{\alpha_1,\cdots,\alpha_n\}} P'}{P+Q \xRightarrow{\{\alpha_1,\cdots,\alpha_n\}} P'}$ |
| **WRes₁** | $\dfrac{P \xRightarrow{\alpha} P'}{P \setminus L \xRightarrow{\alpha} P' \setminus L} \quad (\alpha, \overline{\alpha} \notin L)$ |
| **WRes₂** | $\dfrac{P \xRightarrow{\{\alpha_1,\cdots,\alpha_n\}} P'}{P \setminus L \xRightarrow{\{\alpha_1,\cdots,\alpha_n\}} P' \setminus L} \quad (\alpha_1, \overline{\alpha_1}, \cdots, \alpha_n, \overline{\alpha_n} \notin L)$ |
| **WRel₁** | $\dfrac{P \xRightarrow{\alpha} P'}{P[f] \xRightarrow{f(\alpha)} P'[f]}$ |
| **WRel₂** | $\dfrac{P \xRightarrow{\{\alpha_1,\cdots,\alpha_n\}} P'}{P[f] \xRightarrow{\{f(\alpha_1),\cdots,f(\alpha_n)\}} P'[f]}$ |
| **WCon₁** | $\dfrac{P \xRightarrow{\alpha} P'}{A \xRightarrow{\alpha} P'} \quad (A \overset{\text{def}}{=} P)$ |
| **WCon₂** | $\dfrac{P \xRightarrow{\{\alpha_1,\cdots,\alpha_n\}} P'}{A \xRightarrow{\{\alpha_1,\cdots,\alpha_n\}} P'} \quad (A \overset{\text{def}}{=} P)$ |

**Proposition 10.31** ($\tau$ laws for weakly probabilistic pomset bisimulation). *The $\tau$ laws for weakly probabilistic pomset bisimulation are as follows:*

1. $P \approx_{pp} \tau.P$;
2. $\alpha.\tau.P \approx_{pp} \alpha.P$;
3. $(\alpha_1 \parallel \cdots \parallel \alpha_n).\tau.P \approx_{pp} (\alpha_1 \parallel \cdots \parallel \alpha_n).P$;
4. $P + \tau.P \approx_{pp} \tau.P$;
5. $P \cdot ((Q + \tau \cdot (Q + R)) \boxplus_\pi S) \approx_{pp} P \cdot ((Q + R) \boxplus_\pi S)$;
6. $P \approx_{pp} \tau \parallel P$.

*Proof.* According to the definition of weakly probabilistic pomset bisimulation, we can easily prove the above equations, however, we omit the proof here. □

**Proposition 10.32** ($\tau$ laws for weakly probabilistic step bisimulation). *The $\tau$ laws for weakly probabilistic step bisimulation is as follows:*

1. $P \approx_{ps} \tau.P$;
2. $\alpha.\tau.P \approx_{ps} \alpha.P$;
3. $(\alpha_1 \parallel \cdots \parallel \alpha_n).\tau.P \approx_{ps} (\alpha_1 \parallel \cdots \parallel \alpha_n).P$;

4. $P + \tau.P \approx_{ps} \tau.P$;
5. $P \cdot ((Q + \tau \cdot (Q + R)) \boxplus_\pi S) \approx_{ps} P \cdot ((Q + R) \boxplus_\pi S)$;
6. $P \approx_{ps} \tau \parallel P$.

*Proof.* According to the definition of weakly probabilistic step bisimulation, we can easily prove the above equations, however, we omit the proof here. □

**Proposition 10.33** ($\tau$ laws for weakly probabilistic hp-bisimulation). *The $\tau$ laws for weakly probabilistic hp-bisimulation are as follows:*

1. $P \approx_{php} \tau.P$;
2. $\alpha.\tau.P \approx_{php} \alpha.P$;
3. $(\alpha_1 \parallel \cdots \parallel \alpha_n).\tau.P \approx_{php} (\alpha_1 \parallel \cdots \parallel \alpha_n).P$;
4. $P + \tau.P \approx_{php} \tau.P$;
5. $P \cdot ((Q + \tau \cdot (Q + R)) \boxplus_\pi S) \approx_{php} P \cdot ((Q + R) \boxplus_\pi S)$;
6. $P \approx_{php} \tau \parallel P$.

*Proof.* According to the definition of weakly probabilistic hp-bisimulation, we can easily prove the above equations, however, we omit the proof here. □

**Proposition 10.34** ($\tau$ laws for weakly probabilistic hhp-bisimulation). *The $\tau$ laws for weakly probabilistic hhp-bisimulation are as follows:*

1. $P \approx_{phhp} \tau.P$;
2. $\alpha.\tau.P \approx_{phhp} \alpha.P$;
3. $(\alpha_1 \parallel \cdots \parallel \alpha_n).\tau.P \approx_{phhp} (\alpha_1 \parallel \cdots \parallel \alpha_n).P$;
4. $P + \tau.P \approx_{phhp} \tau.P$;
5. $P \cdot ((Q + \tau \cdot (Q + R)) \boxplus_\pi S) \approx_{phhp} P \cdot ((Q + R) \boxplus_\pi S)$;
6. $P \approx_{phhp} \tau \parallel P$.

*Proof.* According to the definition of weakly probabilistic hhp-bisimulation, we can easily prove the above equations, however, we omit the proof here. □

### 10.3.2 Recursion

**Definition 10.35** (Sequential). $X$ is sequential in $E$ if every subexpression of $E$ that contains $X$, apart from $X$ itself, is of the form $\alpha.F$, or $(\alpha_1 \parallel \cdots \parallel \alpha_n).F$, or $\sum \widetilde{F}$.

**Definition 10.36** (Guarded recursive expression). $X$ is guarded in $E$ if each occurrence of $X$ is with some subexpression $l.F$ or $(l_1 \parallel \cdots \parallel l_n).F$ of $E$.

**Lemma 10.37.** *Let $G$ be guarded and sequential, $Vars(G) \subseteq \widetilde{X}$, and let $G\{\widetilde{P}/\widetilde{X}\} \rightsquigarrow \xrightarrow{\{\alpha_1,\cdots,\alpha_n\}} P'$. Then, there is an expression $H$ such that $G \rightsquigarrow \xrightarrow{\{\alpha_1,\cdots,\alpha_n\}} H$, $P' \equiv H\{\widetilde{P}/\widetilde{X}\}$, and for any $\widetilde{Q}$, $G\{\widetilde{Q}/\widetilde{X}\} \rightsquigarrow \xrightarrow{\{\alpha_1,\cdots,\alpha_n\}} H\{\widetilde{Q}/\widetilde{X}\}$. Moreover, $H$ is sequential, $Vars(H) \subseteq \widetilde{X}$, and if $\alpha_1 = \cdots = \alpha_n = \tau$, then $H$ is also guarded.*

*Proof.* We need to induct on the structure of $G$.

If $G$ is a Constant, a Composition, a Restriction or a Relabeling then it contains no variables, since $G$ is sequential and guarded, then $G \rightsquigarrow \xrightarrow{\{\alpha_1,\cdots,\alpha_n\}} P'$, then let $H \equiv P'$, as desired.

$G$ cannot be a variable, since it is guarded.

If $G \equiv G_1 + G_2$. Then, either $G_1\{\widetilde{P}/\widetilde{X}\} \rightsquigarrow \xrightarrow{\{\alpha_1,\cdots,\alpha_n\}} P'$ or $G_2\{\widetilde{P}/\widetilde{X}\} \rightsquigarrow \xrightarrow{\{\alpha_1,\cdots,\alpha_n\}} P'$, then, we can apply this lemma in either case, as desired.

If $G \equiv G_1 \boxplus_\pi G_2$. Then, either $G_1\{\widetilde{P}/\widetilde{X}\} \rightsquigarrow P'$ or $G_2\{\widetilde{P}/\widetilde{X}\} \rightsquigarrow P'$, then, we can apply this lemma in either case, as desired.

If $G \equiv \beta.H$. Then, we must have $\alpha = \beta$, and $P' \equiv H\{\widetilde{P}/\widetilde{X}\}$, and $G\{\widetilde{Q}/\widetilde{X}\} \equiv \beta.H\{\widetilde{Q}/\widetilde{X}\} \rightsquigarrow \xrightarrow{\beta} H\{\widetilde{Q}/\widetilde{X}\}$, then, let $G'$ be $H$, as desired.

If $G \equiv (\beta_1 \parallel \cdots \parallel \beta_n).H$. Then, we must have $\alpha_i = \beta_i$ for $1 \le i \le n$, and $P' \equiv H\{\widetilde{P}/\widetilde{X}\}$, and $G\{\widetilde{Q}/\widetilde{X}\} \equiv (\beta_1 \parallel \cdots \parallel \beta_n).H\{\widetilde{Q}/\widetilde{X}\} \rightsquigarrow \xrightarrow{\{\beta_1,\cdots,\beta_n\}} H\{\widetilde{Q}/\widetilde{X}\}$, then, let $G'$ be $H$, as desired.

If $G \equiv \tau.H$. Then, we must have $\tau = \tau$, and $P' \equiv H\{\widetilde{P}/\widetilde{X}\}$, and $G\{\widetilde{Q}/\widetilde{X}\} \equiv \tau.H\{\widetilde{Q}/\widetilde{X}\} \rightsquigarrow \xrightarrow{\tau} H\{\widetilde{Q}/\widetilde{X}\}$, then, let $G'$ be $H$, as desired. □

**Theorem 10.38** (Unique solution of equations for weakly probabilistic step bisimulation). *Let the guarded and sequential expressions $\widetilde{E}$ contain free variables $\subseteq \widetilde{X}$, then,*
*If $\widetilde{P} \approx_{ps} \widetilde{E}\{\widetilde{P}/\widetilde{X}\}$ and $\widetilde{Q} \approx_{ps} \widetilde{E}\{\widetilde{Q}/\widetilde{X}\}$, then $\widetilde{P} \approx_{ps} \widetilde{Q}$.*

*Proof.* Like the corresponding theorem in CCS, without loss of generality, we only consider a single equation $X = E$. Hence, we assume $P \approx_{ps} E(P)$, $Q \approx_{ps} E(Q)$, then $P \approx_{ps} Q$.

We will prove $\{(H(P), H(Q)) : H\}$ sequential, if $H(P) \rightsquigarrow \xrightarrow{\{\alpha_1, \cdots, \alpha_n\}} P'$, then, for some $Q'$, $H(Q) \rightsquigarrow \xRightarrow{\{\alpha_1, \cdots, \alpha_n\}} Q'$ and $P' \approx_{ps} Q'$.

Let $H(P) \rightsquigarrow \xrightarrow{\{\alpha_1, \cdot, \alpha_n\}} P'$, then $H(E(P)) \rightsquigarrow \xRightarrow{\{\alpha_1, \cdots, \alpha_n\}} P''$ and $P' \approx_{ps} P''$.

By Lemma 10.37, we know there is a sequential $H'$ such that $H(E(P)) \rightsquigarrow \xRightarrow{\{\alpha_1, \cdots, \alpha_n\}} H'(P) \Rightarrow P'' \approx_{ps} P'$.

Also, $H(E(Q)) \rightsquigarrow \xRightarrow{\{\alpha_1, \cdots, \alpha_n\}} H'(Q) \Rightarrow Q''$ and $P'' \approx_{ps} Q''$. Also, $H(Q) \rightsquigarrow \xrightarrow{\{\alpha_1, \cdots, \alpha_n\}} Q' \approx_{ps} Q''$. Hence, $P' \approx_{ps} Q'$, as desired. □

**Theorem 10.39** (Unique solution of equations for weakly probabilistic pomset bisimulation). *Let the guarded and sequential expressions $\widetilde{E}$ contain free variables $\subseteq \widetilde{X}$, then*
*If $\widetilde{P} \approx_{pp} \widetilde{E}\{\widetilde{P}/\widetilde{X}\}$ and $\widetilde{Q} \approx_{pp} \widetilde{E}\{\widetilde{Q}/\widetilde{X}\}$, then $\widetilde{P} \approx_{pp} \widetilde{Q}$.*

*Proof.* Similarly to the proof of the unique solution of equations for weakly probabilistic step bisimulation $\approx_{ps}$ (Theorem 10.38), we can prove that the unique solution of equations holds for weakly probabilistic pomset bisimulation $\approx_{pp}$, however, we omit the proof here. □

**Theorem 10.40** (Unique solution of equations for weakly probabilistic hp-bisimulation). *Let the guarded and sequential expressions $\widetilde{E}$ contain free variables $\subseteq \widetilde{X}$, then*
*If $\widetilde{P} \approx_{php} \widetilde{E}\{\widetilde{P}/\widetilde{X}\}$ and $\widetilde{Q} \approx_{php} \widetilde{E}\{\widetilde{Q}/\widetilde{X}\}$, then $\widetilde{P} \approx_{php} \widetilde{Q}$.*

*Proof.* Similarly to the proof of the unique solution of equations for weakly probabilistic pomset bisimulation (Theorem 10.39), we can prove that the unique solution of equations holds for weakly probabilistic hp-bisimulation, we just need additionally to check the above conditions on weakly probabilistic hp-bisimulation, however, we omit these here. □

**Theorem 10.41** (Unique solution of equations for weakly probabilistic hhp-bisimulation). *Let the guarded and sequential expressions $\widetilde{E}$ contain free variables $\subseteq \widetilde{X}$, then*
*If $\widetilde{P} \approx_{phhp} \widetilde{E}\{\widetilde{P}/\widetilde{X}\}$ and $\widetilde{Q} \approx_{phhp} \widetilde{E}\{\widetilde{Q}/\widetilde{X}\}$, then $\widetilde{P} \approx_{phhp} \widetilde{Q}$.*

*Proof.* Similarly to the proof of unique solution of equations for weakly probabilistic hp-bisimulation (see Theorem 10.40), we can prove that the unique solution of equations holds for weakly probabilistic hhp-bisimulation, however, we omit the proof here. □

Chapter 11

# Algebraic laws for probabilistic true concurrency

The theory $APPTC$ (Algebra of Probabilistic Processes for True Concurrency) has four modules: $BAPTC$ (Basic Algebra for Probabilistic True Concurrency), $APPTC$ (Algebra for Parallelism in Probabilistic True Concurrency), recursion, and abstraction.

This chapter is organized as follows. We introduce $BAPTC$ in Section 11.1, $APPTC$ in Section 11.2, recursion in Section 11.3, and abstraction in Section 11.4.

## 11.1 Basic algebra for probabilistic true concurrency

In this section, we will discuss the algebraic laws for prime event structure $\mathcal{E}$, exactly for causality $\leq$, conflict $\sharp$, and probabilistic conflict $\sharp_\pi$. We will follow the conventions of process algebra, using $\cdot$ instead of $\leq$, $+$ instead of $\sharp$, and $\boxplus_\pi$ instead of $\sharp_\pi$. The resulted algebra is called Basic Algebra for Probabilistic True Concurrency, abbreviated $BAPTC$.

### 11.1.1 Axiom system of $BAPTC$

In the following, let $e_1, e_2, e'_1, e'_2 \in \mathbb{E}$, and let variables $x, y, z$ range over the set of terms for true concurrency, $p, q, s$ range over the set of closed terms. The set of axioms of $BAPTC$ consists of the laws given in Table 11.1.

Intuitively, the axiom $A1$ says that the binary operator $+$ satisfies the commutative law. The axiom $A2$ says that $+$ satisfies associativity. $A3$ says that $+$ satisfies idempotency. The axiom $A4$ is the right distributivity of the binary operator $\cdot$ to $+$. The axiom $A5$ is the associativity of $\cdot$. The axiom $PA1$ is the commutativity of $\boxplus_\pi$. The axiom $PA2$ is the associativity of $\boxplus_\pi$. The axiom $PA3$ says that $\boxplus_\pi$ satisfies idempotency. The axiom $PA4$ is the right distributivity of $\cdot$ to $\boxplus_\pi$. Also, the axiom $PA5$ is the right distributivity of $+$ to $\boxplus_\pi$.

### 11.1.2 Properties of $BAPTC$

**Definition 11.1** (Basic terms of $BAPTC$). The set of basic terms of $BAPTC$, $\mathcal{B}(BAPTC)$, is inductively defined as follows:

1. $\mathbb{E} \subset \mathcal{B}(BAPTC)$;

**TABLE 11.1** Axioms of $BAPTC$.

| No. | Axiom |
|---|---|
| $A1$ | $x + y = y + x$ |
| $A2$ | $(x + y) + z = x + (y + z)$ |
| $A3$ | $x + x = x$ |
| $A4$ | $(x + y) \cdot z = x \cdot z + y \cdot z$ |
| $A5$ | $(x \cdot y) \cdot z = x \cdot (y \cdot z)$ |
| $PA1$ | $x \boxplus_\pi y = y \boxplus_{1-\pi} x$ |
| $PA2$ | $x \boxplus_\pi (y \boxplus_\rho z) = (x \boxplus_{\frac{\pi}{\pi+\rho-\pi\rho}} y) \boxplus_{\pi+\rho-\pi\rho} z$ |
| $PA3$ | $x \boxplus_\pi x = x$ |
| $PA4$ | $(x \boxplus_\pi y) \cdot z = x \cdot z \boxplus_\pi y \cdot z$ |
| $PA5$ | $(x \boxplus_\pi y) + z = (x + z) \boxplus_\pi (y + z)$ |

Handbook of Truly Concurrent Process Algebra. https://doi.org/10.1016/B978-0-44-321515-5.00015-9

**TABLE 11.2** Term rewrite system of $BAPTC$.

| No. | Rewriting Rule |
|---|---|
| $RA3$ | $x + x \rightarrow x$ |
| $RA4$ | $(x + y) \cdot z \rightarrow x \cdot z + y \cdot z$ |
| $RA5$ | $(x \cdot y) \cdot z \rightarrow x \cdot (y \cdot z)$ |
| $RPA1$ | $x \boxplus_{\pi} y \rightarrow y \boxplus_{1-\pi} x$ |
| $RPA2$ | $x \boxplus_{\pi} (y \boxplus_{\rho} z) \rightarrow (x \boxplus_{\frac{\pi}{\pi+\rho-\pi\rho}} y) \boxplus_{\pi+\rho-\pi\rho} z$ |
| $RPA3$ | $x \boxplus_{\pi} x \rightarrow x$ |
| $RPA4$ | $(x \boxplus_{\pi} y) \cdot z \rightarrow x \cdot z \boxplus_{\pi} y \cdot z$ |
| $RPA5$ | $(x \boxplus_{\pi} y) + z \rightarrow (x + z) \boxplus_{\pi} (y + z)$ |

2. if $e \in \mathbb{E}, t \in \mathcal{B}(BAPTC)$, then $e \cdot t \in \mathcal{B}(BAPTC)$;
3. if $t, s \in \mathcal{B}(BAPTC)$, then $t + s \in \mathcal{B}(BAPTC)$;
4. if $t, s \in \mathcal{B}(BAPTC)$, then $t \boxplus_{\pi} s \in \mathcal{B}(BAPTC)$.

**Theorem 11.2** (Elimination theorem of $BAPTC$). *Let $p$ be a closed $BAPTC$ term. Then, there is a basic $BAPTC$ term $q$ such that $BAPTC \vdash p = q$.*

*Proof.* (1) First, suppose that the following ordering on the signature of $BAPTC$ is defined: $\cdot > + > \boxplus_{\pi}$ and the symbol $\cdot$ is given the lexicographical status for the first argument, then for each rewrite rule $p \rightarrow q$ in Table 11.2 relation $p >_{lpo} q$ can easily be proved. We obtain that the term rewrite system shown in Table 11.2 is strongly normalizing, for it has finitely many rewriting rules, and $>$ is a well-founded ordering on the signature of $BAPTC$, and if $s >_{lpo} t$, for each rewriting rule $s \rightarrow t$ is in Table 11.2 (see Theorem 2.9).

(2) Then, we prove that the normal forms of closed $BAPTC$ terms are basic $BAPTC$ terms.

Suppose that $p$ is a normal form of some closed $BAPTC$ term and suppose that $p$ is not a basic term. Let $p'$ denote the smallest sub-term of $p$ that is not a basic term. This implies that each sub-term of $p'$ is a basic term. Then, we prove that $p$ is not a term in normal form. It is sufficient to induct on the structure of $p'$:

- Case $p' \equiv e, e \in \mathbb{E}$. $p'$ is a basic term, which contradicts the assumption that $p'$ is not a basic term, hence, this case should not occur.
- Case $p' \equiv p_1 \cdot p_2$. By induction on the structure of the basic term $p_1$:
  - Subcase $p_1 \in \mathbb{E}$. $p'$ would be a basic term, which contradicts the assumption that $p'$ is not a basic term;
  - Subcase $p_1 \equiv e \cdot p_1'$. $RA5$ rewriting rule can be applied. Hence, $p$ is not a normal form;
  - Subcase $p_1 \equiv p_1' + p_1''$. $RA4$ rewriting rule can be applied. Hence, $p$ is not a normal form.
- Case $p' \equiv p_1 + p_2$. By induction on the structure of the basic terms both $p_1$ and $p_2$, all subcases will lead to that $p'$ would be a basic term, which contradicts the assumption that $p'$ is not a basic term;
- Case $p' \equiv p_1 \boxplus_{\pi} p_2$. By induction on the structure of the basic terms both $p_1$ and $p_2$, all subcases will lead to that $p'$ would be a basic term, which contradicts the assumption that $p'$ is not a basic term. □

### 11.1.3 Structured operational semantics of $BAPTC$

In this section, we will define a term-deduction system that gives the operational semantics of $BAPTC$. As in [10], we also introduce the counterpart $\breve{e}$ of the event $e$, and also the set $\breve{\mathbb{E}} = \{\breve{e} | e \in \mathbb{E}\}$.

First, we give the definition of PDFs in Table 11.3.

We give the operational transition rules for operators $\cdot$, $+$, and $\boxplus_{\pi}$ as Table 11.4 shows. Also, the predicate $\xrightarrow{e} \surd$ represents successful termination after execution of the event $e$.

The pomset transition rules are shown in Table 11.5, which are different from the single event transition rules in Table 11.4, the pomset transition rules are labeled by pomsets, which are defined by causality $\cdot$, conflict $+$, and $\boxplus_{\pi}$.

**Theorem 11.3** (Congruence of $BAPTC$ with respect to probabilistic pomset bisimulation equivalence). *Probabilistic pomset bisimulation equivalence $\sim_{pp}$ is a congruence with respect to $BAPTC$.*

**TABLE 11.3** PDF definitions of $BAPTC$.

| |
|---|
| $\mu(e, \breve{e}) = 1$ |
| $\mu(x \cdot y, x' \cdot y) = \mu(x, x')$ |
| $\mu(x + y, x' + y') = \mu(x, x') \cdot \mu(y, y')$ |
| $\mu(x \boxplus_\pi y, z) = \pi\mu(x, z) + (1 - \pi)\mu(y, z)$ |
| $\mu(x, y) = 0$, otherwise |

**TABLE 11.4** Single event transition rules of $BAPTC$.

| |
|---|
| $\frac{}{e \rightsquigarrow \breve{e}}$ |
| $\frac{x \rightsquigarrow x'}{x \cdot y \rightsquigarrow x' \cdot y}$ |
| $\frac{x \rightsquigarrow x' \quad y \rightsquigarrow y'}{x + y \rightsquigarrow x' + y'}$ |
| $\frac{x \rightsquigarrow x'}{x \boxplus_\pi y \rightsquigarrow x'} \quad \frac{y \rightsquigarrow y'}{x \boxplus_\pi y \rightsquigarrow y'}$ |
| $\frac{}{\breve{e} \xrightarrow{e} \surd}$ |
| $\frac{x \xrightarrow{e} \surd}{x + y \xrightarrow{e} \surd} \quad \frac{x \xrightarrow{e} x'}{x + y \xrightarrow{e} x'} \quad \frac{y \xrightarrow{e} \surd}{x + y \xrightarrow{e} \surd} \quad \frac{y \xrightarrow{e} y'}{x + y \xrightarrow{e} y'}$ |
| $\frac{x \xrightarrow{e} \surd}{x \cdot y \xrightarrow{e} y} \quad \frac{x \xrightarrow{e} x'}{x \cdot y \xrightarrow{e} x' \cdot y}$ |

**TABLE 11.5** Pomset transition rules of $BAPTC$.

| |
|---|
| $\frac{}{X \rightsquigarrow \breve{X}}$ |
| $\frac{x \rightsquigarrow x'}{x \cdot y \rightsquigarrow x' \cdot y}$ |
| $\frac{x \rightsquigarrow x' \quad y \rightsquigarrow y'}{x + y \rightsquigarrow x' + y'}$ |
| $\frac{x \rightsquigarrow x'}{x \boxplus_\pi y \rightsquigarrow x'} \quad \frac{y \rightsquigarrow y'}{x \boxplus_\pi y \rightsquigarrow y'}$ |
| $\frac{}{\breve{X} \xrightarrow{X} \surd}$ |
| $\frac{x \xrightarrow{X} \surd}{x + y \xrightarrow{X} \surd}(X \subseteq x) \quad \frac{x \xrightarrow{X} x'}{x + y \xrightarrow{X} x'}(X \subseteq x) \quad \frac{y \xrightarrow{Y} \surd}{x + y \xrightarrow{Y} \surd}(Y \subseteq y) \quad \frac{y \xrightarrow{Y} y'}{x + y \xrightarrow{Y} y'}(Y \subseteq y)$ |
| $\frac{x \xrightarrow{X} \surd}{x \cdot y \xrightarrow{X} y}(X \subseteq x) \quad \frac{x \xrightarrow{X} x'}{x \cdot y \xrightarrow{X} x' \cdot y}(X \subseteq x)$ |

*Proof.* It is easy to see that probabilistic pomset bisimulation is an equivalent relation on $BAPTC$ terms, we only need to prove that $\sim_{pp}$ is preserved by the operators $\cdot$, $+$, and $\boxplus_\pi$. That is, if $x \sim_{pp} x'$ and $y \sim_{pp} y'$, we need to prove that $x \cdot y \sim_{pp} x' \cdot y'$, $x + y \sim_{pp} x' + y'$, and $x \boxplus_\pi y \sim_{pp} x' \boxplus_\pi y'$. The proof is quite trivial, hence, we omit it here. □

**Theorem 11.4** (Soundness of $BAPTC$ modulo probabilistic pomset bisimulation equivalence). *Let $x$ and $y$ be $BAPTC$ terms. If $BAPTC \vdash x = y$, then $x \sim_{pp} y$.*

*Proof.* Since probabilistic pomset bisimulation $\sim_{pp}$ is both an equivalent and a congruent relation, we only need to check if each axiom in Table 11.1 is sound modulo probabilistic pomset bisimulation equivalence. The proof is quite trivial, hence, we omit it here. □

**Theorem 11.5** (Completeness of $BAPTC$ modulo probabilistic pomset bisimulation equivalence). *Let p and q be closed $BAPTC$ terms, if $p \sim_{pp} q$, then $p = q$.*

*Proof.* First, by the elimination theorem of $BAPTC$, we know that for each closed $BAPTC$ term $p$, there exists a closed basic $BAPTC$ term $p'$, such that $BAPTC \vdash p = p'$, hence, we only need to consider closed basic $BAPTC$ terms.

The basic terms (see Definition 11.1) modulo associativity and commutativity (AC) of conflict $+$ (defined by axioms $A1$ and $A2$ in Table 11.1), and this equivalence is denoted by $=_{AC}$. Then, each equivalence class $s$ modulo AC of $+$ has the following normal form

$$s_1 \boxplus_{\pi_1} \cdots \boxplus_{\pi_{k-1}} s_k$$

with each $s_i$ has the following form

$$t_1 + \cdots + t_l$$

with each $t_j$ either an atomic event or of the form $u_1 \cdot u_2$, and each $t_j$ is called the summand of $s$.

Now, we prove that for normal forms $n$ and $n'$, if $n \sim_{pp} n'$, then $n =_{AC} n'$. It is sufficient to induct on the sizes of $n$ and $n'$.

- Consider a summand $e$ of $n$. Then, $n \rightsquigarrow \breve{e} \xrightarrow{e} \surd$, hence, $n \sim_{pp} n'$ implies $n' \rightsquigarrow \breve{e} \xrightarrow{e} \surd$, meaning that $n'$ also contains the summand $e$.
- Consider a summand $t_1 \cdot t_2$ of $n$. Then, $n \rightsquigarrow \breve{t_1} \xrightarrow{t_1} t_2$, hence, $n \sim_{pp} n'$ implies $n' \rightsquigarrow \breve{t_1} \xrightarrow{t_1} t_2'$ with $t_2 \sim_{pp} t_2'$, meaning that $n'$ contains a summand $t_1 \cdot t_2'$. Since $t_2$ and $t_2'$ are normal forms and have sizes smaller than $n$ and $n'$, by the induction hypotheses $t_2 \sim_{pp} t_2'$ implies $t_2 =_{AC} t_2'$.

Hence, we obtain $n =_{AC} n'$.

Finally, let $s$ and $t$ be basic terms, and $s \sim_{pp} t$, there are normal forms $n$ and $n'$, such that $s = n$ and $t = n'$. The soundness theorem of $BAPTC$ modulo probabilistic pomset bisimulation equivalence (see Theorem 11.4) yields $s \sim_{pp} n$ and $t \sim_{pp} n'$, hence, $n \sim_{pp} s \sim_{pp} t \sim_{pp} n'$. Since, if $n \sim_{pp} n'$, then $n =_{AC} n'$, $s = n =_{AC} n' = t$, as desired. □

The step transition rules are similar in Table 11.5, which are different from the pomset transition rules, the step transition rules are labeled by steps, in which every event is pairwise concurrent.

**Theorem 11.6** (Congruence of $BAPTC$ with respect to probabilistic step bisimulation equivalence). *Probabilistic step bisimulation equivalence $\sim_{ps}$ is a congruence with respect to $BAPTC$.*

*Proof.* It is easy to see that probabilistic step bisimulation is an equivalent relation on $BAPTC$ terms, we only need to prove that $\sim_{ps}$ is preserved by the operators $\cdot$, $+$, and $\boxplus_\pi$. That is, if $x \sim_{ps} x'$ and $y \sim_{ps} y'$, we need to prove that $x \cdot y \sim_{ps} x' \cdot y'$, $x + y \sim_{ps} x' + y'$, and $x \boxplus_\pi y \sim_{ps} x' \boxplus_\pi y'$. The proof is quite trivial, hence, we omit it here □

**Theorem 11.7** (Soundness of $BAPTC$ modulo probabilistic step bisimulation equivalence). *Let x and y be $BAPTC$ terms. If $BAPTC \vdash x = y$, then $x \sim_{ps} y$.*

*Proof.* Since probabilistic step bisimulation $\sim_{ps}$ is both an equivalent and a congruent relation, we only need to check if each axiom in Table 11.1 is sound modulo probabilistic step bisimulation equivalence. The proof is quite trivial, hence, we omit it here. □

**Theorem 11.8** (Completeness of $BAPTC$ modulo probabilistic step bisimulation equivalence). *Let p and q be closed $BAPTC$ terms, if $p \sim_{ps} q$, then $p = q$.*

*Proof.* First, by the elimination theorem of $BAPTC$, we know that for each closed $BAPTC$ term $p$, there exists a closed basic $BAPTC$ term $p'$, such that $BAPTC \vdash p = p'$, hence, we only need to consider closed basic $BAPTC$ terms.

The basic terms (see Definition 11.1) modulo associativity and commutativity (AC) of conflict $+$ (defined by axioms $A1$ and $A2$ in Table 11.1), and this equivalence is denoted by $=_{AC}$. Then, each equivalence class $s$ modulo AC of $+$ has the following normal form

$$s_1 \boxplus_{\pi_1} \cdots \boxplus_{\pi_{k-1}} s_k$$

with each $s_i$ having the following form

$$t_1 + \cdots + t_l$$

with each $t_j$ either an atomic event or of the form $u_1 \cdot u_2$, and each $t_j$ is called the summand of $s$.

Now, we prove that for normal forms $n$ and $n'$, if $n \sim_{ps} n'$, then $n =_{AC} n'$. It is sufficient to induct on the sizes of $n$ and $n'$.

- Consider a summand $e$ of $n$. Then, $n \rightsquigarrow \breve{e} \xrightarrow{e} \surd$, hence, $n \sim_{ps} n'$ implies $n' \rightsquigarrow \breve{e} \xrightarrow{e} \surd$, meaning that $n'$ also contains the summand $e$.
- Consider a summand $t_1 \cdot t_2$ of $n$. Then, $n \rightsquigarrow \breve{t_1} \xrightarrow{t_1} t_2$, hence, $n \sim_{ps} n'$ implies $n' \rightsquigarrow \breve{t_1} \xrightarrow{t_1} t_2'$ with $t_2 \sim_{ps} t_2'$, meaning that $n'$ contains a summand $t_1 \cdot t_2'$. Since $t_2$ and $t_2'$ are normal forms and have sizes smaller than $n$ and $n'$, by the induction hypotheses $t_2 \sim_{ps} t_2'$ implies $t_2 =_{AC} t_2'$.

Hence, we obtain $n =_{AC} n'$.

Finally, let $s$ and $t$ be basic terms, and $s \sim_{ps} t$, there are normal forms $n$ and $n'$, such that $s = n$ and $t = n'$. The soundness theorem of $BAPTC$ modulo probabilistic pomset bisimulation equivalence (see Theorem 11.7) yields $s \sim_{ps} n$ and $t \sim_{ps} n'$, hence, $n \sim_{ps} s \sim_{ps} t \sim_{ps} n'$. Since if $n \sim_{ps} n'$, then $n =_{AC} n'$, $s = n =_{AC} n' = t$, as desired. □

The transition rules for (hereditary) hp-bisimulation of $BAPTC$ are the same as the single event transition rules in Table 11.4.

**Theorem 11.9** (Congruence of $BAPTC$ with respect to probabilistic hp-bisimulation equivalence). *Probabilistic hp-bisimulation equivalence $\sim_{php}$ is a congruence with respect to $BAPTC$.*

*Proof.* It is easy to see that probabilistic hp-bisimulation is an equivalent relation on $BAPTC$ terms, we only need to prove that $\sim_{php}$ is preserved by the operators $\cdot$, $+$, and $\boxplus_\pi$. That is, if $x \sim_{php} x'$ and $y \sim_{php} y'$, we need to prove that $x \cdot y \sim_{php} x' \cdot y'$, $x + y \sim_{php} x' + y'$, and $x \boxplus_\pi y \sim_{php} x' \boxplus_\pi y'$. The proof is quite trivial, hence, we omit it here. □

**Theorem 11.10** (Soundness of $BAPTC$ modulo probabilistic hp-bisimulation equivalence). *Let $x$ and $y$ be $BAPTC$ terms. If $BAPTC \vdash x = y$, then $x \sim_{php} y$.*

*Proof.* Since probabilistic hp-bisimulation $\sim_{php}$ is both an equivalent and a congruent relation, we only need to check if each axiom in Table 11.1 is sound modulo probabilistic hp-bisimulation equivalence. The proof is quite trivial, hence, we omit it here. □

**Theorem 11.11** (Completeness of $BAPTC$ modulo probabilistic hp-bisimulation equivalence). *Let $p$ and $q$ be closed $BAPTC$ terms, if $p \sim_{php} q$, then $p = q$.*

*Proof.* First, by the elimination theorem of $BAPTC$, we know that for each closed $BAPTC$ term $p$, there exists a closed basic $BAPTC$ term $p'$, such that $BAPTC \vdash p = p'$, hence, we only need to consider closed basic $BAPTC$ terms.

The basic terms (see Definition 11.1) modulo associativity and commutativity (AC) of conflict $+$ (defined by axioms $A1$ and $A2$ in Table 11.1), and this equivalence is denoted by $=_{AC}$. Then, each equivalence class $s$ modulo AC of $+$ has the following normal form

$$s_1 \boxplus_{\pi_1} \cdots \boxplus_{\pi_{k-1}} s_k$$

with each $s_i$ having the following form

$$t_1 + \cdots + t_l$$

with each $t_j$ either an atomic event or of the form $u_1 \cdot u_2$, and each $t_j$ is called the summand of $s$.

Now, we prove that for normal forms $n$ and $n'$, if $n \sim_{php} n'$, then $n =_{AC} n'$. It is sufficient to induct on the sizes of $n$ and $n'$.

- Consider a summand $e$ of $n$. Then, $n \rightsquigarrow \breve{e} \xrightarrow{e} \surd$, hence, $n \sim_{php} n'$ implies $n' \rightsquigarrow \breve{e} \xrightarrow{e} \surd$, meaning that $n'$ also contains the summand $e$.
- Consider a summand $t_1 \cdot t_2$ of $n$. Then, $n \rightsquigarrow \breve{t_1} \xrightarrow{t_1} t_2$, hence, $n \sim_{php} n'$ implies $n' \rightsquigarrow \breve{t_1} \xrightarrow{t_1} t_2'$ with $t_2 \sim_{php} t_2'$, meaning that $n'$ contains a summand $t_1 \cdot t_2'$. Since $t_2$ and $t_2'$ are normal forms and have sizes smaller than $n$ and $n'$, by the induction hypotheses $t_2 \sim_{php} t_2'$ implies $t_2 =_{AC} t_2'$.

Hence, we obtain $n =_{AC} n'$.

Finally, let $s$ and $t$ be basic terms, and $s \sim_{php} t$, there are normal forms $n$ and $n'$, such that $s = n$ and $t = n'$. The soundness theorem of $BAPTC$ modulo probabilistic pomset bisimulation equivalence (see Theorem 11.10) yields $s \sim_{php} n$ and $t \sim_{php} n'$, hence, $n \sim_{php} s \sim_{php} t \sim_{php} n'$. Since if $n \sim_{php} n'$, then $n =_{AC} n'$, $s = n =_{AC} n' = t$, as desired. □

**Theorem 11.12** (Congruence of $BAPTC$ with respect to probabilistic hhp-bisimulation equivalence). *Probabilistic hhp-bisimulation equivalence $\sim_{phhp}$ is a congruence with respect to $BAPTC$.*

*Proof.* It is easy to see that probabilistic hhp-bisimulation is an equivalent relation on $BAPTC$ terms, we only need to prove that $\sim_{phhp}$ is preserved by the operators $\cdot$, $+$, and $\boxplus_\pi$. That is, if $x \sim_{phhp} x'$ and $y \sim_{phhp} y'$, we need to prove that $x \cdot y \sim_{phhp} x' \cdot y'$, $x + y \sim_{phhp} x' + y'$, and $x \boxplus_\pi y \sim_{phhp} x' \boxplus_\pi y'$. The proof is quite trivial, hence, we omit it here. □

**Theorem 11.13** (Soundness of $BAPTC$ modulo probabilistic hhp-bisimulation equivalence). *Let $x$ and $y$ be $BAPTC$ terms. If $BAPTC \vdash x = y$, then $x \sim_{phhp} y$.*

*Proof.* Since probabilistic hhp-bisimulation $\sim_{phhp}$ is both an equivalent and a congruent relation, we only need to check if each axiom in Table 11.1 is sound modulo probabilistic hhp-bisimulation equivalence. This is quite trivial, hence, we omit it here. □

**Theorem 11.14** (Completeness of $BAPTC$ modulo probabilistic hhp-bisimulation equivalence). *Let $p$ and $q$ be closed $BAPTC$ terms, if $p \sim_{phhp} q$, then $p = q$.*

*Proof.* First, by the elimination theorem of $BAPTC$, we know that for each closed $BAPTC$ term $p$, there exists a closed basic $BAPTC$ term $p'$, such that $BAPTC \vdash p = p'$, hence, we only need to consider closed basic $BAPTC$ terms.

The basic terms (see Definition 11.1) modulo associativity and commutativity (AC) of conflict $+$ (defined by axioms $A1$ and $A2$ in Table 11.1), and this equivalence is denoted by $=_{AC}$. Then, each equivalence class $s$ modulo AC of $+$ has the following normal form

$$s_1 \boxplus_{\pi_1} \cdots \boxplus_{\pi_{k-1}} s_k$$

with each $s_i$ having the following form

$$t_1 + \cdots + t_l$$

with each $t_j$ either an atomic event or of the form $u_1 \cdot u_2$, and each $t_j$ is called the summand of $s$.

Now, we prove that for normal forms $n$ and $n'$, if $n \sim_{phhp} n'$, then $n =_{AC} n'$. It is sufficient to induct on the sizes of $n$ and $n'$.

- Consider a summand $e$ of $n$. Then, $n \rightsquigarrow \breve{e} \xrightarrow{e} \surd$, hence, $n \sim_{phhp} n'$ implies $n' \rightsquigarrow \breve{e} \xrightarrow{e} \surd$, meaning that $n'$ also contains the summand $e$.
- Consider a summand $t_1 \cdot t_2$ of $n$. Then, $n \rightsquigarrow \breve{t_1} \xrightarrow{t_1} t_2$, hence, $n \sim_{phhp} n'$ implies $n' \rightsquigarrow \breve{t_1} \xrightarrow{t_1} t_2'$ with $t_2 \sim_{phhp} t_2'$, meaning that $n'$ contains a summand $t_1 \cdot t_2'$. Since $t_2$ and $t_2'$ are normal forms and have sizes smaller than $n$ and $n'$, by the induction hypotheses $t_2 \sim_{phhp} t_2'$ implies $t_2 =_{AC} t_2'$.

Hence, we obtain $n =_{AC} n'$.

Finally, let $s$ and $t$ be basic terms, and $s \sim_{phhp} t$, there are normal forms $n$ and $n'$, such that $s = n$ and $t = n'$. The soundness theorem of $BAPTC$ modulo probabilistic pomset bisimulation equivalence (see Theorem 11.13) yields $s \sim_{phhp} n$ and $t \sim_{phhp} n'$, hence, $n \sim_{phhp} s \sim_{phhp} t \sim_{phhp} n'$. Since if $n \sim_{phhp} n'$, then $n =_{AC} n'$, $s = n =_{AC} n' = t$, as desired. □

## 11.2 Algebra for parallelism in probabilistic true concurrency

In this section, we will discuss parallelism in probabilistic true concurrency. The resulted algebra is called Algebra for Parallelism in Probabilistic True Concurrency, abbreviated $APPTC$.

### 11.2.1 Axiom system of parallelism

We design the axioms of parallelism in Table 11.6, including algebraic laws for parallel operator $\parallel$, communication operator $\mid$, conflict elimination operator $\Theta$, unless operator $\triangleleft$, and also the whole parallel operator $\between$. Since the communication between two communicating events in different parallel branches may cause deadlock (a state of inactivity), which is caused by mismatch of two communicating events or the imperfectness of the communication channel, we introduce a new constant $\delta$ to denote the deadlock, and let the atomic event $e \in \mathbb{E} \cup \{\delta\}$.

We explain the intuitions of the axioms of parallelism in Table 11.6 in the following. The axiom $A6$ says that the deadlock $\delta$ is redundant in the process term $t + \delta$. $A7$ says that the deadlock blocks all behaviors of the process term $\delta \cdot t$.

**TABLE 11.6** Axioms of parallelism.

| No. | Axiom |
|---|---|
| $A3$ | $e+e=e$ |
| $A6$ | $x+\delta=x$ |
| $A7$ | $\delta\cdot x=\delta$ |
| $P1$ | $(x+x=x, y+y=y) \quad x\between y=x\parallel y+x\mid y$ |
| $P2$ | $x\parallel y=y\parallel x$ |
| $P3$ | $(x\parallel y)\parallel z=x\parallel(y\parallel z)$ |
| $P4$ | $(x+x=x, y+y=y) \quad x\parallel y=x\leftmerge y+y\leftmerge x$ |
| $P5$ | $(e_1\leq e_2) \quad e_1\leftmerge(e_2\cdot y)=(e_1\leftmerge e_2)\cdot y$ |
| $P6$ | $(e_1\leq e_2) \quad (e_1\cdot x)\leftmerge e_2=(e_1\leftmerge e_2)\cdot x$ |
| $P7$ | $(e_1\leq e_2) \quad (e_1\cdot x)\leftmerge(e_2\cdot y)=(e_1\leftmerge e_2)\cdot(x\between y)$ |
| $P8$ | $(x+y)\leftmerge z=(x\leftmerge z)+(y\leftmerge z)$ |
| $P9$ | $\delta\leftmerge x=\delta$ |
| $C10$ | $e_1\mid e_2=\gamma(e_1,e_2)$ |
| $C11$ | $e_1\mid(e_2\cdot y)=\gamma(e_1,e_2)\cdot y$ |
| $C12$ | $(e_1\cdot x)\mid e_2=\gamma(e_1,e_2)\cdot x$ |
| $C13$ | $(e_1\cdot x)\mid(e_2\cdot y)=\gamma(e_1,e_2)\cdot(x\between y)$ |
| $C14$ | $(x+y)\mid z=(x\mid z)+(y\mid z)$ |
| $C15$ | $x\mid(y+z)=(x\mid y)+(x\mid z)$ |
| $C16$ | $\delta\mid x=\delta$ |
| $C17$ | $x\mid\delta=\delta$ |
| $PM1$ | $x\parallel(y\boxplus_\pi z)=(x\parallel y)\boxplus_\pi(x\parallel z)$ |
| $PM2$ | $(x\boxplus_\pi y)\parallel z=(x\parallel z)\boxplus_\pi(y\parallel z)$ |
| $PM3$ | $x\mid(y\boxplus_\pi z)=(x\mid y)\boxplus_\pi(x\mid z)$ |
| $PM4$ | $(x\boxplus_\pi y)\mid z=(x\mid z)\boxplus_\pi(y\mid z)$ |
| $CE18$ | $\Theta(e)=e$ |
| $CE19$ | $\Theta(\delta)=\delta$ |
| $CE20$ | $\Theta(x+y)=\Theta(x)+\Theta(y)$ |
| $PCE1$ | $\Theta(x\boxplus_\pi y)=\Theta(x)\boxplus_\pi\Theta(y)$ |
| $CE21$ | $\Theta(x\cdot y)=\Theta(x)\cdot\Theta(y)$ |
| $CE22$ | $\Theta(x\leftmerge y)=((\Theta(x)\triangleleft y)\leftmerge y)+((\Theta(y)\triangleleft x)\leftmerge x)$ |
| $CE23$ | $\Theta(x\mid y)=((\Theta(x)\triangleleft y)\mid y)+((\Theta(y)\triangleleft x)\mid x)$ |
| $U24$ | $(\sharp(e_1,e_2)) \quad e_1\triangleleft e_2=\tau$ |
| $U25$ | $(\sharp(e_1,e_2),e_2\leq e_3) \quad e_1\triangleleft e_3=\tau$ |
| $U26$ | $(\sharp(e_1,e_2),e_2\leq e_3) \quad e3\triangleleft e_1=\tau$ |
| $PU1$ | $(\sharp_\pi(e_1,e_2)) \quad e_1\triangleleft e_2=\tau$ |
| $PU2$ | $(\sharp_\pi(e_1,e_2),e_2\leq e_3) \quad e_1\triangleleft e_3=\tau$ |
| $PU3$ | $(\sharp_\pi(e_1,e_2),e_2\leq e_3) \quad e_3\triangleleft e_1=\tau$ |
| $U27$ | $e\triangleleft\delta=e$ |
| $U28$ | $\delta\triangleleft e=\delta$ |
| $U29$ | $(x+y)\triangleleft z=(x\triangleleft z)+(y\triangleleft z)$ |
| $PU4$ | $(x\boxplus_\pi y)\triangleleft z=(x\triangleleft z)\boxplus_\pi(y\triangleleft z)$ |
| $U30$ | $(x\cdot y)\triangleleft z=(x\triangleleft z)\cdot(y\triangleleft z)$ |
| $U31$ | $(x\leftmerge y)\triangleleft z=(x\triangleleft z)\leftmerge(y\triangleleft z)$ |

*continued on next page*

**TABLE 11.6** (*continued*)

| No. | Axiom |
|---|---|
| $U32$ | $(x \mid y) \triangleleft z = (x \triangleleft z) \mid (y \triangleleft z)$ |
| $U33$ | $x \triangleleft (y + z) = (x \triangleleft y) \triangleleft z$ |
| $PU5$ | $x \triangleleft (y \boxplus_\pi z) = (x \triangleleft y) \triangleleft z$ |
| $U34$ | $x \triangleleft (y \cdot z) = (x \triangleleft y) \triangleleft z$ |
| $U35$ | $x \triangleleft (y \between z) = (x \triangleleft y) \triangleleft z$ |
| $U36$ | $x \triangleleft (y \mid z) = (x \triangleleft y) \triangleleft z$ |

The axiom $P1$ is the definition of the whole parallelism $\between$, which says that $s \between t$ either is the form of $s \parallel t$ or $s \mid t$. $P2$ says that $\parallel$ satisfies commutative law, while $P3$ says that $\parallel$ satisfies associativity. $P4$, $P5$, and $P6$ are the defining axioms of $\parallel$, say the $s \parallel t$ executes $s$ and $t$ concurrently. $P8$ is the right of $\between$ to $+$. $P9$ says that $\delta \between t$ blocks any event.

$C10$, $C11$, $C12$, and $C13$ are the defining axioms of the communication operator $\mid$, which say that $s \mid t$ makes a communication between $s$ and $t$. $C14$ and $C15$ are the right and left distributivity of $\mid$ to $+$. $C16$ and $C17$ say that both $\delta \mid t$ and $t \mid \delta$ block any event.

$CE18$ and $CE19$ say that the conflict elimination operator $\Theta$ leaves atomic events and the deadlock unchanged. $CE20$–$CE23$ are the functions of $\Theta$ acting on the operators $+$, $\cdot$, $\parallel$, and $\mid$. $U24$, $U25$, and $U26$ are the defining laws of the unless operator $\triangleleft$, in $U24$ and $U26$, there is a new constant $\tau$, the silent step, we will discuss $\tau$ in detail in Section 11.4, in these two axioms, we just need to remember that $\tau$ really remains silent. $U27$ says that the deadlock $\delta$ cannot block any event in the process term $e \triangleleft \delta$, while $U28$ says that $\delta \triangleleft e$ does not exhibit any behavior. $U29$ – $U36$ are the disguised right and left distributivity of $\triangleleft$ to the operators $+$, $\cdot$, $\parallel$, and $\mid$.

The axiom $A3$ in the above section is replaced by the new one, because of the introduction of $\delta$. The axioms $PM1$, $PM2$, $PM3$, and $PM4$ are the distributivity of $\parallel$ and $\mid$ to $\boxplus_\pi$. $PCE1$ is the function of $\Theta$ acting on the operator $\boxplus_\pi$. $PU1$, $PU2$, and $PU3$ are the defining laws of the unless operator $\triangleleft$ for $\sharp_\pi$. $PU4$ and $PU5$ are the disguised right and left distributivity of $\triangleleft$ to the operators $\boxplus_\pi$.

**Definition 11.15** (Basic terms of $APPTC$). The set of basic terms of $APPTC$, $\mathcal{B}(APPTC)$, are inductively defined as follows:

1. $\mathbb{E} \subset \mathcal{B}(APPTC)$;
2. if $e \in \mathbb{E}, t \in \mathcal{B}(APPTC)$, then $e \cdot t \in \mathcal{B}(APPTC)$;
3. if $t, s \in \mathcal{B}(APPTC)$, then $t + s \in \mathcal{B}(APPTC)$;
4. if $t, s \in \mathcal{B}(APPTC)$, then $t \boxplus_\pi s \in \mathcal{B}(APPTC)$;
5. if $t, s \in \mathcal{B}(APPTC)$, then $t \between s \in \mathcal{B}(APPTC)$.

Based on the definition of basic terms for $APPTC$ (see Definition 11.15) and axioms of parallelism (see Table 11.6), we can prove the elimination theorem of parallelism.

**Theorem 11.16** (Elimination theorem of parallelism). *Let $p$ be a closed $APPTC$ term. Then, there is a basic $APPTC$ term $q$ such that $APPTC \vdash p = q$.*

*Proof.* (1) First, suppose that the following ordering on the signature of $APPTC$ is defined: $\between > \cdot > + > \boxplus_\pi$ and the symbol $\cdot$ is given the lexicographical status for the first argument, then for each rewrite rule $p \rightarrow q$ in Table 11.7 relation $p >_{lpo} q$ can easily be proved. We obtain that the term rewrite system shown in Table 11.7 is strongly normalizing, for it has finitely many rewriting rules, and $>$ is a well-founded ordering on the signature of $APPTC$, and if $s >_{lpo} t$, for each rewriting rule $s \rightarrow t$ is in Table 11.7 (see Theorem 2.9).

(2) Then, we prove that the normal forms of closed $APPTC$ terms are basic $APPTC$ terms.

Suppose that $p$ is a normal form of some closed $APPTC$ term and suppose that $p$ is not a basic $APPTC$ term. Let $p'$ denote the smallest sub-term of $p$ that is not a basic $APPTC$ term. This implies that each sub-term of $p'$ is a basic $APPTC$ term. Then, we prove that $p$ is not a term in normal form. It is sufficient to induct on the structure of $p'$:

- Case $p' \equiv e, e \in \mathbb{E}$. $p'$ is a basic $APPTC$ term, which contradicts the assumption that $p'$ is not a basic $APPTC$ term, hence, this case should not occur.
- Case $p' \equiv p_1 \cdot p_2$. By induction on the structure of the basic $APPTC$ term $p_1$:

**TABLE 11.7** Term rewrite system of $APPTC$.

| No. | Rewriting Rule |
|---|---|
| $RA3$ | $e + e \rightarrow e$ |
| $RA6$ | $x + \delta \rightarrow x$ |
| $RA7$ | $\delta \cdot x \rightarrow \delta$ |
| $RP1$ | $(x + x = x, y + y = y) \quad x \between y \rightarrow x \parallel y + x \mid y$ |
| $RP2$ | $x \parallel y \rightarrow y \parallel x$ |
| $RP3$ | $(x \parallel y) \parallel z \rightarrow x \parallel (y \parallel z)$ |
| $RP4$ | $(x + x = x, y + y = y) \quad x \parallel y \rightarrow x \mathbin{\lfloor\!\lfloor} y + y \mathbin{\lfloor\!\lfloor} x$ |
| $RP5$ | $(e_1 \leq e_2) \quad e_1 \mathbin{\lfloor\!\lfloor} (e_2 \cdot y) \rightarrow (e_1 \mathbin{\lfloor\!\lfloor} e_2) \cdot y$ |
| $RP6$ | $(e_1 \leq e_2) \quad (e_1 \cdot x) \mathbin{\lfloor\!\lfloor} e_2 \rightarrow (e_1 \mathbin{\lfloor\!\lfloor} e_2) \cdot x$ |
| $RP7$ | $(e_1 \leq e_2) \quad (e_1 \cdot x) \mathbin{\lfloor\!\lfloor} (e_2 \cdot y) \rightarrow (e_1 \mathbin{\lfloor\!\lfloor} e_2) \cdot (x \between y)$ |
| $RP8$ | $(x + y) \mathbin{\lfloor\!\lfloor} z \rightarrow (x \mathbin{\lfloor\!\lfloor} z) + (y \mathbin{\lfloor\!\lfloor} z)$ |
| $RP9$ | $\delta \mathbin{\lfloor\!\lfloor} x \rightarrow \delta$ |
| $RC10$ | $e_1 \mid e_2 \rightarrow \gamma(e_1, e_2)$ |
| $RC11$ | $e_1 \mid (e_2 \cdot y) \rightarrow \gamma(e_1, e_2) \cdot y$ |
| $RC12$ | $(e_1 \cdot x) \mid e_2 \rightarrow \gamma(e_1, e_2) \cdot x$ |
| $RC13$ | $(e_1 \cdot x) \mid (e_2 \cdot y) \rightarrow \gamma(e_1, e_2) \cdot (x \between y)$ |
| $RC14$ | $(x + y) \mid z \rightarrow (x \mid z) + (y \mid z)$ |
| $RC15$ | $x \mid (y + z) \rightarrow (x \mid y) + (x \mid z)$ |
| $RC16$ | $\delta \mid x \rightarrow \delta$ |
| $RC17$ | $x \mid \delta \rightarrow \delta$ |
| $RPM1$ | $x \parallel (y \boxplus_\pi z) \rightarrow (x \parallel y) \boxplus_\pi (x \parallel z)$ |
| $RPM2$ | $(x \boxplus_\pi y) \parallel z \rightarrow (x \parallel z) \boxplus_\pi (y \parallel z)$ |
| $RPM3$ | $x \mid (y \boxplus_\pi z) \rightarrow (x \mid y) \boxplus_\pi (x \mid z)$ |
| $RPM4$ | $(x \boxplus_\pi y) \mid z \rightarrow (x \mid z) \boxplus_\pi (y \mid z)$ |
| $RCE18$ | $\Theta(e) \rightarrow e$ |
| $RCE19$ | $\Theta(\delta) \rightarrow \delta$ |
| $RCE20$ | $\Theta(x + y) \rightarrow \Theta(x) + \Theta(y)$ |
| $RPCE1$ | $\Theta(x \boxplus_\pi y) \rightarrow \Theta(x) \boxplus_\pi \Theta(y)$ |
| $RCE21$ | $\Theta(x \cdot y) \rightarrow \Theta(x) \cdot \Theta(y)$ |
| $RCE22$ | $\Theta(x \mathbin{\lfloor\!\lfloor} y) \rightarrow ((\Theta(x) \triangleleft y) \mathbin{\lfloor\!\lfloor} y) + ((\Theta(y) \triangleleft x) \mathbin{\lfloor\!\lfloor} x)$ |
| $RCE23$ | $\Theta(x \mid y) \rightarrow ((\Theta(x) \triangleleft y) \mid y) + ((\Theta(y) \triangleleft x) \mid x)$ |
| $RU24$ | $(\sharp(e_1, e_2)) \quad e_1 \triangleleft e_2 \rightarrow \tau$ |
| $RU25$ | $(\sharp(e_1, e_2), e_2 \leq e_3) \quad e_1 \triangleleft e_3 \rightarrow \tau$ |
| $RU26$ | $(\sharp(e_1, e_2), e_2 \leq e_3) \quad e3 \triangleleft e_1 \rightarrow \tau$ |
| $RPU1$ | $(\sharp_\pi(e_1, e_2)) \quad e_1 \triangleleft e_2 \rightarrow \tau$ |
| $RPU2$ | $(\sharp_\pi(e_1, e_2), e_2 \leq e_3) \quad e_1 \triangleleft e_3 \rightarrow \tau$ |
| $RPU3$ | $(\sharp_\pi(e_1, e_2), e_2 \leq e_3) \quad e_3 \triangleleft e_1 \rightarrow \tau$ |
| $RU27$ | $e \triangleleft \delta \rightarrow e$ |
| $RU28$ | $\delta \triangleleft e \rightarrow \delta$ |
| $RU29$ | $(x + y) \triangleleft z \rightarrow (x \triangleleft z) + (y \triangleleft z)$ |
| $RPU4$ | $(x \boxplus_\pi y) \triangleleft z \rightarrow (x \triangleleft z) \boxplus_\pi (y \triangleleft z)$ |
| $RU30$ | $(x \cdot y) \triangleleft z \rightarrow (x \triangleleft z) \cdot (y \triangleleft z)$ |
| $RU31$ | $(x \mathbin{\lfloor\!\lfloor} y) \triangleleft z \rightarrow (x \triangleleft z) \mathbin{\lfloor\!\lfloor} (y \triangleleft z)$ |
| $RU32$ | $(x \mid y) \triangleleft z \rightarrow (x \triangleleft z) \mid (y \triangleleft z)$ |

*continued on next page*

**TABLE 11.7** *(continued)*

| No. | Rewriting Rule |
|---|---|
| $RU33$ | $x \triangleleft (y+z) \rightarrow (x \triangleleft y) \triangleleft z$ |
| $RPU5$ | $x \triangleleft (y \boxplus_\pi z) \rightarrow (x \triangleleft y) \triangleleft z$ |
| $RU34$ | $x \triangleleft (y \cdot z) \rightarrow (x \triangleleft y) \triangleleft z$ |
| $RU35$ | $x \triangleleft (y \between z) \rightarrow (x \triangleleft y) \triangleleft z$ |
| $RU36$ | $x \triangleleft (y \mid z) \rightarrow (x \triangleleft y) \triangleleft z$ |

**TABLE 11.8** PDF definitions of $APPTC$.

| |
|---|
| $\mu(\delta, \breve{\delta}) = 1$ |
| $\mu(x \between y, x' \parallel y' + x' \mid y') = \mu(x, x') \cdot \mu(y, y')$ |
| $\mu(x \parallel y, x' \between y + y' \between x) = \mu(x, x') \cdot \mu(y, y')$ |
| $\mu(x \between y, x' \between y) = \mu(x, x')$ |
| $\mu(x \mid y, x' \mid y') = \mu(x, x') \cdot \mu(y, y')$ |
| $\mu(\Theta(x), \Theta(x')) = \mu(x, x')$ |
| $\mu(x \triangleleft y, x' \triangleleft y) = \mu(x, x')$ |
| $\mu(x, y) = 0$, otherwise |

- Subcase $p_1 \in \mathbb{E}$. $p'$ would be a basic $APPTC$ term, which contradicts the assumption that $p'$ is not a basic $APPTC$ term;
- Subcase $p_1 \equiv e \cdot p_1'$. $RA5$ rewriting rule in Table 11.2 can be applied. Hence, $p$ is not a normal form;
- Subcase $p_1 \equiv p_1' + p_1''$. $RA4$ rewriting rule in Table 11.2 can be applied. Hence, $p$ is not a normal form;
- Subcase $p_1 \equiv p_1' \boxplus_\pi p_1''$. $RPA4$ and $RPA5$ rewriting rules in Table 11.2 can be applied. Hence, $p$ is not a normal form;
- Subcase $p_1 \equiv p_1' \between p_1''$. $p'$ would be a basic $APPTC$ term, which contradicts the assumption that $p'$ is not a basic $APPTC$ term;
- Subcase $p_1 \equiv p_1' \mid p_1''$. $RC11$ rewrite rule in Table 11.7 can be applied. Hence, $p$ is not a normal form;
- Subcase $p_1 \equiv \Theta(p_1')$. $RCE19$ and $RCE20$ rewrite rules in Table 11.7 can be applied. Hence, $p$ is not a normal form.

- Case $p' \equiv p_1 + p_2$. By induction on the structure of the basic $APPTC$ terms both $p_1$ and $p_2$, all subcases will lead to that $p'$ would be a basic $APPTC$ term, which contradicts the assumption that $p'$ is not a basic $APPTC$ term.
- Case $p' \equiv p_1 \boxplus_\pi p_2$. By induction on the structure of the basic $APPTC$ terms both $p_1$ and $p_2$, all subcases will lead to that $p'$ would be a basic $APPTC$ term, which contradicts the assumption that $p'$ is not a basic $APPTC$ term.
- Case $p' \equiv p_1 \between p_2$. By induction on the structure of the basic $APPTC$ terms both $p_1$ and $p_2$, all subcases will lead to that $p'$ would be a basic $APPTC$ term, which contradicts the assumption that $p'$ is not a basic $APPTC$ term.
- Case $p' \equiv p_1 \mid p_2$. By induction on the structure of the basic $APPTC$ terms both $p_1$ and $p_2$, all subcases will lead to that $p'$ would be a basic $APPTC$ term, which contradicts the assumption that $p'$ is not a basic $APPTC$ term.
- Case $p' \equiv \Theta(p_1)$. By induction on the structure of the basic $APPTC$ term $p_1$, $RCE19 - RCE24$ rewrite rules in Table 11.7 can be applied. Hence, $p$ is not a normal form.
- Case $p' \equiv p_1 \triangleleft p_2$. By induction on the structure of the basic $APPTC$ terms both $p_1$ and $p_2$, all subcases will lead to that $p'$ would be a basic $APPTC$ term, which contradicts the assumption that $p'$ is not a basic $APPTC$ term. □

### 11.2.2 Structured operational semantics of parallelism

First, we give the definition of PDFs in Table 11.8.

We give the transition rules of APTC in Tables 11.9 and 11.10, which are suitable for all truly concurrent behavioral equivalence, including probabilistic pomset bisimulation, probabilistic step bisimulation, probabilistic hp-bisimulation, and probabilistic hhp-bisimulation.

**Theorem 11.17** (Generalization of the algebra for parallelism with respect to $BAPTC$). *The algebra for parallelism is a generalization of $BAPTC$.*

**TABLE 11.9 Probabilistic transition rules of APPTC.**

| |
|---|
| $\dfrac{x \rightsquigarrow x' \quad y \rightsquigarrow y'}{x \between y \rightsquigarrow x' \parallel y' + x' \mid y'}$ |
| $\dfrac{x \rightsquigarrow x' \quad y \rightsquigarrow y'}{x \parallel y \rightsquigarrow x' \leftmerge y + y' \leftmerge x}$ |
| $\dfrac{x \rightsquigarrow x'}{x \leftmerge y \rightsquigarrow x' \leftmerge y}$ |
| $\dfrac{x \rightsquigarrow x' \quad y \rightsquigarrow y'}{x \mid y \rightsquigarrow x' \mid y'}$ |
| $\dfrac{x \rightsquigarrow x'}{\Theta(x) \rightsquigarrow \Theta(x')}$ |
| $\dfrac{x \rightsquigarrow x'}{x \triangleleft y \rightsquigarrow x' \triangleleft y}$ |

*Proof.* It follows from the following three facts:

1. The transition rules of $BAPTC$ in Section 11.1 are all source dependent;
2. The sources of the transition rules for the algebra for parallelism contain an occurrence of $\between$, or $\parallel$, or $\leftmerge$, or $\mid$, or $\Theta$, or $\triangleleft$;
3. The transition rules of $APPTC$ are all source dependent.

Hence, the algebra for parallelism is a generalization of $BAPTC$, that is, $BAPTC$ is an embedding of the algebra for parallelism, as desired. □

**Theorem 11.18** (Congruence of $APPTC$ with respect to probabilistic pomset bisimulation equivalence). *Probabilistic pomset bisimulation equivalence $\sim_{pp}$ is a congruence with respect to $APPTC$.*

*Proof.* It is easy to see that probabilistic pomset bisimulation is an equivalent relation on $APPTC$ terms, we only need to prove that $\sim_{pp}$ is preserved by the operators $\between$, $\parallel$, $\leftmerge$, $\mid$, $\Theta$, and $\triangleleft$. That is, if $x \sim_{pp} x'$ and $y \sim_{pp} y'$, we need to prove that $x \between y \sim_{pp} x' \between y'$, $x \parallel y \sim_{pp} x' \parallel y'$, $x \leftmerge y \sim_{pp} x' \leftmerge y'$, $x \mid y \sim_{pp} x' \mid y'$, $x \triangleleft y \sim_{pp} x' \triangleleft y'$, and $\Theta(x) \sim_{pp} \Theta(x')$. The proof is quite trivial, hence, we omit it here. □

**Theorem 11.19** (Soundness of parallelism modulo probabilistic pomset bisimulation equivalence). *Let $x$ and $y$ be $APPTC$ terms. If $APPTC \vdash x = y$, then $x \sim_{pp} y$.*

*Proof.* Since probabilistic pomset bisimulation $\sim_{pp}$ is both an equivalent and a congruent relation with respect to the operators $\between$, $\parallel$, $\leftmerge$, $\mid$, $\Theta$, and $\triangleleft$, we only need to check if each axiom in Table 11.6 is sound modulo probabilistic pomset bisimulation equivalence. The proof is quite trivial, hence, we omit it here. □

**Theorem 11.20** (Completeness of parallelism modulo probabilistic pomset bisimulation equivalence). *Let $p$ and $q$ be closed $APPTC$ terms, if $p \sim_{pp} q$, then $p = q$.*

*Proof.* First, by the elimination theorem of $APPTC$ (see Theorem 11.16), we know that for each closed $APPTC$ term $p$, there exists a closed basic $APPTC$ term $p'$, such that $APPTC \vdash p = p'$, hence, we only need to consider closed basic $APPTC$ terms.

The basic terms (see Definition 11.15) modulo associativity and commutativity (AC) of conflict $+$ (defined by axioms $A1$ and $A2$ in Table 11.1) and these equivalences is denoted by $=_{AC}$. Then, each equivalence class $s$ modulo AC of $+$ has the following normal form

$$s_1 \boxplus_{\pi_1} \cdots \boxplus_{\pi_{k-1}} s_k$$

with each $s_i$ having the following form

$$t_1 + \cdots + t_l$$

with each $t_j$ either an atomic event or of the form

$$u_1 \cdot \cdots \cdot u_m$$

**TABLE 11.10 Action transition rules of APPTC.**

| | |
|---|---|
| $\frac{x\xrightarrow{e_1}\surd\quad y\xrightarrow{e_2}\surd}{x\parallel y\xrightarrow{\{e_1,e_2\}}\surd}$ | $\frac{x\xrightarrow{e_1}x'\quad y\xrightarrow{e_2}\surd}{x\parallel y\xrightarrow{\{e_1,e_2\}}x'}$ |
| $\frac{x\xrightarrow{e_1}\surd\quad y\xrightarrow{e_2}y'}{x\parallel y\xrightarrow{\{e_1,e_2\}}y'}$ | $\frac{x\xrightarrow{e_1}x'\quad y\xrightarrow{e_2}y'}{x\parallel y\xrightarrow{\{e_1,e_2\}}x'\between y'}$ |
| $\frac{x\xrightarrow{e_1}\surd\quad y\xrightarrow{e_2}\surd\quad(e_1\leq e_2)}{x\leftmerge y\xrightarrow{\{e_1,e_2\}}\surd}$ | $\frac{x\xrightarrow{e_1}x'\quad y\xrightarrow{e_2}\surd\quad(e_1\leq e_2)}{x\leftmerge y\xrightarrow{\{e_1,e_2\}}x'}$ |
| $\frac{x\xrightarrow{e_1}\surd\quad y\xrightarrow{e_2}y'\quad(e_1\leq e_2)}{x\leftmerge y\xrightarrow{\{e_1,e_2\}}y'}$ | $\frac{x\xrightarrow{e_1}x'\quad y\xrightarrow{e_2}y'\quad(e_1\leq e_2)}{x\leftmerge y\xrightarrow{\{e_1,e_2\}}x'\between y'}$ |
| $\frac{x\xrightarrow{e_1}\surd\quad y\xrightarrow{e_2}\surd}{x\mid y\xrightarrow{\gamma(e_1,e_2)}\surd}$ | $\frac{x\xrightarrow{e_1}x'\quad y\xrightarrow{e_2}\surd}{x\mid y\xrightarrow{\gamma(e_1,e_2)}x'}$ |
| $\frac{x\xrightarrow{e_1}\surd\quad y\xrightarrow{e_2}y'}{x\mid y\xrightarrow{\gamma(e_1,e_2)}y'}$ | $\frac{x\xrightarrow{e_1}x'\quad y\xrightarrow{e_2}y'}{x\mid y\xrightarrow{\gamma(e_1,e_2)}x'\between y'}$ |
| $\frac{x\xrightarrow{e_1}\surd\quad(\sharp(e_1,e_2))}{\Theta(x)\xrightarrow{e_1}\surd}$ | $\frac{x\xrightarrow{e_2}\surd\quad(\sharp(e_1,e_2))}{\Theta(x)\xrightarrow{e_2}\surd}$ |
| $\frac{x\xrightarrow{e_1}x'\quad(\sharp(e_1,e_2))}{\Theta(x)\xrightarrow{e_1}\Theta(x')}$ | $\frac{x\xrightarrow{e_2}x'\quad(\sharp(e_1,e_2))}{\Theta(x)\xrightarrow{e_2}\Theta(x')}$ |
| $\frac{x\xrightarrow{e_1}\surd\quad(\sharp_\pi(e_1,e_2))}{\Theta(x)\xrightarrow{e_1}\surd}$ | $\frac{x\xrightarrow{e_2}\surd\quad(\sharp_\pi(e_1,e_2))}{\Theta(x)\xrightarrow{e_2}\surd}$ |
| $\frac{x\xrightarrow{e_1}x'\quad(\sharp_\pi(e_1,e_2))}{\Theta(x)\xrightarrow{e_1}\Theta(x')}$ | $\frac{x\xrightarrow{e_2}x'\quad(\sharp_\pi(e_1,e_2))}{\Theta(x)\xrightarrow{e_2}\Theta(x')}$ |
| $\frac{x\xrightarrow{e_1}\surd\quad y\nrightarrow^{e_2}\quad(\sharp(e_1,e_2))}{x\triangleleft y\xrightarrow{\tau}\surd}$ | $\frac{x\xrightarrow{e_1}x'\quad y\nrightarrow^{e_2}\quad(\sharp(e_1,e_2))}{x\triangleleft y\xrightarrow{\tau}x'}$ |
| $\frac{x\xrightarrow{e_1}\surd\quad y\nrightarrow^{e_3}\quad(\sharp(e_1,e_2),e_2\leq e_3)}{x\triangleleft y\xrightarrow{\tau}\surd}$ | $\frac{x\xrightarrow{e_1}x'\quad y\nrightarrow^{e_3}\quad(\sharp(e_1,e_2),e_2\leq e_3)}{x\triangleleft y\xrightarrow{\tau}x'}$ |
| $\frac{x\xrightarrow{e_3}\surd\quad y\nrightarrow^{e_2}\quad(\sharp(e_1,e_2),e_1\leq e_3)}{x\triangleleft y\xrightarrow{\tau}\surd}$ | $\frac{x\xrightarrow{e_3}x'\quad y\nrightarrow^{e_2}\quad(\sharp(e_1,e_2),e_1\leq e_3)}{x\triangleleft y\xrightarrow{\tau}x'}$ |
| $\frac{x\xrightarrow{e_1}\surd\quad y\nrightarrow^{e_2}\quad(\sharp_\pi(e_1,e_2))}{x\triangleleft y\xrightarrow{\tau}\surd}$ | $\frac{x\xrightarrow{e_1}x'\quad y\nrightarrow^{e_2}\quad(\sharp_\pi(e_1,e_2))}{x\triangleleft y\xrightarrow{\tau}x'}$ |
| $\frac{x\xrightarrow{e_1}\surd\quad y\nrightarrow^{e_3}\quad(\sharp_\pi(e_1,e_2),e_2\leq e_3)}{x\triangleleft y\xrightarrow{\tau}\surd}$ | $\frac{x\xrightarrow{e_1}x'\quad y\nrightarrow^{e_3}\quad(\sharp_\pi(e_1,e_2),e_2\leq e_3)}{x\triangleleft y\xrightarrow{\tau}x'}$ |
| $\frac{x\xrightarrow{e_3}\surd\quad y\nrightarrow^{e_2}\quad(\sharp_\pi(e_1,e_2),e_1\leq e_3)}{x\triangleleft y\xrightarrow{\tau}\surd}$ | $\frac{x\xrightarrow{e_3}x'\quad y\nrightarrow^{e_2}\quad(\sharp_\pi(e_1,e_2),e_1\leq e_3)}{x\triangleleft y\xrightarrow{\tau}x'}$ |

with each $u_l$ either an atomic event or of the form

$$v_1\leftmerge\cdots\leftmerge v_m$$

with each $v_m$ an atomic event, and each $t_j$ is called the summand of $s$.

Now, we prove that for normal forms $n$ and $n'$, if $n\sim_{pp}n'$, then $n=_{AC}n'$. It is sufficient to induct on the sizes of $n$ and $n'$.

- Consider a summand $e$ of $n$. Then, $n\rightsquigarrow\breve{e}\xrightarrow{e}\surd$, hence, $n\sim_{pp}n'$ implies $n'\rightsquigarrow\breve{e}\xrightarrow{e}\surd$, meaning that $n'$ also contains the summand $e$.
- Consider a summand $t_1\cdot t_2$ of $n$,

- if $t_1 \equiv e'$, then $n \rightsquigarrow \breve{e}' \xrightarrow{e'} t_2$, hence, $n \sim_{pp} n'$ implies $n' \rightsquigarrow \breve{e}' \xrightarrow{e'} t_2'$ with $t_2 \sim_{pp} t_2'$, meaning that $n'$ contains a summand $e' \cdot t_2'$. Since $t_2$ and $t_2'$ are normal forms and have sizes smaller than $n$ and $n'$, by the induction hypotheses if $t_2 \sim_{pp} t_2'$, then $t_2 =_{AC} t_2'$;
- if $t_1 \equiv e_1 \leftmerge \cdots \leftmerge e_m$, then $n \rightsquigarrow \breve{e}_1 \leftmerge \cdots \leftmerge \breve{e}_m \xrightarrow{\{e_1,\cdots,e_m\}} t_2$, hence, $n \sim_{pp} n'$ implies $n' \rightsquigarrow \breve{e}_1 \leftmerge \cdots \leftmerge \breve{e}_m \xrightarrow{\{e_1,\cdots,e_m\}} t_2'$ with $t_2 \sim_{pp} t_2'$, meaning that $n'$ contains a summand $(e_1 \leftmerge \cdots \leftmerge e_m) \cdot t_2'$. Since $t_2$ and $t_2'$ are normal forms and have sizes smaller than $n$ and $n'$, by the induction hypotheses if $t_2 \sim_{pp} t_2'$, then $t_2 =_{AC} t_2'$.

Hence, we obtain $n =_{AC} n'$.

Finally, let $s$ and $t$ be basic $APPTC$ terms, and $s \sim_{pp} t$, there are normal forms $n$ and $n'$, such that $s = n$ and $t = n'$. The soundness theorem of parallelism modulo probabilistic pomset bisimulation equivalence (see Theorem 11.19) yields $s \sim_{pp} n$ and $t \sim_{pp} n'$, hence, $n \sim_{pp} s \sim_{pp} t \sim_{pp} n'$. Since if $n \sim_{pp} n'$, then $n =_{AC} n'$, $s = n =_{AC} n' = t$, as desired. □

**Theorem 11.21** (Congruence of $APPTC$ with respect to probabilistic step bisimulation equivalence). *Probabilistic step bisimulation equivalence $\sim_{ps}$ is a congruence with respect to $APPTC$.*

*Proof.* It is easy to see that probabilistic step bisimulation is an equivalent relation on $APPTC$ terms, we only need to prove that $\sim_{ps}$ is preserved by the operators $\between$, $\parallel$, $\leftmerge$, $\mid$, $\Theta$, and $\triangleleft$. That is, if $x \sim_{ps} x'$ and $y \sim_{ps} y'$, we need to prove that $x \between y \sim_{ps} x' \between y'$, $x \parallel y \sim_{ps} x' \parallel y'$, $x \leftmerge y \sim_{ps} x' \leftmerge y'$, $x \mid y \sim_{ps} x' \mid y'$, $x \triangleleft y \sim_{ps} x' \triangleleft y'$, and $\Theta(x) \sim_{ps} \Theta(x')$. The proof is quite trivial, hence, we omit it here. □

**Theorem 11.22** (Soundness of parallelism modulo probabilistic step bisimulation equivalence). *Let $x$ and $y$ be $APPTC$ terms. If $APPTC \vdash x = y$, then $x \sim_{ps} y$.*

*Proof.* Since probabilistic step bisimulation $\sim_{ps}$ is both an equivalent and a congruent relation with respect to the operators $\between$, $\parallel$, $\leftmerge$, $\mid$, $\Theta$, and $\triangleleft$, we only need to check if each axiom in Table 11.6 is sound modulo probabilistic step bisimulation equivalence. The proof is quite trivial, hence, we omit it here. □

**Theorem 11.23** (Completeness of parallelism modulo probabilistic step bisimulation equivalence). *Let $p$ and $q$ be closed $APPTC$ terms, if $p \sim_{ps} q$, then $p = q$.*

*Proof.* First, by the elimination theorem of $APPTC$ (see Theorem 11.16), we know that for each closed $APPTC$ term $p$, there exists a closed basic $APPTC$ term $p'$, such that $APPTC \vdash p = p'$, hence, we only need to consider closed basic $APPTC$ terms.

The basic terms (see Definition 11.15) modulo associativity and commutativity (AC) of conflict $+$ (defined by axioms $A1$ and $A2$ in Table 11.1 and these equivalences is denoted by $=_{AC}$. Then, each equivalence class $s$ modulo AC of $+$ has the following normal form

$$s_1 \boxplus_{\pi_1} \cdots \boxplus_{\pi_{k-1}} s_k$$

with each $s_i$ having the following form

$$t_1 + \cdots + t_l$$

with each $t_j$ either an atomic event or of the form

$$u_1 \cdot \cdots \cdot u_m$$

with each $u_l$ either an atomic event or of the form

$$v_1 \leftmerge \cdots \leftmerge v_m$$

with each $v_m$ an atomic event, and each $t_j$ is called the summand of $s$.

Now, we prove that for normal forms $n$ and $n'$, if $n \sim_{ps} n'$, then $n =_{AC} n'$. It is sufficient to induct on the sizes of $n$ and $n'$.

- Consider a summand $e$ of $n$. Then, $n \rightsquigarrow \breve{e} \xrightarrow{e} \surd$, hence, $n \sim_{ps} n'$ implies $n' \rightsquigarrow \breve{e} \xrightarrow{e} \surd$, meaning that $n'$ also contains the summand $e$.
- Consider a summand $t_1 \cdot t_2$ of $n$,
  - if $t_1 \equiv e'$, then $n \rightsquigarrow \breve{e}' \xrightarrow{e'} t_2$, hence, $n \sim_{ps} n'$ implies $n' \rightsquigarrow \breve{e}' \xrightarrow{e'} t_2'$ with $t_2 \sim_{ps} t_2'$, meaning that $n'$ contains a summand $e' \cdot t_2'$. Since $t_2$ and $t_2'$ are normal forms and have sizes smaller than $n$ and $n'$, by the induction hypotheses if $t_2 \sim_{ps} t_2'$, then $t_2 =_{AC} t_2'$;

– if $t_1 \equiv e_1 \| \cdots \| e_m$, then $n \rightsquigarrow \breve{e_1} \| \cdots \| \breve{e_m} \xrightarrow{\{e_1,\cdots,e_m\}} t_2$, hence, $n \sim_{ps} n'$ implies $n' \rightsquigarrow \breve{e_1} \| \cdots \| \breve{e_m} \xrightarrow{\{e_1,\cdots,e_m\}} t_2'$ with $t_2 \sim_{ps} t_2'$, meaning that $n'$ contains a summand $(e_1 \| \cdots \| e_m) \cdot t_2'$. Since $t_2$ and $t_2'$ are normal forms and have sizes smaller than $n$ and $n'$, by the induction hypotheses if $t_2 \sim_{ps} t_2'$, then $t_2 =_{AC} t_2'$.

Hence, we obtain $n =_{AC} n'$.

Finally, let $s$ and $t$ be basic $APPTC$ terms, and $s \sim_{ps} t$, there are normal forms $n$ and $n'$, such that $s = n$ and $t = n'$. The soundness theorem of parallelism modulo probabilistic step bisimulation equivalence (see Theorem 11.22) yields $s \sim_{ps} n$ and $t \sim_{ps} n'$, hence, $n \sim_{ps} s \sim_{ps} t \sim_{ps} n'$. Since if $n \sim_{ps} n'$, then $n =_{AC} n'$, $s = n =_{AC} n' = t$, as desired. □

**Theorem 11.24** (Congruence of $APPTC$ with respect to probabilistic hp-bisimulation equivalence). *Probabilistic hp-bisimulation equivalence $\sim_{php}$ is a congruence with respect to $APPTC$.*

*Proof.* It is easy to see that probabilistic hp-bisimulation is an equivalent relation on $APPTC$ terms, we only need to prove that $\sim_{php}$ is preserved by the operators $\between$, $\|$, $\leftmerge$, $\mid$, $\Theta$, and $\triangleleft$. That is, if $x \sim_{php} x'$ and $y \sim_{php} y'$, we need to prove that $x \between y \sim_{php} x' \between y'$, $x \parallel y \sim_{php} x' \parallel y'$, $x \leftmerge y \sim_{php} x' \leftmerge y'$, $x \mid y \sim_{php} x' \mid y'$, $x \triangleleft y \sim_{php} x' \triangleleft y'$, and $\Theta(x) \sim_{php} \Theta(x')$. The proof is quite trivial, hence, we omit it here. □

**Theorem 11.25** (Soundness of parallelism modulo probabilistic hp-bisimulation equivalence). *Let $x$ and $y$ be $APPTC$ terms. If $APPTC \vdash x = y$, then $x \sim_{php} y$.*

*Proof.* Since probabilistic hp-bisimulation $\sim_{php}$ is both an equivalent and a congruent relation with respect to the operators $\between$, $\|$, $\leftmerge$, $\mid$, $\Theta$, and $\triangleleft$, we only need to check if each axiom in Table 11.6 is sound modulo probabilistic hp-bisimulation equivalence. The proof is quite trivial, hence, we omit it here. □

**Theorem 11.26** (Completeness of parallelism modulo probabilistic hp-bisimulation equivalence). *Let $p$ and $q$ be closed $APPTC$ terms, if $p \sim_{php} q$, then $p = q$.*

*Proof.* First, by the elimination theorem of $APPTC$ (see Theorem 11.16), we know that for each closed $APPTC$ term $p$, there exists a closed basic $APPTC$ term $p'$, such that $APPTC \vdash p = p'$, hence, we only need to consider closed basic $APPTC$ terms.

The basic terms (see Definition 11.15) modulo associativity and commutativity (AC) of conflict $+$ (defined by axioms $A1$ and $A2$ in Table 11.1 and these equivalences is denoted by $=_{AC}$. Then, each equivalence class $s$ modulo AC of $+$ has the following normal form

$$s_1 \boxplus_{\pi_1} \cdots \boxplus_{\pi_{k-1}} s_k$$

with each $s_i$ having the following form

$$t_1 + \cdots + t_l$$

with each $t_j$ either an atomic event or of the form

$$u_1 \cdot \cdots \cdot u_m$$

with each $u_l$ either an atomic event or of the form

$$v_1 \leftmerge \cdots \leftmerge v_m$$

with each $v_m$ an atomic event, and each $t_j$ is called the summand of $s$.

Now, we prove that for normal forms $n$ and $n'$, if $n \sim_{php} n'$, then $n =_{AC} n'$. It is sufficient to induct on the sizes of $n$ and $n'$.

- Consider a summand $e$ of $n$. Then, $n \rightsquigarrow \breve{e} \xrightarrow{e} \surd$, hence, $n \sim_{php} n'$ implies $n' \rightsquigarrow \breve{e} \xrightarrow{e} \surd$, meaning that $n'$ also contains the summand $e$.
- Consider a summand $t_1 \cdot t_2$ of $n$,
  - if $t_1 \equiv e'$, then $n \rightsquigarrow \breve{e'} \xrightarrow{e'} t_2$, hence, $n \sim_{php} n'$ implies $n' \rightsquigarrow \breve{e'} \xrightarrow{e'} t_2'$ with $t_2 \sim_{php} t_2'$, meaning that $n'$ contains a summand $e' \cdot t_2'$. Since $t_2$ and $t_2'$ are normal forms and have sizes smaller than $n$ and $n'$, by the induction hypotheses if $t_2 \sim_{php} t_2'$, then $t_2 =_{AC} t_2'$;
  - if $t_1 \equiv e_1 \leftmerge \cdots \leftmerge e_m$, then $n \rightsquigarrow \breve{e_1} \leftmerge \cdots \leftmerge \breve{e_m} \xrightarrow{\{e_1,\cdots,e_m\}} t_2$, hence, $n \sim_{php} n'$ implies $n' \rightsquigarrow \breve{e_1} \leftmerge \cdots \leftmerge \breve{e_m} \xrightarrow{\{e_1,\cdots,e_m\}} t_2'$ with $t_2 \sim_{php} t_2'$, meaning that $n'$ contains a summand $(e_1 \leftmerge \cdots \leftmerge e_m) \cdot t_2'$. Since $t_2$ and $t_2'$ are normal forms and have sizes smaller than $n$ and $n'$, by the induction hypotheses if $t_2 \sim_{php} t_2'$, then $t_2 =_{AC} t_2'$.

Hence, we obtain $n =_{AC} n'$.

Finally, let $s$ and $t$ be basic $APPTC$ terms, and $s \sim_{php} t$, there are normal forms $n$ and $n'$, such that $s = n$ and $t = n'$. The soundness theorem of parallelism modulo probabilistic hp-bisimulation equivalence (see Theorem 11.25) yields $s \sim_{php} n$ and $t \sim_{php} n'$, hence, $n \sim_{php} s \sim_{php} t \sim_{php} n'$. Since if $n \sim_{php} n'$, then $n =_{AC} n'$, $s = n =_{AC} n' = t$, as desired. □

**Theorem 11.27** (Congruence of $APPTC$ with respect to probabilistic hhp-bisimulation equivalence). *Probabilistic hhp-bisimulation equivalence $\sim_{phhp}$ is a congruence with respect to $APPTC$.*

*Proof.* It is easy to see that probabilistic hhp-bisimulation is an equivalent relation on $APPTC$ terms, we only need to prove that $\sim_{phhp}$ is preserved by the operators $\between$, $\parallel$, $\Lleftarrow$, $\mid$, $\Theta$, and $\triangleleft$. That is, if $x \sim_{phhp} x'$ and $y \sim_{phhp} y'$, we need to prove that $x \between y \sim_{phhp} x' \between y'$, $x \parallel y \sim_{phhp} x' \parallel y'$, $x \Lleftarrow y \sim_{phhp} x' \Lleftarrow y'$, $x \mid y \sim_{phhp} x' \mid y'$, $x \triangleleft y \sim_{phhp} x' \triangleleft y'$, and $\Theta(x) \sim_{phhp} \Theta(x')$. The proof is quite trivial, hence, we omit it here. □

**Theorem 11.28** (Soundness of parallelism modulo probabilistic hhp-bisimulation equivalence). *Let $x$ and $y$ be $APPTC$ terms. If $APPTC \vdash x = y$, then $x \sim_{phhp} y$.*

*Proof.* Since probabilistic hhp-bisimulation $\sim_{phhp}$ is both an equivalent and a congruent relation with respect to the operators $\between$, $\parallel$, $\Lleftarrow$, $\mid$, $\Theta$, and $\triangleleft$, we only need to check if each axiom in Table 11.6 is sound modulo probabilistic hhp-bisimulation equivalence. The proof is quite trivial, hence, we omit it here. □

**Theorem 11.29** (Completeness of parallelism modulo probabilistic hhp-bisimulation equivalence). *Let $p$ and $q$ be closed $APPTC$ terms, if $p \sim_{phhp} q$, then $p = q$.*

*Proof.* First, by the elimination theorem of $APPTC$ (see Theorem 11.16), we know that for each closed $APPTC$ term $p$, there exists a closed basic $APPTC$ term $p'$, such that $APPTC \vdash p = p'$, hence, we only need to consider closed basic $APPTC$ terms.

The basic terms (see Definition 11.15) modulo associativity and commutativity (AC) of conflict $+$ (defined by axioms $A1$ and $A2$ in Table 11.1 and these equivalences is denoted by $=_{AC}$. Then, each equivalence class $s$ modulo AC of $+$ has the following normal form

$$s_1 \boxplus_{\pi_1} \cdots \boxplus_{\pi_{k-1}} s_k$$

with each $s_i$ having the following form

$$t_1 + \cdots + t_l$$

with each $t_j$ either an atomic event or of the form

$$u_1 \cdot \cdots \cdot u_m$$

with each $u_l$ either an atomic event or of the form

$$v_1 \Lleftarrow \cdots \Lleftarrow v_m$$

with each $v_m$ an atomic event, and each $t_j$ is called the summand of $s$.

Now, we prove that for normal forms $n$ and $n'$, if $n \sim_{phhp} n'$, then $n =_{AC} n'$. It is sufficient to induct on the sizes of $n$ and $n'$.

- Consider a summand $e$ of $n$. Then, $n \rightsquigarrow \breve{e} \xrightarrow{e} \surd$, hence, $n \sim_{phhp} n'$ implies $n' \rightsquigarrow \breve{e} \xrightarrow{e} \surd$, meaning that $n'$ also contains the summand $e$.
- Consider a summand $t_1 \cdot t_2$ of $n$,
  - if $t_1 \equiv e'$, then $n \rightsquigarrow \breve{e'} \xrightarrow{e'} t_2$, hence, $n \sim_{phhp} n'$ implies $n' \rightsquigarrow \breve{e'} \xrightarrow{e'} t_2'$ with $t_2 \sim_{phhp} t_2'$, meaning that $n'$ contains a summand $e' \cdot t_2'$. Since $t_2$ and $t_2'$ are normal forms and have sizes smaller than $n$ and $n'$, by the induction hypotheses if $t_2 \sim_{phhp} t_2'$, then $t_2 =_{AC} t_2'$;
  - if $t_1 \equiv e_1 \Lleftarrow \cdots \Lleftarrow e_m$, then $n \rightsquigarrow \breve{e}_1 \Lleftarrow \cdots \Lleftarrow \breve{e}_m \xrightarrow{\{e_1,\cdots,e_m\}} t_2$, hence, $n \sim_{phhp} n'$ implies $n' \rightsquigarrow \breve{e}_1 \Lleftarrow \cdots \Lleftarrow \breve{e}_m \xrightarrow{\{e_1,\cdots,e_m\}} t_2'$ with $t_2 \sim_{phhp} t_2'$, meaning that $n'$ contains a summand $(e_1 \Lleftarrow \cdots \Lleftarrow e_m) \cdot t_2'$. Since $t_2$ and $t_2'$ are normal forms and have sizes smaller than $n$ and $n'$, by the induction hypotheses if $t_2 \sim_{phhp} t_2'$, then $t_2 =_{AC} t_2'$.

**TABLE 11.11 PDF definitions of $\partial_H$.**

| |
|---|
| $\mu(\partial_H(x), \partial_H(x')) = \mu(x, x')$ |
| $\mu(x, y) = 0$, otherwise |

**TABLE 11.12 Transition rules of encapsulation operator $\partial_H$.**

$$\frac{x \rightsquigarrow x'}{\partial_H(x) \rightsquigarrow \partial_H(x')}$$

$$\frac{x \xrightarrow{e} \surd}{\partial_H(x) \xrightarrow{e} \surd} \quad (e \notin H) \qquad \frac{x \xrightarrow{e} x'}{\partial_H(x) \xrightarrow{e} \partial_H(x')} \quad (e \notin H)$$

**TABLE 11.13 Axioms of encapsulation operator.**

| No. | Axiom |
|---|---|
| $D1$ | $e \notin H \quad \partial_H(e) = e$ |
| $D2$ | $e \in H \quad \partial_H(e) = \delta$ |
| $D3$ | $\partial_H(\delta) = \delta$ |
| $D4$ | $\partial_H(x + y) = \partial_H(x) + \partial_H(y)$ |
| $D5$ | $\partial_H(x \cdot y) = \partial_H(x) \cdot \partial_H(y)$ |
| $D6$ | $\partial_H(x \between y) = \partial_H(x) \between \partial_H(y)$ |
| $PD1$ | $\partial_H(x \boxplus_\pi y) = \partial_H(x) \boxplus_\pi \partial_H(y)$ |

Hence, we obtain $n =_{AC} n'$.

Finally, let $s$ and $t$ be basic $APPTC$ terms, and $s \sim_{phhp} t$, there are normal forms $n$ and $n'$, such that $s = n$ and $t = n'$. The soundness theorem of parallelism modulo probabilistic hhp-bisimulation equivalence (see Theorem 11.28) yields $s \sim_{phhp} n$ and $t \sim_{phhp} n'$, hence, $n \sim_{phhp} s \sim_{phhp} t \sim_{phhp} n'$. Since if $n \sim_{phhp} n'$, then $n =_{AC} n'$, $s = n =_{AC} n' = t$, as desired. □

### 11.2.3 Encapsulation

The mismatch of two communicating events in different parallel branches can cause deadlock, hence, the deadlocks in the concurrent processes should be eliminated. Like $ACP$ [3], we also introduce the unary encapsulation operator $\partial_H$ for set $H$ of atomic events, which renames all atomic events in $H$ into $\delta$. The whole algebra including parallelism for true concurrency in the above sections, deadlock $\delta$ and encapsulation operator $\partial_H$, is also called Algebra for Parallelism in Probabilistic True Concurrency, abbreviated $APPTC$.

First, we give the definition of PDFs in Table 11.11.

The transition rules of encapsulation operator $\partial_H$ are shown in Table 11.12.

Based on the transition rules for encapsulation operator $\partial_H$ in Table 11.12, we design the axioms as Table 11.13 shows.

The axioms $D1 - D3$ are the defining laws for the encapsulation operator $\partial_H$, $D1$ leaves atomic events outside $H$ unchanged, $D2$ renames atomic events in $H$ into $\delta$, and $D3$ says that it leaves $\delta$ unchanged. $D4 - D6$ and $PD1$ say that in term $\partial_H(t)$, all transitions of $t$ labeled with atomic events in $H$ are blocked.

**Theorem 11.30** (Conservativity of $APPTC$ with respect to the algebra for parallelism). *$APPTC$ is a conservative extension of the algebra for parallelism.*

*Proof.* It follows from the following two facts (see Theorem 2.5):

1. The transition rules of the algebra for parallelism in the above sections are all source dependent;
2. The sources of the transition rules for the encapsulation operator contain an occurrence of $\partial_H$.

Hence, $APPTC$ is a conservative extension of the algebra for parallelism, as desired. □

**TABLE 11.14** Term rewrite system of encapsulation operator $\partial_H$.

| No. | Rewriting Rule |
|---|---|
| $RD1$ | $e \notin H \quad \partial_H(e) \rightarrow e$ |
| $RD2$ | $e \in H \quad \partial_H(e) \rightarrow \delta$ |
| $RD3$ | $\partial_H(\delta) \rightarrow \delta$ |
| $RD4$ | $\partial_H(x+y) \rightarrow \partial_H(x) + \partial_H(y)$ |
| $RD5$ | $\partial_H(x \cdot y) \rightarrow \partial_H(x) \cdot \partial_H(y)$ |
| $RD6$ | $\partial_H(x \between y) \rightarrow \partial_H(x) \between \partial_H(y)$ |
| $RPD1$ | $\partial_H(x \boxplus_\pi y) \rightarrow \partial_H(x) \boxplus_\pi \partial_H(y)$ |

**Theorem 11.31** (Elimination theorem of $APPTC$). *Let $p$ be a closed $APPTC$ term including the encapsulation operator $\partial_H$. Then, there is a basic $APPTC$ term $q$ such that $APPTC \vdash p = q$.*

*Proof.* (1) First, suppose that the following ordering on the signature of $APPTC$ is defined: $\between > \cdot > + > \boxplus_\pi$ and the symbol $\cdot$ is given the lexicographical status for the first argument, then for each rewrite rule $p \rightarrow q$ in Table 11.14 relation $p >_{lpo} q$ can easily be proved. We obtain that the term rewrite system shown in Table 11.14 is strongly normalizing, for it has finitely many rewriting rules, and $>$ is a well-founded ordering on the signature of $APPTC$, and if $s >_{lpo} t$, for each rewriting rule $s \rightarrow t$ is in Table 11.14 (see Theorem 2.9).

(2) Then, we prove that the normal forms of closed $APPTC$ terms including encapsulation operator $\partial_H$ are basic $APPTC$ terms.

Suppose that $p$ is a normal form of some closed $APPTC$ term and suppose that $p$ is not a basic $APPTC$ term. Let $p'$ denote the smallest sub-term of $p$ that is not a basic $APPTC$ term. It implies that each sub-term of $p'$ is a basic $APPTC$ term. Then, we prove that $p$ is not a term in normal form. It is sufficient to induct on the structure of $p'$, following from Theorem 11.16, we only prove the new case $p' \equiv \partial_H(p_1)$:

- Case $p_1 \equiv e$. The transition rules $RD1$ or $RD2$ can be applied, hence, $p$ is not a normal form;
- Case $p_1 \equiv \delta$. The transition rule $RD3$ can be applied, hence, $p$ is not a normal form;
- Case $p_1 \equiv p_1' + p_1''$. The transition rule $RD4$ can be applied, hence, $p$ is not a normal form;
- Case $p_1 \equiv p_1' \cdot p_1''$. The transition rule $RD5$ can be applied, hence, $p$ is not a normal form;
- Case $p_1 \equiv p_1' \between p_1''$. The transition rule $RD6$ can be applied, hence, $p$ is not a normal form;
- Case $p_1 \equiv p_1' \boxplus_\pi p_1''$. The transition rule $RPD1$ can be applied, hence, $p$ is not a normal form. □

**Theorem 11.32** (Congruence theorem of encapsulation operator $\partial_H$ with respect to probabilistic pomset bisimulation equivalence). *Probabilistic pomset bisimulation equivalence $\sim_{pp}$ is a congruence with respect to encapsulation operator $\partial_H$.*

*Proof.* It is easy to see that probabilistic pomset bisimulation is an equivalent relation on $APPTC$ terms, we only need to prove that $\sim_{pp}$ is preserved by the operators $\partial_H$. That is, if $x \sim_{pp} x'$, we need to prove that $\partial_H(x) \sim_{pp} \partial_H(x')$. The proof is quite trivial, hence, we omit it here. □

**Theorem 11.33** (Soundness of $APPTC$ modulo probabilistic pomset bisimulation equivalence). *Let $x$ and $y$ be $APPTC$ terms including encapsulation operator $\partial_H$. If $APPTC \vdash x = y$, then $x \sim_{pp} y$.*

*Proof.* Since probabilistic pomset bisimulation $\sim_{pp}$ is both an equivalent and a congruent relation with respect to the operator $\partial_H$, we only need to check if each axiom in Table 11.13 is sound modulo probabilistic pomset bisimulation equivalence. The proof is quite trivial, hence, we omit it here. □

**Theorem 11.34** (Completeness of $APPTC$ modulo probabilistic pomset bisimulation equivalence). *Let $p$ and $q$ be closed $APPTC$ terms including encapsulation operator $\partial_H$, if $p \sim_{pp} q$, then $p = q$.*

*Proof.* First, by the elimination theorem of $APPTC$ (see Theorem 11.31), we know that the normal form of $APPTC$ does not contain $\partial_H$, and for each closed $APPTC$ term $p$, there exists a closed basic $APPTC$ term $p'$, such that $APPTC \vdash p = p'$, hence, we only need to consider closed basic $APPTC$ terms.

Similarly to Theorem 11.20, we can prove that for normal forms $n$ and $n'$, if $n \sim_{pp} n'$, then $n =_{AC} n'$.

Finally, let $s$ and $t$ be basic $APPTC$ terms, and $s \sim_{pp} t$, there are normal forms $n$ and $n'$, such that $s = n$ and $t = n'$. The soundness theorem of $APPTC$ modulo probabilistic pomset bisimulation equivalence (see Theorem 11.33) yields $s \sim_{pp} n$ and $t \sim_{pp} n'$, hence, $n \sim_{pp} s \sim_{pp} t \sim_{pp} n'$. Since if $n \sim_{pp} n'$, then $n =_{AC} n'$, $s = n =_{AC} n' = t$, as desired. □

**Theorem 11.35** (Congruence theorem of encapsulation operator $\partial_H$ with respect to probabilistic step bisimulation equivalence). *Probabilistic step bisimulation equivalence $\sim_{ps}$ is a congruence with respect to encapsulation operator $\partial_H$.*

*Proof.* It is easy to see that probabilistic step bisimulation is an equivalent relation on $APPTC$ terms, we only need to prove that $\sim_{ps}$ is preserved by the operators $\partial_H$. That is, if $x \sim_{ps} x'$, we need to prove that $\partial_H(x) \sim_{ps} \partial_H(x')$. The proof is quite trivial, hence, we omit it here. □

**Theorem 11.36** (Soundness of $APPTC$ modulo probabilistic step bisimulation equivalence). *Let $x$ and $y$ be $APPTC$ terms including encapsulation operator $\partial_H$. If $APPTC \vdash x = y$, then $x \sim_{ps} y$.*

*Proof.* Since probabilistic step bisimulation $\sim_{ps}$ is both an equivalent and a congruent relation with respect to the operator $\partial_H$, we only need to check if each axiom in Table 11.13 is sound modulo probabilistic step bisimulation equivalence. The proof is quite trivial, hence, we omit it here. □

**Theorem 11.37** (Completeness of $APPTC$ modulo probabilistic step bisimulation equivalence). *Let $p$ and $q$ be closed $APPTC$ terms including encapsulation operator $\partial_H$, if $p \sim_{ps} q$, then $p = q$.*

*Proof.* First, by the elimination theorem of $APPTC$ (see Theorem 11.31), we know that the normal form of $APPTC$ does not contain $\partial_H$, and for each closed $APPTC$ term $p$, there exists a closed basic $APPTC$ term $p'$, such that $APPTC \vdash p = p'$, hence, we only need to consider closed basic $APPTC$ terms.

Similarly to Theorem 11.23, we can prove that for normal forms $n$ and $n'$, if $n \sim_{ps} n'$, then $n =_{AC} n'$.

Finally, let $s$ and $t$ be basic $APPTC$ terms, and $s \sim_{ps} t$, there are normal forms $n$ and $n'$, such that $s = n$ and $t = n'$. The soundness theorem of $APPTC$ modulo probabilistic step bisimulation equivalence (see Theorem 11.36) yields $s \sim_{ps} n$ and $t \sim_{ps} n'$, hence, $n \sim_{ps} s \sim_{ps} t \sim_{ps} n'$. Since if $n \sim_{ps} n'$, then $n =_{AC} n'$, $s = n =_{AC} n' = t$, as desired. □

**Theorem 11.38** (Congruence theorem of encapsulation operator $\partial_H$ with respect to probabilistic hp-bisimulation equivalence). *Probabilistic hp-bisimulation equivalence $\sim_{php}$ is a congruence with respect to encapsulation operator $\partial_H$.*

*Proof.* It is easy to see that probabilistic hp-bisimulation is an equivalent relation on $APPTC$ terms, we only need to prove that $\sim_{php}$ is preserved by the operators $\partial_H$. That is, if $x \sim_{php} x'$, we need to prove that $\partial_H(x) \sim_{php} \partial_H(x')$. The proof is quite trivial, hence, we omit it here. □

**Theorem 11.39** (Soundness of $APPTC$ modulo probabilistic hp-bisimulation equivalence). *Let $x$ and $y$ be $APPTC$ terms including encapsulation operator $\partial_H$. If $APPTC \vdash x = y$, then $x \sim_{php} y$.*

*Proof.* Since probabilistic hp-bisimulation $\sim_{php}$ is both an equivalent and a congruent relation with respect to the operator $\partial_H$, we only need to check if each axiom in Table 11.13 is sound modulo probabilistic hp-bisimulation equivalence. The proof is quite trivial, hence, we omit it here. □

**Theorem 11.40** (Completeness of $APPTC$ modulo probabilistic hp-bisimulation equivalence). *Let $p$ and $q$ be closed $APPTC$ terms including encapsulation operator $\partial_H$, if $p \sim_{php} q$, then $p = q$.*

*Proof.* First, by the elimination theorem of $APPTC$ (see Theorem 11.31), we know that the normal form of $APPTC$ does not contain $\partial_H$, and for each closed $APPTC$ term $p$, there exists a closed basic $APPTC$ term $p'$, such that $APPTC \vdash p = p'$, hence, we only need to consider closed basic $APPTC$ terms.

Similarly to Theorem 11.26, we can prove that for normal forms $n$ and $n'$, if $n \sim_{php} n'$, then $n =_{AC} n'$.

Finally, let $s$ and $t$ be basic $APPTC$ terms, and $s \sim_{php} t$, there are normal forms $n$ and $n'$, such that $s = n$ and $t = n'$. The soundness theorem of $APPTC$ modulo probabilistic hp-bisimulation equivalence (see Theorem 11.39) yields $s \sim_{php} n$ and $t \sim_{php} n'$, hence, $n \sim_{php} s \sim_{php} t \sim_{php} n'$. Since if $n \sim_{php} n'$, then $n =_{AC} n'$, $s = n =_{AC} n' = t$, as desired. □

**Theorem 11.41** (Congruence theorem of encapsulation operator $\partial_H$ with respect to probabilistic hhp-bisimulation equivalence). *Probabilistic hhp-bisimulation equivalence $\sim_{phhp}$ is a congruence with respect to encapsulation operator $\partial_H$.*

*Proof.* It is easy to see that probabilistic hhp-bisimulation is an equivalent relation on $APPTC$ terms, we only need to prove that $\sim_{phhp}$ is preserved by the operators $\partial_H$. That is, if $x \sim_{phhp} x'$, we need to prove that $\partial_H(x) \sim_{phhp} \partial_H(x')$. The proof is quite trivial, hence, we omit it here. □

**Theorem 11.42** (Soundness of $APPTC$ modulo probabilistic hhp-bisimulation equivalence). *Let $x$ and $y$ be $APPTC$ terms including encapsulation operator $\partial_H$. If $APPTC \vdash x = y$, then $x \sim_{phhp} y$.*

*Proof.* Since probabilistic hhp-bisimulation $\sim_{phhp}$ is both an equivalent and a congruent relation with respect to the operator $\partial_H$, we only need to check if each axiom in Table 11.13 is sound modulo probabilistic hhp-bisimulation equivalence. The proof is quite trivial, hence, we omit it here. □

**Theorem 11.43** (Completeness of $APPTC$ modulo probabilistic hhp-bisimulation equivalence). *Let $p$ and $q$ be closed $APPTC$ terms including encapsulation operator $\partial_H$, if $p \sim_{phhp} q$, then $p = q$.*

*Proof.* First, by the elimination theorem of $APPTC$ (see Theorem 11.31), we know that the normal form of $APPTC$ does not contain $\partial_H$, and for each closed $APPTC$ term $p$, there exists a closed basic $APPTC$ term $p'$, such that $APPTC \vdash p = p'$, hence, we only need to consider closed basic $APPTC$ terms.

Similarly to Theorem 11.29, we can prove that for normal forms $n$ and $n'$, if $n \sim_{phhp} n'$, then $n =_{AC} n'$.

Finally, let $s$ and $t$ be basic $APPTC$ terms, and $s \sim_{phhp} t$, there are normal forms $n$ and $n'$, such that $s = n$ and $t = n'$. The soundness theorem of $APPTC$ modulo probabilistic hhp-bisimulation equivalence (see Theorem 11.42) yields $s \sim_{phhp} n$ and $t \sim_{phhp} n'$, hence, $n \sim_{phhp} s \sim_{phhp} t \sim_{phhp} n'$. Since if $n \sim_{phhp} n'$, then $n =_{AC} n'$, $s = n =_{AC} n' = t$, as desired. □

## 11.3 Recursion

In this section, we introduce recursion to capture infinite processes based on $APPTC$. Since in $APPTC$, there are four basic operators $\cdot$, $+$, $\boxplus_\pi$, and $\between$, the recursion must be adapted this situation to include $\boxplus_\pi$ and $\between$.

In the following, $E, F, G$ are recursion specifications, $X, Y, Z$ are recursive variables.

### 11.3.1 Guarded recursive specifications

**Definition 11.44** (Guarded recursive specification). A recursive specification

$$X_1 = t_1(X_1, \cdots, X_n)$$

$$\cdots$$

$$X_n = t_n(X_1, \cdots, X_n)$$

is guarded if the right-hand sides of its recursive equations can be adapted to the form by applications of the axioms in $APPTC$ and replacing recursion variables by the right-hand sides of their recursive equations,

$((a_{111} \between \cdots \between a_{11i_1}) \cdot s_1(X_1, \cdots, X_n) + \cdots + (a_{1k1} \between \cdots \between a_{1ki_k}) \cdot s_k(X_1, \cdots, X_n) + (b_{111} \between \cdots \between b_{11j_1}) + \cdots + (b_{11j_1} \between \cdots \between b_{1lj_l})) \boxplus_{\pi_1} \cdots \boxplus_{\pi_{m-1}} ((a_{m11} \between \cdots \between a_{m1i_1}) \cdot s_1(X_1, \cdots, X_n) + \cdots + (a_{mk1} \between \cdots \between a_{mki_k}) \cdot s_k(X_1, \cdots, X_n) + (b_{m11} \between \cdots \between b_{m1j_1}) + \cdots + (b_{m1j_1} \between \cdots \between b_{mlj_l}))$

where $a_{111}, \cdots, a_{11i_1}, a_{1k1}, \cdots, a_{1ki_k}, b_{111}, \cdots, b_{11j_1}, b_{11j_1}, \cdots, b_{1lj_l}, \cdots, a_{m11}, \cdots, a_{m1i_1}, a_{mk1}, \cdots, a_{mki_k}, b_{m11}, \cdots, b_{m1j_1}, b_{m1j_1}, \cdots, b_{mlj_l} \in \mathbb{E}$, and the sum above is allowed to be empty, in which case it represents the deadlock $\delta$.

**Definition 11.45** (Linear recursive specification). A recursive specification is linear if its recursive equations are of the form

$((a_{111} \between \cdots \between a_{11i_1})X_1 + \cdots + (a_{1k1} \between \cdots \between a_{1ki_k})X_k + (b_{111} \between \cdots \between b_{11j_1}) + \cdots + (b_{11j_1} \between \cdots \between b_{1lj_l})) \boxplus_{\pi_1} \cdots \boxplus_{\pi_{m-1}} ((a_{m11} \between \cdots \between a_{m1i_1})X_1 + \cdots + (a_{mk1} \between \cdots \between a_{mki_k})X_k + (b_{m11} \between \cdots \between b_{m1j_1}) + \cdots + (b_{m1j_1} \between \cdots \between b_{mlj_l}))$

where $a_{111}, \cdots, a_{11i_1}, a_{1k1}, \cdots, a_{1ki_k}, b_{111}, \cdots, b_{11j_1}, b_{11j_1}, \cdots, b_{1lj_l}, \cdots, a_{m11}, \cdots, a_{m1i_1}, a_{mk1}, \cdots, a_{mki_k}, b_{m11}, \cdots, b_{m1j_1}, b_{m1j_1}, \cdots, b_{mlj_l} \in \mathbb{E}$, and the sum above is allowed to be empty, in which case it represents the deadlock $\delta$.

First, we give the definition of PDFs in Table 11.15.

For a guarded recursive specifications $E$ with the form

$$X_1 = t_1(X_1, \cdots, X_n)$$

$$\cdots$$

**TABLE 11.15 PDF definitions of recursion.**

| |
|---|
| $\mu(\langle X\|E\rangle, y) = \mu(\langle t_X\|E\rangle, y)$ |
| $\mu(x, y) = 0$, otherwise |

**TABLE 11.16 Transition rules of guarded recursion.**

| |
|---|
| $\dfrac{t_i(\langle X_1\|E\rangle, \cdots, \langle X_n\|E\rangle) \rightsquigarrow y}{\langle X_i\|E\rangle \rightsquigarrow y}$ |
| $\dfrac{t_i(\langle X_1\|E\rangle, \cdots, \langle X_n\|E\rangle) \xrightarrow{\{e_1,\cdots,e_k\}} \surd}{\langle X_i\|E\rangle \xrightarrow{\{e_1,\cdots,e_k\}} \surd}$ |
| $\dfrac{t_i(\langle X_1\|E\rangle, \cdots, \langle X_n\|E\rangle) \xrightarrow{\{e_1,\cdots,e_k\}} y}{\langle X_i\|E\rangle \xrightarrow{\{e_1,\cdots,e_k\}} y}$ |

**TABLE 11.17 Recursive definition and specification principle.**

| No. | Axiom |
|---|---|
| $RDP$ | $\langle X_i\|E\rangle = t_i(\langle X_1\|E\rangle, \cdots, \langle X_n\|E\rangle) \quad (i \in \{1, \cdots, n\})$ |
| $RSP$ | if $y_i = t_i(y_1, \cdots, y_n)$ for $i \in \{1, \cdots, n\}$, then $y_i = \langle X_i\|E\rangle \quad (i \in \{1, \cdots, n\})$ |

$$X_n = t_n(X_1, \cdots, X_n)$$

the behavior of the solution $\langle X_i|E\rangle$ for the recursion variable $X_i$ in $E$, where $i \in \{1, \cdots, n\}$, is exactly the behavior of their right-hand sides $t_i(X_1, \cdots, X_n)$, which is captured by the two transition rules in Table 11.16.

**Theorem 11.46** (Conservativity of $APPTC$ with guarded recursion). *$APPTC$ with guarded recursion is a conservative extension of $APPTC$.*

*Proof.* Since the transition rules of $APPTC$ are source dependent, and the transition rules for guarded recursion in Table 11.16 contain only a fresh constant in their source, hence, the transition rules of $APPTC$ with guarded recursion are a conservative extension of those of $APPTC$. □

**Theorem 11.47** (Congruence theorem of $APPTC$ with guarded recursion). *Probabilistic truly concurrent bisimulation equivalences $\sim_{pp}$, $\sim_{ps}$, $\sim_{php}$, and $\sim_{phhp}$ are all congruences with respect to $APPTC$ with guarded recursion.*

*Proof.* It follows from the following two facts:

1. In a guarded recursive specification, the right-hand sides of its recursive equations can be adapted to the form by applications of the axioms in $APPTC$ and replacing recursion variables by the right-hand sides of their recursive equations;
2. Truly concurrent bisimulation equivalences $\sim_{pp}$, $\sim_{ps}$, $\sim_{php}$, and $\sim_{phhp}$ are all congruences with respect to all operators of $APPTC$. □

### 11.3.2 Recursive definition and specification principles

The $RDP$ (Recursive Definition Principle) and the $RSP$ (Recursive Specification Principle) are shown in Table 11.17.

$RDP$ follows immediately from the two transition rules for guarded recursion, which express that $\langle X_i|E\rangle$ and $t_i(\langle X_1|E\rangle, \cdots, \langle X_n|E\rangle)$ have the same initial transitions for $i \in \{1, \cdots, n\}$. $RSP$ follows from the fact that guarded recursive specifications have only one solution.

**Theorem 11.48** (Elimination theorem of $APPTC$ with linear recursion). *Each process term in $APPTC$ with linear recursion is equal to a process term $\langle X_1|E\rangle$ with $E$ a linear recursive specification.*

*Proof.* By applying structural induction with respect to term size, each process term $t_1$ in $APPTC$ with linear recursion generates a process that can be expressed in the form of equations

$t_i = ((a_{1i11} \leftmerge \cdots \leftmerge a_{1i1i_1})t_{i1} + \cdots + (a_{1ik_i1} \leftmerge \cdots \leftmerge a_{1ik_ii_k})t_{ik_i} + (b_{1i11} \leftmerge \cdots \leftmerge b_{1i1i_1}) + \cdots + (b_{1il_i1} \leftmerge \cdots \leftmerge b_{1il_ii_l})) \boxplus_{\pi_1} \cdots \boxplus_{\pi_{m-1}} ((a_{mi11} \leftmerge \cdots \leftmerge a_{mi1i_1})t_{i1} + \cdots + (a_{mik_i1} \leftmerge \cdots \leftmerge a_{mik_ii_k})t_{ik_i} + (b_{mi11} \leftmerge \cdots \leftmerge b_{mi1i_1}) + \cdots + (b_{mil_i1} \leftmerge \cdots \leftmerge b_{mil_ii_l}))$

for $i \in \{1, \cdots, n\}$. Let the linear recursive specification $E$ consist of the recursive equations

$X_i = ((a_{1i11} \leftmerge \cdots \leftmerge a_{1i1i_1})X_{i1} + \cdots + (a_{1ik_i1} \leftmerge \cdots \leftmerge a_{1ik_ii_k})X_{ik_i} + (b_{1i11} \leftmerge \cdots \leftmerge b_{1i1i_1}) + \cdots + (b_{1il_i1} \leftmerge \cdots \leftmerge b_{1il_ii_l})) \boxplus_{\pi_1} \cdots \boxplus_{\pi_{m-1}} ((a_{mi11} \leftmerge \cdots \leftmerge a_{mi1i_1})X_{i1} + \cdots + (a_{mik_i1} \leftmerge \cdots \leftmerge a_{mik_ii_k})X_{ik_i} + (b_{mi11} \leftmerge \cdots \leftmerge b_{mi1i_1}) + \cdots + (b_{mil_i1} \leftmerge \cdots \leftmerge b_{mil_ii_l}))$

for $i \in \{1, \cdots, n\}$. Replacing $X_i$ by $t_i$ for $i \in \{1, \cdots, n\}$ is a solution for $E$, $RSP$ yields $t_1 = \langle X_1|E\rangle$. □

**Theorem 11.49** (Soundness of $APPTC$ with guarded recursion). *Let $x$ and $y$ be $APPTC$ with guarded recursion terms. If $APPTC$ with guarded recursion $\vdash x = y$, then*

1. $x \sim_{pp} y$;
2. $x \sim_{ps} y$;
3. $x \sim_{php} y$;
4. $x \sim_{phhp} y$.

*Proof.* Since $\sim_{pp}$, $\sim_{ps}$, $\sim_{php}$, and $\sim_{phhp}$ are all both an equivalent and a congruent relation with respect to $APPTC$ with guarded recursion, we only need to check if each axiom in Table 11.17 is sound modulo $\sim_{pp}$, $\sim_{ps}$, $\sim_{php}$, and $\sim_{phhp}$. The proof is quite trivial, hence, we omit it here. □

**Theorem 11.50** (Completeness of $APPTC$ with linear recursion). *Let $p$ and $q$ be closed $APPTC$ with linear recursion terms, then*

1. *if $p \sim_{pp} q$, then $p = q$;*
2. *if $p \sim_{ps} q$, then $p = q$;*
3. *if $p \sim_{php} q$, then $p = q$;*
4. *if $p \sim_{phhp} q$, then $p = q$.*

*Proof.* First, by the elimination theorem of $APPTC$ with guarded recursion (see Theorem 11.48), we know that each process term in $APPTC$ with linear recursion is equal to a process term $\langle X_1|E\rangle$ with $E$ a linear recursive specification.

It remains to prove the following cases.

(1) If $\langle X_1|E_1\rangle \sim_{pp} \langle Y_1|E_2\rangle$ for linear recursive specification $E_1$ and $E_2$, then $\langle X_1|E_1\rangle = \langle Y_1|E_2\rangle$.

Let $E_1$ consist of recursive equations $X = t_X$ for $X \in \mathcal{X}$ and $E_2$ consists of recursion equations $Y = t_Y$ for $Y \in \mathcal{Y}$. Let the linear recursive specification $E$ consist of recursion equations $Z_{XY} = t_{XY}$, and $\langle X|E_1\rangle \sim_{pp} \langle Y|E_2\rangle$, and $t_{XY}$ consists of the following summands:

1. $t_{XY}$ contains a summand $(a_1 \leftmerge \cdots \leftmerge a_m)Z_{X'Y'}$ iff $t_X$ contains the summand $(a_1 \leftmerge \cdots \leftmerge a_m)X'$ and $t_Y$ contains the summand $(a_1 \leftmerge \cdots \leftmerge a_m)Y'$ such that $\langle X'|E_1\rangle \sim_{pp} \langle Y'|E_2\rangle$;
2. $t_{XY}$ contains a summand $b_1 \leftmerge \cdots \leftmerge b_n$ iff $t_X$ contains the summand $b_1 \leftmerge \cdots \leftmerge b_n$ and $t_Y$ contains the summand $b_1 \leftmerge \cdots \leftmerge b_n$.

Let $\sigma$ map recursion variable $X$ in $E_1$ to $\langle X|E_1\rangle$, and let $\psi$ map recursion variable $Z_{XY}$ in $E$ to $\langle X|E_1\rangle$. Hence, $\sigma((a_1 \leftmerge \cdots \leftmerge a_m)X') \equiv (a_1 \leftmerge \cdots \leftmerge a_m)\langle X'|E_1\rangle \equiv \psi((a_1 \leftmerge \cdots \leftmerge a_m)Z_{X'Y'})$, hence, by $RDP$, we obtain $\langle X|E_1\rangle = \sigma(t_X) = \psi(t_{XY})$. Then, by $RSP$, $\langle X|E_1\rangle = \langle Z_{XY}|E\rangle$, particularly, $\langle X_1|E_1\rangle = \langle Z_{X_1Y_1}|E\rangle$. Similarly, we can obtain $\langle Y_1|E_2\rangle = \langle Z_{X_1Y_1}|E\rangle$. Finally, $\langle X_1|E_1\rangle = \langle Z_{X_1Y_1}|E\rangle = \langle Y_1|E_2\rangle$, as desired.

(2) If $\langle X_1|E_1\rangle \sim_{ps} \langle Y_1|E_2\rangle$ for linear recursive specification $E_1$ and $E_2$, then $\langle X_1|E_1\rangle = \langle Y_1|E_2\rangle$.

It can be proven similarly to (1), hence, we omit it here.

(3) If $\langle X_1|E_1\rangle \sim_{php} \langle Y_1|E_2\rangle$ for linear recursive specification $E_1$ and $E_2$, then $\langle X_1|E_1\rangle = \langle Y_1|E_2\rangle$.

It can be proven similarly to (1), hence, we omit it here.

(4) If $\langle X_1|E_1\rangle \sim_{phhp} \langle Y_1|E_2\rangle$ for linear recursive specification $E_1$ and $E_2$, then $\langle X_1|E_1\rangle = \langle Y_1|E_2\rangle$.

It can be proven similarly to (1), hence, we omit it here. □

### 11.3.3 Approximation induction principle

In this section, we introduce the approximation induction principle ($AIP$) and try to explain that $AIP$ is still valid in probabilistic true concurrency. $AIP$ can be used to try and equate probabilistic truly concurrent bisimilar guarded recursive specifications. $AIP$ says that if two process terms are probabilistic truly concurrent bisimilar up to any finite depth, then they are probabilistic truly concurrent bisimilar.

**TABLE 11.18** PDF definitions of approximation induction principle.

| |
|---|
| $\mu(\Pi_n(x), \Pi_n(x')) = \mu(x, x') \quad n \geq 1$ |
| $\mu(x, y) = 0$, otherwise |

**TABLE 11.19** Transition rules of projection operator $\Pi_n$.

$$\frac{x \rightsquigarrow x'}{\Pi_n(x) \rightsquigarrow \Pi_n(x')}$$

$$\frac{x \xrightarrow{\{e_1,\cdots,e_k\}} \surd}{\Pi_{n+1}(x) \xrightarrow{\{e_1,\cdots,e_k\}} \surd} \qquad \frac{x \xrightarrow{\{e_1,\cdots,e_k\}} x'}{\Pi_{n+1}(x) \xrightarrow{\{e_1,\cdots,e_k\}} \Pi_n(x')}$$

**TABLE 11.20** Axioms of projection operator.

| No. | Axiom |
|---|---|
| $PR1$ | $\Pi_n(x + y) = \Pi_n(x) + \Pi_n(y)$ |
| $PPR1$ | $\Pi_n(x \boxplus_\rho y) = \Pi_n(x) \boxplus_\rho \Pi_n(y)$ |
| $PR2$ | $\Pi_n(x \between y) = \Pi_n(x) \between \Pi_n(y)$ |
| $PR3$ | $\Pi_{n+1}(e_1 \between \cdots \between e_k) = e_1 \between \cdots \between e_k$ |
| $PR4$ | $\Pi_{n+1}((e_1 \between \cdots \between e_k) \cdot x) = (e_1 \between \cdots \between e_k) \cdot \Pi_n(x)$ |
| $PR5$ | $\Pi_0(x) = \delta$ |
| $PR6$ | $\Pi_n(\delta) = \delta$ |

Also, we need the auxiliary unary projection operator $\Pi_n$ for $n \in \mathbb{N}$ and $\mathbb{N} \triangleq \{0, 1, 2, \cdots\}$.

First, we give the definition of PDFs in Table 11.18.

The transition rules of $\Pi_n$ are expressed in Table 11.19.

Based on the transition rules for projection operator $\Pi_n$ in Table 11.19, we design the axioms as Table 11.20 shows.

The axioms $PR1 - PR2$ and $PPR1$ say that $\Pi_n(s + t)$, $\Pi_n(s \between t)$ and $\Pi_n(s \boxplus_\rho t)$ can execute transitions of $s$ and $t$ up to depth $n$. $PR3$ says that $\Pi_{n+1}(e_1 \between \cdots \between e_k)$ executes $\{e_1, \cdots, e_k\}$ and terminates successfully. $PR4$ says that $\Pi_{n+1}((e_1 \between \cdots \between e_k) \cdot t)$ executes $\{e_1, \cdots, e_k\}$ and then executes transitions of $t$ up to depth $n$. $PR5$ and $PR6$ say that $\Pi_0(t)$ and $\Pi_n(\delta)$ exhibit no actions.

**Theorem 11.51** (Conservativity of $APPTC$ with projection operator and guarded recursion). *$APPTC$ with projection operator and guarded recursion is a conservative extension of $APPTC$ with guarded recursion.*

*Proof.* It follows from the following two facts (see Theorem 2.5):

1. The transition rules of $APPTC$ with guarded recursion are all source dependent;
2. The sources of the transition rules for the projection operator contain an occurrence of $\Pi_n$. □

**Theorem 11.52** (Congruence theorem of projection operator $\Pi_n$). *Probabilistic truly concurrent bisimulation equivalences $\sim_{pp}$, $\sim_{ps}$, $\sim_{php}$, and $\sim_{phhp}$ are all congruences with respect to projection operator $\Pi_n$.*

*Proof.* It is easy to see that $\sim_{pp}$, $\sim_{ps}$, $\sim_{php}$, and $\sim_{phhp}$ are all an equivalent relation with respect to projection operator $\Pi_n$, we only need to prove that $\sim_{pp}$, $\sim_{ps}$, $\sim_{php}$, and $\sim_{phhp}$ are preserved by the operators $\Pi_n$. That is, if $x \sim_{pp} x'$, $x \sim_{ps} x'$, $x \sim_{php} x'$, and $x \sim_{phhp} x'$, we need to prove that $\Pi_n(x) \sim_{pp} \Pi_n(x')$, $\Pi_n(x) \sim_{ps} \Pi_n(x')$, $\Pi_n(x) \sim_{php} \Pi_n(x')$, and $\Pi_n(x) \sim_{phhp} \Pi_n(x')$. The proof is quite trivial, hence, hence, we omit it here. □

**Theorem 11.53** (Elimination theorem of $APPTC$ with linear recursion and projection operator). *Each process term in $APPTC$ with linear recursion and projection operator is equal to a process term $\langle X_1|E\rangle$ with $E$ a linear recursive specification.*

**TABLE 11.21** $AIP$.

| No. | Axiom |
|---|---|
| $AIP$ | if $\Pi_n(x) = \Pi_n(y)$ for $n \in \mathbb{N}$, then $x = y$ |

*Proof.* By applying structural induction with respect to term size, each process term $t_1$ in $APPTC$ with linear recursion and projection operator $\Pi_n$ generates a process that can be expressed in the form of equations

$t_i = ((a_{1i11} \Vert \cdots \Vert a_{1i1i_1})t_{i1} + \cdots + (a_{1ik_i1} \Vert \cdots \Vert a_{1ik_ii_k})t_{ik_i} + (b_{1i11} \Vert \cdots \Vert b_{1i1i_1}) + \cdots + (b_{1il_i1} \Vert \cdots \Vert b_{1il_ii_l})) \boxplus_{\pi_1} \cdots \boxplus_{\pi_{m-1}} ((a_{mi11} \Vert \cdots \Vert a_{mi1i_1})t_{i1} + \cdots + (a_{mik_i1} \Vert \cdots \Vert a_{mik_ii_k})t_{ik_i} + (b_{mi11} \Vert \cdots \Vert b_{mi1i_1}) + \cdots + (b_{mil_i1} \Vert \cdots \Vert b_{mil_ii_l}))$

for $i \in \{1, \cdots, n\}$. Let the linear recursive specification $E$ consist of the recursive equations

$X_i = ((a_{1i11} \Vert \cdots \Vert a_{1i1i_1})X_{i1} + \cdots + (a_{1ik_i1} \Vert \cdots \Vert a_{1ik_ii_k})X_{ik_i} + (b_{1i11} \Vert \cdots \Vert b_{1i1i_1}) + \cdots + (b_{1il_i1} \Vert \cdots \Vert b_{1il_ii_l})) \boxplus_{\pi_1} \cdots \boxplus_{\pi_{m-1}} ((a_{mi11} \Vert \cdots \Vert a_{mi1i_1})X_{i1} + \cdots + (a_{mik_i1} \Vert \cdots \Vert a_{mik_ii_k})X_{ik_i} + (b_{mi11} \Vert \cdots \Vert b_{mi1i_1}) + \cdots + (b_{mil_i1} \Vert \cdots \Vert b_{mil_ii_l}))$

for $i \in \{1, \cdots, n\}$. Replacing $X_i$ by $t_i$ for $i \in \{1, \cdots, n\}$ is a solution for $E$, $RSP$ yields $t_1 = \langle X_1 | E \rangle$.

That is, in $E$, there is no occurrence of projection operator $\Pi_n$. □

**Theorem 11.54** (Soundness of $APPTC$ with projection operator and guarded recursion). *Let $x$ and $y$ be $APPTC$ with projection operator and guarded recursion terms. If $APPTC$ with projection operator and guarded recursion $\vdash x = y$, then*

1. $x \sim_{pp} y$;
2. $x \sim_{ps} y$;
3. $x \sim_{php} y$;
4. $x \sim_{phhp} y$.

*Proof.* Since $\sim_{pp}$, $\sim_{ps}$, $\sim_{php}$, and $\sim_{phhp}$ are all both an equivalent and a congruent relation with respect to $APPTC$ with guarded recursion, we only need to check if each axiom in Table 11.19 is sound modulo $\sim_{pp}$, $\sim_{ps}$, $\sim_{php}$, and $\sim_{phhp}$. The proof is quite trivial, hence, we omit it here. □

Then, $AIP$ is given in Table 11.21.

**Theorem 11.55** (Soundness of $AIP$). *Let $x$ and $y$ be $APPTC$ with projection operator and guarded recursion terms:*

1. *If $\Pi_n(x) \sim_{pp} \Pi_n(y)$ for $n \in \mathbb{N}$, then $x \sim_{pp} y$;*
2. *If $\Pi_n(x) \sim_{ps} \Pi_n(y)$ for $n \in \mathbb{N}$, then $x \sim_{ps} y$;*
3. *If $\Pi_n(x) \sim_{php} \Pi_n(y)$ for $n \in \mathbb{N}$, then $x \sim_{php} y$;*
4. *If $\Pi_n(x) \sim_{phhp} \Pi_n(y)$ for $n \in \mathbb{N}$, then $x \sim_{phhp} y$.*

*Proof.* (1) If $\Pi_n(x) \sim_{pp} \Pi_n(y)$ for $n \in \mathbb{N}$, then $x \sim_{pp} y$.

Since $\sim_{pp}$ is both an equivalent and a congruent relation with respect to $APPTC$ with guarded recursion and projection operator, we only need to check if $AIP$ in Table 11.21 is sound modulo $\sim_{pp}$.

Let $p, p_0$ and $q, q_0$ be closed $APPTC$ with projection operator and guarded recursion terms such that $\Pi_n(p_0) \sim_{pp} \Pi_n(q_0)$ for $n \in \mathbb{N}$. We define a relation $R$ such that $pRq$ iff $\Pi_n(p) \sim_{pp} \Pi_n(q)$. Obviously, $p_0 R q_0$, next, we prove that $R \in \sim_{pp}$.

Let $pRq$ and $p \rightsquigarrow \xrightarrow{\{e_1,\cdots,e_k\}} \surd$, then $\Pi_1(p) \rightsquigarrow \xrightarrow{\{e_1,\cdots,e_k\}} \surd$, $\Pi_1(p) \sim_{pp} \Pi_1(q)$ yields $\Pi_1(q) \rightsquigarrow \xrightarrow{\{e_1,\cdots,e_k\}} \surd$. Similarly, $q \rightsquigarrow \xrightarrow{\{e_1,\cdots,e_k\}} \surd$ implies $p \rightsquigarrow \xrightarrow{\{e_1,\cdots,e_k\}} \surd$.

Let $pRq$ and $p \rightsquigarrow \xrightarrow{\{e_1,\cdots,e_k\}} p'$. We define the set of process terms

$$S_n \triangleq \{q' | q \rightsquigarrow \xrightarrow{\{e_1,\cdots,e_k\}} q' \text{ and } \Pi_n(p') \sim_{pp} \Pi_n(q')\}$$

1. Since $\Pi_{n+1}(p) \sim_{pp} \Pi_{n+1}(q)$ and $\Pi_{n+1}(p) \rightsquigarrow \xrightarrow{\{e_1,\cdots,e_k\}} \Pi_n(p')$, there exist $q'$ such that $\Pi_{n+1}(q) \rightsquigarrow \xrightarrow{\{e_1,\cdots,e_k\}} \Pi_n(q')$ and $\Pi_n(p') \sim_{pp} \Pi_n(q')$. Hence, $S_n$ is not empty.
2. There are only finitely many $q'$ such that $q \rightsquigarrow \xrightarrow{\{e_1,\cdots,e_k\}} q'$, hence, $S_n$ is finite.
3. $\Pi_{n+1}(p) \sim_{pp} \Pi_{n+1}(q)$ implies $\Pi_n(p') \sim_{pp} \Pi_n(q')$, hence, $S_n \supseteq S_{n+1}$.

Hence, $S_n$ has a non-empty intersection, and let $q'$ be in this intersection, then $q \rightsquigarrow \xrightarrow{\{e_1,\cdots,e_k\}} q'$ and $\Pi_n(p') \sim_{pp} \Pi_n(q')$, hence, $p'Rq'$. Similarly, let $p \amalg q$, we can obtain that $q \rightsquigarrow \xrightarrow{\{e_1,\cdots,e_k\}} q'$ implies $p \rightsquigarrow \xrightarrow{\{e_1,\cdots,e_k\}} p'$ such that $p'Rq'$.

Finally, $R \in \sim_{pp}$ and $p_0 \sim_{pp} q_0$, as desired.

(2) If $\Pi_n(x) \sim_{ps} \Pi_n(y)$ for $n \in \mathbb{N}$, then $x \sim_{ps} y$.

It can be proven similarly to (1).

(3) If $\Pi_n(x) \sim_{php} \Pi_n(y)$ for $n \in \mathbb{N}$, then $x \sim_{php} y$.

It can be proven similarly to (1).

(4) If $\Pi_n(x) \sim_{phhp} \Pi_n(y)$ for $n \in \mathbb{N}$, then $x \sim_{phhp} y$.

It can be proven similarly to (1). □

**Theorem 11.56** (Completeness of $AIP$). *Let $p$ and $q$ be closed $APPTC$ with linear recursion and projection operator terms, then*

1. *if $p \sim_{pp} q$, then $\Pi_n(p) = \Pi_n(q)$;*
2. *if $p \sim_{ps} q$, then $\Pi_n(p) = \Pi_n(q)$;*
3. *if $p \sim_{php} q$, then $\Pi_n(p) = \Pi_n(q)$;*
4. *if $p \sim_{phhp} q$, then $\Pi_n(p) = \Pi_n(q)$.*

*Proof.* First, by the elimination theorem of $APPTC$ with guarded recursion and projection operator (see Theorem 11.53), we know that each process term in $APPTC$ with linear recursion and projection operator is equal to a process term $\langle X_1|E\rangle$ with $E$ a linear recursive specification:

$X_i = ((a_{1i11} \between \cdots \between a_{1i1i_1})X_{i1} + \cdots + (a_{1ik_i1} \between \cdots \between a_{1ik_ii_k})X_{ik_i} + (b_{1i11} \between \cdots \between b_{1i1i_1}) + \cdots + (b_{1il_i1} \between \cdots \between b_{1il_ii_l})) \boxplus_{\pi_1} \cdots \boxplus_{\pi_{m-1}} ((a_{mi11} \between \cdots \between a_{mi1i_1})X_{i1} + \cdots + (a_{mik_i1} \between \cdots \between a_{mik_ii_k})X_{ik_i} + (b_{mi11} \between \cdots \between b_{mi1i_1}) + \cdots + (b_{mil_i1} \between \cdots \between b_{mil_ii_l}))$

for $i \in \{1, \cdots, n\}$.

It remains to prove the following cases:

(1) if $p \sim_{pp} q$, then $\Pi_n(p) = \Pi_n(q)$.

Let $p \sim_{pp} q$, and fix an $n \in \mathbb{N}$, there are $p', q'$ in basic $APPTC$ terms such that $p' = \Pi_n(p)$ and $q' = \Pi_n(q)$. Since $\sim_{pp}$ is a congruence with respect to $APPTC$, if $p \sim_{pp} q$, then $\Pi_n(p) \sim_{pp} \Pi_n(q)$. The soundness theorem yields $p' \sim_{pp} \Pi_n(p) \sim_{pp} \Pi_n(q) \sim_{pp} q'$. Finally, the completeness of $APPTC$ modulo $\sim_{pp}$ (see Theorem 11.33) ensures $p' = q'$, and $\Pi_n(p) = p' = q' = \Pi_n(q)$, as desired.

(2) if $p \sim_{ps} q$, then $\Pi_n(p) = \Pi_n(q)$.

Let $p \sim_{ps} q$, and fix an $n \in \mathbb{N}$, there are $p', q'$ in basic $APPTC$ terms such that $p' = \Pi_n(p)$ and $q' = \Pi_n(q)$. Since $\sim_{ps}$ is a congruence with respect to $APPTC$, if $p \sim_{ps} q$, then $\Pi_n(p) \sim_{ps} \Pi_n(q)$. The soundness theorem yields $p' \sim_{ps} \Pi_n(p) \sim_{ps} \Pi_n(q) \sim_{ps} q'$. Finally, the completeness of $APPTC$ modulo $\sim_{ps}$ (see Theorem 11.36) ensures $p' = q'$, and $\Pi_n(p) = p' = q' = \Pi_n(q)$, as desired.

(3) if $p \sim_{php} q$, then $\Pi_n(p) = \Pi_n(q)$.

Let $p \sim_{php} q$, and fix an $n \in \mathbb{N}$, there are $p', q'$ in basic $APPTC$ terms such that $p' = \Pi_n(p)$ and $q' = \Pi_n(q)$. Since $\sim_{php}$ is a congruence with respect to $APPTC$, if $p \sim_{php} q$, then $\Pi_n(p) \sim_{php} \Pi_n(q)$. The soundness theorem yields $p' \sim_{php} \Pi_n(p) \sim_{php} \Pi_n(q) \sim_{php} q'$. Finally, the completeness of $APPTC$ modulo $\sim_{php}$ (see Theorem 11.39) ensures $p' = q'$, and $\Pi_n(p) = p' = q' = \Pi_n(q)$, as desired.

(4) if $p \sim_{phhp} q$, then $\Pi_n(p) = \Pi_n(q)$.

Let $p \sim_{phhp} q$, and fix an $n \in \mathbb{N}$, there are $p', q'$ in basic $APPTC$ terms such that $p' = \Pi_n(p)$ and $q' = \Pi_n(q)$. Since $\sim_{phhp}$ is a congruence with respect to $APPTC$, if $p \sim_{phhp} q$, then $\Pi_n(p) \sim_{phhp} \Pi_n(q)$. The soundness theorem yields $p' \sim_{phhp} \Pi_n(p) \sim_{phhp} \Pi_n(q) \sim_{phhp} q'$. Finally, the completeness of $APPTC$ modulo $\sim_{phhp}$ (see Theorem 11.42) ensures $p' = q'$, and $\Pi_n(p) = p' = q' = \Pi_n(q)$, as desired. □

## 11.4 Abstraction

To abstract away from the internal implementations of a program, and verify that the program exhibits the desired external behaviors, the silent step $\tau$ and abstraction operator $\tau_I$ are introduced, where $I \subseteq \mathbb{E}$ denotes the internal events. The silent step $\tau$ represents the internal events, when we consider the external behaviors of a process, $\tau$ events can be removed, that is, $\tau$ events must remain silent. The transition rule of $\tau$ is shown in Table 11.22. In the following, let the atomic event $e$ range over $\mathbb{E} \cup \{\delta\} \cup \{\tau\}$, and let the communication function $\gamma : \mathbb{E} \cup \{\tau\} \times \mathbb{E} \cup \{\tau\} \to \mathbb{E} \cup \{\delta\}$, with each communication involved $\tau$ resulting into $\delta$.

In this section, we try to find the algebraic laws of $\tau$ and $\tau_I$ in probabilistic true concurrency.

**TABLE 11.22 Transition rules of the silent step.**

| |
|---|
| $\frac{}{\tau \rightsquigarrow \breve{\tau}}$ |
| $\frac{}{\tau \xrightarrow{\tau} \surd}$ |

### 11.4.1 Guarded linear recursion

The silent step $\tau$ as an atomic event, is introduced into $E$. Considering the recursive specification $X = \tau X$, $\tau s$, $\tau\tau s$, and $\tau \cdots s$ are all its solutions, that is, the solutions make the existence of $\tau$-loops that cause unfairness. To prevent $\tau$-loops, we extend the definition of linear recursive specification to the guarded one.

**Definition 11.57** (Guarded linear recursive specification). A recursive specification is linear if its recursive equations are of the form

$((a_{111} \parallel \cdots \parallel a_{11i_1})X_1 + \cdots + (a_{1k1} \parallel \cdots \parallel a_{1ki_k})X_k + (b_{111} \parallel \cdots \parallel b_{11j_1}) + \cdots + (b_{11j_1} \parallel \cdots \parallel b_{1lj_l})) \boxplus_{\pi_1} \cdots \boxplus_{\pi_{m-1}}$
$((a_{m11} \parallel \cdots \parallel a_{m1i_1})X_1 + \cdots + (a_{mk1} \parallel \cdots \parallel a_{mki_k})X_k + (b_{m11} \parallel \cdots \parallel b_{m1j_1}) + \cdots + (b_{m1j_1} \parallel \cdots \parallel b_{mlj_l}))$
where $a_{111}, \cdots, a_{11i_1}, a_{1k1}, \cdots, a_{1ki_k}, b_{111}, \cdots, b_{11j_1}, b_{11j_1}, \cdots, b_{1lj_l} \cdots$
$a_{m11}, \cdots, a_{m1i_1}, a_{mk1}, \cdots, a_{mki_k}, b_{m11}, \cdots, b_{m1j_1}, b_{m1j_1}, \cdots, b_{mlj_l} \in \mathbb{E} \cup \{\tau\}$, and the sum above is allowed to be empty, in which case it represents the deadlock $\delta$.

A linear recursive specification $E$ is guarded if there does not exist an infinite sequence of $\tau$-transitions $\langle X|E\rangle \rightsquigarrow\xrightarrow{\tau} \langle X'|E\rangle \rightsquigarrow\xrightarrow{\tau} \langle X''|E\rangle \rightsquigarrow\xrightarrow{\tau} \cdots$.

**Theorem 11.58** (Conservativity of $APPTC$ with silent step and guarded linear recursion). *$APPTC$ with silent step and guarded linear recursion is a conservative extension of $APPTC$ with linear recursion.*

*Proof.* Since the transition rules of $APPTC$ with linear recursion are source dependent, and the transition rules for silent step in Table 11.22 contain only a fresh constant $\tau$ in their source, hence, the transition rules of $APPTC$ with silent step and guarded linear recursion is a conservative extension of those of $APPTC$ with linear recursion. □

**Theorem 11.59** (Congruence theorem of $APPTC$ with silent step and guarded linear recursion). *Probabilistic rooted branching truly concurrent bisimulation equivalences $\approx_{prbp}$, $\approx_{prbs}$, $\approx_{prbhp}$, and $\approx_{prbhhp}$ are all congruences with respect to $APPTC$ with silent step and guarded linear recursion.*

*Proof.* It follows from the following three facts:

1. In a guarded linear recursive specification, the right-hand sides of its recursive equations can be adapted to the form by applications of the axioms in $APPTC$ and replacing recursion variables by the right-hand sides of their recursive equations;
2. Probabilistic truly concurrent bisimulation equivalences $\sim_{pp}$, $\sim_{ps}$, $\sim_{php}$, and $\sim_{phhp}$ are all congruences with respect to all operators of $APPTC$, while probabilistic truly concurrent bisimulation equivalences $\sim_{pp}$, $\sim_{ps}$, $\sim_{php}$, and $\sim_{phhp}$ imply the corresponding probabilistic rooted branching truly concurrent bisimulations $\approx_{prbp}$, $\approx_{prbs}$, $\approx_{prbhp}$, and $\approx_{prbhhp}$, hence, probabilistic rooted branching truly concurrent bisimulations $\approx_{prbp}$, $\approx_{prbs}$, $\approx_{prbhp}$, and $\approx_{prbhhp}$ are all congruences with respect to all operators of $APPTC$;
3. While $\mathbb{E}$ is extended to $\mathbb{E} \cup \{\tau\}$, it can be proved that probabilistic rooted branching truly concurrent bisimulations $\approx_{prbp}$, $\approx_{prbs}$, $\approx_{prbhp}$, and $\approx_{prbhhp}$ are all congruences with respect to all operators of $APPTC$, however, we omit this proof here. □

### 11.4.2 Algebraic laws for the silent step

We design the axioms for the silent step $\tau$ in Table 11.23.

The axioms $B1$ and $B2$ are the conditions in which $\tau$ really remains silent to act with the operators $\cdot$, $+$, $\boxplus_\pi$, and $\between$.

**Theorem 11.60** (Elimination theorem of $APPTC$ with silent step and guarded linear recursion). *Each process term in $APPTC$ with silent step and guarded linear recursion is equal to a process term $\langle X_1|E\rangle$ with $E$ a guarded linear recursive specification.*

**TABLE 11.23** Axioms of silent step.

| No. | Axiom |
|---|---|
| $B1$ | $(y=y+y, z=z+z) \quad x\cdot((y+\tau\cdot(y+z))\boxplus_{\pi} w)=x\cdot((y+z)\boxplus_{\pi} w)$ |
| $B2$ | $(y=y+y, z=z+z) \quad x\between((y+\tau\between(y+z))\boxplus_{\pi} w)=x\between((y+z)\boxplus_{\pi} w)$ |

*Proof.* By applying structural induction with respect to term size, each process term $t_1$ in $APPTC$ with silent step and guarded linear recursion generates a process that can be expressed in the form of equations

$t_i=((a_{1i11}\between\cdots\between a_{1i1i_1})t_{i1}+\cdots+(a_{1ik_i1}\between\cdots\between a_{1ik_ii_k})t_{ik_i}+(b_{1i11}\between\cdots\between b_{1i1i_1})+\cdots+(b_{1il_i1}\between\cdots\between b_{1il_ii_l}))\boxplus_{\pi_1}\cdots\boxplus_{\pi_{m-1}}((a_{mi11}\between\cdots\between a_{mi1i_1})t_{i1}+\cdots+(a_{mik_i1}\between\cdots\between a_{mik_ii_k})t_{ik_i}+(b_{mi11}\between\cdots\between b_{mi1i_1})+\cdots+(b_{mil_i1}\between\cdots\between b_{m1il_ii_l}))$

for $i\in\{1,\cdots,n\}$. Let the linear recursive specification $E$ consist of the recursive equations

$X_i=((a_{1i11}\between\cdots\between a_{1i1i_1})X_{i1}+\cdots+(a_{1ik_i1}\between\cdots\between a_{1ik_ii_k})X_{ik_i}+(b_{1i11}\between\cdots\between b_{1i1i_1})+\cdots+(b_{1il_i1}\between\cdots\between b_{1il_ii_l}))\boxplus_{\pi_1}\cdots\boxplus_{\pi_{m-1}}((a_{mi11}\between\cdots\between a_{mi1i_1})X_{i1}+\cdots+(a_{mik_i1}\between\cdots\between a_{mik_ii_k})X_{ik_i}+(b_{mi11}\between\cdots\between b_{mi1i_1})+\cdots+(b_{mil_i1}\between\cdots\between b_{mil_ii_l}))$

for $i\in\{1,\cdots,n\}$. Replacing $X_i$ by $t_i$ for $i\in\{1,\cdots,n\}$ is a solution for $E$, $RSP$ yields $t_1=\langle X_1|E\rangle$. □

**Theorem 11.61** (Soundness of $APPTC$ with silent step and guarded linear recursion). *Let $x$ and $y$ be $APPTC$ with silent step and guarded linear recursion terms. If $APPTC$ with silent step and guarded linear recursion $\vdash x=y$, then*

1. $x\approx_{prbp} y$;
2. $x\approx_{prbs} y$;
3. $x\approx_{prbhp} y$;
4. $x\approx_{prbhhp} y$.

*Proof.* Since probabilistic truly concurrent rooted branching bisimulation $\approx_{prbp}$, $\approx_{prbs}$, $\approx_{prbhp}$, and $\approx_{prbhhp}$ are all both an equivalent and a congruent relation with respect to $APPTC$ with silent step and guarded linear recursion, we only need to check if each axiom in Table 11.23 is sound modulo probabilistic truly concurrent rooted branching bisimulation $\approx_{prbp}$, $\approx_{prbs}$, $\approx_{prbhp}$, and $\approx_{prbhhp}$. The proof is quite trivial, hence, we omit it here. □

**Theorem 11.62** (Completeness of $APPTC$ with silent step and guarded linear recursion). *Let $p$ and $q$ be closed $APPTC$ with silent step and guarded linear recursion terms, then*

1. *if $p\approx_{prbp} q$, then $p=q$;*
2. *if $p\approx_{prbs} q$, then $p=q$;*
3. *if $p\approx_{prbhp} q$, then $p=q$;*
4. *if $p\approx_{prbhhp} q$, then $p=q$.*

*Proof.* First, by the elimination theorem of $APPTC$ with silent step and guarded linear recursion (see Theorem 11.60), we know that each process term in $APPTC$ with silent step and guarded linear recursion is equal to a process term $\langle X_1|E\rangle$ with $E$ a guarded linear recursive specification.

It remains to prove the following cases:

(1) If $\langle X_1|E_1\rangle\approx_{prbp}\langle Y_1|E_2\rangle$ for guarded linear recursive specification $E_1$ and $E_2$, then $\langle X_1|E_1\rangle=\langle Y_1|E_2\rangle$.

First, the recursive equation $W=\tau+\cdots+\tau$ with $W\not\equiv X_1$ in $E_1$ and $E_2$, can be removed, and the corresponding summands $aW$ are replaced by $a$, to obtain $E_1'$ and $E_2'$, by use of the axioms $RDP$, $A3$, and $B1-2$, and $\langle X|E_1\rangle=\langle X|E_1'\rangle$, $\langle Y|E_2\rangle=\langle Y|E_2'\rangle$.

Let $E_1$ consist of recursive equations $X=t_X$ for $X\in\mathcal{X}$ and $E_2$ consists of recursion equations $Y=t_Y$ for $Y\in\mathcal{Y}$, and are not the form $\tau+\cdots+\tau$. Let the guarded linear recursive specification $E$ consists of recursion equations $Z_{XY}=t_{XY}$, and $\langle X|E_1\rangle\approx_{prbp}\langle Y|E_2\rangle$, and $t_{XY}$ consists of the following summands:

1. $t_{XY}$ contains a summand $(a_1\between\cdots\between a_m)Z_{X'Y'}$ iff $t_X$ contains the summand $(a_1\between\cdots\between a_m)X'$ and $t_Y$ contains the summand $(a_1\between\cdots\between a_m)Y'$ such that $\langle X'|E_1\rangle\approx_{prbp}\langle Y'|E_2\rangle$;
2. $t_{XY}$ contains a summand $b_1\between\cdots\between b_n$ iff $t_X$ contains the summand $b_1\between\cdots\between b_n$ and $t_Y$ contains the summand $b_1\between\cdots\between b_n$;
3. $t_{XY}$ contains a summand $\tau Z_{X'Y}$ iff $XY\not\equiv X_1Y_1$, $t_X$ contains the summand $\tau X'$, and $\langle X'|E_1\rangle\approx_{prbp}\langle Y|E_2\rangle$;
4. $t_{XY}$ contains a summand $\tau Z_{XY'}$ iff $XY\not\equiv X_1Y_1$, $t_Y$ contains the summand $\tau Y'$, and $\langle X|E_1\rangle\approx_{prbp}\langle Y'|E_2\rangle$.

Since $E_1$ and $E_2$ are guarded, $E$ is guarded. Constructing the process term $u_{XY}$ consists of the following summands:

1. $u_{XY}$ contains a summand $(a_1\between\cdots\between a_m)\langle X'|E_1\rangle$ iff $t_X$ contains the summand $(a_1\between\cdots\between a_m)X'$ and $t_Y$ contains the summand $(a_1\between\cdots\between a_m)Y'$ such that $\langle X'|E_1\rangle\approx_{prbp}\langle Y'|E_2\rangle$;

**TABLE 11.24** Transition rules of the abstraction operator.

$$\frac{x \rightsquigarrow x'}{\tau_I(x) \rightsquigarrow \tau_I(x')}$$

$$\frac{x \xrightarrow{e} \surd}{\tau_I(x) \xrightarrow{e} \surd} \quad e \notin I \qquad \frac{x \xrightarrow{e} x'}{\tau_I(x) \xrightarrow{e} \tau_I(x')} \quad e \notin I$$

$$\frac{x \xrightarrow{e} \surd}{\tau_I(x) \xrightarrow{\tau} \surd} \quad e \in I \qquad \frac{x \xrightarrow{e} x'}{\tau_I(x) \xrightarrow{\tau} \tau_I(x')} \quad e \in I$$

**2.** $u_{XY}$ contains a summand $b_1 \parallel \cdots \parallel b_n$ iff $t_X$ contains the summand $b_1 \parallel \cdots \parallel b_n$ and $t_Y$ contains the summand $b_1 \parallel \cdots \parallel b_n$;

**3.** $u_{XY}$ contains a summand $\tau \langle X'|E_1 \rangle$ iff $XY \not\equiv X_1 Y_1$, $t_X$ contains the summand $\tau X'$, and $\langle X'|E_1 \rangle \approx_{rbs} \langle Y|E_2 \rangle$.

Let the process term $s_{XY}$ be defined as follows:

**1.** $s_{XY} \triangleq \tau \langle X|E_1 \rangle + u_{XY}$ iff $XY \not\equiv X_1 Y_1$, $t_Y$ contains the summand $\tau Y'$, and $\langle X|E_1 \rangle \approx_{prbp} \langle Y'|E_2 \rangle$;

**2.** $s_{XY} \triangleq \langle X|E_1 \rangle$, otherwise.

Hence, $\langle X|E_1 \rangle = \langle X|E_1 \rangle + u_{XY}$, and $(a_1 \parallel \cdots \parallel a_m)(\tau \langle X|E_1 \rangle + u_{XY}) = (a_1 \parallel \cdots \parallel a_m)((\tau \langle X|E_1 \rangle + u_{XY}) + u_{XY}) = (a_1 \parallel \cdots \parallel a_m)(\langle X|E_1 \rangle + u_{XY}) = (a_1 \parallel \cdots \parallel a_m)\langle X|E_1 \rangle$, hence, $(a_1 \parallel \cdots \parallel a_m)s_{XY} = (a_1 \parallel \cdots \parallel a_m)\langle X|E_1 \rangle$.

Let $\sigma$ map recursion variable $X$ in $E_1$ to $\langle X|E_1 \rangle$, and let $\psi$ map recursion variable $Z_{XY}$ in $E$ to $s_{XY}$. It is sufficient to prove $s_{XY} = \psi(t_{XY})$ for recursion variables $Z_{XY}$ in $E$. Either $XY \equiv X_1 Y_1$ or $XY \not\equiv X_1 Y_1$, we can obtain $s_{XY} = \psi(t_{XY})$. Hence, $s_{XY} = \langle Z_{XY}|E \rangle$ for recursive variables $Z_{XY}$ in $E$ is a solution for $E$. Then, by $RSP$, particularly, $\langle X_1|E_1 \rangle = \langle Z_{X_1 Y_1}|E \rangle$. Similarly, we can obtain $\langle Y_1|E_2 \rangle = \langle Z_{X_1 Y_1}|E \rangle$. Finally, $\langle X_1|E_1 \rangle = \langle Z_{X_1 Y_1}|E \rangle = \langle Y_1|E_2 \rangle$, as desired.

(2) If $\langle X_1|E_1 \rangle \approx_{prbs} \langle Y_1|E_2 \rangle$ for guarded linear recursive specifications $E_1$ and $E_2$, then $\langle X_1|E_1 \rangle = \langle Y_1|E_2 \rangle$. It can be proven similarly to (1), hence, we omit it here.

(3) If $\langle X_1|E_1 \rangle \approx_{prbhp} \langle Y_1|E_2 \rangle$ for guarded linear recursive specifications $E_1$ and $E_2$, then $\langle X_1|E_1 \rangle = \langle Y_1|E_2 \rangle$. It can be proven similarly to (1), hence, we omit it here.

(4) If $\langle X_1|E_1 \rangle \approx_{prbhhp} \langle Y_1|E_2 \rangle$ for guarded linear recursive specifications $E_1$ and $E_2$, then $\langle X_1|E_1 \rangle = \langle Y_1|E_2 \rangle$. It can be proven similarly to (1), hence, we omit it here. $\quad\square$

### 11.4.3 Abstraction

The unary abstraction operator $\tau_I$ ($I \subseteq \mathbb{E}$) renames all atomic events in $I$ into $\tau$. $APPTC$ with silent step and abstraction operator is called $APPTC_\tau$. The transition rules of operator $\tau_I$ are shown in Table 11.24.

**Theorem 11.63** (Conservativity of $APPTC_\tau$ with guarded linear recursion). *$APPTC_\tau$ with guarded linear recursion is a conservative extension of $APPTC$ with silent step and guarded linear recursion.*

*Proof.* Since the transition rules of $APPTC$ with silent step and guarded linear recursion are source dependent, and the transition rules for abstraction operator in Table 11.24 contain only a fresh operator $\tau_I$ in their source, hence, the transition rules of $APPTC_\tau$ with guarded linear recursion is a conservative extension of those of $APPTC$ with silent step and guarded linear recursion. $\quad\square$

**Theorem 11.64** (Congruence theorem of $APPTC_\tau$ with guarded linear recursion). *Probabilistic rooted branching truly concurrent bisimulation equivalences $\approx_{prbp}$, $\approx_{prbs}$, $\approx_{prbhp}$, and $\approx_{prbhhp}$ are all congruences with respect to $APPTC_\tau$ with guarded linear recursion.*

*Proof.* It is easy to see that probabilistic rooted branching truly concurrent bisimulations $\approx_{prbp}$, $\approx_{prbs}$, $\approx_{prbhp}$, and $\approx_{prbhhp}$ are all equivalent relations on $APPTC$ terms, we only need to prove that $\approx_{prbp}$, $\approx_{prbs}$, $\approx_{prbhp}$, and $\approx_{prbhhp}$ are preserved by the operators $\tau_I$. That is, if $x \approx_{prbp} x'$, $x \approx_{prbs} x'$, $x \approx_{prbhp} x'$, and $x \approx_{prbhhp} x'$, we need to prove that $\tau_I(x) \approx_{prbp} \tau_I(x')$, $\tau_I(x) \approx_{prbs} \tau_I(x')$, $\tau_I(x) \approx_{prbhp} \tau_I(x')$, and $\tau_I(x) \approx_{prbhhp} \tau_I(x')$. The proof is quite trivial, hence, we omit it here. $\quad\square$

We design the axioms for the abstraction operator $\tau_I$ in Table 11.25.

The axioms $TI1 - TI3$ are the defining laws for the abstraction operator $\tau_I$; $TI4 - TI6$ and $PTI1$ say that in process term $\tau_I(t)$, all transitions of $t$ labeled with atomic events from $I$ are renamed into $\tau$.

**TABLE 11.25** Axioms of abstraction operator.

| No. | Axiom |
|---|---|
| $TI1$ | $e \notin I \quad \tau_I(e) = e$ |
| $TI2$ | $e \in I \quad \tau_I(e) = \tau$ |
| $TI3$ | $\tau_I(\delta) = \delta$ |
| $TI4$ | $\tau_I(x + y) = \tau_I(x) + \tau_I(y)$ |
| $PTI1$ | $\tau_I(x \boxplus_\pi y) = \tau_I(x) \boxplus_\pi \tau_I(y)$ |
| $TI5$ | $\tau_I(x \cdot y) = \tau_I(x) \cdot \tau_I(y)$ |
| $TI6$ | $\tau_I(x \between y) = \tau_I(x) \between \tau_I(y)$ |

**TABLE 11.26** Recursive verification rules.

| | |
|---|---|
| $VR_1$ | $\dfrac{x = y + (i_1 \between \cdots \between i_m) \cdot x, y = y + y}{\tau \cdot \tau_I(x) = \tau \cdot \tau_I(y)}$ |
| $VR_2$ | $\dfrac{x = z \boxplus_\pi (u + (i_1 \between \cdots \between i_m) \cdot x), z = z + u, z = z + z}{\tau \cdot \tau_I(x) = \tau \cdot \tau_I(z)}$ |
| $VR_3$ | $\dfrac{x = z + (i_1 \between \cdots \between i_m) \cdot y, y = z \boxplus_\pi (u + (j_1 \between \cdots \between j_n) \cdot x), z = z + u, z = z + z}{\tau \cdot \tau_I(x) = \tau \cdot \tau_I(y') \text{ for } y' = z \boxplus_\pi (u + (i_1 \between \cdots \between i_m) \cdot y')}$ |

**Theorem 11.65** (Soundness of $APPTC_\tau$ with guarded linear recursion). *Let $x$ and $y$ be $APPTC_\tau$ with guarded linear recursion terms. If $APPTC_\tau$ with guarded linear recursion $\vdash x = y$, then*

1. $x \approx_{prbp} y$;
2. $x \approx_{prbs} y$;
3. $x \approx_{prbhp} y$;
4. $x \approx_{prbhhp} y$.

*Proof.* Since probabilistic rooted branching step bisimulations $\approx_{prbp}$, $\approx_{prbs}$, $\approx_{prbhp}$, and $\approx_{prbhhp}$ are all both equivalent and congruent relations with respect to $APPTC_\tau$ with guarded linear recursion, we only need to check if each axiom in Table 11.25 is sound modulo $\approx_{prbp}$, $\approx_{prbs}$, $\approx_{prbhp}$, and $\approx_{prbhhp}$. The proof is quite trivial, hence, we omit it here. □

Although $\tau$-loops are prohibited in guarded linear recursive specifications in a specifiable way, they can be constructed using the abstraction operator, for example, there exist $\tau$-loops in the process term $\tau_{\{a\}}(\langle X|X = aX\rangle)$. To avoid $\tau$-loops caused by $\tau_I$ and ensure fairness, we introduce the following recursive verification rules as Table 11.26 shows, note that $i_1, \cdots, i_m, j_1, \cdots, j_n \in I \subseteq \mathbb{E} \setminus \{\tau\}$.

**Theorem 11.66** (Soundness of $VR_1$, $VR_2$, $VR_3$). *$VR_1$, $VR_2$, and $VR_3$ are sound modulo probabilistic rooted branching truly concurrent bisimulation equivalences $\approx_{prbp}$, $\approx_{prbs}$, $\approx_{prbhp}$, and $\approx_{prbhhp}$.*

Chapter 12

# Mobility

In this chapter, we design a calculus of probabilistic truly concurrent mobile processes ($\pi_{ptc}$). This chapter is organized as follows. We introduce the syntax and operational semantics of $\pi_{ptc}$ in Section 12.1, its properties for strongly probabilistic truly concurrent bisimulations in Section 12.2, and its axiomatization in Section 12.3.

## 12.1 Syntax and operational semantics

We assume an infinite set $\mathcal{N}$ of (action or event) names, and use $a, b, c, \cdots$ to range over $\mathcal{N}$, use $x, y, z, w, u, v$ as meta-variables over names. We denote by $\overline{\mathcal{N}}$ the set of co-names and let $\overline{a}, \overline{b}, \overline{c}, \cdots$ range over $\overline{\mathcal{N}}$. Then, we set $\mathcal{L} = \mathcal{N} \cup \overline{\mathcal{N}}$ as the set of labels, and use $l, \overline{l}$ to range over $\mathcal{L}$. We extend complementation to $\mathcal{L}$ such that $\overline{\overline{a}} = a$. Let $\tau$ denote the silent step (internal action or event) and define $Act = \mathcal{L} \cup \{\tau\}$ to be the set of actions, $\alpha, \beta$ range over $Act$. Also, $K, L$ are used to stand for subsets of $\mathcal{L}$ and $\overline{L}$ is used for the set of complements of labels in $L$.

Further, we introduce a set $\mathcal{X}$ of process variables, and a set $\mathcal{K}$ of process constants, and let $X, Y, \cdots$ range over $\mathcal{X}$, and $A, B, \cdots$ range over $\mathcal{K}$. For each process constant $A$, a nonnegative arity $ar(A)$ is assigned to it. Let $\widetilde{x} = x_1, \cdots, x_{ar(A)}$ be a tuple of distinct name variables, then $A(\widetilde{x})$ is called a process constant. $\widetilde{X}$ is a tuple of distinct process variables, and $E, F, \cdots$ range over the recursive expressions. We write $\mathcal{P}$ for the set of processes. Sometimes, we use $I, J$ to stand for an indexing set, and we write $E_i : i \in I$ for a family of expressions indexed by $I$. $Id_D$ is the identity function or relation over set $D$. The symbol $\equiv_\alpha$ denotes equality under standard alpha-convertibility, note that the subscript $\alpha$ has no relation to the action $\alpha$.

### 12.1.1 Syntax

We use the Prefix . to model the causality relation $\leq$ in true concurrency, the Summation $+$ to model the conflict relation $\sharp$, $\boxplus_\pi$ to model the probabilistic conflict relation $\sharp_\pi$ in probabilistic true concurrency, and the Composition $\parallel$ to explicitly model concurrent relation in true concurrency. Also, we follow the conventions of process algebra.

**Definition 12.1** (Syntax). A truly concurrent process $P$ is defined inductively by the following formation rules:

1. $A(\widetilde{x}) \in \mathcal{P}$;
2. $\mathbf{nil} \in \mathcal{P}$;
3. if $P \in \mathcal{P}$, then the Prefix $\tau.P \in \mathcal{P}$, for $\tau \in Act$ is the silent action;
4. if $P \in \mathcal{P}$, then the Output $\overline{x}y.P \in \mathcal{P}$, for $x, y \in Act$;
5. if $P \in \mathcal{P}$, then the Input $x(y).P \in \mathcal{P}$, for $x, y \in Act$;
6. if $P \in \mathcal{P}$, then the Restriction $(x)P \in \mathcal{P}$, for $x \in Act$;
7. if $P, Q \in \mathcal{P}$, then the Summation $P + Q \in \mathcal{P}$;
8. if $P, Q \in \mathcal{P}$, then the Summation $P \boxplus_\pi Q \in \mathcal{P}$;
9. if $P, Q \in \mathcal{P}$, then the Composition $P \parallel Q \in \mathcal{P}$;

The standard BNF grammar of syntax of $\pi_{tc}$ can be summarized as follows:

$$P ::= A(\widetilde{x}) \quad | \quad \mathbf{nil} \quad | \quad \tau.P \quad | \quad \overline{x}y.P \quad | \quad x(y).P \quad | \quad (x)P \quad | \quad P + P \quad | \quad P \boxplus_\pi P \quad | \quad P \parallel P$$

In $\overline{x}y$, $x(y)$ and $\overline{x}(y)$, $x$ is called the subject, $y$ is called the object and it may be free or bound.

**Definition 12.2** (Free variables). The free names of a process $P$, $fn(P)$, are defined as follows:

1. $fn(A(\widetilde{x})) \subseteq \{\widetilde{x}\}$;
2. $fn(\mathbf{nil}) = \emptyset$;
3. $fn(\tau.P) = fn(P)$;

Handbook of Truly Concurrent Process Algebra. https://doi.org/10.1016/B978-0-44-321515-5.00016-0

**TABLE 12.1** Probabilistic transition rules of $\pi_{ptc}$.

| Rule | Definition |
|---|---|
| PTAU-ACT | $\frac{}{\tau.P \rightsquigarrow \breve{\tau}.P}$ |
| POUTPUT-ACT | $\frac{}{\overline{x}y.P \rightsquigarrow \breve{\overline{x}y}.P}$ |
| PINPUT-ACT | $\frac{}{x(z).P \rightsquigarrow \breve{x(z)}.P}$ |
| PPAR | $\frac{P \rightsquigarrow P' \quad Q \rightsquigarrow Q'}{P \parallel Q \rightsquigarrow P' \parallel Q'}$ |
| PSUM | $\frac{P \rightsquigarrow P' \quad Q \rightsquigarrow Q'}{P + Q \rightsquigarrow P' + Q'}$ |
| PBOX-SUM | $\frac{P \rightsquigarrow P'}{P \boxplus_{\pi} Q \rightsquigarrow P'}$ |
| PIDE | $\frac{P\{\widetilde{y}/\widetilde{x}\} \rightsquigarrow P'}{A(\widetilde{y}) \rightsquigarrow P'} \quad (A(\widetilde{x}) \overset{\text{def}}{=} P)$ |
| PRES | $\frac{P \rightsquigarrow P'}{(y)P \rightsquigarrow (y)P'} \quad (y \notin n(\alpha))$ |

4. $fn(\overline{x}y.P) = fn(P) \cup \{x\} \cup \{y\}$;
5. $fn(x(y).P) = fn(P) \cup \{x\} - \{y\}$;
6. $fn((x)P) = fn(P) - \{x\}$;
7. $fn(P + Q) = fn(P) \cup fn(Q)$;
8. $fn(P \boxplus_{\pi} Q) = fn(P) \cup fn(Q)$;
9. $fn(P \parallel Q) = fn(P) \cup fn(Q)$.

**Definition 12.3** (Bound variables). Let $n(P)$ be the names of a process $P$, then the bound names $bn(P) = n(P) - fn(P)$.

For each process constant schema $A(\widetilde{x})$, a defining equation of the form

$$A(\widetilde{x}) \overset{\text{def}}{=} P$$

is assumed, where $P$ is a process with $fn(P) \subseteq \{\widetilde{x}\}$.

**Definition 12.4** (Substitutions). A substitution is a function $\sigma : \mathcal{N} \to \mathcal{N}$. For $x_i\sigma = y_i$ with $1 \leq i \leq n$, we write $\{y_1/x_1, \cdots, y_n/x_n\}$ or $\{\widetilde{y}/\widetilde{x}\}$ for $\sigma$. For a process $P \in \mathcal{P}$, $P\sigma$ is defined inductively as follows:

1. if $P$ is a process constant $A(\widetilde{x}) = A(x_1, \cdots, x_n)$, then $P\sigma = A(x_1\sigma, \cdots, x_n\sigma)$;
2. if $P = \mathbf{nil}$, then $P\sigma = \mathbf{nil}$;
3. if $P = \tau.P'$, then $P\sigma = \tau.P'\sigma$;
4. if $P = \overline{x}y.P'$, then $P\sigma = \overline{x\sigma}y\sigma.P'\sigma$;
5. if $P = x(y).P'$, then $P\sigma = x\sigma(y).P'\sigma$;
6. if $P = (x)P'$, then $P\sigma = (x\sigma)P'\sigma$;
7. if $P = P_1 + P_2$, then $P\sigma = P_1\sigma + P_2\sigma$;
8. if $P = P_1 \boxplus_{\pi} P_2$, then $P\sigma = P_1\sigma \boxplus_{\pi} P_2\sigma$;
9. if $P = P_1 \parallel P_2$, then $P\sigma = P_1\sigma \parallel P_2\sigma$.

### 12.1.2 Operational semantics

The operational semantics is defined by LTSs (labeled transition systems), and it is detailed by the following definition.

**Definition 12.5** (Semantics). The operational semantics of $\pi_{tc}$ corresponding to the syntax in Definition 12.1 is defined by a series of transition rules, named **PACT**, **PSUM**, **PBOX-SUM PIDE**, **PPAR**, **PRES** and named **ACT**, **SUM**, **IDE**, **PAR**, **COM**, **CLOSE**, **RES**, **OPEN** that indicate that the rules are associated, respectively, with Prefix, Summation, Box-Summation, Identity, Parallel Composition, Communication, and Restriction in Definition 12.1. They are shown in Tables 12.1 and 12.2.

### 12.1.3 Properties of transitions

**Proposition 12.6.** **1.** *If* $P \xrightarrow{\alpha} P'$, *then*

**TABLE 12.2** Action transition rules of $\pi_{ptc}$.

$$\text{TAU-ACT}\ \frac{}{\tau.P \xrightarrow{\tau} P} \qquad \text{OUTPUT-ACT}\ \frac{}{\overline{x}y.P \xrightarrow{\overline{x}y} P}$$

$$\text{INPUT-ACT}\ \frac{}{x(z).P \xrightarrow{x(w)} P\{w/z\}}\ (w \notin fn((z)P))$$

$$\text{PAR}_1\ \frac{P \xrightarrow{\alpha} P' \quad Q \nrightarrow}{P \parallel Q \xrightarrow{\alpha} P' \parallel Q}\ (bn(\alpha) \cap fn(Q) = \emptyset) \qquad \text{PAR}_2\ \frac{Q \xrightarrow{\alpha} Q' \quad P \nrightarrow}{P \parallel Q \xrightarrow{\alpha} P \parallel Q'}\ (bn(\alpha) \cap fn(P) = \emptyset)$$

$$\text{PAR}_3\ \frac{P \xrightarrow{\alpha} P' \quad Q \xrightarrow{\beta} Q'}{P \parallel Q \xrightarrow{\{\alpha,\beta\}} P' \parallel Q'}\ (\beta \neq \overline{\alpha}, bn(\alpha) \cap bn(\beta) = \emptyset, bn(\alpha) \cap fn(Q) = \emptyset, bn(\beta) \cap fn(P) = \emptyset)$$

$$\text{PAR}_4\ \frac{P \xrightarrow{x_1(z)} P' \quad Q \xrightarrow{x_2(z)} Q'}{P \parallel Q \xrightarrow{\{x_1(w), x_2(w)\}} P'\{w/z\} \parallel Q'\{w/z\}}\ (w \notin fn((z)P) \cup fn((z)Q))$$

$$\text{COM}\ \frac{P \xrightarrow{\overline{x}y} P' \quad Q \xrightarrow{x(z)} Q'}{P \parallel Q \xrightarrow{\tau} P' \parallel Q'\{y/z\}}$$

$$\text{CLOSE}\ \frac{P \xrightarrow{\overline{x}(w)} P' \quad Q \xrightarrow{x(w)} Q'}{P \parallel Q \xrightarrow{\tau} (w)(P' \parallel Q')}$$

$$\text{SUM}_1\ \frac{P \xrightarrow{\alpha} P'}{P + Q \xrightarrow{\alpha} P'} \qquad \text{SUM}_2\ \frac{P \xrightarrow{\{\alpha_1,\cdots,\alpha_n\}} P'}{P + Q \xrightarrow{\{\alpha_1,\cdots,\alpha_n\}} P'}$$

$$\text{IDE}_1\ \frac{P\{\widetilde{y}/\widetilde{x}\} \xrightarrow{\alpha} P'}{A(\widetilde{y}) \xrightarrow{\alpha} P'}\ (A(\widetilde{x}) \overset{\text{def}}{=} P) \qquad \text{IDE}_2\ \frac{P\{\widetilde{y}/\widetilde{x}\} \xrightarrow{\{\alpha_1,\cdots,\alpha_n\}} P'}{A(\widetilde{y}) \xrightarrow{\{\alpha_1,\cdots,\alpha_n\}} P'}\ (A(\widetilde{x}) \overset{\text{def}}{=} P)$$

$$\text{RES}_1\ \frac{P \xrightarrow{\alpha} P'}{(y)P \xrightarrow{\alpha} (y)P'}\ (y \notin n(\alpha)) \qquad \text{RES}_2\ \frac{P \xrightarrow{\{\alpha_1,\cdots,\alpha_n\}} P'}{(y)P \xrightarrow{\{\alpha_1,\cdots,\alpha_n\}} (y)P'}\ (y \notin n(\alpha_1) \cup \cdots \cup n(\alpha_n))$$

$$\text{OPEN}_1\ \frac{P \xrightarrow{\overline{x}y} P'}{(y)P \xrightarrow{\overline{x}(w)} P'\{w/y\}}\ (y \neq x, w \notin fn((y)P'))$$

$$\text{OPEN}_2\ \frac{P \xrightarrow{\{\overline{x}_1 y,\cdots,\overline{x}_n y\}} P'}{(y)P \xrightarrow{\{\overline{x}_1(w),\cdots,\overline{x}_n(w)\}} P'\{w/y\}}\ (y \neq x_1 \neq \cdots \neq x_n, w \notin fn((y)P'))$$

   (a) $fn(\alpha) \subseteq fn(P)$;
   (b) $fn(P') \subseteq fn(P) \cup bn(\alpha)$;

2. *If* $P \xrightarrow{\{\alpha_1,\cdots,\alpha_n\}} P'$, *then*
   (a) $fn(\alpha_1) \cup \cdots \cup fn(\alpha_n) \subseteq fn(P)$;
   (b) $fn(P') \subseteq fn(P) \cup bn(\alpha_1) \cup \cdots \cup bn(\alpha_n)$.

*Proof.* By induction on the depth of inference. □

**Proposition 12.7.** *Suppose that* $P \xrightarrow{\alpha(y)} P'$, *where* $\alpha = x$ *or* $\alpha = \overline{x}$, *and* $x \notin n(P)$, *then there exists some* $P'' \equiv_\alpha P'\{z/y\}$, $P \xrightarrow{\alpha(z)} P''$.

*Proof.* By induction on the depth of inference. □

**Proposition 12.8.** *If* $P \to P'$, $bn(\alpha) \cap fn(P'\sigma) = \emptyset$, *and* $\sigma \lceil bn(\alpha) = id$, *then there exists some* $P'' \equiv_\alpha P'\sigma$, $P\sigma \xrightarrow{\alpha\sigma} P''$.

*Proof.* By the definition of substitution (Definition 12.4) and induction on the depth of inference. □

**Proposition 12.9.** 1. *If* $P\{w/z\} \xrightarrow{\alpha} P'$, *where* $w \notin fn(P)$ *and* $bn(\alpha) \cap fn(P, w) = \emptyset$, *then there exist some* $Q$ *and* $\beta$ *with* $Q\{w/z\} \equiv_\alpha P'$ *and* $\beta\sigma = \alpha$, $P \xrightarrow{\beta} Q$;

2. *If* $P\{w/z\} \xrightarrow{\{\alpha_1,\cdots,\alpha_n\}} P'$, *where* $w \notin fn(P)$ *and* $bn(\alpha_1) \cap \cdots \cap bn(\alpha_n) \cap fn(P,w) = \emptyset$, *then there exist some* $Q$ *and* $\beta_1, \cdots, \beta_n$ *with* $Q\{w/z\} \equiv_\alpha P'$ *and* $\beta_1\sigma = \alpha_1, \cdots, \beta_n\sigma = \alpha_n$, $P \xrightarrow{\{\beta_1,\cdots,\beta_n\}} Q$.

*Proof.* By the definition of substitution (Definition 12.4) and induction on the depth of inference. □

## 12.2 Strongly probabilistic truly concurrent bisimilarities

### 12.2.1 Basic definitions

First, in this section, we introduce concepts of (strongly) probabilistic truly concurrent bisimilarities, including probabilistic pomset bisimilarity, probabilistic step bisimilarity, probabilistic history-preserving (hp-)bisimilarity, and probabilistic hereditary history-preserving (hhp-)bisimilarity. In contrast to traditional probabilistic truly concurrent bisimilarities in Chapter 2, these versions in $\pi_{ptc}$ must take care of actions with bound objects. Note that these probabilistic truly concurrent bisimilarities are defined as late bisimilarities, but not early bisimilarities, as defined in $\pi$-calculus [4] [5]. Note that here, a PES $\mathcal{E}$ is deemed to be a process.

**Definition 12.10** (Strongly probabilistic pomset, step bisimilarity 2). Let $\mathcal{E}_1, \mathcal{E}_2$ be PESs. A strongly probabilistic pomset bisimulation is a relation $R \subseteq \mathcal{C}(\mathcal{E}_1) \times \mathcal{C}(\mathcal{E}_2)$, such that (1) if $(C_1, C_2) \in R$, and $C_1 \xrightarrow{X_1} C_1'$ (with $\mathcal{E}_1 \xrightarrow{X_1} \mathcal{E}_1'$), then $C_2 \xrightarrow{X_2} C_2'$ (with $\mathcal{E}_2 \xrightarrow{X_2} \mathcal{E}_2'$), with $X_1 \subseteq \mathbb{E}_1$, $X_2 \subseteq \mathbb{E}_2$, $X_1 \sim X_2$ and $(C_1', C_2') \in R$:

1. for each fresh action $\alpha \in X_1$, if $C_1'' \xrightarrow{\alpha} C_1'''$ (with $\mathcal{E}_1'' \xrightarrow{\alpha} \mathcal{E}_1'''$), then for some $C_2''$ and $C_2'''$, $C_2'' \xrightarrow{\alpha} C_2'''$ (with $\mathcal{E}_2'' \xrightarrow{\alpha} \mathcal{E}_2'''$), such that if $(C_1'', C_2'') \in R$ then $(C_1''', C_2''') \in R$;
2. for each $x(y) \in X_1$ with $(y \notin n(\mathcal{E}_1, \mathcal{E}_2))$, if $C_1'' \xrightarrow{x(y)} C_1'''$ (with $\mathcal{E}_1'' \xrightarrow{x(y)} \mathcal{E}_1'''\{w/y\}$) for all $w$, then for some $C_2''$ and $C_2'''$, $C_2'' \xrightarrow{x(y)} C_2'''$ (with $\mathcal{E}_2'' \xrightarrow{x(y)} \mathcal{E}_2'''\{w/y\}$) for all $w$, such that if $(C_1'', C_2'') \in R$, then $(C_1''', C_2''') \in R$;
3. for each two $x_1(y), x_2(y) \in X_1$ with $(y \notin n(\mathcal{E}_1, \mathcal{E}_2))$, if $C_1'' \xrightarrow{\{x_1(y), x_2(y)\}} C_1'''$ (with $\mathcal{E}_1'' \xrightarrow{\{x_1(y), x_2(y)\}} \mathcal{E}_1'''\{w/y\}$) for all $w$, then for some $C_2''$ and $C_2'''$, $C_2'' \xrightarrow{\{x_1(y), x_2(y)\}} C_2'''$ (with $\mathcal{E}_2'' \xrightarrow{\{x_1(y), x_2(y)\}} \mathcal{E}_2'''\{w/y\}$) for all $w$, such that if $(C_1'', C_2'') \in R$ then $(C_1''', C_2''') \in R$;
4. for each $\overline{x}(y) \in X_1$ with $y \notin n(\mathcal{E}_1, \mathcal{E}_2)$, if $C_1'' \xrightarrow{\overline{x}(y)} C_1'''$ (with $\mathcal{E}_1'' \xrightarrow{\overline{x}(y)} \mathcal{E}_1'''$), then for some $C_2''$ and $C_2'''$, $C_2'' \xrightarrow{\overline{x}(y)} C_2'''$ (with $\mathcal{E}_2'' \xrightarrow{\overline{x}(y)} \mathcal{E}_2'''$), such that if $(C_1'', C_2'') \in R$ then $(C_1''', C_2''') \in R$,

and vice versa; (2) if $(C_1, C_2) \in R$, and $C_1 \xrightarrow{\pi} C_1^\pi$, then $C_2 \xrightarrow{\pi} C_2^\pi$ and $(C_1^\pi, C_2^\pi) \in R$, and vice versa; (3) if $(C_1, C_2) \in R$, then $\mu(C_1, C) = \mu(C_2, C)$ for each $C \in \mathcal{C}(\mathcal{E})/R$; (4) $[\surd]_R = \{\surd\}$.

We say that $\mathcal{E}_1, \mathcal{E}_2$ are strongly probabilistic pomset bisimilar, written $\mathcal{E}_1 \sim_{pp} \mathcal{E}_2$, if there exists a strongly probabilistic pomset bisimulation $R$, such that $(\emptyset, \emptyset) \in R$. By replacing probabilistic pomset transitions with steps, we can obtain the definition of strongly probabilistic step bisimulation. When PESs $\mathcal{E}_1$ and $\mathcal{E}_2$ are strongly probabilistic step bisimilar, we write $\mathcal{E}_1 \sim_{ps} \mathcal{E}_2$.

**Definition 12.11** (Strongly probabilistic (hereditary) history-preserving bisimilarity 2). A strongly probabilistic history-preserving (hp-) bisimulation is a posetal relation $R \subseteq \mathcal{C}(\mathcal{E}_1)\overline{\times}\mathcal{C}(\mathcal{E}_2)$ such that (1) if $(C_1, f, C_2) \in R$, and

1. for $e_1 = \alpha$ a fresh action, if $C_1 \xrightarrow{\alpha} C_1'$ (with $\mathcal{E}_1 \xrightarrow{\alpha} \mathcal{E}_1'$), then for some $C_2'$ and $e_2 = \alpha$, $C_2 \xrightarrow{\alpha} C_2'$ (with $\mathcal{E}_2 \xrightarrow{\alpha} \mathcal{E}_2'$), such that $(C_1', f[e_1 \mapsto e_2], C_2') \in R$;
2. for $e_1 = x(y)$ with $(y \notin n(\mathcal{E}_1, \mathcal{E}_2))$, if $C_1 \xrightarrow{x(y)} C_1'$ (with $\mathcal{E}_1 \xrightarrow{x(y)} \mathcal{E}_1'\{w/y\}$) for all $w$, then for some $C_2'$ and $e_2 = x(y)$, $C_2 \xrightarrow{x(y)} C_2'$ (with $\mathcal{E}_2 \xrightarrow{x(y)} \mathcal{E}_2'\{w/y\}$) for all $w$, such that $(C_1', f[e_1 \mapsto e_2], C_2') \in R$;
3. for $e_1 = \overline{x}(y)$ with $y \notin n(\mathcal{E}_1, \mathcal{E}_2)$, if $C_1 \xrightarrow{\overline{x}(y)} C_1'$ (with $\mathcal{E}_1 \xrightarrow{\overline{x}(y)} \mathcal{E}_1'$), then for some $C_2'$ and $e_2 = \overline{x}(y)$, $C_2 \xrightarrow{\overline{x}(y)} C_2'$ (with $\mathcal{E}_2 \xrightarrow{\overline{x}(y)} \mathcal{E}_2'$), such that $(C_1', f[e_1 \mapsto e_2], C_2') \in R$,

and vice versa; (2) if $(C_1, f, C_2) \in R$, and $C_1 \xrightarrow{\pi} C_1^\pi$, then $C_2 \xrightarrow{\pi} C_2^\pi$ and $(C_1^\pi, f, C_2^\pi) \in R$, and vice versa; (3) if $(C_1, f, C_2) \in R$, then $\mu(C_1, C) = \mu(C_2, C)$ for each $C \in \mathcal{C}(\mathcal{E})/R$; (4) $[\surd]_R = \{\surd\}$. $\mathcal{E}_1, \mathcal{E}_2$ are strongly probabilistic history-preserving (hp-)bisimilar and are written $\mathcal{E}_1 \sim_{php} \mathcal{E}_2$ if there exists a strongly probabilistic hp-bisimulation $R$ such that $(\emptyset, \emptyset, \emptyset) \in R$.

A strongly probabilistic hereditary history-preserving (hhp-)bisimulation is a downward closed strongly probabilistic hp-bisimulation. $\mathcal{E}_1, \mathcal{E}_2$ are strongly probabilistic hereditary history-preserving (hhp-)bisimilar and are written $\mathcal{E}_1 \sim_{phhp} \mathcal{E}_2$.

**Theorem 12.12.** *$\equiv_\alpha$ are strongly probabilistic truly concurrent bisimulations. That is, if $P \equiv_\alpha Q$, then*

1. $P \sim_{pp} Q$;
2. $P \sim_{ps} Q$;
3. $P \sim_{php} Q$;
4. $P \sim_{phhp} Q$.

*Proof.* By induction on the depth of inference (see Table 12.2), we can obtain the following facts:

1. If $\alpha$ is a free action and $P \rightsquigarrow\xrightarrow{\alpha} P'$, then equally for some $Q'$ with $P' \equiv_\alpha Q'$, $Q \rightsquigarrow\xrightarrow{\alpha} Q'$;
2. If $P \rightsquigarrow\xrightarrow{a(y)} P'$ with $a = x$ or $a = \overline{x}$ and $z \notin n(Q)$, then equally for some $Q'$ with $P'\{z/y\} \equiv_\alpha Q'$, $Q \rightsquigarrow\xrightarrow{a(z)} Q'$.

   Then, we can obtain:

1. by the definition of strongly probabilistic pomset bisimilarity (Definition 12.10), $P \sim_{pp} Q$;
2. by the definition of strongly probabilistic step bisimilarity (Definition 12.10), $P \sim_{ps} Q$;
3. by the definition of strongly probabilistic hp-bisimilarity (Definition 12.11), $P \sim_{php} Q$;
4. by the definition of strongly probabilistic hhp-bisimilarity (Definition 12.11), $P \sim_{phhp} Q$. □

### 12.2.2 Laws and congruence

Similarly to CPTC, we can obtain the following laws with respect to probabilistic truly concurrent bisimilarities.

**Theorem 12.13** (Summation laws for strongly probabilistic pomset bisimilarity). *The summation laws for strongly probabilistic pomset bisimilarity are as follows:*

1. $P + \mathbf{nil} \sim_{pp} P$;
2. $P + P \sim_{pp} P$;
3. $P_1 + P_2 \sim_{pp} P_2 + P_1$;
4. $P_1 + (P_2 + P_3) \sim_{pp} (P_1 + P_2) + P_3$.

*Proof.* According to the definition of strongly probabilistic pomset bisimulation, we can easily prove the above equations, however, we omit the proof here. □

**Theorem 12.14** (Summation laws for strongly probabilistic step bisimilarity). *The summation laws for strongly probabilistic step bisimilarity are as follows:*

1. $P + \mathbf{nil} \sim_{ps} P$;
2. $P + P \sim_{ps} P$;
3. $P_1 + P_2 \sim_{ps} P_2 + P_1$;
4. $P_1 + (P_2 + P_3) \sim_{ps} (P_1 + P_2) + P_3$.

*Proof.* According to the definition of strongly probabilistic step bisimulation, we can easily prove the above equations, however, we omit the proof here. □

**Theorem 12.15** (Summation laws for strongly probabilistic hp-bisimilarity). *The summation laws for strongly probabilistic hp-bisimilarity are as follows:*

1. $P + \mathbf{nil} \sim_{php} P$;
2. $P + P \sim_{php} P$;
3. $P_1 + P_2 \sim_{php} P_2 + P_1$;
4. $P_1 + (P_2 + P_3) \sim_{php} (P_1 + P_2) + P_3$.

*Proof.* According to the definition of strongly probabilistic hp-bisimulation, we can easily prove the above equations, however, we omit the proof here. □

**Theorem 12.16** (Summation laws for strongly probabilistic hhp-bisimilarity). *The summation laws for strongly probabilistic hhp-bisimilarity are as follows:*

1. $P + \textbf{nil} \sim_{phhp} P$;
2. $P + P \sim_{phhp} P$;
3. $P_1 + P_2 \sim_{phhp} P_2 + P_1$;
4. $P_1 + (P_2 + P_3) \sim_{phhp} (P_1 + P_2) + P_3$.

*Proof.* According to the definition of strongly probabilistic hhp-bisimulation, we can easily prove the above equations, however, we omit the proof here. □

**Proposition 12.17** (Box-Summation laws for strongly probabilistic pomset bisimulation). *The box-summation laws for strongly probabilistic pomset bisimulation are as follows:*

1. $P \boxplus_\pi \textbf{nil} \sim_{pp} P$;
2. $P \boxplus_\pi P \sim_{pp} P$;
3. $P_1 \boxplus_\pi P_2 \sim_{pp} P_2 \boxplus_{1-\pi} P_1$;
4. $P_1 \boxplus_\pi (P_2 \boxplus_\rho P_3) \sim_{pp} (P_1 \boxplus_{\frac{\pi}{\pi+\rho-\pi\rho}} P_2) \boxplus_{\pi+\rho-\pi\rho} P_3$.

*Proof.* According to the definition of strongly probabilistic pomset bisimulation, we can easily prove the above equations, however, we omit the proof here. □

**Proposition 12.18** (Box-Summation laws for strongly probabilistic step bisimulation). *The box-summation laws for strongly probabilistic step bisimulation are as follows:*

1. $P \boxplus_\pi \textbf{nil} \sim_{ps} P$;
2. $P \boxplus_\pi P \sim_{ps} P$;
3. $P_1 \boxplus_\pi P_2 \sim_{ps} P_2 \boxplus_{1-\pi} P_1$;
4. $P_1 \boxplus_\pi (P_2 \boxplus_\rho P_3) \sim_{ps} (P_1 \boxplus_{\frac{\pi}{\pi+\rho-\pi\rho}} P_2) \boxplus_{\pi+\rho-\pi\rho} P_3$.

*Proof.* According to the definition of strongly probabilistic step bisimulation, we can easily prove the above equations, however, we omit the proof here. □

**Proposition 12.19** (Box-Summation laws for strongly probabilistic hp-bisimulation). *The box-summation laws for strongly probabilistic hp-bisimulation are as follows:*

1. $P \boxplus_\pi \textbf{nil} \sim_{php} P$;
2. $P \boxplus_\pi P \sim_{php} P$;
3. $P_1 \boxplus_\pi P_2 \sim_{php} P_2 \boxplus_{1-\pi} P_1$;
4. $P_1 \boxplus_\pi (P_2 \boxplus_\rho P_3) \sim_{php} (P_1 \boxplus_{\frac{\pi}{\pi+\rho-\pi\rho}} P_2) \boxplus_{\pi+\rho-\pi\rho} P_3$.

*Proof.* According to the definition of strongly probabilistic hp-bisimulation, we can easily prove the above equations, however, we omit the proof here. □

**Proposition 12.20** (Box-Summation laws for strongly probabilistic hhp-bisimulation). *The box-summation laws for strongly probabilistic hhp-bisimulation are as follows:*

1. $P \boxplus_\pi \textbf{nil} \sim_{phhp} P$;
2. $P \boxplus_\pi P \sim_{phhp} P$;
3. $P_1 \boxplus_\pi P_2 \sim_{phhp} P_2 \boxplus_{1-\pi} P_1$;
4. $P_1 \boxplus_\pi (P_2 \boxplus_\rho P_3) \sim_{phhp} (P_1 \boxplus_{\frac{\pi}{\pi+\rho-\pi\rho}} P_2) \boxplus_{\pi+\rho-\pi\rho} P_3$.

*Proof.* According to the definition of strongly probabilistic hhp-bisimulation, we can easily prove the above equations, however, we omit the proof here. □

**Theorem 12.21** (Identity law for probabilistic truly concurrent bisimilarities). *If* $A(\widetilde{x}) \stackrel{def}{=} P$, *then*

1. $A(\widetilde{y}) \sim_{pp} P\{\widetilde{y}/\widetilde{x}\}$;
2. $A(\widetilde{y}) \sim_{ps} P\{\widetilde{y}/\widetilde{x}\}$;
3. $A(\widetilde{y}) \sim_{php} P\{\widetilde{y}/\widetilde{x}\}$;
4. $A(\widetilde{y}) \sim_{phhp} P\{\widetilde{y}/\widetilde{x}\}$.

*Proof.* According to the definition of strongly probabilistic truly concurrent bisimulations, we can easily prove the above equations, however, we omit the proof here. □

**Theorem 12.22** (Restriction laws for strongly probabilistic pomset bisimilarity). *The restriction laws for strongly probabilistic pomset bisimilarity are as follows:*

1. $(y)P \sim_{pp} P$, *if* $y \notin fn(P)$;
2. $(y)(z)P \sim_{pp} (z)(y)P$;
3. $(y)(P + Q) \sim_{pp} (y)P + (y)Q$;
4. $(y)(P \boxplus_\pi Q) \sim_{pp} (y)P \boxplus_\pi (y)Q$;
5. $(y)\alpha.P \sim_{pp} \alpha.(y)P$ *if* $y \notin n(\alpha)$;
6. $(y)\alpha.P \sim_{pp}$ ***nil*** *if* $y$ *is the subject of* $\alpha$.

*Proof.* According to the definition of strongly probabilistic pomset bisimulation, we can easily prove the above equations, however, we omit the proof here. □

**Theorem 12.23** (Restriction laws for strongly probabilistic step bisimilarity). *The restriction laws for strongly probabilistic step bisimilarity are as follows:*

1. $(y)P \sim_{ps} P$, *if* $y \notin fn(P)$;
2. $(y)(z)P \sim_{ps} (z)(y)P$;
3. $(y)(P + Q) \sim_{ps} (y)P + (y)Q$;
4. $(y)(P \boxplus_\pi Q) \sim_{ps} (y)P \boxplus_\pi (y)Q$;
5. $(y)\alpha.P \sim_{ps} \alpha.(y)P$ *if* $y \notin n(\alpha)$;
6. $(y)\alpha.P \sim_{ps}$ ***nil*** *if* $y$ *is the subject of* $\alpha$.

*Proof.* According to the definition of strongly probabilistic step bisimulation, we can easily prove the above equations, however, we omit the proof here. □

**Theorem 12.24** (Restriction laws for strongly probabilistic hp-bisimilarity). *The restriction laws for strongly probabilistic hp-bisimilarity are as follows:*

1. $(y)P \sim_{php} P$, *if* $y \notin fn(P)$;
2. $(y)(z)P \sim_{php} (z)(y)P$;
3. $(y)(P + Q) \sim_{php} (y)P + (y)Q$;
4. $(y)(P \boxplus_\pi Q) \sim_{php} (y)P \boxplus_\pi (y)Q$;
5. $(y)\alpha.P \sim_{php} \alpha.(y)P$ *if* $y \notin n(\alpha)$;
6. $(y)\alpha.P \sim_{php}$ ***nil*** *if* $y$ *is the subject of* $\alpha$.

*Proof.* According to the definition of strongly probabilistic hp-bisimulation, we can easily prove the above equations, however, we omit the proof here. □

**Theorem 12.25** (Restriction laws for strongly probabilistic hhp-bisimilarity). *The restriction laws for strongly probabilistic hhp-bisimilarity are as follows:*

1. $(y)P \sim_{phhp} P$, *if* $y \notin fn(P)$;
2. $(y)(z)P \sim_{phhp} (z)(y)P$;
3. $(y)(P + Q) \sim_{phhp} (y)P + (y)Q$;
4. $(y)(P \boxplus_\pi Q) \sim_{phhp} (y)P \boxplus_\pi (y)Q$;
5. $(y)\alpha.P \sim_{phhp} \alpha.(y)P$ *if* $y \notin n(\alpha)$;
6. $(y)\alpha.P \sim_{phhp}$ ***nil*** *if* $y$ *is the subject of* $\alpha$.

*Proof.* According to the definition of strongly probabilistic hhp-bisimulation, we can easily prove the above equations, however, we omit the proof here. □

**Theorem 12.26** (Parallel laws for strongly probabilistic pomset bisimilarity). *The parallel laws for strongly probabilistic pomset bisimilarity are as follows:*

1. $P \parallel$ ***nil*** $\sim_{pp} P$;
2. $P_1 \parallel P_2 \sim_{pp} P_2 \parallel P_1$;
3. $(P_1 \parallel P_2) \parallel P_3 \sim_{pp} P_1 \parallel (P_2 \parallel P_3)$;
4. $(y)(P_1 \parallel P_2) \sim_{pp} (y)P_1 \parallel (y)P_2$, *if* $y \notin fn(P_1) \cap fn(P_2)$.

*Proof.* According to the definition of strongly probabilistic pomset bisimulation, we can easily prove the above equations, however, we omit the proof here. □

**Theorem 12.27** (Parallel laws for strongly probabilistic step bisimilarity). *The parallel laws for strongly probabilistic step bisimilarity are as follows:*

1. $P \parallel \mathbf{nil} \sim_{ps} P$;
2. $P_1 \parallel P_2 \sim_{ps} P_2 \parallel P_1$;
3. $(P_1 \parallel P_2) \parallel P_3 \sim_{ps} P_1 \parallel (P_2 \parallel P_3)$;
4. $(y)(P_1 \parallel P_2) \sim_{ps} (y)P_1 \parallel (y)P_2$, *if* $y \notin fn(P_1) \cap fn(P_2)$.

*Proof.* According to the definition of strongly probabilistic step bisimulation, we can easily prove the above equations, however, we omit the proof here. □

**Theorem 12.28** (Parallel laws for strongly probabilistic hp-bisimilarity). *The parallel laws for strongly probabilistic hp-bisimilarity are as follows:*

1. $P \parallel \mathbf{nil} \sim_{php} P$;
2. $P_1 \parallel P_2 \sim_{php} P_2 \parallel P_1$;
3. $(P_1 \parallel P_2) \parallel P_3 \sim_{php} P_1 \parallel (P_2 \parallel P_3)$;
4. $(y)(P_1 \parallel P_2) \sim_{php} (y)P_1 \parallel (y)P_2$, *if* $y \notin fn(P_1) \cap fn(P_2)$.

*Proof.* According to the definition of strongly probabilistic hp-bisimulation, we can easily prove the above equations, however, we omit the proof here. □

**Theorem 12.29** (Parallel laws for strongly probabilistic hhp-bisimilarity). *The parallel laws for strongly probabilistic hhp-bisimilarity are as follows:*

1. $P \parallel \mathbf{nil} \sim_{phhp} P$;
2. $P_1 \parallel P_2 \sim_{phhp} P_2 \parallel P_1$;
3. $(P_1 \parallel P_2) \parallel P_3 \sim_{phhp} P_1 \parallel (P_2 \parallel P_3)$;
4. $(y)(P_1 \parallel P_2) \sim_{phhp} (y)P_1 \parallel (y)P_2$, *if* $y \notin fn(P_1) \cap fn(P_2)$.

*Proof.* According to the definition of strongly probabilistic hhp-bisimulation, we can easily prove the above equations, however, we omit the proof here. □

**Theorem 12.30** (Expansion law for probabilistic truly concurrent bisimilarities). *Let* $P \equiv \boxplus_i \sum_j \alpha_{ij}.P_{ij}$ *and* $Q \equiv \boxplus_k \sum_l \beta_{kl}.Q_{kl}$, *where* $bn(\alpha_{ij}) \cap fn(Q) = \emptyset$ *for all* $i, j$, *and* $bn(\beta_{kl}) \cap fn(P) = \emptyset$ *for all* $k, l$. *Then,*

1. $P \parallel Q \sim_{pp} \boxplus_i \boxplus_k \sum_j \sum_l (\alpha_{ij} \parallel \beta_{kl}).(P_{ij} \parallel Q_{kl}) + \boxplus_i \boxplus_k \sum_{\alpha_{ij}\ comp\ \beta_{kl}} \tau.R_{ijkl}$;
2. $P \parallel Q \sim_{ps} \boxplus_i \boxplus_k \sum_j \sum_l (\alpha_{ij} \parallel \beta_{kl}).(P_{ij} \parallel Q_{kl}) + \boxplus_i \boxplus_k \sum_{\alpha_{ij}\ comp\ \beta_{kl}} \tau.R_{ijkl}$;
3. $P \parallel Q \sim_{php} \boxplus_i \boxplus_k \sum_j \sum_l (\alpha_{ij} \parallel \beta_{kl}).(P_{ij} \parallel Q_{kl}) + \boxplus_i \boxplus_k \sum_{\alpha_{ij}\ comp\ \beta_{kl}} \tau.R_{ijkl}$;
4. $P \parallel Q \nsim_{phhp} \boxplus_i \boxplus_k \sum_j \sum_l (\alpha_{ij} \parallel \beta_{kl}).(P_{ij} \parallel Q_{kl}) + \boxplus_i \boxplus_k \sum_{\alpha_{ij}\ comp\ \beta_{kl}} \tau.R_{ijkl}$,

*where* $\alpha_{ij}\ comp\ \beta_{kl}$ *and* $R_{ijkl}$ *are defined as follows:*

1. $\alpha_{ij}$ *is* $\overline{x}u$ *and* $\beta_{kl}$ *is* $x(v)$, *then* $R_{ijkl} = P_{ij} \parallel Q_{kl}\{u/v\}$;
2. $\alpha_{ij}$ *is* $\overline{x}(u)$ *and* $\beta_{kl}$ *is* $x(v)$, *then* $R_{ijkl} = (w)(P_{ij}\{w/u\} \parallel Q_{kl}\{w/v\})$, *if* $w \notin fn((u)P_{ij}) \cup fn((v)Q_{kl})$;
3. $\alpha_{ij}$ *is* $x(v)$ *and* $\beta_{kl}$ *is* $\overline{x}u$, *then* $R_{ijkl} = P_{ij}\{u/v\} \parallel Q_{kl}$;
4. $\alpha_{ij}$ *is* $x(v)$ *and* $\beta_{kl}$ *is* $\overline{x}(u)$, *then* $R_{ijkl} = (w)(P_{ij}\{w/v\} \parallel Q_{kl}\{w/u\})$, *if* $w \notin fn((v)P_{ij}) \cup fn((u)Q_{kl})$.

*Proof.* According to the definition of strongly probabilistic truly concurrent bisimulations, we can easily prove the above equations, however, we omit the proof here. □

**Theorem 12.31** (Equivalence and congruence for strongly probabilistic pomset bisimilarity). **1.** $\sim_{pp}$ *is an equivalence relation;*

2. *If* $P \sim_{pp} Q$, *then*
   (a) $\alpha.P \sim_{pp} \alpha.Q$, $\alpha$ *is a free action;*
   (b) $P + R \sim_{pp} Q + R$;
   (c) $P \boxplus_\pi R \sim_{pp} Q \boxplus_\pi R$;
   (d) $P \parallel R \sim_{pp} Q \parallel R$;

(e) $(w)P \sim_{pp} (w)Q$;
(f) $x(y).P \sim_{pp} x(y).Q$.

*Proof.* According to the definition of strongly probabilistic pomset bisimulation, we can easily prove the above equations, however, we omit the proof here. □

**Theorem 12.32** (Equivalence and congruence for strongly probabilistic step bisimilarity). **1.** $\sim_{ps}$ *is an equivalence relation;*

**2.** *If* $P \sim_{ps} Q$, *then*
(a) $\alpha.P \sim_{ps} \alpha.Q$, $\alpha$ *is a free action;*
(b) $P + R \sim_{ps} Q + R$;
(c) $P \boxplus_\pi R \sim_{ps} Q \boxplus_\pi R$;
(d) $P \parallel R \sim_{ps} Q \parallel R$;
(e) $(w)P \sim_{ps} (w)Q$;
(f) $x(y).P \sim_{ps} x(y).Q$.

*Proof.* According to the definition of strongly probabilistic step bisimulation, we can easily prove the above equations, however, we omit the proof here. □

**Theorem 12.33** (Equivalence and congruence for strongly probabilistic hp-bisimilarity). **1.** $\sim_{php}$ *is an equivalence relation;*

**2.** *If* $P \sim_{php} Q$, *then*
(a) $\alpha.P \sim_{php} \alpha.Q$, $\alpha$ *is a free action;*
(b) $P + R \sim_{php} Q + R$;
(c) $P \boxplus_\pi R \sim_{php} Q \boxplus_\pi R$;
(d) $P \parallel R \sim_{php} Q \parallel R$;
(e) $(w)P \sim_{php} (w)Q$;
(f) $x(y).P \sim_{php} x(y).Q$.

*Proof.* According to the definition of strongly probabilistic hp-bisimulation, we can easily prove the above equations, however, we omit the proof here. □

**Theorem 12.34** (Equivalence and congruence for strongly probabilistic hhp-bisimilarity). **1.** $\sim_{phhp}$ *is an equivalence relation;*

**2.** *If* $P \sim_{phhp} Q$, *then*
(a) $\alpha.P \sim_{phhp} \alpha.Q$, $\alpha$ *is a free action;*
(b) $P + R \sim_{phhp} Q + R$;
(c) $P \boxplus_\pi R \sim_{phhp} Q \boxplus_\pi R$;
(d) $P \parallel R \sim_{phhp} Q \parallel R$;
(e) $(w)P \sim_{phhp} (w)Q$;
(f) $x(y).P \sim_{phhp} x(y).Q$.

*Proof.* According to the definition of strongly probabilistic hhp-bisimulation, we can easily prove the above equations, however, we omit the proof here. □

### 12.2.3 Recursion

**Definition 12.35.** Let $X$ have arity $n$, and let $\widetilde{x} = x_1, \cdots, x_n$ be distinct names, and $fn(P) \subseteq \{x_1, \cdots, x_n\}$. The replacement of $X(\widetilde{x})$ by $P$ in $E$, written $E\{X(\widetilde{x}) := P\}$, means the result of replacing each sub-term $X(\widetilde{y})$ in $E$ by $P\{\widetilde{y}/\widetilde{x}\}$.

**Definition 12.36.** Let $E$ and $F$ be two process expressions containing only $X_1, \cdots, X_m$ with associated name sequences $\widetilde{x}_1, \cdots, \widetilde{x}_m$. Then,

1. $E \sim_{pp} F$ means $E(\widetilde{P}) \sim_{pp} F(\widetilde{P})$;
2. $E \sim_{ps} F$ means $E(\widetilde{P}) \sim_{ps} F(\widetilde{P})$;
3. $E \sim_{php} F$ means $E(\widetilde{P}) \sim_{php} F(\widetilde{P})$;
4. $E \sim_{phhp} F$ means $E(\widetilde{P}) \sim_{phhp} F(\widetilde{P})$;

for all $\widetilde{P}$ such that $fn(P_i) \subseteq \widetilde{x}_i$ for each $i$.

**Definition 12.37.** A term or identifier is weakly guarded in $P$ if it lies within some sub-term $\alpha.Q$ or $(\alpha_1 \parallel \cdots \parallel \alpha_n).Q$ of $P$.

**Theorem 12.38.** *Assume that $\widetilde{E}$ and $\widetilde{F}$ are expressions containing only $X_i$ with $\widetilde{x}_i$, and $\widetilde{A}$ and $\widetilde{B}$ are identifiers with $A_i$, $B_i$. Then, for all $i$,*

1. $E_i \sim_{ps} F_i$, $A_i(\widetilde{x}_i) \overset{def}{=} E_i(\widetilde{A})$, $B_i(\widetilde{x}_i) \overset{def}{=} F_i(\widetilde{B})$, *then* $A_i(\widetilde{x}_i) \sim_{ps} B_i(\widetilde{x}_i)$;
2. $E_i \sim_{pp} F_i$, $A_i(\widetilde{x}_i) \overset{def}{=} E_i(\widetilde{A})$, $B_i(\widetilde{x}_i) \overset{def}{=} F_i(\widetilde{B})$, *then* $A_i(\widetilde{x}_i) \sim_{pp} B_i(\widetilde{x}_i)$;
3. $E_i \sim_{php} F_i$, $A_i(\widetilde{x}_i) \overset{def}{=} E_i(\widetilde{A})$, $B_i(\widetilde{x}_i) \overset{def}{=} F_i(\widetilde{B})$, *then* $A_i(\widetilde{x}_i) \sim_{php} B_i(\widetilde{x}_i)$;
4. $E_i \sim_{phhp} F_i$, $A_i(\widetilde{x}_i) \overset{def}{=} E_i(\widetilde{A})$, $B_i(\widetilde{x}_i) \overset{def}{=} F_i(\widetilde{B})$, *then* $A_i(\widetilde{x}_i) \sim_{phhp} B_i(\widetilde{x}_i)$.

*Proof.* 1. $E_i \sim_{ps} F_i$, $A_i(\widetilde{x}_i) \overset{def}{=} E_i(\widetilde{A})$, $B_i(\widetilde{x}_i) \overset{def}{=} F_i(\widetilde{B})$, then $A_i(\widetilde{x}_i) \sim_{ps} B_i(\widetilde{x}_i)$.
We will consider the case $I = \{1\}$ with loss of generality, and show the following relation $R$ is a strongly probabilistic step bisimulation.

$$R = \{(G(A), G(B)) : G \text{ has only identifier } X\}$$

By choosing $G \equiv X(\widetilde{y})$, it follows that $A(\widetilde{y}) \sim_{ps} B(\widetilde{y})$. It is sufficient to prove the following:

**(a)** If $G(A) \rightsquigarrow \xrightarrow{\{\alpha_1,\cdots,\alpha_n\}} P'$, where $\alpha_i (1 \leq i \leq n)$ is a free action or bound output action with $bn(\alpha_1) \cap \cdots \cap bn(\alpha_n) \cap n(G(A), G(B)) = \emptyset$, then $G(B) \rightsquigarrow \xrightarrow{\{\alpha_1,\cdots,\alpha_n\}} Q''$ such that $P' \sim_{ps} Q''$;

**(b)** If $G(A) \rightsquigarrow \xrightarrow{x(y)} P'$ with $x \notin n(G(A), G(B))$, then $G(B) \rightsquigarrow \xrightarrow{x(y)} Q''$, such that for all $u$, $P'\{u/y\} \sim_{ps} Q''\{u/y\}$.

To prove the above properties, it is sufficient to induct on the depth of inference and quite routine, however, we omit the proof here.

2. $E_i \sim_{pp} F_i$, $A_i(\widetilde{x}_i) \overset{def}{=} E_i(\widetilde{A})$, $B_i(\widetilde{x}_i) \overset{def}{=} F_i(\widetilde{B})$, then $A_i(\widetilde{x}_i) \sim_{pp} B_i(\widetilde{x}_i)$. This can be proven similarly to the above case.
3. $E_i \sim_{php} F_i$, $A_i(\widetilde{x}_i) \overset{def}{=} E_i(\widetilde{A})$, $B_i(\widetilde{x}_i) \overset{def}{=} F_i(\widetilde{B})$, then $A_i(\widetilde{x}_i) \sim_{php} B_i(\widetilde{x}_i)$. This can be proven similarly to the above case.
4. $E_i \sim_{phhp} F_i$, $A_i(\widetilde{x}_i) \overset{def}{=} E_i(\widetilde{A})$, $B_i(\widetilde{x}_i) \overset{def}{=} F_i(\widetilde{B})$, then $A_i(\widetilde{x}_i) \sim_{phhp} B_i(\widetilde{x}_i)$. This can be proven similarly to the above case. □

**Theorem 12.39** (Unique solution of equations). *Assume $\widetilde{E}$ are expressions containing only $X_i$ with $\widetilde{x}_i$, and each $X_i$ is weakly guarded in each $E_j$. Assume that $\widetilde{P}$ and $\widetilde{Q}$ are processes such that $fn(P_i) \subseteq \widetilde{x}_i$ and $fn(Q_i) \subseteq \widetilde{x}_i$. Then, for all $i$,*

1. *if* $P_i \sim_{pp} E_i(\widetilde{P})$, $Q_i \sim_{pp} E_i(\widetilde{Q})$, *then* $P_i \sim_{pp} Q_i$;
2. *if* $P_i \sim_{ps} E_i(\widetilde{P})$, $Q_i \sim_{ps} E_i(\widetilde{Q})$, *then* $P_i \sim_{ps} Q_i$;
3. *if* $P_i \sim_{php} E_i(\widetilde{P})$, $Q_i \sim_{php} E_i(\widetilde{Q})$, *then* $P_i \sim_{php} Q_i$;
4. *if* $P_i \sim_{phhp} E_i(\widetilde{P})$, $Q_i \sim_{phhp} E_i(\widetilde{Q})$, *then* $P_i \sim_{phhp} Q_i$.

*Proof.* 1. It is similar to the proof of the unique solution of equations for strongly probabilistic pomset bisimulation in CPTC, however, we omit it here;
2. It is similar to the proof of the unique solution of equations for strongly probabilistic step bisimulation in CPTC, however, we omit it here;
3. It is similar to the proof of the unique solution of equations for strongly probabilistic hp-bisimulation in CPTC, however, we omit it here;
4. It is similar to the proof of the unique solution of equations for strongly probabilistic hhp-bisimulation in CPTC, however, we omit it here. □

## 12.3 Algebraic theory

In this section, we will try to axiomatize $\pi_{ptc}$, the theory is **SPTC** (for strongly probabilistic true concurrency).

**Definition 12.40** (SPTC). The theory **SPTC** consists of the following axioms and inference rules:

1. Alpha-conversion **A**.

$$\text{if } P \equiv Q, \text{ then } P = Q;$$

**2.** Congruence **C**. If $P = Q$, then

$$\tau.P = \tau.Q \quad \overline{x}y.P = \overline{x}y.Q$$
$$P + R = Q + R \quad P \parallel R = Q \parallel R$$
$$(x)P = (x)Q \quad x(y).P = x(y).Q;$$

**3.** Summation **S**.

$$\textbf{S0} \quad P + \textbf{nil} = P$$
$$\textbf{S1} \quad P + P = P$$
$$\textbf{S2} \quad P + Q = Q + P$$
$$\textbf{S3} \quad P + (Q + R) = (P + Q) + R;$$

**4.** Box-Summation ($BS$).

$$\textbf{BS0} \quad P \boxplus_\pi \textbf{nil} = P$$
$$\textbf{BS1} \quad P \boxplus_\pi P = P$$
$$\textbf{BS2} \quad P \boxplus_\pi Q = Q \boxplus_{1-\pi} P$$
$$\textbf{BS3} \quad P \boxplus_\pi (Q \boxplus_\rho R) = (P \boxplus_{\frac{\pi}{\pi+\rho-\pi\rho}} Q) \boxplus_{\pi+\rho-\pi\rho} R;$$

**5.** Restriction **R**.

$$\textbf{R0} \quad (x)P = P \quad \text{if } x \notin fn(P)$$
$$\textbf{R1} \quad (x)(y)P = (y)(x)P$$
$$\textbf{R2} \quad (x)(P + Q) = (x)P + (x)Q$$
$$\textbf{R3} \quad (x)\alpha.P = \alpha.(x)P \quad \text{if } x \notin n(\alpha)$$
$$\textbf{R4} \quad (x)\alpha.P = \textbf{nil} \quad \text{if } x \text{is the subject of } \alpha;$$

**6.** Expansion **E**. Let $P \equiv \boxplus_i \sum_j \alpha_{ij}.P_{ij}$ and $Q \equiv \boxplus_k \sum_l \beta_{kl}.Q_{kl}$, where $bn(\alpha_{ij}) \cap fn(Q) = \emptyset$ for all $i, j$, and $bn(\beta_{kl}) \cap fn(P) = \emptyset$ for all $k, l$. Then,
**(a)** $P \parallel Q \sim_{pp} \boxplus_i \boxplus_k \sum_j \sum_l (\alpha_{ij} \parallel \beta_{kl}).(P_{ij} \parallel Q_{kl}) + \boxplus_i \boxplus_k \sum_{\alpha_{ij} \text{ comp } \beta_{kl}} \tau.R_{ijkl}$;
**(b)** $P \parallel Q \sim_{ps} \boxplus_i \boxplus_k \sum_j \sum_l (\alpha_{ij} \parallel \beta_{kl}).(P_{ij} \parallel Q_{kl}) + \boxplus_i \boxplus_k \sum_{\alpha_{ij} \text{ comp } \beta_{kl}} \tau.R_{ijkl}$;
**(c)** $P \parallel Q \sim_{php} \boxplus_i \boxplus_k \sum_j \sum_l (\alpha_{ij} \parallel \beta_{kl}).(P_{ij} \parallel Q_{kl}) + \boxplus_i \boxplus_k \sum_{\alpha_{ij} \text{ comp } \beta_{kl}} \tau.R_{ijkl}$;
**(d)** $P \parallel Q \nsim_{phhp} \boxplus_i \boxplus_k \sum_j \sum_l (\alpha_{ij} \parallel \beta_{kl}).(P_{ij} \parallel Q_{kl}) + \boxplus_i \boxplus_k \sum_{\alpha_{ij} \text{ comp } \beta_{kl}} \tau.R_{ijkl}$,
where $\alpha_{ij}$ comp $\beta_{kl}$ and $R_{ijkl}$ are defined as follows:
**(a)** $\alpha_{ij}$ is $\overline{x}u$ and $\beta_{kl}$ is $x(v)$, then $R_{ijkl} = P_{ij} \parallel Q_{kl}\{u/v\}$;
**(b)** $\alpha_{ij}$ is $\overline{x}(u)$ and $\beta_{kl}$ is $x(v)$, then $R_{ijkl} = (w)(P_{ij}\{w/u\} \parallel Q_{kl}\{w/v\})$, if $w \notin fn((u)P_{ij}) \cup fn((v)Q_{kl})$;
**(c)** $\alpha_{ij}$ is $x(v)$ and $\beta_{kl}$ is $\overline{x}u$, then $R_{ijkl} = P_{ij}\{u/v\} \parallel Q_{kl}$;
**(d)** $\alpha_{ij}$ is $x(v)$ and $\beta_{kl}$ is $\overline{x}(u)$, then $R_{ijkl} = (w)(P_{ij}\{w/v\} \parallel Q_{kl}\{w/u\})$, if $w \notin fn((v)P_{ij}) \cup fn((u)Q_{kl})$;

**7.** Identifier **I**.

$$\text{If } A(\widetilde{x}) \overset{\text{def}}{=} P, \text{ then } A(\widetilde{y}) = P\{\widetilde{y}/\widetilde{x}\}.$$

**Theorem 12.41** (Soundness). *If* $STC \vdash P = Q$, *then*

1. $P \sim_{pp} Q$;
2. $P \sim_{ps} Q$;
3. $P \sim_{php} Q$;
4. $P \sim_{phhp} Q$.

*Proof.* The soundness of these laws modulo strongly truly concurrent bisimilarities was already proven in Section 12.2. □

**Definition 12.42.** The agent identifier $A$ is weakly guardedly defined if every agent identifier is weakly guarded in the right-hand side of the definition of $A$.

**Definition 12.43** (Head normal form). A Process $P$ is in head normal form if it is a sum of the prefixes:

$$P \equiv \boxplus_i \sum_j (\alpha_{ij1} \parallel \cdots \parallel \alpha_{ijn}).P_{ij}.$$

**Proposition 12.44.** *If every agent identifier is weakly guardedly defined, then for any process $P$, there is a head normal form $H$ such that*

$$\mathbf{STC} \vdash P = H$$

*Proof.* It is sufficient to induct on the structure of $P$ and quite obvious. □

**Theorem 12.45** (Completeness). *For all processes $P$ and $Q$,*

1. *if $P \sim_{pp} Q$, then $\mathbf{STC} \vdash P = Q$;*
2. *if $P \sim_{ps} Q$, then $\mathbf{STC} \vdash P = Q$;*
3. *if $P \sim_{php} Q$, then $\mathbf{STC} \vdash P = Q$.*

*Proof.* **1.** if $P \sim_{ps} Q$, then $\mathbf{STC} \vdash P = Q$. Since $P$ and $Q$ all have head normal forms, let $P \equiv \boxplus_{j=1}^{l} \sum_{i=1}^{k} \alpha_{ji}.P_{ji}$ and $Q \equiv \boxplus_{j=1}^{l} \sum_{i=1}^{k} \beta_{ji}.Q_{ji}$. Then, the depth of $P$, denoted as $d(P) = 0$, if $k = 0$; $d(P) = 1 + max\{d(P_{ji})\}$ for $1 \leq j, i \leq k$. The depth $d(Q)$ can be defined similarly.
It is sufficient to induct on $d = d(P) + d(Q)$. When $d = 0$, $P \equiv \mathbf{nil}$ and $Q \equiv \mathbf{nil}$, $P = Q$, as desired.
Suppose $d > 0$.

- If $(\alpha_1 \parallel \cdots \parallel \alpha_n).M$ with $\alpha_{ji}(1 \leq j, i \leq n)$ free actions is a summand of $P$, then $P \xrightarrow{\{\alpha_1,\cdots,\alpha_n\}} M$. Since $Q$ is in head normal form and has a summand $(\alpha_1 \parallel \cdots \parallel \alpha_n).N$ such that $M \sim_{ps} N$, by the induction hypothesis $\mathbf{STC} \vdash M = N$, $\mathbf{STC} \vdash (\alpha_1 \parallel \cdots \parallel \alpha_n).M = (\alpha_1 \parallel \cdots \parallel \alpha_n).N$;
- If $x(y).M$ is a summand of $P$, then for $z \notin n(P, Q)$, $P \xrightarrow{x(z)} M' \equiv M\{z/y\}$. Since $Q$ is in head normal form and has a summand $x(w).N$ such that for all $v$, $M'\{v/z\} \sim_{ps} N'\{v/z\}$, where $N' \equiv N\{z/w\}$, by the induction hypothesis $\mathbf{STC} \vdash M'\{v/z\} = N'\{v/z\}$, by the axioms **C** and **A**, $\mathbf{STC} \vdash x(y).M = x(w).N$;
- If $\overline{x}(y).M$ is a summand of $P$, then for $z \notin n(P, Q)$, $P \xrightarrow{\overline{x}(z)} M' \equiv M\{z/y\}$. Since $Q$ is in head normal form and has a summand $\overline{x}(w).N$ such that $M' \sim_{ps} N'$, where $N' \equiv N\{z/w\}$, by the induction hypothesis $\mathbf{STC} \vdash M' = N'$, by the axioms **A** and **C**, $\mathbf{STC} \vdash \overline{x}(y).M = \overline{x}(w).N$;

**2.** if $P \sim_{pp} Q$, then $\mathbf{STC} \vdash P = Q$. It can be proven similarly to the above case.
**3.** if $P \sim_{php} Q$, then $\mathbf{STC} \vdash P = Q$. It can be proven similarly to the above case. □

Chapter 13

# Guards

In this chapter, we introduce guards into probabilistic process algebra in Chapter 4 based on the work on guards for process algebra [14]. This chapter is organized as follows. We introduce the operational semantics of guards in Section 13.1, $BAPTC$ with Guards in Section 13.2, $APPTC$ with Guards in Section 13.3, recursion in Section 13.4, abstraction in Section 13.5, and Hoare Logic for $APPTC_G$ in Section 13.6. Note that all the definitions of PDFs are the same as those in Chapter 4, and we do not repeat these here.

## 13.1 Operational semantics

In this section, we extend probabilistic truly concurrent bisimilarities to those containing data states.

**Definition 13.1** (Prime event structure with silent event and empty event). Let $\Lambda$ be a fixed set of labels, ranged over $a, b, c, \cdots$ and $\tau, \epsilon$. A ($\Lambda$-labeled) prime event structure with silent event $\tau$ and empty event $\epsilon$ is a tuple $\mathcal{E} = \langle \mathbb{E}, \leq, \sharp, \sharp_\pi \lambda \rangle$, where $\mathbb{E}$ is a denumerable set of events, including the silent event $\tau$ and empty event $\epsilon$. Let $\hat{\mathbb{E}} = \mathbb{E} \backslash \{\tau, \epsilon\}$, exactly excluding $\tau$ and $\epsilon$, it is obvious that $\hat{\tau^*} = \epsilon$. Let $\lambda : \mathbb{E} \to \Lambda$ be a labeling function and let $\lambda(\tau) = \tau$ and $\lambda(\epsilon) = \epsilon$. Also, $\leq$, $\sharp$, $\sharp_\pi$ are binary relations on $\mathbb{E}$, called causality, conflict, and probabilistic conflict, respectively, such that:

1. $\leq$ is a partial order and $\lceil e \rceil = \{e' \in \mathbb{E} | e' \leq e\}$ is finite for all $e \in \mathbb{E}$. It is easy to see that $e \leq \tau^* \leq e' = e \leq \tau \leq \cdots \leq \tau \leq e'$, then $e \leq e'$.
2. $\sharp$ is irreflexive, symmetric, and hereditary with respect to $\leq$, that is, for all $e, e', e'' \in \mathbb{E}$, if $e \sharp e' \leq e''$, then $e \sharp e''$;
3. $\sharp_\pi$ is irreflexive, symmetric, and hereditary with respect to $\leq$, that is, for all $e, e', e'' \in \mathbb{E}$, if $e \sharp_\pi e' \leq e''$, then $e \sharp_\pi e''$.

Then, the concepts of consistency and concurrency can be drawn from the above definition:

1. $e, e' \in \mathbb{E}$ are consistent, denoted as $e \frown e'$, if $\neg(e \sharp e')$ and $\neg(e \sharp_\pi e')$. A subset $X \subseteq \mathbb{E}$ is called consistent, if $e \frown e'$ for all $e, e' \in X$.
2. $e, e' \in \mathbb{E}$ are concurrent, denoted as $e \parallel e'$, if $\neg(e \leq e')$, $\neg(e' \leq e)$, and $\neg(e \sharp e')$ and $\neg(e \sharp_\pi e')$.

**Definition 13.2** (Configuration). Let $\mathcal{E}$ be a PES. A (finite) configuration in $\mathcal{E}$ is a (finite) consistent subset of events $C \subseteq \mathcal{E}$, closed with respect to causality (i.e., $\lceil C \rceil = C$), and a data state $s \in S$ with $S$ the set of all data states, denoted $\langle C, s \rangle$. The set of finite configurations of $\mathcal{E}$ is denoted by $\langle \mathcal{C}(\mathcal{E}), S \rangle$. We let $\hat{C} = C \backslash \{\tau\} \cup \{\epsilon\}$.

A consistent subset of $X \subseteq \mathbb{E}$ of events can be seen as a pomset. Given $X, Y \subseteq \mathbb{E}$, $\hat{X} \sim \hat{Y}$ if $\hat{X}$ and $\hat{Y}$ are isomorphic as pomsets. In the remainder of the chapter, when we say $C_1 \sim C_2$, we mean $\hat{C_1} \sim \hat{C_2}$.

**Definition 13.3** (Pomset transitions and step). Let $\mathcal{E}$ be a PES and let $C \in \mathcal{C}(\mathcal{E})$, and $\emptyset \neq X \subseteq \mathbb{E}$, if $C \cap X = \emptyset$ and $C' = C \cup X \in \mathcal{C}(\mathcal{E})$, then $\langle C, s \rangle \xrightarrow{X} \langle C', s' \rangle$ is called a pomset transition from $\langle C, s \rangle$ to $\langle C', s' \rangle$. When the events in $X$ are pairwise concurrent, we say that $\langle C, s \rangle \xrightarrow{X} \langle C', s' \rangle$ is a step. It is obvious that $\to^* \xrightarrow{X} \to^* = \xrightarrow{X}$ and $\to^* \xrightarrow{e} \to^* = \xrightarrow{e}$ for any $e \in \mathbb{E}$ and $X \subseteq \mathbb{E}$.

**Definition 13.4** (Probabilistic transitions). Let $\mathcal{E}$ be a PES and let $C \in \mathcal{C}(\mathcal{E})$, the transition $\langle C, s \rangle \xrightarrow{\pi} \langle C^\pi, s \rangle$ is called a probabilistic transition from $\langle C, s \rangle$ to $\langle C^\pi, s \rangle$.

**Definition 13.5** (Weak pomset transitions and weak step). Let $\mathcal{E}$ be a PES and let $C \in \mathcal{C}(\mathcal{E})$, and $\emptyset \neq X \subseteq \hat{\mathbb{E}}$, if $C \cap X = \emptyset$ and $\hat{C'} = \hat{C} \cup X \in \mathcal{C}(\mathcal{E})$, then $\langle C, s \rangle \xRightarrow{X} \langle C', s' \rangle$ is called a weak pomset transition from $\langle C, s \rangle$ to $\langle C', s' \rangle$, where we define $\xRightarrow{e} \triangleq \xrightarrow{\tau^*} \xrightarrow{e} \xrightarrow{\tau^*}$. Also, $\xRightarrow{X} \triangleq \xrightarrow{\tau^*} \xrightarrow{e} \xrightarrow{\tau^*}$, for every $e \in X$. When the events in $X$ are pairwise concurrent, we say that $\langle C, s \rangle \xRightarrow{X} \langle C', s' \rangle$ is a weak step.

We will also suppose that all the PESs in this chapter are image finite, that is, for any PES $\mathcal{E}$ and $C \in \mathcal{C}(\mathcal{E})$ and $a \in \Lambda$, $\{\langle C, s \rangle \xrightarrow{\pi} \langle C^\pi, s \rangle\}$, $\{e \in \mathbb{E} | \langle C, s \rangle \xrightarrow{e} \langle C', s' \rangle \wedge \lambda(e) = a\}$ and $\{e \in \hat{\mathbb{E}} | \langle C, s \rangle \xRightarrow{e} \langle C', s' \rangle \wedge \lambda(e) = a\}$ is finite.

Handbook of Truly Concurrent Process Algebra. https://doi.org/10.1016/B978-0-44-321515-5.00017-2

**Definition 13.6** (Probabilistic pomset, step bisimulation). Let $\mathcal{E}_1$, $\mathcal{E}_2$ be PESs. A probabilistic pomset bisimulation is a relation $R \subseteq \langle \mathcal{C}(\mathcal{E}_1), S\rangle \times \langle \mathcal{C}(\mathcal{E}_2), S\rangle$, such that (1) if $(\langle C_1, s\rangle, \langle C_2, s\rangle) \in R$, and $\langle C_1, s\rangle \xrightarrow{X_1} \langle C_1', s'\rangle$, then $\langle C_2, s\rangle \xrightarrow{X_2} \langle C_2', s'\rangle$, with $X_1 \subseteq \mathbb{E}_1$, $X_2 \subseteq \mathbb{E}_2$, $X_1 \sim X_2$ and $(\langle C_1', s'\rangle, \langle C_2', s'\rangle) \in R$ for all $s, s' \in S$, and vice versa; (2) if $(\langle C_1, s\rangle, \langle C_2, s\rangle) \in R$, and $\langle C_1, s\rangle \xrightarrow{\pi} \langle C_1^\pi, s\rangle$, then $\langle C_2, s\rangle \xrightarrow{\pi} \langle C_2^\pi, s\rangle$ and $(\langle C_1^\pi, s\rangle, \langle C_2^\pi, s\rangle) \in R$, and vice versa; (3) if $(\langle C_1, s\rangle, \langle C_2, s\rangle) \in R$, then $\mu(C_1, C) = \mu(C_2, C)$ for each $C \in \mathcal{C}(\mathcal{E})/R$; (4) $[\surd]_R = \{\surd\}$. We say that $\mathcal{E}_1$, $\mathcal{E}_2$ are probabilistic pomset bisimilar, written $\mathcal{E}_1 \sim_{pp} \mathcal{E}_2$, if there exists a probabilistic pomset bisimulation $R$, such that $(\langle\emptyset, \emptyset\rangle, \langle\emptyset, \emptyset\rangle) \in R$. By replacing probabilistic pomset transitions with probabilistic steps, we can obtain the definition of probabilistic step bisimulation. When PESs $\mathcal{E}_1$ and $\mathcal{E}_2$ are probabilistic step bisimilar, we write $\mathcal{E}_1 \sim_{ps} \mathcal{E}_2$.

**Definition 13.7** (Weakly probabilistic pomset, step bisimulation). Let $\mathcal{E}_1$, $\mathcal{E}_2$ be PESs. A weakly probabilistic pomset bisimulation is a relation $R \subseteq \langle \mathcal{C}(\mathcal{E}_1), S\rangle \times \langle \mathcal{C}(\mathcal{E}_2), S\rangle$, such that (1) if $(\langle C_1, s\rangle, \langle C_2, s\rangle) \in R$, and $\langle C_1, s\rangle \xRightarrow{X_1} \langle C_1', s'\rangle$, then $\langle C_2, s\rangle \xRightarrow{X_2} \langle C_2', s'\rangle$, with $X_1 \subseteq \hat{\mathbb{E}}_1$, $X_2 \subseteq \hat{\mathbb{E}}_2$, $X_1 \sim X_2$ and $(\langle C_1', s'\rangle, \langle C_2', s'\rangle) \in R$ for all $s, s' \in S$, and vice versa; (2) if $(\langle C_1, s\rangle, \langle C_2, s\rangle) \in R$, and $\langle C_1, s\rangle \xrightarrow{\pi} \langle C_1^\pi, s\rangle$, then $\langle C_2, s\rangle \xrightarrow{\pi} \langle C_2^\pi, s\rangle$ and $(\langle C_1^\pi, s\rangle, \langle C_2^\pi, s\rangle) \in R$, and vice versa; (3) if $(\langle C_1, s\rangle, \langle C_2, s\rangle) \in R$, then $\mu(C_1, C) = \mu(C_2, C)$ for each $C \in \mathcal{C}(\mathcal{E})/R$; (4) $[\surd]_R = \{\surd\}$. We say that $\mathcal{E}_1$, $\mathcal{E}_2$ are weakly probabilistic pomset bisimilar, written $\mathcal{E}_1 \approx_{pp} \mathcal{E}_2$, if there exists a weakly probabilistic pomset bisimulation $R$, such that $(\langle\emptyset, \emptyset\rangle, \langle\emptyset, \emptyset\rangle) \in R$. By replacing weakly probabilistic pomset transitions with weakly probabilistic steps, we can obtain the definition of weakly probabilistic step bisimulation. When PESs $\mathcal{E}_1$ and $\mathcal{E}_2$ are weakly probabilistic step bisimilar, we write $\mathcal{E}_1 \approx_{ps} \mathcal{E}_2$.

**Definition 13.8** (Posetal product). Given two PESs $\mathcal{E}_1$, $\mathcal{E}_2$, the posetal product of their configurations, denoted $\langle \mathcal{C}(\mathcal{E}_1), S\rangle \overline{\times} \langle \mathcal{C}(\mathcal{E}_2), S\rangle$, is defined as

$$\{(\langle C_1, s\rangle, f, \langle C_2, s\rangle) | C_1 \in \mathcal{C}(\mathcal{E}_1), C_2 \in \mathcal{C}(\mathcal{E}_2), f : C_1 \to C_2 \text{ isomorphism}\}$$

A subset $R \subseteq \langle \mathcal{C}(\mathcal{E}_1), S\rangle \overline{\times} \langle \mathcal{C}(\mathcal{E}_2), S\rangle$ is called a posetal relation. We say that $R$ is downward closed when for any $(\langle C_1, s\rangle, f, \langle C_2, s\rangle), (\langle C_1', s'\rangle, f', \langle C_2', s'\rangle) \in \langle \mathcal{C}(\mathcal{E}_1), S\rangle \overline{\times} \langle \mathcal{C}(\mathcal{E}_2), S\rangle$, if $(\langle C_1, s\rangle, f, \langle C_2, s\rangle) \subseteq (\langle C_1', s'\rangle, f', \langle C_2', s'\rangle)$ pointwise and $(\langle C_1', s'\rangle, f', \langle C_2', s'\rangle) \in R$, then $(\langle C_1, s\rangle, f, \langle C_2, s\rangle) \in R$.

For $f : X_1 \to X_2$, we define $f[x_1 \mapsto x_2] : X_1 \cup \{x_1\} \to X_2 \cup \{x_2\}$, $z \in X_1 \cup \{x_1\}$, (1) $f[x_1 \mapsto x_2](z) = x_2$, if $z = x_1$; (2) $f[x_1 \mapsto x_2](z) = f(z)$, otherwise, where $X_1 \subseteq \mathbb{E}_1$, $X_2 \subseteq \mathbb{E}_2$, $x_1 \in \mathbb{E}_1$, $x_2 \in \mathbb{E}_2$.

**Definition 13.9** (Weakly posetal product). Given two PESs $\mathcal{E}_1$, $\mathcal{E}_2$, the weakly posetal product of their configurations, denoted $\langle \mathcal{C}(\mathcal{E}_1), S\rangle \overline{\times} \langle \mathcal{C}(\mathcal{E}_2), S\rangle$, is defined as

$$\{(\langle C_1, s\rangle, f, \langle C_2, s\rangle) | C_1 \in \mathcal{C}(\mathcal{E}_1), C_2 \in \mathcal{C}(\mathcal{E}_2), f : \hat{C}_1 \to \hat{C}_2 \text{ isomorphism}\}$$

A subset $R \subseteq \langle \mathcal{C}(\mathcal{E}_1), S\rangle \overline{\times} \langle \mathcal{C}(\mathcal{E}_2), S\rangle$ is called a weakly posetal relation. We say that $R$ is downward closed when for any $(\langle C_1, s\rangle, f, \langle C_2, s\rangle), (\langle C_1', s'\rangle, f, \langle C_2', s'\rangle) \in \langle \mathcal{C}(\mathcal{E}_1), S\rangle \overline{\times} \langle \mathcal{C}(\mathcal{E}_2), S\rangle$, if $(\langle C_1, s\rangle, f, \langle C_2, s\rangle) \subseteq (\langle C_1', s'\rangle, f', \langle C_2', s'\rangle)$ pointwise and $(\langle C_1', s'\rangle, f', \langle C_2', s'\rangle) \in R$, then $(\langle C_1, s\rangle, f, \langle C_2, s\rangle) \in R$.

For $f : X_1 \to X_2$, we define $f[x_1 \mapsto x_2] : X_1 \cup \{x_1\} \to X_2 \cup \{x_2\}$, $z \in X_1 \cup \{x_1\}$, (1) $f[x_1 \mapsto x_2](z) = x_2$, if $z = x_1$; (2) $f[x_1 \mapsto x_2](z) = f(z)$, otherwise, where $X_1 \subseteq \hat{\mathbb{E}}_1$, $X_2 \subseteq \hat{\mathbb{E}}_2$, $x_1 \in \hat{\mathbb{E}}_1$, $x_2 \in \hat{\mathbb{E}}_2$. Also, we define $f(\tau^*) = f(\tau^*)$.

**Definition 13.10** (Probabilistic (hereditary) history-preserving bisimulation). A probabilistic history-preserving (hp-) bisimulation is a posetal relation $R \subseteq \langle \mathcal{C}(\mathcal{E}_1), S\rangle \overline{\times} \langle \mathcal{C}(\mathcal{E}_2), S\rangle$ such that (1) if $(\langle C_1, s\rangle, f, \langle C_2, s\rangle) \in R$, and $\langle C_1, s\rangle \xrightarrow{e_1} \langle C_1', s'\rangle$, then $\langle C_2, s\rangle \xrightarrow{e_2} \langle C_2', s'\rangle$, with $(\langle C_1', s'\rangle, f[e_1 \mapsto e_2], \langle C_2', s'\rangle) \in R$ for all $s, s' \in S$, and vice versa; (2) if $(\langle C_1, s\rangle, f, \langle C_2, s\rangle) \in R$, and $\langle C_1, s\rangle \xrightarrow{\pi} \langle C_1^\pi, s\rangle$, then $\langle C_2, s\rangle \xrightarrow{\pi} \langle C_2^\pi, s\rangle$ and $(\langle C_1^\pi, s\rangle, f, \langle C_2^\pi, s\rangle) \in R$, and vice versa; (3) if $(C_1, f, C_2) \in R$, then $\mu(C_1, C) = \mu(C_2, C)$ for each $C \in \mathcal{C}(\mathcal{E})/R$; (4) $[\surd]_R = \{\surd\}$. $\mathcal{E}_1, \mathcal{E}_2$ are probabilistic history-preserving (hp-)bisimilar and are written $\mathcal{E}_1 \sim_{php} \mathcal{E}_2$ if there exists a probabilistic hp-bisimulation $R$ such that $(\langle\emptyset, \emptyset\rangle, \emptyset, \langle\emptyset, \emptyset\rangle) \in R$.

A probabilistic hereditary history-preserving (hhp-)bisimulation is a downward closed probabilistic hp-bisimulation. $\mathcal{E}_1, \mathcal{E}_2$ are probabilistic hereditary history-preserving (hhp-)bisimilar and are written $\mathcal{E}_1 \sim_{phhp} \mathcal{E}_2$.

**Definition 13.11** (Weakly probabilistic (hereditary) history-preserving bisimulation). A weakly probabilistic history-preserving (hp-) bisimulation is a weakly posetal relation $R \subseteq \langle \mathcal{C}(\mathcal{E}_1), S\rangle \overline{\times} \langle \mathcal{C}(\mathcal{E}_2), S\rangle$ such that (1) if $(\langle C_1, s\rangle, f, \langle C_2, s\rangle) \in R$, and $\langle C_1, s\rangle \xRightarrow{e_1} \langle C_1', s'\rangle$, then $\langle C_2, s\rangle \xRightarrow{e_2} \langle C_2', s'\rangle$, with $(\langle C_1', s'\rangle, f[e_1 \mapsto e_2], \langle C_2', s'\rangle) \in R$ for all $s, s' \in S$, and vice versa;

(2) if $(\langle C_1, s\rangle, f, \langle C_2, s\rangle) \in R$, and $\langle C_1, s\rangle \xrightarrow{\pi} \langle C_1^{\pi}, s\rangle$, then $\langle C_2, s\rangle \xrightarrow{\pi} \langle C_2^{\pi}, s\rangle$ and $(\langle C_1^{\pi}, s\rangle, f, \langle C_2^{\pi}, s\rangle) \in R$, and vice versa; (3) if $(C_1, f, C_2) \in R$, then $\mu(C_1, C) = \mu(C_2, C)$ for each $C \in \mathcal{C}(\mathcal{E})/R$; (4) $[\surd]_R = \{\surd\}$. $\mathcal{E}_1, \mathcal{E}_2$ are weakly probabilistic history-preserving (hp-)bisimilar and are written $\mathcal{E}_1 \approx_{php} \mathcal{E}_2$ if there exists a weakly probabilistic hp-bisimulation $R$ such that $(\langle\emptyset, \emptyset\rangle, \emptyset, \langle\emptyset, \emptyset\rangle) \in R$.

A weakly probabilistic hereditary history-preserving (hhp-)bisimulation is a downward closed weakly probabilistic hp-bisimulation. $\mathcal{E}_1, \mathcal{E}_2$ are weakly probabilistic hereditary history-preserving (hhp-)bisimilar and are written $\mathcal{E}_1 \approx_{phhp} \mathcal{E}_2$.

**Definition 13.12** (Probabilistic branching pomset, step bisimulation). Assume a special termination predicate $\downarrow$, and let $\surd$ represent a state with $\surd \downarrow$. Let $\mathcal{E}_1, \mathcal{E}_2$ be PESs. A probabilistic branching pomset bisimulation is a relation $R \subseteq \langle \mathcal{C}(\mathcal{E}_1), S\rangle \times \langle \mathcal{C}(\mathcal{E}_2), S\rangle$, such that:

1. if $(\langle C_1, s\rangle, \langle C_2, s\rangle) \in R$, and $\langle C_1, s\rangle \xrightarrow{X} \langle C_1', s'\rangle$, then
   - either $X \equiv \tau^*$, and $(\langle C_1', s'\rangle, \langle C_2, s\rangle) \in R$ with $s' \in \tau(s)$;
   - or there is a sequence of (zero or more) probabilistic transitions and $\tau$-transitions $\langle C_2, s\rangle \rightsquigarrow^* \xrightarrow{\tau^*} \langle C_2^0, s^0\rangle$, such that $(\langle C_1, s\rangle, \langle C_2^0, s^0\rangle) \in R$ and $\langle C_2^0, s^0\rangle \xRightarrow{X} \langle C_2', s'\rangle$ with $(\langle C_1', s'\rangle, \langle C_2', s'\rangle) \in R$;
2. if $(\langle C_1, s\rangle, \langle C_2, s\rangle) \in R$, and $\langle C_2, s\rangle \xrightarrow{X} \langle C_2', s'\rangle$, then
   - either $X \equiv \tau^*$, and $(\langle C_1, s\rangle, \langle C_2', s'\rangle) \in R$;
   - or there is a sequence of (zero or more) probabilistic transitions and $\tau$-transitions $\langle C_1, s\rangle \rightsquigarrow^* \xrightarrow{\tau^*} \langle C_1^0, s^0\rangle$, such that $(\langle C_1^0, s^0\rangle, \langle C_2, s\rangle) \in R$ and $\langle C_1^0, s^0\rangle \xRightarrow{X} \langle C_1', s'\rangle$ with $(\langle C_1', s'\rangle, \langle C_2', s'\rangle) \in R$;
3. if $(\langle C_1, s\rangle, \langle C_2, s\rangle) \in R$ and $\langle C_1, s\rangle \downarrow$, then there is a sequence of (zero or more) probabilistic transitions and $\tau$-transitions $\langle C_2, s\rangle \rightsquigarrow^* \xrightarrow{\tau^*} \langle C_2^0, s^0\rangle$ such that $(\langle C_1, s\rangle, \langle C_2^0, s^0\rangle) \in R$ and $\langle C_2^0, s^0\rangle \downarrow$;
4. if $(\langle C_1, s\rangle, \langle C_2, s\rangle) \in R$ and $\langle C_2, s\rangle \downarrow$, then there is a sequence of (zero or more) probabilistic transitions and $\tau$-transitions $\langle C_1, s\rangle \rightsquigarrow^* \xrightarrow{\tau^*} \langle C_1^0, s^0\rangle$ such that $(\langle C_1^0, s^0\rangle, \langle C_2, s\rangle) \in R$ and $\langle C_1^0, s^0\rangle \downarrow$;
5. if $(C_1, C_2) \in R$, then $\mu(C_1, C) = \mu(C_2, C)$ for each $C \in \mathcal{C}(\mathcal{E})/R$;
6. $[\surd]_R = \{\surd\}$.

We say that $\mathcal{E}_1$, $\mathcal{E}_2$ are probabilistic branching pomset bisimilar, written $\mathcal{E}_1 \approx_{pbp} \mathcal{E}_2$, if there exists a probabilistic branching pomset bisimulation $R$, such that $(\langle\emptyset, \emptyset\rangle, \langle\emptyset, \emptyset\rangle) \in R$.

By replacing probabilistic pomset transitions with steps, we can obtain the definition of probabilistic branching step bisimulation. When PESs $\mathcal{E}_1$ and $\mathcal{E}_2$ are probabilistic branching step bisimilar, we write $\mathcal{E}_1 \approx_{pbs} \mathcal{E}_2$.

**Definition 13.13** (Probabilistic rooted branching pomset, step bisimulation). Assume a special termination predicate $\downarrow$, and let $\surd$ represent a state with $\surd \downarrow$. Let $\mathcal{E}_1, \mathcal{E}_2$ be PESs. A probabilistic rooted branching pomset bisimulation is a relation $R \subseteq \langle \mathcal{C}(\mathcal{E}_1), S\rangle \times \langle \mathcal{C}(\mathcal{E}_2), S\rangle$, such that:

1. if $(\langle C_1, s\rangle, \langle C_2, s\rangle) \in R$ and $\langle C_1, s\rangle \rightsquigarrow\xrightarrow{X} \langle C_1', s'\rangle$, then $\langle C_2, s\rangle \rightsquigarrow\xrightarrow{X} \langle C_2', s'\rangle$ with $\langle C_1', s'\rangle \approx_{pbp} \langle C_2', s'\rangle$;
2. if $(\langle C_1, s\rangle, \langle C_2, s\rangle) \in R$ and $\langle C_2, s\rangle \rightsquigarrow\xrightarrow{X} \langle C_2', s'\rangle$, then $\langle C_1, s\rangle \rightsquigarrow\xrightarrow{X} \langle C_1', s'\rangle$ with $\langle C_1', s'\rangle \approx_{pbp} \langle C_2', s'\rangle$;
3. if $(\langle C_1, s\rangle, \langle C_2, s\rangle) \in R$ and $\langle C_1, s\rangle \downarrow$, then $\langle C_2, s\rangle \downarrow$;
4. if $(\langle C_1, s\rangle, \langle C_2, s\rangle) \in R$ and $\langle C_2, s\rangle \downarrow$, then $\langle C_1, s\rangle \downarrow$.

We say that $\mathcal{E}_1, \mathcal{E}_2$ are probabilistic rooted branching pomset bisimilar, written $\mathcal{E}_1 \approx_{prbp} \mathcal{E}_2$, if there exists a probabilistic rooted branching pomset bisimulation $R$, such that $(\langle\emptyset, \emptyset\rangle, \langle\emptyset, \emptyset\rangle) \in R$.

By replacing pomset transitions with steps, we can obtain the definition of probabilistic rooted branching step bisimulation. When PESs $\mathcal{E}_1$ and $\mathcal{E}_2$ are probabilistic rooted branching step bisimilar, we write $\mathcal{E}_1 \approx_{prbs} \mathcal{E}_2$.

**Definition 13.14** (Probabilistic branching (hereditary) history-preserving bisimulation). Assume a special termination predicate $\downarrow$, and let $\surd$ represent a state with $\surd \downarrow$. A probabilistic branching history-preserving (hp-) bisimulation is a weakly posetal relation $R \subseteq \langle \mathcal{C}(\mathcal{E}_1), S\rangle \overline{\times} \langle \mathcal{C}(\mathcal{E}_2), S\rangle$ such that:

1. if $(\langle C_1, s\rangle, f, \langle C_2, s\rangle) \in R$, and $\langle C_1, s\rangle \xrightarrow{e_1} \langle C_1', s'\rangle$, then
   - either $e_1 \equiv \tau$, and $(\langle C_1', s'\rangle, f[e_1 \mapsto \tau], \langle C_2, s\rangle) \in R$;
   - or there is a sequence of (zero or more) probabilistic transitions and $\tau$-transitions $\langle C_2, s\rangle \rightsquigarrow^* \xrightarrow{\tau^*} \langle C_2^0, s^0\rangle$, such that $(\langle C_1, s\rangle, f, \langle C_2^0, s^0\rangle) \in R$ and $\langle C_2^0, s^0\rangle \xrightarrow{e_2} \langle C_2', s'\rangle$ with $(\langle C_1', s'\rangle, f[e_1 \mapsto e_2], \langle C_2', s'\rangle) \in R$;

2. if $(\langle C_1, s\rangle, f, \langle C_2, s\rangle) \in R$, and $\langle C_2, s\rangle \xrightarrow{e_2} \langle C_2', s'\rangle$, then
   - either $e_2 \equiv \tau$, and $(\langle C_1, s\rangle, f[e_2 \mapsto \tau], \langle C_2', s'\rangle) \in R$;
   - or there is a sequence of (zero or more) probabilistic transitions and $\tau$-transitions $\langle C_1, s\rangle \rightsquigarrow^* \xrightarrow{\tau^*} \langle C_1^0, s^0\rangle$, such that $(\langle C_1^0, s^0\rangle, f, \langle C_2, s\rangle) \in R$ and $\langle C_1^0, s^0\rangle \xrightarrow{e_1} \langle C_1', s'\rangle$ with $(\langle C_1', s'\rangle, f[e_2 \mapsto e_1], \langle C_2', s'\rangle) \in R$;
3. if $(\langle C_1, s\rangle, f, \langle C_2, s\rangle) \in R$ and $\langle C_1, s\rangle \downarrow$, then there is a sequence of (zero or more) probabilistic transitions and $\tau$-transitions $\langle C_2, s\rangle \rightsquigarrow^* \xrightarrow{\tau^*} \langle C_2^0, s^0\rangle$ such that $(\langle C_1, s\rangle, f, \langle C_2^0, s^0\rangle) \in R$ and $\langle C_2^0, s^0\rangle \downarrow$;
4. if $(\langle C_1, s\rangle, f, \langle C_2, s\rangle) \in R$ and $\langle C_2, s\rangle \downarrow$, then there is a sequence of (zero or more) probabilistic transitions and $\tau$-transitions $\langle C_1, s\rangle \rightsquigarrow^* \xrightarrow{\tau^*} \langle C_1^0, s^0\rangle$ such that $(\langle C_1^0, s^0\rangle, f, \langle C_2, s\rangle) \in R$ and $\langle C_1^0, s^0\rangle \downarrow$;
5. if $(C_1, C_2) \in R$, then $\mu(C_1, C) = \mu(C_2, C)$ for each $C \in \mathcal{C}(\mathcal{E})/R$;
6. $[\surd]_R = \{\surd\}$.

$\mathcal{E}_1, \mathcal{E}_2$ are probabilistic branching history-preserving (hp-)bisimilar and are written $\mathcal{E}_1 \approx_{pbhp} \mathcal{E}_2$ if there exists a probabilistic branching hp-bisimulation $R$ such that $(\langle\emptyset, \emptyset\rangle, \emptyset, \langle\emptyset, \emptyset\rangle) \in R$.

A probabilistic branching hereditary history-preserving (hhp-)bisimulation is a downward closed probabilistic branching hp-bisimulation. $\mathcal{E}_1, \mathcal{E}_2$ are probabilistic branching hereditary history-preserving (hhp-)bisimilar and are written $\mathcal{E}_1 \approx_{pbhhp} \mathcal{E}_2$.

**Definition 13.15** (Probabilistic rooted branching (hereditary) history-preserving bisimulation). Assume a special termination predicate $\downarrow$, and let $\surd$ represent a state with $\surd \downarrow$. A probabilistic rooted branching history-preserving (hp-) bisimulation is a weakly posetal relation $R \subseteq \langle \mathcal{C}(\mathcal{E}_1), S\rangle \overline{\times} \langle \mathcal{C}(\mathcal{E}_2), S\rangle$ such that:

1. if $(\langle C_1, s\rangle, f, \langle C_2, s\rangle) \in R$ and $\langle C_1, s\rangle \rightsquigarrow\xrightarrow{e_1} \langle C_1', s'\rangle$, then $\langle C_2, s\rangle \rightsquigarrow\xrightarrow{e_2} \langle C_2', s'\rangle$ with $\langle C_1', s'\rangle \approx_{pbhp} \langle C_2', s'\rangle$;
2. if $(\langle C_1, s\rangle, f, \langle C_2, s\rangle) \in R$ and $\langle C_2, s\rangle \rightsquigarrow\xrightarrow{e_2} \langle C_2', s'\rangle$, then $\langle C_1, s\rangle \rightsquigarrow\xrightarrow{e_1} \langle C_1', s'\rangle$ with $\langle C_1', s'\rangle \approx_{pbhp} \langle C_2', s'\rangle$;
3. if $(\langle C_1, s\rangle, f, \langle C_2, s\rangle) \in R$ and $\langle C_1, s\rangle \downarrow$, then $\langle C_2, s\rangle \downarrow$;
4. if $(\langle C_1, s\rangle, f, \langle C_2, s\rangle) \in R$ and $\langle C_2, s\rangle \downarrow$, then $\langle C_1, s\rangle \downarrow$.

$\mathcal{E}_1, \mathcal{E}_2$ are probabilistic rooted branching history-preserving (hp-)bisimilar and are written $\mathcal{E}_1 \approx_{prbhp} \mathcal{E}_2$ if there exists a probabilistic rooted branching hp-bisimulation $R$ such that $(\langle\emptyset, \emptyset\rangle, \emptyset, \langle\emptyset, \emptyset\rangle) \in R$.

A probabilistic rooted branching hereditary history-preserving (hhp-)bisimulation is a downward closed probabilistic rooted branching hp-bisimulation. $\mathcal{E}_1, \mathcal{E}_2$ are probabilistic rooted branching hereditary history-preserving (hhp-)bisimilar and are written $\mathcal{E}_1 \approx_{prbhhp} \mathcal{E}_2$.

## 13.2 *BAPTC* with guards

In this section, we will discuss the guards for $BAPTC$, which is denoted as $BAPTC_G$. Let $\mathbb{E}$ be the set of atomic events (actions), $G_{at}$ be the set of atomic guards, $\delta$ be the deadlock constant, and $\epsilon$ be the empty event. We extend $G_{at}$ to the set of basic guards $G$ with element $\phi, \psi, \cdots$, which is generated by the following formation rules:

$$\phi ::= \delta|\epsilon|\neg\phi|\psi \in G_{at}|\phi + \psi|\phi \boxplus_\pi \psi|\phi \cdot \psi$$

In the following, let $e_1, e_2, e_1', e_2' \in \mathbb{E}$, $\phi, \psi \in G$ and let variables $x, y, z$ range over the set of terms for true concurrency, $p, q, s$ range over the set of closed terms. The predicate $test(\phi, s)$ represents that $\phi$ holds in the state $s$, and $test(\epsilon, s)$ holds and $test(\delta, s)$ does not hold. $effect(e, s) \in S$ denotes $s'$ in $s \xrightarrow{e} s'$. The predicate weakest precondition $wp(e, \phi)$ denotes that $\forall s, s' \in S, test(\phi, effect(e, s))$ holds.

The set of axioms of $BAPTC_G$ consists of the laws given in Table 13.1.

Note that by eliminating atomic event from the process terms, the axioms in Table 13.1 will lead to a Boolean Algebra. Also, $G8$ and $G9$ are preconditions of $e$ and $\phi$, $G10$ is the weakest precondition of $e$ and $\phi$. A data environment with $effect$ function is sufficiently deterministic, and it is obvious that if the weakest precondition is expressible and $G10$, $G11$ are sound, then the related data environment is sufficiently deterministic.

**Definition 13.16** (Basic terms of $BAPTC_G$). The set of basic terms of $BAPTC_G$, $\mathcal{B}(BAPTC_G)$, is inductively defined as follows:

1. $\mathbb{E} \subset \mathcal{B}(BAPTC_G)$;
2. $G \subset \mathcal{B}(BAPTC_G)$;

**TABLE 13.1** Axioms of $BAPTC_G$.

| No. | Axiom |
|---|---|
| $A1$ | $x + y = y + x$ |
| $A2$ | $(x + y) + z = x + (y + z)$ |
| $A3$ | $e + e = e$ |
| $A4$ | $(x + y) \cdot z = x \cdot z + y \cdot z$ |
| $A5$ | $(x \cdot y) \cdot z = x \cdot (y \cdot z)$ |
| $A6$ | $x + \delta = x$ |
| $A7$ | $\delta \cdot x = \delta$ |
| $A8$ | $\epsilon \cdot x = x$ |
| $A9$ | $x \cdot \epsilon = x$ |
| $PA1$ | $x \boxplus_\pi y = y \boxplus_{1-\pi} x$ |
| $PA2$ | $x \boxplus_\pi (y \boxplus_\rho z) = (x \boxplus_{\frac{\pi}{\pi+\rho-\pi\rho}} y) \boxplus_{\pi+\rho-\pi\rho} z$ |
| $PA3$ | $x \boxplus_\pi x = x$ |
| $PA4$ | $(x \boxplus_\pi y) \cdot z = x \cdot z \boxplus_\pi y \cdot z$ |
| $PA5$ | $(x \boxplus_\pi y) + z = (x + z) \boxplus_\pi (y + z)$ |
| $G1$ | $\phi \cdot \neg\phi = \delta$ |
| $G2$ | $\phi + \neg\phi = \epsilon$ |
| $PG1$ | $\phi \boxplus_\pi \neg\phi = \epsilon$ |
| $G3$ | $\phi\delta = \delta$ |
| $G4$ | $\phi(x + y) = \phi x + \phi y$ |
| $PG2$ | $\phi(x \boxplus_\pi y) = \phi x \boxplus_\pi \phi y$ |
| $G5$ | $\phi(x \cdot y) = \phi x \cdot y$ |
| $G6$ | $(\phi + \psi)x = \phi x + \psi x$ |
| $PG3$ | $(\phi \boxplus_\pi \psi)x = \phi x \boxplus_\pi \psi x$ |
| $G7$ | $(\phi \cdot \psi) \cdot x = \phi \cdot (\psi \cdot x)$ |
| $G8$ | $\phi = \epsilon$ if $\forall s \in S.test(\phi, s)$ |
| $G9$ | $\phi_0 \cdot \cdots \cdot \phi_n = \delta$ if $\forall s \in S, \exists i \leq n.test(\neg\phi_i, s)$ |
| $G10$ | $wp(e, \phi)e\phi = wp(e, \phi)e$ |
| $G11$ | $\neg wp(e, \phi)e\neg\phi = \neg wp(e, \phi)e$ |

3. if $e \in \mathbb{E}, t \in \mathcal{B}(BAPTC_G)$, then $e \cdot t \in \mathcal{B}(BAPTC_G)$;
4. if $\phi \in G, t \in \mathcal{B}(BAPTC_G)$, then $\phi \cdot t \in \mathcal{B}(BAPTC_G)$;
5. if $t, s \in \mathcal{B}(BAPTC_G)$, then $t + s \in \mathcal{B}(BAPTC_G)$;
6. if $t, s \in \mathcal{B}(BAPTC_G)$, then $t \boxplus_\pi s \in \mathcal{B}(BAPTC_G)$.

**Theorem 13.17** (Elimination theorem of $BAPTC_G$). *Let $p$ be a closed $BAPTC_G$ term. Then, there is a basic $BAPTC_G$ term $q$ such that $BAPTC_G \vdash p = q$.*

*Proof.* (1) First, suppose that the following ordering on the signature of $BAPTC_G$ is defined: $\cdot > + > \boxplus_\pi$ and the symbol $\cdot$ is given the lexicographical status for the first argument, then for each rewrite rule $p \to q$ in Table 13.2 relation $p >_{lpo} q$ can easily be proved. We obtain that the term rewrite system shown in Table 13.2 is strongly normalizing, for it has finitely many rewriting rules, and $>$ is a well-founded ordering on the signature of $BAPTC_G$, and if $s >_{lpo} t$, for each rewriting rule $s \to t$ is in Table 13.2 (see Theorem 2.9).

(2) Then, we prove that the normal forms of closed $BAPTC_G$ terms are basic $BAPTC_G$ terms.

Suppose that $p$ is a normal form of some closed $BAPTC_G$ term and suppose that $p$ is not a basic term. Let $p'$ denote the smallest sub-term of $p$ that is not a basic term. This implies that each sub-term of $p'$ is a basic term. Then, we prove that $p$ is not a term in normal form. It is sufficient to induct on the structure of $p'$:

- Case $p' \equiv e, e \in \mathbb{E}$. $p'$ is a basic term, which contradicts the assumption that $p'$ is not a basic term, hence, this case should not occur.

**TABLE 13.2** Term rewrite system of $BAPTC_G$.

| No. | Rewriting Rule |
|---|---|
| $RA3$ | $e + e \rightarrow e$ |
| $RA4$ | $(x + y) \cdot z \rightarrow x \cdot z + y \cdot z$ |
| $RA5$ | $(x \cdot y) \cdot z \rightarrow x \cdot (y \cdot z)$ |
| $RA6$ | $x + \delta \rightarrow x$ |
| $RA7$ | $\delta \cdot x \rightarrow \delta$ |
| $RA8$ | $\epsilon \cdot x \rightarrow x$ |
| $RA9$ | $x \cdot \epsilon \rightarrow x$ |
| $RPA1$ | $x \boxplus_\pi y \rightarrow y \boxplus_{1-\pi} x$ |
| $RPA2$ | $x \boxplus_\pi (y \boxplus_\rho z) \rightarrow (x \boxplus_{\frac{\pi}{\pi+\rho-\pi\rho}} y) \boxplus_{\pi+\rho-\pi\rho} z$ |
| $RPA3$ | $x \boxplus_\pi x \rightarrow x$ |
| $RPA4$ | $(x \boxplus_\pi y) \cdot z \rightarrow x \cdot z \boxplus_\pi y \cdot z$ |
| $RPA5$ | $(x \boxplus_\pi y) + z \rightarrow (x + z) \boxplus_\pi (y + z)$ |
| $RG1$ | $\phi \cdot \neg\phi \rightarrow \delta$ |
| $RG2$ | $\phi + \neg\phi \rightarrow \epsilon$ |
| $RPG1$ | $\phi \boxplus_\pi \neg\phi \rightarrow \epsilon$ |
| $RG3$ | $\phi\delta \rightarrow \delta$ |
| $RG4$ | $\phi(x + y) \rightarrow \phi x + \phi y$ |
| $RPG2$ | $\phi(x \boxplus_\pi y) \rightarrow \phi x \boxplus_\pi \phi y$ |
| $RG5$ | $\phi(x \cdot y) \rightarrow \phi x \cdot y$ |
| $RG6$ | $(\phi + \psi)x \rightarrow \phi x + \psi x$ |
| $RPG3$ | $(\phi \boxplus_\pi \psi)x \rightarrow \phi x \boxplus_\pi \psi x$ |
| $RG7$ | $(\phi \cdot \psi) \cdot x \rightarrow \phi \cdot (\psi \cdot x)$ |
| $RG8$ | $\phi \rightarrow \epsilon$ if $\forall s \in S.test(\phi, s)$ |
| $RG9$ | $\phi_0 \cdot \cdots \cdot \phi_n \rightarrow \delta$ if $\forall s \in S, \exists i \leq n.test(\neg\phi_i, s)$ |
| $RG10$ | $wp(e, \phi)e\phi \rightarrow wp(e, \phi)e$ |
| $RG11$ | $\neg wp(e, \phi)e\neg\phi \rightarrow \neg wp(e, \phi)e$ |

- Case $p' \equiv \phi, \phi \in G$. $p'$ is a basic term, which contradicts the assumption that $p'$ is not a basic term, hence, this case should not occur.
- Case $p' \equiv p_1 \cdot p_2$. By induction on the structure of the basic term $p_1$:
  - Subcase $p_1 \in \mathbb{E}$. $p'$ would be a basic term, which contradicts the assumption that $p'$ is not a basic term;
  - Subcase $p_1 \in G$. $p'$ would be a basic term, which contradicts the assumption that $p'$ is not a basic term;
  - Subcase $p_1 \equiv e \cdot p_1'$. $RA5$ or $RA9$ rewriting rules can be applied. Hence, $p$ is not a normal form;
  - Subcase $p_1 \equiv \phi \cdot p_1'$. $RG1$, $RG3$, $RG4$, $RG5$, $RG7$, or $RG8-9$ rewriting rules can be applied. Hence, $p$ is not a normal form;
  - Subcase $p_1 \equiv p_1' + p_1''$. $RA4$, $RA6$, $RG2$, or $RG6$ rewriting rules can be applied. Hence, $p$ is not a normal form;
  - Subcase $p_1 \equiv p_1' \boxplus_\pi p_1''$. $RPG3$ rewriting rule can be applied. Hence, $p$ is not a normal form.
- Case $p' \equiv p_1 + p_2$. By induction on the structure of the basic terms both $p_1$ and $p_2$, all subcases will lead to that $p'$ would be a basic term, which contradicts the assumption that $p'$ is not a basic term.
- Case $p' \equiv p_1 \boxplus_\pi p_2$. By induction on the structure of the basic terms both $p_1$ and $p_2$, all subcases will lead to that $p'$ would be a basic term, which contradicts the assumption that $p'$ is not a basic term. □

We will define a term-deduction system that gives the operational semantics of $BAPTC_G$. We give the operational transition rules for $\epsilon$, atomic guard $\phi \in G_{at}$, atomic event $e \in \mathbb{E}$, operators $\cdot$ and $+$ as Table 13.3 shows. Also, the predicate $\xrightarrow{e} \surd$ represents successful termination after execution of the event $e$.

Note that we replace the single atomic event $e \in \mathbb{E}$ by $X \subseteq \mathbb{E}$, we can obtain the pomset transition rules of $BAPTC_G$, however, we omit them here.

**TABLE 13.3** Single event transition rules of $BAPTC_G$.

$$\frac{}{\langle \epsilon, s\rangle \rightsquigarrow \langle \breve{\epsilon}, s\rangle}$$

$$\frac{}{\langle e, s\rangle \rightsquigarrow \langle \breve{e}, s\rangle}$$

$$\frac{}{\langle \phi, s\rangle \rightsquigarrow \langle \breve{\phi}, s\rangle}$$

$$\frac{\langle x, s\rangle \rightsquigarrow \langle x', s\rangle}{\langle x \cdot y, s\rangle \rightsquigarrow \langle x' \cdot y, s\rangle}$$

$$\frac{\langle x, s\rangle \rightsquigarrow \langle x', s\rangle \quad \langle y, s\rangle \rightsquigarrow \langle y', s\rangle}{\langle x + y, s\rangle \rightsquigarrow \langle x' + y', s\rangle}$$

$$\frac{\langle x, s\rangle \rightsquigarrow \langle x', s\rangle}{\langle x \boxplus_\pi y, s\rangle \rightsquigarrow \langle x', s\rangle} \quad \frac{\langle y, s\rangle \rightsquigarrow \langle y', s\rangle}{\langle x \boxplus_\pi y, s\rangle \rightsquigarrow \langle y', s\rangle}$$

$$\frac{}{\langle \breve{\epsilon}, s\rangle \rightarrow \langle \surd, s\rangle}$$

$$\frac{}{\langle \breve{e}, s\rangle \xrightarrow{e} \langle \surd, s'\rangle} \text{ if } s' \in effect(e, s)$$

$$\frac{}{\langle \breve{\phi}, s\rangle \rightarrow \langle \surd, s\rangle} \text{ if } test(\phi, s)$$

$$\frac{\langle x, s\rangle \xrightarrow{e} \langle \surd, s'\rangle}{\langle x + y, s\rangle \xrightarrow{e} \langle \surd, s'\rangle} \quad \frac{\langle x, s\rangle \xrightarrow{e} \langle x', s'\rangle}{\langle x + y, s\rangle \xrightarrow{e} \langle x', s'\rangle}$$

$$\frac{\langle y, s\rangle \xrightarrow{e} \langle \surd, s'\rangle}{\langle x + y, s\rangle \xrightarrow{e} \langle \surd, s'\rangle} \quad \frac{\langle y, s\rangle \xrightarrow{e} \langle y', s'\rangle}{\langle x + y, s\rangle \xrightarrow{e} \langle y', s'\rangle}$$

$$\frac{\langle x, s\rangle \xrightarrow{e} \langle \surd, s'\rangle}{\langle x \cdot y, s\rangle \xrightarrow{e} \langle y, s'\rangle} \quad \frac{\langle x, s\rangle \xrightarrow{e} \langle x', s'\rangle}{\langle x \cdot y, s\rangle \xrightarrow{e} \langle x' \cdot y, s'\rangle}$$

**Theorem 13.18** (Congruence of $BAPTC_G$ with respect to probabilistic truly concurrent bisimulation equivalences). *(1) Probabilistic pomset bisimulation equivalence $\sim_{pp}$ is a congruence with respect to $BAPTC_G$.*

*(2) Probabilistic step bisimulation equivalence $\sim_{ps}$ is a congruence with respect to $BAPTC_G$.*

*(3) Probabilistic hp-bisimulation equivalence $\sim_{php}$ is a congruence with respect to $BAPTC_G$.*

*(4) Probabilistic hhp-bisimulation equivalence $\sim_{phhp}$ is a congruence with respect to $BAPTC_G$.*

*Proof.* (1) It is easy to see that probabilistic pomset bisimulation is an equivalent relation on $BAPTC_G$ terms, we only need to prove that $\sim_{pp}$ is preserved by the operators $\cdot$, $+$, and $\boxplus_\pi$. It is trivial and we leave the proof as an exercise for the reader.

(2) It is easy to see that probabilistic step bisimulation is an equivalent relation on $BAPTC_G$ terms, we only need to prove that $\sim_{ps}$ is preserved by the operators $\cdot$, $+$, and $\boxplus_\pi$. It is trivial and we leave the proof as an exercise for the reader.

(3) It is easy to see that probabilistic hp-bisimulation is an equivalent relation on $BAPTC_G$ terms, we only need to prove that $\sim_{php}$ is preserved by the operators $\cdot$, $+$, and $\boxplus_\pi$. It is trivial and we leave the proof as an exercise for the reader.

(4) It is easy to see that probabilistic hhp-bisimulation is an equivalent relation on $BAPTC_G$ terms, we only need to prove that $\sim_{phhp}$ is preserved by the operators $\cdot$, $+$, and $\boxplus_\pi$. It is trivial and we leave the proof as an exercise for the reader. □

**Theorem 13.19** (Soundness of $BAPTC_G$ modulo probabilistic truly concurrent bisimulation equivalences). *(1) Let $x$ and $y$ be $BAPTC_G$ terms. If $BAPTC_G \vdash x = y$, then $x \sim_{pp} y$.*

*(2) Let $x$ and $y$ be $BAPTC_G$ terms. If $BAPTC_G \vdash x = y$, then $x \sim_{ps} y$.*

*(3) Let $x$ and $y$ be $BAPTC_G$ terms. If $BAPTC_G \vdash x = y$, then $x \sim_{php} y$.*

*(4) Let $x$ and $y$ be $BAPTC_G$ terms. If $BAPTC_G \vdash x = y$, then $x \sim_{phhp} y$.*

*Proof.* (1) Since probabilistic pomset bisimulation $\sim_{pp}$ is both an equivalent and a congruent relation, we only need to check if each axiom in Table 13.1 is sound modulo probabilistic pomset bisimulation equivalence. We leave the proof as an exercise for the reader.

(2) Since probabilistic step bisimulation $\sim_{ps}$ is both an equivalent and a congruent relation, we only need to check if each axiom in Table 13.1 is sound modulo probabilistic step bisimulation equivalence. We leave the proof as an exercise for the reader.

(3) Since probabilistic hp-bisimulation $\sim_{php}$ is both an equivalent and a congruent relation, we only need to check if each axiom in Table 13.1 is sound modulo probabilistic hp-bisimulation equivalence. We leave the proof as an exercise for the reader.

(4) Since probabilistic hhp-bisimulation $\sim_{phhp}$ is both an equivalent and a congruent relation, we only need to check if each axiom in Table 13.1 is sound modulo probabilistic hhp-bisimulation equivalence. We leave the proof as an exercise for the reader. □

**Theorem 13.20** (Completeness of $BAPTC_G$ modulo probabilistic truly concurrent bisimulation equivalences). *(1) Let p and q be closed $BAPTC_G$ terms, if $p \sim_{pp} q$, then $p = q$.*

*(2) Let p and q be closed $BAPTC_G$ terms, if $p \sim_{ps} q$, then $p = q$.*

*(3) Let p and q be closed $BAPTC_G$ terms, if $p \sim_{php} q$, then $p = q$.*

*(4) Let p and q be closed $BAPTC_G$ terms, if $p \sim_{phhp} q$, then $p = q$.*

*Proof.* (1) First, by the elimination theorem of $BAPTC_G$, we know that for each closed $BAPTC_G$ term $p$, there exists a closed basic $BAPTC_G$ term $p'$, such that $BAPTC_G \vdash p = p'$, so, we only need to consider closed basic $BAPTC_G$ terms.

The basic terms (see Definition 13.16) modulo associativity and commutativity (AC) of conflict $+$ (defined by axioms $A1$ and $A2$ in Table 13.1), and this equivalence is denoted by $=_{AC}$. Then, each equivalence class $s$ modulo AC of $+$ has the following normal form

$$s_1 \boxplus_{\pi_1} \cdots \boxplus_{\pi_{k-1}} s_k$$

with each $s_i$ having the following form

$$t_1 + \cdots + t_l$$

with each $t_j$ either an atomic event or of the form $u_1 \cdot u_2$, and each $t_j$ is called the summand of $s$.

Now, we prove that for normal forms $n$ and $n'$, if $n \sim_{pp} n'$, then $n =_{AC} n'$. It is sufficient to induct on the sizes of $n$ and $n'$.

- Consider a summand $e$ of $n$. Then, $\langle n, s\rangle \rightsquigarrow \xrightarrow{e} \langle \surd, s'\rangle$, hence, $n \sim_{pp} n'$ implies $\langle n', s\rangle \rightsquigarrow \xrightarrow{e} \langle \surd, s\rangle$, meaning that $n'$ also contains the summand $e$.
- Consider a summand $\phi$ of $n$. Then, $\langle n, s\rangle \rightsquigarrow \rightarrow \langle \surd, s\rangle$, if $test(\phi, s)$ holds, hence, $n \sim_{pp} n'$ implies $\langle n', s\rangle \rightsquigarrow \rightarrow \langle \surd, s\rangle$, if $test(\phi, s)$ holds, meaning that $n'$ also contains the summand $\phi$.
- Consider a summand $t_1 \cdot t_2$ of $n$. Then, $\langle n, s\rangle \rightsquigarrow \xrightarrow{t_1} \langle t_2, s'\rangle$, hence, $n \sim_{pp} n'$ implies $\langle n', s\rangle \rightsquigarrow \xrightarrow{t_1} \langle t_2', s'\rangle$ with $t_2 \sim_{pp} t_2'$, meaning that $n'$ contains a summand $t_1 \cdot t_2'$. Since $t_2$ and $t_2'$ are normal forms and have sizes smaller than $n$ and $n'$, by the induction hypotheses $t_2 \sim_{pp} t_2'$ implies $t_2 =_{AC} t_2'$.

Hence, we obtain $n =_{AC} n'$.

Finally, let $s$ and $t$ be basic terms, and $s \sim_{pp} t$, there are normal forms $n$ and $n'$, such that $s = n$ and $t = n'$. The soundness theorem of $BAPTC_G$ modulo probabilistic pomset bisimulation equivalence (see Theorem 13.19) yields $s \sim_{pp} n$ and $t \sim_{pp} n'$, hence, $n \sim_{pp} s \sim_{pp} t \sim_{pp} n'$, since if $n \sim_{pp} n'$, then $n =_{AC} n'$, $s = n =_{AC} n' = t$, as desired.

(2) It can be proven similarly as (1).

(3) It can be proven similarly as (1).

(4) It can be proven similarly as (1). □

**Theorem 13.21** (Sufficient determinism). *All related data environments with respect to $BAPTC_G$ can be sufficiently deterministic.*

*Proof.* It only needs to be checked that $effect(t, s)$ function is deterministic, and is sufficient to induct on the structure of term $t$. The only matter are the cases $t = t_1 + t_2$ and $t = t_1 \boxplus_\pi t_2$, with the help of guards, we can make $t_1 = \phi_1 \cdot t_1'$ and $t_2 = \phi_2 \cdot t_2'$, and $effect(t)$ is sufficiently deterministic. □

## 13.3 *APPTC* with guards

In this section, we will extend $APPTC$ with guards, which is abbreviated $APPTC_G$. The set of basic guards $G$ with element $\phi, \psi, \cdots$, which is extended by the following formation rules:

$$\phi ::= \delta|\epsilon|\neg\phi|\psi \in G_{at}|\phi + \psi|\phi \boxplus_\pi \psi|\phi \cdot \psi|\phi \between \psi$$

The set of axioms of $APPTC_G$ include the axioms of $BAPTC_G$ in Table 13.1 and the axioms are shown in Table 13.4.

**TABLE 13.4** Axioms of $APPTC_G$.

| No. | Axiom |
|---|---|
| $P1$ | $(x+x=x, y+y=y) \quad x \between y = x \parallel y + x \mid y$ |
| $P2$ | $x \parallel y = y \parallel x$ |
| $P3$ | $(x \parallel y) \parallel z = x \parallel (y \parallel z)$ |
| $P4$ | $(x+x=x, y+y=y) \quad x \parallel y = x \mathbin{\lfloor\!\lfloor} y + y \mathbin{\lfloor\!\lfloor} x$ |
| $P5$ | $(e_1 \leq e_2) \quad e_1 \mathbin{\lfloor\!\lfloor} (e_2 \cdot y) = (e_1 \mathbin{\lfloor\!\lfloor} e_2) \cdot y$ |
| $P6$ | $(e_1 \leq e_2) \quad (e_1 \cdot x) \mathbin{\lfloor\!\lfloor} e_2 = (e_1 \mathbin{\lfloor\!\lfloor} e_2) \cdot x$ |
| $P7$ | $(e_1 \leq e_2) \quad (e_1 \cdot x) \mathbin{\lfloor\!\lfloor} (e_2 \cdot y) = (e_1 \mathbin{\lfloor\!\lfloor} e_2) \cdot (x \between y)$ |
| $P8$ | $(x+y) \mathbin{\lfloor\!\lfloor} z = (x \mathbin{\lfloor\!\lfloor} z) + (y \mathbin{\lfloor\!\lfloor} z)$ |
| $P9$ | $\delta \mathbin{\lfloor\!\lfloor} x = \delta$ |
| $P10$ | $\epsilon \mathbin{\lfloor\!\lfloor} x = x$ |
| $P11$ | $x \mathbin{\lfloor\!\lfloor} \epsilon = x$ |
| $C1$ | $e_1 \mid e_2 = \gamma(e_1, e_2)$ |
| $C2$ | $e_1 \mid (e_2 \cdot y) = \gamma(e_1, e_2) \cdot y$ |
| $C3$ | $(e_1 \cdot x) \mid e_2 = \gamma(e_1, e_2) \cdot x$ |
| $C4$ | $(e_1 \cdot x) \mid (e_2 \cdot y) = \gamma(e_1, e_2) \cdot (x \between y)$ |
| $C5$ | $(x+y) \mid z = (x \mid z) + (y \mid z)$ |
| $C6$ | $x \mid (y+z) = (x \mid y) + (x \mid z)$ |
| $C7$ | $\delta \mid x = \delta$ |
| $C8$ | $x \mid \delta = \delta$ |
| $C9$ | $\epsilon \mid x = \delta$ |
| $C10$ | $x \mid \epsilon = \delta$ |
| $PM1$ | $x \parallel (y \boxplus_\pi z) = (x \parallel y) \boxplus_\pi (x \parallel z)$ |
| $PM2$ | $(x \boxplus_\pi y) \parallel z = (x \parallel z) \boxplus_\pi (y \parallel z)$ |
| $PM3$ | $x \mid (y \boxplus_\pi z) = (x \mid y) \boxplus_\pi (x \mid z)$ |
| $PM4$ | $(x \boxplus_\pi y) \mid z = (x \mid z) \boxplus_\pi (y \mid z)$ |
| $CE1$ | $\Theta(e) = e$ |
| $CE2$ | $\Theta(\delta) = \delta$ |
| $CE3$ | $\Theta(\epsilon) = \epsilon$ |
| $CE4$ | $\Theta(x+y) = \Theta(x) + \Theta(y)$ |
| $PCE1$ | $\Theta(x \boxplus_\pi y) = \Theta(x) \boxplus_\pi \Theta(y)$ |
| $CE5$ | $\Theta(x \cdot y) = \Theta(x) \cdot \Theta(y)$ |
| $CE6$ | $\Theta(x \mathbin{\lfloor\!\lfloor} y) = ((\Theta(x) \triangleleft y) \mathbin{\lfloor\!\lfloor} y) + ((\Theta(y) \triangleleft x) \mathbin{\lfloor\!\lfloor} x)$ |
| $CE7$ | $\Theta(x \mid y) = ((\Theta(x) \triangleleft y) \mid y) + ((\Theta(y) \triangleleft x) \mid x)$ |
| $U1$ | $(\sharp(e_1, e_2)) \quad e_1 \triangleleft e_2 = \tau$ |
| $U2$ | $(\sharp(e_1, e_2), e_2 \leq e_3) \quad e_1 \triangleleft e_3 = \tau$ |
| $U3$ | $(\sharp(e_1, e_2), e_2 \leq e_3) \quad e3 \triangleleft e_1 = \tau$ |
| $PU1$ | $(\sharp_\pi(e_1, e_2)) \quad e_1 \triangleleft e_2 = \tau$ |
| $PU2$ | $(\sharp_\pi(e_1, e_2), e_2 \leq e_3) \quad e_1 \triangleleft e_3 = \tau$ |
| $PU3$ | $(\sharp_\pi(e_1, e_2), e_2 \leq e_3) \quad e_3 \triangleleft e_1 = \tau$ |
| $U4$ | $e \triangleleft \delta = e$ |
| $U5$ | $\delta \triangleleft e = \delta$ |
| $U6$ | $e \triangleleft \epsilon = e$ |
| $U7$ | $\epsilon \triangleleft e = e$ |
| $U8$ | $(x+y) \triangleleft z = (x \triangleleft z) + (y \triangleleft z)$ |

*continued on next page*

**TABLE 13.4 (continued)**

| No. | Axiom |
|---|---|
| $PU4$ | $(x \boxplus_\pi y) \triangleleft z = (x \triangleleft z) \boxplus_\pi (y \triangleleft z)$ |
| $U9$ | $(x \cdot y) \triangleleft z = (x \triangleleft z) \cdot (y \triangleleft z)$ |
| $U10$ | $(x \between y) \triangleleft z = (x \triangleleft z) \between (y \triangleleft z)$ |
| $U11$ | $(x \mid y) \triangleleft z = (x \triangleleft z) \mid (y \triangleleft z)$ |
| $U12$ | $x \triangleleft (y + z) = (x \triangleleft y) \triangleleft z$ |
| $PU5$ | $x \triangleleft (y \boxplus_\pi z) = (x \triangleleft y) \triangleleft z$ |
| $U13$ | $x \triangleleft (y \cdot z) = (x \triangleleft y) \triangleleft z$ |
| $U14$ | $x \triangleleft (y \between z) = (x \triangleleft y) \triangleleft z$ |
| $U15$ | $x \triangleleft (y \mid z) = (x \triangleleft y) \triangleleft z$ |
| $D1$ | $e \notin H \quad \partial_H(e) = e$ |
| $D2$ | $e \in H \quad \partial_H(e) = \delta$ |
| $D3$ | $\partial_H(\delta) = \delta$ |
| $D4$ | $\partial_H(x + y) = \partial_H(x) + \partial_H(y)$ |
| $D5$ | $\partial_H(x \cdot y) = \partial_H(x) \cdot \partial_H(y)$ |
| $D6$ | $\partial_H(x \between y) = \partial_H(x) \between \partial_H(y)$ |
| $PD1$ | $\partial_H(x \boxplus_\pi y) = \partial_H(x) \boxplus_\pi \partial_H(y)$ |
| $G12$ | $\phi(x \between y) = \phi x \between \phi y$ |
| $G13$ | $\phi(x \mid y) = \phi x \mid \phi y$ |
| $G14$ | $\delta \between \phi = \delta$ |
| $G15$ | $\phi \mid \delta = \delta$ |
| $G16$ | $\delta \mid \phi = \delta$ |
| $G17$ | $\phi \between \epsilon = \phi$ |
| $G18$ | $\epsilon \between \phi = \phi$ |
| $G19$ | $\phi \mid \epsilon = \delta$ |
| $G20$ | $\epsilon \mid \phi = \delta$ |
| $G21$ | $\phi \between \neg\phi = \delta$ |
| $G22$ | $\Theta(\phi) = \phi$ |
| $G23$ | $\partial_H(\phi) = \phi$ |
| $G24$ | $\phi_0 \between \cdots \between \phi_n = \delta$ if $\forall s_0, \cdots, s_n \in S, \exists i \leq n.test(\neg\phi_i, s_0 \cup \cdots \cup s_n)$ |

**Definition 13.22** (Basic terms of $APPTC_G$). The set of basic terms of $APPTC_G$, $\mathcal{B}(APPTC_G)$, is inductively defined as follows:

1. $\mathbb{E} \subset \mathcal{B}(APPTC_G)$;
2. $G \subset \mathcal{B}(APPTC_G)$;
3. if $e \in \mathbb{E}, t \in \mathcal{B}(APPTC_G)$, then $e \cdot t \in \mathcal{B}(APPTC_G)$;
4. if $\phi \in G, t \in \mathcal{B}(APPTC_G)$, then $\phi \cdot t \in \mathcal{B}(APPTC_G)$;
5. if $t, s \in \mathcal{B}(APPTC_G)$, then $t + s \in \mathcal{B}(APPTC_G)$;
6. if $t, s \in \mathcal{B}(APPTC_G)$, then $t \boxplus_\pi s \in \mathcal{B}(APPTC_G)$;
7. if $t, s \in \mathcal{B}(APPTC_G)$, then $t \between s \in \mathcal{B}(APPTC_G)$.

Based on the definition of basic terms for $APPTC_G$ (see Definition 13.22) and the axioms of $APPTC_G$, we can prove the elimination theorem of $APPTC_G$.

**Theorem 13.23** (Elimination theorem of $APPTC_G$). *Let $p$ be a closed $APPTC_G$ term. Then, there is a basic $APPTC_G$ term $q$ such that $APPTC_G \vdash p = q$.*

*Proof.* (1) First, suppose that the following ordering on the signature of $APPTC_G$ is defined: $\between > \cdot > + > \boxplus_\pi$ and the symbol $\cdot$ is given the lexicographical status for the first argument, then for each rewrite rule $p \to q$ in Table 13.5 relation

$p >_{lpo} q$ can easily be proved. We obtain that the term rewrite system shown in Table 13.5 is strongly normalizing, for it has finitely many rewriting rules, and $>$ is a well-founded ordering on the signature of $APPTC_G$, and if $s >_{lpo} t$, for each rewriting rule $s \to t$ is in Table 13.5 (see Theorem 2.9).

(2) Then, we prove that the normal forms of closed $APPTC_G$ terms are basic $APPTC_G$ terms.

Suppose that $p$ is a normal form of some closed $APPTC_G$ term and suppose that $p$ is not a basic $APPTC_G$ term. Let $p'$ denote the smallest sub-term of $p$ that is not a basic $APPTC_G$ term. This implies that each sub-term of $p'$ is a basic $APPTC_G$ term. Then, we prove that $p$ is not a term in normal form. It is sufficient to induct on the structure of $p'$:

- Case $p' \equiv e, e \in \mathbb{E}$. $p'$ is a basic $APPTC_G$ term, which contradicts the assumption that $p'$ is not a basic $APPTC_G$ term, hence, this case should not occur.
- Case $p' \equiv \phi, \phi \in G$. $p'$ is a basic term, which contradicts the assumption that $p'$ is not a basic term, hence, this case should not occur.
- Case $p' \equiv p_1 \cdot p_2$. By induction on the structure of the basic $APPTC_G$ term $p_1$:
  - Subcase $p_1 \in \mathbb{E}$. $p'$ would be a basic $APPTC_G$ term, which contradicts the assumption that $p'$ is not a basic $APPTC_G$ term;
  - Subcase $p_1 \in G$. $p'$ would be a basic term, which contradicts the assumption that $p'$ is not a basic term;
  - Subcase $p_1 \equiv e \cdot p_1'$. $RA5$ or $RA9$ rewriting rules in Table 13.2 can be applied. Hence, $p$ is not a normal form;
  - Subcase $p_1 \equiv \phi \cdot p_1'$. $RG1$, $RG3$, $RG4$, $RG5$, $RG7$, or $RG8-9$ rewriting rules can be applied. Hence, $p$ is not a normal form;
  - Subcase $p_1 \equiv p_1' + p_1''$. $RA4$, $RA6$, $RG2$, or $RG6$ rewriting rules in Table 13.2 can be applied. Hence, $p$ is not a normal form;
  - Subcase $p_1 \equiv p_1' \boxplus_\pi p_1''$. $RRA1-5$ rewriting rules in Table 13.2 can be applied. Hence, $p$ is not a normal form;
  - Subcase $p_1 \equiv p_1' \leftmerge p_1''$. $RP5-RP11$ rewrite rules in Table 13.5 can be applied. Hence, $p$ is not a normal form;
  - Subcase $p_1 \equiv p_1' \mid p_1''$. $RC1-RC10$ rewrite rules in Table 13.5 can be applied. Hence, $p$ is not a normal form;
  - Subcase $p_1 \equiv \Theta(p_1')$. $RCE1-RCE7$ rewrite rules in Table 13.5 can be applied. Hence, $p$ is not a normal form;
  - Subcase $p_1 \equiv \partial_H(p_1')$. $RD1-RD7$ rewrite rules in Table 13.5 can be applied. Hence, $p$ is not a normal form.
- Case $p' \equiv p_1 + p_2$. By induction on the structure of the basic $APPTC_G$ terms both $p_1$ and $p_2$, all subcases will lead to that $p'$ would be a basic $APPTC_G$ term, which contradicts the assumption that $p'$ is not a basic $APPTC_G$ term.
- Case $p' \equiv p_1 \boxplus_\pi p_2$. By induction on the structure of the basic $APPTC_G$ terms both $p_1$ and $p_2$, all subcases will lead to that $p'$ would be a basic $APPTC_G$ term, which contradicts the assumption that $p'$ is not a basic $APPTC_G$ term.
- Case $p' \equiv p_1 \leftmerge p_2$. By induction on the structure of the basic $APPTC_G$ terms both $p_1$ and $p_2$, all subcases will lead to that $p'$ would be a basic $APPTC_G$ term, which contradicts the assumption that $p'$ is not a basic $APPTC_G$ term.
- Case $p' \equiv p_1 \mid p_2$. By induction on the structure of the basic $APPTC_G$ terms both $p_1$ and $p_2$, all subcases will lead to that $p'$ would be a basic $APPTC_G$ term, which contradicts the assumption that $p'$ is not a basic $APPTC_G$ term.
- Case $p' \equiv \Theta(p_1)$. By induction on the structure of the basic $APPTC_G$ term $p_1$, $RCE1-RCE7$ rewrite rules in Table 13.5 can be applied. Hence, $p$ is not a normal form.
- Case $p' \equiv p_1 \triangleleft p_2$. By induction on the structure of the basic $APPTC_G$ terms both $p_1$ and $p_2$, all subcases will lead to that $p'$ would be a basic $APPTC_G$ term, which contradicts the assumption that $p'$ is not a basic $APPTC_G$ term.
- Case $p' \equiv \partial_H(p_1)$. By induction on the structure of the basic $APPTC_G$ terms of $p_1$, all subcases will lead to that $p'$ would be a basic $APPTC_G$ term, which contradicts the assumption that $p'$ is not a basic $APPTC_G$ term. □

We will define a term-deduction system that gives the operational semantics of $APPTC_G$. Two atomic events $e_1$ and $e_2$ are in race condition, which are denoted $e_1\%e_2$. See Tables 13.6 and 13.7.

**Theorem 13.24** (Generalization of $APPTC_G$ with respect to $BAPTC_G$). *$APPTC_G$ is a generalization of $BAPTC_G$.*

*Proof.* It follows from the following three facts:

1. The transition rules of $BAPTC_G$ in Section 13.2 are all source dependent;
2. The sources of the transition rules $APPTC_G$ contain an occurrence of $\between$, or $\parallel$, or $\leftmerge$, or $\mid$, or $\Theta$, or $\triangleleft$;
3. The transition rules of $APPTC_G$ are all source dependent.

Hence, $APPTC_G$ is a generalization of $BAPTC_G$, that is, $BAPTC_G$ is an embedding of $APPTC_G$, as desired. □

**Theorem 13.25** (Congruence of $APPTC_G$ with respect to probabilistic truly concurrent bisimulation equivalences). *(1) Probabilistic pomset bisimulation equivalence $\sim_{pp}$ is a congruence with respect to $APPTC_G$.*

*(2) Probabilistic step bisimulation equivalence $\sim_{ps}$ is a congruence with respect to $APPTC_G$.*

**TABLE 13.5** Term rewrite system of $APPTC_G$.

| No. | Rewriting Rule |
|---|---|
| $RP1$ | $(x+x=x, y+y=y) \quad x \between y \to x \parallel y + x \mid y$ |
| $RP2$ | $x \parallel y \to y \parallel x$ |
| $RP3$ | $(x \parallel y) \parallel z \to x \parallel (y \parallel z)$ |
| $RP4$ | $(x+x=x, y+y=y) \quad x \parallel y \to x \LeftFootline y + y \LeftFootline x$ |
| $RP5$ | $(e_1 \leq e_2) \quad e_1 \LeftFootline (e_2 \cdot y) \to (e_1 \LeftFootline e_2) \cdot y$ |
| $RP6$ | $(e_1 \leq e_2) \quad (e_1 \cdot x) \LeftFootline e_2 \to (e_1 \LeftFootline e_2) \cdot x$ |
| $RP7$ | $(e_1 \leq e_2) \quad (e_1 \cdot x) \LeftFootline (e_2 \cdot y) \to (e_1 \LeftFootline e_2) \cdot (x \between y)$ |
| $RP8$ | $(x+y) \LeftFootline z \to (x \LeftFootline z) + (y \LeftFootline z)$ |
| $RP9$ | $\delta \LeftFootline x \to \delta$ |
| $RP10$ | $\epsilon \LeftFootline x \to x$ |
| $RP11$ | $x \LeftFootline \epsilon \to x$ |
| $RC1$ | $e_1 \mid e_2 \to \gamma(e_1, e_2)$ |
| $RC2$ | $e_1 \mid (e_2 \cdot y) \to \gamma(e_1, e_2) \cdot y$ |
| $RC3$ | $(e_1 \cdot x) \mid e_2 \to \gamma(e_1, e_2) \cdot x$ |
| $RC4$ | $(e_1 \cdot x) \mid (e_2 \cdot y) \to \gamma(e_1, e_2) \cdot (x \between y)$ |
| $RC5$ | $(x+y) \mid z \to (x \mid z) + (y \mid z)$ |
| $RC6$ | $x \mid (y+z) \to (x \mid y) + (x \mid z)$ |
| $RC7$ | $\delta \mid x \to \delta$ |
| $RC8$ | $x \mid \delta \to \delta$ |
| $RC9$ | $\epsilon \mid x \to \delta$ |
| $RC10$ | $x \mid \epsilon \to \delta$ |
| $RPM1$ | $x \parallel (y \boxplus_\pi z) \to (x \parallel y) \boxplus_\pi (x \parallel z)$ |
| $RPM2$ | $(x \boxplus_\pi y) \parallel z \to (x \parallel z) \boxplus_\pi (y \parallel z)$ |
| $RPM3$ | $x \mid (y \boxplus_\pi z) \to (x \mid y) \boxplus_\pi (x \mid z)$ |
| $RPM4$ | $(x \boxplus_\pi y) \mid z \to (x \mid z) \boxplus_\pi (y \mid z)$ |
| $RCE1$ | $\Theta(e) \to e$ |
| $RCE2$ | $\Theta(\delta) \to \delta$ |
| $RCE3$ | $\Theta(\epsilon) \to \epsilon$ |
| $RCE4$ | $\Theta(x+y) \to \Theta(x) + \Theta(y)$ |
| $RPCE1$ | $\Theta(x \boxplus_\pi y) \to \Theta(x) \boxplus_\pi \Theta(y)$ |
| $RCE5$ | $\Theta(x \cdot y) \to \Theta(x) \cdot \Theta(y)$ |
| $RCE6$ | $\Theta(x \LeftFootline y) \to ((\Theta(x) \triangleleft y) \LeftFootline y) + ((\Theta(y) \triangleleft x) \LeftFootline x)$ |
| $RCE7$ | $\Theta(x \mid y) \to ((\Theta(x) \triangleleft y) \mid y) + ((\Theta(y) \triangleleft x) \mid x)$ |
| $RU1$ | $(\sharp(e_1, e_2)) \quad e_1 \triangleleft e_2 \to \tau$ |
| $RU2$ | $(\sharp(e_1, e_2), e_2 \leq e_3) \quad e_1 \triangleleft e_3 \to \tau$ |
| $RU3$ | $(\sharp(e_1, e_2), e_2 \leq e_3) \quad e3 \triangleleft e_1 \to \tau$ |
| $RPU1$ | $(\sharp_\pi(e_1, e_2)) \quad e_1 \triangleleft e_2 \to \tau$ |
| $RPU2$ | $(\sharp_\pi(e_1, e_2), e_2 \leq e_3) \quad e_1 \triangleleft e_3 \to \tau$ |
| $RPU3$ | $(\sharp_\pi(e_1, e_2), e_2 \leq e_3) \quad e_3 \triangleleft e_1 \to \tau$ |
| $RU4$ | $e \triangleleft \delta \to e$ |
| $RU5$ | $\delta \triangleleft e \to \delta$ |
| $RU6$ | $e \triangleleft \epsilon \to e$ |
| $RU7$ | $\epsilon \triangleleft e \to e$ |
| $RU8$ | $(x+y) \triangleleft z \to (x \triangleleft z) + (y \triangleleft z)$ |

continued on next page

**TABLE 13.5** *(continued)*

| No. | Rewriting Rule |
|---|---|
| $RPU4$ | $(x \boxplus_\pi y) \triangleleft z \to (x \triangleleft z) \boxplus_\pi (y \triangleleft z)$ |
| $RU9$ | $(x \cdot y) \triangleleft z \to (x \triangleleft z) \cdot (y \triangleleft z)$ |
| $RU10$ | $(x \between y) \triangleleft z \to (x \triangleleft z) \between (y \triangleleft z)$ |
| $RU11$ | $(x \mid y) \triangleleft z \to (x \triangleleft z) \mid (y \triangleleft z)$ |
| $RU12$ | $x \triangleleft (y + z) \to (x \triangleleft y) \triangleleft z$ |
| $RPU5$ | $x \triangleleft (y \boxplus_\pi z) \to (x \triangleleft y) \triangleleft z$ |
| $RU13$ | $x \triangleleft (y \cdot z) \to (x \triangleleft y) \triangleleft z$ |
| $RU14$ | $x \triangleleft (y \between z) \to (x \triangleleft y) \triangleleft z$ |
| $RU15$ | $x \triangleleft (y \mid z) \to (x \triangleleft y) \triangleleft z$ |
| $RD1$ | $e \notin H \quad \partial_H(e) \to e$ |
| $RD2$ | $e \in H \quad \partial_H(e) \to \delta$ |
| $RD3$ | $\partial_H(\delta) \to \delta$ |
| $RD4$ | $\partial_H(x + y) \to \partial_H(x) + \partial_H(y)$ |
| $RD5$ | $\partial_H(x \cdot y) \to \partial_H(x) \cdot \partial_H(y)$ |
| $RD6$ | $\partial_H(x \between y) \to \partial_H(x) \between \partial_H(y)$ |
| $RPD1$ | $\partial_H(x \boxplus_\pi y) \to \partial_H(x) \boxplus_\pi \partial_H(y)$ |
| $RG12$ | $\phi(x \between y) \to \phi x \between \phi y$ |
| $RG13$ | $\phi(x \mid y) \to \phi x \mid \phi y$ |
| $RG14$ | $\delta \between \phi \to \delta$ |
| $RG15$ | $\phi \mid \delta \to \delta$ |
| $RG16$ | $\delta \mid \phi \to \delta$ |
| $RG17$ | $\phi \between \epsilon \to \phi$ |
| $RG18$ | $\epsilon \between \phi \to \phi$ |
| $RG19$ | $\phi \mid \epsilon \to \delta$ |
| $RG20$ | $\epsilon \mid \phi \to \delta$ |
| $RG21$ | $\phi \between \neg\phi \to \delta$ |
| $RG22$ | $\Theta(\phi) \to \phi$ |
| $RG23$ | $\partial_H(\phi) \to \phi$ |
| $RG24$ | $\phi_0 \between \cdots \between \phi_n \to \delta$ if $\forall s_0, \cdots, s_n \in S, \exists i \leq n.test(\neg\phi_i, s_0 \cup \cdots \cup s_n)$ |

**TABLE 13.6** Probabilistic transition rules of $APPTC_G$.

$$\frac{x \rightsquigarrow x' \quad y \rightsquigarrow y'}{x \between y \rightsquigarrow x' \parallel y' + x' \mid y'}$$

$$\frac{x \rightsquigarrow x' \quad y \rightsquigarrow y'}{x \parallel y \rightsquigarrow x' \between y + y' \between x}$$

$$\frac{x \rightsquigarrow x'}{x \between y \rightsquigarrow x' \between y}$$

$$\frac{x \rightsquigarrow x' \quad y \rightsquigarrow y'}{x \mid y \rightsquigarrow x' \mid y'}$$

$$\frac{x \rightsquigarrow x'}{\Theta(x) \rightsquigarrow \Theta(x')}$$

$$\frac{x \rightsquigarrow x'}{x \triangleleft y \rightsquigarrow x' \triangleleft y}$$

**TABLE 13.7 Action transition rules of $APPTC_G$.**

$$\frac{}{\langle \breve{e}_1 \parallel \cdots \parallel \breve{e}_n, s\rangle \xrightarrow{\{e_1,\cdots,e_n\}} \langle \surd, s'\rangle} \text{ if } s' \in effect(e_1, s) \cup \cdots \cup effect(e_n, s)$$

$$\frac{}{\langle \breve{\phi}_1 \parallel \cdots \parallel \breve{\phi}_n, s\rangle \rightarrow \langle \surd, s\rangle} \text{ if } test(\phi_1, s), \cdots, test(\phi_n, s)$$

$$\frac{\langle x, s\rangle \xrightarrow{e_1} \langle \surd, s'\rangle \quad \langle y, s\rangle \xrightarrow{e_2} \langle \surd, s''\rangle}{\langle x \parallel y, s\rangle \xrightarrow{\{e_1,e_2\}} \langle \surd, s' \cup s''\rangle} \qquad \frac{\langle x, s\rangle \xrightarrow{e_1} \langle x', s'\rangle \quad \langle y, s\rangle \xrightarrow{e_2} \langle \surd, s''\rangle}{\langle x \parallel y, s\rangle \xrightarrow{\{e_1,e_2\}} \langle x', s' \cup s''\rangle}$$

$$\frac{\langle x, s\rangle \xrightarrow{e_1} \langle \surd, s'\rangle \quad \langle y, s\rangle \xrightarrow{e_2} \langle y', s''\rangle}{\langle x \parallel y, s\rangle \xrightarrow{\{e_1,e_2\}} \langle y', s' \cup s''\rangle} \qquad \frac{\langle x, s\rangle \xrightarrow{e_1} \langle x', s'\rangle \quad \langle y, s\rangle \xrightarrow{e_2} \langle y', s''\rangle}{\langle x \parallel y, s\rangle \xrightarrow{\{e_1,e_2\}} \langle x' \between y', s' \cup s''\rangle}$$

$$\frac{\langle x, s\rangle \xrightarrow{e_1} \langle \surd, s'\rangle \quad \langle y, s\rangle \stackrel{e_2}{\nrightarrow} \quad (e_1 \% e_2)}{\langle x \parallel y, s\rangle \xrightarrow{e_1} \langle y, s'\rangle} \qquad \frac{\langle x, s\rangle \xrightarrow{e_1} \langle x', s'\rangle \quad \langle y, s\rangle \stackrel{e_2}{\nrightarrow} \quad (e_1 \% e_2)}{\langle x \parallel y, s\rangle \xrightarrow{e_1} \langle x' \between y, s'\rangle}$$

$$\frac{\langle x, s\rangle \stackrel{e_1}{\nrightarrow} \quad \langle y, s\rangle \xrightarrow{e_2} \langle \surd, s''\rangle \quad (e_1 \% e_2)}{\langle x \parallel y, s\rangle \xrightarrow{e_2} \langle x, s''\rangle} \qquad \frac{\langle x, s\rangle \stackrel{e_1}{\nrightarrow} \quad \langle y, s\rangle \xrightarrow{e_2} \langle y', s''\rangle \quad (e_1 \% e_2)}{\langle x \parallel y, s\rangle \xrightarrow{e_2} \langle x \between y', s''\rangle}$$

$$\frac{\langle x, s\rangle \xrightarrow{e_1} \langle \surd, s'\rangle \quad \langle y, s\rangle \xrightarrow{e_2} \langle \surd, s''\rangle \quad (e_1 \leq e_2)}{\langle x \lfloor\!\lfloor y, s\rangle \xrightarrow{\{e_1,e_2\}} \langle \surd, s' \cup s''\rangle} \qquad \frac{\langle x, s\rangle \xrightarrow{e_1} \langle x', s'\rangle \quad \langle y, s\rangle \xrightarrow{e_2} \langle \surd, s''\rangle \quad (e_1 \leq e_2)}{\langle x \lfloor\!\lfloor y, s\rangle \xrightarrow{\{e_1,e_2\}} \langle x', s' \cup s''\rangle}$$

$$\frac{\langle x, s\rangle \xrightarrow{e_1} \langle \surd, s'\rangle \quad \langle y, s\rangle \xrightarrow{e_2} \langle y', s''\rangle \quad (e_1 \leq e_2)}{\langle x \lfloor\!\lfloor y, s\rangle \xrightarrow{\{e_1,e_2\}} \langle y', s' \cup s''\rangle} \qquad \frac{\langle x, s\rangle \xrightarrow{e_1} \langle x', s'\rangle \quad \langle y, s\rangle \xrightarrow{e_2} \langle y', s''\rangle \quad (e_1 \leq e_2)}{\langle x \lfloor\!\lfloor y, s\rangle \xrightarrow{\{e_1,e_2\}} \langle x' \between y', s' \cup s''\rangle}$$

$$\frac{\langle x, s\rangle \xrightarrow{e_1} \langle \surd, s'\rangle \quad \langle y, s\rangle \xrightarrow{e_2} \langle \surd, s''\rangle}{\langle x \mid y, s\rangle \xrightarrow{\gamma(e_1,e_2)} \langle \surd, effect(\gamma(e_1, e_2), s)\rangle} \qquad \frac{\langle x, s\rangle \xrightarrow{e_1} \langle x', s'\rangle \quad \langle y, s\rangle \xrightarrow{e_2} \langle \surd, s''\rangle}{\langle x \mid y, s\rangle \xrightarrow{\gamma(e_1,e_2)} \langle x', effect(\gamma(e_1, e_2), s)\rangle}$$

$$\frac{\langle x, s\rangle \xrightarrow{e_1} \langle \surd, s'\rangle \quad \langle y, s\rangle \xrightarrow{e_2} \langle y', s''\rangle}{\langle x \mid y, s\rangle \xrightarrow{\gamma(e_1,e_2)} \langle y', effect(\gamma(e_1, e_2), s)\rangle} \qquad \frac{\langle x, s\rangle \xrightarrow{e_1} \langle x', s'\rangle \quad \langle y, s\rangle \xrightarrow{e_2} \langle y', s''\rangle}{\langle x \mid y, s\rangle \xrightarrow{\gamma(e_1,e_2)} \langle x' \between y', effect(\gamma(e_1, e_2), s)\rangle}$$

$$\frac{\langle x, s\rangle \xrightarrow{e_1} \langle \surd, s'\rangle \quad (\sharp(e_1, e_2))}{\langle \Theta(x), s\rangle \xrightarrow{e_1} \langle \surd, s'\rangle} \qquad \frac{\langle x, s\rangle \xrightarrow{e_2} \langle \surd, s''\rangle \quad (\sharp(e_1, e_2))}{\langle \Theta(x), s\rangle \xrightarrow{e_2} \langle \surd, s''\rangle}$$

$$\frac{\langle x, s\rangle \xrightarrow{e_1} \langle x', s'\rangle \quad (\sharp(e_1, e_2))}{\langle \Theta(x), s\rangle \xrightarrow{e_1} \langle \Theta(x'), s'\rangle} \qquad \frac{\langle x, s\rangle \xrightarrow{e_2} \langle x'', s''\rangle \quad (\sharp(e_1, e_2))}{\langle \Theta(x), s\rangle \xrightarrow{e_2} \langle \Theta(x''), s''\rangle}$$

$$\frac{\langle x, s\rangle \xrightarrow{e_1} \langle \surd, s'\rangle \quad \langle y, s\rangle \nrightarrow^{e_2} \quad (\sharp(e_1, e_2))}{\langle x \triangleleft y, s\rangle \xrightarrow{\tau} \langle \surd, s'\rangle} \qquad \frac{\langle x, s\rangle \xrightarrow{e_1} \langle x', s'\rangle \quad \langle y, s\rangle \nrightarrow^{e_2} \quad (\sharp(e_1, e_2))}{\langle x \triangleleft y, s\rangle \xrightarrow{\tau} \langle x', s'\rangle}$$

$$\frac{\langle x, s\rangle \xrightarrow{e_1} \langle \surd, s\rangle \quad \langle y, s\rangle \nrightarrow^{e_3} \quad (\sharp(e_1, e_2), e_2 \leq e_3)}{\langle x \triangleleft y, s\rangle \xrightarrow{\tau} \langle \surd, s'\rangle} \qquad \frac{\langle x, s\rangle \xrightarrow{e_1} \langle x', s'\rangle \quad \langle y, s\rangle \nrightarrow^{e_3} \quad (\sharp(e_1, e_2), e_2 \leq e_3)}{\langle x \triangleleft y, s\rangle \xrightarrow{\tau} \langle x', s'\rangle}$$

$$\frac{\langle x, s\rangle \xrightarrow{e_3} \langle \surd, s'\rangle \quad \langle y, s\rangle \nrightarrow^{e_2} \quad (\sharp(e_1, e_2), e_1 \leq e_3)}{\langle x \triangleleft y, s\rangle \xrightarrow{\tau} \langle \surd, s'\rangle} \qquad \frac{\langle x, s\rangle \xrightarrow{e_3} \langle x', s'\rangle \quad \langle y, s\rangle \nrightarrow^{e_2} \quad (\sharp(e_1, e_2), e_1 \leq e_3)}{\langle x \triangleleft y, s\rangle \xrightarrow{\tau} \langle x', s'\rangle}$$

$$\frac{\langle x, s\rangle \xrightarrow{e_1} \langle \surd, s'\rangle \quad (\sharp_\pi(e_1, e_2))}{\langle \Theta(x), s\rangle \xrightarrow{e_1} \langle \surd, s'\rangle} \qquad \frac{\langle x, s\rangle \xrightarrow{e_2} \langle \surd, s''\rangle \quad (\sharp_\pi(e_1, e_2))}{\langle \Theta(x), s\rangle \xrightarrow{e_2} \langle \surd, s''\rangle}$$

$$\frac{\langle x, s\rangle \xrightarrow{e_1} \langle x', s'\rangle \quad (\sharp_\pi(e_1, e_2))}{\langle \Theta(x), s\rangle \xrightarrow{e_1} \langle \Theta(x'), s'\rangle} \qquad \frac{\langle x, s\rangle \xrightarrow{e_2} \langle x'', s''\rangle \quad (\sharp_\pi(e_1, e_2))}{\langle \Theta(x), s\rangle \xrightarrow{e_2} \langle \Theta(x''), s''\rangle}$$

$$\frac{\langle x, s\rangle \xrightarrow{e_1} \langle \surd, s'\rangle \quad \langle y, s\rangle \nrightarrow^{e_2} \quad (\sharp_\pi(e_1, e_2))}{\langle x \triangleleft y, s\rangle \xrightarrow{\tau} \langle \surd, s'\rangle} \qquad \frac{\langle x, s\rangle \xrightarrow{e_1} \langle x', s'\rangle \quad \langle y, s\rangle \nrightarrow^{e_2} \quad (\sharp_\pi(e_1, e_2))}{\langle x \triangleleft y, s\rangle \xrightarrow{\tau} \langle x', s'\rangle}$$

$$\frac{\langle x, s\rangle \xrightarrow{e_1} \langle \surd, s\rangle \quad \langle y, s\rangle \nrightarrow^{e_3} \quad (\sharp_\pi(e_1, e_2), e_2 \leq e_3)}{\langle x \triangleleft y, s\rangle \xrightarrow{\tau} \langle \surd, s'\rangle} \qquad \frac{\langle x, s\rangle \xrightarrow{e_1} \langle x', s'\rangle \quad \langle y, s\rangle \nrightarrow^{e_3} \quad (\sharp_\pi(e_1, e_2), e_2 \leq e_3)}{\langle x \triangleleft y, s\rangle \xrightarrow{\tau} \langle x', s'\rangle}$$

$$\frac{\langle x, s\rangle \xrightarrow{e_3} \langle \surd, s'\rangle \quad \langle y, s\rangle \nrightarrow^{e_2} \quad (\sharp_\pi(e_1, e_2), e_1 \leq e_3)}{\langle x \triangleleft y, s\rangle \xrightarrow{\tau} \langle \surd, s'\rangle} \qquad \frac{\langle x, s\rangle \xrightarrow{e_3} \langle x', s'\rangle \quad \langle y, s\rangle \nrightarrow^{e_2} \quad (\sharp_\pi(e_1, e_2), e_1 \leq e_3)}{\langle x \triangleleft y, s\rangle \xrightarrow{\tau} \langle x', s'\rangle}$$

$$\frac{\langle x, s\rangle \xrightarrow{e} \langle \surd, s'\rangle}{\langle \partial_H(x), s\rangle \xrightarrow{e} \langle \surd, s'\rangle} \quad (e \notin H) \qquad \frac{\langle x, s\rangle \xrightarrow{e} \langle x', s'\rangle}{\langle \partial_H(x), s\rangle \xrightarrow{e} \langle \partial_H(x'), s'\rangle} \quad (e \notin H)$$

*(3) Probabilistic hp-bisimulation equivalence $\sim_{php}$ is a congruence with respect to $APPTC_G$.*

*(4) Probabilistic hhp-bisimulation equivalence $\sim_{phhp}$ is a congruence with respect to $APPTC_G$.*

*Proof.* (1) It is easy to see that probabilistic pomset bisimulation is an equivalent relation on $APPTC_G$ terms, we only need to prove that $\sim_{pp}$ is preserved by the operators $\parallel$, $\llcorner\!\llcorner$, $|$, $\Theta$, $\triangleleft$, $\partial_H$. It is trivial and we leave the proof as an exercise for the reader.

(2) It is easy to see that probabilistic step bisimulation is an equivalent relation on $APPTC_G$ terms, we only need to prove that $\sim_{ps}$ is preserved by the operators $\parallel$, $\llcorner\!\llcorner$, $|$, $\Theta$, $\triangleleft$, $\partial_H$. It is trivial and we leave the proof as an exercise for the reader.

(3) It is easy to see that probabilistic hp-bisimulation is an equivalent relation on $APPTC_G$ terms, we only need to prove that $\sim_{php}$ is preserved by the operators $\parallel$, $\llcorner\!\llcorner$, $|$, $\Theta$, $\triangleleft$, $\partial_H$. It is trivial and we leave the proof as an exercise for the reader.

(4) It is easy to see that probabilistic hhp-bisimulation is an equivalent relation on $APPTC_G$ terms, we only need to prove that $\sim_{phhp}$ is preserved by the operators $\parallel$, $\llcorner\!\llcorner$, $|$, $\Theta$, $\triangleleft$, $\partial_H$. It is trivial and we leave the proof as an exercise for the reader. □

**Theorem 13.26** (Soundness of $APPTC_G$ modulo probabilistic truly concurrent bisimulation equivalences). *(1) Let $x$ and $y$ be $APPTC_G$ terms. If $APPTC_G \vdash x = y$, then $x \sim_{pp} y$.*

*(2) Let $x$ and $y$ be $APPTC_G$ terms. If $APPTC_G \vdash x = y$, then $x \sim_{ps} y$.*

*(3) Let $x$ and $y$ be $APPTC_G$ terms. If $APPTC_G \vdash x = y$, then $x \sim_{php} y$;*

*(3) Let $x$ and $y$ be $APPTC_G$ terms. If $APPTC_G \vdash x = y$, then $x \sim_{phhp} y$.*

*Proof.* (1) Since probabilistic pomset bisimulation $\sim_{pp}$ is both an equivalent and a congruent relation, we only need to check if each axiom in Table 13.4 is sound modulo probabilistic pomset bisimulation equivalence. We leave the proof as an exercise for the reader.

(2) Since probabilistic step bisimulation $\sim_{ps}$ is both an equivalent and a congruent relation, we only need to check if each axiom in Table 13.4 is sound modulo probabilistic step bisimulation equivalence. We leave the proof as an exercise for the reader.

(3) Since probabilistic hp-bisimulation $\sim_{php}$ is both an equivalent and a congruent relation, we only need to check if each axiom in Table 13.4 is sound modulo probabilistic hp-bisimulation equivalence. We leave the proof as an exercise for the reader.

(4) Since probabilistic hhp-bisimulation $\sim_{phhp}$ is both an equivalent and a congruent relation, we only need to check if each axiom in Table 13.4 is sound modulo probabilistic hhp-bisimulation equivalence. We leave the proof as an exercise for the reader. □

**Theorem 13.27** (Completeness of $APPTC_G$ modulo probabilistic truly concurrent bisimulation equivalences). *(1) Let $p$ and $q$ be closed $APPTC_G$ terms, if $p \sim_{pp} q$, then $p = q$.*

*(2) Let $p$ and $q$ be closed $APPTC_G$ terms, if $p \sim_{ps} q$, then $p = q$.*

*(3) Let $p$ and $q$ be closed $APPTC_G$ terms, if $p \sim_{php} q$, then $p = q$.*

*(3) Let $p$ and $q$ be closed $APPTC_G$ terms, if $p \sim_{phhp} q$, then $p = q$.*

*Proof.* (1) First, by the elimination theorem of $APPTC_G$ (see Theorem 13.23), we know that for each closed $APPTC_G$ term $p$, there exists a closed basic $APPTC_G$ term $p'$, such that $APPTC \vdash p = p'$, hence, we only need to consider closed basic $APPTC_G$ terms.

The basic terms (see Definition 13.22) modulo associativity and commutativity (AC) of conflict $+$ (defined by axioms $A1$ and $A2$ in Table 13.1), and these equivalences is denoted by $=_{AC}$. Then, each equivalence class $s$ modulo AC of $+$ has the following normal form

$$s_1 \boxplus_{\pi_1} \cdots \boxplus_{\pi_{k-1}} s_k$$

with each $s_i$ having the following form

$$t_1 + \cdots + t_l$$

with each $t_j$ either an atomic event or of the form

$$u_1 \cdot \cdots \cdot u_m$$

with each $u_l$ either an atomic event or of the form

$$v_1 \leftmerge \cdots \leftmerge v_m$$

with each $v_m$ an atomic event, and each $t_j$ is called the summand of $s$.

Now, we prove that for normal forms $n$ and $n'$, if $n \sim_{pp} n'$, then $n =_{AC} n'$. It is sufficient to induct on the sizes of $n$ and $n'$.

- Consider a summand $e$ of $n$. Then, $\langle n, s\rangle \rightsquigarrow \xrightarrow{e} \langle \surd, s'\rangle$, hence, $n \sim_{pp} n'$ implies $\langle n', s\rangle \rightsquigarrow \xrightarrow{e} \langle \surd, s\rangle$, meaning that $n'$ also contains the summand $e$.
- Consider a summand $\phi$ of $n$. Then, $\langle n, s\rangle \rightsquigarrow \rightarrow \langle \surd, s\rangle$, if $test(\phi, s)$ holds, hence, $n \sim_{pp} n'$ implies $\langle n', s\rangle \rightsquigarrow \rightarrow \langle \surd, s\rangle$, if $test(\phi, s)$ holds, meaning that $n'$ also contains the summand $\phi$.
- Consider a summand $t_1 \cdot t_2$ of $n$,
  - if $t_1 \equiv e'$, then $\langle n, s\rangle \rightsquigarrow \xrightarrow{e'} \langle t_2, s'\rangle$, hence, $n \sim_{pp} n'$ implies $\langle n', s\rangle \rightsquigarrow \xrightarrow{e'} \langle t_2', s'\rangle$ with $t_2 \sim_{pp} t_2'$, meaning that $n'$ contains a summand $e' \cdot t_2'$. Since $t_2$ and $t_2'$ are normal forms and have sizes smaller than $n$ and $n'$, by the induction hypotheses if $t_2 \sim_{pp} t_2'$, then $t_2 =_{AC} t_2'$;
  - if $t_1 \equiv \phi'$, then $\langle n, s\rangle \rightsquigarrow \rightarrow \langle t_2, s\rangle$, if $test(\phi', s)$ holds, hence, $n \sim_{pp} n'$ implies $\langle n', s\rangle \rightsquigarrow \rightarrow \langle t_2', s\rangle$ with $t_2 \sim_{pp} t_2'$, if $test(\phi', s)$ holds, meaning that $n'$ contains a summand $\phi' \cdot t_2'$. Since $t_2$ and $t_2'$ are normal forms and have sizes smaller than $n$ and $n'$, by the induction hypotheses if $t_2 \sim_{pp} t_2'$, then $t_2 =_{AC} t_2'$;
  - if $t_1 \equiv e_1 \leftmerge \cdots \leftmerge e_l$, then $\langle n, s\rangle \rightsquigarrow \xrightarrow{\{e_1, \cdots, e_l\}} \langle t_2, s'\rangle$, hence, $n \sim_{pp} n'$ implies $\langle n', s\rangle \rightsquigarrow \xrightarrow{\{e_1, \cdots, e_l\}} \langle t_2', s'\rangle$ with $t_2 \sim_{pp} t_2'$, meaning that $n'$ contains a summand $(e_1 \leftmerge \cdots \leftmerge e_l) \cdot t_2'$. Since $t_2$ and $t_2'$ are normal forms and have sizes smaller than $n$ and $n'$, by the induction hypotheses if $t_2 \sim_{pp} t_2'$, then $t_2 =_{AC} t_2'$;
  - if $t_1 \equiv \phi_1 \leftmerge \cdots \leftmerge \phi_l$, then $\langle n, s\rangle \rightsquigarrow \rightarrow \langle t_2, s\rangle$, if $test(\phi_1, s), \cdots, test(\phi_l, s)$ hold, hence, $n \sim_p n'$ implies $\langle n', s\rangle \rightsquigarrow \rightarrow \langle t_2', s\rangle$ with $t_2 \sim_{pp} t_2'$, if $test(\phi_1, s), \cdots, test(\phi_l, s)$ hold, meaning that $n'$ contains a summand $(\phi_1 \leftmerge \cdots \leftmerge \phi_l) \cdot t_2'$. Since $t_2$ and $t_2'$ are normal forms and have sizes smaller than $n$ and $n'$, by the induction hypotheses if $t_2 \sim_{pp} t_2'$ then $t_2 =_{AC} t_2'$.

Hence, we obtain $n =_{AC} n'$.

Finally, let $s$ and $t$ be basic $APPTC_G$ terms, and $s \sim_{pp} t$, there are normal forms $n$ and $n'$, such that $s = n$ and $t = n'$. The soundness theorem of $APPTC_G$ modulo probabilistic pomset bisimulation equivalence (see Theorem 13.26) yields $s \sim_{pp} n$ and $t \sim_{pp} n'$, hence, $n \sim_{pp} s \sim_{pp} t \sim_{pp} n'$. Since if $n \sim_{pp} n'$, then $n =_{AC} n'$, $s = n =_{AC} n' = t$, as desired.

(2) It can be proven similarly as (1).

(3) It can be proven similarly as (1).

(4) It can be proven similarly as (1). □

**Theorem 13.28** (Sufficient determinism). *All related data environments with respect to $APPTC_G$ can be sufficiently deterministic.*

*Proof.* It only needs to check $effect(t, s)$ function is deterministic, and is sufficient to induct on the structure of term $t$. The new matter is the case $t = t_1 \between t_2$, the whole thing is $t_1 \between t_2 = t_1 \leftmerge t_2 + t_2 \leftmerge t_1 + t_1 \mid t_2$. We can make $effect(t)$ sufficiently deterministic: eliminating non-determinism during the modeling time by use of empty event $\epsilon$. We can make $t = t_1 \parallel t_2$ become $t = (\epsilon \cdot t_1) \parallel t_2$ or $t = t_1 \parallel (\epsilon \cdot t_2)$ during the modeling phase, and then $effect(t, s)$ becomes sufficiently deterministic. □

## 13.4 Recursion

In this section, we introduce recursion to capture infinite processes based on $APPTC_G$. In the following, $E, F, G$ are recursion specifications, $X, Y, Z$ are recursive variables.

**Definition 13.29** (Guarded recursive specification). A recursive specification

$$X_1 = t_1(X_1, \cdots, X_n)$$

$$\cdots$$

$$X_n = t_n(X_1, \cdots, X_n)$$

is guarded if the right-hand sides of its recursive equations can be adapted to the form by applications of the axioms in $APPTC$ and replacing recursion variables by the right-hand sides of their recursive equations,

**TABLE 13.8 Transition rules of guarded recursion.**

$$\frac{\langle t_i(\langle X_1|E\rangle,\cdots,\langle X_n|E\rangle),s\rangle \rightsquigarrow \langle y,s\rangle}{\langle\langle X_i|E\rangle,s\rangle \rightsquigarrow \langle y,s\rangle}$$

$$\frac{\langle t_i(\langle X_1|E\rangle,\cdots,\langle X_n|E\rangle),s\rangle \xrightarrow{\{e_1,\cdots,e_k\}} \langle\surd,s'\rangle}{\langle\langle X_i|E\rangle,s\rangle \xrightarrow{\{e_1,\cdots,e_k\}} \langle\surd,s'\rangle}$$

$$\frac{\langle t_i(\langle X_1|E\rangle,\cdots,\langle X_n|E\rangle),s\rangle \xrightarrow{\{e_1,\cdots,e_k\}} \langle y,s'\rangle}{\langle\langle X_i|E\rangle,s\rangle \xrightarrow{\{e_1,\cdots,e_k\}} \langle y,s'\rangle}$$

$((a_{111}\leftmerge\cdots\leftmerge a_{11i_1})\cdot s_1(X_1,\cdots,X_n)+\cdots+(a_{1k1}\leftmerge\cdots\leftmerge a_{1ki_k})\cdot s_k(X_1,\cdots,X_n)+(b_{111}\leftmerge\cdots\leftmerge b_{11j_1})+\cdots+(b_{11j_1}\leftmerge\cdots\leftmerge b_{1lj_l}))\boxplus_{\pi_1}\cdots\boxplus_{\pi_{m-1}}((a_{m11}\leftmerge\cdots\leftmerge a_{m1i_1})\cdot s_1(X_1,\cdots,X_n)+\cdots+(a_{mk1}\leftmerge\cdots\leftmerge a_{mki_k})\cdot s_k(X_1,\cdots,X_n)+(b_{m11}\leftmerge\cdots\leftmerge b_{m1j_1})+\cdots+(b_{m1j_1}\leftmerge\cdots\leftmerge b_{mlj_l}))$
where $a_{111},\cdots,a_{11i_1},a_{1k1},\cdots,a_{1ki_k},b_{111},\cdots,b_{11j_1},b_{11j_1},\cdots,b_{1lj_l},\cdots,a_{m11},\cdots,a_{m1i_1},a_{1k1},\cdots,a_{mki_k},$
$b_{111},\cdots,b_{m1j_1},b_{m1j_1},\cdots,b_{mlj_l}\in\mathbb{E}$, and the sum above is allowed to be empty, in which case it represents the deadlock $\delta$. Also, there does not exist an infinite sequence of $\epsilon$-transitions $\langle X|E\rangle\rightarrow\langle X'|E\rangle\rightarrow\langle X''|E\rangle\rightarrow\cdots$.

**Theorem 13.30** (Conservativity of $APPTC_G$ with guarded recursion). *$APPTC_G$ with guarded recursion is a conservative extension of $APPTC_G$.*

*Proof.* Since the transition rules of $APPTC_G$ are source dependent, and the transition rules for guarded recursion in Table 13.8 contain only a fresh constant in their source, hence, the transition rules of $APPTC_G$ with guarded recursion are a conservative extension of those of $APPTC_G$. □

**Theorem 13.31** (Congruence theorem of $APPTC_G$ with guarded recursion). *Probabilistic truly concurrent bisimulation equivalences $\sim_{pp}$, $\sim_p$, $\sim_{php}$, and $\sim_{phhp}$ are all congruences with respect to $APPTC_G$ with guarded recursion.*

*Proof.* It follows from the following two facts:

1. In a guarded recursive specification, the right-hand sides of its recursive equations can be adapted to the form by applications of the axioms in $APPTC_G$ and replacing recursion variables by the right-hand sides of their recursive equations;
2. Probabilistic truly concurrent bisimulation equivalences $\sim_{pp}$, $\sim_{ps}$, $\sim_{php}$, and $\sim_{phhp}$ are all congruences with respect to all operators of $APPTC_G$. □

**Theorem 13.32** (Elimination theorem of $APPTC_G$ with linear recursion). *Each process term in $APPTC_G$ with linear recursion is equal to a process term $\langle X_1|E\rangle$ with $E$ a linear recursive specification.*

*Proof.* By applying structural induction with respect to term size, each process term $t_1$ in $APPTC$ with linear recursion generates a process that can be expressed in the form of equations
$t_i=((a_{1i11}\leftmerge\cdots\leftmerge a_{1i1i_1})t_{i1}+\cdots+(a_{1ik_i1}\leftmerge\cdots\leftmerge a_{1ik_ii_k})t_{ik_i}+(b_{1i11}\leftmerge\cdots\leftmerge b_{1i1i_1})+\cdots+(b_{1il_i1}\leftmerge\cdots\leftmerge b_{1il_ii_l}))\boxplus_{\pi_1}\cdots\boxplus_{\pi_{m-1}}((a_{mi11}\leftmerge\cdots\leftmerge a_{mi1i_1})t_{i1}+\cdots+(a_{mik_i1}\leftmerge\cdots\leftmerge a_{mik_ii_k})t_{ik_i}+(b_{mi11}\leftmerge\cdots\leftmerge b_{mi1i_1})+\cdots+(b_{mil_i1}\leftmerge\cdots\leftmerge b_{mil_ii_l}))$
for $i\in\{1,\cdots,n\}$. Let the linear recursive specification $E$ consist of the recursive equations
$X_i=((a_{1i11}\leftmerge\cdots\leftmerge a_{1i1i_1})X_{i1}+\cdots+(a_{1ik_i1}\leftmerge\cdots\leftmerge a_{1ik_ii_k})X_{ik_i}+(b_{1i11}\leftmerge\cdots\leftmerge b_{1i1i_1})+\cdots+(b_{1il_i1}\leftmerge\cdots\leftmerge b_{1il_ii_l}))\boxplus_{\pi_1}\cdots\boxplus_{\pi_{m-1}}((a_{mi11}\leftmerge\cdots\leftmerge a_{mi1i_1})X_{i1}+\cdots+(a_{mik_i1}\leftmerge\cdots\leftmerge a_{mik_ii_k})X_{ik_i}+(b_{mi11}\leftmerge\cdots\leftmerge b_{mi1i_1})+\cdots+(b_{mil_i1}\leftmerge\cdots\leftmerge b_{mil_ii_l}))$
for $i\in\{1,\cdots,n\}$. Replacing $X_i$ by $t_i$ for $i\in\{1,\cdots,n\}$ is a solution for $E$, $RSP$ yields $t_1=\langle X_1|E\rangle$. □

**Theorem 13.33** (Soundness of $APPTC_G$ with guarded recursion). *Let $x$ and $y$ be $APPTC_G$ with guarded recursion terms. If $APPTC_G$ with guarded recursion $\vdash x=y$, then*

*(1) $x\sim_{ps}y$;*

*(2) $x\sim_{pp}y$;*

*(3) $x\sim_{php}y$;*

*(4) $x\sim_{phhp}y$.*

*Proof.* (1) Since probabilistic step bisimulation $\sim_{ps}$ is both an equivalent and a congruent relation with respect to $APPTC_G$ with guarded recursion, we only need to check if each axiom in Table 2.6 is sound modulo probabilistic step bisimulation equivalence. We leave them as exercises for the reader.

(2) Since probabilistic pomset bisimulation $\sim_{pp}$ is both an equivalent and a congruent relation with respect to the guarded recursion, we only need to check if each axiom in Table 2.6 is sound modulo probabilistic pomset bisimulation equivalence. We leave them as exercises for the reader.

(3) Since probabilistic hp-bisimulation $\sim_{php}$ is both an equivalent and a congruent relation with respect to guarded recursion, we only need to check if each axiom in Table 2.6 is sound modulo probabilistic hp-bisimulation equivalence. We leave them as exercises for the reader.

(4) Since probabilistic hhp-bisimulation $\sim_{phhp}$ is both an equivalent and a congruent relation with respect to guarded recursion, we only need to check if each axiom in Table 2.6 is sound modulo probabilistic hhp-bisimulation equivalence. We leave them as exercises for the reader. □

**Theorem 13.34** (Completeness of $APPTC_G$ with linear recursion). *Let p and q be closed $APPTC_G$ with linear recursion terms, then*

*(1) if $p \sim_{ps} q$, then $p = q$.*
*(2) if $p \sim_{pp} q$, then $p = q$.*
*(3) if $p \sim_{php} q$, then $p = q$.*
*(4) if $p \sim_{phhp} q$, then $p = q$.*

*Proof.* First, by the elimination theorem of $APPTC_G$ with guarded recursion (see Theorem 13.32), we know that each process term in $APPTC_G$ with linear recursion is equal to a process term $\langle X_1|E\rangle$ with $E$ a linear recursive specification. Also, for simplicity, without loss of generalization, we do not consider empty event $\epsilon$, just because recursion with $\epsilon$ is similar to that with silent event $\tau$.

It remains to prove the following cases:

(1) If $\langle X_1|E_1\rangle \sim_{ps} \langle Y_1|E_2\rangle$ for linear recursive specification $E_1$ and $E_2$, then $\langle X_1|E_1\rangle = \langle Y_1|E_2\rangle$.

Let $E_1$ consist of recursive equations $X = t_X$ for $X \in \mathcal{X}$ and $E_2$ consists of recursion equations $Y = t_Y$ for $Y \in \mathcal{Y}$. Let the linear recursive specification $E$ consist of recursion equations $Z_{XY} = t_{XY}$, and $\langle X|E_1\rangle \sim_s \langle Y|E_2\rangle$, and $t_{XY}$ consists of the following summands:

1. $t_{XY}$ contains a summand $(a_1 \between \cdots \between a_m)Z_{X'Y'}$ iff $t_X$ contains the summand $(a_1 \between \cdots \between a_m)X'$ and $t_Y$ contains the summand $(a_1 \between \cdots \between a_m)Y'$ such that $\langle X'|E_1\rangle \sim_s \langle Y'|E_2\rangle$;
2. $t_{XY}$ contains a summand $b_1 \between \cdots \between b_n$ iff $t_X$ contains the summand $b_1 \between \cdots \between b_n$ and $t_Y$ contains the summand $b_1 \between \cdots \between b_n$.

Let $\sigma$ map recursion variable $X$ in $E_1$ to $\langle X|E_1\rangle$, and let $\pi$ map recursion variable $Z_{XY}$ in $E$ to $\langle X|E_1\rangle$. Hence, $\sigma((a_1 \between \cdots \between a_m)X') \equiv (a_1 \between \cdots \between a_m)\langle X'|E_1\rangle \equiv \pi((a_1 \between \cdots \between a_m)Z_{X'Y'})$, hence, by $RDP$, we obtain $\langle X|E_1\rangle = \sigma(t_X) = \pi(t_{XY})$. Then, by $RSP$, $\langle X|E_1\rangle = \langle Z_{XY}|E\rangle$, particularly, $\langle X_1|E_1\rangle = \langle Z_{X_1Y_1}|E\rangle$. Similarly, we can obtain $\langle Y_1|E_2\rangle = \langle Z_{X_1Y_1}|E\rangle$. Finally, $\langle X_1|E_1\rangle = \langle Z_{X_1Y_1}|E\rangle = \langle Y_1|E_2\rangle$, as desired.

(2) If $\langle X_1|E_1\rangle \sim_{pp} \langle Y_1|E_2\rangle$ for linear recursive specifications $E_1$ and $E_2$, then $\langle X_1|E_1\rangle = \langle Y_1|E_2\rangle$.

It can be proven similarly to (1), hence, we omit it here.

(3) If $\langle X_1|E_1\rangle \sim_{php} \langle Y_1|E_2\rangle$ for linear recursive specifications $E_1$ and $E_2$, then $\langle X_1|E_1\rangle = \langle Y_1|E_2\rangle$.

It can be proven similarly to (1), hence, we omit it here.

(4) If $\langle X_1|E_1\rangle \sim_{phhp} \langle Y_1|E_2\rangle$ for linear recursive specification s$E_1$ and $E_2$, then $\langle X_1|E_1\rangle = \langle Y_1|E_2\rangle$.

It can be proven similarly to (1), hence, we omit it here. □

## 13.5 Abstraction

To abstract away from the internal implementations of a program, and verify that the program exhibits the desired external behaviors, the silent step $\tau$ and abstraction operator $\tau_I$ are introduced, where $I \subseteq \mathbb{E} \cup G_{at}$ denotes the internal events or guards. The silent step $\tau$ represents the internal events, and $\tau_\phi$ for internal guards, when we consider the external behaviors of a process, $\tau$ steps can be removed, that is, $\tau$ steps must remain silent. The transition rule of $\tau$ is shown in Table 13.9. In the following, let the atomic event $e$ range over $\mathbb{E} \cup \{\epsilon\} \cup \{\delta\} \cup \{\tau\}$, and $\phi$ range over $G \cup \{\tau\}$, and let the communication function $\gamma : \mathbb{E} \cup \{\tau\} \times \mathbb{E} \cup \{\tau\} \rightarrow \mathbb{E} \cup \{\delta\}$, with each communication involved $\tau$ resulting in $\delta$. We use $\tau(s)$ to denote $effect(\tau, s)$, for the fact that $\tau$ only changes the state of internal data environment, that is, for the external data environments, $s = \tau(s)$.

**TABLE 13.9** Transition rule of the silent step.

$$\frac{}{\tau \rightsquigarrow \breve{\tau}}$$

$$\frac{}{\langle \tau_\phi, s\rangle \rightarrow \langle \surd, s\rangle} \text{ if } test(\tau_\phi, s)$$

$$\frac{}{\langle \tau, s\rangle \xrightarrow{\tau} \langle \surd, \tau(s)\rangle}$$

**TABLE 13.10** Axioms of silent step.

| No. | Axiom |
|---|---|
| $B1$ | $(y = y + y, z = z + z) \quad x \cdot ((y + \tau \cdot (y + z)) \boxplus_\pi w) = x \cdot ((y + z) \boxplus_\pi w)$ |
| $B2$ | $(y = y + y, z = z + z) \quad x \between ((y + \tau \between (y + z)) \boxplus_\pi w) = x \between ((y + z) \boxplus_\pi w)$ |

**Definition 13.35** (Guarded linear recursive specification). A linear recursive specification $E$ is guarded if there does not exist an infinite sequence of $\tau$-transitions $\langle X|E\rangle \xrightarrow{\tau} \langle X'|E\rangle \xrightarrow{\tau} \langle X''|E\rangle \xrightarrow{\tau} \cdots$, and there does not exist an infinite sequence of $\epsilon$-transitions $\langle X|E\rangle \rightarrow \langle X'|E\rangle \rightarrow \langle X''|E\rangle \rightarrow \cdots$.

**Theorem 13.36** (Conservativity of $APPTC_G$ with silent step and guarded linear recursion). *$APPTC_G$ with silent step and guarded linear recursion is a conservative extension of $APPTC_G$ with linear recursion.*

*Proof.* Since the transition rules of $APPTC_G$ with linear recursion are source dependent, and the transition rules for silent step in Table 13.9 contain only a fresh constant $\tau$ in their source, hence, the transition rules of $APPTC_G$ with silent step and guarded linear recursion is a conservative extension of those of $APPTC_G$ with linear recursion. □

**Theorem 13.37** (Congruence theorem of $APPTC_G$ with silent step and guarded linear recursion). *Probabilistic rooted branching truly concurrent bisimulation equivalences $\approx_{prbp}$, $\approx_{prbs}$, $\approx_{prbhp}$, and $\approx_{rbhhp}$ are all congruences with respect to $APPTC_G$ with silent step and guarded linear recursion.*

*Proof.* It follows from the following three facts:

1. In a guarded linear recursive specification, the right-hand sides of its recursive equations can be adapted to the form by applications of the axioms in $APPTC_G$ and replacing recursion variables by the right-hand sides of their recursive equations;
2. Probabilistic truly concurrent bisimulation equivalences $\sim_{pp}$, $\sim_{ps}$, $\sim_{php}$, and $\sim_{phhp}$ are all congruences with respect to all operators of $APPTC_G$, while probabilistic truly concurrent bisimulation equivalences $\sim_{pp}$, $\sim_{ps}$, $\sim_{php}$, and $\sim_{phhp}$ imply the corresponding probabilistic rooted branching truly concurrent bisimulations $\approx_{prbp}$, $\approx_{prbs}$, $\approx_{prbhp}$, and $\approx_{prbhhp}$, hence, probabilistic rooted branching truly concurrent bisimulations $\approx_{prbp}$, $\approx_{prbs}$, $\approx_{prbhp}$, and $\approx_{prbhhp}$ are all congruences with respect to all operators of $APPTC_G$;
3. While $\mathbb{E}$ is extended to $\mathbb{E} \cup \{\tau\}$, and $G$ is extended to $G \cup \{\tau\}$, it can be proved that probabilistic rooted branching truly concurrent bisimulations $\approx_{prbp}$, $\approx_{prbs}$, $\approx_{prbhp}$, and $\approx_{prbhhp}$ are all congruences with respect to all operators of $APPTC_G$, however, we omit the proof here. □

We design the axioms for the silent step $\tau$ in Table 13.10.

**Theorem 13.38** (Elimination theorem of $APPTC_G$ with silent step and guarded linear recursion). *Each process term in $APPTC_G$ with silent step and guarded linear recursion is equal to a process term $\langle X_1|E\rangle$ with $E$ a guarded linear recursive specification.*

*Proof.* By applying structural induction with respect to term size, each process term $t_1$ in $APPTC_G$ with silent step and guarded linear recursion generates a process that can be expressed in the form of equations

$t_i = ((a_{1i11} \between \cdots \between a_{1i1i_1})t_{i1} + \cdots + (a_{1ik_i1} \between \cdots \between a_{1ik_ii_k})t_{ik_i} + (b_{1i11} \between \cdots \between b_{1i1i_1}) + \cdots + (b_{1il_i1} \between \cdots \between b_{1il_ii_l})) \boxplus_{\pi_1} \cdots \boxplus_{\pi_{m-1}} ((a_{mi11} \between \cdots \between a_{mi1i_1})t_{i1} + \cdots + (a_{mik_i1} \between \cdots \between a_{mik_ii_k})t_{ik_i} + (b_{mi11} \between \cdots \between b_{mi1i_1}) + \cdots + (b_{mil_i1} \between \cdots \between b_{m1il_ii_l}))$

for $i \in \{1, \cdots, n\}$. Let the linear recursive specification $E$ consist of the recursive equations

$X_i = ((a_{1i11} \between \cdots \between a_{1i1i_1})X_{i1} + \cdots + (a_{1ik_i1} \between \cdots \between a_{1ik_ii_k})X_{ik_i} + (b_{1i11} \between \cdots \between b_{1i1i_1}) + \cdots + (b_{1il_i1} \between \cdots \between b_{1il_ii_l})) \boxplus_{\pi_1} \cdots \boxplus_{\pi_{m-1}} ((a_{mi11} \between \cdots \between a_{mi1i_1})X_{i1} + \cdots + (a_{mik_i1} \between \cdots \between a_{mik_ii_k})X_{ik_i} + (b_{mi11} \between \cdots \between b_{mi1i_1}) + \cdots + (b_{mil_i1} \between \cdots \between b_{mil_ii_l}))$

for $i \in \{1, \cdots, n\}$. Replacing $X_i$ by $t_i$ for $i \in \{1, \cdots, n\}$ is a solution for $E$, $RSP$ yields $t_1 = \langle X_1|E\rangle$. □

**Theorem 13.39** (Soundness of $APPTC_G$ with silent step and guarded linear recursion). *Let $x$ and $y$ be $APPTC_G$ with silent step and guarded linear recursion terms. If $APPTC_G$ with silent step and guarded linear recursion $\vdash x = y$, then*

*(1) $x \approx_{prbs} y$;*

*(2) $x \approx_{prbp} y$;*

*(3) $x \approx_{prbhp} y$;*

*(4) $x \approx_{prbhhp} y$.*

*Proof.* (1) Since probabilistic rooted branching step bisimulation $\approx_{prbs}$ is both an equivalent and a congruent relation with respect to $APPTC_G$ with silent step and guarded linear recursion, we only need to check if each axiom in Table 13.10 is sound modulo probabilistic rooted branching step bisimulation $\approx_{prbs}$. We leave them as exercises for the reader.

(2) Since probabilistic rooted branching pomset bisimulation $\approx_{prbp}$ is both an equivalent and a congruent relation with respect to $APPTC_G$ with silent step and guarded linear recursion, we only need to check if each axiom in Table 13.10 is sound modulo probabilistic rooted branching pomset bisimulation $\approx_{prbp}$. We leave them as exercises for the reader.

(3) Since probabilistic rooted branching hp-bisimulation $\approx_{prbhp}$ is both an equivalent and a congruent relation with respect to $APPTC_G$ with silent step and guarded linear recursion, we only need to check if each axiom in Table 13.10 is sound modulo probabilistic rooted branching hp-bisimulation $\approx_{prbhp}$. We leave them as exercises for the reader.

(4) Since probabilistic rooted branching hhp-bisimulation $\approx_{prbhhp}$ is both an equivalent and a congruent relation with respect to $APPTC_G$ with silent step and guarded linear recursion, we only need to check if each axiom in Table 13.10 is sound modulo probabilistic rooted branching hhp-bisimulation $\approx_{prbhhp}$. We leave them as exercises for the reader. □

**Theorem 13.40** (Completeness of $APPTC_G$ with silent step and guarded linear recursion). *Let $p$ and $q$ be closed $APPTC_G$ with silent step and guarded linear recursion terms, then,*

*(1) if $p \approx_{prbs} q$, then $p = q$;*

*(2) if $p \approx_{prbp} q$, then $p = q$;*

*(3) if $p \approx_{prbhp} q$, then $p = q$;*

*(3) if $p \approx_{prbhhp} q$, then $p = q$.*

*Proof.* First, by the elimination theorem of $APPTC_G$ with silent step and guarded linear recursion (see Theorem 13.38), we know that each process term in $APPTC_G$ with silent step and guarded linear recursion is equal to a process term $\langle X_1|E\rangle$ with $E$ a guarded linear recursive specification.

It remains to prove the following cases:

(1) If $\langle X_1|E_1\rangle \approx_{prbs} \langle Y_1|E_2\rangle$ for guarded linear recursive specification $E_1$ and $E_2$, then $\langle X_1|E_1\rangle = \langle Y_1|E_2\rangle$.

First, the recursive equation $W = \tau + \cdots + \tau$ with $W \not\equiv X_1$ in $E_1$ and $E_2$, can be removed, and the corresponding summands $aW$ are replaced by $a$, to obtain $E_1'$ and $E_2'$, by use of the axioms $RDP$, $A3$, and $B1$, and $\langle X|E_1\rangle = \langle X|E_1'\rangle$, $\langle Y|E_2\rangle = \langle Y|E_2'\rangle$.

Let $E_1$ consist of recursive equations $X = t_X$ for $X \in \mathcal{X}$ and $E_2$ consist of recursion equations $Y = t_Y$ for $Y \in \mathcal{Y}$, and are not of the form $\tau + \cdots + \tau$. Let the guarded linear recursive specification $E$ consist of recursion equations $Z_{XY} = t_{XY}$, and $\langle X|E_1\rangle \approx_{rbs} \langle Y|E_2\rangle$, and $t_{XY}$ consist of the following summands:

1. $t_{XY}$ contains a summand $(a_1 \between \cdots \between a_m)Z_{X'Y'}$ iff $t_X$ contains the summand $(a_1 \between \cdots \between a_m)X'$ and $t_Y$ contains the summand $(a_1 \between \cdots \between a_m)Y'$ such that $\langle X'|E_1\rangle \approx_{rbs} \langle Y'|E_2\rangle$;
2. $t_{XY}$ contains a summand $b_1 \between \cdots \between b_n$ iff $t_X$ contains the summand $b_1 \between \cdots \between b_n$ and $t_Y$ contains the summand $b_1 \between \cdots \between b_n$;
3. $t_{XY}$ contains a summand $\tau Z_{X'Y}$ iff $XY \not\equiv X_1Y_1$, $t_X$ contains the summand $\tau X'$, and $\langle X'|E_1\rangle \approx_{prbs} \langle Y|E_2\rangle$;
4. $t_{XY}$ contains a summand $\tau Z_{XY'}$ iff $XY \not\equiv X_1Y_1$, $t_Y$ contains the summand $\tau Y'$, and $\langle X|E_1\rangle \approx_{prbs} \langle Y'|E_2\rangle$.

Since $E_1$ and $E_2$ are guarded, $E$ is guarded. Constructing the process term $u_{XY}$ consists of the following summands:

1. $u_{XY}$ contains a summand $(a_1 \between \cdots \between a_m)\langle X'|E_1\rangle$ iff $t_X$ contains the summand $(a_1 \between \cdots \between a_m)X'$ and $t_Y$ contains the summand $(a_1 \between \cdots \between a_m)Y'$ such that $\langle X'|E_1\rangle \approx_{prbs} \langle Y'|E_2\rangle$;
2. $u_{XY}$ contains a summand $b_1 \between \cdots \between b_n$ iff $t_X$ contains the summand $b_1 \between \cdots \between b_n$ and $t_Y$ contains the summand $b_1 \between \cdots \between b_n$;
3. $u_{XY}$ contains a summand $\tau\langle X'|E_1\rangle$ iff $XY \not\equiv X_1Y_1$, $t_X$ contains the summand $\tau X'$, and $\langle X'|E_1\rangle \approx_{prbs} \langle Y|E_2\rangle$.

Let the process term $s_{XY}$ be defined as follows:

1. $s_{XY} \triangleq \tau\langle X|E_1\rangle + u_{XY}$ iff $XY \not\equiv X_1Y_1$, $t_Y$ contains the summand $\tau Y'$, and $\langle X|E_1\rangle \approx_{prbs} \langle Y'|E_2\rangle$;
2. $s_{XY} \triangleq \langle X|E_1\rangle$, otherwise.

Hence, $\langle X|E_1\rangle = \langle X|E_1\rangle + u_{XY}$, and $(a_1 \between \cdots \between a_m)(\tau\langle X|E_1\rangle + u_{XY}) = (a_1 \between \cdots \between a_m)((\tau\langle X|E_1\rangle + u_{XY}) + u_{XY}) = (a_1 \between \cdots \between a_m)(\langle X|E_1\rangle + u_{XY}) = (a_1 \between \cdots \between a_m)\langle X|E_1\rangle$, hence, $(a_1 \between \cdots \between a_m)s_{XY} = (a_1 \between \cdots \between a_m)\langle X|E_1\rangle$.

**TABLE 13.11 Transition rule of the abstraction operator.**

| | |
|---|---|
| $\frac{\langle x,s\rangle \rightsquigarrow \langle x',s\rangle}{\langle \tau_I(x),s\rangle \rightsquigarrow \langle \tau_I(x'),s\rangle}$ | |
| $\frac{\langle x,s\rangle \xrightarrow{e} \langle \surd,s'\rangle}{\langle \tau_I(x),s\rangle \xrightarrow{e} \langle \surd,s'\rangle}\quad e\notin I$ | $\frac{\langle x,s\rangle \xrightarrow{e} \langle x',s'\rangle}{\langle \tau_I(x),s\rangle \xrightarrow{e} \langle \tau_I(x'),s'\rangle}\quad e\notin I$ |
| $\frac{\langle x,s\rangle \xrightarrow{e} \langle \surd,s'\rangle}{\langle \tau_I(x),s\rangle \xrightarrow{\tau} \langle \surd,\tau(s)\rangle}\quad e\in I$ | $\frac{\langle x,s\rangle \xrightarrow{e} \langle x',s'\rangle}{\langle \tau_I(x),s\rangle \xrightarrow{\tau} \langle \tau_I(x'),\tau(s)\rangle}\quad e\in I$ |

Let $\sigma$ map recursion variable $X$ in $E_1$ to $\langle X|E_1\rangle$, and let $\pi$ map recursion variable $Z_{XY}$ in $E$ to $s_{XY}$. It is sufficient to prove $s_{XY} = \pi(t_{XY})$ for recursion variables $Z_{XY}$ in $E$. Either $XY \equiv X_1Y_1$ or $XY \not\equiv X_1Y_1$, we can obtain $s_{XY} = \pi(t_{XY})$. Hence, $s_{XY} = \langle Z_{XY}|E\rangle$ for recursive variables $Z_{XY}$ in $E$ is a solution for $E$. Then, by $RSP$, particularly, $\langle X_1|E_1\rangle = \langle Z_{X_1Y_1}|E\rangle$. Similarly, we can obtain $\langle Y_1|E_2\rangle = \langle Z_{X_1Y_1}|E\rangle$. Finally, $\langle X_1|E_1\rangle = \langle Z_{X_1Y_1}|E\rangle = \langle Y_1|E_2\rangle$, as desired.

(2) If $\langle X_1|E_1\rangle \approx_{prbp} \langle Y_1|E_2\rangle$ for guarded linear recursive specification s$E_1$ and $E_2$, then $\langle X_1|E_1\rangle = \langle Y_1|E_2\rangle$.
It can be proven similarly to (1), hence, we omit it here.

(3) If $\langle X_1|E_1\rangle \approx_{prbhb} \langle Y_1|E_2\rangle$ for guarded linear recursive specifications $E_1$ and $E_2$, then $\langle X_1|E_1\rangle = \langle Y_1|E_2\rangle$.
It can be proven similarly to (1), hence, we omit it here.

(4) If $\langle X_1|E_1\rangle \approx_{prbhhb} \langle Y_1|E_2\rangle$ for guarded linear recursive specifications $E_1$ and $E_2$, then $\langle X_1|E_1\rangle = \langle Y_1|E_2\rangle$.
It can be proven similarly to (1), hence, we omit it here. □

The unary abstraction operator $\tau_I$ ($I \subseteq \mathbb{E} \cup G_{at}$) renames all atomic events or atomic guards in $I$ into $\tau$. $APPTC_G$ with silent step and abstraction operator is called $APPTC_{G_\tau}$. The transition rules of operator $\tau_I$ are shown in Table 13.11.

**Theorem 13.41** (Conservativity of $APPTC_{G_\tau}$ with guarded linear recursion). *$APPTC_{G_\tau}$ with guarded linear recursion is a conservative extension of $APPTC_G$ with silent step and guarded linear recursion.*

*Proof.* Since the transition rules of $APPTC_G$ with silent step and guarded linear recursion are source dependent, and the transition rules for abstraction operator in Table 13.11 contain only a fresh operator $\tau_I$ in their source, the transition rules of $APPTC_{G_\tau}$ with guarded linear recursion is a conservative extension of those of $APPTC_G$ with silent step and guarded linear recursion. □

**Theorem 13.42** (Congruence theorem of $APPTC_{G_\tau}$ with guarded linear recursion). *Probabilistic rooted branching truly concurrent bisimulation equivalences $\approx_{prbp}$, $\approx_{prbs}$, $\approx_{prbhp}$, and $\approx_{prbhhp}$ are all congruences with respect to $APPTC_{G_\tau}$ with guarded linear recursion.*

*Proof.* (1) It is easy to see that probabilistic rooted branching pomset bisimulation is an equivalent relation on $APPTC_{G_\tau}$ with guarded linear recursion terms, we only need to prove that $\approx_{prbp}$ is preserved by the operators $\tau_I$. It is trivial and we leave the proof as an exercise for the reader.

(2) It is easy to see that probabilistic rooted branching step bisimulation is an equivalent relation on $APPTC_{G_\tau}$ with guarded linear recursion terms, we only need to prove that $\approx_{prbs}$ is preserved by the operators $\tau_I$. It is trivial and we leave the proof as an exercise for the reader.

(3) It is easy to see that probabilistic rooted branching hp-bisimulation is an equivalent relation on $APPTC_{G_\tau}$ with guarded linear recursion terms, we only need to prove that $\approx_{prbhp}$ is preserved by the operators $\tau_I$. It is trivial and we leave the proof as an exercise for the reader.

(4) It is easy to see that probabilistic rooted branching hhp-bisimulation is an equivalent relation on $APPTC_{G_\tau}$ with guarded linear recursion terms, we only need to prove that $\approx_{prbhhp}$ is preserved by the operators $\tau_I$. It is trivial and we leave the proof as an exercise for the reader. □

We design the axioms for the abstraction operator $\tau_I$ in Table 13.12.

**Theorem 13.43** (Soundness of $APPTC_{G_\tau}$ with guarded linear recursion). *Let $x$ and $y$ be $APPTC_{G_\tau}$ with guarded linear recursion terms. If $APPTC_{G_\tau}$ with guarded linear recursion $\vdash x = y$, then*

*(1) $x \approx_{prbs} y$;*
*(2) $x \approx_{prbp} y$;*
*(3) $x \approx_{prbhp} y$;*
*(4) $x \approx_{prbhhp} y$.*

**TABLE 13.12** Axioms of abstraction operator.

| No. | Axiom |
|---|---|
| $TI1$ | $e \notin I \quad \tau_I(e) = e$ |
| $TI2$ | $e \in I \quad \tau_I(e) = \tau$ |
| $TI3$ | $\tau_I(\delta) = \delta$ |
| $TI4$ | $\tau_I(x + y) = \tau_I(x) + \tau_I(y)$ |
| $PTI1$ | $\tau_I(x \boxplus_\pi y) = \tau_I(x) \boxplus_\pi \tau_I(y)$ |
| $TI5$ | $\tau_I(x \cdot y) = \tau_I(x) \cdot \tau_I(y)$ |
| $TI6$ | $\tau_I(x \between y) = \tau_I(x) \between \tau_I(y)$ |
| $G28$ | $\tau_\phi \cdot x = x$ |
| $G29$ | $x \cdot \tau_\phi = x$ |
| $G30$ | $x \parallel \tau_\phi = x$ |
| $G31$ | $\phi \notin I \quad \tau_I(\phi) = \phi$ |
| $G32$ | $\phi \in I \quad \tau_I(\phi) = \tau_\phi$ |

**TABLE 13.13** Recursive verification rules.

| | |
|---|---|
| $VR_1$ | $\dfrac{x = y + (i_1 \between \cdots \between i_m) \cdot x, y = y + y}{\tau \cdot \tau_I(x) = \tau \cdot \tau_I(y)}$ |
| $VR_2$ | $\dfrac{x = z \boxplus_\pi (u + (i_1 \between \cdots \between i_m) \cdot x), z = z + u, z = z + z}{\tau \cdot \tau_I(x) = \tau \cdot \tau_I(z)}$ |
| $VR_3$ | $\dfrac{x = z + (i_1 \between \cdots \between i_m) \cdot y, y = z \boxplus_\pi (u + (j_1 \between \cdots \between j_n) \cdot x), z = z + u, z = z + z}{\tau \cdot \tau_I(x) = \tau \cdot \tau_I(y') \text{ for } y' = z \boxplus_\pi (u + (i_1 \between \cdots \between i_m) \cdot y')}$ |

*Proof.* (1) Since probabilistic rooted branching step bisimulation $\approx_{prbs}$ is both an equivalent and a congruent relation with respect to $APPTC_{G_\tau}$ with guarded linear recursion, we only need to check if each axiom in Table 13.12 is sound modulo probabilistic rooted branching step bisimulation $\approx_{prbs}$. We leave them as exercises for the reader.

(2) Since probabilistic rooted branching pomset bisimulation $\approx_{prbp}$ is both an equivalent and a congruent relation with respect to $APPTC_{G_\tau}$ with guarded linear recursion, we only need to check if each axiom in Table 13.12 is sound modulo probabilistic rooted branching pomset bisimulation $\approx_{prbp}$. We leave them as exercises for the reader.

(3) Since probabilistic rooted branching hp-bisimulation $\approx_{prbhp}$ is both an equivalent and a congruent relation with respect to $APPTC_{G_\tau}$ with guarded linear recursion, we only need to check if each axiom in Table 13.12 is sound modulo probabilistic rooted branching hp-bisimulation $\approx_{prbhp}$. We leave them as exercises for the reader.

(4) Since probabilistic rooted branching hhp-bisimulation $\approx_{prbhhp}$ is both an equivalent and a congruent relation with respect to $APPTC_{G_\tau}$ with guarded linear recursion, we only need to check if each axiom in Table 13.12 is sound modulo probabilistic rooted branching hhp-bisimulation $\approx_{prbhhp}$. We leave them as exercises for the reader. □

Although $\tau$-loops are prohibited in guarded linear recursive specifications in a specifiable way, they can be constructed using the abstraction operator, for example, there exist $\tau$-loops in the process term $\tau_{\{a\}}(\langle X|X = aX\rangle)$. To avoid $\tau$-loops caused by $\tau_I$ and ensure fairness, we introduce the following recursive verification rules as Table 13.13 shows, note that $i_1, \cdots, i_m, j_1, \cdots, j_n \in I \subseteq \mathbb{E} \setminus \{\tau\}$.

**Theorem 13.44** (Soundness of $VR_1, VR_2, VR_3$). *$VR_1$, $VR_2$, and $VR_3$ are sound modulo probabilistic rooted branching truly concurrent bisimulation equivalences $\approx_{prbp}$, $\approx_{prbs}$, $\approx_{prbhp}$, and $\approx_{prbhhp}$.*

## 13.6 Hoare logic for $APPTC_G$

In this section, we introduce Hoare logic for $APPTC_G$. We do not introduce the preliminaries of Hoare logic, please refer to [13] for details.

A partial correct formula has the form

$$\{pre\}P\{post\}$$

**TABLE 13.14** The proof system $H$.

| | |
|---|---|
| $(H1)$ | $\{wp(e,\alpha)\}e\{\alpha\}$ if $e \in \mathbb{E}$ |
| $(H2)$ | $\{\alpha\}\phi\{\alpha \cdot \phi\}$ if $\phi \in G$ |
| $(H3)$ | $\frac{\{\alpha\}t\{\beta\} \quad \{\alpha\}t'\{\beta\}}{\{\alpha\}t+t'\{\beta\}}$ |
| $(PH1)$ | $\frac{\{\alpha\}t\{\beta\} \quad \{\alpha\}t'\{\beta\}}{\{\alpha\}t\boxplus_{\pi}t'\{\beta\}}$ |
| $(H4)$ | $\frac{\{\alpha\}t\{\alpha'\} \quad \{\alpha'\}t'\{\beta\}}{\{\alpha\}t\cdot t'\{\beta\}}$ |
| $(H5)$ | $\frac{\{\alpha\}t\{\alpha'\} \quad \{\beta\}t'\{\beta'\}}{\{\alpha\leftmerge\beta\}t\between t'\{\alpha'\leftmerge\beta'\}}$ |
| $(H6)$ | $\frac{\{\alpha\}t\{\beta\}}{\{\alpha\}\Theta(t)\{\beta\}}$ |
| $(H7)$ | $\frac{\{\alpha\}t\{\beta\}}{\{\alpha\}\partial_H(t)\{\beta\}}$ |
| $(H8)$ | $\frac{\{\alpha\}t\{\beta\}}{\{\alpha\}\tau_I(t)\{\beta\}}$ |
| $(H9)$ | $\frac{\alpha\rightarrow\alpha' \quad \{\alpha'\}t\{\beta'\} \quad \beta'\rightarrow\beta}{\{\alpha\}t\{\beta\}}$ |
| $(H10)$ | For $E=\{x=t_x \mid x\in V_E\}$ a guarded linear recursive specification $\forall y\in V_E$ and $z\in V_E$: $\frac{\{\alpha_x\}t_x\{\beta_x\} \quad \cdots \quad \{\alpha_y\}t_y\{\beta_y\}}{\{\alpha_z\}\langle z \mid E\rangle\{\beta_z\}}$ |
| $(H10')$ | For $E=\{x=t_x \mid x\in V_E\}$ a guarded linear recursive specification $\forall y\in V_E$ and $z\in V_E$: $\frac{\{\alpha_{x_1}\leftmerge\cdots\leftmerge\alpha_{x_{nx}}\}t_x\{\beta_{x_1}\leftmerge\cdots\leftmerge\beta_{x_{nx}}\} \quad \cdots \quad \{\alpha_{y_1}\leftmerge\cdots\leftmerge\alpha_{y_{ny}}\}t_y\{\beta_{y_1}\leftmerge\cdots\leftmerge\beta_{y_{ny}}\}}{\{\alpha_{z_1}\leftmerge\cdots\leftmerge\alpha_{z_{nz}}\}\langle z \mid E\rangle\{\beta_{z_1}\leftmerge\cdots\leftmerge\beta_{z_{nz}}\}}$ |

where $pre$ are preconditions, $post$ are postconditions, and $P$ are programs. $\{pre\}P\{post\}$ means that $pre$ hold, then $P$ are executed and $post$ hold. We take the guards $G$ of $APPTC_G$ as the language of conditions, and closed terms of $APPTC_G$ as programs. For some condition $\alpha \in G$ and some data state $s \in S$, we denote $S \models \alpha[s]$ for $\langle\alpha, s\rangle \rightarrow \langle\surd, s\rangle$, and $S \models \alpha$ for $\forall s \in S, S \models \alpha[s]$, $S \models \{\alpha\}p\{\beta\}$ for all $s \in S$, $v \subseteq \mathbb{E} \cup G$, $S \models \alpha[s]$, $\langle p, s\rangle \xrightarrow{v} \langle p', s'\rangle$, $S \models \beta[s']$ with $s' \in S$. It is obvious that $S \models \{\alpha\}p\{\beta\} \Leftrightarrow \alpha p \approx_{prbp} (\approx_{prbs}, \approx_{prbhp}, \approx_{prbhhp})\alpha p\beta$.

We design a proof system $H$ to deriving partial correct formulas over terms of $APPTC_G$ as Table 13.14 shows. Let $\Gamma$ be a set of conditions and partial correct formulas, we denote $\Gamma \vdash \{\alpha\}t\{\beta\}$ iff we can derive $\{\alpha\}t\{\beta\}$ in $H$, note that $t$ does not need to be closed terms. Also, we write $\alpha \rightarrow \beta$ for $S \models \alpha \Rightarrow S \models \beta$.

**Theorem 13.45** (Soundness of $H$). *Let $Tr_S$ be the set of conditions that hold in S. Let p be a closed term of $APPTC_{G_\tau}$ with guarded linear recursion and $VR_1, VR_2, VR_3$, and $\alpha, \beta \in G$ be guards. Then,*

$$\begin{aligned}
Tr_S \vdash \{\alpha\}p\{\beta\} &\Rightarrow APPTC_{G_\tau} \textit{ with guarded linear recursion and } VR_1, VR_2, VR_3 \vdash \alpha p = \alpha p\beta \\
&\Leftrightarrow \alpha p \approx_{prbs} (\approx_{prbp}, \approx_{prbhp}, \approx_{prbhhp})\alpha p\beta \\
&\Leftrightarrow S \models \{\alpha\}p\{\beta\}
\end{aligned}$$

*Proof.* We only need to prove

$$Tr_S \vdash \{\alpha\}p\{\beta\} \Rightarrow APPTC_{G_\tau} \text{ with guarded linear recursion and } VR_1, VR_2, VR_3 \vdash \alpha p = \alpha p\beta$$

For $H1$–$H10$, by induction on the length of derivation, the soundness of $H1$–$H10$ are straightforward. We only prove the soundness of $H10'$.

Let $E = \{x_i = t_i(x_1, \cdots, x_n) \mid i = 1, \cdots, n\}$ be a guarded linear recursive specification. Assume that

$$Tr_S, \{\{\alpha_1\leftmerge\cdots\leftmerge\alpha_{n_i}\}x_i\{\beta_1\leftmerge\cdots\leftmerge\beta_{n_i}\} \mid i = 1, \cdots, n\} \vdash \{\alpha_1\leftmerge\cdots\leftmerge\alpha_{n_j}\}t_j(x_1, \cdots, x_n)\{\beta_1\leftmerge\cdots\leftmerge\beta_{n_j}\}$$

for $j = 1, \cdots, n$. We would show that $APPTC_{G_\tau}$ with guarded linear recursion and $VR_1, VR_2, VR_3 \vdash (\alpha_1 \leftmerge \cdots \leftmerge \alpha_{n_j})X_j = (\alpha_1 \leftmerge \cdots \leftmerge \alpha_{n_j})X_j(\beta_1 \leftmerge \cdots \leftmerge \beta_{n_j})$.

We write recursive specifications $E'$ and $E''$ for

$$E' = \{y_i = (\alpha_1 \leftmerge \cdots \leftmerge \alpha_{n_i})t_i(y_1, \cdots, y_n)|i = 1, \cdots, n\}$$
$$E'' = \{z_i = (\alpha_1 \leftmerge \cdots \leftmerge \alpha_{n_i})t_i(z_1(\beta_1 \leftmerge \cdots \leftmerge \beta_{n_1}), \cdots, z_n(\beta_1 \leftmerge \cdots \leftmerge \beta_{n_n}))|i = 1, \cdots, n\}$$

and would show that for $j = 1, \cdots, n$,

(1) $(\alpha_1 \leftmerge \cdots \leftmerge \alpha_{n_j})X_j = Y_j$;

(2) $Z_j(\beta_1 \leftmerge \cdots \leftmerge \beta_{n_j}) = Z_j$;

(3) $Z_j = Y_j$.

For (1), we have

$$\begin{aligned}(\alpha_1 \leftmerge \cdots \leftmerge \alpha_{n_j})X_j &= (\alpha_1 \leftmerge \cdots \leftmerge \alpha_{n_j})t_j(X_1, \cdots, X_n)\\ &= (\alpha_1 \leftmerge \cdots \leftmerge \alpha_{n_j})t_j((\alpha_1 \leftmerge \cdots \leftmerge \alpha_{n_1})X_1, \cdots, (\alpha_1 \leftmerge \cdots \leftmerge \alpha_{n_n})X_n)\end{aligned}$$

by RDP, we have $(\alpha_1 \leftmerge \cdots \leftmerge \alpha_{n_j})X_j = Y_j$.

For (2), we have

$$\begin{aligned}Z_j(\beta_1 \leftmerge \cdots \leftmerge \beta_{n_j}) &= (\alpha_1 \leftmerge \cdots \leftmerge \alpha_{n_j})t_j(Z_1(\beta_1 \leftmerge \cdots \leftmerge \beta_{n_1}), \cdots, Z_n(\beta_1 \leftmerge \cdots \leftmerge \beta_{n_n}))(\beta_1 \leftmerge \cdots \leftmerge \beta_{n_j})\\ &= (\alpha_1 \leftmerge \cdots \leftmerge \alpha_{n_j})t_j(Z_1(\beta_1 \leftmerge \cdots \leftmerge \beta_{n_1}), \cdots, Z_n(\beta_1 \leftmerge \cdots \leftmerge \beta_{n_n}))\\ &= (\alpha_1 \leftmerge \cdots \leftmerge \alpha_{n_j})t_j((Z_1(\beta_1 \leftmerge \cdots \leftmerge \beta_{n_1}))(\beta_1 \leftmerge \cdots \leftmerge \beta_{n_1}), \cdots, (Z_n(\beta_1 \leftmerge \cdots \leftmerge \beta_{n_n}))\\ &\quad (\beta_1 \leftmerge \cdots \leftmerge \beta_{n_n}))\end{aligned}$$

by RDP, we have $Z_j(\beta_1 \leftmerge \cdots \leftmerge \beta_{n_j}) = Z_j$.

For (3), we have

$$\begin{aligned}Z_j &= (\alpha_1 \leftmerge \cdots \leftmerge \alpha_{n_j})t_j(Z_1(\beta_1 \leftmerge \cdots \leftmerge \beta_{n_1}), \cdots, Z_n(\beta_1 \leftmerge \cdots \leftmerge \beta_{n_n}))\\ &= (\alpha_1 \leftmerge \cdots \leftmerge \alpha_{n_j})t_j(Z_1, \cdots, Z_n)\end{aligned}$$

by RDP, we have $Z_j = Y_j$. □

Chapter 14

# CTC with probability and guards

In this chapter, we design the calculus CTC with probability and guards. This chapter is organized as follows. We introduce the operational semantics in Section 14.1, its syntax and operational semantics in Section 14.2, its properties for strong bisimulations in Section 14.3, and its properties for weak bisimulations in Section 14.4.

## 14.1 Operational semantics

**Definition 14.1** (Prime event structure with silent event and empty event). Let $\Lambda$ be a fixed set of labels, ranged over $a, b, c, \cdots$ and $\tau, \epsilon$. A ($\Lambda$-labeled) prime event structure with silent event $\tau$ and empty event $\epsilon$ is a tuple $\mathcal{E} = \langle \mathbb{E}, \leq, \sharp, \sharp_\pi \lambda \rangle$, where $\mathbb{E}$ is a denumerable set of events, including the silent event $\tau$ and empty event $\epsilon$. Let $\hat{\mathbb{E}} = \mathbb{E} \backslash \{\tau, \epsilon\}$, exactly excluding $\tau$ and $\epsilon$, it is obvious that $\hat{\tau^*} = \epsilon$. Let $\lambda : \mathbb{E} \to \Lambda$ be a labeling function and let $\lambda(\tau) = \tau$ and $\lambda(\epsilon) = \epsilon$. Also, $\leq$, $\sharp$, $\sharp_\pi$ are binary relations on $\mathbb{E}$, called causality, conflict, and probabilistic conflict, respectively, such that:

1. $\leq$ is a partial order and $\lceil e \rceil = \{e' \in \mathbb{E} | e' \leq e\}$ is finite for all $e \in \mathbb{E}$. It is easy to see that $e \leq \tau^* \leq e' = e \leq \tau \leq \cdots \leq \tau \leq e'$, then $e \leq e'$.
2. $\sharp$ is irreflexive, symmetric, and hereditary with respect to $\leq$, that is, for all $e, e', e'' \in \mathbb{E}$, if $e \sharp e' \leq e''$, then $e \sharp e''$;
3. $\sharp_\pi$ is irreflexive, symmetric, and hereditary with respect to $\leq$, that is, for all $e, e', e'' \in \mathbb{E}$, if $e \sharp_\pi e' \leq e''$, then $e \sharp_\pi e''$.

Then, the concepts of consistency and concurrency can be drawn from the above definition:

1. $e, e' \in \mathbb{E}$ are consistent, denoted as $e \frown e'$, if $\neg(e \sharp e')$ and $\neg(e \sharp_\pi e')$. A subset $X \subseteq \mathbb{E}$ is called consistent, if $e \frown e'$ for all $e, e' \in X$.
2. $e, e' \in \mathbb{E}$ are concurrent, denoted as $e \parallel e'$, if $\neg(e \leq e')$, $\neg(e' \leq e)$, and $\neg(e \sharp e')$ and $\neg(e \sharp_\pi e')$.

**Definition 14.2** (Configuration). Let $\mathcal{E}$ be a PES. A (finite) configuration in $\mathcal{E}$ is a (finite) consistent subset of events $C \subseteq \mathcal{E}$, closed with respect to causality (i.e., $\lceil C \rceil = C$), and a data state $s \in S$ with $S$ the set of all data states, denoted $\langle C, s \rangle$. The set of finite configurations of $\mathcal{E}$ is denoted by $\langle \mathcal{C}(\mathcal{E}), S \rangle$. We let $\hat{C} = C \backslash \{\tau\} \cup \{\epsilon\}$.

A consistent subset of $X \subseteq \mathbb{E}$ of events can be seen as a pomset. Given $X, Y \subseteq \mathbb{E}$, $\hat{X} \sim \hat{Y}$ if $\hat{X}$ and $\hat{Y}$ are isomorphic as pomsets. In the remainder of the chapter, when we say $C_1 \sim C_2$, we mean $\hat{C_1} \sim \hat{C_2}$.

**Definition 14.3** (Pomset transitions and step). Let $\mathcal{E}$ be a PES and let $C \in \mathcal{C}(\mathcal{E})$, and $\emptyset \neq X \subseteq \mathbb{E}$, if $C \cap X = \emptyset$ and $C' = C \cup X \in \mathcal{C}(\mathcal{E})$, then $\langle C, s \rangle \xrightarrow{X} \langle C', s' \rangle$ is called a pomset transition from $\langle C, s \rangle$ to $\langle C', s' \rangle$. When the events in $X$ are pairwise concurrent, we say that $\langle C, s \rangle \xrightarrow{X} \langle C', s' \rangle$ is a step. It is obvious that $\to^* \xrightarrow{X} \to^* = \xrightarrow{X}$ and $\to^* \xrightarrow{e} \to^* = \xrightarrow{e}$ for any $e \in \mathbb{E}$ and $X \subseteq \mathbb{E}$.

**Definition 14.4** (Probabilistic transitions). Let $\mathcal{E}$ be a PES and let $C \in \mathcal{C}(\mathcal{E})$, the transition $\langle C, s \rangle \overset{\pi}{\rightsquigarrow} \langle C^\pi, s \rangle$ is called a probabilistic transition from $\langle C, s \rangle$ to $\langle C^\pi, s \rangle$.

**Definition 14.5** (Weak pomset transitions and weak step). Let $\mathcal{E}$ be a PES and let $C \in \mathcal{C}(\mathcal{E})$, and $\emptyset \neq X \subseteq \hat{\mathbb{E}}$, if $C \cap X = \emptyset$ and $\hat{C'} = \hat{C} \cup X \in \mathcal{C}(\mathcal{E})$, then $\langle C, s \rangle \xRightarrow{X} \langle C', s' \rangle$ is called a weak pomset transition from $\langle C, s \rangle$ to $\langle C', s' \rangle$, where we define $\xRightarrow{e} \triangleq \xrightarrow{\tau^*} \xrightarrow{e} \xrightarrow{\tau^*}$. Also, $\xRightarrow{X} \triangleq \xrightarrow{\tau^*} \xrightarrow{e} \xrightarrow{\tau^*}$, for every $e \in X$. When the events in $X$ are pairwise concurrent, we say that $\langle C, s \rangle \xRightarrow{X} \langle C', s' \rangle$ is a weak step.

We will also suppose that all the PESs in this chapter are image finite, that is, for any PES $\mathcal{E}$ and $C \in \mathcal{C}(\mathcal{E})$ and $a \in \Lambda$, $\{\langle C, s \rangle \overset{\pi}{\rightsquigarrow} \langle C^\pi, s \rangle\}$, $\{e \in \mathbb{E} | \langle C, s \rangle \xrightarrow{e} \langle C', s' \rangle \wedge \lambda(e) = a\}$ and $\{e \in \hat{\mathbb{E}} | \langle C, s \rangle \xRightarrow{e} \langle C', s' \rangle \wedge \lambda(e) = a\}$ is finite.

**Definition 14.6** (Probabilistic pomset, step bisimulation). Let $\mathcal{E}_1, \mathcal{E}_2$ be PESs. A probabilistic pomset bisimulation is a relation $R \subseteq \langle \mathcal{C}(\mathcal{E}_1), S \rangle \times \langle \mathcal{C}(\mathcal{E}_2), S \rangle$, such that (1) if $(\langle C_1, s \rangle, \langle C_2, s \rangle) \in R$, and $\langle C_1, s \rangle \xrightarrow{X_1} \langle C_1', s' \rangle$, then $\langle C_2, s \rangle \xrightarrow{X_2} \langle C_2', s' \rangle$,

Handbook of Truly Concurrent Process Algebra. https://doi.org/10.1016/B978-0-44-321515-5.00018-4

with $X_1 \subseteq \mathbb{E}_1$, $X_2 \subseteq \mathbb{E}_2$, $X_1 \sim X_2$ and $(\langle C_1', s'\rangle, \langle C_2', s'\rangle) \in R$ for all $s, s' \in S$, and vice versa; (2) if $(\langle C_1, s\rangle, \langle C_2, s\rangle) \in R$, and $\langle C_1, s\rangle \overset{\pi}{\rightsquigarrow} \langle C_1^{\pi}, s\rangle$, then $\langle C_2, s\rangle \overset{\pi}{\rightsquigarrow} \langle C_2^{\pi}, s\rangle$ and $(\langle C_1^{\pi}, s\rangle, \langle C_2^{\pi}, s\rangle) \in R$, and vice versa; (3) if $(\langle C_1, s\rangle, \langle C_2, s\rangle) \in R$, then $\mu(C_1, C) = \mu(C_2, C)$ for each $C \in \mathcal{C}(\mathcal{E})/R$; (4) $[\surd]_R = \{\surd\}$. We say that $\mathcal{E}_1$, $\mathcal{E}_2$ are probabilistic pomset bisimilar, written $\mathcal{E}_1 \sim_{pp} \mathcal{E}_2$, if there exists a probabilistic pomset bisimulation $R$, such that $(\langle\emptyset, \emptyset\rangle, \langle\emptyset, \emptyset\rangle) \in R$. By replacing probabilistic pomset transitions with probabilistic steps, we can obtain the definition of probabilistic step bisimulation. When PESs $\mathcal{E}_1$ and $\mathcal{E}_2$ are probabilistic step bisimilar, we write $\mathcal{E}_1 \sim_{ps} \mathcal{E}_2$.

**Definition 14.7** (Weakly probabilistic pomset, step bisimulation). Let $\mathcal{E}_1$, $\mathcal{E}_2$ be PESs. A weakly probabilistic pomset bisimulation is a relation $R \subseteq \langle \mathcal{C}(\mathcal{E}_1), S\rangle \times \langle \mathcal{C}(\mathcal{E}_2), S\rangle$, such that (1) if $(\langle C_1, s\rangle, \langle C_2, s\rangle) \in R$, and $\langle C_1, s\rangle \overset{X_1}{\Longrightarrow} \langle C_1', s'\rangle$, then $\langle C_2, s\rangle \overset{X_2}{\Longrightarrow} \langle C_2', s'\rangle$, with $X_1 \subseteq \hat{\mathbb{E}}_1$, $X_2 \subseteq \hat{\mathbb{E}}_2$, $X_1 \sim X_2$ and $(\langle C_1', s'\rangle, \langle C_2', s'\rangle) \in R$ for all $s, s' \in S$, and vice versa; (2) if $(\langle C_1, s\rangle, \langle C_2, s\rangle) \in R$, and $\langle C_1, s\rangle \overset{\pi}{\rightsquigarrow} \langle C_1^{\pi}, s\rangle$, then $\langle C_2, s\rangle \overset{\pi}{\rightsquigarrow} \langle C_2^{\pi}, s\rangle$ and $(\langle C_1^{\pi}, s\rangle, \langle C_2^{\pi}, s\rangle) \in R$, and vice versa; (3) if $(\langle C_1, s\rangle, \langle C_2, s\rangle) \in R$, then $\mu(C_1, C) = \mu(C_2, C)$ for each $C \in \mathcal{C}(\mathcal{E})/R$; (4) $[\surd]_R = \{\surd\}$. We say that $\mathcal{E}_1$, $\mathcal{E}_2$ are weakly probabilistic pomset bisimilar, written $\mathcal{E}_1 \approx_{pp} \mathcal{E}_2$, if there exists a weakly probabilistic pomset bisimulation $R$, such that $(\langle\emptyset, \emptyset\rangle, \langle\emptyset, \emptyset\rangle) \in R$. By replacing weakly probabilistic pomset transitions with weakly probabilistic steps, we can obtain the definition of weakly probabilistic step bisimulation. When PESs $\mathcal{E}_1$ and $\mathcal{E}_2$ are weakly probabilistic step bisimilar, we write $\mathcal{E}_1 \approx_{ps} \mathcal{E}_2$.

**Definition 14.8** (Posetal product). Given two PESs $\mathcal{E}_1$, $\mathcal{E}_2$, the posetal product of their configurations, denoted $\langle \mathcal{C}(\mathcal{E}_1), S\rangle \overline{\times} \langle \mathcal{C}(\mathcal{E}_2), S\rangle$, is defined as

$$\{(\langle C_1, s\rangle, f, \langle C_2, s\rangle) | C_1 \in \mathcal{C}(\mathcal{E}_1), C_2 \in \mathcal{C}(\mathcal{E}_2), f : C_1 \to C_2 \text{ isomorphism}\}$$

A subset $R \subseteq \langle \mathcal{C}(\mathcal{E}_1), S\rangle \overline{\times} \langle \mathcal{C}(\mathcal{E}_2), S\rangle$ is called a posetal relation. We say that $R$ is downward closed when for any $(\langle C_1, s\rangle, f, \langle C_2, s\rangle), (\langle C_1', s'\rangle, f', \langle C_2', s'\rangle) \in \langle \mathcal{C}(\mathcal{E}_1), S\rangle \overline{\times} \langle \mathcal{C}(\mathcal{E}_2), S\rangle$, if $(\langle C_1, s\rangle, f, \langle C_2, s\rangle) \subseteq (\langle C_1', s'\rangle, f', \langle C_2', s'\rangle)$ pointwise and $(\langle C_1', s'\rangle, f', \langle C_2', s'\rangle) \in R$, then $(\langle C_1, s\rangle, f, \langle C_2, s\rangle) \in R$.

For $f : X_1 \to X_2$, we define $f[x_1 \mapsto x_2] : X_1 \cup \{x_1\} \to X_2 \cup \{x_2\}$, $z \in X_1 \cup \{x_1\}$, (1) $f[x_1 \mapsto x_2](z) = x_2$, if $z = x_1$; (2) $f[x_1 \mapsto x_2](z) = f(z)$, otherwise, where $X_1 \subseteq \mathbb{E}_1$, $X_2 \subseteq \mathbb{E}_2$, $x_1 \in \mathbb{E}_1$, $x_2 \in \mathbb{E}_2$.

**Definition 14.9** (Weakly posetal product). Given two PESs $\mathcal{E}_1$, $\mathcal{E}_2$, the weakly posetal product of their configurations, denoted $\langle \mathcal{C}(\mathcal{E}_1), S\rangle \overline{\times} \langle \mathcal{C}(\mathcal{E}_2), S\rangle$, is defined as

$$\{(\langle C_1, s\rangle, f, \langle C_2, s\rangle) | C_1 \in \mathcal{C}(\mathcal{E}_1), C_2 \in \mathcal{C}(\mathcal{E}_2), f : \hat{C_1} \to \hat{C_2} \text{ isomorphism}\}$$

A subset $R \subseteq \langle \mathcal{C}(\mathcal{E}_1), S\rangle \overline{\times} \langle \mathcal{C}(\mathcal{E}_2), S\rangle$ is called a weakly posetal relation. We say that $R$ is downward closed when for any $(\langle C_1, s\rangle, f, \langle C_2, s\rangle), (\langle C_1', s'\rangle, f, \langle C_2', s'\rangle) \in \langle \mathcal{C}(\mathcal{E}_1), S\rangle \overline{\times} \langle \mathcal{C}(\mathcal{E}_2), S\rangle$, if $(\langle C_1, s\rangle, f, \langle C_2, s\rangle) \subseteq (\langle C_1', s'\rangle, f', \langle C_2', s'\rangle)$ pointwise and $(\langle C_1', s'\rangle, f', \langle C_2', s'\rangle) \in R$, then $(\langle C_1, s\rangle, f, \langle C_2, s\rangle) \in R$.

For $f : X_1 \to X_2$, we define $f[x_1 \mapsto x_2] : X_1 \cup \{x_1\} \to X_2 \cup \{x_2\}$, $z \in X_1 \cup \{x_1\}$, (1) $f[x_1 \mapsto x_2](z) = x_2$, if $z = x_1$; (2) $f[x_1 \mapsto x_2](z) = f(z)$, otherwise, where $X_1 \subseteq \hat{\mathbb{E}}_1$, $X_2 \subseteq \hat{\mathbb{E}}_2$, $x_1 \in \hat{\mathbb{E}}_1$, $x_2 \in \hat{\mathbb{E}}_2$. Also, we define $f(\tau^*) = f(\tau^*)$.

**Definition 14.10** (Probabilistic (hereditary) history-preserving bisimulation). A probabilistic history-preserving (hp-) bisimulation is a posetal relation $R \subseteq \langle \mathcal{C}(\mathcal{E}_1), S\rangle \overline{\times} \langle \mathcal{C}(\mathcal{E}_2), S\rangle$ such that (1) if $(\langle C_1, s\rangle, f, \langle C_2, s\rangle) \in R$, and $\langle C_1, s\rangle \xrightarrow{e_1} \langle C_1', s'\rangle$, then $\langle C_2, s\rangle \xrightarrow{e_2} \langle C_2', s'\rangle$, with $(\langle C_1', s'\rangle, f[e_1 \mapsto e_2], \langle C_2', s'\rangle) \in R$ for all $s, s' \in S$, and vice versa; (2) if $(\langle C_1, s\rangle, f, \langle C_2, s\rangle) \in R$, and $\langle C_1, s\rangle \overset{\pi}{\rightsquigarrow} \langle C_1^{\pi}, s\rangle$, then $\langle C_2, s\rangle \overset{\pi}{\rightsquigarrow} \langle C_2^{\pi}, s\rangle$ and $(\langle C_1^{\pi}, s\rangle, f, \langle C_2^{\pi}, s\rangle) \in R$, and vice versa; (3) if $(C_1, f, C_2) \in R$, then $\mu(C_1, C) = \mu(C_2, C)$ for each $C \in \mathcal{C}(\mathcal{E})/R$; (4) $[\surd]_R = \{\surd\}$. $\mathcal{E}_1, \mathcal{E}_2$ are probabilistic history-preserving (hp-)bisimilar and are written $\mathcal{E}_1 \sim_{php} \mathcal{E}_2$ if there exists a probabilistic hp-bisimulation $R$ such that $(\langle\emptyset, \emptyset\rangle, \emptyset, \langle\emptyset, \emptyset\rangle) \in R$.

A probabilistic hereditary history-preserving (hhp-)bisimulation is a downward closed probabilistic hp-bisimulation. $\mathcal{E}_1, \mathcal{E}_2$ are probabilistic hereditary history-preserving (hhp-)bisimilar and are written $\mathcal{E}_1 \sim_{phhp} \mathcal{E}_2$.

**Definition 14.11** (Weakly probabilistic (hereditary) history-preserving bisimulation). A weakly probabilistic history-preserving (hp-) bisimulation is a weakly posetal relation $R \subseteq \langle \mathcal{C}(\mathcal{E}_1), S\rangle \overline{\times} \langle \mathcal{C}(\mathcal{E}_2), S\rangle$ such that (1) if $(\langle C_1, s\rangle, f, \langle C_2, s\rangle) \in R$, and $\langle C_1, s\rangle \overset{e_1}{\Longrightarrow} \langle C_1', s'\rangle$, then $\langle C_2, s\rangle \overset{e_2}{\Longrightarrow} \langle C_2', s'\rangle$, with $(\langle C_1', s'\rangle, f[e_1 \mapsto e_2], \langle C_2', s'\rangle) \in R$ for all $s, s' \in S$, and vice versa; (2) if $(\langle C_1, s\rangle, f, \langle C_2, s\rangle) \in R$, and $\langle C_1, s\rangle \overset{\pi}{\rightsquigarrow} \langle C_1^{\pi}, s\rangle$, then $\langle C_2, s\rangle \overset{\pi}{\rightsquigarrow} \langle C_2^{\pi}, s\rangle$ and $(\langle C_1^{\pi}, s\rangle, f, \langle C_2^{\pi}, s\rangle) \in R$, and vice versa; (3) if $(C_1, f, C_2) \in R$, then $\mu(C_1, C) = \mu(C_2, C)$ for each $C \in \mathcal{C}(\mathcal{E})/R$; (4) $[\surd]_R = \{\surd\}$. $\mathcal{E}_1, \mathcal{E}_2$ are weakly probabilistic history-preserving (hp-)bisimilar and are written $\mathcal{E}_1 \approx_{php} \mathcal{E}_2$ if there exists a weakly probabilistic hp-bisimulation $R$ such that $(\langle\emptyset,\emptyset\rangle,\emptyset,\langle\emptyset,\emptyset\rangle) \in R$.

A weakly probabilistic hereditary history-preserving (hhp-)bisimulation is a downward closed weakly probabilistic hp-bisimulation. $\mathcal{E}_1, \mathcal{E}_2$ are weakly probabilistic hereditary history-preserving (hhp-)bisimilar and are written $\mathcal{E}_1 \approx_{phhp} \mathcal{E}_2$.

## 14.2 Syntax and operational semantics

We assume an infinite set $\mathcal{N}$ of (action or event) names, and use $a, b, c, \cdots$ to range over $\mathcal{N}$. We denote by $\overline{\mathcal{N}}$ the set of co-names and let $\overline{a}, \overline{b}, \overline{c}, \cdots$ range over $\overline{\mathcal{N}}$. Then, we set $\mathcal{L} = \mathcal{N} \cup \overline{\mathcal{N}}$ as the set of labels, and use $l, \overline{l}$ to range over $\mathcal{L}$. We extend complementation to $\mathcal{L}$ such that $\overline{\overline{a}} = a$. Let $\tau$ denote the silent step (internal action or event) and define $Act = \mathcal{L} \cup \{\tau\}$ to be the set of actions, $\alpha, \beta$ range over $Act$. Also, $K, L$ are used to stand for subsets of $\mathcal{L}$ and $\overline{L}$ is used for the set of complements of labels in $L$. A relabeling function $f$ is a function from $\mathcal{L}$ to $\mathcal{L}$ such that $f(\overline{l}) = \overline{f(l)}$. By defining $f(\tau) = \tau$, we extend $f$ to $Act$.

Further, we introduce a set $\mathcal{X}$ of process variables, and a set $\mathcal{K}$ of process constants, and let $X, Y, \cdots$ range over $\mathcal{X}$, and $A, B, \cdots$ range over $\mathcal{K}$, $\widetilde{X}$ is a tuple of distinct process variables, and also $E, F, \cdots$ range over the recursive expressions. We write $\mathcal{P}$ for the set of processes. Sometimes, we use $I, J$ to stand for an indexing set, and we write $E_i : i \in I$ for a family of expressions indexed by $I$. $Id_D$ is the identity function or relation over set $D$.

For each process constant schema $A$, a defining equation of the form

$$A \overset{\text{def}}{=} P$$

is assumed, where $P$ is a process.

Let $G_{at}$ be the set of atomic guards, $\delta$ be the deadlock constant, and $\epsilon$ be the empty action, and extend $Act$ to $Act \cup \{\epsilon\} \cup \{\delta\}$. We extend $G_{at}$ to the set of basic guards $G$ with element $\phi, \psi, \cdots$, which is generated by the following formation rules:

$$\phi ::= \delta|\epsilon|\neg\phi|\psi \in G_{at}|\phi + \psi|\phi \cdot \psi$$

The predicate $test(\phi, s)$ represents that $\phi$ holds in the state $s$, and $test(\epsilon, s)$ holds and $test(\delta, s)$ does not hold. $effect(e, s) \in S$ denotes $s'$ in $s \xrightarrow{e} s'$. The predicate weakest precondition $wp(e, \phi)$ denotes that $\forall s, s' \in S, test(\phi, effect(e, s))$ holds.

### 14.2.1 Syntax

We use the Prefix . to model the causality relation $\leq$ in true concurrency, the Summation $+$ to model the conflict relation $\sharp$, the Box-Summation $\boxplus_\pi$ to model the probabilistic conflict relation $\sharp_\pi$ in true concurrency, and the Composition $\parallel$ to explicitly model concurrent relation in true concurrency. Also, we follow the conventions of process algebra.

**Definition 14.12** (Syntax). Truly concurrent processes CTC with probabilism and guards are defined inductively by the following formation rules:

1. $A \in \mathcal{P}$;
2. $\phi \in \mathcal{P}$;
3. $\mathbf{nil} \in \mathcal{P}$;
4. if $P \in \mathcal{P}$, then the Prefix $\alpha.P \in \mathcal{P}$, for $\alpha \in Act$;
5. if $P \in \mathcal{P}$, then the Prefix $\phi.P \in \mathcal{P}$, for $\phi \in G_{at}$;
6. if $P, Q \in \mathcal{P}$, then the Summation $P + Q \in \mathcal{P}$;
7. if $P, Q \in \mathcal{P}$, then the Box-Summation $P \boxplus_\pi Q \in \mathcal{P}$;
8. if $P, Q \in \mathcal{P}$, then the Composition $P \parallel Q \in \mathcal{P}$;
9. if $P \in \mathcal{P}$, then the Prefix $(\alpha_1 \parallel \cdots \parallel \alpha_n).P \in \mathcal{P} \quad (n \in I)$, for $\alpha_1, \cdots, \alpha_n \in Act$;
10. if $P \in \mathcal{P}$, then the Restriction $P \setminus L \in \mathcal{P}$ with $L \in \mathcal{L}$;
11. if $P \in \mathcal{P}$, then the Relabeling $P[f] \in \mathcal{P}$.

The standard BNF grammar of syntax of CTC with probabilism and guards can be summarized as follows:

$$P ::= A|\mathbf{nil}|\alpha.P|\phi.P|P + P|P \boxplus_\pi P|P \parallel P|(\alpha_1 \parallel \cdots \parallel \alpha_n).P|P \setminus L|P[f]$$

**TABLE 14.1** Probabilistic transition rules of CTC with probabilism and guards.

| | |
|---|---|
| **PAct**$_1$ | $\frac{}{\langle\alpha.P,s\rangle\rightsquigarrow\langle\breve{\alpha}.P,s\rangle}$ |
| **PSum** | $\frac{\langle P,s\rangle\rightsquigarrow\langle P',s\rangle\quad Q\rightsquigarrow Q'}{\langle P+Q,s\rangle\rightsquigarrow\langle P'+Q',s\rangle}$ |
| **PBox-Sum** | $\frac{\langle P,s\rangle\rightsquigarrow\langle P',s\rangle}{\langle P\boxplus_\pi Q,s\rangle\rightsquigarrow\langle P',s\rangle}\quad\frac{\langle Q,s\rangle\rightsquigarrow\langle Q',s\rangle}{\langle P\boxplus_\pi Q,s\rangle\rightsquigarrow\langle Q',s\rangle}$ |
| **PCom** | $\frac{\langle P,s\rangle\rightsquigarrow\langle P',s\rangle\quad\langle Q,s\rangle\rightsquigarrow\langle Q',s\rangle}{\langle P\parallel Q,s\rangle\rightsquigarrow\langle P'+Q',s\rangle}$ |
| **PAct**$_2$ | $\frac{}{\langle(\alpha_1\parallel\cdots\parallel\alpha_n).P,s\rangle\rightsquigarrow\langle(\breve{\alpha}_1\parallel\cdots\parallel\breve{\alpha}_n).P,s\rangle}$ |
| **PRes** | $\frac{\langle P,s\rangle\rightsquigarrow\langle P',s\rangle}{\langle P\setminus L,s\rangle\rightsquigarrow\langle P'\setminus L,s\rangle}$ |
| **PRel** | $\frac{\langle P,s\rangle\rightsquigarrow\langle P',s\rangle}{\langle P[f],s\rangle\rightsquigarrow\langle P'[f],s\rangle}$ |
| **PCon** | $\frac{\langle P,s\rangle\rightsquigarrow\langle P',s\rangle}{\langle A,s\rangle\rightsquigarrow\langle P',s\rangle}\quad(A\overset{\text{def}}{=}P)$ |

### 14.2.2 Operational semantics

The operational semantics is defined by LTSs (labeled transition systems), and it is detailed by the following definition.

**Definition 14.13** (Semantics). The operational semantics of CTC with probabilism and guards corresponding to the syntax in Definition 14.12 is defined by a series of transition rules, named **PAct**, **PSum**, **PBox-Sum**, **Com**, **Res**, **Rel**, and **Con** and named **Act**, **Sum**, **Com**, **Res**, **Rel**, and **Con** indicate that the rules are associated, respectively, with Prefix, Summation, Composition, Restriction, Relabeling, and Constants in Definition 14.12. They are shown in Tables 14.1 and 14.2.

### 14.2.3 Properties of transitions

**Definition 14.14** (Sorts). Given the sorts $\mathcal{L}(A)$ and $\mathcal{L}(X)$ of constants and variables, we define $\mathcal{L}(P)$ inductively as follows.

1. $\mathcal{L}(l.P)=\{l\}\cup\mathcal{L}(P)$;
2. $\mathcal{L}((l_1\parallel\cdots\parallel l_n).P)=\{l_1,\cdots,l_n\}\cup\mathcal{L}(P)$;
3. $\mathcal{L}(\tau.P)=\mathcal{L}(P)$;
4. $\mathcal{L}(\epsilon.P)=\mathcal{L}(P)$;
5. $\mathcal{L}(\phi.P)=\mathcal{L}(P)$;
6. $\mathcal{L}(P+Q)=\mathcal{L}(P)\cup\mathcal{L}(Q)$;
7. $\mathcal{L}(P\boxplus_\pi Q)=\mathcal{L}(P)\cup\mathcal{L}(Q)$;
8. $\mathcal{L}(P\parallel Q)=\mathcal{L}(P)\cup\mathcal{L}(Q)$;
9. $\mathcal{L}(P\setminus L)=\mathcal{L}(P)-(L\cup\overline{L})$;
10. $\mathcal{L}(P[f])=\{f(l):l\in\mathcal{L}(P)\}$;
11. for $A\overset{\text{def}}{=}P$, $\mathcal{L}(P)\subseteq\mathcal{L}(A)$.

Now, we present some properties of the transition rules defined in Table 14.2.

**Proposition 14.15.** *If $P\xrightarrow{\alpha}P'$, then*

1. $\alpha\in\mathcal{L}(P)\cup\{\tau\}\cup\{\epsilon\}$;
2. $\mathcal{L}(P')\subseteq\mathcal{L}(P)$.

*If $P\xrightarrow{\{\alpha_1,\cdots,\alpha_n\}}P'$, then*

1. $\alpha_1,\cdots,\alpha_n\in\mathcal{L}(P)\cup\{\tau\}\cup\{\epsilon\}$;
2. $\mathcal{L}(P')\subseteq\mathcal{L}(P)$.

*Proof.* By induction on the inference of $P\xrightarrow{\alpha}P'$ and $P\xrightarrow{\{\alpha_1,\cdots,\alpha_n\}}P'$, there are several cases corresponding to the transition rules named **Act**$_{1,2,3}$, **Gur**, **Sum**$_{1,2}$, **Com**$_{1,2,3,4}$, **Res**$_{1,2}$, **Rel**$_{1,2}$, and **Con**$_{1,2}$ in Table 14.2, we just prove the one case **Act**$_1$ and **Act**$_3$, and omit the others here.

**TABLE 14.2** Action transition rules of CTC with probabilism and guards.

| | |
|---|---|
| **Act₁** | $\dfrac{}{\langle \breve{\alpha}.P, s\rangle \xrightarrow{\alpha} \langle P, s'\rangle}$ |
| **Act₂** | $\dfrac{}{\langle \epsilon, s\rangle \rightarrow \langle \surd, s\rangle}$ |
| **Gur** | $\dfrac{}{\langle \phi, s\rangle \rightarrow \langle \surd, s\rangle}$ if $test(\phi, s)$ |
| **Sum₁** | $\dfrac{\langle P, s\rangle \xrightarrow{\alpha} \langle P', s'\rangle}{\langle P+Q, s\rangle \xrightarrow{\alpha} \langle P', s'\rangle}$ |
| **Com₁** | $\dfrac{\langle P, s\rangle \xrightarrow{\alpha} \langle P', s'\rangle \quad Q \nrightarrow}{\langle P \parallel Q, s\rangle \xrightarrow{\alpha} \langle P' \parallel Q, s'\rangle}$ |
| **Com₂** | $\dfrac{\langle Q, s\rangle \xrightarrow{\alpha} \langle Q', s'\rangle \quad P \nrightarrow}{\langle P \parallel Q, s\rangle \xrightarrow{\alpha} \langle P \parallel Q', s'\rangle}$ |
| **Com₃** | $\dfrac{\langle P, s\rangle \xrightarrow{\alpha} \langle P', s'\rangle \quad \langle Q, s\rangle \xrightarrow{\beta} \langle Q', s''\rangle}{\langle P \parallel Q, s\rangle \xrightarrow{\{\alpha,\beta\}} \langle P' \parallel Q', s' \cup s''\rangle} \quad (\beta \neq \overline{\alpha})$ |
| **Com₄** | $\dfrac{\langle P, s\rangle \xrightarrow{l} \langle P', s'\rangle \quad \langle Q, s\rangle \xrightarrow{\overline{l}} \langle Q', s''\rangle}{\langle P \parallel Q, s\rangle \xrightarrow{\tau} \langle P' \parallel Q', s' \cup s''\rangle}$ |
| **Act₃** | $\dfrac{}{\langle (\breve{\alpha}_1 \parallel \cdots \parallel \breve{\alpha}_n).P, s\rangle \xrightarrow{\{\alpha_1,\cdots,\alpha_n\}} \langle P, s'\rangle} \quad (\alpha_i \neq \overline{\alpha_j} \quad i, j \in \{1, \cdots, n\})$ |
| **Sum₂** | $\dfrac{\langle P, s\rangle \xrightarrow{\{\alpha_1,\cdots,\alpha_n\}} \langle P', s'\rangle}{\langle P+Q, s\rangle \xrightarrow{\{\alpha_1,\cdots,\alpha_n\}} \langle P', s'\rangle}$ |
| **Res₁** | $\dfrac{\langle P, s\rangle \xrightarrow{\alpha} \langle P', s'\rangle}{\langle P \setminus L, s\rangle \xrightarrow{\alpha} \langle P' \setminus L, s'\rangle} \quad (\alpha, \overline{\alpha} \notin L)$ |
| **Res₂** | $\dfrac{\langle P, s\rangle \xrightarrow{\{\alpha_1,\cdots,\alpha_n\}} \langle P', s'\rangle}{\langle P \setminus L, s\rangle \xrightarrow{\{\alpha_1,\cdots,\alpha_n\}} \langle P' \setminus L, s'\rangle} \quad (\alpha_1, \overline{\alpha_1}, \cdots, \alpha_n, \overline{\alpha_n} \notin L)$ |
| **Rel₁** | $\dfrac{\langle P, s\rangle \xrightarrow{\alpha} \langle P', s'\rangle}{\langle P[f], s\rangle \xrightarrow{\langle f(\alpha)} P'[f], s'\rangle}$ |
| **Rel₂** | $\dfrac{\langle P, s\rangle \xrightarrow{\{\alpha_1,\cdots,\alpha_n\}} \langle P', s'\rangle}{\langle P[f], s\rangle \xrightarrow{\{f(\alpha_1),\cdots,f(\alpha_n)\}} \langle P'[f], s'\rangle}$ |
| **Con₁** | $\dfrac{\langle P, s\rangle \xrightarrow{\alpha} \langle P', s'\rangle}{\langle A, s\rangle \xrightarrow{\alpha} \langle P', s'\rangle} \quad (A \overset{\text{def}}{=} P)$ |
| **Con₂** | $\dfrac{\langle P, s\rangle \xrightarrow{\{\alpha_1,\cdots,\alpha_n\}} \langle P', s'\rangle}{\langle A, s\rangle \xrightarrow{\{\alpha_1,\cdots,\alpha_n\}} \langle P', s'\rangle} \quad (A \overset{\text{def}}{=} P)$ |

Case **Act₁**: by **Act₁**, with $P \equiv \alpha.P'$. Then, by Definition 14.14, we have (1) $\mathcal{L}(P) = \{\alpha\} \cup \mathcal{L}(P')$ if $\alpha \neq \tau$; (2) $\mathcal{L}(P) = \mathcal{L}(P')$ if $\alpha = \tau$ or $\alpha = \epsilon$. Hence, $\alpha \in \mathcal{L}(P) \cup \{\tau\} \cup \{\epsilon\}$, and $\mathcal{L}(P') \subseteq \mathcal{L}(P)$, as desired.

Case **Act₃**: by **Act₃**, with $P \equiv (\alpha_1 \parallel \cdots \parallel \alpha_n).P'$. Then, by Definition 14.14, we have (1) $\mathcal{L}(P) = \{\alpha_1, \cdots, \alpha_n\} \cup \mathcal{L}(P')$ if $\alpha_i \neq \tau$ for $i \leq n$; (2) $\mathcal{L}(P) = \mathcal{L}(P')$ if $\alpha_1, \cdots, \alpha_n = \tau$ or $\alpha_1, \cdots, \alpha_n = \epsilon$. Hence, $\alpha_1, \cdots, \alpha_n \in \mathcal{L}(P) \cup \{\tau\} \cup \{\epsilon\}$, and $\mathcal{L}(P') \subseteq \mathcal{L}(P)$, as desired. □

## 14.3 Strong bisimulations

### 14.3.1 Laws and congruence

Based on the concepts of strongly probabilistic truly concurrent bisimulation equivalences, we obtain the following laws.

**Proposition 14.16** (Monoid laws for strongly probabilistic pomset bisimulation). *The monoid laws for strongly probabilistic pomset bisimulation are as follows:*

1. $P+Q \sim_{pp} Q+P$;
2. $P+(Q+R) \sim_{pp} (P+Q)+R$;
3. $P+P \sim_{pp} P$;
4. $P+\textbf{\textit{nil}} \sim_{pp} P$.

*Proof.* 1. $P+Q \sim_{pp} Q+P$. It is sufficient to prove the relation $R = \{(P+Q, Q+P)\} \cup \textbf{Id}$ is a strongly probabilistic pomset bisimulation, however, we omit the proof here;
2. $P+(Q+R) \sim_{pp} (P+Q)+R$. It is sufficient to prove the relation $R = \{(P+(Q+R), (P+Q)+R)\} \cup \textbf{Id}$ is a strongly probabilistic pomset bisimulation, however, we omit the proof here;
3. $P+P \sim_{pp} P$. It is sufficient to prove the relation $R = \{(P+P, P)\} \cup \textbf{Id}$ is a strongly probabilistic pomset bisimulation, however, we omit the proof here;
4. $P+\textbf{nil} \sim_{pp} P$. It is sufficient to prove the relation $R = \{(P+\textbf{nil}, P)\} \cup \textbf{Id}$ is a strongly probabilistic pomset bisimulation, however, we omit the proof here. □

**Proposition 14.17** (Monoid laws for strongly probabilistic step bisimulation). *The monoid laws for strongly probabilistic step bisimulation are as follows:*

1. $P+Q \sim_{ps} Q+P$;
2. $P+(Q+R) \sim_{ps} (P+Q)+R$;
3. $P+P \sim_{ps} P$;
4. $P+\textbf{\textit{nil}} \sim_{ps} P$.

*Proof.* 1. $P+Q \sim_{ps} Q+P$. It is sufficient to prove the relation $R = \{(P+Q, Q+P)\} \cup \textbf{Id}$ is a strongly probabilistic step bisimulation, however, we omit the proof here;
2. $P+(Q+R) \sim_{ps} (P+Q)+R$. It is sufficient to prove the relation $R = \{(P+(Q+R), (P+Q)+R)\} \cup \textbf{Id}$ is a strongly probabilistic step bisimulation, however, we omit the proof here;
3. $P+P \sim_{ps} P$. It is sufficient to prove the relation $R = \{(P+P, P)\} \cup \textbf{Id}$ is a strongly probabilistic step bisimulation, however, we omit the proof here;
4. $P+\textbf{nil} \sim_{ps} P$. It is sufficient to prove the relation $R = \{(P+\textbf{nil}, P)\} \cup \textbf{Id}$ is a strongly probabilistic step bisimulation, however, we omit the proof here. □

**Proposition 14.18** (Monoid laws for strongly probabilistic hp-bisimulation). *The monoid laws for strongly probabilistic hp-bisimulation are as follows:*

1. $P+Q \sim_{php} Q+P$;
2. $P+(Q+R) \sim_{php} (P+Q)+R$;
3. $P+P \sim_{php} P$;
4. $P+\textbf{\textit{nil}} \sim_{php} P$.

*Proof.* 1. $P+Q \sim_{php} Q+P$. It is sufficient to prove the relation $R = \{(P+Q, Q+P)\} \cup \textbf{Id}$ is a strongly probabilistic hp-bisimulation, however, we omit the proof here;
2. $P+(Q+R) \sim_{php} (P+Q)+R$. It is sufficient to prove the relation $R = \{(P+(Q+R), (P+Q)+R)\} \cup \textbf{Id}$ is a strongly probabilistic hp-bisimulation, however, we omit the proof here;
3. $P+P \sim_{php} P$. It is sufficient to prove the relation $R = \{(P+P, P)\} \cup \textbf{Id}$ is a strongly probabilistic hp-bisimulation, however, we omit the proof here;
4. $P+\textbf{nil} \sim_{php} P$. It is sufficient to prove the relation $R = \{(P+\textbf{nil}, P)\} \cup \textbf{Id}$ is a strongly probabilistic hp-bisimulation, however, we omit the proof here. □

**Proposition 14.19** (Monoid laws for strongly probabilistic hhp-bisimulation). *The monoid laws for strongly probabilistic hhp-bisimulation are as follows:*

1. $P+Q \sim_{phhp} Q+P$;
2. $P+(Q+R) \sim_{phhp} (P+Q)+R$;
3. $P+P \sim_{phhp} P$;
4. $P+\textbf{\textit{nil}} \sim_{phhp} P$.

*Proof.* **1.** $P + Q \sim_{phhp} Q + P$. It is sufficient to prove the relation $R = \{(P + Q, Q + P)\} \cup \mathbf{Id}$ is a strongly probabilistic hhp-bisimulation, however, we omit the proof here;

**2.** $P + (Q + R) \sim_{phhp} (P + Q) + R$. It is sufficient to prove the relation $R = \{(P + (Q + R), (P + Q) + R)\} \cup \mathbf{Id}$ is a strongly probabilistic hhp-bisimulation, however, we omit the proof here;

**3.** $P + P \sim_{phhp} P$. It is sufficient to prove the relation $R = \{(P + P, P)\} \cup \mathbf{Id}$ is a strongly probabilistic hhp-bisimulation, however, we omit the proof here;

**4.** $P + \mathbf{nil} \sim_{phhp} P$. It is sufficient to prove the relation $R = \{(P + \mathbf{nil}, P)\} \cup \mathbf{Id}$ is a strongly probabilistic hhp-bisimulation, however, we omit the proof here. □

**Proposition 14.20** (Monoid laws 2 for strongly probabilistic pomset bisimulation). *The monoid laws 2 for strongly probabilistic pomset bisimulation are as follows:*

**1.** $P \boxplus_\pi Q \sim_{pp} Q \boxplus_{1-\pi} P$;

**2.** $P \boxplus_\pi (Q \boxplus_\rho R) \sim_{pp} (P \boxplus_{\frac{\pi}{\pi+\rho-\pi\rho}} Q) \boxplus_{\pi+\rho-\pi\rho} R$;

**3.** $P \boxplus_\pi P \sim_{pp} P$;

**4.** $P \boxplus_\pi \mathit{nil} \sim_{pp} P$.

*Proof.* **1.** $P \boxplus_\pi Q \sim_{pp} Q \boxplus_{1-\pi} P$. It is sufficient to prove the relation $R = \{(P \boxplus_\pi Q, Q \boxplus_{1-\pi} P)\} \cup \mathbf{Id}$ is a strongly probabilistic pomset bisimulation, however, we omit the proof here;

**2.** $P \boxplus_\pi (Q \boxplus_\rho R) \sim_{pp} (P \boxplus_{\frac{\pi}{\pi+\rho-\pi\rho}} Q) \boxplus_{\pi+\rho-\pi\rho} R$. It is sufficient to prove the relation $R = \{(P \boxplus_\pi (Q \boxplus_\rho R), (P \boxplus_{\frac{\pi}{\pi+\rho-\pi\rho}} Q) \boxplus_{\pi+\rho-\pi\rho} R)\} \cup \mathbf{Id}$ is a strongly probabilistic pomset bisimulation, however, we omit the proof here;

**3.** $P \boxplus_\pi P \sim_{pp} P$. It is sufficient to prove the relation $R = \{(P \boxplus_\pi P, P)\} \cup \mathbf{Id}$ is a strongly probabilistic pomset bisimulation, however, we omit the proof here;

**4.** $P \boxplus_\pi \mathbf{nil} \sim_{pp} P$. It is sufficient to prove the relation $R = \{(P \boxplus_\pi \mathbf{nil}, P)\} \cup \mathbf{Id}$ is a strongly probabilistic pomset bisimulation, however, we omit the proof here. □

**Proposition 14.21** (Monoid laws 2 for strongly probabilistic step bisimulation). *The monoid laws 2 for strongly probabilistic step bisimulation are as follows:*

**1.** $P \boxplus_\pi Q \sim_{ps} Q \boxplus_{1-\pi} P$;

**2.** $P \boxplus_\pi (Q \boxplus_\rho R) \sim_{ps} (P \boxplus_{\frac{\pi}{\pi+\rho-\pi\rho}} Q) \boxplus_{\pi+\rho-\pi\rho} R$;

**3.** $P \boxplus_\pi P \sim_{ps} P$;

**4.** $P \boxplus_\pi \mathit{nil} \sim_{ps} P$.

*Proof.* **1.** $P \boxplus_\pi Q \sim_{ps} Q \boxplus_{1-\pi} P$. It is sufficient to prove the relation $R = \{(P \boxplus_\pi Q, Q \boxplus_{1-\pi} P)\} \cup \mathbf{Id}$ is a strongly probabilistic step bisimulation, however, we omit the proof here;

**2.** $P \boxplus_\pi (Q \boxplus_\rho R) \sim_{ps} (P \boxplus_{\frac{\pi}{\pi+\rho-\pi\rho}} Q) \boxplus_{\pi+\rho-\pi\rho} R$. It is sufficient to prove the relation $R = \{(P \boxplus_\pi (Q \boxplus_\rho R), (P \boxplus_{\frac{\pi}{\pi+\rho-\pi\rho}} Q) \boxplus_{\pi+\rho-\pi\rho} R)\} \cup \mathbf{Id}$ is a strongly probabilistic step bisimulation, however, we omit the proof here;

**3.** $P \boxplus_\pi P \sim_{ps} P$. It is sufficient to prove the relation $R = \{(P \boxplus_\pi P, P)\} \cup \mathbf{Id}$ is a strongly probabilistic step bisimulation, however, we omit the proof here;

**4.** $P \boxplus_\pi \mathbf{nil} \sim_{ps} P$. It is sufficient to prove the relation $R = \{(P \boxplus_\pi \mathbf{nil}, P)\} \cup \mathbf{Id}$ is a strongly probabilistic step bisimulation, however, we omit the proof here. □

**Proposition 14.22** (Monoid laws 2 for strongly probabilistic hp-bisimulation). *The monoid laws 2 for strongly probabilistic hp-bisimulation are as follows:*

**1.** $P \boxplus_\pi Q \sim_{php} Q \boxplus_{1-\pi} P$;

**2.** $P \boxplus_\pi (Q \boxplus_\rho R) \sim_{php} (P \boxplus_{\frac{\pi}{\pi+\rho-\pi\rho}} Q) \boxplus_{\pi+\rho-\pi\rho} R$;

**3.** $P \boxplus_\pi P \sim_{php} P$;

**4.** $P \boxplus_\pi \mathit{nil} \sim_{php} P$.

*Proof.* **1.** $P \boxplus_\pi Q \sim_{php} Q \boxplus_{1-\pi} P$. It is sufficient to prove the relation $R = \{(P \boxplus_\pi Q, Q \boxplus_{1-\pi} P)\} \cup \mathbf{Id}$ is a strongly probabilistic hp-bisimulation, however, we omit the proof here;

**2.** $P \boxplus_\pi (Q \boxplus_\rho R) \sim_{php} (P \boxplus_{\frac{\pi}{\pi+\rho-\pi\rho}} Q) \boxplus_{\pi+\rho-\pi\rho} R$. It is sufficient to prove the relation $R = \{(P \boxplus_\pi (Q \boxplus_\rho R), (P \boxplus_{\frac{\pi}{\pi+\rho-\pi\rho}} Q) \boxplus_{\pi+\rho-\pi\rho} R)\} \cup \mathbf{Id}$ is a strongly probabilistic hp-bisimulation, however, we omit the proof here;

**3.** $P \boxplus_\pi P \sim_{php} P$. It is sufficient to prove the relation $R = \{(P \boxplus_\pi P, P)\} \cup \mathbf{Id}$ is a strongly probabilistic hp-bisimulation, however, we omit the proof here;

4. $P \boxplus_\pi \mathbf{nil} \sim_{php} P$. It is sufficient to prove the relation $R = \{(P \boxplus_\pi \mathbf{nil}, P)\} \cup \mathbf{Id}$ is a strongly probabilistic hp-bisimulation, however, we omit the proof here. □

**Proposition 14.23** (Monoid laws 2 for strongly probabilistic hhp-bisimulation). *The monoid laws 2 for strongly probabilistic hhp-bisimulation are as follows:*

1. $P \boxplus_\pi Q \sim_{phhp} Q \boxplus_{1-\pi} P$;
2. $P \boxplus_\pi (Q \boxplus_\rho R) \sim_{phhp} (P \boxplus_{\frac{\pi}{\pi+\rho-\pi\rho}} Q) \boxplus_{\pi+\rho-\pi\rho} R$;
3. $P \boxplus_\pi P \sim_{phhp} P$;
4. $P \boxplus_\pi nil \sim_{phhp} P$.

*Proof.* 1. $P \boxplus_\pi Q \sim_{phhp} Q \boxplus_{1-\pi} P$. It is sufficient to prove the relation $R = \{(P \boxplus_\pi Q, Q \boxplus_{1-\pi} P)\} \cup \mathbf{Id}$ is a strongly probabilistic hhp-bisimulation, however, we omit the proof here;
2. $P \boxplus_\pi (Q \boxplus_\rho R) \sim_{phhp} (P \boxplus_{\frac{\pi}{\pi+\rho-\pi\rho}} Q) \boxplus_{\pi+\rho-\pi\rho} R$. It is sufficient to prove the relation $R = \{(P \boxplus_\pi (Q \boxplus_\rho R), (P \boxplus_{\frac{\pi}{\pi+\rho-\pi\rho}} Q) \boxplus_{\pi+\rho-\pi\rho} R)\} \cup \mathbf{Id}$ is a strongly probabilistic hhp-bisimulation, however, we omit the proof here;
3. $P \boxplus_\pi P \sim_{phhp} P$. It is sufficient to prove the relation $R = \{(P \boxplus_\pi P, P)\} \cup \mathbf{Id}$ is a strongly probabilistic hhp-bisimulation, however, we omit the proof here;
4. $P \boxplus_\pi \mathbf{nil} \sim_{phhp} P$. It is sufficient to prove the relation $R = \{(P \boxplus_\pi \mathbf{nil}, P)\} \cup \mathbf{Id}$ is a strongly probabilistic hhp-bisimulation, however, we omit the proof here. □

**Proposition 14.24** (Static laws for strongly probabilistic pomset bisimulation). *The static laws for strongly probabilistic pomset bisimulation are as follows:*

1. $P \parallel Q \sim_{pp} Q \parallel P$;
2. $P \parallel (Q \parallel R) \sim_{pp} (P \parallel Q) \parallel R$;
3. $P \parallel nil \sim_{pp} P$;
4. $P \setminus L \sim_{pp} P$, *if* $\mathcal{L}(P) \cap (L \cup \overline{L}) = \emptyset$;
5. $P \setminus K \setminus L \sim_{pp} P \setminus (K \cup L)$;
6. $P[f] \setminus L \sim_{pp} P \setminus f^{-1}(L)[f]$;
7. $(P \parallel Q) \setminus L \sim_{pp} P \setminus L \parallel Q \setminus L$, *if* $\mathcal{L}(P) \cap \overline{\mathcal{L}(Q)} \cap (L \cup \overline{L}) = \emptyset$;
8. $P[Id] \sim_{pp} P$;
9. $P[f] \sim_{pp} P[f']$, *if* $f \upharpoonright \mathcal{L}(P) = f' \upharpoonright \mathcal{L}(P)$;
10. $P[f][f'] \sim_{pp} P[f' \circ f]$;
11. $(P \parallel Q)[f] \sim_{pp} P[f] \parallel Q[f]$, *if* $f \upharpoonright (L \cup \overline{L})$ *is one-to-one, where* $L = \mathcal{L}(P) \cup \mathcal{L}(Q)$.

*Proof.* 1. $P \parallel Q \sim_{pp} Q \parallel P$. It is sufficient to prove the relation $R = \{(P \parallel Q, Q \parallel P)\} \cup \mathbf{Id}$ is a strongly probabilistic pomset bisimulation, however, we omit the proof here;
2. $P \parallel (Q \parallel R) \sim_{pp} (P \parallel Q) \parallel R$. It is sufficient to prove the relation $R = \{(P \parallel (Q \parallel R), (P \parallel Q) \parallel R)\} \cup \mathbf{Id}$ is a strongly probabilistic pomset bisimulation, however, we omit the proof here;
3. $P \parallel \mathbf{nil} \sim_{pp} P$. It is sufficient to prove the relation $R = \{(P \parallel \mathbf{nil}, P)\} \cup \mathbf{Id}$ is a strongly probabilistic pomset bisimulation, however, we omit the proof here;
4. $P \setminus L \sim_{pp} P$, if $\mathcal{L}(P) \cap (L \cup \overline{L}) = \emptyset$. It is sufficient to prove the relation $R = \{(P \setminus L, P)\} \cup \mathbf{Id}$, if $\mathcal{L}(P) \cap (L \cup \overline{L}) = \emptyset$, is a strongly probabilistic pomset bisimulation, however, we omit the proof here;
5. $P \setminus K \setminus L \sim_{pp} P \setminus (K \cup L)$. It is sufficient to prove the relation $R = \{(P \setminus K \setminus L, P \setminus (K \cup L))\} \cup \mathbf{Id}$ is a strongly probabilistic pomset bisimulation, however, we omit the proof here;
6. $P[f] \setminus L \sim_{pp} P \setminus f^{-1}(L)[f]$. It is sufficient to prove the relation $R = \{(P[f] \setminus L, P \setminus f^{-1}(L)[f])\} \cup \mathbf{Id}$ is a strongly probabilistic pomset bisimulation, however, we omit the proof here;
7. $(P \parallel Q) \setminus L \sim_{pp} P \setminus L \parallel Q \setminus L$, if $\mathcal{L}(P) \cap \overline{\mathcal{L}(Q)} \cap (L \cup \overline{L}) = \emptyset$. It is sufficient to prove the relation $R = \{(P + Q, Q + P)\} \cup \mathbf{Id}$ is a strongly probabilistic pomset bisimulation, however, we omit the proof here;
8. $P[Id] \sim_{pp} P$. It is sufficient to prove the relation $R = \{(P[Id], P)\} \cup \mathbf{Id}$ is a strongly probabilistic pomset bisimulation, however, we omit the proof here;
9. $P[f] \sim_{pp} P[f']$, if $f \upharpoonright \mathcal{L}(P) = f' \upharpoonright \mathcal{L}(P)$. It is sufficient to prove the relation $R = \{(P[f], P[f'])\} \cup \mathbf{Id}$, if $f \upharpoonright \mathcal{L}(P) = f' \upharpoonright \mathcal{L}(P)$, is a strongly probabilistic pomset bisimulation, however, we omit the proof here;
10. $P[f][f'] \sim_{pp} P[f' \circ f]$. It is sufficient to prove the relation $R = \{(P[f][f'], P[f' \circ f])\} \cup \mathbf{Id}$ is a strongly probabilistic pomset bisimulation, however, we omit the proof here;

11. $(P \parallel Q)[f] \sim_{pp} P[f] \parallel Q[f]$, if $f \upharpoonright (L \cup \overline{L})$ is one-to-one, where $L = \mathcal{L}(P) \cup \mathcal{L}(Q)$. It is sufficient to prove the relation $R = \{((P \parallel Q)[f], P[f] \parallel Q[f])\} \cup \mathbf{Id}$, if $f \upharpoonright (L \cup \overline{L})$ is one-to-one, where $L = \mathcal{L}(P) \cup \mathcal{L}(Q)$, is a strongly probabilistic pomset bisimulation, however, we omit the proof here. □

**Proposition 14.25** (Static laws for strongly probabilistic step bisimulation). *The static laws for strongly probabilistic step bisimulation are as follows:*

1. $P \parallel Q \sim_{ps} Q \parallel P$;
2. $P \parallel (Q \parallel R) \sim_{ps} (P \parallel Q) \parallel R$;
3. $P \parallel \mathbf{nil} \sim_{ps} P$;
4. $P \setminus L \sim_{ps} P$, *if* $\mathcal{L}(P) \cap (L \cup \overline{L}) = \emptyset$;
5. $P \setminus K \setminus L \sim_{ps} P \setminus (K \cup L)$;
6. $P[f] \setminus L \sim_{ps} P \setminus f^{-1}(L)[f]$;
7. $(P \parallel Q) \setminus L \sim_{ps} P \setminus L \parallel Q \setminus L$, *if* $\mathcal{L}(P) \cap \overline{\mathcal{L}(Q)} \cap (L \cup \overline{L}) = \emptyset$;
8. $P[Id] \sim_{ps} P$;
9. $P[f] \sim_{ps} P[f']$, *if* $f \upharpoonright \mathcal{L}(P) = f' \upharpoonright \mathcal{L}(P)$;
10. $P[f][f'] \sim_{ps} P[f' \circ f]$;
11. $(P \parallel Q)[f] \sim_{ps} P[f] \parallel Q[f]$, *if* $f \upharpoonright (L \cup \overline{L})$ *is one-to-one, where* $L = \mathcal{L}(P) \cup \mathcal{L}(Q)$.

*Proof.* 1. $P \parallel Q \sim_{ps} Q \parallel P$. It is sufficient to prove the relation $R = \{(P \parallel Q, Q \parallel P)\} \cup \mathbf{Id}$ is a strongly probabilistic step bisimulation, however, we omit the proof here;

2. $P \parallel (Q \parallel R) \sim_{ps} (P \parallel Q) \parallel R$. It is sufficient to prove the relation $R = \{(P \parallel (Q \parallel R), (P \parallel Q) \parallel R)\} \cup \mathbf{Id}$ is a strongly probabilistic step bisimulation, however, we omit the proof here;
3. $P \parallel \mathbf{nil} \sim_{ps} P$. It is sufficient to prove the relation $R = \{(P \parallel \mathbf{nil}, P)\} \cup \mathbf{Id}$ is a strongly probabilistic step bisimulation, however, we omit the proof here;
4. $P \setminus L \sim_{ps} P$, if $\mathcal{L}(P) \cap (L \cup \overline{L}) = \emptyset$. It is sufficient to prove the relation $R = \{(P \setminus L, P)\} \cup \mathbf{Id}$, if $\mathcal{L}(P) \cap (L \cup \overline{L}) = \emptyset$, is a strongly probabilistic step bisimulation, however, we omit the proof here;
5. $P \setminus K \setminus L \sim_{ps} P \setminus (K \cup L)$. It is sufficient to prove the relation $R = \{(P \setminus K \setminus L, P \setminus (K \cup L))\} \cup \mathbf{Id}$ is a strongly probabilistic step bisimulation, however, we omit the proof here;
6. $P[f] \setminus L \sim_{ps} P \setminus f^{-1}(L)[f]$. It is sufficient to prove the relation $R = \{(P[f] \setminus L, P \setminus f^{-1}(L)[f])\} \cup \mathbf{Id}$ is a strongly probabilistic step bisimulation, however, we omit the proof here;
7. $(P \parallel Q) \setminus L \sim_{ps} P \setminus L \parallel Q \setminus L$, if $\mathcal{L}(P) \cap \overline{\mathcal{L}(Q)} \cap (L \cup \overline{L}) = \emptyset$. It is sufficient to prove the relation $R = \{(P + Q, Q + P)\} \cup \mathbf{Id}$ is a strongly probabilistic step bisimulation, however, we omit the proof here;
8. $P[Id] \sim_{ps} P$. It is sufficient to prove the relation $R = \{(P[Id], P)\} \cup \mathbf{Id}$ is a strongly probabilistic step bisimulation, however, we omit the proof here;
9. $P[f] \sim_{ps} P[f']$, if $f \upharpoonright \mathcal{L}(P) = f' \upharpoonright \mathcal{L}(P)$. It is sufficient to prove the relation $R = \{(P[f], P[f'])\} \cup \mathbf{Id}$, if $f \upharpoonright \mathcal{L}(P) = f' \upharpoonright \mathcal{L}(P)$, is a strongly probabilistic step bisimulation, however, we omit the proof here;
10. $P[f][f'] \sim_{ps} P[f' \circ f]$. It is sufficient to prove the relation $R = \{(P[f][f'], P[f' \circ f])\} \cup \mathbf{Id}$ is a strongly probabilistic step bisimulation, however, we omit the proof here;
11. $(P \parallel Q)[f] \sim_{ps} P[f] \parallel Q[f]$, if $f \upharpoonright (L \cup \overline{L})$ is one-to-one, where $L = \mathcal{L}(P) \cup \mathcal{L}(Q)$. It is sufficient to prove the relation $R = \{((P \parallel Q)[f], P[f] \parallel Q[f])\} \cup \mathbf{Id}$, if $f \upharpoonright (L \cup \overline{L})$ is one-to-one, where $L = \mathcal{L}(P) \cup \mathcal{L}(Q)$, is a strongly probabilistic step bisimulation, however, we omit the proof here. □

**Proposition 14.26** (Static laws for strongly probabilistic hp-bisimulation). *The static laws for strongly probabilistic hp-bisimulation are as follows:*

1. $P \parallel Q \sim_{php} Q \parallel P$;
2. $P \parallel (Q \parallel R) \sim_{php} (P \parallel Q) \parallel R$;
3. $P \parallel \mathbf{nil} \sim_{php} P$;
4. $P \setminus L \sim_{php} P$, *if* $\mathcal{L}(P) \cap (L \cup \overline{L}) = \emptyset$;
5. $P \setminus K \setminus L \sim_{php} P \setminus (K \cup L)$;
6. $P[f] \setminus L \sim_{php} P \setminus f^{-1}(L)[f]$;
7. $(P \parallel Q) \setminus L \sim_{php} P \setminus L \parallel Q \setminus L$, *if* $\mathcal{L}(P) \cap \overline{\mathcal{L}(Q)} \cap (L \cup \overline{L}) = \emptyset$;
8. $P[Id] \sim_{php} P$;
9. $P[f] \sim_{php} P[f']$, *if* $f \upharpoonright \mathcal{L}(P) = f' \upharpoonright \mathcal{L}(P)$;
10. $P[f][f'] \sim_{php} P[f' \circ f]$;

11. $(P \parallel Q)[f] \sim_{php} P[f] \parallel Q[f]$, *if* $f \upharpoonright (L \cup \overline{L})$ *is one-to-one, where* $L = \mathcal{L}(P) \cup \mathcal{L}(Q)$.

*Proof.* 1. $P \parallel Q \sim_{php} Q \parallel P$. It is sufficient to prove the relation $R = \{(P \parallel Q, Q \parallel P)\} \cup \mathbf{Id}$ is a strongly probabilistic hp-bisimulation, however, we omit the proof here;
2. $P \parallel (Q \parallel R) \sim_{php} (P \parallel Q) \parallel R$. It is sufficient to prove the relation $R = \{(P \parallel (Q \parallel R), (P \parallel Q) \parallel R)\} \cup \mathbf{Id}$ is a strongly probabilistic hp-bisimulation, however, we omit the proof here;
3. $P \parallel \mathbf{nil} \sim_{php} P$. It is sufficient to prove the relation $R = \{(P \parallel \mathbf{nil}, P)\} \cup \mathbf{Id}$ is a strongly probabilistic hp-bisimulation, however, we omit the proof here;
4. $P \setminus L \sim_{php} P$, if $\mathcal{L}(P) \cap (L \cup \overline{L}) = \emptyset$. It is sufficient to prove the relation $R = \{(P \setminus L, P)\} \cup \mathbf{Id}$, if $\mathcal{L}(P) \cap (L \cup \overline{L}) = \emptyset$, is a strongly probabilistic hp-bisimulation, however, we omit the proof here;
5. $P \setminus K \setminus L \sim_{php} P \setminus (K \cup L)$. It is sufficient to prove the relation $R = \{(P \setminus K \setminus L, P \setminus (K \cup L))\} \cup \mathbf{Id}$ is a strongly probabilistic hp-bisimulation, however, we omit the proof here;
6. $P[f] \setminus L \sim_{php} P \setminus f^{-1}(L)[f]$. It is sufficient to prove the relation $R = \{(P[f] \setminus L, P \setminus f^{-1}(L)[f])\} \cup \mathbf{Id}$ is a strongly probabilistic hp-bisimulation, however, we omit the proof here;
7. $(P \parallel Q) \setminus L \sim_{php} P \setminus L \parallel Q \setminus L$, if $\mathcal{L}(P) \cap \overline{\mathcal{L}(Q)} \cap (L \cup \overline{L}) = \emptyset$. It is sufficient to prove the relation $R = \{(P + Q, Q + P)\} \cup \mathbf{Id}$ is a strongly probabilistic hp-bisimulation, however, we omit the proof here;
8. $P[Id] \sim_{php} P$. It is sufficient to prove the relation $R = \{(P[Id], P)\} \cup \mathbf{Id}$ is a strongly probabilistic hp-bisimulation, however, we omit the proof here;
9. $P[f] \sim_{php} P[f']$, if $f \upharpoonright \mathcal{L}(P) = f' \upharpoonright \mathcal{L}(P)$. It is sufficient to prove the relation $R = \{(P[f], P[f'])\} \cup \mathbf{Id}$, if $f \upharpoonright \mathcal{L}(P) = f' \upharpoonright \mathcal{L}(P)$, is a strongly probabilistic hp-bisimulation, however, we omit the proof here;
10. $P[f][f'] \sim_{php} P[f' \circ f]$. It is sufficient to prove the relation $R = \{(P[f][f'], P[f' \circ f])\} \cup \mathbf{Id}$ is a strongly probabilistic hp-bisimulation, however, we omit the proof here;
11. $(P \parallel Q)[f] \sim_{php} P[f] \parallel Q[f]$, if $f \upharpoonright (L \cup \overline{L})$ is one-to-one, where $L = \mathcal{L}(P) \cup \mathcal{L}(Q)$. It is sufficient to prove the relation $R = \{((P \parallel Q)[f], P[f] \parallel Q[f])\} \cup \mathbf{Id}$, if $f \upharpoonright (L \cup \overline{L})$ is one-to-one, where $L = \mathcal{L}(P) \cup \mathcal{L}(Q)$, is a strongly probabilistic hp-bisimulation, however, we omit the proof here. □

**Proposition 14.27** (Static laws for strongly probabilistic hhp-bisimulation). *The static laws for strongly probabilistic hhp-bisimulation are as follows:*

1. $P \parallel Q \sim_{phhp} Q \parallel P$;
2. $P \parallel (Q \parallel R) \sim_{phhp} (P \parallel Q) \parallel R$;
3. $P \parallel \mathbf{nil} \sim_{phhp} P$;
4. $P \setminus L \sim_{phhp} P$, *if* $\mathcal{L}(P) \cap (L \cup \overline{L}) = \emptyset$;
5. $P \setminus K \setminus L \sim_{phhp} P \setminus (K \cup L)$;
6. $P[f] \setminus L \sim_{phhp} P \setminus f^{-1}(L)[f]$;
7. $(P \parallel Q) \setminus L \sim_{phhp} P \setminus L \parallel Q \setminus L$, *if* $\mathcal{L}(P) \cap \overline{\mathcal{L}(Q)} \cap (L \cup \overline{L}) = \emptyset$;
8. $P[Id] \sim_{phhp} P$;
9. $P[f] \sim_{phhp} P[f']$, *if* $f \upharpoonright \mathcal{L}(P) = f' \upharpoonright \mathcal{L}(P)$;
10. $P[f][f'] \sim_{phhp} P[f' \circ f]$;
11. $(P \parallel Q)[f] \sim_{phhp} P[f] \parallel Q[f]$, *if* $f \upharpoonright (L \cup \overline{L})$ *is one-to-one, where* $L = \mathcal{L}(P) \cup \mathcal{L}(Q)$.

*Proof.* 1. $P \parallel Q \sim_{phhp} Q \parallel P$. It is sufficient to prove the relation $R = \{(P \parallel Q, Q \parallel P)\} \cup \mathbf{Id}$ is a strongly probabilistic hhp-bisimulation, however, we omit the proof here;
2. $P \parallel (Q \parallel R) \sim_{phhp} (P \parallel Q) \parallel R$. It is sufficient to prove the relation $R = \{(P \parallel (Q \parallel R), (P \parallel Q) \parallel R)\} \cup \mathbf{Id}$ is a strongly probabilistic hhp-bisimulation, however, we omit the proof here;
3. $P \parallel \mathbf{nil} \sim_{phhp} P$. It is sufficient to prove the relation $R = \{(P \parallel \mathbf{nil}, P)\} \cup \mathbf{Id}$ is a strongly probabilistic hhp-bisimulation, however, we omit the proof here;
4. $P \setminus L \sim_{phhp} P$, if $\mathcal{L}(P) \cap (L \cup \overline{L}) = \emptyset$. It is sufficient to prove the relation $R = \{(P \setminus L, P)\} \cup \mathbf{Id}$, if $\mathcal{L}(P) \cap (L \cup \overline{L}) = \emptyset$, is a strongly probabilistic hhp-bisimulation, however, we omit the proof here;
5. $P \setminus K \setminus L \sim_{phhp} P \setminus (K \cup L)$. It is sufficient to prove the relation $R = \{(P \setminus K \setminus L, P \setminus (K \cup L))\} \cup \mathbf{Id}$ is a strongly probabilistic hhp-bisimulation, however, we omit the proof here;
6. $P[f] \setminus L \sim_{phhp} P \setminus f^{-1}(L)[f]$. It is sufficient to prove the relation $R = \{(P[f] \setminus L, P \setminus f^{-1}(L)[f])\} \cup \mathbf{Id}$ is a strongly probabilistic hhp-bisimulation, however, we omit the proof here;
7. $(P \parallel Q) \setminus L \sim_{phhp} P \setminus L \parallel Q \setminus L$, if $\mathcal{L}(P) \cap \overline{\mathcal{L}(Q)} \cap (L \cup \overline{L}) = \emptyset$. It is sufficient to prove the relation $R = \{(P + Q, Q + P)\} \cup \mathbf{Id}$ is a strongly probabilistic hhp-bisimulation, however, we omit the proof here;

8. $P[Id] \sim_{phhp} P$. It is sufficient to prove the relation $R = \{(P[Id], P)\} \cup \mathbf{Id}$ is a strongly probabilistic hhp-bisimulation, however, we omit the proof here;
9. $P[f] \sim_{phhp} P[f']$, if $f \upharpoonright \mathcal{L}(P) = f' \upharpoonright \mathcal{L}(P)$. It is sufficient to prove the relation $R = \{(P[f], P[f'])\} \cup \mathbf{Id}$, if $f \upharpoonright \mathcal{L}(P) = f' \upharpoonright \mathcal{L}(P)$, is a strongly probabilistic hhp-bisimulation, however, we omit the proof here;
10. $P[f][f'] \sim_{phhp} P[f' \circ f]$. It is sufficient to prove the relation $R = \{(P[f][f'], P[f' \circ f])\} \cup \mathbf{Id}$ is a strongly probabilistic hhp-bisimulation, however, we omit the proof here;
11. $(P \parallel Q)[f] \sim_{phhp} P[f] \parallel Q[f]$, if $f \upharpoonright (L \cup \overline{L})$ is one-to-one, where $L = \mathcal{L}(P) \cup \mathcal{L}(Q)$. It is sufficient to prove the relation $R = \{((P \parallel Q)[f], P[f] \parallel Q[f])\} \cup \mathbf{Id}$, if $f \upharpoonright (L \cup \overline{L})$ is one-to-one, where $L = \mathcal{L}(P) \cup \mathcal{L}(Q)$, is a strongly probabilistic hhp-bisimulation, however, we omit the proof here. □

**Proposition 14.28** (Guards laws for strongly probabilistic pomset bisimulation). *The guards laws for strongly probabilistic pomset bisimulation are as follows:*

1. $P + \delta \sim_{pp} P$;
2. $\delta.P \sim_{pp} \delta$;
3. $\epsilon.P \sim_{pp} P$;
4. $P.\epsilon \sim_{pp} P$;
5. $\phi.\neg\phi \sim_{pp} \delta$;
6. $\phi + \neg\phi \sim_{pp} \epsilon$;
7. $\phi.\delta \sim_{pp} \delta$;
8. $\phi.(P + Q) \sim_{pp} \phi.P + \phi.Q$;
9. $\phi.(P.Q) \sim_{pp} \phi.P.Q$;
10. $(\phi + \psi).P \sim_{pp} \phi.P + \psi.P$;
11. $(\phi.\psi).P \sim_{pp} \phi.(\psi.P)$;
12. $\phi \sim_{pp} \epsilon$ *if* $\forall s \in S.test(\phi, s)$;
13. $\phi_0.\cdots.\phi_n \sim_{pp} \delta$ *if* $\forall s \in S, \exists i \leq n.test(\neg\phi_i, s)$;
14. $wp(\alpha, \phi).\alpha.\phi \sim_{pp} wp(\alpha, \phi).\alpha$;
15. $\neg wp(\alpha, \phi).\alpha.\neg\phi \sim_{pp} \neg wp(\alpha, \phi).\alpha$;
16. $\delta \parallel P \sim_{pp} \delta$;
17. $P \parallel \delta \sim_{pp} \delta$;
18. $\epsilon \parallel P \sim_{pp} P$;
19. $P \parallel \epsilon \sim_{pp} P$;
20. $\phi.(P \parallel Q) \sim_{pp} \phi.P \parallel \phi.Q$;
21. $\phi \parallel \delta \sim_{pp} \delta$;
22. $\delta \parallel \phi \sim_{pp} \delta$;
23. $\phi \parallel \epsilon \sim_{pp} \phi$;
24. $\epsilon \parallel \phi \sim_{pp} \phi$;
25. $\phi \parallel \neg\phi \sim_{pp} \delta$;
26. $\phi_0 \parallel \cdots \parallel \phi_n \sim_{pp} \delta$ *if* $\forall s_0, \cdots, s_n \in S, \exists i \leq n.test(\neg\phi_i, s_0 \cup \cdots \cup s_n)$.

*Proof.* 1. $P + \delta \sim_{pp} P$. It is sufficient to prove the relation $R = \{(P + \delta, P)\} \cup \mathbf{Id}$ is a strongly probabilistic pomset bisimulation, however, we omit the proof here;
2. $\delta.P \sim_{pp} \delta$. It is sufficient to prove the relation $R = \{(\delta.P, \delta)\} \cup \mathbf{Id}$ is a strongly probabilistic pomset bisimulation, however, we omit the proof here;
3. $\epsilon.P \sim_{pp} P$. It is sufficient to prove the relation $R = \{(\epsilon.P, P)\} \cup \mathbf{Id}$ is a strongly probabilistic pomset bisimulation, however, we omit the proof here;
4. $P.\epsilon \sim_{pp} P$. It is sufficient to prove the relation $R = \{(P.\epsilon, P)\} \cup \mathbf{Id}$ is a strongly probabilistic pomset bisimulation, however, we omit the proof here;
5. $\phi.\neg\phi \sim_{pp} \delta$. It is sufficient to prove the relation $R = \{(\phi.\neg\phi, \delta)\} \cup \mathbf{Id}$ is a strongly probabilistic pomset bisimulation, however, we omit the proof here;
6. $\phi + \neg\phi \sim_{pp} \epsilon$. It is sufficient to prove the relation $R = \{(\phi + \neg\phi, \epsilon)\} \cup \mathbf{Id}$ is a strongly probabilistic pomset bisimulation, however, we omit the proof here;
7. $\phi.\delta \sim_{pp} \delta$. It is sufficient to prove the relation $R = \{(\phi.\delta, \delta)\} \cup \mathbf{Id}$ is a strongly probabilistic pomset bisimulation, however, we omit the proof here;
8. $\phi.(P + Q) \sim_{pp} \phi.P + \phi.Q$. It is sufficient to prove the relation $R = \{(\phi.(P + Q), \phi.P + \phi.Q)\} \cup \mathbf{Id}$ is a strongly probabilistic pomset bisimulation, however, we omit the proof here;

9. $\phi.(P.Q) \sim_{pp} \phi.P.Q$. It is sufficient to prove the relation $R = \{(\phi.(P.Q), \phi.P.Q)\} \cup \mathbf{Id}$ is a strongly probabilistic pomset bisimulation, however, we omit the proof here;
10. $(\phi + \psi).P \sim_{pp} \phi.P + \psi.P$. It is sufficient to prove the relation $R = \{((\phi + \psi).P, \phi.P + \psi.P)\} \cup \mathbf{Id}$ is a strongly probabilistic pomset bisimulation, however, we omit the proof here;
11. $(\phi.\psi).P \sim_{pp} \phi.(\psi.P)$. It is sufficient to prove the relation $R = \{((\phi.\psi).P, \phi.(\psi.P))\} \cup \mathbf{Id}$ is a strongly probabilistic pomset bisimulation, however, we omit the proof here;
12. $\phi \sim_{pp} \epsilon$ if $\forall s \in S.test(\phi, s)$. It is sufficient to prove the relation $R = \{(\phi, \epsilon)\} \cup \mathbf{Id}$, if $\forall s \in S.test(\phi, s)$, is a strongly probabilistic pomset bisimulation, however, we omit the proof here;
13. $\phi_0.\cdots.\phi_n \sim_{pp} \delta$ if $\forall s \in S, \exists i \leq n.test(\neg\phi_i, s)$. It is sufficient to prove the relation $R = \{(\phi_0.\cdots.\phi_n, \delta)\} \cup \mathbf{Id}$, if $\forall s \in S, \exists i \leq n.test(\neg\phi_i, s)$, is a strongly probabilistic pomset bisimulation, however, we omit the proof here;
14. $wp(\alpha, \phi).\alpha.\phi \sim_{pp} wp(\alpha, \phi).\alpha$. It is sufficient to prove the relation $R = \{(wp(\alpha, \phi).\alpha.\phi, wp(\alpha, \phi).\alpha)\} \cup \mathbf{Id}$ is a strongly probabilistic pomset bisimulation, however, we omit the proof here;
15. $\neg wp(\alpha, \phi).\alpha.\neg\phi \sim_{pp} \neg wp(\alpha, \phi).e$. It is sufficient to prove the relation $R = \{(\neg wp(\alpha, \phi).\alpha.\neg\phi, \neg wp(\alpha, \phi).\alpha)\} \cup \mathbf{Id}$ is a strongly probabilistic pomset bisimulation, however, we omit the proof here;
16. $\delta \parallel P \sim_{pp} \delta$. It is sufficient to prove the relation $R = \{(\delta \parallel P, \delta)\} \cup \mathbf{Id}$ is a strongly probabilistic pomset bisimulation, however, we omit the proof here;
17. $P \parallel \delta \sim_{pp} \delta$. It is sufficient to prove the relation $R = \{(P \parallel \delta, \delta)\} \cup \mathbf{Id}$ is a strongly probabilistic pomset bisimulation, however, we omit the proof here;
18. $\epsilon \parallel P \sim_{pp} P$. It is sufficient to prove the relation $R = \{(\epsilon \parallel P, P)\} \cup \mathbf{Id}$ is a strongly probabilistic pomset bisimulation, however, we omit the proof here;
19. $P \parallel \epsilon \sim_{pp} P$. It is sufficient to prove the relation $R = \{(P \parallel \epsilon, P)\} \cup \mathbf{Id}$ is a strongly probabilistic pomset bisimulation, however, we omit the proof here;
20. $\phi.(P \parallel Q) \sim_{pp} \phi.P \parallel \phi.Q$. It is sufficient to prove the relation $R = \{(\phi.(P \parallel Q), \phi.P \parallel \phi.Q)\} \cup \mathbf{Id}$ is a strongly probabilistic pomset bisimulation, however, we omit the proof here;
21. $\phi \parallel \delta \sim_{pp} \delta$. It is sufficient to prove the relation $R = \{(\phi \parallel \delta, \delta)\} \cup \mathbf{Id}$ is a strongly probabilistic pomset bisimulation, however, we omit the proof here;
22. $\delta \parallel \phi \sim_{pp} \delta$. It is sufficient to prove the relation $R = \{(\delta \parallel \phi, \delta)\} \cup \mathbf{Id}$ is a strongly probabilistic pomset bisimulation, however, we omit the proof here;
23. $\phi \parallel \epsilon \sim_{pp} \phi$. It is sufficient to prove the relation $R = \{(\phi \parallel \epsilon, \phi)\} \cup \mathbf{Id}$ is a strongly probabilistic pomset bisimulation, however, we omit the proof here;
24. $\epsilon \parallel \phi \sim_{pp} \phi$. It is sufficient to prove the relation $R = \{(\epsilon \parallel \phi, \phi)\} \cup \mathbf{Id}$ is a strongly probabilistic pomset bisimulation, however, we omit the proof here;
25. $\phi \parallel \neg\phi \sim_{pp} \delta$. It is sufficient to prove the relation $R = \{(\phi \parallel \neg\phi, \delta)\} \cup \mathbf{Id}$ is a strongly probabilistic pomset bisimulation, however, we omit the proof here;
26. $\phi_0 \parallel \cdots \parallel \phi_n \sim_{pp} \delta$ if $\forall s_0, \cdots, s_n \in S, \exists i \leq n.test(\neg\phi_i, s_0 \cup \cdots \cup s_n)$. It is sufficient to prove the relation $R = \{(\phi_0 \parallel \cdots \parallel \phi_n, \delta)\} \cup \mathbf{Id}$, if $\forall s_0, \cdots, s_n \in S, \exists i \leq n.test(\neg\phi_i, s_0 \cup \cdots \cup s_n)$, is a strongly probabilistic pomset bisimulation, however, we omit the proof here. □

**Proposition 14.29** (Guards laws for strongly probabilistic step bisimulation). *The guards laws for strongly probabilistic step bisimulation are as follows:*

1. $P + \delta \sim_{ps} P$;
2. $\delta.P \sim_{ps} \delta$;
3. $\epsilon.P \sim_{ps} P$;
4. $P.\epsilon \sim_{ps} P$;
5. $\phi.\neg\phi \sim_{ps} \delta$;
6. $\phi + \neg\phi \sim_{ps} \epsilon$;
7. $\phi.\delta \sim_{ps} \delta$;
8. $\phi.(P + Q) \sim_{ps} \phi.P + \phi.Q$;
9. $\phi.(P.Q) \sim_{ps} \phi.P.Q$;
10. $(\phi + \psi).P \sim_{ps} \phi.P + \psi.P$;
11. $(\phi.\psi).P \sim_{ps} \phi.(\psi.P)$;
12. $\phi \sim_{ps} \epsilon$ *if* $\forall s \in S.test(\phi, s)$;
13. $\phi_0.\cdots.\phi_n \sim_{ps} \delta$ *if* $\forall s \in S, \exists i \leq n.test(\neg\phi_i, s)$;
14. $wp(\alpha, \phi).\alpha.\phi \sim_{ps} wp(\alpha, \phi).\alpha$;

15. $\neg wp(\alpha, \phi).\alpha.\neg\phi \sim_{ps} \neg wp(\alpha, \phi).\alpha$;
16. $\delta \parallel P \sim_{ps} \delta$;
17. $P \parallel \delta \sim_{ps} \delta$;
18. $\epsilon \parallel P \sim_{ps} P$;
19. $P \parallel \epsilon \sim_{ps} P$;
20. $\phi.(P \parallel Q) \sim_{ps} \phi.P \parallel \phi.Q$;
21. $\phi \parallel \delta \sim_{ps} \delta$;
22. $\delta \parallel \phi \sim_{ps} \delta$;
23. $\phi \parallel \epsilon \sim_{ps} \phi$;
24. $\epsilon \parallel \phi \sim_{ps} \phi$;
25. $\phi \parallel \neg\phi \sim_{ps} \delta$;
26. $\phi_0 \parallel \cdots \parallel \phi_n \sim_{ps} \delta$ *if* $\forall s_0, \cdots, s_n \in S, \exists i \leq n.test(\neg\phi_i, s_0 \cup \cdots \cup s_n)$.

*Proof.* 1. $P + \delta \sim_{ps} P$. It is sufficient to prove the relation $R = \{(P + \delta, P)\} \cup \mathbf{Id}$ is a strongly probabilistic step bisimulation, however, we omit the proof here;

2. $\delta.P \sim_{ps} \delta$. It is sufficient to prove the relation $R = \{(\delta.P, \delta)\} \cup \mathbf{Id}$ is a strongly probabilistic step bisimulation, however, we omit the proof here;
3. $\epsilon.P \sim_{ps} P$. It is sufficient to prove the relation $R = \{(\epsilon.P, P)\} \cup \mathbf{Id}$ is a strongly probabilistic step bisimulation, however, we omit the proof here;
4. $P.\epsilon \sim_{ps} P$. It is sufficient to prove the relation $R = \{(P.\epsilon, P)\} \cup \mathbf{Id}$ is a strongly probabilistic step bisimulation, however, we omit the proof here;
5. $\phi.\neg\phi \sim_{ps} \delta$. It is sufficient to prove the relation $R = \{(\phi.\neg\phi, \delta)\} \cup \mathbf{Id}$ is a strongly probabilistic step bisimulation, however, we omit the proof here;
6. $\phi + \neg\phi \sim_{ps} \epsilon$. It is sufficient to prove the relation $R = \{(\phi + \neg\phi, \epsilon)\} \cup \mathbf{Id}$ is a strongly probabilistic step bisimulation, however, we omit the proof here;
7. $\phi.\delta \sim_{ps} \delta$. It is sufficient to prove the relation $R = \{(\phi.\delta, \delta)\} \cup \mathbf{Id}$ is a strongly probabilistic step bisimulation, however, we omit the proof here;
8. $\phi.(P + Q) \sim_{ps} \phi.P + \phi.Q$. It is sufficient to prove the relation $R = \{(\phi.(P + Q), \phi.P + \phi.Q)\} \cup \mathbf{Id}$ is a strongly probabilistic step bisimulation, however, we omit the proof here;
9. $\phi.(P.Q) \sim_{ps} \phi.P.Q$. It is sufficient to prove the relation $R = \{(\phi.(P.Q), \phi.P.Q)\} \cup \mathbf{Id}$ is a strongly probabilistic step bisimulation, however, we omit the proof here;
10. $(\phi + \psi).P \sim_{ps} \phi.P + \psi.P$. It is sufficient to prove the relation $R = \{((\phi + \psi).P, \phi.P + \psi.P)\} \cup \mathbf{Id}$ is a strongly probabilistic step bisimulation, however, we omit the proof here;
11. $(\phi.\psi).P \sim_{ps} \phi.(\psi.P)$. It is sufficient to prove the relation $R = \{((\phi.\psi).P, \phi.(\psi.P))\} \cup \mathbf{Id}$ is a strongly probabilistic step bisimulation, however, we omit the proof here;
12. $\phi \sim_{ps} \epsilon$ if $\forall s \in S.test(\phi, s)$. It is sufficient to prove the relation $R = \{(\phi, \epsilon)\} \cup \mathbf{Id}$, if $\forall s \in S.test(\phi, s)$, is a strongly probabilistic step bisimulation, however, we omit the proof here;
13. $\phi_0. \cdots .\phi_n \sim_{ps} \delta$ if $\forall s \in S, \exists i \leq n.test(\neg\phi_i, s)$. It is sufficient to prove the relation $R = \{(\phi_0. \cdots .\phi_n, \delta)\} \cup \mathbf{Id}$, if $\forall s \in S, \exists i \leq n.test(\neg\phi_i, s)$, is a strongly probabilistic step bisimulation, however, we omit the proof here;
14. $wp(\alpha, \phi).\alpha.\phi \sim_{ps} wp(\alpha, \phi).\alpha$. It is sufficient to prove the relation $R = \{(wp(\alpha, \phi).\alpha.\phi, wp(\alpha, \phi).\alpha)\} \cup \mathbf{Id}$ is a strongly probabilistic step bisimulation, however, we omit the proof here;
15. $\neg wp(\alpha, \phi).\alpha.\neg\phi \sim_{ps} \neg wp(\alpha, \phi).\alpha$. It is sufficient to prove the relation $R = \{(\neg wp(\alpha, \phi).\alpha.\neg\phi, \neg wp(\alpha, \phi).\alpha)\} \cup \mathbf{Id}$ is a strongly probabilistic step bisimulation, however, we omit the proof here;
16. $\delta \parallel P \sim_{ps} \delta$. It is sufficient to prove the relation $R = \{(\delta \parallel P, \delta)\} \cup \mathbf{Id}$ is a strongly probabilistic step bisimulation, however, we omit the proof here;
17. $P \parallel \delta \sim_{ps} \delta$. It is sufficient to prove the relation $R = \{(P \parallel \delta, \delta)\} \cup \mathbf{Id}$ is a strongly probabilistic step bisimulation, however, we omit the proof here;
18. $\epsilon \parallel P \sim_{ps} P$. It is sufficient to prove the relation $R = \{(\epsilon \parallel P, P)\} \cup \mathbf{Id}$ is a strongly probabilistic step bisimulation, however, we omit the proof here;
19. $P \parallel \epsilon \sim_{ps} P$. It is sufficient to prove the relation $R = \{(P \parallel \epsilon, P)\} \cup \mathbf{Id}$ is a strongly probabilistic step bisimulation, however, we omit the proof here;
20. $\phi.(P \parallel Q) \sim_{ps} \phi.P \parallel \phi.Q$. It is sufficient to prove the relation $R = \{(\phi.(P \parallel Q), \phi.P \parallel \phi.Q)\} \cup \mathbf{Id}$ is a strongly probabilistic step bisimulation, however, we omit the proof here;
21. $\phi \parallel \delta \sim_{ps} \delta$. It is sufficient to prove the relation $R = \{(\phi \parallel \delta, \delta)\} \cup \mathbf{Id}$ is a strongly probabilistic step bisimulation, however, we omit the proof here;

22. $\delta \parallel \phi \sim_{ps} \delta$. It is sufficient to prove the relation $R = \{(\delta \parallel \phi, \delta)\} \cup \mathbf{Id}$ is a strongly probabilistic step bisimulation, however, we omit the proof here;
23. $\phi \parallel \epsilon \sim_{ps} \phi$. It is sufficient to prove the relation $R = \{(\phi \parallel \epsilon, \phi)\} \cup \mathbf{Id}$ is a strongly probabilistic step bisimulation, however, we omit the proof here;
24. $\epsilon \parallel \phi \sim_{ps} \phi$. It is sufficient to prove the relation $R = \{(\epsilon \parallel \phi, \phi)\} \cup \mathbf{Id}$ is a strongly probabilistic step bisimulation, however, we omit the proof here;
25. $\phi \parallel \neg\phi \sim_{ps} \delta$. It is sufficient to prove the relation $R = \{(\phi \parallel \neg\phi, \delta)\} \cup \mathbf{Id}$ is a strongly probabilistic step bisimulation, however, we omit the proof here;
26. $\phi_0 \parallel \cdots \parallel \phi_n \sim_{ps} \delta$ if $\forall s_0, \cdots, s_n \in S, \exists i \leq n.test(\neg\phi_i, s_0 \cup \cdots \cup s_n)$. It is sufficient to prove the relation $R = \{(\phi_0 \parallel \cdots \parallel \phi_n, \delta)\} \cup \mathbf{Id}$, if $\forall s_0, \cdots, s_n \in S, \exists i \leq n.test(\neg\phi_i, s_0 \cup \cdots \cup s_n)$, is a strongly probabilistic step bisimulation, however, we omit the proof here. □

**Proposition 14.30** (Guards laws for strongly probabilistic hp-bisimulation). *The guards laws for strongly probabilistic hp-bisimulation are as follows:*

1. $P + \delta \sim_{php} P$;
2. $\delta.P \sim_{php} \delta$;
3. $\epsilon.P \sim_{php} P$;
4. $P.\epsilon \sim_{php} P$;
5. $\phi.\neg\phi \sim_{php} \delta$;
6. $\phi + \neg\phi \sim_{php} \epsilon$;
7. $\phi.\delta \sim_{php} \delta$;
8. $\phi.(P + Q) \sim_{php} \phi.P + \phi.Q$;
9. $\phi.(P.Q) \sim_{php} \phi.P.Q$;
10. $(\phi + \psi).P \sim_{php} \phi.P + \psi.P$;
11. $(\phi.\psi).P \sim_{php} \phi.(\psi.P)$;
12. $\phi \sim_{php} \epsilon$ *if* $\forall s \in S.test(\phi, s)$;
13. $\phi_0. \cdots .\phi_n \sim_{php} \delta$ *if* $\forall s \in S, \exists i \leq n.test(\neg\phi_i, s)$;
14. $wp(\alpha, \phi).\alpha.\phi \sim_{php} wp(\alpha, \phi).\alpha$;
15. $\neg wp(\alpha, \phi).\alpha.\neg\phi \sim_{php} \neg wp(\alpha, \phi).\alpha$;
16. $\delta \parallel P \sim_{php} \delta$;
17. $P \parallel \delta \sim_{php} \delta$;
18. $\epsilon \parallel P \sim_{php} P$;
19. $P \parallel \epsilon \sim_{php} P$;
20. $\phi.(P \parallel Q) \sim_{php} \phi.P \parallel \phi.Q$;
21. $\phi \parallel \delta \sim_{php} \delta$;
22. $\delta \parallel \phi \sim_{php} \delta$;
23. $\phi \parallel \epsilon \sim_{php} \phi$;
24. $\epsilon \parallel \phi \sim_{php} \phi$;
25. $\phi \parallel \neg\phi \sim_{php} \delta$;
26. $\phi_0 \parallel \cdots \parallel \phi_n \sim_{php} \delta$ *if* $\forall s_0, \cdots, s_n \in S, \exists i \leq n.test(\neg\phi_i, s_0 \cup \cdots \cup s_n)$.

*Proof.* 1. $P + \delta \sim_{php} P$. It is sufficient to prove the relation $R = \{(P + \delta, P)\} \cup \mathbf{Id}$ is a strongly probabilistic hp-bisimulation, however, we omit the proof here;
2. $\delta.P \sim_{php} \delta$. It is sufficient to prove the relation $R = \{(\delta.P, \delta)\} \cup \mathbf{Id}$ is a strongly probabilistic hp-bisimulation, however, we omit the proof here;
3. $\epsilon.P \sim_{php} P$. It is sufficient to prove the relation $R = \{(\epsilon.P, P)\} \cup \mathbf{Id}$ is a strongly probabilistic hp-bisimulation, however, we omit the proof here;
4. $P.\epsilon \sim_{php} P$. It is sufficient to prove the relation $R = \{(P.\epsilon, P)\} \cup \mathbf{Id}$ is a strongly probabilistic hp-bisimulation, however, we omit the proof here;
5. $\phi.\neg\phi \sim_{php} \delta$. It is sufficient to prove the relation $R = \{(\phi.\neg\phi, \delta)\} \cup \mathbf{Id}$ is a strongly probabilistic hp-bisimulation, however, we omit the proof here;
6. $\phi + \neg\phi \sim_{php} \epsilon$. It is sufficient to prove the relation $R = \{(\phi + \neg\phi, \epsilon)\} \cup \mathbf{Id}$ is a strongly probabilistic hp-bisimulation, however, we omit the proof here;
7. $\phi.\delta \sim_{php} \delta$. It is sufficient to prove the relation $R = \{(\phi.\delta, \delta)\} \cup \mathbf{Id}$ is a strongly probabilistic hp-bisimulation, however, we omit the proof here;

8. $\phi.(P+Q) \sim_{php} \phi.P+\phi.Q$. It is sufficient to prove the relation $R=\{(\phi.(P+Q), \phi.P+\phi.Q)\} \cup \mathbf{Id}$ is a strongly probabilistic hp-bisimulation, however, we omit the proof here;
9. $\phi.(P.Q) \sim_{php} \phi.P.Q$. It is sufficient to prove the relation $R=\{(\phi.(P.Q), \phi.P.Q)\} \cup \mathbf{Id}$ is a strongly probabilistic hp-bisimulation, however, we omit the proof here;
10. $(\phi+\psi).P \sim_{php} \phi.P+\psi.P$. It is sufficient to prove the relation $R=\{((\phi+\psi).P, \phi.P+\psi.P)\} \cup \mathbf{Id}$ is a strongly probabilistic hp-bisimulation, however, we omit the proof here;
11. $(\phi.\psi).P \sim_{php} \phi.(\psi.P)$. It is sufficient to prove the relation $R=\{((\phi.\psi).P, \phi.(\psi.P))\} \cup \mathbf{Id}$ is a strongly probabilistic hp-bisimulation, however, we omit the proof here;
12. $\phi \sim_{php} \epsilon$ if $\forall s \in S.test(\phi, s)$. It is sufficient to prove the relation $R=\{(\phi, \epsilon)\} \cup \mathbf{Id}$, if $\forall s \in S.test(\phi, s)$, is a strongly probabilistic hp-bisimulation, however, we omit the proof here;
13. $\phi_0.\cdots.\phi_n \sim_{php} \delta$ if $\forall s \in S, \exists i \leq n.test(\neg\phi_i, s)$. It is sufficient to prove the relation $R=\{(\phi_0.\cdots.\phi_n, \delta)\} \cup \mathbf{Id}$, if $\forall s \in S, \exists i \leq n.test(\neg\phi_i, s)$, is a strongly probabilistic hp-bisimulation, however, we omit the proof here;
14. $wp(\alpha, \phi).\alpha.\phi \sim_{php} wp(\alpha, \phi).\alpha$. It is sufficient to prove the relation $R=\{(wp(\alpha, \phi).\alpha.\phi, wp(\alpha, \phi).\alpha)\} \cup \mathbf{Id}$ is a strongly probabilistic hp-bisimulation, however, we omit the proof here;
15. $\neg wp(\alpha, \phi).\alpha.\neg\phi \sim_{php} \neg wp(\alpha, \phi).\alpha$. It is sufficient to prove the relation $R=\{(\neg wp(\alpha, \phi).\alpha.\neg\phi, \neg wp(\alpha, \phi).\alpha)\} \cup \mathbf{Id}$ is a strongly probabilistic hp-bisimulation, however, we omit the proof here;
16. $\delta \parallel P \sim_{php} \delta$. It is sufficient to prove the relation $R=\{(\delta \parallel P, \delta)\} \cup \mathbf{Id}$ is a strongly probabilistic hp-bisimulation, however, we omit the proof here;
17. $P \parallel \delta \sim_{php} \delta$. It is sufficient to prove the relation $R=\{(P \parallel \delta, \delta)\} \cup \mathbf{Id}$ is a strongly probabilistic hp-bisimulation, however, we omit the proof here;
18. $\epsilon \parallel P \sim_{php} P$. It is sufficient to prove the relation $R=\{(\epsilon \parallel P, P)\} \cup \mathbf{Id}$ is a strongly probabilistic hp-bisimulation, however, we omit the proof here;
19. $P \parallel \epsilon \sim_{php} P$. It is sufficient to prove the relation $R=\{(P \parallel \epsilon, P)\} \cup \mathbf{Id}$ is a strongly probabilistic hp-bisimulation, however, we omit the proof here;
20. $\phi.(P \parallel Q) \sim_{php} \phi.P \parallel \phi.Q$. It is sufficient to prove the relation $R=\{(\phi.(P \parallel Q), \phi.P \parallel \phi.Q)\} \cup \mathbf{Id}$ is a strongly probabilistic hp-bisimulation, however, we omit the proof here;
21. $\phi \parallel \delta \sim_{php} \delta$. It is sufficient to prove the relation $R=\{(\phi \parallel \delta, \delta)\} \cup \mathbf{Id}$ is a strongly probabilistic hp-bisimulation, however, we omit the proof here;
22. $\delta \parallel \phi \sim_{php} \delta$. It is sufficient to prove the relation $R=\{(\delta \parallel \phi, \delta)\} \cup \mathbf{Id}$ is a strongly probabilistic hp-bisimulation, however, we omit the proof here;
23. $\phi \parallel \epsilon \sim_{php} \phi$. It is sufficient to prove the relation $R=\{(\phi \parallel \epsilon, \phi)\} \cup \mathbf{Id}$ is a strongly probabilistic hp-bisimulation, however, we omit the proof here;
24. $\epsilon \parallel \phi \sim_{php} \phi$. It is sufficient to prove the relation $R=\{(\epsilon \parallel \phi, \phi)\} \cup \mathbf{Id}$ is a strongly probabilistic hp-bisimulation, however, we omit the proof here;
25. $\phi \parallel \neg\phi \sim_{php} \delta$. It is sufficient to prove the relation $R=\{(\phi \parallel \neg\phi, \delta)\} \cup \mathbf{Id}$ is a strongly probabilistic hp-bisimulation, however, we omit the proof here;
26. $\phi_0 \parallel \cdots \parallel \phi_n \sim_{php} \delta$ if $\forall s_0, \cdots, s_n \in S, \exists i \leq n.test(\neg\phi_i, s_0 \cup \cdots \cup s_n)$. It is sufficient to prove the relation $R=\{(\phi_0 \parallel \cdots \parallel \phi_n, \delta)\} \cup \mathbf{Id}$, if $\forall s_0, \cdots, s_n \in S, \exists i \leq n.test(\neg\phi_i, s_0 \cup \cdots \cup s_n)$, is a strongly probabilistic hp-bisimulation, however, we omit the proof here. □

**Proposition 14.31** (Guards laws for strongly probabilistic hhp-bisimulation). *The guards laws for strongly probabilistic hhp-bisimulation are as follows:*

1. $P+\delta \sim_{phhp} P$;
2. $\delta.P \sim_{phhp} \delta$;
3. $\epsilon.P \sim_{phhp} P$;
4. $P.\epsilon \sim_{phhp} P$;
5. $\phi.\neg\phi \sim_{phhp} \delta$;
6. $\phi+\neg\phi \sim_{phhp} \epsilon$;
7. $\phi.\delta \sim_{phhp} \delta$;
8. $\phi.(P+Q) \sim_{phhp} \phi.P+\phi.Q$;
9. $\phi.(P.Q) \sim_{phhp} \phi.P.Q$;
10. $(\phi+\psi).P \sim_{phhp} \phi.P+\psi.P$;
11. $(\phi.\psi).P \sim_{phhp} \phi.(\psi.P)$;
12. $\phi \sim_{phhp} \epsilon$ *if* $\forall s \in S.test(\phi, s)$;

**13.** $\phi_0.\cdots.\phi_n \sim_{phhp} \delta$ *if* $\forall s \in S, \exists i \leq n.test(\neg\phi_i, s)$;
**14.** $wp(\alpha,\phi).\alpha.\phi \sim_{phhp} wp(\alpha,\phi).\alpha$;
**15.** $\neg wp(\alpha,\phi).\alpha.\neg\phi \sim_{phhp} \neg wp(\alpha,\phi).\alpha$;
**16.** $\delta \parallel P \sim_{phhp} \delta$;
**17.** $P \parallel \delta \sim_{phhp} \delta$;
**18.** $\epsilon \parallel P \sim_{phhp} P$;
**19.** $P \parallel \epsilon \sim_{phhp} P$;
**20.** $\phi.(P \parallel Q) \sim_{phhp} \phi.P \parallel \phi.Q$;
**21.** $\phi \parallel \delta \sim_{phhp} \delta$;
**22.** $\delta \parallel \phi \sim_{phhp} \delta$;
**23.** $\phi \parallel \epsilon \sim_{phhp} \phi$;
**24.** $\epsilon \parallel \phi \sim_{phhp} \phi$;
**25.** $\phi \parallel \neg\phi \sim_{phhp} \delta$;
**26.** $\phi_0 \parallel \cdots \parallel \phi_n \sim_{phhp} \delta$ *if* $\forall s_0,\cdots,s_n \in S, \exists i \leq n.test(\neg\phi_i, s_0 \cup \cdots \cup s_n)$.

*Proof.* **1.** $P + \delta \sim_{phhp} P$. It is sufficient to prove the relation $R = \{(P + \delta, P)\} \cup \mathbf{Id}$ is a strongly probabilistic hhp-bisimulation, however, we omit the proof here;
**2.** $\delta.P \sim_{phhp} \delta$. It is sufficient to prove the relation $R = \{(\delta.P, \delta)\} \cup \mathbf{Id}$ is a strongly probabilistic hhp-bisimulation, however, we omit the proof here;
**3.** $\epsilon.P \sim_{phhp} P$. It is sufficient to prove the relation $R = \{(\epsilon.P, P)\} \cup \mathbf{Id}$ is a strongly probabilistic hhp-bisimulation, however, we omit the proof here;
**4.** $P.\epsilon \sim_{phhp} P$. It is sufficient to prove the relation $R = \{(P.\epsilon, P)\} \cup \mathbf{Id}$ is a strongly probabilistic hhp-bisimulation, however, we omit the proof here;
**5.** $\phi.\neg\phi \sim_{phhp} \delta$. It is sufficient to prove the relation $R = \{(\phi.\neg\phi, \delta)\} \cup \mathbf{Id}$ is a strongly probabilistic hhp-bisimulation, however, we omit the proof here;
**6.** $\phi + \neg\phi \sim_{phhp} \epsilon$. It is sufficient to prove the relation $R = \{(\phi + \neg\phi, \epsilon)\} \cup \mathbf{Id}$ is a strongly probabilistic hhp-bisimulation, however, we omit the proof here;
**7.** $\phi.\delta \sim_{phhp} \delta$. It is sufficient to prove the relation $R = \{(\phi.\delta, \delta)\} \cup \mathbf{Id}$ is a strongly probabilistic hhp-bisimulation, however, we omit the proof here;
**8.** $\phi.(P + Q) \sim_{phhp} \phi.P + \phi.Q$. It is sufficient to prove the relation $R = \{(\phi.(P + Q), \phi.P + \phi.Q)\} \cup \mathbf{Id}$ is a strongly probabilistic hhp-bisimulation, however, we omit the proof here;
**9.** $\phi.(P.Q) \sim_{phhp} \phi.P.Q$. It is sufficient to prove the relation $R = \{(\phi.(P.Q), \phi.P.Q)\} \cup \mathbf{Id}$ is a strongly probabilistic hhp-bisimulation, however, we omit the proof here;
**10.** $(\phi + \psi).P \sim_{phhp} \phi.P + \psi.P$. It is sufficient to prove the relation $R = \{((\phi + \psi).P, \phi.P + \psi.P)\} \cup \mathbf{Id}$ is a strongly probabilistic hhp-bisimulation, however, we omit the proof here;
**11.** $(\phi.\psi).P \sim_{phhp} \phi.(\psi.P)$. It is sufficient to prove the relation $R = \{((\phi.\psi).P, \phi.(\psi.P))\} \cup \mathbf{Id}$ is a strongly probabilistic hhp-bisimulation, however, we omit the proof here;
**12.** $\phi \sim_{phhp} \epsilon$ if $\forall s \in S.test(\phi, s)$. It is sufficient to prove the relation $R = \{(\phi, \epsilon)\} \cup \mathbf{Id}$, if $\forall s \in S.test(\phi, s)$, is a strongly probabilistic hhp-bisimulation, however, we omit the proof here;
**13.** $\phi_0.\cdots.\phi_n \sim_{phhp} \delta$ if $\forall s \in S, \exists i \leq n.test(\neg\phi_i, s)$. It is sufficient to prove the relation $R = \{(\phi_0.\cdots.\phi_n, \delta)\} \cup \mathbf{Id}$, if $\forall s \in S, \exists i \leq n.test(\neg\phi_i, s)$, is a strongly probabilistic hhp-bisimulation, however, we omit the proof here;
**14.** $wp(\alpha,\phi).\alpha.\phi \sim_{phhp} wp(\alpha,\phi).\alpha$. It is sufficient to prove the relation $R = \{(wp(\alpha,\phi).\alpha.\phi, wp(\alpha,\phi).\alpha)\} \cup \mathbf{Id}$ is a strongly probabilistic hhp-bisimulation, however, we omit the proof here;
**15.** $\neg wp(\alpha,\phi).\alpha.\neg\phi \sim_{phhp} \neg wp(\alpha,\phi).\alpha$. It is sufficient to prove the relation $R = \{(\neg wp(\alpha,\phi).\alpha.\neg\phi, \neg wp(\alpha,\phi).\alpha)\} \cup \mathbf{Id}$ is a strongly probabilistic hhp-bisimulation, however, we omit the proof here;
**16.** $\delta \parallel P \sim_{phhp} \delta$. It is sufficient to prove the relation $R = \{(\delta \parallel P, \delta)\} \cup \mathbf{Id}$ is a strongly probabilistic hhp-bisimulation, however, we omit the proof here;
**17.** $P \parallel \delta \sim_{phhp} \delta$. It is sufficient to prove the relation $R = \{(P \parallel \delta, \delta)\} \cup \mathbf{Id}$ is a strongly probabilistic hhp-bisimulation, however, we omit the proof here;
**18.** $\epsilon \parallel P \sim_{phhp} P$. It is sufficient to prove the relation $R = \{(\epsilon \parallel P, P)\} \cup \mathbf{Id}$ is a strongly probabilistic hhp-bisimulation, however, we omit the proof here;
**19.** $P \parallel \epsilon \sim_{phhp} P$. It is sufficient to prove the relation $R = \{(P \parallel \epsilon, P)\} \cup \mathbf{Id}$ is a strongly probabilistic hhp-bisimulation, however, we omit the proof here;

**20.** $\phi.(P \parallel Q) \sim_{phhp} \phi.P \parallel \phi.Q$. It is sufficient to prove the relation $R = \{(\phi.(P \parallel Q), \phi.P \parallel \phi.Q)\} \cup \mathbf{Id}$ is a strongly probabilistic hhp-bisimulation, however, we omit the proof here;
**21.** $\phi \parallel \delta \sim_{phhp} \delta$. It is sufficient to prove the relation $R = \{(\phi \parallel \delta, \delta)\} \cup \mathbf{Id}$ is a strongly probabilistic hhp-bisimulation, however, we omit the proof here;
**22.** $\delta \parallel \phi \sim_{phhp} \delta$. It is sufficient to prove the relation $R = \{(\delta \parallel \phi, \delta)\} \cup \mathbf{Id}$ is a strongly probabilistic hhp-bisimulation, however, we omit the proof here;
**23.** $\phi \parallel \epsilon \sim_{phhp} \phi$. It is sufficient to prove the relation $R = \{(\phi \parallel \epsilon, \phi)\} \cup \mathbf{Id}$ is a strongly probabilistic hhp-bisimulation, however, we omit the proof here;
**24.** $\epsilon \parallel \phi \sim_{phhp} \phi$. It is sufficient to prove the relation $R = \{(\epsilon \parallel \phi, \phi)\} \cup \mathbf{Id}$ is a strongly probabilistic hhp-bisimulation, however, we omit the proof here;
**25.** $\phi \parallel \neg\phi \sim_{phhp} \delta$. It is sufficient to prove the relation $R = \{(\phi \parallel \neg\phi, \delta)\} \cup \mathbf{Id}$ is a strongly probabilistic hhp-bisimulation, however, we omit the proof here;
**26.** $\phi_0 \parallel \cdots \parallel \phi_n \sim_{phhp} \delta$ if $\forall s_0, \cdots, s_n \in S, \exists i \leq n.test(\neg\phi_i, s_0 \cup \cdots \cup s_n)$. It is sufficient to prove the relation $R = \{(\phi_0 \parallel \cdots \parallel \phi_n, \delta)\} \cup \mathbf{Id}$, if $\forall s_0, \cdots, s_n \in S, \exists i \leq n.test(\neg\phi_i, s_0 \cup \cdots \cup s_n)$, is a strongly probabilistic hhp-bisimulation, however, we omit the proof here. □

**Proposition 14.32** (Expansion law for strongly probabilistic pomset bisimulation). *Let* $P \equiv (P_1[f_1] \parallel \cdots \parallel P_n[f_n]) \setminus L$, *with* $n \geq 1$. *Then,*

$$P \sim_{pp} \{(f_1(\alpha_1) \parallel \cdots \parallel f_n(\alpha_n)).(P_1'[f_1] \parallel \cdots \parallel P_n'[f_n]) \setminus L : \langle P_i, s_i\rangle \rightsquigarrow\xrightarrow{\alpha_i} \langle P_i', s_i'\rangle, i \in \{1, \cdots, n\}, f_i(\alpha_i) \notin L \cup \overline{L}\}$$
$$+\sum\{\tau.(P_1[f_1] \parallel \cdots \parallel P_i'[f_i] \parallel \cdots \parallel P_j'[f_j] \parallel \cdots \parallel P_n[f_n]) \setminus L : \langle P_i, s_i\rangle \rightsquigarrow\xrightarrow{l_1} \langle P_i', s_i'\rangle, \langle P_j, s_j\rangle \rightsquigarrow\xrightarrow{l_2} \langle P_j', s_j'\rangle, f_i(l_1) = \overline{f_j(l_2)}, i < j\}$$

*Proof.* First, we consider the case without Restriction and Relabeling. That is, we suffice to prove the following case by induction on the size $n$.

For $P \equiv P_1 \parallel \cdots \parallel P_n$, with $n \geq 1$, we need to prove

$$P \sim_{pp} \{(\alpha_1 \parallel \cdots \parallel \alpha_n).(P_1' \parallel \cdots \parallel P_n') : \langle P_i, s_i\rangle \rightsquigarrow\xrightarrow{\alpha_i} \langle P_i', s_i'\rangle, i \in \{1, \cdots, n\}$$
$$+\sum\{\tau.(P_1 \parallel \cdots \parallel P_i' \parallel \cdots \parallel P_j' \parallel \cdots \parallel P_n) : \langle P_i, s_i\rangle \rightsquigarrow\xrightarrow{l} \langle P_i', s_i'\rangle, \langle P_j, s_j\rangle \rightsquigarrow\xrightarrow{\overline{l}} \langle P_j', s_j'\rangle, i < j\}$$

For $n = 1$, $P_1 \sim_{pp} \alpha_1.P_1' : \langle P_1, s_1\rangle \rightsquigarrow\xrightarrow{\alpha_1} \langle P_1', s_1'\rangle$ is obvious. Then, with a hypothesis $n$, we consider $R \equiv P \parallel P_{n+1}$. By the transition rules $\mathbf{Com}_{1,2,3,4}$, we can obtain

$$R \sim_{pp} \{(p \parallel \alpha_{n+1}).(P' \parallel P_{n+1}') : \langle P, s\rangle \rightsquigarrow\xrightarrow{p} \langle P', s'\rangle, \langle P_{n+1}, s\rangle \rightsquigarrow\xrightarrow{\alpha_{n+1}} \langle P_{n+1}', s''\rangle, p \subseteq P\}$$
$$+\sum\{\tau.(P' \parallel P_{n+1}') : \langle P, s\rangle \rightsquigarrow\xrightarrow{l} \langle P', s'\rangle, \langle P_{n+1}, s\rangle \rightsquigarrow\xrightarrow{\overline{l}} \langle P_{n+1}', s''\rangle\}$$

Now, with the induction assumption $P \equiv P_1 \parallel \cdots \parallel P_n$, the right-hand side can be reformulated as follows:

$$\{(\alpha_1 \parallel \cdots \parallel \alpha_n \parallel \alpha_{n+1}).(P_1' \parallel \cdots \parallel P_n' \parallel P_{n+1}') : \langle P_i, s_i\rangle \rightsquigarrow\xrightarrow{\alpha_i} \langle P_i', s_i'\rangle, i \in \{1, \cdots, n+1\}$$
$$+\sum\{\tau.(P_1 \parallel \cdots \parallel P_i' \parallel \cdots \parallel P_j' \parallel \cdots \parallel P_n \parallel P_{n+1}) : \langle P_i, s_i\rangle \rightsquigarrow\xrightarrow{l} \langle P_i', s_i'\rangle, \langle P_j, s_j\rangle \rightsquigarrow\xrightarrow{\overline{l}} \langle P_j', s_j'\rangle, i < j\}$$
$$+\sum\{\tau.(P_1 \parallel \cdots \parallel P_i' \parallel \cdots \parallel P_j \parallel \cdots \parallel P_n \parallel P_{n+1}') : \langle P_i, s_i\rangle \rightsquigarrow\xrightarrow{l} \langle P_i', s_i'\rangle, \langle P_{n+1}, s_{n+1}\rangle \rightsquigarrow\xrightarrow{\overline{l}} \langle P_{n+1}', s_{n+1}'\rangle, i \in \{1, \cdots, n\}\}$$

Hence,

$$R \sim_{pp} \{(\alpha_1 \parallel \cdots \parallel \alpha_n \parallel \alpha_{n+1}).(P_1' \parallel \cdots \parallel P_n' \parallel P_{n+1}') :$$

$$P_i \overset{\alpha_i}{\rightsquigarrow} P_i', i \in \{1, \cdots, n+1\}$$

$$+\sum\{\tau.(P_1 \parallel \cdots \parallel P_i' \parallel \cdots \parallel P_j' \parallel \cdots \parallel P_n):$$

$$P_i \overset{l}{\rightsquigarrow} P_i', P_j \overset{\bar{l}}{\rightsquigarrow} P_j', 1 \leq i < j \geq n+1\}$$

Then, we can easily add the full conditions with Restriction and Relabeling. □

**Proposition 14.33** (Expansion law for strongly probabilistic step bisimulation). *Let $P \equiv (P_1[f_1] \parallel \cdots \parallel P_n[f_n]) \setminus L$, with $n \geq 1$. Then,*

$$P \sim_{ps} \{(f_1(\alpha_1) \parallel \cdots \parallel f_n(\alpha_n)).(P_1'[f_1] \parallel \cdots \parallel P_n'[f_n]) \setminus L :$$

$$\langle P_i, s_i\rangle \overset{\alpha_i}{\rightsquigarrow} \langle P_i', s_i'\rangle, i \in \{1, \cdots, n\}, f_i(\alpha_i) \notin L \cup \overline{L}\}$$

$$+\sum\{\tau.(P_1[f_1] \parallel \cdots \parallel P_i'[f_i] \parallel \cdots \parallel P_j'[f_j] \parallel \cdots \parallel P_n[f_n]) \setminus L :$$

$$\langle P_i, s_i\rangle \overset{l_1}{\rightsquigarrow} \langle P_i', s_i'\rangle, \langle P_j, s_j\rangle \overset{l_2}{\rightsquigarrow} \langle P_j', s_j'\rangle, f_i(l_1) = \overline{f_j(l_2)}, i < j\}$$

*Proof.* First, we consider the case without Restriction and Relabeling. That is, we suffice to prove the following case by induction on the size $n$.

For $P \equiv P_1 \parallel \cdots \parallel P_n$, with $n \geq 1$, we need to prove

$$P \sim_{ps} \{(\alpha_1 \parallel \cdots \parallel \alpha_n).(P_1' \parallel \cdots \parallel P_n') : \langle P_i, s_i\rangle \overset{\alpha_i}{\rightsquigarrow} \langle P_i', s_i'\rangle, i \in \{1, \cdots, n\}$$

$$+\sum\{\tau.(P_1 \parallel \cdots \parallel P_i' \parallel \cdots \parallel P_j' \parallel \cdots \parallel P_n) : \langle P_i, s_i\rangle \overset{l}{\rightsquigarrow} \langle P_i', s_i'\rangle, \langle P_j, s_j\rangle \overset{\bar{l}}{\rightsquigarrow} \langle P_j', s_j'\rangle, i < j\}$$

For $n = 1$, $P_1 \sim_{ps} \alpha_1.P_1' : \langle P_1, s_1\rangle \overset{\alpha_1}{\rightsquigarrow} \langle P_1', s_1'\rangle$ is obvious. Then, with a hypothesis $n$, we consider $R \equiv P \parallel P_{n+1}$. By the transition rules $\mathbf{Com}_{1,2,3,4}$, we can obtain

$$R \sim_{ps} \{(p \parallel \alpha_{n+1}).(P' \parallel P_{n+1}') : \langle P, s\rangle \overset{p}{\rightsquigarrow} \langle P', s'\rangle, \langle P_{n+1}, s\rangle \overset{\alpha_{n+1}}{\rightsquigarrow} \langle P_{n+1}', s''\rangle, p \subseteq P\}$$

$$+\sum\{\tau.(P' \parallel P_{n+1}') : \langle P, s\rangle \overset{l}{\rightsquigarrow} \langle P', s'\rangle, \langle P_{n+1}, s\rangle \overset{\bar{l}}{\rightsquigarrow} \langle P_{n+1}', s''\rangle\}$$

Now, with the induction assumption $P \equiv P_1 \parallel \cdots \parallel P_n$, the right-hand side can be reformulated as follows:

$$\{(\alpha_1 \parallel \cdots \parallel \alpha_n \parallel \alpha_{n+1}).(P_1' \parallel \cdots \parallel P_n' \parallel P_{n+1}') :$$

$$\langle P_i, s_i\rangle \overset{\alpha_i}{\rightsquigarrow} \langle P_i', s_i'\rangle, i \in \{1, \cdots, n+1\}$$

$$+\sum\{\tau.(P_1 \parallel \cdots \parallel P_i' \parallel \cdots \parallel P_j' \parallel \cdots \parallel P_n \parallel P_{n+1}) :$$

$$\langle P_i, s_i\rangle \overset{l}{\rightsquigarrow} \langle P_i', s_i'\rangle, \langle P_j, s_j\rangle \overset{\bar{l}}{\rightsquigarrow} \langle P_j', s_j'\rangle, i < j\}$$

$$+\sum\{\tau.(P_1 \parallel \cdots \parallel P_i' \parallel \cdots \parallel P_j \parallel \cdots \parallel P_n \parallel P_{n+1}') :$$

$$\langle P_i, s_i\rangle \overset{l}{\rightsquigarrow} \langle P_i', s_i'\rangle, \langle P_{n+1}, s_{n+1}\rangle \overset{\bar{l}}{\rightsquigarrow} \langle P_{n+1}', s_{n+1}'\rangle, i \in \{1, \cdots, n\}\}$$

Hence,

$$R \sim_{ps} \{(\alpha_1 \parallel \cdots \parallel \alpha_n \parallel \alpha_{n+1}).(P_1' \parallel \cdots \parallel P_n' \parallel P_{n+1}') :$$

$$P_i \overset{\alpha_i}{\rightsquigarrow} P_i', i \in \{1, \cdots, n+1\}$$

$$+\sum\{\tau.(P_1 \parallel \cdots \parallel P_i' \parallel \cdots \parallel P_j' \parallel \cdots \parallel P_n) :$$

$$P_i \overset{l}{\rightsquigarrow} P_i', P_j \overset{\bar{l}}{\rightsquigarrow} P_j', 1 \leq i < j \geq n+1\}$$

Then, we can easily add the full conditions with Restriction and Relabeling. □

**Proposition 14.34** (Expansion law for strongly probabilistic hp-bisimulation). *Let $P \equiv (P_1[f_1] \parallel \cdots \parallel P_n[f_n]) \setminus L$, with $n \geq 1$. Then,*

$$P \sim_{php} \{(f_1(\alpha_1) \parallel \cdots \parallel f_n(\alpha_n)).(P_1'[f_1] \parallel \cdots \parallel P_n'[f_n]) \setminus L :$$
$$\langle P_i, s_i\rangle \rightsquigarrow\xrightarrow{\alpha_i} \langle P_i', s_i'\rangle, i \in \{1, \cdots, n\}, f_i(\alpha_i) \notin L \cup \overline{L}\}$$
$$+\sum\{\tau.(P_1[f_1] \parallel \cdots \parallel P_i'[f_i] \parallel \cdots \parallel P_j'[f_j] \parallel \cdots \parallel P_n[f_n]) \setminus L :$$
$$\langle P_i, s_i\rangle \rightsquigarrow\xrightarrow{l_1} \langle P_i', s_i'\rangle, \langle P_j, s_j\rangle \rightsquigarrow\xrightarrow{l_2} \langle P_j', s_j'\rangle, f_i(l_1) = \overline{f_j(l_2)}, i < j\}$$

*Proof.* First, we consider the case without Restriction and Relabeling. That is, we suffice to prove the following case by induction on the size $n$.

For $P \equiv P_1 \parallel \cdots \parallel P_n$, with $n \geq 1$, we need to prove

$$P \sim_{php} \{(\alpha_1 \parallel \cdots \parallel \alpha_n).(P_1' \parallel \cdots \parallel P_n') : \langle P_i, s_i\rangle \rightsquigarrow\xrightarrow{\alpha_i} \langle P_i', s_i'\rangle, i \in \{1, \cdots, n\}$$
$$+\sum\{\tau.(P_1 \parallel \cdots \parallel P_i' \parallel \cdots \parallel P_j' \parallel \cdots \parallel P_n) : \langle P_i, s_i\rangle \rightsquigarrow\xrightarrow{l} \langle P_i', s_i'\rangle, \langle P_j, s_j\rangle \rightsquigarrow\xrightarrow{\overline{l}} \langle P_j', s_j'\rangle, i < j\}$$

For $n = 1$, $P_1 \sim_{php} \alpha_1.P_1' : \langle P_1, s_1\rangle \rightsquigarrow\xrightarrow{\alpha_1} \langle P_1', s_1'\rangle$ is obvious. Then, with a hypothesis $n$, we consider $R \equiv P \parallel P_{n+1}$. By the transition rules $\mathbf{Com}_{1,2,3,4}$, we can obtain

$$R \sim_{php} \{(p \parallel \alpha_{n+1}).(P' \parallel P_{n+1}') : \langle P, s\rangle \rightsquigarrow\xrightarrow{p} \langle P', s'\rangle, \langle P_{n+1}, s\rangle \rightsquigarrow\xrightarrow{\alpha_{n+1}} \langle P_{n+1}', s''\rangle, p \subseteq P\}$$
$$+\sum\{\tau.(P' \parallel P_{n+1}') : \langle P, s\rangle \rightsquigarrow\xrightarrow{l} \langle P', s'\rangle, \langle P_{n+1}, s\rangle \rightsquigarrow\xrightarrow{\overline{l}} \langle P_{n+1}', s''\rangle\}$$

Now, with the induction assumption $P \equiv P_1 \parallel \cdots \parallel P_n$, the right-hand side can be reformulated as follows:

$$\{(\alpha_1 \parallel \cdots \parallel \alpha_n \parallel \alpha_{n+1}).(P_1' \parallel \cdots \parallel P_n' \parallel P_{n+1}') :$$
$$\langle P_i, s_i\rangle \rightsquigarrow\xrightarrow{\alpha_i} \langle P_i', s_i'\rangle, i \in \{1, \cdots, n+1\}$$
$$+\sum\{\tau.(P_1 \parallel \cdots \parallel P_i' \parallel \cdots \parallel P_j' \parallel \cdots \parallel P_n \parallel P_{n+1}) :$$
$$\langle P_i, s_i\rangle \rightsquigarrow\xrightarrow{l} \langle P_i', s_i'\rangle, \langle P_j, s_j\rangle \rightsquigarrow\xrightarrow{\overline{l}} \langle P_j', s_j'\rangle, i < j\}$$
$$+\sum\{\tau.(P_1 \parallel \cdots \parallel P_i' \parallel \cdots \parallel P_j \parallel \cdots \parallel P_n \parallel P_{n+1}') :$$
$$\langle P_i, s_i\rangle \rightsquigarrow\xrightarrow{l} \langle P_i', s_i'\rangle, \langle P_{n+1}, s_{n+1}\rangle \rightsquigarrow\xrightarrow{\overline{l}} \langle P_{n+1}', s_{n+1}'\rangle, i \in \{1, \cdots, n\}\}$$

Hence,

$$R \sim_{php} \{(\alpha_1 \parallel \cdots \parallel \alpha_n \parallel \alpha_{n+1}).(P_1' \parallel \cdots \parallel P_n' \parallel P_{n+1}') :$$
$$P_i \rightsquigarrow\xrightarrow{\alpha_i} P_i', i \in \{1, \cdots, n+1\}$$
$$+\sum\{\tau.(P_1 \parallel \cdots \parallel P_i' \parallel \cdots \parallel P_j' \parallel \cdots \parallel P_n) :$$
$$P_i \rightsquigarrow\xrightarrow{l} P_i', P_j \rightsquigarrow\xrightarrow{\overline{l}} P_j', 1 \leq i < j \geq n+1\}$$

Then, we can easily add the full conditions with Restriction and Relabeling. □

**Proposition 14.35** (Expansion law for strongly probabilistic hhp-bisimulation). *Let $P \equiv (P_1[f_1] \parallel \cdots \parallel P_n[f_n]) \setminus L$, with $n \geq 1$. Then,*

$$P \sim_{phhp} \{(f_1(\alpha_1) \parallel \cdots \parallel f_n(\alpha_n)).(P_1'[f_1] \parallel \cdots \parallel P_n'[f_n]) \setminus L :$$
$$\langle P_i, s_i\rangle \rightsquigarrow\xrightarrow{\alpha_i} \langle P_i', s_i'\rangle, i \in \{1, \cdots, n\}, f_i(\alpha_i) \notin L \cup \overline{L}\}$$
$$+\sum\{\tau.(P_1[f_1] \parallel \cdots \parallel P_i'[f_i] \parallel \cdots \parallel P_j'[f_j] \parallel \cdots \parallel P_n[f_n]) \setminus L :$$

$$\langle P_i, s_i\rangle \rightsquigarrow\xrightarrow{l_1} \langle P_i', s_i'\rangle, \langle P_j, s_j\rangle \rightsquigarrow\xrightarrow{l_2} \langle P_j', s_j'\rangle, f_i(l_1) = \overline{f_j(l_2)}, i < j\}$$

*Proof.* First, we consider the case without Restriction and Relabeling. That is, we suffice to prove the following case by induction on the size $n$.

For $P \equiv P_1 \parallel \cdots \parallel P_n$, with $n \geq 1$, we need to prove

$$P \sim_{phhp} \{(\alpha_1 \parallel \cdots \parallel \alpha_n).(P_1' \parallel \cdots \parallel P_n') : \langle P_i, s_i\rangle \rightsquigarrow\xrightarrow{\alpha_i} \langle P_i', s_i'\rangle, i \in \{1, \cdots, n\}$$

$$+\sum\{\tau.(P_1 \parallel \cdots \parallel P_i' \parallel \cdots \parallel P_j' \parallel \cdots \parallel P_n) : \langle P_i, s_i\rangle \rightsquigarrow\xrightarrow{l} \langle P_i', s_i'\rangle, \langle P_j, s_j\rangle \rightsquigarrow\xrightarrow{\bar{l}} \langle P_j', s_j'\rangle, i < j\}$$

For $n = 1$, $P_1 \sim_{phhp} \alpha_1.P_1' : \langle P_1, s_1\rangle \rightsquigarrow\xrightarrow{\alpha_1} \langle P_1', s_1'\rangle$ is obvious. Then, with a hypothesis $n$, we consider $R \equiv P \parallel P_{n+1}$. By the transition rules $\mathbf{Com}_{1,2,3,4}$, we can obtain

$$R \sim_{phhp} \{(p \parallel \alpha_{n+1}).(P' \parallel P_{n+1}') : \langle P, s\rangle \rightsquigarrow\xrightarrow{p} \langle P', s'\rangle, \langle P_{n+1}, s\rangle \rightsquigarrow\xrightarrow{\alpha_{n+1}} \langle P_{n+1}', s''\rangle, p \subseteq P\}$$

$$+\sum\{\tau.(P' \parallel P_{n+1}') : \langle P, s\rangle \rightsquigarrow\xrightarrow{l} \langle P', s'\rangle, \langle P_{n+1}, s\rangle \rightsquigarrow\xrightarrow{\bar{l}} \langle P_{n+1}', s''\rangle\}$$

Now, with the induction assumption $P \equiv P_1 \parallel \cdots \parallel P_n$, the right-hand side can be reformulated as follows:

$$\{(\alpha_1 \parallel \cdots \parallel \alpha_n \parallel \alpha_{n+1}).(P_1' \parallel \cdots \parallel P_n' \parallel P_{n+1}') :$$
$$\langle P_i, s_i\rangle \rightsquigarrow\xrightarrow{\alpha_i} \langle P_i', s_i'\rangle, i \in \{1, \cdots, n+1\}$$
$$+\sum\{\tau.(P_1 \parallel \cdots \parallel P_i' \parallel \cdots \parallel P_j' \parallel \cdots \parallel P_n \parallel P_{n+1}) :$$
$$\langle P_i, s_i\rangle \rightsquigarrow\xrightarrow{l} \langle P_i', s_i'\rangle, \langle P_j, s_j\rangle \rightsquigarrow\xrightarrow{\bar{l}} \langle P_j', s_j'\rangle, i < j\}$$
$$+\sum\{\tau.(P_1 \parallel \cdots \parallel P_i' \parallel \cdots \parallel P_j \parallel \cdots \parallel P_n \parallel P_{n+1}') :$$
$$\langle P_i, s_i\rangle \rightsquigarrow\xrightarrow{l} \langle P_i', s_i'\rangle, \langle P_{n+1}, s_{n+1}\rangle \rightsquigarrow\xrightarrow{\bar{l}} \langle P_{n+1}', s_{n+1}'\rangle, i \in \{1, \cdots, n\}\}$$

Hence,

$$R \sim_{phhp} \{(\alpha_1 \parallel \cdots \parallel \alpha_n \parallel \alpha_{n+1}).(P_1' \parallel \cdots \parallel P_n' \parallel P_{n+1}') :$$
$$P_i \rightsquigarrow\xrightarrow{\alpha_i} P_i', i \in \{1, \cdots, n+1\}$$
$$+\sum\{\tau.(P_1 \parallel \cdots \parallel P_i' \parallel \cdots \parallel P_j' \parallel \cdots \parallel P_n) :$$
$$P_i \rightsquigarrow\xrightarrow{l} P_i', P_j \rightsquigarrow\xrightarrow{\bar{l}} P_j', 1 \leq i < j \geq n+1\}$$

Then, we can easily add the full conditions with Restriction and Relabeling. □

**Theorem 14.36** (Congruence for strongly probabilistic pomset bisimulation). *We can enjoy the full congruence for strongly probabilistic pomset bisimulation as follows:*

1. *If $A \stackrel{def}{=} P$, then $A \sim_{pp} P$;*
2. *Let $P_1 \sim_{pp} P_2$. Then,*
   (a) $\alpha.P_1 \sim_{pp} \alpha.P_2$;
   (b) $\phi.P_1 \sim_{pp} \phi.P_2$;
   (c) $(\alpha_1 \parallel \cdots \parallel \alpha_n).P_1 \sim_{pp} (\alpha_1 \parallel \cdots \parallel \alpha_n).P_2$;
   (d) $P_1 + Q \sim_{pp} P_2 + Q$;
   (e) $P_1 \boxplus_{pi} Q \sim_{pp} P_2 \boxplus_{\pi} Q$;
   (f) $P_1 \parallel Q \sim_{pp} P_2 \parallel Q$;
   (g) $P_1 \setminus L \sim_{pp} P_2 \setminus L$;
   (h) $P_1[f] \sim_{pp} P_2[f]$.

*Proof.* **1.** If $A \stackrel{\text{def}}{=} P$, then $A \sim_{pp} P$. It is obvious.

2. Let $P_1 \sim_{pp} P_2$. Then,
   (a) $\alpha.P_1 \sim_{pp} \alpha.P_2$. It is sufficient to prove the relation $R = \{(\alpha.P_1, \alpha.P_2)\} \cup \mathbf{Id}$ is a strongly probabilistic pomset bisimulation, however, we omit the proof here;
   (b) $\phi.P_1 \sim_{pp} \phi.P_2$. It is sufficient to prove the relation $R = \{(\phi.P_1, \phi.P_2)\} \cup \mathbf{Id}$ is a strongly probabilistic pomset bisimulation, however, we omit the proof here;
   (c) $(\alpha_1 \parallel \cdots \parallel \alpha_n).P_1 \sim_{pp} (\alpha_1 \parallel \cdots \parallel \alpha_n).P_2$. It is sufficient to prove the relation $R = \{((\alpha_1 \parallel \cdots \parallel \alpha_n).P_1, (\alpha_1 \parallel \cdots \parallel \alpha_n).P_2)\} \cup \mathbf{Id}$ is a strongly probabilistic pomset bisimulation, however, we omit the proof here;
   (d) $P_1 + Q \sim_{pp} P_2 + Q$. It is sufficient to prove the relation $R = \{(P_1 + Q, P_2 + Q)\} \cup \mathbf{Id}$ is a strongly probabilistic pomset bisimulation, however, we omit the proof here;
   (e) $P_1 \boxplus_{pi} Q \sim_{pp} P_2 \boxplus_{pi} Q$. It is sufficient to prove the relation $R = \{(P_1 \boxplus_{pi} Q, P_2 \boxplus_{pi} Q)\} \cup \mathbf{Id}$ is a strongly probabilistic pomset bisimulation, however, we omit the proof here;
   (f) $P_1 \parallel Q \sim_{pp} P_2 \parallel Q$. It is sufficient to prove the relation $R = \{(P_1 \parallel Q, P_2 \parallel Q)\} \cup \mathbf{Id}$ is a strongly probabilistic pomset bisimulation, however, we omit the proof here;
   (g) $P_1 \setminus L \sim_{pp} P_2 \setminus L$. It is sufficient to prove the relation $R = \{(P_1 \setminus L, P_2 \setminus L)\} \cup \mathbf{Id}$ is a strongly probabilistic pomset bisimulation, however, we omit the proof here;
   (h) $P_1[f] \sim_{pp} P_2[f]$. It is sufficient to prove the relation $R = \{(P_1[f], P_2[f])\} \cup \mathbf{Id}$ is a strongly probabilistic pomset bisimulation, however, we omit the proof here. □

**Theorem 14.37** (Congruence for strongly probabilistic step bisimulation). *We can enjoy the full congruence for strongly probabilistic step bisimulation as follows:*

1. *If $A \stackrel{def}{=} P$, then $A \sim_{ps} P$;*
2. *Let $P_1 \sim_{ps} P_2$. Then,*
   (a) $\alpha.P_1 \sim_{ps} \alpha.P_2$;
   (b) $\phi.P_1 \sim_{ps} \phi.P_2$;
   (c) $(\alpha_1 \parallel \cdots \parallel \alpha_n).P_1 \sim_{ps} (\alpha_1 \parallel \cdots \parallel \alpha_n).P_2$;
   (d) $P_1 + Q \sim_{ps} P_2 + Q$;
   (e) $P_1 \boxplus_{pi} Q \sim_{ps} P_2 \boxplus_{\pi} Q$;
   (f) $P_1 \parallel Q \sim_{ps} P_2 \parallel Q$;
   (g) $P_1 \setminus L \sim_{ps} P_2 \setminus L$;
   (h) $P_1[f] \sim_{ps} P_2[f]$.

*Proof.* 1. If $A \stackrel{\text{def}}{=} P$, then $A \sim_{ps} P$. It is obvious.
2. Let $P_1 \sim_{ps} P_2$. Then,
   (a) $\alpha.P_1 \sim_{ps} \alpha.P_2$. It is sufficient to prove the relation $R = \{(\alpha.P_1, \alpha.P_2)\} \cup \mathbf{Id}$ is a strongly probabilistic step bisimulation, however, we omit the proof here;
   (b) $\phi.P_1 \sim_{ps} \phi.P_2$. It is sufficient to prove the relation $R = \{(\phi.P_1, \phi.P_2)\} \cup \mathbf{Id}$ is a strongly probabilistic step bisimulation, however, we omit the proof here;
   (c) $(\alpha_1 \parallel \cdots \parallel \alpha_n).P_1 \sim_{ps} (\alpha_1 \parallel \cdots \parallel \alpha_n).P_2$. It is sufficient to prove the relation $R = \{((\alpha_1 \parallel \cdots \parallel \alpha_n).P_1, (\alpha_1 \parallel \cdots \parallel \alpha_n).P_2)\} \cup \mathbf{Id}$ is a strongly probabilistic step bisimulation, however, we omit the proof here;
   (d) $P_1 + Q \sim_{ps} P_2 + Q$. It is sufficient to prove the relation $R = \{(P_1 + Q, P_2 + Q)\} \cup \mathbf{Id}$ is a strongly probabilistic step bisimulation, however, we omit the proof here;
   (e) $P_1 \boxplus_{pi} Q \sim_{ps} P_2 \boxplus_{pi} Q$. It is sufficient to prove the relation $R = \{(P_1 \boxplus_{pi} Q, P_2 \boxplus_{pi} Q)\} \cup \mathbf{Id}$ is a strongly probabilistic step bisimulation, however, we omit the proof here;
   (f) $P_1 \parallel Q \sim_{ps} P_2 \parallel Q$. It is sufficient to prove the relation $R = \{(P_1 \parallel Q, P_2 \parallel Q)\} \cup \mathbf{Id}$ is a strongly probabilistic step bisimulation, however, we omit the proof here;
   (g) $P_1 \setminus L \sim_{ps} P_2 \setminus L$. It is sufficient to prove the relation $R = \{(P_1 \setminus L, P_2 \setminus L)\} \cup \mathbf{Id}$ is a strongly probabilistic step bisimulation, however, we omit the proof here;
   (h) $P_1[f] \sim_{ps} P_2[f]$. It is sufficient to prove the relation $R = \{(P_1[f], P_2[f])\} \cup \mathbf{Id}$ is a strongly probabilistic step bisimulation, however, we omit the proof here. □

**Theorem 14.38** (Congruence for strongly probabilistic hp-bisimulation). *We can enjoy the full congruence for strongly probabilistic hp-bisimulation as follows:*

1. *If $A \stackrel{def}{=} P$, then $A \sim_{php} P$;*

2. *Let $P_1 \sim_{php} P_2$. Then,*
   (a) $\alpha.P_1 \sim_{php} \alpha.P_2$;
   (b) $\phi.P_1 \sim_{php} \phi.P_2$;
   (c) $(\alpha_1 \parallel \cdots \parallel \alpha_n).P_1 \sim_{php} (\alpha_1 \parallel \cdots \parallel \alpha_n).P_2$;
   (d) $P_1 + Q \sim_{php} P_2 + Q$;
   (e) $P_1 \boxplus_{pi} Q \sim_{php} P_2 \boxplus_{\pi} Q$;
   (f) $P_1 \parallel Q \sim_{php} P_2 \parallel Q$;
   (g) $P_1 \setminus L \sim_{php} P_2 \setminus L$;
   (h) $P_1[f] \sim_{php} P_2[f]$.

*Proof.* **1.** If $A \stackrel{\text{def}}{=} P$, then $A \sim_{php} P$. It is obvious.
**2.** Let $P_1 \sim_{php} P_2$. Then,
   (a) $\alpha.P_1 \sim_{php} \alpha.P_2$. It is sufficient to prove the relation $R = \{(\alpha.P_1, \alpha.P_2)\} \cup \mathbf{Id}$ is a strongly probabilistic hp-bisimulation, however, we omit the proof here;
   (b) $\phi.P_1 \sim_{php} \phi.P_2$. It is sufficient to prove the relation $R = \{(\phi.P_1, \phi.P_2)\} \cup \mathbf{Id}$ is a strongly probabilistic hp-bisimulation, however, we omit the proof here;
   (c) $(\alpha_1 \parallel \cdots \parallel \alpha_n).P_1 \sim_{php} (\alpha_1 \parallel \cdots \parallel \alpha_n).P_2$. It is sufficient to prove the relation $R = \{((\alpha_1 \parallel \cdots \parallel \alpha_n).P_1, (\alpha_1 \parallel \cdots \parallel \alpha_n).P_2)\} \cup \mathbf{Id}$ is a strongly probabilistic hp-bisimulation, however, we omit the proof here;
   (d) $P_1 + Q \sim_{php} P_2 + Q$. It is sufficient to prove the relation $R = \{(P_1 + Q, P_2 + Q)\} \cup \mathbf{Id}$ is a strongly probabilistic hp-bisimulation, however, we omit the proof here;
   (e) $P_1 \boxplus_{pi} Q \sim_{php} P_2 \boxplus_{pi} Q$. It is sufficient to prove the relation $R = \{(P_1 \boxplus_{pi} Q, P_2 \boxplus_{pi} Q)\} \cup \mathbf{Id}$ is a strongly probabilistic hp-bisimulation, however, we omit the proof here;
   (f) $P_1 \parallel Q \sim_{php} P_2 \parallel Q$. It is sufficient to prove the relation $R = \{(P_1 \parallel Q, P_2 \parallel Q)\} \cup \mathbf{Id}$ is a strongly probabilistic hp-bisimulation, however, we omit the proof here;
   (g) $P_1 \setminus L \sim_{php} P_2 \setminus L$. It is sufficient to prove the relation $R = \{(P_1 \setminus L, P_2 \setminus L)\} \cup \mathbf{Id}$ is a strongly probabilistic hp-bisimulation, however, we omit the proof here;
   (h) $P_1[f] \sim_{php} P_2[f]$. It is sufficient to prove the relation $R = \{(P_1[f], P_2[f])\} \cup \mathbf{Id}$ is a strongly probabilistic hp-bisimulation, however, we omit the proof here. □

**Theorem 14.39** (Congruence for strongly probabilistic hhp-bisimulation). *We can enjoy the full congruence for strongly probabilistic hhp-bisimulation as follows:*

1. *If $A \stackrel{def}{=} P$, then $A \sim_{phhp} P$;*
2. *Let $P_1 \sim_{phhp} P_2$. Then,*
   (a) $\alpha.P_1 \sim_{phhp} \alpha.P_2$;
   (b) $\phi.P_1 \sim_{phhp} \phi.P_2$;
   (c) $(\alpha_1 \parallel \cdots \parallel \alpha_n).P_1 \sim_{phhp} (\alpha_1 \parallel \cdots \parallel \alpha_n).P_2$;
   (d) $P_1 + Q \sim_{phhp} P_2 + Q$;
   (e) $P_1 \boxplus_{pi} Q \sim_{phhp} P_2 \boxplus_{\pi} Q$;
   (f) $P_1 \parallel Q \sim_{phhp} P_2 \parallel Q$;
   (g) $P_1 \setminus L \sim_{phhp} P_2 \setminus L$;
   (h) $P_1[f] \sim_{phhp} P_2[f]$.

*Proof.* **1.** If $A \stackrel{\text{def}}{=} P$, then $A \sim_{phhp} P$. It is obvious.
**2.** Let $P_1 \sim_{phhp} P_2$. Then,
   (a) $\alpha.P_1 \sim_{phhp} \alpha.P_2$. It is sufficient to prove the relation $R = \{(\alpha.P_1, \alpha.P_2)\} \cup \mathbf{Id}$ is a strongly probabilistic hhp-bisimulation, however, we omit the proof here;
   (b) $\phi.P_1 \sim_{phhp} \phi.P_2$. It is sufficient to prove the relation $R = \{(\phi.P_1, \phi.P_2)\} \cup \mathbf{Id}$ is a strongly probabilistic hhp-bisimulation, however, we omit the proof here;
   (c) $(\alpha_1 \parallel \cdots \parallel \alpha_n).P_1 \sim_{phhp} (\alpha_1 \parallel \cdots \parallel \alpha_n).P_2$. It is sufficient to prove the relation $R = \{((\alpha_1 \parallel \cdots \parallel \alpha_n).P_1, (\alpha_1 \parallel \cdots \parallel \alpha_n).P_2)\} \cup \mathbf{Id}$ is a strongly probabilistic hhp-bisimulation, however, we omit the proof here;
   (d) $P_1 + Q \sim_{phhp} P_2 + Q$. It is sufficient to prove the relation $R = \{(P_1 + Q, P_2 + Q)\} \cup \mathbf{Id}$ is a strongly probabilistic hhp-bisimulation, however, we omit the proof here;
   (e) $P_1 \boxplus_{pi} Q \sim_{phhp} P_2 \boxplus_{pi} Q$. It is sufficient to prove the relation $R = \{(P_1 \boxplus_{pi} Q, P_2 \boxplus_{pi} Q)\} \cup \mathbf{Id}$ is a strongly probabilistic hhp-bisimulation, however, we omit the proof here;

(f) $P_1 \parallel Q \sim_{phhp} P_2 \parallel Q$. It is sufficient to prove the relation $R = \{(P_1 \parallel Q, P_2 \parallel Q)\} \cup \mathbf{Id}$ is a strongly probabilistic hhp-bisimulation, however, we omit the proof here;

(g) $P_1 \setminus L \sim_{phhp} P_2 \setminus L$. It is sufficient to prove the relation $R = \{(P_1 \setminus L, P_2 \setminus L)\} \cup \mathbf{Id}$ is a strongly probabilistic hhp-bisimulation, however, we omit the proof here;

(h) $P_1[f] \sim_{phhp} P_2[f]$. It is sufficient to prove the relation $R = \{(P_1[f], P_2[f])\} \cup \mathbf{Id}$ is a strongly probabilistic hhp-bisimulation, however, we omit the proof here. □

### 14.3.2 Recursion

**Definition 14.40** (Weakly guarded recursive expression). $X$ is weakly guarded in $E$ if each occurrence of $X$ is with some subexpression $\alpha.F$ or $(\alpha_1 \parallel \cdots \parallel \alpha_n).F$ of $E$.

**Lemma 14.41.** *If the variables $\widetilde{X}$ are weakly guarded in E, and $\langle E\{\widetilde{P}/\widetilde{X}\}, s\rangle \rightsquigarrow \xrightarrow{\{\alpha_1,\cdots,\alpha_n\}} \langle P', s'\rangle$, then $P'$ takes the form $E'\{\widetilde{P}/\widetilde{X}\}$ for some expression $E'$, and moreover, for any $\widetilde{Q}$, $\langle E\{\widetilde{Q}/\widetilde{X}\}, s\rangle \rightsquigarrow \xrightarrow{\{\alpha_1,\cdots,\alpha_n\}} \langle E'\{\widetilde{Q}/\widetilde{X}\}, s'\rangle$.*

*Proof.* It needs to induct on the depth of the inference of $\langle E\{\widetilde{P}/\widetilde{X}\}, s\rangle \rightsquigarrow \xrightarrow{\{\alpha_1,\cdots,\alpha_n\}} \langle P', s'\rangle$.

1. Case $E \equiv Y$, a variable. Then, $Y \notin \widetilde{X}$. Since $\widetilde{X}$ are weakly guarded, $\langle Y\{\widetilde{P}/\widetilde{X} \equiv Y\}, s\rangle \not\rightsquigarrow$, this case is impossible.
2. Case $E \equiv \beta.F$. Then, we must have $\alpha = \beta$, and $P' \equiv F\{\widetilde{P}/\widetilde{X}\}$, and $\langle E\{\widetilde{Q}/\widetilde{X}\}, s\rangle \equiv \langle \beta.F\{\widetilde{Q}/\widetilde{X}\}, s\rangle \rightsquigarrow \xrightarrow{\beta} \langle F\{\widetilde{Q}/\widetilde{X}\}, s'\rangle$, then, let $E'$ be $F$, as desired.
3. Case $E \equiv (\beta_1 \parallel \cdots \parallel \beta_n).F$. Then, we must have $\alpha_i = \beta_i$ for $1 \le i \le n$, and $P' \equiv F\{\widetilde{P}/\widetilde{X}\}$, and $\langle E\{\widetilde{Q}/\widetilde{X}\}, s\rangle \equiv \langle (\beta_1 \parallel \cdots \parallel \beta_n).F\{\widetilde{Q}/\widetilde{X}\}, s\rangle \rightsquigarrow \xrightarrow{\{\beta_1,\cdots,\beta_n\}} \langle F\{\widetilde{Q}/\widetilde{X}\}, s'\rangle$, then, let $E'$ be $F$, as desired.
4. Case $E \equiv E_1 + E_2$. Then, either $\langle E_1\{\widetilde{P}/\widetilde{X}\}, s\rangle \rightsquigarrow \xrightarrow{\{\alpha_1,\cdots,\alpha_n\}} \langle P', s'\rangle$ or $\langle E_2\{\widetilde{P}/\widetilde{X}\}, s\rangle \rightsquigarrow \xrightarrow{\{\alpha_1,\cdots,\alpha_n\}} \langle P', s'\rangle$, then, we can apply this lemma in either case, as desired.
5. Case $E \equiv E_1 \parallel E_2$. There are four possibilities:
   (a) We may have $\langle E_1\{\widetilde{P}/\widetilde{X}\}, s\rangle \rightsquigarrow \xrightarrow{\alpha} \langle P_1', s'\rangle$ and $\langle E_2\{\widetilde{P}/\widetilde{X}\}, s\rangle \not\rightsquigarrow$ with $P' \equiv P_1' \parallel (E_2\{\widetilde{P}/\widetilde{X}\})$, then by applying this lemma, $P_1'$ is of the form $E_1'\{\widetilde{P}/\widetilde{X}\}$, and for any $Q$, $\langle E_1\{\widetilde{Q}/\widetilde{X}\}, s\rangle \rightsquigarrow \xrightarrow{\alpha} \langle E_1'\{\widetilde{Q}/\widetilde{X}\}, s'\rangle$. Hence, $P'$ is of the form $E_1' \parallel E_2\{\widetilde{P}/\widetilde{X}\}$, and for any $Q$, $\langle E\{\widetilde{Q}/\widetilde{X}\} \equiv E_1\{\widetilde{Q}/\widetilde{X}\} \parallel E_2\{\widetilde{Q}/\widetilde{X}\}, s\rangle \rightsquigarrow \xrightarrow{\alpha} \langle (E_1' \parallel E_2)\{\widetilde{Q}/\widetilde{X}\}, s'\rangle$, then let $E'$ be $E_1' \parallel E_2$, as desired.
   (b) We may have $\langle E_2\{\widetilde{P}/\widetilde{X}\}, s\rangle \rightsquigarrow \xrightarrow{\alpha} \langle P_2', s'\rangle$ and $\langle E_1\{\widetilde{P}/\widetilde{X}\}, s\rangle \not\rightsquigarrow$ with $P' \equiv P_2' \parallel (E_1\{\widetilde{P}/\widetilde{X}\})$, this case can be prove similarly to the above subcase, as desired.
   (c) We may have $\langle E_1\{\widetilde{P}/\widetilde{X}\}, s\rangle \rightsquigarrow \xrightarrow{\alpha} \langle P_1', s'\rangle$ and $\langle E_2\{\widetilde{P}/\widetilde{X}\}, s\rangle \rightsquigarrow \xrightarrow{\beta} \langle P_2', s''\rangle$ with $\alpha \neq \overline{\beta}$ and $P' \equiv P_1' \parallel P_2'$, then by applying this lemma, $P_1'$ is of the form $E_1'\{\widetilde{P}/\widetilde{X}\}$, and for any $Q$, $\langle E_1\{\widetilde{Q}/\widetilde{X}\}, s\rangle \rightsquigarrow \xrightarrow{\alpha} \langle E_1'\{\widetilde{Q}/\widetilde{X}\}, s'\rangle$; $P_2'$ is of the form $E_2'\{\widetilde{P}/\widetilde{X}\}$, and for any $Q$, $\langle E_2\{\widetilde{Q}/\widetilde{X}\}, s\rangle \rightsquigarrow \xrightarrow{\alpha} \langle E_2'\{\widetilde{Q}/\widetilde{X}\}, s''\rangle$. Hence, $P'$ is of the form $E_1' \parallel E_2'\{\widetilde{P}/\widetilde{X}\}$, and for any $Q$, $\langle E\{\widetilde{Q}/\widetilde{X}\} \equiv E_1\{\widetilde{Q}/\widetilde{X}\} \parallel E_2\{\widetilde{Q}/\widetilde{X}\}, s\rangle \rightsquigarrow \xrightarrow{\{\alpha,\beta\}} \langle (E_1' \parallel E_2')\{\widetilde{Q}/\widetilde{X}\}, s' \cup s''\rangle$, then let $E'$ be $E_1' \parallel E_2'$, as desired.
   (d) We may have $\langle E_1\{\widetilde{P}/\widetilde{X}\}, s\rangle \rightsquigarrow \xrightarrow{l} \langle P_1', s'\rangle$ and $\langle E_2\{\widetilde{P}/\widetilde{X}\}, s\rangle \rightsquigarrow \xrightarrow{\overline{l}} \langle P_2', s''\rangle$ with $P' \equiv P_1' \parallel P_2'$, then by applying this lemma, $P_1'$ is of the form $E_1'\{\widetilde{P}/\widetilde{X}\}$, and for any $Q$, $\langle E_1\{\widetilde{Q}/\widetilde{X}\}, s\rangle \rightsquigarrow \xrightarrow{l} \langle E_1'\{\widetilde{Q}/\widetilde{X}\}, s'\rangle$; $P_2'$ is of the form $E_2'\{\widetilde{P}/\widetilde{X}\}$, and for any $Q$, $\langle E_2\{\widetilde{Q}/\widetilde{X}\}, s\rangle \rightsquigarrow \xrightarrow{\overline{l}} \langle E_2'\{\widetilde{Q}/\widetilde{X}\}, s''\rangle$. Hence, $P'$ is of the form $E_1' \parallel E_2'\{\widetilde{P}/\widetilde{X}\}$, and for any $Q$, $\langle E\{\widetilde{Q}/\widetilde{X}\} \equiv E_1\{\widetilde{Q}/\widetilde{X}\} \parallel E_2\{\widetilde{Q}/\widetilde{X}\}, s\rangle \rightsquigarrow \xrightarrow{\tau} \langle (E_1' \parallel E_2')\{\widetilde{Q}/\widetilde{X}\}, s' \cup s''\rangle$, then let $E'$ be $E_1' \parallel E_2'$, as desired.
6. Case $E \equiv F[R]$ and $E \equiv F \setminus L$. These cases can be proven similarly to the above case.
7. Case $E \equiv C$, an agent constant defined by $C \stackrel{\text{def}}{=} R$. Then, there is no $X \in \widetilde{X}$ occurring in $E$, so $\langle C, s\rangle \rightsquigarrow \xrightarrow{\{\alpha_1,\cdots,\alpha_n\}} \langle P', s'\rangle$, let $E'$ be $P'$, as desired. □

**Theorem 14.42** (Unique solution of equations for strongly probabilistic pomset bisimulation). *Let the recursive expressions $E_i (i \in I)$ contain at most the variables $X_i (i \in I)$, and let each $X_j (j \in I)$ be weakly guarded in each $E_i$. Then,*
*If $\widetilde{P} \sim_{pp} \widetilde{E}\{\widetilde{P}/\widetilde{X}\}$ and $\widetilde{Q} \sim_{pp} \widetilde{E}\{\widetilde{Q}/\widetilde{X}\}$, then $\widetilde{P} \sim_{pp} \widetilde{Q}$.*

*Proof.* It is sufficient to induct on the depth of the inference of $\langle E\{\widetilde{P}/\widetilde{X}\}, s\rangle \rightsquigarrow \xrightarrow{\{\alpha_1,\cdots,\alpha_n\}} \langle P', s'\rangle$.

1. Case $E \equiv X_i$. Then, we have $\langle E\{\widetilde{P}/\widetilde{X}\}, s\rangle \equiv \langle P_i, s\rangle \xrightarrow{\{\alpha_1,\cdots,\alpha_n\}} \langle P', s'\rangle$, since $P_i \sim_{pp} E_i\{\widetilde{P}/\widetilde{X}\}$, we have $\langle E_i\{\widetilde{P}/\widetilde{X}\}, s\rangle \xrightarrow{\{\alpha_1,\cdots,\alpha_n\}} \langle P'', s'\rangle \sim_{pp} \langle P', s'\rangle$. Since $\widetilde{X}$ are weakly guarded in $E_i$, by Lemma 14.41, $P'' \equiv E'\{\widetilde{P}/\widetilde{X}\}$ and $\langle E_i\{\widetilde{P}/\widetilde{X}\}, s\rangle \xrightarrow{\{\alpha_1,\cdots,\alpha_n\}} \langle E'\{\widetilde{P}/\widetilde{X}\}, s'\rangle$. Since $E\{\widetilde{Q}/\widetilde{X}\} \equiv X_i\{\widetilde{Q}/\widetilde{X}\} \equiv Q_i \sim_{pp} E_i\{\widetilde{Q}/\widetilde{X}\}$, $\langle E\{\widetilde{Q}/\widetilde{X}\}, s\rangle \xrightarrow{\{\alpha_1,\cdots,\alpha_n\}} \langle Q', s'\rangle \sim_{pp} \langle E'\{\widetilde{Q}/\widetilde{X}\}, s'\rangle$. Hence, $P' \sim_{pp} Q'$, as desired.
2. Case $E \equiv \alpha.F$. This case can be proven similarly.
3. Case $E \equiv (\alpha_1 \parallel \cdots \parallel \alpha_n).F$. This case can be proven similarly.
4. Case $E \equiv E_1 + E_2$. We have $\langle E_i\{\widetilde{P}/\widetilde{X}\}, s\rangle \xrightarrow{\{\alpha_1,\cdots,\alpha_n\}} \langle P', s'\rangle$, $\langle E_i\{\widetilde{Q}/\widetilde{X}\}, s\rangle \xrightarrow{\{\alpha_1,\cdots,\alpha_n\}} \langle Q', s'\rangle$, then, $P' \sim_{pp} Q'$, as desired.
5. Case $E \equiv E_1 \parallel E_2$, $E \equiv F[R]$ and $E \equiv F \setminus L$, $E \equiv C$. These cases can be proven similarly to the above case. □

**Theorem 14.43** (Unique solution of equations for strongly probabilistic step bisimulation). *Let the recursive expressions $E_i(i \in I)$ contain at most the variables $X_i(i \in I)$, and let each $X_j(j \in I)$ be weakly guarded in each $E_i$. Then, If $\widetilde{P} \sim_{ps} \widetilde{E}\{\widetilde{P}/\widetilde{X}\}$ and $\widetilde{Q} \sim_{ps} \widetilde{E}\{\widetilde{Q}/\widetilde{X}\}$, then $\widetilde{P} \sim_{ps} \widetilde{Q}$.*

*Proof.* It is sufficient to induct on the depth of the inference of $\langle E\{\widetilde{P}/\widetilde{X}\}, s\rangle \xrightarrow{\{\alpha_1,\cdots,\alpha_n\}} \langle P', s'\rangle$.

1. Case $E \equiv X_i$. Then, we have $\langle E\{\widetilde{P}/\widetilde{X}\}, s\rangle \equiv \langle P_i, s\rangle \xrightarrow{\{\alpha_1,\cdots,\alpha_n\}} \langle P', s'\rangle$, since $P_i \sim_{ps} E_i\{\widetilde{P}/\widetilde{X}\}$, we have $\langle E_i\{\widetilde{P}/\widetilde{X}\}, s\rangle \xrightarrow{\{\alpha_1,\cdots,\alpha_n\}} \langle P'', s'\rangle \sim_{ps} \langle P', s'\rangle$. Since $\widetilde{X}$ are weakly guarded in $E_i$, by Lemma 14.41, $P'' \equiv E'\{\widetilde{P}/\widetilde{X}\}$ and $\langle E_i\{\widetilde{P}/\widetilde{X}\}, s\rangle \xrightarrow{\{\alpha_1,\cdots,\alpha_n\}} \langle E'\{\widetilde{P}/\widetilde{X}\}, s'\rangle$. Since $E\{\widetilde{Q}/\widetilde{X}\} \equiv X_i\{\widetilde{Q}/\widetilde{X}\} \equiv Q_i \sim_{ps} E_i\{\widetilde{Q}/\widetilde{X}\}$, $\langle E\{\widetilde{Q}/\widetilde{X}\}, s\rangle \xrightarrow{\{\alpha_1,\cdots,\alpha_n\}} \langle Q', s'\rangle \sim_{ps} \langle E'\{\widetilde{Q}/\widetilde{X}\}, s'\rangle$. Hence, $P' \sim_{ps} Q'$, as desired.
2. Case $E \equiv \alpha.F$. This case can be proven similarly.
3. Case $E \equiv (\alpha_1 \parallel \cdots \parallel \alpha_n).F$. This case can be proven similarly.
4. Case $E \equiv E_1 + E_2$. We have $\langle E_i\{\widetilde{P}/\widetilde{X}\}, s\rangle \xrightarrow{\{\alpha_1,\cdots,\alpha_n\}} \langle P', s'\rangle$, $\langle E_i\{\widetilde{Q}/\widetilde{X}\}, s\rangle \xrightarrow{\{\alpha_1,\cdots,\alpha_n\}} \langle Q', s'\rangle$, then, $P' \sim_{ps} Q'$, as desired.
5. Case $E \equiv E_1 \parallel E_2$, $E \equiv F[R]$ and $E \equiv F \setminus L$, $E \equiv C$. These cases can be proven similarly to the above case. □

**Theorem 14.44** (Unique solution of equations for strongly probabilistic hp-bisimulation). *Let the recursive expressions $E_i(i \in I)$ contain at most the variables $X_i(i \in I)$, and let each $X_j(j \in I)$ be weakly guarded in each $E_i$. Then, If $\widetilde{P} \sim_{php} \widetilde{E}\{\widetilde{P}/\widetilde{X}\}$ and $\widetilde{Q} \sim_{php} \widetilde{E}\{\widetilde{Q}/\widetilde{X}\}$, then $\widetilde{P} \sim_{php} \widetilde{Q}$.*

*Proof.* It is sufficient to induct on the depth of the inference of $\langle E\{\widetilde{P}/\widetilde{X}\}, s\rangle \xrightarrow{\{\alpha_1,\cdots,\alpha_n\}} \langle P', s'\rangle$.

1. Case $E \equiv X_i$. Then, we have $\langle E\{\widetilde{P}/\widetilde{X}\}, s\rangle \equiv \langle P_i, s\rangle \xrightarrow{\{\alpha_1,\cdots,\alpha_n\}} \langle P', s'\rangle$, since $P_i \sim_{php} E_i\{\widetilde{P}/\widetilde{X}\}$, we have $\langle E_i\{\widetilde{P}/\widetilde{X}\}, s\rangle \xrightarrow{\{\alpha_1,\cdots,\alpha_n\}} \langle P'', s'\rangle \sim_{php} \langle P', s'\rangle$. Since $\widetilde{X}$ are weakly guarded in $E_i$, by Lemma 14.41, $P'' \equiv E'\{\widetilde{P}/\widetilde{X}\}$ and $\langle E_i\{\widetilde{P}/\widetilde{X}\}, s\rangle \xrightarrow{\{\alpha_1,\cdots,\alpha_n\}} \langle E'\{\widetilde{P}/\widetilde{X}\}, s'\rangle$. Since $E\{\widetilde{Q}/\widetilde{X}\} \equiv X_i\{\widetilde{Q}/\widetilde{X}\} \equiv Q_i \sim_{php} E_i\{\widetilde{Q}/\widetilde{X}\}$, $\langle E\{\widetilde{Q}/\widetilde{X}\}, s\rangle \xrightarrow{\{\alpha_1,\cdots,\alpha_n\}} \langle Q', s'\rangle \sim_{php} \langle E'\{\widetilde{Q}/\widetilde{X}\}, s'\rangle$. Hence, $P' \sim_{php} Q'$, as desired.
2. Case $E \equiv \alpha.F$. This case can be proven similarly.
3. Case $E \equiv (\alpha_1 \parallel \cdots \parallel \alpha_n).F$. This case can be proven similarly.
4. Case $E \equiv E_1 + E_2$. We have $\langle E_i\{\widetilde{P}/\widetilde{X}\}, s\rangle \xrightarrow{\{\alpha_1,\cdots,\alpha_n\}} \langle P', s'\rangle$, $\langle E_i\{\widetilde{Q}/\widetilde{X}\}, s\rangle \xrightarrow{\{\alpha_1,\cdots,\alpha_n\}} \langle Q', s'\rangle$, then, $P' \sim_{php} Q'$, as desired.
5. Case $E \equiv E_1 \parallel E_2$, $E \equiv F[R]$ and $E \equiv F \setminus L$, $E \equiv C$. These cases can be proven similarly to the above case. □

**Theorem 14.45** (Unique solution of equations for strongly probabilistic hhp-bisimulation). *Let the recursive expressions $E_i(i \in I)$ contain at most the variables $X_i(i \in I)$, and let each $X_j(j \in I)$ be weakly guarded in each $E_i$. Then, If $\widetilde{P} \sim_{phhp} \widetilde{E}\{\widetilde{P}/\widetilde{X}\}$ and $\widetilde{Q} \sim_{phhp} \widetilde{E}\{\widetilde{Q}/\widetilde{X}\}$, then $\widetilde{P} \sim_{phhp} \widetilde{Q}$.*

*Proof.* It is sufficient to induct on the depth of the inference of $\langle E\{\widetilde{P}/\widetilde{X}\}, s\rangle \xrightarrow{\{\alpha_1,\cdots,\alpha_n\}} \langle P', s'\rangle$.

1. Case $E \equiv X_i$. Then, we have $\langle E\{\widetilde{P}/\widetilde{X}\}, s\rangle \equiv \langle P_i, s\rangle \xrightarrow{\{\alpha_1,\cdots,\alpha_n\}} \langle P', s'\rangle$, since $P_i \sim_{phhp} E_i\{\widetilde{P}/\widetilde{X}\}$, we have $\langle E_i\{\widetilde{P}/\widetilde{X}\}, s\rangle \xrightarrow{\{\alpha_1,\cdots,\alpha_n\}} \langle P'', s'\rangle \sim_{phhp} \langle P', s'\rangle$. Since $\widetilde{X}$ are weakly guarded in $E_i$, by Lemma 14.41, $P'' \equiv$

$E'\{\widetilde{P}/\widetilde{X}\}$ and $\langle E_i\{\widetilde{P}/\widetilde{X}\}, s\rangle \rightsquigarrow \xrightarrow{\{\alpha_1,\cdots,\alpha_n\}} \langle E'\{\widetilde{P}/\widetilde{X}\}, s'\rangle$. Since $E\{\widetilde{Q}/\widetilde{X}\} \equiv X_i\{\widetilde{Q}/\widetilde{X}\} \equiv Q_i \sim_{phhp} E_i\{\widetilde{Q}/\widetilde{X}\}$, $\langle E\{\widetilde{Q}/\widetilde{X}\}, s\rangle \rightsquigarrow \xrightarrow{\{\alpha_1,\cdots,\alpha_n\}} \langle Q', s'\rangle \sim_{phhp} \langle E'\{\widetilde{Q}/\widetilde{X}\}, s'\rangle$. Hence, $P' \sim_{phhp} Q'$, as desired.

2. Case $E \equiv \alpha.F$. This case can be proven similarly.
3. Case $E \equiv (\alpha_1 \parallel \cdots \parallel \alpha_n).F$. This case can be proven similarly.
4. Case $E \equiv E_1 + E_2$. We have $\langle E_i\{\widetilde{P}/\widetilde{X}\}, s\rangle \rightsquigarrow \xrightarrow{\{\alpha_1,\cdots,\alpha_n\}} \langle P', s'\rangle$, $\langle E_i\{\widetilde{Q}/\widetilde{X}\}, s\rangle \rightsquigarrow \xrightarrow{\{\alpha_1,\cdots,\alpha_n\}} \langle Q', s'\rangle$, then, $P' \sim_{phhp} Q'$, as desired.
5. Case $E \equiv E_1 \parallel E_2$, $E \equiv F[R]$ and $E \equiv F \setminus L$, $E \equiv C$. These cases can be proven similarly to the above case. □

## 14.4 Weak bisimulations

The weak probabilistic transition rules of CTC with probabilism and guards are the same as the strong one in Table 14.1. Also, the weak action transition rules of CTC with probabilism and guards are listed in Table 14.3.

### 14.4.1 Laws and congruence

Remembering that $\tau$ can neither be restricted nor relabeled, we know that the monoid laws, the monoid laws 2, the static laws, and the expansion law in Section 14.3 still hold with respect to the corresponding weakly probabilistic truly concurrent bisimulations. Also, we can enjoy the full congruence of Prefix, Summation, Composition, Restriction, Relabeling, and Constants with respect to corresponding weakly probabilistic truly concurrent bisimulations. We will not retype these laws here, instead we just give the $\tau$-specific laws.

**Proposition 14.46** ($\tau$ laws for weakly probabilistic pomset bisimulation). *The $\tau$ laws for weakly probabilistic pomset bisimulation are as follows:*

1. $P \approx_{pp} \tau.P$;
2. $\alpha.\tau.P \approx_{pp} \alpha.P$;
3. $(\alpha_1 \parallel \cdots \parallel \alpha_n).\tau.P \approx_{pp} (\alpha_1 \parallel \cdots \parallel \alpha_n).P$;
4. $P + \tau.P \approx_{pp} \tau.P$;
5. $P \cdot ((Q + \tau \cdot (Q + R)) \boxplus_\pi S) \approx_{pp} P \cdot ((Q + R) \boxplus_\pi S)$;
6. $P \approx_{pp} \tau \parallel P$.

*Proof.* 1. $P \approx_{pp} \tau.P$. It is sufficient to prove the relation $R = \{(P, \tau.P)\} \cup \mathbf{Id}$ is a weakly probabilistic pomset bisimulation, however, we omit the proof here;
2. $\alpha.\tau.P \approx_{pp} \alpha.P$. It is sufficient to prove the relation $R = \{(\alpha.\tau.P, \alpha.P)\} \cup \mathbf{Id}$ is a weakly probabilistic pomset bisimulation, however, we omit the proof here;
3. $(\alpha_1 \parallel \cdots \parallel \alpha_n).\tau.P \approx_{pp} (\alpha_1 \parallel \cdots \parallel \alpha_n).P$. It is sufficient to prove the relation $R = \{((\alpha_1 \parallel \cdots \parallel \alpha_n).\tau.P, (\alpha_1 \parallel \cdots \parallel \alpha_n).P)\} \cup \mathbf{Id}$ is a weakly probabilistic pomset bisimulation, however, we omit the proof here;
4. $P + \tau.P \approx_{pp} \tau.P$. It is sufficient to prove the relation $R = \{(P + \tau.P, \tau.P)\} \cup \mathbf{Id}$ is a weakly probabilistic pomset bisimulation, however, we omit the proof here;
5. $P \cdot ((Q + \tau \cdot (Q + R)) \boxplus_\pi S) \approx_{pp} P \cdot ((Q + R) \boxplus_\pi S)$. It is sufficient to prove the relation $R = \{(P \cdot ((Q + \tau \cdot (Q + R)) \boxplus_\pi S), P \cdot ((Q + R) \boxplus_\pi S))\} \cup \mathbf{Id}$ is a weakly probabilistic pomset bisimulation, however, we omit the proof here;
6. $P \approx_{pp} \tau \parallel P$. It is sufficient to prove the relation $R = \{(P, \tau \parallel P)\} \cup \mathbf{Id}$ is a weakly probabilistic pomset bisimulation, however, we omit the proof here. □

**Proposition 14.47** ($\tau$ laws for weakly probabilistic step bisimulation). *The $\tau$ laws for weakly probabilistic step bisimulation are as follows:*

1. $P \approx_{ps} \tau.P$;
2. $\alpha.\tau.P \approx_{ps} \alpha.P$;
3. $(\alpha_1 \parallel \cdots \parallel \alpha_n).\tau.P \approx_{ps} (\alpha_1 \parallel \cdots \parallel \alpha_n).P$;
4. $P + \tau.P \approx_{ps} \tau.P$;
5. $P \cdot ((Q + \tau \cdot (Q + R)) \boxplus_\pi S) \approx_{ps} P \cdot ((Q + R) \boxplus_\pi S)$;
6. $P \approx_{ps} \tau \parallel P$.

*Proof.* 1. $P \approx_{ps} \tau.P$. It is sufficient to prove the relation $R = \{(P, \tau.P)\} \cup \mathbf{Id}$ is a weakly probabilistic step bisimulation, however, we omit the proof here;

**TABLE 14.3 Weak action transition rules of CTC with probabilism and guards.**

| Rule | Transition rule | Side condition |
|---|---|---|
| Act$_1$ | $\dfrac{}{\langle \breve{\alpha}.P, s\rangle \xRightarrow{\alpha} \langle P, s'\rangle}$ | |
| Act$_2$ | $\dfrac{}{\langle \epsilon, s\rangle \Rightarrow \langle \surd, s\rangle}$ | |
| Gur | $\dfrac{}{\langle \phi, s\rangle \Rightarrow \langle \surd, s\rangle}$ | if $test(\phi, s)$ |
| Sum$_1$ | $\dfrac{\langle P, s\rangle \xRightarrow{\alpha} \langle P', s'\rangle}{\langle P+Q, s\rangle \xRightarrow{\alpha} \langle P', s'\rangle}$ | |
| Com$_1$ | $\dfrac{\langle P, s\rangle \xRightarrow{\alpha} \langle P', s'\rangle \quad Q \nrightarrow}{\langle P \parallel Q, s\rangle \xRightarrow{\alpha} \langle P' \parallel Q, s'\rangle}$ | |
| Com$_2$ | $\dfrac{\langle Q, s\rangle \xRightarrow{\alpha} \langle Q', s'\rangle \quad P \nrightarrow}{\langle P \parallel Q, s\rangle \xRightarrow{\alpha} \langle P \parallel Q', s'\rangle}$ | |
| Com$_3$ | $\dfrac{\langle P, s\rangle \xRightarrow{\alpha} \langle P', s'\rangle \quad \langle Q, s\rangle \xRightarrow{\beta} \langle Q', s''\rangle}{\langle P \parallel Q, s\rangle \xRightarrow{\{\alpha,\beta\}} \langle P' \parallel Q', s' \cup s''\rangle}$ | $(\beta \neq \overline{\alpha})$ |
| Com$_4$ | $\dfrac{\langle P, s\rangle \xRightarrow{l} \langle P', s'\rangle \quad \langle Q, s\rangle \xRightarrow{\overline{l}} \langle Q', s''\rangle}{\langle P \parallel Q, s\rangle \xRightarrow{\tau} \langle P' \parallel Q', s' \cup s''\rangle}$ | |
| Act$_3$ | $\dfrac{}{\langle(\breve{\alpha}_1 \parallel \cdots \parallel \breve{\alpha}_n).P, s\rangle \xRightarrow{\{\alpha_1,\cdots,\alpha_n\}} \langle P, s'\rangle}$ | $(\alpha_i \neq \overline{\alpha_j} \quad i, j \in \{1, \cdots, n\})$ |
| Sum$_2$ | $\dfrac{\langle P, s\rangle \xRightarrow{\{\alpha_1,\cdots,\alpha_n\}} \langle P', s'\rangle}{\langle P+Q, s\rangle \xRightarrow{\{\alpha_1,\cdots,\alpha_n\}} \langle P', s'\rangle}$ | |
| Res$_1$ | $\dfrac{\langle P, s\rangle \xRightarrow{\alpha} \langle P', s'\rangle}{\langle P \setminus L, s\rangle \xRightarrow{\alpha} \langle P' \setminus L, s'\rangle}$ | $(\alpha, \overline{\alpha} \notin L)$ |
| Res$_2$ | $\dfrac{\langle P, s\rangle \xRightarrow{\{\alpha_1,\cdots,\alpha_n\}} \langle P', s'\rangle}{\langle P \setminus L, s\rangle \xRightarrow{\{\alpha_1,\cdots,\alpha_n\}} \langle P' \setminus L, s'\rangle}$ | $(\alpha_1, \overline{\alpha_1}, \cdots, \alpha_n, \overline{\alpha_n} \notin L)$ |
| Rel$_1$ | $\dfrac{\langle P, s\rangle \xRightarrow{\alpha} \langle P', s'\rangle}{\langle P[f], s\rangle \xRightarrow{f(\alpha)} \langle P'[f], s'\rangle}$ | |
| Rel$_2$ | $\dfrac{\langle P, s\rangle \xRightarrow{\{\alpha_1,\cdots,\alpha_n\}} \langle P', s'\rangle}{\langle P[f], s\rangle \xRightarrow{\{f(\alpha_1),\cdots,f(\alpha_n)\}} \langle P'[f], s'\rangle}$ | |
| Con$_1$ | $\dfrac{\langle P, s\rangle \xRightarrow{\alpha} \langle P', s'\rangle}{\langle A, s\rangle \xRightarrow{\alpha} \langle P', s'\rangle}$ | $(A \overset{\text{def}}{=} P)$ |
| Con$_2$ | $\dfrac{\langle P, s\rangle \xRightarrow{\{\alpha_1,\cdots,\alpha_n\}} \langle P', s'\rangle}{\langle A, s\rangle \xRightarrow{\{\alpha_1,\cdots,\alpha_n\}} \langle P', s'\rangle}$ | $(A \overset{\text{def}}{=} P)$ |

2. $\alpha.\tau.P \approx_{ps} \alpha.P$. It is sufficient to prove the relation $R = \{(\alpha.\tau.P, \alpha.P)\} \cup \mathbf{Id}$ is a weakly probabilistic step bisimulation, however, we omit the proof here;
3. $(\alpha_1 \parallel \cdots \parallel \alpha_n).\tau.P \approx_{ps} (\alpha_1 \parallel \cdots \parallel \alpha_n).P$. It is sufficient to prove the relation $R = \{((\alpha_1 \parallel \cdots \parallel \alpha_n).\tau.P, (\alpha_1 \parallel \cdots \parallel \alpha_n).P)\} \cup \mathbf{Id}$ is a weakly probabilistic step bisimulation, however, we omit the proof here;
4. $P + \tau.P \approx_{ps} \tau.P$. It is sufficient to prove the relation $R = \{(P + \tau.P, \tau.P)\} \cup \mathbf{Id}$ is a weakly probabilistic step bisimulation, however, we omit the proof here;
5. $P \cdot ((Q + \tau \cdot (Q + R)) \boxplus_\pi S) \approx_{ps} P \cdot ((Q + R) \boxplus_\pi S)$. It is sufficient to prove the relation $R = \{(P \cdot ((Q + \tau \cdot (Q + R)) \boxplus_\pi S), P \cdot ((Q + R) \boxplus_\pi S))\} \cup \mathbf{Id}$ is a weakly probabilistic step bisimulation, however, we omit the proof here;
6. $P \approx_{ps} \tau \parallel P$. It is sufficient to prove the relation $R = \{(P, \tau \parallel P)\} \cup \mathbf{Id}$ is a weakly probabilistic step bisimulation, however, we omit the proof here. □

**Proposition 14.48** ($\tau$ laws for weakly probabilistic hp-bisimulation). *The $\tau$ laws for weakly probabilistic hp-bisimulation are as follows:*

1. $P \approx_{php} \tau.P$;
2. $\alpha.\tau.P \approx_{php} \alpha.P$;
3. $(\alpha_1 \parallel \cdots \parallel \alpha_n).\tau.P \approx_{php} (\alpha_1 \parallel \cdots \parallel \alpha_n).P$;
4. $P + \tau.P \approx_{php} \tau.P$;
5. $P \cdot ((Q + \tau \cdot (Q + R)) \boxplus_\pi S) \approx_{php} P \cdot ((Q + R) \boxplus_\pi S)$;
6. $P \approx_{php} \tau \parallel P$.

*Proof.* **1.** $P \approx_{php} \tau.P$. It is sufficient to prove the relation $R = \{(P, \tau.P)\} \cup \mathbf{Id}$ is a weakly probabilistic hp-bisimulation, however, we omit the proof here;

2. $\alpha.\tau.P \approx_{php} \alpha.P$. It is sufficient to prove the relation $R = \{(\alpha.\tau.P, \alpha.P)\} \cup \mathbf{Id}$ is a weakly probabilistic hp-bisimulation, however, we omit the proof here;
3. $(\alpha_1 \parallel \cdots \parallel \alpha_n).\tau.P \approx_{php} (\alpha_1 \parallel \cdots \parallel \alpha_n).P$. It is sufficient to prove the relation $R = \{((\alpha_1 \parallel \cdots \parallel \alpha_n).\tau.P, (\alpha_1 \parallel \cdots \parallel \alpha_n).P)\} \cup \mathbf{Id}$ is a weakly probabilistic hp-bisimulation, however, we omit the proof here;
4. $P + \tau.P \approx_{php} \tau.P$. It is sufficient to prove the relation $R = \{(P + \tau.P, \tau.P)\} \cup \mathbf{Id}$ is a weakly probabilistic hp-bisimulation, however, we omit the proof here;
5. $P \cdot ((Q + \tau \cdot (Q + R)) \boxplus_\pi S) \approx_{php} P \cdot ((Q + R) \boxplus_\pi S)$. It is sufficient to prove the relation $R = \{(P \cdot ((Q + \tau \cdot (Q + R)) \boxplus_\pi S), P \cdot ((Q + R) \boxplus_\pi S))\} \cup \mathbf{Id}$ is a weakly probabilistic hp-bisimulation, however, we omit the proof here;
6. $P \approx_{php} \tau \parallel P$. It is sufficient to prove the relation $R = \{(P, \tau \parallel P)\} \cup \mathbf{Id}$ is a weakly probabilistic hp-bisimulation, however, we omit the proof here. □

**Proposition 14.49** ($\tau$ laws for weakly probabilistic hhp-bisimulation). *The $\tau$ laws for weakly probabilistic hhp-bisimulation are as follows:*

1. $P \approx_{phhp} \tau.P$;
2. $\alpha.\tau.P \approx_{phhp} \alpha.P$;
3. $(\alpha_1 \parallel \cdots \parallel \alpha_n).\tau.P \approx_{phhp} (\alpha_1 \parallel \cdots \parallel \alpha_n).P$;
4. $P + \tau.P \approx_{phhp} \tau.P$;
5. $P \cdot ((Q + \tau \cdot (Q + R)) \boxplus_\pi S) \approx_{phhp} P \cdot ((Q + R) \boxplus_\pi S)$;
6. $P \approx_{phhp} \tau \parallel P$.

*Proof.* **1.** $P \approx_{phhp} \tau.P$. It is sufficient to prove the relation $R = \{(P, \tau.P)\} \cup \mathbf{Id}$ is a weakly probabilistic hhp-bisimulation, however, we omit the proof here;

2. $\alpha.\tau.P \approx_{phhp} \alpha.P$. It is sufficient to prove the relation $R = \{(\alpha.\tau.P, \alpha.P)\} \cup \mathbf{Id}$ is a weakly probabilistic hhp-bisimulation, however, we omit the proof here;
3. $(\alpha_1 \parallel \cdots \parallel \alpha_n).\tau.P \approx_{phhp} (\alpha_1 \parallel \cdots \parallel \alpha_n).P$. It is sufficient to prove the relation $R = \{((\alpha_1 \parallel \cdots \parallel \alpha_n).\tau.P, (\alpha_1 \parallel \cdots \parallel \alpha_n).P)\} \cup \mathbf{Id}$ is a weakly probabilistic hhp-bisimulation, however, we omit the proof here;
4. $P + \tau.P \approx_{phhp} \tau.P$. It is sufficient to prove the relation $R = \{(P + \tau.P, \tau.P)\} \cup \mathbf{Id}$ is a weakly probabilistic hhp-bisimulation, however, we omit the proof here;
5. $P \cdot ((Q + \tau \cdot (Q + R)) \boxplus_\pi S) \approx_{phhp} P \cdot ((Q + R) \boxplus_\pi S)$. It is sufficient to prove the relation $R = \{(P \cdot ((Q + \tau \cdot (Q + R)) \boxplus_\pi S), P \cdot ((Q + R) \boxplus_\pi S))\} \cup \mathbf{Id}$ is a weakly probabilistic hhp-bisimulation, however, we omit the proof here;
6. $P \approx_{phhp} \tau \parallel P$. It is sufficient to prove the relation $R = \{(P, \tau \parallel P)\} \cup \mathbf{Id}$ is a weakly probabilistic hhp-bisimulation, however, we omit the proof here. □

### 14.4.2 Recursion

**Definition 14.50** (Sequential). $X$ is sequential in $E$ if every subexpression of $E$ that contains $X$, apart from $X$ itself, is of the form $\alpha.F$, or $(\alpha_1 \parallel \cdots \parallel \alpha_n).F$, or $\sum \widetilde{F}$.

**Definition 14.51** (Guarded recursive expression). $X$ is guarded in $E$ if each occurrence of $X$ is with some subexpression $l.F$ or $(l_1 \parallel \cdots \parallel l_n).F$ of $E$.

**Lemma 14.52.** *Let $G$ be guarded and sequential, $Vars(G) \subseteq \widetilde{X}$, and let $\langle G\{\widetilde{P}/\widetilde{X}\}, s\rangle \rightsquigarrow \xrightarrow{\{\alpha_1, \cdots, \alpha_n\}} \langle P', s'\rangle$. Then, there is an expression $H$ such that $\langle G, s\rangle \rightsquigarrow \xrightarrow{\{\alpha_1, \cdots, \alpha_n\}} \langle H, s'\rangle$, $P' \equiv H\{\widetilde{P}/\widetilde{X}\}$, and for any $\widetilde{Q}$, $\langle G\{\widetilde{Q}/\widetilde{X}\}, s\rangle \rightsquigarrow \xrightarrow{\{\alpha_1, \cdots, \alpha_n\}} \langle H\{\widetilde{Q}/\widetilde{X}\}, s'\rangle$. Moreover, $H$ is sequential, $Vars(H) \subseteq \widetilde{X}$, and if $\alpha_1 = \cdots = \alpha_n = \tau$, then $H$ is also guarded.*

*Proof.* We need to induct on the structure of $G$.

If $G$ is a Constant, a Composition, a Restriction or a Relabeling then it contains no variables, since $G$ is sequential and guarded, then $\langle G, s\rangle \rightsquigarrow\xrightarrow{\{\alpha_1,\cdots,\alpha_n\}} \langle P', s'\rangle$, then let $H \equiv P'$, as desired.

$G$ cannot be a variable, since it is guarded.

If $G \equiv G_1 + G_2$. Then, either $\langle G_1\{\widetilde{P}/\widetilde{X}\}, s\rangle \rightsquigarrow\xrightarrow{\{\alpha_1,\cdots,\alpha_n\}} \langle P', s'\rangle$ or $\langle G_2\{\widetilde{P}/\widetilde{X}\}, s\rangle \rightsquigarrow\xrightarrow{\{\alpha_1,\cdots,\alpha_n\}} \langle P', s'\rangle$, then, we can apply this lemma in either case, as desired.

If $G \equiv \beta.H$. Then, we must have $\alpha = \beta$, and $P' \equiv H\{\widetilde{P}/\widetilde{X}\}$, and $\langle G\{\widetilde{Q}/\widetilde{X}\}, s\rangle \equiv \langle \beta.H\{\widetilde{Q}/\widetilde{X}\}, s\rangle \rightsquigarrow\xrightarrow{\beta} \langle H\{\widetilde{Q}/\widetilde{X}\}, s'\rangle$, then, let $G'$ be $H$, as desired.

If $G \equiv (\beta_1 \parallel \cdots \parallel \beta_n).H$. Then, we must have $\alpha_i = \beta_i$ for $1 \le i \le n$, and $P' \equiv H\{\widetilde{P}/\widetilde{X}\}$, and $\langle G\{\widetilde{Q}/\widetilde{X}\}, s\rangle \equiv \langle(\beta_1 \parallel \cdots \parallel \beta_n).H\{\widetilde{Q}/\widetilde{X}\}, s\rangle \rightsquigarrow\xrightarrow{\{\beta_1,\cdots,\beta_n\}} \langle H\{\widetilde{Q}/\widetilde{X}\}, s'\rangle$, then, let $G'$ be $H$, as desired.

If $G \equiv \tau.H$. Then, we must have $\tau = \tau$, and $P' \equiv H\{\widetilde{P}/\widetilde{X}\}$, and $\langle G\{\widetilde{Q}/\widetilde{X}\}, s\rangle \equiv \langle \tau.H\{\widetilde{Q}/\widetilde{X}\}, s\rangle \rightsquigarrow\xrightarrow{\tau} \langle H\{\widetilde{Q}/\widetilde{X}\}, s'\rangle$, then, let $G'$ be $H$, as desired. □

**Theorem 14.53** (Unique solution of equations for weakly probabilistic pomset bisimulation). *Let the guarded and sequential expressions $\widetilde{E}$ contain free variables $\subseteq \widetilde{X}$, then*

*If $\widetilde{P} \approx_{pp} \widetilde{E}\{\widetilde{P}/\widetilde{X}\}$ and $\widetilde{Q} \approx_{pp} \widetilde{E}\{\widetilde{Q}/\widetilde{X}\}$, then $\widetilde{P} \approx_{pp} \widetilde{Q}$.*

*Proof.* Like the corresponding theorem in CCS, without loss of generality, we only consider a single equation $X = E$. Hence, we assume $P \approx_{pp} E(P)$, $Q \approx_{pp} E(Q)$, then $P \approx_{pp} Q$.

We will prove $\{(H(P), H(Q)) : H\}$ sequential, if $\langle H(P), s\rangle \rightsquigarrow\xrightarrow{\{\alpha_1,\cdots,\alpha_n\}} \langle P', s'\rangle$, then, for some $Q'$, $\langle H(Q), s\rangle \rightsquigarrow\xRightarrow{\{\alpha_1,\cdots,\alpha_n\}} \langle Q', s'\rangle$ and $P' \approx_{pp} Q'$.

Let $\langle H(P), s\rangle \rightsquigarrow\xrightarrow{\{\alpha_1,\cdot,\alpha_n\}} \langle P', s'\rangle$, then $\langle H(E(P)), s\rangle \rightsquigarrow\xRightarrow{\{\alpha_1,\cdots,\alpha_n\}} \langle P'', s''\rangle$ and $P' \approx_{pp} P''$.

By Lemma 14.52, we know there is a sequential $H'$ such that $\langle H(E(P)), s\rangle \rightsquigarrow\xRightarrow{\{\alpha_1,\cdots,\alpha_n\}} \langle H'(P), s'\rangle \Rightarrow P'' \approx_{pp} P'$.

Also, $\langle H(E(Q)), s\rangle \rightsquigarrow\xRightarrow{\{\alpha_1,\cdots,\alpha_n\}} \langle H'(Q), s'\rangle \Rightarrow Q''$ and $P'' \approx_{pp} Q''$. Also, $\langle H(Q), s\rangle \rightsquigarrow\xrightarrow{\{\alpha_1,\cdots,\alpha_n\}} \langle Q', s'\rangle \approx_{pp}\Rightarrow Q' \approx_{pp} Q''$. Hence, $P' \approx_{pp} Q'$, as desired. □

**Theorem 14.54** (Unique solution of equations for weakly probabilistic step bisimulation). *Let the guarded and sequential expressions $\widetilde{E}$ contain free variables $\subseteq \widetilde{X}$, then*

*If $\widetilde{P} \approx_{ps} \widetilde{E}\{\widetilde{P}/\widetilde{X}\}$ and $\widetilde{Q} \approx_{ps} \widetilde{E}\{\widetilde{Q}/\widetilde{X}\}$, then $\widetilde{P} \approx_{ps} \widetilde{Q}$.*

*Proof.* Like the corresponding theorem in CCS, without loss of generality, we only consider a single equation $X = E$. Hence, we assume $P \approx_{ps} E(P)$, $Q \approx_{ps} E(Q)$, then $P \approx_{ps} Q$.

We will prove $\{(H(P), H(Q)) : H\}$ sequential, if $\langle H(P), s\rangle \rightsquigarrow\xrightarrow{\{\alpha_1,\cdots,\alpha_n\}} \langle P', s'\rangle$, then, for some $Q'$, $\langle H(Q), s\rangle \rightsquigarrow\xRightarrow{\{\alpha_1,\cdots,\alpha_n\}} \langle Q', s'\rangle$ and $P' \approx_{ps} Q'$.

Let $\langle H(P), s\rangle \rightsquigarrow\xrightarrow{\{\alpha_1,\cdot,\alpha_n\}} \langle P', s'\rangle$, then $\langle H(E(P)), s\rangle \rightsquigarrow\xRightarrow{\{\alpha_1,\cdots,\alpha_n\}} \langle P'', s''\rangle$ and $P' \approx_{ps} P''$.

By Lemma 14.52, we know there is a sequential $H'$ such that $\langle H(E(P)), s\rangle \rightsquigarrow\xRightarrow{\{\alpha_1,\cdots,\alpha_n\}} \langle H'(P), s'\rangle \Rightarrow P'' \approx_{ps} P'$.

Also, $\langle H(E(Q)), s\rangle \rightsquigarrow\xRightarrow{\{\alpha_1,\cdots,\alpha_n\}} \langle H'(Q), s'\rangle \Rightarrow Q''$ and $P'' \approx_{ps} Q''$. Also, $\langle H(Q), s\rangle \rightsquigarrow\xrightarrow{\{\alpha_1,\cdots,\alpha_n\}} \langle Q', s'\rangle \approx_{ps}\Rightarrow Q' \approx_{ps} Q''$. Hence, $P' \approx_{ps} Q'$, as desired. □

**Theorem 14.55** (Unique solution of equations for weakly probabilistic hp-bisimulation). *Let the guarded and sequential expressions $\widetilde{E}$ contain free variables $\subseteq \widetilde{X}$, then*

*If $\widetilde{P} \approx_{php} \widetilde{E}\{\widetilde{P}/\widetilde{X}\}$ and $\widetilde{Q} \approx_{php} \widetilde{E}\{\widetilde{Q}/\widetilde{X}\}$, then $\widetilde{P} \approx_{php} \widetilde{Q}$.*

*Proof.* Like the corresponding theorem in CCS, without loss of generality, we only consider a single equation $X = E$. Hence, we assume $P \approx_{php} E(P)$, $Q \approx_{php} E(Q)$, then $P \approx_{php} Q$.

We will prove $\{(H(P), H(Q)) : H\}$ sequential, if $\langle H(P), s\rangle \rightsquigarrow\xrightarrow{\{\alpha_1,\cdots,\alpha_n\}} \langle P', s'\rangle$, then, for some $Q'$, $\langle H(Q), s\rangle \rightsquigarrow\xRightarrow{\{\alpha_1,\cdots,\alpha_n\}} \langle Q', s'\rangle$ and $P' \approx_{php} Q'$.

Let $\langle H(P), s\rangle \rightsquigarrow\xrightarrow{\{\alpha_1,\cdot,\alpha_n\}} \langle P', s'\rangle$, then $\langle H(E(P)), s\rangle \rightsquigarrow\xRightarrow{\{\alpha_1,\cdots,\alpha_n\}} \langle P'', s''\rangle$ and $P' \approx_{php} P''$.

By Lemma 14.52, we know there is a sequential $H'$ such that $\langle H(E(P)), s\rangle \rightsquigarrow\xRightarrow{\{\alpha_1,\cdots,\alpha_n\}} \langle H'(P), s'\rangle \Rightarrow P'' \approx_{php} P'$.

Also, $\langle H(E(Q)), s\rangle \rightsquigarrow\xRightarrow{\{\alpha_1,\cdots,\alpha_n\}} \langle H'(Q), s'\rangle \Rightarrow Q''$ and $P'' \approx_{php} Q''$. Also, $\langle H(Q), s\rangle \rightsquigarrow\xrightarrow{\{\alpha_1,\cdots,\alpha_n\}} \langle Q', s'\rangle \approx_{php}\Rightarrow Q' \approx_{php} Q''$. Hence, $P' \approx_{php} Q'$, as desired. □

**Theorem 14.56** (Unique solution of equations for weakly probabilistic hhp-bisimulation). *Let the guarded and sequential expressions $\widetilde{E}$ contain free variables $\subseteq \widetilde{X}$, then*

*If $\widetilde{P} \approx_{phhp} \widetilde{E}\{\widetilde{P}/\widetilde{X}\}$ and $\widetilde{Q} \approx_{phhp} \widetilde{E}\{\widetilde{Q}/\widetilde{X}\}$, then $\widetilde{P} \approx_{phhp} \widetilde{Q}$.*

*Proof.* Like the corresponding theorem in CCS, without loss of generality, we only consider a single equation $X = E$. Hence, we assume $P \approx_{phhp} E(P)$, $Q \approx_{phhp} E(Q)$, then $P \approx_{phhp} Q$.

We will prove $\{(H(P), H(Q)) : H\}$ sequential, if $\langle H(P), s\rangle \rightsquigarrow \xrightarrow{\{\alpha_1,\cdots,\alpha_n\}} \langle P', s'\rangle$, then, for some $Q'$, $\langle H(Q), s\rangle \rightsquigarrow \xRightarrow{\{\alpha_1,\cdots,\alpha_n\}} \langle Q', s'\rangle$ and $P' \approx_{phhp} Q'$.

Let $\langle H(P), s\rangle \rightsquigarrow \xrightarrow{\{\alpha_1,\cdot,\alpha_n\}} \langle P', s'\rangle$, then $\langle H(E(P)), s\rangle \rightsquigarrow \xRightarrow{\{\alpha_1,\cdots,\alpha_n\}} \langle P'', s''\rangle$ and $P' \approx_{phhp} P''$.

By Lemma 14.52, we know there is a sequential $H'$ such that $\langle H(E(P)), s\rangle \rightsquigarrow \xRightarrow{\{\alpha_1,\cdots,\alpha_n\}} \langle H'(P), s'\rangle \Rightarrow P'' \approx_{phhp} P'$.

Also, $\langle H(E(Q)), s\rangle \rightsquigarrow \xRightarrow{\{\alpha_1,\cdots,\alpha_n\}} \langle H'(Q), s'\rangle \Rightarrow Q''$ and $P'' \approx_{phhp} Q''$. Also, $\langle H(Q), s\rangle \rightsquigarrow \xrightarrow{\{\alpha_1,\cdots,\alpha_n\}} \langle Q', s'\rangle \approx_{phhp} \Rightarrow Q' \approx_{phhp} Q''$. Hence, $P' \approx_{phhp} Q'$, as desired. □

# Chapter 15

# $\pi_{tc}$ with probability and guards

In this chapter, we design $\pi_{tc}$ with probability and guards. This chapter is organized as follows. In Section 15.1, we introduce the truly concurrent operational semantics. Then, we introduce the syntax and operational semantics, laws modulo strongly truly concurrent bisimulations, and algebraic theory of $\pi_{tc}$ with probabilism and guards in Sections 15.2, 15.3, and 15.4, respectively.

## 15.1 Operational semantics

First, in this section, we introduce concepts of (strongly) probabilistic truly concurrent bisimilarities, including probabilistic pomset bisimilarity, probabilistic step bisimilarity, probabilistic history-preserving (hp-)bisimilarity, and probabilistic hereditary history-preserving (hhp-)bisimilarity. In contrast to traditional probabilistic truly concurrent bisimilarities in Chapter 2, these versions in $\pi_{ptc}$ must take care of actions with bound objects. Note that these probabilistic truly concurrent bisimilarities are defined as late bisimilarities, but not early bisimilarities, as defined in $\pi$-calculus [4] [5]. Note that here, a PES $\mathcal{E}$ is deemed as a process.

**Definition 15.1** (Prime event structure with silent event and empty event). Let $\Lambda$ be a fixed set of labels, ranged over $a, b, c, \cdots$ and $\tau, \epsilon$. A ($\Lambda$-labeled) prime event structure with silent event $\tau$ and empty event $\epsilon$ is a tuple $\mathcal{E} = \langle \mathbb{E}, \leq, \sharp, \lambda \rangle$, where $\mathbb{E}$ is a denumerable set of events, including the silent event $\tau$ and empty event $\epsilon$. Let $\hat{\mathbb{E}} = \mathbb{E} \backslash \{\tau, \epsilon\}$, exactly excluding $\tau$ and $\epsilon$, it is obvious that $\hat{\tau^*} = \epsilon$. Let $\lambda : \mathbb{E} \to \Lambda$ be a labeling function and let $\lambda(\tau) = \tau$ and $\lambda(\epsilon) = \epsilon$. Also, $\leq$, $\sharp$ are binary relations on $\mathbb{E}$, called causality and conflict, respectively, such that:

1. $\leq$ is a partial order and $\lceil e \rceil = \{e' \in \mathbb{E} | e' \leq e\}$ is finite for all $e \in \mathbb{E}$. It is easy to see that $e \leq \tau^* \leq e' = e \leq \tau \leq \cdots \leq \tau \leq e'$, then $e \leq e'$.
2. $\sharp$ is irreflexive, symmetric, and hereditary with respect to $\leq$, that is, for all $e, e', e'' \in \mathbb{E}$, if $e \sharp e' \leq e''$, then $e \sharp e''$.

Then, the concepts of consistency and concurrency can be drawn from the above definition:

1. $e, e' \in \mathbb{E}$ are consistent, denoted as $e \frown e'$, if $\neg(e \sharp e')$. A subset $X \subseteq \mathbb{E}$ is called consistent, if $e \frown e'$ for all $e, e' \in X$.
2. $e, e' \in \mathbb{E}$ are concurrent, denoted as $e \parallel e'$, if $\neg(e \leq e')$, $\neg(e' \leq e)$, and $\neg(e \sharp e')$.

**Definition 15.2** (Configuration). Let $\mathcal{E}$ be a PES. A (finite) configuration in $\mathcal{E}$ is a (finite) consistent subset of events $C \subseteq \mathcal{E}$, closed with respect to causality (i.e., $\lceil C \rceil = C$), and a data state $s \in S$ with $S$ the set of all data states, denoted $\langle C, s \rangle$. The set of finite configurations of $\mathcal{E}$ is denoted by $\langle \mathcal{C}(\mathcal{E}), S \rangle$. We let $\hat{C} = C \backslash \{\tau\} \cup \{\epsilon\}$.

A consistent subset of $X \subseteq \mathbb{E}$ of events can be seen as a pomset. Given $X, Y \subseteq \mathbb{E}$, $\hat{X} \sim \hat{Y}$ if $\hat{X}$ and $\hat{Y}$ are isomorphic as pomsets. In the remainder of the chapter, when we say $C_1 \sim C_2$, we mean $\hat{C_1} \sim \hat{C_2}$.

**Definition 15.3** (Probabilistic transitions). Let $\mathcal{E}$ be a PES and let $C \in \mathcal{C}(\mathcal{E})$, the transition $\langle C, s \rangle \xrightarrow{\pi} \langle C^\pi, s \rangle$ is called a probabilistic transition from $\langle C, s \rangle$ to $\langle C^\pi, s \rangle$.

A probability distribution function (PDF) $\mu$ is a map $\mu : \mathcal{C} \times \mathcal{C} \to [0, 1]$ and $\mu^*$ is the cumulative probability distribution function (cPDF).

**Definition 15.4** (Strongly probabilistic pomset, step bisimilarity). Let $\mathcal{E}_1$, $\mathcal{E}_2$ be PESs. A strongly probabilistic pomset bisimulation is a relation $R \subseteq \langle \mathcal{C}(\mathcal{E}_1), s \rangle \times \langle \mathcal{C}(\mathcal{E}_2), s \rangle$, such that (1) if $(\langle C_1, s \rangle, \langle C_2, s \rangle) \in R$, and $\langle C_1, s \rangle \xrightarrow{X_1} \langle C_1', s' \rangle$ (with $\mathcal{E}_1 \xrightarrow{X_1} \mathcal{E}_1'$) then $\langle C_2, s \rangle \xrightarrow{X_2} \langle C_2', s' \rangle$ (with $\mathcal{E}_2 \xrightarrow{X_2} \mathcal{E}_2'$), with $X_1 \subseteq \mathbb{E}_1$, $X_2 \subseteq \mathbb{E}_2$, $X_1 \sim X_2$ and $(\langle C_1', s' \rangle, \langle C_2', s' \rangle) \in R$:

1. for each fresh action $\alpha \in X_1$, if $\langle C_1'', s'' \rangle \xrightarrow{\alpha} \langle C_1''', s''' \rangle$ (with $\mathcal{E}_1'' \xrightarrow{\alpha} \mathcal{E}_1'''$), then for some $C_2''$ and $\langle C_2''', s''' \rangle$, $\langle C_2'', s'' \rangle \xrightarrow{\alpha} \langle C_2''', s''' \rangle$ (with $\mathcal{E}_2'' \xrightarrow{\alpha} \mathcal{E}_2'''$), such that if $(\langle C_1'', s'' \rangle, \langle C_2'', s'' \rangle) \in R$, then $(\langle C_1''', s''' \rangle, \langle C_2''', s''' \rangle) \in R$;

Handbook of Truly Concurrent Process Algebra. https://doi.org/10.1016/B978-0-44-321515-5.00019-6

2. for each $x(y) \in X_1$ with $(y \notin n(\mathcal{E}_1, \mathcal{E}_2))$, if $\langle C_1'', s''\rangle \xrightarrow{x(y)} \langle C_1''', s'''\rangle$ (with $\mathcal{E}_1'' \xrightarrow{x(y)} \mathcal{E}_1'''\{w/y\}$) for all $w$, then for some $C_2''$ and $C_2'''$, $\langle C_2'', s''\rangle \xrightarrow{x(y)} \langle C_2''', s'''\rangle$ (with $\mathcal{E}_2'' \xrightarrow{x(y)} \mathcal{E}_2'''\{w/y\}$) for all $w$, such that if $(\langle C_1'', s''\rangle, \langle C_2'', s''\rangle) \in R$, then $(\langle C_1''', s'''\rangle, \langle C_2''', s'''\rangle) \in R$;
3. for each two $x_1(y), x_2(y) \in X_1$ with $(y \notin n(\mathcal{E}_1, \mathcal{E}_2))$, if $\langle C_1'', s''\rangle \xrightarrow{\{x_1(y), x_2(y)\}} \langle C_1''', s'''\rangle$ (with $\mathcal{E}_1'' \xrightarrow{\{x_1(y), x_2(y)\}} \mathcal{E}_1'''\{w/y\}$) for all $w$, then for some $C_2''$ and $C_2'''$, $\langle C_2'', s''\rangle \xrightarrow{\{x_1(y), x_2(y)\}} \langle C_2''', s'''\rangle$ (with $\mathcal{E}_2'' \xrightarrow{\{x_1(y), x_2(y)\}} \mathcal{E}_2'''\{w/y\}$) for all $w$, such that if $(\langle C_1'', s''\rangle, \langle C_2'', s''\rangle) \in R$, then $(\langle C_1''', s'''\rangle, \langle C_2''', s'''\rangle) \in R$;
4. for each $\overline{x}(y) \in X_1$ with $y \notin n(\mathcal{E}_1, \mathcal{E}_2)$, if $\langle C_1'', s''\rangle \xrightarrow{\overline{x}(y)} \langle C_1''', s'''\rangle$ (with $\mathcal{E}_1'' \xrightarrow{\overline{x}(y)} \mathcal{E}_1'''$), then for some $C_2''$ and $C_2'''$, $\langle C_2'', s''\rangle \xrightarrow{\overline{x}(y)} \langle C_2''', s'''\rangle$ (with $\mathcal{E}_2'' \xrightarrow{\overline{x}(y)} \mathcal{E}_2'''$), such that if $(\langle C_1'', s''\rangle, \langle C_2'', s''\rangle) \in R$, then $(\langle C_1''', s'''\rangle, \langle C_2''', s'''\rangle) \in R$,

and vice versa; (2) if $(\langle C_1, s\rangle, \langle C_2, s\rangle) \in R$, and $\langle C_1, s\rangle \overset{\pi}{\rightsquigarrow} \langle C_1^{\pi}, s\rangle$, then $\langle C_2, s\rangle \overset{\pi}{\rightsquigarrow} \langle C_2^{\pi}, s\rangle$ and $(\langle C_1^{\pi}, s\rangle, \langle C_2^{\pi}, s\rangle) \in R$, and vice versa; (3) if $(\langle C_1, s\rangle, \langle C_2, s\rangle) \in R$, then $\mu(C_1, C) = \mu(C_2, C)$ for each $C \in \mathcal{C}(\mathcal{E})/R$; (4) $[\surd]_R = \{\surd\}$.

We say that $\mathcal{E}_1, \mathcal{E}_2$ are strongly probabilistic pomset bisimilar, written $\mathcal{E}_1 \sim_{pp} \mathcal{E}_2$, if there exists a strongly probabilistic pomset bisimulation $R$, such that $(\emptyset, \emptyset) \in R$. By replacing probabilistic pomset transitions with steps, we can obtain the definition of strongly probabilistic step bisimulation. When PESs $\mathcal{E}_1$ and $\mathcal{E}_2$ are strongly probabilistic step bisimilar, we write $\mathcal{E}_1 \sim_{ps} \mathcal{E}_2$.

**Definition 15.5** (Posetal product). Given two PESs $\mathcal{E}_1$, $\mathcal{E}_2$, the posetal product of their configurations, denoted $\langle \mathcal{C}(\mathcal{E}_1), S\rangle \overline{\times} \langle \mathcal{C}(\mathcal{E}_2), S\rangle$, is defined as

$$\{(\langle C_1, s\rangle, f, \langle C_2, s\rangle) | C_1 \in \mathcal{C}(\mathcal{E}_1), C_2 \in \mathcal{C}(\mathcal{E}_2), f : C_1 \to C_2 \text{ isomorphism}\}$$

A subset $R \subseteq \langle \mathcal{C}(\mathcal{E}_1), S\rangle \overline{\times} \langle \mathcal{C}(\mathcal{E}_2), S\rangle$ is called a posetal relation. We say that $R$ is downward closed when for any $(\langle C_1, s\rangle, f, \langle C_2, s\rangle), (\langle C_1', s'\rangle, f', \langle C_2', s'\rangle) \in \langle \mathcal{C}(\mathcal{E}_1), S\rangle \overline{\times} \langle \mathcal{C}(\mathcal{E}_2), S\rangle$, if $(\langle C_1, s\rangle, f, \langle C_2, s\rangle) \subseteq (\langle C_1', s'\rangle, f', \langle C_2', s'\rangle)$ pointwise and $(\langle C_1', s'\rangle, f', \langle C_2', s'\rangle) \in R$, then $(\langle C_1, s\rangle, f, \langle C_2, s\rangle) \in R$.

For $f : X_1 \to X_2$, we define $f[x_1 \mapsto x_2] : X_1 \cup \{x_1\} \to X_2 \cup \{x_2\}$, $z \in X_1 \cup \{x_1\}$, (1) $f[x_1 \mapsto x_2](z) = x_2$, if $z = x_1$; (2) $f[x_1 \mapsto x_2](z) = f(z)$, otherwise, where $X_1 \subseteq \mathbb{E}_1$, $X_2 \subseteq \mathbb{E}_2$, $x_1 \in \mathbb{E}_1$, $x_2 \in \mathbb{E}_2$.

**Definition 15.6** (Strongly probabilistic (hereditary) history-preserving bisimilarity). A strongly probabilistic history-preserving (hp-) bisimulation is a posetal relation $R \subseteq \mathcal{C}(\mathcal{E}_1) \overline{\times} \mathcal{C}(\mathcal{E}_2)$ such that (1) if $(\langle C_1, s\rangle, f, \langle C_2, s\rangle) \in R$, and

1. for $e_1 = \alpha$ a fresh action, if $\langle C_1, s\rangle \xrightarrow{\alpha} \langle C_1', s'\rangle$ (with $\mathcal{E}_1 \xrightarrow{\alpha} \mathcal{E}_1'$), then for some $C_2'$ and $e_2 = \alpha$, $\langle C_2, s\rangle \xrightarrow{\alpha} \langle C_2', s'\rangle$ (with $\mathcal{E}_2 \xrightarrow{\alpha} \mathcal{E}_2'$), such that $(\langle C_1', s'\rangle, f[e_1 \mapsto e_2], \langle C_2', s'\rangle) \in R$;
2. for $e_1 = x(y)$ with $(y \notin n(\mathcal{E}_1, \mathcal{E}_2))$, if $\langle C_1, s\rangle \xrightarrow{x(y)} \langle C_1', s'\rangle$ (with $\mathcal{E}_1 \xrightarrow{x(y)} \mathcal{E}_1'\{w/y\}$) for all $w$, then for some $C_2'$ and $e_2 = x(y)$, $\langle C_2, s\rangle \xrightarrow{x(y)} \langle C_2', s'\rangle$ (with $\mathcal{E}_2 \xrightarrow{x(y)} \mathcal{E}_2'\{w/y\}$) for all $w$, such that $(\langle C_1', s'\rangle, f[e_1 \mapsto e_2], \langle C_2', s'\rangle) \in R$;
3. for $e_1 = \overline{x}(y)$ with $y \notin n(\mathcal{E}_1, \mathcal{E}_2)$, if $\langle C_1, s\rangle \xrightarrow{\overline{x}(y)} \langle C_1', s'\rangle$ (with $\mathcal{E}_1 \xrightarrow{\overline{x}(y)} \mathcal{E}_1'$), then for some $C_2'$ and $e_2 = \overline{x}(y)$, $\langle C_2, s\rangle \xrightarrow{\overline{x}(y)} \langle C_2', s'\rangle$ (with $\mathcal{E}_2 \xrightarrow{\overline{x}(y)} \mathcal{E}_2'$), such that $(\langle C_1', s'\rangle, f[e_1 \mapsto e_2], \langle C_2', s'\rangle) \in R$,

and vice versa; (2) if $(\langle C_1, s\rangle, f, \langle C_2, s\rangle) \in R$, and $\langle C_1, s\rangle \overset{\pi}{\rightsquigarrow} \langle C_1^{\pi}, s\rangle$, then $\langle C_2, s\rangle \overset{\pi}{\rightsquigarrow} \langle C_2^{\pi}, s\rangle$ and $(\langle C_1^{\pi}, s\rangle, f, \langle C_2^{\pi}, s\rangle) \in R$, and vice versa; (3) if $(\langle C_1, s\rangle, f, \langle C_2, s\rangle) \in R$, then $\mu(C_1, C) = \mu(C_2, C)$ for each $C \in \mathcal{C}(\mathcal{E})/R$; (4) $[\surd]_R = \{\surd\}$. $\mathcal{E}_1, \mathcal{E}_2$ are strongly probabilistic history-preserving (hp-)bisimilar and are written $\mathcal{E}_1 \sim_{php} \mathcal{E}_2$ if there exists a strongly probabilistic hp-bisimulation $R$ such that $(\emptyset, \emptyset, \emptyset) \in R$.

A strongly probabilistic hereditary history-preserving (hhp-)bisimulation is a downward closed strongly probabilistic hp-bisimulation. $\mathcal{E}_1, \mathcal{E}_2$ are FR strongly probabilistic hereditary history-preserving (hhp-)bisimilar and are written $\mathcal{E}_1 \sim_{phhp} \mathcal{E}_2$.

## 15.2 Syntax and operational semantics

We assume an infinite set $\mathcal{N}$ of (action or event) names, and use $a, b, c, \cdots$ to range over $\mathcal{N}$, and use $x, y, z, w, u, v$ as meta-variables over names. We denote by $\overline{\mathcal{N}}$ the set of co-names and let $\overline{a}, \overline{b}, \overline{c}, \cdots$ range over $\overline{\mathcal{N}}$. Then, we set $\mathcal{L} = \mathcal{N} \cup \overline{\mathcal{N}}$ as the set of labels, and use $l, \overline{l}$ to range over $\mathcal{L}$. We extend complementation to $\mathcal{L}$ such that $\overline{\overline{a}} = a$. Let $\tau$ denote the silent step (internal action or event) and define $Act = \mathcal{L} \cup \{\tau\}$ to be the set of actions, $\alpha, \beta$ range over $Act$. Also, $K, L$ are used to stand for subsets of $\mathcal{L}$ and $\overline{L}$ is used for the set of complements of labels in $L$.

Further, we introduce a set $\mathcal{X}$ of process variables, and a set $\mathcal{K}$ of process constants, and let $X, Y, \cdots$ range over $\mathcal{X}$, and $A, B, \cdots$ range over $\mathcal{K}$. For each process constant $A$, a nonnegative arity $ar(A)$ is assigned to it. Let $\widetilde{x} = x_1, \cdots, x_{ar(A)}$ be a tuple of distinct name variables, then $A(\widetilde{x})$ is called a process constant. $\widetilde{X}$ is a tuple of distinct process variables, also $E, F, \cdots$ range over the recursive expressions. We write $\mathcal{P}$ for the set of processes. Sometimes, we use $I, J$ to stand for an indexing set, and we write $E_i : i \in I$ for a family of expressions indexed by $I$. $Id_D$ is the identity function or relation over set $D$. The symbol $\equiv_\alpha$ denotes equality under standard alpha-convertibility, note that the subscript $\alpha$ has no relation to the action $\alpha$.

Let $G_{at}$ be the set of atomic guards, $\delta$ be the deadlock constant, and $\epsilon$ be the empty action, and extend $Act$ to $Act \cup \{\epsilon\} \cup \{\delta\}$. We extend $G_{at}$ to the set of basic guards $G$ with element $\phi, \psi, \cdots$, which is generated by the following formation rules:

$$\phi ::= \delta|\epsilon|\neg\phi|\psi \in G_{at}|\phi + \psi|\phi \cdot \psi$$

The predicate $test(\phi, s)$ represents that $\phi$ holds in the state $s$, and $test(\epsilon, s)$ holds and $test(\delta, s)$ does not hold. $effect(e, s) \in S$ denotes $s'$ in $s \xrightarrow{e} s'$. The predicate weakest precondition $wp(e, \phi)$ denotes that $\forall s, s' \in S, test(\phi, effect(e, s))$ holds.

### 15.2.1 Syntax

We use the Prefix . to model the causality relation $\leq$ in true concurrency, the Summation $+$ to model the conflict relation $\sharp$, $\boxplus_\pi$ to model the probabilistic conflict relation $\sharp_\pi$ in probabilistic true concurrency, and the Composition $\parallel$ to explicitly model concurrent relation in true concurrency. Also, we follow the conventions of process algebra.

**Definition 15.7** (Syntax). A truly concurrent process $\pi_{tc}$ with probabilism and guards is defined inductively by the following formation rules:

1. $A(\widetilde{x}) \in \mathcal{P}$;
2. $\phi \in \mathcal{P}$;
3. $\mathbf{nil} \in \mathcal{P}$;
4. if $P \in \mathcal{P}$, then the Prefix $\tau.P \in \mathcal{P}$, for $\tau \in Act$ is the silent action;
5. if $P \in \mathcal{P}$, then the Prefix $\phi.P \in \mathcal{P}$, for $\phi \in G_{at}$;
6. if $P \in \mathcal{P}$, then the Output $\overline{x}y.P \in \mathcal{P}$, for $x, y \in Act$;
7. if $P \in \mathcal{P}$, then the Input $x(y).P \in \mathcal{P}$, for $x, y \in Act$;
8. if $P \in \mathcal{P}$, then the Restriction $(x)P \in \mathcal{P}$, for $x \in Act$;
9. if $P, Q \in \mathcal{P}$, then the Summation $P + Q \in \mathcal{P}$;
10. if $P, Q \in \mathcal{P}$, then the Summation $P \boxplus_\pi Q \in \mathcal{P}$;
11. if $P, Q \in \mathcal{P}$, then the Composition $P \parallel Q \in \mathcal{P}$.

The standard BNF grammar of syntax of $\pi_{tc}$ with probabilism and guards can be summarized as follows:

$$P ::= A(\widetilde{x})|\mathbf{nil}|\tau.P|\overline{x}y.P|x(y).P|(x)P|\phi.P|P + P|P \boxplus_\pi P|P \parallel P$$

In $\overline{x}y$, $x(y)$ and $\overline{x}(y)$, $x$ is called the subject, $y$ is called the object and it may be free or bound.

**Definition 15.8** (Free variables). The free names of a process $P$, $fn(P)$, are defined as follows:

1. $fn(A(\widetilde{x})) \subseteq \{\widetilde{x}\}$;
2. $fn(\mathbf{nil}) = \emptyset$;
3. $fn(\tau.P) = fn(P)$;
4. $fn(\phi.P) = fn(P)$;
5. $fn(\overline{x}y.P) = fn(P) \cup \{x\} \cup \{y\}$;
6. $fn(x(y).P) = fn(P) \cup \{x\} - \{y\}$;
7. $fn((x)P) = fn(P) - \{x\}$;
8. $fn(P + Q) = fn(P) \cup fn(Q)$;
9. $fn(P \boxplus_\pi Q) = fn(P) \cup fn(Q)$;
10. $fn(P \parallel Q) = fn(P) \cup fn(Q)$.

**Definition 15.9** (Bound variables). Let $n(P)$ be the names of a process $P$, then the bound names $bn(P) = n(P) - fn(P)$.

**TABLE 15.1 Probabilistic transition rules.**

| | | |
|---|---|---|
| **PTAU-ACT** | $\dfrac{}{\langle \tau.P, s\rangle \rightsquigarrow \langle \breve{\tau}.P, s\rangle}$ | |
| **POUTPUT-ACT** | $\dfrac{}{\langle \overline{x}y.P, s\rangle \rightsquigarrow \langle \breve{\overline{x}y}.P, s\rangle}$ | |
| **PINPUT-ACT** | $\dfrac{}{\langle x(z).P, s\rangle \rightsquigarrow \langle x(\breve{z}).P, s\rangle}$ | |
| **PPAR** | $\dfrac{\langle P, s\rangle \rightsquigarrow \langle P', s\rangle \quad \langle Q, s\rangle \rightsquigarrow \langle Q', s\rangle}{\langle P \parallel Q, s\rangle \rightsquigarrow \langle P' \parallel Q', s\rangle}$ | |
| **PSUM** | $\dfrac{\langle P, s\rangle \rightsquigarrow \langle P', s\rangle \quad \langle Q, s\rangle \rightsquigarrow \langle Q', s\rangle}{\langle P + Q, s\rangle \rightsquigarrow \langle P' + Q', s\rangle}$ | |
| **PBOX-SUM** | $\dfrac{\langle P, s\rangle \rightsquigarrow \langle P', s\rangle}{\langle P \boxplus_\pi Q, s\rangle \rightsquigarrow \langle P', s\rangle}$ | |
| **PIDE** | $\dfrac{\langle P\{\widetilde{y}/\widetilde{x}\}, s\rangle \rightsquigarrow \langle P', s\rangle}{\langle A(\widetilde{y}), s\rangle \rightsquigarrow \langle P', s\rangle}$ | $(A(\widetilde{x}) \overset{\text{def}}{=} P)$ |
| **PRES** | $\dfrac{\langle P, s\rangle \rightsquigarrow \langle P', s\rangle}{\langle (y)P, s\rangle \rightsquigarrow \langle (y)P', s\rangle}$ | $(y \notin n(\alpha))$ |

For each process constant schema $A(\widetilde{x})$, a defining equation of the form

$$A(\widetilde{x}) \overset{\text{def}}{=} P$$

is assumed, where $P$ is a process with $fn(P) \subseteq \{\widetilde{x}\}$.

**Definition 15.10** (Substitutions). A substitution is a function $\sigma : \mathcal{N} \to \mathcal{N}$. For $x_i\sigma = y_i$ with $1 \le i \le n$, we write $\{y_1/x_1, \cdots, y_n/x_n\}$ or $\{\widetilde{y}/\widetilde{x}\}$ for $\sigma$. For a process $P \in \mathcal{P}$, $P\sigma$ is defined inductively as follows:

1. if $P$ is a process constant $A(\widetilde{x}) = A(x_1, \cdots, x_n)$, then $P\sigma = A(x_1\sigma, \cdots, x_n\sigma)$;
2. if $P = \mathbf{nil}$, then $P\sigma = \mathbf{nil}$;
3. if $P = \tau.P'$, then $P\sigma = \tau.P'\sigma$;
4. if $P = \phi.P'$, then $P\sigma = \phi.P'\sigma$;
5. if $P = \overline{x}y.P'$, then $P\sigma = \overline{x\sigma}y\sigma.P'\sigma$;
6. if $P = x(y).P'$, then $P\sigma = x\sigma(y).P'\sigma$;
7. if $P = (x)P'$, then $P\sigma = (x\sigma)P'\sigma$;
8. if $P = P_1 + P_2$, then $P\sigma = P_1\sigma + P_2\sigma$;
9. if $P = P_1 \boxplus_\pi P_2$, then $P\sigma = P_1\sigma \boxplus_\pi P_2\sigma$;
10. if $P = P_1 \parallel P_2$, then $P\sigma = P_1\sigma \parallel P_2\sigma$.

### 15.2.2 Operational semantics

The operational semantics is defined by LTSs (labeled transition systems), and it is detailed by the following definition.

**Definition 15.11** (Semantics). The operational semantics of $\pi_{tc}$ with probabilism and guards corresponding to the syntax in Definition 15.7 is defined by a series of transition rules, named **PACT**, **PSUM**, **PBOX-SUM**, **PIDE**, **PPAR**, **PRES** and named **ACT**, **SUM**, **IDE**, **PAR**, **COM**, **CLOSE**, **RES**, **OPEN** indicate that the rules are associated, respectively, with Prefix, Summation, Box-Summation, Identity, Parallel Composition, Communication, and Restriction in Definition 15.7. They are shown in Tables 15.1 and 15.2.

### 15.2.3 Properties of transitions

**Proposition 15.12.** **1.** *If* $\langle P, s\rangle \xrightarrow{\alpha} \langle P', s'\rangle$, *then*

(a) $fn(\alpha) \subseteq fn(P)$;

(b) $fn(P') \subseteq fn(P) \cup bn(\alpha)$;

**2.** *If* $\langle P, s\rangle \xrightarrow{\{\alpha_1, \cdots, \alpha_n\}} \langle P', s\rangle$, *then*

(a) $fn(\alpha_1) \cup \cdots \cup fn(\alpha_n) \subseteq fn(P)$;

**TABLE 15.2 Action transition rules.**

| | | |
|---|---|---|
| TAU-ACT | $\dfrac{}{\langle \check{\tau}.P,s\rangle \xrightarrow{\tau} \langle P,\tau(s)\rangle}$ | |
| OUTPUT-ACT | $\dfrac{}{\langle \check{\overline{x}y}.P,s\rangle \xrightarrow{\overline{x}y} \langle P,s'\rangle}$ | |
| INPUT-ACT | $\dfrac{}{\langle x(\check{z}).P,s\rangle \xrightarrow{x(w)} \langle P\{w/z\},s'\rangle}$ | $(w\notin fn((z)P))$ |
| PAR$_1$ | $\dfrac{\langle P,s\rangle \xrightarrow{\alpha} \langle P',s'\rangle \quad \langle Q,s\rangle \nrightarrow}{\langle P\parallel Q,s\rangle \xrightarrow{\alpha} \langle P'\parallel Q,s'\rangle}$ | $(bn(\alpha)\cap fn(Q)=\emptyset)$ |
| PAR$_2$ | $\dfrac{\langle Q,s\rangle \xrightarrow{\alpha} \langle Q',s'\rangle \quad \langle P,s\rangle \nrightarrow}{\langle P\parallel Q,s\rangle \xrightarrow{\alpha} \langle P\parallel Q',s'\rangle}$ | $(bn(\alpha)\cap fn(P)=\emptyset)$ |
| PAR$_3$ | $\dfrac{\langle P,s\rangle \xrightarrow{\alpha} \langle P',s'\rangle \quad \langle Q,s\rangle \xrightarrow{\beta} \langle Q',s''\rangle}{\langle P\parallel Q,s\rangle \xrightarrow{\{\alpha,\beta\}} \langle P'\parallel Q',s'\cup s''\rangle}$ $(\beta\neq\overline{\alpha}, bn(\alpha)\cap bn(\beta)=\emptyset, bn(\alpha)\cap fn(Q)=\emptyset, bn(\beta)\cap fn(P)=\emptyset)$ | |
| PAR$_4$ | $\dfrac{\langle P,s\rangle \xrightarrow{x_1(z)} \langle P',s'\rangle \quad \langle Q,s\rangle \xrightarrow{x_2(z)} \langle Q',s''\rangle}{\langle P\parallel Q,s\rangle \xrightarrow{\{x_1(w),x_2(w)\}} \langle P'\{w/z\}\parallel Q'\{w/z\},s'\cup s''\rangle}$ | $(w\notin fn((z)P)\cup fn((z)Q))$ |
| COM | $\dfrac{\langle P,s\rangle \xrightarrow{\overline{x}y} \langle P',s'\rangle \quad \langle Q,s\rangle \xrightarrow{x(z)} \langle Q',s''\rangle}{\langle P\parallel Q,s\rangle \xrightarrow{\tau} \langle P'\parallel Q'\{y/z\},s'\cup s''\rangle}$ | |
| CLOSE | $\dfrac{\langle P,s\rangle \xrightarrow{\overline{x}(w)} \langle P',s'\rangle \quad \langle Q,s\rangle \xrightarrow{x(w)} \langle Q',s''\rangle}{\langle P\parallel Q,s\rangle \xrightarrow{\tau} \langle (w)(P'\parallel Q'),s'\cup s''\rangle}$ | |
| SUM$_1$ | $\dfrac{\langle P,s\rangle \xrightarrow{\alpha} \langle P',s'\rangle}{\langle P+Q,s\rangle \xrightarrow{\alpha} \langle P',s'\rangle}$ | |
| SUM$_2$ | $\dfrac{\langle P,s\rangle \xrightarrow{\{\alpha_1,\cdots,\alpha_n\}} \langle P',s'\rangle}{\langle P+Q,s\rangle \xrightarrow{\{\alpha_1,\cdots,\alpha_n\}} \langle P',s'\rangle}$ | |
| IDE$_1$ | $\dfrac{\langle P\{\widetilde{y}/\widetilde{x}\},s\rangle \xrightarrow{\alpha} \langle P',s'\rangle}{\langle A(\widetilde{y}),s\rangle \xrightarrow{\alpha} \langle P',s'\rangle}$ | $(A(\widetilde{x})\overset{\text{def}}{=}P)$ |
| IDE$_2$ | $\dfrac{\langle P\{\widetilde{y}/\widetilde{x}\},s\rangle \xrightarrow{\{\alpha_1,\cdots,\alpha_n\}} \langle P',s'\rangle}{\langle A(\widetilde{y}),s\rangle \xrightarrow{\{\alpha_1,\cdots,\alpha_n\}} \langle P',s'\rangle}$ | $(A(\widetilde{x})\overset{\text{def}}{=}P)$ |
| RES$_1$ | $\dfrac{\langle P,s\rangle \xrightarrow{\alpha} \langle P',s'\rangle}{\langle (y)P,s\rangle \xrightarrow{\alpha} \langle (y)P',s'\rangle}$ | $(y\notin n(\alpha))$ |
| RES$_2$ | $\dfrac{\langle P,s\rangle \xrightarrow{\{\alpha_1,\cdots,\alpha_n\}} \langle P',s'\rangle}{\langle (y)P,s\rangle \xrightarrow{\{\alpha_1,\cdots,\alpha_n\}} \langle (y)P',s'\rangle}$ | $(y\notin n(\alpha_1)\cup\cdots\cup n(\alpha_n))$ |
| OPEN$_1$ | $\dfrac{\langle P,s\rangle \xrightarrow{\overline{x}y} \langle P',s'\rangle}{\langle (y)P,s\rangle \xrightarrow{\overline{x}(w)} \langle P'\{w/y\},s'\rangle}$ | $(y\neq x, w\notin fn((y)P'))$ |
| OPEN$_2$ | $\dfrac{\langle P,s\rangle \xrightarrow{\{\overline{x}_1 y,\cdots,\overline{x}_n y\}} \langle P',s'\rangle}{\langle (y)P,s\rangle \xrightarrow{\{\overline{x}_1(w),\cdots,\overline{x}_n(w)\}} \langle P'\{w/y\},s'\rangle}$ | $(y\neq x_1\neq\cdots\neq x_n, w\notin fn((y)P'))$ |

**(b)** $fn(P')\subseteq fn(P)\cup bn(\alpha_1)\cup\cdots\cup bn(\alpha_n)$.

*Proof.* By induction on the depth of inference. □

**Proposition 15.13.** *Suppose that* $\langle P,s\rangle \xrightarrow{\alpha(y)} \langle P',s'\rangle$, *where* $\alpha=x$ *or* $\alpha=\overline{x}$, *and* $x\notin n(P)$, *then there exists some* $P''\equiv_\alpha P'\{z/y\}$, $\langle P,s\rangle \xrightarrow{\alpha(z)} \langle P'',s''\rangle$.

*Proof.* By induction on the depth of inference. □

**Proposition 15.14.** *If* $\langle P, s\rangle \xrightarrow{\alpha} \langle P', s'\rangle$, $bn(\alpha) \cap fn(P'\sigma) = \emptyset$, *and* $\sigma \lceil bn(\alpha) = id$, *then there exists some* $P'' \equiv_\alpha P'\sigma$, $\langle P, s\rangle\sigma \xrightarrow{\alpha\sigma} \langle P'', s''\rangle$.

*Proof.* By the definition of substitution (Definition 15.10) and induction on the depth of inference. □

**Proposition 15.15.** **1.** *If* $\langle P\{w/z\}, s\rangle \xrightarrow{\alpha} \langle P', s'\rangle$, *where* $w \notin fn(P)$ *and* $bn(\alpha) \cap fn(P, w) = \emptyset$, *then there exist some* $Q$ *and* $\beta$ *with* $Q\{w/z\} \equiv_\alpha P'$ *and* $\beta\sigma = \alpha$, $\langle P, s\rangle \xrightarrow{\beta} \langle Q, s'\rangle$;

**2.** *If* $\langle P\{w/z\}, s\rangle \xrightarrow{\{\alpha_1,\cdots,\alpha_n\}} \langle P', s'\rangle$, *where* $w \notin fn(P)$ *and* $bn(\alpha_1) \cap \cdots \cap bn(\alpha_n) \cap fn(P, w) = \emptyset$, *then there exist some* $Q$ *and* $\beta_1, \cdots, \beta_n$ *with* $Q\{w/z\} \equiv_\alpha P'$ *and* $\beta_1\sigma = \alpha_1, \cdots, \beta_n\sigma = \alpha_n$, $\langle P, s\rangle \xrightarrow{\{\beta_1,\cdots,\beta_n\}} \langle Q, s'\rangle$.

*Proof.* By the definition of substitution (Definition 15.10) and induction on the depth of inference. □

## 15.3 Strong bisimilarities

### 15.3.1 Laws and congruence

**Theorem 15.16.** $\equiv_\alpha$ *are strongly probabilistic truly concurrent bisimulations. That is, if* $P \equiv_\alpha Q$, *then*

1. $P \sim_{pp} Q$;
2. $P \sim_{ps} Q$;
3. $P \sim_{php} Q$;
4. $P \sim_{phhp} Q$.

*Proof.* By induction on the depth of inference, we can obtain the following facts:

1. If $\alpha$ is a free action and $\langle P, s\rangle \rightsquigarrow\xrightarrow{\alpha} \langle P', s'\rangle$, then equally for some $Q'$ with $P' \equiv_\alpha Q'$, $\langle Q, s\rangle \rightsquigarrow\xrightarrow{\alpha} \langle Q', s'\rangle$;
2. If $\langle P, s\rangle \rightsquigarrow\xrightarrow{a(y)} \langle P', s'\rangle$ with $a = x$ or $a = \overline{x}$ and $z \notin n(Q)$, then equally for some $Q'$ with $P'\{z/y\} \equiv_\alpha Q'$, $\langle Q, s\rangle \rightsquigarrow\xrightarrow{a(z)} \langle Q', s'\rangle$.

   Then, we can obtain:

1. by the definition of strongly probabilistic pomset bisimilarity, $P \sim_{pp} Q$;
2. by the definition of strongly probabilistic step bisimilarity, $P \sim_{ps} Q$;
3. by the definition of strongly probabilistic hp-bisimilarity, $P \sim_{php} Q$;
4. by the definition of strongly probabilistic hhp-bisimilarity, $P \sim_{phhp} Q$. □

**Proposition 15.17** (Summation laws for strongly probabilistic pomset bisimulation). *The Summation laws for strongly probabilistic pomset bisimulation are as follows:*

1. $P + Q \sim_{pp} Q + P$;
2. $P + (Q + R) \sim_{pp} (P + Q) + R$;
3. $P + P \sim_{pp} P$;
4. $P + \mathbf{nil} \sim_{pp} P$.

*Proof.* **1.** $P + Q \sim_{pp} Q + P$. It is sufficient to prove the relation $R = \{(P + Q, Q + P)\} \cup \mathbf{Id}$ is a strongly probabilistic pomset bisimulation, however, we omit the proof here;

2. $P + (Q + R) \sim_{pp} (P + Q) + R$. It is sufficient to prove the relation $R = \{(P + (Q + R), (P + Q) + R)\} \cup \mathbf{Id}$ is a strongly probabilistic pomset bisimulation, however, we omit the proof here;
3. $P + P \sim_{pp} P$. It is sufficient to prove the relation $R = \{(P + P, P)\} \cup \mathbf{Id}$ is a strongly probabilistic pomset bisimulation, however, we omit the proof here;
4. $P + \mathbf{nil} \sim_{pp} P$. It is sufficient to prove the relation $R = \{(P + \mathbf{nil}, P)\} \cup \mathbf{Id}$ is a strongly probabilistic pomset bisimulation, however, we omit the proof here. □

**Proposition 15.18** (Summation laws for strongly probabilistic step bisimulation). *The Summation laws for strongly probabilistic step bisimulation are as follows:*

1. $P + Q \sim_{ps} Q + P$;
2. $P + (Q + R) \sim_{ps} (P + Q) + R$;
3. $P + P \sim_{ps} P$;

4. $P + \boldsymbol{nil} \sim_{ps} P$.

*Proof.* 1. $P + Q \sim_{ps} Q + P$. It is sufficient to prove the relation $R = \{(P + Q, Q + P)\} \cup \mathbf{Id}$ is a strongly probabilistic step bisimulation, however, we omit the proof here;
2. $P + (Q + R) \sim_{ps} (P + Q) + R$. It is sufficient to prove the relation $R = \{(P + (Q + R), (P + Q) + R)\} \cup \mathbf{Id}$ is a strongly probabilistic step bisimulation, however, we omit the proof here;
3. $P + P \sim_{ps} P$. It is sufficient to prove the relation $R = \{(P + P, P)\} \cup \mathbf{Id}$ is a strongly probabilistic step bisimulation, however, we omit the proof here;
4. $P + \mathbf{nil} \sim_{ps} P$. It is sufficient to prove the relation $R = \{(P + \mathbf{nil}, P)\} \cup \mathbf{Id}$ is a strongly probabilistic step bisimulation, however, we omit the proof here. □

**Proposition 15.19** (Summation laws for strongly probabilistic hp-bisimulation). *The Summation laws for strongly probabilistic hp-bisimulation are as follows:*

1. $P + Q \sim_{php} Q + P$;
2. $P + (Q + R) \sim_{php} (P + Q) + R$;
3. $P + P \sim_{php} P$;
4. $P + \boldsymbol{nil} \sim_{php} P$.

*Proof.* 1. $P + Q \sim_{php} Q + P$. It is sufficient to prove the relation $R = \{(P + Q, Q + P)\} \cup \mathbf{Id}$ is a strongly probabilistic hp-bisimulation, however, we omit the proof here;
2. $P + (Q + R) \sim_{php} (P + Q) + R$. It is sufficient to prove the relation $R = \{(P + (Q + R), (P + Q) + R)\} \cup \mathbf{Id}$ is a strongly probabilistic hp-bisimulation, however, we omit the proof here;
3. $P + P \sim_{php} P$. It is sufficient to prove the relation $R = \{(P + P, P)\} \cup \mathbf{Id}$ is a strongly probabilistic hp-bisimulation, however, we omit the proof here;
4. $P + \mathbf{nil} \sim_{php} P$. It is sufficient to prove the relation $R = \{(P + \mathbf{nil}, P)\} \cup \mathbf{Id}$ is a strongly probabilistic hp-bisimulation, however, we omit the proof here. □

**Proposition 15.20** (Summation laws for strongly probabilistic hhp-bisimulation). *The Summation laws for strongly probabilistic hhp-bisimulation are as follows:*

1. $P + Q \sim_{phhp} Q + P$;
2. $P + (Q + R) \sim_{phhp} (P + Q) + R$;
3. $P + P \sim_{phhp} P$;
4. $P + \boldsymbol{nil} \sim_{phhp} P$.

*Proof.* 1. $P + Q \sim_{phhp} Q + P$. It is sufficient to prove the relation $R = \{(P + Q, Q + P)\} \cup \mathbf{Id}$ is a strongly probabilistic hhp-bisimulation, however, we omit the proof here;
2. $P + (Q + R) \sim_{phhp} (P + Q) + R$. It is sufficient to prove the relation $R = \{(P + (Q + R), (P + Q) + R)\} \cup \mathbf{Id}$ is a strongly probabilistic hhp-bisimulation, however, we omit the proof here;
3. $P + P \sim_{phhp} P$. It is sufficient to prove the relation $R = \{(P + P, P)\} \cup \mathbf{Id}$ is a strongly probabilistic hhp-bisimulation, however, we omit the proof here;
4. $P + \mathbf{nil} \sim_{phhp} P$. It is sufficient to prove the relation $R = \{(P + \mathbf{nil}, P)\} \cup \mathbf{Id}$ is a strongly probabilistic hhp-bisimulation, however, we omit the proof here. □

**Proposition 15.21** (Box-Summation laws for strongly probabilistic pomset bisimulation). *The Box-Summation laws for strongly probabilistic pomset bisimulation are as follows:*

1. $P \boxplus_{\pi} Q \sim_{pp} Q \boxplus_{1-\pi} P$;
2. $P \boxplus_{\pi} (Q \boxplus_{\rho} R) \sim_{pp} (P \boxplus_{\frac{\pi}{\pi+\rho-\pi\rho}} Q) \boxplus_{\pi+\rho-\pi\rho} R$;
3. $P \boxplus_{\pi} P \sim_{pp} P$;
4. $P \boxplus_{\pi} \boldsymbol{nil} \sim_{pp} P$.

*Proof.* 1. $P \boxplus_{\pi} Q \sim_{pp} Q \boxplus_{1-\pi} P$. It is sufficient to prove the relation $R = \{(P \boxplus_{\pi} Q, Q \boxplus_{1-\pi} P)\} \cup \mathbf{Id}$ is a strongly probabilistic pomset bisimulation, however, we omit the proof here;
2. $P \boxplus_{\pi} (Q \boxplus_{\rho} R) \sim_{pp} (P \boxplus_{\frac{\pi}{\pi+\rho-\pi\rho}} Q) \boxplus_{\pi+\rho-\pi\rho} R$. It is sufficient to prove the relation $R = \{(P \boxplus_{\pi} (Q \boxplus_{\rho} R), (P \boxplus_{\frac{\pi}{\pi+\rho-\pi\rho}} Q) \boxplus_{\pi+\rho-\pi\rho} R)\} \cup \mathbf{Id}$ is a strongly probabilistic pomset bisimulation, however, we omit the proof here;
3. $P \boxplus_{\pi} P \sim_{pp} P$. It is sufficient to prove the relation $R = \{(P \boxplus_{\pi} P, P)\} \cup \mathbf{Id}$ is a strongly probabilistic pomset bisimulation, however, we omit the proof here;

4. $P \boxplus_\pi \mathbf{nil} \sim_{pp} P$. It is sufficient to prove the relation $R = \{(P \boxplus_\pi \mathbf{nil}, P)\} \cup \mathbf{Id}$ is a strongly probabilistic pomset bisimulation, however, we omit the proof here. □

**Proposition 15.22** (Box-Summation laws for strongly probabilistic step bisimulation). *The Box-Summation laws for strongly probabilistic step bisimulation are as follows:*

1. $P \boxplus_\pi Q \sim_{ps} Q \boxplus_{1-\pi} P$;
2. $P \boxplus_\pi (Q \boxplus_\rho R) \sim_{ps} (P \boxplus_{\frac{\pi}{\pi+\rho-\pi\rho}} Q) \boxplus_{\pi+\rho-\pi\rho} R$;
3. $P \boxplus_\pi P \sim_{ps} P$;
4. $P \boxplus_\pi \mathbf{nil} \sim_{ps} P$.

*Proof.* 1. $P \boxplus_\pi Q \sim_{ps} Q \boxplus_{1-\pi} P$. It is sufficient to prove the relation $R = \{(P \boxplus_\pi Q, Q \boxplus_{1-\pi} P)\} \cup \mathbf{Id}$ is a strongly probabilistic step bisimulation, however, we omit the proof here;

2. $P \boxplus_\pi (Q \boxplus_\rho R) \sim_{ps} (P \boxplus_{\frac{\pi}{\pi+\rho-\pi\rho}} Q) \boxplus_{\pi+\rho-\pi\rho} R$. It is sufficient to prove the relation $R = \{(P \boxplus_\pi (Q \boxplus_\rho R), (P \boxplus_{\frac{\pi}{\pi+\rho-\pi\rho}} Q) \boxplus_{\pi+\rho-\pi\rho} R)\} \cup \mathbf{Id}$ is a strongly probabilistic step bisimulation, however, we omit the proof here;
3. $P \boxplus_\pi P \sim_{ps} P$. It is sufficient to prove the relation $R = \{(P \boxplus_\pi P, P)\} \cup \mathbf{Id}$ is a strongly probabilistic step bisimulation, however, we omit the proof here;
4. $P \boxplus_\pi \mathbf{nil} \sim_{ps} P$. It is sufficient to prove the relation $R = \{(P \boxplus_\pi \mathbf{nil}, P)\} \cup \mathbf{Id}$ is a strongly probabilistic step bisimulation, however, we omit the proof here. □

**Proposition 15.23** (Box-Summation laws for strongly probabilistic hp-bisimulation). *The Box-Summation laws for strongly probabilistic hp-bisimulation are as follows:*

1. $P \boxplus_\pi Q \sim_{php} Q \boxplus_{1-\pi} P$;
2. $P \boxplus_\pi (Q \boxplus_\rho R) \sim_{php} (P \boxplus_{\frac{\pi}{\pi+\rho-\pi\rho}} Q) \boxplus_{\pi+\rho-\pi\rho} R$;
3. $P \boxplus_\pi P \sim_{php} P$;
4. $P \boxplus_\pi \mathbf{nil} \sim_{php} P$.

*Proof.* 1. $P \boxplus_\pi Q \sim_{php} Q \boxplus_{1-\pi} P$. It is sufficient to prove the relation $R = \{(P \boxplus_\pi Q, Q \boxplus_{1-\pi} P)\} \cup \mathbf{Id}$ is a strongly probabilistic hp-bisimulation, however, we omit the proof here;

2. $P \boxplus_\pi (Q \boxplus_\rho R) \sim_{php} (P \boxplus_{\frac{\pi}{\pi+\rho-\pi\rho}} Q) \boxplus_{\pi+\rho-\pi\rho} R$. It is sufficient to prove the relation $R = \{(P \boxplus_\pi (Q \boxplus_\rho R), (P \boxplus_{\frac{\pi}{\pi+\rho-\pi\rho}} Q) \boxplus_{\pi+\rho-\pi\rho} R)\} \cup \mathbf{Id}$ is a strongly probabilistic hp-bisimulation, however, we omit the proof here;
3. $P \boxplus_\pi P \sim_{php} P$. It is sufficient to prove the relation $R = \{(P \boxplus_\pi P, P)\} \cup \mathbf{Id}$ is a strongly probabilistic hp-bisimulation, however, we omit the proof here;
4. $P \boxplus_\pi \mathbf{nil} \sim_{php} P$. It is sufficient to prove the relation $R = \{(P \boxplus_\pi \mathbf{nil}, P)\} \cup \mathbf{Id}$ is a strongly probabilistic hp-bisimulation, however, we omit the proof here. □

**Proposition 15.24** (Box-Summation laws for strongly probabilistic hhp-bisimulation). *The Box-Summation laws for strongly probabilistic hhp-bisimulation are as follows:*

1. $P \boxplus_\pi Q \sim_{phhp} Q \boxplus_{1-\pi} P$;
2. $P \boxplus_\pi (Q \boxplus_\rho R) \sim_{phhp} (P \boxplus_{\frac{\pi}{\pi+\rho-\pi\rho}} Q) \boxplus_{\pi+\rho-\pi\rho} R$;
3. $P \boxplus_\pi P \sim_{phhp} P$;
4. $P \boxplus_\pi \mathbf{nil} \sim_{phhp} P$.

*Proof.* 1. $P \boxplus_\pi Q \sim_{phhp} Q \boxplus_{1-\pi} P$. It is sufficient to prove the relation $R = \{(P \boxplus_\pi Q, Q \boxplus_{1-\pi} P)\} \cup \mathbf{Id}$ is a strongly probabilistic hhp-bisimulation, however, we omit the proof here;

2. $P \boxplus_\pi (Q \boxplus_\rho R) \sim_{phhp} (P \boxplus_{\frac{\pi}{\pi+\rho-\pi\rho}} Q) \boxplus_{\pi+\rho-\pi\rho} R$. It is sufficient to prove the relation $R = \{(P \boxplus_\pi (Q \boxplus_\rho R), (P \boxplus_{\frac{\pi}{\pi+\rho-\pi\rho}} Q) \boxplus_{\pi+\rho-\pi\rho} R)\} \cup \mathbf{Id}$ is a strongly probabilistic hhp-bisimulation, however, we omit the proof here;
3. $P \boxplus_\pi P \sim_{phhp} P$. It is sufficient to prove the relation $R = \{(P \boxplus_\pi P, P)\} \cup \mathbf{Id}$ is a strongly probabilistic hhp-bisimulation, however, we omit the proof here;
4. $P \boxplus_\pi \mathbf{nil} \sim_{phhp} P$. It is sufficient to prove the relation $R = \{(P \boxplus_\pi \mathbf{nil}, P)\} \cup \mathbf{Id}$ is a strongly probabilistic hhp-bisimulation, however, we omit the proof here. □

**Theorem 15.25** (Identity law for strongly probabilistic truly concurrent bisimilarities). *If $A(\widetilde{x}) \stackrel{def}{=} P$, then*

1. $A(\widetilde{y}) \sim_{pp} P\{\widetilde{y}/\widetilde{x}\}$;
2. $A(\widetilde{y}) \sim_{ps} P\{\widetilde{y}/\widetilde{x}\}$;

3. $A(\widetilde{y}) \sim_{php} P\{\widetilde{y}/\widetilde{x}\}$;
4. $A(\widetilde{y}) \sim_{phhp} P\{\widetilde{y}/\widetilde{x}\}$.

*Proof.* 1. $A(\widetilde{y}) \sim_{pp} P\{\widetilde{y}/\widetilde{x}\}$. It is sufficient to prove the relation $R = \{(A(\widetilde{y}), P\{\widetilde{y}/\widetilde{x}\})\} \cup \mathbf{Id}$ is a strongly probabilistic pomset bisimulation, however, we omit the proof here;
2. $A(\widetilde{y}) \sim_{ps} P\{\widetilde{y}/\widetilde{x}\}$. It is sufficient to prove the relation $R = \{(A(\widetilde{y}), P\{\widetilde{y}/\widetilde{x}\})\} \cup \mathbf{Id}$ is a strongly probabilistic step bisimulation, however, we omit the proof here;
3. $A(\widetilde{y}) \sim_{php} P\{\widetilde{y}/\widetilde{x}\}$. It is sufficient to prove the relation $R = \{(A(\widetilde{y}), P\{\widetilde{y}/\widetilde{x}\})\} \cup \mathbf{Id}$ is a strongly probabilistic hp-bisimulation, however, we omit the proof here;
4. $A(\widetilde{y}) \sim_{phhp} P\{\widetilde{y}/\widetilde{x}\}$. It is sufficient to prove the relation $R = \{(A(\widetilde{y}), P\{\widetilde{y}/\widetilde{x}\})\} \cup \mathbf{Id}$ is a strongly probabilistic hhp-bisimulation, however, we omit the proof here. □

**Theorem 15.26** (Restriction laws for strongly probabilistic pomset bisimilarity). *The restriction laws for strongly probabilistic pomset bisimilarity are as follows:*

1. $(y)P \sim_{pp} P$, *if* $y \notin fn(P)$;
2. $(y)(z)P \sim_{pp} (z)(y)P$;
3. $(y)(P + Q) \sim_{pp} (y)P + (y)Q$;
4. $(y)(P \boxplus_\pi Q) \sim_{pp} (y)P \boxplus_\pi (y)Q$;
5. $(y)\alpha.P \sim_{pp} \alpha.(y)P$ *if* $y \notin n(\alpha)$;
6. $(y)\alpha.P \sim_{pp}$ **nil** *if y is the subject of* $\alpha$.

*Proof.* 1. $(y)P \sim_{pp} P$, if $y \notin fn(P)$. It is sufficient to prove the relation $R = \{((y)P, P)\} \cup \mathbf{Id}$, if $y \notin fn(P)$, is a strongly probabilistic pomset bisimulation, however, we omit the proof here;
2. $(y)(z)P \sim_{pp} (z)(y)P$. It is sufficient to prove the relation $R = \{((y)(z)P, (z)(y)P)\} \cup \mathbf{Id}$ is a strongly probabilistic pomset bisimulation, however, we omit the proof here;
3. $(y)(P + Q) \sim_{pp} (y)P + (y)Q$. It is sufficient to prove the relation $R = \{((y)(P + Q), (y)P + (y)Q)\} \cup \mathbf{Id}$ is a strongly probabilistic pomset bisimulation, however, we omit the proof here;
4. $(y)(P \boxplus_\pi Q) \sim_{pp} (y)P \boxplus_\pi (y)Q$. It is sufficient to prove the relation $R = \{((y)(P \boxplus_\pi Q), (y)P \boxplus_\pi (y)Q)\} \cup \mathbf{Id}$ is a strongly probabilistic pomset bisimulation, however, we omit the proof here;
5. $(y)\alpha.P \sim_{pp} \alpha.(y)P$ if $y \notin n(\alpha)$. It is sufficient to prove the relation $R = \{((y)\alpha.P, \alpha.(y)P)\} \cup \mathbf{Id}$, if $y \notin n(\alpha)$, is a strongly probabilistic pomset bisimulation, however, we omit the proof here;
6. $(y)\alpha.P \sim_{pp}$ **nil** if $y$ is the subject of $\alpha$. It is sufficient to prove the relation $R = \{((y)\alpha.P, \mathbf{nil})\} \cup \mathbf{Id}$, if $y$ is the subject of $\alpha$, is a strongly probabilistic pomset bisimulation, however, we omit the proof here. □

**Theorem 15.27** (Restriction laws for strongly probabilistic step bisimilarity). *The restriction laws for strongly probabilistic step bisimilarity are as follows:*

1. $(y)P \sim_{ps} P$, *if* $y \notin fn(P)$;
2. $(y)(z)P \sim_{ps} (z)(y)P$;
3. $(y)(P + Q) \sim_{ps} (y)P + (y)Q$;
4. $(y)(P \boxplus_\pi Q) \sim_{ps} (y)P \boxplus_\pi (y)Q$;
5. $(y)\alpha.P \sim_{ps} \alpha.(y)P$ *if* $y \notin n(\alpha)$;
6. $(y)\alpha.P \sim_{ps}$ **nil** *if y is the subject of* $\alpha$.

*Proof.* 1. $(y)P \sim_{ps} P$, if $y \notin fn(P)$. It is sufficient to prove the relation $R = \{((y)P, P)\} \cup \mathbf{Id}$, if $y \notin fn(P)$, is a strongly probabilistic step bisimulation, however, we omit the proof here;
2. $(y)(z)P \sim_{ps} (z)(y)P$. It is sufficient to prove the relation $R = \{((y)(z)P, (z)(y)P)\} \cup \mathbf{Id}$ is a strongly probabilistic step bisimulation, however, we omit the proof here;
3. $(y)(P + Q) \sim_{ps} (y)P + (y)Q$. It is sufficient to prove the relation $R = \{((y)(P + Q), (y)P + (y)Q)\} \cup \mathbf{Id}$ is a strongly probabilistic step bisimulation, however, we omit the proof here;
4. $(y)(P \boxplus_\pi Q) \sim_{ps} (y)P \boxplus_\pi (y)Q$. It is sufficient to prove the relation $R = \{((y)(P \boxplus_\pi Q), (y)P \boxplus_\pi (y)Q)\} \cup \mathbf{Id}$ is a strongly probabilistic step bisimulation, however, we omit the proof here;
5. $(y)\alpha.P \sim_{ps} \alpha.(y)P$ if $y \notin n(\alpha)$. It is sufficient to prove the relation $R = \{((y)\alpha.P, \alpha.(y)P)\} \cup \mathbf{Id}$, if $y \notin n(\alpha)$, is a strongly probabilistic step bisimulation, however, we omit the proof here;
6. $(y)\alpha.P \sim_{ps}$ **nil** if $y$ is the subject of $\alpha$. It is sufficient to prove the relation $R = \{((y)\alpha.P, \mathbf{nil})\} \cup \mathbf{Id}$, if $y$ is the subject of $\alpha$, is a strongly probabilistic step bisimulation, however, we omit the proof here. □

**Theorem 15.28** (Restriction laws for strongly probabilistic hp-bisimilarity). *The restriction laws for strongly probabilistic hp-bisimilarity are as follows:*

1. $(y)P \sim_{php} P$, *if* $y \notin fn(P)$;
2. $(y)(z)P \sim_{php} (z)(y)P$;
3. $(y)(P+Q) \sim_{php} (y)P+(y)Q$;
4. $(y)(P \boxplus_\pi Q) \sim_{php} (y)P \boxplus_\pi (y)Q$;
5. $(y)\alpha.P \sim_{php} \alpha.(y)P$ *if* $y \notin n(\alpha)$;
6. $(y)\alpha.P \sim_{php}$ **nil** *if y is the subject of* $\alpha$.

*Proof.* 1. $(y)P \sim_{php} P$, if $y \notin fn(P)$. It is sufficient to prove the relation $R = \{((y)P, P)\} \cup \mathbf{Id}$, if $y \notin fn(P)$, is a strongly probabilistic hp-bisimulation, however, we omit the proof here;
2. $(y)(z)P \sim_{php} (z)(y)P$. It is sufficient to prove the relation $R = \{((y)(z)P, (z)(y)P)\} \cup \mathbf{Id}$ is a strongly probabilistic hp-bisimulation, however, we omit the proof here;
3. $(y)(P+Q) \sim_{php} (y)P+(y)Q$. It is sufficient to prove the relation $R = \{((y)(P+Q), (y)P+(y)Q)\} \cup \mathbf{Id}$ is a strongly probabilistic hp-bisimulation, however, we omit the proof here;
4. $(y)(P \boxplus_\pi Q) \sim_{php} (y)P \boxplus_\pi (y)Q$. It is sufficient to prove the relation $R = \{((y)(P \boxplus_\pi Q), (y)P \boxplus_\pi (y)Q)\} \cup \mathbf{Id}$ is a strongly probabilistic hp-bisimulation, however, we omit the proof here;
5. $(y)\alpha.P \sim_{php} \alpha.(y)P$ if $y \notin n(\alpha)$. It is sufficient to prove the relation $R = \{((y)\alpha.P, \alpha.(y)P)\} \cup \mathbf{Id}$, if $y \notin n(\alpha)$, is a strongly probabilistic hp-bisimulation, however, we omit the proof here;
6. $(y)\alpha.P \sim_{php}$ **nil** if $y$ is the subject of $\alpha$. It is sufficient to prove the relation $R = \{((y)\alpha.P, \mathbf{nil})\} \cup \mathbf{Id}$, if $y$ is the subject of $\alpha$, is a strongly probabilistic hp-bisimulation, however, we omit the proof here. □

**Theorem 15.29** (Restriction laws for strongly probabilistic hhp-bisimilarity). *The restriction laws for strongly probabilistic hhp-bisimilarity are as follows:*

1. $(y)P \sim_{phhp} P$, *if* $y \notin fn(P)$;
2. $(y)(z)P \sim_{phhp} (z)(y)P$;
3. $(y)(P+Q) \sim_{phhp} (y)P+(y)Q$;
4. $(y)(P \boxplus_\pi Q) \sim_{phhp} (y)P \boxplus_\pi (y)Q$;
5. $(y)\alpha.P \sim_{phhp} \alpha.(y)P$ *if* $y \notin n(\alpha)$;
6. $(y)\alpha.P \sim_{phhp}$ **nil** *if y is the subject of* $\alpha$.

*Proof.* 1. $(y)P \sim_{phhp} P$, if $y \notin fn(P)$. It is sufficient to prove the relation $R = \{((y)P, P)\} \cup \mathbf{Id}$, if $y \notin fn(P)$, is a strongly probabilistic hhp-bisimulation, however, we omit the proof here;
2. $(y)(z)P \sim_{phhp} (z)(y)P$. It is sufficient to prove the relation $R = \{((y)(z)P, (z)(y)P)\} \cup \mathbf{Id}$ is a strongly probabilistic hhp-bisimulation, however, we omit the proof here;
3. $(y)(P+Q) \sim_{phhp} (y)P+(y)Q$. It is sufficient to prove the relation $R = \{((y)(P+Q), (y)P+(y)Q)\} \cup \mathbf{Id}$ is a strongly probabilistic hhp-bisimulation, however, we omit the proof here;
4. $(y)(P \boxplus_\pi Q) \sim_{phhp} (y)P \boxplus_\pi (y)Q$. It is sufficient to prove the relation $R = \{((y)(P \boxplus_\pi Q), (y)P \boxplus_\pi (y)Q)\} \cup \mathbf{Id}$ is a strongly probabilistic hhp-bisimulation, however, we omit the proof here;
5. $(y)\alpha.P \sim_{phhp} \alpha.(y)P$ if $y \notin n(\alpha)$. It is sufficient to prove the relation $R = \{((y)\alpha.P, \alpha.(y)P)\} \cup \mathbf{Id}$, if $y \notin n(\alpha)$, is a strongly probabilistic hhp-bisimulation, however, we omit the proof here;
6. $(y)\alpha.P \sim_{phhp}$ **nil** if $y$ is the subject of $\alpha$. It is sufficient to prove the relation $R = \{((y)\alpha.P, \mathbf{nil})\} \cup \mathbf{Id}$, if $y$ is the subject of $\alpha$, is a strongly probabilistic hhp-bisimulation, however, we omit the proof here. □

**Theorem 15.30** (Parallel laws for strongly probabilistic pomset bisimilarity). *The parallel laws for strongly probabilistic pomset bisimilarity are as follows:*

1. $P \parallel \mathbf{nil} \sim_{pp} P$;
2. $P_1 \parallel P_2 \sim_{pp} P_2 \parallel P_1$;
3. $(P_1 \parallel P_2) \parallel P_3 \sim_{pp} P_1 \parallel (P_2 \parallel P_3)$;
4. $(y)(P_1 \parallel P_2) \sim_{pp} (y)P_1 \parallel (y)P_2$, *if* $y \notin fn(P_1) \cap fn(P_2)$.

*Proof.* 1. $P \parallel \mathbf{nil} \sim_{pp} P$. It is sufficient to prove the relation $R = \{(P \parallel \mathbf{nil}, P)\} \cup \mathbf{Id}$ is a strongly probabilistic pomset bisimulation, however, we omit the proof here;
2. $P_1 \parallel P_2 \sim_{pp} P_2 \parallel P_1$. It is sufficient to prove the relation $R = \{(P_1 \parallel P_2, P_2 \parallel P_1)\} \cup \mathbf{Id}$ is a strongly probabilistic pomset bisimulation, however, we omit the proof here;

3. $(P_1 \parallel P_2) \parallel P_3 \sim_{pp} P_1 \parallel (P_2 \parallel P_3)$. It is sufficient to prove the relation $R = \{((P_1 \parallel P_2) \parallel P_3, P_1 \parallel (P_2 \parallel P_3))\} \cup \mathbf{Id}$ is a strongly probabilistic pomset bisimulation, however, we omit the proof here;
4. $(y)(P_1 \parallel P_2) \sim_{pp} (y)P_1 \parallel (y)P_2$, if $y \notin fn(P_1) \cap fn(P_2)$. It is sufficient to prove the relation $R = \{((y)(P_1 \parallel P_2), (y)P_1 \parallel (y)P_2)\} \cup \mathbf{Id}$, if $y \notin fn(P_1) \cap fn(P_2)$, is a strongly probabilistic pomset bisimulation, however, we omit the proof here. □

**Theorem 15.31** (Parallel laws for strongly probabilistic step bisimilarity). *The parallel laws for strongly probabilistic step bisimilarity are as follows:*

1. $P \parallel \mathbf{nil} \sim_{ps} P$;
2. $P_1 \parallel P_2 \sim_{ps} P_2 \parallel P_1$;
3. $(P_1 \parallel P_2) \parallel P_3 \sim_{ps} P_1 \parallel (P_2 \parallel P_3)$;
4. $(y)(P_1 \parallel P_2) \sim_{ps} (y)P_1 \parallel (y)P_2$, *if* $y \notin fn(P_1) \cap fn(P_2)$.

*Proof.* 1. $P \parallel \mathbf{nil} \sim_{ps} P$. It is sufficient to prove the relation $R = \{(P \parallel \mathbf{nil}, P)\} \cup \mathbf{Id}$ is a strongly probabilistic step bisimulation, however, we omit the proof here;
2. $P_1 \parallel P_2 \sim_{ps} P_2 \parallel P_1$. It is sufficient to prove the relation $R = \{(P_1 \parallel P_2, P_2 \parallel P_1)\} \cup \mathbf{Id}$ is a strongly probabilistic step bisimulation, however, we omit the proof here;
3. $(P_1 \parallel P_2) \parallel P_3 \sim_{ps} P_1 \parallel (P_2 \parallel P_3)$. It is sufficient to prove the relation $R = \{((P_1 \parallel P_2) \parallel P_3, P_1 \parallel (P_2 \parallel P_3))\} \cup \mathbf{Id}$ is a strongly probabilistic step bisimulation, however, we omit the proof here;
4. $(y)(P_1 \parallel P_2) \sim_{ps} (y)P_1 \parallel (y)P_2$, if $y \notin fn(P_1) \cap fn(P_2)$. It is sufficient to prove the relation $R = \{((y)(P_1 \parallel P_2), (y)P_1 \parallel (y)P_2)\} \cup \mathbf{Id}$, if $y \notin fn(P_1) \cap fn(P_2)$, is a strongly probabilistic step bisimulation, however, we omit the proof here. □

**Theorem 15.32** (Parallel laws for strongly probabilistic hp-bisimilarity). *The parallel laws for strongly probabilistic hp-bisimilarity are as follows:*

1. $P \parallel \mathbf{nil} \sim_{php} P$;
2. $P_1 \parallel P_2 \sim_{php} P_2 \parallel P_1$;
3. $(P_1 \parallel P_2) \parallel P_3 \sim_{php} P_1 \parallel (P_2 \parallel P_3)$;
4. $(y)(P_1 \parallel P_2) \sim_{php} (y)P_1 \parallel (y)P_2$, *if* $y \notin fn(P_1) \cap fn(P_2)$.

*Proof.* 1. $P \parallel \mathbf{nil} \sim_{php} P$. It is sufficient to prove the relation $R = \{(P \parallel \mathbf{nil}, P)\} \cup \mathbf{Id}$ is a strongly probabilistic hp-bisimulation, however, we omit the proof here;
2. $P_1 \parallel P_2 \sim_{php} P_2 \parallel P_1$. It is sufficient to prove the relation $R = \{(P_1 \parallel P_2, P_2 \parallel P_1)\} \cup \mathbf{Id}$ is a strongly probabilistic hp-bisimulation, however, we omit the proof here;
3. $(P_1 \parallel P_2) \parallel P_3 \sim_{php} P_1 \parallel (P_2 \parallel P_3)$. It is sufficient to prove the relation $R = \{((P_1 \parallel P_2) \parallel P_3, P_1 \parallel (P_2 \parallel P_3))\} \cup \mathbf{Id}$ is a strongly probabilistic hp-bisimulation, however, we omit the proof here;
4. $(y)(P_1 \parallel P_2) \sim_{php} (y)P_1 \parallel (y)P_2$, if $y \notin fn(P_1) \cap fn(P_2)$. It is sufficient to prove the relation $R = \{((y)(P_1 \parallel P_2), (y)P_1 \parallel (y)P_2)\} \cup \mathbf{Id}$, if $y \notin fn(P_1) \cap fn(P_2)$, is a strongly probabilistic hp-bisimulation, however, we omit the proof here. □

**Theorem 15.33** (Parallel laws for strongly probabilistic hhp-bisimilarity). *The parallel laws for strongly probabilistic hhp-bisimilarity are as follows:*

1. $P \parallel \mathbf{nil} \sim_{phhp} P$;
2. $P_1 \parallel P_2 \sim_{phhp} P_2 \parallel P_1$;
3. $(P_1 \parallel P_2) \parallel P_3 \sim_{phhp} P_1 \parallel (P_2 \parallel P_3)$;
4. $(y)(P_1 \parallel P_2) \sim_{phhp} (y)P_1 \parallel (y)P_2$, *if* $y \notin fn(P_1) \cap fn(P_2)$.

*Proof.* 1. $P \parallel \mathbf{nil} \sim_{phhp} P$. It is sufficient to prove the relation $R = \{(P \parallel \mathbf{nil}, P)\} \cup \mathbf{Id}$ is a strongly probabilistic hhp-bisimulation, however, we omit the proof here;
2. $P_1 \parallel P_2 \sim_{phhp} P_2 \parallel P_1$. It is sufficient to prove the relation $R = \{(P_1 \parallel P_2, P_2 \parallel P_1)\} \cup \mathbf{Id}$ is a strongly probabilistic hhp-bisimulation, however, we omit the proof here;
3. $(P_1 \parallel P_2) \parallel P_3 \sim_{phhp} P_1 \parallel (P_2 \parallel P_3)$. It is sufficient to prove the relation $R = \{((P_1 \parallel P_2) \parallel P_3, P_1 \parallel (P_2 \parallel P_3))\} \cup \mathbf{Id}$ is a strongly probabilistic hhp-bisimulation, however, we omit the proof here;
4. $(y)(P_1 \parallel P_2) \sim_{phhp} (y)P_1 \parallel (y)P_2$, if $y \notin fn(P_1) \cap fn(P_2)$. It is sufficient to prove the relation $R = \{((y)(P_1 \parallel P_2), (y)P_1 \parallel (y)P_2)\} \cup \mathbf{Id}$, if $y \notin fn(P_1) \cap fn(P_2)$, is a strongly probabilistic hhp-bisimulation, however, we omit the proof here. □

**Theorem 15.34** (Expansion law for truly concurrent bisimilarities). *Let* $P \equiv \boxplus_i \sum_j \alpha_{ij}.P_{ij}$ *and* $Q \equiv \boxplus_k \sum_l \beta_{kl}.Q_{kl}$, *where* $bn(\alpha_{ij}) \cap fn(Q) = \emptyset$ *for all* $i, j$, *and* $bn(\beta_{kl}) \cap fn(P) = \emptyset$ *for all* $k, l$. *Then,*

1. $P \parallel Q \sim_{pp} \boxplus_i \boxplus_k \sum_j \sum_l (\alpha_{ij} \parallel \beta_{kl}).(P_{ij} \parallel Q_{kl}) + \boxplus_i \boxplus_k \sum_{\alpha_{ij}\ comp\ \beta_{kl}} \tau.R_{ijkl}$;
2. $P \parallel Q \sim_{ps} \boxplus_i \boxplus_k \sum_j \sum_l (\alpha_{ij} \parallel \beta_{kl}).(P_{ij} \parallel Q_{kl}) + \boxplus_i \boxplus_k \sum_{\alpha_{ij}\ comp\ \beta_{kl}} \tau.R_{ijkl}$;
3. $P \parallel Q \sim_{php} \boxplus_i \boxplus_k \sum_j \sum_l (\alpha_{ij} \parallel \beta_{kl}).(P_{ij} \parallel Q_{kl}) + \boxplus_i \boxplus_k \sum_{\alpha_{ij}\ comp\ \beta_{kl}} \tau.R_{ijkl}$;
4. $P \parallel Q \nsim_{phhp} \boxplus_i \boxplus_k \sum_j \sum_l (\alpha_{ij} \parallel \beta_{kl}).(P_{ij} \parallel Q_{kl}) + \boxplus_i \boxplus_k \sum_{\alpha_{ij}\ comp\ \beta_{kl}} \tau.R_{ijkl}$,

*where* $\alpha_{ij}\ comp\ \beta_{kl}$ *and* $R_{ijkl}$ *are defined as follows:*

1. $\alpha_{ij}$ *is* $\overline{x}u$ *and* $\beta_{kl}$ *is* $x(v)$, *then* $R_{ijkl} = P_{ij} \parallel Q_{kl}\{u/v\}$;
2. $\alpha_{ij}$ *is* $\overline{x}(u)$ *and* $\beta_{kl}$ *is* $x(v)$, *then* $R_{ijkl} = (w)(P_{ij}\{w/u\} \parallel Q_{kl}\{w/v\})$, *if* $w \notin fn((u)P_{ij}) \cup fn((v)Q_{kl})$;
3. $\alpha_{ij}$ *is* $x(v)$ *and* $\beta_{kl}$ *is* $\overline{x}u$, *then* $R_{ijkl} = P_{ij}\{u/v\} \parallel Q_{kl}$;
4. $\alpha_{ij}$ *is* $x(v)$ *and* $\beta_{kl}$ *is* $\overline{x}(u)$, *then* $R_{ijkl} = (w)(P_{ij}\{w/v\} \parallel Q_{kl}\{w/u\})$, *if* $w \notin fn((v)P_{ij}) \cup fn((u)Q_{kl})$.

*Proof.* According to the definition of strongly probabilistic truly concurrent bisimulations, we can easily prove the above equations, hence, we omit the proof here. □

**Theorem 15.35** (Equivalence and congruence for strongly probabilistic pomset bisimilarity). *We can enjoy the full congruence modulo strongly probabilistic pomset bisimilarity:*

1. $\sim_{pp}$ *is an equivalence relation;*
2. *If* $P \sim_{pp} Q$, *then*
   (a) $\alpha.P \sim_{pp} \alpha.Q$, $\alpha$ *is a free action;*
   (b) $\phi.P \sim_{pp} \phi.Q$;
   (c) $P + R \sim_{pp} Q + R$;
   (d) $P \boxplus_\pi R \sim_{pp} Q \boxplus_\pi R$;
   (e) $P \parallel R \sim_{pp} Q \parallel R$;
   (f) $(w)P \sim_{pp} (w)Q$;
   (g) $x(y).P \sim_{pp} x(y).Q$.

*Proof.* **1.** $\sim_{pp}$ is an equivalence relation, it is obvious that;
**2.** If $P \sim_{pp} Q$, then
   (a) $\alpha.P \sim_{pp} \alpha.Q$, $\alpha$ is a free action. It is sufficient to prove the relation $R = \{(\alpha.P, \alpha.Q)\} \cup \mathbf{Id}$ is a strongly probabilistic pomset bisimulation, however, we omit the proof here;
   (b) $\phi.P \sim_{pp} \phi.Q$. It is sufficient to prove the relation $R = \{(\phi.P, \phi.Q)\} \cup \mathbf{Id}$ is a strongly probabilistic pomset bisimulation, however, we omit the proof here;
   (c) $P + R \sim_{pp} Q + R$. It is sufficient to prove the relation $R = \{(P + R, Q + R)\} \cup \mathbf{Id}$ is a strongly probabilistic pomset bisimulation, however, we omit the proof here;
   (d) $P \boxplus_\pi R \sim_{pp} Q \boxplus_\pi R$. It is sufficient to prove the relation $R = \{(P \boxplus_\pi R, Q \boxplus_\pi R)\} \cup \mathbf{Id}$ is a strongly probabilistic pomset bisimulation, however, we omit the proof here;
   (e) $P \parallel R \sim_{pp} Q \parallel R$. It is sufficient to prove the relation $R = \{(P \parallel R, Q \parallel R)\} \cup \mathbf{Id}$ is a strongly probabilistic pomset bisimulation, however, we omit the proof here;
   (f) $(w)P \sim_{pp} (w)Q$. It is sufficient to prove the relation $R = \{((w)P, (w)Q)\} \cup \mathbf{Id}$ is a strongly probabilistic pomset bisimulation, however, we omit the proof here;
   (g) $x(y).P \sim_{pp} x(y).Q$. It is sufficient to prove the relation $R = \{(x(y).P, x(y).Q)\} \cup \mathbf{Id}$ is a strongly probabilistic pomset bisimulation, however, we omit the proof here. □

**Theorem 15.36** (Equivalence and congruence for strongly probabilistic step bisimilarity). *We can enjoy the full congruence modulo strongly probabilistic step bisimilarity:*

1. $\sim_{ps}$ *is an equivalence relation;*
2. *If* $P \sim_{ps} Q$, *then*
   (a) $\alpha.P \sim_{ps} \alpha.Q$, $\alpha$ *is a free action;*
   (b) $\phi.P \sim_{ps} \phi.Q$;
   (c) $P + R \sim_{ps} Q + R$;
   (d) $P \boxplus_\pi R \sim_{ps} Q \boxplus_\pi R$;
   (e) $P \parallel R \sim_{ps} Q \parallel R$;

(f) $(w)P \sim_{ps} (w)Q$;
(g) $x(y).P \sim_{ps} x(y).Q$.

*Proof.* **1.** $\sim_{ps}$ is an equivalence relation, it is obvious that;

**2.** If $P \sim_{ps} Q$, then

(a) $\alpha.P \sim_{ps} \alpha.Q$, $\alpha$ is a free action. It is sufficient to prove the relation $R = \{(\alpha.P, \alpha.Q)\} \cup \mathbf{Id}$ is a strongly probabilistic step bisimulation, however, we omit the proof here;
(b) $\phi.P \sim_{ps} \phi.Q$. It is sufficient to prove the relation $R = \{(\phi.P, \phi.Q)\} \cup \mathbf{Id}$ is a strongly probabilistic step bisimulation, however, we omit the proof here;
(c) $P + R \sim_{ps} Q + R$. It is sufficient to prove the relation $R = \{(P + R, Q + R)\} \cup \mathbf{Id}$ is a strongly probabilistic step bisimulation, however, we omit the proof here;
(d) $P \boxplus_{\pi} R \sim_{ps} Q \boxplus_{\pi} R$. It is sufficient to prove the relation $R = \{(P \boxplus_{\pi} R, Q \boxplus_{\pi} R)\} \cup \mathbf{Id}$ is a strongly probabilistic step bisimulation, however, we omit the proof here;
(e) $P \parallel R \sim_{ps} Q \parallel R$. It is sufficient to prove the relation $R = \{(P \parallel R, Q \parallel R)\} \cup \mathbf{Id}$ is a strongly probabilistic step bisimulation, however, we omit the proof here;
(f) $(w)P \sim_{ps} (w)Q$. It is sufficient to prove the relation $R = \{((w)P, (w)Q)\} \cup \mathbf{Id}$ is a strongly probabilistic step bisimulation, however, we omit the proof here;
(g) $x(y).P \sim_{ps} x(y).Q$. It is sufficient to prove the relation $R = \{(x(y).P, x(y).Q)\} \cup \mathbf{Id}$ is a strongly probabilistic step bisimulation, however, we omit the proof here. □

**Theorem 15.37** (Equivalence and congruence for strongly probabilistic hp-bisimilarity). *We can enjoy the full congruence modulo strongly probabilistic hp-bisimilarity:*

**1.** $\sim_{php}$ *is an equivalence relation;*

**2.** *If* $P \sim_{php} Q$, *then*

(a) $\alpha.P \sim_{php} \alpha.Q$, *$\alpha$ is a free action;*
(b) $\phi.P \sim_{php} \phi.Q$;
(c) $P + R \sim_{php} Q + R$;
(d) $P \boxplus_{\pi} R \sim_{php} Q \boxplus_{\pi} R$;
(e) $P \parallel R \sim_{php} Q \parallel R$;
(f) $(w)P \sim_{php} (w)Q$;
(g) $x(y).P \sim_{php} x(y).Q$.

*Proof.* **1.** $\sim_{php}$ is an equivalence relation, it is obvious that;

**2.** If $P \sim_{php} Q$, then

(a) $\alpha.P \sim_{php} \alpha.Q$, $\alpha$ is a free action. It is sufficient to prove the relation $R = \{(\alpha.P, \alpha.Q)\} \cup \mathbf{Id}$ is a strongly probabilistic hp-bisimulation, however, we omit the proof here;
(b) $\phi.P \sim_{php} \phi.Q$. It is sufficient to prove the relation $R = \{(\phi.P, \phi.Q)\} \cup \mathbf{Id}$ is a strongly probabilistic hp-bisimulation, however, we omit the proof here;
(c) $P + R \sim_{php} Q + R$. It is sufficient to prove the relation $R = \{(P + R, Q + R)\} \cup \mathbf{Id}$ is a strongly probabilistic hp-bisimulation, however, we omit the proof here;
(d) $P \boxplus_{\pi} R \sim_{php} Q \boxplus_{\pi} R$. It is sufficient to prove the relation $R = \{(P \boxplus_{\pi} R, Q \boxplus_{\pi} R)\} \cup \mathbf{Id}$ is a strongly probabilistic hp-bisimulation, however, we omit the proof here;
(e) $P \parallel R \sim_{php} Q \parallel R$. It is sufficient to prove the relation $R = \{(P \parallel R, Q \parallel R)\} \cup \mathbf{Id}$ is a strongly probabilistic hp-bisimulation, however, we omit the proof here;
(f) $(w)P \sim_{php} (w)Q$. It is sufficient to prove the relation $R = \{((w)P, (w)Q)\} \cup \mathbf{Id}$ is a strongly probabilistic hp-bisimulation, however, we omit the proof here;
(g) $x(y).P \sim_{php} x(y).Q$. It is sufficient to prove the relation $R = \{(x(y).P, x(y).Q)\} \cup \mathbf{Id}$ is a strongly probabilistic hp-bisimulation, however, we omit the proof here. □

**Theorem 15.38** (Equivalence and congruence for strongly probabilistic hhp-bisimilarity). *We can enjoy the full congruence modulo strongly probabilistic hhp-bisimilarity:*

**1.** $\sim_{phhp}$ *is an equivalence relation;*

**2.** *If* $P \sim_{phhp} Q$, *then*

(a) $\alpha.P \sim_{phhp} \alpha.Q$, *$\alpha$ is a free action;*
(b) $\phi.P \sim_{phhp} \phi.Q$;

(c) $P + R \sim_{phhp} Q + R$;
(d) $P \boxplus_\pi R \sim_{phhp} Q \boxplus_\pi R$;
(e) $P \parallel R \sim_{phhp} Q \parallel R$;
(f) $(w)P \sim_{phhp} (w)Q$;
(g) $x(y).P \sim_{phhp} x(y).Q$.

*Proof.* **1.** $\sim_{phhp}$ is an equivalence relation, it is obvious that;
**2.** If $P \sim_{phhp} Q$, then
(a) $\alpha.P \sim_{phhp} \alpha.Q$, $\alpha$ is a free action. It is sufficient to prove the relation $R = \{(\alpha.P, \alpha.Q)\} \cup \mathbf{Id}$ is a strongly probabilistic hhp-bisimulation, however, we omit the proof here;
(b) $\phi.P \sim_{phhp} \phi.Q$. It is sufficient to prove the relation $R = \{(\phi.P, \phi.Q)\} \cup \mathbf{Id}$ is a strongly probabilistic hhp-bisimulation, however, we omit the proof here;
(c) $P + R \sim_{phhp} Q + R$. It is sufficient to prove the relation $R = \{(P + R, Q + R)\} \cup \mathbf{Id}$ is a strongly probabilistic hhp-bisimulation, however, we omit the proof here;
(d) $P \boxplus_\pi R \sim_{phhp} Q \boxplus_\pi R$. It is sufficient to prove the relation $R = \{(P \boxplus_\pi R, Q \boxplus_\pi R)\} \cup \mathbf{Id}$ is a strongly probabilistic hhp-bisimulation, however, we omit the proof here;
(e) $P \parallel R \sim_{phhp} Q \parallel R$. It is sufficient to prove the relation $R = \{(P \parallel R, Q \parallel R)\} \cup \mathbf{Id}$ is a strongly probabilistic hhp-bisimulation, however, we omit the proof here;
(f) $(w)P \sim_{phhp} (w)Q$. It is sufficient to prove the relation $R = \{((w)P, (w)Q)\} \cup \mathbf{Id}$ is a strongly probabilistic hhp-bisimulation, however, we omit the proof here;
(g) $x(y).P \sim_{phhp} x(y).Q$. It is sufficient to prove the relation $R = \{(x(y).P, x(y).Q)\} \cup \mathbf{Id}$ is a strongly probabilistic hhp-bisimulation, however, we omit the proof here. □

### 15.3.2 Recursion

**Definition 15.39.** Let $X$ have arity $n$, and let $\widetilde{x} = x_1, \cdots, x_n$ be distinct names, and $fn(P) \subseteq \{x_1, \cdots, x_n\}$. The replacement of $X(\widetilde{x})$ by $P$ in $E$, written $E\{X(\widetilde{x}) := P\}$, means the result of replacing each subterm $X(\widetilde{y})$ in $E$ by $P\{\widetilde{y}/\widetilde{x}\}$.

**Definition 15.40.** Let $E$ and $F$ be two process expressions containing only $X_1, \cdots, X_m$ with associated name sequences $\widetilde{x}_1, \cdots, \widetilde{x}_m$. Then,

1. $E \sim_{pp} F$ means $E(\widetilde{P}) \sim_{pp} F(\widetilde{P})$;
2. $E \sim_{ps} F$ means $E(\widetilde{P}) \sim_{ps} F(\widetilde{P})$;
3. $E \sim_{php} F$ means $E(\widetilde{P}) \sim_{php} F(\widetilde{P})$;
4. $E \sim_{phhp} F$ means $E(\widetilde{P}) \sim_{phhp} F(\widetilde{P})$;

for all $\widetilde{P}$ such that $fn(P_i) \subseteq \widetilde{x}_i$ for each $i$.

**Definition 15.41.** A term or identifier is weakly guarded in $P$ if it lies within some sub-term $\alpha.Q$ or $(\alpha_1 \parallel \cdots \parallel \alpha_n).Q$ of $P$.

**Theorem 15.42.** *Assume that $\widetilde{E}$ and $\widetilde{F}$ are expressions containing only $X_i$ with $\widetilde{x}_i$, and $\widetilde{A}$ and $\widetilde{B}$ are identifiers with $A_i$, $B_i$. Then, for all $i$,*

1. $E_i \sim_{ps} F_i$, $A_i(\widetilde{x}_i) \overset{\text{def}}{=} E_i(\widetilde{A})$, $B_i(\widetilde{x}_i) \overset{\text{def}}{=} F_i(\widetilde{B})$, *then* $A_i(\widetilde{x}_i) \sim_{ps} B_i(\widetilde{x}_i)$;
2. $E_i \sim_{pp} F_i$, $A_i(\widetilde{x}_i) \overset{\text{def}}{=} E_i(\widetilde{A})$, $B_i(\widetilde{x}_i) \overset{\text{def}}{=} F_i(\widetilde{B})$, *then* $A_i(\widetilde{x}_i) \sim_{pp} B_i(\widetilde{x}_i)$;
3. $E_i \sim_{php} F_i$, $A_i(\widetilde{x}_i) \overset{\text{def}}{=} E_i(\widetilde{A})$, $B_i(\widetilde{x}_i) \overset{\text{def}}{=} F_i(\widetilde{B})$, *then* $A_i(\widetilde{x}_i) \sim_{php} B_i(\widetilde{x}_i)$;
4. $E_i \sim_{phhp} F_i$, $A_i(\widetilde{x}_i) \overset{\text{def}}{=} E_i(\widetilde{A})$, $B_i(\widetilde{x}_i) \overset{\text{def}}{=} F_i(\widetilde{B})$, *then* $A_i(\widetilde{x}_i) \sim_{phhp} B_i(\widetilde{x}_i)$.

*Proof.* **1.** $E_i \sim_{ps} F_i$, $A_i(\widetilde{x}_i) \overset{\text{def}}{=} E_i(\widetilde{A})$, $B_i(\widetilde{x}_i) \overset{\text{def}}{=} F_i(\widetilde{B})$, then $A_i(\widetilde{x}_i) \sim_{ps} B_i(\widetilde{x}_i)$.
We will consider the case $I = \{1\}$ with loss of generality, and show the following relation $R$ is a strongly probabilistic step bisimulation.

$$R = \{(G(A), G(B)) : G \text{ has only identifier } X\}$$

By choosing $G \equiv X(\widetilde{y})$, it follows that $A(\widetilde{y}) \sim_{ps} B(\widetilde{y})$. It is sufficient to prove the following:

(a) If $\langle G(A), s\rangle \rightsquigarrow \xrightarrow{\{\alpha_1,\cdots,\alpha_n\}} \langle P', s'\rangle$, where $\alpha_i (1 \le i \le n)$ is a free action or bound output action with $bn(\alpha_1) \cap \cdots \cap bn(\alpha_n) \cap n(G(A), G(B)) = \emptyset$, then $\langle G(B), s\rangle \rightsquigarrow \xrightarrow{\{\alpha_1,\cdots,\alpha_n\}} \langle Q'', s''\rangle$ such that $P' \sim_{ps} Q''$;

(b) If $\langle G(A), s\rangle \rightsquigarrow \xrightarrow{x(y)} \langle P', s'\rangle$ with $x \notin n(G(A), G(B))$, then $\langle G(B), s\rangle \rightsquigarrow \xrightarrow{x(y)} \langle Q'', s''\rangle$, such that for all $u$, $\langle P', s'\rangle\{u/y\} \sim_{ps} \langle Q''\{u/y\}, s''\rangle$.

To prove the above properties, it is sufficient to induct on the depth of inference and it is quite routine, however, we omit the proof here.

2. $E_i \sim_{pp} F_i$, $A_i(\widetilde{x_i}) \overset{\text{def}}{=} E_i(\widetilde{A})$, $B_i(\widetilde{x_i}) \overset{\text{def}}{=} F_i(\widetilde{B})$, then $A_i(\widetilde{x_i}) \sim_{pp} B_i(\widetilde{x_i})$. It can be proven similarly to the above case.
3. $E_i \sim_{php} F_i$, $A_i(\widetilde{x_i}) \overset{\text{def}}{=} E_i(\widetilde{A})$, $B_i(\widetilde{x_i}) \overset{\text{def}}{=} F_i(\widetilde{B})$, then $A_i(\widetilde{x_i}) \sim_{php} B_i(\widetilde{x_i})$. It can be proven similarly to the above case.
4. $E_i \sim_{phhp} F_i$, $A_i(\widetilde{x_i}) \overset{\text{def}}{=} E_i(\widetilde{A})$, $B_i(\widetilde{x_i}) \overset{\text{def}}{=} F_i(\widetilde{B})$, then $A_i(\widetilde{x_i}) \sim_{phhp} B_i(\widetilde{x_i})$. It can be proven similarly to the above case. □

**Theorem 15.43** (Unique solution of equations). *Assume $\widetilde{E}$ are expressions containing only $X_i$ with $\widetilde{x_i}$, and each $X_i$ is weakly guarded in each $E_j$. Assume that $\widetilde{P}$ and $\widetilde{Q}$ are processes such that $fn(P_i) \subseteq \widetilde{x_i}$ and $fn(Q_i) \subseteq \widetilde{x_i}$. Then, for all i,*

1. *if $P_i \sim_{pp} E_i(\widetilde{P})$, $Q_i \sim_{pp} E_i(\widetilde{Q})$, then $P_i \sim_{pp} Q_i$;*
2. *if $P_i \sim_{ps} E_i(\widetilde{P})$, $Q_i \sim_{ps} E_i(\widetilde{Q})$, then $P_i \sim_{ps} Q_i$;*
3. *if $P_i \sim_{php} E_i(\widetilde{P})$, $Q_i \sim_{php} E_i(\widetilde{Q})$, then $P_i \sim_{php} Q_i$;*
4. *if $P_i \sim_{phhp} E_i(\widetilde{P})$, $Q_i \sim_{phhp} E_i(\widetilde{Q})$, then $P_i \sim_{phhp} Q_i$.*

*Proof.* 1. It is similar to the proof of the unique solution of equations for strongly probabilistic pomset bisimulation in CTC, please refer to Chapter 10 for details, however, we omit the proof here;
2. It is similar to the proof of unique solution of equations for strongly probabilistic step bisimulation in CTC, please refer to Chapter 10 for details, however, we omit the proof here;
3. It is similar to the proof of unique solution of equations for strongly probabilistic hp-bisimulation in CTC, please refer to Chapter 10 for details, however, we omit the proof here;
4. It is similar to the proof of unique solution of equations for strongly probabilistic hhp-bisimulation in CTC, please refer to Chapter 10 for details, however, we omit the proof here. □

## 15.4 Algebraic theory

**Definition 15.44** (STC). The theory **STC** consists of the following axioms and inference rules:

1. Alpha-conversion **A**.

$$\text{if } P \equiv Q, \text{ then } P = Q;$$

2. Congruence **C**. If $P = Q$, then

$$\tau.P = \tau.Q \quad \overline{x}y.P = \overline{x}y.Q$$
$$P + R = Q + R \quad P \parallel R = Q \parallel R$$
$$(x)P = (x)Q \quad x(y).P = x(y).Q;$$

3. Summation **S**.

$$\textbf{S0} \quad P + \textbf{nil} = P$$
$$\textbf{S1} \quad P + P = P$$
$$\textbf{S2} \quad P + Q = Q + P$$
$$\textbf{S3} \quad P + (Q + R) = (P + Q) + R;$$

4. Box-Summation $(BS)$.

$$\textbf{BS0} \quad P \boxplus_\pi \textbf{nil} = P$$
$$\textbf{BS1} \quad P \boxplus_\pi P = P$$
$$\textbf{BS2} \quad P \boxplus_\pi Q = Q \boxplus_{1-\pi} P$$

$$\textbf{BS3} \quad P \boxplus_{\pi} (Q \boxplus_{\rho} R) = (P \boxplus_{\frac{\pi}{\pi+\rho-\pi\rho}} Q) \boxplus_{\pi+\rho-\pi\rho} R;$$

5. Restriction **R**.

$$\textbf{R0} \quad (x)P = P \quad \text{if } x \notin fn(P)$$

$$\textbf{R1} \quad (x)(y)P = (y)(x)P$$

$$\textbf{R2} \quad (x)(P+Q) = (x)P + (x)Q$$

$$\textbf{R3} \quad (x)\alpha.P = \alpha.(x)P \quad \text{if } x \notin n(\alpha)$$

$$\textbf{R4} \quad (x)\alpha.P = \textbf{nil} \quad \text{if } x \text{is the subject of } \alpha;$$

6. Expansion **E**. Let $P \equiv \boxplus_i \sum_j \alpha_{ij}.P_{ij}$ and $Q \equiv \boxplus_k \sum_l \beta_{kl}.Q_{kl}$, where $bn(\alpha_{ij}) \cap fn(Q) = \emptyset$ for all $i, j$, and $bn(\beta_{kl}) \cap fn(P) = \emptyset$ for all $k, l$. Then,
   (a) $P \parallel Q \sim_{pp} \boxplus_i \boxplus_k \sum_j \sum_l (\alpha_{ij} \parallel \beta_{kl}).(P_{ij} \parallel Q_{kl}) + \boxplus_i \boxplus_k \sum_{\alpha_{ij} \text{ comp } \beta_{kl}} \tau.R_{ijkl}$;
   (b) $P \parallel Q \sim_{ps} \boxplus_i \boxplus_k \sum_j \sum_l (\alpha_{ij} \parallel \beta_{kl}).(P_{ij} \parallel Q_{kl}) + \boxplus_i \boxplus_k \sum_{\alpha_{ij} \text{ comp } \beta_{kl}} \tau.R_{ijkl}$;
   (c) $P \parallel Q \sim_{php} \boxplus_i \boxplus_k \sum_j \sum_l (\alpha_{ij} \parallel \beta_{kl}).(P_{ij} \parallel Q_{kl}) + \boxplus_i \boxplus_k \sum_{\alpha_{ij} \text{ comp } \beta_{kl}} \tau.R_{ijkl}$;
   (d) $P \parallel Q \nsim_{phhp} \boxplus_i \boxplus_k \sum_j \sum_l (\alpha_{ij} \parallel \beta_{kl}).(P_{ij} \parallel Q_{kl}) + \boxplus_i \boxplus_k \sum_{\alpha_{ij} \text{ comp } \beta_{kl}} \tau.R_{ijkl}$,
   where $\alpha_{ij}$ comp $\beta_{kl}$ and $R_{ijkl}$ are defined as follows:
   (a) $\alpha_{ij}$ is $\overline{x}u$ and $\beta_{kl}$ is $x(v)$, then $R_{ijkl} = P_{ij} \parallel Q_{kl}\{u/v\}$;
   (b) $\alpha_{ij}$ is $\overline{x}(u)$ and $\beta_{kl}$ is $x(v)$, then $R_{ijkl} = (w)(P_{ij}\{w/u\} \parallel Q_{kl}\{w/v\})$, if $w \notin fn((u)P_{ij}) \cup fn((v)Q_{kl})$;
   (c) $\alpha_{ij}$ is $x(v)$ and $\beta_{kl}$ is $\overline{x}u$, then $R_{ijkl} = P_{ij}\{u/v\} \parallel Q_{kl}$;
   (d) $\alpha_{ij}$ is $x(v)$ and $\beta_{kl}$ is $\overline{x}(u)$, then $R_{ijkl} = (w)(P_{ij}\{w/v\} \parallel Q_{kl}\{w/u\})$, if $w \notin fn((v)P_{ij}) \cup fn((u)Q_{kl})$.
7. Identifier **I**.

$$\text{If } A(\widetilde{x}) \overset{\text{def}}{=} P, \text{ then } A(\widetilde{y}) = P\{\widetilde{y}/\widetilde{x}\}.$$

**Theorem 15.45** (Soundness). *If* $STC \vdash P = Q$*, then*

1. $P \sim_{pp} Q$;
2. $P \sim_{pp} Q$;
3. $P \sim_{php} Q$;
4. $P \sim_{phhp} Q$.

*Proof.* The soundness of these laws modulo strongly truly concurrent bisimilarities was already proven in Section 15.3. □

**Definition 15.46.** The agent identifier $A$ is weakly guardedly defined if every agent identifier is weakly guarded in the right-hand side of the definition of $A$.

**Definition 15.47** (Head normal form). A Process $P$ is in head normal form if it is a sum of the prefixes:

$$P \equiv \boxplus_i \sum_j (\alpha_{ij1} \parallel \cdots \parallel \alpha_{ijn}).P_{ij}$$

**Proposition 15.48.** *If every agent identifier is weakly guardedly defined, then for any process P, there is a head normal form H such that*

$$STC \vdash P = H$$

*Proof.* It is sufficient to induct on the structure of $P$ and it is quite obvious. □

**Theorem 15.49** (Completeness). *For all processes P and Q,*

1. *if* $P \sim_{pp} Q$*, then* $STC \vdash P = Q$;
2. *if* $P \sim_{pp} Q$*, then* $STC \vdash P = Q$;
3. *if* $P \sim_{php} Q$*, then* $STC \vdash P = Q$.

*Proof.* 1. if $P \sim_{pp} Q$, then $\textbf{STC} \vdash P = Q$.

Since $P$ and $Q$ all have head normal forms, let $P \equiv \boxplus_{j=1}^{l} \sum_{i=1}^{k} \alpha_{ji}.P_{ji}$ and $Q \equiv \boxplus_{j=1}^{l} \sum_{i=1}^{k} \beta_{ji}.Q_{ji}$. Then, the depth of $P$, denoted as $d(P) = 0$, if $k = 0$; $d(P) = 1 + max\{d(P_{ji})\}$ for $1 \leq j, i \leq k$. The depth $d(Q)$ can be defined similarly. It is sufficient to induct on $d = d(P) + d(Q)$. When $d = 0$, $P \equiv \mathbf{nil}$ and $Q \equiv \mathbf{nil}$, $P = Q$, as desired.
Suppose $d > 0$.

- If $(\alpha_1 \parallel \cdots \parallel \alpha_n).M$ with $\alpha_{ji}(1 \leq j, i \leq n)$ free actions is a summand of $P$, then $\langle P, s\rangle \rightsquigarrow \xrightarrow{\{\alpha_1, \cdots, \alpha_n\}} \langle M, s'\rangle$. Since $Q$ is in head normal form and has a summand $(\alpha_1 \parallel \cdots \parallel \alpha_n).N$ such that $M \sim_{pp} N$, by the induction hypothesis $\mathbf{STC} \vdash M = N$, $\mathbf{STC} \vdash (\alpha_1 \parallel \cdots \parallel \alpha_n).M = (\alpha_1 \parallel \cdots \parallel \alpha_n).N$;
- If $x(y).M$ is a summand of $P$, then for $z \notin n(P, Q)$, $\langle P, s\rangle \rightsquigarrow \xrightarrow{x(z)} \langle M', s'\rangle \equiv \langle M\{z/y\}, s'\rangle$. Since $Q$ is in head normal form and has a summand $x(w).N$ such that for all $v$, $M'\{v/z\} \sim_{pp} N'\{v/z\}$, where $N' \equiv N\{z/w\}$, by the induction hypothesis $\mathbf{STC} \vdash M'\{v/z\} = N'\{v/z\}$, by the axioms $\mathbf{C}$ and $\mathbf{A}$, $\mathbf{STC} \vdash x(y).M = x(w).N$;
- If $\overline{x}(y).M$ is a summand of $P$, then for $z \notin n(P, Q)$, $\langle P, s\rangle \rightsquigarrow \xrightarrow{\overline{x}(z)} \langle M', s'\rangle \equiv \langle M\{z/y\}, s'\rangle$. Since $Q$ is in head normal form and has a summand $\overline{x}(w).N$ such that $M' \sim_{pp} N'$, where $N' \equiv N\{z/w\}$, by the induction hypothesis $\mathbf{STC} \vdash M' = N'$, by the axioms $\mathbf{A}$ and $\mathbf{C}$, $\mathbf{STC} \vdash \overline{x}(y).M = \overline{x}(w).N$.

**2.** if $P \sim_{pp} Q$, then $\mathbf{STC} \vdash P = Q$. It can be proven similarly to the above case.
**3.** if $P \sim_{php} Q$, then $\mathbf{STC} \vdash P = Q$. It can be proven similarly to the above case. □

Chapter 16

# Introduction to actors

There are many studies on the formalization for concurrency, such as process algebra [1] [2] [3] and actors [19] [20] [21] [22]. Traditionally, process algebras model the interleaving concurrency and actors capture the so-called true concurrency.

An actor [19] [20] [21] acts as an atomic function unit of concurrency and encapsulates a set of states, a control thread, and a set of local computations. It has a unique mail address and maintains a mail box to accept messages sent by other actors. Actors do local computations by means of processing the messages stored in the mail box sequentially and block when their mail boxes are empty.

During processing a message in a mail box, an actor may perform three candidate actions:

1. **send** action: sending messages asynchronously to other actors by their mail box addresses;
2. **create** action: creating new actors with new behaviors;
3. **ready** action: becoming ready to process the next message from the mail box or block if the mail box is empty.

The illustration of an actor model is shown in Fig. 16.1.

The work A$\pi$ of Agha [22] gives actors an algebraic model based on $\pi$-calculus [4] [5]. In this work, Agha pointed out that it must satisfy the following characteristics as an actor:

1. Concurrency: all actors execute concurrently;
2. Asynchrony: an actor receives and sends messages asynchronously;
3. Uniqueness: an actor has a unique name and the associated unique mail box name;
4. Concentration: an actor focuses on the processing messages, including some local computations, creations of some new actors, and sending some messages to other actors;
5. Communication Dependency: the only way of affecting an actor is sending a message to it;
6. Abstraction: except for the receiving and sending message, and creating new actors, the local computations are abstracted;

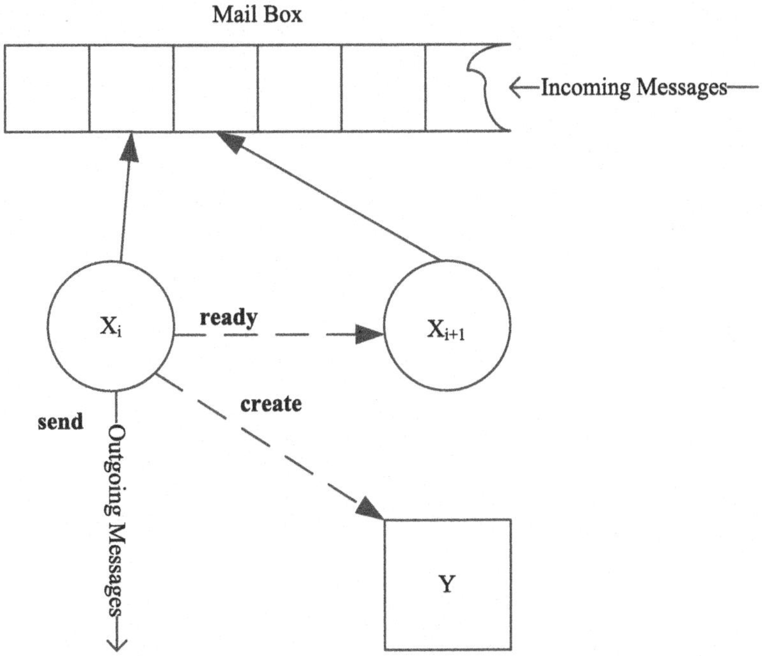


**FIGURE 16.1** Model of an actor.

Handbook of Truly Concurrent Process Algebra. https://doi.org/10.1016/B978-0-44-321515-5.00020-2

7. Persistence: an actor does not disappear after processing a message.

We have done some work on truly concurrent process algebra [9], which is proven to be a generalization of traditional process algebra for true concurrency. Now, for actors and truly concurrent process algebra that are all models for true concurrency, can we model actors based on truly concurrent process algebra? This is a natural problem, in this book, we try to do this work.

We capture the actor model in the following characteristics:

1. Concurrency: all actors are modeled as encapsulated processes and execute concurrently;
2. Asynchrony: an actor receives and sends messages asynchronously modeled by the asynchronous communication mechanisms in truly concurrent process algebra;
3. Uniqueness: the uniqueness of an actor is ensured by a unique process;
4. Concentration: the local computations, creations of some new actors, and sending some messages to other actors are modeled by the computational properties of truly concurrent process algebra;
5. Communication Dependency: the communications are modeled by the communication mechanisms of truly concurrent process algebra;
6. Abstraction: the local computations are abstracted by the abstraction mechanisms of truly concurrent process algebra;
7. Persistence: an actor is modeled as a persistent process.

Truly concurrent process algebra has a rich expressive ability to model the above characteristics of actors, and more importantly, they are all models for true concurrency. Compared with other models of actors, the truly concurrent process algebra based model has the following advantages:

1. The truly concurrent process algebra has rich expressive abilities to describe almost all characteristics of actors, especially for asynchronous communication, actor creation, recursion, abstraction, etc.;
2. The truly concurrent process algebra and actors are all models for true concurrency, and have inborn intimacy;
3. The truly concurrent process algebra has a firm semantics foundation and a powerful proof theory, the correctness of an actor system can be proven easily.

This part of the book is organized as follows. In Chapter 17, for self-sufficiency, we introduce some preliminaries on truly concurrent process algebra, for more details, please refer to Chapters 2 and 9. We give the model of actors based on truly concurrent process algebra in Chapter 18. We use the truly concurrent process algebra based actor model to model some applications and systems, in Chapters 19, 20, 21, 22, and 23, we model Map-Reduce, the Google File System, cloud resource management, Web Service composition, and the QoS-aware Web Service orchestration engine, respectively.

Chapter 17

# Truly concurrent process algebra

In this chapter, to make this book self-sufficient, we introduce the preliminaries on truly concurrent process algebra [9], which is based on truly concurrent operational semantics. For the basic knowledge of truly concurrent process algebras, please refer to Chapter 2.

## 17.1 Process creation

To model process creation, we introduce a unity operator **new** inspired by Baeten's work on process creation [16].

The transition rules of **new** are as Table 17.1 shows.

Also, the transition rules of the sequential composition $\cdot$ are adjusted to the followings, as Table 17.2 shows.

We design the axioms for the **new** operator in Table 17.3. The predicate $isParallelizable(x, y)$ means that $x$ and $y$ can be parallelized.

**Theorem 17.1** (Soundness of the **new** operator). *Let $x$ and $y$ be $APTC_\tau$ with guarded linear recursion and the **new** operator terms. If $APTC_\tau$ with guarded linear recursion and **new** operator $\vdash x = y$, then*

1. $x \approx_{rbs} y$;
2. $x \approx_{rbp} y$;
3. $x \approx_{rbhp} y$.

**Theorem 17.2** (Completeness of the **new** operator). *Let $p$ and $q$ be closed $APTC_\tau$ with guarded linear recursion and $CFAR$ and the **new** operator terms, then*

1. *if $p \approx_{rbs} q$, then $p = q$;*

**TABLE 17.1** Transition rule of the new operator.

| | |
|---|---|
| $\frac{}{\mathbf{new}(x) \rightarrow x}$ | $\frac{x \xrightarrow{e} x'}{\mathbf{new}(x) \xrightarrow{e} \mathbf{new}(x')}$ |

**TABLE 17.2** New transition rule of the $\cdot$ operator.

| | |
|---|---|
| $\frac{x \xrightarrow{e} \surd}{x \cdot y \xrightarrow{e} y}$ | $\frac{x \xrightarrow{e} x'}{x \cdot y \xrightarrow{e} x' \cdot y}$ |
| $\frac{x \rightarrow x' \quad y \xrightarrow{e} y'}{x \cdot y \xrightarrow{e} x' \between y'}$ | $\frac{x \xrightarrow{e} x' \quad y \rightarrow y'}{x \cdot y \xrightarrow{e} x' \between y'}$ |
| $\frac{x \xrightarrow{e_1} x' \quad y \xrightarrow{e_2} y' \quad \gamma(e_1, e_2) \text{ does not exist}}{x \cdot y \xrightarrow{\{e_1, e_2\}} x' \between y'}$ | $\frac{x \xrightarrow{e_1} x' \quad y \xrightarrow{e_2} y' \quad \gamma(e_1, e_2) \text{ exists}}{x \cdot y \xrightarrow{\gamma(e_1, e_2)} x' \between y'}$ |

**TABLE 17.3** Axioms of new operator.

| No. | Axiom |
|---|---|
| $PC1$ | if $isParallelizable(x, y)$, then $\mathbf{new}(x) \cdot y = x \between y$ |
| $PC2$ | $\mathbf{new}(x) \between y = x \between y$ |
| $PC3$ | $x \between \mathbf{new}(y) = x \between y$ |

Handbook of Truly Concurrent Process Algebra. https://doi.org/10.1016/B978-0-44-321515-5.00021-4

2. *if* $p \approx_{rbp} q$*, then* $p = q$*;*
3. *if* $p \approx_{rbhp} q$*, then* $p = q$*.*

## 17.2 Asynchronous communication

The communication in APTC is synchronous, for two atomic actions $a, b \in A$, if there exists a communication between $a$ and $b$, then they merge into a new communication action $\gamma(a, b)$; otherwise let $\gamma(a, b) = \delta$.

Asynchronous communication between actions $a, b \in A$ does not exist in a merged $\gamma(a, b)$, and it is only explicitly defined by the causality relation $a \leq b$ to ensure that the send action $a$ is executed before the receive action $b$.

APTC naturally support asynchronous communication to be adapted to the following aspects:

1. remove the communication merge operator $|$, just because there does not exist a communication merger $\gamma(a, b)$ between two asynchronous communicating action $a, b \in A$;
2. remove the asynchronous communicating actions $a, b \in A$ from $H$ of the encapsulation operator $\partial_H$;
3. ensure the send action $a$ is executed before the receive action $b$, by inserting appropriate numbers of placeholders during the modeling time; or by adding a causality constraint between the communicating actions $a \leq b$, all process terms violate this constraint will cause deadlocks.

## 17.3 Applications

$APTC$ provides a formal framework based on truly concurrent behavioral semantics, which can be used to verify the correctness of system behaviors. In this section, we tend to choose alternating bit protocol (ABP) [18].

The ABP protocol is used to ensure successful transmission of data through a corrupted channel. This success is based on the assumption that data can be resent an unlimited number of times, which is illustrated in Fig. 17.1, we alter it into the true concurrency situation.

1. Data elements $d_1, d_2, d_3, \cdots$ from a finite set $\Delta$ are communicated between a Sender and a Receiver;
2. If the Sender reads a datum from channel $A_1$, then this datum is sent to the Receiver in parallel through channel $A_2$;
3. The Sender processes the data in $\Delta$, forms new data, and sends them to the Receiver through channel $B$;
4. The Receiver sends the datum into channel $C_2$;
5. If channel $B$ is corrupted, the message communicated through $B$ can be turned into an error message $\bot$;
6. Every time the Receiver receives a message via channel $B$, it sends an acknowledgment to the Sender via channel $D$, which is also corrupted;
7. Finally, the Sender and the Receiver send out their outputs in parallel through channels $C_1$ and $C_2$.

In the truly concurrent ABP, the Sender sends its data to the Receiver; and the Receiver can also send its data to the Sender, for simplicity and without loss of generality, we assume that only the Sender sends its data and the Receiver only receives the data from the Sender. The Sender attaches a bit 0 to data elements $d_{2k-1}$ and a bit 1 to data elements $d_{2k}$, when they are sent into channel $B$. When the Receiver reads a datum, it sends back the attached bit via channel $D$. If the Receiver receives a corrupted message, then it sends back the previous acknowledgment to the Sender.

Then, the state transition of the Sender can be described by $APTC$ as follows:

$$S_b = \sum_{d\in\Delta} r_{A_1}(d) \cdot T_{db}$$

$$T_{db} = (\sum_{d'\in\Delta} (s_B(d', b) \cdot s_{C_1}(d')) + s_B(\bot)) \cdot U_{db}$$

$$U_{db} = r_D(b) \cdot S_{1-b} + (r_D(1-b) + r_D(\bot)) \cdot T_{db}$$

where $s_B$ denotes sending data through channel $B$, $r_D$ denotes receiving data through channel $D$, similarly, $r_{A_1}$ means receiving data via channel $A_1$, $s_{C_1}$ denotes sending data via channel $C_1$, and $b \in \{0, 1\}$.

Also, the state transition of the Receiver can be described by $APTC$ as follows:

$$R_b = \sum_{d\in\Delta} r_{A_2}(d) \cdot R'_b$$

$$R'_b = \sum_{d'\in\Delta} \{r_B(d', b) \cdot s_{C_2}(d') \cdot Q_b + r_B(d', 1-b) \cdot Q_{1-b}\} + r_B(\bot) \cdot Q_{1-b}$$

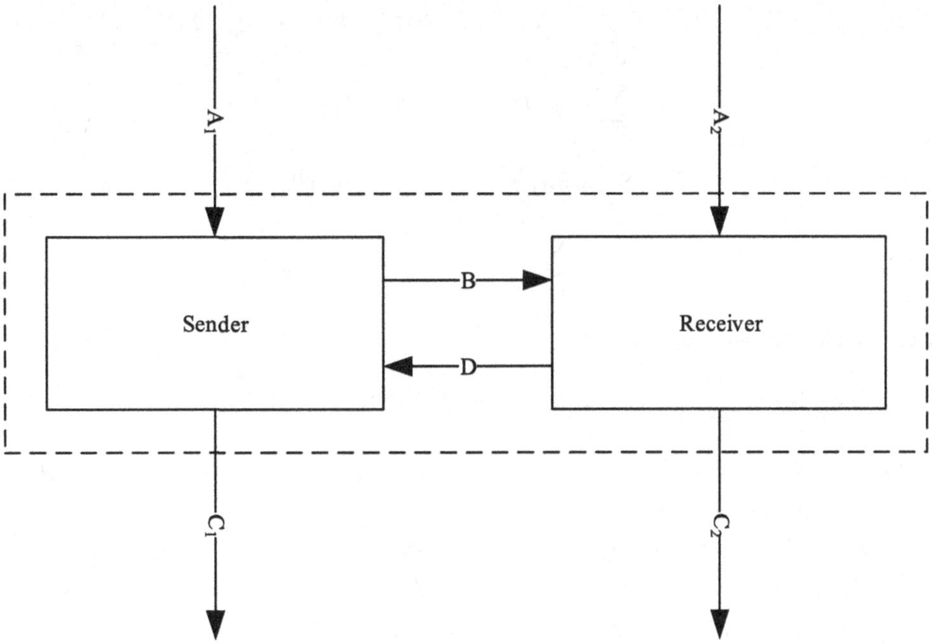


**FIGURE 17.1** Alternating bit protocol.

$$Q_b = (s_D(b) + s_D(\perp)) \cdot R_{1-b}$$

where $r_{A_2}$ denotes receiving data via channel $A_2$, $r_B$ denotes receiving data via channel $B$, $s_{C_2}$ denotes sending data via channel $C_2$, $s_D$ denotes sending data via channel $D$, and $b \in \{0, 1\}$.

The send action and receive action of the same data through the same channel can communicate with each other, otherwise, a deadlock $\delta$ will be caused. We define the following communication functions:

$$\gamma(s_B(d', b), r_B(d', b)) \triangleq c_B(d', b)$$
$$\gamma(s_B(\perp), r_B(\perp)) \triangleq c_B(\perp)$$
$$\gamma(s_D(b), r_D(b)) \triangleq c_D(b)$$
$$\gamma(s_D(\perp), r_D(\perp)) \triangleq c_D(\perp)$$

Let $R_0$ and $S_0$ be in parallel, then the system $R_0 S_0$ can be represented by the following process term:

$$\tau_I(\partial_H(\Theta(R_0 \between S_0))) = \tau_I(\partial_H(R_0 \between S_0))$$

where $H = \{s_B(d', b), r_B(d', b), s_D(b), r_D(b) | d' \in \Delta, b \in \{0, 1\}\}$
$\{s_B(\perp), r_B(\perp), s_D(\perp), r_D(\perp)\}$

$I = \{c_B(d', b), c_D(b) | d' \in \Delta, b \in \{0, 1\}\} \cup \{c_B(\perp), c_D(\perp)\}$

Then, we obtain the following conclusion.

**Theorem 17.3** (Correctness of the ABP protocol). *The ABP protocol $\tau_I(\partial_H(R_0 \between S_0))$ can exhibit the desired external behaviors.*

*Proof.* By use of the algebraic laws of $APTC$, we have the following expansions:

$$R_0 \between S_0 \overset{\text{P1}}{=} R_0 \parallel S_0 + R_0 \mid S_0$$
$$\overset{\text{RDP}}{=} (\sum_{d\in\Delta} r_{A_2}(d) \cdot R'_0) \parallel (\sum_{d\in\Delta} r_{A_1}(d) T_{d0})$$
$$+ (\sum_{d\in\Delta} r_{A_2}(d) \cdot R'_0) \mid (\sum_{d\in\Delta} r_{A_1}(d) T_{d0})$$

$$\overset{\text{P6,C14}}{=} \sum_{d\in\Delta}(r_{A_2}(d) \parallel r_{A_1}(d))R'_0 \between T_{d0} + \delta \cdot R'_0 \between T_{d0}$$

$$\overset{\text{A6,A7}}{=} \sum_{d\in\Delta}(r_{A_2}(d) \parallel r_{A_1}(d))R'_0 \between T_{d0}$$

$$\partial_H(R_0 \between S_0) = \partial_H(\sum_{d\in\Delta}(r_{A_2}(d) \parallel r_{A_1}(d))R'_0 \between T_{d0})$$

$$= \sum_{d\in\Delta}(r_{A_2}(d) \parallel r_{A_1}(d))\partial_H(R'_0 \between T_{d0})$$

Similarly, we can obtain the following equations:

$$\partial_H(R_0 \between S_0) = \sum_{d\in\Delta}(r_{A_2}(d) \parallel r_{A_1}(d)) \cdot \partial_H(T_{d0} \between R'_0)$$
$$\partial_H(T_{d0} \between R'_0) = c_B(d',0) \cdot (s_{C_1}(d') \parallel s_{C_2}(d')) \cdot \partial_H(U_{d0} \between Q_0) + c_B(\perp) \cdot \partial_H(U_{d0} \between Q_1)$$
$$\partial_H(U_{d0} \between Q_1) = (c_D(1) + c_D(\perp)) \cdot \partial_H(T_{d0} \between R'_0)$$
$$\partial_H(Q_0 \between U_{d0}) = c_D(0) \cdot \partial_H(R_1 \between S_1) + c_D(\perp) \cdot \partial_H(R'_1 \between T_{d0})$$
$$\partial_H(R'_1 \between T_{d0}) = (c_B(d',0) + c_B(\perp)) \cdot \partial_H(Q_0 \between U_{d0})$$
$$\partial_H(R_1 \between S_1) = \sum_{d\in\Delta}(r_{A_2}(d) \parallel r_{A_1}(d)) \cdot \partial_H(T_{d1} \between R'_1)$$
$$\partial_H(T_{d1} \between R'_1) = c_B(d',1) \cdot (s_{C_1}(d') \parallel s_{C_2}(d')) \cdot \partial_H(U_{d1} \between Q_1) + c_B(\perp) \cdot \partial_H(U_{d1} \between Q'_0)$$
$$\partial_H(U_{d1} \between Q'_0) = (c_D(0) + c_D(\perp)) \cdot \partial_H(T_{d1} \between R'_1)$$
$$\partial_H(Q_1 \between U_{d1}) = c_D(1) \cdot \partial_H(R_0 \between S_0) + c_D(\perp) \cdot \partial_H(R'_0 \between T_{d1})$$
$$\partial_H(R'_0 \between T_{d1}) = (c_B(d',1) + c_B(\perp)) \cdot \partial_H(Q_1 \between U_{d1})$$

Let $\partial_H(R_0 \between S_0) = \langle X_1|E\rangle$, where E is the following guarded linear recursion specification:

$$\{X_1 = \sum_{d\in\Delta}(r_{A_2}(d) \parallel r_{A_1}(d)) \cdot X_{2d}, Y_1 = \sum_{d\in\Delta}(r_{A_2}(d) \parallel r_{A_1}(d)) \cdot Y_{2d},$$
$$X_{2d} = c_B(d',0) \cdot X_{4d} + c_B(\perp) \cdot X_{3d}, Y_{2d} = c_B(d',1) \cdot Y_{4d} + c_B(\perp) \cdot Y_{3d},$$
$$X_{3d} = (c_D(1) + c_D(\perp)) \cdot X_{2d}, Y_{3d} = (c_D(0) + c_D(\perp)) \cdot Y_{2d},$$
$$X_{4d} = (s_{C_1}(d') \parallel s_{C_2}(d')) \cdot X_{5d}, Y_{4d} = (s_{C_1}(d') \parallel s_{C_2}(d')) \cdot Y_{5d},$$
$$X_{5d} = c_D(0) \cdot Y_1 + c_D(\perp) \cdot X_{6d}, Y_{5d} = c_D(1) \cdot X_1 + c_D(\perp) \cdot Y_{6d},$$
$$X_{6d} = (c_B(d,0) + c_B(\perp)) \cdot X_{5d}, Y_{6d} = (c_B(d,1) + c_B(\perp)) \cdot Y_{5d}$$
$$|d, d' \in \Delta\}$$

Then, we apply abstraction operator $\tau_I$ into $\langle X_1|E\rangle$.

$$\tau_I(\langle X_1|E\rangle) = \sum_{d\in\Delta}(r_{A_1}(d) \parallel r_{A_2}(d)) \cdot \tau_I(\langle X_{2d}|E\rangle)$$
$$= \sum_{d\in\Delta}(r_{A_1}(d) \parallel r_{A_2}(d)) \cdot \tau_I(\langle X_{4d}|E\rangle)$$
$$= \sum_{d,d'\in\Delta}(r_{A_1}(d) \parallel r_{A_2}(d)) \cdot (s_{C_1}(d') \parallel s_{C_2}(d')) \cdot \tau_I(\langle X_{5d}|E\rangle)$$
$$= \sum_{d,d'\in\Delta}(r_{A_1}(d) \parallel r_{A_2}(d)) \cdot (s_{C_1}(d') \parallel s_{C_2}(d')) \cdot \tau_I(\langle Y_1|E\rangle)$$

Similarly, we can obtain $\tau_I(\langle Y_1|E\rangle) = \sum_{d,d'\in\Delta}(r_{A_1}(d) \parallel r_{A_2}(d)) \cdot (s_{C_1}(d') \parallel s_{C_2}(d')) \cdot \tau_I(\langle X_1|E\rangle)$.

We obtain $\tau_I(\partial_H(R_0 \between S_0)) = \sum_{d,d'\in\Delta}(r_{A_1}(d) \parallel r_{A_2}(d)) \cdot (s_{C_1}(d') \parallel s_{C_2}(d')) \cdot \tau_I(\partial_H(R_0 \between S_0))$. Hence, the ABP protocol $\tau_I(\partial_H(R_0 \between S_0))$ can exhibit the desired external behaviors. □

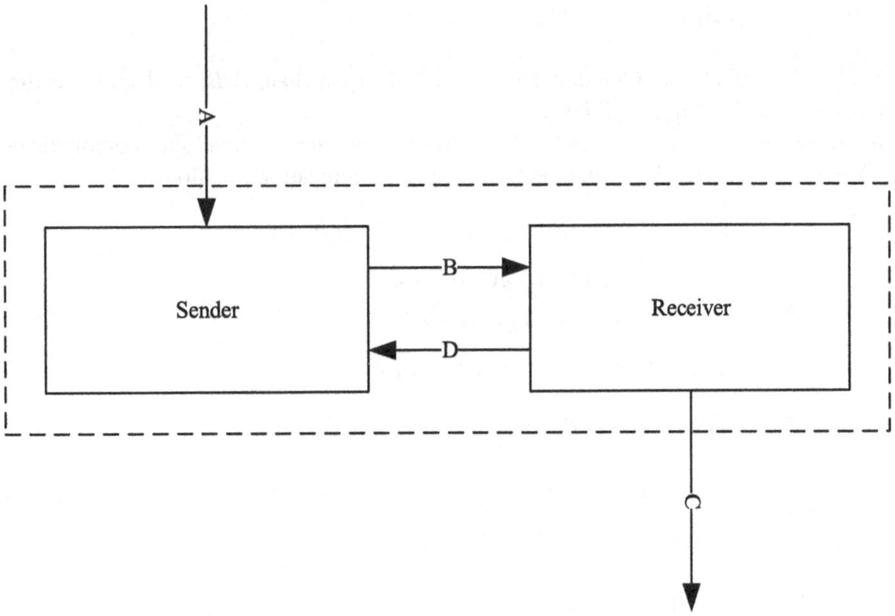


FIGURE 17.2 Alternating bit protocol.

With the help of a shadow constant, now we can verify the traditional alternating bit protocol (ABP) [18].

The ABP protocol is used to ensure successful transmission of data through a corrupted channel. This success is based on the assumption that data can be resent an unlimited number of times, which is illustrated in Fig. 17.2, we alter it into the true concurrency situation.

1. Data elements $d_1, d_2, d_3, \cdots$ from a finite set $\Delta$ are communicated between a Sender and a Receiver.
2. If the Sender reads a datum from channel $A$.
3. The Sender processes the data in $\Delta$, forms new data, and sends them to the Receiver through channel $B$.
4. Also, the Receiver sends the datum into channel $C$.
5. If channel $B$ is corrupted, the message communicated through $B$ can be turned into an error message $\perp$.
6. Every time the Receiver receives a message via channel $B$, it sends an acknowledgment to the Sender via channel $D$, which is also corrupted.

The Sender attaches a bit 0 to data elements $d_{2k-1}$ and a bit 1 to data elements $d_{2k}$, when they are sent into channel $B$. When the Receiver reads a datum, it sends back the attached bit via channel $D$. If the Receiver receives a corrupted message, then it sends back the previous acknowledgment to the Sender.

Then, the state transition of the Sender can be described by $APTC$ as follows:

$$S_b = \sum_{d\in\Delta} r_A(d) \cdot T_{db}$$

$$T_{db} = (\sum_{d'\in\Delta} (s_B(d', b) \cdot \circledS^{s_C(d')}) + s_B(\perp)) \cdot U_{db}$$

$$U_{db} = r_D(b) \cdot S_{1-b} + (r_D(1-b) + r_D(\perp)) \cdot T_{db}$$

where $s_B$ denotes sending data through channel $B$, $r_D$ denotes receiving data through channel $D$, similarly, $r_A$ means receiving data via channel $A$, $\circledS^{s_C(d')}$ denotes the shadow of $s_C(d')$.

Also, the state transition of the Receiver can be described by $APTC$ as follows:

$$R_b = \sum_{d\in\Delta} \circledS^{r_A(d)} \cdot R'_b$$

$$R'_b = \sum_{d'\in\Delta} \{r_B(d', b) \cdot s_C(d') \cdot Q_b + r_B(d', 1-b) \cdot Q_{1-b}\} + r_B(\perp) \cdot Q_{1-b}$$

$$Q_b = (s_D(b) + s_D(\bot)) \cdot R_{1-b}$$

where $Ⓢ^{r_A(d)}$ denotes the shadow of $r_A(d)$, $r_B$ denotes receiving data via channel $B$, $s_C$ denotes sending data via channel $C$, $s_D$ denotes sending data via channel $D$, and $b \in \{0, 1\}$.

The send action and receive action of the same data through the same channel can communicate with each other, otherwise, a deadlock $\delta$ will be caused. We define the following communication functions:

$$\gamma(s_B(d', b), r_B(d', b)) \triangleq c_B(d', b)$$
$$\gamma(s_B(\bot), r_B(\bot)) \triangleq c_B(\bot)$$
$$\gamma(s_D(b), r_D(b)) \triangleq c_D(b)$$
$$\gamma(s_D(\bot), r_D(\bot)) \triangleq c_D(\bot)$$

Let $R_0$ and $S_0$ be in parallel, then the system $R_0 S_0$ can be represented by the following process term:

$$\tau_I(\partial_H(\Theta(R_0 \between S_0))) = \tau_I(\partial_H(R_0 \between S_0))$$

where $H = \{s_B(d', b), r_B(d', b), s_D(b), r_D(b) | d' \in \Delta, b \in \{0, 1\}\}$
$\{s_B(\bot), r_B(\bot), s_D(\bot), r_D(\bot)\}$

$I = \{c_B(d', b), c_D(b) | d' \in \Delta, b \in \{0, 1\}\} \cup \{c_B(\bot), c_D(\bot)\}$

Then, we obtain the following conclusion.

**Theorem 17.4** (Correctness of the ABP protocol). *The ABP protocol $\tau_I(\partial_H(R_0 \between S_0))$ can exhibit the desired external behaviors.*

*Proof.* Similarly, we can obtain $\tau_I(\langle X_1|E\rangle) = \sum_{d,d' \in \Delta} r_A(d) \cdot s_C(d') \cdot \tau_I(\langle Y_1|E\rangle)$ and $\tau_I(\langle Y_1|E\rangle) = \sum_{d,d' \in \Delta} r_A(d) \cdot s_C(d') \cdot \tau_I(\langle X_1|E\rangle)$.

Hence, the ABP protocol $\tau_I(\partial_H(R_0 \between S_0))$ can exhibit the desired external behaviors. □

Chapter 18

# Process algebra based actor model

In this chapter, we introduce an actor model described by the truly concurrent process algebra. First, we introduce the model based on truly concurrent process algebra, and then analyze the advantages of this model.

## 18.1 Modeling characteristics of an actor

The characteristics of an actor are modeled as follows:

1. Computations: the computations are modeled as atomic actions, and the computational logics are captured by sequential composition $\cdot$, alternative composition $+$, parallel composition $\between$, and the conditional guards (see Section 9.1 for details) of truly concurrent process algebra;
2. Asynchronous Communications: a communications are composed of a pair of sending/receiving actions, the asynchrony of communication only requires that the sending action occurs before the receiving action, see Section 17.2 for details;
3. Uniqueness: for simplicity, the unique name of an actor and the unique name of its mail box are combined into its one unique name;
4. Abstraction: the local computations are encapsulated and abstracted as internal steps $\tau$, see abstraction of truly concurrent process algebra;
5. Actor Creations: by use of process creations in Section 17.1, we can create new actors;
6. Concurrency: all the actors are executed in parallel that can be captured by the parallel composition $\between$ of truly concurrent process algebra;
7. Persistence: once an actor has been created, it will receive and process messages continuously, this infinite computation can be captured by recursion of truly concurrent process algebra.

## 18.2 Combining all the elements into a whole

Based on the modeling elements of an actor, we can model a whole actor computational system consisted of a set of actors as follows:

1. According to the requirements of the system, design the system (including the inputs/outputs and functions) and divide it into a set of actors by the modular methodology;
2. Determine the interfaces of all actors, including receiving messages, sending messages, and creating other actors;
3. Determine the interactions among all actors, mainly including the causal relations of the sending/receiving actions for each interaction;
4. Implement the functions of each actor by programming its state transitions based on truly concurrent process algebra, and the program consists of a set of atomic actions and the computational logics among them, including $\cdot$, $+$, $\between$, and guards;
5. Apply recursion to the program of each actor to capture the persistence property of each actor;
6. Apply abstraction to the program of each actor to encapsulate it;
7. Prove that each actor has the desired external behaviors;
8. Put all actors in parallel and insert the interactions among them to implement the whole actor system;
9. Apply recursion to the whole system to capture the persistence property of the whole system;
10. Apply abstraction to the whole system by abstracting the interactions among actors as internal actions;
11. Finally, prove that the whole system has the desired external behaviors.

Compared with other models of actors, the truly concurrent process algebra based model has the following advantages:

1. The truly concurrent process algebra has rich expressive abilities to describe almost all characteristics of actors, especially for asynchronous communication, actor creation, recursion, abstraction, etc.;
2. The truly concurrent process algebra and actors are all models for true concurrency, and have inborn intimacy;

Handbook of Truly Concurrent Process Algebra. https://doi.org/10.1016/B978-0-44-321515-5.00022-6

3. The truly concurrent process algebra has a firm semantics foundation and a powerful proof theory, the correctness of an actor system can be proven easily.

In the following chapters, we will apply this new actor model to model and verify different computational systems, and show the advantages of this new model together.

Chapter 19

# Process algebra based actor model of Map–Reduce

In this chapter, we will use the process algebra based actor model to model and verify map–reduce. In Section 19.1, we introduce the requirements of map–reduce; we model the map–reduce by use of the new actor model in Section 19.2.

## 19.1 Requirements of Map–Reduce

Map–Reduce is a programming model and system aiming at a large-scale data set, which uses the thinking of functional programming language. It includes two programs: Map and Reduce, and also a framework to execute the program instances on a computer cluster. Map program reads the data set from the inputting files, executes some filters and transformations, and then outputs the data set as the form of $(key, value)$, while the Reduce program combines the outputs of the Map program according to the rules defined by the user. The architecture and the execution process are shown in Fig. 19.1.

As shown in Fig. 19.1, the execution process is as follows:

1. The lib of Map–Reduce in the user program divides the input files into 16–64 MB size of file segments;
2. Then, the Master program receives the requests from the user including the addresses of the input files, then creates $m$ map worker programs, and allocates a map task for each map worker including the addresses of the input files;

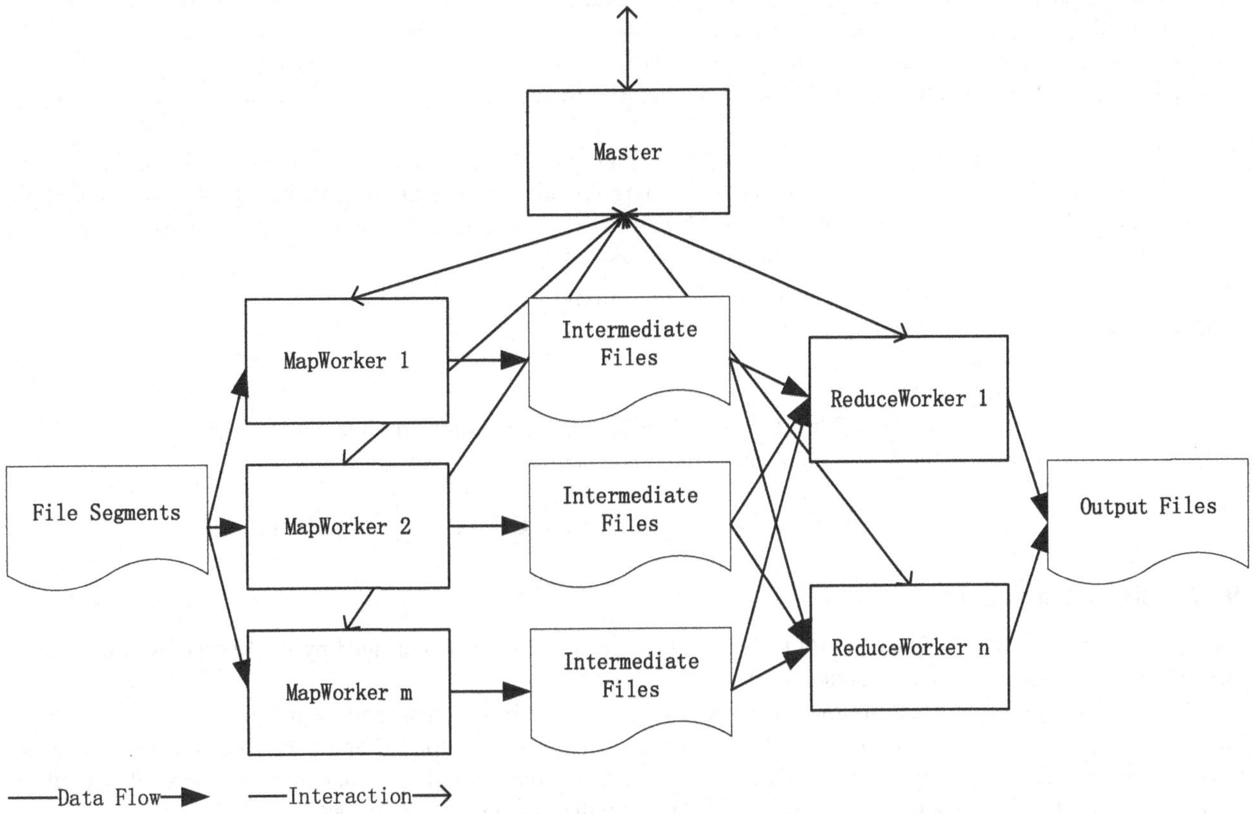


FIGURE 19.1 An architecture of Map–Reduce.

Handbook of Truly Concurrent Process Algebra. https://doi.org/10.1016/B978-0-44-321515-5.00023-8

3. The map workers receive the tasks from the Master and obtain the addresses of the input files, read the corresponding input file segments, execute some filters and transformations, and generate the $(key, value)$ pairs form intermediate files, and also notify the Master when their map tasks are finished;
4. The Master receives the task finished notifications from the map workers, including the addresses of the intermediate files, then creates $n$ reduce workers, and sends the reduce tasks to the reduce workers (also including the addresses of the intermediate files);
5. The reduce workers receive the tasks from the Master and obtain the addresses of the intermediate files, read the corresponding intermediate files, execute some reduce actions, and generate the output files, and also notify the Master when their reduce tasks are finished;
6. The Master receives the task finished notifications from the reduce workers, including the addresses of the output files, then generates the output responses to the user.

## 19.2 The new actor model of Map–Reduce

According to the architecture of Map–Reduce, the whole actors system implemented by actors can be divided into three kinds of actors: the Map actors (MapAs), the Reduce actors (RAs), and the Master actor (Mas).

### 19.2.1 Map actor, MapA

A Map worker is an atomic function unit to execute the map tasks and is managed by the Master. We use an actor called Map actor (MapA) to model a Map worker.

A MapA has a unique name, local information, and variables to contain its states, and local computation procedures to manipulate the information and variables. A MapA is always managed by the Master and it receives messages from the Master, sends messages to the Master, and is created by the Master. Note that a MapA can not create new MapAs, it can only be created by the Master. That is, a MapA is an actor with a constraint that is without a **create** action.

After a MapA is created, the typical process is as follows:

1. The MapA receives the map tasks $DI_{MapA}$ (including the addresses of the input files) from the Master through its mail box denoted by its name $MapA$ (the corresponding reading action is denoted $r_{MapA}(DI_{MapA})$);
2. Then, it does some local computations mixed some atomic filter and transformation actions by computation logics, including $\cdot$, $+$, $\between$, and guards, the whole local computations are denoted $I_{MapA}$, which is the set of all local atomic actions;
3. When the local computations are finished, the MapA generates the intermediate files containing a series of $(key, value)$ pairs, generates the output message $DO_{MapA}$ (containing the addresses of the intermediate files), and sends to the Master's mail box denoted by the Master's name $Mas$ (the corresponding sending action is denoted $s_{Mas}(DO_{MapA})$), and then processes the next message from the Master recursively.

The above process is described as the following state transitions by APTC:

$MapA = r_{MapA}(DI_{MapA}) \cdot MapA_1$

$MapA_1 = I_{MapA} \cdot MapA_2$

$MapA_2 = s_{Mas}(DO_{MapA}) \cdot MapA$

By use of the algebraic laws of APTC, the MapA may be proven to exhibit the desired external behaviors. If it exhibits the desired external behaviors, the MapA should have the following form:

$$\tau_{I_{MapA}}(\partial_{\emptyset}(MapA)) = r_{MapA}(DI_{MapA}) \cdot s_{Mas}(DO_{MapA}) \cdot \tau_{I_{MapA}}(\partial_{\emptyset}(MapA))$$

### 19.2.2 Reduce actor, RA

A Reduce worker is an atomic function unit to execute the reduce tasks and is managed by the Master. We use an actor called Reduce actor (RA) to model a Reduce worker.

A RA has a unique name, local information, and variables to contain its states, and local computation procedures to manipulate the information and variables. A RA is always managed by the Master and it receives messages from the Master, sends messages to the Master, and is created by the Master. Note that a RA can not create new RAs, it can only be created by the Master. That is, a RA is an actor with a constraint that is without a **create** action.

After a RA is created, the typical process is as follows:

1. The RA receives the reduce tasks $DI_{RA}$ (including the addresses of the intermediate files) from the Master through its mail box denoted by its name $RA$ (the corresponding reading action is denoted $r_{RA}(DI_{RA})$);
2. Then, it does some local computations mixed with some atomic reduce actions by computation logics, including $\cdot$, $+$, $\between$, and guards, the whole local computations are denoted $I_{RA}$, which is the set of all local atomic actions;
3. When the local computations are finished, the RA generates the output files, generates the output message $DO_{RA}$ (containing the addresses of the output files), and sends to the Master's mail box denoted by the Master's name $Mas$ (the corresponding sending action is denoted $s_{Mas}(DO_{RA})$), and then processes the next message from the Master recursively.

The above process is described as the following state transitions by APTC:

$RA = r_{RA}(DI_{RA}) \cdot RA_1$

$RA_1 = I_{RA} \cdot RA_2$

$RA_2 = s_{Mas}(DO_{RA}) \cdot RA$

By use of the algebraic laws of APTC, the RA may be proven to exhibit the desired external behaviors. If it exhibits the desired external behaviors, the RA should have the following form:

$$\tau_{I_{RA}}(\partial_\emptyset(RA)) = r_{RA}(DI_{RA}) \cdot s_{Mas}(DO_{RA}) \cdot \tau_{I_{RA}}(\partial_\emptyset(RA))$$

### 19.2.3 Master actor, Mas

The Master receives the requests from the user, manages the Map actors and the Reduce actors, and returns the responses to the user. We use an actor called Master actor (Mas) to model the Master.

After the Master actor is created, the typical process is as follows:

1. The Mas receives the requests $DI_{Mas}$ from the user through its mail box denoted by its name $Mas$ (the corresponding reading action is denoted $r_{Mas}(DI_{Mas})$);
2. Then, it does some local computations mixed with some atomic division actions to divide the input files into file segments by computation logics, including $\cdot$, $+$, $\between$, and guards, the whole local computations are denoted and included into $I_{Mas}$, which is the set of all local atomic actions;
3. The Mas creates $m$ Map actors $MapA_i$ for $1 \leq i \leq m$ in parallel through actions $\textbf{new}(MapA_1) \parallel \cdots \parallel \textbf{new}(MapA_m)$;
4. When the local computations are finished, the Mas generates the map tasks $DI_{MapA_i}$ containing the addresses of the corresponding file segments for each $MapA_i$ with $1 \leq i \leq m$, sends them to the MapAs' mail box denoted by the MapAs' name $MapA_i$ (the corresponding sending actions are denoted $s_{MapA_1}(DI_{MapA_1}) \parallel \cdots \parallel s_{MapA_m}(DI_{MapA_m})$);
5. The Mas receives the responses $DO_{MapA_i}$ (containing the addresses of the intermediate files) from $MapA_i$ for $1 \leq i \leq m$ through its mail box denoted by its name $Mas$ (the corresponding reading actions are denoted $r_{Mas}(DO_{MapA_1}) \parallel \cdots \parallel r_{Mas}(DO_{MapA_m})$);
6. Then, it does some local computations mixed with some atomic division actions by computation logics, including $\cdot$, $+$, $\between$, and guards, the whole local computations are denoted and included into $I_{Mas}$, which is the set of all local atomic actions;
7. The Mas creates $n$ Reduce actors $RA_j$ for $1 \leq j \leq n$ in parallel through actions $\textbf{new}(RA_1) \parallel \cdots \parallel \textbf{new}(RA_n)$;
8. When the local computations are finished, the Mas generates the reduce tasks $DI_{RA_j}$ containing the addresses of the corresponding intermediate files for each $RA_j$ with $1 \leq j \leq n$, sends them to the RAs' mail box denoted by the RAs' name $RA_j$ (the corresponding sending actions are denoted $s_{RA_1}(DI_{RA_1}) \parallel \cdots \parallel s_{RA_n}(DI_{RA_n})$);
9. The Mas receives the responses $DO_{RA_j}$ (containing the addresses of the output files) from $RA_j$ for $1 \leq j \leq n$ through its mail box denoted by its name $Mas$ (the corresponding reading actions are denoted $r_{Mas}(DO_{RA_1}) \parallel \cdots \parallel r_{Mas}(DO_{RA_n})$);
10. Then, it does some local computations mixed with some atomic actions by computation logics, including $\cdot$, $+$, $\between$, and guards, the whole local computations are denoted and included into $I_{Mas}$, which is the set of all local atomic actions;
11. When the local computations are finished, the Mas generates the output responses $DO_{Mas}$ containing the addresses of the output files, sends them to users (the corresponding sending action is denoted $s_O(DO_{Mas})$), and then processes the next message from the user recursively.

The above process is described as the following state transitions by APTC:

$Mas = r_{Mas}(DI_{Mas}) \cdot Mas_1$

$Mas_1 = I_{Mas} \cdot Mas_2$

$Mas_2 = \textbf{new}(MapA_1) \parallel \cdots \parallel \textbf{new}(MapA_m) \cdot Mas_3$

$Mas_3 = s_{MapA_1}(DI_{MapA_1}) \parallel \cdots \parallel s_{MapA_m}(DI_{MapA_m}) \cdot Mas_4$
$Mas_4 = r_{Mas}(DO_{MapA_1}) \parallel \cdots \parallel r_{Mas}(DO_{MapA_m}) \cdot Mas_5$
$Mas_5 = I_{Mas} \cdot Mas_6$
$Mas_6 = \mathbf{new}(RA_1) \parallel \cdots \parallel \mathbf{new}(RA_n) \cdot Mas_7$
$Mas_7 = s_{RA_1}(DI_{RA_1}) \parallel \cdots \parallel s_{RA_n}(DI_{RA_n}) \cdot Mas_8$
$Mas_8 = r_{Mas}(DO_{RA_1}) \parallel \cdots \parallel r_{Mas}(DO_{RA_n}) \cdot Mas_9$
$Mas_9 = I_{Mas} \cdot Mas_{10}$
$Mas_{10} = s_O(DO_{Mas}) \cdot Mas$

By use of the algebraic laws of APTC, the Mas may be proven to exhibit the desired external behaviors. If it exhibits the desired external behaviors, the Mas should have the following form:

$\tau_{I_{Mas}}(\partial_\emptyset(Mas)) = r_{Mas}(DI_{Mas}) \cdot (s_{MapA_1}(DI_{MapA_1}) \parallel \cdots \parallel s_{MapA_m}(DI_{MapA_m})) \cdot$
$(r_{Mas}(DO_{MapA_1}) \parallel \cdots \parallel r_{Mas}(DO_{MapA_m})) \cdot (s_{RA_1}(DI_{RA_1}) \parallel \cdots \parallel s_{RA_n}(DI_{RA_n})) \cdot$
$(r_{Mas}(DO_{RA_1}) \parallel \cdots \parallel r_{Mas}(DO_{RA_n})) \cdot s_{Mas}(DO_{Mas}) \cdot \tau_{I_{Mas}}(\partial_\emptyset(Mas))$

### 19.2.4 Putting all actors together into a whole

We put all actors together into a whole, including all MapAs, RAs, and Mas, according to the architecture as illustrated in Fig. 19.1. The whole actor system $Mas = Mas \quad MapA_1 \quad \cdots \quad MapA_m$ $RA_1 \quad \cdots \quad RA_n$ can be represented by the following process term of APTC.

$$\tau_I(\partial_H(Mas)) = \tau_I(\partial_H(Mas \between MapA_1 \between \cdots \between MapA_m \between RA_1 \between \cdots \between RA_n))$$

Among all the actors, there are synchronous communications. The actor's reading and the same actor's sending actions with the same type messages may cause communications. If the actor's sending action occurs before the same actor's reading action, an asynchronous communication will occur; otherwise, a deadlock $\delta$ will be caused.

There are two kinds of asynchronous communications as follows:

(1) The communications between a MapA and Mas with the following constraints:

$s_{MapA}(DI_{MapA}) \leq r_{MapA}(DI_{MapA})$
$s_{Mas}(DO_{MapA}) \leq r_{Mas}(DO_{MapA})$

(2) The communications between a RA and Mas with the following constraints:

$s_{RA}(DI_{RA}) \leq r_{RA}(DI_{RA})$
$s_{Mas}(DO_{RA}) \leq r_{Mas}(DO_{RA})$

Hence, the set $H$ and $I$ can be defined as follows:

$H = \{s_{MapA_1}(DI_{MapA_1}), r_{MapA_1}(DI_{MapA_1}), \cdots, s_{MapA_m}(DI_{MapA_m}), r_{MapA_m}(DI_{MapA_m}),$
$s_{Mas}(DO_{MapA_1}), r_{Mas}(DO_{MapA_1}), \cdots, s_{Mas}(DO_{MapA_m}), r_{Mas}(DO_{MapA_m}),$
$s_{RA_1}(DI_{RA_1}), r_{RA_1}(DI_{RA_1}), \cdots, s_{RA_n}(DI_{RA_n}), r_{RA_n}(DI_{RA_n}),$
$s_{Mas}(DO_{RA_1}), r_{Mas}(DO_{RA_1}), \cdot, s_{Mas}(DO_{RA_n}), r_{Mas}(DO_{RA_n})$
$|s_{MapA_1}(DI_{MapA_1}) \nleq r_{MapA_1}(DI_{MapA_1}), \cdots, s_{MapA_m}(DI_{MapA_m}) \nleq r_{MapA_m}(DI_{MapA_m}),$
$s_{Mas}(DO_{MapA_1}) \nleq r_{Mas}(DO_{MapA_1}), \cdots, s_{Mas}(DO_{MapA_m}) \nleq r_{Mas}(DO_{MapA_m}),$
$s_{RA_1}(DI_{RA_1}) \nleq r_{RA_1}(DI_{RA_1}), \cdots, s_{RA_n}(DI_{RA_n}) \nleq r_{RA_n}(DI_{RA_n}),$
$s_{Mas}(DO_{RA_1}) \nleq r_{Mas}(DO_{RA_1}), \cdot, s_{Mas}(DO_{RA_n}) \nleq r_{Mas}(DO_{RA_n})\}$

$I = \{s_{MapA_1}(DI_{MapA_1}), r_{MapA_1}(DI_{MapA_1}), \cdots, s_{MapA_m}(DI_{MapA_m}), r_{MapA_m}(DI_{MapA_m}),$
$s_{Mas}(DO_{MapA_1}), r_{Mas}(DO_{MapA_1}), \cdots, s_{Mas}(DO_{MapA_m}), r_{Mas}(DO_{MapA_m}),$
$s_{RA_1}(DI_{RA_1}), r_{RA_1}(DI_{RA_1}), \cdots, s_{RA_n}(DI_{RA_n}), r_{RA_n}(DI_{RA_n}),$
$s_{Mas}(DO_{RA_1}), r_{Mas}(DO_{RA_1}), \cdot, s_{Mas}(DO_{RA_n}), r_{Mas}(DO_{RA_n})$
$|s_{MapA_1}(DI_{MapA_1}) \leq r_{MapA_1}(DI_{MapA_1}), \cdots, s_{MapA_m}(DI_{MapA_m}) \leq r_{MapA_m}(DI_{MapA_m}),$
$s_{Mas}(DO_{MapA_1}) \leq r_{Mas}(DO_{MapA_1}), \cdots, s_{Mas}(DO_{MapA_m}) \leq r_{Mas}(DO_{MapA_m}),$
$s_{RA_1}(DI_{RA_1}) \leq r_{RA_1}(DI_{RA_1}), \cdots, s_{RA_n}(DI_{RA_n}) \leq r_{RA_n}(DI_{RA_n}),$
$s_{Mas}(DO_{RA_1}) \leq r_{Mas}(DO_{RA_1}), \cdot, s_{Mas}(DO_{RA_n}) \leq r_{Mas}(DO_{RA_n})\}$
$\cup I_{MapA_1} \cup \cdots \cup I_{MapA_m} \cup I_{RA_1} \cup \cdots \cup I_{RA_n} \cup I_{Mas}$

Then, we can obtain the following conclusion.

**Theorem 19.1.** *The whole actor system of Map–Reduce illustrated in Fig. 19.1 exhibits the desired external behaviors.*

*Proof.* By use of the algebraic laws of APTC, we can prove the following equation:

$\tau_I(\partial_H(Mas)) = \tau_I(\partial_H(Mas \between MapA_1 \between \cdots \between MapA_m \between RA_1 \between \cdots \between RA_n))$
$= r_{Mas}(DI_{Mas}) \cdot s_O(DO_{Mas}) \cdot \tau_I(\partial_H(Mas \between MapA_1 \between \cdots \between MapA_m \between RA_1 \between \cdots \between RA_n))$
$= r_{Mas}(DI_{Mas}) \cdot s_O(DO_{Mas}) \cdot \tau_I(\partial_H(Mas))$

For the details of the proof, as we omit them here, please refer to Section 17.3. □

Chapter 20

# Process algebra based actor model of the Google File System

In this chapter, we will use the process algebra based actor model to model and verify the Google File System. In Section 20.1, we introduce the requirements of the Google File System; we model the Google File System by use of the new actor model in Section 20.2.

## 20.1 Requirements of the Google File System

The Google File System (GFS) is a distributed file system used to deal with large-scale data-density applications. GFS has some design goals that are the same as the other traditional distributed file systems, such as performance, scalability, reliability, and usability. However, GFS has some other advantages, such as fault-tolerance, the huge size of files, appended writing of files, and also the flexibility caused by the cooperative design of the APIs of GFS and the applications.

A GFS cluster includes a Master and some chunk server, and can be accessed by multiple clients, as Fig. 20.1 illustrates. A file is divided into the fixed size of chunks with a global unique identity allocated by the Master; each chunk is saved on the disk of a chunk server as a Linux file, and can be accessed by the identity and the byte boundary through the chunk server. To improve the reliability, each chunk has three copies located on different chunk servers.

The Master manages the meta data of all file system, including name space, accessing information, mapping information from a file to chunks, and the locations of chunks.

A client implementing APIs of GFS, can interact with the Master to exchange the meta information of files and interact with the chunk servers to exchange the actual chunks.

As shown in Fig. 20.1, the execution process is as follows:

1. The client receives the file accessing requests from the outside, including the meta information of the files. The client processes the requests, and generates the file information, and sends to the Master;

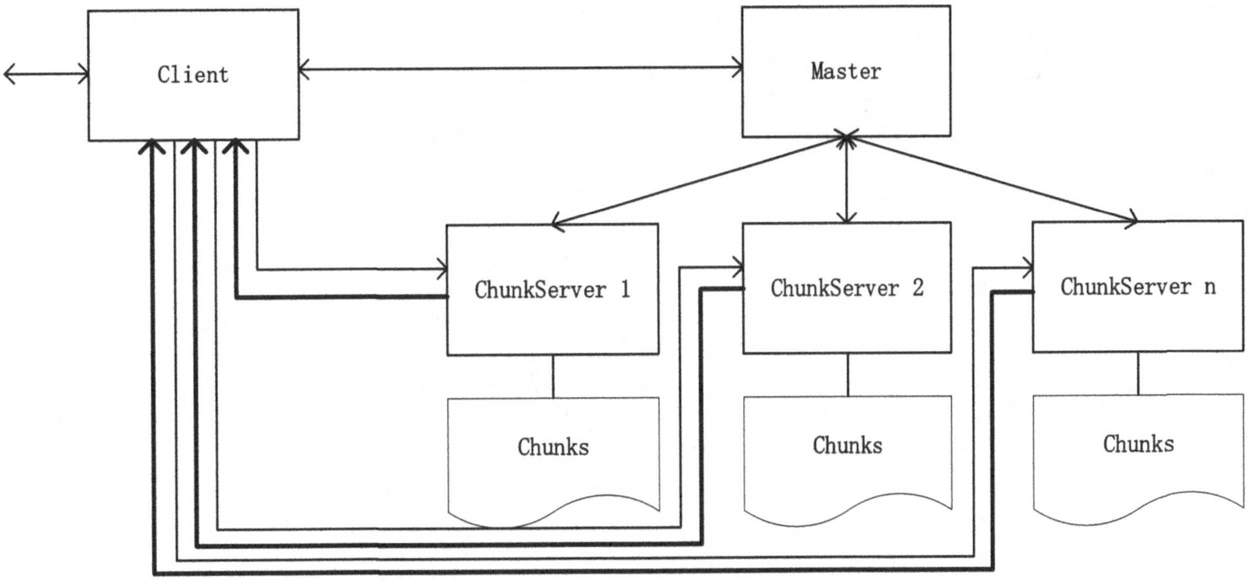


**FIGURE 20.1** An architecture of the Google File System.

Handbook of Truly Concurrent Process Algebra. https://doi.org/10.1016/B978-0-44-321515-5.00024-X

2. The Master receives the file information requests, creates some chunk servers according to the meta information of the files and the locations of the chunks, generates the file requests (including the address of the client) for each chunk server, and sends the requests to each chunk server, respectively;
3. The chunk server receives the requests, obtains the related chunks, and sends them to the client.

## 20.2 The new actor model of the Google File System

According to the architecture of GFS, the whole actors system implemented by actors can be divided into three kinds of actors: the client actor (CA), the chunk server actors (CSAs), and the Master actor (Mas).

### 20.2.1 Client actor, CA

We use an actor called Client actor (CA) to model the client.

After the CA is created, the typical process is as follows:

1. The CA receives the requests $DI_{CA}$ (including the meta information of the request files) from the outside through its mail box denoted by its name $CA$ (the corresponding reading action is denoted $r_{CA}(DI_{CA})$);
2. Then, it does some local computations mixed with some atomic actions by computation logics, including $\cdot$, $+$, $\between$, and guards, the whole local computations are denoted and included into $I_{CA}$, which is the set of all local atomic actions;
3. When the local computations are finished, the CA generates the output message $DI_{Mas}$ (containing the meta information of the request files and the address of the client), and sends to the Master's mail box denoted by the Master's name $Mas$ (the corresponding sending action is denoted $s_{Mas}(DI_{Mas})$);
4. The CA receives the chunks from the $n$ chunk servers $CSA_i$ with $1 \leq i \leq n$ through its mail box denoted by its name $CA$ (the corresponding reading actions are denoted $r_{CA}(DO_{CSA_1}) \parallel \cdots \parallel r_{CA}(DO_{CSA_n})$);
5. Then, it does some local computations mixed with some atomic combination actions to combine the chunks by computation logics, including $\cdot$, $+$, $\between$, and guards, the whole local computations are denoted and included into $I_{CA}$, which is the set of all local atomic actions;
6. When the local computations are finished, the CA generates the output message $DO_{CA}$ (containing the files), and sends to the outside (the corresponding sending action is denoted $s_O(DO_{CA})$), and then processes the next message from the outside recursively.

The above process is described as the following state transitions by APTC:

$CA = r_{CA}(DI_{CA}) \cdot CA_1$

$CA_1 = I_{CA} \cdot CA_2$

$CA_2 = s_{Mas}(DI_{Mas}) \cdot CA_3$

$CA_3 = r_{CA}(DO_{CSA_1}) \parallel \cdots \parallel r_{CA}(DO_{CSA_n}) \cdot CA_4$

$CA_4 = I_{CA} \cdot CA_5$

$CA_5 = s_O(DO_{CA}) \cdot CA$

By use of the algebraic laws of APTC, the CA may be proven to exhibit the desired external behaviors. If it can exhibit the desired external behaviors, the CA should have the following form:

$$\tau_{I_{CA}}(\partial_\emptyset(CA)) = r_{CA}(DI_{CA}) \cdot s_{Mas}(DI_{Mas}) \cdot (r_{CA}(DO_{CSA_1}) \parallel \cdots \parallel r_{CA}(DO_{CSA_n})) \cdot s_O(DO_{CA}) \cdot \tau_{I_{CA}}(\partial_\emptyset(CA))$$

### 20.2.2 Chunk Server actor, CSA

A chunk server is an atomic function unit to access the chunks and is managed by the Master. We use an actor called chunk server actor (CSA) to model a chunk server.

A CSA has a unique name, local information and variables to contain its states, and local computation procedures to manipulate the information and variables. A CSA is always managed by the Master and it receives messages from the Master, sends messages to the Master and the client, and is created by the Master. Note that a CSA cannot create new CSAs, it can only be created by the Master. That is, a CSA is an actor with a constraint that is without a **create** action.

After a CSA is created, the typical process is as follows:

1. The CSA receives the chunks' requests $DI_{CSA}$ (including the information of the chunks and the address of the client) from the Master through its mail box denoted by its name $CSA$ (the corresponding reading action is denoted $r_{CSA}(DI_{CSA})$);
2. Then, it does some local computations mixed with some atomic actions by computation logics, including $\cdot$, $+$, $\between$, and guards, the whole local computations are denoted $I_{CSA}$, which is the set of all local atomic actions;

3. When the local computations are finished, it generates the output message $DO_{CSA}$ (containing the chunks and their meta information), and sends it to the client's mail box denoted by the client's name $CA$ (the corresponding sending action is denoted $s_{CA}(DO_{CSA})$), and then processes the next message from the Master recursively.

The above process is described as the following state transitions by APTC:
$CSA = r_{CSA}(DI_{CSA}) \cdot CSA_1$
$CSA_1 = I_{CSA} \cdot CSA_2$
$CSA_2 = s_{CA}(DO_{CSA}) \cdot CSA$
By use of the algebraic laws of APTC, the CSA may be proven to exhibit the desired external behaviors. If it can exhibit the desired external behaviors, the CSA should have the following form:

$$\tau_{I_{CSA}}(\partial_{\emptyset}(CSA)) = r_{CSA}(DI_{CSA}) \cdot s_{CA}(DO_{CSA}) \cdot \tau_{I_{CSA}}(\partial_{\emptyset}(CSA))$$

### 20.2.3 Master actor, Mas

The Master receives the requests from the client, and manages the chunk server actors. We use an actor called Master actor (Mas) to model the Master.

After the Master actor is created, the typical process is as follows:

1. The Mas receives the requests $DI_{Mas}$ from the client through its mail box denoted by its name $Mas$ (the corresponding reading action is denoted $r_{Mas}(DI_{Mas})$);
2. Then, it does some local computations mixed with some atomic actions by computation logics, including $\cdot$, $+$, $\between$, and guards, the whole local computations are denoted and included into $I_{Mas}$, which is the set of all local atomic actions;
3. The Mas creates $n$ chunk server actors $CSA_i$ for $1 \leq i \leq n$ in parallel through actions $\mathbf{new}(CSA_1) \parallel \cdots \parallel \mathbf{new}(CSA_n)$;
4. When the local computations are finished, the Mas generates the request $DI_{CSA_i}$ containing the meta information of chunks and the address of the client for each $CSA_i$ with $1 \leq i \leq n$, sends them to the CSAs' mail box denoted by the CSAs' name $CSA_i$ (the corresponding sending actions are denoted $s_{CSA_1}(DI_{CSA_1}) \parallel \cdots \parallel s_{CSA_n}(DI_{CSA_n})$), and then processes the next message from the client recursively.

The above process is described as the following state transitions by APTC:
$Mas = r_{Mas}(DI_{Mas}) \cdot Mas_1$
$Mas_1 = I_{Mas} \cdot Mas_2$
$Mas_2 = \mathbf{new}(CSA_1) \parallel \cdots \parallel \mathbf{new}(CSA_n) \cdot Mas_3$
$Mas_3 = s_{CSA_1}(DI_{CSA_1}) \parallel \cdots \parallel s_{CSA_n}(DI_{CSA_n}) \cdot Mas$
By use of the algebraic laws of APTC, the Mas may be proven to exhibit the desired external behaviors. If it can exhibit the desired external behaviors, the Mas should have the following form:
$\tau_{I_{Mas}}(\partial_{\emptyset}(Mas)) = r_{Mas}(DI_{Mas}) \cdot (s_{CSA_1}(DI_{CSA_1}) \parallel \cdots \parallel s_{CSA_n}(DI_{CSA_n})) \cdot \tau_{I_{Mas}}(\partial_{\emptyset}(Mas))$

### 20.2.4 Putting all actors together into a whole

We put all actors together into a whole, including all CA, CSAs, and Mas, according to the architecture as illustrated in Fig. 20.1. The whole actor system $CA \quad Mas = CA \quad Mas \quad CSA_1 \quad \cdots \quad CSA_n$ can be represented by the following process term of APTC:

$$\tau_I(\partial_H(CA \between Mas)) = \tau_I(\partial_H(CA \between Mas \between CSA_1 \between \cdots \between CSA_n))$$

Among all the actors, there are synchronous communications. The actor's reading and the same actor's sending actions with the same type of messages may cause communications. If the actor's sending action occurs before the same actor's reading action, an asynchronous communication will occur; otherwise, a deadlock $\delta$ will be caused.

There are three kinds of asynchronous communications as follows:
(1) The communications between a CSA and Mas with the following constraint:
$s_{CSA}(DI_{CSA}) \leq r_{CSA}(DI_{CSA})$
(2) The communications between a CSA and CA with the following constraint:
$s_{CA}(DO_{CSA}) \leq r_{CA}(DO_{CSA})$
(3) The communications between CA and Mas with the following constraint:
$s_{Mas}(DI_{Mas}) \leq r_{Mas}(DI_{Mas})$
Hence, the set $H$ and $I$ can be defined as follows:

$H = \{s_{CSA_1}(DI_{CSA_1}), r_{CSA_1}(DI_{CSA_1}), \cdots, s_{CSA_n}(DI_{CSA_n}), r_{CSA_n}(DI_{CSA_n}),$
$s_{CA}(DO_{CSA_1}), r_{CA}(DO_{CSA_1}), \cdots, s_{CA}(DO_{CSA_n}), r_{CA}(DO_{CSA_n}),$
$s_{Mas}(DI_{Mas}), r_{Mas}(DI_{Mas})$
$|s_{CSA_1}(DI_{CSA_1}) \nleq r_{CSA_1}(DI_{CSA_1}), \cdots, s_{CSA_n}(DI_{CSA_n}) \nleq r_{CSA_n}(DI_{CSA_n}),$
$s_{CA}(DO_{CSA_1}) \nleq r_{CA}(DO_{CSA_1}), \cdots, s_{CA}(DO_{CSA_n}) \nleq r_{CA}(DO_{CSA_n}),$
$s_{Mas}(DI_{Mas}) \nleq r_{Mas}(DI_{Mas})\}$

$I = \{s_{CSA_1}(DI_{CSA_1}), r_{CSA_1}(DI_{CSA_1}), \cdots, s_{CSA_n}(DI_{CSA_n}), r_{CSA_n}(DI_{CSA_n}),$
$s_{CA}(DO_{CSA_1}), r_{CA}(DO_{CSA_1}), \cdots, s_{CA}(DO_{CSA_n}), r_{CA}(DO_{CSA_n}),$
$s_{Mas}(DI_{Mas}), r_{Mas}(DI_{Mas})$
$|s_{CSA_1}(DI_{CSA_1}) \leq r_{CSA_1}(DI_{CSA_1}), \cdots, s_{CSA_n}(DI_{CSA_n}) \leq r_{CSA_n}(DI_{CSA_n}),$
$s_{CA}(DO_{CSA_1}) \leq r_{CA}(DO_{CSA_1}), \cdots, s_{CA}(DO_{CSA_n}) \leq r_{CA}(DO_{CSA_n}),$
$s_{Mas}(DI_{Mas}) \leq r_{Mas}(DI_{Mas})\} \cup I_{CA} \cup I_{CSA_1} \cup \cdots \cup I_{CSA_n} \cup I_{Mas}$

Then, we can obtain the following conclusion.

**Theorem 20.1.** *The whole actor system of GFS illustrated in Fig. 20.1 can exhibit the desired external behaviors.*

*Proof.* By use of the algebraic laws of APTC, we can prove the following equation:
$\tau_I(\partial_H(CA \between Mas)) = \tau_I(\partial_H(CA \between Mas \between CSA_1 \between \cdots \between CSA_n))$
$= r_{CA}(DI_{CA}) \cdot s_O(DO_{CA}) \cdot \tau_I(\partial_H(CA \between Mas \between CSA_1 \between \cdots \between CSA_n))$
$= r_{CA}(DI_{CA}) \cdot s_O(DO_{CA}) \cdot \tau_I(\partial_H(CA \between Mas))$

For the details of the proof, as we omit them here, please refer to Section 17.3. □

Chapter 21

# Process algebra based actor model of cloud resource management

In this chapter, we will use the process algebra based actor model to model and verify cloud resource management. In Section 21.1, we introduce the requirements of cloud resource management; we model the cloud resource management by use of the new actor model in Section 21.2.

## 21.1 Requirements of cloud resource management

There are various kinds of resources in cloud computing, such as computational ability, storage ability, operation system platform, middle-ware platform, development platform, and various common and specific softwares. Such various kinds of resources should be managed uniformly, in the forms of uniform lifetime management, uniform execution and monitoring, and also uniform utilization and accessing.

The way of uniform management of various resources is the adoption of virtualization. Each resource is encapsulated as a virtual resource, which provides accessing of the actual resource downward, and uniform management and accessing interface upward. Hence, the core architecture of cloud resource management is illustrated in Fig. 21.1. In this architecture, there are four main kinds of components:

1. The Client: it receives the resource accessing requests, sends to the Resource manager, and obtains the running states and execution results from the Resource Manager, and sends them out;
2. The Resource Manager: it receives the requests from the Client, creates, accesses, and manages the virtual resources;
3. The State Collector: it collects the states of the involved running virtual resources;
4. The Virtual Resources: they encapsulate various kinds of resources as a uniform management interface.

As shown in Fig. 21.1, the typical execution process of cloud resource management is as follows:

1. The Client receives the resource accessing requests, and sends them to the Resource Manager;
2. The Resource Manager receives the requests from the Client, divides the computational tasks, creates the related virtual resources, and sends the divided tasks to the involved virtual resources;
3. The created virtual resources receive their tasks from the Resource Manager, accesses the actual resources to run the computational tasks, during the running they report their running states to the State Collector;
4. The State Collector receives the running states from the virtual resources, stores the states into the State Base, and sends the running states of the involved virtual resources to the Resource Manager;
5. The Resource Manager receives the running states, after an inner processing, sends the states to the Client;
6. The Client receives the states and sends them to the outside;
7. When the running of virtual resources are finished, they send the results to the Resource Manager;
8. The Resource Manager receives the computational results, and after an inner combination sends the combined results to the Client;
9. The Client receives the results and sends them to the outside.

## 21.2 The new actor model of cloud resource management

According to the architecture of cloud resource management, the whole actors system implemented by actors can be divided into four kinds of actors: the client actor (CA), the Virtual Resource actors (VAs), the Resource Manager actor (RA), and the State Collector actor (SA).

Handbook of Truly Concurrent Process Algebra. https://doi.org/10.1016/B978-0-44-321515-5.00025-1

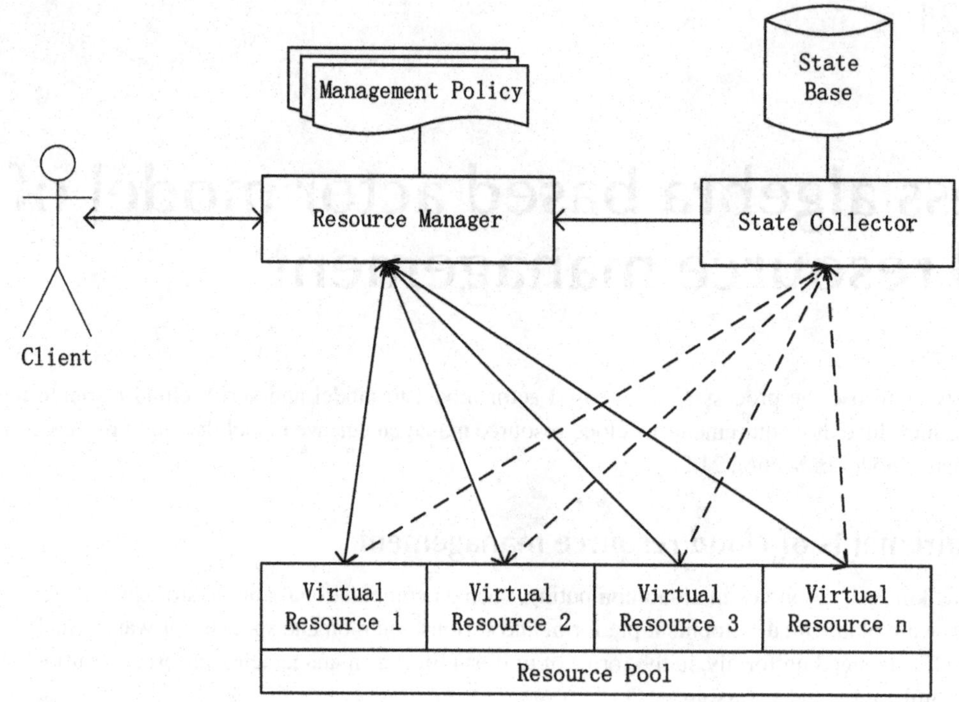


**FIGURE 21.1** An architecture of cloud resource management.

### 21.2.1 Client actor, CA

We use an actor called Client Actor (CA) to model the Client.

After the CA is created, the typical process is as follows:

1. The CA receives the requests $DI_{CA}$ from the outside through its mail box denoted by its name $CA$ (the corresponding reading action is denoted $r_{CA}(DI_{CA})$);
2. Then, it does some local computations mixed with some atomic actions by computation logics, including $\cdot$, $+$, $\between$, and guards, the whole local computations are denoted and included into $I_{CA}$, which is the set of all local atomic actions;
3. When the local computations are finished, the CA generates the output requests $DI_{RA}$, and sends to the RA's mail box denoted by the RA's name $RA$ (the corresponding sending action is denoted $s_{RA}(DI_{RA})$);
4. The CA receives the running states (we assume just one time) from RA through its mail box denoted by its name $CA$ (the corresponding reading actions are denoted $r_{CA}(RS_{RA})$);
5. Then, it does some local computations mixed with some atomic actions by computation logics, including $\cdot$, $+$, $\between$, and guards, the whole local computations are denoted and included into $I_{CA}$, which is the set of all local atomic actions;
6. When the local computations are finished, the CA generates the output states $RS_{CA}$ (containing the files), and sends them to the outside (the corresponding sending action is denoted $s_O(RS_{CA})$);
7. The CA receives the computational results from RA through its mail box denoted by its name $CA$ (the corresponding reading actions are denoted $r_{CA}(CR_{RA})$);
8. Then, it does some local computations mixed with some atomic actions by computation logics, including $\cdot$, $+$, $\between$, and guards, the whole local computations are denoted and included into $I_{CA}$, which is the set of all local atomic actions;
9. When the local computations are finished, the CA generates the output message $DO_{CA}$, and sends it to the outside (the corresponding sending action is denoted $s_O(DO_{CA})$), and then processes the next message from the outside recursively.

The above process is described as the following state transitions by APTC:

$CA = r_{CA}(DI_{CA}) \cdot CA_1$

$CA_1 = I_{CA} \cdot CA_2$

$CA_2 = s_{RA}(DI_{RA}) \cdot CA_3$

$CA_3 = r_{CA}(RS_{RA}) \cdot CA_4$

$CA_4 = I_{CA} \cdot CA_5$

$CA_5 = s_O(RS_{CA}) \cdot CA_6$

$CA_6 = r_{CA}(CR_{RA}) \cdot CA_7$
$CA_7 = I_{CA} \cdot CA_8$
$CA_8 = s_O(DO_{CA}) \cdot CA$

By use of the algebraic laws of APTC, the CA may be proven to exhibit the desired external behaviors. If it can exhibit the desired external behaviors, the CA should have the following form:

$\tau_{I_{CA}}(\partial_\emptyset(CA)) = r_{CA}(DI_{CA}) \cdot s_{RA}(DI_{RA}) \cdot r_{CA}(RS_{RA}) \cdot s_O(RS_{CA}) \cdot r_{CA}(CR_{RA}) \cdot s_O(DO_{CA}) \cdot \tau_{I_{CA}}(\partial_\emptyset(CA))$

### 21.2.2 Virtual Resource actor, VA

A Virtual Resource is an atomic function unit to access actual resource and is managed by the RA. We use an actor called Virtual Resource actor (VA) to model a Virtual Resource.

A VA has a unique name, local information and variables to contain its states, and local computation procedures to manipulate the information and variables. A VA is always managed by the Master and it receives messages from the Master, sends messages to the Master and the client, and is created by the Master. Note that a VA can not create new VAs, it can only be created by the Master. That is, a VA is an actor with a constraint that is without a **create** action.

After a VA is created, the typical process is as follows:

1. The VA receives the computational tasks $DI_{VA}$ from RA through its mail box denoted by its name $VA$ (the corresponding reading action is denoted $r_{VA}(DI_{VA})$);
2. Then, it does some local computations mixed with some atomic actions by computation logics, including $\cdot$, $+$, $\between$, and guards, the whole local computations are denoted and included into $I_{VA}$, which is the set of all local atomic actions;
3. During the local computations, it generates the running states $RS_{VA}$, and sends them (we assume just one time) to the SA's mail box denoted by the SA's name $SA$ (the corresponding sending action is denoted $s_{SA}(RS_{VA})$);
4. Then, it does some local computations mixed with some atomic actions by computation logics, including $\cdot$, $+$, $\between$, and guards, the whole local computations are denoted and included into $I_{VA}$, which is the set of all local atomic actions;
5. If the local computations are finished, VA generates the computational results $CR_{VA}$, and sends them to the RA's mail box denoted by the RA's name $RA$ (the corresponding sending action is denoted $s_{RA}(CR_{VA})$), and then processes the next task from RA recursively.

The above process is described as the following state transitions by APTC:

$VA = r_{VA}(DI_{VA}) \cdot VA_1$
$VA_1 = I_{VA} \cdot VA_2$
$VA_2 = s_{SA}(RS_{VA}) \cdot VA_3$
$VA_3 = I_{VA} \cdot VA_4$
$VA_4 = s_{RA}(CR_{VA}) \cdot VA$

By use of the algebraic laws of APTC, the VA may be proven to exhibit the desired external behaviors. If it can exhibit the desired external behaviors, the VA should have the following form:

$$\tau_{I_{VA}}(\partial_\emptyset(VA)) = r_{VA}(DI_{VA}) \cdot s_{SA}(RS_{VA}) \cdot s_{RA}(CR_{VA}) \cdot \tau_{I_{VA}}(\partial_\emptyset(VA))$$

### 21.2.3 Resource Manager actor, RA

RA receives the requests from the client, and manages the VAs. We use an actor called Resource Manager actor (RA) to model the Resource Manager.

After RA is created, the typical process is as follows:

1. The RA receives the requests $DI_{RA}$ from the Client through its mail box denoted by its name $RA$ (the corresponding reading action is denoted $r_{RA}(DI_{RA})$);
2. Then, it does some local computations mixed with some atomic actions by computation logics, including $\cdot$, $+$, $\between$, and guards, the whole local computations are denoted and included into $I_{RA}$, which is the set of all local atomic actions;
3. The RA creates $n$ VAs $VA_i$ for $1 \leq i \leq n$ in parallel through actions $\textbf{new}(VA_1) \parallel \cdots \parallel \textbf{new}(VA_n)$;
4. When the local computations are finished, the RA generates the computational tasks $DI_{VA_i}$ for each $VA_i$ with $1 \leq i \leq n$, sends them to the VAs' mail box denoted by the VAs' name $VA_i$ (the corresponding sending actions are denoted $s_{VA_1}(DI_{VA_1}) \parallel \cdots \parallel s_{VA_n}(DI_{VA_n})$);
5. The RA receives the running states $RS_{SA}$ (we assume just one time) from the SA through its mail box denoted by its name $RA$ (the corresponding reading action is denoted $r_{RA}(RS_{SA})$);

6. Then, it does some local computations mixed with some atomic actions by computation logics, including $\cdot$, $+$, $\between$, and guards, the whole local computations are denoted and included into $I_{RA}$, which is the set of all local atomic actions;
7. When the local computations are finished, the RA generates running states $RS_{RA}$, sends them to the CA's mail box denoted by the CA's name $CA$ (the corresponding sending actions are denoted $s_{CA}(RS_{RA})$);
8. The RA receives the computational results $CR_{VA_i}$ from the $VA_i$ for $1 \leq i \leq n$ through its mail box denoted by its name $RA$ (the corresponding reading action is denoted $r_{RA}(CR_{VA_1}) \parallel \cdots \parallel r_{RA}(CR_{VA_n})$);
9. Then, it does some local computations mixed with some atomic actions by computation logics, including $\cdot$, $+$, $\between$, and guards, the whole local computations are denoted and included into $I_{RA}$, which is the set of all local atomic actions;
10. When the local computations are finished, the RA generates results $CR_{RA}$, sends them to the CAs' mail box denoted by the CA's name $CA$ (the corresponding sending actions are denoted $s_{CA}(CR_{RA})$), and then processes the next message from the client recursively.

The above process is described as the following state transitions by APTC:

$RA = r_{RA}(DI_{RA}) \cdot RA_1$

$RA_1 = I_{RA} \cdot RA_2$

$RA_2 = \mathbf{new}(VA_1) \parallel \cdots \parallel \mathbf{new}(VA_n) \cdot RA_3$

$RA_3 = s_{VA_1}(DI_{VA_1}) \parallel \cdots \parallel s_{VA_n}(DI_{VA_n}) \cdot RA_4$

$RA_4 = r_{RA}(RS_{SA}) \cdot RA_5$

$RA_5 = I_{RA} \cdot RA_6$

$RA_6 = s_{CA}(RS_{RA}) \cdot RA_7$

$RA_7 = r_{RA}(CR_{VA_1}) \parallel \cdots \parallel r_{RA}(CR_{VA_n}) \cdot RA_8$

$RA_8 = I_{RA} \cdot RA_9$

$RA_9 = s_{CA}(CR_{RA}) \cdot RA$

By use of the algebraic laws of APTC, the RA may be proven to exhibit the desired external behaviors. If it can exhibit the desired external behaviors, the RA should have the following form:

$\tau_{I_{RA}}(\partial_\emptyset(RA)) = r_{RA}(DI_{RA}) \cdot (s_{VA_1}(DI_{VA_1}) \parallel \cdots \parallel s_{VA_n}(DI_{VA_n})) \cdot r_{RA}(RS_{SA}) \cdot s_{CA}(RS_{RA}) \cdot (r_{RA}(CR_{VA_1}) \parallel \cdots \parallel r_{RA}(CR_{VA_n})) \cdot s_{CA}(CR_{RA}) \cdot \tau_{I_{RA}}(\partial_\emptyset(RA))$

### 21.2.4 State Collector actor, SA

We use an actor called State Collector actor (SA) to model the State Collector.

After the SA is created, the typical process is as follows:

1. The SA receives the running states $RS_{VA_i}$ from $VA_i$ (we assume just one time) for $1 \leq i \leq n$ through its mail box denoted by its name $SA$ (the corresponding reading action is denoted $r_{SA}(RS_{VA_1}) \parallel \cdots \parallel r_{SA}(RS_{VA_n})$);
2. Then, it does some local computations mixed with some atomic actions by computation logics, including $\cdot$, $+$, $\between$, and guards, the whole local computations are denoted and included into $I_{SA}$, which is the set of all local atomic actions;
3. When the local computations are finished, SA generates the running states $RS_{SA}$, and sends them to the RA's mail box denoted by the RA's name $RA$ (the corresponding sending action is denoted $s_{RA}(RS_{SA})$), and then processes the next task from RA recursively.

The above process is described as the following state transitions by APTC:

$SA = r_{SA}(RS_{VA_1}) \parallel \cdots \parallel r_{SA}(RS_{VA_n}) \cdot SA_1$

$SA_1 = I_{SA} \cdot SA_2$

$SA_2 = s_{RA}(RS_{SA}) \cdot SA$

By use of the algebraic laws of APTC, the SA may be proven to exhibit the desired external behaviors. If it can exhibit the desired external behaviors, the SA should have the following form:

$\tau_{I_{SA}}(\partial_\emptyset(SA)) = (r_{SA}(RS_{VA_1}) \parallel \cdots \parallel r_{SA}(RS_{VA_n})) \cdot s_{RA}(RS_{SA})) \cdot \tau_{I_{SA}}(\partial_\emptyset(SA))$

### 21.2.5 Putting all actors together into a whole

We put all the actors together into a whole, including all CA, VAs, RA, and SA, according to the architecture as illustrated in Fig. 21.1. The whole actor system $CA \quad RA \quad SA = CA \quad RA \quad SA \quad VA_1 \quad \cdots \quad VA_n$ can be represented by the following process term of APTC:

$$\tau_I(\partial_H(CA \between RA \between SA)) = \tau_I(\partial_H(CA \between RA \between SA \between VA_1 \between \cdots \between VA_n))$$

Among all the actors, there are synchronous communications. The actor's reading and the same actor's sending actions with the same type messages may cause communications. If the actor's sending action occurs before the same actor's reading action, an asynchronous communication will occur; otherwise, a deadlock $\delta$ will be caused.

There are four kinds of asynchronous communications as follows:

(1) The communications between a VA and RA with the following constraints:

$s_{VA}(DI_{VA}) \leq r_{VA}(DI_{VA})$

$s_{RA}(CR_{VA}) \leq r_{RA}(CR_{VA})$

(2) The communications between a VA and SA with the following constraint:

$s_{SA}(RS_{VA}) \leq r_{SA}(RS_{VA})$

(3) The communications between CA and RA with the following constraints:

$s_{RA}(DI_{RA}) \leq r_{RA}(DI_{RA})$

$s_{CA}(RS_{RA}) \leq r_{CA}(RS_{RA})$

$s_{CA}(CR_{RA}) \leq r_{CA}(CR_{RA})$

(4) The communications between RA and SA with the following constraint:

$s_{RA}(RS_{SA}) \leq r_{RA}(RS_{SA})$

Hence, the set $H$ and $I$ can be defined as follows:

$$H = \{s_{VA_1}(DI_{VA_1}), r_{VA_1}(DI_{VA_1}), \cdots, s_{VA_n}(DI_{VA_n}), r_{VA_n}(DI_{VA_n}),$$
$$s_{RA}(CR_{VA_1}), r_{RA}(CR_{VA_1}), \cdots, s_{RA}(CR_{VA_n}), r_{RA}(CR_{VA_n}),$$
$$s_{SA}(RS_{VA_1}), r_{SA}(RS_{VA_1}), \cdots, s_{SA}(RS_{VA_n}), r_{SA}(RS_{VA_n}),$$
$$s_{RA}(DI_{RA}), r_{RA}(DI_{RA}), s_{CA}(RS_{RA}), r_{CA}(RS_{RA}),$$
$$s_{CA}(CR_{RA}), r_{CA}(CR_{RA}), s_{RA}(RS_{SA}), r_{RA}(RS_{SA})$$
$$|s_{VA_1}(DI_{VA_1}) \nleq r_{VA_1}(DI_{VA_1}), \cdots, s_{VA_n}(DI_{VA_n}) \nleq r_{VA_n}(DI_{VA_n}),$$
$$s_{RA}(CR_{VA_1}) \nleq r_{RA}(CR_{VA_1}), \cdots, s_{RA}(CR_{VA_n}) \nleq r_{RA}(CR_{VA_n}),$$
$$s_{SA}(RS_{VA_1}) \nleq r_{SA}(RS_{VA_1}), \cdots, s_{SA}(RS_{VA_n}) \nleq r_{SA}(RS_{VA_n}),$$
$$s_{RA}(DI_{RA}) \nleq r_{RA}(DI_{RA}), s_{CA}(RS_{RA}) \nleq r_{CA}(RS_{RA}),$$
$$s_{CA}(CR_{RA}) \nleq r_{CA}(CR_{RA}), s_{RA}(RS_{SA}) \nleq r_{RA}(RS_{SA})\}$$

$$I = \{s_{VA_1}(DI_{VA_1}), r_{VA_1}(DI_{VA_1}), \cdots, s_{VA_n}(DI_{VA_n}), r_{VA_n}(DI_{VA_n}),$$
$$s_{RA}(CR_{VA_1}), r_{RA}(CR_{VA_1}), \cdots, s_{RA}(CR_{VA_n}), r_{RA}(CR_{VA_n}),$$
$$s_{SA}(RS_{VA_1}), r_{SA}(RS_{VA_1}), \cdots, s_{SA}(RS_{VA_n}), r_{SA}(RS_{VA_n}),$$
$$s_{RA}(DI_{RA}), r_{RA}(DI_{RA}), s_{CA}(RS_{RA}), r_{CA}(RS_{RA}),$$
$$s_{CA}(CR_{RA}), r_{CA}(CR_{RA}), s_{RA}(RS_{SA}), r_{RA}(RS_{SA})$$
$$|s_{VA_1}(DI_{VA_1}) \leq r_{VA_1}(DI_{VA_1}), \cdots, s_{VA_n}(DI_{VA_n}) \leq r_{VA_n}(DI_{VA_n}),$$
$$s_{RA}(CR_{VA_1}) \leq r_{RA}(CR_{VA_1}), \cdots, s_{RA}(CR_{VA_n}) \leq r_{RA}(CR_{VA_n}),$$
$$s_{SA}(RS_{VA_1}) \leq r_{SA}(RS_{VA_1}), \cdots, s_{SA}(RS_{VA_n}) \leq r_{SA}(RS_{VA_n}),$$
$$s_{RA}(DI_{RA}) \leq r_{RA}(DI_{RA}), s_{CA}(RS_{RA}) \leq r_{CA}(RS_{RA}),$$
$$s_{CA}(CR_{RA}) \leq r_{CA}(CR_{RA}), s_{RA}(RS_{SA}) \leq r_{RA}(RS_{SA})\}$$
$$\cup I_{CA} \cup I_{VA_1} \cup \cdots \cup I_{VA_n} \cup I_{RA} \cup I_{SA}$$

Then, we can obtain the following conclusion.

**Theorem 21.1.** *The whole actor system of cloud resource management illustrated in Fig. 21.1 can exhibit the desired external behaviors.*

*Proof.* By use of the algebraic laws of APTC, we can prove the following equation:

$$\tau_I(\partial_H(CA \between RA \between SA)) = \tau_I(\partial_H(CA \between RA \between SA \between VA_1 \between \cdots \between VA_n))$$
$$= r_{CA}(DI_{CA}) \cdot s_O(RS_{CA}) \cdot s_O(CR_{CA}) \cdot \tau_I(\partial_H(CA \between RA \between SA \between VA_1 \between \cdots \between VA_n))$$
$$= r_{CA}(DI_{CA}) \cdot s_O(RS_{CA}) \cdot s_O(CR_{CA}) \cdot \tau_I(\partial_H(CA \between RA \between SA))$$

For the details of the proof, as we omit them here, please refer to Section 17.3. □

Chapter 22

# Process algebra based actor model of the Web Service composition

In this chapter, we will use the process algebra based actor model to model and verify Web Service composition based on the previous work [25]. In Section 22.1, we introduce the requirements of the Web Service composition runtime system; we model the Web Service composition runtime by use of the new actor model in Section 22.2; finally, we take an example to show the usage of the model in Section 22.3.

## 22.1 Requirements of the Web Service composition

Web Service (WS) is a distributed software component that emerged about ten years ago to utilize the most widely used Internet application protocol – HTTP as its base transport protocol. As a component, a WS has similar ingredients as the others, such as DCOM, EJB, CORBA, and so on. That is, a WS uses HTTP-based SOAP as its transport protocol, WSDL as its interface description language, and UDDI as its name and directory service.

WS Composition creates new composite WSs using different composition patterns from the collection of existing WSs. Due to the advantages of WS to solve cross-organizational application integrations, two composition patterns are dominant. One is called the Web Service Orchestration (WSO) [23], which uses a workflow-like composition pattern to orchestrate business activities (implemented as WS Operations) and models cross-organizational business processes or other kind of processes. The other is called Web Service Choreography (WSC) [24], which has an aggregate composition pattern to capture the external interaction behaviors of WSs and acts as a contract or a protocol among WSs.

We now take a simple example of buying books from a book store to illustrate some concepts of WS composition. Although this example is quite simple and only includes the sequence control flow (that is, each business activity in a business process is executed in sequence), it is enough to explain the concepts and ideas of this chapter and avoids unnecessary complexity without loss of generality. We use this example throughout this chapter. The requirements of this example are as Fig. 22.1 shows.

A customer buys books from a book store through a user agent. In this example, we ignore interactions between the customer and the user agent, and focus on those between the user agent and the book store. Either the user agent or book store has business processes to interact with each other.

We give the process of user agent as follows. The process of the book store can be obtained from that of user agent as contrasts.

1. The user agent requests a list of all books to the book store;
2. It receives the book list from the book store;
3. It selects the books by the customer and sends the list of selected books to the book store;
4. It receives the prices of selected books from the book store;
5. It accepts the prices and pays for the selected book to the book store. Then, the process terminates.

Since the business activities, such as the book store accepting request for a list of books from the user agent, are implemented as WSs (exactly WS operations), such buyer agent and book store business processes are called WSOs. These WSOs are published as WSs, called their interface WSs for interacting among each other. The interaction behaviors among WSs described by some contracts or protocols are called WSCs.

There are many efforts for WS Composition, including its specifications, design methods, verifications, simulations, and runtime supports. Different methods and tools are used in WS Composition research, such as XML-based WSO description specifications and WSC description specifications, formal verification techniques based on Process Algebra and Petri-Net, and runtime implementations using programming languages. Some of these works mainly focus on WSO, others mainly on WSC, and also a few works attempt to establish a relationship between WSO and WSC.

Can a WS interact with another one? Also, can a WSO interact with another one via their interfaces? Is the definition of a WSC compatible with its partner WSs or partner WSOs? To solve these problems, a correct relationship between WSO

Handbook of Truly Concurrent Process Algebra. https://doi.org/10.1016/B978-0-44-321515-5.00026-3

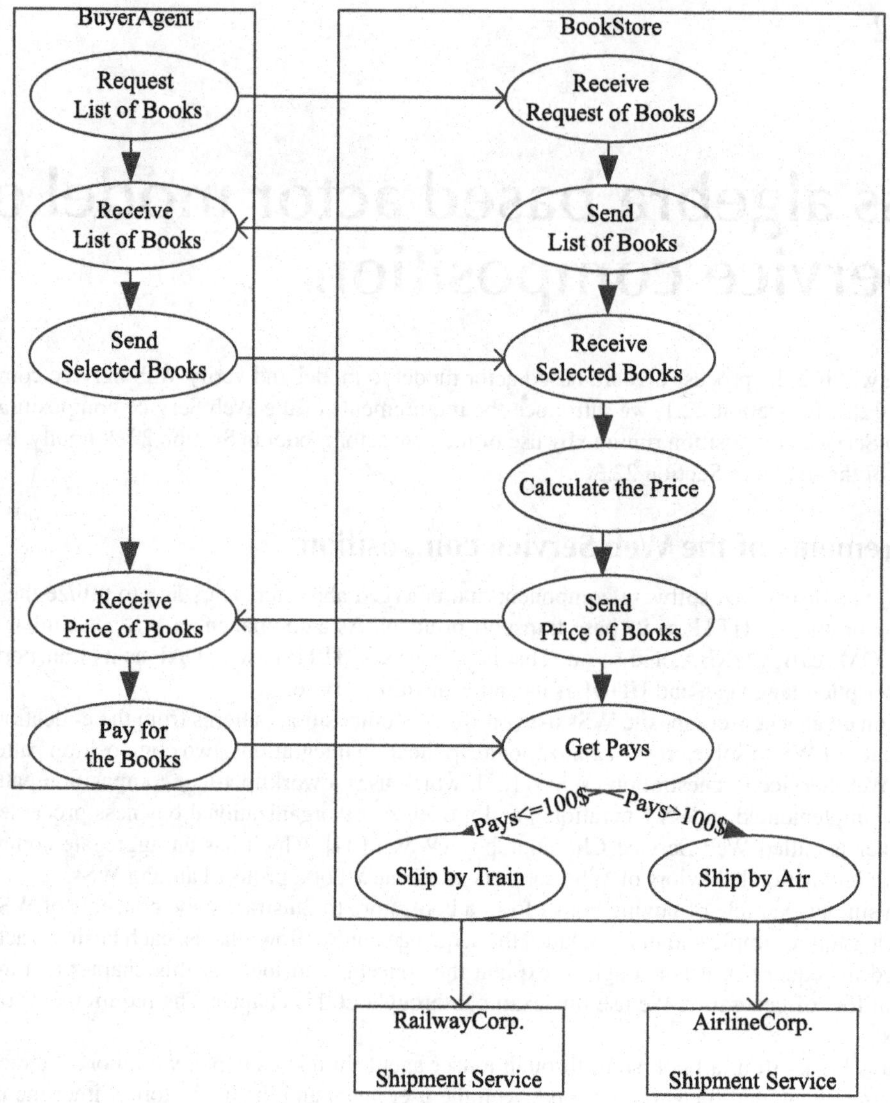


**FIGURE 22.1** Requirements of an example.

and WSC must be established. A WS Composition system combining WSO and WSC, with a natural relationship between the two, is an attractive direction. In a systematic viewpoint, WS, WSO, and WSC are organized with a natural relationship under the whole environment of cross-organizational business integration. More importantly, such a system should have a firm theoretical foundation.

In this chapter, we try to make such a system based on the new actor model.

### 22.1.1 WSO and WSC

A WS is a distributed software component with transport protocol – SOAP, interface description by WSDL, and can be registered into UDDI to be searched and discovered by its customers.

A WSO orchestrates WSs existing on the Web into a process through the so-called control flow constructs. That is, within a WSO, there are a collection of atomic function units called activities with control flows to manipulate them. Hence, the main ingredients of a WSO are the following:

- Inputs and Outputs: At the start time of a WSO, it accepts some inputs, and it sends out outcomes at the end of its execution;

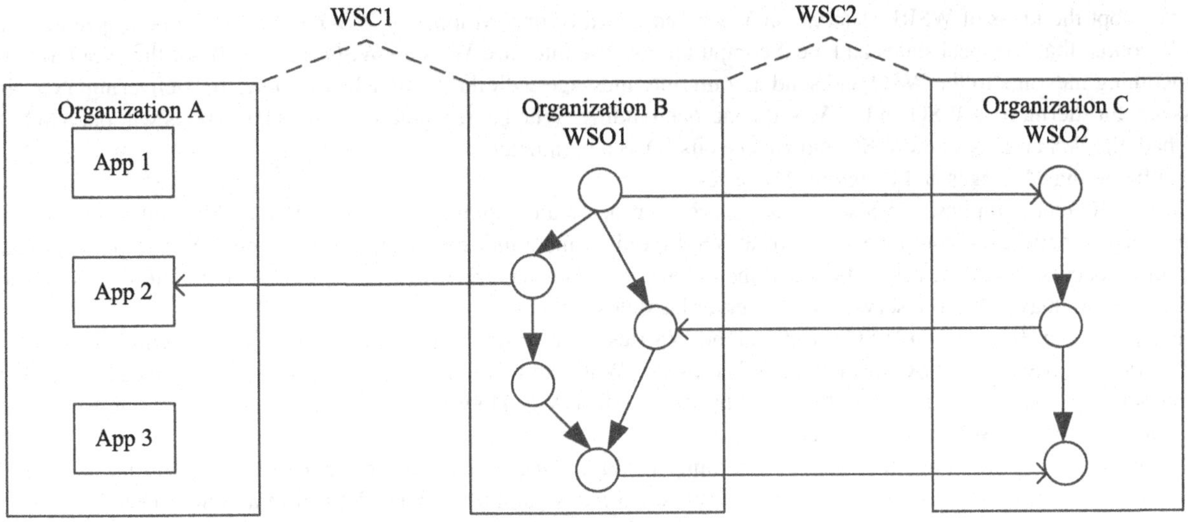


**FIGURE 22.2** Relationship between WSO and WSC.

- Information and Variable Definitions: A WSO has local states that may be transferred among activities. Finally, the local states are sent to WSs outside by activities in the form of messages. In turn, activities receiving message outside can alter the local states;
- Activity Definitions: An activity is an atomic unit with several pre-defined function kinds, such as invoking a WS outside, invoking an application inside, receiving a request from a customer inside/outside, local variable assignments, etc.;
- Control Flow Definitions: Control flow definitions give activities an execution order. In terms of structural model based control flow definitions, control flows are the so-called structural activities that can be sequence activity, choice activity, loop activity, parallel activity, and their variants;
- Binding WS Information: Added values of WS Composition are the so-called recursive composition, that is, a WSO orchestrating existing WSs is published as a new WS itself too. A WSO interacts with other WSs outside through this new WS.

In Fig. 22.1, the user agent business process is modeled as UserAgent WSO described by WS-BPEL.

The interface WS for UserAgent WSO is called UserAgent WS described by WSDL.

A WSC defines the external interaction behaviors and serves as a contract or a protocol among WSs. The main ingredients of a WSC are as follows:

- Partner Definitions: They define the partners within a WSC including the role of a partner acting as and relationships among partners;
- Information and Variable Definitions: A WSC may also have local states exchanged among the interacting WSs;
- Interactions among Partners: Interaction points and interaction behaviors are defined as the core contents in a WSC.

In the buying books example, the WSC between user agent and bookstore (exactly UserAgentWS and BookStoreWS) called BuyingBookWSC is described by WS-CDL.

The WSO and the WSC define two different aspects of WS Composition. Their relationships are as Fig. 22.2 illustrates. Note that a WSO may require at least a WSC, but a WSC does not need to depend on a WSO.

### 22.1.2 Design decisions on Web Service composition runtime

(1) Stateless WS or Stateful WS

In the viewpoint of W3C, a WS itself is an interface or a wrapper of an application inside the boundary of an organization that has a willingness to interact with applications outside. That is, a W3C WS has no independent programming model like other component models and has no need of containing local states for local computations. Indeed, there are different means of developing WS to be a full sense component, such as OGSI. Incompatibility between W3C WS and OGSI-like WS leads to WSRF as a compromised solution that reserves the W3C WS and develops a notion of WS Resource to model states.

We adopt the ideas of WSRF. That is, let WS be an interface or a wrapper of WSO and let WSO be a special kind of WS Resource that has local states and local computations. The interface WS of a WSO reserves ID of the WSO to deliver an incoming message to the WSO and send an outgoing message with the ID attached in order for delivering a call-back message. Furthermore, a WSO and its WS are one-one binding. When a new incoming message arrives without a WSO ID attached, the WS creates a new WSO and attaches its ID as a parameter.

(2) Incoming Messages and Outgoing Messages

Just as the name implies, a WS serves as a server to process an incoming message within a C/S framework. However, an interaction between a component WS or a WSO requires incoming message and outgoing message pairs. When an interaction occurs, one serves as a client and the other serves as a server. However, in the next interaction, the one served as client before may serve as a server and the server becomes a client.

The problem is that when a WSO (or other kind WS Resource) inside interacts with WSs outside, who is willing to act as the bridge between the WSO inside and WSs outside? When an incoming message arrives, it is easily understood that the incoming message is delivered to the WSO by the interface WS. However, how is an outgoing message from a WSO inside to a component WS outside delivered?

In fact, there are two ways to solve the outgoing message. One is the way of WS-BPEL [23], and the other is that of an early version of WSDL. The former uses a so-called *invoke* atomic activity defined in a WSO to send an outgoing message directly without the assistance of its interface WS. In contrast, the latter specifies that everything exchanged between resources inside and functions outside must go via the interface WS of the resource inside. Furthermore, in an early edition of WSDL, there are four kinds of WS operations defined, including an **In** operation, an **In-Out** operation, an **Out** operation, and an **Out-In** operation. An **In** operation and an **In-Out** operation receive the incoming messages, while an **Out** operation and an **Out-In** operation deliver the outgoing messages. An **Out** operation and an **Out-In** operation are somewhat strange because a WS is a kind of server in nature. Hence, in the later versions of WSDL, an **Out** operation and an **Out-In** operation are removed. However, the problem of how to process the outgoing message remains.

The way of WS-BPEL will cause some confusion in the WS Composition runtime architecture design, and the method of the early edition of WSDL looks somewhat strange. Hence, our way of processing outgoing messages is a compromise of the above ones. That is, the outgoing messages from an internal WSO to an external resource, must go via the WS of the internal WSO. However, the WS does not need to declare operations for processing the outgoing messages in the WSDL definitions.

(3) Functions and Enablements of WSC

A WSC acts as a contract or a protocol between interacting WSs. From a viewpoint of business integration requirements, a WSC serves as a business contract to constrain the rights and obligations of business partners. Also, from the view of utilized technologies, a WSC can be deemed as a communication protocol that coordinates the interaction behaviors of the involved WSs.

About the enablements of a WSC, there are also two different enable patterns. One is a concentrated architecture and the other is a distributed one.

The concentrated way considers that the enablements of a WSC must be under supervision of a third authorized party or all involved partners. An absolutely concentrated way may require that any operation about interacting WSs must be done via a supervisor. This way may cause the supervisor to become a performance bottleneck when many interactions occur, but it can bring trustworthiness of interaction results if the supervisor is trustworthy itself.

The distributed way argues that each WS interacts among others with constraints of a WSC and there is no need for a supervisor. It is regarded that WSs just behave *correctly* to obey a WSC and maybe take an example of enablements of open Internet protocols. However, there are cheating business behaviors of an intendedly *incorrect* WS, that are unlike almost purely technical motivations of open Internet protocols.

We use a hybrid enablements of WSC. That is, when a WSC is contracted (either contracted dynamically at runtime or contracted with human interventions at design time) among WSs and enabled, the WSC creates the partner WSs at the beginning of enablements. Also, then the WSs interact with each other.

### 22.1.3 A WS composition runtime architecture

Based on the above introductions and discussions, we design an architecture of WS Composition runtime as Fig. 22.3 shows. Fig. 22.3 illustrates the typical architecture of a WS Composition runtime. We explain the compositions and their relationships in the following. There are four components: WSO, WS, WSC, and applications inside.

The functions and ingredients of a WSO usually it have a collection of activities that may interact with partner WSs outside or applications inside. Enablements of a WSO require a runtime environment that is not illustrated in Fig. 22.3. For

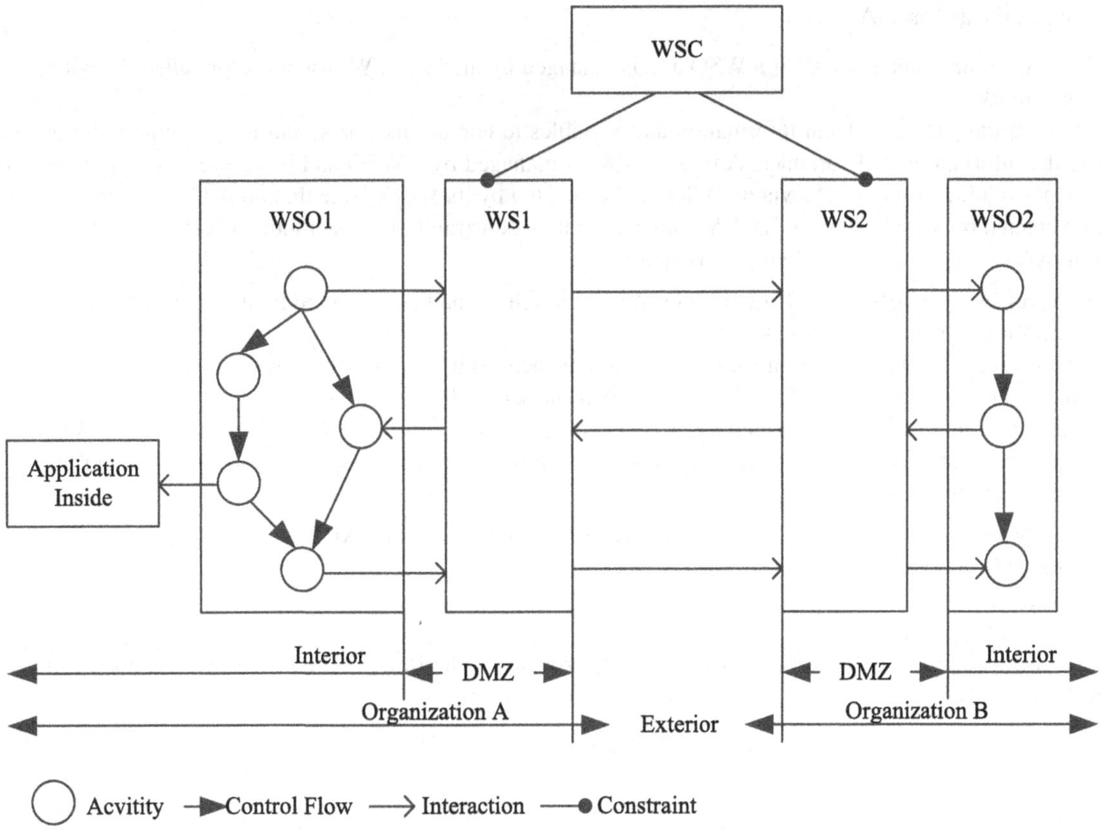


FIGURE 22.3 An architecture of WS composition runtime.

example, execution of a WSO described by WS-BPEL needs a WS-BPEL interpreter (also called a WSO engine). A WSO locates in the interior of an organization. It interacts with applications inside with private exchanging mechanisms and with other partner WSOs outside via its interface WS.

Applications inside may be any legacy application or any newly developed application within the interior of a organization. These applications can be implemented in any technical framework and provide interfaces to interact with other applications inside, including a WSO. Interactions between a WSO and an application inside may be based on any private communication mechanism, such as local object method call, RPC, RMI, etc., which depends on the technical framework adopted by the application.

An interface WS acts as an interface of a WSO to interact with partner WSs outside. A WSO is with a one-to-one binding to its interface WS and is created by its interface WS at the time of first interaction with the exterior. Enablements of a WS also require a runtime support usually called a SOAP engine that implies a HTTP server installed to couple with HTTP requests. A WS and its runtime support locate at the demilitarized zone (DMZ) of an organization that has different management policies and different security policies to the interior of an organization.

A WSC acts as a contract or a protocol of partner WSs. When a WSC is enabled, it creates all partner WSs at their accurate positions. Enablements of a WSC also require a runtime support to interpret the WSC description language like WS-CDL. A WSC and its support environment can be located at a third authorized party or other places negotiated by the partners.

## 22.2 The new actor model of Web Service composition

According to the architecture of WS composition runtime, the whole actors system implemented by actors can be divided into four kinds of actors: the activity actors, the WS actors, the WSO actors, and the WSC actor.

### 22.2.1 Activity actor, AA

An activity is an atomic function unit of a WSO and is managed by the WSO. We use an actor called Activity Actor (AA) to model an activity.

An AA has a unique name, local information and variables to contain its states, and local computation procedures to manipulate the information and variables. An AA is always managed by a WSO and it receives messages from its WSO, sends messages to other AAs or WSs via its WSO, and is created by its WSO. Note that an AA cannot create new AAs, it can only be created by a WSO. That is, an AA is an actor with a constraint that is without a **create** action.

After an AA is created, the typical process is as follows:

1. The AA receives some messages $DI_{AA}$ from its WSO through its mail box denoted by its name $AA$ (the corresponding reading action is denoted $r_{AA}(DI_{AA})$);
2. Then, it does some local computations mixed some atomic actions by computation logics, including $\cdot$, $+$, $\between$, and guards, the whole local computations are denoted $I_{AA}$, which is the set of all local atomic actions;
3. When the local computations are finished, the AA generates the output message $DO_{AA}$ and sends to its WSO's mail box denoted by the WSO's name $WSO$ (the corresponding sending action is denoted $s_{WSO}(DO_{AA})$), and then processes the next message from its WSO recursively.

The above process is described as the following state transition skeletons by APTC:

$AA = r_{AA}(DI_{AA}) \cdot AA_1$

$AA_1 = I_{AA} \cdot AA_2$

$AA_2 = s_{WSO}(DO_{AA}) \cdot AA$

By use of the algebraic laws of APTC, the AA may be proven to exhibit the desired external behaviors: If it can exhibit the desired external behaviors, the AA should have the following form:

$$\tau_{I_{AA}}(\partial_{\emptyset}(AA)) = r_{AA}(DI_{AA}) \cdot s_{WSO}(DO_{AA}) \cdot \tau_{I_{AA}}(\partial_{\emptyset}(AA))$$

### 22.2.2 Web Service orchestration, WSO

A WSO includes a set of AAs and acts as the manager of the AAs. The management operations may be creating a member AA, or acting as a bridge between AAs and WSs outside.

After a WSO is created, the typical process is as follows:

1. The WSO receives the initialization message $DI_{WSO}$ from its interface WS through its mail box by its name $WSO$ (the corresponding reading action is denoted $r_{WSO}(DI_{WSO})$);
2. The WSO may create its AAs in parallel through actions $\mathbf{new}(AA_1) \parallel \cdots \parallel \mathbf{new}(AA_n)$ if it is not initialized;
3. The WSO may receive messages from its interface WS or its AAs through its mail box by its name $WSO$ (the corresponding reading actions are distinct by the message names);
4. The WSO does some local computations mixed some atomic actions by computation logics, including $\cdot$, $+$, $\between$, and guards, the local computations are included into $I_{WSO}$, which is the set of all local atomic actions;
5. When the local computations are finished, the WSO generates the output messages and may send to its AAs or its interface WS (the corresponding sending actions are distinct by the names of AAs and WS, and also the names of messages), and then processes the next message from its AAs or the interface WS.

The above process is described as the following state transition skeletons by APTC:

$WSO = r_{WSO}(DI_{WSO}) \cdot WSO_1$

$WSO_1 = (\{isInitialed(WSO) = FLALSE\} \cdot (\mathbf{new}(AA_1) \parallel \cdots \parallel \mathbf{new}(AA_n)) + \{isInitialed(WSO) = TRUE\}) \cdot WSO_2$

$WSO_2 = r_{WSO}(DI_{AAs}, DI_{WS}) \cdot WSO_3$

$WSO_3 = I_{WSO} \cdot WSO_4$

$WSO_4 = s_{AAs,WS}(DO_{WSO}) \cdot WSO$

By use of the algebraic laws of APTC, the WSO may be proven to exhibit the desired external behaviors: If it can exhibit the desired external behaviors, the WSO should have the following form:

$$\tau_{I_{WSO}}(\partial_{\emptyset}(WSO)) = r_{WSO}(DI_{WSO}) \cdot \cdots \cdot s_{WS}(DO_{WSO}) \cdot \tau_{I_{WSO}}(\partial_{\emptyset}(WSO))$$

with $I_{WSO}$ extended to $I_{WSO} \cup \{\{isInitialed(WSO) = FLALSE\}, \{isInitialed(WSO) = TRUE\}\}$.

### 22.2.3 Web Service, WS

A WS is an actor that has the characteristics of an ordinary actor. It acts as a communication bridge between the inner WSO and the external partner WS and creates a new WSO when it receives a new incoming message.

After A WS is created, the typical process is as follows:

1. The WS receives the initialization message $DI_{WS}$ from its WSC actor through its mail box by its name $WS$ (the corresponding reading action is denoted $r_{WS}(DI_{WS})$);
2. The WS may create its WSO through actions $\mathbf{new}(WSO)$ if it is not initialized;
3. The WS may receive messages from its partner WS or its WSO through its mail box by its name $WSO$ (the corresponding reading actions are distinct by the message names);
4. The WS does some local computations mixed with some atomic actions by computation logics, including $\cdot$, $+$, $\between$, and guards, the local computations are included into $I_{WS}$, which is the set of all local atomic actions;
5. When the local computations are finished, the WS generates the output messages and may send to its WSO or its partner WS (the corresponding sending actions are distinct by the names of WSO and the partner WS, and also the names of messages), and then processes the next message from its WSO or the partner WS.

The above process is described as the following state transition skeletons by APTC:

$WS = r_{WS}(DI_{WS}) \cdot WS_1$

$WS_1 = (\{isInitialed(WS) = FLALSE\} \cdot \mathbf{new}(WSO) + \{isInitialed(WS) = TRUE\}) \cdot WS_2$

$WS_2 = r_{WS}(DI_{WSO}, DI_{WS'}) \cdot WS_3$

$WS_3 = I_{WS} \cdot WS_4$

$WS_4 = s_{WSO,WS'}(DO_{WS}) \cdot WS$

By use of the algebraic laws of APTC, the WS may be proven to exhibit the desired external behaviors: If it can exhibit the desired external behaviors, the WS should have the following form:

$$\tau_{I_{WS}}(\partial_\emptyset(WS)) = r_{WS}(DI_{WS}) \cdot \cdots \cdot s_{WS'}(DO_{WS}) \cdot \tau_{I_{WS}}(\partial_\emptyset(WS))$$

with $I_{WS}$ extended to $I_{WS} \cup \{\{isInitialed(WS) = FLALSE\}, \{isInitialed(WS) = TRUE\}\}$.

### 22.2.4 Web Service choreography, WSC

A WSC actor creates partner WSs as some kind of roles and sets each WS to the other one as their partner WSs.

After a WSC is created, the typical process is as follows:

1. The WSC receives the initialization message $DI_{WSC}$ from the outside through its mail box by its name $WSC$ (the corresponding reading action is denoted $r_{WSC}(DI_{WSC})$);
2. The WSC may create its WSs through actions $\mathbf{new}(WS_1) \parallel \mathbf{new}(WS_2)$ if it is not initialized;
3. The WSC does some local computations mixed with some atomic actions by computation logics, including $\cdot$, $+$, $\between$, and guards, the local computations are included into $I_{WSC}$, which is the set of all local atomic actions;
4. When the local computations are finished, the WSC generates the output messages and sends to its WSs, or the outside (the corresponding sending actions are distinct by the names of WSs, and also the names of messages), and then processes the next message from the outside.

The above process is described as the following state transition skeletons by APTC:

$WSC = r_{WSC}(DI_{WSC}) \cdot WSC_1$

$WSC_1 = (\{isInitialed(WSC) = FLALSE\} \cdot (\mathbf{new}(WS_1) \parallel \mathbf{new}(WS_2)) + \{isInitialed(WSC) = TRUE\}) \cdot WSC_2$

$WSC_2 = I_{WSC} \cdot WSC_3$

$WSC_3 = s_{WS_1,WS_2,O}(DO_{WSC}) \cdot WSC$

By use of the algebraic laws of APTC, the WSC may be proven to exhibit the desired external behaviors: If it can exhibit the desired external behaviors, the WSC should have the following form:

$$\tau_{I_{WSC}}(\partial_\emptyset(WSC)) = r_{WSC}(DI_{WSC}) \cdot s_{WS_1,WS_2,O}(DO_{WSC}) \cdot \tau_{I_{WSC}}(\partial_\emptyset(WSC))$$

with $I_{WSC}$ extended to $I_{WSC} \cup \{\{isInitialed(WSC) = FLALSE\}, \{isInitialed(WSC) = TRUE\}\}$.

### 22.2.5 Putting all actors together into a whole

We put all the actors together into a whole, including all AAs, WSOs, WSs, and WSC, according to the architecture as illustrated in Fig. 22.3. The whole actor system $WSC = WSC \quad WSs \quad WSOs \quad AAs$ can be represented by the following process term of APTC.

$$\tau_I(\partial_H(WSC)) = \tau_I(\partial_H(WSC \between WSs \between WSOs \between AAs))$$

Among all the actors, there are synchronous communications. The actor's reading and the same actor's sending actions with the same type messages may cause communications. If the actor's sending action occurs before the same actor's reading action, an asynchronous communication will occur; otherwise, a deadlock $\delta$ will be caused.

There are four pairs kinds of asynchronous communications as follows:

(1) The communications between an AA and its WSO with the following constraints:

$s_{AA}(DI_{AA-WSO}) \leq r_{AA}(DI_{AA-WSO})$

$s_{WSO}(DI_{WSO-AA}) \leq r_{WSO}(DI_{WSO-AA})$

Note that the messages $DI_{AA-WSO}$ and $DO_{WSO-AA}$, $DI_{WSO-AA}$ and $DO_{AA-WSO}$ are the same messages.

(2) The communications between a WSO and its interface WS with the following constraints:

$s_{WSO}(DI_{WSO-WS}) \leq r_{WSO}(DI_{WSO-WS})$

$s_{WS}(DI_{WS-WSO}) \leq r_{WS}(DI_{WS-WSO})$

Note that the messages $DI_{WSO-WS}$ and $DO_{WS-WSO}$, $DI_{WS-WSO}$ and $DO_{WSO-WS}$ are the same messages.

(3) The communications between a WS and its partner WS with the following constraints:

$s_{WS_1}(DI_{WS_1-WS_2}) \leq r_{WS_1}(DI_{WS_1-WS_2})$

$s_{WS_2}(DI_{WS_2-WS_1}) \leq r_{WS_2}(DI_{WS_2-WS_1})$

Note that the messages $DI_{WS_1-WS_2}$ and $DO_{WS_2-WS_1}$, $DI_{WS_2-WS_1}$ and $DO_{WS_1-WS_2}$ are the same messages.

(4) The communications between a WS and its WSC with the following constraints:

$s_{WSC}(DI_{WSC-WS}) \leq r_{WSC}(DI_{WSC-WS})$

$s_{WS}(DI_{WS-WSC}) \leq r_{WS}(DI_{WS-WSC})$

Note that the messages $DI_{WSC-WS}$ and $DO_{WS-WSC}$, $DI_{WS-WSC}$ and $DO_{WSC-WS}$ are the same messages.

Hence, the set $H$ and $I$ can be defined as follows:

$H = \{s_{AA}(DI_{AA-WSO}), r_{AA}(DI_{AA-WSO}), s_{WSO}(DI_{WSO-AA}), r_{WSO}(DI_{WSO-AA}),$
$s_{WSO}(DI_{WSO-WS}), r_{WSO}(DI_{WSO-WS}), s_{WS}(DI_{WS-WSO}), r_{WS}(DI_{WS-WSO}),$
$s_{WS_1}(DI_{WS_1-WS_2}), r_{WS_1}(DI_{WS_1-WS_2}), s_{WS_2}(DI_{WS_2-WS_1}), r_{WS_2}(DI_{WS_2-WS_1}),$
$s_{WSC}(DI_{WSC-WS}), r_{WSC}(DI_{WSC-WS}), s_{WS}(DI_{WS-WSC}), r_{WS}(DI_{WS-WSC})$
$|s_{AA}(DI_{AA-WSO}) \nleq r_{AA}(DI_{AA-WSO}), s_{WSO}(DI_{WSO-AA}) \nleq r_{WSO}(DI_{WSO-AA}),$
$s_{WSO}(DI_{WSO-WS}) \nleq r_{WSO}(DI_{WSO-WS}), s_{WS}(DI_{WS-WSO}) \nleq r_{WS}(DI_{WS-WSO}),$
$s_{WS_1}(DI_{WS_1-WS_2}) \nleq r_{WS_1}(DI_{WS_1-WS_2}), s_{WS_2}(DI_{WS_2-WS_1}) \nleq r_{WS_2}(DI_{WS_2-WS_1}),$
$s_{WSC}(DI_{WSC-WS}) \nleq r_{WSC}(DI_{WSC-WS}), s_{WS}(DI_{WS-WSC}) \nleq r_{WS}(DI_{WS-WSC})\}$

$I = \{s_{AA}(DI_{AA-WSO}), r_{AA}(DI_{AA-WSO}), s_{WSO}(DI_{WSO-AA}), r_{WSO}(DI_{WSO-AA}),$
$s_{WSO}(DI_{WSO-WS}), r_{WSO}(DI_{WSO-WS}), s_{WS}(DI_{WS-WSO}), r_{WS}(DI_{WS-WSO}),$
$s_{WS_1}(DI_{WS_1-WS_2}), r_{WS_1}(DI_{WS_1-WS_2}), s_{WS_2}(DI_{WS_2-WS_1}), r_{WS_2}(DI_{WS_2-WS_1}),$
$s_{WSC}(DI_{WSC-WS}), r_{WSC}(DI_{WSC-WS}), s_{WS}(DI_{WS-WSC}), r_{WS}(DI_{WS-WSC})$
$|s_{AA}(DI_{AA-WSO}) \leq r_{AA}(DI_{AA-WSO}), s_{WSO}(DI_{WSO-AA}) \leq r_{WSO}(DI_{WSO-AA}),$
$s_{WSO}(DI_{WSO-WS}) \leq r_{WSO}(DI_{WSO-WS}), s_{WS}(DI_{WS-WSO}) \leq r_{WS}(DI_{WS-WSO}),$
$s_{WS_1}(DI_{WS_1-WS_2}) \leq r_{WS_1}(DI_{WS_1-WS_2}), s_{WS_2}(DI_{WS_2-WS_1}) \leq r_{WS_2}(DI_{WS_2-WS_1}),$
$s_{WSC}(DI_{WSC-WS}) \leq r_{WSC}(DI_{WSC-WS}), s_{WS}(DI_{WS-WSC}) \leq r_{WS}(DI_{WS-WSC})\}$
$\cup I_{AAs} \cup I_{WSOs} \cup I_{WSs} \cup I_{WSC}$

If the whole actor system of WS composition runtime can exhibit the desired external behaviors, the system should have the following form:

$\tau_I(\partial_H(WSC)) = \tau_I(\partial_H(WSC \between WSs \between WSOs \between AAs))$
$= r_{WSC}(DI_{WSC}) \cdot s_O(DO_{WSC}) \cdot \tau_I(\partial_H(WSC \between WSs \between WSOs \between AAs))$
$= r_{WSC}(DI_{WSC}) \cdot s_O(DO_{WSC}) \cdot \tau_I(\partial_H(WSC))$

## 22.3 An example

Using the architecture in Fig. 22.3, we obtain an implementation of the buying books example as shown in Fig. 22.4. In this implementation, there are one WSC (named BuyingBookWSC, denoted $WSC$), two WSs (one is named UserAgentWS and

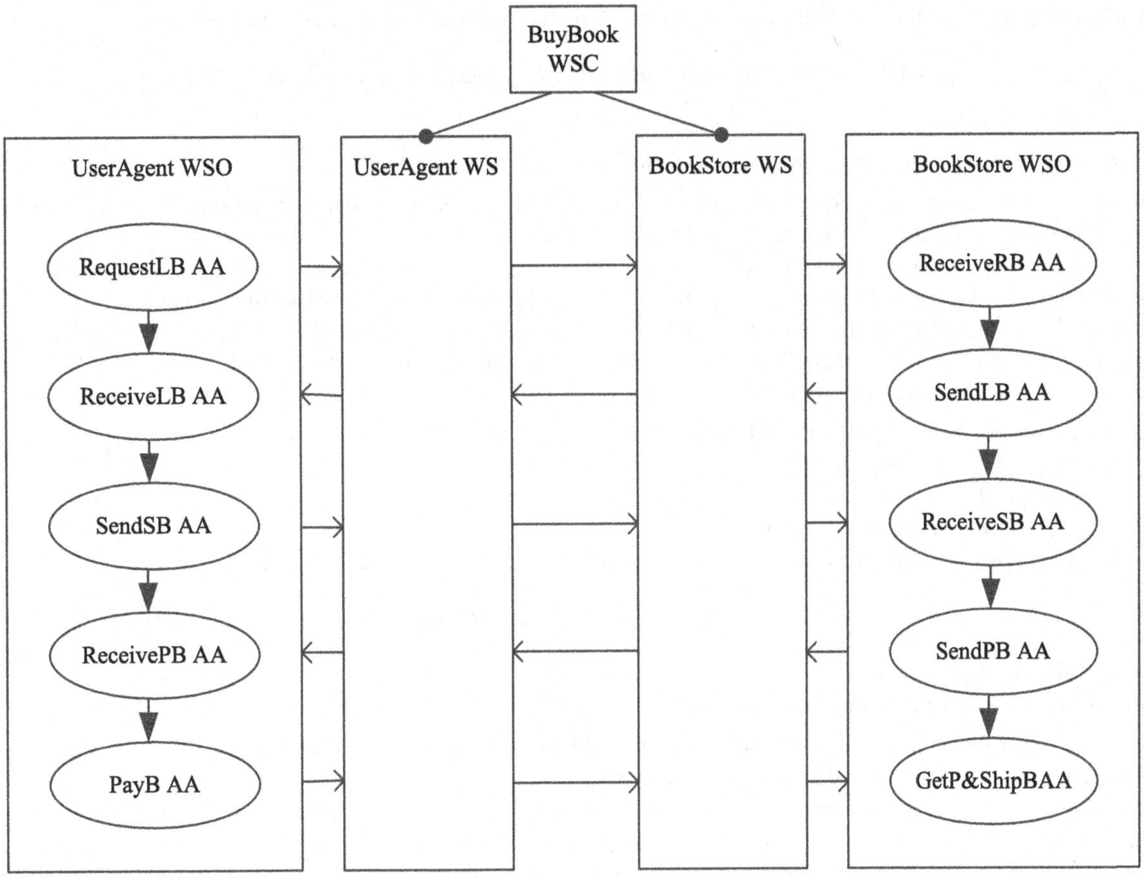


**FIGURE 22.4** Implementation of the buying books example.

denoted $WS_1$, the other is named BookStoreWS and denoted $WS_2$), two WSOs (one is named UserAgentWSO and denoted $WSO_1$, the other is named BookStoreWSO and denoted $WSO_2$), and two set of AAs denoted $AA_{1i}$ and $AA_{2j}$. The set of AAs belong to UserAgentWSO including RequstLBAA denoted $AA_{11}$, ReceiveLBAA denoted $AA_{12}$, SendSBAA denoted $AA_{13}$, ReceivePBAA denoted $AA_{14}$ PayBAA denoted $AA_{15}$, and the other set of AAs belong to BookStoreWSO including ReceiveRBAA denoted $AA_{21}$, SendLBAA denoted $AA_{22}$, ReceiveSBAA denoted $AA_{23}$, SendPBAA denoted $AA_{24}$, and GetP&ShipBAA denoted $AA_{25}$.

The detailed implementations of actors in Fig. 22.4 is the following:

### 22.3.1 User Agent AAs

(1) RequstLBAA ($AA_{11}$)

After $AA_{11}$ is created, the typical process is as follows:

1. The $AA_{11}$ receives some messages $RequestLB_{WA_1}$ from $WSO_1$ through its mail box denoted by its name $AA_{11}$ (the corresponding reading action is denoted $r_{AA_{11}}(RequestLB_{WA_1})$);
2. Then, it does some local computations mixed with some atomic actions by computation logics, including $\cdot$, $+$, $\between$, and guards, the whole local computations are denoted $I_{AA_{11}}$, which is the set of all local atomic actions;
3. When the local computations are finished, the $AA_{11}$ generates the output message $RequestLB_{AW_1}$ and sends to $WSO_1$'s mail box denoted by $WSO_1$'s name $WSO_1$ (the corresponding sending action is denoted $s_{WSO_1}(RequestLB_{AW_1})$), and then processes the next message from $WSO_1$ recursively.

The above process is described as the following state transitions by APTC:

$AA_{11} = r_{AA_{11}}(RequestLB_{WA_1}) \cdot AA_{11_1}$

$AA_{11_1} = I_{AA_{11}} \cdot AA_{11_2}$

$AA_{11_2} = s_{WSO_1}(RequestLB_{AW_1}) \cdot AA_{11}$

By use of the algebraic laws of APTC, $AA_{11}$ can be proven to exhibit the desired external behaviors:

$$\tau_{I_{AA_{11}}}(\partial_\emptyset(AA_{11})) = r_{AA_{11}}(RequestLB_{WA_1}) \cdot s_{WSO_1}(RequestLB_{AW_1}) \cdot \tau_{I_{AA_{11}}}(\partial_\emptyset(AA_{11}))$$

(2) ReceiveLBAA ($AA_{12}$)

After $AA_{12}$ is created, the typical process is as follows:

1. The $AA_{12}$ receives some messages $ReceiveLB_{WA_1}$ from $WSO_1$ through its mail box denoted by its name $AA_{12}$ (the corresponding reading action is denoted $r_{AA_{12}}(ReceiveLB_{WA_1})$);
2. Then, it does some local computations mixed with some atomic actions by computation logics, including $\cdot$, $+$, $\between$, and guards, the whole local computations are denoted $I_{AA_{12}}$, which is the set of all local atomic actions;
3. When the local computations are finished, the $AA_{12}$ generates the output message $ReceiveLB_{AW_1}$ and sends to $WSO_1$'s mail box denoted by $WSO_1$'s name $WSO_1$ (the corresponding sending action is denoted $s_{WSO_1}(ReceiveLB_{AW_1})$), and then processes the next message from $WSO_1$ recursively.

The above process is described as the following state transitions by APTC:

$AA_{12} = r_{AA_{12}}(ReceiveLB_{WA_1}) \cdot AA_{12_1}$

$AA_{12_1} = I_{AA_{12}} \cdot AA_{12_2}$

$AA_{12_2} = s_{WSO_1}(ReceiveLB_{AW_1}) \cdot AA_{12}$

By use of the algebraic laws of APTC, $AA_{12}$ can be proven to exhibit the desired external behaviors:

$$\tau_{I_{AA_{12}}}(\partial_\emptyset(AA_{12})) = r_{AA_{12}}(ReceiveLB_{WA_1}) \cdot s_{WSO_1}(ReceiveLB_{AW_1}) \cdot \tau_{I_{AA_{12}}}(\partial_\emptyset(AA_{12}))$$

(3) SendSBAA ($AA_{13}$)

After $AA_{13}$ is created, the typical process is as follows:

1. The $AA_{13}$ receives some messages $SendSB_{WA_1}$ from $WSO_1$ through its mail box denoted by its name $AA_{13}$ (the corresponding reading action is denoted $r_{AA_{13}}(SendSB_{WA_1})$);
2. Then, it does some local computations mixed with some atomic actions by computation logics, including $\cdot$, $+$, $\between$, and guards, the whole local computations are denoted $I_{AA_{13}}$, which is the set of all local atomic actions;
3. When the local computations are finished, the $AA_{13}$ generates the output message $SendSB_{AW_1}$ and sends to $WSO_1$'s mail box denoted by $WSO_1$'s name $WSO_1$ (the corresponding sending action is denoted $s_{WSO_1}(SendSB_{AW_1})$), and then processes the next message from $WSO_1$ recursively.

The above process is described as the following state transitions by APTC:

$AA_{13} = r_{AA_{13}}(SendSB_{WA_1}) \cdot AA_{13_1}$

$AA_{13_1} = I_{AA_{13}} \cdot AA_{13_2}$

$AA_{13_2} = s_{WSO_1}(SendSB_{AW_1}) \cdot AA_{13}$

By use of the algebraic laws of APTC, $AA_{13}$ can be proven to exhibit the desired external behaviors:

$$\tau_{I_{AA_{13}}}(\partial_\emptyset(AA_{13})) = r_{AA_{13}}(SendSB_{WA_1}) \cdot s_{WSO_1}(SendSB_{AW_1}) \cdot \tau_{I_{AA_{13}}}(\partial_\emptyset(AA_{13}))$$

(4) ReceivePBAA ($AA_{14}$)

After $AA_{14}$ is created, the typical process is as follows:

1. The $AA_{14}$ receives some messages $ReceivePB_{WA_1}$ from $WSO_1$ through its mail box denoted by its name $AA_{14}$ (the corresponding reading action is denoted $r_{AA_{14}}(ReceivePB_{WA_1})$);
2. Then, it does some local computations mixed with some atomic actions by computation logics, including $\cdot$, $+$, $\between$, and guards, the whole local computations are denoted $I_{AA_{14}}$, which is the set of all local atomic actions;
3. When the local computations are finished, the $AA_{14}$ generates the output message $ReceivePB_{AW_1}$ and sends to $WSO_1$'s mail box denoted by $WSO_1$'s name $WSO_1$ (the corresponding sending action is denoted $s_{WSO_1}(ReceivePB_{AW_1})$), and then processes the next message from $WSO_1$ recursively.

The above process is described as the following state transitions by APTC:

$AA_{14} = r_{AA_{14}}(ReceivePB_{WA_1}) \cdot AA_{14_1}$

$AA_{14_1} = I_{AA_{14}} \cdot AA_{14_2}$

$AA_{14_2} = s_{WSO_1}(ReceivePB_{AW_1}) \cdot AA_{14}$

By use of the algebraic laws of APTC, $AA_{14}$ can be proven to exhibit the desired external behaviors:

$$\tau_{I_{AA_{14}}}(\partial_\emptyset(AA_{14})) = r_{AA_{14}}(ReceivePB_{WA_1}) \cdot s_{WSO_1}(ReceivePB_{AW_1}) \cdot \tau_{I_{AA_{14}}}(\partial_\emptyset(AA_{14}))$$

(5) PayBAA ($AA_{15}$)

After $AA_{15}$ is created, the typical process is as follows:

1. The $AA_{15}$ receives some messages $PayB_{WA_1}$ from $WSO_1$ through its mail box denoted by its name $AA_{15}$ (the corresponding reading action is denoted $r_{AA_{15}}(PayB_{WA_1})$);
2. Then, it does some local computations mixed with some atomic actions by computation logics, including $\cdot$, $+$, $\between$, and guards, the whole local computations are denoted $I_{AA_{15}}$, which is the set of all local atomic actions;
3. When the local computations are finished, the $AA_{15}$ generates the output message $PayB_{AW_1}$ and sends to $WSO_1$'s mail box denoted by $WSO_1$'s name $WSO_1$ (the corresponding sending action is denoted $s_{WSO_1}(PayB_{AW_1})$), and then processes the next message from $WSO_1$ recursively.

The above process is described as the following state transitions by APTC:

$AA_{15} = r_{AA_{15}}(PayB_{WA_1}) \cdot AA_{15_1}$

$AA_{15_1} = I_{AA_{15}} \cdot AA_{15_2}$

$AA_{15_2} = s_{WSO_1}(PayB_{AW_1}) \cdot AA_{15}$

By use of the algebraic laws of APTC, $AA_{15}$ can be proven to exhibit the desired external behaviors:

$$\tau_{I_{AA_{15}}}(\partial_\emptyset(AA_{15})) = r_{AA_{15}}(PayB_{WA_1}) \cdot s_{WSO_1}(PayB_{AW_1}) \cdot \tau_{I_{AA_{15}}}(\partial_\emptyset(AA_{15}))$$

## 22.3.2 UserAgent WSO

After UserAgent WSO ($WSO_1$) is created, the typical process is as follows:

1. The $WSO_1$ receives the initialization message $ReBuyingBooks_{WW_1}$ from its interface WS through its mail box by its name $WSO_1$ (the corresponding reading action is denoted $r_{WSO_1}(ReBuyingBooks_{WW_1})$);
2. The $WSO_1$ may create its AAs in parallel through actions **new**($AA_{11}$) $\parallel \cdots \parallel$ **new**($AA_{15}$) if it is not initialized;
3. The $WSO_1$ does some local computations mixed with some atomic actions by computation logics, including $\cdot$, $+$, $\between$, and guards, the local computations are included into $I_{WSO_1}$, which is the set of all local atomic actions;
4. When the local computations are finished, the $WSO_1$ generates the output messages $RequestLB_{WA_1}$ and sends to $AA_{11}$ (the corresponding sending action is denoted $s_{AA_{11}}(RequestLB_{WA_1})$);
5. The $WSO_1$ receives the response message $RequestLB_{AW_1}$ from $AA_{11}$ through its mail box by its name $WSO_1$ (the corresponding reading action is denoted $r_{WSO_1}(RequestLB_{AW_1})$);
6. The $WSO_1$ does some local computations mixed with some atomic actions by computation logics, including $\cdot$, $+$, $\between$, and guards, the local computations are included into $I_{WSO_1}$, which is the set of all local atomic actions;
7. When the local computations are finished, the $WSO_1$ generates the output messages $RequestLB_{WW_1}$ and sends to $WS_1$ (the corresponding sending action is denoted $s_{WS_1}(RequestLB_{WW_1})$);
8. The $WSO_1$ receives the response message $ReceiveLB_{WW_1}$ from $WS_1$ through its mail box by its name $WSO_1$ (the corresponding reading action is denoted $r_{WSO_1}(ReceiveLB_{WW_1})$);
9. The $WSO_1$ does some local computations mixed with some atomic actions by computation logics, including $\cdot$, $+$, $\between$, and guards, the local computations are included into $I_{WSO_1}$, which is the set of all local atomic actions;
10. When the local computations are finished, the $WSO_1$ generates the output messages $ReceiveLB_{WA_1}$ and sends to $AA_{12}$ (the corresponding sending action is denoted $s_{AA_{12}}(ReceiveLB_{WA_1})$);
11. The $WSO_1$ receives the response message $ReceiveLB_{AW_1}$ from $AA_{12}$ through its mail box by its name $WSO_1$ (the corresponding reading action is denoted $r_{WSO_1}(ReceiveLB_{AW_1})$);
12. The $WSO_1$ does some local computations mixed with some atomic actions by computation logics, including $\cdot$, $+$, $\between$, and guards, the local computations are included into $I_{WSO_1}$, which is the set of all local atomic actions;
13. When the local computations are finished, the $WSO_1$ generates the output messages $SendSB_{WA_1}$ and sends to $AA_{13}$ (the corresponding sending action is denoted $s_{AA_13}(SendSB_{WA_1})$);
14. The $WSO_1$ receives the response message $SendSB_{AW_1}$ from $AA_{13}$ through its mail box by its name $WSO_1$ (the corresponding reading action is denoted $r_{WSO_1}(SendSB_{AW_1})$);
15. The $WSO_1$ does some local computations mixed with some atomic actions by computation logics, including $\cdot$, $+$, $\between$, and guards, the local computations are included into $I_{WSO_1}$, which is the set of all local atomic actions;
16. When the local computations are finished, the $WSO_1$ generates the output messages $SendSB_{WW_1}$ and sends to $WS_1$ (the corresponding sending action is denoted $s_{WS_1}(SendSB_{WW_1})$);
17. The $WSO_1$ receives the response message $ReceivePB_{WW_1}$ from $WS_1$ through its mail box by its name $WSO_1$ (the corresponding reading action is denoted $r_{WSO_1}(ReceivePB_{WW_1})$);

18. The $WSO_1$ does some local computations mixed with some atomic actions by computation logics, including $\cdot$, $+$, $\between$, and guards, the local computations are included into $I_{WSO_1}$, which is the set of all local atomic actions;
19. When the local computations are finished, the $WSO_1$ generates the output messages $ReceivePB_{WA_1}$ and sends to $AA_{14}$ (the corresponding sending action is denoted $s_{AA_{14}}(ReceivePB_{WA_1})$);
20. The $WSO_1$ receives the response message $ReceivePB_{AW_1}$ from $AA_{14}$ through its mail box by its name $WSO_1$ (the corresponding reading action is denoted $r_{WSO_1}(ReceivePB_{AW_1})$);
21. The $WSO_1$ does some local computations mixed with some atomic actions by computation logics, including $\cdot$, $+$, $\between$, and guards, the local computations are included into $I_{WSO_1}$, which is the set of all local atomic actions;
22. When the local computations are finished, the $WSO_1$ generates the output messages $PayB_{WA_1}$ and sends to $AA_{15}$ (the corresponding sending action is denoted $s_{AA_{15}}(PayB_{WA_1})$);
23. The $WSO_1$ receives the response message $PayB_{AW_1}$ from $AA_{15}$ through its mail box by its name $WSO_1$ (the corresponding reading action is denoted $r_{WSO_1}(PayB_{AW_1})$);
24. The $WSO_1$ does some local computations mixed with some atomic actions by computation logics, including $\cdot$, $+$, $\between$, and guards, the local computations are included into $I_{WSO_1}$, which is the set of all local atomic actions;
25. When the local computations are finished, the $WSO_1$ generates the output messages $PayB_{WW_1}$ and sends to $WS_1$ (the corresponding sending action is denoted $s_{WS_1}(PayB_{WW_1})$), and then processes the messages from $WS_1$ recursively.

The above process is described as the following state transitions by APTC:

$WSO_1 = r_{WSO_1}(ReBuyingBooks_{WW_1}) \cdot WSO_{1_1}$

$WSO_{1_1} = (\{isInitialed(WSO_1) = FLALSE\} \cdot (\mathbf{new}(AA_{11}) \parallel \cdots \parallel \mathbf{new}(AA_{15})) + \{isInitialed(WSO_1) = TRUE\}) \cdot WSO_{1_2}$

$WSO_{1_2} = I_{WSO_1} \cdot WSO_{1_3}$

$WSO_{1_3} = s_{AA_{11}}(RequestLB_{WA_1}) \cdot WSO_{1_4}$

$WSO_{1_4} = r_{WSO_1}(RequestLB_{AW_1}) \cdot WSO_{1_5}$

$WSO_{1_5} = I_{WSO_1} \cdot WSO_{1_6}$

$WSO_{1_6} = s_{WS_1}(RequestLB_{WW_1}) \cdot WSO_{1_7}$

$WSO_{1_7} = r_{WSO_1}(ReceiveLB_{WW_1}) \cdot WSO_{1_8}$

$WSO_{1_8} = I_{WSO_1} \cdot WSO_{1_9}$

$WSO_{1_9} = s_{AA_{12}}(ReceiveLB_{WA_1}) \cdot WSO_{1_{10}}$

$WSO_{1_{10}} = r_{WSO_1}(ReceiveLB_{AW_1}) \cdot WSO_{1_{11}}$

$WSO_{1_{11}} = I_{WSO_1} \cdot WSO_{1_{12}}$

$WSO_{1_{12}} = s_{AA_13}(SendSB_{WA_1}) \cdot WSO_{1_{13}}$

$WSO_{1_{13}} = r_{WSO_1}(SendSB_{AW_1}) \cdot WSO_{1_{14}}$

$WSO_{1_{14}} = I_{WSO_1} \cdot WSO_{1_{15}}$

$WSO_{1_{15}} = s_{WS_1}(SendSB_{WW_1}) \cdot WSO_{1_{16}}$

$WSO_{1_{16}} = r_{WSO_1}(ReceivePB_{WW_1}) \cdot WSO_{1_{17}}$

$WSO_{1_{17}} = I_{WSO_1} \cdot WSO_{1_{18}}$

$WSO_{1_{18}} = s_{AA_{14}}(ReceivePB_{WA_1}) \cdot WSO_{1_{19}}$

$WSO_{1_{19}} = r_{WSO_1}(ReceivePB_{AW_1}) \cdot WSO_{1_{20}}$

$WSO_{1_{20}} = I_{WSO_1} \cdot WSO_{1_{21}}$

$WSO_{1_{21}} = s_{AA_{15}}(PayB_{WA_1}) \cdot WSO_{1_{22}}$

$WSO_{1_{22}} = r_{WSO_1}(PayB_{AW_1}) \cdot WSO_{1_{23}}$

$WSO_{1_{23}} = I_{WSO_1} \cdot WSO_{1_{24}}$

$WSO_{1_{24}} = s_{WS_1}(PayB_{WW_1}) \cdot WSO_1$

By use of the algebraic laws of APTC, the $WSO_1$ can be proven to exhibit the desired external behaviors:

$$\tau_{I_{WSO_1}}(\partial_\emptyset(WSO_1)) = r_{WSO_1}(ReBuyingBooks_{WW_1}) \cdot s_{AA_{11}}(RequestLB_{WA_1}) \cdot r_{WSO_1}(RequestLB_{AW_1}) \cdot s_{WS_1}(RequestLB_{WW_1}) \cdot r_{WSO_1}(ReceiveLB_{WW_1}) \cdot s_{AA_{12}}(ReceiveLB_{WA_1}) \cdot r_{WSO_1}(ReceiveLB_{AW_1}) \cdot s_{AA_13}(SendSB_{WA_1}) \cdot r_{WSO_1}(SendSB_{AW_1}) \cdot s_{WS_1}(SendSB_{WW_1}) \cdot r_{WSO_1}(ReceivePB_{WW_1}) \cdot s_{AA_{14}}(ReceivePB_{WA_1}) \cdot r_{WSO_1}(ReceivePB_{AW_1}) \cdot s_{AA_{15}}(PayB_{WA_1}) \cdot r_{WSO_1}(PayB_{AW_1}) \cdot s_{WS_1}(PayB_{WW_1}) \cdot \tau_{I_{WSO_1}}(\partial_\emptyset(WSO_1))$$

with $I_{WSO_1}$ extended to $I_{WSO_1} \cup \{\{isInitialed(WSO_1) = FLALSE\}, \{isInitialed(WSO_1) = TRUE\}\}$.

### 22.3.3 UserAgent WS

After UserAgent WS ($WS_1$) is created, the typical process is as follows:

1. The $WS_1$ receives the initialization message $ReBuyingBooks_{WC_1}$ from the buying books WSC $WSC$ through its mail box by its name $WS_1$ (the corresponding reading action is denoted $r_{WS_1}(ReBuyingBooks_{WC_1})$);
2. The $WS_1$ may create its $WSO_1$ through an action $\mathbf{new}(WSO_1)$ if it is not initialized;
3. The $WS_1$ does some local computations mixed with some atomic actions by computation logics, including $\cdot$, $+$, $\between$, and guards, the local computations are included into $I_{WS_1}$, which is the set of all local atomic actions;
4. When the local computations are finished, the $WS_1$ generates the output messages $ReBuyingBooks_{WW_1}$ and sends to $WSO_1$ (the corresponding sending action is denoted $s_{WSO_1}(ReBuyingBooks_{WW_1})$);
5. The $WS_1$ receives the response message $RequestLB_{WW_1}$ from $WSO_1$ through its mail box by its name $WS_1$ (the corresponding reading action is denoted $r_{WS_1}(RequestLB_{WW_1})$);
6. The $WS_1$ does some local computations mixed with some atomic actions by computation logics, including $\cdot$, $+$, $\between$, and guards, the local computations are included into $I_{WS_1}$, which is the set of all local atomic actions;
7. When the local computations are finished, the $WS_1$ generates the output messages $RequestLB_{WW_{12}}$ and sends to $WS_2$ (the corresponding sending action is denoted $s_{WS_2}(RequestLB_{WW_{12}})$);
8. The $WS_1$ receives the response message $SendLB_{WW_{21}}$ from $WS_2$ through its mail box by its name $WS_1$ (the corresponding reading action is denoted $r_{WS_1}(SendLB_{WW_{21}})$);
9. The $WS_1$ does some local computations mixed with some atomic actions by computation logics, including $\cdot$, $+$, $\between$, and guards, the local computations are included into $I_{WS_1}$, which is the set of all local atomic actions;
10. When the local computations are finished, the $WS_1$ generates the output messages $ReceiveLB_{WW_1}$ and sends to $WSO_1$ (the corresponding sending action is denoted $s_{WSO_1}(ReceiveLB_{WW_1})$);
11. The $WS_1$ receives the response message $SendSB_{WW_1}$ from $WSO_1$ through its mail box by its name $WS_1$ (the corresponding reading action is denoted $r_{WS_1}(SendSB_{WW_1})$);
12. The $WS_1$ does some local computations mixed with some atomic actions by computation logics, including $\cdot$, $+$, $\between$, and guards, the local computations are included into $I_{WS_1}$, which is the set of all local atomic actions;
13. When the local computations are finished, the $WS_1$ generates the output messages $SendSB_{WW_{12}}$ and sends to $WS_2$ (the corresponding sending action is denoted $s_{WS_2}(SendSB_{WW_{12}})$);
14. The $WS_1$ receives the response message $SendPB_{WW_{21}}$ from $WS_2$ through its mail box by its name $WS_1$ (the corresponding reading action is denoted $r_{WS_1}(SendPB_{WW_{21}})$);
15. The $WS_1$ does some local computations mixed with some atomic actions by computation logics, including $\cdot$, $+$, $\between$, and guards, the local computations are included into $I_{WS_1}$, which is the set of all local atomic actions;
16. When the local computations are finished, the $WS_1$ generates the output messages $ReceivePB_{WW_1}$ and sends to $WSO_1$ (the corresponding sending action is denoted $s_{WSO_1}(ReceivePB_{WW_1})$);
17. The $WS_1$ receives the response message $PayB_{WW_1}$ from $WSO_1$ through its mail box by its name $WS_1$ (the corresponding reading action is denoted $r_{WS_1}(PayB_{WW_1})$);
18. The $WS_1$ does some local computations mixed with some atomic actions by computation logics, including $\cdot$, $+$, $\between$, and guards, the local computations are included into $I_{WS_1}$, which is the set of all local atomic actions;
19. When the local computations are finished, the $WS_1$ generates the output messages $PayB_{WW_{12}}$ and sends to $WS_2$ (the corresponding sending action is denoted $s_{WS_2}(PayB_{WW_{12}})$), and then processes the messages from $WSC$ recursively.

The above process is described as the following state transitions by APTC:

$WS_1 = r_{WS_1}(ReBuyingBooks_{WC_1}) \cdot WS_{1_1}$
$WS_{1_1} = (\{isInitialed(WS_1) = FLALSE\} \cdot \mathbf{new}(WSO_1) + \{isInitialed(WS_1) = TRUE\}) \cdot WS_{1_2}$
$WS_{1_2} = I_{WS_1} \cdot WS_{1_3}$
$WS_{1_3} = s_{WSO_1}(ReBuyingBooks_{WW_1}) \cdot WS_{1_4}$
$WS_{1_4} = r_{WS_1}(RequestLB_{WW_1}) \cdot WS_{1_5}$
$WS_{1_5} = I_{WS_1} \cdot WS_{1_6}$
$WS_{1_6} = s_{WS_2}(RequestLB_{WW_{12}}) \cdot WS_{1_7}$
$WS_{1_7} = r_{WS_1}(SendLB_{WW_{21}}) \cdot WS_{1_8}$
$WS_{1_8} = I_{WS_1} \cdot WS_{1_9}$
$WS_{1_9} = s_{WSO_1}(ReceiveLB_{WW_1}) \cdot WS_{1_{10}}$
$WS_{1_{10}} = r_{WS_1}(SendSB_{WW_1}) \cdot WS_{1_{11}}$
$WS_{1_{11}} = I_{WS_1} \cdot WS_{1_{12}}$
$WS_{1_{12}} = s_{WS_2}(SendSB_{WW_{12}}) \cdot WS_{1_{13}}$
$WS_{1_{13}} = r_{WS_1}(SendPB_{WW_{21}}) \cdot WS_{1_{14}}$
$WS_{1_{14}} = I_{WS_1} \cdot WS_{1_{15}}$
$WS_{1_{15}} = s_{WSO_1}(ReceivePB_{WW_1}) \cdot WS_{1_{16}}$

$WS1_{16} = r_{WS_1}(PayB_{WW_1}) \cdot WS1_{17}$
$WS1_{17} = I_{WS_1} \cdot WS1_{18}$
$WS1_{18} = s_{WS_2}(PayB_{WW_{12}}) \cdot WS_1$
By use of the algebraic laws of APTC, the $WS_1$ can be proven to exhibit the desired external behaviors:
$\tau_{I_{WS_1}}(\partial_\emptyset(WS_1)) = r_{WSO_1}(r_{WS_1}(ReBuyingBooks_{WC_1}) \cdot s_{WSO_1}(ReBuyingBooks_{WW_1})$
$\cdot r_{WS_1}(RequestLB_{WW_1}) \cdot s_{WS_2}(RequestLB_{WW_{12}}) \cdot r_{WS_1}(SendLB_{WW_{21}})$
$\cdot s_{WSO_1}(ReceiveLB_{WW_1}) \cdot r_{WS_1}(SendSB_{WW_1}) \cdot s_{WS_2}(SendSB_{WW_{12}})$
$\cdot r_{WS_1}(SendPB_{WW_{21}}) \cdot s_{WSO_1}(ReceivePB_{WW_1}) \cdot r_{WS_1}(PayB_{WW_1})$
$\cdot s_{WS_2}(PayB_{WW_{12}}) \cdot \tau_{I_{WS_1}}(\partial_\emptyset(WS_1))$
with $I_{WS_1}$ extended to $I_{WS_1} \cup \{\{isInitialed(WS_1) = FLALSE\}, \{isInitialed(WS_1) = TRUE\}\}$.

### 22.3.4 BookStore AAs

(1) ReceiveRBAA ($AA_{21}$)
After $AA_{21}$ is created, the typical process is as follows:

1. The $AA_{21}$ receives some messages $ReceiveRB_{WA_2}$ from $WSO_2$ through its mail box denoted by its name $AA_{21}$ (the corresponding reading action is denoted $r_{AA_{21}}(ReceiveRB_{WA_2})$);
2. Then, it does some local computations mixed with some atomic actions by computation logics, including $\cdot$, $+$, $\between$, and guards, the whole local computations are denoted $I_{AA_{21}}$, which is the set of all local atomic actions;
3. When the local computations are finished, the $AA_{21}$ generates the output message $ReceiveRB_{AW_2}$ and sends to $WSO_2$'s mail box denoted by $WSO_2$'s name $WSO_2$ (the corresponding sending action is denoted $s_{WSO_2}(ReceiveRB_{AW_2})$), and then processes the next message from $WSO_2$ recursively.

The above process is described as the following state transitions by APTC:
$AA_{21} = r_{AA_{21}}(ReceiveRB_{WA_2}) \cdot AA21_1$
$AA21_1 = I_{AA_{21}} \cdot AA21_2$
$AA21_2 = s_{WSO_2}(ReceiveRB_{AW_2}) \cdot AA21$
By use of the algebraic laws of APTC, $AA_{21}$ can be proven to exhibit the desired external behaviors:

$$\tau_{I_{AA_{21}}}(\partial_\emptyset(AA21)) = r_{AA_{21}}(RequestLB_{WA_2}) \cdot s_{WSO_2}(RequestLB_{AW_2}) \cdot \tau_{I_{AA_{21}}}(\partial_\emptyset(AA21))$$

(2) SendLBAA ($AA_{22}$)
After $AA_{22}$ is created, the typical process is as follows:

1. The $AA_{22}$ receives some messages $SendLB_{WA_2}$ from $WSO_2$ through its mail box denoted by its name $AA_{22}$ (the corresponding reading action is denoted $r_{AA_{22}}(ReceiveLB_{WA_2})$);
2. Then, it does some local computations mixed with some atomic actions by computation logics, including $\cdot$, $+$, $\between$, and guards, the whole local computations are denoted $I_{AA_{22}}$, which is the set of all local atomic actions;
3. When the local computations are finished, the $AA_{22}$ generates the output message $SendLB_{AW_2}$ and sends to $WSO_2$'s mail box denoted by $WSO_2$'s name $WSO_2$ (the corresponding sending action is denoted $s_{WSO_2}(SendLB_{AW_2})$), and then processes the next message from $WSO_2$ recursively.

The above process is described as the following state transitions by APTC:
$AA22 = r_{AA_{22}}(SendLB_{WA_2}) \cdot AA22_1$
$AA22_1 = I_{AA_{22}} \cdot AA22_2$
$AA22_2 = s_{WSO_2}(SendLB_{AW_2}) \cdot AA22$
By use of the algebraic laws of APTC, $AA_{22}$ can be proven to exhibit the desired external behaviors:

$$\tau_{I_{AA_{22}}}(\partial_\emptyset(AA22)) = r_{AA_{22}}(SendLB_{WA_2}) \cdot s_{WSO_2}(SendLB_{AW_2}) \cdot \tau_{I_{AA_{22}}}(\partial_\emptyset(AA22))$$

(3) ReceiveSBAA ($AA_{23}$)
After $AA_{23}$ is created, the typical process is as follows:

1. The $AA_{23}$ receives some messages $ReceiveSB_{WA_2}$ from $WSO_2$ through its mail box denoted by its name $AA_{23}$ (the corresponding reading action is denoted $r_{AA_{23}}(ReceiveSB_{WA_2})$);
2. Then, it does some local computations mixed with some atomic actions by computation logics, including $\cdot$, $+$, $\between$, and guards, the whole local computations are denoted $I_{AA_{23}}$, which is the set of all local atomic actions;

3. When the local computations are finished, the $AA_{23}$ generates the output message $ReceiveSB_{AW_2}$ and sends to $WSO_2$'s mail box denoted by $WSO_2$'s name $WSO_2$ (the corresponding sending action is denoted $s_{WSO_2}(ReceiveSB_{AW_2})$), and then processes the next message from $WSO_2$ recursively.

The above process is described as the following state transitions by APTC:
$AA_{23} = r_{AA_{23}}(ReceiveSB_{WA_2}) \cdot AA_{23_1}$
$AA_{23_1} = I_{AA_{23}} \cdot AA_{23_2}$
$AA_{23_2} = s_{WSO_2}(ReceiveSB_{AW_2}) \cdot AA_{23}$
By use of the algebraic laws of APTC, $AA_{23}$ can be proven to exhibit the desired external behaviors:

$$\tau_{I_{AA_{23}}}(\partial_\emptyset(AA_{23})) = r_{AA_{23}}(ReceiveSB_{WA_1}) \cdot s_{WSO_2}(ReceiveSB_{AW_2}) \cdot \tau_{I_{AA_{23}}}(\partial_\emptyset(AA_{23}))$$

(4) SendPBAA ($AA_{24}$)
After $AA_{24}$ is created, the typical process is as follows:

1. The $AA_{24}$ receives some messages $SendPB_{WA_2}$ from $WSO_2$ through its mail box denoted by its name $AA_{24}$ (the corresponding reading action is denoted $r_{AA_{24}}(SendPB_{WA_2})$);
2. Then, it does some local computations mixed with some atomic actions by computation logics, including $\cdot$, $+$, $\between$, and guards, the whole local computations are denoted $I_{AA_{24}}$, which is the set of all local atomic actions;
3. When the local computations are finished, the $AA_{24}$ generates the output message $SendPB_{AW_2}$ and sends to $WSO_2$'s mail box denoted by $WSO_2$'s name $WSO_2$ (the corresponding sending action is denoted $s_{WSO_2}(SendPB_{AW_2})$), and then processes the next message from $WSO_2$ recursively.

The above process is described as the following state transitions by APTC:
$AA_{24} = r_{AA_{24}}(SendPB_{WA_2}) \cdot AA_{24_1}$
$AA_{24_1} = I_{AA_{24}} \cdot AA_{24_2}$
$AA_{24_2} = s_{WSO_2}(SendPB_{AW_2}) \cdot AA_{24}$
By use of the algebraic laws of APTC, $AA_{24}$ can be proven to exhibit the desired external behaviors:

$$\tau_{I_{AA_{24}}}(\partial_\emptyset(AA_{24})) = r_{AA_{24}}(SendPB_{WA_2}) \cdot s_{WSO_2}(SendPB_{AW_2}) \cdot \tau_{I_{AA_{24}}}(\partial_\emptyset(AA_{24}))$$

(5) GetP&ShipBAA ($AA_{25}$)
After $AA_{25}$ is created, the typical process is as follows:

1. The $AA_{25}$ receives some messages $GetP\&ShipB_{WA_2}$ from $WSO_2$ through its mail box denoted by its name $AA_{25}$ (the corresponding reading action is denoted $r_{AA_{25}}(GetP\&ShipB_{WA_2})$);
2. Then, it does some local computations mixed with some atomic actions by computation logics, including $\cdot$, $+$, $\between$, and guards, the whole local computations are denoted $I_{AA_{25}}$, which is the set of all local atomic actions;
3. When the local computations are finished, the $AA_{25}$ generates the output message $GetP\&ShipB_{AW_2}$ and sends to $WSO_2$'s mail box denoted by $WSO_2$'s name $WSO_2$ (the corresponding sending action is denoted $s_{WSO_2}(GetP\&ShipB_{AW_2})$), and then processes the next message from $WSO_2$ recursively.

The above process is described as the following state transitions by APTC:
$AA_{25} = r_{AA_{25}}(GetP\&ShipB_{WA_2}) \cdot AA_{25_1}$
$AA_{25_1} = I_{AA_{25}} \cdot AA_{25_2}$
$AA_{25_2} = s_{WSO_2}(GetP\&ShipB_{AW_2}) \cdot AA_{25}$
By use of the algebraic laws of APTC, $AA_{25}$ can be proven to exhibit the desired external behaviors:

$$\tau_{I_{AA_{25}}}(\partial_\emptyset(AA_{25})) = r_{AA_{25}}(GetP\&ShipB_{WA_1}) \cdot s_{WSO_2}(GetP\&ShipB_{AW_2}) \cdot \tau_{I_{AA_{25}}}(\partial_\emptyset(AA_{25}))$$

### 22.3.5 BookStore WSO

After BookStore WSO ($WSO_2$) is created, the typical process is as follows:

1. The $WSO_2$ receives the initialization message $ReceiveRB_{WW_2}$ from its interface WS $WS_2$ through its mail box by its name $WSO_2$ (the corresponding reading action is denoted $r_{WSO_2}(ReceiveRB_{WW_2})$);
2. The $WSO_2$ may create its AAs in parallel through actions $\mathbf{new}(AA_{21}) \parallel \cdots \parallel \mathbf{new}(AA_{25})$ if it is not initialized;
3. The $WSO_2$ does some local computations mixed with some atomic actions by computation logics, including $\cdot$, $+$, $\between$, and guards, the local computations are included into $I_{WSO_2}$, which is the set of all local atomic actions;

4. When the local computations are finished, the $WSO_2$ generates the output messages $ReceiveRB_{WA_2}$ and sends to $AA_{21}$ (the corresponding sending action is denoted $s_{AA_{21}}(ReceiveRB_{WA_2})$);
5. The $WSO_2$ receives the response message $ReceiveRB_{AW_2}$ from $AA_{21}$ through its mail box by its name $WSO_2$ (the corresponding reading action is denoted $r_{WSO_2}(ReceiveRB_{AW_2})$);
6. The $WSO_2$ does some local computations mixed with some atomic actions by computation logics, including $\cdot$, $+$, $\between$, and guards, the local computations are included into $I_{WSO_2}$, which is the set of all local atomic actions;
7. When the local computations are finished, the $WSO_2$ generates the output messages $SendLB_{WA_2}$ and sends to $AA_{22}$ (the corresponding sending action is denoted $s_{AA_{22}}(SendLB_{WA_2})$);
8. The $WSO_2$ receives the response message $SendLB_{AW_2}$ from $AA_{22}$ through its mail box by its name $WSO_2$ (the corresponding reading action is denoted $r_{WSO_2}(SendLB_{AW_2})$);
9. The $WSO_2$ does some local computations mixed with some atomic actions by computation logics, including $\cdot$, $+$, $\between$, and guards, the local computations are included into $I_{WSO_2}$, which is the set of all local atomic actions;
10. When the local computations are finished, the $WSO_2$ generates the output messages $SendLB_{WW_2}$ and sends to $WS_2$ (the corresponding sending action is denoted $s_{WS_2}(SendLB_{WW_2})$);
11. The $WSO_2$ receives the response message $ReceiveSB_{WW_2}$ from $WS_2$ through its mail box by its name $WSO_2$ (the corresponding reading action is denoted $r_{WSO_2}(ReceiveSB_{WW_2})$);
12. The $WSO_2$ does some local computations mixed with some atomic actions by computation logics, including $\cdot$, $+$, $\between$, and guards, the local computations are included into $I_{WSO_2}$, which is the set of all local atomic actions;
13. When the local computations are finished, the $WSO_2$ generates the output messages $ReceiveSB_{WA_2}$ and sends to $AA_{23}$ (the corresponding sending action is denoted $s_{AA_{23}}(ReceiveSB_{WA_2})$);
14. The $WSO_2$ receives the response message $ReceiveSB_{AW_2}$ from $AA_{23}$ through its mail box by its name $WSO_2$ (the corresponding reading action is denoted $r_{WSO_2}(ReceiveSB_{AW_2})$);
15. The $WSO_2$ does some local computations mixed with some atomic actions by computation logics, including $\cdot$, $+$, $\between$, and guards, the local computations are included into $I_{WSO_2}$, which is the set of all local atomic actions;
16. When the local computations are finished, the $WSO_2$ generates the output messages $SendPB_{WA_2}$ and sends to $AA_{24}$ (the corresponding sending action is denoted $s_{AA_{24}}(SendPB_{WA_2})$);
17. The $WSO_2$ receives the response message $SendPB_{AW_2}$ from $AA_{24}$ through its mail box by its name $WSO_2$ (the corresponding reading action is denoted $r_{WSO_2}(SendPB_{AW_2})$);
18. The $WSO_2$ does some local computations mixed with some atomic actions by computation logics, including $\cdot$, $+$, $\between$, and guards, the local computations are included into $I_{WSO_2}$, which is the set of all local atomic actions;
19. When the local computations are finished, the $WSO_2$ generates the output messages $SendPB_{WW_2}$ and sends to $WS_2$ (the corresponding sending action is denoted $s_{WS_2}(SendPB_{WW_2})$);
20. The $WSO_2$ receives the response message $SendPB_{WW_2}$ from $WS_2$ through its mail box by its name $WSO_2$ (the corresponding reading action is denoted $r_{WSO_2}(SendPB_{WW_2})$);
21. The $WSO_2$ does some local computations mixed with some atomic actions by computation logics, including $\cdot$, $+$, $\between$, and guards, the local computations are included into $I_{WSO_2}$, which is the set of all local atomic actions;
22. When the local computations are finished, the $WSO_2$ generates the output messages $GetP\&ShipB_{WA_2}$ and sends to $AA_{25}$ (the corresponding sending action is denoted $s_{AA_{25}}(GetP\&ShipB_{WA_2})$);
23. The $WSO_2$ receives the response message $GetP\&ShipB_{AW_2}$ from $AA_{25}$ through its mail box by its name $WSO_2$ (the corresponding reading action is denoted $r_{WSO_2}(GetP\&ShipB_{AW_2})$);
24. The $WSO_2$ does some local computations mixed with some atomic actions by computation logics, including $\cdot$, $+$, $\between$, and guards, the local computations are included into $I_{WSO_2}$, which is the set of all local atomic actions;
25. When the local computations are finished, the $WSO_2$ generates the output messages $GetP\&ShipB_{WW_2}$ and sends to $WS_2$ (the corresponding sending action is denoted $s_{WS_2}(GetP\&ShipB_{WW_2})$), and then processes the messages from $WS_2$ recursively.

The above process is described as the following state transitions by APTC:

$WSO_2 = r_{WSO_2}(ReceiveRB_{WW_2}) \cdot WSO_{2_1}$

$WSO_{2_1} = (\{isInitialed(WSO_2) = FLALSE\} \cdot (\mathbf{new}(AA_{21}) \parallel \cdots \parallel \mathbf{new}(AA_{25})) + \{isInitialed(WSO_2) = TRUE\}) \cdot WSO_{2_2}$

$WSO_{2_2} = I_{WSO_2} \cdot WSO_{2_3}$

$WSO_{2_3} = s_{AA_{21}}(ReceiveRB_{WA_2}) \cdot WSO_{2_4}$

$WSO_{2_4} = r_{WSO_2}(ReceiveRB_{AW_2}) \cdot WSO_{2_5}$

$WSO_{2_5} = I_{WSO_2} \cdot WSO_{2_6}$

$WSO_{2_6} = s_{AA_{22}}(SendLB_{WA_2}) \cdot WSO_{2_7}$

$WSO2_7 = r_{WSO_2}(SendLB_{AW_2}) \cdot WSO2_8$
$WSO2_8 = I_{WSO_2} \cdot WSO2_9$
$WSO2_9 = s_{WS_2}(SendLB_{WW_2}) \cdot WSO2_{10}$
$WSO2_{10} = r_{WSO_2}(ReceiveSB_{WW_2}) \cdot WSO2_{11}$
$WSO2_{11} = I_{WSO_2} \cdot WSO2_{12}$
$WSO2_{12} = s_{AA_{23}}(ReceiveSB_{WA_2}) \cdot WSO2_{13}$
$WSO2_{13} = r_{WSO_2}(ReceiveSB_{AW_2}) \cdot WSO2_{14}$
$WSO2_{14} = I_{WSO_2} \cdot WSO2_{15}$
$WSO2_{15} = s_{AA_{24}}(SendPB_{WA_2}) \cdot WSO2_{16}$
$WSO2_{16} = r_{WSO_2}(SendPB_{AW_2}) \cdot WSO2_{17}$
$WSO2_{17} = I_{WSO_2} \cdot WSO2_{18}$
$WSO2_{18} = s_{WS_2}(SendPB_{WW_2}) \cdot WSO2_{19}$
$WSO2_{19} = r_{WSO_2}(SendPB_{WW_2}) \cdot WSO2_{20}$
$WSO2_{20} = I_{WSO_2} \cdot WSO2_{21}$
$WSO2_{21} = s_{AA_{25}}(GetP\&ShipB_{WA_2}) \cdot WSO2_{22}$
$WSO2_{22} = r_{WSO_2}(GetP\&ShipB_{AW_2}) \cdot WSO2_{23}$
$WSO2_{23} = I_{WSO_2} \cdot WSO2_{24}$
$WSO2_{24} = s_{WS_2}(GetP\&ShipB_{WW_2}) \cdot WSO_2$

By use of the algebraic laws of APTC, the $WSO_2$ can be proven to exhibit the desired external behaviors:

$$\tau_{I_{WSO_2}}(\partial_\emptyset(WSO_2)) = r_{WSO_2}(ReceiveRB_{WW_2}) \cdot s_{AA_{21}}(ReceiveRB_{WA_2}) \cdot r_{WSO_2}(ReceiveRB_{AW_2})$$
$$\cdot s_{AA_{22}}(SendLB_{WA_2}) \cdot r_{WSO_2}(SendLB_{AW_2}) \cdot s_{WS_2}(SendLB_{WW_2}) \cdot r_{WSO_2}(ReceiveSB_{WW_2})$$
$$\cdot s_{AA_{23}}(ReceiveSB_{WA_2}) \cdot r_{WSO_2}(ReceiveSB_{AW_2}) \cdot s_{AA_{24}}(SendPB_{WA_2}) \cdot r_{WSO_2}(SendPB_{AW_2})$$
$$\cdot s_{WS_2}(SendPB_{WW_2}) \cdot r_{WSO_2}(SendPB_{WW_2}) \cdot s_{AA_{25}}(GetP\&ShipB_{WA_2}) \cdot r_{WSO_2}(GetP\&ShipB_{AW_2})$$
$$\cdot s_{WS_2}(GetP\&ShipB_{WW_2}) \cdot \tau_{I_{WSO_2}}(\partial_\emptyset(WSO_2))$$

with $I_{WSO_2}$ extended to $I_{WSO_2} \cup \{\{isInitialed(WSO_2) = FLALSE\}, \{isInitialed(WSO_2) = TRUE\}\}$.

### 22.3.6 BookStore WS

After BookStore WS ($WS_2$) is created, the typical process is as follows:

1. The $WS_2$ receives the initialization message $RequestLB_{WW_{12}}$ from its interface WS $WS_1$ through its mail box by its name $WS_2$ (the corresponding reading action is denoted $r_{WS_2}(RequestLB_{WW_{12}})$);
2. The $WS_2$ may create its $WSO_2$ through actions **new**($WSO_2$) if it is not initialized;
3. The $WS_2$ does some local computations mixed with some atomic actions by computation logics, including $\cdot$, $+$, $\between$, and guards, the local computations are included into $I_{WS_2}$, which is the set of all local atomic actions;
4. When the local computations are finished, the $WS_2$ generates the output messages $ReceiveRB_{WW_2}$ and sends to $WSO_2$ (the corresponding sending action is denoted $s_{WSO_2}(ReceiveRB_{WW_2})$);
5. The $WS_2$ receives the response message $SendLB_{WW_2}$ from $WSO_2$ through its mail box by its name $WS_2$ (the corresponding reading action is denoted $r_{WS_2}(SendLB_{WW_2})$);
6. The $WS_2$ does some local computations mixed with some atomic actions by computation logics, including $\cdot$, $+$, $\between$, and guards, the local computations are included into $I_{WS_2}$, which is the set of all local atomic actions;
7. When the local computations are finished, the $WS_2$ generates the output messages $SendLB_{WW_{21}}$ and sends to $WS_1$ (the corresponding sending action is denoted $s_{WS_1}(SendLB_{WW_{21}})$);
8. The $WS_2$ receives the response message $SendSB_{WW_{12}}$ from $WS_1$ through its mail box by its name $WS_2$ (the corresponding reading action is denoted $r_{WS_2}(SendSB_{WW_{12}})$);
9. The $WS_2$ does some local computations mixed with some atomic actions by computation logics, including $\cdot$, $+$, $\between$, and guards, the local computations are included into $I_{WS_2}$, which is the set of all local atomic actions;
10. When the local computations are finished, the $WS_2$ generates the output messages $ReceiveSB_{WW_2}$ and sends to $WSO_2$ (the corresponding sending action is denoted $s_{WSO_2}(ReceiveSB_{WW_2})$);
11. The $WS_2$ receives the response message $SendPB_{WW_2}$ from $WSO_2$ through its mail box by its name $WS_2$ (the corresponding reading action is denoted $r_{WS_2}(SendPB_{WW_2})$);
12. The $WS_2$ does some local computations mixed with some atomic actions by computation logics, including $\cdot$, $+$, $\between$, and guards, the local computations are included into $I_{WS_2}$, which is the set of all local atomic actions;
13. When the local computations are finished, the $WS_2$ generates the output messages $SendPB_{WW_{21}}$ and sends to $WS_1$ (the corresponding sending action is denoted $s_{WS_1}(SendPB_{WW_{21}})$);

14. The $WS_2$ receives the response message $PayB_{WW_{21}}$ from $WS_1$ through its mail box by its name $WS_2$ (the corresponding reading action is denoted $r_{WS_2}(PayB_{WW_{21}})$);
15. The $WS_2$ does some local computations mixed with some atomic actions by computation logics, including $\cdot$, $+$, $\between$, and guards, the local computations are included into $I_{WS_2}$, which is the set of all local atomic actions;
16. When the local computations are finished, the $WS_2$ generates the output messages $GetP\&ShipB_{WA_2}$ and sends to $WSO_2$ (the corresponding sending action is denoted $s_{WSO_2}(GetP\&ShipB_{WW_2})$);
17. The $WS_2$ receives the response message $GetP\&ShipB_{WW_2}$ from $WSO_2$ through its mail box by its name $WS_2$ (the corresponding reading action is denoted $r_{WS_2}(GetP\&ShipB_{WW_2})$);
18. The $WS_2$ does some local computations mixed with some atomic actions by computation logics, including $\cdot$, $+$, $\between$, and guards, the local computations are included into $I_{WS_2}$, which is the set of all local atomic actions;
19. When the local computations are finished, the $WS_2$ generates the output messages $GetP\&ShipB_{WC_2}$ and sends to $WSC$ (the corresponding sending action is denoted $s_{WSC}(GetP\&ShipB_{WC_2})$), and then processes the messages from $WS_1$ recursively.

The above process is described as the following state transitions by APTC:

$WS_2 = r_{WS_2}(RequestLB_{WW_{12}}) \cdot WS_{2_1}$

$WS_{2_1} = (\{isInitialed(WS_2) = FLALSE\} \cdot \mathbf{new}(WSO_2) + \{isInitialed(WS_2) = TRUE\}) \cdot WS_{2_2}$

$WS_{2_2} = I_{WS_2} \cdot WS_{2_3}$

$WS_{2_3} = s_{WSO_2}(ReceiveRB_{WW_2}) \cdot WS_{2_4}$

$WS_{2_4} = r_{WS_2}(SendLB_{WW_2}) \cdot WS_{2_5}$

$WS_{2_5} = I_{WS_2} \cdot WS_{2_6}$

$WS_{2_6} = s_{WS_1}(SendLB_{WW_{21}}) \cdot WS_{2_7}$

$WS_{2_7} = r_{WS_2}(SendSB_{WW_{12}}) \cdot WS_{2_8}$

$WS_{2_8} = I_{WS_2} \cdot WS_{2_9}$

$WS_{2_9} = s_{WSO_2}(ReceiveSB_{WW_2}) \cdot WS_{2_{10}}$

$WS_{2_{10}} = r_{WS_2}(SendPB_{WW_2}) \cdot WS_{2_{11}}$

$WS_{2_{11}} = I_{WS_2} \cdot WS_{2_{12}}$

$WS_{2_{12}} = s_{WS_1}(SendPB_{WW_{21}}) \cdot WS_{2_{13}}$

$WS_{2_{13}} = r_{WS_2}(PayB_{WW_{21}}) \cdot WS_{2_{14}}$

$WS_{2_{14}} = I_{WS_2} \cdot WS_{2_{15}}$

$WS_{2_{15}} = s_{WSO_2}(GetP\&ShipB_{WW_2}) \cdot WS_{2_{16}}$

$WS_{2_{16}} = r_{WS_2}(GetP\&ShipB_{WW_2}) \cdot WS_{2_{17}}$

$WS_{2_{17}} = I_{WS_2} \cdot WS_{2_{18}}$

$WS_{2_{18}} = s_{WSC}(GetP\&ShipB_{WC_2}) \cdot WS_2$

By use of the algebraic laws of APTC, the $WS_2$ can be proven to exhibit the desired external behaviors:

$\tau_{I_{WS_2}}(\partial_\emptyset(WS_2)) = r_{WS_2}(RequestLB_{WW_{12}}) \cdot s_{WSO_2}(ReceiveRB_{WW_2})$
$\cdot r_{WS_2}(SendLB_{WW_2}) \cdot s_{WS_1}(SendLB_{WW_{21}}) \cdot r_{WS_2}(SendSB_{WW_{12}})$
$\cdot s_{WSO_2}(ReceiveSB_{WW_2}) \cdot r_{WS_2}(SendPB_{WW_2}) \cdot s_{WS_1}(SendPB_{WW_{21}})$
$\cdot r_{WS_2}(PayB_{WW_{21}}) \cdot s_{WSO_2}(GetP\&ShipB_{WW_2}) \cdot r_{WS_2}(GetP\&ShipB_{WW_2})$
$\cdot s_{WSC}(GetP\&ShipB_{WC}) \cdot \tau_{I_{WS_2}}(\partial_\emptyset(WS_2))$
with $I_{WS_2}$ extended to $I_{WS_2} \cup \{\{isInitialed(WS_2) = FLALSE\}, \{isInitialed(WS_2) = TRUE\}\}$.

### 22.3.7 BuyingBooks WSC

After $WSC$ is created, the typical process is as follows:

1. The WSC receives the initialization message $DI_{WSC}$ from the outside through its mail box by its name $WSC$ (the corresponding reading action is denoted $r_{WSC}(DI_{WSC})$);
2. The WSC may create its WSs through actions $\mathbf{new}(WS_1) \parallel \mathbf{new}(WS_2)$ if it is not initialized;
3. The WSC does some local computations mixed with some atomic actions by computation logics, including $\cdot$, $+$, $\between$, and guards, the local computations are included into $I_{WSC}$, which is the set of all local atomic actions;
4. When the local computations are finished, the WSC generates the output messages $ReBuyingBooks_{WC_1}$ and sends to $WS_1$ (the corresponding sending action is denoted $s_{WS_1}(ReBuyingBooks_{WC_1})$);
5. The WSC receives the result message $GetP\&ShipB_{WC_2}$ from $WS_2$ through its mail box by its name $WSC$ (the corresponding reading action is denoted $r_{WSC}(GetP\&ShipB_{WC_2})$);

6. The WSC does some local computations mixed with some atomic actions by computation logics, including $\cdot$, $+$, $\between$, and guards, the local computations are included into $I_{WSC}$, which is the set of all local atomic actions;
7. When the local computations are finished, the WSC generates the output messages $DO_{WSC}$ and sends to the outside (the corresponding sending action is denoted $s_O(DO_{WSC})$), and then processes the next message from the outside.

The above process is described as the following state transitions by APTC:
$WSC = r_{WSC}(DI_{WSC}) \cdot WSC_1$
$WSC_1 = (\{isInitialed(WSC) = FLALSE\} \cdot (\mathbf{new}(WS_1) \parallel \mathbf{new}(WS_2)) + \{isInitialed(WSC) = TRUE\}) \cdot WSC_2$
$WSC_2 = I_{WSC} \cdot WSC_3$
$WSC_3 = s_{WS_1}(ReBuyingBooks_{WC_1}) \cdot WSC_4$
$WSC_4 = r_{WSC}(GetP\&ShipB_{WC_2}) \cdot WSC_5$
$WSC_5 = I_{WSC} \cdot WSC_6$
$WSC_6 = s_O(DO_{WSC}) \cdot WSC$
By use of the algebraic laws of APTC, the WSC can be proven to exhibit the desired external behaviors:
$\tau_{I_{WSC}}(\partial_\emptyset(WSC)) = r_{WSC}(DI_{WSC}) \cdot s_{WS_1}(ReBuyingBooks_{WC_1}) \cdot r_{WSC}(GetP\&ShipB_{WC_2}) \cdot s_O(DO_{WSC}) \cdot \tau_{I_{WSC}}(\partial_\emptyset(WSC))$
with $I_{WSC}$ extended to $I_{WSC} \cup \{\{isInitialed(WSC) = FLALSE\}, \{isInitialed(WSC) = TRUE\}\}$.

### 22.3.8 Putting all actors together into a whole

Now, we can put all actors together into a whole, including all AAs, WSOs, WSs, and WSC, according to the buying books example as illustrated in Fig. 22.4. The whole actor system
$WSC = WSC \quad WS_1 \quad WS_2 \quad WSO_1 \quad WSO_2 \quad AA_{11} \quad AA_{12} \quad AA_{13} \quad AA_{14} \quad AA_{15}$
$AA_{21} \quad AA_{22} \quad AA_{23} \quad AA_{24} \quad AA_{25}$ can be represented by the following process term of APTC:
$\tau_I(\partial_H(WSC)) = \tau_I(\partial_H(WSC \between WS_1 \between WS_2 \between WSO_1 \between WSO_2 \between AA_{11} \between AA_{12} \between AA_{13} \between AA_{14} \between AA_{15} \between AA_{21} \between AA_{22} \between AA_{23} \between AA_{24} \between AA_{25}))$

Among all the actors, there are synchronous communications. The actor's reading and the same actor's sending actions with the same type messages may cause communications. If the actor's sending action occurs before the same actor's reading action, an asynchronous communication will occur; otherwise, a deadlock $\delta$ will be caused.

There are seven kinds of asynchronous communications as follows:

(1) The communications between $WSO_1$ and its AAs with the following constraints:
$s_{AA_{11}}(RequestLB_{WA_1}) \leq r_{AA_{11}}(RequestLB_{WA_1})$
$s_{WSO_1}(RequestLB_{AW_1}) \leq r_{WSO_1}(RequestLB_{AW_1})$
$s_{AA_{12}}(ReceiveLB_{WA_1}) \leq r_{AA_{12}}(ReceiveLB_{WA_1})$
$s_{WSO_1}(ReceiveLB_{AW_1}) \leq r_{WSO_1}(ReceiveLB_{AW_1})$
$s_{AA_13}(SendSB_{WA_1}) \leq r_{AA_13}(SendSB_{WA_1})$
$s_{WSO_1}(SendSB_{AW_1}) \leq r_{WSO_1}(SendSB_{AW_1})$
$s_{AA_{14}}(ReceivePB_{WA_1}) \leq r_{AA_{14}}(ReceivePB_{WA_1})$
$s_{WSO_1}(ReceivePB_{AW_1}) \leq r_{WSO_1}(ReceivePB_{AW_1})$
$s_{AA_{15}}(PayB_{WA_1}) \leq r_{AA_{15}}(PayB_{WA_1})$
$s_{WSO_1}(PayB_{AW_1}) \leq r_{WSO_1}(PayB_{AW_1})$

(2) The communications between $WSO_1$ and its interface WS $WS_1$ with the following constraints:
$s_{WSO_1}(ReBuyingBooks_{WW_1}) \leq r_{WSO_1}(ReBuyingBooks_{WW_1})$
$s_{WS_1}(RequestLB_{WW_1}) \leq r_{WS_1}(RequestLB_{WW_1})$
$s_{WSO_1}(ReceiveLB_{WW_1}) \leq r_{WSO_1}(ReceiveLB_{WW_1})$
$s_{WS_1}(SendSB_{WW_1}) \leq r_{WS_1}(SendSB_{WW_1})$
$s_{WSO_1}(ReceivePB_{WW_1}) \leq r_{WSO_1}(ReceivePB_{WW_1})$
$s_{WS_1}(PayB_{WW_1}) \leq r_{WS_1}(PayB_{WW_1})$

(3) The communications between $WSO_2$ and its AAs with the following constraints:
$s_{AA_{21}}(ReceiveRB_{WA_2}) \leq r_{AA_{21}}(ReceiveRB_{WA_2})$
$s_{WSO_2}(ReceiveRB_{AW_2}) \leq r_{WSO_2}(ReceiveRB_{AW_2})$
$s_{AA_{22}}(SendLB_{WA_2}) \leq r_{AA_{22}}(SendLB_{WA_2})$
$s_{WSO_2}(SendLB_{AW_2}) \leq r_{WSO_2}(SendLB_{AW_2})$
$s_{AA_{23}}(ReceiveSB_{WA_2}) \leq r_{AA_{23}}(ReceiveSB_{WA_2})$
$s_{WSO_2}(ReceiveSB_{AW_2}) \leq r_{WSO_2}(ReceiveSB_{AW_2})$

$s_{AA_{24}}(SendPB_{WA_2}) \leq r_{AA_{24}}(SendPB_{WA_2})$
$s_{WSO_2}(SendPB_{AW_2}) \leq r_{WSO_2}(SendPB_{AW_2})$
$s_{AA_{25}}(GetP\&ShipB_{WA_2}) \leq r_{AA_{25}}(GetP\&ShipB_{WA_2})$
$s_{WSO_2}(GetP\&ShipB_{AW_2}) \leq r_{WSO_2}(GetP\&ShipB_{AW_2})$

(4) The communications between $WSO_2$ and its interface WS $WS_2$ with the following constraints:

$s_{WSO_2}(ReceiveRB_{WW_2}) \leq r_{WSO_2}(ReceiveRB_{WW_2})$
$s_{WS_2}(SendLB_{WW_2}) \leq r_{WS_2}(SendLB_{WW_2})$
$s_{WSO_2}(ReceiveSB_{WW_2}) \leq r_{WSO_2}(ReceiveSB_{WW_2})$
$s_{WS_2}(SendPB_{WW_2}) \leq r_{WS_2}(SendPB_{WW_2})$
$s_{WSO_2}(SendPB_{WW_2}) \leq r_{WSO_2}(SendPB_{WW_2})$
$s_{WS_2}(GetP\&ShipB_{WW_2}) \leq r_{WS_2}(GetP\&ShipB_{WW_2})$

(5) The communications between $WS_1$ and $WS_2$ with the following constraints:

$s_{WS_2}(RequestLB_{WW_{12}}) \leq r_{WS_2}(RequestLB_{WW_{12}})$
$s_{WS_1}(SendLB_{WW_{21}}) \leq r_{WS_1}(SendLB_{WW_{21}})$
$s_{WS_2}(SendSB_{WW_{12}}) \leq r_{WS_2}(SendSB_{WW_{12}})$
$s_{WS_1}(SendPB_{WW_{21}}) \leq r_{WS_1}(SendPB_{WW_{21}})$
$s_{WS_2}(PayB_{WW_{12}}) \leq r_{WS_2}(PayB_{WW_{12}})$

(6) The communications between $WS_1$ and its WSC $WSC$ with the following constraint:

$s_{WS_1}(ReBuyingBooks_{WC_1}) \leq r_{WS_1}(ReBuyingBooks_{WC_1})$

(7) The communications between $WS_2$ and its WSC $WSC$ with the following constraint:

$s_{WSC}(GetP\&ShipB_{WC}) \leq r_{WSC}(GetP\&ShipB_{WC})$

Hence, the set $H$ and $I$ can be defined as follows:

$H = \{s_{AA_{11}}(RequestLB_{WA_1}), r_{AA_{11}}(RequestLB_{WA_1}),$
$s_{WSO_1}(RequestLB_{AW_1}), r_{WSO_1}(RequestLB_{AW_1}),$
$s_{AA_{12}}(ReceiveLB_{WA_1}), r_{AA_{12}}(ReceiveLB_{WA_1}),$
$s_{WSO_1}(ReceiveLB_{AW_1}), r_{WSO_1}(ReceiveLB_{AW_1}),$
$s_{AA_13}(SendSB_{WA_1}), r_{AA_13}(SendSB_{WA_1}),$
$s_{WSO_1}(SendSB_{AW_1}), r_{WSO_1}(SendSB_{AW_1}),$
$s_{AA_{14}}(ReceivePB_{WA_1}), r_{AA_{14}}(ReceivePB_{WA_1}),$
$s_{WSO_1}(ReceivePB_{AW_1}), r_{WSO_1}(ReceivePB_{AW_1}),$
$s_{AA_{15}}(PayB_{WA_1}), r_{AA_{15}}(PayB_{WA_1}),$
$s_{WSO_1}(PayB_{AW_1}), r_{WSO_1}(PayB_{AW_1}),$
$s_{WSO_1}(ReBuyingBooks_{WW_1}), r_{WSO_1}(ReBuyingBooks_{WW_1}),$
$s_{WS_1}(RequestLB_{WW_1}), r_{WS_1}(RequestLB_{WW_1}),$
$s_{WSO_1}(ReceiveLB_{WW_1}), r_{WSO_1}(ReceiveLB_{WW_1}),$
$s_{WS_1}(SendSB_{WW_1}), r_{WS_1}(SendSB_{WW_1}),$
$s_{WSO_1}(ReceivePB_{WW_1}), r_{WSO_1}(ReceivePB_{WW_1}),$
$s_{WS_1}(PayB_{WW_1}), r_{WS_1}(PayB_{WW_1}),$
$s_{AA_{21}}(ReceiveRB_{WA_2}), r_{AA_{21}}(ReceiveRB_{WA_2}),$
$s_{WSO_2}(ReceiveRB_{AW_2}), r_{WSO_2}(ReceiveRB_{AW_2}),$
$s_{AA_{22}}(SendLB_{WA_2}), r_{AA_{22}}(SendLB_{WA_2}),$
$s_{WSO_2}(SendLB_{AW_2}), r_{WSO_2}(SendLB_{AW_2}),$
$s_{AA_{23}}(ReceiveSB_{WA_2}), r_{AA_{23}}(ReceiveSB_{WA_2}),$
$s_{WSO_2}(ReceiveSB_{AW_2}), r_{WSO_2}(ReceiveSB_{AW_2}),$
$s_{AA_{24}}(SendPB_{WA_2}), r_{AA_{24}}(SendPB_{WA_2}),$
$s_{WSO_2}(SendPB_{AW_2}), r_{WSO_2}(SendPB_{AW_2}),$
$s_{AA_{25}}(GetP\&ShipB_{WA_2}), r_{AA_{25}}(GetP\&ShipB_{WA_2}),$
$s_{WSO_2}(GetP\&ShipB_{AW_2}), r_{WSO_2}(GetP\&ShipB_{AW_2}),$
$s_{WSO_2}(ReceiveRB_{WW_2}), r_{WSO_2}(ReceiveRB_{WW_2}),$
$s_{WS_2}(SendLB_{WW_2}), r_{WS_2}(SendLB_{WW_2}),$
$s_{WSO_2}(ReceiveSB_{WW_2}), r_{WSO_2}(ReceiveSB_{WW_2}),$
$s_{WS_2}(SendPB_{WW_2}), r_{WS_2}(SendPB_{WW_2}),$
$s_{WSO_2}(SendPB_{WW_2}), r_{WSO_2}(SendPB_{WW_2}),$
$s_{WS_2}(GetP\&ShipB_{WW_2}), r_{WS_2}(GetP\&ShipB_{WW_2}),$

$s_{WS_2}(RequestLB_{WW_{12}}), r_{WS_2}(RequestLB_{WW_{12}}),$
$s_{WS_1}(SendLB_{WW_{21}}), r_{WS_1}(SendLB_{WW_{21}}),$
$s_{WS_2}(SendSB_{WW_{12}}), r_{WS_2}(SendSB_{WW_{12}}),$
$s_{WS_1}(SendPB_{WW_{21}}), r_{WS_1}(SendPB_{WW_{21}}),$
$s_{WS_2}(PayB_{WW_{12}}), r_{WS_2}(PayB_{WW_{12}}),$
$s_{WS_1}(ReBuyingBooks_{WC_1}), r_{WS_1}(ReBuyingBooks_{WC_1}),$
$s_{WSC}(GetP\&ShipB_{WC}), r_{WSC}(GetP\&ShipB_{WC})$
$|s_{AA_{11}}(RequestLB_{WA_1}) \nleq r_{AA_{11}}(RequestLB_{WA_1}),$
$s_{WSO_1}(RequestLB_{AW_1}) \nleq r_{WSO_1}(RequestLB_{AW_1}),$
$s_{AA_{12}}(ReceiveLB_{WA_1}) \nleq r_{AA_{12}}(ReceiveLB_{WA_1}),$
$s_{WSO_1}(ReceiveLB_{AW_1}) \nleq r_{WSO_1}(ReceiveLB_{AW_1}),$
$s_{AA_13}(SendSB_{WA_1}) \nleq r_{AA_13}(SendSB_{WA_1}),$
$s_{WSO_1}(SendSB_{AW_1}) \nleq r_{WSO_1}(SendSB_{AW_1}),$
$s_{AA_{14}}(ReceivePB_{WA_1}) \nleq r_{AA_{14}}(ReceivePB_{WA_1}),$
$s_{WSO_1}(ReceivePB_{AW_1}) \nleq r_{WSO_1}(ReceivePB_{AW_1}),$
$s_{AA_{15}}(PayB_{WA_1}) \nleq r_{AA_{15}}(PayB_{WA_1}),$
$s_{WSO_1}(PayB_{AW_1}) \nleq r_{WSO_1}(PayB_{AW_1}),$
$s_{WSO_1}(ReBuyingBooks_{WW_1}) \nleq r_{WSO_1}(ReBuyingBooks_{WW_1}),$
$s_{WS_1}(RequestLB_{WW_1}) \nleq r_{WS_1}(RequestLB_{WW_1}),$
$s_{WSO_1}(ReceiveLB_{WW_1}) \nleq r_{WSO_1}(ReceiveLB_{WW_1}),$
$s_{WS_1}(SendSB_{WW_1}) \nleq r_{WS_1}(SendSB_{WW_1}),$
$s_{WSO_1}(ReceivePB_{WW_1}) \nleq r_{WSO_1}(ReceivePB_{WW_1}),$
$s_{WS_1}(PayB_{WW_1}) \nleq r_{WS_1}(PayB_{WW_1}),$
$s_{AA_{21}}(ReceiveRB_{WA_2}) \nleq r_{AA_{21}}(ReceiveRB_{WA_2}),$
$s_{WSO_2}(ReceiveRB_{AW_2}) \nleq r_{WSO_2}(ReceiveRB_{AW_2}),$
$s_{AA_{22}}(SendLB_{WA_2}) \nleq r_{AA_{22}}(SendLB_{WA_2}),$
$s_{WSO_2}(SendLB_{AW_2}) \nleq r_{WSO_2}(SendLB_{AW_2}),$
$s_{AA_{23}}(ReceiveSB_{WA_2}) \nleq r_{AA_{23}}(ReceiveSB_{WA_2}),$
$s_{WSO_2}(ReceiveSB_{AW_2}) \nleq r_{WSO_2}(ReceiveSB_{AW_2}),$
$s_{AA_{24}}(SendPB_{WA_2}) \nleq r_{AA_{24}}(SendPB_{WA_2}),$
$s_{WSO_2}(SendPB_{AW_2}) \nleq r_{WSO_2}(SendPB_{AW_2}),$
$s_{AA_{25}}(GetP\&ShipB_{WA_2}) \nleq r_{AA_{25}}(GetP\&ShipB_{WA_2}),$
$s_{WSO_2}(GetP\&ShipB_{AW_2}) \nleq r_{WSO_2}(GetP\&ShipB_{AW_2}),$
$s_{WSO_2}(ReceiveRB_{WW_2}) \nleq r_{WSO_2}(ReceiveRB_{WW_2}),$
$s_{WS_2}(SendLB_{WW_2}) \nleq r_{WS_2}(SendLB_{WW_2}),$
$s_{WSO_2}(ReceiveSB_{WW_2}) \nleq r_{WSO_2}(ReceiveSB_{WW_2}),$
$s_{WS_2}(SendPB_{WW_2}) \nleq r_{WS_2}(SendPB_{WW_2}),$
$s_{WSO_2}(SendPB_{WW_2}) \nleq r_{WSO_2}(SendPB_{WW_2}),$
$s_{WS_2}(GetP\&ShipB_{WW_2}) \nleq r_{WS_2}(GetP\&ShipB_{WW_2}),$
$s_{WS_2}(RequestLB_{WW_{12}}) \nleq r_{WS_2}(RequestLB_{WW_{12}}),$
$s_{WS_1}(SendLB_{WW_{21}}) \nleq r_{WS_1}(SendLB_{WW_{21}}),$
$s_{WS_2}(SendSB_{WW_{12}}) \nleq r_{WS_2}(SendSB_{WW_{12}}),$
$s_{WS_1}(SendPB_{WW_{21}}) \nleq r_{WS_1}(SendPB_{WW_{21}}),$
$s_{WS_2}(PayB_{WW_{12}}) \nleq r_{WS_2}(PayB_{WW_{12}}),$
$s_{WS_1}(ReBuyingBooks_{WC_1}) \nleq r_{WS_1}(ReBuyingBooks_{WC_1}),$
$s_{WSC}(GetP\&ShipB_{WC}) \nleq r_{WSC}(GetP\&ShipB_{WC})\}$

$I = \{s_{AA_{11}}(RequestLB_{WA_1}), r_{AA_{11}}(RequestLB_{WA_1}),$
$s_{WSO_1}(RequestLB_{AW_1}), r_{WSO_1}(RequestLB_{AW_1}),$
$s_{AA_{12}}(ReceiveLB_{WA_1}), r_{AA_{12}}(ReceiveLB_{WA_1}),$
$s_{WSO_1}(ReceiveLB_{AW_1}), r_{WSO_1}(ReceiveLB_{AW_1}),$
$s_{AA_13}(SendSB_{WA_1}), r_{AA_13}(SendSB_{WA_1}),$
$s_{WSO_1}(SendSB_{AW_1}), r_{WSO_1}(SendSB_{AW_1}),$
$s_{AA_{14}}(ReceivePB_{WA_1}), r_{AA_{14}}(ReceivePB_{WA_1}),$
$s_{WSO_1}(ReceivePB_{AW_1}), r_{WSO_1}(ReceivePB_{AW_1}),$

$s_{AA_{15}}(PayB_{WA_1}), r_{AA_{15}}(PayB_{WA_1}),$
$s_{WSO_1}(PayB_{AW_1}), r_{WSO_1}(PayB_{AW_1}),$
$s_{WSO_1}(ReBuyingBooks_{WW_1}), r_{WSO_1}(ReBuyingBooks_{WW_1}),$
$s_{WS_1}(RequestLB_{WW_1}), r_{WS_1}(RequestLB_{WW_1}),$
$s_{WSO_1}(ReceiveLB_{WW_1}), r_{WSO_1}(ReceiveLB_{WW_1}),$
$s_{WS_1}(SendSB_{WW_1}), r_{WS_1}(SendSB_{WW_1}),$
$s_{WSO_1}(ReceivePB_{WW_1}), r_{WSO_1}(ReceivePB_{WW_1}),$
$s_{WS_1}(PayB_{WW_1}), r_{WS_1}(PayB_{WW_1}),$
$s_{AA_{21}}(ReceiveRB_{WA_2}), r_{AA_{21}}(ReceiveRB_{WA_2}),$
$s_{WSO_2}(ReceiveRB_{AW_2}), r_{WSO_2}(ReceiveRB_{AW_2}),$
$s_{AA_{22}}(SendLB_{WA_2}), r_{AA_{22}}(SendLB_{WA_2}),$
$s_{WSO_2}(SendLB_{AW_2}), r_{WSO_2}(SendLB_{AW_2}),$
$s_{AA_{23}}(ReceiveSB_{WA_2}), r_{AA_{23}}(ReceiveSB_{WA_2}),$
$s_{WSO_2}(ReceiveSB_{AW_2}), r_{WSO_2}(ReceiveSB_{AW_2}),$
$s_{AA_{24}}(SendPB_{WA_2}), r_{AA_{24}}(SendPB_{WA_2}),$
$s_{WSO_2}(SendPB_{AW_2}), r_{WSO_2}(SendPB_{AW_2}),$
$s_{AA_{25}}(GetP\&ShipB_{WA_2}), r_{AA_{25}}(GetP\&ShipB_{WA_2}),$
$s_{WSO_2}(GetP\&ShipB_{AW_2}), r_{WSO_2}(GetP\&ShipB_{AW_2}),$
$s_{WSO_2}(ReceiveRB_{WW_2}), r_{WSO_2}(ReceiveRB_{WW_2}),$
$s_{WS_2}(SendLB_{WW_2}), r_{WS_2}(SendLB_{WW_2}),$
$s_{WSO_2}(ReceiveSB_{WW_2}), r_{WSO_2}(ReceiveSB_{WW_2}),$
$s_{WS_2}(SendPB_{WW_2}), r_{WS_2}(SendPB_{WW_2}),$
$s_{WSO_2}(SendPB_{WW_2}), r_{WSO_2}(SendPB_{WW_2}),$
$s_{WS_2}(GetP\&ShipB_{WW_2}), r_{WS_2}(GetP\&ShipB_{WW_2}),$
$s_{WS_2}(RequestLB_{WW_{12}}), r_{WS_2}(RequestLB_{WW_{12}}),$
$s_{WS_1}(SendLB_{WW_{21}}), r_{WS_1}(SendLB_{WW_{21}}),$
$s_{WS_2}(SendSB_{WW_{12}}), r_{WS_2}(SendSB_{WW_{12}}),$
$s_{WS_1}(SendPB_{WW_{21}}), r_{WS_1}(SendPB_{WW_{21}}),$
$s_{WS_2}(PayB_{WW_{12}}), r_{WS_2}(PayB_{WW_{12}}),$
$s_{WS_1}(ReBuyingBooks_{WC_1}), r_{WS_1}(ReBuyingBooks_{WC_1}),$
$s_{WSC}(GetP\&ShipB_{WC}), r_{WSC}(GetP\&ShipB_{WC})$
$|s_{AA_{11}}(RequestLB_{WA_1}) \leq r_{AA_{11}}(RequestLB_{WA_1}),$
$s_{WSO_1}(RequestLB_{AW_1}) \leq r_{WSO_1}(RequestLB_{AW_1}),$
$s_{AA_{12}}(ReceiveLB_{WA_1}) \leq r_{AA_{12}}(ReceiveLB_{WA_1}),$
$s_{WSO_1}(ReceiveLB_{AW_1}) \leq r_{WSO_1}(ReceiveLB_{AW_1}),$
$s_{AA_13}(SendSB_{WA_1}) \leq r_{AA_13}(SendSB_{WA_1}),$
$s_{WSO_1}(SendSB_{AW_1}) \leq r_{WSO_1}(SendSB_{AW_1}),$
$s_{AA_{14}}(ReceivePB_{WA_1}) \leq r_{AA_{14}}(ReceivePB_{WA_1}),$
$s_{WSO_1}(ReceivePB_{AW_1}) \leq r_{WSO_1}(ReceivePB_{AW_1}),$
$s_{AA_{15}}(PayB_{WA_1}) \leq r_{AA_{15}}(PayB_{WA_1}),$
$s_{WSO_1}(PayB_{AW_1}) \leq r_{WSO_1}(PayB_{AW_1}),$
$s_{WSO_1}(ReBuyingBooks_{WW_1}) \leq r_{WSO_1}(ReBuyingBooks_{WW_1}),$
$s_{WS_1}(RequestLB_{WW_1}) \leq r_{WS_1}(RequestLB_{WW_1}),$
$s_{WSO_1}(ReceiveLB_{WW_1}) \leq r_{WSO_1}(ReceiveLB_{WW_1}),$
$s_{WS_1}(SendSB_{WW_1}) \leq r_{WS_1}(SendSB_{WW_1}),$
$s_{WSO_1}(ReceivePB_{WW_1}) \leq r_{WSO_1}(ReceivePB_{WW_1}),$
$s_{WS_1}(PayB_{WW_1}) \leq r_{WS_1}(PayB_{WW_1}),$
$s_{AA_{21}}(ReceiveRB_{WA_2}) \leq r_{AA_{21}}(ReceiveRB_{WA_2}),$
$s_{WSO_2}(ReceiveRB_{AW_2}) \leq r_{WSO_2}(ReceiveRB_{AW_2}),$
$s_{AA_{22}}(SendLB_{WA_2}) \leq r_{AA_{22}}(SendLB_{WA_2}),$
$s_{WSO_2}(SendLB_{AW_2}) \leq r_{WSO_2}(SendLB_{AW_2}),$
$s_{AA_{23}}(ReceiveSB_{WA_2}) \leq r_{AA_{23}}(ReceiveSB_{WA_2}),$
$s_{WSO_2}(ReceiveSB_{AW_2}) \leq r_{WSO_2}(ReceiveSB_{AW_2}),$
$s_{AA_{24}}(SendPB_{WA_2}) \leq r_{AA_{24}}(SendPB_{WA_2}),$

$s_{WSO_2}(SendPB_{AW_2}) \leq r_{WSO_2}(SendPB_{AW_2})$,
$s_{AA_{25}}(GetP\&ShipB_{WA_2}) \leq r_{AA_{25}}(GetP\&ShipB_{WA_2})$,
$s_{WSO_2}(GetP\&ShipB_{AW_2}) \leq r_{WSO_2}(GetP\&ShipB_{AW_2})$,
$s_{WSO_2}(ReceiveRB_{WW_2}) \leq r_{WSO_2}(ReceiveRB_{WW_2})$,
$s_{WS_2}(SendLB_{WW_2}) \leq r_{WS_2}(SendLB_{WW_2})$,
$s_{WSO_2}(ReceiveSB_{WW_2}) \leq r_{WSO_2}(ReceiveSB_{WW_2})$,
$s_{WS_2}(SendPB_{WW_2}) \leq r_{WS_2}(SendPB_{WW_2})$,
$s_{WSO_2}(SendPB_{WW_2}) \leq r_{WSO_2}(SendPB_{WW_2})$,
$s_{WS_2}(GetP\&ShipB_{WW_2}) \leq r_{WS_2}(GetP\&ShipB_{WW_2})$,
$s_{WS_2}(RequestLB_{WW_{12}}) \leq r_{WS_2}(RequestLB_{WW_{12}})$,
$s_{WS_1}(SendLB_{WW_{21}}) \leq r_{WS_1}(SendLB_{WW_{21}})$,
$s_{WS_2}(SendSB_{WW_{12}}) \leq r_{WS_2}(SendSB_{WW_{12}})$,
$s_{WS_1}(SendPB_{WW_{21}}) \leq r_{WS_1}(SendPB_{WW_{21}})$,
$s_{WS_2}(PayB_{WW_{12}}) \leq r_{WS_2}(PayB_{WW_{12}})$,
$s_{WS_1}(ReBuyingBooks_{WC_1}) \leq r_{WS_1}(ReBuyingBooks_{WC_1})$,
$s_{WSC}(GetP\&ShipB_{WC}) \leq r_{WSC}(GetP\&ShipB_{WC})\}$
$\cup I_{AA_{11}} \cup I_{AA_{12}} \cup I_{AA_{13}} \cup I_{AA_{14}} \cup I_{AA_{15}} \cup I_{AA_{21}} \cup I_{AA_{22}} \cup I_{AA_{23}} \cup I_{AA_{24}} \cup I_{AA_{25}} \cup I_{WSO_1} \cup I_{WSO_2} \cup I_{WS_1} \cup I_{WS_2} \cup I_{WSC}$

Then, we can obtain the following conclusion.

**Theorem 22.1.** *The whole actor system of buying books example illustrated in Fig. 22.4 can exhibit the desired external behaviors.*

*Proof.* By use of the algebraic laws of APTC, we can prove the following equation:

$\tau_I(\partial_H(WSC))$
$= \tau_I(\partial_H(WSC \between WS_1 \between WS_2 \between WSO_1 \between WSO_2 \between AA_{11} \between AA_{12} \between AA_{13} \between AA_{14} \between AA_{15} \between AA_{21} \between AA_{22} \between AA_{23} \between AA_{24} \between AA_{25}))$
$= r_{WSC}(DI_{WSC}) \cdot s_O(DO_{WSC}) \cdot \tau_I(\partial_H(WSC \between WS_1 \between WS_2 \between WSO_1 \between WSO_2 \between AA_{11} \between AA_{12} \between AA_{13} \between AA_{14} \between AA_{15} \between AA_{21} \between AA_{22} \between AA_{23} \between AA_{24} \between AA_{25}))$
$= r_{WSC}(DI_{WSC}) \cdot s_O(DO_{WSC}) \cdot \tau_I(\partial_H(WSC))$

For the details of the proof, as we omit them here, please refer to Section 17.3. □

Chapter 23

# Process algebra based actor model of the QoS-aware Web Service orchestration engine

In this chapter, we will use the process algebra based actor model to model and verify the QoS-aware Web Service orchestration engine based on the previous work [26]. In Section 23.1, we introduce the requirements of the QoS-aware Web Service orchestration engine; we model the QoS-aware Web Service orchestration engine by use of the new actor model in Section 23.2; finally, we take an example to show the usage of the model in Section 23.3.

## 23.1 Requirements of the QoS-aware Web Service orchestration engine

Web Service (WS) is a distributed component that emerged about ten years ago, which uses WSDL as its interface description language, SOAP as its communication protocol, and UDDI as its directory service. Because WS uses the Web as its provision platform, it is suitable to be used to develop cross-organizational business integrations.

Cross-organizational business processes are usual forms in e-commerce that orchestrate some business activities into a workflow. WS Orchestration (WSO) provides a solution for such business process based on WS technologies, hereby representing a business process where business activities are modeled as component WSs (a component WS corresponds to a business activity, it may be an atomic WS or another composite WS).

From the WS viewpoint, WSO provides a workflow-like pattern to orchestrate existing WSs to create a new composite WS, and embodies the added values of WS. In particular, we use the term WSO, rather than another term – WS Composition, because there are also other WS composition patterns, such as WS Choreography (WSC) [24]. However, about WSC and the relationship of WSO and WSC, we do not explain more here, as it is not the focus of this chapter, please see Chapter 22 for details.

In this chapter, we focus on WSO, in fact, the QoS-aware WSO engine (runtime of WSO) and its formal model. A QoS-aware WSO enables the customers to be satisfied with not only their functional requirements, but also their QoS requirements, such as performance requirements, reliability requirements, security requirements, etc. A single execution of a WSO is called a WSO instance (WSOI). A QoS-aware WSO engine provides runtime supports for WSOs with the assurance of QoS implementations. These runtime supports include lifetime operations on a WSO instance, queue processing for requests from the customers and incoming message delivery to a WSO instance.

WS and WSO are within a continuously changing and evolving environment. The customers, the requirements of the customers, and the component WSs are all changing dynamically. To assure safe adaptation to dynamically changing and evolving requirements, it is important to have a rigorous semantic model of the system: the component WSs, the WSO engine that provides WSO instance management and invocation of the component WSs, the customer accesses, and the interactions among these elements. Using such a model, designs can be analyzed to clarify assumptions that must be met for correct operation.

We give a so-called BuyingBooks example for the scenario of cross-organizational business process integration and use a so-called BookStore WSO to illustrate some related concepts, such as WSO, activity, etc. Also, we use the BookStore WSO to explain the formal model we established in the following.

An example is BuyingBooks as Fig. 23.1 shows. We use this BuyingBooks example throughout this chapter to illustrate concepts and mechanisms in WS Composition.

In Fig. 23.1, there are four organizations: BuyerAgent, BookStore, RailwayCorp, and AirlineCorp. Also, each organization has one business process. In fact, there are two business processes, the business processes in RailwayCorp and AirlineCorp are simplified as just WSs for simpleness without loss of generality. We introduce the business process of BookStore as follows, and the process of BuyerAgent can be understood as contrasts:

Handbook of Truly Concurrent Process Algebra. https://doi.org/10.1016/B978-0-44-321515-5.00027-5

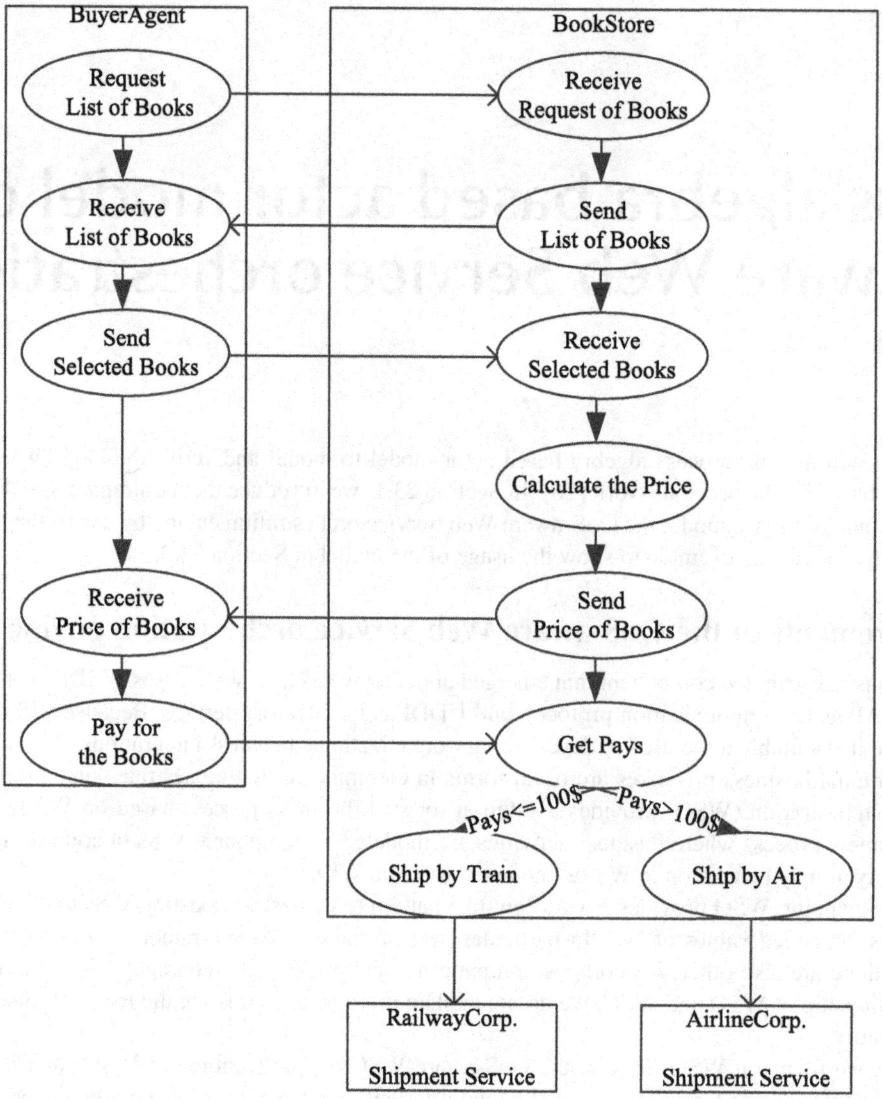


FIGURE 23.1 The BuyingBooks example.

1. The BookStore receives a request of list of books from the buyer through BuyerAgent;
2. It sends the list of books to the buyer via BuyerAgent;
3. It receives the selected book list by the buyer via BuyerAgent;
4. It calculates the price of the selected books;
5. It sends the price of the selected books to the buyer via BuyerAgent;
6. It receives payments for the books from the buyer via BuyerAgent;
7. If the payments are greater than 100$, then the BookStore calls the shipment service of AirlineCorp for the shipment of books;
8. Otherwise, the BookStore calls the shipment service of RailwayCorp for the shipment of books. Then, the process is ended.

Each business process is implemented by a WSO, for example, the BookStore WSO and BuyerAgent WSO implement BookStore process and BuyerAgent process, respectively. Each WSO invokes external WSs through its activities directly. Also, each WSO is published as a WS to receive the incoming messages.

### 23.1.1 The BookStore WSO

The flow of BookStore WSO is as Fig. 23.1 shows. There are several receive–reply activity pairs and several invoke activities in the BookStore WSO. The QoS requirements are not included in the WS-BPEL description, because these need an extension of WS-BPEL and are out of the scope. In the request message from the BuyerAgent WSO, the QoS requirements, such as the whole execution time threshold and the additional charges, can also be attached, not only the functional parameters.

Another related specification is the WSDL description of the interface WS for BuyingBooks WSO. As we focus on WS composition, this WSDL specification is omitted.

### 23.1.2 Architecture of a typical QoS-aware WSO engine, QoS-WSOE

In this section, we first analyze the requirements of a WSO Engine. Then, we discuss problems about QoS management of WS and define the QoS aspects used in this chapter. Finally, we give the architecture of QoS-WSOE and discuss the state transition of a WSO instance.

As the introduction above says, a WSO description language, such as WS-BPEL, has:

- basic constructs called atomic activities to model invocation to an external WS, receiving invocation from an external WS and reply to that WS, and other inner basic functions;
- information and variables exchanged between WSs;
- control flows called structural activities to orchestrate activities;
- other inner transaction processing mechanisms, such as exception definitions and throwing mechanisms, event definitions, and response mechanisms.

Therefore, a WSO described by WS-BPEL is a program with WSs as its basic function units and must be enabled by a WSO engine. An execution of a WSO is called an instance of that WSO. The WSO engine can create a new WSO instance according to information included in a request of a customer via the interface WS (note that a WSO is encapsulated as a WS also) of the WSO. Once a WSO instance is created, it has a thread of control to execute independently according to its definition described by a kind of description language, such as WS-BPEL. During its execution, it may create activities to interact with WSs outside and also may do inner processings, such as local variable assignments. When it ends execution, it replies to the customer with its execution outcomes.

In order to provide the adaptability of a WSO, the bindings between its activities and WSs outside are not direct and static. That is, WSs are classified according to the ontologies of specific domains and the WSs belonging to the same ontology have the same functions and interfaces, and different access points and different QoS. To make this possible, from a system viewpoint, a name and directory service – UDDI is necessary. All WSs with access information and QoS information are registered into a UDDI that classifies WSs by their ontologies to be discovered and invoked in future. UDDI should provide multi-interfaces to search WSs registered in for its users, for example, a user can obtain information of a specific set of WSs by providing a service ontology and specific QoS requirements via an interface of the UDDI.

The above mechanisms make QoS-aware service selection possible. In a QoS-aware WSO engine, after a new WSO instance is created, the new WSO instance first selects its component WSs according to the QoS requirements provided by the customer and ontologies of component WSs defined in the description file of the WSO by WS-BPEL (the description of QoS and ontologies of component WSs by WS-BPEL, needs an extension of WS-BPEL, but this is out of the scope).

About QoS of a WS, there are various QoS aspects, such as performance QoS, security QoS, reliability QoS, availability QoS, and so on. In this chapter, we use a cost-effective QoS approach. That is, cost QoS is used to measure the costs of one invocation of a WS while response time QoS is used to capture the effectiveness of one invocation of a WS. In the following, we assume all WSs are aware of cost-effective QoS.

According to the requirements of a WSO engine discussed above, the architecture of QoS-WSOE is given as Fig. 23.2 shows.

In the architecture of QoS-WSOE, there are external components, such as Client, WS of a WSO, UDDI and component WSs, and inner components, including WSO Instance Manager, WSO Instances, Activities, and Service Selector. Among them, WS of a WSO, UDDI, WSO Instance Manager, and Service Selector are permanent components and Client, component WSs, WSO Instances, Activities are transient components. Component WSs are transient components since they are determined after a service selection process is executed by Service Selector.

Through a typical requirement process, we illustrate the functions and relationships of these components:

1. A Client submits its requests including the WSO ontology, input parameters, and QoS requirements to the WS of a WSO through SOAP protocol;

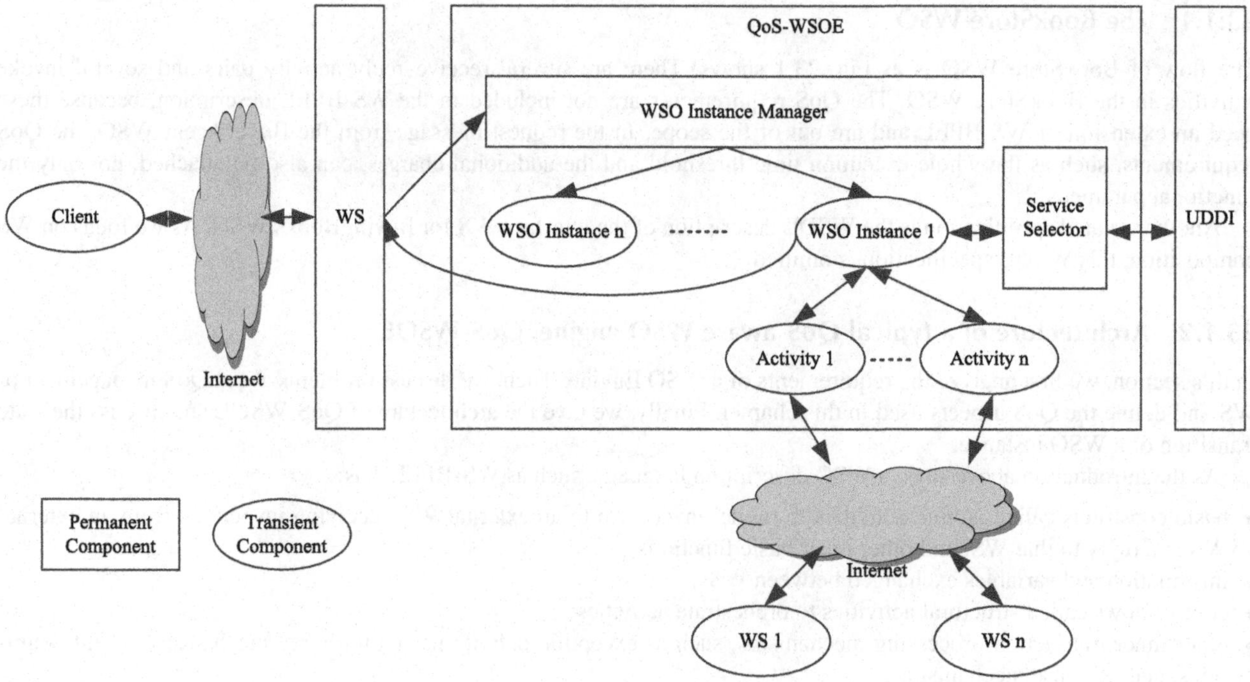


**FIGURE 23.2** Architecture of QoS-WSOE.

2. The WS transmits the requirements from a SOAP message sent by the Client to the WSO Instance Manager using private communication mechanisms;
3. The WSO Instance Manager creates a new WSO Instance including its Activities and transmits the input parameters and the QoS requirements to the new instance;
4. The instance transmits ontologies of its component WSs and the QoS requirements to the Service Selector to perform a service selection process via interactions with a UDDI. If the QoS requirements cannot be satisfied, the instance replies to the Client to deny this time service;
5. If the QoS requirements can be satisfied, each activity in the WSO Instance is bound to an external WS;
6. The WSO Instance transmits input parameters to each activity for an invocation to its binding WS;
7. After the WSO Instance ends its execution, that is, every invocation to its component WSs by activities in the WSO Instance is returned, the WSO Instance returns the execution outcomes to the Client.

An execution of a WSO is called a WSO instance (WSOI). A WSOI is created when the WSO Instance Manager receives a new request (including the functional parameters and the QoS requirements).

## 23.2 The new actor model of the QoS-aware Web Service orchestration engine

According to the architecture of the QoS-aware Web Service Orchestration Engine, the whole actors system implemented by actors can be divided into five kinds of actors: the WS actors, the Web Service Orchestration Instance Manager actor, the WSO actors, the activity actors, and the service selector actor.

### 23.2.1 Web Service, WS

A WS is an actor that has the characteristics of an ordinary actor. It acts as a communication bridge between the inner WSO and the outside, and the outside and the inner implementations.

After a WS is created, the typical process is as follows:

1. The WS receives the incoming message $DI_{WS}$ from the outside through its mail box by its name $WS$ (the corresponding reading action is denoted $r_{WS}(DI_{WS})$);

2. The WS may invokes the inner implementations, and does some local computations mixed with some atomic actions by computation logics, including $\cdot$, $+$, $\between$, and guards, the local computations are included into $I_{WS}$, which is the set of all local atomic actions;
3. When the local computations are finished, the WS generates the output messages and may send to the outside (the corresponding sending actions are distinct by the names of the outside actors, and also the names of messages), and then processes the next message from the outside.

The above process is described as the following state transition skeletons by APTC:
$WS = r_{WS}(DI_{WS}) \cdot WS_1$
$WS_1 = I_{WS} \cdot WS_2$
$WS_2 = s_O(DO_{WS}) \cdot WS$
By use of the algebraic laws of APTC, the WS may be proven to exhibit the desired external behaviors. If it can exhibit the desired external behaviors, the WS should have the following form:

$$\tau_{I_{WS}}(\partial_{\emptyset}(WS)) = r_{WS}(DI_{WS}) \cdot s_O(DO_{WS}) \cdot \tau_{I_{WS}}(\partial_{\emptyset}(WS))$$

### 23.2.2 Web Service orchestration instance manager, WSOIM

The WSOIM manages a set of WSO actors. The management operations may be creating a WSO actor.

After the WSOIM is created, the typical process is as follows:

1. The WSOIM receives the incoming message $DI_{WSOIM}$ from the interface WS through its mail box by its name $WSOIM$ (the corresponding reading action is denoted $r_{WSOIM}(DI_{WSO})$);
2. The WSOIM may create a WSO actor through an action **new**$(WSO)$ if it is not initialized;
3. The WSOIM does some local computations mixed with some atomic actions by computation logics, including $\cdot$, $+$, $\between$, and guards, the local computations are included into $I_{WSOIM}$, which is the set of all local atomic actions;
4. When the local computations are finished, the WSOIM generates the output messages $DO_{WSOIM}$ and sends them to the WSO (the corresponding sending action is denoted $s_{WSO}(DO_{WSOIM})$), and then processes the next message from the interface WS.

The above process is described as the following state transition skeletons by APTC:
$WSOIM = r_{WSOIM}(DI_{WSOIM}) \cdot WSOIM_1$
$WSOIM_1 = (\{isInitialed(WSO) = FLALSE\} \cdot \mathbf{new}(WSO) + \{isInitialed(WSO) = TRUE\}) \cdot WSOIM_2$
$WSOIM_2 = I_{WSOIM} \cdot WSOIM_3$
$WSOIM_3 = s_{WSO}(DO_{WSOIM}) \cdot WSOIM$
By use of the algebraic laws of APTC, the WSOIM may be proven to exhibit the desired external behaviors. If it can exhibit the desired external behaviors, the WSOIM should have the following form:

$$\tau_{I_{WSOIM}}(\partial_{\emptyset}(WSOIM)) = r_{WSOIM}(DI_{WSOIM}) \cdot s_{WSO}(DO_{WSOIM}) \cdot \tau_{I_{WSOIM}}(\partial_{\emptyset}(WSOIM))$$

with $I_{WSOIM}$ extended to $I_{WSOIM} \cup \{\{isInitialed(WSO) = FLALSE\}, \{isInitialed(WSO) = TRUE\}\}$.

### 23.2.3 Web Service orchestration (instance), WSO

A WSO includes a set of AAs and acts as the manager of the AAs. The management operations may be creating a member AA.

After a WSO is created, the typical process is as follows:

1. The WSO receives the incoming message $DI_{WSO}$ from the WSOIM through its mail box by its name $WSO$ (the corresponding reading action is denoted $r_{WSO}(DI_{WSO})$);
2. The WSO may create its AAs in parallel through actions **new**$(AA_1) \parallel \cdots \parallel$ **new**$(AA_n)$ if it is not initialized;
3. The WSO may receive messages from its AAs through its mail box by its name $WSO$ (the corresponding reading actions are distinct by the message names);
4. The WSO does some local computations mixed with some atomic actions by computation logics, including $\cdot$, $+$, $\between$, and guards, the local computations are included into $I_{WSO}$, which is the set of all local atomic actions;
5. When the local computations are finished, the WSO generates the output messages and may send to its AAs or the interface WS (the corresponding sending actions are distinct by the names of AAs and WS, and also the names of messages), and then processes the next message from its AAs or the interface WS.

The above process is described as the following state transition skeletons by APTC:
$WSO = r_{WSO}(DI_{WSO}) \cdot WSO_1$
$WSO_1 = (\{isInitialed(WSO) = FLALSE\} \cdot (\mathbf{new}(AA_1) \parallel \cdots \parallel \mathbf{new}(AA_n)) + \{isInitialed(WSO) = TRUE\}) \cdot WSO_2$
$WSO_2 = r_{WSO}(DI_{AAs}) \cdot WSO_3$
$WSO_3 = I_{WSO} \cdot WSO_4$
$WSO_4 = s_{AAs,WS}(DO_{WSO}) \cdot WSO$

By use of the algebraic laws of APTC, the WSO may be proven to exhibit the desired external behaviors. If it can exhibit the desired external behaviors, the WSO should have the following form:

$$\tau_{I_{WSO}}(\partial_\emptyset(WSO)) = r_{WSO}(DI_{WSO}) \cdot \cdots \cdot s_{WS}(DO_{WSO}) \cdot \tau_{I_{WSO}}(\partial_\emptyset(WSO))$$

with $I_{WSO}$ extended to $I_{WSO} \cup \{\{isInitialed(WSO) = FLALSE\}, \{isInitialed(WSO) = TRUE\}\}$.

### 23.2.4 Activity actor, AA

An activity is an atomic function unit of a WSO and is managed by the WSO. We use an actor called activity actor (AA) to model an activity.

An AA has a unique name, local information and variables to contain its states, and local computation procedures to manipulate the information and variables. An AA is always managed by a WSO and it receives messages from its WSO, sends messages to other AAs or WSs via its WSO, and is created by its WSO. Note that an AA cannot create new AAs, it can only be created by a WSO. That is, an AA is an actor with a constraint that is without a **create** action.

After an AA is created, the typical process is as follows:

1. The AA receives some messages $DI_{AA}$ from its WSO through its mail box denoted by its name $AA$ (the corresponding reading action is denoted $r_{AA}(DI_{AA})$);
2. Then, it does some local computations mixed with some atomic actions by computation logics, including $\cdot$, $+$, $\between$, and guards, the whole local computations are denoted $I_{AA}$, which is the set of all local atomic actions;
3. When the local computations are finished, the AA generates the output message $DO_{AA}$ and sends to its WSO's mail box denoted by the WSO's name $WSO$ (the corresponding sending action is denoted $s_{WSO}(DO_{AA})$), and then processes the next message from its WSO recursively.

The above process is described as the following state transition skeletons by APTC:
$AA = r_{AA}(DI_{AA}) \cdot AA_1$
$AA_1 = I_{AA} \cdot AA_2$
$AA_2 = s_{WSO}(DO_{AA}) \cdot AA$

By use of the algebraic laws of APTC, the AA may be proven to exhibit the desired external behaviors. If it can exhibit the desired external behaviors, the AA should have the following form:

$$\tau_{I_{AA}}(\partial_\emptyset(AA)) = r_{AA}(DI_{AA}) \cdot s_{WSO}(DO_{AA}) \cdot \tau_{I_{AA}}(\partial_\emptyset(AA))$$

### 23.2.5 Service selector, SS

The service selector (SS) is an actor accepting the request (including the WSO definitions and the QoS requirements) from the WSO, and returning the WS selection response:

1. The SS receives the request $DI_{SS}$ from the WSO through its mail box denoted by its name $SS$ (the corresponding reading action is denoted $r_{SS}(DI_{SS})$);
2. Then, it does some local computations mixed with some atomic actions by computation logics, including $\cdot$, $+$, $\between$, and guards, the whole local computations are denoted $I_{SS}$, which is the set of all local atomic actions. For simplicity, we assume that the interaction with the UDDI is also an internal action and included into $I_{SS}$;
3. When the local computations are finished, the SS generates the WS selection results $DO_{SS}$ and sends to the WSO's mail box denoted by the WSO's name $WSO$ (the corresponding sending action is denoted $s_{WSO}(DO_{SS})$), and then processes the next message from the WSO recursively.

The above process is described as the following state transition skeletons by APTC:
$SS = r_{SS}(DI_{SS}) \cdot SS_1$
$SS_1 = I_{SS} \cdot SS_2$

$SS_2 = s_{WSO}(DO_{SS}) \cdot SS$

By use of the algebraic laws of APTC, the AA may be proven to exhibit the desired external behaviors. If it can exhibit the desired external behaviors, the AA should have the following form:

$$\tau_{I_{SS}}(\partial_\emptyset(SS)) = r_{SS}(DI_{SS}) \cdot s_{WSO}(DO_{SS}) \cdot \tau_{I_{SS}}(\partial_\emptyset(SS))$$

### 23.2.6 Putting all actors together into a whole

We put all actors together into a whole, including all WSOIM, SS, AAs, WSOs, and WSs, according to the architecture as illustrated in Fig. 23.2. The whole actor system $WSs \quad WSOIM \quad SS = WSs \quad WSOIM \quad SS \quad WSOs \quad AAs$ can be represented by the following process term of APTC:

$$\tau_I(\partial_H(WSs \between WSOIM \between SS)) = \tau_I(\partial_H(WSs \between WSOIM \between SS \between WSOs \between AAs))$$

Among all the actors, there are synchronous communications. The actor's reading and the same actor's sending actions with the same type messages may cause communications. If to the actor's sending action occurs before the same actor's reading action, an asynchronous communication will occur; otherwise, a deadlock $\delta$ will be caused.

There are four pairs kinds of asynchronous communications as follows:

(1) The communications between an AA and its WSO with the following constraints:

$s_{AA}(DI_{AA-WSO}) \leq r_{AA}(DI_{AA-WSO})$

$s_{WSO}(DI_{WSO-AA}) \leq r_{WSO}(DI_{WSO-AA})$

Note that the messages $DI_{AA-WSO}$ and $DO_{WSO-AA}$, $DI_{WSO-AA}$ and $DO_{AA-WSO}$ are the same messages.

(2) The communications between a WSO and its interface WS with the following constraint:

$s_{WS}(DI_{WS-WSO}) \leq r_{WS}(DI_{WS-WSO})$

Note that $DI_{WS-WSO}$ and $DO_{WSO-WS}$ are the same messages.

(3) The communications between the interface WS and the WSOIM with the following constraint:

$s_{WSOIM}(DI_{WSOIM-WS}) \leq r_{WSOIM}(DI_{WSOIM-WS})$

Note that the messages $DI_{WSOIM-WS}$ and $DO_{WS-WSOIM}$ are the same messages.

(4) The communications between the WSO and the WSOIM with the following constraint:

$s_{WSO}(DI_{WSO-WSOIM}) \leq r_{WSO}(DI_{WSO-WSOIM})$

Note that the messages $DI_{WSO-WSOIM}$ and $DO_{WSOIM-WSO}$ are the same messages.

(5) The communications between a WS and a WSO with the following constraints:

$s_{WS}(DI_{WS-WSO}) \leq r_{WS}(DI_{WS-WSO})$

$s_{WSO}(DI_{WSO-WS}) \leq r_{WSO}(DI_{WSO-WS})$

Note that the messages $DI_{WS-WSO}$ and $DO_{WSO-WS}$, $DI_{WSO-WS}$ and $DO_{WS-WSO}$ are the same messages.

(6) The communications between a WSO and SS with the following constraints:

$s_{SS}(DI_{SS-WSO}) \leq r_{SS}(DI_{SS-WSO})$

$s_{WSO}(DI_{WSO-SS}) \leq r_{WSO}(DI_{WSO-SS})$

Note that the messages $DI_{SS-WSO}$ and $DO_{WSO-SS}$, $DI_{WSO-SS}$ and $DO_{SS-WSO}$ are the same messages.

(7) The communications between a WS and its partner WS with the following constraints:

$s_{WS_1}(DI_{WS_1-WS_2}) \leq r_{WS_1}(DI_{WS_1-WS_2})$

$s_{WS_2}(DI_{WS_2-WS_1}) \leq r_{WS_2}(DI_{WS_2-WS_1})$

Note that the messages $DI_{WS_1-WS_2}$ and $DO_{WS_2-WS_1}$, $DI_{WS_2-WS_1}$, and $DO_{WS_1-WS_2}$ are the same messages.

Hence, the set $H$ and $I$ can be defined as follows:

$H = \{s_{AA}(DI_{AA-WSO}), r_{AA}(DI_{AA-WSO}), s_{WSO}(DI_{WSO-AA}), r_{WSO}(DI_{WSO-AA}),$
$s_{WS}(DI_{WS-WSO}), r_{WS}(DI_{WS-WSO}), s_{WSOIM}(DI_{WSOIM-WS}), r_{WSOIM}(DI_{WSOIM-WS}),$
$s_{WSO}(DI_{WSO-WSOIM}), r_{WSO}(DI_{WSO-WSOIM}), s_{WS}(DI_{WS-WSO}), r_{WS}(DI_{WS-WSO}),$
$s_{WSO}(DI_{WSO-WS}), r_{WSO}(DI_{WSO-WS}), s_{SS}(DI_{SS-WSO}), r_{SS}(DI_{SS-WSO}),$
$s_{WSO}(DI_{WSO-SS}), r_{WSO}(DI_{WSO-SS}), s_{WS_1}(DI_{WS_1-WS_2}), r_{WS_1}(DI_{WS_1-WS_2}),$
$s_{WS_2}(DI_{WS_2-WS_1}), r_{WS_2}(DI_{WS_2-WS_1})$
$|s_{AA}(DI_{AA-WSO}) \nleq r_{AA}(DI_{AA-WSO}), s_{WSO}(DI_{WSO-AA}) \nleq r_{WSO}(DI_{WSO-AA}),$
$s_{WS}(DI_{WS-WSO}) \nleq r_{WS}(DI_{WS-WSO}), s_{WSOIM}(DI_{WSOIM-WS}) \nleq r_{WSOIM}(DI_{WSOIM-WS}),$
$s_{WSO}(DI_{WSO-WSOIM}) \nleq r_{WSO}(DI_{WSO-WSOIM}), s_{WS}(DI_{WS-WSO}) \nleq r_{WS}(DI_{WS-WSO}),$
$s_{WSO}(DI_{WSO-WS}) \nleq r_{WSO}(DI_{WSO-WS}), s_{SS}(DI_{SS-WSO}) \nleq r_{SS}(DI_{SS-WSO}),$

$s_{WSO}(DI_{WSO-SS}) \nleq r_{WSO}(DI_{WSO-SS}), s_{WS_1}(DI_{WS_1-WS_2}) \nleq r_{WS_1}(DI_{WS_1-WS_2}),$
$s_{WS_2}(DI_{WS_2-WS_1}) \nleq r_{WS_2}(DI_{WS_2-WS_1})\}$

$I = \{s_{AA}(DI_{AA-WSO}), r_{AA}(DI_{AA-WSO}), s_{WSO}(DI_{WSO-AA}), r_{WSO}(DI_{WSO-AA}),$
$s_{WS}(DI_{WS-WSO}), r_{WS}(DI_{WS-WSO}), s_{WSOIM}(DI_{WSOIM-WS}), r_{WSOIM}(DI_{WSOIM-WS}),$
$s_{WSO}(DI_{WSO-WSOIM}), r_{WSO}(DI_{WSO-WSOIM}), s_{WS}(DI_{WS-WSO}), r_{WS}(DI_{WS-WSO}),$
$s_{WSO}(DI_{WSO-WS}), r_{WSO}(DI_{WSO-WS}), s_{SS}(DI_{SS-WSO}), r_{SS}(DI_{SS-WSO}),$
$s_{WSO}(DI_{WSO-SS}), r_{WSO}(DI_{WSO-SS}), s_{WS_1}(DI_{WS_1-WS_2}), r_{WS_1}(DI_{WS_1-WS_2}),$
$s_{WS_2}(DI_{WS_2-WS_1}), r_{WS_2}(DI_{WS_2-WS_1})$
$|s_{AA}(DI_{AA-WSO}) \leq r_{AA}(DI_{AA-WSO}), s_{WSO}(DI_{WSO-AA}) \leq r_{WSO}(DI_{WSO-AA}),$
$s_{WS}(DI_{WS-WSO}) \leq r_{WS}(DI_{WS-WSO}), s_{WSOIM}(DI_{WSOIM-WS}) \leq r_{WSOIM}(DI_{WSOIM-WS}),$
$s_{WSO}(DI_{WSO-WSOIM}) \leq r_{WSO}(DI_{WSO-WSOIM}), s_{WS}(DI_{WS-WSO}) \leq r_{WS}(DI_{WS-WSO}),$
$s_{WSO}(DI_{WSO-WS}) \leq r_{WSO}(DI_{WSO-WS}), s_{SS}(DI_{SS-WSO}) \leq r_{SS}(DI_{SS-WSO}),$
$s_{WSO}(DI_{WSO-SS}) \leq r_{WSO}(DI_{WSO-SS}), s_{WS_1}(DI_{WS_1-WS_2}) \leq r_{WS_1}(DI_{WS_1-WS_2}),$
$s_{WS_2}(DI_{WS_2-WS_1}) \leq r_{WS_2}(DI_{WS_2-WS_1})\} \cup I_{AAs} \cup I_{WSO} \cup I_{WSs} \cup I_{SS} \cup I_{WSOIM}$

If the whole actor system of QoS-aware WS orchestration engine can exhibit the desired external behaviors, the system should have the following form:

$\tau_I(\partial_H(WSs \between WSOIM \between SS)) = \tau_I(\partial_H(WSs \between WSOIM \between SS \between WSOs \between AAs))$
$= r_{WS}(DI_{WS}) \cdot s_O(DO_{WS}) \cdot \tau_I(\partial_H(WSs \between WSOIM \between SS \between WSOs \between AAs))$
$= r_{WS}(DI_{WS}) \cdot s_O(DO_{WS}) \cdot \tau_I(\partial_H(WSs \between WSOIM \between SS))$

## 23.3 An example

Using the architecture in Fig. 23.2, we obtain an implementation of the buying books example as shown in Fig. 23.1. In this implementation, there are four WSs (BuyerAgentWS denoted $WS_1$, BookStoreWS denoted $WS_2$, RailwayWS denoted $WS_3$, and AirlineWS denoted $WS_4$), the focused BookStore WSO denoted $WSO$, and the focused set of AAs (ReceiveRBAA denoted $AA_1$, SendLBAA denoted $AA_2$, ReceiveSBAA denoted $AA_3$, CalculatePAA denoted $AA_4$, SendPAA denoted $AA_5$, GetPaysAA denoted $AA_6$, ShipByTAA denoted $AA_7$, and ShipByAAA denoted $AA_8$), one WSOIM denoted $WSOIM$, one service selector denoted $SS$.

The detailed implementations of actors in Fig. 23.1 is the following:

### 23.3.1 BookStore AAs

(1) ReceiveRBAA ($AA_1$)

After $AA_1$ is created, the typical process is as follows:

1. The $AA_1$ receives some messages $ReceiveRB_{WA}$ from $WSO$ through its mail box denoted by its name $AA_1$ (the corresponding reading action is denoted $r_{AA_1}(ReceiveRB_{WA})$);
2. Then, it does some local computations mixed with some atomic actions by computation logics, including $\cdot$, $+$, $\between$, and guards, the whole local computations are denoted $I_{AA_1}$, which is the set of all local atomic actions;
3. When the local computations are finished, the $AA_1$ generates the output message $ReceiveRB_{AW}$ and sends to $WSO$'s mail box denoted by $WSO$'s name $WSO$ (the corresponding sending action is denoted $s_{WSO}(ReceiveRB_{AW})$), and then processes the next message from $WSO$ recursively.

The above process is described as the following state transitions by APTC:

$AA_1 = r_{AA_1}(ReceiveRB_{WA}) \cdot AA_{1_1}$
$AA_{1_1} = I_{AA_1} \cdot AA_{1_2}$
$AA_{1_2} = s_{WSO}(ReceiveRB_{AW}) \cdot AA_1$

By use of the algebraic laws of APTC, $AA_1$ can be proven to exhibit the desired external behaviors:

$$\tau_{I_{AA_1}}(\partial_\emptyset(AA_1)) = r_{AA_1}(RequestLB_{WA}) \cdot s_{WSO}(RequestLB_{AW}) \cdot \tau_{I_{AA_1}}(\partial_\emptyset(AA_1))$$

(2) SendLBAA ($AA_2$)

After $AA_2$ is created, the typical process is as follows:

1. The $AA_2$ receives some messages $SendLB_{WA}$ from $WSO$ through its mail box denoted by its name $AA_2$ (the corresponding reading action is denoted $r_{AA_2}(ReceiveLB_{WA})$);

2. Then, it does some local computations mixed with some atomic actions by computation logics, including $\cdot$, $+$, $\between$, and guards, the whole local computations are denoted $I_{AA_2}$, which is the set of all local atomic actions;
3. When the local computations are finished, the $AA_2$ generates the output message $SendLB_{AW}$ and sends to $WSO$'s mail box denoted by $WSO$'s name $WSO$ (the corresponding sending action is denoted $s_{WSO}(SendLB_{AW})$), and then processes the next message from $WSO$ recursively.

The above process is described as the following state transitions by APTC:
$AA_2 = r_{AA_2}(SendLB_{WA}) \cdot AA_{2_1}$
$AA_{2_1} = I_{AA_2} \cdot AA_{2_2}$
$AA_{2_2} = s_{WSO}(SendLB_{AW}) \cdot AA_2$
By use of the algebraic laws of APTC, $AA_2$ can be proven to exhibit the desired external behaviors:

$$\tau_{I_{AA_2}}(\partial_{\emptyset}(AA_2)) = r_{AA_2}(SendLB_{WA}) \cdot s_{WSO}(SendLB_{AW}) \cdot \tau_{I_{AA_2}}(\partial_{\emptyset}(AA_2))$$

(3) ReceiveSBAA ($AA_3$)
After $AA_3$ is created, the typical process is as follows:

1. The $AA_3$ receives some messages $ReceiveSB_{WA_2}$ from $WSO$ through its mail box denoted by its name $AA_3$ (the corresponding reading action is denoted $r_{AA_3}(ReceiveSB_{WA})$);
2. Then, it does some local computations mixed with some atomic actions by computation logics, including $\cdot$, $+$, $\between$, and guards, the whole local computations are denoted $I_{AA_3}$, which is the set of all local atomic actions;
3. When the local computations are finished, the $AA_3$ generates the output message $ReceiveSB_{AW}$ and sends to $WSO$'s mail box denoted by $WSO$'s name $WSO$ (the corresponding sending action is denoted $s_{WSO}(ReceiveSB_{AW})$), and then processes the next message from $WSO$ recursively.

The above process is described as the following state transitions by APTC:
$AA_3 = r_{AA_3}(ReceiveSB_{WA}) \cdot AA_{3_1}$
$AA_{3_1} = I_{AA_3} \cdot AA_{3_2}$
$AA_{3_2} = s_{WSO}(ReceiveSB_{AW}) \cdot AA_3$
By use of the algebraic laws of APTC, $AA_3$ can be proven to exhibit the desired external behaviors:

$$\tau_{I_{AA_3}}(\partial_{\emptyset}(AA_3)) = r_{AA_3}(ReceiveSB_{WA}) \cdot s_{WSO}(ReceiveSB_{AW}) \cdot \tau_{I_{AA_3}}(\partial_{\emptyset}(AA_3))$$

(4) CalculatePAA ($AA_4$)
After $AA_4$ is created, the typical process is as follows:

1. The $AA_4$ receives some messages $CalculateP_{WA}$ from $WSO$ through its mail box denoted by its name $AA_4$ (the corresponding reading action is denoted $r_{AA_4}(CalculateP_{WA})$);
2. Then, it does some local computations mixed with some atomic actions by computation logics, including $\cdot$, $+$, $\between$, and guards, the whole local computations are denoted $I_{AA_4}$, which is the set of all local atomic actions;
3. When the local computations are finished, the $AA_4$ generates the output message $CalculateP_{AW}$ and sends to $WSO$'s mail box denoted by $WSO$'s name $WSO$ (the corresponding sending action is denoted $s_{WSO}(CalculateP_{AW})$), and then processes the next message from $WSO$ recursively.

The above process is described as the following state transitions by APTC:
$AA_4 = r_{AA_4}(CalculateP_{WA}) \cdot AA_{4_1}$
$AA_{4_1} = I_{AA_4} \cdot AA_{4_2}$
$AA_{4_2} = s_{WSO}(CalculateP_{AW}) \cdot AA_4$
By use of the algebraic laws of APTC, $AA_4$ can be proven to exhibit the desired external behaviors:

$$\tau_{I_{AA_4}}(\partial_{\emptyset}(AA_4)) = r_{AA_4}(CalculateP_{WA}) \cdot s_{WSO}(CalculateP_{AW}) \cdot \tau_{I_{AA_4}}(\partial_{\emptyset}(AA_4))$$

(5) SendPAA ($AA_5$)
After $AA_5$ is created, the typical process is as follows:

1. The $AA_5$ receives some messages $SendP_{WA}$ from $WSO$ through its mail box denoted by its name $AA_5$ (the corresponding reading action is denoted $r_{AA_5}(SendP_{WA})$);
2. Then, it does some local computations mixed with some atomic actions by computation logics, including $\cdot$, $+$, $\between$, and guards, the whole local computations are denoted $I_{AA_5}$, which is the set of all local atomic actions;

3. When the local computations are finished, the $AA_5$ generates the output message $SendP_{AW}$ and sends to $WSO$'s mail box denoted by $WSO$'s name $WSO$ (the corresponding sending action is denoted $s_{WSO}(SendP_{AW})$), and then processes the next message from $WSO$ recursively.

   The above process is described as the following state transitions by APTC:
   $AA_5 = r_{AA_5}(SendP_{WA}) \cdot AA_{5_1}$
   $AA_{5_1} = I_{AA_5} \cdot AA_{5_2}$
   $AA_{5_2} = s_{WSO}(SendP_{AW}) \cdot AA_5$
   By use of the algebraic laws of APTC, $AA_5$ can be proven to exhibit the desired external behaviors:

$$\tau_{I_{AA_5}}(\partial_\emptyset(AA_5)) = r_{AA_5}(SendP_{WA}) \cdot s_{WSO}(SendP_{AW}) \cdot \tau_{I_{AA_5}}(\partial_\emptyset(AA_5))$$

(6) ShipByTAA ($AA_6$)
After $AA_6$ is created, the typical process is as follows:

1. The $AA_6$ receives some messages $ShipByT_{WA}$ from $WSO$ through its mail box denoted by its name $AA_6$ (the corresponding reading action is denoted $r_{AA_6}(ShipByT_{WA})$);
2. Then, it does some local computations mixed with some atomic actions by computation logics, including $\cdot$, $+$, $\between$, and guards, the whole local computations are denoted $I_{AA_6}$, which is the set of all local atomic actions;
3. When the local computations are finished, the $AA_6$ generates the output message $ShipByT_{AW}$ and sends to $WSO$'s mail box denoted by $WSO$'s name $WSO$ (the corresponding sending action is denoted $s_{WSO}(ShipByT_{AW})$), and then processes the next message from $WSO$ recursively.

   The above process is described as the following state transitions by APTC:
   $AA_6 = r_{AA_6}(ShipByT_{WA}) \cdot AA_{6_1}$
   $AA_{6_1} = I_{AA_6} \cdot AA_{6_2}$
   $AA_{6_2} = s_{WSO}(ShipByT_{AW}) \cdot AA_6$
   By use of the algebraic laws of APTC, $AA_6$ can be proven to exhibit the desired external behaviors:

$$\tau_{I_{AA_6}}(\partial_\emptyset(AA_6)) = r_{AA_6}(ShipByT_{WA}) \cdot s_{WSO}(ShipByT_{AW}) \cdot \tau_{I_{AA_6}}(\partial_\emptyset(AA_6))$$

(7) ShipByAAA ($AA_7$)
After $AA_7$ is created, the typical process is as follows:

1. The $AA_7$ receives some messages $ShipByA_{WA}$ from $WSO$ through its mail box denoted by its name $AA_7$ (the corresponding reading action is denoted $r_{AA_7}(ShipByA_{WA})$);
2. Then, it does some local computations mixed with some atomic actions by computation logics, including $\cdot$, $+$, $\between$, and guards, the whole local computations are denoted $I_{AA_7}$, which is the set of all local atomic actions;
3. When the local computations are finished, the $AA_7$ generates the output message $ShipByA_{AW}$ and sends to $WSO$'s mail box denoted by $WSO$'s name $WSO$ (the corresponding sending action is denoted $s_{WSO}(ShipByA_{AW})$), and then processes the next message from $WSO$ recursively.

   The above process is described as the following state transitions by APTC:
   $AA_7 = r_{AA_7}(ShipByA_{WA}) \cdot AA_{7_1}$
   $AA_{7_1} = I_{AA_7} \cdot AA_{7_2}$
   $AA_{7_2} = s_{WSO}(ShipByA_{AW}) \cdot AA_7$
   By use of the algebraic laws of APTC, $AA_7$ can be proven to exhibit the desired external behaviors:

$$\tau_{I_{AA_7}}(\partial_\emptyset(AA_7)) = r_{AA_7}(ShipByA_{WA}) \cdot s_{WSO}(ShipByA_{AW}) \cdot \tau_{I_{AA_7}}(\partial_\emptyset(AA_7))$$

### 23.3.2 WSOIM

After $WSOIM$ is created, the typical process is as follows:

1. The $WSOIM$ receives some messages $DI_{WSOIM}$ from $WS_2$ through its mail box denoted by its name $WSOIM$ (the corresponding reading action is denoted $r_{WSOIM}(DI_{WSOIM})$);
2. The $WSOIM$ may create a $WSO$ through action **new**($WSO$) if it is not initialized;
3. Then, it does some local computations mixed with some atomic actions by computation logics, including $\cdot$, $+$, $\between$, and guards, the whole local computations are denoted $I_{WSOIM}$, which is the set of all local atomic actions;

4. When the local computations are finished, the $WSOIM$ generates the output message $DO_{WSOIM}$ and sends to $WSO$'s mail box denoted by $WSO$'s name $WSO$ (the corresponding sending action is denoted $s_{WSO}(DO_{WSOIM})$), and then processes the next message from $WS_2$ recursively.

   The above process is described as the following state transitions by APTC:
   $WSOIM = r_{WSOIM}(DI_{WSOIM}) \cdot WSOIM_1$
   $WSOIM_1 = (\{isInitialed(WSO) = FLALSE\} \cdot \mathbf{new}(WSO) + \{isInitialed(WSO) = TRUE\}) \cdot WSOIM_2$
   $WSOIM_2 = I_{WSOIM} \cdot WSOIM_3$
   $WSOIM_3 = s_{WSO}(DO_{WSOIM}) \cdot WSOIM$
   By use of the algebraic laws of APTC, $WSOIM$ can be proven to exhibit the desired external behaviors:

$$\tau_{I_{WSOIM}}(\partial_{\emptyset}(WSOIM)) = r_{WSOIM}(DI_{WSOIM}) \cdot s_{WSO}(DO_{WSOIM}) \cdot \tau_{I_{WSOIM}}(\partial_{\emptyset}(WSOIM))$$

### 23.3.3 BookStore WSO

After BookStore WSO ($WSO$) is created, the typical process is as follows:

1. The $WSO$ receives the requests $ReceiveRB_{MW}$ from $WSOIM$ through its mail box by its name $WSO$ (the corresponding reading action is denoted $r_{WSO}(ReceiveRB_{MW})$);
2. The $WSO$ may create its AAs in parallel through actions $\mathbf{new}(AA_1) \parallel \cdots \parallel \mathbf{new}(AA_7)$ if it is not initialized;
3. The $WSO$ does some local computations mixed with some atomic actions by computation logics, including $\cdot$, $+$, $\between$, and guards, the local computations are included into $I_{WSO}$, which is the set of all local atomic actions;
4. When the local computations are finished, the $WSO$ generates the output messages $ReceiveRB_{WA}$ and sends to $AA_1$ (the corresponding sending action is denoted $s_{AA_1}(ReceiveRB_{WA})$);
5. The $WSO$ receives the response message $ReceiveRB_{AW}$ from $AA_1$ through its mail box by its name $WSO$ (the corresponding reading action is denoted $r_{WSO}(ReceiveRB_{AW})$);
6. The $WSO$ does some local computations mixed with some atomic actions by computation logics, including $\cdot$, $+$, $\between$, and guards, the local computations are included into $I_{WSO}$, which is the set of all local atomic actions;
7. When the local computations are finished, the $WSO$ generates the output messages $SendLB_{WA}$ and sends to $AA_2$ (the corresponding sending action is denoted $s_{AA_2}(SendLB_{WA})$);
8. The $WSO$ receives the response message $SendLB_{AW}$ from $AA_2$ through its mail box by its name $WSO$ (the corresponding reading action is denoted $r_{WSO}(SendLB_{AW})$);
9. The $WSO$ does some local computations mixed with some atomic actions by computation logics, including $\cdot$, $+$, $\between$, and guards, the local computations are included into $I_{WSO}$, which is the set of all local atomic actions;
10. When the local computations are finished, the $WSO$ generates the output messages $SendLB_{WW_1}$ and sends to $WS_1$ (the corresponding sending action is denoted $s_{WS_1}(SendLB_{WW_1})$);
11. The $WSO$ receives the response message $ReceiveSB_{WW_1}$ from $WS_1$ through its mail box by its name $WSO$ (the corresponding reading action is denoted $r_{WSO}(ReceiveSB_{WW_1})$);
12. The $WSO$ does some local computations mixed with some atomic actions by computation logics, including $\cdot$, $+$, $\between$, and guards, the local computations are included into $I_{WSO}$, which is the set of all local atomic actions;
13. When the local computations are finished, the $WSO$ generates the output messages $ReceiveSB_{WA}$ and sends to $AA_3$ (the corresponding sending action is denoted $s_{AA_3}(ReceiveSB_{WA})$);
14. The $WSO$ receives the response message $ReceiveSB_{AW}$ from $AA_3$ through its mail box by its name $WSO$ (the corresponding reading action is denoted $r_{WSO}(ReceiveSB_{AW})$);
15. The $WSO$ does some local computations mixed with some atomic actions by computation logics, including $\cdot$, $+$, $\between$, and guards, the local computations are included into $I_{WSO}$, which is the set of all local atomic actions;
16. When the local computations are finished, the $WSO$ generates the output messages $CalculateP_{WA}$ and sends to $AA_4$ (the corresponding sending action is denoted $s_{AA_4}(CalculateP_{WA})$);
17. The $WSO$ receives the response message $CalculateP_{AW}$ from $AA_4$ through its mail box by its name $WSO$ (the corresponding reading action is denoted $r_{WSO}(CalculateP_{AW})$);
18. The $WSO$ does some local computations mixed with some atomic actions by computation logics, including $\cdot$, $+$, $\between$, and guards, the local computations are included into $I_{WSO}$, which is the set of all local atomic actions;
19. When the local computations are finished, the $WSO$ generates the output messages $SendP_{WA}$ and sends to $AA_5$ (the corresponding sending action is denoted $s_{AA_5}(SendP_{WA})$);
20. The $WSO$ receives the response message $sendP_{AW}$ from $AA_5$ through its mail box by its name $WSO$ (the corresponding reading action is denoted $r_{WSO}(sendP_{AW})$);

21. The $WSO$ does some local computations mixed with some atomic actions by computation logics, including $\cdot$, $+$, $\between$, and guards, the local computations are included into $I_{WSO}$, which is the set of all local atomic actions;
22. When the local computations are finished, the $WSO$ generates the output messages $SendP_{WW_1}$ and sends to $WS_1$ (the corresponding sending action is denoted $s_{WS_1}(SendP_{WW_1})$);
23. The $WSO$ receives the response message $GetPays_{WW_1}$ from $WS_1$ through its mail box by its name $WSO$ (the corresponding reading action is denoted $r_{WSO}(GetPays_{WW_1})$);
24. The $WSO$ does some local computations mixed with some atomic actions by computation logics, including $\cdot$, $+$, $\between$, and guards, the local computations are included into $I_{WSO}$, which is the set of all local atomic actions;
25. When the local computations are finished, the $WSO$ generates the output messages $GetPays_{WA}$ and sends to $AA_6$ (the corresponding sending action is denoted $s_{AA_6}(GetPays_{WA})$);
26. The $WSO$ receives the response message $GetPays_{AW}$ from $AA_6$ through its mail box by its name $WSO$ (the corresponding reading action is denoted $r_{WSO}(GetPays_{AW})$);
27. The $WSO$ does some local computations mixed with some atomic actions by computation logics, including $\cdot$, $+$, $\between$, and guards, the local computations are included into $I_{WSO}$, which is the set of all local atomic actions;
28. When the local computations are finished, the $WSO$ generates the WS selection request messages $DI_{SS}$ and sends to $SS$ (the corresponding sending action is denoted $s_{SS}(DI_{SS})$);
29. The $WSO$ receives the response message $DO_{SS}$ from $SS$ through its mail box by its name $WSO$ (the corresponding reading action is denoted $r_{WSO}(DO_{SS})$);
30. The $WSO$ selects $WS_3$ and $WS_4$, does some local computations mixed with some atomic actions by computation logics, including $\cdot$, $+$, $\between$, and guards, the local computations are included into $I_{WSO}$, which is the set of all local atomic actions;
31. When the local computations are finished, if $Pays <= 100\$$, the $WSO$ generates the output messages $ShipByT_{WW_3}$ and sends to $WS_3$ (the corresponding sending action is denoted $s_{WS_3}(ShipByT_{WW_3})$); if $Pays > 100\$$, the $WSO$ generates the output message $ShipByA_{WW_4}$ and sends to $WS_4$ (the corresponding sending action is denoted $s_{WS_4}(ShipByA_{WW_4})$);
32. The $WSO$ receives the response message $ShipFinish_{WW_3}$ from $WS_3$ through its mail box by its name $WSO$ (the corresponding reading action is denoted $r_{WSO}(ShipFinish_{WW_3})$), or the response message $ShipFinish_{WW_4}$ from $WS_4$ through its mail box by its name $WSO$ (the corresponding reading action is denoted $r_{WSO}(ShipFinish_{WW_4})$);
33. The $WSO$ does some local computations mixed with some atomic actions by computation logics, including $\cdot$, $+$, $\between$, and guards, the local computations are included into $I_{WSO}$, which is the set of all local atomic actions;
34. When the local computations are finished, the $WSO$ generates the output messages $BBFinish_{WW_2}$ and sends to $WS_2$ (the corresponding sending action is denoted $s_{WS_2}(BBFinish_{WW_2})$), and then processes the messages from $WS_2$ recursively.

The above process is described as the following state transitions by APTC:

$WSO = r_{WSO}(ReceiveRB_{MW}) \cdot WSO_1$

$WSO_1 = (\{isInitialed(WSO) = FLALSE\} \cdot (\mathbf{new}(AA_1) \parallel \cdots \parallel \mathbf{new}(AA_7)) + \{isInitialed(WSO) = TRUE\}) \cdot WSO_2$

$WSO_2 = I_{WSO} \cdot WSO_3$

$WSO_3 = s_{AA_1}(ReceiveRB_{WA}) \cdot WSO_4$

$WSO_4 = r_{WSO}(ReceiveRB_{AW}) \cdot WSO_5$

$WSO_5 = I_{WSO} \cdot WSO_6$

$WSO_6 = s_{AA_2}(SendLB_{WA}) \cdot WSO_7$

$WSO_7 = r_{WSO}(SendLB_{AW}) \cdot WSO_8$

$WSO_8 = I_{WSO} \cdot WSO_9$

$WSO_9 = s_{WS_1}(SendLB_{WW_1}) \cdot WSO_{10}$

$WSO_{10} = r_{WSO}(ReceiveSB_{WW_1}) \cdot WSO_{11}$

$WSO_{11} = I_{WSO} \cdot WSO_{12}$

$WSO_{12} = s_{AA_3}(ReceiveSB_{WA}) \cdot WSO_{13}$

$WSO_{13} = r_{WSO}(ReceiveSB_{AW}) \cdot WSO_{14}$

$WSO_{14} = I_{WSO} \cdot WSO_{15}$

$WSO_{15} = s_{AA_4}(CalculteP_{WA}) \cdot WSO_{16}$

$WSO_{16} = r_{WSO}(CalculateP_{AW}) \cdot WSO_{17}$

$WSO_{17} = I_{WSO} \cdot WSO_{18}$

$WSO_{18} = s_{AA_5}(SendP_{WA}) \cdot WSO_{19}$

$WSO_{19} = r_{WSO}(SendP_{AW}) \cdot WSO_{20}$
$WSO_{20} = I_{WSO} \cdot WSO_{21}$
$WSO_{21} = s_{WS_1}(SendP_{WW_1}) \cdot WSO_{22}$
$WSO_{22} = r_{WSO}(GetPays_{WW_1}) \cdot WSO_{23}$
$WSO_{23} = I_{WSO} \cdot WSO_{24}$
$WSO_{24} = s_{AA_6}(GetPays_{WA}) \cdot WSO_{25}$
$WSO_{25} = r_{WSO}(GetPays_{AW}) \cdot WSO_{26}$
$WSO_{26} = I_{WSO} \cdot WSO_{27}$
$WSO_{27} = s_{SS}(DI_{SS}) \cdot WSO_{28}$
$WSO_{28} = r_{WSO}(DO_{SS}) \cdot WSO_{29}$
$WSO_{29} = I_{WSO} \cdot WSO_{30}$
$WSO_{30} = (\{Pays <= 100\$\} \cdot s_{WS_3}(ShipByT_{WW_3}) \cdot r_{WSO}(ShipFinish_{WW_3}) + \{Pays > 100\$\} \cdot s_{WS_4}(ShipByA_{WW_4}) \cdot r_{WSO}(ShipFinish_{WW_4})) \cdot WSO_{31}$
$WSO_{31} = I_{WSO} \cdot WSO_{32}$
$WSO_{32} = s_{WS_2}(BBFinish_{WW_2}) \cdot WSO$

By use of the algebraic laws of APTC, the $WSO_2$ can be proven to exhibit the desired external behaviors:

$\tau_{I_{WSO}}(\partial_{\emptyset}(WSO)) = r_{WSO}(ReceiveRB_{MW}) \cdot s_{AA_1}(ReceiveRB_{WA}) \cdot r_{WSO}(ReceiveRB_{AW})$
$s_{AA_2}(SendLB_{WA}) \cdot r_{WSO}(SendLB_{AW}) \cdot s_{WS_1}(SendLB_{WW_1}) \cdot r_{WSO}(ReceiveSB_{WW_1}) \cdot$
$s_{AA_3}(ReceiveSB_{WA}) \cdot r_{WSO}(ReceiveSB_{AW}) \cdot s_{AA_4}(CalcultePA_{WA}) \cdot r_{WSO}(CalculateP_{AW}) \cdot$
$s_{AA_5}(SendP_{WA}) \cdot r_{WSO}(SendP_{AW}) \cdot s_{WS_1}(SendP_{WW_1}) \cdot r_{WSO}(GetPays_{WW_1}) \cdot$
$s_{AA_6}(GetPays_{WA}) \cdot r_{WSO}(GetPays_{AW}) \cdot s_{SS}(DI_{SS}) \cdot r_{WSO}(DO_{SS}) \cdot$
$(s_{WS_3}(ShipByT_{WW_3}) \cdot r_{WSO}(ShipFinish_{WW_3}) + s_{WS_4}(ShipByA_{WW_4}) \cdot r_{WSO}(ShipFinish_{WW_4})) \cdot$
$s_{WS_2}(BBFinish_{WW_2}) \cdot \tau_{I_{WSO}}(\partial_{\emptyset}(WSO))$

with $I_{WSO}$ extended to $I_{WSO} \cup \{\{isInitialed(WSO) = FLALSE\}, \{isInitialed(WSO) = TRUE\}, \{Pays <= 100\$\}, \{Pays > 100\$\}\}$.

### 23.3.4 BuyerAgent WS

After BuyerAgent WS ($WS_1$) is created, the typical process is as follows:

1. The $WS_1$ receives the message $SendLB_{WW_1}$ from the $WSO$ through its mail box by its name $WS_1$ (the corresponding reading action is denoted $r_{WS_1}(SendLB_{WW_1})$);
2. The $WS_1$ does some local computations mixed with some atomic actions by computation logics, including $\cdot$, $+$, $\between$, and guards, the local computations are included into $I_{WS_1}$, which is the set of all local atomic actions;
3. When the local computations are finished, the $WS_1$ generates the output messages $ReceiveSB_{WW_1}$ and sends to the $WSO$ (the corresponding sending action is denoted $s_{WSO}(ReceiveSB_{WW_1})$);
4. The $WS_1$ receives the response message $SendP_{WW_1}$ from $WSO$ through its mail box by its name $WS_1$ (the corresponding reading action is denoted $r_{WS_1}(SendP_{WW_1})$);
5. The $WS_1$ does some local computations mixed with some atomic actions by computation logics, including $\cdot$, $+$, $\between$, and guards, the local computations are included into $I_{WS_1}$, which is the set of all local atomic actions;
6. When the local computations are finished, the $WS_1$ generates the output messages $GetPays_{WW_1}$ and sends to $WSO$ (the corresponding sending action is denoted $s_{WSO}(GetPays_{WW_1})$), and then processes the messages from $WSO$ recursively.

The above process is described as the following state transitions by APTC:

$WS_1 = r_{WS_1}(SendLB_{WW_1}) \cdot WS_{1_1}$
$WS_{1_1} = I_{WS_1} \cdot WS_{1_2}$
$WS_{1_2} = s_{WSO}(ReceiveSB_{WW_1}) \cdot WS_{1_3}$
$WS_{1_3} = r_{WS_1}(SendP_{WW_1}) \cdot WS_{1_4}$
$WS_{1_4} = I_{WS_1} \cdot WS_{1_5}$
$WS_{1_5} = s_{WSO}(GetPays_{WW_1}) \cdot WS_1$

By use of the algebraic laws of APTC, the $WS_1$ can be proven to exhibit the desired external behaviors:

$\tau_{I_{WS_1}}(\partial_{\emptyset}(WS_1)) = r_{WS_1}(SendLB_{WW_1}) \cdot s_{WSO}(ReceiveSB_{WW_1}) \cdot r_{WS_1}(SendP_{WW_1}) \cdot s_{WSO}(GetPays_{WW_1}) \cdot \tau_{I_{WS_1}}(\partial_{\emptyset}(WS_1))$

### 23.3.5 BookStore WS

After BookStore WS ($WS_2$) is created, the typical process is as follows:

1. The $WS_2$ receives the request message $RequestLB_{WS_2}$ from the outside through its mail box by its name $WS_2$ (the corresponding reading action is denoted $r_{WS_2}(RequestLB_{WS_2})$);
2. The $WS_2$ does some local computations mixed with some atomic actions by computation logics, including $\cdot$, $+$, $\between$, and guards, the local computations are included into $I_{WS_2}$, which is the set of all local atomic actions;
3. When the local computations are finished, the $WS_2$ generates the output messages $ReceiveRB_{WM}$ and sends to $WSOIM$ (the corresponding sending action is denoted $s_{WSOIM}(ReceiveRB_{WM})$);
4. The $WS_2$ receives the response message $BBFinish_{WW_2}$ from $WSO$ through its mail box by its name $WS_2$ (the corresponding reading action is denoted $r_{WS_2}(BBFinish_{WW_2})$);
5. The $WS_2$ does some local computations mixed with some atomic actions by computation logics, including $\cdot$, $+$, $\between$, and guards, the local computations are included into $I_{WS_2}$, which is the set of all local atomic actions;
6. When the local computations are finished, the $WS_2$ generates the output messages $BBFinish_O$ and sends to the outside (the corresponding sending action is denoted $s_O(BBFinish_O)$), and then processes the messages from the outside recursively.

The above process is described as the following state transitions by APTC:

$WS_2 = r_{WS_2}(RequestLB_{WS_2}) \cdot WS_{2_1}$

$WS_{2_1} = I_{WS_2} \cdot WS_{2_2}$

$WS_{2_2} = s_{WSOIM}(ReceiveRB_{WM}) \cdot WS_{2_3}$

$WS_{2_3} = r_{WS_2}(BBFinish_{WW_2}) \cdot WS_{2_4}$

$WS_{2_4} = I_{WS_2} \cdot WS_{2_5}$

$WS_{2_5} = s_O(BBFinish_O) \cdot WS_2$

By use of the algebraic laws of APTC, the $WS_2$ can be proven to exhibit the desired external behaviors:

$\tau_{I_{WS_2}}(\partial_\emptyset(WS_2)) = r_{WS_2}(RequestLB_{WS_2}) \cdot s_{WSOIM}(ReceiveRB_{WM}) \cdot r_{WS_2}(BBFinish_{WW_2}) \cdot s_O(BBFinish_O) \cdot \tau_{I_{WS_2}}(\partial_\emptyset(WS_2))$

### 23.3.6 Railway WS

After Railway WS ($WS_3$) is created, the typical process is as follows:

1. The $WS_3$ receives the message $ShipByT_{WW_3}$ from the $WSO$ through its mail box by its name $WS_3$ (the corresponding reading action is denoted $r_{WS_3}(ShipByT_{WW_3})$);
2. The $WS_3$ does some local computations mixed with some atomic actions by computation logics, including $\cdot$, $+$, $\between$, and guards, the local computations are included into $I_{WS_3}$, which is the set of all local atomic actions;
3. When the local computations are finished, the $WS_3$ generates the output messages $ShipFinish_{WW_3}$ and sends to the $WSO$ (the corresponding sending action is denoted $s_{WSO}(ShipFinish_{WW_3})$), and then processes the messages from $WSO$ recursively.

The above process is described as the following state transitions by APTC.

$WS_3 = r_{WS_3}(ShipByT_{WW_3}) \cdot WS_{3_1}$

$WS_{3_1} = I_{WS_3} \cdot WS_{3_2}$

$WS_{3_2} = s_{WSO}(ShipFinish_{WW_3}) \cdot WS_3$

By use of the algebraic laws of APTC, the $WS_3$ can be proven to exhibit the desired external behaviors:

$\tau_{I_{WS_3}}(\partial_\emptyset(WS_3)) = r_{WS_3}(ShipByT_{WW_3}) \cdot s_{WSO}(ShipFinish_{WW_3}) \cdot \tau_{I_{WS_3}}(\partial_\emptyset(WS_3))$

### 23.3.7 Airline WS

After Airline WS ($WS_4$) is created, the typical process is as follows:

1. The $WS_4$ receives the message $ShipByA_{WW_4}$ from the $WSO$ through its mail box by its name $WS_4$ (the corresponding reading action is denoted $r_{WS_4}(ShipByA_{WW_4})$);
2. The $WS_4$ does some local computations mixed with some atomic actions by computation logics, including $\cdot$, $+$, $\between$, and guards, the local computations are included into $I_{WS_4}$, which is the set of all local atomic actions;
3. When the local computations are finished, the $WS_4$ generates the output messages $ShipFinish_{WW_4}$ and sends to the $WSO$ (the corresponding sending action is denoted $s_{WSO}(ShipFinish_{WW_4})$), and then processes the messages from $WSO$ recursively.

The above process is described as the following state transitions by APTC:

$WS_4 = r_{WS_4}(ShipByA_{WW_4}) \cdot WS_{4_1}$

$WS_{4_1} = I_{WS_4} \cdot WS_{4_2}$

$WS_{4_2} = s_{WSO}(ShipFinish_{WW_4}) \cdot WS_4$

By use of the algebraic laws of APTC, the $WS_4$ can be proven to exhibit the desired external behaviors:

$\tau_{I_{WS_4}}(\partial_\emptyset(WS_4)) = r_{WS_4}(ShipByA_{WW_4}) \cdot s_{WSO}(ShipFinish_{WW_4}) \cdot \tau_{I_{WS_4}}(\partial_\emptyset(WS_4))$

### 23.3.8 Service selector

After $SS$ is created, the typical process is as follows:

1. The $SS$ receives the QoS-based WS selection request message $DI_{SS}$ from $WSO$ through its mail box by its name $SS$ (the corresponding reading action is denoted $r_{SS}(DI_{SS})$);
2. The $SS$ does some local computations mixed with some atomic actions and interactions with UDDI by computation logics, including $\cdot$, $+$, $\between$, and guards, the local computations are included into $I_{SS}$, which is the set of all local atomic actions;
3. When the local computations are finished, the $SS$ generates the output messages $DO_{SS}$ and sends to $WSO$ (the corresponding sending action is denoted $s_{WSO}(DO_{SS})$), and then processes the next message from the $WSO$s recursively.

The above process is described as the following state transitions by APTC:

$SS = r_{SS}(DI_{SS}) \cdot SS_1$

$SS_1 = I_{SS} \cdot SS_2$

$SS_2 = s_{WSO}(DO_{SS}) \cdot SS$

By use of the algebraic laws of APTC, the $SS$ can be proven to exhibit the desired external behaviors:

$\tau_{I_{SS}}(\partial_\emptyset(SS)) = r_{SS}(DI_{SS}) \cdot s_{WSO}(DO_{SS}) \cdot \tau_{I_{SS}}(\partial_\emptyset(SS))$

### 23.3.9 Putting all actors together into a whole

Now, we can put all actors together into a whole, including all AAs, WSOIM, WSO, WSs, and SS, according to the buying books example as illustrated in Fig. 23.1. The whole actor system
$WS_1 \quad WS_2 \quad WS_3 \quad WS_4 \quad WSOIM \quad SS = WS_1 \quad WS_2 \quad WS_3 \quad WS_4 \quad WSOIM \quad SS \quad WSO$
$AA_1 \quad AA_2 \quad AA_3 \quad AA_4 \quad AA_5 \quad AA_6 \quad AA_7$ can be represented by the following process term of APTC:

$\tau_I(\partial_H(WS_1 \between WS_2 \between WS_3 \between WS_4 \between WSOIM \between SS)) = \tau_I(\partial_H(WS_1 \between WS_2 \between WS_3 \between WS_4 \between WSOIM \between SS \between WSO \between AA_1 \between AA_2 \between AA_3 \between AA_4 \between AA_5 \between AA_6 \between AA_7))$

Among all the actors, there are synchronous communications. The actor's reading and the same actor's sending actions with the same type messages may cause communications. If the actor's sending action occurs before the same actor's reading action, an asynchronous communication will occur; otherwise, a deadlock $\delta$ will be caused.

There are eight kinds of asynchronous communications as follows:

(1) The communications between $WSO$ and its AAs with the following constraints:

$s_{AA_1}(ReceiveRB_{WA}) \leq r_{AA_1}(ReceiveRB_{WA})$

$s_{WSO}(ReceiveRB_{AW}) \leq r_{WSO}(ReceiveRB_{AW})$

$s_{AA_2}(SendLB_{WA}) \leq r_{AA_2}(SendLB_{WA})$

$s_{WSO}(SendLB_{AW}) \leq r_{WSO}(SendLB_{AW})$

$s_{AA_3}(ReceiveSB_{WA}) \leq r_{AA_3}(ReceiveSB_{WA})$

$s_{WSO}(ReceiveSB_{AW}) \leq r_{WSO}(ReceiveSB_{AW})$

$s_{AA_4}(CalcultePs_{WA}) \leq r_{AA_4}(CalcultePs_{WA})$

$s_{WSO}(CalculatePs_{AW}) \leq r_{WSO}(CalculatePs_{AW})$

$s_{AA_5}(SendPs_{WA}) \leq r_{AA_5}(SendPs_{WA})$

$s_{WSO}(SendPs_{AW}) \leq r_{WSO}(SendPs_{AW})$

$s_{AA_6}(GetPays_{WA}) \leq r_{AA_6}(GetPays_{WA})$

$s_{WSO}(GetPays_{AW}) \leq r_{WSO}(GetPays_{AW})$

(2) The communications between $WSO$ and $WS_1$ with the following constraints:

$s_{WS_1}(SendLB_{WW_1}) \leq r_{WS_1}(SendLB_{WW_1})$

$s_{WSO}(ReceiveSB_{WW_1}) \leq r_{WSO}(ReceiveSB_{WW_1})$

$s_{WS_1}(SendPs_{WW_1}) \leq r_{WS_1}(SendPs_{WW_1})$

$s_{WSO}(GetPays_{WW_1}) \leq r_{WSO}(GetPays_{WW_1})$

(3) The communications between $WSO$ and $WS_2$ with the following constraint:
$s_{WS_2}(BBFinish_{WW_2}) \leq r_{WS_2}(BBFinish_{WW_2})$
(4) The communications between $WSO$ and $WS_3$ with the following constraints:
$s_{WS_3}(ShipByT_{WW_3}) \leq r_{WS_3}(ShipByT_{WW_3})$
$s_{WSO}(ShipFinish_{WW_3}) \leq r_{WSO}(ShipFinish_{WW_3})$
(5) The communications between $WSO$ and $WS_4$ with the following constraints:
$s_{WS_4}(ShipByA_{WW_4}) \leq r_{WS_4}(ShipByA_{WW_4})$
$s_{WSO}(ShipFinish_{WW_4}) \leq r_{WSO}(ShipFinish_{WW_4})$
(6) The communications between $WSO$ and $WSOIM$ with the following constraint:
$s_{WSO}(ReceiveRB_{MW}) \leq r_{WSO}(ReceiveRB_{MW})$
(7) The communications between $WSO$ and $SS$ with the following constraints:
$s_{SS}(DI_{SS}) \leq r_{SS}(DI_{SS})$
$s_{WSO}(DO_{SS}) \leq r_{WSO}(DO_{SS})$
(8) The communications between $WS_2$ and $WSOIM$ with the following constraint:
$s_{WSOIM}(ReceiveRB_{WM}) \leq r_{WSOIM}(ReceiveRB_{WM})$
Hence, the set $H$ and $I$ can be defined as follows:

$H = \{s_{AA_1}(ReceiveRB_{WA}), r_{AA_1}(ReceiveRB_{WA}),$
$s_{WSO}(ReceiveRB_{AW}), r_{WSO}(ReceiveRB_{AW}),$
$s_{AA_2}(SendLB_{WA}), r_{AA_2}(SendLB_{WA}),$
$s_{WSO}(SendLB_{AW}), r_{WSO}(SendLB_{AW}),$
$s_{AA_3}(ReceiveSB_{WA}), r_{AA_3}(ReceiveSB_{WA}),$
$s_{WSO}(ReceiveSB_{AW}), r_{WSO}(ReceiveSB_{AW}),$
$s_{AA_4}(CalculteP_{WA}), r_{AA_4}(CalculteP_{WA}),$
$s_{WSO}(CalculateP_{AW}), r_{WSO}(CalculateP_{AW}),$
$s_{AA_5}(SendP_{WA}), r_{AA_5}(SendP_{WA}),$
$s_{WSO}(SendP_{AW}), r_{WSO}(SendP_{AW}),$
$s_{AA_6}(GetPays_{WA}), r_{AA_6}(GetPays_{WA}),$
$s_{WSO}(GetPays_{AW}), r_{WSO}(GetPays_{AW}),$
$s_{WS_1}(SendLB_{WW_1}), r_{WS_1}(SendLB_{WW_1}),$
$s_{WSO}(ReceiveSB_{WW_1}), r_{WSO}(ReceiveSB_{WW_1}),$
$s_{WS_1}(SendP_{WW_1}), r_{WS_1}(SendP_{WW_1}),$
$s_{WSO}(GetPays_{WW_1}), r_{WSO}(GetPays_{WW_1}),$
$s_{WS_2}(BBFinish_{WW_2}), r_{WS_2}(BBFinish_{WW_2}),$
$s_{WS_3}(ShipByT_{WW_3}), r_{WS_3}(ShipByT_{WW_3}),$
$s_{WSO}(ShipFinish_{WW_3}), r_{WSO}(ShipFinish_{WW_3}),$
$s_{WS_4}(ShipByA_{WW_4}), r_{WS_4}(ShipByA_{WW_4}),$
$s_{WSO}(ShipFinish_{WW_4}), r_{WSO}(ShipFinish_{WW_4}),$
$s_{WSO}(ReceiveRB_{MW}), r_{WSO}(ReceiveRB_{MW}),$
$s_{SS}(DI_{SS}), r_{SS}(DI_{SS}),$
$s_{WSO}(DO_{SS}), r_{WSO}(DO_{SS}),$
$s_{WSOIM}(ReceiveRB_{WM}), r_{WSOIM}(ReceiveRB_{WM})$
$|s_{AA_1}(ReceiveRB_{WA}) \nleq r_{AA_1}(ReceiveRB_{WA}),$
$s_{WSO}(ReceiveRB_{AW}) \nleq r_{WSO}(ReceiveRB_{AW}),$
$s_{AA_2}(SendLB_{WA}) \nleq r_{AA_2}(SendLB_{WA}),$
$s_{WSO}(SendLB_{AW}) \nleq r_{WSO}(SendLB_{AW}),$
$s_{AA_3}(ReceiveSB_{WA}) \nleq r_{AA_3}(ReceiveSB_{WA}),$
$s_{WSO}(ReceiveSB_{AW}) \nleq r_{WSO}(ReceiveSB_{AW}),$
$s_{AA_4}(CalculteP_{WA}) \nleq r_{AA_4}(CalculteP_{WA}),$
$s_{WSO}(CalculateP_{AW}) \nleq r_{WSO}(CalculateP_{AW}),$
$s_{AA_5}(SendP_{WA}) \nleq r_{AA_5}(SendP_{WA}),$
$s_{WSO}(SendP_{AW}) \nleq r_{WSO}(SendP_{AW}),$
$s_{AA_6}(GetPays_{WA}) \nleq r_{AA_6}(GetPays_{WA}),$
$s_{WSO}(GetPays_{AW}) \nleq r_{WSO}(GetPays_{AW}),$
$s_{WS_1}(SendLB_{WW_1}) \nleq r_{WS_1}(SendLB_{WW_1}),$

$$s_{WSO}(ReceiveSB_{WW_1}) \nleq r_{WSO}(ReceiveSB_{WW_1}),$$
$$s_{WS_1}(SendP_{WW_1}) \nleq r_{WS_1}(SendP_{WW_1}),$$
$$s_{WSO}(GetPays_{WW_1}) \nleq r_{WSO}(GetPays_{WW_1}),$$
$$s_{WS_2}(BBFinish_{WW_2}) \nleq r_{WS_2}(BBFinish_{WW_2}),$$
$$s_{WS_3}(ShipByT_{WW_3}) \nleq r_{WS_3}(ShipByT_{WW_3}),$$
$$s_{WSO}(ShipFinish_{WW_3}) \nleq r_{WSO}(ShipFinish_{WW_3}),$$
$$s_{WS_4}(ShipByA_{WW_4}) \nleq r_{WS_4}(ShipByA_{WW_4}),$$
$$s_{WSO}(ShipFinish_{WW_4}) \nleq r_{WSO}(ShipFinish_{WW_4}),$$
$$s_{WSO}(ReceiveRB_{MW}) \nleq r_{WSO}(ReceiveRB_{MW}),$$
$$s_{SS}(DI_{SS}) \nleq r_{SS}(DI_{SS}),$$
$$s_{WSO}(DO_{SS}) \nleq r_{WSO}(DO_{SS}),$$
$$s_{WSOIM}(ReceiveRB_{WM}) \nleq r_{WSOIM}(ReceiveRB_{WM})\}$$

$$I = \{s_{AA_1}(ReceiveRB_{WA}), r_{AA_1}(ReceiveRB_{WA}),$$
$$s_{WSO}(ReceiveRB_{AW}), r_{WSO}(ReceiveRB_{AW}),$$
$$s_{AA_2}(SendLB_{WA}), r_{AA_2}(SendLB_{WA}),$$
$$s_{WSO}(SendLB_{AW}), r_{WSO}(SendLB_{AW}),$$
$$s_{AA_3}(ReceiveSB_{WA}), r_{AA_3}(ReceiveSB_{WA}),$$
$$s_{WSO}(ReceiveSB_{AW}), r_{WSO}(ReceiveSB_{AW}),$$
$$s_{AA_4}(CalculteP_{WA}), r_{AA_4}(CalculteP_{WA}),$$
$$s_{WSO}(CalculateP_{AW}), r_{WSO}(CalculateP_{AW}),$$
$$s_{AA_5}(SendP_{WA}), r_{AA_5}(SendP_{WA}),$$
$$s_{WSO}(SendP_{AW}), r_{WSO}(SendP_{AW}),$$
$$s_{AA_6}(GetPays_{WA}), r_{AA_6}(GetPays_{WA}),$$
$$s_{WSO}(GetPays_{AW}), r_{WSO}(GetPays_{AW}),$$
$$s_{WS_1}(SendLB_{WW_1}), r_{WS_1}(SendLB_{WW_1}),$$
$$s_{WSO}(ReceiveSB_{WW_1}), r_{WSO}(ReceiveSB_{WW_1}),$$
$$s_{WS_1}(SendP_{WW_1}), r_{WS_1}(SendP_{WW_1}),$$
$$s_{WSO}(GetPays_{WW_1}), r_{WSO}(GetPays_{WW_1}),$$
$$s_{WS_2}(BBFinish_{WW_2}), r_{WS_2}(BBFinish_{WW_2}),$$
$$s_{WS_3}(ShipByT_{WW_3}), r_{WS_3}(ShipByT_{WW_3}),$$
$$s_{WSO}(ShipFinish_{WW_3}), r_{WSO}(ShipFinish_{WW_3}),$$
$$s_{WS_4}(ShipByA_{WW_4}), r_{WS_4}(ShipByA_{WW_4}),$$
$$s_{WSO}(ShipFinish_{WW_4}), r_{WSO}(ShipFinish_{WW_4}),$$
$$s_{WSO}(ReceiveRB_{MW}), r_{WSO}(ReceiveRB_{MW}),$$
$$s_{SS}(DI_{SS}), r_{SS}(DI_{SS}),$$
$$s_{WSO}(DO_{SS}), r_{WSO}(DO_{SS}),$$
$$s_{WSOIM}(ReceiveRB_{WM}), r_{WSOIM}(ReceiveRB_{WM})$$
$$|s_{AA_1}(ReceiveRB_{WA}) \leq r_{AA_1}(ReceiveRB_{WA}),$$
$$s_{WSO}(ReceiveRB_{AW}) \leq r_{WSO}(ReceiveRB_{AW}),$$
$$s_{AA_2}(SendLB_{WA}) \leq r_{AA_2}(SendLB_{WA}),$$
$$s_{WSO}(SendLB_{AW}) \leq r_{WSO}(SendLB_{AW}),$$
$$s_{AA_3}(ReceiveSB_{WA}) \leq r_{AA_3}(ReceiveSB_{WA}),$$
$$s_{WSO}(ReceiveSB_{AW}) \leq r_{WSO}(ReceiveSB_{AW}),$$
$$s_{AA_4}(CalculteP_{WA}) \leq r_{AA_4}(CalculteP_{WA}),$$
$$s_{WSO}(CalculateP_{AW}) \leq r_{WSO}(CalculateP_{AW}),$$
$$s_{AA_5}(SendP_{WA}) \leq r_{AA_5}(SendP_{WA}),$$
$$s_{WSO}(SendP_{AW}) \leq r_{WSO}(SendP_{AW}),$$
$$s_{AA_6}(GetPays_{WA}) \leq r_{AA_6}(GetPays_{WA}),$$
$$s_{WSO}(GetPays_{AW}) \leq r_{WSO}(GetPays_{AW}),$$
$$s_{WS_1}(SendLB_{WW_1}) \leq r_{WS_1}(SendLB_{WW_1}),$$
$$s_{WSO}(ReceiveSB_{WW_1}) \leq r_{WSO}(ReceiveSB_{WW_1}),$$
$$s_{WS_1}(SendP_{WW_1}) \leq r_{WS_1}(SendP_{WW_1}),$$
$$s_{WSO}(GetPays_{WW_1}) \leq r_{WSO}(GetPays_{WW_1}),$$
$$s_{WS_2}(BBFinish_{WW_2}) \leq r_{WS_2}(BBFinish_{WW_2}),$$

$s_{WS_3}(ShipByTw_{W_3}) \leq r_{WS_3}(ShipByTw_{W_3})$,
$s_{WSO}(ShipFinish_{WW_3}) \leq r_{WSO}(ShipFinish_{WW_3})$,
$s_{WS_4}(ShipByAw_{W_4}) \leq r_{WS_4}(ShipByAw_{W_4})$,
$s_{WSO}(ShipFinish_{WW_4}) \leq r_{WSO}(ShipFinish_{WW_4})$,
$s_{WSO}(ReceiveRB_{MW}) \leq r_{WSO}(ReceiveRB_{MW})$,
$s_{SS}(DI_{SS}) \leq r_{SS}(DI_{SS})$,
$s_{WSO}(DO_{SS}) \leq r_{WSO}(DO_{SS})$,
$s_{WSOIM}(ReceiveRB_{WM}) \leq r_{WSOIM}(ReceiveRB_{WM})\}$
$\cup I_{AA_1} \cup I_{AA_2} \cup I_{AA_3} \cup I_{AA_4} \cup I_{AA_5} \cup I_{AA_6} \cup I_{AA_7} \cup I_{WSOIM} \cup I_{WSO} \cup I_{WS_1} \cup I_{WS_2} \cup I_{WS_3} \cup I_{WS_4} \cup I_{SS}$
Then, we can obtain the following conclusion.

**Theorem 23.1.** *The whole actor system of buying books example illustrated in Fig. 23.1 can exhibit the desired external behaviors.*

*Proof.* By use of the algebraic laws of APTC, we can prove the following equation:
$\tau_I(\partial_H(WS_1 \between WS_2 \between WS_3 \between WS_4 \between WSOIM \between SS))$
$= \tau_I(\partial_H(WS_1 \between WS_2 \between WS_3 \between WS_4 \between WSOIM \between SS \between WSO \between AA_1 \between AA_2 \between AA_3 \between AA_4 \between AA_5 \between AA_6 \between AA_7))$
$= r_{WS_2}(RequestLB_{WS_2}) \cdot s_O(BBFinish_O) \cdot \tau_I(\partial_H(WS_1 \between WS_2 \between WS_3 \between WS_4 \between WSOIM \between SS \between WSO \between AA_1 \between AA_2 \between AA_3 \between AA_4 \between AA_5 \between AA_6 \between AA_7))$
$= r_{WS_2}(RequestLB_{WS_2}) \cdot s_O(BBFinish_O) \cdot \tau_I(\partial_H(WS_1 \between WS_2 \between WS_3 \between WS_4 \between WSOIM \between SS))$
For the details of the proof, as we omit them here, please refer to Section 17.3. □

Chapter 24

# Introduction to secure process algebra

A security protocol [29] includes some computational operations, some cryptographic operations (for examples, symmetric encryption/decryption, asymmetric encryption/decryption, hash function, digital signatures, message authentication codes, random sequence generations, XOR operations, etc.), some communication operations to exchange data, and also the computational logics among these operations.

Designing a perfectly practical security protocol is quite a complex task, because of the open network environments and the complex security requirements against various known and unknown attacks. How to design a security protocol usually heavily depends on the experiences of security engineering. From experience, formal verifications can be used in the design of security protocols to satisfy the main goal of the security protocol.

There are many formal verification tools to support the verifications of security protocols, such as BAN logic [30] and those works based on process algebra. In the work based on process algebra, there are works based on pi-calculus, such as spi-calculus [27] and the applied pi-calculus [28]. The work based on process algebra has some advantages: they describe the security protocols in a programming style, and have firm theoretical foundations.

Based on our previous work on truly concurrent process algebras APTC [9], we use it to verify the security protocols. This work (called Secure APTC, abbreviated SAPTC) has the following advantages in verifying security protocols:

1. It has a firm theoretical foundation, including equational logics, structured operational semantics, and axiomatizations between them;
2. It has rich expressive powers to describe security protocols. Cryptographic operations are modeled as atomic actions and can be extended, explicit parallelism and communication mechanism to model communication operations and principals, rich computational properties to describing computational logics in the security protocols, including conditional guards, alternative composition, sequential composition, parallelism and communication, encapsulation and deadlock, recursion, and abstraction.
3. Especially by abstraction, it is convenient and obvious to observe the relations between the inputs and outputs of a security protocol, including the relations without any attack, the relations under each known attack, and the relations under unknown attacks if the unknown attacks can be described.

This part of the book is organized as follows. About truly concurrent process algebra APTC, including operational semantics, proof techniques, BATC (Basic Algebra for True Concurrency), APTC (Algebra of Parallelism for True Concurrency), recursion and abstraction, an important extension of APTC – placeholder, axiomatization for hhp (hereditary history-preserving)- bisimulation, and also the application of APTC through an example, please refer to Chapters 2, 9, and 17.

We extend APTC to SAPTC to describe cryptographic properties in Chapter 25, including the description of symmetric and asymmetric encryption/decryption, hash, digital signatures, message authentication codes, random sequence generation, and XOR, etc.

Then, we introduce the cases of verifying security protocols, including key exchange related protocols in Chapter 26, including key exchange protocols with symmetric cryptography and public key cryptography, interlock protocol against man-in-the-middle attack, key exchange protocol with digital signature, key and message transmission protocol, and key and message broadcast protocol.

Also, authentication protocols are introduced in Chapter 27, including mutual authentication using the interlock protocol against man-in-the-middle attack, and SKID.

Also, key exchange and authentication mixed protocols are introduced in Chapter 28, including Wide-Mouth Frog protocol, Yahalom protocol, Needham–Schroeder protocol, Otway–Rees protocol, Kerberos protocol, Neuman–Stubblebine protocol, Denning–Sacco protocol, DASS protocol, and Woo–Lam protocol.

Also, other protocols are introduced in Chapter 29, including secret splitting protocols, bit commitment protocols, anonymous key distribution protocols.

Also, digital cash protocols are introduced in Chapter 30, including four digital cash protocols.

Finally, secure elections protocols are introduced in Chapter 31, including six secure elections protocols.

**Handbook of Truly Concurrent Process Algebra. https://doi.org/10.1016/B978-0-44-321515-5.00028-7**

Chapter 25

# Secure APTC

Cryptography mainly includes two aspects: the cryptographic operations and security protocols. The former includes symmetric and asymmetric encryption/decryption, hash, digital signatures, message authentication codes, random sequence generation, and XOR, etc. The latter includes the computational logic driven by the security application logics among the cryptographic operations.

In this chapter, we model the above two cryptographic properties by APTC ($APTC_G$). In Section 25.1, we model symmetric encryption/decryption by APTC. Also, we model asymmetric encryption/decryption, hash, digital signatures, message authentication codes, random sequence generation, blind signatures, and XOR in Sections 25.2, 25.3, 25.4, 25.5, 25.6, 25.7, and 25.8. In Section 25.9, we extended the communication merge to support data substitution. Finally, in Section 25.10, we show how to analyze the security protocols by use of APTC ($APTC_G$).

## 25.1 Symmetric encryption

In the symmetric encryption and decryption, only one key $k$ is used. The inputs of symmetric encryption are the key $k$ and the plaintext $D$ and the output is the ciphertext, so we treat the symmetric encryption as an atomic action denoted $enc_k(D)$. We also use $ENC_k(D)$ to denote the ciphertext output. The inputs of symmetric decryption are the same key $k$ and the ciphertext $ENC_k(D)$ and output is the plaintext $D$, we also treat the symmetric decryption as an atomic action $dec_k(ENC_k(D))$. We also use $DEC_k(ENC_k(D))$ to denote the output of the corresponding decryption.

As $D$ is plaintext, it is obvious that $DEC_k(ENC_k(D)) = D$ and $enc_k(D) \leq dec_k(ENC_k(D))$, where $\leq$ is the causal relation; and for $D$ is the ciphertext, $ENC_k(DEC_k(D)) = D$ and $dec_k(D) \leq enc_k(DEC_k(D))$ hold.

## 25.2 Asymmetric encryption

In the asymmetric encryption and decryption, two keys are used: the public key $pk_s$ and the private key $sk_s$ generated from the same seed $s$. The inputs of asymmetric encryption are the key $pk_s$ or $sk_s$ and the plaintext $D$ and the output is the ciphertext, so we treat the asymmetric encryption as an atomic action denoted $enc_{pk_s}(D)$ or $enc_{sk_s}(D)$. We also use $ENC_{pk_s}(D)$ and $ENC_{sk_s}(D)$ to denote the ciphertext outputs. The inputs of asymmetric decryption are the corresponding key $sk_s$ or $pk_s$ and the ciphertext $ENC_{pk_s}(D)$ or $ENC_{sk_s}(D)$, and output is the plaintext $D$, we also treat the asymmetric decryption as an atomic action $dec_{sk_s}(ENC_{pk_s}(D))$ and $dec_{pk_s}(ENC_{sk_s}(D))$. We also use $DEC_{sk_s}(ENC_{pk_s}(D))$ and $DEC_{pk_s}(ENC_{sk_s}(D))$ to denote the corresponding decryption outputs.

As $D$ is plaintext, it is obvious that $DEC_{sk_s}(ENC_{pk_s}(D)) = D$ and $DEC_{pk_s}(ENC_{sk_s}(D)) = D$, and $enc_{pk_s}(D) \leq dec_{sk_s}(ENC_{pk_s}(D))$ and $enc_{sk_s}(D) \leq dec_{pk_s}(ENC_{sk_s}(D))$, where $\leq$ is the causal relation; and for $D$ is the ciphertext, $ENC_{sk_s}(DEC_{pk_s}(D)) = D$ and $ENC_{pk_s}(DEC_{sk_s}(D)) = D$, and $dec_{pk_s}(D) \leq enc_{sk_s}(DEC_{pk_s}(D))$ and $dec_{sk_s}(D) \leq enc_{pk_s}(DEC_{sk_s}(D))$.

## 25.3 Hash

The hash function is used to generate the digest of the data. The input of the hash function $hash$ is the data $D$ and the output is the digest of the data. We treat the hash function as an atomic action denoted $hash(D)$, and we also use $HASH(D)$ to denote the output digest.

As $D_1 = D_2$, it is obvious that $HASH(D_1) = HASH(D_2)$.

## 25.4 Digital signatures

Digital signature uses the private key $sk_s$ to encrypt some data and the public key $pk_s$ to decrypt the encrypted data to implement the so-called non-repudiation. The inputs of the sign function are some data $D$ and the private key $sk_s$ and the output is the signature. We treat the signing function as an atomic action $sign_{sk_s}(D)$, and also use $SIGN_{sk_s}(D)$ to denote

**Handbook of Truly Concurrent Process Algebra. https://doi.org/10.1016/B978-0-44-321515-5.00029-9**

the signature. The inputs of the de-sign function are the public key $pk_s$ and the signature $SIGN_{sk_s}(D)$, and the output is the original data $D$. We also treat the de-sign function as an atomic action $de\text{-}sign_{pk_s}(SIGN_{sk_s}(D))$, and also we use $DE\text{-}SIGN_{pk_s}(SIGN_{sk_s}(D))$ to denote the output of the de-sign action.

It is obvious that $DE\text{-}SIGN_{pk_s}(SIGN_{sk_s}(D)) = D$.

## 25.5 Message authentication codes

MAC (Message Authentication Code) is used to authenticate data by symmetric keys $k$ and it is often assumed that $k$ is privately shared only between two principals $A$ and $B$. The inputs of the MAC function are the key $k$ and some data $D$, and the output is the MACs. We treat the MAC function as an atomic action $mac_k(D)$, and use $MAC_k(D)$ to denote the output MACs.

The MACs $MAC_k(D)$ are generated by one principal $A$ and with $D$ together sent to the other principal $B$. The other principal $B$ regenerate the MACs $MAC_k(D)'$, if $MAC_k(D) = MAC_k(D)'$, then the data $D$ are from $A$.

## 25.6 Random sequence generation

Random sequence generation is used to generate a random sequence, which may be a symmetric key $k$, a pair of public key $pk_s$ and $sk_s$, or a nonce $nonce$ (usually used to resist replay attacks). We treat the random sequence generation function as an atomic action $rsg_k$ for symmetric key generation, $rsg_{pk_s,sk_s}$ for asymmetric key pair generation, and $rsg_N$ for nonce generation, and the corresponding outputs are $k$, $pk_s$ and $sk_s$, $N$, respectively.

## 25.7 Blind signatures

In the blind signatures, only one key $k$ is used. The inputs of blind function are the key $k$ and the plaintext $D$ and the output is the ciphertext, so we treat the blind function as an atomic action denoted $blind_k(D)$. We also use $BLIND_k(D)$ to denote the ciphertext output. The inputs of unblind function are the same key $k$ and the ciphertext $BLIND_k(D)$ and output is the plaintext $D$, we also treat the unblind function as an atomic action $unblind_k(BLIND_k(D))$. We also use $UNBLIND_k(BLIND_k(D))$ to denote the output of the corresponding unblind function.

As $D$ is plaintext, it is obvious that $UNBLIND_k(BLIND_k(D)) = D$ and $blind_k(D) \leq unblind_k(BLIND_k(D))$, where $\leq$ is the causal relation; and for $D$ is the ciphertext. Also, $UNBLIND_k(SIGN_{sk}(BLIND_k(D))) = SIGN_{sk}(D)$.

## 25.8 XOR

The inputs of the XOR function are two data $D_1$ and $D_2$, and the output is the XOR result. We treat the XOR function as an atomic action $xor(D_1, D_2)$, and we also use $XOR(D_1, D_2)$ to denote the XOR result.

It is obvious that the following equations hold:

1. $XOR(XOR(D_1, D_2), D_3) = XOR(D_1, XOR(D_2, D_3))$;
2. $XOR(D_1, D_2) = XOR(D_2, D_1)$;
3. $XOR(D, 0) = D$;
4. $XOR(D, D) = 0$;
5. $XOR(D_2, XOR(D_1, D_2)) = D_1$.

## 25.9 Extended communications

In APTC ($APTC_G$), the communication between two parallel processes is modeled as the communication merge of two communicating actions. One communicating action is the sending data ($D_1, \cdots, D_n \in \Delta$) action through certain channel $A$, which is denoted $s_A(D_1, \cdots, D_n)$, the other communicating action is the receiving data ($d_1, \cdots, d_n$ range over $\Delta$) action through the corresponding channel $A$ that is denoted $r_A(d_1, \cdots, d_n)$, note that $d_i$ and $D_i$ for $1 \leq i \leq n$ have the same data type.

We extend communication merge to this situation. The axioms of the extended communication merge are shown in Table 25.1, and the transition rules are shown in Table 25.2.

Obviously, the conclusions of the theories of $APTC$ and $APTC_G$ still hold without any alternation.

**TABLE 25.1** Axioms of the Extended Communication Merge.

| No. | Axiom |
|---|---|
| $C1$ | $e_1(D_1,\cdots,D_n) \mid e_2(d_1,\cdots,d_n) = \gamma(e_1(D_1,\cdots,D_n),e_2(d_1,\cdots,d_n))$ |
| $C2$ | $e_1(D_1,\cdots,D_n) \mid (e_2(d_1,\cdots,d_n)\cdot y) = \gamma(e_1(D_1,\cdots,D_n),e_2(d_1,\cdots,d_n))\cdot y[D_1/d_1,\cdots,D_n/d_n]$ |
| $C3$ | $(e_1(D_1,\cdots,D_n)\cdot x) \mid e_2(d_1,\cdots,d_n) = \gamma(e_1(D_1,\cdots,D_n),e_2(d_1,\cdots,d_n))\cdot x$ |
| $C4$ | $(e_1(D_1,\cdots,D_n)\cdot x) \mid (e_2(d_1,\cdots,d_n)\cdot y) = \gamma(e_1(D_1,\cdots,D_n),e_2(d_1,\cdots,d_n))\cdot (x \between y[D_1/d_1,\cdots,D_n/d_n])$ |

**TABLE 25.2** Transition Rules of the Extended Communication Merge.

$$\frac{\langle x,s\rangle \xrightarrow{e_1(D_1,\cdots,D_n)} \langle\surd,s'\rangle \quad \langle y,s\rangle \xrightarrow{e_2(d_1,\cdots,d_n)} \langle\surd,s''\rangle}{\langle x\mid y,s\rangle \xrightarrow{\gamma(e_1(D_1,\cdots,D_n),e_2(d_1,\cdots,d_n))} \langle\surd, effect(\gamma(e_1(D_1,\cdots,D_n),e_2(d_1,\cdots,d_n)),s)\rangle}$$

$$\frac{\langle x,s\rangle \xrightarrow{e_1(D_1,\cdots,D_n)} \langle x',s'\rangle \quad \langle y,s\rangle \xrightarrow{e_2(d_1,\cdots,d_n)} \langle\surd,s''\rangle}{\langle x\mid y,s\rangle \xrightarrow{\gamma(e_1(D_1,\cdots,D_n),e_2(d_1,\cdots,d_n))} \langle x', effect(\gamma(e_1(D_1,\cdots,D_n),e_2(d_1,\cdots,d_n)),s)\rangle}$$

$$\frac{\langle x,s\rangle \xrightarrow{e_1(D_1,\cdots,D_n)} \langle\surd,s'\rangle \quad \langle y,s\rangle \xrightarrow{e_2(d_1,\cdots,d_n)} \langle y',s''\rangle}{\langle x\mid y,s\rangle \xrightarrow{\gamma(e_1(D_1,\cdots,D_n),e_2(d_1,\cdots,d_n))} \langle y'[D_1/d_1,\cdots,D_n/d_n], effect(\gamma(e_1(D_1,\cdots,D_n),e_2(d_1,\cdots,d_n)),s)\rangle}$$

$$\frac{\langle x,s\rangle \xrightarrow{e_1(D_1,\cdots,D_n)} \langle x',s'\rangle \quad \langle y,s\rangle \xrightarrow{e_2(d_1,\cdots,d_n)} \langle y',s''\rangle}{\langle x\mid y,s\rangle \xrightarrow{\gamma(e_1(D_1,\cdots,D_n),e_2(d_1,\cdots,d_n))} \langle x'\between y'[D_1/d_1,\cdots,D_n/d_n], effect(\gamma(e_1(D_1,\cdots,D_n),e_2(d_1,\cdots,d_n)),s)\rangle}$$

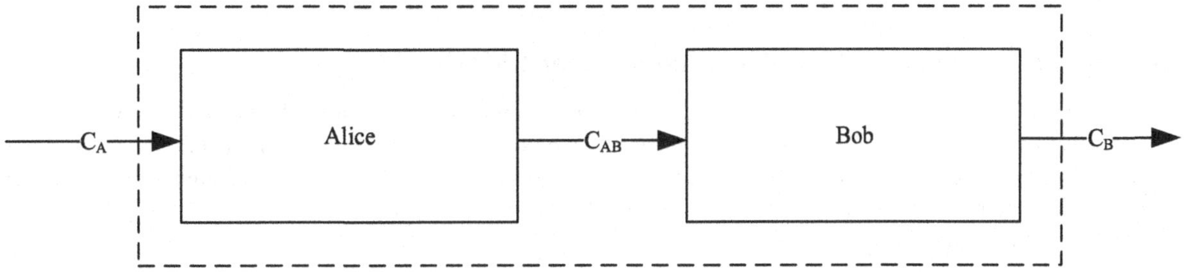


**FIGURE 25.1** A protocol using private channels.

## 25.10 Analyses of security protocols

In this section, we will show the application of analyzing security protocols by APTC ($APTC_G$) via several examples.

### 25.10.1 A protocol using private channels

The protocol shown in Fig. 25.1 uses private channels, that is, the channel $C_{AB}$ between Alice and Bob is private to Alice and Bob, there is no one else who can use this channel.

The process of the protocol is as follows:

1. Alice receives some messages $D$ from the outside through the channel $C_A$ (the corresponding reading action is denoted $r_{C_A}(D)$), after an internal processing $af$, she sends $D$ to Bob through the private channel $C_{AB}$ (the corresponding sending action is denoted $s_{C_{AB}}(D)$);
2. Bob receives the message $D$ through the private channel $C_{AB}$ (the corresponding reading action is denoted $r_{C_{AB}}(D)$), after an internal processing $bf$, he sends $D$ to the outside through the channel $C_B$ (the corresponding sending action is denoted $s_{C_B}(D)$),

where $D \in \Delta$, $\Delta$ is the set of data.

Alice's state transitions described by $APTC_G$ are as follows:

$A = \sum_{D\in\Delta} r_{C_A}(D)\cdot A_2$

$A_2 = af \cdot A_3$

$A_3 = s_{C_{AB}}(D)\cdot A$

Bob's state transitions described by $APTC_G$ are as follows:
$B = r_{C_{AB}}(D) \cdot B_2$
$B_2 = bf \cdot B_3$
$B_3 = s_{C_B}(D) \cdot B$
The sending action and the reading action of the same type data through the same channel can communicate with each other, otherwise, they will cause a deadlock $\delta$. We define the following communication functions:
$\gamma(r_{C_{AB}}(D), s_{C_{AB}}(D) \triangleq c_{C_{AB}}(D)$
Let all modules be in parallel, then the protocol $A \quad B$ can be presented by the following process term:

$$\tau_I(\partial_H(\Theta(A \between B))) = \tau_I(\partial_H(A \between B))$$

where $H = \{r_{C_{AB}}(D), s_{C_{AB}}(D)|D \in \Delta\}$, $I = \{c_{C_{AB}}(D), af, bf|D \in \Delta\}$.
Then, we obtain the following conclusion on the protocol.

**Theorem 25.1.** *The protocol using private channels in Fig. 25.1 is secure.*

*Proof.* Based on the above state transitions of the above modules, by use of the algebraic laws of $APTC_G$, we can prove that
$\tau_I(\partial_H(A \between B)) = \sum_{D\in\Delta}(r_{C_A}(D) \cdot s_{C_B}(D)) \cdot \tau_I(\partial_H(A \between B))$
For the details of the proof, please refer to Section 17.3, as we omit it here.
That is, the protocol in Fig. 25.1 $\tau_I(\partial_H(A \between B))$ can exhibit the desired external behaviors, and because the channel $C_{AB}$ is private, there is no attack.
Hence, the protocol using private channels in Fig. 25.1 is secure. □

### 25.10.2 Secure communication protocols using symmetric keys

The protocol shown in Fig. 25.2 uses symmetric keys for secure communication, that is, the key $k_{AB}$ between Alice and Bob is privately shared to Alice and Bob, there is no one else who can use this key. For secure communication, the main challenge is the information leakage to against the confidentiality. Since all channels in Fig. 25.2 are public, there may be an Eve to intercept the messages sent from Alice to Bob, and try to crack the secrets.
The process of the protocol is as follows:

1. Alice receives some messages $D$ from the outside through the channel $C_A$ (the corresponding reading action is denoted $r_{C_A}(D)$), after an encryption processing $enc_{k_{AB}}(D)$, she sends $ENC_{k_{AB}}(D)$ to Bob through the channel $C_{AB}$ (the corresponding sending action is denoted $s_{C_{AB}}(ENC_{k_{AB}}(D))$). She also sends $ENC_{k_{AB}}(D)$ to Eve through the channel $C_{AE}$ (the corresponding sending action is denoted $s_{C_{AE}}(ENC_{k_{AB}}(D))$);
2. Bob receives the message $ENC_{k_{AB}}(D)$ through the channel $C_{AB}$ (the corresponding reading action is denoted $r_{C_{AB}}(ENC_{k_{AB}}(D))$), after a decryption processing $dec_{k_{AB}}(ENC_{k_{AB}}(D))$, he sends $D$ to the outside through the channel $C_B$ (the corresponding sending action is denoted $s_{C_B}(D)$);
3. Eve receives the message $ENC_{k_{AB}}(D)$ through the channel $C_{AE}$ (the corresponding reading action is denoted $r_{C_{AE}}(ENC_{k_{AB}}(D))$), after a decryption processing $dec_{k_E}(ENC_{k_{AB}}(D))$, he sends $DEC_{k_E}(ENC_{k_{AB}}(D))$ to the outside through the channel $C_E$ (the corresponding sending action is denoted $s_{C_E}(DEC_{k_E}(ENC_{k_{AB}}(D)))$),

where $D \in \Delta$, $\Delta$ is the set of data.
Alice's state transitions described by $APTC_G$ are as follows:
$A = \sum_{D\in\Delta} r_{C_A}(D) \cdot A_2$
$A_2 = enc_{k_{AB}}(D) \cdot A_3$
$A_3 = (s_{C_{AB}}(ENC_{k_{AB}}(D)) \parallel s_{C_{AE}}(ENC_{k_{AB}}(D))) \cdot A$
Bob's state transitions described by $APTC_G$ are as follows:
$B = r_{C_{AB}}(ENC_{k_{AB}}(D)) \cdot B_2$
$B_2 = dec_{k_{AB}}(ENC_{k_{AB}}(D)) \cdot B_3$
$B_3 = s_{C_B}(D) \cdot B$
Eve's state transitions described by $APTC_G$ are as follows:
$E = r_{C_{AE}}(ENC_{k_{AB}}(D)) \cdot E_2$
$E_2 = dec_{k_E}(ENC_{k_{AB}}(D)) \cdot E_3$
$E_3 = (\{k_E \neq k_{AB}\} \cdot s_{C_E}(DEC_{k_E}(ENC_{k_{AB}}(D))) + \{k_E = k_{AB}\} \cdot s_{C_E}(D)) \cdot E$

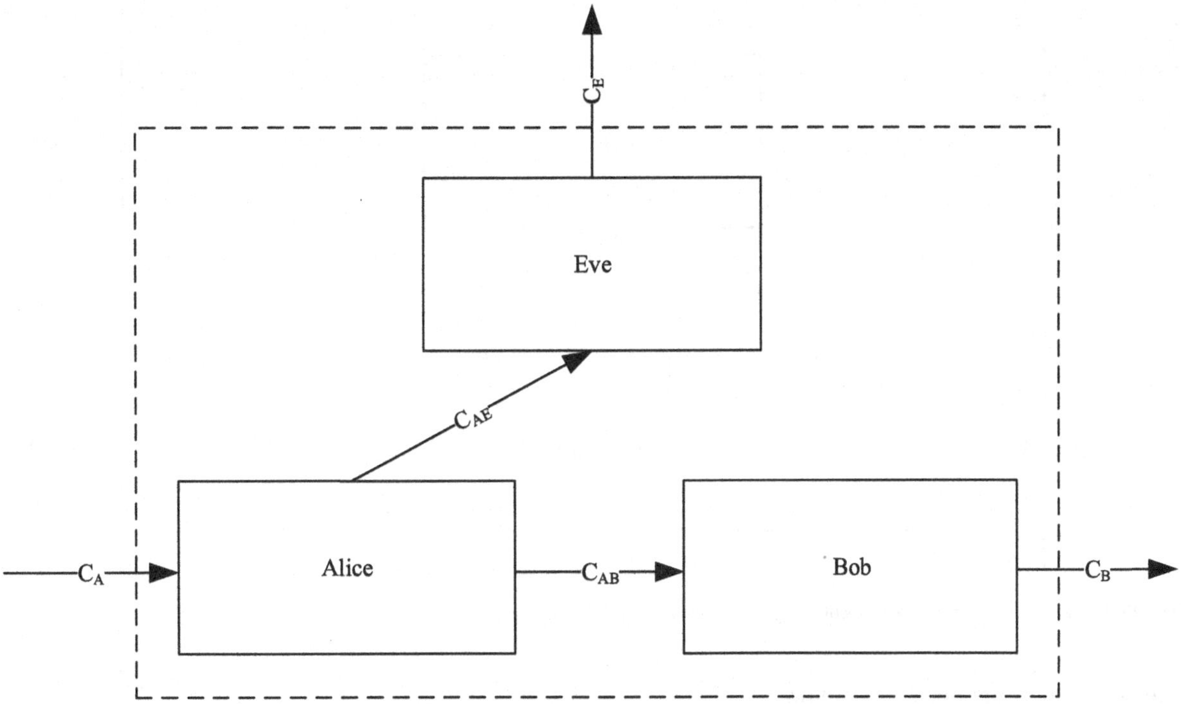


**FIGURE 25.2** Secure communication protocol using symmetric keys.

The sending action and the reading action of the same type data through the same channel can communicate with each other, otherwise, they will cause a deadlock $\delta$. We define the following communication functions:

$\gamma(r_{C_{AB}}(ENC_{k_{AB}}(D)), s_{C_{AB}}(ENC_{k_{AB}}(D)) \triangleq c_{C_{AB}}(ENC_{k_{AB}}(D))$

$\gamma(r_{C_{AE}}(ENC_{k_{AB}}(D)), s_{C_{AE}}(ENC_{k_{AB}}(D)) \triangleq c_{C_{AE}}(ENC_{k_{AB}}(D))$

Let all modules be in parallel, then the protocol $A \quad B \quad E$ can be presented by the following process term:

$$\tau_I(\partial_H(\Theta(A \between B \between E))) = \tau_I(\partial_H(A \between B \between E))$$

where $H = \{r_{C_{AB}}(ENC_{k_{AB}}(D)), s_{C_{AB}}(ENC_{k_{AB}}(D)), r_{C_{AE}}(ENC_{k_{AB}}(D)), s_{C_{AE}}(ENC_{k_{AB}}(D)) | D \in \Delta\}$,

$I = \{c_{C_{AB}}(ENC_{k_{AB}}(D)), c_{C_{AE}}(ENC_{k_{AB}}(D)), enc_{k_{AB}}(D), dec_{k_{AB}}(ENC_{k_{AB}}(D)), dec_{k_E}(ENC_{k_{AB}}(D)),$
$\{k_E \neq k_{AB}\}, \{k_E = k_{AB}\} | D \in \Delta\}$

Then, we obtain the following conclusion on the protocol.

**Theorem 25.2.** *The protocol using symmetric keys for secure communication in Fig. 25.2 is confidential.*

*Proof.* Based on the above state transitions of the above modules, by use of the algebraic laws of $APTC_G$, we can prove that

$\tau_I(\partial_H(A \between B \between E)) = \sum_{D\in\Delta}(r_{C_A}(D) \cdot (s_{C_B}(D) \parallel s_{C_E}(DEC_{k_E}(ENC_{k_{AB}}(D))))) \cdot \tau_I(\partial_H(A \between B \between E))$

For the details of the proof, please refer to Section 17.3, as we omit them here.

That is, the protocol in Fig. 25.2 $\tau_I(\partial_H(A \between B \between E))$ can exhibit the desired external behaviors, and because the key $k_{AB}$ is private, $DEC_{k_E}(ENC_{k_{AB}}(D)) \neq D$ (for $k_E \neq k_{AB}$).

Hence, the protocol using symmetric keys in Fig. 25.2 is confidential. □

## 25.10.3 Discussion

From the above section, we can see the process of analysis of security protocols, that is, through abstract away the internal series of cryptographic operations, we can see the relation between the inputs and the outputs of the whole protocol, then we can obtain the conclusions of whether or not the protocol is secure.

A security protocol is designed for one or several goals. For example, the secure communication protocol using symmetric keys in Fig. 25.2 is designed for the confidentiality of the communication. Hence, we only verify if the protocol is confidential. In fact, the protocol in Fig. 25.2 cannot resist other attacks, for example, the replay attack, as Fig. 25.3 shows.

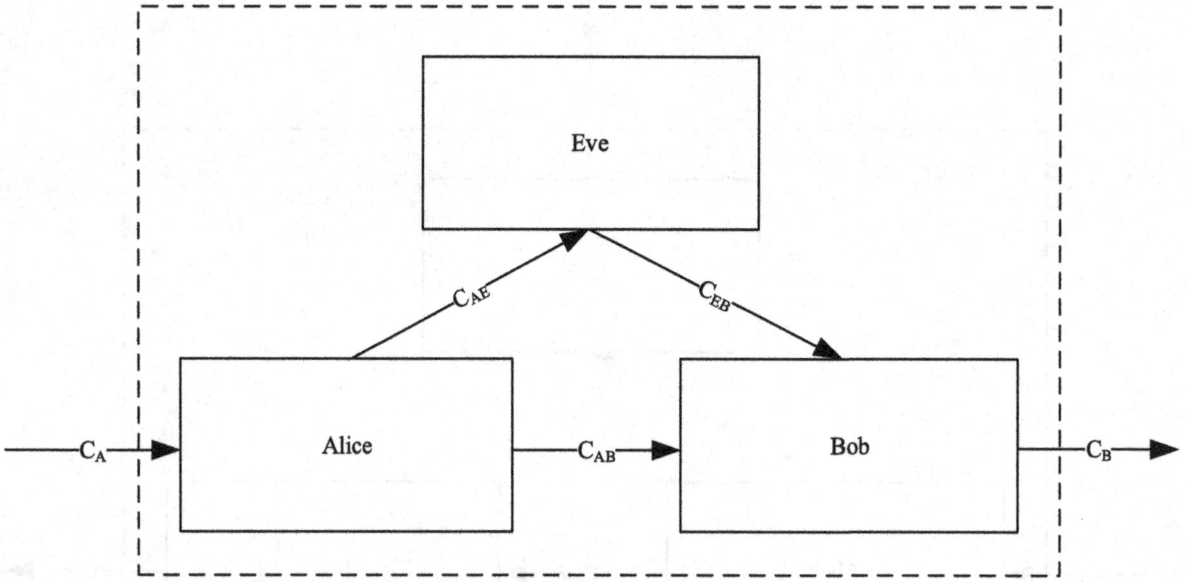


**FIGURE 25.3** Secure communication protocol using symmetric keys with Replay Attack.

The process of the protocol is as follows:

1. Alice receives some messages $D$ from the outside through the channel $C_A$ (the corresponding reading action is denoted $r_{C_A}(D)$), after an encryption processing $enc_{k_{AB}}(D)$, she sends $ENC_{k_{AB}}(D)$ to Bob through the channel $C_{AB}$ (the corresponding sending action is denoted $s_{C_{AB}}(ENC_{k_{AB}}(D))$). She also sends $ENC_{k_{AB}}(D)$ to Eve through the channel $C_{AE}$ (the corresponding sending action is denoted $s_{C_{AE}}(ENC_{k_{AB}}(D))$);
2. Eve receives the message $ENC_{k_{AB}}(D)$ through the channel $C_{AE}$ (the corresponding reading action is denoted $r_{C_{AE}}(ENC_{k_{AB}}(D))$), without an internal processing, he sends $ENC_{k_{AB}}(D)$ to the outside through the channel $C_{EB}$ (the corresponding sending action is denoted $s_{C_{EB}}(ENC_{k_{AB}}(D))$);
3. Bob receives the message $ENC_{k_{AB}}(D)$ through the channel $C_{AB}$ (the corresponding reading action is denoted $r_{C_{AB}}(ENC_{k_{AB}}(D))$), after a decryption processing $dec_{k_{AB}}(ENC_{k_{AB}}(D))$, he sends $D$ to the outside through the channel $C_B$ (the corresponding sending action is denoted $s_{C_B}(D)$); Bob receives the message $ENC_{k_{AB}}(D)$ through the channel $C_{EB}$ (the corresponding reading action is denoted $r_{C_{EB}}(ENC_{k_{AB}}(D))$), after a decryption processing $dec_{k_{AB}}(ENC_{k_{AB}}(D))$, he sends $D$ to the outside through the channel $C_B$ (the corresponding sending action is denoted $s_{C_B}(D)$),

where $D \in \Delta$, $\Delta$ is the set of data.

Alice's state transitions described by $APTC_G$ are as follows:

$A = \sum_{D\in\Delta} r_{C_A}(D) \cdot A_2$

$A_2 = enc_{k_{AB}}(D) \cdot A_3$

$A_3 = (s_{C_{AB}}(ENC_{k_{AB}}(D)) \parallel s_{C_{AE}}(ENC_{k_{AB}}(D))) \cdot A$

Bob's state transitions described by $APTC_G$ are as follows:

$B = (r_{C_{AB}}(ENC_{k_{AB}}(D)) \parallel r_{C_{AB}}(ENC_{k_{EB}}(D))) \cdot B_2$

$B_2 = (dec_{k_{AB}}(ENC_{k_{AB}}(D)) \parallel dec_{k_{AB}}(ENC_{k_{AB}}(D))) \cdot B_3$

$B_3 = (s_{C_B}(D) \parallel s_{C_B}(D)) \cdot B$

Eve's state transitions described by $APTC_G$ are as follows:

$E = r_{C_{AE}}(ENC_{k_{AB}}(D)) \cdot E_2$

$E_2 = dec_{k_E}(ENC_{k_{AB}}(D)) \cdot E_3$

$E_3 = s_{C_{EB}}(ENC_{k_{AB}}(D)) \cdot E$

The sending action and the reading action of the same type data through the same channel can communicate with each other, otherwise, they will cause a deadlock $\delta$. We define the following communication functions:

$\gamma(r_{C_{AB}}(ENC_{k_{AB}}(D)), s_{C_{AB}}(ENC_{k_{AB}}(D)) \triangleq c_{C_{AB}}(ENC_{k_{AB}}(D))$

$\gamma(r_{C_{AE}}(ENC_{k_{AB}}(D)), s_{C_{AE}}(ENC_{k_{AB}}(D)) \triangleq c_{C_{AE}}(ENC_{k_{AB}}(D))$

$\gamma(r_{C_{BE}}(ENC_{k_{AB}}(D)), s_{C_{BE}}(ENC_{k_{AB}}(D)) \triangleq c_{C_{BE}}(ENC_{k_{AB}}(D))$

Let all modules be in parallel, then the protocol $A \quad B \quad E$ can be presented by the following process term:

$$\tau_I(\partial_H(\Theta(A \between B \between E))) = \tau_I(\partial_H(A \between B \between E))$$

where $H = \{r_{C_{AB}}(ENC_{k_{AB}}(D)), s_{C_{AB}}(ENC_{k_{AB}}(D)), r_{C_{AE}}(ENC_{k_{AB}}(D)), s_{C_{AE}}(ENC_{k_{AB}}(D)),$
$r_{C_{BE}}(ENC_{k_{AB}}(D)), s_{C_{BE}}(ENC_{k_{AB}}(D))|D \in \Delta\},$

$I = \{c_{C_{AB}}(ENC_{k_{AB}}(D)), c_{C_{AE}}(ENC_{k_{AB}}(D)), c_{C_{BE}}(ENC_{k_{AB}}(D)), enc_{k_{AB}}(D), dec_{k_{AB}}(ENC_{k_{AB}}(D))|D \in \Delta\}$

Then, we obtain the following conclusion on the protocol.

**Theorem 25.3.** *The protocol using symmetric keys for secure communication in Fig. 25.2 is not secure for replay attack.*

*Proof.* Based on the above state transitions of the above modules, by use of the algebraic laws of $APTC_G$, we can prove that

$\tau_I(\partial_H(A \between B \between E)) = \sum_{D\in\Delta}(r_{C_A}(D) \cdot (s_{C_B}(D) \parallel s_{C_B}(D))) \cdot \tau_I(\partial_H(A \between B \between E))$

For the details of the proof, please refer to Section 17.3, as we omit it here.

That is, the protocol in Fig. 25.2 $\tau_I(\partial_H(A \between B \between E))$ can exhibit undesired external behaviors ($D$ is outputted twice).

Hence, the protocol using symmetric keys in Fig. 25.2 is not secure for replay attack. □

Generally, in the following chapters, when we introduce the analysis of a security protocol, we will mainly analyze the secure properties related to its design goal.

Chapter 26

# Analyses of key exchange protocols

In this chapter, we will introduce several key exchange protocols, including key exchange protocols with symmetric cryptography in Section 26.1 and public key cryptography in Section 26.2, interlock protocol against man-in-the-middle attack in Section 26.3, key exchange protocol with digital signature in Section 26.4, key and message transmission protocol in Section 26.5, and key and message broadcast protocol in Section 26.6.

## 26.1 Key exchange with symmetric cryptography

The protocol shown in Fig. 26.1 uses symmetric keys for secure communication, that is, the key $k_{AB}$ between Alice and Bob is privately shared to Alice and Bob, and $k_{AB}$ is generated by Trent, Alice and Bob have shared keys $k_{AT}$ and $k_{BT}$ already. For secure communication, the main challenge is the information leakage against the confidentiality.

The process of the protocol is as follows:

1. Alice receives some messages $D$ from the outside through the channel $C_A$ (the corresponding reading action is denoted $r_{C_A}(D)$), if $k_{AB}$ is not established, she sends a key request message $M$ to Trent through the channel $C_{AT}$ (the corresponding sending action is denoted $s_{C_{AT}}(M)$);
2. Trent receives the message $M$ through the channel $C_{AT}$ (the corresponding reading action is denoted $r_{C_{AT}}(M)$), generates a session key $k_{AB}$ through an action $rsg_{k_{AB}}$, and encrypts it for Alice and Bob through an action $enc_{k_{AT}}(k_{AB})$ and action $enc_{k_{BT}}(k_{AB})$ respectively, he sends $ENC_{k_{AT}}(k_{AB}), ENC_{k_{BT}}(k_{AB})$ to Alice through the channel $C_{TA}$ (the corresponding sending action is denoted $s_{C_{TA}}(ENC_{k_{AT}}(k_{AB}), ENC_{k_{BT}}(k_{AB}))$);
3. Alice receives $ENC_{k_{AT}}(k_{AB}), ENC_{k_{BT}}(k_{AB})$ from Trent through the channel $C_{TA}$ (the corresponding reading action is denoted $r_{C_{TA}}(ENC_{k_{AT}}(k_{AB}), ENC_{k_{BT}}(k_{AB}))$), she decrypts $ENC_{k_{AT}}(k_{AB})$ through an action $dec_{k_{AT}}(ENC_{k_{AT}}(k_{AB}))$ and obtains $k_{AB}$, and sends $ENC_{k_{BT}}(k_{AB})$ to Bob through the channel $C_{AB}$ (the corresponding sending action is denoted $s_{C_{AB}}(ENC_{k_{BT}}(k_{AB}))$);

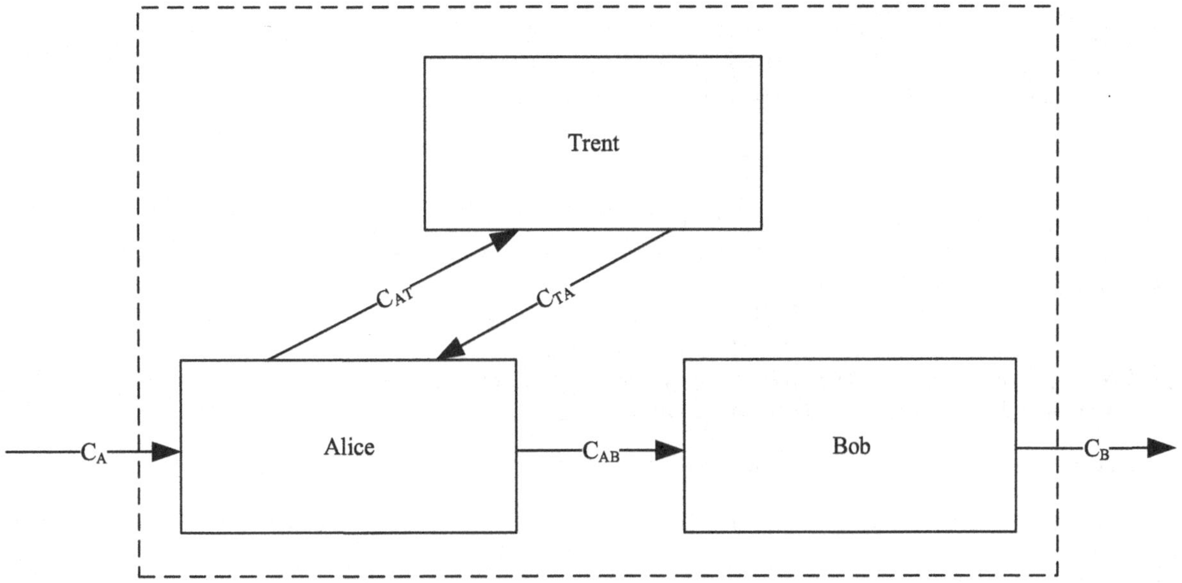


**FIGURE 26.1** Key exchange protocol with symmetric cryptography.

Handbook of Truly Concurrent Process Algebra. https://doi.org/10.1016/B978-0-44-321515-5.00030-5

4. Bob receives $ENC_{k_{BT}}(k_{AB})$ from Alice through the channel $C_{AB}$ (the corresponding reading action is denoted $r_{C_{AB}}(ENC_{k_{BT}}(k_{AB}))$), he decrypts $ENC_{k_{AT}}(k_{AB})$ through an action $dec_{k_{AT}}(ENC_{k_{AT}}(k_{AB}))$ and obtains $k_{AB}$, then $k_{AB}$ is established;
5. If $k_{AB}$ is established, after an encryption processing $enc_{k_{AB}}(D)$, Alice sends $ENC_{k_{AB}}(D)$ to Bob through the channel $C_{AB}$ (the corresponding sending action is denoted $s_{C_{AB}}(ENC_{k_{AB}}(D))$);
6. Bob receives the message $ENC_{k_{AB}}(D)$ through the channel $C_{AB}$ (the corresponding reading action is denoted $r_{C_{AB}}(ENC_{k_{AB}}(D))$), after a decryption processing $dec_{k_{AB}}(ENC_{k_{AB}}(D))$, he sends $D$ to the outside through the channel $C_B$ (the corresponding sending action is denoted $s_{C_B}(D)$),

where $D \in \Delta$, $\Delta$ is the set of data.

Alice's state transitions described by $APTC_G$ are as follows:

$A = \sum_{D\in\Delta} r_{C_A}(D) \cdot A_2$

$A_2 = \{k_{AB} = NULL\} \cdot s_{C_{AT}}(M) \cdot A_3 + \{k_{AB} \neq NULL\} \cdot A_6$

$A_3 = r_{C_{TA}}(ENC_{k_{AT}}(k_{AB}), ENC_{k_{BT}}(k_{AB})) \cdot A_4$

$A_4 = dec_{k_{AT}}(ENC_{k_{AT}}(k_{AB})) \cdot A_5$

$A_5 = s_{C_{AB}}(ENC_{k_{BT}}(k_{AB})) \cdot A_6$

$A_6 = enc_{k_{AB}}(D) \cdot A_7$

$A_7 = s_{C_{AB}}(ENC_{k_{AB}}(D)) \cdot A$

Bob's state transitions described by $APTC_G$ are as follows:

$B = \{k_{AB} = NULL\} \cdot B_1 + \{k_{AB} \neq NULL\} \cdot B_3$

$B_1 = r_{C_{AB}}(ENC_{k_{BT}}(k_{AB})) \cdot B_2$

$B_2 = dec_{k_{BT}}(ENC_{k_{AB}}(k_{AB})) \cdot B_3$

$B_3 = r_{C_{AB}}(ENC_{k_{AB}}(D)) \cdot B_4$

$B_4 = dec_{k_{AB}}(ENC_{k_{AB}}(D)) \cdot B_5$

$B_5 = s_{C_B}(D) \cdot B$

Trent's state transitions described by $APTC_G$ are as follows:

$T = r_{C_{AT}}(M) \cdot T_2$

$T_2 = rsg_{k_{AB}} \cdot T_3$

$T_3 = (enc_{k_{AT}}(k_{AB}) \parallel enc_{k_{BT}}(k_{AB})) \cdot T_4$

$T_4 = s_{C_{TA}}(ENC_{k_{AT}}(k_{AB}), ENC_{k_{BT}}(k_{AB})) \cdot T$

The sending action and the reading action of the same type of data through the same channel can communicate with each other, otherwise, they will cause a deadlock $\delta$. We define the following communication functions:

$\gamma(r_{C_{AT}}(M), s_{C_{AT}}(M)) \triangleq c_{C_{AT}}(M)$

$\gamma(r_{C_{TA}}(ENC_{k_{AT}}(k_{AB}), ENC_{k_{BT}}(k_{AB})), s_{C_{TA}}(ENC_{k_{AT}}(k_{AB}), ENC_{k_{BT}}(k_{AB})))$
$\triangleq c_{C_{TA}}(ENC_{k_{AT}}(k_{AB}), ENC_{k_{BT}}(k_{AB}))$

$\gamma(r_{C_{AB}}(ENC_{k_{BT}}(k_{AB})), s_{C_{AB}}(ENC_{k_{BT}}(k_{AB}))) \triangleq c_{C_{AB}}(ENC_{k_{BT}}(k_{AB}))$

$\gamma(r_{C_{AB}}(ENC_{k_{AB}}(D)), s_{C_{AB}}(ENC_{k_{AB}}(D)) \triangleq c_{C_{AB}}(ENC_{k_{AB}}(D))$

Let all modules be in parallel, then the protocol $A \quad B \quad T$ can be presented by the following process term:

$$\tau_I(\partial_H(\Theta(A \between B \between T))) = \tau_I(\partial_H(A \between B \between T))$$

where $H = \{r_{C_{AT}}(M), s_{C_{AT}}(M), r_{C_{TA}}(ENC_{k_{AT}}(k_{AB}), ENC_{k_{BT}}(k_{AB})),$
$s_{C_{TA}}(ENC_{k_{AT}}(k_{AB}), ENC_{k_{BT}}(k_{AB})), r_{C_{AB}}(ENC_{k_{BT}}(k_{AB})), s_{C_{AB}}(ENC_{k_{BT}}(k_{AB})),$
$r_{C_{AB}}(ENC_{k_{AB}}(D)), s_{C_{AB}}(ENC_{k_{AB}}(D)|D \in \Delta\},$

$I = \{c_{C_{AT}}(M), c_{C_{TA}}(ENC_{k_{AT}}(k_{AB}), ENC_{k_{BT}}(k_{AB})), c_{C_{AB}}(ENC_{k_{BT}}(k_{AB})), c_{C_{AB}}(ENC_{k_{AB}}(D)),$
$\{k_{AB} = NULL\}, \{k_{AB} \neq NULL\}, dec_{k_{AT}}(ENC_{k_{AT}}(k_{AB})), enc_{k_{AB}}(D),$
$dec_{k_{BT}}(ENC_{k_{AB}}(k_{AB})), dec_{k_{AB}}(ENC_{k_{AB}}(D)), rsg_{k_{AB}}, enc_{k_{AT}}(k_{AB}), enc_{k_{BT}}(k_{AB})|D \in \Delta\}$

Then, we obtain the following conclusion on the protocol.

**Theorem 26.1.** *The key exchange protocol with symmetric cryptography in Fig. 26.1 is confidential.*

*Proof.* Based on the above state transitions of the above modules, by use of the algebraic laws of $APTC_G$, we can prove that

$\tau_I(\partial_H(A \between B \between T)) = \sum_{D\in\Delta}(r_{C_A}(D) \cdot (s_{C_B}(D) \parallel s_{C_E}(DEC_{k_E}(ENC_{k_{AB}}(D))))) \cdot \tau_I(\partial_H(A \between B \between T))$

For the details of the proof, please refer to Section 17.3, as we omit them here.

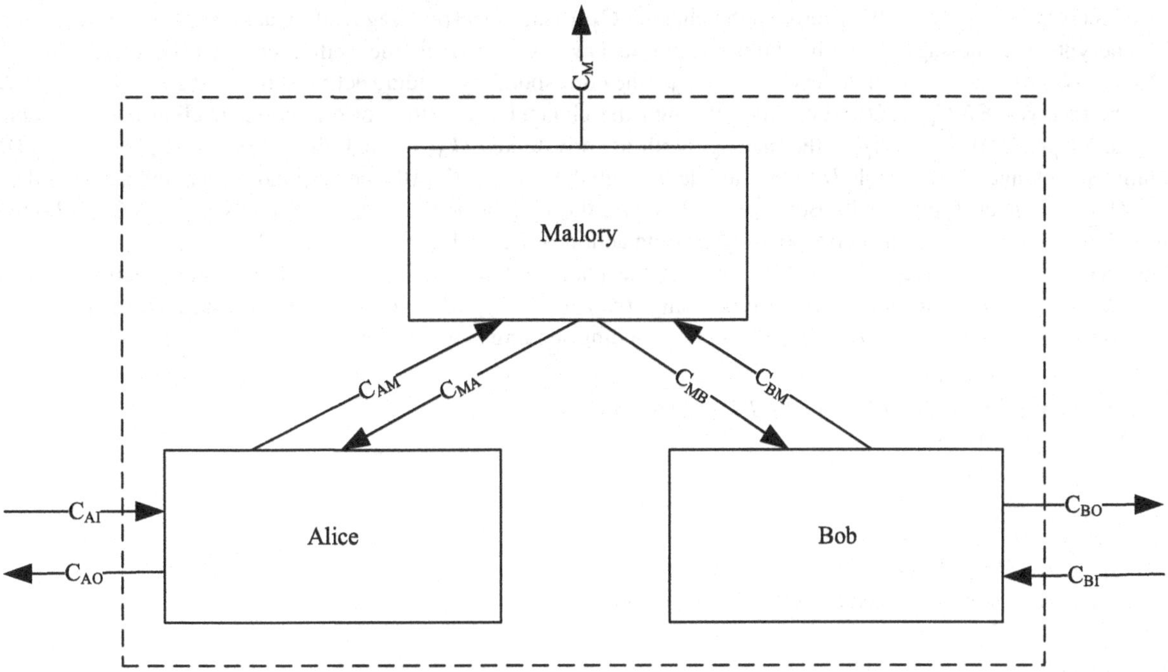


**FIGURE 26.2** Key exchange protocol with public key cryptography and man-in-the-middle attack.

That is, the protocol in Fig. 26.1 $\tau_I(\partial_H(A \between B \between T))$ can exhibit the desired external behaviors, and because the key $k_{AB}$ is private. The protocol using symmetric keys in Fig. 26.1 is confidential and similar to the protocol in Section 25.10.2, and we do not model the information leakage attack. □

## 26.2 Key exchange with public key cryptography

The protocol shown in Fig. 26.2 uses public keys for secure communication with man-in-the-middle attack, that is, Alice and Bob have shared their public keys $pk_A$ and $pk_B$ already. For secure communication, the main challenge is the information leakage against the confidentiality.

The process of key exchange protocol with public key cryptography is:

1. Alice obtains Bob's public key from Trent;
2. Alice generates a random session key, encrypts it using Bob's public key, and sends it to Bob;
3. Bob receives the encrypted session key, decrypted by his private key, and obtains the session key;
4. Alice and Bob can communicate by use of the session key.

We do not verify the above protocols, but we verify the above protocols with man-in-the-middle attack as Fig. 26.2 shows.

The process of the protocol with man-in-the-middle attack is as follows, and we only consider the message in one direction: from Alice to Bob.

1. Alice receives some messages $D$ from the outside through the channel $C_{AI}$ (the corresponding reading action is denoted $r_{C_{AI}}(D)$), she sends a key request message $Me$ to Mallory through the channel $C_{AM}$ (the corresponding sending action is denoted $s_{C_{AM}}(Me)$);
2. Mallory receives the message $Me$ through the channel $C_{AM}$ (the corresponding reading action is denoted $r_{C_{AM}}(Me)$), he sends $Me$ to the Bob through the channel $C_{MB}$ (the corresponding sending action is denoted $s_{C_{MB}}(Me)$);
3. Bob receives the message $Me$ from Mallory through the channel $C_{MB}$ (the corresponding reading action is denoted $r_{C_{MB}}(Me)$), and sends his public key $pk_B$ to Mallory through the channel $C_{BM}$ (the corresponding sending action is denoted $s_{C_{BM}}(pk_B)$);
4. Mallory receives $pk_B$ from Bob through the channel $C_{BM}$ (the corresponding reading action is denoted $r_{C_{BM}}(pk_B)$), then he stores $pk_B$, and sends his public key $pk_M$ to Alice through the channel $C_{MA}$ (the corresponding sending action is denoted $s_{C_{MA}}(pk_M)$);

5. Alice receives $pk_M$ from Mallory through the channel $C_{MA}$ (the corresponding reading action is denoted $r_{C_{MA}}(pk_M)$), she encrypts the message $D$ with Mallory's public key $pk_M$ through the action $enc_{pk_M}(D)$, then Alice sends $ENC_{pk_M}(D)$ to Mallory through the channel $C_{AM}$ (the corresponding sending action is denoted $s_{C_{AM}}(ENC_{pk_M}(D))$);
6. Mallory receives $ENC_{pk_M}(D)$ from Alice through the channel $C_{AM}$ (the corresponding reading action is denoted $r_{C_{AM}}(ENC_{pk_M}(D))$), he decrypts the message with his private key $sk_M$ through the action $dec_{sk_M}(ENC_{pk_M}(D))$ to obtain the message $D$, and sends $D$ to the outside through the channel $C_M$ (the corresponding sending action is denoted $s_{C_M}(D)$), then he encrypts $D$ with Bob's public key $pk_B$ through the action $enc_{pk_B}(D)$ and sends $ENC_{pk_B}(D)$ to Bob through the channel $C_{MB}$ (the corresponding sending action is denoted $s_{C_{MB}}(ENC_{pk_B}(D))$);
7. Bob receives the message $ENC_{pk_B}(D)$ through the channel $C_{MB}$ (the corresponding reading action is denoted $r_{C_{MB}}(ENC_{pk_B}(D))$), after a decryption processing $dec_{sk_B}(ENC_{pk_B}(D))$ to obtain the message $D$, then he sends $D$ to the outside through the channel $C_{BO}$ (the corresponding sending action is denoted $s_{C_{BO}}(D)$),

where $D \in \Delta$, $\Delta$ is the set of data.

Alice's state transitions described by $APTC_G$ are as follows:

$A = \sum_{D\in\Delta} r_{C_{AI}}(D) \cdot A_2$

$A_2 = s_{C_{AM}}(Me) \cdot A_3$

$A_3 = r_{C_{MA}}(pk_M) \cdot A_4$

$A_4 = enc_{pk_M}(D) \cdot A_5$

$A_5 = s_{C_{AM}}(ENC_{pk_M}(D)) \cdot A$

Bob's state transitions described by $APTC_G$ are as follows:

$B = r_{C_{MB}}(Me) \cdot B_2$

$B_2 = s_{C_{BM}}(pk_B) \cdot B_3$

$B_3 = r_{C_{MB}}(ENC_{pk_B}(D)) \cdot B_4$

$B_4 = dec_{sk_B}(ENC_{pk_B}(D)) \cdot B_5$

$B_5 = s_{C_{BO}}(D) \cdot B$

Mallory's state transitions described by $APTC_G$ are as follows:

$Ma = r_{C_{AM}}(Me) \cdot Ma_2$

$Ma_2 = s_{C_{MB}}(Me) \cdot Ma_3$

$Ma_3 = r_{C_{BM}}(pk_B) \cdot Ma_4$

$Ma_4 = s_{C_{MA}}(pk_M) \cdot Ma_5$

$Ma_5 = r_{C_{AM}}(ENC_{pk_M}(D)) \cdot Ma_6$

$Ma_6 = dec_{sk_M}(ENC_{pk_M}(D)) \cdot Ma_7$

$Ma_7 = s_{C_M}(D) \cdot Ma_8$

$Ma_8 = enc_{pk_B}(D) \cdot Ma_9$

$Ma_9 = s_{C_{MB}}(ENC_{pk_B}(D)) \cdot Ma$

The sending action and the reading action of the same type of data through the same channel can communicate with each other, otherwise, they will cause a deadlock $\delta$. We define the following communication functions:

$\gamma(r_{C_{AM}}(Me), s_{C_{AM}}(Me)) \triangleq c_{C_{AM}}(Me)$

$\gamma(r_{C_{MB}}(Me), s_{C_{MB}}(Me)) \triangleq c_{C_{MB}}(Me)$

$\gamma(r_{C_{BM}}(pk_B), s_{C_{BM}}(pk_B)) \triangleq c_{C_{BM}}(pk_B)$

$\gamma(r_{C_{MA}}(pk_M), s_{C_{MA}}(pk_M)) \triangleq c_{C_{MA}}(pk_M)$

$\gamma(r_{C_{AM}}(ENC_{pk_M}(D)), s_{C_{AM}}(ENC_{pk_M}(D))) \triangleq c_{C_{AM}}(ENC_{pk_M}(D))$

$\gamma(r_{C_{MB}}(ENC_{pk_B}(D)), s_{C_{MB}}(ENC_{pk_B}(D))) \triangleq c_{C_{MB}}(ENC_{pk_B}(D))$

Let all modules be in parallel, then the protocol $A \quad B \quad Ma$ can be presented by the following process term:

$$\tau_I(\partial_H(\Theta(A \between B \between Ma))) = \tau_I(\partial_H(A \between B \between Ma))$$

where $H = \{r_{C_{AM}}(Me), s_{C_{AM}}(Me), r_{C_{MB}}(Me), s_{C_{MB}}(Me), r_{C_{BM}}(pk_B), s_{C_{BM}}(pk_B),$
$r_{C_{MA}}(pk_M), s_{C_{MA}}(pk_M), r_{C_{AM}}(ENC_{pk_M}(D)), s_{C_{AM}}(ENC_{pk_M}(D)),$
$r_{C_{MB}}(ENC_{pk_B}(D)), s_{C_{MB}}(ENC_{pk_B}(D))|D \in \Delta\}$,

$I = \{c_{C_{AM}}(Me), c_{C_{MB}}(Me), c_{C_{BM}}(pk_B), c_{C_{MA}}(pk_M), c_{C_{AM}}(ENC_{pk_M}(D)),$
$c_{C_{MB}}(ENC_{pk_B}(D)), enc_{pk_M}(D), dec_{sk_B}(ENC_{pk_B}(D)), dec_{sk_M}(ENC_{pk_M}(D)), enc_{pk_B}(D)|D \in \Delta\}$

Then, we obtain the following conclusion on the protocol.

**Theorem 26.2.** *The key exchange protocol with public key cryptography in Fig. 26.2 is insecure.*

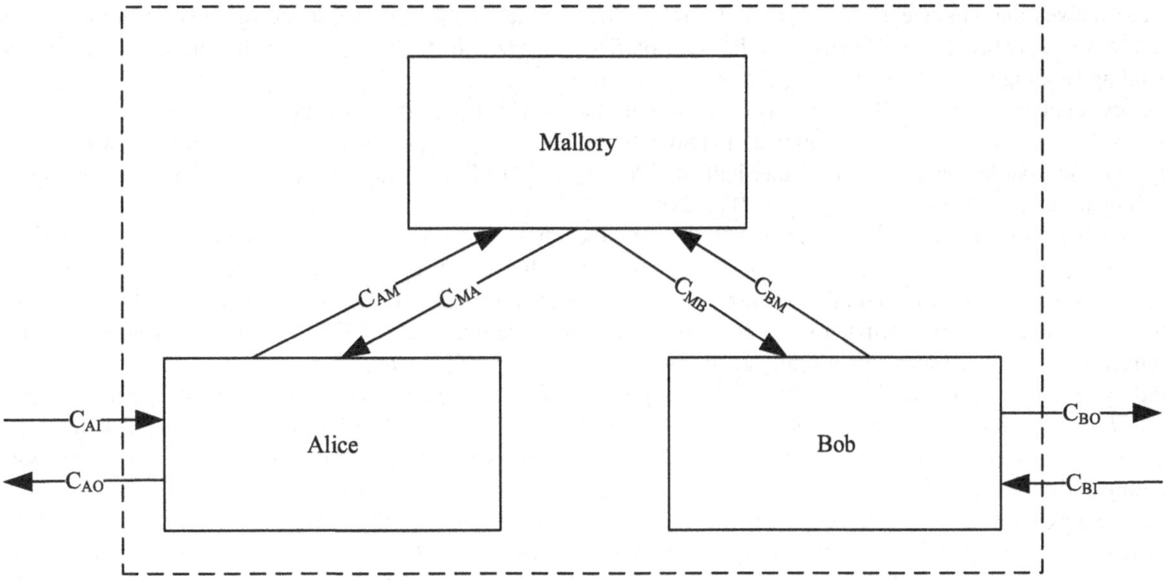


**FIGURE 26.3** Interlock protocol with man-in-the-middle attack.

*Proof.* Based on the above state transitions of the above modules, by use of the algebraic laws of $APTC_G$, we can prove that

$\tau_I(\partial_H(A \between B \between Ma)) = \sum_{D\in\Delta}(r_{C_{AI}}(D) \cdot s_{C_M}(D) \cdot s_{C_{BO}}(D)) \cdot \tau_I(\partial_H(A \between B \between Ma))$

For the details of the proof, please refer to Section 17.3, as we omit them here.

That is, the protocol in Fig. 26.2 $\tau_I(\partial_H(A \between B \between Ma))$ can exhibit undesired external behaviors, that is, there is an external action $s_{C_M}(D)$ while Alice and Bob are not aware. □

## 26.3 Interlock protocol

The interlock protocol shown in Fig. 26.3 also uses public keys for secure communication with man-in-the-middle attack, that is, Alice and Bob have shared their public keys $pk_A$ and $pk_B$ already. However, the interlock protocol can resist man-in-the-middle attack, that is, Alice and Bob can be aware of the existence of the man in the middle.

The process of the interlock protocol with man-in-the-middle attack is as follows, we assume that Alice has "Bob's" public key $pk_M$, Bob has "Alice's" public key $pk_M$, and Mallory has Alice's public key $pk_A$ and Bob's public key $pk_B$.

1. Alice receives some messages $D_A$ from the outside through the channel $C_{AI}$ (the corresponding reading action is denoted $r_{C_{AI}}(D_A)$), she encrypts the message $D_A$ with Mallory's public key $pk_M$ through the action $enc_{pk_M}(D_A)$, then Alice sends the half of $ENC_{pk_M}(D_A)$ to Mallory through the channel $C_{AM}$ (the corresponding sending action is denoted $s_{C_{AM}}(ENC_{pk_M}(D_A)/2)$);
2. Mallory receives $ENC_{pk_M}(D_A)/2$ from Alice through the channel $C_{AM}$ (the corresponding reading action is denoted $r_{C_{AM}}(ENC_{pk_M}(D_A)/2)$), he can not decrypt the message with his private key $sk_M$, and has to make another message $D'_A$ and encrypt $D'_A$ with Bob's public key $pk_B$ through the action $enc_{pk_B}(D'_A)$, and sends the half of $ENC_{pk_B}(D'_A)$ to Bob through the channel $C_{MB}$ (the corresponding sending action is denoted $s_{C_{MB}}(ENC_{pk_B}(D'_A)/2)$);
3. Bob receives the message $ENC_{pk_B}(D'_A)/2$ through the channel $C_{MB}$ (the corresponding reading action is denoted $r_{C_{MB}}(ENC_{pk_B}(D'_A)/2)$), and receives some message $D_B$ from the outside through the channel $C_{BI}$ (the corresponding reading action is denoted $r_{BI}(D_B)$), after an encryption processing $enc_{pk_M}(D_B)$ to obtain the message $ENC_{pk_M}(D_B)$, then he sends the half of $ENC_{pk_M}(D_B)$ to Mallory through the channel $C_{BM}$ (the corresponding sending action is denoted $s_{C_{BM}}(ENC_{pk_M}(D_B)/2)$);
4. Mallory receives $ENC_{pk_M}(D_B)/2$ from Bob through the channel $C_{BM}$ (the corresponding reading action is denoted $r_{C_{BM}}(ENC_{pk_M}(D_B)/2)$), he can not decrypt the message with his private key $sk_M$, and has to make another message $D'_B$ and encrypt $D'_B$ with Alice's public key $pk_A$ through the action $enc_{pk_A}(D'_B)$, and sends the half of $ENC_{pk_A}(D'_B)$ to Alice through the channel $C_{MA}$ (the corresponding sending action is denoted $s_{C_{MA}}(ENC_{pk_A}(D'_B)/2)$);

5. Alice receives the message $ENC_{pk_A}(D'_B)/2$ through the channel $C_{MA}$ (the corresponding reading action is denoted $r_{C_{MA}}(ENC_{pk_A}(D'_B)/2))$, and sends the other half of $ENC_{pk_M}(D_A)$ to Mallory through the channel $C_{AM}$ (the corresponding sending action is denoted $s_{C_{AM}}(ENC_{pk_M}(D_A)/2))$;
6. Mallory receives $ENC_{pk_M}(D_A)/2$ from Alice through the channel $C_{AM}$ (the corresponding reading action is denoted $r_{C_{AM}}(ENC_{pk_M}(D_A)/2))$, he can combine the two halves of $ENC_{pk_M}(D_A)/2$ and decrypt the message with his private key $sk_M$, and but he has to send the other half of $ENC_{pk_B}(D'_A)$ to Bob through the channel $C_{MB}$ (the corresponding sending action is denoted $s_{C_{MB}}(ENC_{pk_B}(D'_A)/2))$;
7. Bob receives the message $ENC_{pk_B}(D'_A)/2$ through the channel $C_{MB}$ (the corresponding reading action is denoted $r_{C_{MB}}(ENC_{pk_B}(D'_A)/2))$, after a combination of two halves of $ENC_{pk_B}(D'_A)$ and a decryption processing $dec_{sk_B}(ENC_{pk_B}(D'_A))$ to obtain the message $D'_A$, then he sends it to the outside through the channel $C_{BO}$ (the corresponding sending action is denoted $s_{C_{BO}}(D'_A))$. Then, he sends the other half of $ENC_{pk_M}(D_B)$ to Mallory through the channel $C_{BM}$ (the corresponding sending action is denoted $s_{C_{BM}}(ENC_{pk_M}(D_B)/2))$;
8. Mallory receives $ENC_{pk_M}(D_B)/2$ from Bob through the channel $C_{BM}$ (the corresponding reading action is denoted $r_{C_{BM}}(ENC_{pk_M}(D_B)/2))$, he can combine the two halves of $ENC_{pk_M}(D_B)/2$ and decrypt the message with his private key $sk_M$, but he has to send the other half of $ENC_{pk_A}(D'_B)$ to Alice through the channel $C_{MA}$ (the corresponding sending action is denoted $s_{C_{MA}}(ENC_{pk_A}(D'_B)/2))$;
9. Alice receives the message $ENC_{pk_A}(D'_B)/2$ through the channel $C_{MA}$ (the corresponding reading action is denoted $r_{C_{MA}}(ENC_{pk_A}(D'_B)/2))$, after a combination of two halves of $ENC_{pk_A}(D'_B)$ and a decryption processing $dec_{sk_A}(ENC_{pk_A}(D'_B))$ to obtain the message $D'_B$, then she sends it to the outside through the channel $C_{AO}$ (the corresponding sending action is denoted $s_{C_{AO}}(D'_B))$,

where $D \in \Delta$, $\Delta$ is the set of data.

Alice's state transitions described by $APTC_G$ are as follows:

$A = \sum_{D_A\in\Delta} r_{C_{AI}}(D_A) \cdot A_2$
$A_2 = enc_{pk_M}(D_A) \cdot A_3$
$A_3 = s_{C_{AM}}(ENC_{pk_M}(D_A)/2) \cdot A_4$
$A_4 = r_{C_{MA}}(ENC_{pk_A}(D'_B)/2) \cdot A_5$
$A_5 = s_{C_{AM}}(ENC_{pk_M}(D_A)/2) \cdot A_6$
$A_6 = r_{C_{MA}}(ENC_{pk_A}(D'_B)/2) \cdot A_7$
$A_7 = dec_{sk_A}(ENC_{pk_A}(D'_B)) \cdot A_8$
$A_8 = s_{C_{AO}}(D'_B) \cdot A$

Bob's state transitions described by $APTC_G$ are as follows:

$B = r_{C_{MB}}(ENC_{pk_B}(D'_A)/2) \cdot B_2$
$B_2 = \sum_{D_B\in\Delta} r_{BI}(D_B) \cdot B_3$
$B_3 = enc_{pk_M}(D_B) \cdot B_4$
$B_4 = s_{C_{BM}}(ENC_{pk_M}(D_B)/2) \cdot B_5$
$B_5 = r_{C_{MB}}(ENC_{pk_B}(D'_A)/2) \cdot B_6$
$B_6 = dec_{sk_B}(ENC_{pk_B}(D'_A)) \cdot B_7$
$B_7 = s_{C_{BO}}(D'_A) \cdot B_8$
$B_8 = s_{C_{BM}}(ENC_{pk_M}(D_B)/2) \cdot B$

Mallory's state transitions described by $APTC_G$ are as follows:

$Ma = r_{C_{AM}}(ENC_{pk_M}(D_A)/2) \cdot Ma_2$
$Ma_2 = enc_{pk_B}(D'_A) \cdot Ma_3$
$Ma_3 = s_{C_{MB}}(ENC_{pk_B}(D'_A)/2) \cdot Ma_4$
$Ma_4 = r_{C_{BM}}(ENC_{pk_M}(D_B)/2) \cdot Ma_5$
$Ma_5 = enc_{pk_A}(D'_B) \cdot Ma_6$
$Ma_6 = s_{C_{MA}}(ENC_{pk_A}(D'_B)/2) \cdot Ma_7$
$Ma_7 = r_{C_{AM}}(ENC_{pk_M}(D_A)/2) \cdot Ma_8$
$Ma_8 = s_{C_{MB}}(ENC_{pk_B}(D'_A)/2) \cdot Ma_9$
$Ma_9 = r_{C_{BM}}(ENC_{pk_M}(D_B)/2) \cdot Ma_{10}$
$Ma_{10} = s_{C_{MA}}(ENC_{pk_A}(D'_B)/2) \cdot Ma$

The sending action and the reading action of the same type of data through the same channel can communicate with each other, otherwise, they will cause a deadlock $\delta$. We define the following communication functions:

$\gamma(r_{C_{AM}}(ENC_{pk_M}(D_A)/2), s_{C_{AM}}(ENC_{pk_M}(D_A)/2)) \triangleq c_{C_{AM}}(ENC_{pk_M}(D_A)/2)$
$\gamma(r_{C_{MB}}(ENC_{pk_B}(D'_A)/2), s_{C_{MB}}(ENC_{pk_B}(D'_A)/2)) \triangleq c_{C_{MB}}(ENC_{pk_B}(D'_A)/2)$

$\gamma(r_{C_{BM}}(ENC_{pk_M}(D_B)/2), s_{C_{BM}}(ENC_{pk_M}(D_B)/2)) \triangleq c_{C_{BM}}(ENC_{pk_M}(D_B)/2)$
$\gamma(r_{C_{MA}}(ENC_{pk_A}(D'_B)/2), s_{C_{MA}}(ENC_{pk_A}(D'_B)/2)) \triangleq c_{C_{MA}}(ENC_{pk_A}(D'_B)/2)$
$\gamma(r_{C_{AM}}(ENC_{pk_M}(D_A)/2), s_{C_{AM}}(ENC_{pk_M}(D_A)/2)) \triangleq c_{C_{AM}}(ENC_{pk_M}(D_A)/2)$
$\gamma(r_{C_{MB}}(ENC_{pk_B}(D'_A)/2), s_{C_{MB}}(ENC_{pk_B}(D'_A)/2)) \triangleq c_{C_{MB}}(ENC_{pk_B}(D'_A)/2)$
$\gamma(r_{C_{BM}}(ENC_{pk_M}(D_B)/2), s_{C_{BM}}(ENC_{pk_M}(D_B)/2)) \triangleq c_{C_{BM}}(ENC_{pk_M}(D_B)/2)$
$\gamma(r_{C_{MA}}(ENC_{pk_A}(D'_B)/2), s_{C_{MA}}(ENC_{pk_A}(D'_B)/2)) \triangleq c_{C_{MA}}(ENC_{pk_A}(D'_B)/2)$

Let all modules be in parallel, then the protocol $A$ $B$ $Ma$ can be presented by the following process term:

$$\tau_I(\partial_H(\Theta(A \between B \between Ma))) = \tau_I(\partial_H(A \between B \between Ma))$$

where $H = \{r_{C_{AM}}(ENC_{pk_M}(D_A)/2), s_{C_{AM}}(ENC_{pk_M}(D_A)/2), r_{C_{MB}}(ENC_{pk_B}(D'_A)/2),$
$s_{C_{MB}}(ENC_{pk_B}(D'_A)/2), r_{C_{BM}}(ENC_{pk_M}(D_B)/2), s_{C_{BM}}(ENC_{pk_M}(D_B)/2),$
$r_{C_{MA}}(ENC_{pk_A}(D'_B)/2), s_{C_{MA}}(ENC_{pk_A}(D'_B)/2), r_{C_{AM}}(ENC_{pk_M}(D_A)/2),$
$s_{C_{AM}}(ENC_{pk_M}(D_A)/2), r_{C_{MB}}(ENC_{pk_B}(D'_A)/2), s_{C_{MB}}(ENC_{pk_B}(D'_A)/2),$
$r_{C_{BM}}(ENC_{pk_M}(D_B)/2), s_{C_{BM}}(ENC_{pk_M}(D_B)/2), r_{C_{MA}}(ENC_{pk_A}(D'_B)/2),$
$s_{C_{MA}}(ENC_{pk_A}(D'_B)/2) | D_A, D_B, D'_A, D'_B \in \Delta\},$

$I = \{c_{C_{AM}}(ENC_{pk_M}(D_A)/2), c_{C_{MB}}(ENC_{pk_B}(D'_A)/2), c_{C_{BM}}(ENC_{pk_M}(D_B)/2), c_{C_{MA}}(ENC_{pk_A}(D'_B)/2),$
$c_{C_{AM}}(ENC_{pk_M}(D_A)/2), c_{C_{MB}}(ENC_{pk_B}(D'_A)/2), c_{C_{BM}}(ENC_{pk_M}(D_B)/2), c_{C_{MA}}(ENC_{pk_A}(D'_B)/2),$
$enc_{pk_M}(D_A), dec_{sk_A}(ENC_{pk_A}(D'_B)), enc_{pk_M}(D_B), dec_{sk_B}(ENC_{pk_B}(D'_A)), enc_{pk_B}(D'_A),$
$enc_{pk_A}(D'_B) | D_A, D_B, D'_A, D'_B \in \Delta\}$

Then, we obtain the following conclusion on the protocol.

**Theorem 26.3.** *The interlock protocol with public key cryptography in Fig. 26.3 is secure.*

*Proof.* Based on the above state transitions of the above modules, by use of the algebraic laws of $APTC_G$, we can prove that

$\tau_I(\partial_H(A \between B \between Ma)) = \sum_{D_A, D_B, D'_A, D'_B \in \Delta}(r_{C_{AI}}(D_A) \cdot r_{C_{BI}}(D_B) \cdot s_{C_{BO}}(D'_A) \cdot s_{C_{AO}}(D'_B)) \cdot \tau_I(\partial_H(A \between B \between Ma))$

For the details of the proof, please refer to Section 17.3, as we omit them here.

That is, the interlock protocol in Fig. 26.3 $\tau_I(\partial_H(A \between B \between Ma))$ can exhibit the desired external behaviors, that is, Alice and Bob can be aware of the existence of the man in the middle. □

## 26.4 Key exchange with digital signatures

The protocol shown in Fig. 26.4 uses digital signature for secure communication with man-in-the-middle attack, that is, Alice and Bob have shared their public keys $pk_A$ and $pk_B$, and the public keys are signed by Trent: $SIGN_{sk_T}(A, pk_A)$, $SIGN_{sk_T}(B, pk_B)$ and $SIGN_{sk_T}(M, pk_M)$. Note that Trent's public key $pk_T$ is well known. Also, the key exchange protocol with digital signature can resist man-in-the-middle attack, that is, Alice and Bob can be aware of the existence of the man in the middle.

The process of the protocol with man-in-the-middle attack is as follows, and we only consider the message in one direction: from Alice to Bob.

1. Alice receives some messages $D$ from the outside through the channel $C_{AI}$ (the corresponding reading action is denoted $r_{C_{AI}}(D)$), she sends a key request message $Me$ to Mallory through the channel $C_{AM}$ (the corresponding sending action is denoted $s_{C_{AM}}(Me)$);
2. Mallory receives the message $Me$ through the channel $C_{AM}$ (the corresponding reading action is denoted $r_{C_{AM}}(Me)$), he sends $Me$ to the Bob through the channel $C_{MB}$ (the corresponding sending action is denoted $s_{C_{MB}}(Me)$);
3. Bob receives the message $Me$ from Mallory through the channel $C_{MB}$ (the corresponding reading action is denoted $r_{C_{MB}}(Me)$), and sends his signed public key $SIGN_{sk_T}(B, pk_B)$ to Mallory through the channel $C_{BM}$ (the corresponding sending action is denoted $s_{C_{BM}}(SIGN_{sk_T}(B, pk_B))$);
4. Mallory receives $SIGN_{sk_T}(B, pk_B)$ from Bob through the channel $C_{BM}$ (the corresponding reading action is denoted $r_{C_{BM}}(SIGN_{sk_T}(B, pk_B))$), he can obtain $pk_B$, then he sends his signed public key $SIGN_{sk_T}(M, pk_M)$ or $SIGN_{sk_T}(B, pk_B)$ to Alice through the channel $C_{MA}$ (the corresponding sending action is denoted $s_{C_{MA}}(SIGN_{sk_T}(M, pk_M))$);
5. Alice receives $SIGN_{sk_T}(M, pk_M)$ or $SIGN_{sk_T}(B, pk_B)$ from Mallory through the channel $C_{MA}$ (the corresponding reading action is denoted $r_{C_{MA}}(SIGN_{sk_T}(d_1, d_2))$), she de-signs this message using Trent's public key $pk_T$ through

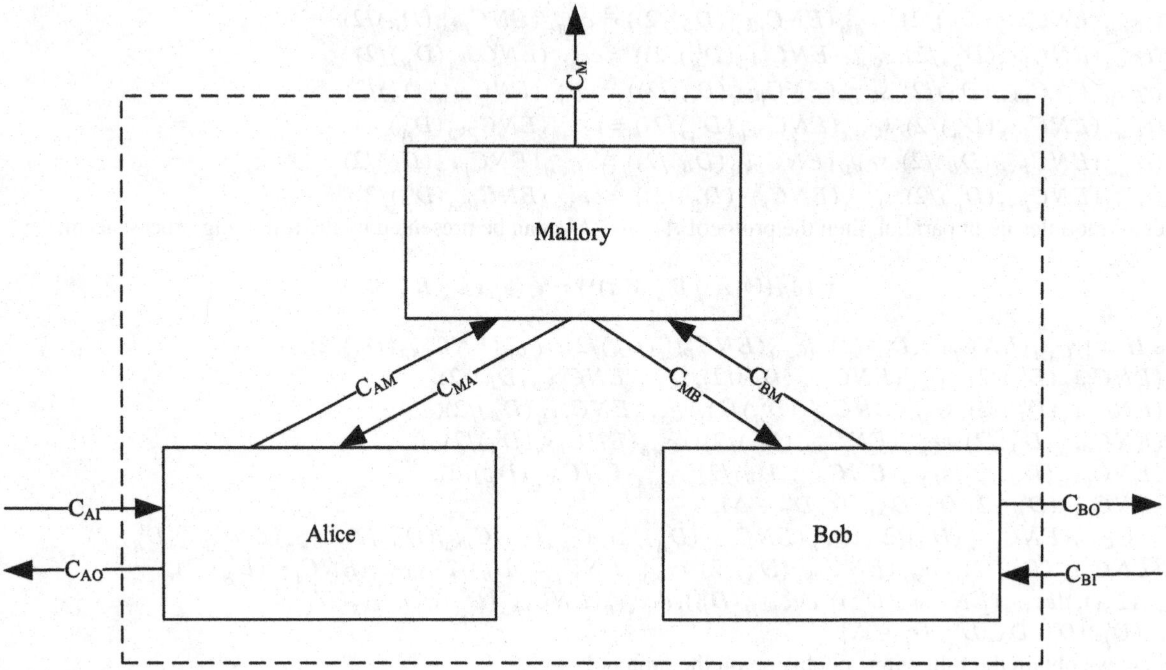


**FIGURE 26.4** Key exchange protocol with digital signature and man-in-the-middle attack.

the action $de\text{-}sign_{pk_T}(SIGN_{sk_T}(d_1, d_2))$, if $d_1 = B$: she encrypts the message $D$ with Bob's public key $pk_B$ through the action $enc_{pk_B}(D)$, then Alice sends $ENC_{pk_B}(D)$ to Mallory through the channel $C_{AM}$ (the corresponding sending action is denoted $s_{C_{AM}}(ENC_{pk_B}(D))$); if $d_1 \neq B$, she encrypts the message $\perp$ (a special meaningless message) with Mallory's public key $pk_M$ through the action $enc_{pk_M}(\perp)$, then Alice sends $ENC_{pk_M}(\perp)$ to Mallory through the channel $C_{AM}$ (the corresponding sending action is denoted $s_{C_{AM}}(ENC_{pk_M}(\perp))$);

6. Mallory receives $ENC_{pk_M}(d_3)$ from Alice through the channel $C_{AM}$ (the corresponding reading action is denoted $r_{C_{AM}}(ENC_{pk_M}(d_3))$), he decrypts the message with his private key $sk_M$ through the action $dec_{sk_M}(ENC_{pk_M}(d_3))$ to obtain the message $d_3$ (maybe $\perp$ or another meaningless data, all denoted $\perp$), and sends $\perp$ to the outside through the channel $C_M$ (the corresponding sending action is denoted $s_{C_M}(\perp)$), then he encrypts $\perp$ with Bob's public key $pk_B$ through the action $enc_{pk_B}(\perp)$ and sends $ENC_{pk_B}(\perp)$ to Bob through the channel $C_{MB}$ (the corresponding sending action is denoted $s_{C_{MB}}(ENC_{pk_B}(\perp))$);
7. Bob receives the message $ENC_{pk_B}(\perp)$ through the channel $C_{MB}$ (the corresponding reading action is denoted $r_{C_{MB}}(ENC_{pk_B}(\perp))$), after a decryption processing $dec_{sk_B}(ENC_{pk_B}(\perp))$ to obtain the message $\perp$, then he sends $\perp$ to the outside through the channel $C_{BO}$ (the corresponding sending action is denoted $s_{C_{BO}}(\perp)$),

where $D \in \Delta$, $\Delta$ is the set of data.

Alice's state transitions described by $APTC_G$ are as follows:

$A = \sum_{D\in\Delta} r_{C_{AI}}(D) \cdot A_2$

$A_2 = s_{C_{AM}}(Me) \cdot A_3$

$A_3 = r_{C_{MA}}(SIGN_{sk_T}(d_1, d_2)) \cdot A_4$

$A_4 = de\text{-}sign_{pk_T}(SIGN_{sk_T}(d_1, d_2)) \cdot A_5$

$A_5 = (\{d_1 = B\} \cdot enc_{pk_B}(D) \cdot s_{C_{AM}}(ENC_{pk_B}(D)) + \{d_1 \neq B\} \cdot enc_{pk_M}(\perp) \cdot s_{C_{AM}}(ENC_{pk_M}(\perp))) \cdot A$

Bob's state transitions described by $APTC_G$ are as follows:

$B = r_{C_{MB}}(Me) \cdot B_2$

$B_2 = s_{C_{BM}}(SIGN_{sk_T}(B, pk_B)) \cdot B_3$

$B_3 = r_{C_{MB}}(ENC_{pk_B}(\perp)) \cdot B_4$

$B_4 = dec_{sk_B}(ENC_{pk_B}(\perp)) \cdot B_5$

$B_5 = s_{C_{BO}}(\perp) \cdot B$

Mallory's state transitions described by $APTC_G$ are as follows:

$Ma = r_{C_{AM}}(Me) \cdot Ma_2$

$Ma_2 = s_{C_{MB}}(Me) \cdot Ma_3$

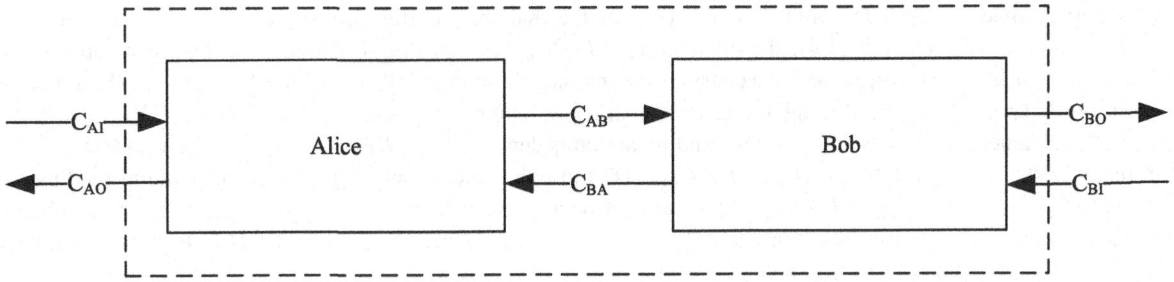


FIGURE 26.5 Key and message transmission protocol.

$Ma_3 = r_{C_{BM}}(SIGN_{sk_T}(B, pk_B)) \cdot Ma_4$
$Ma_4 = s_{C_{MA}}(SIGN_{sk_T}(M, pk_M)) \cdot Ma_5$
$Ma_5 = r_{C_{AM}}(ENC_{pk_M}(d_3)) \cdot Ma_6$
$Ma_6 = dec_{sk_M}(ENC_{pk_M}(d_3)) \cdot Ma_7$
$Ma_7 = s_{C_M}(\perp) \cdot Ma_8$
$Ma_8 = enc_{pk_B}(\perp) \cdot Ma_9$
$Ma_9 = s_{C_{MB}}(ENC_{pk_B}(\perp)) \cdot Ma$

The sending action and the reading action of the same type of data through the same channel can communicate with each other, otherwise, they will cause a deadlock $\delta$. We define the following communication functions:

$\gamma(r_{C_{AM}}(Me), s_{C_{AM}}(Me)) \triangleq c_{C_{AM}}(Me)$
$\gamma(r_{C_{MB}}(Me), s_{C_{MB}}(Me)) \triangleq c_{C_{MB}}(Me)$
$\gamma(r_{C_{BM}}(SIGN_{sk_T}(B, pk_B)), s_{C_{BM}}(SIGN_{sk_T}(B, pk_B))) \triangleq c_{C_{BM}}(SIGN_{sk_T}(B, pk_B))$
$\gamma(r_{C_{MA}}(SIGN_{sk_T}(M, pk_M)), s_{C_{MA}}(SIGN_{sk_T}(M, pk_M))) \triangleq c_{C_{MA}}(SIGN_{sk_T}(M, pk_M))$
$\gamma(r_{C_{AM}}(ENC_{pk_M}(d_3)), s_{C_{AM}}(ENC_{pk_M}(d_3))) \triangleq c_{C_{AM}}(ENC_{pk_M}(d_3))$
$\gamma(r_{C_{MB}}(ENC_{pk_B}(\perp)), s_{C_{MB}}(ENC_{pk_B}(\perp))) \triangleq c_{C_{MB}}(ENC_{pk_B}(\perp))$

Let all modules be in parallel, then the protocol $A \quad B \quad Ma$ can be presented by the following process term:

$$\tau_I(\partial_H(\Theta(A \between B \between Ma))) = \tau_I(\partial_H(A \between B \between Ma))$$

where $H = \{r_{C_{AM}}(Me), s_{C_{AM}}(Me), r_{C_{MB}}(Me), s_{C_{MB}}(Me), r_{C_{BM}}(SIGN_{sk_T}(B, pk_B)),$
$s_{C_{BM}}(SIGN_{sk_T}(B, pk_B)), r_{C_{MA}}(SIGN_{sk_T}(M, pk_M)), s_{C_{MA}}(SIGN_{sk_T}(M, pk_M)),$
$r_{C_{AM}}(ENC_{pk_M}(d_3)), s_{C_{AM}}(ENC_{pk_M}(d_3)), r_{C_{MB}}(ENC_{pk_B}(\perp)), s_{C_{MB}}(ENC_{pk_B}(\perp)) | D \in \Delta\},$
$I = \{c_{C_{AM}}(Me), c_{C_{MB}}(Me), c_{C_{BM}}(SIGN_{sk_T}(B, pk_B)), c_{C_{MA}}(SIGN_{sk_T}(M, pk_M)),$
$c_{C_{AM}}(ENC_{pk_M}(d_3)), c_{C_{MB}}(ENC_{pk_B}(\perp)), de\text{-}sign_{pk_T}(SIGN_{sk_T}(d_1, d_2)),$
$\{d_1 = B\}, ENC_{pk_B}(D), \{d_1 \neq B\}, enc_{pk_M}(\perp), dec_{sk_B}(ENC_{pk_B}(\perp)),$
$dec_{sk_M}(ENC_{pk_M}(d_3)), enc_{pk_B}(\perp) | D \in \Delta\}$

Then, we obtain the following conclusion on the protocol.

**Theorem 26.4.** *The key exchange protocol with digital signature in Fig. 26.4 is secure.*

*Proof.* Based on the above state transitions of the above modules, by use of the algebraic laws of $APTC_G$, we can prove that

$\tau_I(\partial_H(A \between B \between Ma)) = \sum_{D\in\Delta}(r_{C_{AI}}(D) \cdot s_{C_M}(\perp) \cdot s_{C_{BO}}(\perp)) \cdot \tau_I(\partial_H(A \between B \between Ma))$

For the details of the proof, please refer to Section 17.3, as we omit them here.

That is, the protocol in Fig. 26.4 $\tau_I(\partial_H(A \between B \between Ma))$ can exhibit the desired external behaviors, that is, Alice and Bob can be aware of the existence of the man in the middle. □

## 26.5 Key and message transmission

The protocol shown in Fig. 26.5 uses digital signature for secure communication, that is, Alice and Bob have shared their public keys $pk_A$ and $pk_B$, and the public keys are signed by Trent: $SIGN_{sk_T}(A, pk_A)$, $SIGN_{sk_T}(B, pk_B)$. Note that Trent's public key $pk_T$ is well known. There is no session key exchange process before the message is transferred.

The process of the protocol is as follows, and we only consider the message in one direction: from Alice to Bob:

1. Alice receives some messages $D$ from the outside through the channel $C_{AI}$ (the corresponding reading action is denoted $r_{C_{AI}}(D)$), she has $SIGN_{sk_T}(B, pk_B)$, she $de\text{-}sign_{pk_T}(SIGN_{sk_T}(B, pk_B))$ and obtains $pk_B$, then generates a session key $k_{AB}$ through an action $rsg_{k_AB}$, and she encrypts the message $D$ with $k_{AB}$ through an action $enc_{k_{AB}}(D)$ and encrypts $k_{AB}$ with Bob's public key $pk_B$ through the action $enc_{pk_B}(k_{AB})$, then Alice sends $ENC_{pk_B}(k_{AB}), ENC_{k_{AB}}(D)$ to Bob through the channel $C_{AB}$ (the corresponding sending action is denoted $s_{C_{AB}}(ENC_{pk_B}(k_{AB}), ENC_{k_{AB}}(D))$);
2. Bob receives the message $ENC_{pk_B}(k_{AB}), ENC_{k_{AB}}(D)$ through the channel $C_{AB}$ (the corresponding reading action is denoted $r_{C_{AB}}(ENC_{pk_B}(k_{AB}), ENC_{k_{AB}}(D))$), after a decryption processing $dec_{sk_B}(ENC_{pk_B}(k_{AB}))$ to obtain the message $k_{AB}$ and a decryption processing $dec_{k_{AB}}(ENC_{k_{AB}}(D))$ to obtain $D$, then he sends $D$ to the outside through the channel $C_{BO}$ (the corresponding sending action is denoted $s_{C_{BO}}(D)$),

where $D \in \Delta$, $\Delta$ is the set of data.

Alice's state transitions described by $APTC_G$ are as follows:

$A = \sum_{D\in\Delta} r_{C_{AI}}(D) \cdot A_2$

$A_2 = de\text{-}sign_{pk_T}(SIGN_{sk_T}(B, pk_B)) \cdot A_3$

$A_3 = rsg_{k_{AB}} \cdot A_4$

$A_4 = enc_{k_{AB}}(D) \cdot A_5$

$A_5 = enc_{pk_B}(k_{AB}) \cdot A_6$

$A_6 = s_{C_{AB}}(ENC_{pk_B}(k_{AB}), ENC_{k_{AB}}(D)) \cdot A$

Bob's state transitions described by $APTC_G$ are as follows:

$B = r_{C_{AB}}(ENC_{pk_B}(k_{AB}), ENC_{k_{AB}}(D)) \cdot B_2$

$B_2 = dec_{sk_B}(ENC_{pk_B}(k_{AB})) \cdot B_3$

$B_3 = dec_{k_{AB}}(ENC_{k_{AB}}(D)) \cdot B_4$

$B_4 = s_{C_{BO}}(D) \cdot B$

The sending action and the reading action of the same type of data through the same channel can communicate with each other, otherwise, they will cause a deadlock $\delta$. We define the following communication functions:

$\gamma(r_{C_{AB}}(ENC_{pk_B}(k_{AB}), ENC_{k_{AB}}(D)), s_{C_{AB}}(ENC_{pk_B}(k_{AB}), ENC_{k_{AB}}(D)))$
$\triangleq c_{C_{AB}}(ENC_{pk_B}(k_{AB}), ENC_{k_{AB}}(D))$

Let all modules be in parallel, then the protocol $A$ $B$ can be presented by the following process term:

$$\tau_I(\partial_H(\Theta(A \between B))) = \tau_I(\partial_H(A \between B))$$

where $H = \{r_{C_{AB}}(ENC_{pk_B}(k_{AB}), ENC_{k_{AB}}(D)), s_{C_{AB}}(ENC_{pk_B}(k_{AB}), ENC_{k_{AB}}(D))|D \in \Delta\}$,

$I = \{c_{C_{AB}}(ENC_{pk_B}(k_{AB}), ENC_{k_{AB}}(D)), de\text{-}sign_{pk_T}(SIGN_{sk_T}(B, pk_B)), rsg_{k_{AB}},$
$enc_{k_{AB}}(D), enc_{pk_B}(k_{AB}), dec_{sk_B}(ENC_{pk_B}(k_{AB})), dec_{k_{AB}}(ENC_{k_{AB}}(D))|D \in \Delta\}$

Then, we obtain the following conclusion on the protocol.

**Theorem 26.5.** *The key and message transmission protocol with digital signature in Fig. 26.5 is secure.*

*Proof.* Based on the above state transitions of the above modules, by use of the algebraic laws of $APTC_G$, we can prove that

$\tau_I(\partial_H(A \between B)) = \sum_{D\in\Delta}(r_{C_{AI}}(D) \cdot s_{C_{BO}}(D)) \cdot \tau_I(\partial_H(A \between B))$

For the details of the proof, please refer to Section 17.3, as we omit them here.

That is, the protocol in Fig. 26.5 $\tau_I(\partial_H(A \between B))$ can exhibit the desired external behaviors, and similarly to the protocol in Section 26.4, this protocol can resist the man-in-the-middle attack. □

## 26.6 Key and message broadcast

The protocol shown in Fig. 26.6 uses digital signature for secure broadcast communication, that is, Alice, Bob, Carol, and Dave have shared their public keys $pk_A$, $pk_B$, $pk_C$, and $pk_{Da}$ and the public keys are signed by the Trent: $SIGN_{sk_T}(A, pk_A)$, $SIGN_{sk_T}(B, pk_B)$, $SIGN_{sk_T}(C, pk_C)$, and $SIGN_{sk_T}(Da, pk_{Da})$. Note that Trent's public key $pk_T$ is well known. There is no session key exchange process before the message is transferred.

The process of the protocol is as follows, and we only consider the message in one direction: from Alice to Bob, Carol, and Dave.

1. Alice receives some messages $D$ from the outside through the channel $C_{AI}$ (the corresponding reading action is denoted $r_{C_{AI}}(D)$), she has $SIGN_{sk_T}(B,pk_B)$, $SIGN_{sk_T}(C,pk_C)$, and $SIGN_{sk_T}(Da,pk_{Da})$, she $de\text{-}sign_{pk_T}(SIGN_{sk_T}(B,pk_B))$

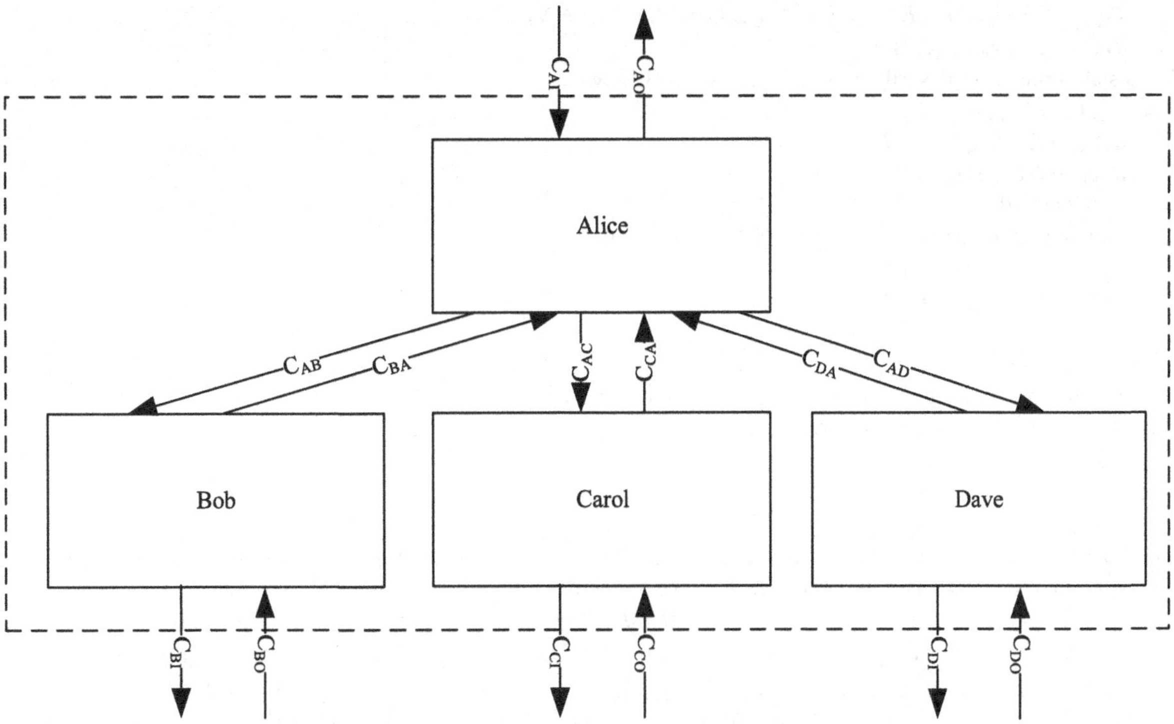


**FIGURE 26.6** Key and message broadcast protocol.

and obtains $pk_B$, $de\text{-}sign_{pk_T}(SIGN_{sk_T}(C, pk_C))$ and obtains $pk_C$, $de\text{-}sign_{pk_T}(SIGN_{sk_T}(Da, pk_{Da}))$ and obtains $pk_{Da}$, then generates a session key $k$ through an action $rsg_k$, and she encrypts the message $D$ with $k$ through an action $enc_k(D)$ and encrypts $k$ with Bob's public key $pk_B$ through the action $enc_{pk_B}(k_{AB})$, Carol's public key $pk_C$ through the action $enc_{pk_C}(k)$, Dave's public key $pk_{Da}$ through the action $enc_{pk_{Da}}(k)$, then Alice sends $ENC_{pk_B}(k), ENC_k(D)$ to Bob through the channel $C_{AB}$ (the corresponding sending action is denoted $s_{C_{AB}}(ENC_{pk_B}(k), ENC_k(D))$), sends $ENC_{pk_C}(k), ENC_k(D)$ to Bob through the channel $C_{AC}$ (the corresponding sending action is denoted $s_{C_{AC}}(ENC_{pk_C}(k), ENC_k(D))$), sends $ENC_{pk_{Da}}(k), ENC_k(D)$ to Bob through the channel $C_{AD}$ (the corresponding sending action is denoted $s_{C_{AD}}(ENC_{pk_{Da}}(k), ENC_k(D))$);

2. Bob receives the message $ENC_{pk_B}(k), ENC_k(D)$ through the channel $C_{AB}$ (the corresponding reading action is denoted $r_{C_{AB}}(ENC_{pk_B}(k), ENC_k(D))$), after a decryption processing $dec_{sk_B}(ENC_{pk_B}(k))$ to obtain the key $k$ and a decryption processing $dec_k(ENC_k(D))$ to obtain $D$, then he sends $D$ to the outside through the channel $C_{BO}$ (the corresponding sending action is denoted $s_{C_{BO}}(D)$);
3. Carol receives the message $ENC_{pk_C}(k), ENC_k(D)$ through the channel $C_{AC}$ (the corresponding reading action is denoted $r_{C_{AC}}(ENC_{pk_C}(k), ENC_k(D))$), after a decryption processing $dec_{sk_C}(ENC_{pk_C}(k))$ to obtain the message $k$ and a decryption processing $dec_k(ENC_k(D))$ to obtain $D$, then she sends $D$ to the outside through the channel $C_{CO}$ (the corresponding sending action is denoted $s_{C_{CO}}(D)$);
4. Dave receives the message $ENC_{pk_{Da}}(k), ENC_k(D)$ through the channel $C_{AD}$ (the corresponding reading action is denoted $r_{C_{AD}}(ENC_{pk_{Da}}(k), ENC_k(D))$), after a decryption processing $dec_{sk_{Da}}(ENC_{pk_{Da}}(k))$ to obtain the message $k$ and a decryption processing $dec_k(ENC_k(D))$ to obtain $D$, then he sends $D$ to the outside through the channel $C_{DO}$ (the corresponding sending action is denoted $s_{C_{DO}}(D)$),

where $D \in \Delta$, $\Delta$ is the set of data.

Alice's state transitions described by $APTC_G$ are as follows:

$A = \sum_{D\in\Delta} r_{C_{AI}}(D) \cdot A_2$

$A_2 = (de\text{-}sign_{pk_T}(SIGN_{sk_T}(B, pk_B)) \parallel de\text{-}sign_{pk_T}(SIGN_{sk_T}(C, pk_C))$
$\parallel de\text{-}sign_{pk_T}(SIGN_{sk_T}(D, pk_D))) \cdot A_3$

$A_3 = rsg_k \cdot A_4$

$A_4 = enc_k(D) \cdot A_5$

$A_5 = (enc_{pk_B}(k) \parallel enc_{pk_C}(k) \parallel enc_{pk_D}(k)) \cdot A_6$

$A_6 = (s_{C_{AB}}(ENC_{pk_B}(k), ENC_k(D)) \parallel s_{C_{AC}}(ENC_{pk_C}(k), ENC_k(D))$
$\parallel s_{C_{AD}}(ENC_{pk_D}(k), ENC_k(D))) \cdot A$

Bob's state transitions described by $APTC_G$ are as follows:

$B = r_{C_{AB}}(ENC_{pk_B}(k), ENC_k(D)) \cdot B_2$
$B_2 = dec_{sk_B}(ENC_{pk_B}(k)) \cdot B_3$
$B_3 = dec_k(ENC_k(D)) \cdot B_4$
$B_4 = s_{C_{BO}}(D) \cdot B$

Carol's state transitions described by $APTC_G$ are as follows:

$C = r_{C_{AC}}(ENC_{pk_C}(k), ENC_k(D)) \cdot C_2$
$C_2 = dec_{sk_C}(ENC_{pk_C}(k)) \cdot C_3$
$C_3 = dec_k(ENC_k(D)) \cdot C_4$
$C_4 = s_{C_{CO}}(D) \cdot C$

Dave's state transitions described by $APTC_G$ are as follows:

$Da = r_{C_{AD}}(ENC_{pk_{Da}}(k), ENC_k(D)) \cdot Da_2$
$Da_2 = dec_{sk_{Da}}(ENC_{pk_{Da}}(k)) \cdot Da_3$
$Da_3 = dec_k(ENC_k(D)) \cdot Da_4$
$Da_4 = s_{C_{DO}}(D) \cdot Da$

The sending action and the reading action of the same type of data through the same channel can communicate with each other, otherwise, they will cause a deadlock $\delta$. We define the following communication functions:

$\gamma(r_{C_{AB}}(ENC_{pk_B}(k), ENC_k(D)), s_{C_{AB}}(ENC_{pk_B}(k), ENC_k(D))) \triangleq c_{C_{AB}}(ENC_{pk_B}(k), ENC_k(D))$
$\gamma(r_{C_{AC}}(ENC_{pk_C}(k), ENC_k(D)), s_{C_{AC}}(ENC_{pk_C}(k), ENC_k(D))) \triangleq c_{C_{AC}}(ENC_{pk_C}(k), ENC_k(D))$
$\gamma(r_{C_{AD}}(ENC_{pk_{Da}}(k), ENC_k(D)), s_{C_{AD}}(ENC_{pk_{Da}}(k), ENC_k(D))) \triangleq c_{C_{AD}}(ENC_{pk_{Da}}(k), ENC_k(D))$

Let all modules be in parallel, then the protocol $A \quad B \quad C \quad Da$ can be presented by the following process term:

$$\tau_I(\partial_H(\Theta(A \between B \between C \between Da))) = \tau_I(\partial_H(A \between B \between C \between Da))$$

where $H = \{r_{C_{AB}}(ENC_{pk_B}(k), ENC_k(D)), s_{C_{AB}}(ENC_{pk_B}(k), ENC_k(D)), r_{C_{AC}}(ENC_{pk_C}(k), ENC_k(D)),$
$s_{C_{AC}}(ENC_{pk_C}(k), ENC_k(D)), r_{C_{AD}}(ENC_{pk_{Da}}(k), ENC_k(D)), s_{C_{AD}}(ENC_{pk_{Da}}(k), ENC_k(D))|D \in \Delta\}$,

$I = \{c_{C_{AB}}(ENC_{pk_B}(k), ENC_k(D)), c_{C_{AC}}(ENC_{pk_C}(k), ENC_k(D)), c_{C_{AD}}(ENC_{pk_{Da}}(k), ENC_k(D)),$
$de\text{-}sign_{pk_T}(SIGN_{sk_T}(C, pk_C)), de\text{-}sign_{pk_T}(SIGN_{sk_T}(Da, pk_{Da})), de\text{-}sign_{pk_T}(SIGN_{sk_T}(B, pk_B)),$
$rsg_k, enc_k(D), enc_{pk_B}(k), enc_{pk_C}(k), enc_{pk_{Da}}(k), dec_k(ENC_k(D)),$
$dec_{sk_B}(ENC_{pk_B}(k)), dec_{sk_C}(ENC_{pk_C}(k)), dec_{sk_{Da}}(ENC_{pk_{Da}}(k))|D \in \Delta\}$

Then, we obtain the following conclusion on the protocol.

**Theorem 26.6.** *The key and message broadcast protocol with digital signature in Fig. 26.6 is secure.*

*Proof.* Based on the above state transitions of the above modules, by use of the algebraic laws of $APTC_G$, we can prove that

$\tau_I(\partial_H(A \between B \between C \between Da)) = \sum_{D\in\Delta}(r_{C_{AI}}(D) \cdot (s_{C_{BO}}(D) \parallel s_{C_{CO}}(D) \parallel s_{C_{DO}}(D))) \cdot \tau_I(\partial_H(A \between B \between C \between Da))$

For the details of the proof, please refer to Section 17.3, as we omit them here.

That is, the protocol in Fig. 26.6 $\tau_I(\partial_H(A \between B \between C \between Da))$ can exhibit the desired external behaviors, and similarly to the protocol in Section 26.4, this protocol can resist the man-in-the-middle attack. □

Chapter 27

# Analyses of authentication protocols

An authentication protocol is used to verify the principal's identity, including verification of one principal's identity and mutual verifications of more that two principals' identities. We omit some quite simple authentication protocols, including authentication using one-way functions, etc. We will analyze mutual authentication using the interlock protocol against man-in-the-middle attack in Section 27.1, and SKID in Section 27.2.

## 27.1 Mutual authentication using the interlock protocol

The mutual authentication using the interlock protocol shown in Fig. 27.1 also uses public keys for secure communication with man-in-the-middle attack, that is, Alice and Bob have shared their public keys $pk_A$ and $pk_B$ already. However, the interlock protocol can resist man-in-the-middle attack, that is, Alice and Bob can be aware of the existence of the man in the middle.

The process of the mutual authentication using the interlock protocol with man-in-the-middle attack is as follows, we assume that Alice has "Bob's" public key $pk_M$, Bob has "Alice's" public key $pk_M$, and Mallory has Alice's public key $pk_A$ and Bob's public key $pk_B$.

1. Alice receives some password $P_A$ from the outside through the channel $C_{AI}$ (the corresponding reading action is denoted $r_{C_{AI}}(P_A)$), she encrypts the password $P_A$ with Mallory's public key $pk_M$ through the action $enc_{pk_M}(P_A)$, then Alice sends the half of $ENC_{pk_M}(P_A)$ to Mallory through the channel $C_{AM}$ (the corresponding sending action is denoted $s_{C_{AM}}(ENC_{pk_M}(P_A)/2)$);
2. Mallory receives $ENC_{pk_M}(P_A)/2$ from Alice through the channel $C_{AM}$ (the corresponding reading action is denoted $r_{C_{AM}}(ENC_{pk_M}(P_A)/2)$), he cannot decrypt the password with his private key $sk_M$, and has to make another password $P'_A$ and encrypt $P'_A$ with Bob's public key $pk_B$ through the action $enc_{pk_B}(P'_A)$, and sends the half of $ENC_{pk_B}(P'_A)$ to Bob through the channel $C_{MB}$ (the corresponding sending action is denoted $s_{C_{MB}}(ENC_{pk_B}(P'_A)/2)$);

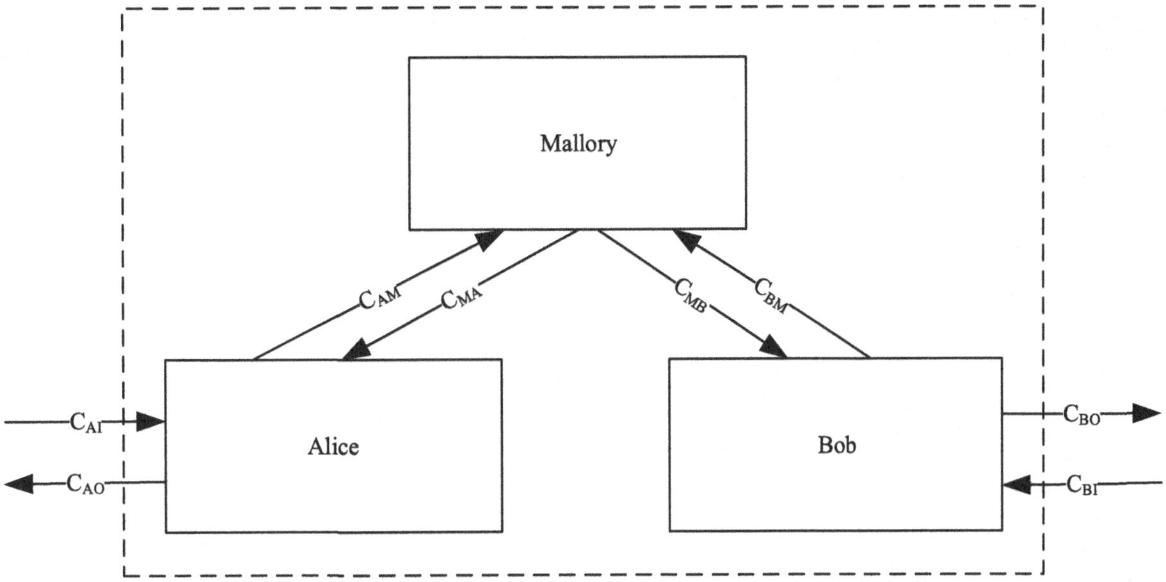


**FIGURE 27.1** Mutual authentication using the interlock protocol with man-in-the-middle attack.

Handbook of Truly Concurrent Process Algebra. https://doi.org/10.1016/B978-0-44-321515-5.00031-7

3. Bob receives the password $ENC_{pk_B}(P'_A)/2$ through the channel $C_{MB}$ (the corresponding reading action is denoted $r_{C_{MB}}(ENC_{pk_B}(P'_A)/2)$), and receives some password $P_B$ from the outside through the channel $C_{BI}$ (the corresponding reading action is denoted $r_{BI}(P_B)$), after an encryption processing $enc_{pk_M}(P_B)$ to obtain the password $ENC_{pk_M}(P_B)$, then he sends the half of $ENC_{pk_M}(P_B)$ to Mallory through the channel $C_{BM}$ (the corresponding sending action is denoted $s_{C_{BM}}(ENC_{pk_M}(P_B)/2)$);
4. Mallory receives $ENC_{pk_M}(P_B)/2$ from Bob through the channel $C_{BM}$ (the corresponding reading action is denoted $r_{C_{BM}}(ENC_{pk_M}(P_B)/2)$), he cannot decrypt the password with his private key $sk_M$, and has to make another password $P'_B$ and encrypt $P'_B$ with Alice's public key $pk_A$ through the action $enc_{pk_A}(P'_B)$, and sends the half of $ENC_{pk_A}(P'_B)$ to Alice through the channel $C_{MA}$ (the corresponding sending action is denoted $s_{C_{MA}}(ENC_{pk_A}(P'_B)/2)$);
5. Alice receives the password $ENC_{pk_A}(P'_B)/2$ through the channel $C_{MA}$ (the corresponding reading action is denoted $r_{C_{MA}}(ENC_{pk_A}(P'_B)/2)$), and sends the other half of $ENC_{pk_M}(P_A)$ to Mallory through the channel $C_{AM}$ (the corresponding sending action is denoted $s_{C_{AM}}(ENC_{pk_M}(P_A)/2)$);
6. Mallory receives $ENC_{pk_M}(P_A)/2$ from Alice through the channel $C_{AM}$ (the corresponding reading action is denoted $r_{C_{AM}}(ENC_{pk_M}(P_A)/2)$), he can combine the two halves of $ENC_{pk_M}(P_A)/2$ and decrypt the password with his private key $sk_M$, and but he has to send the other half of $ENC_{pk_B}(P'_A)$ to Bob through the channel $C_{MB}$ (the corresponding sending action is denoted $s_{C_{MB}}(ENC_{pk_B}(P'_A)/2)$);
7. Bob receives the password $ENC_{pk_B}(P'_A)/2$ through the channel $C_{MB}$ (the corresponding reading action is denoted $r_{C_{MB}}(ENC_{pk_B}(P'_A)/2)$), after a combination of two halves of $ENC_{pk_B}(P'_A)$ and a decryption processing $dec_{sk_B}(ENC_{pk_B}(P'_A))$ to obtain the password $P'_A$, then he sends it to the outside through the channel $C_{BO}$ (the corresponding sending action is denoted $s_{C_{BO}}(P'_A)$). Then, he sends the other half of $ENC_{pk_M}(P_B)$ to Mallory through the channel $C_{BM}$ (the corresponding sending action is denoted $s_{C_{BM}}(ENC_{pk_M}(P_B)/2)$);
8. Mallory receives $ENC_{pk_M}(P_B)/2$ from Bob through the channel $C_{BM}$ (the corresponding reading action is denoted $r_{C_{BM}}(ENC_{pk_M}(P_B)/2)$), he can combine the two halves of $ENC_{pk_M}(P_B)/2$ and decrypt the password with his private key $sk_M$, and but he has to send the other half of $ENC_{pk_A}(P'_B)$ to Alice through the channel $C_{MA}$ (the corresponding sending action is denoted $s_{C_{MA}}(ENC_{pk_A}(P'_B)/2)$);
9. Alice receives the password $ENC_{pk_A}(P'_B)/2$ through the channel $C_{MA}$ (the corresponding reading action is denoted $r_{C_{MA}}(ENC_{pk_A}(P'_B)/2)$), after a combination of two halves of $ENC_{pk_A}(P'_B)$ and a decryption processing $dec_{sk_A}(ENC_{pk_A}(P'_B))$ to obtain the password $P'_B$, then she sends it to the outside through the channel $C_{AO}$ (the corresponding sending action is denoted $s_{C_{AO}}(P'_B)$),

where $P_A, P_B, P'_A, P'_B \in \Delta$, $\Delta$ is the set of data.

Alice's state transitions described by $APTC_G$ are as follows:

$A = \sum_{P_A\in\Delta} r_{C_{AI}}(P_A) \cdot A_2$
$A_2 = enc_{pk_M}(P_A) \cdot A_3$
$A_3 = s_{C_{AM}}(ENC_{pk_M}(P_A)/2) \cdot A_4$
$A_4 = r_{C_{MA}}(ENC_{pk_A}(P'_B)/2) \cdot A_5$
$A_5 = s_{C_{AM}}(ENC_{pk_M}(P_A)/2) \cdot A_6$
$A_6 = r_{C_{MA}}(ENC_{pk_A}(P'_B)/2) \cdot A_7$
$A_7 = dec_{sk_A}(ENC_{pk_A}(P'_B)) \cdot A_8$
$A_8 = s_{C_{AO}}(P'_B) \cdot A$

Bob's state transitions described by $APTC_G$ are as follows:

$B = r_{C_{MB}}(ENC_{pk_B}(P'_A)/2) \cdot B_2$
$B_2 = \sum_{P_B\in\Delta} r_{BI}(P_B) \cdot B_3$
$B_3 = enc_{pk_M}(P_B) \cdot B_4$
$B_4 = s_{C_{BM}}(ENC_{pk_M}(P_B)/2) \cdot B_5$
$B_5 = r_{C_{MB}}(ENC_{pk_B}(P'_A)/2) \cdot B_6$
$B_6 = dec_{sk_B}(ENC_{pk_B}(P'_A)) \cdot B_7$
$B_7 = s_{C_{BO}}(P'_A) \cdot B_8$
$B_8 = s_{C_{BM}}(ENC_{pk_M}(P_B)/2) \cdot B$

Mallory's state transitions described by $APTC_G$ are as follows:

$Ma = r_{C_{AM}}(ENC_{pk_M}(P_A)/2) \cdot Ma_2$
$Ma_2 = enc_{pk_B}(P'_A) \cdot Ma_3$
$Ma_3 = s_{C_{MB}}(ENC_{pk_B}(P'_A)/2) \cdot Ma_4$
$Ma_4 = r_{C_{BM}}(ENC_{pk_M}(P_B)/2) \cdot Ma_5$
$Ma_5 = enc_{pk_A}(P'_B) \cdot Ma_6$

$Ma_6 = s_{C_{MA}}(ENC_{pk_A}(P'_B)/2) \cdot Ma_7$
$Ma_7 = r_{C_{AM}}(ENC_{pk_M}(P_A)/2) \cdot Ma_8$
$Ma_8 = s_{C_{MB}}(ENC_{pk_B}(P'_A)/2) \cdot Ma_9$
$Ma_9 = r_{C_{BM}}(ENC_{pk_M}(P_B)/2) \cdot Ma_{10}$
$Ma_{10} = s_{C_{MA}}(ENC_{pk_A}(P'_B)/2) \cdot Ma$

The sending action and the reading action of the same type data through the same channel can communicate with each other, otherwise, they will cause a deadlock $\delta$. We define the following communication functions:

$\gamma(r_{C_{AM}}(ENC_{pk_M}(P_A)/2), s_{C_{AM}}(ENC_{pk_M}(P_A)/2)) \triangleq c_{C_{AM}}(ENC_{pk_M}(P_A)/2)$
$\gamma(r_{C_{MB}}(ENC_{pk_B}(P'_A)/2), s_{C_{MB}}(ENC_{pk_B}(P'_A)/2)) \triangleq c_{C_{MB}}(ENC_{pk_B}(P'_A)/2)$
$\gamma(r_{C_{BM}}(ENC_{pk_M}(P_B)/2), s_{C_{BM}}(ENC_{pk_M}(P_B)/2)) \triangleq c_{C_{BM}}(ENC_{pk_M}(P_B)/2)$
$\gamma(r_{C_{MA}}(ENC_{pk_A}(P'_B)/2), s_{C_{MA}}(ENC_{pk_A}(P'_B)/2)) \triangleq c_{C_{MA}}(ENC_{pk_A}(P'_B)/2)$
$\gamma(r_{C_{AM}}(ENC_{pk_M}(P_A)/2), s_{C_{AM}}(ENC_{pk_M}(P_A)/2)) \triangleq c_{C_{AM}}(ENC_{pk_M}(P_A)/2)$
$\gamma(r_{C_{MB}}(ENC_{pk_B}(P'_A)/2), s_{C_{MB}}(ENC_{pk_B}(P'_A)/2)) \triangleq c_{C_{MB}}(ENC_{pk_B}(P'_A)/2)$
$\gamma(r_{C_{BM}}(ENC_{pk_M}(P_B)/2), s_{C_{BM}}(ENC_{pk_M}(P_B)/2)) \triangleq c_{C_{BM}}(ENC_{pk_M}(P_B)/2)$
$\gamma(r_{C_{MA}}(ENC_{pk_A}(P'_B)/2), s_{C_{MA}}(ENC_{pk_A}(P'_B)/2)) \triangleq c_{C_{MA}}(ENC_{pk_A}(P'_B)/2)$

Let all modules be in parallel, then the protocol $A$ $B$ $Ma$ can be presented by the following process term:

$$\tau_I(\partial_H(\Theta(A \between B \between Ma))) = \tau_I(\partial_H(A \between B \between Ma))$$

where $H = \{r_{C_{AM}}(ENC_{pk_M}(P_A)/2), s_{C_{AM}}(ENC_{pk_M}(P_A)/2), r_{C_{MB}}(ENC_{pk_B}(P'_A)/2),$
$s_{C_{MB}}(ENC_{pk_B}(P'_A)/2), r_{C_{BM}}(ENC_{pk_M}(P_B)/2), s_{C_{BM}}(ENC_{pk_M}(P_B)/2),$
$r_{C_{MA}}(ENC_{pk_A}(P'_B)/2), s_{C_{MA}}(ENC_{pk_A}(P'_B)/2), r_{C_{AM}}(ENC_{pk_M}(P_A)/2),$
$s_{C_{AM}}(ENC_{pk_M}(P_A)/2), r_{C_{MB}}(ENC_{pk_B}(P'_A)/2), s_{C_{MB}}(ENC_{pk_B}(P'_A)/2),$
$r_{C_{BM}}(ENC_{pk_M}(P_B)/2), s_{C_{BM}}(ENC_{pk_M}(P_B)/2), r_{C_{MA}}(ENC_{pk_A}(P'_B)/2),$
$s_{C_{MA}}(ENC_{pk_A}(P'_B)/2) | P_A, P_B, P'_A, P'_B \in \Delta\},$

$I = \{c_{C_{AM}}(ENC_{pk_M}(P_A)/2), c_{C_{MB}}(ENC_{pk_B}(P'_A)/2), c_{C_{BM}}(ENC_{pk_M}(P_B)/2), c_{C_{MA}}(ENC_{pk_A}(P'_B)/2),$
$c_{C_{AM}}(ENC_{pk_M}(P_A)/2), c_{C_{MB}}(ENC_{pk_B}(P'_A)/2), c_{C_{BM}}(ENC_{pk_M}(P_B)/2), c_{C_{MA}}(ENC_{pk_A}(P'_B)/2),$
$enc_{pk_M}(P_A), dec_{sk_A}(ENC_{pk_A}(P'_B)), enc_{pk_M}(P_B), dec_{sk_B}(ENC_{pk_B}(P'_A)), enc_{pk_B}(P'_A),$
$enc_{pk_A}(P'_B) | P_A, P_B, P'_A, P'_B \in \Delta\}$

Then, we obtain the following conclusion on the protocol.

**Theorem 27.1.** *The mutual authentication using the interlock protocol in Fig. 27.1 is secure.*

*Proof.* Based on the above state transitions of the above modules, by use of the algebraic laws of $APTC_G$, we can prove that

$\tau_I(\partial_H(A \between B \between Ma)) = \sum_{P_A, P_B, P'_A, P'_B \in \Delta}(r_{C_{AI}}(P_A) \cdot r_{C_{BI}}(P_B) \cdot s_{C_{BO}}(P'_A) \cdot s_{C_{AO}}(P'_B)) \cdot \tau_I(\partial_H(A \between B \between Ma))$

For the details of the proof, please refer to Section 17.3, as we omit them here.

That is, the mutual authentication using the interlock protocol in Fig. 27.1 $\tau_I(\partial_H(A \between B \between Ma))$ can exhibit the desired external behaviors, that is, Alice and Bob can be aware of the existence of the man in the middle. □

## 27.2 SKID

The SKID protocol shown in Fig. 27.2 uses symmetric cryptography to authenticate each other, that is, Alice and Bob have shared their key $k_{AB}$.

The process of the protocol is as follows, and we only consider the message in one direction: from Alice to Bob.

1. Alice receives some messages $D$ from the outside through the channel $C_{AI}$ (the corresponding reading action is denoted $r_{C_{AI}}(D)$), she generates a random number $R_A$ through an action $rsg_{R_A}$, she sends $R_A$ to Bob through the channel $C_{AB}$ (the corresponding sending action is denoted $s_{C_{AB}}(R_A)$);
2. Bob receives the number $R_A$ through the channel $C_{AB}$ (the corresponding reading action is denoted $r_{C_{AB}}(R_A)$), he generates a random number $R_B$ through an action $rsg_{R_B}$, and generates a MAC (Message Authentication Code) through an action $mac_{k_{AB}}(R_A, R_B, B)$, then he sends $B, R_B, MAC_{k_{AB}}(R_A, R_B, B)$ to Alice through the channel $C_{BA}$ (the corresponding sending action is denoted $s_{C_{BA}}(B, R_B, MAC_{k_{AB}}(R_A, R_B, B)))$;
3. Alice receives $d_B, d_{R_B}, d_{MAC_{k_{AB}}(R_A, R_B, B)}$ from Bob through the channel $C_{BA}$ (the corresponding reading action is denoted $r_{C_{BA}}(d_B, d_{R_B}, d_{MAC_{k_{AB}}(R_A, R_B, B)})$), she generates a MAC through an action $mac_{k_{AB}}(R_A, d_{R_B}, d_B)$, if

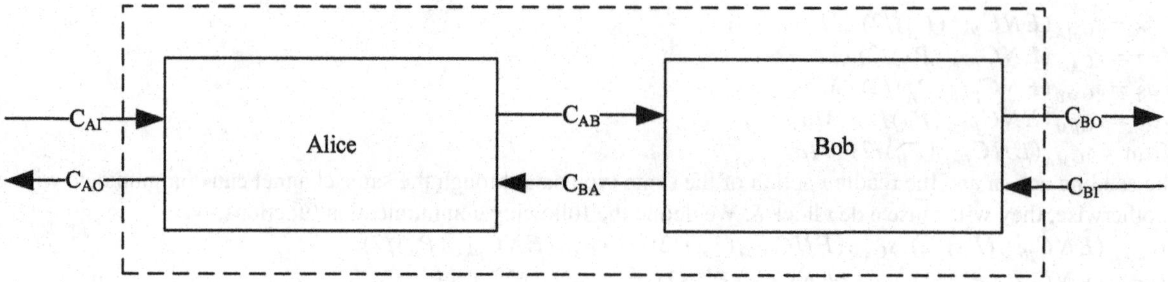


**FIGURE 27.2** SKID protocol.

$MAC_{k_{AB}}(R_A, d_{R_B}, d_B) = d_{MAC_{k_{AB}}(R_A,R_B,B)}$, she generates a MAC through an action $mac_{k_{AB}}(R_B, A)$ and encrypts $D$ by $k_{AB}$ through an action $enc_{k_{AB}}(D)$, then she sends $A, MAC_{k_{AB}}(R_B, A), ENC_{k_{AB}}(D)$ to Bob through the channel $C_{AB}$ (the corresponding sending action is denoted $s_{C_{AB}}(A, MAC_{k_{AB}}(R_B, A), ENC_{k_{AB}}(D))$);

4. Bob receives the data $d_A, d_{MAC_{k_{AB}}(R_B,A)}, ENC_{k_{AB}}(D)$ from Alice through the channel $C_{AB}$ (the corresponding reading action is denoted $r_{C_{AB}}(d_A, d_{MAC_{k_{AB}}(R_B,A)}, ENC_{k_{AB}}(D))$), he generates a MAC through an action $mac_{k_{AB}}(R_B, d_A)$, if $MAC_{k_{AB}}(R_B, d_A) = d_{MAC_{k_{AB}}(R_B,A)}$, he decrypts $ENC_{k_{AB}}(D)$ by $k_{AB}$ through an action $dec_{k_{AB}}(ENC_{k_{AB}}(D))$ to obtain $D$, then he sends $D$ to the outside through the channel $C_{BO}$ (the corresponding sending action is denoted $s_{C_{BO}}(D)$),

where $D \in \Delta$, $\Delta$ is the set of data.

Alice's state transitions described by $APTC_G$ are as follows:

$A = \sum_{D\in\Delta} r_{C_{AI}}(D) \cdot A_2$

$A_2 = rsg_{R_A} \cdot A_3$

$A_3 = s_{C_{AB}}(R_A) \cdot A_4$

$A_4 = r_{C_{BA}}(d_B, d_{R_B}, d_{MAC_{k_{AB}}(R_A,R_B,B)}) \cdot A_5$

$A_5 = mac_{k_{AB}}(R_A, d_{R_B}, d_B) \cdot A_6$

$A_6 = \{MAC_{k_{AB}}(R_A, d_{R_B}, d_B) = d_{MAC_{k_{AB}}(R_A,R_B,B)}\} \cdot mac_{k_{AB}}(R_B, A) \cdot enc_{k_{AB}}(D)$
$\cdot s_{C_{AB}}(A, MAC_{k_{AB}}(R_B, A), ENC_{k_{AB}}(D)) \cdot A$

Bob's state transitions described by $APTC_G$ are as follows:

$B = r_{C_{AB}}(R_A) \cdot B_2$

$B_2 = rsg_{R_B} \cdot B_3$

$B_3 = mac_{k_{AB}}(R_A, R_B, B) \cdot B_4$

$B_4 = s_{C_{BA}}(B, R_B, MAC_{k_{AB}}(R_A, R_B, B)) \cdot B_5$

$B_5 = r_{C_{AB}}(d_A, d_{MAC_{k_{AB}}(R_B,A)}, ENC_{k_{AB}}(D)) \cdot B_6$

$B_6 = mac_{k_{AB}}(R_B, d_A) \cdot B_7$

$B_7 = \{MAC_{k_{AB}}(R_B, d_A) = d_{MAC_{k_{AB}}(R_B,A)}\} \cdot dec_{k_{AB}}(ENC_{k_{AB}}(D)) \cdot s_{C_{BO}}(D) \cdot B$

The sending action and the reading action of the same type data through the same channel can communicate with each other, otherwise, they will cause a deadlock $\delta$. We define the following communication functions:

$\gamma(r_{C_{AB}}(R_A), s_{C_{AB}}(R_A)) \triangleq c_{C_{AB}}(R_A)$

$\gamma(r_{C_{BA}}(B, R_B, MAC_{k_{AB}}(R_A, R_B, B)), s_{C_{BA}}(B, R_B, MAC_{k_{AB}}(R_A, R_B, B)))$
$\triangleq c_{C_{BA}}(B, R_B, MAC_{k_{AB}}(R_A, R_B, B))$

$\gamma(r_{C_{AB}}(d_A, d_{MAC_{k_{AB}}(R_B,A)}, ENC_{k_{AB}}(D)), s_{C_{AB}}(d_A, d_{MAC_{k_{AB}}(R_B,A)}, ENC_{k_{AB}}(D)))$
$\triangleq c_{C_{AB}}(d_A, d_{MAC_{k_{AB}}(R_B,A)}, ENC_{k_{AB}}(D))$

Let all modules be in parallel, then the protocol $A \quad B$ can be presented by the following process term:

$$\tau_I(\partial_H(\Theta(A \between B))) = \tau_I(\partial_H(A \between B))$$

where $H = \{r_{C_{AB}}(R_A), s_{C_{AB}}(R_A), r_{C_{BA}}(B, R_B, MAC_{k_{AB}}(R_A, R_B, B)),$
$s_{C_{BA}}(B, R_B, MAC_{k_{AB}}(R_A, R_B, B)), r_{C_{AB}}(d_A, d_{MAC_{k_{AB}}(R_B,A)}, ENC_{k_{AB}}(D)),$
$s_{C_{AB}}(d_A, d_{MAC_{k_{AB}}(R_B,A)}, ENC_{k_{AB}}(D))|D \in \Delta\}$,

$I = \{c_{C_{AB}}(R_A), c_{C_{BA}}(B, R_B, MAC_{k_{AB}}(R_A, R_B, B)), c_{C_{AB}}(d_A, d_{MAC_{k_{AB}}(R_B,A)}, ENC_{k_{AB}}(D)),$
$rsg_{R_A}, mac_{k_{AB}}(R_A, d_{R_B}, d_B), \{MAC_{k_{AB}}(R_A, d_{R_B}, d_B) = d_{MAC_{k_{AB}}(R_A,R_B,B)}\}, mac_{k_{AB}}(R_B, A),$
$enc_{k_{AB}}(D), rsg_{R_B}, mac_{k_{AB}}(R_A, R_B, B), mac_{k_{AB}}(R_B, d_A), \{MAC_{k_{AB}}(R_B, d_A) = d_{MAC_{k_{AB}}(R_B,A)}\},$
$dec_{k_{AB}}(ENC_{k_{AB}}(D))|D \in \Delta\}$

Then, we obtain the following conclusion on the protocol.

**Theorem 27.2.** *The key and message transmission protocol with digital signature in Fig. 27.2 is secure.*

*Proof.* Based on the above state transitions of the above modules, by use of the algebraic laws of $APTC_G$, we can prove that

$\tau_I(\partial_H(A \between B)) = \sum_{D\in\Delta}(r_{C_{AI}}(D) \cdot s_{C_{BO}}(D)) \cdot \tau_I(\partial_H(A \between B))$

For the details of the proof, please refer to Section 17.3, as we omit them here.

That is, the protocol in Fig. 27.2 $\tau_I(\partial_H(A \between B))$ can exhibit the desired external behaviors, and similarly to the protocol in Section 26.4, without leasing of $k_{AB}$, this protocol can resist the man-in-the-middle attack. □

Chapter 28

# Analyses of practical protocols

In this chapter, we will introduce analyses of some practical authentication and key exchange protocols. For a perfectly practical security protocol, it should resist any kind of attack. There are many kinds of attacks, it is difficult to model all known attacks, for simplicity, we only analyze the protocols with several kinds of main attacks.

We introduce analyses of Wide-Mouth Frog protocol in Section 28.1, Yahalom protocol in Section 28.2, Needham–Schroeder protocol in Section 28.3, Otway–Rees protocol in Section 28.4, Kerberos protocol in Section 28.5, Neuman–Stubblebine protocol in Section 28.6, Denning–Sacco protocol in Section 28.7, DASS protocol in Section 28.8 and Woo–Lam protocol in Section 28.9.

## 28.1 Wide-Mouth Frog protocol

The Wide-Mouth Frog protocol shown in Fig. 28.1 uses symmetric keys for secure communication, that is, the key $k_{AB}$ between Alice and Bob is privately shared to Alice and Bob, Alice and Bob have shared keys with Trent $k_{AT}$ and $k_{BT}$ already.

The process of the protocol is as follows:

1. Alice receives some messages $D$ from the outside through the channel $C_{AI}$ (the corresponding reading action is denoted $r_{C_{AI}}(D)$), if $k_{AB}$ is not established, she generates a random session key $k_{AB}$ through an action $rsg_{k_{AB}}$, encrypts the key request message $T_A, B, k_{AB}$ with $k_{AT}$ through an action $enc_{k_{AT}}(T_A, B, k_{AB})$, where $T_A$ is Alice's time stamp, and sends $A, ENC_{k_{AT}}(T_A, B, k_{AB})$ to Trent through the channel $C_{AT}$ (the corresponding sending action is denoted $s_{C_{AT}}(A, ENC_{k_{AT}}(T_A, B, k_{AB}))$);
2. Trent receives the message $A, ENC_{k_{AT}}(T_A, B, k_{AB})$ through the channel $C_{AT}$ (the corresponding reading action is denoted $r_{C_{AT}}(A, ENC_{k_{AT}}(T_A, B, k_{AB}))$), he decrypts the message through an action $dec_{k_{AT}}(ENC_{k_{AT}}(T_A, B, k_{AB}))$. If $isFresh(T_A) = TRUE$, where $isFresh$ is a function to decide whether a time stamp is fresh, he encrypts $T_B, A, K_{AB}$ with $k_{BT}$ through an action $enc_{k_{BT}}(T_B, A, K_{AB})$, sends $\top$ to Alice through the channel $C_{TA}$ (the corresponding sending

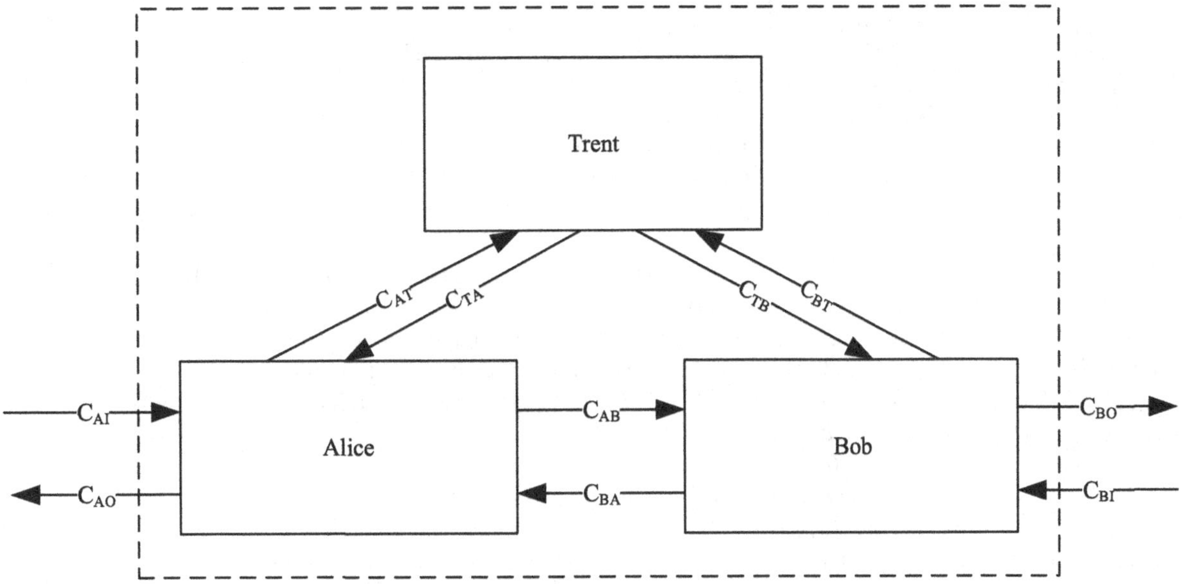


**FIGURE 28.1** Wide-Mouth Frog protocol.

Handbook of Truly Concurrent Process Algebra. https://doi.org/10.1016/B978-0-44-321515-5.00032-9

action is denoted $s_{C_{TA}}(\top)$) and $ENC_{k_{BT}}(T_B, A, K_{AB})$ to Bob through the channel $C_{TB}$ (the corresponding sending action is denoted $s_{C_{TB}}(ENC_{k_{BT}}(T_B, A, K_{AB}))$); else if $isFresh(T_A) = FALSE$, he sends $\bot$ to Alice and Bob (the corresponding sending actions are denoted $s_{C_{TA}}(\bot)$ and $s_{C_{TB}}(\bot)$, respectively);

3. Bob receives $d_{TB}$ from Trent through the channel $C_{TB}$ (the corresponding reading action is denoted $r_{C_{TB}}(d_{TB})$). If $d_{TB} = \bot$, he sends $\bot$ to Alice through the channel $C_{BA}$ (the corresponding sending action is denoted $s_{C_{BA}}(\bot)$); if $d_{TB} \neq \bot$, he decrypts $ENC_{k_{BT}}(T_B, A, K_{AB})$ through an action $dec_{k_{BT}}(ENC_{k_{BT}}(T_B, A, K_{AB}))$. If $isFresh(T_B) = TRUE$, he obtains $k_{AB}$, and sends $\top$ to Alice (the corresponding sending action is denoted $s_{C_{BA}}(\top)$); if $isFresh(T_B) = FALSE$, he sends $\bot$ to Alice through the channel $C_{BA}$ (the corresponding sending action is denoted $s_{C_{BA}}(\bot)$);
4. Alice receives $d_{TA}$ from Trent through the channel $C_{TA}$ (the corresponding reading action is denoted $r_{C_{TA}}(d_{TA})$), receives $d_{BA}$ from Bob through the channel $C_{BA}$ (the corresponding reading action is denoted $r_{C_{BA}}(d_{BA})$). If $d_{TA} = \top \cdot d_{BA} = \top$, after an encryption processing $enc_{k_{AB}}(T_{A_D}, D)$, Alice sends $ENC_{k_{AB}}(T_{A_D}, D)$ to Bob through the channel $C_{AB}$ (the corresponding sending action is denoted $s_{C_{AB}}(T_{A_D}, ENC_{k_{AB}}(D))$); else if $d_{TA} = \bot + d_{BA} = \bot$, Alice sends $\bot$ to the outside through the channel $C_{AO}$ (the corresponding sending action is denoted $s_{C_{AO}}(\bot)$);
5. Bob receives the message $ENC_{k_{AB}}(T_{A_D}, D)$ through the channel $C_{AB}$ (the corresponding reading action is denoted $r_{C_{AB}}(T_{A_D}, ENC_{k_{AB}}(D))$), after a decryption processing $dec_{k_{AB}}(ENC_{k_{AB}}(T_{A_D}, D))$, if $isFresh(T_{A_D}) = TRUE$, he sends $D$ to the outside through the channel $C_{BO}$ (the corresponding sending action is denoted $s_{C_{BO}}(D)$), if $isFresh(T_{A_D}) = FALSE$, he sends $\bot$ to the outside through the channel $C_{BO}$ (the corresponding sending action is denoted $s_{C_{BO}}(\bot)$),

where $D \in \Delta$, $\Delta$ is the set of data.

Alice's state transitions described by $APTC_G$ are as follows:

$A = \sum_{D\in\Delta} r_{C_{AI}}(D) \cdot A_2$

$A_2 = \{k_{AB} = NULL\} \cdot rsg_{k_{AB}} \cdot A_3 + \{k_{AB} \neq NULL\} \cdot A_7$

$A_3 = enc_{k_{AT}}(T_A, B, k_{AB}) \cdot A_4$

$A_4 = s_{C_{AT}}(A, ENC_{k_{AT}}(T_A, B, k_{AB})) \cdot A_5$

$A_5 = (r_{C_{TA}}(d_{TA}) \parallel r_{C_{BA}}(d_{BA})) \cdot A_6$

$A_6 = \{d_{TA} = \top \cdot d_{BA} = \top\} \cdot A_7 + \{d_{TA} = \bot + d_{BA} = \bot\} \cdot A_9$

$A_7 = enc_{k_{AB}}(T_{A_D}, D) \cdot A_8$

$A_8 = s_{C_{AB}}(T_{A_D}, ENC_{k_{AB}}(D)) \cdot A$

$A_9 = s_{C_{AO}}(\bot) \cdot A$

Bob's state transitions described by $APTC_G$ are as follows:

$B = \{k_{AB} = NULL\} \cdot B_1 + \{k_{AB} \neq NULL\} \cdot B_5$

$B_1 = r_{C_{TB}}(d_{TB}) \cdot B_2$

$B_2 = \{d_{TB} \neq \bot\} \cdot B_3 + \{d_{TB} = \bot\} \cdot s_{C_{BA}}(\bot) \cdot B$

$B_3 = dec_{k_{BT}}(ENC_{k_{BT}}(T_B, A, K_{AB})) \cdot B_4$

$B_4 = \{isFresh(T_B) = TRUE\} \cdot s_{C_{BA}}(\top) \cdot B_5 + \{isFresh(T_B) = FALSE\} \cdot s_{C_{BA}}(\bot) \cdot B$

$B_5 = r_{C_{AB}}(T_{A_D}, ENC_{k_{AB}}(D)) \cdot B_6$

$B_6 = dec_{k_{AB}}(ENC_{k_{AB}}(T_{A_D}, D)) \cdot B_7$

$B_7 = \{isFresh(T_{A_D}) = TRUE\} \cdot s_{C_{BO}}(D) \cdot B + \{isFresh(T_{A_D}) = FALSE\} \cdot s_{C_{BO}}(\bot) \cdot B$

Trent's state transitions described by $APTC_G$ are as follows:

$T = r_{C_{AT}}(A, ENC_{k_{AT}}(T_A, B, k_{AB})) \cdot T_2$

$T_2 = dec_{k_{AT}}(ENC_{k_{AT}}(T_A, B, k_{AB})) \cdot T_3$

$T_3 = \{isFresh(T_A) = TRUE\} \cdot enc_{k_{BT}}(T_B, A, K_{AB}) \cdot (s_{C_{TA}}(\top) \parallel s_{C_{TB}}(ENC_{k_{BT}}(T_B, A, K_{AB})))T$
$+ \{isFresh(T_A) = FALSE\} \cdot (s_{C_{TA}}(\bot) \parallel s_{C_{TB}}(\bot)) \cdot T$

The sending action and the reading action of the same type data through the same channel can communicate with each other, otherwise, they will cause a deadlock $\delta$. We define the following communication functions:

$\gamma(r_{C_{AT}}(A, ENC_{k_{AT}}(T_A, B, k_{AB})), s_{C_{AT}}(A, ENC_{k_{AT}}(T_A, B, k_{AB}))) \triangleq c_{C_{AT}}(A, ENC_{k_{AT}}(T_A, B, k_{AB}))$

$\gamma(r_{C_{TA}}(d_{TA}), s_{C_{TA}}(d_{TA})) \triangleq c_{C_{TA}}(d_{TA})$

$\gamma(r_{C_{BA}}(d_{BA}), s_{C_{BA}}(d_{BA})) \triangleq c_{C_{BA}}(d_{BA})$

$\gamma(r_{C_{AB}}(T_{A_D}, ENC_{k_{AB}}(D)), s_{C_{AB}}(T_{A_D}, ENC_{k_{AB}}(D))) \triangleq c_{C_{AB}}(T_{A_D}, ENC_{k_{AB}}(D))$

$\gamma(r_{C_{TB}}(d_{TB}), s_{C_{TB}}(d_{TB})) \triangleq c_{C_{TB}}(d_{TB})$

Let all modules be in parallel, then the protocol $A \quad B \quad T$ can be presented by the following process term:

$$\tau_I(\partial_H(\Theta(A \between B \between T))) = \tau_I(\partial_H(A \between B \between T))$$

where $H = \{r_{C_{AT}}(A, ENC_{k_{AT}}(T_A, B, k_{AB})), s_{C_{AT}}(A, ENC_{k_{AT}}(T_A, B, k_{AB})),$
$r_{C_{TA}}(d_{TA}), s_{C_{TA}}(d_{TA}), r_{C_{BA}}(d_{BA}), s_{C_{BA}}(d_{BA}),$
$r_{C_{AB}}(T_{A_D}, ENC_{k_{AB}}(D)), s_{C_{AB}}(T_{A_D}, ENC_{k_{AB}}(D)), r_{C_{TB}}(d_{TB}), s_{C_{TB}}(d_{TB})|D \in \Delta\},$
$I = \{c_{C_{AT}}(A, ENC_{k_{AT}}(T_A, B, k_{AB})), c_{C_{TA}}(d_{TA}), c_{C_{BA}}(d_{BA}),$
$c_{C_{AB}}(T_{A_D}, ENC_{k_{AB}}(D)), c_{C_{TB}}(d_{TB}), \{k_{AB} = NULL\}, rsg_{k_{AB}},$
$\{k_{AB} \neq NULL\}, enc_{k_{AT}}(T_A, B, k_{AB}), \{d_{TA} = \top \cdot d_{BA} = \top\}, \{d_{TA} = \bot + d_{BA} = \bot\},$
$enc_{k_{AB}}(T_{A_D}, D), \{d_{TB} \neq \bot\}, \{d_{TB} = \bot\}, dec_{k_{BT}}(ENC_{k_{BT}}(T_B, A, K_{AB})),$
$\{isFresh(T_B) = TRUE\}, \{isFresh(T_B) = FALSE\}, dec_{k_{AB}}(ENC_{k_{AB}}(T_{A_D}, D)),$
$\{isFresh(T_{A_D}) = TRUE\}, \{isFresh(T_{A_D}) = FALSE\}, dec_{k_{AT}}(ENC_{k_{AT}}(T_A, B, k_{AB})),$
$\{isFresh(T_A) = TRUE\}, enc_{k_{BT}}(T_B, A, K_{AB}), \{isFresh(T_A) = FALSE\}|D \in \Delta\}$

Then, we obtain the following conclusion on the protocol.

**Theorem 28.1.** *The Wide-Mouth Frog protocol in Fig. 28.1 is secure.*

*Proof.* Based on the above state transitions of the above modules, by use of the algebraic laws of $APTC_G$, we can prove that

$\tau_I(\partial_H(A \between B \between T)) = \sum_{D\in\Delta}(r_{C_{AI}}(D) \cdot ((s_{C_{AO}}(\bot) \parallel s_{C_{BO}}(\bot)) + s_{C_{BO}}(D))) \cdot \tau_I(\partial_H(A \between B \between T))$

For the details of the proof, please refer to Section 17.3, as we omit them here.

That is, the Wide-Mouth Frog protocol in Fig. 28.1 $\tau_I(\partial_H(A \between B \between T))$ can exhibit the desired external behaviors:

1. For information leakage, because $k_{AT}$ is privately shared only between Alice and Trent, $k_{BT}$ is privately shared only between Bob and Trent, $k_{AB}$ is privately shared only among Trent, Alice, and Bob. For the modeling of confidentiality, it is similar to the protocol in Section 25.10.2, the Wide-Mouth Frog protocol is confidential;
2. For replay attack, the use of time stamps $T_A$, $T_B$, and $T_{A_D}$ makes it that $\tau_I(\partial_H(A \between B \between T)) = \sum_{D\in\Delta}(r_{C_{AI}}(D) \cdot (s_{C_{AO}}(\bot) \parallel s_{C_{BO}}(\bot))) \cdot \tau_I(\partial_H(A \between B \between T))$, as desired;
3. Without replay attack, the protocol would be $\tau_I(\partial_H(A \between B \between T)) = \sum_{D\in\Delta}(r_{C_{AI}}(D) \cdot s_{C_{BO}}(D)) \cdot \tau_I(\partial_H(A \between B \between T))$, as desired;
4. For the man-in-the-middle attack, because $k_{AT}$ is privately shared only between Alice and Trent, $k_{BT}$ is privately shared only between Bob and Trent, $k_{AB}$ is privately shared only among Trent, Alice, and Bob. For the modeling of the man-in-the-middle attack, it is similar to the protocol in Section 26.4, the Wide-Mouth Frog protocol can be used against the man-in-the-middle attack;
5. For the unexpected and non-technical leaking of $k_{AT}$, $k_{BT}$, $k_{AB}$, or they are not strong enough, or Trent is dishonest, they are out of the scope of analyses of security protocols;
6. For malicious tampering and transmission errors, they are out of the scope of analyses of security protocols. □

## 28.2 Yahalom protocol

The Yahalom protocol shown in Fig. 28.2 uses symmetric keys for secure communication, that is, the key $k_{AB}$ between Alice and Bob is privately shared to Alice and Bob, Alice and Bob have shared keys with Trent $k_{AT}$ and $k_{BT}$ already.

The process of the protocol is as follows:

1. Alice receives some messages $D$ from the outside through the channel $C_{AI}$ (the corresponding reading action is denoted $r_{C_{AI}}(D)$), if $k_{AB}$ is not established, she generates a random number $R_A$ through an action $rsg_{R_A}$, and sends $A, R_A$ to Bob through the channel $C_{AB}$ (the corresponding sending action is denoted $s_{C_{AB}}(A, R_A)$);
2. Bob receives $A, R_A$ from Alice through the channel $C_{AB}$ (the corresponding reading action is denoted $r_{C_{AB}}(A, R_A)$), he generates a random number $R_B$ through an action $rsg_{R_B}$, encrypts $A, R_A, R_B$ by $k_{BT}$ through an action $enc_{k_{BT}}(A, R_A, R_B)$, and sends $ENC_{k_{BT}}(A, R_A, R_B)$ to Trent through the channel $C_{BT}$ (the corresponding sending action is denoted $s_{C_{BT}}(ENC_{k_{BT}}(A, R_A, R_B)))$;
3. Trent receives $ENC_{k_{BT}}(A, R_A, R_B)$ through the channel $C_{BT}$ (the corresponding reading action is denoted $r_{C_{BT}}(ENC_{k_{BT}}(A, R_A, R_B)))$, he decrypts the message through an action $dec_{k_{BT}}(ENC_{k_{BT}}(A, R_A, R_B))$, generates a random session key $k_{AB}$ through an action $rsg_{k_{AB}}$, then he encrypts $B, k_{AB}, R_A, R_B$ by $k_{AT}$ through an action $enc_{k_{AT}}(B, k_{AB}, R_A, R_B)$, encrypts $A, k_{AB}$ by $k_{BT}$ through an action $enc_{k_{BT}}(A, k_{AB})$, and sends them to Alice through the channel $C_{TA}$ (the corresponding sending action is denoted $s_{C_{TA}}(ENC_{k_{AT}}(B, k_{AB}, R_A, R_B), ENC_{k_{BT}}(A, k_{AB})))$;
4. Alice receives the message from Trent through the channel $C_{TA}$ (the corresponding reading action is denoted $r_{C_{TA}}(ENC_{k_{AT}}(B, k_{AB}, d_{R_A}, R_B), ENC_{k_{BT}}(A, k_{AB})))$, she decrypts $ENC_{k_{AT}}(B, k_{AB}, d_{R_A}, R_B)$ by $k_{AT}$ through an action $dec_{k_{AT}}(ENC_{k_{AT}}(B, k_{AB}, d_{R_A}, R_B))$, if $d_{R_A} = R_A$, she encrypts $R_B, D$ by $k_{AB}$ through an action $enc_{k_{AB}}(R_B, D)$,

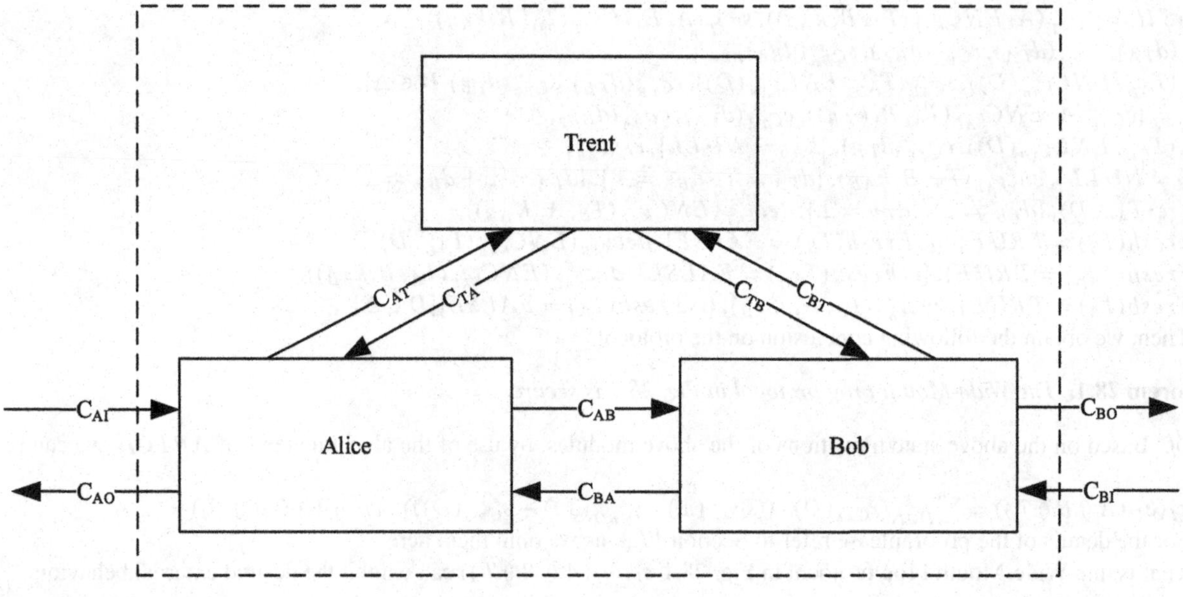


**FIGURE 28.2** Yahalom protocol.

and sends $ENC_{k_{BT}}(A, k_{AB}), ENC_{k_{AB}}(R_B, D)$ to Bob through the channel $C_{AB}$ (the corresponding sending action is denoted $s_{C_{AB}}(ENC_{k_{BT}}(A, k_{AB}), ENC_{k_{AB}}(R_B, D)))$; else if $d_{R_A} \neq R_A$, she sends $\perp$ to Bob through the channel $C_{AB}$ (the corresponding sending action is denoted $s_{C_{AB}}(\perp)$);

5. Bob receives $d_{AB}$ from Alice (the corresponding reading action is denoted $r_{C_{AB}}(d_{AB})$), if $d_{AB} = \perp$, he sends $\perp$ to the outside through the channel $C_{BO}$ (the corresponding sending action is denoted $s_{C_{BO}}(\perp)$); else if $d_{AB} \neq \perp$, $d_{AB}$ must be of the form of $ENC_{k_{BT}}(A, k_{AB}), ENC_{k_{AB}}(d_{R_B}, D)$ (without considering the malicious tampering and transmission errors), he decrypts $ENC_{k_{BT}}(A, k_{AB})$ by $k_{BT}$ through an action $dec_{k_{BT}}(ENC_{k_{BT}}(A, k_{AB}))$ to ensure the message is from Alice and obtain $k_{AB}$, then he decrypts $ENC_{k_{AB}}(d_{R_B}, D)$ by $k_{AB}$ through an action $dec_{k_{AB}}(ENC_{k_{AB}}(d_{R_B}, D))$, if $d_{R_B} = R_B$, he sends $D$ to the outside through the channel $C_{BO}$ (the corresponding sending action is denoted $s_{C_{BO}}(D)$), else if $d_{R_B} \neq R_B$, he sends $\perp$ to the outside through the channel $C_{BO}$ (the corresponding sending action is denoted $s_{C_{BO}}(\perp)$),

where $D \in \Delta$, $\Delta$ is the set of data.

Alice's state transitions described by $APTC_G$ are as follows:

$A = \sum_{D\in\Delta} r_{C_{AI}}(D) \cdot A_2$

$A_2 = \{k_{AB} = NULL\} \cdot rsg_{R_A} \cdot A_3 + \{k_{AB} \neq NULL\} \cdot A_7$

$A_3 = s_{C_{AB}}(A, R_A) \cdot A_4$

$A_4 = r_{C_{TA}}(ENC_{k_{AT}}(B, k_{AB}, d_{R_A}, R_B), ENC_{k_{BT}}(A, k_{AB})) \cdot A_5$

$A_5 = dec_{k_{AT}}(ENC_{k_{AT}}(B, k_{AB}, d_{R_A}, R_B)) \cdot A_6$

$A_6 = \{d_{R_A} = R_A\} \cdot A_7 + \{d_{R_A} \neq R_A\} \cdot A_9$

$A_7 = enc_{k_{AB}}(R_B, D) \cdot A_8$

$A_8 = s_{C_{AB}}(ENC_{k_{BT}}(A, k_{AB}), ENC_{k_{AB}}(R_B, D)) \cdot A$

$A_9 = s_{C_{AB}}(\perp) \cdot A$

Bob's state transitions described by $APTC_G$ are as follows:

$B = \{k_{AB} = NULL\} \cdot B_1 + \{k_{AB} \neq NULL\} \cdot B_5$

$B_1 = r_{C_{AB}}(A, R_A) \cdot B_2$

$B_2 = rsg_{R_B} \cdot B_3$

$B_3 = enc_{k_{BT}}(A, R_A, R_B) \cdot B_4$

$B_4 = s_{C_{BT}}(ENC_{k_{BT}}(A, R_A, R_B)) \cdot B_5$

$B_5 = r_{C_{AB}}(d_{AB}) \cdot B_6$

$B_6 = \{d_{AB} = \perp\} \cdot s_{C_{BO}}(\perp) \cdot B + \{d_{AB} \neq \perp\} \cdot B_7$

$B_7 = dec_{k_{BT}}(ENC_{k_{BT}}(A, k_{AB})) \cdot B_8$

$B_8 = dec_{k_{AB}}(ENC_{k_{AB}}(d_{R_B}, D)) \cdot B_9$

$B_9 = \{d_{R_B} = R_B\} \cdot s_{C_{BO}}(D) \cdot B + \{d_{R_B} \neq R_B\} \cdot s_{C_{BO}}(\perp) \cdot B$

Trent's state transitions described by $APTC_G$ are as follows:

$T = r_{C_{BT}}(ENC_{k_{BT}}(A, R_A, R_B)) \cdot T_2$

$T_2 = dec_{k_{BT}}(ENC_{k_{BT}}(A, R_A, R_B)) \cdot T_3$

$T_3 = rsg_{k_{AB}} \cdot T_4$

$T_4 = enc_{k_{AT}}(B, k_{AB}, R_A, R_B) \cdot T_5$

$T_5 = enc_{k_{BT}}(A, k_{AB}) \cdot T_6$

$T_6 = s_{C_{TA}}(ENC_{k_{AT}}(B, k_{AB}, R_A, R_B), ENC_{k_{BT}}(A, k_{AB})) \cdot T$

The sending action and the reading action of the same type data through the same channel can communicate with each other, otherwise, they will cause a deadlock $\delta$. We define the following communication functions:

$\gamma(r_{C_{AB}}(A, R_A), s_{C_{AB}}(A, R_A)) \triangleq c_{C_{AB}}(A, R_A)$

$\gamma(r_{C_{TA}}(ENC_{k_{AT}}(B, k_{AB}, d_{R_A}, R_B), ENC_{k_{BT}}(A, k_{AB})),$
$s_{C_{TA}}(ENC_{k_{AT}}(B, k_{AB}, d_{R_A}, R_B), ENC_{k_{BT}}(A, k_{AB})))$
$\triangleq c_{C_{TA}}(ENC_{k_{AT}}(B, k_{AB}, d_{R_A}, R_B), ENC_{k_{BT}}(A, k_{AB}))$

$\gamma(r_{C_{BT}}(ENC_{k_{BT}}(A, R_A, R_B)), s_{C_{BT}}(ENC_{k_{BT}}(A, R_A, R_B))) \triangleq c_{C_{BT}}(ENC_{k_{BT}}(A, R_A, R_B))$

$\gamma(r_{C_{AB}}(d_{AB}), s_{C_{AB}}(d_{AB})) \triangleq c_{C_{AB}}(d_{AB})$

Let all modules be in parallel, then the protocol $A \quad B \quad T$ can be presented by the following process term:

$$\tau_I(\partial_H(\Theta(A \between B \between T))) = \tau_I(\partial_H(A \between B \between T))$$

where $H = \{r_{C_{AB}}(A, R_A), s_{C_{AB}}(A, R_A), r_{C_{AB}}(d_{AB}), s_{C_{AB}}(d_{AB}),$
$r_{C_{TA}}(ENC_{k_{AT}}(B, k_{AB}, d_{R_A}, R_B), ENC_{k_{BT}}(A, k_{AB})), s_{C_{TA}}(ENC_{k_{AT}}(B, k_{AB}, d_{R_A}, R_B),$
$ENC_{k_{BT}}(A, k_{AB})), r_{C_{BT}}(ENC_{k_{BT}}(A, R_A, R_B)), s_{C_{BT}}(ENC_{k_{BT}}(A, R_A, R_B))|D \in \Delta\},$

$I = \{c_{C_{AB}}(A, R_A), c_{C_{TA}}(ENC_{k_{AT}}(B, k_{AB}, d_{R_A}, R_B), ENC_{k_{BT}}(A, k_{AB})),$
$c_{C_{BT}}(ENC_{k_{BT}}(A, R_A, R_B)), c_{C_{AB}}(d_{AB}), \{k_{AB} = NULL\}, rsg_{R_A}, \{k_{AB} \neq NULL\},$
$dec_{k_{AT}}(ENC_{k_{AT}}(B, k_{AB}, d_{R_A}, R_B)), \{d_{R_A} = R_A\}, \{d_{R_A} \neq R_A\},$
$enc_{k_{AB}}(R_B, D), rsg_{R_B}, enc_{k_{BT}}(A, R_A, R_B), \{d_{AB} = \perp\}, \{d_{AB} \neq \perp\},$
$dec_{k_{BT}}(ENC_{k_{BT}}(A, k_{AB})), dec_{k_{AB}}(ENC_{k_{AB}}(d_{R_B}, D)),$
$\{d_{R_B} = R_B\}, \{d_{R_B} \neq R_B\}, dec_{k_{BT}}(ENC_{k_{BT}}(A, R_A, R_B)), rsg_{k_{AB}},$
$enc_{k_{AT}}(B, k_{AB}, R_A, R_B), enc_{k_{BT}}(A, k_{AB})|D \in \Delta\}$

Then, we obtain the following conclusion on the protocol.

**Theorem 28.2.** *The Yahalom protocol in Fig. 28.2 is secure.*

*Proof.* Based on the above state transitions of the above modules, by use of the algebraic laws of $APTC_G$, we can prove that

$\tau_I(\partial_H(A \between B \between T)) = \sum_{D\in\Delta}(r_{C_{AI}}(D) \cdot (s_{C_{BO}}(\perp) + s_{C_{BO}}(D))) \cdot \tau_I(\partial_H(A \between B \between T))$

For the details of the proof, please refer to Section 17.3, as we omit them here.

That is, the Yahalom protocol in Fig. 28.2 $\tau_I(\partial_H(A \between B \between T))$ can exhibit the desired external behaviors:

1. For information leakage, because $k_{AT}$ is privately shared only between Alice and Trent, $k_{BT}$ is privately shared only between Bob and Trent, $k_{AB}$ is privately shared only among Trent, Alice, and Bob. For the modeling of confidentiality, it is similar to the protocol in Section 25.10.2, the Yahalom protocol is confidential;
2. For the man-in-the-middle attack, because $k_{AT}$ is privately shared only between Alice and Trent, $k_{BT}$ is privately shared only between Bob and Trent, $k_{AB}$ is privately shared only among Trent, Alice, and Bob, and the use of the random numbers $R_A$ and $R_B$, the protocol would be $\tau_I(\partial_H(A \between B \between T)) = \sum_{D\in\Delta}(r_{C_{AI}}(D) \cdot s_{C_{BO}}(\perp)) \cdot \tau_I(\partial_H(A \between B \between T))$, as desired, the Yahalom protocol can be used against the man-in-the-middle attack;
3. Without man-in-the-middle attack, the protocol would be $\tau_I(\partial_H(A \between B \between T)) = \sum_{D\in\Delta}(r_{C_{AI}}(D) \cdot s_{C_{BO}}(D)) \cdot \tau_I(\partial_H(A \between B \between T))$, as desired;
4. For the unexpected and non-technical leaking of $k_{AT}$, $k_{BT}$, $k_{AB}$, or they are not strong enough, or Trent is dishonest, they are out of the scope of analyses of security protocols;
5. For malicious tampering and transmission errors, they are out of the scope of analyses of security protocols. □

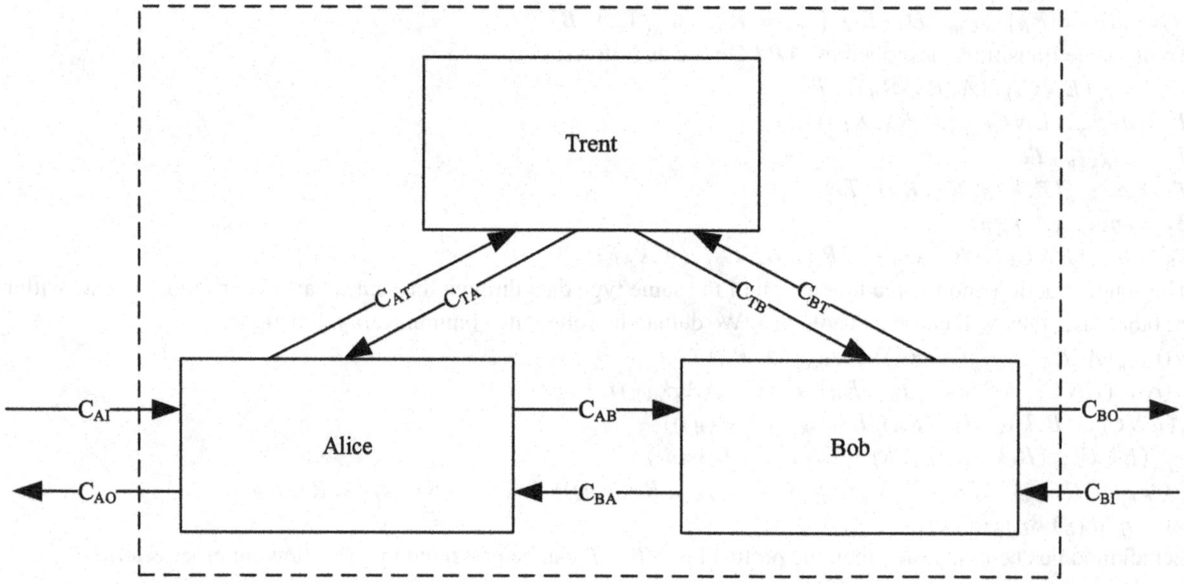


**FIGURE 28.3** Needham–Schroeder protocol.

## 28.3 Needham–Schroeder protocol

The Needham–Schroeder protocol shown in Fig. 28.3 uses symmetric keys for secure communication, that is, the key $k_{AB}$ between Alice and Bob is privately shared to Alice and Bob, Alice and Bob have shared keys with Trent $k_{AT}$ and $k_{BT}$ already.

The process of the protocol is as follows:

1. Alice receives some messages $D$ from the outside through the channel $C_{AI}$ (the corresponding reading action is denoted $r_{C_{AI}}(D)$), if $k_{AB}$ is not established, she generates a random number $R_A$ through an action $rsg_{R_A}$, and sends $A, B, R_A$ to Trent through the channel $C_{AT}$ (the corresponding sending action is denoted $s_{C_{AT}}(A, B, R_A)$);
2. Trent receives $A, B, R_A$ from Alice through the channel $C_{AT}$ (the corresponding reading action is denoted $r_{C_{AT}}(A, B, R_A)$), he generates a random session key $k_{AB}$ through an action $rsg_{k_{AB}}$, then he encrypts $A, k_{AB}$ by $k_{BT}$ through an action $enc_{k_{BT}}(A, k_{AB})$, encrypts $R_A, B, k_{AB}, ENC_{k_{BT}}(A, k_{AB})$ by $k_{AT}$ through an action $enc_{k_{AT}}(R_A, B, k_{AB}, ENC_{k_{BT}}(A, k_{AB}))$, and sends them to Alice through the channel $C_{TA}$ (the corresponding sending action is denoted $s_{C_{TA}}(ENC_{k_{AT}}(R_A, B, k_{AB}, ENC_{k_{BT}}(A, k_{AB}))))$;
3. Alice receives the message from Trent through the channel $C_{TA}$ (the corresponding reading action is denoted $r_{C_{TA}}(ENC_{k_{AT}}(d_{R_A}, B, k_{AB}, ENC_{k_{BT}}(A, k_{AB}))))$, she decrypts $ENC_{k_{AT}}(d_{R_A}, B, k_{AB}, ENC_{k_{BT}}(A, k_{AB}))$ by $k_{AT}$ through an action $dec_{k_{AT}}(ENC_{k_{AT}}(d_{R_A}, B, k_{AB}, ENC_{k_{BT}}(A, k_{AB})))$, if $d_{R_A} = R_A$, she sends $ENC_{k_{BT}}(A, k_{AB})$ to Bob through the channel $C_{AB}$ (the corresponding sending action is denoted $s_{C_{AB}}(ENC_{k_{BT}}(A, k_{AB})))$; else if $d_{R_A} \neq R_A$, she sends $\perp$ to Bob through the channel $C_{AB}$ (the corresponding sending action is denoted $s_{C_{AB}}(\perp)$);
4. Bob receives $d_{AB}$ from Alice (the corresponding reading action is denoted $r_{C_{AB}}(d_{AB})$), if $d_{AB} = \perp$, he sends $\perp$ to the outside through the channel $C_{BO}$ (the corresponding sending action is denoted $s_{C_{BO}}(\perp)$), and sends $\perp$ to Alice through the channel $C_{AB}$ (the corresponding sending action is denoted $s_{C_{AB}}(\perp)$); else if $d_{AB} \neq \perp$, $d_{AB}$ must be of the form of $ENC_{k_{BT}}(A, k_{AB})$ (without considering the malicious tampering and transmission errors), he decrypts $ENC_{k_{BT}}(A, k_{AB})$ by $k_{BT}$ through an action $dec_{k_{BT}}(ENC_{k_{BT}}(A, k_{AB}))$ to ensure the message is from Alice and obtain $k_{AB}$, then he generates a random number $R_B$ through an action $rsg_{R_B}$, encrypts $R_B$ by $k_{AB}$ through an action $enc_{k_{AB}}(R_B)$, and sends $ENC_{k_{AB}}(R_B)$ to Alice through the channel $C_{BA}$ (the corresponding sending action is denoted $s_{C_{BA}}(ENC_{k_{AB}}(R_B)))$;
5. Alice receives $d_{BA}$ from Bob through the channel $C_{BA}$ (the corresponding reading action is denoted $r_{C_{BA}}(d_{BA})$), if $d_{BA} \neq \perp$, she decrypts $ENC_{k_{AB}}(R_B)$ to obtain $R_B$ by $k_{AB}$ through an action $dec_{k_{AB}}(ENC_{k_{AB}}(R_B))$, encrypts $R_B - 1, D$ through an action $enc_{k_{AB}}(R_B - 1, D)$, and sends $ENC_{k_{AB}}(R_B - 1, D)$ to Bob through the channel $C_{AB}$ (the corresponding sending action is denoted $s_{C_{AB}}(ENC_{k_{AB}}(R_B - 1, D)))$; else if $d_{BA} = \perp$, she sends $\perp$ to Bob through the channel $C_{AB}$ (the corresponding sending action is denoted $s_{C_{AB}}(\perp)$);
6. Bob receives $d'_{AB}$ from Alice through the channel $C_{AB}$ (the corresponding reading action is denoted $r_{C_{AB}}(d'_{AB})$), if $d'_{AB} \neq \perp$, he decrypts $ENC_{k_{AB}}(d_{R_B-1}, D)$ by $k_{AB}$ through an action $dec_{k_{AB}}(ENC_{k_{AB}}(d_{R_B-1}, D))$, if $d_{R_B-1} = R_B - 1$,

he sends $D$ to the outside through the channel $C_{BO}$ (the corresponding sending action is denoted $s_{C_{BO}}(D)$), else if $d_{R_B-1} \neq R_B - 1$, he sends $\perp$ to the outside through the channel $C_{BO}$ (the corresponding sending action is denoted $s_{C_{BO}}(\perp)$); else if $d'_{AB} = \perp$, he sends $\perp$ to the outside through the channel $C_{BO}$ (the corresponding sending action is denoted $s_{C_{BO}}(\perp)$),

where $D \in \Delta$, $\Delta$ is the set of data.

Alice's state transitions described by $APTC_G$ are as follows:

$A = \sum_{D\in\Delta} r_{C_{AI}}(D) \cdot A_2$

$A_2 = \{k_{AB} = NULL\} \cdot rsg_{R_A} \cdot A_3 + \{k_{AB} \neq NULL\} \cdot A_{11}$

$A_3 = s_{C_{AT}}(A, B, R_A) \cdot A_4$

$A_4 = r_{C_{TA}}(ENC_{k_{AT}}(d_{R_A}, B, k_{AB}, ENC_{k_{BT}}(A, k_{AB}))) \cdot A_5$

$A_5 = dec_{k_{AT}}(enc_{k_{AT}}(d_{R_A}, B, k_{AB}, ENC_{k_{BT}}(A, k_{AB}))) \cdot A_6$

$A_6 = \{d_{R_A} = R_A\} \cdot A_7 + \{d_{R_A} \neq R_A\} \cdot s_{C_{AB}}(\perp) \cdot A_8$

$A_7 = s_{C_{AB}}(ENC_{k_{BT}}(A, k_{AB})) \cdot A_8$

$A_8 = r_{C_{BA}}(d_{BA}) \cdot A_9$

$A_9 = \{d_{BA} \neq \perp\} \cdot A_{10} + \{d_{BA} = \perp\} \cdot A_{13}$

$A_{10} = dec_{k_{AB}}(ENC_{k_{AB}}(R_B)) \cdot A_{11}$

$A_{11} = enc_{k_{AB}}(R_B - 1, D) \cdot A_{12}$

$A_{12} = s_{C_{AB}}(ENC_{k_{AB}}(R_B - 1, D)) \cdot A$

$A_{13} = s_{C_{AB}}(\perp) \cdot A$

Bob's state transitions described by $APTC_G$ are as follows:

$B = \{k_{AB} = NULL\} \cdot B_1 + \{k_{AB} \neq NULL\} \cdot B_7$

$B_1 = r_{C_{AB}}(d_{AB}) \cdot B_2$

$B_2 = \{d_{AB} = \perp\} \cdot (s_{C_{BO}}(\perp) \parallel s_{C_{AB}}(\perp)) \cdot B_7 + \{d_{AB} \neq \perp\} \cdot B_3$

$B_3 = dec_{k_{BT}}(ENC_{k_{BT}}(A, k_{AB})) \cdot B_4$

$B_4 = rsg_{R_B} \cdot B_5$

$B_5 = enc_{k_{AB}}(R_B) \cdot B_6$

$B_6 = s_{C_{BA}}(ENC_{k_{AB}}(R_B)) \cdot B_7$

$B_7 = r_{C_{AB}}(d'_{AB}) \cdot B_8$

$B_8 = \{d'_{AB} \neq \perp\} \cdot B_9 + \{d'_{AB} = \perp\} \cdot s_{C_{BO}}(\perp) \cdot B$

$B_9 = dec_{k_{AB}}(ENC_{k_{AB}}(d_{R_B-1}, D)) \cdot B_{10}$

$B_{10} = \{d_{R_B-1} = R_B - 1\} \cdot s_{C_{BO}}(D) \cdot B + \{d_{R_B-1} \neq R_B - 1\} \cdot s_{C_{BO}}(\perp) \cdot B$

Trent's state transitions described by $APTC_G$ are as follows:

$T = r_{C_{AT}}(A, B, R_A) \cdot T_2$

$T_2 = rsg_{k_{AB}} \cdot T_3$

$T_3 = enc_{k_{BT}}(A, k_{AB}) \cdot T_4$

$T_4 = enc_{k_{AT}}(R_A, B, k_{AB}, ENC_{k_{BT}}(A, k_{AB})) \cdot T_5$

$T_5 = s_{C_{TA}}(ENC_{k_{AT}}(R_A, B, k_{AB}, ENC_{k_{BT}}(A, k_{AB}))) \cdot T$

The sending action and the reading action of the same type data through the same channel can communicate with each other, otherwise, they will cause a deadlock $\delta$. We define the following communication functions:

$\gamma(r_{C_{AT}}(A, B, R_A), s_{C_{AT}}(A, B, R_A)) \triangleq c_{C_{AT}}(A, B, R_A)$

$\gamma(r_{C_{TA}}(ENC_{k_{AT}}(d_{R_A}, B, k_{AB}, ENC_{k_{BT}}(A, k_{AB}))),$
$s_{C_{TA}}(ENC_{k_{AT}}(d_{R_A}, B, k_{AB}, ENC_{k_{BT}}(A, k_{AB}))))$
$\triangleq c_{C_{TA}}(ENC_{k_{AT}}(d_{R_A}, B, k_{AB}, ENC_{k_{BT}}(A, k_{AB})))$

$\gamma(r_{C_{AB}}(d_{AB}), s_{C_{AB}}(d_{AB})) \triangleq c_{C_{AB}}(d_{AB})$

$\gamma(r_{C_{BA}}(d_{BA}), s_{C_{BA}}(d_{BA})) \triangleq c_{C_{BA}}(d_{BA})$

$\gamma(r_{C_{AB}}(d'_{AB}), s_{C_{AB}}(d'_{AB})) \triangleq c_{C_{AB}}(d'_{AB})$

Let all modules be in parallel, then the protocol $A \quad B \quad T$ can be presented by the following process term:

$$\tau_I(\partial_H(\Theta(A \between B \between T))) = \tau_I(\partial_H(A \between B \between T))$$

where $H = \{r_{C_{AT}}(A, B, R_A), s_{C_{AT}}(A, B, R_A), r_{C_{AB}}(d_{AB}), s_{C_{AB}}(d_{AB}),$
$r_{C_{BA}}(d_{BA}), s_{C_{BA}}(d_{BA}), r_{C_{AB}}(d'_{AB}), s_{C_{AB}}(d'_{AB}),$
$r_{C_{TA}}(ENC_{k_{AT}}(d_{R_A}, B, k_{AB}, ENC_{k_{BT}}(A, k_{AB}))),$
$s_{C_{TA}}(ENC_{k_{AT}}(d_{R_A}, B, k_{AB}, ENC_{k_{BT}}(A, k_{AB})))|D \in \Delta\},$

$I = \{c_{C_{AT}}(A, B, R_A), c_{C_{AB}}(d_{AB}), c_{C_{BA}}(d_{BA}), c_{C_{AB}}(d'_{AB}),$
$c_{C_{TA}}(ENC_{k_{AT}}(d_{R_A}, B, k_{AB}, ENC_{k_{BT}}(A, k_{AB}))),$
$\{k_{AB} = NULL\}, rsg_{R_A}, \{k_{AB} \neq NULL\},$
$dec_{k_{AT}}(enc_{k_{AT}}(d_{R_A}, B, k_{AB}, ENC_{k_{BT}}(A, k_{AB}))),$
$\{d_{R_A} = R_A\}, \{d_{R_A} \neq R_A\}, \{d_{BA} \neq \perp\}, \{d_{BA} = \perp\},$
$dec_{k_{AB}}(ENC_{k_{AB}}(R_B)), enc_{k_{AB}}(R_B - 1, D), \{d_{AB} = \perp\}, \{d_{AB} \neq \perp\},$
$dec_{k_{BT}}(ENC_{k_{BT}}(A, k_{AB})), rsg_{R_B}, enc_{k_{AB}}(R_B),$
$\{d'_{AB} = \perp\}, \{d'_{AB} \neq \perp\}, dec_{k_{AB}}(ENC_{k_{AB}}(d_{R_B-1}, D)),$
$\{d_{R_B-1} = R_B - 1\}, \{d_{R_B-1} \neq R_B - 1\}, rsg_{k_{AB}}, enc_{k_{BT}}(A, k_{AB}),$
$enc_{k_{AT}}(R_A, B, k_{AB}, ENC_{k_{BT}}(A, k_{AB})) | D \in \Delta\}$

Then, we obtain the following conclusion on the protocol.

**Theorem 28.3.** *The Needham–Schroeder protocol in Fig. 28.3 is secure.*

*Proof.* Based on the above state transitions of the above modules, by use of the algebraic laws of $APTC_G$, we can prove that

$\tau_I(\partial_H(A \between B \between T)) = \sum_{D\in\Delta}(r_{C_{AI}}(D) \cdot (s_{C_{BO}}(\perp) + s_{C_{BO}}(D))) \cdot \tau_I(\partial_H(A \between B \between T))$

For the details of the proof, please refer to Section 17.3, as we omit them here.

That is, the Needham–Schroeder protocol in Fig. 28.3 $\tau_I(\partial_H(A \between B \between T))$ can exhibit the desired external behaviors:

1. For information leakage, because $k_{AT}$ is privately shared only between Alice and Trent, $k_{BT}$ is privately shared only between Bob and Trent, $k_{AB}$ is privately shared only among Trent, Alice, and Bob. For the modeling of confidentiality, it is similar to the protocol in Section 25.10.2, the Needham–Schroeder protocol is confidential;
2. For replay attack, the use of random numbers $R_A$, $R_B$, makes it that $\tau_I(\partial_H(A \between B \between T)) = \sum_{D\in\Delta}(r_{C_{AI}}(D) \cdot s_{C_{BO}}(\perp)) \cdot \tau_I(\partial_H(A \between B \between T))$, as desired;
3. Without replay attack, the protocol would be $\tau_I(\partial_H(A \between B \between T)) = \sum_{D\in\Delta}(r_{C_{AI}}(D) \cdot s_{C_{BO}}(D)) \cdot \tau_I(\partial_H(A \between B \between T))$, as desired;
4. For the man-in-the-middle attack, because $k_{AT}$ is privately shared only between Alice and Trent, $k_{BT}$ is privately shared only between Bob and Trent, $k_{AB}$ is privately shared only among Trent, Alice, and Bob. For the modeling of the man-in-the-middle attack, it is similar to the protocol in Section 26.4, the Needham–Schroeder protocol can be used against the man-in-the-middle attack;
5. For the unexpected and non-technical leaking of $k_{AT}$, $k_{BT}$, $k_{AB}$, or they are not strong enough, or Trent is dishonest, they are out of the scope of analyses of security protocols;
6. For malicious tampering and transmission errors, they are out of the scope of analyses of security protocols. □

## 28.4 Otway–Rees protocol

The Otway–Rees protocol shown in Fig. 28.4 uses symmetric keys for secure communication, that is, the key $k_{AB}$ between Alice and Bob is privately shared to Alice and Bob, Alice and Bob have shared keys with Trent $k_{AT}$ and $k_{BT}$ already.

The process of the protocol is as follows:

1. Alice receives some messages $D$ from the outside through the channel $C_{AI}$ (the corresponding reading action is denoted $r_{C_{AI}}(D)$), if $k_{AB}$ is not established, she generates the random numbers $I$, $R_A$ through the actions $rsg_I$ and $rsg_{R_A}$, encrypts $R_A, I, A, B$ by $k_{AT}$ through an action $enc_{k_{AT}}(R_A, I, A, B)$, and sends $I, A, B, ENC_{k_{AT}}(R_A, I, A, B)$ to Bob through the channel $C_{AB}$ (the corresponding sending action is denoted$s_{C_{AB}}(I, A, B, ENC_{k_{AT}}(R_A, I, A, B))$);
2. Bob receives $I, A, B, ENC_{k_{AT}}(R_A, I, A, B)$ from Alice through the channel $C_{AB}$ (the corresponding reading action is denoted $r_{C_{AB}}(I, A, B, ENC_{k_{AT}}(R_A, I, A, B))$), he generates a random number $R_B$ through an action $rsg_{R_B}$, encrypts $R_B, I, A, B$ by $k_{BT}$ through an action $enc_{k_{BT}}(R_B, I, A, B)$, and sends $I, A, B, ENC_{k_{AT}}(R_A, I, A, B), ENC_{k_{BT}}(R_B, I, A, B)$ to Trent through the channel $C_{BT}$ (the corresponding sending action is denoted $s_{C_{BT}}(I, A, B, ENC_{k_{AT}}(R_A, I, A, B), ENC_{k_{BT}}(R_B, I, A, B))$);
3. Trent receives $I, A, B, ENC_{k_{AT}}(R_A, I, A, B), ENC_{k_{BT}}(R_B, I, A, B)$ through the channel $C_{BT}$ (the corresponding reading action is denoted$r_{C_{BT}}(I, A, B, ENC_{k_{AT}}(R_A, I, A, B), ENC_{k_{BT}}(R_B, I, A, B))$), he decrypts the message $ENC_{k_{AT}}(R_A, I, A, B)$ through an action $dec_{k_{AT}}(ENC_{k_{AT}}(R_A, I, A, B))$ and the message $ENC_{k_{BT}}(R_B, I, A, B)$ through an action $dec_{k_{BT}}(ENC_{k_{BT}}(R_B, I, A, B))$, generates a random session key $k_{AB}$ through an action $rsg_{k_{AB}}$, then he encrypts $R_A, k_{AB}$ by $k_{AT}$ through an action $enc_{k_{AT}}(R_A, k_{AB})$, encrypts $R_B, k_{AB}$ by $k_{BT}$ through an action $enc_{k_{BT}}(R_B, k_{AB})$, and sends them to Bob through the channel $C_{TB}$ (the corresponding sending action is denoted $s_{C_{TB}}(I, ENC_{k_{AT}}(R_A, k_{AB}), ENC_{k_{BT}}(R_B, k_{AB}))$);

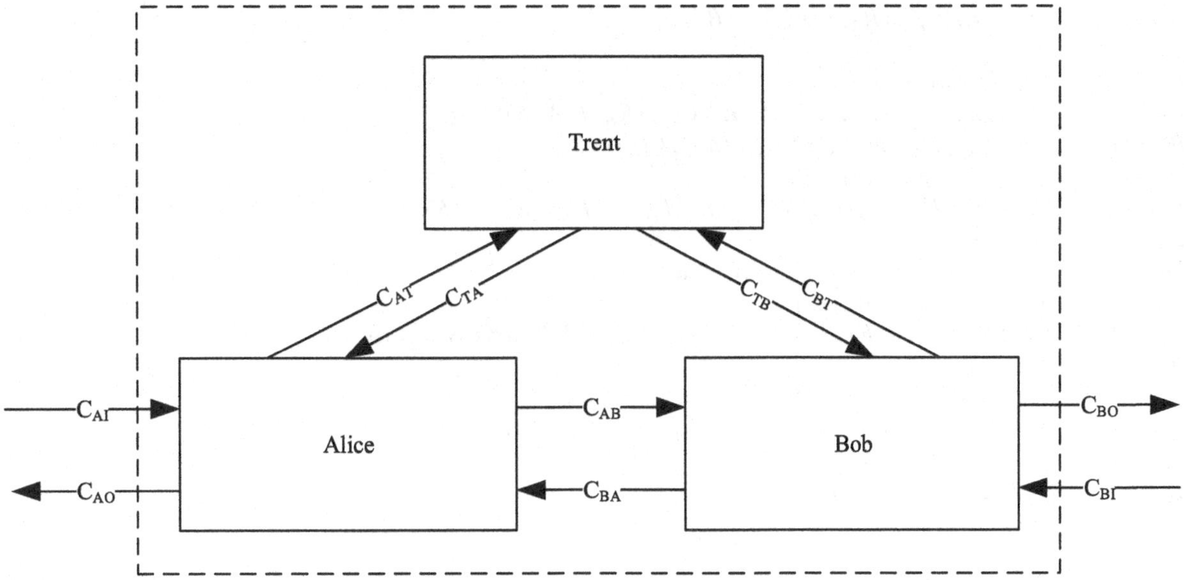


**FIGURE 28.4** Otway–Rees protocol.

4. Bob receives the message from Trent through the channel $C_{TB}$ (the corresponding reading action is denoted $r_{C_{TB}}(d_I, ENC_{k_{AT}}(R_A, k_{AB}), ENC_{k_{BT}}(d_{R_B}, k_{AB})))$, he decrypts $ENC_{k_{BT}}(d_{R_B}, k_{AB})$ by $k_{BT}$ through an action $dec_{k_{BT}}(ENC_{k_{BT}}(d_{R_B}, k_{AB}))$, if $d_{R_B} = R_B$ and $d_I = I$, he sends $I, ENC_{k_{AT}}(R_A, k_{AB})$ to Alice through the channel $C_{BA}$ (the corresponding sending action is denoted $s_{C_{BA}}(I, ENC_{k_{AB}}(R_A, k_{AB})))$; else if $d_{R_B} \neq R_B$ or $D_I \neq I$, he sends $\perp$ to Alice through the channel $C_{BA}$ (the corresponding sending action is denoted $s_{C_{BA}}(\perp)$);
5. Alice receives $d_{BA}$ from Bob (the corresponding reading action is denoted $r_{C_{BA}}(d_{BA})$), if $d_{BA} = \perp$, she sends $\perp$ to Bob through the channel $C_{AB}$ (the corresponding sending action is denoted $s_{C_{AB}}(\perp)$); else if $d_{BA} \neq \perp$, she decrypts $ENC_{k_{AT}}(R_A, k_{AB})$ by $k_{AT}$ through an action $dec_{k_{AT}}(ENC_{k_{AT}}(R_A, k_{AB}))$, if $d_{R_A} = R_A$ and $d_I = I$, she generates a random number $R_D$ through an action $rsg_{R_D}$, encrypts $R_D, D$ by $k_{AB}$ through an action $enc_{k_{AB}}(R_D, D)$, and sends it to Bob through the channel $C_{AB}$ (the corresponding sending action is denoted $s_{C_{AB}}(ENC_{k_{AB}}(R_D, D)))$, else if $d_{R_A} \neq R_A$ or $d_I \neq I$, she sends $\perp$ to Bob through the channel $C_{AB}$ (the corresponding sending action is denoted $s_{C_{AB}}(\perp)$);
6. Bob receives $d_{AB}$ from Alice (the corresponding reading action is denoted $r_{C_{AB}}(d_{AB})$), if $d_{AB} = \perp$, he sends $\perp$ to the outside through the channel $C_{BO}$ (the corresponding sending action is denoted $s_{C_{BO}}(\perp)$); else if $d_{AB} \neq \perp$, he decrypts $ENC_{k_{AB}}(R_D, D)$ by $k_{AB}$ through an action $dec_{k_{AB}}(ENC_{k_{AB}}(R_D, D))$, if $isFresh(R_D) = TRUE$, he sends $D$ to the outside through the channel $C_{BO}$ (the corresponding sending action is denoted $s_{C_{BO}}(D)$), else if $isFresh(d_{R_D}) = FALSE$, he sends $\perp$ to the outside through the channel $C_{BO}$ (the corresponding sending action is denoted $s_{C_{BO}}(\perp)$),

where $D \in \Delta$, $\Delta$ is the set of data.

Alice's state transitions described by $APTC_G$ are as follows:

$A = \sum_{D\in\Delta} r_{C_{AI}}(D) \cdot A_2$

$A_2 = \{k_{AB} = NULL\} \cdot rsg_I \cdot rsg_{R_A} \cdot A_3 + \{k_{AB} \neq NULL\} \cdot A_9$

$A_3 = enc_{k_{AT}}(R_A, I, A, B) \cdot A_4$

$A_4 = s_{C_{AB}}(I, A, B, ENC_{k_{AT}}(R_A, I, A, B)) \cdot A_5$

$A_5 = r_{C_{BA}}(d_{BA}) \cdot A_6$

$A_6 = \{d_{BA} \neq \perp\} \cdot A_7 + \{d_{BA} = \perp\} \cdot s_{C_{AB}}(\perp) \cdot A$

$A_7 = dec_{k_{AT}}(ENC_{k_{AT}}(R_A, k_{AB})) \cdot A_8$

$A_8 = \{d_{R_A} = R_A \cdot d_I = I\} \cdot A_9 + \{d_{R_A} \neq R_A + d_I \neq I\} \cdot A_{12}$

$A_9 = rsg_{R_D} \cdot A_{10}$

$A_{10} = enc_{k_{AB}}(R_D, D) \cdot A_{11}$

$A_{11} = s_{C_{AB}}(ENC_{k_{AB}}(R_D, D)) \cdot A$

$A_{12} = s_{C_{AB}}(\perp) \cdot A$

Bob's state transitions described by $APTC_G$ are as follows:

$B = \{k_{AB} = NULL\} \cdot B_1 + \{k_{AB} \neq NULL\} \cdot B_8$

$B_1 = r_{C_{AB}}(I, A, B, ENC_{k_{AT}}(R_A, I, A, B)) \cdot B_2$
$B_2 = rsg_{R_B} \cdot B_3$
$B_3 = enc_{k_{BT}}(R_B, I, A, B) \cdot B_4$
$B_4 = s_{C_{BT}}(I, A, B, ENC_{k_{AT}}(R_A, I, A, B), ENC_{k_{BT}}(R_B, I, A, B)) \cdot B_5$
$B_5 = r_{C_{TB}}(d_I, ENC_{k_{AT}}(R_A, k_{AB}), ENC_{k_{BT}}(d_{R_B}, k_{AB})) \cdot B_6$
$B_6 = dec_{k_{BT}}(ENC_{k_{BT}}(d_{R_B}, k_{AB})) \cdot B_7$
$B_7 = \{d_{R_B} = R_B \cdot d_I = I\} \cdot s_{C_{BA}}(I, ENC_{k_{AB}}(R_A, k_{AB})) \cdot B_8 + \{d_{R_B} \neq R_B + d_I \neq I\} \cdot s_{C_{AB}}(\bot) \cdot B_8$
$B_8 = r_{C_{AB}}(d_{AB}) \cdot B_9$
$B_9 = \{d_{AB} = \bot\} \cdot s_{C_{BO}}(\bot) \cdot B + \{d_{AB} \neq \bot\} \cdot B_{10}$
$B_{10} = dec_{k_{AB}}(ENC_{k_{AB}}(R_D, D)) \cdot B_{11}$
$B_{11} = \{isFresh(R_D) = TRUE\} \cdot B_{12} + \{isFresh(R_D) = FALSE\} \cdot s_{C_{BO}}(\bot) \cdot B$
$B_{12} = s_{C_{BO}}(D) \cdot B$

Trent's state transitions described by $APTC_G$ are as follows:

$T = r_{C_{BT}}(I, A, B, ENC_{k_{AT}}(R_A, I, A, B), ENC_{k_{BT}}(R_B, I, A, B)) \cdot T_2$
$T_2 = dec_{k_{AT}}(ENC_{k_{AT}}(R_A, I, A, B)) \cdot T_3$
$T_3 = dec_{k_{BT}}(ENC_{k_{BT}}(R_B, I, A, B)) \cdot T_4$
$T_4 = rsg_{k_{AB}} \cdot T_5$
$T_5 = enc_{k_{AT}}(R_A, k_{AB}) \cdot T_6$
$T_6 = enc_{k_{BT}}(R_B, k_{AB}) \cdot T_7$
$T_7 = s_{C_{TB}}(I, ENC_{k_{AT}}(R_A, k_{AB}), ENC_{k_{BT}}(R_B, k_{AB})) \cdot T$

The sending action and the reading action of the same type data through the same channel can communicate with each other, otherwise, they will cause a deadlock $\delta$. We define the following communication functions:

$\gamma(r_{C_{AB}}(I, A, B, ENC_{k_{AT}}(R_A, I, A, B)), s_{C_{AB}}(I, A, B, ENC_{k_{AT}}(R_A, I, A, B)))$
$\triangleq c_{C_{AB}}(I, A, B, ENC_{k_{AT}}(R_A, I, A, B))$
$\gamma(r_{C_{BA}}(d_{BA}), s_{C_{BA}}(d_{BA})) \triangleq c_{C_{BA}}(d_{BA})$
$\gamma(r_{C_{BT}}(I, A, B, ENC_{k_{AT}}(R_A, I, A, B), ENC_{k_{BT}}(R_B, I, A, B)),$
$s_{C_{BT}}(I, A, B, ENC_{k_{AT}}(R_A, I, A, B), ENC_{k_{BT}}(R_B, I, A, B)))$
$\triangleq c_{C_{BT}}(I, A, B, ENC_{k_{AT}}(R_A, I, A, B), ENC_{k_{BT}}(R_B, I, A, B))$
$\gamma(r_{C_{TB}}(d_I, ENC_{k_{AT}}(R_A, k_{AB}), ENC_{k_{BT}}(d_{R_B}, k_{AB})),$
$s_{C_{TB}}(d_I, ENC_{k_{AT}}(R_A, k_{AB}), ENC_{k_{BT}}(d_{R_B}, k_{AB})))$
$\triangleq c_{C_{TB}}(d_I, ENC_{k_{AT}}(R_A, k_{AB}), ENC_{k_{BT}}(d_{R_B}, k_{AB}))$
$\gamma(r_{C_{AB}}(d_{AB}), s_{C_{AB}}(d_{AB})) \triangleq c_{C_{AB}}(d_{AB})$

Let all modules be in parallel, then the protocol $A \quad B \quad T$ can be presented by the following process term:

$$\tau_I(\partial_H(\Theta(A \between B \between T))) = \tau_I(\partial_H(A \between B \between T))$$

where $H = \{r_{C_{AB}}(I, A, B, ENC_{k_{AT}}(R_A, I, A, B)), s_{C_{AB}}(I, A, B, ENC_{k_{AT}}(R_A, I, A, B)),$
$r_{C_{BA}}(d_{BA}), s_{C_{BA}}(d_{BA}), r_{C_{AB}}(d_{AB}), s_{C_{AB}}(d_{AB}),$
$r_{C_{BT}}(I, A, B, ENC_{k_{AT}}(R_A, I, A, B), ENC_{k_{BT}}(R_B, I, A, B)),$
$s_{C_{BT}}(I, A, B, ENC_{k_{AT}}(R_A, I, A, B), ENC_{k_{BT}}(R_B, I, A, B)),$
$r_{C_{TB}}(d_I, ENC_{k_{AT}}(R_A, k_{AB}), ENC_{k_{BT}}(d_{R_B}, k_{AB})),$
$s_{C_{TB}}(d_I, ENC_{k_{AT}}(R_A, k_{AB}), ENC_{k_{BT}}(d_{R_B}, k_{AB}))|D \in \Delta\},$
$I = \{c_{C_{AB}}(I, A, B, ENC_{k_{AT}}(R_A, I, A, B)), c_{C_{BA}}(d_{BA}), c_{C_{AB}}(d_{AB}),$
$c_{C_{BT}}(I, A, B, ENC_{k_{AT}}(R_A, I, A, B), ENC_{k_{BT}}(R_B, I, A, B)),$
$c_{C_{TB}}(d_I, ENC_{k_{AT}}(R_A, k_{AB}), ENC_{k_{BT}}(d_{R_B}, k_{AB})),$
$\{k_{AB} = NULL\}, rsg_I, rsg_{R_A}, \{k_{AB} \neq NULL\}, enc_{k_{AT}}(R_A, I, A, B),$
$\{d_{BA} \neq \bot\}, \{d_{BA} = \bot\}, dec_{k_{AT}}(ENC_{k_{AT}}(R_A, k_{AB})),$
$\{d_{R_A} = R_A \cdot d_I = I\}, \{d_{R_A} \neq R_A + d_I \neq I\}, rsg_{R_D},$
$enc_{k_{AB}}(R_D, D), rsg_{R_B}, enc_{k_{BT}}(R_B, I, A, B),$
$dec_{k_{BT}}(ENC_{k_{BT}}(d_{R_B}, k_{AB})), \{d_{R_B} = R_B \cdot d_I = I\},$
$\{d_{R_B} \neq R_B + d_I \neq I\}, \{d_{AB} = \bot\}, \{d_{AB} \neq \bot\},$
$dec_{k_{AB}}(ENC_{k_{AB}}(R_D, D)), \{isFresh(R_D) = TRUE\}, \{isFresh(R_D) = FALSE\},$
$dec_{k_{AT}}(ENC_{k_{AT}}(R_A, I, A, B)), dec_{k_{BT}}(ENC_{k_{BT}}(R_B, I, A, B)),$
$rsg_{k_{AB}}, enc_{k_{AT}}(R_A, k_{AB}), enc_{k_{BT}}(R_B, k_{AB})|D \in \Delta\}$

Then, we obtain the following conclusion on the protocol.

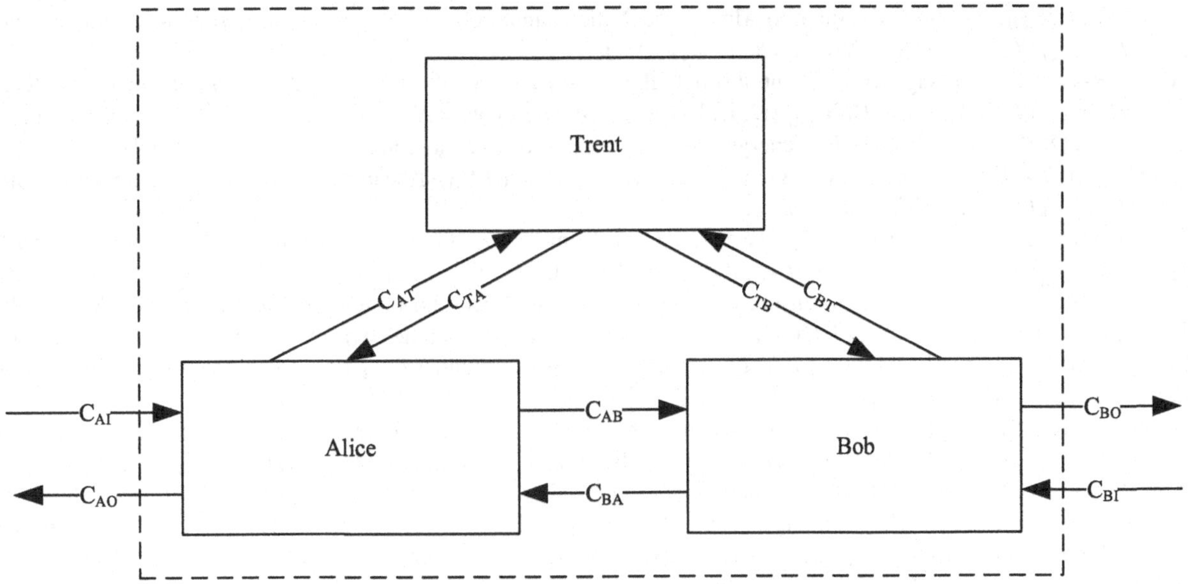


**FIGURE 28.5** Kerberos protocol.

**Theorem 28.4.** *The Otway–Rees protocol in Fig. 28.4 is secure.*

*Proof.* Based on the above state transitions of the above modules, by use of the algebraic laws of $APTC_G$, we can prove that

$\tau_I(\partial_H(A \between B \between T)) = \sum_{D\in\Delta}(r_{C_{AI}}(D) \cdot (s_{C_{BO}}(\perp) + s_{C_{BO}}(D))) \cdot \tau_I(\partial_H(A \between B \between T))$

For the details of the proof, please refer to Section 17.3, as we omit them here.

That is, the Otway–Rees protocol in Fig. 28.4 $\tau_I(\partial_H(A \between B \between T))$ can exhibit the desired external behaviors:

1. For information leakage, because $k_{AT}$ is privately shared only between Alice and Trent, $k_{BT}$ is privately shared only between Bob and Trent, $k_{AB}$ is privately shared only among Trent, Alice, and Bob. For the modeling of confidentiality, it is similar to the protocol in Section 25.10.2, the Otway–Rees protocol is confidential;
2. For the man-in-the-middle attack, because $k_{AT}$ is privately shared only between Alice and Trent, $k_{BT}$ is privately shared only between Bob and Trent, $k_{AB}$ is privately shared only among Trent, Alice, and Bob, and the use of the random numbers $I$, $R_A$ and $R_B$, the protocol would be $\tau_I(\partial_H(A \between B \between T)) = \sum_{D\in\Delta}(r_{C_{AI}}(D) \cdot s_{C_{BO}}(\perp)) \cdot \tau_I(\partial_H(A \between B \between T))$, as desired, the Otway–Rees protocol can be used against the man-in-the-middle attack;
3. For replay attack, the use of the random numbers $I$, $R_A$ and $R_B$, makes it that $\tau_I(\partial_H(A \between B \between T)) = \sum_{D\in\Delta}(r_{C_{AI}}(D) \cdot s_{C_{BO}}(\perp)) \cdot \tau_I(\partial_H(A \between B \between T))$, as desired;
4. Without man-in-the-middle and replay attack, the protocol would be $\tau_I(\partial_H(A \between B \between T)) = \sum_{D\in\Delta}(r_{C_{AI}}(D) \cdot s_{C_{BO}}(D)) \cdot \tau_I(\partial_H(A \between B \between T))$, as desired;
5. For the unexpected and non-technical leaking of $k_{AT}$, $k_{BT}$, $k_{AB}$, or they are not strong enough, or Trent is dishonest, they are out of the scope of analyses of security protocols;
6. For malicious tampering and transmission errors, they are out of the scope of analyses of security protocols. □

## 28.5 Kerberos protocol

The Kerberos protocol shown in Fig. 28.5 uses symmetric keys for secure communication, that is, the key $k_{AB}$ between Alice and Bob is privately shared to Alice and Bob, Alice and Bob have shared keys with Trent $k_{AT}$ and $k_{BT}$ already.

The process of the protocol is as follows:

1. Alice receives some messages $D$ from the outside through the channel $C_{AI}$ (the corresponding reading action is denoted $r_{C_{AI}}(D)$), if $k_{AB}$ is not established, she sends $A, B$ to Trent through the channel $C_{AT}$ (the corresponding sending action is denoted $s_{C_{AT}}(A, B)$);
2. Trent receives $A, B$ from Alice through the channel $C_{AT}$ (the corresponding reading action is denoted $r_{C_{AT}}(A, B)$), he generates a random session key $k_{AB}$ through an action $rsg_{k_{AB}}$, obtains time stamp $T$ and lifetime $L$, then he encrypts $T, L, k_{AB}, B$ by $k_{AT}$ through an action $enc_{k_{AT}}(T, L, k_{AB}, B)$, encrypts $T, L, k_{AB}, A$ by $k_{BT}$ through an action

$enc_{k_{BT}}(T, L, k_{AB}, A)$, and sends them to Alice through the channel $C_{TA}$ (the corresponding sending action is denoted $s_{C_{TA}}(ENC_{k_{AT}}(T, L, k_{AB}, B), ENC_{k_{BT}}(K, L, k_{AB}, A)))$;

3. Alice receives the message from Trent through the channel $C_{TA}$ (the corresponding reading action is denoted $r_{C_{TA}}(ENC_{k_{AT}}(T, L, k_{AB}, B), ENC_{k_{BT}}(K, L, k_{AB}, A)))$, she decrypts $ENC_{k_{AT}}(T, L, k_{AB}, B)$ by $k_{AT}$ through an action $dec_{k_{AT}}(ENC_{k_{AT}}(T, L, k_{AB}, B))$, encrypts $A, T$ by $k_{AB}$ through an action $enc_{k_{AB}}(A, T)$, she sends $ENC_{k_{AB}}(A, T), ENC_{k_{BT}}(T, L, k_{AB}, A)$ to Bob through the channel $C_{AB}$ (the corresponding sending action is denoted $s_{C_{AB}}(ENC_{k_{AB}}(A, T), ENC_{k_{BT}}(T, L, k_{AB}, A)))$;
4. Bob receives $ENC_{k_{AB}}(A, T), ENC_{k_{BT}}(T, L, k_{AB}, A)$ from Alice (the corresponding reading action is denoted $r_{C_{AB}}(ENC_{k_{AB}}(A, T), ENC_{k_{BT}}(T, L, k_{AB}, A)))$, he decrypts $ENC_{k_{BT}}(T, L, k_{AB}, A)$ by $k_{BT}$ through an action $dec_{k_{BT}}(ENC_{k_{BT}}(T, L, k_{AB}, A))$ to ensure the message is from Alice and obtain $k_{AB}$, and decrypts $ENC_{k_{AB}}(A, T)$ by $k_{AB}$ through an action $dec_{k_{AB}}(ENC_{k_{AB}}(A, T))$ to obtain $A$ and $T$, then he encrypts $T + 1$ by $k_{AB}$ through an action $enc_{k_{AB}}(T + 1)$, and sends $ENC_{k_{AB}}(T + 1)$ to Alice through the channel $C_{BA}$ (the corresponding sending action is denoted $s_{C_{BA}}(ENC_{k_{AB}}(T + 1)))$;
5. Alice receives $ENC_{k_{AB}}(d_{T+1})$ from Bob through the channel $C_{BA}$ (the corresponding reading action is denoted $r_{C_{BA}}(ENC_{k_{AB}}(d_{T+1})))$, he decrypts $ENC_{k_{AB}}(d_{T+1})$ by $k_{AB}$ through an action $dec_{k_{AB}}(ENC_{k_{AB}}(d_{T+1}))$, if $d_{T+1} = T + 1$, she encrypts $T + 2, D$ through an action $enc_{k_{AB}}(T + 2, D)$, and sends $ENC_{k_{AB}}(T + 2, D)$ to Bob through the channel $C_{AB}$ (the corresponding sending action is denoted $s_{C_{AB}}(ENC_{k_{AB}}(T + 2, D)))$; else if $d_{T+1} \neq T + 1$, she sends $\perp$ to Bob through the channel $C_{AB}$ (the corresponding sending action is denoted $s_{C_{AB}}(\perp)$);
6. Bob receives $d_{AB}$ from Alice through the channel $C_{AB}$ (the corresponding reading action is denoted $r_{C_{AB}}(d_{AB})$), if $d_{AB} \neq \perp$, he decrypts $ENC_{k_{AB}}(d_{T+2}, D)$ by $k_{AB}$ through an action $dec_{k_{AB}}(ENC_{k_{AB}}(d_{T+2}, D))$, if $d_{T+2} = T + 2$, he sends $D$ to the outside through the channel $C_{BO}$ (the corresponding sending action is denoted $s_{C_{BO}}(D)$), else if $d_{T+2} \neq T + 2$, he sends $\perp$ to the outside through the channel $C_{BO}$ (the corresponding sending action is denoted $s_{C_{BO}}(\perp)$); else if $d_{AB} = \perp$, he sends $\perp$ to the outside through the channel $C_{BO}$ (the corresponding sending action is denoted $s_{C_{BO}}(\perp)$),

where $D \in \Delta$, $\Delta$ is the set of data.

Alice's state transitions described by $APTC_G$ are as follows:

$A = \sum_{D\in\Delta} r_{C_{AI}}(D) \cdot A_2$
$A_2 = \{k_{AB} = NULL\} \cdot A_3 + \{k_{AB} \neq NULL\} \cdot A_{12}$
$A_3 = s_{C_{AT}}(A, B) \cdot A_4$
$A_4 = r_{C_{TA}}(ENC_{k_{AT}}(T, L, k_{AB}, B), ENC_{k_{BT}}(K, L, k_{AB}, A)) \cdot A_5$
$A_5 = dec_{k_{AT}}(ENC_{k_{AT}}(T, L, k_{AB}, B)) \cdot A_6$
$A_6 = enc_{k_{AB}}(A, T) \cdot A_7$
$A_7 = s_{C_{AB}}(ENC_{k_{AB}}(A, T), ENC_{k_{BT}}(T, L, k_{AB}, A)) \cdot A_8$
$A_8 = r_{C_{BA}}(ENC_{k_{AB}}(d_{T+1})) \cdot A_9$
$A_9 = dec_{k_{AB}}(ENC_{k_{AB}}(d_{T+1})) \cdot A_{10}$
$A_{10} = dec_{k_{AB}}(ENC_{k_{AB}}(R_B)) \cdot A_{11}$
$A_{11} = \{d_{T+1} = T + 1\} \cdot A_{12} + \{d_{T+1} \neq T + 1\} \cdot A_{14}$
$A_{12} = enc_{k_{AB}}(T + 2, D) \cdot A_{13}$
$A_{13} = s_{C_{AB}}(ENC_{k_{AB}}(R_B - 1, D)) \cdot A$
$A_{14} = s_{C_{AB}}(\perp) \cdot A$

Bob's state transitions described by $APTC_G$ are as follows:

$B = \{k_{AB} = NULL\} \cdot B_1 + \{k_{AB} \neq NULL\} \cdot B_6$
$B_1 = r_{C_{AB}}(ENC_{k_{AB}}(A, T), ENC_{k_{BT}}(T, L, k_{AB}, A)) \cdot B_2$
$B_2 = dec_{k_{BT}}(ENC_{k_{BT}}(T, L, k_{AB}, A)) \cdot B_3$
$B_3 = dec_{k_{AB}}(ENC_{k_{AB}}(A, T)) \cdot B_4$
$B_4 = enc_{k_{AB}}(T + 1) \cdot B_5$
$B_5 = s_{C_{BA}}(ENC_{k_{AB}}(T + 1)) \cdot B_6$
$B_6 = r_{C_{AB}}(d_{AB}) \cdot B_7$
$B_7 = \{d_{AB} \neq \perp\} \cdot B_8 + \{d_{AB} = \perp\} \cdot s_{C_{BO}}(\perp) \cdot B$
$B_8 = dec_{k_{AB}}(ENC_{k_{AB}}(d_{T+2}, D)) \cdot B_9$
$B_9 = \{d_{T+2} = T + 2\} \cdot B_{10} + \{d_{T+2} \neq T + 2\} \cdot s_{C_{BO}}(\perp) \cdot B$
$B_{10} = s_{C_{BO}}(D) \cdot B$

Trent's state transitions described by $APTC_G$ are as follows:

$T = r_{C_{AT}}(A, B) \cdot T_2$
$T_2 = rsg_{k_{AB}} \cdot T_3$

$T_3 = enc_{k_{AT}}(T, L, k_{AB}, B) \cdot T_4$
$T_4 = enc_{k_{BT}}(T, L, k_{AB}, A) \cdot T_5$
$T_5 = s_{C_{TA}}(ENC_{k_{AT}}(T, L, k_{AB}, B), ENC_{k_{BT}}(K, L, k_{AB}, A)) \cdot T$

The sending action and the reading action of the same type data through the same channel can communicate with each other, otherwise, they will cause a deadlock $\delta$. We define the following communication functions:

$\gamma(r_{C_{AT}}(A, B), s_{C_{AT}}(A, B)) \triangleq c_{C_{AT}}(A, B)$
$\gamma(r_{C_{TA}}(ENC_{k_{AT}}(T, L, k_{AB}, B), ENC_{k_{BT}}(K, L, k_{AB}, A)),$
$s_{C_{TA}}(ENC_{k_{AT}}(T, L, k_{AB}, B), ENC_{k_{BT}}(K, L, k_{AB}, A)))$
$\triangleq c_{C_{TA}}(ENC_{k_{AT}}(T, L, k_{AB}, B), ENC_{k_{BT}}(K, L, k_{AB}, A))$
$\gamma(r_{C_{AB}}(ENC_{k_{AB}}(A, T), ENC_{k_{BT}}(T, L, k_{AB}, A)), s_{C_{AB}}(ENC_{k_{AB}}(A, T), ENC_{k_{BT}}(T, L, k_{AB}, A)))$
$\triangleq c_{C_{AB}}(ENC_{k_{AB}}(A, T), ENC_{k_{BT}}(T, L, k_{AB}, A))$
$\gamma(r_{C_{BA}}(ENC_{k_{AB}}(d_{T+1})), s_{C_{BA}}(ENC_{k_{AB}}(d_{T+1}))) \triangleq c_{C_{BA}}(ENC_{k_{AB}}(d_{T+1}))$
$\gamma(r_{C_{AB}}(d_{AB}), s_{C_{AB}}(d_{AB})) \triangleq c_{C_{AB}}(d_{AB})$

Let all modules be in parallel, then the protocol $A \quad B \quad T$ can be presented by the following process term:

$$\tau_I(\partial_H(\Theta(A \between B \between T))) = \tau_I(\partial_H(A \between B \between T))$$

where $H = \{r_{C_{AT}}(A, B), s_{C_{AT}}(A, B), r_{C_{AB}}(d_{AB}), s_{C_{AB}}(d_{AB}),$
$r_{C_{TA}}(ENC_{k_{AT}}(T, L, k_{AB}, B), ENC_{k_{BT}}(K, L, k_{AB}, A)),$
$s_{C_{TA}}(ENC_{k_{AT}}(T, L, k_{AB}, B), ENC_{k_{BT}}(K, L, k_{AB}, A)),$
$r_{C_{AB}}(ENC_{k_{AB}}(A, T), ENC_{k_{BT}}(T, L, k_{AB}, A)),$
$s_{C_{AB}}(ENC_{k_{AB}}(A, T), ENC_{k_{BT}}(T, L, k_{AB}, A)),$
$r_{C_{BA}}(ENC_{k_{AB}}(d_{T+1})), s_{C_{BA}}(ENC_{k_{AB}}(d_{T+1}))|D \in \Delta\},$

$I = \{c_{C_{AT}}(A, B), c_{C_{BA}}(ENC_{k_{AB}}(d_{T+1})), c_{C_{AB}}(d_{AB}),$
$c_{C_{TA}}(ENC_{k_{AT}}(T, L, k_{AB}, B), ENC_{k_{BT}}(K, L, k_{AB}, A)),$
$c_{C_{AB}}(ENC_{k_{AB}}(A, T), ENC_{k_{BT}}(T, L, k_{AB}, A)),$
$\{k_{AB} = NULL\}, \{k_{AB} \neq NULL\}, dec_{k_{AT}}(ENC_{k_{AT}}(T, L, k_{AB}, B)),$
$enc_{k_{AB}}(A, T), dec_{k_{AB}}(ENC_{k_{AB}}(d_{T+1})), dec_{k_{AB}}(ENC_{k_{AB}}(R_B)),$
$\{d_{T+1} = T + 1\}, \{d_{T+1} \neq T + 1\}, enc_{k_{AB}}(T + 2, D),$
$dec_{k_{BT}}(ENC_{k_{BT}}(T, L, k_{AB}, A)), dec_{k_{AB}}(ENC_{k_{AB}}(A, T)),$
$enc_{k_{AB}}(T + 1), \{d_{AB} \neq \bot\}, \{d_{AB} = \bot\}, dec_{k_{AB}}(ENC_{k_{AB}}(d_{T+2}, D)),$
$\{d_{T+2} = T + 2\}, \{d_{T+2} \neq T + 2\}, rsg_{k_{AB}}, enc_{k_{AT}}(T, L, k_{AB}, B),$
$enc_{k_{BT}}(T, L, k_{AB}, A)|D \in \Delta\}$

Then, we obtain the following conclusion on the protocol.

**Theorem 28.5.** *The Kerberos protocol in Fig. 28.5 is secure.*

*Proof.* Based on the above state transitions of the above modules, by use of the algebraic laws of $APTC_G$, we can prove that

$\tau_I(\partial_H(A \between B \between T)) = \sum_{D\in\Delta}(r_{C_{AI}}(D) \cdot (s_{C_{BO}}(\bot) + s_{C_{BO}}(D))) \cdot \tau_I(\partial_H(A \between B \between T))$

For the details of the proof, please refer to Section 17.3, as we omit them here.

That is, the Kerberos protocol in Fig. 28.5 $\tau_I(\partial_H(A \between B \between T))$ can exhibit the desired external behaviors:

1. For information leakage, because $k_{AT}$ is privately shared only between Alice and Trent, $k_{BT}$ is privately shared only between Bob and Trent, $k_{AB}$ is privately shared only among Trent, Alice, and Bob. For the modeling of confidentiality, it is similar to the protocol in Section 25.10.2, the Kerberos protocol is confidential;
2. For replay attack, the use of the time stamp $T$, makes it that $\tau_I(\partial_H(A \between B \between T)) = \sum_{D\in\Delta}(r_{C_{AI}}(D) \cdot s_{C_{BO}}(\bot)) \cdot \tau_I(\partial_H(A \between B \between T))$, as desired;
3. Without replay attack, the protocol would be $\tau_I(\partial_H(A \between B \between T)) = \sum_{D\in\Delta}(r_{C_{AI}}(D) \cdot s_{C_{BO}}(D)) \cdot \tau_I(\partial_H(A \between B \between T))$, as desired;
4. For the man-in-the-middle attack, because $k_{AT}$ is privately shared only between Alice and Trent, $k_{BT}$ is privately shared only between Bob and Trent, $k_{AB}$ is privately shared only among Trent, Alice, and Bob. For the modeling of the man-in-the-middle attack, it is similar to the protocol in Section 26.4, the Kerberos protocol can be used against the man-in-the-middle attack;
5. For the unexpected and non-technical leaking of $k_{AT}$, $k_{BT}$, $k_{AB}$, or they are not strong enough, or Trent is dishonest, they are out of the scope of analyses of security protocols;
6. For malicious tampering and transmission errors, they are out of the scope of analyses of security protocols. □

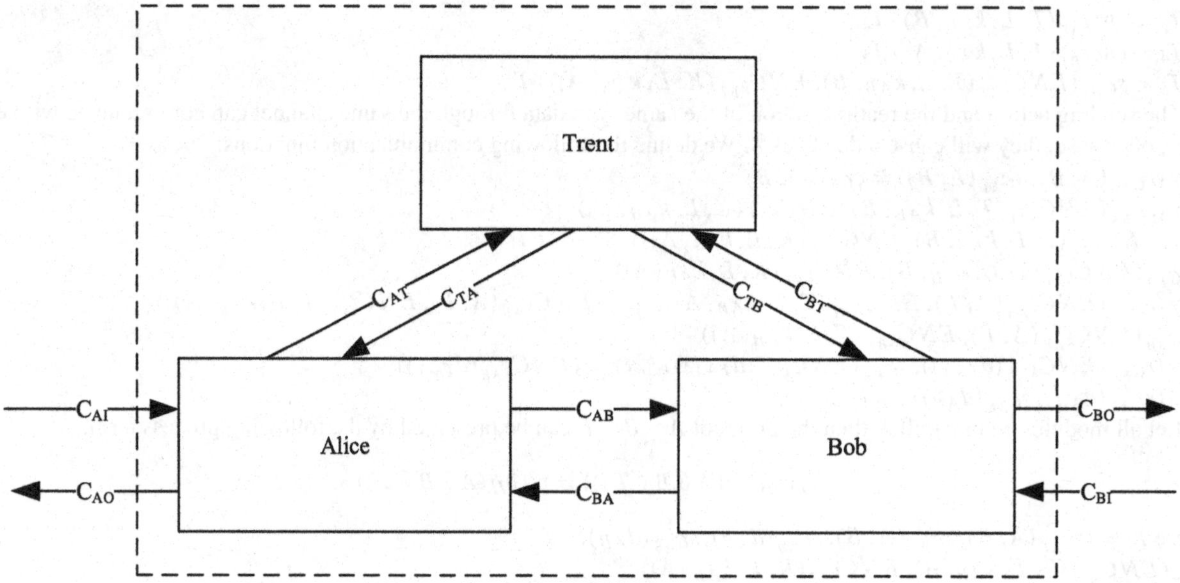


**FIGURE 28.6** Neuman–Stubblebine protocol.

## 28.6 Neuman–Stubblebine protocol

The Neuman–Stubblebine protocol shown in Fig. 28.6 uses symmetric keys for secure communication, that is, the key $k_{AB}$ between Alice and Bob is privately shared to Alice and Bob, Alice and Bob have shared keys with Trent $k_{AT}$ and $k_{BT}$ already.

The process of the protocol is as follows:

1. Alice receives some messages $D$ from the outside through the channel $C_{AI}$ (the corresponding reading action is denoted $r_{C_{AI}}(D)$), if $k_{AB}$ is not established, she generates a random number $R_A$ through the action $rsg_{R_A}$, sends $R_A, A$ to Bob through the channel $C_{AB}$ (the corresponding sending action is denoted $s_{C_{AB}}(R_A, A)$);
2. Bob receives $R_A, A$ from Alice through the channel $C_{AB}$ (the corresponding reading action is denoted $r_{C_{AB}}(R_A, A)$), he generates a random number $R_B$ through an action $rsg_{R_B}$, encrypts $R_A, A, T_B$ by $k_{BT}$ through an action $enc_{k_{BT}}(R_A, A, T_B)$, and sends $R_B, B, ENC_{k_{BT}}(R_A, A, T_B)$ to Trent through the channel $C_{BT}$ (the corresponding sending action is denoted $s_{C_{BT}}(R_B, B, ENC_{k_{BT}}(R_A, A, T_B))$);
3. Trent receives $R_B, B, ENC_{k_{BT}}(R_A, A, T_B)$ through the channel $C_{BT}$ (the corresponding reading action is denoted $r_{C_{BT}}(R_B, B, ENC_{k_{BT}}(R_A, A, T_B))$), he decrypts the message $ENC_{k_{BT}}(R_A, A, T_B)$ through an action $dec_{k_{BT}}(ENC_{k_{BT}}(R_A, A, T_B))$, generates a random session key $k_{AB}$ through an action $rsg_{k_{AB}}$, then he encrypts $B, R_A, k_{AB}, T_B$ by $k_{AT}$ through an action $enc_{k_{AT}}(B, R_A, k_{AB}, T_B)$, encrypts $A, k_{AB}, T_B$ by $k_{BT}$ through an action $enc_{k_{BT}}(A, k_{AB}, T_B)$, and sends them to Alice through the channel $C_{TA}$ (the corresponding sending action is denoted $s_{C_{TA}}(ENC_{k_{AT}}(B, R_A, k_{AB}, T_B), ENC_{k_{BT}}(A, k_{AB}, T_B), R_B)$);
4. Alice receives the message from Trent through the channel $C_{TA}$ (the corresponding reading action is denoted $r_{C_{TA}}(ENC_{k_{AT}}(B, d_{R_A}, k_{AB}, T_B), ENC_{k_{BT}}(A, k_{AB}, T_B), R_B)$), she decrypts $ENC_{k_{AT}}(B, d_{R_A}, k_{AB}, T_B)$ by $k_{AT}$ through an action $dec_{k_{AT}}(ENC_{k_{AT}}(B, d_{R_A}, k_{AB}, T_B))$, if $d_{R_A} = R_A$, she encrypts $R_B$ by $k_{AB}$ through an action $enc_{k_{AB}}(R_B)$, and sends $ENC_{k_{BT}}(A, k_{AB}, T_B), ENC_{k_{AB}}(R_B)$ to Bob through the channel $C_{AB}$ (the corresponding sending action is denoted $s_{C_{AB}}(ENC_{k_{BT}}(A, k_{AB}, T_B), ENC_{k_{AB}}(R_B))$); else if $d_{R_A} \neq R_A$, she sends $\perp$ to Bob through the channel $C_{AB}$ (the corresponding sending action is denoted $s_{C_{AB}}(\perp)$);
5. Bob receives $d_{AB}$ from Alice (the corresponding reading action is denoted $r_{C_{AB}}(d_{AB})$), if $d_{AB} = \perp$, he sends $\perp$ to Alice through the channel $C_{BA}$ (the corresponding sending action is denoted $s_{C_{BA}}(\perp)$); else if $d_{AB} \neq \perp$, she decrypts $ENC_{k_{BT}}(A, k_{AB}, T_B)$ by $k_{BT}$ through an action $dec_{k_{BT}}(ENC_{k_{BT}}(A, k_{AB}, T_B))$, decrypts $ENC_{k_{AB}}(d_{R_B})$ by $k_{AB}$ through an action $dec_{k_{AB}}(ENC_{K_{AB}}(d_{R_B}))$, if $d_{R_B} = R_B$, he generates a random number $R_D$ through an action $rsg_{R_D}$, encrypts $R_D$ by $k_{AB}$ through an action $enc_{k_{AB}}(R_D)$, and sends it to Alice through the channel $C_{BA}$ (the corresponding sending action is denoted $s_{C_{BA}}(ENC_{k_{AB}}(R_D))$), else if $d_{R_B} \neq R_B$, he sends $\perp$ to Alice through the channel $C_{BA}$ (the corresponding sending action is denoted $s_{C_{BA}}(\perp)$);

6. Alice receives $d_{BA}$ from Bob (the corresponding reading action is denoted $r_{C_{BA}}(d_{BA})$), if $d_{BA} = \perp$, she sends $\perp$ to Bob through the channel $C_{AB}$ (the corresponding sending action is denoted $s_{C_{AB}}(\perp)$); else if $d_{BA} \neq \perp$, she decrypts $ENC_{k_{AB}}(R_D)$ by $k_{AB}$ through an action $dec_{k_{AB}}(ENC_{k_{AB}}(R_D))$, if $isFresh(R_D) = TRUE$, she generates a random number $R'_D$ through an action $rsg_{R'_D}$, encrypts $R'_D, D$ by $k_{AB}$ through an action $enc_{k_{AB}}(R'_D, D)$, and sends it to Bob through the channel $C_{AB}$ (the corresponding sending action is denoted $s_{C_{AB}}(ENC_{k_{AB}}(R'_D, D))$), else if $isFresh(R_D) = FALSE$, she sends $\perp$ to Bob through the channel $C_{AB}$ (the corresponding sending action is denoted $s_{C_{AB}}(\perp)$);
7. Bob receives $d'_{AB}$ from Alice (the corresponding reading action is denoted $r_{C_{AB}}(d'_{AB})$), if $d'_{AB} = \perp$, he sends $\perp$ to the outside through the channel $C_{BO}$ (the corresponding sending action is denoted $s_{C_{BO}}(\perp)$); else if $d'_{AB} \neq \perp$, she decrypts $ENC_{k_{AB}}(R'_D, D)$ by $k_{AB}$ through an action $dec_{k_{AB}}(ENC_{k_{AB}}(R'_D, D))$, if $isFresh(d_{R'_D}) = TRUE$, she sends $D$ to the outside through the channel $C_{BO}$ (the corresponding sending action is denoted $s_{C_{BO}}(D)$), else if $isFresh(d'_{R_D}) = FALSE$, he sends $\perp$ to the outside through the channel $C_{BO}$ (the corresponding sending action is denoted $s_{C_{BO}}(\perp)$),

where $D \in \Delta$, $\Delta$ is the set of data.

Alice's state transitions described by $APTC_G$ are as follows:

$A = \sum_{D\in\Delta} r_{C_{AI}}(D) \cdot A_2$
$A_2 = \{k_{AB} = NULL\} \cdot rsg_{R_A} \cdot A_3 + \{k_{AB} \neq NULL\} \cdot A_13$
$A_3 = s_{C_{AB}}(R_A, A) \cdot A_4$
$A_4 = r_{C_{TA}}(ENC_{k_{AT}}(B, d_{R_A}, k_{AB}, T_B), ENC_{k_{BT}}(A, k_{AB}, T_B), R_B) \cdot A_5$
$A_5 = dec_{k_{AT}}(ENC_{k_{AT}}(B, d_{R_A}, k_{AB}, T_B)) \cdot A_6$
$A_6 = \{d_{R_A} = R_A\} \cdot A_7 + \{d_{R_A} \neq R_A\} \cdot s_{C_{AB}}(\perp) \cdot A_9$
$A_7 = enc_{k_{AB}}(R_B) \cdot A_8$
$A_8 = s_{C_{AB}}(ENC_{k_{BT}}(A, k_{AB}, T_B), ENC_{k_{AB}}(R_B)) \cdot A_9$
$A_9 = r_{C_{BA}}(d_{BA}) \cdot A_{10}$
$A_{10} = \{d_{BA} \neq \perp\} \cdot A_{11} + \{d_{BA} = \perp\} \cdot s_{C_{AB}}(\perp) \cdot A$
$A_{11} = dec_{k_{AB}}(ENC_{k_{AB}}(R_D)) \cdot A_{12}$
$A_{12} = \{isFresh(d_{R_D}) = TRUE\} \cdot A_{13} + \{isFresh(d_{R_D}) = FALSE\} \cdot s_{C_{AB}}(\perp) \cdot A$
$A_{13} = rsg_{R'_D} \cdot A_{14}$
$A_{14} = enc_{k_{AB}}(R'_D, D) \cdot A_{15}$
$A_{15} = s_{C_{AB}}(ENC_{k_{AB}}(R'_D, D)) \cdot A$

Bob's state transitions described by $APTC_G$ are as follows:

$B = \{k_{AB} = NULL\} \cdot B_1 + \{k_{AB} \neq NULL\} \cdot B_{13}$
$B_1 = r_{C_{AB}}(R_A, A) \cdot B_2$
$B_2 = rsg_{R_B} \cdot B_3$
$B_3 = enc_{k_{BT}}(R_A, A, T_B) \cdot B_4$
$B_4 = s_{C_{BT}}(R_B, B, ENC_{k_{BT}}(R_A, A, T_B)) \cdot B_5$
$B_5 = r_{C_{AB}}(d_{AB}) \cdot B_6$
$B_6 = \{d_{AB} \neq \perp\} \cdot B_7 + \{d_{AB} = \perp\} \cdot s_{C_{BA}}(\perp) \cdot B_{13}$
$B_7 = dec_{k_{BT}}(ENC_{k_{BT}}(A, k_{AB}, T_B)) \cdot B_8$
$B_8 = dec_{k_{AB}}(ENC_{K_{AB}}(d_{R_B})) \cdot B_9$
$B_9 = \{d_{R_B} \neq R_B\} \cdot s_{C_{BO}}(\perp) \cdot B_{13} + \{d_{R_B} = R_B\} \cdot B_{10}$
$B_{10} = rsg_{R_D} \cdot B_{11}$
$B_{11} = enc_{k_{AB}}(R_D) \cdot B_{12}$
$B_{12} = s_{C_{BA}}(ENC_{k_{AB}}(R_D)) \cdot B_{13}$
$B_{13} = r_{C_{AB}}(d'_{AB}) \cdot B_{14}$
$B_{14} = \{d'_{AB} = \perp\} \cdot s_{C_{BO}}(\perp) \cdot B + \{d'_{AB} \neq \perp\} \cdot B_{15}$
$B_{15} = dec_{k_{AB}}(ENC_{k_{AB}}(R'_D, D)) \cdot B_{16}$
$B_{16} = \{isFresh(R'_D) = FALSE\} \cdot s_{C_{BO}}(\perp) \cdot B + \{isFresh(R'_D) = TRUE\} \cdot B_{17}$
$B_{17} = s_{C_{BO}}(D) \cdot B$

Trent's state transitions described by $APTC_G$ are as follows:

$T = r_{C_{BT}}(R_B, B, ENC_{k_{BT}}(R_A, A, T_B)) \cdot T_2$
$T_2 = dec_{k_{BT}}(ENC_{k_{BT}}(R_A, A, T_B)) \cdot T_3$
$T_3 = rsg_{k_{AB}} \cdot T_4$
$T_4 = enc_{k_{AT}}(B, R_A, k_{AB}, T_B) \cdot T_5$
$T_5 = enc_{k_{BT}}(A, k_{AB}, T_B) \cdot T_6$

$T_6 = s_{C_{TA}}(ENC_{k_{AT}}(B, R_A, k_{AB}, T_B), ENC_{k_{BT}}(A, k_{AB}, T_B), R_B) \cdot T$

The sending action and the reading action of the same type data through the same channel can communicate with each other, otherwise, they will cause a deadlock $\delta$. We define the following communication functions:

$\gamma(r_{C_{AB}}(R_A, A), s_{C_{AB}}(R_A, A)) \triangleq c_{C_{AB}}(R_A, A)$

$\gamma(r_{C_{BT}}(R_B, B, ENC_{k_{BT}}(R_A, A, T_B)), s_{C_{BT}}(R_B, B, ENC_{k_{BT}}(R_A, A, T_B)))$
$\triangleq c_{C_{BT}}(R_B, B, ENC_{k_{BT}}(R_A, A, T_B))$

$\gamma(r_{C_{TA}}(ENC_{k_{AT}}(B, d_{R_A}, k_{AB}, T_B), ENC_{k_{BT}}(A, k_{AB}, T_B), R_B),$
$s_{C_{TA}}(ENC_{k_{AT}}(B, d_{R_A}, k_{AB}, T_B), ENC_{k_{BT}}(A, k_{AB}, T_B), R_B)) \triangleq$
$c_{C_{TA}}(ENC_{k_{AT}}(B, d_{R_A}, k_{AB}, T_B), ENC_{k_{BT}}(A, k_{AB}, T_B), R_B)$

$\gamma(r_{C_{AB}}(d_{AB}), s_{C_{AB}}(d_{AB})) \triangleq c_{C_{AB}}(d_{AB})$

$\gamma(r_{C_{BA}}(d_{BA}), s_{C_{BA}}(d_{BA})) \triangleq c_{C_{BA}}(d_{BA})$

$\gamma(r_{C_{AB}}(d'_{AB}), s_{C_{AB}}(d'_{AB})) \triangleq c_{C_{AB}}(d'_{AB})$

Let all modules be in parallel, then the protocol $A \quad B \quad T$ can be presented by the following process term:

$$\tau_I(\partial_H(\Theta(A \between B \between T))) = \tau_I(\partial_H(A \between B \between T))$$

where $H = \{r_{C_{AB}}(R_A, A), s_{C_{AB}}(R_A, A), r_{C_{AB}}(d_{AB}), s_{C_{AB}}(d_{AB}),$
$r_{C_{BA}}(d_{BA}), s_{C_{BA}}(d_{BA}), r_{C_{AB}}(d'_{AB}), s_{C_{AB}}(d'_{AB})'$
$r_{C_{BT}}(R_B, B, ENC_{k_{BT}}(R_A, A, T_B)), s_{C_{BT}}(R_B, B, ENC_{k_{BT}}(R_A, A, T_B)),$
$r_{C_{TA}}(ENC_{k_{AT}}(B, d_{R_A}, k_{AB}, T_B), ENC_{k_{BT}}(A, k_{AB}, T_B), R_B),$
$s_{C_{TA}}(ENC_{k_{AT}}(B, d_{R_A}, k_{AB}, T_B), ENC_{k_{BT}}(A, k_{AB}, T_B), R_B)|D \in \Delta\},$

$I = \{c_{C_{AB}}(R_A, A), c_{C_{BT}}(R_B, B, ENC_{k_{BT}}(R_A, A, T_B)),$
$c_{C_{TA}}(ENC_{k_{AT}}(B, d_{R_A}, k_{AB}, T_B), ENC_{k_{BT}}(A, k_{AB}, T_B), R_B),$
$c_{C_{AB}}(d_{AB}), c_{C_{BA}}(d_{BA}), c_{C_{AB}}(d'_{AB}),$
$\{k_{AB} = NULL\}, rsg_{R_A}, \{k_{AB} \neq NULL\}, dec_{k_{AT}}(ENC_{k_{AT}}(B, d_{R_A}, k_{AB}, T_B)),$
$\{d_{R_A} = R_A\}, \{d_{R_A} \neq R_A\}, enc_{k_{AB}}(R_B), \{d_{BA} = \perp\}, \{d_{BA} \neq \perp\},$
$dec_{k_{AB}}(ENC_{k_{AB}}(R_D)), \{isFresh(R_D) = TRUE\}, \{isFresh(R_D) = FALSE\},$
$rsg_{R'_D}, enc_{k_{AB}}(R'_D, D), rsg_{R_B}, enc_{k_{BT}}(R_A, A, T_B), \{d_{AB} \neq \perp\},$
$\{d_{AB} = \perp\}, dec_{k_{BT}}(ENC_{k_{BT}}(A, k_{AB}, T_B)), dec_{k_{AB}}(ENC_{K_{AB}}(d_{R_B})),$
$\{d_{R_B} \neq R_B\}, \{d_{R_B} = R_B\}, rsg_{R_D}, enc_{k_{AB}}(R_D), \{d'_{AB} = \perp\}, \{d'_{AB} \neq \perp\},$
$dec_{k_{AB}}(ENC_{k_{AB}}(R'_D, D)), \{isFresh(d_{R'_D}) = FALSE\}, \{isFresh(d_{R'_D}) = TRUE\},$
$dec_{k_{BT}}(ENC_{k_{BT}}(R_A, A, T_B)), rsg_{k_{AB}}, enc_{k_{AT}}(B, R_A, k_{AB}, T_B),$
$enc_{k_{BT}}(A, k_{AB}, T_B)|D \in \Delta\}$

Then, we obtain the following conclusion on the protocol.

**Theorem 28.6.** *The Neuman–Stubblebine protocol in Fig. 28.6 is secure.*

*Proof.* Based on the above state transitions of the above modules, by use of the algebraic laws of $APTC_G$, we can prove that

$\tau_I(\partial_H(A \between B \between T)) = \sum_{D\in\Delta}(r_{C_{AI}}(D) \cdot (s_{C_{BO}}(\perp) + s_{C_{BO}}(D))) \cdot \tau_I(\partial_H(A \between B \between T))$

For the details of the proof, please refer to Section 17.3, as we omit them here.

That is, the Neuman–Stubblebine protocol in Fig. 28.6 $\tau_I(\partial_H(A \between B \between T))$ can exhibit the desired external behaviors:

1. For information leakage, because $k_{AT}$ is privately shared only between Alice and Trent, $k_{BT}$ is privately shared only between Bob and Trent, $k_{AB}$ is privately shared only among Trent, Alice, and Bob. For the modeling of confidentiality, it is similar to the protocol in Section 25.10.2, the Neuman–Stubblebine protocol is confidential;
2. For the man-in-the-middle attack, because $k_{AT}$ is privately shared only between Alice and Trent, $k_{BT}$ is privately shared only between Bob and Trent, $k_{AB}$ is privately shared only among Trent, Alice, and Bob, and the use of the random numbers $R_A$, $R_B$, $R_D$, and $R'_D$, the protocol would be $\tau_I(\partial_H(A \between B \between T)) = \sum_{D\in\Delta}(r_{C_{AI}}(D) \cdot s_{C_{BO}}(\perp)) \cdot \tau_I(\partial_H(A \between B \between T))$, as desired, the Neuman–Stubblebine protocol can be used against the man-in-the-middle attack;
3. For replay attack, the use of the random numbers $T$, $R_A$, $R_B$, $R_D$, and $R'_D$, makes it that $\tau_I(\partial_H(A \between B \between T)) = \sum_{D\in\Delta}(r_{C_{AI}}(D) \cdot s_{C_{BO}}(\perp)) \cdot \tau_I(\partial_H(A \between B \between T))$, as desired;
4. Without man-in-the-middle and replay attack, the protocol would be $\tau_I(\partial_H(A \between B \between T)) = \sum_{D\in\Delta}(r_{C_{AI}}(D) \cdot s_{C_{BO}}(D)) \cdot \tau_I(\partial_H(A \between B \between T))$, as desired;
5. For the unexpected and non-technical leaking of $k_{AT}$, $k_{BT}$, $k_{AB}$, or they are not strong enough, or Trent is dishonest, they are out of the scope of analyses of security protocols;
6. For malicious tampering and transmission errors, they are out of the scope of analyses of security protocols. □

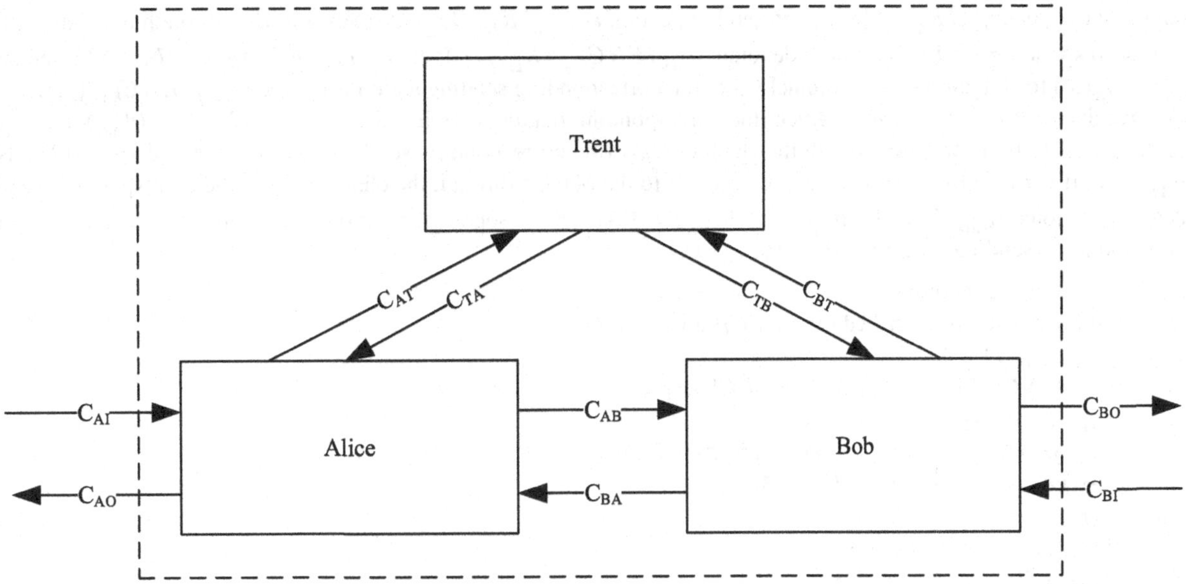


FIGURE 28.7 Denning–Sacco protocol.

## 28.7 Denning–Sacco protocol

The Denning–Sacco protocol shown in Fig. 28.7 uses asymmetric keys and symmetric keys for secure communication, that is, the key $k_{AB}$ between Alice and Bob is privately shared to Alice and Bob, Alice's, Bob's and Trent's public keys $pk_A$, $pk_B$, and $pk_T$ can be publicly obtained.

The process of the protocol is as follows:

1. Alice receives some messages $D$ from the outside through the channel $C_{AI}$ (the corresponding reading action is denoted $r_{C_{AI}}(D)$), if $k_{AB}$ is not established, she sends $A, B$ to Trent through the channel $C_{AT}$ (the corresponding sending action is denoted $s_{C_{AT}}(A, B)$);
2. Trent receives $A, B$ through the channel $C_{AT}$ (the corresponding reading action is denoted $r_{C_{AT}}(A, B)$), he signs Alice's and Bob's public keys $pk_A$ and $pk_B$ through the actions $sign_{sk_T}(A, pk_A)$ and $sign_{sk_T}(B, pk_B)$, and sends the signatures to Alice through the channel $C_{TA}$ (the corresponding sending action is denoted $s_{C_{TA}}(SIGN_{sk_T}(A, pk_A), SIGN_{sk_T}(B, pk_B))$);
3. Alice receives the message from Trent through the channel $C_{TA}$ (the corresponding reading action is denoted $r_{C_{TA}}(SIGN_{sk_T}(A, pk_A), SIGN_{sk_T}(B, pk_B))$), she de-signs $SIGN_{sk_T}(B, pk_B)$ through an action $de\text{-}sign_{pk_T}(SIGN_{sk_T}(B, pk_B))$ to obtain $pk_B$, generates a random session key $k_{AB}$ through an action $rsg_{k_{AB}}$, signs $A, B, k_{AB}, T_A$ through an action $sign_{sk_A}(A, B, k_{AB}, T_A)$, and encrypts the signature by $pk_B$ through an action $enc_{pk_B}(SIGN_{sk_A}(A, B, k_{AB}, T_A))$, then sends $ENC_{pk_B}(SIGN_{sk_A}(A, B, k_{AB}, T_A)), SIGN_{sk_T}(A, pk_A), SIGN_{sk_T}(B, pk_B)$ to Bob through the channel $C_{AB}$ (the corresponding sending action is denoted $s_{C_{AB}}(ENC_{pk_B}(SIGN_{sk_A}(A, B, k_{AB}, T_A)), SIGN_{sk_T}(A, pk_A), SIGN_{sk_T}(B, pk_B))$);
4. Bob receives $ENC_{pk_B}(SIGN_{sk_A}(A, B, k_{AB}, T_A)), SIGN_{sk_T}(A, pk_A), SIGN_{sk_T}(B, pk_B)$ from Alice (the corresponding reading action is denoted$r_{C_{AB}}(ENC_{pk_B}(SIGN_{sk_A}(A, B, k_{AB}, T_A)), SIGN_{sk_T}(A, pk_A), SIGN_{sk_T}(B, pk_B))$), he de-signs $SIGN_{sk_T}(A, pk_A)$ through an action $de\text{-}sign_{pk_T}(SIGN_{sk_T}(A, pk_A))$ to obtain $pk_A$, decrypts $ENC_{pk_B}(SIGN_{sk_A}(A, B, k_{AB}, T_A))$ through an action $dec_{sk_B}(ENC_{pk_B}(SIGN_{sk_A}(A, B, k_{AB}, T_A)))$ and de-signs $SIGN_{sk_A}(A, B, k_{AB}, T_A)$ through an action $de\text{-}sign_{pk_A}(SIGN_{sk_A}(A, B, k_{AB}, T_A))$ to obtain $k_{AB}$ and $T_A$, if $isValid(T_A) = TRUE$, he generates a random number $R_D$ through an action $rsg_{R_D}$, encrypts $R_D$ by $k_{AB}$ through an action $enc_{k_{AB}}(R_D)$, and sends it to Alice through the channel $C_{BA}$ (the corresponding sending action is denoted $s_{C_{BA}}(ENC_{k_{AB}}(R_D))$), else if $isValid(T_A) = FALSE$, he sends $ENC_{k_{AB}}(\perp)$ to Alice through the channel $C_{BA}$ (the corresponding sending action is denoted $s_{C_{BA}}(ENC_{k_{AB}}(\perp))$);
5. Alice receives $ENC_{k_{AB}}(d_{BA})$ from Bob (the corresponding reading action is denoted $r_{C_{BA}}(ENC_{k_{AB}}(d_{BA}))$), if $d_{BA} = \perp$, she sends $ENC_{k_{AB}}(\perp)$ to Bob through the channel $C_{AB}$ (the corresponding sending action is denoted $s_{C_{AB}}(ENC_{k_{AB}}(\perp))$); else if $d_{BA} \neq \perp$, if $isFresh(d_{BA}) = TRUE$, she generates a random number $R'_D$ through an

action $rsg_{R'_D}$, encrypts $R'_D, D$ by $k_{AB}$ through an action $enc_{k_{AB}}(R'_D, D)$, and sends it to Bob through the channel $C_{AB}$ (the corresponding sending action is denoted $s_{C_{AB}}(ENC_{k_{AB}}(R'_D, D))$), else if $isFresh(d_{BA}) = FALSE$, she sends $ENC_{k_{AB}}(\perp)$ to Bob through the channel $C_{AB}$ (the corresponding sending action is denoted $s_{C_{AB}}(ENC_{k_{AB}}(\perp))$);

**6.** Bob receives $ENC_{k_{AB}}(d'_{AB})$ from Alice (the corresponding reading action is denoted $r_{C_{AB}}(ENC_{k_{AB}}(d'_{AB}))$), if $d'_{AB} = \perp$, he sends $\perp$ to the outside through the channel $C_{BO}$ (the corresponding sending action is denoted $s_{C_{BO}}(\perp)$); else if $d'_{AB} \neq \perp$, if $isFresh(d_{R'_D}) = TRUE$, he sends $D$ to the outside through the channel $C_{BO}$ (the corresponding sending action is denoted $s_{C_{BO}}(D)$), else if $isFresh(d'_{R_D}) = FALSE$, he sends $\perp$ to the outside through the channel $C_{BO}$ (the corresponding sending action is denoted $s_{C_{BO}}(\perp)$),

where $D \in \Delta$, $\Delta$ is the set of data.

Alice's state transitions described by $APTC_G$ are as follows:

$A = \sum_{D\in\Delta} r_{C_{AI}}(D) \cdot A_2$

$A_2 = \{k_{AB} = NULL\} \cdot A_3 + \{k_{AB} \neq NULL\} \cdot A_{13}$

$A_3 = s_{C_{AT}}(A, B) \cdot A_4$

$A_4 = r_{C_{TA}}(SIGN_{sk_T}(A, pk_A), SIGN_{sk_T}(B, pk_B)) \cdot A_5$

$A_5 = de\text{-}sign_{pk_T}(SIGN_{sk_T}(B, pk_B)) \cdot A_6$

$A_6 = rsg_{k_{AB}} \cdot A_7$

$A_7 = sign_{sk_A}(A, B, k_{AB}, T_A) \cdot A_8$

$A_8 = enc_{pk_B}(SIGN_{sk_A}(A, B, k_{AB}, T_A)) \cdot A_9$

$A_9 = s_{C_{AB}}(ENC_{pk_B}(SIGN_{sk_A}(A, B, k_{AB}, T_A)), SIGN_{sk_T}(A, pk_A), SIGN_{sk_T}(B, pk_B)) \cdot A_{10}$

$A_{10} = r_{C_{BA}}(ENC_{k_{AB}}(d_{BA})) \cdot A_{11}$

$A_{11} = \{d_{BA} \neq \perp\} \cdot A_{12} + \{d_{BA} = \perp\} \cdot s_{C_{AB}}(ENC_{k_{AB}}(\perp)) \cdot A$

$A_{12} = \{isFresh(d_{BA}) = TRUE\} \cdot A_{13} + \{isFresh(d_{BA}) = FALSE\} \cdot s_{C_{AB}}(ENC_{k_{AB}}(\perp)) \cdot A$

$A_{13} = rsg_{R'_D} \cdot A_{14}$

$A_{14} = enc_{k_{AB}}(R'_D, D) \cdot A_{15}$

$A_{15} = s_{C_{AB}}(ENC_{k_{AB}}(R'_D, D)) \cdot A$

Bob's state transitions described by $APTC_G$ are as follows:

$B = \{k_{AB} = NULL\} \cdot B_1 + \{k_{AB} \neq NULL\} \cdot B_9$

$B_1 = r_{C_{AB}}(ENC_{pk_B}(SIGN_{sk_A}(A, B, k_{AB}, T_A)), SIGN_{sk_T}(A, pk_A), SIGN_{sk_T}(B, pk_B)) \cdot B_2$

$B_2 = de\text{-}sign_{pk_T}(SIGN_{sk_T}(A, pk_A)) \cdot B_3$

$B_3 = dec_{sk_B}(ENC_{pk_B}(SIGN_{sk_A}(A, B, k_{AB}, T_A))) \cdot B_4$

$B_4 = de\text{-}sign_{pk_A}(SIGN_{sk_A}(A, B, k_{AB}, T_A)) \cdot B_5$

$B_5 = \{isValid(T_A) = TRUE\} \cdot B_6 + \{isValid(T_A) = FALSE\} \cdot s_{C_{BA}}(ENC_{k_{AB}}(\perp)) \cdot B_9$

$B_6 = rsg_{R_D} \cdot B_7$

$B_7 = enc_{k_{AB}}(R_D) \cdot B_8$

$B_8 = s_{C_{BA}}(ENC_{k_{AB}}(d_{BA})) \cdot B_9$

$B_9 = r_{C_{AB}}(ENC_{k_{AB}}(d'_{AB})) \cdot B_{10}$

$B_{10} = dec_{k_{AB}}(ENC_{k_{AB}}(d'_{AB})) \cdot B_{11}$

$B_{11} = \{d'_{AB} = \perp\} \cdot s_{C_{BO}}(\perp) \cdot B + \{d'_{AB} \neq \perp\} \cdot B_{12}$

$B_{12} = \{isFresh(d_{R'_D}) = FALSE\} \cdot s_{C_{BO}}(\perp) B + \{isFresh(d_{R'_D}) = TRUE\} \cdot B_{13}$

$B_{13} = s_{C_{BO}}(D) \cdot B$

Trent's state transitions described by $APTC_G$ are as follows:

$T = r_{C_{AT}}(A, B) \cdot T_2$

$T_2 = sign_{sk_T}(A, pk_A) \cdot T_3$

$T_3 = sign_{sk_T}(B, pk_B) \cdot T_4$

$T_4 = s_{C_{TA}}(SIGN_{sk_T}(A, pk_A), SIGN_{sk_T}(B, pk_B)) \cdot T$

The sending action and the reading action of the same type data through the same channel can communicate with each other, otherwise, they will cause a deadlock $\delta$. We define the following communication functions:

$\gamma(r_{C_{AT}}(A, B), s_{C_{AT}}(A, B)) \triangleq c_{C_{AT}}(A, B)$

$\gamma(r_{C_{TA}}(SIGN_{sk_T}(A, pk_A), SIGN_{sk_T}(B, pk_B)), s_{C_{TA}}(SIGN_{sk_T}(A, pk_A), SIGN_{sk_T}(B, pk_B)))$
$\triangleq c_{C_{TA}}(SIGN_{sk_T}(A, pk_A), SIGN_{sk_T}(B, pk_B))$

$\gamma(r_{C_{AB}}(ENC_{pk_B}(SIGN_{sk_A}(A, B, k_{AB}, T_A)), SIGN_{sk_T}(A, pk_A), SIGN_{sk_T}(B, pk_B)),$
$s_{C_{AB}}(ENC_{pk_B}(SIGN_{sk_A}(A, B, k_{AB}, T_A)), SIGN_{sk_T}(A, pk_A), SIGN_{sk_T}(B, pk_B)))$
$\triangleq c_{C_{AB}}(ENC_{pk_B}(SIGN_{sk_A}(A, B, k_{AB}, T_A)), SIGN_{sk_T}(A, pk_A), SIGN_{sk_T}(B, pk_B))$

$\gamma(r_{C_{BA}}(ENC_{k_{AB}}(d_{BA})), s_{C_{BA}}(ENC_{k_{AB}}(d_{BA}))) \triangleq c_{C_{BA}}(ENC_{k_{AB}}(d_{BA}))$
$\gamma(r_{C_{AB}}(ENC_{k_{AB}}(d'_{AB})), s_{C_{AB}}(ENC_{k_{AB}}(d'_{AB}))) \triangleq c_{C_{AB}}(ENC_{k_{AB}}(d'_{AB}))$

Let all modules be in parallel, then the protocol $A \quad B \quad T$ can be presented by the following process term:

$$\tau_I(\partial_H(\Theta(A \between B \between T))) = \tau_I(\partial_H(A \between B \between T))$$

where $H = \{r_{C_{AT}}(A, B), s_{C_{AT}}(A, B), r_{C_{BA}}(ENC_{k_{AB}}(d_{BA})), s_{C_{BA}}(ENC_{k_{AB}}(d_{BA})),$
$r_{C_{AB}}(ENC_{k_{AB}}(d'_{AB})), s_{C_{AB}}(ENC_{k_{AB}}(d'_{AB})),$
$r_{C_{TA}}(SIGN_{sk_T}(A, pk_A), SIGN_{sk_T}(B, pk_B)), s_{C_{TA}}(SIGN_{sk_T}(A, pk_A), SIGN_{sk_T}(B, pk_B)),$
$r_{C_{AB}}(ENC_{pk_B}(SIGN_{sk_A}(A, B, k_{AB}, T_A)), SIGN_{sk_T}(A, pk_A), SIGN_{sk_T}(B, pk_B)),$
$s_{C_{AB}}(ENC_{pk_B}(SIGN_{sk_A}(A, B, k_{AB}, T_A)), SIGN_{sk_T}(A, pk_A), SIGN_{sk_T}(B, pk_B))|D \in \Delta\},$
$I = \{c_{C_{AT}}(A, B), c_{C_{BA}}(ENC_{k_{AB}}(d_{BA})), c_{C_{AB}}(ENC_{k_{AB}}(d'_{AB})),$
$c_{C_{TA}}(SIGN_{sk_T}(A, pk_A), SIGN_{sk_T}(B, pk_B)),$
$c_{C_{AB}}(ENC_{pk_B}(SIGN_{sk_A}(A, B, k_{AB}, T_A)), SIGN_{sk_T}(A, pk_A), SIGN_{sk_T}(B, pk_B)),$
$\{k_{AB} = NULL\}, \{k_{AB} \neq NULL\}, de\text{-}sign_{pk_T}(SIGN_{sk_T}(B, pk_B)),$
$rsg_{k_{AB}}, sign_{sk_A}(A, B, k_{AB}, T_A), enc_{pk_B}(SIGN_{sk_A}(A, B, k_{AB}, T_A)),$
$\{isFresh(d_{BA}) = TRUE\}, \{isFresh(d_{BA}) = FALSE\}, \{d_{BA} \neq \perp\}, \{d_{BA} = \perp\},$
$rsg_{R'_D}, enc_{k_{AB}}(R'_D, D), de\text{-}sign_{pk_T}(SIGN_{sk_T}(A, pk_A)),$
$dec_{sk_B}(ENC_{pk_B}(SIGN_{sk_A}(A, B, k_{AB}, T_A))), de\text{-}sign_{pk_A}(SIGN_{sk_A}(A, B, k_{AB}, T_A)),$
$\{isValid(T_A) = TRUE\}, \{isValid(T_A) = FALSE\}, rsg_{R_D}, enc_{k_{AB}}(R_D),$
$dec_{k_{AB}}(ENC_{k_{AB}}(d'_{AB})), \{d'_{AB} = \perp\}, \{d'_{AB} \neq \perp\},$
$\{isFresh(d_{R'_D}) = TRUE\}, \{isFresh(d_{R'_D}) = FALSE\}, sign_{sk_T}(A, pk_A), sign_{sk_T}(B, pk_B)|D \in \Delta\}$

Then, we obtain the following conclusion on the protocol.

**Theorem 28.7.** *The Denning–Sacco protocol in Fig. 28.7 is secure.*

*Proof.* Based on the above state transitions of the above modules, by use of the algebraic laws of $APTC_G$, we can prove that

$\tau_I(\partial_H(A \between B \between T)) = \sum_{D\in\Delta}(r_{C_{AI}}(D) \cdot (s_{C_{BO}}(\perp) + s_{C_{BO}}(D))) \cdot \tau_I(\partial_H(A \between B \between T))$

For the details of the proof, please refer to Section 17.3, as we omit them here.

That is, the Denning–Sacco protocol in Fig. 28.7 $\tau_I(\partial_H(A \between B \between T))$ can exhibit the desired external behaviors:

1. For the modeling of confidentiality, it is similar to the protocol in Section 25.10.2, the Denning–Sacco protocol is confidential;
2. For the man-in-the-middle attack, because $pk_A$ and $pk_B$ are signed by Trent, the protocol would be $\tau_I(\partial_H(A \between B \between T)) = \sum_{D\in\Delta}(r_{C_{AI}}(D) \cdot s_{C_{BO}}(\perp)) \cdot \tau_I(\partial_H(A \between B \between T))$, as desired, the Denning–Sacco protocol can be used against the man-in-the-middle attack;
3. For replay attack, the use of the time stamp $T_A$, random numbers $R_D$ and $R'_D$, makes it that $\tau_I(\partial_H(A \between B \between T)) = \sum_{D\in\Delta}(r_{C_{AI}}(D) \cdot s_{C_{BO}}(\perp)) \cdot \tau_I(\partial_H(A \between B \between T))$, as desired;
4. Without man-in-the-middle and replay attack, the protocol would be $\tau_I(\partial_H(A \between B \between T)) = \sum_{D\in\Delta}(r_{C_{AI}}(D) \cdot s_{C_{BO}}(D)) \cdot \tau_I(\partial_H(A \between B \between T))$, as desired;
5. For the unexpected and non-technical leaking of $sk_A$, $sk_B$, $k_{AB}$, or they are not strong enough, or Trent is dishonest, they are out of the scope of analyses of security protocols;
6. For malicious tampering and transmission errors, they are out of the scope of analyses of security protocols. □

## 28.8 DASS protocol

The DASS (Distributed Authentication Security Service) protocol shown in Fig. 28.8 uses asymmetric keys and symmetric keys for secure communication, that is, the key $k_{AB}$ between Alice and Bob is privately shared to Alice and Bob, Alice's, Bob's, and Trent's public keys $pk_A$, $pk_B$, and $pk_T$ can be publicly obtained.

The process of the protocol is as follows:

1. Alice receives some messages $D$ from the outside through the channel $C_{AI}$ (the corresponding reading action is denoted $r_{C_{AI}}(D)$), if $k_{AB}$ is not established, she sends $B$ to Trent through the channel $C_{AT}$ (the corresponding sending action is denoted $s_{C_{AT}}(B)$);
2. Trent receives $B$ through the channel $C_{AT}$ (the corresponding reading action is denoted $r_{C_{AT}}(B)$), he signs Bob's public key $pk_B$ through the action $sign_{sk_T}(B, pk_B)$, and sends the signature to Alice through the channel $C_{TA}$ (the corresponding sending action is denoted $s_{C_{TA}}(SIGN_{sk_T}(B, pk_B))$);

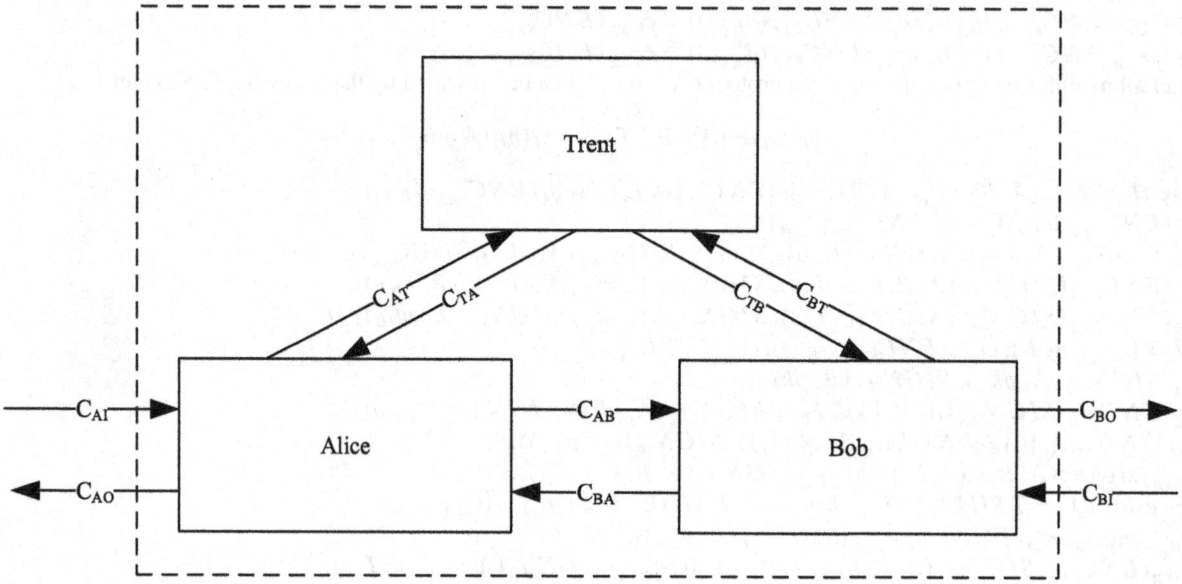


FIGURE 28.8 DASS protocol.

3. Alice receives the message from Trent through the channel $C_{TA}$ (the corresponding reading action is denoted $r_{C_{TA}}(SIGN_{sk_T}(B, pk_B)))$, she de-signs $SIGN_{sk_T}(B, pk_B)$ through an action $de\text{-}sign_{pk_T}(SIGN_{sk_T}(B, pk_B))$ to obtain $pk_B$, generates a random session key $k_{AB}$ through an action $rsg_{k_{AB}}$, generates a public key $pk_P$ through an action $rsg_{pk_P}$ and generates a private key $sk_P$ through an action $rsg_{sk_P}$, signs $L, A, k_{AB}, sk_P, pk_P$ through an action $sign_{sk_A}(L, A, k_{AB}, sk_P, pk_P)$, where $L$ is the life cycle of $k_{AB}$, and encrypts the time stamp $T_A$ by $k_{AB}$ through an action $enc_{k_{AB}}(T_A)$, encrypts $k_{AB}$ by $pk_B$ through an action $enc_{pk_B}(k_{AB})$ and then re-encrypts it by $sk_P$ through an action $enc_{sk_P}(ENC_{pk_B}(k_{AB}))$, then sends $ENC_{k_{AB}}(T_A), SIGN_{sk_A}(L, A, k_{AB}, sk_P, pk_P), ENC_{sk_P}(ENC_{pk_B}(k_{AB}))$ to Bob through the channel $C_{AB}$ (the corresponding sending action is denoted $s_{C_{AB}}(ENC_{k_{AB}}(T_A), SIGN_{sk_A}(L, A, k_{AB}, sk_P, pk_P), ENC_{sk_P}(ENC_{pk_B}(k_{AB}))))$;
4. Bob receives $ENC_{k_{AB}}(T_A), SIGN_{sk_A}(L, A, k_{AB}, sk_P, pk_P), ENC_{sk_P}(ENC_{pk_B}(k_{AB}))$ from Alice (the corresponding reading action is denoted $r_{C_{AB}}(ENC_{k_{AB}}(T_A), SIGN_{sk_A}(L, A, k_{AB}, sk_P, pk_P), ENC_{sk_P}(ENC_{pk_B}(k_{AB}))))$, he sends the name of Alice $A$ to Trent through the channel $C_{BT}$ (the corresponding sending action is denoted $s_{C_{BT}}(A)$);
5. Trent receives the name of Alice $A$ from Bob through the channel $C_{BT}$ (the corresponding reading action is denoted $r_{C_{BT}}(A)$), signs $A$ and $pk_A$ through an action $sign_{sk_T}(A, pk_A)$, and sends the signature $SIGN_{sk_T}(A, pk_A)$ to Bob through the channel $C_{TB}$ (the corresponding sending action is denoted $s_{C_{TB}}(SIGN_{sk_T}(A, pk_A)))$;
6. Bob receives the signature from Trent through the channel $C_{TB}$ (the corresponding reading action is denoted $r_{C_{TB}}(SIGN_{sk_T}(A, pk_A)))$, he de-signs $SIGN_{sk_T}(A, pk_A)$ through an action $de\text{-}sign_{pk_T}(SIGN_{sk_T}(A, pk_A))$ to obtain $pk_A$, de-signs $SIGN_{sk_A}(L, A, k_{AB}, sk_P, pk_P)$ through an action $de\text{-}signpk_A(SIGN_{sk_A}(L, A, k_{AB}, sk_P, pk_P))$ and decrypts $ENC_{sk_P}(ENC_{pk_B}(k_{AB}))$ and $ENC_{k_{AB}}(T_A)$ through an action $dec_{pk_P}(ENC_{sk_P}(ENC_{pk_B}(k_{AB})))$ and an action $dec_{sk_B}(ENC_{pk_B}(k_{AB}))$ and an action $dec_{k_{AB}}(T_A)$ to obtain $k_{AB}$ and $T_A$, if $isValid(T_A) = TRUE$, he encrypts the time stamp $T_B$ by $k_{AB}$ through an action $enc_{k_{AB}}(T_B)$, and sends it to Alice through the channel $C_{BA}$ (the corresponding sending action is denoted $s_{C_{BA}}(ENC_{k_{AB}}(T_B)))$, else if $isValid(T_A) = FALSE$, he sends $ENC_{k_{AB}}(\bot)$ to Alice through the channel $C_{BA}$ (the corresponding sending action is denoted $s_{C_{BA}}(ENC_{k_{AB}}(\bot)))$;
7. Alice receives $ENC_{k_{AB}}(d_{BA})$ from Bob (the corresponding reading action is denoted $r_{C_{BA}}(ENC_{k_{AB}}(d_{BA})))$, if $d_{BA} = \bot$, she sends $ENC_{k_{AB}}(\bot)$ to Bob through the channel $C_{AB}$ (the corresponding sending action is denoted $s_{C_{AB}}(ENC_{k_{AB}}(\bot)))$; else if $d_{BA} \neq \bot$, if $isFresh(d_{BA}) = TRUE$, she generates a random number $R_D$ through an action $rsg_{R_D}$, encrypts $R_D, D$ by $k_{AB}$ through an action $enc_{k_{AB}}(R_D, D)$, and sends it to Bob through the channel $C_{AB}$ (the corresponding sending action is denoted $s_{C_{AB}}(ENC_{k_{AB}}(R_D, D)))$, else if $isFresh(d_{BA}) = FALSE$, he sends $ENC_{k_{AB}}(\bot)$ to Bob through the channel $C_{AB}$ (the corresponding sending action is denoted $s_{C_{AB}}(ENC_{k_{AB}}(\bot)))$;
8. Bob receives $ENC_{k_{AB}}(d_{AB})$ from Alice (the corresponding reading action is denoted $r_{C_{AB}}(ENC_{k_{AB}}(d_{AB})))$, if $d_{AB} = \bot$, he sends $\bot$ to the outside through the channel $C_{BO}$ (the corresponding sending action is denoted $s_{C_{BO}}(\bot)$); else if $d_{AB} \neq \bot$, if $isFresh(d_{R_D}) = TRUE$, he sends $D$ to the outside through the channel $C_{BO}$ (the corresponding sending

action is denoted $s_{C_{BO}}(D)$), else if $isFresh(d_{R_D}) = FALSE$, he sends $\perp$ to the outside through the channel $C_{BO}$ (the corresponding sending action is denoted $s_{C_{BO}}(\perp)$),

where $D \in \Delta$, $\Delta$ is the set of data.

Alice's state transitions described by $APTC_G$ are as follows:

$A = \sum_{D\in\Delta} r_{C_{AI}}(D) \cdot A_2$

$A_2 = \{k_{AB} = NULL\} \cdot A_3 + \{k_{AB} \neq NULL\} \cdot A_{13}$

$A_3 = s_{C_{AT}}(B) \cdot A_4$

$A_4 = r_{C_{TA}}(SIGN_{sk_T}(B, pk_B)) \cdot A_5$

$A_5 = de\text{-}sign_{pk_T}(SIGN_{sk_T}(B, pk_B)) \cdot A_6$

$A_6 = (rsg_{k_{AB}} \parallel rsg_{pk_P} \parallel rsg_{sk_P}) \cdot A_7$

$A_7 = sign_{sk_A}(L, A, k_{AB}, sk_P, pk_P) \cdot A_8$

$A_8 = (enc_{sk_P}(ENC_{pk_B}(k_{AB})) \parallel enc_{k_{AB}}(T_A)) \cdot A_9$

$A_9 = s_{C_{AB}}(ENC_{k_{AB}}(T_A), SIGN_{sk_A}(L, A, k_{AB}, sk_P, pk_P), ENC_{sk_P}(ENC_{pk_B}(k_{AB}))) \cdot A_{10}$

$A_{10} = r_{C_{BA}}(ENC_{k_{AB}}(d_{BA})) \cdot A_{11}$

$A_{11} = \{d_{BA} \neq \perp\} \cdot A_{12} + \{d_{BA} = \perp\} \cdot s_{C_{AB}}(ENC_{k_{AB}}(\perp)) \cdot A$

$A_{12} = \{isFresh(d_{BA}) = TRUE\} \cdot A_{13} + \{isFresh(d_{BA}) = FALSE\} \cdot s_{C_{AB}}(ENC_{k_{AB}}(\perp)) \cdot A$

$A_{13} = rsg_{R_D} \cdot A_{14}$

$A_{14} = enc_{k_{AB}}(R_D, D) \cdot A_{15}$

$A_{15} = s_{C_{AB}}(ENC_{k_{AB}}(R_D, D)) \cdot A$

Bob's state transitions described by $APTC_G$ are as follows:

$B = \{k_{AB} = NULL\} \cdot B_1 + \{k_{AB} \neq NULL\} \cdot B_{11}$

$B_1 = r_{C_{AB}}(ENC_{k_{AB}}(T_A), SIGN_{sk_A}(L, A, k_{AB}, sk_P, pk_P), ENC_{sk_P}(ENC_{pk_B}(k_{AB}))) \cdot B_2$

$B_2 = s_{C_{BT}}(A) \cdot B_3$

$B_3 = r_{C_{TB}}(SIGN_{sk_T}(A, pk_A)) \cdot B_4$

$B_4 = de\text{-}sign_{pk_T}(SIGN_{sk_T}(A, pk_A)) \cdot B_5$

$B_5 = de\text{-}sign pk_A(SIGN_{sk_A}(L, A, k_{AB}, sk_P, pk_P)) \cdot B_6$

$B_6 = (dec_{pk_P}(ENC_{sk_P}(ENC_{pk_B}(k_{AB}))) \parallel dec_{sk_B}(ENC_{pk_B}(k_{AB})) \parallel dec_{k_{AB}}(T_A)) \cdot B_7$

$B_7 = \{isValid(T_A) = TRUE\} \cdot B_8 + \{isValid(T_A) = FALSE\} \cdot s_{C_{BA}}(ENC_{k_{AB}}(\perp)) \cdot B_{11}$

$B_8 = rsg_{T_B} \cdot B_9$

$B_9 = enc_{k_{AB}}(T_B) \cdot B_{10}$

$B_{10} = s_{C_{BA}}(ENC_{k_{AB}}(d_{BA})) \cdot B_{11}$

$B_{11} = r_{C_{AB}}(ENC_{k_{AB}}(d_{AB})) \cdot B_{12}$

$B_{12} = dec_{k_{AB}}(ENC_{k_{AB}}(d_{AB})) \cdot B_{13}$

$B_{13} = \{d_{AB} = \perp\} \cdot s_{C_{BO}}(\perp) \cdot B + \{d_{AB} \neq \perp\} \cdot B_{14}$

$B_{14} = \{isFresh(d_{R_D}) = FALSE\} \cdot s_{C_{BO}}(\perp)B + \{isFresh(d_{R_D}) = TRUE\} \cdot B_{15}$

$B_{15} = s_{C_{BO}}(D) \cdot B$

Trent's state transitions described by $APTC_G$ are as follows:

$T = r_{C_{AT}}(B) \cdot T_2$

$T_2 = sign_{sk_T}(B, pk_B) \cdot T_3$

$T_3 = s_{C_{TA}}(SIGN_{sk_T}(B, pk_B)) \cdot T_4$

$T_4 = r_{C_{BT}}(A) \cdot T_5$

$T_5 = sign_{sk_T}(A, pk_A) \cdot T_6$

$T_6 = s_{C_{TA}}(SIGN_{sk_T}(A, pk_A)) \cdot T$

The sending action and the reading action of the same type data through the same channel can communicate with each other, otherwise, they will cause a deadlock $\delta$. We define the following communication functions:

$\gamma(r_{C_{AT}}(B), s_{C_{AT}}(B)) \triangleq c_{C_{AT}}(B)$

$\gamma(r_{C_{BT}}(A), s_{C_{AT}}(A)) \triangleq c_{C_{AT}}(A)$

$\gamma(r_{C_{TA}}(SIGN_{sk_T}(B, pk_B)), s_{C_{TA}}(SIGN_{sk_T}(B, pk_B)))$
$\triangleq c_{C_{TA}}(SIGN_{sk_T}(B, pk_B))$

$\gamma(r_{C_{TB}}(SIGN_{sk_T}(A, pk_A)), s_{C_{TB}}(SIGN_{sk_T}(A, pk_A)))$
$\triangleq c_{C_{TB}}(SIGN_{sk_T}(A, pk_A))$

$\gamma(r_{C_{AB}}(ENC_{k_{AB}}(T_A), SIGN_{sk_A}(L, A, k_{AB}, sk_P, pk_P), ENC_{sk_P}(ENC_{pk_B}(k_{AB}))),$
$s_{C_{AB}}(ENC_{k_{AB}}(T_A), SIGN_{sk_A}(L, A, k_{AB}, sk_P, pk_P), ENC_{sk_P}(ENC_{pk_B}(k_{AB}))))$
$\triangleq c_{C_{AB}}(ENC_{k_{AB}}(T_A), SIGN_{sk_A}(L, A, k_{AB}, sk_P, pk_P), ENC_{sk_P}(ENC_{pk_B}(k_{AB})))$

$\gamma(r_{C_{BA}}(ENC_{k_{AB}}(d_{BA})), s_{C_{BA}}(ENC_{k_{AB}}(d_{BA}))) \triangleq c_{C_{BA}}(ENC_{k_{AB}}(d_{BA}))$
$\gamma(r_{C_{AB}}(ENC_{k_{AB}}(d_{AB})), s_{C_{AB}}(ENC_{k_{AB}}(d_{AB}))) \triangleq c_{C_{AB}}(ENC_{k_{AB}}(d_{AB}))$

Let all modules be in parallel, then the protocol $A \quad B \quad T$ can be presented by the following process term:

$$\tau_I(\partial_H(\Theta(A \between B \between T))) = \tau_I(\partial_H(A \between B \between T))$$

where $H = \{r_{C_{AT}}(B), s_{C_{AT}}(B), r_{C_{BT}}(A), s_{C_{BT}}(A), r_{C_{BA}}(ENC_{k_{AB}}(d_{BA})), s_{C_{BA}}(ENC_{k_{AB}}(d_{BA})),$
$r_{C_{AB}}(ENC_{k_{AB}}(d_{AB})), s_{C_{AB}}(ENC_{k_{AB}}(d_{AB})),$
$r_{C_{TA}}(SIGN_{sk_T}(B, pk_B)), s_{C_{TA}}(SIGN_{sk_T}(B, pk_B)),$
$r_{C_{TB}}(SIGN_{sk_T}(A, pk_A)), s_{C_{TB}}(SIGN_{sk_T}(A, pk_A)),$
$r_{C_{AB}}(ENC_{k_{AB}}(T_A), SIGN_{sk_A}(L, A, k_{AB}, sk_P, pk_P), ENC_{sk_P}(ENC_{pk_B}(k_{AB}))),$
$s_{C_{AB}}(ENC_{k_{AB}}(T_A), SIGN_{sk_A}(L, A, k_{AB}, sk_P, pk_P), ENC_{sk_P}(ENC_{pk_B}(k_{AB})))|D \in \Delta\},$
$I = \{c_{C_{AT}}(B), c_{C_{BT}}(A), c_{C_{BA}}(ENC_{k_{AB}}(d_{BA})), c_{C_{AB}}(ENC_{k_{AB}}(d_{AB})),$
$c_{C_{TA}}(SIGN_{sk_T}(B, pk_B)), c_{C_{TB}}(SIGN_{sk_T}(A, pk_A))$
$c_{C_{AB}}(ENC_{k_{AB}}(T_A), SIGN_{sk_A}(L, A, k_{AB}, sk_P, pk_P), ENC_{sk_P}(ENC_{pk_B}(k_{AB}))),$
$\{k_{AB} = NULL\}, \{k_{AB} \neq NULL\}, de\text{-}sign_{pk_T}(SIGN_{sk_T}(B, pk_B)),$
$rsg_{k_{AB}}, rsg_{pk_P}, rsg_{sk_P}, sign_{sk_A}(L, A, k_{AB}, sk_P, pk_P), enc_{sk_P}(ENC_{pk_B}(k_{AB})), enc_{k_{AB}}(T_A),$
$\{isFresh(d_{BA}) = TRUE\}, \{isFresh(d_{BA}) = FALSE\}, \{d_{BA} \neq \perp\}, \{d_{BA} = \perp\},$
$rsg_{R_D}, enc_{k_{AB}}(R_D, D), de\text{-}sign_{pk_T}(SIGN_{sk_T}(A, pk_A)),$
$dec_{sk_B}(ENC_{pk_B}(SIGN_{sk_A}(L, A, k_{AB}, sk_P, pk_P))), dec_{pk_P}(ENC_{sk_P}(ENC_{pk_B}(k_{AB}))),$
$dec_{sk_B}(ENC_{pk_B}(k_{AB})), dec_{k_{AB}}(T_A), \{isValid(T_A) = TRUE\},$
$\{isValid(T_A) = FALSE\}, rsg_{R_D}, enc_{k_{AB}}(R_D),$
$dec_{k_{AB}}(ENC_{k_{AB}}(d_{AB})), \{d'_{AB} = \perp\}, \{d_{AB} \neq \perp\},$
$\{isFresh(d_{R_D}) = TRUE\}, \{isFresh(d_{R_D}) = FALSE\}, sign_{sk_T}(A, pk_A), sign_{sk_T}(B, pk_B)|D \in \Delta\}$

Then, we obtain the following conclusion on the protocol.

**Theorem 28.8.** *The DASS protocol in Fig. 28.8 is secure.*

*Proof.* Based on the above state transitions of the above modules, by use of the algebraic laws of $APTC_G$, we can prove that

$\tau_I(\partial_H(A \between B \between T)) = \sum_{D\in\Delta}(r_{C_{AI}}(D) \cdot (s_{C_{BO}}(\perp) + s_{C_{BO}}(D))) \cdot \tau_I(\partial_H(A \between B \between T))$

For the details of the proof, please refer to Section 17.3, as we omit them here.

That is, the DASS protocol in Fig. 28.8 $\tau_I(\partial_H(A \between B \between T))$ can exhibit the desired external behaviors:

1. For the modeling of confidentiality, it is similar to the protocol in Section 25.10.2, the DASS protocol is confidential;
2. For the man-in-the-middle attack, because $pk_A$ and $pk_B$ are signed by Trent, the protocol would be $\tau_I(\partial_H(A \between B \between T)) = \sum_{D\in\Delta}(r_{C_{AI}}(D) \cdot s_{C_{BO}}(\perp)) \cdot \tau_I(\partial_H(A \between B \between T))$, as desired, the DASS protocol can be used against the man-in-the-middle attack;
3. For replay attack, the use of the time stamp $T_A$, $T_B$, and random number $R_D$, makes it that $\tau_I(\partial_H(A \between B \between T)) = \sum_{D\in\Delta}(r_{C_{AI}}(D) \cdot s_{C_{BO}}(\perp)) \cdot \tau_I(\partial_H(A \between B \between T))$, as desired;
4. Without man-in-the-middle and replay attack, the protocol would be $\tau_I(\partial_H(A \between B \between T)) = \sum_{D\in\Delta}(r_{C_{AI}}(D) \cdot s_{C_{BO}}(D)) \cdot \tau_I(\partial_H(A \between B \between T))$, as desired;
5. For the unexpected and non-technical leaking of $sk_A$, $sk_B$, $k_{AB}$, or they are not strong enough, or Trent is dishonest, they are out of the scope of analyses of security protocols;
6. For malicious tampering and transmission errors, they are out of the scope of analyses of security protocols. □

## 28.9 Woo–Lam protocol

The Woo–Lam protocol shown in Fig. 28.9 uses asymmetric keys and symmetric keys for secure communication, that is, the key $k_{AB}$ between Alice and Bob is privately shared to Alice and Bob, Alice's, Bob's, and Trent's public keys $pk_A$, $pk_B$, and $pk_T$ can be publicly obtained.

The process of the protocol is as follows:

1. Alice receives some messages $D$ from the outside through the channel $C_{AI}$ (the corresponding reading action is denoted $r_{C_{AI}}(D)$), if $k_{AB}$ is not established, she sends $A, B$ to Trent through the channel $C_{AT}$ (the corresponding sending action is denoted $s_{C_{AT}}(A, B)$);

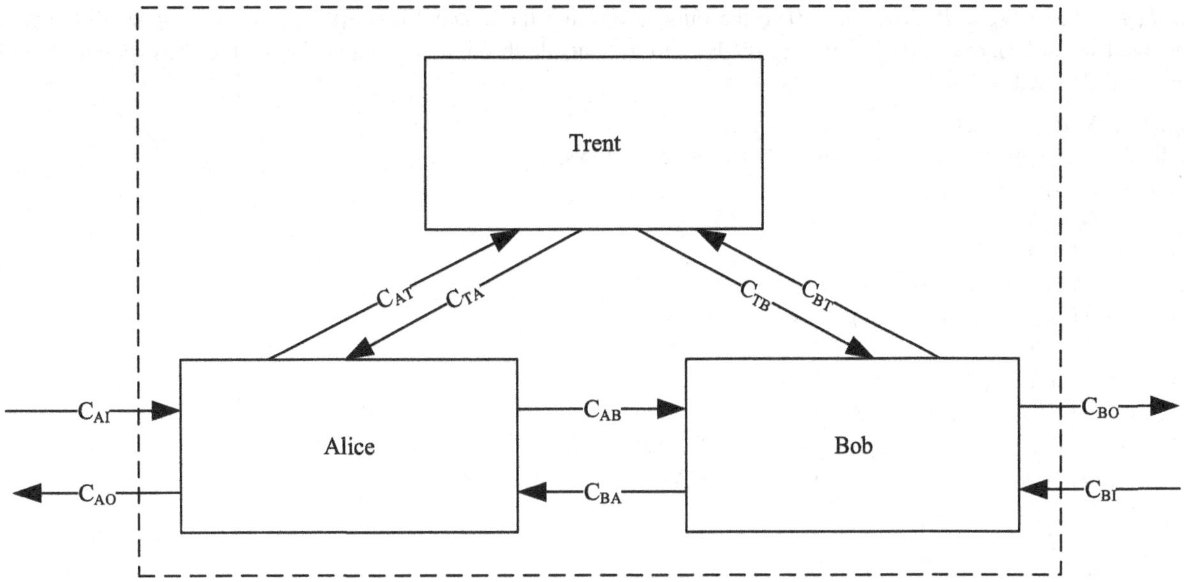


**FIGURE 28.9** Woo–Lam protocol.

2. Trent receives $A, B$ through the channel $C_{AT}$ (the corresponding reading action is denoted $r_{C_{AT}}(A, B)$), he signs Bob's public key $pk_B$ through the action $sign_{sk_T}(pk_B)$, and sends the signature to Alice through the channel $C_{TA}$ (the corresponding sending action is denoted $s_{C_{TA}}(SIGN_{sk_T}(pk_B))$);
3. Alice receives the message from Trent through the channel $C_{TA}$ (the corresponding reading action is denoted $r_{C_{TA}}(SIGN_{sk_T}(pk_B))$), she de-signs $SIGN_{sk_T}(pk_B)$ through an action $de\text{-}sign_{pk_T}(SIGN_{sk_T}(pk_B))$ to obtain $pk_B$, generates a random number $R_A$ through an action $rsg_{R_A}$ and encrypts $A, R_A$ by $pk_B$ through an action $enc_{pk_B}(A, R_A)$, and sends $ENC_{pk_B}(A, R_A)$ to Bob through the channel $C_{AB}$ (the corresponding sending action is denoted $s_{C_{AB}}(ENC_{pk_B}(A, R_A))$);
4. Bob receives $ENC_{pk_B}(A, R_A)$ from Alice (the corresponding reading action is denoted $r_{C_{AB}}(ENC_{pk_B}(A, R_A))$), he decrypts $ENC_{pk_B}(A, R_A)$ through an action $dec_{sk_B}(ENC_{pk_B}(A, R_A))$ to obtain $A$ and $R_A$, encrypts $R_A$ by $pk_T$ through an action $enc_{pk_T}(R_A)$, then sends $A, B, ENC_{pk_T}(R_A)$ to Trent through the channel $C_{BT}$ (the corresponding sending action is denoted $s_{C_{BT}}(A, B, ENC_{pk_T}(R_A))$);
5. Trent receives $A, B, ENC_{pk_T}(R_A)$ from Bob through the channel $C_{BT}$ (the corresponding reading action is denoted $r_{C_{BT}}(A, B, ENC_{pk_T}(R_A))$), he decrypts the message through an action $dec_{sk_T}(ENC_{pk_T}(R_A))$, signs $pk_A$ through an action $sign_{sk_T}(pk_A)$, generates a random session key $k_{AB}$ through an action $rsg_{k_{AB}}$ and signs $R_A, k_{AB}, A, B$ through an action $sign_{sk_T}(R_A, k_{AB}, A, B)$, encrypts $SIGN_{sk_T}(R_A, k_{AB}, A, B)$ through an action $enc_{pk_B}(SIGN_{sk_T}(R_A, k_{AB}, A, B))$ and sends them to Bob through the channel $C_{TB}$ (the corresponding sending action is denoted $s_{C_{TB}}(SIGN_{sk_T}(pk_A), ENC_{pk_B}(SIGN_{sk_T}(R_A, k_{AB}, A, B)))$);
6. Bob receives the signatures from Trent through the channel $C_{TB}$ (the corresponding reading action is denoted $r_{C_{TB}}(SIGN_{sk_T}(pk_A), ENC_{pk_B}(SIGN_{sk_T}(R_A, k_{AB}, A, B)))$), he de-signs $SIGN_{sk_T}(pk_A)$ through an action $de\text{-}sign_{pk_T}(SIGN_{sk_T}(pk_A))$ to obtain $pk_A$, decrypts $ENC_{pk_B}(SIGN_{sk_T}(R_A, k_{AB}, A, B))$ through an action $dec_{sk_B}(ENC_{pk_B}(SIGN_{sk_T}(R_A, k_{AB}, A, B)))$, generates a random number $R_B$ through an action $rsg_{R_B}$, encrypts them through an action $enc_{pk_A}(SIGN_{sk_T}(R_A, k_{AB}, A, B), R_B)$ and sends $ENC_{pk_A}(SIGN_{sk_T}(R_A, k_{AB}, A, B), R_B)$ to Alice through the channel $C_{BA}$ (the corresponding sending action is denoted $s_{C_{BA}}(ENC_{pk_A}(SIGN_{sk_T}(R_A, k_{AB}, A, B), R_B))$);
7. Alice receives $ENC_{pk_A}(SIGN_{sk_T}(d_{R_A}, k_{AB}, A, B), R_B)$ from Bob (the corresponding reading action is denoted $r_{C_{BA}}(ENC_{pk_A}(SIGN_{sk_T}(d_{R_A}, k_{AB}, A, B), R_B))$), she decrypts the message through an action $dec_{sk_A}(ENC_{pk_A}(SIGN_{sk_T}(R_A, k_{AB}, A, B), R_B))$, de-signs $SIGN_{sk_T}(R_A, k_{AB}, A, B)$ through an action $de\text{-}sign_{pk_T}(SIGN_{sk_T}(R_A, k_{AB}, A, B))$, if $d_{R_A} \neq R_A$, she sends $ENC_{k_{AB}}(\perp)$ to Bob through the channel $C_{AB}$ (the corresponding sending action is denoted $s_{C_{AB}}(ENC_{k_{AB}}(\perp))$); else if $d_{R_A} = R_A$, encrypts $R_B, D$ by $k_{AB}$ through an action $enc_{k_{AB}}(R_B, D)$, and sends it to Bob through the channel $C_{AB}$ (the corresponding sending action is denoted $s_{C_{AB}}(ENC_{k_{AB}}(R_B, D))$);
8. Bob receives $ENC_{k_{AB}}(d_{AB})$ from Alice (the corresponding reading action is denoted $r_{C_{AB}}(ENC_{k_{AB}}(d_{AB}))$), if $d_{AB} = \perp$, he sends $\perp$ to the outside through the channel $C_{BO}$ (the corresponding sending action is denoted $s_{C_{BO}}(\perp)$); else

if $d_{AB} \neq \perp$, if $d_{R_B} = R_B$, he sends $D$ to the outside through the channel $C_{BO}$ (the corresponding sending action is denoted $s_{C_{BO}}(D)$), else if $d_{R_B} \neq R_B$, he sends $\perp$ to the outside through the channel $C_{BO}$ (the corresponding sending action is denoted $s_{C_{BO}}(\perp)$),

where $D \in \Delta$, $\Delta$ is the set of data.

Alice's state transitions described by $APTC_G$ are as follows:

$A = \sum_{D\in\Delta} r_{C_{AI}}(D) \cdot A_2$

$A_2 = \{k_{AB} = NULL\} \cdot A_3 + \{k_{AB} \neq NULL\} \cdot A_9$

$A_3 = s_{C_{AT}}(A, B) \cdot A_4$

$A_4 = r_{C_{TA}}(SIGN_{sk_T}(pk_B)) \cdot A_5$

$A_5 = de\text{-}sign_{pk_T}(SIGN_{sk_T}(pk_B)) \cdot A_6$

$A_6 = rsg_{R_A} \cdot A_7$

$A_7 = enc_{sk_P}(A, R_A) \cdot A_8$

$A_8 = s_{C_{AB}}(ENC_{sk_P}(A, R_A)) \cdot A_9$

$A_9 = r_{C_{BA}}(ENC_{pk_A}(SIGN_{sk_T}(d_{R_A}, k_{AB}, A, B), R_B)) \cdot A_{10}$

$A_{10} = \{d_{R_A} = R_A\} \cdot A_{11} + \{d_{R_A} \neq R_A\} \cdot s_{C_{AB}}(ENC_{k_{AB}}(\perp)) \cdot A$

$A_{11} = enc_{k_{AB}}(R_B, D) \cdot A_{12}$

$A_{12} = s_{C_{AB}}(ENC_{k_{AB}}(R_B, D)) \cdot A$

Bob's state transitions described by $APTC_G$ are as follows:

$B = \{k_{AB} = NULL\} \cdot B_1 + \{k_{AB} \neq NULL\} \cdot B_{10}$

$B_1 = r_{C_{AB}}(ENC_{sk_P}(A, R_A)) \cdot B_2$

$B_2 = dec_{sk_B}(ENC_{pk_B}(A, R_A)) \cdot B_3$

$B_3 = s_{C_{BT}}(A, B, ENC_{pk_T}(R_A)) \cdot B_4$

$B_4 = r_{C_{TB}}(SIGN_{sk_T}(pk_A), ENC_{pk_B}(SIGN_{sk_T}(R_A, k_{AB}, A, B))) \cdot B_5$

$B_5 = de\text{-}sign_{pk_T}(SIGN_{sk_T}(pk_A)) \cdot B_6$

$B_6 = dec_{sk_B}(ENC_{pk_B}(SIGN_{sk_T}(R_A, k_{AB}, A, B))) \cdot B_7$

$B_7 = rsg_{R_B} \cdot B_8$

$B_8 = enc_{pk_A}(SIGN_{sk_T}(R_A, k_{AB}, A, B), R_B) \cdot B_9$

$B_9 = s_{C_{BA}}(ENC_{pk_A}(SIGN_{sk_T}(R_A, k_{AB}, A, B), R_B)) \cdot B_{10}$

$B_{10} = r_{C_{AB}}(ENC_{k_{AB}}(d_{AB})) \cdot B_{11}$

$B_{11} = dec_{k_{AB}}(ENC_{k_{AB}}(d_{AB})) \cdot B_{12}$

$B_{12} = \{d_{AB} = \perp\} \cdot s_{C_{BO}}(\perp) \cdot B + \{d_{AB} \neq \perp\} \cdot B_{13}$

$B_{13} = \{d_{R_B} \neq R_B\} \cdot s_{C_{BO}}(\perp) B + \{d_{R_B} = R_B\} \cdot B_{14}$

$B_{14} = s_{C_{BO}}(D) \cdot B$

Trent's state transitions described by $APTC_G$ are as follows:

$T = r_{C_{AT}}(A, B) \cdot T_2$

$T_2 = sign_{sk_T}(pk_B) \cdot T_3$

$T_3 = s_{C_{TA}}(SIGN_{sk_T}(pk_B)) \cdot T_4$

$T_4 = r_{C_{BT}}(A, B, ENC_{pk_T}(R_A)) \cdot T_5$

$T_5 = dec_{sk_T}(ENC_{pk_T}(R_A)) \cdot T_6$

$T_6 = sign_{sk_T}(pk_A) \cdot T_7$

$T_7 = rsg_{k_{AB}} \cdot T_8$

$T_8 = sign_{sk_T}(R_A, k_{AB}, A, B) \cdot T_9$

$T_9 = enc_{pk_B}(SIGN_{sk_T}(R_A, k_{AB}, A, B)) \cdot T_{10}$

$T_{10} = s_{C_{TB}}(SIGN_{sk_T}(pk_A), ENC_{pk_B}(SIGN_{sk_T}(R_A, k_{AB}, A, B))) \cdot T$

The sending action and the reading action of the same type data through the same channel can communicate with each other, otherwise, they will cause a deadlock $\delta$. We define the following communication functions:

$\gamma(r_{C_{AT}}(A, B), s_{C_{AT}}(A, B)) \triangleq c_{C_{AT}}(A, B)$

$\gamma(r_{C_{BT}}(A, B, ENC_{pk_T}(R_A)), s_{C_{AT}}(A, B, ENC_{pk_T}(R_A))) \triangleq c_{C_{AT}}(A, B, ENC_{pk_T}(R_A))$

$\gamma(r_{C_{TA}}(SIGN_{sk_T}(pk_B)), s_{C_{TA}}(SIGN_{sk_T}(pk_B)))$
$\triangleq c_{C_{TA}}(SIGN_{sk_T}(pk_B))$

$\gamma(r_{C_{TB}}(SIGN_{sk_T}(pk_A), ENC_{pk_B}(SIGN_{sk_T}(R_A, k_{AB}, A, B))),$
$s_{C_{TB}}(SIGN_{sk_T}(pk_A), ENC_{pk_B}(SIGN_{sk_T}(R_A, k_{AB}, A, B))))$
$\triangleq c_{C_{TB}}(SIGN_{sk_T}(A, pk_A))$

$\gamma(r_{C_{AB}}(ENC_{sk_P}(A, R_A)), s_{C_{AB}}(ENC_{sk_P}(A, R_A)))$
$\triangleq c_{C_{AB}}(ENC_{sk_P}(A, R_A))$
$\gamma(r_{C_{BA}}(ENC_{pk_A}(SIGN_{sk_T}(d_{R_A}, k_{AB}, A, B), R_B)),$
$s_{C_{BA}}(ENC_{pk_A}(SIGN_{sk_T}(d_{R_A}, k_{AB}, A, B), R_B)))$
$\triangleq c_{C_{BA}}(ENC_{pk_A}(SIGN_{sk_T}(d_{R_A}, k_{AB}, A, B), R_B))$
$\gamma(r_{C_{AB}}(ENC_{k_{AB}}(R_B, D)), s_{C_{AB}}(ENC_{k_{AB}}(R_B, D))) \triangleq c_{C_{AB}}(ENC_{k_{AB}}(R_B, D))$

Let all modules be in parallel, then the protocol $A \quad B \quad T$ can be presented by the following process term:

$$\tau_I(\partial_H(\Theta(A \between B \between T))) = \tau_I(\partial_H(A \between B \between T))$$

where $H = \{r_{C_{AT}}(A, B), s_{C_{AT}}(A, B), r_{C_{BT}}(A, B, ENC_{pk_T}(R_A)), s_{C_{AT}}(A, B, ENC_{pk_T}(R_A)),$
$r_{C_{TA}}(SIGN_{sk_T}(pk_B)), s_{C_{TA}}(SIGN_{sk_T}(pk_B)),$
$r_{C_{TB}}(SIGN_{sk_T}(pk_A), ENC_{pk_B}(SIGN_{sk_T}(R_A, k_{AB}, A, B))),$
$s_{C_{TB}}(SIGN_{sk_T}(pk_A), ENC_{pk_B}(SIGN_{sk_T}(R_A, k_{AB}, A, B))),$
$r_{C_{AB}}(ENC_{sk_P}(A, R_A)), s_{C_{AB}}(ENC_{sk_P}(A, R_A)),$
$r_{C_{BA}}(ENC_{pk_A}(SIGN_{sk_T}(d_{R_A}, k_{AB}, A, B), R_B)),$
$s_{C_{BA}}(ENC_{pk_A}(SIGN_{sk_T}(d_{R_A}, k_{AB}, A, B), R_B)),$
$r_{C_{AB}}(ENC_{k_{AB}}(R_B, D)), s_{C_{AB}}(ENC_{k_{AB}}(R_B, D))|D \in \Delta\},$
$I = \{c_{C_{AT}}(A, B), c_{C_{AT}}(A, B, ENC_{pk_T}(R_A)), c_{C_{TA}}(SIGN_{sk_T}(pk_B)),$
$c_{C_{TB}}(SIGN_{sk_T}(A, pk_A)), c_{C_{AB}}(ENC_{sk_P}(A, R_A)),$
$c_{C_{BA}}(ENC_{pk_A}(SIGN_{sk_T}(d_{R_A}, k_{AB}, A, B), R_B)), c_{C_{AB}}(ENC_{k_{AB}}(R_B, D)),$
$\{k_{AB} = NULL\}, \{k_{AB} \neq NULL\}, de\text{-}sign_{pk_T}(SIGN_{sk_T}(pk_B)),$
$rsg_{R_A}, enc_{sk_P}(A, R_A), \{d_{R_A} = R_A\}, \{d_{R_A} \neq R_A\}, enc_{k_{AB}}(R_B, D),$
$dec_{sk_B}(ENC_{pk_B}(A, R_A)), dec_{sk_B}(ENC_{pk_B}(A, R_A)), de\text{-}sign_{pk_T}(SIGN_{sk_T}(pk_A)),$
$dec_{sk_B}(ENC_{pk_B}(SIGN_{sk_T}(R_A, k_{AB}, A, B))), rsg_{R_B}, enc_{pk_A}(SIGN_{sk_T}(R_A, k_{AB}, A, B), R_B),$
$dec_{k_{AB}}(ENC_{k_{AB}}(d_{AB})), \{d_{AB} = \perp\}, \{d_{AB} \neq \perp\},$
$\{d_{R_B} = R_B\}, \{d_{R_B} \neq R_B\}, sign_{sk_T}(pk_B), dec_{sk_T}(ENC_{pk_T}(R_A)),$
$sign_{sk_T}(pk_A), rsg_{k_{AB}}, sign_{sk_T}(R_A, k_{AB}, A, B), enc_{pk_B}(SIGN_{sk_T}(R_A, k_{AB}, A, B))|D \in \Delta\}$

Then, we obtain the following conclusion on the protocol.

**Theorem 28.9.** *The Woo–Lam protocol in Fig. 28.9 is secure.*

*Proof.* Based on the above state transitions of the above modules, by use of the algebraic laws of $APTC_G$, we can prove that

$\tau_I(\partial_H(A \between B \between T)) = \sum_{D\in\Delta}(r_{C_{AI}}(D) \cdot (s_{C_{BO}}(\perp) + s_{C_{BO}}(D))) \cdot \tau_I(\partial_H(A \between B \between T))$

For the details of the proof, please refer to Section 17.3, as we omit them here.

That is, the Woo–Lam protocol in Fig. 28.9 $\tau_I(\partial_H(A \between B \between T))$ can exhibit the desired external behaviors:

1. For the modeling of confidentiality, it is similar to the protocol in Section 25.10.2, the Woo–Lam protocol is confidential;
2. For the man-in-the-middle attack, because $pk_A$ and $pk_B$ are signed by Trent, the protocol would be $\tau_I(\partial_H(A \between B \between T)) = \sum_{D\in\Delta}(r_{C_{AI}}(D) \cdot s_{C_{BO}}(\perp)) \cdot \tau_I(\partial_H(A \between B \between T))$, as desired, the Woo–Lam protocol can be used against the man-in-the-middle attack;
3. For replay attack, the use of the random number $R_A$, $R_B$, makes it that $\tau_I(\partial_H(A \between B \between T)) = \sum_{D\in\Delta}(r_{C_{AI}}(D) \cdot s_{C_{BO}}(\perp)) \cdot \tau_I(\partial_H(A \between B \between T))$, as desired;
4. Without man-in-the-middle and replay attack, the protocol would be $\tau_I(\partial_H(A \between B \between T)) = \sum_{D\in\Delta}(r_{C_{AI}}(D) \cdot s_{C_{BO}}(D)) \cdot \tau_I(\partial_H(A \between B \between T))$, as desired;
5. For the unexpected and non-technical leaking of $sk_A$, $sk_B$, $k_{AB}$, or they are not strong enough, or Trent is dishonest, they are out of the scope of analyses of security protocols;
6. For malicious tampering and transmission errors, they are out of the scope of analyses of security protocols. □

Chapter 29

# Analyses of other protocols

In this chapter, we will introduce some other useful security protocols, including secret splitting protocols in Section 29.1, bit commitment protocols in Section 29.2, and anonymous key distribution protocols in Section 29.3.

## 29.1 Analyses of secret splitting protocols

The hypothetical secret splitting protocol is shown in Fig. 29.1. Trent receives a message, splits this into four parts, and each part is sent to Alice, Bob, Carol, and Dave. Then, Trent gathers the four parts from Alice, Bob, Carol, and Dave, and combines them into a message. If the combined message is the original message, he then sends out the message.

The process of the protocol is as follows:

1. Trent receives some messages $D$ from the outside through the channel $C_{TI}$ (the corresponding reading action is denoted $r_{C_{TI}}(D)$), he generates three random numbers $R_1, R_2, R_3$ of equal lengths to $D$ through three actions $rsg_{R_1}$, $rsg_{R_2}$, and $rsg_{R_3}$ respectively. Then, he does an XOR operation to the data $D$, $R_1$, $R_2$, and $R_3$ through an XOR action $xor(R_1, R_2, R_3, D)$ to obtain $R_4 = XOR(R_1, R_2, R_3, D)$, he sends $R_1$, $R_2$, $R_3$, $R_4$ to Alice, Bob, Carol, and Dave through the channels $C_{TA}$, $C_{TB}$, $C_{TC}$, and $C_{TD}$, respectively (the corresponding sending actions are denoted $s_{C_{TA}}(R_1)$, $s_{C_{TB}}(R_2)$, $s_{C_{TC}}(R_3)$, $s_{C_{TD}}(R_4)$);
2. Alice receives $R_1$ from Trent through the channel $C_{TA}$ (the corresponding reading action is denoted $r_{C_{TA}}(R_1)$), she may store $R_1$, we assume that she sends $R_1$ to Trent immediately through the channel $C_{AT}$ (the corresponding sending action is denoted $s_{C_{AT}}(R_1)$);
3. Bob receives $R_2$ from Trent through the channel $C_{TB}$ (the corresponding reading action is denoted $r_{C_{TB}}(R_2)$), he may store $R_2$, we assume that he sends $R_2$ to Trent immediately through the channel $C_{BT}$ (the corresponding sending action is denoted $s_{C_{BT}}(R_2)$);

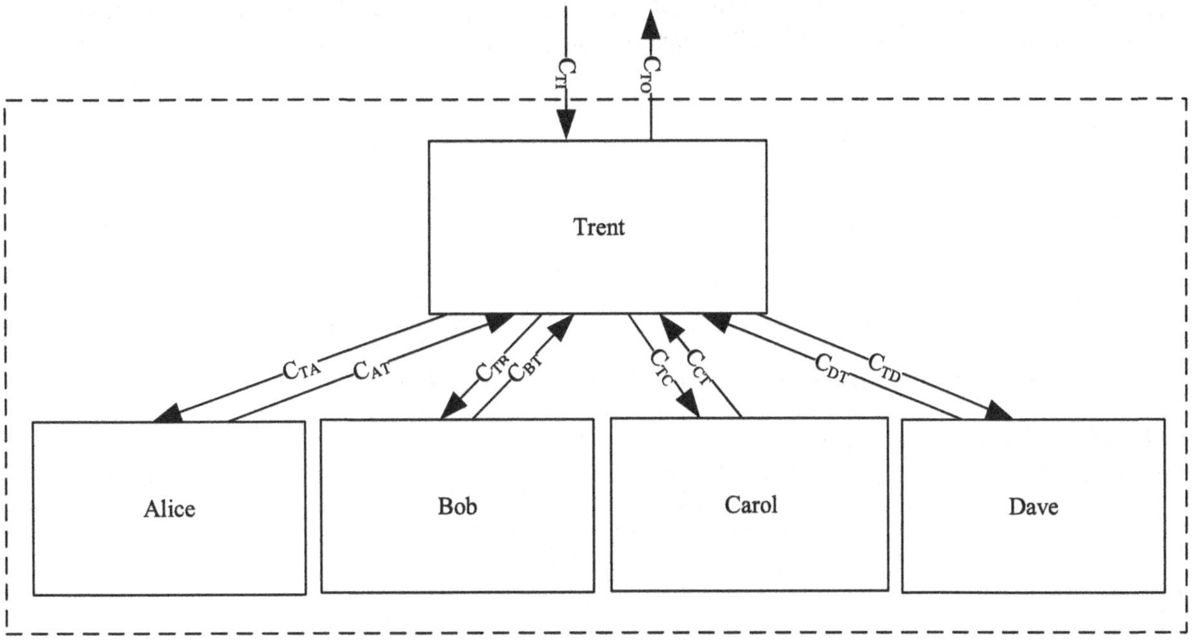


**FIGURE 29.1** Secret splitting protocol.

Handbook of Truly Concurrent Process Algebra. https://doi.org/10.1016/B978-0-44-321515-5.00033-0

4. Carol receives $R_3$ from Trent through the channel $C_{TC}$ (the corresponding reading action is denoted $r_{C_{TC}}(R_3)$), she may store $R_3$, we assume that she sends $R_3$ to Trent immediately through the channel $C_{CT}$ (the corresponding sending action is denoted $s_{C_{CT}}(R_3)$);
5. Dave receives $R_4$ from Trent through the channel $C_{TD}$ (the corresponding reading action is denoted $r_{C_{TD}}(R_4)$), he may store $R_4$, we assume that he sends $R_4$ to Trent immediately through the channel $C_{DT}$ (the corresponding sending action is denoted $s_{C_{DT}}(R_4)$);
6. Trent receives $d_{R_1}$, $d_{R_3}$, $d_{R_3}$, and $d_{R_4}$ from Alice, Bob, Carol, and Dave through the channel $C_{AT}$, $C_{BT}$, $C_{CT}$, and $C_{DT}$, respectively (the corresponding reading actions are denoted $r_{C_{AT}}(d_{R_1})$, $r_{C_{BT}}(d_{R_2})$, $r_{C_{CT}}(d_{R_3})$, $r_{C_{DT}}(d_{R_4})$), he does an XOR operation on the data $d_{R_1}, d_{R_2}, d_{R_3}$, and $d_{R_4}$ through an XOR action $xor(d_{R_1}, d_{R_2}, d_{R_3}, d_{R_4})$ to obtain $D' = XOR(d_{R_1}, d_{R_2}, d_{R_3}, d_{R_4})$, if $D = D'$, he sends $D$ to the outside through the channel $C_{TO}$ (the corresponding sending action is denoted $s_{C_{TO}}(D)$),

where $D \in \Delta$, $\Delta$ is the set of data.

Trent's state transitions described by $APTC_G$ are as follows:

$T = \sum_{D\in\Delta} r_{C_{TI}}(D) \cdot T_2$

$T_2 = (rsg_{R_1} \parallel rsg_{R_2} \parallel rsg_{R_3}) \cdot T_3$

$T_3 = xor(R_1, R_2, R_3, D) \cdot T_4$

$T_4 = (s_{C_{TA}}(R_1) \parallel s_{C_{TB}}(R_2) \parallel s_{C_{TC}}(R_3) \parallel s_{C_{TD}}(R_4)) \cdot T_5$

$T_5 = (r_{C_{AT}}(d_{R_1}) \parallel r_{C_{BT}}(d_{R_2}) \parallel r_{C_{CT}}(d_{R_3}) \parallel r_{C_{DT}}(d_{R_4})) \cdot T_6$

$T_6 = xor(d_{R_1}, d_{R_2}, d_{R_3}, d_{R_4}) \cdot T_7$

$T_7 = \{D = D'\} \cdot s_{C_{TO}}(D) \cdot T$

Alice's state transitions described by $APTC_G$ are as follows:

$A = r_{C_{TB}}(R_2) \cdot A_2$

$A_2 = s_{C_{BT}}(R_2) \cdot A$

Bob's state transitions described by $APTC_G$ are as follows:

$B = r_{C_{TB}}(R_2) \cdot B_2$

$B_2 = s_{C_{BT}}(R_2) \cdot B$

Carol's state transitions described by $APTC_G$ are as follows:

$C = r_{C_{TB}}(R_2) \cdot C_2$

$C_2 = s_{C_{BT}}(R_2) \cdot C$

Dave's state transitions described by $APTC_G$ are as follows:

$Da = r_{C_{TB}}(R_2) \cdot Da_2$

$Da_2 = s_{C_{BT}}(R_2) \cdot Da$

The sending action and the reading action of the same type data through the same channel can communicate with each other, otherwise, they will cause a deadlock $\delta$. We define the following communication functions:

$\gamma(r_{C_{TA}}(R_1), s_{C_{TA}}(R_1)) \triangleq c_{C_{TA}}(R_1)$

$\gamma(r_{C_{TB}}(R_2), s_{C_{TB}}(R_2)) \triangleq c_{C_{TB}}(R_2)$

$\gamma(r_{C_{TC}}(R_3), s_{C_{TC}}(R_3)) \triangleq c_{C_{TC}}(R_3)$

$\gamma(r_{C_{TD}}(R_4), s_{C_{TD}}(R_4)) \triangleq c_{C_{TD}}(R_4)$

$\gamma(r_{C_{AT}}(d_{R_1}), s_{C_{AT}}(d_{R_1})) \triangleq c_{C_{AT}}(d_{R_1})$

$\gamma(r_{C_{BT}}(d_{R_2}), s_{C_{BT}}(d_{R_2})) \triangleq c_{C_{BT}}(d_{R_2})$

$\gamma(r_{C_{CT}}(d_{R_3}), s_{C_{CT}}(d_{R_3})) \triangleq c_{C_{CT}}(d_{R_3})$

$\gamma(r_{C_{DT}}(d_{R_4}), s_{C_{DT}}(d_{R_4})) \triangleq c_{C_{DT}}(d_{R_4})$

Let all modules be in parallel, then the protocol $A \quad B \quad C \quad Da \quad T$ can be presented by the following process term:

$$\tau_I(\partial_H(\Theta(A \between B \between C \between Da \between T))) = \tau_I(\partial_H(A \between B \between C \between Da \between T))$$

where $H = \{r_{C_{TA}}(R_1), s_{C_{TA}}(R_1), r_{C_{TB}}(R_2), s_{C_{TB}}(R_2), r_{C_{TC}}(R_3), s_{C_{TC}}(R_3),$
$r_{C_{TD}}(R_4), s_{C_{TD}}(R_4), r_{C_{AT}}(d_{R_1}), s_{C_{AT}}(d_{R_1}), r_{C_{BT}}(d_{R_2}), s_{C_{BT}}(d_{R_2}),$
$r_{C_{CT}}(d_{R_3}), s_{C_{CT}}(d_{R_3}), r_{C_{DT}}(d_{R_4}), s_{C_{DT}}(d_{R_4}) | D \in \Delta\}$,

$I = \{c_{C_{TA}}(R_1), c_{C_{TB}}(R_2), c_{C_{TC}}(R_3), c_{C_{TD}}(R_4), c_{C_{AT}}(d_{R_1}), c_{C_{BT}}(d_{R_2}),$
$c_{C_{CT}}(d_{R_3}), c_{C_{DT}}(d_{R_4}), rsg_{R_1}, rsg_{R_2}, rsg_{R_3}, xor(R_1, R_2, R_3, D),$
$xor(d_{R_1}, d_{R_2}, d_{R_3}, d_{R_4}), \{D = D'\} | D \in \Delta\}$

Then, we obtain the following conclusion on the protocol.

**Theorem 29.1.** *The secret splitting protocol in Fig. 29.1 is secure.*

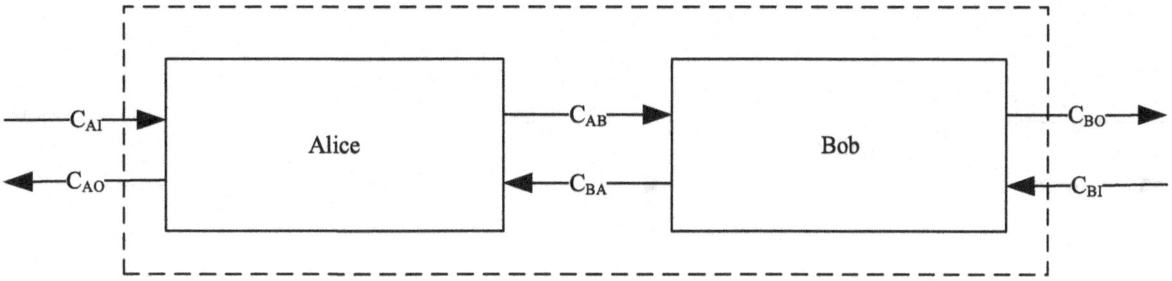


**FIGURE 29.2** Bit commitment protocol 1.

*Proof.* Based on the above state transitions of the above modules, by use of the algebraic laws of $APTC_G$, we can prove that

$\tau_I(\partial_H(A \between B \between C \between Da \between T)) = \sum_{D\in\Delta}(r_{C_{TI}}(D) \cdot s_{C_{TO}}(D)) \cdot \tau_I(\partial_H(A \between B \between C \between Da \between T))$

For the details of the proof, please refer to Section 17.3, as we omit it here.

That is, the protocol in Fig. 29.1 $\tau_I(\partial_H(A \between B \between C \between Da \between T))$ can exhibit the desired external behaviors, and satisfies the main goal of secret splitting. It must be noted that the distribution and gathering of $R_1, R_2, R_3, R_4$ have no cryptographic assurance, they can be made into information leakage. □

## 29.2 Analyses of bit commitment protocols

In this section, we will introduce analyses of bit commitment protocols. We introduce analyses of bit commitment protocol based on symmetric cryptography in Section 29.2.1, and bit commitment protocol based on one-way function in Section 29.2.2.

### 29.2.1 Bit commitment protocol 1

The protocol shown in Fig. 29.2 uses symmetric cryptography to implement bit commitment.

The process of the protocol is as follows:

1. Bob receives some requests $D$ from the outside through the channel $C_{BI}$ (the corresponding reading action is denoted $r_{C_{BI}}(D)$), he generates a random sequence $R$ through an action $rsg_R$, then Bob sends $R$ to Alice through the channel $C_{BA}$ (the corresponding sending action is denoted $s_{C_{BA}}(R)$);
2. Alice receives $R$ from Bob through the channel $C_{BA}$ (the corresponding reading action is denoted $r_{C_{BA}}(R)$), she generates the commitment $b$ through an action $rsg_b$ and generates a random key $k$ through an action $rsg_k$, encrypts $b$ and $R$ by $k$ through an action $enc_k(R, b)$, and sends $ENC_k(R, b)$ to Bob through the channel $C_{AB}$ (the corresponding sending action is denoted $s_{C_{AB}}(ENC_k(R, b))$);
3. Bob receives the message $ENC_k(R, b)$ from Alice through the channel $C_{AB}$ (the corresponding reading action is denoted $r_{C_{AB}}(ENC_k(R, b))$), he cannot decrypt the message due to the absence of $k$; after some time, he sends a commitment release request $r$ to Alice through the channel $C_{BA}$ (the corresponding sending action is denoted $s_{C_{BA}}(r)$);
4. Alice receives $r$ from Bob through the channel $C_{BA}$ (the corresponding reading action is denoted $r_{C_{BA}}(r)$), she sends $k$ to Bob through the channel $C_{AB}$ (the corresponding sending action is denoted $s_{C_{AB}}(k)$);
5. Bob receives $k$ from Alice through the channel $C_{AB}$ (the corresponding reading action is denoted $r_{C_{AB}}(k)$), he decrypts $ENC_k(d_R, b)$ through an action $dec_k(ENC_k(d_R, b))$, if $d_R = R$, he sends $b$ to the outside through the channel $C_{BO}$ (the corresponding sending action is denoted $s_{C_{BO}}(b)$); else if $d_R \neq R$, he sends $\perp$ to the outside through the channel $C_{BO}$ (the corresponding sending action is denoted $s_{C_{BO}}(\perp)$),

where $D \in \Delta$, $\Delta$ is the set of data.

Alice's state transitions described by $APTC_G$ are as follows:

$A = r_{C_{BA}}(R) \cdot A_2$

$A_2 = rsg_b \cdot A_3$

$A_3 = rsg_k \cdot A_4$

$A_4 = s_{C_{AB}}(ENC_k(R, b)) \cdot A_5$

$A_5 = r_{C_{BA}}(r) \cdot A_6$

$A_6 = s_{C_{AB}}(k) \cdot A$

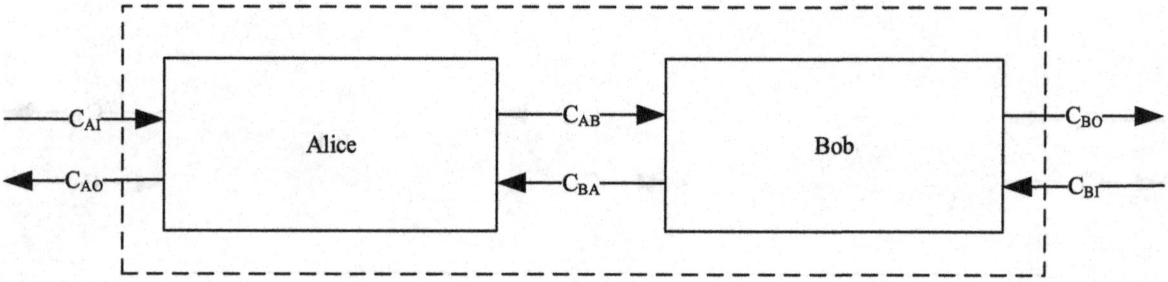


**FIGURE 29.3** Bit commitment protocol 2.

Bob's state transitions described by $APTC_G$ are as follows:

$B = \sum_{D\in\Delta} r_{C_{BI}}(D) \cdot B_2$

$B_2 = rsg_R \cdot B_3$

$B_3 = s_{C_{BA}}(R) \cdot B_4$

$B_4 = r_{C_{AB}}(ENC_k(R,b)) \cdot B_5$

$B_5 = s_{C_{BA}}(r) \cdot B_6$

$B_6 = r_{C_{AB}}(k) \cdot B_7$

$B_7 = dec_k(ENC_k(d_R,b)) \cdot B_8$

$B_8 = \{d_R = R\} \cdot B_9 + \{d_R \neq R\} \cdot B_{10}$

$B_9 = s_{C_{BO}}(b) \cdot B$

$B_{10} = s_{C_{BO}}(\perp) \cdot B$

The sending action and the reading action of the same type data through the same channel can communicate with each other, otherwise, they will cause a deadlock $\delta$. We define the following communication functions:

$\gamma(r_{C_{BA}}(R), s_{C_{BA}}(R)) \triangleq c_{C_{BA}}(R)$

$\gamma(r_{C_{AB}}(ENC_k(R,b)), s_{C_{AB}}(ENC_k(R,b))) \triangleq c_{C_{AB}}(ENC_k(R,b))$

$\gamma(r_{C_{BA}}(r), s_{C_{BA}}(r)) \triangleq c_{C_{BA}}(r)$

$\gamma(r_{C_{AB}}(k), s_{C_{AB}}(k)) \triangleq c_{C_{AB}}(k)$

Let all modules be in parallel, then the protocol $A \quad B$ can be presented by the following process term:

$$\tau_I(\partial_H(\Theta(A \between B))) = \tau_I(\partial_H(A \between B))$$

where $H = \{r_{C_{BA}}(R), s_{C_{BA}}(R), r_{C_{AB}}(ENC_k(R,b)), s_{C_{AB}}(ENC_k(R,b)),$
$r_{C_{BA}}(r), s_{C_{BA}}(r), r_{C_{AB}}(k), s_{C_{AB}}(k) | D \in \Delta\}$,

$I = \{c_{C_{BA}}(R), c_{C_{AB}}(ENC_k(R,b)), c_{C_{BA}}(r), c_{C_{AB}}(k),$
$rsg_b, rsg_k, rsg_R, dec_k(ENC_k(d_R,b)), \{d_R = R\}, \{d_R \neq R\} | D \in \Delta\}$

Then, we obtain the following conclusion on the protocol.

**Theorem 29.2.** *The bit commitment protocol 1 in Fig. 29.2 is secure.*

*Proof.* Based on the above state transitions of the above modules, by use of the algebraic laws of $APTC_G$, we can prove that

$\tau_I(\partial_H(A \between B)) = \sum_{D\in\Delta}(r_{C_{BI}}(D) \cdot (s_{C_{BO}}(b) + s_{C_{BO}}(\perp))) \cdot \tau_I(\partial_H(A \between B))$

For the details of the proof, please refer to Section 17.3, as we omit them here.

That is, the protocol in Fig. 29.2 $\tau_I(\partial_H(A \between B))$ can exhibit the desired external behaviors, that is, if the bits are committed, the system would be $\tau_I(\partial_H(A \between B)) = \sum_{D\in\Delta}(r_{C_{BI}}(D) \cdot s_{C_{BO}}(b)) \cdot \tau_I(\partial_H(A \between B))$; otherwise, the system would be $\tau_I(\partial_H(A \between B)) = \sum_{D\in\Delta}(r_{C_{BI}}(D) \cdot s_{C_{BO}}(\perp)) \cdot \tau_I(\partial_H(A \between B))$.

Note that the main security goals are bit commitment, the protocol in Fig. 29.2 cannot satisfy other security goals, such as confidentiality. □

### 29.2.2 Bit commitment protocol 2

The protocol shown in Fig. 29.3 uses a one-way function to implement bit commitment.

The process of the protocol is as follows:

1. Alice receives some requests $D$ from the outside through the channel $C_{AI}$ (the corresponding reading action is denoted $r_{C_{AI}}(D)$), she generates a random sequence $R_1$ through an action $rsg_{R_1}$, and a random sequence $R_2$ through an action $rsg_{R_2}$, generates the commitment $b$ through an action $rsg_b$, computes the hash of $R_1, R_2, b$ through an action $hash(R_1, R_2, b)$, and sends $HASH(R_1, R_2, b), R_1$ to Bob through the channel $C_{AB}$ (the corresponding sending action is denoted $s_{C_{AB}}(HASH(R_1, R_2, b), R_1)$);
2. Bob receives the message $HASH(R_1, R_2, b), R_1)$ from Alice through the channel $C_{AB}$ (the corresponding reading action is denoted $r_{C_{AB}}(HASH(R_1, R_2, b), R_1)$), after some time, he sends a commitment release request $r$ to Alice through the channel $C_{BA}$ (the corresponding sending action is denoted $s_{C_{BA}}(r)$);
3. Alice receives $r$ from Bob through the channel $C_{BA}$ (the corresponding reading action is denoted $r_{C_{BA}}(r)$), she sends $R_1, R_2, b$ to Bob through the channel $C_{AB}$ (the corresponding sending action is denoted $s_{C_{AB}}(R_1, R_2, b)$);
4. Bob receives $d_{R_1}, R_2, b$ from Alice through the channel $C_{AB}$ (the corresponding reading action is denoted $r_{C_{AB}}(d_{R_1}, R_2, b)$), if $d_{R_1} = R_1$ and $HASH(R_1, R_2, b) = HASH(d_{R_1}, R_2, b)$, he sends $b$ to the outside through the channel $C_{BO}$ (the corresponding sending action is denoted $s_{C_{BO}}(b)$); else if $d_{R_1} \neq R_1$ or $HASH(R_1, R_2, b) \neq HASH(d_{R_1}, R_2, b)$, he sends $\perp$ to the outside through the channel $C_{BO}$ (the corresponding sending action is denoted $s_{C_{BO}}(\perp)$),

where $D \in \Delta$, $\Delta$ is the set of data.

Alice's state transitions described by $APTC_G$ are as follows:

$A = \sum_{D\in\Delta} r_{C_{BI}}(D) \cdot A_2$
$A_2 = rsg_{R_1} \cdot A_3$
$A_3 = rsg_{R_2} \cdot A_4$
$A_4 = rsg_b \cdot A_5$
$A_5 = hash(R_1, R_2, b) \cdot A_6$
$A_6 = s_{C_{AB}}(HASH(R_1, R_2, b), R_1) \cdot A_7$
$A_7 = r_{C_{BA}}(r) \cdot A_8$
$A_8 = s_{C_{AB}}(R_1, R_2, b) \cdot A$

Bob's state transitions described by $APTC_G$ are as follows:

$B = r_{C_{AB}}(HASH(R_1, R_2, b), R_1) \cdot B_2$
$B_2 = s_{C_{BA}}(r) \cdot B_3$
$B_3 = r_{C_{AB}}(d_{R_1}, R_2, b) \cdot B_4$
$B_4 = \{d_{R_1} = R_1\} \cdot \{HASH(R_1, R_2, b) = HASH(d_{R_1}, R_2, b)\} \cdot B_5 + (\{d_{R_1} \neq R_1\} + \{HASH(R_1, R_2, b) \neq HASH(d_{R_1}, R_2, b)\}) \cdot B_6$
$B_5 = s_{C_{BO}}(b) \cdot B$
$B_6 = s_{C_{BO}}(\perp) \cdot B$

The sending action and the reading action of the same type data through the same channel can communicate with each other, otherwise, they will cause a deadlock $\delta$. We define the following communication functions:

$\gamma(r_{C_{AB}}(HASH(R_1, R_2, b), R_1), s_{C_{AB}}(HASH(R_1, R_2, b), R_1)) \triangleq c_{C_{AB}}(HASH(R_1, R_2, b), R_1)$
$\gamma(r_{C_{BA}}(r), s_{C_{BA}}(r)) \triangleq c_{C_{BA}}(r)$
$\gamma(r_{C_{AB}}(d_{R_1}, R_2, b), s_{C_{AB}}(d_{R_1}, R_2, b)) \triangleq c_{C_{AB}}(d_{R_1}, R_2, b)$

Let all modules be in parallel, then the protocol $A \quad B$ can be presented by the following process term:

$$\tau_I(\partial_H(\Theta(A \between B))) = \tau_I(\partial_H(A \between B))$$

where $H = \{r_{C_{AB}}(HASH(R_1, R_2, b), R_1), s_{C_{AB}}(HASH(R_1, R_2, b), R_1), r_{C_{BA}}(r), s_{C_{BA}}(r), r_{C_{AB}}(d_{R_1}, R_2, b), s_{C_{AB}}(d_{R_1}, R_2, b) | D \in \Delta\}$,

$I = \{c_{C_{AB}}(HASH(R_1, R_2, b), R_1), c_{C_{BA}}(r), c_{C_{AB}}(d_{R_1}, R_2, b), rsg_{R_1}, rsg_{R_2}, rsg_b, hash(R_1, R_2, b), \{d_{R_1} = R_1\}, \{HASH(R_1, R_2, b) = HASH(d_{R_1}, R_2, b), \{d_{R_1} \neq R_1\}, \{HASH(R_1, R_2, b) \neq HASH(d_{R_1}, R_2, b)\}\} | D \in \Delta\}$

Then, we obtain the following conclusion on the protocol.

**Theorem 29.3.** *The bit commitment protocol 2 in Fig. 29.3 is secure.*

*Proof.* Based on the above state transitions of the above modules, by use of the algebraic laws of $APTC_G$, we can prove that

$\tau_I(\partial_H(A \between B)) = \sum_{D\in\Delta}(r_{C_{AI}}(D) \cdot (s_{C_{BO}}(b) + s_{C_{BO}}(\perp))) \cdot \tau_I(\partial_H(A \between B))$

For the details of the proof, please refer to Section 17.3, as we omit them here.

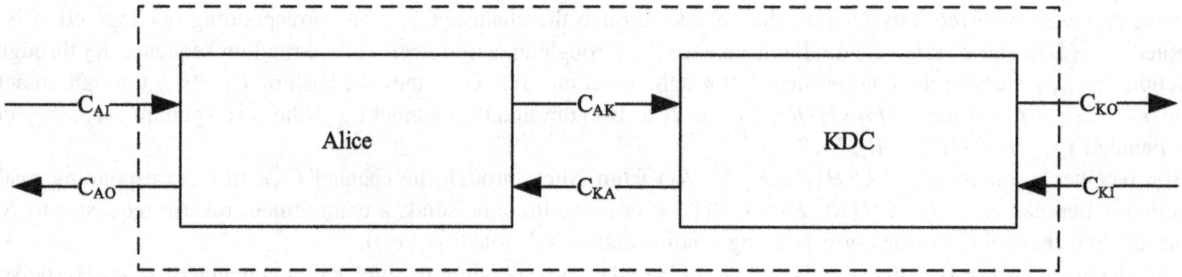


**FIGURE 29.4** Anonymous key distribution protocol.

That is, the protocol in Fig. 29.3 $\tau_I(\partial_H(A \between B))$ can exhibit the desired external behaviors, that is, if the bits are committed, the system would be $\tau_I(\partial_H(A \between B)) = \sum_{D\in\Delta}(r_{C_{BI}}(D) \cdot s_{C_{BO}}(b)) \cdot \tau_I(\partial_H(A \between B))$; otherwise, the system would be $\tau_I(\partial_H(A \between B)) = \sum_{D\in\Delta}(r_{C_{BI}}(D) \cdot s_{C_{BO}}(\bot)) \cdot \tau_I(\partial_H(A \between B))$.

Note that the main security goals are bit commitment, the protocol in Fig. 29.3 cannot satisfy other security goals, such as confidentiality. □

## 29.3 Analyses of anonymous key distribution protocols

The protocol shown in Fig. 29.4 uses asymmetric cryptography to implement anonymous key distribution.

The process of the protocol is as follows:

1. Alice receives some requests $D$ from the outside through the channel $C_{BI}$ (the corresponding reading action is denoted $r_{C_{BI}}(D)$), she generates a public/private key pair through an action $rsg_{pk_A,sk_A}$, and sends the key request $r$ to KDC through the channel $C_{AK}$ (the corresponding sending action is denoted $s_{C_{AK}}(r)$);
2. The KDC receives the key request $r$ from Alice through the channel $C_{AK}$ (the corresponding reading action is denoted $r_{C_{AK}}(r)$), he generates a series of keys $k_i$ through actions $rsg_{k_i}$ for $1 \le i \le n$, and encrypts these keys by his public key $pk_K$ through actions $enc_{pk_K}(k_i)$ for $1 \le i \le n$, then sends these encrypted keys to Alice through the channel $C_{KA}$ (the corresponding sending action is denoted $s_{C_{KA}}(ENC_{pk_K}(k_1), \cdots, ENC_{pk_K}(k_n))$);
3. Alice receives the encrypted keys from the KDC through the channel $C_{KA}$ (the corresponding reading action is denoted $r_{C_{KA}}(ENC_{pk_K}(k_1), \cdots, ENC_{pk_K}(k_n))$), she randomly selects one $ENC_{pk_K}(k_j)$ for $1 \le j \le n$, encrypts it by her public key $pk_A$ through an action $enc_{pk_A}(ENC_{pk_K}(k_j))$, and sends the doubly encrypted key $ENC_{pk_A}(ENC_{pk_K}(k_j))$ to the KDC through the channel $C_{AK}$ (the corresponding sending action is denoted $s_{C_{AK}}(ENC_{pk_A}(ENC_{pk_K}(k_j)))$);
4. The KDC receives the doubly encrypted key from Alice through the channel $C_{AK}$ (the corresponding reading action is denoted $r_{C_{AK}}(ENC_{pk_A}(ENC_{pk_K}(k_j)))$), he decrypts it by his private key $sk_K$ through an action $dec_{sk_K}(ENC_{pk_A}(ENC_{pk_K}(k_j)))$ to obtain $ENC_{pk_A}(k_j)$, and sends $ENC_{pk_A}(k_j)$ to Alice through the channel $C_{KA}$ (the corresponding sending action is denoted $s_{C_{KA}}(ENC_{pk_A}(k_j))$);
5. Alice receives $ENC_{pk_A}(k_j)$ from the KDC through the channel $C_{KA}$ (the corresponding reading action is denoted $r_{C_{KA}}(ENC_{pk_A}(k_j))$), she decrypts it by her private key $sk_A$ through an action $dec_{sk_A}(ENC_{pk_A}(k_j))$ to obtain $k_j$, and sends $k_j$ to the outside through the channel $C_{AO}$ (the corresponding sending action is denoted $s_{C_{AO}}(k_j)$),

where $D \in \Delta$, $\Delta$ is the set of data.

Alice's state transitions described by $APTC_G$ are as follows:

$A = \sum_{D\in\Delta} r_{C_{AI}}(D) \cdot A_2$
$A_2 = rsg_{pk_A,sk_A} \cdot A_3$
$A_3 = s_{C_{AK}}(r) \cdot A_4$
$A_4 = r_{C_{KA}}(ENC_{pk_K}(k_1), \cdots, ENC_{pk_K}(k_n)) \cdot A_5$
$A_5 = enc_{pk_A}(ENC_{pk_K}(k_j)) \cdot A_6$
$A_6 = s_{C_{AK}}(ENC_{pk_A}(ENC_{pk_K}(k_j))) \cdot A_7$
$A_7 = r_{C_{KA}}(ENC_{pk_A}(k_j)) \cdot A_8$
$A_8 = dec_{sk_A}(ENC_{pk_A}(k_j)) \cdot A_9$
$A_9 = s_{C_{AO}}(k_j) \cdot A$

The KDC's state transitions described by $APTC_G$ are as follows:

$K = r_{C_{AK}}(r) \cdot K_2$
$K_2 = rsg_{k_1} \parallel \cdots \parallel rsg_{k_n} \cdot K_3$

$K_3 = enc_{pk_K}(k_1) \parallel \cdots \parallel enc_{pk_K}(k_n) \cdot K_4$
$K_4 = s_{C_{KA}}(ENC_{pk_K}(k_1), \cdots, ENC_{pk_K}(k_n)) \cdot K_5$
$K_5 = r_{C_{AK}}(ENC_{pk_A}(ENC_{pk_K}(k_j))) \cdot K_6$
$K_6 = dec_{sk_K}(ENC_{pk_A}(ENC_{pk_K}(k_j))) \cdot K_7$
$K_7 = s_{C_{KA}}(ENC_{pk_A}(k_j)) \cdot K$

The sending action and the reading action of the same type data through the same channel can communicate with each other, otherwise, they will cause a deadlock $\delta$. We define the following communication functions:

$\gamma(r_{C_{AK}}(r), s_{C_{AK}}(r)) \triangleq c_{C_{AK}}(r)$
$\gamma(r_{C_{KA}}(ENC_{pk_K}(k_1), \cdots, ENC_{pk_K}(k_n)), s_{C_{KA}}(ENC_{pk_K}(k_1), \cdots, ENC_{pk_K}(k_n)))$
$\triangleq c_{C_{KA}}(ENC_{pk_K}(k_1), \cdots, ENC_{pk_K}(k_n))$
$\gamma(r_{C_{AK}}(ENC_{pk_A}(ENC_{pk_K}(k_j))), s_{C_{AK}}(ENC_{pk_A}(ENC_{pk_K}(k_j)))) \triangleq c_{C_{AK}}(ENC_{pk_A}(ENC_{pk_K}(k_j)))$
$\gamma(r_{C_{KA}}(ENC_{pk_A}(k_j)), s_{C_{KA}}(ENC_{pk_A}(k_j))) \triangleq c_{C_{KA}}(ENC_{pk_A}(k_j))$

Let all modules be in parallel, then the protocol $A \quad K$ can be presented by the following process term:

$$\tau_I(\partial_H(\Theta(A \between K))) = \tau_I(\partial_H(A \between K))$$

where $H = \{r_{C_{AK}}(r), s_{C_{AK}}(r), r_{C_{KA}}(ENC_{pk_A}(k_j)), s_{C_{KA}}(ENC_{pk_A}(k_j)),$
$r_{C_{KA}}(ENC_{pk_K}(k_1), \cdots, ENC_{pk_K}(k_n)), s_{C_{KA}}(ENC_{pk_K}(k_1), \cdots, ENC_{pk_K}(k_n)),$
$r_{C_{AK}}(ENC_{pk_A}(ENC_{pk_K}(k_j))), s_{C_{AK}}(ENC_{pk_A}(ENC_{pk_K}(k_j)))|D \in \Delta\},$

$I = \{c_{C_{AK}}(r), c_{C_{KA}}(ENC_{pk_A}(k_j)), c_{C_{AK}}(ENC_{pk_A}(ENC_{pk_K}(k_j))),$
$c_{C_{KA}}(ENC_{pk_K}(k_1), \cdots, ENC_{pk_K}(k_n)), rsg_{pk_A, sk_A}, enc_{pk_A}(ENC_{pk_K}(k_j)),$
$dec_{sk_A}(ENC_{pk_A}(k_j)), rsg_{k_1}, \cdots, rsg_{k_n}, enc_{pk_K}(k_1), \cdots, enc_{pk_K}(k_n),$
$dec_{sk_K}(ENC_{pk_A}(ENC_{pk_K}(k_j)))|D \in \Delta\}$

Then, we obtain the following conclusion on the protocol.

**Theorem 29.4.** *The anonymous key distribution protocol in Fig. 29.4 is secure.*

*Proof.* Based on the above state transitions of the above modules, by use of the algebraic laws of $APTC_G$, we can prove that

$\tau_I(\partial_H(A \between K)) = \sum_{D\in\Delta}(r_{C_{AI}}(D) \cdot s_{C_{AO}}(k_j)) \cdot \tau_I(\partial_H(A \between K))$

For the details of the proof, please refer to Section 17.3, as we omit them here.

That is, the protocol in Fig. 29.4 $\tau_I(\partial_H(A \between K))$ can exhibit the desired external behaviors, and is secure. □

Chapter 30

# Analyses of digital cash protocols

Digital cash makes it possible to use cash digitally. Digital cash may have the following six properties:

1. Independence. The digital cash is independent of the location, and can be used through the network;
2. Security. The digital cash cannot be copied and reused;
3. Privacy. The privacy of the owner of the digital cash is protected;
4. Off-line payment. The digital cash can be used off-line;
5. Transferability. The digital cash can be transferred to the other users;
6. Divisibility. The digital cash can be divided into small pieces of digital cash.

In this chapter, we will introduce four digital cash protocols in the following sections. In the analyses of these four protocols, we will mainly analyze the security and privacy properties.

## 30.1 Digital cash protocol 1

The Digital Cash Protocol 1 shown in Fig. 30.1 is the basic digital cash protocol to ensure anonymity.

The process of the protocol is as follows:

1. Alice receives some requests $D$ from the outside through the channel $C_{AI}$ (the corresponding reading action is denoted $r_{C_{AI}}(D)$), she generates $n$ $\$m_i$ orders with each order encrypted by the bank's public key $pk_B$ through actions $enc_{pk_B}(m_i)$ for $1 \leq i \leq n$, and sends them to the bank through the channel $C_{AB}$ (the corresponding sending action is denoted $s_{C_{AB}}(ENC_{pk_B}(m_1), \cdots, ENC_{pk_B}(m_n))$);
2. The bank receives these orders from Alice through the channel $C_{AB}$ (the corresponding reading action is denoted $r_{C_{AB}}(ENC_{pk_B}(m_1), \cdots, ENC_{pk_B}(m_n))$), he randomly selects $n-1$ orders and decrypts them through actions $dec_{sk_B}(ENC_{pk_B}(m_j))$ for $1 \leq j \leq n-1$ to ensure that each $m_j = m$. Then, he signs the left $ENC_{pk_B}(m_k)$ through an action $sign_{sk_B}(ENC_{pk_B}(m_k))$, checks the identity of Alice and deducts $\$m$ from Alice's account through an action

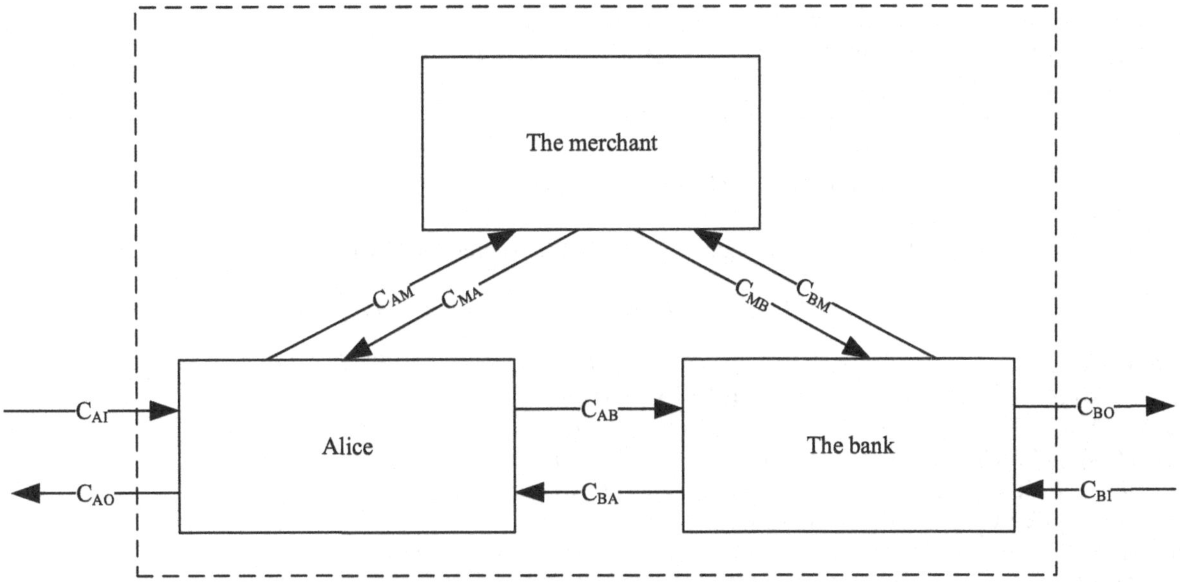


**FIGURE 30.1** Digital Cash Protocol 1.

Handbook of Truly Concurrent Process Algebra. https://doi.org/10.1016/B978-0-44-321515-5.00034-2

$s_{C_{BO}}(-m)$, then sends $SIGN_{sk_B}(ENC_{pk_B}(m_k))$ to Alice through the channel $C_{BA}$ (the corresponding sending action is denoted $s_{C_{BA}}(SIGN_{sk_B}(ENC_{pk_B}(m_k)))$);

3. Alice receives the signed order $SIGN_{sk_B}(ENC_{pk_B}(m_k))$ from the bank through the channel $C_{BA}$ (the corresponding reading action is denoted $r_{C_{BA}}(SIGN_{sk_B}(ENC_{pk_B}(m_k)))$), she may send the signed order to some merchant through the channel $C_{AM}$ (the corresponding sending action is denoted $s_{C_{AM}}(SIGN_{sk_B}(ENC_{pk_B}(m_k)))$);
4. The merchant receives the signed cash from Alice through the channel $C_{AM}$ (the corresponding reading action is denoted $r_{C_{AM}}(SIGN_{sk_B}(ENC_{pk_B}(m_k)))$), he sends it to the bank through the channel $C_{MB}$ (the corresponding sending action is denoted $s_{C_{MB}}(SIGN_{sk_B}(ENC_{pk_B}(m_k)))$);
5. The bank receives the signed cash from the merchant through the channel $C_{MB}$ (the corresponding reading action is denoted $r_{C_{MB}}(SIGN_{sk_B}(ENC_{pk_B}(m_k)))$), he de-signs the cash through an action $de\text{-}sign(SIGN_{sk_B}(ENC_{pk_B}(m_k)))$, then decrypts it through an action $dec_{sk_B}(ENC_{pk_B}(m_k))$, checks the identity of the merchant and credits \$$m$ to the merchant's account through an action $s_{C_{BO}}(+m)$,

where $D \in \Delta$, $\Delta$ is the set of data.

Alice's state transitions described by $APTC_G$ are as follows:

$A = \sum_{D\in\Delta} r_{C_{AI}}(D) \cdot A_2$

$A_2 = enc_{pk_B}(m_1) \parallel \cdots \parallel enc_{pk_B}(m_n) \cdot A_3$

$A_3 = s_{C_{AB}}(ENC_{pk_B}(m_1), \cdots, ENC_{pk_B}(m_n)) \cdot A_4$

$A_4 = r_{C_{BA}}(SIGN_{sk_B}(ENC_{pk_B}(m_k))) \cdot A_5$

$A_5 = s_{C_{AM}}(SIGN_{sk_B}(ENC_{pk_B}(m_k))) \cdot A$

The bank's state transitions described by $APTC_G$ are as follows:

$B = r_{C_{AB}}(ENC_{pk_B}(m_1), \cdots, ENC_{pk_B}(m_n)) \cdot B_2$

$B_2 = dec_{sk_B}(ENC_{pk_B}(m_1)) \parallel \cdots \parallel dec_{sk_B}(ENC_{pk_B}(m_{n-1})) \cdot B_3$

$B_3 = sign_{sk_B}(ENC_{pk_B}(m_k)) \cdot B_4$

$B_4 = s_{C_{BO}}(-m) \cdot B_5$

$B_5 = s_{C_{BA}}(SIGN_{sk_B}(ENC_{pk_B}(m_k))) \cdot B_6$

$B_6 = r_{C_{MB}}(SIGN_{sk_B}(ENC_{pk_B}(m_k))) \cdot B_7$

$B_7 = de\text{-}sign(SIGN_{sk_B}(ENC_{pk_B}(m_k))) \cdot B_8$

$B_8 = s_{C_{BO}}(+m) \cdot B$

The merchant's state transitions described by $APTC_G$ are as follows:

$M = r_{C_{AM}}(SIGN_{sk_B}(ENC_{pk_B}(m_k))) \cdot M_2$

$M_2 = s_{C_{MB}}(SIGN_{sk_B}(ENC_{pk_B}(m_k))) \cdot M$

The sending action and the reading action of the same type data through the same channel can communicate with each other, otherwise, they will cause a deadlock $\delta$. We define the following communication functions:

$\gamma(r_{C_{AB}}(ENC_{pk_B}(m_1), \cdots, ENC_{pk_B}(m_n)), s_{C_{AB}}(ENC_{pk_B}(m_1), \cdots, ENC_{pk_B}(m_n)))$
$\triangleq c_{C_{AB}}(ENC_{pk_B}(m_1), \cdots, ENC_{pk_B}(m_n))$

$\gamma(r_{C_{BA}}(SIGN_{sk_B}(ENC_{pk_B}(m_k))), s_{C_{BA}}(SIGN_{sk_B}(ENC_{pk_B}(m_k))))$
$\triangleq c_{C_{BA}}(SIGN_{sk_B}(ENC_{pk_B}(m_k)))$

$\gamma(r_{C_{MB}}(SIGN_{sk_B}(ENC_{pk_B}(m_k))), s_{C_{MB}}(SIGN_{sk_B}(ENC_{pk_B}(m_k))))$
$\triangleq c_{C_{MB}}(SIGN_{sk_B}(ENC_{pk_B}(m_k)))$

$\gamma(r_{C_{AM}}(SIGN_{sk_B}(ENC_{pk_B}(m_k))), s_{C_{AM}}(SIGN_{sk_B}(ENC_{pk_B}(m_k))))$
$\triangleq c_{C_{AM}}(SIGN_{sk_B}(ENC_{pk_B}(m_k)))$

Let all modules be in parallel, then the protocol $A \quad B \quad M$ can be presented by the following process term.

$$\tau_I(\partial_H(\Theta(A \between B \between M))) = \tau_I(\partial_H(A \between B \between M))$$

where $H = \{r_{C_{AB}}(ENC_{pk_B}(m_1), \cdots, ENC_{pk_B}(m_n)), s_{C_{AB}}(ENC_{pk_B}(m_1), \cdots, ENC_{pk_B}(m_n)),$
$r_{C_{BA}}(SIGN_{sk_B}(ENC_{pk_B}(m_k))), s_{C_{BA}}(SIGN_{sk_B}(ENC_{pk_B}(m_k))),$
$r_{C_{MB}}(SIGN_{sk_B}(ENC_{pk_B}(m_k))), s_{C_{MB}}(SIGN_{sk_B}(ENC_{pk_B}(m_k))),$
$r_{C_{AM}}(SIGN_{sk_B}(ENC_{pk_B}(m_k))), s_{C_{AM}}(SIGN_{sk_B}(ENC_{pk_B}(m_k)))|D \in \Delta\},$

$I = \{c_{C_{AB}}(ENC_{pk_B}(m_1), \cdots, ENC_{pk_B}(m_n)), c_{C_{BA}}(SIGN_{sk_B}(ENC_{pk_B}(m_k))),$
$c_{C_{MB}}(SIGN_{sk_B}(ENC_{pk_B}(m_k))), c_{C_{AM}}(SIGN_{sk_B}(ENC_{pk_B}(m_k))),$
$enc_{pk_B}(m_1), \cdots, enc_{pk_B}(m_n), dec_{sk_B}(ENC_{pk_B}(m_1)), \cdots, dec_{sk_B}(ENC_{pk_B}(m_{n-1})),$
$sign_{sk_B}(ENC_{pk_B}(m_k)), de\text{-}sign(SIGN_{sk_B}(ENC_{pk_B}(m_k)))|D \in \Delta\}$

Then, we obtain the following conclusion on the protocol.

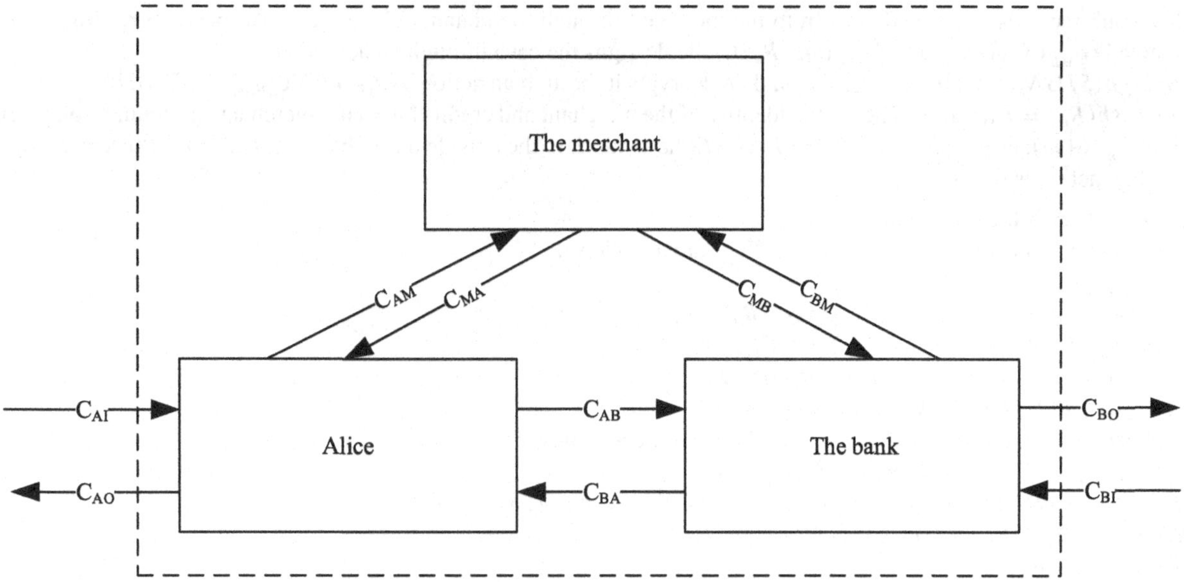


**FIGURE 30.2** Digital Cash Protocol 2.

**Theorem 30.1.** *The Digital Cash Protocol 1 in Fig. 30.1 is anonymous.*

*Proof.* Based on the above state transitions of the above modules, by use of the algebraic laws of $APTC_G$, we can prove that

$\tau_I(\partial_H(A \between B \between M)) = \sum_{D\in\Delta}(r_{C_{AI}}(D)\cdot s_{C_{BO}}(-m)\cdot s_{C_{BO}}(+m))\cdot\tau_I(\partial_H(A \between B \between M))$

For the details of the proof, please refer to Section 17.3, as we omit them here.

That is, the Digital Cash Protocol 1 in Fig. 30.1 $\tau_I(\partial_H(A \between B \between M))$ can exhibit the desired external behaviors:

1. The digital cash of Alice $SIGN_{sk_B}(ENC_{pk_B}(m_k))$ is anonymous for the merchant and the bank;
2. The protocol cannot resist replay attack, for digital cash, this is the so-called double spending problem, either for Alice or the merchant. The system would be $\tau_I(\partial_H(A \between B \between M)) = \sum_{D\in\Delta}(r_{C_{AI}}(D)\cdot s_{C_{BO}}(-m)\cdot s_{C_{BO}}(+m)\cdot s_{C_{BO}}(+m))\cdot \tau_I(\partial_H(A \between B \between M))$. □

## 30.2 Digital cash protocol 2

The Digital Cash Protocol 2 shown in Fig. 30.2 is the basic digital cash protocol to ensure anonymity and resist replay attacks.

The process of the protocol is as follows:

1. Alice receives some requests $D$ from the outside through the channel $C_{AI}$ (the corresponding reading action is denoted $r_{C_{AI}}(D)$), she generates $n$ $\$m_i$ orders containing a random number $R_i$ with each order encrypted by the bank's public key $pk_B$ through actions $enc_{pk_B}(m_i, R_i)$ for $1 \le i \le n$, and sends them to the bank through the channel $C_{AB}$ (the corresponding sending action is denoted $s_{C_{AB}}(ENC_{pk_B}(m_1, R_1), \cdots, ENC_{pk_B}(m_n, R_n)))$;
2. The bank receives these orders from Alice through the channel $C_{AB}$ (the corresponding reading action is denoted $r_{C_{AB}}(ENC_{pk_B}(m_1, R_1), \cdots, ENC_{pk_B}(m_n, R_n)))$, he randomly selects $n-1$ orders and decrypts them through actions $dec_{sk_B}(ENC_{pk_B}(m_j, R_j))$ for $1 \le j \le n-1$ to ensure that each $m_j = m$ and $R_j$ is fresh. Then, he signs the left $ENC_{pk_B}(m_k, R_k)$ through an action $sign_{sk_B}(ENC_{pk_B}(m_k, R_k))$, checks the identity of Alice and deducts $\$m$ from Alice's account through an action $s_{C_{BO}}(-m)$, then sends $SIGN_{sk_B}(ENC_{pk_B}(m_k, R_k))$ to Alice through the channel $C_{BA}$ (the corresponding sending action is denoted $s_{C_{BA}}(SIGN_{sk_B}(ENC_{pk_B}(m_k, R_k))))$;
3. Alice receives the signed order $SIGN_{sk_B}(ENC_{pk_B}(m_k, R_k))$ from the bank through the channel $C_{BA}$ (the corresponding reading action is denoted $r_{C_{BA}}(SIGN_{sk_B}(ENC_{pk_B}(m_k, R_k))))$, she may send the signed order to some merchant through the channel $C_{AM}$ (the corresponding sending action is denoted $s_{C_{AM}}(SIGN_{sk_B}(ENC_{pk_B}(m_k, R_k))))$;
4. The merchant receives the signed cash from Alice through the channel $C_{AM}$ (the corresponding reading action is denoted $r_{C_{AM}}(SIGN_{sk_B}(ENC_{pk_B}(m_k, R_k))))$, he sends it to the bank through the channel $C_{MB}$ (the corresponding sending action is denoted $s_{C_{MB}}(SIGN_{sk_B}(ENC_{pk_B}(m_k, R_k))))$;

5. The bank receives the signed cash from the merchant through the channel $C_{MB}$ (the corresponding reading action is denoted $r_{C_{MB}}(SIGN_{sk_B}(ENC_{pk_B}(m_k, R_k))))$, he de-signs the cash through an action $de\text{-}sign(SIGN_{sk_B}(ENC_{pk_B}(m_k, R_k)))$, then decrypts it through an action $dec_{sk_B}(ENC_{pk_B}(m_k, R_k))$, if $isFresh(R_k) = TRUE$, he checks the identity of the merchant and credits \$$m$ to the merchant's account through an action $s_{C_{BO}}(+m)$; else if $isFresh(R_k) = FALSE$, he sends $\perp$ to the outside through the channel $C_{BO}$ (the corresponding sending action is denoted $s_{C_{BO}}(\perp)$),

where $D \in \Delta$, $\Delta$ is the set of data.

Alice's state transitions described by $APTC_G$ are as follows:

$A = \sum_{D\in\Delta} r_{C_{AI}}(D) \cdot A_2$

$A_2 = enc_{pk_B}(m_1, R_1) \parallel \cdots \parallel enc_{pk_B}(m_n, R_n) \cdot A_3$

$A_3 = s_{C_{AB}}(ENC_{pk_B}(m_1, R_1), \cdots, ENC_{pk_B}(m_n, R_n)) \cdot A_4$

$A_4 = r_{C_{BA}}(SIGN_{sk_B}(ENC_{pk_B}(m_k, R_k))) \cdot A_5$

$A_5 = s_{C_{AM}}(SIGN_{sk_B}(ENC_{pk_B}(m_k, R_k))) \cdot A$

The bank's state transitions described by $APTC_G$ are as follows:

$B = r_{C_{AB}}(ENC_{pk_B}(m_1, R_1), \cdots, ENC_{pk_B}(m_n, R_n)) \cdot B_2$

$B_2 = dec_{sk_B}(ENC_{pk_B}(m_1, R_1)) \parallel \cdots \parallel dec_{sk_B}(ENC_{pk_B}(m_{n-1}, R_{n-1})) \cdot B_3$

$B_3 = sign_{sk_B}(ENC_{pk_B}(m_k, R_k)) \cdot B_4$

$B_4 = s_{C_{BO}}(-m) \cdot B_5$

$B_5 = s_{C_{BA}}(SIGN_{sk_B}(ENC_{pk_B}(m_k, R_k))) \cdot B_6$

$B_6 = r_{C_{MB}}(SIGN_{sk_B}(ENC_{pk_B}(m_k, R_k))) \cdot B_7$

$B_7 = de\text{-}sign(SIGN_{sk_B}(ENC_{pk_B}(m_k, R_k))) \cdot B_8$

$B_8 = \{isFresh(R_k) = TRUE\} \cdot s_{C_{BO}}(+m) \cdot B + \{isFresh(R_k) = FALSE\} \cdot s_{C_{BO}}(\perp) \cdot B$

The merchant's state transitions described by $APTC_G$ are as follows:

$M = r_{C_{AM}}(SIGN_{sk_B}(ENC_{pk_B}(m_k, R_k))) \cdot M_2$

$M_2 = s_{C_{MB}}(SIGN_{sk_B}(ENC_{pk_B}(m_k, R_k))) \cdot M$

The sending action and the reading action of the same type data through the same channel can communicate with each other, otherwise, they will cause a deadlock $\delta$. We define the following communication functions:

$\gamma(r_{C_{AB}}(ENC_{pk_B}(m_1, R_1), \cdots, ENC_{pk_B}(m_n, R_n)), s_{C_{AB}}(ENC_{pk_B}(m_1, R_1), \cdots, ENC_{pk_B}(m_n, R_n)))$
$\triangleq c_{C_{AB}}(ENC_{pk_B}(m_1, R_1), \cdots, ENC_{pk_B}(m_n, R_n))$

$\gamma(r_{C_{BA}}(SIGN_{sk_B}(ENC_{pk_B}(m_k, R_k))), s_{C_{BA}}(SIGN_{sk_B}(ENC_{pk_B}(m_k, R_k))))$
$\triangleq c_{C_{BA}}(SIGN_{sk_B}(ENC_{pk_B}(m_k, R_k)))$

$\gamma(r_{C_{MB}}(SIGN_{sk_B}(ENC_{pk_B}(m_k, R_k))), s_{C_{MB}}(SIGN_{sk_B}(ENC_{pk_B}(m_k, R_k))))$
$\triangleq c_{C_{MB}}(SIGN_{sk_B}(ENC_{pk_B}(m_k, R_k)))$

$\gamma(r_{C_{AM}}(SIGN_{sk_B}(ENC_{pk_B}(m_k, R_k))), s_{C_{AM}}(SIGN_{sk_B}(ENC_{pk_B}(m_k, R_k))))$
$\triangleq c_{C_{AM}}(SIGN_{sk_B}(ENC_{pk_B}(m_k, R_k)))$

Let all modules be in parallel, then the protocol $A \quad B \quad M$ can be presented by the following process term.

$$\tau_I(\partial_H(\Theta(A \between B \between M))) = \tau_I(\partial_H(A \between B \between M))$$

where $H = \{r_{C_{AB}}(ENC_{pk_B}(m_1, R_1), \cdots, ENC_{pk_B}(m_n, R_n)),$
$s_{C_{AB}}(ENC_{pk_B}(m_1, R_1), \cdots, ENC_{pk_B}(m_n, R_n)),$
$r_{C_{BA}}(SIGN_{sk_B}(ENC_{pk_B}(m_k, R_k))), s_{C_{BA}}(SIGN_{sk_B}(ENC_{pk_B}(m_k, R_k))),$
$r_{C_{MB}}(SIGN_{sk_B}(ENC_{pk_B}(m_k, R_k))), s_{C_{MB}}(SIGN_{sk_B}(ENC_{pk_B}(m_k, R_k))),$
$r_{C_{AM}}(SIGN_{sk_B}(ENC_{pk_B}(m_k, R_k))), s_{C_{AM}}(SIGN_{sk_B}(ENC_{pk_B}(m_k, R_k)))|D \in \Delta\},$

$I = \{c_{C_{AB}}(ENC_{pk_B}(m_1, R_1), \cdots, ENC_{pk_B}(m_n, R_n)), c_{C_{BA}}(SIGN_{sk_B}(ENC_{pk_B}(m_k, R_k))),$
$c_{C_{MB}}(SIGN_{sk_B}(ENC_{pk_B}(m_k, R_k))), c_{C_{AM}}(SIGN_{sk_B}(ENC_{pk_B}(m_k, R_k))),$
$enc_{pk_B}(m_1, R_1), \cdots, enc_{pk_B}(m_n, R_n), dec_{sk_B}(ENC_{pk_B}(m_1, R_1)), \cdots, dec_{sk_B}(ENC_{pk_B}(m_{n-1}, R_{n-1})),$
$sign_{sk_B}(ENC_{pk_B}(m_k, R_k)), de\text{-}sign(SIGN_{sk_B}(ENC_{pk_B}(m_k, R_k))),$
$\{isFresh(R_k) = TRUE\}, \{isFresh(R_k) = FALSE\}|D \in \Delta\}$

Then, we obtain the following conclusion on the protocol.

**Theorem 30.2.** *The Digital Cash Protocol 2 in Fig. 30.2 is anonymous and resists replaying.*

*Proof.* Based on the above state transitions of the above modules, by use of the algebraic laws of $APTC_G$, we can prove that

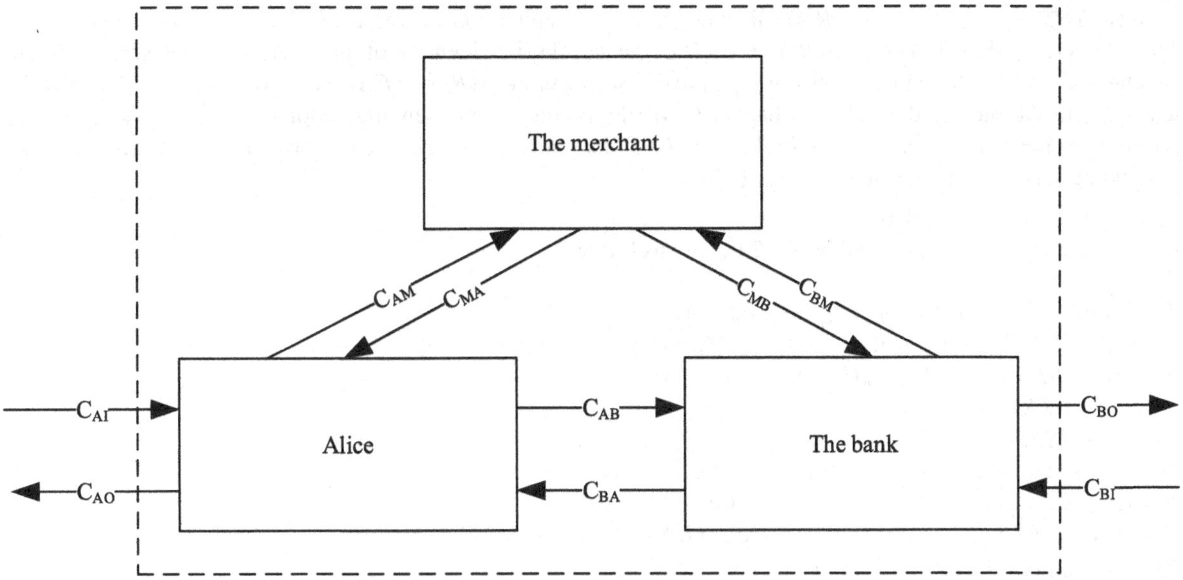


**FIGURE 30.3** Digital Cash Protocol 3.

$\tau_I(\partial_H(A \between B \between M)) = \sum_{D\in\Delta}(r_{C_{AI}}(D) \cdot s_{C_{BO}}(-m) \cdot (s_{C_{BO}}(+m) + s_{C_{BO}}(\bot))) \cdot \tau_I(\partial_H(A \between B \between M))$

For the details of the proof, please refer to Section 17.3, as we omit them here.

That is, the Digital Cash Protocol 2 in Fig. 30.2 $\tau_I(\partial_H(A \between B \between M))$ can exhibit the desired external behaviors:

1. The digital cash of Alice $SIGN_{sk_B}(ENC_{pk_B}(m_k))$ is anonymous for the merchant and the bank;
2. The protocol can resist replay attacks, for the use of the random number in each digital cash;
3. The bank does not know who cheats him when the double spending problem occurs, either the owner of the cash or the merchant. □

## 30.3 Digital cash protocol 3

The Digital Cash Protocol 3 shown in Fig. 30.3 is the basic digital cash protocol to ensure the anonymity, resist replay attacks, and know partly who the attacker was.

The process of the protocol is as follows:

1. Alice receives some requests $D$ from the outside through the channel $C_{AI}$ (the corresponding reading action is denoted $r_{C_{AI}}(D)$), she generates $n$ $\$m_i$ orders containing a random number $R_i$ with each order encrypted by the bank's public key $pk_B$ through actions $enc_{pk_B}(m_i, R_i)$ for $1 \leq i \leq n$, and sends them to the bank through the channel $C_{AB}$ (the corresponding sending action is denoted $s_{C_{AB}}(ENC_{pk_B}(m_1, R_1), \cdots, ENC_{pk_B}(m_n, R_n))$);
2. The bank receives these orders from Alice through the channel $C_{AB}$ (the corresponding reading action is denoted $r_{C_{AB}}(ENC_{pk_B}(m_1, R_1), \cdots, ENC_{pk_B}(m_n, R_n))$), he randomly selects $n-1$ orders and decrypts them through actions $dec_{sk_B}(ENC_{pk_B}(m_j, R_j))$ for $1 \leq j \leq n-1$ to ensure that each $m_j = m$ and $R_j$ is fresh. Then, he signs the left $ENC_{pk_B}(m_k, R_k)$ through an action $sign_{sk_B}(ENC_{pk_B}(m_k, R_k))$, checks the identity of Alice and deducts $\$m$ from Alice's account through an action $s_{C_{BO}}(-m)$, then sends $SIGN_{sk_B}(ENC_{pk_B}(m_k, R_k))$ to Alice through the channel $C_{BA}$ (the corresponding sending action is denoted $s_{C_{BA}}(SIGN_{sk_B}(ENC_{pk_B}(m_k, R_k)))$);
3. Alice receives the signed order $SIGN_{sk_B}(ENC_{pk_B}(m_k, R_k))$ from the bank through the channel $C_{BA}$ (the corresponding reading action is denoted $r_{C_{BA}}(SIGN_{sk_B}(ENC_{pk_B}(m_k, R_k)))$), she generates a random string $R$ and encrypts it through an action $enc_{pk_B}(R)$, she may send the signed order to some merchant through the channel $C_{AM}$ (the corresponding sending action is denoted $s_{C_{AM}}(SIGN_{sk_B}(ENC_{pk_B}(m_k, R_k)), ENC_{pk_B}(R))$);
4. The merchant receives the signed cash from Alice through the channel $C_{AM}$ (the corresponding reading action is denoted $r_{C_{AM}}(SIGN_{sk_B}(ENC_{pk_B}(m_k, R_k)), ENC_{pk_B}(R))$), he sends it to the bank through the channel $C_{MB}$ (the corresponding sending action is denoted $s_{C_{MB}}(SIGN_{sk_B}(ENC_{pk_B}(m_k, R_k)))$);
5. The bank receives the signed cash from the merchant through the channel $C_{MB}$ (the corresponding reading action is denoted $r_{C_{MB}}(SIGN_{sk_B}(ENC_{pk_B}(m_k, R_k)), ENC_{pk_B}(R))$), he de-signs the cash through an action

$de\text{-}sign(SIGN_{sk_B}(ENC_{pk_B}(m_k, R_k)))$, then decrypts it through an action $dec_{sk_B}(ENC_{pk_B}(m_k, R_k))$ and $dec_{sk_B}(ENC_{pk_B}(R))$, if $isFresh(R_k) = TRUE$, he checks the identity of the merchant and credits \$$m$ to the merchant's account through an action $s_{C_{BO}}(+m)$; else if $isFresh(R_k) = FALSE$ and $isFresh(R) = TRUE$, he sends $\bot_A$ to the outside through the channel $C_{BO}$ (the corresponding sending action is denoted $s_{C_{BO}}(\bot_A)$); else if $isFresh(R_k) = FALSE$ and $isFresh(R) = FALSE$, he sends $\bot_M$ to the outside through the channel $C_{BO}$ (the corresponding sending action is denoted $s_{C_{BO}}(\bot_M)$),

where $D \in \Delta$, $\Delta$ is the set of data.

Alice's state transitions described by $APTC_G$ are as follows:

$A = \sum_{D\in\Delta} r_{C_{AI}}(D) \cdot A_2$

$A_2 = enc_{pk_B}(m_1, R_1) \parallel \cdots \parallel enc_{pk_B}(m_n, R_n) \cdot A_3$

$A_3 = s_{C_{AB}}(ENC_{pk_B}(m_1, R_1), \cdots, ENC_{pk_B}(m_n, R_n)) \cdot A_4$

$A_4 = r_{C_{BA}}(SIGN_{sk_B}(ENC_{pk_B}(m_k, R_k))) \cdot A_5$

$A_5 = enc_{pk_B}(R) \cdot A_6$

$A_6 = s_{C_{AM}}(SIGN_{sk_B}(ENC_{pk_B}(m_k, R_k)), ENC_{pk_B}(R)) \cdot A$

The bank's state transitions described by $APTC_G$ are as follows:

$B = r_{C_{AB}}(ENC_{pk_B}(m_1, R_1), \cdots, ENC_{pk_B}(m_n, R_n)) \cdot B_2$

$B_2 = dec_{sk_B}(ENC_{pk_B}(m_1, R_1)) \parallel \cdots \parallel dec_{sk_B}(ENC_{pk_B}(m_{n-1}, R_{n-1})) \cdot B_3$

$B_3 = sign_{sk_B}(ENC_{pk_B}(m_k, R_k)) \cdot B_4$

$B_4 = s_{C_{BO}}(-m) \cdot B_5$

$B_5 = s_{C_{BA}}(SIGN_{sk_B}(ENC_{pk_B}(m_k, R_k))) \cdot B_6$

$B_6 = r_{C_{MB}}(SIGN_{sk_B}(ENC_{pk_B}(m_k, R_k)), ENC_{pk_B}(R)) \cdot B_7$

$B_7 = de\text{-}sign(SIGN_{sk_B}(ENC_{pk_B}(m_k, R_k))) \cdot B_8$

$B_8 = dec_{sk_B}(ENC_{pk_B}(R)) \cdot B_9$

$B_9 = \{isFresh(R_k) = TRUE\} \cdot s_{C_{BO}}(+m) \cdot B + \{isFresh(R_k) = FALSE\} \cdot \{isFresh(R) = TRUE\} \cdot s_{C_{BO}}(\bot_A) \cdot B$
$+ \{isFresh(R_k) = FALSE\} \cdot \{isFresh(R) = FALSE\} \cdot s_{C_{BO}}(\bot_M) \cdot B$

The merchant's state transitions described by $APTC_G$ are as follows:

$M = r_{C_{AM}}(SIGN_{sk_B}(ENC_{pk_B}(m_k, R_k)), ENC_{pk_B}(R)) \cdot M_2$

$M_2 = s_{C_{MB}}(SIGN_{sk_B}(ENC_{pk_B}(m_k, R_k)), ENC_{pk_B}(R)) \cdot M$

The sending action and the reading action of the same type data through the same channel can communicate with each other, otherwise, they will cause a deadlock $\delta$. We define the following communication functions:

$\gamma(r_{C_{AB}}(ENC_{pk_B}(m_1, R_1), \cdots, ENC_{pk_B}(m_n, R_n)), s_{C_{AB}}(ENC_{pk_B}(m_1, R_1), \cdots, ENC_{pk_B}(m_n, R_n)))$
$\triangleq c_{C_{AB}}(ENC_{pk_B}(m_1, R_1), \cdots, ENC_{pk_B}(m_n, R_n))$

$\gamma(r_{C_{BA}}(SIGN_{sk_B}(ENC_{pk_B}(m_k, R_k))), s_{C_{BA}}(SIGN_{sk_B}(ENC_{pk_B}(m_k, R_k))))$
$\triangleq c_{C_{BA}}(SIGN_{sk_B}(ENC_{pk_B}(m_k, R_k)))$

$\gamma(r_{C_{MB}}(SIGN_{sk_B}(ENC_{pk_B}(m_k, R_k)), ENC_{pk_B}(R)),$
$s_{C_{MB}}(SIGN_{sk_B}(ENC_{pk_B}(m_k, R_k)), ENC_{pk_B}(R)))$
$\triangleq c_{C_{MB}}(SIGN_{sk_B}(ENC_{pk_B}(m_k, R_k)), ENC_{pk_B}(R))$

$\gamma(r_{C_{AM}}(SIGN_{sk_B}(ENC_{pk_B}(m_k, R_k)), ENC_{pk_B}(R)),$
$s_{C_{AM}}(SIGN_{sk_B}(ENC_{pk_B}(m_k, R_k)), ENC_{pk_B}(R)))$
$\triangleq c_{C_{AM}}(SIGN_{sk_B}(ENC_{pk_B}(m_k, R_k)), ENC_{pk_B}(R))$

Let all modules be in parallel, then the protocol $A \quad B \quad M$ can be presented by the following process term.

$$\tau_I(\partial_H(\Theta(A \between B \between M))) = \tau_I(\partial_H(A \between B \between M))$$

where $H = \{r_{C_{AB}}(ENC_{pk_B}(m_1, R_1), \cdots, ENC_{pk_B}(m_n, R_n)),$
$s_{C_{AB}}(ENC_{pk_B}(m_1, R_1), \cdots, ENC_{pk_B}(m_n, R_n)),$
$r_{C_{BA}}(SIGN_{sk_B}(ENC_{pk_B}(m_k, R_k))), s_{C_{BA}}(SIGN_{sk_B}(ENC_{pk_B}(m_k, R_k))),$
$r_{C_{MB}}(SIGN_{sk_B}(ENC_{pk_B}(m_k, R_k)), ENC_{pk_B}(R)),$
$s_{C_{MB}}(SIGN_{sk_B}(ENC_{pk_B}(m_k, R_k)), ENC_{pk_B}(R)),$
$r_{C_{AM}}(SIGN_{sk_B}(ENC_{pk_B}(m_k, R_k)), ENC_{pk_B}(R)),$
$s_{C_{AM}}(SIGN_{sk_B}(ENC_{pk_B}(m_k, R_k)), ENC_{pk_B}(R))|D \in \Delta\},$

$I = \{c_{C_{AB}}(ENC_{pk_B}(m_1, R_1), \cdots, ENC_{pk_B}(m_n, R_n)), c_{C_{BA}}(SIGN_{sk_B}(ENC_{pk_B}(m_k, R_k))),$
$c_{C_{MB}}(SIGN_{sk_B}(ENC_{pk_B}(m_k, R_k)), ENC_{pk_B}(R)), c_{C_{AM}}(SIGN_{sk_B}(ENC_{pk_B}(m_k, R_k)), ENC_{pk_B}(R)),$
$enc_{pk_B}(m_1, R_1), \cdots, enc_{pk_B}(m_n, R_n), dec_{sk_B}(ENC_{pk_B}(m_1, R_1)), \cdots, dec_{sk_B}(ENC_{pk_B}(m_{n-1}, R_{n-1})),$

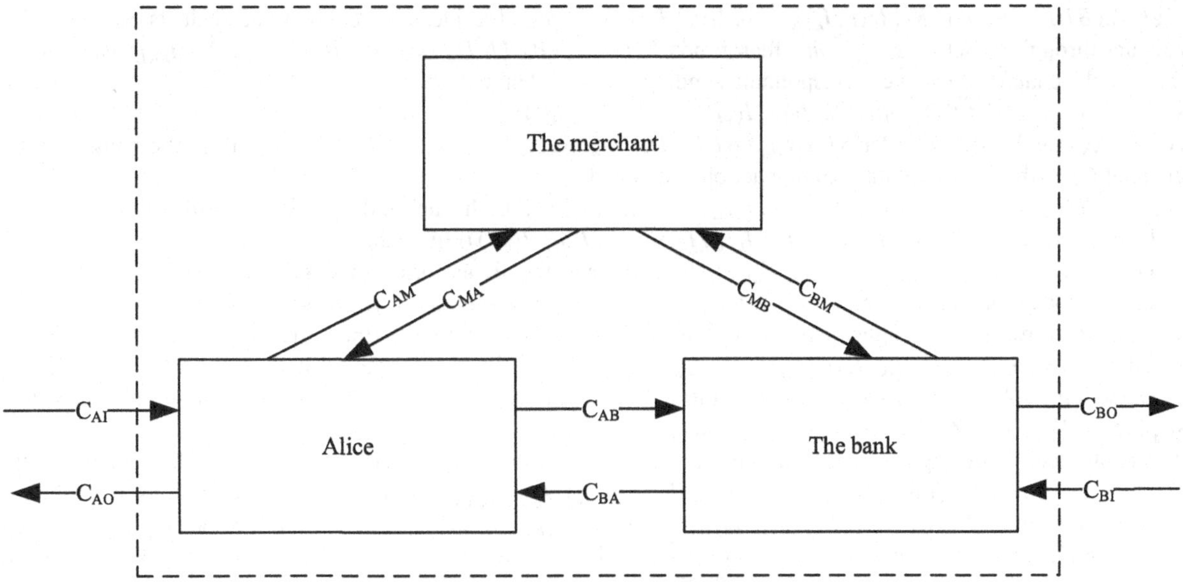


**FIGURE 30.4** Digital Cash Protocol 4.

$sign_{sk_B}(ENC_{pk_B}(m_k, R_k)), de\text{-}sign(SIGN_{sk_B}(ENC_{pk_B}(m_k, R_k))), enc_{pk_B}(R), dec_{sk_B}(ENC_{pk_B}(R)),$
$\{isFresh(R_k) = TRUE\}, \{isFresh(R_k) = FALSE\}, \{isFresh(R) = TRUE\}, \{isFresh(R) = FALSE\} | D \in \Delta\}$

Then, we obtain the following conclusion on the protocol.

**Theorem 30.3.** *The Digital Cash Protocol 3 in Fig. 30.3 is anonymous, resists replaying, and knows partly who the attacker was.*

*Proof.* Based on the above state transitions of the above modules, by use of the algebraic laws of $APTC_G$, we can prove that

$\tau_I(\partial_H(A \between B \between M)) = \sum_{D\in\Delta}(r_{C_{AI}}(D) \cdot s_{C_{BO}}(-m) \cdot (s_{C_{BO}}(+m) + s_{C_{BO}}(\bot_A) + s_{C_{BO}}(\bot_M))) \cdot \tau_I(\partial_H(A \between B \between M))$

For the details of the proof, please refer to Section 17.3, as we omit them here.

That is, the Digital Cash Protocol 3 in Fig. 30.3 $\tau_I(\partial_H(A \between B \between M))$ can exhibit the desired external behaviors:

1. The digital cash of Alice $SIGN_{sk_B}(ENC_{pk_B}(m_k))$ is anonymous for the merchant and the bank;
2. The protocol can resist replay attack, for the use of the random number in each digital cash;
3. The bank knows who cheats him when the double spending problem occurs, either the owner of the cash or the merchant. However, he does not know exactly the identity of the person. □

## 30.4 Digital cash protocol 4

The Digital Cash Protocol 4 shown in Fig. 30.4 is the basic digital cash protocol to ensure anonymity, resist replay attacks, and know who exactly the cheat is.

The process of the protocol is as follows:

1. Alice receives some requests $D$ from the outside through the channel $C_{AI}$ (the corresponding reading action is denoted $r_{C_{AI}}(D)$), she generates $n$ \$$m_i$ orders containing $m$, a random number $R_i$, and $n$ pair of string $I_{i1L}, I_{i1R}, \cdots, I_{inL}, I_{inR}$, with each order blinded through actions $blind_{k_1}(m_i, R_i, I_{i1L}, I_{i1R}, \cdots, I_{inL}, I_{inR})$ for $1 \le i \le n$, and sends them to the bank through the channel $C_{AB}$ (the corresponding sending action is denoted $s_{C_{AB}}(BLIND_{k_1}(m_1, R_1, I_{1L}, I_{1R}), \cdots, BLIND_{k_1}(m_n, R_n, I_{nL}, I_{nR})))$;
2. The bank receives these orders from Alice through the channel $C_{AB}$ (the corresponding reading action is denoted $r_{C_{AB}}(BLIND_{k_1}(m_1, R_1, I_{11L}, I_{11R}, \cdots, I_{1nL}, I_{1nR}), \cdots, BLIND_{k_1}(m_n, R_n, I_{1nL}, I_{1nR}, \cdots, I_{nnL}, I_{nnR})))$, he randomly selects $n-1$ orders and asks Alice to unblind them through actions $unblind_{k_1}(BLIND_{k_1}(m_j, R_j, I_{j1L}, I_{j1R}, \cdots, I_{jnL}, I_{jnR}))$ and to reveal $I_{j1L}, I_{j1R}, \cdots, I_{jnL}, I_{jnR}$ (see Section 29.2) for $1 \le j \le n-1$ to ensure that each $m_j = m$ and $R_j$ is fresh. Then, he signs the left $BLIND_{k_1}(m_k, R_k, I_{k1L}, I_{k1R}, \cdots, I_{knL}, I_{knR})$ through an action

$sign_{sk_B}(BLIND_{k_1}(m_k, R_k, I_{k1L}, I_{k1R}, \cdots, I_{knL}, I_{knR}))$, checks the identity of Alice and deducts \$m from Alice's account through an action $s_{C_{BO}}(-m)$, then sends $SIGN_{sk_B}(BLIND_{k_1}(m_k, R_k, I_{k1L}, I_{k1R}, \cdots, I_{knL}, I_{knR}))$ to Alice through the channel $C_{BA}$ (the corresponding sending action is denoted $s_{C_{BA}}(SIGN_{sk_B}(BLIND_{k_1}(m_k, R_k, I_{k1L}, I_{k1R}, \cdots, I_{knL}, I_{knR}))))$;

**3.** Alice receives the signed order $SIGN_{sk_B}(BLIND_{k_1}(m_k, R_k, I_{k1L}, I_{k1R}, \cdots, I_{knL}, I_{knR}))$ from the bank through the channel $C_{BA}$ (the corresponding reading action is denoted $r_{C_{BA}}(SIGN_{sk_B}(BLIND_{k_1}(m_k, R_k, I_{k1L}, I_{k1R}, \cdots, I_{knL}, I_{knR}))))$, she unblinds the signed order through an action $unblind_{k_1}(SIGN_{sk_B}(BLIND_{k_1}(m_k, R_k, I_{k1L}, I_{k1R}, \cdots, I_{knL}, I_{knR})))$ to obtain $SIGN_{sk_B}(m_k, R_k, I_{k1L}, I_{k1R}, \cdots, I_{knL}, I_{knR})$ she may send the signed order to some merchant through the channel $C_{AM}$ (the corresponding sending action is denoted $s_{C_{AM}}(SIGN_{sk_B}(m_k, R_k, I_{k1L}, I_{k1R}, \cdots, I_{knL}, I_{knR})))$;

**4.** The merchant receives the signed cash from Alice through the channel $C_{AM}$ (the corresponding reading action is denoted $r_{C_{AM}}(SIGN_{sk_B}(m_k, R_k, I_{k1L}, I_{k1R}, \cdots, I_{knL}, I_{knR})))$, he asks Alice to reveal half of $I_{j1L}, I_{j1R}, \cdots, I_{jnL}, I_{jnR}$ (see Section 29.2), and sends it to the bank through the channel $C_{MB}$ (the corresponding sending action is denoted $s_{C_{MB}}(SIGN_{sk_B}(m_k, R_k, I_{k1L}, I_{k1R}, \cdots, I_{knL}, I_{knR}), I'_{k1L}, I'_{k1R}, \cdots, I'_{knL}, I'_{knR}))$;

**5.** The bank receives the signed cash from the merchant through the channel $C_{MB}$ (the corresponding reading action is denoted $r_{C_{MB}}(SIGN_{sk_B}(m_k, R_k, I_{k1L}, I_{k1R}, \cdots, I_{knL}, I_{knR}), I'_{k1L}, I'_{k1R}, \cdots, I'_{knL}, I'_{knR}))$, he de-signs the cash through an action $de\text{-}sign(SIGN_{sk_B}(m_k, R_k, I_{k1L}, I_{k1R}, \cdots, I_{knL}, I_{knR}))$, if $isFresh(R_k) = TRUE$, he checks the identity of the merchant and credits \$m to the merchant's account through an action $s_{C_{BO}}(+m)$; else if $isFresh(R_k) = FALSE$ and $isFresh(I'_{k1L}, I'_{k1R}, \cdots, I'_{knL}, I'_{knR}) = TRUE$, he obtains the identity of Alice and sends $\perp_A$ to the outside through the channel $C_{BO}$ (the corresponding sending action is denoted $s_{C_{BO}}(\perp_A)$); else if $isFresh(R_k) = FALSE$ and $isFresh(I'_{k1L}, I'_{k1R}, \cdots, I'_{knL}, I'_{knR}) = FALSE$, he sends $\perp_M$ to the outside through the channel $C_{BO}$ (the corresponding sending action is denoted $s_{C_{BO}}(\perp_M)$),

where $D \in \Delta$, $\Delta$ is the set of data.

Alice's state transitions described by $APTC_G$ are as follows:

$A = \sum_{D\in\Delta} r_{C_{AI}}(D) \cdot A_2$

$A_2 = blind_{k_1}(m_1, R_i, I_{11L}, I_{11R}, \cdots, I_{1nL}, I_{1nR}) \parallel \cdots \parallel blind_{k_1}(m_n, R_n, I_{n1L}, I_{n1R}, \cdots, I_{nnL}, I_{nnR}) \cdot A_3$

$A_3 = s_{C_{AB}}(BLIND_{k_1}(m_1, R_1, I_{1L}, I_{1R}), \cdots, BLIND_{k_1}(m_n, R_n, I_{nL}, I_{nR})) \cdot A_4$

$A_4 = r_{C_{BA}}(SIGN_{sk_B}(BLIND_{k_1}(m_k, R_k, I_{k1L}, I_{k1R}, \cdots, I_{knL}, I_{knR}))) \cdot A_5$

$A_5 = unblind_{k_1}(SIGN_{sk_B}(BLIND_{k_1}(m_k, R_k, I_{k1L}, I_{k1R}, \cdots, I_{knL}, I_{knR}))) \cdot A_6$

$A_6 = s_{C_{AM}}(SIGN_{sk_B}(m_k, R_k, I_{k1L}, I_{k1R}, \cdots, I_{knL}, I_{knR})) \cdot A$

The bank's state transitions described by $APTC_G$ are as follows:

$B = r_{C_{AB}}(BLIND_{k_1}(m_1, R_1, I_{11L}, I_{11R}, \cdots, I_{1nL}, I_{1nR}), \cdots,$
$BLIND_{k_1}(m_n, R_n, I_{1nL}, I_{1nR}, \cdots, I_{nnL}, I_{nnR})) \cdot B_2$

$B_2 = unblind_{k_1}(BLIND_{k_1}(m_1, R_1, I_{11L}, I_{11R}, \cdots, I_{1nL}, I_{1nR})) \parallel \cdots$
$\parallel unblind_{k_1}(BLIND_{k_1}(m_{n-1}, R_{n-1}, I_{n-11L}, I_{n-11R}, \cdots, I_{n-1nL}, I_{n-1nR})) \cdot B_3$

$B_3 = sign_{sk_B}(BLIND_{k_1}(m_k, R_k, I_{k1L}, I_{k1R}, \cdots, I_{knL}, I_{knR})) \cdot B_4$

$B_4 = s_{C_{BO}}(-m) \cdot B_5$

$B_5 = s_{C_{BA}}(SIGN_{sk_B}(BLIND_{k_1}(m_k, R_k, I_{k1L}, I_{k1R}, \cdots, I_{knL}, I_{knR}))) \cdot B_6$

$B_6 = r_{C_{MB}}(SIGN_{sk_B}(m_k, R_k, I_{k1L}, I_{k1R}, \cdots, I_{knL}, I_{knR}), I'_{k1L}, I'_{k1R}, \cdots, I'_{knL}, I'_{knR}) \cdot B_7$

$B_7 = de\text{-}sign(SIGN_{sk_B}(m_k, R_k, I_{k1L}, I_{k1R}, \cdots, I_{knL}, I_{knR})) \cdot B_8$

$B_8 = \{isFresh(R_k) = TRUE\} \cdot s_{C_{BO}}(+m) \cdot B$
$+ \{isFresh(R_k) = FALSE\} \cdot \{isFresh(I'_{k1L}, I'_{k1R}, \cdots, I'_{knL}, I'_{knR}) = TRUE\} \cdot s_{C_{BO}}(\perp_A) \cdot B$
$+ \{isFresh(R_k) = FALSE\} \cdot \{isFresh(I'_{k1L}, I'_{k1R}, \cdots, I'_{knL}, I'_{knR}) = FALSE\} \cdot s_{C_{BO}}(\perp_M) \cdot B$

The merchant's state transitions described by $APTC_G$ are as follows:

$M = r_{C_{AM}}(SIGN_{sk_B}(m_k, R_k, I_{k1L}, I_{k1R}, \cdots, I_{knL}, I_{knR})) \cdot M_2$

$M_2 = s_{C_{MB}}(SIGN_{sk_B}(m_k, R_k, I_{k1L}, I_{k1R}, \cdots, I_{knL}, I_{knR}), I'_{k1L}, I'_{k1R}, \cdots, I'_{knL}, I'_{knR}) \cdot M$

The sending action and the reading action of the same type data through the same channel can communicate with each other, otherwise, they will cause a deadlock $\delta$. We define the following communication functions:

$\gamma(r_{C_{AB}}(BLIND_{k_1}(m_1, R_1, I_{11L}, I_{11R}, \cdots, I_{1nL}, I_{1nR}), \cdots, BLIND_{k_1}(m_n, R_n, I_{1nL}, I_{1nR}, \cdots, I_{nnL}, I_{nnR})),$
$s_{C_{AB}}(BLIND_{k_1}(m_1, R_1, I_{11L}, I_{11R}, \cdots, I_{1nL}, I_{1nR}), \cdots, BLIND_{k_1}(m_n, R_n, I_{1nL}, I_{1nR}, \cdots, I_{nnL}, I_{nnR})))$
$\triangleq c_{C_{AB}}(BLIND_{k_1}(m_1, R_1, I_{11L}, I_{11R}, \cdots, I_{1nL}, I_{1nR}), \cdots, BLIND_{k_1}(m_n, R_n, I_{1nL}, I_{1nR}, \cdots, I_{nnL}, I_{nnR}))$

$\gamma(r_{C_{BA}}(SIGN_{sk_B}(BLIND_{k_1}(m_k, R_k, I_{k1L}, I_{k1R}, \cdots, I_{knL}, I_{knR}))),$
$s_{C_{BA}}(SIGN_{sk_B}(BLIND_{k_1}(m_k, R_k, I_{k1L}, I_{k1R}, \cdots, I_{knL}, I_{knR}))))$
$\triangleq c_{C_{BA}}(SIGN_{sk_B}(BLIND_{k_1}(m_k, R_k, I_{k1L}, I_{k1R}, \cdots, I_{knL}, I_{knR})))$

$\gamma(r_{C_{MB}}(SIGN_{sk_B}(m_k, R_k, I_{k1L}, I_{k1R}, \cdots, I_{knL}, I_{knR}), I'_{k1L}, I'_{k1R}, \cdots, I'_{knL}, I'_{knR}),$
$s_{C_{MB}}(SIGN_{sk_B}(m_k, R_k, I_{k1L}, I_{k1R}, \cdots, I_{knL}, I_{knR}), I'_{k1L}, I'_{k1R}, \cdots, I'_{knL}, I'_{knR}))$
$\triangleq c_{C_{MB}}(SIGN_{sk_B}(m_k, R_k, I_{k1L}, I_{k1R}, \cdots, I_{knL}, I_{knR}), I'_{k1L}, I'_{k1R}, \cdots, I'_{knL}, I'_{knR})$
$\gamma(r_{C_{AM}}(SIGN_{sk_B}(m_k, R_k, I_{k1L}, I_{k1R}, \cdots, I_{knL}, I_{knR})),$
$s_{C_{AM}}(SIGN_{sk_B}(m_k, R_k, I_{k1L}, I_{k1R}, \cdots, I_{knL}, I_{knR})))$
$\triangleq c_{C_{AM}}(SIGN_{sk_B}(m_k, R_k, I_{k1L}, I_{k1R}, \cdots, I_{knL}, I_{knR}))$

Let all modules be in parallel, then the protocol $A \quad B \quad M$ can be presented by the following process term.

$$\tau_I(\partial_H(\Theta(A \between B \between M))) = \tau_I(\partial_H(A \between B \between M))$$

where $H = \{r_{C_{AB}}(BLIND_{k_1}(m_1, R_1, I_{11L}, I_{11R}, \cdots, I_{1nL}, I_{1nR}), \cdots,$
$BLIND_{k_1}(m_n, R_n, I_{1nL}, I_{1nR}, \cdots, I_{nnL}, I_{nnR})),$
$s_{C_{AB}}(BLIND_{k_1}(m_1, R_1, I_{11L}, I_{11R}, \cdots, I_{1nL}, I_{1nR}), \cdots, BLIND_{k_1}(m_n, R_n, I_{1nL}, I_{1nR}, \cdots, I_{nnL}, I_{nnR})),$
$r_{C_{BA}}(SIGN_{sk_B}(BLIND_{k_1}(m_k, R_k, I_{k1L}, I_{k1R}, \cdots, I_{knL}, I_{knR}))),$
$s_{C_{BA}}(SIGN_{sk_B}(BLIND_{k_1}(m_k, R_k, I_{k1L}, I_{k1R}, \cdots, I_{knL}, I_{knR}))),$
$r_{C_{MB}}(SIGN_{sk_B}(m_k, R_k, I_{k1L}, I_{k1R}, \cdots, I_{knL}, I_{knR}), I'_{k1L}, I'_{k1R}, \cdots, I'_{knL}, I'_{knR}),$
$s_{C_{MB}}(SIGN_{sk_B}(m_k, R_k, I_{k1L}, I_{k1R}, \cdots, I_{knL}, I_{knR}), I'_{k1L}, I'_{k1R}, \cdots, I'_{knL}, I'_{knR}),$
$r_{C_{AM}}(SIGN_{sk_B}(m_k, R_k, I_{k1L}, I_{k1R}, \cdots, I_{knL}, I_{knR})),$
$s_{C_{AM}}(SIGN_{sk_B}(m_k, R_k, I_{k1L}, I_{k1R}, \cdots, I_{knL}, I_{knR}))|D \in \Delta\},$

$I = \{c_{C_{AB}}(BLIND_{k_1}(m_1, R_1, I_{11L}, I_{11R}, \cdots, I_{1nL}, I_{1nR}), \cdots, BLIND_{k_1}(m_n, R_n, I_{1nL}, I_{1nR}, \cdots, I_{nnL}, I_{nnR})),$
$c_{C_{BA}}(SIGN_{sk_B}(BLIND_{k_1}(m_k, R_k, I_{k1L}, I_{k1R}, \cdots, I_{knL}, I_{knR}))),$
$c_{C_{MB}}(SIGN_{sk_B}(m_k, R_k, I_{k1L}, I_{k1R}, \cdots, I_{knL}, I_{knR}), I'_{k1L}, I'_{k1R}, \cdots, I'_{knL}, I'_{knR}),$
$c_{C_{AM}}(SIGN_{sk_B}(m_k, R_k, I_{k1L}, I_{k1R}, \cdots, I_{knL}, I_{knR})),$
$blind_{k_1}(m_1, R_i, I_{11L}, I_{11R}, \cdots, I_{1nL}, I_{1nR}), \cdots, blind_{k_1}(m_n, R_n, I_{n1L}, I_{n1R}, \cdots, I_{nnL}, I_{nnR}),$
$unblind_{k_1}(SIGN_{sk_B}(BLIND_{k_1}(m_k, R_k, I_{k1L}, I_{k1R}, \cdots, I_{knL}, I_{knR}))),$
$unblind_{k_1}(BLIND_{k_1}(m_1, R_1, I_{11L}, I_{11R}, \cdots, I_{1nL}, I_{1nR})),$
$\cdots, unblind_{k_1}(BLIND_{k_1}(m_{n-1}, R_{n-1}, I_{n-11L}, I_{n-11R}, \cdots, I_{n-1nL}, I_{n-1nR})),$
$sign_{sk_B}(BLIND_{k_1}(m_k, R_k, I_{k1L}, I_{k1R}, \cdots, I_{knL}, I_{knR})),$
$de\text{-}sign(SIGN_{sk_B}(m_k, R_k, I_{k1L}, I_{k1R}, \cdots, I_{knL}, I_{knR})),$
$\{isFresh(R_k) = TRUE\}, \{isFresh(R_k) = FALSE\},$
$\{isFresh(I'_{k1L}, I'_{k1R}, \cdots, I'_{knL}, I'_{knR}) = TRUE\}, \{isFresh(I'_{k1L}, I'_{k1R}, \cdots, I'_{knL}, I'_{knR}) = FALSE\}|D \in \Delta\}$

Then, we obtain the following conclusion on the protocol.

**Theorem 30.4.** *The Digital Cash Protocol 4 in Fig. 30.4 is anonymous, resists replaying, and knows who exactly is the cheat.*

*Proof.* Based on the above state transitions of the above modules, by use of the algebraic laws of $APTC_G$, we can prove that

$\tau_I(\partial_H(A \between B \between M)) = \sum_{D\in\Delta}(r_{C_{AI}}(D) \cdot s_{C_{BO}}(-m) \cdot (s_{C_{BO}}(+m) + s_{C_{BO}}(\perp_A) + s_{C_{BO}}(\perp_M))) \cdot \tau_I(\partial_H(A \between B \between M))$

For the details of the proof, please refer to Section 17.3, as we omit them here.

That is, the Digital Cash Protocol 4 in Fig. 30.4 $\tau_I(\partial_H(A \between B \between M))$ can exhibit the desired external behaviors:

1. The digital cash of Alice $SIGN_{sk_B}(ENC_{pk_B}(m_k))$ is anonymous for the merchant and the bank;
2. The protocol can resist replay attacks, for the use of the random number in each digital cash;
3. The bank knows who cheats him when the double spending problem occurs, either the owner of the cash or the merchant, and he knows exactly the identity of the person. □

Chapter 31

# Analyses of secure elections protocols

Secure elections protocols should be able to prevent cheating and maintain the voter's privacy. An ideal secure election protocol should have the following properties:

1. Legitimacy: only authorized voters can vote;
2. Oneness: no one can vote more than once;
3. Privacy: no one can determine for whom anyone else voted;
4. Non-replicability: no one can duplicate anyone else's vote;
5. Non-changeability: no one can change anyone else's vote;
6. Validness: every voter can make sure that his vote has been taken into account in the final tabulation.

In this chapter, we will introduce six secure elections protocols in the following sections. In the analyses of these six protocols, we will mainly analyze the security and privacy properties.

## 31.1 Secure elections protocol 1

The secure elections protocol 1 is shown in Fig. 31.1, which is a basic one to implement the basic voting function. In this protocol, there are a CTF (Central Tabulating Facility), to collect the votes, and four voters: Alice, Bob, Carol, and Dave.

The process of the protocol is as follows:

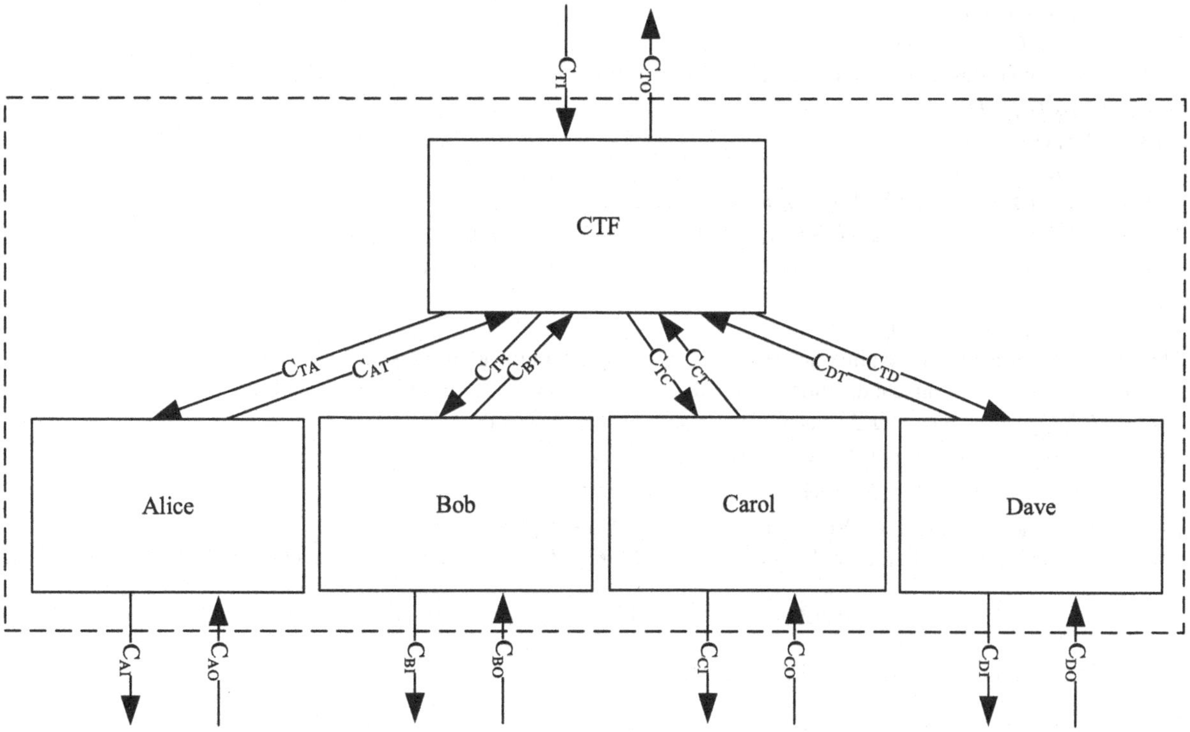


**FIGURE 31.1** Secure elections protocol 1.

Handbook of Truly Concurrent Process Algebra. https://doi.org/10.1016/B978-0-44-321515-5.00035-4

1. Alice receives some voting request $D_A$ from the outside through the channel $C_{AI}$ (the corresponding reading action is denoted $r_{C_{AI}}(D_A)$), she generates the votes $v_A$, encrypts $v_A$ by CTF's public key $pk_T$ through an action $enc_{pk_T}(v_A)$, and sends it to CTF through the channel $C_{AT}$ (the corresponding sending action is denoted $s_{C_{AT}}(ENC_{pk_T}(v_A))$);
2. Bob receives some voting request $D_B$ from the outside through the channel $C_{BI}$ (the corresponding reading action is denoted $r_{C_{BI}}(D_B)$), he generates the votes $v_B$, encrypts $v_B$ by CTF's public key $pk_T$ through an action $enc_{pk_T}(v_B)$, and sends it to CTF through the channel $C_{BT}$ (the corresponding sending action is denoted $s_{C_{BT}}(ENC_{pk_T}(v_B))$);
3. Carol receives some voting request $D_C$ from the outside through the channel $C_{CI}$ (the corresponding reading action is denoted $r_{C_{CI}}(D_C)$), she generates the votes $v_C$, encrypts $v_C$ by CTF's public key $pk_T$ through an action $enc_{pk_T}(v_C)$, and sends it to CTF through the channel $C_{CT}$ (the corresponding sending action is denoted $s_{C_{CT}}(ENC_{pk_T}(v_C))$);
4. Dave receives some voting request $D_D$ from the outside through the channel $C_{DI}$ (the corresponding reading action is denoted $r_{C_{DI}}(D_D)$), he generates the votes $v_D$, encrypts $v_D$ by CTF's public key $pk_T$ through an action $enc_{pk_T}(v_D)$, and sends it to CTF through the channel $C_{DT}$ (the corresponding sending action is denoted $s_{C_{DT}}(ENC_{pk_T}(v_D))$);
5. CTF receives encrypted votes from Alice, Bob, Carol, and Dave through the channels $C_{AT}$, $C_BT$, $C_{CT}$, and $C_{DT}$ (the corresponding reading actions are denoted $r_{C_{AT}}(ENC_{pk_T}(v_A))$, $r_{C_{BT}}(ENC_{pk_T}(v_B))$, $r_{C_{CT}}(ENC_{pk_T}(v_C))$ and $r_{C_{DT}}(ENC_{pk_T}(v_D))$, respectively), decrypts the encrypted votes through actions $dec_{sk_T}(ENC_{pk_T}(v_A))$, $dec_{sk_T}(ENC_{pk_T}(v_B))$, $dec_{sk_T}(ENC_{pk_T}(v_C))$, and $dec_{sk_T}(ENC_{pk_T}(v_D))$ to obtain $v_A$, $v_B$, $v_C$, and $v_D$, then sends $v_A+v_B+v_C+v_D$ to the outside through the channel $C_{TO}$ (the corresponding sending action is denoted $s_{C_{TO}}(v_A+v_B+v_C+v_D)$),

where $D_A, D_B, D_C, D_D \in \Delta$, $\Delta$ is the set of data.

Alice's state transitions described by $APTC_G$ are as follows:

$A = \sum_{D_A\in\Delta} r_{C_{AI}}(D_A)\cdot A_2$

$A_2 = enc_{pk_T}(v_A)\cdot A_3$

$A_3 = s_{C_{AT}}(ENC_{pk_T}(v_A))\cdot A$

Bob's state transitions described by $APTC_G$ are as follows:

$B = \sum_{D_B\in\Delta} r_{C_{BI}}(D_B)\cdot B_2$

$B_2 = enc_{pk_T}(v_B)\cdot B_3$

$B_3 = s_{C_{BT}}(ENC_{pk_T}(v_B))\cdot B$

Carol's state transitions described by $APTC_G$ are as follows:

$C = \sum_{D_C\in\Delta} r_{C_{CI}}(D_C)\cdot C_2$

$C_2 = enc_{pk_T}(v_C)\cdot C_3$

$C_3 = s_{C_{CT}}(ENC_{pk_T}(v_C))\cdot C$

Dave's state transitions described by $APTC_G$ are as follows:

$D = \sum_{D_D\in\Delta} r_{C_{DI}}(D_D)\cdot D_2$

$D_2 = enc_{pk_T}(v_D)\cdot D_3$

$D_3 = s_{C_{DT}}(ENC_{pk_T}(v_D))\cdot D$

CTF's state transitions described by $APTC_G$ are as follows:

$T = r_{C_{AT}}(ENC_{pk_T}(v_A)) \parallel r_{C_{BT}}(ENC_{pk_T}(v_B)) \parallel r_{C_{CT}}(ENC_{pk_T}(v_C)) \parallel r_{C_{DT}}(ENC_{pk_T}(v_D))\cdot T_2$

$T_2 = dec_{sk_T}(ENC_{pk_T}(v_A)) \parallel dec_{sk_T}(ENC_{pk_T}(v_B)) \parallel dec_{sk_T}(ENC_{pk_T}(v_C))$
$\parallel dec_{sk_T}(ENC_{pk_T}(v_D))\cdot T_3$

$T_3 = s_{C_{TO}}(v_A+v_B+v_C+v_D)\cdot T$

The sending action and the reading action of the same type data through the same channel can communicate with each other, otherwise, they will cause a deadlock $\delta$. We define the following communication functions:

$\gamma(r_{C_{AT}}(ENC_{pk_T}(v_A)), s_{C_{AT}}(ENC_{pk_T}(v_A))) \triangleq c_{C_{AT}}(ENC_{pk_T}(v_A))$

$\gamma(r_{C_{BT}}(ENC_{pk_T}(v_B)), s_{C_{BT}}(ENC_{pk_T}(v_B))) \triangleq c_{C_{BT}}(ENC_{pk_T}(v_B))$

$\gamma(r_{C_{CT}}(ENC_{pk_T}(v_C)), s_{C_{CT}}(ENC_{pk_T}(v_C))) \triangleq c_{C_{CT}}(ENC_{pk_T}(v_C))$

$\gamma(r_{C_{DT}}(ENC_{pk_T}(v_D)), s_{C_{DT}}(ENC_{pk_T}(v_D))) \triangleq c_{C_{DT}}(ENC_{pk_T}(v_D))$

Let all modules be in parallel, then the protocol $A \quad B \quad C \quad D \quad T$ can be presented by the following process term:

$$\tau_I(\partial_H(\Theta(A \between B \between C \between D \between T))) = \tau_I(\partial_H(A \between B \between C \between D \between T))$$

where $H = \{r_{C_{AT}}(ENC_{pk_T}(v_A)), s_{C_{AT}}(ENC_{pk_T}(v_A)),$
$r_{C_{BT}}(ENC_{pk_T}(v_B)), s_{C_{BT}}(ENC_{pk_T}(v_B)),$
$r_{C_{CT}}(ENC_{pk_T}(v_C)), s_{C_{CT}}(ENC_{pk_T}(v_C)),$
$r_{C_{DT}}(ENC_{pk_T}(v_D)), s_{C_{DT}}(ENC_{pk_T}(v_D)) | D_A, D_B, D_C, D_D \in \Delta\},$

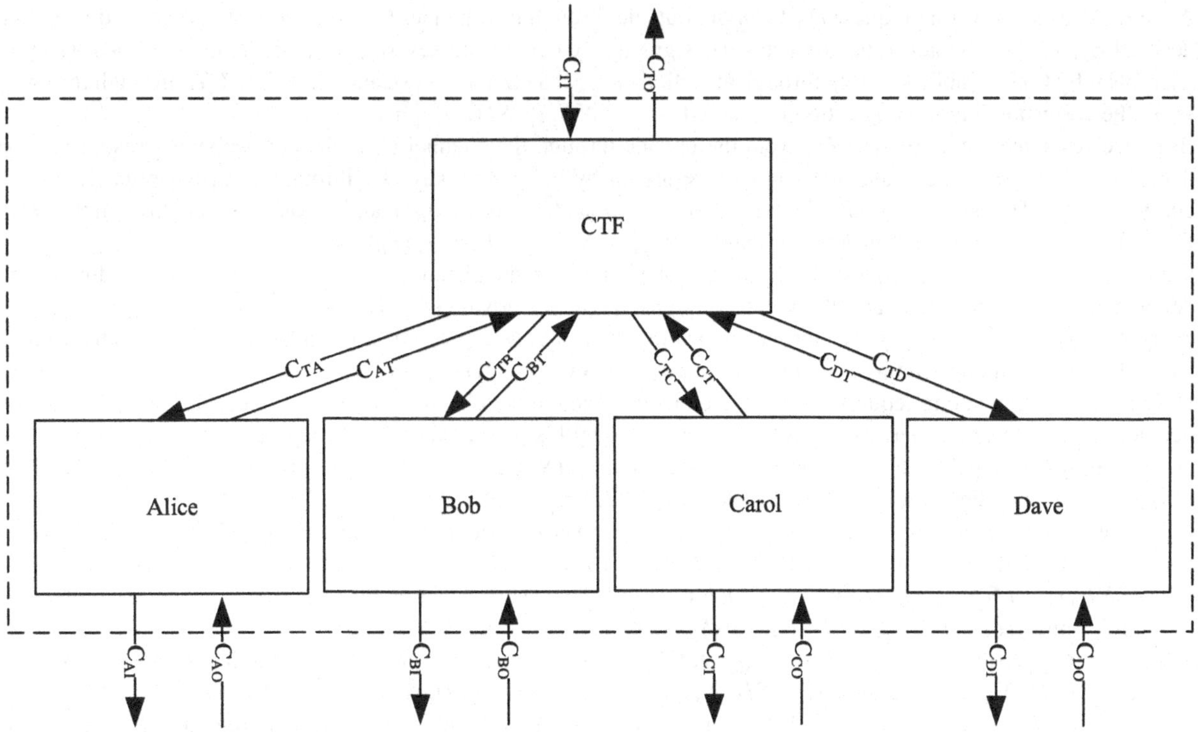


**FIGURE 31.2** Secure elections protocol 2.

$I = \{c_{C_{AT}}(ENC_{pk_T}(v_A)), c_{C_{BT}}(ENC_{pk_T}(v_B)), c_{C_{CT}}(ENC_{pk_T}(v_C)),$
$c_{C_{DT}}(ENC_{pk_T}(v_D)), enc_{pk_T}(v_A), enc_{pk_T}(v_B), enc_{pk_T}(v_C), enc_{pk_T}(v_D),$
$dec_{sk_T}(ENC_{pk_T}(v_A)), dec_{sk_T}(ENC_{pk_T}(v_B)), dec_{sk_T}(ENC_{pk_T}(v_C)),$
$dec_{sk_T}(ENC_{pk_T}(v_D)) | D_A, D_B, D_C, D_D \in \Delta\}$

Then, we obtain the following conclusion on the protocol.

**Theorem 31.1.** *The secure elections protocol 1 in Fig. 31.1 is secure, but basic.*

*Proof.* Based on the above state transitions of the above modules, by use of the algebraic laws of $APTC_G$, we can prove that

$\tau_I(\partial_H(A \between B \between C \between D \between T)) = \sum_{D_A, D_B, D_C, D_D \in \Delta}(r_{C_{AI}}(D_A) \parallel r_{C_{BI}}(D_B) \parallel r_{C_{CI}}(D_C) \parallel r_{C_{DI}}(D_D) \cdot s_{C_{TO}}(v_A + v_B + v_C + v_D)) \cdot \tau_I(\partial_H(A \between B \between C \between D \between T))$

For the details of the proof, please refer to Section 17.3, as we omit them here.

That is, the protocol in Fig. 31.1 $\tau_I(\partial_H(A \between B \between C \between D \between T))$ can exhibit the desired external behaviors, and is secure. However, for the properties of secure elections protocols:

1. Legitimacy: all voters can vote;
2. Oneness: anyone can vote more than once;
3. Privacy: no one can determine for whom anyone else voted;
4. Non-replicability: CTF can duplicate anyone else's vote;
5. Non-changeability: CTF can change anyone else's vote;
6. Validness: every voter cannot make sure that his vote has been taken into account in the final tabulation. □

## 31.2 Secure elections protocol 2

The secure elections protocol 2 is shown in Fig. 31.2, which is an improved one based on the secure elections protocol 1 in Section 31.1. In this protocol, there are a CTF (Central Tabulating Facility), to check the identity of voters and collect the votes, and four voters: Alice, Bob, Carol, and Dave.

The process of the protocol is as follows:

1. Alice receives some voting request $D_A$ from the outside through the channel $C_{AI}$ (the corresponding reading action is denoted $r_{C_{AI}}(D_A)$), she generates the votes $v_A$, signs $v_A$ by her private key $sk_A$ through an action $sign_{sk_A}(v_A)$, then encrypts it by CTF's public key $pk_T$ through an action $enc_{pk_T}(SIGN_{sk_A}(v_A))$, and sends it to CTF through the channel $C_{AT}$ (the corresponding sending action is denoted $s_{C_{AT}}(ENC_{pk_T}(SIGN_{sk_A}(v_A)))$);
2. Bob receives some voting request $D_B$ from the outside through the channel $C_{BI}$ (the corresponding reading action is denoted $r_{C_{BI}}(D_B)$), he generates the votes $v_B$, signs $v_B$ by his private key $sk_B$ through an action $sign_{sk_B}(v_B)$, then encrypts it by CTF's public key $pk_T$ through an action $enc_{pk_T}(SIGN_{sk_B}(v_B))$, and sends it to CTF through the channel $C_{BT}$ (the corresponding sending action is denoted $s_{C_{BT}}(ENC_{pk_T}(SIGN_{sk_B}(v_B)))$);
3. Carol receives some voting request $D_C$ from the outside through the channel $C_{CI}$ (the corresponding reading action is denoted $r_{C_{CI}}(D_C)$), she generates the votes $v_C$, signs $v_C$ by her private key $sk_C$ through an action $sign_{sk_C}(v_C)$, then encrypts it by CTF's public key $pk_T$ through an action $enc_{pk_T}(SIGN_{sk_C}(v_C))$, and sends it to CTF through the channel $C_{CT}$ (the corresponding sending action is denoted $s_{C_{CT}}(ENC_{pk_T}(SIGN_{sk_C}(v_C)))$);
4. Dave receives some voting request $D_D$ from the outside through the channel $C_{DI}$ (the corresponding reading action is denoted $r_{C_{DI}}(D_D)$), he generates the votes $v_D$, signs $v_D$ by his private key $sk_D$ through an action $sign_{sk_D}(v_D)$, then encrypts it by CTF's public key $pk_T$ through an action $enc_{pk_T}(SIGN_{sk_D}(v_D))$, and sends it to CTF through the channel $C_{DT}$ (the corresponding sending action is denoted $s_{C_{DT}}(ENC_{pk_T}(SIGN_{sk_D}(v_D)))$);
5. CTF receives encrypted votes from Alice, Bob, Carol, and Dave through the channels $C_{AT}$, $C_BT$, $C_{CT}$ and $C_{DT}$ (the corresponding reading actions are denoted$r_{C_{AT}}(ENC_{pk_T}(SIGN_{sk_A}(v_A)))$, $r_{C_{BT}}(ENC_{pk_T}(SIGN_{sk_B}(v_B)))$, $r_{C_{CT}}(ENC_{pk_T}(SIGN_{sk_C}(v_C)))$, and $r_{C_{DT}}(ENC_{pk_T}(SIGN_{sk_D}(v_D)))$, respectively), decrypts the encrypted votes through actions $dec_{sk_T}(ENC_{pk_T}(SIGN_{sk_A}(v_A)))$, $dec_{sk_T}(ENC_{pk_T}(SIGN_{sk_B}(v_B)))$, $dec_{sk_T}(ENC_{pk_T}(SIGN_{sk_C}(v_C)))$, and $dec_{sk_T}(ENC_{pk_T}(SIGN_{sk_D}(v_D)))$, then de-signs them through actions $de\text{-}sign_{pk_A}(SIGN_{sk_A}(v_A))$, $de\text{-}sign_{pk_B}(SIGN_{sk_B}(v_B))$, $de\text{-}sign_{pk_C}(SIGN_{sk_C}(v_C))$, and $de\text{-}sign_{pk_D}(SIGN_{sk_D}(v_D))$ to obtain $v_A$, $v_B$, $v_C$, and $v_D$, then sends $v_A + v_B + v_C + v_D, A, B, C, D$ to the outside through the channel $C_{TO}$ (the corresponding sending action is denoted $s_{C_{TO}}(v_A + v_B + v_C + v_D, A, B, C, D)$),

where $D_A, D_B, D_C, D_D \in \Delta$, $\Delta$ is the set of data.

Alice's state transitions described by $APTC_G$ are as follows:

$A = \sum_{D_A \in \Delta} r_{C_{AI}}(D_A) \cdot A_2$

$A_2 = sign_{sk_A}(v_A) \cdot A_3$

$A_3 = enc_{pk_T}(SIGN_{sk_A}(v_A)) \cdot A_4$

$A_4 = s_{C_{AT}}(ENC_{pk_T}(SIGN_{sk_A}(v_A))) \cdot A$

Bob's state transitions described by $APTC_G$ are as follows:

$B = \sum_{D_B \in \Delta} r_{C_{BI}}(D_B) \cdot B_2$

$B_2 = sign_{sk_B}(v_B) \cdot B_3$

$B_3 = enc_{pk_T}(SIGN_{sk_B}(v_B)) \cdot B_4$

$B_4 = s_{C_{BT}}(ENC_{pk_T}(SIGN_{sk_B}(v_B))) \cdot B$

Carol's state transitions described by $APTC_G$ are as follows:

$C = \sum_{D_C \in \Delta} r_{C_{CI}}(D_C) \cdot C_2$

$C_2 = sign_{sk_C}(v_C) \cdot C_3$

$C_3 = enc_{pk_T}(SIGN_{sk_C}(v_C)) \cdot C_4$

$C_4 = s_{C_{CT}}(ENC_{pk_T}(SIGN_{sk_C}(v_C))) \cdot C$

Dave's state transitions described by $APTC_G$ are as follows:

$D = \sum_{D_D \in \Delta} r_{C_{DI}}(D_D) \cdot D_2$

$D_2 = sign_{sk_D}(v_D) \cdot D_3$

$D_3 = enc_{pk_T}(SIGN_{sk_D}(v_D)) \cdot D_4$

$D_4 = s_{C_{DT}}(ENC_{pk_T}(SIGN_{sk_D}(v_D))) \cdot D$

CTF's state transitions described by $APTC_G$ are as follows:

$T = r_{C_{AT}}(ENC_{pk_T}(SIGN_{sk_A}(v_A))) \parallel r_{C_{BT}}(ENC_{pk_T}(SIGN_{sk_B}(v_B)))$
$\parallel r_{C_{CT}}(ENC_{pk_T}(SIGN_{sk_C}(v_C))) \parallel r_{C_{DT}}(ENC_{pk_T}(SIGN_{sk_D}(v_D))) \cdot T_2$

$T_2 = dec_{sk_T}(ENC_{pk_T}(SIGN_{sk_A}(v_A))) \parallel dec_{sk_T}(ENC_{pk_T}(SIGN_{sk_B}(v_B)))$
$\parallel dec_{sk_T}(ENC_{pk_T}(SIGN_{sk_C}(v_C))) \parallel dec_{sk_T}(ENC_{pk_T}(SIGN_{sk_D}(v_D))) \cdot T_3$

$T_3 = de\text{-}sign_{pk_A}(SIGN_{sk_A}(v_A)) \parallel de\text{-}sign_{pk_B}(SIGN_{sk_B}(v_B))$
$\parallel de\text{-}sign_{pk_C}(SIGN_{sk_C}(v_C)) \parallel de\text{-}sign_{pk_D}(SIGN_{sk_D}(v_D)) \cdot T_4$

$T_4 = s_{C_{TO}}(v_A + v_B + v_C + v_D, A, B, C, D) \cdot T$

The sending action and the reading action of the same type data through the same channel can communicate with each other, otherwise, they will cause a deadlock $\delta$. We define the following communication functions:

$\gamma(r_{C_{AT}}(ENC_{pk_T}(SIGN_{sk_A}(v_A))), s_{C_{AT}}(ENC_{pk_T}(SIGN_{sk_A}(v_A))))$
$\triangleq c_{C_{AT}}(ENC_{pk_T}(SIGN_{sk_A}(v_A)))$

$\gamma(r_{C_{BT}}(ENC_{pk_T}(SIGN_{sk_B}(v_B))), s_{C_{BT}}(ENC_{pk_T}(SIGN_{sk_B}(v_B))))$
$\triangleq c_{C_{BT}}(ENC_{pk_T}(SIGN_{sk_B}(v_B)))$

$\gamma(r_{C_{CT}}(ENC_{pk_T}(SIGN_{sk_C}(v_C))), s_{C_{CT}}(ENC_{pk_T}(SIGN_{sk_C}(v_C))))$
$\triangleq c_{C_{CT}}(ENC_{pk_T}(SIGN_{sk_C}(v_C)))$

$\gamma(r_{C_{DT}}(ENC_{pk_T}(SIGN_{sk_D}(v_D))), s_{C_{DT}}(ENC_{pk_T}(SIGN_{sk_D}(v_D))))$
$\triangleq c_{C_{DT}}(ENC_{pk_T}(SIGN_{sk_D}(v_D)))$

Let all modules be in parallel, then the protocol $A \quad B \quad C \quad D \quad T$ can be presented by the following process term:

$$\tau_I(\partial_H(\Theta(A \between B \between C \between D \between T))) = \tau_I(\partial_H(A \between B \between C \between D \between T))$$

where $H = \{r_{C_{AT}}(ENC_{pk_T}(SIGN_{sk_A}(v_A))), s_{C_{AT}}(ENC_{pk_T}(SIGN_{sk_A}(v_A))),$
$r_{C_{BT}}(ENC_{pk_T}(SIGN_{sk_B}(v_B))), s_{C_{BT}}(ENC_{pk_T}(SIGN_{sk_B}(v_B))),$
$r_{C_{CT}}(ENC_{pk_T}(SIGN_{sk_C}(v_C))), s_{C_{CT}}(ENC_{pk_T}(SIGN_{sk_C}(v_C))),$
$r_{C_{DT}}(ENC_{pk_T}(SIGN_{sk_D}(v_D))), s_{C_{DT}}(ENC_{pk_T}(SIGN_{sk_D}(v_D)))|D_A, D_B, D_C, D_D \in \Delta\},$

$I = \{c_{C_{AT}}(ENC_{pk_T}(SIGN_{sk_A}(v_A))), c_{C_{BT}}(ENC_{pk_T}(SIGN_{sk_B}(v_B))),$
$c_{C_{CT}}(ENC_{pk_T}(SIGN_{sk_C}(v_C))), c_{C_{DT}}(ENC_{pk_T}(SIGN_{sk_D}(v_D))),$
$sign_{sk_A}(v_A), sign_{sk_B}(v_B), sign_{sk_C}(v_C), sign_{sk_D}(v_D),$
$enc_{pk_T}(SIGN_{sk_A}(v_A)), enc_{pk_T}(SIGN_{sk_B}(v_B)), enc_{pk_T}(SIGN_{sk_C}(v_C)),$
$enc_{pk_T}(SIGN_{sk_D}(v_D)), dec_{sk_T}(ENC_{pk_T}(SIGN_{sk_A}(v_A))),$
$dec_{sk_T}(ENC_{pk_T}(SIGN_{sk_B}(v_B))), dec_{sk_T}(ENC_{pk_T}(SIGN_{sk_C}(v_C))),$
$dec_{sk_T}(ENC_{pk_T}(SIGN_{sk_D}(v_D))), de\text{-}sign_{pk_A}(SIGN_{sk_A}(v_A)),$
$de\text{-}sign_{pk_B}(SIGN_{sk_B}(v_B)), de\text{-}sign_{pk_C}(SIGN_{sk_C}(v_C)),$
$de\text{-}sign_{pk_D}(SIGN_{sk_D}(v_D))|D_A, D_B, D_C, D_D \in \Delta\}$

Then, we obtain the following conclusion on the protocol.

**Theorem 31.2.** *The secure elections protocol 2 in Fig. 31.2 is improved based on the secure elections protocol 1.*

*Proof.* Based on the above state transitions of the above modules, by use of the algebraic laws of $APTC_G$, we can prove that

$\tau_I(\partial_H(A \between B \between C \between D \between T)) = \sum_{D_A, D_B, D_C, D_D \in \Delta}(r_{C_{AI}}(D_A) \parallel r_{C_{BI}}(D_B) \parallel r_{C_{CI}}(D_C) \parallel r_{C_{DI}}(D_D) \cdot s_{C_{TO}}(v_A + v_B + v_C + v_D, A, B, C, D)) \cdot \tau_I(\partial_H(A \between B \between C \between D \between T))$

For the details of the proof, please refer to Section 17.3, as we omit them here.

That is, the protocol in Fig. 31.2 $\tau_I(\partial_H(A \between B \between C \between D \between T))$ can exhibit the desired external behaviors, and is secure. However, for the properties of secure elections protocols:

1. Legitimacy: only authorized voters can vote;
2. Oneness: no one can vote more than once;
3. Privacy: CTF can determine for whom anyone else voted;
4. Non-replicability: CTF can duplicate anyone else's vote;
5. Non-changeability: CTF can change anyone else's vote;
6. Validness: every voter cannot make sure that his vote has been taken into account in the final tabulation. □

## 31.3 Secure elections protocol 3

The secure elections protocol 3 is shown in Fig. 31.3, which is an improved one based on the secure elections protocol 2 in Section 31.2. In this protocol, there are a CTF (Central Tabulating Facility), to check the identity of voters and collect the votes, and four voters: Alice, Bob, Carol, and Dave.

The process of the protocol is as follows:

1. Alice receives some voting request $D_A$ from the outside through the channel $C_{AI}$ (the corresponding reading action is denoted $r_{C_{AI}}(D_A)$), she generates a message containing all possible voting results $V_A$ and a random number $R_A$, blinds this message through an action $blind_{k_A}(V_A, R_A)$, totally there are 10 such messages generated,

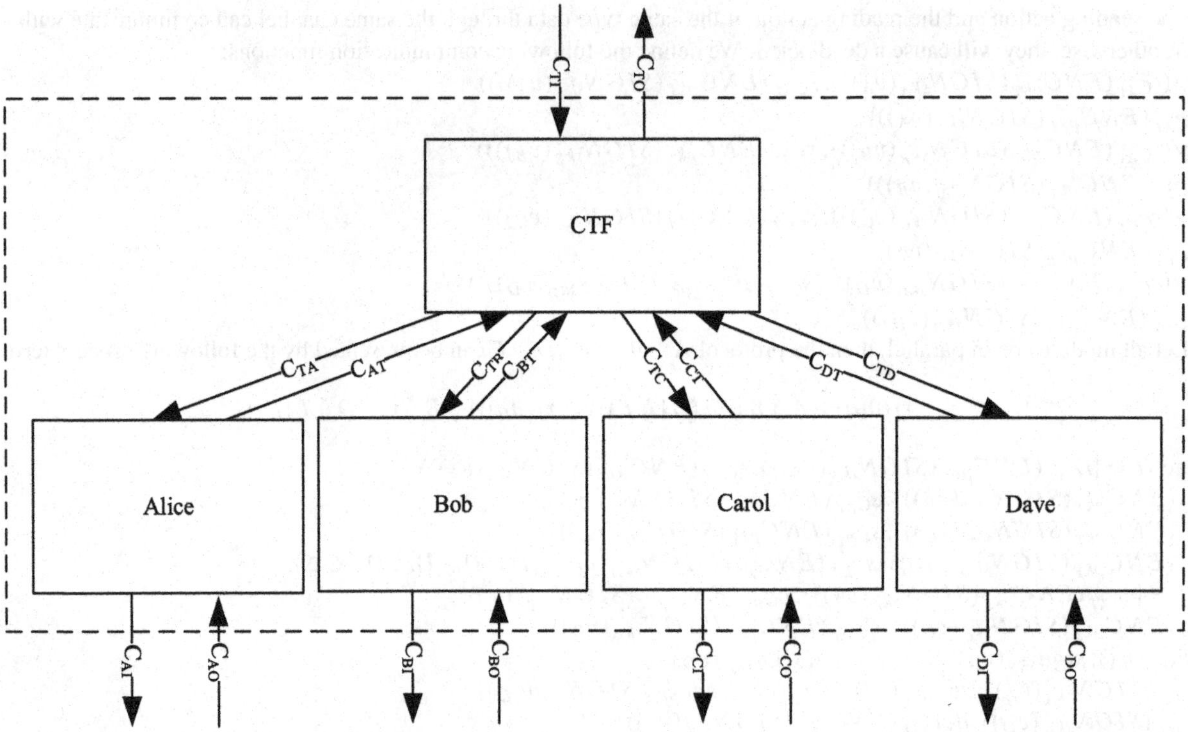


**FIGURE 31.3** Secure elections protocol 3.

then she sends these 10 messages to CTF through the channel $C_{AT}$ (the corresponding sending action is denoted $s_{C_{AT}}(10 \times BLIND_{k_A}(V_A, R_A), k_A, A)$);

2. Bob receives some voting request $D_B$ from the outside through the channel $C_{BI}$ (the corresponding reading action is denoted $r_{C_{BI}}(D_B)$), he generates a message containing all possible voting results $V_B$ and a random number $R_B$, blinds this message through an action $blind_{k_B}(V_B, R_B)$, totally there are 10 such messages generated, then he sends these 10 messages to CTF through the channel $C_{BT}$ (the corresponding sending action is denoted $s_{C_{BT}}(10 \times BLIND_{k_B}(V_B, R_B), k_B, B)$);
3. Carol receives some voting request $D_C$ from the outside through the channel $C_{CI}$ (the corresponding reading action is denoted $r_{C_{CI}}(D_C)$), she generates a message containing all possible voting results $V_C$ and a random number $R_C$, blinds this message through an action $blind_{k_C}(V_C, R_C)$, totally there are 10 such messages generated, then she sends these 10 messages to CTF through the channel $C_{CT}$ (the corresponding sending action is denoted $s_{C_{CT}}(10 \times BLIND_{k_C}(V_C, R_C), k_C, C)$);
4. Dave receives some voting request $D_D$ from the outside through the channel $C_{DI}$ (the corresponding reading action is denoted $r_{C_{DI}}(D_D)$), he generates a message containing all possible voting results $V_D$ and a random number $R_D$, blinds this message through an action $blind_{k_D}(V_D, R_D)$, totally there are 10 such messages generated, then he sends these 10 messages to CTF through the channel $C_{DT}$ (the corresponding sending action is denoted $s_{C_{DT}}(10 \times BLIND_{k_D}(V_D, R_D), k_D, D)$);
5. CTF receives the messages from Alice, Bob, Carol, and Dave through the channels $C_{AT}$, $C_{BT}$, $C_{CT}$, and $C_{DT}$ (the corresponding reading actions are denoted $r_{C_{AT}}(10 \times BLIND_{k_A}(V_A, R_A), k_A, A)$, $s_{C_{BT}}(10 \times BLIND_{k_B}(V_B, R_B), k_B, B)$, $s_{C_{CT}}(10 \times BLIND_{k_C}(V_C, R_C), k_C, C)$, and $s_{C_{DT}}(10 \times BLIND_{k_D}(V_D, R_D), k_D, D)$, respectively), he checks the names of Alice, Bob, Carol, and Dave to make sure that they submit the blinded messages in the first time and stores their names through the actions $check(A)$, $check(B)$, $check(C)$, and $check(D)$, then he unblinds randomly their 9 sets of messages to make sure that they are formed correctly through the actions $9 \times unblind_{k_A}(BLIND_{k_A}(V_A, R_A))$, $9 \times unblind_{k_B}(BLIND_{k_B}(V_B, R_B))$, $9 \times unblind_{k_C}(BLIND_{k_C}(V_C, R_C))$, and $9 \times unblind_{k_D}(BLIND_{k_D}(V_D, R_D))$. Then, he signs their left message through the actions $sign_{sk_T}(BLIND_{k_A}(V_A, R_A))$, $sign_{sk_T}(BLIND_{k_B}(V_B, R_B))$,$sign_{sk_T}(BLIND_{k_C}(V_C, R_C))$, and $sign_{sk_T}(BLIND_{k_D}(V_D, R_D))$, and sends them to Alice, Bob, Carol, and Dave through the channels $C_{TA}$, $C_{TB}$, $C_{TC}$ and $C_{TD}$ (the corresponding sending actions is denoted $s_{C_{TA}}(SIGN_{sk_T}(BLIND_{k_A}(V_A, R_A)))$,

$s_{C_{TB}}(SIGN_{sk_T}(BLIND_{k_B}(V_B, R_B)))$, $s_{C_{TC}}(SIGN_{sk_T}(BLIND_{k_C}(V_C, R_C)))$, and $s_{C_{TD}}(SIGN_{sk_T}(BLIND_{k_D}(V_D, R_D))))$;

6. Alice receives the signed message from CTF through the channel $C_{TA}$ (the corresponding reading action is denoted $r_{C_{TA}}(SIGN_{sk_T}(BLIND_{k_A}(V_A, R_A))))$, she unblinds the message through the action $unblind_{k_A}(SIGN_{sk_T}(BLIND_{k_A}(V_A, R_A)))$, selects her vote $v_A$ from $V_A$, encrypts the vote through an action $enc_{pk_T}(SIGN_{sk_T}(v_A, R_A))$, and sends her encrypted vote to CTF through the channel $C_{AT}$ (the corresponding sending action is denoted $s_{C_{AT}}(ENC_{pk_T}(SIGN_{sk_T}(v_A, R_A))))$;
7. Bob receives the signed message from CTF through the channel $C_{TB}$ (the corresponding reading action is denoted $r_{C_{TB}}(SIGN_{sk_T}(BLIND_{k_B}(V_B, R_B))))$, he unblinds the message through the action $unblind_{k_B}(SIGN_{sk_T}(BLIND_{k_B}(V_B, R_B)))$, selects his vote $v_B$ from $V_B$, encrypts the vote through an action $enc_{pk_T}(SIGN_{sk_T}(v_B, R_B))$, and sends his encrypted vote to CTF through the channel $C_{BT}$ (the corresponding sending action is denoted $s_{C_{BT}}(ENC_{pk_T}(SIGN_{sk_T}(v_B, R_B))))$;
8. Carol receives the signed message from CTF through the channel $C_{TC}$ (the corresponding reading action is denoted $r_{C_{TC}}(SIGN_{sk_T}(BLIND_{k_C}(V_C, R_C))))$, she unblinds the message through the action $unblind_{k_C}(SIGN_{sk_T}(BLIND_{k_C}(V_C, R_C)))$, selects her vote $v_C$ from $V_C$, encrypts the vote through an action $enc_{pk_T}(SIGN_{sk_T}(v_C, R_C))$, and sends her encrypted vote to CTF through the channel $C_{CT}$ (the corresponding sending action is denoted $s_{C_{CT}}(ENC_{pk_T}(SIGN_{sk_T}(v_C, R_C))))$;
9. Dave receives the signed message from CTF through the channel $C_{TD}$ (the corresponding reading action is denoted $r_{C_{TD}}(SIGN_{sk_T}(BLIND_{k_D}(V_D, R_D))))$, he unblinds the message through the action $unblind_{k_D}(SIGN_{sk_T}(BLIND_{k_D}(V_D, R_D)))$, selects his vote $v_D$ from $V_D$, encrypts the vote through an action $enc_{pk_T}(SIGN_{sk_T}(v_D, R_D))$, and sends his encrypted vote to CTF through the channel $C_{DT}$ (the corresponding sending action is denoted $s_{C_{DT}}(ENC_{pk_T}(SIGN_{sk_T}(v_D, R_D))))$;
10. CTF receives the votes from Alice, Bob, Carol, and Dave through the channels $C_{AT}$, $C_{BT}$, $C_{CT}$ and $C_{DT}$ (the corresponding reading actions are denoted $r_{C_{AT}}(ENC_{pk_T}(SIGN_{sk_T}(v_A, R_A)))$, $r_{C_{BT}}(ENC_{pk_T}(SIGN_{sk_T}(v_B, R_B)))$, $r_{C_{CT}}(ENC_{pk_T}(SIGN_{sk_T}(v_C, R_C)))$, and $r_{C_{DT}}(ENC_{pk_T}(SIGN_{sk_T}(v_D, R_D)))$, respectively), he decrypts and de-signs these votes through the actions $dec_{sk_T}(ENC_{pk_T}(SIGN_{sk_T}(v_A, R_A)))$, $dec_{sk_T}(ENC_{pk_T}(SIGN_{sk_T}(v_B, R_B)))$, $dec_{sk_T}(ENC_{pk_T}(SIGN_{sk_T}(v_C, R_C)))$, $dec_{sk_T}(ENC_{pk_T}(SIGN_{sk_T}(v_D, R_D)))$, and $de\text{-}sign_{pk_T}(SIGN_{sk_T}(v_A, R_A))$, $de\text{-}sign_{pk_T}(SIGN_{sk_T}(v_B, R_B))$, $de\text{-}sign_{pk_T}(SIGN_{sk_T}(v_C, R_C))$, $de\text{-}sign_{pk_T}(SIGN_{sk_T}(v_D, R_D))$. If $isFresh(R_A) = TRUE$, he tabulates $v_A$ through an action $tab(v_A)$, else $tab(0)$; if $isFresh(R_B) = TRUE$, he tabulates $v_B$ through an action $tab(v_B)$, else $tab(0)$; if $isFresh(R_C) = TRUE$, he tabulates $v_C$ through an action $tab(v_C)$, else $tab(0)$; if $isFresh(R_D) = TRUE$, he tabulates $v_D$ through an action $tab(v_D)$, else $tab(0)$. Finally, he sends the voting results $TAB$ to the outside through the channel $C_{TO}$ (the corresponding sending action is denoted $s_{C_{TO}}(TAB)$),

where $D_A, D_B, D_C, D_D \in \Delta$, $\Delta$ is the set of data.

Alice's state transitions described by $APTC_G$ are as follows:

$A = \sum_{D_A \in \Delta} r_{C_{AI}}(D_A) \cdot A_2$
$A_2 = blind_{k_A}(V_A, R_A) \cdot A_3$
$A_3 = s_{C_{AT}}(10 \times BLIND_{k_A}(V_A, R_A), k_A, A) \cdot A_4$
$A_4 = r_{C_{TA}}(SIGN_{sk_T}(BLIND_{k_A}(V_A, R_A))) \cdot A_5$
$A_5 = unblind_{k_A}(SIGN_{sk_T}(BLIND_{k_A}(V_A, R_A))) \cdot A_6$
$A_6 = enc_{pk_T}(SIGN_{sk_T}(v_A, R_A)) \cdot A_7$
$A_7 = s_{C_{AT}}(ENC_{pk_T}(SIGN_{sk_T}(v_A, R_A))) \cdot A$

Bob's state transitions described by $APTC_G$ are as follows:

$B = \sum_{D_B \in \Delta} r_{C_{BI}}(D_B) \cdot B_2$
$B_2 = blind_{k_B}(V_B, R_B) \cdot B_3$
$B_3 = s_{C_{BT}}(10 \times BLIND_{k_B}(V_B, R_B), k_B, B) \cdot B_4$
$B_4 = r_{C_{TB}}(SIGN_{sk_T}(BLIND_{k_B}(V_B, R_B))) \cdot B_5$
$B_5 = unblind_{k_B}(SIGN_{sk_T}(BLIND_{k_B}(V_B, R_B))) \cdot B_6$
$B_6 = enc_{pk_T}(SIGN_{sk_T}(v_B, R_B)) \cdot B_7$
$B_7 = s_{C_{BT}}(ENC_{pk_T}(SIGN_{sk_T}(v_B, R_B))) \cdot B$

Carol's state transitions described by $APTC_G$ are as follows:

$C = \sum_{D_C \in \Delta} r_{C_{CI}}(D_C) \cdot C_2$
$C_2 = blind_{k_C}(V_C, R_C) \cdot C_3$
$C_3 = s_{C_{CT}}(10 \times BLIND_{k_C}(V_C, R_C), k_C, C) \cdot C_4$

$C_4 = r_{C_{TC}}(SIGN_{sk_T}(BLIND_{k_C}(V_C, R_C))) \cdot C_5$
$C_5 = unblind_{k_C}(SIGN_{sk_T}(BLIND_{k_C}(V_C, R_C))) \cdot C_6$
$C_6 = enc_{pk_T}(SIGN_{sk_T}(v_C, R_C)) \cdot C_7$
$C_7 = s_{C_{CT}}(ENC_{pk_T}(SIGN_{sk_T}(v_C, R_C))) \cdot C$

Dave's state transitions described by $APTC_G$ are as follows:

$D = \sum_{D_D \in \Delta} r_{C_{DI}}(D_D) \cdot D_2$
$D_2 = blind_{k_D}(V_D, R_D) \cdot D_3$
$D_3 = s_{C_{DT}}(10 \times BLIND_{k_D}(V_D, R_D), k_D, D) \cdot D_4$
$D_4 = r_{C_{TD}}(SIGN_{sk_T}(BLIND_{k_D}(V_D, R_D))) \cdot D_5$
$D_5 = unblind_{k_D}(SIGN_{sk_T}(BLIND_{k_D}(V_D, R_D))) \cdot D_6$
$D_6 = enc_{pk_T}(SIGN_{sk_T}(v_D, R_D)) \cdot D_7$
$D_7 = s_{C_{DT}}(ENC_{pk_T}(SIGN_{sk_T}(v_D, R_D))) \cdot D$

CTF's state transitions described by $APTC_G$ are as follows:

$T = r_{C_{AT}}(10 \times BLIND_{k_A}(V_A, R_A), k_A, A) \parallel s_{C_{BT}}(10 \times BLIND_{k_B}(V_B, R_B), k_B, B)$
$\parallel s_{C_{CT}}(10 \times BLIND_{k_C}(V_C, R_C), k_C, C) \parallel s_{C_{DT}}(10 \times BLIND_{k_D}(V_D, R_D), k_D, D) \cdot T_2$
$T_2 = check(A) \parallel check(B) \parallel check(C) \parallel check(D) \cdot T_3$
$T_3 = 9 \times unblind_{k_A}(BLIND_{k_A}(V_A, R_A)) \parallel 9 \times unblind_{k_B}(BLIND_{k_B}(V_B, R_B))$
$\parallel 9 \times unblind_{k_C}(BLIND_{k_C}(V_C, R_C)) \parallel 9 \times unblind_{k_D}(BLIND_{k_D}(V_D, R_D)) \cdot T_4$
$T_4 = sign_{sk_T}(BLIND_{k_A}(V_A, R_A)) \parallel sign_{sk_T}(BLIND_{k_B}(V_B, R_B))$
$\parallel sign_{sk_T}(BLIND_{k_C}(V_C, R_C)) \parallel sign_{sk_T}(BLIND_{k_D}(V_D, R_D)) \cdot T_5$
$T_5 = s_{C_{TA}}(SIGN_{sk_T}(BLIND_{k_A}(V_A, R_A))) \parallel s_{C_{TB}}(SIGN_{sk_T}(BLIND_{k_B}(V_B, R_B)))$
$\parallel s_{C_{TC}}(SIGN_{sk_T}(BLIND_{k_C}(V_C, R_C))) \parallel s_{C_{TD}}(SIGN_{sk_T}(BLIND_{k_D}(V_D, R_D))) \cdot T_6$
$T_6 = r_{C_{AT}}(ENC_{pk_T}(SIGN_{sk_T}(v_A, R_A))) \parallel r_{C_{BT}}(ENC_{pk_T}(SIGN_{sk_T}(v_B, R_B)))$
$\parallel r_{C_{CT}}(ENC_{pk_T}(SIGN_{sk_T}(v_C, R_C))) \parallel r_{C_{DT}}(ENC_{pk_T}(SIGN_{sk_T}(v_D, R_D))) \cdot T_7$
$T_7 = dec_{sk_T}(ENC_{pk_T}(SIGN_{sk_T}(v_A, R_A))) \parallel dec_{sk_T}(ENC_{pk_T}(SIGN_{sk_T}(v_B, R_B)))$
$\parallel dec_{sk_T}(ENC_{pk_T}(SIGN_{sk_T}(v_C, R_C))) \parallel dec_{sk_T}(ENC_{pk_T}(SIGN_{sk_T}(v_D, R_D))) \cdot T_8$
$T_8 = de\text{-}sign_{pk_T}(SIGN_{sk_T}(v_A, R_A)) \parallel de\text{-}sign_{pk_T}(SIGN_{sk_T}(v_B, R_B))$
$\parallel de\text{-}sign_{pk_T}(SIGN_{sk_T}(v_C, R_C)) \parallel de\text{-}sign_{pk_T}(SIGN_{sk_T}(v_D, R_D)) \cdot T_9$
$T_9 = ((\{isFresh(R_A) = TRUE\} \cdot tab(v_A) + \{isFresh(R_A) = FALSE\} \cdot tab(0))$
$\parallel (\{isFresh(R_B) = TRUE\} \cdot tab(v_B) + \{isFresh(R_B) = FALSE\} \cdot tab(0))$
$\parallel (\{isFresh(R_C) = TRUE\} \cdot tab(v_C) + \{isFresh(R_C) = FALSE\} \cdot tab(0))$
$\parallel (\{isFresh(R_D) = TRUE\} \cdot tab(v_D) + \{isFresh(R_D) = FALSE\} \cdot tab(0))) \cdot T_{10}$
$T_{10} = s_{C_{TO}}(TAB) \cdot T$

The sending action and the reading action of the same type data through the same channel can communicate with each other, otherwise, they will cause a deadlock $\delta$. We define the following communication functions:

$\gamma(r_{C_{AT}}(10 \times BLIND_{k_A}(V_A, R_A), k_A, A), s_{C_{AT}}(10 \times BLIND_{k_A}(V_A, R_A), k_A, A))$
$\triangleq c_{C_{AT}}(10 \times BLIND_{k_A}(V_A, R_A), k_A, A)$
$\gamma(r_{C_{TA}}(SIGN_{sk_T}(BLIND_{k_A}(V_A, R_A))), s_{C_{TA}}(SIGN_{sk_T}(BLIND_{k_A}(V_A, R_A))))$
$\triangleq c_{C_{TA}}(SIGN_{sk_T}(BLIND_{k_A}(V_A, R_A)))$
$\gamma(r_{C_{AT}}(ENC_{pk_T}(SIGN_{sk_T}(v_A, R_A))), s_{C_{AT}}(ENC_{pk_T}(SIGN_{sk_T}(v_A, R_A))))$
$\triangleq c_{C_{AT}}(ENC_{pk_T}(SIGN_{sk_T}(v_A, R_A)))$
$\gamma(r_{C_{BT}}(10 \times BLIND_{k_B}(V_B, R_B), k_B, B), s_{C_{BT}}(10 \times BLIND_{k_B}(V_B, R_B), k_B, B))$
$\triangleq c_{C_{BT}}(10 \times BLIND_{k_B}(V_B, R_B), k_B, B)$
$\gamma(r_{C_{TB}}(SIGN_{sk_T}(BLIND_{k_B}(V_B, R_B))), s_{C_{TB}}(SIGN_{sk_T}(BLIND_{k_B}(V_B, R_B))))$
$\triangleq c_{C_{TB}}(SIGN_{sk_T}(BLIND_{k_B}(V_B, R_B)))$
$\gamma(r_{C_{BT}}(ENC_{pk_T}(SIGN_{sk_T}(v_B, R_B))), s_{C_{BT}}(ENC_{pk_T}(SIGN_{sk_T}(v_B, R_B))))$
$\triangleq c_{C_{BT}}(ENC_{pk_T}(SIGN_{sk_T}(v_B, R_B)))$
$\gamma(r_{C_{CT}}(10 \times BLIND_{k_C}(V_C, R_C), k_C, C), s_{C_{CT}}(10 \times BLIND_{k_C}(V_C, R_C), k_C, C))$
$\triangleq c_{C_{CT}}(10 \times BLIND_{k_C}(V_C, R_C), k_C, C)$
$\gamma(r_{C_{TC}}(SIGN_{sk_T}(BLIND_{k_C}(V_C, R_C))), s_{C_{TC}}(SIGN_{sk_T}(BLIND_{k_C}(V_C, R_C))))$
$\triangleq c_{C_{TC}}(SIGN_{sk_T}(BLIND_{k_C}(V_C, R_C)))$
$\gamma(r_{C_{CT}}(ENC_{pk_T}(SIGN_{sk_T}(v_C, R_C))), s_{C_{CT}}(ENC_{pk_T}(SIGN_{sk_T}(v_C, R_C))))$
$\triangleq c_{C_{CT}}(ENC_{pk_T}(SIGN_{sk_T}(v_C, R_C)))$

$\gamma(r_{C_{DT}}(10 \times BLIND_{k_D}(V_D, R_D), k_D, D), s_{C_{DT}}(10 \times BLIND_{k_D}(V_D, R_D), k_D, D))$
$\triangleq c_{C_{DT}}(10 \times BLIND_{k_D}(V_D, R_D), k_D, D)$
$\gamma(r_{C_{TD}}(SIGN_{sk_T}(BLIND_{k_D}(V_D, R_D))), s_{C_{TD}}(SIGN_{sk_T}(BLIND_{k_D}(V_D, R_D))))$
$\triangleq c_{C_{TD}}(SIGN_{sk_T}(BLIND_{k_D}(V_D, R_D)))$
$\gamma(r_{C_{DT}}(ENC_{pk_T}(SIGN_{sk_T}(v_D, R_D))), s_{C_{DT}}(ENC_{pk_T}(SIGN_{sk_T}(v_D, R_D))))$
$\triangleq c_{C_{DT}}(ENC_{pk_T}(SIGN_{sk_T}(v_D, R_D)))$

Let all modules be in parallel, then the protocol $A$ $B$ $C$ $D$ $T$ can be presented by the following process term:

$$\tau_I(\partial_H(\Theta(A \between B \between C \between D \between T))) = \tau_I(\partial_H(A \between B \between C \between D \between T))$$

where $H = \{r_{C_{AT}}(10 \times BLIND_{k_A}(V_A, R_A), k_A, A), s_{C_{AT}}(10 \times BLIND_{k_A}(V_A, R_A), k_A, A),$
$r_{C_{TA}}(SIGN_{sk_T}(BLIND_{k_A}(V_A, R_A))), s_{C_{TA}}(SIGN_{sk_T}(BLIND_{k_A}(V_A, R_A))),$
$r_{C_{AT}}(ENC_{pk_T}(SIGN_{sk_T}(v_A, R_A))), s_{C_{AT}}(ENC_{pk_T}(SIGN_{sk_T}(v_A, R_A))),$
$r_{C_{BT}}(10 \times BLIND_{k_B}(V_B, R_B), k_B, B), s_{C_{BT}}(10 \times BLIND_{k_B}(V_B, R_B), k_B, B),$
$r_{C_{TB}}(SIGN_{sk_T}(BLIND_{k_B}(V_B, R_B))), s_{C_{TB}}(SIGN_{sk_T}(BLIND_{k_B}(V_B, R_B))),$
$r_{C_{BT}}(ENC_{pk_T}(SIGN_{sk_T}(v_B, R_B))), s_{C_{BT}}(ENC_{pk_T}(SIGN_{sk_T}(v_B, R_B))),$
$r_{C_{CT}}(10 \times BLIND_{k_C}(V_C, R_C), k_C, C), s_{C_{CT}}(10 \times BLIND_{k_C}(V_C, R_C), k_C, C),$
$r_{C_{TC}}(SIGN_{sk_T}(BLIND_{k_C}(V_C, R_C))), s_{C_{TC}}(SIGN_{sk_T}(BLIND_{k_C}(V_C, R_C))),$
$r_{C_{CT}}(ENC_{pk_T}(SIGN_{sk_T}(v_C, R_C))), s_{C_{CT}}(ENC_{pk_T}(SIGN_{sk_T}(v_C, R_C))),$
$r_{C_{DT}}(10 \times BLIND_{k_D}(V_D, R_D), k_D, D), s_{C_{DT}}(10 \times BLIND_{k_D}(V_D, R_D), k_D, D),$
$r_{C_{TD}}(SIGN_{sk_T}(BLIND_{k_D}(V_D, R_D))), s_{C_{TD}}(SIGN_{sk_T}(BLIND_{k_D}(V_D, R_D))),$
$r_{C_{DT}}(ENC_{pk_T}(SIGN_{sk_T}(v_D, R_D))), s_{C_{DT}}(ENC_{pk_T}(SIGN_{sk_T}(v_D, R_D)))|D_A, D_B, D_C, D_D \in \Delta\},$

$I = \{c_{C_{AT}}(10 \times BLIND_{k_A}(V_A, R_A), k_A, A), c_{C_{TA}}(SIGN_{sk_T}(BLIND_{k_A}(V_A, R_A))),$
$c_{C_{AT}}(ENC_{pk_T}(SIGN_{sk_T}(v_A, R_A))), c_{C_{BT}}(10 \times BLIND_{k_B}(V_B, R_B), k_B, B),$
$c_{C_{TB}}(SIGN_{sk_T}(BLIND_{k_B}(V_B, R_B))), c_{C_{BT}}(ENC_{pk_T}(SIGN_{sk_T}(v_B, R_B))),$
$c_{C_{CT}}(10 \times BLIND_{k_C}(V_C, R_C), k_C, C), c_{C_{TC}}(SIGN_{sk_T}(BLIND_{k_C}(V_C, R_C))),$
$c_{C_{CT}}(ENC_{pk_T}(SIGN_{sk_T}(v_C, R_C))), c_{C_{DT}}(10 \times BLIND_{k_D}(V_D, R_D), k_D, D),$
$c_{C_{TD}}(SIGN_{sk_T}(BLIND_{k_D}(V_D, R_D))), c_{C_{DT}}(ENC_{pk_T}(SIGN_{sk_T}(v_D, R_D))),$
$blind_{k_A}(V_A, R_A), blind_{k_B}(V_B, R_B), blind_{k_C}(V_C, R_C), blind_{k_D}(V_D, R_D),$
$unblind_{k_A}(SIGN_{sk_T}(BLIND_{k_A}(V_A, R_A))), unblind_{k_B}(SIGN_{sk_T}(BLIND_{k_B}(V_B, R_B))),$
$unblind_{k_C}(SIGN_{sk_T}(BLIND_{k_C}(V_C, R_C))), unblind_{k_D}(SIGN_{sk_T}(BLIND_{k_D}(V_D, R_D))),$
$enc_{pk_T}(SIGN_{sk_T}(v_A, R_A)), enc_{pk_T}(SIGN_{sk_T}(v_B, R_B)), enc_{pk_T}(SIGN_{sk_T}(v_C, R_C)),$
$enc_{pk_T}(SIGN_{sk_T}(v_D, R_D)), check(A), check(B), check(C), check(D),$
$9 \times unblind_{k_A}(BLIND_{k_A}(V_A, R_A)), 9 \times unblind_{k_B}(BLIND_{k_B}(V_B, R_B)),$
$9 \times unblind_{k_C}(BLIND_{k_C}(V_C, R_C)), 9 \times unblind_{k_D}(BLIND_{k_D}(V_D, R_D)),$
$sign_{sk_T}(BLIND_{k_A}(V_A, R_A)), sign_{sk_T}(BLIND_{k_B}(V_B, R_B)),$
$sign_{sk_T}(BLIND_{k_C}(V_C, R_C)), sign_{sk_T}(BLIND_{k_D}(V_D, R_D)),$
$dec_{sk_T}(ENC_{pk_T}(SIGN_{sk_T}(v_A, R_A))), dec_{sk_T}(ENC_{pk_T}(SIGN_{sk_T}(v_B, R_B))),$
$dec_{sk_T}(ENC_{pk_T}(SIGN_{sk_T}(v_C, R_C))), dec_{sk_T}(ENC_{pk_T}(SIGN_{sk_T}(v_D, R_D))),$
$de\text{-}sign_{pk_T}(SIGN_{sk_T}(v_A, R_A)), de\text{-}sign_{pk_T}(SIGN_{sk_T}(v_B, R_B)),$
$de\text{-}sign_{pk_T}(SIGN_{sk_T}(v_C, R_C)), de\text{-}sign_{pk_T}(SIGN_{sk_T}(v_D, R_D)),$
$\{isFresh(R_A) = TRUE\}, tab(v_A), \{isFresh(R_A) = FALSE\}, tab(0),$
$\{isFresh(R_B) = TRUE\}, tab(v_B), \{isFresh(R_B) = FALSE\},$
$\{isFresh(R_C) = TRUE\}, tab(v_C), \{isFresh(R_C) = FALSE\},$
$\{isFresh(R_D) = TRUE\}, tab(v_D), \{isFresh(R_D) = FALSE\}|D_A, D_B, D_C, D_D \in \Delta\}$

Then, we obtain the following conclusion on the protocol.

**Theorem 31.3.** *The secure elections protocol 3 in Fig. 31.3 is improved based on the secure elections protocol 2.*

*Proof.* Based on the above state transitions of the above modules, by use of the algebraic laws of $APTC_G$, we can prove that

$\tau_I(\partial_H(A \between B \between C \between D \between T)) = \sum_{D_A, D_B, D_C, D_D \in \Delta}(r_{C_{AI}}(D_A) \parallel r_{C_{BI}}(D_B) \parallel r_{C_{CI}}(D_C) \parallel r_{C_{DI}}(D_D) \cdot s_{C_{TO}}(TAB)) \cdot \tau_I(\partial_H(A \between B \between C \between D \between T))$

For the details of the proof, please refer to Section 17.3, as we omit them here.

That is, the protocol in Fig. 31.3 $\tau_I(\partial_H(A \between B \between C \between D \between T))$ can exhibit the desired external behaviors, and is secure. However, for the properties of secure elections protocols:

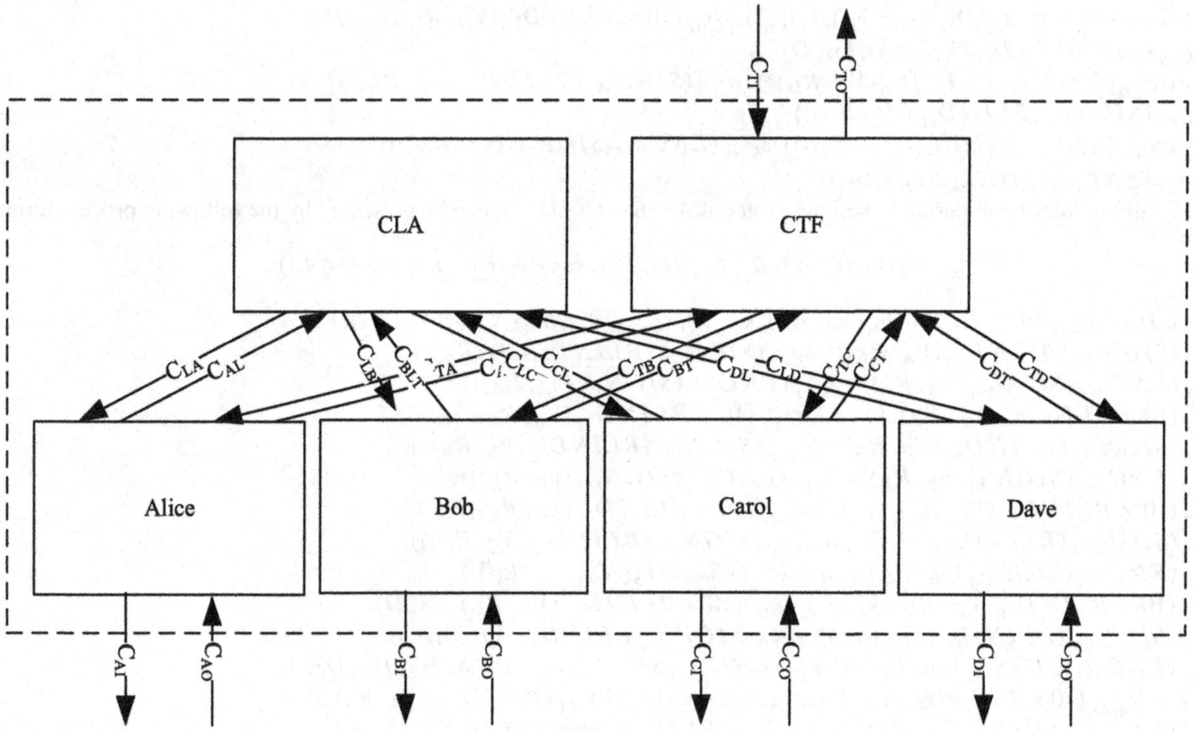


**FIGURE 31.4** Secure elections protocol 4.

1. Legitimacy: only authorized voters can vote;
2. Oneness: no one can vote more than once;
3. Privacy: no one can determine for whom anyone else voted;
4. Non-replicability: no one can duplicate anyone else's vote;
5. Non-changeability: no one can change anyone else's vote;
6. Validness: every voter can make sure that his vote has been taken into account in the final tabulation, if CTF is trustworthy.

However, CTF still can make valid signatures to cheat. □

## 31.4 Secure elections protocol 4

The secure elections protocol 4 is shown in Fig. 31.4, which is an improved one based on the secure elections protocol 3 in Section 31.3. In this protocol, there are a CLA (Central Legitimization Agency) to check the identity of voters, a CTF (Central Tabulating Facility) to collect the votes, and four voters: Alice, Bob, Carol, and Dave.

The process of the protocol is as follows:

1. Alice receives some voting request $D_A$ from the outside through the channel $C_{AI}$ (the corresponding reading action is denoted $r_{C_{AI}}(D_A)$), she generates a request $r_A$, encrypts it by CLA's public key through an action $enc_{pk_L}(r_A)$, and sends it to CLA through the channel $C_{AL}$ (the corresponding sending action is denoted $s_{C_{AL}}(ENC_{pk_L}(r_A))$);
2. Bob receives some voting request $D_B$ from the outside through the channel $C_{BI}$ (the corresponding reading action is denoted $r_{C_{BI}}(D_B)$), he generates a request $r_B$, encrypts it by CLA's public key through an action $enc_{pk_L}(r_B)$, and sends it to CLA through the channel $C_{BL}$ (the corresponding sending action is denoted $s_{C_{BL}}(ENC_{pk_L}(r_B))$);
3. Carol receives some voting request $D_C$ from the outside through the channel $C_{CI}$ (the corresponding reading action is denoted $r_{C_{CI}}(D_C)$), she generates a request $r_C$, encrypts it by CLA's public key through an action $enc_{pk_L}(r_C)$, and sends it to CLA through the channel $C_{CL}$ (the corresponding sending action is denoted $s_{C_{CL}}(ENC_{pk_L}(r_C))$);
4. Dave receives some voting request $D_D$ from the outside through the channel $C_{DI}$ (the corresponding reading action is denoted $r_{C_{DI}}(D_D)$), he generates a request $r_D$, encrypts it by CLA's public key through an action $enc_{pk_L}(r_D)$, and sends it to CLA through the channel $C_{DL}$ (the corresponding sending action is denoted $s_{C_{DL}}(ENC_{pk_L}(r_D))$);

5. CLA receives the requests from Alice, Bob, Carol, and Dave through the channels $C_{AL}$, $C_{BL}$, $C_{CL}$, and $C_{DL}$ (the corresponding reading actions are denoted $s_{C_{AL}}(ENC_{pk_L}(r_A))$, $s_{C_{BL}}(ENC_{pk_L}(r_B))$, $s_{C_{CL}}(ENC_{pk_L}(r_C))$, and $s_{C_{DL}}(ENC_{pk_L}(r_D))$, respectively), he decrypts these encrypted requests through the actions $dec_{sk_L}(ENC_{pk_L}(r_A))$, $dec_{sk_L}(ENC_{pk_L}(r_B))$, $dec_{sk_L}(ENC_{pk_L}(r_C))$, and $dec_{sk_L}(ENC_{pk_L}(r_D))$ to obtain $r_A$, $r_B$, $r_C$, and $r_D$, records the names of Alice, Bob, Carol, and Dave through actions $rec(A)$, $rec(B)$, $rec(C)$, and $rec(D)$; Both CLA and CTF maintain a table of valid numbers, and the table of CTF is obtained from that of CLA; then CLA randomly selects numbers $R_A$, $R_B$, $R_C$, and $R_D$, encrypts them through actions $enc_{pk_A}(R_A)$, $enc_{pk_B}(R_B)$, $enc_{pk_C}(R_C)$, and $enc_{pk_D}(R_D)$ and sends them to Alice, Bob, Carol, and Dave through the channels $C_{LA}$, $C_{LB}$, $C_{LC}$, and $C_{LD}$, respectively (the corresponding sending actions are denoted $s_{C_{LA}}(ENC_{pk_A}(R_A))$, $s_{C_{LB}}(ENC_{pk_B}(R_B))$, $s_{C_{LC}}(ENC_{pk_C}(R_C))$, and $s_{C_{LD}}(ENC_{pk_D}(R_D)))$;
6. Alice receives the encrypted number from CLA through the channel $C_{LA}$ (the corresponding reading action is denoted $r_{C_{LA}}(ENC_{pk_A}(R_A)))$, she decrypts the encrypted number through an action $dec_{sk_A}(ENC_{pk_A}(R_A))$ to obtain $R_A$, generates a random identity number $I_A$ through an action $rsg_{I_A}$ and her vote $v_A$, encrypted $I_A, R_A, v_A$ by CTF's public key through an action $enc_{pk_T}(I_A, R_A, v_A)$ and sends the encrypted message to CTF through the channel $C_{AT}$ (the corresponding sending action is denoted $s_{C_{AT}}(ENC_{pk_T}(I_A, R_A, v_A)))$;
7. Bob receives the encrypted number from CLA through the channel $C_{LB}$ (the corresponding reading action is denoted $r_{C_{LB}}(ENC_{pk_B}(R_B)))$, he decrypts the encrypted number through an action $dec_{sk_B}(ENC_{pk_B}(R_B))$ to obtain $R_B$, generates a random identity number $I_B$ through an action $rsg_{I_B}$ and his vote $v_B$, encrypted $I_B, R_B, v_B$ by CTF's public key through an action $enc_{pk_T}(I_B, R_B, v_B)$ and sends the encrypted message to CTF through the channel $C_{BT}$ (the corresponding sending action is denoted $s_{C_{BT}}(ENC_{pk_T}(I_B, R_B, v_B)))$;
8. Carol receives the encrypted number from CLA through the channel $C_{LC}$ (the corresponding reading action is denoted $r_{C_{LC}}(ENC_{pk_C}(R_C)))$, she decrypts the encrypted number through an action $dec_{sk_C}(ENC_{pk_C}(R_C))$ to obtain $R_C$, generates a random identity number $I_C$ through an action $rsg_{I_C}$ and her vote $v_C$, encrypted $I_C, R_C, v_C$ by CTF's public key through an action $enc_{pk_T}(I_C, R_C, v_C)$ and sends the encrypted message to CTF through the channel $C_{CT}$ (the corresponding sending action is denoted $s_{C_{CT}}(ENC_{pk_T}(I_C, R_C, v_C)))$;
9. Dave receives the encrypted number from CLA through the channel $C_{LD}$ (the corresponding reading action is denoted $r_{C_{LD}}(ENC_{pk_A}(R_D)))$, he decrypts the encrypted number through an action $dec_{sk_D}(ENC_{pk_D}(R_D))$ to obtain $R_D$, generates a random identity number $I_D$ through an action $rsg_{I_D}$ and his vote $v_D$, encrypted $I_D, R_D, v_D$ by CTF's public key through an action $enc_{pk_T}(I_D, R_D, v_D)$ and sends the encrypted message to CTF through the channel $C_{DT}$ (the corresponding sending action is denoted $s_{C_{DT}}(ENC_{pk_T}(I_D, R_D, v_D)))$;
10. CTF receives the encrypted messages from Alice, Bob, Carol, and Dave through the channels $C_{AT}$, $C_{BT}$, $C_{CT}$, and $C_{DT}$ (the corresponding reading actions are denoted $r_{C_{AT}}(ENC_{pk_T}(I_A, R_A, v_A))$, $r_{C_{BT}}(ENC_{pk_T}(I_B, R_B, v_B))$, $r_{C_{CT}}(ENC_{pk_T}(I_C, R_C, v_C))$, and $r_{C_{DT}}(ENC_{pk_T}(I_D, R_D, v_D))$, respectively), he decrypts these encrypted messages through actions $dec_{sk_T}(ENC_{pk_T}(I_A, R_A, v_A))$, $dec_{sk_T}(ENC_{pk_T}(I_B, R_B, v_B))$, $dec_{sk_T}(ENC_{pk_T}(I_C, R_C, v_C))$, and $dec_{sk_T}(ENC_{pk_T}(I_D, R_D, v_D))$. If $isExisted(R_A) = TRUE$, he removes $R_A$ from its table through an action $remove(R_A)$, records the vote $v_A$ and the pair of $I_A$ and $v_A$ into the voting results $TAB$ through an action $rec(I_A, v_A)$, else he does nothing; if $isExisted(R_B) = TRUE$, he removes $R_B$ from its table through an action $remove(R_B)$, records the vote $v_B$ and the pair of $I_B$ and $v_B$ into the voting results $TAB$ through an action $rec(I_B, v_B)$, else he does nothing; if $isExisted(R_C) = TRUE$, he removes $R_C$ from its table through an action $remove(R_C)$, records the vote $v_C$ and the pair of $I_C$ and $v_C$ into the voting results $TAB$ through an action $rec(I_C, v_C)$, else he does nothing; if $isExisted(R_D) = TRUE$, he removes $R_D$ from its table through an action $remove(R_D)$, records the vote $v_D$ and the pair of $I_D$ and $v_D$ into the voting results $TAB$ through an action $rec(I_D, v_D)$, else he does nothing. Finally, he sends the voting results $TAB$ to the outside through the channel $C_{TO}$ (the corresponding sending action is denoted $s_{C_{TO}}(TAB)$),

where $D_A, D_B, D_C, D_D \in \Delta$, $\Delta$ is the set of data.

Alice's state transitions described by $APTC_G$ are as follows:

$A = \sum_{D_A \in \Delta} r_{C_{AI}}(D_A) \cdot A_2$

$A_2 = enc_{pk_L}(r_A) \cdot A_3$

$A_3 = s_{C_{AL}}(ENC_{pk_L}(r_A)) \cdot A_4$

$A_4 = r_{C_{LA}}(ENC_{pk_A}(R_A)) \cdot A_5$

$A_5 = dec_{sk_A}(ENC_{pk_A}(R_A)) \cdot A_6$

$A_6 = rsg_{I_A} \cdot A_7$

$A_7 = enc_{pk_T}(I_A, R_A, v_A) \cdot A_8$

$A_8 = s_{C_{AT}}(ENC_{pk_T}(I_A, R_A, v_A)) \cdot A$

Bob's state transitions described by $APTC_G$ are as follows:

$B = \sum_{D_B\in\Delta} r_{C_{BI}}(D_B) \cdot B_2$

$B_2 = enc_{pk_L}(r_B) \cdot B_3$

$B_3 = s_{C_{BL}}(ENC_{pk_L}(r_B)) \cdot B_4$

$B_4 = r_{C_{LB}}(ENC_{pk_B}(R_B)) \cdot B_5$

$B_5 = dec_{sk_B}(ENC_{pk_B}(R_B)) \cdot B_6$

$B_6 = rsg_{I_B} \cdot B_7$

$B_7 = enc_{pk_T}(I_B, R_B, v_B) \cdot B_8$

$B_8 = s_{C_{BT}}(ENC_{pk_T}(I_B, R_B, v_B)) \cdot B$

Carol's state transitions described by $APTC_G$ are as follows:

$C = \sum_{D_C\in\Delta} r_{C_{CI}}(D_C) \cdot C_2$

$C_2 = enc_{pk_L}(r_C) \cdot C_3$

$C_3 = s_{C_{CL}}(ENC_{pk_L}(r_C)) \cdot C_4$

$C_4 = r_{C_{LC}}(ENC_{pk_C}(R_C)) \cdot C_5$

$C_5 = dec_{sk_C}(ENC_{pk_C}(R_C)) \cdot C_6$

$C_6 = rsg_{I_C} \cdot C_7$

$C_7 = enc_{pk_T}(I_C, R_C, v_C) \cdot C_8$

$C_8 = s_{C_{CT}}(ENC_{pk_T}(I_C, R_C, v_C)) \cdot C$

Dave's state transitions described by $APTC_G$ are as follows:

$D = \sum_{D_D\in\Delta} r_{C_{DI}}(D_D) \cdot D_2$

$D_2 = enc_{pk_L}(r_D) \cdot D_3$

$D_3 = s_{C_{DL}}(ENC_{pk_L}(r_D)) \cdot D_4$

$D_4 = r_{C_{LD}}(ENC_{pk_D}(R_D)) \cdot D_5$

$D_5 = dec_{sk_D}(ENC_{pk_D}(R_D)) \cdot D_6$

$D_6 = rsg_{I_D} \cdot D_7$

$D_7 = enc_{pk_T}(I_D, R_D, v_D) \cdot D_8$

$D_8 = s_{C_{DT}}(ENC_{pk_T}(I_D, R_D, v_D)) \cdot D$

CLA's state transitions described by $APTC_G$ are as follows:

$L = s_{C_{AL}}(ENC_{pk_L}(r_A)) \parallel s_{C_{BL}}(ENC_{pk_L}(r_B))$
$\parallel s_{C_{CL}}(ENC_{pk_L}(r_C)) \parallel s_{C_{DL}}(ENC_{pk_L}(r_D)) \cdot L_2$

$L_2 = dec_{sk_L}(ENC_{pk_L}(r_A)) \parallel dec_{sk_L}(ENC_{pk_L}(r_B))$
$\parallel dec_{sk_L}(ENC_{pk_L}(r_C)) \parallel dec_{sk_L}(ENC_{pk_L}(r_D)) \cdot L_3$

$L_3 = rec(A) \parallel rec(B) \parallel rec(C) \parallel rec(D) \cdot L_4$

$L_4 = enc_{pk_A}(R_A) \parallel enc_{pk_B}(R_B) \parallel enc_{pk_C}(R_C) \parallel enc_{pk_D}(R_D) \cdot L_5$

$L_5 = s_{C_{LA}}(ENC_{pk_A}(R_A)) \parallel s_{C_{LB}}(ENC_{pk_B}(R_B))$
$\parallel s_{C_{LC}}(ENC_{pk_C}(R_C)) \parallel s_{C_{LD}}(ENC_{pk_D}(R_D)) \cdot L$

CTF's state transitions described by $APTC_G$ are as follows:

$T = r_{C_{AT}}(ENC_{pk_T}(I_A, R_A, v_A)) \parallel r_{C_{BT}}(ENC_{pk_T}(I_B, R_B, v_B))$
$\parallel r_{C_{CT}}(ENC_{pk_T}(I_C, R_C, v_C)) \parallel r_{C_{DT}}(ENC_{pk_T}(I_D, R_D, v_D)) \cdot T_2$

$T_2 = dec_{sk_T}(ENC_{pk_T}(I_A, R_A, v_A)) \parallel dec_{sk_T}(ENC_{pk_T}(I_B, R_B, v_B))$
$\parallel dec_{sk_T}(ENC_{pk_T}(I_C, R_C, v_C)) \parallel dec_{sk_T}(ENC_{pk_T}(I_D, R_D, v_D)) \cdot T_3$

$T_3 = ((\{isExisted(R_A) = TRUE\} \cdot remove(R_A) \cdot rec(I_A, v_A) + \{isExisted(R_A) = FALSE\})$
$\parallel (\{isExisted(R_B) = TRUE\} \cdot remove(R_B) \cdot rec(I_B, v_B) + \{isExisted(R_B) = FALSE\})$
$\parallel (\{isExisted(R_C) = TRUE\} \cdot remove(R_C) \cdot rec(I_C, v_C) + \{isExisted(R_C) = FALSE\})$
$\parallel (\{isExisted(R_D) = TRUE\} \cdot remove(R_D) \cdot rec(I_D, v_D) + \{isExisted(R_D) = FALSE\})) \cdot T_4$

$T_4 = s_{C_{TO}}(TAB) \cdot T$

The sending action and the reading action of the same type data through the same channel can communicate with each other, otherwise, they will cause a deadlock $\delta$. We define the following communication functions:

$\gamma(r_{C_{AL}}(ENC_{pk_L}(r_A)), s_{C_{AL}}(ENC_{pk_L}(r_A))) \triangleq c_{C_{AL}}(ENC_{pk_L}(r_A))$

$\gamma(r_{C_{LA}}(ENC_{pk_A}(R_A)), s_{C_{LA}}(ENC_{pk_A}(R_A))) \triangleq c_{C_{LA}}(ENC_{pk_A}(R_A))$

$\gamma(r_{C_{AT}}(ENC_{pk_T}(I_A, R_A, v_A)), s_{C_{AT}}(ENC_{pk_T}(I_A, R_A, v_A))) \triangleq c_{C_{AT}}(ENC_{pk_T}(I_A, R_A, v_A))$

$\gamma(r_{C_{BL}}(ENC_{pk_L}(r_B)), s_{C_{BL}}(ENC_{pk_L}(r_B))) \triangleq c_{C_{BL}}(ENC_{pk_L}(r_B))$

$\gamma(r_{C_{LB}}(ENC_{pk_B}(R_B)), s_{C_{LB}}(ENC_{pk_B}(R_B))) \triangleq c_{C_{LB}}(ENC_{pk_B}(R_B))$

$\gamma(r_{C_{BT}}(ENC_{pk_T}(I_B, R_B, v_B)), s_{C_{BT}}(ENC_{pk_T}(I_B, R_B, v_B))) \triangleq c_{C_{BT}}(ENC_{pk_T}(I_B, R_B, v_B))$
$\gamma(r_{C_{CL}}(ENC_{pk_L}(r_C)), s_{C_{CL}}(ENC_{pk_L}(r_C))) \triangleq c_{C_{CL}}(ENC_{pk_L}(r_C))$
$\gamma(r_{C_{LC}}(ENC_{pk_C}(R_C)), s_{C_{LC}}(ENC_{pk_C}(R_C))) \triangleq c_{C_{LC}}(ENC_{pk_C}(R_C))$
$\gamma(r_{C_{CT}}(ENC_{pk_T}(I_C, R_C, v_C)), s_{C_{CT}}(ENC_{pk_T}(I_C, R_C, v_C))) \triangleq c_{C_{CT}}(ENC_{pk_T}(I_C, R_C, v_C))$
$\gamma(r_{C_{DL}}(ENC_{pk_L}(r_D)), s_{C_{DL}}(ENC_{pk_L}(r_D))) \triangleq c_{C_{DL}}(ENC_{pk_L}(r_D))$
$\gamma(r_{C_{LD}}(ENC_{pk_D}(R_D)), s_{C_{LD}}(ENC_{pk_D}(R_D))) \triangleq c_{C_{LD}}(ENC_{pk_D}(R_D))$
$\gamma(r_{C_{DT}}(ENC_{pk_T}(I_D, R_D, v_D)), s_{C_{DT}}(ENC_{pk_T}(I_D, R_D, v_D))) \triangleq c_{C_{DT}}(ENC_{pk_T}(I_D, R_D, v_D))$

Let all modules be in parallel, then the protocol $A \quad B \quad C \quad D \quad L \quad T$ can be presented by the following process term:

$$\tau_I(\partial_H(\Theta(A \between B \between C \between D \between L \between T))) = \tau_I(\partial_H(A \between B \between C \between D \between L \between T))$$

where $H = \{r_{C_{AL}}(ENC_{pk_L}(r_A)), s_{C_{AL}}(ENC_{pk_L}(r_A)),$
$r_{C_{LA}}(ENC_{pk_A}(R_A)), s_{C_{LA}}(ENC_{pk_A}(R_A)),$
$r_{C_{AT}}(ENC_{pk_T}(I_A, R_A, v_A)), s_{C_{AT}}(ENC_{pk_T}(I_A, R_A, v_A)),$
$r_{C_{BL}}(ENC_{pk_L}(r_B)), s_{C_{BL}}(ENC_{pk_L}(r_B)),$
$r_{C_{LB}}(ENC_{pk_B}(R_B)), s_{C_{LB}}(ENC_{pk_B}(R_B)),$
$r_{C_{BT}}(ENC_{pk_T}(I_B, R_B, v_B)), s_{C_{BT}}(ENC_{pk_T}(I_B, R_B, v_B)),$
$r_{C_{CL}}(ENC_{pk_L}(r_C)), s_{C_{CL}}(ENC_{pk_L}(r_C)),$
$r_{C_{LC}}(ENC_{pk_C}(R_C)), s_{C_{LC}}(ENC_{pk_C}(R_C)),$
$r_{C_{CT}}(ENC_{pk_T}(I_C, R_C, v_C)), s_{C_{CT}}(ENC_{pk_T}(I_C, R_C, v_C)),$
$r_{C_{DL}}(ENC_{pk_L}(r_D)), s_{C_{DL}}(ENC_{pk_L}(r_D)),$
$r_{C_{LD}}(ENC_{pk_D}(R_D)), s_{C_{LD}}(ENC_{pk_D}(R_D)),$
$r_{C_{DT}}(ENC_{pk_T}(I_D, R_D, v_D)), s_{C_{DT}}(ENC_{pk_T}(I_D, R_D, v_D)) | D_A, D_B, D_C, D_D \in \Delta\},$
$I = \{c_{C_{AL}}(ENC_{pk_L}(r_A)), c_{C_{LA}}(ENC_{pk_A}(R_A)),$
$c_{C_{AT}}(ENC_{pk_T}(I_A, R_A, v_A)), c_{C_{BL}}(ENC_{pk_L}(r_B)),$
$c_{C_{LB}}(ENC_{pk_B}(R_B)), c_{C_{BT}}(ENC_{pk_T}(I_B, R_B, v_B)),$
$c_{C_{CL}}(ENC_{pk_L}(r_C)), c_{C_{LC}}(ENC_{pk_C}(R_C)),$
$c_{C_{CT}}(ENC_{pk_T}(I_C, R_C, v_C)), c_{C_{DL}}(ENC_{pk_L}(r_D)),$
$c_{C_{LD}}(ENC_{pk_D}(R_D)), c_{C_{DT}}(ENC_{pk_T}(I_D, R_D, v_D)),$
$enc_{pk_L}(r_A), enc_{pk_L}(r_B), enc_{pk_L}(r_C), enc_{pk_L}(r_D),$
$dec_{sk_A}(ENC_{pk_A}(R_A)), dec_{sk_B}(ENC_{pk_B}(R_B)), dec_{sk_C}(ENC_{pk_C}(R_C)),$
$dec_{sk_D}(ENC_{pk_D}(R_D)), rsg_{I_A}, rsg_{I_B}, rsg_{I_C}, rsg_{I_D},$
$enc_{pk_T}(I_A, R_A, v_A), enc_{pk_T}(I_B, R_B, v_B), enc_{pk_T}(I_C, R_C, v_C), enc_{pk_T}(I_D, R_D, v_D),$
$dec_{sk_L}(ENC_{pk_L}(r_A)), dec_{sk_L}(ENC_{pk_L}(r_B)),$
$dec_{sk_L}(ENC_{pk_L}(r_C)), dec_{sk_L}(ENC_{pk_L}(r_D)),$
$rec(A), rec(B), rec(C), rec(D), enc_{pk_A}(R_A), enc_{pk_B}(R_B),$
$enc_{pk_C}(R_C), enc_{pk_D}(R_D), dec_{sk_T}(ENC_{pk_T}(I_A, R_A, v_A)),$
$dec_{sk_T}(ENC_{pk_T}(I_B, R_B, v_B)), dec_{sk_T}(ENC_{pk_T}(I_C, R_C, v_C)), dec_{sk_T}(ENC_{pk_T}(I_D, R_D, v_D)),$
$\{isExisted(R_A) = TRUE\}, remove(R_A), rec(I_A, v_A), \{isExisted(R_A) = FALSE\},$
$\{isExisted(R_B) = TRUE\}, remove(R_B), rec(I_B, v_B), \{isExisted(R_B) = FALSE\},$
$\{isExisted(R_C) = TRUE\}, remove(R_C), rec(I_C, v_C), \{isExisted(R_C) = FALSE\},$
$\{isExisted(R_D) = TRUE\}, remove(R_D), rec(I_D, v_D), \{isExisted(R_D) = FALSE\}$
$| D_A, D_B, D_C, D_D \in \Delta\}$

Then, we obtain the following conclusion on the protocol.

**Theorem 31.4.** *The secure elections protocol 4 in Fig. 31.4 is improved based on the secure elections protocol 3.*

*Proof.* Based on the above state transitions of the above modules, by use of the algebraic laws of $APTC_G$, we can prove that

$\tau_I(\partial_H(A \between B \between C \between D \between L \between T)) = \sum_{D_A, D_B, D_C, D_D \in \Delta}(r_{C_{AI}}(D_A) \parallel r_{C_{BI}}(D_B) \parallel r_{C_{CI}}(D_C) \parallel r_{C_{DI}}(D_D) \cdot s_{C_{TO}}(TAB)) \cdot \tau_I(\partial_H(A \between B \between C \between D \between L \between T))$

For the details of the proof, please refer to Section 17.3, as we omit them here.

That is, the protocol in Fig. 31.4 $\tau_I(\partial_H(A \between B \between C \between D \between L \between T))$ can exhibit the desired external behaviors, and is secure. However, for the properties of secure elections protocols:

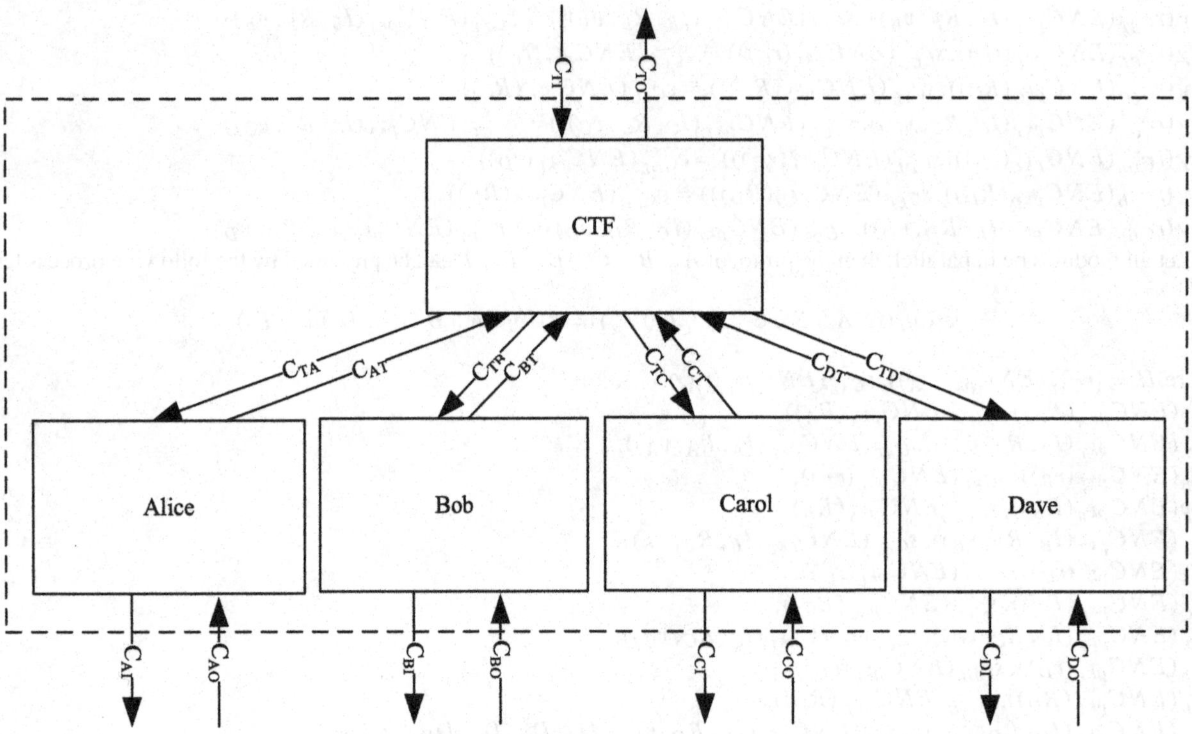


**FIGURE 31.5** Secure elections protocol 5.

1. Legitimacy: only authorized voters can vote;
2. Oneness: no one can vote more than once;
3. Privacy: no one can determine for whom anyone else voted;
4. Non-replicability: no one can duplicate anyone else's vote;
5. Non-changeability: no one can change anyone else's vote;
6. Validness: every voter can make sure that his vote has been taken into account in the final tabulation, if CLA and CTF are trustworthy.

However, CLA and CTF still can conspire to distribute valid numbers to illegal voters. □

## 31.5 Secure elections protocol 5

The secure elections protocol 5 is shown in Fig. 31.5, which is an improved one based on the secure elections protocol 4 in Section 31.4. In this protocol, there are a CTF (Central Tabulating Facility) to check the identity of voters and collect the votes, and four voters: Alice, Bob, Carol, and Dave.

The process of the protocol is as follows:

1. Alice receives some voting request $D_A$ from the outside through the channel $C_{AI}$ (the corresponding reading action is denoted $r_{C_{AI}}(D_A)$), she generates a request $r_A$, encrypts it by CTF's public key through an action $enc_{pk_T}(r_A)$, and sends it to CTF through the channel $C_{AT}$ (the corresponding sending action is denoted $s_{C_{AT}}(ENC_{pk_T}(r_A))$);
2. Bob receives some voting request $D_B$ from the outside through the channel $C_{BI}$ (the corresponding reading action is denoted $r_{C_{BI}}(D_B)$), he generates a request $r_B$, encrypts it by CTF's public key through an action $enc_{pk_T}(r_B)$, and sends it to CTF through the channel $C_{BT}$ (the corresponding sending action is denoted $s_{C_{BT}}(ENC_{pk_T}(r_B))$);
3. Carol receives some voting request $D_C$ from the outside through the channel $C_{CI}$ (the corresponding reading action is denoted $r_{C_{CI}}(D_C)$), she generates a request $r_C$, encrypts it by CTF's public key through an action $enc_{pk_T}(r_C)$, and sends it to CTF through the channel $C_{CT}$ (the corresponding sending action is denoted $s_{C_{CT}}(ENC_{pk_T}(r_C))$);
4. Dave receives some voting request $D_D$ from the outside through the channel $C_{DI}$ (the corresponding reading action is denoted $r_{C_{DI}}(D_D)$), he generates a request $r_D$, encrypts it by CTF's public key through an action $enc_{pk_T}(r_D)$, and sends it to CTF through the channel $C_{DT}$ (the corresponding sending action is denoted $s_{C_{DT}}(ENC_{pk_T}(r_D))$);

5. CTF receives the requests from Alice, Bob, Carol, and Dave through the channels $C_{AT}$, $C_{BT}$, $C_{CT}$, and $C_{DT}$ (the corresponding reading actions are denoted $s_{C_{AT}}(ENC_{pk_T}(r_A))$, $s_{C_{BT}}(ENC_{pk_T}(r_B))$, $s_{C_{CT}}(ENC_{pk_T}(r_C))$, and $s_{C_{DT}}(ENC_{pk_T}(r_D))$, respectively), he decrypts these encrypted requests through the actions $dec_{sk_T}(ENC_{pk_T}(r_A))$, $dec_{sk_T}(ENC_{pk_T}(r_B))$, $dec_{sk_T}(ENC_{pk_T}(r_C))$, and $dec_{sk_T}(ENC_{pk_T}(r_D))$ to obtain $r_A$, $r_B$, $r_C$, and $r_D$, records the names of Alice, Bob, Carol, and Dave through actions $rec(A)$, $rec(B)$, $rec(C)$, and $rec(D)$; CTF maintains a table of valid numbers; then CTF encrypts all numbers $R$ through actions $enc_{pk_A}(R)$, $enc_{pk_B}(R)$, $enc_{pk_C}(R)$, and $enc_{pk_D}(R)$ and sends them to Alice, Bob, Carol, and Dave through the channels $C_{TA}$, $C_{TB}$, $C_{TC}$, and $C_{TD}$, respectively (the corresponding sending actions are denoted $s_{C_{TA}}(ENC_{pk_A}(R))$, $s_{C_{TB}}(ENC_{pk_B}(R))$, $s_{C_{TC}}(ENC_{pk_C}(R))$, and $s_{C_{TD}}(ENC_{pk_D}(R))$);
6. Alice receives the encrypted number from CTF through the channel $C_{TA}$ (the corresponding reading action is denoted $r_{C_{TA}}(ENC_{pk_A}(R))$), she decrypts the encrypted number through an action $dec_{sk_A}(ENC_{pk_A}(R))$ to randomly select one $R_A$, generates a random identity number $I_A$ through an action $rsg_{I_A}$ and her vote $v_A$, encrypted $I_A, R_A, v_A$ by CTF's public key through an action $enc_{pk_T}(I_A, R_A, v_A)$ and sends the encrypted message to CTF through the channel $C_{AT}$ (the corresponding sending action is denoted $s_{C_{AT}}(ENC_{pk_T}(I_A, R_A, v_A))$);
7. Bob receives the encrypted number from CTF through the channel $C_{TB}$ (the corresponding reading action is denoted $r_{C_{TB}}(ENC_{pk_B}(R))$), he decrypts the encrypted number through an action $dec_{sk_B}(ENC_{pk_B}(R))$ to randomly select one $R_B$, generates a random identity number $I_B$ through an action $rsg_{I_B}$ and his vote $v_B$, encrypted $I_B, R_B, v_B$ by CTF's public key through an action $enc_{pk_T}(I_B, R_B, v_B)$ and sends the encrypted message to CTF through the channel $C_{BT}$ (the corresponding sending action is denoted $s_{C_{BT}}(ENC_{pk_T}(I_B, R_B, v_B))$);
8. Carol receives the encrypted number from CTF through the channel $C_{TC}$ (the corresponding reading action is denoted $r_{C_{TC}}(ENC_{pk_C}(R))$), she decrypts the encrypted number through an action $dec_{sk_C}(ENC_{pk_C}(R))$ to randomly select one $R_C$, generates a random identity number $I_C$ through an action $rsg_{I_C}$ and her vote $v_C$, encrypted $I_C, R_C, v_C$ by CTF's public key through an action $enc_{pk_T}(I_C, R_C, v_C)$ and sends the encrypted message to CTF through the channel $C_{CT}$ (the corresponding sending action is denoted $s_{C_{CT}}(ENC_{pk_T}(I_C, R_C, v_C))$);
9. Dave receives the encrypted number from CTF through the channel $C_{TD}$ (the corresponding reading action is denoted $r_{C_{TD}}(ENC_{pk_A}(R))$), he decrypts the encrypted number through an action $dec_{sk_D}(ENC_{pk_D}(R))$ to randomly select one $R_D$, generates a random identity number $I_D$ through an action $rsg_{I_D}$ and his vote $v_D$, encrypted $I_D, R_D, v_D$ by CTF's public key through an action $enc_{pk_T}(I_D, R_D, v_D)$ and sends the encrypted message to CTF through the channel $C_{DT}$ (the corresponding sending action is denoted $s_{C_{DT}}(ENC_{pk_T}(I_D, R_D, v_D))$);
10. CTF receives the encrypted messages from Alice, Bob, Carol, and Dave through the channels $C_{AT}$, $C_{BT}$, $C_{CT}$, and $C_{DT}$ (the corresponding reading actions are denoted $r_{C_{AT}}(ENC_{pk_T}(I_A, R_A, v_A))$, $r_{C_{BT}}(ENC_{pk_T}(I_B, R_B, v_B))$, $r_{C_{CT}}(ENC_{pk_T}(I_C, R_C, v_C))$, and $r_{C_{DT}}(ENC_{pk_T}(I_D, R_D, v_D))$, respectively), he decrypts these encrypted messages through actions $dec_{sk_T}(ENC_{pk_T}(I_A, R_A, v_A))$, $dec_{sk_T}(ENC_{pk_T}(I_B, R_B, v_B))$, $dec_{sk_T}(ENC_{pk_T}(I_C, R_C, v_C))$, and $dec_{sk_T}(ENC_{pk_T}(I_D, R_D, v_D))$. If $isExisted(R_A) = TRUE$, he removes $R_A$ from its table through an action $remove(R_A)$, records the vote $v_A$ and the pair of $I_A$ and $v_A$ into the voting results $TAB$ through an action $rec(I_A, v_A)$, else he does nothing; if $isExisted(R_B) = TRUE$, he removes $R_B$ from its table through an action $remove(R_B)$, records the vote $v_B$ and the pair of $I_B$ and $v_B$ into the voting results $TAB$ through an action $rec(I_B, v_B)$, else he does nothing; if $isExisted(R_C) = TRUE$, he removes $R_C$ from its table through an action $remove(R_C)$, records the vote $v_C$ and the pair of $I_C$ and $v_C$ into the voting results $TAB$ through an action $rec(I_C, v_C)$, else he does nothing; if $isExisted(R_D) = TRUE$, he removes $R_D$ from its table through an action $remove(R_D)$, records the vote $v_D$ and the pair of $I_D$ and $v_D$ into the voting results $TAB$ through an action $rec(I_D, v_D)$, else he does nothing. Finally, he sends the voting results $TAB$ to the outside through the channel $C_{TO}$ (the corresponding sending action is denoted $s_{C_{TO}}(TAB)$),

where $D_A, D_B, D_C, D_D \in \Delta$, $\Delta$ is the set of data.

Alice's state transitions described by $APTC_G$ are as follows:

$A = \sum_{D_A \in \Delta} r_{C_{AI}}(D_A) \cdot A_2$

$A_2 = enc_{pk_T}(r_A) \cdot A_3$

$A_3 = s_{C_{AT}}(ENC_{pk_T}(r_A)) \cdot A_4$

$A_4 = r_{C_{TA}}(ENC_{pk_A}(R)) \cdot A_5$

$A_5 = dec_{sk_A}(ENC_{pk_A}(R)) \cdot A_6$

$A_6 = rsg_{I_A} \cdot A_7$

$A_7 = enc_{pk_T}(I_A, R_A, v_A) \cdot A_8$

$A_8 = s_{C_{AT}}(ENC_{pk_T}(I_A, R_A, v_A)) \cdot A$

Bob's state transitions described by $APTC_G$ are as follows:

$B = \sum_{D_B \in \Delta} r_{C_{BI}}(D_B) \cdot B_2$
$B_2 = enc_{pk_T}(r_B) \cdot B_3$
$B_3 = s_{C_{BT}}(ENC_{pk_T}(r_B)) \cdot B_4$
$B_4 = r_{C_{TB}}(ENC_{pk_B}(R)) \cdot B_5$
$B_5 = dec_{sk_B}(ENC_{pk_B}(R)) \cdot B_6$
$B_6 = rsg_{I_B} \cdot B_7$
$B_7 = enc_{pk_T}(I_B, R_B, v_B) \cdot B_8$
$B_8 = s_{C_{BT}}(ENC_{pk_T}(I_B, R_B, v_B)) \cdot B$

Carol's state transitions described by $APTC_G$ are as follows:

$C = \sum_{D_C \in \Delta} r_{C_{CI}}(D_C) \cdot C_2$
$C_2 = enc_{pk_T}(r_C) \cdot C_3$
$C_3 = s_{C_{CT}}(ENC_{pk_T}(r_C)) \cdot C_4$
$C_4 = r_{C_{TC}}(ENC_{pk_C}(R)) \cdot C_5$
$C_5 = dec_{sk_C}(ENC_{pk_C}(R)) \cdot C_6$
$C_6 = rsg_{I_C} \cdot C_7$
$C_7 = enc_{pk_T}(I_C, R_C, v_C) \cdot C_8$
$C_8 = s_{C_{CT}}(ENC_{pk_T}(I_C, R_C, v_C)) \cdot C$

Dave's state transitions described by $APTC_G$ are as follows:

$D = \sum_{D_D \in \Delta} r_{C_{DI}}(D_D) \cdot D_2$
$D_2 = enc_{pk_T}(r_D) \cdot D_3$
$D_3 = s_{C_{DT}}(ENC_{pk_T}(r_D)) \cdot D_4$
$D_4 = r_{C_{TD}}(ENC_{pk_D}(R)) \cdot D_5$
$D_5 = dec_{sk_D}(ENC_{pk_D}(R)) \cdot D_6$
$D_6 = rsg_{I_D} \cdot D_7$
$D_7 = enc_{pk_T}(I_D, R_D, v_D) \cdot D_8$
$D_8 = s_{C_{DT}}(ENC_{pk_T}(I_D, R_D, v_D)) \cdot D$

CTF's state transitions described by $APTC_G$ are as follows:

$T = s_{C_{AT}}(ENC_{pk_T}(r_A)) \parallel s_{C_{BT}}(ENC_{pk_T}(r_B))$
$\parallel s_{C_{CT}}(ENC_{pk_T}(r_C)) \parallel s_{C_{DT}}(ENC_{pk_T}(r_D)) \cdot T_2$
$T_2 = dec_{sk_T}(ENC_{pk_T}(r_A)) \parallel dec_{sk_T}(ENC_{pk_T}(r_B))$
$\parallel dec_{sk_T}(ENC_{pk_T}(r_C)) \parallel dec_{sk_T}(ENC_{pk_T}(r_D)) \cdot T_3$
$T_3 = rec(A) \parallel rec(B) \parallel rec(C) \parallel rec(D) \cdot T_4$
$T_4 = enc_{pk_A}(R) \parallel enc_{pk_B}(R) \parallel enc_{pk_C}(R) \parallel enc_{pk_D}(R) \cdot T_5$
$T_5 = s_{C_{TA}}(ENC_{pk_A}(R)) \parallel s_{C_{TB}}(ENC_{pk_B}(R))$
$\parallel s_{C_{TC}}(ENC_{pk_C}(R)) \parallel s_{C_{TD}}(ENC_{pk_D}(R)) \cdot T_6$
$T_6 = r_{C_{AT}}(ENC_{pk_T}(I_A, R_A, v_A)) \parallel r_{C_{BT}}(ENC_{pk_T}(I_B, R_B, v_B))$
$\parallel r_{C_{CT}}(ENC_{pk_T}(I_C, R_C, v_C)) \parallel r_{C_{DT}}(ENC_{pk_T}(I_D, R_D, v_D)) \cdot T_7$
$T_7 = dec_{sk_T}(ENC_{pk_T}(I_A, R_A, v_A)) \parallel dec_{sk_T}(ENC_{pk_T}(I_B, R_B, v_B))$
$\parallel dec_{sk_T}(ENC_{pk_T}(I_C, R_C, v_C)) \parallel dec_{sk_T}(ENC_{pk_T}(I_D, R_D, v_D)) \cdot T_8$
$T_8 = ((\{isExisted(R_A) = TRUE\} \cdot remove(R_A) \cdot rec(I_A, v_A) + \{isExisted(R_A) = FALSE\})$
$\parallel (\{isExisted(R_B) = TRUE\} \cdot remove(R_B) \cdot rec(I_B, v_B) + \{isExisted(R_B) = FALSE\})$
$\parallel (\{isExisted(R_C) = TRUE\} \cdot remove(R_C) \cdot rec(I_C, v_C) + \{isExisted(R_C) = FALSE\})$
$\parallel (\{isExisted(R_D) = TRUE\} \cdot remove(R_D) \cdot rec(I_D, v_D) + \{isExisted(R_D) = FALSE\})) \cdot T_9$
$T_9 = s_{C_{TO}}(TAB) \cdot T$

The sending action and the reading action of the same type data through the same channel can communicate with each other, otherwise, they will cause a deadlock $\delta$. We define the following communication functions:

$\gamma(r_{C_{AT}}(ENC_{pk_T}(r_A)), s_{C_{AT}}(ENC_{pk_T}(r_A))) \triangleq c_{C_{AT}}(ENC_{pk_T}(r_A))$
$\gamma(r_{C_{TA}}(ENC_{pk_A}(R)), s_{C_{TA}}(ENC_{pk_A}(R))) \triangleq c_{C_{TA}}(ENC_{pk_A}(R))$
$\gamma(r_{C_{AT}}(ENC_{pk_T}(I_A, R_A, v_A)), s_{C_{AT}}(ENC_{pk_T}(I_A, R_A, v_A))) \triangleq c_{C_{AT}}(ENC_{pk_T}(I_A, R_A, v_A))$
$\gamma(r_{C_{BT}}(ENC_{pk_T}(r_B)), s_{C_{BT}}(ENC_{pk_T}(r_B))) \triangleq c_{C_{BT}}(ENC_{pk_T}(r_B))$
$\gamma(r_{C_{TB}}(ENC_{pk_B}(R)), s_{C_{TB}}(ENC_{pk_B}(R))) \triangleq c_{C_{TB}}(ENC_{pk_B}(R))$
$\gamma(r_{C_{BT}}(ENC_{pk_T}(I_B, R_B, v_B)), s_{C_{BT}}(ENC_{pk_T}(I_B, R_B, v_B))) \triangleq c_{C_{BT}}(ENC_{pk_T}(I_B, R_B, v_B))$
$\gamma(r_{C_{CT}}(ENC_{pk_T}(r_C)), s_{C_{CT}}(ENC_{pk_T}(r_C))) \triangleq c_{C_{CT}}(ENC_{pk_T}(r_C))$

$\gamma(r_{C_{TC}}(ENC_{pk_C}(R)), s_{C_{TC}}(ENC_{pk_C}(R))) \triangleq c_{C_{TC}}(ENC_{pk_C}(R))$
$\gamma(r_{C_{CT}}(ENC_{pk_T}(I_C, R_C, v_C)), s_{C_{CT}}(ENC_{pk_T}(I_C, R_C, v_C))) \triangleq c_{C_{CT}}(ENC_{pk_T}(I_C, R_C, v_C))$
$\gamma(r_{C_{DT}}(ENC_{pk_T}(r_D)), s_{C_{DT}}(ENC_{pk_T}(r_D))) \triangleq c_{C_{DT}}(ENC_{pk_T}(r_D))$
$\gamma(r_{C_{TD}}(ENC_{pk_D}(R)), s_{C_{TD}}(ENC_{pk_D}(R))) \triangleq c_{C_{TD}}(ENC_{pk_D}(R))$
$\gamma(r_{C_{DT}}(ENC_{pk_T}(I_D, R_D, v_D)), s_{C_{DT}}(ENC_{pk_T}(I_D, R_D, v_D))) \triangleq c_{C_{DT}}(ENC_{pk_T}(I_D, R_D, v_D))$

Let all modules be in parallel, then the protocol $A \quad B \quad C \quad D \quad T$ can be presented by the following process term:

$$\tau_I(\partial_H(\Theta(A \between B \between C \between D \between T))) = \tau_I(\partial_H(A \between B \between C \between D \between T))$$

where $H = \{r_{C_{AT}}(ENC_{pk_T}(r_A)), s_{C_{AT}}(ENC_{pk_T}(r_A)),$
$r_{C_{TA}}(ENC_{pk_A}(R)), s_{C_{TA}}(ENC_{pk_A}(R)),$
$r_{C_{AT}}(ENC_{pk_T}(I_A, R_A, v_A)), s_{C_{AT}}(ENC_{pk_T}(I_A, R_A, v_A)),$
$r_{C_{BT}}(ENC_{pk_T}(r_B)), s_{C_{BT}}(ENC_{pk_T}(r_B)),$
$r_{C_{TB}}(ENC_{pk_B}(R)), s_{C_{TB}}(ENC_{pk_B}(R)),$
$r_{C_{BT}}(ENC_{pk_T}(I_B, R_B, v_B)), s_{C_{BT}}(ENC_{pk_T}(I_B, R_B, v_B)),$
$r_{C_{CT}}(ENC_{pk_T}(r_C)), s_{C_{CT}}(ENC_{pk_T}(r_C)),$
$r_{C_{TC}}(ENC_{pk_C}(R)), s_{C_{TC}}(ENC_{pk_C}(R)),$
$r_{C_{CT}}(ENC_{pk_T}(I_C, R_C, v_C)), s_{C_{CT}}(ENC_{pk_T}(I_C, R_C, v_C)),$
$r_{C_{DT}}(ENC_{pk_T}(r_D)), s_{C_{DT}}(ENC_{pk_T}(r_D)),$
$r_{C_{TD}}(ENC_{pk_D}(R)), s_{C_{TD}}(ENC_{pk_D}(R)),$
$r_{C_{DT}}(ENC_{pk_T}(I_D, R_D, v_D)), s_{C_{DT}}(ENC_{pk_T}(I_D, R_D, v_D)) | D_A, D_B, D_C, D_D \in \Delta\},$
$I = \{c_{C_{AT}}(ENC_{pk_T}(r_A)), c_{C_{TA}}(ENC_{pk_A}(R)),$
$c_{C_{AT}}(ENC_{pk_T}(I_A, R_A, v_A)), c_{C_{BT}}(ENC_{pk_T}(r_B)),$
$c_{C_{TB}}(ENC_{pk_B}(R)), c_{C_{BT}}(ENC_{pk_T}(I_B, R_B, v_B)),$
$c_{C_{CT}}(ENC_{pk_T}(r_C)), c_{C_{TC}}(ENC_{pk_C}(R)),$
$c_{C_{CT}}(ENC_{pk_T}(I_C, R_C, v_C)), c_{C_{DT}}(ENC_{pk_T}(r_D)),$
$c_{C_{TD}}(ENC_{pk_D}(R)), c_{C_{DT}}(ENC_{pk_T}(I_D, R_D, v_D)),$
$enc_{pk_T}(r_A), enc_{pk_T}(r_B), enc_{pk_T}(r_C), enc_{pk_T}(r_D),$
$dec_{sk_A}(ENC_{pk_A}(R)), dec_{sk_B}(ENC_{pk_B}(R)), dec_{sk_C}(ENC_{pk_C}(R)),$
$dec_{sk_D}(ENC_{pk_D}(R)), rsg_{I_A}, rsg_{I_B}, rsg_{I_C}, rsg_{I_D},$
$enc_{pk_T}(I_A, R_A, v_A), enc_{pk_T}(I_B, R_B, v_B), enc_{pk_T}(I_C, R_C, v_C), enc_{pk_T}(I_D, R_D, v_D),$
$dec_{sk_T}(ENC_{pk_T}(r_A)), dec_{sk_T}(ENC_{pk_T}(r_B)),$
$dec_{sk_T}(ENC_{pk_T}(r_C)), dec_{sk_T}(ENC_{pk_T}(r_D)),$
$rec(A), rec(B), rec(C), rec(D), enc_{pk_A}(R), enc_{pk_B}(R),$
$enc_{pk_C}(R), enc_{pk_D}(R), dec_{sk_T}(ENC_{pk_T}(I_A, R_A, v_A)),$
$dec_{sk_T}(ENC_{pk_T}(I_B, R_B, v_B)), dec_{sk_T}(ENC_{pk_T}(I_C, R_C, v_C)), dec_{sk_T}(ENC_{pk_T}(I_D, R_D, v_D)),$
$\{isExisted(R_A) = TRUE\}, remove(R_A), rec(I_A, v_A), \{isExisted(R_A) = FALSE\},$
$\{isExisted(R_B) = TRUE\}, remove(R_B), rec(I_B, v_B), \{isExisted(R_B) = FALSE\},$
$\{isExisted(R_C) = TRUE\}, remove(R_C), rec(I_C, v_C), \{isExisted(R_C) = FALSE\},$
$\{isExisted(R_D) = TRUE\}, remove(R_D), rec(I_D, v_D), \{isExisted(R_D) = FALSE\}$
$| D_A, D_B, D_C, D_D \in \Delta\}$

Then, we obtain the following conclusion on the protocol.

**Theorem 31.5.** *The secure elections protocol 5 in Fig. 31.5 is improved based on the secure elections protocol 4.*

*Proof.* Based on the above state transitions of the above modules, by use of the algebraic laws of $APTC_G$, we can prove that

$\tau_I(\partial_H(A \between B \between C \between D \between T)) = \sum_{D_A, D_B, D_C, D_D \in \Delta}(r_{C_{AI}}(D_A) \parallel r_{C_{BI}}(D_B) \parallel r_{C_{CI}}(D_C) \parallel r_{C_{DI}}(D_D) \cdot s_{C_{TO}}(TAB)) \cdot \tau_I(\partial_H(A \between B \between C \between D \between T))$

For the details of the proof, please refer to Section 17.3, as we omit them here.

That is, the protocol in Fig. 31.5 $\tau_I(\partial_H(A \between B \between C \between D \between T))$ can exhibit the desired external behaviors, and is secure. However, for the properties of secure elections protocols:

1. Legitimacy: only authorized voters can vote;
2. Oneness: no one can vote more than once;

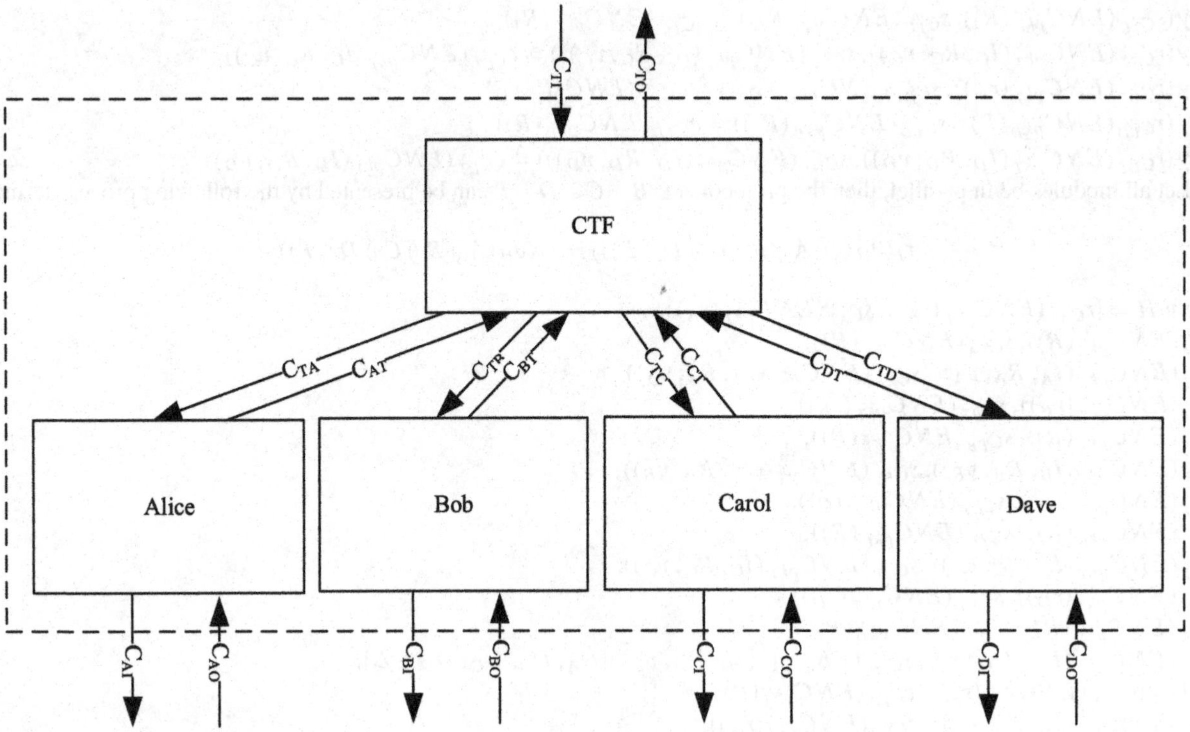


**FIGURE 31.6** Secure elections protocol 6.

3. Privacy: no one can determine for whom anyone else voted;
4. Non-replicability: no one can duplicate anyone else's vote;
5. Non-changeability: no one can change anyone else's vote;
6. Validness: every voter can make sure that his vote has been taken into account in the final tabulation, if CTF is trustworthy.

   The anonymous valid numbers distribution can avoid the distribution of valid numbers to illegal voters. □

## 31.6 Secure elections protocol 6

The secure elections protocol 6 is shown in Fig. 31.6, which is an improved one based on the secure elections protocol 5 in Section 31.5. In this protocol, there are a CTF (Central Tabulating Facility) to check the identity of voters and collect the votes, and four voters: Alice, Bob, Carol, and Dave.

The process of the protocol is as follows:

1. Alice receives some voting request $D_A$ from the outside through the channel $C_{AI}$ (the corresponding reading action is denoted $r_{C_{AI}}(D_A)$), she generates a request $r_A$, encrypts it by CTF's public key through an action $enc_{pk_T}(r_A)$, and sends it to CTF through the channel $C_{AT}$ (the corresponding sending action is denoted $s_{C_{AT}}(ENC_{pk_T}(r_A))$);
2. Bob receives some voting request $D_B$ from the outside through the channel $C_{BI}$ (the corresponding reading action is denoted $r_{C_{BI}}(D_B)$), he generates a request $r_B$, encrypts it by CTF's public key through an action $enc_{pk_T}(r_B)$, and sends it to CTF through the channel $C_{BT}$ (the corresponding sending action is denoted $s_{C_{BT}}(ENC_{pk_T}(r_B))$);
3. Carol receives some voting request $D_C$ from the outside through the channel $C_{CI}$ (the corresponding reading action is denoted $r_{C_{CI}}(D_C)$), she generates a request $r_C$, encrypts it by CTF's public key through an action $enc_{pk_T}(r_C)$, and sends it to CTF through the channel $C_{CT}$ (the corresponding sending action is denoted $s_{C_{CT}}(ENC_{pk_T}(r_C))$);
4. Dave receives some voting request $D_D$ from the outside through the channel $C_{DI}$ (the corresponding reading action is denoted $r_{C_{DI}}(D_D)$), he generates a request $r_D$, encrypts it by CTF's public key through an action $enc_{pk_T}(r_D)$, and sends it to CTF through the channel $C_{DT}$ (the corresponding sending action is denoted $s_{C_{DT}}(ENC_{pk_T}(r_D))$);
5. CTF receives the requests from Alice, Bob, Carol, and Dave through the channels $C_{AT}$, $C_{BT}$, $C_{CT}$, and $C_{DT}$ (the corresponding reading actions are denoted $s_{C_{AT}}(ENC_{pk_T}(r_A))$, $s_{C_{BT}}(ENC_{pk_T}(r_B))$, $s_{C_{CT}}(ENC_{pk_T}(r_C))$, and

$s_{C_{DT}}(ENC_{pk_T}(r_D))$, respectively), he decrypts these encrypted requests through the actions $dec_{sk_T}(ENC_{pk_T}(r_A))$, $dec_{sk_T}(ENC_{pk_T}(r_B))$, $dec_{sk_T}(ENC_{pk_T}(r_C))$, and $dec_{sk_T}(ENC_{pk_T}(r_D))$ to obtain $r_A$, $r_B$, $r_C$, and $r_D$, records the names of Alice, Bob, Carol, and Dave through actions $rec(A)$, $rec(B)$, $rec(C)$, and $rec(D)$; CTF maintains a table of valid numbers; then CTF encrypts all numbers $R$ through actions $enc_{pk_A}(R)$, $enc_{pk_B}(R)$, $enc_{pk_C}(R)$, and $enc_{pk_D}(R)$ and sends them to Alice, Bob, Carol, and Dave through the channels $C_{TA}$, $C_{TB}$, $C_{TC}$, and $C_{TD}$, respectively (the corresponding sending actions are denoted $s_{C_{TA}}(ENC_{pk_A}(R))$, $s_{C_{TB}}(ENC_{pk_B}(R))$, $s_{C_{TC}}(ENC_{pk_C}(R))$, and $s_{C_{TD}}(ENC_{pk_D}(R))$);

6. Alice receives the encrypted number from CTF through the channel $C_{TA}$ (the corresponding reading action is denoted $r_{C_{TA}}(ENC_{pk_A}(R))$), she decrypts the encrypted number through an action $dec_{sk_A}(ENC_{pk_A}(R))$ to randomly select one $R_A$, generates a random identity number $I_A$ through an action $rsg_{I_A}$ and her vote $v_A$, generates a pair of public/private keys through an action $rsg_{pk'_A,sk'_A}$, encrypted $I_A, R_A, v_A$ through an action $enc_{pk'_A}(I_A, R_A, v_A)$, and sends the encrypted message to CTF through the channel $C_{AT}$ (the corresponding sending action is denoted $s_{C_{AT}}(ENC_{pk'_A}(I_A, R_A, v_A))$);
7. Bob receives the encrypted number from CTF through the channel $C_{TB}$ (the corresponding reading action is denoted $r_{C_{TB}}(ENC_{pk_B}(R))$), he decrypts the encrypted number through an action $dec_{sk_B}(ENC_{pk_B}(R))$ to randomly select one $R_B$, generates a random identity number $I_B$ through an action $rsg_{I_B}$ and his vote $v_B$, generates a pair of public/private keys through an action $rsg_{pk'_B,sk'_B}$, encrypted $I_A, R_A, v_A$ through an action $enc_{pk'_B}(I_A, R_A, v_A)$, and sends the encrypted message to CTF through the channel $C_{BT}$ (the corresponding sending action is denoted $s_{C_{BT}}(ENC_{pk'_B}(I_B, R_B, v_B))$);
8. Carol receives the encrypted number from CTF through the channel $C_{TC}$ (the corresponding reading action is denoted $r_{C_{TC}}(ENC_{pk_C}(R))$), she decrypts the encrypted number through an action $dec_{sk_C}(ENC_{pk_C}(R))$ to randomly select one $R_C$, generates a random identity number $I_C$ through an action $rsg_{I_C}$ and her vote $v_C$, generates a pair of public/private keys through an action $rsg_{pk'_C,sk'_C}$, encrypted $I_A, R_A, v_A$ through an action $enc_{pk'_C}(I_A, R_A, v_A)$, and sends the encrypted message to CTF through the channel $C_{CT}$ (the corresponding sending action is denoted $s_{C_{CT}}(ENC_{pk'_C}(I_C, R_C, v_C))$);
9. Dave receives the encrypted number from CTF through the channel $C_{TD}$ (the corresponding reading action is denoted $r_{C_{TD}}(ENC_{pk_A}(R))$), he decrypts the encrypted number through an action $dec_{sk_D}(ENC_{pk_D}(R))$ to randomly select one $R_D$, generates a random identity number $I_D$ through an action $rsg_{I_D}$ and his vote $v_D$, generates a pair of public/private keys through an action $rsg_{pk'_D,sk'_D}$, encrypted $I_A, R_A, v_A$ through an action $enc_{pk'_D}(I_A, R_A, v_A)$, and sends the encrypted message to CTF through the channel $C_{DT}$ (the corresponding sending action is denoted $s_{C_{DT}}(ENC_{pk'_D}(I_D, R_D, v_D))$);
10. CTF receives the encrypted messages from Alice, Bob, Carol, and Dave through the channels $C_{AT}$, $C_{BT}$, $C_{CT}$, and $C_{DT}$ (the corresponding reading actions are denoted $r_{C_{AT}}(ENC_{pk'_A}(I_A, R_A, v_A))$, $r_{C_{BT}}(ENC_{pk'_B}(I_B, R_B, v_B))$, $r_{C_{CT}}(ENC_{pk'_C}(I_C, R_C, v_C))$, and $r_{C_{DT}}(ENC_{pk'_D}(I_D, R_D, v_D))$, respectively), he sends them to the outside through the channel $C_{TO}$ (the corresponding sending actions are denoted $s_{C_{TO}}(ENC_{pk'_A}(I_A, R_A, v_A))$, $s_{C_{TO}}(ENC_{pk'_B}(I_B, R_B, v_B))$, $s_{C_{TO}}(ENC_{pk'_C}(I_C, R_C, v_C))$, and $s_{C_{TO}}(ENC_{pk'_D}(I_D, R_D, v_D))$, respectively); then he sends the request $r_T$ to request the voter to reveal their votes through the channels $C_{TA}$, $C_{TB}$, $C_{TC}$, and $C_{TD}$ (the corresponding sending actions are denoted $s_{C_{TA}}(r_T)$, $s_{C_{TB}}(r_T)$, $s_{C_{TC}}(r_T)$, and $s_{C_{TD}}(r_T)$);
11. Alice receives the request $r_T$ from CTF through the channel $C_{TA}$ (the corresponding reading action is denoted $r_{C_{TA}}(r_T)$), she sends $R_A, I_A.sk'_A$ to CTF through the channel $C_{AT}$ (the corresponding sending action is denoted $s_{C_{AT}}(R_A, I_A, sk'_A)$);
12. Bob receives the request $r_T$ from CTF through the channel $C_{TB}$ (the corresponding reading action is denoted $r_{C_{TB}}(r_T)$), he sends $R_B, I_B.sk'_B$ to CTF through the channel $C_{BT}$ (the corresponding sending action is denoted $s_{C_{BT}}(R_B, I_B, sk'_B)$);
13. Carol receives the request $r_T$ from CTF through the channel $C_{TC}$ (the corresponding reading action is denoted $r_{C_{TC}}(r_T)$), she sends $R_C, I_C.sk'_C$ to CTF through the channel $C_{CT}$ (the corresponding sending action is denoted $s_{C_{CT}}(R_C, I_C, sk'_C)$);
14. Dave receives the request $r_T$ from CTF through the channel $C_{TD}$ (the corresponding reading action is denoted $r_{C_{TD}}(r_T)$), he sends $R_D, I_D.sk'_D$ to CTF through the channel $C_{DT}$ (the corresponding sending action is denoted $s_{C_{DT}}(R_D, I_D, sk'_D)$);
15. CTF receives the message from Alice, Bob, Carol, and Dave through the channels $C_{AT}$, $C_{BT}$, $C_{CT}$, and $C_{DT}$ (the corresponding reading actions are denoted $r_{C_{AT}}(R_A, I_A, sk'_A)$, $r_{C_{BT}}(R_B, I_B, sk'_B)$, $r_{C_{CT}}(R_C, I_C, sk'_C)$, and $r_{C_{DT}}(R_D, I_D, sk'_D)$, respectively), he decrypts the above encrypted messages through actions $dec_{sk'_A}(ENC_{pk'_A}(I_A, R_A, v_A))$, $dec_{sk'_B}(ENC_{pk'_B}(I_B, R_B, v_B))$, $dec_{sk'_C}(ENC_{pk'_C}(I_C, R_C, v_C))$, and

$dec_{sk'_D}(ENC_{pk'_D}(I_D, R_D, v_D))$. If $isExisted(R_A) = TRUE$, he removes $R_A$ from its table through an action $remove(R_A)$, records the vote $v_A$ and the pair of $I_A$ and $v_A$ into the voting results $TAB$ through an action $rec(I_A, v_A)$, else he does nothing; if $isExisted(R_B) = TRUE$, he removes $R_B$ from its table through an action $remove(R_B)$, records the vote $v_B$ and the pair of $I_B$ and $v_B$ into the voting results $TAB$ through an action $rec(I_B, v_B)$, else he does nothing; if $isExisted(R_C) = TRUE$, he removes $R_C$ from its table through an action $remove(R_C)$, records the vote $v_C$ and the pair of $I_C$ and $v_C$ into the voting results $TAB$ through an action $rec(I_C, v_C)$, else he does nothing; if $isExisted(R_D) = TRUE$, he removes $R_D$ from its table through an action $remove(R_D)$, records the vote $v_D$ and the pair of $I_D$ and $v_D$ into the voting results $TAB$ through an action $rec(I_D, v_D)$, else he does nothing. Finally, he sends the voting results $TAB$ to the outside through the channel $C_{TO}$ (the corresponding sending action is denoted $s_{C_{TO}}(TAB)$),

where $D_A, D_B, D_C, D_D \in \Delta$, $\Delta$ is the set of data.

Alice's state transitions described by $APTC_G$ are as follows:

$A = \sum_{D_A \in \Delta} r_{C_{AI}}(D_A) \cdot A_2$
$A_2 = enc_{pk_T}(r_A) \cdot A_3$
$A_3 = s_{C_{AT}}(ENC_{pk_T}(r_A)) \cdot A_4$
$A_4 = r_{C_{TA}}(ENC_{pk_A}(R)) \cdot A_5$
$A_5 = dec_{sk_A}(ENC_{pk_A}(R)) \cdot A_6$
$A_6 = rsg_{I_A} \cdot A_7$
$A_7 = rsg_{pk'_A, sk'_A} \cdot A_8$
$A_8 = enc_{pk'_A}(I_A, R_A, v_A) \cdot A_9$
$A_9 = s_{C_{AT}}(ENC_{pk'_A}(I_A, R_A, v_A)) \cdot A_{10}$
$A_{10} = r_{C_{TA}}(r_T) \cdot A_{11}$
$A_{11} = s_{C_{AT}}(R_A, I_A, sk'_A) \cdot A$

Bob's state transitions described by $APTC_G$ are as follows:

$B = \sum_{D_B \in \Delta} r_{C_{BI}}(D_B) \cdot B_2$
$B_2 = enc_{pk_T}(r_B) \cdot B_3$
$B_3 = s_{C_{BT}}(ENC_{pk_T}(r_B)) \cdot B_4$
$B_4 = r_{C_{TB}}(ENC_{pk_B}(R)) \cdot B_5$
$B_5 = dec_{sk_B}(ENC_{pk_B}(R)) \cdot B_6$
$B_6 = rsg_{I_B} \cdot B_7$
$B_7 = rsg_{pk'_B, sk'_B} \cdot B_8$
$B_8 = enc_{pk'_B}(I_B, R_B, v_B) \cdot B_9$
$B_9 = s_{C_{BT}}(ENC_{pk'_B}(I_B, R_B, v_B)) \cdot B_{10}$
$B_{10} = r_{C_{TB}}(r_T) \cdot B_{11}$
$B_{11} = s_{C_{BT}}(R_B, I_B, sk'_B) \cdot B$

Carol's state transitions described by $APTC_G$ are as follows:

$C = \sum_{D_C \in \Delta} r_{C_{CI}}(D_C) \cdot C_2$
$C_2 = enc_{pk_T}(r_C) \cdot C_3$
$C_3 = s_{C_{CT}}(ENC_{pk_T}(r_C)) \cdot C_4$
$C_4 = r_{C_{TC}}(ENC_{pk_C}(R)) \cdot C_5$
$C_5 = dec_{sk_C}(ENC_{pk_C}(R)) \cdot C_6$
$C_6 = rsg_{I_C} \cdot C_7$
$C_7 = rsg_{pk'_C, sk'_C} \cdot C_8$
$C_8 = enc_{pk'_C}(I_C, R_C, v_C) \cdot C_9$
$C_9 = s_{C_{CT}}(ENC_{pk'_C}(I_C, R_C, v_C)) \cdot C_{10}$
$C_{10} = r_{C_{TC}}(r_T) \cdot C_{11}$
$C_{11} = s_{C_{CT}}(R_C, I_C, sk'_C) \cdot C$

Dave's state transitions described by $APTC_G$ are as follows:

$D = \sum_{D_D \in \Delta} r_{C_{DI}}(D_D) \cdot D_2$
$D_2 = enc_{pk_T}(r_D) \cdot D_3$
$D_3 = s_{C_{DT}}(ENC_{pk_T}(r_D)) \cdot D_4$
$D_4 = r_{C_{TD}}(ENC_{pk_D}(R)) \cdot D_5$
$D_5 = dec_{sk_D}(ENC_{pk_D}(R)) \cdot D_6$
$D_6 = rsg_{I_D} \cdot D_7$

$D_7 = rsg_{pk'_D, sk'_D} \cdot D_8$

$D_8 = enc_{pk'_D}(I_D, R_D, v_D) \cdot D_9$

$D_9 = s_{C_{DT}}(ENC_{pk'_D}(I_D, R_D, v_D)) \cdot D_{10}$

$D_{10} = r_{C_{TD}}(r_T) \cdot D_{11}$

$D_{11} = s_{C_{DT}}(R_D, I_D, sk'_D) \cdot D$

CTF's state transitions described by $APTC_G$ are as follows:

$T = s_{C_{AT}}(ENC_{pk_T}(r_A)) \parallel s_{C_{BT}}(ENC_{pk_T}(r_B))$
$\parallel s_{C_{CT}}(ENC_{pk_T}(r_C)) \parallel s_{C_{DT}}(ENC_{pk_T}(r_D)) \cdot T_2$

$T_2 = dec_{sk_T}(ENC_{pk_T}(r_A)) \parallel dec_{sk_T}(ENC_{pk_T}(r_B))$
$\parallel dec_{sk_T}(ENC_{pk_T}(r_C)) \parallel dec_{sk_T}(ENC_{pk_T}(r_D)) \cdot T_3$

$T_3 = rec(A) \parallel rec(B) \parallel rec(C) \parallel rec(D) \cdot T_4$

$T_4 = enc_{pk_A}(R) \parallel enc_{pk_B}(R) \parallel enc_{pk_C}(R) \parallel enc_{pk_D}(R) \cdot T_5$

$T_5 = s_{C_{TA}}(ENC_{pk_A}(R)) \parallel s_{C_{TB}}(ENC_{pk_B}(R))$
$\parallel s_{C_{TC}}(ENC_{pk_C}(R)) \parallel s_{C_{TD}}(ENC_{pk_D}(R)) \cdot T_6$

$T_6 = r_{C_{AT}}(ENC_{pk'_A}(I_A, R_A, v_A)) \parallel r_{C_{BT}}(ENC_{pk'_B}(I_B, R_B, v_B))$
$\parallel r_{C_{CT}}(ENC_{pk'_C}(I_C, R_C, v_C)) \parallel r_{C_{DT}}(ENC_{pk'_D}(I_D, R_D, v_D)) \cdot T_7$

$T_7 = s_{C_{TO}}(ENC_{pk'_A}(I_A, R_A, v_A)) \parallel s_{C_{TO}}(ENC_{pk'_B}(I_B, R_B, v_B))$
$\parallel s_{C_{TO}}(ENC_{pk'_C}(I_C, R_C, v_C)) \parallel s_{C_{TO}}(ENC_{pk'_D}(I_D, R_D, v_D)) \cdot T_8$

$T_8 = s_{C_{TA}}(r_T) \parallel s_{C_{TB}}(r_T) \parallel s_{C_{TC}}(r_T) \parallel s_{C_{TD}}(r_T) \cdot T_9$

$T_9 = r_{C_{AT}}(R_A, I_A, sk'_A) \parallel r_{C_{BT}}(R_B, I_B, sk'_B)$
$\parallel r_{C_{CT}}(R_C, I_C, sk'_C) \parallel r_{C_{DT}}(R_D, I_D, sk'_D) \cdot T_{10}$

$T_{10} = dec_{sk'_A}(ENC_{pk'_A}(I_A, R_A, v_A)) \parallel dec_{sk'_b}(ENC_{pk'_B}(I_B, R_B, v_B))$
$\parallel dec_{sk'_C}(ENC_{pk'_C}(I_C, R_C, v_C)) \parallel dec_{sk'_D}(ENC_{pk'_D}(I_D, R_D, v_D)) \cdot T_{11}$

$T_{11} = ((\{isExisted(R_A) = TRUE\} \cdot remove(R_A) \cdot rec(I_A, v_A) + \{isExisted(R_A) = FALSE\})$
$\parallel (\{isExisted(R_B) = TRUE\} \cdot remove(R_B) \cdot rec(I_B, v_B) + \{isExisted(R_B) = FALSE\})$
$\parallel (\{isExisted(R_C) = TRUE\} \cdot remove(R_C) \cdot rec(I_C, v_C) + \{isExisted(R_C) = FALSE\})$
$\parallel (\{isExisted(R_D) = TRUE\} \cdot remove(R_D) \cdot rec(I_D, v_D) + \{isExisted(R_D) = FALSE\})) \cdot T_{12}$

$T_{12} = s_{C_{TO}}(TAB) \cdot T$

The sending action and the reading action of the same type data through the same channel can communicate with each other, otherwise, they will cause a deadlock $\delta$. We define the following communication functions:

$\gamma(r_{C_{AT}}(ENC_{pk_T}(r_A)), s_{C_{AT}}(ENC_{pk_T}(r_A))) \triangleq c_{C_{AT}}(ENC_{pk_T}(r_A))$

$\gamma(r_{C_{TA}}(ENC_{pk_A}(R)), s_{C_{TA}}(ENC_{pk_A}(R))) \triangleq c_{C_{TA}}(ENC_{pk_A}(R))$

$\gamma(r_{C_{AT}}(ENC_{pk'_A}(I_A, R_A, v_A)), s_{C_{AT}}(ENC_{pk'_A}(I_A, R_A, v_A))) \triangleq c_{C_{AT}}(ENC_{pk'_A}(I_A, R_A, v_A))$

$\gamma(r_{C_{TA}}(r_T), s_{C_{TA}}(r_T)) \triangleq c_{C_{TA}}(r_T)$

$\gamma(r_{C_{AT}}(R_A, I_A, sk'_A), s_{C_{AT}}(R_A, I_A, sk'_A)) \triangleq c_{C_{AT}}(R_A, I_A, sk'_A)$

$\gamma(r_{C_{BT}}(ENC_{pk_T}(r_B)), s_{C_{BT}}(ENC_{pk_T}(r_B))) \triangleq c_{C_{BT}}(ENC_{pk_T}(r_B))$

$\gamma(r_{C_{TB}}(ENC_{pk_B}(R)), s_{C_{TB}}(ENC_{pk_B}(R))) \triangleq c_{C_{TB}}(ENC_{pk_B}(R))$

$\gamma(r_{C_{BT}}(ENC_{pk'_B}(I_B, R_B, v_B)), s_{C_{BT}}(ENC_{pk'_B}(I_B, R_B, v_B))) \triangleq c_{C_{BT}}(ENC_{pk'_B}(I_B, R_B, v_B))$

$\gamma(r_{C_{TB}}(r_T), s_{C_{TB}}(r_T)) \triangleq c_{C_{TB}}(r_T)$

$\gamma(r_{C_{BT}}(R_B, I_B, sk'_B), s_{C_{BT}}(R_B, I_B, sk'_B)) \triangleq c_{C_{BT}}(R_B, I_B, sk'_B)$

$\gamma(r_{C_{CT}}(ENC_{pk_T}(r_C)), s_{C_{CT}}(ENC_{pk_T}(r_C))) \triangleq c_{C_{CT}}(ENC_{pk_T}(r_C))$

$\gamma(r_{C_{TC}}(ENC_{pk_C}(R)), s_{C_{TC}}(ENC_{pk_C}(R))) \triangleq c_{C_{TC}}(ENC_{pk_C}(R))$

$\gamma(r_{C_{CT}}(ENC_{pk'_C}(I_C, R_C, v_C)), s_{C_{CT}}(ENC_{pk'_C}(I_C, R_C, v_C))) \triangleq c_{C_{CT}}(ENC_{pk'_C}(I_C, R_C, v_C))$

$\gamma(r_{C_{TC}}(r_T), s_{C_{TC}}(r_T)) \triangleq c_{C_{TC}}(r_T)$

$\gamma(r_{C_{CT}}(R_C, I_C, sk'_C), s_{C_{CT}}(R_C, I_C, sk'_C)) \triangleq c_{C_{CT}}(R_C, I_C, sk'_C)$

$\gamma(r_{C_{DT}}(ENC_{pk_T}(r_D)), s_{C_{DT}}(ENC_{pk_T}(r_D))) \triangleq c_{C_{DT}}(ENC_{pk_T}(r_D))$

$\gamma(r_{C_{TD}}(ENC_{pk_D}(R)), s_{C_{TD}}(ENC_{pk_D}(R))) \triangleq c_{C_{TD}}(ENC_{pk_D}(R))$

$\gamma(r_{C_{DT}}(ENC_{pk'_D}(I_D, R_D, v_D)), s_{C_{DT}}(ENC_{pk'_D}(I_D, R_D, v_D))) \triangleq c_{C_{DT}}(ENC_{pk'_D}(I_D, R_D, v_D))$

$\gamma(r_{C_{TD}}(r_T), s_{C_{TD}}(r_T)) \triangleq c_{C_{TD}}(r_T)$

$\gamma(r_{C_{DT}}(R_D, I_D, sk'_D), s_{C_{DT}}(R_D, I_D, sk'_D)) \triangleq c_{C_{DT}}(R_D, I_D, sk'_D)$

Let all modules be in parallel, then the protocol $A \quad B \quad C \quad D \quad T$ can be presented by the following process term:

$$\tau_I(\partial_H(\Theta(A \between B \between C \between D \between T))) = \tau_I(\partial_H(A \between B \between C \between D \between T))$$

where $H = \{r_{C_{AT}}(ENC_{pk_T}(r_A)), s_{C_{AT}}(ENC_{pk_T}(r_A)),$
$r_{C_{TA}}(ENC_{pk_A}(R)), s_{C_{TA}}(ENC_{pk_A}(R)),$
$r_{C_{AT}}(ENC_{pk'_A}(I_A, R_A, v_A)), s_{C_{AT}}(ENC_{pk'_A}(I_A, R_A, v_A)),$
$r_{C_{TA}}(r_T), s_{C_{TA}}(r_T),$
$r_{C_{AT}}(R_A, I_A, sk'_A), s_{C_{AT}}(R_A, I_A, sk'_A),$
$r_{C_{BT}}(ENC_{pk_T}(r_B)), s_{C_{BT}}(ENC_{pk_T}(r_B)),$
$r_{C_{TB}}(ENC_{pk_B}(R)), s_{C_{TB}}(ENC_{pk_B}(R)),$
$r_{C_{BT}}(ENC_{pk'_B}(I_B, R_B, v_B)), s_{C_{BT}}(ENC_{pk'_B}(I_B, R_B, v_B)),$
$r_{C_{TB}}(r_T), s_{C_{TB}}(r_T),$
$r_{C_{BT}}(R_B, I_B, sk'_B), s_{C_{BT}}(R_B, I_B, sk'_B),$
$r_{C_{CT}}(ENC_{pk_T}(r_C)), s_{C_{CT}}(ENC_{pk_T}(r_C)),$
$r_{C_{TC}}(ENC_{pk_C}(R)), s_{C_{TC}}(ENC_{pk_C}(R)),$
$r_{C_{CT}}(ENC_{pk'_C}(I_C, R_C, v_C)), s_{C_{CT}}(ENC_{pk'_C}(I_C, R_C, v_C)),$
$r_{C_{TC}}(r_T), s_{C_{TC}}(r_T),$
$r_{C_{CT}}(R_C, I_C, sk'_C), s_{C_{CT}}(R_C, I_C, sk'_C),$
$r_{C_{DT}}(ENC_{pk_T}(r_D)), s_{C_{DT}}(ENC_{pk_T}(r_D)),$
$r_{C_{TD}}(ENC_{pk_D}(R)), s_{C_{TD}}(ENC_{pk_D}(R)),$
$r_{C_{DT}}(ENC_{pk'_D}(I_D, R_D, v_D)), s_{C_{DT}}(ENC_{pk'_D}(I_D, R_D, v_D)),$
$r_{C_{TD}}(r_T), s_{C_{TD}}(r_T),$
$r_{C_{DT}}(R_D, I_D, sk'_D), s_{C_{DT}}(R_D, I_D, sk'_D)|D_A, D_B, D_C, D_D \in \Delta\},$
$I = \{c_{C_{AT}}(ENC_{pk_T}(r_A)), c_{C_{TA}}(ENC_{pk_A}(R)),$
$c_{C_{AT}}(ENC_{pk'_A}(I_A, R_A, v_A)), c_{C_{BT}}(ENC_{pk_T}(r_B)),$
$c_{C_{TB}}(ENC_{pk_B}(R)), c_{C_{BT}}(ENC_{pk'_B}(I_B, R_B, v_B)),$
$c_{C_{CT}}(ENC_{pk_T}(r_C)), c_{C_{TC}}(ENC_{pk_C}(R)),$
$c_{C_{CT}}(ENC_{pk'_C}(I_C, R_C, v_C)), c_{C_{DT}}(ENC_{pk_T}(r_D)),$
$c_{C_{TD}}(ENC_{pk_D}(R)), c_{C_{DT}}(ENC_{pk'_D}(I_D, R_D, v_D)),$
$c_{C_{TA}}(r_T), c_{C_{AT}}(R_A, I_A, sk'_A), c_{C_{TB}}(r_T), c_{C_{BT}}(R_B, I_B, sk'_B),$
$c_{C_{TC}}(r_T), c_{C_{CT}}(R_C, I_C, sk'_C), c_{C_{TD}}(r_T), c_{C_{DT}}(R_D, I_D, sk'_D),$
$enc_{pk_T}(r_A), enc_{pk_T}(r_B), enc_{pk_T}(r_C), enc_{pk_T}(r_D),$
$dec_{sk_A}(ENC_{pk_A}(R)), dec_{sk_B}(ENC_{pk_B}(R)), dec_{sk_C}(ENC_{pk_C}(R)),$
$dec_{sk_D}(ENC_{pk_D}(R)), rsg_{I_A}, rsg_{I_B}, rsg_{I_C}, rsg_{I_D},$
$rsg_{pk'_A, sk'_A}, rsg_{pk'_B, sk'_B}, rsg_{pk'_C, sk'_C}, rsg_{pk'_D, sk'_D},$
$enc_{pk'_A}(I_A, R_A, v_A), enc_{pk'_B}(I_B, R_B, v_B), enc_{pk'_C}(I_C, R_C, v_C), enc_{pk'_D}(I_D, R_D, v_D),$
$dec_{sk_T}(ENC_{pk_T}(r_A)), dec_{sk_T}(ENC_{pk_T}(r_B)),$
$dec_{sk_T}(ENC_{pk_T}(r_C)), dec_{sk_T}(ENC_{pk_T}(r_D)),$
$rec(A), rec(B), rec(C), rec(D), enc_{pk_A}(R), enc_{pk_B}(R),$
$enc_{pk_C}(R), enc_{pk_D}(R), dec_{sk'_A}(ENC_{pk'_A}(I_A, R_A, v_A)),$
$dec_{sk'_B}(ENC_{pk'_B}(I_B, R_B, v_B)), dec_{sk'_C}(ENC_{pk'_C}(I_C, R_C, v_C)), dec_{sk'_D}(ENC_{pk'_D}(I_D, R_D, v_D)),$
$\{isExisted(R_A) = TRUE\}, remove(R_A), rec(I_A, v_A), \{isExisted(R_A) = FALSE\},$
$\{isExisted(R_B) = TRUE\}, remove(R_B), rec(I_B, v_B), \{isExisted(R_B) = FALSE\},$
$\{isExisted(R_C) = TRUE\}, remove(R_C), rec(I_C, v_C), \{isExisted(R_C) = FALSE\},$
$\{isExisted(R_D) = TRUE\}, remove(R_D), rec(I_D, v_D), \{isExisted(R_D) = FALSE\}$
$|D_A, D_B, D_C, D_D \in \Delta\}$

Then, we obtain the following conclusion on the protocol.

**Theorem 31.6.** *The secure elections protocol 6 in Fig. 31.6 is improved based on the secure elections protocol 5.*

*Proof.* Based on the above state transitions of the above modules, by use of the algebraic laws of $APTC_G$, we can prove that

$\tau_I(\partial_H(A \between B \between C \between D \between T)) = \sum_{D_A, D_B, D_C, D_D \in \Delta}((r_{C_{AI}}(D_A) \parallel r_{C_{BI}}(D_B) \parallel r_{C_{CI}}(D_C) \parallel r_{C_{DI}}(D_D)) \cdot$
$(s_{C_{TO}}(ENC_{pk'_A}(I_A, R_A, v_A)) \parallel s_{C_{TO}}(ENC_{pk'_B}(I_B, R_B, v_B)) \parallel s_{C_{TO}}(ENC_{pk'_C}(I_C, R_C, v_C)) \parallel$
$s_{C_{TO}}(ENC_{pk'_D}(I_D, R_D, v_D))) \cdot s_{C_{TO}}(TAB)) \cdot \tau_I(\partial_H(A \between B \between C \between D \between T))$

For the details of the proof, please refer to Section 17.3, as we omit them here.

That is, the protocol in Fig. 31.6 $\tau_I(\partial_H(A \between B \between C \between D \between T))$ can exhibit the desired external behaviors, and is secure. However, for the properties of secure elections protocols:

1. Legitimacy: only authorized voters can vote;
2. Oneness: no one can vote more than once;
3. Privacy: no one can determine for whom anyone else voted;
4. Non-replicability: no one can duplicate anyone else's vote;
5. Non-changeability: no one can change anyone else's vote;
6. Validness: every voter can make sure that his vote has been taken into account in the final tabulation, if CTF is trustworthy.

Additionally, (1) If a voter observes that his vote is not properly counted, he can protest; (2) A voter can change his vote later. □

Chapter 32

# Introduction to verification of patterns

The software patterns provide building blocks to the design and implementation of a software system, and try to make the software engineering progress from experience to science. The software patterns were made famous because of the introduction of the design patterns [31]. After that, patterns have been researched and developed widely and rapidly.

The series of books of pattern-oriented software architecture [32] [33] [34] [35] [36] should be marked in the development of software patterns. In these books, patterns are detailed in the following aspects:

1. Patterns are categorized from great granularity to tiny granularity. The greatest granularity is called architecture patterns, the medium granularity is called design patterns, and the tiniest granularity is called idioms. In each granularity, patterns are detailed and classified according to their functionalities.
2. Every pattern is detailed according to a regular format to be understood and utilized easily, which includes introduction to a pattern on example, context, problem, solution, structure, dynamics, implementation, example resolved, and variants.
3. Except for the general patterns, patterns of the vertical domains are also involved, including the domains of networked objects and resource management.
4. To make the development and utilization of patterns scientific, the pattern languages are discussed.

As mentioned in these books, formalization of patterns and an intermediate pattern language are needed and should be developed in the future of patterns. Hence, in this part of the book, we formalize software patterns according to the categories of the series of books of pattern-oriented software architecture, and verify the correctness of patterns based on truly concurrent process algebra [9]. In one aspect, patterns are formalized and verified; in the other aspect, truly concurrent process algebra can play a role of an intermediate pattern language for its rigorous theory.

This part of the book is organized as follows:

About the preliminaries of truly concurrent process algebra, including its whole theory, modeling of race condition and asynchronous communication, and applications, please refer to Chapters 2, 9, and 17.

In Chapter 33, we formalize and verify the architectural patterns, including structural patterns, such as the Layers pattern, the Pipes and Filters pattern, and the Blackboard pattern, then patterns considering distribution aspects, patterns that feature human–computer interaction, and patterns supporting extension of applications.

In Chapter 34, we formalize and verify the design patterns, including the patterns related to structural decomposition, the patterns related to organization of work, the patterns related to access control, management oriented patterns, and the communication oriented patterns.

In Chapter 35, we formalize and verify the idioms, including the Singleton pattern and the Counted Pointer pattern.

In Chapter 36, we formalize and verify the patterns for concurrent and networked objects, including service access and configuration patterns, patterns related to event handling, synchronization patterns, and concurrency patterns.

In Chapter 37, we formalize and verify the patterns for resource management, including patterns for resource management, patterns related to resource acquisition, patterns for resource life cycle, and patterns for resource release.

In Chapter 38, we show the formalization and verification of composition of patterns, including the composition of the Layers patterns, the composition of Presentation–Abstraction–Control (PAC) patterns, and the composition of patterns for resource management.

**Handbook of Truly Concurrent Process Algebra. https://doi.org/10.1016/B978-0-44-321515-5.00036-6**

Chapter 33

# Verification of architectural patterns

Architecture patterns are highest-level patterns that present structural organizations for software systems and contain a set of subsystems and the relationships among them.

In this chapter, we verify four categories of architectural patterns, in Section 33.1, we verify structural patterns including the Layers pattern, the Pipes and Filters pattern, and the Blackboard pattern. In Section 33.2, we verify patterns considering distribution aspects. We verify patterns that feature human–computer interaction in Section 33.3. In Section 33.4, we verify patterns supporting extension of applications.

## 33.1 From mud to structure

In this section, we verify structural patterns including the Layers pattern, the Pipes and Filters pattern, and the Blackboard pattern.

### 33.1.1 Verification of the Layers pattern

The Layers pattern contains several layers with each layer being a particular level of abstraction of subtasks. In the Layers pattern, there are only communications between the adjacent layers. That is, for layer $i$, it receives data (the data denoted $d_{U_i}$) from layer $i+1$, processes the data (the processing function denoted $UF_i$), and sends the processed data (the processed data denoted $UF_i(d_{U_i})$) to layer $i-1$; in the other direction, it receives data (the data denoted $d_{L_i}$) from layer $i-1$, processes the data (the processing function denoted $LF_i$), and sends the processed data (the processed data denoted $LF_i(d_{L_i})$) to layer $i+1$, as Fig. 33.1 illustrates. The four channels are denoted $UI_i$ (the Upper Input of layer $i$), $LO_i$ (the Lower Output of layer $i$), $LI_i$ (the Lower Input of layer $i$), and $UO_i$ (the Upper Output of layer $i$), respectively.

The whole Layers pattern containing $n$ layers is illustrated in Fig. 33.2. Note that the numbering of layers is in reverse order, that is, the highest layer is called layer $n$ and the lowest layer is called layer 1.

There exist two typical processes in the Layers pattern corresponding to two directions of data processing as Fig. 33.3 illustrates. One process is as follows:

1. The highest layer $n$ receives data from the application that is denoted $d_{U_n}$ through channel $UI_n$ (the corresponding reading action is denoted $r_{UI_n}(d_{U_n})$), then processes the data, and sends the processed data to layer $n-1$ that is denoted $UF_n(d_{U_n})$ through channel $LO_n$ (the corresponding sending action is denoted $s_{LO_n}(UF_n(d_{U_n}))$);
2. The layer $i$ receives data from the layer $i+1$ that is denoted $d_{U_i}$ through channel $UI_i$ (the corresponding reading action is denoted $r_{UI_i}(d_{U_i})$), then processes the data, and sends the processed data to layer $i-1$ that is denoted $UF_i(d_{U_i})$ through channel $LO_i$ (the corresponding sending action is denoted $s_{LO_i}(UF_i(d_{U_i}))$);

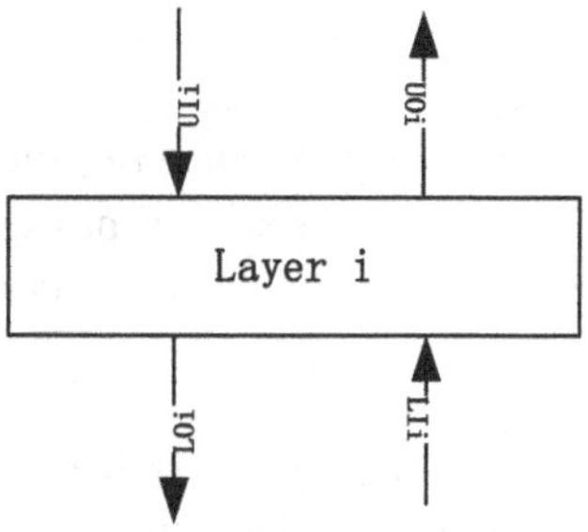


FIGURE 33.1 Layer i.

Handbook of Truly Concurrent Process Algebra. https://doi.org/10.1016/B978-0-44-321515-5.00037-8

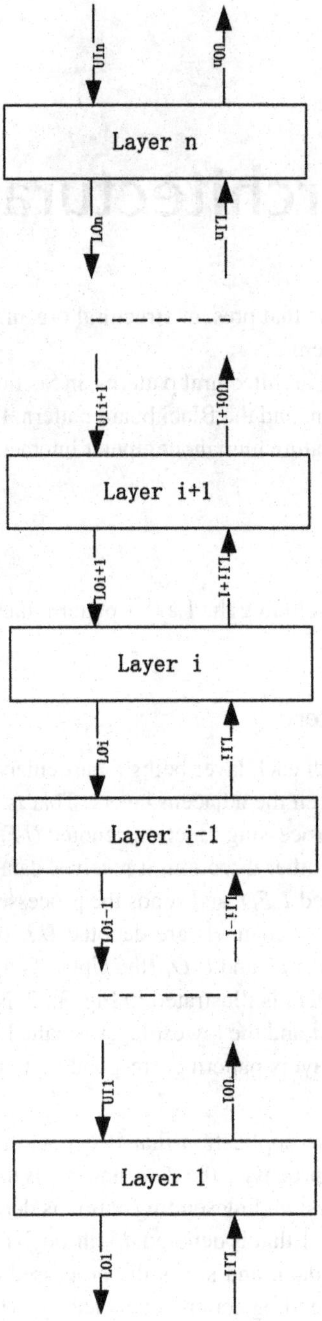


FIGURE 33.2 Layers pattern.

3. The lowest layer 1 receives data from the layer 2 that is denoted $d_{U_1}$ through channel $UI_1$ (the corresponding reading action is denoted $r_{UI_1}(d_{U_1})$), then processes the data, and sends the processed data to another layer's peer that is denoted $UF_1(d_{U_1})$ through channel $LO_1$ (the corresponding sending action is denoted $s_{LO_1}(UF_1(d_{U_1}))$).

The other process is the following:

1. The lowest layer 1 receives data from another layer's peer that is denoted $d_{L_1}$ through channel $LI_1$ (the corresponding reading action is denoted $r_{LI_1}(d_{L_1})$), then processes the data, and sends the processed data to layer 2 that is denoted $LF_1(d_{L_1})$ through channel $UO_1$ (the corresponding sending action is denoted $s_{UO_1}(LF_1(d_{L_1}))$);

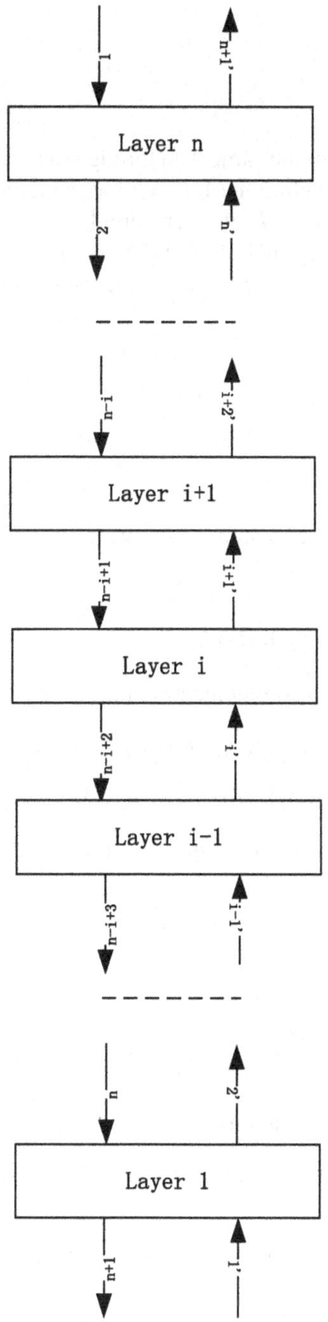


**FIGURE 33.3** Typical process of Layers pattern.

2. The layer $i$ receives data from the layer $i-1$ that is denoted $d_{L_i}$ through channel $LI_i$ (the corresponding reading action is denoted $r_{LI_i}(d_{L_i})$), then processes the data, and sends the processed data to layer $i+1$ that is denoted $LF_i(d_{L_i})$ through channel $UO_i$ (the corresponding sending action is denoted $s_{UO_i}(LF_i(d_{L_i}))$);
3. The highest layer $n$ receives data from layer $n-1$ that is denoted $d_{L_n}$ through channel $LI_n$ (the corresponding reading action is denoted $r_{LI_n}(d_{L_n})$), then processes the data, and sends the processed data to the application that is denoted $LF_n(d_{L_n})$ through channel $UO_n$ (the corresponding sending action is denoted $s_{UO_n}(LF_n(d_{L_n}))$).

We begin to verify the Layers pattern. We assume all data elements $d_{U_i}$ and $d_{L_i}$ (for $1 \leq i \leq n$) are from a finite set $\Delta$. The state transitions of layer $i$ (for $1 \leq i \leq n$) described by APTC are as follows:

$L_i = \sum_{d_{U_i}, d_{L_i} \in \Delta}(r_{UI_i}(d_{U_i}) \cdot L_{i_2} \between r_{LI_i}(d_{L_i}) \cdot L_{i_3})$

$L_{i_2} = UF_i \cdot L_{i_4}$
$L_{i_3} = LF_i \cdot L_{i_5}$
$L_{i_4} = \sum_{d_{U_i} \in \Delta}(s_{LO_i}(UF_i(d_{U_i})) \cdot L_i)$
$L_{i_5} = \sum_{d_{L_i} \in \Delta}(s_{UO_i}(LF_i(d_{L_i})) \cdot L_i)$

The sending action and the reading action of the same data through the same channel can communicate with each other, otherwise, they will cause a deadlock $\delta$. We define the following communication functions for $1 \leq i \leq n$. Note that the channel of $LO_{i+1}$ of layer $i+1$ and the channel $UI_i$ of layer $i$ are the same channel, and the channel $LI_{i+1}$ of layer $i+1$ and the channel $UO_i$ of layer $i$ are the same channel. Also, the data $d_{L_{i+1}}$ of layer $i+1$ and the data $LF_i(d_{L_i})$ of layer $i$ are the same data, and the data $UF_{i+1}(d_{U_{i+1}})$ of layer $i+1$ and the data $d_{U_i}$ of layer $i$ are the same data:

$$\gamma(r_{UI_i}(d_{U_i}), s_{LO_{i+1}}(UF_{i+1}(d_{U_{i+1}}))) \triangleq c_{UI_i}(d_{U_i})$$
$$\gamma(r_{LI_i}(d_{L_i}), s_{UO_{i-1}}(LF_{i-1}(d_{L_{i-1}}))) \triangleq c_{LI_i}(d_{L_i})$$
$$\gamma(r_{UI_{i-1}}(d_{U_{i-1}}), s_{LO_i}(UF_i(d_{U_i}))) \triangleq c_{UI_{i-1}}(d_{U_{i-1}})$$
$$\gamma(r_{LI_{i+1}}(d_{L_{i+1}}), s_{UO_i}(LF_i(d_{L_i}))) \triangleq c_{LI_{i+1}}(d_{L_{i+1}})$$

Note that for the layer $n$, there are only two communication functions as follows:

$$\gamma(r_{LI_n}(d_{L_n}), s_{UO_{n-1}}(LF_{n-1}(d_{L_{n-1}}))) \triangleq c_{LI_n}(d_{L_n})$$
$$\gamma(r_{UI_{n-1}}(d_{U_{n-1}}), s_{LO_n}(UF_n(d_{U_n}))) \triangleq c_{UI_{n-1}}(d_{U_{n-1}})$$

Also, for the layer 1, there are also only two communication functions as follows:

$$\gamma(r_{UI_1}(d_{U_1}), s_{LO_2}(UF_2(d_{U_2}))) \triangleq c_{UI_1}(d_{U_1})$$
$$\gamma(r_{LI_2}(d_{L_2}), s_{UO_1}(LF_1(d_{L_1}))) \triangleq c_{LI_2}(d_{L_2})$$

Let all layers from layer $n$ to layer 1 be in parallel, then the Layers pattern $L_n \cdots L_i \cdots L_1$ can be presented by the following process term:

$$\tau_I(\partial_H(\Theta(L_n \between \cdots \between L_i \between \cdots \between L_1))) = \tau_I(\partial_H(L_n \between \cdots \between L_i \between \cdots \between L_1))$$

where $H = \{r_{UI_1}(d_{U_1}), s_{UO_1}(LF_1(d_{L_1})), \cdots, r_{UI_i}(d_{U_i}), r_{LI_i}(d_{L_i}), s_{LO_i}(UF_i(d_{U_i})), s_{UO_i}(LF_i(d_{L_i})), \cdots, r_{LI_n}(d_{L_n}), s_{LO_n}(UF_n(d_{U_n})) | d_{U_1}, d_{L_1}, \cdots, d_{U_i}, d_{L_i} \cdots, d_{U_n}, d_{L_n} \in \Delta\}$,

$I = \{c_{UI_1}(d_{U_1}), c_{LI_2}(d_{L_2}), \cdots, c_{UI_i}(d_{U_i}), c_{LI_i}(d_{L_i}), c_{UI_{i-1}}(d_{U_{i-1}}), c_{LI_{i+1}}(d_{L_{i+1}}), \cdots, c_{LI_n}(d_{L_n}), c_{UI_{n-1}}(d_{U_{n-1}}), LF_1, UF_1, \cdots, LF_i, UF_i, \cdots, LF_n, UF_n | d_{U_1}, d_{L_1}, \cdots, d_{U_i}, d_{L_i} \cdots, d_{U_n}, d_{L_n} \in \Delta\}$

Then, we obtain the following conclusion on the Layers pattern.

**Theorem 33.1** (Correctness of the Layers pattern). *The Layers pattern $\tau_I(\partial_H(L_n \between \cdots \between L_i \between \cdots \between L_1))$ can exhibit the desired external behaviors.*

*Proof.* Based on the above state transitions of layer $i$ (for $1 \leq i \leq n$), by use of the algebraic laws of APTC, we can prove that

$\tau_I(\partial_H(L_n \between \cdots \between L_i \between \cdots \between L_1)) = \sum_{d_{U_1}, d_{L_1}, d_{U_n}, d_{L_n} \in \Delta}((r_{UI_n}(d_{U_n}) \parallel r_{LI_1}(d_{L_1})) \cdot (s_{UO_n}(LF_n(d_{L_n})) \parallel s_{LO_1}(UF_1(d_{U_1})))) \cdot \tau_I(\partial_H(L_n \between \cdots \between L_i \between \cdots \between L_1))$

That is, the Layers pattern $\tau_I(\partial_H(L_n \between \cdots \between L_i \between \cdots \between L_1))$ can exhibit the desired external behaviors.

For the details of the proof, please refer to Section 17.3, as we omit them here. □

Two Layers pattern peers can be composed together, just by linking the lower output of layer 1 of one peer together with the lower input of layer 1 of the other peer, and vice versa. As Fig. 33.4 illustrated.

There are also two typical data processing processes in the composition of two layers peers, as Fig. 33.5 shows. One process is data transferred from peer $P$ to another peer $P'$ as follows:

1. The highest layer $n$ of peer $P$ receives data from the application of peer $P$ that is denoted $d_{U_n}$ through channel $UI_n$ (the corresponding reading action is denoted $r_{UI_n}(d_{U_n})$), then processes the data, and sends the processed data to layer $n-1$ of peer $P$ that is denoted $UF_n(d_{U_n})$ through channel $LO_n$ (the corresponding sending action is denoted $s_{LO_n}(UF_n(d_{U_n}))$);

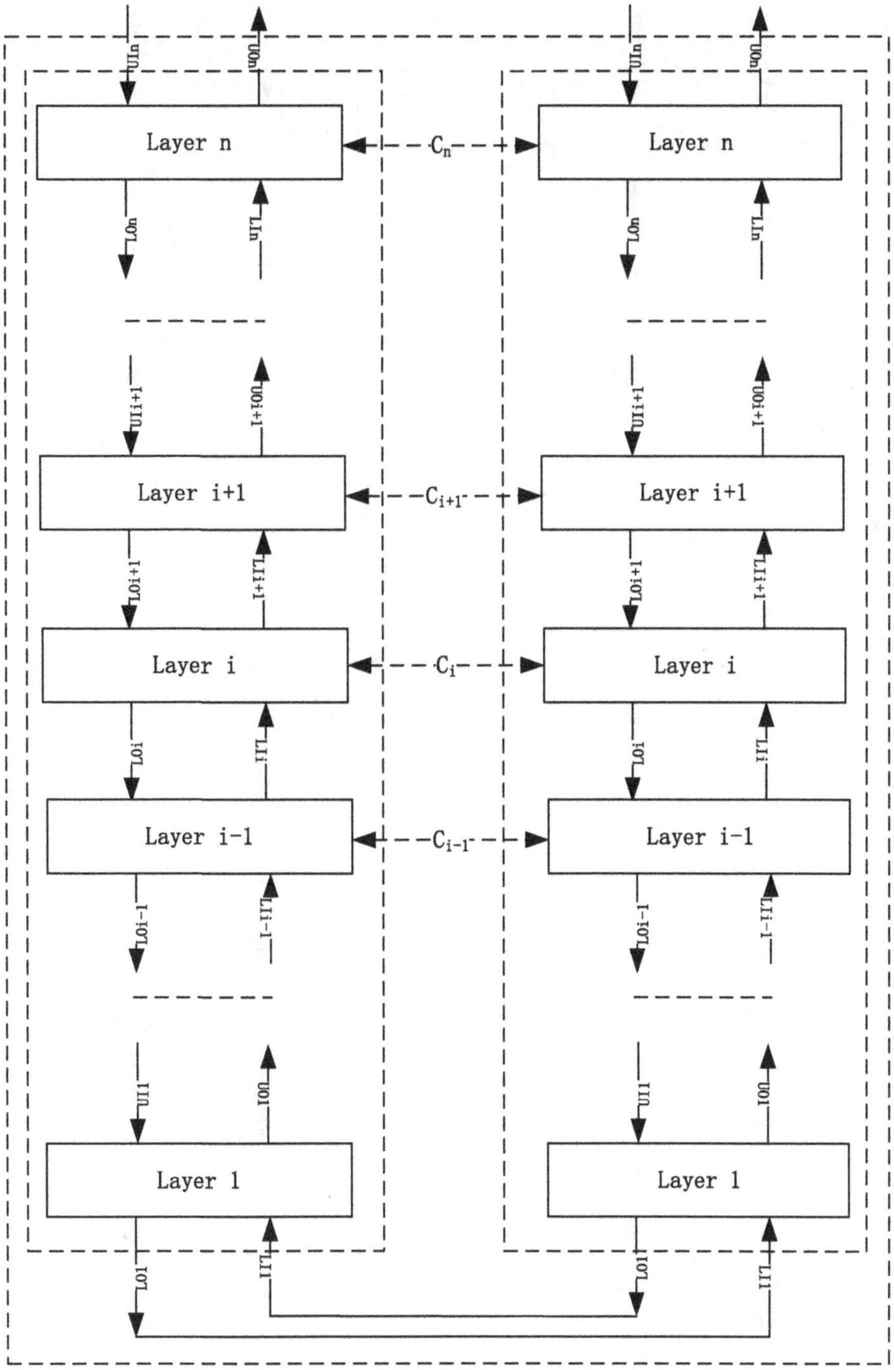


**FIGURE 33.4** Two layers peers.

2. The layer $i$ of peer $P$ receives data from the layer $i+1$ of peer $P$ that is denoted $d_{U_i}$ through channel $UI_i$ (the corresponding reading action is denoted $r_{UI_i}(d_{U_i})$), then processes the data, and sends the processed data to layer $i-1$ that is denoted $UF_i(d_{U_i})$ through channel $LO_i$ (the corresponding sending action is denoted $s_{LO_i}(UF_i(d_{U_i}))$);
3. The lowest layer 1 of peer $P$ receives data from the layer 2 of peer $P$ that is denoted $d_{U_1}$ through channel $UI_1$ (the corresponding reading action is denoted $r_{UI_1}(d_{U_1})$), then processes the data, and sends the processed data to another layer's peer $P'$ that is denoted $UF_1(d_{U_1})$ through channel $LO_1$ (the corresponding sending action is denoted $s_{LO_1}(UF_1(d_{U_1}))$);

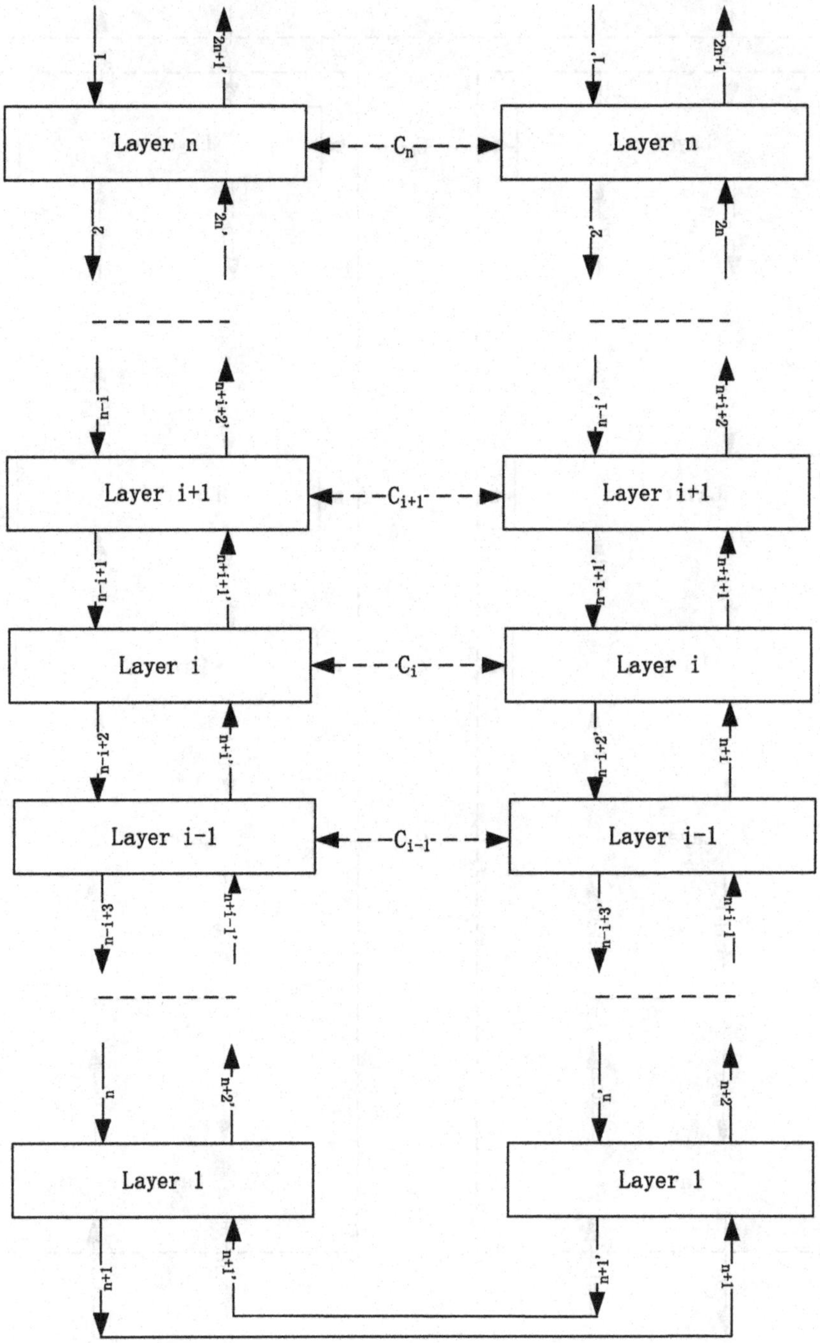


**FIGURE 33.5** Typical process 1 of two layers peers.

4. The lowest layer $1'$ of $P'$ receives data from the another layers peer $P$ that is denoted $d_{L_{1'}}$ through channel $LI_{1'}$ (the corresponding reading action is denoted $r_{LI_{1'}}(d_{L_{1'}})$), then processes the data, and sends the processed data to layer 2 of $P'$ that is denoted $LF_{1'}(d_{L_{1'}})$ through channel $UO_{1'}$ (the corresponding sending action is denoted $s_{UO_{1'}}(LF_{1'}(d_{L_{1'}}))$);
5. The layer $i'$ of peer $P'$ receives data from the layer $i' - 1$ of peer $P'$ that is denoted $d_{L_{i'}}$ through channel $LI_{i'}$ (the corresponding reading action is denoted $r_{LI_{i'}}(d_{L_{i'}})$), then processes the data, and sends the processed data to layer $i' + 1$ of peer $P'$ that is denoted $LF_{i'}(d_{L_{i'}})$ through channel $UO_{i'}$ (the corresponding sending action is denoted $s_{UO_{i'}}(LF_{i'}(d_{L_{i'}}))$);

6. The highest layer $n'$ of peer $P'$ receives data from layer $n'-1$ of peer $P'$ that is denoted $d_{L_{n'}}$ through channel $LI_{n'}$ (the corresponding reading action is denoted $r_{LI_{n'}}(d_{L_{n'}})$), then processes the data, and sends the processed data to the application of peer $P'$ that is denoted $LF_{n'}(d_{L_{n'}})$ through channel $UO_{n'}$ (the corresponding sending action is denoted $s_{UO_{n'}}(LF_{n'}(d_{L_{n'}}))$).

The other similar process is data transferred from peer $P'$ to peer $P$, hence, we do not repeat this here.

The verification of two layers peers is as follows:

We also assume all data elements $d_{U_i}$, $d_{L_i}$, $d_{U_{i'}}$, and $d_{L_{i'}}$ (for $1 \leq i, i' \leq n$) are from a finite set $\Delta$. The state transitions of layer $i$ (for $1 \leq i \leq n$) described by APTC are as follows:

$L_i = \sum_{d_{U_i}, d_{L_i} \in \Delta}(r_{UI_i}(d_{U_i}) \cdot L_{i_2} \between r_{LI_i}(d_{L_i}) \cdot L_{i_3})$

$L_{i_2} = UF_i \cdot L_{i_4}$

$L_{i_3} = LF_i \cdot L_{i_5}$

$L_{i_4} = \sum_{d_{U_i} \in \Delta}(s_{LO_i}(UF_i(d_{U_i})) \cdot L_i)$

$L_{i_5} = \sum_{d_{L_i} \in \Delta}(s_{UO_i}(LF_i(d_{L_i})) \cdot L_i)$

The state transitions of layer $i'$ (for $1 \leq i' \leq n$) described by APTC are as follows:

$L_{i'} = \sum_{d_{U_{i'}}, d_{L_{i'}} \in \Delta}(r_{UI_{i'}}(d_{U_{i'}}) \cdot L_{i'_2} \between r_{LI_{i'}}(d_{L_{i'}}) \cdot L_{i'_3})$

$L_{i'_2} = UF_{i'} \cdot L_{i'_4}$

$L_{i'_3} = LF_{i'} \cdot L_{i'_5}$

$L_{i'_4} = \sum_{d_{U_{i'}} \in \Delta}(s_{LO_{i'}}(UF_{i'}(d_{U_{i'}})) \cdot L_{i'})$

$L_{i'_5} = \sum_{d_{L_{i'}} \in \Delta}(s_{UO_{i'}}(LF_{i'}(d_{L_{i'}})) \cdot L_{i'})$

The sending action and the reading action of the same data through the same channel can communicate with each other, otherwise, they will cause a deadlock $\delta$. We define the following communication functions for $1 \leq i \leq n$ and $1 \leq i' \leq n$. Note that the channel of $LO_{i+1}$ of layer $i+1$ and the channel $UI_i$ of layer $i$ are the same channel, and the channel $LI_{i+1}$ of layer $i+1$ and the channel $UO_i$ of layer $i$ are the same channel; the channel of $LO_{i'+1}$ of layer $i'+1$ and the channel $UI_{i'}$ of layer $i'$ are the same channel, and the channel $LI_{i'+1}$ of layer $i'+1$ and the channel $UO_{i'}$ of layer $i'$ are the same channel. Also, the data $d_{L_{i+1}}$ of layer $i+1$ and the data $LF_i(d_{L_i})$ of layer $i$ are the same data, and the data $UF_{i+1}(d_{U_{i+1}})$ of layer $i+1$ and the data $d_{U_i}$ of layer $i$ are the same data; the data $d_{L_{i'+1}}$ of layer $i'+1$ and the data $LF_{i'}(d_{L_{i'}})$ of layer $i'$ are the same data, and the data $UF_{i'+1}(d_{U_{i'+1}})$ of layer $i'+1$ and the data $d_{U_{i'}}$ of layer $i'$ are the same data:

$$\gamma(r_{UI_i}(d_{U_i}), s_{LO_{i+1}}(UF_{i+1}(d_{U_{i+1}}))) \triangleq c_{UI_i}(d_{U_i})$$

$$\gamma(r_{LI_i}(d_{L_i}), s_{UO_{i-1}}(LF_{i-1}(d_{L_{i-1}}))) \triangleq c_{LI_i}(d_{L_i})$$

$$\gamma(r_{UI_{i-1}}(d_{U_{i-1}}), s_{LO_i}(UF_i(d_{U_i}))) \triangleq c_{UI_{i-1}}(d_{U_{i-1}})$$

$$\gamma(r_{LI_{i+1}}(d_{L_{i+1}}), s_{UO_i}(LF_i(d_{L_i}))) \triangleq c_{LI_{i+1}}(d_{L_{i+1}})$$

$$\gamma(r_{UI_{i'}}(d_{U_{i'}}), s_{LO_{i'+1}}(UF_{i'+1}(d_{U_{i'+1}}))) \triangleq c_{UI_{i'}}(d_{U_{i'}})$$

$$\gamma(r_{LI_{i'}}(d_{L_{i'}}), s_{UO_{i'-1}}(LF_{i'-1}(d_{L_{i'-1}}))) \triangleq c_{LI_{i'}}(d_{L_{i'}})$$

$$\gamma(r_{UI_{i'-1}}(d_{U_{i'-1}}), s_{LO_{i'}}(UF_{i'}(d_{U_{i'}}))) \triangleq c_{UI_{i'-1}}(d_{U_{i'-1}})$$

$$\gamma(r_{LI_{i'+1}}(d_{L_{i'+1}}), s_{UO_{i'}}(LF_i(d_{L_{i'}}))) \triangleq c_{LI_{i'+1}}(d_{L_{i'+1}})$$

Note that for the layer $n$, there are only two communication functions as follows:

$$\gamma(r_{LI_n}(d_{L_n}), s_{UO_{n-1}}(LF_{n-1}(d_{L_{n-1}}))) \triangleq c_{LI_n}(d_{L_n})$$

$$\gamma(r_{UI_{n-1}}(d_{U_{n-1}}), s_{LO_n}(UF_n(d_{U_n}))) \triangleq c_{UI_{n-1}}(d_{U_{n-1}})$$

For the layer $n'$, there are only two communication functions as follows:

$$\gamma(r_{LI_{n'}}(d_{L_{n'}}), s_{UO_{n'-1}}(LF_{n'-1}(d_{L_{n'-1}}))) \triangleq c_{LI_{n'}}(d_{L_{n'}})$$

$$\gamma(r_{UI_{n'-1}}(d_{U_{n'-1}}), s_{LO_{n'}}(UF_{n'}(d_{U_{n'}}))) \triangleq c_{UI_{n'-1}}(d_{U_{n'-1}})$$

For the layer 1, there are four communication functions as follows:

$$\gamma(r_{UI_1}(d_{U_1}), s_{LO_2}(UF_2(d_{U_2}))) \triangleq c_{UI_1}(d_{U_1})$$

$$\gamma(r_{LI_2}(d_{L_2}), s_{UO_1}(LF_1(d_{L_1}))) \triangleq c_{LI_2}(d_{L_2})$$
$$\gamma(r_{LI_1}(d_{L_1}), s_{LO_{1'}}(UF_{1'}(d_{U_{1'}}))) \triangleq c_{LI_1}(d_{L_1})$$
$$\gamma(r_{LI_{1'}}(d_{L_{1'}}), s_{LO_1}(UF_1(d_{U_1}))) \triangleq c_{LI_{1'}}(d_{L_{1'}})$$

Also, for the layer $1'$, there are four communication functions as follows:

$$\gamma(r_{UI_{1'}}(d_{U_{1'}}), s_{LO_{2'}}(UF_{2'}(d_{U_{2'}}))) \triangleq c_{UI_{1'}}(d_{U_{1'}})$$
$$\gamma(r_{LI_{2'}}(d_{L_{2'}}), s_{UO_{1'}}(LF_{1'}(d_{L_{1'}}))) \triangleq c_{LI_{2'}}(d_{L_{2'}})$$
$$\gamma(r_{LI_1}(d_{L_1}), s_{LO_{1'}}(UF_{1'}(d_{U_{1'}}))) \triangleq c_{LI_1}(d_{L_1})$$
$$\gamma(r_{LI_{1'}}(d_{L_{1'}}), s_{LO_1}(UF_1(d_{U_1}))) \triangleq c_{LI_{1'}}(d_{L_{1'}})$$

Let all layers from layer $n$ to layer 1 and from layer $1'$ to $n'$ be in parallel, then the Layers pattern $L_n \cdots L_i \cdots L_1 L_{1'} \cdots L_{i'} \cdots L_{n'}$ can be presented by the following process term:

$\tau_I(\partial_H(\Theta(L_n \between \cdots \between L_i \between \cdots \between L_1 \between L_{1'} \between \cdots \between L_{i'} \between \cdots \between L_{n'}))) = \tau_I(\partial_H(L_n \between \cdots \between L_i \between \cdots \between L_1 \between L_{1'} \between \cdots \between L_{i'} \between \cdots \between L_{n'}))$

where $H = \{r_{LI_1}(d_{L_1}), s_{LO_1}(UF_1(d_{U_1})), r_{UI_1}(d_{U_1}), s_{UO_1}(LF_1(d_{L_1})), \cdots, r_{UI_i}(d_{U_i}), r_{LI_i}(d_{L_i}), s_{LO_i}(UF_i(d_{U_i})), s_{UO_i}(LF_i(d_{L_i})), \cdots, r_{LI_n}(d_{L_n}), s_{LO_n}(UF_n(d_{U_n})), r_{LI_{1'}}(d_{L_{1'}}), s_{LO_{1'}}(UF_1(d_{U_{1'}})), r_{UI_{1'}}(d_{U_{1'}}), s_{UO_{1'}}(LF_{1'}(d_{L_{1'}})), \cdots, r_{UI_{i'}}(d_{U_{i'}}), r_{LI_{i'}}(d_{L_{i'}}), s_{LO_{i'}}(UF_{i'}(d_{U_{i'}})), s_{UO_{i'}}(LF_{i'}(d_{L_{i'}})), \cdots, r_{LI_{n'}}(d_{L_{n'}}), s_{LO_{n'}}(UF_{n'}(d_{U_{n'}})) |d_{U_1}, d_{L_1}, \cdots, d_{U_i}, d_{L_i} \cdots, d_{U_n}, d_{L_n}, d_{U_{1'}}, d_{L_{1'}}, \cdots, d_{U_{i'}}, d_{L_{i'}} \cdots, d_{U_{n'}}, d_{L_{n'}} \in \Delta\}$,

$I = \{c_{UI_1}(d_{U_1}), c_{LI_1}(d_{L_1}), c_{LI_2}(d_{L_2}), \cdots, c_{UI_i}(d_{U_i}), c_{LI_i}(d_{L_i}), c_{UI_{i-1}}(d_{U_{i-1}}), c_{LI_{i+1}}(d_{L_{i+1}}), \cdots, c_{LI_n}(d_{L_n}), c_{UI_{n-1}}(d_{U_{n-1}}), LF_1, UF_1, \cdots, LF_i, UF_i, \cdots, LF_n, UF_n, c_{UI_{1'}}(d_{U_{1'}}), c_{LI_{1'}}(d_{L_{1'}}), c_{LI_{2'}}(d_{L_{2'}}), \cdots, c_{UI_{i'}}(d_{U_{i'}}), c_{LI_{i'}}(d_{L_{i'}}), c_{UI_{i'-1}}(d_{U_{i'-1}}), c_{LI_{i'+1}}(d_{L_{i'+1}}), \cdots, c_{LI_{n'}}(d_{L_{n'}}), c_{UI_{n'-1}}(d_{U_{n'-1}}), LF_{1'}, UF_{1'}, \cdots, LF_{i'}, UF_{i'}, \cdots, LF_{n'}, UF_{n'} |d_{U_1}, d_{L_1}, \cdots, d_{U_i}, d_{L_i} \cdots, d_{U_n}, d_{L_n}, d_{U_{1'}}, d_{L_{1'}}, \cdots, d_{U_{i'}}, d_{L_{i'}} \cdots, d_{U_{n'}}, d_{L_{n'}} \in \Delta\}$

Then, we obtain the following conclusion on the Layers pattern.

**Theorem 33.2** (Correctness of two layers peers). *The two layers peers $\tau_I(\partial_H(L_n \between \cdots \between L_i \between \cdots \between L_1 \between L_{1'} \between \cdots \between L_{i'} \between \cdots \between L_{n'}))$ can exhibit the desired external behaviors.*

*Proof.* Based on the above state transitions of layer $i$ and $i'$ (for $1 \leq i, i' \leq n$), by use of the algebraic laws of APTC, we can prove that

$\tau_I(\partial_H(L_n \between \cdots \between L_i \between \cdots \between L_1 \between L_{1'} \between \cdots \between L_{i'} \between \cdots \between L_{n'})) = \sum_{d_{U_n}, d_{L_n}, d_{U_{n'}}, d_{L_{n'}} \in \Delta}((r_{UI_n}(d_{U_n}) \parallel r_{UI_{n'}}(d_{U_{n'}})) \cdot (s_{UO_n}(LF_n(d_{L_n})) \parallel s_{UO_{n'}}(LF_{n'}(d_{L_{n'}})))) \cdot \tau_I(\partial_H(L_n \between \cdots \between L_i \between \cdots \between L_1 \between L_{1'} \between \cdots \between L_{i'} \between \cdots \between L_{n'}))$

That is, the two layers peers $\tau_I(\partial_H(L_n \between \cdots \between L_i \between \cdots \between L_1 \between L_{1'} \between \cdots \between L_{i'} \between \cdots \between L_{n'}))$ can exhibit the desired external behaviors.

For the details of the proof, please refer to Section 17.3, as we omit them here. □

There exists another composition of two layers peers. There are communications between two peer layers that are called virtual communication. Virtual communications are specified by communication protocols, and we assume data transferred between layer $i$ through virtual communications. Also, the two typical processes are illustrated in Fig. 33.6. The process from peer $P$ to peer $P'$ is as follows:

1. The highest layer $n$ of peer $P$ receives data from the application of peer $P$ that is denoted $d_{U_n}$ through channel $UI_n$ (the corresponding reading action is denoted $r_{UI_n}(d_{U_n})$), then processes the data, and sends the processed data to layer $n-1$ of peer $P$ that is denoted $UF_n(d_{U_n})$ through channel $LO_n$ (the corresponding sending action is denoted $s_{LO_n}(UF_n(d_{U_n}))$);
2. The layer $i$ of peer $P$ receives data from the layer $i+1$ of peer $P$ that is denoted $d_{U_i}$ through channel $UI_i$ (the corresponding reading action is denoted $r_{UI_i}(d_{U_i})$), then processes the data, and sends the processed data to layer $i$ of peer $P'$ that is denoted $UF_i(d_{U_i})$ through channel $LO_i$ (the corresponding sending action is denoted $s_{LO_i}(UF_i(d_{U_i}))$);
3. The layer $i'$ of peer $P'$ receives data from the layer $i$ of peer $P$ that is denoted $d_{L_{i'}}$ through channel $LI_{i'}$ (the corresponding reading action is denoted $r_{LI_{i'}}(d_{L_{i'}})$), then processes the data, and sends the processed data to layer $i'+1$ of peer $P'$ that is denoted $LF_{i'}(d_{L_{i'}})$ through channel $UO_{i'}$ (the corresponding sending action is denoted $s_{UO_{i'}}(LF_{i'}(d_{L_{i'}}))$);

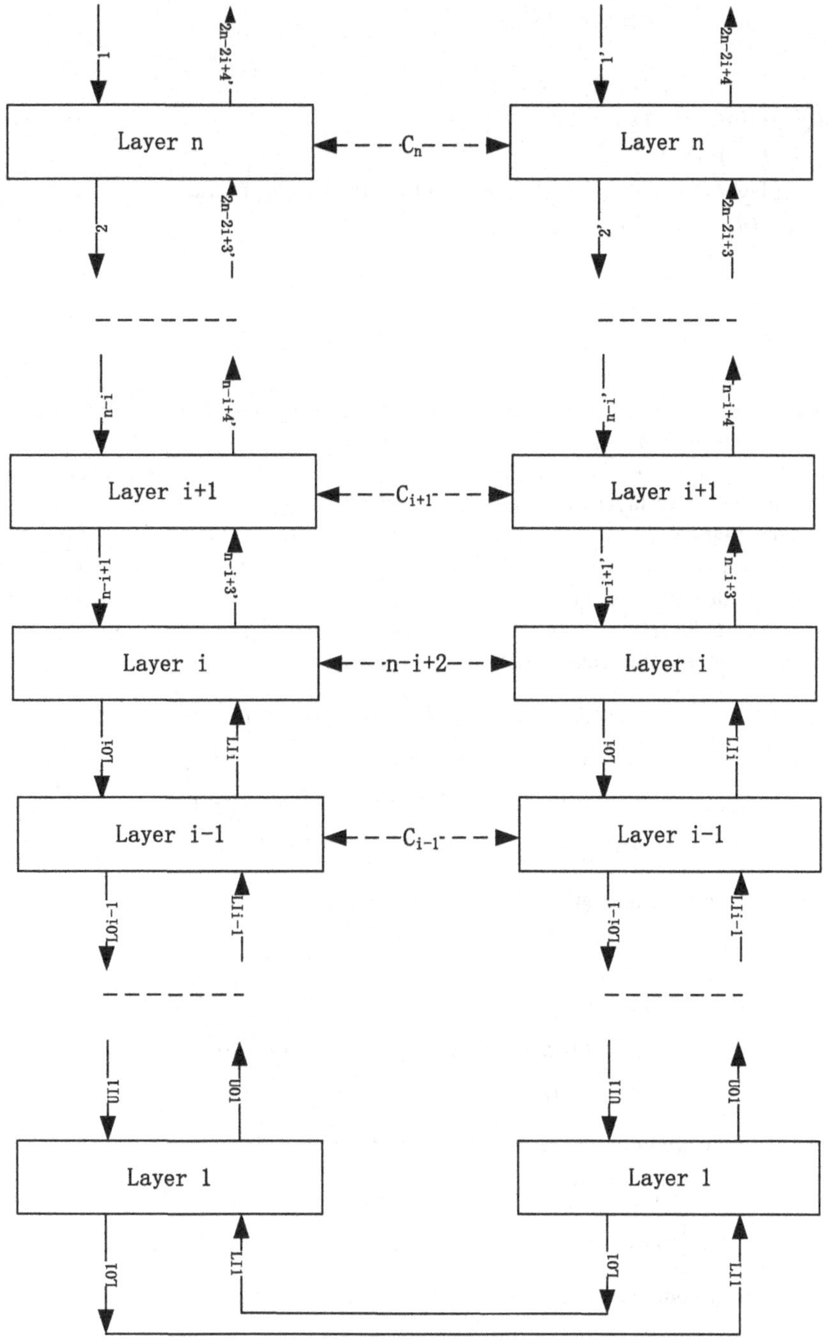


FIGURE 33.6 Typical process 2 of two layers peers.

4. The highest layer $n'$ of peer $P'$ receives data from layer $n'-1$ of peer $P'$ that is denoted $d_{L_{n'}}$ through channel $LI_{n'}$ (the corresponding reading action is denoted $r_{LI_{n'}}(d_{L_{n'}})$), then processes the data, and sends the processed data to the application of peer $P'$ that is denoted $LF_{n'}(d_{L_{n'}})$ through channel $UO_{n'}$ (the corresponding sending action is denoted $s_{UO_{n'}}(LF_{n'}(d_{L_{n'}}))$).

The other similar process is data transferred from $P'$ to $P$, hence, we do not repeat this here.

The verification of two layers peers' communication through virtual communication is as follows:

We also assume all data elements $d_{U_i}$, $d_{L_i}$, $d_{U_{i'}}$, and $d_{L_{i'}}$ (for $1 \leq i, i' \leq n$) are from a finite set $\Delta$. The state transitions of layer $i$ (for $1 \leq i \leq n$) described by APTC are as follows:

$L_i = \sum_{d_{U_i}, d_{L_i} \in \Delta}(r_{UI_i}(d_{U_i}) \cdot L_{i_2} \between r_{LI_i}(d_{L_i}) \cdot L_{i_3})$
$L_{i_2} = UF_i \cdot L_{i_4}$
$L_{i_3} = LF_i \cdot L_{i_5}$
$L_{i_4} = \sum_{d_{U_i} \in \Delta}(s_{LO_i}(UF_i(d_{U_i})) \cdot L_i)$
$L_{i_5} = \sum_{d_{L_i} \in \Delta}(s_{UO_i}(LF_i(d_{L_i})) \cdot L_i)$
The state transitions of layer $i'$ (for $1 \leq i' \leq n$) described by APTC are as follows:
$L_{i'} = \sum_{d_{U_{i'}}, d_{L_{i'}} \in \Delta}(r_{UI_{i'}}(d_{U_{i'}}) \cdot L_{i'_2} \between r_{LI_{i'}}(d_{L_{i'}}) \cdot L_{i'_3})$
$L_{i'_2} = UF_{i'} \cdot L_{i'_4}$
$L_{i'_3} = LF_{i'} \cdot L_{i'_5}$
$L_{i'_4} = \sum_{d_{U_{i'}} \in \Delta}(s_{LO_{i'}}(UF_{i'}(d_{U_{i'}})) \cdot L_{i'})$
$L_{i'_5} = \sum_{d_{L_{i'}} \in \Delta}(s_{UO_{i'}}(LF_{i'}(d_{L_{i'}})) \cdot L_{i'})$

The sending action and the reading action of the same data through the same channel can communicate with each other, otherwise, they will cause a deadlock $\delta$. We define the following communication functions for $1 \leq i \leq n$ and $1 \leq i' \leq n$. Note that the channel of $LO_{i+1}$ of layer $i+1$ and the channel $UI_i$ of layer $i$ are the same channel, and the channel $LI_{i+1}$ of layer $i+1$ and the channel $UO_i$ of layer $i$ are the same channel; the channel of $LO_{i'+1}$ of layer $i'+1$ and the channel $UI_{i'}$ of layer $i'$ are the same channel, and the channel $LI_{i'+1}$ of layer $i'+1$ and the channel $UO_{i'}$ of layer $i'$ are the same channel. Also, the data $d_{L_{i+1}}$ of layer $i+1$ and the data $LF_i(d_{L_i})$ of layer $i$ are the same data, and the data $UF_{i+1}(d_{U_{i+1}})$ of layer $i+1$ and the data $d_{U_i}$ of layer $i$ are the same data; the data $d_{L_{i'+1}}$ of layer $i'+1$ and the data $LF_{i'}(d_{L_{i'}})$ of layer $i'$ are the same data, and the data $UF_{i'+1}(d_{U_{i'+1}})$ of layer $i'+1$ and the data $d_{U_{i'}}$ of layer $i'$ are the same data:

For the layer $i$, there are four communication functions as follows:

$$\gamma(r_{UI_i}(d_{U_i}), s_{LO_{i+1}}(UF_{i+1}(d_{U_{i+1}}))) \triangleq c_{UI_i}(d_{U_i})$$
$$\gamma(r_{LI_{i+1}}(d_{L_{i+1}}), s_{UO_i}(LF_i(d_{L_i}))) \triangleq c_{LI_{i+1}}(d_{L_{i+1}})$$
$$\gamma(r_{LI_i}(d_{L_i}), s_{LO_{i'}}(UF_{i'}(d_{U_{i'}}))) \triangleq c_{LI_i}(d_{L_i})$$
$$\gamma(r_{LI_{i'}}(d_{L_{i'}}), s_{LO_i}(UF_i(d_{U_i}))) \triangleq c_{LI_{i'}}(d_{L_{i'}})$$

For the layer $i'$, there are four communication functions as follows:

$$\gamma(r_{UI_{i'}}(d_{U_{i'}}), s_{LO_{i'+1}}(UF_{i'+1}(d_{U_{i'+1}}))) \triangleq c_{UI_{i'}}(d_{U_{i'}})$$
$$\gamma(r_{LI_{i'+1}}(d_{L_{i'+1}}), s_{UO_{i'}}(LF_{i'}(d_{L_{i'}}))) \triangleq c_{LI_{i'+1}}(d_{L_{i'+1}})$$
$$\gamma(r_{LI_i}(d_{L_i}), s_{LO_{i'}}(UF_{i'}(d_{U_{i'}}))) \triangleq c_{LI_i}(d_{L_i})$$
$$\gamma(r_{LI_{i'}}(d_{L_{i'}}), s_{LO_i}(UF_i(d_{U_i}))) \triangleq c_{LI_{i'}}(d_{L_{i'}})$$

Note that for the layer $n$, there are only two communication functions as follows:

$$\gamma(r_{LI_n}(d_{L_n}), s_{UO_{n-1}}(LF_{n-1}(d_{L_{n-1}}))) \triangleq c_{LI_n}(d_{L_n})$$
$$\gamma(r_{UI_{n-1}}(d_{U_{n-1}}), s_{LO_n}(UF_n(d_{U_n}))) \triangleq c_{UI_{n-1}}(d_{U_{n-1}})$$

Also, for the layer $n'$, there are only two communication functions as follows:

$$\gamma(r_{LI_{n'}}(d_{L_{n'}}), s_{UO_{n'-1}}(LF_{n'-1}(d_{L_{n'-1}}))) \triangleq c_{LI_{n'}}(d_{L_{n'}})$$
$$\gamma(r_{UI_{n'-1}}(d_{U_{n'-1}}), s_{LO_{n'}}(UF_{n'}(d_{U_{n'}}))) \triangleq c_{UI_{n'-1}}(d_{U_{n'-1}})$$

Let all layers from layer $n$ to layer $i$ be in parallel, then the Layers pattern $L_n \cdots L_i L_{i'} \cdots L_{n'}$ can be presented by the following process term:

$\tau_I(\partial_H(\Theta(L_n \between \cdots \between L_i \between L_{1'} \between \cdots \between L_{i'} \between \cdots \between L_{n'}))) = \tau_I(\partial_H(L_n \between \cdots \between L_i \between L_{i'} \between \cdots \between L_{n'}))$
where $H = \{r_{UI_i}(d_{U_i}), r_{LI_i}(d_{L_i}), s_{LO_i}(UF_i(d_{U_i})), s_{UO_i}(LF_i(d_{L_i})), \cdots, r_{LI_n}(d_{L_n}), s_{LO_n}(UF_n(d_{U_n})),$
$r_{UI_{i'}}(d_{U_{i'}}), r_{LI_{i'}}(d_{L_{i'}}), s_{LO_{i'}}(UF_{i'}(d_{U_{i'}})), s_{UO_{i'}}(LF_{i'}(d_{L_{i'}})), \cdots, r_{LI_{n'}}(d_{L_{n'}}), s_{LO_{n'}}(UF_{n'}(d_{U_{n'}}))$
$|d_{U_i}, d_{L_i} \cdots, d_{U_n}, d_{L_n}, d_{U_{i'}}, d_{L_{i'}} \cdots, d_{U_{n'}}, d_{L_{n'}} \in \Delta\}$,

$I = \{c_{UI_i}(d_{U_i}), c_{LI_i}(d_{L_i}), c_{LI_{i+1}}(d_{L_{i+1}}), \cdots, c_{LI_n}(d_{L_n}), c_{UI_{n-1}}(d_{U_{n-1}}), LF_i, UF_i, \cdots, LF_n, UF_n,$
$c_{UI_{i'}}(d_{U_{i'}}), c_{LI_{i'}}(d_{L_{i'}}), c_{LI_{i'+1}}(d_{L_{i'+1}}), \cdots, c_{LI_{n'}}(d_{L_{n'}}), c_{UI_{n'-1}}(d_{U_{n'-1}}), LF_{i'}, UF_{i'}, \cdots, LF_{n'}, UF_{n'}$
$|d_{U_i}, d_{L_i} \cdots, d_{U_n}, d_{L_n}, d_{U_{i'}}, d_{L_{i'}} \cdots, d_{U_{n'}}, d_{L_{n'}} \in \Delta\}$

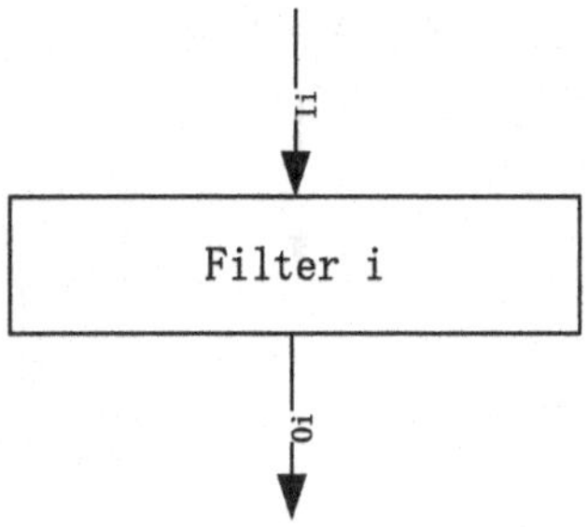


FIGURE 33.7 Filter i.

Then, we obtain the following conclusion on the Layers pattern.

**Theorem 33.3** (Correctness of two layers peers via virtual communication). *The two layers peers via virtual communication* $\tau_I(\partial_H(L_n \between \cdots \between L_i \between L_{i'} \between \cdots \between L_{n'}))$ *can exhibit the desired external behaviors.*

*Proof.* Based on the above state transitions of layer $i$ and $i'$ (for $1 \leq i, i' \leq n$), by use of the algebraic laws of APTC, we can prove that

$\tau_I(\partial_H(L_n \between \cdots \between L_i \between L_{i'} \between \cdots \between L_{n'})) =$

$\sum_{d_{U_n}, d_{L_n}, d_{U_{n'}}, d_{L_{n'}} \in \Delta}((r_{UI_n}(d_{U_n}) \parallel r_{UI_{n'}}(d_{U_{n'}})) \cdot (s_{UO_n}(LF_n(d_{L_n})) \parallel s_{UO_{n'}}(LF_{n'}(d_{L_{n'}})))) \cdot \tau_I(\partial_H(L_n \between \cdots \between L_i \between L_{i'} \between \cdots \between L_{n'}))$

That is, the two layers peers via virtual communication $\tau_I(\partial_H(L_n \between \cdots \between L_i \between L_{i'} \between \cdots \between L_{n'}))$ can exhibit the desired external behaviors.

For the details of the proof, please refer to Section 17.3, as we omit them here. □

### 33.1.2 Verification of the Pipes and Filters pattern

The Pipes and Filters pattern is used to process a stream of data with each processing step being encapsulated in a filter component. The data stream flows out of the data source, and into the first filter; the first filter processes the data, and sends the processed data to the next filter; eventually, the data stream flows out of the pipes of filters and into the data sink, as Fig. 33.8 illustrates, there are $n$ filters in the pipes. In particular, for filter $i$ ($1 \leq i \leq n$), as illustrated in Fig. 33.7, it has an input channel $I_i$ to read the data $d_i$, then processes the data via a processing function $FF_i$, finally sends the processed data to the next filter through an output channel $O_i$.

There is one typical process in the Pipes and Filters pattern as illustrated in Fig. 33.9 and is as follows:

1. The filter 1 receives the data from the data source that is denoted $d_1$ through the channel $I_1$ (the corresponding reading action is denoted $r_{I_1}(d_1)$), then processes the data through a processing function $FF_1$, and sends the processed data to the filter 2 that is denoted $FF_1(d_1)$ through the channel $O_1$ (the corresponding sending action is denoted $s_{O_1}(FF_1(d_1))$);
2. The filter $i$ receives the data from filter $i-1$ that is denoted $d_i$ through the channel $I_i$ (the corresponding reading action is denoted $r_{I_i}(d_i)$), then processes the data through a processing function $FF_i$, and sends the processed data to the filter $i+1$ that is denoted $FF_i(d_i)$ through the channel $O_i$ (the corresponding sending action is denoted $s_{O_i}(FF_i(d_i))$);
3. The filter $n$ receives the data from filter $n-1$ that is denoted $d_n$ through the channel $I_n$ (the corresponding reading action is denoted $r_{I_n}(d_n)$), then processes the data through a processing function $FF_n$, and sends the processed data to the data sink that is denoted $FF_n(d_n)$ through the channel $O_n$ (the corresponding sending action is denoted $s_{O_n}(FF_1(d_n))$).

In the following, we verify the Pipes and Filters pattern. We assume all data elements $d_i$ (for $1 \leq i \leq n$) are from a finite set $\Delta$. The state transitions of filter $i$ (for $1 \leq i \leq n$) described by APTC are as follows:

$F_i = \sum_{d_i, \in \Delta}(r_{I_i}(d_i) \cdot F_{i_2})$

$F_{i_2} = FF_i \cdot F_{i_3}$

$F_{i_3} = \sum_{d_i \in \Delta}(s_{O_i}(FF_i(d_i)) \cdot F_i)$

The sending action and the reading action of the same data through the same channel can communicate with each other, otherwise, they will cause a deadlock $\delta$. We define the following communication functions for $1 \leq i \leq n$. Note that the channel of $I_{i+1}$ of filter $i+1$ and the channel $O_i$ of filter $i$ are the same channel. Also, the data $d_{i+1}$ of filter $i+1$ and the data $FF_i(d_i)$ of filter $i$ are the same data:

$$\gamma(r_{I_i}(d_i), s_{O_{i-1}}(FF_{i-1}(d_{i-1}))) \triangleq c_{I_i}(d_i)$$

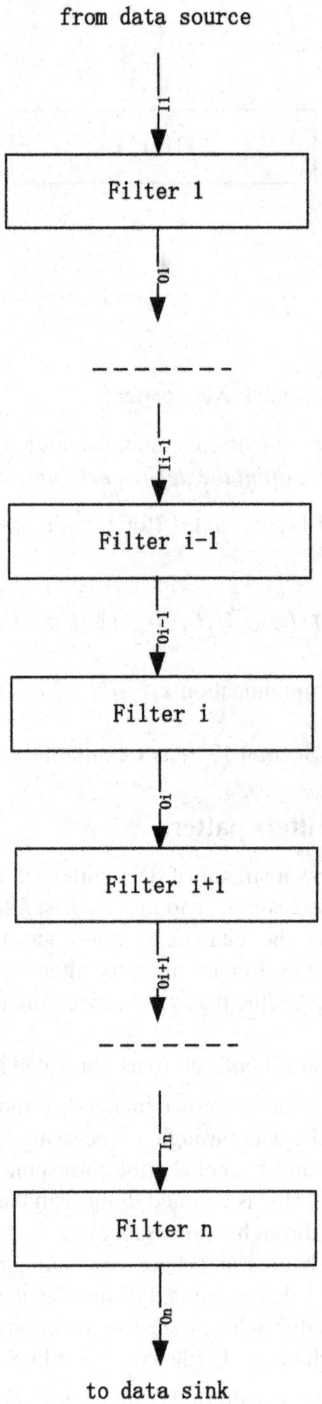


**FIGURE 33.8** Pipes and Filters pattern.

$$\gamma(r_{I_{i+1}}(d_{i+1}), s_{O_i}(FF_i(d_i))) \triangleq c_{I_{i+1}}(d_{i+1})$$

Note that for the filter $n$, there is only one communication function as follows:

$$\gamma(r_{I_n}(d_n), s_{O_{n-1}}(FF_{n-1}(d_{n-1}))) \triangleq c_{I_n}(d_n)$$

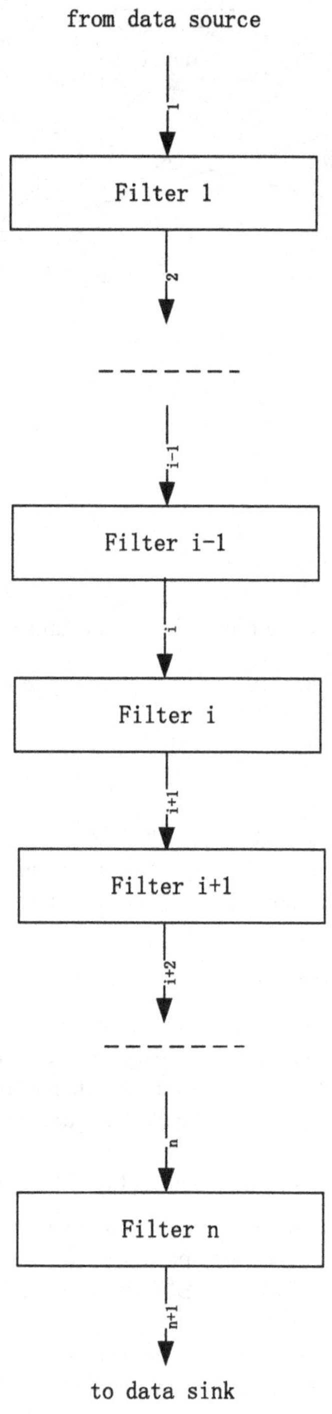


**FIGURE 33.9** Typical process of Pipes and Filters pattern.

Also, for the filter 1, there is also only one communication function as follows:

$$\gamma(r_{I_2}(d_2), s_{O_1}(FF_1(d_1))) \triangleq c_{I_2}(d_2)$$

Let all filters from filter 1 to filter $n$ be in parallel, then the Pipes and Filters pattern $F_1 \cdots F_i \cdots F_n$ can be presented by the following process term:

$$\tau_I(\partial_H(\Theta(F_1 \between \cdots \between F_i \between \cdots \between F_n))) = \tau_I(\partial_H(F_1 \between \cdots \between F_i \between \cdots \between F_n))$$

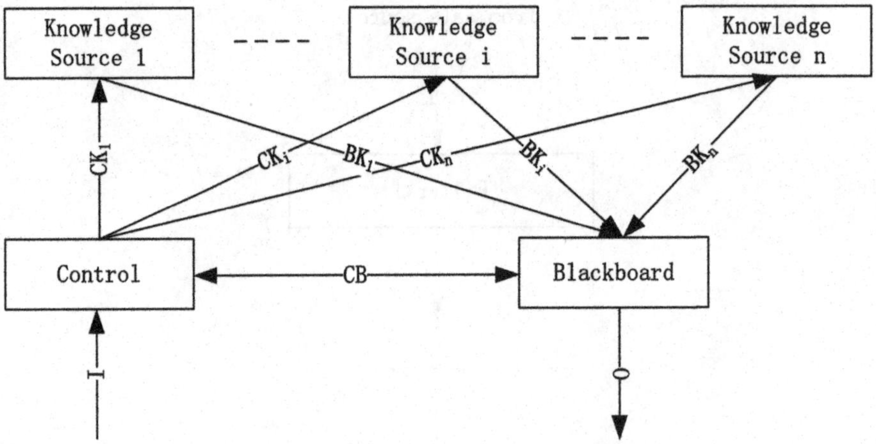


**FIGURE 33.10** Blackboard pattern.

where $H=\{s_{O_1}(FF_1(d_1)),\cdots,r_{I_i}(d_i),s_{O_i}(FF_i(d_i)),\cdots,r_{I_n}(d_n)|d_1,\cdots,d_i,\cdots,d_n\in\Delta\}$,
$I=\{c_{I_2}(d_2),\cdots,c_{I_i}(d_i),\cdots,c_{I_n}(d_n),FF_1,\cdots,FF_i\cdots,FF_n|d_1,\cdots,d_i,\cdots,d_n\in\Delta\}$
Then, we obtain the following conclusion on the Pipes and Filters pattern.

**Theorem 33.4** (Correctness of the Pipes and Filters pattern). *The Pipes and Filters pattern* $\tau_I(\partial_H(F_1\between\cdots\between F_i\between\cdots\between F_n))$ *can exhibit the desired external behaviors.*

*Proof.* Based on the above state transitions of filter $i$ (for $1\leq i\leq n$), by use of the algebraic laws of APTC, we can prove that

$\tau_I(\partial_H(F_1\between\cdots\between F_i\between\cdots\between F_n))=\sum_{d_1,d_n\in\Delta}(r_{I_1}(d_1)\cdot s_{O_n}(FF_n(d_n)))\cdot\tau_I(\partial_H(F_1\between\cdots\between F_i\between\cdots\between F_n))$

That is, the Pipes and Filters pattern $\tau_I(\partial_H(F_1\between\cdots\between F_i\between\cdots\between F_n))$ can exhibit the desired external behaviors.

For the details of the proof, please refer to Section 17.3, as we omit them here. □

### 33.1.3 Verification of the Blackboard pattern

The Blackboard pattern is used to solve problems with no deterministic solutions. In the Blackboard pattern, there are one Control module, one Blackboard module, and several Knowledge Source modules. When the Control module receives a request, it queries the Blackboard module about the involved Knowledge Sources; then the Control module invokes the related Knowledge Sources; finally, the related Knowledge Sources update the Blackboard with the invoked results, as illustrated in Fig. 33.10.

The typical process of the Blackboard pattern is illustrated in Fig. 33.11 and is as follows:

1. The Control module receives the request from outside applications that is denoted $d_I$, through the input channel $I$ (the corresponding reading action is denoted $r_I(d_I)$), then processes the request through a processing function $CF_1$, and sends the processed data that is denoted $CF_1(d_I)$ to the Blackboard module through the channel $CB$ (the corresponding sending action is denoted $s_{CB}(CF_1(d_I))$);
2. The Blackboard module receives the request (information of the involved Knowledge Sources) from the Control module through the channel $CB$ (the corresponding reading action is denoted $r_{CB}(CF_1(d_I))$), then processes the request through a processing function $BF_1$, and generates and sends the response that is denoted $d_B$ to the Control module through the channel $CB$ (the corresponding sending action is denoted $s_{CB}(d_B)$);
3. The Control module receives the data from the Blackboard module through the channel $CB$ (the corresponding reading action is denoted $r_{CB}(d_B)$), then processes the data through another processing function $CF_2$, and generates and sends the requests to the related Knowledge Sources that are denoted $d_{C_i}$ through the channels $CK_i$ (the corresponding sending action is denoted $s_{CK_i}(d_{C_i})$) with $1\leq i\leq n$;
4. The Knowledge Source $i$ receives the request from the Control module through the channel $CK_i$ (the corresponding reading action is denoted $r_{CK_i}(d_{C_i})$), then processes the request through a processing function $KF_i$, and generates and sends the processed data $d_{K_i}$ to the Blackboard module through the channel $BK_i$ (the corresponding sending action is denoted $s_{BK_i}(d_{K_i})$);

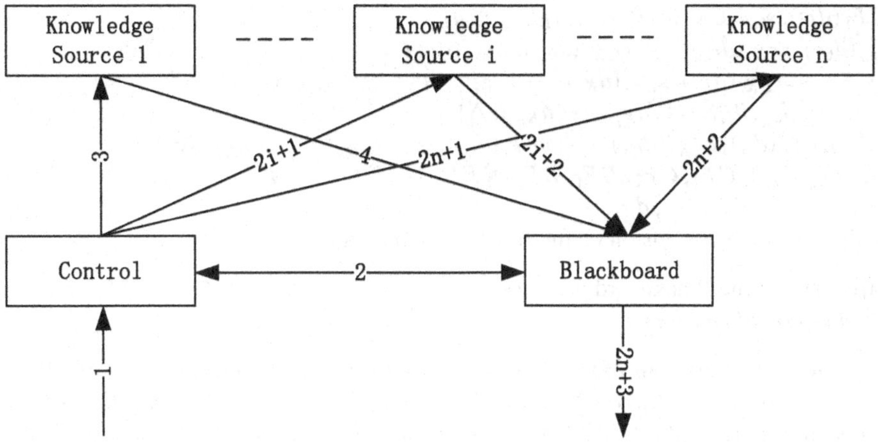


**FIGURE 33.11** Typical process of Blackboard pattern.

5. The Blackboard module receives the invoked results from Knowledge Source $i$ through the channel $BK_i$ (the corresponding reading action is denoted $r_{BK_i}(d_{K_i})$) $(1 \leq i \leq n)$, then processes the results through another processing function $BF_2$, generates and sends the output $d_O$ through the channel $O$ (the corresponding sending action is denoted $s_O(d_O)$).

In the following, we verify the Blackboard pattern. We assume all data elements $d_I, d_B, d_{C_i}, d_{K_i}, d_O$ (for $1 \leq i \leq n$) are from a finite set $\Delta$. The state transitions of the Control module described by APTC are as follows:

$C = \sum_{d_I \in \Delta}(r_I(d_I) \cdot C_2)$
$C_2 = CF_1 \cdot C_3$
$C_3 = \sum_{d_I \in \Delta}(s_{CB}(CF_1(d_I)) \cdot C_4)$
$C_4 = \sum_{d_B \in \Delta}(r_{CB}(d_B) \cdot C_5)$
$C_5 = CF_2 \cdot C_6$
$C_6 = \sum_{d_{C_1},\cdots,d_{C_i},\cdots,d_{C_n} \in \Delta}(s_{CK_1}(d_{C_1}) \between \cdots \between s_{CK_i}(d_{C_i}) \between \cdots \between s_{CK_n}(d_{C_n}) \cdot C)$

The state transitions of the Blackboard module described by APTC are as follows:

$B = \sum_{d_I \in \Delta}(r_{CB}(CF_1(d_I)) \cdot B_2)$
$B_2 = BF_1 \cdot B_3$
$B_3 = \sum_{d_B \in \Delta}(s_{CB}(d_B) \cdot B_4)$
$B_4 = \sum_{d_{K_1},\cdots,d_{K_i},\cdots,d_{K_n} \in \Delta}(r_{BK_1}(d_{K_1}) \between \cdots \between r_{BK_i}(d_{K_i}) \between \cdots \between r_{BK_n}(d_{K_n}) \cdot B_5)$
$B_5 = BF_2 \cdot B_6$
$B_6 = \sum_{d_O \in \Delta}(s_O(d_O) \cdot B)$

The state transitions of the Knowledge Source $i$ described by APTC are as follows:

$K_i = \sum_{d_{C_i} \in \Delta}(r_{CK_i}(d_{C_i}) \cdot K_{i_2})$
$K_{i_2} = KF_i \cdot K_{i_3}$
$K_{i_3} = \sum_{d_{K_i} \in \Delta}(s_{BK_i}(d_{K_i}) \cdot K_i)$

The sending action and the reading action of the same data through the same channel can communicate with each other, otherwise, they will cause a deadlock $\delta$. We define the following communication functions for $1 \leq i \leq n$:

$$\gamma(r_{CB}(CF_1(d_I)), s_{CB}(CF_1(d_I))) \triangleq c_{CB}(CF_1(d_I))$$
$$\gamma(r_{CB}(d_B), s_{CB}(d_B)) \triangleq c_{CB}(d_B)$$
$$\gamma(r_{CK_i}(d_{C_i}), s_{CK_i}(d_{C_i})) \triangleq c_{CK_i}(d_{C_i})$$
$$\gamma(r_{BK_i}(d_{K_i}), s_{BK_i}(d_{K_i})) \triangleq c_{BK_i}(d_{K_i})$$

Let all modules be in parallel, then the Blackboard pattern $C \quad B \quad K_1 \cdots K_i \cdots K_n$ can be presented by the following process term:

$$\tau_I(\partial_H(\Theta(C \between B \between K_1 \between \cdots \between K_i \between \cdots \between K_n))) = \tau_I(\partial_H(C \between B \between K_1 \between \cdots \between K_i \between \cdots \between K_n))$$

where $H = \{r_{CB}(CF_1(d_I)), s_{CB}(CF_1(d_I)), r_{CB}(d_B), s_{CB}(d_B), r_{CK_1}(d_{C_1}),$
$s_{CK_1}(d_{C_1}), \cdots, r_{CK_i}(d_{C_i}), s_{CK_i}(d_{C_i}), \cdots, r_{CK_n}(d_{C_n}), s_{CK_n}(d_{C_n}),$
$r_{BK_1}(d_{K_1}), s_{BK_1}(d_{K_1}), \cdots, r_{BK_i}(d_{K_i}), s_{BK_i}(d_{K_i}), \cdots, r_{BK_n}(d_{K_n}), s_{BK_n}(d_{K_n})$
$|d_I, d_B, d_{C_1}, \cdots, d_{C_i}, \cdots, d_{C_n}, d_{K_1}, \cdots, d_{K_i}, \cdots, d_{K_n} \in \Delta\}$,
$I = \{c_{CB}(CF_1(d_I)), c_{CB}(d_B), c_{CK_1}(d_{C_1}), \cdots, c_{CK_i}(d_{C_i}), \cdots, c_{CK_n}(d_{C_n}), c_{BK_1}(d_{K_1}),$
$\cdots, c_{BK_i}(d_{K_i}), \cdots, c_{BK_n}(d_{K_n}), CF_1, CF_2, BF_1, BF_2, KF1, \cdots, KF_i, \cdots, KF_n$
$|d_I, d_B, d_{C_1}, \cdots, d_{C_i}, \cdots, d_{C_n}, d_{K_1}, \cdots, d_{K_i}, \cdots, d_{K_n} \in \Delta\}$

Then, we obtain the following conclusion on the Blackboard pattern.

**Theorem 33.5** (Correctness of the Blackboard pattern). *The Blackboard pattern $\tau_I(\partial_H(C \between B \between K_1 \between \cdots \between K_i \between \cdots \between K_n))$ can exhibit the desired external behaviors.*

*Proof.* Based on the above state transitions of the above modules, by use of the algebraic laws of APTC, we can prove that

$$\tau_I(\partial_H(C \between B \between K_1 \between \cdots \between K_i \between \cdots \between K_n)) = \sum_{d_I, d_O \in \Delta}(r_I(d_I) \cdot s_O(d_O)) \cdot \tau_I(\partial_H(C \between B \between K_1 \between \cdots \between K_i \between \cdots \between K_n))$$

That is, the Blackboard pattern $\tau_I(\partial_H(C \between B \between K_1 \between \cdots \between K_i \between \cdots \between K_n))$ can exhibit the desired external behaviors.

For the details of the proof, please refer to Section 17.3, as we omit them here. □

## 33.2 Distributed systems

In this section, we verify the distributed systems oriented patterns, including the Broker pattern, the Pipes and Filters pattern in Section 33.1, and the Microkernel Pattern in Section 33.4.

### 33.2.1 Verification of the Broker pattern

The Broker pattern decouples the invocation process between the Client and the Server. There are five types of modules in the Broker pattern: the Client, the Client-side Proxy, the Brokers, the Server-side Proxy, and the Server. The Client receives the request from the user and passes it to the Client-side Proxy, then to the first broker and the next one, the last broker passes it the Server-side Proxy, finally leads to the invocation to the Server; the Server processes the request and generates the response, then the response is returned to the user in a reverse way, as illustrated in Fig. 33.12.

The typical process of the Broker pattern is illustrated in Fig. 33.13 and is as follows:

1. The Client receives the request $d_I$ through the channel $I$ (the corresponding reading action is denoted $r_I(d_I)$), then processes the request through a processing function that is denoted $CF_1$, then sends the processed request $CF_1(d_I)$ to the Client-side Proxy through the channel $I_{CP}$ (the corresponding sending action is denoted $s_{I_{CP}}(CF_1(d_I))$);
2. The Client-side Proxy receives the request $d_{I_{CP}}$ from the Client through the channel $I_{CP}$ (the corresponding reading action is denoted $r_{I_{CP}}(d_{I_{CP}})$), then processes the request through a processing function $CPF_1$, and then sends the processed request $CPF_1(d_{I_{CP}})$ to the first broker 1 through the channel $I_{CB}$ (the corresponding sending action is denoted $s_{I_{CB}}(CPF_1(d_{I_{CP}}))$);
3. The broker $i$ (for $1 \leq i \leq n$) receives the request $d_{I_{B_i}}$ from the broker $i-1$ through the channel $I_{BB_i}$ (the corresponding reading action is denoted $r_{I_{BB_i}}(d_{I_{B_i}})$), then processes the request through a processing function $BF_{i_1}$, and then sends the processes request $BF_{i_1}(d_{I_{B_i}})$ to the broker $i+1$ through the channel $I_{BB_{i+1}}$ (the corresponding sending action is denoted $s_{I_{BB_{i+1}}}(BF_{i_1}(d_{I_{B_i}}))$);
4. The Server-side Proxy receives the request $d_{I_{SP}}$ from the last broker $n$ through the channel $I_{BS}$ (the corresponding reading action is denoted $r_{I_{BS}}(d_{I_{SP}})$), then processes the request through a processing function $SPF_1$, and then sends the processed request $SPF_1(d_{I_{SP}})$ to the Server through the channel $I_{PS}$ (the corresponding sending action is denoted $s_{I_{PS}}(SPF_1(d_{I_{SP}}))$);
5. The Server receives the request $d_{I_S}$ from the Server-side Proxy through the channel $I_{PS}$ (the corresponding reading action is denoted $r_{I_{PS}}(d_{I_S})$), then processes the request and generates the response $d_{O_S}$ through a processing function $SF$, and then sends the response to the Server-side Proxy through the channel $O_{PS}$ (the corresponding sending action is denoted $s_{O_{PS}}(d_{O_S})$);
6. The Server-side Proxy receives the response $d_{O_{SP}}$ from the Server through the channel $O_{PS}$ (the corresponding reading action is denoted $r_{O_{PS}}(d_{O_{SP}})$), then processes the response through a processing function $SPF_2$, and sends the processed response $SPF_2(d_{O_{SP}})$ to the last broker $n$ through the channel $O_{BS}$ (the corresponding sending action is denoted $s_{O_{BS}}(SPF_2(d_{O_{SP}}))$);
7. the broker $i$ receives the response $d_{O_{B_i}}$ from the broker $i+1$ through the channel $O_{BB_{i+1}}$ (the corresponding reading action is denoted $r_{O_{BB_{i+1}}}(d_{O_{B_i}})$), then processes the response through a processing function $BF_{i_2}$, and then sends the

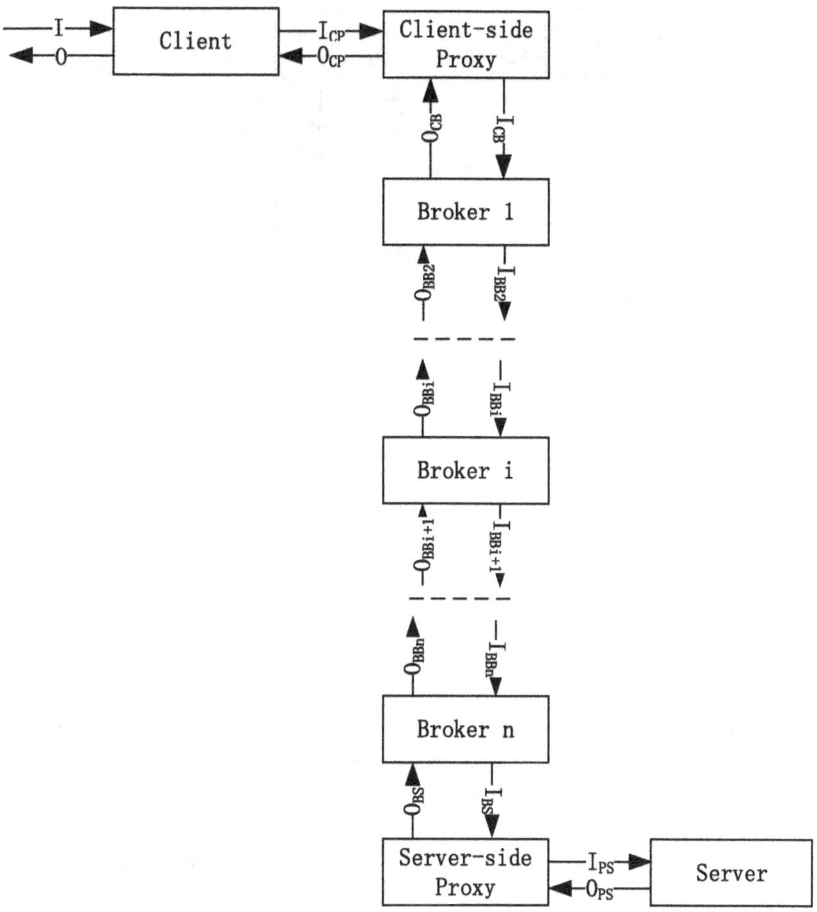


**FIGURE 33.12** Broker pattern.

processed response $BF_{i_2}(d_{O_{B_i}})$ to the broker $i-1$ through the channel $O_{BB_i}$ (the corresponding sending action is denoted $s_{O_{BB_i}}(BF_{i_2}(d_{O_{B_i}}))$);

8. The Client-side Proxy receives the response $d_{O_{CP}}$ from the first broker 1 through the channel $O_{CB}$ (the corresponding reading action is denoted $r_{O_{CB}}(d_{O_{CP}})$), then processes the response through a processing function $CPF_2$, and sends the processed response $CPF_2(d_{O_{CP}})$ to the Client through the channel $O_{CP}$ (the corresponding sending action is denoted $s_{O_{CP}}(CPF_2(d_{O_{CP}}))$);
9. The Client receives the response $d_{O_C}$ from the Client-side Proxy through the channel $O_{CP}$ (the corresponding reading action is denoted $r_{O_{CP}}(d_{O_C})$), then processes the response through a processing function $CF_2$ and generates the response $d_O$, and then sends the response out through the channel $O$ (the corresponding sending action is denoted $s_O(d_O)$).

In the following, we verify the Broker pattern. We assume all data elements $d_I$, $d_{I_{CP}}$, $d_{I_{B_i}}$, $d_{I_{SP}}$, $d_{I_S}$, $d_{O_S}$, $d_{O_{B_i}}$, $d_{O_{CP}}$, $d_{O_C}$, $d_O$ (for $1 \leq i \leq n$) are from a finite set $\Delta$. Note that the channels $I_{BB_1}$ and $I_{CB}$ are the same channel; the channels $O_{BB_1}$ and $O_{CB}$ are the same channel; the channels $I_{BB_{n+1}}$ and $I_{BS}$ are the same channel; the channels $O_{BB_{n+1}}$ and $O_{BS}$ are the same channel. Also, the data $CF_1(d_I)$ and $d_{I_{CP}}$ are the same data; the data $CPF_1(d_{I_{CP}})$ and $d_{I_{B_1}}$ are the same data; the data $BF_{i_1}(d_{I_{B_i}})$ and $d_{I_{B_{i+1}}}$ are the same data; the data $BF_{n_1}(d_{I_{B_n}})$ and the data $d_{I_{SP}}$ are the same data; the data $SPF_1(d_{I_{SP}})$ and $d_{I_S}$ are the same data; the data $SPF_2(d_{O_S})$ and $d_{O_{B_n}}$ are the same data; the data $BF_{i_2}(d_{O_{B_i}})$ and the data $d_{O_{B_{i-1}}}$ are the same data; the data $BF_{1_2}(d_{O_{B_1}})$ and $d_{O_{CP}}$ are the same data; the data $CPF_2(d_{O_{CP}})$ and $d_{O_C}$ are the same data; the data $CF_2(d_{O_C})$ and the data $d_O$ are the same data.

The state transitions of the Client module described by APTC are as follows:

$C = \sum_{d_I \in \Delta}(r_I(d_I) \cdot C_2)$
$C_2 = CF_1 \cdot C_3$
$C_3 = \sum_{d_I \in \Delta}(s_{I_{CP}}(CF_1(d_I)) \cdot C_4)$
$C_4 = \sum_{d_{O_C} \in \Delta}(r_{O_{CP}}(d_{O_C}) \cdot C_5)$

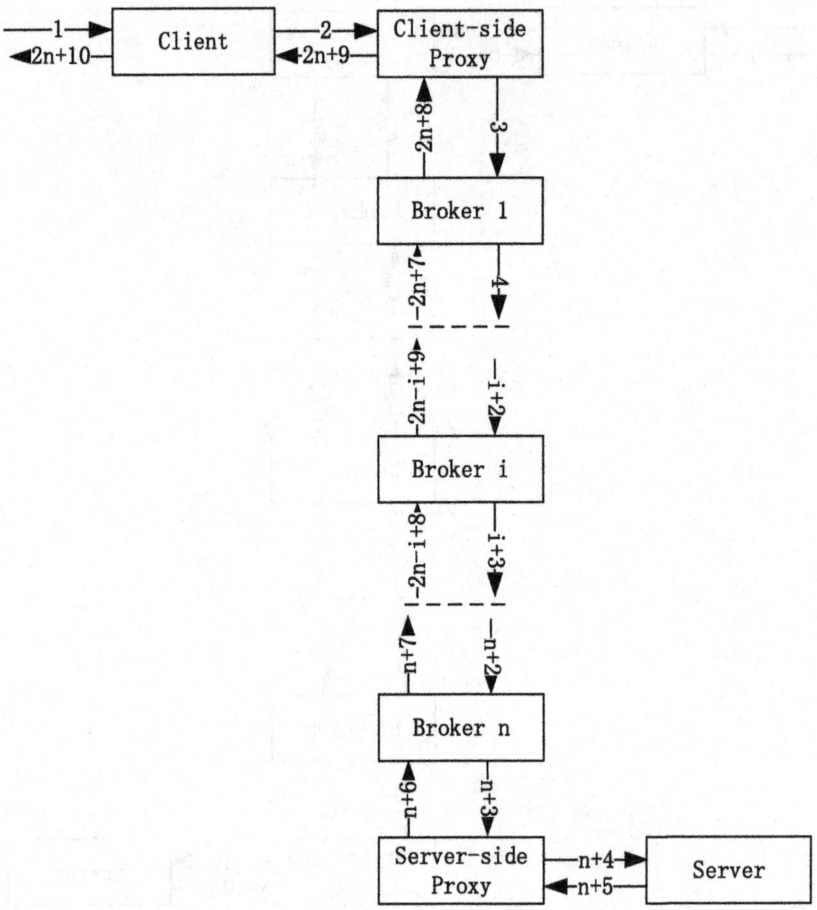


FIGURE 33.13 Typical process of Broker pattern.

$C_5 = CF_2 \cdot C_6$

$C_6 = \sum_{d_O \in \Delta}(s_O(d_O) \cdot C)$

The state transitions of the Client-side Proxy module described by APTC are as follows:

$CP = \sum_{d_{I_{CP}} \in \Delta}(r_{I_{CP}}(d_{I_{CP}}) \cdot CP_2)$

$CP_2 = CPF_1 \cdot CP_3$

$CP_3 = \sum_{d_{I_{CP}} \in \Delta}(s_{I_{CB}}(CPF_1(d_{I_{CP}})) \cdot CP_4)$

$CP_4 = \sum_{d_{O_{CP}} \in \Delta}(r_{O_{CB}}(d_{O_{CP}}) \cdot CP_5)$

$CP_5 = CPF_2 \cdot CP_6$

$CP_6 = \sum_{d_{O_{CP}} \in \Delta}(s_{O_{CP}}(CPF_2(d_{O_{CP}})) \cdot CP)$

The state transitions of the Broker $i$ described by APTC are as follows:

$B_i = \sum_{d_{I_{B_i}} \in \Delta}(r_{I_{BB_i}}(d_{I_{B_i}}) \cdot B_{i_2})$

$B_{i_2} = BF_{i_1} \cdot B_{i_3}$

$B_{i_3} = \sum_{d_{I_{B_i}} \in \Delta}(s_{I_{BB_{i+1}}}(BF_{i_1}(d_{I_{B_i}})) \cdot B_{i_4})$

$B_{i_4} = \sum_{d_{O_{B_i}} \in \Delta}(r_{O_{BB_{i+1}}}(d_{O_{B_i}}) \cdot B_{i_5})$

$B_{i_5} = BF_{i_2} \cdot B_{i_6}$

$B_{i_6} = \sum_{d_{O_{B_i}} \in \Delta}(s_{O_{BB_i}}(BF_{i_2}(d_{O_{B_i}})) \cdot B_i)$

The state transitions of the Server-side Proxy described by APTC are as follows:

$SP = \sum_{d_{I_{SP}} \in \Delta}(r_{I_{BS}}(d_{I_{SP}}) \cdot SP_2)$

$SP_2 = SPF_1 \cdot SP_3$

$SP_3 = \sum_{d_{I_{SP}} \in \Delta}(s_{I_{PS}}(SPF_1(d_{I_{SP}})) \cdot SP_4)$

$SP_4 = \sum_{d_{O_{SP}} \in \Delta}(r_{O_{PS}}(d_{O_{SP}}) \cdot SP_5)$
$SP_5 = SPF_2 \cdot SP_6$
$SP_6 = \sum_{d_{O_{SP}} \in \Delta}(s_{O_{BS}}(SPF_2(d_{O_{SP}})) \cdot SP)$
The state transitions of the Server described by APTC are as follows:
$S = \sum_{d_{I_S} \in \Delta}(r_I(d_I) \cdot S_2)$
$S_2 = SF \cdot S_3$
$S_3 = \sum_{d_{O_S} \in \Delta}(s_{O_{PS}}(d_{O_S}) \cdot S)$

The sending action and the reading action of the same data through the same channel can communicate with each other, otherwise, they will cause a deadlock $\delta$. We define the following communication functions of the broker $i$ for $1 \leq i \leq n$:

$$\gamma(r_{I_{BB_i}}(d_{I_{B_i}}), s_{I_{BB_i}}(BF_{i-1_1}(d_{I_{B_{i-1}}}))) \triangleq c_{I_{BB_i}}(d_{I_{B_i}})$$
$$\gamma(r_{O_{BB_{i+1}}}(d_{O_{B_i}}), s_{O_{BB_{i+1}}}(BF_{i+1_2}(d_{O_{B_{i+1}}}))) \triangleq c_{O_{BB_{i+1}}}(d_{O_{B_i}})$$

There are two communication functions between the Client and the Client-side Proxy as follows:

$$\gamma(r_{I_{CP}}(d_{I_{CP}}), s_{I_{CP}}(CF_1(d_I))) \triangleq c_{I_{CP}}(d_{I_{CP}})$$
$$\gamma(r_{O_{CP}}(d_{O_C}), s_{O_{CP}}(CPF_2(d_{O_{CP}}))) \triangleq c_{O_{CP}}(d_{O_C})$$

There are two communication functions between the Broker 1 and the Client-side Proxy as follows:

$$\gamma(r_{I_{BB_1}}(d_{I_{B_1}}), s_{I_{CB}}(CPF_1(d_{I_{CP}}))) \triangleq c_{I_{BB_1}}(d_{I_{B_1}})$$
$$\gamma(r_{O_{CB}}(d_{O_{CP}}), s_{O_{BB_i}}(BF_{i_2}(d_{O_{B_i}}))) \triangleq c_{O_{CB}}(d_{O_{CP}})$$

There are two communication functions between the Broker $n$ and the Server-side Proxy as follows:

$$\gamma(r_{I_{BS}}(d_{I_{SP}}), s_{I_{BB_{n+1}}}(BF_{n_1}(d_{I_{B_n}}))) \triangleq c_{I_{BS}}(d_{I_{SP}})$$
$$\gamma(r_{O_{BB_{n+1}}}(d_{O_{B_n}}), s_{O_{BS}}(SPF_2(d_{O_{SP}}))) \triangleq c_{O_{BB_{n+1}}}(d_{O_{B_n}})$$

There are two communication functions between the Server and the Server-side Proxy as follows:

$$\gamma(r_{I_{PS}}(d_{I_S}), s_{I_{PS}}(SPF_1(d_{I_{SP}}))) \triangleq c_{I_{PS}}(d_{I_S})$$
$$\gamma(r_{O_{PS}}(d_{O_{SP}}), s_{O_{PS}}(d_{O_S})) \triangleq c_{O_{PS}}(d_{O_{SP}})$$

Let all modules be in parallel, then the Broker pattern $C \quad CP \quad SP \quad S \quad B_1 \cdots B_i \cdots B_n$ can be presented by the following process term:

$\tau_I(\partial_H(\Theta(C \between CP \between SP \between S \between B_1 \between \cdots \between B_i \between \cdots \between B_n))) = \tau_I(\partial_H(C \between CP \between SP \between S \between B_1 \between \cdots \between B_i \between \cdots \between B_n))$
where $H = \{r_{I_{BB_i}}(d_{I_{B_i}}), s_{I_{BB_i}}(BF_{i-1_1}(d_{I_{B_{i-1}}})), r_{O_{BB_{i+1}}}(d_{O_{B_i}}), s_{O_{BB_{i+1}}}(BF_{i+1_2}(d_{O_{B_{i+1}}})),$
$r_{I_{CP}}(d_{I_{CP}}), s_{I_{CP}}(CF_1(d_I)), r_{O_{CP}}(d_{O_C}), s_{O_{CP}}(CPF_2(d_{O_{CP}})), r_{I_{BB_1}}(d_{I_{B_1}}), s_{I_{CB}}(CPF_1(d_{I_{CP}})),$
$r_{O_{CB}}(d_{O_{CP}}), s_{O_{BB_i}}(BF_{i_2}(d_{O_{B_i}})), r_{I_{BS}}(d_{I_{SP}}), s_{I_{BB_{n+1}}}(BF_{n_1}(d_{I_{B_n}})), r_{O_{BB_{n+1}}}(d_{O_{B_n}}), s_{O_{BS}}(SPF_2(d_{O_{SP}})),$
$r_{I_{PS}}(d_{I_S}), s_{I_{PS}}(SPF_1(d_{I_{SP}})), r_{O_{PS}}(d_{O_{SP}}), s_{O_{PS}}(d_{O_S})$
$|d_I, d_{I_{CP}}, d_{O_{CP}}, d_{I_{SP}}, d_{O_{SP}}, d_{I_S}, d_{O_S}, d_{O_C}, d_{I_{CP}}, d_{O_{CP}}, d_O, d_{I_{B_1}}, \cdots, d_{I_{B_i}}, \cdots, d_{I_{B_n}}, d_{O_{B_1}}, \cdots, d_{O_{B_i}}, \cdots, d_{O_{B_n}} \in \Delta\},$
$I = \{c_{I_{BB_1}}(d_{I_{B_1}}), c_{O_{BB_2}}(d_{O_{B_1}}), \cdots, c_{I_{BB_i}}(d_{I_{B_i}}), c_{O_{BB_{i+1}}}(d_{O_{B_i}}), \cdots, c_{I_{BB_n}}(d_{I_{B_n}}), c_{O_{BB_{n+1}}}(d_{O_{B_n}}),$
$c_{I_{CP}}(d_{I_{CP}}), c_{O_{CP}}(d_{O_C}), c_{I_{BB_1}}(d_{I_{B_1}}), c_{O_{CB}}(d_{O_{CP}}), c_{I_{BS}}(d_{I_{SP}}), c_{O_{BB_{n+1}}}(d_{O_{B_n}}), c_{I_{PS}}(d_{I_S}), c_{O_{PS}}(d_{O_{SP}}),$
$CF_1, CF_2, CPF_1, CPF_2, BF_{1_1}, BF_{1_2}, \cdots, BF_{i_1}, BF_{i_2}, \cdots, BF_{n_1}, BF_{n_2}, SPF_1, SPF_2, SF$
$|d_I, d_{I_{CP}}, d_{O_{CP}}, d_{I_{SP}}, d_{O_{SP}}, d_{I_S}, d_{O_S}, d_{O_C}, d_{I_{CP}}, d_{O_{CP}}, d_O, d_{I_{B_1}}, \cdots, d_{I_{B_i}}, \cdots, d_{I_{B_n}}, d_{O_{B_1}}, \cdots, d_{O_{B_i}}, \cdots, d_{O_{B_n}} \in \Delta\}$

Then, we obtain the following conclusion on the Broker pattern.

**Theorem 33.6** (Correctness of the Broker pattern). *The Broker pattern* $\tau_I(\partial_H(C \between CP \between SP \between S \between B_1 \between \cdots \between B_i \between \cdots \between B_n))$ *can exhibit the desired external behaviors.*

*Proof.* Based on the above state transitions of the above modules, by use of the algebraic laws of APTC, we can prove that
$\tau_I(\partial_H(C \between CP \between SP \between S \between B_1 \between \cdots \between B_i \between \cdots \between B_n)) = \sum_{d_I, d_O \in \Delta}(r_I(d_I) \cdot s_O(d_O)) \cdot \tau_I(\partial_H(C \between CP \between SP \between S \between B_1 \between \cdots \between B_i \between \cdots \between B_n))$

That is, the Broker pattern $\tau_I(\partial_H(C \between CP \between SP \between S \between B_1 \between \cdots \between B_i \between \cdots \between B_n))$ can exhibit the desired external behaviors.

For the details of the proof, please refer to Section 17.3, as we omit them here. □

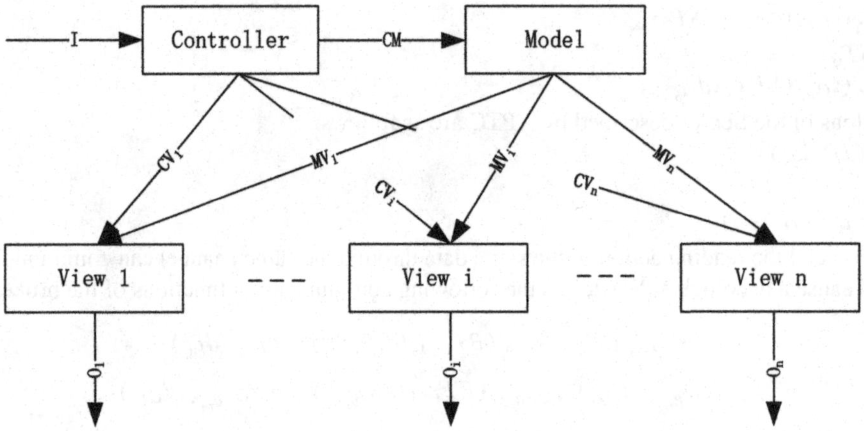


FIGURE 33.14 MVC pattern.

## 33.3 Interactive systems

In this section, we verify interactive systems oriented patterns, including the Model–View–Controller (MVC) pattern and the Presentation–Abstraction–Control (PAC) pattern.

### 33.3.1 Verification of the MVC pattern

The MVC pattern is used to model the interactive systems, which has three components: the Model, the Views, and the Controller. The Model is used to contain the data and encapsulate the core functionalities; the Views is to show the computational results to the user; and the Controller interacts between the system and user, accepts the instructions and controls the Model and the Views. The Controller receives the instructions from the user through the channel $I$, then it sends the instructions to the Model through the channel $CM$ and to the View $i$ through the channel $CV_i$ for $1 \leq i \leq n$; the model receives the instructions from the Controller, updates the data and computes the results, and sends the results to the View $i$ through the channel $MV_i$ for $1 \leq i \leq n$; When the View $i$ receives the results from the Model, it generates or updates the view to the user. This is illustrated in Fig. 33.14.

The typical process of the MVC pattern is shown in Fig. 33.15 and is the following:

1. The Controller receives the instructions $d_I$ from the user through the channel $I$ (the corresponding reading action is denoted $r_I(D_I)$), processes the instructions through a processing function $CF$, and generates the instructions to the Model $d_{I_M}$ and those to the View $i$ $d_{I_{V_i}}$ for $1 \leq i \leq n$; it sends $d_{I_M}$ to the Model through the channel $CM$ (the corresponding sending action is denoted $s_{CM}(d_{I_M})$) and sends $d_{I_{V_i}}$ to the View $i$ through the channel $CV_i$ (the corresponding sending action is denoted $s_{CV_i}(d_{I_{V_i}})$);
2. The Model receives the instructions from the Controller through the channel $CM$ (the corresponding reading action is denoted $r_{CM}(d_{I_M})$), processes the instructions through a processing function $MF$, generates the computational results to the View $i$ (for $1 \leq i \leq n$) that is denoted $d_{O_{M_i}}$; then sends the results to the View $i$ through the channel $MV_i$ (the corresponding sending action is denoted $s_{MV_i}(d_{O_{M_i}})$);
3. The View $i$ (for $1 \leq i \leq n$) receives the instructions from the Controller through the channel $CV_i$ (the corresponding reading action is denoted $r_{CV_i}(d_{I_{V_i}})$), processes the instructions through a processing function $VF_{i_1}$ to make ready to receive the computational results from the Model; then it receives the computational results from the Model through the channel $MV_i$ (the corresponding reading action is denoted $r_{MV_i}(d_{O_{M_i}})$), processes the results through a processing function $VF_{i_2}$, generates the output $d_{O_i}$, then sends the output through the channel $O_i$ (the corresponding sending action is denoted $s_{O_i}(d_{O_i})$).

In the following, we verify the MVC pattern. We assume all data elements $d_I$, $d_{I_M}$, $d_{I_{V_i}}$, $d_{O_{M_i}}$, $d_{O_i}$ (for $1 \leq i \leq n$) are from a finite set $\Delta$.

The state transitions of the Controller module described by APTC are as follows:

$C = \sum_{d_I \in \Delta}(r_I(d_I) \cdot C_2)$

$C_2 = CF \cdot C_3$

$C_3 = \sum_{d_{I_M} \in \Delta}(s_{CM}(d_{I_M}) \cdot C_4)$

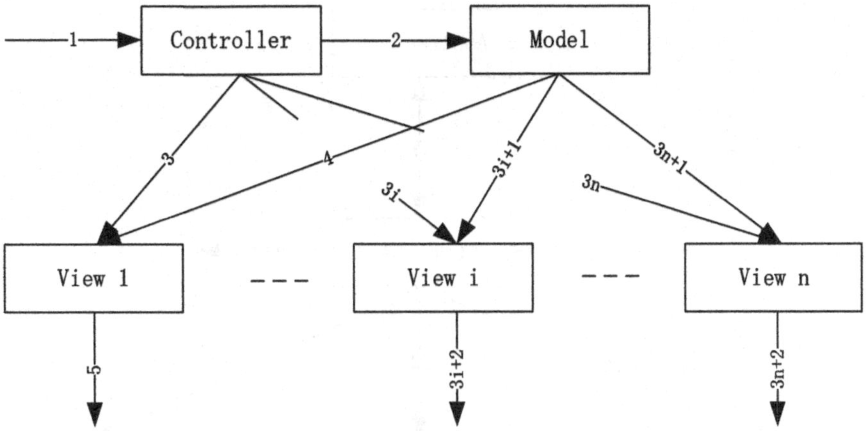


FIGURE 33.15 Typical process of MVC pattern.

$C_4 = \sum_{d_{I_{V_1}},\cdots,d_{I_{V_n}} \in \Delta}(s_{CV_1}(d_{I_{V_1}}) \between \cdots \between s_{CV_n}(d_{I_{V_n}}) \cdot C)$

The state transitions of the Model described by APTC are as follows:

$M = \sum_{d_{I_M} \in \Delta}(r_{CM}(d_{I_M}) \cdot M_2)$

$M_2 = MF \cdot M_3$

$M_3 = \sum_{d_{O_{M_1}},\cdots,d_{O_{M_n}} \in \Delta}(s_{MV_1}(d_{O_{M_1}}) \between \cdots \between s_{MV_n}(d_{O_{M_n}}) \cdot M)$

The state transitions of the View $i$ described by APTC are as follows:

$V_i = \sum_{d_{I_{V_i}} \in \Delta}(r_{CV_i}(d_{I_{V_i}}) \cdot V_{i_2})$

$V_{i_2} = VF_{i_1} \cdot V_{i_3}$

$V_{i_3} = \sum_{d_{O_{M_i}} \in \Delta}(r_{MV_i}(d_{O_{M_i}}) \cdot V_{i_4})$

$V_{i_4} = VF_{i_2} \cdot V_{i_5}$

$V_{i_5} = \sum_{d_{O_i} \in \Delta}(s_{O_i}(d_{O_i}) \cdot V_i)$

The sending action and the reading action of the same data through the same channel can communicate with each other, otherwise, they will cause a deadlock $\delta$. We define the following communication functions of the View $i$ for $1 \leq i \leq n$:

$$\gamma(r_{CV_i}(d_{I_{V_i}}), s_{CV_i}(d_{I_{V_i}})) \triangleq c_{CV_i}(d_{I_{V_i}})$$

$$\gamma(r_{MV_i}(d_{O_{M_i}}), s_{MV_i}(d_{O_{M_i}})) \triangleq c_{MV_i}(d_{O_{M_i}})$$

There is one communication function between the Controller and the Model as follows:

$$\gamma(r_{CM}(d_{I_M}), s_{CM}(d_{I_M})) \triangleq c_{CM}(d_{I_M})$$

Let all modules be in parallel, then the MVC pattern $C \quad M \quad V_1 \cdots V_i \cdots V_n$ can be presented by the following process term:

$\tau_I(\partial_H(\Theta(C \between M \between V_1 \between \cdots \between V_i \between \cdots \between V_n))) = \tau_I(\partial_H(C \between M \between V_1 \between \cdots \between V_i \between \cdots \between V_n))$

where $H = \{r_{CV_i}(d_{I_{V_i}}), s_{CV_i}(d_{I_{V_i}}), r_{MV_i}(d_{O_{M_i}}), s_{MV_i}(d_{O_{M_i}}), r_{CM}(d_{I_M}), s_{CM}(d_{I_M})$
$|d_I, d_{I_M}, d_{I_{V_i}}, d_{O_{M_i}}, d_{O_i} \in \Delta\}$ for $1 \leq i \leq n$,

$I = \{c_{CV_i}(d_{I_{V_i}}), c_{MV_i}(d_{O_{M_i}}), c_{CM}(d_{I_M}), CF, MF, VF_{1_1}, VF_{1_2}, \cdots, VF_{n_1}, VF_{n_2}$
$|d_I, d_{I_M}, d_{I_{V_i}}, d_{O_{M_i}}, d_{O_i} \in \Delta\}$ for $1 \leq i \leq n$

Then, we obtain the following conclusion on the MVC pattern.

**Theorem 33.7** (Correctness of the MVC pattern). *The MVC pattern $\tau_I(\partial_H(C \between M \between V_1 \between \cdots \between V_i \between \cdots \between V_n))$ can exhibit the desired external behaviors.*

*Proof.* Based on the above state transitions of the above modules, by use of the algebraic laws of APTC, we can prove that

$\tau_I(\partial_H(C \between M \between V_1 \between \cdots \between V_i \between \cdots \between V_n)) = \sum_{d_I,d_{O_1},\cdots,d_{O_n} \in \Delta}(r_I(d_I) \cdot s_{O_1}(d_{O_1}) \parallel \cdots \parallel s_{O_i}(d_{O_i}) \parallel \cdots \parallel s_{O_n}(d_{O_n})) \cdot \tau_I(\partial_H(C \between M \between V_1 \between \cdots \between V_i \between \cdots \between V_n))$

That is, the MVC pattern $\tau_I(\partial_H(C \between M \between V_1 \between \cdots \between V_i \between \cdots \between V_n))$ can exhibit the desired external behaviors.

For the details of the proof, please refer to Section 17.3, as we omit them here. □

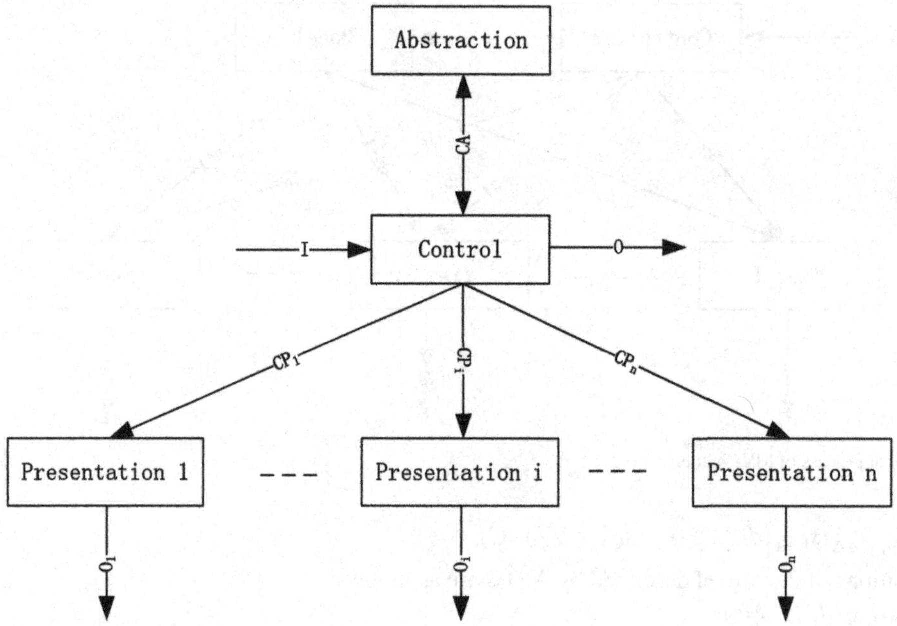


FIGURE 33.16 PAC pattern.

### 33.3.2 Verification of the PAC pattern

The PAC pattern is also used to model the interactive systems, which has three components: the Abstraction, the Presentations, and the Control. The Abstraction is used to contain the data and encapsulate the core functionalities; the Presentations is to show the computational results to the user; and the Control interacts between the system and user, accepts the instructions, and controls the Abstraction and the Presentations, and also other PACs. The Control receives the instructions from the user through the channel $I$, then it sends the instructions to the Abstraction through the channel $CA$ and receives the results through the same channel. Then, the Control sends the results to the Presentation $i$ through the channel $CP_i$ for $1 \leq i \leq n$, and also sends the unprocessed instructions to other PACs through the channel $O$; when the Presentation $i$ receives the results from the Control, it generates or updates the presentation to the user. This is illustrated in Fig. 33.16.

The typical process of the PAC pattern is shown in Fig. 33.17 and is the following:

1. The Control receives the instructions $d_I$ from the user through the channel $I$ (the corresponding reading action is denoted $r_I(D_I)$), processes the instructions through a processing function $CF_1$, and generates the instructions to the Abstraction $d_{I_A}$ and the remaining instructions $d_O$; it sends $d_{I_A}$ to the Abstraction through the channel $CA$ (the corresponding sending action is denoted $s_{CA}(d_{I_A})$), sends $d_O$ to the other PAC through the channel $O$ (the corresponding sending action is denoted $s_O(d_O)$);
2. The Abstraction receives the instructions from the Control through the channel $CA$ (the corresponding reading action is denoted $r_{CA}(d_{I_A})$), processes the instructions through a processing function $AF$, generates the computational results to Control that is denoted $d_{O_A}$, and sends the results to the Control through the channel $CA$ (the corresponding sending action is denoted $s_{CA}(d_{O_A})$);
3. The Control receives the computational results from the Abstraction through channel $CA$ (the corresponding reading action is denoted $r_{CA}(d_{O_A})$), processes the results through a processing function $CF_2$ to generate the results to the Presentation $i$ (for $1 \leq i \leq n$) that is denoted $d_{O_{C_i}}$; then sends the results to the Presentation $i$ through the channel $CP_i$ (the corresponding sending action is denoted $s_{CP_i}(d_{O_{C_i}})$);
4. The Presentation $i$ (for $1 \leq i \leq n$) receives the computational results from the Control through the channel $CP_i$ (the corresponding reading action is denoted $r_{CP_i}(d_{O_{C_i}})$), processes the results through a processing function $PF_i$, generates the output $d_{O_i}$, then sends the output through the channel $O_i$ (the corresponding sending action is denoted $s_{O_i}(d_{O_i})$).

In the following, we verify the PAC pattern. We assume all data elements $d_I$, $d_{I_A}$, $d_{O_A}$, $d_O$, $d_{O_{C_i}}$, $d_{O_i}$ (for $1 \leq i \leq n$) are from a finite set $\Delta$.

The state transitions of the Control module described by APTC are as follows:

$C = \sum_{d_I \in \Delta}(r_I(d_I) \cdot C_2)$

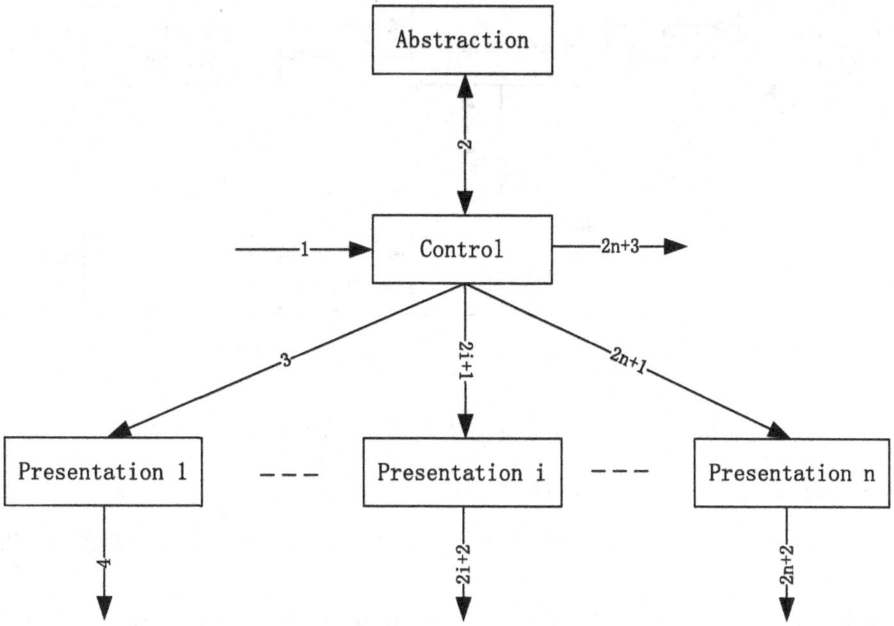


**FIGURE 33.17** Typical process of PAC pattern.

$C_2 = CF_1 \cdot C_3$
$C_3 = \sum_{d_{I_A} \in \Delta}(s_{CA}(d_{I_A}) \cdot C_4)$
$C_4 = \sum_{d_O \in \Delta}(s_O(d_O) \cdot C_5)$
$C_5 = \sum_{d_{O_A} \in \Delta}(r_{CA}(d_{O_A}) \cdot C_6)$
$C_6 = CF_2 \cdot C_7$
$C_7 = \sum_{d_{O_{C_1}}, \cdots, d_{O_{C_n}} \in \Delta}(s_{CP_1}(d_{O_{C_1}}) \between \cdots \between s_{CP_n}(d_{O_{C_n}}) \cdot C)$
The state transitions of the Abstraction described by APTC are as follows:
$A = \sum_{d_{I_A} \in \Delta}(r_{CA}(d_{I_A}) \cdot A_2)$
$A_2 = AF \cdot A_3$
$A_3 = \sum_{d_{O_A} \in \Delta}(s_{CA}(d_{O_A}) \cdot A)$
The state transitions of the Presentation $i$ described by APTC are as follows:
$P_i = \sum_{d_{O_{C_i}} \in \Delta}(r_{CP_i}(d_{O_{C_i}}) \cdot P_{i_2})$
$P_{i_2} = PF_i \cdot P_{i_3}$
$P_{i_3} = \sum_{d_{O_i} \in \Delta}(s_{O_i}(d_{O_i}) \cdot P_i)$
The sending action and the reading action of the same data through the same channel can communicate with each other, otherwise, they will cause a deadlock $\delta$. We define the following communication function of the Presentation $i$ for $1 \le i \le n$:

$$\gamma(r_{CP_i}(d_{O_{C_i}}), s_{CP_i}(d_{O_{C_i}})) \triangleq c_{CP_i}(d_{O_{C_i}})$$

There are two communication functions between the Control and the Abstraction as follows:

$$\gamma(r_{CA}(d_{I_A}), s_{CA}(d_{I_A})) \triangleq c_{CA}(d_{I_A})$$
$$\gamma(r_{CA}(d_{O_A}), s_{CA}(d_{O_A})) \triangleq c_{CA}(d_{O_A})$$

Let all modules be in parallel, then the PAC pattern $C \quad A \quad P_1 \cdots P_i \cdots P_n$ can be presented by the following process term:
$\tau_I(\partial_H(\Theta(C \between A \between P_1 \between \cdots \between P_i \between \cdots \between P_n))) = \tau_I(\partial_H(C \between A \between P_1 \between \cdots \between P_i \between \cdots \between P_n))$
where $H = \{r_{CP_i}(d_{O_{C_i}}), s_{CP_i}(d_{O_{C_i}}), r_{CA}(d_{I_A}), s_{CA}(d_{I_A}), r_{CA}(d_{O_A}), s_{CA}(d_{O_A})$
$|d_I, d_{I_A}, d_{O_A}, d_O, d_{O_{C_i}}, d_{O_i} \in \Delta\}$ for $1 \le i \le n$,

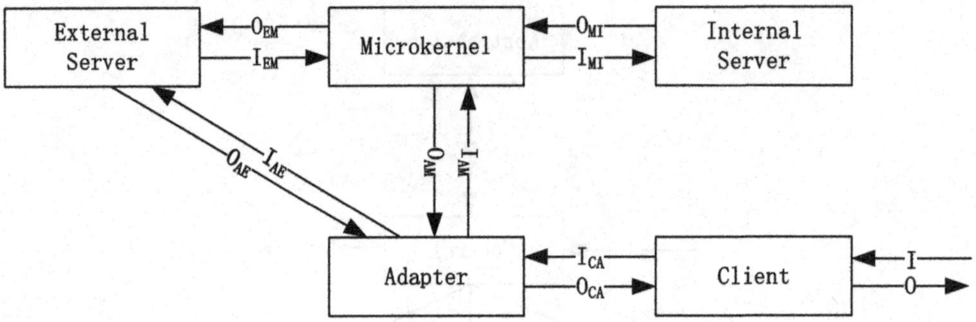


**FIGURE 33.18** Microkernel pattern.

$I = \{c_{CA}(d_{I_A}), c_{CA}(d_{O_A}), c_{CP_1}(d_{O_{C_1}}), \cdots, c_{CP_n}(d_{O_{C_n}}), CF_1, CF_2, AF, PF_1, \cdots, PF_n$
$|d_I, d_{I_A}, d_{O_A}, d_O, d_{O_{C_i}}, d_{O_i} \in \Delta\}$ for $1 \leq i \leq n$

Then, we obtain the following conclusion on the PAC pattern.

**Theorem 33.8** (Correctness of the PAC pattern). *The PAC pattern $\tau_I(\partial_H(C \between A \between P_1 \between \cdots \between P_i \between \cdots \between P_n))$ can exhibit the desired external behaviors.*

*Proof.* Based on the above state transitions of the above modules, by use of the algebraic laws of APTC, we can prove that

$\tau_I(\partial_H(C \between A \between P_1 \between \cdots \between P_i \between \cdots \between P_n)) = \sum_{d_I, d_O, d_{O_1}, \cdots, d_{O_n} \in \Delta}(r_I(d_I) \cdot s_O(d_O) \cdot s_{O_1}(d_{O_1}) \parallel \cdots \parallel s_{O_i}(d_{O_i}) \parallel \cdots \parallel s_{O_n}(d_{O_n})) \cdot \tau_I(\partial_H(C \between A \between P_1 \between \cdots \between P_i \between \cdots \between P_n))$

That is, the PAC pattern $\tau_I(\partial_H(C \between A \between P_1 \between \cdots \between P_i \between \cdots \between P_n))$ can exhibit the desired external behaviors.

For the details of the proof, please refer to Section 17.3, as we omit them here. □

## 33.4 Adaptable systems

In this section, we verify adaptive systems oriented patterns, including the Microkernel pattern and the Reflection pattern.

### 33.4.1 Verification of the Microkernel pattern

The Microkernel pattern adapts the changing of the requirements by implementing the unchangeable requirements as a minimal functional kernel and changeable requirements as external functionalities. There are five modules in the Microkernel pattern: the Microkernel, the Internal Server, the External Server, the Adapter, and the Client. The Client interacts with the user through the channels $I$ and $O$; The Adapter interacts with the Microkernel through the channels $I_{AM}$ and $O_{AM}$, and with the External Server through the channels $I_{AE}$ and $O_{AE}$; The Microkernel interacts with the Internal Server through the channels $I_{MI}$ and $O_{MI}$ and with the External Server through the channels $I_{EM}$ and $O_{EM}$. This is illustrated in Fig. 33.18.

The typical process of the Microkernel pattern is shown in Fig. 33.19 and is as follows:

1. The Client receives the request $d_I$ from the user through the channel $I$ (the corresponding reading action is denoted $r_I(d_I)$), then processes the request $d_I$ through a processing function $CF_1$, and sends the processed request $d_{I_C}$ to the Adapter through the channel $I_{CA}$ (the corresponding sending action is denoted $s_{I_{CA}}(d_{I_C})$);
2. The Adapter receives $d_{I_C}$ from the Client through the channel $I_{CA}$ (the corresponding reading action is denoted $r_{I_{CA}}(d_{I_C})$), then processes the request through a processing function $AF_1$, generates and sends the processed request $d_{I_A}$ to the Microkernel through the channel $I_{AM}$ (the corresponding sending action is denoted $s_{I_{AM}}(d_{I_A})$);
3. The Microkernel receives the request $d_{I_A}$ from the Adapter through the channel $I_{AM}$ (the corresponding reading action is denoted $r_{I_{AM}}(d_{I_A})$), then processes the request through a processing function $MF_1$, generates and sends the processed request $d_{I_M}$ to the Internal Server through the channel $I_{MI}$ (the corresponding sending action is denoted $s_{I_{MI}}(d_{I_M})$), and to the External Server through the channel $I_{EM}$ (the corresponding sending action is denoted $s_{I_{EM}}(d_{I_M})$);
4. The Internal Server receives the request $d_{I_M}$ from the Microkernel through the channel $I_{MI}$ (the corresponding reading action is denoted $r_{I_{MI}}(d_{I_M})$), then processes the request and generates the response $d_{O_I}$ through a processing function $IF$, and sends the response to the Microkernel through the channel $O_{MI}$ (the corresponding sending action is denoted $s_{O_{MI}}(d_{O_I})$);

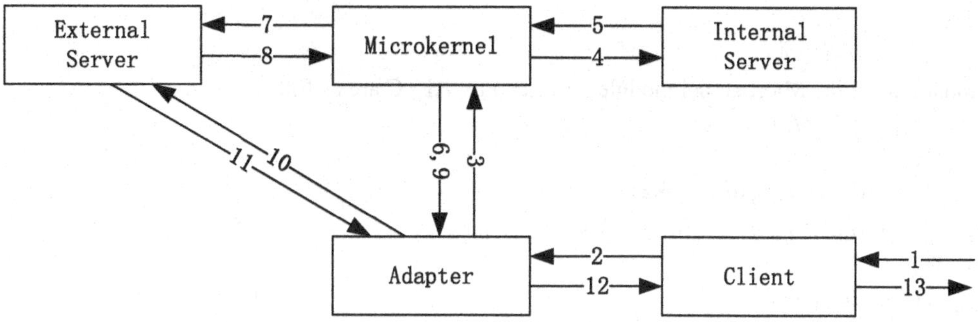


**FIGURE 33.19** Typical process of Microkernel pattern.

5. The External Server receives the request $d_{I_M}$ from the Microkernel through the channel $I_{EM}$ (the corresponding reading action is denoted $r_{I_{EM}}(d_{I_M})$), then processes the request and generates the response $d_{O_E}$ through a processing function $EF_1$, and sends the response to the Microkernel through the channel $O_{EM}$ (the corresponding sending action is denoted $s_{O_{EM}}(d_{O_E})$);
6. The Microkernel receives the response $d_{O_I}$ from the Internal Server through the channel $O_{MI}$ (the corresponding reading action is denoted $r_{O_{MI}}(d_{O_I})$) and the response $d_{O_E}$ from the External Server through the channel $O_{EM}$ (the corresponding reading action is denoted $r_{O_{EM}}(d_{O_E})$), then processes the responses and generates the response $d_{O_M}$ through a processing function $MF_2$, and sends $d_{O_M}$ to the Adapter through the channel $O_{AM}$ (the corresponding sending action is denoted $s_{O_{AM}}(d_{O_M})$);
7. The Adapter receives the response $d_{O_M}$ from the Microkernel through the channel $O_{AM}$ (the corresponding reading action is denoted $r_{O_{AM}}(d_{O_M})$), it may send $d_{I_{A'}}$ to the External Server through the channel $I_{AE}$ (the corresponding sending action is denoted $s_{I_{AE}}(d_{I_{A'}})$);
8. The External Server receives the request $d_{I_{A'}}$ from the Adapter through the channel $I_{AE}$ (the corresponding reading action is denoted $r_{I_{AE}}(d_{I_{A'}})$), then processes the request and generates the response $d_{O_{E'}}$ through a processing function $EF_2$, and sends $d_{O_{E'}}$ to the Adapter through the channel $O_{AE}$ (the corresponding sending action is denoted $s_{O_{AE}}(d_{O_{E'}})$);
9. The Adapter receives the response from the External Server through the channel $O_{AE}$ (the corresponding reading action is denoted $r_{O_AE}(d_{O_{E'}})$), then processes $d_{O_M}$ and $d_{O_{E'}}$ through a processing function $AF_2$ and generates the response $d_{O_A}$, and sends $d_{O_A}$ to the Client through the channel $O_{CA}$ (the corresponding sending action is denoted $s_{O_{CA}}(d_{O_A})$);
10. The Client receives the response $d_{O_A}$ from the Adapter through the channel $O_{CA}$ (the corresponding reading action is denoted $r_{O_{CA}}(d_{O_A})$), then processes $d_{O_A}$ through a processing function $CF_2$ and generates the response $d_O$, and sends $d_O$ to the user through the channel $O$ (the corresponding sending action is denoted $s_O(d_O)$).

In the following, we verify the Microkernel pattern. We assume all data elements $d_I, d_{I_C}, d_{I_A}, d_{I_{A'}}, d_{I_M}, d_{O_I}, d_{O_E}, d_{O_{E'}}$, $d_{O_M}, d_O, d_{O_A}$ (for $1 \leq i \leq n$) are from a finite set $\Delta$.

The state transitions of the Client module described by APTC are as follows:

$C = \sum_{d_I \in \Delta}(r_I(d_I) \cdot C_2)$
$C_2 = CF_1 \cdot C_3$
$C_3 = \sum_{d_{I_C} \in \Delta}(s_{I_{CA}}(d_{I_C}) \cdot C_4)$
$C_4 = \sum_{d_{O_A} \in \Delta}(r_{O_{CA}}(d_{O_A}) \cdot C_5)$
$C_5 = CF_2 \cdot C_6$
$C_6 = \sum_{d_O \in \Delta}(s_O(d_O) \cdot C)$

The state transitions of the Adapter module described by APTC are as follows:

$A = \sum_{d_{I_C} \in \Delta}(r_{I_{CA}}(d_{I_C}) \cdot A_2)$
$A_2 = AF_1 \cdot A_3$
$A_3 = \sum_{d_{I_A} \in \Delta}(s_{I_{AM}}(d_{I_A}) \cdot A_4)$
$A_4 = \sum_{d_{O_M} \in \Delta}(r_{O_{AM}}(d_{O_M}) \cdot A_5)$
$A_5 = \sum_{d_{I_{A'}} \in \Delta}(s_{I_{AE}}(d_{I_{A'}}) \cdot A_6)$
$A_6 = \sum_{d_{O_{E'}} \in \Delta}(r_{O_{AE}}(d_{O_{E'}}) \cdot A_7)$

$A_7 = AF_2 \cdot A_8$
$A_8 = \sum_{d_{O_A} \in \Delta}(s_{O_{CA}}(d_{O_A}) \cdot A)$
The state transitions of the Microkernel module described by APTC are as follows:
$M = \sum_{d_{I_A} \in \Delta}(r_{I_{AM}}(d_{I_A}) \cdot M_2)$
$M_2 = MF_1 \cdot M_3$
$M_3 = \sum_{d_{I_M} \in \Delta}(s_{I_{MI}}(d_{I_M}) \between s_{I_{EM}}(d_{I_M}) \cdot M_4)$
$M_4 = \sum_{d_{O_I}, d_{O_E} \in \Delta}(r_{O_{MI}}(d_{O_I}) \between r_{O_{EM}}(d_{O_E}) \cdot M_5)$
$M_5 = MF_2 \cdot M_6)$
$M_6 = \sum_{d_{O_M} \in \Delta}(s_{O_{AM}}(d_{O_M}) \cdot M)$
The state transitions of the Internal Server described by APTC are as follows:
$I = \sum_{d_{I_M} \in \Delta}(r_{I_{MI}}(d_{I_M}) \cdot I_2)$
$I_2 = IF \cdot I_3$
$I_3 = \sum_{d_{O_I} \in \Delta}(s_{O_{MI}}(d_{O_I}) \cdot I)$
The state transitions of the External Server described by APTC are as follows:
$E = \sum_{d_{I_M} \in \Delta}(r_{I_{EM}}(d_{I_M}) \cdot E_2)$
$E_2 = EF_1 \cdot E_3$
$E_3 = \sum_{d_{O_E} \in \Delta}(s_{O_{EM}}(d_{O_E}) \cdot E_4)$
$E_4 = \sum_{d_{I_{A'}} \in \Delta}(r_{I_{AE}}(d_{I_{A'}}) \cdot E_5)$
$E_5 = EF_2 \cdot E_6$
$E_6 = \sum_{d_{O_{E'}} \in \Delta}(s_{O_{AE}}(d_{O_{E'}}) \cdot E)$

The sending action and the reading action of the same data through the same channel can communicate with each other, otherwise, they will cause a deadlock $\delta$. We define the following communication functions between the Client and the Adapter:

$$\gamma(r_{I_{CA}}(d_{I_C}), s_{I_{CA}}(d_{I_C})) \triangleq c_{I_{CA}}(d_{I_C})$$
$$\gamma(r_{O_{CA}}(d_{O_A}), s_{O_{CA}}(d_{O_A})) \triangleq c_{O_{CA}}(d_{O_A})$$

There are two communication functions between the Adapter and the Microkernel as follows:

$$\gamma(r_{I_{AM}}(d_{I_A}), s_{I_{AM}}(d_{I_A})) \triangleq c_{I_{AM}}(d_{I_A})$$
$$\gamma(r_{O_{AM}}(d_{O_M}), s_{O_{AM}}(d_{O_M})) \triangleq c_{O_{AM}}(d_{O_M})$$

There are two communication functions between the Adapter and the External Server as follows:

$$\gamma(r_{I_{AE}}(d_{I_{A'}}), s_{I_{AE}}(d_{I_{A'}})) \triangleq c_{I_{AE}}(d_{I_{A'}})$$
$$\gamma(r_{O_{AE}}(d_{O_{E'}}), s_{O_{AE}}(d_{O_{E'}})) \triangleq c_{O_{AE}}(d_{O_{E'}})$$

There are two communication functions between the Internal Server and the Microkernel as follows:

$$\gamma(r_{I_{MI}}(d_{I_M}), s_{I_{MI}}(d_{I_M})) \triangleq c_{I_{MI}}(d_{I_M})$$
$$\gamma(r_{O_{MI}}(d_{O_I}), s_{O_{MI}}(d_{O_I})) \triangleq c_{O_{MI}}(d_{O_I})$$

There are two communication functions between the External Server and the Microkernel as follows:

$$\gamma(r_{I_{EM}}(d_{I_M}), s_{I_{EM}}(d_{I_M})) \triangleq c_{I_{EM}}(d_{I_M})$$
$$\gamma(r_{O_{EM}}(d_{O_E}), s_{O_{EM}}(d_{O_E})) \triangleq c_{O_{EM}}(d_{O_E})$$

Let all modules be in parallel, then the Microkernel pattern $C \quad A \quad M \quad I \quad E$ can be presented by the following process term:

$\tau_I(\partial_H(\Theta(C \between A \between M \between I \between E))) = \tau_I(\partial_H(C \between A \between M \between I \between E))$
where $H = \{r_{I_{CA}}(d_{I_C}), s_{I_{CA}}(d_{I_C}), r_{O_{CA}}(d_{O_A}), s_{O_{CA}}(d_{O_A}), r_{I_{AM}}(d_{I_A}), s_{I_{AM}}(d_{I_A}),$
$r_{O_{AM}}(d_{O_M}), s_{O_{AM}}(d_{O_M}), r_{I_{AE}}(d_{I_{A'}}), s_{I_{AE}}(d_{I_{A'}}), r_{O_{AE}}(d_{O_{E'}}), s_{O_{AE}}(d_{O_{E'}}),$

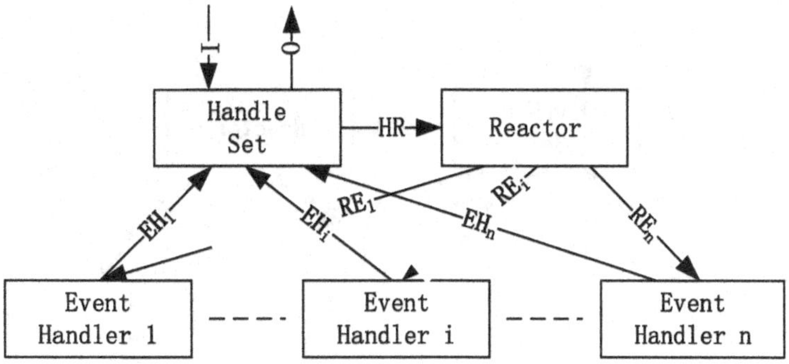


FIGURE 33.20 Reflection pattern.

$r_{I_{MI}}(d_{I_M}), s_{I_{MI}}(d_{I_M}), r_{O_{MI}}(d_{O_I}), s_{O_{MI}}(d_{O_I}), r_{I_{EM}}(d_{I_M}), s_{I_{EM}}(d_{I_M}),$
$r_{O_{EM}}(d_{O_E}), s_{O_{EM}}(d_{O_E}) | d_I, d_{I_C}, d_{I_A}, d_{I_{A'}}, d_{I_M}, d_{O_I}, d_{O_E}, d_{O_{E'}}, d_{O_M}, d_O, d_{O_A} \in \Delta\}$,

$I = \{c_{I_{CA}}(d_{I_C}), c_{O_{CA}}(d_{O_A}), c_{I_{AM}}(d_{I_A}), c_{O_{AM}}(d_{O_M}), c_{I_{AE}}(d_{I_{A'}}), c_{O_{AE}}(d_{O_{E'}}),$
$c_{I_{MI}}(d_{I_M}), c_{O_{MI}}(d_{O_I}), c_{I_{EM}}(d_{I_M}), c_{O_{EM}}(d_{O_E}), CF_1, CF_2, AF_1, AF_2, MF_1, MF_2, IF, EF_1, EF_2$
$| d_I, d_{I_C}, d_{I_A}, d_{I_{A'}}, d_{I_M}, d_{O_I}, d_{O_E}, d_{O_{E'}}, d_{O_M}, d_O, d_{O_A} \in \Delta\}$

Then, we obtain the following conclusion on the Microkernel pattern.

**Theorem 33.9** (Correctness of the Microkernel pattern). *The Microkernel pattern $\tau_I(\partial_H(C \between A \between M \between I \between E))$ can exhibit the desired external behaviors.*

*Proof.* Based on the above state transitions of the above modules, by use of the algebraic laws of APTC, we can prove that

$\tau_I(\partial_H(C \between A \between M \between I \between E)) = \sum_{d_I, d_O \in \Delta}(r_I(d_I) \cdot s_O(d_O)) \cdot \tau_I(\partial_H(C \between A \between M \between I \between E))$

That is, the Microkernel pattern $\tau_I(\partial_H(C \between A \between M \between I \between E))$ can exhibit the desired external behaviors.

For the details of the proof, please refer to Section 17.3, as we omit them here. □

## 33.4.2 Verification of the Reflection pattern

The Reflection pattern makes the system able to change its structure and behaviors dynamically. There are two levels in the Reflection pattern: one is the meta level to encapsulate the information of system properties and make the system self-aware; the other is the base level to implement the concrete application logic. The meta level modules include the Metaobject Protocol and $n$ Metaobject. The Metaobject Protocol is used to configure the Metaobjects, and it interacts with Metaobject $i$ through the channels $I_{MP_i}$ and $O_{MP_i}$, and it exchanges configuration information with the outside through the input channel $I_M$ and $O_M$. The Metaobject encapsulate the system properties, and it interacts with the Metaobject Protocol, and with the Component through the channels $I_{MC_i}$ and $O_{MC_i}$. The base level modules including concrete Components, which interact with the Metaobject, and with the outside through the input channel $I_C$ and $O_C$. This is illustrated in Fig. 33.20.

The typical process of the Reflection pattern is shown in Fig. 33.21 and is as follows:

1. The Metaobject Protocol receives the configuration information from the user through the channel $I_M$ (the corresponding reading action is denoted $r_{I_M}(d_{I_M})$), then processes the information through a processing function $PF_1$ and generates the configuration $d_{I_P}$, and sends $d_{I_P}$ to the Metaobject $i$ (for $1 \leq i \leq n$) through the channel $I_{MP_i}$ (the corresponding sending action is denoted $s_{I_{MP_i}}(d_{I_P})$);
2. The Metaobject $i$ receives the configuration $d_{I_P}$ from the Metaobject Protocol through the channel $I_{MP_i}$ (the corresponding reading action is denoted $r_{I_{MP_i}}(d_{I_P})$), then configures the properties through a configuration function $MF_{i1}$, and sends the configuration results $d_{O_{M_{i1}}}$ to the Metaobject Protocol through the channel $O_{MP_i}$ (the corresponding sending action is denoted $s_{O_{MP_i}}(d_{O_{M_{i1}}})$);
3. The Metaobject Protocol receives the configuration results from the Metaobject $i$ through the channel $O_{MP_i}$ (the corresponding reading action is denoted $r_{O_{MP_i}}(d_{O_{M_{i1}}})$), then processes the results through a processing function $PF_2$ and generates the result $d_{O_M}$, and sends $d_{O_M}$ to the outside through the channel $O_M$ (the corresponding sending action is denoted $s_{O_M}(d_{O_M})$);
4. The Component receives the invocation from the user through the channel $I_C$ (the corresponding reading action is denoted $r_{I_C}(d_{I_C})$), then processes the invocation through a processing function $CF_1$ and generates the configuration $d_{I_C}$, and sends $d_{I_C}$ to the Metaobject $i$ (for $1 \leq i \leq n$) through the channel $I_{MC_i}$ (the corresponding sending action is denoted $s_{I_{MC_i}}(d_{I_C})$);

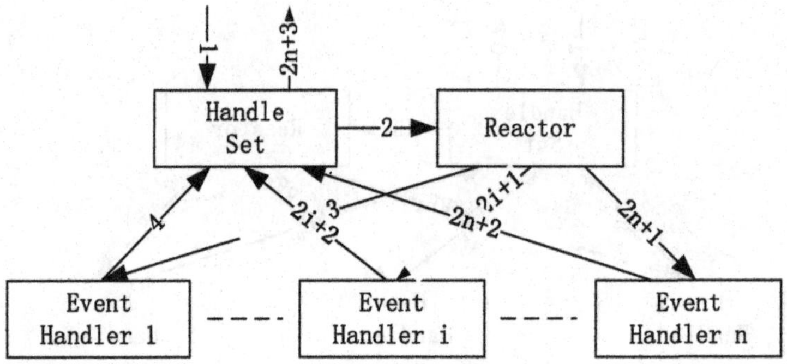


FIGURE 33.21 Typical process of Reflection pattern.

5. The Metaobject $i$ receives the invocation $d_{I_C}$ from the Component through the channel $I_{MC_i}$ (the corresponding reading action is denoted $r_{I_{MC_i}}(d_{I_C})$), then computes through a computational function $MF_{i2}$, and sends the computational results $d_{O_{M_{i2}}}$ to the Component through the channel $O_{MC_i}$ (the corresponding sending action is denoted $s_{O_{MC_i}}(d_{O_{M_{i2}}})$);
6. The Component receives the computational results from the Metaobject $i$ through the channel $O_{MC_i}$ (the corresponding reading action is denoted $r_{O_{MC_i}}(d_{O_{M_{i2}}})$), then processes the results through a processing function $CF_2$ and generates the result $d_{O_C}$, then sends $d_{O_C}$ to the outside through the channel $O_C$ (the corresponding sending action is denoted $s_{O_C}(d_{O_C})$).

In the following, we verify the Reflection pattern. We assume all data elements $d_{I_M}, d_{I_C}, d_{I_P}, d_{O_{M_{i1}}}, d_{O_{M_{i2}}}, d_{O_M}, d_{O_C}$ (for $1 \leq i \leq n$) are from a finite set $\Delta$.

The state transitions of the Metaobject Protocol module described by APTC are as follows:

$P = \sum_{d_{I_M} \in \Delta}(r_{I_M}(d_{I_M}) \cdot P_2)$

$P_2 = PF_1 \cdot P_3$

$P_3 = \sum_{d_{I_P} \in \Delta}(s_{I_{MP_1}}(d_{I_P}) \between \cdots \between s_{I_{MP_n}}(d_{I_P}) \cdot P_4)$

$P_4 = \sum_{d_{O_{M_{11}}}, \cdots, d_{O_{M_{n1}}} \in \Delta}(r_{O_{MP_1}}(d_{O_{M_{11}}}) \between \cdots \between r_{O_{MP_n}}(d_{O_{M_{n1}}}) \cdot P_5)$

$P_5 = PF_2 \cdot P_6$

$P_6 = \sum_{d_{O_M} \in \Delta}(s_{O_M}(d_{O_M}) \cdot P)$

The state transitions of the Component described by APTC are as follows:

$C = \sum_{d_{I_C} \in \Delta}(r_{I_C}(d_{I_C}) \cdot C_2)$

$C_2 = CF_1 \cdot C_3$

$C_3 = \sum_{d_{I_P} \in \Delta}(s_{I_{MC_1}}(d_{I_C}) \between \cdots \between s_{I_{MC_n}}(d_{I_C}) \cdot C_4)$

$C_4 = \sum_{d_{O_{M_{12}}}, \cdots, d_{O_{M_{n2}}} \in \Delta}(r_{O_{MC_1}}(d_{O_{M_{12}}}) \between \cdots \between r_{O_{MC_n}}(d_{O_{M_{n2}}}) \cdot C_5)$

$C_5 = CF_2 \cdot C_6$

$C_6 = \sum_{d_{O_C} \in \Delta}(s_{O_C}(d_{O_C}) \cdot C)$

The state transitions of the Metaobject $i$ described by APTC are as follows:

$M_i = \sum_{d_{I_P} \in \Delta}(r_{I_{MP_i}}(d_{I_P}) \cdot M_{i_2})$

$M_{i_2} = MF_{i1} \cdot M_{i_3}$

$M_{i_3} = \sum_{d_{O_{M_{i1}}} \in \Delta}(s_{O_{MP_i}}(d_{O_{M_{i1}}}) \cdot M_{i_4})$

$M_{i_4} = \sum_{d_{I_C} \in \Delta}(r_{I_{MC_i}}(d_{I_C}) \cdot M_{i_5})$

$M_{i_5} = MF_{i2} \cdot M_{i_6}$

$M_{i_6} = \sum_{d_{O_{M_{i2}}} \in \Delta}(s_{O_{MC_i}}(d_{O_{M_{i2}}}) \cdot M_i)$

The sending action and the reading action of the same data through the same channel can communicate with each other, otherwise, they will cause a deadlock $\delta$. We define the following communication functions of the Metaobject $i$ for $1 \leq i \leq n$:

$$\gamma(r_{I_{MP_i}}(d_{I_P}), s_{I_{MP_i}}(d_{I_P})) \triangleq c_{I_{MP_i}}(d_{I_P})$$

$$\gamma(r_{O_{MP_i}}(d_{O_{M_{i1}}}), s_{O_{MP_i}}(d_{O_{M_{i1}}})) \triangleq c_{O_{MP_i}}(d_{O_{M_{i1}}})$$

$$\gamma(r_{I_{MC_i}}(d_{I_C}), s_{I_{MC_i}}(d_{I_C})) \triangleq c_{I_{MC_i}}(d_{I_C})$$

$$\gamma(r_{OMC_i}(d_{O_{M_{i2}}}), s_{OMC_i}(d_{O_{M_{i2}}})) \triangleq c_{OMC_i}(d_{O_{M_{i2}}})$$

Let all modules be in parallel, then the Reflection pattern $C \quad P \quad M_1 \cdots M_i \cdots M_n$ can be presented by the following process term:

$\tau_I(\partial_H(\Theta(C \between P \between M_1 \between \cdots \between M_i \between \cdots \between M_n))) = \tau_I(\partial_H(C \between P \between M_1 \between \cdots \between M_i \between \cdots \between M_n))$

where $H = \{r_{IMP_i}(d_{I_P}), s_{IMP_i}(d_{I_P})), r_{OMP_i}(d_{O_{M_{i1}}}), s_{OMP_i}(d_{O_{M_{i1}}}),$
$r_{IMC_i}(d_{I_C}), s_{IMC_i}(d_{I_C}), r_{OMC_i}(d_{O_{M_{i2}}}), s_{OMC_i}(d_{O_{M_{i2}}}) | d_{I_M}, d_{I_C}, d_{I_P}, d_{O_{M_{i1}}}, d_{O_{M_{i2}}}, d_{O_M}, d_{O_C} \in \Delta\}$ for $1 \le i \le n$,

$I = \{c_{IMP_i}(d_{I_P}), c_{OMP_i}(d_{O_{M_{i1}}}), c_{IMC_i}(d_{I_C}), c_{OMC_i}(d_{O_{M_{i2}}}), PF_1, PF_2, CF_1, CF_2, MF_{i1}, MF_{i2}$
$| d_{I_M}, d_{I_C}, d_{I_P}, d_{O_{M_{i1}}}, d_{O_{M_{i2}}}, d_{O_M}, d_{O_C} \in \Delta\}$ for $1 \le i \le n$

Then, we obtain the following conclusion on the Reflection pattern.

**Theorem 33.10** (Correctness of the Reflection pattern). *The Reflection pattern $\tau_I(\partial_H(C \between P \between M_1 \between \cdots \between M_i \between \cdots \between M_n))$ can exhibit the desired external behaviors.*

*Proof.* Based on the above state transitions of the above modules, by use of the algebraic laws of APTC, we can prove that

$\tau_I(\partial_H(C \between P \between M_1 \between \cdots \between M_i \between \cdots \between M_n)) = \sum_{d_{I_M}, d_{I_C}, d_{O_M}, d_{O_C} \in \Delta}(r_{I_M}(d_{I_M}) \parallel r_{I_C}(d_{I_C}) \cdot s_{O_M}(d_{O_M}) \parallel s_{O_C}(d_{O_C})) \cdot \tau_I(\partial_H(C \between P \between M_1 \between \cdots \between M_i \between \cdots \between M_n))$

That is, the Reflection pattern $\tau_I(\partial_H(C \between P \between M_1 \between \cdots \between M_i \between \cdots \between M_n))$ can exhibit the desired external behaviors.

For the details of the proof, please refer to Section 17.3, as we omit them here. □

Chapter 34

# Verification of design patterns

Design patterns are middle-level patterns, which are lower than the architecture patterns and higher than the programming language-specific idioms. Design patterns describe the architecture of the subsystems.

In this chapter, we verify the five categories of design patterns. In Section 34.1, we verify the patterns related to structural decomposition. In Section 34.2, we verify the patterns related to organization of work. We verify the patterns related to access control in Section 34.3 and verify management oriented patterns in Section 34.4. Finally, we verify the communication oriented patterns in Section 34.5.

## 34.1 Structural decomposition

In this section, we verify structural decomposition related patterns, including the Whole-Part pattern.

### 34.1.1 Verification the Whole-Part pattern

The Whole-Part pattern is used to divide application logics into Parts and aggregate the Parts into a Whole. In this pattern, there are a Whole module and $n$ Part modules. The Whole module interacts with the outside through the channels $I$ and $O$, and with Part $i$ (for $1 \leq i \leq n$) through the channels $I_{WP_i}$ and $O_{WP_i}$, as illustrated in Fig. 34.1.

The typical process of the Whole-Part pattern is shown in Fig. 34.2 and is as follows:

1. The Whole receives the request $d_I$ from the outside through the channel $I$ (the corresponding reading action is denoted $r_I(d_I)$), then processes the request through a processing function $WF_1$ and generates the request $d_{I_W}$, and sends the $d_{I_W}$ to the Part $i$ through the channel $I_{WP_i}$ (the corresponding sending action is denoted $s_{I_{WP_i}}(d_{I_W})$);
2. The Part $i$ receives the request $d_{I_W}$ from the Whole through the channel $I_{WP_i}$ (the corresponding reading action is denoted $r_{I_{WP_i}}(d_{I_W})$), then processes the request through a processing function $PF_i$ and generates the response $d_{O_{P_i}}$, and sends the response to the Whole through the channel $O_{WP_i}$ (the corresponding sending action is denoted $s_{O_{WP_i}}(d_{O_{P_i}})$);
3. The Whole receives the response $d_{O_{P_i}}$ from the Part $i$ through the channel $O_{WP_i}$ (the corresponding reading action is denoted $r_{O_{WP_i}}(d_{O_{P_i}})$), then processes the request through a processing function $WF_2$ and generates the request $d_O$, and sends the $d_O$ to the outside through the channel $O$ (the corresponding sending action is denoted $s_O(d_O)$).

In the following, we verify the Whole-Part pattern. We assume all data elements $d_I$, $d_{I_W}$, $d_{O_{P_i}}$, $d_O$ (for $1 \leq i \leq n$) are from a finite set $\Delta$.

The state transitions of the Whole module described by APTC are as follows:

$W = \sum_{d_I \in \Delta}(r_I(d_I) \cdot W_2)$

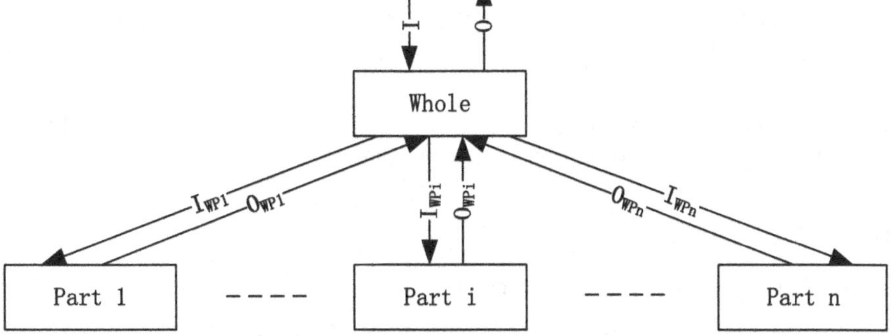


FIGURE 34.1 Whole-Part pattern.

Handbook of Truly Concurrent Process Algebra. https://doi.org/10.1016/B978-0-44-321515-5.00038-X

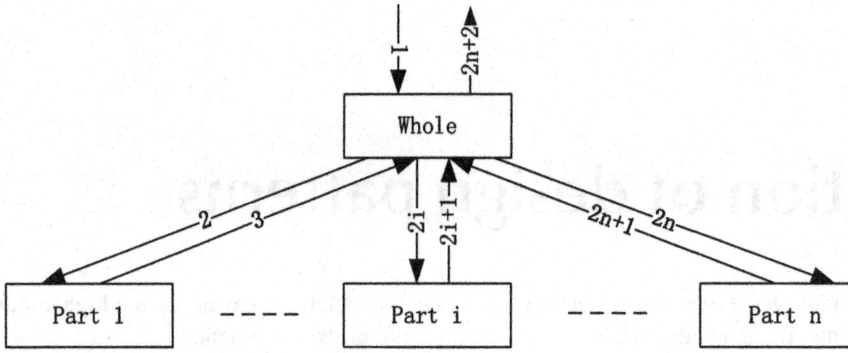

**FIGURE 34.2** Typical process of Whole-Part pattern.

$W_2 = WF_1 \cdot W_3$

$W_3 = \sum_{d_{I_W} \in \Delta} (s_{I_{WP_1}}(d_{I_W}) \between \cdots \between s_{I_{WP_n}}(d_{I_W}) \cdot W_4)$

$W_4 = \sum_{d_{O_{P_1}}, \cdots, d_{O_{P_n}} \in \Delta} (r_{O_{WP_1}}(d_{O_{P_1}}) \between \cdots \between r_{O_{WP_n}}(d_{O_{P_n}}) \cdot W_5)$

$W_5 = WF_2 \cdot W_6$

$W_6 = \sum_{d_O \in \Delta} (s_O(d_O) \cdot W)$

The state transitions of the Part $i$ described by APTC are as follows:

$P_i = \sum_{d_{I_W} \in \Delta} (r_{I_{WP_i}}(d_{I_W}) \cdot P_{i_2})$

$P_{i_2} = PF_i \cdot P_{i_3}$

$P_{i_3} = \sum_{d_{O_{P_i}} \in \Delta} (s_{O_{WP_i}}(d_{O_{P_i}}) \cdot P)$

The sending action and the reading action of the same data through the same channel can communicate with each other, otherwise, they will cause a deadlock $\delta$. We define the following communication functions of the Part $i$ for $1 \le i \le n$:

$$\gamma(r_{I_{WP_i}}(d_{I_W}), s_{I_{WP_i}}(d_{I_W})) \triangleq c_{I_{WP_i}}(d_{I_W})$$

$$\gamma(r_{O_{WP_i}}(d_{O_{P_i}}), s_{O_{WP_i}}(d_{O_{P_i}})) \triangleq c_{O_{WP_i}}(d_{O_{P_i}})$$

Let all modules be in parallel, then the Whole-Part pattern $Q \quad P_1 \cdots P_i \cdots P_n$ can be presented by the following process term:

$\tau_I(\partial_H(\Theta(W \between P_1 \between \cdots \between P_i \between \cdots \between P_n))) = \tau_I(\partial_H(W \between P_1 \between \cdots \between P_i \between \cdots \between P_n))$

where $H = \{r_{I_{WP_i}}(d_{I_W}), s_{I_{WP_i}}(d_{I_W}), r_{O_{WP_i}}(d_{O_{P_i}}), s_{O_{WP_i}}(d_{O_{P_i}})$

$|d_I, d_{I_W}, d_{O_{P_i}}, d_O \in \Delta\}$ for $1 \le i \le n$,

$I = \{c_{I_{WP_i}}(d_{I_W}), c_{O_{WP_i}}(d_{O_{P_i}}), WF_1, WF_2, PF_i$

$|d_I, d_{I_W}, d_{O_{P_i}}, d_O \in \Delta\}$ for $1 \le i \le n$

Then, we obtain the following conclusion on the Whole-Part pattern.

**Theorem 34.1** (Correctness of the Whole-Part pattern). *The Whole-Part pattern $\tau_I(\partial_H(W \between P_1 \between \cdots \between P_i \between \cdots \between P_n))$ can exhibit the desired external behaviors.*

*Proof.* Based on the above state transitions of the above modules, by use of the algebraic laws of APTC, we can prove that

$\tau_I(\partial_H(W \between P_1 \between \cdots \between P_i \between \cdots \between P_n)) = \sum_{d_{I_M}, d_{I_C}, d_{O_M}, d_{O_C} \in \Delta} (r_{I_M}(d_{I_M}) \parallel r_{I_C}(d_{I_C}) \cdot s_{O_M}(d_{O_M}) \parallel s_{O_C}(d_{O_C})) \cdot \tau_I(\partial_H(W \between P_1 \between \cdots \between P_i \between \cdots \between P_n))$

That is, the Whole-Part pattern $\tau_I(\partial_H(W \between P_1 \between \cdots \between P_i \between \cdots \between P_n))$ can exhibit the desired external behaviors.

For the details of the proof, please refer to Section 17.3, as we omit them here. □

## 34.2 Organization of work

### 34.2.1 Verification of the Master–Slave pattern

The Master–Slave pattern is used to implement large-scale computation. In this pattern, there are a Master module and $n$ Slave modules. The Slaves are used to implement concrete computations and the Master is used to distribute computational

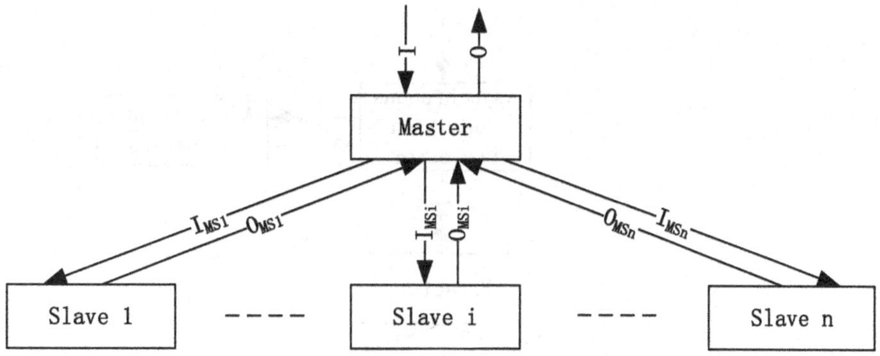


**FIGURE 34.3** Master–Slave pattern.

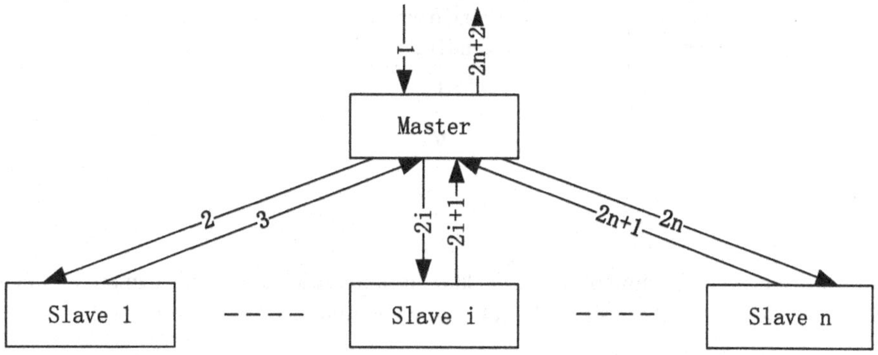


**FIGURE 34.4** Typical process of Master–Slave pattern.

tasks and collect the computational results. The Master module interacts with the outside through the channels $I$ and $O$, and with Slave $i$ (for $1 \leq i \leq n$) through the channels $I_{MS_i}$ and $O_{MS_i}$, as illustrated in Fig. 34.3.

The typical process of the Master–Slave pattern is shown in Fig. 34.4 and is as follows:

1. The Master receives the request $d_I$ from the outside through the channel $I$ (the corresponding reading action is denoted $r_I(d_I)$), then processes the request through a processing function $MF_1$ and generates the request $d_{I_M}$, and sends the $d_{I_M}$ to the Slave $i$ through the channel $I_{MS_i}$ (the corresponding sending action is denoted $s_{I_{MS_i}}(d_{I_M})$);
2. The Slave $i$ receives the request $d_{I_M}$ from the Master through the channel $I_{MS_i}$ (the corresponding reading action is denoted $r_{I_{MS_i}}(d_{I_M})$), then processes the request through a processing function $SF_i$ and generates the response $d_{O_{S_i}}$, and sends the response to the Master through the channel $O_{MS_i}$ (the corresponding sending action is denoted $s_{O_{MS_i}}(d_{O_{S_i}})$);
3. The Master receives the response $d_{O_{S_i}}$ from the Slave $i$ through the channel $O_{MS_i}$ (the corresponding reading action is denoted $r_{O_{MS_i}}(d_{O_{S_i}})$), then processes the request through a processing function $MF_2$ and generates the request $d_O$, and sends the $d_O$ to the outside through the channel $O$ (the corresponding sending action is denoted $s_O(d_O)$).

In the following, we verify the Master–Slave pattern. We assume all data elements $d_I, d_{I_M}, d_{O_{S_i}}, d_O$ (for $1 \leq i \leq n$) are from a finite set $\Delta$.

The state transitions of the Master module described by APTC are as follows:

$M = \sum_{d_I \in \Delta}(r_I(d_I) \cdot M_2)$

$M_2 = MF_1 \cdot M_3$

$M_3 = \sum_{d_{I_M} \in \Delta}(s_{I_{MS_1}}(d_{I_M}) \between \cdots \between s_{I_{MS_n}}(d_{I_M}) \cdot M_4)$

$M_4 = \sum_{d_{O_{S_1}}, \cdots, d_{O_{S_n}} \in \Delta}(r_{O_{MS_1}}(d_{O_{S_1}}) \between \cdots \between r_{O_{MS_n}}(d_{O_{S_n}}) \cdot M_5)$

$M_5 = MF_2 \cdot M_6$

$M_6 = \sum_{d_O \in \Delta}(s_O(d_O) \cdot M)$

The state transitions of the Slave $i$ described by APTC are as follows:

$S_i = \sum_{d_{I_M} \in \Delta}(r_{I_{MS_i}}(d_{I_M}) \cdot S_{i_2})$

$S_{i_2} = SF_i \cdot S_{i_3}$

$S_{i_3} = \sum_{d_{O_{S_i}} \in \Delta}(s_{O_{MS_i}}(d_{O_{S_i}}) \cdot S)$

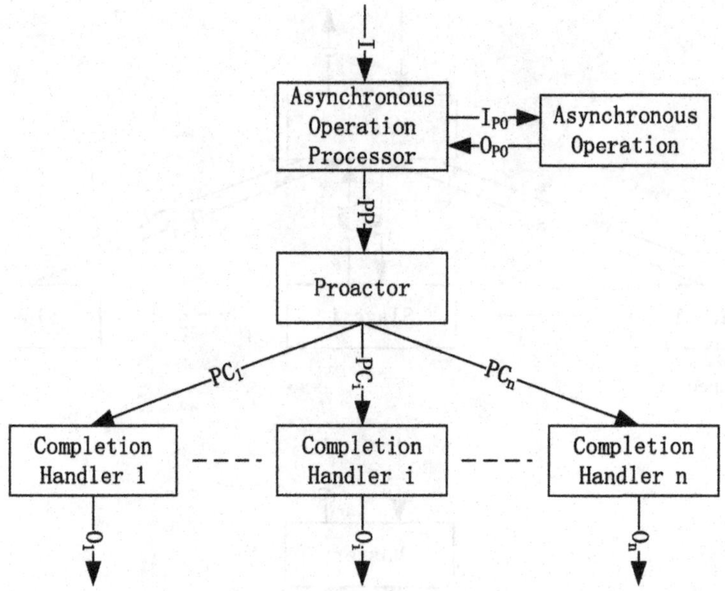


**FIGURE 34.5** Proxy pattern.

The sending action and the reading action of the same data through the same channel can communicate with each other, otherwise, they will cause a deadlock $\delta$. We define the following communication functions of the Slave $i$ for $1 \leq i \leq n$:

$$\gamma(r_{IMS_i}(d_{IM}), s_{IMS_i}(d_{IM})) \triangleq c_{IMS_i}(d_{IM})$$

$$\gamma(r_{OMS_i}(d_{OS_i}), s_{OMS_i}(d_{OS_i})) \triangleq c_{OMS_i}(d_{OS_i})$$

Let all modules be in parallel, then the Master–Slave pattern $M \quad S_1 \cdots S_i \cdots S_n$ can be presented by the following process term:

$\tau_I(\partial_H(\Theta(M \between S_1 \between \cdots \between S_i \between \cdots \between S_n))) = \tau_I(\partial_H(M \between S_1 \between \cdots \between S_i \between \cdots \between S_n))$

where $H = \{r_{IMS_i}(d_{IM}), s_{IMS_i}(d_{IM}), r_{OMS_i}(d_{OS_i}), s_{OMS_i}(d_{OS_i})$
$|d_I, d_{IM}, d_{OS_i}, d_O \in \Delta\}$ for $1 \leq i \leq n$,

$I = \{c_{IMS_i}(d_{IM}), c_{OMS_i}(d_{OS_i}), MF_1, MF_2, SF_i$
$|d_I, d_{IM}, d_{OS_i}, d_O \in \Delta\}$ for $1 \leq i \leq n$

Then, we obtain the following conclusion on the Master–Slave pattern.

**Theorem 34.2** (Correctness of the Master–Slave pattern). *The Master–Slave pattern $\tau_I(\partial_H(M \between S_1 \between \cdots \between S_i \between \cdots \between S_n))$ can exhibit the desired external behaviors.*

*Proof.* Based on the above state transitions of the above modules, by use of the algebraic laws of APTC, we can prove that

$\tau_I(\partial_H(M \between S_1 \between \cdots \between S_i \between \cdots \between S_n)) = \sum_{d_I, d_O \in \Delta}(r_I(d_I) \cdot s_O(d_O)) \cdot \tau_I(\partial_H(M \between S_1 \between \cdots \between S_i \between \cdots \between S_n))$

That is, the Master–Slave pattern $\tau_I(\partial_H(M \between S_1 \between \cdots \between S_i \between \cdots \between S_n))$ can exhibit the desired external behaviors.

For the details of the proof, please refer to Section 17.3, as we omit them here. □

## 34.3 Access control

### 34.3.1 Verification of the Proxy pattern

The Proxy pattern is used to decouple the access of original components through a proxy. In this pattern, there are a Proxy module and $n$ Original modules. The Originals are used to implement concrete computations and the Proxy is used to decouple the access to the Originals. The Proxy module interacts with the outside through the channels $I$ and $O$, and with Original $i$ (for $1 \leq i \leq n$) through the channels $I_{PO_i}$ and $O_{PO_i}$, as illustrated in Fig. 34.5.

The typical process of the Proxy pattern is shown in Fig. 34.6 and is as follows:

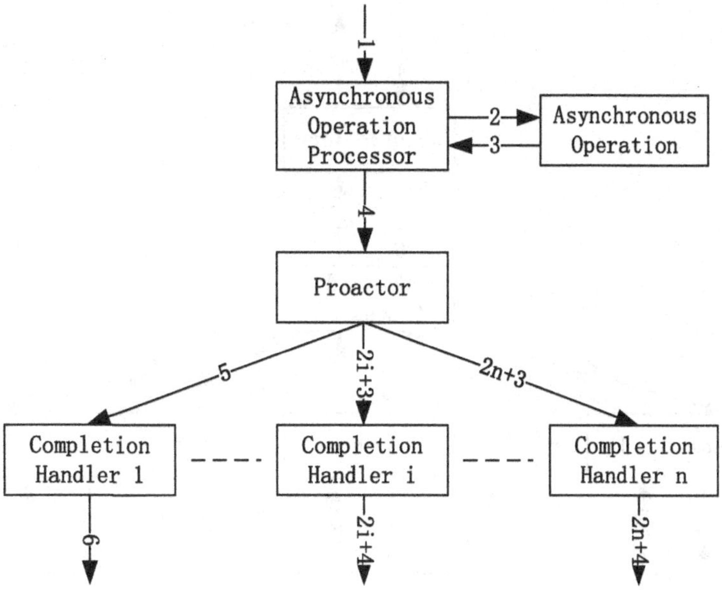


**FIGURE 34.6** Typical process of Proxy pattern.

1. The Proxy receives the request $d_I$ from the outside through the channel $I$ (the corresponding reading action is denoted $r_I(d_I)$), then processes the request through a processing function $PF_1$ and generates the request $d_{I_P}$, and sends the $d_{I_P}$ to the Original $i$ through the channel $I_{PO_i}$ (the corresponding sending action is denoted $s_{I_{PO_i}}(d_{I_P})$);
2. The Original $i$ receives the request $d_{I_P}$ from the Proxy through the channel $I_{PO_i}$ (the corresponding reading action is denoted $r_{I_{PO_i}}(d_{I_P})$), then processes the request through a processing function $OF_i$ and generates the response $d_{O_{O_i}}$, and sends the response to the Proxy through the channel $O_{PO_i}$ (the corresponding sending action is denoted $s_{O_{PO_i}}(d_{O_{O_i}})$);
3. The Proxy receives the response $d_{O_{O_i}}$ from the Original $i$ through the channel $O_{PO_i}$ (the corresponding reading action is denoted $r_{O_{PO_i}}(d_{O_{O_i}})$), then processes the request through a processing function $PF_2$ and generates the request $d_O$, and sends the $d_O$ to the outside through the channel $O$ (the corresponding sending action is denoted $s_O(d_O)$).

In the following, we verify the Proxy pattern. We assume all data elements $d_I, d_{I_P}, d_{O_{O_i}}, d_O$ (for $1 \le i \le n$) are from a finite set $\Delta$.

The state transitions of the Proxy module described by APTC are as follows:

$P = \sum_{d_I \in \Delta}(r_I(d_I) \cdot P_2)$

$P_2 = PF_1 \cdot P_3$

$P_3 = \sum_{d_{I_P} \in \Delta}(s_{I_{PO_1}}(d_{I_P}) \between \cdots \between s_{I_{PO_n}}(d_{I_P}) \cdot P_4)$

$P_4 = \sum_{d_{O_{O_1}}, \cdots, d_{O_{O_n}} \in \Delta}(r_{O_{PO_1}}(d_{O_{O_1}}) \between \cdots \between r_{O_{PO_n}}(d_{O_{O_n}}) \cdot P_5)$

$P_5 = PF_2 \cdot P_6$

$P_6 = \sum_{d_O \in \Delta}(s_O(d_O) \cdot P)$

The state transitions of the Original $i$ described by APTC are as follows:

$O_i = \sum_{d_{I_P} \in \Delta}(r_{I_{PO_i}}(d_{I_P}) \cdot O_{i_2})$

$O_{i_2} = OF_i \cdot O_{i_3}$

$O_{i_3} = \sum_{d_{O_{O_i}} \in \Delta}(s_{O_{PO_i}}(d_{O_{O_i}}) \cdot O_i)$

The sending action and the reading action of the same data through the same channel can communicate with each other, otherwise, they will cause a deadlock $\delta$. We define the following communication functions of the Original $i$ for $1 \le i \le n$:

$$\gamma(r_{I_{PO_i}}(d_{I_P}), s_{I_{PO_i}}(d_{I_P})) \triangleq c_{I_{PO_i}}(d_{I_P})$$

$$\gamma(r_{O_{PO_i}}(d_{O_{O_i}}), s_{O_{PO_i}}(d_{O_{O_i}})) \triangleq c_{O_{PO_i}}(d_{O_{O_i}})$$

Let all modules be in parallel, then the Proxy pattern $P \quad O_1 \cdots O_i \cdots O_n$ can be presented by the following process term:

$\tau_I(\partial_H(\Theta(P \between O_1 \between \cdots \between O_i \between \cdots \between O_n))) = \tau_I(\partial_H(P \between O_1 \between \cdots \between O_i \between \cdots \between O_n))$

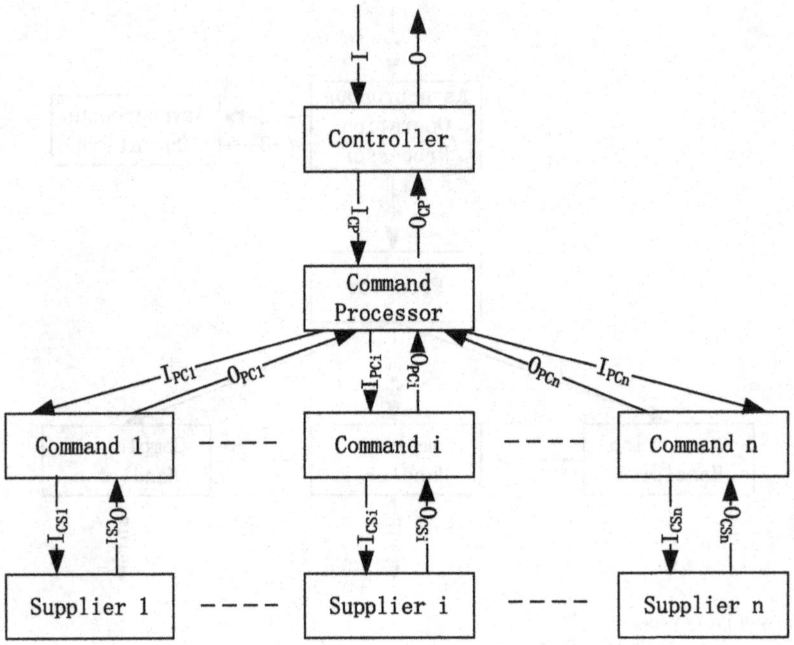


**FIGURE 34.7** Command Processor pattern.

where $H = \{r_{I_{PO_i}}(d_{I_P}), s_{I_{PO_i}}(d_{I_P}), r_{O_{PO_i}}(d_{O_{O_i}}), s_{O_{PO_i}}(d_{O_{O_i}})$
$|d_I, d_{I_P}, d_{O_{O_i}}, d_O \in \Delta\}$ for $1 \leq i \leq n$,
$I = \{c_{I_{PO_i}}(d_{I_P}), c_{O_{PO_i}}(d_{O_{O_i}}), PF_1, PF_2, OF_i$
$|d_I, d_{I_P}, d_{O_{O_i}}, d_O \in \Delta\}$ for $1 \leq i \leq n$

Then, we obtain the following conclusion on the Proxy pattern.

**Theorem 34.3** (Correctness of the Proxy pattern). *The Proxy pattern $\tau_I(\partial_H(P \between O_1 \between \cdots \between O_i \between \cdots \between O_n))$ can exhibit the desired external behaviors.*

*Proof.* Based on the above state transitions of the above modules, by use of the algebraic laws of APTC, we can prove that
$\tau_I(\partial_H(P \between O_1 \between \cdots \between O_i \between \cdots \between O_n)) = \sum_{d_I, d_O \in \Delta}(r_I(d_I) \cdot s_O(d_O)) \cdot \tau_I(\partial_H(P \between O_1 \between \cdots \between O_i \between \cdots \between O_n))$

That is, the Proxy pattern $\tau_I(\partial_H(P \between O_1 \between \cdots \between O_i \between \cdots \between O_n))$ can exhibit the desired external behaviors.

For the details of the proof, please refer to Section 17.3, as we omit them here. □

## 34.4 Management

### 34.4.1 Verification of the Command Processor pattern

The Command Processor pattern is used to decouple the request and execution of a service. In this pattern, there are a Controller module, a Command Processor module, $n$ Command modules, and $n$ Supplier modules. The Supplier is used to implement concrete computation, the Command is used to encapsulate a Supplier into a command, and the Command Processor is used to manage Commands. The Controller module interacts with the outside through the channels $I$ and $O$, and with the Command Processor through the channels $I_{CP}$ and $O_{CP}$. The Command Processor interacts with Command $i$ (for $1 \leq i \leq n$) through the channels $I_{PC_i}$ and $O_{PC_i}$, and the Command $i$ interacts with the Supplier $i$ through the channels $I_{CS_i}$ and $O_{CS_i}$, as illustrated in Fig. 34.7.

The typical process of the Command Processor pattern is shown in Fig. 34.8 and is as follows:

1. The Controller receives the request $d_I$ from the outside through the channel $I$ (the corresponding reading action is denoted $r_I(d_I)$), then processes the request through a processing function $CF_1$ and generates the request $d_{I_P}$, and sends the $d_{I_P}$ to the Command Processor through the channel $I_{CP}$ (the corresponding sending action is denoted $s_{I_{CP}}(d_{I_P})$);
2. The Command Processor receives the request $d_{I_P}$ from the Controller through the channel $I_{CP}$ (the corresponding reading action is denoted $r_{I_{CP}}(d_{I_P})$), then processes the request through a processing function $PF_1$ and generates the

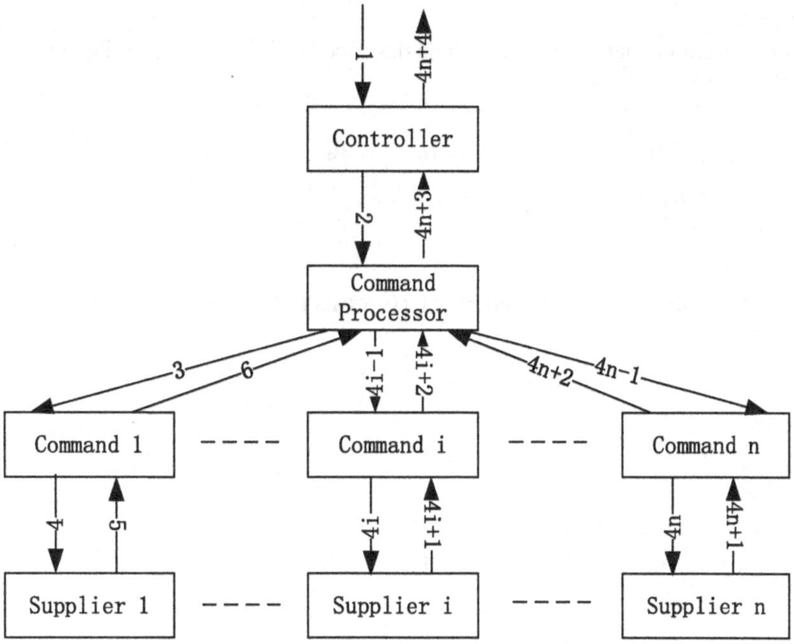


**FIGURE 34.8** Typical process of Command Processor pattern.

request $d_{I_{Com_i}}$, and sends the $d_{I_{Com_i}}$ to the Command $i$ through the channel $I_{PC_i}$ (the corresponding sending action is denoted $s_{I_{PC_i}}(d_{I_{Com_i}})$);

3. The Command $i$ receives the request $d_{I_{Com_i}}$ from the Command Processor through the channel $I_{PC_i}$ (the corresponding reading action is denoted $r_{I_{PC_i}}(d_{I_{Com_i}})$), then processes the request through a processing function $ComF_{i1}$ and generates the request $d_{I_{S_i}}$, and sends the request to the Supplier $i$ through the channel $I_{CS_i}$ (the corresponding sending action is denoted $s_{I_{CS_i}}(d_{I_{S_i}})$);
4. The Supplier $i$ receives the request $d_{I_{S_i}}$ from the Command $i$ through the channel $I_{CS_i}$ (the corresponding reading action is denoted $r_{I_{CS_i}}(d_{I_{S_i}})$), then processes the request through a processing function $SF_i$ and generates the response $d_{O_{S_i}}$, and sends the response to the Command through the channel $O_{CS_i}$ (the corresponding sending action is denoted $s_{O_{CS_i}}(d_{O_{S_i}})$);
5. The Command $i$ receives the response $d_{O_{S_i}}$ from the Supplier $i$ through the channel $O_{CS_i}$ (the corresponding reading action is denoted $r_{O_{CS_i}}(d_{O_{S_i}})$), then processes the request through a processing function $ComF_{i2}$ and generates the response $d_{O_{Com_i}}$, and sends the response to the Command Processor through the channel $O_{PC_i}$ (the corresponding sending action is denoted $s_{O_{PC_i}}(d_{O_{Com_i}})$);
6. The Command Processor receives the response $d_{O_{Com_i}}$ from the Command $i$ through the channel $O_{PC_i}$ (the corresponding reading action is denoted $r_{O_{PC_i}}(d_{O_{Com_i}})$), then processes the response and generates the response $d_{O_P}$ through a processing function $PF_2$, and sends $d_{O_P}$ to the Controller through the channel $O_{CP}$ (the corresponding sending action is denoted $s_{O_{CP}}(d_{O_P})$);
7. The Controller receives the response $d_{O_P}$ from the Command Processor through the channel $O_{CP}$ (the corresponding reading action is denoted $r_{O_{CP}}(d_{O_P})$), then processes the request through a processing function $CF_2$ and generates the request $d_O$, and sends the $d_O$ to the outside through the channel $O$ (the corresponding sending action is denoted $s_O(d_O)$).

In the following, we verify the Command Processor pattern. We assume all data elements $d_I$, $d_{I_P}$, $d_{I_{Com_i}}$, $d_{I_{S_i}}$, $d_{O_{S_i}}$, $d_{O_{Com_i}}$, $d_{O_P}$, $d_O$ (for $1 \leq i \leq n$) are from a finite set $\Delta$.

The state transitions of the Controller module described by APTC are as follows:

$C = \sum_{d_I \in \Delta}(r_I(d_I) \cdot C_2)$

$C_2 = CF_1 \cdot C_3$

$C_3 = \sum_{d_{I_P} \in \Delta}(s_{I_{CP}}(d_{I_P}) \cdot C_4)$

$C_4 = \sum_{d_{O_P} \in \Delta}(r_{O_{CP}}(d_{O_P}) \cdot C_5)$

$C_5 = CF_2 \cdot C_6$

$C_6 = \sum_{d_O \in \Delta}(s_O(d_O) \cdot C)$

The state transitions of the Command Processor module described by APTC are as follows:

$P = \sum_{d_{I_P} \in \Delta}(r_{ICP}(d_{I_P}) \cdot P_2)$

$P_2 = PF_1 \cdot P_3$

$P_3 = \sum_{d_{I_{Com_1}}, \cdots, d_{I_{Com_n}} \in \Delta}(s_{IPC_1}(d_{I_{Com_1}}) \between \cdots \between s_{IPC_n}(d_{I_{Com_n}}) \cdot P_4)$

$P_4 = \sum_{d_{O_{Com_1}}, \cdots, d_{O_{Com_n}} \in \Delta}(r_{OPC_1}(d_{O_{Com_1}}) \between \cdots \between r_{OPC_n}(d_{O_{Com_n}}) \cdot P_5)$

$P_5 = PF_2 \cdot P_6$

$P_6 = \sum_{d_{O_P} \in \Delta}(s_{OCP}(d_{O_P}) \cdot P)$

The state transitions of the Command $i$ described by APTC are as follows:

$Com_i = \sum_{d_{I_{Com_i}} \in \Delta}(r_{IPC_i}(d_{I_{Com_i}}) \cdot Com_{i_2})$

$Com_{i_2} = ComF_{i1} \cdot Com_{i_3}$

$Com_{i_3} = \sum_{d_{I_{S_i}} \in \Delta}(s_{ICS_i}(d_{I_{S_i}}) \cdot Com_{i_4})$

$Com_{i_4} = \sum_{d_{O_{S_i}} \in \Delta}(r_{OCS_i}(d_{O_{S_i}}) \cdot Com_{i_5})$

$Com_{i_5} = ComF_{i2} \cdot Com_{i_6}$

$Com_{i_6} = \sum_{d_{O_{Com_i}} \in \Delta}(s_{OPC_i}(d_{O_{Com_i}}) \cdot Com_i)$

The state transitions of the Supplier $i$ described by APTC are as follows:

$S_i = \sum_{d_{I_{S_i}} \in \Delta}(r_{ICS_i}(d_{I_{S_i}}) \cdot S_{i_2})$

$S_{i_2} = SF_i \cdot S_{i_3}$

$S_{i_3} = \sum_{d_{O_{S_i}} \in \Delta}(s_{OCS_i}(d_{O_{S_i}}) \cdot S_i)$

The sending action and the reading action of the same data through the same channel can communicate with each other, otherwise, they will cause a deadlock $\delta$. We define the following communication functions between the Controller the Command Processor:

$$\gamma(r_{ICP}(d_{I_P}), s_{ICP}(d_{I_P})) \triangleq c_{ICP}(d_{I_P})$$

$$\gamma(r_{OCP}(d_{O_P}), s_{OCP}(d_{O_P})) \triangleq c_{OCP}(d_{O_P})$$

There are two communication functions between the Command Processor and the Command $i$ for $1 \le i \le n$:

$$\gamma(r_{IPC_i}(d_{I_{Com_i}}), s_{IPC_i}(d_{I_{Com_i}})) \triangleq c_{IPC_i}(d_{I_{Com_i}})$$

$$\gamma(r_{OPC_i}(d_{O_{Com_i}}), s_{OPC_i}(d_{O_{Com_i}})) \triangleq c_{OPC_i}(d_{O_{Com_i}})$$

There are two communication functions between the Supplier $i$ and the Command $i$ for $1 \le i \le n$:

$$\gamma(r_{ICS_i}(d_{I_{S_i}}), s_{ICS_i}(d_{I_{S_i}})) \triangleq c_{ICS_i}(d_{I_{S_i}})$$

$$\gamma(r_{OCS_i}(d_{O_{S_i}}), s_{OCS_i}(d_{O_{S_i}})) \triangleq c_{OCS_i}(d_{O_{S_i}})$$

Let all modules be in parallel, then the Command Processor pattern

$C \quad P \quad Com_1 \cdots \quad Com_i \quad \cdots Com_n \quad S_1 \cdots S_i \cdots S_n$

can be presented by the following process term:

$\tau_I(\partial_H(\Theta(C \between P \between Com_1 \between \cdots \between Com_i \between \cdots \between Com_n \between S_1 \between \cdots \between S_i \between \cdots \between S_n))) = \tau_I(\partial_H(C \between P \between Com_1 \between \cdots \between Com_i \between \cdots \between Com_n \between S_1 \between \cdots \between S_i \between \cdots \between S_n))$

where $H = \{r_{ICP}(d_{I_P}), s_{ICP}(d_{I_P}), r_{OCP}(d_{O_P}), s_{OCP}(d_{O_P}), r_{IPC_i}(d_{I_{Com_i}}), s_{IPC_i}(d_{I_{Com_i}}), r_{OPC_i}(d_{O_{Com_i}}), s_{OPC_i}(d_{O_{Com_i}}), r_{ICS_i}(d_{I_{S_i}}), s_{ICS_i}(d_{I_{S_i}}), r_{OCS_i}(d_{O_{S_i}}), s_{OCS_i}(d_{O_{S_i}}) |d_I, d_{I_P}, d_{I_{Com_i}}, d_{I_{S_i}}, d_{O_{S_i}}, d_{O_{Com_i}}, d_{O_P}, d_O \in \Delta\}$ for $1 \le i \le n$,

$I = \{c_{ICP}(d_{I_P}), c_{OCP}(d_{O_P}), c_{IPC_i}(d_{I_{Com_i}}), c_{OPC_i}(d_{O_{Com_i}}), c_{ICS_i}(d_{I_{S_i}}), c_{OCS_i}(d_{O_{S_i}}), CF_1, CF_2, PF_1, PF_2, ComF_{i1}, ComF_{i2}, SF_i |d_I, d_{I_P}, d_{I_{Com_i}}, d_{I_{S_i}}, d_{O_{S_i}}, d_{O_{Com_i}}, d_{O_P}, d_O \in \Delta\}$ for $1 \le i \le n$

Then, we obtain the following conclusion on the Command Processor pattern.

**Theorem 34.4** (Correctness of the Command Processor pattern). *The Command Processor pattern* $\tau_I(\partial_H(C \between P \between Com_1 \between \cdots \between Com_i \between \cdots \between Com_n \between S_1 \between \cdots \between S_i \between \cdots \between S_n))$ *can exhibit the desired external behaviors.*

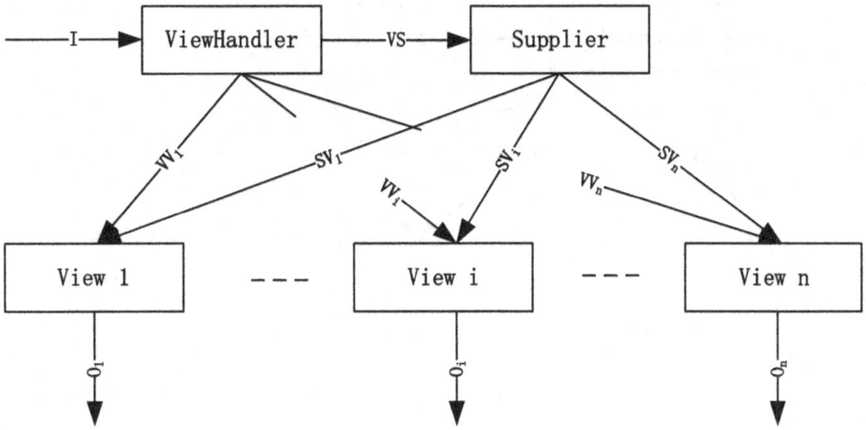


**FIGURE 34.9** View Handler pattern.

*Proof.* Based on the above state transitions of the above modules, by use of the algebraic laws of APTC, we can prove that $\tau_I(\partial_H(C \between P \between Com_1 \between \cdots \between Com_i \between \cdots \between Com_n \between S_1 \between \cdots \between S_i \between \cdots \between S_n)) = \sum_{d_I, d_O \in \Delta}(r_I(d_I) \cdot s_O(d_O)) \cdot \tau_I(\partial_H(C \between P \between Com_1 \between \cdots \between Com_i \between \cdots \between Com_n \between S_1 \between \cdots \between S_i \between \cdots \between S_n))$

That is, the Command Processor pattern $\tau_I(\partial_H(C \between P \between Com_1 \between \cdots \between Com_i \between \cdots \between Com_n \between S_1 \between \cdots \between S_i \between \cdots \between S_n))$ can exhibit the desired external behaviors.

For the details of the proof, please refer to Section 17.3, as we omit them here. □

### 34.4.2 Verification of the View Handler pattern

The View Handler pattern is used to manage all views of the system, which has three components: the Supplier, the Views, and the ViewHandler. The Supplier is used to contain the data and encapsulate the core functionalities; the Views is to show the computational results to the user; and the ViewHandler interacts between the system and user, accepts the instructions, and controls the Supplier and the Views. The ViewHandler receives the instructions from the user through the channel $I$, then it sends the instructions to the Supplier through the channel $VS$ and to the View $i$ through the channel $VV_i$ for $1 \leq i \leq n$; the model receives the instructions from the ViewHandler, updates the data and computes the results, and sends the results to the View $i$ through the channel $SV_i$ for $1 \leq i \leq n$; When the View $i$ receives the results from the Supplier, it generates or updates the view to the user. This is illustrated in Fig. 34.9.

The typical process of the View Handler pattern is shown in Fig. 34.10 and is the following:

1. The ViewHandler receives the instructions $d_I$ from the user through the channel $I$ (the corresponding reading action is denoted $r_I(D_I)$), processes the instructions through a processing function $VHF$, and generates the instructions to the Supplier $d_{I_S}$ and those to the View $i$ $d_{I_{V_i}}$ for $1 \leq i \leq n$; it sends $d_{I_S}$ to the Supplier through the channel $CM$ (the corresponding sending action is denoted $s_{VS}(d_{I_M})$) and sends $d_{I_{V_i}}$ to the View $i$ through the channel $VV_i$ (the corresponding sending action is denoted $s_{VV_i}(d_{I_{V_i}})$);
2. The Supplier receives the instructions from the ViewHandler through the channel $VS$ (the corresponding reading action is denoted $r_{VS}(d_{I_S})$), processes the instructions through a processing function $SF$, generates the computational results to the View $i$ (for $1 \leq i \leq n$) that is denoted $d_{O_{S_i}}$; then sends the results to the View $i$ through the channel $SV_i$ (the corresponding sending action is denoted $s_{MV_i}(d_{O_{M_i}})$);
3. The View $i$ (for $1 \leq i \leq n$) receives the instructions from the ViewHandler through the channel $VV_i$ (the corresponding reading action is denoted $r_{VV_i}(d_{I_{V_i}})$), processes the instructions through a processing function $VF_{i_1}$ to make ready to receive the computational results from the Supplier; then it receives the computational results from the Supplier through the channel $SV_i$ (the corresponding reading action is denoted $r_{SV_i}(d_{O_{S_i}})$), processes the results through a processing function $VF_{i_2}$, generates the output $d_{O_i}$, then sends the output through the channel $O_i$ (the corresponding sending action is denoted $s_{O_i}(d_{O_i})$).

In the following, we verify the View Handler pattern. We assume all data elements $d_I, d_{I_S}, d_{I_{V_i}}, d_{O_{S_i}}, d_{O_i}$ (for $1 \leq i \leq n$) are from a finite set $\Delta$.

The state transitions of the ViewHandler module described by APTC are as follows:

$VH = \sum_{d_I \in \Delta}(r_I(d_I) \cdot VH_2)$

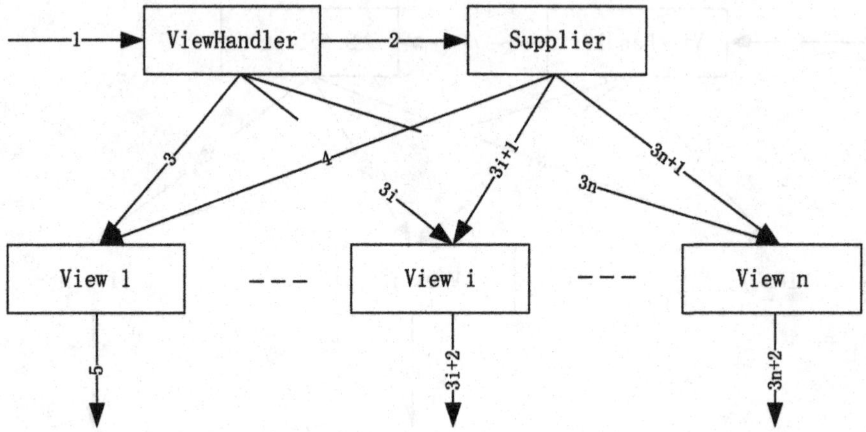


FIGURE 34.10 Typical process of View Handler pattern.

$VH_2 = VHF \cdot VH_3$
$VH_3 = \sum_{d_{I_S} \in \Delta}(s_{VS}(d_{I_S}) \cdot VH_4)$
$VH_4 = \sum_{d_{I_{V_1}}, \cdots, d_{I_{V_n}} \in \Delta}(s_{VV_1}(d_{I_{V_1}}) \between \cdots \between s_{VV_n}(d_{I_{V_n}}) \cdot VH)$
The state transitions of the Supplier described by APTC are as follows:
$S = \sum_{d_{I_M} \in \Delta}(r_{VS}(d_{I_S}) \cdot S_2)$
$S_2 = SF \cdot S_3$
$S_3 = \sum_{d_{O_{S_1}}, \cdots, d_{O_{S_n}} \in \Delta}(s_{SV_1}(d_{O_{S_1}}) \between \cdots \between s_{SV_n}(d_{O_{S_n}}) \cdot S)$
The state transitions of the View $i$ described by APTC are as follows:
$V_i = \sum_{d_{I_{V_i}} \in \Delta}(r_{VV_i}(d_{I_{V_i}}) \cdot V_{i_2})$
$V_{i_2} = VF_{i_1} \cdot V_{i_3}$
$V_{i_3} = \sum_{d_{O_{S_i}} \in \Delta}(r_{SV_i}(d_{O_{S_i}}) \cdot V_{i_4})$
$V_{i_4} = VF_{i_2} \cdot V_{i_5}$
$V_{i_5} = \sum_{d_{O_i} \in \Delta}(s_{O_i}(d_{O_i}) \cdot V_i)$
The sending action and the reading action of the same data through the same channel can communicate with each other, otherwise, they will cause a deadlock $\delta$. We define the following communication functions of the View $i$ for $1 \leq i \leq n$:

$$\gamma(r_{VV_i}(d_{I_{V_i}}), s_{VV_i}(d_{I_{V_i}})) \triangleq c_{VV_i}(d_{I_{V_i}})$$
$$\gamma(r_{SV_i}(d_{O_{S_i}}), s_{SV_i}(d_{O_{S_i}})) \triangleq c_{SV_i}(d_{O_{S_i}})$$

There is one communication function between the ViewHandler and the Supplier as follows:

$$\gamma(r_{VS}(d_{I_S}), s_{VS}(d_{I_S})) \triangleq c_{VS}(d_{I_S})$$

Let all modules be in parallel, then the View Handler pattern $VH \quad S \quad V_1 \cdots V_i \cdots V_n$ can be presented by the following process term:
$\tau_I(\partial_H(\Theta(VH \between S \between V_1 \between \cdots \between V_i \between \cdots \between V_n))) = \tau_I(\partial_H(VH \between S \between V_1 \between \cdots \between V_i \between \cdots \between V_n))$
where $H = \{r_{VV_i}(d_{I_{V_i}}), s_{VV_i}(d_{I_{V_i}}), r_{SV_i}(d_{O_{S_i}}), s_{SV_i}(d_{O_{S_i}}), r_{VS}(d_{I_S}), s_{VS}(d_{I_S})$
$|d_I, d_{I_S}, d_{I_{V_i}}, d_{O_{S_i}}, d_{O_i} \in \Delta\}$ for $1 \leq i \leq n$,
$I = \{c_{VV_i}(d_{I_{V_i}}), c_{VV_i}(d_{O_{M_i}}), c_{VS}(d_{I_S}), VHF, SF, VF_{1_1}, VF_{1_2}, \cdots, VF_{n_1}, VF_{n_2}$
$|d_I, d_{I_S}, d_{I_{V_i}}, d_{O_{S_i}}, d_{O_i} \in \Delta\}$ for $1 \leq i \leq n$
Then, we obtain the following conclusion on the View Handler pattern.

**Theorem 34.5** (Correctness of the View Handler pattern). *The View Handler pattern $\tau_I(\partial_H(VH \between S \between V_1 \between \cdots \between V_i \between \cdots \between V_n))$ can exhibit the desired external behaviors.*

*Proof.* Based on the above state transitions of the above modules, by use of the algebraic laws of APTC, we can prove that

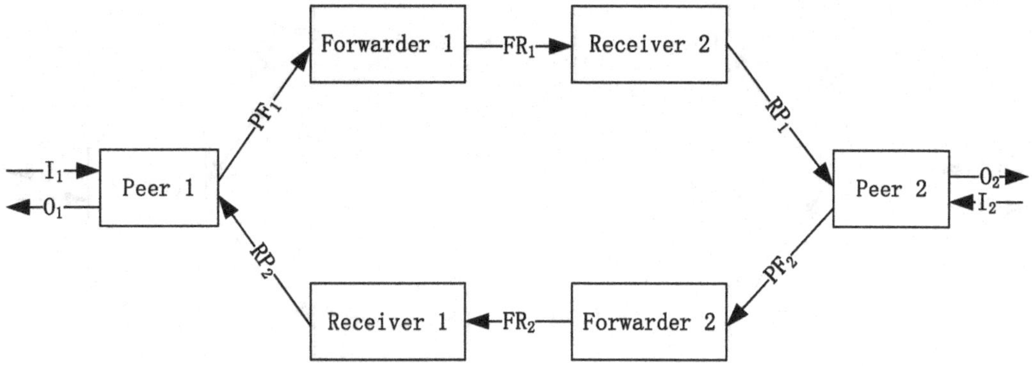


**FIGURE 34.11** Forwarder–Receiver pattern.

$\tau_I(\partial_H(VH \between S \between V_1 \between \cdots \between V_i \between \cdots \between V_n)) = \sum_{d_I, d_{O_1}, \cdots, d_{O_n} \in \Delta}(r_I(d_I) \cdot s_{O_1}(d_{O_1}) \parallel \cdots \parallel s_{O_i}(d_{O_i}) \parallel \cdots \parallel s_{O_n}(d_{O_n})) \cdot$
$\tau_I(\partial_H(VH \between S \between V_1 \between \cdots \between V_i \between \cdots \between V_n))$

That is, the View Handler pattern $\tau_I(\partial_H(VH \between S \between V_1 \between \cdots \between V_i \between \cdots \between V_n))$ can exhibit the desired external behaviors.

For the details of the proof, please refer to Section 17.3, as we omit them here. □

## 34.5 Communication

### 34.5.1 Verification of the Forwarder–Receiver pattern

The Forwarder–Receiver pattern decouples the communication of two communicating peers. There are six modules in the Forwarder–Receiver pattern: the two Peers, the two Forwarders, and the two Receivers. The Peers interact with the user through the channels $I_1$, $I_2$, and $O_1$, $O_2$; and with the Forwarder through the channels $PF_1$ and $PF_2$. The Receivers interact with Forwarders through the channels $FR_1$ and $FR_2$, and with the Peers through the channels $RP_1$ and $RP_2$. This is illustrated in Fig. 34.11.

The typical process of the Forwarder–Receiver pattern is shown in Fig. 34.12 and is as follows:

1. The Peer 1 receives the request $d_{I_1}$ from the user through the channel $I_1$ (the corresponding reading action is denoted $r_{I_1}(d_{I_1})$), then processes the request $d_{I_1}$ through a processing function $P1F_1$, and sends the processed request $d_{I_{F_1}}$ to the Forwarder 1 through the channel $PF_1$ (the corresponding sending action is denoted $s_{PF_1}(d_{I_{F_1}})$);
2. The Forwarder 1 receives $d_{I_{F_1}}$ from the Peer 1 through the channel $PF_1$ (the corresponding reading action is denoted $r_{PF_1}(d_{I_{F_1}})$), then processes the request through a processing function $F1F$, generates and sends the processed request $d_{I_{R_2}}$ to the Receiver 2 through the channel $FR_1$ (the corresponding sending action is denoted $s_{FR_1}(d_{I_{R_2}})$);
3. The Receiver 2 receives the request $d_{I_{R_2}}$ from the Forwarder 1 through the channel $FR_1$ (the corresponding reading action is denoted $r_{FR_1}(d_{I_{R_2}})$), then processes the request through a processing function $R2F$, generates and sends the processed request $d_{I_{P_2}}$ to the Peer 2 through the channel $RP_1$ (the corresponding sending action is denoted $s_{RP_1}(d_{I_{P_2}})$);
4. The Peer 2 receives the request $d_{I_{P_2}}$ from the Receiver 2 through the channel $RP_1$ (the corresponding reading action is denoted $r_{RP_1}(d_{I_{P_2}})$), then processes the request and generates the response $d_{O_2}$ through a processing function $P2F_2$, and sends the response to the outside through the channel $O_2$ (the corresponding sending action is denoted $s_{O_2}(d_{O_2})$).

There is another symmetric process from Peer 2 to Peer 1, however, we omit it here.

In the following, we verify the Forwarder–Receiver pattern. We assume all data elements $d_{I_1}, d_{I_2}, d_{I_{F_1}}, d_{I_{F_2}}, d_{I_{R_1}}, d_{I_{R_2}}, d_{I(P_1)}, d_{I_{P_2}}, d_{O_1}, d_{O_2}$ are from a finite set $\Delta$. We only give the transitions of the first process.

The state transitions of the Peer 1 module described by APTC are as follows:

$P1 = \sum_{d_{I_1} \in \Delta}(r_{I_1}(d_{I_1}) \cdot P1_2)$

$P1_2 = P1F_1 \cdot P1_3$

$P1_3 = \sum_{d_{I_{F_1}} \in \Delta}(s_{PF_1}(d_{I_{F_1}}) \cdot P1)$

The state transitions of the Forwarder 1 module described by APTC are as follows:

$F1 = \sum_{d_{I_{F_1}} \in \Delta}(r_{PF_1}(d_{I_{F_1}}) \cdot F1_2)$

$F1_2 = F1F \cdot F1_3$

$F1_3 = \sum_{d_{I_{R_2}} \in \Delta}(s_{FR_1}(d_{I_{R_2}}) \cdot F1)$

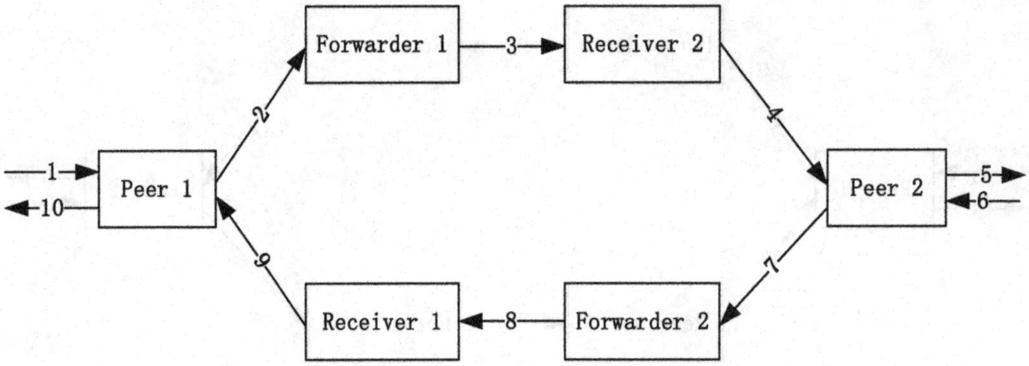


**FIGURE 34.12** Typical process of Forwarder–Receiver pattern.

The state transitions of the Receiver 2 module described by APTC are as follows:

$R2 = \sum_{d_{I_{R_2}} \in \Delta}(r_{FR_1}(d_{I_{R_2}}) \cdot R2_2)$

$R2_2 = R2F \cdot R2_3$

$R2_3 = \sum_{d_{I_{P_2}} \in \Delta}(s_{RP_1}(d_{I_{P_2}}) \cdot R2)$

The state transitions of the Peer 2 module described by APTC are as follows:

$P2 = \sum_{d_{I_{P_2}} \in \Delta}(r_{RP_1}(d_{I_{P_2}}) \cdot P2_2)$

$P2_2 = P2F_2 \cdot P2_3$

$P2_3 = \sum_{d_{O_2} \in \Delta}(s_{O_2}(d_{O_2}) \cdot P2)$

The sending action and the reading action of the same data through the same channel can communicate with each other, otherwise, they will cause a deadlock $\delta$. We define the following communication function between the Peer 1 and the Forwarder 1:

$$\gamma(r_{PF_1}(d_{I_{F_1}}), s_{PF_1}(d_{I_{F_1}})) \triangleq c_{PF_1}(d_{I_{F_1}})$$

There is one communication function between the Forwarder 1 and the Receiver 2 as follows:

$$\gamma(r_{FR_1}(d_{I_{R_2}}), s_{FR_1}(d_{I_{R_2}})) \triangleq c_{FR_1}(d_{I_{R_2}})$$

There is one communication function between the Receiver 2 and the Peer 2 as follows:

$$\gamma(r_{RP_1}(d_{I_{P_2}}), s_{RP_1}(d_{I_{P_2}})) \triangleq c_{RP_1}(d_{I_{P_2}})$$

We define the following communication function between the Peer 2 and the Forwarder 2:

$$\gamma(r_{PF_2}(d_{I_{F_2}}), s_{PF_2}(d_{I_{F_2}})) \triangleq c_{PF_2}(d_{I_{F_2}})$$

There is one communication function between the Forwarder 2 and the Receiver 1 as follows:

$$\gamma(r_{FR_2}(d_{I_{R_1}}), s_{FR_2}(d_{I_{R_1}})) \triangleq c_{FR_2}(d_{I_{R_1}})$$

There is one communication function between the Receiver 1 and the Peer 1 as follows:

$$\gamma(r_{RP_2}(d_{I_{P_1}}), s_{RP_2}(d_{I_{P_1}})) \triangleq c_{RP_2}(d_{I_{P_1}})$$

Let all modules be in parallel, then the Forwarder–Receiver pattern $P1 \quad F1 \quad R1 \quad R2 \quad F2 \quad P2$ can be presented by the following process term:

$\tau_I(\partial_H(\Theta(P1 \between F1 \between R1 \between R2 \between F2 \between P2))) = \tau_I(\partial_H(P1 \between F1 \between R1 \between R2 \between F2 \between P2))$

where $H = \{r_{PF_1}(d_{I_{F_1}}), s_{PF_1}(d_{I_{F_1}}), r_{FR_1}(d_{I_{R_2}}), s_{FR_1}(d_{I_{R_2}}), r_{RP_1}(d_{I_{P_2}}), s_{RP_1}(d_{I_{P_2}}),$
$r_{PF_2}(d_{I_{F_2}}), s_{PF_2}(d_{I_{F_2}}), r_{FR_2}(d_{I_{R_1}}), s_{FR_2}(d_{I_{R_1}}), r_{RP_2}(d_{I_{P_1}}), s_{RP_2}(d_{I_{P_1}})$
$|d_{I_1}, d_{I_2}, d_{I_{F_1}}, d_{I_{F_2}}, d_{I_{R_1}}, d_{I_{R_2}}, d_{I(P_1)}, d_{I_{P_2}}, d_{O_1}, d_{O_2} \in \Delta\}$,

$I = \{c_{PF_1}(d_{I_{F_1}}), c_{FR_1}(d_{I_{R_2}}), c_{RP_1}(d_{I_{P_2}}), c_{PF_2}(d_{I_{F_2}}), c_{FR_2}(d_{I_{R_1}}), c_{RP_2}(d_{I_{P_1}}),$
$P1F_1, P1F_2, P2F_1, P2F_2, F1F, F2F, R1F, R2F|d_{I_1}, d_{I_2}, d_{I_{F_1}}, d_{I_{F_2}}, d_{I_{R_1}}, d_{I_{R_2}}, d_{I(P_1)}, d_{I_{P_2}}, d_{O_1}, d_{O_2} \in \Delta\}$

Then, we obtain the following conclusion on the Forwarder–Receiver pattern.

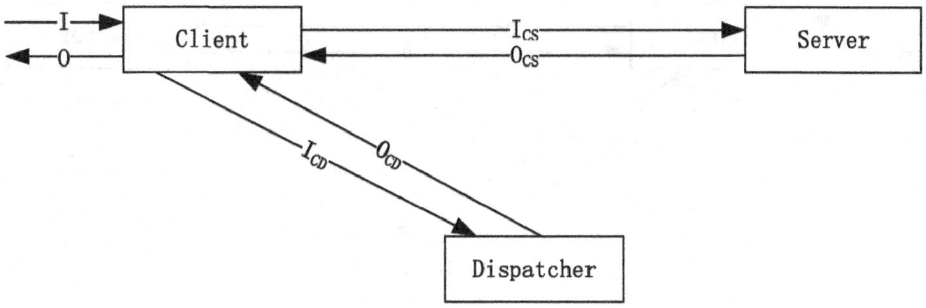


FIGURE 34.13 Client–Dispatcher–Server pattern.

**Theorem 34.6** (Correctness of the Forwarder–Receiver pattern). *The Forwarder–Receiver pattern* $\tau_I(\partial_H(P1 \between F1 \between R1 \between R2 \between F2 \between P2))$ *can exhibit the desired external behaviors.*

*Proof.* Based on the above state transitions of the above modules, by use of the algebraic laws of APTC, we can prove that $\tau_I(\partial_H(P1 \between F1 \between R1 \between R2 \between F2 \between P2)) = \sum_{d_{I_1},d_{I_2},d_{O_1},d_{O_2}\in\Delta}((r_{I_1}(d_{I_1}) \cdot s_{O_2}(d_{O_2})) \parallel (r_{I_2}(d_{I_2}) \cdot s_{O_1}(d_{O_1}))) \cdot \tau_I(\partial_H(P1 \between F1 \between R1 \between R2 \between F2 \between P2))$

That is, the Forwarder–Receiver pattern $\tau_I(\partial_H(P1 \between F1 \between R1 \between R2 \between F2 \between P2))$ can exhibit the desired external behaviors.

For the details of the proof, please refer to Section 17.3, as we omit them here. □

### 34.5.2 Verification of the Client–Dispatcher–Server pattern

The Client–Dispatcher–Server pattern decouples the invocation of the client and the server to introduce an intermediate dispatcher. There are three modules in the Client–Dispatcher–Server pattern: the Client, the Dispatcher, and the Server. The Client interacts with the user through the channels $I$ and $O$; with the Dispatcher through the channels $I_{CD}$ and $O_{CD}$; and with the Server through the channels $I_{CS}$ and $O_{CS}$. This is illustrated in Fig. 34.13.

The typical process of the Client–Dispatcher–Server pattern is shown in Fig. 34.14 and is as follows:

1. The Client receives the request $d_I$ from the user through the channel $I$ (the corresponding reading action is denoted $r_I(d_I)$), then processes the request $d_I$ through a processing function $CF_1$, and sends the processed request $d_{I_D}$ to the Dispatcher through the channel $I_{CD}$ (the corresponding sending action is denoted $s_{I_CD}(d_{I_D})$);
2. The Dispatcher receives $d_{I_D}$ from the Client through the channel $I_{CD}$ (the corresponding reading action is denoted $r_{I_{CD}}(d_{I_D})$), then processes the request through a processing function $DF$, generates and sends the processed response $d_{O_D}$ to the Client through the channel $O_{CD}$ (the corresponding sending action is denoted $s_{O_{CD}}(d_{O_D})$);
3. The Client receives the response $d_{O_D}$ from the Dispatcher through the channel $O_{CD}$ (the corresponding reading action is denoted $r_{O_{CD}}(d_{O_D})$), then processes the request through a processing function $CF_2$, generates and sends the processed request $d_{I_S}$ to the Server through the channel $I_{CS}$ (the corresponding sending action is denoted $s_{I_{CS}}(d_{I_S})$);
4. The Server receives the request $d_{I_S}$ from the Client through the channel $I_{CS}$ (the corresponding reading action is denoted $r_{I_{CS}}(d_{I_S})$), then processes the request and generates the response $d_{O_S}$ through a processing function $SF$, and sends the response to the outside through the channel $O_{CS}$ (the corresponding sending action is denoted $s_{O_{CS}}(d_{O_S})$);
5. The Client receives the response $d_{O_S}$ from the Server through the channel $O_{CS}$ (the corresponding reading action is denoted $r_{O_{CS}}(d_{O_S})$), then processes the request through a processing function $CF_3$, generates and sends the processed response $d_O$ to the user through the channel $O$ (the corresponding sending action is denoted $s_O(d_O)$).

In the following, we verify the Client–Dispatcher–Server pattern. We assume all data elements $d_I$, $d_{I_D}$, $d_{I_S}$, $d_{O_D}$, $d_{O_S}$, $d_O$ are from a finite set $\Delta$.

The state transitions of the Client module described by APTC are as follows:

$C = \sum_{d_I\in\Delta}(r_I(d_I) \cdot C_2)$

$C_2 = CF_1 \cdot C_3$

$C_3 = \sum_{d_{I_D}\in\Delta}(s_{I_{CD}}(d_{I_D}) \cdot C_4)$

$C_4 = \sum_{d_{O_D}\in\Delta}(r_{O_{CD}}(d_{O_D}) \cdot C_5)$

$C_5 = CF_2 \cdot C_6$

$C_6 = \sum_{d_{I_S}\in\Delta}(s_{I_{CS}}(d_{I_S}) \cdot C_7)$

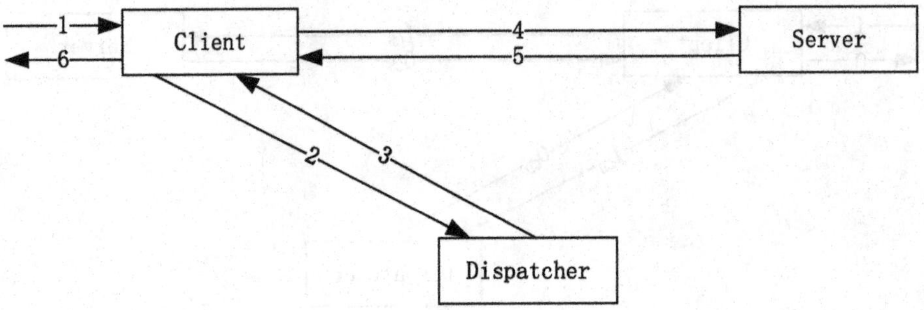


**FIGURE 34.14** Typical process of Client–Dispatcher–Server pattern.

$C_7 = \sum_{d_{O_S} \in \Delta}(r_{O_{CS}}(d_{O_S}) \cdot C_8)$
$C_8 = CF_3 \cdot C_9$
$C_9 = \sum_{d_O \in \Delta}(s_O(d_O) \cdot C)$

The state transitions of the Dispatcher module described by APTC are as follows:

$D = \sum_{d_{I_D} \in \Delta}(r_{I_{CD}}(d_{I_D}) \cdot D_2)$
$D_2 = DF \cdot D_3$
$D_3 = \sum_{d_{O_D} \in \Delta}(s_{O_{CD}}(d_{O_D}) \cdot D)$

The state transitions of the Server module described by APTC are as follows:

$S = \sum_{d_{I_S} \in \Delta}(r_{I_{CS}}(d_{I_S}) \cdot S_2)$
$S_2 = SF \cdot S_3$
$S_3 = \sum_{d_{O_S} \in \Delta}(s_{O_{CS}}(d_{O_S}) \cdot S)$

The sending action and the reading action of the same data through the same channel can communicate with each other, otherwise, they will cause a deadlock $\delta$. We define the following communication functions between the Client and the Dispatcher:

$$\gamma(r_{I_{CD}}(d_{I_D}), s_{I_{CD}}(d_{I_D})) \triangleq c_{I_{CD}}(d_{I_D})$$
$$\gamma(r_{O_{CD}}(d_{O_D}), s_{O_{CD}}(d_{O_D})) \triangleq c_{O_{CD}}(d_{O_D})$$

There are two communication functions between the Client and the Server as follows:

$$\gamma(r_{I_{CS}}(d_{I_S}), s_{I_{CS}}(d_{I_S})) \triangleq c_{I_{CS}}(d_{I_S})$$
$$\gamma(r_{O_{CS}}(d_{O_S}), s_{O_{CS}}(d_{O_S})) \triangleq c_{O_{CS}}(d_{O_S})$$

Let all modules be in parallel, then the Client–Dispatcher–Server pattern $C \quad D \quad S$ can be presented by the following process term:

$\tau_I(\partial_H(\Theta(C \between D \between S))) = \tau_I(\partial_H(C \between D \between S))$

where $H = \{r_{I_{CD}}(d_{I_D}), s_{I_{CD}}(d_{I_D}), r_{O_{CD}}(d_{O_D}), s_{O_{CD}}(d_{O_D}), r_{I_{CS}}(d_{I_S}), s_{I_{CS}}(d_{I_S}),$
$r_{O_{CS}}(d_{O_S}), s_{O_{CS}}(d_{O_S}) | d_I, d_{I_D}, d_{I_S}, d_{O_D}, d_{O_S}, d_O \in \Delta\}$,

$I = \{c_{I_{CD}}(d_{I_D}), c_{O_{CD}}(d_{O_D}), c_{I_{CS}}(d_{I_S}), c_{O_{CS}}(d_{O_S}), CF_1, CF_2, CF_3, DF, SF$
$| d_I, d_{I_D}, d_{I_S}, d_{O_D}, d_{O_S}, d_O \in \Delta\}$

Then, we obtain the following conclusion on the Client–Dispatcher–Server pattern.

**Theorem 34.7** (Correctness of the Client–Dispatcher–Server pattern). *The Client–Dispatcher–Server pattern $\tau_I(\partial_H(C \between D \between S))$ can exhibit the desired external behaviors.*

*Proof.* Based on the above state transitions of the above modules, by use of the algebraic laws of APTC, we can prove that

$\tau_I(\partial_H(C \between D \between S)) = \sum_{d_I, d_O \in \Delta}(r_I(d_I) \cdot s_O(d_O)) \cdot \tau_I(\partial_H(C \between D \between S))$

That is, the Client–Dispatcher–Server pattern $\tau_I(\partial_H(C \between D \between S))$ can exhibit the desired external behaviors.

For the details of the proof, please refer to Section 17.3, as we omit them here. □

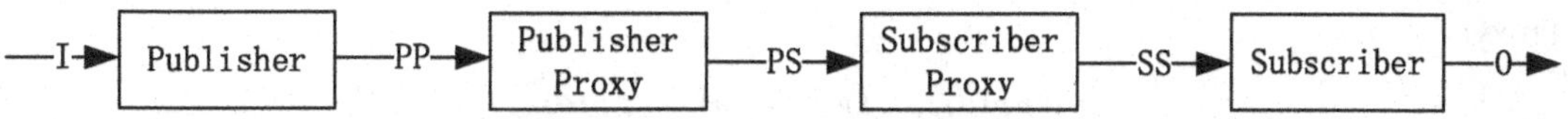


**FIGURE 34.15** Publisher–Subscriber pattern.

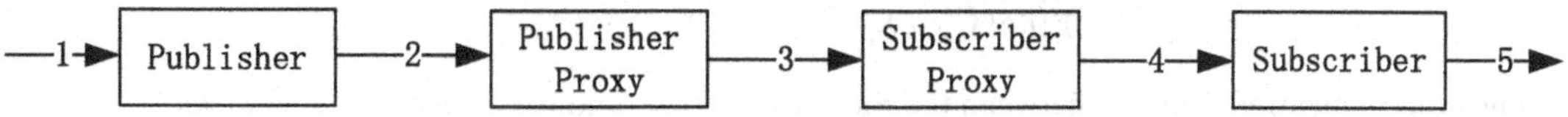


**FIGURE 34.16** Typical process of Publisher–Subscriber pattern.

### 34.5.3 Verification of the Publisher–Subscriber pattern

The Publisher–Subscriber pattern decouples the communication of the publisher and subscriber. There are four modules in the Publisher–Subscriber pattern: the Publisher, the Publisher Proxy, the Subscriber Proxy, and the Subscriber. The Publisher interacts with the outside through the channel $I$, and with the Publisher Proxy through the channel $PP$. The Publisher Proxy interacts with the Subscriber Proxy through the channel $PS$. The Subscriber interacts with the Subscriber Proxy through the channel $SS$, and with the outside through the channels $O$. This is illustrated in Fig. 34.15.

The typical process of the Publisher–Subscriber pattern is shown in Fig. 34.16 and is as follows:

1. The Publisher receives the input $d_I$ from the outside through the channel $I$ (the corresponding reading action is denoted $r_I(d_I)$), then processes the input $d_I$ through a processing function $PF$, and sends the processed input $d_{I_{PP}}$ to the Publisher Proxy through the channel $PP$ (the corresponding sending action is denoted $s_{PP}(d_{I_{PP}})$);
2. The Publisher Proxy receives $d_{I_{PP}}$ from the Publisher through the channel $PP$ (the corresponding reading action is denoted $r_{PP}(d_{I_{PP}})$), then processes the request through a processing function $PPF$, generates and sends the processed input $d_{I_{SP}}$ to the Subscriber Proxy through the channel $PS$ (the corresponding sending action is denoted $s_{PS}(d_{I_{SP}})$);
3. The Subscriber Proxy receives the input $d_{I_{SP}}$ from the Publisher Proxy through the channel $PS$ (the corresponding reading action is denoted $r_{PS}(d_{I_{SP}})$), then processes the request through a processing function $SPF$, generates and sends the processed input $d_{I_S}$ to the Subscriber through the channel $SS$ (the corresponding sending action is denoted $s_{SS}(d_{I_S})$);
4. The Subscriber receives the input $d_{I_S}$ from the Subscriber Proxy through the channel $SS$ (the corresponding reading action is denoted $r_{SS}(d_{I_S})$), then processes the request and generates the response $d_O$ through a processing function $SF$, and sends the response to the outside through the channel $O$ (the corresponding sending action is denoted $s_O(d_O)$).

In the following, we verify the Publisher–Subscriber pattern. We assume all data elements $d_I$, $d_{I_{PP}}$, $d_{I_{SP}}$, $d_{I_S}$, $d_O$ are from a finite set $\Delta$.

The state transitions of the Publisher module described by APTC are as follows:

$P = \sum_{d_I \in \Delta}(r_I(d_I) \cdot P_2)$

$P_2 = PF \cdot P_3$

$P_3 = \sum_{d_{I_{PP}} \in \Delta}(s_{PP}(d_{I_{PP}}) \cdot P)$

The state transitions of the Publisher Proxy module described by APTC are as follows:

$PP = \sum_{d_{I_{PP}} \in \Delta}(r_{PP}(d_{I_{PP}}) \cdot PP_2)$

$PP_2 = PPF \cdot PP_3$

$PP_3 = \sum_{d_{I_{SP}} \in \Delta}(s_{PS}(d_{I_{SP}}) \cdot PP)$

The state transitions of the Subscriber Proxy module described by APTC are as follows:

$SP = \sum_{d_{I_{SP}} \in \Delta}(r_{PS}(d_{I_{SP}}) \cdot SP_2)$

$SP_2 = SPF \cdot SP_3$

$SP_3 = \sum_{d_{I_S} \in \Delta}(s_{SS}(d_{I_S}) \cdot SP)$

The state transitions of the Subscriber module described by APTC are as follows:

$S = \sum_{d_{I_S} \in \Delta}(r_{SS}(d_{I_S}) \cdot S_2)$

$S_2 = SF \cdot S_3$

$S_3 = \sum_{d_O \in \Delta}(s_O(d_O) \cdot S)$

The sending action and the reading action of the same data through the same channel can communicate with each other, otherwise, they will cause a deadlock $\delta$. We define the following communication function between the Publisher and the

Publisher Proxy:

$$\gamma(r_{PP}(d_{I_{PP}}), s_{PP}(d_{I_{PP}})) \triangleq c_{PP}(d_{I_{PP}})$$

There is one communication function between the Publisher Proxy and the Subscriber Proxy as follows:

$$\gamma(r_{PS}(d_{I_{SP}}), s_{PS}(d_{I_{SP}})) \triangleq c_{PS}(d_{I_{SP}})$$

There is one communication function between the Subscriber Proxy and the Subscriber as follows:

$$\gamma(r_{SS}(d_{I_S}), s_{SS}(d_{I_S})) \triangleq c_{SS}(d_{I_S})$$

Let all modules be in parallel, then the Publisher–Subscriber pattern $P \quad PP \quad SP \quad S$ can be presented by the following process term:

$\tau_I(\partial_H(\Theta(P \between PP \between SP \between S))) = \tau_I(\partial_H(P \between PP \between SP \between S))$

where $H = \{r_{PP}(d_{I_{PP}}), s_{PP}(d_{I_{PP}}), r_{PS}(d_{I_{SP}}), s_{PS}(d_{I_{SP}}), r_{SS}(d_{I_S}), s_{SS}(d_{I_S})$
$|d_I, d_{I_{PP}}, d_{I_{SP}}, d_{I_S}, d_O \in \Delta\}$,

$I = \{c_{PP}(d_{I_{PP}}), c_{PS}(d_{I_{SP}}), c_{SS}(d_{I_S}), PF, PPF, SPF, SF|d_I, d_{I_{PP}}, d_{I_{SP}}, d_{I_S}, d_O \in \Delta\}$

Then, we obtain the following conclusion on the Publisher–Subscriber pattern.

**Theorem 34.8** (Correctness of the Publisher–Subscriber pattern). *The Publisher–Subscriber pattern $\tau_I(\partial_H(P \between PP \between SP \between S))$ can exhibit the desired external behaviors.*

*Proof.* Based on the above state transitions of the above modules, by use of the algebraic laws of APTC, we can prove that $\tau_I(\partial_H(P \between PP \between SP \between S)) = \sum_{d_I, d_O \in \Delta}(r_I(d_I) \cdot s_O(d_O)) \cdot \tau_I(\partial_H(P \between PP \between SP \between S))$

That is, the Publisher–Subscriber pattern $\tau_I(\partial_H(P \between PP \between SP \between S))$ can exhibit the desired external behaviors.

For the details of the proof, please refer to Section 17.3, as we omit them here. □

Chapter 35

# Verification of idioms

Idioms are the lowest-level patterns that are programming language-specific and implement some specific concrete problems.

There are numerous language-specific idioms, in this chapter, we only verify two idioms called the Singleton pattern and the Counted Pointer pattern.

## 35.1 Verification of the Singleton pattern

The Singleton pattern ensures that there is only one instance in runtime for an object. In the Singleton pattern, there is only one module: The Singleton. The Singleton interacts with the outside through the input channels $I_i$ and the output channels $O_i$ for $1 \leq i \leq n$, as illustrated in Fig. 35.1.

The typical process is shown in Fig. 35.2 and is as follows:

1. The Singleton receives the input $d_{I_i}$ from the outside through the channel $I_i$ (the corresponding reading action is denoted $r_{I_i}(d_{I_i})$);
2. Then, it processes the input and generates the output $d_{O_i}$ through a processing function $SF_i$;
3. Then, it sends the output to the outside through the channel $O_i$ (the corresponding sending action is denoted $s_{O_i}(d_{O_i})$).

In the following, we verify the Singleton pattern. We assume all data elements $d_{I_i}$, $d_{O_i}$ for $1 \leq i \leq n$ are from a finite set $\Delta$.

The state transitions of the Singleton module described by APTC are as follows:

$S = \sum_{d_{I_1},\cdots,d_{I_n}\in\Delta}(r_{I_1}(d_{I_1}) \between \cdots \between r_{I_n}(d_{I_n}) \cdot S_2)$

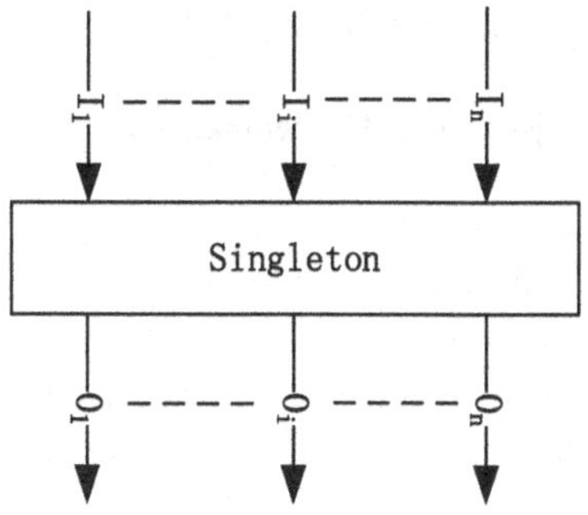


**FIGURE 35.1** Singleton pattern.

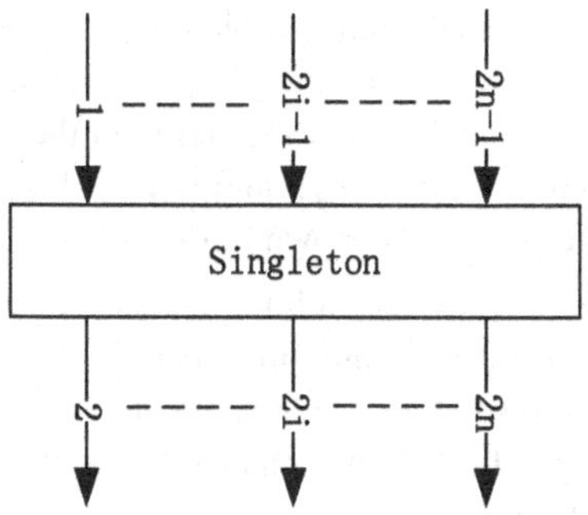


**FIGURE 35.2** Typical process of Singleton pattern.

**Handbook of Truly Concurrent Process Algebra. https://doi.org/10.1016/B978-0-44-321515-5.00039-1**

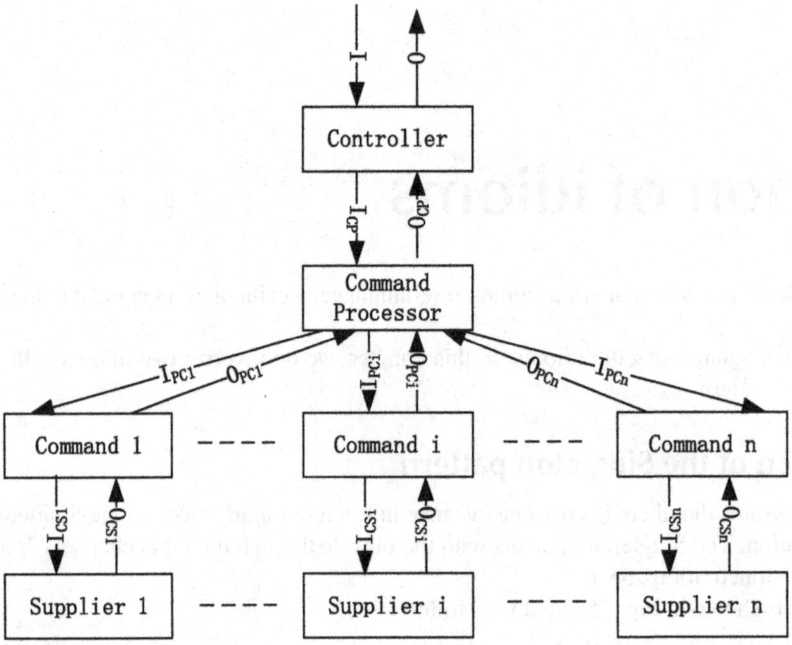


**FIGURE 35.3** Counted Pointer pattern.

$S_2 = SF_1 \between \cdots \between SF_n \cdot S_3$

$S_3 = \sum_{d_{O_1},\cdots,d_{O_n}\in\Delta}(s_{O_1}(d_{O_1}) \between \cdots \between s_{O_n}(d_{O_n}) \cdot S)$

There are no communications in the Singleton pattern.

Let all modules be in parallel, then the Singleton pattern $S$ can be presented by the following process term:

$\tau_I(\partial_H(\Theta(S))) = \tau_I(\partial_H(S))$

where $H = \emptyset$, $I = \{SF_i\}$ for $1 \leq i \leq n$.

Then, we obtain the following conclusion on the Singleton pattern.

**Theorem 35.1** (Correctness of the Singleton pattern). *The Singleton pattern $\tau_I(\partial_H(S))$ can exhibit the desired external behaviors.*

*Proof.* Based on the above state transitions of the above modules, by use of the algebraic laws of APTC, we can prove that

$\tau_I(\partial_H(S)) = \sum_{d_{I_1},d_{O_1},\cdots,d_{I_n},d_{O_n}\in\Delta}(r_{I_1}(d_{I_1}) \parallel \cdots \parallel r_{I_n}(d_{I_n}) \cdot s_{O_1}(d_{O_1}) \parallel \cdots \parallel s_{O_n}(d_{O_n})) \cdot \tau_I(\partial_H(S))$

That is, the Singleton pattern $\tau_I(\partial_H(S))$ can exhibit the desired external behaviors.

For the details of the proof, please refer to Section 17.3, as we omit them here. □

## 35.2 Verification of the Counted Pointer pattern

The Counted Pointer pattern makes memory management (implemented as Handle) of shared objects (implemented as Bodys) easier in C++. There are three modules in the Counted Pointer pattern: the Client, the Handle, and the Body. The Client interacts with the outside through the channels $I$ and $O$, and with the Handle through the channel $I_{CH}$ and $O_{CH}$. The Handle interacts with the Body through the channels $I_{HB}$ and $O_{HB}$. This is illustrated in Fig. 35.3.

The typical process of the Counted Pointer pattern is shown in Fig. 35.4 and is as follows:

1. The Client receives the input $d_I$ from the outside through the channel $I$ (the corresponding reading action is denoted $r_I(d_I)$), then processes the input $d_I$ through a processing function $CF_1$, and sends the processed input $d_{I_H}$ to the Handle through the channel $I_{CH}$ (the corresponding sending action is denoted $s_{I_{CH}}(d_{I_H})$);
2. The Handle receives $d_{I_H}$ from the Client through the channel $I_{CH}$ (the corresponding reading action is denoted $r_{I_{CH}}(d_{I_H})$), then processes the request through a processing function $HF_1$, generates and sends the processed input $d_{I_B}$ to the Body through the channel $I_{HB}$ (the corresponding sending action is denoted $s_{I_{HB}}(d_{I_B})$);

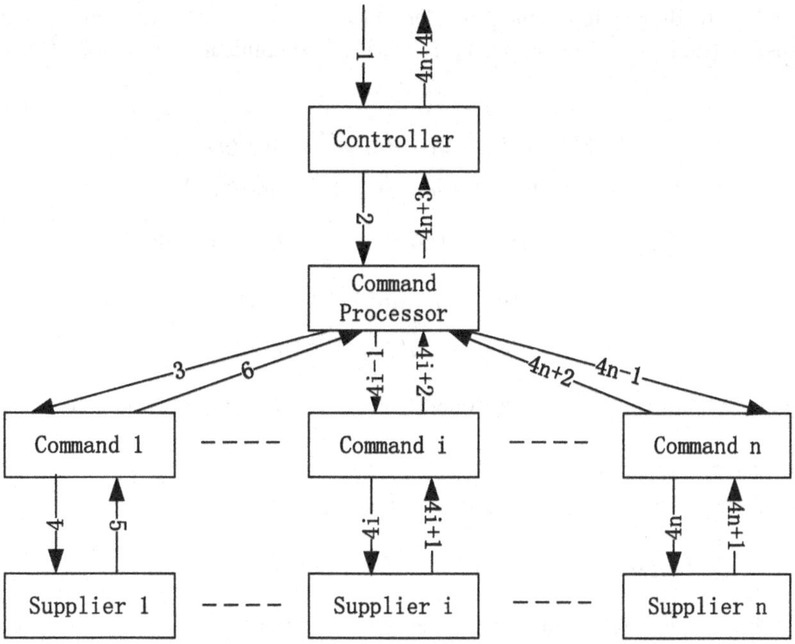


**FIGURE 35.4** Typical process of Counted Pointer pattern.

3. The Body receives the input $d_{I_B}$ from the Handle through the channel $I_{HB}$ (the corresponding reading action is denoted $r_{I_{HB}}(d_{I_B})$), then processes the input through a processing function $BF$, generates and sends the response $d_{O_B}$ to the Handle through the channel $O_{HB}$ (the corresponding sending action is denoted $s_{O_{HB}}(d_{O_B})$);
4. The Handle receives the response $d_{O_B}$ from the Body through the channel $O_{HB}$ (the corresponding reading action is denoted $r_{O_{HB}}(d_{O_B})$), then processes the response through a processing function $HF_2$, generates and sends the response $d_{O_H}$ (the corresponding sending action is denoted $s_{O_{CH}}(d_{O_H})$);
5. The Client receives the response $d_{O_H}$ from the Handle through the channel $O_{CH}$ (the corresponding reading action is denoted $r_{O_{CH}}(d_{O_H})$), then processes the request and generates the response $d_O$ through a processing function $CF_2$, and sends the response to the outside through the channel $O$ (the corresponding sending action is denoted $s_O(d_O)$).

In the following, we verify the Counted Pointer pattern. We assume all data elements $d_I, d_{I_H}, d_{I_B}, d_{O_B}, d_{O_H}, d_O$ are from a finite set $\Delta$.

The state transitions of the Client module described by APTC are as follows:

$C = \sum_{d_I \in \Delta}(r_I(d_I) \cdot C_2)$

$C_2 = CF_1 \cdot C_3$

$C_3 = \sum_{d_{I_H} \in \Delta}(s_{I_{CH}}(d_{I_H}) \cdot C_4)$

$C_4 = \sum_{d_{O_H} \in \Delta}(r_{O_{CH}}(d_{O_H}) \cdot C_5)$

$C_5 = CF_2 \cdot C_6$

$C_6 = \sum_{d_O \in \Delta}(s_O(d_O) \cdot C)$

The state transitions of the Handle module described by APTC are as follows:

$H = \sum_{d_{I_H} \in \Delta}(r_{I_{CH}}(d_{I_H}) \cdot H_2)$

$H_2 = HF_1 \cdot H_3$

$H_3 = \sum_{d_{I_B} \in \Delta}(s_{I_{HB}}(d_{I_B}) \cdot H_4)$

$H_4 = \sum_{d_{O_B} \in \Delta}(r_{O_{HB}}(d_{O_B}) \cdot H_5)$

$H_5 = HF_2 \cdot H_6$

$H_6 = \sum_{d_{O_H} \in \Delta}(s_{O_{CH}}(d_{O_H}) \cdot H)$

The state transitions of the Body module described by APTC are as follows:

$B = \sum_{d_{I_B} \in \Delta}(r_{I_{HB}}(d_{I_B}) \cdot B_2)$

$B_2 = BF \cdot B_3$

$B_3 = \sum_{d_{O_B} \in \Delta}(s_{O_{HB}}(d_{O_B}) \cdot B)$

The sending action and the reading action of the same data through the same channel can communicate with each other, otherwise, they will cause a deadlock $\delta$. We define the following communication functions between the Client and the Handle Proxy:

$$\gamma(r_{ICH}(d_{I_H}), s_{ICH}(d_{I_H})) \triangleq c_{ICH}(d_{I_H})$$

$$\gamma(r_{OCH}(d_{O_H}), s_{OCH}(d_{O_H})) \triangleq c_{OCH}(d_{O_H})$$

There are two communication functions between the Handle and the Body as follows:

$$\gamma(r_{IHB}(d_{I_B}), s_{IHB}(d_{I_B})) \triangleq c_{IHB}(d_{I_B})$$

$$\gamma(r_{OHB}(d_{O_B}), s_{OHB}(d_{O_B})) \triangleq c_{OHB}(d_{O_B})$$

Let all modules be in parallel, then the Counted Pointer pattern $C \quad H \quad B$ can be presented by the following process term:

$\tau_I(\partial_H(\Theta(C \between H \between B))) = \tau_I(\partial_H(C \between H \between B))$

where $H = \{r_{ICH}(d_{I_H}), s_{ICH}(d_{I_H}), r_{OCH}(d_{O_H}), s_{OCH}(d_{O_H}), r_{IHB}(d_{I_B}), s_{IHB}(d_{I_B}), r_{OHB}(d_{O_B}), s_{OHB}(d_{O_B})|d_I, d_{I_H}, d_{I_B}, d_{O_B}, d_{O_H}, d_O \in \Delta\}$,

$I = \{c_{ICH}(d_{I_H}), c_{OCH}(d_{O_H}), c_{IHB}(d_{I_B}), c_{OHB}(d_{O_B}), CF_1, CF_2, HF_1, HF_2, BF |d_I, d_{I_H}, d_{I_B}, d_{O_B}, d_{O_H}, d_O \in \Delta\}$

Then, we obtain the following conclusion on the Counted Pointer pattern.

**Theorem 35.2** (Correctness of the Counted Pointer pattern). *The Counted Pointer pattern $\tau_I(\partial_H(C \between H \between B))$ can exhibit the desired external behaviors.*

*Proof.* Based on the above state transitions of the above modules, by use of the algebraic laws of APTC, we can prove that

$\tau_I(\partial_H(C \between H \between B)) = \sum_{d_I, d_O \in \Delta}(r_I(d_I) \cdot s_O(d_O)) \cdot \tau_I(\partial_H(C \between H \between B))$

That is, the Counted Pointer pattern $\tau_I(\partial_H(C \between H \between B))$ can exhibit the desired external behaviors.

For the details of the proof, please refer to Section 17.3, as we omit them here. □

Chapter 36

# Verification of patterns for concurrent and networked objects

Patterns for concurrent and networked objects can be used both in higher-level and lower-level systems and applications.

In this chapter, we verify patterns for concurrent and networked objects. In Section 36.1, we verify service access and configuration patterns. In Section 36.2, we verify patterns related to event handling. We verify synchronization patterns in Section 36.3 and concurrency patterns in Section 36.4.

## 36.1 Service access and configuration patterns

In this section, we verify patterns for service access and configuration, including the Wrapper Facade pattern, the Component Configurator pattern, the Interceptor pattern, and the Extension Interface pattern.

### 36.1.1 Verification of the Wrapper Facade pattern

The Wrapper Facade pattern encapsulates the non-object-oriented APIs into the object-oriented ones. There are two classes of modules in the Wrapper Facade pattern: the Wrapper Facade and $n$ API Functions. The Wrapper Facade interacts with API Function $i$ through the channels $I_{WA_i}$ and $O_{WA_i}$, and it exchanges information with the outside through the input channel $I$ and $O$. This is illustrated in Fig. 36.1.

The typical process of the Wrapper Facade pattern is shown in Fig. 36.2 and is as follows:

1. The Wrapper Facade receives the input from the user through the channel $I$ (the corresponding reading action is denoted $r_I(d_I)$), then processes the input through a processing function $WF_1$ and generates the input $d_{I_{A_i}}$, and sends $d_{I_{A_i}}$ to the API Function $i$ (for $1 \leq i \leq n$) through the channel $I_{WA_i}$ (the corresponding sending action is denoted $s_{I_{WA_i}}(d_{I_{A_i}})$);
2. The API Function $i$ receives the input $d_{I_{A_i}}$ from the Wrapper Facade through the channel $I_{WA_i}$ (the corresponding reading action is denoted $r_{I_{WA_i}}(d_{I_{A_i}})$), then processes the input through a processing function $AF_i$, and sends the results $d_{O_{A_i}}$ to the Wrapper Facade through the channel $O_{WA_i}$ (the corresponding sending action is denoted $s_{O_{WA_i}}(d_{O_{A_i}})$);
3. The Wrapper Facade receives the computational results from the API Function $i$ through the channel $O_{WA_i}$ (the corresponding reading action is denoted $r_{O_{WA_i}}(d_{O_{A_i}})$), then processes the results through a processing function $WF_2$ and generates the result $d_O$, and sends $d_O$ to the outside through the channel $O$ (the corresponding sending action is denoted $s_O(d_O)$).

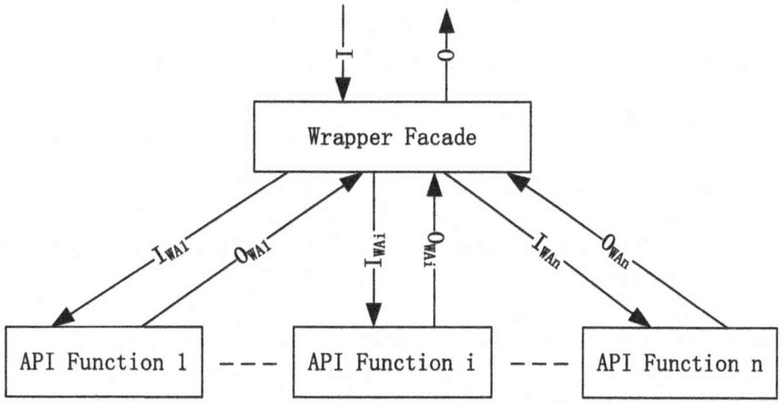


FIGURE 36.1 Wrapper Facade pattern.

Handbook of Truly Concurrent Process Algebra. https://doi.org/10.1016/B978-0-44-321515-5.00040-8

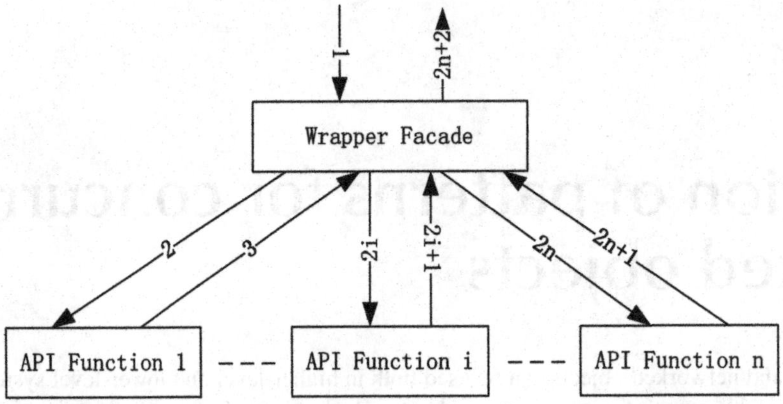


**FIGURE 36.2** Typical process of Wrapper Facade pattern.

In the following, we verify the Wrapper Facade pattern. We assume all data elements $d_I, d_{I_{A_i}}, d_{O_{A_i}}, d_O$ (for $1 \leq i \leq n$) are from a finite set $\Delta$.

The state transitions of the Wrapper Facade module described by APTC are as follows:

$W = \sum_{d_I \in \Delta}(r_I(d_I) \cdot W_2)$

$W_2 = WF_1 \cdot W_3$

$W_3 = \sum_{d_{I_{A_1}}, \cdots, d_{I_{A_n}} \in \Delta}(s_{IWA_1}(d_{I_{A_1}}) \between \cdots \between s_{IWA_n}(d_{I_{A_n}}) \cdot W_4)$

$W_4 = \sum_{d_{O_{A_1}}, \cdots, d_{O_{A_n}} \in \Delta}(r_{OWA_1}(d_{O_{A_1}}) \between \cdots \between r_{OWA_n}(d_{O_{A_n}}) \cdot W_5)$

$W_5 = WF_2 \cdot W_6$

$W_6 = \sum_{d_O \in \Delta}(s_O(d_O) \cdot W)$

The state transitions of the API Function $i$ described by APTC are as follows:

$A_i = \sum_{d_{I_{A_i}} \in \Delta}(r_{IWA_i}(d_{I_{A_i}}) \cdot A_{i_2})$

$A_{i_2} = AF_i \cdot A_{i_3}$

$A_{i_3} = \sum_{d_{O_{A_i}} \in \Delta}(s_{OWA_i}(d_{O_{A_i}}) \cdot A_i)$

The sending action and the reading action of the same data through the same channel can communicate with each other, otherwise, they will cause a deadlock $\delta$. We define the following communication functions of the API Function $i$ for $1 \leq i \leq n$:

$$\gamma(r_{IWA_i}(d_{I_{A_i}}), s_{IWA_i}(d_{I_{A_i}})) \triangleq c_{IWA_i}(d_{I_{A_i}})$$

$$\gamma(r_{OWA_i}(d_{O_{A_i}}), s_{OWA_i}(d_{O_{A_i}})) \triangleq c_{OWA_i}(d_{O_{A_i}})$$

Let all modules be in parallel, then the Wrapper Facade pattern $W$ $A_1 \cdots A_i \cdots A_n$ can be presented by the following process term:

$\tau_I(\partial_H(\Theta(W \between A_1 \between \cdots \between A_i \between \cdots \between A_n))) = \tau_I(\partial_H(W \between A_1 \between \cdots \between A_i \between \cdots \between A_n))$

where $H = \{r_{IWA_i}(d_{I_{A_i}}), s_{IWA_i}(d_{I_{A_i}}), r_{OWA_i}(d_{O_{A_i}}), s_{OWA_i}(d_{O_{A_i}})$
$|d_I, d_{I_{A_i}}, d_{O_{A_i}}, d_O \in \Delta\}$ for $1 \leq i \leq n$,

$I = \{c_{IWA_i}(d_{I_{A_i}}), c_{OWA_i}(d_{O_{A_i}}), WF_1, WF_2, AF_i$
$|d_I, d_{I_{A_i}}, d_{O_{A_i}}, d_O \in \Delta\}$ for $1 \leq i \leq n$

Then, we obtain the following conclusion on the Wrapper Facade pattern.

**Theorem 36.1** (Correctness of the Wrapper Facade pattern). *The Wrapper Facade pattern $\tau_I(\partial_H(W \between A_1 \between \cdots \between A_i \between \cdots \between A_n))$ can exhibit the desired external behaviors.*

*Proof.* Based on the above state transitions of the above modules, by use of the algebraic laws of APTC, we can prove that

$\tau_I(\partial_H(W \between A_1 \between \cdots \between A_i \between \cdots \between A_n)) = \sum_{d_I, d_O \in \Delta}(r_I(d_I) \cdot s_O(d_O)) \cdot \tau_I(\partial_H(W \between A_1 \between \cdots \between A_i \between \cdots \between A_n))$

That is, the Wrapper Facade pattern $\tau_I(\partial_H(W \between A_1 \between \cdots \between A_i \between \cdots \between A_n))$ can exhibit the desired external behaviors.

For the details of the proof, please refer to Section 17.3, as we omit them here. □

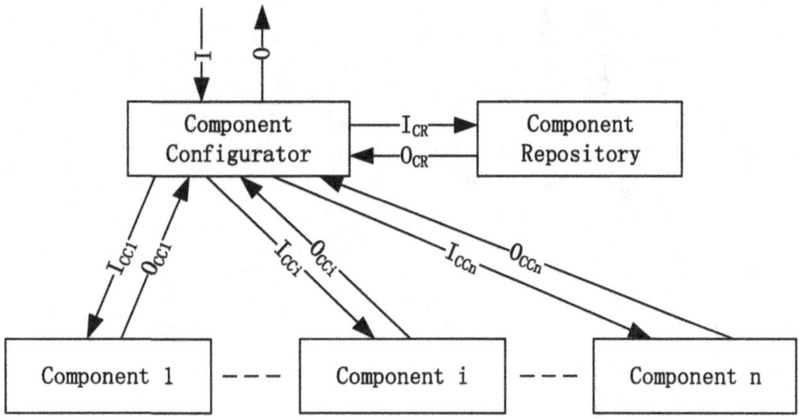


**FIGURE 36.3** Component Configurator pattern.

## 36.1.2 Verification of the Component Configurator pattern

The Component Configurator pattern allows us to configure the components dynamically. There are three classes of modules in the Component Configurator pattern: the Component Configurator, $n$ Components, and the Component Repository. The Component Configurator interacts with Component $i$ through the channels $I_{CC_i}$ and $O_{CC_i}$, and it exchanges information with the outside through the input channel $I$ and $O$, and with the Component Repository through the channels $I_{CR}$ and $O_{CR}$. This is illustrated in Fig. 36.3.

The typical process of the Component Configurator pattern is shown in Fig. 36.4 and is as follows:

1. The Component Configurator receives the input from the user through the channel $I$ (the corresponding reading action is denoted $r_I(d_I)$), then processes the input through a processing function $CCF_1$ and generates the input $d_{I_{C_i}}$, and sends $d_{I_{C_i}}$ to the Component $i$ (for $1 \le i \le n$) through the channel $I_{CC_i}$ (the corresponding sending action is denoted $s_{I_{CC_i}}(d_{I_{C_i}})$);
2. The Component $i$ receives the input $d_{I_{C_i}}$ from the Component Configurator through the channel $I_{CC_i}$ (the corresponding reading action is denoted $r_{I_{CC_i}}(d_{I_{C_i}})$), then processes the input through a processing function $CF_i$, and sends the results $d_{O_{C_i}}$ to the Component Configurator through the channel $O_{CC_i}$ (the corresponding sending action is denoted $s_{O_{CC_i}}(d_{O_{C_i}})$);
3. The Component Configurator receives the configurational results from the Component $i$ through the channel $O_{CC_i}$ (the corresponding reading action is denoted $r_{O_{CC_i}}(d_{O_{C_i}})$), then processes the results through a processing function $CCF_2$ and generates the configurational information $d_{I_R}$, and sends $d_{I_R}$ to the Component Repository through the channel $I_{CR}$ (the corresponding sending action is denoted $s_{I_{CR}}(d_{I_R})$);
4. The Component Repository receives the configurational information $d_{I_R}$ through the channel $I_{CR}$ (the corresponding reading action is denoted $r_{I_{CR}}(d_{I_R})$), then processes the information and generates the results $d_{O_R}$ through a processing function $RF$, and sends the results $d_{O_R}$ to the Component Configurator through the channels $O_{CR}$ (the corresponding sending action is denoted $s_{O_{CR}}(d_{O_R})$);
5. The Component Configurator receives the results $d_{O_R}$ from the Component Repository through the channel $O_{CR}$ (the corresponding reading action is denoted $r_{O_{CR}}(d_{O_R})$), then processes the results and generates the results $d_O$ through a processing function $CCF_3$, and sends $d_O$ to the outside through the channel $O$ (the corresponding sending action is denoted $s_O(d_O)$).

In the following, we verify the Component Configurator pattern. We assume all data elements $d_I$, $d_{I_R}$, $d_{I_{C_i}}$, $d_{O_{C_i}}$, $d_{O_R}$, $d_O$ (for $1 \le i \le n$) are from a finite set $\Delta$.

The state transitions of the Component Configurator module described by APTC are as follows:

$CC = \sum_{d_I \in \Delta}(r_I(d_I) \cdot CC_2)$

$CC_2 = CCF_1 \cdot CC_3$

$CC_3 = \sum_{d_{I_{C_1}},\cdots,d_{I_{C_n}} \in \Delta}(s_{I_{CC_1}}(d_{I_{C_1}}) \between \cdots \between s_{I_{CC_n}}(d_{I_{C_n}}) \cdot CC_4)$

$CC_4 = \sum_{d_{O_{C_1}},\cdots,d_{O_{C_n}} \in \Delta}(r_{O_{CC_1}}(d_{O_{C_1}}) \between \cdots \between r_{O_{CC_n}}(d_{O_{C_n}}) \cdot CC_5)$

$CC_5 = CCF_2 \cdot CC_6$

$CC_6 = \sum_{d_{I_R} \in \Delta}(s_{I_{CR}}(d_{I_R}) \cdot CC_7)$

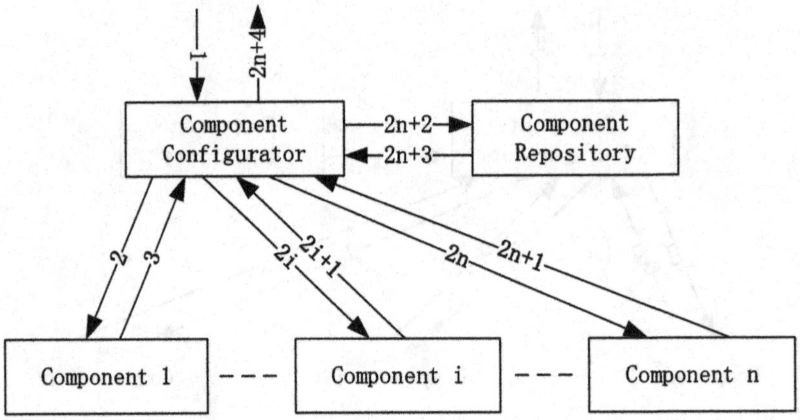

**FIGURE 36.4** Typical process of Component Configurator pattern.

$CC_7 = \sum_{d_{O_R} \in \Delta}(r_{O_{CR}}(d_{O_R}) \cdot CC_8)$

$CC_8 = CCF_3 \cdot CC_9$

$CC_9 = \sum_{d_O \in \Delta}(s_O(d_O) \cdot CC)$

The state transitions of the Component $i$ described by APTC are as follows:

$C_i = \sum_{d_{I_{C_i}} \in \Delta}(r_{I_{CC_i}}(d_{I_{C_i}}) \cdot C_{i_2})$

$C_{i_2} = CF_i \cdot C_{i_3}$

$C_{i_3} = \sum_{d_{O_{C_i}} \in \Delta}(s_{O_{CC_i}}(d_{O_{C_i}}) \cdot C_i)$

The state transitions of the Component Repository described by APTC are as follows:

$R = \sum_{d_{I_R} \in \Delta}(r_{I_{CR}}(d_{I_R}) \cdot R_2)$

$R_2 = RF \cdot R_3$

$R_3 = \sum_{d_{O_R} \in \Delta}(s_{O_{CR}}(d_{O_R}) \cdot R)$

The sending action and the reading action of the same data through the same channel can communicate with each other, otherwise, they will cause a deadlock $\delta$. We define the following communication functions of the Component Configurator for $1 \leq i \leq n$:

$$\gamma(r_{I_{CC_i}}(d_{I_{C_i}}), s_{I_{CC_i}}(d_{I_{C_i}})) \triangleq c_{I_{CC_i}}(d_{I_{C_i}})$$

$$\gamma(r_{O_{CC_i}}(d_{O_{C_i}}), s_{O_{CC_i}}(d_{O_{C_i}})) \triangleq c_{O_{CC_i}}(d_{O_{C_i}})$$

$$\gamma(r_{I_{CR}}(d_{I_R}), s_{I_{CR}}(d_{I_R})) \triangleq c_{I_{CR}}(d_{I_R})$$

$$\gamma(r_{O_{CR}}(d_{O_R}), s_{O_{CR}}(d_{O_R})) \triangleq c_{O_{CR}}(d_{O_R})$$

Let all modules be in parallel, then the Component Configurator pattern $CC \quad R \quad C_1 \cdots C_i \cdots C_n$ can be presented by the following process term:

$\tau_I(\partial_H(\Theta(CC \between R \between C_1 \between \cdots \between C_i \between \cdots \between C_n))) = \tau_I(\partial_H(CC \between R \between C_1 \between \cdots \between C_i \between \cdots \between C_n))$

where $H = \{r_{I_{CC_i}}(d_{I_{C_i}}), s_{I_{CC_i}}(d_{I_{C_i}}), r_{O_{CC_i}}(d_{O_{C_i}}), s_{O_{CC_i}}(d_{O_{C_i}}), r_{I_{CR}}(d_{I_R}), s_{I_{CR}}(d_{I_R}),$
$r_{O_{CR}}(d_{O_R}), s_{O_{CR}}(d_{O_R}) | d_I, d_{I_R}, d_{I_{C_i}}, d_{O_{C_i}}, d_{O_R}, d_O \in \Delta\}$ for $1 \leq i \leq n$,

$I = \{c_{I_{CC_i}}(d_{I_{C_i}}), c_{O_{CC_i}}(d_{O_{C_i}}), c_{I_{CR}}(d_{I_R}), c_{O_{CR}}(d_{O_R}), CCF_1, CCF_2, CCF_3, CF_i, RF$
$| d_I, d_{I_R}, d_{I_{C_i}}, d_{O_{C_i}}, d_{O_R}, d_O \in \Delta\}$ for $1 \leq i \leq n$

Then, we obtain the following conclusion on the Component Configurator pattern.

**Theorem 36.2** (Correctness of the Component Configurator pattern). *The Component Configurator pattern $\tau_I(\partial_H(CC \between R \between C_1 \between \cdots \between C_i \between \cdots \between C_n))$ can exhibit the desired external behaviors.*

*Proof.* Based on the above state transitions of the above modules, by use of the algebraic laws of APTC, we can prove that

$\tau_I(\partial_H(CC \between R \between C_1 \between \cdots \between C_i \between \cdots \between C_n)) = \sum_{d_I, d_O \in \Delta}(r_I(d_I) \cdot s_O(d_O)) \cdot \tau_I(\partial_H(CC \between R \between C_1 \between \cdots \between C_i \between \cdots \between C_n))$

That is, the Component Configurator pattern $\tau_I(\partial_H(CC \between R \between C_1 \between \cdots \between C_i \between \cdots \between C_n))$ can exhibit the desired external behaviors.

For the details of the proof, please refer to Section 17.3, as we omit them here. □

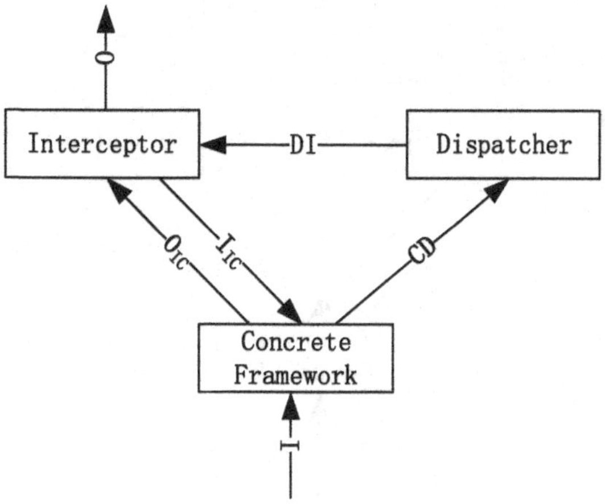


FIGURE 36.5 Interceptor pattern.

### 36.1.3 Verification of the Interceptor pattern

The Interceptor pattern adds functionalities to the concrete framework to introduce an intermediate Dispatcher and an Interceptor. There are three modules in the Interceptor pattern: the Concrete Framework, the Dispatcher, and the Interceptor. The Concrete Framework interacts with the user through the channels $I$ and $O$; with the Dispatcher through the channel $CD$, and with the Interceptor through the channels $I_{IC}$ and $O_{IC}$. The Dispatcher interacts with the Interceptor through the channel $DI$. This is illustrated in Fig. 36.5.

The typical process of the Interceptor pattern is shown in Fig. 36.6 and is as follows:

1. The Concrete Framework receives the request $d_I$ from the user through the channel $I$ (the corresponding reading action is denoted $r_I(d_I)$), then processes the request $d_I$ through a processing function $CF_1$, and sends the processed request $d_{I_D}$ to the Dispatcher through the channel $CD$ (the corresponding sending action is denoted $s_{CD}(d_{I_D})$);
2. The Dispatcher receives $d_{I_D}$ from the Concrete Framework through the channel $CD$ (the corresponding reading action is denoted $r_{CD}(d_{I_D})$), then processes the request through a processing function $DF$, generates and sends the processed request $d_{O_D}$ to the Interceptor through the channel $DI$ (the corresponding sending action is denoted $s_{DI}(d_{O_D})$);
3. The Interceptor receives the request $d_{O_D}$ from the Dispatcher through the channel $DI$ (the corresponding reading action is denoted $r_{DI}(d_{O_D})$), then processes the request and generates the request $d_{I_C}$ through a processing function $IF_1$, and sends the request to the Concrete Framework through the channel $I_{IC}$ (the corresponding sending action is denoted $s_{I_{IC}}(d_{I_C})$);
4. The Concrete Framework receives the request $d_{I_C}$ from the Interceptor through the channel $I_{IC}$ (the corresponding reading action is denoted $r_{I_{IC}}(d_{I_C})$), then processes the request through a processing function $CF_2$, generates and sends the response $d_{O_C}$ to the Interceptor through the channel $O_{IC}$ (the corresponding sending action is denoted $s_{O_{IC}}(d_{O_C})$);
5. The Interceptor receives the response $d_{O_C}$ from the Concrete Framework through the channel $O_{IC}$ (the corresponding reading action is denoted $r_{O_{IC}}(d_{O_C})$), then processes the request through a processing function $IF_2$, generates and sends the processed response $d_O$ to the user through the channel $O$ (the corresponding sending action is denoted $s_O(d_O)$).

In the following, we verify the Interceptor pattern. We assume all data elements $d_I, d_{I_D}, d_{I_C}, d_{O_D}, d_{O_C}, d_O$ are from a finite set $\Delta$.

The state transitions of the Concrete Framework module described by APTC are as follows:

$C = \sum_{d_I\in\Delta}(r_I(d_I) \cdot C_2)$

$C_2 = CF_1 \cdot C_3$

$C_3 = \sum_{d_{I_D}\in\Delta}(s_{CD}(d_{I_D}) \cdot C_4)$

$C_4 = \sum_{d_{I_C}\in\Delta}(r_{I_{IC}}(d_{I_C}) \cdot C_5)$

$C_5 = CF_2 \cdot C_6$

$C_6 = \sum_{d_{O_C}\in\Delta}(s_{O_{IC}}(d_{O_C}) \cdot C)$

The state transitions of the Dispatcher module described by APTC are as follows:

$D = \sum_{d_{I_D}\in\Delta}(r_{CD}(d_{I_D}) \cdot D_2)$

$D_2 = DF \cdot D_3$

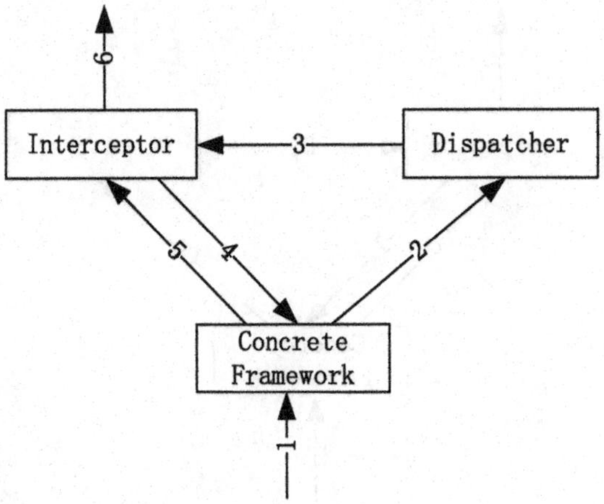


**FIGURE 36.6** Typical process of Interceptor pattern.

$D_3 = \sum_{d_{O_D} \in \Delta}(s_{DI}(d_{O_D}) \cdot D)$

The state transitions of the Interceptor module described by APTC are as follows:

$I = \sum_{d_{O_D} \in \Delta}(r_{DI}(d_{O_D}) \cdot I_2)$

$I_2 = IF_1 \cdot I_3$

$I_3 = \sum_{d_{I_C} \in \Delta}(s_{I_{IC}}(d_{I_C}) \cdot I_4)$

$I_4 = \sum_{d_{O_C} \in \Delta}(r_{O_{IC}}(d_{O_C}) \cdot I_5)$

$I_5 = IF_2 \cdot I_6$

$I_6 = \sum_{d_O \in \Delta}(s_O(d_O) \cdot I)$

The sending action and the reading action of the same data through the same channel can communicate with each other, otherwise, they will cause a deadlock $\delta$. We define the following communication function between the Concrete Framework and the Dispatcher:

$$\gamma(r_{CD}(d_{I_D}), s_{CD}(d_{I_D})) \triangleq c_{CD}(d_{I_D})$$

There are two communication functions between the Concrete Framework and the Interceptor as follows:

$$\gamma(r_{I_{IC}}(d_{I_C}), s_{I_{IC}}(d_{I_C})) \triangleq c_{I_{IC}}(d_{I_C})$$

$$\gamma(r_{O_{IC}}(d_{O_C}), s_{O_{IC}}(d_{O_C})) \triangleq c_{O_{IC}}(d_{O_C})$$

There is one communication function between the Dispatcher and the Interceptor as follows:

$$\gamma(r_{DI}(d_{O_D}), s_{DI}(d_{O_D})) \triangleq c_{DI}(d_{O_D})$$

Let all modules be in parallel, then the Interceptor pattern $C \quad D \quad I$ can be presented by the following process term:

$\tau_I(\partial_H(\Theta(C \between D \between I))) = \tau_I(\partial_H(C \between D \between I))$

where $H = \{r_{CD}(d_{I_D}), s_{CD}(d_{I_D}), r_{I_{IC}}(d_{I_C}), s_{I_{IC}}(d_{I_C}), r_{O_{IC}}(d_{O_C}), s_{O_{IC}}(d_{O_C}), r_{DI}(d_{O_D}), s_{DI}(d_{O_D}) | d_I, d_{I_D}, d_{I_C}, d_{O_D}, d_{O_C}, d_O \in \Delta\}$,

$I = \{c_{CD}(d_{I_D}), c_{I_{IC}}(d_{I_C}), c_{O_{IC}}(d_{O_C}), c_{DI}(d_{O_D}), CF_1, CF_2, DF, IF_1, IF_2 | d_I, d_{I_D}, d_{I_C}, d_{O_D}, d_{O_C}, d_O \in \Delta\}$

Then, we obtain the following conclusion on the Interceptor pattern.

**Theorem 36.3** (Correctness of the Interceptor pattern). *The Interceptor pattern $\tau_I(\partial_H(C \between D \between I))$ can exhibit the desired external behaviors.*

*Proof.* Based on the above state transitions of the above modules, by use of the algebraic laws of APTC, we can prove that

$\tau_I(\partial_H(C \between D \between I)) = \sum_{d_I, d_O \in \Delta}(r_I(d_I) \cdot s_O(d_O)) \cdot \tau_I(\partial_H(C \between D \between I))$

That is, the Interceptor pattern $\tau_I(\partial_H(C \between D \between I))$ can exhibit the desired external behaviors.

For the details of the proof, please refer to Section 17.3, as we omit them here. □

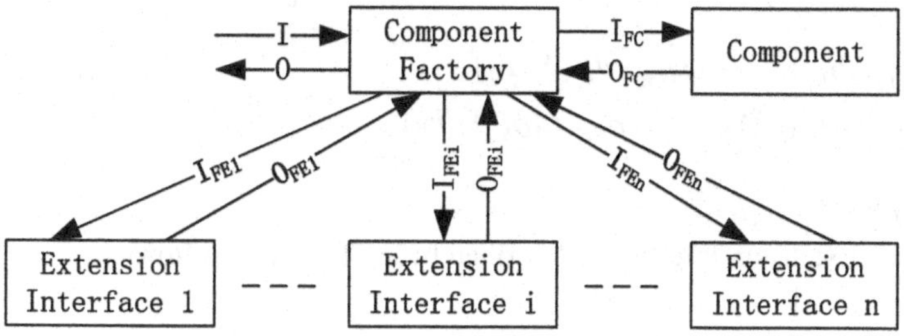


**FIGURE 36.7** Extension Interface pattern.

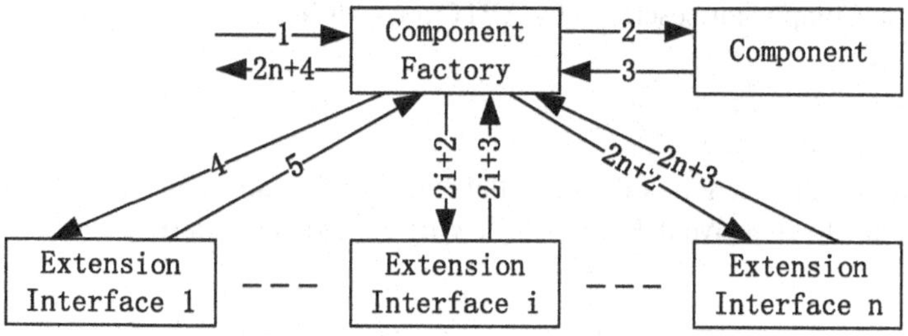


**FIGURE 36.8** Typical process of Extension Interface pattern.

## 36.1.4 Verification of the Extension Interface pattern

The Extension Interface pattern allows us to export multiple interfaces of a component to extend or modify the functionalities of the component. There are three classes of modules in the Extension Interface pattern: the Component Factory, $n$ Extension Interfaces, and the Component. The Component Factory interacts with Extension Interface $i$ through the channels $I_{FE_i}$ and $O_{FE_i}$, and it exchanges information with the outside through the input channel $I$ and the output channel $O$, and with the Component through the channels $I_{FC}$ and $O_{FC}$. This is illustrated in Fig. 36.7.

The typical process of the Extension Interface pattern is shown in Fig. 36.8 and is as follows:

1. The Component Factory receives the input from the user through the channel $I$ (the corresponding reading action is denoted $r_I(d_I)$), then processes the input through a processing function $FF_1$ and generates the input $d_{I_C}$, and sends $d_{I_C}$ to the Component through the channel $I_{FC}$ (the corresponding sending action is denoted $s_{I_{FC}}(d_{I_C})$);
2. The Component receives the input $d_{I_C}$ through the channel $I_{FC}$ (the corresponding reading action is denoted $r_{I_{FC}}(d_{I_C})$), then processes the information and generates the results $d_{O_C}$ through a processing function $CF$, and sends the results $d_{O_C}$ to the Component Factory through the channels $O_{FC}$ (the corresponding sending action is denoted $s_{O_{FC}}(d_{O_C})$);
3. The Component Factory receives the results $d_{O_C}$ from the Component through the channel $O_{FC}$ (the corresponding reading action is denoted $r_{O_{FC}}(d_{O_C})$), then processes the results through a processing function $FF_2$ and generates the input $d_{I_{E_i}}$, and sends $d_{I_{E_i}}$ to the Extension Interface $i$ (for $1 \leq i \leq n$) through the channel $I_{FE_i}$ (the corresponding sending action is denoted $s_{I_{FE_i}}(d_{I_{E_i}})$);
4. The Extension Interface $i$ receives the input $d_{I_{E_i}}$ from the Component Factory through the channel $I_{FE_i}$ (the corresponding reading action is denoted $r_{I_{FE_i}}(d_{I_{E_i}})$), then processes the input through a processing function $EF_i$, and sends the results $d_{O_{E_i}}$ to the Component Factory through the channel $O_{FE_i}$ (the corresponding sending action is denoted $s_{O_{FE_i}}(d_{O_{E_i}})$);
5. The Component Factory receives the results from the Extension Interface $i$ through the channel $O_{FE_i}$ (the corresponding reading action is denoted $r_{O_{FE_i}}(d_{O_{E_i}})$), then processes the results and generates the results $d_O$ through a processing function $FF_3$, and sends $d_O$ to the outside through the channel $O$ (the corresponding sending action is denoted $s_O(d_O)$).

In the following, we verify the Extension Interface pattern. We assume all data elements $d_I$, $d_{I_C}$, $d_{I_{E_i}}$, $d_{O_{E_i}}$, $d_{O_C}$, $d_O$ (for $1 \leq i \leq n$) are from a finite set $\Delta$.

The state transitions of the Component Factory module described by APTC are as follows:

$F = \sum_{d_I \in \Delta}(r_I(d_I) \cdot F_2)$

$F_2 = FF_1 \cdot F_3$

$F_3 = \sum_{d_{I_C} \in \Delta}(s_{I_{FC}}(d_{I_C}) \cdot F_4)$

$F_4 = \sum_{d_{O_C} \in \Delta}(r_{O_{FC}}(d_{O_C}) \cdot F_5)$

$F_5 = FF_2 \cdot F_6$
$F_6 = \sum_{d_{I_{E_1}},\cdots,d_{I_{E_n}}\in\Delta}(s_{IFE_1}(d_{I_{E_1}}) \between \cdots \between s_{IFE_n}(d_{I_{E_n}}) \cdot F_7)$
$F_7 = \sum_{d_{O_{E_1}},\cdots,d_{O_{E_n}}\in\Delta}(r_{OFE_1}(d_{O_{E_1}}) \between \cdots \between r_{OFE_n}(d_{O_{E_n}}) \cdot F_8)$
$F_8 = FF_3 \cdot F_9$
$F_9 = \sum_{d_O\in\Delta}(s_O(d_O) \cdot F)$
The state transitions of the Extension Interface $i$ described by APTC are as follows:
$E_i = \sum_{d_{I_{E_i}}\in\Delta}(r_{IFE_i}(d_{I_{E_i}}) \cdot E_{i_2})$
$E_{i_2} = EF_i \cdot E_{i_3}$
$E_{i_3} = \sum_{d_{O_{E_i}}\in\Delta}(s_{OFE_i}(d_{O_{E_i}}) \cdot E_i)$
The state transitions of the Component described by APTC are as follows:
$C = \sum_{d_{I_C}\in\Delta}(r_{IFC}(d_{I_C}) \cdot C_2)$
$C_2 = CF \cdot C_3$
$C_3 = \sum_{d_{O_C}\in\Delta}(s_{OFC}(d_{O_C}) \cdot C)$
The sending action and the reading action of the same data through the same channel can communicate with each other, otherwise, they will cause a deadlock $\delta$. We define the following communication functions of the Component Factory for $1 \le i \le n$:

$$\gamma(r_{IFE_i}(d_{I_{E_i}}), s_{IFE_i}(d_{I_{E_i}})) \triangleq c_{IFE_i}(d_{I_{E_i}})$$

$$\gamma(r_{OFE_i}(d_{O_{E_i}}), s_{OFE_i}(d_{O_{E_i}})) \triangleq c_{OFE_i}(d_{O_{E_i}})$$

$$\gamma(r_{IFC}(d_{I_C}), s_{IFC}(d_{I_C})) \triangleq c_{IFC}(d_{I_C})$$

$$\gamma(r_{OFC}(d_{O_C}), s_{OFC}(d_{O_C})) \triangleq c_{OFC}(d_{O_C})$$

Let all modules be in parallel, then the Extension Interface pattern $F \quad C \quad E_1 \cdots E_i \cdots E_n$ can be presented by the following process term:
$\tau_I(\partial_H(\Theta(F \between C \between E_1 \between \cdots \between E_i \between \cdots \between E_n))) = \tau_I(\partial_H(F \between C \between E_1 \between \cdots \between E_i \between \cdots \between E_n))$
where $H = \{r_{IFE_i}(d_{I_{E_i}}), s_{IFE_i}(d_{I_{E_i}}), r_{OFE_i}(d_{O_{E_i}}), s_{OFE_i}(d_{O_{E_i}}), r_{IFC}(d_{I_C}), s_{IFC}(d_{I_C}),$
$r_{OFC}(d_{O_C}), s_{OFC}(d_{O_C}) | d_I, d_{I_C}, d_{I_{E_i}}, d_{O_{E_i}}, d_{O_C}, d_O \in \Delta\}$ for $1 \le i \le n$,
$I = \{c_{IFE_i}(d_{I_{E_i}}), c_{OFE_i}(d_{O_{E_i}}), c_{IFC}(d_{I_C}), c_{OFC}(d_{O_C}), FF_1, FF_2, FF_3, EF_i, CF$
$| d_I, d_{I_C}, d_{I_{E_i}}, d_{O_{E_i}}, d_{O_C}, d_O \in \Delta\}$ for $1 \le i \le n$
Then, we obtain the following conclusion on the Extension Interface pattern.

**Theorem 36.4** (Correctness of the Extension Interface pattern). *The Extension Interface pattern $\tau_I(\partial_H(F \between C \between E_1 \between \cdots \between E_i \between \cdots \between E_n))$ can exhibit the desired external behaviors.*

*Proof.* Based on the above state transitions of the above modules, by use of the algebraic laws of APTC, we can prove that
$\tau_I(\partial_H(F \between C \between E_1 \between \cdots \between E_i \between \cdots \between E_n)) = \sum_{d_I,d_O\in\Delta}(r_I(d_I) \cdot s_O(d_O)) \cdot \tau_I(\partial_H(F \between C \between E_1 \between \cdots \between E_i \between \cdots \between E_n))$
That is, the Extension Interface pattern $\tau_I(\partial_H(F \between C \between E_1 \between \cdots \between E_i \between \cdots \between E_n))$ can exhibit the desired external behaviors.

For the details of the proof, please refer to Section 17.3, as we omit them here. □

## 36.2 Event handling patterns

In this section, we verify patterns related to event handling, including the Reactor pattern, the Proactor pattern, the Asynchronous Completion Token pattern, and the Acceptor–Connector pattern.

### 36.2.1 Verification of the Reactor pattern

The Reactor pattern allows us to demultiplex and dispatch the request event to the event-driven applications. There are three classes of modules in the Reactor pattern: the Handle Set, $n$ Event Handlers, and the Reactor. The Handle Set interacts with Event Handler $i$ through the channel $EH_i$, and it exchanges information with the outside through the input channel $I$ and the output channel $O$, and with the Reactor through the channel $HR$. The Reactor interacts with the Event Handler $i$ through the channel $RE_i$. This is illustrated in Fig. 36.9.

The typical process of the Reactor pattern is shown in Fig. 36.10 and is as follows:

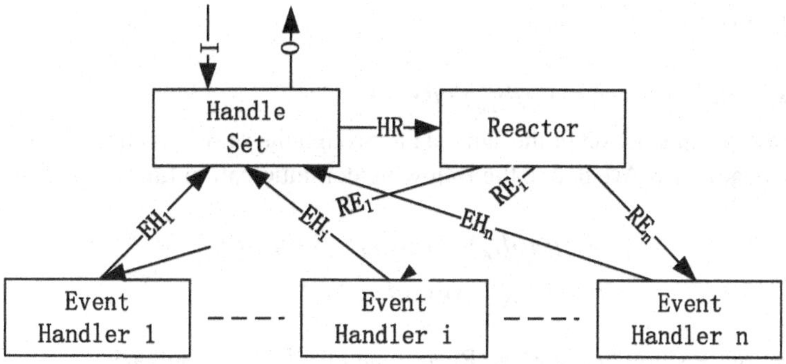


**FIGURE 36.9** Reactor pattern.

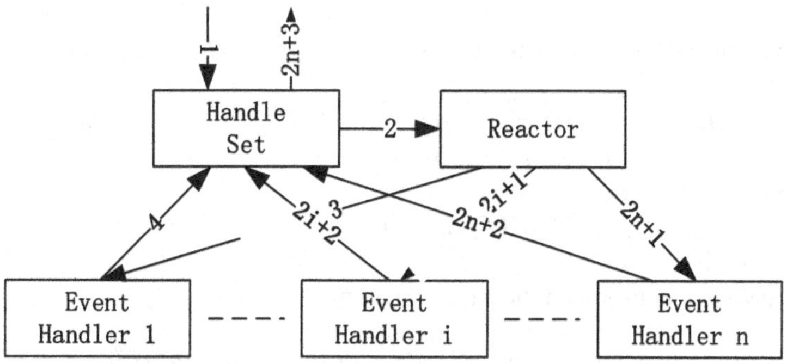


**FIGURE 36.10** Typical process of Reactor pattern.

1. The Handle Set receives the input from the user through the channel $I$ (the corresponding reading action is denoted $r_I(d_I)$), then processes the input through a processing function $HF_1$ and generates the input $d_{I_R}$, and sends $d_{I_R}$ to the Reactor through the channel $HR$ (the corresponding sending action is denoted $s_{HR}(d_{I_R})$);
2. The Reactor receives the input $d_{I_R}$ through the channel $HR$ (the corresponding reading action is denoted $r_{HR}(d_{I_R})$), then processes the information and generates the results $d_{I_{E_i}}$ through a processing function $RF$, and sends the results $d_{I_{E_i}}$ to the Event Handler $i$ through the channels $RE_i$ (the corresponding sending action is denoted $s_{RE_i}(d_{I_{E_i}})$);
3. The Event Handler $i$ receives the input $d_{I_{E_i}}$ from the Reactor through the channel $RE_i$ (the corresponding reading action is denoted $r_{RE_i}(d_{I_{E_i}})$), then processes the input through a processing function $EF_i$, and sends the results $d_{O_{E_i}}$ to the Handle Set through the channel $EH_i$ (the corresponding sending action is denoted $s_{EH_i}(d_{O_{E_i}})$);
4. The Handle Set receives the results from the Event Handler $i$ through the channel $EH_i$ (the corresponding reading action is denoted $r_{EH_i}(d_{O_{E_i}})$), then processes the results and generates the results $d_O$ through a processing function $HF_2$, and sends $d_O$ to the outside through the channel $O$ (the corresponding sending action is denoted $s_O(d_O)$).

In the following, we verify the Reactor pattern. We assume all data elements $d_I, d_{I_R}, d_{I_{E_i}}, d_{O_{E_i}}, d_O$ (for $1 \leq i \leq n$) are from a finite set $\Delta$.

The state transitions of the Handle Set module described by APTC are as follows:

$H = \sum_{d_I\in\Delta}(r_I(d_I)\cdot F_2)$

$H_2 = HF_1 \cdot H_3$

$H_3 = \sum_{d_{I_R}\in\Delta}(s_{HR}(d_{I_R})\cdot H_4)$

$H_4 = \sum_{d_{O_{E_i}}\in\Delta}(r_{EH_i}(d_{O_{E_i}})\cdot H_5)$

$H_5 = HF_2 \cdot H_6$

$H_6 = \sum_{d_O\in\Delta}(s_O(d_O)\cdot H)$

The state transitions of the Event Handler $i$ described by APTC are as follows:

$E_i = \sum_{d_{I_{E_i}}\in\Delta}(r_{RE_i}(d_{I_{E_i}})\cdot E_{i_2})$

$E_{i_2} = EF_i \cdot E_{i_3}$

$E_{i_3} = \sum_{d_{O_{E_i}}\in\Delta}(s_{EH_i}(d_{O_{E_i}})\cdot E_i)$

The state transitions of the Reactor described by APTC are as follows:

$R = \sum_{d_{I_R}\in\Delta}(r_{HR}(d_{I_R}) \cdot R_2)$
$R_2 = RF \cdot R_3$
$R_3 = \sum_{d_{I_{E_1}},\cdots,d_{I_{E_n}}\in\Delta}(s_{RE_1}(d_{I_{E_1}}) \between \cdots \between s_{RE_n}(d_{I_{E_n}}) \cdot R)$

The sending action and the reading action of the same data through the same channel can communicate with each other, otherwise, they will cause a deadlock $\delta$. We define the following communication functions of the Handle Set for $1 \leq i \leq n$:

$$\gamma(r_{HR}(d_{I_R}), s_{HR}(d_{I_R})) \triangleq c_{HR}(d_{I_R})$$

$$\gamma(r_{EH_i}(d_{O_{E_i}}), s_{EH_i}(d_{O_{E_i}})) \triangleq c_{EH_i}(d_{O_{E_i}})$$

There is one communication function between the Reactor and the Event Handler $i$:

$$\gamma(r_{RE_i}(d_{I_{E_i}}), s_{RE_i}(d_{I_{E_i}})) \triangleq c_{RE_i}(d_{I_{E_i}})$$

Let all modules be in parallel, then the Reactor pattern $H \quad R \quad E_1 \cdots E_i \cdots E_n$ can be presented by the following process term:

$\tau_I(\partial_H(\Theta(H \between R \between E_1 \between \cdots \between E_i \between \cdots \between E_n))) = \tau_I(\partial_H(H \between R \between E_1 \between \cdots \between E_i \between \cdots \between E_n))$

where $H = \{r_{HR}(d_{I_R}), s_{HR}(d_{I_R}), r_{EH_i}(d_{O_{E_i}}), s_{EH_i}(d_{O_{E_i}}), r_{RE_i}(d_{I_{E_i}}), s_{RE_i}(d_{I_{E_i}})$
$|d_I, d_{I_R}, d_{I_{E_i}}, d_{O_{E_i}}, d_O \in \Delta\}$ for $1 \leq i \leq n$,

$I = \{c_{HR}(d_{I_R}), c_{EH_i}(d_{O_{E_i}}), c_{RE_i}(d_{I_{E_i}}), HF_1, HF_2, EF_i, RF$
$|d_I, d_{I_R}, d_{I_{E_i}}, d_{O_{E_i}}, d_O \in \Delta\}$ for $1 \leq i \leq n$

Then, we obtain the following conclusion on the Reactor pattern.

**Theorem 36.5** (Correctness of the Reactor pattern). *The Reactor pattern $\tau_I(\partial_H(H \between R \between E_1 \between \cdots \between E_i \between \cdots \between E_n))$ can exhibit the desired external behaviors.*

*Proof.* Based on the above state transitions of the above modules, by use of the algebraic laws of APTC, we can prove that

$\tau_I(\partial_H(H \between R \between E_1 \between \cdots \between E_i \between \cdots \between E_n)) = \sum_{d_I,d_O\in\Delta}(r_I(d_I) \cdot s_O(d_O)) \cdot \tau_I(\partial_H(H \between R \between E_1 \between \cdots \between E_i \between \cdots \between E_n))$

That is, the Reactor pattern $\tau_I(\partial_H(H \between R \between E_1 \between \cdots \between E_i \between \cdots \between E_n))$ can exhibit the desired external behaviors.

For the details of the proof, please refer to Section 17.3, as we omit them here. □

### 36.2.2 Verification of the Proactor pattern

The Proactor pattern also decouples the delivery of the events between the event-driven applications and clients, but the events are triggered by the completion of asynchronous operations, which has four classes of components: the Asynchronous Operation Processor, the Asynchronous Operation, the Proactor, and $n$ Completion Handlers. The Asynchronous Operation Processor interacts with the outside through the channel $I$; with the Asynchronous Operation through the channels $I_{PO}$ and $O_{PO}$, and with the Proactor through the channel $PP$. The Proactor interacts with the Completion Handler $i$ through the channel $PC_i$. The Completion Handler $i$ interacts with the outside through the channel $O_i$. This is illustrated in Fig. 36.11.

The typical process of the Proactor pattern is shown in Fig. 36.12 and is the following:

1. The Asynchronous Operation Processor receives the input $d_I$ from the user through the channel $I$ (the corresponding reading action is denoted $r_I(D_I)$), processes the input through a processing function $AOPF_1$, and generates the input to the Asynchronous Operation $d_{I_{AO}}$ and it sends $d_{I_{AO}}$ to the Asynchronous Operation through the channel $I_{PO}$ (the corresponding sending action is denoted $s_{I_{PO}}(d_{I_{AO}})$);
2. The Asynchronous Operation receives the input from the Asynchronous Operation Processor through the channel $I_{PO}$ (the corresponding reading action is denoted $r_{I_{PO}}(d_{I_{AO}})$), processes the input through a processing function $AOF$, generates the computational results to the Asynchronous Operation Processor that is denoted $d_{O_{AO}}$; then sends the results to the Asynchronous Operation Processor through the channel $O_{PO}$ (the corresponding sending action is denoted $s_{O_{PO}}(d_{O_{AO}})$);
3. The Asynchronous Operation Processor receives the results from the Asynchronous Operation through the channel $O_{PO}$ (the corresponding reading action is denoted $r_{O_{PO}}(d_{O_{AO}})$), then processes the results and generates the events $d_{I_P}$ through a processing function $AOPF_2$, and sends it to the Proactor through the channel $PP$ (the corresponding sending action is denoted $s_{PP}(d_{I_P})$);

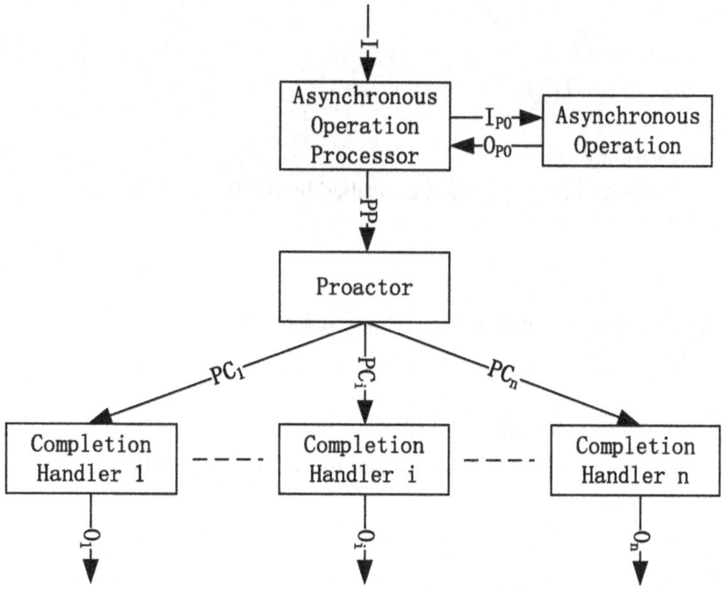


FIGURE 36.11 Proactor pattern.

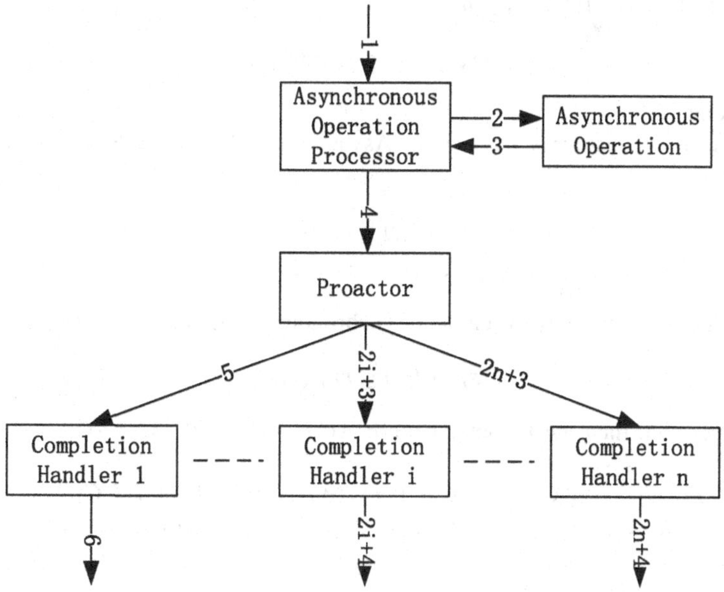


FIGURE 36.12 Typical process of Proactor pattern.

4. The Proactor receives the events $d_{I_P}$ from the Asynchronous Operation Processor through the channel $PP$ (the corresponding reading action is denoted $r_{PP}(d_{I_P})$), then processes the events through a processing function $PF$, and sends the processed events to the Completion Handler $i$ $d_{I_{C_i}}$ for $1 \leq i \leq n$ through the channel $PC_i$ (the corresponding sending action is denoted $s_{PC_i}(d_{I_{C_i}})$);
5. The Completion Handler $i$ (for $1 \leq i \leq n$) receives the events from the Proactor through the channel $PC_i$ (the corresponding reading action is denoted $r_{PC_i}(d_{I_{C_i}})$), processes the events through a processing function $CF_i$, generates the output $d_{O_i}$, then sends the output through the channel $O_i$ (the corresponding sending action is denoted $s_{O_i}(d_{O_i})$).

In the following, we verify the Proactor pattern. We assume all data elements $d_I$, $d_{I_{AO}}$, $d_{I_P}$, $d_{I_{C_i}}$, $d_{O_{AO}}$, $d_{O_i}$ (for $1 \leq i \leq n$) are from a finite set $\Delta$.

The state transitions of the Asynchronous Operation Processor module described by APTC are as follows:

$AOP = \sum_{d_I \in \Delta}(r_I(d_I) \cdot AOP_2)$

$AOP_2 = AOPF_1 \cdot AOP_3$
$AOP_3 = \sum_{d_{I_{AO}} \in \Delta}(s_{I_{PO}}(d_{I_{AO}}) \cdot AOP_4)$
$AOP_4 = \sum_{d_{O_{AO}} \in \Delta}(r_{O_{PO}}(d_{O_{AO}}) \cdot AOP_5)$
$AOP_5 = AOPF_2 \cdot AOP_6$
$AOP_6 = \sum_{d_{I_P} \in \Delta}(s_{PP}(d_{I_P}) \cdot AOP)$

The state transitions of the Asynchronous Operation described by APTC are as follows:

$AO = \sum_{d_{I_{AO}} \in \Delta}(r_{I_{PO}}(d_{I_{AO}}) \cdot AO_2)$
$AO_2 = AOF \cdot AO_3$
$AO_3 = \sum_{d_{O_{AO}} \in \Delta}(s_{O_{PO}}(d_{O_{AO}}) \cdot AO)$

The state transitions of the Proactor described by APTC are as follows:

$P = \sum_{d_{I_P} \in \Delta}(r_{PP}(d_{I_P}) \cdot P_2)$
$P_2 = PF \cdot P_3$
$P_3 = \sum_{d_{I_{C_1}}, \cdots, d_{I_{C_n}} \in \Delta}(s_{PC_1}(d_{I_{C_1}}) \between \cdots \between s_{PC_n}(d_{I_{C_n}}) \cdot P)$

The state transitions of the Completion Handler $i$ described by APTC are as follows:

$C_i = \sum_{d_{I_{C_i}} \in \Delta}(r_{PC_i}(d_{I_{C_i}}) \cdot C_{i_2})$
$C_{i_2} = CF_i \cdot C_{i_3}$
$C_{i_3} = \sum_{d_{O_i} \in \Delta}(s_{O_i}(d_{O_i}) \cdot C_i)$

The sending action must occur before the reading action of the same data through the same channel, then they can asynchronously communicate with each other, otherwise, they will cause a deadlock $\delta$. We define the following communication constraint of the Completion Handler $i$ for $1 \leq i \leq n$:

$$s_{PC_i}(d_{I_{C_i}}) \leq r_{PC_i}(d_{I_{C_i}})$$

Here, $\leq$ is a causality relation.

There are two communication constraints between the Asynchronous Operation Processor and the Asynchronous Operation as follows:

$$s_{I_{PO}}(d_{I_{AO}}) \leq r_{PO}(d_{I_{AO}})$$
$$s_{O_{PO}}(d_{O_{AO}}) \leq r_{O_{PO}}(d_{O_{AO}})$$

There is one communication constraint between the Asynchronous Operation Processor and the Proactor as follows:

$$s_{PP}(d_{I_P}) \leq r_{PP}(d_{I_P})$$

Let all modules be in parallel, then the Proactor pattern $AOP \quad AO \quad P \quad C_1 \cdots C_i \cdots C_n$ can be presented by the following process term:

$\tau_I(\partial_H(\Theta(AOP \between AO \between P \between C_1 \between \cdots \between C_i \between \cdots \between C_n))) = \tau_I(\partial_H(AOP \between AO \between P \between C_1 \between \cdots \between C_i \between \cdots \between C_n))$

where $H = \{s_{PC_i}(d_{I_{C_i}}), r_{PC_i}(d_{I_{C_i}}), s_{I_{PO}}(d_{I_{AO}}), r_{PO}(d_{I_{AO}}), s_{O_{PO}}(d_{O_{AO}}), r_{O_{PO}}(d_{O_{AO}}), s_{PP}(d_{I_P}), r_{PP}(d_{I_P})$
$|s_{PC_i}(d_{I_{C_i}}) \nleq r_{PC_i}(d_{I_{C_i}}), s_{I_{PO}}(d_{I_{AO}}) \nleq r_{PO}(d_{I_{AO}}), s_{O_{PO}}(d_{O_{AO}}) \nleq r_{O_{PO}}(d_{O_{AO}}), s_{PP}(d_{I_P}) \nleq r_{PP}(d_{I_P}),$
$d_I, d_{I_{AO}}, d_{I_P}, d_{I_{C_i}}, d_{O_{AO}}, d_{O_i} \in \Delta\}$ for $1 \leq i \leq n$,

$I = \{s_{PC_i}(d_{I_{C_i}}), r_{PC_i}(d_{I_{C_i}}), s_{I_{PO}}(d_{I_{AO}}), r_{PO}(d_{I_{AO}}), s_{O_{PO}}(d_{O_{AO}}), r_{O_{PO}}(d_{O_{AO}}),$
$s_{PP}(d_{I_P}), r_{PP}(d_{I_P}), AOPF_1, AOPF_2, AOF, CF_i$
$|s_{PC_i}(d_{I_{C_i}}) \leq r_{PC_i}(d_{I_{C_i}}), s_{I_{PO}}(d_{I_{AO}}) \leq r_{PO}(d_{I_{AO}}), s_{O_{PO}}(d_{O_{AO}}) \leq r_{O_{PO}}(d_{O_{AO}}), s_{PP}(d_{I_P}) \leq r_{PP}(d_{I_P}),$
$d_I, d_{I_{AO}}, d_{I_P}, d_{I_{C_i}}, d_{O_{AO}}, d_{O_i} \in \Delta\}$ for $1 \leq i \leq n$

Then, we obtain the following conclusion on the Proactor pattern.

**Theorem 36.6** (Correctness of the Proactor pattern). *The Proactor pattern $\tau_I(\partial_H(AOP \between AO \between P \between C_1 \between \cdots \between C_i \between \cdots \between C_n))$ can exhibit the desired external behaviors.*

*Proof.* Based on the above state transitions of the above modules, by use of the algebraic laws of APTC, we can prove that

$\tau_I(\partial_H(AOP \between AO \between P \between C_1 \between \cdots \between C_i \between \cdots \between C_n)) = \sum_{d_I, d_{O_1}, \cdots, d_{O_n} \in \Delta}(r_I(d_I) \cdot s_{O_1}(d_{O_1}) \parallel \cdots \parallel s_{O_i}(d_{O_i}) \parallel \cdots \parallel s_{O_n}(d_{O_n})) \cdot \tau_I(\partial_H(AOP \between AO \between P \between C_1 \between \cdots \between C_i \between \cdots \between C_n))$

That is, the Proactor pattern $\tau_I(\partial_H(AOP \between AO \between P \between C_1 \between \cdots \between C_i \between \cdots \between C_n))$ can exhibit the desired external behaviors.

For the details of the proof, please refer to Section 17.3, as we omit them here. □

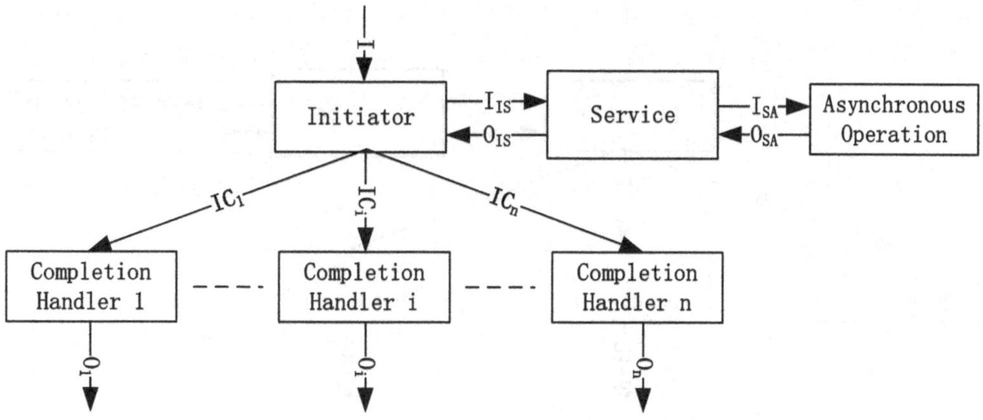


**FIGURE 36.13** Asynchronous Completion Token pattern.

### 36.2.3 Verification of the Asynchronous Completion Token pattern

The Asynchronous Completion Token pattern also decouples the delivery of the events between the event-driven applications and clients, but the events are triggered by the completion of asynchronous operations, which has four classes of components: the Initiator, the Asynchronous Operation, the Service, and $n$ Completion Handlers. The Initiator interacts with the outside through the channel $I$; with the Service through the channels $I_{IS}$ and $O_{IS}$, and with the Completion Handler $i$ with the channel $IC_i$. The Service interacts with the Asynchronous Operation through the channel $I_{SO}$ and $O_{SO}$. The Completion Handler $i$ interacts with the outside through the channel $O_i$. This is illustrated in Fig. 36.13.

The typical process of the Asynchronous Completion Token pattern is shown in Fig. 36.14 and is the following:

1. The Initiator receives the input $d_I$ from the user through the channel $I$ (the corresponding reading action is denoted $r_I(D_I)$), processes the input through a processing function $IF_1$, and generates the input to the Service $d_{I_S}$ and it sends $d_{I_S}$ to the Asynchronous Operation through the channel $I_{IS}$ (the corresponding sending action is denoted $s_{I_{IS}}(d_{I_S})$);
2. The Service receives the input from the Initiator through the channel $I_{IS}$ (the corresponding reading action is denoted $r_{I_{IS}}(d_{I_S})$), processes the input through a processing function $SF_1$, generates the input to the Asynchronous Operation that is denoted $d_{I_A}$; then sends the input to the Asynchronous Operation through the channel $I_{SA}$ (the corresponding sending action is denoted $s_{O_{SA}}(d_{O_A})$);
3. The Asynchronous Operation receives the input from the Service through the channel $I_{SA}$ (the corresponding reading action is denoted $r_{I_{SA}}(d_{I_A})$), then processes the input and generates the results $d_{O_A}$ through a processing function $AF$, and sends the results to the Service through the channel $O_{SA}$ (the corresponding sending action is denoted $s_{O_{SA}}(d_{O_A})$);
4. The Service receives the results $d_{O_A}$ from the Asynchronous Operation through the channel $O_{SA}$ (the corresponding reading action is denoted $r_{O_{SA}}(d_{O_A})$), then processes the results and generates the results $d_{O_S}$ through a processing function $SF_2$, and sends the results to the Initiator through the channel $O_{IS}$ (the corresponding sending action is denoted $s_{O_{IS}}(d_{O_S})$);
5. The Initiator receives the results $d_{O_S}$ from the Service through the channel $O_{IS}$ (the corresponding reading action is denoted $r_{O_{IS}}(d_{O_S})$), then processes the results and generates the events $d_{IC_i}$ through a processing function $IF_2$, and sends the processed events to the Completion Handler $i$ $d_{IC_i}$ for $1 \leq i \leq n$ through the channel $IC_i$ (the corresponding sending action is denoted $s_{IC_i}(d_{IC_i})$);
6. The Completion Handler $i$ (for $1 \leq i \leq n$) receives the events from the Initiator through the channel $IC_i$ (the corresponding reading action is denoted $r_{IC_i}(d_{IC_i})$), processes the events through a processing function $CF_i$, generates the output $d_{O_i}$, then sends the output through the channel $O_i$ (the corresponding sending action is denoted $s_{O_i}(d_{O_i})$).

In the following, we verify the Asynchronous Completion Token pattern. We assume all data elements $d_I, d_{I_S}, d_{I_A}, d_{IC_i}, d_{O_A}, d_{O_S}, d_{O_i}$ (for $1 \leq i \leq n$) are from a finite set $\Delta$.

The state transitions of the Initiator module described by APTC are as follows:

$I = \sum_{d_I \in \Delta}(r_I(d_I) \cdot I_2)$

$I_2 = IF_1 \cdot I_3$

$I_3 = \sum_{d_{I_S} \in \Delta}(s_{I_{IS}}(d_{I_{IS}}) \cdot I_4)$

$I_4 = \sum_{d_{O_S} \in \Delta}(r_{O_{IS}}(d_{O_S}) \cdot I_5)$

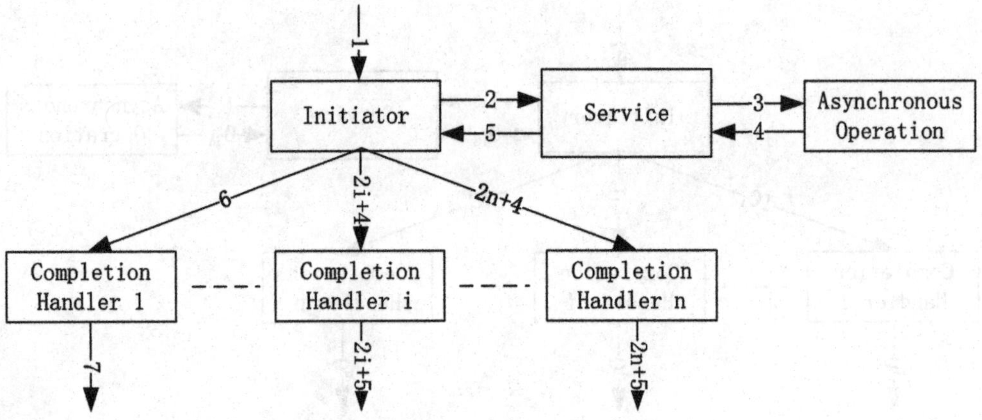

**FIGURE 36.14** Typical process of Asynchronous Completion Token pattern.

$I_5 = IF_2 \cdot I_6$

$I_6 = \sum_{d_{I_{C_1}}, d_{I_{C_n}} \in \Delta}(s_{IC_i}(d_{I_{C_i}}) \cdot I)$

The state transitions of the Service described by APTC are as follows:

$S = \sum_{d_{I_S} \in \Delta}(r_{I_{IS}}(d_{I_S}) \cdot S_2)$

$S_2 = SF_1 \cdot S_3$

$S_3 = \sum_{d_{I_A} \in \Delta}(s_{I_{SA}}(d_{I_A}) \cdot S_4)$

$S_4 = \sum_{d_{O_A} \in \Delta}(r_{O_{SA}}(d_{O_A}) \cdot S_5)$

$S_5 = SF_2 \cdot S_6$

$S_6 = \sum_{d_{O_S} \in \Delta}(s_{O_{IS}}(d_{O_S}) \cdot S)$

The state transitions of the Asynchronous Operation described by APTC are as follows:

$A = \sum_{d_{I_A} \in \Delta}(r_{I_{SA}}(d_{I_A}) \cdot A_2)$

$A_2 = A\bar{F} \cdot A_3$

$A_3 = \sum_{d_{O_A} \in \Delta}(s_{O_{SA}}(d_{O_A}) \cdot A)$

The state transitions of the Completion Handler $i$ described by APTC are as follows:

$C_i = \sum_{d_{I_{C_i}} \in \Delta}(r_{IC_i}(d_{I_{C_i}}) \cdot C_{i_2})$

$C_{i_2} = CF_i \cdot C_{i_3}$

$C_{i_3} = \sum_{d_{O_i} \in \Delta}(s_{O_i}(d_{O_i}) \cdot C_i)$

The sending action must occur before the reading action of the same data through the same channel, then they can asynchronously communicate with each other, otherwise, they will cause a deadlock $\delta$. We define the following communication constraint of the Completion Handler $i$ for $1 \leq i \leq n$:

$$s_{IC_i}(d_{I_{C_i}}) \leq r_{IC_i}(d_{I_{C_i}})$$

Here, $\leq$ is a causality relation.

There are two communication constraints between the Initiator and the Service as follows:

$$s_{I_{IS}}(d_{I_S}) \leq r_{I_{IS}}(d_{I_S})$$
$$s_{O_{IS}}(d_{O_S}) \leq r_{O_{IS}}(d_{O_S})$$

There are two communication constraints between the Service and the Asynchronous Operation as follows:

$$s_{I_{SA}}(d_{I_A}) \leq r_{I_{SA}}(d_{I_A})$$
$$s_{O_{SA}}(d_{O_A}) \leq r_{O_{SA}}(d_{O_A})$$

Let all modules be in parallel, then the Asynchronous Completion Token pattern $I \quad S \quad A \quad C_1 \cdots C_i \cdots C_n$ can be presented by the following process term:

$\tau_I(\partial_H(\Theta(I \between S \between A \between C_1 \between \cdots \between C_i \between \cdots \between C_n))) = \tau_I(\partial_H(I \between S \between A \between C_1 \between \cdots \between C_i \between \cdots \between C_n))$

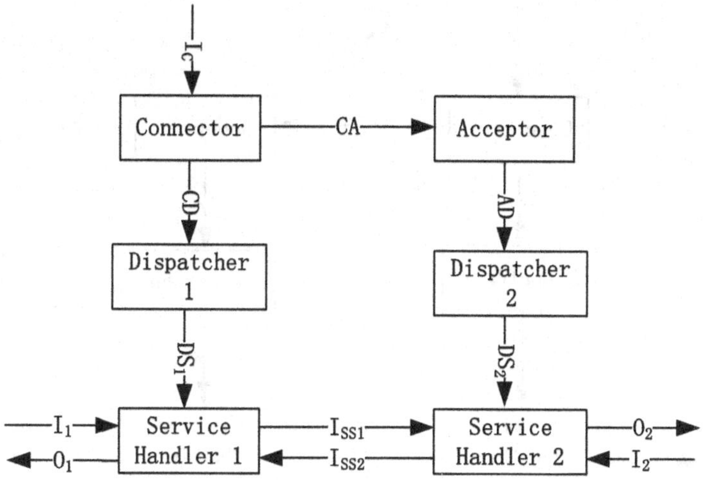


**FIGURE 36.15** Acceptor–Connector pattern.

where $H = \{s_{IC_i}(d_{IC_i}), r_{IC_i}(d_{IC_i}), s_{I_{IS}}(d_{I_S}), r_{I_{IS}}(d_{I_S}), s_{O_{IS}}(d_{O_S}), r_{O_{IS}}(d_{O_S}),$
$s_{I_{SA}}(d_{I_A}), r_{I_{SA}}(d_{I_A}), s_{O_{SA}}(d_{O_A}), r_{O_{SA}}(d_{O_A})$
$|s_{IC_i}(d_{IC_i}) \nleq r_{IC_i}(d_{IC_i}), s_{I_{IS}}(d_{I_S}) \nleq r_{I_{IS}}(d_{I_S}), s_{O_{IS}}(d_{O_S}) \nleq r_{O_{IS}}(d_{O_S}), s_{I_{SA}}(d_{I_A}) \nleq r_{I_{SA}}(d_{I_A}),$
$s_{O_{SA}}(d_{O_A}) \nleq r_{O_{SA}}(d_{O_A}), d_I, d_{I_S}, d_{I_A}, d_{IC_i}, d_{O_A}, d_{O_S}, d_{O_i} \in \Delta\}$ for $1 \leq i \leq n$,
$I = \{s_{IC_i}(d_{IC_i}), r_{IC_i}(d_{IC_i}), s_{I_{IS}}(d_{I_S}), r_{I_{IS}}(d_{I_S}), s_{O_{IS}}(d_{O_S}), r_{O_{IS}}(d_{O_S}),$
$s_{I_{SA}}(d_{I_A}), r_{I_{SA}}(d_{I_A}), s_{O_{SA}}(d_{O_A}), r_{O_{SA}}(d_{O_A}), IF_1, IF_2, SF_1, SF_2, AF, CF_i$
$|s_{IC_i}(d_{IC_i}) \leq r_{IC_i}(d_{IC_i}), s_{I_{IS}}(d_{I_S}) \leq r_{I_{IS}}(d_{I_S}), s_{O_{IS}}(d_{O_S}) \leq r_{O_{IS}}(d_{O_S}), s_{I_{SA}}(d_{I_A}) \leq r_{I_{SA}}(d_{I_A}),$
$s_{O_{SA}}(d_{O_A}) \leq r_{O_{SA}}(d_{O_A}), d_I, d_{I_S}, d_{I_A}, d_{IC_i}, d_{O_A}, d_{O_S}, d_{O_i} \in \Delta\}$ for $1 \leq i \leq n$

Then, we obtain the following conclusion on the Asynchronous Completion Token pattern.

**Theorem 36.7** (Correctness of the Asynchronous Completion Token pattern). *The Asynchronous Completion Token pattern $\tau_I(\partial_H(I \between S \between A \between C_1 \between \cdots \between C_i \between \cdots \between C_n))$ can exhibit the desired external behaviors.*

*Proof.* Based on the above state transitions of the above modules, by use of the algebraic laws of APTC, we can prove that
$\tau_I(\partial_H(I \between S \between A \between C_1 \between \cdots \between C_i \between \cdots \between C_n)) = \sum_{d_I, d_{O_1}, \cdots, d_{O_n} \in \Delta}(r_I(d_I) \cdot s_{O_1}(d_{O_1}) \parallel \cdots \parallel s_{O_i}(d_{O_i}) \parallel \cdots \parallel s_{O_n}(d_{O_n})) \cdot$
$\tau_I(\partial_H(I \between S \between A \between C_1 \between \cdots \between C_i \between \cdots \between C_n))$

That is, the Asynchronous Completion Token pattern $\tau_I(\partial_H(I \between S \between A \between C_1 \between \cdots \between C_i \between \cdots \between C_n))$ can exhibit the desired external behaviors.

For the details of the proof, please refer to Section 17.3, as we omit them here. □

### 36.2.4 Verification of the Acceptor–Connector pattern

The Acceptor–Connector pattern decouples the connection and initialization of two cooperating peers. There are six modules in the Acceptor–Connector pattern: the two Service Handlers, the two Dispatchers, and the two initiators: the Connector and the Acceptor. The Service Handlers interact with the user through the channels $I_1$, $I_2$ and $O_1$, $O_2$; with the Dispatcher through the channels $DS_1$ and $DS_2$, and with each other through the channels $I_{SS_1}$ and $I_{SS_2}$. The Connector interacts with Dispatcher 1 through the channels $CD$, and with the outside through the channels $I_C$. The Acceptor interacts with the Dispatcher 2 through the channel $AD$, and with the outside through the channel $I_A$. The Dispatchers interact with the Service Handlers through the channels $DS_1$ and $DS_2$. This is illustrated in Fig. 36.15.

The typical process of the Acceptor–Connector pattern is shown in Fig. 36.16 and is as follows:

1. The Connector receives the request $d_{I_C}$ from the outside through the channel $I_C$ (the corresponding reading action is denoted $r_{I_C}(d_{I_C})$), then processes the request and generates the request $d_{I_{D_1}}$ and $d_{I_A}$ through a processing function $CF$, and sends the request to the Dispatcher 1 through the channel $CD$ (the corresponding sending action is denoted $s_{CD}(d_{I_{D_1}})$) and sends the request to the Acceptor through the channel $CA$ (the corresponding sending action is denoted $s_{CA}(d_{I_A})$);
2. The Dispatcher 1 receives the request $d_{I_{D_1}}$ from the Connector through the channel $CD$ (the corresponding reading action is denoted $r_{CD}(d_{I_{D_1}})$), then processes the request and generates the request $d_{I_{S_1}}$ through a processing function

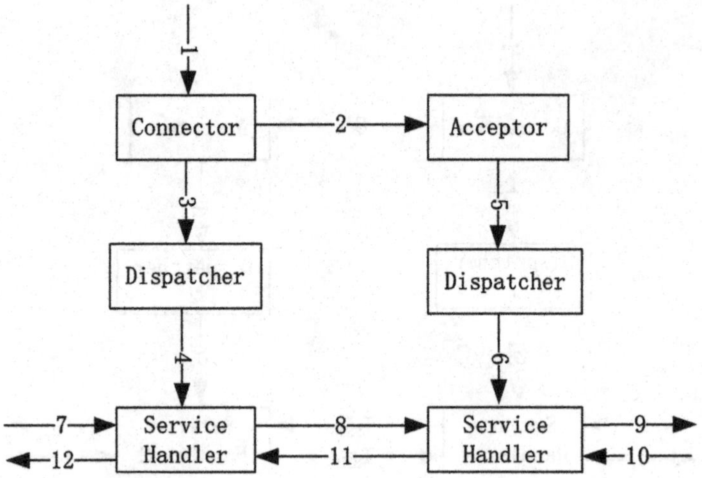


**FIGURE 36.16** Typical process of Acceptor–Connector pattern.

$D1F$, and sends the request to the Service Handler 1 through the channel $DS_1$ (the corresponding sending action is denoted $s_{DS_1}(d_{I_{S_1}})$);

3. The Service Handler 1 receives the request $d_{I_{S_1}}$ from the Dispatcher 1 through the channel $DS_1$ (the corresponding reading action is denoted $r_{DS_1}(d_{I_{S_1}})$), then processes the request through a processing function $S1F_1$ and makes ready to accept the request from the outside;
4. The Acceptor receives the request $d_{I_A}$ from the Connector through the channel $CA$ (the corresponding reading action is denoted $r_{CA}(d_{I_A})$), then processes the request and generates the request $d_{I_{D_2}}$ through a processing function $AF$, and sends the request to the Dispatcher 1 through the channel $AD$ (the corresponding sending action is denoted $s_{AD}(d_{I_{D_2}})$);
5. The Dispatcher 2 receives the request $d_{I_{D_2}}$ from the Acceptor through the channel $AD$ (the corresponding reading action is denoted $r_{AD}(d_{I_{D_2}})$), then processes the request and generates the request $d_{I_{S_2}}$ through a processing function $D2F$, and sends the request to the Service Handler 2 through the channel $DS_2$ (the corresponding sending action is denoted $s_{DS_2}(d_{I_{S_2}})$);
6. The Service Handler 2 receives the request $d_{I_{S_2}}$ from the Dispatcher 2 through the channel $DS_2$ (the corresponding reading action is denoted $r_{DS_2}(d_{I_{S_2}})$), then processes the request through a processing function $S2F_1$ and makes ready to accept the request from the outside;
7. The Service Handler 1 receives the request $d_{I_1}$ from the user through the channel $I_1$ (the corresponding reading action is denoted $r_{I_1}(d_{I_1})$), then processes the request $d_{I_1}$ through a processing function $S1F_2$, and sends the processed request $d_{I_{SS_2}}$ to the Service Handler 2 through the channel $I_{SS_1}$ (the corresponding sending action is denoted $s_{I_{SS_1}}(d_{I_{SS_2}})$);
8. The Service Handler 2 receives the request $d_{I_{SS_2}}$ from the Service Handler 1 through the channel $I_{SS_1}$ (the corresponding reading action is denoted $r_{I_{SS_1}}(d_{I_{SS_2}})$), then processes the request and generates the response $d_{O_2}$ through a processing function $S2F_3$, and sends the response to the outside through the channel $O_2$ (the corresponding sending action is denoted $s_{O_2}(d_{O_2})$);
9. The Service Handler 2 receives the request $d_{I_2}$ from the user through the channel $I_2$ (the corresponding reading action is denoted $r_{I_2}(d_{I_2})$), then processes the request $d_{I_2}$ through a processing function $S2F_2$, and sends the processed request $d_{I_{SS_1}}$ to the Service Handler 1 through the channel $I_{SS_2}$ (the corresponding sending action is denoted $s_{I_{SS_2}}(d_{I_{SS_1}})$);
10. The Service Handler 1 receives the request $d_{I_{SS_1}}$ from the Service Handler 2 through the channel $I_{SS_2}$ (the corresponding reading action is denoted $r_{I_{SS_2}}(d_{I_{SS_1}})$), then processes the request and generates the response $d_{O_1}$ through a processing function $S1F_3$, and sends the response to the outside through the channel $O_1$ (the corresponding sending action is denoted $s_{O_1}(d_{O_1})$).

In the following, we verify the Acceptor–Connector pattern. We assume all data elements $d_{I_1}, d_{I_2}, d_{I_C}, d_{I_A}, d_{I_{D_1}}, d_{I_{D_2}}, d_{I_{S_1}}, d_{I_{S_2}}, d_{I_{SS_1}}, d_{I_{SS_2}}, d_{O_1}, d_{O_2}$ are from a finite set $\Delta$. We only give the transitions of the first process.

The state transitions of the Connector module described by APTC are as follows:

$C = \sum_{d_{I_C}\in\Delta}(r_{I_C}(d_{I_C}) \cdot C_2)$

$C_2 = CF \cdot C_3$

$C_3 = \sum_{d_{I_A}, d_{I_{D_1}}\in\Delta}(s_{CA}(d_{I_A}) \between s_{CD}(d_{I_{D_1}}) \cdot C)$

The state transitions of the Dispatcher 1 module described by APTC are as follows:
$D1 = \sum_{d_{I_{D_1}} \in \Delta}(r_{CD}(d_{I_{D_1}}) \cdot D1_2)$
$D1_2 = D1F \cdot D1_3$
$D1_3 = \sum_{d_{I_{S_1}} \in \Delta}(s_{DS_1}(d_{I_{S_1}}) \cdot D1)$
The state transitions of the Service Handler 1 module described by APTC are as follows:
$S1 = \sum_{d_{I_1}, d_{I_{S_1}}, d_{I_{SS_1}} \in \Delta}(r_{I_1}(d_{I_1}) \between r_{DS_1}(d_{I_{S_1}}) \between r_{I_{SS_2}}(d_{I_{SS_1}}) \cdot S1_2)$
$S1_2 = S1F_1 \between S1F_2 \between S1F_3 \cdot S1_3$
$S1_3 = \sum_{d_{I_{SS_2}}, d_{O_1} \in \Delta}(s_{I_{SS_1}}(d_{I_{SS_2}}) \between s_{O_1}(d_{O_1}) \cdot S1)$
The state transitions of the Acceptor module described by APTC are as follows:
$A = \sum_{d_{I_A} \in \Delta}(r_{CA}(d_{I_A}) \cdot A_2)$
$A_2 = AF \cdot A_3$
$A_3 = \sum_{d_{I_{D_2}} \in \Delta}(s_{AD}(d_{I_{D_2}}) \cdot A)$
The state transitions of the Dispatcher 2 module described by APTC are as follows:
$D2 = \sum_{d_{I_{D_2}} \in \Delta}(r_{AD}(d_{I_{D_2}}) \cdot D2_2)$
$D2_2 = D2F \cdot D2_3$
$D2_3 = \sum_{d_{I_{S_2}} \in \Delta}(s_{DS_2}(d_{I_{S_2}}) \cdot D2)$
The state transitions of the Service Handler 2 module described by APTC are as follows:
$S2 = \sum_{d_{I_2}, d_{I_{S_2}}, d_{I_{SS_2}} \in \Delta}(r_{I_2}(d_{I_2}) \between r_{DS_2}(d_{I_{S_2}}) \between r_{I_{SS_1}}(d_{I_{SS_2}}) \cdot S2_2)$
$S2_2 = S2F_1 \between S2F_2 \between S2F_3 \cdot S2_3$
$S2_3 = \sum_{d_{I_{SS_1}}, d_{O_2} \in \Delta}(s_{I_{SS_2}}(d_{I_{SS_1}}) \between s_{O_2}(d_{O_2}) \cdot S2)$
The sending action and the reading action of the same data through the same channel can communicate with each other, otherwise, they will cause a deadlock $\delta$. We define the following communication function between the Connector and the Acceptor:

$$\gamma(r_{CA}(d_{I_A}), s_{CA}(d_{I_A})) \triangleq c_{CA}(d_{I_A})$$

There is one communication function between the Connector and the Dispatcher 1 as follows:

$$\gamma(r_{CD}(d_{I_{D_1}}), s_{CD}(d_{I_{D_1}})) \triangleq c_{CD}(d_{I_{D_1}})$$

There is one communication function between the Dispatcher 1 and the Service Handler 1 as follows:

$$\gamma(r_{DS_1}(d_{I_{S_1}}), s_{DS_1}(d_{I_{S_1}})) \triangleq c_{DS_1}(d_{I_{S_1}})$$

We define the following communication function between the Acceptor and the Dispatcher 2:

$$\gamma(r_{AD}(d_{I_{D_2}}), s_{AD}(d_{I_{D_2}})) \triangleq c_{AD}(d_{I_{D_2}})$$

There is one communication function between the Dispatcher 2 and the Service Handler 2 as follows:

$$\gamma(r_{DS_2}(d_{I_{S_2}}), s_{DS_2}(d_{I_{S_2}})) \triangleq c_{DS_2}(d_{I_{S_2}})$$

There are two communication functions between the Service Handler 1 and the Service Handler 2 as follows:

$$\gamma(r_{I_{SS_1}}(d_{I_{SS_2}}), s_{I_{SS_1}}(d_{I_{SS_2}})) \triangleq c_{I_{SS_1}}(d_{I_{SS_2}})$$
$$\gamma(r_{I_{SS_2}}(d_{I_{SS_1}}), s_{I_{SS_2}}(d_{I_{SS_1}})) \triangleq c_{I_{SS_2}}(d_{I_{SS_1}})$$

Let all modules be in parallel, then the Acceptor–Connector pattern $C \quad D1 \quad S1 \quad A \quad D2 \quad S2$ can be presented by the following process term:
$\tau_I(\partial_H(\Theta(C \between D1 \between S1 \between A \between D2 \between S2))) = \tau_I(\partial_H(C \between D1 \between S1 \between A \between D2 \between S2))$
where $H = \{r_{CA}(d_{I_A}), s_{CA}(d_{I_A}), r_{CD}(d_{I_{D_1}}), s_{CD}(d_{I_{D_1}}), r_{DS_1}(d_{I_{S_1}}), s_{DS_1}(d_{I_{S_1}}), r_{AD}(d_{I_{D_2}}), s_{AD}(d_{I_{D_2}}),$
$r_{DS_2}(d_{I_{S_2}}), s_{DS_2}(d_{I_{S_2}}), r_{I_{SS_1}}(d_{I_{SS_2}}), s_{I_{SS_1}}(d_{I_{SS_2}}), r_{I_{SS_2}}(d_{I_{SS_1}}), s_{I_{SS_2}}(d_{I_{SS_1}})$
$|d_{I_1}, d_{I_2}, d_{I_C}, d_{I_A}, d_{I_{D_1}}, d_{I_{D_2}}, d_{I_{S_1}}, d_{I_{S_2}}, d_{I_{SS_1}}, d_{I_{SS_2}}, d_{O_1}, d_{O_2} \in \Delta\},$

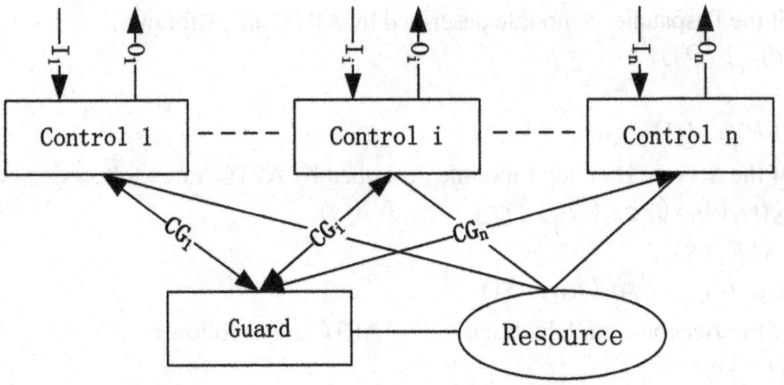


**FIGURE 36.17** Scoped Locking pattern.

$I = \{c_{CA}(d_{I_A}), c_{CD}(d_{I_{D_1}}), c_{DS_1}(d_{I_{S_1}}), c_{AD}(d_{I_{D_2}}), c_{DS_2}(d_{I_{S_2}}), c_{ISS_1}(d_{I_{SS_2}}), c_{ISS_2}(d_{I_{SS_1}}),$
$CF, AF, D1F, D2F, S1F_1, S1F_2, S1F_3, S2F_1, S2F_2, S2F_3$
$|d_{I_1}, d_{I_2}, d_{I_C}, d_{I_A}, d_{I_{D_1}}, d_{I_{D_2}}, d_{I_{S_1}}, d_{I_{S_2}}, d_{I_{SS_1}}, d_{I_{SS_2}}, d_{O_1}, d_{O_2} \in \Delta\}$

Then, we obtain the following conclusion on the Acceptor–Connector pattern.

**Theorem 36.8** (Correctness of the Acceptor–Connector pattern). *The Acceptor–Connector pattern $\tau_I(\partial_H(C \between D1 \between S1 \between A \between D2 \between S2))$ can exhibit the desired external behaviors.*

*Proof.* Based on the above state transitions of the above modules, by use of the algebraic laws of APTC, we can prove that
$\tau_I(\partial_H(C \between D1 \between S1 \between A \between D2 \between S2)) = \sum_{d_{I_C}, d_{I_1}, d_{I_2}, d_{O_1}, d_{O_2} \in \Delta}(r_{I_C}(d_{I_C}) \parallel (r_{I_1}(d_{I_1}) \cdot s_{O_2}(d_{O_2})) \parallel (r_{I_2}(d_{I_2}) \cdot s_{O_1}(d_{O_1}))) \cdot \tau_I(\partial_H(C \between D1 \between S1 \between A \between D2 \between S2))$

That is, the Acceptor–Connector pattern $\tau_I(\partial_H(C \between D1 \between S1 \between A \between D2 \between S2))$ can exhibit the desired external behaviors.

For the details of the proof, please refer to Section 17.3, as we omit them here. □

## 36.3 Synchronization patterns

In this section, we verify the synchronization patterns, including the Scoped Locking pattern, the Strategized Locking pattern, the Thread-Safe Interface pattern, and the Double-Checked Locking Optimization pattern.

### 36.3.1 Verification of the Scoped Locking pattern

The Scoped Locking pattern ensures that a lock is acquired automatically when control enters a scope and released when control leaves the scope. In the Scoped Locking pattern, there are two classes of modules: The $n$ Controls and the Guard. The Control $i$ interacts with the outside through the input channel $I_i$ and the output channel $O_i$, and with the Guard through the channel $CG_i$ for $1 \leq i \leq n$, as illustrated in Fig. 36.17.

The typical process is shown in Fig. 36.18 and is as follows:

1. The Control $i$ receives the input $d_{I_i}$ from the outside through the channel $I_i$ (the corresponding reading action is denoted $r_{I_i}(d_{I_i})$), then it processes the input and generates the input $d_{I_{G_i}}$ through a processing function $CF_{i1}$, and it sends the input to the Guard through the channel $CG_i$ (the corresponding sending action is denoted $s_{CG_i}(d_{I_{G_i}})$);
2. The Guard receives the input $d_{I_{G_i}}$ from the Control $i$ through the channel $CG_i$ (the corresponding reading action is denoted $r_{CG_i}(d_{I_{G_i}})$) for $1 \leq i \leq n$, then processes the request and generates the output $d_{O_{G_i}}$ through a processing function $GF_i$, (note that after the processing, a lock is acquired), and sends the output to the Control $i$ through the channel $CG_i$ (the corresponding sending action is denoted $s_{CG_i}(d_{O_{G_i}})$);
3. The Control $i$ receives the output from the Guard through the channel $CG_i$ (the corresponding reading action is denoted $r_{CG_i}(d_{O_{G_i}})$), then processes the output and generates the output $d_{O_i}$ through a processing function $CF_{i2}$ (accessing the resource), and sends the output to the outside through the channel $O_i$ (the corresponding sending action is denoted $s_{O_i}(d_{O_i})$).

In the following, we verify the Scoped Locking pattern. We assume all data elements $d_{I_i}, d_{O_i}, d_{I_{G_i}}, d_{O_{G_i}}$ for $1 \leq i \leq n$ are from a finite set $\Delta$.

The state transitions of the Control $i$ module described by APTC are as follows:

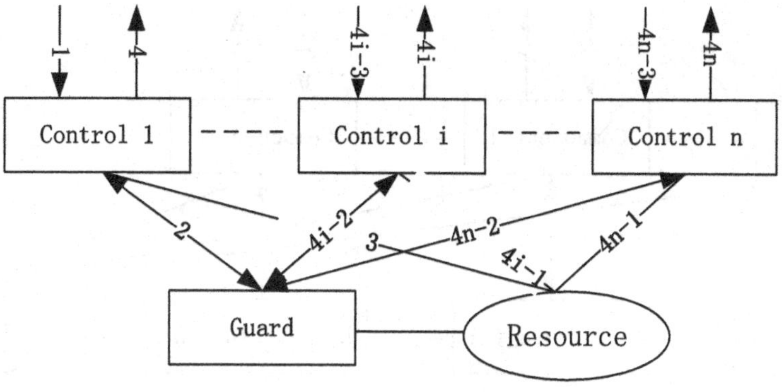


**FIGURE 36.18** Typical process of Scoped Locking pattern.

$C_i = \sum_{d_{I_i} \in \Delta}(r_{I_i}(d_{I_i}) \cdot C_{i_2})$
$C_{i_2} = CF_{i1} \cdot C_{i_3}$
$C_{i_3} = \sum_{d_{IG_i} \in \Delta}(s_{CG_i}(d_{IG_i}) \cdot C_{i_4})$
$C_{i_4} = \sum_{d_{OG_i} \in \Delta}(r_{CG_i}(d_{OG_i}) \cdot C_{i_5})$
$C_{i_5} = CF_{i2} \cdot C_{i_6} \quad CF_{12}\% \cdots \%CF_{n2}$
$C_{i_6} = \sum_{d_{O_i} \in \Delta}(s_{O_i}(d_{O_i}) \cdot C_i)$
The state transitions of the Guard module described by APTC are as follows:
$G = \sum_{d_{IG_1},\cdots,d_{IG_n} \in \Delta}(r_{CG_1}(d_{IG_1}) \between \cdots \between r_{CG_n}(d_{IG_n}) \cdot G_2)$
$G_2 = GF_1 \between \cdots \between GF_n \cdot G_3 \quad (GF_1\% \cdots \%GF_n)$
$G_3 = \sum_{d_{OG_1},\cdots,d_{OG_n} \in \Delta}(s_{CG_1}(d_{OG_1}) \between \cdots \between s_{CG_n}(d_{OG_n}) \cdot G)$
The sending action and the reading action of the same data through the same channel can communicate with each other, otherwise, they will cause a deadlock $\delta$. We define the following communication functions between the Control $i$ and the Guard:

$$\gamma(r_{CG_i}(d_{IG_i}), s_{CG_i}(d_{IG_i})) \triangleq c_{CG_i}(d_{IG_i})$$
$$\gamma(r_{CG_i}(d_{OG_i}), s_{CG_i}(d_{OG_i})) \triangleq c_{CG_i}(d_{OG_i})$$

Let all modules be in parallel, then the Scoped Locking pattern $C_1 \cdots C_n \quad G$ can be presented by the following process term:
$\tau_I(\partial_H(\Theta(C_1 \between \cdots \between C_n \between G))) = \tau_I(\partial_H(C_1 \between \cdots \between C_n \between G))$
where $H = \{r_{CG_i}(d_{IG_i}), s_{CG_i}(d_{IG_i}), r_{CG_i}(d_{OG_i}), s_{CG_i}(d_{OG_i}) | d_{I_i}, d_{O_i}, d_{IG_i}, d_{OG_i} \in \Delta\}$,
$I = \{c_{CG_i}(d_{IG_i}), c_{CG_i}(d_{OG_i}), CF_{i1}, CF_{i2}, GF_i | d_{I_i}, d_{O_i}, d_{IG_i}, d_{OG_i} \in \Delta\}$ for $1 \le i \le n$
Then, we obtain the following conclusion on the Scoped Locking pattern.

**Theorem 36.9** (Correctness of the Scoped Locking pattern). *The Scoped Locking pattern $\tau_I(\partial_H(C_1 \between \cdots \between C_n \between G))$ can exhibit the desired external behaviors.*

*Proof.* Based on the above state transitions of the above modules, by use of the algebraic laws of APTC, we can prove that
$\tau_I(\partial_H(C_1 \between \cdots \between C_n \between G)) = \sum_{d_{I_1},d_{O_1},\cdots,d_{I_n},d_{O_n} \in \Delta}(r_{I_1}(d_{I_1}) \parallel \cdots \parallel r_{I_n}(d_{I_n}) \cdot s_{O_1}(d_{O_1}) \parallel \cdots \parallel s_{O_n}(d_{O_n})) \cdot \tau_I(\partial_H(C_1 \between \cdots \between C_n \between G))$
That is, the Scoped Locking pattern $\tau_I(\partial_H(C_1 \between \cdots \between C_n \between G))$ can exhibit the desired external behaviors.
For the details of the proof, please refer to Section 17.3, as we omit them here. □

### 36.3.2 Verification of the Strategized Locking pattern

The Strategized Locking pattern uses a component (the LockStrategy) to parameterize the synchronization for protecting the concurrent access to the critical section. In the Strategized Locking pattern, there are two classes of modules: The $n$ Components and the $n$ LockStrategies. The Component $i$ interacts with the outside through the input channel $I_i$ and the output channel $O_i$, and with the LockStrategy $i$ through the channel $CL_i$ for $1 \le i \le n$, as illustrated in Fig. 36.19.

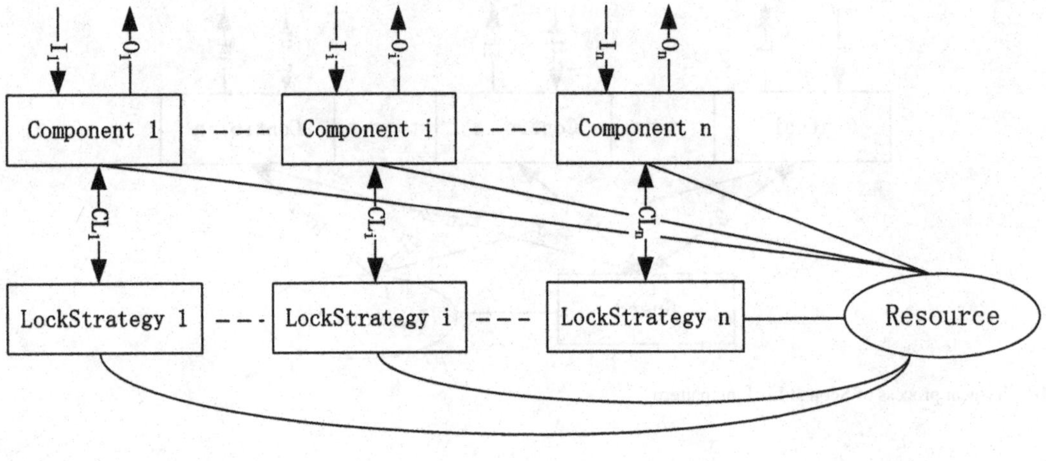


**FIGURE 36.19** Strategized Locking pattern.

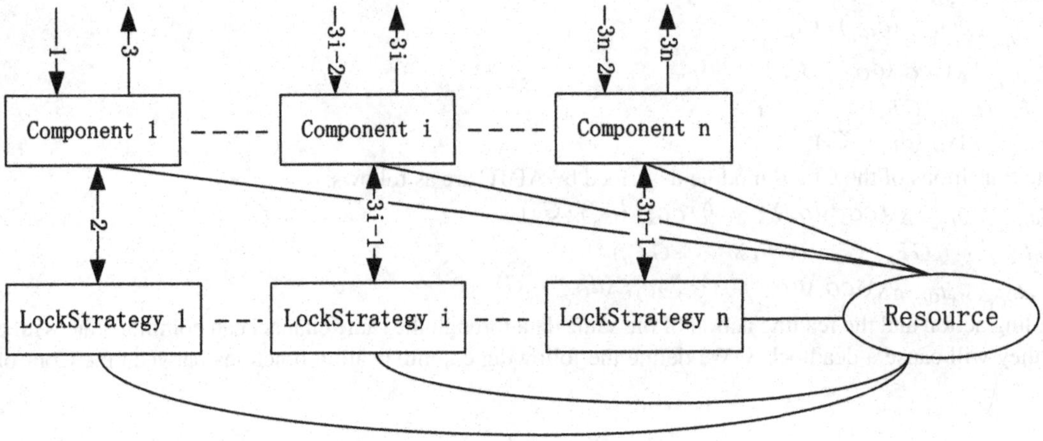


**FIGURE 36.20** Typical process of Strategized Locking pattern.

The typical process is shown in Fig. 36.20 and is as follows:

1. The Component $i$ receives the input $d_{I_i}$ from the outside through the channel $I_i$ (the corresponding reading action is denoted $r_{I_i}(d_{I_i})$), then it processes the input and generates the input $d_{I_{L_i}}$ through a processing function $CF_{i1}$, and it sends the input to the LockStrategy through the channel $CL_i$ (the corresponding sending action is denoted $s_{CL_i}(d_{I_{L_i}})$);
2. The LockStrategy receives the input $d_{I_{L_i}}$ from the Component $i$ through the channel $CL_i$ (the corresponding reading action is denoted $r_{CL_i}(d_{I_{L_i}})$) for $1 \leq i \leq n$, then processes the request and generates the output $d_{O_{L_i}}$ through a processing function $LF_i$, (note that after the processing, a lock is acquired), and sends the output to the Component $i$ through the channel $CL_i$ (the corresponding sending action is denoted $s_{CL_i}(d_{O_{L_i}})$);
3. The Component $i$ receives the output from the LockStrategy through the channel $CL_i$ (the corresponding reading action is denoted $r_{CL_i}(d_{O_{L_i}})$), then processes the output and generates the output $d_{O_i}$ through a processing function $CF_{i2}$, and sends the output to the outside through the channel $O_i$ (the corresponding sending action is denoted $s_{O_i}(d_{O_i})$).

In the following, we verify the Strategized Locking pattern. We assume all data elements $d_{I_i}, d_{O_i}, d_{I_{L_i}}, d_{O_{L_i}}$ for $1 \leq i \leq n$ are from a finite set $\Delta$.

The state transitions of the Component $i$ module described by APTC are as follows:

$C_i = \sum_{d_{I_i} \in \Delta}(r_{I_i}(d_{I_i}) \cdot C_{i_2})$

$C_{i_2} = CF_{i1} \cdot C_{i_3}$

$C_{i_3} = \sum_{d_{I_{L_i}} \in \Delta}(s_{CL_i}(d_{I_{L_i}}) \cdot C_{i_4})$

$C_{i_4} = \sum_{d_{O_{L_i}} \in \Delta}(r_{CL_i}(d_{O_{L_i}}) \cdot C_{i_5})$

$C_{i_5} = CF_{i2} \cdot C_{i_6} \quad (CF_{12}\% \cdots \% CF_{n2})$

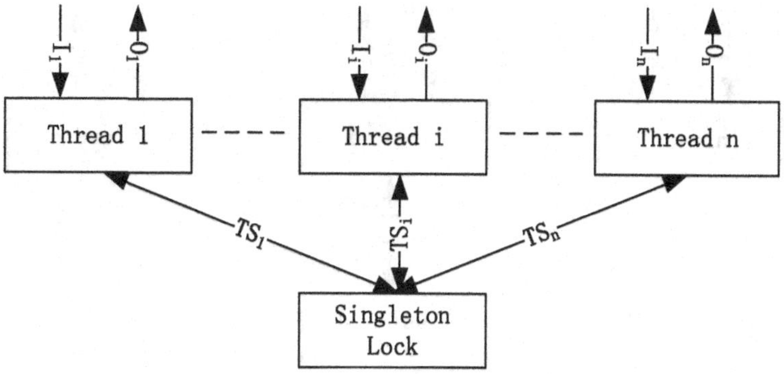


**FIGURE 36.21** Double-Checked Locking Optimization pattern.

$C_{i_6} = \sum_{d_{O_i} \in \Delta} (s_{O_i}(d_{O_i}) \cdot C_i)$

The state transitions of the LockStrategy $i$ module described by APTC are as follows:

$L_i = \sum_{d_{I_{L_i}} \in \Delta} (r_{CL_i}(d_{I_{L_i}}) \cdot L_{i_2})$

$L_{i_2} = LF_i \cdot L_{i_3} \quad (LF_1 \% \cdots \% LF_n)$

$L_{i_3} = \sum_{d_{O_{L_i}} \in \Delta} (s_{CL_i}(d_{O_{L_i}}) \cdot L_i)$

The sending action and the reading action of the same data through the same channel can communicate with each other, otherwise, they will cause a deadlock $\delta$. We define the following communication functions between the Component $i$ and the LockStrategy $i$:

$$\gamma(r_{CL_i}(d_{I_{L_i}}), s_{CL_i}(d_{I_{L_i}})) \triangleq c_{CL_i}(d_{I_{L_i}})$$

$$\gamma(r_{CL_i}(d_{O_{L_i}}), s_{CL_i}(d_{O_{L_i}})) \triangleq c_{CL_i}(d_{O_{L_i}})$$

Let all modules be in parallel, then the Strategized Locking pattern $C_1 \cdots C_n \quad L_1 \cdots L_n$ can be presented by the following process term:

$\tau_I(\partial_H(\Theta(C_1 \between \cdots \between C_n \between L_1 \between \cdots \between L_n))) = \tau_I(\partial_H(C_1 \between \cdots \between C_n \between L_1 \between \cdots \between L_n))$

where $H = \{r_{CL_i}(d_{I_{L_i}}), s_{CL_i}(d_{I_{L_i}}), r_{CL_i}(d_{O_{L_i}}), s_{CL_i}(d_{O_{L_i}}) | d_{I_i}, d_{O_i}, d_{I_{L_i}}, d_{O_{L_i}} \in \Delta\}$,

$I = \{c_{CL_i}(d_{I_{L_i}}), c_{CL_i}(d_{O_{L_i}}), CF_{i1}, CF_{i2}, LF_i | d_{I_i}, d_{O_i}, d_{I_{L_i}}, d_{O_{L_i}} \in \Delta\}$ for $1 \leq i \leq n$

Then, we obtain the following conclusion on the Strategized Locking pattern.

**Theorem 36.10** (Correctness of the Strategized Locking pattern). *The Strategized Locking pattern $\tau_I(\partial_H(C_1 \between \cdots \between C_n \between L_1 \between \cdots \between L_n))$ can exhibit the desired external behaviors.*

*Proof.* Based on the above state transitions of the above modules, by use of the algebraic laws of APTC, we can prove that

$\tau_I(\partial_H(C_1 \between \cdots \between C_n \between L_1 \between \cdots \between L_n)) = \sum_{d_{I_1}, d_{O_1}, \cdots, d_{I_n}, d_{O_n} \in \Delta} (r_{I_1}(d_{I_1}) \parallel \cdots \parallel r_{I_n}(d_{I_n}) \cdot s_{O_1}(d_{O_1}) \parallel \cdots \parallel s_{O_n}(d_{O_n})) \cdot \tau_I(\partial_H(C_1 \between \cdots \between C_n \between L_1 \between \cdots \between L_n))$

That is, the Strategized Locking pattern $\tau_I(\partial_H(C_1 \between \cdots \between C_n \between L_1 \between \cdots \between L_n))$ can exhibit the desired external behaviors.

For the details of the proof, please refer to Section 17.3, as we omit them here. □

### 36.3.3 Verification of the Double-Checked Locking Optimization pattern

The Double-Checked Locking Optimization pattern ensures that a lock is acquired in a thread-safe manner. In the Double-Checked Locking Optimization pattern, there are two classes of modules: The $n$ Threads and the Singleton Lock. The Thread $i$ interacts with the outside through the input channel $I_i$ and the output channel $O_i$, and with the Singleton Lock through the channel $TS_i$ for $1 \leq i \leq n$, as illustrated in Fig. 36.21.

The typical process is shown in Fig. 36.22 and is as follows:

1. The Thread $i$ receives the input $d_{I_i}$ from the outside through the channel $I_i$ (the corresponding reading action is denoted $r_{I_i}(d_{I_i})$), then it processes the input and generates the input $d_{IS_i}$ through a processing function $TF_{i1}$, and it sends the input to the Singleton Lock through the channel $TS_i$ (the corresponding sending action is denoted $s_{TS_i}(d_{IS_i})$);
2. The Singleton Lock receives the input $d_{IS_i}$ from the Thread $i$ through the channel $TS_i$ (the corresponding reading action is denoted $r_{TS_i}(d_{IS_i})$) for $1 \leq i \leq n$, then processes the request and generates the output $d_{OS_i}$ through a processing

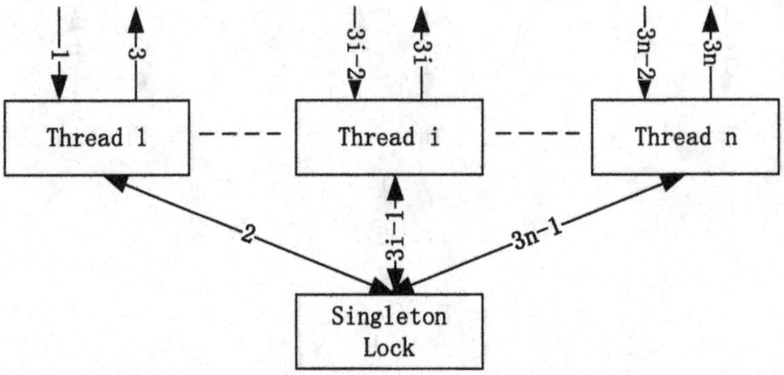


**FIGURE 36.22** Typical process of Double-Checked Locking Optimization pattern.

function $SF_i$, (note that after the processing, a lock is acquired), and sends the output to the Thread $i$ through the channel $TS_i$ (the corresponding sending action is denoted $s_{TS_i}(d_{O_{S_i}})$);

3. The Thread $i$ receives the output from the Singleton Lock through the channel $TS_i$ (the corresponding reading action is denoted $r_{TS_i}(d_{O_{S_i}})$), then processes the output and generates the output $d_{O_i}$ through a processing function $TF_{i2}$ (accessing the resource), and sends the output to the outside through the channel $O_i$ (the corresponding sending action is denoted $s_{O_i}(d_{O_i})$).

In the following, we verify the Double-Checked Locking Optimization pattern. We assume all data elements $d_{I_i}$, $d_{O_i}$, $d_{I_{S_i}}$, $d_{O_{S_i}}$ for $1 \leq i \leq n$ are from a finite set $\Delta$.

The state transitions of the Thread $i$ module described by APTC are as follows:

$T_i = \sum_{d_{I_i} \in \Delta}(r_{I_i}(d_{I_i}) \cdot T_{i_2})$

$T_{i_2} = TF_{i1} \cdot T_{i_3}$

$T_{i_3} = \sum_{d_{I_{S_i}} \in \Delta}(s_{TS_i}(d_{I_{S_i}}) \cdot T_{i_4})$

$T_{i_4} = \sum_{d_{O_{S_i}} \in \Delta}(r_{TS_i}(d_{O_{S_i}}) \cdot T_{i_5})$

$T_{i_5} = TF_{i2} \cdot T_{i_6} \quad (TF_{12}\% \cdots \% TF_{n2})$

$T_{i_6} = \sum_{d_{O_i} \in \Delta}(s_{O_i}(d_{O_i}) \cdot T_i)$

The state transitions of the Singleton Lock module described by APTC are as follows:

$S = \sum_{d_{I_{S_1}}, \cdots, d_{I_{S_n}} \in \Delta}(r_{TS_1}(d_{I_{S_1}}) \between \cdots \between r_{TS_n}(d_{I_{S_n}}) \cdot S_2)$

$S_2 = SF_1 \between \cdots \between SF_n \cdot S_3 \quad (SF_1\% \cdots \% SF_n)$

$S_3 = \sum_{d_{O_{S_1}}, \cdots, d_{O_{S_n}} \in \Delta}(s_{TS_1}(d_{O_{S_1}}) \between \cdots \between s_{TS_n}(d_{O_{S_n}}) \cdot S)$

The sending action and the reading action of the same data through the same channel can communicate with each other, otherwise, they will cause a deadlock $\delta$. We define the following communication functions between the Thread $i$ and the Singleton Lock:

$$\gamma(r_{TS_i}(d_{I_{S_i}}), s_{TS_i}(d_{I_{S_i}})) \triangleq c_{TS_i}(d_{I_{S_i}})$$

$$\gamma(r_{TS_i}(d_{O_{S_i}}), s_{TS_i}(d_{O_{S_i}})) \triangleq c_{TS_i}(d_{O_{S_i}})$$

Let all modules be in parallel, then the Double-Checked Locking Optimization pattern $T_1 \cdots T_n \quad S$ can be presented by the following process term:

$\tau_I(\partial_H(\Theta(T_1 \between \cdots \between T_n \between S))) = \tau_I(\partial_H(T_1 \between \cdots \between T_n \between S))$

where $H = \{r_{TS_i}(d_{I_{S_i}}), s_{TS_i}(d_{I_{S_i}}), r_{TS_i}(d_{O_{S_i}}), s_{TS_i}(d_{O_{S_i}}) | d_{I_i}, d_{O_i}, d_{I_{S_i}}, d_{O_{S_i}} \in \Delta\}$,

$I = \{c_{TS_i}(d_{I_{S_i}}), c_{TS_i}(d_{O_{S_i}}), TF_{i1}, TF_{i2}, SF_i | d_{I_i}, d_{O_i}, d_{I_{S_i}}, d_{O_{S_i}} \in \Delta\}$ for $1 \leq i \leq n$

Then, we obtain the following conclusion on the Double-Checked Locking Optimization pattern.

**Theorem 36.11** (Correctness of the Double-Checked Locking Optimization pattern). *The Double-Checked Locking Optimization pattern $\tau_I(\partial_H(T_1 \between \cdots \between T_n \between S))$ can exhibit the desired external behaviors.*

*Proof.* Based on the above state transitions of the above modules, by use of the algebraic laws of APTC, we can prove that $\tau_I(\partial_H(T_1 \between \cdots \between T_n \between S)) = \sum_{d_{I_1}, d_{O_1}, \cdots, d_{I_n}, d_{O_n} \in \Delta}(r_{I_1}(d_{I_1}) \parallel \cdots \parallel r_{I_n}(d_{I_n}) \cdot s_{O_1}(d_{O_1}) \parallel \cdots \parallel s_{O_n}(d_{O_n})) \cdot \tau_I(\partial_H(T_1 \between \cdots \between T_n \between S))$

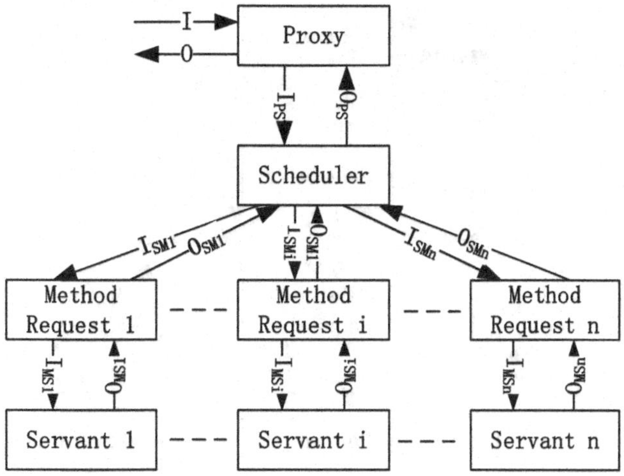


**FIGURE 36.23** Active Object pattern.

That is, the Double-Checked Locking Optimization pattern $\tau_I(\partial_H(T_1 \between \cdots \between T_n \between S))$ can exhibit the desired external behaviors.

For the details of the proof, please refer to Section 17.3, as we omit them here. □

## 36.4 Concurrency patterns

In this section, we verify concurrency-related patterns, including the Active Object pattern, the Monitor Object pattern, the Half-Sync/Half-Async pattern, the Leader/Followers pattern, and the Thread-Specific Storage pattern.

### 36.4.1 Verification of the Active Object pattern

The Active Object pattern is used to decouple the method request and method execution of an object. In this pattern, there are a Proxy module, a Scheduler module, $n$ Method Request modules, and $n$ Servant modules. The Servant is used to implement concrete computation, the Method Request is used to encapsulate a Servant, and the Scheduler is used to manage Method Requests. The Proxy module interacts with the outside through the channels $I$ and $O$, and with the Scheduler through the channels $I_{PS}$ and $O_{PS}$. The Scheduler interacts with Method Request $i$ (for $1 \leq i \leq n$) through the channels $I_{SM_i}$ and $O_{SM_i}$, and the Method Request $i$ interacts with the Servant $i$ through the channels $I_{MS_i}$ and $O_{CS_i}$, as illustrated in Fig. 36.23.

The typical process of the Active Object pattern is shown in Fig. 36.24 and is as follows:

1. The Proxy receives the request $d_I$ from the outside through the channel $I$ (the corresponding reading action is denoted $r_I(d_I)$), then processes the request through a processing function $PF_1$ and generates the request $d_{I_{Sh}}$, and sends the $d_{I_{Sh}}$ to the Scheduler through the channel $I_{PS}$ (the corresponding sending action is denoted $s_{I_{PS}}(d_{I_{Sh}})$);
2. The Scheduler receives the request $d_{I_{Sh}}$ from the Proxy through the channel $I_{PS}$ (the corresponding reading action is denoted $r_{I_{PS}}(d_{I_{Sh}})$), then processes the request through a processing function $ShF_1$ and generates the request $d_{I_{M_i}}$, and sends the $d_{I_{M_i}}$ to the Method Request $i$ through the channel $I_{SM_i}$ (the corresponding sending action is denoted $s_{I_{SM_i}}(d_{I_{M_i}})$);
3. The Method Request $i$ receives the request $d_{I_{M_i}}$ from the Scheduler through the channel $I_{SM_i}$ (the corresponding reading action is denoted $r_{I_{SM_i}}(d_{I_{M_i}})$), then processes the request through a processing function $MF_{i1}$ and generates the request $d_{I_{S_i}}$, and sends the request to the Servant $i$ through the channel $I_{MS_i}$ (the corresponding sending action is denoted $s_{I_{MS_i}}(d_{I_{S_i}})$);
4. The Servant $i$ receives the request $d_{I_{S_i}}$ from the Method Request $i$ through the channel $I_{MS_i}$ (the corresponding reading action is denoted $r_{I_{MS_i}}(d_{I_{S_i}})$), then processes the request through a processing function $SF_i$ and generates the response $d_{O_{S_i}}$, and sends the response to the Method Request through the channel $O_{MS_i}$ (the corresponding sending action is denoted $s_{O_{MS_i}}(d_{O_{S_i}})$);
5. The Method Request $i$ receives the response $d_{O_{S_i}}$ from the Servant $i$ through the channel $O_{MS_i}$ (the corresponding reading action is denoted $r_{O_{MS_i}}(d_{O_{S_i}})$), then processes the request through a processing function $MF_{i2}$ and generates

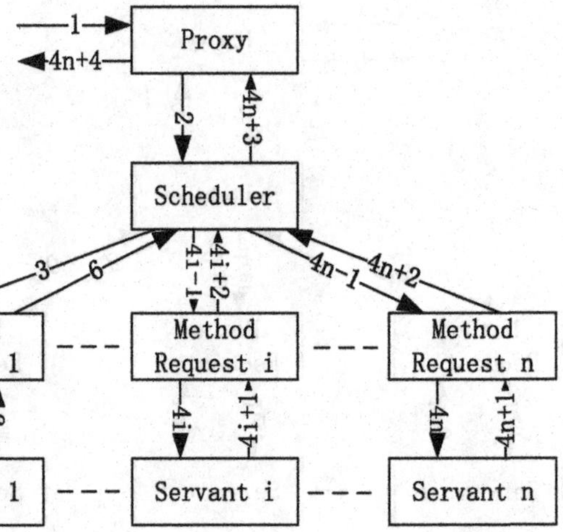


**FIGURE 36.24** Typical process of Active Object pattern.

the response $d_{O_{M_i}}$, and sends the response to the Scheduler through the channel $O_{SM_i}$ (the corresponding sending action is denoted $s_{O_{SM_i}}(d_{O_{M_i}})$);

6. The Scheduler receives the response $d_{O_{M_i}}$ from the Method Request $i$ through the channel $O_{SM_i}$ (the corresponding reading action is denoted $r_{O_{SM_i}}(d_{O_{M_i}})$), then processes the response and generates the response $d_{O_{Sh}}$ through a processing function $ShF_2$, and sends $d_{O_{Sh}}$ to the Proxy through the channel $O_{PS}$ (the corresponding sending action is denoted $s_{O_{PS}}(d_{O_{Sh}})$);
7. The Proxy receives the response $d_{O_{Sh}}$ from the Scheduler through the channel $O_{PS}$ (the corresponding reading action is denoted $r_{O_{PS}}(d_{O_{Sh}})$), then processes the request through a processing function $PF_2$ and generates the request $d_O$, and sends the $d_O$ to the outside through the channel $O$ (the corresponding sending action is denoted $s_O(d_O)$).

In the following, we verify the Active Object pattern. We assume all data elements $d_I, d_{I_{Sh}}, d_{I_{M_i}}, d_{I_{S_i}}, d_{O_{S_i}}, d_{O_{M_i}}, d_{O_{Sh}}, d_O$ (for $1 \leq i \leq n$) are from a finite set $\Delta$.

The state transitions of the Proxy module described by APTC are as follows:

$P = \sum_{d_I\in\Delta}(r_I(d_I) \cdot P_2)$

$P_2 = PF_1 \cdot P_3$

$P_3 = \sum_{d_{I_{Sh}}\in\Delta}(s_{I_{PS}}(d_{I_{Sh}}) \cdot P_4)$

$P_4 = \sum_{d_{O_{Sh}}\in\Delta}(r_{O_{PS}}(d_{O_{Sh}}) \cdot P_5)$

$P_5 = PF_2 \cdot P_6$

$P_6 = \sum_{d_O\in\Delta}(s_O(d_O) \cdot P)$

The state transitions of the Scheduler module described by APTC are as follows:

$Sh = \sum_{d_{I_{Sh}}\in\Delta}(r_{I_{PS}}(d_{I_{Sh}}) \cdot Sh_2)$

$Sh_2 = ShF_1 \cdot Sh_3$

$Sh_3 = \sum_{d_{I_{M_1}},\cdots,d_{I_{M_n}}\in\Delta}(s_{I_{SM_1}}(d_{I_{M_1}}) \between \cdots \between s_{I_{SM_n}}(d_{I_{M_n}}) \cdot Sh_4)$

$Sh_4 = \sum_{d_{O_{M_1}},\cdots,d_{O_{M_n}}\in\Delta}(r_{O_{SM_1}}(d_{O_{M_1}}) \between \cdots \between r_{O_{SM_n}}(d_{O_{M_n}}) \cdot Sh_5)$

$Sh_5 = ShF_2 \cdot Sh_6$

$Sh_6 = \sum_{d_{O_{Sh}}\in\Delta}(s_{O_{PS}}(d_{O_{Sh}}) \cdot Sh)$

The state transitions of the Method Request $i$ described by APTC are as follows:

$M_i = \sum_{d_{I_{M_i}}\in\Delta}(r_{I_{SM_i}}(d_{I_{M_i}}) \cdot M_{i_2})$

$M_{i_2} = MF_{i1} \cdot M_{i_3}$

$M_{i_3} = \sum_{d_{I_{S_i}}\in\Delta}(s_{I_{MS_i}}(d_{I_{S_i}}) \cdot M_{i_4})$

$M_{i_4} = \sum_{d_{O_{S_i}}\in\Delta}(r_{O_{MS_i}}(d_{O_{S_i}}) \cdot M_{i_5})$

$M_{i_5} = MF_{i2} \cdot M_{i_6}$

$M_{i_6} = \sum_{d_{O_{M_i}}\in\Delta}(s_{O_{SM_i}}(d_{O_{M_i}}) \cdot M_i)$

The state transitions of the Servant $i$ described by APTC are as follows:

$S_i = \sum_{d_{IS_i} \in \Delta}(r_{ICS_i}(d_{IS_i}) \cdot S_{i2})$
$S_{i2} = SF_i \cdot S_{i3}$
$S_{i3} = \sum_{d_{OS_i} \in \Delta}(s_{OCS_i}(d_{OS_i}) \cdot S_i)$

The sending action and the reading action of the same data through the same channel can communicate with each other, otherwise, they will cause a deadlock $\delta$. We define the following communication functions between the Proxy the Scheduler:

$$\gamma(r_{IPS}(d_{ISh}), s_{IPS}(d_{ISh})) \triangleq c_{IPS}(d_{ISh})$$
$$\gamma(r_{OPS}(d_{OSh}), s_{OPS}(d_{OSh})) \triangleq c_{OPS}(d_{OSh})$$

There are two communication functions between the Scheduler and the Method Request $i$ for $1 \le i \le n$:

$$\gamma(r_{ISM_i}(d_{IM_i}), s_{ISM_i}(d_{IM_i})) \triangleq c_{ISM_i}(d_{IM_i})$$
$$\gamma(r_{OSM_i}(d_{OM_i}), s_{OSM_i}(d_{OM_i})) \triangleq c_{OSM_i}(d_{OM_i})$$

There are two communication functions between the Servant $i$ and the Method Request $i$ for $1 \le i \le n$:

$$\gamma(r_{IMS_i}(d_{IS_i}), s_{IMS_i}(d_{IS_i})) \triangleq c_{IMS_i}(d_{IS_i})$$
$$\gamma(r_{OMS_i}(d_{OS_i}), s_{OMS_i}(d_{OS_i})) \triangleq c_{OMS_i}(d_{OS_i})$$

Let all modules be in parallel, then the Active Object pattern
$P \quad Sh \quad M_1 \cdots \quad M_i \quad \cdots M_n \quad S_1 \cdots S_i \cdots S_n$
can be presented by the following process term:
$\tau_I(\partial_H(\Theta(P \between Sh \between M_1 \between \cdots \between M_i \between \cdots \between M_n \between S_1 \between \cdots \between S_i \between \cdots \between S_n))) = \tau_I(\partial_H(P \between Sh \between M_1 \between \cdots \between M_i \between \cdots \between M_n \between S_1 \between \cdots \between S_i \between \cdots \between S_n))$
where $H = \{r_{IPS}(d_{ISh}), s_{IPS}(d_{ISh}), r_{OPS}(d_{OSh}), s_{OPS}(d_{OSh}), r_{ISM_i}(d_{IM_i}), s_{ISM_i}(d_{IM_i}),$
$r_{OSM_i}(d_{OM_i}), s_{OSM_i}(d_{OM_i}), r_{IMS_i}(d_{IS_i}), s_{IMS_i}(d_{IS_i}), r_{OMS_i}(d_{OS_i}), s_{OMS_i}(d_{OS_i})$
$|d_I, d_{ISh}, d_{IM_i}, d_{IS_i}, d_{OS_i}, d_{OM_i}, d_{OSh}, d_O \in \Delta\}$ for $1 \le i \le n$,
$I = \{c_{IPS}(d_{ISh}), c_{OPS}(d_{OSh}), c_{ISM_i}(d_{IM_i}), c_{OSM_i}(d_{OM_i}), c_{IMS_i}(d_{IS_i}),$
$c_{OMS_i}(d_{OS_i}), PF_1, PF_2, ShF_1, ShF_2, MF_{i1}, MF_{i2}, SF_i$
$|d_I, d_{ISh}, d_{IM_i}, d_{IS_i}, d_{OS_i}, d_{OM_i}, d_{OSh}, d_O \in \Delta\}$ for $1 \le i \le n$
Then, we obtain the following conclusion on the Active Object pattern.

**Theorem 36.12** (Correctness of the Active Object pattern). *The Active Object pattern $\tau_I(\partial_H(P \between Sh \between M_1 \between \cdots \between M_i \between \cdots \between M_n \between S_1 \between \cdots \between S_i \between \cdots \between S_n))$ can exhibit the desired external behaviors.*

*Proof.* Based on the above state transitions of the above modules, by use of the algebraic laws of APTC, we can prove that
$\tau_I(\partial_H(P \between Sh \between M_1 \between \cdots \between M_i \between \cdots \between M_n \between S_1 \between \cdots \between S_i \between \cdots \between S_n)) = \sum_{d_I, d_O \in \Delta}(r_I(d_I) \cdot s_O(d_O)) \cdot \tau_I(\partial_H(P \between Sh \between M_1 \between \cdots \between M_i \between \cdots \between M_n \between S_1 \between \cdots \between S_i \between \cdots \between S_n))$

That is, the Active Object pattern $\tau_I(\partial_H(P \between Sh \between M_1 \between \cdots \between M_i \between \cdots \between M_n \between S_1 \between \cdots \between S_i \between \cdots \between S_n))$ can exhibit the desired external behaviors.

For the details of the proof, please refer to Section 17.3, as we omit them here. □

### 36.4.2 Verification of the Monitor Object pattern

The Monitor Object pattern synchronizes concurrent method execution to ensure that only one method runs at a time. In the Monitor Object pattern, there are two classes of modules: The $n$ Client Threads and the Monitor Object. The Client Thread $i$ interacts with the outside through the input channel $I_i$ and the output channel $O_i$, and with the Monitor Object through the channel $CM_i$ for $1 \le i \le n$, as illustrated in Fig. 36.25.

The typical process is shown in Fig. 36.26 and is as follows:

1. The Client Thread $i$ receives the input $d_{I_i}$ from the outside through the channel $I_i$ (the corresponding reading action is denoted $r_{I_i}(d_{I_i})$), then it processes the input and generates the input $d_{IM_i}$ through a processing function $CF_{i1}$, and it sends the input to the Monitor Object through the channel $CM_i$ (the corresponding sending action is denoted $s_{CM_i}(d_{IM_i})$);

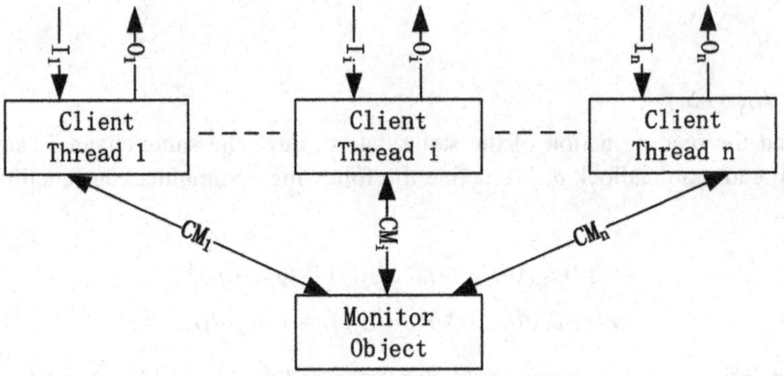


**FIGURE 36.25** Monitor Object pattern.

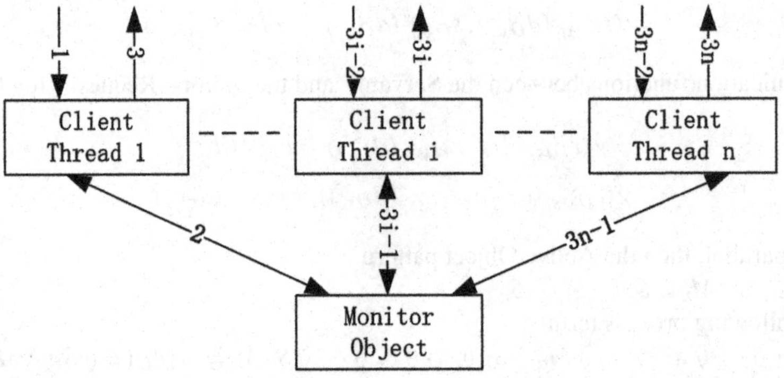


**FIGURE 36.26** Typical process of Monitor Object pattern.

2. The Monitor Object receives the input $d_{I_{M_i}}$ from the Client Thread $i$ through the channel $CM_i$ (the corresponding reading action is denoted $r_{CM_i}(d_{I_{M_i}})$) for $1 \leq i \leq n$, then processes the request and generates the output $d_{O_{M_i}}$ through a processing function $MF_i$, and sends the output to the Client Thread $i$ through the channel $CM_i$ (the corresponding sending action is denoted $s_{CM_i}(d_{O_{M_i}})$);
3. The Client Thread $i$ receives the output from the Monitor Object through the channel $CM_i$ (the corresponding reading action is denoted $r_{CM_i}(d_{O_{M_i}})$), then processes the output and generates the output $d_{O_i}$ through a processing function $CF_{i2}$ (accessing the resource), and sends the output to the outside through the channel $O_i$ (the corresponding sending action is denoted $s_{O_i}(d_{O_i})$).

In the following, we verify the Monitor Object pattern. We assume all data elements $d_{I_i}, d_{O_i}, d_{I_{M_i}}, d_{O_{M_i}}$ for $1 \leq i \leq n$ are from a finite set $\Delta$.

The state transitions of the Client Thread $i$ module described by APTC are as follows:

$C_i = \sum_{d_{I_i} \in \Delta}(r_{I_i}(d_{I_i}) \cdot C_{i_2})$

$C_{i_2} = CF_{i1} \cdot C_{i_3}$

$C_{i_3} = \sum_{d_{I_{M_i}} \in \Delta}(s_{CM_i}(d_{I_{M_i}}) \cdot C_{i_4})$

$C_{i_4} = \sum_{d_{O_{M_i}} \in \Delta}(r_{CM_i}(d_{O_{M_i}}) \cdot C_{i_5})$

$C_{i_5} = CF_{i2} \cdot C_{i_6} \quad (CF_{12}\% \cdots \%CF_{n2})$

$C_{i_6} = \sum_{d_{O_i} \in \Delta}(s_{O_i}(d_{O_i}) \cdot C_i)$

The state transitions of the Monitor Object module described by APTC are as follows:

$M = \sum_{d_{I_{M_1}}, \cdots, d_{I_{M_n}} \in \Delta}(r_{CM_1}(d_{I_{M_1}}) \between \cdots \between r_{CM_n}(d_{I_{M_n}}) \cdot M_2)$

$M_2 = MF_1 \between \cdots \between MF_n \cdot M_3 \quad (MF_1\% \cdots \%MF_n)$

$M_3 = \sum_{d_{O_{M_1}}, \cdots, d_{O_{M_n}} \in \Delta}(s_{CM_1}(d_{O_{M_1}}) \between \cdots \between s_{CM_n}(d_{O_{M_n}}) \cdot M)$

The sending action and the reading action of the same data through the same channel can communicate with each other, otherwise, they will cause a deadlock $\delta$. We define the following communication functions between the Client Thread $i$ and

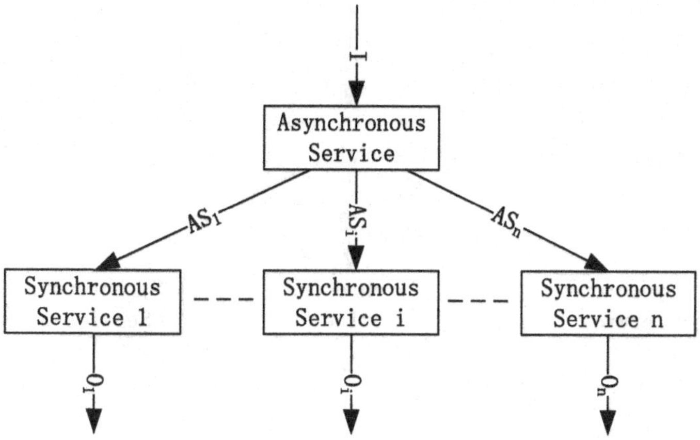


**FIGURE 36.27** Half-Sync/Half-Async pattern.

the Monitor Object:

$$\gamma(r_{CM_i}(d_{I_{M_i}}), s_{CM_i}(d_{I_{M_i}})) \triangleq c_{CM_i}(d_{I_{M_i}})$$
$$\gamma(r_{CM_i}(d_{O_{M_i}}), s_{CM_i}(d_{O_{M_i}})) \triangleq c_{CM_i}(d_{O_{M_i}})$$

Let all modules be in parallel, then the Monitor Object pattern $C_1 \cdots C_n \quad M$ can be presented by the following process term:

$\tau_I(\partial_H(\Theta(C_1 \between \cdots \between C_n \between M))) = \tau_I(\partial_H(C_1 \between \cdots \between C_n \between M))$

where $H = \{r_{CM_i}(d_{I_{M_i}}), s_{CM_i}(d_{I_{M_i}}), r_{CM_i}(d_{O_{M_i}}), s_{CM_i}(d_{O_{M_i}})|d_{I_i}, d_{O_i}, d_{I_{M_i}}, d_{O_{M_i}} \in \Delta\}$,

$I = \{c_{CM_i}(d_{I_{M_i}}), c_{CM_i}(d_{O_{M_i}}), CF_{i1}, CF_{i2}, MF_i|d_{I_i}, d_{O_i}, d_{I_{M_i}}, d_{O_{M_i}} \in \Delta\}$ for $1 \leq i \leq n$

Then, we obtain the following conclusion on the Monitor Object pattern.

**Theorem 36.13** (Correctness of the Monitor Object pattern). *The Monitor Object pattern $\tau_I(\partial_H(C_1 \between \cdots \between C_n \between M))$ can exhibit the desired external behaviors.*

*Proof.* Based on the above state transitions of the above modules, by use of the algebraic laws of APTC, we can prove that

$\tau_I(\partial_H(C_1 \between \cdots \between C_n \between M)) = \sum_{d_{I_1}, d_{O_1}, \cdots, d_{I_n}, d_{O_n} \in \Delta}(r_{I_1}(d_{I_1}) \parallel \cdots \parallel r_{I_n}(d_{I_n}) \cdot s_{O_1}(d_{O_1}) \parallel \cdots \parallel s_{O_n}(d_{O_n})) \cdot \tau_I(\partial_H(C_1 \between \cdots \between C_n \between M))$

That is, the Monitor Object pattern $\tau_I(\partial_H(C_1 \between \cdots \between C_n \between M))$ can exhibit the desired external behaviors.

For the details of the proof, please refer to Section 17.3, as we omit them here. □

### 36.4.3 Verification of the Half-Sync/Half-Async pattern

The Half-Sync/Half-Async pattern decouples the asynchronous and synchronous processings, which has two classes of components: $n$ Synchronous Services and the Asynchronous Service. The Asynchronous Service receives the inputs asynchronously from the user through the channel $I$, then the Asynchronous Service sends the results to the Synchronous Service $i$ through the channel $AS_i$ synchronously for $1 \leq i \leq n$; When the Synchronous Service $i$ receives the input from the Asynchronous Service, it generates and sends the results out to the user through the channel $O_i$. This is illustrated in Fig. 36.27.

The typical process of the Half-Sync/Half-Async pattern is shown in Fig. 36.28 and is the following:

1. The Asynchronous Service receives the input $d_I$ from the user through the channel $I$ (the corresponding reading action is denoted $r_I(D_I)$), processes the input through a processing function $AF$, and generates the input to the Synchronous Service $i$ (for $1 \leq i \leq n$) that is denoted $d_{I_{S_i}}$; then sends the input to the Synchronous Service $i$ through the channel $AS_i$ (the corresponding sending action is denoted $s_{AS_i}(d_{I_{S_i}})$);
2. The Synchronous Service $i$ (for $1 \leq i \leq n$) receives the input from the Asynchronous Service through the channel $AS_i$ (the corresponding reading action is denoted $r_{AS_i}(d_{I_{S_i}})$), processes the results through a processing function $SF_i$, generates the output $d_{O_i}$, then sends the output through the channel $O_i$ (the corresponding sending action is denoted $s_{O_i}(d_{O_i})$).

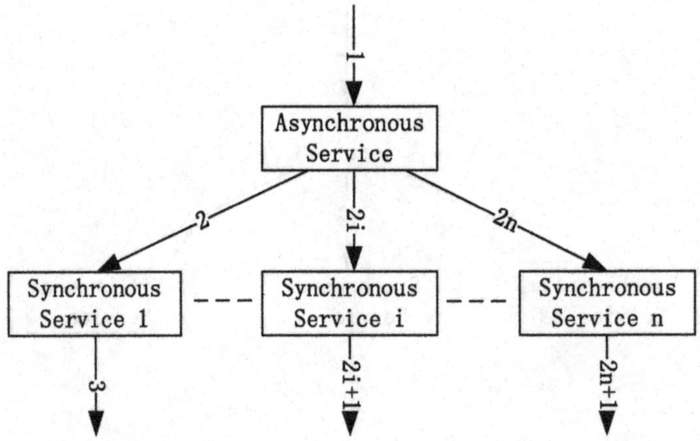


**FIGURE 36.28** Typical process of Half-Sync/Half-Async pattern.

In the following, we verify the Half-Sync/Half-Async pattern. We assume all data elements $d_I, d_{IS_i}, d_{O_i}$ (for $1 \leq i \leq n$) are from a finite set $\Delta$.

The state transitions of the Asynchronous Service module described by APTC are as follows:

$A = \sum_{d_I \in \Delta}(r_I(d_I) \cdot A_2)$

$A_2 = AF \cdot A_3$

$A_3 = \sum_{d_{IS_1}, \cdots, d_{IS_n} \in \Delta}(s_{AS_1}(d_{IS_1}) \between \cdots \between s_{AS_n}(d_{IS_n}) \cdot A)$

The state transitions of the Synchronous Service $i$ described by APTC are as follows:

$S_i = \sum_{d_{IS_i} \in \Delta}(r_{AS_i}(d_{IS_i}) \cdot S_{i_2})$

$S_{i_2} = SF_i \cdot S_{i_3}$

$S_{i_3} = \sum_{d_{O_i} \in \Delta}(s_{O_i}(d_{O_i}) \cdot S_i)$

The sending action and the reading action of the same data through the same channel can communicate with each other, otherwise, they will cause a deadlock $\delta$. We define the following communication function of the Synchronous Service $i$ for $1 \leq i \leq n$:

$$\gamma(r_{AS_i}(d_{IS_i}), s_{AS_i}(d_{IS_i})) \triangleq c_{AS_i}(d_{IS_i})$$

Let all modules be in parallel, then the Half-Sync/Half-Async pattern $A \quad S_1 \cdots S_i \cdots S_n$ can be presented by the following process term:

$\tau_I(\partial_H(\Theta(A \between S_1 \between \cdots \between S_i \between \cdots \between S_n))) = \tau_I(\partial_H(A \between S_1 \between \cdots \between S_i \between \cdots \between S_n))$

where $H = \{r_{AS_i}(d_{OS_i}), s_{AS_i}(d_{OS_i}) | d_I, d_{IS_i}, d_{O_i} \in \Delta\}$ for $1 \leq i \leq n$,

$I = \{c_{AS_i}(d_{IS_i}), AF, SF_i | d_I, d_{IS_i}, d_{O_i} \in \Delta\}$ for $1 \leq i \leq n$

Then, we obtain the following conclusion on the Half-Sync/Half-Async pattern.

**Theorem 36.14** (Correctness of the Half-Sync/Half-Async pattern). *The Half-Sync/Half-Async pattern* $\tau_I(\partial_H(A \between S_1 \between \cdots \between S_i \between \cdots \between S_n))$ *can exhibit the desired external behaviors.*

*Proof.* Based on the above state transitions of the above modules, by use of the algebraic laws of APTC, we can prove that

$\tau_I(\partial_H(A \between S_1 \between \cdots \between S_i \between \cdots \between S_n)) = \sum_{d_I, d_{O_1}, \cdots, d_{O_n} \in \Delta}(r_I(d_I) \cdot s_{O_1}(d_{O_1}) \parallel \cdots \parallel s_{O_i}(d_{O_i}) \parallel \cdots \parallel s_{O_n}(d_{O_n})) \cdot \tau_I(\partial_H(A \between S_1 \between \cdots \between S_i \between \cdots \between S_n))$

That is, the Half-Sync/Half-Async pattern $\tau_I(\partial_H(A \between S_1 \between \cdots \between S_i \between \cdots \between S_n))$ can exhibit the desired external behaviors.

For the details of the proof, please refer to Section 17.3, as we omit them here. □

### 36.4.4 Verification of the Leader/Followers pattern

The Leader/Followers pattern decouples the event delivery between the event source and event handler. There are four modules in the Leader/Followers pattern: the Handle Set, the Leader, the Follower, and the Event Handler. The Handle Set interacts with the outside through the channel $I$, and with the Leader through the channel $HL$. The Leader interacts with the Follower through the channel $LF$. The Event Handler interacts with the Follower through the channel $FE$, and with the outside through the channels $O$. This is illustrated in Fig. 36.29.

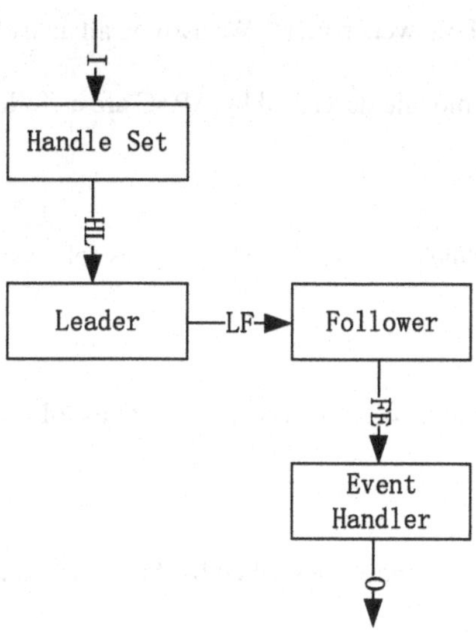


**FIGURE 36.29** Leader/Followers pattern.

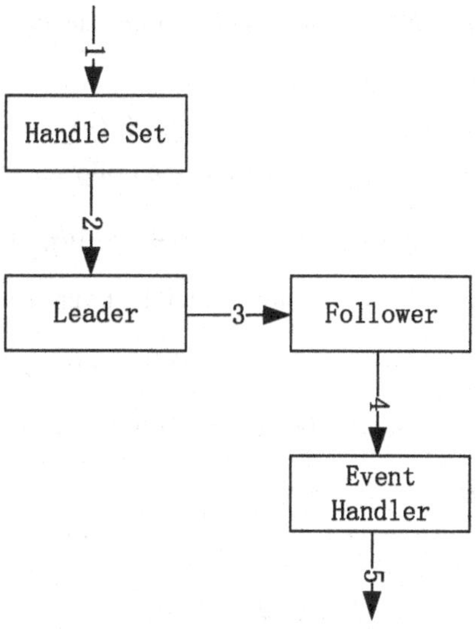


**FIGURE 36.30** Typical process of Leader/Followers pattern.

The typical process of the Leader/Followers pattern is shown in Fig. 36.30 and is as follows:

1. The Handle Set receives the input $d_I$ from the outside through the channel $I$ (the corresponding reading action is denoted $r_I(d_I)$), then processes the input $d_I$ through a processing function $HF$, and sends the processed input $d_{I_{HL}}$ to the Leader through the channel $HL$ (the corresponding sending action is denoted $s_{PP}(d_{I_{HL}})$);
2. The Leader receives $d_{I_{HL}}$ from the Handle Set through the channel $HL$ (the corresponding reading action is denoted $r_{HL}(d_{I_{HL}})$), then processes the request through a processing function $LF$, generates and sends the processed input $d_{I_{LF}}$ to the Follower through the channel $LF$ (the corresponding sending action is denoted $s_{LF}(d_{I_{LF}})$);
3. The Follower receives the input $d_{I_{LF}}$ from the Leader through the channel $LF$ (the corresponding reading action is denoted $r_{LF}(d_{I_{LF}})$), then processes the request through a processing function $FF$, generates and sends the processed input $d_{I_{FE}}$ to the Event Handler through the channel $FE$ (the corresponding sending action is denoted $s_{FE}(d_{I_{FE}})$);
4. The Event Handler receives the input $d_{I_{FE}}$ from the Follower through the channel $FE$ (the corresponding reading action is denoted $r_{FE}(d_{I_{FE}})$), then processes the request and generates the response $d_O$ through a processing function $EF$, and sends the response to the outside through the channel $O$ (the corresponding sending action is denoted $s_O(d_O)$).

In the following, we verify the Leader/Followers pattern. We assume all data elements $d_I, d_{I_{HL}}, d_{I_{LF}}, d_{I_{FE}}, d_O$ are from a finite set $\Delta$.

The state transitions of the Handle Set module described by APTC are as follows:

$H = \sum_{d_I \in \Delta}(r_I(d_I) \cdot H_2)$

$H_2 = HF \cdot H_3$

$H_3 = \sum_{d_{I_{HL}} \in \Delta}(s_{HL}(d_{I_{HL}}) \cdot H)$

The state transitions of the Leader module described by APTC are as follows:

$L = \sum_{d_{I_{HL}} \in \Delta}(r_{HL}(d_{I_{HL}}) \cdot L_2)$

$L_2 = LF \cdot L_3$

$L_3 = \sum_{d_{I_{LF}} \in \Delta}(s_{LF}(d_{I_{LF}}) \cdot L)$

The state transitions of the Follower module described by APTC are as follows:

$F = \sum_{d_{I_{LF}} \in \Delta}(r_{LF}(d_{I_{LF}}) \cdot F_2)$

$F_2 = FF \cdot F_3$

$F_3 = \sum_{d_{I_{FE}} \in \Delta}(s_{FE}(d_{I_{FE}}) \cdot F)$

The state transitions of the Event Handler module described by APTC are as follows:

$E = \sum_{d_{I_{FE}} \in \Delta}(r_{FE}(d_{I_{FE}}) \cdot E_2)$

$E_2 = EF \cdot E_3$

$E_3 = \sum_{d_O \in \Delta}(s_O(d_O) \cdot E)$

The sending action and the reading action of the same data through the same channel can communicate with each other, otherwise, they will cause a deadlock $\delta$. We define the following communication function between the Handle Set and the Leader:

$$\gamma(r_{HL}(d_{I_{HL}}), s_{HL}(d_{I_{HL}})) \triangleq c_{HL}(d_{I_{HL}})$$

There is one communication function between the Leader and the Follower as follows:

$$\gamma(r_{LF}(d_{I_{LF}}), s_{LF}(d_{I_{LF}})) \triangleq c_{LF}(d_{I_{LF}})$$

There is one communication function between the Follower and the Event Handler as follows:

$$\gamma(r_{FE}(d_{I_{FE}}), s_{FE}(d_{I_{FE}})) \triangleq c_{FE}(d_{I_{FE}})$$

Let all modules be in parallel, then the Leader/Followers pattern $H \quad L \quad F \quad E$ can be presented by the following process term:

$\tau_I(\partial_H(\Theta(H \between L \between F \between E))) = \tau_I(\partial_H(H \between L \between F \between E))$

where $H = \{r_{HL}(d_{I_{HL}}), s_{HL}(d_{I_{HL}}), r_{LF}(d_{I_{LF}}), s_{LF}(d_{I_{LF}}), r_{FE}(d_{I_{FE}}), s_{FE}(d_{I_{FE}})$
$|d_I, d_{I_{HL}}, d_{I_{LF}}, d_{I_{FE}}, d_O \in \Delta\}$,

$I = \{c_{HL}(d_{I_{HL}}), c_{LF}(d_{I_{LF}}), c_{FE}(d_{I_{FE}}), HF, LF, FF, EF | d_I, d_{I_{HL}}, d_{I_{LF}}, d_{I_{FE}}, d_O \in \Delta\}$

Then, we obtain the following conclusion on the Leader/Followers pattern.

**Theorem 36.15** (Correctness of the Leader/Followers pattern). *The Leader/Followers pattern $\tau_I(\partial_H(H \between L \between F \between E))$ can exhibit the desired external behaviors.*

*Proof.* Based on the above state transitions of the above modules, by use of the algebraic laws of APTC, we can prove that

$\tau_I(\partial_H(H \between L \between F \between E)) = \sum_{d_I, d_O \in \Delta}(r_I(d_I) \cdot s_O(d_O)) \cdot \tau_I(\partial_H(H \between L \between F \between E))$

That is, the Leader/Followers pattern $\tau_I(\partial_H(H \between L \between F \between E))$ can exhibit the desired external behaviors.

For the details of the proof, please refer to Section 17.3, as we omit them here. □

### 36.4.5 Verification of the Thread-Specific Storage pattern

The Thread-Specific Storage pattern allows the application threads to obtain a global access point to a local object. There are four modules in the Thread-Specific Storage pattern: the Application Thread, the Thread-Specific Object Proxy, the Key Factory, and the Thread-Specific Object. The Application Thread interacts with the outside through the channels $I$ and $O$, and with the Thread-Specific Object Proxy through the channels $I_{AP}$ and $O_{AP}$. The Thread-Specific Object Proxy interacts with the Thread-Specific Object through the channels $I_{PO}$ and $O_{PO}$, and with the Key Factory through the channels $I_{PF}$ and $O_{PF}$. This is illustrated in Fig. 36.31.

The typical process of the Thread-Specific Storage pattern is shown in Fig. 36.32 and is as follows:

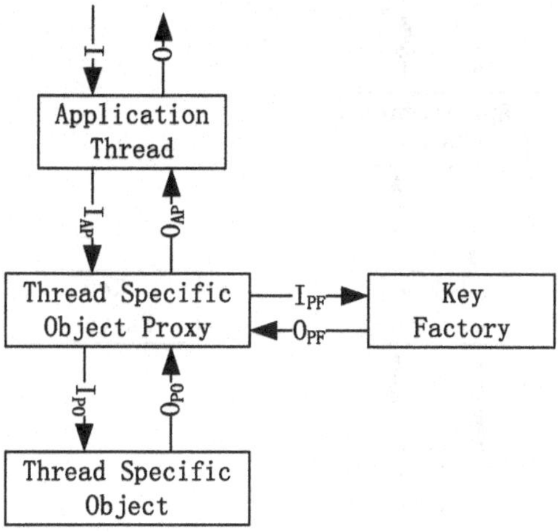


**FIGURE 36.31** Thread-Specific Storage pattern.

1. The Application Thread receives the input $d_I$ from the outside through the channel $I$ (the corresponding reading action is denoted $r_I(d_I)$), then processes the input $d_I$ through a processing function $AF_1$, and sends the processed input $d_{I_P}$ to the Thread-Specific Object Proxy through the channel $I_{AP}$ (the corresponding sending action is denoted $s_{I_{AP}}(d_{I_P})$);
2. The Thread-Specific Object Proxy receives $d_{I_P}$ from the Application Thread through the channel $I_{AP}$ (the corresponding reading action is denoted $r_{I_{AP}}(d_{I_P})$), then processes the request through a processing function $PF_1$, generates and sends the processed input $d_{I_F}$ to the Key Factory through the channel $I_{PF}$ (the corresponding sending action is denoted $s_{I_{PF}}(d_{I_F})$);
3. The Key Factory receives the input $d_{I_F}$ from the Thread-Specific Object Proxy through the channel $I_{PF}$ (the corresponding reading action is denoted $r_{I_{PF}}(d_{I_F})$), then processes the request and generates the result $d_{O_F}$ through a processing function $FF$, and sends the result to the Thread-Specific Object Proxy through the channel $O_{PF}$ (the corresponding sending action is denoted $s_{O_{PF}}(d_{O_F})$);
4. The Thread-Specific Object Proxy receives the result from the Key Factory through the channel $O_{PF}$ (the corresponding reading action is denoted $r_{O_{PF}}(d_{O_F})$), then processes the results and generates the request $d_{I_O}$ to the Thread-Specific Object through a processing function $PF_2$, and sends the request to the Thread-Specific Object through the channel $I_{PO}$ (the corresponding sending action is denoted $s_{I_{PO}}(d_{I_O})$);
5. The Thread-Specific Object receives the input $d_{I_O}$ from the Thread-Specific Object Proxy through the channel $I_{PO}$ (the corresponding reading action is denoted $r_{I_{PO}}(d_{I_O})$), then processes the input through a processing function $OF$, generates and sends the response $d_{O_O}$ to the Thread-Specific Object Proxy through the channel $O_{PO}$ (the corresponding sending action is denoted $s_{O_{PO}}(d_{O_O})$);
6. The Thread-Specific Object Proxy receives the response $d_{O_O}$ from the Thread-Specific Object through the channel $O_{PO}$ (the corresponding reading action is denoted $r_{O_{PO}}(d_{O_O})$), then processes the response through a processing function $PF_3$, generates and sends the response $d_{O_P}$ (the corresponding sending action is denoted $s_{O_{AP}}(d_{O_P})$);
7. The Application Thread receives the response $d_{O_P}$ from the Thread-Specific Object Proxy through the channel $O_{AP}$ (the corresponding reading action is denoted $r_{O_{AP}}(d_{O_P})$), then processes the request and generates the response $d_O$ through a processing function $AF_2$, and sends the response to the outside through the channel $O$ (the corresponding sending action is denoted $s_O(d_O)$).

In the following, we verify the Thread-Specific Storage pattern. We assume all data elements $d_I$, $d_{I_P}$, $d_{I_F}$, $d_{I_O}$, $d_{O_P}$, $d_{O_F}$, $d_{O_O}$, $d_O$ are from a finite set $\Delta$.

The state transitions of the Application Thread module described by APTC are as follows:

$A = \sum_{d_I \in \Delta}(r_I(d_I) \cdot A_2)$

$A_2 = AF_1 \cdot A_3$

$A_3 = \sum_{d_{I_P} \in \Delta}(s_{I_{AP}}(d_{I_P}) \cdot A_4)$

$A_4 = \sum_{d_{O_P} \in \Delta}(r_{O_{AP}}(d_{O_P}) \cdot A_5)$

$A_5 = AF_2 \cdot A_6$

$A_6 = \sum_{d_O \in \Delta}(s_O(d_O) \cdot A)$

The state transitions of the Thread-Specific Object Proxy module described by APTC are as follows:

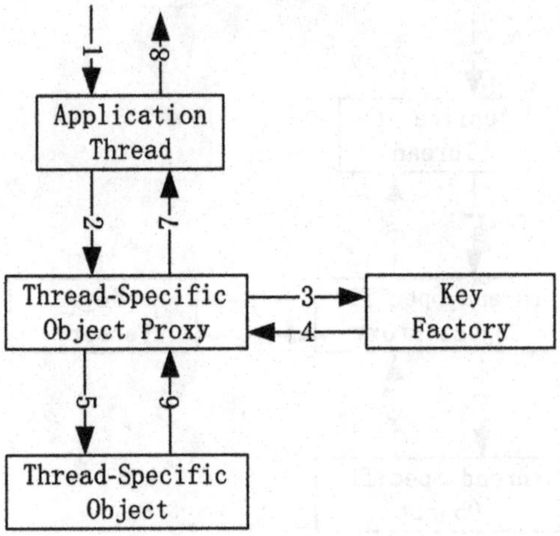


FIGURE 36.32 Typical process of Thread-Specific Storage pattern.

$P=\sum_{d_{I_P}\in\Delta}(r_{I_{AP}}(d_{I_P})\cdot P_2)$
$P_2=PF_1\cdot P_3$
$P_3=\sum_{d_{I_F}\in\Delta}(s_{I_{PF}}(d_{I_F})\cdot P_4)$
$P_4=\sum_{d_{O_F}\in\Delta}(r_{O_{PF}}(d_{O_F})\cdot P_5)$
$P_5=PF_2\cdot P_6$
$P_6=\sum_{d_{I_O}\in\Delta}(s_{I_{PO}}(d_{I_O})\cdot P_7)$
$P_7=\sum_{d_{O_O}\in\Delta}(r_{O_{PO}}(d_{O_O})\cdot P_8)$
$P_8=PF_3\cdot P_9$
$P_9=\sum_{d_{O_P}\in\Delta}(s_{O_{AP}}(d_{O_P})\cdot P)$

The state transitions of the Key Factory module described by APTC are as follows:

$F=\sum_{d_{I_F}\in\Delta}(r_{I_{PF}}(d_{I_F})\cdot F_2)$
$F_2=FF\cdot F_3$
$F_3=\sum_{d_{O_F}\in\Delta}(s_{O_{PF}}(d_{O_F})\cdot F)$

The state transitions of the Thread-Specific Object module described by APTC are as follows:

$O=\sum_{d_{I_O}\in\Delta}(r_{I_{PO}}(d_{I_O})\cdot O_2)$
$O_2=OF\cdot O_3$
$O_3=\sum_{d_{O_O}\in\Delta}(s_{O_{PO}}(d_{O_O})\cdot O)$

The sending action and the reading action of the same data through the same channel can communicate with each other, otherwise, they will cause a deadlock $\delta$. We define the following communication functions between the Application Thread and the Thread-Specific Object Proxy:

$$\gamma(r_{I_{AP}}(d_{I_P}),s_{I_{AP}}(d_{I_P}))\triangleq c_{I_{AP}}(d_{I_P})$$
$$\gamma(r_{O_{AP}}(d_{O_P}),s_{O_{AP}}(d_{O_P}))\triangleq c_{O_{AP}}(d_{O_P})$$

There are two communication functions between the Thread-Specific Object Proxy and the Key Factory as follows:

$$\gamma(r_{I_{PF}}(d_{I_F}),s_{I_{PF}}(d_{I_F}))\triangleq c_{I_{PF}}(d_{I_F})$$
$$\gamma(r_{O_{PF}}(d_{O_F}),s_{O_{PF}}(d_{O_F}))\triangleq c_{O_{PF}}(d_{O_F})$$

There are two communication functions between the Thread-Specific Object Proxy and the Thread-Specific Object as follows:

$$\gamma(r_{I_{PO}}(d_{I_O}),s_{I_{PO}}(d_{I_O}))\triangleq c_{I_{PO}}(d_{I_O})$$
$$\gamma(r_{O_{PO}}(d_{O_O}),s_{O_{PO}}(d_{O_O}))\triangleq c_{O_{PO}}(d_{O_O})$$

Let all modules be in parallel, then the Thread-Specific Storage pattern $A \quad P \quad F \quad O$ can be presented by the following process term:

$\tau_I(\partial_H(\Theta(A \between P \between F \between O))) = \tau_I(\partial_H(A \between P \between F \between O))$

where $H = \{r_{I_{AP}}(d_{I_P}), s_{I_{AP}}(d_{I_P}), r_{O_{AP}}(d_{O_P}), s_{O_{AP}}(d_{O_P}), r_{I_{PF}}(d_{I_F}), s_{I_{PF}}(d_{I_F}),$
$r_{O_{PF}}(d_{O_F}), s_{O_{PF}}(d_{O_F}), r_{I_{PO}}(d_{I_O}), s_{I_{PO}}(d_{I_O}), r_{O_{PO}}(d_{O_O}), s_{O_{PO}}(d_{O_O})$
$|d_I, d_{I_P}, d_{I_F}, d_{I_O}, d_{O_P}, d_{O_F}, d_{O_O}, d_O \in \Delta\}$,

$I = \{c_{I_{AP}}(d_{I_P}), c_{O_{AP}}(d_{O_P}), c_{I_{PF}}(d_{I_F}), c_{O_{PF}}(d_{O_F}), c_{I_{PO}}(d_{I_O}), c_{O_{PO}}(d_{O_O}),$
$AF_1, AF_2, PF_1, PF_2, PF_3, FF, OF|d_I, d_{I_P}, d_{I_F}, d_{I_O}, d_{O_P}, d_{O_F}, d_{O_O}, d_O \in \Delta\}$

Then, we obtain the following conclusion on the Thread-Specific Storage pattern.

**Theorem 36.16** (Correctness of the Thread-Specific Storage pattern). *The Thread-Specific Storage pattern $\tau_I(\partial_H(A \between P \between F \between O))$ can exhibit the desired external behaviors.*

*Proof.* Based on the above state transitions of the above modules, by use of the algebraic laws of APTC, we can prove that

$\tau_I(\partial_H(A \between P \between F \between O)) = \sum_{d_I, d_O \in \Delta}(r_I(d_I) \cdot s_O(d_O)) \cdot \tau_I(\partial_H(A \between P \between F \between O))$

That is, the Thread-Specific Storage pattern $\tau_I(\partial_H(A \between P \between F \between O))$ can exhibit the desired external behaviors.

For the details of the proof, please refer to Section 17.3, as we omit them here. □

Chapter 37

# Verification of patterns for resource management

Patterns for resource management are patterns related to resource management, and can be used in higher-level and lower-level systems and applications.

In this chapter, we verify patterns for resource management. In Section 37.1, we verify patterns related to resource acquisition. We verify patterns for resource life cycle in Section 37.2 and patterns for resource release in Section 37.3.

## 37.1 Resource acquisition

In this section, we verify patterns for resource acquisition, including the Lookup pattern, the Lazy Acquisition pattern, the Eager Acquisition pattern, and the Partial Acquisition pattern.

### 37.1.1 Verification of the Lookup pattern

The Lookup pattern uses a mediating lookup service to find and access resources. There are four modules in the Lookup pattern: the Resource User, the Resource Provider, the Lookup Service, and the Resource. The Resource User interacts with the outside through the channels $I$ and $O$; with the Resource Provider through the channel $I_{UP}$ and $O_{UP}$; with the Resource through the channels $I_{UR}$ and $O_{UR}$, and with the Lookup Service through the channels $I_{US}$ and $O_{US}$. This is illustrated in Fig. 37.1.

The typical process of the Lookup pattern is shown in Fig. 37.2 and is as follows:

1. The Resource User receives the input $d_I$ from the outside through the channel $I$ (the corresponding reading action is denoted $r_I(d_I)$), then processes the input $d_I$ through a processing function $UF_1$, and sends the processed input $d_{I_S}$ to the Lookup Service through the channel $I_{US}$ (the corresponding sending action is denoted $s_{I_{US}}(d_{I_S})$);
2. The Lookup Service receives $d_{I_S}$ from the Resource User through the channel $I_{US}$ (the corresponding reading action is denoted $r_{I_{US}}(d_{I_S})$), then processes the request through a processing function $SF$, generates and sends the processed output $d_{O_S}$ to the Resource User through the channel $O_{US}$ (the corresponding sending action is denoted $s_{O_{US}}(d_{O_S})$);
3. The Resource User receives the output $d_{O_S}$ from the Lookup Service through the channel $O_{US}$ (the corresponding reading action is denoted $r_{O_{US}}(d_{O_S})$), then processes the output and generates the input $d_{I_P}$ through a processing function $UF_2$, and sends the input to the Resource Provider through the channel $I_{UP}$ (the corresponding sending action is denoted $s_{I_{UP}}(d_{I_P})$);
4. The Resource Provider receives the input from the Resource User through the channel $I_{UP}$ (the corresponding reading action is denoted $r_{I_{UP}}(d_{I_P})$), then processes the input and generates the output $d_{O_P}$ to the Resource User through

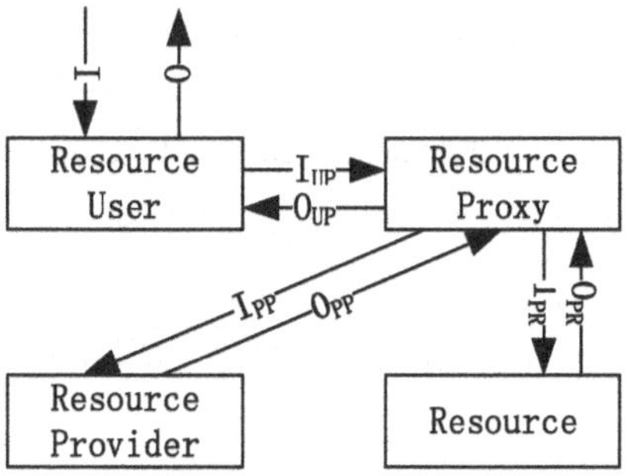


**FIGURE 37.1** Lookup pattern.

Handbook of Truly Concurrent Process Algebra. https://doi.org/10.1016/B978-0-44-321515-5.00041-X

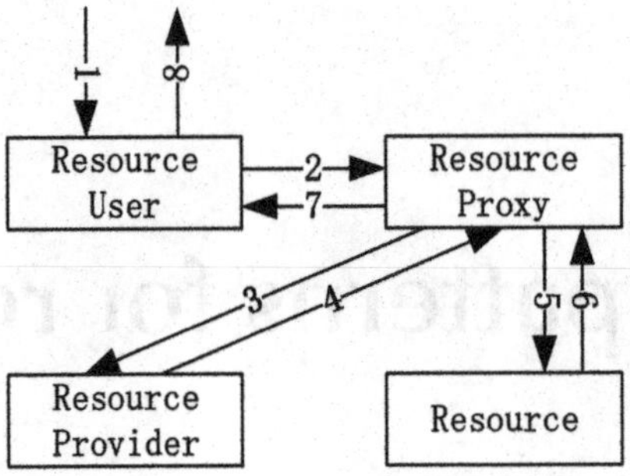


FIGURE 37.2 Typical process of Lookup pattern.

a processing function $PF$, and sends the output to the Resource User through the channel $O_{UP}$ (the corresponding sending action is denoted $s_{O_{UP}}(d_{O_P})$);

5. The Resource User receives the output $d_{O_P}$ from the Resource Provider through the channel $O_{UP}$ (the corresponding reading action is denoted $r_{O_{UP}}(d_{O_P})$), then processes the input through a processing function $UF_3$, generates and sends the input $d_{I_R}$ to the Resource through the channel $I_{UR}$ (the corresponding sending action is denoted $s_{I_{UR}}(d_{I_R})$);
6. The Resource receives the input $d_{I_R}$ from the Resource User through the channel $I_{UR}$ (the corresponding reading action is denoted $r_{I_{UR}}(d_{I_R})$), then processes the input through a processing function $RF$, generates and sends the response $d_{O_R}$ (the corresponding sending action is denoted $s_{O_{UR}}(d_{O_R})$);
7. The Resource User receives the response $d_{O_R}$ from the Resource through the channel $O_{UR}$ (the corresponding reading action is denoted $r_{O_{UR}}(d_{O_R})$), then processes the response and generates the response $d_O$ through a processing function $UF_4$, and sends the response to the outside through the channel $O$ (the corresponding sending action is denoted $s_O(d_O)$).

In the following, we verify the Lookup pattern. We assume all data elements $d_I$, $d_{I_S}$, $d_{I_P}$, $d_{I_R}$, $d_{O_S}$, $d_{O_P}$, $d_{O_R}$, $d_O$ are from a finite set $\Delta$.

The state transitions of the Resource User module described by APTC are as follows:

$U = \sum_{d_I\in\Delta}(r_I(d_I)\cdot U_2)$

$U_2 = UF_1\cdot U_3$

$U_3 = \sum_{d_{I_S}\in\Delta}(s_{I_{US}}(d_{I_S})\cdot U_4)$

$U_4 = \sum_{d_{O_S}\in\Delta}(r_{O_{US}}(d_{O_S})\cdot U_5)$

$U_5 = UF_2\cdot U_6$

$U_6 = \sum_{d_{I_P}\in\Delta}(s_{I_{UP}}(d_{I_P})\cdot U_7)$

$U_7 = \sum_{d_{O_P}\in\Delta}(r_{O_{UP}}(d_{O_P})\cdot U_8)$

$U_8 = UF_3\cdot U_9$

$U_9 = \sum_{d_{I_R}\in\Delta}(s_{I_{UR}}(d_{I_R})\cdot U_{10})$

$U_{10} = \sum_{d_{O_R}\in\Delta}(r_{O_{UR}}(d_{O_R})\cdot U_{11})$

$U_{11} = UF_4\cdot U_{12}$

$U_{12} = \sum_{d_O\in\Delta}(s_O(d_O)\cdot U)$

The state transitions of the Resource Provider module described by APTC are as follows:

$P = \sum_{d_{I_P}\in\Delta}(r_{I_{UP}}(d_{I_P})\cdot P_2)$

$P_2 = PF\cdot P_3$

$P_3 = \sum_{d_{O_P}\in\Delta}(s_{O_{UP}}(d_{O_P})\cdot P)$

The state transitions of the Lookup Service module described by APTC are as follows:

$S = \sum_{d_{I_S}\in\Delta}(r_{I_{US}}(d_{I_S})\cdot S_2)$

$S_2 = SF\cdot S_3$

$S_3 = \sum_{d_{O_S}\in\Delta}(s_{O_{US}}(d_{O_S})\cdot S)$

The state transitions of the Resource module described by APTC are as follows:

$R = \sum_{d_{I_R}\in\Delta}(r_{I_{UR}}(d_{I_R})\cdot R_2)$

$R_2 = RF\cdot R_3$

$R_3 = \sum_{d_{O_R}\in\Delta}(s_{O_{UR}}(d_{O_R})\cdot R)$

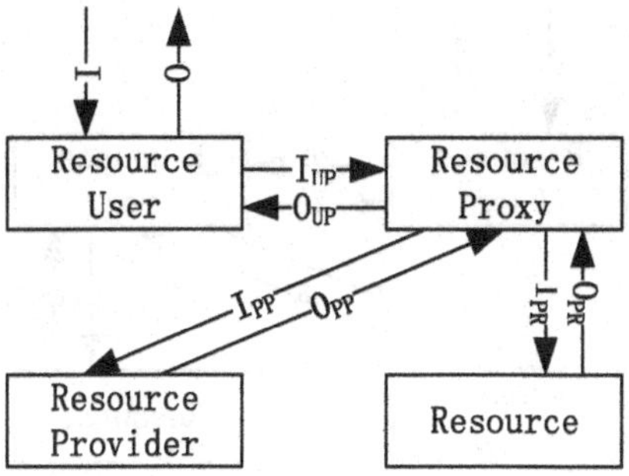


**FIGURE 37.3** Lazy Acquisition pattern.

The sending action and the reading action of the same data through the same channel can communicate with each other, otherwise, they will cause a deadlock $\delta$. We define the following communication functions between the Resource User and the Resource Provider Proxy:

$$\gamma(r_{I_{UP}}(d_{I_P}), s_{I_{UP}}(d_{I_P})) \triangleq c_{I_{UP}}(d_{I_P})$$
$$\gamma(r_{O_{UP}}(d_{O_P}), s_{O_{UP}}(d_{O_P})) \triangleq c_{O_{UP}}(d_{O_P})$$

There are two communication functions between the Resource User and the Lookup Service as follows:

$$\gamma(r_{I_{US}}(d_{I_S}), s_{I_{US}}(d_{I_S})) \triangleq c_{I_{US}}(d_{I_S})$$
$$\gamma(r_{O_{US}}(d_{O_S}), s_{O_{US}}(d_{O_S})) \triangleq c_{O_{US}}(d_{O_S})$$

There are two communication functions between the Resource User and the Resource as follows:

$$\gamma(r_{I_{UR}}(d_{I_R}), s_{I_{UR}}(d_{I_R})) \triangleq c_{I_{UR}}(d_{I_R})$$
$$\gamma(r_{O_{UR}}(d_{O_R}), s_{O_{UR}}(d_{O_R})) \triangleq c_{O_{UR}}(d_{O_R})$$

Let all modules be in parallel, then the Lookup pattern $U \quad S \quad P \quad R$ can be presented by the following process term:
$\tau_I(\partial_H(\Theta(U \between S \between P \between R))) = \tau_I(\partial_H(U \between S \between P \between R))$
where $H = \{r_{I_{UP}}(d_{I_P}), s_{I_{UP}}(d_{I_P}), r_{O_{UP}}(d_{O_P}), s_{O_{UP}}(d_{O_P}), r_{I_{US}}(d_{I_S}), s_{I_{US}}(d_{I_S}),$
$r_{O_{US}}(d_{O_S}), s_{O_{US}}(d_{O_S}), r_{I_{UR}}(d_{I_R}), s_{I_{UR}}(d_{I_R}), r_{O_{UR}}(d_{O_R}), s_{O_{UR}}(d_{O_R})$
$|d_I, d_{I_P}, d_{I_S}, d_{I_R}, d_{O_P}, d_{O_S}, d_{O_R}, d_O \in \Delta\},$
$I = \{c_{I_{UP}}(d_{I_P}), c_{O_{UP}}(d_{O_P}), c_{I_{US}}(d_{I_S}), c_{O_{US}}(d_{O_S}), c_{I_{UR}}(d_{I_R}), c_{O_{UR}}(d_{O_R}),$
$UF_1, UF_2, UF_3, UF_4, PF, SF, RF|d_I, d_{I_P}, d_{I_S}, d_{I_R}, d_{O_P}, d_{O_S}, d_{O_R}, d_O \in \Delta\}$
Then, we obtain the following conclusion on the Lookup pattern.

**Theorem 37.1** (Correctness of the Lookup pattern). *The Lookup pattern $\tau_I(\partial_H(U \between S \between P \between R))$ can exhibit the desired external behaviors.*

*Proof.* Based on the above state transitions of the above modules, by use of the algebraic laws of APTC, we can prove that
$\tau_I(\partial_H(U \between S \between P \between R)) = \sum_{d_I, d_O \in \Delta}(r_I(d_I) \cdot s_O(d_O)) \cdot \tau_I(\partial_H(U \between S \between P \between R))$
That is, the Lookup pattern $\tau_I(\partial_H(U \between S \between P \between R))$ can exhibit the desired external behaviors.
For the details of the proof, please refer to Section 17.3, as we omit them here. □

### 37.1.2 Verification of the Lazy Acquisition pattern

The Lazy Acquisition pattern defers the acquisitions of resources to the latest possible time. There are four modules in the Lazy Acquisition pattern: the Resource User, the Resource Provider, the Resource Proxy, and the Resource. The Resource User interacts with the outside through the channels $I$ and $O$, and with the Resource Proxy through the channels $I_{UP}$ and $O_{UP}$. The Resource Proxy interacts with the Resource through the channels $I_{PR}$ and $O_{PR}$, and with the Resource Provider through the channels $I_{PP}$ and $O_{PP}$. This is illustrated in Fig. 37.3.

The typical process of the Lazy Acquisition pattern is shown in Fig. 37.4 and is as follows:

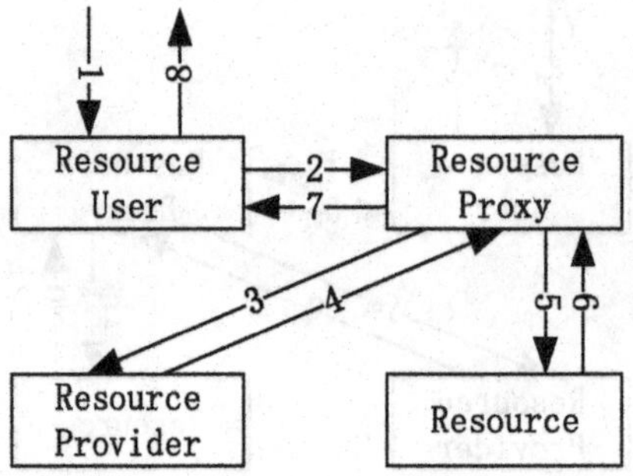


FIGURE 37.4 Typical process of Lazy Acquisition pattern.

1. The Resource User receives the input $d_I$ from the outside through the channel $I$ (the corresponding reading action is denoted $r_I(d_I)$), then processes the input $d_I$ through a processing function $UF_1$, and sends the processed input $d_{I_P}$ to the Resource Proxy through the channel $I_{UP}$ (the corresponding sending action is denoted $s_{I_{UP}}(d_{I_P})$);
2. The Resource Proxy receives $d_{I_P}$ from the Resource User through the channel $I_{UP}$ (the corresponding reading action is denoted $r_{I_{UP}}(d_{I_P})$), then processes the request through a processing function $PF_1$, generates and sends the processed input $d_{I_{RP}}$ to the Resource Provider through the channel $I_{PP}$ (the corresponding sending action is denoted $s_{I_{PP}}(d_{I_{RP}})$);
3. The Resource Provider receives the input $d_{I_{RP}}$ from the Resource Proxy through the channel $I_{PP}$ (the corresponding reading action is denoted $r_{I_{PP}}(d_{I_{RP}})$), then processes the input and generates the output $d_{O_{RP}}$ through a processing function $RPF$, and sends the output to the Resource Proxy through the channel $O_{PP}$ (the corresponding sending action is denoted $s_{O_{PP}}(d_{O_{RP}})$);
4. The Resource Proxy receives the output from the Resource Provider through the channel $O_{PP}$ (the corresponding reading action is denoted $r_{O_{PP}}(d_{O_{RP}})$), then processes the results and generates the input $d_{I_R}$ to the Resource through a processing function $PF_2$, and sends the input to the Resource through the channel $I_{PR}$ (the corresponding sending action is denoted $s_{I_{PR}}(d_{I_R})$);
5. The Resource receives the input $d_{I_R}$ from the Resource Proxy through the channel $I_{PR}$ (the corresponding reading action is denoted $r_{I_{PR}}(d_{I_R})$), then processes the input through a processing function $RF$, generates and sends the output $d_{O_R}$ to the Resource Proxy through the channel $O_{PR}$ (the corresponding sending action is denoted $s_{O_{PR}}(d_{O_R})$);
6. The Resource Proxy receives the output $d_{O_R}$ from the Resource through the channel $O_{PR}$ (the corresponding reading action is denoted $r_{O_{PR}}(d_{O_R})$), then processes the response through a processing function $PF_3$, generates and sends the response $d_{O_P}$ (the corresponding sending action is denoted $s_{O_{UP}}(d_{O_P})$);
7. The Resource User receives the response $d_{O_P}$ from the Resource Proxy through the channel $O_{UP}$ (the corresponding reading action is denoted $r_{O_{UP}}(d_{O_P})$), then processes the response and generates the response $d_O$ through a processing function $UF_2$, and sends the response to the outside through the channel $O$ (the corresponding sending action is denoted $s_O(d_O)$).

In the following, we verify the Lazy Acquisition pattern. We assume all data elements $d_I, d_{I_S}, d_{I_P}, d_{I_R}, d_{O_S}, d_{O_P}, d_{O_R}, d_O$ are from a finite set $\Delta$.

The state transitions of the Resource User module described by APTC are as follows:

$U = \sum_{d_I \in \Delta}(r_I(d_I) \cdot U_2)$

$U_2 = UF_1 \cdot U_3$

$U_3 = \sum_{d_{I_P} \in \Delta}(s_{I_{UP}}(d_{I_P}) \cdot U_4)$

$U_4 = \sum_{d_{O_P} \in \Delta}(r_{O_{UP}}(d_{O_P}) \cdot U_5)$

$U_5 = UF_2 \cdot U_6$

$U_6 = \sum_{d_O \in \Delta}(s_O(d_O) \cdot U)$

The state transitions of the Resource Proxy module described by APTC are as follows:

$P = \sum_{d_{I_P} \in \Delta}(r_{I_{UP}}(d_{I_P}) \cdot P_2)$

$P_2 = PF_1 \cdot P_3$

$P_3 = \sum_{d_{I_{RP}} \in \Delta}(s_{I_{PP}}(d_{I_{RP}}) \cdot P_4)$

$P_4 = \sum_{d_{O_{RP}} \in \Delta}(r_{O_{PP}}(d_{O_{RP}}) \cdot P_5)$

$P_5 = PF_2 \cdot P_6$

$P_6 = \sum_{d_{I_R} \in \Delta}(s_{I_{PR}}(d_{I_R}) \cdot P_7)$

$P_7 = \sum_{d_{O_R} \in \Delta}(r_{O_{PR}}(d_{O_R}) \cdot P_8)$
$P_8 = PF_3 \cdot P_9$
$P_9 = \sum_{d_{O_P} \in \Delta}(s_{O_{UP}}(d_{O_P}) \cdot P)$
The state transitions of the Resource Provider module described by APTC are as follows:
$RP = \sum_{d_{I_{RP}} \in \Delta}(r_{I_{PP}}(d_{I_{RP}}) \cdot RP_2)$
$RP_2 = RPF \cdot RP_3$
$RP_3 = \sum_{d_{O_{RP}} \in \Delta}(s_{O_{PP}}(d_{O_{RP}}) \cdot RP)$
The state transitions of the Resource module described by APTC are as follows:
$R = \sum_{d_{I_R} \in \Delta}(r_{I_{PR}}(d_{I_R}) \cdot R_2)$
$R_2 = RF \cdot R_3$
$R_3 = \sum_{d_{O_R} \in \Delta}(s_{O_{PR}}(d_{O_R}) \cdot R)$
The sending action and the reading action of the same data through the same channel can communicate with each other, otherwise, they will cause a deadlock $\delta$. We define the following communication functions between the Resource User and the Resource Proxy:

$$\gamma(r_{I_{UP}}(d_{I_P}), s_{I_{UP}}(d_{I_P})) \triangleq c_{I_{UP}}(d_{I_P})$$
$$\gamma(r_{O_{UP}}(d_{O_P}), s_{O_{UP}}(d_{O_P})) \triangleq c_{O_{UP}}(d_{O_P})$$

There are two communication functions between the Resource Provider and the Resource Proxy as follows:

$$\gamma(r_{I_{PP}}(d_{I_{RP}}), s_{I_{PP}}(d_{I_{RP}})) \triangleq c_{I_{PP}}(d_{I_{RP}})$$
$$\gamma(r_{O_{PP}}(d_{O_{RP}}), s_{O_{PP}}(d_{O_{RP}})) \triangleq c_{O_{PP}}(d_{O_{RP}})$$

There are two communication functions between the Resource Proxy and the Resource as follows:

$$\gamma(r_{I_{PR}}(d_{I_R}), s_{I_{PR}}(d_{I_R})) \triangleq c_{I_{PR}}(d_{I_R})$$
$$\gamma(r_{O_{PR}}(d_{O_R}), s_{O_{PR}}(d_{O_R})) \triangleq c_{O_{PR}}(d_{O_R})$$

Let all modules be in parallel, then the Lazy Acquisition pattern $U \quad P \quad RP \quad R$ can be presented by the following process term:
$\tau_I(\partial_H(\Theta(U \between P \between RP \between R))) = \tau_I(\partial_H(U \between P \between RP \between R))$
where $H = \{r_{I_{UP}}(d_{I_P}), s_{I_{UP}}(d_{I_P}), r_{O_{UP}}(d_{O_P}), s_{O_{UP}}(d_{O_P}), r_{I_{PP}}(d_{I_{RP}}), s_{I_{PP}}(d_{I_{RP}}),$
$r_{O_{PP}}(d_{O_{RP}}), s_{O_{PP}}(d_{O_{RP}}), r_{I_{PR}}(d_{I_R}), s_{I_{PR}}(d_{I_R}), r_{O_{PR}}(d_{O_R}), s_{O_{PR}}(d_{O_R})$
$|d_I, d_{I_P}, d_{I_{RP}}, d_{I_R}, d_{O_P}, d_{O_{RP}}, d_{O_R}, d_O \in \Delta\},$
$I = \{c_{I_{UP}}(d_{I_P}), c_{O_{UP}}(d_{O_P}), c_{I_{PP}}(d_{I_{RP}}), c_{O_{PP}}(d_{O_{RP}}), c_{I_{PR}}(d_{I_R}), c_{O_{PR}}(d_{O_R}),$
$UF_1, UF_2, PF_1, PF_2, PF_3, RPF, RF|d_I, d_{I_P}, d_{I_{RP}}, d_{I_R}, d_{O_P}, d_{O_{RP}}, d_{O_R}, d_O \in \Delta\}$
Then, we obtain the following conclusion on the Lazy Acquisition pattern.

**Theorem 37.2** (Correctness of the Lazy Acquisition pattern). *The Lazy Acquisition pattern $\tau_I(\partial_H(U \between P \between RP \between R))$ can exhibit the desired external behaviors.*

*Proof.* Based on the above state transitions of the above modules, by use of the algebraic laws of APTC, we can prove that
$\tau_I(\partial_H(U \between P \between RP \between R)) = \sum_{d_I, d_O \in \Delta}(r_I(d_I) \cdot s_O(d_O)) \cdot \tau_I(\partial_H(U \between P \between RP \between R))$
That is, the Lazy Acquisition pattern $\tau_I(\partial_H(U \between P \between RP \between R))$ can exhibit the desired external behaviors.
For the details of the proof, please refer to Section 17.3, as we omit them here. □

### 37.1.3 Verification of the Eager Acquisition pattern

The Eager Acquisition pattern acquires the resources eagerly. There are four modules in the Eager Acquisition pattern: the Resource User, the Resource Provider, the Resource Proxy, and the Resource. The Resource User interacts with the outside through the channels $I$ and $O$, and with the Resource Proxy through the channels $I_{UP}$ and $O_{UP}$. The Resource Proxy interacts with the Resource through the channels $I_{PR}$ and $O_{PR}$, and with the Resource Provider through the channels $I_{PP}$ and $O_{PP}$. This is illustrated in Fig. 37.5.

The typical process of the Eager Acquisition pattern is shown in Fig. 37.6 and is as follows:

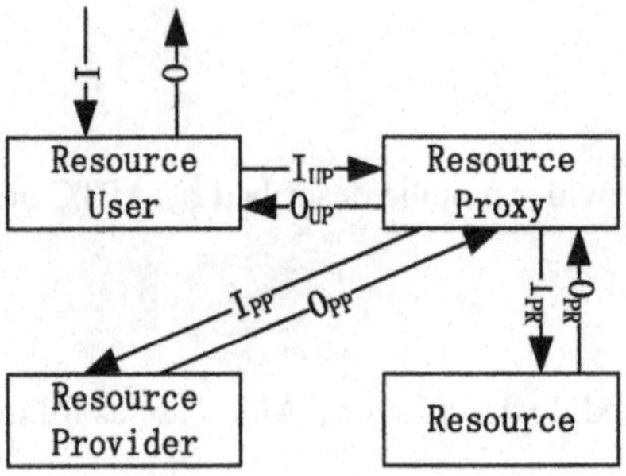


FIGURE 37.5 Eager Acquisition pattern.

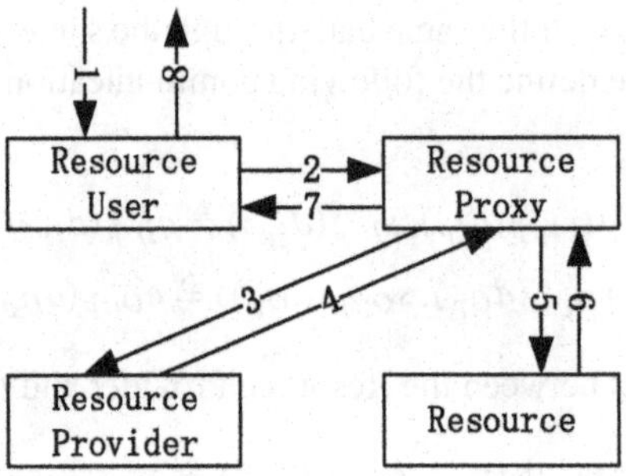


FIGURE 37.6 Typical process of Eager Acquisition pattern.

1. The Resource User receives the input $d_I$ from the outside through the channel $I$ (the corresponding reading action is denoted $r_I(d_I)$), then processes the input $d_I$ through a processing function $UF_1$, and sends the processed input $d_{I_P}$ to the Resource Proxy through the channel $I_{UP}$ (the corresponding sending action is denoted $s_{I_{UP}}(d_{I_P})$);
2. The Resource Proxy receives $d_{I_P}$ from the Resource User through the channel $I_{UP}$ (the corresponding reading action is denoted $r_{I_{UP}}(d_{I_P})$), then processes the request through a processing function $PF_1$, generates and sends the processed input $d_{I_{RP}}$ to the Resource Provider through the channel $I_{PP}$ (the corresponding sending action is denoted $s_{I_{PP}}(d_{I_{RP}})$);
3. The Resource Provider receives the input $d_{I_{RP}}$ from the Resource Proxy through the channel $I_{PP}$ (the corresponding reading action is denoted $r_{I_{PP}}(d_{I_{RP}})$), then processes the input and generates the output $d_{O_{RP}}$ through a processing function $RPF$, and sends the output to the Resource Proxy through the channel $O_{PP}$ (the corresponding sending action is denoted $s_{O_{PP}}(d_{O_{RP}})$);
4. The Resource Proxy receives the output from the Resource Provider through the channel $O_{PP}$ (the corresponding reading action is denoted $r_{O_{PP}}(d_{O_{RP}})$), then processes the results and generates the input $d_{I_R}$ to the Resource through a processing function $PF_2$, and sends the input to the Resource through the channel $I_{PR}$ (the corresponding sending action is denoted $s_{I_{PR}}(d_{I_R})$);
5. The Resource receives the input $d_{I_R}$ from the Resource Proxy through the channel $I_{PR}$ (the corresponding reading action is denoted $r_{I_{PR}}(d_{I_R})$), then processes the input through a processing function $RF$, generates and sends the output $d_{O_R}$ to the Resource Proxy through the channel $O_{PR}$ (the corresponding sending action is denoted $s_{O_{PR}}(d_{O_R})$);
6. The Resource Proxy receives the output $d_{O_R}$ from the Resource through the channel $O_{PR}$ (the corresponding reading action is denoted $r_{O_{PR}}(d_{O_R})$), then processes the response through a processing function $PF_3$, generates and sends the response $d_{O_P}$ (the corresponding sending action is denoted $s_{O_{UP}}(d_{O_P})$);
7. The Resource User receives the response $d_{O_P}$ from the Resource Proxy through the channel $O_{UP}$ (the corresponding reading action is denoted $r_{O_{UP}}(d_{O_P})$), then processes the response and generates the response $d_O$ through a processing function $UF_2$, and sends the response to the outside through the channel $O$ (the corresponding sending action is denoted $s_O(d_O)$).

In the following, we verify the Eager Acquisition pattern. We assume all data elements $d_I, d_{I_S}, d_{I_P}, d_{I_R}, d_{O_S}, d_{O_P}, d_{O_R}, d_O$ are from a finite set $\Delta$.

The state transitions of the Resource User module described by APTC are as follows:

$U = \sum_{d_I \in \Delta}(r_I(d_I) \cdot U_2)$

$U_2 = UF_1 \cdot U_3$

$U_3 = \sum_{d_{I_P} \in \Delta}(s_{I_{UP}}(d_{I_P}) \cdot U_4)$
$U_4 = \sum_{d_{O_P} \in \Delta}(r_{O_{UP}}(d_{O_P}) \cdot U_5)$
$U_5 = UF_2 \cdot U_6$
$U_6 = \sum_{d_O \in \Delta}(s_O(d_O) \cdot U)$
The state transitions of the Resource Proxy module described by APTC are as follows:
$P = \sum_{d_{I_P} \in \Delta}(r_{I_{UP}}(d_{I_P}) \cdot P_2)$
$P_2 = PF_1 \cdot P_3$
$P_3 = \sum_{d_{I_{RP}} \in \Delta}(s_{I_{PP}}(d_{I_{RP}}) \cdot P_4)$
$P_4 = \sum_{d_{O_{RP}} \in \Delta}(r_{O_{PP}}(d_{O_{RP}}) \cdot P_5)$
$P_5 = PF_2 \cdot P_6$
$P_6 = \sum_{d_{I_R} \in \Delta}(s_{I_{PR}}(d_{I_R}) \cdot P_7)$
$P_7 = \sum_{d_{O_R} \in \Delta}(r_{O_{PR}}(d_{O_R}) \cdot P_8)$
$P_8 = PF_3 \cdot P_9$
$P_9 = \sum_{d_{O_P} \in \Delta}(s_{O_{UP}}(d_{O_P}) \cdot P)$
The state transitions of the Resource Provider module described by APTC are as follows:
$RP = \sum_{d_{I_{RP}} \in \Delta}(r_{I_{PP}}(d_{I_{RP}}) \cdot RP_2)$
$RP_2 = RPF \cdot RP_3$
$RP_3 = \sum_{d_{O_{RP}} \in \Delta}(s_{O_{PP}}(d_{O_{RP}}) \cdot RP)$
The state transitions of the Resource module described by APTC are as follows:
$R = \sum_{d_{I_R} \in \Delta}(r_{I_{PR}}(d_{I_R}) \cdot R_2)$
$R_2 = RF \cdot R_3$
$R_3 = \sum_{d_{O_R} \in \Delta}(s_{O_{PR}}(d_{O_R}) \cdot R)$

The sending action and the reading action of the same data through the same channel can communicate with each other, otherwise, they will cause a deadlock $\delta$. We define the following communication functions between the Resource User and the Resource Proxy:

$$\gamma(r_{I_{UP}}(d_{I_P}), s_{I_{UP}}(d_{I_P})) \triangleq c_{I_{UP}}(d_{I_P})$$
$$\gamma(r_{O_{UP}}(d_{O_P}), s_{O_{UP}}(d_{O_P})) \triangleq c_{O_{UP}}(d_{O_P})$$

There are two communication functions between the Resource Provider and the Resource Proxy as follows:

$$\gamma(r_{I_{PP}}(d_{I_{RP}}), s_{I_{PP}}(d_{I_{RP}})) \triangleq c_{I_{PP}}(d_{I_{RP}})$$
$$\gamma(r_{O_{PP}}(d_{O_{RP}}), s_{O_{PP}}(d_{O_{RP}})) \triangleq c_{O_{PP}}(d_{O_{RP}})$$

There are two communication functions between the Resource Proxy and the Resource as follows:

$$\gamma(r_{I_{PR}}(d_{I_R}), s_{I_{PR}}(d_{I_R})) \triangleq c_{I_{PR}}(d_{I_R})$$
$$\gamma(r_{O_{PR}}(d_{O_R}), s_{O_{PR}}(d_{O_R})) \triangleq c_{O_{PR}}(d_{O_R})$$

Let all modules be in parallel, then the Eager Acquisition pattern $U \quad P \quad RP \quad R$ can be presented by the following process term:

$\tau_I(\partial_H(\Theta(U \between P \between RP \between R))) = \tau_I(\partial_H(U \between P \between RP \between R))$
where $H = \{r_{I_{UP}}(d_{I_P}), s_{I_{UP}}(d_{I_P}), r_{O_{UP}}(d_{O_P}), s_{O_{UP}}(d_{O_P}), r_{I_{PP}}(d_{I_{RP}}), s_{I_{PP}}(d_{I_{RP}}),$
$r_{O_{PP}}(d_{O_{RP}}), s_{O_{PP}}(d_{O_{RP}}), r_{I_{PR}}(d_{I_R}), s_{I_{PR}}(d_{I_R}), r_{O_{PR}}(d_{O_R}), s_{O_{PR}}(d_{O_R})$
$|d_I, d_{I_P}, d_{I_{RP}}, d_{I_R}, d_{O_P}, d_{O_{RP}}, d_{O_R}, d_O \in \Delta\}$,
$I = \{c_{I_{UP}}(d_{I_P}), c_{O_{UP}}(d_{O_P}), c_{I_{PP}}(d_{I_{RP}}), c_{O_{PP}}(d_{O_{RP}}), c_{I_{PR}}(d_{I_R}), c_{O_{PR}}(d_{O_R}),$
$UF_1, UF_2, PF_1, PF_2, PF_3, RPF, RF|d_I, d_{I_P}, d_{I_{RP}}, d_{I_R}, d_{O_P}, d_{O_{RP}}, d_{O_R}, d_O \in \Delta\}$

Then, we obtain the following conclusion on the Eager Acquisition pattern.

**Theorem 37.3** (Correctness of the Eager Acquisition pattern). *The Eager Acquisition pattern* $\tau_I(\partial_H(U \between P \between RP \between R))$ *can exhibit the desired external behaviors.*

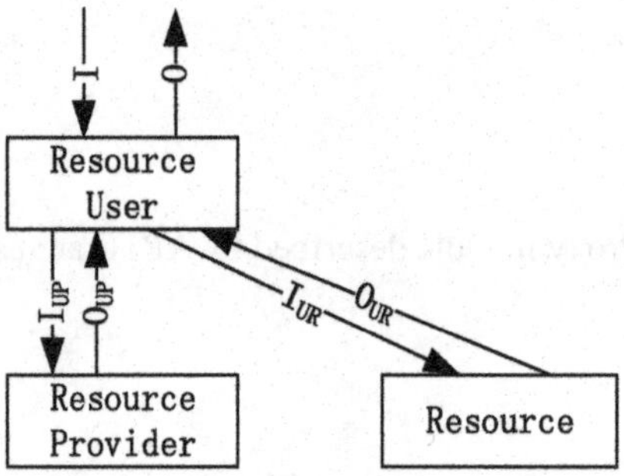


FIGURE 37.7 Partial Acquisition pattern.

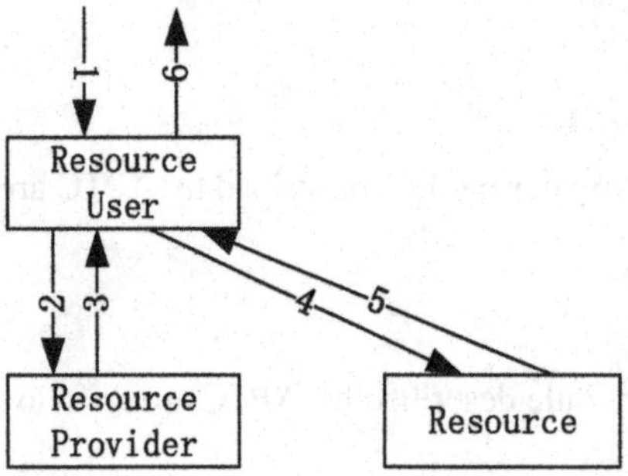


FIGURE 37.8 Typical process of Partial Acquisition pattern.

*Proof.* Based on the above state transitions of the above modules, by use of the algebraic laws of APTC, we can prove that
$\tau_I(\partial_H(U \between P \between RP \between R)) = \sum_{d_I, d_O \in \Delta}(r_I(d_I) \cdot s_O(d_O)) \cdot \tau_I(\partial_H(U \between P \between RP \between R))$
That is, the Eager Acquisition pattern $\tau_I(\partial_H(U \between P \between RP \between R))$ can exhibit the desired external behaviors.
For the details of the proof, please refer to Section 17.3, as we omit them here. □

### 37.1.4 Verification of the Partial Acquisition pattern

The Partial Acquisition pattern partially acquires the resources into multi-stages to optimize resource management. There are three modules in the Partial Acquisition pattern: the Resource User, the Resource Provider, and the Resource. The Resource User interacts with the outside through the channels $I$ and $O$; with the Resource Provider through the channels $I_{UP}$ and $O_{UP}$, and with the Resource through the channels $I_{UR}$ and $O_{UR}$. This is illustrated in Fig. 37.7.

The typical process of the Partial Acquisition pattern is shown in Fig. 37.8 and is as follows:

1. The Resource User receives the input $d_I$ from the outside through the channel $I$ (the corresponding reading action is denoted $r_I(d_I)$), then processes the input $d_I$ through a processing function $UF_1$, and generates the input $d_{I_P}$, and sends the input to the Resource Provider through the channel $I_{UP}$ (the corresponding sending action is denoted $s_{I_{UP}}(d_{I_P})$);
2. The Resource Provider receives the input from the Resource User through the channel $I_{UP}$ (the corresponding reading action is denoted $r_{I_{UP}}(d_{I_P})$), then processes the input and generates the output $d_{O_P}$ to the Resource User through a processing function $PF$, and sends the output to the Resource User through the channel $O_{UP}$ (the corresponding sending action is denoted $s_{O_{UP}}(d_{O_P})$);
3. The Resource User receives the output $d_{O_P}$ from the Resource Provider through the channel $O_{UP}$ (the corresponding reading action is denoted $r_{O_{UP}}(d_{O_P})$), then processes the output through a processing function $UF_2$, generates and sends the input $d_{I_R}$ to the Resource through the channel $I_{UR}$ (the corresponding sending action is denoted $s_{I_{UR}}(d_{I_R})$);
4. The Resource receives the input $d_{I_R}$ from the Resource User through the channel $I_{UR}$ (the corresponding reading action is denoted $r_{I_{UR}}(d_{I_R})$), then processes the input through a processing function $RF$, generates and sends the response $d_{O_R}$ (the corresponding sending action is denoted $s_{O_{UR}}(d_{O_R})$);
5. The Resource User receives the response $d_{O_R}$ from the Resource through the channel $O_{UR}$ (the corresponding reading action is denoted $r_{O_{UR}}(d_{O_R})$), then processes the response and generates the response $d_O$ through a processing function $UF_3$, and sends the response to the outside through the channel $O$ (the corresponding sending action is denoted $s_O(d_O)$).

In the following, we verify the Partial Acquisition pattern. We assume all data elements $d_I, d_{I_P}, d_{I_R}, d_{O_P}, d_{O_R}, d_O$ are from a finite set $\Delta$.

The state transitions of the Resource User module described by APTC are as follows:

$U = \sum_{d_I\in\Delta}(r_I(d_I)\cdot U_2)$

$U_2 = UF_1 \cdot U_3$

$U_3 = \sum_{d_{I_P}\in\Delta}(s_{IUP}(d_{I_P})\cdot U_4)$

$U_4 = \sum_{d_{O_P}\in\Delta}(r_{OUP}(d_{O_P})\cdot U_5)$

$U_5 = UF_2 \cdot U_6$

$U_6 = \sum_{d_{I_R}\in\Delta}(s_{IUR}(d_{I_R})\cdot U_7)$

$U_7 = \sum_{d_{O_R}\in\Delta}(r_{OUR}(d_{O_R})\cdot U_8)$

$U_8 = UF_3 \cdot U_9$

$U_9 = \sum_{d_O\in\Delta}(s_O(d_O)\cdot U)$

The state transitions of the Resource Provider module described by APTC are as follows:

$P = \sum_{d_{I_P}\in\Delta}(r_{IUP}(d_{I_P})\cdot P_2)$

$P_2 = PF \cdot P_3$

$P_3 = \sum_{d_{O_P}\in\Delta}(s_{OUP}(d_{O_P})\cdot P)$

The state transitions of the Resource module described by APTC are as follows:

$R = \sum_{d_{I_R}\in\Delta}(r_{IUR}(d_{I_R})\cdot R_2)$

$R_2 = RF \cdot R_3$

$R_3 = \sum_{d_{O_R}\in\Delta}(s_{OUR}(d_{O_R})\cdot R)$

The sending action and the reading action of the same data through the same channel can communicate with each other, otherwise, they will cause a deadlock $\delta$. We define the following communication functions between the Resource User and the Resource Provider:

$$\gamma(r_{IUP}(d_{I_P}), s_{IUP}(d_{I_P})) \triangleq c_{IUP}(d_{I_P})$$

$$\gamma(r_{OUP}(d_{O_P}), s_{OUP}(d_{O_P})) \triangleq c_{OUP}(d_{O_P})$$

There are two communication functions between the Resource User and the Resource as follows:

$$\gamma(r_{IUR}(d_{I_R}), s_{IUR}(d_{I_R})) \triangleq c_{IUR}(d_{I_R})$$

$$\gamma(r_{OUR}(d_{O_R}), s_{OUR}(d_{O_R})) \triangleq c_{OUR}(d_{O_R})$$

Let all modules be in parallel, then the Partial Acquisition pattern $U \quad P \quad R$ can be presented by the following process term:

$\tau_I(\partial_H(\Theta(U \between P \between R))) = \tau_I(\partial_H(U \between P \between R))$

where $H = \{r_{IUP}(d_{I_P}), s_{IUP}(d_{I_P}), r_{OUP}(d_{O_P}), s_{OUP}(d_{O_P}), r_{IUR}(d_{I_R}), s_{IUR}(d_{I_R}),$
$r_{OUR}(d_{O_R}), s_{OUR}(d_{O_R})|d_I, d_{I_P}, d_{I_R}, d_{O_P}, d_{O_R}, d_O \in \Delta\}$,

$I = \{c_{IUP}(d_{I_P}), c_{OUP}(d_{O_P}), c_{IUR}(d_{I_R}), c_{OUR}(d_{O_R}),$
$UF_1, UF_2, UF_3, PF, RF|d_I, d_{I_P}, d_{I_R}, d_{O_P}, d_{O_R}, d_O \in \Delta\}$

Then, we obtain the following conclusion on the Partial Acquisition pattern.

**Theorem 37.4** (Correctness of the Partial Acquisition pattern). *The Partial Acquisition pattern $\tau_I(\partial_H(U \between P \between R))$ can exhibit the desired external behaviors.*

*Proof.* Based on the above state transitions of the above modules, by use of the algebraic laws of APTC, we can prove that
$\tau_I(\partial_H(U \between P \between R)) = \sum_{d_I,d_O\in\Delta}(r_I(d_I)\cdot s_O(d_O))\cdot \tau_I(\partial_H(U \between P \between R))$

That is, the Partial Acquisition pattern $\tau_I(\partial_H(U \between P \between R))$ can exhibit the desired external behaviors.

For the details of the proof, please refer to Section 17.3, as we omit them here. □

## 37.2 Resource Life cycle

In this section, we verify patterns related to resource lifecycle, including the Caching pattern, the Pooling pattern, the Coordinator pattern, and the Resource Life cycle Manager pattern.

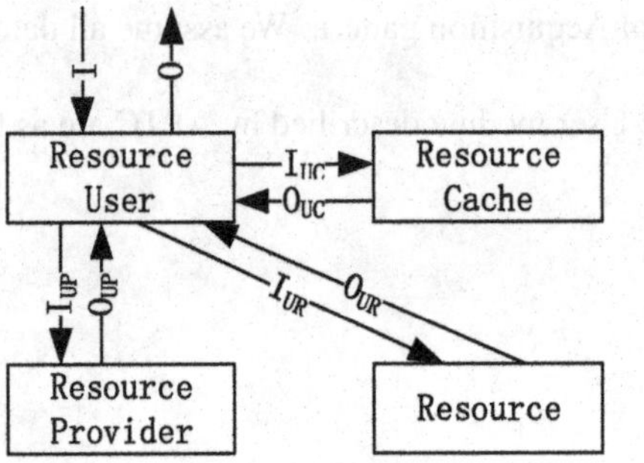


FIGURE 37.9 Caching pattern.

### 37.2.1 Verification of the Caching pattern

The Caching pattern allows us to cache the resources to avoid re-acquisitions of the resources. There are four modules in the Caching pattern: the Resource User, the Resource Provider, the Resource Cache, and the Resource. The Resource User interacts with the outside through the channels $I$ and $O$; with the Resource Provider through the channels $I_{UP}$ and $O_{UP}$; with the Resource through the channels $I_{UR}$ and $O_{UR}$, and with the Resource Cache through the channels $I_{UC}$ and $O_{UC}$. This is illustrated in Fig. 37.9.

The typical process of the Caching pattern is shown in Fig. 37.10 and is as follows:

1. The Resource User receives the input $d_I$ from the outside through the channel $I$ (the corresponding reading action is denoted $r_I(d_I)$), then processes the input and generates the input $d_{I_P}$ through a processing function $UF_1$, and sends the input to the Resource Provider through the channel $I_{UP}$ (the corresponding sending action is denoted $s_{I_{UP}}(d_{I_P})$);
2. The Resource Provider receives the input from the Resource User through the channel $I_{UP}$ (the corresponding reading action is denoted $r_{I_{UP}}(d_{I_P})$), then processes the input and generates the output $d_{O_P}$ to the Resource User through a processing function $PF$, and sends the output to the Resource User through the channel $O_{UP}$ (the corresponding sending action is denoted $s_{O_{UP}}(d_{O_P})$);
3. The Resource User receives the output $d_{O_P}$ from the Resource Provider through the channel $O_{UP}$ (the corresponding reading action is denoted $r_{O_{UP}}(d_{O_P})$), then processes the input through a processing function $UF_2$, generates and sends the input $d_{I_R}$ to the Resource through the channel $I_{UR}$ (the corresponding sending action is denoted $s_{I_{UR}}(d_{I_R})$);
4. The Resource receives the input $d_{I_R}$ from the Resource User through the channel $I_{UR}$ (the corresponding reading action is denoted $r_{I_{UR}}(d_{I_R})$), then processes the input through a processing function $RF$, generates and sends the response $d_{O_R}$ (the corresponding sending action is denoted $s_{O_{UR}}(d_{O_R})$);
5. The Resource User receives the response $d_{O_R}$ from the Resource through the channel $O_{UR}$ (the corresponding reading action is denoted $r_{O_{UR}}(d_{O_R})$), then processes the input $d_{O_R}$ through a processing function $UF_3$, and sends the processed input $d_{I_C}$ to the Resource Cache through the channel $I_{UC}$ (the corresponding sending action is denoted $s_{I_{UC}}(d_{I_C})$);
6. The Resource Cache receives $d_{I_C}$ from the Resource User through the channel $I_{UC}$ (the corresponding reading action is denoted $r_{I_{UC}}(d_{I_C})$), then processes the request through a processing function $CF$, generates and sends the processed output $d_{O_C}$ to the Resource User through the channel $O_{UC}$ (the corresponding sending action is denoted $s_{O_{UC}}(d_{O_C})$);
7. The Resource User receives the output $d_{O_C}$ from the Resource Cache through the channel $O_{UC}$ (the corresponding reading action is denoted $r_{O_{UC}}(d_{O_C})$), then processes the response and generates the response $d_O$ through a processing function $UF_4$, and sends the response to the outside through the channel $O$ (the corresponding sending action is denoted $s_O(d_O)$).

In the following, we verify the Caching pattern. We assume all data elements $d_I, d_{I_C}, d_{I_P}, d_{I_R}, d_{O_C}, d_{O_P}, d_{O_R}, d_O$ are from a finite set $\Delta$.

The state transitions of the Resource User module described by APTC are as follows:

$U = \sum_{d_I\in\Delta}(r_I(d_I)\cdot U_2)$

$U_2 = UF_1\cdot U_3$

$U_3 = \sum_{d_{I_P}\in\Delta}(s_{I_{UP}}(d_{I_P})\cdot U_4)$

$U_4 = \sum_{d_{O_P}\in\Delta}(r_{O_{UP}}(d_{O_P})\cdot U_5)$

$U_5 = UF_2\cdot U_6$

$U_6 = \sum_{d_{I_R}\in\Delta}(s_{I_{UR}}(d_{I_R})\cdot U_7)$

$U_7 = \sum_{d_{O_R}\in\Delta}(r_{O_{UR}}(d_{O_R})\cdot U_8)$

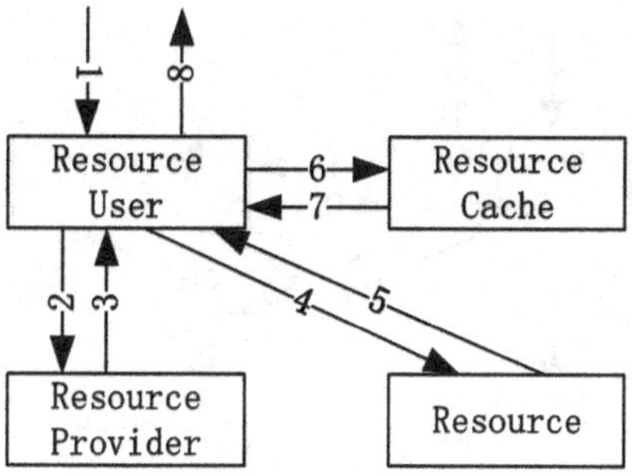


**FIGURE 37.10** Typical process of Caching pattern.

$U_8 = UF_3 \cdot U_9$
$U_9 = \sum_{d_{I_C} \in \Delta}(s_{IUC}(d_{I_C}) \cdot U_{10})$
$U_{10} = \sum_{d_{O_C} \in \Delta}(r_{OUC}(d_{O_C}) \cdot U_{11})$
$U_{11} = UF_4 \cdot U_{12}$
$U_{12} = \sum_{d_O \in \Delta}(s_O(d_O) \cdot U)$
The state transitions of the Resource Provider module described by APTC are as follows:
$P = \sum_{d_{I_P} \in \Delta}(r_{IUP}(d_{I_P}) \cdot P_2)$
$P_2 = PF \cdot P_3$
$P_3 = \sum_{d_{O_P} \in \Delta}(s_{OUP}(d_{O_P}) \cdot P)$
The state transitions of the Resource Cache module described by APTC are as follows:
$C = \sum_{d_{I_C} \in \Delta}(r_{IUC}(d_{I_C}) \cdot C_2)$
$C_2 = CF \cdot C_3$
$C_3 = \sum_{d_{O_C} \in \Delta}(s_{OUC}(d_{O_C}) \cdot C)$
The state transitions of the Resource module described by APTC are as follows:
$R = \sum_{d_{I_R} \in \Delta}(r_{IUR}(d_{I_R}) \cdot R_2)$
$R_2 = RF \cdot R_3$
$R_3 = \sum_{d_{O_R} \in \Delta}(s_{OUR}(d_{O_R}) \cdot R)$

The sending action and the reading action of the same data through the same channel can communicate with each other, otherwise, they will cause a deadlock $\delta$. We define the following communication functions between the Resource User and the Resource Provider:

$$\gamma(r_{IUP}(d_{I_P}), s_{IUP}(d_{I_P})) \triangleq c_{IUP}(d_{I_P})$$
$$\gamma(r_{OUP}(d_{O_P}), s_{OUP}(d_{O_P})) \triangleq c_{OUP}(d_{O_P})$$

There are two communication functions between the Resource User and the Resource Cache as follows:

$$\gamma(r_{IUC}(d_{I_C}), s_{IUC}(d_{I_C})) \triangleq c_{IUC}(d_{I_C})$$
$$\gamma(r_{OUC}(d_{O_C}), s_{OUC}(d_{O_C})) \triangleq c_{OUC}(d_{O_C})$$

There are two communication functions between the Resource User and the Resource as follows:

$$\gamma(r_{IUR}(d_{I_R}), s_{IUR}(d_{I_R})) \triangleq c_{IUR}(d_{I_R})$$
$$\gamma(r_{OUR}(d_{O_R}), s_{OUR}(d_{O_R})) \triangleq c_{OUR}(d_{O_R})$$

Let all modules be in parallel, then the Caching pattern $U \quad C \quad P \quad R$ can be presented by the following process term:
$\tau_I(\partial_H(\Theta(U \between C \between P \between R))) = \tau_I(\partial_H(U \between C \between P \between R))$
where $H = \{r_{IUP}(d_{I_P}), s_{IUP}(d_{I_P}), r_{OUP}(d_{O_P}), s_{OUP}(d_{O_P}), r_{IUC}(d_{I_C}), s_{IUC}(d_{I_C}),$
$r_{OUC}(d_{O_C}), s_{OUC}(d_{O_C}), r_{IUR}(d_{I_R}), s_{IUR}(d_{I_R}), r_{OUR}(d_{O_R}), s_{OUR}(d_{O_R})$
$|d_I, d_{I_P}, d_{I_C}, d_{I_R}, d_{O_P}, d_{O_C}, d_{O_R}, d_O \in \Delta\},$

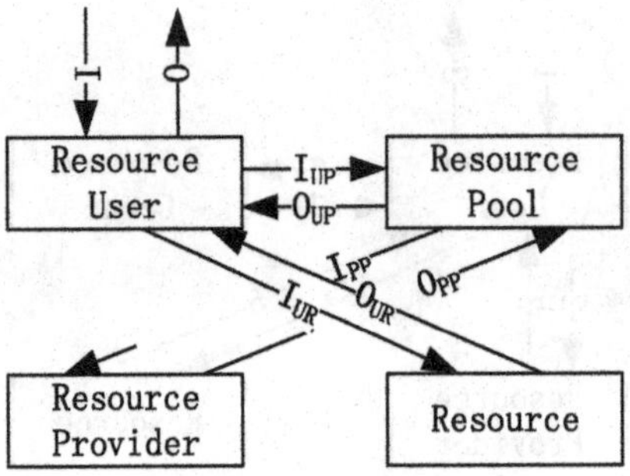


**FIGURE 37.11** Pooling pattern.

$I = \{c_{I_{UP}}(d_{I_P}), c_{O_{UP}}(d_{O_P}), c_{I_{UC}}(d_{I_C}), c_{O_{UC}}(d_{O_C}), c_{I_{UR}}(d_{I_R}), c_{O_{UR}}(d_{O_R}),$
$UF_1, UF_2, UF_3, UF_4, PF, CF, RF | d_I, d_{I_P}, d_{I_C}, d_{I_R}, d_{O_P}, d_{O_C}, d_{O_R}, d_O \in \Delta\}$

Then, we obtain the following conclusion on the Caching pattern.

**Theorem 37.5** (Correctness of the Caching pattern). *The Caching pattern $\tau_I(\partial_H(U \between C \between P \between R))$ can exhibit the desired external behaviors.*

*Proof.* Based on the above state transitions of the above modules, by use of the algebraic laws of APTC, we can prove that
$\tau_I(\partial_H(U \between C \between P \between R)) = \sum_{d_I, d_O \in \Delta}(r_I(d_I) \cdot s_O(d_O)) \cdot \tau_I(\partial_H(U \between C \between P \between R))$

That is, the Caching pattern $\tau_I(\partial_H(U \between C \between P \between R))$ can exhibit the desired external behaviors.

For the details of the proof, please refer to Section 17.3, as we omit them here. □

### 37.2.2 Verification of the Pooling pattern

The Pooling pattern allows us to recycle the resources to avoid re-acquisitions of the resources. There are four modules in the Pooling pattern: the Resource User, the Resource Provider, the Resource Pool, and the Resource. The Resource User interacts with the outside through the channels $I$ and $O$; with the Resource Provider through the channels $I_{UP}$ and $O_{UP}$, and with the Resource through the channels $I_{UR}$ and $O_{UR}$. The Resource Pool interacts with the Resource Provider through the channels $I_{PP}$ and $O_{PP}$. This is illustrated in Fig. 37.11.

The typical process of the Pooling pattern is shown in Fig. 37.12 and is as follows:

1. The Resource User receives the input $d_I$ from the outside through the channel $I$ (the corresponding reading action is denoted $r_I(d_I)$), then processes the input and generates the input $d_{I_P}$ through a processing function $UF_1$, and sends the input to the Resource Pool through the channel $I_{UP}$ (the corresponding sending action is denoted $s_{I_{UP}}(d_{I_P})$);
2. The Resource Pool receives the input from the Resource User through the channel $I_{UP}$ (the corresponding reading action is denoted $r_{I_{UP}}(d_{I_P})$), then processes the input and generates the input $d_{I_{RP}}$ to the Resource Provider through a processing function $PF_1$, and sends the input to the Resource Provider through the channel $I_{PP}$ (the corresponding sending action is denoted $s_{I_{PP}}(d_{I_{RP}})$);
3. The Resource Provider receives the output $d_{I_{RP}}$ from the Resource Pool through the channel $I_{PP}$ (the corresponding reading action is denoted $r_{I_{PP}}(d_{I_{RP}})$), then processes the input through a processing function $RPF$, generates and sends the output $d_{O_{RP}}$ to the Resource Pool through the channel $O_{PP}$ (the corresponding sending action is denoted $s_{O_{PP}}(d_{O_{RP}})$);
4. The Resource Pool receives the input $d_{O_{RP}}$ from the Resource Provider through the channel $O_{PP}$ (the corresponding reading action is denoted $r_{O_{PP}}(d_{O_{RP}})$), then processes the output through a processing function $PF_2$, generates and sends the response $d_{O_P}$ (the corresponding sending action is denoted $s_{O_{UP}}(d_{O_P})$);
5. The Resource User receives the response $d_{O_P}$ from the Resource Pool through the channel $O_{UP}$ (the corresponding reading action is denoted $r_{O_{UP}}(d_{O_P})$), then processes the output $d_{O_P}$ through a processing function $UF_2$, and sends the processed input $d_{I_R}$ to the Resource through the channel $I_{UR}$ (the corresponding sending action is denoted $s_{I_{UR}}(d_{I_R})$);
6. The Resource receives $d_{I_R}$ from the Resource User through the channel $I_{UR}$ (the corresponding reading action is denoted $r_{I_{UR}}(d_{I_R})$), then processes the request through a processing function $RF$, generates and sends the processed output $d_{O_R}$ to the Resource User through the channel $O_{UR}$ (the corresponding sending action is denoted $s_{O_{UR}}(d_{O_R})$);
7. The Resource User receives the output $d_{O_R}$ from the Resource through the channel $O_{UR}$ (the corresponding reading action is denoted $r_{O_{UR}}(d_{O_R})$), then processes the response and generates the response $d_O$ through a processing function $UF_3$, and sends the response to the outside through the channel $O$ (the corresponding sending action is denoted $s_O(d_O)$).

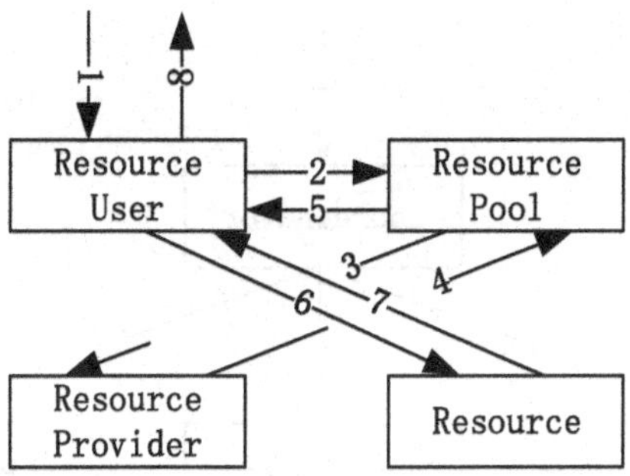


**FIGURE 37.12** Typical process of Pooling pattern.

In the following, we verify the Pooling pattern. We assume all data elements $d_I, d_{I_P}, d_{I_{RP}}, d_{I_R}, d_{O_{RP}}, d_{O_P}, d_{O_R}, d_O$ are from a finite set $\Delta$.

The state transitions of the Resource User module described by APTC are as follows:

$U = \sum_{d_I\in\Delta}(r_I(d_I) \cdot U_2)$

$U_2 = UF_1 \cdot U_3$

$U_3 = \sum_{d_{I_P}\in\Delta}(s_{IUP}(d_{I_P}) \cdot U_4)$

$U_4 = \sum_{d_{O_P}\in\Delta}(r_{OUP}(d_{O_P}) \cdot U_5)$

$U_5 = UF_2 \cdot U_6$

$U_6 = \sum_{d_{I_R}\in\Delta}(s_{IUR}(d_{I_R}) \cdot U_7)$

$U_7 = \sum_{d_{O_R}\in\Delta}(r_{OUR}(d_{O_R}) \cdot U_8)$

$U_8 = UF_3 \cdot U_9$

$U_9 = \sum_{d_O\in\Delta}(s_O(d_O) \cdot U)$

The state transitions of the Resource Provider module described by APTC are as follows:

$RP = \sum_{d_{I_{RP}}\in\Delta}(r_{IPP}(d_{I_{RP}}) \cdot RP_2)$

$RP_2 = RPF \cdot RP_3$

$RP_3 = \sum_{d_{O_{RP}}\in\Delta}(s_{OPP}(d_{O_{RP}}) \cdot RP)$

The state transitions of the Resource Pool module described by APTC are as follows:

$P = \sum_{d_{I_P}\in\Delta}(r_{IUP}(d_{I_P}) \cdot P_2)$

$P_2 = PF_1 \cdot P_3$

$P_3 = \sum_{d_{I_{RP}}\in\Delta}(s_{IPP}(d_{I_{RP}}) \cdot P_4)$

$P_4 = \sum_{d_{O_{RP}}\in\Delta}(r_{OPP}(d_{O_{RP}}) \cdot P_5)$

$P_5 = PF_2 \cdot P_6$

$P_6 = \sum_{d_{O_P}\in\Delta}(s_{OUP}(d_{O_P}) \cdot P)$

The state transitions of the Resource module described by APTC are as follows:

$R = \sum_{d_{I_R}\in\Delta}(r_{IUR}(d_{I_R}) \cdot R_2)$

$R_2 = RF \cdot R_3$

$R_3 = \sum_{d_{O_R}\in\Delta}(s_{OUR}(d_{O_R}) \cdot R)$

The sending action and the reading action of the same data through the same channel can communicate with each other, otherwise, they will cause a deadlock $\delta$. We define the following communication functions between the Resource User and the Resource Pool:

$$\gamma(r_{IUP}(d_{I_P}), s_{IUP}(d_{I_P})) \triangleq c_{IUP}(d_{I_P})$$
$$\gamma(r_{OUP}(d_{O_P}), s_{OUP}(d_{O_P})) \triangleq c_{OUP}(d_{O_P})$$

There are two communication functions between the Resource Provider and the Resource Pool as follows:

$$\gamma(r_{IPP}(d_{I_{RP}}), s_{IPP}(d_{I_{RP}})) \triangleq c_{IPP}(d_{I_{RP}})$$
$$\gamma(r_{OPP}(d_{O_{RP}}), s_{OPP}(d_{O_{RP}})) \triangleq c_{OPP}(d_{O_{RP}})$$

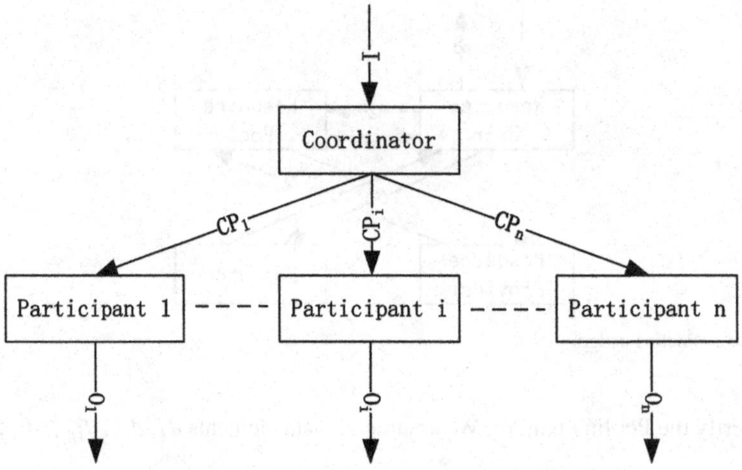


**FIGURE 37.13** Coordinator pattern.

There are two communication functions between the Resource User and the Resource as follows:

$$\gamma(r_{IUR}(d_{IR}), s_{IUR}(d_{IR})) \triangleq c_{IUR}(d_{IR})$$
$$\gamma(r_{OUR}(d_{OR}), s_{OUR}(d_{OR})) \triangleq c_{OUR}(d_{OR})$$

Let all modules be in parallel, then the Pooling pattern $U \quad RP \quad P \quad R$ can be presented by the following process term:
$\tau_I(\partial_H(\Theta(U \between RP \between P \between R))) = \tau_I(\partial_H(U \between RP \between P \between R))$
where $H = \{r_{IUP}(d_{IP}), s_{IUP}(d_{IP}), r_{OUP}(d_{OP}), s_{OUP}(d_{OP}), r_{IPP}(d_{IRP}), s_{IPP}(d_{IRP}),$
$r_{OPP}(d_{ORP}), s_{OPP}(d_{ORP}), r_{IUR}(d_{IR}), s_{IUR}(d_{IR}), r_{OUR}(d_{OR}), s_{OUR}(d_{OR})$
$|d_I, d_{IP}, d_{IRP}, d_{IR}, d_{OP}, d_{ORP}, d_{OR}, d_O \in \Delta\}$,
$I = \{c_{IUP}(d_{IP}), c_{OUP}(d_{OP}), c_{IPP}(d_{IRP}), c_{OPP}(d_{ORP}), c_{IUR}(d_{IR}), c_{OUR}(d_{OR}),$
$UF_1, UF_2, UF_3, PF_1, PF_2, RPF, RF|d_I, d_{IP}, d_{IRP}, d_{IR}, d_{OP}, d_{ORP}, d_{OR}, d_O \in \Delta\}$
Then, we obtain the following conclusion on the Pooling pattern.

**Theorem 37.6** (Correctness of the Pooling pattern). *The Pooling pattern $\tau_I(\partial_H(U \between RP \between P \between R))$ can exhibit the desired external behaviors.*

*Proof.* Based on the above state transitions of the above modules, by use of the algebraic laws of APTC, we can prove that
$\tau_I(\partial_H(U \between RP \between P \between R)) = \sum_{d_I, d_O \in \Delta}(r_I(d_I) \cdot s_O(d_O)) \cdot \tau_I(\partial_H(U \between RP \between P \between R))$
That is, the Pooling pattern $\tau_I(\partial_H(U \between RP \between P \between R))$ can exhibit the desired external behaviors.
For the details of the proof, please refer to Section 17.3, as we omit them here. □

### 37.2.3 Verification of the Coordinator pattern

The Coordinator pattern gives a solution to maintain the consistency by coordinating the completion of tasks involving multiple participants, which has two classes of components: $n$ Synchronous Services and the Coordinator. The Coordinator receives the inputs from the user through the channel $I$, then the Coordinator sends the results to the Participant $i$ through the channels $CP_i$ for $1 \leq i \leq n$; When the Participant $i$ receives the input from the Coordinator, it generates and sends the results out to the user through the channel $O_i$. This is illustrated in Fig. 37.13.

The typical process of the Coordinator pattern is shown in Fig. 37.14 and is the following:

1. The Coordinator receives the input $d_I$ from the user through the channel $I$ (the corresponding reading action is denoted $r_I(D_I)$), processes the input through a processing function $CF$, and generates the input to the Participant $i$ (for $1 \leq i \leq n$) that is denoted $d_{IP_i}$; then sends the input to the Participant $i$ through the channel $CP_i$ (the corresponding sending action is denoted $s_{CP_i}(d_{IP_i})$);
2. The Participant $i$ (for $1 \leq i \leq n$) receives the input from the Coordinator through the channel $CP_i$ (the corresponding reading action is denoted $r_{CP_i}(d_{IP_i})$), processes the results through a processing function $PF_i$, generates the output $d_{O_i}$, then sends the output through the channel $O_i$ (the corresponding sending action is denoted $s_{O_i}(d_{O_i})$).

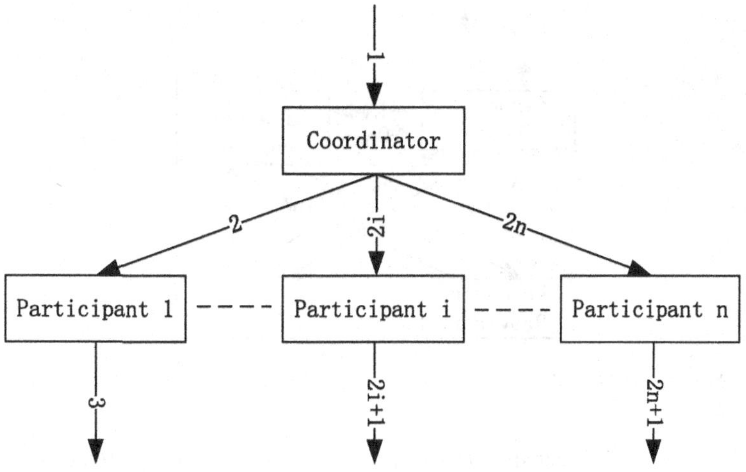


**FIGURE 37.14** Typical process of Coordinator pattern.

In the following, we verify the Coordinator pattern. We assume all data elements $d_I, d_{IP_i}, d_{O_i}$ (for $1 \leq i \leq n$) are from a finite set $\Delta$.

The state transitions of the Coordinator module described by APTC are as follows:

$C = \sum_{d_I \in \Delta}(r_I(d_I) \cdot C_2)$

$C_2 = CF \cdot C_3$

$C_3 = \sum_{d_{IP_1},\cdots,d_{IP_n} \in \Delta}(s_{CP_1}(d_{IP_1}) \between \cdots \between s_{CP_n}(d_{IP_n}) \cdot C)$

The state transitions of the Participant $i$ described by APTC are as follows:

$P_i = \sum_{d_{IP_i} \in \Delta}(r_{CP_i}(d_{IP_i}) \cdot P_{i_2})$

$P_{i_2} = PF_i \cdot P_{i_3}$

$P_{i_3} = \sum_{d_{O_i} \in \Delta}(s_{O_i}(d_{O_i}) \cdot P_i)$

The sending action and the reading action of the same data through the same channel can communicate with each other, otherwise, they will cause a deadlock $\delta$. We define the following communication function of the Participant $i$ for $1 \leq i \leq n$:

$$\gamma(r_{CP_i}(d_{IP_i}), s_{CP_i}(d_{IP_i})) \triangleq c_{CP_i}(d_{IP_i})$$

Let all modules be in parallel, then the Coordinator pattern $C \quad P_1 \cdots P_i \cdots P_n$ can be presented by the following process term:

$\tau_I(\partial_H(\Theta(C \between P_1 \between \cdots \between P_i \between \cdots \between P_n))) = \tau_I(\partial_H(C \between P_1 \between \cdots \between P_i \between \cdots \between P_n))$

where $H = \{r_{CP_i}(d_{OP_i}), s_{CP_i}(d_{OP_i}) | d_I, d_{IP_i}, d_{O_i} \in \Delta\}$ for $1 \leq i \leq n$,

$I = \{c_{CP_i}(d_{IP_i}), CF, PF_i | d_I, d_{IP_i}, d_{O_i} \in \Delta\}$ for $1 \leq i \leq n$

Then, we obtain the following conclusion on the Coordinator pattern.

**Theorem 37.7** (Correctness of the Coordinator pattern). *The Coordinator pattern $\tau_I(\partial_H(C \between P_1 \between \cdots \between P_i \between \cdots \between P_n))$ can exhibit the desired external behaviors.*

*Proof.* Based on the above state transitions of the above modules, by use of the algebraic laws of APTC, we can prove that

$\tau_I(\partial_H(C \between P_1 \between \cdots \between P_i \between \cdots \between P_n)) = \sum_{d_I, d_{O_1}, \cdots, d_{O_n} \in \Delta}(r_I(d_I) \cdot s_{O_1}(d_{O_1}) \parallel \cdots \parallel s_{O_i}(d_{O_i}) \parallel \cdots \parallel s_{O_n}(d_{O_n})) \cdot \tau_I(\partial_H(C \between P_1 \between \cdots \between P_i \between \cdots \between P_n))$

That is, the Coordinator pattern $\tau_I(\partial_H(A \between S_1 \between \cdots \between S_i \between \cdots \between S_n))$ can exhibit the desired external behaviors.

For the details of the proof, please refer to Section 17.3, as we omit them here. □

### 37.2.4 Verification of the Resource Life cycle Manager pattern

The Resource Life cycle Manager pattern decouples the life cycle management by introducing a Resource Life cycle Manager. There are four modules in the Resource Life cycle Manager pattern: the Resource User, the Resource Provider, the Resource Life cycle Manager, and the Resource. The Resource User interacts with the outside through the channels $I$ and $O$; with the Resource Provider through the channels $I_{UM}$ and $O_{UM}$, and with the Resource through the channels $I_{UR}$

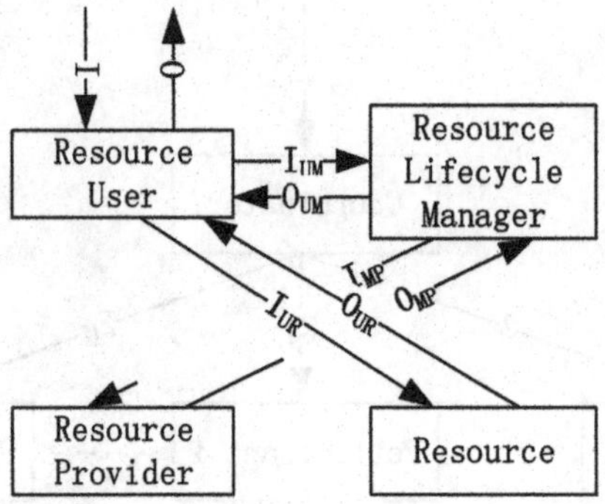


FIGURE 37.15 Resource Life cycle Manager pattern.

and $O_{UR}$. The Resource Life cycle Manager interacts with the Resource Provider through the channels $I_{MP}$ and $O_{MP}$. This is illustrated in Fig. 37.15.

The typical process of the Resource Life cycle Manager pattern is shown in Fig. 37.16 and is as follows:

1. The Resource User receives the input $d_I$ from the outside through the channel $I$ (the corresponding reading action is denoted $r_I(d_I)$), then processes the input and generates the input $d_{I_M}$ through a processing function $UF_1$, and sends the input to the Resource Life cycle Manager through the channel $I_{UM}$ (the corresponding sending action is denoted $s_{I_{UM}}(d_{I_M})$);
2. The Resource Life cycle Manager receives the input from the Resource User through the channel $I_{UM}$ (the corresponding reading action is denoted $r_{I_{UM}}(d_{I_M})$), then processes the input and generates the input $d_{I_P}$ to the Resource Provider through a processing function $MF_1$, and sends the input to the Resource Provider through the channel $I_{MP}$ (the corresponding sending action is denoted $s_{I_{MP}}(d_{I_P})$);
3. The Resource Provider receives the output $d_{I_P}$ from the Resource Life Cycle Manager through the channel $I_{MP}$ (the corresponding reading action is denoted $r_{I_{MP}}(d_{I_P})$), then processes the input through a processing function $PF$, generates and sends the output $d_{O_P}$ to the Resource Life Cycle Manager through the channel $O_{MP}$ (the corresponding sending action is denoted $s_{O_{MP}}(d_{O_P})$);
4. The Resource Life Cycle Manager receives the input $d_{O_P}$ from the Resource Provider through the channel $O_{MP}$ (the corresponding reading action is denoted $r_{O_{MP}}(d_{O_P})$), then processes the output through a processing function $MF_2$, generates and sends the response $d_{O_M}$ (the corresponding sending action is denoted $s_{O_{UM}}(d_{O_M})$);
5. The Resource User receives the response $d_{O_M}$ from the Resource Life Cycle Manager through the channel $O_{UM}$ (the corresponding reading action is denoted $r_{O_{UM}}(d_{O_M})$), then processes the output $d_{O_M}$ through a processing function $UF_2$, and sends the processed input $d_{I_R}$ to the Resource through the channel $I_{UR}$ (the corresponding sending action is denoted $s_{I_{UR}}(d_{I_R})$);
6. The Resource receives $d_{I_R}$ from the Resource User through the channel $I_{UR}$ (the corresponding reading action is denoted $r_{I_{UR}}(d_{I_R})$), then processes the request through a processing function $RF$, generates and sends the processed output $d_{O_R}$ to the Resource User through the channel $O_{UR}$ (the corresponding sending action is denoted $s_{O_{UR}}(d_{O_R})$);
7. The Resource User receives the output $d_{O_R}$ from the Resource through the channel $O_{UR}$ (the corresponding reading action is denoted $r_{O_{UR}}(d_{O_R})$), then processes the response and generates the response $d_O$ through a processing function $UF_3$, and sends the response to the outside through the channel $O$ (the corresponding sending action is denoted $s_O(d_O)$).

In the following, we verify the Resource Life Cycle Manager pattern. We assume all data elements $d_I$, $d_{I_P}$, $d_{I_M}$, $d_{I_R}$, $d_{O_M}$, $d_{O_P}$, $d_{O_R}$, $d_O$ are from a finite set $\Delta$.

The state transitions of the Resource User module described by APTC are as follows:

$U = \sum_{d_I\in\Delta}(r_I(d_I)\cdot U_2)$

$U_2 = UF_1\cdot U_3$

$U_3 = \sum_{d_{I_M}\in\Delta}(s_{I_{UM}}(d_{I_M})\cdot U_4)$

$U_4 = \sum_{d_{O_M}\in\Delta}(r_{O_{UM}}(d_{O_M})\cdot U_5)$

$U_5 = UF_2\cdot U_6$

$U_6 = \sum_{d_{I_R}\in\Delta}(s_{I_{UR}}(d_{I_R})\cdot U_7)$

$U_7 = \sum_{d_{O_R}\in\Delta}(r_{O_{UR}}(d_{O_R})\cdot U_8)$

$U_8 = UF_3\cdot U_9$

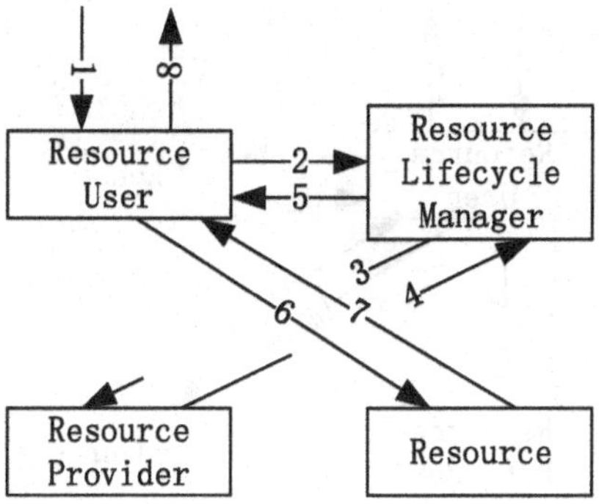


**FIGURE 37.16** Typical process of Resource Life Cycle Manager pattern.

$U_9 = \sum_{d_O \in \Delta}(s_O(d_O) \cdot U)$

The state transitions of the Resource Provider module described by APTC are as follows:

$P = \sum_{d_{I_P} \in \Delta}(r_{I_{MP}}(d_{I_P}) \cdot P_2)$

$P_2 = PF \cdot P_3$

$P_3 = \sum_{d_{O_P} \in \Delta}(s_{O_{MP}}(d_{O_P}) \cdot P)$

The state transitions of the Resource Life Cycle Manager module described by APTC are as follows:

$M = \sum_{d_{I_M} \in \Delta}(r_{I_{UM}}(d_{I_M}) \cdot M_2)$

$M_2 = MF_1 \cdot M_3$

$M_3 = \sum_{d_{I_P} \in \Delta}(s_{I_{MP}}(d_{I_P}) \cdot M_4)$

$M_4 = \sum_{d_{O_P} \in \Delta}(r_{O_{MP}}(d_{O_P}) \cdot M_5)$

$M_5 = MF_2 \cdot M_6$

$M_6 = \sum_{d_{O_M} \in \Delta}(s_{O_{UM}}(d_{O_M}) \cdot M)$

The state transitions of the Resource module described by APTC are as follows:

$R = \sum_{d_{I_R} \in \Delta}(r_{I_{UR}}(d_{I_R}) \cdot R_2)$

$R_2 = RF \cdot R_3$

$R_3 = \sum_{d_{O_R} \in \Delta}(s_{O_{UR}}(d_{O_R}) \cdot R)$

The sending action and the reading action of the same data through the same channel can communicate with each other, otherwise, they will cause a deadlock $\delta$. We define the following communication functions between the Resource User and the Resource Life Cycle Manager:

$$\gamma(r_{I_{UM}}(d_{I_M}), s_{I_{UM}}(d_{I_M})) \triangleq c_{I_{UM}}(d_{I_M})$$

$$\gamma(r_{O_{UM}}(d_{O_M}), s_{O_{UM}}(d_{O_M})) \triangleq c_{O_{UM}}(d_{O_M})$$

There are two communication functions between the Resource Provider and the Resource Life Cycle Manager as follows:

$$\gamma(r_{I_{MP}}(d_{I_P}), s_{I_{MP}}(d_{I_P})) \triangleq c_{I_{MP}}(d_{I_P})$$

$$\gamma(r_{O_{MP}}(d_{O_P}), s_{O_{MP}}(d_{O_P})) \triangleq c_{O_{MP}}(d_{O_P})$$

There are two communication functions between the Resource User and the Resource as follows:

$$\gamma(r_{I_{UR}}(d_{I_R}), s_{I_{UR}}(d_{I_R})) \triangleq c_{I_{UR}}(d_{I_R})$$

$$\gamma(r_{O_{UR}}(d_{O_R}), s_{O_{UR}}(d_{O_R})) \triangleq c_{O_{UR}}(d_{O_R})$$

Let all modules be in parallel, then the Resource Life Cycle Manager pattern $U \quad M \quad P \quad R$ can be presented by the following process term:

$\tau_I(\partial_H(\Theta(U \between M \between P \between R))) = \tau_I(\partial_H(U \between M \between P \between R))$

where $H = \{r_{I_{UM}}(d_{I_M}), s_{I_{UM}}(d_{I_M}), r_{O_{UM}}(d_{O_M}), s_{O_{UM}}(d_{O_M}), r_{I_{MP}}(d_{I_P}), s_{I_{MP}}(d_{I_P}),$
$r_{O_{MP}}(d_{O_P}), s_{O_{MP}}(d_{O_P}), r_{I_{UR}}(d_{I_R}), s_{I_{UR}}(d_{I_R}), r_{O_{UR}}(d_{O_R}), s_{O_{UR}}(d_{O_R})$
$|d_I, d_{I_P}, d_{I_M}, d_{I_R}, d_{O_P}, d_{O_M}, d_{O_R}, d_O \in \Delta\},$

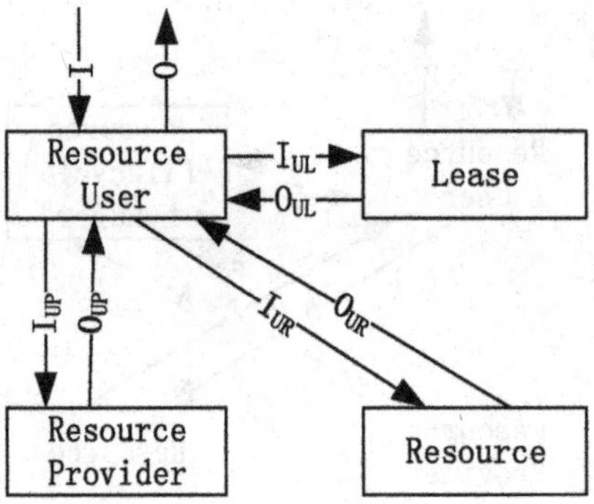


FIGURE 37.17 Leasing pattern.

$I = \{c_{I_{UM}}(d_{I_M}), c_{O_{UM}}(d_{O_M}), c_{I_{MP}}(d_{I_P}), c_{O_{MP}}(d_{O_P}), c_{I_{UR}}(d_{I_R}), c_{O_{UR}}(d_{O_R}),$
$UF_1, UF_2, UF_3, MF_1, MF_2, PF, RF | d_I, d_{I_P}, d_{I_M}, d_{I_R}, d_{O_P}, d_{O_M}, d_{O_R}, d_O \in \Delta\}$

Then, we obtain the following conclusion on the Resource Life Cycle Manager pattern.

**Theorem 37.8** (Correctness of the Resource Life Cycle Manager pattern). *The Resource Life Cycle Manager pattern $\tau_I(\partial_H(U \between M \between P \between R))$ can exhibit the desired external behaviors.*

*Proof.* Based on the above state transitions of the above modules, by use of the algebraic laws of APTC, we can prove that

$\tau_I(\partial_H(U \between M \between P \between R)) = \sum_{d_I, d_O \in \Delta}(r_I(d_I) \cdot s_O(d_O)) \cdot \tau_I(\partial_H(U \between M \between P \between R))$

That is, the Resource Life Cycle Manager pattern $\tau_I(\partial_H(U \between RP \between P \between R))$ can exhibit the desired external behaviors.

For the details of the proof, please refer to Section 17.3, as we omit them here. □

## 37.3 Resource release

In this section, we verify patterns for resource release, including the Leasing pattern and the Evictor pattern.

### 37.3.1 Verification of the Leasing pattern

The Leasing pattern uses a mediating lookup service to find and access resources. There are four modules in the Leasing pattern: the Resource User, the Resource Provider, the Lease, and the Resource. The Resource User interacts with the outside through the channels $I$ and $O$; with the Resource Provider through the channels $I_{UP}$ and $O_{UP}$; with the Resource through the channels $I_{UR}$ and $O_{UR}$, and with the Lease through the channels $I_{UL}$ and $O_{UL}$. This is illustrated in Fig. 37.17.

The typical process of the Leasing pattern is shown in Fig. 37.18 and is as follows:

1. The Resource User receives the input $d_I$ from the outside through the channel $I$ (the corresponding reading action is denoted $r_I(d_I)$), then processes the input $d_I$ through a processing function $UF_1$, and sends the input $d_{I_P}$ to the Resource Provider through the channel $I_{UP}$ (the corresponding sending action is denoted $s_{I_{UP}}(d_{I_P})$);
2. The Resource Provider receives the input from the Resource User through the channel $I_{UP}$ (the corresponding reading action is denoted $r_{I_{UP}}(d_{I_P})$), then processes the input and generates the output $d_{O_P}$ to the Resource User through a processing function $PF$, and sends the output to the Resource User through the channel $O_{UP}$ (the corresponding sending action is denoted $s_{O_{UP}}(d_{O_P})$);
3. The Resource User receives the output $d_{O_P}$ from the Resource Provider through the channel $O_{UP}$ (the corresponding reading action is denoted $r_{O_{UP}}(d_{O_P})$), then processes the output through a processing function $UF_2$, generates and sends the input $d_{I_R}$ to the Resource through the channel $I_{UR}$ (the corresponding sending action is denoted $s_{I_{UR}}(d_{I_R})$);
4. The Resource receives the input $d_{I_R}$ from the Resource User through the channel $I_{UR}$ (the corresponding reading action is denoted $r_{I_{UR}}(d_{I_R})$), then processes the input through a processing function $RF$, generates and sends the response $d_{O_R}$ (the corresponding sending action is denoted $s_{O_{UR}}(d_{O_R})$);
5. The Resource User receives the response $d_{O_R}$ from the Resource through the channel $O_{UR}$ (the corresponding reading action is denoted $r_{O_{UR}}(d_{O_R})$), then processes the response and generates the response $d_O$ through a processing function $UF_3$, and sends the processed input $d_{I_L}$ to the Lease through the channel $I_{UL}$ (the corresponding sending action is denoted $s_{I_{UL}}(d_{I_L})$);

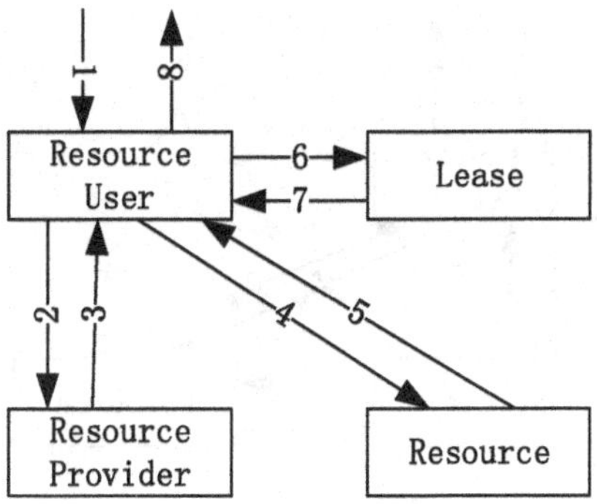


**FIGURE 37.18** Typical process of Leasing pattern.

6. The Lease receives $d_{I_L}$ from the Resource User through the channel $I_{UL}$ (the corresponding reading action is denoted $r_{I_{UL}}(d_{I_L})$), then processes the request through a processing function $LF$, generates and sends the processed output $d_{O_L}$ to the Resource User through the channel $O_{UL}$ (the corresponding sending action is denoted $s_{O_{UL}}(d_{O_L})$);
7. The Resource User receives the output $d_{O_L}$ from the Lease through the channel $O_{UL}$ (the corresponding reading action is denoted $r_{O_{UL}}(d_{O_L})$), then processes the output and generates $d_O$ through a processing function $UF_4$, and sends the response to the outside through the channel $O$ (the corresponding sending action is denoted $s_O(d_O)$).

In the following, we verify the Leasing pattern. We assume all data elements $d_I, d_{I_L}, d_{I_P}, d_{I_R}, d_{O_L}, d_{O_P}, d_{O_R}, d_O$ are from a finite set $\Delta$.

The state transitions of the Resource User module described by APTC are as follows:

$U = \sum_{d_I\in\Delta}(r_I(d_I)\cdot U_2)$
$U_2 = UF_1\cdot U_3$
$U_3 = \sum_{d_{I_P}\in\Delta}(s_{I_{UP}}(d_{I_P})\cdot U_4)$
$U_4 = \sum_{d_{O_P}\in\Delta}(r_{O_{UP}}(d_{O_P})\cdot U_5)$
$U_5 = UF_2\cdot U_6$
$U_6 = \sum_{d_{I_R}\in\Delta}(s_{I_{UR}}(d_{I_R})\cdot U_7)$
$U_7 = \sum_{d_{O_R}\in\Delta}(r_{O_{UR}}(d_{O_R})\cdot U_8)$
$U_8 = UF_3\cdot U_9$
$U_9 = \sum_{d_{I_L}\in\Delta}(s_{I_{UL}}(d_{I_L})\cdot U_{10})$
$U_{10} = \sum_{d_{O_L}\in\Delta}(r_{O_{UL}}(d_{O_L})\cdot U_{11})$
$U_{11} = UF_4\cdot U_{12}$
$U_{12} = \sum_{d_O\in\Delta}(s_O(d_O)\cdot U)$

The state transitions of the Resource Provider module described by APTC are as follows:

$P = \sum_{d_{I_P}\in\Delta}(r_{I_{UP}}(d_{I_P})\cdot P_2)$
$P_2 = PF\cdot P_3$
$P_3 = \sum_{d_{O_P}\in\Delta}(s_{O_{UP}}(d_{O_P})\cdot P)$

The state transitions of the Lease module described by APTC are as follows:

$L = \sum_{d_{I_L}\in\Delta}(r_{I_{UL}}(d_{I_L})\cdot L_2)$
$L_2 = LF\cdot L_3$
$L_3 = \sum_{d_{O_L}\in\Delta}(s_{O_{UL}}(d_{O_L})\cdot L)$

The state transitions of the Resource module described by APTC are as follows:

$R = \sum_{d_{I_R}\in\Delta}(r_{I_{UR}}(d_{I_R})\cdot R_2)$
$R_2 = RF\cdot R_3$
$R_3 = \sum_{d_{O_R}\in\Delta}(s_{O_{UR}}(d_{O_R})\cdot R)$

The sending action and the reading action of the same data through the same channel can communicate with each other, otherwise, they will cause a deadlock $\delta$. We define the following communication functions between the Resource User and the Resource Provider Proxy:

$$\gamma(r_{I_{UP}}(d_{I_P}), s_{I_{UP}}(d_{I_P})) \triangleq c_{I_{UP}}(d_{I_P})$$

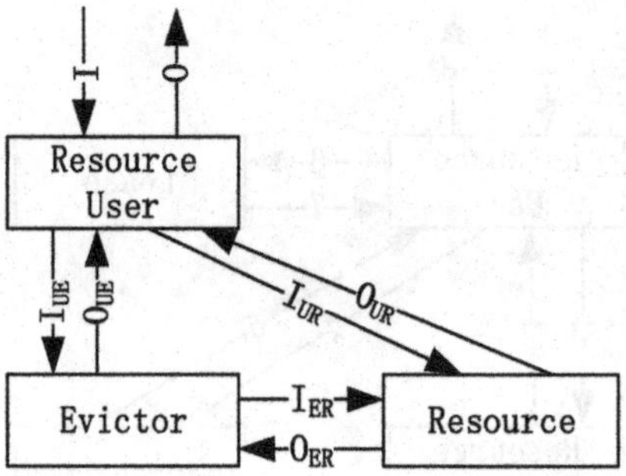


**FIGURE 37.19** Evictor pattern.

$$\gamma(r_{O_{UP}}(d_{O_P}), s_{O_{UP}}(d_{O_P})) \triangleq c_{O_{UP}}(d_{O_P})$$

There are two communication functions between the Resource User and the Lease as follows:

$$\gamma(r_{I_{UL}}(d_{I_L}), s_{I_{UL}}(d_{I_L})) \triangleq c_{I_{UL}}(d_{I_L})$$
$$\gamma(r_{O_{UL}}(d_{O_L}), s_{O_{UL}}(d_{O_L})) \triangleq c_{O_{UL}}(d_{O_L})$$

There are two communication functions between the Resource User and the Resource as follows:

$$\gamma(r_{I_{UR}}(d_{I_R}), s_{I_{UR}}(d_{I_R})) \triangleq c_{I_{UR}}(d_{I_R})$$
$$\gamma(r_{O_{UR}}(d_{O_R}), s_{O_{UR}}(d_{O_R})) \triangleq c_{O_{UR}}(d_{O_R})$$

Let all modules be in parallel, then the Leasing pattern $U \quad L \quad P \quad R$ can be presented by the following process term:

$\tau_I(\partial_H(\Theta(U \between L \between P \between R))) = \tau_I(\partial_H(U \between L \between P \between R))$

where $H = \{r_{I_{UP}}(d_{I_P}), s_{I_{UP}}(d_{I_P}), r_{O_{UP}}(d_{O_P}), s_{O_{UP}}(d_{O_P}), r_{I_{UL}}(d_{I_L}), s_{I_{UL}}(d_{I_L}),$
$r_{O_{UL}}(d_{O_L}), s_{O_{UL}}(d_{O_L}), r_{I_{UR}}(d_{I_R}), s_{I_{UR}}(d_{I_R}), r_{O_{UR}}(d_{O_R}), s_{O_{UR}}(d_{O_R})$
$|d_I, d_{I_P}, d_{I_L}, d_{I_R}, d_{O_P}, d_{O_L}, d_{O_R}, d_O \in \Delta\}$,

$I = \{c_{I_{UP}}(d_{I_P}), c_{O_{UP}}(d_{O_P}), c_{I_{UL}}(d_{I_L}), c_{O_{UL}}(d_{O_L}), c_{I_{UR}}(d_{I_R}), c_{O_{UR}}(d_{O_R}),$
$UF_1, UF_2, UF_3, UF_4, PF, LF, RF | d_I, d_{I_P}, d_{I_L}, d_{I_R}, d_{O_P}, d_{O_L}, d_{O_R}, d_O \in \Delta\}$

Then, we obtain the following conclusion on the Leasing pattern.

**Theorem 37.9** (Correctness of the Leasing pattern). *The Leasing pattern $\tau_I(\partial_H(U \between L \between P \between R))$ can exhibit the desired external behaviors.*

*Proof.* Based on the above state transitions of the above modules, by use of the algebraic laws of APTC, we can prove that

$\tau_I(\partial_H(U \between L \between P \between R)) = \sum_{d_I, d_O \in \Delta}(r_I(d_I) \cdot s_O(d_O)) \cdot \tau_I(\partial_H(U \between L \between P \between R))$

That is, the Leasing pattern $\tau_I(\partial_H(U \between L \between P \between R))$ can exhibit the desired external behaviors.

For the details of the proof, please refer to Section 17.3, as we omit them here. □

### 37.3.2 Verification of the Evictor pattern

The Evictor pattern allows different strategies to release the resources. There are three modules in the Evictor pattern: the Resource User, the Evictor, and the Resource. The Resource User interacts with the outside through the channels $I$ and $O$; with the Evictor through the channels $I_{UE}$ and $O_{UE}$, and with the Resource through the channels $I_{UR}$ and $O_{UR}$. The Evictor interacts with the Resource through the channels $I_{ER}$ and $O_{ER}$. This is illustrated in Fig. 37.19.

The typical process of the Evictor pattern is shown in Fig. 37.20 and is as follows:

1. The Resource User receives the input $d_I$ from the outside through the channel $I$ (the corresponding reading action is denoted $r_I(d_I)$), then processes the input $d_I$ through a processing function $UF_1$, and generates the input $d_{I_R}$, and sends the input to the Resource through the channel $I_{UR}$ (the corresponding sending action is denoted $s_{I_{UR}}(d_{I_R})$);
2. The Resource receives the input from the Resource User through the channel $I_{UR}$ (the corresponding reading action is denoted $r_{I_{UR}}(d_{I_R})$), then processes the input and generates the output $d_{O_R}$ to the Resource User through a processing function $RF_1$, and sends the output to the Resource User through the channel $O_{UR}$ (the corresponding sending action is denoted $s_{O_{UR}}(d_{O_R})$);

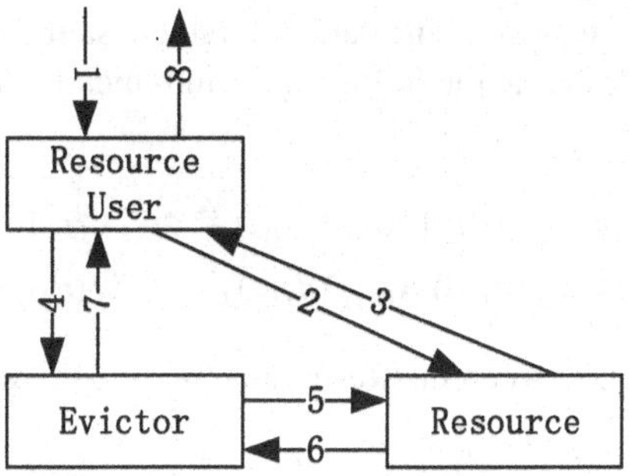


**FIGURE 37.20** Typical process of Evictor pattern.

3. The Resource User receives the output $d_{O_R}$ from the Resource through the channel $O_{UR}$ (the corresponding reading action is denoted $r_{O_{UR}}(d_{O_R})$), then processes the output through a processing function $UF_2$, generates and sends the input $d_{I_E}$ to the Evictor through the channel $I_{UE}$ (the corresponding sending action is denoted $s_{I_{UE}}(d_{I_E})$);
4. The Evictor receives the input $d_{I_E}$ from the Resource User through the channel $I_{UE}$ (the corresponding reading action is denoted $r_{I_{UE}}(d_{I_E})$), then processes the input through a processing function $EF_1$, generates and sends the input $d_{I_{R'}}$ (the corresponding sending action is denoted $s_{I_{ER}}(d_{I_{R'}})$);
5. The Resource receives the input from the Evictor through the channel $I_{ER}$ (the corresponding reading action is denoted $r_{I_{ER}}(d_{I_{R'}})$), then processes the input and generates the output $d_{O_{R'}}$ to the Evictor through a processing function $RF_2$, and sends the output to the Evictor through the channel $O_{ER}$ (the corresponding sending action is denoted $s_{O_{ER}}(d_{O_{R'}})$);
6. The Evictor receives $d_{O_{R'}}$ from the Resource through the channel $O_{ER}$ (the corresponding reading action is denoted $r_{O_{ER}}(d_{O_{R'}})$), then processes the input through a processing function $EF_2$, generates and sends the output $d_{O_E}$ (the corresponding sending action is denoted $s_{O_{UE}}(d_{O_E})$);
7. The Resource User receives the response $d_{O_E}$ from the Evictor through the channel $O_{UE}$ (the corresponding reading action is denoted $r_{O_{UE}}(d_{O_E})$), then processes the response and generates the response $d_O$ through a processing function $UF_3$, and sends the response to the outside through the channel $O$ (the corresponding sending action is denoted $s_O(d_O)$).

In the following, we verify the Evictor pattern. We assume all data elements $d_I, d_{I_E}, d_{I_R}, d_{I_{R'}}, d_{O_E}, d_{O_R}, d_{O_{R'}}, d_O$ are from a finite set $\Delta$.

The state transitions of the Resource User module described by APTC are as follows:

$U = \sum_{d_I\in\Delta}(r_I(d_I)\cdot U_2)$

$U_2 = UF_1 \cdot U_3$

$U_3 = \sum_{d_{I_R}\in\Delta}(s_{I_{UR}}(d_{I_R})\cdot U_4)$

$U_4 = \sum_{d_{O_R}\in\Delta}(r_{O_{UR}}(d_{O_R})\cdot U_5)$

$U_5 = UF_2 \cdot U_6$

$U_6 = \sum_{d_{I_E}\in\Delta}(s_{I_{UE}}(d_{I_E})\cdot U_7)$

$U_7 = \sum_{d_{O_E}\in\Delta}(r_{O_{UE}}(d_{O_E})\cdot U_8)$

$U_8 = UF_3 \cdot U_9$

$U_9 = \sum_{d_O\in\Delta}(s_O(d_O)\cdot U)$

The state transitions of the Evictor module described by APTC are as follows:

$E = \sum_{d_{I_E}\in\Delta}(r_{I_{UE}}(d_{I_E})\cdot E_2)$

$E_2 = EF_1 \cdot E_3$

$E_3 = \sum_{d_{I_{R'}}\in\Delta}(s_{I_{ER}}(d_{I_{R'}})\cdot E_4)$

$E_4 = \sum_{d_{O_{R'}}\in\Delta}(r_{O_{ER}}(d_{O_{R'}})\cdot E_5)$

$E_5 = EF_2 \cdot E_6$

$E_6 = \sum_{d_{O_E}\in\Delta}(s_{O_{UE}}(d_{O_E})\cdot E)$

The state transitions of the Resource module described by APTC are as follows:

$R = \sum_{d_{I_R}\in\Delta}(r_{I_{UR}}(d_{I_R})\cdot R_2)$

$R_2 = RF \cdot R_3$

$R_3 = \sum_{d_{O_R}\in\Delta}(s_{O_{UR}}(d_{O_R})\cdot R)$

The sending action and the reading action of the same data through the same channel can communicate with each other, otherwise, they will cause a deadlock $\delta$. We define the following communication functions between the Resource User and the Evictor:

$$\gamma(r_{I_{UE}}(d_{I_E}), s_{I_{UE}}(d_{I_E})) \triangleq c_{I_{UE}}(d_{I_E})$$

$$\gamma(r_{O_{UE}}(d_{O_E}), s_{O_{UE}}(d_{O_E})) \triangleq c_{O_{UE}}(d_{O_E})$$

There are two communication functions between the Resource User and the Resource as follows:

$$\gamma(r_{I_{UR}}(d_{I_R}), s_{I_{UR}}(d_{I_R})) \triangleq c_{I_{UR}}(d_{I_R})$$

$$\gamma(r_{O_{UR}}(d_{O_R}), s_{O_{UR}}(d_{O_R})) \triangleq c_{O_{UR}}(d_{O_R})$$

There are two communication functions between the Evictor and the Resource as follows:

$$\gamma(r_{I_{ER}}(d_{I_{R'}}), s_{I_{ER}}(d_{I_{R'}})) \triangleq c_{I_{ER}}(d_{I_{R'}})$$

$$\gamma(r_{O_{ER}}(d_{O_{R'}}), s_{O_{ER}}(d_{O_{R'}})) \triangleq c_{O_{ER}}(d_{O_{R'}})$$

Let all modules be in parallel, then the Evictor pattern $U \quad E \quad R$ can be presented by the following process term:
$\tau_I(\partial_H(\Theta(U \between E \between R))) = \tau_I(\partial_H(U \between E \between R))$
where $H = \{r_{I_{UE}}(d_{I_E}), s_{I_{UE}}(d_{I_E}), r_{O_{UE}}(d_{O_E}), s_{O_{UE}}(d_{O_E}), r_{I_{UR}}(d_{I_R}), s_{I_{UR}}(d_{I_R}),$
$r_{O_{UR}}(d_{O_R}), s_{O_{UR}}(d_{O_R}), r_{I_{ER}}(d_{I_{R'}}), s_{I_{ER}}(d_{I_{R'}}), r_{O_{ER}}(d_{O_{R'}}), s_{O_{ER}}(d_{O_{R'}})$
$|d_I, d_{I_E}, d_{I_R}, d_{I_{R'}}, d_{O_{R'}}, d_{O_E}, d_{O_R}, d_O \in \Delta\}$,
$I = \{c_{I_{UE}}(d_{I_E}), c_{O_{UE}}(d_{O_E}), c_{I_{UR}}(d_{I_R}), c_{O_{UR}}(d_{O_R}), c_{I_{ER}}(d_{I_{R'}}), c_{O_{ER}}(d_{O_{R'}}),$
$UF_1, UF_2, UF_3, EF_1, EF_2, RF | d_I, d_{I_E}, d_{I_R}, d_{I_{R'}}, d_{O_{R'}}, d_{O_E}, d_{O_R}, d_O \in \Delta\}$
Then, we obtain the following conclusion on the Evictor pattern.

**Theorem 37.10** (Correctness of the Evictor pattern). *The Evictor pattern $\tau_I(\partial_H(U \between E \between R))$ can exhibit the desired external behaviors.*

*Proof.* Based on the above state transitions of the above modules, by use of the algebraic laws of APTC, we can prove that
$\tau_I(\partial_H(U \between E \between R)) = \sum_{d_I, d_O \in \Delta}(r_I(d_I) \cdot s_O(d_O)) \cdot \tau_I(\partial_H(U \between E \between R))$
That is, the Evictor pattern $\tau_I(\partial_H(U \between E \between R))$ can exhibit the desired external behaviors.
For the details of the proof, please refer to Section 17.3, as we omit them here. □

Chapter 38

# Composition of patterns

Patterns can be composed to satisfy the actual requirements freely, once the syntax and semantics of the output of one pattern can be inserted into the syntax and semantics of the input of another pattern.

In this chapter, we show the composition of patterns. In Section 38.1, we verify the composition of the Layers patterns. In Section 38.2, we show the composition of Presentation–Abstraction–Control (PAC) patterns. We compose patterns for resource management in Section 38.3.

## 38.1 Composition of the Layers patterns

In this section, we show the composition of the Layers patterns, and verify the correctness of the composition. We have already verified the correctness of the Layers pattern and its composition in Section 33.1.1, here we verify the correctness of the composition of the Layers patterns based on the correctness result of the Layers pattern.

The composition of two layer's peers is illustrated in Fig. 38.1. Each layer's peer is abstracted as a module, and the composition of two layer's peers is also abstracted as a new module, as the dotted rectangles illustrate in Fig. 38.1.

There are two typical processes in the composition of two layer's peers: one is the direction from peer $P$ to peer $P'$, the other is the direction from $P'$ to $P$. However, we omit them here, hence, please refer to Section 33.1.1 for details.

In the following, we verify the correctness of the inserting of two layer's peers. We assume all data elements $d_{L_1}, d_{L_{1'}}, d_{U_n}, d_{U_{n'}}$ are from a finite set $\Delta$. Note that the channel $LO_1$ and the channel $LI_{1'}$ are the same channel; the channel $LO_{1'}$ and the channel $LI_1$ are the same channel. Also, the data $d_{L_{1'}}$ and the data $PUF(d_{U_n})$ are the same data; the data $d_{L_1}$ and the data $P'UF(d_{U_{n'}})$ are the same data.

The state transitions of the $P$ described by APTC are as follows:

$P = \sum_{d_{U_n}, d_{L_1} \in \Delta}(r_{UI_n}(d_{U_n}) \cdot P_2 \between r_{LI_1}(d_{L_1}) \cdot P_3)$

$P_2 = PUF \cdot P_4$

$P_3 = PLF \cdot P_5$

$P_4 = \sum_{d_{U_n} \in \Delta}(s_{LO_1}(PUF(d_{U_n})) \cdot P)$

$P_5 = \sum_{d_{L_1} \in \Delta}(s_{UO_n}(PLF(d_{L_1})) \cdot P)$

The state transitions of the $P'$ described by APTC are as follows:

$P' = \sum_{d_{U_{n'}}, d_{L_{1'}} \in \Delta}(r_{UI_{n'}}(d_{U_{n'}}) \cdot P_2' \between r_{LI_{1'}}(d_{L_{1'}}) \cdot P_3')$

$P_2' = P'UF \cdot P_4'$

$P_3' = P'LF \cdot P_5'$

$P_4' = \sum_{d_{U_{n'}} \in \Delta}(s_{LO_{1'}}(PUF(d_{U_{n'}})) \cdot P')$

$P_5' = \sum_{d_{L_{1'}} \in \Delta}(s_{UO_{n'}}(PLF(d_{L_{1'}})) \cdot P')$

The sending action and the reading action of the same data through the same channel can communicate with each other, otherwise, they will cause a deadlock $\delta$. We define the following communication functions:

$$\gamma(r_{LI_1}(d_{L_1}), s_{LO_{1'}}(P'UF(d_{U_{n'}}))) \triangleq c_{LI_1}(d_{L_1})$$

$$\gamma(r_{LI_{1'}}(d_{L_{1'}}), s_{LO_1}(PUF(d_{U_n}))) \triangleq c_{LI_{1'}}(d_{L_{1'}})$$

Let all modules be in parallel, then the two layer's peers $P$ $P'$ can be presented by the following process term:

$\tau_I(\partial_H(\Theta(\tau_{I_1}(\partial_{H_1}(P)) \between \tau_{I_2}(\partial_{H_2}(P'))))) = \tau_I(\partial_H(\tau_{I_1}(\partial_{H_1}(P)) \between \tau_{I_2}(\partial_{H_2}(P'))))$

where $H = \{r_{LI_1}(d_{L_1}), s_{LO_{1'}}(P'UF(d_{U_{n'}})), r_{LI_{1'}}(d_{L_{1'}}), s_{LO_1}(PUF(d_{U_n}))$
$|d_{L_1}, d_{L_{1'}}, d_{U_n}, d_{U_{n'}} \in \Delta\}$,

$I = \{c_{LI_1}(d_{L_1}), c_{LI_{1'}}(d_{L_{1'}}), PUF, PLF, P'UF, P'LF | d_{L_1}, d_{L_{1'}}, d_{U_n}, d_{U_{n'}} \in \Delta\}$

Also, about the definitions of $H_1$ and $I_1$, $H_2$ and $I_2$, please see Section 33.1.1.

Then, we obtain the following conclusion on the inserting of two layer's peers.

**Handbook of Truly Concurrent Process Algebra. https://doi.org/10.1016/B978-0-44-321515-5.00042-1**

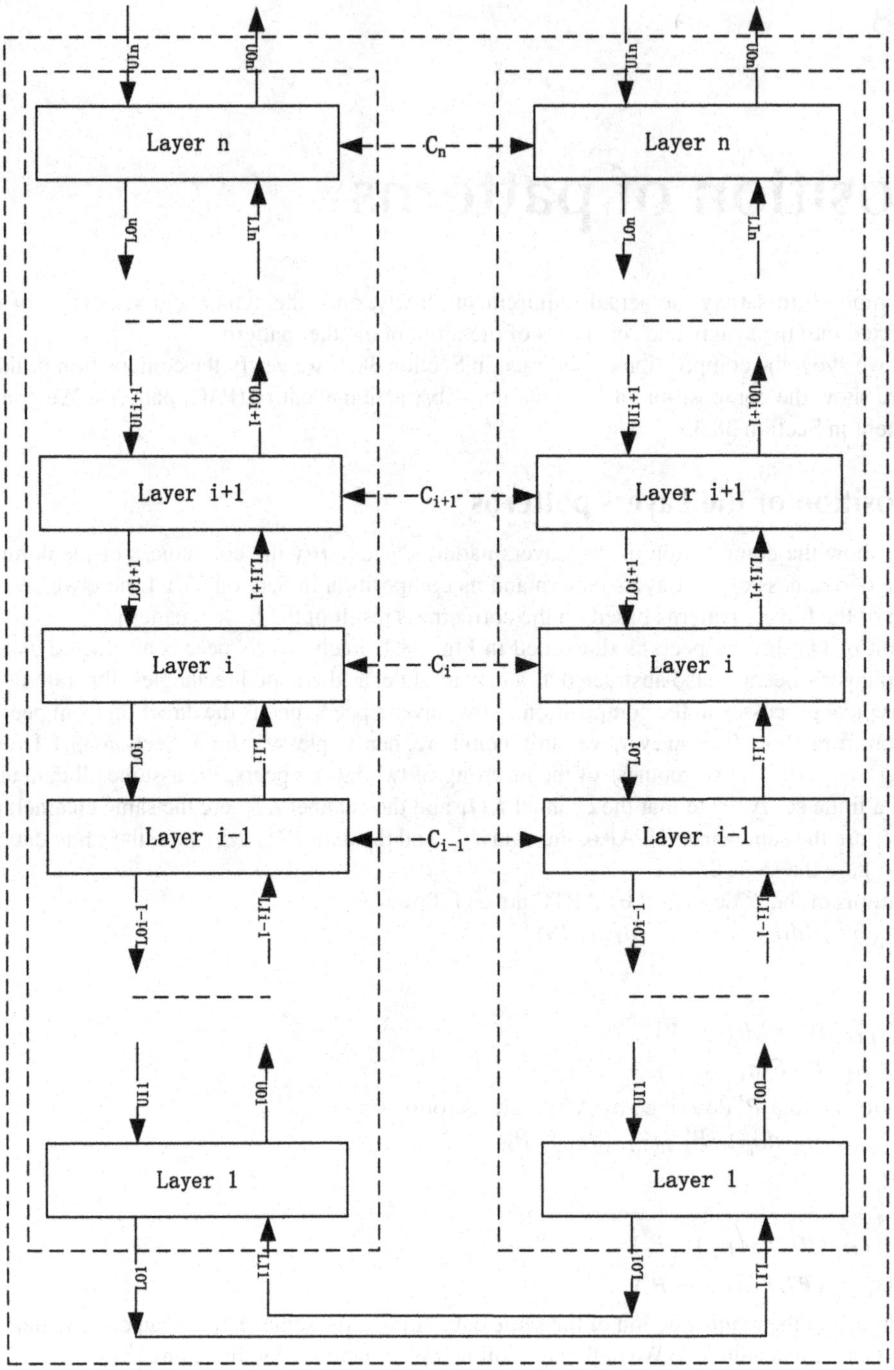


FIGURE 38.1 Composition of the Layers patterns.

**Theorem 38.1** (Correctness of the inserting of two layer's peers). *The inserting of two layer's peers*
$\tau_I(\partial_H(\tau_{I_1}(\partial_{H_1}(P)) \between \tau_{I_2}(\partial_{H_2}(P'))))$
*can exhibit the desired external behaviors.*

*Proof.* Based on the above state transitions of the above modules, by use of the algebraic laws of APTC, we can prove that
$\tau_I(\partial_H(\tau_{I_1}(\partial_{H_1}(P)) \between \tau_{I_2}(\partial_{H_2}(P')))) = \sum_{d_{U_n}, d_{U_{n'}} \in \Delta}((r_{UI_n}(d_{U_n}) \parallel r_{UI_{n'}}(d_{U_{n'}}))$
$\cdot (s_{UO_n}(PLF(P'UF(d_{U_{n'}}))) \parallel s_{UO_{n'}}(P'LF(PUF(d_{U_n}))))) \cdot \tau_I(\partial_H(\tau_{I_1}(\partial_{H_1}(P)) \between \tau_{I_2}(\partial_{H_2}(P'))))$
That is, the inserting of two layer's peers $\tau_I(\partial_H(\tau_{I_1}(\partial_{H_1}(P)) \between \tau_{I_2}(\partial_{H_2}(P'))))$ can exhibit the desired external behaviors.

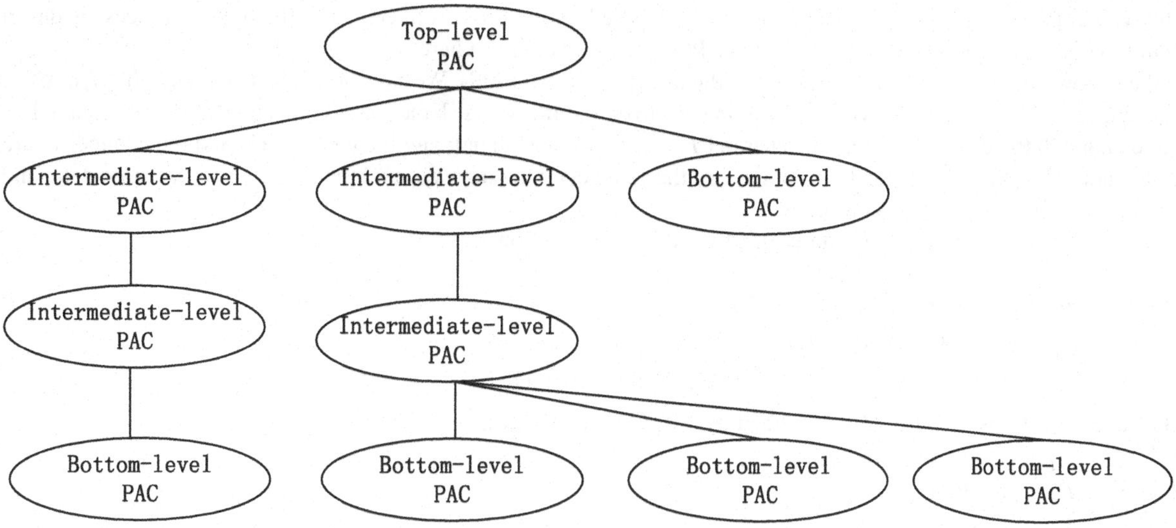


**FIGURE 38.2** Levels of the PAC pattern.

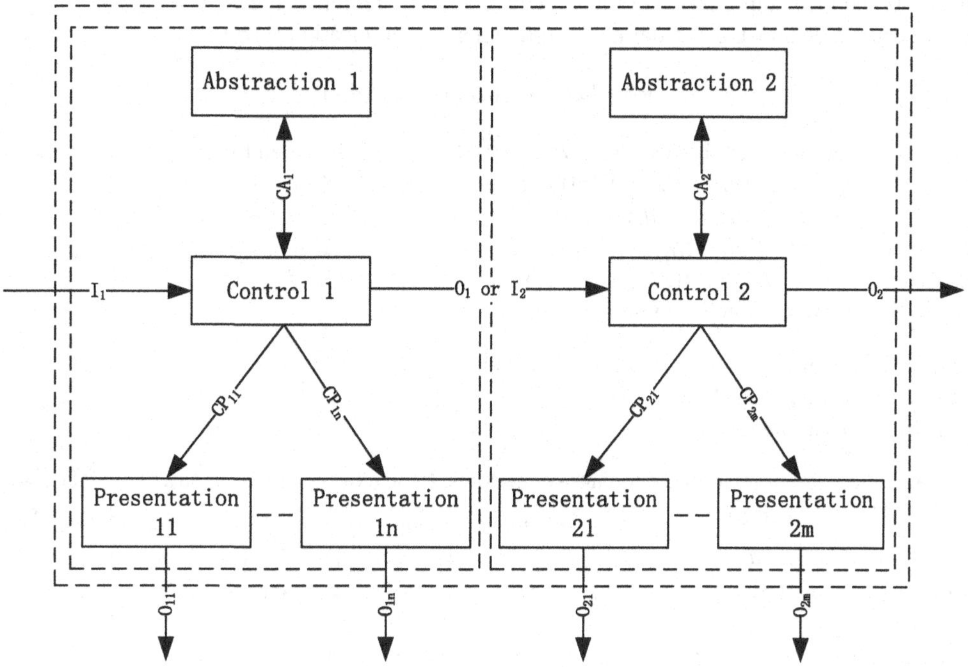


**FIGURE 38.3** Inserting of two PACs.

For the details of the proof, please refer to Section 17.3, as we omit them here. □

## 38.2 Composition of the PAC patterns

In this section, we show the composition of Presentation–Abstraction–Control (PAC) patterns (we already verified its correctness in Section 33.3.2) and verify the correctness of the composition.

The PAC patterns can be composed into levels of PACs, as illustrated in Fig. 38.2.

If the syntax and semantics of the output of one PAC match the syntax and semantics of the input of another PAC, then they can be composed. For simplicity and without loss of generality, we show the inserting of only two PACs, as illustrated in Fig. 38.3. Each PAC is abstracted as a module, and the composition of two PACs is also abstracted as a new module, as the dotted rectangles illustrate in Fig. 38.3.

The typical process of inserting of two PACs is composed by the process of one PAC (the typical process is described in Section 33.3.2) follows the process of the other PAC, hence, we omit it here.

In the following, we verify the correctness of the inserting of two PACs. We assume all data elements $d_{I_1}, d_{O_1}, d_{I_2}, d_{O_2}, d_I, d_O, d_{O_{1i}}$ (for $1 \leq i \leq n$), $d_{O_{2j}}$ (for $1 \leq j \leq m$) are from a finite set $\Delta$. Note that the channel $I$ and the channel $I_1$ are the same channel; the channel $O_1$ and the channel $I_2$ are the same channel; and the channel $O_2$ and the channel $O$ are the same channel. Also, the data $d_{I_1}$ and the data $d_I$ are the same data; the data $d_{O_1}$ and the data $d_{I_2}$ are the same data; and the data $d_{O_2}$ and the data $d_O$ are the same data.

The state transitions of the $PAC_1$ described by APTC are as follows:

$PAC_1 = \sum_{d_{I_1} \in \Delta}(r_{I_1}(d_{I_1}) \cdot PAC_{1_2})$

$PAC_{1_2} = PAC_1F \cdot PAC_{1_3}$

$PAC_{1_3} = \sum_{d_{O_1} \in \Delta}(s_{O_1}(d_{O_1}) \cdot PAC_{1_4})$

$PAC_{1_4} = \sum_{d_{O_{1_1}},\cdots,d_{O_{1_n}} \in \Delta}(s_{O_{1_1}}(d_{O_{1_1}}) \between \cdots \between s_{O_{1_n}}(d_{O_{1_n}}) \cdot PAC_1)$

The state transitions of the $PAC_2$ described by APTC are as follows:

$PAC_2 = \sum_{d_{I_2} \in \Delta}(r_{I_2}(d_{I_2}) \cdot PAC_{2_2})$

$PAC_{2_2} = PAC_2F \cdot PAC_{2_3}$

$PAC_{2_3} = \sum_{d_{O_2} \in \Delta}(s_{O_2}(d_{O_2}) \cdot PAC_{2_4})$

$PAC_{2_4} = \sum_{d_{O_{2_1}},\cdots,d_{O_{2m}} \in \Delta}(s_{O_{2_1}}(d_{O_{2_1}}) \between \cdots \between s_{O_{2m}}(d_{O_{2m}}) \cdot PAC_2)$

The sending action and the reading action of the same data through the same channel can communicate with each other, otherwise, they will cause a deadlock $\delta$. We define the following communication function:

$$\gamma(r_{I_2}(d_{I_2}), s_{O_1}(d_{O_1})) \triangleq c_{I_2}(d_{I_2})$$

Let all modules be in parallel, then the two PACs $PAC_1 PAC_2$ can be presented by the following process term:

$\tau_I(\partial_H(\Theta(\tau_{I_1}(\partial_{H_1}(PAC_1)) \between \tau_{I_2}(\partial_{H_2}((PAC_2)))))) = \tau_I(\partial_H(\tau_{I_1}(\partial_{H_1}(PAC_1)) \between \tau_{I_2}(\partial_{H_2}((PAC_2)))))$

where $H = \{r_{I_2}(d_{I_2}), s_{O_1}(d_{O_1}) | d_{I_1}, d_{O_1}, d_{I_2}, d_{O_2}, d_I, d_O, d_{O_{1i}}, d_{O_{2j}} \in \Delta\}$ for $1 \leq i \leq n$ and $1 \leq j \leq m$,

$I = \{c_{I_2}(d_{I_2}), PAC_1F, PAC_2F | d_{I_1}, d_{O_1}, d_{I_2}, d_{O_2}, d_I, d_O, d_{O_{1i}}, d_{O_{2j}} \in \Delta\}$ for $1 \leq i \leq n$ and $1 \leq j \leq m$

Also, about the definitions of $H_1$ and $I_1$, $H_2$ and $I_2$, please see Section 33.3.2.

Then, we obtain the following conclusion on the inserting of two PACs.

**Theorem 38.2** (Correctness of the inserting of two PACs). *The inserting of two PACs*

$\tau_I(\partial_H(\tau_{I_1}(\partial_{H_1}(PAC_1)) \between \tau_{I_2}(\partial_{H_2}(PAC_2))))$

*can exhibit the desired external behaviors.*

*Proof.* Based on the above state transitions of the above modules, by use of the algebraic laws of APTC, we can prove that

$\tau_I(\partial_H(\tau_{I_1}(\partial_{H_1}(PAC_1)) \between \tau_{I_2}(\partial_{H_2}(PAC_2)))) = \sum_{d_I, d_O, d_{O_{1_1}},\cdots,d_{O_{1_n}}, d_{O_{2_1}},\cdots,d_{O_{2m}} \in \Delta}(r_I(d_I) \cdot s_O(d_O) \cdot s_{O_{1_1}}(d_{O_{1_1}}) \parallel \cdots \parallel s_{O_{1_i}}(d_{O_{1_i}}) \parallel \cdots \parallel s_{O_{1_n}}(d_{O_{1_n}}) \cdot s_{O_{2_1}}(d_{O_{2_1}}) \parallel \cdots \parallel s_{O_{2_j}}(d_{O_{2_j}}) \parallel \cdots \parallel s_{O_{2m}}(d_{O_{2m}})) \cdot \tau_I(\partial_H(\tau_{I_1}(\partial_{H_1}(PAC_1)) \between \tau_{I_2}(\partial_{H_2}(PAC_2))))$

That is, the inserting of two PACs $\tau_I(\partial_H(\tau_{I_1}(\partial_{H_1}(PAC_1)) \between \tau_{I_2}(\partial_{H_2}(PAC_2))))$ can exhibit the desired external behaviors.

For the details of the proof, please refer to Section 17.3, as we omit them here. □

## 38.3 Composition of resource management patterns

In this section, we show the composition of resource management patterns (we have already verified the correctness of patterns for resource management in Chapter 37), and verify the correctness of the composition.

The whole process of resource management involves resource acquisition first, resource utilization and life cycle management secondly, and resource release lastly, as Fig. 38.4 illustrates. For resource acquisition, we take an example of the Lookup pattern, and the Life Cycle Manager pattern for resource life cycle management, and the Leasing pattern for resource release. The whole process of resource management is composed of the typical processes of the Lookup pattern, the Life Cycle Manager pattern, and the Leasing pattern, however, we do not repeat these here, please refer to the details of these three patterns in Chapter 37. Also, we can verify the correctness of the whole resource management system shown in Fig. 38.4, just like the work we discussed many times for concrete patterns in the above chapters.

However, we do not verify the correctness of the whole resource management system like in the previous work. The whole resource management system in Fig. 38.4 contains the full functions of the Lookup pattern, the Life Cycle Manager

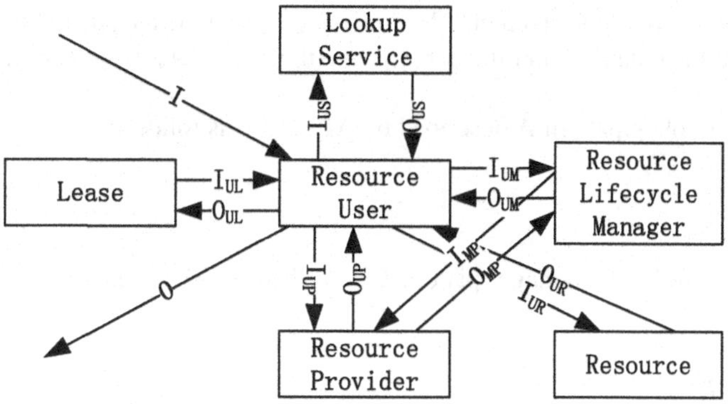


FIGURE 38.4 The whole resource management process.

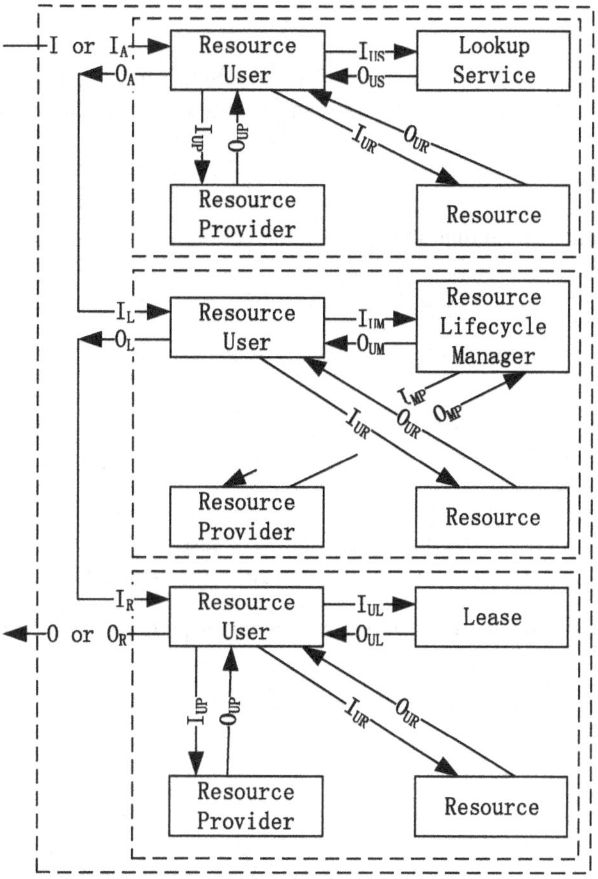


FIGURE 38.5 Inserting of resource management patterns.

pattern, and the Leasing pattern, and actually can be implemented by the composition of these three patterns, as Fig. 38.5 illustrates. For the whole process of resource management, first the Lookup pattern works, then the Life Cycle Manager pattern, and lastly the Leasing pattern. That is, the output of the Lookup pattern is inserted into the input of the Life Cycle Manager pattern, and the output of the Life Cycle Manager pattern is inserted into the Leasing pattern. Each pattern is abstracted as a module, and the composition of these three patterns is also abstracted as a new module, as the dotted rectangles illustrate in Fig. 38.5.

In the following, we verify the correctness of inserting of resource management patterns. We assume all data elements $d_I, d_O, d_{I_A}, d_{O_A}, d_{I_L}, d_{O_L}, d_{I_R}, d_{O_R}$ are from a finite set $\Delta$. Note that the channel $I$ and the channel $I_A$ are the same channel; the channel $O_A$ and the channel $I_L$ are the same channel; the channel $O_L$ and the channel $I_R$ are the same channel; and the channel $O_R$ and the channel $O$ are the same channel. Also, the data $d_{I_A}$ and the data $d_I$ are the same data; the data $d_{O_A}$ and the data $d_{I_L}$ are the same data; the data $d_{O_L}$ and the data $d_{I_R}$ are the same data; and the data $d_{O_R}$ and the data $d_O$ are the same data.

The state transitions of the Lookup pattern $A$ described by APTC are as follows:

$A = \sum_{d_{I_A} \in \Delta}(r_{I_A}(d_{I_A}) \cdot A_2)$

$A_2 = AF \cdot A_3$

$A_3 = \sum_{d_{O_A} \in \Delta}(s_{O_A}(d_{O_A}) \cdot A)$

The state transitions of the Life Cycle Manager pattern $L$ described by APTC are as follows:

$L = \sum_{d_{I_L} \in \Delta}(r_{I_L}(d_{I_L}) \cdot L_2)$

$L_2 = LF \cdot L_3$

$L_3 = \sum_{d_{O_L} \in \Delta}(s_{O_L}(d_{O_L}) \cdot L)$

The state transitions of the Leasing pattern $R$ described by APTC are as follows:

$R = \sum_{d_{I_R} \in \Delta}(r_{I_R}(d_{I_R}) \cdot R_2)$

$R_2 = RF \cdot R_3$

$R_3 = \sum_{d_{O_R} \in \Delta}(s_{O_R}(d_{O_R}) \cdot R)$

The sending action and the reading action of the same data through the same channel can communicate with each other, otherwise, they will cause a deadlock $\delta$. We define the following communication function between the Lookup pattern and the Life Cycle Manager pattern:

$$\gamma(r_{I_L}(d_{I_L}), s_{O_A}(d_{O_A})) \triangleq c_{I_L}(d_{I_L})$$

We define the following communication function between the Life Cycle Manager pattern and the Leasing pattern:

$$\gamma(r_{I_R}(d_{I_R}), s_{O_L}(d_{O_L})) \triangleq c_{I_R}(d_{I_R})$$

Let all modules be in parallel, then the resource management patterns $A \quad L \quad R$ can be presented by the following process term:

$\tau_I(\partial_H(\Theta(\tau_{I_1}(\partial_{H_1}(A)) \between \tau_{I_2}(\partial_{H_2}(L)) \between \tau_{I_3}(\partial_{H_3}(R))))) = \tau_I(\partial_H(\tau_{I_1}(\partial_{H_1}(A)) \between \tau_{I_2}(\partial_{H_2}(L)) \between \tau_{I_3}(\partial_{H_3}(R))))$

where $H = \{r_{I_L}(d_{I_L}), s_{O_A}(d_{O_A}), r_{I_R}(d_{I_R}), s_{O_L}(d_{O_L})$
$|d_I, d_O, d_{I_A}, d_{O_A}, d_{I_L}, d_{O_L}, d_{I_R}, d_{O_R} \in \Delta\}$,

$I = \{c_{I_L}(d_{I_L}), c_{I_R}(d_{I_R}), AF, LF, RF | d_I, d_O, d_{I_A}, d_{O_A}, d_{I_L}, d_{O_L}, d_{I_R}, d_{O_R} \in \Delta\}$

Also, about the definitions of $H_1$ and $I_1$, $H_2$ and $I_2$, $H_3$ and $I_3$, please see Chapter 37.

Then, we obtain the following conclusion on the inserting of resource management patterns.

**Theorem 38.3** (Correctness of the inserting of resource management patterns). *The inserting of resource management patterns*

$\tau_I(\partial_H(\tau_{I_1}(\partial_{H_1}(A)) \between \tau_{I_2}(\partial_{H_2}(L)) \between \tau_{I_3}(\partial_{H_3}(R))))$

*can exhibit the desired external behaviors.*

*Proof.* Based on the above state transitions of the above modules, by use of the algebraic laws of APTC, we can prove that

$\tau_I(\partial_H(\tau_{I_1}(\partial_{H_1}(A)) \between \tau_{I_2}(\partial_{H_2}(L)) \between \tau_{I_3}(\partial_{H_3}(R)))) = \sum_{d_I, d_O \in \Delta}(r_I(d_I) \cdot s_O(d_O)) \cdot$
$\tau_I(\partial_H(\tau_{I_1}(\partial_{H_1}(A)) \between \tau_{I_2}(\partial_{H_2}(L)) \between \tau_{I_3}(\partial_{H_3}(R))))$

That is, the inserting of resource management patterns $\tau_I(\partial_H(\tau_{I_1}(\partial_{H_1}(A)) \between \tau_{I_2}(\partial_{H_2}(L)) \between \tau_{I_3}(\partial_{H_3}(R))))$ can exhibit the desired external behaviors.

For the details of the proof, please refer to Section 17.3, as we omit them here. □

# References


[1] R. Milner, Communication and Concurrency, Prentice Hall, 1989.

[2] R. Milner, A Calculus of Communicating Systems, LNCS, vol. 92, Springer, 1980.

[3] W. Fokkink, Introduction to Process Algebra, 2nd ed., Springer-Verlag, 2007.

[4] R. Milner, J. Parrow, D. Walker, A calculus of mobile processes, part I, Information and Computation 100 (1) (1992) 1–40.

[5] R. Milner, J. Parrow, D. Walker, A calculus of mobile processes, part II, Information and Computation 100 (1) (1992) 41–77.

[6] I. Phillips, I. Ulidowski, Reversing algebraic process calculi, Journal of Logic and Algebraic Programming 73 (1–2) (2007) 70–96.

[7] I. Phillips, I. Ulidowski, True concurrency semantics via reversibility, http://www.researchgate.net/publication/266891384, 2014.

[8] I. Ulidowski, I. Phillips, S. Yuen, Concurrency and reversibility, in: RC, in: LNCS, vol. 8507, Springer, 2014, pp. 1–14.

[9] Y. Wang, Algebraic Theory for True Concurrency, Elsevier, ISBN 978-0-443-18912-8, 2023, 2023-1-1.

[10] S. Andova, Probabilistic process algebra, Annals of Operations Research 128 (2002) 204–219.

[11] S. Andova, J. Baeten, T. Willemse, A complete axiomatisation of branching bisimulation for probabilistic systems with an application in protocol verification, in: International Conference on Concurrency Theory, Springer Berlin Heidelberg, 2006.

[12] S. Andova, S. Georgievska, On compositionality, efficiency, and applicability of abstraction in probabilistic systems, in: Conference on Current Trends in Theory and Practice of Computer Science, Springer-Verlag, 2009.

[13] C.A.R. Hoare, An axiomatic basis for computer programming, Communications of the ACM 12 (10) (1969).

[14] J.F. Groote, A. Ponse, Process algebra with guards: combining Hoare logic with process algebra, Formal Aspects of Computing 6 (2) (1994) 115–164.

[15] F. Moller, The importance of the left merge operator in process algebras, in: M.S. Paterson (Ed.), Proceedings 17th Colloquium on Automata, Languages and Programming (ICALP'90), Warwick, in: LNCS, vol. 443, Springer, 1990, pp. 752–764.

[16] J. Baeten, F.W. Vaandrager, An algebra for process creation, Acta Informatica 29 (4) (1992) 303–334.

[17] F.W. Vaandrager, Verification of two communication protocols by means of process algebra, Report CS-R8608, CWI, Amsterdam, 1986.

[18] K.A. Bartlett, R.A. Scantlebury, P.T. Wilkinson, A note on reliable full-duplex transmission over half-duplex links, Communications of the ACM 12 (5) (1969) 260–261.

[19] C. Hewitt, View control structures as patterns of passing messages, Journal of Artificial Intelligence 8 (3) (1977) 323–346.

[20] G. Agha, Actors: a model of concurrent computation in distributed systems, Ph.D. thesis, MIT, 1986.

[21] G. Agha, I. Mason, S. Smith, C. Talcott, A foundation for actor computation, Journal of Functional Programming (1993).

[22] G. Agha, P. Thati, An algebraic theory of actors and its application to a simple object-based language, in: From Object-Orientation to Formal Methods, Essays in Memory of Ole-Johan Dahl DBLP, 2004, pp. 26–57.

[23] D. Jordan, J. Evdemon, Web Services Business Process Execution Language Version 2.0. OASIS Standard, 2007.

[24] N. Kavantzas, D. Burdett, G. Ritzinger, et al., Web Services Choreography Description Language Version 1.0. W3C Candidate Recommendation, 2005.

[25] Y. Wang, Formal model of Web Service composition: an actor-based approach to unifying orchestration and choreography, arXiv:1312.0677, 2013.

[26] Y. Wang, A formal model of QoS-aware Web Service orchestration engine, IEEE Transactions on Network and Service Management 13 (1) (2016) 113–125.

[27] M. Abadi, A.D. Gordon, A calculus for cryptographic protocols: the spi calculus, Information and Computation 148 (1) (1999) 1–70.

[28] M. Abadi, B. Blanchet, C. Fournet, The applied pi calculus: mobile values, new names, and secure communication, Journal of the ACM 65 (1) (2017) 1–41.

[29] B. Schneier, Applied Cryptography: Protocols, Algorithms, and Source Code in C, 2nd ed., Government Information Quarterly, vol. 13(3), 1996, p. 336.

[30] M. Burrows, M. Abadi, R.M. Needham, A logic of authentication, Proceedings of the Royal Society of London A 426 (1989) 233–271. A preliminary version appeared as Digital Equipment Corporation Systems Research Center report No. 39.

[31] E. Gamma, R. Helm, R. Johnson, J. Vlissides, Design patterns: elements of reusable object-oriented software, 1995.

[32] F. Buschmann, R. Meunier, H. Rohnert, P. Sommerlad, M. Stal, Pattern-Oriented Software Architecture - volume 1: A System of Patterns, Wiley Publishing, 1996.

[33] D.C. Schmidt, M. Stal, H. Rohnert, F. Buschmann, Pattern-Oriented Software Architecture - volume 2: Patterns for Concurrent and Networked Objects, Wiley Publishing, 2000.

[34] M. Kircher, P. Jain, Pattern-Oriented Software Architecture - volume 3: Patterns for Resource Management, Wiley Publishing, 2004.

[35] F. Buschmann, K. Henney, D.C. Schmidt, Pattern-Oriented Software Architecture - volume 4: A Pattern Language for Distributed Computing, Wiley Publishing, 2007.

[36] F. Buschmann, K. Henney, D.C. Schmidt, Pattern-Oriented Software Architecture - volume 5: On Patterns and Pattern Languages, Wiley Publishing, 2007.

# Index

## Symbols



## A



## B

## C



## D



## E

## F



## G



## H



## I



## K

## Q



## R



## S

Printed in the United States
by Baker & Taylor Publisher Services